Succeed in Calculus

Use a wide range of valuable resources to excel in calculus.

■ *Step-by-step solutions help you review and prepare.*

Study and Solutions Guide, Volumes I (0-395-88767-4) and II (0-395-88768-2)

• Detailed solutions to selected odd-numbered text exercises
• Study strategies and algebra review

■ *Use the power of technology to apply calculus to real-world settings.*

Lab Manual Series

• Available in five versions—*Maple* (Windows: 0-395-90054-9; Mac: 0-395-90053-0), *Mathematica* (Windows: 0-395-90059-X; Mac: 0-395-90058-1), *Derive* (Windows: 0-395-90052-2), *Mathcad* (Windows: 0-395-90056-5; Mac: 0-395-90055-7), and the *TI-92* graphing calculator (Windows: 0-395-90062-X; Mac: 0-395-90061-1)
• Challenging, real-world projects using technology
• Includes data disks
• See pp. xxiv–xxv for more details.

■ *Graphing is easy when you have the right tools.*

Graphing Technology Guide (0-395-88773-9)

• Keystroke instructions for a wide variety of *Texas Instruments, Casio, Sharp,* and *Hewlett-Packard* graphing calculators, including the most current models
• Examples with step-by-step solutions
• Extensive graphics screen output and technology tips

■ *Brush up on precalculus to succeed in calculus.*

The Algebra of Calculus (0-669-21885-5)

• Review of the algebra, trigonometry, and analytic geometry required for calculus
• Over 200 examples with solutions
• Pretests and exercises with answers

■ *Learn calculus using innovative technology.*

Interactive Calculus, Version 2.0 (0-395-91102-8)

• Internet accessible
• Multimedia, CD-ROM format
• Includes all of the content of the Sixth Edition
• Active mathematics, including editable 2D graphs and rotatable 3D graphs
• New explorations and simulations
• See pp. xxvi–xxvii for more details.

Look for these resources in your bookstore. If you don't find them, check with your bookstore manager or call Houghton Mifflin toll free at **1-800-225-1464** to place an order.

Two ways to get the advantage in your *calculus* class

Sixth Edition

Calculus
with Analytic Geometry

Roland E. Larson
Robert P. Hostetler

The Pennsylvania State University
The Behrend College

Bruce H. Edwards

University of Florida

with the assistance of
David E. Heyd

The Pennsylvania State University
The Behrend College

Houghton Mifflin Company   Boston   New York

Editor in Chief, Mathematics: Charles Hartford
Managing Editor: Cathy Cantin
Senior Associate Editor: Maureen Brooks
Associate Editor: Michael Richards
Assistant Editor: Carolyn Johnson
Supervising Editor: Karen Carter
Art Supervisor: Gary Crespo
Marketing Manager: Sara Whittern
Associate Marketing Manager: Ros Kane
Marketing Assistant: Carrie Lipscomb
Design: Henry Rachlin
Composition and Art: Meridian Creative Group

We have included examples and exercises that use real-life data as well as technology output from a variety of software. This would not have been possible without the help of many people and organizations. Our wholehearted thanks goes to all for their time and effort.

Trademark Acknowledgments: TI is a registered trademark of Texas Instruments, Inc. Mathcad is a registered trademark of MathSoft, Inc. Windows, Microsoft, and MS-DOS are registered trademarks of Microsoft, Inc. Mathematica is a registered trademark of Wolfram Research, Inc. DERIVE is a registered trademark of Soft Warehouse, Inc. IBM is a registered trademark of International Business Machines Corporation. Maple is a registered trademark of the University of Waterloo.

Printed in the U.S.A.

Library of Congress Catalog Card Number: 97-72511

ISBN: 0-395-86974-9

3456789–VH–01 00 99

Contents

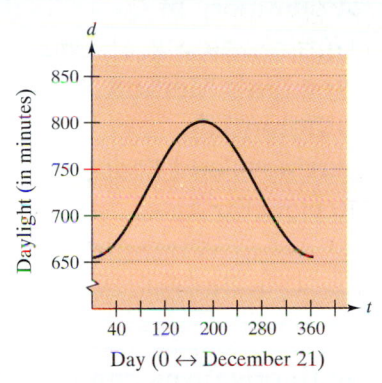

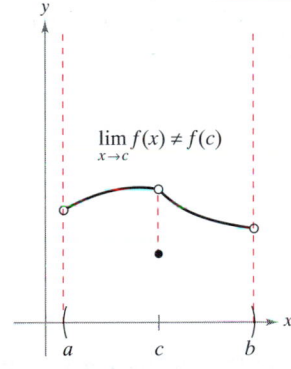

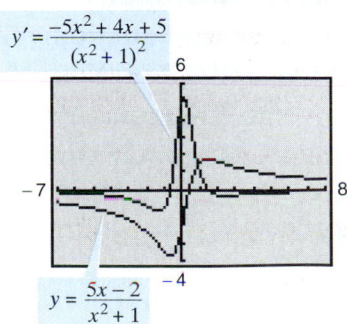

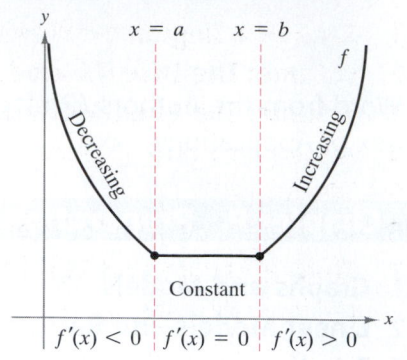

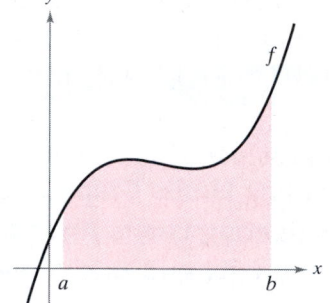

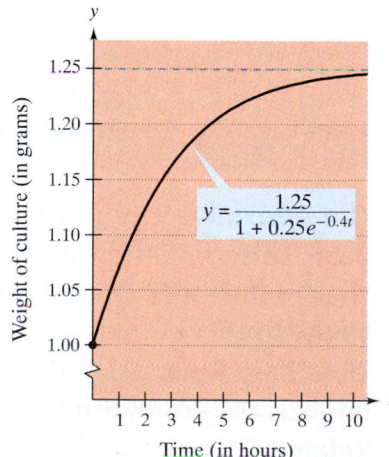

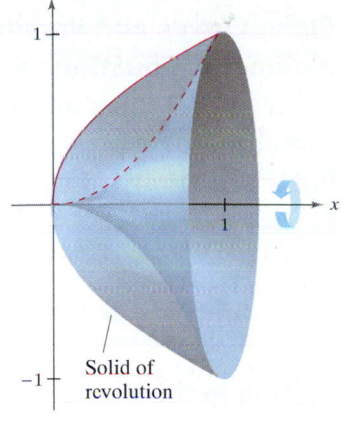

Solid of
revolution

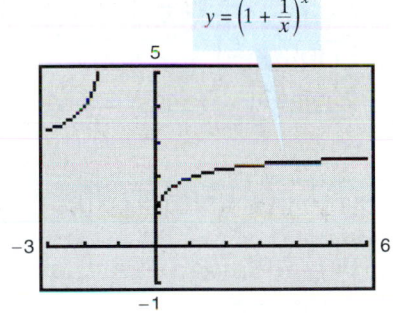

$y = \left(1 + \dfrac{1}{x}\right)^x$

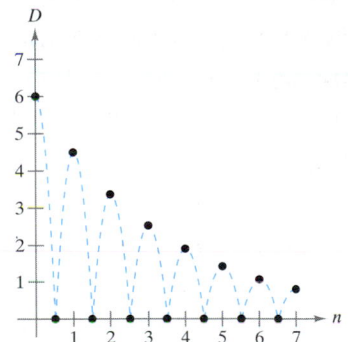

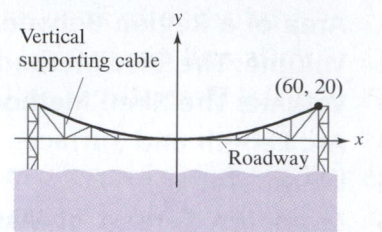

Chapter 10   Vectors and the Geometry of Space   699

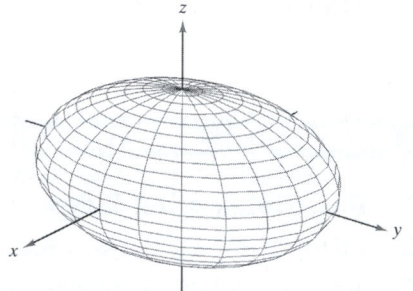

Chapter 11   Vector-Valued Functions   767

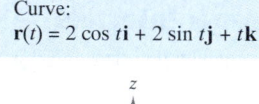

Curve:
$\mathbf{r}(t) = 2\cos t\,\mathbf{i} + 2\sin t\,\mathbf{j} + t\,\mathbf{k}$

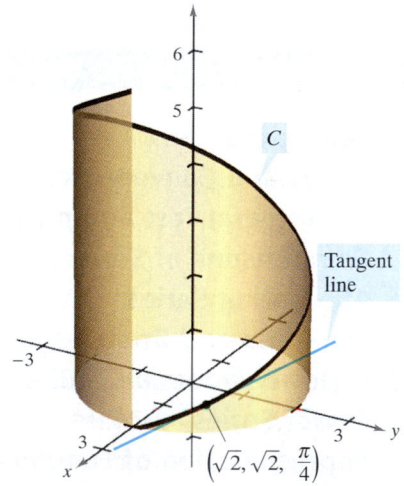

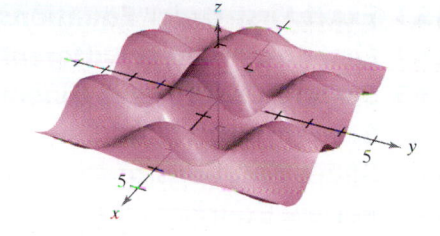

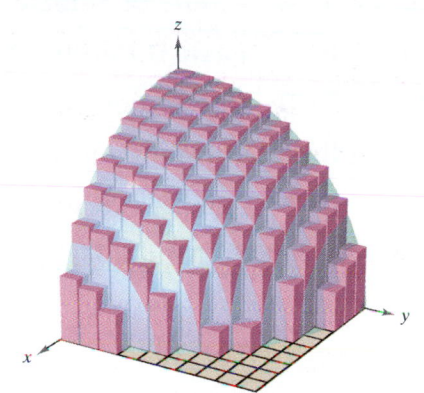

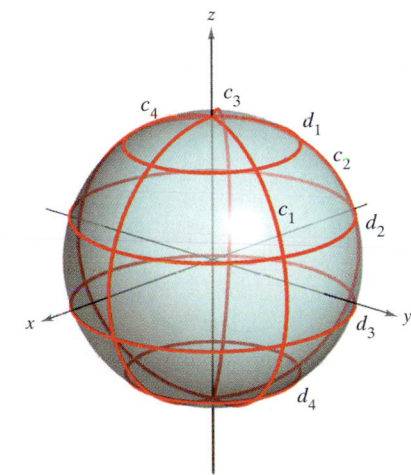

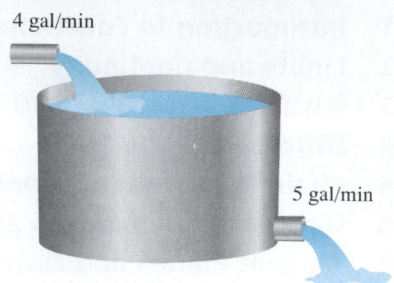

4 gal/min

5 gal/min

A word from the authors…

Welcome to *Calculus with Analytic Geometry,* Sixth Edition! We are excited about the Sixth Edition and hope you will be too after you hear about why we wrote it, what's new about it, and how it will carry you and your calculus students into the twenty-first century!

Reform in Mathematics Education

As you know, the current reform movement in math education started about 15 years ago and has involved all levels of mathematics education from kindergarten through college. There may be some who say reform was not needed. However, the vast majority of math educators agree that reform was essential. In fact, the new math of the 1960s, 1970s, and early 1980s was a national disaster. It was far too abstract and far too removed from the real-life applications that were the actual foundation of mathematics. The result was evident to everyone—math phobia, falling test scores, high drop-out rates, and a general sense that students were not learning to be creative problem solvers.

So what were the proposed solutions? There were many—more real-life connections, more incorporation of technology, curriculum revisions, and the development of alternative forms of teaching, assessment, and learning.

Where This Text Stands with Respect to Reform

"The text has definitely benefited from reform. More than that, over the years our calculus text has actually led the way in developing many innovative learning techniques."

Where does the Sixth Edition stand with respect to reform? This is one of the most common questions we are asked. Our answer: The text has definitely benefited from reform. More than that, over the years our calculus text has actually led the way in developing many innovative learning techniques. From its first edition, the text stressed the importance of graphical learning—much more than other texts in use in the late 1970s and early 1980s. This text was one of the first to incorporate computer-generated art—both two-dimensional and three-dimensional—to aid in the visualization of complex mathematical concepts.

We have always paid careful attention to the presentation—using precise mathematical language, innovative full-color designs for emphasis and clarity, and a level of exposition that appeals to students—to create an effective teaching and learning tool. Although difficult to quantify, this feature has been praised by thousands of students and their instructors over the past 20 years. With each edition, we have continued to incorporate the best strategies for teaching calculus, using the pedagogy we have developed as a result of our teaching experiences, as well as many suggestions from thoughtful users.

This Sixth Edition might best be described as fitting midway between texts that define themselves as traditional and those that are considered reform texts. Our approach is like that of a traditional text in that we firmly believe in the importance of carefully developed theory, correct statements of theorems, inclusion of proofs, and mastery of traditional calculus skills. We have found no evidence that it is somehow possible to apply calculus in real-life situations without first being able to understand and "do" calculus.

On the other hand, we wholeheartedly embrace many of the features of calculus reform. For instance, this edition features additional opportunities to use technology, an increased emphasis on real-life applications and modeling data, new motivational features, many more conceptual exercises and multi-part exercises, new explorations in many sections, and a myriad of student and teacher aids.

We believe in giving teachers options to teach calculus the way they want. Because of this, you will find a wide variety of approaches and features in the text. By choosing the options that best fit you as a teacher, you can customize the book in dozens of different ways: lecture or discovery approaches, pencil and paper skills or technology, formal or informal, theoretical or intuitive, analytic or graphical, mathematics-centered or applied, classroom presentation or distance learning—they are all here as options for you.

What We Changed in the Sixth Edition

"In the Sixth Edition, we continue to lead the way in incorporating the best aspects of reform in a meaningful yet easy-to-use manner."

In the Sixth Edition, we continue to lead the way in incorporating the best aspects of reform in a meaningful yet easy-to-use manner. Here are some of the most significant new features.

New *Explorations*   For instructors and students who benefit from a frequent or occasional "discovery" mode, we have incorporated this option into the Sixth Edition. By discovering concepts that are new to them, students get a taste of what it is like to be a real mathematician. Also, some students find it easier to remember concepts they have "discovered." Truly effective explorations are difficult to create. The challenge is finding a good balance between what is given and what is expected to be discovered. Our Explorations represent a combination of suggestions by users and the results of our own classroom experience.

"As all calculus teachers know, one of the difficulties in teaching calculus is that we are often answering questions that students have not yet asked. The motivators address this dilemma by presenting real-life situations with exploratory questions."

New *Motivating the Chapter*   Each chapter now begins with a full-page chapter motivator. As all calculus teachers know, one of the difficulties in teaching calculus is that we are often answering questions that students have not yet asked. The motivators address this dilemma by presenting real-life situations with exploratory questions. As students attempt to use the techniques of their current skill set to answer the questions, they will learn to appreciate the power and efficiency of the new calculus techniques presented in the chapter. We spent a lot of time researching effective settings and developing meaningful questions, and we found that the ones selected for this text sparked interest among the students in our classes.

New *Lab Manuals*   The real-life application that is introduced in each *Motivating the Chapter* forms the basis for the extended technology lab projects in the supplementary Lab Manuals. In each *Motivating the Chapter*, students are asked to solve a problem using the techniques of their current skill set. As the chapter progresses, students can be assigned projects from the Lab Manuals that ask them to take a new look at these problems using concepts learned in that chapter as well as technology. The Lab Manuals come in five versions (each with data disks) designed for use with *Maple, Mathematica, Derive, Mathcad,* and the *TI-92* graphing calculator.

New *Art Program*   Visualization is a problem-solving skill that is critical for the understanding of complex calculus theory and concepts. To help students develop this skill, the Sixth Edition features a completely new art program that was created using state-of-the-art computer technology for accuracy, clarity, and realism. The effect of the new three-dimensional art is particularly striking. Using computer graphics software, we adjusted the color and transparency, view, light sources, and shadows on the solids and surfaces until we found the optimal combination of features to show true perspective. The result, we believe, is the most accurate representation of three-dimensional calculus surfaces ever! For instance, the image on the book's cover was produced with this computer graphics system and the illusion of three-dimensionality is amazing.

Revised Exercise Sets   All of the exercise sets of the previous edition were considered for revision, and many new exercises were added to the Sixth Edition. We focused on adding problems that were technology-oriented, thought-provoking, conceptual, creative, real, and engaging. There are many more opportunities for writing, for individual and group projects, and for solving problems with graphical, numerical, and analytical approaches. We expanded on the wide variety of applications, which were already distinctive for their relevance and originality. Also new to the Sixth Edition are many modeling data exercises that ask students to find and interpret mathematical models from the real-life data that are given. To the many who helped us in our search for such exercises, we offer our thanks.

Revised *Interactive Calculus*   The Fifth Edition of this text was available in an interactive CD-ROM format that was innovative and well received. Now, four years later, we have continued to push the limits of what technology can do as a medium for teaching calculus. Among the new and enhanced features of the CD-ROM version of the Sixth Edition are: all of the content of the revised text, more active mathematics, editable two-dimensional graphs, dazzling three-dimensional graphs, additional explorations and simulations, and a syllabus builder for instructors.

Table of Contents   Some of the text was revised to enhance the flexibility of the presentation and to reflect our perspective on reform. We rewrote Chapter P, moving much of the precalculus review to Appendix A. What remains of precalculus in Chapter P was restructured to introduce students to new ways of thinking about math, including graphical, numerical, and analytical approaches; modeling; problem solving; and data analysis. Instructors who skipped this chapter in previous editions might want to reconsider that decision.

Chapter 1 now begins with a new section, "A Preview of Calculus," which introduces students to the distinctions between precalculus and calculus, emphasing the necessity of calculus in our dynamic, everyday surroundings. One section on differential equations was moved from Chapter 15 to Chapter 5 to allow students to make more use of this material, and to better prepare them for their courses in other disciplines such as physics and chemistry. In Chapter 6, we repositioned the section on moments, centers of mass, and centroids so that it precedes the section on fluid pressure and fluid force, to address changing teaching styles. We reduced the coverage of conic sections in Chapter 9, because this material is mostly a review for students at this level. This edition discusses conics in the same chapter that covers parametric equations and the polar coordinate system. Finally, the section on rotation and the general second-degree equation was moved to Appendix E, leaving it optional for those instructors who wish to cover it.

"… the most accurate representation of three-dimensional calculus surfaces ever!"

"We focused on adding problems that were technology-oriented, thought-provoking, conceptual, creative, real, and engaging."

*"Although we carefully and thoroughly revised the text ..., we **didn't change** many of the things that our colleagues and the 1,500,000 students who have used the book have told us worked for them."*

What We Didn't Change

Although we carefully and thoroughly revised the text by enhancing the usefulness of some features and topics and adding others, we *didn't* change many of the things that our colleagues and the 1,500,000 students who have used this book have told us worked for them. We still offer comprehensive coverage of the material required by students in a three-semester calculus course, including carefully stated theory and proofs. Additionally, as we do with all our books, we painstakingly formatted every page of the Sixth Edition to achieve a clear presentation of the material. Finally, the text was carefully written in a style that is mathematically precise, as well as engaging, direct, and readable.

We hope you will enjoy the Sixth Edition. We are proud to have it as our calculus entry for the next century!

Roland E. Larson

Robert P. Hostetler

Bruce H. Edwards

Features

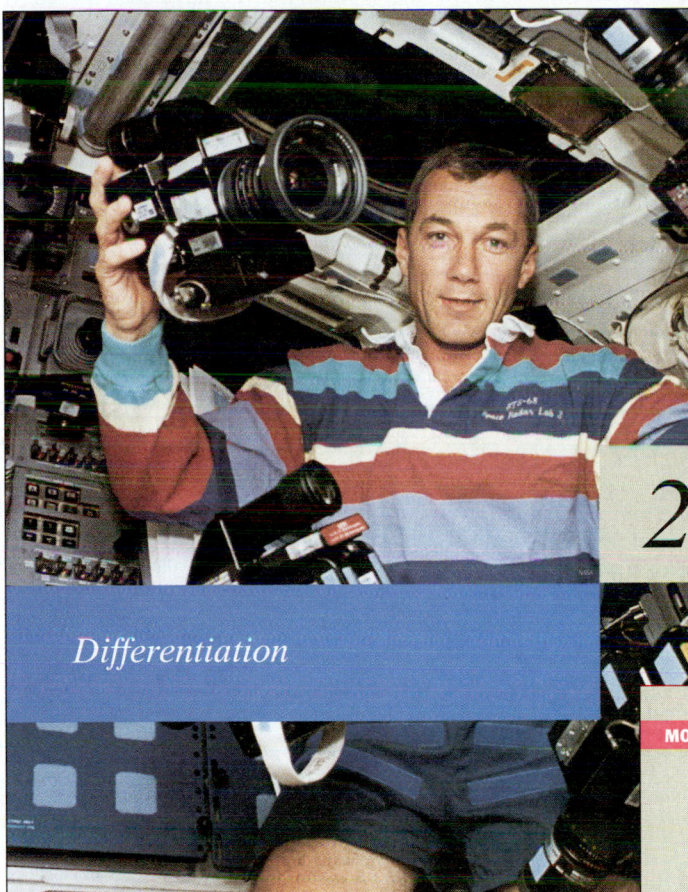

Differentiation

2

Chapter Openers

Each chapter opens with a photograph that corresponds to the mathematical application in *Motivating the Chapter.*

Motivating the Chapter

Each *Motivating the Chapter* explores the concepts to be covered in the chapter using a real-world setting. Following a short introduction, open-ended questions guide students through an introduction to the main themes of the chapter.

Kathryn C. Thorton, Ph.D., Payload Commander on the USML-2 mission, has been a NASA astronaut since July 1985. The USML-2 mission was her fourth space flight.

MOTIVATING THE CHAPTER   **Gravity: Finding It Experimentally**

SPACE SHUTTLE EXPERIMENTS
On October 20, 1995, the Second United States Microgravity Laboratory (USML-2) was launched aboard the space shuttle Columbia. The USML-2 was built to take advantage of the low-gravity conditions in order to research how a near-weightless environment influences the behavior of fluids, combustion, material structure, and protein crystals.

The study of dynamics dates back to the sixteenth century. As the Dark Ages gave way to the Renaissance, Galileo Galilei (1564–1642) was one of the first to take steps toward understanding the motion of objects under the influence of gravity.

Up until Galileo's time, it was recognized that a falling object moved faster and faster as it fell, but what mathematical law governed this accelerating motion was unknown. Free-falling objects move too fast to have been measured with any of the equipment available at that time. Galileo solved this problem with a rather ingenious setup. He reasoned that gravity could be "diluted" by rolling a ball down an inclined plane. He used a water clock, which kept track of time by measuring the amount of water that poured through a small opening at the bottom.

We now have relatively inexpensive instruments, such as the *Texas Instruments Calculator-Based Laboratory (CBL) System*, that allow accurate position data to be gathered on a free-falling object. A CBL System was used to track the positions of a falling ball at time intervals of 0.02 second. The results are shown below.

Time (sec)	Height (meters)	Velocity (meters/sec)
0.00	0.290864	−0.16405
0.02	0.284279	−0.32857
0.04	0.274400	−0.49403
0.06	0.260131	−0.71322
0.08	0.241472	−0.93309
0.10	0.219520	−1.09409
0.12	0.189885	−1.47655
0.14	0.160250	−1.47891
0.16	0.126224	−1.69994
0.18	0.086711	−1.96997
0.20	0.045002	−2.07747
0.22	0.000000	−2.25010

QUESTIONS

1. Use a graphing utility to sketch a scatter plot of the positions of the falling ball. What type of model seems to be the best fit? Use the regression features of the graphing utility to find the best-fitting model.

2. Repeat the procedure in Question 1 for the velocities of the falling ball. Describe any relationships between the two models.

3. In theory, the position of a free-falling object in a vacuum is given by $s = \frac{1}{2}gt^2 + v_0 t + s_0$, where g is the acceleration due to gravity (meters per second per second), t is the time (seconds), v_0 is the initial velocity (meters per second), and s_0 is the initial height (meters). From this experiment, estimate the value of g. Do you think your estimate is too great or too small? Explain your reasoning.

The concepts presented here will be explored further in this chapter. For an extension of this application, see the lab series that accompanies this text.

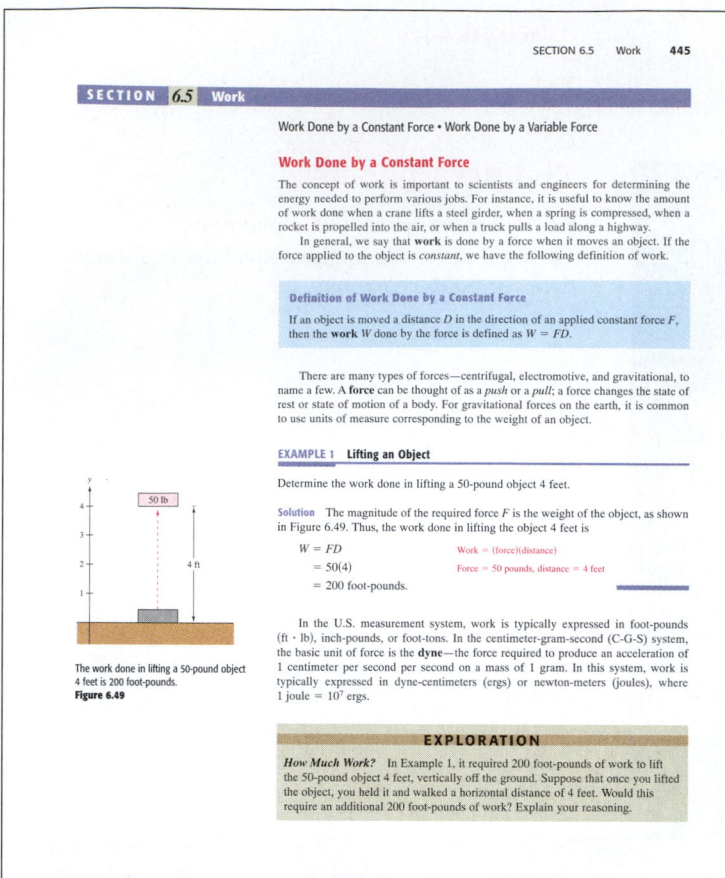

Section Topics

Each section begins with a list of subsection topics. This outline will help instructors with class planning and help students with studying and synthesizing the material in the section.

Explorations

Before students are exposed to selected topics, exploratory projects allow them to discover concepts on their own, making them more likely to remember the results. These optional boxed features can be omitted, if the instructor so desires, with no loss of continuity in the coverage of material.

Examples

To increase the usefulness of the text as a study tool, the Sixth Edition contains over 1000 examples, each titled for easy reference. Many of these detailed examples display solutions that are presented graphically, analytically, and/or numerically to provide further insight into mathematical concepts. Side comments clarify the steps of the solution as necessary.

Graphics

The Sixth Edition has over 3500 figures. Computer-generated for accuracy, clarity, and realism, this new art program will help students visualize mathematical concepts more easily. The surfaces and solids of complex, three-dimensional figures were created using the optimal combination of color and transparency, view, perspective, light sources, and shadows to show true perspective.

CHAPTER 13   Multiple Integration

EXAMPLE 1   Approximating the Volume of a Solid

Approximate the volume of the solid lying between the paraboloid

$$f(x, y) = 1 - \frac{1}{2}x^2 - \frac{1}{2}y^2$$

and the square region R given by $0 \le x \le 1$, $0 \le y \le 1$. Use a partition made up of squares whose edges have a length of $\frac{1}{4}$.

Solution Begin by forming the specified partition of R. For this partition, it is convenient to choose the centers of the subregions as the points at which to evaluate $f(x, y)$.

$$\left(\tfrac{1}{8}, \tfrac{1}{8}\right) \quad \left(\tfrac{1}{8}, \tfrac{3}{8}\right) \quad \left(\tfrac{1}{8}, \tfrac{5}{8}\right) \quad \left(\tfrac{1}{8}, \tfrac{7}{8}\right)$$
$$\left(\tfrac{3}{8}, \tfrac{1}{8}\right) \quad \left(\tfrac{3}{8}, \tfrac{3}{8}\right) \quad \left(\tfrac{3}{8}, \tfrac{5}{8}\right) \quad \left(\tfrac{3}{8}, \tfrac{7}{8}\right)$$
$$\left(\tfrac{5}{8}, \tfrac{1}{8}\right) \quad \left(\tfrac{5}{8}, \tfrac{3}{8}\right) \quad \left(\tfrac{5}{8}, \tfrac{5}{8}\right) \quad \left(\tfrac{5}{8}, \tfrac{7}{8}\right)$$
$$\left(\tfrac{7}{8}, \tfrac{1}{8}\right) \quad \left(\tfrac{7}{8}, \tfrac{3}{8}\right) \quad \left(\tfrac{7}{8}, \tfrac{5}{8}\right) \quad \left(\tfrac{7}{8}, \tfrac{7}{8}\right)$$

Because the area of each square is $\Delta x_i \Delta y_i = \frac{1}{16}$, you can approximate the volume by the sum

$$\sum_{i=1}^{16} f(x_i, y_i)\Delta x_i \Delta y_i = \sum_{i=1}^{16}\left(1 - \frac{1}{2}x_i^2 - \frac{1}{2}y_i^2\right)\left(\frac{1}{16}\right) \approx 0.672.$$

This approximation is shown graphically in Figure 13.12. The exact volume of the solid is $\frac{2}{3}$ (see Example 2). You can obtain a better approximation by using a finer partition. For example, with a partition of squares with sides of length $\frac{1}{10}$, the approximation is 0.668.

TECHNOLOGY Some three-dimensional graphing utilities are capable of sketching figures such as that shown in Figure 13.12. For instance, the sketch shown in Figure 13.13 was drawn with a computer program. In this sketch, note that each of the rectangular prisms lies within the solid region.

In Example 1, note that by using finer partitions, you can obtain better approximations of the volume. This observation suggests that you could obtain the exact volume by taking a limit. That is

$$\text{Volume} = \lim_{\|\Delta\|\to 0}\sum_{i=1}^{n} f(x_i, y_i)\Delta x_i \Delta y_i.$$

The precise meaning of this limit is that the limit is equal to L if for every $\varepsilon > 0$ there exists a $\delta > 0$ such that

$$\left| L - \sum_{i=1}^{n} f(x_i, y_i)\Delta x_i \Delta y_i \right| < \varepsilon$$

for all partitions Δ of the plane region R (that satisfy $\|\Delta\| < \delta$) and for all possible choices of x_i and y_i in the ith region.

Using the limit of a Riemann sum to define volume is a special case of using the limit to define a **double integral**. The general case, however, does not require that the function be positive or continuous.

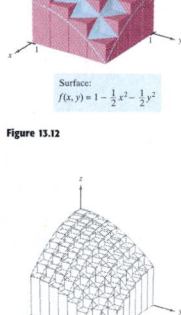

Surface:
$f(x, y) = 1 - \frac{1}{2}x^2 - \frac{1}{2}y^2$

Figure 13.12

Figure 13.13

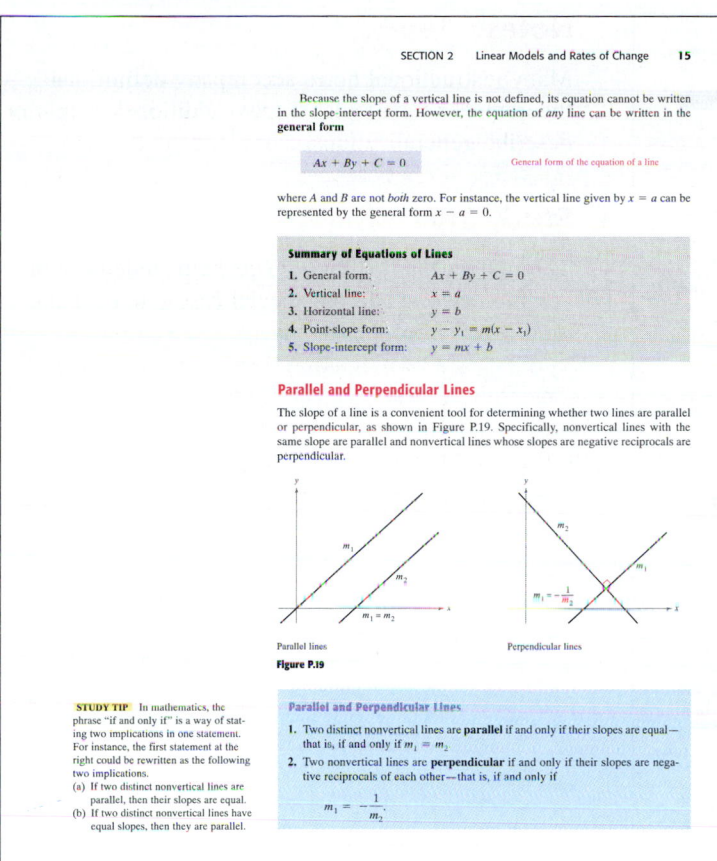

SECTION 2   Linear Models and Rates of Change   15

Because the slope of a vertical line is not defined, its equation cannot be written in the slope-intercept form. However, the equation of *any* line can be written in the **general form**

$$Ax + By + C = 0 \qquad \text{General form of the equation of a line}$$

where A and B are not *both* zero. For instance, the vertical line given by $x = a$ can be represented by the general form $x - a = 0$.

Summary of Equations of Lines

1. General form: $Ax + By + C = 0$
2. Vertical line: $x = a$
3. Horizontal line: $y = b$
4. Point-slope form: $y - y_1 = m(x - x_1)$
5. Slope-intercept form: $y = mx + b$

Parallel and Perpendicular Lines

The slope of a line is a convenient tool for determining whether two lines are parallel or perpendicular, as shown in Figure P.19. Specifically, nonvertical lines with the same slope are parallel and nonvertical lines whose slopes are negative reciprocals are perpendicular.

Parallel lines   Perpendicular lines
Figure P.19

STUDY TIP In mathematics, the phrase "if and only if" is a way of stating two implications in one statement. For instance, the first statement at the right could be rewritten as the following two implications.
(a) If two distinct nonvertical lines are parallel, then their slopes are equal.
(b) If two distinct nonvertical lines have equal slopes, then they are parallel.

Parallel and Perpendicular Lines

1. Two distinct nonvertical lines are **parallel** if and only if their slopes are equal—that is, if and only if $m_1 = m_2$.
2. Two nonvertical lines are **perpendicular** if and only if their slopes are negative reciprocals of each other—that is, if and only if

$$m_1 = -\frac{1}{m_2}.$$

Summaries

Many sections have summaries that identify core ideas and procedures. In some instances, an entire section summarizes the preceding topics.

Definitions and Theorems

All definitions and theorems are highlighted for emphasis and easy reference.

Technology

Students are encouraged to use a graphing utility or computer algebra system as a tool for exploration, discovery, and problem solving. Many opportunities to execute complicated computations, to visualize theoretical concepts, to discover alternative approaches, and to verify the results of other solution methods using technology are presented. However, students are not required to have access to a graphing utility to use this text effectively. In addition to describing the benefits of using technology, the text also pays special attention to its possible misuse or misinterpretation.

Historical Notes

Historical notes are integrated throughout the text to help students understand that calculus has a past.

CHAPTER 1   Limits and Their Properties

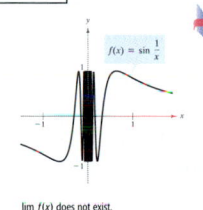

$f(x) = \sin \dfrac{1}{x}$

$\lim_{x \to 0} f(x)$ does not exist.
Figure 1.10

EXAMPLE 5   Oscillating Behavior

Discuss the existence of the limit $\displaystyle\lim_{x \to 0} \sin \frac{1}{x}$.

Solution Let $f(x) = \sin(1/x)$. In Figure 1.10, you can see that as x approaches 0, $f(x)$ oscillates between -1 and 1. Therefore, the limit does not exist because no matter how small you choose δ, it is possible to choose x_1 and x_2 within δ units of 0 such that $\sin(1/x_1) = 1$ and $\sin(1/x_2) = -1$, as indicated in the table.

x	$\dfrac{2}{\pi}$	$\dfrac{2}{3\pi}$	$\dfrac{2}{5\pi}$	$\dfrac{2}{7\pi}$	$\dfrac{2}{9\pi}$	$\dfrac{2}{11\pi}$	$x \to 0$
$\sin \dfrac{1}{x}$	1	-1	1	-1	1	-1	Limit does not exist.

Peter Gustav Dirichlet (1805–1859)
In the early development of calculus, the definition of a function was much more restricted than it is today, and "functions" such as the Dirichlet function would not have been considered. The modern definition of a function was given by the German mathematician Peter Gustav Dirichlet.

Common Types of Behavior Associated with the Nonexistence of a Limit

1. $f(x)$ approaches a different number from the right side of c than it approaches from the left side.
2. $f(x)$ increases or decreases without bound as x approaches c.
3. $f(x)$ oscillates between two fixed values as x approaches c.

There are many other interesting functions that have unusual limit behavior. An often cited one is the *Dirichlet function*

$$f(x) = \begin{cases} 0, & \text{if } x \text{ is rational.} \\ 1, & \text{if } x \text{ is irrational.} \end{cases}$$

This function has *no limit* at any real number c.

TECHNOLOGY When you use a graphing utility to investigate the behavior of a function near the x-value at which you are trying to evaluate a limit, remember that you can't always trust the pictures that graphing utilities draw. For instance, if you use a graphing utility to sketch the graph of the function in Example 5 over an interval containing 0, you will most likely obtain an incorrect graph—such as that shown in Figure 1.11. The reason that a graphing utility can't show the correct graph is that the graph has infinitely many oscillations over any interval that contains 0.

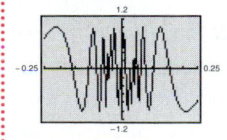

Incorrect graph of $f(x) = \sin(1/x)$
Figure 1.11

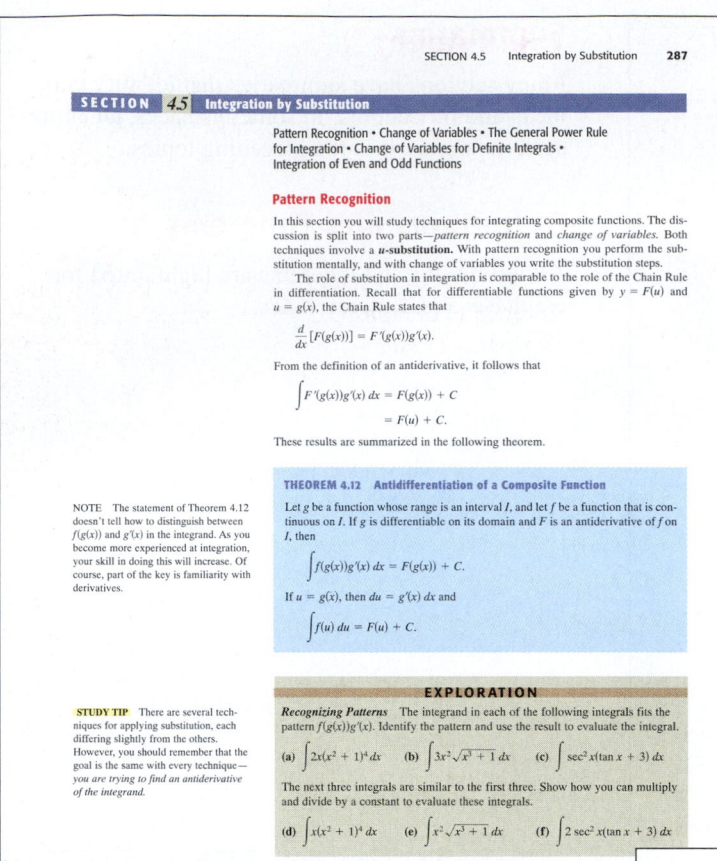

SECTION 4.5   Integration by Substitution   **287**

SECTION 4.5   Integration by Substitution

Pattern Recognition • Change of Variables • The General Power Rule for Integration • Change of Variables for Definite Integrals • Integration of Even and Odd Functions

Pattern Recognition

In this section you will study techniques for integrating composite functions. The discussion is split into two parts—*pattern recognition* and *change of variables*. Both techniques involve a *u-substitution*. With pattern recognition you perform the substitution mentally, and with change of variables you write the substitution steps.

The role of substitution in integration is comparable to the role of the Chain Rule in differentiation. Recall that for differentiable functions given by $y = F(u)$ and $u = g(x)$, the Chain Rule states that

$$\frac{d}{dx}[F(g(x))] = F'(g(x))g'(x).$$

From the definition of an antiderivative, it follows that

$$\int F'(g(x))g'(x)\,dx = F(g(x)) + C$$
$$= F(u) + C.$$

These results are summarized in the following theorem.

NOTE   The statement of Theorem 4.12 doesn't tell how to distinguish between $f(g(x))$ and $g'(x)$ in the integrand. As you become more experienced at integration, your skill in doing this will increase. Of course, part of the key is familiarity with derivatives.

THEOREM 4.12   Antidifferentiation of a Composite Function

Let g be a function whose range is an interval I, and let f be a function that is continuous on I. If g is differentiable on its domain and F is an antiderivative of f on I, then

$$\int f(g(x))g'(x)\,dx = F(g(x)) + C.$$

If $u = g(x)$, then $du = g'(x)\,dx$ and

$$\int f(u)\,du = F(u) + C.$$

STUDY TIP   There are several techniques for applying substitution, each differing slightly from the others. However, you should remember that the goal is the same with every technique—*you are trying to find an antiderivative of the integrand.*

EXPLORATION

Recognizing Patterns   The integrand in each of the following integrals fits the pattern $f(g(x))g'(x)$. Identify the pattern and use the result to evaluate the integral.

(a) $\int 2x(x^2 + 1)^4\,dx$ (b) $\int 3x^2\sqrt{x^3 + 1}\,dx$ (c) $\int \sec^2 x(\tan x + 3)\,dx$

The next three integrals are similar to the first three. Show how you can multiply and divide by a constant to evaluate these integrals.

(d) $\int x(x^2 + 1)^4\,dx$ (e) $\int x^2\sqrt{x^3 + 1}\,dx$ (f) $\int 2\sec^2 x(\tan x + 3)\,dx$

Notes

Many instructional notes accompany definitions, theorems, and examples to give additional insight or describe generalizations.

Study Tips

Throughout the text, *Study Tips* help students avoid common errors, address special cases, and expand on theoretical concepts.

Exercises

The text contains nearly 10,000 exercises. Each exercise set is graded, progressing from skill-development problems to more challenging problems involving applications and proofs. The wide variety of exercises includes many real, technology-oriented, thought-provoking, and engaging problems. Review exercises are included at the end of each chapter. Answers to all odd-numbered exercises are included in the back of the text. Red (blue, in the appendix) exercise numbers indicate selected exercises that can be found in the *Study and Solutions Guide*. To help instructors make homework assignments, many of the exercises in the text are labeled to indicate the area of application (e.g., *Break-Even Point*) or the type of exercise (e.g., *Writing* or *Approximation*).

Lab Series

The real-life application introduced in each *Motivating the Chapter* forms the basis for extended lab projects in the supplementary Lab Manuals using *Maple, Mathematica, Derive, Mathcad,* and the *TI-92* graphing calculator. The labs are referenced in the text where it seemed most appropriate to assign them.

34   CHAPTER P   Preparation for Calculus

EXERCISES FOR SECTION P.4   **LAB SERIES   Lab P.1**

In Exercises 1–4, a scatter plot of data is given. Determine whether the data can be modeled by a linear function, a quadratic function, or a trigonometric function, or that there appears to be no relationship between x and y.

1.

2.

3.

4.

5. *Carcinogens*   The ordered pairs give the exposure index x of a carcinogenic substance and the cancer mortality y per 100,000 people in the population.

(3.50, 150.1), (3.58, 133.1), (4.42, 132.9),
(2.26, 116.7), (2.63, 140.7), (4.85, 165.5),
(12.65, 210.7), (7.42, 181.0), (9.35, 213.4)

(a) Plot the data. From the graph, do the data appear to be approximately linear?

(b) Visually find a linear model for the data. Graph the model.

(c) Use the model to approximate y if $x = 3$.

6. *Quiz Scores*   The ordered pairs give the scores of two consecutive 15-point quizzes for a class of 18 students.

(7, 13), (9, 7), (14, 14), (15, 15), (10, 15), (9, 7),
(14, 11), (14, 15), (8, 10), (15, 9), (10, 11) (9, 10),
(11, 14), (7, 14), (11, 10), (14, 11), (10, 15), (9, 6)

(a) Plot the data. From the graph, does the relationship between consecutive scores appear approximately linear?

(b) If the data appear approximately linear, find a linear model for the data. If not, give some possible explanations.

7. *Hooke's Law*   Hooke's Law states that the force F required to compress or stretch a spring (within its elastic limits) is proportional to the distance d that the spring is compressed or stretched from its original length. That is, $F = kd$, where k is a measure of the stiffness of the spring and is called the *spring constant*. The table gives the elongation d in centimeters of a spring when a force of F kilograms is applied.

F	20	40	60	80	100
d	1.4	2.5	4.0	5.3	6.6

(a) Use the regression capabilities of a graphing utility to find a linear model for the data.

(b) Use a graphing utility to plot the data and graph the model. How well does the model fit the data? Explain your reasoning.

(c) Use the model to estimate the elongation of the spring when a force of 55 kilograms is applied.

8. *Falling Object*   In an experiment, students measured the speed s (in meters per second) of a falling object t seconds after it was released. The results are given in the table

t	0	1	2	3	4
s	0	11.0	19.4	29.2	39.4

(a) Use the regression capabilities of a graphing utility to find a linear model for the data.

(b) Use a graphing utility to plot the data and graph the model. How well does the model fit the data? Explain your reasoning.

(c) Use the model to estimate the speed of the object after 2.5 seconds.

9. *Energy Consumption*   The data give the per capita energy usage (in thousands of kilograms of coal equivalent) and the per capita gross national product (in thousands of U.S. dollars) for a sample of countries in 1990. (*Source: Statistical Office of the United Nations*)

Argentina	(1.83, 3.7)	Bangladesh	(0.07, 0.2)
Brazil	(0.77, 2.6)	Canada	(10.51, 21.5)
Denmark	(4.70, 24.0)	Finland	(5.93, 26.0)
France	(3.87, 20.8)	Greece	(3.05, 6.8)
India	(0.31, 0.3)	Italy	(3.86, 19.4)
Japan	(4.21, 26.2)	Mexico	(1.75, 3.0)
Pakistan	(0.28, 0.4)	South Korea	(2.47, 6.0)
Tanzania	(0.04, 0.1)	United States	(10.32, 23.0)

(a) Use the regression capabilities of a graphing utility to find a linear model for the data.

(b) Use a graphing utility to plot the data and graph the model.

(c) Interpret the graph in part (b). Use the graph to identify any countries that appear to differ from the linear model.

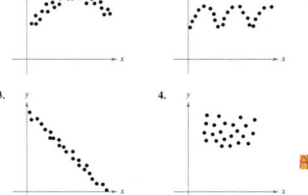

30. Use integration to confirm your results in Exercise 29, where the region is bounded by the graphs of $y = x^{2/3}$, $y = 0$, and $x = 5$.

Think About It In Exercises 31 and 32, determine which value best approximates the volume of the solid generated by revolving the region bounded by the graphs of the equations about the *y*-axis. (Make your selection on the basis of a sketch of the solid and *not* by performing any calculations.)

31. $y = 2e^{-x}$, $y = 0$, $x = 0$, $x = 2$
   (a) $\frac{3}{2}$ (b) -2 (c) 4   (d) 7.5   (e) 15

32. $y = \tan x$, $y = 0$, $x = 0$, $x = \frac{\pi}{4}$
   (a) 3.5   (b) $-\frac{9}{4}$ (c) 8   (d) 10   (e) 1

33. *Machine Part* A solid is generated by revolving the region bounded by $y = \frac{1}{2}x^2$ and $y = 2$ about the *y*-axis. A hole, centered along the axis of revolution, is drilled through this solid so that one-fourth of the volume is removed. Find the diameter of the hole.

34. *Machine Part* A solid is generated by revolving the region bounded by $y = \sqrt{9 - x^2}$ and $y = 0$ about the *y*-axis. A hole, centered along the axis of revolution, is drilled through this solid so that one-third of the volume is removed. Find the diameter of the hole.

35. A hole is cut through the center of a sphere of radius *r*. The height of the remaining spherical ring is *h*, as shown in the figure. Show that the volume of the ring is $V = \pi h^3/6$. (*Note:* The volume is independent of *r*.)

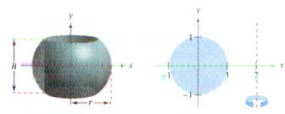

Figure for 35   **Figure for 36**

36. *Volume of a Torus* A torus is formed by revolving the region bounded by the circle $x^2 + y^2 = 1$ about the line $x = 2$, as shown in the figure. Find the volume of this "doughnut-shaped" solid. (*Hint:* The integral $\int_{-1}^{1} \sqrt{1 - x^2}\, dx$ represents the area of a semicircle.)

37. *Volume of a Torus* Repeat Exercise 36 for a torus formed by revolving the region bounded by the circle $x^2 + y^2 = r^2$ about the line $x = R$, where $r < R$.

38. *Volume of a Segment of a Sphere* Let a sphere of radius *r* be cut by a plane, thus forming a segment of height *h*. Show that the volume of this segment is $\frac{1}{3}\pi h^2(3r - h)$.

Think About It In Exercises 39 and 40, give a geometric argument that explains why the integrals have equal values.

39. $\pi \int_{1}^{5} (x - 1)\, dx, \qquad 2\pi \int_{0}^{2} y[5 - (y^2 + 1)]\, dy$

40. $\pi \int_{0}^{2} [16 - (2y)^2]\, dy, \qquad 2\pi \int_{0}^{4} x\left(\frac{x}{2}\right) dx$

41. *Think About It* Match each of the integrals with the solid whose volume it represents, and give the dimensions of each solid.
   (a) Right circular cone   (b) Torus   (c) Sphere
   (d) Right circular cylinder   (e) Ellipsoid

   (i) $2\pi \int_{0}^{h} hx\, dx$

   (ii) $2\pi \int_{0}^{h} hx\left(1 - \frac{x}{r}\right) dx$

   (iii) $2\pi \int_{0}^{r} 2x\sqrt{r^2 - x^2}\, dx$

   (iv) $2\pi \int_{0}^{b} 2ax\sqrt{1 - \frac{x^2}{b^2}}\, dx$

   (v) $2\pi \int_{-r}^{r} (R - x)(2\sqrt{r^2 - x^2})\, dx$

42. *Volume of a Storage Shed* A storage shed has a circular base of diameter 80 feet (see figure). Starting at the center, the interior height is measured every 10 feet and recorded in the table.

x	0	10	20	30	40
Height	50	45	40	20	0

   (a) Use Simpson's Rule to approximate the volume of the building.
   (b) Note that the roof line consists of two line segments. Find the equations of the line segments and use integration to find the volume of the shed.

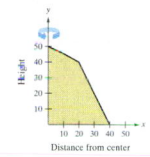

Distance from center

Think About It

These exercises are thought-provoking, conceptual problems that help students grasp the underlying theories.

Applications

A wide variety of relevant application problems are included to demonstrate clearly the real-world usage of mathematics.

Modeling Data

These new multipart questions ask students to find and interpret mathematical models from real-life data. Often the questions are enhanced by the use of a graphing utility.

Section Projects

New to the Sixth Edition, *Section Projects* appear at the end of selected exercise sets. These extended applications can be assigned to individual students or to groups of students in a peer-assisted learning environment.

In Exercises 39–42, use the result of Exercise 37 to find the least squares regression quadratic for the given points. Use the regression capabilities of a graphing utility to confirm your results. Use the graphing utility to plot the points and graph the least squares regression quadratic.

39. $(-2, 0), (-1, 0), (0, 1), (1, 2), (2, 5)$
40. $(-4, 5), (-2, 6), (2, 6), (4, 2)$
41. $(0, 0), (2, 2), (3, 6), (4, 12)$
42. $(0, 10), (1, 9), (2, 6), (3, 0)$

43. *Modeling Data* After a new turbocharger for an automobile engine was developed, the following experimental data were obtained for speed in miles per hour at 2-second intervals.

Time (x)	0	2	4	6	8	10
Speed (y)	0	15	30	50	65	70

   (a) Find a least squares regression quadratic for the data. Use a graphing utility to confirm your results.
   (b) Use a graphing utility to plot the points and graph the model.

44. *Modeling Data* The table gives the world population (in billions) for five different years. (*Source: U.S. Bureau of the Census*)

Year (x)	1960	1970	1980	1990	1996
Population (y)	3.0	3.7	4.5	5.3	5.8

   Let $x = 0$ represent the year 1960.
   (a) Use the regression capabilities of a graphing utility to find the least squares regression line for the data.
   (b) Use the regression capabilities of a graphing utility to find the least squares regression quadratic for the data.
   (c) Use a graphing utility to plot the data and graph the models.
   (d) Use both models to forecast the world population for the year 2010. How do the two models differ as you extrapolate into the future?

45. *Modeling Data* A meteorologist measures the atmospheric pressure *P* (in kilograms per square meter) at altitude *h* (in kilometers). The data are shown below.

h	0	5	10	15	20
P	10,332	5583	2376	1240	517

   (a) Use the regression capabilities of a graphing utility to find a least squares regression line for the points $(h, \ln P)$.
   (b) The result in part (a) is an equation of the form $\ln P = ah + b$. Write this logarithmic form in exponential form.
   (c) Use a graphing utility to plot the original data and graph the exponential model in part (b).

46. *Modeling Data* The endpoints of the interval over which distinct vision is possible are called the *near point* and *far point* of the eye. With increasing age, these points normally change. The table gives the approximate near point *y* in centimeters for various ages *x*.

x	10	20	30	40	50
y	7	10	14	22	40

   (a) Find a rational model for the data by taking the reciprocal of the near points to generate the points $(x, 1/y)$. Use the regression capabilities of a graphing utility to find a least squares regression line for the revised data. The resulting line has the form

   $$\frac{1}{y} = ax + b.$$

   Solve for *y*.
   (b) Use a graphing utility to plot the data and graph the model.
   (c) Do you think the model can be used to predict the near point for a person who is 60 years old? Explain.

Building a Pipeline An oil company wishes to construct a pipeline from its offshore facility *A* to its refinery *B*. The offshore facility is 2 miles from shore, and the refinery is 1 mile inland. Furthermore, *A* and *B* are 5 miles apart, as indicated in the figure.

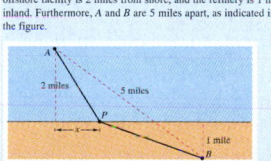

   The cost of building the pipeline is $3 million per mile in the water, and $4 million per mile on land. Hence, the cost of the pipeline depends on the location of point *P*, where it meets the shore. What would be the most economical route of the pipeline?
   Imagine that you are to write a report to the oil company about this problem. Let *x* be the distance indicated in the figure. Determine the cost of building the pipeline from *A* to *P*, and the cost from *P* to *B*. Analyze some sample pipeline routes and their corresponding costs. For instance, what is the cost of the most direct route? Then use calculus to determine the route of the pipeline that minimizes the cost. Explain all steps of your development and include any relevant graphs.

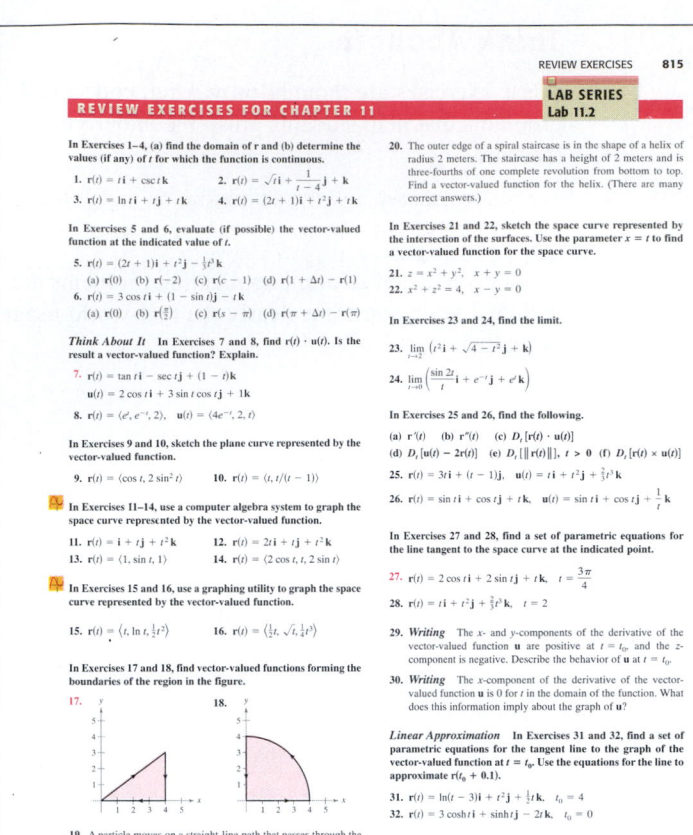

Writing

To develop students' reasoning skills and make them comfortable discussing mathematical concepts, the text now contains many *Writing* exercises.

Graphing Utilities

Many exercises in the text can be solved using technology; however, the symbol identifies all exercises in which students are specifically instructed to use a graphing utility or a computer algebra system.

True or False?

To help students understand the logical structure of calculus, a set of "true or false" questions is included toward the ends of many exercise sets. These questions help students focus on concepts, common errors, and the correct statements of definitions and theorems.

Journal References

References to articles in readily available journals help students understand that calculus is a current, dynamic field.

CHAPTER 3   Applications of Differentiation

48. *Inventory Cost* A retailer has determined that the cost C of ordering and storing x units of a certain product is

$$C = 2x + \frac{300{,}000}{x}, \qquad 1 \le x \le 300.$$

The delivery truck can bring at most 300 units per order. Find the order size that will minimize cost. Could the cost be decreased if the truck were replaced with one that could bring at most 400 units? Explain.

49. *Lawn Sprinkler* A lawn sprinkler is constructed in such a way that $d\theta/dt$ is constant, where θ ranges between 45° and 135° (see figure). The distance the water travels horizontally is

$$x = \frac{v^2 \sin 2\theta}{32}, \qquad \frac{\pi}{4} \le \theta \le \frac{3\pi}{4}$$

where v is the speed of the water. Find dx/dt and explain why this lawn sprinkler does not water evenly. What part of the lawn receives the most water?

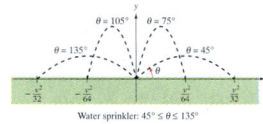

Water sprinkler: 45° ≤ θ ≤ 135°

FOR FURTHER INFORMATION For more information on the "calculus of lawn sprinklers," see the article "Design of an Oscillating Sprinkler" by Bart Braden in the January 1985 issue of *Mathematics Magazine.*

50. *Modeling Data* The defense outlays as percents of the gross domestic product for the years 1976 through 1995 are as follows. (*Source: U.S. Office of Management and Budget*)

1976: (5.3%); 1977: (5.1%); 1978: (4.8%); 1979: (4.8%); 1980: (5.1%); 1981: (5.3%); 1982: (5.9%); 1983: (6.3%); 1984: (6.2%); 1985: (6.4%); 1986: (6.5%); 1987: (6.3%); 1988: (6.0%); 1989: (5.9%); 1990: (5.5%); 1991: (4.8%); 1992: (5.0%); 1993: (4.7%); 1994: (4.2%); 1995: (3.9%).

(a) Use the regression capabilities of a graphing utility to find a model of the form $y = at^4 + bt^3 + ct^2 + dt + e$ for the data. (Let t represent the time in years, with $t = 0$ corresponding to 1980.)

(b) Use a graphing utility to plot the data and graph the model.

(c) Locate the absolute extrema of the model on the interval $[-4, 15]$.

51. *Honeycomb* The surface area of a cell in a honeycomb is

$$S = 6hs + \frac{3s^2}{2}\left(\frac{\sqrt{3} - \cos\theta}{\sin\theta}\right)$$

where h and s are positive constants and θ is the angle at which the upper faces meet the altitude of the cell. Find the angle θ ($\pi/6 \le \theta \le \pi/2$) that minimizes the surface area S.

FOR FURTHER INFORMATION For more information on the geometric structure of a honeycomb cell, see the article "The Design of Honeycombs" by Anthony L. Peressini in UMAP Module 502, published by COMAP, Inc., Suite 210, 57 Bedford Street, Lexington, MA.

52. *Highway Design* In order to build a highway it is necessary to fill a section of a valley where the grades (slopes) of the sides are 6% and 9% (see figure). The top of the filled region will have the shape of a parabolic arc that is tangent to the two slopes at the points A and B. The horizontal distance between the points A and B is 1000 feet.

(a) Find a quadratic function $y = ax^2 + bx + c$, $-500 \le x \le 500$, that describes the top of the filled region.

(b) Complete the table giving the depths d of the fill at the specified values of x.

x	-500	-400	-300	-200	-100
d					

x	0	100	200	300	400	500
d						

(c) What will be the lowest point on the completed highway? Will it be directly over the point where the two hillsides come together?

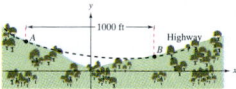

True or False? In Exercises 53–56, determine whether the statement is true or false. If it is false, explain why or give an example that shows it is false.

53. The maximum of a function that is continuous on a closed interval can occur at two different values in the interval.

54. If a function is continuous on a closed interval, then it must have a minimum on the interval.

55. If $x = c$ is a critical number of the function f, then it is also a critical number of the function $g(x) = f(x) + k$, where k is a constant.

56. If $x = c$ is a critical number of the function f, then it is also a critical number of the function $g(x) = f(x - k)$, where k is a constant.

57. Find all critical numbers of the greatest integer function $f(x) = [\![x]\!]$.

Supplements

Calculus with Analytic Geometry, Sixth Edition, by Larson, Hostetler, and Edwards, is accompanied by a comprehensive supplements package with ancillaries for students, for instructors, and for classroom resource. Most items are keyed directly to the book.

Printed Resources

For the Instructor

Instructor's Resource Guide by Ann R. Kraus, The Pennsylvania State University, The Behrend College

- Notes to the instructor
- Chapter summaries
- Ready-made chapter tests and finals
- Gateway tests
- Teaching strategies
- Sample syllabi
- Suggested solutions for *Exploration, Technology,* and *Motivating the Chapter* features
- Solutions to *Section Projects*

Complete Solutions Guide, Volumes I, II, and III, by Bruce H. Edwards, University of Florida

- Detailed solutions to all text exercises

Test Item File by Ann R. Kraus, The Pennsylvania State University, The Behrend College

- Printed test bank
- Approximately 3000 test items
- Multiple-choice and open-ended questions coded by level of difficulty
- Technology required to solve some test questions
- Also available as test-generating software

Graphing Calculator Demonstration Problems by August J. Zarcone and Russell Lundstrom, College of DuPage

- Classroom demonstration problems using the *TI-85* graphing calculator

Transparency Package

- 120 color transparencies of figures from the text

For the Student

Study and Solutions Guide, Volumes I and II, by David E. Heyd, The Pennsylvania State University, The Behrend College

- Detailed solutions to selected odd-numbered text exercises, which are indicated by a red exercise number in the text (blue exercise number in the appendix)
- Algebra review
- Study strategies

Lab Manual Series

- Available in five versions—*Maple, Mathematica, Derive, Mathcad,* and the *TI-92* graphing calculator
- See pages xxiv–xxv for more details.

Graphing Technology Guide by Benjamin N. Levy and Laurel Technical Services

- Keystroke instructions for a wide variety of Texas Instruments, Casio, Sharp, and Hewlett-Packard graphing calculators, including the most current models
- Examples with step-by-step solutions
- Extensive graphics screen output
- Technology tips

The Algebra of Calculus by Eric J. Braude

- Review of algebra, trigonometry, and analytic geometry required for calculus
- Over 200 examples with solutions
- Pretests and exercises

Calculus Applications in Engineering and Science by Stuart Goldenberg and Harvey Greenwald, California Polytechnic State University—San Luis Obispo

- Extended applications

Computer Projects in Calculus by Bruce H. Edwards, University of Florida

- Guided exploration labs
- Brief introductions to *Mathematica, Maple,* and *Derive*

Insights into Calculus with the Graphics Calculator by Herbert A. Hollister, Bowling Green State University

- Examples with step-by-step solutions
- Exercises with answers to odd-numbered exercises
- Graphics screen displays
- Graphing calculator programs

Media Resources

For the Instructor

3-D Demonstration Software

- Rotatable three-dimensional figures from the text
- Available on disk for demonstration in classroom

Computerized Testing

- Test-generating software
- Approximately 3000 test items
- Also available as a printed test bank

For the Student

Interactive Calculus, Version 2.0

- Internet accessible
- Multimedia, CD-ROM format
- Includes all of the content of the Sixth Edition
- Active mathematics, including editable two-dimensional graphs and rotatable three-dimensional graphs
- New explorations and simulations
- See pages xxvi–xxvii for more details.

Topics in Calculus Videotapes   by Dana Mosely

- Comprehensive coverage of key topics
- Explanation of concepts, sample problems, and applications

Graphing Calculator Instructional Videotape   by Dana Mosely

- Introduces students to a wide variety of Texas Instruments, Casio, Sharp, and Hewlett-Packard graphing calculators, including the most current models
- Covers basic capabilities shared by many graphing utilities

BestGrapher   by George Best

- Function grapher
- Simultaneous display of equation, graph, and table of values
- Zoom and print features

Workout!   by Joe Mazur

- Self-study tutorial software
- On-line hints and guidance
- Generates problems at four levels of difficulty

Note: Visit the Houghton Mifflin home page at http://www.hmco.com

SUPPLEMENTS

Lab Manual Series

Using Technology to Apply Calculus

Each *Motivating the Chapter* introduces the student to an application that explores the concepts to be covered in the chapter. The real-world settings and meaningful questions spark students' interest. These applications are extended in technology lab projects in the supplementary lab series. The labs are available in five versions—*Maple, Mathematica, Derive, Mathcad,* and the *TI-92* graphing calculator.

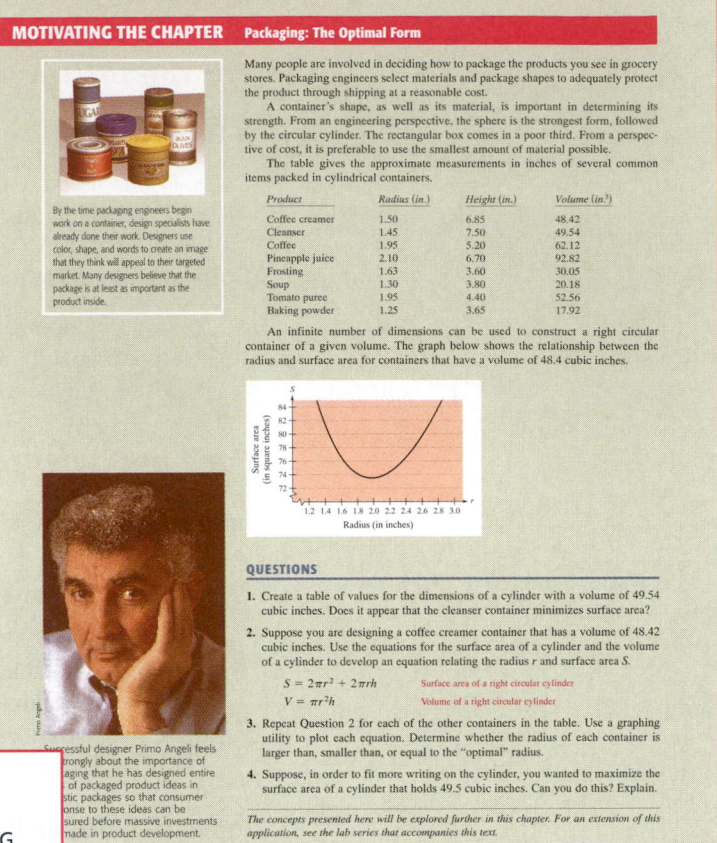

MOTIVATING THE CHAPTER   Packaging: The Optimal Form

By the time packaging engineers begin work on a container, design specialists have already done their work. Designers use color, shape, and words to create an image that they think will appeal to their targeted market. Many designers believe that the package is at least as important as the product inside.

Many people are involved in deciding how to package the products you see in grocery stores. Packaging engineers select materials and package shapes to adequately protect the product through shipping at a reasonable cost.

A container's shape, as well as its material, is important in determining its strength. From an engineering perspective, the sphere is the strongest form, followed by the circular cylinder. The rectangular box comes in a poor third. From a perspective of cost, it is preferable to use the smallest amount of material possible.

The table gives the approximate measurements in inches of several common items packed in cylindrical containers.

Product	Radius (in.)	Height (in.)	Volume (in.³)
Coffee creamer	1.50	6.85	48.42
Cleanser	1.45	7.50	49.54
Coffee	1.95	5.20	62.12
Pineapple juice	2.10	6.70	92.82
Frosting	1.63	3.60	30.05
Soup	1.30	3.80	20.18
Tomato puree	1.95	4.40	52.56
Baking powder	1.25	3.65	17.92

An infinite number of dimensions can be used to construct a right circular container of a given volume. The graph below shows the relationship between the radius and surface area for containers that have a volume of 48.4 cubic inches.

Successful designer Primo Angeli feels strongly about the importance of packaging that he has designed entire [...] of packaged product ideas in [...]stic packages so that consumer [...]onse to these ideas can be [...]sured before massive investments [...] made in product development.

QUESTIONS

1. Create a table of values for the dimensions of a cylinder with a volume of 49.54 cubic inches. Does it appear that the cleanser container minimizes surface area?

2. Suppose you are designing a coffee creamer container that has a volume of 48.42 cubic inches. Use the equations for the surface area of a cylinder and the volume of a cylinder to develop an equation relating the radius r and surface area S.

$$S = 2\pi r^2 + 2\pi rh \qquad \text{Surface area of a right circular cylinder}$$
$$V = \pi r^2 h \qquad \text{Volume of a right circular cylinder}$$

3. Repeat Question 2 for each of the other containers in the table. Use a graphing utility to plot each equation. Determine whether the radius of each container is larger than, smaller than, or equal to the "optimal" radius.

4. Suppose, in order to fit more writing on the cylinder, you wanted to maximize the surface area of a cylinder that holds 49.5 cubic inches. Can you do this? Explain.

The concepts presented here will be explored further in this chapter. For an extension of this application, see the lab series that accompanies this text.

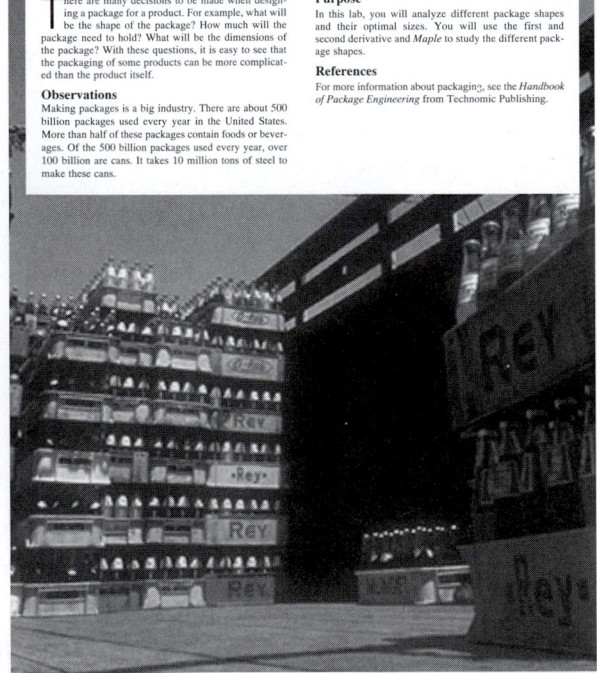

LAB 3.1

PACKAGING

Optimization

There are many decisions to be made when designing a package for a product. For example, what will be the shape of the package? How much will the package need to hold? What will be the dimensions of the package? With these questions, it is easy to see that the packaging of some products can be more complicated than the product itself.

Observations

Making packages is a big industry. There are about 500 billion packages used every year in the United States. More than half of these packages contain foods or beverages. Of the 500 billion packages used every year, over 100 billion are cans. It takes 10 million tons of steel to make these cans.

Purpose

In this lab, you will analyze different package shapes and their optimal sizes. You will use the first and second derivative and *Maple* to study the different package shapes.

References

For more information about packaging, see the *Handbook of Package Engineering* from Technomic Publishing.

LAB 3.1 PACKAGING 25

Each lab begins with background information and observations, as well as a statement of the purpose of the lab, in order to motivate students' interest and place the application in context. Data, graphs, and equations are available on the accompanying data disk for ease of use enabling students to focus on problem solving and modeling data.

A variety of open-ended exercises encourages students to deepen their understanding of calculus. Challenging, analytical, thought-provoking, comparative, and exploratory questions are included.

Through the use of technology, the labs expand on concepts discussed in class and challenge students' knowledge above and beyond what classtime allows. The labs give students the opportunity to appreciate the applicability of calculus and practice their problem-solving and computer algebra skills—important skills for students to carry into the future.

Each lab manual contains at least one lab per chapter, corresponding to the *Motivating the Chapter* opener. The labs are referenced in the text where it seems most appropriate to assign them. However, they are flexible enough to be assigned at the instructor's discretion.

---

6. A Different Container Shape. Design a rectangular container with a square base for each of the following volumes and heights. The volume of a rectangular container with a square base is

$$V = x^2h$$

and the surface area is

$$S = x^2 + 4xh.$$

Volume (in.³)	Height (in.)	Side of Base (in.)	Surface Area (in.²)
17.92	3.65		
49.54	7.50		
62.12	5.20		
48.42	6.85		
30.05	3.60		
92.82	6.70		
20.18	3.80		
52.56	4.40		

Compare the rectangular containers with square bases to the cylinders that have the same volume in Exercise 1. What advantages do the rectangular containers have over the cylinders? What disadvantages do they have?

LAB 3.1 PACKAGING   29

LAB MANUALS

Interactive Calculus, Version 2.0

To accommodate a variety of teaching and learning styles, *Calculus* is also available on the internet or in a multimedia, CD-ROM format. *Interactive Calculus,* Version 2.0, is networkable and cross-platform, and incorporates live mathematics throughout the entire program. Live mathematics help students visualize and explore—leading to a deeper understanding of calculus concepts than has ever before been possible.

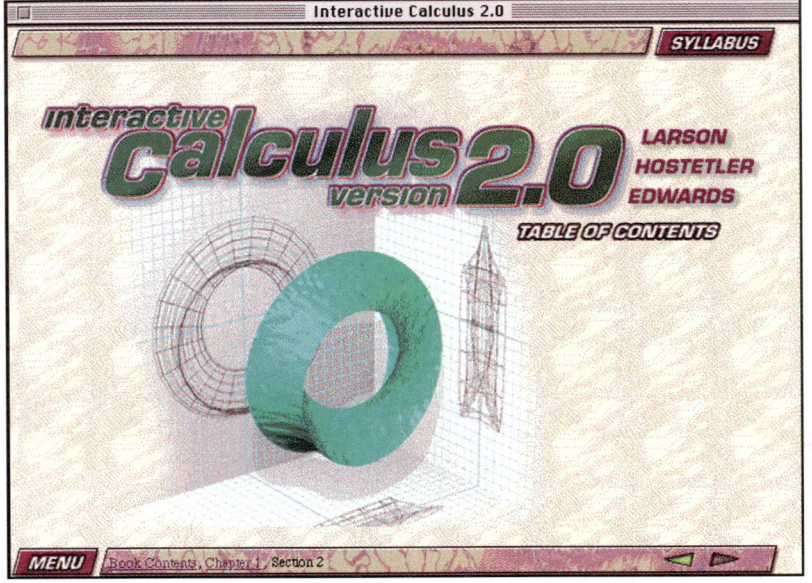

Live mathematics throughout

- Open explorations give students the opportunity to explore using computer algebra systems.
- Editable two-dimensional graphs, featured in examples throughout the entire program, provide additional opportunities to explore and investigate.
- Rotatable three-dimensional graphs allow for a whole new level of visualization.
- New and enhanced explorations, simulations, and animations make concepts come alive.

Syllabus management tool

All of the content of the Sixth Edition text—a wealth of applications, exercises, worked examples, and detailed explanations—is included in *Interactive Calculus,* Version 2.0, as a database. Instructors have the flexibility of customizing content and interactive features for students as desired. Instructors may simply add dates to a default syllabus or may modify the order of topics. Either way, a customized syllabus is easy to distribute electronically and update instantly. This tool is particularly useful for managing distance learning courses.

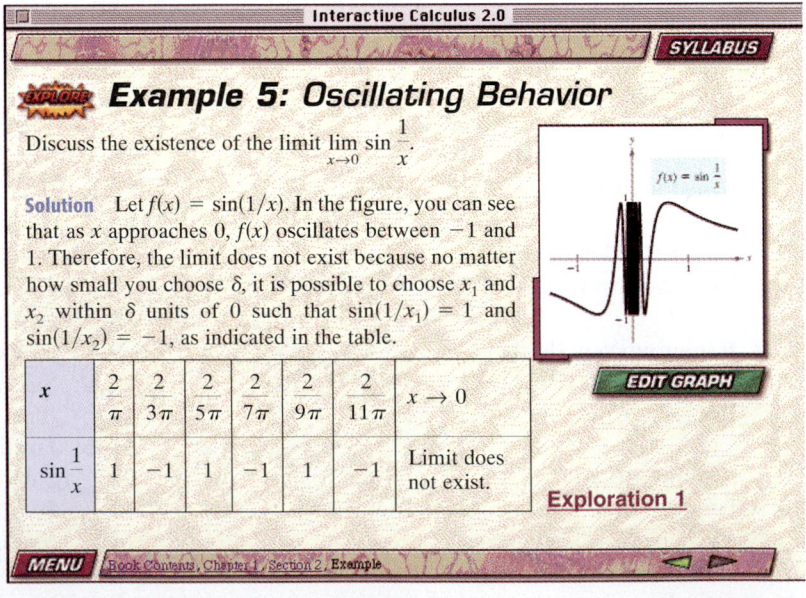

Other new features include:

- Enhanced three-dimensional art for clarity, accuracy, and realism
- Searchable text
- Theorem index
- Features index
- Integrated on-line help
- Bookmarks

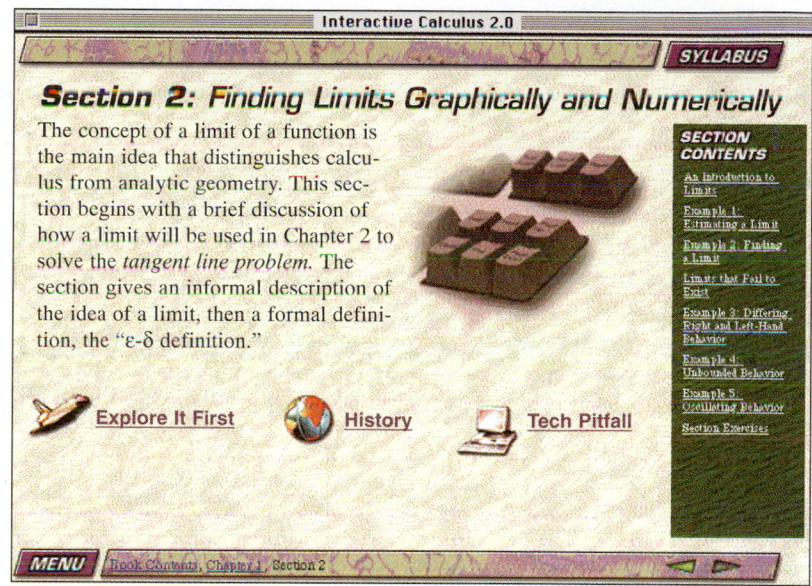

New design with intuitive navigation

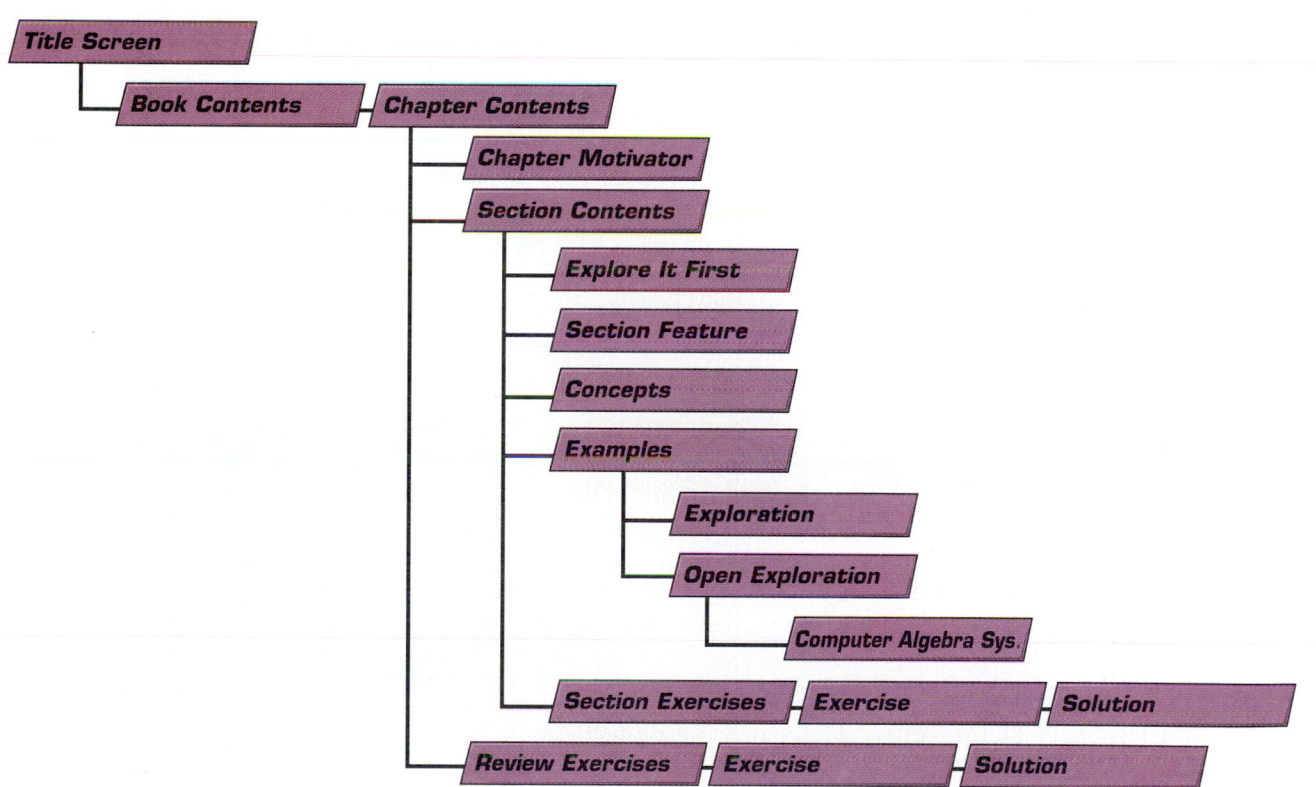

About the Cover

Helaman Ferguson's *Umbilic Torus NC* was shown on the cover of the
Fifth Edition of this text. This commanding bronze sculpture stands
about 27 inches high and has a single edge. Tracing this edge carries
you three times around the ring before returning you to the starting
point. A cross section of the surface corresponds to a hypocycloid.
Compelling and mysterious, the torus epitomizes the elegance that
mathematicians find when using calculus to describe the world.

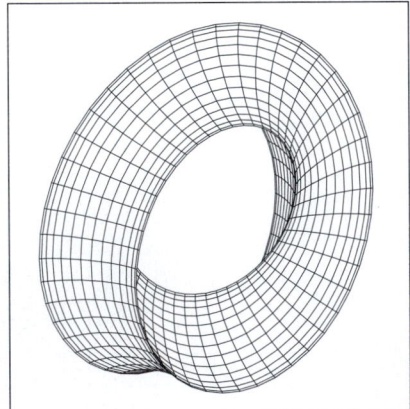

For the cover of the Sixth Edition, we decided to
use the torus again. Instead of a photo, however, we
took a step into virtual reality and created the torus
as a computer image. We started the same way
Ferguson did—we used a set of parametric equations to create a wire-frame image of the torus. To
recreate this image on your own computer, try
using the equations for a parametric surface on
page 984.

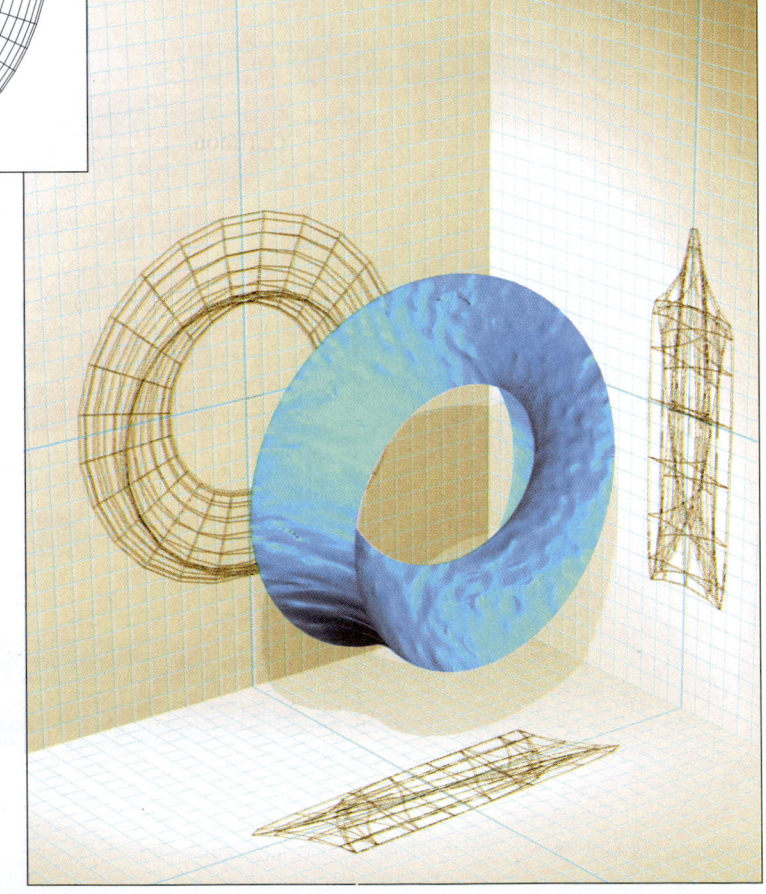

Next, we imported the wire-frame image into a
three-dimensional rendering program called
SoftImage. In that program, the torus was given a
skin, texture, and color. Then the torus was placed
in the first octant of a three-dimensional coordinate
system, as shown at the right. Finally, we projected
the three standard engineering graphics views of the
wire-frame image onto the three coordinate planes.

Acknowledgements

We would like to thank the many people who have helped us at various stages of this project during the past 21 years. Their encouragement, criticisms, and suggestions have been invaluable to us.

Sixth Edition Consultants

Kerry D. Bailey
Laramie County Community College

Michael Montaño
Riverside Community College

Mark A. Manchester
SUNY College of Agriculture and
Technology at Morrisville

Carol Urban
College of DuPage

Sixth Edition Reviewers

Andrew A. Bulleri
Howard Community College

Ruth A. Hartman
Black Hawk College

Donna E. Nordstrom
Pasadena City College

Paula Castagna
Fresno City College

Irvin Roy Hentzel
Iowa State University

Darrell J. Peterson
Santa Monica College

Charles L. Cope
Morehouse College

Howard E. Holcomb
Monroe Community College

John E. Santomas
Villanova University

Jack Courtney
Michigan State University

Robert Kowalczyk
University of Massachusetts—
Dartmouth

Mark Stevenson
Oakland Community College

James Daniels
Palomar College

Maita Levine
University of Cincinnati

Anthony D. Thomas
University of Wisconsin—Platteville

Luz M. DeAlba
Drake University

Darrell Minor
Columbus State Community College

Robert J. Vojack
Ridgewood High School, Ridgewood,
NJ

Dewey Furness
Ricks College

David C. Morency
University of Vermont

John R. Watret
Embry-Riddle Aeronautical University

Javier Garza
Tarleton State University

Gerald Mueller
Columbus State Community College

Previous Editions Reviewers

Dennis Alber, Palm Beach Junior College; James Angelos, Central Michigan University; Harry L. Baldwin, Jr., San Diego City College; Homer F. Bechtell, University of New Hampshire; Keith Bergeron, United States Air Force Academy; Norman Biernes, University of Regina; Brian Blank, Washington University; Jack Ceder, University of California—Santa Barbara; Jorge Cossio, Miami-Dade Community College; Kathy Davis, University of Texas; Paul W. Davis, Worcester Polytechnic Institute; Rosario Diprizio, Oakton Community College; Garret J. Etgen, University of Houston; Russell Euler, Northwest Missouri State University; Phillip A. Ferguson, Fresno City College; Li Fong, Johnson County Community College; Michael Frantz, University of La Verne; William R. Fuller, Purdue University; K. Elayn Gay, University of New Orleans; Thomas M. Green, Contra Costa College; Ali

Hajjafar, University of Akron; Eric R. Immel, Georgia Institute of Technology; Arnold J. Insel, Illinois State University; Elgin Johnston, Iowa State University; Hideaki Kaneko, Old Dominion University; Toni Kasper, Borough of Manhattan Community College; William J. Keane, Boston College; Timothy J. Kearns, Boston College; Ronnie Khuri, University of Florida; Frank T. Kocher, Jr., Pennsylvania State University; Joseph F. Krebs, Boston College; David C. Lantz, Colgate University; Norbert Lerner, State University of New York at Cortland; Murray Lieb, New Jersey Institute of Technology; Ransom Van B. Lynch, Phillips Exeter Academy; Bennet Manvel, Colorado State University; Mauricio Marroquin, Los Angeles Valley College; Robert L. Maynard, Tidewater Community College; Robert McMaster, John Abbott College; Maurice L. Monahan, South Dakota State University; Philip Montgomery, University of Kansas; Duff A. Muir, United States Air Force Academy; Charlotte J. Newsom, Tidewater Community College; Terry J. Newton, United States Air Force Academy; Robert A. Nowlan, Southern Connecticut State University; Luis Ortiz-Franco, Chapman University; Barbara L. Osofsky, Rutgers University; Judith A. Palagallo, University of Akron; Wayne J. Peeples, University of Texas; Jorge A. Perez, LaGuardia Community College; Donald Poulson, Mesa Community College; Jean L. Rubin, Purdue University; Barry J. Sarnacki, United States Air Force Academy; N. James Schoonmaker, University of Vermont; George W. Schultz, St. Petersburg Junior College; Richard E. Shermoen, Washburn University; Thomas W. Shilgalis, Illinois State University; J. Philip Smith, Southern Connecticut State University; Frank Soler, De Anza College; Enid Steinbart, University of New Orleans; Michael Steuer, Nassau Community College; Lawrence A. Trivieri, Mohawk Valley Community College; John Tweed, Old Dominion University; Marjorie Valentine, North Side ISD, San Antonio; Bert K. Waits, Ohio State University; Florence A. Warfel, University of Pittsburgh; Carroll G. Wells, Western Kentucky University; Jay Wiestling, Palomar College; Paul D. Zahn, Borough of Manhattan Community College; August J. Zarcone, College of DuPage.

We would like to express our special thanks to Nicolae Dinculeanu, University of Florida, for several suggestions that we have incorporated into the Sixth Edition.

A special thanks to all the people at Houghton Mifflin Company who worked with us is the development of the Sixth Edition, especially Chris Hoag, Cathy Cantin, Maureen Brooks, Michael Richards, Carolyn Johnson, Karen Carter, Rachel Wimberly, Carrie Lipscomb, Gary Crespo, Henry Rachlin, Sara Whittern, and Ros Kane.

We would also like to thank the staff at Larson Texts, Inc., who assisted with proofreading the manuscript; preparing and proofreading the art package; and checking and typesetting the supplements.

A special note of thanks goes to the over 1,500,000 students who have used earlier editions of the text.

On a personal level, we are grateful to our wives, Deanna Gilbert Larson, Eloise Hostetler, and Consuelo Edwards, for their love, patience, and support. Also, a special thanks goes to R. Scott O'Neil.

If you have suggestions for improving this text, please feel free to write to us. Over the past two decades we have received many useful comments from both instructors and students, and we value these very much.

Roland E. Larson
Robert P. Hostetler
Bruce H. Edwards

Index of Applications

Engineering and the Physical Sciences

Absolute zero, 71
Acceleration, 123, 250, 251, 306
Acceleration due to gravity, 32, 120
Acceleration on the moon, 123
Acid rain, 830
Adiabatic expansion, 148
Aircraft control, 802
Aircraft glide rate, 185
Aircraft separation, 251
Airplane ascent, A26
Air pressure, 404
Air traffic control, 670
Altitude of a plane, 149
Anamorphic art, 680
Angle of elevation, 148, 149
Angle of elevation of a camera, 144
Angular rate of change, 384
Angular speed, A25
Annual snowfall, 932
Annual temperature range, 830
Antenna radiation, 687
Apparent temperature, 848
Architecture, 648
Area, 30, 38, 113, 146, 210, 211, 212
Area of end of a log, 226
Area of a lot, 263, 305
Area of a spiral ramp, 1037
Asteroid Apollo, 693
Atmospheric pressure, 320, 346, 364, 901
Auditorium lights, 719
Autocatalytic chemical reaction, 210
Automobile aerodynamics, 29
Automobile engine, 901
Average displacement, 488
Average field strength, 507
Average speed, 85
Average velocity of a falling object, 108
Bacterial culture growth, 364
Barge towing, 711
Beam deflection, 186, 647
Beam strength, 35, 212
Bessel function, 614
Boiling temperature, 36
Bouncing ball, 563, 633
Boyle's Law, 122, 453
Braking load, 728

Bridge design, 648
Brinell hardness, 35
Building construction, 719
Building design, 415, 522, 956, 1008, A16
Building a pipeline, 901
Bulb design, 443
Buoyant force, 469
Camera surveillance design, 149
Capillary action, 956
Carbon dating, 363
Car performance, 35, 36
Cavalieri's Theorem, 426
Center of mass of a conversion van window, 463
Center of mass of a section of a boat's hull, 464
Center of pressure on a sail, 949
Centripetal acceleration, 802, 855
Centripetal force, 802, 816
Centroid, 472
Changing shadow length, 148
Charles's Law, 71
Chemical mixtures, 1080, 1084
Chemical reaction, 376, 401, 516, 912
Chemical release from a storage tank, 347
Circular motion, 792, 793, 801
Climb rate for an airplane, 403
Comparing fluid forces, 504
Constant flow rate, 29
Constructing an arch dam, 406
Construction cost, 829
Construction of the Gateway Arch in St. Louis, 402
Construction of a semielliptical arch, 649
Conveyer design, 17
Cornu spiral, 814
Cross section of a canal, 214
Curtate cycloid, 660
Cycloidal motion, 792, 801
Daily temperature, 36, 133
Deceleration, 250
Depth, 146, 147, 152
Dimensions of a box, 853, 909
Distance between cities, 764
Distance between planes, 147
Distance traveled, 285, 565
Doppler effect, 132
Drag force, 912
Earthquake intensity, 365
Electrical charge, 1049

APPLICATIONS

APPLICATIONS

Business and Economics

The Social and Behavioral Sciences

The Life Sciences

APPLICATIONS

General

Calculus Options

To accommodate the various methods of teaching calculus, Houghton Mifflin offers the programs described below. All texts and interactive versions are written by Larson, Hostetler, and Edwards.

Calculus with Analytic Geometry, Sixth Edition

This core text covers the entire three-semester course in 16 chapters.

Calculus of a Single Variable, Sixth Edition

This single-variable text is designed for a two-semester course in calculus, presented in ten chapters.

Multivariable Calculus, Sixth Edition

This multivariable text is designed for a third-semester course in calculus, presented in six chapters.

Calculus with Analytic Geometry, Alternate Sixth Edition

This text, with trigonometry introduced in Chapter 8, covers the entire three-semester course in 18 chapters. It also offers alternative treatments of the following topics: limits, applications of integration, exponential and logarithmic functions, and vectors.

Calculus: Early Transcendental Functions

This three-semester text integrates coverage of transcendental functions from the beginning of the text, providing a variety of interesting functions and applications beginning early in the course.

Calculus of a Single Variable: Early Transcendental Functions

This single-variable text is designed for a two-semester course in calculus, presented in ten chapters, with coverage of transcendental functions integrated from the beginning of the text.

Interactive Calculus, Version 2.0

This multimedia, interactive version of *Calculus with Analytic Geometry,* Sixth Edition, contains all of the content of the text itself, and provides an opportunity to use editable graphs, active mathematics, and many types of exploratory activities.

Interactive Calculus: Early Transcendental Functions

This multimedia, interactive version of *Calculus: Early Transcendental Functions* contains all of the content of the text itself, and provides an opportunity to use editable graphs, active mathematics, and many types of exploratory activities.

Preparation for Calculus

P

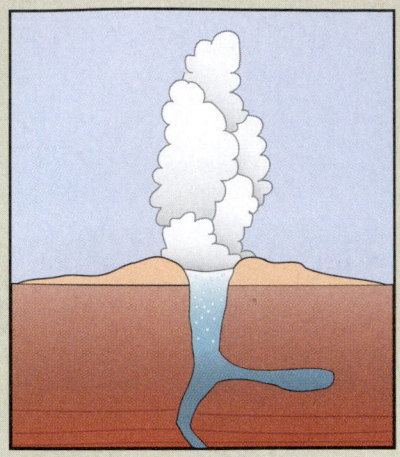

For a geyser to exist, three conditions must exist: a supply of water, an intense heat source, and pressure-tight plumbing.

Volcanically heated water enters a geyser's plumbing system at great depths, boiling and heating cooler water that is flowing in from the surface. Eventually the steam from the boiling water creates enough pressure to explosively force great volumes of water high into the air. Once an eruption is over, the entire filling, heating, and boiling process begins again.

Yellowstone National Park, located in the northwest corner of Wyoming and adjacent parts of Montana and Idaho, contains over half of all known geysers on earth. Few of these hydrothermal spectacles are regular enough to be anticipated accurately. Of the more than 400 geysers in Yellowstone National Park, predictions are posted for only seven, one of which is Old Faithful.

Old Faithful geyser was named in 1870. In 1938 geologist Harry M. Woodward noticed that there was a correlation between the duration of Old Faithful's eruption and the length of time (interval) before the next eruption.

More than 137,000 eruptions of Old Faithful have been observed and recorded. The durations have varied between 1.5 and 5.5 minutes, and the intervals have varied between 30 and 120 minutes. Since Woodward's observations, the intervals have tended to increase. It is speculated that major earthquakes in the region have shifted the circulation of hot water away from the geyser, resulting in a longer "fill time."

The data below show 35 eruptions as ordered pairs of the form (x, y), where x is the duration in minutes and y is the interval in minutes. *(Source: Yellowstone National Park)*

(1.80, 56),   (1.82, 58),   (1.88, 60),   (1.90, 62),   (1.92, 60),
(1.93, 56),   (1.98, 59),   (2.03, 60),   (2.05, 57),   (2.13, 60),
(2.30, 57),   (2.35, 57),   (2.37, 61),   (2.82, 73),   (3.13, 76),
(3.27, 77),   (3.65, 77),   (3.70, 82),   (3.78, 79),   (3.83, 85),
(3.87, 81),   (3.88, 80),   (4.10, 89),   (4.27, 90),   (4.30, 84),
(4.30, 89),   (4.43, 84),   (4.43, 89),   (4.47, 80),   (4.47, 86),
(4.53, 89),   (4.55, 86),   (4.60, 88),   (4.60, 92),   (4.63, 91)

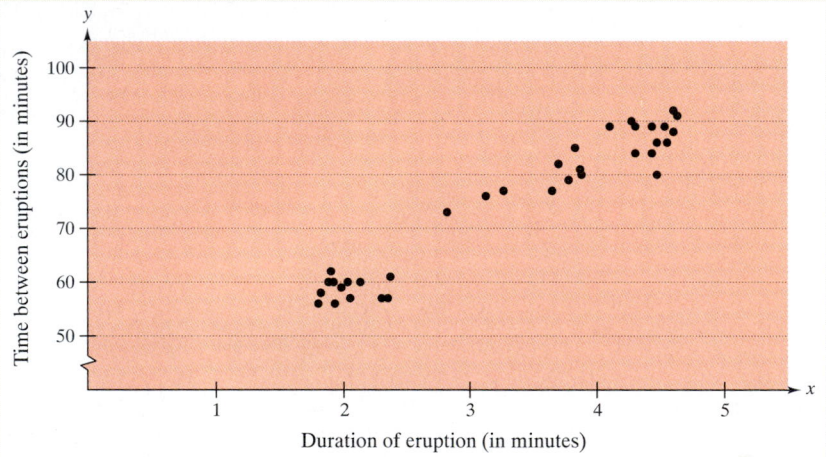

Harry M. Woodward was the first to describe a relationship between the durations and intervals of Old Faithful's eruptions.

QUESTIONS

1. Describe the relationship between the durations and the intervals of Old Faithful's eruptions. Is there a correlation between the two measures?

2. Suppose that you observe an eruption that lasts for 2 minutes and 40 seconds. When would you expect the next eruption?

3. Write a model (an equation involving x and y) that can be used to predict the length of time until the next eruption. Explain how you obtained the model. Did you use technology or did you obtain the model using only hand calculations?

The concepts presented here will be explored further in this chapter. For an extension of this application, see the lab series that accompanies this text.

SECTION *P.1*   **Graphs and Models**

RENÉ DESCARTES (1596–1650)

Descartes made many contributions to philosophy, science, and mathematics. The idea of representing points in the plane by pairs of real numbers and representing curves in the plane by equations was described by Descartes in his book *La Géométrie*, published in 1637.

The Graph of an Equation • Intercepts of a Graph • Symmetry of a Graph • Points of Intersection • Mathematical Models

The Graph of an Equation

In 1637 the French mathematician René Descartes revolutionized the study of mathematics by joining its two major fields—algebra and geometry. With Descartes's coordinate plane, geometric concepts could be formulated analytically and algebraic concepts could be viewed graphically. Such is the power of this approach that within a century, much of calculus had been developed.

You can enhance your chance of success in calculus by following the same approach. That is, by viewing calculus from multiple perspectives—*graphically*, *analytically*, and *numerically*, you will increase your understanding of core concepts.

Consider the equation $3x + y = 7$. The point $(2, 1)$ is a **solution point** of the equation because the equation is satisfied (is true) when 2 is substituted for x and 1 is substituted for y. This equation has many other solutions, such as $(1, 4)$ and $(0, 7)$. To systematically find other solutions, solve the original equation for y.

$$y = 7 - 3x \qquad \text{Analytic approach}$$

Then construct a **table of values** by substituting several values of x.

x	0	1	2	3	4
y	7	4	1	-2	-5

Numerical approach

From the table, you can see that $(0, 7)$, $(1, 4)$, $(2, 1)$, $(3, -2)$, and $(4, -5)$ are solutions of the original equation

$$3x + y = 7. \qquad \text{Original equation}$$

Like many equations, this equation has an infinite number of solutions. The set of all solution points is the **graph** of the equation, as shown in Figure P.1.

NOTE   Even though we refer to the sketch shown in Figure P.1 as the graph of $3x + y = 7$, it really represents only a *portion* of the graph. The entire graph would extend beyond the page.

In this course, you will study many sketching aids. The simplest is point plotting—that is, you plot points until the basic shape of the graph seems apparent.

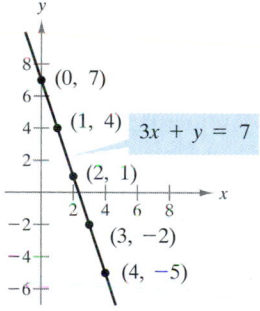

Graphical approach: $3x + y = 7$
Figure P.1

EXAMPLE 1   Sketching a Graph by Point Plotting

Sketch the graph of $y = x^2 - 2$.

Solution   First construct a table of values. Then plot the points given in the table.

x	-2	-1	0	1	2	3
y	2	-1	-2	-1	2	7

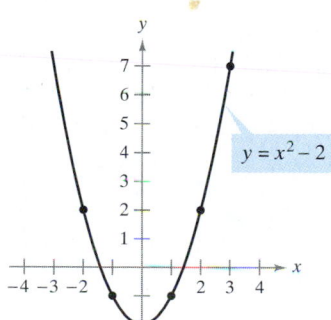

The parabola $y = x^2 - 2$
Figure P.2

Finally, connect the points with a *smooth curve*, as shown in Figure P.2. This graph is a **parabola.** It is one of the conics you will study in Chapter 9.

NOTE   For more information on how to use a graphing utility, refer to the Graphing Technology Guide and the Graphing Calculator Training Videotape that accompany this text.

One disadvantage of point plotting is that to get a good idea about the shape of a graph, you may need to plot many points. With only a few points, you could badly misrepresent the graph. For instance, suppose that to sketch the graph of

$$y = \tfrac{1}{30}x(39 - 10x^2 + x^4)$$

you plotted only five points:

$$(-3, -3), (-1, -1), (0, 0), (1, 1), \text{ and } (3, 3),$$

as shown in Figure P.3(a). From these five points, you might conclude that the graph is a line. This, however, is not correct. By plotting several more points, you can see that the graph is more complicated, as shown in Figure P.3(b).

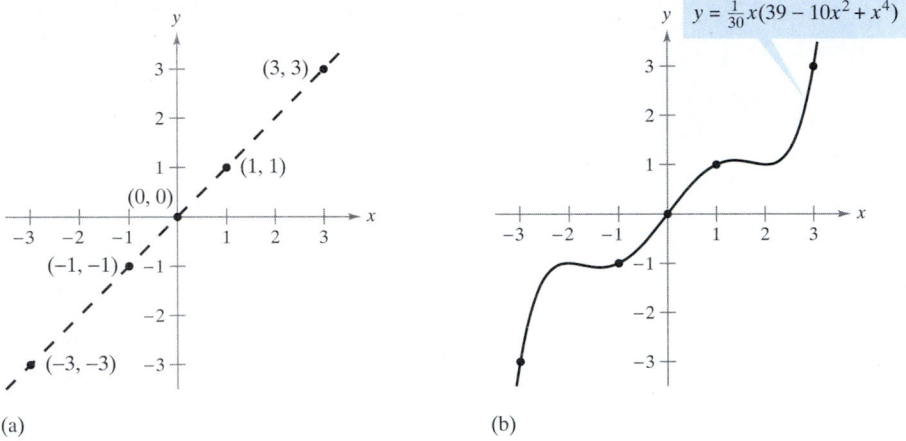

(a)   (b)

Plotting only a few points can misrepresent a graph.
Figure P.3

EXPLORATION

Comparing Graphical and Analytic Approaches   Use a graphing utility to graph each of the following. In each case, find a viewing rectangle that shows the important characteristics of the graph.

(a) $y = x^3 - 3x^2 + 2x + 5$

(b) $y = x^3 - 3x^2 + 2x + 25$

(c) $y = -x^3 - 3x^2 + 20x + 5$

(d) $y = 3x^3 - 40x^2 + 50x - 45$

(e) $y = -(x + 12)^3$

(f) $y = (x - 2)(x - 4)(x - 6)$

A pure graphical approach to this problem would involve a simple "guess, check, and revise" strategy. What types of things do you think an analytic approach might involve? For instance, does the graph have symmetry? Does the graph have turns? If so, where are they?

As you proceed through Chapters 1, 2, and 3 of this text, you will study many new analytic tools that will help you analyze graphs of equations such as these.

TECHNOLOGY   Technology has made sketching of graphs easier. Even with technology, however, it is possible to badly misrepresent a graph. For instance, each of the graphing utility screens shown in Figure P.4 shows a portion of the graph of

$$y = x^3 - x^2 - 25.$$

From the screen on the left, you might assume that the graph is a line. From the screen on the right, however, you can see that the graph is not a line. Thus, whether you are sketching a graph by hand or using a graphing utility, you must realize that different "viewing rectangles" can produce very different views of a graph. In choosing a viewing rectangle, your goal is to show a view of the graph that fits well in the context of the problem.

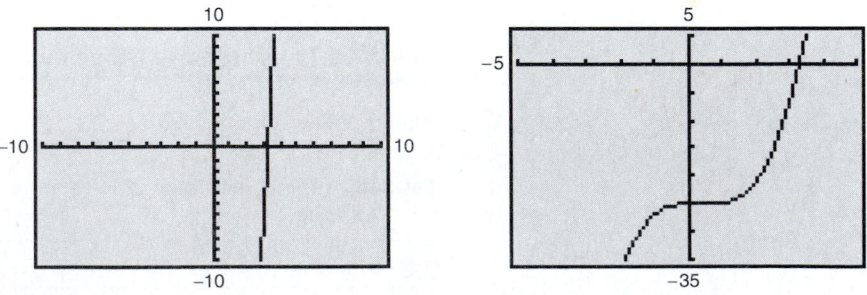

Graphing utility screens of $y = x^3 - x^2 - 25$
Figure P.4

NOTE   In this text, we use the term *graphing utility* to mean either a *graphing calculator* or computer graphing software such as *Maple, Mathematica, Derive, Mathcad,* or the *TI-92*.

Intercepts of a Graph

Two types of solution points that are especially useful are those having zero as their x- or y-coordinate. Such points are called **intercepts** because they are the points at which the graph intersects the x- or y-axis. The point $(a, 0)$ is an **x-intercept** of the graph of an equation if it is a solution point of the equation. To find the x-intercepts of a graph, let y be zero and solve the equation for x. The point $(0, b)$ is a **y-intercept** of the graph of an equation if it is a solution point of the equation. To find the y-intercepts of a graph, let x be zero and solve the equation for y.

NOTE   Some texts denote the x-intercept as the x-coordinate of the point $(a, 0)$ rather than the point itself. Unless it is necessary to make a distinction, we will use the term *intercept* to mean either the point or the coordinate.

It is possible for a graph to have no intercepts, or it might have several. For instance, consider the four graphs shown in Figure P.5.

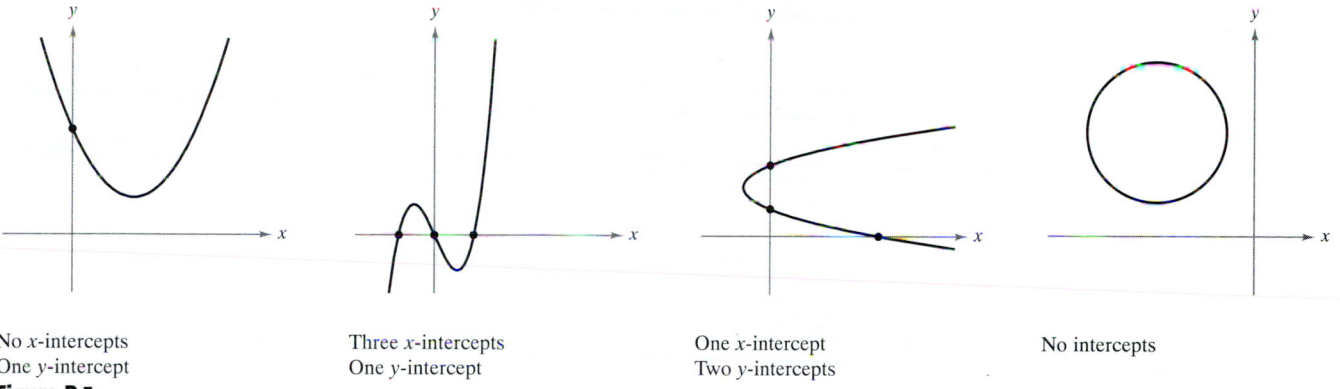

No x-intercepts
One y-intercept
Figure P.5

Three x-intercepts
One y-intercept

One x-intercept
Two y-intercepts

No intercepts

EXAMPLE 2   Finding x- and y-intercepts

Find the x- and y-intercepts of the graph of $y = x^3 - 4x$.

Solution   To find the x-intercepts, let y be zero and solve for x.

$$x^3 - 4x = 0 \qquad \text{Let } y \text{ be zero.}$$
$$x(x - 2)(x + 2) = 0 \qquad \text{Factor.}$$
$$x = 0, 2, \text{ or } -2 \qquad \text{Solve for } x.$$

Because this equation has three solutions, you can conclude that the graph has three x-intercepts:

$$(0, 0), (2, 0), \text{ and } (-2, 0). \qquad x\text{-intercepts}$$

To find the y-intercepts, let x be zero. Doing this produces $y = 0$. Thus, the y-intercept is

$$(0, 0). \qquad y\text{-intercept}$$

(See Figure P.6.)

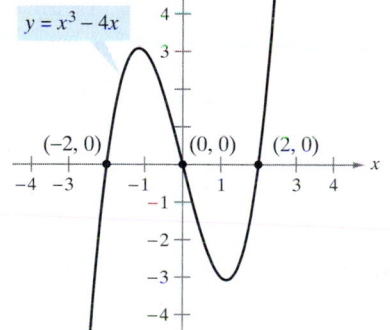

$y = x^3 - 4x$
$(-2, 0)$ $(0, 0)$ $(2, 0)$

Intercepts of a graph
Figure P.6

TECHNOLOGY   Example 2 uses an analytic approach to finding intercepts. When an analytic approach is not possible, you can use a graphical approach by finding the points where the graph intersects the axes. Try using a graphing utility to approximate the intercepts.

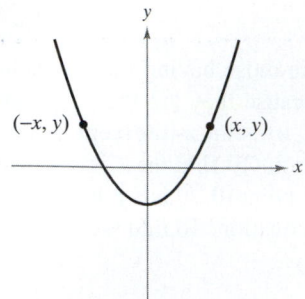

y-Axis symmetry

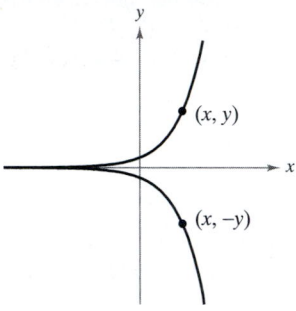

x-Axis symmetry

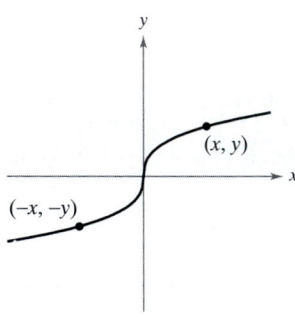

Origin symmetry
Figure P.7

Symmetry of a Graph

The following three types of symmetry can be used to help sketch the graph of an equation (see Figure P.7).

1. A graph is **symmetric with respect to the *y*-axis** if, whenever (x, y) is a point on the graph, $(-x, y)$ is also a point on the graph. This means that the portion of the graph to the left of the *y*-axis is a mirror image of the portion to the right of the *y*-axis.

2. A graph is **symmetric with respect to the *x*-axis** if, whenever (x, y) is a point on the graph, $(x, -y)$ is also a point on the graph. This means that the portion of the graph above the *x*-axis is a mirror image of the portion below the *x*-axis.

3. A graph is **symmetric with respect to the origin** if, whenever (x, y) is a point on the graph, $(-x, -y)$ is also a point on the graph. This means that the graph is unchanged by a rotation of 180° about the origin.

   Knowing that a graph has symmetry *before* attempting to sketch it is useful because it helps you determine an appropriate viewing rectangle.

Tests for Symmetry

1. The graph of an equation in *x* and *y* is symmetric with respect to the *y*-axis if replacing *x* by $-x$ yields an equivalent equation.
2. The graph of an equation in *x* and *y* is symmetric with respect to the *x*-axis if replacing *y* by $-y$ yields an equivalent equation.
3. The graph of an equation in *x* and *y* is symmetric with respect to the origin if replacing *x* by $-x$ and *y* by $-y$ yields an equivalent equation.

The graph of a polynomial has symmetry with respect to the *y*-axis if each term has an even exponent (or is a constant). For instance, the graph of

$$y = 2x^4 - x^2 + 2 \qquad \text{\textcolor{red}{*y*-axis symmetry}}$$

has symmetry with respect to the *y*-axis. Similarly, the graph of a polynomial has symmetry with respect to the origin if each term has an odd exponent, as illustrated in Example 3.

EXAMPLE 3 Testing for Origin Symmetry

Show that the graph of

$$y = 2x^3 - x$$

is symmetric with respect to the origin.

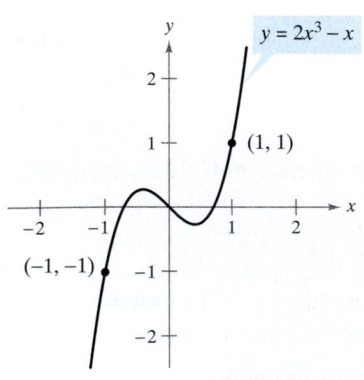

Origin symmetry
Figure P.8

Solution

$$\begin{aligned}
y &= 2x^3 - x && \text{Original equation} \\
-y &= 2(-x)^3 - (-x) && \text{Replace } x \text{ by } -x \text{ and } y \text{ by } -y. \\
-y &= -2x^3 + x && \text{Simplify.} \\
y &= 2x^3 - x && \text{Equivalent equation}
\end{aligned}$$

Because the replacement produces an equivalent equation, you can conclude that the graph of $y = 2x^3 - x$ is symmetric with respect to the origin, as shown in Figure P.8.

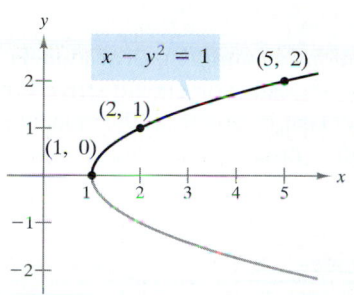

First plot the points above the *x*-axis. Then use symmetry to complete the graph.
Figure P.9

EXAMPLE 4   Using Symmetry to Sketch a Graph

Sketch the graph of $x - y^2 = 1$.

Solution   The graph is symmetric with respect to the *x*-axis because replacing *y* by $-y$ yields an equivalent equation.

$$x - y^2 = 1 \qquad \text{Original equation}$$
$$x - (-y)^2 = 1 \qquad \text{Replace } y \text{ by } -y.$$
$$x - y^2 = 1 \qquad \text{Equivalent equation}$$

This means that the portion of the graph below the *x*-axis is a mirror image of the portion above the *x*-axis. To sketch the graph, first sketch the portion above the *x*-axis. Then reflect in the *x*-axis to obtain the entire graph, as shown in Figure P.9.

TECHNOLOGY   Graphing utilities are designed so that they most easily graph equations in which *y* is a function of *x* (see Section P.3 for a definition of **function**). To graph other types of equations, you need to split the graph into two or more parts *or* you need to use a different graphing mode. For instance, to sketch the graph of the equation in Example 4, you can split it into two parts.

$$y_1 = \sqrt{x - 1} \qquad \text{Top portion of graph}$$
$$y_2 = -\sqrt{x - 1} \qquad \text{Bottom portion of graph}$$

Points of Intersection

A **point of intersection** of the graphs of two equations is a point that satisfies both equations. You can find the points of intersection of two graphs by solving their equations simultaneously.

EXAMPLE 5   Finding Points of Intersection

Find all points of intersection of the graphs of $x^2 - y = 3$ and $x - y = 1$.

Solution   It is helpful to begin by sketching the graphs of both equations on the *same* rectangular coordinate system, as shown in Figure P.10. Having done this, it appears that the two graphs have two points of intersection. To find these two points, you can use the following steps.

$$y = x^2 - 3 \qquad \text{Solve first equation for } y.$$
$$y = x - 1 \qquad \text{Solve second equation for } y.$$
$$x^2 - 3 = x - 1 \qquad \text{Equate } y\text{-values.}$$
$$x^2 - x - 2 = 0 \qquad \text{Write in general form.}$$
$$(x - 2)(x + 1) = 0 \qquad \text{Factor.}$$
$$x = 2 \text{ or } -1 \qquad \text{Solve for } x.$$

The corresponding values of *y* are obtained by substituting $x = 2$ and $x = -1$ into either of the original equations. Doing this produces two points of intersection:

$$(2, 1) \quad \text{and} \quad (-1, -2). \qquad \text{Points of intersection}$$

You can check these points by substituting into *both* of the original equations or by using the trace feature of a graphing utility.

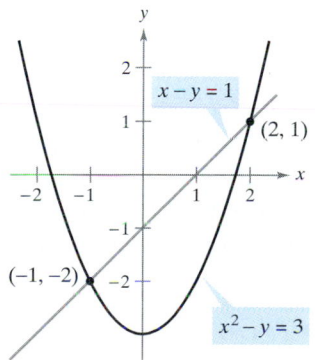

Two points of intersection
Figure P.10

From 1960 through the mid 1990s, Mauna Loa Observatory in Hawaii has measured the increasing concentration of carbon dioxide in earth's atmosphere.

Mathematical Models

Real-life applications of mathematics often use equations as **mathematical models.** In developing a mathematical model to represent actual data, you should strive for two (often conflicting) goals: accuracy and simplicity. That is, you want the model to be simple enough to be workable, yet accurate enough to produce meaningful results. Section P.4 explores these goals more completely.

EXAMPLE 6   The Rise in Atmospheric Carbon Dioxide

Between 1960 and 1990, the Mauna Loa Observatory in Hawaii recorded the carbon dioxide concentration y (in parts per million) in earth's atmosphere. The January readings for each year are shown in Figure P.11. In Figure P.11(a), a linear model has been fitted to the data

$$y = 313.6 + 1.24t, \qquad \text{Linear model}$$

where $t = 0$ represents 1960. In Figure P.11(b), a quadratic model has been fitted to the data

$$y = 316.2 + 0.70t + 0.018t^2. \qquad \text{Quadratic model}$$

Which model better represents the data? In the July 1990 issue of *Scientific American,* these data were used to predict the carbon dioxide level in earth's atmosphere in the year 2035. The prediction was 470 parts per million. Which, if either, of the two models could have been used to make this prediction?

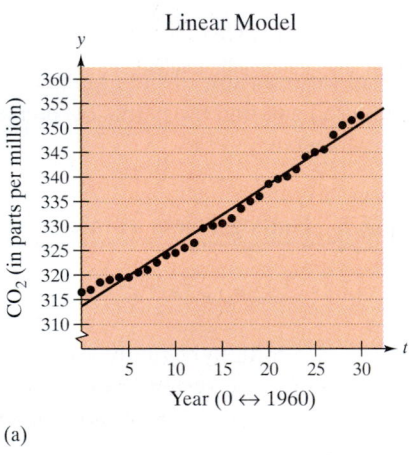
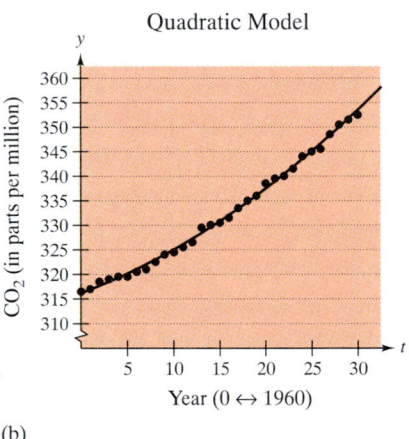

Graphs of linear and quadratic models
Figure P.11

NOTE   The models in Example 6 were developed using a procedure called least squares regression (see Section 12.9). The linear model has a correlation given by $r^2 = 0.984$. The quadratic model has a correlation given by $r^2 = 0.997$. The closer r^2 is to 1, the "better" it is.

Solution   To answer the first question, you need to define what "better" means. Using a statistical definition involving the squares of the differences between the actual y-values and the y-values given by the model, the quadratic model is better because its values are closer to the actual values. This conclusion is shared by the author of the *Scientific American* article. Using the linear model with $t = 75$ (for 2035), the prediction would be

$$y = 313.6 + 1.24(75) = 406.6. \qquad \text{Linear model}$$

Using the quadratic model, the prediction for 2035 would be

$$y = 316.2 + 0.70(75) + 0.018(75)^2 = 469.95. \qquad \text{Quadratic model}$$

EXERCISES FOR SECTION P.1

In Exercises 1–4, match the equation with its graph. [Graphs are labeled (a), (b), (c), and (d).]

(a)
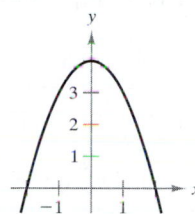

(b)
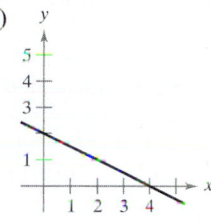

(c)
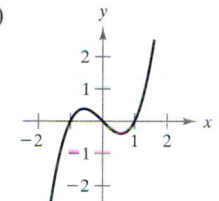

(d)
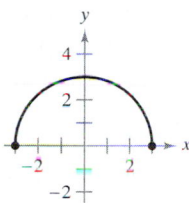

1. $y = -\dfrac{1}{2}x + 2$

2. $y = \sqrt{9 - x^2}$

3. $y = 4 - x^2$

4. $y = x^3 - x$

In Exercises 5–10, find any intercepts.

5. $y = x^2 + x - 2$

6. $y^2 = x^3 - 4x$

7. $y = x^2\sqrt{9 - x^2}$

8. $y = \dfrac{x^2 + 3x}{(3x + 1)^2}$

9. $x^2 y - x^2 + 4y = 0$

10. $y = 2x - \sqrt{x^2 + 1}$

In Exercises 11–18, check for symmetry with respect to each axis and to the origin.

11. $y = x^2 - 2$

12. $xy - \sqrt{4 - x^2} = 0$

13. $y^2 = x^3 - 4x$

14. $xy^2 = -10$

15. $y = x^3 + x$

16. $xy = 1$

17. $y = \dfrac{x}{x^2 + 1}$

18. $y = x^3 + x - 3$

In Exercises 19–30, sketch the graph of the equation. Identify any intercepts and test for symmetry.

19. $y = -3x + 2$

20. $y = -\dfrac{1}{2}x + 2$

21. $y = \dfrac{1}{2}x - 4$

22. $y = x^2 + 3$

23. $y = 1 - x^2$

24. $y = 2x^2 + x$

25. $y = x^3 + 2$

26. $y = \sqrt{9 - x^2}$

27. $y = (x + 2)^2$

28. $x = y^2 - 4$

29. $y = \dfrac{1}{x}$

30. $y = 2x^4$

In Exercises 31 and 32, describe the viewing rectangle of a graphing utility that yields the figure.

31. $y = x^3 - 3x^2 + 4$

32. $y = |x| + |x - 10|$

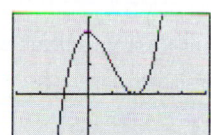

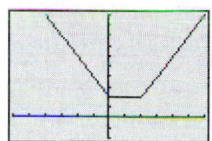

In Exercises 33–36, use a graphing utility to graph the equation. (See Technology note, page **7**.) Identify any intercepts and test for symmetry.

33. $y = -2x^2 + x + 1$

34. $y = x\sqrt{25 - x^2}$

35. $y = \dfrac{5}{x^2 + 1} - 1$

36. $x^2 + 4y^2 = 4$

In Exercises 37 and 38, use a graphing utility to graph the equation. Move the cursor along the curve to approximate the unknown coordinate of each given solution point accurate to two decimal places. (*Hint:* You may need to use the zoom feature to obtain the required accuracy.)

37. $y = \sqrt{5 - x}$ (a) $(2, y)$ (b) $(x, 3)$

38. $y = x^5 - 5x$ (a) $(-0.5, y)$ (b) $(x, -4)$

In Exercises 39–42, write an equation whose graph has the indicated property. (There may be more than one correct answer.)

39. The graph has intercepts at $x = -2$, $x = 4$, and $x = 6$.

40. The graph has intercepts at $x = -\dfrac{5}{2}$, $x = 2$, and $x = \dfrac{3}{2}$.

41. The graph is symmetric with respect to the origin.

42. The graph is symmetric with respect to the x-axis.

In Exercises 43–50, find the points of intersection of the graphs of the equations, and check your results analytically.

43. $x + y = 2$
         $2x - y = 1$

44. $2x - 3y = 13$
         $5x + 3y = 1$

45. $x + y = 7$
         $3x - 2y = 11$

46. $x^2 + y^2 = 25$
         $2x + y = 10$

47. $x^2 + y^2 = 5$
         $x - y = 1$

48. $x^2 + y = 4$
         $2x - y = 1$

49. $y = x^3$
         $y = x$

50. $x = 3 - y^2$
         $y = x - 1$

The symbol *indicates an exercise in which you are instructed to use graphing technology or a symbolic computer algebra system. The solutions of other exercises may also be facilitated by use of appropriate technology.*

A red number indicates that a detailed solution can be found in the Study and Solutions Guide.

In Exercises 51 and 52, use a graphing utility to graph the equations and find the points of intersection of the graphs.

51. $y = x^3 - 2x^2 + x - 1$

$y = -x^2 + 3x - 1$

52. $y = x^4 - 2x^2 + 1$

$y = 1 - x^2$

53. *Break-Even Point* Find the sales necessary to break even $(R = C)$ if the cost C of producing x units is

$C = 5.5\sqrt{x} + 10,000$ Cost equation

and the revenue R for selling x units is

$R = 3.29x$. Revenue equation

54. *Think About It* Each table gives solution points for one of the following equations.

(i) $y = kx + 5$

(ii) $y = x^2 + k$

(iii) $y = kx^{3/2}$

(iv) $xy = k$

For each equation, match the equation with the correct table and find k.

(a)

x	1	4	9
y	3	24	81

(b)

x	1	4	9
y	7	13	23

(c)

x	1	4	9
y	36	9	4

(d)

x	1	4	9
y	-9	6	71

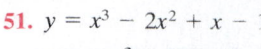 **55.** *Modeling Data* The table shows the consumer price index (1982–84 = 100) for selected years. *(Source: Bureau of Labor Statistics)*

Year	1970	1975	1980	1985	1990	1994
CPI	38.8	53.8	82.4	107.6	130.7	148.2

(a) Use the regression capabilities of a graphing utility to find a mathematical model of the form

$y = at^2 + bt + c$

for the data. In the model, y represents the consumer price index and t represents the year, with $t = 0$ corresponding to 1970.

(b) Use a graphing utility to graph the model and compare the data with the model.

(c) Use the model to predict the CPI for the year 2000.

56. *Modeling Data* The table shows the average number of acres per farm in the United States for selected years. *(Source: U.S. Department of Agriculture)*

Year	1950	1960	1970	1980	1990	1994
Acreage	213	297	374	426	461	478

(a) Use the regression capabilities of a graphing utility to find a mathematical model of the form

$y = at^2 + bt + c$

for the data. In the model, y represents the average acreage and t represents the year, with $t = 0$ corresponding to 1950.

(b) Use a graphing utility to graph the model and compare the data with the model.

(c) Use the model to predict the average number of acres per farm in the United States in the year 2000.

57. *Copper Wire* The resistance y in ohms of 1000 feet of solid copper wire at 77°F can be approximated by the mathematical model

$$y = \frac{10,770}{x^2} - 0.37, \qquad 5 \le x \le 100$$

where x is the diameter of the wire in mils (0.001 in.). Use a graphing utility to graph the model. If the diameter of the wire is doubled, the resistance is changed by approximately what factor?

58. (a) Prove that if a graph is symmetric with respect to the x-axis and to the y-axis, then it is symmetric with respect to the origin. Give an example to show that the converse is not true.

(b) Prove that if a graph is symmetric with respect to one axis and to the origin, then it is symmetric with respect to the other axis.

True or False? **In Exercises 59–62, determine whether the statement is true or false. If it is false, explain why or give an example that shows it is false.**

59. If $(1, -2)$ is a point on a graph that is symmetric with respect to the x-axis, then $(-1, -2)$ is also a point on the graph.

60. If $(1, -2)$ is a point on a graph that is symmetric with respect to the y-axis, then $(-1, -2)$ is also a point on the graph.

61. If $b^2 - 4ac > 0$ and $a \ne 0$, then the graph of $y = ax^2 + bx + c$ has two x-intercepts.

62. If $b^2 - 4ac = 0$ and $a \ne 0$, then the graph of $y = ax^2 + bx + c$ has only one x-intercept.

63. Find an equation of the graph that consists of all points (x, y) whose distance from the origin is K times $(K \ne 1)$ the distance from $(2, 0)$. (For a review on the distance formula, see the appendix.)

The Slope of a Line • Equations of Lines • Ratios and Rates of Change • Graphing Linear Models • Parallel and Perpendicular Lines

The Slope of a Line

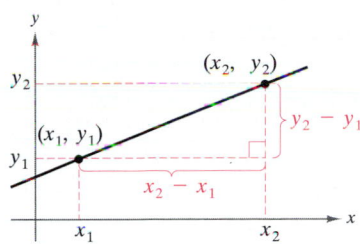

$\Delta y = y_2 - y_1 =$ change in y
$\Delta x = x_2 - x_1 =$ change in x

Figure P.12

The **slope** of a nonvertical line is a measure of the number of units the line rises (or falls) vertically for each unit of horizontal change from left to right. Consider the two points (x_1, y_1) and (x_2, y_2) on the line in Figure P.12. As you move from left to right along this line, a vertical change of

$$\Delta y = y_2 - y_1 \qquad \text{Change in } y$$

units corresponds to a horizontal change of

$$\Delta x = x_2 - x_1 \qquad \text{Change in } x$$

units. (Δ is a Greek uppercase letter *delta*, and the symbols Δy and Δx are read "delta y" and "delta x.")

> **Definition of the Slope of a Line**
>
> The **slope** m of a nonvertical line passing through the points (x_1, y_1) and (x_2, y_2) is
>
> $$m = \frac{\Delta y}{\Delta x} = \frac{y_2 - y_1}{x_2 - x_1}, \qquad x_1 \neq x_2.$$

SYMBOL FOR SLOPE

The use of the letter m to represent the slope of a line comes from *monter*, the French verb meaning to mount, to climb, or to rise.

NOTE   When using the formula for slope, note that

$$\frac{y_2 - y_1}{x_2 - x_1} = \frac{-(y_1 - y_2)}{-(x_1 - x_2)} = \frac{y_1 - y_2}{x_1 - x_2}.$$

Thus, it does not matter in which order you subtract *as long as* you are consistent and both "subtracted coordinates" come from the same point.

Figure P.13 shows four lines: one has a positive slope, one has a slope of zero, one has a negative slope, and one has an "undefined slope." In general, the greater the absolute value of the slope of a line, the steeper the line is. For instance, in Figure P.13, the line with a slope of -5 is steeper than the line with a slope of $\frac{1}{5}$.

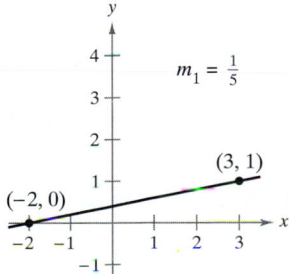

If m is positive, then the line rises from left to right.

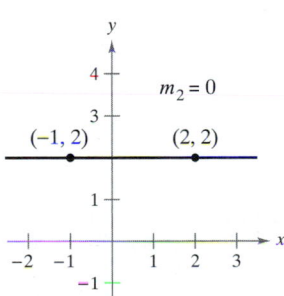

If m is zero, then the line is horizontal.

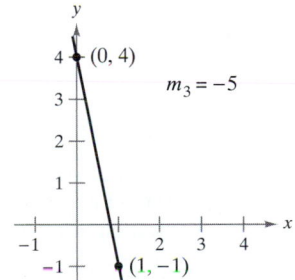

If m is negative, then the line falls from left to right.

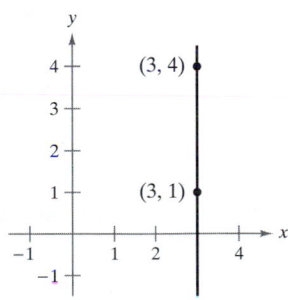

If m is undefined, then the line is vertical.

Figure P.13

EXPLORATION

Investigating Slopes of Lines
Use a graphing utility to graph each of the linear equations. Which point is common to all seven lines? Which value in the equation determines the slope of each line?

(a) $y - 4 = -2(x + 1)$

(b) $y - 4 = -1(x + 1)$

(c) $y - 4 = -\frac{1}{2}(x + 1)$

(d) $y - 4 = 0(x + 1)$

(e) $y - 4 = \frac{1}{2}(x + 1)$

(f) $y - 4 = 1(x + 1)$

(g) $y - 4 = 2(x + 1)$

Equations of Lines

Any two points on a nonvertical line can be used to calculate its slope. This can be verified from the similar triangles shown in Figure P.14. (Recall that the ratios of corresponding sides of similar triangles are equal.)

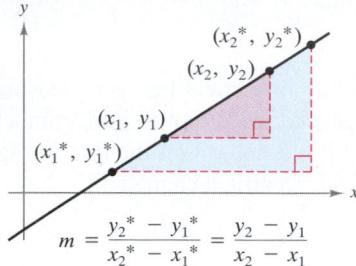

$$m = \frac{y_2^* - y_1^*}{x_2^* - x_1^*} = \frac{y_2 - y_1}{x_2 - x_1}$$

Any two points on a line can be used to determine its slope.
Figure P.14

You can write an equation of a line if you know the slope of the line and the coordinates of one point on the line. Suppose the slope is m and the point is (x_1, y_1). If (x, y) is any other point on the line, then

$$\frac{y - y_1}{x - x_1} = m.$$

This equation, involving the two variables x and y, can be rewritten in the form $y - y_1 = m(x - x_1)$, which is called the **point-slope equation of a line.**

> **Point-Slope Equation of a Line**
>
> An equation of the line with slope m passing through the point (x_1, y_1) is given by
>
> $$y - y_1 = m(x - x_1).$$

EXAMPLE 1 Finding an Equation of a Line

Find an equation of the line that has a slope of 3 and passes through the point $(1, -2)$.

Solution

$y - y_1 = m(x - x_1)$	Point-slope form
$y - (-2) = 3(x - 1)$	Substitute -2 for y_1, 1 for x_1, and 3 for m.
$y + 2 = 3x - 3$	Simplify.
$y = 3x - 5$	Solve for y.

(See Figure P.15.)

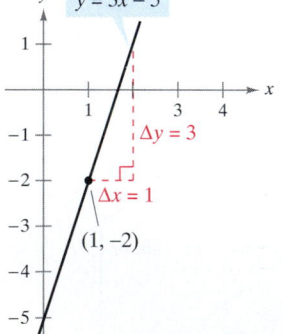

The line with a slope of 3 passing through point $(1, -2)$
Figure P.15

NOTE   Remember that slope can be used to describe only a line that is not vertical. Thus, vertical lines cannot be written in point-slope form. For instance, the equation of the vertical line passing through the point $(1, -2)$ is $x = 1$.

Ratios and Rates of Change

The slope of a line can be interpreted as either a *ratio* or a *rate*. If the x- and y-axes have the same unit of measure, the slope has no units and is a **ratio.** If the x- and y-axes have different units of measure, the slope is a rate or **rate of change.** In your study of calculus, you will encounter applications involving both interpretations of slope.

EXAMPLE 2   Population Growth and Engineering Design

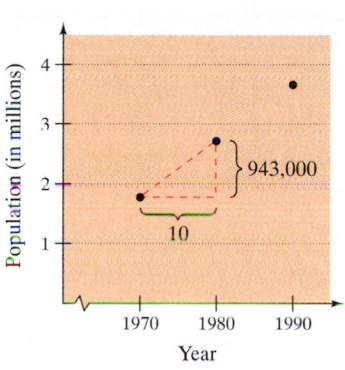

Population of Arizona in census years
Figure P.16

a. The population of Arizona was 1,775,000 in 1970 and 2,718,000 in 1980. Over this 10-year period, the average rate of change of the population was

$$\text{Rate of change} = \frac{\text{change in population}}{\text{change in years}}$$

$$= \frac{2,718,000 - 1,775,000}{1980 - 1970}$$

$$= 94,300 \text{ people per year.}$$

If Arizona's population had continued to increase at this same rate for the next 10 years, it would have had a 1990 population of 3,661,000. In the 1990 census, however, Arizona's population was determined to be 3,665,000, so the population's rate of change from 1980 to 1990 was a little greater than in the previous decade (see Figure P.16).

b. In tournament water-ski jumping, the ramp rises to a height of 6 feet on a raft that is 21 feet long, as shown in Figure P.17. The slope of the ski ramp is the ratio of its height (the rise) to the length of its base (the run).

$$\text{Slope of ramp} = \frac{\text{rise}}{\text{run}}$$ Rise is vertical change, run is horizontal change.

$$= \frac{6 \text{ feet}}{21 \text{ feet}}$$

$$= \frac{2}{7}$$

In this case, note that the slope is a ratio and has no units.

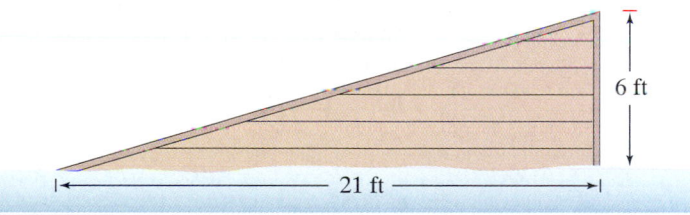

Dimensions of a water-ski ramp
Figure P.17

NOTE   The rate of change found in Example 2a is an **average rate of change.** An average rate of change is always calculated over an interval. In this case, the interval is [1970, 1980]. In Chapter 2 you will study another type of rate of change called an *instantaneous rate of change.*

Graphing Linear Models

Many problems in analytic geometry can be classified in two basic categories: (1) Given a graph, what is its equation? and (2) Given an equation, what is its graph? The point-slope equation of a line can be used to solve problems in the first category. However, this form is not especially useful for solving problems in the second category. The form that is better suited to sketching the graph of a line is the **slope-intercept** form of the equation of a line.

> ### The Slope-Intercept Equation of a Line
>
> The graph of the linear equation
>
> $$y = mx + b$$
>
> is a line having a *slope* of m and a *y-intercept* at $(0, b)$.

EXAMPLE 3 Sketching Lines in the Plane

Sketch the graph of each equation.

a. $y = 2x + 1$ **b.** $y = 2$ **c.** $3y + x - 6 = 0$

Solution

a. Because $b = 1$, the y-intercept is $(0, 1)$. Because the slope is $m = 2$, you know that the line rises two units for each unit it moves to the right, as shown in Figure P.18(a).

b. Because $b = 2$, the y-intercept is $(2, 0)$. Because the slope is $m = 0$, you know that the line is horizontal, as shown in Figure P.18(b).

c. Begin by writing the equation in slope-intercept form.

$$3y + x - 6 = 0 \qquad \text{Original equation}$$
$$3y = -x + 6 \qquad \text{Isolate y-term on the left.}$$
$$y = -\frac{1}{3}x + 2 \qquad \text{Slope-intercept form}$$

In this form, you can see that the y-intercept is $(0, 2)$ and the slope is $m = -\frac{1}{3}$. This means that the line falls one unit for every three units it moves to the right, as shown in Figure P.18(c).

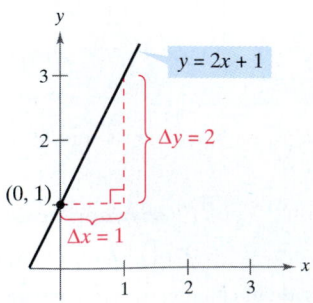

(a) $m = 2$; line rises

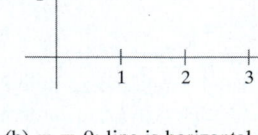

(b) $m = 0$; line is horizontal

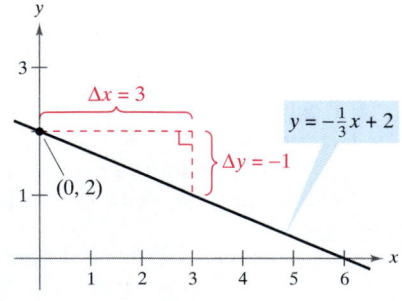

(c) $m = -\frac{1}{3}$; line falls

Figure P.18

Because the slope of a vertical line is not defined, its equation cannot be written in the slope-intercept form. However, the equation of *any* line can be written in the **general form**

$$Ax + By + C = 0$$ General form of the equation of a line

where A and B are not *both* zero. For instance, the vertical line given by $x = a$ can be represented by the general form $x - a = 0$.

Summary of Equations of Lines

1. General form: $Ax + By + C = 0$
2. Vertical line: $x = a$
3. Horizontal line: $y = b$
4. Point-slope form: $y - y_1 = m(x - x_1)$
5. Slope-intercept form: $y = mx + b$

Parallel and Perpendicular Lines

The slope of a line is a convenient tool for determining whether two lines are parallel or perpendicular, as shown in Figure P.19. Specifically, nonvertical lines with the same slope are parallel and nonvertical lines whose slopes are negative reciprocals are perpendicular.

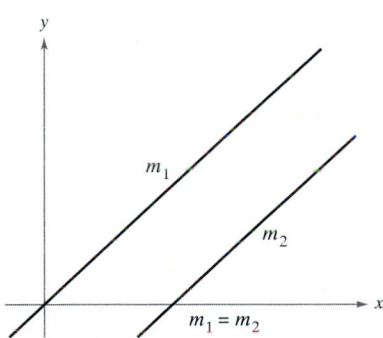

Parallel lines

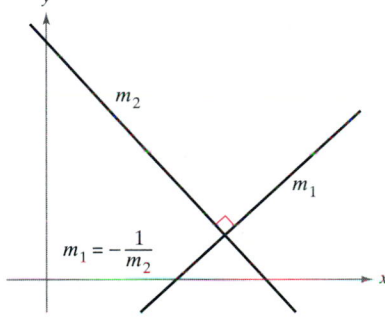

Perpendicular lines

Figure P.19

STUDY TIP In mathematics, the phrase "if and only if" is a way of stating two implications in one statement. For instance, the first statement at the right could be rewritten as the following two implications.

(a) If two distinct nonvertical lines are parallel, then their slopes are equal.
(b) If two distinct nonvertical lines have equal slopes, then they are parallel.

Parallel and Perpendicular Lines

1. Two distinct nonvertical lines are **parallel** if and only if their slopes are equal—that is, if and only if $m_1 = m_2$.

2. Two nonvertical lines are **perpendicular** if and only if their slopes are negative reciprocals of each other—that is, if and only if

$$m_1 = -\frac{1}{m_2}.$$

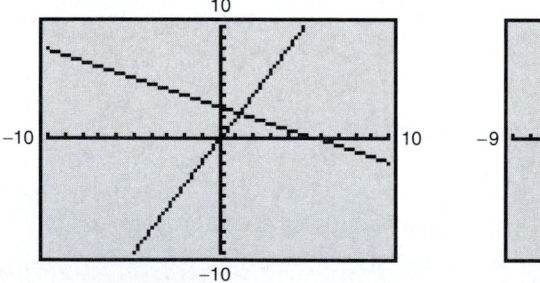

$3x + 2y = 4$

$2x - 3y = 5$

$(2, -1)$

$2x - 3y = 7$

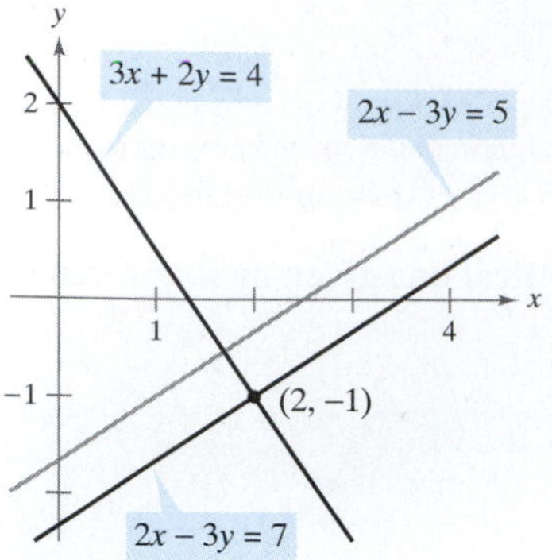

Lines parallel and perpendicular to
$2x - 3y = 5$
Figure P.20

EXAMPLE 4   Finding Parallel and Perpendicular Lines

Find the general forms of the equations of the lines that pass through the point $(2, -1)$ and are

a. parallel to the line $2x - 3y = 5$ **b.** perpendicular to the line $2x - 3y = 5$

as shown in Figure P.20.

Solution   By writing the linear equation $2x - 3y = 5$ in slope-intercept form, $y = \frac{2}{3}x - \frac{5}{3}$, you can see that the given line has a slope of $m = \frac{2}{3}$.

a. The line through $(2, -1)$ that is parallel to the given line also has a slope of $\frac{2}{3}$.

$$y - y_1 = m(x - x_1) \qquad \text{Point-slope form}$$
$$y - (-1) = \tfrac{2}{3}(x - 2) \qquad \text{Substitute.}$$
$$3(y + 1) = 2(x - 2) \qquad \text{Simplify.}$$
$$2x - 3y - 7 = 0 \qquad \text{General form}$$

Note the similarity to the original equation.

b. Using the negative reciprocal of the slope of the given line, you can determine that the slope of a line perpendicular to the given line is $-\frac{3}{2}$. Therefore, the line through the point $(2, -1)$ that is perpendicular to the given line has the following equation.

$$y - y_1 = m(x - x_1) \qquad \text{Point-slope form}$$
$$y - (-1) = -\tfrac{3}{2}(x - 2) \qquad \text{Substitute.}$$
$$2(y + 1) = -3(x - 2) \qquad \text{Simplify.}$$
$$3x + 2y - 4 = 0 \qquad \text{General form}$$

TECHNOLOGY   The apparent slope of a line will be distorted if you use different tick-mark spacing on the x- and y-axes. For instance, the graphing calculator screens in Figures P.21(a) and P.21(b) both show the lines given by $y = 2x$ and $y = -\frac{1}{2}x + 3$. Because these lines have slopes that are negative reciprocals, they must be perpendicular. In Figure P.21(a), however, the lines don't appear to be perpendicular because the tick-mark spacing on the x-axis is not the same as that on the y-axis.

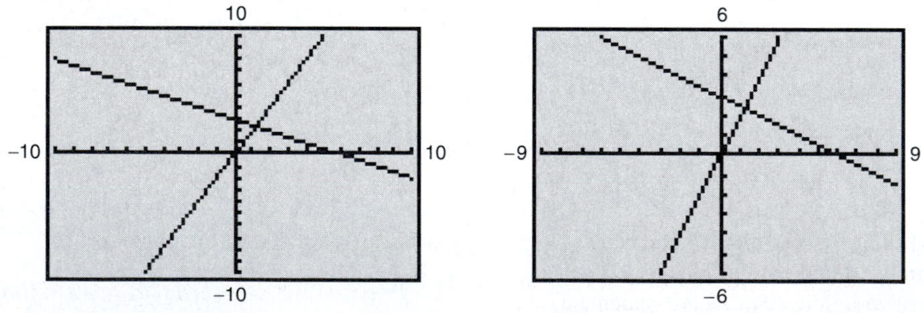

(a) Tick-mark spacing on the x-axis is not the same as tick-mark spacing on the y-axis.

(b) Tick-mark spacing on the x-axis is the same as tick-mark spacing on the y-axis.

Figure P.21

EXERCISES FOR SECTION P.2

In Exercises 1–6, estimate the slope of the line from its graph.

1.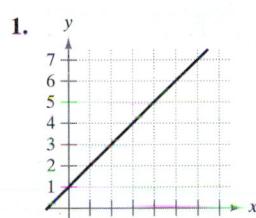

2.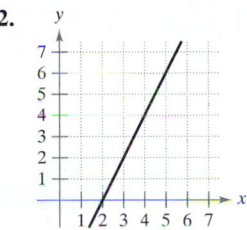

3.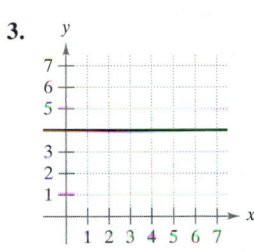

4.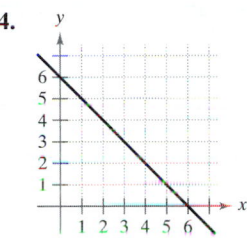

5.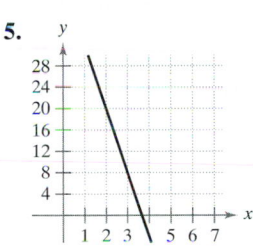

6.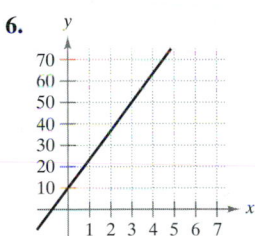

In Exercises 7 and 8, sketch the graph of the lines through the point with the indicated slopes. Make the sketches on the same set of coordinate axes.

Point	Slopes			
7. (2, 3)	(a) 0	(b) 1	(c) -2	(d) Undefined
8. $(-4, 1)$	(a) 3	(b) -3	(c) $\frac{1}{3}$	(d) 0

In Exercises 9–12, plot the pair of points and find the slope of the line passing through them.

9. $(3, -4), (5, 2)$

10. $(2, 1), (2, 5)$

11. $(1, 2), (-2, 4)$

12. $\left(\frac{7}{8}, \frac{3}{4}\right), \left(\frac{5}{4}, -\frac{1}{4}\right)$

In Exercises 13–16, use the point on the line and the slope of the line to find three additional points that the line passes through. (There is more than one correct answer.)

Point	Slope		Point	Slope
13. (2, 1)	$m = 0$		14. $(-3, 4)$	m undefined
15. (1, 7)	$m = -3$		16. $(-2, -2)$	$m = 2$

17. *Modeling Data* The table gives the earnings per share of common stock for General Mills for the years 1987 through 1994. Time in years is represented by t, with $t = 0$ corresponding to 1990, and the earnings per share are represented by y. (*Source: General Mills 1994 Annual Report*)

t	-3	-2	-1	0
y	\$1.25	\$1.63	\$2.53	\$2.32

t	1	2	3	4
y	\$2.87	\$2.99	\$3.10	\$2.95

(a) Plot the data by hand and connect adjacent points with a line segment.

(b) Use the slope to determine the years when earnings decreased most rapidly and increased most rapidly.

18. *Conveyor Design* A moving conveyor is built to rise 1 meter for each 3 meters of horizontal change.

(a) Find the slope of the conveyor.

(b) Suppose the conveyor runs between two floors in a factory. Find the length of the conveyor if the vertical distance between floors is 10 feet.

19. *Writing* Write a short paragraph explaining whether or not any two points on a line can be used to calculate the slope of the line.

20. *Rate of Change* Each of the following is the slope of a line representing daily revenue y in terms of time x in days. Use the slope to interpret any change in daily revenue for a 1-day increase in time.

(a) $m = 400$ (b) $m = 100$ (c) $m = 0$

In Exercises 21–24, find the slope and the y-intercept (if possible) of the line.

21. $x + 5y = 20$

22. $6x - 5y = 15$

23. $x = 4$

24. $y = -1$

In Exercises 25–30, find an equation of the line that passes through the points, and sketch the line.

25. $(2, 1), (0, -3)$

26. $(-3, -4), (1, 4)$

27. $(0, 0), (-1, 3)$

28. $(-3, 6), (1, 2)$

29. $(1, -2), (3, -2)$

30. $\left(\frac{7}{8}, \frac{3}{4}\right), \left(\frac{5}{4}, -\frac{1}{4}\right)$

In Exercises 31–36, find an equation of the line that passes through the point and has the indicated slope. Sketch the line.

Point	Slope		Point	Slope
31. (0, 3)	$m = \frac{3}{4}$		32. $(-1, 2)$	m undefined
33. (0, 0)	$m = \frac{2}{3}$		34. $(-2, 4)$	$m = -\frac{3}{5}$
35. (0, 2)	$m = 4$		36. (0, 4)	$m = 0$

37. Find an equation of the vertical line with x-intercept at 3.

38. Show that the line with intercepts $(a, 0)$ and $(0, b)$ has the following equation.

$$\frac{x}{a} + \frac{y}{b} = 1, \qquad a \neq 0, b \neq 0$$

In Exercises 39–42, use the result of Exercise 38 to write an equation of the line.

39. x-intercept: $(2, 0)$
y-intercept: $(0, 3)$

40. x-intercept: $\left(-\frac{2}{3}, 0\right)$
y-intercept: $(0, -2)$

41. Point on line: $(1, 2)$
x-intercept: $(a, 0)$
y-intercept: $(0, a)$
$(a \neq 0)$

42. Point on line: $(-3, 4)$
x-intercept: $(a, 0)$
y-intercept: $(0, a)$
$(a \neq 0)$

In Exercises 43–48, write an equation of the line through the point (a) parallel to the given line and (b) perpendicular to the given line.

	Point	*Line*
43.	$(2, 1)$	$4x - 2y = 3$
44.	$(-3, 2)$	$x + y = 7$
45.	$\left(\frac{7}{8}, \frac{3}{4}\right)$	$5x + 3y = 0$
46.	$(-6, 4)$	$3x + 4y = 7$
47.	$(2, 5)$	$x = 4$
48.	$(-1, 0)$	$y = -3$

In Exercises 49–54, sketch a graph of the equation.

49. $y = -3$

50. $x = 4$

51. $2x - y - 3 = 0$

52. $x + 2y + 6 = 0$

53. $y = -2x + 1$

54. $y - 1 = 3(x + 4)$

 Writing **In Exercises 55 and 56, use a graphing utility to graph the equation using each of the viewing rectangles. Describe the difference between the two views.**

55. $y = 0.5x - 3$

Xmin = -5	Xmin = -2
Xmax = 10	Xmax = 10
Xscl = 1	Xscl = 1
Ymin = -1	Ymin = -4
Ymax = 10	Ymax = 1
Yscl = 1	Yscl = 1

56. $y = -8x + 5$

Xmin = -2	Xmin = -5
Xmax = 2	Xmax = 10
Xscl = 1	Xscl = 1
Ymin = -5	Ymin = -80
Ymax = 5	Ymax = 80
Yscl = 1	Yscl = 20

Rate of Change **In Exercises 57–60, you are given the dollar value of a product in 1998 *and* the rate at which the value of the product is expected to change during the next 5 years. Use this information to write a linear equation that gives the dollar value V of the product in terms of the year t. (Let $t = 8$ represent 1998.)**

	1998 Value	*Rate*
57.	$2540	$125 increase per year
58.	$156	$4.50 increase per year
59.	$20,400	$2000 decrease per year
60.	$245,000	$5600 decrease per year

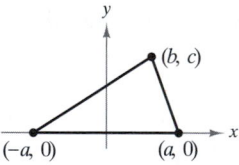 **In Exercises 61 and 62, use a graphing utility to graph the parabolas and find their points of intersection. Find an equation of the line through the points of intersection and sketch its graph in the same viewing rectangle.**

61. $y = x^2$
$y = 4x - x^2$

62. $y = x^2 - 4x + 3$
$y = -x^2 + 2x + 3$

In Exercises 63 and 64, determine whether the points are collinear. (Three points are *collinear* if they lie on the same line.)

63. $(-2, 1), (-1, 0), (2, -2)$

64. $(0, 4), (7, -6), (-5, 11)$

In Exercises 65–68, refer to the triangle in the figure.

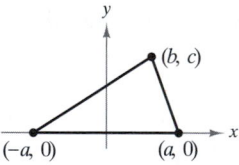

65. Find the coordinates of the point of intersection of the perpendicular bisectors of the sides.

66. Find the coordinates of the point of intersection of the medians.

67. Find the coordinates of the point of intersection of the altitudes.

68. Show that the points of intersection in Exercises 65, 66, and 67 are collinear.

69. *Temperature Conversion* Find the linear equation that expresses the relationship between the temperature in degrees Celsius C and degrees Fahrenheit F. Use the fact that water freezes at 0°C (32°F) and boils at 100°C (212°F) to convert 72°F to degrees Celsius.

70. *Reimbursed Expenses* A company reimburses its sales representatives $150 per day for lodging and meals plus 30¢ per mile driven. Write a linear equation giving the daily cost C to the company in terms of x, the number of miles driven.

71. *Career Choice* An employee has two options for positions in a large corporation. One position pays $12.50 per hour *plus* an additional unit rate of $0.75 per unit produced. The other pays $9.20 per hour *plus* a unit rate of $1.30.

(a) Find linear equations for the hourly wages W in terms of x, the number of units produced per hour, for each of the options.

(b) Use a graphing utility to graph the linear equations and find the point of intersection.

(c) Interpret the meaning of the point of intersection of the graphs in part (b). How would you use this information to select the correct option if the goal were to obtain the highest hourly wage?

72. *Straight-Line Depreciation* A small business purchases a piece of equipment for $875. After 5 years the equipment will be outdated, having no value.

(a) Write a linear equation giving the value y of the equipment in terms of the time x, $0 \le x \le 5$.

(b) Use a graphing utility to graph the equation.

(c) Use the trace feature to estimate (to two-decimal-place accuracy) the value of the equipment when $x = 2$.

(d) Use the trace feature to estimate (to two-decimal-place accuracy) the time when the value of the equipment is $200.

73. *Apartment Rental* A real estate office handles an apartment complex with 50 units. When the rent is $580 per month, all 50 units are occupied. However, when the rent is $625, the average number of occupied units drops to 47. Assume that the relationship between the monthly rent p and the demand x is linear. (*Note:* Here we use the term *demand* to refer to the number of occupied units.)

(a) Write a linear equation giving the demand x in terms of the rent p.

(b) *Linear extrapolation* Use a graphing utility to graph the demand equation and use the trace feature to predict the number of units occupied if the rent is raised to $655.

(c) *Linear interpolation* Predict the number of units occupied if the rent is lowered to $595. Verify graphically.

74. *Modeling Data* An instructor gives regular 20-point quizzes and 100-point exams in a mathematics course. Average scores for six students, given as ordered pairs (x, y) where x is the average quiz score and y is the average test score, are $(18, 87)$, $(10, 55)$, $(19, 96)$, $(16, 79)$, $(13, 76)$, and $(15, 82)$.

(a) Use the regression capabilities of a graphing utility to find the least squares regression line for the data.

(b) Use a graphing utility to plot the points and graph the regression line in the same viewing rectangle.

(c) Use the regression line to predict the average exam score for a student with an average quiz score of 17.

(d) Interpret the meaning of the slope of the regression line.

(e) If the instructor added 4 points to the average test score of everyone in the class, describe the change in the position of the plotted points and the change in the equation of the line.

Distance **In Exercises 75–80, find the distance between the point and line, or between the lines, using the formula for the distance between the point (x_1, y_1) and the line $Ax + By + C = 0$.**

$$\text{Distance} = \frac{|Ax_1 + By_1 + C|}{\sqrt{A^2 + B^2}}$$

75. Point: $(0, 0)$
Line: $4x + 3y = 10$

76. Point: $(2, 3)$
Line: $4x + 3y = 10$

77. Point: $(-2, 1)$
Line: $x - y - 2 = 0$

78. Point: $(6, 2)$
Line: $x = -1$

79. Line: $x + y = 1$
Line: $x + y = 5$

80. Line: $3x - 4y = 1$
Line: $3x - 4y = 10$

81. Show that the distance between the point (x_1, y_1) and the line $Ax + By + C = 0$ is

$$\text{Distance} = \frac{|Ax_1 + By_1 + C|}{\sqrt{A^2 + B^2}}$$

82. Write the distance d between the point $(3, 1)$ and the line $y = mx + 4$ in terms of m. Use a graphing utility to graph the equation. When is the distance 0? Explain the result geometrically.

83. Prove that the diagonals of a rhombus intersect at right angles.

84. Prove that the figure formed by connecting consecutive midpoints of the sides of any quadrilateral is a parallelogram.

85. Prove that if the points (x_1, y_1) and (x_2, y_2) lie on the same line as $(x_1{}^*, y_1{}^*)$ and $(x_2{}^*, y_2{}^*)$, then

$$\frac{y_2{}^* - y_1{}^*}{x_2{}^* - x_1{}^*} = \frac{y_2 - y_1}{x_2 - x_1}.$$

Assume $x_1 \ne x_2$ and $x_1{}^* \ne x_2{}^*$.

86. Prove that if the slopes of two nonvertical lines are negative reciprocals of each other, then the lines are perpendicular.

True or False? **In Exercises 87 and 88, determine whether the statement is true or false. If it is false, explain why or give an example that shows it is false.**

87. The lines represented by $ax + by = c_1$ and $bx - ay = c_2$ are perpendicular. Assume $a \ne 0$ and $b \ne 0$.

88. It is possible for two lines with positive slopes to be perpendicular to each other.

SECTION $P.3$ **Functions and Their Graphs**

Functions and Function Notation • The Domain and Range of a Function •
The Graph of a Function • Transformations of Functions •
Classifications and Combinations of Functions

Functions and Function Notation

A **relation** between two sets X and Y is a set of ordered pairs, each of the form (x, y) where x is a member of X and y is a member of Y. A **function** from X to Y is a relation between X and Y that has the property that any two ordered pairs with the same x-value also have the same y-value. The variable x is the **independent variable,** and the variable y is the **dependent variable.**

Many real-life situations can be modeled by functions. For instance, the area A of a circle is a function of the circle's radius r.

$$A = \pi r^2 \qquad \text{\textcolor{red}{A is a function of r.}}$$

In this case r is the independent variable and A is the dependent variable.

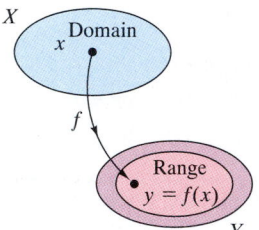

A real-valued function f of a real variable
Figure P.22

> ### Definition of a Real-Valued Function of a Real Variable
>
> Let X and Y be sets of real numbers. A **real-valued function f of a real variable** x from X to Y is a correspondence that assigns to each number x in X exactly one number y in Y.
>
> The **domain** of f is the set X. The number y is the **image** of x under f and is denoted by $f(x)$. The **range** of f is a subset of Y and consists of all images of numbers in X (see Figure P.22).

Functions can be specified in a variety of ways. In this text, however, we will concentrate primarily on functions that are given by equations involving the dependent and independent variables. For instance, the equation

$$x^2 + 2y = 1 \qquad \text{\textcolor{red}{Equation in implicit form}}$$

defines y, the dependent variable, as a function of x, the independent variable. To **evaluate** this function (that is, to find the y-value that corresponds to a given x-value), it is convenient to isolate y on the left side of the equation.

$$y = \tfrac{1}{2}(1 - x^2) \qquad \text{\textcolor{red}{Equation in explicit form}}$$

Using f as the name of the function, you can write this equation as

$$f(x) = \tfrac{1}{2}(1 - x^2). \qquad \text{\textcolor{red}{Function notation}}$$

The original equation, $x^2 + 2y = 1$, **implicitly** defines y as a function of x. When you solve the equation for y, you are writing the equation in **explicit** form.

Function notation has the advantage of clearly identifying the dependent variable as $f(x)$ while at the same time telling you that x is the independent variable and that the function itself is "f." The symbol $f(x)$ is read "f of x." Function notation allows you to be less wordy. Instead of asking "What is the value of y that corresponds to $x = 3$?" you can ask "What is $f(3)$?"

FUNCTION NOTATION

The word *function* was first used by Gottfried Wilhelm Leibniz in 1694 as a term to denote any quantity connected with a curve, such as the coordinates of a point on a curve or the slope of a curve. Forty years later, Leonhard Euler used the word function to describe any expression made up of a variable and some constants. He introduced the notation

$$y = f(x).$$

In an equation that defines a function, the role of the variable x is simply that of a placeholder. For instance, the function given by

$$f(x) = 2x^2 - 4x + 1$$

can be described by the form

$$f(\;\;\;\;) = 2(\;\;\;\;)^2 - 4(\;\;\;\;) + 1$$

where parentheses are used instead of x. To evaluate $f(-2)$, simply place -2 in each set of parentheses.

$$f(-2) = 2(-2)^2 - 4(-2) + 1 \qquad \text{Substitute } -2 \text{ for } x.$$
$$= 2(4) + 8 + 1 \qquad \text{Simplify.}$$
$$= 17 \qquad \text{Simplify.}$$

NOTE   Although f is often used as a convenient function name and x as the independent variable, you can use other symbols. For instance, the following equations all define the same function.

$$f(x) = x^2 - 4x + 7 \qquad \text{Function name is } f, \text{ independent variable is } x.$$
$$f(t) = t^2 - 4t + 7 \qquad \text{Function name is } f, \text{ independent variable is } t.$$
$$g(s) = s^2 - 4s + 7 \qquad \text{Function name is } g, \text{ independent variable is } s.$$

EXAMPLE 1   Evaluating a Function

For the function f defined by $f(x) = x^2 + 7$, evaluate each of the following.

a. $f(3a)$ **b.** $f(b - 1)$ **c.** $\dfrac{f(x + \Delta x) - f(x)}{\Delta x}, \quad \Delta x \neq 0$

Solution

a. $f(3a) = (3a)^2 + 7$ 　　　　　　　Substitute $3a$ for x.

$\qquad = 9a^2 + 7$ 　　　　　　　Simplify.

b. $f(b - 1) = (b - 1)^2 + 7$ 　　　　Substitute $b - 1$ for x.

$\qquad\qquad = b^2 - 2b + 1 + 7$ 　　　Expand binomial.

$\qquad\qquad = b^2 - 2b + 8$ 　　　　　Simplify.

c. $\dfrac{f(x + \Delta x) - f(x)}{\Delta x} = \dfrac{[(x + \Delta x)^2 + 7] - (x^2 + 7)}{\Delta x}$

$$= \frac{x^2 + 2x\Delta x + (\Delta x)^2 + 7 - x^2 - 7}{\Delta x}$$

$$= \frac{2x\Delta x + (\Delta x)^2}{\Delta x}$$

$$= \frac{\cancel{\Delta x}(2x + \Delta x)}{\cancel{\Delta x}}$$

$$= 2x + \Delta x, \qquad \Delta x \neq 0$$

NOTE   The expression in Example 1(c) is called a *difference quotient* and has a special significance in calculus. We will say more about this in Chapter 2.

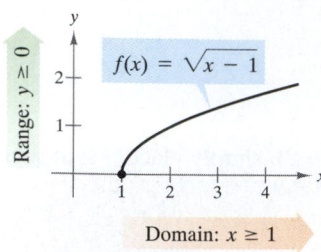

(a) The domain of f is $[1, \infty)$ and the range is $[0, \infty)$.

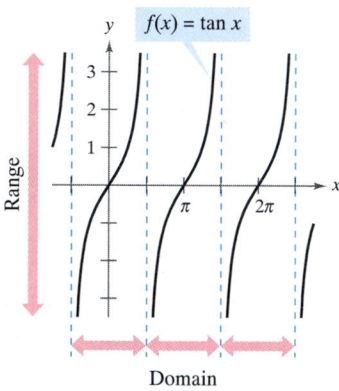

(b) The domain of f is all x-values such that $x \neq \dfrac{\pi}{2} + n\pi$ and the range is $(-\infty, \infty)$.

Figure P.23

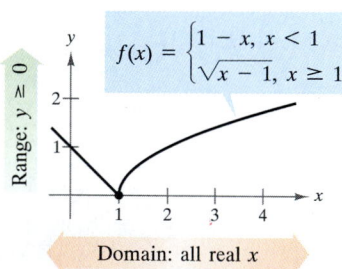

The domain of f is $(-\infty, \infty)$ and the range is $[0, \infty)$.

Figure P.24

The Domain and Range of a Function

The domain of a function can be described explicitly, or it may be described *implicitly* by an equation used to define the function. The implied domain is the set of all real numbers for which the equation is defined, whereas an explicitly defined domain is one that is given along with the function. For example, the function given by

$$f(x) = \frac{1}{x^2 - 4}, \qquad 4 \leq x \leq 5$$

has an explicitly defined domain given by $\{x: 4 \leq x \leq 5\}$. On the other hand, the function given by

$$g(x) = \frac{1}{x^2 - 4}$$

has an implied domain that is the set $\{x: x \neq \pm 2\}$.

EXAMPLE 2   Finding the Domain and Range of a Function

a. The domain of the function

$$f(x) = \sqrt{x - 1}$$

is the set of all x-values for which $x - 1 \geq 0$, which is the interval $[1, \infty)$. To find the range observe that $f(x) = \sqrt{x - 1}$ is never negative. Thus, the range is the interval $[0, \infty)$, as indicated in Figure P.23(a).

b. The domain of the tangent function, as shown in Figure P.23(b),

$$f(x) = \tan x$$

is the set of all x-values such that

$$x \neq \frac{\pi}{2} + n\pi, \qquad n \text{ is an integer.} \qquad \text{\color{red}Domain of tangent function}$$

The range of this function is the set of all real numbers. For a review of the characteristics of this and other trigonometric functions, see the appendix.

EXAMPLE 3   A Function Defined by More than One Equation

Determine the domain and range of the function.

$$f(x) = \begin{cases} 1 - x, & \text{if } x < 1 \\ \sqrt{x - 1}, & \text{if } x \geq 1 \end{cases}$$

Solution Because f is defined for $x < 1$ and $x \geq 1$, the domain is the entire set of real numbers. On the portion of the domain for which $x \geq 1$, the function behaves as in Example 2(a). For $x < 1$, the values of $1 - x$ are positive. Therefore, the range of the function is the interval $[0, \infty)$. (See Figure P.24.)

A function from X to Y is **one-to-one** if to each y-value in the range there corresponds exactly one x-value in the domain. For instance, the function given in Example 2(a) is one-to-one, whereas the functions given in Examples 2(b) and 3 are not one-to-one. A function is **onto** if its range consists of all of Y.

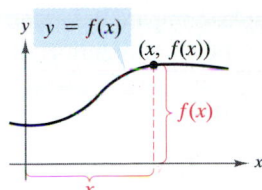

The graph of a function
Figure P.25

The Graph of a Function

The graph of the function $y = f(x)$ consists of all points $(x, f(x))$, where x is in the domain of f. In Figure P.25, note that

$$x = \text{the directed distance from the } y\text{-axis}$$

$$f(x) = \text{the directed distance from the } x\text{-axis.}$$

A vertical line can intersect the graph of a function of x at most *once*. This observation provides a convenient visual test (called the **vertical line test**) for functions of x. For example, in Figure P.26(a), you can see that the graph does not define y as a function of x because a vertical line intersects the graph twice, whereas in Figures P.26(b) and (c), the graphs do define y as a function of x.

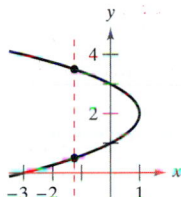

(a) Not a function of x

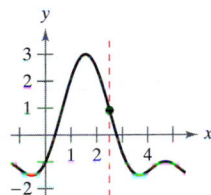

(b) A function of x

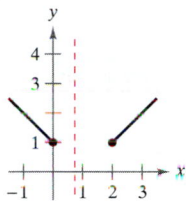

(c) A function of x

Figure P.26

Figure P.27 shows the graphs of eight basic functions. You should be able to recognize these graphs. (Graphs of the other four basic trigonometric functions are shown in the appendix.)

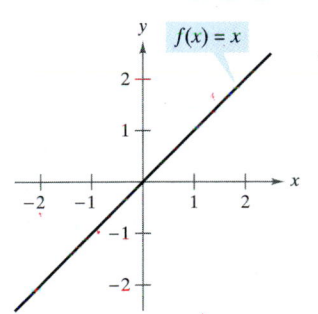

Identity function

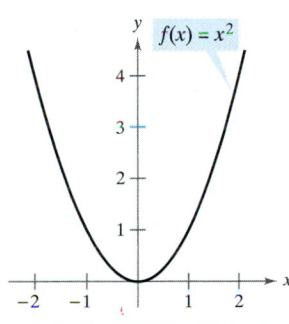

Squaring function

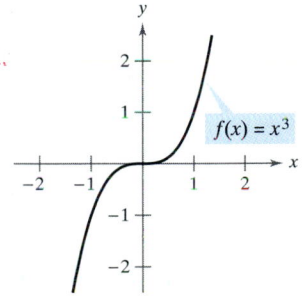

Cubing function

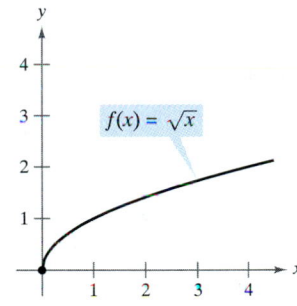

Square root function

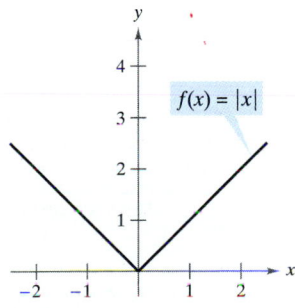

Absolute value function

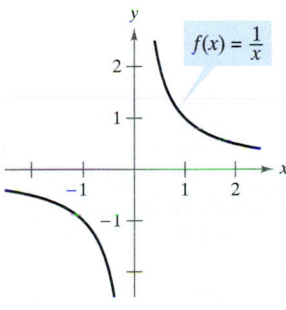

Rational function

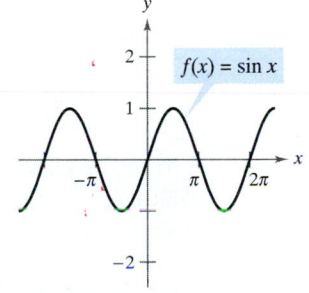

Sine function

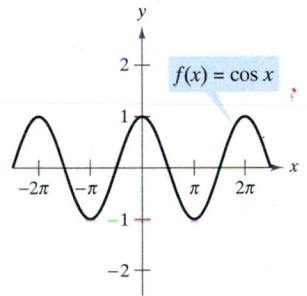

Cosine function

The graphs of eight basic functions
Figure P.27

Writing Equations for Functions
Each of the graphing utility screens shown below has the graph of one of the eight basic functions shown on page 23. Each screen also has a transformation of the graph. Describe the transformation. Then use your description to write an equation for the transformation.

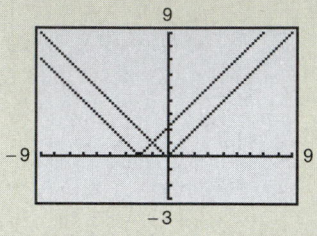

(a)

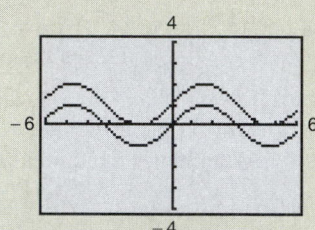

(b)

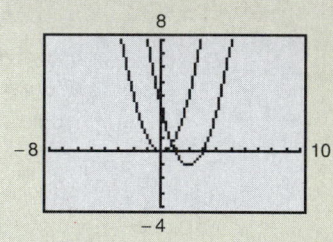

(c)

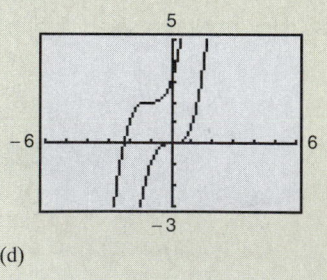

(d)

Transformations of Functions

Some families of graphs have the same basic shape. For example, compare the graph of $y = x^2$ with the graphs of the four other quadratic functions shown in Figure P.28.

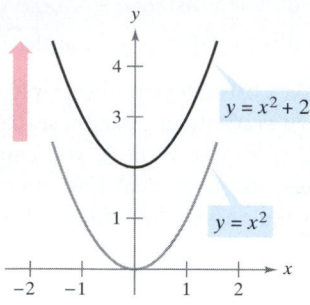
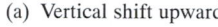

(a) Vertical shift upward

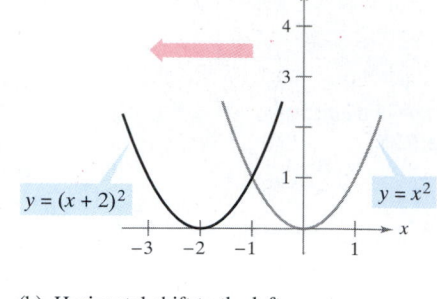

(b) Horizontal shift to the left

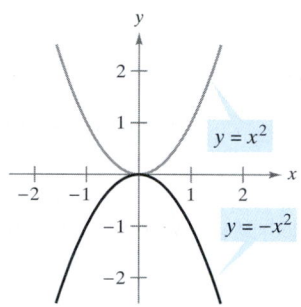

(c) Reflection

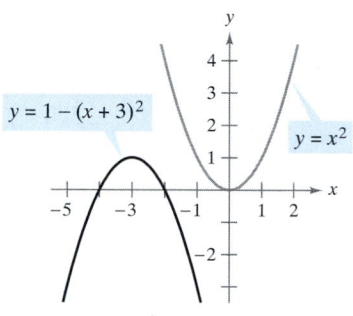

(d) Shift left, reflect, and shift upward

Figure P.28

Each of the graphs in Figure P.28 is a **transformation** of the graph of $y = x^2$. The three basic types of transformations illustrated by these graphs are vertical shifts, horizontal shifts, and reflections. Function notation lends itself well to describing transformations of graphs in the plane. For instance, if $f(x) = x^2$ is considered to be the original function in Figure P.28, the transformations shown can be represented by the following equations.

$y = f(x) + 2$	Vertical shift up 2 units
$y = f(x + 2)$	Horizontal shift to the left 2 units
$y = -f(x)$	Reflection about the x-axis
$y = -f(x + 3) + 1$	Shift left 3 units, reflect about x-axis, and shift up 1 unit

Basic Types of Transformations ($c > 0$)

Original graph:	$y = f(x)$
Horizontal shift c units to the **right**:	$y = f(x - c)$
Horizontal shift c units to the **left**:	$y = f(x + c)$
Vertical shift c units **downward**:	$y = f(x) - c$
Vertical shift c units **upward**:	$y = f(x) + c$
Reflection (about the x-axis):	$y = -f(x)$
Reflection (about the y-axis):	$y = f(-x)$
Reflection (about the origin):	$y = -f(-x)$

Leonhard Euler (1707–1783)

In addition to making major contributions to almost every branch of mathematics, Euler was one of the first to apply calculus to real-life problems in physics. His extensive published writings include such topics as shipbuilding, acoustics, optics, astronomy, mechanics, and magnetism.

FOR FURTHER INFORMATION For more on the history of the concept of a function, see the article "Evolution of the Function Concept: A Brief Survey" by Israel Kleiner in the September 1989 issue of *The College Mathematics Journal.*

Classifications and Combinations of Functions

The modern notion of a function is derived from the efforts of many seventeenth- and eighteenth-century mathematicians. Of particular note was Leonhard Euler, to whom we are indebted for the function notation $y = f(x)$. By the end of the eighteenth century, mathematicians and scientists had concluded that most real-world phenomena could be represented by mathematical models taken from a collection of functions called **elementary functions.** Elementary functions fall into three categories.

1. Algebraic functions (polynomial, radical, rational)
2. Trigonometric functions (sine, cosine, tangent, and so on)
3. Exponential and logarithmic functions

You can review the trigonometric functions in the appendix. The other nonalgebraic functions, such as the inverse trigonometric functions and the exponential and logarithmic functions, are introduced in Chapter 5.

The most common type of algebraic function is a **polynomial function**

$$f(x) = a_n x^n + a_{n-1} x^{n-1} + \cdots + a_2 x^2 + a_1 x + a_0, \qquad a_n \neq 0$$

where the positive integer n is the **degree** of the polynomial function. The numbers a_i are **coefficients,** with a_n the **leading coefficient** and a_0 the **constant term** of the polynomial function. It is common practice to use subscript notation for coefficients of general polynomial functions, but for polynomial functions of low degree, the following simpler forms are often used.

Zeroth degree:	$f(x) = a$	Constant function
First degree:	$f(x) = ax + b$	Linear function
Second degree:	$f(x) = ax^2 + bx + c$	Quadratic function
Third degree:	$f(x) = ax^3 + bx^2 + cx + d$	Cubic function

Although the graph of a polynomial function can have several turns, eventually the graph will rise or fall without bound as x moves to the right or left. Whether the graph of

$$f(x) = a_n x^n + a_{n-1} x^{n-1} + \cdots + a_2 x^2 + a_1 x + a_0$$

eventually rises or falls can be determined by the function's degree (odd or even) and by the leading coefficient a_n, as indicated in Figure P.29. Note that the dashed portions of the graphs indicate that the **leading coefficient test** determines *only* the right and left behavior of the graph.

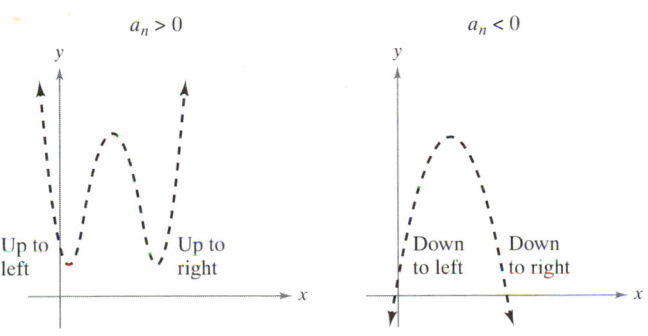

Graphs of polynomial functions of even degree

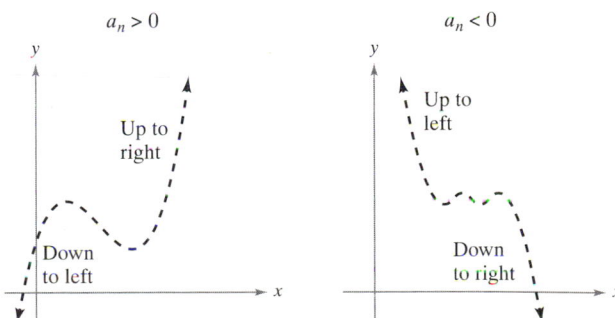

Graphs of polynomial functions of odd degree

The leading coefficient test for polynomial functions
Figure P.29

EXERCISES FOR SECTION P.3

In Exercises 1–10, evaluate (if possible) the function at the given value(s) of the independent variable. Simplify the results.

1. $f(x) = 2x - 3$
  (a) $f(0)$
  (b) $f(-3)$
  (c) $f(b)$
  (d) $f(x - 1)$

2. $f(x) = \sqrt{x + 3}$
  (a) $f(-2)$
  (b) $f(6)$
  (c) $f(c)$
  (d) $f(x + \Delta x)$

3. $f(x) = \begin{cases} 2x + 1, & x < 0 \\ 2x + 2, & x \ge 0 \end{cases}$
  (a) $f(-1)$ (b) $f(0)$ (c) $f(2)$ (d) $f(t^2 + 1)$

4. $f(x) = \begin{cases} x^2 + 2, & x \le 1 \\ 2x^2 + 2, & x > 1 \end{cases}$
  (a) $f(-2)$ (b) $f(0)$ (c) $f(1)$ (d) $f(s^2 + 2)$

5. $f(x) = \cos 2x$
  (a) $f(0)$ (b) $f(-\pi/4)$ (c) $f(\pi/3)$

6. $f(x) = \sin x$
  (a) $f(\pi)$ (b) $f(5\pi/4)$ (c) $f(2\pi/3)$

7. $f(x) = x^3$
$$\frac{f(x + \Delta x) - f(x)}{\Delta x}$$

8. $f(x) = 3x - 1$
$$\frac{f(x) - f(1)}{x - 1}$$

9. $f(x) = \dfrac{1}{\sqrt{x - 1}}$
$$\frac{f(x) - f(2)}{x - 2}$$

10. $f(x) = x^3 - x$
$$\frac{f(x) - f(1)}{x - 1}$$

In Exercises 11–18, sketch a graph of the function and find its domain and range. You can use a graphing utility to verify your graph.

11. $f(x) = 4 - x$
12. $g(x) = \dfrac{4}{x}$

13. $h(x) = \sqrt{x - 1}$
14. $f(x) = \frac{1}{2}x^3 + 2$

15. $f(x) = \sqrt{9 - x^2}$
16. $f(x) = x + \sqrt{4 - x^2}$

17. $g(t) = 2 \sin \pi t$
18. $h(\theta) = -5 \cos \dfrac{\theta}{2}$

In Exercises 19 and 20, use the vertical line test to determine whether y is a function of x.

19. $x - y^2 = 0$
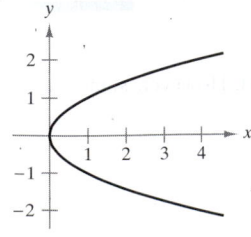

20. $\sqrt{x^2 - 4} - y = 0$
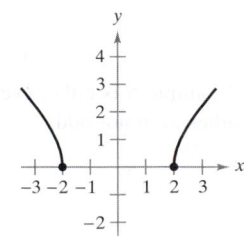

21. *Think About It* Express the function
$$f(x) = |x| + |x - 2|$$
without using absolute value signs. (For a review of absolute value, see the appendix.)

22. *Writing* Use a graphing utility to graph the polynomial functions $p_1(x) = x^3 - x + 1$ and $p_2(x) = x^3 - x$. How many zeros does each function have? Is there a cubic polynomial that has no zeros? Explain.

In Exercises 23–26, determine whether y is a function of x.

23. $x^2 + y^2 = 4$
24. $x^2 + y = 4$
25. $y^2 = x^2 - 1$
26. $x^2y - x^2 + 4y = 0$

In Exercises 27 and 28, determine whether the relation is a function.

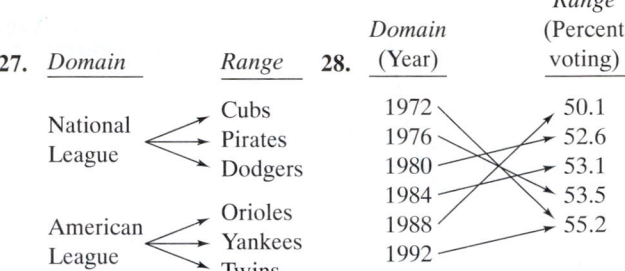

Modeling Data In Exercises 29–32, match the data with a function from the following list.

(i) $f(x) = cx$ (ii) $g(x) = cx^2$
(iii) $h(x) = c\sqrt{|x|}$ (iv) $r(x) = c/x$

Determine the value of the constant c for each function so the function fits the data given in the table.

29.

x	-4	-1	0	1	4
y	-32	-2	0	-2	-32

30.

x	-4	-1	0	1	4
y	-1	$-\frac{1}{4}$	0	$\frac{1}{4}$	1

31.

x	-4	-1	0	1	4
y	-8	-32	Undef.	32	8

32.

x	-4	-1	0	1	4
y	6	3	0	3	6

33. *Think About It*   Water runs into a vase of height 30 centimeters at a constant rate. The vase is full after 5 seconds. Use this information and the shape of the vase shown in the figure to answer the questions if d is the depth of the water in centimeters and t is the time in seconds.

(a) Explain why d is a function of t.

(b) Determine the domain and range of the function.

(c) Sketch a possible graph of the function.

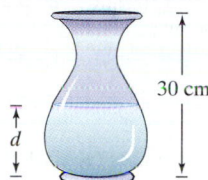

30 cm

d

34. *Think About It*   A student who commutes 27 miles to attend college remembers, after driving a few minutes, that a term paper that is due has been forgotten. Driving faster than usual, the student returns home, picks up the paper, and once again starts toward school. Sketch a possible graph of the student's distance from home as a function of time.

35. Select the viewing rectangle on a graphing utility that shows the most complete graph of the function $f(x) = 10x\sqrt{400 - x^2}$.

Xmin = -5	Xmin = -20	Xmin = -25
Xmax = 50	Xmax = 20	Xmax = 25
Xscl = 5	Xscl = 2	Xscl = 5
Ymin = -5000	Ymin = -500	Ymin = -2000
Ymax = 5000	Ymax = 500	Ymax = 2000
Yscl = 500	Yscl = 50	Yscl = 200

36. Use the graph of f shown in the figure to sketch the graph of each function.

(a) $f(x - 4)$ (b) $f(x + 2)$ (c) $f(x) + 4$

(d) $f(x) - 1$ (e) $2f(x)$ (f) $\frac{1}{2}f(x)$

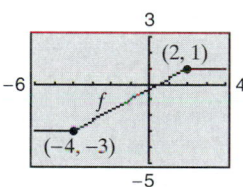

37. Use the graph of $f(x) = \sqrt{x}$ to sketch the graph of each function. In each case, describe the transformation.

(a) $y = \sqrt{x} + 2$ (b) $y = -\sqrt{x}$ (c) $y = \sqrt{x - 2}$

38. Specify the sequence of transformations that will yield each graph of h from the graph of the function $f(x) = \sin x$.

(a) $h(x) = \sin\left(x + \dfrac{\pi}{2}\right) + 1$ (b) $h(x) = -\sin(x - 1)$

39. *Graphical Reasoning*   An electronically controlled thermostat in a home is programmed to automatically lower the temperature during the night (see figure). The temperature T in degrees Celsius is given in terms of t, the time in hours on a 24-hour clock.

(a) Approximate $T(4)$ and $T(15)$.

(b) Suppose the thermostat were reprogrammed to produce a temperature $H(t) = T(t - 1)$. How would this change the temperature in the house? Explain.

(c) Suppose the thermostat were reprogrammed to produce a temperature $H(t) = T(t) - 1$. How would this change the temperature in the house? Explain.

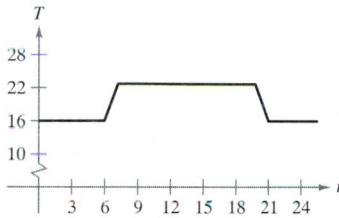

40. Given $f(x) = \sqrt{x}$ and $g(x) = x^2 - 1$, find the following.

(a) $f(g(1))$ (b) $g(f(1))$ (c) $g(f(0))$

(d) $f(g(-4))$ (e) $f(g(x))$ (f) $g(f(x))$

In Exercises 41–44, find the composite functions $(f \circ g)$ and $(g \circ f)$. What is the domain of each composite function? Are the two composite functions equal?

41. $f(x) = x^2$

$g(x) = \sqrt{x}$

42. $f(x) = x^2 - 1$

$g(x) = \cos x$

43. $f(x) = \dfrac{1}{x}$

$g(x) = x^2 + 1$

44. $f(x) = \dfrac{1}{x}$

$g(x) = \sqrt{x + 2}$

45. *Ripples*   A pebble is dropped into a calm pond, causing ripples in the form of concentric circles. The radius (in feet) of the outer ripple is given by $r(t) = 0.6t$, where t is the time in seconds after the pebble strikes the water. The area of the circle is given by the function $A(r) = \pi r^2$. Find and interpret $(A \circ r)(t)$.

46. *Automobile Aerodynamics*   The number of horsepower H required to overcome wind drag on a certain automobile is approximated by

$$H(x) = 0.002x^2 + 0.005x - 0.029, \quad 10 \le x \le 100$$

where x is the speed of the car in miles per hour.

(a) Use a graphing utility to graph H.

(b) Rewrite the power function so that x represents the speed in kilometers per hour. [Find $H(1.6x)$.]

In Exercises 47–50, determine whether the function is even, odd, or neither. Use a graphing utility to verify your result.

47. $f(x) = 4 - x^2$

48. $f(x) = \sqrt[3]{x}$

49. $f(x) = x \cos x$

50. $f(x) = \sin^2 x$

Think About It **In Exercises 51 and 52, find the coordinates of a second point on the graph of a function *f* if the given point is on the graph and the function is (a) even and (b) odd.**

51. $\left(-\frac{3}{2}, 4\right)$ **52.** $(4, 9)$

53. Prove that the function is odd.

$$f(x) = a_{2n+1}x^{2n+1} + \cdots + a_3 x^3 + a_1 x$$

54. Prove that the function is even.

$$f(x) = a_{2n}x^{2n} + a_{2n-2}x^{2n-2} + \cdots + a_2 x^2 + a_0$$

55. Prove that the product of two even (or two odd) functions is even.

56. Prove that the product of an odd function and an even function is odd.

57. Use a graphing utility to graphically demonstrate the results of Exercises 55 and 56. Use functions of your choice.

58. What can be said about the sum or difference of (a) two even functions, (b) two odd functions, and (c) an odd function and an even function? Demonstrate your conclusions graphically.

59. ***Area*** A rectangle has a perimeter of 100 meters.

(a) Make a sketch that gives a visual representation of the rectangle where *x* represents its length.

(b) Express the area *A* of the rectangle as a function of *x*.

(c) Use a graphing utility to graph the area function. What is the domain of the function?

(d) Use the graph in part (c) to determine the dimensions that yield a maximum area.

60. ***Time*** You are in a boat 2 miles from the nearest point on the coast. You are to go to a point *Q*, 3 miles down the coast and 1 mile inland (see figure). You can row at 2 miles per hour and walk at 4 miles per hour. Express the total time *T* of the trip as a function of *x*.

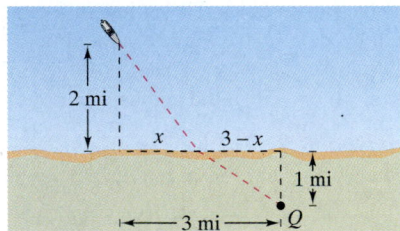

61. ***Volume*** An open box of maximum volume is to be made from a square piece of material 24 centimeters on a side by cutting equal squares from the corners and turning up the sides (see figure).

(a) Use the table feature of a graphing utility to complete six rows of a table. (The first two rows are shown.) Use the result to guess the maximum volume.

Height, *x*	Length and Width	Volume, *V*
1	$24 - 2(1)$	$1[24 - 2(1)]^2 = 484$
2	$24 - 2(2)$	$2[24 - 2(2)]^2 = 800$

(b) Use a graphing utility to plot the points (x, V). Is *V* a function of *x*?

(c) If yes, write *V* as a function of *x*, and determine its domain.

(d) Use a graphing utility to graph the volume function and approximate the dimensions of the box that yield a maximum volume.

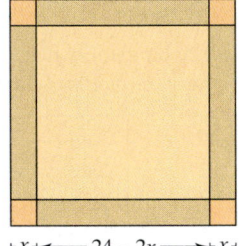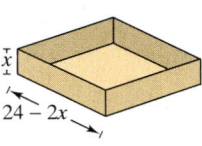

62. ***Length*** A right triangle is formed in the first quadrant by the *x*- and *y*-axes and the line through the point $(3, 2)$ (see figure). Write the length *L* of the hypotenuse as a function of *x*.

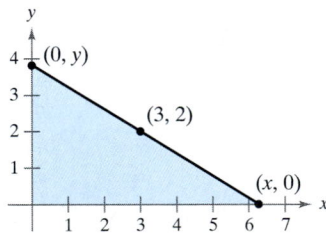

True or False? **In Exercises 63–66, determine whether the statement is true or false. If it is false, explain why or give an example that shows it is false.**

63. If $f(a) = f(b)$, then $a = b$.

64. A vertical line can intersect the graph of a function at most once.

65. If $f(x) = f(-x)$ for all *x* in the domain of *f*, then the graph of *f* is symmetric with respect to the *y*-axis.

66. If *f* is a function, then $f(ax) = af(x)$.

Fitting a Linear Model to Data • Fitting a Quadratic Model to Data •
Fitting a Trigonometric Model to Data

Fitting a Linear Model to Data

A basic premise of science is that much of the physical world can be described mathematically and that many physical phenomena are predictable. This scientific outlook was part of the scientific revolution that took place in Europe during the late 1500s. Two early publications that are connected with this revolution were *On the Revolutions of the Heavenly Spheres* by the Polish astronomer Nicolaus Copernicus and *On the Structure of the Human Body* by the Belgian anatomist Andreas Vesalius. Each of these books was published in 1543 and each broke with prior tradition by suggesting the use of a scientific method rather than unquestioned reliance on authority.

One characteristic of modern science is gathering data and then describing the data with a mathematical model. For instance, the data given in Example 1 are inspired by Leonardo da Vinci's famous drawing that indicates that a person's height and arm span are equal.

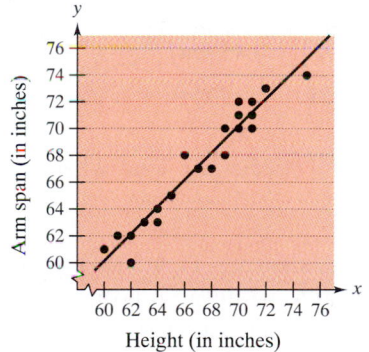

A computer graphics drawing based on the pen and ink drawing of Leonardo da Vinci's famous study of human proportions, called *Vitruvian Man*

EXAMPLE 1   Fitting a Linear Model to Data

A class of 28 people collected the following data, which represents their heights x and arm spans y (rounded to the nearest inch).

(60, 61), (65, 65), (68, 67), (72, 73), (61, 62), (63, 63), (70, 71),

(75, 74), (71, 72), (62, 60), (65, 65), (66, 68), (62, 62), (72, 73),

(70, 70), (69, 68), (69, 70), (60, 61), (63, 63), (64, 64), (71, 71),

(68, 67), (69, 70), (70, 72), (65, 65), (64, 63), (71, 70), (67, 67)

Find a linear model to represent these data.

Solution   There are different ways to model these data with an equation. The simplest would be to observe that x and y are about the same and list the model as simply $y = x$. A more careful analysis would be to use a procedure from statistics called linear regression. (You will study this procedure in Section 12.9.) The least squares regression line for these data is

$$y = 1.006x - 0.225.$$ Least squares regression line

The graph of the model and the data are shown in Figure P.32. From this model, you can see that a person's arm span tends to be about the same as his or her height.

Linear model and data
Figure P.32

TECHNOLOGY   Many scientific and graphing calculators have built-in least squares regression programs. Typically, you enter the data into the calculator and then run the linear regression program. The program usually displays the slope and y-intercept of the best-fitting line *and* the correlation coefficient r. The closer $|r|$ is to 1, the better the model fits the data. For instance, in Example 1, the value of r is 0.97, which indicates that the model is a good fit for the data. If the r-value is positive, the variables have a positive correlation, as in Example 1. If the r-value is negative, the variables have a negative correlation.

Fitting a Quadratic Model to Data

A function that gives the height s of a falling object in terms of the time t is called a position function. If air resistance is not considered, the position of a falling object can be modeled by

$$s(t) = \tfrac{1}{2}gt^2 + v_0t + s_0$$

where g is the acceleration due to gravity, v_0 is the initial velocity, and s_0 is the initial height. The value of g depends on where the object is dropped. On earth, g is approximately -32 feet per second per second, or -9.8 meters per second per second.

To discover the value of g experimentally, you could record the heights of a falling object at several increments, as shown in Example 2.

EXAMPLE 2   Fitting a Quadratic Model to Data

A basketball is dropped from a height of about $5\tfrac{1}{4}$ feet, as shown in Figure P.33. The height of the basketball is recorded 23 times at intervals of about 0.02 second*. The results are shown in the table.

Time	0.0	0.02	0.04	0.06	0.08	0.099996
Height	5.23594	5.20353	5.16031	5.0991	5.02707	4.95146

Time	0.119996	0.139992	0.159988	0.179988	0.199984	0.219984
Height	4.85062	4.74979	4.63096	4.50132	4.35728	4.19523

Time	0.23998	0.25993	0.27998	0.299976	0.319972	0.339961
Height	4.02958	3.84593	3.65507	3.44981	3.23375	3.01048

Time	0.359961	0.379951	0.399941	0.419941	0.439941
Height	2.76921	2.52074	2.25786	1.98058	1.63488

Find a model to fit these data. Then use the model to predict the time when the basketball will hit the ground.

Solution   Begin by drawing a scatter plot of the data, as shown in Figure P.34. From the scatter plot, you can see that the data do not appear to be linear. It does appear, however, that they might be quadratic. To check this, enter the data into a calculator or computer that has a quadratic regression program. You should obtain the model

$$s = -15.45t^2 - 1.30t + 5.234.$$ Least squares regression quadratic

Using this model, you can predict the time when the basketball hits the ground by substituting 0 for s and solving the resulting equation for t.

$$0 = -15.45t^2 - 1.30t + 5.234$$ Let $s = 0$.

$$t = \frac{1.30 \pm \sqrt{(-1.30)^2 - 4(-15.45)(5.234)}}{2(-15.45)}$$ Quadratic Formula

$$t \approx 0.54$$ Choose positive solution.

The solution is about 0.54 second. In other words, the basketball will continue to fall for about 0.1 second more before hitting the ground.

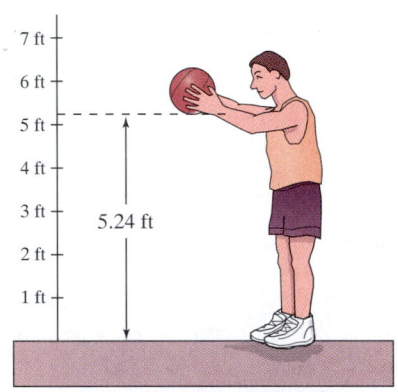

A basketball is dropped from a height of about 5.24 feet. Using a CBL System, the ball's height is recorded every 0.02 second.
Figure P.33

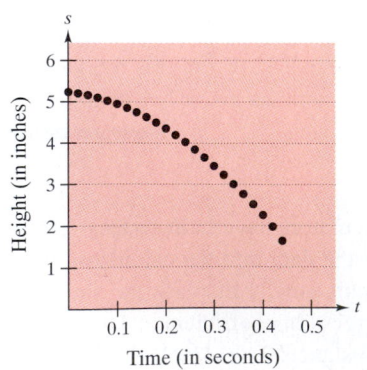

Scatter plot of data
Figure P.34

---

*Data were collected with a Texas Instruments CBL (Calculator-Based Laboratory) System.

Fitting a Trigonometric Model to Data

What is mathematical modeling? This is one of the questions that is asked in the book *Guide to Mathematical Modelling*. Here is part of the answer.*

1. Mathematical modeling consists of applying your mathematical skills to obtain useful answers to real problems.

2. Learning to apply mathematical skills is very different from learning mathematics itself.

3. Models are used in a very wide range of applications, some of which do not appear initially to be mathematical in nature.

4. Models often allow quick and cheap evaluation of alternatives, leading to optimal solutions that are not otherwise obvious.

5. There are no precise rules in mathematical modeling and no "correct" answers.

6. Modeling can be learned only by *doing*.

EXAMPLE 3   Fitting a Trigonometric Model to Data

The number of hours of daylight on earth depends on two variables: the latitude and the time of year. Here are the numbers of minutes of daylight at a location of 20° latitude on the longest and shortest days of the year: June 21 (summer solstice), 801 minutes; December 21 (winter solstice), 655 minutes. Use this data to write a model for the number of minutes of daylight d on each day of the year at a location of 20° latitude. How could you check the accuracy of your model.

Solution   Here is one way to create a model. You can hypothesize that the model is a sine function whose period is 365 days. Using the given data, you can conclude that the amplitude of the graph is $(801 - 655)/2$, or 73. Thus, one possible model is

$$d = 728 - 73 \sin\left(\frac{2\pi t}{365} + \frac{\pi}{2}\right).$$

In this model, t represents the number of each day of the year, with December 21 represented by $t = 0$. A graph of this model is shown in Figure P.35. To check the accuracy of this model, we used a weather almanac to find the numbers of minutes of daylight on different days of the year at the location of 20° latitude.

Date	Value of t	Actual Daylight	Daylight Given by Model
Dec 21	0	655 min	655 min
Jan 1	11	657 min	656 min
Feb 1	42	676 min	673 min
Mar 1	70	705 min	702 min
Apr 1	101	740 min	740 min
May 1	131	772 min	774 min
Jun 1	162	796 min	797 min
Jun 21	182	801 min	801 min
Jul 1	192	799 min	800 min
Aug 1	223	782 min	784 min
Sep 1	254	752 min	752 min
Oct 1	284	718 min	715 min
Nov 1	315	685 min	680 min
Dec 1	345	661 min	659 min

You can see that the model is fairly accurate.

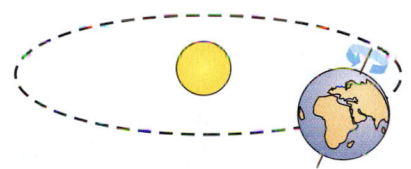

The plane of earth's orbit about the sun and its axis of rotation are not perpendicular. Instead, earth's axis is tilted with respect to its orbit. The result is that the amount of daylight received by locations on earth varies with the time of year. That is, it varies with the position of earth in its orbit.

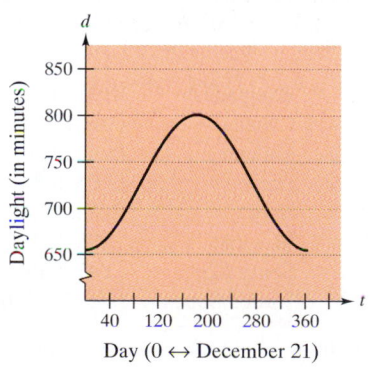

Graph of model
Figure P.35

Text from Dilwyn Edwards and Mike Hamson, GUIDE TO MATHEMATICAL MODELLING (Boca Raton: CRC Press, 1990). Used by permission of the authors

In Exercises 1–4, a scatter plot of data is given. Determine whether the data can be modeled by a linear function, a quadratic function, or a trigonometric function, or that there appears to be no relationship between *x* and *y*.

1.   *y*

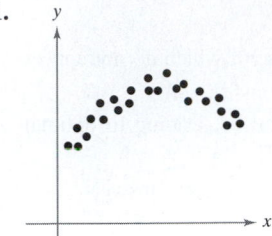

2.   *y*

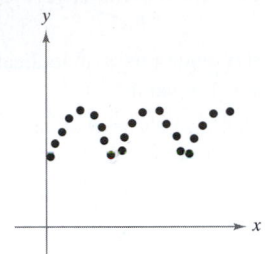

3.   *y*

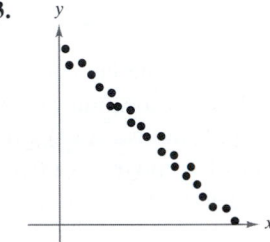

4.   *y*

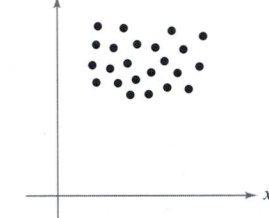

5. Carcinogens   The ordered pairs give the exposure index *x* of a carcinogenic substance and the cancer mortality *y* per 100,000 people in the population.

(3.50, 150.1), (3.58, 133.1), (4.42, 132.9),

(2.26, 116.7), (2.63, 140.7), (4.85, 165.5),

(12.65, 210.7), (7.42, 181.0), (9.35, 213.4)

(a) Plot the data. From the graph, do the data appear to be approximately linear?

(b) Visually find a linear model for the data. Graph the model.

(c) Use the model to approximate *y* if *x* = 3.

6. Quiz Scores   The ordered pairs give the scores of two consecutive 15-point quizzes for a class of 18 students.

(7, 13), (9, 7), (14, 14), (15, 15), (10, 15), (9, 7),

(14, 11), (14, 15), (8, 10), (15, 9), (10, 11) (9, 10),

(11, 14), (7, 14), (11, 10), (14, 11), (10, 15), (9, 6)

(a) Plot the data. From the graph, does the relationship between consecutive scores appear approximately linear?

(b) If the data appear approximately linear, find a linear model for the data. If not, give some possible explanations.

 7. Hooke's Law   Hooke's Law states that the force *F* required to compress or stretch a spring (within its elastic limits) is proportional to the distance *d* that the spring is compressed or stretched from its original length. That is, *F* = *kd*, where *k* is a measure of the stiffness of the spring and is called the *spring constant.* The table gives the elongation *d* in centimeters of a spring when a force of *F* kilograms is applied.

F	20	40	60	80	100
d	1.4	2.5	4.0	5.3	6.6

(a) Use the regression capabilities of a graphing utility to find a linear model for the data.

(b) Use a graphing utility to plot the data and graph the model. How well does the model fit the data? Explain your reasoning.

(c) Use the model to estimate the elongation of the spring when a force of 55 kilograms is applied.

 8. Falling Object   In an experiment, students measured the speed *s* (in meters per second) of a falling object *t* seconds after it was released. The results are given in the table

t	0	1	2	3	4
s	0	11.0	19.4	29.2	39.4

(a) Use the regression capabilities of a graphing utility to find a linear model for the data.

(b) Use a graphing utility to plot the data and graph the model. How well does the model fit the data? Explain your reasoning.

(c) Use the model to estimate the speed of the object after 2.5 seconds.

 9. Energy Consumption   The data give the per capita energy usage (in thousands of kilograms of coal equivalent) and the per capita gross national product (in thousands of U.S. dollars) for a sample of countries in 1990.   *(Source: Statistical Office of the United Nations)*

Argentina	(1.83, 3.7)	Bangladesh	(0.07, 0.2)
Brazil	(0.77, 2.6)	Canada	(10.51, 21.5)
Denmark	(4.70, 24.0)	Finland	(5.93, 26.0)
France	(3.87, 20.8)	Greece	(3.05, 6.8)
India	(0.31, 0.3)	Italy	(3.86, 19.4)
Japan	(4.21, 26.2)	Mexico	(1.75, 3.0)
Pakistan	(0.28, 0.4)	South Korea	(2.47, 6.0)
Tanzania	(0.04, 0.1)	United States	(10.32, 23.0)

(a) Use the regression capabilities of a graphing utility to find a linear model for the data.

(b) Use a graphing utility to plot the data and graph the model.

(c) Interpret the graph in part (b). Use the graph to identify any countries that appear to differ from the linear model.

10. Brinell Hardness The data in the table give the Brinell hardness H of 0.35 carbon steel when hardened and tempered at temperature t (degrees Fahrenheit). (*Source: Standard Handbook for Mechanical Engineers*)

t	200	400	600	800	1000	1200
H	534	495	415	352	269	217

(a) Use the regression capabilities of a graphing utility to find a linear model for the data.

(b) Use a graphing utility to plot the data and graph the model. How well does the model fit the data? Explain your reasoning.

(c) Use the model to estimate the hardness when t is 500°F.

11. Automobile Costs The data in the table give the variable costs for operating an automobile in the United States for the years 1985 through 1991. The functions y_1, y_2, and y_3 represent the costs in cents per mile for gas and oil, maintenance, and tires. (*Source: American Automobile Manufacturers Association*)

Year	y_1	y_2	y_3
1985	6.16	1.23	0.65
1986	4.48	1.37	0.67
1987	4.80	1.60	0.80
1988	5.20	1.60	0.80
1989	5.20	1.90	0.80
1990	5.40	2.10	0.90
1991	6.70	2.20	0.90

(a) Let t be the time in years, with $t = 5$ corresponding to 1985. Use the regression capabilities of a graphing utility to find a quadratic model for y_1 and linear models for y_2 and y_3.

(b) Use a graphing utility to graph y_1, y_2, y_3, and $y_1 + y_2 + y_3$ in the same viewing rectangle. Use the model to estimate the total variable cost per mile in 1998.

12. Beam Strength Students in a lab measured the breaking strength S (in pounds) of wood 2 inches thick, x inches high, and 12 inches long. The results are given in the table.

x	4	6	8	10	12
S	2370	5460	10,310	16,250	23,860

(a) Use the regression capabilities of a graphing utility to fit a quadratic model to the data.

(b) Use a graphing utility to plot the data and graph the model.

(c) Use the model to approximate the breaking strength when $x = 2$.

13. Health Maintenance Organization The bar graph gives the number of people N (in millions) receiving their care in HMOs for the years 1986 through 1995. (*Source: GHHA's National Directory of HMOs*)

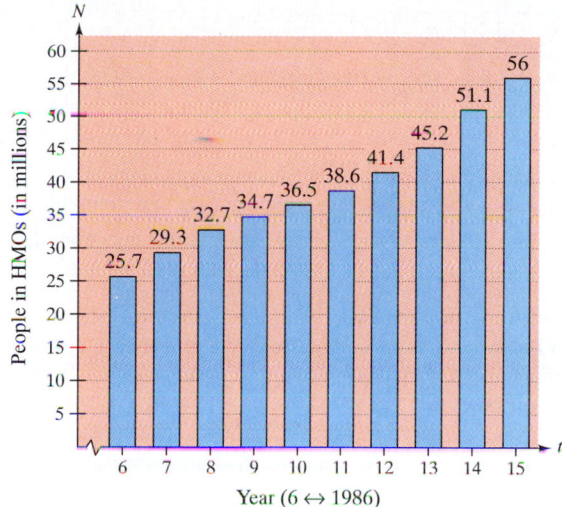

(a) Let t be the time in years, with $t = 6$ corresponding to 1986. Use the regression capabilities of a graphing utility to find linear and cubic models for the data.

(b) Use a graphing utility to graph the data and the linear and cubic models.

(c) Use the graphs in part (b) to determine which is the better model.

(d) Use a graphing utility to find and graph a quadratic model for the data.

(e) Interpret the slope of the linear model in terms of the context of the data.

(f) Use the linear and cubic models to predict the number of people receiving care in HMOs in the year 2000.

14. Car Performance The time t (in seconds) required to attain a speed of s miles per hour from a standing start for a 1995 Dodge Avenger is given in the table. (*Source: Road & Track, March 1995*)

s	30	40	50	60	70	80	90
t	3.4	5.0	7.0	9.3	12.0	15.8	20.0

(a) Use the regression capabilities of a graphing utility to find a quadratic model for the data.

(b) Use a graphing utility to plot the data and graph the model.

(c) Use the graph in part (b) to state why the model is not appropriate for determining the time required to attain speeds less than 20 miles per hour.

(d) Because the test began from a standing start, add the point $(0, 0)$ to the data. Fit a quadratic model to the revised data and graph the new model. Does it more accurately model the behavior of the car for low speeds? Explain.

15. Car Performance   A V8 car engine is coupled to a dynamometer and the horsepower y is measured at different engine speeds x (in thousands of revolutions per minute). The results are shown in the table.

x	1	2	3	4	5	6
y	40	85	140	200	225	245

(a) Use the regression capabilities of a graphing utility to find a cubic model for the data.

(b) Use a graphing utility to plot the data and graph the model.

(c) Use the model to approximate the horsepower when the engine is running at 4500 revolutions per minute.

16. Boiling Temperature   The table gives the temperature T (°F) at which water boils at selected pressures p (pounds per square inch). *(Source: Standard Handbook for Mechanical Engineers)*

p	5	10	14.696 (1 atmosphere)	20
T	162.24°	193.21°	212.00°	227.96°

p	30	40	60	80	100
T	250.33°	267.25°	292.71°	312.03°	327.81°

(a) Use the regression capabilities of a graphing utility to find a cubic model for the data.

(b) Use a graphing utility to plot the data and graph the model.

(c) Use the graph to estimate the pressure required for the boiling point of water to exceed 300°F.

(d) Explain why the model would not be correct for pressures exceeding 100 pounds per square inch.

17. Harmonic Motion   The motion of an oscillating weight suspended by a spring was measured by a motion detector. The data collected and the approximate maximum (positive and negative) displacements from equilibrium are shown in the figure. The displacement y is measured in centimeters and the time t is measured in seconds.

(a) Is y a function of t? Explain.

(b) Approximate the amplitude and period of the oscillations.

(c) Find a model for the data.

(d) Use a graphing utility to graph the model in part (c). Compare the result with the data in the figure.

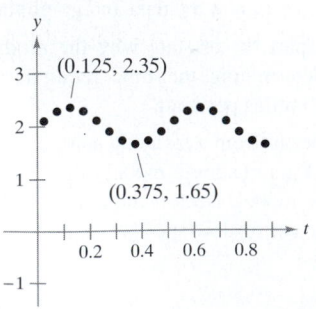

18. Temperature   The table gives the normal daily high temperatures for Honolulu H and Chicago C (in degrees Fahrenheit) for month t, with $t = 1$ corresponding to January. *(Source: NOAA)*

t	1	2	3	4	5	6
H	80.1	80.5	81.6	82.8	84.7	86.5
C	29.0	33.5	45.8	58.6	70.1	79.6

t	7	8	9	10	11	12
H	87.5	88.7	88.5	86.9	84.1	81.2
C	83.7	81.8	74.8	63.3	48.4	34.0

(a) A model for Honolulu is

$$H(t) = 84.40 + 4.28 \sin\left(\frac{\pi t}{6} + 3.86\right).$$

Find a model for Chicago.

(b) Use a graphing utility to graph the data and the model for the temperatures in Honolulu. How well does the model fit?

(c) Use a graphing utility to graph the data and the model for the temperatures in Chicago. How well does the model fit?

(d) Use the models to estimate the average annual temperature in each city. What term of the model did you use? Explain.

(e) What is the period of each model? Is it what you expected? Explain.

(f) Which city has a greater variability of temperatures throughout the year? Which factor of the models determines this variability? Explain.

19. Individual Project   Search for real-life data in a newspaper or magazine article and fit the data to a model. What does your model imply about the data?

REVIEW EXERCISES FOR CHAPTER P

In Exercises 1–4, find the intercepts (if any).

1. $y = 2x - 3$

2. $y = (x - 1)(x - 3)$

3. $y = \dfrac{x - 1}{x - 2}$

4. $xy = 4$

In Exercises 5 and 6, check for symmetry with respect to both axes and to the origin.

5. $x^2 y - x^2 + 4y = 0$

6. $y = x^4 - x^2 + 3$

In Exercises 7–14, sketch the graph of the equation.

7. $y = \frac{1}{2}(-x + 3)$

8. $4x - 2y = 6$

9. $-\frac{1}{3}x + \frac{5}{6}y = 1$

10. $0.02x + 0.15y = 0.25$

11. $y = 7 - 6x - x^2$

12. $y = 6x - x^2$

13. $y = \sqrt{5 - x}$

14. $y = |x - 4| - 4$

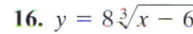 **In Exercises 15 and 16, describe the viewing rectangle of a graphing utility that yields the figure.**

15. $y = 4x^2 - 25$

16. $y = 8\sqrt[3]{x - 6}$

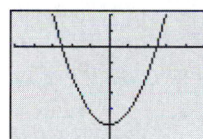

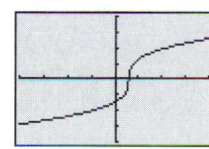

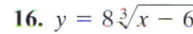 **In Exercises 17 and 18, use a graphing utility to find the point(s) of intersection of the graphs of the equations.**

17. $3x - 4y = 8$
    $x + \ y = 5$

18. $x - y + 1 = 0$
    $y - x^2 = 7$

19. **Think About It** Write an equation whose graph has intercepts at $x = -2$ and $x = 2$ and is symmetric with respect to the origin.

20. **Think About It** For what value of k does the graph of $y = kx^3$ pass through the point?

    (a) $(1, 4)$ (b) $(-2, 1)$
    (c) $(0, 0)$ (d) $(-1, -1)$

In Exercises 21 and 22, plot the points and find the slope of the line passing through the points.

21. $\left(\frac{3}{2}, 1\right), \left(5, \frac{5}{2}\right)$

22. $(7, -1), (7, 12)$

In Exercises 23 and 24, use the concept of slope to find t such that the three points are collinear.

23. $(-2, 5), (0, t), (1, 1)$

24. $(-3, 3), (t, -1), (8, 6)$

25. Find the equations of the lines passing through $(-2, 4)$ and having the following characteristics.

    (a) Slope of $\frac{7}{16}$

    (b) Parallel to the line $5x - 3y = 3$

    (c) Passing through the origin

    (d) Parallel to the y-axis

26. Find the equations of the lines passing through $(1, 3)$ and having the following characteristics.

    (a) Slope of $-\frac{2}{3}$

    (b) Perpendicular to the line $x + y = 0$

    (c) Passing through the point $(2, 4)$

    (d) Parallel to the x-axis

27. **Rate of Change** The purchase price of a new machine is \$12,500, and its value will decrease by \$850 per year. Use this information to write a linear equation that gives the value V of the machine t years after it is purchased. Find its value at the end of 3 years.

28. **Break-Even Analysis** A contractor purchases a piece of equipment for \$36,500 that costs an average of \$9.25 per hour for fuel and maintenance. The equipment operator is paid \$13.50 per hour, and customers are charged \$30 per hour.

    (a) Write an equation for the cost C of operating this equipment for t hours.

    (b) Write an equation for the revenue R derived from t hours of use.

    (c) Find the break-even point for this equipment by finding the time at which $R = C$.

In Exercises 29–32, sketch the graph of the equation and use the vertical line test to determine whether the equation expresses y as a function of x.

29. $x - y^2 = 0$

30. $x^2 - y = 0$

31. $y = x^2 - 2x$

32. $x = 9 - y^2$

33. Evaluate (if possible) the function $f(x) = 1/x$ at the specified values of the independent variable, and simplify the results.

    (a) $f(0)$

    (b) $\dfrac{f(1 + \Delta x) - f(1)}{\Delta x}$

34. Given $f(x) = 1 - x^2$ and $g(x) = 2x + 1$, find the following.

    (a) $f(x) - g(x)$

    (b) $f(x)g(x)$

    (c) $g(f(x))$

35. Sketch (on the same set of coordinate axes) a graph of f for $c = -2, 0,$ and 2.

(a) $f(x) = x^3 + c$ (b) $f(x) = (x - c)^3$

(c) $f(x) = (x - 2)^3 + c$ (d) $f(x) = cx^3$

 36. Use a graphing utility to graph $f(x) = x^3 - 3x^2$. Use the graph to write a formula for the function g shown in the figure.

(a) 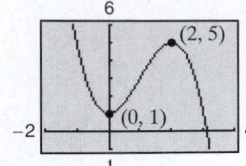   (b)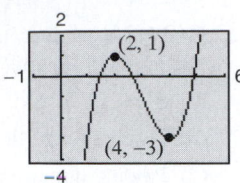

37. Conjecture

(a) Use a graphing utility to graph the functions f, g, and h in the same viewing rectangle. Write a description of any similarities and differences you observe among the graphs.

Odd powers: $f(x) = x$, $g(x) = x^3$, $h(x) = x^5$

Even powers: $f(x) = x^2$, $g(x) = x^4$, $h(x) = x^6$

(b) Use the result in part (a) to make a conjecture about the graphs of the functions $y = x^7$ and $y = x^8$. Use a graphing utility to verify your conjecture.

38. Think About It Use the result of Exercise 37 to guess the shapes of the graphs of the functions f, g, and h. Then use a graphing utility to graph each function and compare the result with your guess.

(a) $f(x) = x^2(x - 6)^2$

(b) $g(x) = x^3(x - 6)^2$

(c) $h(x) = x^3(x - 6)^3$

39. Area A wire 24 inches long is to be cut into four pieces to form a rectangle whose shortest side has a length of x.

(a) Express the area A of the rectangle as a function of x.

(b) Determine the domain of the function and use a graphing utility to graph the function over that domain.

(c) Use the graph of the function to approximate the maximum area of the rectangle. Make a conjecture about the dimensions that yield a maximum area.

40. Writing The following graphs give the profits P for two small companies over a period of 2 years. Create a story to describe the behavior of each profit function for some hypothetical product the company produces.

(a) 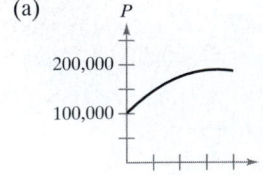   (b)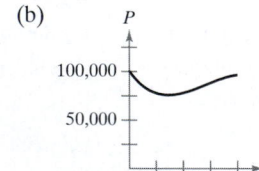

41. Think About It What is the minimum degree of the polynomial function whose graph approximates the given graph? What sign must the leading coefficient have?

(a) 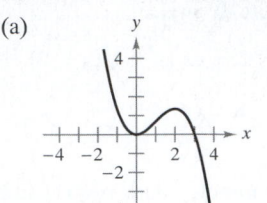   (b)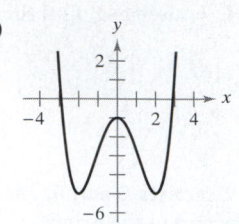

(c) 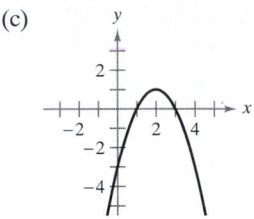   (d)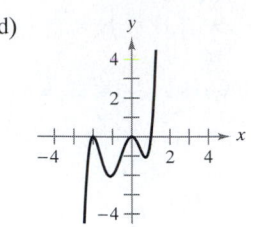

42. Stress Test A machine part was tested by bending it x centimeters ten times per minute until the time y (in hours) of failure. The results are recorded in the table.

x	3	6	9	12	15	18	21	24	27	30
y	61	56	53	55	48	35	36	33	44	23

(a) Use the regression capabilities of a graphing utility to find a linear model for the data.

(b) Use a graphing utility to plot the data and graph the model.

(c) Use the graph to determine whether there may have been an error made in conducting one of the tests or in recording the results. If so, eliminate the erroneous point and find the model for the revised data.

43. Harmonic Motion The motion of an oscillating weight suspended by a spring was measured by a motion detector. The data collected and the approximate maximum (positive and negative) displacements from equilibrium are shown in the figure. The displacement y is measured in feet and the time t is measured in seconds.

(a) Is y a function of t? Explain.

(b) Approximate the amplitude and period of the oscillations.

(c) Find a model for the data.

(d) Use a graphing utility to graph the model in part (c). Compare the result with the data in the figure.

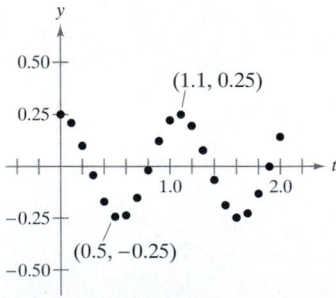

1

Limits and Their Properties

How High? How Fast?

Look more closely at the march of winning times and record distances, of gold-medal weights and precedent-setting heights. The law of diminishing returns has set in. The world-record time in the women's 400-meter freestyle, for example, dropped more than two minutes— a full 33 percent—from 1921 (6:16.6) to 1976 (4:11.69). In the 20 years since, it has fallen just eight seconds, to Janet Evans's 4:03.85 at the 1988 Seoul Olympics. If you were to plot world records on graph paper, you would get curves that seem to approach a limit asymptotically, coming tantalizingly closer but never quite reaching it. It is as if the curves were little south-pole magnets and the limit an imposing bar of north polarity. But what is the limit?

In a story written for *Newsweek,* Sharon Begley and Adam Rogers questioned the limits of human endurance as evidenced by world records in various sporting events.

A look at records set in various sports over the past century shows that humans continue to run faster, jump higher, and throw farther than ever before. What is allowing this to occur?

One factor is training. Physiologists are working to identify which systems in the human body limit performance, and to create training techniques that develop those systems. Similarly, sports psychologists work with individuals and team members to help them develop the mental "flow" that will allow them to deliver peak performances. Moreover, trainers have developed devices to monitor athletes' bodies and provide them with more feedback on their performance than was available even 20 years ago.

Equipment has also improved vastly over the years. In some sports, the advancement is obvious. Bicycles are lighter and more aerodynamic than ever before. Improved track surfaces boosted runners' speeds and aluminum poles drastically increased vault heights. Even sports such as swimming, with no obvious equipment, have benefited from technology. Shaving body hair cut a full second from male swimmers' times in the 100-meter freestyle, and new styles of swimsuits are expected to reduce drag and improve time even more. The two scatter plots below show the successive world records (in seconds) for two men's swimming events.

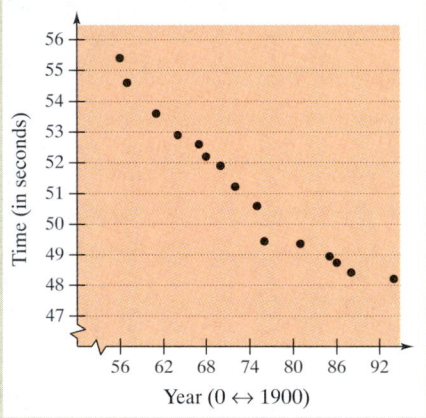

Men's 100-meter freestyle

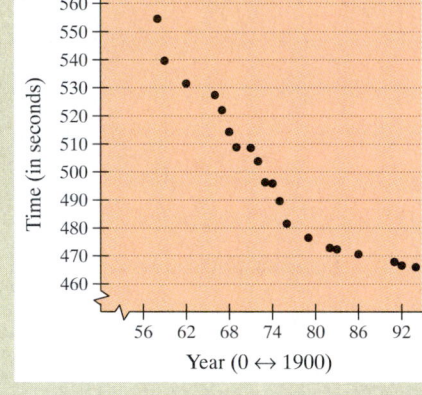

Men's 800-meter freestyle

QUESTIONS

1. From the scatter plots shown above, can you determine which year body shaving was started? Explain your reasoning.

2. In which other years do you think there may have been technological advances in swimming? Explain your reasoning.

3. What does the lower limit appear to be for a man to swim 100 meters? To swim 800 meters? How did you determine this?

4. Copy the two scatter plots and draw a curve that seems to fit the data best. What type of equation do you think would produce the curve you have drawn?

5. Read the excerpt from *Newsweek* at the left. What do the authors mean by the phrase "approach a limit asymptotically"?

The concepts presented here will be explored further in this chapter. For an extension of this application, see the lab series that accompanies this text.

SECTION 1.1 A Preview of Calculus

What Is Calculus? • The Tangent Line Problem • The Area Problem

What Is Calculus?

Calculus is the mathematics of change—velocities and accelerations. Calculus is also the mathematics of tangent lines, slopes, areas, volumes, arc lengths, centroids, curvatures, and a variety of other concepts that have enabled scientists, engineers, and economists to model real-life situations.

Although precalculus mathematics also deals with velocities, accelerations, tangent lines, slopes, and so on, here is a fundamental difference between precalculus mathematics and calculus. Precalculus mathematics is more static, whereas calculus is more dynamic. Here are some examples.

- An object traveling at a constant velocity can be modeled with precalculus mathematics. To model the velocity of an accelerating object, you need calculus.
- The slope of a line can be modeled with precalculus mathematics. To model the slope of a curve, you need calculus.
- A tangent line to a circle can be modeled with precalculus mathematics. To model a tangent line of a general graph, you need calculus.
- The area of a rectangle can be modeled with precalculus mathematics. To model the area under a general curve, you need calculus.

Each of these situations involves the same general strategy—the reformulation of precalculus mathematics through the use of a limit process. Thus, one way to answer the question "What is calculus?" is to say that calculus is a "limit machine" that involves three stages. The first stage is precalculus mathematics, such as the slope of a line or the area of a rectangle. The second stage is the limit process, and the third stage is a new calculus formulation, such as a derivative or integral.

| Precalculus Mathematics | → | Limit Process | → | Calculus |

STUDY TIP As you progress through this course, remember that learning calculus is just one of your goals. Your most important goal is to learn how to use calculus to model and solve real-life problems. Here are a few problem-solving strategies that may help you.

- Be sure you understand the question. What is given? What are you asked to find?
- Outline a plan. There are many approaches you could use: look for a pattern, solve a simpler problem, work backwards, draw a diagram, use technology, or any of many other approaches.
- Complete your plan. Be sure to answer the question. Verbalize your answer. For example, rather than writing the answer as $x = 4.6$, it would be better to write the answer as "The area of the region is 4.6 meters."
- Look back at your work. Does your answer make sense? Is there a way you can check the reasonableness of your answer?

GRACE CHISHOLM YOUNG (1868–1944)

Grace Chisholm Young received her degree in mathematics from Girton College in Cambridge, England. Her early work was published under the name of William Young, her husband. Between 1914 and 1916, Grace Young published work on the foundations of calculus that won her the Gamble Prize from Girton College.

Some students try to learn calculus as if it were simply a collection of new formulas. This is unfortunate. If you reduce calculus to the memorization of differentiation and integration formulas, you will miss a great deal of understanding, self-confidence, and satisfaction.

On the following two pages we have listed some familiar precalculus concepts coupled with their calculus counterparts. Throughout the text, your goal should be to learn how precalculus formulas and techniques are used as building blocks to produce the more general calculus formulas and techniques. Don't worry if you are unfamiliar with some of the "old formulas" listed on the following two pages—we will be reviewing all of them.

As you proceed through this text, we suggest that you come back to this discussion repeatedly. Try to keep track of where you are relative to the three stages involved in the study of calculus. For example, the first three chapters break down as follows.

Chapter P: Preparation for Calculus	Precalculus
Chapter 1: Limits and Their Properties	The limit process
Chapter 2: Differentiation	The new calculus formula

WITHOUT CALCULUS	WITH DIFFERENTIAL CALCULUS
Value of $f(x)$ when $x = c$	Limit of $f(x)$ as x approaches c
Slope of a line	Slope of a curve
Secant line to a curve	Tangent line to a curve
Average rate of change between $t = a$ and $t = b$	Instantaneous rate of change at $t = c$
Curvature of a circle	Curvature of a curve
Height of a curve when $x = c$	Maximum height of a curve on an interval
Tangent plane to a sphere	Tangent plane to a surface
Direction of motion along a straight line	Direction of motion along a curved line

WITHOUT CALCULUS		WITH INTEGRAL CALCULUS	
Area of a rectangle		Area under a curve	
Work done by a constant force		Work done by a variable force	
Center of a rectangle		Centroid of a region	
Length of a line segment		Length of an arc	
Surface area of a cylinder		Surface area of a solid of revolution	
Mass of a solid of constant density		Mass of a solid of variable density	
Volume of a rectangular solid		Volume of a region under a surface	
Sum of a finite number of terms	$a_1 + a_2 + \cdots + a_n = S$	Sum of an infinite number of terms	$a_1 + a_2 + a_3 + \cdots = S$

The Tangent Line Problem

The notion of a limit is fundamental to the study of calculus. The following brief descriptions of two classic problems in calculus—*the tangent line problem* and *the area problem*—should give you some idea of the way limits are used in calculus.

In the tangent line problem, you are given a function f and a point P on its graph and are asked to find an equation of the tangent line to the graph at point P, as shown in Figure 1.1.

Except for cases involving a vertical tangent line, the problem of finding the **tangent line** at a point P is equivalent to finding the *slope* of the tangent line at P. You can approximate this slope by using a line through the point of tangency and a second point on the curve, as shown in Figure 1.2(a). Such a line is called a **secant line.** If $P(c, f(c))$ is the point of tangency and

$$Q(c + \Delta x, f(c + \Delta x))$$

is a second point on the graph of f, the slope of the secant line through these two points is given by

$$m_{sec} = \frac{f(c + \Delta x) - f(c)}{c + \Delta x - c} = \frac{f(c + \Delta x) - f(c)}{\Delta x}.$$

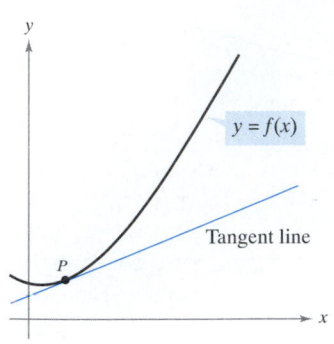

The tangent line to the graph of f at P
Figure 1.1

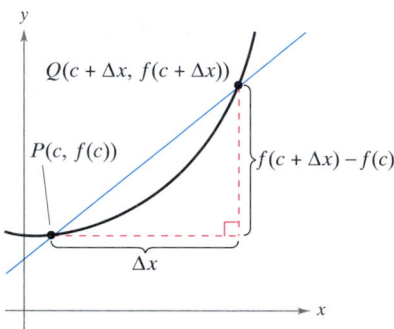

(a) The secant line through $(c, f(c))$ and $(c + \Delta x, f(c + \Delta x))$

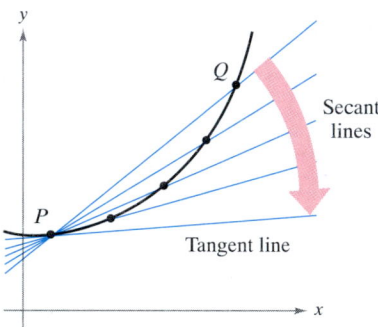

(b) As Q approaches P, the secant lines approach the tangent line.

Figure 1.2

As point Q approaches point P, the slope of the secant line approaches the slope of the tangent line, as shown in Figure 1.2(b). When such a "limiting position" exists, the slope of the tangent line is said to be the **limit** of the slope of the secant line. (Much more will be said about this important problem in Chapter 2.)

EXPLORATION

The following points lie on the graph of $f(x) = x^2$. Each successive point gets closer to the point $P(1, 1)$. Find the slope of the secant line through Q_1 and P, Q_2 and P, and so on. Graph these secant lines on a graphing utility. Then use your results to estimate the slope of the tangent line to the graph of f at the point P.

$$Q_1(1.5, f(1.5)), \quad Q_2(1.1, f(1.1)), \quad Q_3(1.01, f(1.01)),$$

$$Q_4(1.001, f(1.001)), \quad Q_5(1.0001, f(1.0001))$$

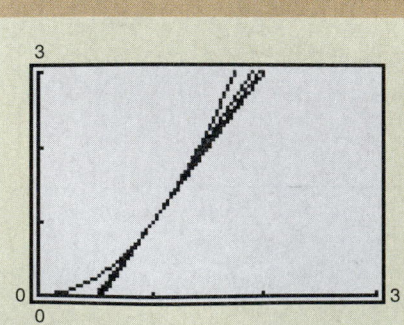

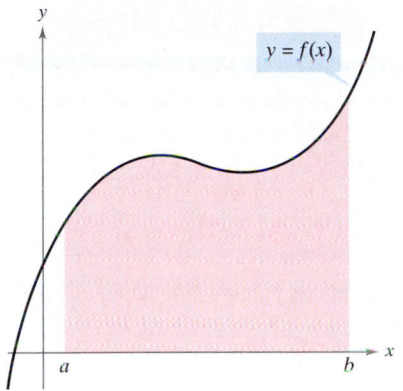

Area under a curve
Figure 1.3

The Area Problem

In the tangent line problem, you saw how the limit process can be applied to the slope of a line to find the slope of a general curve. A second classic problem in calculus is finding the area of a plane region that is bounded by the graphs of functions. This problem can also be solved with a limit process. In this case, the limit process is applied to the area of a rectangle to find the area of a general region.

As a simple example, consider the region bounded by the graph of the function $y = f(x)$, the x-axis, and the vertical lines $x = a$ and $x = b$, as shown in Figure 1.3. You can approximate the area of the region with several rectangular regions, as shown in Figure 1.4. As you increase the number of rectangles, the approximation tends to become better and better because the amount of area missed by the rectangles decreases. Your goal is to determine the limit of the sum of the areas of the rectangles as the number of rectangles increases without bound.

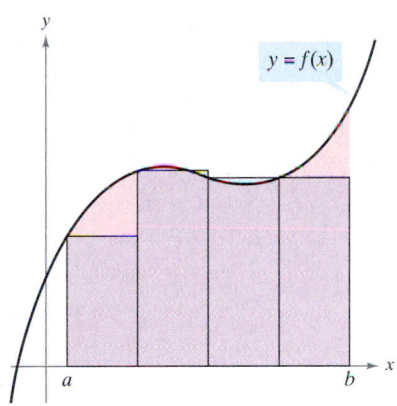

Approximation using four rectangles
Figure 1.4

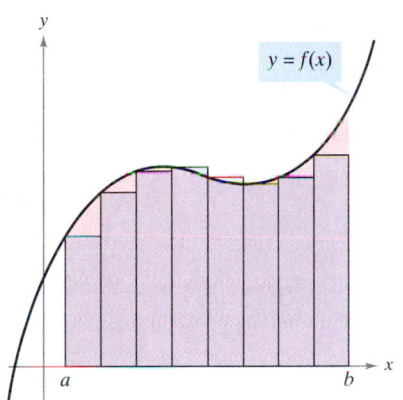

Approximation using eight rectangles

EXPLORATION

Consider the region bounded by the graphs of $f(x) = x^2$, $y = 0$, and $x = 1$, as shown in part (a) of the figure. The area of the region can be approximated by two sets of rectangles—one set inscribed in the region and the other set circumscribed over the region, as shown in parts (b) and (c). Find the sum of the areas of each set of rectangles. Then use your results to approximate the area of the region.

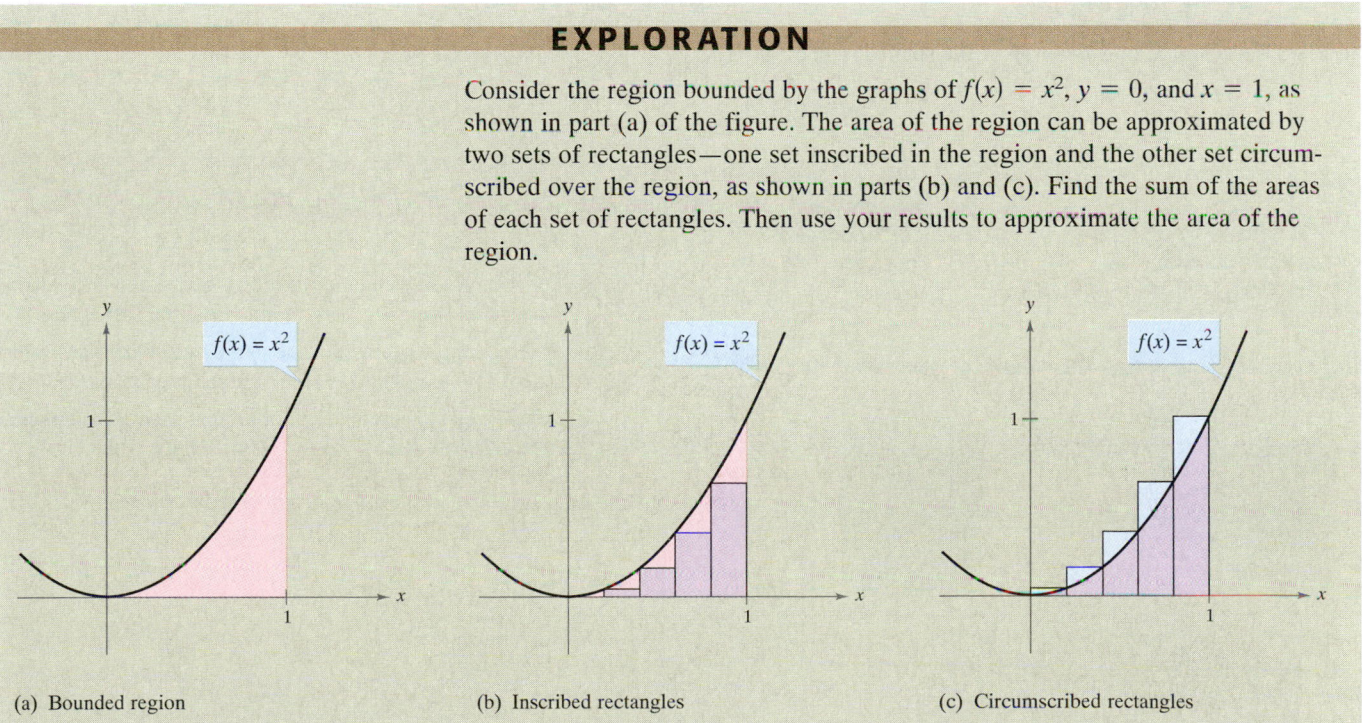

(a) Bounded region   (b) Inscribed rectangles   (c) Circumscribed rectangles

In Exercises 1–4, decide whether the problem can be solved using precalculus, or whether calculus is required. If the problem can be solved using precalculus, solve it. If the problem seems to require calculus, explain your reasoning and use a graphical or numerical approach to estimate the solution.

1. Find the distance traveled in 15 seconds by an object traveling at a constant velocity of 20 feet per second.

2. Find the distance traveled in 15 seconds by an object moving with a velocity of $v(t) = 20 + 7 \cos t$ feet per second.

3. A bicyclist is riding on a path modeled by the equation $y = 0.04(8x - x^2)$, where x and y are measured in miles. Find the rate of change of elevation when $x = 2$.

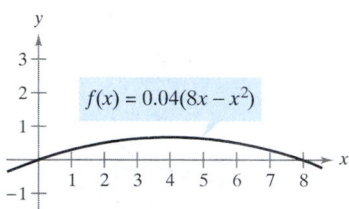

4. A bicyclist is riding on a path modeled by the equation $y = 0.08x$, where x and y are measured in miles. Find the rate of change of elevation when $x = 2$.

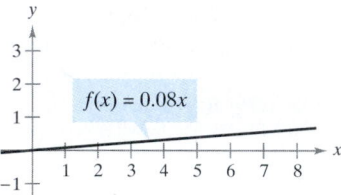

In Exercises 5 and 6, find the area of the shaded region.

5.
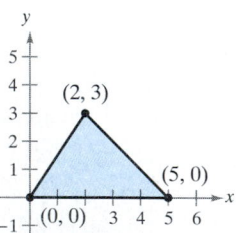

6.
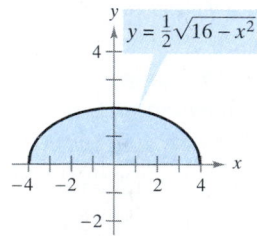

In Exercises 7 and 8, find the volume of the solid shown.

7.
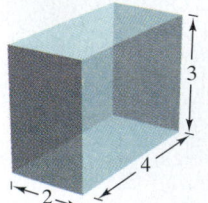

8.
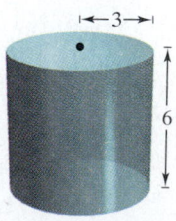

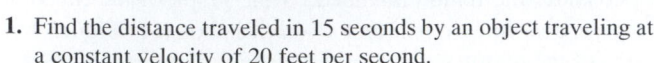
9. (a) Use the list feature of a graphing utility to graph the following.

$$y_1 = 4x - x^2$$

$$y_2 = \left[\frac{y_1(1 + \{2, 1.5, 1, 0.5\}) - y_1(1)}{\{2, 1.5, 1, 0.5\}} \right](x - 1) + y_1(1)$$

(*Note:* If you cannot use lists on your graphing utility, graph y_2 four times using 2, 1.5, 1, and 0.5.)

(b) Give a written description of the graphs of y_2 relative to the graph of y_1.

(c) Use the results in part (a) to estimate the slope of the tangent line to the graph of y_1 at $(1, 3)$. If you want to improve your approximation of the slope, how could you change the list in the formula for y_2?

10. (a) Use the rectangles to approximate the area of the region bounded by $y = 5/x$, $y = 0$, $x = 1$, and $x = 5$.

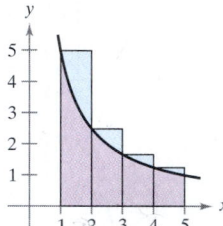 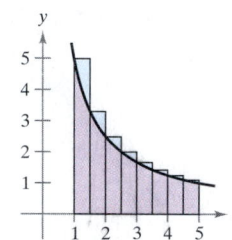

(b) Describe how you could continue this process to obtain a more accurate approximation of the area.

11. You are asked to find the length of the graph of $f(x) = 5/x$ from $(1, 5)$ to $(5, 1)$.

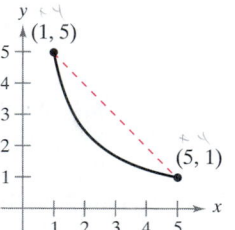 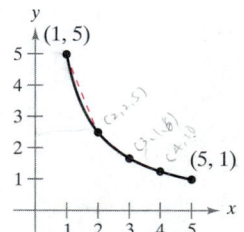

(a) Approximate the length of the curve by finding the distance between its two endpoints, as shown in the first figure.

(b) Approximate the length of the curve by finding the lengths of four line segments, as shown in the second figure.

(c) Describe how you could continue this process to obtain a more accurate approximation of the length of the curve.

The symbol *indicates an exercise in which you are instructed to use graphing technology or a symbolic computer algebra system. The solutions of other exercises may also be facilitated by use of appropriate technology.*

A red number indicates that a detailed solution can be found in the Study and Solutions Guide.

An Introduction to Limits • Limits That Fail to Exist • A Formal Definition of a Limit

An Introduction to Limits

Suppose you are asked to sketch the graph of the function f given by

$$f(x) = \frac{x^3 - 1}{x - 1}, \qquad x \neq 1.$$

For all values other than $x = 1$, you can use standard curve-sketching techniques. However, at $x = 1$, it is not clear what to expect. To get an idea of the behavior of the graph of f near $x = 1$, you can use two sets of x-values—one set that approaches 1 from the left and one that approaches 1 from the right, as shown in the table.

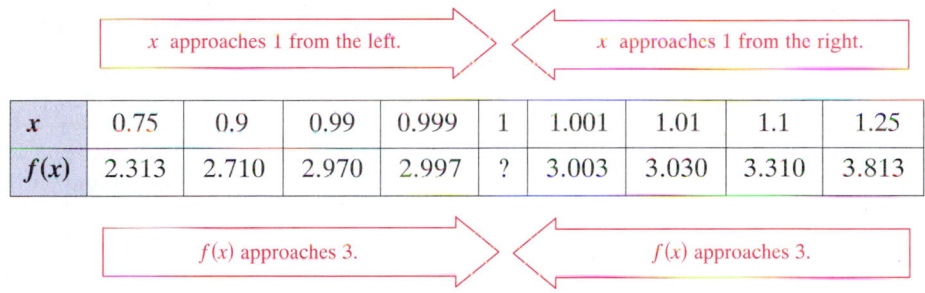

	x approaches 1 from the left.					x approaches 1 from the right.			
x	0.75	0.9	0.99	0.999	1	1.001	1.01	1.1	1.25
$f(x)$	2.313	2.710	2.970	2.997	?	3.003	3.030	3.310	3.813

	$f(x)$ approaches 3.			$f(x)$ approaches 3.	

When you graph the function, it appears that the graph of f is a parabola that has a gap at the point $(1, 3)$, as shown in Figure 1.5. Although x cannot equal 1, you can move arbitrarily close to 1, and as a result $f(x)$ moves arbitrarily close to 3. Using limit notation, you can write

$$\lim_{x \to 1} f(x) = 3. \qquad \text{This is read as "the limit of } f(x) \text{ as } x \text{ approaches 1 is 3."}$$

This discussion leads to an informal description of a limit. If $f(x)$ becomes arbitrarily close to a single number L as x approaches c from either side, the **limit** of $f(x)$, as x approaches c, is L. This limit is written as

$$\lim_{x \to c} f(x) = L.$$

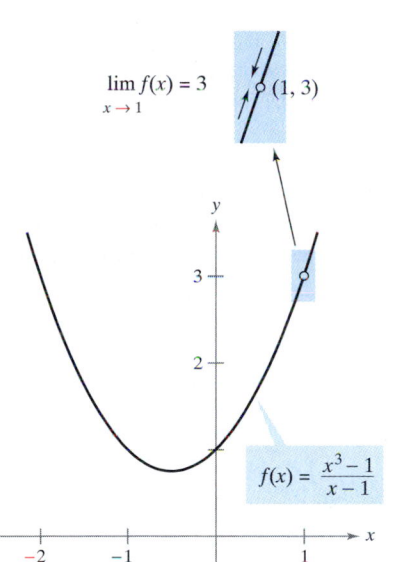

$\lim\limits_{x \to 1} f(x) = 3$ $(1, 3)$

$$f(x) = \frac{x^3 - 1}{x - 1}$$

The limit of $f(x)$ as x approaches 1 is 3.
Figure 1.5

EXPLORATION

The discussion above gives an example of how you can approximate a limit *numerically* by constructing a table and *graphically* by drawing a graph. Try approximating the limit

$$\lim_{x \to 2} \frac{x^2 - 3x + 2}{x - 2}$$

numerically by completing the table.

x	1.75	1.9	1.99	1.999	2	2.001	2.01	2.1	2.25
$f(x)$	?	?	?	?	?	?	?	?	?

Then use a graphing utility to approximate the limit graphically.

EXAMPLE 1 Estimating a Limit Numerically

Evaluate the function $f(x) = x/(\sqrt{x + 1} - 1)$ at several points near $x = 0$ and use the result to estimate the limit

$$\lim_{x \to 0} \frac{x}{\sqrt{x + 1} - 1}.$$

Solution The table lists the values of $f(x)$ for several x-values near 0.

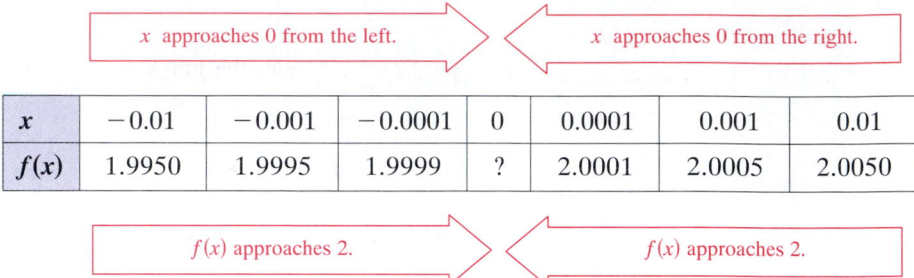

x	-0.01	-0.001	-0.0001	0	0.0001	0.001	0.01
$f(x)$	1.9950	1.9995	1.9999	?	2.0001	2.0005	2.0050

$f(x)$ approaches 2. $f(x)$ approaches 2.

From the results shown in the table, you can estimate the limit to be 2. This limit is reinforced by the graph of f (see Figure 1.6).

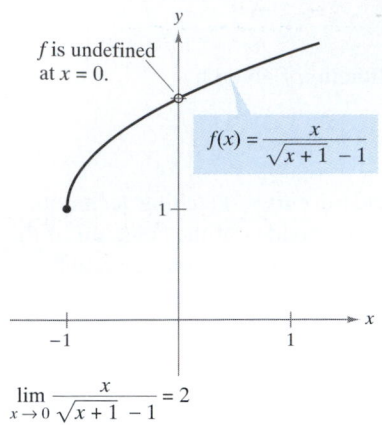

f is undefined at $x = 0$.

$f(x) = \dfrac{x}{\sqrt{x + 1} - 1}$

$\lim_{x \to 0} \dfrac{x}{\sqrt{x + 1} - 1} = 2$

The limit of $f(x)$ as x approaches 0 is 2.
Figure 1.6

So far in this section, you have been approximating limits numerically and graphically. Each of these approaches produces an approximation of the limit. In Section 1.3, you will study analytic techniques for evaluating limits. Throughout the course, try to develop a habit of using this three-pronged approach to problem solving.

1. Numerical approach   Construct a table of values.
2. Graphical approach   Draw a graph by hand or using technology.
3. Analytic approach   Use algebra or calculus.

NOTE In Example 1, note that the function is undefined at $x = 0$ and yet $f(x)$ appears to be approaching a limit as x approaches 0. This often happens, and it is important to realize that *the existence or nonexistence of $f(x)$ at $x = c$ has no bearing on the existence of the limit of $f(x)$ as x approaches c.*

EXAMPLE 2 Finding a Limit

Find the limit of $f(x)$ as x approaches 2 where f is defined as

$$f(x) = \begin{cases} 1, & x \neq 2 \\ 0, & x = 2. \end{cases}$$

Solution Because $f(x) = 1$ for all x other than $x = 2$, you can conclude that the limit is 1, as shown in Figure 1.7. Thus, you can write

$$\lim_{x \to 2} f(x) = 1.$$

The fact that $f(2) = 0$ has no bearing on the existence or value of the limit as x approaches 2. For instance, if the function were defined as

$$f(x) = \begin{cases} 1, & x \neq 2 \\ 2, & x = 2 \end{cases}$$

the limit would be the same.

$f(x) = \begin{cases} 1, x \neq 2 \\ 0, x = 2 \end{cases}$

$\lim_{x \to 2} f(x) = 1$

The limit of $f(x)$ as x approaches 2 is 1.
Figure 1.7

Limits That Fail to Exist

In the next three examples you will examine some limits that fail to exist.

EXAMPLE 3   Behavior That Differs from the Right and Left

Show that the limit does not exist.

$$\lim_{x \to 0} \frac{|x|}{x}$$

Solution   Consider the graph of the function $f(x) = |x|/x$. From Figure 1.8, you can see that for positive x-values

$$\frac{|x|}{x} = 1, \quad x > 0$$

and for negative x-values

$$\frac{|x|}{x} = -1, \quad x < 0.$$

This means that no matter how close x gets to 0, there will be both positive and negative x-values that yield $f(x) = 1$ and $f(x) = -1$. Specifically, if δ (the lowercase Greek letter *delta*) is a positive number, then for x-values satisfying the inequality $0 < |x| < \delta$, you can classify the values of $|x|/x$ as follows.

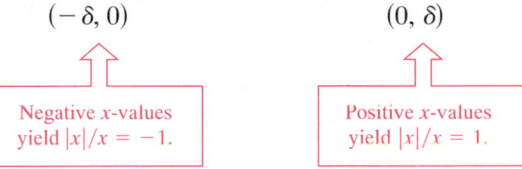

$(-\delta, 0)$ $(0, \delta)$

Negative x-values yield $|x|/x = -1$.   Positive x-values yield $|x|/x = 1$.

This implies that the limit does not exist. ▬▬▬

EXAMPLE 4   Unbounded Behavior

Discuss the existence of the limit

$$\lim_{x \to 0} \frac{1}{x^2}.$$

Solution   Let $f(x) = 1/x^2$. In Figure 1.9, you can see that as x approaches 0 from either the right or the left, $f(x)$ increases without bound. This means that by choosing x close enough to 0, you can force $f(x)$ to be as large as you want. For instance, $f(x)$ will be larger than 100 if you choose x that is within $\frac{1}{10}$ of 0. That is,

$$0 < |x| < \frac{1}{10} \implies f(x) = \frac{1}{x^2} > 100.$$

Similarly, you can force $f(x)$ to be larger than 1,000,000, as follows.

$$0 < |x| < \frac{1}{1000} \implies f(x) = \frac{1}{x^2} > 1,000,000$$

Because $f(x)$ is not approaching a real number L as x approaches 0, you can conclude that the limit does not exist. ▬▬▬

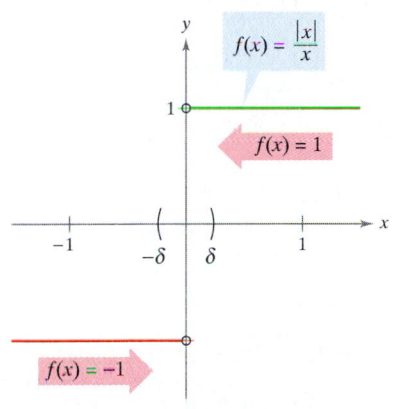

$\lim_{x \to 0} f(x)$ does not exist.

Figure 1.8

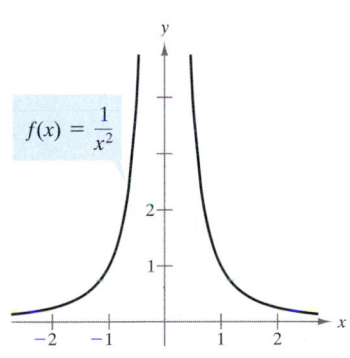

$\lim_{x \to 0} f(x)$ does not exist.

Figure 1.9

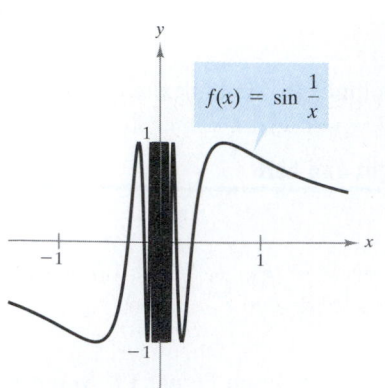

$$f(x) = \sin \frac{1}{x}$$

$\lim\limits_{x \to 0} f(x)$ does not exist.

Figure 1.10

EXAMPLE 5   **Oscillating Behavior**

Discuss the existence of the limit $\lim\limits_{x \to 0} \sin \dfrac{1}{x}$.

Solution   Let $f(x) = \sin(1/x)$. In Figure 1.10, you can see that as x approaches 0, $f(x)$ oscillates between -1 and 1. Therefore, the limit does not exist because no matter how small you choose δ, it is possible to choose x_1 and x_2 within δ units of 0 such that $\sin(1/x_1) = 1$ and $\sin(1/x_2) = -1$, as indicated in the table.

x	$\dfrac{2}{\pi}$	$\dfrac{2}{3\pi}$	$\dfrac{2}{5\pi}$	$\dfrac{2}{7\pi}$	$\dfrac{2}{9\pi}$	$\dfrac{2}{11\pi}$	$x \to 0$
$\sin \dfrac{1}{x}$	1	-1	1	-1	1	-1	Limit does not exist.

PETER GUSTAV DIRICHLET (1805–1859)

In the early development of calculus, the definition of a function was much more restricted than it is today, and "functions" such as the Dirichlet function would not have been considered. The modern definition of a function was given by the German mathematician Peter Gustav Dirichlet.

Common Types of Behavior Associated with the Nonexistence of a Limit

1. $f(x)$ approaches a different number from the right side of c than it approaches from the left side.

2. $f(x)$ increases or decreases without bound as x approaches c.

3. $f(x)$ oscillates between two fixed values as x approaches c.

There are many other interesting functions that have unusual limit behavior. An often cited one is the *Dirichlet function*

$$f(x) = \begin{cases} 0, & \text{if } x \text{ is rational.} \\ 1, & \text{if } x \text{ is irrational.} \end{cases}$$

This function has *no limit* at any real number c.

TECHNOLOGY   When you use a graphing utility to investigate the behavior of a function near the x-value at which you are trying to evaluate a limit, remember that you can't always trust the pictures that graphing utilities draw. For instance, if you use a graphing utility to sketch the graph of the function in Example 5 over an interval containing 0, you will most likely obtain an incorrect graph—such as that shown in Figure 1.11. The reason that a graphing utility can't show the correct graph is that the graph has infinitely many oscillations over any interval that contains 0.

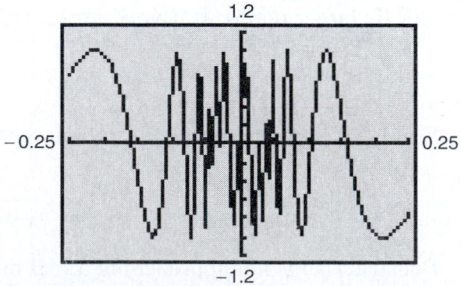

Incorrect graph of $f(x) = \sin(1/x)$

Figure 1.11

A Formal Definition of a Limit

Let's take another look at the informal description of a limit. If $f(x)$ becomes arbitrarily close to a single number L as x approaches c from either side, we say that the limit of $f(x)$ as x approaches c is L, written as

$$\lim_{x \to c} f(x) = L.$$

At first glance, this description looks fairly technical. Even so, we call it informal because we have yet to give exact meanings to the two phrases

"$f(x)$ becomes arbitrarily close to L"

and

"x approaches c."

The first person to assign mathematical rigorous meanings to these two phrases was Augustin-Louis Cauchy. His **ε-δ definition of a limit** is the standard used today.

In Figure 1.12, let ε (the lowercase Greek letter *epsilon*) represent a (small) positive number. Then the phrase "$f(x)$ becomes arbitrarily close to L" means that $f(x)$ lies in the interval $(L - \varepsilon, L + \varepsilon)$. Using absolute value, you can write this as

$$|f(x) - L| < \varepsilon.$$

Similarly, the phrase "x approaches c" means that there exists a positive number δ such that x lies in either the interval $(c - \delta, c)$ or the interval $(c, c + \delta)$. This fact can be concisely expressed by the double inequality

$$0 < |x - c| < \delta.$$

The first inequality

$$0 < |x - c| \qquad\qquad \text{The difference between } x \text{ and } c \text{ is more than 0.}$$

expresses the fact that $x \neq c$. The second inequality

$$|x - c| < \delta \qquad\qquad x \text{ is within } \delta \text{ units of } c.$$

says that x is within a distance δ of c.

Augustin-Louis Cauchy (1789–1857)

Cauchy began his mathematical work at age 22. During his brilliant career, he wrote almost 800 papers on subjects ranging from mechanics to astronomy.

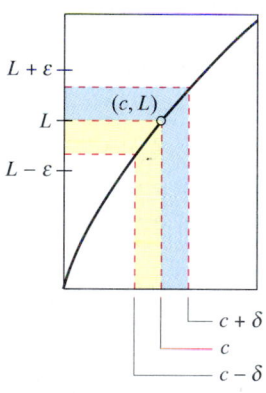

The ε-δ definition of the limit of $f(x)$ as x approaches c.
Figure 1.12

FOR FURTHER INFORMATION For more on the introduction of rigor to calculus, see "Who Gave You the Epsilon? Cauchy and the Origins of Rigorous Calculus" by Judith V. Grabiner in the March 1983 issue of *The American Mathematical Monthly.*

Definition of Limit

Let f be a function defined on an open interval containing c (except possibly at c) and let L be a real number. The statement

$$\lim_{x \to c} f(x) = L$$

means that for each $\varepsilon > 0$ there exists a $\delta > 0$ such that if

$$0 < |x - c| < \delta, \quad \text{then} \quad |f(x) - L| < \varepsilon.$$

NOTE   Throughout this text, when we write

$$\lim_{x \to c} f(x) = L$$

we imply two statements—the limit **exists** *and* the limit is L.

Some functions do not have limits as $x \to c$, but those that do cannot have two different limits as $x \to c$. That is, *if the limit of a function exists, it is unique* (see Exercise 49).

The next three examples should help you develop a better understanding of the ε-δ definition of a limit.

EXAMPLE 6 Finding a δ for a Given ε

Given the limit

$$\lim_{x \to 3} (2x - 5) = 1$$

find δ such that $|(2x - 5) - 1| < 0.01$ whenever $0 < |x - 3| < \delta$.

Solution In this problem, you are working with a given value of ε—namely, $\varepsilon = 0.01$. To find an appropriate δ, notice that

$$|(2x - 5) - 1| = |2x - 6| = 2|x - 3|.$$

Because the inequality $|(2x - 5) - 1| < 0.01$ is equivalent to $2|x - 3| < 0.01$, you can choose $\delta = \frac{1}{2}(0.01) = 0.005$. This choice works because

$$0 < |x - 3| < 0.005$$

implies that

$$|(2x - 5) - 1| = 2|x - 3| < 2(0.005) = 0.01$$

as shown in Figure 1.13.

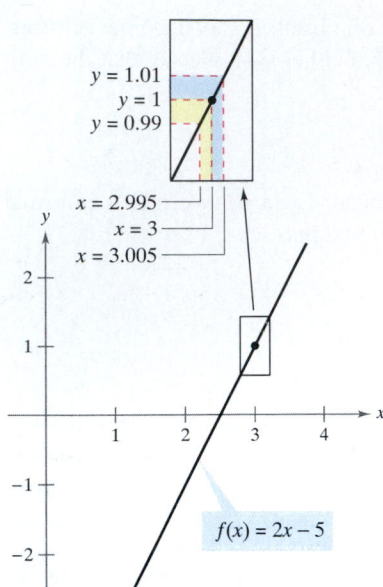

The limit of $f(x)$ as x approaches 3 is 1.
Figure 1.13

NOTE In Example 6, note that 0.005 is the *largest* value of δ that will guarantee $|(2x - 5) - 1| < 0.01$ whenever $0 < |x - 3| < \delta$. Any *smaller* positive value of δ would, of course, also work.

In Example 6, you found a δ-value for a *given* ε. This does not prove the existence of the limit. To do that, you must prove that you can find a δ for any ε, as demonstrated in the next example.

EXAMPLE 7 Using the ε-δ Definition of a Limit

Use the ε-δ definition of a limit to prove that

$$\lim_{x \to 2} (3x - 2) = 4.$$

Solution You must show that for each $\varepsilon > 0$, there exists a $\delta > 0$ such that $|(3x - 2) - 4| < \varepsilon$ whenever $0 < |x - 2| < \delta$. Because your choice of δ depends on ε, you need to establish a connection between the absolute values $|(3x - 2) - 4|$ and $|x - 2|$.

$$|(3x - 2) - 4| = |3x - 6| = 3|x - 2|$$

Thus, for a given $\varepsilon > 0$ you can choose $\delta = \varepsilon/3$. This choice works because

$$0 < |x - 2| < \delta = \frac{\varepsilon}{3}$$

implies that

$$|(3x - 2) - 4| = 3|x - 2| < 3\left(\frac{\varepsilon}{3}\right) = \varepsilon$$

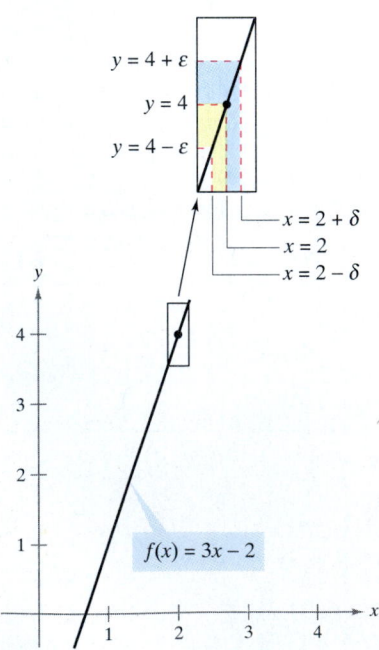

The limit of $f(x)$ as x approaches 2 is 4.
Figure 1.14

as shown in Figure 1.14.

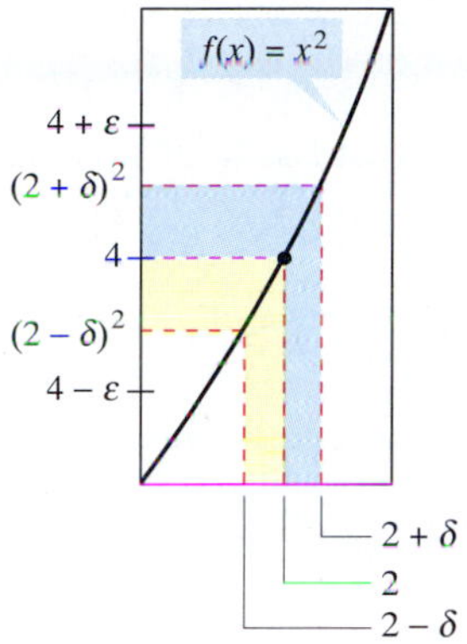

EXAMPLE 8   Using the ε-δ Definition of a Limit

Use the ε-δ definition of a limit to prove that

$$\lim_{x \to 2} x^2 = 4.$$

Solution   You must show that for each $\varepsilon > 0$, there exists a $\delta > 0$ such that

$$|x^2 - 4| < \varepsilon \quad \text{when} \quad 0 < |x - 2| < \delta.$$

To find an appropriate δ, begin by writing $|x^2 - 4| = |x - 2||x + 2|$. For all x in the interval $(1, 3)$, you know that $|x + 2| < 5$. Thus, letting δ be the minimum of $\varepsilon/5$ and 1, it follows that, whenever $0 < |x - 2| < \delta$, you have

$$|x^2 - 4| = |x - 2||x + 2| < \left(\frac{\varepsilon}{5}\right)(5) = \varepsilon$$

as shown in Figure 1.15.

Throughout this chapter you will use the ε-δ definition of a limit primarily to prove theorems about limits and to establish the existence or nonexistence of particular types of limits. For *finding* limits, you will learn techniques that are easier to use than the ε-δ definition of a limit.

The limit of $f(x)$ as x approaches 2 is 4.
Figure 1.15

EXERCISES FOR SECTION 1.2

1. *Modeling Data*   The cost of a telephone call between two cities is \$0.75 for the first minute and \$0.50 for each additional minute. A formula for the cost is given by

$$C(t) = 0.75 - 0.50 [\![-(t - 1)]\!]$$

where t is the time in minutes.

(*Note:* $[\![x]\!]$ = greatest integer n such that $n \le x$. For example, $[\![3.2]\!] = 3$ and $[\![-1.6]\!] = -2$.)

(a) Use a graphing utility to graph the cost function for $0 < t \le 5$.

(b) Use the graph to complete the table and observe the behavior of the function as t approaches 3.5. Use the graph and the table to find

$$\lim_{t \to 3.5} C(t).$$

t	3	3.3	3.4	3.5	3.6	3.7	4
C				?			

(c) Use the graph to complete the table and observe the behavior of the function as t approaches 3.

t	2	2.5	2.9	3	3.1	3.5	4
C				?			

Does the limit of $C(t)$ as t approaches 3 exist? Explain.

2. *Writing*   Write a brief description of the meaning of the notation

$$\lim_{x \to 8} f(x) = 25.$$

In Exercises 3–10, complete the table and use the result to estimate the limit. If desired, use a graphing utility to graph the function to confirm your result.

3. $\displaystyle \lim_{x \to 2} \frac{x - 2}{x^2 - x - 2}$

x	1.9	1.99	1.999	2.001	2.01	2.1
$f(x)$						

4. $\displaystyle \lim_{x \to 2} \frac{x - 2}{x^2 - 4}$

x	1.9	1.99	1.999	2.001	2.01	2.1
$f(x)$						

5. $\displaystyle \lim_{x \to 0} \frac{\sqrt{x + 3} - \sqrt{3}}{x}$

x	-0.1	-0.01	-0.001	0.001	0.01	1.0
$f(x)$						

6. $\lim\limits_{x \to -3} \dfrac{\sqrt{1-x}-2}{x+3}$

x	-3.1	-3.01	-3.001	-2.999	-2.99	-2.9
$f(x)$						

7. $\lim\limits_{x \to 3} \dfrac{[1/(x+1)]-(1/4)}{x-3}$

x	2.9	2.99	2.999	3.001	3.01	3.1
$f(x)$						

8. $\lim\limits_{x \to 4} \dfrac{[x/(x+1)]-(4/5)}{x-4}$

x	3.9	3.99	3.999	4.001	4.01	4.1
$f(x)$						

9. $\lim\limits_{x \to 0} \dfrac{\sin x}{x}$

x	-0.1	-0.01	-0.001	0.001	0.01	0.1
$f(x)$						

10. $\lim\limits_{x \to 0} \dfrac{\cos x - 1}{x}$

x	-0.1	-0.01	-0.001	0.001	0.01	0.1
$f(x)$						

In Exercises 11–20, use the graph to find the limit (if it exists).

11. $\lim\limits_{x \to 3} (4-x)$

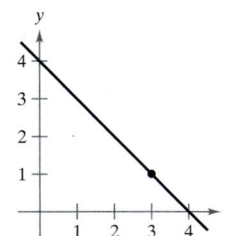

12. $\lim\limits_{x \to 1} (x^2+2)$

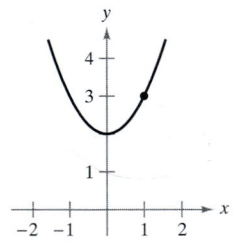

13. $\lim\limits_{x \to 2} f(x)$

$$f(x) = \begin{cases} 4-x, & x \neq 2 \\ 0, & x = 2 \end{cases}$$

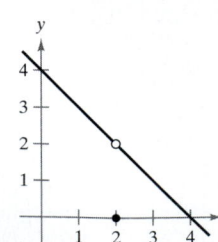

14. $\lim\limits_{x \to 1} f(x)$

$$f(x) = \begin{cases} x^2+2, & x \neq 1 \\ 1, & x = 1 \end{cases}$$

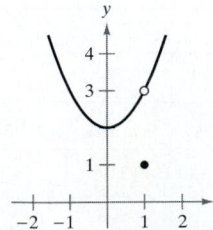

15. $\lim\limits_{x \to 5} \dfrac{|x-5|}{x-5}$

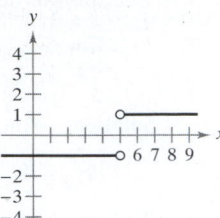

16. $\lim\limits_{x \to 3} \dfrac{1}{x-3}$

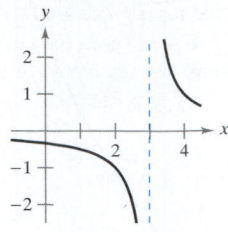

17. $\lim\limits_{x \to \pi/2} \tan x$

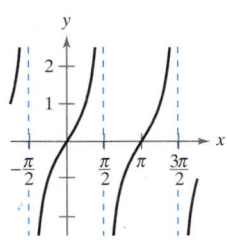

18. $\lim\limits_{x \to 0} \sec x$

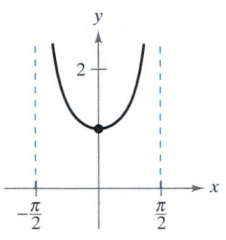

19. $\lim\limits_{x \to 0} \cos \dfrac{1}{x}$

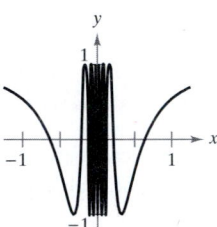

20. $\lim\limits_{x \to 1} \sin \pi x$

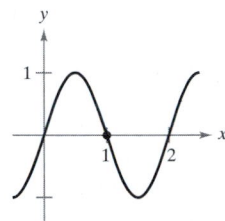

In Exercises 21–24, find the limit L. Then find $\delta > 0$ such that $|f(x) - L| < 0.01$ whenever $0 < |x - c| < \delta$.

21. $\lim\limits_{x \to 2} (3x+2)$

22. $\lim\limits_{x \to 4} \left(4 - \dfrac{x}{2}\right)$

23. $\lim\limits_{x \to 2} (x^2-3)$

24. $\lim\limits_{x \to 5} (x^2+4)$

In Exercises 25–36, find the limit L. Then use the ε-δ definition to prove that the limit is L.

25. $\lim\limits_{x \to 2} (x+3)$

26. $\lim\limits_{x \to -3} (2x+5)$

27. $\lim\limits_{x \to -4} \left(\tfrac{1}{2}x - 1\right)$

28. $\lim\limits_{x \to 1} \left(\tfrac{2}{3}x + 9\right)$

29. $\lim\limits_{x \to 6} 3$

30. $\lim\limits_{x \to 2} (-1)$

31. $\lim\limits_{x \to 0} \sqrt[3]{x}$

32. $\lim\limits_{x \to 4} \sqrt{x}$

33. $\lim\limits_{x \to -2} |x-2|$

34. $\lim\limits_{x \to 3} |x-3|$

35. $\lim\limits_{x \to 1} (x^2+1)$

36. $\lim\limits_{x \to -3} (x^2+3x)$

Writing In Exercises 37–40, use a graphing utility to graph the function and estimate the limit (if it exists). What is the domain of the function? Can you detect a possible danger in determining the domain of a function solely by analyzing the graph generated by a graphing utility? Write a short paragraph about the importance of examining a function analytically as well as graphically.

37. $f(x) = \dfrac{\sqrt{x+5}-3}{x-4}$ **38.** $f(x) = \dfrac{x-3}{x^2 - 4x + 3}$

$\lim\limits_{x \to 4} f(x)$ $\lim\limits_{x \to 3} f(x)$

39. $f(x) = \dfrac{x-9}{\sqrt{x}-3}$ **40.** $f(x) = \dfrac{x-3}{x^2 - 9}$

$\lim\limits_{x \to 9} f(x)$ $\lim\limits_{x \to 3} f(x)$

 41. Graphical Analysis   The statement

$$\lim_{x \to 2} \frac{x^2 - 4}{x - 2} = 4$$

means that for each $\varepsilon > 0$ there corresponds a $\delta > 0$ such that if $0 < |x - 2| < \delta$, then

$$\left| \frac{x^2 - 4}{x - 2} - 4 \right| < \varepsilon.$$

If $\varepsilon = 0.001$, then

$$\left| \frac{x^2 - 4}{x - 2} - 4 \right| < 0.001.$$

Use a graphing utility to graph each side of this inequality. Use the zoom feature to find an interval $(2 - \delta, 2 + \delta)$ such that the graph of the left side is below the graph of the right side of the inequality.

42. Think About It   Answer each of the following and give reasons for your answer.

(a) If $f(2) = 4$, can you conclude anything about the limit of $f(x)$ as x approaches 2?

(b) If the limit of $f(x)$ as x approaches 2 is 4, can you conclude anything about $f(2)$?

True or False?   In Exercises 43–46, determine whether the statement is true or false. If it is false, explain why or give an example that shows it is false.

43. If f is undefined at $x = c$, then the limit of $f(x)$ as x approaches c does not exist.

44. If the limit of $f(x)$ as x approaches c is 0, then there must exist a number k such that $f(k) < 0.001$.

45. If $f(c) = L$, then $\lim\limits_{x \to c} f(x) = L$.

46. If $\lim\limits_{x \to c} f(x) = L$, then $f(c) = L$.

 47. Programming   Use the programming capabilities of a graphing utility to write a program for approximating

$$\lim_{x \to c} f(x).$$

Assume the program will be applied only to functions whose limits exists as x approaches c. [*Hint:* Let $y_1 = f(x)$ and generate two lists whose entries form the ordered pairs

$$(c \pm [0.1]^n, \ f(c \pm [0.1]^n))$$

for $n = 0, 1, 2, 3,$ and 4.]

48. Use the program you created in Exercise 47 to approximate the limit

$$\lim_{x \to 4} \frac{x^2 - x - 12}{x - 4}.$$

49. Prove that if the limit of $f(x)$ as $x \to c$ exists, then the limit must be unique. [*Hint:* Let

$$\lim_{x \to c} f(x) = L_1 \quad \text{and} \quad \lim_{x \to c} f(x) = L_2$$

and prove that $L_1 = L_2$.]

50. Consider the line $f(x) = mx + b$, where $m \neq 0$. Use the ε-δ definition of the limit to prove that

$$\lim_{x \to c} f(x) = mc + b.$$

51. Prove that

$$\lim_{x \to c} f(x) = L$$

is equivalent to

$$\lim_{x \to c} [\, f(x) - L\,] = 0.$$

52. Consider the function $f(x) = (1 + x)^{1/x}$. Estimate the limit $\lim\limits_{x \to 0} (1 + x)^{1/x}$ by evaluating f at x-values near 0. Sketch the graph of f.

SECTION *1.3* **Evaluating Limits Analytically**

Properties of Limits • A Strategy for Finding Limits •
Cancellation and Rationalization Techniques • The Squeeze Theorem

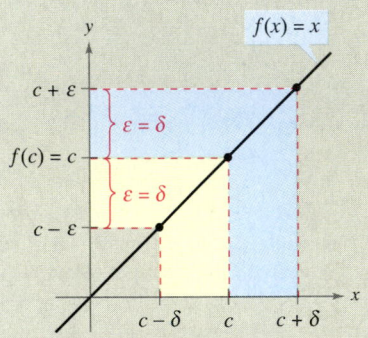
Properties of Limits

In Section 1.2, you learned that the limit of $f(x)$ as x approaches c does not depend on the value of f at $x = c$. It may happen, however, that the limit is precisely $f(c)$. In such cases, the limit can be evaluated by **direct substitution.** That is,

$$\lim_{x \to c} f(x) = f(c). \qquad \text{Substitute } c \text{ for } x.$$

Such *well-behaved* functions are **continuous at c.** You will examine this concept more closely in Section 1.4.

THEOREM 1.1 Some Basic Limits

Let b and c be real numbers and let n be a positive integer.

1. $\lim\limits_{x \to c} b = b$ **2.** $\lim\limits_{x \to c} x = c$ **3.** $\lim\limits_{x \to c} x^n = c^n$

EXAMPLE 1 Evaluating Basic Limits

a. $\lim\limits_{x \to 2} 3 = 3$

b. $\lim\limits_{x \to -4} x = -4$

c. $\lim\limits_{x \to 2} x^2 = 2^2 = 4$

NOTE When you encounter new notations or symbols in mathematics, be sure you know how the notations are read. For instance, the limit in Example 1c is read as "the limit of x^2 as x approaches 2 is 4."

THEOREM 1.2 Properties of Limits

Let b and c be real numbers, let n be a positive integer, and let f and g be functions with the following limits.

$$\lim_{x \to c} f(x) = L \qquad \text{and} \qquad \lim_{x \to c} g(x) = K$$

1. Scalar multiple: $\lim\limits_{x \to c} [b f(x)] = bL$

2. Sum or difference: $\lim\limits_{x \to c} [f(x) \pm g(x)] = L \pm K$

3. Product: $\lim\limits_{x \to c} [f(x)g(x)] = LK$

4. Quotient: $\lim\limits_{x \to c} \dfrac{f(x)}{g(x)} = \dfrac{L}{K}, \qquad$ provided $K \neq 0$

5. Power: $\lim\limits_{x \to c} [f(x)]^n = L^n$

EXAMPLE 2   The Limit of a Polynomial

$$\lim_{x \to 2} (4x^2 + 3) = \lim_{x \to 2} 4x^2 + \lim_{x \to 2} 3 \qquad \text{Property 2}$$

$$= 4\left(\lim_{x \to 2} x^2\right) + \lim_{x \to 2} 3 \qquad \text{Property 1}$$

$$= 4(2^2) + 3 \qquad \text{Example 1}$$

$$= 19 \qquad \text{Simplify.}$$

In Example 2, note that the limit (as $x \to 2$) of the *polynomial function* $p(x) = 4x^2 + 3$ is simply the value of p at $x = 2$.

$$\lim_{x \to 2} p(x) = p(2) = 4(2^2) + 3 = 19$$

This *direct substitution* property is valid for all polynomial and rational functions with nonzero denominators.

THEOREM 1.3   Limits of Polynomial and Rational Functions

If p is a polynomial function and c is a real number, then

$$\lim_{x \to c} p(x) = p(c).$$

If r is a rational function given by $r(x) = p(x)/q(x)$ and c is a real number such that $q(c) \neq 0$, then

$$\lim_{x \to c} r(x) = r(c) = \frac{p(c)}{q(c)}.$$

EXAMPLE 3   The Limit of a Rational Function

Find the limit: $\lim\limits_{x \to 1} \dfrac{x^2 + x + 2}{x + 1}$.

Solution   Because the denominator is not 0 when $x = 1$, you can apply Theorem 1.3 to obtain

$$\lim_{x \to 1} \frac{x^2 + x + 2}{x + 1} = \frac{1^2 + 1 + 2}{1 + 1} = \frac{4}{2} = 2.$$

Polynomial functions and rational functions are two of the three basic types of algebraic functions. The following theorem deals with the limit of the third type of algebraic function—one that involves a radical.

THEOREM 1.4   The Limit of a Function Involving a Radical

Let n be a positive integer. The following limit is valid for all c if n is odd, and is valid for $c > 0$ if n is even.

$$\lim_{x \to c} \sqrt[n]{x} = \sqrt[n]{c}$$

The following theorem greatly expands your ability to evaluate limits because it shows how to analyze the limit of a composite function.

> **THEOREM 1.5   The Limit of a Composite Function**
>
> If f and g are functions such that $\lim\limits_{x \to c} g(x) = L$ and $\lim\limits_{x \to L} f(x) = f(L)$, then
>
> $$\lim_{x \to c} f(g(x)) = f(L).$$

EXAMPLE 4   The Limit of a Composite Function

a. Because

$$\lim_{x \to 0} (x^2 + 4) = 0^2 + 4 = 4 \qquad \text{and} \qquad \lim_{x \to 4} \sqrt{x} = 2$$

it follows that

$$\lim_{x \to 0} \sqrt{x^2 + 4} = \sqrt{4} = 2.$$

b. Because

$$\lim_{x \to 3} (2x^2 - 10) = 2(3^2) - 10 = 8 \quad \text{and} \quad \lim_{x \to 8} \sqrt[3]{x} = 2$$

it follows that

$$\lim_{x \to 3} \sqrt[3]{2x^2 - 10} = \sqrt[3]{8} = 2.$$

You have seen that the limits of many algebraic functions can be evaluated by direct substitution. Each of the six basic trigonometric functions also possesses this desirable quality, as shown in the next theorem (presented without proof).

> **THEOREM 1.6   Limits of Trigonometric Functions**
>
> Let c be a real number in the domain of the given trigonometric function.
>
> **1.** $\lim\limits_{x \to c} \sin x = \sin c$ **2.** $\lim\limits_{x \to c} \cos x = \cos c$
>
> **3.** $\lim\limits_{x \to c} \tan x = \tan c$ **4.** $\lim\limits_{x \to c} \cot x = \cot c$
>
> **5.** $\lim\limits_{x \to c} \sec x = \sec c$ **6.** $\lim\limits_{x \to c} \csc x = \csc c$

EXAMPLE 5   Limits of Trigonometric Functions

a. $\lim\limits_{x \to 0} \tan x = \tan (0) = 0$

b. $\lim\limits_{x \to \pi} (x \cos x) = \left(\lim\limits_{x \to \pi} x \right) \left(\lim\limits_{x \to \pi} \cos x \right) = \pi \cos (\pi) = -\pi$

c. $\lim\limits_{x \to 0} \sin^2 x = \lim\limits_{x \to 0} (\sin x)^2 = 0^2 = 0$

A Strategy for Finding Limits

On the previous three pages, you studied several types of functions whose limits can be evaluated by direct substitution. This knowledge, together with the following theorem, can be used to develop a strategy for finding limits.

> **THEOREM 1.7   Functions That Agree at All But One Point**
>
> Let c be a real number and let $f(x) = g(x)$ for all $x \neq c$ in an open interval containing c. If the limit of $g(x)$ as x approaches c exists, then the limit of $f(x)$ also exists and
>
> $$\lim_{x \to c} f(x) = \lim_{x \to c} g(x).$$

EXAMPLE 6   Finding the Limit of a Function

Find the limit: $\displaystyle \lim_{x \to 1} \frac{x^3 - 1}{x - 1}$.

Solution   Let $f(x) = (x^3 - 1)/(x - 1)$. By factoring and canceling, you can rewrite f as

$$f(x) = \frac{(x - 1)(x^2 + x + 1)}{(x - 1)} = x^2 + x + 1 = g(x), \qquad x \neq 1.$$

Thus, for all x-values other than $x = 1$, the functions f and g agree, as shown in Figure 1.16. Because $\lim_{x \to 1} g(x)$ exists, you can apply Theorem 1.7 to conclude that f and g have the same limit at $x = 1$.

$$\lim_{x \to 1} \frac{x^3 - 1}{x - 1} = \lim_{x \to 1} \frac{(x - 1)(x^2 + x + 1)}{x - 1} \qquad \text{Factor.}$$

$$= \lim_{x \to 1} \frac{(x - 1)(x^2 + x + 1)}{x - 1} \qquad \text{Cancel like factors.}$$

$$= \lim_{x \to 1} (x^2 + x + 1) \qquad \text{Apply Theorem 1.7.}$$

$$= 1^2 + 1 + 1 \qquad \text{Use direct substitution.}$$

$$= 3 \qquad \text{Simplify.}$$

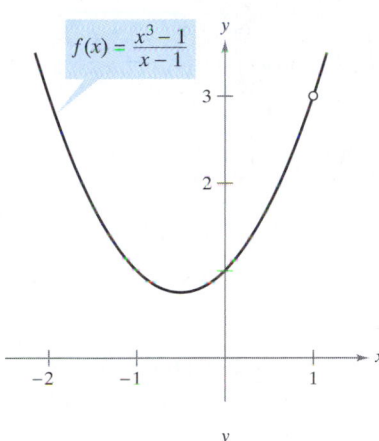

$f(x) = \dfrac{x^3 - 1}{x - 1}$

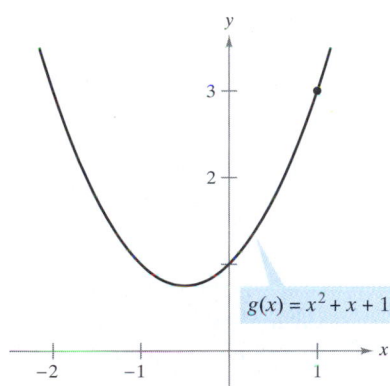

$g(x) = x^2 + x + 1$

f and g agree at all but one point.
Figure 1.16

STUDY TIP   When applying this strategy for finding limits, remember that some functions do not have limits (as x approaches c). For instance, the following limit does not exist.

$$\lim_{x \to 1} \frac{x^3 + 1}{x - 1}$$

A Strategy for Finding Limits

1. Learn to recognize which limits can be evaluated by direct substitution. (These limits are listed in Theorems 1.1 through 1.6)

2. If the limit of $f(x)$ as x approaches c *cannot* be evaluated by direct substitution, try to find a function g that agrees with f for all x other than $x = c$. [Choose g such that the limit of $g(x)$ *can* be evaluated by direct substitution.]

3. Apply Theorem 1.7 to conclude *analytically* that

$$\lim_{x \to c} f(x) = \lim_{x \to c} g(x) = g(c).$$

4. Use a *graph* or *table* to reinforce your conclusion.

Cancellation and Rationalization Techniques

Two techniques for finding limits analytically are shown in Examples 7 and 8. The first technique involves canceling common factors, and the second technique involves rationalizing the numerator of a fractional expression.

EXAMPLE 7   Cancellation Technique

Find the limit: $\displaystyle\lim_{x \to -3} \frac{x^2 + x - 6}{x + 3}$.

Solution   Although you are taking the limit of a rational function, you *cannot* apply Theorem 1.3 because the limit of the denominator is 0.

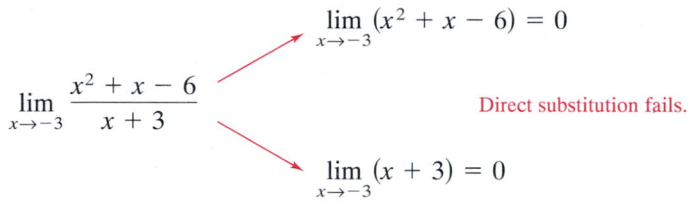

$$\lim_{x \to -3} (x^2 + x - 6) = 0$$

$$\lim_{x \to -3} \frac{x^2 + x - 6}{x + 3}$$ Direct substitution fails.

$$\lim_{x \to -3} (x + 3) = 0$$

Because the limit of the numerator is also 0, the numerator and denominator have a *common factor* of $(x + 3)$. Thus, for all $x \neq -3$, you can cancel this factor to obtain

$$\frac{x^2 + x - 6}{x + 3} = \frac{\cancel{(x + 3)}(x - 2)}{\cancel{x + 3}} = x - 2, \qquad x \neq -3.$$

Using Theorem 1.7, it follows that

$$\lim_{x \to -3} \frac{x^2 + x - 6}{x + 3} = \lim_{x \to -3} (x - 2)$$ Apply Theorem 1.7.

$$= -5.$$ Use direct substitution.

This result is shown graphically in Figure 1.17. Note that the graph of the function f coincides with the graph of the function $g(x) = x - 2$, except that the graph of f has a gap at the point $(-3, -5)$.

In Example 7, direct substitution produced the meaningless fractional form $0/0$. Such an expression is called an **indeterminate form** because you cannot (from the form alone) determine the limit. When you try to evaluate a limit and encounter this form, remember that you must rewrite the fraction so that the new denominator does not have 0 as its limit. One way to do this is to *cancel like factors*, as shown in Example 7. A second way is to *rationalize the numerator*, as shown in Example 8.

TECHNOLOGY   Because the graphs of

$$f(x) = \frac{x^2 + x - 6}{x + 3} \qquad \text{and} \qquad g(x) = x - 2$$

differ only at the point $(-3, -5)$ a graphing utility may not be able to distinguish visually between these graphs. However, because of the pixel configuration and rounding error of a graphing utility, it may be possible to find screen settings that distinguish between the graphs. Specifically, by repeatedly zooming in near the point $(-3, -5)$ on the graph of f, your graphing utility may show glitches or irregularities that do not exist on the actual graph. (See Figure 1.18.) By changing the screen settings on your graphing utility you may obtain the correct graph of f.

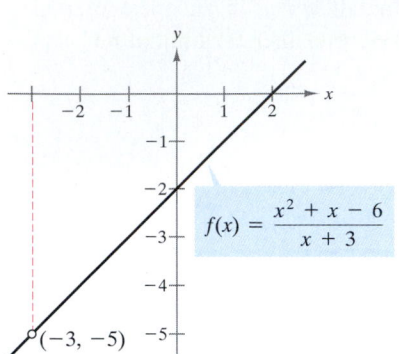

f is undefined when $x = -3$.

Figure 1.17

NOTE   In the solution of Example 7, be sure you see the usefulness of the Factor Theorem of Algebra. This theorem states that if c is a zero of a polynomial function, $(x - c)$ is a factor of the polynomial. Thus, if you apply direct substitution to a rational function and obtain

$$r(c) = \frac{p(c)}{q(c)} = \frac{0}{0},$$

you can conclude that $(x - c)$ must be a common factor to both $p(x)$ and $q(x)$.

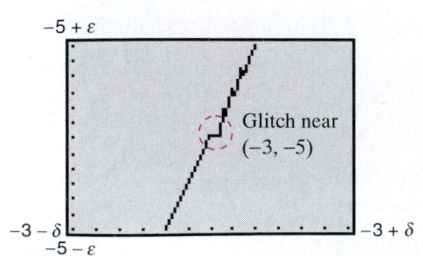

Incorrect graph of f

Figure 1.18

EXAMPLE 8   Rationalization Technique

Find the limit: $\displaystyle\lim_{x \to 0} \frac{\sqrt{x + 1} - 1}{x}$.

Solution   By direct substitution, you obtain the indeterminate form $0/0$.

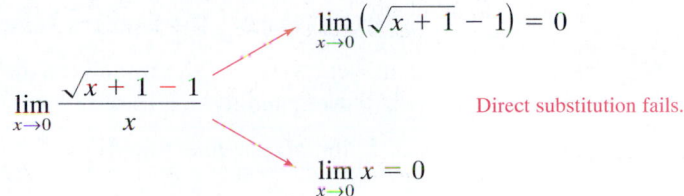

$$\lim_{x \to 0} \left(\sqrt{x + 1} - 1\right) = 0$$

$$\lim_{x \to 0} \frac{\sqrt{x + 1} - 1}{x}$$

Direct substitution fails.

$$\lim_{x \to 0} x = 0$$

In this case, you can rewrite the fraction by rationalizing the numerator.

$$\frac{\sqrt{x + 1} - 1}{x} = \left(\frac{\sqrt{x + 1} - 1}{x}\right)\left(\frac{\sqrt{x + 1} + 1}{\sqrt{x + 1} + 1}\right)$$

$$= \frac{(x + 1) - 1}{x\left(\sqrt{x + 1} + 1\right)}$$

$$= \frac{x}{x\left(\sqrt{x + 1} + 1\right)}$$

$$= \frac{1}{\sqrt{x + 1} + 1}, \quad x \neq 0$$

Now, using Theorem 1.7, you can evaluate the limit as follows.

$$\lim_{x \to 0} \frac{\sqrt{x + 1} - 1}{x} = \lim_{x \to 0} \frac{1}{\sqrt{x + 1} + 1}$$

$$= \frac{1}{1 + 1}$$

$$= \frac{1}{2}$$

A table or a graph can reinforce your conclusion that the limit is $\frac{1}{2}$. (See Figure 1.19.)

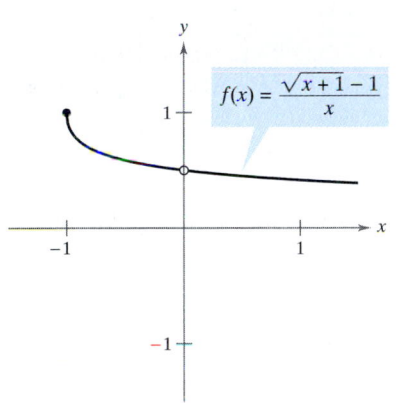

$f(x) = \dfrac{\sqrt{x+1} - 1}{x}$

The limit of $f(x)$ as x approaches 0 is $\frac{1}{2}$.
Figure 1.19

x approaches 0 from the left. ⟶   ⟵ x approaches 0 from the right.

x	-0.25	-0.1	-0.01	-0.001	0	0.001	0.01	0.1	0.25
$f(x)$	0.5359	0.5132	0.5013	0.5001	?	0.4999	0.4988	0.4881	0.4721

$f(x)$ approaches 0.5. ⟶   ⟵ $f(x)$ approaches 0.5.

NOTE   The rationalization technique for evaluating limits is based on multiplication by a convenient form of 1. In Example 8, the convenient form is

$$1 = \frac{\sqrt{x + 1} + 1}{\sqrt{x + 1} + 1}.$$

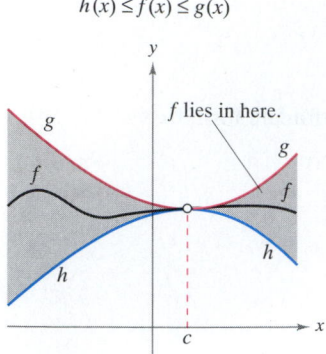

$h(x) \leq f(x) \leq g(x)$

f lies in here.

The Squeeze Theorem
Figure 1.20

The Squeeze Theorem

The next theorem concerns the limit of a function that is squeezed between two other functions, each of which has the same limit at a given *x*-value, as shown in Figure 1.20. (The proof of this theorem is given in the appendix.)

THEOREM 1.8   The Squeeze Theorem

If $h(x) \leq f(x) \leq g(x)$ for all *x* in an open interval containing *c*, except possibly at *c* itself, and if

$$\lim_{x \to c} h(x) = L = \lim_{x \to c} g(x)$$

then $\lim_{x \to c} f(x)$ exists and is equal to *L*.

You can see the usefulness of the Squeeze Theorem in the proof of Theorem 1.9.

THEOREM 1.9   Two Special Trigonometric Limits

1. $\lim_{x \to 0} \dfrac{\sin x}{x} = 1$ **2.** $\lim_{x \to 0} \dfrac{1 - \cos x}{x} = 0$

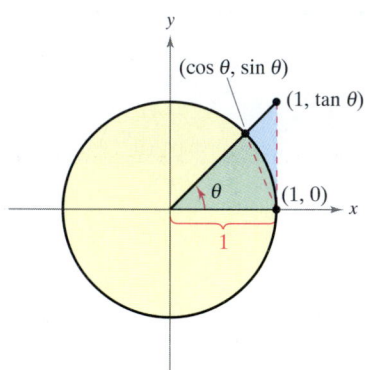

$(\cos \theta, \sin \theta)$

$(1, \tan \theta)$

$(1, 0)$

A circular sector is used to prove Theorem 1.9.
Figure 1.21

Proof   To avoid the confusion of two different uses of *x*, the proof is presented using the variable θ, where θ is an acute positive angle *measured in radians*. Figure 1.21 shows a circular sector that is squeezed between two triangles.

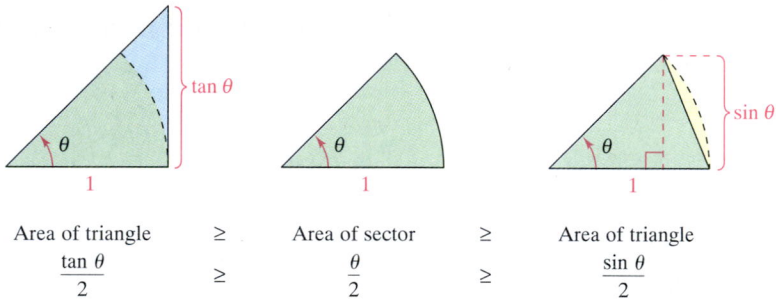

Area of triangle	$\geq$	Area of sector	$\geq$	Area of triangle
$\dfrac{\tan \theta}{2}$	$\geq$	$\dfrac{\theta}{2}$	$\geq$	$\dfrac{\sin \theta}{2}$

Multiplying each expression by $2/\sin \theta$ produces

$$\frac{1}{\cos \theta} \geq \frac{\theta}{\sin \theta} \geq 1$$

and taking reciprocals and reversing the inequalities yields

$$\cos \theta \leq \frac{\sin \theta}{\theta} \leq 1.$$

Because $\cos \theta = \cos(-\theta)$ and $(\sin \theta)/\theta = [\sin(-\theta)]/(-\theta)$, we can conclude that this inequality is valid for *all* nonzero θ in the open interval $(-\pi/2, \pi/2)$. Finally, because $\lim_{\theta \to 0} \cos \theta = 1$ and $\lim_{\theta \to 0} 1 = 1$, you can apply the Squeeze Theorem to conclude that $\lim_{\theta \to 0} (\sin \theta)/\theta = 1$. The proof of the second limit is left as an exercise (see Exercise 103).

FOR FURTHER INFORMATION
For more information on the function $f(x) = (\sin x)/x$, see the article "The Function $(\sin x)/x$" by William B. Gearhart and Harris S. Shultz in the March 1990 issue of *The College Mathematics Journal.*

EXAMPLE 9   A Limit Involving a Trigonometric Function

Find the limit: $\lim\limits_{x \to 0} \dfrac{\tan x}{x}$.

Solution   Direct substitution yields the indeterminate form 0/0. To solve this problem, you can write $\tan x$ as $(\sin x)/(\cos x)$ and obtain

$$\lim_{x \to 0} \frac{\tan x}{x} = \lim_{x \to 0} \left(\frac{\sin x}{x} \right) \left(\frac{1}{\cos x} \right).$$

Now, because

$$\lim_{x \to 0} \frac{\sin x}{x} = 1 \qquad \text{and} \qquad \lim_{x \to 0} \frac{1}{\cos x} = 1$$

you can obtain

$$\lim_{x \to 0} \frac{\tan x}{x} = \left(\lim_{x \to 0} \frac{\sin x}{x} \right) \left(\lim_{x \to 0} \frac{1}{\cos x} \right)$$
$$= (1)(1)$$
$$= 1.$$

(See Figure 1.22.)

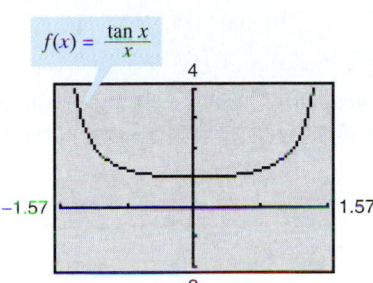

$f(x) = \dfrac{\tan x}{x}$

The limit of $f(x)$ as x approaches 0 is 1.
Figure 1.22

EXAMPLE 10   A Limit Involving a Trigonometric Function

Find the limit: $\lim\limits_{x \to 0} \dfrac{\sin 4x}{x}$.

Solution   Direct substitution yields the indeterminate form 0/0. To solve this problem, you can rewrite the limit as

$$\lim_{x \to 0} \frac{\sin 4x}{x} = 4 \left(\lim_{x \to 0} \frac{\sin 4x}{4x} \right).$$

Now, by letting $y = 4x$ and observing that $x \to 0$ if and only if $y \to 0$, you can write

$$\lim_{x \to 0} \frac{\sin 4x}{x} = 4 \left(\lim_{x \to 0} \frac{\sin 4x}{4x} \right)$$
$$= 4 \left(\lim_{y \to 0} \frac{\sin y}{y} \right)$$
$$= 4(1)$$
$$= 4.$$

(See Figure 1.23.)

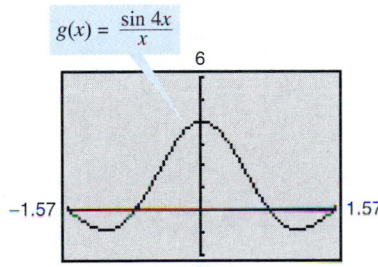

$g(x) = \dfrac{\sin 4x}{x}$

The limit of $g(x)$ as x approaches 0 is 4.
Figure 1.23

TECHNOLOGY   Try using a graphing utility to confirm the limits in the examples and exercise set. For instance, Figures 1.22 and 1.23 show the graphs of

$$f(x) = \frac{\tan x}{x} \qquad \text{and} \qquad g(x) = \frac{\sin 4x}{x}.$$

Note that the first graph appears to contain the point (0, 1) and the second graph appears to contain the point (0, 4), which lends support to the conclusions obtained in Examples 9 and 10.

EXERCISES FOR SECTION 1.3

In Exercises 1–4, use a graphing utility to graph the function and visually estimate the limits.

1. $h(x) = x^2 - 5x$

  (a) $\lim\limits_{x \to 5} h(x)$

  (b) $\lim\limits_{x \to -1} h(x)$

2. $g(x) = \dfrac{12(\sqrt{x} - 3)}{x - 9}$

  (a) $\lim\limits_{x \to 4} g(x)$

  (b) $\lim\limits_{x \to 0} g(x)$

3. $f(x) = x \cos x$

  (a) $\lim\limits_{x \to 0} f(x)$

  (b) $\lim\limits_{x \to \pi/3} f(x)$

4. $f(t) = t|t - 4|$

  (a) $\lim\limits_{t \to 4} f(t)$

  (b) $\lim\limits_{t \to -1} f(t)$

In Exercises 5–28, find the limit.

5. $\lim\limits_{x \to 4} x^2$

6. $\lim\limits_{x \to -3} (3x + 2)$

7. $\lim\limits_{x \to 0} (2x - 1)$

8. $\lim\limits_{x \to 1} (-x^2 + 1)$

9. $\lim\limits_{x \to 2} (-x^2 + x - 2)$

10. $\lim\limits_{x \to 1} (3x^3 - 2x^2 + 4)$

11. $\lim\limits_{x \to 3} \sqrt{x + 1}$

12. $\lim\limits_{x \to 4} \sqrt[3]{x + 4}$

13. $\lim\limits_{x \to -4} (x + 3)^2$

14. $\lim\limits_{x \to 0} (2x - 1)^3$

15. $\lim\limits_{x \to 2} \dfrac{1}{x}$

16. $\lim\limits_{x \to -3} \dfrac{2}{x + 2}$

17. $\lim\limits_{x \to -1} \dfrac{x^2 + 1}{x}$

18. $\lim\limits_{x \to 3} \dfrac{\sqrt{x + 1}}{x - 4}$

19. $\lim\limits_{x \to \pi/2} \sin x$

20. $\lim\limits_{x \to \pi} \tan x$

21. $\lim\limits_{x \to 1} \cos \pi x$

22. $\lim\limits_{x \to 1} \sin \dfrac{\pi x}{2}$

23. $\lim\limits_{x \to 0} \sec 2x$

24. $\lim\limits_{x \to \pi} \cos 3x$

25. $\lim\limits_{x \to 5\pi/6} \sin x$

26. $\lim\limits_{x \to 5\pi/3} \cos x$

27. $\lim\limits_{x \to 3} \tan\left(\dfrac{\pi x}{4}\right)$

28. $\lim\limits_{x \to 7} \sec\left(\dfrac{\pi x}{6}\right)$

In Exercises 29–32, use the information to evaluate the limits.

29. $\lim\limits_{x \to c} f(x) = 2$

  $\lim\limits_{x \to c} g(x) = 3$

  (a) $\lim\limits_{x \to c} [5g(x)]$

  (b) $\lim\limits_{x \to c} [f(x) + g(x)]$

  (c) $\lim\limits_{x \to c} [f(x)g(x)]$

  (d) $\lim\limits_{x \to c} \dfrac{f(x)}{g(x)}$

30. $\lim\limits_{x \to c} f(x) = \tfrac{3}{2}$

  $\lim\limits_{x \to c} g(x) = \tfrac{1}{2}$

  (a) $\lim\limits_{x \to c} [4f(x)]$

  (b) $\lim\limits_{x \to c} [f(x) + g(x)]$

  (c) $\lim\limits_{x \to c} [f(x)g(x)]$

  (d) $\lim\limits_{x \to c} \dfrac{f(x)}{g(x)}$

31. $\lim\limits_{x \to c} f(x) = 4$

  (a) $\lim\limits_{x \to c} [f(x)]^3$

  (b) $\lim\limits_{x \to c} \sqrt{f(x)}$

  (c) $\lim\limits_{x \to c} [3f(x)]$

  (d) $\lim\limits_{x \to c} [f(x)]^{3/2}$

32. $\lim\limits_{x \to c} f(x) = 27$

  (a) $\lim\limits_{x \to c} \sqrt[3]{f(x)}$

  (b) $\lim\limits_{x \to c} \dfrac{f(x)}{18}$

  (c) $\lim\limits_{x \to c} [f(x)]^2$

  (d) $\lim\limits_{x \to c} [f(x)]^{2/3}$

In Exercises 33–36, use the graph to determine the limit visually (if it exists). When possible, identify two functions that agree at all but one point.

33. $g(x) = \dfrac{-2x^2 + x}{x}$

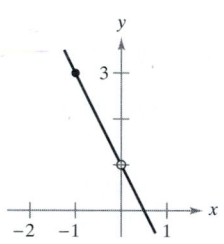

  (a) $\lim\limits_{x \to 0} g(x)$

  (b) $\lim\limits_{x \to -1} g(x)$

34. $h(x) = \dfrac{x^2 - 3x}{x}$

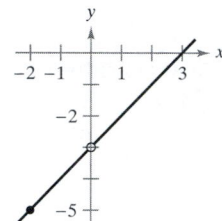

  (a) $\lim\limits_{x \to -2} h(x)$

  (b) $\lim\limits_{x \to 0} h(x)$

35. $g(x) = \dfrac{x^3 - x}{x - 1}$

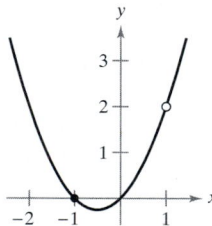

  (a) $\lim\limits_{x \to 1} g(x)$

  (b) $\lim\limits_{x \to -1} g(x)$

36. $f(x) = \dfrac{x}{x^2 - x}$

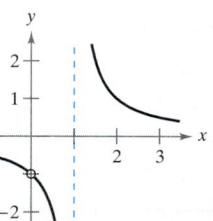

  (a) $\lim\limits_{x \to 1} f(x)$

  (b) $\lim\limits_{x \to 0} f(x)$

 In Exercises 37–40, find the limit of the function (if it exists). Identify two functions that agree at all but one point and use a graphing utility to graph the function.

37. $\lim\limits_{x \to -1} \dfrac{x^2 - 1}{x + 1}$

38. $\lim\limits_{x \to -1} \dfrac{2x^2 - x - 3}{x + 1}$

39. $\lim\limits_{x \to -2} \dfrac{x^3 + 8}{x + 2}$

40. $\lim\limits_{x \to -1} \dfrac{x^3 + 1}{x + 1}$

In Section P.1, an *x*-intercept of a graph was defined to be a point $(a, 0)$ at which the graph crosses the *x*-axis. If the graph represents a function *f*, the number *a* is a **zero** of *f*. In other words, *the zeros of a function f are the solutions of the equation* $f(x) = 0$. For example, the function $f(x) = x - 4$ has a zero at $x = 4$ because $f(4) = 0$.

In Section P.1 we also discussed different types of symmetry. In the terminology of functions, a function is **even** if its graph is symmetric with respect to the *y*-axis, and is **odd** if its graph is symmetric with respect to the origin. The symmetry tests in Section P.1 yield the following test for even and odd functions.

> **Test for Even and Odd Functions**
>
> The function $y = f(x)$ is **even** if $f(-x) = f(x)$.
> The function $y = f(x)$ is **odd** if $f(-x) = -f(x)$.

NOTE   Except for the constant function $f(x) = 0$, the graph of a function of *x* cannot have symmetry with respect to the *x*-axis because it then would fail the vertical line test for the graph of the function.

EXAMPLE 5   Even and Odd Functions and Zeros of Functions

Determine whether each function is even, odd, or neither. Then find the zeros of the function.

a. $f(x) = x^3 - x$ **b.** $g(x) = 1 + \cos x$

Solution

a. This function is odd because

$$f(-x) = (-x)^3 - (-x) = -x^3 + x = -(x^3 - x) = -f(x).$$

The zeros of *f* are found as follows.

$$x^3 - x = 0 \qquad \text{Let } f(x) = 0.$$
$$x(x^2 - 1) = x(x - 1)(x + 1) = 0 \qquad \text{Factor.}$$
$$x = 0, 1, -1 \qquad \text{Zeros of } f$$

See Figure P.31(a).

b. This function is even because

$$g(-x) = 1 + \cos(-x) = 1 + \cos x = g(x). \qquad \cos(-x) = \cos(x)$$

The zeros of *f* are found as follows.

$$1 + \cos x = 0 \qquad \text{Let } g(x) = 0.$$
$$\cos x = -1 \qquad \text{Subtract 1 from each side.}$$
$$x = (2n + 1)\pi, \ n \text{ is an integer.} \qquad \text{Zeros of } g$$

See Figure P.31(b).

NOTE   Each of the functions in Example 5 is either even or odd. However, some functions, such as $f(x) = x^2 + x + 1$, are neither even nor odd.

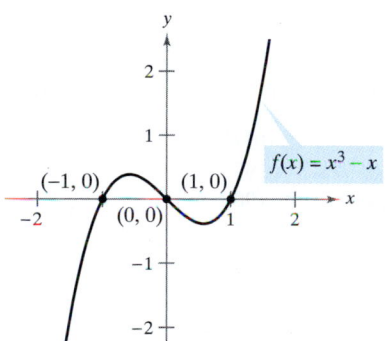

(a) Odd function

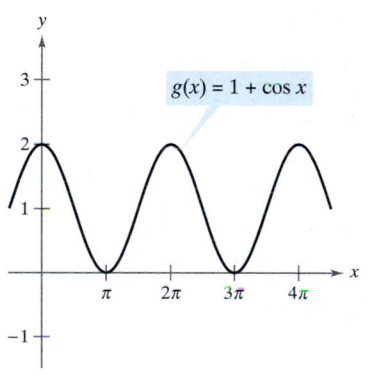

(b) Even function

Figure P.31

EXERCISES FOR SECTION P.3

In Exercises 1–10, evaluate (if possible) the function at the given value(s) of the independent variable. Simplify the results.

1. $f(x) = 2x - 3$

  (a) $f(0)$

  (b) $f(-3)$

  (c) $f(b)$

  (d) $f(x - 1)$

2. $f(x) = \sqrt{x + 3}$

  (a) $f(-2)$

  (b) $f(6)$

  (c) $f(c)$

  (d) $f(x + \Delta x)$

3. $f(x) = \begin{cases} 2x + 1, & x < 0 \\ 2x + 2, & x \geq 0 \end{cases}$

  (a) $f(-1)$ (b) $f(0)$ (c) $f(2)$ (d) $f(t^2 + 1)$

4. $f(x) = \begin{cases} x^2 + 2, & x \leq 1 \\ 2x^2 + 2, & x > 1 \end{cases}$

  (a) $f(-2)$ (b) $f(0)$ (c) $f(1)$ (d) $f(s^2 + 2)$

5. $f(x) = \cos 2x$

  (a) $f(0)$ (b) $f(-\pi/4)$ (c) $f(\pi/3)$

6. $f(x) = \sin x$

  (a) $f(\pi)$ (b) $f(5\pi/4)$ (c) $f(2\pi/3)$

7. $f(x) = x^3$

$$\frac{f(x + \Delta x) - f(x)}{\Delta x}$$

8. $f(x) = 3x - 1$

$$\frac{f(x) - f(1)}{x - 1}$$

9. $f(x) = \dfrac{1}{\sqrt{x - 1}}$

$$\frac{f(x) - f(2)}{x - 2}$$

10. $f(x) = x^3 - x$

$$\frac{f(x) - f(1)}{x - 1}$$

In Exercises 11–18, sketch a graph of the function and find its domain and range. You can use a graphing utility to verify your graph.

11. $f(x) = 4 - x$

12. $g(x) = \dfrac{4}{x}$

13. $h(x) = \sqrt{x - 1}$

14. $f(x) = \frac{1}{2}x^3 + 2$

15. $f(x) = \sqrt{9 - x^2}$

16. $f(x) = x + \sqrt{4 - x^2}$

17. $g(t) = 2 \sin \pi t$

18. $h(\theta) = -5 \cos \dfrac{\theta}{2}$

In Exercises 19 and 20, use the vertical line test to determine whether y is a function of x.

19. $x - y^2 = 0$

20. $\sqrt{x^2 - 4} - y = 0$

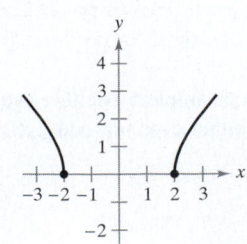

21. *Think About It*   Express the function

$$f(x) = |x| + |x - 2|$$

without using absolute value signs. (For a review of absolute value, see the appendix.)

22. *Writing*   Use a graphing utility to graph the polynomial functions $p_1(x) = x^3 - x + 1$ and $p_2(x) = x^3 - x$. How many zeros does each function have? Is there a cubic polynomial that has no zeros? Explain.

In Exercises 23–26, determine whether y is a function of x.

23. $x^2 + y^2 = 4$

24. $x^2 + y = 4$

25. $y^2 = x^2 - 1$

26. $x^2 y - x^2 + 4y = 0$

In Exercises 27 and 28, determine whether the relation is a function.

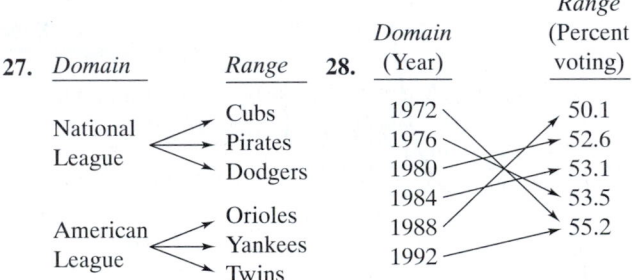

27.

Domain	Range
National League	Cubs, Pirates, Dodgers
American League	Orioles, Yankees, Twins

28.

Domain (Year)	Range (Percent voting)
1972	50.1
1976	52.6
1980	53.1
1984	53.5
1988	55.2
1992	

Modeling Data   **In Exercises 29–32, match the data with a function from the following list.**

  (i) $f(x) = cx$ (ii) $g(x) = cx^2$

  (iii) $h(x) = c\sqrt{|x|}$ (iv) $r(x) = c/x$

Determine the value of the constant c for each function so the function fits the data given in the table.

29.

x	-4	-1	0	1	4
y	-32	-2	0	-2	-32

30.

x	-4	-1	0	1	4
y	-1	$-\frac{1}{4}$	0	$\frac{1}{4}$	1

31.

x	-4	-1	0	1	4
y	-8	-32	Undef.	32	8

32.

x	-4	-1	0	1	4
y	6	3	0	3	6

In Exercises 41–52, find the limit (if it exists).

41. $\lim\limits_{x \to 5} \dfrac{x - 5}{x^2 - 25}$

42. $\lim\limits_{x \to 2} \dfrac{2 - x}{x^2 - 4}$

43. $\lim\limits_{x \to 1} \dfrac{x^2 + x - 2}{x^2 - 1}$

44. $\lim\limits_{x \to 0} \dfrac{\sqrt{2 + x} - \sqrt{2}}{x}$

45. $\lim\limits_{x \to 0} \dfrac{\sqrt{3 + x} - \sqrt{3}}{x}$

46. $\lim\limits_{x \to 0} \dfrac{[1/(x + 4)] - (1/4)}{x}$

47. $\lim\limits_{x \to 0} \dfrac{[1/(2 + x)] - (1/2)}{x}$

48. $\lim\limits_{x \to 3} \dfrac{\sqrt{x + 1} - 2}{x - 3}$

49. $\lim\limits_{\Delta x \to 0} \dfrac{2(x + \Delta x) - 2x}{\Delta x}$

50. $\lim\limits_{\Delta x \to 0} \dfrac{(x + \Delta x)^2 - x^2}{\Delta x}$

51. $\lim\limits_{\Delta x \to 0} \dfrac{(x + \Delta x)^2 - 2(x + \Delta x) + 1 - (x^2 - 2x + 1)}{\Delta x}$

52. $\lim\limits_{\Delta x \to 0} \dfrac{(x + \Delta x)^3 - x^3}{\Delta x}$

Graphical, Numerical, and Analytic Analysis In Exercises 53–56, use a graphing utility to graph the function and estimate the limit. Use a table to reinforce your conclusion. Then find the limit by analytic methods.

53. $\lim\limits_{x \to 0} \dfrac{\sqrt{x + 2} - \sqrt{2}}{x}$

54. $\lim\limits_{x \to 16} \dfrac{4 - \sqrt{x}}{x - 16}$

55. $\lim\limits_{x \to 0} \dfrac{[1/(2 + x)] - (1/2)}{x}$

56. $\lim\limits_{x \to 2} \dfrac{x^5 - 32}{x - 2}$

In Exercises 57–68, determine the limit of the trigonometric function (if it exists).

57. $\lim\limits_{x \to 0} \dfrac{\sin x}{5x}$

58. $\lim\limits_{x \to 0} \dfrac{3(1 - \cos x)}{x}$

59. $\lim\limits_{\theta \to 0} \dfrac{\sec \theta - 1}{\theta \sec \theta}$

60. $\lim\limits_{\theta \to 0} \dfrac{\cos \theta \tan \theta}{\theta}$

61. $\lim\limits_{x \to 0} \dfrac{\sin^2 x}{x}$

62. $\lim\limits_{x \to 0} \dfrac{\tan^2 x}{x}$

63. $\lim\limits_{h \to 0} \dfrac{(1 - \cos h)^2}{h}$

64. $\lim\limits_{\phi \to \pi} \phi \sec \phi$

65. $\lim\limits_{x \to \pi/2} \dfrac{\cos x}{\cot x}$

66. $\lim\limits_{x \to \pi/4} \dfrac{1 - \tan x}{\sin x - \cos x}$

67. $\lim\limits_{t \to 0} \dfrac{\sin^2 t}{t^2}$

$\left[\textit{Hint: Find } \lim\limits_{t \to 0}\left(\dfrac{\sin t}{t} \right)^2 . \right]$

68. $\lim\limits_{x \to 0} \dfrac{\sin 2x}{\sin 3x}$

$\left[\textit{Hint: Find } \lim\limits_{x \to 0}\left(\dfrac{2 \sin 2x}{2x} \right)\left(\dfrac{3x}{3 \sin 3x} \right) . \right]$

Graphical, Numerical, and Analytic Analysis In Exercises 69–72, use a graphing utility to graph the function and estimate the limit. Use a table to reinforce your conclusion. Then find the limit by analytic methods.

69. $\lim\limits_{t \to 0} \dfrac{\sin 3t}{t}$

70. $\lim\limits_{h \to 0} (1 + \cos 2h)$

71. $\lim\limits_{x \to 0} \dfrac{\sin x^2}{x}$

72. $\lim\limits_{x \to 0} \dfrac{\sin x}{\sqrt[3]{x}}$

In Exercises 73–76, find $\lim\limits_{h \to 0} \dfrac{f(x + h) - f(x)}{h}$.

73. $f(x) = 2x + 3$

74. $f(x) = \sqrt{x}$

75. $f(x) = \dfrac{4}{x}$

76. $f(x) = x^2 - 4x$

In Exercises 77 and 78, use the Squeeze Theorem to find $\lim\limits_{x \to c} f(x)$.

77. $c = 0$

$4 - x^2 \leq f(x) \leq 4 + x^2$

78. $c = a$

$b - |x - a| \leq f(x) \leq b + |x - a|$

In Exercises 79–84, use a graphing utility to graph the given function and the equations $y = |x|$ and $y = -|x|$ in the same viewing rectangle. Using the graphs to visually observe the Squeeze Theorem, find $\lim\limits_{x \to 0} f(x)$.

79. $f(x) = x \cos x$

80. $f(x) = |x \sin x|$

81. $f(x) = |x| \sin x$

82. $f(x) = |x| \cos x$

83. $f(x) = x \sin \dfrac{1}{x}$

84. $h(x) = x \cos \dfrac{1}{x}$

85. Writing   Use a graphing utility to graph

$$f(x) = x, \ g(x) = \sin x, \text{ and } h(x) = \dfrac{\sin x}{x}$$

in the same viewing rectangle. Compare the magnitudes of $f(x)$ and $g(x)$ when x is "close to" 0. Use the comparison to write a short paragraph explaining why

$$\lim\limits_{x \to 0} h(x) = 1.$$

86. Writing   Use a graphing utility to graph

$$f(x) = x, \ g(x) = \sin^2 x, \text{ and } h(x) = \dfrac{\sin^2 x}{x}$$

in the same viewing rectangle. Compare the magnitudes of $f(x)$ and $g(x)$ when x is "close to" 0. Use the comparison to write a short paragraph explaining why

$$\lim\limits_{x \to 0} h(x) = 0.$$

Free-Falling Object   In Exercises 87 and 88, use the position function $s(t) = -16t^2 + 1000$, which gives the height (in feet) of an object that has fallen for t seconds from a height of 1000 feet. The velocity at time $t = a$ seconds is given by

$$\lim_{t \to a} \frac{s(a) - s(t)}{a - t}.$$

87. If a construction worker drops a wrench from a height of 1000 feet, how fast will the wrench be falling after 5 seconds?

88. If a construction worker drops a wrench from a height of 1000 feet, when will the wrench hit the ground? At what velocity will the wrench impact the ground?

If a construction worker drops a wrench from a height of 1000 feet and yells "Look out below!", how long will a person on the ground have to get out of the way? Exercise 88 gives part of the answer.

Free-Falling Object   In Exercises 89 and 90, use the position function $s(t) = -4.9t^2 + 150$, which gives the height (in meters) of an object that has fallen from a height of 150 meters. The velocity at time $t = a$ seconds is given by

$$\lim_{t \to a} \frac{s(a) - s(t)}{a - t}.$$

89. Find the velocity of the object when $t = 3$.

90. At what velocity will the object impact the ground?

91. ***Modeling Data***   The average typing speed S of a typing student after t weeks of lessons is given in the table.

t	5	10	15	20	25	30
S	28	56	79	90	93	94

(a) Create a line graph of the data.

(b) Does there appear to be a limiting typing speed? Explain.

92. Find two functions f and g such that $\lim_{x \to 0} f(x)$ and $\lim_{x \to 0} g(x)$ do not exist, but $\lim_{x \to 0} [f(x) + g(x)]$ does exist.

93. Prove that if $\lim_{x \to c} f(x)$ exists and $\lim_{x \to c} [f(x) + g(x)]$ does not exist, then $\lim_{x \to c} g(x)$ does not exist.

94. Prove Property 1 of Theorem 1.1.

95. Prove Property 3 of Theorem 1.1. (You may use Property 3 of Theorem 1.2.)

96. Prove Property 1 of Theorem 1.2.

97. ***True or False***   If $f(x) < g(x)$ for all $x \neq a$, is it true that

$$\lim_{x \to a} f(x) < \lim_{x \to a} g(x)?$$

If it is false, explain why or give an example that shows it is false.

98. Prove that if $\lim_{x \to c} f(x) = 0$, then $\lim_{x \to c} |f(x)| = 0$.

99. Prove that if $\lim_{x \to c} f(x) = 0$ and $|g(x)| \leq M$ for a fixed number M and all $x \neq c$, then $\lim_{x \to c} f(x)g(x) = 0$.

100. Prove that if $\lim_{x \to c} |f(x)| = 0$, then $\lim_{x \to c} f(x) = 0$.

     (*Note:* This is the converse of Exercise 98.)

101. Prove that if $\lim_{x \to c} f(x) = L$, then $\lim_{x \to c} |f(x)| = |L|$.

     [*Hint:* Use the inequality $\big| |f(x)| - |L| \big| \leq |f(x) - L|$.]

102. ***Think About It***   Find a function f to show that the converse of Exercise 101 is not true. [*Hint:* Find a function f such that $\lim_{x \to c} |f(x)| = |L|$ but $\lim_{x \to c} f(x)$ does not exist.]

103. Prove the second part of Theorem 1.9 by proving that

$$\lim_{x \to 0} \frac{1 - \cos x}{x} = 0.$$

104. Let $f(x) = \begin{cases} 0, & \text{if } x \text{ is rational} \\ 1, & \text{if } x \text{ is irrational} \end{cases}$

     and

     $g(x) = \begin{cases} 0, & \text{if } x \text{ is rational} \\ x, & \text{if } x \text{ is irrational.} \end{cases}$

     Find (if possible) $\lim_{x \to 0} f(x)$ and $\lim_{x \to 0} g(x)$.

105. ***Graphical Reasoning***   Consider the function $f(x) = \dfrac{\sec x - 1}{x^2}$.

     (a) Find the domain of f.

     (b) Use a graphing utility to graph f. Is the domain of f obvious from the graph? If not, explain.

     (c) Use the graph of f to approximate $\lim_{x \to 0} f(x)$.

     (d) Confirm the answer in part (c) analytically.

106. ***Approximation***

     (a) Find $\lim_{x \to 0} \dfrac{1 - \cos x}{x^2}$.

     (b) Use the result in part (a) to derive the approximation $\cos x \approx 1 - \frac{1}{2}x^2$ for x near 0.

     (c) Use the result in part (b) to approximate $\cos(0.1)$.

     (d) Use a calculator to approximate $\cos(0.1)$ to four decimal places. Compare the result with the approximation in part (c).

107. ***Writing***   In the context of finding limits, discuss what is meant by two functions that agree at all but one point.

Tony Stone Images

SECTION *1.4* **Continuity and One-Sided Limits**

Continuity at a Point and on an Open Interval •
One-Sided Limits and Continuity on a Closed Interval •
Properties of Continuity • The Intermediate Value Theorem

Continuity at a Point and on an Open Interval

In mathematics, the term *continuous* has much the same meaning as it has in everyday usage. To say that a function is continuous at $x = c$ means that there is no interruption in the graph of f at c. That is, its graph is unbroken at c and there are no holes, jumps, or gaps. Figure 1.24 identifies three values of x at which the graph of f is *not* continuous. At all other points in the interval (a, b), the graph of f is uninterrupted and **continuous.**

EXPLORATION

Informally, you might say that a function is *continuous* on an open interval if its graph can be drawn with a pencil without lifting the pencil from the paper. Use a graphing calculator to graph each of the following functions on the indicated interval. From the graphs, which functions would you say are continuous on the interval? Do you think you can trust the results you obtained graphically? Explain your reasoning.

Function	Interval
(a) $y = x^2 + 1$	$(-3, 3)$
(b) $y = \dfrac{1}{x - 2}$	$(-3, 3)$
(c) $y = \dfrac{\sin x}{x}$	$(-\pi, \pi)$
(d) $y = \dfrac{x^2 - 4}{x + 2}$	$(-3, 3)$
(e) $y = \begin{cases} 2x - 4, & x \leq 0 \\ x + 1, & x > 0 \end{cases}$	$(-3, 3)$

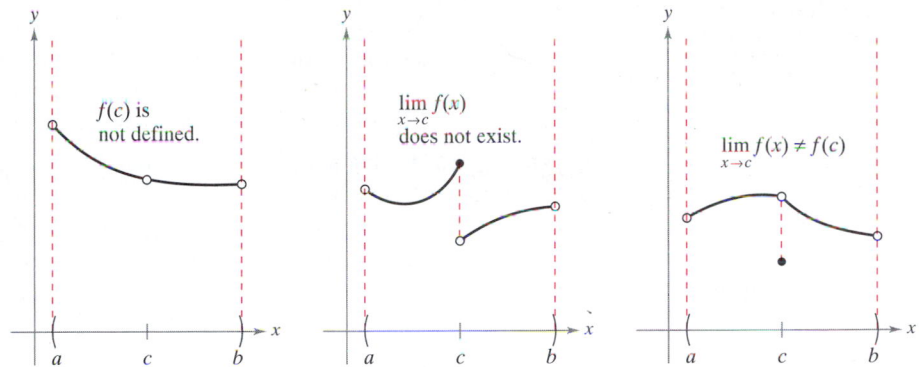

Three conditions exist for which the graph of f is not continuous at $x = c$.
Figure 1.24

In Figure 1.24, it appears that continuity at $x = c$ can be destroyed by any one of the following conditions.

1. The function is not defined at $x = c$.

2. The limit of $f(x)$ does not exist at $x = c$.

3. The limit of $f(x)$ exists at $x = c$, but it is not equal to $f(c)$.

If *none* of the above three conditions is true, the function f is called **continuous at c,** as indicated in the following important definition.

FOR FURTHER INFORMATION For more information on the concept of continuity, see the article "Leibniz and the Spell of the Continuous" by Hardy Grant in the September 1994 issue of *The College Mathematics Journal.*

Definition of Continuity

Continuity at a Point: A function f is **continuous at c** if the following three conditions are met.

1. $f(c)$ is defined.

2. $\displaystyle\lim_{x \to c} f(x)$ exists.

3. $\displaystyle\lim_{x \to c} f(x) = f(c)$.

Continuity on an Open Interval: A function is **continuous on an open interval (a, b)** if it is continuous at each point in the interval. A function that is continuous on the entire real line $(-\infty, \infty)$ is **everywhere continuous.**

Consider an open interval I that contains a real number c. If a function f is defined on I (except possibly at c), and f is not continuous at c, then f is said to have a **discontinuity** at c. Discontinuities fall into two categories: **removable** and **nonremovable**. A discontinuity at c is called removable if f can be made continuous by appropriately defining (or redefining) $f(c)$. For instance, the functions shown in Figure 1.25(a) and (c) have removable discontinuities at c and the function shown in Figure 1.25(b) has a nonremovable discontinuity at c.

EXAMPLE 1   Continuity of a Function

Discuss the continuity of each function.

a. $f(x) = \dfrac{1}{x}$ **b.** $g(x) = \dfrac{x^2 - 1}{x - 1}$ **c.** $h(x) = \begin{cases} x + 1, & x \le 0 \\ x^2 + 1, & x > 0 \end{cases}$ **d.** $y = \sin x$

Solution

a. The domain of f is all nonzero real numbers. From Theorem 1.3, you can conclude that f is continuous at every x-value in its domain. At $x = 0$, f has a nonremovable discontinuity, as shown if Figure 1.26(a). In other words, there is no way to define $f(0)$ so as to make the function continuous at $x = 0$.

b. The domain of g is all real numbers except $x = 1$. From Theorem 1.3, you can conclude that g is continuous at every x-value in its domain. At $x = 1$, the function has a removable discontinuity, as shown in Figure 1.26(b). If $g(1)$ is defined as 2, the "newly defined" function is continuous for all real numbers.

c. The domain of h is all real numbers. The function h is continuous on $(-\infty, 0)$ and $(0, \infty)$, and, because $\lim_{x \to 0} h(x) = 1$, h is continuous on the entire real line, as shown in Figure 1.26(c).

d. The domain of y is all real numbers. From Theorem 1.6, you can conclude that the function is continuous on its entire domain, $(-\infty, \infty)$, as shown in Figure 1.26(d).

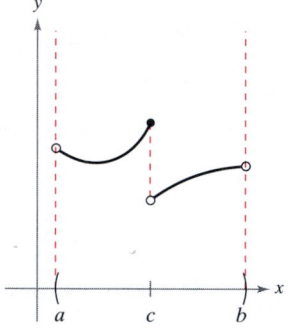

(a) Removable discontinuity

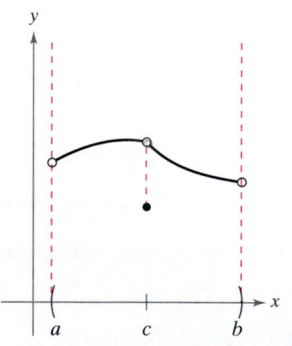

(b) Nonremovable discontinuity

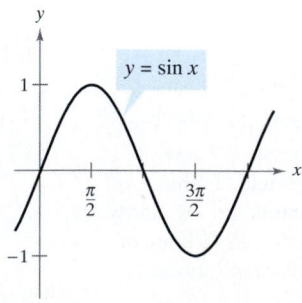

(c) Removable discontinuity

Figure 1.25

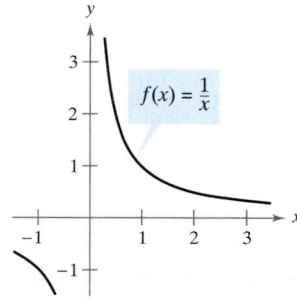

(a) Nonremovable discontinuity at $x = 0$

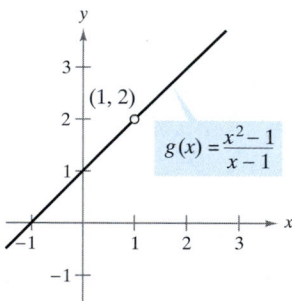

(b) Removable discontinuity at $x = 1$

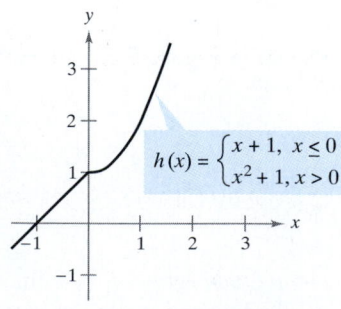

(c) Continuous on entire real line

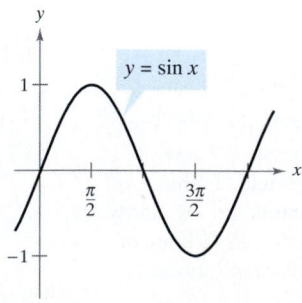

(d) Continuous on entire real line

Figure 1.26

One-Sided Limits and Continuity on a Closed Interval

To understand continuity on a closed interval, you first need to look at a different type of limit called a **one-sided limit.** For example, the **limit from the right** means that x approaches c from values greater then c. This limit is denoted as

$$\lim_{x \to c^+} f(x) = L. \qquad \text{Limit from the right}$$

Similarly, the **limit from the left** means that x approaches c from values less than c. This limit is denoted as

$$\lim_{x \to c^-} f(x) = L. \qquad \text{Limit from the left}$$

One-sided limits are useful in taking limits of functions involving radicals. For instance, if n is an even integer,

$$\lim_{x \to 0^+} \sqrt[n]{x} = 0.$$

EXAMPLE 2   A One-Sided Limit

Find the limit of $f(x) = \sqrt{4 - x^2}$ as x approaches -2 from the right.

Solution   As indicated in Figure 1.27, the limit as x approaches -2 from the right is

$$\lim_{x \to -2^+} \sqrt{4 - x^2} = 0.$$

One-sided limits can be used to investigate the behavior of **step functions.** One common type of step function is the **greatest integer function** $[\![x]\!]$, defined by

$$[\![x]\!] = \text{greatest integer } n \text{ such that } n \leq x. \qquad \text{Greatest integer function}$$

For instance, $[\![2.5]\!] = 2$ and $[\![-2.5]\!] = -3$.

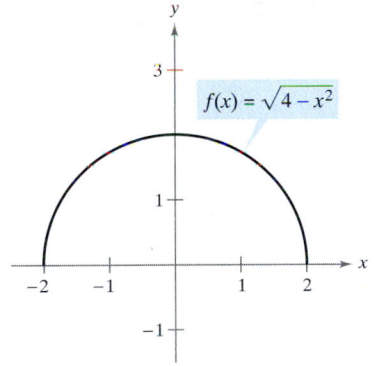

The limit of $f(x)$ as x approaches -2 from the right is 0.
Figure 1.27

EXAMPLE 3   The Greatest Integer Function

Find the limit of the greatest integer function $f(x) = [\![x]\!]$ as x approaches 0 from the left and from the right.

Solution   As shown in Figure 1.28, the limit as x approaches 0 *from the left* is given by

$$\lim_{x \to 0^-} [\![x]\!] = -1$$

and the limit as x approaches 0 *from the right* is given by

$$\lim_{x \to 0^+} [\![x]\!] = 0.$$

The greatest integer function is not continuous at 0 because the left and right limits at zero are different. By similar reasoning, you can see that the greatest integer function is not continuous at any integer n.

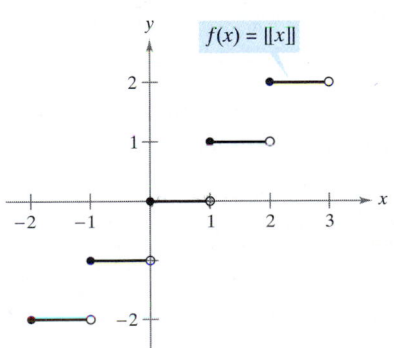

Greatest integer function
Figure 1.28

When the limit from the left is not equal to the limit from the right, the (two-sided) limit *does not exist*. The next theorem makes this more explicit. The proof of this theorem follows directly from the definition of a one-sided limit.

THEOREM 1.10   The Existence of a Limit

Let f be a function and let c and L be real numbers. The limit of $f(x)$ as x approaches c is L if and only if

$$\lim_{x \to c^-} f(x) = L \quad \text{and} \quad \lim_{x \to c^+} f(x) = L.$$

The concept of a one-sided limit allows you to extend the definition of continuity to closed intervals. Basically, a function is continuous on a closed interval if it is continuous in the interior of the interval and possesses one-sided continuity at the endpoints. We state this formally, as follows.

Definition of Continuity on a Closed Interval

A function f is **continuous on the closed interval** $[a, b]$ if it is continuous on the open interval (a, b) and

$$\lim_{x \to a^+} f(x) = f(a) \quad \text{and} \quad \lim_{x \to b^-} f(x) = f(b).$$

The function f is **continuous from the right** at a and **continuous from the left** at b (see Figure 1.29).

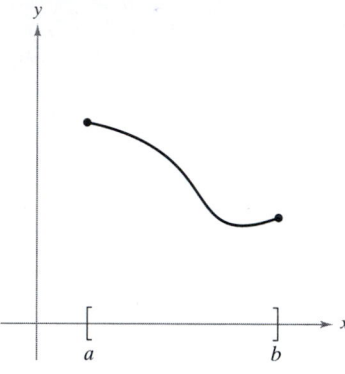

Continuous function on a closed interval
Figure 1.29

Similar definitions can be made to cover continuity on intervals of the form $(a, b]$ and $[a, b)$ that are neither open nor closed, or on infinite intervals. For example, the function

$$f(x) = \sqrt{x}$$

is continuous on the infinite interval $[0, \infty)$, and the function

$$g(x) = \sqrt{2 - x}$$

is continuous on the infinite interval $(-\infty, 2]$.

EXAMPLE 4   Continuity on a Closed Interval

Discuss the continuity of $f(x) = \sqrt{1 - x^2}$.

Solution   The domain of f is the closed interval $[-1, 1]$. At all points in the open interval $(-1, 1)$, the continuity of f follows from Theorems 1.4 and 1.5. Moreover, because

$$\lim_{x \to -1^+} \sqrt{1 - x^2} = 0 = f(-1) \qquad \text{\textcolor{red}{Continuous from the right}}$$

and

$$\lim_{x \to 1^-} \sqrt{1 - x^2} = 0 = f(1) \qquad \text{\textcolor{red}{Continuous from the left}}$$

you can conclude that f is continuous on the closed interval $[-1, 1]$, as shown in Figure 1.30.

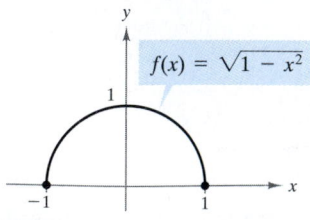

Continuous function on $[-1, 1]$
Figure 1.30

The next example shows how a one-sided limit can be used to determine the value of absolute zero on the Kelvin scale.

EXAMPLE 5   Charles's Law and Absolute Zero

On the Kelvin scale, *absolute zero* is the temperature 0 K. Although temperatures of approximately 0.0001 K have been produced in laboratories, absolute zero has never been attained. In fact, evidence suggests that absolute zero *cannot* be attained. How did scientists determine that 0 K is the "lower limit" of the temperature of matter? What is absolute zero of the Celsius scale?

Solution   The determination of absolute zero stems from the work of the French physicist Jacques Charles (1746–1823). Charles discovered that the volume of gas at a constant pressure increases linearly with the temperature of the gas. The table illustrates this relationship between volume and temperature. In the table, one mole of hydrogen is held at a constant pressure of one atmosphere. The volume V is measured in liters and the temperature T is measured in degrees Celsius.

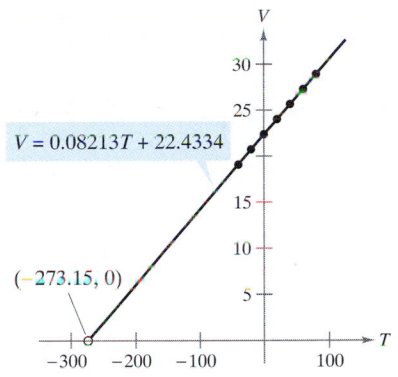

$V = 0.08213T + 22.4334$

$(-273.15, 0)$

The volume of hydrogen gas depends on its temperature.
Figure 1.31

T	-40	-20	0	20	40	60	80
V	19.1482	20.7908	22.4334	24.0760	25.7186	27.3612	29.0038

The points represented by the table are shown in Figure 1.31. Moreover, by using the points in the table, you can determine that T and V are related by the linear equation

$$V = 0.08213T + 22.4334 \qquad \text{or} \qquad T = \frac{V - 22.4334}{0.08213}.$$

By reasoning that the volume of the gas can approach 0 (but never go below 0) you can determine that the "least possible temperature" is given by

$$\lim_{V \to 0^+} T = \lim_{V \to 0^+} \frac{V - 22.4334}{0.08213}$$

$$= \frac{0 - 22.4334}{0.08213} \qquad \text{Use direct substitution.}$$

$$\approx -273.15.$$

Thus, absolute zero on the Kelvin scale (0 K) is approximately $-273.15°$ on the Celsius scale.

The following table shows the temperatures in Example 5, converted to the Fahrenheit scale. Try repeating the solution shown in Example 5 using these temperatures and volumes. Use the result to find the value of absolute zero on the Fahrenheit scale.

T	-40	-4	32	68	104	140	176
V	19.1482	20.7908	22.4334	24.0760	25.7186	27.3612	29.0038

NOTE   Charles's Law for gases (assuming constant pressure) can be stated as

$$V = RT \qquad \text{Charles's Law}$$

where V is volume, R is constant, and T is temperature. In the statement of this law, what property must the temperature scale have?

The space shuttle uses liquid hydrogen and liquid oxygen as fuel. The liquid hydrogen must be stored at a temperature between $-262°C$ and $-253°C$. At lower temperatures the hydrogen would be a solid, and at higher temperatures it would be a gas.

NASA

Properties of Continuity

In Section 1.3, you studied several properties of limits. Each of those properties yields a corresponding property pertaining to the continuity of a function. For instance, Theorem 1.11 follows directly from Theorem 1.2.

THEOREM 1.11   Properties of Continuity

If b is a real number and f and g are continuous at $x = c$, then the following functions are also continuous at c.

1. Scalar multiple: bf 2. Sum and difference: $f \pm g$

3. Product: fg 4. Quotient: $\dfrac{f}{g}$, if $g(c) \neq 0$

The following types of functions are continuous at every point in their domains.

1. Polynomial functions: $p(x) = a_n x^n + a_{n-1} x^{n-1} + \cdots + a_1 x + a_0$

2. Rational functions: $r(x) = \dfrac{p(x)}{q(x)}, \quad q(x) \neq 0$

3. Radical functions: $f(x) = \sqrt[n]{x}$

4. Trigonometric functions: $\sin x, \cos x, \tan x, \cot x, \sec x, \csc x$

By combining Theorem 1.11 with this summary, you can conclude that a wide variety of elementary functions are continuous at every point in their domains.

 EXAMPLE 6   Applying Properties of Continuity

By Theorem 1.11, it follows that each of the following functions is continuous at every point in its domain.

$$f(x) = x + \sin x, \quad f(x) = 3 \tan x, \quad f(x) = \frac{x^2 + 1}{\cos x}$$

The next theorem, which is a consequence of Theorem 1.5, allows you to determine the continuity of *composite* functions such as

$$f(x) = \sin 3x, \quad f(x) = \sqrt{x^2 + 1}, \quad f(x) = \tan \frac{1}{x}.$$

THEOREM 1.12   Continuity of a Composite Function

If g is continuous at c and f is continuous at $g(c)$, then the composite function given by $(f \circ g)(x) = f(g(x))$ is continuous at c.

NOTE   One consequence of Theorem 1.12 is that if f and g satisfy the given conditions, you can determine the limit of $f(g(x))$ as x approaches c to be

$$\lim_{x \to c} f(g(x)) = f(g(c)).$$

EXAMPLE 7   Testing for Continuity

Describe the interval(s) on which each function is continuous.

a. $f(x) = \tan x$ **b.** $g(x) = \begin{cases} \sin \dfrac{1}{x}, & x \neq 0 \\ 0, & x = 0 \end{cases}$ **c.** $h(x) = \begin{cases} x \sin \dfrac{1}{x}, & x \neq 0 \\ 0, & x = 0 \end{cases}$

Solution

a. The tangent function $f(x) = \tan x$ is undefined at

$$x = \frac{\pi}{2} + n\pi, \qquad n \text{ is an integer.}$$

At all other points it is continuous. Thus, $f(x) = \tan x$ is continuous on the open intervals

$$\ldots, \left(-\frac{3\pi}{2}, -\frac{\pi}{2}\right), \left(-\frac{\pi}{2}, \frac{\pi}{2}\right), \left(\frac{\pi}{2}, \frac{3\pi}{2}\right), \ldots$$

as shown in Figure 1.32(a).

b. Because $y = 1/x$ is continuous except at $x = 0$ and the sine function is continuous for all real values of x, it follows that $y = \sin(1/x)$ is continuous at all real values except $x = 0$. At $x = 0$, the limit of $g(x)$ does not exist (see Example 5, Section 1.2). Therefore, g is continuous on the intervals $(-\infty, 0)$ and $(0, \infty)$, as indicated in Figure 1.32(b).

c. This function is similar to that in part (b) except that the oscillations are damped by the factor x. Using the Squeeze Theorem, you obtain

$$-|x| \leq x \sin \frac{1}{x} \leq |x|, \qquad x \neq 0$$

and you can conclude that

$$\lim_{x \to 0} h(x) = 0.$$

Thus, h is continuous on the entire real line, as indicated in Figure 1.32(c).

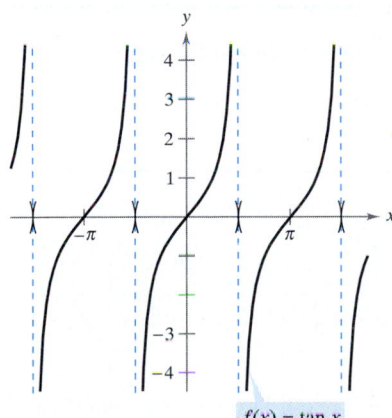

$f(x) = \tan x$

(a) f is continuous on each open interval in its domain.

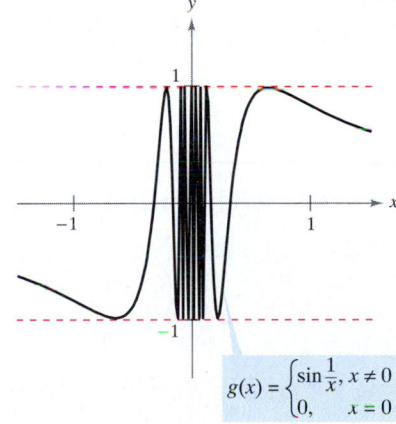

$g(x) = \begin{cases} \sin \dfrac{1}{x}, & x \neq 0 \\ 0, & x = 0 \end{cases}$

(b) g is continuous on $(-\infty, 0)$ and $(0, \infty)$.

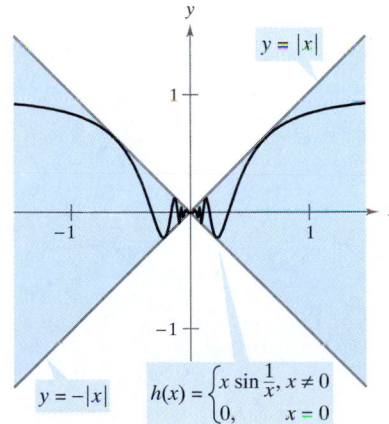

$y = |x|$

$y = -|x|$

$h(x) = \begin{cases} x \sin \dfrac{1}{x}, & x \neq 0 \\ 0, & x = 0 \end{cases}$

(c) h is continuous on the entire real line.

Figure 1.32

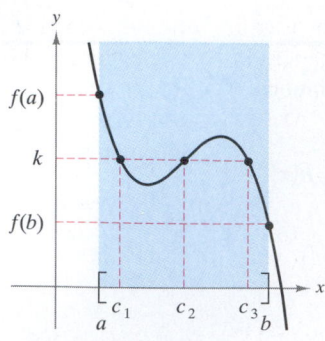

f is continuous on $[a, b]$.
[There exist 3 c's such that $f(c) = k$.]
Figure 1.33

The Intermediate Value Theorem

We conclude this section with an important theorem concerning the behavior of functions that are continuous on a closed interval.

> **THEOREM 1.13 Intermediate Value Theorem**
>
> If f is continuous on the closed interval $[a, b]$ and k is any number between $f(a)$ and $f(b)$, then there is at least one number c in $[a, b]$ such that $f(c) = k$.

NOTE   The Intermediate Value Theorem tells you that at least one c exists, but it does not give a method for finding c. Such theorems are called **existence theorems**.

By referring to a text on advanced calculus, you will find that a proof of this theorem is based on a property of real numbers called *completeness*. The Intermediate Value Theorem states that for a continuous function f, if x takes on all values between a and b, $f(x)$ must take on all values between $f(a)$ and $f(b)$.

As a simple example of this theorem, consider a person's height. Suppose that a girl is 5 feet tall on her thirteenth birthday and 5 feet 7 inches tall on her fourteenth birthday. Then, for any height h between 5 feet and 5 feet 7 inches, there must have been a time t when her height was exactly h. This seems reasonable because human growth is continuous and a person's height does not abruptly change from one value to another.

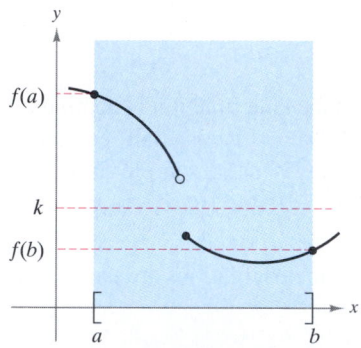

f is not continuous on $[a, b]$.
[There are no c's such that $f(c) = k$.]
Figure 1.34

The Intermediate Value Theorem guarantees the existence of *at least one* number c in the closed interval $[a, b]$. There may, of course, be more than one number c such that $f(c) = k$, as shown in Figure 1.33. A function that is not continuous does not necessarily possess the intermediate value property. For example, the graph of the function shown in Figure 1.34 jumps over the horizontal line given by $y = k$, and for this function there is no value of c in $[a, b]$ such that $f(c) = k$.

The Intermediate Value Theorem often can be used to locate the zeros of a function that is continuous on a closed interval. Specifically, if f is continuous on $[a, b]$ and $f(a)$ and $f(b)$ differ in sign, the Intermediate Value Theorem guarantees the existence of at least one zero of f in the closed interval $[a, b]$.

EXAMPLE 8   An Application of the Intermediate Value Theorem

Use the Intermediate Value Theorem to show that the polynomial function

$$f(x) = x^3 + 2x - 1$$

has a zero in the interval $[0, 1]$.

Solution   Note that f is continuous on the closed interval $[0, 1]$. Because

$$f(0) = 0^3 + 2(0) - 1 = -1$$

and

$$f(1) = 1^3 + 2(1) - 1 = 2$$

it follows that $f(0) < 0$ and $f(1) > 0$. You can therefore apply the Intermediate Value Theorem to conclude that there must be some c in $[0, 1]$ such that

$$f(c) = 0 \qquad \text{\small{\textit{f} has a zero in the closed interval } [0, 1].}$$

as shown in Figure 1.35.

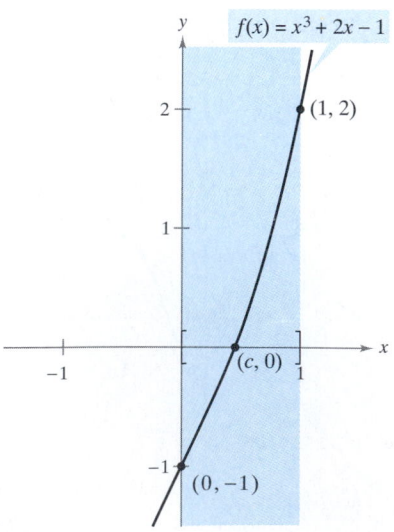

$f(x) = x^3 + 2x - 1$

f is continuous on $[0, 1]$ with $f(0) < 0$ and $f(1) > 0$.
Figure 1.35

The **bisection method** for approximating the real zeros of a continuous function is similar to the method used in Example 8. If you know that a zero exists in the closed interval $[a, b]$, the zero must lie in the interval

$$\left[a, \frac{a + b}{2} \right] \quad \text{or} \quad \left[\frac{a + b}{2}, b \right].$$

From the sign of $f([a + b]/2)$, you can determine which interval contains the zero. By repeatedly bisecting the interval, you can "close in" on the zero of the function.

TECHNOLOGY   You can also use the zoom feature of a graphing utility to approximate the real zeros of a continuous function. By repeatedly zooming in on the point where the graph crosses the x-axis, and adjusting the x-axis scale, you can approximate the zero of the function to any desired accuracy. The zero of $x^3 + 2x - 1$ is approximately 0.453, as shown in Figure 1.36.

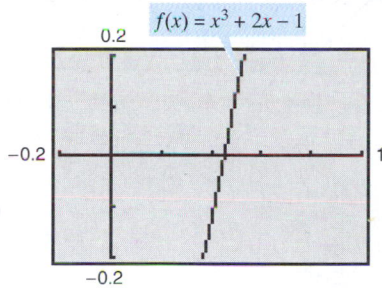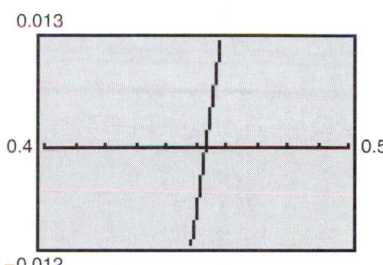

Zooming in on the zero of f

Figure 1.36

EXERCISES FOR SECTION 1.4

1. ***Think About It***   State how continuity is destroyed at $x = c$ for each of the following.

(a)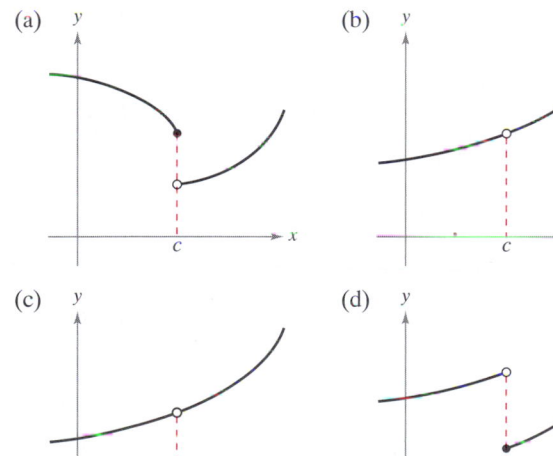

(b)

(c)

(d)

2. ***Writing***   Describe the difference between a discontinuity that is removable and one that is nonremovable. In your explanation, give examples of the following.

(a) A function with a nonremovable discontinuity at $x = 2$.

(b) A function with a removable discontinuity at $x = -2$.

(c) A function that has both of the characteristics described in parts (a) and (b).

3. ***Think About It***   Sketch the graph of any function f such that

$$\lim_{x \to 3^+} f(x) = 1$$

and

$$\lim_{x \to 3^-} f(x) = 0.$$

Is the function continuous at $x = 3$? Explain.

4. ***Think About It***   If the functions f and g are continuous for all real x, is $f + g$ always continuous for all real x? Is f/g always continuous for all real x? If either is not continuous, give an example to verify your conclusion.

In Exercises 5–10, use the graph to determine the limit.

(a) $\lim_{x \to c^+} f(x)$ (b) $\lim_{x \to c^-} f(x)$ (c) $\lim_{x \to c} f(x)$

5.

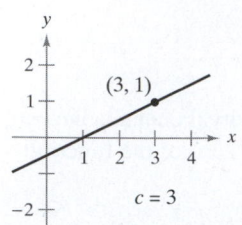

6.

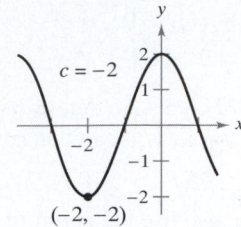

7.

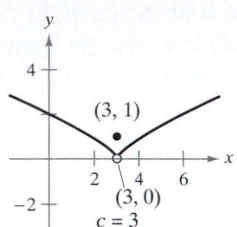

8.

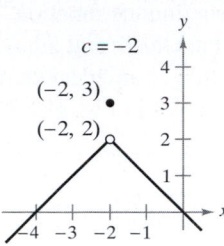

9.

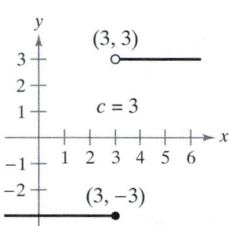

10.

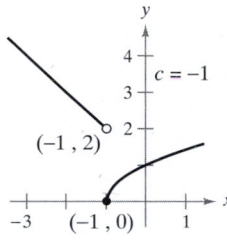

In Exercises, 11–26, find the limit (if it exists).

11. $\lim_{x \to 5^+} \dfrac{x - 5}{x^2 - 25}$

12. $\lim_{x \to 2^+} \dfrac{2 - x}{x^2 - 4}$

13. $\lim_{x \to 2^+} \dfrac{x}{\sqrt{x^2 - 4}}$

14. $\lim_{x \to 4^-} \dfrac{\sqrt{x} - 2}{x - 4}$

15. $\lim_{x \to 0} \dfrac{|x|}{x}$

16. $\lim_{x \to 2} \dfrac{|x - 2|}{x - 2}$

17. $\lim_{\Delta x \to 0^-} \dfrac{\dfrac{1}{x + \Delta x} - \dfrac{1}{x}}{\Delta x}$

18. $\lim_{\Delta x \to 0^+} \dfrac{(x + \Delta x)^2 + x + \Delta x - (x^2 + x)}{\Delta x}$

19. $\lim_{x \to 3} f(x)$, where $f(x) = \begin{cases} \dfrac{x + 2}{2}, & x \le 3 \\ \dfrac{12 - 2x}{3}, & x > 3 \end{cases}$

20. $\lim_{x \to 2} f(x)$, where $f(x) = \begin{cases} x^2 - 4x + 6, & x < 2 \\ -x^2 + 4x - 2, & x \ge 2 \end{cases}$

21. $\lim_{x \to 1} f(x)$, where $f(x) = \begin{cases} x^3 + 1, & x < 1 \\ x + 1, & x \ge 1 \end{cases}$

22. $\lim_{x \to 1} f(x)$, where $f(x) = \begin{cases} x, & x \le 1 \\ 1 - x, & x > 1 \end{cases}$

23. $\lim_{x \to \pi} \cot x$

24. $\lim_{x \to \pi/2} \sec x$

25. $\lim_{x \to 3^-} (2[\![x]\!] - 1)$

26. $\lim_{x \to 2^+} (2x - [\![x]\!])$

In Exercises 27–30, find the x-values (if any) at which f is not continuous.

27. $f(x) = \dfrac{1}{x^2 - 4}$

28. $f(x) = \dfrac{x^2 - 1}{x + 1}$

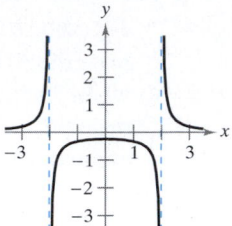

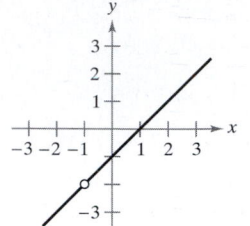

29. $f(x) = \frac{1}{2}[\![x]\!] + x$

30. $f(x) = \begin{cases} x, & x < 1 \\ 2, & x = 1 \\ 2x - 1, & x > 1 \end{cases}$

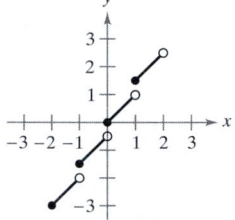

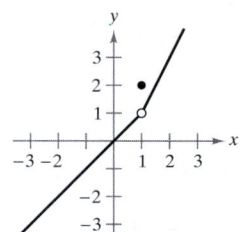

In Exercises 31–52, find the x-values (if any) at which f is not continuous. Which of the discontinuities are removable?

31. $f(x) = x^2 - 2x + 1$

32. $f(x) = \dfrac{1}{x^2 + 1}$

33. $f(x) = x + \sin x$

34. $f(x) = \cos \dfrac{\pi x}{2}$

35. $f(x) = \dfrac{1}{x - 1}$

36. $f(x) = \dfrac{x}{x^2 - 1}$

37. $f(x) = \dfrac{x}{x^2 + 1}$

38. $f(x) = \dfrac{x - 3}{x^2 - 9}$

39. $f(x) = \dfrac{x + 2}{x^2 - 3x - 10}$

40. $f(x) = \dfrac{x - 1}{x^2 + x - 2}$

41. $f(x) = \dfrac{|x + 2|}{x + 2}$

42. $f(x) = \dfrac{|x - 3|}{x - 3}$

43. $f(x) = \begin{cases} x, & x \le 1 \\ x^2, & x > 1 \end{cases}$

44. $f(x) = \begin{cases} -2x + 3, & x < 1 \\ x^2, & x \ge 1 \end{cases}$

45. $f(x) = \begin{cases} \frac{1}{2}x + 1, & x \le 2 \\ 3 - x, & x > 2 \end{cases}$

46. $f(x) = \begin{cases} -2x, & x \le 2 \\ x^2 - 4x + 1, & x > 2 \end{cases}$

47. $f(x) = \begin{cases} \csc \frac{\pi x}{6}, & |x - 3| \le 2 \\ 2, & |x - 3| > 2 \end{cases}$

48. $f(x) = \begin{cases} \tan \frac{\pi x}{4}, & |x| < 1 \\ x, & |x| \ge 1 \end{cases}$

49. $f(x) = \csc 2x$

50. $f(x) = \tan \frac{\pi x}{2}$

51. $f(x) = [\![x - 1]\!]$

52. $f(x) = x - [\![x]\!]$

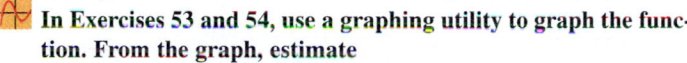 In Exercises 53 and 54, use a graphing utility to graph the function. From the graph, estimate

$$\lim_{x \to 0^+} f(x) \quad \text{and} \quad \lim_{x \to 0^-} f(x).$$

Is the function continuous on the entire real line?

53. $f(x) = \dfrac{|x^2 - 4|x}{x + 2}$

54. $f(x) = \dfrac{|x^2 + 4x|(x + 2)}{x + 4}$

In Exercises 55–58, find the constants a and b such that the function is continuous on the entire real line.

55. $f(x) = \begin{cases} x^3, & x \le 2 \\ ax^2, & x > 2 \end{cases}$

56. $g(x) = \begin{cases} \dfrac{4 \sin x}{x}, & x < 0 \\ a - 2x, & x \ge 0 \end{cases}$

57. $f(x) = \begin{cases} 2, & x \le -1 \\ ax + b, & -1 < x < 3 \\ -2, & x \ge 3 \end{cases}$

58. $g(x) = \begin{cases} \dfrac{x^2 - a^2}{x - a}, & x \ne a \\ 8, & x = a \end{cases}$

In Exercises 59–62, discuss the continuity of the composite function $h(x) = f(g(x))$.

59. $f(x) = x^2$

$g(x) = x - 1$

60. $f(x) = \dfrac{1}{\sqrt{x}}$

$g(x) = x - 1$

61. $f(x) = \dfrac{1}{(x - 6)}$

$g(x) = x^2 + 5$

62. $f(x) = \sin x$

$g(x) = x^2$

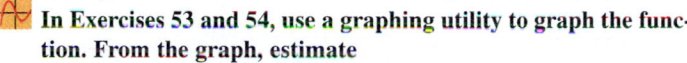 In Exercises 63–66, use a graphing utility to graph the function. Use the graph to determine any x-values at which the function is not continuous.

63. $f(x) = [\![x]\!] - x$

64. $h(x) = \dfrac{1}{x^2 - x - 2}$

65. $g(x) = \begin{cases} 2x - 4, & x \le 3 \\ x^2 - 2x, & x > 3 \end{cases}$

66. $f(x) = \begin{cases} \dfrac{\cos x - 1}{x}, & x < 0 \\ 5x, & x \ge 0 \end{cases}$

In Exercises 67–70, find the interval(s) on which the function is continuous.

67. $f(x) = \dfrac{x}{x^2 + 1}$

68. $f(x) = x\sqrt{x + 3}$

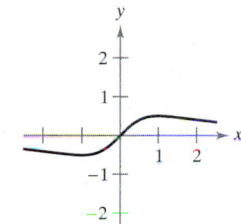

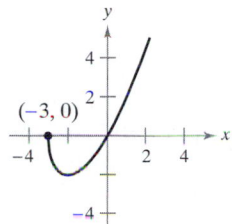

69. $f(x) = \csc \dfrac{x}{2}$

70. $f(x) = \dfrac{x + 1}{\sqrt{x}}$

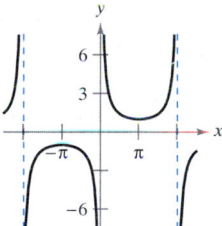

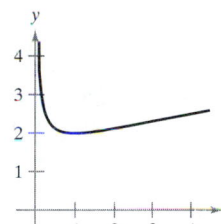

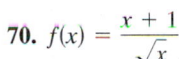 *Writing*   In Exercises 71 and 72, use a graphing utility to graph the function on the interval $[-4, 4]$. Does the graph of the function *appear* continuous on this interval? *Is* the function continuous on $[-4, 4]$? Write a short paragraph about the importance of examining a function analytically as well as graphically.

71. $f(x) = \dfrac{\sin x}{x}$

72. $f(x) = \dfrac{x^3 - 8}{x - 2}$

Writing   In Exercises 73 and 74, give a written explanation of why the function has a zero in the specified interval.

73. $f(x) = x^2 - 4x + 3, \quad [2, 4]$

74. $f(x) = x^3 + 3x - 2, \quad [0, 1]$

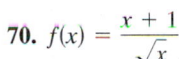 In Exercises 75–78, use the Intermediate Value Theorem and a graphing utility to approximate the zero of the function in the interval $[0, 1]$. Repeatedly "zoom in" on the graph of the function to approximate the zero accurate to two decimal places. Use the root-finding capabilities of the graphing utility to approximate the zero accurate to four decimal places.

75. $f(x) = x^3 + x - 1$

76. $f(x) = x^3 + 3x - 2$

77. $g(t) = 2 \cos t - 3t$

78. $h(\theta) = 1 + \theta - 3 \tan \theta$

In Exercises 79–82, verify that the Intermediate Value Theorem applies to the indicated interval and find the value of c guaranteed by the theorem.

79. $f(x) = x^2 + x - 1$, $\quad [0, 5]$, $\quad f(c) = 11$

80. $f(x) = x^2 - 6x + 8$, $\quad [0, 3]$, $\quad f(c) = 0$

81. $f(x) = x^3 - x^2 + x - 2$, $\quad [0, 3]$, $\quad f(c) = 4$

82. $f(x) = \dfrac{x^2 + x}{x - 1}$, $\quad \left[\dfrac{5}{2}, 4\right]$, $\quad f(c) = 6$

83. *Volume* Use the Intermediate Value Theorem to show that for all spheres with radii in the interval $[0, 5]$, there is one with a volume of 275 cubic centimeters.

84. *Telephone Charges* A dial-direct long distance call between two cities costs $1.04 for the first 2 minutes and $0.36 for each additional minute or fraction thereof. Use the greatest integer function to write the cost C of a call in terms of the time t (in minutes). Sketch a graph of this function and discuss its continuity.

85. *Inventory Management* The number of units in inventory in a small company is given by

$$N(t) = 25\left(2\left[\!\left[\dfrac{t + 2}{2}\right]\!\right] - t\right)$$

where t is the time in months. Sketch the graph of this function and discuss its continuity. How often must this company replenish its inventory?

86. *Déjà Vu* At 8:00 A.M. on Saturday a man begins running up the side of a mountain to his weekend campsite (see figure). On Sunday morning at 8:00 A.M. he runs back down the mountain. It takes him 20 minutes to run up, but only 10 minutes to run down. At some point on the way down, he realizes that he passed the same place at exactly the same time on Saturday. Prove that he is correct. [*Hint:* Let $s(t)$ and $r(t)$ be the position functions for the runs up and down, and apply the Intermediate Value Theorem to the function $f(t) = s(t) - r(t)$.]

Saturday 8:00 A.M. Sunday 8:00 A.M.

87. Prove that if f is continuous and has no zeros on $[a, b]$, then either

$f(x) > 0$ for all x in $[a, b]$

or

$f(x) < 0$ for all x in $[a, b]$.

88. Show that the Dirichlet function

$$f(x) = \begin{cases} 0, & \text{if } x \text{ is rational} \\ 1, & \text{if } x \text{ is irrational} \end{cases}$$

is not continuous at any real number.

89. Show that the function

$$f(x) = \begin{cases} 0, & \text{if } x \text{ is rational} \\ kx, & \text{if } x \text{ is irrational} \end{cases}$$

is continuous only at $x = 0$. (Assume that k is any nonzero real number.)

90. The **signum function** is defined by

$$\text{sgn}(x) = \begin{cases} -1, & x < 0 \\ 0, & x = 0 \\ 1, & x > 0. \end{cases}$$

Sketch a graph of $\text{sgn}(x)$ and find the following limits (if possible).

(a) $\lim\limits_{x \to 0^-} \text{sgn}(x)$ $\qquad$ (b) $\lim\limits_{x \to 0^+} \text{sgn}(x)$ $\qquad$ (c) $\lim\limits_{x \to 0} \text{sgn}(x)$

True or False? **In Exercises 91–94, determine whether the statement is true or false. If it is false, explain why or give an example that shows it is false.**

91. If $\lim\limits_{x \to c} f(x) = L$ and $f(c) = L$, then f is continuous at c.

92. If $f(x) = g(x)$ for $x \neq c$ and $f(c) \neq g(c)$, then either f or g is not continuous at c.

93. A rational function can have infinitely many x-values at which it is not continuous.

94. The function $f(x) = |x - 1|/(x - 1)$ is continuous on $(-\infty, \infty)$.

95. *Modeling Data* After an object falls for t seconds, the speed S (in feet per second) of the object is recorded in the the table.

t	0	5	10	15	20	25	30
S	0	48.2	53.5	55.2	55.9	56.2	56.3

(a) Create a line graph of the data.

(b) Does there appear to be a limiting speed of the object? If there is a limiting speed, identify a possible cause.

96. Prove that if $\lim\limits_{\Delta x \to 0} f(x_0 + \Delta x) = f(x_0)$, then f is continuous at x_0.

97. Prove that for any real number y there exists x in $(-\pi/2, \pi/2)$ such that $\tan x = y$.

98. Discuss the continuity of the function $h(x) = x[\![x]\!]$.

99. Let $f(x) = (\sqrt{x + c^2} - c)/x$, $c > 0$. What is the domain of f? How can you define f at $x = 0$ in order for f to be continuous there?

100. Let $f_1(x)$ and $f_2(x)$ be continuous on the closed interval $[a, b]$. If $f_1(a) < f_2(a)$ and $f_1(b) > f_2(b)$, prove that there exists c between a and b such that $f_1(c) = f_2(c)$.

Infinite Limits • Vertical Asymptotes

Infinite Limits

Let f be the function given by

$$f(x) = \frac{3}{x - 2}.$$

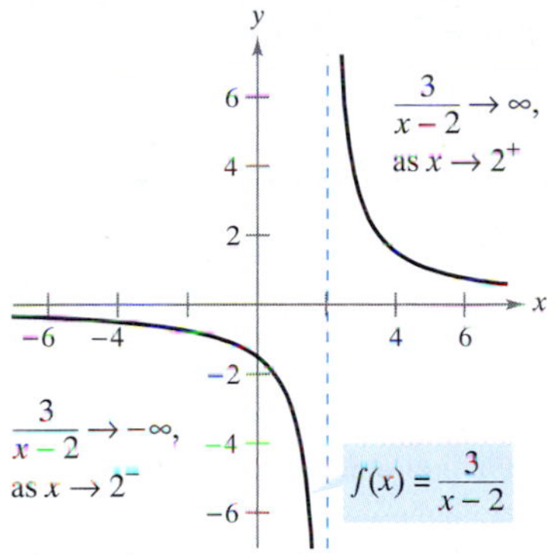

From Figure 1.37 and the table, you can see that $f(x)$ *decreases without bound* as x approaches 2 from the left, and $f(x)$ *increases without bound* as x approaches 2 from the right. This behavior is denoted as

$$\lim_{x \to 2^-} \frac{3}{x - 2} = -\infty \qquad \text{\textit{f(x) decreases without bound as x approaches 2 from the left.}}$$

and

$$\lim_{x \to 2^+} \frac{3}{x - 2} = \infty \qquad \text{\textit{f(x) increases without bound as x approaches 2 from the right.}}$$

| | x approaches 2 from the left. | | | | | x approaches 2 from the right. | | | |
x	1.5	1.9	1.99	1.999	2	2.001	2.01	2.1	2.5
$f(x)$	-6	-30	-300	-3000	?	3000	300	30	6
	$f(x)$ decreases without bound.					$f(x)$ increases without bound.			

A limit in which $f(x)$ increases or decreases without bound as x approaches c is called an **infinite limit.**

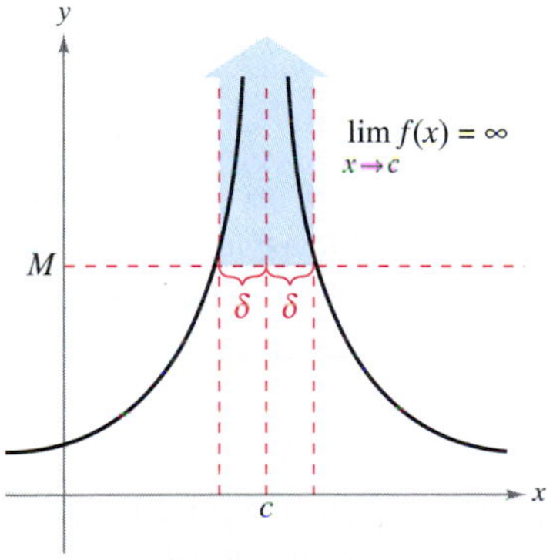

$\dfrac{3}{x-2} \to \infty$, as $x \to 2^+$

$\dfrac{3}{x-2} \to -\infty$, as $x \to 2^-$

$f(x) = \dfrac{3}{x - 2}$

$f(x)$ increases and decreases without bound
as x approaches 2.
Figure 1.37

$\lim_{x \to c} f(x) = \infty$

Infinite limits
Figure 1.38

Definition of Infinite Limits

Let f be a function that is defined at every real number in some open interval containing c (except possibly at c itself). The statement

$$\lim_{x \to c} f(x) = \infty$$

means that for each $M > 0$ there exists a $\delta > 0$ such that $f(x) > M$ whenever $0 < |x - c| < \delta$ (see Figure 1.38). Similarly, the statement

$$\lim_{x \to c} f(x) = -\infty$$

means that for each $N < 0$ there exists a $\delta > 0$ such that $f(x) < N$ whenever $0 < |x - c| < \delta$. To define the **infinite limit from the left,** replace $0 < |x - c| < \delta$ by $c - \delta < x < c$. To define the **infinite limit from the right,** replace $0 < |x - c| < \delta$ by $c < x < c + \delta$.

Be sure you see that the equal sign in the statement $\lim f(x) = \infty$ does not mean that the limit exists! On the contrary, it tells you how the limit *fails to exist* by denoting the unbounded behavior of $f(x)$ as x approaches c.

EXAMPLE 1 Determining Infinite Limits from a Graph

Use Figure 1.39 to determine the limit of each function as x approaches 1 from the left and from the right.

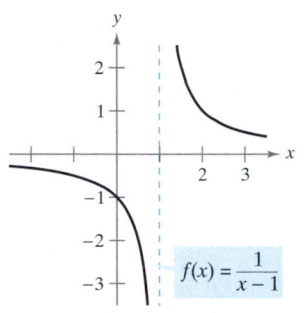

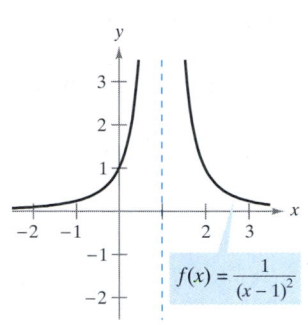

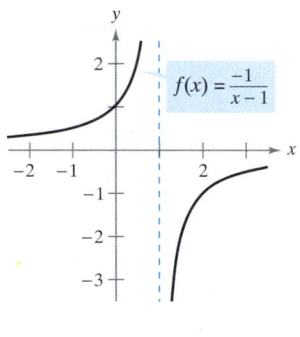

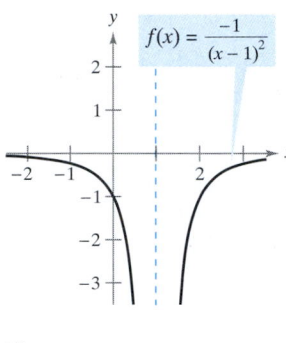

(a)   (b)   (c)   (d)

Each graph has an asymptote at $x = 1$.
Figure 1.39

Solution

a. $\displaystyle\lim_{x \to 1^-} \frac{1}{x - 1} = -\infty$ and $\displaystyle\lim_{x \to 1^+} \frac{1}{x - 1} = \infty$

b. $\displaystyle\lim_{x \to 1} \frac{1}{(x - 1)^2} = \infty$ Limit from each side is ∞.

c. $\displaystyle\lim_{x \to 1^-} \frac{-1}{x - 1} = \infty$ and $\displaystyle\lim_{x \to 1^+} \frac{-1}{x - 1} = -\infty$

d. $\displaystyle\lim_{x \to 1} \frac{-1}{(x - 1)^2} = -\infty$ Limit from each side is $-\infty$.

Vertical Asymptotes

If it were possible to extend the graphs in Figure 1.39 toward positive and negative infinity, you would see that each graph becomes arbitrarily close to the vertical line $x = 1$. This line is a **vertical asymptote** of the graph of f.

NOTE   If a function f has a vertical asymptote at $x = c$, then f is *not continuous* at c.

Definition of Vertical Asymptote

If $f(x)$ approaches infinity (or negative infinity) as x approaches c from the right or the left, then the line $x = c$ is a **vertical asymptote** of the graph of f.

In Example 1, note that each of the functions is a *quotient* and that the vertical asymptote occurs at a number where the denominator is 0 (and the numerator is not 0). The next theorem generalizes this observation. (A proof of this theorem is given in the appendix.)

> ### THEOREM 1.14   Vertical Asymptotes
>
> Let f and g be continuous on an open interval containing c. If $f(c) \neq 0$, $g(c) = 0$, and there exists an open interval containing c such that $g(x) \neq 0$ for all $x \neq c$ in the interval, then the graph of the function given by
>
> $$h(x) = \frac{f(x)}{g(x)}$$
>
> has a vertical asymptote at $x = c$.

EXAMPLE 2   Finding Vertical Asymptotes

Determine all vertical asymptotes of the graph of each function.

a. $f(x) = \dfrac{1}{2(x + 1)}$ **b.** $f(x) = \dfrac{x^2 + 1}{x^2 - 1}$ **c.** $f(x) = \cot x$

Solution

a. When $x = -1$, the denominator of

$$f(x) = \frac{1}{2(x + 1)}$$

is 0 and the numerator is not 0. Hence, by Theorem 1.14, you can conclude that $x = -1$ is a vertical asymptote, as shown in Figure 1.40(a).

b. By factoring the denominator as

$$f(x) = \frac{x^2 + 1}{x^2 - 1} = \frac{x^2 + 1}{(x - 1)(x + 1)}$$

you can see that the denominator is 0 at $x = -1$ and $x = 1$. Moreover, because the numerator is not 0 at these two points, you can apply Theorem 1.14 to conclude that the graph of f has two vertical asymptotes, as shown in Figure 1.40(b).

c. By writing the cotangent function in the form

$$f(x) = \cot x = \frac{\cos x}{\sin x}$$

you can apply Theorem 1.14 to conclude that vertical asymptotes occur at all values of x such that $\cos x \neq 0$ and $\sin x = 0$, as shown in Figure 1.40(c). Hence, the graph of this function has infinitely many vertical asymptotes. These asymptotes occur when $x = n\pi$, where n is an integer.

Theorem 1.14 requires that the value of the numerator at $x = c$ be nonzero. If both the numerator and the denominator are 0 at $x = c$, you obtain the *indeterminate form* $0/0$, and you cannot determine the limit behavior at $x = c$ without further investigation.

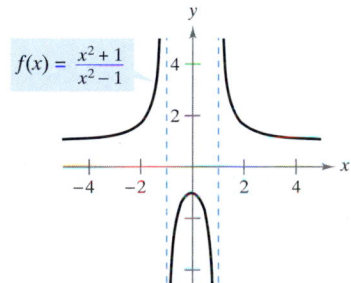

$f(x) = \dfrac{1}{2(x + 1)}$

(a)

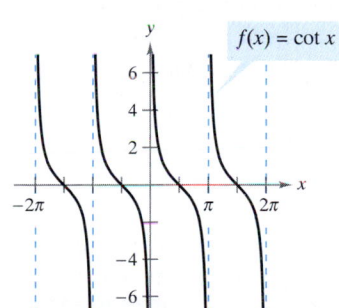

$f(x) = \dfrac{x^2 + 1}{x^2 - 1}$

(b)

$f(x) = \cot x$

(c)

Functions with vertical asymptotes
Figure 1.40

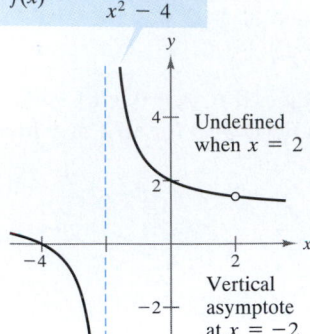

$$f(x) = \frac{x^2 + 2x - 8}{x^2 - 4}$$

Undefined when $x = 2$

Vertical asymptote at $x = -2$

$f(x)$ increases and decreases without bound as x approaches -2.
Figure 1.41

EXAMPLE 3 A Rational Function with Common Factors

Determine all vertical asymptotes of the graph of

$$f(x) = \frac{x^2 + 2x - 8}{x^2 - 4}.$$

Solution Begin by simplifying the expression, as follows.

$$f(x) = \frac{x^2 + 2x - 8}{x^2 - 4}$$

$$= \frac{(x + 4)(x - 2)}{(x + 2)(x - 2)}$$

$$= \frac{x + 4}{x + 2}, \quad x \neq 2$$

At all x-values other than $x = 2$, the graph of f coincides with the graph of $g(x) = (x + 4)/(x + 2)$. Thus, you can apply Theorem 1.14 to g to conclude that there is a vertical asymptote at $x = -2$, as shown in Figure 1.41. From the graph, you can see that

$$\lim_{x \to -2^-} \frac{x^2 + 2x - 8}{x^2 - 4} = -\infty \quad \text{and} \quad \lim_{x \to -2^+} \frac{x^2 + 2x - 8}{x^2 - 4} = \infty.$$

Note that $x = 2$ is *not* a vertical asymptote.

EXAMPLE 4 Determining Infinite Limits

Find each limit.

$$\lim_{x \to 1^-} \frac{x^2 - 3x}{x - 1} \quad \text{and} \quad \lim_{x \to 1^+} \frac{x^2 - 3x}{x - 1}$$

Solution Because the denominator is 0 when $x = 1$ (and the numerator is not zero), you know that the graph of

$$f(x) = \frac{x^2 - 3x}{x - 1}$$

has a vertical asymptote at $x = 1$. This means that each of the given limits is either ∞ or $-\infty$. A graphing utility can help determine the result. From the graph of f shown in Figure 1.42, you can see that the graph approaches ∞ from the left of $x = 1$ and approaches $-\infty$ from the right of $x = 1$. Thus, you can conclude that

$$\lim_{x \to 1^-} \frac{x^2 - 3x}{x - 1} = \infty \qquad \text{The limit from the left is infinity.}$$

and

$$\lim_{x \to 1^+} \frac{x^2 - 3x}{x - 1} = -\infty. \qquad \text{The limit from the right is negative infinity.}$$

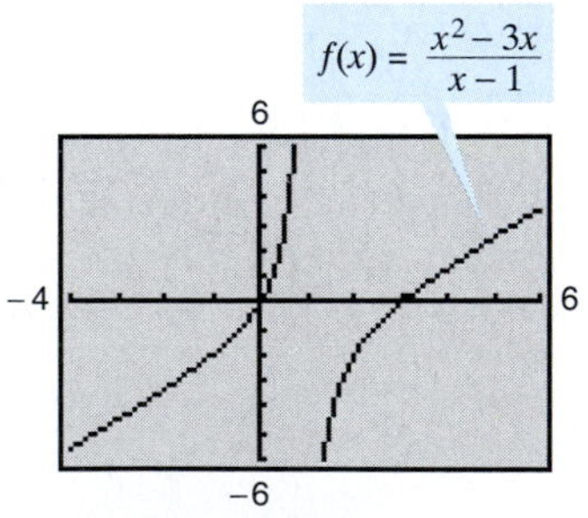

$$f(x) = \frac{x^2 - 3x}{x - 1}$$

f has a vertical asymptote at $x = 1$.
Figure 1.42

STUDY TIP When using a graphing calculator or graphing software, be careful to correctly interpret the graph of a function with a vertical asymptote—graphing utilities often have difficulty drawing this type of graph.

THEOREM 1.15   Properties of Infinite Limits

Let c and L be real numbers and let f and g be functions such that

$$\lim_{x \to c} f(x) = \infty \quad \text{and} \quad \lim_{x \to c} g(x) = L.$$

1. Sum or difference: $\quad \lim_{x \to c} [f(x) \pm g(x)] = \infty$

2. Product: $\quad \lim_{x \to c} [f(x)g(x)] = \infty, \quad L > 0$

$\qquad\qquad\qquad\qquad \lim_{x \to c} [f(x)g(x)] = -\infty, \quad L < 0$

3. Quotient: $\quad \lim_{x \to c} \dfrac{g(x)}{f(x)} = 0$

Similar properties hold for one-sided limits and for functions for which the limit of $f(x)$ as x approaches c is $-\infty$.

Proof   To show that the limit of $f(x) + g(x)$ is infinite, choose $M > 0$. You then need to find $\delta > 0$ such that

$$[f(x) + g(x)] > M$$

whenever $0 < |x - c| < \delta$. For simplicity's sake, you can assume L is positive and let $M_1 = M + 1$. Because the limit of $f(x)$ is infinite, there exists δ_1 such that $f(x) > M_1$ whenever $0 < |x - c| < \delta_1$. Also because the limit of $g(x)$ is L, there exists δ_2 such that $|g(x) - L| < 1$ whenever $0 < |x - c| < \delta_2$. By letting δ be the smaller of δ_1 and δ_2, you can conclude that $0 < |x - c| < \delta$ implies $f(x) > M + 1$ and $|g(x) - L| < 1$. The second of these two inequalities implies that $g(x) > L - 1$, and, adding this to the first inequality, you can write

$$f(x) + g(x) > (M + 1) + (L - 1) = M + L > M.$$

Thus, you can conclude that

$$\lim_{x \to c} [f(x) + g(x)] = \infty.$$

We leave the proofs of the remaining properties as exercises (see Exercise 60).

EXAMPLE 5   Determining Limits

a. Because $\lim\limits_{x \to 0} 1 = 1$ and $\lim\limits_{x \to 0} \dfrac{1}{x^2} = \infty$, you can write

$$\lim_{x \to 0} \left(1 + \frac{1}{x^2}\right) = \infty. \qquad \textcolor{red}{\text{Property 1, Theorem 1.15}}$$

b. Because $\lim\limits_{x \to 1^-} (x^2 + 1) = 2$ and $\lim\limits_{x \to 1^-} \dfrac{1}{x - 1} = -\infty$, you can write

$$\lim_{x \to 1^-} \frac{x^2 + 1}{1/(x - 1)} = 0. \qquad \textcolor{red}{\text{Property 3, Theorem 1.15}}$$

c. Because $\lim\limits_{x \to 0^+} 3 = 3$ and $\lim\limits_{x \to 0^+} \cot x = \infty$, you can write

$$\lim_{x \to 0^+} 3 \cot x = \infty. \qquad \textcolor{red}{\text{Property 2, Theorem 1.15}}$$

EXERCISES FOR SECTION 1.5

In Exercises 1–4, determine whether $f(x)$ approaches ∞ or $-\infty$ as x approaches -2 from the left and from the right.

1. $f(x) = \dfrac{1}{(x + 2)^2}$

2. $f(x) = \dfrac{1}{x + 2}$

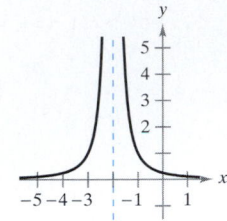

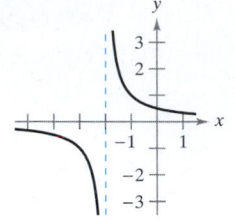

3. $f(x) = \tan \dfrac{\pi x}{4}$

4. $f(x) = \sec \dfrac{\pi x}{4}$

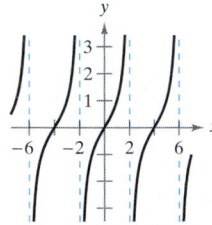

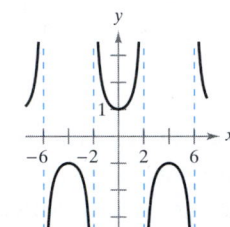

Numerical and Graphical Analysis In Exercises 5–8, determine whether $f(x)$ approaches ∞ or $-\infty$ as x approaches -3 from the left and from the right by completing the table. Use a graphing utility to graph the function and confirm your answer.

x	-3.5	-3.1	-3.01	-3.001
$f(x)$				

x	-2.999	-2.99	-2.9	-2.5
$f(x)$				

5. $f(x) = \dfrac{1}{x^2 - 9}$

6. $f(x) = \dfrac{x}{x^2 - 9}$

7. $f(x) = \dfrac{x^2}{x^2 - 9}$

8. $f(x) = \sec \dfrac{\pi x}{6}$

In Exercises 9–24, find the vertical asymptotes (if any) of the function.

9. $f(x) = \dfrac{1}{x^2}$

10. $f(x) = \dfrac{4}{(x - 2)^3}$

11. $h(x) = \dfrac{x^2 - 2}{x^2 - x - 2}$

12. $g(x) = \dfrac{2 + x}{1 - x}$

13. $f(x) = \dfrac{x^3}{x^2 - 1}$

14. $f(x) = \dfrac{-4x}{x^2 + 4}$

15. $f(x) = \tan 2x$

16. $f(x) = \sec \pi x$

17. $T(t) = 1 - \dfrac{4}{t^2}$

18. $V(s) = \dfrac{-2}{(s - 2)^2}$

19. $f(x) = \dfrac{x}{x^2 + x - 2}$

20. $f(x) = \dfrac{1}{(x + 3)^4}$

21. $g(x) = \dfrac{x^3 + 1}{x + 1}$

22. $h(x) = \dfrac{x^2 - 4}{x^3 + 2x^2 + x + 2}$

23. $s(t) = \dfrac{t}{\sin t}$

24. $g(\theta) = \dfrac{\tan \theta}{\theta}$

In Exercises 25–28, determine whether the function has a vertical asymptote or a removable discontinuity at $x = -1$. Graph the function using a graphing utility to confirm your answer.

25. $f(x) = \dfrac{x^2 - 1}{x + 1}$

26. $f(x) = \dfrac{x^2 - 6x - 7}{x + 1}$

27. $f(x) = \dfrac{x^2 + 1}{x + 1}$

28. $f(x) = \dfrac{\sin(x + 1)}{x + 1}$

In Exercises 29–40, find the limit.

29. $\displaystyle\lim_{x \to 2^+} \dfrac{x - 3}{x - 2}$

30. $\displaystyle\lim_{x \to 1^+} \dfrac{2 + x}{1 - x}$

31. $\displaystyle\lim_{x \to 4^+} \dfrac{x^2}{x^2 - 16}$

32. $\displaystyle\lim_{x \to 4^-} \dfrac{x^2}{x^2 + 16}$

33. $\displaystyle\lim_{x \to -3^-} \dfrac{x^2 + 2x - 3}{x^2 + x - 6}$

34. $\displaystyle\lim_{x \to (-1/2)^+} \dfrac{6x^2 + x - 1}{4x^2 - 4x - 3}$

35. $\displaystyle\lim_{x \to 0^-} \left(1 + \dfrac{1}{x}\right)$

36. $\displaystyle\lim_{x \to 0^-} \left(x^2 - \dfrac{1}{x}\right)$

37. $\displaystyle\lim_{x \to 0^+} \dfrac{2}{\sin x}$

38. $\displaystyle\lim_{x \to (\pi/2)^+} \dfrac{-2}{\cos x}$

39. $\displaystyle\lim_{x \to 1} \dfrac{x^2 - x}{(x^2 + 1)(x - 1)}$

40. $\displaystyle\lim_{x \to 3} \dfrac{x - 2}{x^2}$

In Exercises 41–44, use a graphing utility to graph the function and determine the one-sided limit.

41. $f(x) = \dfrac{x^2 + x + 1}{x^3 - 1}$

$\displaystyle\lim_{x \to 1^+} f(x)$

42. $f(x) = \dfrac{x^3 - 1}{x^2 + x + 1}$

$\displaystyle\lim_{x \to 1^-} f(x)$

43. $f(x) = \dfrac{1}{x^2 - 25}$

$\displaystyle\lim_{x \to 5^-} f(x)$

44. $f(x) = \sec \dfrac{\pi x}{6}$

$\displaystyle\lim_{x \to 3^+} f(x)$

45. A given sum S is inversely proportional to $1 - r$, where $0 < |r| < 1$. Find the limit of S as $r \to 1^-$.

46. *Boyle's Law* For a quantity of gas at a constant temperature, the pressure P is inversely proportional to the volume V. Find the limit of P as $V \to 0^+$.

47. Rate of Change A 25-foot ladder is leaning against a house (see figure). If the base of the ladder is pulled away from the house at a rate of 2 feet per second, the top will move down the wall at a rate of

$$r = \frac{2x}{\sqrt{625 - x^2}} \text{ ft/sec}$$

where x is the distance between the base of the ladder and the house.

(a) Find the rate r when x is 7 feet.

(b) Find the rate r when x is 15 feet.

(c) Find the limit of r as $x \to 25^-$.

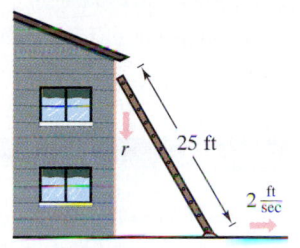

Figure for 47

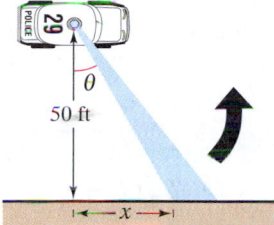

Figure for 48

48. Rate of Change A patrol car is parked 50 feet from a long warehouse (see figure). The revolving light on top of the car turns at a rate of $\frac{1}{2}$ revolution per second. The rate at which the light beam moves along the wall is

$$r = 50\pi \sec^2 \theta \text{ ft/sec.}$$

(a) Find the rate r when θ is $\pi/6$.

(b) Find the rate r when θ is $\pi/3$.

(c) Find the limit of r as $\theta \to (\pi/2)^-$.

49. Illegal Drugs The cost in millions of dollars for a governmental agency to seize $x\%$ of an illegal drug is

$$C = \frac{528x}{100 - x}, \quad 0 \le x < 100.$$

(a) Find the cost of seizing 25%.

(b) Find the cost of seizing 50%.

(c) Find the cost of seizing 75%.

(d) Find the limit of C as $x \to 100^-$.

50. Average Speed On a trip of d miles to another city, a truck driver's average speed was x miles per hour. On the return trip the average speed was y miles per hour. The average speed for the round trip was 50 miles per hour.

(a) Verify that $y = \dfrac{25x}{x - 25}$. What is the domain?

(b) Complete the table.

x	30	40	50	60
y				

Are the values of y different than expected? Explain.

(c) Find the limit of y as $x \to 25^+$.

51. Relativity According to the theory of relativity, the mass m of a particle depends on its velocity v. That is,

$$m = \frac{m_0}{\sqrt{1 - (v^2/c^2)}}$$

where m_0 is the mass when the particle is at rest and c is the speed of light. Find the limit of the mass as v approaches c^-.

52. Think About It Use the graph of the function f (see figure) to sketch the graph of $g(x) = 1/f(x)$ on the interval $[-2, 3]$.

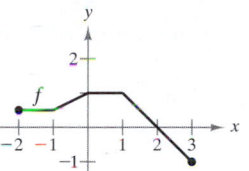

53. Numerical and Graphical Reasoning A crossed belt connects a 20-centimeter pulley (10-cm radius) on an electric motor with a 40-centimeter pulley (20-cm radius) on a saw arbor (see figure). The electric motor runs at 1700 revolutions per minute.

(a) Determine the number of revolutions per minute of the saw.

(b) How does crossing the belt affect the saw in relation to the motor?

(c) Let L be the total length of the belt. Write L as a function of ϕ, where ϕ is measured in radians. What is the domain of the function? [Hint: Add the lengths of the straight sections of the belt and the length of the belt around each pulley.]

(d) Use a graphing utility to complete the table.

ϕ	0.3	0.6	0.9	1.2	1.5
L					

(e) Use a graphing utility to graph the function over the appropriate domain.

(f) Find

$$\lim_{\phi \to (\pi/2)^-} L.$$

Use a geometric argument as the basis of a second method of finding this limit.

(g) Find $\displaystyle\lim_{\phi \to 0^+} L$.

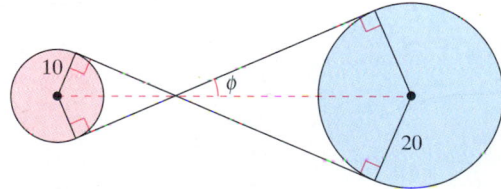

54. *Numerical and Graphical Analysis* Consider the shaded region outside the sector of a circle of radius 10 meters and inside a right triangle (see figure).

(a) Write the area $A = f(\theta)$ of the region as a function of θ. Determine the domain of the function.

(b) Use a graphing utility to complete the table.

θ	0.3	0.6	0.9	1.2	1.5
$f(\theta)$					

(c) Use a graphing utility to graph the function over the appropriate domain.

(d) Find the limit of A as $\theta \to \pi/2^-$.

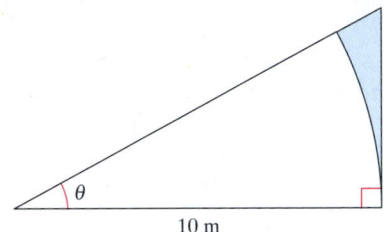

10 m

True or False? **In Exercises 55–58, determine whether the statement is true or false. If it is false, explain why or give an example that shows it is false.**

55. If $p(x)$ is a polynomial, then the function given by

$$f(x) = \frac{p(x)}{x - 1}$$

has a vertical asymptote at $x = 1$.

56. A rational function has at least one vertical asymptote.

57. Polynomial functions have no vertical asymptotes.

58. If f has a vertical asymptote at $x = 0$, then f is undefined at $x = 0$.

59. Find functions f and g such that

$$\lim_{x \to c} f(x) = \infty \quad \text{and} \quad \lim_{x \to c} g(x) = \infty$$

but $\lim_{x \to c} [f(x) - g(x)] \neq 0$.

60. Prove the remaining properties of Theorem 1.15.

61. Prove that if

$$\lim_{x \to c} f(x) = \infty$$

then

$$\lim_{x \to c} \frac{1}{f(x)} = 0.$$

62. Prove that if

$$\lim_{x \to c} \frac{1}{f(x)} = 0$$

then $\lim_{x \to c} f(x)$ does not exist.

63. *Writing* Because

$$\lim_{x \to 1} \frac{x + 2}{(x - 1)^2} = \infty,$$

for each $M > 0$ there exists a $\delta > 0$ such that

$$\frac{x + 2}{(x - 1)^2} > M$$

whenever $0 < |x - 1| < \delta$.

(a) Let $M = 100$. Use a graphing utility to graph the function and the line $y = 100$. Use the graphs to estimate δ.

(b) Let $M = 1000$. Use a graphing utility to graph the function and the line $y = 1000$. Use the graphs to estimate δ.

(c) Write a short paragraph describing the change in δ for increasing M.

SECTION PROJECT

Recall from Theorem 1.9 that the limit of $f(x) = \dfrac{\sin x}{x}$ as x approaches 0 is

$$\lim_{x \to 0} \frac{\sin x}{x} = 1.$$

(a) Use a graphing utility to graph the function f on the interval $-\pi \le 0 \le \pi$. Explain how this graph helps confirm that

$$\lim_{x \to 0} \frac{\sin x}{x} = 1.$$

(b) Explain how you could use a table of values to confirm the value of this limit numerically.

(c) Graph the function $g(x) = \sin x$ by hand. Sketch a tangent line at the point $(0, 0)$ and visually estimate the slope of this tangent line.

(d) Let $(x, \sin x)$ be a point on the graph of g near $(0, 0)$, and write a formula for the slope of the secant line joining $(x, \sin x)$ and $(0, 0)$. Evaluate this formula for $x = 0.1$ and $x = 0.01$. Then determine the exact slope of the tangent line to g at the point $(0, 0)$.

(e) Sketch the graph of the cosine function $h(x) = \cos x$. What is the slope of the tangent line at the point $(0, 1)$? Use limits to find this slope analytically. [*Hint:* See the second part of Theorem 1.9.]

(f) Find the slope of the tangent line to $k(x) = \tan x$ at the point $(0, 0)$.

REVIEW EXERCISES FOR CHAPTER 1

In Exercises 1 and 2, determine whether the problem can be solved using precalculus or if calculus is required. If the problem can be solved using precalculus, solve it. If the problem seems to require calculus, explain your reasoning. Use a graphical or numerical approach to estimate the solution.

1. Find the distance between the points $(1, 1)$ and $(3, 9)$ along the curve $y = x^2$.

2. Find the distance between the points $(1, 1)$ and $(3, 9)$ along the line $y = 4x - 3$.

In Exercises 3 and 4, complete the table and use the result to estimate the limit. Use a graphing utility to graph the function to confirm your result.

x	-0.1	-0.01	-0.001	0.001	0.01	0.1
$f(x)$						

3. $\lim\limits_{x \to 0} \dfrac{[1/(x + 2)] - (1/2)}{x}$

4. $\lim\limits_{x \to 0} \dfrac{\ln(x + 5) - \ln 5}{x}$

In Exercises 5 and 6, use the graph to determine the limit.

5. $h(x) = \dfrac{x^2 - 2x}{x}$

6. $g(x) = \dfrac{3x}{x - 2}$

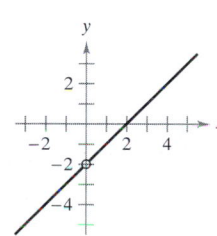

(a) $\lim\limits_{x \to 0} h(x)$

(b) $\lim\limits_{x \to -1} h(x)$

(a) $\lim\limits_{x \to 2^+} g(x)$

(b) $\lim\limits_{x \to 0} g(x)$

In Exercises 7–22, find the limit (if it exists).

7. $\lim\limits_{x \to 2} (5x - 3)$

8. $\lim\limits_{x \to 2} (3x + 5)$

9. $\lim\limits_{x \to 2} (5x - 3)(3x + 5)$

10. $\lim\limits_{x \to 2} \dfrac{3x + 5}{5x - 3}$

11. $\lim\limits_{t \to 3} \dfrac{t^2 + 1}{t}$

12. $\lim\limits_{t \to 3} \dfrac{t^2 - 9}{t - 3}$

13. $\lim\limits_{t \to -2} \dfrac{t + 2}{t^2 - 4}$

14. $\lim\limits_{x \to 0} \dfrac{\sqrt{4 + x} - 2}{x}$

15. $\lim\limits_{x \to 0} \dfrac{[1/(x + 1)] - 1}{x}$

16. $\lim\limits_{s \to 0} \dfrac{(1/\sqrt{1 + s}) - 1}{s}$

17. $\lim\limits_{x \to -5} \dfrac{x^3 + 125}{x + 5}$

18. $\lim\limits_{x \to -2} \dfrac{x^2 - 4}{x^3 + 8}$

19. $\lim\limits_{x \to 0^+} \left(x - \dfrac{1}{x^3} \right)$

20. $\lim\limits_{x \to 2} \dfrac{1}{\sqrt[3]{x^2 - 4}}$

21. $\lim\limits_{\Delta x \to 0} \dfrac{\sin[(\pi/6) + \Delta x] - (1/2)}{\Delta x}$

[*Hint:* $\sin(\theta + \phi) = \sin\theta\cos\phi + \cos\theta\sin\phi$]

22. $\lim\limits_{\Delta x \to 0} \dfrac{\cos(\pi + \Delta x) + 1}{\Delta x}$

[*Hint:* $\cos(\theta + \phi) = \cos\theta\cos\phi - \sin\theta\sin\phi$]

In Exercises 23–32, find the one-sided limit.

23. $\lim\limits_{x \to -2^-} \dfrac{2x^2 + x + 1}{x + 2}$

24. $\lim\limits_{x \to (1/2)^+} \dfrac{x}{2x - 1}$

25. $\lim\limits_{x \to -1^+} \dfrac{x + 1}{x^3 + 1}$

26. $\lim\limits_{x \to -1^-} \dfrac{x + 1}{x^4 - 1}$

27. $\lim\limits_{x \to 1^-} \dfrac{x^2 + 2x + 1}{x - 1}$

28. $\lim\limits_{x \to -1^+} \dfrac{x^2 - 2x + 1}{x + 1}$

will
be
unkno

29. $\lim\limits_{x \to 0^+} \dfrac{\sin 4x}{5x}$

30. $\lim\limits_{x \to 0^+} \dfrac{\sec x}{x}$

31. $\lim\limits_{x \to 0^+} \dfrac{\csc 2x}{x}$

32. $\lim\limits_{x \to 0^-} \dfrac{\cos^2 x}{x}$

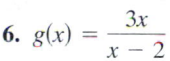 *Numerical, Graphical, and Analytic Analysis*   **In Exercises 33 and 34, consider** $\lim\limits_{x \to 1^+} f(x)$.

(a) Complete the table to approximate the limit.

(b) Use a graphing utility to graph the function and use the graph to approximate the limit.

(c) Rationalize the numerator to find the exact value of the limit analytically.

x	1.1	1.01	1.001	1.0001
$f(x)$				

33. $f(x) = \dfrac{\sqrt{2x + 1} - \sqrt{3}}{x - 1}$

34. $f(x) = \dfrac{1 - \sqrt[3]{x}}{x - 1}$

[*Hint:* $a^3 - b^3 = (a - b)(a^2 + ab + b^2)$]

In Exercises 35–44, determine the intervals on which the function is continuous.

35. $f(x) = [\![x + 3]\!]$

36. $f(x) = \dfrac{3x^2 - x - 2}{x - 1}$

37. $f(x) = \begin{cases} \dfrac{3x^2 - x - 2}{x - 1}, & x \neq 1 \\ 0, & x = 1 \end{cases}$

38. $f(x) = \begin{cases} 5 - x, & x \leq 2 \\ 2x - 3, & x > 2 \end{cases}$

39. $f(x) = \dfrac{1}{(x - 2)^2}$

40. $f(x) = \sqrt{\dfrac{x + 1}{x}}$

41. $f(x) = \dfrac{3}{x + 1}$

42. $f(x) = \dfrac{x + 1}{2x + 2}$

43. $f(x) = \csc \dfrac{\pi x}{2}$

44. $f(x) = \tan 2x$

45. Determine the value of c such that the function is continuous on the entire real line.

$$f(x) = \begin{cases} x + 3, & x \le 2 \\ cx + 6, & x > 2 \end{cases}$$

46. Determine the values of b and c such that the function is continuous on the entire real line.

$$f(x) = \begin{cases} x + 1, & 1 < x < 3 \\ x^2 + bx + c, & |x - 2| \ge 1 \end{cases}$$

47. *Compound Interest* A sum of $5000 is deposited in a savings plan that pays 12% interest compounded semiannually. The account balance after t years is given by $A = 5000(1.06)^{\llbracket 2t \rrbracket}$. Use a graphing utility to graph the function and discuss its continuity.

48. *Cost of Overnight Delivery* The cost of sending an overnight package from New York to Atlanta is $9.80 for the first pound and $2.50 for each additional pound. Use the greatest integer function to create a model for the cost C of overnight delivery of a package weighing x pounds. Use a graphing utility to graph the function and discuss its continuity.

In Exercises 49–52, find the vertical asymptotes (if any) of the function.

49. $g(x) = 1 + \dfrac{2}{x}$

50. $h(x) = \dfrac{4x}{4 - x^2}$

51. $f(x) = \dfrac{8}{(x - 10)^2}$

52. $f(x) = \csc \pi x$

53. *Cost of Clean Air* A utility company burns coal to generate electricity. The cost C in dollars of removing $p\%$ of the air pollutants in the stack emissions is

$$C = \dfrac{80,000p}{100 - p}, \qquad 0 \le p < 100.$$

Find the cost of removing (a) 15%, (b) 50%, and (c) 90%. (d) Find the limit of C as $p \to 100^-$.

54. The function f is defined as follows.

$$f(x) = \dfrac{\tan 2x}{x}, \qquad x \ne 0$$

(a) Find $\lim\limits_{x \to 0} \dfrac{\tan 2x}{x}$ (if it exists).

(b) Can the function f be defined such that it is continuous at $x = 0$?

Free-Falling Object In Exercises 55 and 56, use the position function $s(t) = -4.9t^2 + 200$, which gives the height (in meters) of an object that has fallen from a height of 200 meters. The velocity at time $t = a$ seconds is given by

$$\lim_{t \to a} \dfrac{s(a) - s(t)}{a - t}.$$

55. Find the velocity of the object when $t = 4$.

56. At what velocity will the object impact the ground?

True or False? In Exercises 57–63, determine whether the statement is true or false. If it is false, explain why or give an example that shows it is false.

57. $\lim\limits_{x \to 0} \dfrac{|x|}{x} = 1$

58. $\lim\limits_{x \to 0} x^3 = 0$

59. If $f(x) = g(x)$ for all real numbers other than $x = 0$, and

$$\lim_{x \to 0} f(x) = L$$

then

$$\lim_{x \to 0} g(x) = L.$$

60. If $\lim\limits_{x \to c} f(x) = L$, then $f(c) = L$.

61. For polynomial functions, the limits from the right and the left must exist and must be equal.

62. $\lim\limits_{x \to 2} f(x) = 3$, where

$$f(x) = \begin{cases} 3, & x \le 2 \\ 0, & x > 2 \end{cases}$$

63. $\lim\limits_{x \to 3} f(x) = 1$, where

$$f(x) = \begin{cases} x - 2, & x \le 3 \\ -x^2 + 8x - 14, & x > 3 \end{cases}$$

64. Let $f(x) = \dfrac{x^2 - 4}{|x - 2|}$. Find each limit (if possible).

(a) $\lim\limits_{x \to 2^-} f(x)$

(b) $\lim\limits_{x \to 2^+} f(x)$

(c) $\lim\limits_{x \to 2} f(x)$

65. Let $f(x) = \sqrt{x(x - 1)}$.

(a) Find the domain of f.

(b) Find $\lim\limits_{x \to 0^-} f(x)$.

(c) Find $\lim\limits_{x \to 1^+} f(x)$.

2

Differentiation

The study of dynamics dates back to the sixteenth century. As the Dark Ages gave way to the Renaissance, Galileo Galilei (1564–1642) was one of the first to take steps toward understanding the motion of objects under the influence of gravity.

Up until Galileo's time, it was recognized that a falling object moved faster and faster as it fell, but what mathematical law governed this accelerating motion was unknown. Free-falling objects move too fast to have been measured with any of the equipment available at that time. Galileo solved this problem with a rather ingenious setup. He reasoned that gravity could be "diluted" by rolling a ball down an inclined plane. He used a water clock, which kept track of time by measuring the amount of water that poured through a small opening at the bottom.

We now have relatively inexpensive instruments, such as the *Texas Instruments Calculator-Based Laboratory (CBL) System,* that allow accurate position data to be gathered on a free-falling object. A CBL System was used to track the positions of a falling ball at time intervals of 0.02 second. The results are shown below.

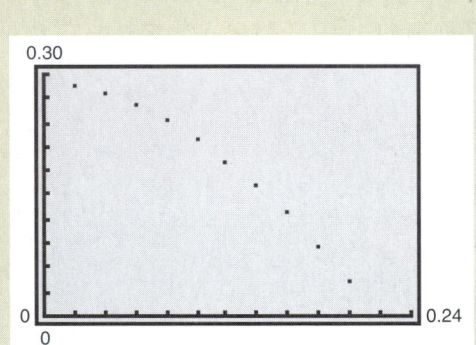

Time (sec)	Height (meters)	Velocity (meters/sec)
0.00	0.290864	-0.16405
0.02	0.284279	-0.32857
0.04	0.274400	-0.49403
0.06	0.260131	-0.71322
0.08	0.241472	-0.93309
0.10	0.219520	-1.09409
0.12	0.189885	-1.47655
0.14	0.160250	-1.47891
0.16	0.126224	-1.69994
0.18	0.086711	-1.96997
0.20	0.045002	-2.07747
0.22	0.000000	-2.25010

Kathryn C. Thorton, Ph.D., Payload Commander on the USML-2 mission, has been a NASA astronaut since July 1985. The USML-2 mission was her fourth space flight.

QUESTIONS

1. Use a graphing utility to sketch a scatter plot of the positions of the falling ball. What type of model seems to be the best fit? Use the regression features of the graphing utility to find the best-fitting model.

2. Repeat the procedure in Question 1 for the velocities of the falling ball. Describe any relationships between the two models.

3. In theory, the position of a free-falling object in a vacuum is given by $s = \frac{1}{2}gt^2 + v_0 t + s_0$, where g is the acceleration due to gravity (meters per second per second), t is the time (seconds), v_0 is the initial velocity (meters per second), and s_0 is the initial height (meters). From this experiment, estimate the value of g. Do you think your estimate is too great or too small? Explain your reasoning.

The concepts presented here will be explored further in this chapter. For an extension of this application, see the lab series that accompanies this text.

The Tangent Line Problem • The Derivative of a Function •
Differentiability and Continuity

The Tangent Line Problem

Calculus grew out of four major problems that European mathematicians were working on during the seventeenth century.

1. The tangent line problem (Section 1.1 and this section)
2. The velocity and acceleration problem (Sections 2.2 and 2.3)
3. The minimum and maximum problem (Section 3.1)
4. The area problem (Sections 1.1 and 4.2)

Each problem involves the notion of a limit, and we could introduce calculus with any of the four problems.

We gave a brief introduction to the tangent line problem in Section 1.1. Although partial solutions to this problem were given by Pierre de Fermat (1601–1665), René Descartes (1596–1650), Christian Huygens (1629–1695), and Isaac Barrow (1630–1677), credit for the first general solution is usually given to Isaac Newton (1642–1727) and Gottfried Leibniz (1646–1716). Newton's work on this problem stemmed from his interest in optics and light refraction.

What does it mean to say that a line is tangent to a curve at a point? For a circle, the tangent line at a point P is the line that is perpendicular to the radial line at point P, as shown in Figure 2.1.

For a general curve, however, the problem is more difficult. For example, how would you define the tangent lines shown in Figure 2.2? You might say that a line is tangent to a curve at a point P if it touches, but does not cross, the curve at point P. This definition would work for the first curve shown in Figure 2.2, but not for the second. *Or* you might say that a line is tangent to a curve if the line touches or intersects the curve at exactly one point. This definition would work for a circle but not for more general curves, as the third curve in Figure 2.2 shows.

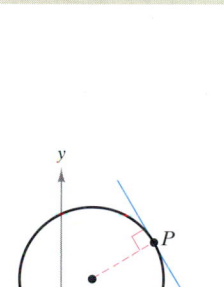

ISAAC NEWTON (1642–1727)

In addition to his work in calculus, Newton made revolutionary contributions to physics, including the Universal Law of Gravitation and his three laws of motion.

Mary Evans Picture Library

Tangent line to a circle
Figure 2.1

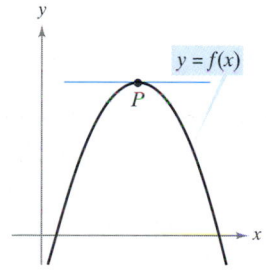 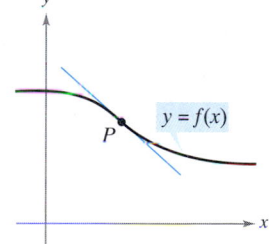 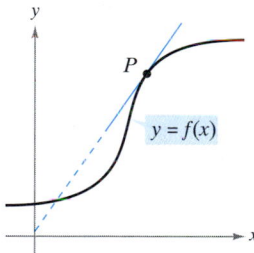

Tangent line to a curve at a point
Figure 2.2

EXPLORATION

Identifying a Tangent Line   Use a graphing utility to sketch the graph of $f(x) = 2x^3 - 4x^2 + 3x - 5$. On the same screen, sketch the graphs of $y = x - 5$, $y = 2x - 5$, and $y = 3x - 5$. Which of these lines, if any, appears to be tangent to the graph of f at the point $(0, -5)$? Explain your reasoning.

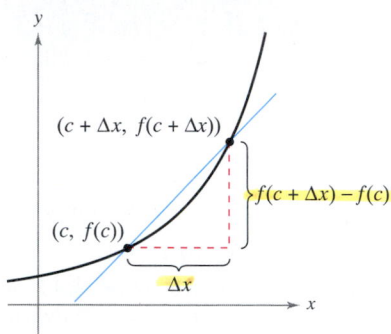

The secant line through $(c, f(c))$ and $(c + \Delta x, f(c + \Delta x))$

Figure 2.3

Essentially, the problem of finding the tangent line at a point P boils down to the problem of finding the *slope* of the tangent line at point P. You can approximate this slope using a **secant line*** through the point of tangency and a second point on the curve, as shown in Figure 2.3. If $(c, f(c))$ is the point of tangency and $(c + \Delta x, f(c + \Delta x))$ is a second point on the graph of f, the slope of the secant line through the two points is given by substitution into the formula

$$m = \frac{y_2 - y_1}{x_2 - x_1}$$

$$
m_{sec} = \frac{f(c + \Delta x) - f(c)}{(c + \Delta x) - c} \qquad \text{Change in } y \atop \text{Change in } x
$$

$$
= \frac{f(c + \Delta x) - f(c)}{\Delta x}. \qquad \text{Slope of secant line}
$$

The right-hand side of this equation is a **different quotient.** The denominator Δx is the **change in x,** and the numerator $\Delta y = f(c + \Delta x) - f(c)$ is the **change in y.**

The beauty of this procedure is that you can obtain more and more accurate approximations to the slope of the tangent line by choosing points closer and closer to the point of tangency, as shown in Figure 2.4.

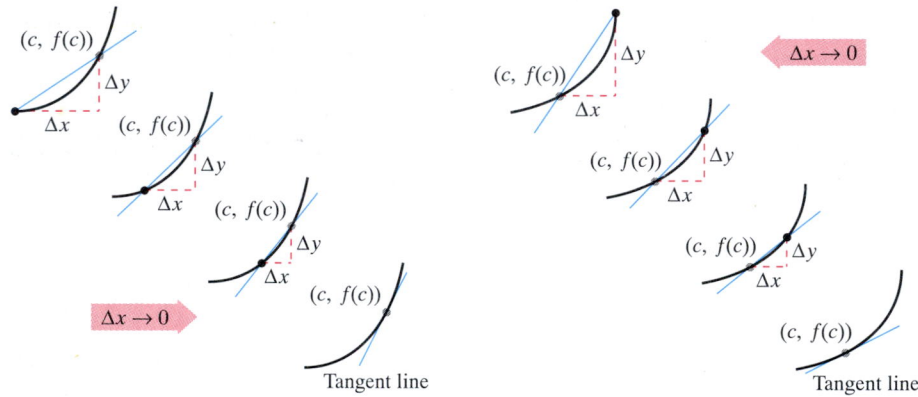

Tangent line approximations

Figure 2.4

Definition of Tangent Line with Slope m

If f is defined on an open interval containing c, and if the limit

$$
\lim_{\Delta x \to 0} \frac{\Delta y}{\Delta x} = \lim_{\Delta x \to 0} \frac{f(c + \Delta x) - f(c)}{\Delta x} = m
$$

exists, then the line passing through $(c, f(c))$ with slope m is the **tangent line** to the graph of f at the point $(c, f(c))$.

The slope of the tangent line to the graph of f at the point $(c, f(c))$ is also called the **slope of the graph of f at $x = c$.**

This use of the word secant comes from the Latin secare, meaning to cut, and is not a reference to the trigonometric function of the same name.

EXAMPLE 1   The Slope of the Graph of a Linear Function

Find the slope of the graph of $f(x) = 2x - 3$ at the point $(2, 1)$.

Solution   To find the slope of the graph of f when $c = 2$, you can apply the definition of the slope of a tangent line, as follows.

$$\lim_{\Delta x \to 0} \frac{f(2 + \Delta x) - f(2)}{\Delta x} = \lim_{\Delta x \to 0} \frac{[2(2 + \Delta x) - 3] - [2(2) - 3]}{\Delta x}$$

$$= \lim_{\Delta x \to 0} \frac{4 + 2\Delta x - 3 - 4 + 3}{\Delta x}$$

$$= \lim_{\Delta x \to 0} \frac{2\Delta x}{\Delta x}$$

$$= \lim_{\Delta x \to 0} 2$$

$$= 2$$

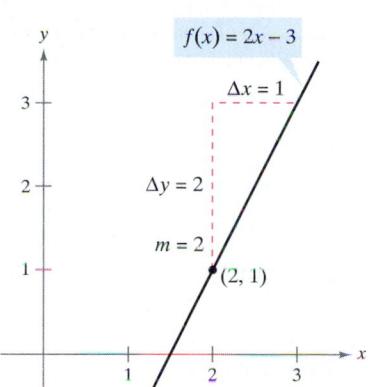

$f(x) = 2x - 3$

$\Delta x = 1$

$\Delta y = 2$

$m = 2$

$(2, 1)$

The slope of f at $(2, 1)$ is $m = 2$.
Figure 2.5

The slope of f at $(2, 1)$ is $m = 2$, as shown in Figure 2.5.

NOTE   In Example 1, the limit definition of the slope of f agrees with the definition of the slope of a line as discussed in Section P.2.

The graph of a linear function has the same slope at any point. This is not true of nonlinear functions, as can be seen in the following example.

EXAMPLE 2   Tangent Lines to the Graph of a Nonlinear Function

Find the slopes of the tangent lines to the graph of

$$f(x) = x^2 + 1$$

at the points $(0, 1)$ and $(-1, 2)$, as shown in Figure 2.6.

Solution   Let $(x, f(x))$ represent an arbitrary point on the graph of f. Then the slope of the tangent line at $(x, f(x))$ is given by

$$\lim_{\Delta x \to 0} \frac{f(x + \Delta x) - f(x)}{\Delta x} = \lim_{\Delta x \to 0} \frac{(x + \Delta x)^2 + 1 - (x^2 + 1)}{\Delta x}$$

$$= \lim_{\Delta x \to 0} \frac{x^2 + 2x(\Delta x) + (\Delta x)^2 + 1 - x^2 - 1}{\Delta x}$$

$$= \lim_{\Delta x \to 0} \frac{2x(\Delta x) + (\Delta x)^2}{\Delta x}$$

$$= \lim_{\Delta x \to 0} (2x + \Delta x)$$

$$= 2x.$$

(handwritten annotations:) whatever x gives a y on the line ; whatever $\frac{y_2 - y_1}{\Delta x}$ will be taken by limit so 2x apply to all point on line ; y_2 ; y_1

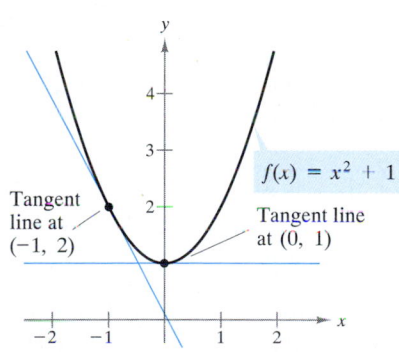

$f(x) = x^2 + 1$

Tangent line at $(-1, 2)$

Tangent line at $(0, 1)$

The slope of f at any point $(x, f(x))$ is $m = 2x$.
Figure 2.6

Therefore, the slope at *any* point $(x, f(x))$ on the graph of f is $m = 2x$. At the point $(0, 1)$, the slope is $m = 2(0) = 0$, and at $(-1, 2)$, the slope is $m = 2(-1) = -2$.

NOTE   In Example 2, note that x is held constant in the limit process (as $\Delta x \to 0$).

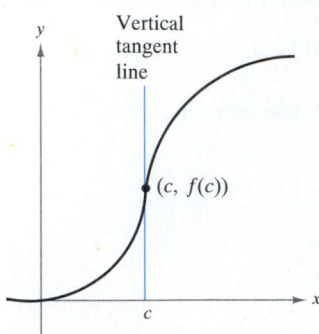

The graph of *f* has a vertical tangent line at $(c, f(c))$.

Figure 2.7

The definition of a tangent line to a curve does not cover the possibility of a vertical tangent line. For vertical tangent lines, you can use the following definition. If *f* is continuous at *c* and

$$\lim_{\Delta x \to 0} \frac{f(c + \Delta x) - f(c)}{\Delta x} = \infty$$

or

$$\lim_{\Delta x \to 0} \frac{f(c + \Delta x) - f(c)}{\Delta x} = -\infty$$

the vertical line, $x = c$, passing through $(c, f(c))$ is a **vertical tangent line** to the graph of *f*. For example, the function shown in Figure 2.7 has a vertical tangent line at $(c, f(c))$. If the domain of *f* is the closed interval $[a, b]$, you can extend the definition of a vertical tangent line to include the endpoints by considering continuity and limits from the right (for $x = a$) and from the left (for $x = b$).

The Derivative of a Function

You have now arrived at a crucial point in the study of calculus. The limit used to define the slope of a tangent line is also used to define one of the two fundamental operations of calculus—**differentiation.**

Definition of the Derivative of a Function

The **derivative** of *f* at *x* is given by

$$f'(x) = \lim_{\Delta x \to 0} \frac{f(x + \Delta x) - f(x)}{\Delta x}$$

provided the limit exists.

The process of finding the derivative of a function is called **differentiation.** A function is **differentiable** at *x* if its derivative exists at *x* and **differentiable on an open interval** (a, b) if it is differentiable at every point in the interval.

In addition to $f'(x)$, which is read as "*f* prime of *x*," other notations are used to denote the derivative of $y = f(x)$. The most common are

$$f'(x), \quad \frac{dy}{dx}, \quad y', \quad \frac{d}{dx}[f(x)], \quad D_x[y].$$ Notation for derivatives

The notation dy/dx is read as "the derivative of *y with respect to x.*" Using limit notation, you can write

$$\frac{dy}{dx} = \lim_{\Delta x \to 0} \frac{\Delta y}{\Delta x}$$
$$= \lim_{\Delta x \to 0} \frac{f(x + \Delta x) - f(x)}{\Delta x}$$
$$= f'(x).$$

EXAMPLE 3   Finding the Derivative by the Limit Process

Find the derivative of $f(x) = x^3 + 2x$.

Solution

$$
\begin{aligned}
f'(x) &= \lim_{\Delta x \to 0} \frac{f(x + \Delta x) - f(x)}{\Delta x} \\[2mm]
&= \lim_{\Delta x \to 0} \frac{(x + \Delta x)^3 + 2(x + \Delta x) - (x^3 + 2x)}{\Delta x} \\[2mm]
&= \lim_{\Delta x \to 0} \frac{x^3 + 3x^2\Delta x + 3x(\Delta x)^2 + (\Delta x)^3 + 2x + 2\Delta x - x^3 - 2x}{\Delta x} \\[2mm]
&= \lim_{\Delta x \to 0} \frac{3x^2\Delta x + 3x(\Delta x)^2 + (\Delta x)^3 + 2\Delta x}{\Delta x} \\[2mm]
&= \lim_{\Delta x \to 0} \frac{\Delta x\left[3x^2 + 3x\Delta x + (\Delta x)^2 + 2\right]}{\Delta x} \\[2mm]
&= \lim_{\Delta x \to 0} \left[3x^2 + 3x\Delta x + (\Delta x)^2 + 2\right] \\[2mm]
&= 3x^2 + 2
\end{aligned}
$$

STUDY TIP Note that in Examples 1, 2, and 3, the key to finding the derivative of a function is to rewrite the difference quotient so that Δx does not occur as a factor of the denominator.

Remember that the derivative of a function f is itself a function, which can be used to find the slope of the tangent line at the point $(x, f(x))$ on the graph of f.

EXAMPLE 4   Using the Derivative to Find the Slope at a Point

Find $f'(x)$ for $f(x) = \sqrt{x}$. Then find the slope of the graph of f at the points $(1, 1)$ and $(4, 2)$. Discuss the behavior of f at $(0, 0)$.

Solution Use the procedure for rationalizing numerators, as discussed in Section 1.3.

$$
\begin{aligned}
f'(x) &= \lim_{\Delta x \to 0} \frac{f(x + \Delta x) - f(x)}{\Delta x} \\[2mm]
&= \lim_{\Delta x \to 0} \frac{\sqrt{x + \Delta x} - \sqrt{x}}{\Delta x} \\[2mm]
&= \lim_{\Delta x \to 0} \left(\frac{\sqrt{x + \Delta x} - \sqrt{x}}{\Delta x}\right)\left(\frac{\sqrt{x + \Delta x} + \sqrt{x}}{\sqrt{x + \Delta x} + \sqrt{x}}\right) \\[2mm]
&= \lim_{\Delta x \to 0} \frac{(x + \Delta x) - x}{\Delta x\left(\sqrt{x + \Delta x} + \sqrt{x}\right)} \\[2mm]
&= \lim_{\Delta x \to 0} \frac{\Delta x}{\Delta x\left(\sqrt{x + \Delta x} + \sqrt{x}\right)} \\[2mm]
&= \lim_{\Delta x \to 0} \frac{1}{\sqrt{x + \Delta x} + \sqrt{x}} \\[2mm]
&= \frac{1}{2\sqrt{x}}
\end{aligned}
$$

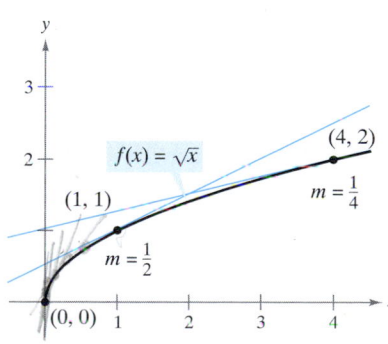

The slope of f at $(x, f(x))$, $x > 0$, is $m = 1/(2\sqrt{x})$.

Figure 2.8

At the point $(1, 1)$, the slope is $f'(1) = \frac{1}{2}$. At the point $(4, 2)$, the slope is $f'(4) = \frac{1}{4}$. (See Figure 2.8.) At the point $(0, 0)$ the slope is undefined. Moreover, because the limit of $f'(x)$ as $x \to 0$ from the right is infinite, the graph of f has a vertical tangent line at $(0, 0)$.

In many applications, it is convenient to use a variable other than x as the independent variable, as shown in Example 5.

EXAMPLE 5 Finding the Derivative of a Function

Find the derivative with respect to t for the function $y = 2/t$.

Solution Considering $y = f(t)$, you obtain the following.

$$\frac{dy}{dt} = \lim_{\Delta t \to 0} \frac{f(t + \Delta t) - f(t)}{\Delta t} \qquad \text{Definition of derivative}$$

$$= \lim_{\Delta t \to 0} \frac{\dfrac{2}{t + \Delta t} - \dfrac{2}{t}}{\Delta t} \qquad f(t + \Delta t) = 2/(t + \Delta t) \text{ and } f(t) = 2/t$$

$$= \lim_{\Delta t \to 0} \frac{\dfrac{2t - 2(t + \Delta t)}{t(t + \Delta t)}}{\Delta t} \qquad \text{Combine fractions in numerator.}$$

$$= \lim_{\Delta t \to 0} \frac{-2\Delta t}{\Delta t(t)(t + \Delta t)} \qquad \text{Cancel common factor of } \Delta t.$$

$$= \lim_{\Delta t \to 0} \frac{-2}{t(t + \Delta t)} \qquad \text{Simplify.}$$

$$= -\frac{2}{t^2} \qquad \text{Evaluate limit as } \Delta t \to 0.$$

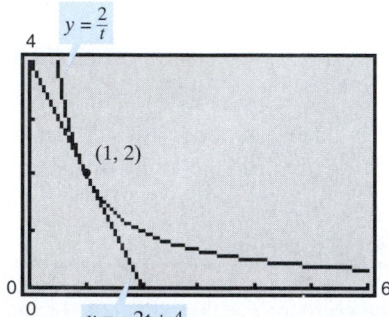

At the point $(1, 2)$ the line $y = -2t + 4$ is tangent to the graph of $y = 2/t$.

Figure 2.9

TECHNOLOGY A graphing utility can be used to reinforce the result given in Example 5. For instance, using the formula $dy/dt = -2/t^2$, you know that the slope of the graph of $y = 2/t$ at the point $(1, 2)$ is $m = -2$. This implies that an equation of the tangent line to the graph at $(1, 2)$ is $y - 2 = -2(t - 1)$ or $y = -2t + 4$, as shown in Figure 2.9.

Differentiability and Continuity

The following alternative limit form of the derivative is useful in investigating the relationship between differentiability and continuity. The derivative of f at c is

$$f'(c) = \lim_{x \to c} \frac{f(x) - f(c)}{x - c} \qquad \text{Alternative form of derivative}$$

provided this limit exists (see Figure 2.10). (A proof of the equivalence of this form is given in the appendix.) Note that the existence of the limit in this alternative form requires that the one-sided limits

$$\lim_{x \to c^-} \frac{f(x) - f(c)}{x - c} \quad \text{and} \quad \lim_{x \to c^+} \frac{f(x) - f(c)}{x - c}$$

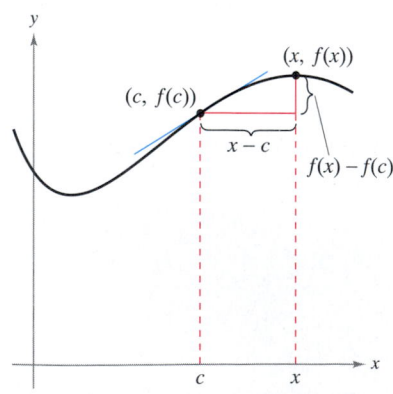

As x approaches c, the secant line approaches the tangent line.

Figure 2.10

exists and are equal. These one-sided limits are called the **derivatives from the left and from the right,** respectively. We say that f is **differentiable on the closed interval** $[a, b]$ if it is differentiable on (a, b) and if the derivative from the right at a and the derivative from the left at b both exist.

If a function is not continuous at $x = c$, it is also not differentiable at $x = c$. For instance, the greatest integer function

$$f(x) = [\![x]\!]$$

is not continuous at $x = 0$—hence, it is not differentiable at $x = 0$ (see Figure 2.11). You can verify this by observing that

$$\lim_{x \to 0^-} \frac{f(x) - f(0)}{x - 0} = \lim_{x \to 0^-} \frac{[\![x]\!] - 0}{x} = \infty \qquad \textcolor{red}{\text{Derivative from the left}}$$

and

$$\lim_{x \to 0^+} \frac{f(x) - f(0)}{x - 0} = \lim_{x \to 0^+} \frac{[\![x]\!] - 0}{x} = 0. \qquad \textcolor{red}{\text{Derivative from the right}}$$

Although it is true that differentiability implies continuity (as we will show in Theorem 2.1), the converse is not true. That is, it is possible for a function to be continuous at $x = c$ and *not* differentiable at $x = c$. Examples 6 and 7 illustrate this possibility.

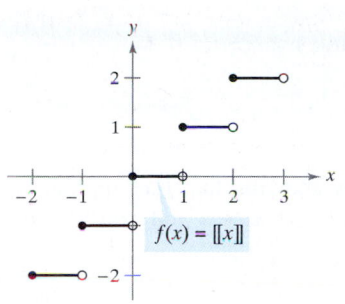

The greatest integer function is not differentiable at $x = 0$, because it is not continuous at $x = 0$.
Figure 2.11

EXAMPLE 6   A Graph with a Sharp Turn

The function

$$f(x) = |x - 2|$$

shown in Figure 2.12 is continuous at $x = 2$. However, the one-sided limits

$$\lim_{x \to 2^-} \frac{f(x) - f(2)}{x - 2} = \lim_{x \to 2^-} \frac{|x - 2| - 0}{x - 2} = -1 \qquad \textcolor{red}{\text{Derivative from the left}}$$

and

$$\lim_{x \to 2^+} \frac{f(x) - f(2)}{x - 2} = \lim_{x \to 2^+} \frac{|x - 2| - 0}{x - 2} = 1 \qquad \textcolor{red}{\text{Derivative from the right}}$$

are not equal. Therefore, f is not differentiable at $x = 2$ and the graph of f does not have a tangent line at the point $(2, 0)$.

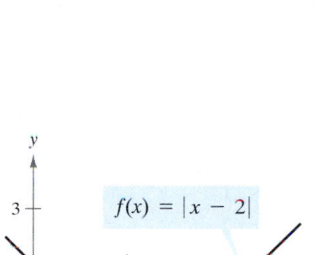

f is not differentiable at $x = 2$, because the derivatives from the left and from the right are not equal.
Figure 2.12

EXAMPLE 7   A Graph with a Vertical Tangent Line

The function

$$f(x) = x^{1/3}$$

is continuous at $x = 0$, as shown in Figure 2.13. However, because the limit

$$\lim_{x \to 0} \frac{f(x) - f(0)}{x - 0} = \lim_{x \to 0} \frac{x^{1/3} - 0}{x}$$

$$= \lim_{x \to 0} \frac{1}{x^{2/3}}$$

$$= \infty$$

is infinite, you can conclude that the tangent line is vertical at $x = 0$. Therefore, f is not differentiable at $x = 0$.

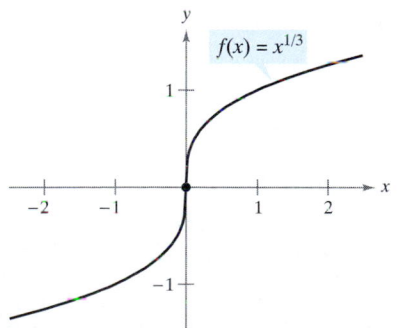

f is not differentiable at $x = 0$, because f has a vertical tangent at $x = 0$.
Figure 2.13

From Examples 6 and 7, you can see that a function is not differentiable at a point at which its graph has a sharp turn *or* a vertical tangent.

TECHNOLOGY Some graphing util-ities such as *Derive, Maple, Mathcad, Mathematica,* and the *TI-92* perform symbolic differentiation. Others per-form *numerical differentiation* by finding values of derivatives using the formula

$$f'(x) \approx \frac{f(x+h) - f(x-h)}{2h}$$

where h is a small number such as 0.001. Can you see any problems with this definition? For instance, using this definition, what is the value of the derivative of $f(x) = |x|$ when $x = 0$?

THEOREM 2.1   Differentiability Implies Continuity

If f is differentiable at $x = c$, then f is continuous at $x = c$.

Proof You can prove that f is continuous at $x = c$ by showing that $f(x)$ approaches $f(c)$ as $x \to c$. To do this, use the differentiability of f at $x = c$ and consider the following limit.

$$\lim_{x \to c} [f(x) - f(c)] = \lim_{x \to c} \left[(x - c)\left(\frac{f(x) - f(c)}{x - c} \right) \right]$$

$$= \left[\lim_{x \to c} (x - c) \right]\left[\lim_{x \to c} \frac{f(x) - f(c)}{x - c} \right]$$

$$= (0)[f'(c)]$$

$$= 0$$

Because the difference $f(x) - f(c)$ approaches zero as $x \to c$, you can conclude that $\lim_{x \to c} f(x) = f(c)$. Therefore, f is continuous at $x = c$. ▬▬▬

You can summarize the relationship between continuity and differentiability as follows.

1. If a function is differentiable at $x = c$, then it is continuous at $x = c$. Thus, differ-entiability implies continuity.

2. It is possible for a function to be continuous at $x = c$ and *not* be differentiable at $x = c$. Thus, continuity does not imply differentiability.

EXERCISES FOR SECTION 2.1

In Exercises 1 and 2, estimate the slope of the curve at the point (x, y).

1. (a)
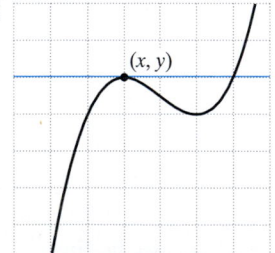
(b)
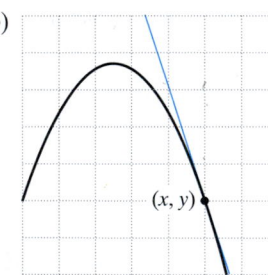

2. (a)
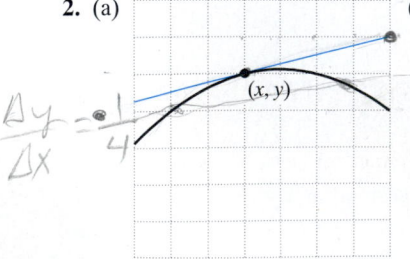
(b)
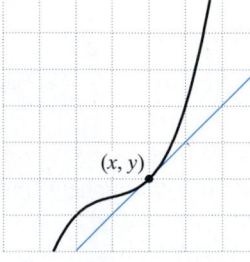

Think About It In Exercises 3 and 4, use the graph shown in the figure.

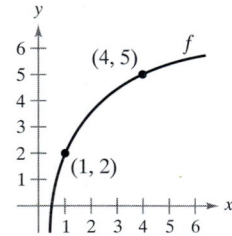

3. Identify or sketch each of the quantities on the figure.

  (a) $f(1)$ and $f(4)$ (b) $f(4) - f(1)$

  (c) $y = \dfrac{f(4) - f(1)}{4 - 1}(x - 1) + f(1)$

4. Insert the proper inequality symbol (< or >) between the given quantities.

  (a) $\dfrac{f(4) - f(1)}{4 - 1}$ ▮ $\dfrac{f(4) - f(3)}{4 - 3}$

  (b) $\dfrac{f(4) - f(1)}{4 - 1}$ ▮ $f'(1)$

In Exercises 5–16, use the definition of the derivative to find $f'(x)$.

5. $f(x) = 3$

6. $f(x) = 3x + 2$

7. $f(x) = -5x$

8. $f(x) = 9 - \frac{1}{2}x$

9. $f(x) = 2x^2 + x - 1$

10. $f(x) = 1 - x^2$

11. $f(x) = x^3 - 12x$

12. $f(x) = x^3 + x^2$

13. $f(x) = \dfrac{1}{x - 1}$

14. $f(x) = \dfrac{1}{x^2}$

15. $f(x) = \sqrt{x - 4}$

16. $f(x) = \dfrac{1}{\sqrt{x}}$

In Exercises 17–22, **(a)** find an equation of the tangent line to the graph of f at the indicated point, **(b)** use a graphing utility to graph the function and its tangent line at the point, and **(c)** use the derivative feature of a graphing utility to confirm your results.

17. $f(x) = x^2 + 1$

(2, 5)

18. $f(x) = x^2 + 2x + 1$

(−3, 4)

19. $f(x) = x^3$

(2, 8)

20. $f(x) = \sqrt{x}$

(1, 1)

21. $f(x) = x + \dfrac{1}{x}$

(1, 2)

22. $f(x) = \dfrac{1}{x + 1}$

(0, 1)

In Exercises 23 and 24, find an equation of the line that is tangent to the graph of f **and** parallel to the given line.

Function	Line
23. $f(x) = x^3$	$3x - y + 1 = 0$
24. $f(x) = \dfrac{1}{\sqrt{x}}$	$x + 2y - 6 = 0$

In Exercises 25 and 26, find the equations of the two tangent lines to the graph of f that pass through the indicated point.

25. $f(x) = 4x - x^2$

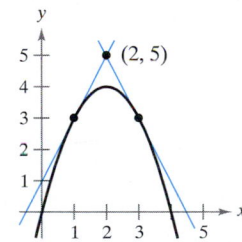

26. $f(x) = x^2$

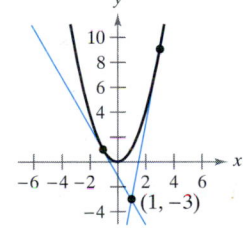

In Exercises 27–32, use your knowledge of the graph of the function and the geometric interpretation of the derivative to match the function with the graph of its derivative. It is not necessary to find the derivative of the function analytically.

(a)

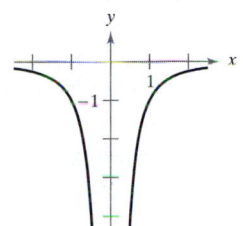

(b)

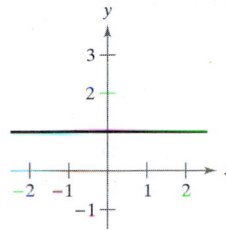

(c)

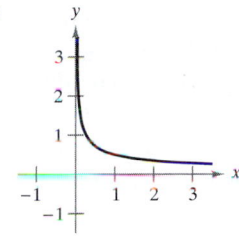

(d)

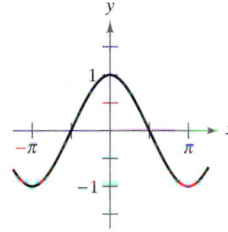

(e)

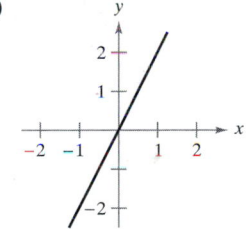

(f)

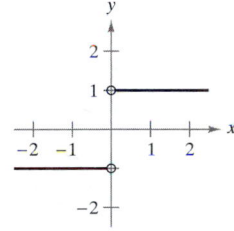

27. $f(x) = x$

28. $f(x) = x^2$

29. $f(x) = \sqrt{x}$

30. $f(x) = \dfrac{1}{x}$

31. $f(x) = |x|$

32. $f(x) = \sin x$

33. *Graphical Reasoning* The figure shows the graph of g'.

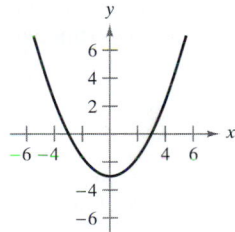

(a) $g'(0) = $ ▨

(b) $g'(3) = $ ▨

(c) What can you conclude about the graph of g knowing that $g'(1) = -\frac{8}{3}$?

(d) What can you conclude about the graph of g knowing that $g'(-4) = \frac{7}{3}$?

(e) Is $g(6) - g(4)$ positive or negative? Explain.

(f) Is it possible to find $g(2)$ from the graph? Explain.

34. *Graphical Reasoning* Use a graphing utility to graph each function and its tangent lines when $x = -1$, $x = 0$, and $x = 1$. Based on the results, determine whether the slope of a tangent line to the graph of a function is always distinct for different values of x.

(a) $f(x) = x^2$ (b) $g(x) = x^3$

35. *Think About It* Sketch a graph of a function whose derivative is always negative.

36. *Think About It* Sketch a graph of a function whose derivative is always positive.

Graphical, Numerical, and Analytic Analysis In Exercises 37 and 38, use a graphing utility to graph f on the interval $[-2, 2]$. Complete the table by *graphically* estimating the slopes of the graph at the indicated points. Then evaluate the slopes *analytically* and compare your results with those obtained graphically.

x	-2	-1.5	-1	-0.5	0	0.5	1	1.5	2
$f(x)$									
$f'(x)$									

37. $f(x) = \frac{1}{4}x^3$ **38.** $f(x) = \frac{1}{2}x^2$

Graphical Reasoning In Exercises 39 and 40, use a graphing utility to graph the functions f and g in the same viewing rectangle where

$$g(x) = \frac{f(x + 0.01) - f(x)}{0.01}.$$

Label the graphs and describe the relationship between them.

39. $f(x) = 2x - x^2$ **40.** $f(x) = 3\sqrt{x}$

Technology In Exercises 41 and 42, use a graphing utility to graph the function and its derivative in the same viewing rectangle. Label the graphs and describe the relationship between them.

41. $f(x) = \dfrac{1}{\sqrt{x}}$ **42.** $f(x) = \dfrac{x^3}{4} - 3x$

Writing In Exercises 43 and 44, consider the functions $f(x)$ and $S_{\Delta x}(x)$ where

$$S_{\Delta x}(x) = \frac{f(2 + \Delta x) - f(2)}{\Delta x}(x - 2) + f(2).$$

(a) Use a graphing utility to graph f and $S_{\Delta x}$ in the same viewing rectangle for $\Delta x = 1$, 0.5, and 0.1.

(b) Give a written description of the graphs of S for the different values of Δx in part (a).

43. $f(x) = 4 - (x - 3)^2$ **44.** $f(x) = x + \dfrac{1}{x}$

In Exercises 45–50, use the alternative form of the derivative to find the derivative at $x = c$ (if it exists).

45. $f(x) = x^2 - 1$, $c = 2$

46. $f(x) = x^3 + 2x$, $c = 1$

47. $f(x) = x^3 + 2x^2 + 1$, $c = -2$

48. $f(x) = 1/x$, $c = 3$

49. $f(x) = (x - 1)^{2/3}$, $c = 1$

50. $f(x) = |x - 2|$, $c = 2$

In Exercises 51–60, describe the x-values at which f is differentiable.

51. $f(x) = |x + 3|$ **52.** $f(x) = |x^2 - 9|$

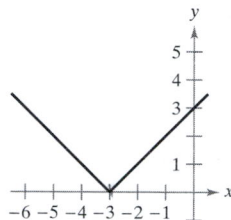 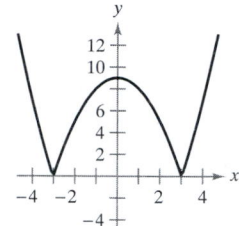

53. $f(x) = \dfrac{1}{x + 1}$ **54.** $f(x) = \dfrac{2x}{x - 1}$

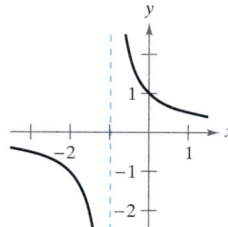 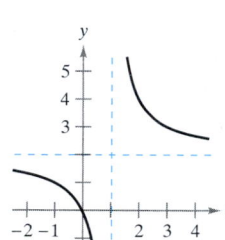

55. $f(x) = (x - 3)^{2/3}$ **56.** $f(x) = x^{2/5}$

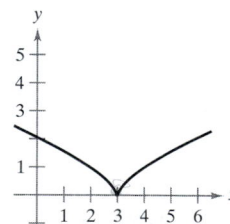 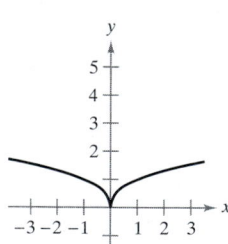

57. $f(x) = \sqrt{x - 1}$ **58.** $f(x) = \dfrac{x^2}{x^2 - 4}$

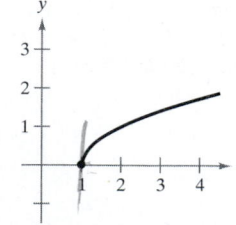 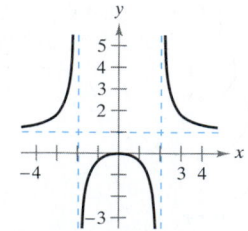

59. $f(x) = \begin{cases} 4 - x^2, & x > 0 \\ x^2 - 4, & x \le 0 \end{cases}$

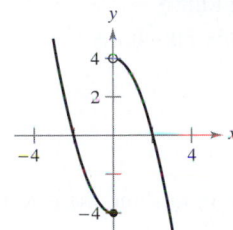

60. $f(x) = \begin{cases} x^2 - 2x, & x > 1 \\ x^3 - 3x^2 + 3x, & x \le 1 \end{cases}$

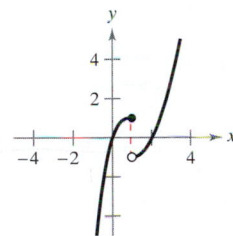

In Exercises 61–64, find the derivatives from the left and from the right at $x = 1$ (if they exist). Is the function differentiable at $x = 1$?

61. $f(x) = |x - 1|$

62. $f(x) = \sqrt{1 - x^2}$

63. $f(x) = \begin{cases} (x - 1)^3, & x \le 1 \\ (x - 1)^2, & x > 1 \end{cases}$

64. $f(x) = \begin{cases} x, & x \le 1 \\ x^2, & x > 1 \end{cases}$

In Exercises 65 and 66, determine whether the function is differentiable at $x = 2$.

65. $f(x) = \begin{cases} x^2 + 1, & x \le 2 \\ 4x - 3, & x > 2 \end{cases}$

66. $f(x) = \begin{cases} \frac{1}{2}x + 1, & x < 2 \\ \sqrt{2x}, & x \ge 2 \end{cases}$

 67. *Graphical Reasoning*   A line with slope m passes through the point $(0, 4)$ and has the equation $y = mx + 4$.

 (a) Write the distance d between the line and the point $(3, 1)$ as a function of m.

 (b) Use a graphing utility to graph the function d in part (a). Based on the graph, is the function differentiable at every value of m? If not, where is it not differentiable?

68. *Think About It*   Assume that $f'(c) = 3$. Find $f'(-c)$ given the following conditions.

 (a) f is an odd function.

 (b) f is an even function.

True or False?   **In Exercises 69–71, determine whether the statement is true or false. If it is false, explain why or give an example that shows it is false.**

69. If a function is continuous at a point, then it is differentiable at that point.

70. If a function has derivatives from both the right and the left at a point, then it is differentiable at that point.

71. If a function is differentiable at a point, then it is continuous at that point.

72. *Conjecture*   Consider the functions $f(x) = x^2$ and $g(x) = x^3$.

 (a) Graph f and f' on the same set of axes.

 (b) Graph g and g' on the same set of axes.

 (c) Identify any pattern between the functions f and g and their respective derivatives. Use the pattern to make a conjecture about $h'(x)$ if $h(x) = x^n$, where n is an integer and $n \ge 2$.

 (d) Find $f'(x)$ if $f(x) = x^4$. Compare the result with the conjecture in part (c). Is this a proof of your conjecture? Explain.

73. Let

$$f(x) = \begin{cases} x \sin \dfrac{1}{x}, & x \ne 0 \\ 0, & x = 0 \end{cases}$$

and

$$g(x) = \begin{cases} x^2 \sin \dfrac{1}{x}, & x \ne 0 \\ 0, & x = 0 \end{cases}$$

Show that f is continuous, but not differentiable, at $x = 0$. Show that g is differentiable at 0, and find $g'(0)$.

 74. *Writing*   Use a graphing utility to graph the two functions $f(x) = x^2 + 1$ and $g(x) = |x| + 1$ in the same viewing rectangle. Use the zoom and trace features to analyze the graphs near the point $(0, 1)$. What do you observe? Which function is differentiable at this point? Write a short paragraph describing the geometric significance of differentiability at a point.

The Constant Rule • The Power Rule • The Constant Multiple Rule • The Sum and Difference Rules • Derivatives of Sine and Cosine Functions • Rates of Change

The Constant Rule

In Section 2.1 you used the limit definition to find derivatives. In this and the next two sections you will be introduced to several "differentiation rules" that allow you to find derivatives without the *direct* use of the limit definition.

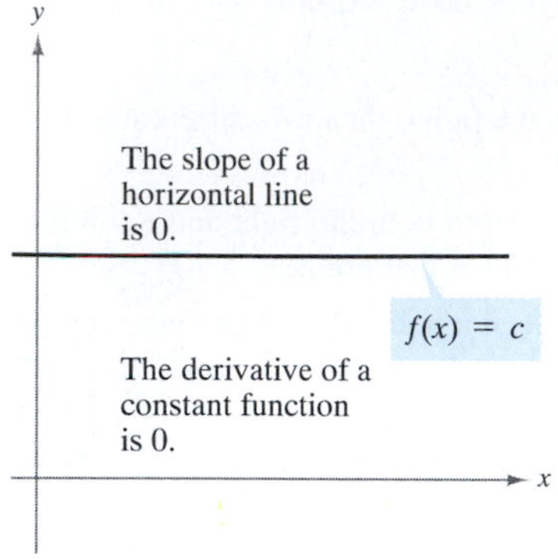

THEOREM 2.2 The Constant Rule

The derivative of a constant function is 0. That is, if c is a real number, then

$$\frac{d}{dx}[c] = 0.$$

The slope of a horizontal line is 0.

$f(x) = c$

The derivative of a constant function is 0.

The Constant Rule
Figure 2.14

NOTE   In Figure 2.14, note that the Constant Rule is equivalent to saying that the slope of a horizontal line is 0. This demonstrates the relationship between slope and derivative.

Proof   Let $f(x) = c$. Then, by the limit definition of the derivative,

$$\frac{d}{dx}[c] = f'(x)$$

$$= \lim_{\Delta x \to 0} \frac{f(x + \Delta x) - f(x)}{\Delta x}$$

$$= \lim_{\Delta x \to 0} \frac{c - c}{\Delta x}$$

$$= 0.$$

EXAMPLE 1   Using the Constant Rule

Function	Derivative
a. $y = 7$	$\dfrac{dy}{dx} = 0$
b. $f(x) = 0$	$f'(x) = 0$
c. $s(t) = -3$	$s'(t) = 0$
d. $y = k\pi^2$, k is constant	$y' = 0$

EXPLORATION

Writing a Conjecture   Use the definition of the derivative given in Section 2.1 to find the derivative of each of the following. What patterns do you see? Use your results to write a conjecture about the derivative of $f(x) = x^n$.

(a) $f(x) = x^1$ **(b)** $f(x) = x^2$ **(c)** $f(x) = x^3$

(d) $f(x) = x^4$ **(e)** $f(x) = x^{1/2}$ **(f)** $f(x) = x^{-1}$

The Power Rule

Before proving the next rule, we review the procedure for expanding a binomial.

$$(x + \Delta x)^2 = x^2 + 2x\Delta x + (\Delta x)^2$$
$$(x + \Delta x)^3 = x^3 + 3x^2\Delta x + 3x(\Delta x)^2 + (\Delta x)^3$$

The general binomial expansion for a positive integer n is

$$(x + \Delta x)^n = x^n + nx^{n-1}(\Delta x) + \underbrace{\frac{n(n-1)x^{n-2}}{2}(\Delta x)^2 + \cdots + (\Delta x)^n}_{(\Delta x)^2 \text{ is a factor of these terms.}}$$

This binomial expansion is used in proving a special case of the Power Rule.

THEOREM 2.3   The Power Rule

If n is a rational number, then the function $f(x) = x^n$ is differentiable and

$$\frac{d}{dx}[x^n] = nx^{n-1}.$$

For f to be differentiable at $x = 0$, n must be a number such that x^{n-1} is defined on an interval containing 0.

Proof   If n is a positive integer greater than 1, then the binomial expansion produces the following.

$$\frac{d}{dx}[x^n] = \lim_{\Delta x \to 0} \frac{(x + \Delta x)^n - x^n}{\Delta x}$$

$$= \lim_{\Delta x \to 0} \frac{x^n + nx^{n-1}(\Delta x) + \dfrac{n(n-1)x^{n-2}}{2}(\Delta x)^2 + \cdots + (\Delta x)^n - x^n}{\Delta x}$$

$$= \lim_{\Delta x \to 0} \left[nx^{n-1} + \frac{n(n-1)x^{n-2}}{2}(\Delta x) + \cdots + (\Delta x)^{n-1} \right]$$

$$= nx^{n-1} + 0 + \cdots + 0$$

$$= nx^{n-1}$$

This proves the case for which n is a positive integer greater than 1. We leave it to you to prove the case for $n = 1$. Example 7 in Section 2.3 proves the case for which n is a negative integer. Exercise 59 in Section 2.5 proves the case for which n is rational. (In Section 5.5, the Power Rule will be extended to cover irrational values of n.)

In the Power Rule, the case for which $n = 1$ is best thought of as a separate differentiation rule. That is,

$$\frac{d}{dx}[x] = 1. \qquad \text{Power Rule when } n = 1$$

This rule is consistent with the fact that the slope of the line $y = x$ is 1, as shown in Figure 2.15.

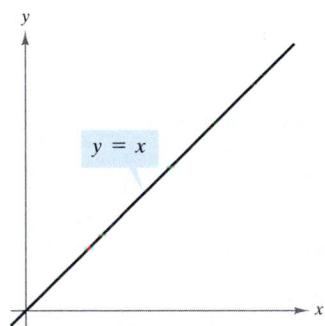

$y = x$

The slope of the line $y = x$ is 1.
Figure 2.15

EXAMPLE 2   Using the Power Rule

Function	Derivative
a. $f(x) = x^3$	$f'(x) = 3x^2$
b. $g(x) = \sqrt[3]{x}$	$g'(x) = \dfrac{d}{dx}\left[x^{1/3}\right] = \dfrac{1}{3}x^{-2/3} = \dfrac{1}{3x^{2/3}}$
c. $y = \dfrac{1}{x^2}$	$\dfrac{dy}{dx} = \dfrac{d}{dx}\left[x^{-2}\right] = (-2)x^{-3} = -\dfrac{2}{x^3}$

In Example 2c, note that *before* differentiating, $1/x^2$ was rewritten as x^{-2}. Rewriting is the first step in *many* differentiation problems.

Given:		Rewrite:		Differentiate:		Simplify:
$y = \dfrac{1}{x^2}$	⇨	$y = x^{-2}$	⇨	$\dfrac{dy}{dx} = (-2)x^{-3}$	⇨	$\dfrac{dy}{dx} = -\dfrac{2}{x^3}$

EXAMPLE 3   Finding the Slope of a Graph

Find the slope of the graph of $f(x) = x^4$ when

a. $x = -1$ **b.** $x = 0$ **c.** $x = 1$.

Solution   The derivative of f is $f'(x) = 4x^3$.

a. When $x = -1$, the slope is $f'(-1) = 4(-1)^3 = -4$.

b. When $x = 0$, the slope is $f'(0) = 4(0)^3 = 0$.

c. When $x = 1$, the slope is $f'(1) = 4(1)^3 = 4$.

In Figure 2.16, note that the slope of the graph is negative at the point $(-1, 1)$, the slope is zero at the point $(0, 0)$, and the slope is positive at the point $(1, 1)$.

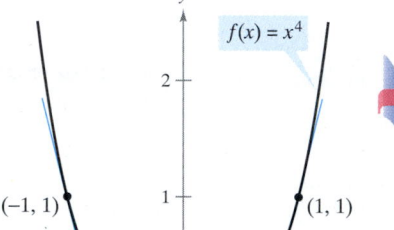

The slope of a graph at a point is the value of the derivative at that point.
Figure 2.16

EXAMPLE 4   Finding an Equation of a Tangent Line

Find an equation of the tangent line to the graph of $f(x) = x^2$ when $x = -2$.

Solution   To find the *point* on the graph of f, evaluate the original function at $x = -2$.

$$(-2, f(-2)) = (-2, 4) \qquad \text{Point on graph}$$

To find the *slope* of the graph when $x = -2$, evaluate the derivative, $f'(x) = 2x$, at $x = -2$.

$$m = f'(-2) = -4 \qquad \text{Slope of graph at } (-2, 4)$$

Now, using the point-slope form of the equation of a line, you can write

$$y - y_1 = m(x - x_1) \qquad \text{Point-slope form}$$
$$y - 4 = -4[x - (-2)] \qquad \text{Substitute for } y_1, m, \text{ and } x_1.$$
$$y = -4x - 4. \qquad \text{Simplify.}$$

(See Figure 2.17.)

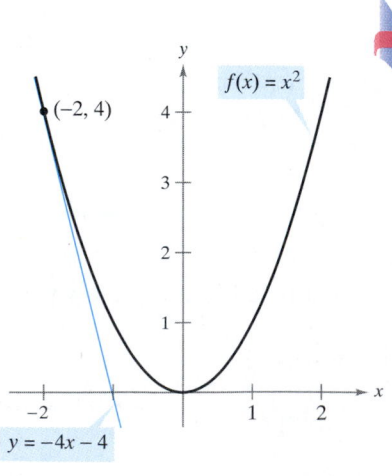

The line $y = -4x - 4$ is tangent to the graph of $f(x) = x^2$ at the point $(-2, 4)$.
Figure 2.17

The Constant Multiple Rule

> **THEOREM 2.4   The Constant Multiple Rule**
>
> If f is a differentiable function and c is a real number, then cf is also differentiable and
>
> $$\frac{d}{dx}[cf(x)] = cf'(x).$$

Proof

$$\frac{d}{dx}[cf(x)] = \lim_{\Delta x \to 0} \frac{cf(x + \Delta x) - cf(x)}{\Delta x}$$

$$= \lim_{\Delta x \to 0} c\left[\frac{f(x + \Delta x) - f(x)}{\Delta x}\right]$$

$$= c\left[\lim_{\Delta x \to 0} \frac{f(x + \Delta x) - f(x)}{\Delta x}\right]$$

$$= cf'(x)$$

Informally, the Constant Multiple Rule states that constants can be factored out of the differentiation process, even if the constants appear in the denominator.

$$\frac{d}{dx}[cf(x)] = c\frac{d}{dx}[f(x)] = cf'(x)$$

$$\frac{d}{dx}\left[\frac{f(x)}{c}\right] = \frac{d}{dx}\left[\left(\frac{1}{c}\right)f(x)\right] = \left(\frac{1}{c}\right)\frac{d}{dx}[f(x)] = \left(\frac{1}{c}\right)f'(x)$$

EXAMPLE 5   Using the Constant Multiple Rule

Function	*Derivative*
a. $y = \dfrac{2}{x}$	$\dfrac{dy}{dx} = \dfrac{d}{dx}[2x^{-1}] = 2\dfrac{d}{dx}[x^{-1}] = 2(-1)x^{-2} = -\dfrac{2}{x^2}$
b. $f(t) = \dfrac{4t^2}{5}$	$f'(t) = \dfrac{d}{dt}\left[\dfrac{4}{5}t^2\right] = \dfrac{4}{5}\dfrac{d}{dt}[t^2] = \dfrac{4}{5}(2t) = \dfrac{8}{5}t$
c. $y = 2\sqrt{x}$	$\dfrac{dy}{dx} = \dfrac{d}{dx}[2x^{1/2}] = 2\left(\dfrac{1}{2}x^{-1/2}\right) = x^{-1/2} = \dfrac{1}{\sqrt{x}}$
d. $y = \dfrac{1}{2\sqrt[3]{x^2}}$	$\dfrac{dy}{dx} = \dfrac{d}{dx}\left[\dfrac{1}{2}x^{-2/3}\right] = \dfrac{1}{2}\left(-\dfrac{2}{3}\right)x^{-5/3} = -\dfrac{1}{3x^{5/3}}$
e. $y = -\dfrac{3x}{2}$	$y' = \dfrac{d}{dx}\left[-\dfrac{3}{2}x\right] = -\dfrac{3}{2}(1) = -\dfrac{3}{2}$

NOTE   The Constant Multiple Rule and the Power Rule can be combined into one rule. The combination rule is $D_x[cx^n] = cnx^{n-1}$.

EXAMPLE 6 **Using Parentheses When Differentiating**

Original Function	Rewrite	Differentiate	Simplify
a. $y = \dfrac{5}{2x^3}$	$y = \dfrac{5}{2}(x^{-3})$	$y' = \dfrac{5}{2}(-3x^{-4})$	$y' = -\dfrac{15}{2x^4}$
b. $y = \dfrac{5}{(2x)^3}$	$y = \dfrac{5}{8}(x^{-3})$	$y' = \dfrac{5}{8}(-3x^{-4})$	$y' = -\dfrac{15}{8x^4}$
c. $y = \dfrac{7}{3x^{-2}}$	$y = \dfrac{7}{3}(x^2)$	$y' = \dfrac{7}{3}(2x)$	$y' = \dfrac{14x}{3}$
d. $y = \dfrac{7}{(3x)^{-2}}$	$y = 63(x^2)$	$y' = 63(2x)$	$y' = 126x$

The Sum and Difference Rules

THEOREM 2.5 **The Sum and Difference Rules**

The derivative of the sum (or difference) of two differentiable functions is differentiable and is the sum (or difference) of their derivatives.

$$\frac{d}{dx}[f(x) + g(x)] = f'(x) + g'(x) \qquad \text{Sum Rule}$$

$$\frac{d}{dx}[f(x) - g(x)] = f'(x) - g'(x) \qquad \text{Difference Rule}$$

Proof A proof of the Sum Rule follows from Theorem 1.2. (The Difference Rule can be proved in a similar way.)

$$\frac{d}{dx}[f(x) + g(x)] = \lim_{\Delta x \to 0} \frac{[f(x + \Delta x) + g(x + \Delta x)] - [f(x) + g(x)]}{\Delta x}$$

$$= \lim_{\Delta x \to 0} \frac{f(x + \Delta x) + g(x + \Delta x) - f(x) - g(x)}{\Delta x}$$

$$= \lim_{\Delta x \to 0} \left[\frac{f(x + \Delta x) - f(x)}{\Delta x} + \frac{g(x + \Delta x) - g(x)}{\Delta x} \right]$$

$$= \lim_{\Delta x \to 0} \frac{f(x + \Delta x) - f(x)}{\Delta x} + \lim_{\Delta x \to 0} \frac{g(x + \Delta x) - g(x)}{\Delta x}$$

$$= f'(x) + g'(x)$$

The Sum and Difference Rules can be extended to any finite number of functions. For instance, if $F(x) = f(x) + g(x) - h(x) - k(x)$, then $F'(x) = f'(x) + g'(x) - h'(x) - k'(x)$.

EXAMPLE 7 **Using the Sum and Difference Rules**

Function	Derivative
a. $f(x) = x^3 - 4x + 5$	$f'(x) = 3x^2 - 4$
b. $g(x) = -\dfrac{x^4}{2} + 3x^3 - 2x$	$g'(x) = -2x^3 + 9x^2 - 2$

FOR FURTHER INFORMATION For the outline of a geometric proof of the derivatives of the sine and cosine functions, see the article "The Spider's Spacewalk Derivation of sin′ and cos′" by Tim Hesterberg in the March 1995 issue of *The College Mathematics Journal.*

Derivatives of Sine and Cosine Functions

In Section 1.3, you studied the following limits.

$$\lim_{\Delta x \to 0} \frac{\sin \Delta x}{\Delta x} = 1 \quad \text{and} \quad \lim_{\Delta x \to 0} \frac{1 - \cos \Delta x}{\Delta x} = 0$$

These two limits can be used to prove differentiation rules for the sine and cosine functions. (The derivatives of the other four trigonometric functions are discussed in Section 2.3.)

THEOREM 2.6   Derivatives of Sine and Cosine Functions

$$\frac{d}{dx}[\sin x] = \cos x \qquad \frac{d}{dx}[\cos x] = -\sin x$$

Proof

$$\frac{d}{dx}[\sin x] = \lim_{\Delta x \to 0} \frac{\sin(x + \Delta x) - \sin x}{\Delta x}$$

$$= \lim_{\Delta x \to 0} \frac{\sin x \cos \Delta x + \cos x \sin \Delta x - \sin x}{\Delta x}$$

$$= \lim_{\Delta x \to 0} \frac{\cos x \sin \Delta x - (\sin x)(1 - \cos \Delta x)}{\Delta x}$$

$$= \lim_{\Delta x \to 0} \left[(\cos x)\left(\frac{\sin \Delta x}{\Delta x}\right) - (\sin x)\left(\frac{1 - \cos \Delta x}{\Delta x}\right) \right]$$

$$= \cos x \left(\lim_{\Delta x \to 0} \frac{\sin \Delta x}{\Delta x} \right) - \sin x \left(\lim_{\Delta x \to 0} \frac{1 - \cos \Delta x}{\Delta x} \right)$$

$$= (\cos x)(1) - (\sin x)(0)$$

$$= \cos x$$

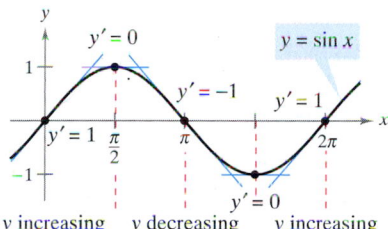

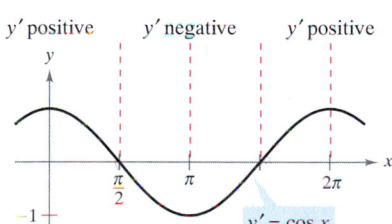

The derivative of the sine function is the cosine function.
Figure 2.18

This differentiation rule is shown graphically in Figure 2.18. Note that for each x, the *slope* of the sine curve is equal to the *value* of the cosine. The proof of the second rule is left as an exercise (see Exercise 93). ▬▬▬

  EXAMPLE 8   Derivatives Involving Sines and Cosines

Function	*Derivative*
a. $y = 2 \sin x$	$y' = 2 \cos x$
b. $y = \dfrac{\sin x}{2} = \dfrac{1}{2} \sin x$	$y' = \dfrac{1}{2} \cos x = \dfrac{\cos x}{2}$
c. $y = x + \cos x$	$y' = 1 - \sin x$

$$\frac{d}{dx}[a \sin x] = a \cos x$$

Figure 2.19

TECHNOLOGY A graphing utility can provide insight into the interpretation of a derivative. For instance, Figure 2.19 shows the graphs of

$$y = a \sin x$$

for $a = \frac{1}{2}$, 1, $\frac{3}{2}$, and 2. Estimate the slope of each graph at the point $(0, 0)$. Then verify your estimates analytically by evaluating the derivative of each function when $x = 0$.

Rates of Change

You have seen how the derivative is used to determine slope. The derivative can also be used to determine the rate of change of one variable with respect to another. Applications involving rates of change occur in a wide variety of fields. A few examples are population growth rates, production rates, water flow rates, velocity, and acceleration.

A common use of rate of change is to describe the motion of an object moving in a straight line. In such problems, it is customary to use either a horizontal or a vertical line with a designated origin to represent the line of motion. On such lines, movement to the right (or upward) is considered to be in the positive direction, and movement to the left (or downward) is considered to be in the negative direction.

The function s that gives the position (relative to the origin) of an object as a function of time t is called a **position function.** If, over a period of time Δt, the object changes its position by the amount $\Delta s = s(t + \Delta t) - s(t)$, then, by the familiar formula

$$\text{Rate} = \frac{\text{distance}}{\text{time}}$$

the **average velocity** is

$$\frac{\text{Change in distance}}{\text{Change in time}} = \frac{\Delta s}{\Delta t}. \qquad \text{Average velocity}$$

EXAMPLE 9   Finding Average Velocity of a Falling Object

If a billiard ball is dropped from a height of 100 feet, its height s at time t is given by the position function

$$s = -16t^2 + 100 \qquad\qquad \text{Position function}$$

where s is measured in feet and t is measured in seconds. Find the average velocity over each of the following time intervals.

a. $[1, 2]$ **b.** $[1, 1.5]$ **c.** $[1, 1.1]$

Solution

a. For the interval $[1, 2]$ the object falls from a height of $s(1) = -16(1)^2 + 100 = 84$ feet to a height of $s(2) = -16(2)^2 + 100 = 36$ feet. The average velocity is

$$\frac{\Delta s}{\Delta t} = \frac{36 - 84}{2 - 1} = \frac{-48}{1} = -48 \text{ feet per second.}$$

b. For the interval $[1, 1.5]$, the object falls from a height of 84 feet to a height of 64 feet. The average velocity is

$$\frac{\Delta s}{\Delta t} = \frac{64 - 84}{1.5 - 1} = \frac{-20}{0.5} = -40 \text{ feet per second.}$$

c. For the interval $[1, 1.1]$, the object falls from a height of 84 feet to a height of 80.64 feet. The average velocity is

$$\frac{\Delta s}{\Delta t} = \frac{80.64 - 84}{1.1 - 1} = \frac{-3.36}{0.1} = -33.6 \text{ feet per second.}$$

Note that the average velocities are *negative*, indicating that the object is moving downward.

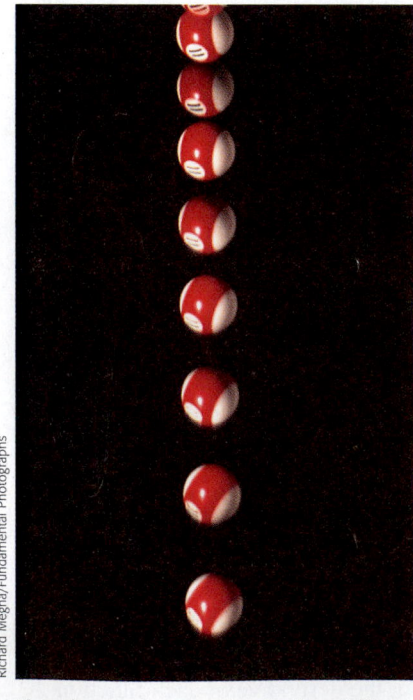

Time-lapse photograph of a free-falling billiard ball.

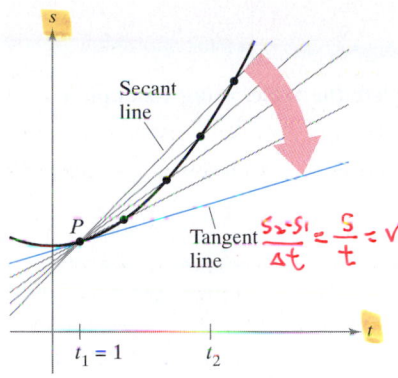

The average velocity between t_1 and t_2 is the slope of the secant line, and the instantaneous velocity at t_1 is the slope of the tangent line.

Figure 2.20

Suppose that in Example 9 you wanted to find the *instantaneous* velocity (or simply the velocity) of the object when $t = 1$. Just as you can approximate the slope of the tangent line by calculating the slope of the secant line, you can approximate the velocity at $t = 1$ by calculating the average velocity over a small interval $[1, 1 + \Delta t]$ (see Figure 2.20). By taking the limit as Δt approaches zero, you obtain the velocity when $t = 1$. Try doing this—you will find that the velocity when $t = 1$ is -32 feet per second.

In general, if $s = s(t)$ is the position function for an object moving along a straight line, the **velocity** of the object at time t is

$$v(t) = \lim_{\Delta t \to 0} \frac{s(t + \Delta t) - s(t)}{\Delta t} = s'(t).$$ Velocity function

In other words, the velocity function is the derivative of the position function. (Velocity can be negative, zero, or positive. The **speed** of an object is the absolute value of its velocity. Speed cannot be negative.)

The position of a free-falling object (neglecting air resistance) under the influence of gravity can be represented by the equation

$$s(t) = \frac{1}{2} g t^2 + v_0 t + s_0$$ Position function

where s_0 is the initial height of the object, v_0 is the initial velocity of the object, and g is the acceleration due to gravity. On earth, the value of g is approximately -32 feet per second per second or -9.8 meters per second per second.

EXAMPLE 10   Using the Derivative to Find Velocity

At time $t = 0$, a diver jumps from a diving board that is 32 feet above the water (see Figure 2.21). The position of the diver is given by

$$s(t) = -16t^2 + 16t + 32$$

where s is measured in feet and t is measured in seconds.

a. When does the diver hit the water?

b. What is the diver's velocity at impact?

Solution

a. To find when the diver hits the water, let $s = 0$ and solve for t.

$$-16t^2 + 16t + 32 = 0$$
$$-16(t + 1)(t - 2) = 0$$
$$t = -1 \text{ or } 2$$

Because $t \geq 0$, choose the positive value to conclude that the diver hits the water at $t = 2$ seconds.

b. The velocity at time t is given by the derivative $s'(t) = -32t + 16$. Therefore, the velocity at time $t = 2$ is

$$s'(2) = -32(2) + 16 = -48 \text{ feet per second.}$$

Velocity is positive when an object is rising, and is negative when an object is falling.

Figure 2.21

NOTE   In Figure 2.21, note that the diver moves upward for the first half-second because the velocity is positive for $0 < t < \frac{1}{2}$. When the velocity is 0, the diver has reached the maximum height of the dive.

EXERCISES FOR SECTION 2.2

In Exercises 1 and 2, find the slope of the tangent line to $y = x^n$ at the point $(1, 1)$.

1. (a) $y = x^{1/2}$

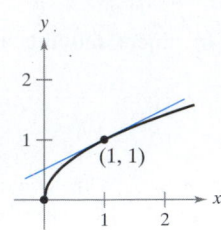

(b) $y = x^{3/2}$

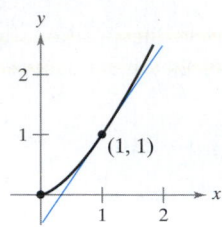

(c) $y = x^2$

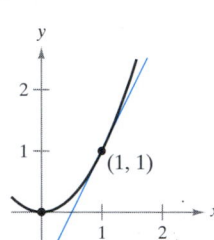

(d) $y = x^3$

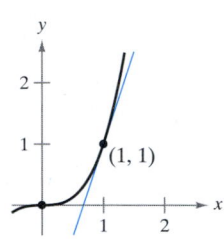

2. (a) $y = x^{-1/2}$

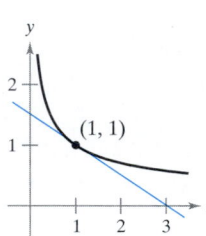

(b) $y = x^{-1}$

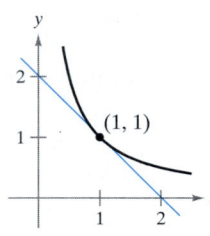

(c) $y = x^{-3/2}$

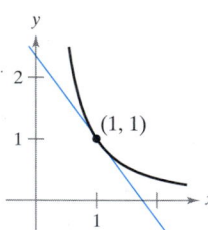

(d) $y = x^{-2}$

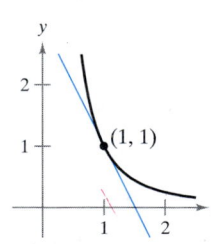

In Exercises 3–16, find the derivative of the function.

3. $y = 3$

4. $f(x) = -2$

5. $f(x) = x + 1$

6. $g(x) = 3x - 1$

7. $g(x) = x^2 + 4$

8. $y = t^2 + 2t - 3$

9. $f(t) = -2t^2 + 3t - 6$

10. $y = x^3 - 9$

11. $s(t) = t^3 - 2t + 4$

12. $f(x) = 2x^3 - x^2 + 3x$

13. $y = x^2 - \frac{1}{2}\cos x$

14. $y = 5 + \sin x$

15. $y = \frac{1}{x} - 3\sin x$

16. $g(t) = \pi \cos t$

In Exercises 17–22, complete the table, using Example 6 as a model.

Original Function	Rewrite	Differentiate	Simplify

17. $y = \dfrac{1}{3x^3}$

18. $y = \dfrac{2}{3x^2}$

19. $y = \dfrac{1}{(3x)^3}$

20. $y = \dfrac{\pi}{(3x)^2}$

21. $y = \dfrac{\sqrt{x}}{x}$

22. $y = \dfrac{4}{x^{-3}}$

In Exercises 23–30, find the value of the derivative of the function at the indicated point. Use the derivative feature of a graphing utility to confirm your results.

Function	Point
23. $f(x) = \dfrac{1}{x}$	$(1, 1)$
24. $f(t) = 3 - \dfrac{3}{5t}$	$\left(\frac{3}{5}, 2\right)$
25. $f(x) = -\frac{1}{2} + \frac{7}{5}x^3$	$\left(0, -\frac{1}{2}\right)$
26. $y = 3x\left(x^2 - \dfrac{2}{x}\right)$	$(2, 18)$
27. $y = (2x + 1)^2$	$(0, 1)$
28. $f(x) = 3(5 - x)^2$	$(5, 0)$
29. $f(\theta) = 4\sin\theta - \theta$	$(0, 0)$
30. $g(t) = 2 + 3\cos t$	$(\pi, -1)$

In Exercises 31–42, find the derivative of the function.

31. $f(x) = x^3 - 3x - 2x^{-4}$

32. $f(x) = x^2 - 3x - 3x^{-2}$

33. $g(t) = t^2 - \dfrac{4}{t}$

34. $f(x) = x + \dfrac{1}{x^2}$

35. $f(x) = \dfrac{x^3 - 3x^2 + 4}{x^2}$

36. $h(x) = \dfrac{2x^2 - 3x + 1}{x}$

37. $y = x(x^2 + 1)$

38. $f(x) = \sqrt[3]{x} + \sqrt[5]{x}$

39. $h(s) = s^{4/5}$

40. $f(t) = t^{1/3} - 1$

41. $f(x) = 4\sqrt{x} + 3\cos x$

42. $f(x) = 2\sin x + 3\cos x$

In Exercises 43–46, (a) find an equation of the tangent line to the graph of f at the indicated point, (b) use a graphing utility to graph the function and its tangent line at the point, and (c) use the derivative feature of a graphing utility to confirm your results.

Function	Point
43. $y = x^4 - 3x^2 + 2$	$(1, 0)$
44. $y = x^3 + x$	$(-1, -2)$
45. $f(x) = \dfrac{1}{\sqrt[3]{x^2}}$	$(8, \frac{1}{4})$
46. $y = (x^2 + 2x)(x + 1)$	$(1, 6)$

In Exercises 47–52, determine the point(s) (if any) at which the function has a horizontal tangent line.

47. $y = x^4 - 8x^2 + 2$ 48. $y = x^3 + x$

49. $y = \dfrac{1}{x^2}$ 50. $y = x^2 + 1$

51. $y = x + \sin x,\ \ 0 \le x < 2\pi$

52. $y = \sqrt{3}x + 2\cos x,\ \ 0 \le x < 2\pi$

Writing   In Exercises 53 and 54, the graphs of a function f and its derivative f' are given on the same set of coordinate axes. Label the graphs as f or f' and write a short paragraph stating the criteria used in making the selection.

53. 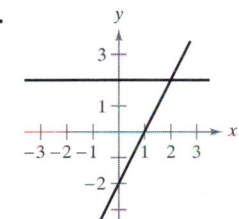   54.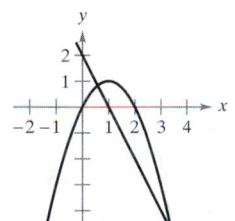

55. Sketch the graphs of the two equations $y = x^2$ and $y = -x^2 + 6x - 5$, and sketch the two lines that are tangent to both graphs. Find the equations of these lines.

56. Show that the graphs of the two equations $y = x$ and $y = 1/x$ have tangent lines that are perpendicular to each other at their point of intersection.

In Exercises 57 and 58, find an equation of the tangent line to the graph of the function f through the point (x_0, y_0) not on the graph. To find the point of tangency (x, y) on the graph of f, solve the equation

$$f'(x) = \frac{y_0 - y}{x_0 - x}.$$

57. $f(x) = \sqrt{x}$ 58. $f(x) = \dfrac{2}{x}$

$(x_0, y_0) = (-4, 0)$ $(x_0, y_0) = (5, 0)$

59. *Linear Approximation*   Consider the function $f(x) = x^{3/2}$ with the solution point $(4, 8)$.

(a) Use a graphing utility to obtain the graph of f. Use the zoom feature to obtain successive magnifications of the graph in the neighborhood of the point $(4, 8)$. After zooming in a few times, the graph should appear nearly linear. Use the trace feature to determine the coordinates of a point "near" $(4, 8)$. Find an equation of the secant line $S(x)$ through the two points.

(b) Find the equation of the line

$$T(x) = f'(4)(x - 4) + f(4)$$

tangent to the graph of f passing through the given point. Why are the linear functions S and T nearly the same?

(c) Use a graphing utility to graph f and T on the same set of coordinate axes. Note that T is a "good" approximation of f when x is "close to" 4. What happens to the accuracy of the approximation as you move farther away from the point of tangency?

(d) Demonstrate the conclusion in part (c) by completing the table.

Δx	-3	-2	-1	-0.5	-0.1	0
$f(x)$						
$T(x)$						

Δx	0.1	0.5	1	2	3
$f(x)$					
$T(x)$					

60. *Linear Approximation*   Repeat Exercise 59 for the function $f(x) = x^3$ where $T(x)$ is the line tangent to the graph at the point $(1, 1)$. Explain why the accuracy of the linear approximation decreases more rapidly than in Exercise 59.

True or False?   In Exercises 61–64, determine whether the statement is true or false. If it is false, explain why or give an example that shows it is false.

61. If $f'(x) = g'(x)$, then $f(x) = g(x)$.

62. If $f(x) = g(x) + c$, then $f'(x) = g'(x)$.

63. If $y = \pi^2$, then $dy/dx = 2\pi$.

64. If $y = x/\pi$, then $dy/dx = 1/\pi$.

In Exercises 65–68, find the average rate of change of the function over the indicated interval. Compare this average rate of change with the instantaneous rates of change at the endpoints of the interval.

Function	Interval
65. $f(t) = 2t + 7$	$[1, 2]$
66. $f(t) = t^2 - 3$	$[2, 2.1]$

Function	Interval
67. $f(x) = \dfrac{-1}{x}$	$[1, 2]$
68. $f(x) = \sin x$	$\left[0, \dfrac{\pi}{6}\right]$

69. ***Think About It***   Use the graph of f to answer each question.

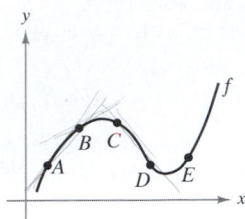

(a) Between which two consecutive points is the average rate of change of the function greatest?

(b) Is the average rate of change of the function between A and B greater than or less than the instantaneous rate of change at B?

(c) Sketch a tangent line to the graph between the points C and D such that the slope of the tangent line is the same as the average rate of change of the function between C and D.

(d) Give any sets of consecutive points for which the average rates of change of the function are approximately equal.

70. ***Think About It***   Sketch the graph of a function f such that $f' > 0$ for all x and the rate of change of the function is decreasing.

Vertical Motion   **In Exercises 71 and 72, use the position function $s(t) = -16t^2 + v_0 t + s_0$ for free-falling objects.**

71. A silver dollar is dropped from the top of the World Trade Center, which is 1362 feet tall.

(a) Determine the position and velocity functions for the coin.

(b) Determine the average velocity on the interval $[1, 2]$.

(c) Find the instantaneous velocities when $t = 1$ and $t = 2$.

(d) Find the time required for the coin to reach ground level.

(e) Find the velocity of the dollar just before it hits the ground.

72. A ball is thrown straight down from the top of a 220-foot building with an initial velocity of -22 feet per second. What is its velocity after 3 seconds? What is its velocity after falling 108 feet?

Vertical Motion   **In Exercises 73 and 74, use the position function $s(t) = -4.9t^2 + v_0 t + s_0$ for free-falling objects.**

73. A projectile is shot upward from the surface of the earth with an initial velocity of 120 meters per second. What is its velocity after 5 seconds? After 10 seconds?

74. To estimate the height of a building, a stone is dropped from the top of the building into a pool of water at ground level. How high is the building if the splash is seen 6.8 seconds after the stone is dropped?

Think About It   **In Exercises 75 and 76, the graph of a position function is given. It represents the distance in miles that a person drives during a 10-minute trip to work. Make a sketch of the corresponding velocity function.**

75. 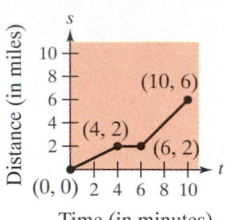   **76.**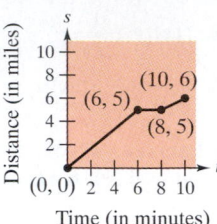

Think About It   **In Exercises 77 and 78, the graph of a velocity function is given. It represents the velocity in miles per hour during a 10-minute drive to work. Make a sketch of the corresponding position function.**

77. 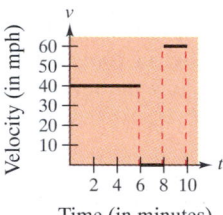   **78.**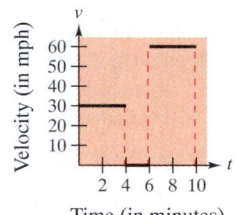

79. ***Modeling Data***   The stopping distance of an automobile traveling at a speed v (kilometers per hour) is the distance R (meters) the car travels during the reaction time of the driver plus the distance B (meters) the car travels after the brakes are applied. The table shows the results of an experiment.

v	20	40	60	80	100
R	3.3	6.7	10.0	13.3	16.7
B	2.3	8.9	20.2	35.9	56.7

(a) Use the regression capabilities of a graphing utility to find a linear model for reaction time.

(b) Use the regression capabilities of a graphing utility to find a quadratic model for braking time.

(c) Determine the polynomial giving the total stopping distance T.

(d) Use a graphing utility to graph the functions R, B, and T in the same viewing rectangle.

(e) Find the derivative of T and the rate of change of the total stopping distance for $v = 40$, $v = 80$, and $v = 100$.

(f) Use the results of this exercise to draw conclusions about the total stopping distance as speed increases.

80. *Velocity* Verify that the average velocity over the time interval $[t_0 - \Delta t, t_0 + \Delta t]$ is the same as the instantaneous velocity at $t = t_0$ for the position function.

$$s(t) = -\tfrac{1}{2}at^2 + c.$$

81. *Area* The area of a square with sides of length s is given by $A = s^2$. Find the rate of change of the area with respect to s when $s = 4$ meters.

82. *Volume* The volume of a cube with sides of length s is given by $V = s^3$. Find the rate of change of the volume with respect to s when $s = 4$ centimeters.

83. *Inventory Management* The annual inventory cost C for a certain manufacturer is

$$C = \frac{1{,}008{,}000}{Q} + 6.3Q$$

where Q is the order size when the inventory is replenished. Find the change in annual cost when Q is increased from 350 to 351, and compare this with the instantaneous rate of change when $Q = 350$.

84. *Fuel Cost* A car is driven 15,000 miles a year and gets x miles per gallon. Assume that the average fuel cost is $1.25 per gallon. Find the annual cost of fuel C as a function of x and use this function to complete the table.

x	10	15	20	25	30	35	40
C							
$\dfrac{dC}{dx}$							

Who would benefit more from a 1-mile-per-gallon increase in fuel efficiency—the driver of a car that gets 15 miles per gallon or the driver of a car that gets 35 miles per gallon? Explain.

85. *Writing* The photo at the bottom of the page was taken at game 4 of the 1996 World Series between the New York Yankees and the Atlanta Braves. Do you think the runner was safe or out? Write a detailed explanation of your reasoning.

86. *Newton's Law of Cooling* This law states that the rate of change of the temperature of an object is proportional to the difference between the object's temperature T and the temperature T_a of the surrounding medium. Write an equation for this law.

87. Find an equation of the parabola $y = ax^2 + bx + c$ that passes through $(0, 1)$ and is tangent to the line $y = x - 1$ at $(1, 0)$.

88. Let (a, b) be an arbitrary point on the graph of $y = 1/x$, $x > 0$. Prove that the area of the triangle formed by the tangent line through (a, b) and the coordinate axes is 2.

89. Find the tangent line(s) to the curve $y = x^3 - 9x$ through the point $(1, -9)$.

90. Find the equation(s) of the tangent line(s) to the parabola $y = x^2$ through the given point.

(a) $(0, a)$ (b) $(a, 0)$

Are there any restrictions on the constant a?

91. Find a and b such that

$$f(x) = \begin{cases} ax^3, & x \le 2 \\ x^2 + b, & x > 2 \end{cases}$$

is differentiable everywhere.

92. Where are the functions $f_1(x) = |\sin x|$ and $f_2(x) = \sin |x|$ differentiable?

93. Prove that $\dfrac{d}{dx}[\cos x] = -\sin x$.

FOR FURTHER INFORMATION For a geometric interpretation of the derivatives of trigonometric functions, see the article "Sines and Cosines of the Times" by Victor J. Katz in the April 1995 issue of *Math Horizons*.

David Burnett/Contact Press Images, Inc.

The Product Rule • The Quotient Rule • Derivatives of Trigonometric Functions • Higher-Order Derivatives

The Product Rule

In Section 2.2 you learned that the derivative of the sum of two functions is simply the sum of their derivatives. The rules for the derivatives of the product and quotient of two functions are not as simple.

THEOREM 2.7   The Product Rule

The derivative of the product of two differentiable functions f and g is itself differentiable. Moreover, the derivative of fg is the first function times the derivative of the second, plus the second function times the derivative of the first.

$$\frac{d}{dx}[f(x)g(x)] = f(x)g'(x) + g(x)f'(x)$$

Proof   Some mathematical proofs, such as the proof of the Sum Rule, are straightforward. Others involve clever steps that may appear unmotivated to a reader. This proof involves such a step—subtracting and adding the same quantity—which is shown in color.

$$\frac{d}{dx}[f(x)g(x)] = \lim_{\Delta x \to 0} \frac{f(x + \Delta x)g(x + \Delta x) - f(x)g(x)}{\Delta x}$$

$$= \lim_{\Delta x \to 0} \frac{f(x + \Delta x)g(x + \Delta x) - f(x + \Delta x)g(x) + f(x + \Delta x)g(x) - f(x)g(x)}{\Delta x}$$

$$= \lim_{\Delta x \to 0} \left[f(x + \Delta x)\frac{g(x + \Delta x) - g(x)}{\Delta x} + g(x)\frac{f(x + \Delta x) - f(x)}{\Delta x} \right]$$

$$= \lim_{\Delta x \to 0} \left[f(x + \Delta x)\frac{g(x + \Delta x) - g(x)}{\Delta x} \right] + \lim_{\Delta x \to 0} \left[g(x)\frac{f(x + \Delta x) - f(x)}{\Delta x} \right]$$

$$= \lim_{\Delta x \to 0} f(x + \Delta x) \cdot \lim_{\Delta x \to 0} \frac{g(x + \Delta x) - g(x)}{\Delta x} + \lim_{\Delta x \to 0} g(x) \cdot \lim_{\Delta x \to 0} \frac{f(x + \Delta x) - f(x)}{\Delta x}$$

$$= f(x)g'(x) + g(x)f'(x)$$

The Product Rule can be extended to cover products involving more than two factors. For example, if f, g, and h are differentiable functions of x, then

$$\frac{d}{dx}[f(x)g(x)h(x)] = f'(x)g(x)h(x) + f(x)g'(x)h(x) + f(x)g(x)h'(x).$$

For instance, the derivative of $y = x^2 \sin x \cos x$ is

$$\frac{dy}{dx} = (2x \sin x \cos x) + (x^2 \cos x \cos x) + (x^2 \sin x (-\sin x))$$

$$= 2x \sin x \cos x + x^2(\cos^2 x - \sin^2 x).$$

The derivative of a product of two functions is not (in general) given by the product of the derivatives of the two functions. To see this, try comparing the product of the derivatives of $f(x) = 3x - 2x^2$ and $g(x) = 5 + 4x$ with the derivative in Example 1.

THE PRODUCT RULE

When Liebniz originally wrote a formula for the Product Rule, he was motivated by the expression

$(x + dx)(y + dy) - xy$

from which he subtracted $dx\, dy$ (as being negligible) and obtained the differential form $x\, dy + y\, dx$. This derivation has resulted in the traditional form of the Product Rule, as stated on page 114. (Source: *The History of Mathematics*, David M. Burton)

A version that some people prefer is

$\dfrac{d}{dx}[f(x)g(x)] = f'(x)g(x) + f(x)g'(x).$

The advantage of this form is that it generalizes easily to products involving 3 or more factors.

EXAMPLE 1   Using the Product Rule

Find the derivative of $h(x) = (3x - 2x^2)(5 + 4x)$.

Solution

$$h'(x) = \overbrace{(3x - 2x^2)}^{\text{First}} \overbrace{\frac{d}{dx}[5 + 4x]}^{\substack{\text{Derivative} \\ \text{of second}}} + \overbrace{(5 + 4x)}^{\text{Second}} \overbrace{\frac{d}{dx}[3x - 2x^2]}^{\substack{\text{Derivative} \\ \text{of first}}}$$

$$= (3x - 2x^2)(4) + (5 + 4x)(3 - 4x)$$

$$= (12x - 8x^2) + (15 - 8x - 16x^2)$$

$$= -24x^2 + 4x + 15$$

In Example 1, you have the option of finding the derivative with or without the Product Rule. To find the derivative without the Product Rule, you can write

$$D_x[(3x - 2x^2)(5 + 4x)] = D_x[-8x^3 + 2x^2 + 15x]$$

$$= -24x^2 + 4x + 15.$$

In the next example, you must use the Product Rule.

EXAMPLE 2   Using the Product Rule

$$\frac{d}{dx}[x \sin x] = x\frac{d}{dx}[\sin x] + \sin x \frac{d}{dx}[x]$$

$$= x \cos x + (\sin x)(1)$$

$$= x \cos x + \sin x$$

EXAMPLE 3   Using the Product Rule

Find the derivative of $y = 2x \cos x - 2 \sin x.$

Solution

$$\frac{dy}{dx} = \overbrace{(2x)\left(\frac{d}{dx}[\cos x]\right) + (\cos x)\left(\frac{d}{dx}[2x]\right)}^{\text{Product Rule}} - \overbrace{2\frac{d}{dx}[\sin x]}^{\text{Constant Multiple Rule}}$$

$$= (2x)(-\sin x) + (\cos x)(2) - 2(\cos x)$$

$$= -2x \sin x$$

NOTE   In Example 3, notice that you use the Product Rule when both factors of the product are variable, and you use the Constant Multiple Rule when one of the factors is a constant.

The Quotient Rule

THEOREM 2.8   The Quotient Rule

The derivative of the quotient f/g of two differentiable functions f and g is itself differentiable at all values of x for which $g(x) \neq 0$. Moreover, the derivative of f/g is given by the denominator times the derivative of the numerator minus the numerator times the derivative of the denominator, all divided by the square of the denominator.

$$\frac{d}{dx}\left[\frac{f(x)}{g(x)}\right] = \frac{g(x)f'(x) - f(x)g'(x)}{[g(x)]^2}, \qquad g(x) \neq 0$$

Proof   As with the proof of Theorem 2.7, the key to this proof is subtracting and adding the same quantity.

$$\frac{d}{dx}\left[\frac{f(x)}{g(x)}\right] = \lim_{\Delta x \to 0} \frac{\dfrac{f(x + \Delta x)}{g(x + \Delta x)} - \dfrac{f(x)}{g(x)}}{\Delta x}$$

$$= \lim_{\Delta x \to 0} \frac{g(x)f(x + \Delta x) - f(x)g(x + \Delta x)}{\Delta x g(x)g(x + \Delta x)}$$

$$= \lim_{\Delta x \to 0} \frac{g(x)f(x + \Delta x) - f(x)g(x) + f(x)g(x) - f(x)g(x + \Delta x)}{\Delta x g(x)g(x + \Delta x)}$$

$$= \frac{\displaystyle\lim_{\Delta x \to 0} \frac{g(x)[f(x + \Delta x) - f(x)]}{\Delta x} - \lim_{\Delta x \to 0} \frac{f(x)[g(x + \Delta x) - g(x)]}{\Delta x}}{\displaystyle\lim_{\Delta x \to 0} [g(x)g(x + \Delta x)]}$$

$$= \frac{g(x)\left[\displaystyle\lim_{\Delta x \to 0} \frac{f(x + \Delta x) - f(x)}{\Delta x}\right] - f(x)\left[\displaystyle\lim_{\Delta x \to 0} \frac{g(x + \Delta x) - g(x)}{\Delta x}\right]}{\displaystyle\lim_{\Delta x \to 0} [g(x)g(x + \Delta x)]}$$

$$= \frac{g(x)f'(x) - f(x)g'(x)}{[g(x)]^2}$$

TECHNOLOGY   Graphing utilities can be used to compare the graph of a function with the graph of its derivative. For instance, in Figure 2.22, the graph of the function in Example 4 appears to have two points that have horizontal tangent lines. What are the values of y' at these two points?

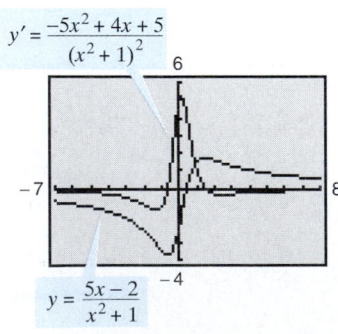

Graphical comparison of a function and its derivative

Figure 2.22

EXAMPLE 4   Using the Quotient Rule

$$\frac{d}{dx}\left[\frac{5x - 2}{x^2 + 1}\right] = \frac{(x^2 + 1)\dfrac{d}{dx}[5x - 2] - (5x - 2)\dfrac{d}{dx}[x^2 + 1]}{(x^2 + 1)^2}$$

$$= \frac{(x^2 + 1)(5) - (5x - 2)(2x)}{(x^2 + 1)^2}$$

$$= \frac{(5x^2 + 5) - (10x^2 - 4x)}{(x^2 + 1)^2}$$

$$= \frac{-5x^2 + 4x + 5}{(x^2 + 1)^2}$$

Note the use of parentheses in Example 4. A liberal use of parentheses is recommended for *all* types of differentiation problems. For instance, with the Quotient Rule, it is a good idea to enclose all factors and derivatives in parentheses, and to pay special attention to the subtraction required in the numerator.

When we introduced differentiation rules in the preceding section, we emphasized the need for rewriting *before* differentiating. The next example illustrates this point with the Quotient Rule.

EXAMPLE 5 Rewriting Before Differentiating

Find the derivative of $y = \dfrac{3 - (1/x)}{x + 5}$.

Solution

$$y = \frac{3 - (1/x)}{x + 5} \qquad \text{Original function}$$

$$= \frac{(3x - 1)/x}{x + 5}$$

$$= \frac{3x - 1}{x(x + 5)}$$

$$= \frac{3x - 1}{x^2 + 5x} \qquad \text{Rewrite.}$$

$$\frac{dy}{dx} = \frac{(x^2 + 5x)(3) - (3x - 1)(2x + 5)}{(x^2 + 5x)^2} \qquad \text{Quotient Rule}$$

$$= \frac{(3x^2 + 15x) - (6x^2 + 13x - 5)}{(x^2 + 5x)^2}$$

$$= \frac{-3x^2 + 2x + 5}{(x^2 + 5x)^2} \qquad \text{Simplify.}$$

Not every quotient needs to be differentiated by the Quotient Rule. For example, each quotient in the next example can be considered as the product of a constant times a function of *x*. In such cases it is more convenient to use the Constant Multiple Rule.

EXAMPLE 6 Using the Constant Multiple Rule

NOTE   To see the benefit of using the Constant Multiple Rule for some quotients, try using the Quotient Rule to differentiate the functions in Example 6—you should obtain the same results, but with more work.

	Original Function	*Rewrite*	*Differentiate*	*Simplify*
a.	$y = \dfrac{x^2 + 3x}{6}$	$y = \dfrac{1}{6}(x^2 + 3x)$	$y' = \dfrac{1}{6}(2x + 3)$	$y' = \dfrac{2x + 3}{6}$
b.	$y = \dfrac{5x^4}{8}$	$y = \dfrac{5}{8}x^4$	$y' = \dfrac{5}{8}(4x^3)$	$y' = \dfrac{5}{2}x^3$
c.	$y = \dfrac{-3(3x - 2x^2)}{7x}$	$y = -\dfrac{3}{7}(3 - 2x)$	$y' = -\dfrac{3}{7}(-2)$	$y' = \dfrac{6}{7}$
d.	$y = \dfrac{9}{5x^2}$	$y = \dfrac{9}{5}(x^{-2})$	$y' = \dfrac{9}{5}(-2x^{-3})$	$y' = -\dfrac{18}{5x^3}$

In Section 2.2, we proved the Power Rule only for the case where the exponent n is a positive integer greater than 1. The next example extends the proof to include negative integer exponents.

EXAMPLE 7   Proof of the Power Rule (Negative Integer Exponents)

If n is a negative integer, there exists a positive integer k such that $n = -k$. Thus, by the Quotient Rule, you can write

$$\frac{d}{dx}[x^n] = \frac{d}{dx}\left[\frac{1}{x^k}\right]$$

$$= \frac{x^k(0) - (1)(kx^{k-1})}{(x^k)^2} \qquad \text{Quotient Rule}$$

$$= \frac{0 - kx^{k-1}}{x^{2k}}$$

$$= -kx^{-k-1}$$

$$= nx^{n-1}. \qquad n = -k$$

Thus, the Power Rule

$$D_x[x^n] = nx^{n-1} \qquad \text{Power Rule}$$

is valid for any integer. (You will be asked to prove that the Power Rule is valid for any rational number in Exercise 59 in Section 2.5.)

Derivatives of Trigonometric Functions

Knowing the derivatives of the sine and cosine functions, you can use the Quotient Rule to find the derivatives of the four remaining trigonometric functions.

> **THEOREM 2.9   Derivatives of Trigonometric Functions**
>
> $$\frac{d}{dx}[\tan x] = \sec^2 x \qquad\qquad \frac{d}{dx}[\cot x] = -\csc^2 x$$
>
> $$\frac{d}{dx}[\sec x] = \sec x \tan x \qquad\qquad \frac{d}{dx}[\csc x] = -\csc x \cot x$$

Proof   Considering $\tan x = (\sin x)/(\cos x)$ and applying the Quotient Rule, you obtain

$$\frac{d}{dx}[\tan x] = \frac{(\cos x)(\cos x) - (\sin x)(-\sin x)}{\cos^2 x}$$

$$= \frac{\cos^2 x + \sin^2 x}{\cos^2 x}$$

$$= \frac{1}{\cos^2 x}$$

$$= \sec^2 x.$$

You are asked to prove the other three parts of the theorem in Exercise 69.

EXAMPLE 8   Differentiating Trigonometric Functions

NOTE   Because of trigonometric identities, the derivative of a trigonometric function can take many forms. This presents a challenge when you are trying to match your answers to those given in the back of the text.

Function	*Derivative*
a. $y = x - \tan x$	$\dfrac{dy}{dx} = 1 - \sec^2 x$
b. $y = x \sec x$	$y' = x(\sec x \tan x) + (\sec x)(1)$
	$\quad = (\sec x)(1 + x \tan x)$

EXAMPLE 9   Different Forms of a Derivative

Differentiate both forms of $y = \dfrac{1 - \cos x}{\sin x} = \csc x - \cot x.$

Solution

First form: $y = \dfrac{1 - \cos x}{\sin x}$

$$y' = \frac{(\sin x)(\sin x) - (1 - \cos x)(\cos x)}{\sin^2 x}$$

$$= \frac{\sin^2 x + \cos^2 x - \cos x}{\sin^2 x}$$

$$= \frac{1 - \cos x}{\sin^2 x}$$

Second form: $y = \csc x - \cot x$

$$y' = -\csc x \cot x + \csc^2 x$$

To show that the two derivatives are equal, you can write

$$\frac{1 - \cos x}{\sin^2 x} = \frac{1}{\sin^2 x} - \left(\frac{1}{\sin x}\right)\left(\frac{\cos x}{\sin x}\right) = \csc^2 x - \csc x \cot x.$$

The following summary shows that much of the work in obtaining a simplified form of a derivative occurs *after* differentiating. Note that two characteristics of a simplified form are the absence of negative exponents and the combining of like terms.

	$f'(x)$ *After Differentiating*	$f'(x)$ *After Simplifying*
Example 1	$(3x - 2x^2)(4) + (5 + 4x)(3 - 4x)$	$-24x^2 + 4x + 15$
Example 3	$(2x)(-\sin x) + (\cos x)(2) - 2(\cos x)$	$-2x \sin x$
Example 4	$\dfrac{(x^2 + 1)(5) - (5x - 2)(2x)}{(x^2 + 1)^2}$	$\dfrac{-5x^2 + 4x + 5}{(x^2 + 1)^2}$
Example 5	$\dfrac{(x^2 + 5x)(3) - (3x - 1)(2x + 5)}{(x^2 + 5x)^2}$	$\dfrac{-3x^2 + 2x + 5}{(x^2 + 5x)^2}$
Example 9	$\dfrac{(\sin x)(\sin x) - (1 - \cos x)(\cos x)}{\sin^2 x}$	$\dfrac{1 - \cos x}{\sin^2 x}$

Higher-Order Derivatives

Just as you can obtain a velocity function by differentiating a position function, you can obtain an **acceleration** function by differentiating a velocity function. Another way of looking at this is that you can obtain an acceleration function by differentiating a position function *twice*.

$$s(t) \qquad \text{Position function}$$
$$v(t) = s'(t) \qquad \text{Velocity function}$$
$$a(t) = v'(t) = s''(t) \qquad \text{Acceleration function}$$

The function given by $a(t)$ is the **second derivative** of $s(t)$ and is denoted by $s''(t)$.

The second derivative is an example of a **higher-order derivative.** You can define derivatives of any positive integer order. For instance, the **third derivative** is the derivative of the second derivative. Higher-order derivatives are denoted as follows.

First derivative:	y',	$f'(x)$,	$\dfrac{dy}{dx}$, $\dfrac{d}{dx}[f(x)]$,	$D_x[y]$
Second derivative:	y'',	$f''(x)$,	$\dfrac{d^2y}{dx^2}$, $\dfrac{d^2}{dx^2}[f(x)]$,	$D_x^{\,2}[y]$
Third derivative:	y''',	$f'''(x)$,	$\dfrac{d^3y}{dx^3}$, $\dfrac{d^3}{dx^3}[f(x)]$,	$D_x^{\,3}[y]$
Fourth derivative:	$y^{(4)}$,	$f^{(4)}(x)$,	$\dfrac{d^4y}{dx^4}$, $\dfrac{d^4}{dx^4}[f(x)]$,	$D_x^{\,4}[y]$
$\vdots$				
nth derivative:	$y^{(n)}$,	$f^{(n)}(x)$,	$\dfrac{d^ny}{dx^n}$, $\dfrac{d^n}{dx^n}[f(x)]$,	$D_x^{\,n}[y]$

THE MOON

The moon's mass is 7.354×10^{22} kilograms, and earth's mass is 5.979×10^{24} kilograms. The moon's radius is 1738 kilometers, and earth's radius is 6371 kilometers. Because the gravitational force on the surface of a planet is directly proportional to its mass and inversely proportional to the square of its radius, the ratio of the gravitational force on earth to the gravitational force on the moon is

$$\frac{(5.979 \times 10^{24})/6371^2}{(7.354 \times 10^{22})/1738^2} \approx 6.05.$$

EXAMPLE 10   Finding the Acceleration Due to Gravity

Because the moon has no atmosphere, a falling object on the moon encounters no air resistance. In 1971, astronaut David Scott demonstrated that a feather and a hammer fall at the same rate on the moon. The position function for each of these falling objects is given by

$$s(t) = -0.81t^2 + 2$$

where $s(t)$ is the height in meters and t is the time in seconds. What is the ratio of earth's gravitational force to the moon's?

Solution   To find the acceleration, differentiate the position function twice.

$$s(t) = -0.81t^2 + 2 \qquad \text{Position function}$$
$$s'(t) = -1.62t \qquad \text{Velocity function}$$
$$s''(t) = -1.62 \qquad \text{Acceleration function}$$

Thus, the acceleration due to gravity on the moon is -1.62 meters per second per second. Because the acceleration due to gravity on earth is -9.8 meters per second per second, the ratio of earth's gravitational force to the moon's is

$$\frac{\text{Earth's gravitational force}}{\text{Moon's gravitational force}} = \frac{-9.8}{-1.62} \approx 6.05.$$

EXERCISES FOR SECTION 2.3

In Exercises 1–6, find $f'(x)$ and $f'(c)$.

Function	Value of c
1. $f(x) = \frac{1}{3}(2x^3 - 4)$	$c = 0$
2. $f(x) = (x^2 - 2x + 1)(x^3 - 1)$	$c = 1$
3. $f(x) = (x^3 - 3x)(2x^2 + 3x + 5)$	$c = 0$
4. $f(x) = \frac{x + 1}{x - 1}$	$c = 2$
5. $f(x) = x \cos x$	$c = \frac{\pi}{4}$
6. $f(x) = \frac{\sin x}{x}$	$c = \frac{\pi}{6}$

In Exercises 7–12, complete the table without using the Quotient Rule (see Example 6).

Function	Rewrite	Differentiate	Simplify
7. $y = \frac{x^2 + 2x}{x}$			
8. $y = \frac{4x^{3/2}}{x}$			
9. $y = \frac{7}{3x^3}$			
10. $y = \frac{4}{5x^2}$			
11. $y = \frac{3x^2 - 5}{7}$			
12. $y = \frac{x^2 - 4}{x + 2}$			

In Exercises 13–26, find the derivative of the algebraic function.

13. $f(x) = \frac{3x - 2}{2x - 3}$

14. $f(x) = \frac{x^3 + 3x + 2}{x^2 - 1}$

15. $f(x) = \frac{3 - 2x - x^2}{x^2 - 1}$

16. $f(x) = x^4\left(1 - \frac{2}{x + 1}\right)$

17. $f(x) = \frac{x + 1}{\sqrt{x}}$

18. $f(x) = \sqrt[3]{x}(\sqrt{x} + 3)$

19. $h(s) = (s^3 - 2)^2$

20. $h(x) = (x^2 - 1)^2$

21. $h(t) = \frac{t + 1}{t^2 + 2t + 2}$

22. $f(x) = \frac{x(x^2 - 1)}{x + 3}$

23. $f(x) = (3x^3 + 4x)(x - 5)(x + 1)$

24. $f(x) = (x^2 - x)(x^2 + 1)(x^2 + x + 1)$

25. $f(x) = \frac{x^2 + c^2}{x^2 - c^2}$, c is a constant

26. $f(x) = \frac{c^2 - x^2}{c^2 + x^2}$, c is a constant

In Exercises 27–42, find the derivative of the trigonometric function.

27. $f(t) = t^2 \sin t$

28. $f(\theta) = (\theta + 1)\cos \theta$

29. $f(t) = \frac{\cos t}{t}$

30. $f(x) = \frac{\sin x}{x}$

31. $f(x) = -x + \tan x$

32. $y = x + \cot x$

33. $g(t) = \sqrt{t} + 4 \sec t$

34. $h(s) = \frac{1}{s} - 10 \csc s$

35. $y = 5x \csc x$

36. $y = \frac{\sec x}{x}$

37. $y = -\csc x - \sin x$

38. $y = x \sin x + \cos x$

39. $y = x^2 \sin x + 2x \cos x$

40. $f(x) = \sin x \cos x$

41. $f(x) = x^2 \tan x$

42. $h(\theta) = 5 \sec \theta + \tan \theta$

In Exercises 43–46, use a symbolic differentiation utility to differentiate the function.

43. $g(x) = \left(\frac{x + 1}{x + 2}\right)(2x - 5)$

44. $f(x) = \left(\frac{x^2 - x - 3}{x^2 + 1}\right)(x^2 + x + 1)$

45. $g(\theta) = \frac{\theta}{1 - \sin \theta}$

46. $f(\theta) = \frac{\sin \theta}{1 - \cos \theta}$

In Exercises 47–50, evaluate the derivative of the function at the indicated point. Use a graphing utility to verify your result.

Function	Point
47. $y = \frac{1 + \csc x}{1 - \csc x}$	$\left(\frac{\pi}{6}, -3\right)$
48. $f(x) = \tan x \cot x$	$(1, 1)$
49. $h(t) = \frac{\sec t}{t}$	$\left(\pi, -\frac{1}{\pi}\right)$
50. $f(x) = \sin x(\sin x + \cos x)$	$\left(\frac{\pi}{4}, 1\right)$

In Exercises 51–56, (a) find an equation of the tangent line to the graph of f at the indicated point, (b) use a graphing utility to graph the function and its tangent line at the point, and (c) use the derivative feature of a graphing utility to confirm your results.

Function	Point
51. $f(x) = \frac{x}{x - 1}$	$(2, 2)$
52. $f(x) = (x - 1)(x^2 - 2)$	$(0, 2)$
53. $f(x) = (x^3 - 3x + 1)(x + 2)$	$(1, -3)$
54. $f(x) = \frac{(x - 1)}{(x + 1)}$	$\left(2, \frac{1}{3}\right)$

Function	Point
55. $f(x) = \tan x$	$\left(\dfrac{\pi}{4}, 1\right)$
56. $f(x) = \sec x$	$\left(\dfrac{\pi}{3}, 2\right)$

In Exercises 57 and 58, determine the point(s) at which the graph of the function has a horizontal tangent.

57. $f(x) = \dfrac{x^2}{x-1}$ **58.** $f(x) = \dfrac{x^2}{x^2+1}$

Think About It **In Exercises 59–62, find $f'(2)$ given the following.**

$g(2) = 3 \quad \text{and} \quad g'(2) = -2$

$h(2) = -1 \quad \text{and} \quad h'(2) = 4$

59. $f(x) = 2g(x) + h(x)$ **60.** $f(x) = 4 - h(x)$

61. $f(x) = \dfrac{g(x)}{h(x)}$ **62.** $f(x) = g(x)h(x)$

In Exercises 63 and 64, find the derivative of the function f for $n = 1, 2, 3,$ and 4. Use the result to write a general rule for $f'(x)$ in terms of n.

63. $f(x) = x^n \sin x$ **64.** $f(x) = \dfrac{\cos x}{x^n}$

65. *Inventory Replenishment* The ordering and transportation cost C of the components used in manufacturing a certain product is

$$C = 100\left(\dfrac{200}{x^2} + \dfrac{x}{x+30}\right), \quad 1 \le x$$

where C is measured in thousands of dollars and x is the order size in hundreds. Find the rate of change of C with respect to x when (a) $x = 10$, (b) $x = 15$, and (c) $x = 20$. What do these rates of change imply about increasing order size?

66. *Boyle's Law* This law states that if the temperature of a gas remains constant, its pressure is inversely proportional to its volume. Use the derivative to show that the rate of change of the pressure is inversely proportional to the square of the volume.

67. *Population Growth* A population of 500 bacteria is introduced into a culture and grows in number according to the equation

$$P(t) = 500\left(1 + \dfrac{4t}{50 + t^2}\right)$$

where t is measured in hours. Find the rate at which the population is growing when $t = 2$.

68. *Rate of Change* Determine whether there exist any values of x in the interval $[0, 2\pi)$ such that the rate of change of $f(x) = \sec x$ and the rate of change of $g(x) = \csc x$ are equal.

69. Prove the following differentiation rules.

(a) $\dfrac{d}{dx}[\sec x] = \sec x \tan x$

(b) $\dfrac{d}{dx}[\csc x] = -\csc x \cot x$

(c) $\dfrac{d}{dx}[\cot x] = -\csc^2 x$

70. *Think About It* Sketch a graph of a differentiable function f such that $f > 0$ and $f' < 0$ for all real numbers x.

71. *Think About It* Sketch a graph of a differentiable function f such that $f(2) = 0$, $f' < 0$ for $-\infty < x < 2$, and $f' > 0$ for $2 < x < \infty$.

72. *Modeling Data* The federal tax burden per capita is given in the table. *(Source: U.S. Internal Revenue Service)*

Year	1984	1985	1986	1987	1988
Tax	$2738	$2982	$3090	$3414	$3598

Year	1989	1990	1991	1992	1993
Tax	$3884	$4026	$4064	$4153	$4382

Year	1994	1995
Tax	$4728	$4996

Let T be the per capita tax burden and let t be the time in years. Then $T = (2{,}231{,}291 + 110{,}636t)/(1000 - 14t)$ is a model for the data, with $t = 4$ corresponding to 1984.

(a) Find T' and use a graphing utility to graph the derivative.

(b) Interpret the graph in part (a), assuming that this model will be used to forecast the per capita tax burden through the year 2000.

(c) Use the model to predict the per capita tax burden in the year 2000.

In Exercises 73–78, find the second derivative of the function.

73. $f(x) = 4x^{3/2}$ **74.** $f(x) = \dfrac{x^2 + 2x - 1}{x}$

75. $f(x) = \dfrac{x}{x-1}$ **76.** $f(x) = x + \dfrac{32}{x^2}$

77. $f(x) = 3 \sin x$ **78.** $f(x) = \sec x$

In Exercises 79–82, find the higher-order derivative.

Given	Find
79. $f'(x) = x^2$	$f''(x)$
80. $f''(x) = 2 - \dfrac{2}{x}$	$f'''(x)$
81. $f'''(x) = 2\sqrt{x}$	$f^{(4)}(x)$
82. $f^{(4)}(x) = 2x + 1$	$f^{(6)}(x)$

Think About It   In Exercises 83 and 84, the graphs of f, f', and f'' are shown on the same set of coordinate axes. Which is which?

83. 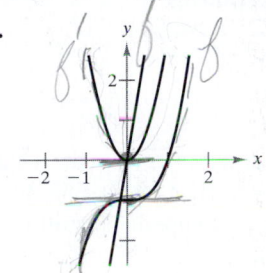   **84.**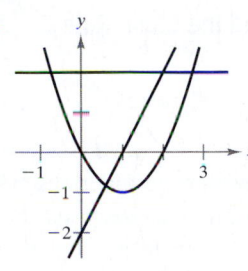

85. *Finding a Pattern*   Consider the function $f(x) = g(x)h(x)$.

(a) Use the product rule to generate rules for finding $f''(x)$, $f'''(x)$, and $f^{(4)}(x)$.

(b) Use the results in part (a) to write a general rule for $f^{(n)}$.

86. *Finding a Pattern*   Develop a general rule for $f^{(n)}(x)$ if

(a) $f(x) = x^n$ and   (b) $f(x) = \dfrac{1}{x}$.

87. *Acceleration*   The velocity of an object in meters per second is

$$v(t) = 36 - t^2, \quad 0 \le t \le 6.$$

Find the velocity and acceleration of the object when $t = 3$. What can be said about the speed of the object when the velocity and acceleration have opposite signs?

88. *Acceleration*   An automobile's velocity starting from rest is

$$v(t) = \frac{100t}{2t + 15}$$

where v is measured in feet per second. Find the acceleration at each of the following times.

(a) 5 seconds   (b) 10 seconds   (c) 20 seconds

89. *Stopping Distance*   A car is traveling at a rate of 66 feet per second (45 miles per hour) when the brakes are applied. The position function for the car is

$$s(t) = -8.25t^2 + 66t$$

where s is measured in feet and t is measured in seconds. Use this function to complete the table, and find the average velocity during each time interval.

t	0	1	2	3	4
$s(t)$					
$v(t)$					
$a(t)$					

90. *Acceleration on the Moon*   An astronaut standing on the moon throws a rock into the air. The height of the rock is

$$s = -\frac{27}{10}t^2 + 27t + 6$$

where s is measured in feet and t is measured in seconds.

(a) Find expressions for the velocity and acceleration of the rock.

(b) Find the time when the rock is at its highest point by finding the time when the velocity is zero. What is its height at this time?

(c) How does the acceleration of the rock compare with the acceleration due to gravity on the earth?

Linear and Quadratic Approximations   The linear and quadratic approximation of a function f at $x = a$ are

$$P_1(x) = f'(a)(x - a) + f(a) \text{ and}$$
$$P_2(x) = \tfrac{1}{2}f''(a)(x - a)^2 + f'(a)(x - a) + f(a).$$

In Exercises 91 and 92, (a) find the specified linear and quadratic approximations of f, (b) use a graphing utility to graph f and the approximations, (c) determine whether P_1 or P_2 is the better approximation, and (d) state how the accuracy changes as you move farther from $x = a$.

91. $f(x) = \cos x$ **92.** $f(x) = \tan x$

$a = \dfrac{\pi}{3}$ $a = \dfrac{\pi}{4}$

True or False?   In Exercises 93–98, determine whether the statement is true or false. If it is false, explain why or give an example that shows it is false.

93. If $y = f(x)g(x)$, then $dy/dx = f'(x)g'(x)$.

94. If $y = (x + 1)(x + 2)(x + 3)(x + 4)$, then $d^5y/dx^5 = 0$.

95. If $f'(c)$ and $g'(c)$ are zero and $h(x) = f(x)g(x)$, then $h'(c) = 0$.

96. If $f(x)$ is an nth-degree polynomial, then $f^{(n+1)}(x) = 0$.

97. The second derivative represents the rate of change of the first derivative.

98. If the velocity of an object is constant, then its acceleration is zero.

99. Find the derivative of the function $f(x) = x|x|$. Does $f''(0)$ exist?

100. *Think About It*   Let f and g be functions whose first and second derivatives exist on an interval I. Which of the following formulas is (are) true?

(a) $fg'' - f''g = (fg' - f'g)'$

(b) $fg'' + f''g = (fg)''$

SECTION 2.4 The Chain Rule

The Chain Rule • The General Power Rule • Simplifying Derivatives • Trigonometric Functions and the Chain Rule

The Chain Rule

We have yet to discuss one of the most powerful differentiation rules—the **Chain Rule.** This rule deals with composite functions and adds a surprising versatility to the rules discussed in the two previous sections. For example, compare the following functions. Those on the left can be differentiated without the Chain Rule, and those on the right are best done with the Chain Rule.

Without the Chain Rule	*With the Chain Rule*
$y = x^2 + 1$	$y = \sqrt{x^2 + 1}$
$y = \sin x$	$y = \sin 6x$
$y = 3x + 2$	$y = (3x + 2)^5$
$y = x + \tan x$	$y = x + \tan x^2$

Basically, the Chain Rule states that if y changes dy/du times as fast as u, and u changes du/dx times as fast as x, then y changes $(dy/du)(du/dx)$ times as fast as x.

EXAMPLE 1   The Derivative of a Composite Function

A set of gears is constructed, as shown in Figure 2.23, such that the second and third gears are on the same axle. As the first axle revolves, it drives the second axle, which in turn drives the third axle. Let y, u, and x represent the numbers of revolutions per minute of the first, second, and third axles. Find dy/du, du/dx, and dy/dx, and show that

$$\frac{dy}{dx} = \frac{dy}{du} \cdot \frac{du}{dx}.$$

Solution Because the circumference of the second gear is three times that of the first, the first axle must make three revolutions to turn the second axle once. Similarly, the second axle must make two revolutions to turn the third axle once, and you can write

$$\frac{dy}{du} = 3 \quad \text{and} \quad \frac{du}{dx} = 2.$$

Combining these two results, you know that the first axle must make six revolutions to turn the third axle once. Thus, you can write

$$\frac{dy}{dx} = \boxed{\begin{array}{c}\text{Rate of change of first axle}\\\text{with respect to second axle}\end{array}} \cdot \boxed{\begin{array}{c}\text{Rate of change of second axle}\\\text{with respect to third axle}\end{array}}$$

$$= \frac{dy}{du} \cdot \frac{du}{dx} = 3 \cdot 2 = 6$$

$$= \boxed{\begin{array}{c}\text{Rate of change of first axle}\\\text{with respect to third axle}\end{array}}.$$

In other words, the rate of change of y with respect to x is the product of the rate of change of y with respect to u and the rate of change of u with respect to x.

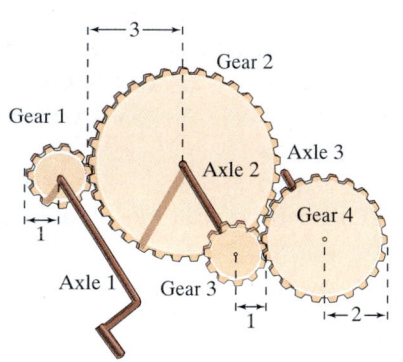

Axle 1: *y* revolutions per minute
Axle 2: *u* revolutions per minute
Axle 3: *x* revolutions per minute
Figure 2.23

Example 1 illustrates a simple case of the Chain Rule. The general rule is stated below.

THEOREM 2.10   The Chain Rule

If $y = f(u)$ is a differentiable function of u and $u = g(x)$ is a differentiable function of x, then $y = f(g(x))$ is a differentiable function of x and

$$\frac{dy}{dx} = \frac{dy}{du} \cdot \frac{du}{dx}$$

or, equivalently,

$$\frac{d}{dx}\left[f(g(x))\right] = f'(g(x))g'(x).$$

Proof   Let $h(x) = f(g(x))$. Then, using the alternative form of the derivative, you need to show that, for $x = c$,

$$h'(c) = f'(g(c))g'(c).$$

An important consideration in this proof is the behavior of g as x approaches c. A problem occurs if there are values of x, other than c, such that $g(x) = g(c)$. In the appendix we show how to use the differentiability of f and g to overcome this problem. For now, assume that $g(x) \neq g(c)$ for values of x other than c. In the proofs of the Product Rule and the Quotient Rule, we added and subtracted the same quantity to obtain the desired form. This proof uses a similar technique—multiplying and dividing by the same (nonzero) quantity. Note that because g is differentiable, it is also continuous, and it follows that $g(x) \rightarrow g(c)$ as $x \rightarrow c$.

$$
\begin{aligned}
h'(c) &= \lim_{x \to c} \frac{f(g(x)) - f(g(c))}{x - c} \\[2mm]
&= \lim_{x \to c} \left[\frac{f(g(x)) - f(g(c))}{g(x) - g(c)} \cdot \frac{g(x) - g(c)}{x - c} \right], \quad g(x) \neq g(c) \\[2mm]
&= \left[\lim_{x \to c} \frac{f(g(x)) - f(g(c))}{g(x) - g(c)} \right] \left[\lim_{x \to c} \frac{g(x) - g(c)}{x - c} \right] \\[2mm]
&= f'(g(c))g'(c)
\end{aligned}
$$

When applying the Chain Rule, it is helpful to think of the composite function $f \circ g$ as having two parts—an inner part and an outer part.

Outer function

$$y = f(g(x)) = f(u)$$

Inner function

The derivative of $y = f(u)$ is the derivative of the outer function (at the inner function u) *times* the derivative of the inner function.

$$y' = f'(u) \cdot u'$$

EXAMPLE 2 Decomposition of a Composite Function

$y = f(g(x))$	$u = g(x)$	$y = f(u)$
a. $y = \dfrac{1}{x + 1}$	$u = x + 1$	$y = \dfrac{1}{u}$
b. $y = \sin 2x$	$u = 2x$	$y = \sin u$
c. $y = \sqrt{3x^2 - x + 1}$	$u = 3x^2 - x + 1$	$y = \sqrt{u}$
d. $y = \tan^2 x$	$u = \tan x$	$y = u^2$

EXAMPLE 3 Using the Chain Rule

Find dy/dx for $y = (x^2 + 1)^3$.

STUDY TIP You could also solve the problem in Example 3 without using the Chain Rule by observing that

$$y = x^6 + 3x^4 + 3x^2 + 1$$

and

$$y' = 6x^5 + 12x^3 + 6x.$$

Verify that this is the same as the derivative in Example 3. Which method would you use to find

$$\frac{d}{dx}(x^2 + 1)^{50}?$$

Solution For this function, you can consider the inside function to be $u = x^2 + 1$. By the Chain Rule, you obtain

$$\frac{dy}{dx} = \underbrace{3(x^2 + 1)^2}_{\frac{dy}{du}}\underbrace{(2x)}_{\frac{du}{dx}} = 6x(x^2 + 1)^2.$$

The General Power Rule

The function in Example 3 is an example of one of the most common types of composite functions, $y = [u(x)]^n$. The rule for differentiating such functions is called the **General Power Rule,** and it is a special case of the Chain Rule.

> **THEOREM 2.11 The General Power Rule**
>
> If $y = [u(x)]^n$, where u is a differentiable function of x and n is a rational number, then
>
> $$\frac{dy}{dx} = n[u(x)]^{n-1}\frac{du}{dx}$$
>
> or, equivalently,
>
> $$\frac{d}{dx}[u^n] = nu^{n-1}u'.$$

Proof Because $y = u^n$, you apply the Chain Rule to obtain

$$\frac{dy}{dx} = \left(\frac{dy}{du}\right)\left(\frac{du}{dx}\right)$$

$$= \frac{d}{du}[u^n]\frac{du}{dx}.$$

By the (simple) Power Rule in Section 2.2, you have $D_u[u^n] = nu^{n-1}$, and it follows that $dy/dx = n[u(x)]^{n-1}(du/dx)$.

EXAMPLE 4   Applying the General Power Rule

Find the derivative of $f(x) = (3x - 2x^2)^3$.

Solution   Let $u = 3x - 2x^2$. Then

$$f(x) = (3x - 2x^2)^3 = u^3$$

and, by the General Power Rule, the derivative is

$$f'(x) = 3(3x - 2x^2)^2 \frac{d}{dx}[3x - 2x^2]$$ Apply General Power Rule.

$$= 3(3x - 2x^2)^2(3 - 4x).$$ Differentiate $3x - 2x^2$.

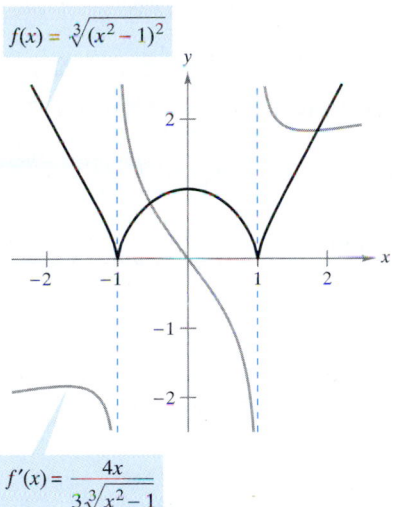

$f(x) = \sqrt[3]{(x^2 - 1)^2}$

$f'(x) = \dfrac{4x}{3\sqrt[3]{x^2 - 1}}$

The derivative of f is 0 at $x = 0$ and is undefined at $x = \pm 1$.
Figure 2.24

EXAMPLE 5   Differentiating Functions Involving Radicals

Find all points on the graph of $f(x) = \sqrt[3]{(x^2 - 1)^2}$ for which $f'(x) = 0$ and those for which $f'(x)$ does not exist.

Solution   Begin by rewriting the function as

$$f(x) = (x^2 - 1)^{2/3}.$$

Then, applying the General Power Rule (with $u = x^2 - 1$) produces

$$f'(x) = \frac{2}{3}(x^2 - 1)^{-1/3}(2x)$$ Apply General Power Rule.

$$= \frac{4x}{3\sqrt[3]{x^2 - 1}}.$$ Write in radical form.

Thus, $f'(x) = 0$ when $x = 0$ and $f'(x)$ does not exist when $x = \pm 1$, as indicated in Figure 2.24.

EXAMPLE 6   Differentiating Quotients with Constant Numerators

Differentiate $g(t) = \dfrac{-7}{(2t - 3)^2}$.

Solution   Begin by rewriting the function as

$$g(t) = -7(2t - 3)^{-2}.$$

Then, applying the General Power Rule produces

$$g'(t) = (-7)(-2)(2t - 3)^{-3}(2)$$ Apply General Power Rule.

Constant Multiple Rule

$$= 28(2t - 3)^{-3}$$ Simplify.

$$= \frac{28}{(2t - 3)^3}.$$ Write with positive exponent.

NOTE   Try differentiating the function in Example 6 using the Quotient Rule. You should obtain the same result, but using the Quotient Rule is less efficient than using the General Power Rule.

Simplifying Derivatives

The next three examples illustrate some techniques for simplifying the "raw derivatives" of functions involving products, quotients, and composites.

EXAMPLE 7 Simplifying by Factoring Out the Least Powers

$$f(x) = x^2\sqrt{1-x^2}$$ Original function

$$= x^2(1-x^2)^{1/2}$$ Rewrite.

$$f'(x) = x^2\frac{d}{dx}\left[(1-x^2)^{1/2}\right] + (1-x^2)^{1/2}\frac{d}{dx}[x^2]$$ Product Rule

$$= x^2\left[\frac{1}{2}(1-x^2)^{-1/2}(-2x)\right] + (1-x^2)^{1/2}(2x)$$ General Power Rule

$$= -x^3(1-x^2)^{-1/2} + 2x(1-x^2)^{1/2}$$ Simplify.

$$= x(1-x^2)^{-1/2}[-x^2(1) + 2(1-x^2)]$$ Factor.

$$= \frac{x(2-3x^2)}{\sqrt{1-x^2}}$$ Simplify.

EXAMPLE 8 Simplifying the Derivative of a Quotient

TECHNOLOGY Symbolic differentiation utilities are capable of differentiating very complicated functions. Often, however, the result is given in unsimplified form. If you have access to such a utility, use it to find the derivatives of the functions given in Examples 7, 8, and 9. Then compare the results with those given on this page.

$$f(x) = \frac{x}{\sqrt[3]{x^2+4}}$$ Original function

$$= \frac{x}{(x^2+4)^{1/3}}$$ Rewrite.

$$f'(x) = \frac{(x^2+4)^{1/3}(1) - x(1/3)(x^2+4)^{-2/3}(2x)}{(x^2+4)^{2/3}}$$ Quotient Rule

$$= \frac{1}{3}(x^2+4)^{-2/3}\left[\frac{3(x^2+4) - (2x^2)(1)}{(x^2+4)^{2/3}}\right]$$ Factor.

$$= \frac{x^2+12}{3(x^2+4)^{4/3}}$$ Simplify.

EXAMPLE 9 Simplifying the Derivative of a Power

$$y = \left(\frac{3x-1}{x^2+3}\right)^2$$ Original function

$$y' = 2\overbrace{\left(\frac{3x-1}{x^2+3}\right)}^{u^{n-1}}\overbrace{\frac{d}{dx}\left[\frac{3x-1}{x^2+3}\right]}^{u'}$$ General Power Rule

$$= \left[\frac{2(3x-1)}{x^2+3}\right]\left[\frac{(x^2+3)(3) - (3x-1)(2x)}{(x^2+3)^2}\right]$$ Quotient Rule

$$= \frac{2(3x-1)(3x^2+9-6x^2+2x)}{(x^2+3)^3}$$ Multiply.

$$= \frac{2(3x-1)(-3x^2+2x+9)}{(x^2+3)^3}$$ Simplify.

Trigonometric Functions and the Chain Rule

The "Chain Rule versions" of the derivatives of the six trigonometric functions are as follows.

$$\frac{d}{dx}[\sin u] = (\cos u)\, u' \qquad \frac{d}{dx}[\cos u] = -(\sin u)\, u'$$

$$\frac{d}{dx}[\tan u] = (\sec^2 u)\, u' \qquad \frac{d}{dx}[\cot u] = -(\csc^2 u)\, u'$$

$$\frac{d}{dx}[\sec u] = (\sec u \tan u)\, u' \qquad \frac{d}{dx}[\csc u] = -(\csc u \cot u)\, u'$$

EXAMPLE 10   Applying the Chain Rule to Trigonometric Functions

a. $y = \sin 2x$ $\qquad$ $y' = \overset{\cos u}{\cos 2x}\ \overset{u'}{\frac{d}{dx}[2x]} = (\cos 2x)(2) = 2\cos 2x$

b. $y = \cos(x - 1)$ $\qquad$ $y' = -\sin(x - 1)$

c. $y = \tan 3x$ $\qquad$ $y' = 3\sec^2 3x$

Be sure that you understand the mathematical conventions regarding parentheses and trigonometric functions. For instance, in Example 10a, $\sin 2x$ is written to mean $\sin(2x)$.

EXAMPLE 11   Parentheses and Trigonometric Functions

a. $y = \cos 3x^2 = \cos(3x^2)$ $\qquad$ $y' = (-\sin 3x^2)(6x) = -6x\sin 3x^2$

b. $y = (\cos 3)x^2$ $\qquad$ $y' = (\cos 3)(2x) = 2x\cos 3$

c. $y = \cos(3x)^2 = \cos(9x^2)$ $\qquad$ $y' = (-\sin 9x^2)(18x) = -18x\sin 9x^2$

d. $y = \cos^2 x = (\cos x)^2$ $\qquad$ $y' = 2(\cos x)(-\sin x)$

$\qquad\qquad\qquad\qquad\qquad\qquad = -2\cos x \sin x$

To find the derivative of a function of the form $k(x) = f(g(h(x)))$, you need to apply the Chain Rule twice, as shown in Example 12.

EXAMPLE 12   Repeated Application of the Chain Rule

$f(t) = \sin^3 4t$ $\qquad\qquad\qquad$ Original function

$\quad = (\sin 4t)^3$ $\qquad\qquad\qquad$ Rewrite.

$f'(t) = 3(\sin 4t)^2 \dfrac{d}{dt}[\sin 4t]$ $\qquad$ Apply Chain Rule once.

$\quad = 3(\sin 4t)^2(\cos 4t)\dfrac{d}{dt}[4t]$ $\qquad$ Apply Chain Rule a second time.

$\quad = 3(\sin 4t)^2(\cos 4t)(4)$

$\quad = 12\sin^2 4t \cos 4t$ $\qquad\qquad$ Simplify.

We conclude this section with a summary of the differentiation rules studied so far. To become skilled at differentiation, you should memorize each rule.

Summary of Differentiation Rules

General Differentiation Rules   Let u and v be differentiable functions of x.

Constant Multiple Rule:

$$\frac{d}{dx}[cu] = cu'$$

Sum or Difference Rule:

$$\frac{d}{dx}[u \pm v] = u' \pm v'$$

Product Rule:

$$\frac{d}{dx}[uv] = uv' + vu'$$

Quotient Rule:

$$\frac{d}{dx}\left[\frac{u}{v}\right] = \frac{vu' - uv'}{v^2}$$

Derivatives of Algebraic Functions

Constant Rule:

$$\frac{d}{dx}[c] = 0$$

Simple Power Rule:

$$\frac{d}{dx}[x^n] = nx^{n-1}, \quad \frac{d}{dx}[x] = 1$$

Derivatives of Trigonometric Functions

$$\frac{d}{dx}[\sin x] = \cos x$$

$$\frac{d}{dx}[\tan x] = \sec^2 x \qquad \frac{d}{dx}[\sec x] = \sec x \tan x$$

$$\frac{d}{dx}[\cos x] = -\sin x$$

$$\frac{d}{dx}[\cot x] = -\csc^2 x \qquad \frac{d}{dx}[\csc x] = -\csc x \cot x$$

Chain Rule

Chain Rule:

$$\frac{d}{dx}[f(u)] = f'(u)\,u'$$

General Power Rule:

$$\frac{d}{dx}[u^n] = nu^{n-1}\,u'$$

STUDY TIP   As an aid to memorization, note that the cofunctions (cosine, cotangent, and cosecant) require a negative sign as part of their derivatives.

EXERCISES FOR SECTION 2.4

In Exercises 1–6, complete the table using Example 2 as a model.

$y = f(g(x))$	$u = g(x)$	$y = f(u)$

1. $y = (6x - 5)^4$

2. $y = \dfrac{1}{\sqrt{x+1}}$

3. $y = \sqrt{x^2 - 1}$

4. $y = \tan(\pi x + 1)$

5. $y = \csc^3 x$

6. $y = \cos\dfrac{3x}{2}$

In Exercises 7–30, find the first derivative of the algebraic function.

7. $y = (2x - 7)^3$

8. $y = (3x^2 + 1)^4$

9. $g(x) = 3(4 - 9x)^4$

10. $f(x) = 2(1 - x^2)^3$

11. $f(x) = (9 - x^2)^{2/3}$

12. $f(t) = (9t + 2)^{2/3}$

13. $f(t) = \sqrt{1 - t}$

14. $g(x) = \sqrt{3 - 2x}$

15. $y = \sqrt[3]{9x^2 + 4}$

16. $g(x) = \sqrt{x^2 - 2x + 1}$

17. $y = 2\sqrt{4 - x^2}$

18. $f(x) = -3\sqrt[4]{2 - 9x}$

19. $y = \dfrac{1}{x - 2}$

20. $s(t) = \dfrac{1}{t^2 + 3t - 1}$

21. $f(t) = \left(\dfrac{1}{t - 3}\right)^2$

22. $y = -\dfrac{4}{(t + 2)^2}$

23. $y = \dfrac{1}{\sqrt{x+2}}$

24. $g(t) = \sqrt{\dfrac{1}{t^2 - 2}}$

25. $f(x) = x^2(x-2)^4$

26. $f(x) = x(3x-9)^3$

27. $y = x\sqrt{1-x^2}$

28. $y = x^2\sqrt{9-x^2}$

29. $y = \dfrac{x}{\sqrt{x^2+1}}$

30. $y = \dfrac{x^2}{\sqrt{x^9+9}}$

In Exercises 31–40, use a symbolic differentiation utility to find the first derivative of the function. Then use the utility to graph the function and its derivative on the same set of coordinate axes. Describe the behavior of the function that corresponds to any zeros of the graph of the derivative.

31. $y = \dfrac{\sqrt{x}+1}{x^2+1}$

32. $y = \sqrt{\dfrac{2x}{x+1}}$

33. $g(t) = \dfrac{3t^2}{\sqrt{t^2+2t-1}}$

34. $f(x) = \sqrt{x}(2-x)^2$

35. $y = \sqrt{\dfrac{x+1}{x}}$

36. $y = (t^2-9)\sqrt{t+2}$

37. $s(t) = \dfrac{-2(2-t)\sqrt{1+t}}{3}$

38. $g(x) = \sqrt{x-1} + \sqrt{x+1}$

39. $y = \dfrac{\cos \pi x + 1}{x}$

40. $y = x^2 \tan \dfrac{1}{x}$

In Exercises 41 and 42, find the slope of the tangent line to the sine function at the origin. Compare this value with the number of complete cycles in the interval $[0, 2\pi]$.

41. (a) $y = \sin x$

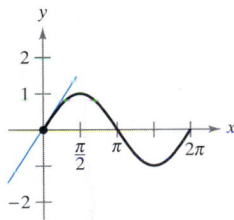

(b) $y = \sin 2x$

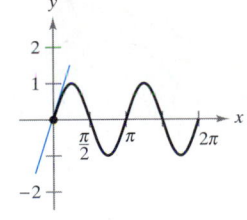

42. (a) $y = \sin 3x$

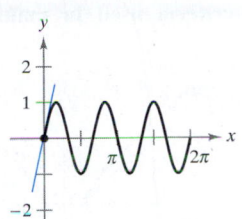

(b) $y = \sin \dfrac{x}{2}$

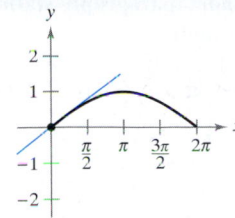

In Exercises 43–52, find the first derivative of the function.

43. $y = \cos 3x$

44. $y = \sin \pi x$

45. $g(x) = 3 \tan 4x$

46. $h(x) = \sec x^2$

47. $f(\theta) = \tfrac{1}{4}\sin^2 2\theta$

48. $g(t) = 5\cos^2 \pi t$

49. $y = \sqrt{x} + \tfrac{1}{4}\sin(2x)^2$

50. $y = 3x - 5\cos(\pi x)^2$

51. $y = \sin(\cos x)$

52. $y = \sin\sqrt{x} + \sqrt{\sin x}$

In Exercises 53–60, evaluate the derivative of the function at the indicated point. You can use a graphing utility to verify your result.

Function	Point
53. $s(t) = \sqrt{t^2 + 2t + 8}$	$(2, 4)$
54. $y = \sqrt[5]{3x^3 + 4x}$	$(2, 2)$
55. $f(x) = \dfrac{3}{x^3 - 4}$	$\left(-1, -\dfrac{3}{5}\right)$
56. $f(x) = \dfrac{1}{(x^2 - 3x)^2}$	$\left(4, \dfrac{1}{16}\right)$
57. $f(t) = \dfrac{3t + 2}{t - 1}$	$(0, -2)$
58. $f(x) = \dfrac{x + 1}{2x - 3}$	$(2, 3)$
59. $y = 37 - \sec^3(2x)$	$(0, 36)$
60. $y = \dfrac{1}{x} + \sqrt{\cos x}$	$\left(\dfrac{\pi}{2}, \dfrac{2}{\pi}\right)$

In Exercises 61–64, (a) find an equation of the tangent line to the graph of f at the indicated point, (b) use a graphing utility to graph the function and its tangent line at the point, and (c) use the derivative feature of a graphing utility to confirm your results.

Function	Point
61. $f(x) = \sqrt{3x^2 - 2}$	$(3, 5)$
62. $f(x) = \tfrac{1}{3}x\sqrt{x^2 + 5}$	$(2, 2)$
63. $f(x) = \sin 2x$	$(\pi, 0)$
64. $f(x) = \tan^2 x$	$\left(\dfrac{\pi}{4}, 1\right)$

Writing In Exercises 65–68, the graphs of a function *f* and its derivative *f′* are given. Label the graphs as *f* or *f′* and write a short paragraph stating the criteria used in making the selection.

65.

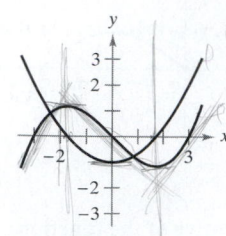

66.

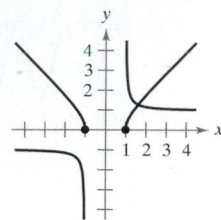

67.

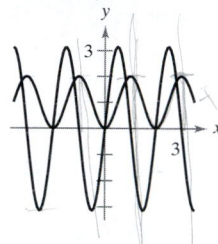

68.

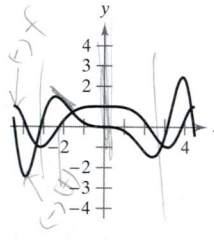

In Exercises 69–72, find the second derivative of the function.

69. $f(x) = 2(x^2 - 1)^3$

70. $f(x) = \dfrac{1}{x - 2}$

71. $f(x) = \sin x^2$

72. $f(x) = \sec^2 \pi x$

73. *Think About It* Given that $g(5) = -3$, $g'(5) = 6$, $h(5) = 3$, and $h'(5) = -2$, find $f'(5)$ (if possible) for each of the following. If it is not possible, state what additional information is required.

(a) $f(x) = g(x)h(x)$ (b) $f(x) = g(h(x))$

(c) $f(x) = \dfrac{g(x)}{h(x)}$ (d) $f(x) = [g(x)]^3$

74. (a) Find the derivative of the function $g(x) = \sin^2 x + \cos^2 x$ in two ways.

(b) For $f(x) = \sec^2 x$ and $g(x) = \tan^2 x$, show that $f'(x) = g'(x)$.

75. *Doppler Effect* The frequency F of a fire truck siren heard by a stationary observer is

$$F = \frac{132{,}400}{331 \pm v}$$

where $\pm v$ represents the velocity of the accelerating fire truck (see figure). Find the rate of change of F with respect to v when

(a) the fire truck is approaching at a velocity of 30 m/s (use $-v$).

(b) the fire truck is moving away at a velocity of 30 m/s (use $+v$).

$$F = \frac{132{,}400}{331 + v} \qquad\qquad F = \frac{132{,}400}{331 - v}$$

Figure for 75

76. *Harmonic Motion* The displacement from equilibrium of an object in harmonic motion on the end of a spring is

$$y = \tfrac{1}{3} \cos 12t - \tfrac{1}{4} \sin 12t$$

where *y* is measured in feet and *t* is the time in seconds. Determine the position and velocity of the object when $t = \pi/8$.

77. *Pendulum* A 15-centimeter pendulum moves according to the equation

$$\theta = 0.2 \cos 8t$$

where θ is the angular displacement from the vertical in radians and *t* is the time in seconds. Determine the maximum angular displacement and the rate of change of θ when $t = 3$ seconds.

78. *Wave Motion* A buoy oscillates in simple harmonic motion

$$y = A \cos \omega t$$

as waves move past it. The buoy moves a total of 3.5 feet (vertically) from its low point to its high point. It returns to its high point every 10 seconds.

(a) Write an equation describing the motion of the buoy if it is at its high point at $t = 0$.

(b) Determine the velocity of the buoy as a function of *t*.

79. *Circulatory System* The speed *S* of blood that is *r* centimeters from the center of an artery is

$$S = C(R^2 - r^2)$$

where *C* is a constant, *R* is the radius of the artery, and *S* is measured in centimeters per second. Suppose a drug is administered and the artery begins to dilate at a rate of dR/dt. At a constant distance *r*, find the rate at which *S* changes with respect to *t* for $C = 1.76 \times 10^5$, $R = 1.2 \times 10^{-2}$, and $dR/dt = 10^{-5}$.

80. *Modeling Data*   The normal daily maximum temperature T (in degrees Fahrenheit) for Denver, Colorado is given in the table. (*Source: National Oceanic and Atmosphere Administration*)

Month	Jan	Feb	Mar	Apr	May	Jun
Temperature	43.2	46.6	52.2	61.8	70.8	81.4

Month	Jul	Aug	Sep	Oct	Nov	Dec
Temperature	88.2	85.8	76.9	66.3	52.5	44.5

(a) Use a graphing utility to plot the data and find a model for the data of the form

$$T(t) = a + b \sin (\pi t/6 - c)$$

where T is the temperature and t is the time in months, with $t = 1$ corresponding to January.

(b) Use a graphing utility to graph the model. How well does the model fit the data?

(c) Find T' and use a graphing utility to graph the derivative.

(d) Based on the graph of the derivative, during what times does the temperature change most rapidly? Most slowly? Do your answers agree with your observations of temperature changes? Explain.

81. *Think About It*   The table gives some values of the derivative of an unknown function f. Complete the table by finding (if possible) the derivative of each of the following transformations of f.

(a) $g(x) = f(x) - 2$ (b) $h(x) = 2 f(x)$

(c) $r(x) = f(-3x)$ (d) $s(x) = f(x + 2)$

x	-2	-1	0	1	2	3
$f'(x)$	4	$\frac{2}{3}$	$-\frac{1}{3}$	-1	-2	-4
$g'(x)$						
$h'(x)$						
$r'(x)$						
$s'(x)$						

82. *Finding a Pattern*   Consider the function $f(x) = \sin \beta x$, where β is a constant.

(a) Find the first-, second-, third-, and fourth-order derivatives of the function.

(b) Verify that the function and its second derivative satisfy the equation $f''(x) + \beta^2 f(x) = 0$.

(c) Use the results in part (a) to write general rules for the even- and odd-order derivatives

$$f^{(2k)}(x) \text{ and } f^{(2k-1)}(x).$$

[*Hint:* $(-1)^k$ is positive if k is even and negative if k is odd.]

83. *Conjecture*   Let f be a differentiable function of period p.

(a) Is the function f' periodic? Verify your answer.

(b) Consider the function $g(x) = f(2x)$. Is the function $g'(x)$ periodic? Verify your answer.

84. Show that the derivative of an odd function is even. That is, if $f(-x) = -f(x)$, then $f'(-x) = f'(x)$.

85. The geometric mean of x and $x + n$ is $g = \sqrt{x(x + n)}$, and the arithmetic mean is $a = [x + (x + n)]/2$. Show that

$$\frac{dg}{dx} = \frac{a}{g}.$$

86. Let u be a differentiable function of x. Use the fact that $|u| = \sqrt{u^2}$ to prove that

$$\frac{d}{dx}[|u|] = u' \frac{u}{|u|}, \quad u \neq 0.$$

In Exercises 87–90, use the result of Exercise 86 to find the derivative of the function.

87. $g(x) = |2x - 3|$ **88.** $f(x) = |x^2 - 4|$

89. $h(x) = |x| \cos x$ **90.** $f(x) = |\sin x|$

Linear and Quadratic Approximations   The linear and quadratic approximations of a function f at $x = a$ are

$$P_1(x) = f'(a)(x - a) + f(a) \text{ and}$$
$$P_2(x) = \tfrac{1}{2} f''(a)(x - a)^2 + f'(a)(x - a) + f(a).$$

In Exercises 91 and 92, (a) find the specified linear and quadratic approximations of f, (b) use a graphing utility to graph f and the approximations, (c) determine whether P_1 or P_2 is the better approximation, and (d) state how the accuracy changes as you move farther from $x = a$.

91. $f(x) = \sin \dfrac{x}{2}$ **92.** $f(x) = \sec 2x$

     $a = \pi$ $a = \dfrac{\pi}{6}$

True or False?   **In Exercises 93–95, determine whether the statement is true or false. If it is false, explain why or give an example that shows it is false.**

93. If $y = (1 - x)^{1/2}$, then $y' = \tfrac{1}{2}(1 - x)^{-1/2}$.

94. If $f(x) = \sin^2(2x)$, then $f'(x) = 2(\sin 2x)(\cos 2x)$.

95. If y is a differentiable function of u, u is a differentiable function of v, and v is a differentiable function of x, then

$$\frac{dy}{dx} = \frac{dy}{du} \frac{du}{dv} \frac{dv}{dx}.$$

SECTION 2.5   Implicit Differentiation

Implicit and Explicit Functions • Implicit Differentiation

Implicit and Explicit Functions

Up to this point in the text, most functions have been expressed in **explicit form.** For example, in the equation

$$y = 3x^2 - 5 \qquad \text{Explicit form}$$

the variable y is explicitly written as a function of x. Some functions, however, are only implied by an equation. For instance, the function $y = 1/x$ is defined **implicitly** by the equation $xy = 1$. Suppose you were asked to find dy/dx for this equation. For this equation, you could begin by writing y explicitly as a function of x and then differentiating.

Implicit Form	*Explicit Form*	*Derivative*
$xy = 1$	$y = \dfrac{1}{x} = x^{-1}$	$\dfrac{dy}{dx} = -x^{-2} = -\dfrac{1}{x^2}$

This strategy works well whenever you can solve for the function explicitly. You cannot, however, use this procedure when you are unable to solve for y as a function of x. For instance, how would you find dy/dx for the equation

$$x^2 - 2y^3 + 4y = 2$$

where it is very difficult to express y as a function of x explicitly? To do this, you can use **implicit differentiation.**

To understand how to find dy/dx implicitly, you must realize that the differentiation is taking place *with respect to x.* This means that when you differentiate terms involving x alone, you can differentiate as usual. However, when you differentiate terms involving y, you must apply the Chain Rule, because you are assuming that y is defined implicitly as a function of x.

EXAMPLE 1   Differentiating with Respect to *x*

a. $\dfrac{d}{dx}[x^3] = 3x^2$ Variables agree: use Simple Power Rule.

Variables agree

b. $\dfrac{d}{dx}[y^3] = 3y^2\dfrac{dy}{dx}$ Variables disagree: use Chain Rule.

$u^n \quad nu^{n-1} \quad u'$

Variables disagree

c. $\dfrac{d}{dx}[x + 3y] = 1 + 3\dfrac{dy}{dx}$ Chain Rule: $\dfrac{d}{dx}[3y] = 3y'$

d. $\dfrac{d}{dx}[xy^2] = x\dfrac{d}{dx}[y^2] + y^2\dfrac{d}{dx}[x]$ Product Rule

$\qquad\qquad = x\left(2y\dfrac{dy}{dx}\right) + y^2(1)$ Chain Rule

$\qquad\qquad = 2xy\dfrac{dy}{dx} + y^2$ Simplify.

Implicit Differentiation

Guidelines for Implicit Differentiation

1. Differentiate both sides of the equation *with respect to x*.
2. Collect all terms involving dy/dx on the left side of the equation and move all other terms to the right side of the equation.
3. Factor dy/dx out of the left side of the equation.
4. Solve for dy/dx by dividing both sides of the equation by the left-hand factor that does not contain dy/dx.

EXAMPLE 2   Implicit Differentiation

Find dy/dx given that $y^3 + y^2 - 5y - x^2 = -4$.

Solution

1. Differentiate both sides of the equation with respect to x.

$$\frac{d}{dx}[y^3 + y^2 - 5y - x^2] = \frac{d}{dx}[-4]$$

$$\frac{d}{dx}[y^3] + \frac{d}{dx}[y^2] - \frac{d}{dx}[5y] - \frac{d}{dx}[x^2] = \frac{d}{dx}[-4]$$

$$3y^2\frac{dy}{dx} + 2y\frac{dy}{dx} - 5\frac{dy}{dx} - 2x = 0$$

2. Collect the dy/dx terms on the left side of the equation.

$$3y^2\frac{dy}{dx} + 2y\frac{dy}{dx} - 5\frac{dy}{dx} = 2x$$

3. Factor dy/dx out of the left side of the equation.

$$\frac{dy}{dx}(3y^2 + 2y - 5) = 2x$$

4. Solve for dy/dx by dividing by $(3y^2 + 2y - 5)$.

$$\frac{dy}{dx} = \frac{2x}{3y^2 + 2y - 5}$$

Note that implicit differentiation can produce an expression for dy/dx that contains both x and y.

To see how you can use an *implicit derivative,* consider the graph shown in Figure 2.25. From the graph, you can see that y is not a function x. Even so, the derivative found in Example 2 gives a formula for the slope of the tangent line at a point on this graph. The slopes at several points on the graph are shown below the graph.

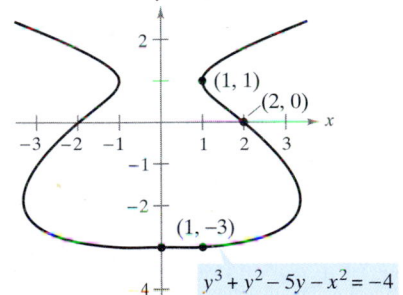

Point on Graph	Slope of Graph
$(2, 0)$	$-\frac{4}{5}$
$(1, -3)$	$\frac{1}{8}$
$x = 0$	0
$(1, 1)$	Undefined

The implicit equation

$$y^3 + y^2 - 5y - x^2 = -4$$

has the derivative

$$\frac{dy}{dx} = \frac{2x}{3y^2 + 2y - 5}.$$

Figure 2.25

TECHNOLOGY With most graphing utilities, it is easy to sketch the graph of an equation that explicitly represents y as a function of x. Sketching graphs of other equations, however, can require some ingenuity. For instance, to sketch the graph of the equation given in Example 2, try using a graphing utility, set in parametric mode, to sketch the graphs given by $x = \sqrt{t^3 + t^2 - 5t + 4}$, $y = t$, and $x = -\sqrt{t^3 + t^2 - 5t + 4}$, $y = t$, for $-5 \leq t \leq 5$. How does the result compare with the graph shown in Figure 2.25?

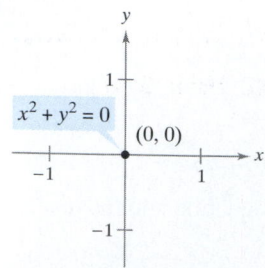

(a)

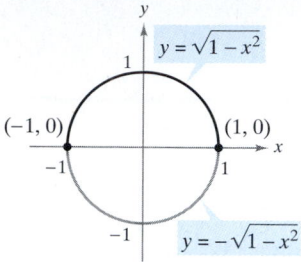

(b)

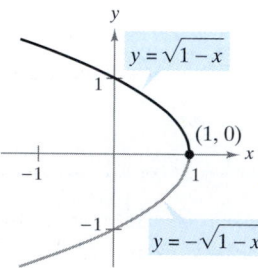

(c)

Some graph segments can be represented by differentiable functions.
Figure 2.26

It is meaningless to solve for dy/dx in an equation that has no solution points. (For example, $x^2 + y^2 = -4$ has no solution points.) If, however, a segment of a graph can be represented by a differentiable function, dy/dx will have meaning as the slope at each point on the segment. Recall that a function is not differentiable at (1) points with vertical tangents and (2) points at which the function is not continuous.

EXAMPLE 3   Representing a Graph by Differentiable Functions

If possible, represent y as a differentiable function of x (see Figure 2.26).

a. $x^2 + y^2 = 0$ **b.** $x^2 + y^2 = 1$ **c.** $x + y^2 = 1$

Solution

a. The graph of this equation is a single point. Therefore, it does not define y as a differentiable function of x.

b. The graph of this equation is the unit circle, centered at $(0, 0)$. The upper semicircle is given by the differentiable function

$$y = \sqrt{1 - x^2}, \; -1 < x < 1$$

and the lower semicircle is given by the differentiable function

$$y = -\sqrt{1 - x^2}, \; -1 < x < 1.$$

At the points $(-1, 0)$ and $(1, 0)$, the slope of the graph is undefined.

c. The upper half of this parabola is given by the differentiable function

$$y = \sqrt{1 - x}, x < 1$$

and the lower semicircle is given by the differentiable function

$$y = -\sqrt{1 - x}, x < 1.$$

At the point $(1, 0)$, the slope of the graph is undefined.

 ### EXAMPLE 4   Finding the Slope of a Graph Implicitly

Determine the slope of the tangent line to the graph of $x^2 + 4y^2 = 4$ at the point $\left(\sqrt{2}, -1/\sqrt{2}\right)$. (See Figure 2.27.)

Solution   Implicit differentiation of the equation $x^2 + 4y^2 = 4$ with respect to x yields

$$2x + 8y\frac{dy}{dx} = 0$$

$$\frac{dy}{dx} = \frac{-2x}{8y} = \frac{-x}{4y}.$$

Therefore, at $\left(\sqrt{2}, -1/\sqrt{2}\right)$, the slope is

$$\frac{dy}{dx} = \frac{-\sqrt{2}}{-4/\sqrt{2}} = \frac{1}{2}.$$

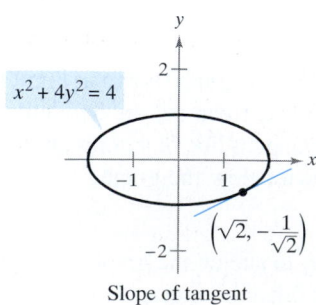

Slope of tangent line is $\frac{1}{2}$.

Figure 2.27

NOTE   To see the benefit of implicit differentiation, try doing Example 4 using the explicit function $y = -\frac{1}{2}\sqrt{4 - x^2}$.

EXAMPLE 5   Finding the Slope of a Graph Implicitly

Determine the slope of the graph of $3(x^2 + y^2)^2 = 100xy$ at the point $(3, 1)$.

Solution

$$\frac{d}{dx}[3(x^2 + y^2)^2] = \frac{d}{dx}[100xy]$$

$$3(2)(x^2 + y^2)\left(2x + 2y\frac{dy}{dx}\right) = 100\left[x\frac{dy}{dx} + y(1)\right]$$

$$12y(x^2 + y^2)\frac{dy}{dx} - 100x\frac{dy}{dx} = 100y - 12x(x^2 + y^2)$$

$$[12y(x^2 + y^2) - 100x]\frac{dy}{dx} = 100y - 12x(x^2 + y^2)$$

$$\frac{dy}{dx} = \frac{100y - 12x(x^2 + y^2)}{-100x + 12y(x^2 + y^2)}$$

$$= \frac{25y - 3x(x^2 + y^2)}{-25x + 3y(x^2 + y^2)}$$

At the point $(3, 1)$, the slope of the graph is

$$\frac{dy}{dx} = \frac{25(1) - 3(3)(3^2 + 1^2)}{-25(3) + 3(1)(3^2 + 1^2)} = \frac{25 - 90}{-75 + 30} = \frac{-65}{-45} = \frac{13}{9}$$

as shown in Figure 2.28. This graph is called a **lemniscate.**

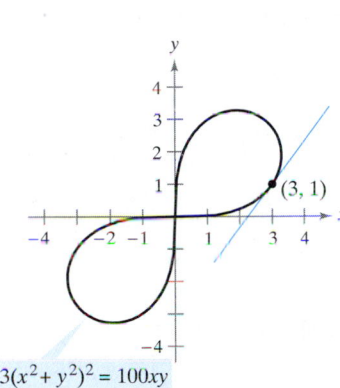

$3(x^2 + y^2)^2 = 100xy$

Lemniscate
Figure 2.28

EXAMPLE 6   Determining a Differentiable Function

Find dy/dx implicitly for the equation $\sin y = x$. Then find the largest interval of the form $-a < y < a$ such that y is a differentiable function of x (see Figure 2.29).

Solution

$$\frac{d}{dx}[\sin y] = \frac{d}{dx}[x]$$

$$\cos y\frac{dy}{dx} = 1$$

$$\frac{dy}{dx} = \frac{1}{\cos y}$$

The largest interval about the origin for which y is a differentiable function of x is $-\pi/2 < y < \pi/2$. To see this, note that $\cos y$ is positive for all y in this interval and is 0 at the endpoints. If you restrict y to the interval $-\pi/2 < y < \pi/2$, you should be able to write dy/dx explicitly as a function of x. To do this, you can use

$$\cos y = \sqrt{1 - \sin^2 y}$$

$$= \sqrt{1 - x^2}, \quad -\frac{\pi}{2} < y < \frac{\pi}{2}$$

and conclude that

$$\frac{dy}{dx} = \frac{1}{\sqrt{1 - x^2}}.$$

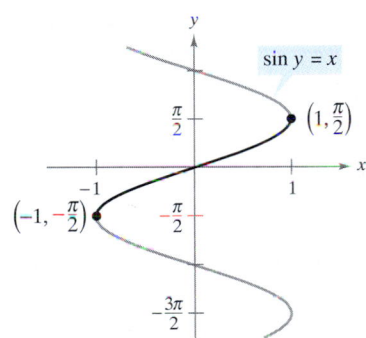

$\sin y = x$

The derivative is $\dfrac{dy}{dx} = \dfrac{1}{\sqrt{1 - x^2}}$.

Figure 2.29

ISAAC BARROW (1630–1677)

The graph in Example 8 is called the **kappa curve** because it resembles the Greek letter kappa, κ. The general solution for the tangent line to this curve was discovered by the English mathematician Isaac Barrow. Newton was Barrow's student and they corresponded frequently regarding their work in the early development of calculus.

With implicit differentiation, the form of the derivative often can be simplified (as in Example 6) by an appropriate use of the *original* equation. A similar technique can be used to find and simplify higher-order derivatives obtained implicitly.

EXAMPLE 7   Finding the Second Derivative Implicitly

Given $x^2 + y^2 = 25$, find $\dfrac{d^2y}{dx^2}$.

Solution   Differentiating each term with respect to x produces

$$2x + 2y\frac{dy}{dx} = 0$$

$$2y\frac{dy}{dx} = -2x$$

$$\frac{dy}{dx} = \frac{-2x}{2y} = -\frac{x}{y}.$$

Differentiating a second time with respect to x yields

$$\frac{d^2y}{dx^2} = -\frac{(y)(1) - (x)(dy/dx)}{y^2} \qquad \text{Quotient Rule}$$

$$= -\frac{y - (x)(-x/y)}{y^2} \qquad \text{Substitute } -x/y \text{ for } \frac{dy}{dx}.$$

$$= -\frac{y^2 + x^2}{y^3}$$

$$= -\frac{25}{y^3}. \qquad \text{Substitute 25 for } x^2 + y^2.$$

EXAMPLE 8   Finding a Tangent Line to a Graph

Find the tangent line to the graph given by $x^2(x^2 + y^2) = y^2$ at the point $\left(\sqrt{2}/2,\ \sqrt{2}/2\right)$, as shown in Figure 2.30.

Solution   By rewriting and differentiating implicitly, you obtain

$$x^4 + x^2y^2 - y^2 = 0$$

$$4x^3 + x^2\left(2y\frac{dy}{dx}\right) + 2xy^2 - 2y\frac{dy}{dx} = 0$$

$$2y(x^2 - 1)\frac{dy}{dx} = -2x(2x^2 + y^2)$$

$$\frac{dy}{dx} = \frac{x(2x^2 + y^2)}{y(1 - x^2)}.$$

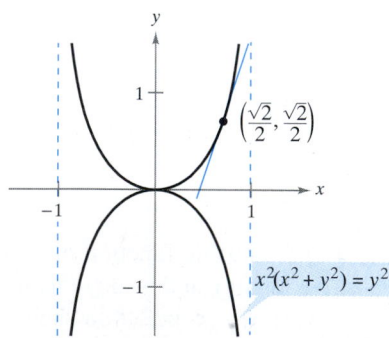

The kappa curve
Figure 2.30

At the point $\left(\sqrt{2}/2,\ \sqrt{2}/2\right)$, the slope is

$$m = \frac{\left(\sqrt{2}/2\right)[2(1/2) + (1/2)]}{\left(\sqrt{2}/2\right)[1 - (1/2)]} = \frac{3/2}{1/2} = 3$$

and the equation of the tangent line at this point is

$$y - \frac{\sqrt{2}}{2} = 3\left(x - \frac{\sqrt{2}}{2}\right)$$

$$y = 3x - \sqrt{2}.$$

EXERCISES FOR SECTION 2.5

In Exercises 1–16, find dy/dx by implicit differentiation.

1. $x^2 + y^2 = 16$

2. $x^2 - y^2 = 16$

3. $x^{1/2} + y^{1/2} = 9$

4. $x^3 + y^3 = 8$

5. $x^3 - xy + y^2 = 4$

6. $x^2 y + y^2 x = -2$

7. $x^3 y^3 - y = x$

8. $\sqrt{xy} = x - 2y$

9. $x^3 - 2x^2 y + 3xy^2 = 38$

10. $2 \sin x \cos y = 1$

11. $\sin x + 2 \cos 2y = 1$

12. $(\sin \pi x + \cos \pi y)^2 = 2$

13. $\sin x = x(1 + \tan y)$

14. $\cot y = x - y$

15. $y = \sin(xy)$

16. $x = \sec \dfrac{1}{y}$

In Exercises 17–24, find dy/dx by implicit differentiation and evaluate the derivative at the indicated point.

Equation	Point
17. $xy = 4$	$(-4, -1)$
18. $x^2 - y^3 = 0$	$(1, 1)$
19. $y^2 = \dfrac{x^2 - 9}{x^2 + 9}$	$(3, 0)$
20. $(x + y)^3 = x^3 + y^3$	$(-1, 1)$
21. $x^{2/3} + y^{2/3} = 5$	$(8, 1)$
22. $x^3 + y^3 = 2xy$	$(1, 1)$
23. $\tan(x + y) = x$	$(0, 0)$
24. $x \cos y = 1$	$\left(2, \dfrac{\pi}{3}\right)$

 In Exercises 25 and 26, use a graphing utility to graph the equation. Find an equation of the tangent line to the graph at the indicated point and sketch its graph.

25. $\sqrt{x} + \sqrt{y} = 3$, $(4, 1)$

26. $y^2 = \dfrac{x - 1}{x^2 + 1}$, $\left(2, \dfrac{\sqrt{5}}{5}\right)$

In Exercises 27–30, find the slope of the tangent line to the graph at the indicated point.

27. Witch of Agnesi:

$(x^2 + 4)y = 8$

Point: $(2, 1)$

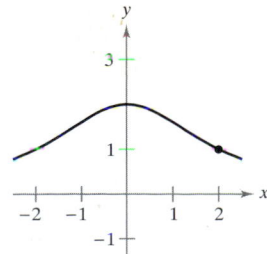

28. Cissoid:

$(4 - x)y^2 = x^3$

Point: $(2, 2)$

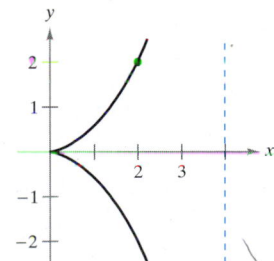

29. Bifolium:

$(x^2 + y^2)^2 = 4x^2 y$

Point: $(1, 1)$

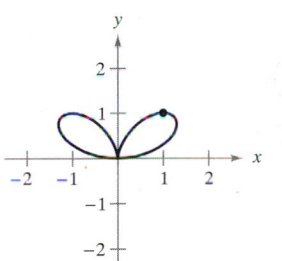

30. Folium of Descartes:

$x^3 + y^3 - 6xy = 0$

Point: $\left(\dfrac{4}{3}, \dfrac{8}{3}\right)$

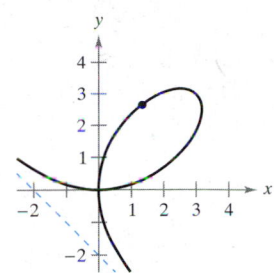

In Exercises 31–34, (a) find two explicit functions by solving the equation for y in terms of x, (b) sketch the graph of the equation and label the parts given by the corresponding explicit functions, (c) differentiate the explicit functions, and (d) find dy/dx implicitly and show that the result is equivalent to that of part (c).

31. $x^2 + y^2 = 16$

32. $x^2 + y^2 - 4x + 6y + 9 = 0$

33. $9x^2 + 16y^2 = 144$

34. $4y^2 - x^2 = 4$

In Exercises 35–40, find $d^2 y/dx^2$ in terms of x and y.

35. $x^2 + xy = 5$

36. $x^2 y^2 - 2x = 3$

37. $x^2 - y^2 = 16$

38. $1 - xy = x - y$

39. $y^2 = x^3$

40. $y^2 = 4x$

In Exercises 41 and 42, find equations for the tangent line and normal line to the circle at the indicated points. (The *normal line* at a point is perpendicular to the tangent line at the point.) Use a graphing utility to graph the equation, tangent line, and normal line.

41. $x^2 + y^2 = 25$

$(4, 3), (-3, 4)$

42. $x^2 + y^2 - 9$

$(0, 3), \left(2, \sqrt{5}\right)$

43. Show that the normal line at any point on the circle $x^2 + y^2 = r^2$ passes through the origin.

44. Two circles of radius 4 are tangent to the graph of $y^2 = 4x$ at the point $(1, 2)$. Find the equations of these two circles.

In Exercises 45 and 46, find the points at which the graph of the equation has a vertical or horizontal tangent line.

45. $25x^2 + 16y^2 + 200x - 160y + 400 = 0$

46. $4x^2 + y^2 - 8x + 4y + 4 = 0$

✦ *Orthogonal Trajectories* In Exercises 47–50, use a graphing utility to sketch the intersecting graphs of the equations and show that they are orthogonal. [Two graphs are *orthogonal* if at their point(s) of intersection their tangent lines are perpendicular to each other.]

47. $2x^2 + y^2 = 6$

$y^2 = 4x$

48. $y^2 = x^3$

$2x^2 + 3y^2 = 5$

49. $x + y = 0$

$x = \sin y$

50. $x^3 = 3(y - 1)$

$x(3y - 29) = 3$

✦ *Orthogonal Trajectories* In Exercises 51 and 52, verify that the two families of curves are orthogonal where C and K are real numbers. Use a graphing utility to graph the two families for two values of C and two values of K.

51. $xy = C$

$x^2 - y^2 = K$

52. $x^2 + y^2 = C^2$

$y = Kx$

In Exercises 53–56, differentiate (a) with respect to x (y is a function of x) and (b) with respect to t (x and y are functions of t).

53. $2y^2 - 3x^4 = 0$ **54.** $x^2 - 3xy^2 + y^3 = 10$

55. $\cos \pi y - 3 \sin \pi x = 1$ **56.** $4 \sin x \cos y = 1$

✦ 57. Consider the equation $x^4 = 4(4x^2 - y^2)$.

    (a) Use a graphing utility to graph the equation.

    (b) Find and graph the four tangent lines to the curve for $y = 3$.

    (c) Find the exact coordinates of the point of intersection of the two tangent lines in the first quadrant.

58. Use Example 6 as a model to find dy/dx implicitly for the equation $\tan y = x$ and find the largest interval of the form $-a < y < a$ such that y is a differentiable function of x. Then express dy/dx as a function of x.

59. Prove (Theorem 2.3) that

$$\frac{d}{dx}[x^n] = nx^{n-1}$$

for the case in which n is a rational number. (*Hint:* Write $y = x^{p/q}$ in the form $y^q = x^p$ and differentiate implicitly. Assume that p and q are integers, where $q > 0$.)

60. Let L be any tangent line to the curve $\sqrt{x} + \sqrt{y} = \sqrt{c}$. Show that the sum of the x- and y-intercepts of L is c.

SECTION PROJECT

Optical Illusions In each graph below, an optical illusion is created by having lines intersect a family of curves. In each case, the lines appear to be curved. Find the value of dy/dx for the indicated values of x and y.

(a) Circles: $x^2 + y^2 = C^2$

$x = 3, y = 4, C = 5$

(b) Hyperbolas: $xy = C$

$x = 1, y = 4, C = 4$

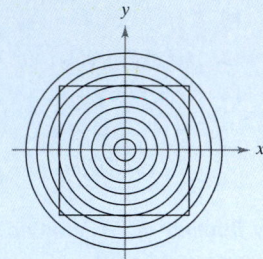 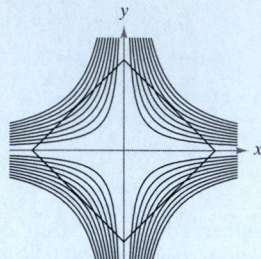

(c) Lines: $ax = by$

$x = \sqrt{3}, y = 3,$

$a = \sqrt{3}, b = 1$

(d) Cosine curves: $y = C \cos x$

$x = \dfrac{\pi}{3}, y = \dfrac{1}{3}, C = \dfrac{2}{3}$

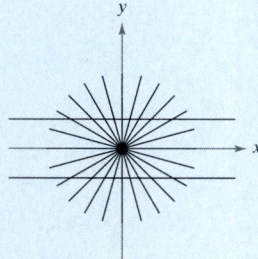 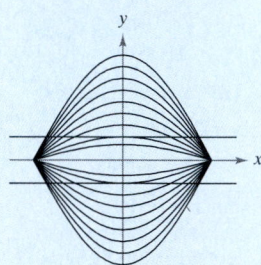

FOR FURTHER INFORMATION For more information on the mathematics of optical illusions, see the article "Descriptive Models for Perception of Optical Illusions" by David A. Smith in the 1985 (Volume 2) issue of the *UMAP Journal.*

Finding Related Rates • Problem Solving with Related Rates

Finding Related Rates

You have seen how the Chain Rule can be used to find dy/dx implicitly. Another important use of the Chain Rule is to find the rates of change of two or more related variables that are changing with respect to *time.*

For example, when water is drained out of a conical tank (see Figure 2.31), the volume V, the radius r, and the height h of the water level are all functions of time t. Knowing that these variables are related by the equation

$$V = \frac{\pi}{3} r^2 h$$

you can differentiate implicitly with respect to t to obtain the **related-rate** equation

$$\frac{dV}{dt} = \frac{\pi}{3} \left[r^2 \frac{dh}{dt} + h \left(2r \frac{dr}{dt} \right) \right]$$

$$= \frac{\pi}{3} \left(r^2 \frac{dh}{dt} + 2rh \frac{dr}{dt} \right).$$

From this equation you can see that the rate of change of V is related to the rates of change of both h and r.

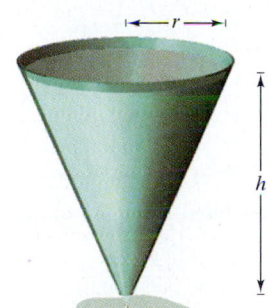

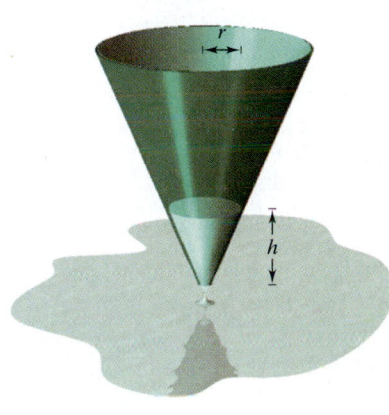

Volume is related to radius and height.
Figure 2.31

EXPLORATION

Finding a Related Rate In the conical tank shown in Figure 2.31, suppose that the height is changing at a rate of -0.2 foot per minute and the radius is changing at a rate of -0.1 foot per minute. What is the rate of change in the volume when the radius is $r = 1$ foot and the height is $h = 2$ feet? Does the rate of change in the volume depend on the values of r and h? Explain.

EXAMPLE 1   Two Rates That Are Related

Suppose x and y are both differentiable functions of t and are related by the equation

$$y = x^2 + 3.$$

Find dy/dt when $x = 1$, given that $dx/dt = 2$ when $x = 1$.

Solution   Using the Chain Rule, you can differentiate both sides of the equation *with respect to t.*

$$y = x^2 + 3 \qquad \text{Original equation}$$

$$\frac{d}{dt}[y] = \frac{d}{dt}[x^2 + 3] \qquad \text{Differentiate with respect to } t.$$

$$\frac{dy}{dt} = 2x \frac{dx}{dt} \qquad \text{Chain Rule}$$

When $x = 1$ and $dx/dt = 2$, you have

$$\frac{dy}{dt} = 2(1)(2) = 4.$$

Problem Solving with Related Rates

In Example 1, you were *given* an equation that related the variables x and y and were asked to find the rate of change of y when $x = 1$.

Equation: $\quad y = x^2 + 3$

Given rate: $\quad \dfrac{dx}{dt} = 2 \quad$ when $\quad x = 1$

Find: $\quad \dfrac{dy}{dt} \quad$ when $\quad x = 1$

In each of the remaining examples in this section, you must *create* a mathematical model from a verbal description.

EXAMPLE 2   Ripples in a Pond

A pebble is dropped into a calm pond, causing ripples in the form of concentric circles, as shown in Figure 2.32. The radius r of the outer ripple is increasing at a constant rate of 1 foot per second. When the radius is 4 feet, at what rate is the total area A of the disturbed water changing?

Solution   The variables r and A are related by $A = \pi r^2$. The rate of change of the radius r is $dr/dt = 1$.

Equation: $\quad A = \pi r^2$

Given rate: $\quad \dfrac{dr}{dt} = 1$

Find: $\quad \dfrac{dA}{dt} \quad$ when $\quad r = 4$

With this information, you can proceed as in Example 1.

$$\frac{d}{dt}[A] = \frac{d}{dt}[\pi r^2] \qquad \text{Differentiate with respect to } t.$$

$$\frac{dA}{dt} = 2\pi r \frac{dr}{dt} \qquad \text{Chain Rule}$$

$$\frac{dA}{dt} = 2\pi(4)(1) = 8\pi \qquad \text{Substitute 1 for } dr/dt \text{ and 4 for } r.$$

When $r = 4$, the area is changing at a rate of 8π square feet per second.

Total area increases as outer radius increases.

Figure 2.32

George Bernard/Science Photo Library/Photo Researchers, Inc.

Guidelines For Solving Related-Rate Problems

1. Identify all *given* quantities and quantities *to be determined*. Make a sketch and label the quantities.
2. Write an equation involving the variables whose rates of change either are given or are to be determined.
3. Using the Chain Rule, implicitly differentiate both sides of the equation *with respect to time t.*
4. *After* completing Step 3, substitute into the resulting equation all known values for the variables and their rates of change. Then solve for the required rate of change.

NOTE   In these guidelines, be sure you perform Step 3 *before* Step 4. Substituting the known values of the variables before differentiating will produce an inappropriate derivative.

The following table lists examples of mathematical models involving rates of change. For instance, the rate of change in the first example is the velocity of a car.

Verbal Statement	Mathematical Model
The velocity of a car after traveling for 1 hour is 50 miles per hour.	x = distance traveled $\dfrac{dx}{dt} = 50$ when $t = 1$
Water is being pumped into a swimming pool at a rate of 10 cubic meters per hour.	V = volume of water in pool $\dfrac{dV}{dt} = 10\text{m}^3/\text{hr}$
A gear is revolving at a rate of 25 revolutions per minute (1 revolution $= 2\pi$ rad).	θ = angle of revolution $\dfrac{d\theta}{dt} = 25(2\pi)$ rad/min

EXAMPLE 3   An Inflating Balloon

Air is being pumped into a spherical balloon (see Figure 2.33) at a rate of 4.5 cubic inches per minute. Find the rate of change of the radius when the radius is 2 inches.

Solution Let V be the volume of the balloon and let r be its radius. Because the volume is increasing at a rate of 4.5 cubic inches per minute, you know that at time t the rate of change of the volume is $dV/dt = \frac{9}{2}$. Thus, the problem can be stated as follows.

Given rate: $\dfrac{dV}{dt} = \dfrac{9}{2}$ (constant rate)

Find: $\dfrac{dr}{dt}$ when $r = 2$

To find the rate of change of the radius, you must find an equation that relates the radius r to the volume V.

$$V = \frac{4}{3}\pi r^3 \qquad \text{Volume of a sphere}$$

Implicit differentiation *with respect to t* produces

$$\frac{dV}{dt} = 4\pi r^2 \frac{dr}{dt} \qquad \text{Differentiate with respect to } t.$$

$$\frac{dr}{dt} = \frac{1}{4\pi r^2}\left(\frac{dV}{dt}\right). \qquad \text{Solve for } dr/dt.$$

Finally, when $r = 2$, the rate of change of the radius is

$$\frac{dr}{dt} = \frac{1}{16\pi}\left(\frac{9}{2}\right) \approx 0.09 \text{ inches per minute.}$$

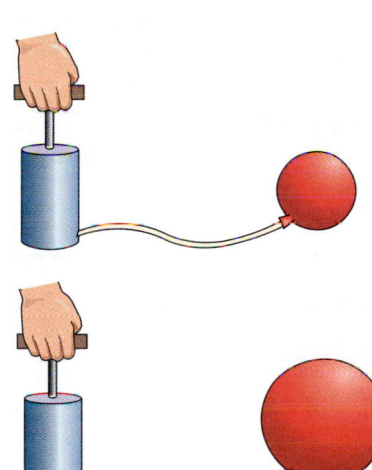

Expanding balloon
Figure 2.33

In Example 3, note that the volume is increasing at a *constant* rate but the radius is increasing at a *variable* rate. Just because two rates are related does not mean that they are proportional. In this particular case, the radius is growing more and more slowly as t increases. Do you see why?

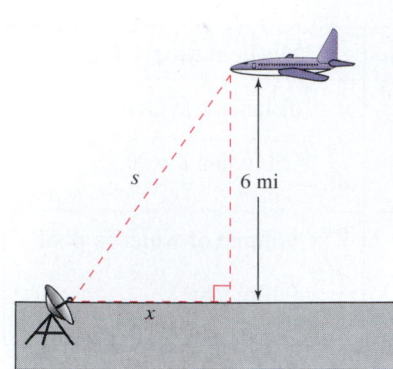

An airplane is flying at an altitude of 6 miles, s miles from the station.
Figure 2.34

EXAMPLE 4   **The Speed of an Airplane Tracked by Radar**

An airplane is flying on a flight path that will take it directly over a radar tracking station, as shown in Figure 2.34. If s is decreasing at a rate of 400 miles per hour when $s = 10$ miles, what is the speed of the plane?

Solution   Let x be the horizontal distance from the station, as shown in Figure 2.34. Notice that when $s = 10$, $x = \sqrt{10^2 - 36} = 8$.

Given rate: $ds/dt = -400$ when $s = 10$
Find: dx/dt when $s = 10$ and $x = 8$

You can find the velocity of the plane as follows.

$$x^2 + 6^2 = s^2 \qquad \text{Pythagorean Theorem}$$

$$2x\frac{dx}{dt} = 2s\frac{ds}{dt} \qquad \text{Differentiate with respect to } t.$$

$$\frac{dx}{dt} = \frac{s}{x}\left(\frac{ds}{dt}\right) \qquad \text{Solve for } dx/dt.$$

$$\frac{dx}{dt} = \frac{10}{8}(-400) = -500 \text{ miles per hour} \qquad \text{Substitute for } s, x, \text{ and } ds/dt.$$

Because the velocity is -500 miles per hour, the *speed* is 500 miles per hour.

EXAMPLE 5   **A Changing Angle of Elevation**

Find the rate of change in the angle of elevation of the camera shown in Figure 2.35 at 10 seconds after lift-off.

Solution   Let θ be the angle of elevation, as shown in Figure 2.35. When $t = 10$, the height s of the rocket is $s = 50t^2 = 50(10^2) = 5000$ feet.

Given rate: $ds/dt = 100t = $ velocity of rocket
Find: $d\theta/dt$ when $t = 10$ and $s = 5000$

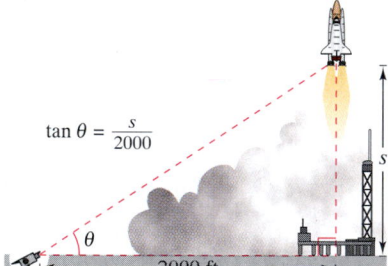

$\tan\theta = \dfrac{s}{2000}$

A television camera at ground level is filming the lift-off of a space shuttle that is rising vertically according to the position equation $s = 50t^2$, where s is measured in feet and t is measured in seconds. The camera is 2000 feet from the launch pad.
Figure 2.35

Using Figure 2.35, you can relate s and θ by the equation $\tan\theta = s/2000$.

$$\tan\theta = \frac{s}{2000} \qquad \text{See Figure 2.35.}$$

$$(\sec^2\theta)\frac{d\theta}{dt} = \frac{1}{2000}\left(\frac{ds}{dt}\right) \qquad \text{Differentiate with respect to } t.$$

$$\frac{d\theta}{dt} = \cos^2\theta\,\frac{100t}{2000} \qquad \text{Substitute } 100t \text{ for } ds/dt.$$

$$= \left(\frac{2000}{\sqrt{s^2 + 2000^2}}\right)^2 \frac{100t}{2000} \qquad \cos\theta = 2000/\sqrt{s^2 + 2000^2}$$

$$\frac{d\theta}{dt} = \frac{2000(100)(10)}{5000^2 + 2000^2} \qquad \text{Substitute for } s \text{ and } t.$$

$$= \frac{2}{29} \text{ radian per second.} \qquad \text{Simplify.}$$

When $t = 10$, θ is changing at a rate of $\frac{2}{29}$ radian per second.

EXAMPLE 6   The Velocity of a Piston

In the engine shown in Figure 2.36, a 7-inch connecting rod is fastened to a crank of radius 3 inches. The crankshaft rotates counterclockwise at a constant rate of 200 revolutions per minute. Find the velocity of the piston when $\theta = \pi/3$.

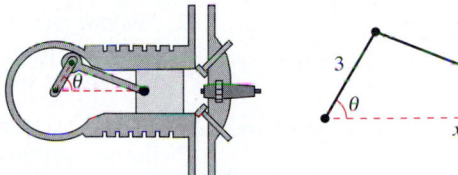

The velocity of a piston is related to the angle of the crankshaft.
Figure 2.36

Solution   Label the distances as shown in Figure 2.36. Because a complete revolution corresponds to 2π radians, it follows that $d\theta/dt = 200(2\pi) = 400\pi$ radians per minute.

Given rate: $\dfrac{d\theta}{dt} = 400\pi$ (constant rate)

Find: $\dfrac{dx}{dt}$ when $\theta = \dfrac{\pi}{3}$

You can use the Law of Cosines (Figure 2.37) to find an equation that relates x and θ.

$$7^2 = 3^2 + x^2 - 2(3)(x) \cos \theta$$

$$0 = 2x \frac{dx}{dt} - 6 \left(-x \sin \theta \frac{d\theta}{dt} + \cos \theta \frac{dx}{dt} \right)$$

$$(6 \cos \theta - 2x) \frac{dx}{dt} = 6x \sin \theta \frac{d\theta}{dt}$$

$$\frac{dx}{dt} = \frac{6x \sin \theta}{6 \cos \theta - 2x} \left(\frac{d\theta}{dt} \right)$$

When $\theta = \pi/3$, you can solve for x as follows.

$$7^2 = 3^2 + x^2 - 2(3)(x) \cos \frac{\pi}{3}$$

$$49 = 9 + x^2 - 6x \left(\frac{1}{2} \right)$$

$$0 = x^2 - 3x - 40$$

$$0 = (x - 8)(x + 5)$$

$$x = 8 \qquad\qquad \text{Choose positive solution.}$$

Thus, when $x = 8$ and $\theta = \pi/3$, the velocity of the piston is

$$\frac{dx}{dt} = \frac{6(8)(\sqrt{3}/2)}{6(1/2) - 16}(400\pi)$$

$$= \frac{9600\pi\sqrt{3}}{-13}$$

$$\approx -4018 \text{ inches per minute.}$$

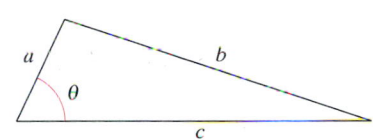

Law of Cosines:
$b^2 = a^2 + c^2 - 2ac \cos \theta$
Figure 2.37

NOTE   Note that the velocity in Example 6 is negative, because x represents a distance that is decreasing.

In Exercises 1–4, assume that x and y are both differentiable functions of t and find the required values of dy/dt and dx/dt.

Equation	Find	Given
1. $y = \sqrt{x}$	(a) $\dfrac{dy}{dt}$ when $x = 4$	$\dfrac{dx}{dt} = 3$
	(b) $\dfrac{dx}{dt}$ when $x = 25$	$\dfrac{dy}{dt} = 2$
2. $y = x^2 - 3x$	(a) $\dfrac{dy}{dt}$ when $x = 3$	$\dfrac{dx}{dt} = 2$
	(b) $\dfrac{dx}{dt}$ when $x = 1$	$\dfrac{dy}{dt} = 5$
3. $xy = 4$	(a) $\dfrac{dy}{dt}$ when $x = 8$	$\dfrac{dx}{dt} = 10$
	(b) $\dfrac{dx}{dt}$ when $x = 1$	$\dfrac{dy}{dt} = -6$
4. $x^2 + y^2 = 25$	(a) $\dfrac{dy}{dt}$ when $x = 3, y = 4$	$\dfrac{dx}{dt} = 8$
	(b) $\dfrac{dx}{dt}$ when $x = 4, y = 3$	$\dfrac{dy}{dt} = -2$

In Exercises 5–8, a point is moving along the graph of the function such that dx/dt is 2 centimeters per second. Find dy/dt for the specified values of x.

Function	Values of x	
5. $y = x^2 + 1$	(a) $x = -1$	(b) $x = 0$
	(c) $x = 1$	(d) $x = 3$
6. $y = \dfrac{1}{1 + x^2}$	(a) $x = -2$	(b) $x = 0$
	(c) $x = 2$	(d) $x = 10$
7. $y = \tan x$	(a) $x = -\dfrac{\pi}{3}$	(b) $x = -\dfrac{\pi}{4}$
	(c) $x = 0$	(d) $x = 1$
8. $y = \sin x$	(a) $x = \dfrac{\pi}{6}$	(b) $x = \dfrac{\pi}{4}$
	(c) $x = \dfrac{\pi}{3}$	(d) $x = \dfrac{\pi}{2}$

Think About It **In Exercises 9 and 10, using the graph of f, (a) determine whether dy/dt increases or decreases for increasing x and constant dx/dt, and (b) determine whether dx/dt increases or decreases for increasing y and constant dy/dt.**

9.

10.

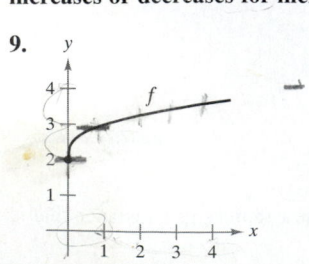

11. Find the rate of change of the distance between the origin and a moving point on the graph of $y = x^2 + 1$ if $dx/dt = 2$ centimeters per second.

12. Find the rate of change of the distance between the origin and a moving point on the graph of $y = \sin x$ if $dx/dt = 2$ centimeters per second.

13. ***Area*** The radius r of a circle is increasing at a rate of 2 centimeters per minute. Find the rate of change of the area when (a) $r = 6$ centimeters and (b) $r = 24$ centimeters.

14. ***Area*** Let A be the area of a circle of radius r that is changing with respect to time. If dr/dt is constant, is dA/dt constant? Explain.

15. ***Area*** The included angle of the two sides of constant equal length s of an isosceles triangle is θ.

   (a) Show that the area of the triangle is given by $A = \frac{1}{2}s^2 \sin \theta$.

   (b) If θ is increasing at the rate of $\frac{1}{2}$ radian per minute, find the rate of change of the area when $\theta = \pi/6$ and $\theta = \pi/3$.

   (c) Explain why the rate of change of the area of the triangle is not constant even though $d\theta/dt$ is constant.

16. ***Volume*** The radius r of a sphere is increasing at a rate of 2 inches per minute.

   (a) Find the rate of change of the volume when $r = 6$ inches and $r = 24$ inches.

   (b) Explain why the rate of change of the volume of the sphere is not constant even though dr/dt is constant.

17. ***Volume*** A spherical balloon is inflated with gas at the rate of 500 cubic centimeters per minute. How fast is the radius of the balloon increasing at the instant the radius is (a) 30 centimeters and (b) 60 centimeters?

18. ***Volume*** All edges of a cube are expanding at a rate of 3 centimeters per second. How fast is the volume changing when each edge is (a) 1 centimeter and (b) 10 centimeters?

19. ***Surface Area*** The conditions are the same as in Exercise 18. Determine how fast the *surface area* is changing when each edge is (a) 1 centimeter and (b) 10 centimeters.

20. ***Volume*** The formula for the volume of a cone is $V = \frac{1}{3}\pi r^2 h$. Find the rate of change of the volume if dr/dt is 2 inches per minute and $h = 3r$ when (a) $r = 6$ inches and (b) $r = 24$ inches.

21. ***Volume*** At a sand and gravel plant, sand is falling off a conveyor and onto a conical pile at a rate of 10 cubic feet per minute. The diameter of the base of the cone is approximately three times the altitude. At what rate is the height of the pile changing when the pile is 15 feet high?

22. ***Depth*** A conical tank (with vertex down) is 10 feet across the top and 12 feet deep. If water is flowing into the tank at a rate of 10 cubic feet per minute, find the rate of change of the depth of the water when the water is 8 feet deep.

23. Depth   A swimming pool is 12 meters long, 6 meters wide, 1 meter deep at the shallow end, and 3 meters deep at the deep end (see figure). Water is being pumped into the pool at $\frac{1}{4}$ cubic meter per minute, and there is 1 meter of water at the deep end.

(a) What percent of the pool is filled?

(b) At what rate is the water level rising?

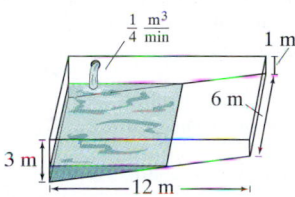
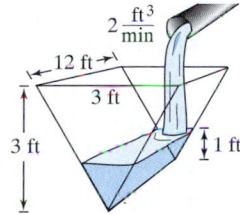

Figure for 23   **Figure for 24**

24. Depth   A trough is 12 feet long and 3 feet across the top (see figure). Its ends are isosceles triangles with altitudes of 3 feet. If water is being pumped into the trough at 2 cubic feet per minute, how fast is the water level rising when the water is 1 foot deep?

25. Moving Ladder   A ladder 25 feet long is leaning against the wall of a house (see figure). The base of the ladder is pulled away from the wall at a rate of 2 feet per second.

(a) How fast is the top moving down the wall when the base of the ladder is 7 feet, 15 feet, and 24 feet from the wall?

(b) Consider the triangle formed by the side of the house, the ladder, and the ground. Find the rate at which the area of the triangle is changing when the base of the ladder is 7 feet from the wall.

(c) Find the rate at which the angle between the ladder and the wall of the house is changing when the base of the ladder is 7 feet from the wall.

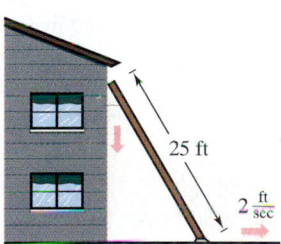
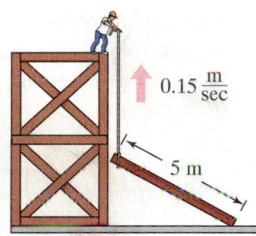

Figure for 25   **Figure for 26**

FOR FURTHER INFORMATION For more information on the mathematics of moving ladders see the article "The Falling Ladder Paradox" by Paul Scholten and Andrew Simoson in the January 1996 issue of *The College Mathematics Journal*.

26. Construction   A construction worker pulls a 5-meter plank up the side of a building under construction by means of a rope tied to one end of a plank (see figure). Assume the opposite end of the plank follows a path perpendicular to the wall of the building and the worker pulls the rope at a rate of 0.15 meter per second. How fast is the end of the plank sliding along the ground when it is 2.5 meters from the wall of the building?

27. Construction   A winch at the top of a 12-meter building pulls a pipe of the same length to a vertical position, as shown in the figure. The winch pulls in rope at a rate of -0.2 meters per second. Find the rate of vertical change and the rate of horizontal change at the end of the pipe when $y = 6$.

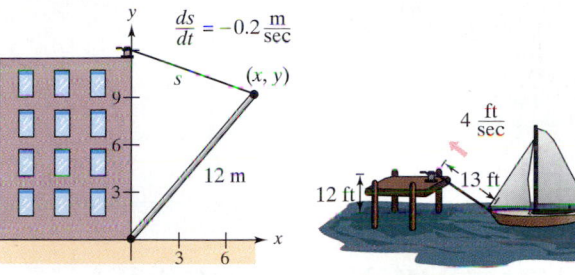

Figure for 27   **Figure for 28**

28. Boating   A boat is pulled into a dock by means of a winch 12 feet above the deck of the boat (see figure). The winch pulls in rope at a rate of 4 feet per second. Determine the speed of the boat when there are 13 feet of rope out. What happens to the speed of the boat as it gets closer to the dock?

29. Air Traffic Control   An air traffic controller spots two planes at the same altitude converging on a point as they fly at right angles to each other (see figure). One plane is 150 miles from the point moving at 450 miles per hour. The other plane is 200 miles from the point moving at 600 miles per hour.

(a) At what rate is the distance between the planes decreasing?

(b) How much time does the air traffic controller have to get one of the planes on a different flight path?

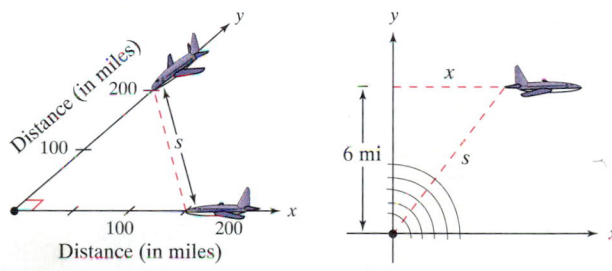

Figure for 29   **Figure for 30**

30. Air Traffic Control   An airplane is flying at an altitude of 6 miles and passes directly over a radar antenna (see figure). When the plane is 10 miles away ($s = 10$), the radar detects that the distance s is changing at a rate of 240 miles per hour. What is the speed of the plane?

31. Baseball   A baseball diamond has the shape of a square with sides 90 feet long (see figure). A player running from second base to third base at a speed of 28 feet per second is 30 feet from third base. At what rate is the player's distance s from home plate changing?

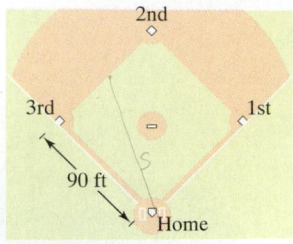 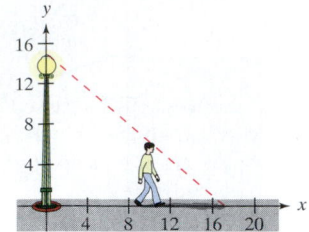

Figure for 31 and 32   **Figure for 33**

32. Baseball   For the baseball diamond in Exercise 31, suppose the player is running from first to second at a speed of 28 feet per second. Find the rate at which the distance from home plate is changing when the player is 30 feet from second base.

33. Shadow Length   A man 6 feet tall walks at a rate of 5 feet per second away from a light that is 15 feet above the ground (see figure). When he is 10 feet from the base of the light,

(a) at what rate is the tip of his shadow moving?

(b) at what rate is the length of his shadow changing?

34. Shadow Length   Repeat Exercise 33 for a man 6 feet tall walking at a rate of 5 feet per second *toward* a light that is 20 feet above the ground (see figure).

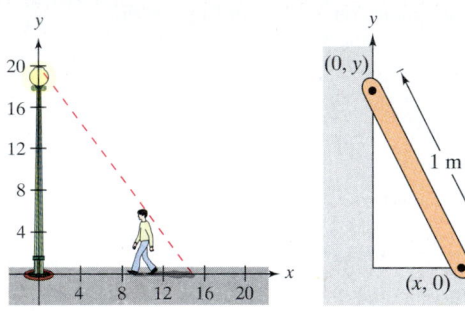

Figure for 34   **Figure for 35**

35. Machine Design   The endpoints of a movable rod of length 1 meter have coordinates $(x, 0)$ and $(0, y)$ (see figure). The position of the end on the x-axis is

$$x(t) = \frac{1}{2} \sin \frac{\pi t}{6}$$

where t is the time in seconds.

(a) Find the time of one complete cycle of the rod.

(b) What is the lowest point reached by the end of the rod on the y-axis?

(c) Find the speed of the y-axis endpoint when the x-axis endpoint is $\left(\frac{1}{4}, 0\right)$.

36. Machine Design   Repeat Exercise 35 for a position function of $x(t) = \frac{3}{5} \sin \pi t$. Use the point $\left(\frac{3}{10}, 0\right)$ for part (c).

37. Evaporation   As a spherical raindrop falls, it reaches a layer of dry air and begins to evaporate at a rate that is proportional to its surface area $(S = 4\pi r^2)$. Show that the radius of the raindrop decreases at a constant rate.

38. Electricity   The combined electrical resistance R of R_1 and R_2, connected in parallel, is given by

$$\frac{1}{R} = \frac{1}{R_1} + \frac{1}{R_2}$$

where R, R_1, and R_2 are measured in ohms. R_1 and R_2 are increasing at rates of 1 and 1.5 ohms per second, respectively. At what rate is R changing when $R_1 = 50$ ohms and $R_2 = 75$ ohms?

39. Adiabatic Expansion   When a certain polyatomic gas undergoes adiabatic expansion, its pressure p and volume v satisfy the equation

$$pv^{1.3} = k$$

where k is a constant. Find the relationship between the related rates dp/dt and dv/dt.

40. Roadway Design   Cars on a certain roadway travel on a circular arc of radius r. In order not to rely on friction alone to overcome the centrifugal force, the road is banked at an angle of magnitude θ from the horizontal. The banking angle must satisfy the equation

$$rg \tan \theta = v^2$$

where v is the velocity of the cars and $g = 32$ feet per second per second is the acceleration due to gravity. Find the relationship between the related rates dv/dt and $d\theta/dt$.

41. Angle of Elevation   A balloon rises at a rate of 3 meters per second from a point on the ground 30 meters from an observer. Find the rate of change of the angle of elevation of the balloon from the observer when the balloon is 30 meters above the ground.

42. Angle of Elevation   A fish is reeled in at a rate of 1 foot per second from a point 15 feet above the water (see figure). At what rate is the angle between the line and the water changing when there are 25 feet of line out?

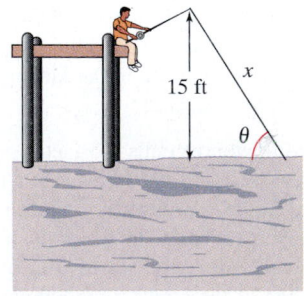

43. *Angle of Elevation*   An airplane flies at an altitude of 5 miles toward a point directly over an observer (see figure). The speed of the plane is 600 miles per hour. Find the rate at which the angle of elevation θ is changing when the angle is (a) $\theta = 30°$, (b) $\theta = 60°$, and (c) $\theta = 75°$.

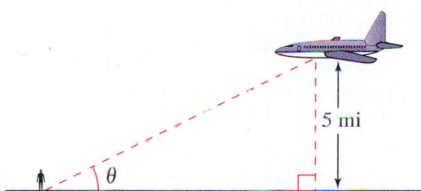

44. *Linear vs. Angular Speed*   A patrol car is parked 50 feet from a long warehouse (see figure). The revolving light on top of the car turns at a rate of 30 revolutions per minute. How fast is the light beam moving along the wall when the beam makes angles of (a) $\theta = 30°$, (b) $\theta = 60°$, and (c) $\theta = 70°$ with the line perpendicular from the light to the wall?

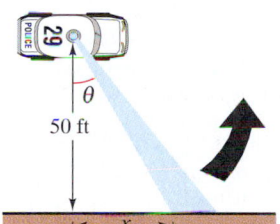   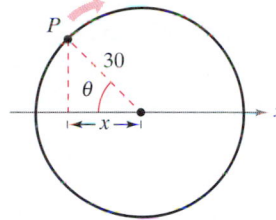

Figure for 44   **Figure for 45**

45. *Linear vs. Angular Speed*   A wheel of radius 30 centimeters revolves at a rate of 10 revolutions per second. A dot is painted at a point P on the rim of the wheel (see figure).

(a) Find dx/dt as a function of θ.

(b) Use a graphing utility to graph the function in part (a).

(c) When is the absolute value of rate of change of x greatest? When is it least?

(d) Find dx/dt when $\theta = 30°$ and $\theta = 60°$.

46. *Flight Control*   An airplane is flying in still air with an airspeed of 240 miles per hour. If it is climbing at an angle of 22°, find the rate at which it is gaining altitude.

47. *Security Camera*   A security camera is centered 50 feet above a 100-foot hallway (see figure). It is easiest to design the camera with a constant angular rate of rotation, but this results in a variable rate at which the images of the surveillance area are recorded. Therefore, it is desirable to design a system with a variable rate of rotation and a constant rate of movement of the scanning beam along the hallway. Find a model for the variable rate of rotation if $|dx/dt| = 2$ feet per second.

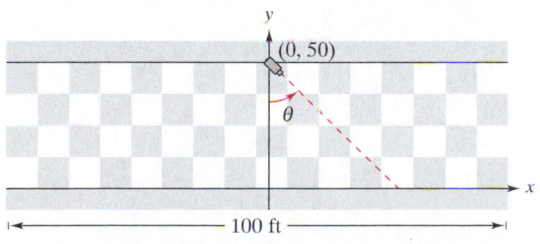

48. *Think About It*   Describe the relationship between the rate of change of y and the rate of change of x in each of the following. Assume all variables and derivatives are positive.

(a) $\dfrac{dy}{dt} = 3\dfrac{dx}{dt}$

(b) $\dfrac{dy}{dt} = x(L - x)\dfrac{dx}{dt}, \quad 0 \le x \le L$

Acceleration   **In Exercises 49 and 50, find the acceleration of the specified object. (*Hint:* Recall that if a variable is changing at a constant rate, its acceleration is zero.)**

49. Find the acceleration of the top of the ladder described in Exercise 25 when the base of the ladder is 7 feet from the wall.

50. Find the acceleration of the boat in Exercise 28 when there are 13 feet of rope out.

51. *Modeling Data*   The table gives the number (in millions) of single women s and married women m in the civilian work force in the United States for the years 1990 through 1994. (*Source: U.S. Bureau of Labor Statistics*)

Year	1990	1991	1992	1993	1994
s	14.2	14.3	14.5	14.6	15.3
m	31.0	31.2	31.7	32.0	32.9

(a) Use the regression capabilities of a graphing utility to find a model of the form $m(s) = as^2 + bs + c$ for the data, where t is the time in years, with $t = 0$ corresponding to 1990.

(b) Find $\dfrac{dm}{dt}$.

(c) Use the model to estimate dm/dt for $t = 5$ if it is predicted that the number of single women in the work force will increase at the rate 1.2 million per year.

52. A ball is dropped from a height of 20 meters, 12 meters away from the top of a 20-meter lamppost (see figure). The ball's shadow, caused by the light at the top of the lamppost, is moving along the level ground. How fast is the shadow moving 1 second after the ball is released? (*Submitted by Dennis Gittinger, St. Philips College, San Antonio, TX*)

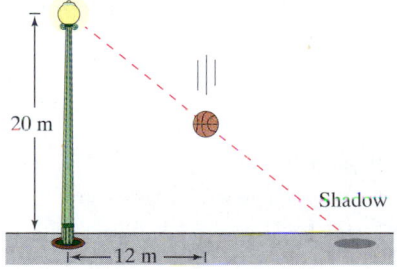

REVIEW EXERCISES FOR CHAPTER 2

In Exercises 1 and 2, find the derivative of the function by using the definition of the derivative.

1. $f(x) = x^2 - 2x + 3$

2. $f(x) = \dfrac{x + 1}{x - 1}$

3. *Writing* Each figure shows the graphs of a function and its derivative. Label the graphs as f or f' and write a short paragraph stating the criteria used in making the selection.

(a)

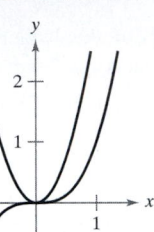

(b)

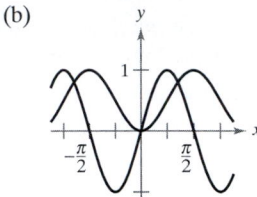

4. *Writing* Use the function $f(x) = x\sqrt{4 - x}$ for each of the following.

(a) Use a graphing utility to graph the function.

(b) Find an equation of the tangent line to the graph of f at the point $(0, 0)$.

(c) Use a graphing utility to complete the table for $x = 0$.

Δx	$f(x + \Delta x)$	$f(x)$	$\dfrac{f(x + \Delta x) - f(x)}{\Delta x}$
2			
1			
0.5			
0.1			

(d) Write a short paragraph giving the geometric interpretation of the last column of the table. How does it relate to the result in part (b)?

In Exercises 5 and 6, describe the x-values at which f is differentiable.

5. $f(x) = (x + 1)^{2/3}$

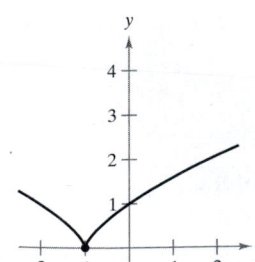

6. $f(x) = \dfrac{4x}{x + 3}$

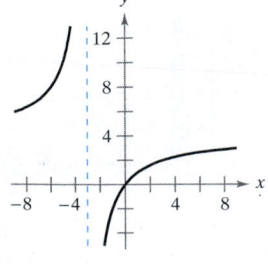

In Exercises 7–22, find the derivative of the algebraic function.

7. $f(x) = x^3 - 3x^2$

8. $f(x) = x^{1/2} - x^{-1/2}$

9. $f(x) = \dfrac{2x^3 - 1}{x^2}$

10. $f(x) = \dfrac{x + 1}{x - 1}$

11. $g(t) = \dfrac{2}{3t^2}$

12. $h(x) = \dfrac{2}{(3x)^2}$

13. $f(x) = \sqrt{1 - x^3}$

14. $f(x) = \sqrt[3]{x^2 - 1}$

15. $f(x) = (3x^2 + 7)(x^2 - 2x + 3)$

16. $f(s) = (s^2 - 1)^{5/2}(s^3 + 5)$

17. $f(x) = \left(x^2 + \dfrac{1}{x}\right)^5$

18. $h(\theta) = \dfrac{\theta}{(1 - \theta)^3}$

19. $f(x) = \dfrac{x^2 + x - 1}{x^2 - 1}$

20. $f(x) = \dfrac{6x - 5}{x^2 + 1}$

21. $f(x) = \dfrac{1}{4 - 3x^2}$

22. $f(x) = \dfrac{9}{3x^2 - 2x}$

In Exercises 23–34, find the derivative of the trigonometric function.

23. $y = 3\cos(3x + 1)$

24. $y = 1 - \cos 2x + 2\cos^2 x$

25. $y = \frac{1}{2}\csc 2x$

26. $y = \csc 3x + \cot 3x$

27. $y = \dfrac{x}{2} - \dfrac{\sin 2x}{4}$

28. $y = \dfrac{1 + \sin x}{1 - \sin x}$

29. $y = \frac{2}{3}\sin^{3/2}x - \frac{2}{7}\sin^{7/2}x$

30. $y = \dfrac{\sec^7 x}{7} - \dfrac{\sec^5 x}{5}$

31. $y = -x\tan x$

32. $y = x\cos x - \sin x$

33. $y = \dfrac{\sin x}{x^2}$

34. $y = \dfrac{\cos(x - 1)}{x - 1}$

In Exercises 35–42, use a symbolic differentiation utility to find the derivative of the function. Use the utility to graph the function and its derivative on the same set of coordinate axes. Describe the behavior of the function that corresponds to any zeros of the graph of the derivative.

35. $f(t) = t^2(t - 1)^5$

36. $f(x) = [(x - 2)(x + 4)]^2$

37. $g(x) = \dfrac{2x}{\sqrt{x + 1}}$

38. $g(x) = x\sqrt{x^2 + 1}$

39. $f(t) = \sqrt{t + 1}\sqrt[3]{t + 1}$

40. $y = \sqrt{3x}(x + 2)^3$

41. $y = \tan\sqrt{1 - x}$

42. $y = 2\csc^3(\sqrt{x})$

In Exercises 43 and 44, find the value of the derivative of the function at the indicated point. Use the derivative feature of a graphing utility to confirm your result.

Function	Point
43. $f(x) = \cos \dfrac{\pi x}{2}$	$\left(\frac{2}{3}, \frac{1}{2}\right)$
44. $f(x) = x + \dfrac{8}{x^2}$	$(2, 4)$

In Exercises 45–48, find the second derivative of the function.

45. $y = 2x^2 + \sin 2x$ **46.** $y = \dfrac{1}{x} + \tan x$

47. $f(x) = \cot x$ **48.** $y = \sin^2 x$

In Exercises 49–52, use a symbolic differentiation utility to find the second derivative of the function.

49. $f(t) = \dfrac{t}{(1 - t)^2}$ **50.** $g(x) = \dfrac{6x - 5}{x^2 + 1}$

51. $g(x) = x \tan x$ **52.** $h(x) = x\sqrt{x^2 - 1}$

In Exercises 53–58, use implicit differentiation to find dy/dx.

53. $x^2 + 3xy + y^3 = 10$ **54.** $x^2 + 9y^2 - 4x + 3y = 0$

55. $y\sqrt{x} - x\sqrt{y} = 16$ **56.** $y^2 = (x - y)(x^2 + y)$

57. $x \sin y = y \cos x$ **58.** $\cos(x + y) = x$

In Exercises 59–64, find the equations of the tangent line and the normal line to the graph of the equation at the indicated point. Use a graphing utility to graph the equation, the tangent line, and the normal line.

59. $y = (x + 3)^3$, $(-2, 1)$ **60.** $y = (x - 2)^2$, $(2, 0)$

61. $x^2 + y^2 = 20$, $(2, 4)$ **62.** $x^2 - y^2 = 16$, $(5, 3)$

63. $y = \sqrt[3]{(x - 2)^2}$, $(3, 1)$ **64.** $y = \dfrac{2x}{1 - x^2}$, $(0, 0)$

65. Find the points on the graph of $f(x) = \frac{1}{3}x^3 + x^2 - x - 1$ when the slope is (a) -1, (b) 2, and (c) 0.

66. Find the points on the graph of $f(x) = x^2 + 1$ when the slope is (a) -1, (b) 0, and (c) 1.

67. Sketch the graph of $f(x) = 4 - |x - 2|$.

(a) Is f continuous at $x = 2$?

(b) Is f differentiable at $x = 2$? Explain.

68. Sketch the graph of

$$f(x) = \begin{cases} x^2 + 4x + 2, & x < -2 \\ 1 - 4x - x^2, & x \geq -2. \end{cases}$$

(a) Is f continuous at $x = -2$?

(b) Is f differentiable at $x = -2$? Explain.

In Exercises 69 and 70, show that the function satisfies the equation.

Function	Equation
69. $y = 2 \sin x + 3 \cos x$	$y'' + y = 0$
70. $y = \dfrac{10 - \cos x}{x}$	$xy' + y = \sin x$

71. *Refrigeration* The temperature T of food put in a freezer is

$$T = \frac{700}{t^2 + 4t + 10}$$

where t is the time in hours. Find the rate of change T with respect to t at each of the following times.

(a) $t = 1$ (b) $t = 3$ (c) $t = 5$ (d) $t = 10$

72. *Fluid Flow* The emergent velocity v of a liquid flowing from a hole in the bottom of a tank is given by $v = \sqrt{2gh}$, where g is the acceleration due to gravity (32 feet per second per second) and h is the depth of the liquid in the tank. Find the rate of change of v with respect to h when (a) $h = 9$ and (b) $h = 4$. (Note that $g = +32$ feet per second per second. The sign of g depends on how a problem is modeled. In this case, letting g be negative would produce an imaginary value for v.)

73. *Vibrating String* When a guitar string is plucked, it vibrates with a frequency of $F = 200\sqrt{T}$, where F is measured in vibrations per second and the tension T is measured in pounds. Find the rate of change in F when (a) $T = 4$ and (b) $T = 9$.

74. *Vertical Motion* A ball is dropped from a height of 100 feet. One second later, another ball is dropped from a height of 75 feet. Which ball hits the ground first?

75. *Vertical Motion* What is the smallest initial velocity that is required to throw a stone 49 feet up to the top of a silo?

76. *Vertical Motion* A bomb is dropped from an airplane at an altitude of 14,400 feet. How long will it take to reach the ground? (Because of the motion of the plane, the fall will not be vertical, but the time will be the same as that for a vertical fall.) The plane is moving at 600 miles per hour. How far will the bomb move horizontally after it is released from the plane?

77. *Projectile Motion* A ball thrown follows a path described by $y = x - 0.02x^2$.

(a) Sketch a graph of the path.

(b) Find the total horizontal distance the ball was thrown.

(c) At what x-value does the ball reach its maximum height? (Use the symmetry of the path.)

(d) Find an equation that gives the instantaneous rate of change of the height of the ball with respect to the horizontal change. Evaluate the equation at $x = 0$, 10, 25, 30, and 50.

(e) What is the instantaneous rate of change of the height when the ball reaches its maximum height?

78. A point moves along the curve $y = \sqrt{x}$ in such a way that the y-value is increasing at a rate of 2 units per second. At what rate is x changing for each of the following values?

(a) $x = \frac{1}{2}$ (b) $x = 1$ (c) $x = 4$

79. The same conditions exist as in Exercise 78. Find the rate of change of the distance between the origin and a point (x, y) on the graph for each of the following.

(a) $x = \frac{1}{2}$ (b) $x = 1$ (c) $x = 4$

80. *Projectile Motion* The path of a projectile thrown at an angle of 45° with level ground is

$$y = x - \frac{32}{v_0^2}(x^2)$$

where the initial velocity is v_0 feet per second.

(a) Find the x-coordinate of the point where the projectile strikes the ground. Use the symmetry of the path of the projectile to locate the x-coordinate of the point where the projectile reaches its maximum height.

(b) What is the instantaneous rate of change of the height when the projectile is at its maximum height?

(c) Show that doubling the initial velocity of the projectile multiplies both the maximum height and the range by a factor of 4.

(d) Find the maximum height and range of a projectile thrown with an initial velocity of 70 feet per second. Use a graphing utility to sketch the path of the projectile.

81. *Changing Depth* The cross section of a 5-meter trough is an isosceles trapezoid with a 2-meter lower base, a 3-meter upper base, and an altitude of 2 meters. Water is running into the trough at a rate of 1 cubic meter per minute. How fast is the water level rising when the water is 1 meter deep?

82. *Linear and Angular Velocity* A rotating beacon is located 1 kilometer off a straight shoreline. If the beacon rotates at a rate of 3 revolutions per minute, how fast (in kilometers per hour) does the beam of light appear to be moving to a viewer who is $\frac{1}{2}$ kilometer down the shoreline?

83. *Moving Shadow* A sandbag is dropped from a balloon at a height of 60 meters when the angle of elevation to the sun is 30° (see figure). Find the rate at which the shadow of the sandbag is traveling along the ground when the sandbag is at a height of 35 meters. (*Hint:* The position of the sandbag is given by $s(t) = 60 - 4.9t^2$.)

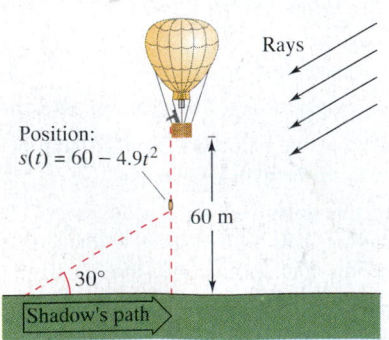

Position:
$s(t) = 60 - 4.9t^2$

Rays

60 m

30°

Shadow's path

84. *Modeling Data* Sales S, in billions of dollars, of recreational vehicles in the United States for the years 1980 through 1994 are given in the table. (*Source: National Sporting Goods Association*)

Year	0	1	2	3	4	5	6	7
Sales	1.2	1.8	1.7	3.4	4.1	3.5	3.9	4.5

Year	8	9	10	11	12	13	14
Sales	4.8	4.5	4.1	3.6	4.4	4.8	4.8

(a) Use the regression capabilities of a graphing utility to find a quadratic model for the data, where t is the time in years, with $t = 0$ corresponding to 1980.

(b) Use a graphing utility to plot the data and graph the model.

(c) Use a graphing utility to graph dS/dt.

(d) The table shows that sales were down from 1989 through 1991. Does the derivative of the model show this decline? Explain. Give a possible explanation for the decline.

(e) Does the model show the full magnitude of the sales slump? Explain.

(f) Use the derivative to determine the interval of time when sales were increasing most rapidly. Explain.

85. *Modeling Data* The speed of a car in miles per hour and the stopping distance in feet are recorded in the table.

Speed (x)	20	30	40	50	60
Stopping Distance (y)	25	55	105	188	300

(a) Use the regression capabilities of a graphing utility to find a quadratic model for the data.

(b) Use a graphing utility to plot the data and graph the model.

(c) Use a graphing utility to graph dy/dx.

(d) Use the model to approximate the stopping distance at a speed of 65 miles per hour.

(e) Use the graphs in parts (b) and (c) to explain the change in stopping distance as the speed is increased.

86. *Horizontal Motion* The position function of a particle moving along the x-axis is

$$x(t) = t^2 - 3t + 2$$

for $-\infty < t < \infty$.

(a) Find the velocity and acceleration of the particle.

(b) Find the open t-interval(s) in which the particle is moving to the left.

(c) Find the position of the particle when the velocity is 0.

(d) Find the speed of the particle when the position is 0.

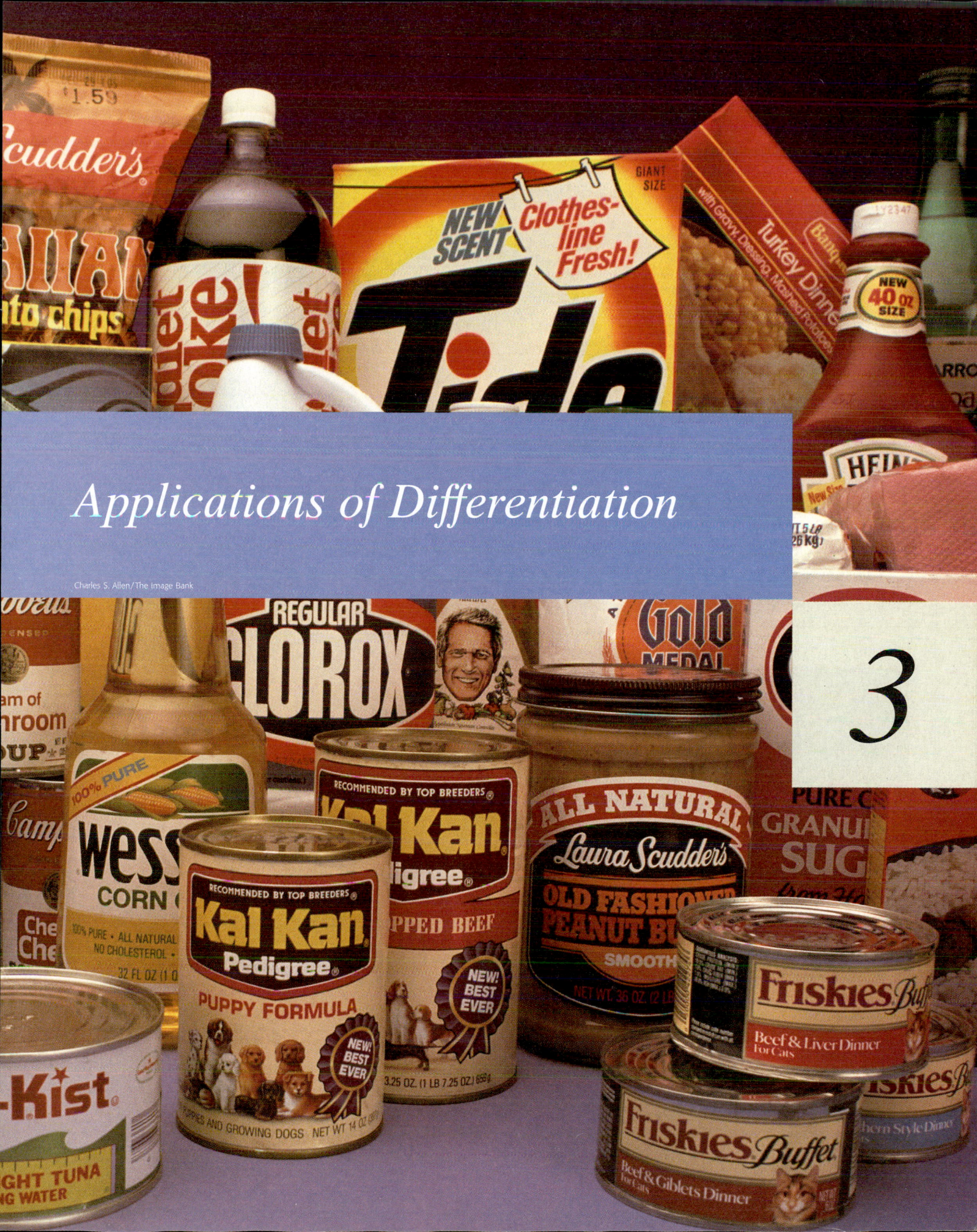

Applications of Differentiation

Charles S. Allen/The Image Bank

3

Many people are involved in deciding how to package the products you see in grocery stores. Packaging engineers select materials and package shapes to adequately protect the product through shipping at a reasonable cost.

A container's shape, as well as its material, is important in determining its strength. From an engineering perspective, the sphere is the strongest form, followed by the circular cylinder. The rectangular box comes in a poor third. From a perspective of cost, it is preferable to use the smallest amount of material possible.

The table gives the approximate measurements in inches of several common items packed in cylindrical containers.

By the time packaging engineers begin work on a container, design specialists have already done their work. Designers use color, shape, and words to create an image that they think will appeal to their targeted market. Many designers believe that the package is at least as important as the product inside.

Product	Radius (in.)	Height (in.)	Volume (in.³)
Coffee creamer	1.50	6.85	48.42
Cleanser	1.45	7.50	49.54
Coffee	1.95	5.20	62.12
Pineapple juice	2.10	6.70	92.82
Frosting	1.63	3.60	30.05
Soup	1.30	3.80	20.18
Tomato puree	1.95	4.40	52.56
Baking powder	1.25	3.65	17.92

An infinite number of dimensions can be used to construct a right circular container of a given volume. The graph below shows the relationship between the radius and surface area for containers that have a volume of 48.4 cubic inches.

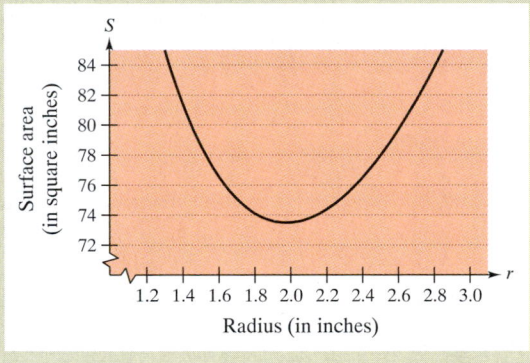

Successful designer Primo Angeli feels so strongly about the importance of packaging that he has designed entire lines of packaged product ideas in realistic packages so that consumer response to these ideas can be measured before massive investments are made in product development.

QUESTIONS

1. Create a table of values for the dimensions of a cylinder with a volume of 49.54 cubic inches. Does it appear that the cleanser container minimizes surface area?

2. Suppose you are designing a coffee creamer container that has a volume of 48.42 cubic inches. Use the equations for the surface area of a cylinder and the volume of a cylinder to develop an equation relating the radius r and surface area S.

   $S = 2\pi r^2 + 2\pi rh$ Surface area of a right circular cylinder

   $V = \pi r^2 h$ Volume of a right circular cylinder

3. Repeat Question 2 for each of the other containers in the table. Use a graphing utility to plot each equation. Determine whether the radius of each container is larger than, smaller than, or equal to the "optimal" radius.

4. Suppose, in order to fit more writing on the cylinder, you wanted to maximize the surface area of a cylinder that holds 49.5 cubic inches. Can you do this? Explain.

The concepts presented here will be explored further in this chapter. For an extension of this application, see the lab series that accompanies this text.

Extrema of a Function • Relative Extrema and Critical Numbers •
Finding Extrema on a Closed Interval

Extrema of a Function

In calculus, much effort is devoted to determining the behavior of a function f on an interval I. Does f have a maximum value on I? Does it have a minimum value? Where is the function increasing? Where is it decreasing? In this chapter you will learn how derivatives can be used to answer these questions. You will also see why these questions are important in real-life applications.

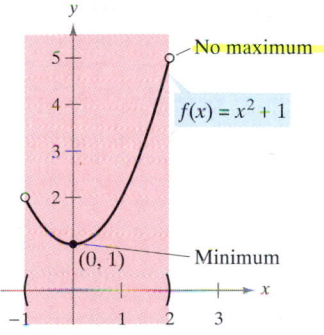

(a) f is continuous, $[-1, 2]$ is closed.

> ### Definition of Extrema
>
> Let f be defined on an interval I containing c.
>
> 1. $f(c)$ is the **minimum of f on I** if $f(c) \leq f(x)$ for all x in I.
> 2. $f(c)$ is the **maximum of f on I** if $f(c) \geq f(x)$ for all x in I.
>
> The minimum and maximum of a function on an interval are the **extreme values,** or **extrema,** of the function on the interval. The minimum and maximum of a function on an interval are also called the **absolute minimum** and **absolute maximum** on the interval.

A function need not have a minimum or a maximum on an interval. For instance, in Figure 3.1(a) and (b), you can see that the function $f(x) = x^2 + 1$ has both a minimum and a maximum on the closed interval $[-1, 2]$, but does not have a maximum on the open interval $(-1, 2)$. Moreover, in Figure 3.1(c), you can see that continuity (or the lack of it) can affect the existence of an extremum on the interval. This suggests the following theorem. (A proof of this theorem is not within the scope of this text.)

(b) f is continuous, $(-1, 2)$ is open.

> ### THEOREM 3.1   The Extreme Value Theorem
>
>  If f is continuous on a closed interval $[a, b]$, then f has both a minimum and a maximum on the interval.

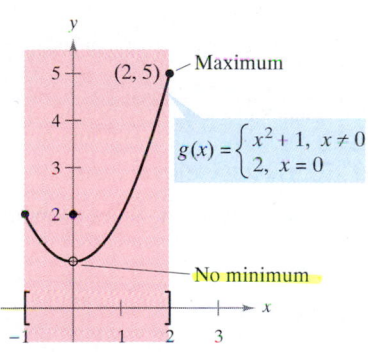

(c) g is not continuous, $[-1, 2]$ is closed.

Extrema can occur at interior points or endpoints of an interval. Extrema that occur at the endpoints are called **endpoint extrema.**

Figure 3.1

EXPLORATION

Finding Minimum and Maximum Values   The Extreme Value Theorem (like the Intermediate Value Theorem) is an *existence theorem* because it tells of the existence of minimum and maximum values but does not show how to find these values. Use the extreme-value capability of a graphing utility to find the minimum and maximum values of each of the following. In each case, do you think the x-values are exact or approximate? Explain your reasoning.

(a) $f(x) = x^2 - 4x + 5$ on the closed interval $[-1, 3]$
(b) $f(x) = x^3 - 2x^2 - 3x - 2$ on the closed interval $[-1, 3]$

Relative Extrema and Critical Numbers

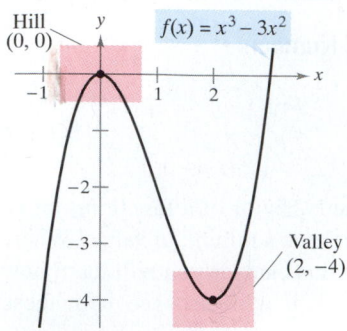

f has a relative maximum at $(0, 0)$ and a relative minimum at $(2, -4)$.
Figure 3.2

In Figure 3.2, the graph of $f(x) = x^3 - 3x^2$ has a **relative maximum** at the point $(0, 0)$ and a **relative minimum** at the point $(2, -4)$. Informally, you can think of a relative maximum as occurring on a "hill" on the graph, and a relative minimum as occurring in a "valley" on the graph. Such a hill and valley can occur in two ways. If the hill (or valley) is smooth and rounded, the graph has a horizontal tangent line at the high point (or low point). If the hill (or valley) is sharp and peaked, the graph represents a function that is not differentiable at the high point (or low point).

> **Definition of Relative Extrema**
>
> **1.** If there is an open interval containing c on which $f(c)$ is a maximum, then $f(c)$ is called a **relative maximum** of f.
> **2.** If there is an open interval containing c on which $f(c)$ is a minimum, then $f(c)$ is called a **relative minimum** of f.
>
> The plural of relative maximum is relative maxima, and the plural of relative minimum is relative minima.

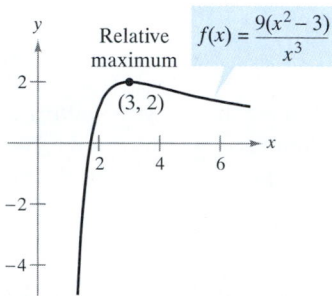

(a)

Example 1 examines the derivatives of functions at *given* relative extrema. (Much more is said about *finding* the relative extrema of a function in Section 3.3.)

EXAMPLE 1   The Value of the Derivative at Relative Extrema

Find the value of the derivative at each of the relative extrema shown in Figure 3.3.

Solution

a. The derivative of

$$f(x) = \frac{9(x^2 - 3)}{x^3}$$

is

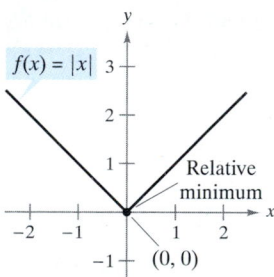

(b)

$$f'(x) = \frac{x^3(18x) - (9)(x^2 - 3)(3x^2)}{(x^3)^2}$$

$$= \frac{9(9 - x^2)}{x^4}.$$

At the point $(3, 2)$, the value of the derivative is $f'(3) = 0$ (see Figure 3.3a).

b. At $x = 0$, the derivative of $f(x) = |x|$ *does not exist* because the following one-sided limits differ (see Figure 3.3b).

$$\lim_{x \to 0^-} \frac{f(x) - f(0)}{x - 0} = \lim_{x \to 0^-} \frac{|x|}{x} = -1 \qquad \text{Limit from the left}$$

$$\lim_{x \to 0^+} \frac{f(x) - f(0)}{x - 0} = \lim_{x \to 0^+} \frac{|x|}{x} = 1 \qquad \text{Limit from the right}$$

c. The derivative of $f(x) = \sin x$ is

$$f'(x) = \cos x.$$

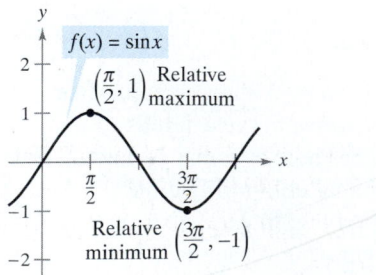

(c)

At the relative extrema, the derivative is zero or is undefined.
Figure 3.3

At the point $(\pi/2, 1)$, the value of the derivative is $f'(\pi/2) = \cos(\pi/2) = 0$. At the point $(3\pi/2, -1)$, the value of the derivative is $f'(3\pi/2) = \cos(3\pi/2) = 0$ (see Figure 3.3c).

Note in Example 1 that at the relative extrema, the derivative is either zero or undefined. The *x*-values at these special points are called **critical numbers.** Figure 3.4 illustrates the two types of critical numbers.

Definition of Critical Number

Let *f* be defined at *c*. If $f'(c) = 0$ or if f' is undefined at *c*, then *c* is a **critical number** of *f*.

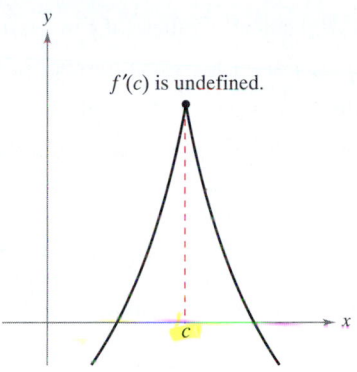

f′(c) is undefined.

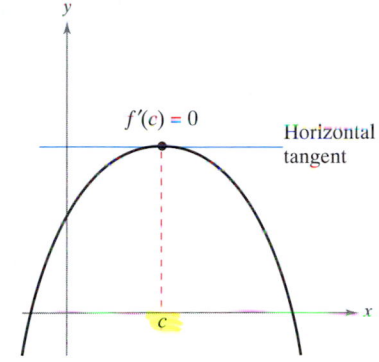

f′(c) = 0

Horizontal tangent

c is a critical number of *f.*
Figure 3.4

THEOREM 3.2   Relative Extrema Occur Only at Critical Numbers

If *f* has a relative minimum or relative maximum at $x = c$, then *c* is a critical number of *f.*

PIERRE DE FERMAT (1601–1665)

For Fermat, who was trained as a lawyer, mathematics was more of a hobby than a profession. Nevertheless, Fermat made many contributions to analytic geometry, number theory, calculus, and probability. In letters to friends, he wrote of many of the fundamental ideas of calculus, long before Newton or Leibniz. For instance, the theorem at the right is sometimes attributed to Fermat.

Proof

Case 1: If *f* is *not* differentiable at $x = c$, then, by definition, *c* is a critical number of *f* and the theorem is valid.

Case 2: If *f* is differentiable at $x = c$, then $f'(c)$ must be positive, negative, or 0. Suppose $f'(c)$ is positive. Then

$$f'(c) = \lim_{x \to c} \frac{f(x) - f(c)}{x - c} > 0$$

which implies that there exists an interval (a, b) containing *c* such that

$$\frac{f(x) - f(c)}{x - c} > 0, \text{ for all } x \neq c \text{ in } (a, b).$$

Because this quotient is positive, the signs of the denominator and numerator must agree. This produces the following inequalities for *x*-values in the interval (a, b).

Left of c: $x < c$ and $f(x) < f(c)$ ⟹ $f(c)$ is not a relative minimum
Right of c: $x > c$ and $f(x) > f(c)$ ⟹ $f(c)$ is not a relative maximum

Thus, the assumption that $f'(c) > 0$ contradicts the hypothesis that $f(c)$ is a relative extremum. Assuming that $f'(c) < 0$ produces a similar contradiction, you are left with only one possibility—namely, $f'(c) = 0$—and so, by definition, *c* is a critical number of *f* and the theorem is valid.

Finding Extrema on a Closed Interval

Theorem 3.2 states that the relative extrema of a function can occur *only* at the critical numbers of the function. Knowing this, you can use the following guidelines to find extrema on a closed interval.

Guidelines for Finding Extrema on a Closed Interval

To find the extrema of a continuous function f on a closed interval $[a, b]$, use the following steps.

1. Find the critical numbers of f in (a, b).
2. Evaluate f at each critical number in (a, b).
3. Evaluate f at each endpoint of $[a, b]$.
4. The least of these values is the minimum. The greatest is the maximum.

The next three examples show how to apply these guidelines. Be sure you see that finding the critical numbers of the function is only part of the procedure. Evaluating the function at the critical numbers *and* the endpoints is the other part.

EXAMPLE 2   Finding Extrema on a Closed Interval

Find the extrema of $f(x) = 3x^4 - 4x^3$ on the interval $[-1, 2]$.

Solution   Begin by differentiating the function.

$$f(x) = 3x^4 - 4x^3 \qquad \text{Original function}$$
$$f'(x) = 12x^3 - 12x^2 \qquad \text{Derivative}$$

To find the critical numbers of f, you must find all x-values for which $f'(x) = 0$ *and* all x-values for which $f'(x)$ is undefined.

$$f'(x) = 12x^3 - 12x^2 = 0 \qquad \text{Set } f'(x) = 0.$$
$$12x^2(x - 1) = 0 \qquad \text{Factor.}$$
$$x = 0, 1 \qquad \text{Critical numbers}$$

Because f' is defined for all x, you can conclude that these are the only critical numbers of f. By evaluating f at these two critical numbers and at the endpoints of $[-1, 2]$, you can determine that the maximum is $f(2) = 16$ and the minimum is $f(1) = -1$, as indicated in the table. The graph of f is shown in Figure 3.5.

Left Endpoint	Critical Number	Critical Number	Right Endpoint
$f(-1) = 7$	$f(0) = 0$	$f(1) = -1$ Minimum	$f(2) = 16$ Maximum

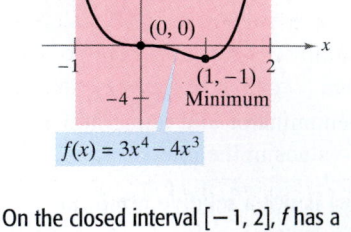

On the closed interval $[-1, 2]$, f has a minimum at $(1, -1)$ and a maximum at $(2, 16)$.

Figure 3.5

NOTE   In Figure 3.5, note that the critical number $x = 0$ does not yield a relative minimum or a relative maximum. This tells you that the converse of Theorem 3.2 is not true. In other words, *the critical numbers of a function need not produce relative extrema.*

EXAMPLE 3   Finding Extrema on a Closed Interval

Find the extrema of

$$f(x) = 2x - 3x^{2/3}$$

on the interval $[-1, 3]$.

Solution   Differentiating produces

$$f'(x) = 2 - \frac{2}{x^{1/3}} = 2\left(\frac{x^{1/3} - 1}{x^{1/3}}\right).$$

From this derivative, you can see that the function has two critical numbers in the interval $[-1, 3]$. The number 1 is a critical number because $f'(1) = 0$, and the number 0 is a critical number because $f'(0)$ is undefined. By evaluating f at these two numbers and at the endpoints of the interval, you can conclude that the minimum is $f(-1) = -5$ and the maximum is $f(0) = 0$, as indicated in the table below. The graph of f is shown in Figure 3.6.

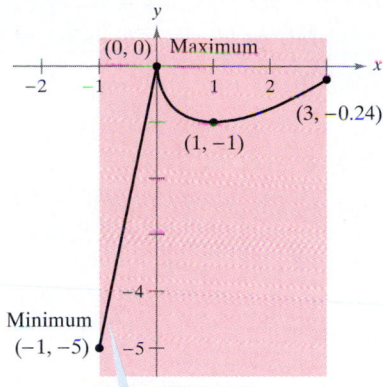

On the closed interval $[-1, 3]$, f has a minimum at $(-1, -5)$ and a maximum at $(0, 0)$.
Figure 3.6

Left Endpoint	Critical Number	Critical Number	Right Endpoint
$f(-1) = -5$ Minimum	$f(0) = 0$ Maximum	$f(1) = -1$	$f(3) = 6 - 3\sqrt[3]{9} \approx -0.24$

EXAMPLE 4   Finding Extrema on a Closed Interval

Find the extrema of

$$f(x) = 2\sin x - \cos 2x$$

on the interval $[0, 2\pi]$.

Solution   This function is differentiable for all real x, so you can find all critical numbers by setting $f'(x)$ equal to zero, as follows.

$$f'(x) = 2\cos x + 2\sin 2x = 0$$
$$2\cos x + 4\cos x \sin x = 0 \qquad \textcolor{red}{\sin 2x = 2\cos x \sin x}$$
$$2(\cos x)(1 + 2\sin x) = 0 \qquad \textcolor{red}{\text{Factor.}}$$

In the interval $[0, 2\pi]$, the factor $\cos x$ is zero when $x = \pi/2$ and when $x = 3\pi/2$. The factor $(1 + 2\sin x)$ is zero when $x = 7\pi/6$ and when $x = 11\pi/6$. By evaluating f at these four critical numbers and at the endpoints of the interval, you can conclude that the maximum is $f(\pi/2) = 3$ and the minimum occurs at *two* points, $f(7\pi/6) = -3/2$ and $f(11\pi/6) = -3/2$, as indicated in the table. The graph is shown in Figure 3.7.

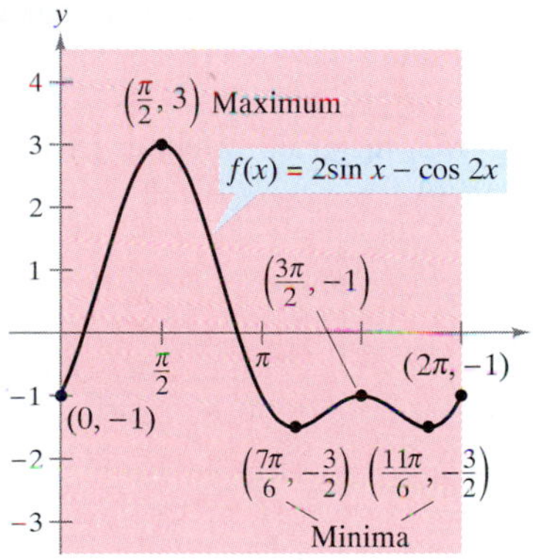

On the closed interval $[0, 2\pi]$, f has two minima at $(7\pi/6, -3/2)$ and $(11\pi/6, -3/2)$ and a maximum at $(\pi/2, 3)$.
Figure 3.7

Left Endpoint	Critical Number	Critical Number	Critical Number	Critical Number	Right Endpoint
$f(0) = -1$	$f\left(\frac{\pi}{2}\right) = 3$ Maximum	$f\left(\frac{7\pi}{6}\right) = -\frac{3}{2}$ Minimum	$f\left(\frac{3\pi}{2}\right) = -1$	$f\left(\frac{11\pi}{6}\right) = -\frac{3}{2}$ Minimum	$f(2\pi) = -1$

EXERCISES FOR SECTION 3.1

In Exercises 1–6, find the value of the derivative (if it exists) at each indicated extremum.

1. $f(x) = \dfrac{x^2}{x^2 + 4}$

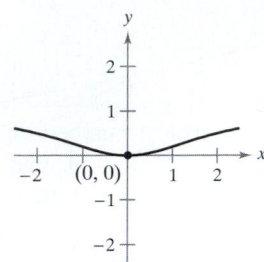

2. $f(x) = \cos \dfrac{\pi x}{2}$

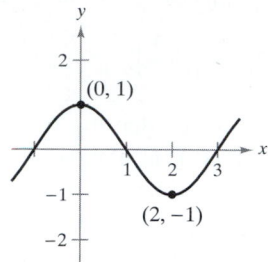

3. $f(x) = x + \dfrac{32}{x^2}$

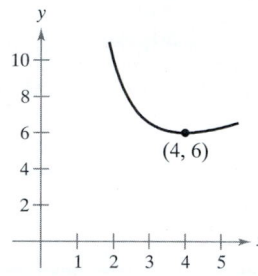

4. $f(x) = -3x\sqrt{x + 1}$

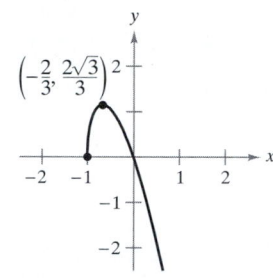

5. $f(x) = (x + 2)^{2/3}$

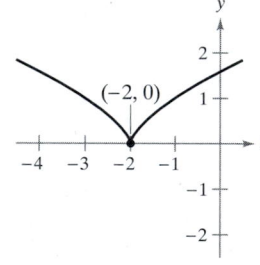

6. $f(x) = 4 - |x|$

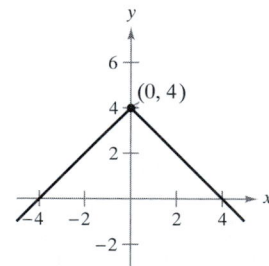

In Exercises 7–12, find any critical numbers of the function.

7. $f(x) = x^2(x - 3)$

8. $g(x) = x^2(x^2 - 4)$

9. $g(t) = t\sqrt{4 - t}$

10. $f(x) = \dfrac{4x}{x^2 + 1}$

11. $h(x) = \sin^2 x + \cos x$
$0 \le x < 2\pi$

12. $f(\theta) = 2 \sec \theta + \tan \theta$
$0 \le \theta < 2\pi$

In Exercises 13–26, determine the absolute extrema of the function and the x-value in the closed interval where it occurs.

Function	Interval
13. $f(x) = 2(3 - x)$	$[-1, 2]$
14. $f(x) = \dfrac{2x + 5}{3}$	$[0, 5]$

Function	Interval
15. $f(x) = -x^2 + 3x$	$[0, 3]$
16. $f(x) = x^2 + 2x - 4$	$[-1, 1]$
17. $f(x) = x^3 - 3x^2$	$[-1, 3]$
18. $f(x) = x^3 - 12x$	$[0, 4]$
19. $f(x) = 3x^{2/3} - 2x$	$[-1, 1]$
20. $g(x) = \sqrt[3]{x}$	$[-1, 1]$
21. $h(t) = 4 - |t - 4|$	$[1, 6]$
22. $g(t) = \dfrac{t^2}{t^2 + 3}$	$[-1, 1]$
23. $h(s) = \dfrac{1}{s - 2}$	$[0, 1]$
24. $h(t) = \dfrac{t}{t - 2}$	$[3, 5]$
25. $f(x) = \cos \pi x$	$\left[0, \dfrac{1}{6}\right]$
26. $g(x) = \csc x$	$\left[\dfrac{\pi}{6}, \dfrac{\pi}{3}\right]$

27. *Writing* Explain why the function $f(x) = \tan x$ has a maximum on $[0, \pi/4]$ but not on $[0, \pi]$.

28. *Writing* Write a short paragraph explaining why a continuous function on an open interval may not have a maximum or minimum. Illustrate your explanation with a sketch of the graph of a function.

In Exercises 29–32, determine from the graph whether f has a minimum in the open interval (a, b).

29. (a)

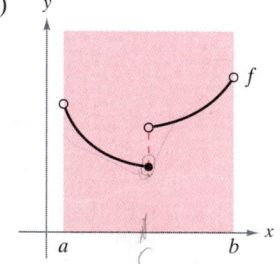

(b)

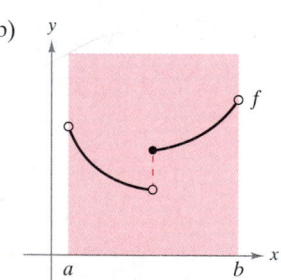

30. (a)

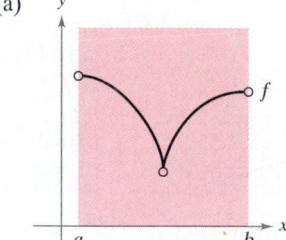

(b)

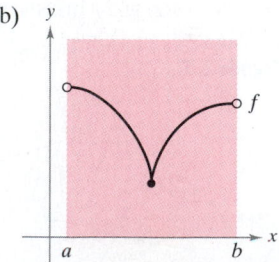

31. (a) 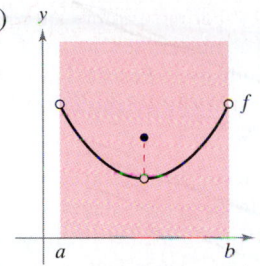   (b)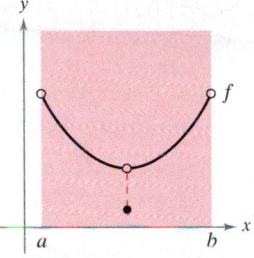

32. (a) 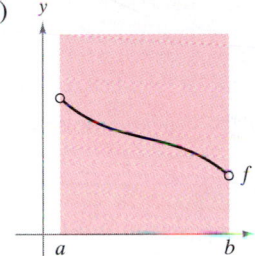   (b)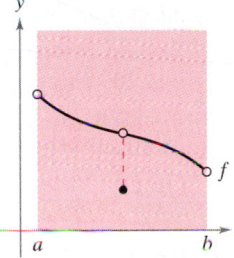

In Exercises 33–36, locate the absolute extrema of the function (if any exists) over the indicated intervals.

33. $f(x) = 2x - 3$

(a) $[0, 2]$

(b) $[0, 2)$

(c) $(0, 2]$

(d) $(0, 2)$

34. $f(x) = 5 - x$

(a) $[1, 4]$

(b) $[1, 4)$

(c) $(1, 4]$

(d) $(1, 4)$

35. $f(x) = x^2 - 2x$

(a) $[-1, 2]$

(b) $(1, 3]$

(c) $(0, 2)$

(d) $[1, 4)$

36. $f(x) = \sqrt{4 - x^2}$

(a) $[-2, 2]$

(b) $[-2, 0)$

(c) $(-2, 2)$

(d) $[1, 2)$

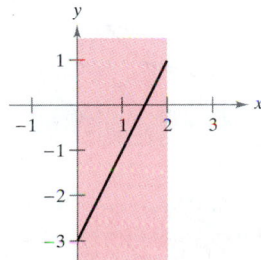

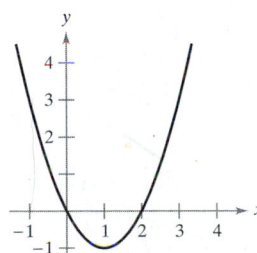

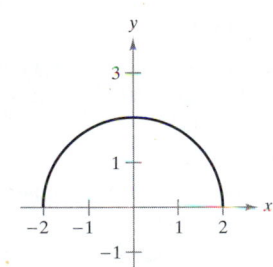

In Exercises 37–40, use a graphing utility to graph the function. Determine the absolute extrema of the function and the x-value in the closed interval where it occurs.

Function	Interval
37. $f(x) = \begin{cases} 2x + 2, & 0 \leq x \leq 1 \\ 4x^2, & 1 < x \leq 3 \end{cases}$	$[0, 3]$
38. $f(x) = \begin{cases} 2 - x^2, & 1 \leq x < 3 \\ 2 - 3x, & 3 \leq x \leq 5 \end{cases}$	$[1, 5]$
39. $f(x) = \dfrac{3}{x - 1}$	$(1, 4]$
40. $f(x) = \dfrac{2}{2 - x}$	$[0, 2)$

In Exercises 41 and 42, (a) use a symbolic differentiation utility to graph the function and approximate any absolute extrema on the indicated interval. (b) Use the utility to find any critical numbers, and use them to find any absolute extrema not located at the endpoints. Compare the results with those in part (a).

Function	Interval
41. $f(x) = 3.2x^5 + 5x^3 - 3.5x$	$[0, 1]$
42. $f(x) = \dfrac{4}{3}x\sqrt{3 - x}$	$[0, 3]$

In Exercises 43 and 44, use a symbolic differentiation utility to find the maximum value of $|f''(x)|$ on the indicated interval. (This value is used in the error estimate for the Trapezoidal Rule, as discussed in Section 4.6.)

Function	Interval
43. $f(x) = \sqrt{1 + x^3}$	$[0, 2]$
44. $f(x) = \dfrac{1}{x^2 + 1}$	$\left[\dfrac{1}{2}, 3\right]$

In Exercises 45 and 46, use a symbolic differentiation utility to find the maximum value of $|f^4(x)|$ on the indicated interval. (This value is used in the error estimate for Simpson's Rule, as discussed in Section 4.6.)

Function	Interval
45. $f(x) = (x + 1)^{2/3}$	$[0, 2]$
46. $f(x) = \dfrac{1}{x^2 + 1}$	$[-1, 1]$

47. *Power* The formula for the power output P of a battery is $P = VI - RI^2$ where V is the electromotive force in volts, R is the resistance, and I is the current. Find the current (measured in amperes) that corresponds to a maximum value of P in a battery for which $V = 12$ volts and $R = 0.5$ ohm. Assume that a 15-ampere fuse bounds the output in the interval $0 \leq I \leq 15$. Could the power output be increased by replacing the 15-ampere fuse with a 20-ampere fuse? Explain.

48. Inventory Cost   A retailer has determined that the cost C of ordering and storing x units of a certain product is

$$C = 2x + \frac{300{,}000}{x}, \qquad 1 \le x \le 300.$$

The delivery truck can bring at most 300 units per order. Find the order size that will minimize cost. Could the cost be decreased if the truck were replaced with one that could bring at most 400 units? Explain.

49. Lawn Sprinkler   A lawn sprinkler is constructed in such a way that $d\theta/dt$ is constant, where θ ranges between 45° and 135° (see figure). The distance the water travels horizontally is

$$x = \frac{v^2 \sin 2\theta}{32}, \qquad \frac{\pi}{4} \le \theta \le \frac{3\pi}{4}$$

where v is the speed of the water. Find dx/dt and explain why this lawn sprinkler does not water evenly. What part of the lawn receives the most water?

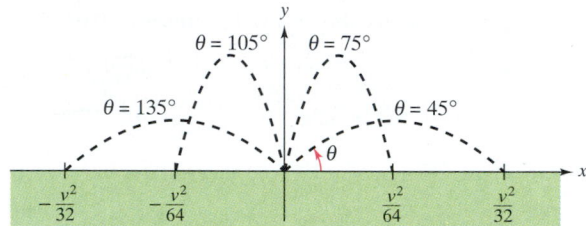

Water sprinkler: $45° \le \theta \le 135°$

FOR FURTHER INFORMATION   For more information on the "calculus of lawn sprinklers," see the article "Design of an Oscillating Sprinkler" by Bart Braden in the January 1985 issue of *Mathematics Magazine*.

50. Modeling Data   The defense outlays as percents of the gross domestic product for the years 1976 through 1995 are as follows.   *(Source: U.S. Office of Management and Budget)*

1976: (5.3%); 1977: (5.1%); 1978: (4.8%); 1979: (4.8%);
1980: (5.1%); 1981: (5.3%); 1982: (5.9%); 1983: (6.3%);
1984: (6.2%); 1985: (6.4%); 1986: (6.5%); 1987: (6.3%);
1988: (6.0%); 1989: (5.9%); 1990: (5.5%); 1991: (4.8%);
1992: (5.0%); 1993: (4.7%); 1994: (4.2%); 1995: (3.9%);

(a) Use the regression capabilities of a graphing utility to find a model of the form $y = at^4 + bt^3 + ct^2 + dt + e$ for the data. (Let t represent the time in years, with $t = 0$ corresponding to 1980.)

(b) Use a graphing utility to plot the data and graph the model.

(c) Locate the absolute extrema of the model on the interval $[-4, 15]$.

51. Honeycomb   The surface area of a cell in a honeycomb is

$$S = 6hs + \frac{3s^2}{2}\left(\frac{\sqrt{3} - \cos\theta}{\sin\theta}\right)$$

where h and s are positive constants and θ is the angle at which the upper faces meet the altitude of the cell. Find the angle θ $(\pi/6 \le \theta \le \pi/2)$ that minimizes the surface area S.

FOR FURTHER INFORMATION   For more information on the geometric structure of a honeycomb cell, see the article "The Design of Honeycombs" by Anthony L. Peressini in UMAP Module 502, published by COMAP, Inc., Suite 210, 57 Bedford Street, Lexington, MA.

52. Highway Design   In order to build a highway it is necessary to fill a section of a valley where the grades (slopes) of the sides are 6% and 9% (see figure). The top of the filled region will have the shape of a parabolic arc that is tangent to the two slopes at the points A and B. The horizontal distance between the points A and B is 1000 feet.

(a) Find a quadratic function $y = ax^2 + bx + c$, $-500 \le x \le 500$, that describes the top of the filled region.

(b) Complete the table giving the depths d of the fill at the specified values of x.

x	-500	-400	-300	-200	-100
d					

x	0	100	200	300	400	500
d						

(c) What will be the lowest point on the completed highway? Will it be directly over the point where the two hillsides come together?

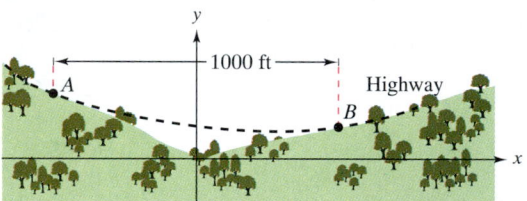

True or False?   In Exercises 53–56, determine whether the statement is true or false. If it is false, explain why or give an example that shows it is false.

53. The maximum of a function that is continuous on a closed interval can occur at two different values in the interval.

54. If a function is continuous on a closed interval, then it must have a minimum on the interval.

55. If $x = c$ is a critical number of the function f, then it is also a critical number of the function $g(x) = f(x) + k$, where k is a constant.

56. If $x = c$ is a critical number of the function f, then it is also a critical number of the function $g(x) = f(x - k)$, where k is a constant.

57. Find all critical numbers of the greatest integer function $f(x) = [\![x]\!]$.

Rolle's Theorem • The Mean Value Theorem

Rolle's Theorem

ROLLE'S THEOREM

French mathematician Michel Rolle first published the theorem that bears his name in 1691. Before this time, however, Rolle was one of the most vocal critics of calculus, stating that the subject gave erroneous results and was based on unsound reasoning. Later in life, Rolle came to see the usefulness of calculus.

The Extreme Value Theorem (Section 3.1) states that a continuous function on a closed interval $[a, b]$ must have both a minimum and a maximum on the interval. Both of these values, however, can occur at the endpoints. **Rolle's Theorem,** named after the French mathematician Michel Rolle (1652–1719), gives conditions that guarantee the existence of an extreme value in the interior of a closed interval.

EXPLORATION

Extreme Values in a Closed Interval   Sketch a rectangular coordinate plane on a piece of paper. Label the points (1, 3) and (5, 3). Using a pencil or pen, draw the graph of a differentiable function f that starts at (1, 3) and ends at (5, 3). Is there at least one point on the graph for which the derivative is zero? Would it be possible to draw the graph so that there *isn't* a point for which the derivative is zero? Explain your reasoning.

THEOREM 3.3   Rolle's Theorem

Let f be continuous on the closed interval $[a, b]$ and differentiable on the open interval (a, b). If

$$f(a) = f(b)$$

then there is at least one number c in (a, b) such that $f'(c) = 0$.

Proof   Let $f(a) = d = f(b)$.

Case 1: If $f(x) = d$ for all x in $[a, b]$, f is constant on the interval and, by Theorem 2.2, $f'(x) = 0$ for all x in (a, b).

Case 2: Suppose $f(x) > d$ for some x in (a, b). By the Extreme Value Theorem, you know that f has a maximum at some c in the interval. Moreover, because $f(c) > d$, this maximum does not occur at either endpoint. Therefore, f has a maximum in the *open* interval (a, b). This implies that $f(c)$ is a *relative* maximum and, by Theorem 3.2, c is a critical number of f. Finally, because f is differentiable at c, you can conclude that $f'(c) = 0$.

Case 3: If $f(x) < d$ for some x in (a, b), you can use an argument similar to that in Case 2, but involving the minimum instead of the maximum.

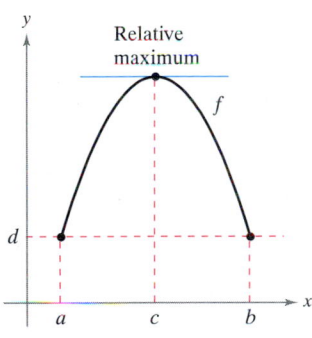

(a) f is continuous on $[a, b]$ and differentiable on (a, b).

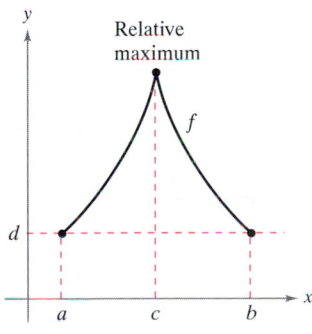

(b) f is continuous on $[a, b]$.

Figure 3.8

From Rolle's Theorem, you can see that if a function f is continuous on $[a, b]$ and differentiable on (a, b), and if $f(a) = f(b)$, there must be at least one x-value between a and b at which the graph of f has a horizontal tangent, as shown in Figure 3.8(a). If the differentiability requirement is dropped from Rolle's Theorem, f will still have a critical number in (a, b), but it may not yield a horizontal tangent. Such a case is shown in Figure 3.8(b).

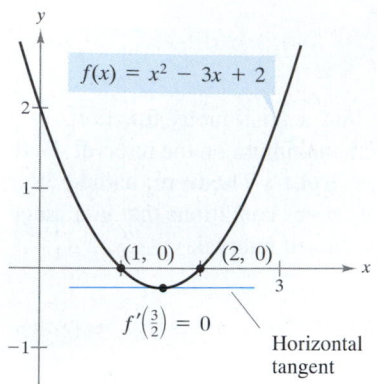

The x-value for which $f'(x) = 0$ is between the two x-intercepts.
Figure 3.9

EXAMPLE 1   Illustrating Rolle's Theorem

Find the two x-intercepts of

$$f(x) = x^2 - 3x + 2$$

and show that $f'(x) = 0$ at some point between the two intercepts.

Solution   Note that f is differentiable on the entire real line. Setting $f(x)$ equal to 0 produces

$$x^2 - 3x + 2 = 0 \qquad \text{Set } f(x) \text{ equal to 0.}$$
$$(x - 1)(x - 2) = 0. \qquad \text{Factor.}$$

Thus, $f(1) = f(2) = 0$, and from Rolle's Theorem you know that there *exists* at least one c in the interval $(1, 2)$ such that $f'(c) = 0$. To *find* such a c, you can solve the equation

$$f'(x) = 2x - 3 = 0 \qquad \text{Set } f'(x) \text{ equal to 0.}$$

and determine that $f'(x) = 0$ when $x = \frac{3}{2}$. Note that the x-value lies in the open interval $(1, 2)$, as shown in Figure 3.9.

Rolle's Theorem states that if f satisfies the conditions of the theorem, there must be *at least* one point between a and b at which the derivative is 0. There may of course be more than one such point, as illustrated in the next example.

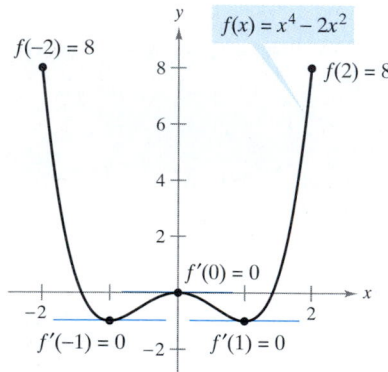

$f'(x) = 0$ for more than one x-value in the interval $(-2, 2)$.
Figure 3.10

EXAMPLE 2   Illustrating Rolle's Theorem

Let $f(x) = x^4 - 2x^2$. Find all values of c in the interval $(-2, 2)$ such that $f'(c) = 0$.

Solution   To begin, note that the function satisfies the conditions of Rolle's Theorem. That is, f is continuous on the interval $[-2, 2]$ and differentiable on the interval $(-2, 2)$. Moreover, because $f(-2) = 8 = f(2)$, you can conclude that there exists at least one c in $(-2, 2)$ such that $f'(c) = 0$. Setting the derivative equal to 0 produces

$$f'(x) = 4x^3 - 4x = 0$$
$$4x(x^2 - 1) = 0$$
$$x = 0, 1, -1.$$

Thus, in the interval $(-2, 2)$, the derivative is zero at three different values of x, as shown in Figure 3.10.

TECHNOLOGY   A graphing utility can be used to indicate whether the points on the graphs in Examples 1 and 2 are relative minima or relative maxima of the functions. When using a graphing utility, however, you should keep in mind that it can give misleading pictures of graphs. For example, try using a graphing utility to graph

$$f(x) = 1 - (x - 1)^2 - \frac{1}{1000(x - 1)^{1/7} + 1}.$$

With most viewing rectangles, it appears that the function has a maximum of 1 when $x = 1$ (see Figure 3.11). By evaluating the function at $x = 1$, however, you can see that $f(1) = 0$. To determine the behavior of this function near $x = 1$, you need to examine the graph analytically to get the complete picture.

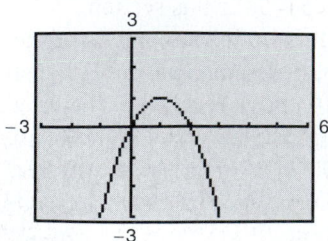

Figure 3.11

The Mean Value Theorem

Rolle's Theorem can be used to prove another theorem—the **Mean Value Theorem.**

THEOREM 3.4   The Mean Value Theorem

If f is continuous on the closed interval $[a, b]$ and differentiable on the open interval (a, b), then there exists a number c in (a, b) such that

$$f'(c) = \frac{f(b) - f(a)}{b - a}.$$

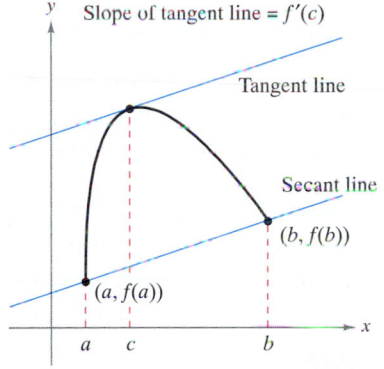

Figure 3.12

Proof   Refer to Figure 3.12. The equation of the secant line containing the points $(a, f(a))$ and $(b, f(b))$ is

$$y = \left[\frac{f(b) - f(a)}{b - a}\right](x - a) + f(a).$$

Let $g(x)$ be the difference between $f(x)$ and y. Then

$$g(x) = f(x) - y$$
$$= f(x) - \left[\frac{f(b) - f(a)}{b - a}\right](x - a) - f(a).$$

By evaluating g at a and b, you can see that $g(a) = 0 = g(b)$. Furthermore, because f is differentiable, g is also differentiable, and you can apply Rolle's Theorem to the function g. Thus, there exists a number c in (a, b) such that $g'(c) = 0$, which implies that

$$0 = g'(c)$$
$$= f'(c) - \frac{f(b) - f(a)}{b - a}.$$

Therefore, there exists a number c in (a, b) such that

$$f'(c) = \frac{f(b) - f(a)}{b - a}.$$

NOTE   The "mean" in the Mean Value Theorem refers to the mean (or average) rate of change of f in the interval $[a, b]$.

JOSEPH-LOUIS LAGRANGE (1736-1813)

The Mean Value Theorem was first proved by the famous mathematician Joseph-Louis Lagrange. Born in Italy, Lagrange held a position in the court of Frederick the Great in Berlin for 20 years. Afterwards, he moved to France, where he met emperor Napoleon Bonaparte, who is quoted as saying, "Lagrange is the lofty pyramid of the mathematical sciences."

Although the Mean Value Theorem can be used directly in problem solving, it is used more often to prove other theorems. In fact, some people consider this to be the most important theorem in calculus—it is closely related to the Fundamental Theorem of Calculus discussed in Chapter 4. For now, you can get an idea of the versatility of this theorem by looking at the results stated in Exercises 51–56 in this section.

The Mean Value Theorem has implications for both basic interpretations of the derivative. Geometrically, the theorem guarantees the existence of a tangent line that is parallel to the secant line through the points $(a, f(a))$ and $(b, f(b))$, as shown in Figure 3.12. Example 3 illustrates this geometric interpretation of the Mean Value Theorem. In terms of rates of change, the Mean Value Theorem implies that there must be a point in the open interval (a, b) at which the instantaneous rate of change is equal to the average rate of change over the interval $[a, b]$. This is illustrated in Example 4.

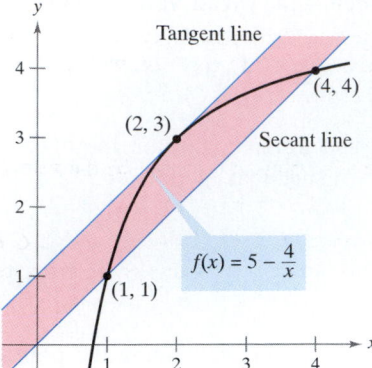

The tangent line at (2, 3) is parallel to the secant line through (1, 1) and (4, 4).
Figure 3.13

EXAMPLE 3   Finding a Tangent Line

Given $f(x) = 5 - (4/x)$, find all values of c in the open interval $(1, 4)$ such that

$$f'(c) = \frac{f(4) - f(1)}{4 - 1}.$$

Solution   The slope of the secant line through $(1, f(1))$ and $(4, f(4))$ is

$$\frac{f(4) - f(1)}{4 - 1} = \frac{4 - 1}{4 - 1} = 1.$$

Because f satisfies the conditions of the Mean Value Theorem, there exists at least one number c in $(1, 4)$ such that $f'(c) = 1$. Solving the equation $f'(x) = 1$ yields

$$f'(x) = \frac{4}{x^2} = 1$$

which implies that $x = \pm 2$. Therefore, in the interval $(1, 4)$, you can conclude that $c = 2$, as shown in Figure 3.13.

EXAMPLE 4   Finding an Instantaneous Rate of Change

Two stationary patrol cars equipped with radar are 5 miles apart on a highway, as shown in Figure 3.14. As a truck passes the first patrol car, its speed is clocked at 55 miles per hour. Four minutes later, when the truck passes the second patrol car, its speed is clocked at 50 miles per hour. Prove that the truck must have exceeded the speed limit (of 55 miles per hour) at some time during the four minutes.

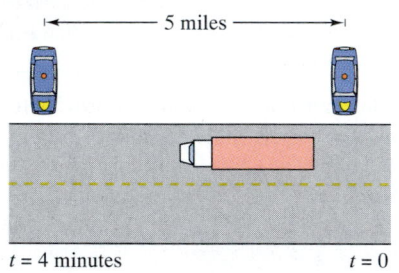

At some time t, the instantaneous velocity is equal to the average velocity over 4 minutes.
Figure 3.14

Solution   Let $t = 0$ be the time (in hours) when the truck passes the first patrol car. The time when the truck passes the second patrol car is

$$t = \frac{4}{60} = \frac{1}{15} \text{ hour.}$$

By letting $s(t)$ represent the distance (in miles) traveled by the truck, you have $s(0) = 0$ and $s\left(\frac{1}{15}\right) = 5$. Therefore, the average velocity of the truck over the 5-mile stretch of highway is

$$\text{Average velocity} = \frac{s(1/15) - s(0)}{(1/15) - 0}$$

$$= \frac{5}{1/15} = 75 \text{ mph.}$$

Assuming that the position function is differentiable, you can apply the Mean Value Theorem to conclude that the truck must have been traveling at a rate of 75 miles per hour sometime during the 4 minutes.

A useful alternative form of the Mean Value Theorem is as follows: If f is continuous on $[a, b]$ and differentiable on (a, b), then there exists a number c in (a, b) such that

$$f(b) = f(a) + (b - a)f'(c).$$ Alternative form of Mean Value Theorem

NOTE   When working the exercises for this section, keep in mind that polynomial functions, rational functions, and trigonometric functions are differentiable at all points in their domains.

EXERCISES FOR SECTION 3.2

Think About It   In Exercises 1 and 2, state why Rolle's Theorem does not apply to the function even though there exist a and b such that $f(a) = f(b)$.

1. $f(x) = 1 - |x - 1|$

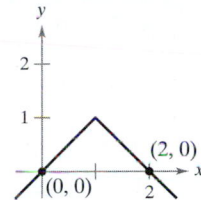

2. $f(x) = \cot \dfrac{x}{2}$

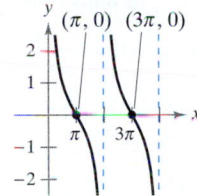

In Exercises 3–16, determine whether Rolle's Theorem can be applied to f on the indicated interval. If Rolle's Theorem can be applied, find all values of c in the interval such that $f'(c) = 0$.

Function	Interval
3. $f(x) = x^2 - 2x$	$[0, 2]$
4. $f(x) = x^2 - 3x + 2$	$[1, 2]$
5. $f(x) = (x - 1)(x - 2)(x - 3)$	$[1, 3]$
6. $f(x) = (x - 3)(x + 1)^2$	$[-1, 3]$
7. $f(x) = x^{2/3} - 1$	$[-8, 8]$
8. $f(x) = 3 - |x - 3|$	$[0, 6]$
9. $f(x) = \dfrac{x^2 - 2x - 3}{x + 2}$	$[-1, 3]$
10. $f(x) = \dfrac{x^2 - 1}{x}$	$[-1, 1]$
11. $f(x) = \sin x$	$[0, 2\pi]$
12. $f(x) = \cos x$	$[0, 2\pi]$
13. $f(x) = \sin 2x$	$\left[\dfrac{\pi}{6}, \dfrac{\pi}{3} \right]$
14. $f(x) = \dfrac{6x}{\pi} - 4 \sin^2 x$	$\left[0, \dfrac{\pi}{6} \right]$
15. $f(x) = \tan x$	$[0, \pi]$
16. $f(x) = \sec x$	$\left[-\dfrac{\pi}{4}, \dfrac{\pi}{4} \right]$

In Exercises 17–20, use a graphing utility to graph the function on the indicated interval. Determine whether Rolle's Theorem can be applied to f on the interval and, if so, find all values of c in the interval such that $f'(c) = 0$.

Function	Interval
17. $f(x) = |x| - 1$	$[-1, 1]$
18. $f(x) = x - x^{1/3}$	$[0, 1]$
19. $f(x) = 4x - \tan \pi x$	$\left[-\frac{1}{4}, \frac{1}{4} \right]$
20. $f(x) = \dfrac{x}{2} - \sin \dfrac{\pi x}{6}$	$[-1, 0]$

21. ***Vertical Motion***   The height of a ball t seconds after it is thrown upward from a height of 32 feet and with an initial velocity of 48 feet per second is $f(t) = -16t^2 + 48t + 32$.

(a) Verify that $f(1) = f(2)$.

(b) According to Rolle's Theorem, what must be the velocity at some time in the interval $[1, 2]$?

22. ***Reorder Costs***   The ordering and transportation cost C of components used in a manufacturing process is approximated by

$$C(x) = 10\left(\frac{1}{x} + \frac{x}{x + 3} \right)$$

where C is measured in thousands of dollars and x is the order size in hundreds.

(a) Verify that $C(3) = C(6)$.

(b) According to Rolle's Theorem, the rate of change of cost must be 0 for some order size in the interval $[3, 6]$. Find this order size.

23. ***Think About It***   Let f be continuous on $[a, b]$ and differentiable on (a, b). If there exists c in (a, b) such that $f'(c) = 0$, does it follow that $f(a) = f(b)$? Explain.

24. ***Think About It***   Let f be continuous on the closed interval $[a, b]$ and differentiable on the open interval (a, b). Also, suppose that $f(a) = f(b)$ and that c is a real number in the interval such that $f'(c) = 0$. Find an interval for the function g over which Rolle's Theorem can be applied, and find the corresponding critical number of g (k is a constant).

(a) $g(x) = f(x) + k$

(b) $g(x) = f(x - k)$

(c) $g(x) = f(kx)$

25. ***Graphical Reasoning***   The figure gives two parts of the graph of a continuous differentiable function f on $[-10, 4]$. The derivative f' is also continuous.

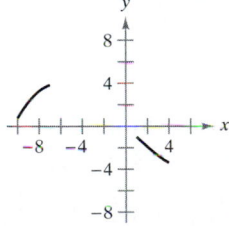

(a) Explain why f must have at least one zero in $[-10, 4]$.

(b) Explain why f' must also have at least one zero in the interval $[-10, 4]$. What are these zeros called?

(c) Make a possible sketch of the function with one zero of f' on the interval $[-10, 4]$.

(d) Make a possible sketch of the function with two zeros of f' on the interval $[-10, 4]$.

(e) Were the conditions of continuity of f and f' necessary to do parts (a) through (d)? Explain.

26. Consider the function $f(x) = 3 \cos^2\left(\dfrac{\pi x}{2}\right)$.

(a) Use a graphing utility to graph f and f'.

(b) Is f a continuous function? Is f'?

(c) Does Rolle's Theorem apply on the interval $[-1, 1]$? Does it apply on the interval $[1, 2]$? Explain.

(d) Evaluate, if possible, $\lim\limits_{x \to 3^-} f'(x)$ and $\lim\limits_{x \to 3^+} f'(x)$.

In Exercises 27–34, apply the Mean Value Theorem to f on the indicated interval. In each case, find all values of c in the interval (a, b) such that

$$f'(c) = \frac{f(b) - f(a)}{b - a}.$$

Function	Interval
27. $f(x) = x^2$	$[-2, 1]$
28. $f(x) = x(x^2 - x - 2)$	$[-1, 1]$
29. $f(x) = x^{2/3}$	$[0, 1]$
30. $f(x) = \dfrac{x + 1}{x}$	$\left[\tfrac{1}{2}, 2\right]$
31. $f(x) = \sqrt{x - 2}$	$[2, 6]$
32. $f(x) = x^3$	$[0, 1]$
33. $f(x) = \sin x$	$[0, \pi]$
34. $f(x) = 2 \sin x + \sin 2x$	$[0, \pi]$

In Exercises 35–38, use a graphing utility to (a) graph the function f on the indicated interval, (b) find and sketch the secant line through points on the graph of f at the endpoints of the indicated interval, and (c) find and sketch any tangent lines to the graph of f that are parallel to the secant line.

Function	Interval
35. $f(x) = \dfrac{x}{x + 1}$	$\left[-\tfrac{1}{2}, 2\right]$
36. $f(x) = x - 2 \sin x$	$[-\pi, \pi]$
37. $f(x) = \sqrt{x}$	$[1, 9]$
38. $f(x) = -x^4 + 4x^3 + 8x^2 + 5$	$[0, 5]$

Writing In Exercises 39 and 40, explain why the Mean Value Theorem does not apply to the function on the interval $[0, 6]$.

39. $f(x) = \dfrac{1}{x - 3}$ **40.** $f(x) = |x - 3|$

Think About It In Exercises 41 and 42, sketch the graph of an arbitrary function f that satisfies the given condition but does not satisfy the conditions of the Mean Value Theorem on the interval $[-5, 5]$.

41. f is continuous on $[-5, 5]$.

42. f is not continuous on $[-5, 5]$.

43. Vertical Motion The height of an object t seconds after it is dropped from a height of 500 meters is $s(t) = -4.9t^2 + 500$.

(a) Find the average velocity of the object during the first 3 seconds.

(b) Use the Mean Value Theorem to verify that at some time during the first 3 seconds of fall the instantaneous velocity equals the average velocity. Find that time.

44. Sales A company introduces a new product for which the number of units sold S is

$$S(t) = 200\left(5 - \frac{9}{2 + t}\right)$$

where t is the time in months.

(a) Find the average value of $S(t)$ during the first year.

(b) During what month does $S'(t)$ equal the average value during the first year.

45. Think About It A plane begins its takeoff at 2:00 P.M. on a 2500-mile flight. The plane arrives at its destination at 7:30 P.M. Explain why there were at least two times during the flight when the speed of the plane was 400 miles per hour.

46. Think About It When an object is removed from a furnace and placed in an environment with a constant temperature of $90°$F, its core temperature is $1500°$F. Five hours later the core temperature is $390°$F. Explain why there must exist a time in the interval when the temperature is decreasing at a rate of $222°$F per hour.

True or False? In Exercises 47–50, determine whether the statement is true or false. If it is false, explain why or give an example that shows it is false.

47. The Mean Value Theorem can be applied to $f(x) = 1/x$ on the interval $[-1, 1]$.

48. If the graph of a function has three x-intercepts, then it must have at least two points at which its tangent line is horizontal.

49. If the graph of a polynomial function has three x-intercepts, then it must have at least two points at which its tangent line is horizontal.

50. If $f'(x) = 0$ for all x in the domain of f, then f is a constant function.

51. Prove that if $a > 0$ and n is any positive integer, then the polynomial function $p(x) = x^{2n+1} + ax + b$ cannot have two real roots.

52. Prove that if $f'(x) = 0$ for all x in an interval (a, b), then f is constant on $[a, b]$.

53. Let $p(x) = Ax^2 + Bx + C$. Prove that for any interval $[a, b]$, the value c guaranteed by the Mean Value Theorem is the midpoint of the interval.

54. Prove that if f is differentiable on $(-\infty, \infty)$ and $f'(x) < 1$ for all real numbers, then f has at most one fixed point. A **fixed point** of a function f is a real number c such that $f(c) = c$.

55. Use the result of Exercise 54 to show that $f(x) = \tfrac{1}{2} \cos x$ has at most one fixed point.

56. Prove that $|\cos x - \cos y| \leq |x - y|$ for all x and y.

Increasing and Decreasing Functions • The First Derivative Test

Increasing and Decreasing Functions

In this section you will learn how derivatives can be used to *classify* relative extrema as either relative minima or relative maxima. We begin by defining increasing and decreasing functions.

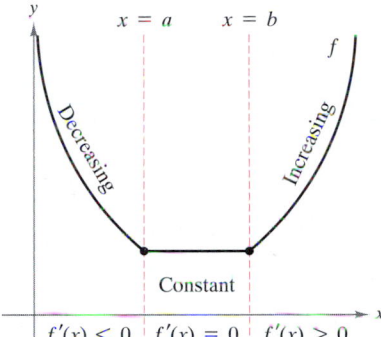

The derivative is related to the slope of a function.

Figure 3.15

> ### Definitions of Increasing and Decreasing Functions
>
> A function f is **increasing** on an interval if for any two numbers x_1 and x_2 in the interval, $x_1 < x_2$ implies $f(x_1) < f(x_2)$.
>
> A function f is **decreasing** on an interval if for any two numbers x_1 and x_2 in the interval, $x_1 < x_2$ implies $f(x_1) > f(x_2)$.

A function is increasing if, *as x moves to the right,* its graph moves up, and is decreasing if its graph moves down. For example, the function in Figure 3.15 is decreasing on the interval $(-\infty, a)$, is constant on the interval (a, b), and is increasing on the interval (b, ∞). As we show in Theorem 3.5 below, a positive derivative implies that the function is increasing; a negative derivative implies that the function is decreasing; and a zero derivative on an entire interval implies that the function is constant on that interval.

> ### THEOREM 3.5   Test for Increasing and Decreasing Functions
>
> Let f be a function that is continuous on the closed interval $[a, b]$ and differentiable on the open interval (a, b).
>
> **1.** If $f'(x) > 0$ for all x in (a, b), then f is increasing on $[a, b]$.
>
> **2.** If $f'(x) < 0$ for all x in (a, b), then f is decreasing on $[a, b]$.
>
> **3.** If $f'(x) = 0$ for all x in (a, b), then f is constant on $[a, b]$.

NOTE   The conclusions in the first two cases of Theorem 3.5 are valid even if $f'(x) = 0$ at a finite number of x-values in (a, b).

Proof   To prove the first case, assume that $f'(x) > 0$ for all x in the interval (a, b) and let $x_1 < x_2$ be any two points in the interval. By the Mean Value Theorem, you know that there exists a number c such that $x_1 < c < x_2$, and

$$f'(c) = \frac{f(x_2) - f(x_1)}{x_2 - x_1}.$$

Because $f'(c) > 0$ and $x_2 - x_1 > 0$, you know that

$$f(x_2) - f(x_1) > 0$$

which implies that $f(x_1) < f(x_2)$. Thus, f is increasing on the interval. The second case has a similar proof (see Exercise 69), and the third case was given as Exercises 52 in Section 3.2.

EXAMPLE 1   Intervals on Which *f* Is Increasing or Decreasing

Find the open intervals on which

$$f(x) = x^3 - \tfrac{3}{2}x^2$$

is increasing or decreasing.

Solution   Note that f is continuous on the entire real line. To determine the critical numbers of f, set $f'(x)$ equal to zero.

$$f'(x) = 3x^2 - 3x = 0 \qquad \text{Let } f'(x) = 0.$$

$$3(x)(x - 1) = 0 \qquad \text{Factor.}$$

$$x = 0, 1 \qquad \text{Critical numbers}$$

Because there are no points for which f' is undefined, you can conclude that $x = 0$ and $x = 1$ are the only critical numbers. The following table summarizes the testing of the three intervals determined by these two critical numbers.

Interval	$-\infty < x < 0$	$0 < x < 1$	$1 < x < \infty$
Test Value	$x = -1$	$x = \tfrac{1}{2}$	$x = 2$
Sign of $f'(x)$	$f'(-1) = 6 > 0$	$f'\left(\tfrac{1}{2}\right) = -\tfrac{3}{4} < 0$	$f'(2) = 6 > 0$
Conclusion	Increasing	Decreasing	Increasing

Thus, f is increasing on the intervals $(-\infty, 0)$ and $(1, \infty)$ and decreasing on the interval $(0, 1)$, as shown in Figure 3.16.

Example 1 gives you one example of how to find intervals on which a function is increasing or decreasing. The guidelines below summarize the steps followed in the example.

Guidelines for Finding Intervals on Which a Function Is Increasing or Decreasing

Let f be continuous on the interval (a, b). To find the open intervals on which f is increasing or decreasing, use the following steps.

1. Locate the critical numbers of f in (a, b), and use these numbers to determine test intervals.
2. Determine the sign of $f'(x)$ at one test value in each of the intervals.
3. Use Theorem 3.5 to decide whether f is increasing or decreasing on each interval.

These guidelines are also valid if the interval (a, b) is replaced by an interval of the form $(-\infty, b)$, (a, ∞), or $(-\infty, \infty)$.

A function is **strictly monotonic** on an interval if it is either increasing on the entire interval or decreasing on the entire interval. For instance, the function $f(x) = x^3$ is strictly monotonic on the entire real line because it is increasing on the entire real line, as shown in Figure 3.17(a). The function shown in Figure 3.17(b) is not strictly monotonic on the entire real line because it is constant on the interval $[0, 1]$.

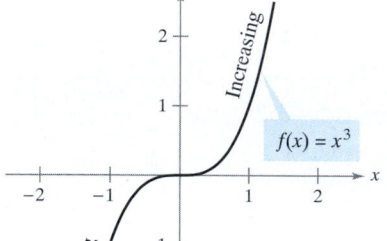

Figure 3.16

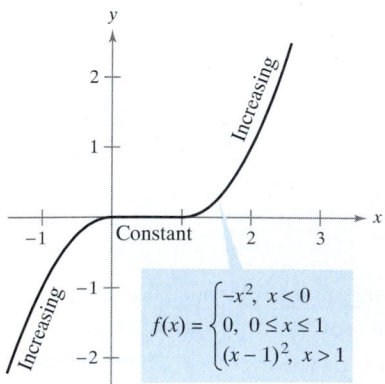

(a) Strictly monotonic function

(b) Not strictly monotonic

Figure 3.17

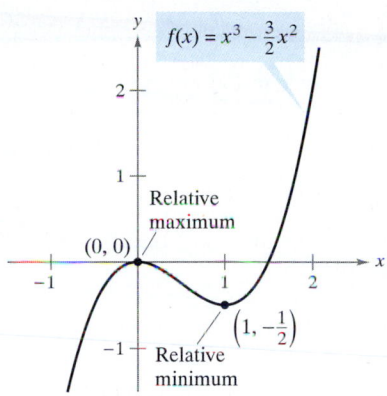

Relative extrema of *f*
Figure 3.18

The First Derivative Test

After you have determined the intervals on which a function is increasing or decreasing, it is not difficult to locate the relative extrema of the function. For instance, in Figure 3.18 (from Example 1), the function

$$f(x) = x^3 - \frac{3}{2}x^2$$

has a relative maximum at the point $(0, 0)$ because f is increasing immediately to the left of $x = 0$ and decreasing immediately to the right of $x = 0$. Similarly, f has a relative minimum at the point $\left(1, -\frac{1}{2}\right)$ because f is decreasing immediately to the left of $x = 1$ and increasing immediately to the right of $x = 1$. The following theorem, called the First Derivative Test, makes this more explicit.

THEOREM 3.6   The First Derivative Test

Let c be a critical number of a function f that is continuous on an open interval I containing c. If f is differentiable on the interval, except possibly at c, then $f(c)$ can be classified as follows.

1. If $f'(x)$ changes from negative to positive at c, then $f(c)$ is a **relative minimum** of f.

2. If $f'(x)$ changes from positive to negative at c, then $f(c)$ is a **relative maximum** of f.

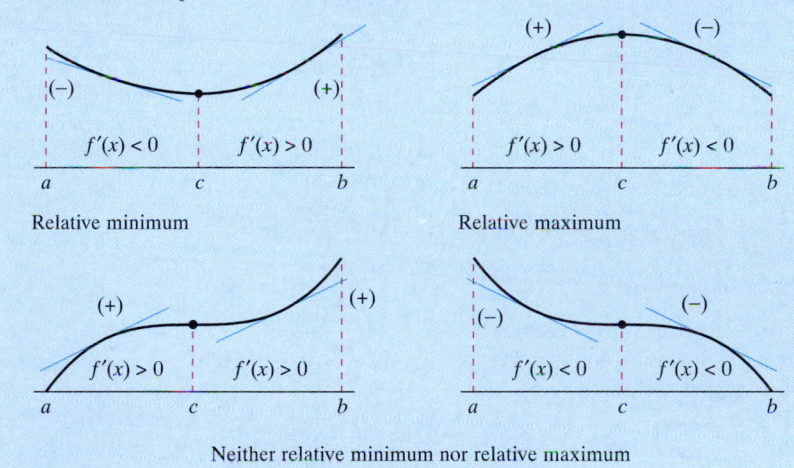

Neither relative minimum nor relative maximum

Proof   Assume that $f'(x)$ changes from negative to positive at c. Then there exist a and b in I such that

$$f'(x) < 0 \text{ for all } x \text{ in } (a, c)$$

and

$$f'(x) > 0 \text{ for all } x \text{ in } (c, b).$$

By Theorem 3.5, f is decreasing on (a, c) and increasing on (c, b). Therefore, $f(c)$ is a minimum of f on the open interval (a, b) and, consequently, a relative minimum of f. This proves the first case of the theorem. The second case can be proved in a similar way (see Exercise 70).

EXAMPLE 2   Applying the First Derivative Test

Find the relative extrema of the function $f(x) = \frac{1}{2}x - \sin x$ in the interval $(0, 2\pi)$.

Solution   Note that f is continuous on the interval $(0, 2\pi)$. To determine the critical numbers of f in this interval, set $f'(x)$ equal to 0.

$$f'(x) = \frac{1}{2} - \cos x = 0 \qquad \text{Let } f'(x) = 0.$$
$$\cos x = \frac{1}{2}$$
$$x = \frac{\pi}{3}, \frac{5\pi}{3} \qquad \text{Critical numbers}$$

Because there are no points for which f' is undefined, you can conclude that $x = \pi/3$ and $x = 5\pi/3$ are the only critical numbers. The following table summarizes the testing of the three intervals determined by these two critical numbers.

Interval	$0 < x < \dfrac{\pi}{3}$	$\dfrac{\pi}{3} < x < \dfrac{5\pi}{3}$	$\dfrac{5\pi}{3} < x < 2\pi$
Test Value	$x = \dfrac{\pi}{4}$	$x = \pi$	$x = \dfrac{7\pi}{4}$
Sign of $f'(x)$	$f'\left(\dfrac{\pi}{4}\right) < 0$	$f'(\pi) > 0$	$f'\left(\dfrac{7\pi}{4}\right) < 0$
Conclusion	Decreasing	Increasing	Decreasing

By applying the First Derivative Test, you can conclude that f has a relative minimum at

$$x = \frac{\pi}{3} \qquad \text{x-value of relative minimum}$$

and a relative maximum at

$$x = \frac{5\pi}{3} \qquad \text{x-value of relative maximum}$$

as shown in Figure 3.19.

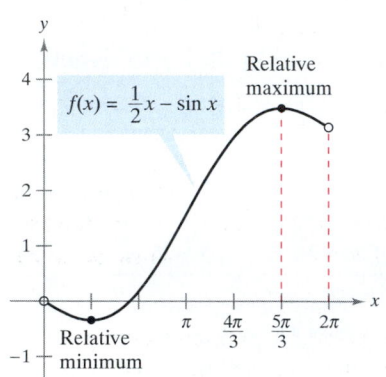

$f(x) = \frac{1}{2}x - \sin x$

Relative maximum

Relative minimum

A relative minimum occurs where f changes from decreasing to increasing, and a relative maximum occurs where f changes from increasing to decreasing.

Figure 3.19

EXPLORATION

Comparing Graphical and Analytical Approaches   In Section 3.2, we pointed out that, *by itself*, a graphing utility can give misleading information about the relative extrema of a graph. *Used in conjunction with an analytical approach*, however, a graphing utility can provide a good way to reinforce your conclusions. Try using a graphing utility to graph the function in Example 2. Then use the zoom and trace features to estimate the relative extrema. How close are your graphical approximations?

Note that in Examples 1 and 2 the given functions are differentiable on the entire real line. For such functions, the only critical numbers are those for which $f'(x) = 0$. Example 3 concerns a function that has two types of critical numbers—those for which $f'(x) = 0$ and those for which f' is undefined.

EXAMPLE 3   Applying the First Derivative Test

Find the relative extrema of

$$f(x) = (x^2 - 4)^{2/3}.$$

Solution   Begin by noting that f is continuous on the entire real line. The derivative of f

$$f'(x) = \frac{2}{3}(x^2 - 4)^{-1/3}(2x) \qquad \text{General Power Rule}$$

$$= \frac{4x}{3(x^2 - 4)^{1/3}} \qquad \text{Simplify.}$$

is 0 when $x = 0$ and undefined when $x = \pm 2$. Thus, the critical numbers are $x = -2$, $x = 0$, and $x = 2$. The table summarizes the testing of the four intervals determined by these three critical numbers.

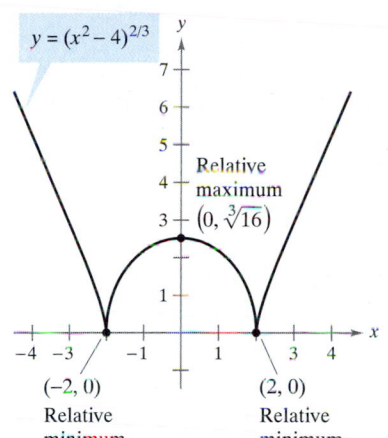

$y = (x^2 - 4)^{2/3}$

Relative maximum $\left(0, \sqrt[3]{16}\right)$

$(-2, 0)$ $(2, 0)$

Relative minimum   Relative minimum

You can apply the First Derivative Test to find relative extrema.
Figure 3.20

Interval	$-\infty < x < -2$	$-2 < x < 0$	$0 < x < 2$	$2 < x < \infty$
Test Value	$x = -3$	$x = -1$	$x = 1$	$x = 3$
Sign of $f'(x)$	$f'(-3) < 0$	$f'(-1) > 0$	$f'(1) < 0$	$f'(3) > 0$
Conclusion	Decreasing	Increasing	Decreasing	Increasing

By applying the First Derivative Test, you can conclude that f has a relative minimum at the point $(-2, 0)$, a relative maximum at the point $\left(0, \sqrt[3]{16}\right)$, and another relative minimum at the point $(2, 0)$, as shown in Figure 3.20.

TECHNOLOGY   When using a graphing utility to graph a function involving radicals or rational exponents, be sure you understand the way the utility evaluates radical expressions. For instance, even though

$$f(x) = (x^2 - 4)^{2/3}$$

and

$$g(x) = [(x^2 - 4)^2]^{1/3}$$

are the same algebraically, many graphing utilities distinguish between these two functions. Which of the graphs shown in Figure 3.21 is incorrect? Why did the graphing utility produce an incorrect graph?

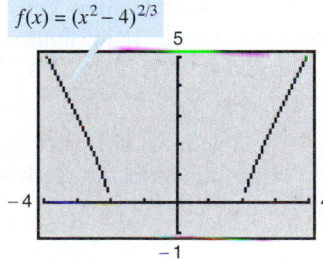

$f(x) = (x^2 - 4)^{2/3}$

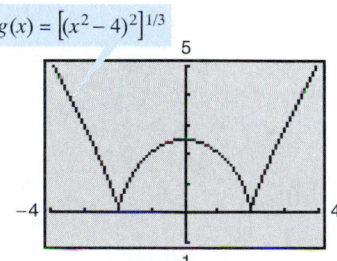

$g(x) = \left[(x^2 - 4)^2\right]^{1/3}$

Which graph is incorrect?
Figure 3.21

When using the First Derivative Test, be sure to consider the domain of the function. For instance, in the next example the function

$$f(x) = \frac{x^4 + 1}{x^2}$$

is not defined when $x = 0$. This x-value must be used with the critical numbers to determine the test intervals.

EXAMPLE 4   Applying the First Derivative Test

Find the relative extrema of $f(x) = \dfrac{x^4 + 1}{x^2}$.

Solution

$f(x) = x^2 + x^{-2}$	Rewrite.
$f'(x) = 2x - 2x^{-3}$	Differentiate.
$\quad = 2x - \dfrac{2}{x^3}$	
$\quad = \dfrac{2(x^4 - 1)}{x^3}$	Simplify.
$\quad = \dfrac{2(x^2 + 1)(x - 1)(x + 1)}{x^3}$	Factor.

Thus, $f'(x)$ is zero at $x = \pm 1$. Moreover, because $x = 0$ is not in the domain of f, you should use this x-value along with the critical numbers to determine the test intervals.

$x = \pm 1$ Critical numbers, $f'(\pm 1) = 0$

$x = 0$ 0 is not in the domain of f.

The table summarizes the testing of the four intervals determined by these three x-values.

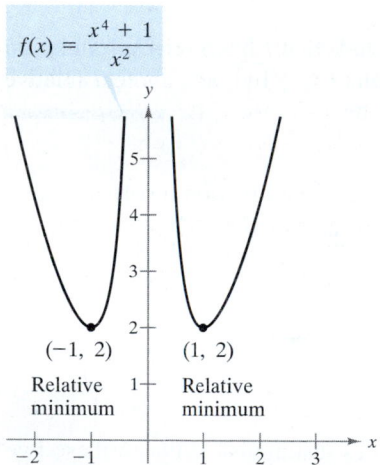

$f(x) = \dfrac{x^4 + 1}{x^2}$

$(-1, 2)$ $(1, 2)$

Relative minimum   Relative minimum

x-values that are not in the domain of *f*, as well as critical numbers, determine test intervals for *f'*.

Figure 3.22

Interval	$-\infty < x < -1$	$-1 < x < 0$	$0 < x < 1$	$1 < x < \infty$
Test Value	$x = -2$	$x = -\frac{1}{2}$	$x = \frac{1}{2}$	$x = 2$
Sign of $f'(x)$	$f'(-2) < 0$	$f'\left(-\frac{1}{2}\right) > 0$	$f'\left(\frac{1}{2}\right) < 0$	$f'(2) > 0$
Conclusion	Decreasing	Increasing	Decreasing	Increasing

By applying the First Derivative Test, you can conclude that f has one relative minimum at the point $(-1, 2)$ and another at the point $(1, 2)$, as shown in Figure 3.22.

TECHNOLOGY   The most difficult step in applying the First Derivative Test is finding the values for which the derivative is equal to 0. For instance, the values of x for which the derivative of

$$f(x) = \frac{x^4 + 1}{x^2 + 1}$$

is equal to 0 are 0 and $\pm \sqrt{\sqrt{2} - 1}$. If you have access to technology that can perform symbolic differentiation and solve equations, try using it to apply the First Derivative Test to this function.

If a projectile is propelled from ground level and air resistance is neglected, the object will travel farthest with an initial angle of 45°. If, however, the projectile is propelled from a point above ground level, the angle that yields a maximum horizontal distance is not 45° (see Example 5).

EXAMPLE 5   The Path of a Projectile

Neglecting air resistance, the path of a projectile that is propelled at an angle θ is

$$y = \frac{g \sec^2 \theta}{2v_0^2} x^2 + (\tan \theta)x + h, \qquad 0 \le \theta \le \frac{\pi}{2}$$

where y is the height, x is the horizontal distance, g is the acceleration due to gravity, v_0 is the initial velocity, and h is the initial height. (This equation is derived in Section 11.3.) Let $g = -32$ feet per second per second, $v_0 = 24$ feet per second, and $h = 9$ feet. What value of θ will produce a maximum horizontal distance?

Solution   To find the distance the projectile travels, let $y = 0$, and use the quadratic formula to solve for x.

$$\frac{g \sec^2 \theta}{2v_0^2} x^2 + (\tan \theta)x + h = 0$$

$$\frac{-32 \sec^2 \theta}{2(24^2)} x^2 + (\tan \theta)x + 9 = 0$$

$$-\frac{\sec^2 \theta}{36} x^2 + (\tan \theta)x + 9 = 0$$

$$x = \frac{-\tan \theta \pm \sqrt{\tan^2 \theta + \sec^2 \theta}}{-\sec^2 \theta/18}$$

$$x = 18 \cos \theta \left(\sin \theta + \sqrt{\sin^2 \theta + 1} \right), \qquad x \ge 0$$

At this point, you need to find the value of θ that produces a maximum value of x. Applying the First Derivative Test by hand would be very tedious. Using technology to solve the equation $dx/d\theta = 0$, however, eliminates most of the messy computations. The result is that the maximum value of x occurs when

$$\theta \approx 0.61548 \text{ radians, or } 35.3°.$$

This conclusion is reinforced by sketching the path of the projectile for different values of θ, as shown in Figure 3.23. Of the three paths shown, note that the distance traveled is greatest for $\theta = 35°$.

STUDY TIP   A computer simulation of this example is given in the interactive CD-ROM version of this text. Using that simulation, you can experimentally discover that the maximum value of x occurs when $\theta \approx 35.3°$.

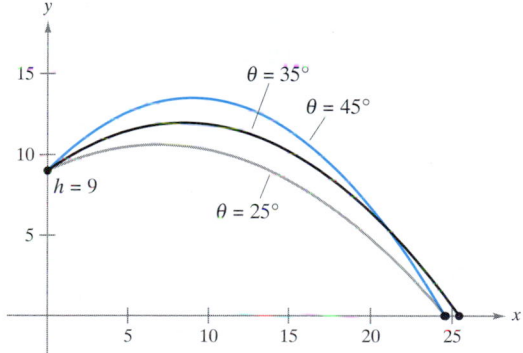

The path of a projectile with initial angle θ
Figure 3.23

EXERCISES FOR SECTION 3.3

In Exercises 1–6, identify the open intervals on which the function is increasing or decreasing.

1. $f(x) = x^2 - 6x + 8$

2. $y = -(x + 1)^2$

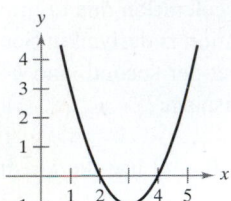

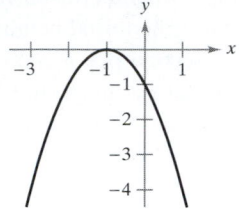

3. $y = \dfrac{x^3}{4} - 3x$

4. $f(x) = x^4 - 2x^2$

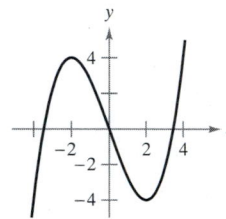

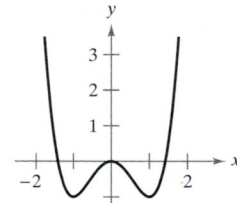

5. $f(x) = \dfrac{1}{x^2}$

6. $y = \dfrac{x^2}{x + 1}$

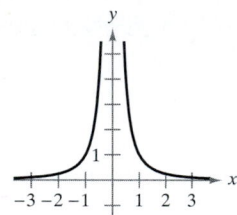

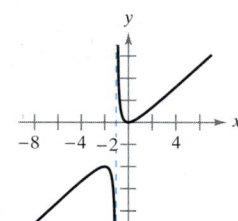

In Exercises 7–20, find the critical numbers of f (if any), find the open intervals on which the algebraic function is increasing or decreasing, and locate all relative extrema.

7. $f(x) = -2x^2 + 4x + 3$

8. $f(x) = x^2 + 8x + 10$

9. $f(x) = x^2 - 6x$

10. $f(x) = (x - 1)^2(x + 2)$

11. $f(x) = 2x^3 + 3x^2 - 12x$

12. $f(x) = (x - 3)^3$

13. $f(x) = \dfrac{x^5 - 5x}{5}$

14. $f(x) = x^4 - 32x + 4$

15. $f(x) = x^{1/3} + 1$

16. $f(x) = x^{2/3}(x - 5)$

17. $f(x) = 5 - |x - 5|$

18. $f(x) = |x + 3| - 1$

19. $f(x) = \dfrac{x^2}{x^2 - 9}$

20. $f(x) = \dfrac{x + 3}{x^2}$

In Exercises 21–28, find the open intervals on which the function is increasing or decreasing, and locate all relative extrema. Use a graphing utility to confirm your results.

21. $f(x) = x^3 - 6x^2 + 15$

22. $f(x) = x^4 - 2x^3$

23. $f(x) = (x - 1)^{2/3}$

24. $f(x) = (x - 1)^{1/3}$

25. $f(x) = x + \dfrac{1}{x}$

26. $f(x) = \dfrac{x}{x + 1}$

27. $f(x) = \dfrac{x^2 - 2x + 1}{x + 1}$

28. $f(x) = \dfrac{x^2 - 3x - 4}{x - 2}$

In Exercises 29–32, consider the trigonometric function on the interval $(0, 2\pi)$. Find the open intervals on which the function is increasing or decreasing, and locate all relative extrema. Use a graphing utility to confirm your results.

29. $f(x) = \dfrac{x}{2} + \cos x$

30. $f(x) = \sin x \cos x$

31. $f(x) = \sin^2 x + \sin x$

32. $f(x) = \dfrac{\cos x}{1 + \sin^2 x}$

In Exercises 33–36, (a) use a symbolic differentiation utility to differentiate the function, (b) sketch the graphs of f and f' on the same set of coordinate axes over the indicated interval, (c) find the critical numbers of f in the open interval, and (d) find the interval(s) on which f' is positive and the interval(s) on which it is negative. Compare the behavior of f and the sign of f'.

Function	Interval
33. $f(x) = 2x\sqrt{9 - x^2}$	$[-3, 3]$
34. $f(x) = 10\left(5 - \sqrt{x^2 - 3x + 16}\right)$	$[0, 5]$
35. $f(t) = t^2 \sin t$	$[0, 2\pi]$
36. $f(x) = \dfrac{x}{2} + \cos \dfrac{x}{2}$	$[0, 4\pi]$

Think About It In Exercises 37–42, assume that f is differentiable for all x. The sign of f' is as follows.

$f'(x) > 0$ on $(-\infty, -4)$

$f'(x) < 0$ on $(-4, 6)$

$f'(x) > 0$ on $(6, \infty)$

Supply the appropriate inequality for the indicated value of c.

Function	Sign of $g'(c)$
37. $g(x) = f(x) + 5$	$g'(0)$ ▮ 0
38. $g(x) = 3f(x) - 3$	$g'(-5)$ ▮ 0
39. $g(x) = -f(x)$	$g'(-6)$ ▮ 0
40. $g(x) = -f(x)$	$g'(0)$ ▮ 0
41. $g(x) = f(x - 10)$	$g'(0)$ ▮ 0
42. $g(x) = f(x - 10)$	$g'(8)$ ▮ 0

Think About It   In Exercises 43–48, a function f is shown in the figure. Sketch a graph of the derivative of f.

43.

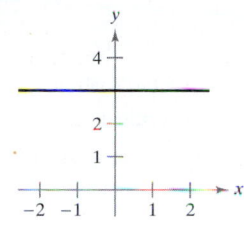

44.

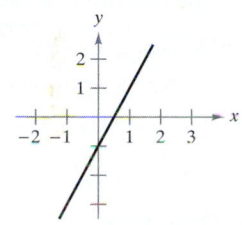

45.

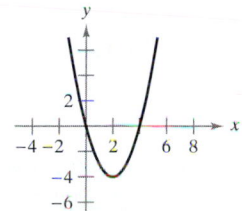

46.

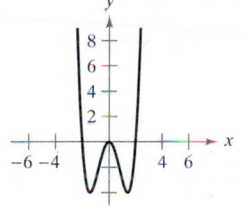

47.

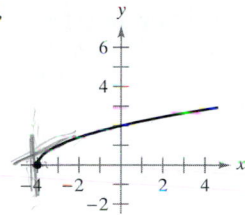

48.

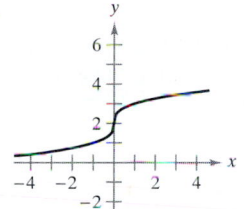

49. *Think About It*   Sketch the graph of an arbitrary function f such that

$$f'(x) \begin{cases} > 0, & x < 4 \\ \text{undefined}, & x = 4 \\ < 0, & x > 4 \end{cases}$$

50. *Think About It*   A differentiable function f has one critical number at $x = 5$. Identify the relative extrema of f at the critical number if $f'(4) = -2.5$ and $f'(6) = 3$.

51. *Think About It*   The function f is differentiable on the interval $[-1, 1]$. The table gives the values of f' for selected values of x. Sketch a graph of f, approximate the critical numbers, and identify the relative extrema.

x	−1	−0.75	−0.50	−0.25
f'(x)	−10	−3.2	−0.5	0.8

x	0	0.25	0.50	0.75	1
f'(x)	5.6	3.6	−0.2	−6.7	−20.1

52. *Rolling a Ball Bearing*   A ball bearing is placed on an inclined plane and begins to roll. The angle of elevation of the plane is θ. The distance (in meters) the ball bearing rolls in t seconds is $s(t) = 4.9(\sin \theta)t^2$.

(a) Determine the speed of the ball bearing.

(b) What value of θ will produce the maximum speed at a particular time?

53. *Numerical, Graphical, and Analytic Analysis*   Consider the functions $f(x) = x$ and $g(x) = \sin x$ on the interval $(0, \pi)$.

(a) Complete the table and make a conjecture about which is the greater function on the interval $(0, \pi)$.

x	0.5	1	1.5	2	2.5	3
f(x)						
g(x)						

(b) Use a graphing utility to graph the functions and use the graphs to make a conjecture about which is the greater function on the interval $(0, \pi)$.

(c) Prove that $f(x) > g(x)$ on the interval $(0, \pi)$. [*Hint:* Show that $h'(x) > 0$ where $h = f - g$.]

54. *Numerical, Graphical, and Analytic Analysis*   The concentration C of a chemical in the bloodstream t hours after injection into muscle tissue is

$$C = \frac{3t}{27 + t^3}, \quad t \geq 0.$$

(a) Complete the table and use the table to approximate the time when the concentration is greatest.

t	0	0.5	1	1.5	2	2.5	3
C(t)							

(b) Use a graphing utility to graph the concentration function and use the graph to approximate the time when the concentration is greatest.

(c) Use calculus to determine analytically the time when the concentration is greatest.

55. *Trachea Contraction*   Coughing forces the trachea (windpipe) to contract, which affects the velocity v of the air passing through the trachea. Suppose the velocity of the air during coughing is

$$v = k(R - r)r^2, \quad 0 \leq r < R$$

where k is constant, R is the normal radius of the trachea, and r is the radius during coughing. What radius will produce the maximum air velocity?

56. *Profit*   The profit P (in dollars) made by a fast-food restaurant selling x hamburgers is

$$P = 2.44x - \frac{x^2}{20,000} - 5000, \quad 0 \leq x \leq 35,000.$$

Find the open intervals on which P is increasing or decreasing.

57. *Power*   The electric power P in watts in a direct-current circuit with two resistors R_1 and R_2 connected in series is

$$P = \frac{vR_1R_2}{(R_1 + R_2)^2}$$

where v is the voltage. If v and R_1 are held constant, what resistance R_2 produces maximum power?

58. *Electrical Resistance*   The resistance R of a certain type of resistor is

$$R = \sqrt{0.001T^4 - 4T + 100}$$

where R is measured in ohms and the temperature T is measured in degrees Celsius.

(a) Use a symbolic differentiation utility to find dR/dT and the critical number of the function. Determine the minimum resistance for this type of resistor.

(b) Use a graphing utility to graph the function R and use the graph to approximate the minimum resistance for this type of resistor.

59. *Modeling Data*   The number of bankruptcies (in thousands) for the years 1981 through 1994 are as follows.

1981: 360.3; 1982: 367.9; 1983: 374.7; 1984: 344.3;

1985: 364.5; 1986: 477.9; 1987: 561.3; 1988: 594.6

1989: 643.0; 1990: 725.5; 1991: 880.4; 1992: 972.5

1993: 918.7; 1994: 845.3

(Source: Administrative Office of the U.S. Courts)

(a) Use the regression capabilities of a graphing utility to find a model of the form

$$B = at^4 + bt^3 + ct^2 + dt + e$$

for the data. (Let $t = 1$ represent 1981.)

(b) Use a graphing utility to plot the data and graph the model.

(c) Analytically find the maximum of the model and compare the result with the actual data.

60. Use a graphing utility to graph $f(x) = 2 \sin 3x + 4 \cos 3x$. Find the maximum value of f. How could you use calculus to estimate the maximum?

Creating Polynomial Functions   **In Exercises 61–64, find a polynomial function**

$$f(x) = a_n x^n + a_{n-1} x^{n-1} + \cdots + a_2 x^2 + a_1 x + a_0$$

that has only the specified extrema. (a) Determine the minimum degree of the function and give the criteria you used in determining the degree. (b) Using the fact that the coordinates of the extrema are solution points of the function, and that the x-coordinates are critical numbers, determine a system of linear equations whose solution yields the coefficients of the required function. (c) Use a graphing utility to solve the system of equations and determine the function. (d) Use a graphing utility to confirm your result graphically.

61. Relative minimum: $(0, 0)$; Relative maximum: $(2, 2)$

62. Relative minimum: $(0, 0)$; Relative maximum: $(4, 1000)$

63. Relative minima: $(0, 0)$, $(4, 0)$

Relative maximum: $(2, 4)$

64. Relative minimum: $(1, 2)$

Relative maxima: $(-1, 4)$, $(3, 4)$

True or False?   **In Exercises 65–68, determine whether the statement is true or false. If it is false, explain why or give an example that shows it is false.**

65. The sum of two increasing functions is increasing.

66. The product of two increasing functions is increasing.

67. Every nth-degree polynomial has $(n - 1)$ critical numbers.

68. An nth-degree polynomial has at most $(n - 1)$ critical numbers.

69. Prove the second case of Theorem 3.5.

70. Prove the second case of Theorem 3.6.

71. Let $x > 0$ and $n > 1$ be real numbers. Prove that $(1 + x)^n > 1 + nx$.

SECTION PROJECT

Rainbows   Rainbows are formed when light strikes raindrops and is reflected and refracted, as shown in the figure. (This figure shows a cross section of a spherical raindrop.) The Law of Refraction states that $(\sin \alpha)/(\sin \beta) = k$, where $k \approx 1.33$ (for water). The angle of deflection is given by $D = \pi + 2\alpha - 4\beta$.

(a) Sketch the graph of D for $0 \leq \alpha \leq \pi/2$. Use a graphing utility with

$$D = \pi + 2\alpha - 4 \sin^{-1}\left(\frac{1}{k} \sin \alpha\right).$$

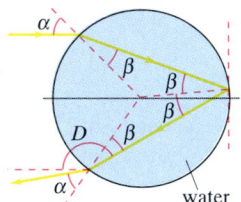

(b) Prove that the minimum angle of deflection occurs when

$$\cos \alpha = \sqrt{\frac{k^2 - 1}{3}}.$$

For water, what is the minimum angle of deflection, D_{min}? (The angle $\pi - D_{min}$ is called the *rainbow angle*.) What value of α produces this minimum angle? (A ray of sunlight that strikes a raindrop at this angle, α, is called a *rainbow ray*.)

FOR FURTHER INFORMATION For more information about the mathematics of rainbows, see the article "Somewhere Within the Rainbow" by Steven Janke in *UMAP Journal*, Volume 13, Number 2, 1992.

Concavity • Points of Inflection • The Second Derivative Test

Concavity

You have already seen that locating the intervals in which a function f increases or decreases helps to describe its graph. In this section, you will see how locating the intervals in which f' increases or decreases can be used to determine where the graph of f is *curving upward* or *curving downward*.

> **Definition of Concavity**
>
> Let f be differentiable on an open interval I. The graph of f is **concave upward** on I if f' is increasing on the interval and **concave downward** on I if f' is decreasing on the interval.

The following graphical interpretation of concavity is useful. (See the appendix for a proof of these results.)

1. Let f be differentiable at c. If the graph of f is concave upward at $(c, f(c))$, the graph of f lies *above* the tangent line at $(c, f(c))$ on some open interval containing c (see Figure 3.24a).

2. Let f be differentiable at c. If the graph of f is concave downward at $(c, f(c))$, the graph of f lies *below* the tangent line at $(c, f(c))$ on some open interval containing c (see Figure 3.24b).

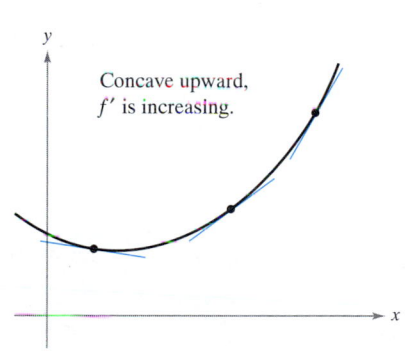

Concave upward,
f' is increasing.

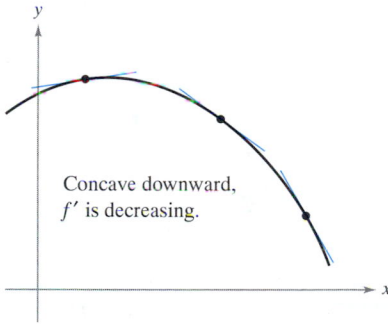

Concave downward,
f' is decreasing.

(a) The graph of f lies above its tangent lines.

(b) The graph of f lies below its tangent lines.

Figure 3.24

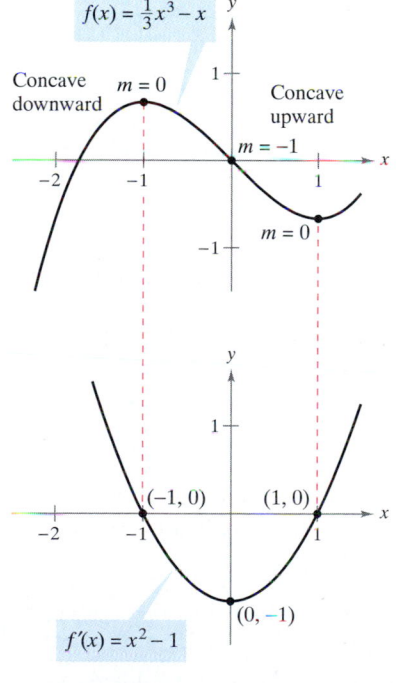

f' is decreasing f' is increasing

The concavity of f is related to the slope of the derivative.

Figure 3.25

To find the open intervals on which the graph of a function f is concave upward or downward, you need to find the intervals on which f' is increasing or decreasing. For instance, the graph of

$$f(x) = \tfrac{1}{3}x^3 - x$$

is concave downward on the open interval $(-\infty, 0)$ because $f'(x) = x^2 - 1$ is decreasing there. (See Figure 3.25.) Similarly, the graph of f is concave upward on the interval $(0, \infty)$ because f' is increasing on $(0, \infty)$.

The following theorem shows how to use the *second* derivative of a function f to determine intervals on which the graph of f is concave upward or downward. A proof of this theorem follows directly from Theorem 3.5 and the definition of concavity.

> **THEOREM 3.7 Test for Concavity**
>
> Let f be a function whose second derivative exists on an open interval I.
>
> **1.** If $f''(x) > 0$ for all x in I, then the graph of f is concave upward in I.
> **2.** If $f''(x) < 0$ for all x in I, then the graph of f is concave downward in I.

NOTE   A third case of Theorem 3.7 could be that if $f''(x) = 0$ for all x in I, then f is linear. Note, however, that concavity is not defined for a line. In other words, a straight line is neither concave upward nor concave downward.

To apply Theorem 3.7, locate the x-values at which $f''(x) = 0$ or f'' is undefined. Second, use these x-values to determine test intervals. Finally, test the sign of $f''(x)$ in each of the test intervals.

EXAMPLE 1   Determining Concavity

Determine the open intervals on which the graph of

$$f(x) = 6(x^2 + 3)^{-1}$$

is concave upward or downward.

Solution   Begin by observing that f is continuous on the entire real line. Next, find the second derivative of f.

$$f(x) = 6(x^2 + 3)^{-1} \qquad \text{Original function}$$
$$f'(x) = (-6)(x^2 + 3)^{-2}(2x)$$
$$= \frac{-12x}{(x^2 + 3)^2} \qquad \text{First derivative}$$
$$f''(x) = \frac{(x^2 + 3)^2(-12) - (-12x)(2)(x^2 + 3)(2x)}{(x^2 + 3)^4}$$
$$= \frac{36(x^2 - 1)}{(x^2 + 3)^3} \qquad \text{Second derivative}$$

Because $f''(x) = 0$ when $x = \pm 1$ and f'' is defined on the entire real line, you should test f'' in the intervals $(-\infty, -1)$, $(-1, 1)$, and $(1, \infty)$. The results are shown in the table and in Figure 3.26.

Interval	$-\infty < x < -1$	$-1 < x < 1$	$1 < x < \infty$
Test value	$x = -2$	$x = 0$	$x = 2$
Sign of $f''(x)$	$f''(-2) > 0$	$f''(0) < 0$	$f''(2) > 0$
Conclusion	Concave upward	Concave downward	Concave upward

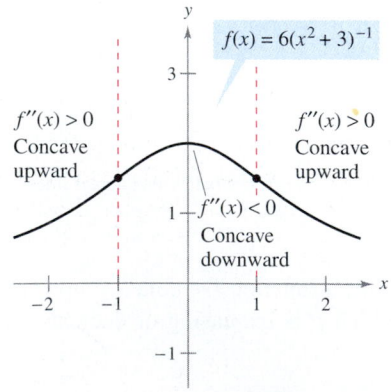

$f''(x) > 0$
Concave upward

$f''(x) > 0$
Concave upward

$f''(x) < 0$
Concave downward

$f(x) = 6(x^2 + 3)^{-1}$

From the sign of f'' you can determine the concavity of the graph of f.
Figure 3.26

The function given in Example 1 is continuous on the entire real line. If there are x-values at which the function is not continuous, these values should be used along with the points at which $f''(x) = 0$ or is undefined to form the test intervals.

EXAMPLE 2   Determining Concavity

Determine the open intervals in which the graph of

$$f(x) = \frac{x^2 + 1}{x^2 - 4}$$
Original function

is concave upward or downward.

Solution

$$f'(x) = \frac{(x^2 - 4)(2x) - (x^2 + 1)(2x)}{(x^2 - 4)^2}$$

$$= \frac{-10x}{(x^2 - 4)^2}$$
First derivative

$$f''(x) = \frac{(x^2 - 4)^2(-10) - (-10x)(2)(x^2 - 4)(2x)}{(x^2 - 4)^4}$$

$$= \frac{10(3x^2 + 4)}{(x^2 - 4)^3}$$
Second derivative

There are no points at which $f''(x) = 0$, but at $x = \pm 2$ the function f is not continuous, so you test for concavity in the intervals $(-\infty, -2)$, $(-2, 2)$, and $(2, \infty)$, as shown in the table. The graph of f is shown in Figure 3.27.

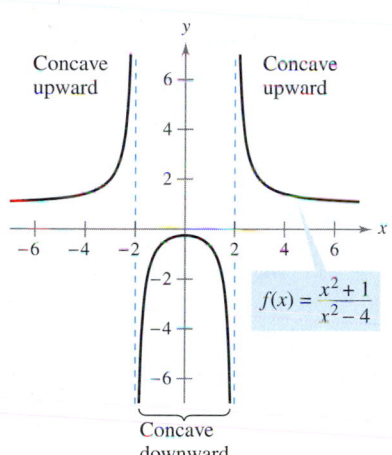

Figure 3.27

Interval	$-\infty < x < -2$	$-2 < x < 2$	$2 < x < \infty$
Test value	$x = -3$	$x = 0$	$x = 3$
Sign of $f''(x)$	$f''(-3) > 0$	$f''(0) < 0$	$f''(3) > 0$
Conclusion	Concave upward	Concave downward	Concave upward

Points of Inflection

The graph in Figure 3.26 has two points at which the concavity changes. If the tangent line to the graph exists at such a point, that point is a **point of inflection.** Three types of points of inflection are shown in Figure 3.28. Note that a graph crosses its tangent line at a point of inflection.

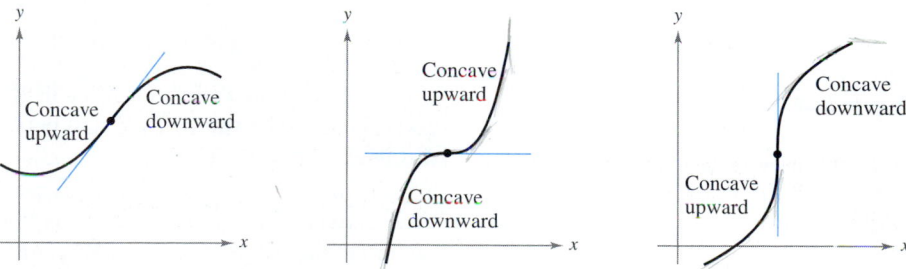

The concavity of f changes at a point of inflection.
Figure 3.28

To locate possible points of inflection, you need only determine the values of x for which $f''(x) = 0$ or for which f'' is undefined. This is similar to the procedure for locating relative extrema of f.

> **THEOREM 3.8 Points of Inflection**
>
> If $(c, f(c))$ is a point of inflection of the graph of f, then either $f''(c) = 0$ or f'' is undefined at $x = c$.

EXAMPLE 3 Finding Points of Inflection

Determine the points of inflection and discuss the concavity of the graph of

$$f(x) = x^4 - 4x^3. \qquad \text{Original function}$$

Solution Differentiating twice produces

$$f'(x) = 4x^3 - 12x^2 \qquad \text{First derivative}$$

$$f''(x) = 12x^2 - 24x = 12x(x - 2). \qquad \text{Second derivative}$$

Possible points of inflection occur at $x = 0$ and $x = 2$. By testing the intervals determined by these x-values, you can conclude that they both yield points of inflection. A summary of this testing is shown in the table, and the graph of f is shown in Figure 3.29.

Interval	$-\infty < x < 0$	$0 < x < 2$	$2 < x < \infty$
Test value	$x = -1$	$x = 1$	$x = 3$
Sign of $f''(x)$	$f''(-1) > 0$	$f''(1) < 0$	$f''(3) > 0$
Conclusion	Concave upward	Concave downward	Concave upward

It is possible for the second derivative to be 0 at a point that is *not* a point of inflection. For instance, the graph of $f(x) = x^4$ is shown in Figure 3.30. The second derivative is 0 when $x = 0$, but the point $(0, 0)$ is not a point of inflection because the graph of f is concave upward in both intervals $-\infty < x < 0$ and $0 < x < \infty$.

EXPLORATION

Consider a general cubic function of the form

$$f(x) = ax^3 + bx^2 + cx + d.$$

You know that the value of d has a bearing on the location of the graph but has no bearing on the value of the first derivative at given values of x. Graphically, this is true because changes in the value of d shift the graph up or down but do not change its basic shape. Use a graphing utility to graph several cubics with different values of c. Then give a graphical explanation of why changes in c do not affect the values of the second derivative.

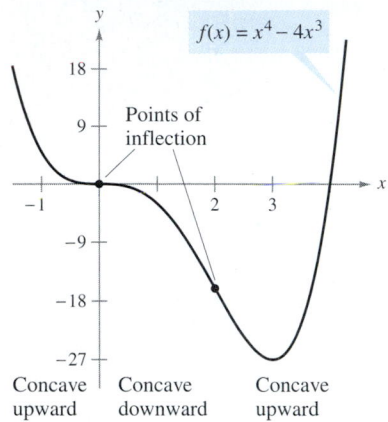

Points of inflection can occur where $f''(x) = 0$ or is undefined.
Figure 3.29

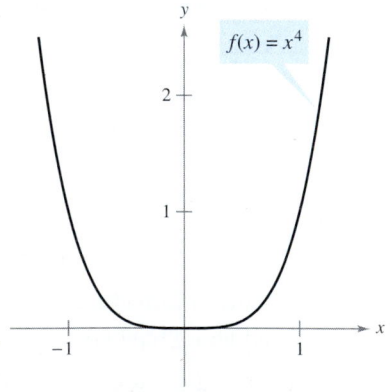

$f''(0) = 0$, but $(0, 0)$ is *not* a point of inflection.
Figure 3.30

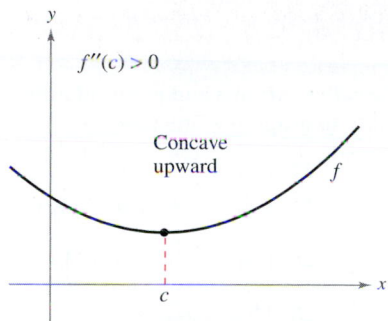

If $f'(c) = 0$ and $f''(c) > 0$, $f(c)$ is a relative minimum.

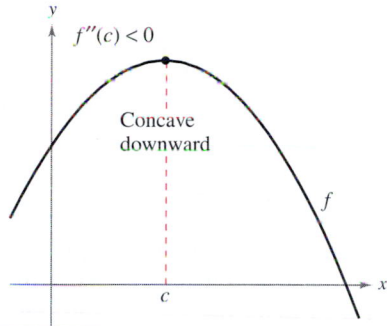

If $f'(c) = 0$ and $f''(c) < 0$, $f(c)$ is a relative maximum.
Figure 3.31

The Second Derivative Test

In addition to testing for concavity, the second derivative can be used to perform a simple test for relative maxima and minima. The test is based on the fact that if the graph of a function f is concave upward on an open interval containing c, and $f'(c) = 0$, $f(c)$ must be a relative minimum of f. Similarly, if the graph of a function f is concave downward on an open interval containing c, and $f'(c) = 0$, $f(c)$ must be a relative maximum of f (see Figure 3.31).

> **THEOREM 3.9   Second Derivative Test**
>
> Let f be a function such that $f'(c) = 0$ and the second derivative of f exists on an open interval containing c.
>
> **1.** If $f''(c) > 0$, then $f(c)$ is a relative minimum.
> **2.** If $f''(c) < 0$, then $f(c)$ is a relative maximum.
>
> If $f''(c) = 0$, the test fails. In such cases, you can use the First Derivative Test.

Proof   If $f'(c) = 0$ and $f''(c) > 0$, there exists an open interval I containing c for which

$$\frac{f'(x) - f'(c)}{x - c} = \frac{f'(x)}{x - c} > 0$$

for all $x \neq c$ in I. If $x < c$, then $x - c < 0$ and $f'(x) < 0$. Also, if $x > c$, then $x - c > 0$ and $f'(x) > 0$. Thus, $f'(x)$ changes from negative to positive at c, and the First Derivative Test implies that $f(c)$ is a relative minimum. A proof of the second case is left to you.

EXAMPLE 4   Using the Second Derivative Test

Find the relative extrema for $f(x) = -3x^5 + 5x^3$.

Solution   Begin by finding the critical numbers of f.

$$f'(x) = -15x^4 + 15x^2 = 15x^2(1 - x^2) = 0 \qquad \text{Set } f'(x) = 0.$$
$$x = -1, 0, 1 \qquad \text{Critical numbers}$$

Using $f''(x) = -60x^3 + 30x = 30(-2x^3 + x)$, you can apply the Second Derivative Test as follows.

Point	Sign of f''		Conclusion
$(-1, -2)$	$f''(-1) = 30 > 0$		Relative minimum
$(1, 2)$	$f''(1) = -30 < 0$		Relative maximum
$(0, 0)$	$f''(0) = 0$		Test fails

Because the Second Derivative Test fails at $(0, 0)$, you can use the First Derivative Test and observe that f increases to the left and right of $x = 0$. Thus, $(0, 0)$ is neither a relative minimum nor a relative maximum (even though the graph has a horizontal tangent line at this point). The graph of f is shown in Figure 3.32.

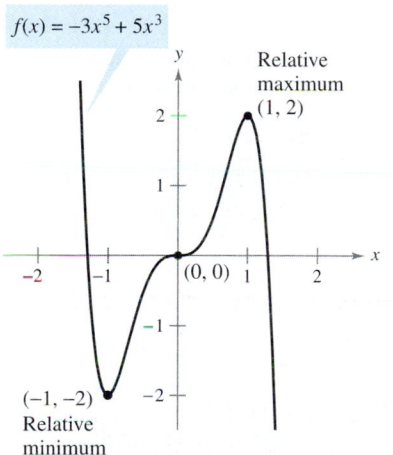

$f(x) = -3x^5 + 5x^3$

$(0, 0)$ is neither a relative minimum nor a relative maximum.
Figure 3.32

EXERCISES FOR SECTION 3.4

In Exercises 1–6, find the open intervals on which the graph is concave upward and those on which it is concave downward.

1. $y = x^2 - x - 2$

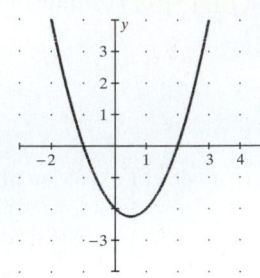

Generated by Derive

2. $y = -x^3 + 3x^2 - 2$

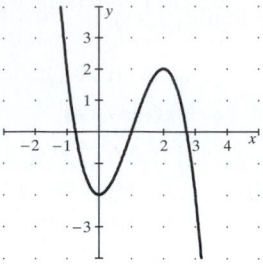

Generated by Derive

3. $f(x) = \dfrac{24}{x^2 + 12}$

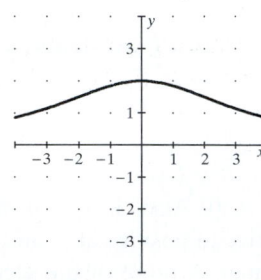

Generated by Derive

4. $f(x) = \dfrac{x^2 - 1}{2x + 1}$

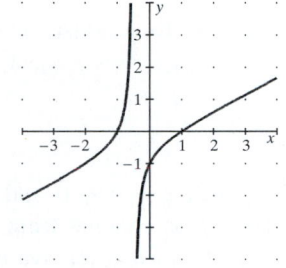

Generated by Derive

5. $f(x) = \dfrac{x^2 + 1}{x^2 - 1}$

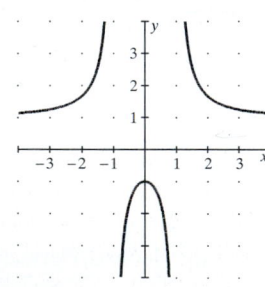

Generated by Derive

6. $y = \dfrac{-3x^5 + 40x^3 + 135x}{270}$

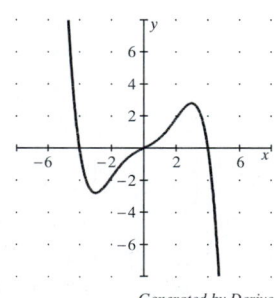

Generated by Derive

In Exercises 7–20, find all relative extrema. Use the Second Derivative Test where applicable.

7. $f(x) = 6x - x^2$

8. $f(x) = x^2 + 3x - 8$

9. $f(x) = (x - 5)^2$

10. $f(x) = -(x - 5)^2$

11. $f(x) = x^3 - 3x^2 + 3$

12. $f(x) = 5 + 3x^2 - x^3$

13. $f(x) = x^4 - 4x^3 + 2$

14. $f(x) = x^3 - 9x^2 + 27x$

15. $f(x) = x^{2/3} - 3$

16. $f(x) = \sqrt{x^2 + 1}$

17. $f(x) = x + \dfrac{4}{x}$

18. $f(x) = \dfrac{x}{x - 1}$

19. $f(x) = \cos x - x$

$0 \le x \le 4\pi$

20. $f(x) = 2 \sin x + \cos 2x$

$0 \le x \le 2\pi$

In Exercises 21–36, find all relative extrema and points of inflection and use a graphing utility to graph the function.

21. $f(x) = x^3 - 12x$

22. $f(x) = x^3 + 1$

23. $f(x) = x^3 - 6x^2 + 12x$

24. $f(x) = 2x^3 - 3x^2 - 12x$

25. $f(x) = \dfrac{1}{4}x^4 - 2x^2$

26. $f(x) = 2x^4 - 8x + 3$

27. $f(x) = x(x - 4)^3$

28. $f(x) = x^3(x - 4)$

29. $f(x) = x\sqrt{x + 3}$

30. $f(x) = x\sqrt{x + 1}$

31. $f(x) = \sin \dfrac{x}{2}$

$0 \le x \le 4\pi$

32. $f(x) = 2 \csc \dfrac{3x}{2}$

$0 < x < 2\pi$

33. $f(x) = \sec\left(x - \dfrac{\pi}{2}\right)$

$0 < x < 4\pi$

34. $f(x) = \sin x + \cos x$

$0 \le x \le 2\pi$

35. $f(x) = 2 \sin x + \sin 2x$

$0 \le x \le 2\pi$

36. $f(x) = x - \sin x$

$0 \le x \le 4\pi$

In Exercises 37–40, use a symbolic differentiation utility to analyze the function over the indicated interval. **(a)** Find the first- and second-order derivatives of the function. **(b)** Find any relative extrema and points of inflection. **(c)** Graph f, f', and f'' on the same set of coordinate axes and state the relationship between the behavior of f and the signs of f' and f''.

37. $f(x) = 0.2x^2(x - 3)^3$, $[-1, 4]$

38. $f(x) = x^2\sqrt{6 - x^2}$, $[-\sqrt{6}, \sqrt{6}]$

39. $f(x) = \sin x - \dfrac{1}{3} \sin 3x + \dfrac{1}{5} \sin 5x$, $[0, \pi]$

40. $f(x) = \sqrt{2x} \sin x$, $[0, 2\pi]$

Think About It In Exercises 41–44, trace the graph of f. On the same set of coordinate axes, sketch the graphs of f' and f''.

41.

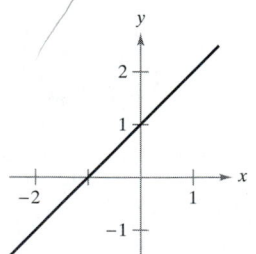

42.

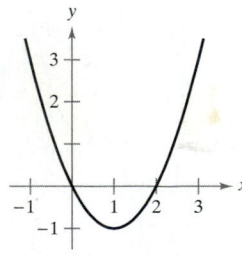

43.

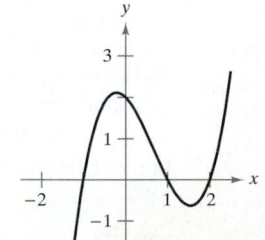

44.

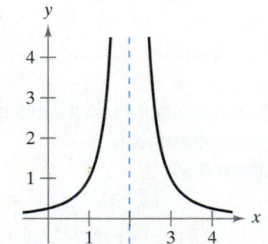

45. ***Think About It***   Consider a function f such that f' is increasing. Sketch graphs of f for (a) $f' < 0$ and (b) $f' > 0$.

46. ***Think About It***   Consider a function f such that f' is decreasing. Sketch graphs of f for (a) $f' < 0$ and (b) $f' > 0$.

Think About It   **In Exercises 47–50, sketch the graph of a function f having the indicated characteristics.**

47. $f(2) = f(4) = 0$

$f(3)$ is defined.

$f'(x) < 0$ if $x < 3$

$f'(3)$ is undefined.

$f'(x) > 0$ if $x > 3$

$f''(x) < 0, x \neq 3$

48. $f(0) = f(2) = 0$

$f'(x) > 0$ if $x < 1$

$f'(1) = 0$

$f'(x) < 0$ if $x > 1$

$f''(x) < 0$

49. $f(2) = f(4) = 0$

$f'(x) > 0$ if $x < 3$

$f'(3)$ is undefined.

$f'(x) < 0$ if $x > 3$

$f''(x) > 0, x \neq 3$

50. $f(0) = f(2) = 0$

$f'(x) < 0$ if $x < 1$

$f'(1) = 0$

$f'(x) > 0$ if $x > 1$

$f''(x) > 0$

In Exercises 51 and 52, find a, b, c, and d such that the cubic $f(x) = ax^3 + bx^2 + cx + d$ satisfies the indicated conditions.

51. Relative maximum: $(3, 3)$

Relative minimum: $(5, 1)$

Inflection point: $(4, 2)$

52. Relative maximum: $(2, 4)$

Relative minimum: $(4, 2)$

Inflection point: $(3, 3)$

53. ***Think About It***   The figure shows the graph of the second derivative of a function f. Sketch a graph of f. (The answer is not unique.)

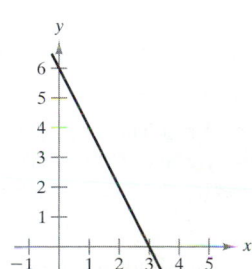

Figure for 53

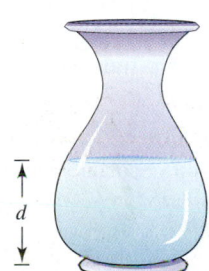

Figure for 54

54. ***Think About It***   Water is running into the vase shown in the figure at a constant rate.

(a) Sketch a graph of the depth d of water in the vase as a function of time.

(b) Does the function have any extrema? Explain.

(c) Give an interpretation of the inflection points of the graph of d.

55. ***Conjecture***   Consider the function $f(x) = (x - 2)^n$.

(a) Use a graphing utility to graph f for $n = 1, 2, 3,$ and 4. Use the graphs to make a conjecture about the relationship between n and any inflection points of the graph of f.

(b) Verify your conjecture in part (a).

56. (a) Graph $f(x) = \sqrt[3]{x}$ and identify the inflection point.

(b) Does $f''(x)$ exist at the inflection point? Explain.

57. ***Think About It*** S represents weekly sales of a product. What can be said of S' and S'' for each of the following?

(a) The rate of change of sales is increasing.

(b) Sales are increasing at a slower rate.

(c) The rate of change of sales is constant.

(d) Sales are steady.

(e) Sales are declining, but at a slower rate.

(f) Sales have bottomed out and have started to rise.

58. ***Think About It***   Sketch the graph of an arbitrary function that does *not* have a point of inflection at $(c, f(c))$ even though $f''(c) = 0$.

59. ***Aircraft Glide Path***   A small aircraft starts its descent from an altitude of 1 mile, 4 miles west of the runway (see figure).

(a) Find the cubic $f(x) = ax^3 + bx^2 + cx + d$ on the interval $[-4, 0]$ that describes a smooth glide path for the landing.

(b) If the glide path of the plane is described by the function in part (a), when would the plane be descending at the most rapid rate?

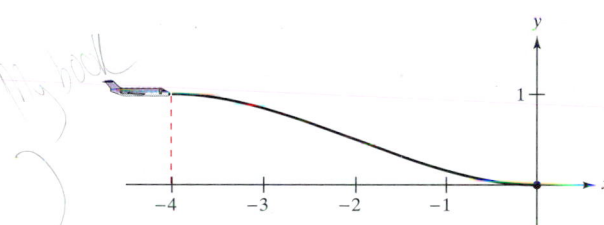

FOR FURTHER INFORMATION   For more information on this type of modeling, see the article "How Not to Land at Lake Tahoe!" by Richard Barshinger in the May 1992 issue of the *The American Mathematical Monthly*.

60. ***Highway Design***   A section of highway connecting two hillsides with grades of 6% and 4% is to be built between two points that are separated by a horizontal distance of 2000 feet (see figure). At the point where the two hillsides come together, there is a 50-foot difference in elevation.

(a) Design a section of highway connecting the hillsides modeled by the function $f(x) = ax^3 + bx^2 + cx + d (-1000 \leq x \leq 1000)$. At the points A and B, the slope of the model must match the grade of the hillside.

(b) Use a graphing utility to graph the model.

(c) Use a graphing utility to graph the derivative of the model.

(d) Determine the grade at the steepest part of the transitional section of the highway.

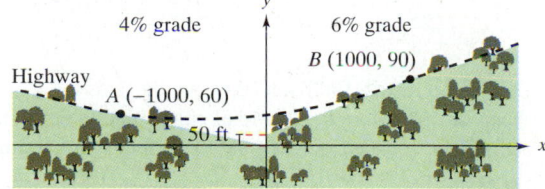

61. Beam Deflection The deflection D of a particular beam of length L is

$$D = 2x^4 - 5Lx^3 + 3L^2x^2$$

where x is the distance from one end of the beam. Find the value of x that yields the maximum deflection.

62. Specific Gravity A model for the specific gravity of water S is

$$S = \frac{5.755}{10^8}T^3 - \frac{8.521}{10^6}T^2 + \frac{6.540}{10^5}T + 0.99987, \quad 0 < T < 25$$

where T is the water temperature in degrees Celsius.

(a) Use a symbolic differentiation utility to find the coordinates of the maximum value of the function.

(b) Sketch a graph of the function over the specified domain. (Use a setting in which $0.996 \leq S \leq 1.001$.)

(c) Estimate the specific gravity of water when $T = 20°$.

63. Average Cost A manufacturer has determined that the total cost C of operating a factory is

$$C = 0.5x^2 + 15x + 5000$$

where x is the number of units produced. At what level of production will the average cost per unit be minimized? (The average cost per unit is C/x.)

64. Inventory Cost The total cost C for ordering and storing x units is

$$C = 2x + \frac{300,000}{x}.$$

What order size will produce a minimum cost?

65. Engine Design In the engine shown in the figure, a connecting rod 18 centimeters long is fastened to a crank of radius 6 centimeters at point P. The crankshaft rotates counterclockwise at a constant rate of 200 revolutions per minute. The horizontal velocity (cm/min) of point P is

$$v = -2400\pi \sin\theta$$

where θ is the central angle of the crankshaft. What values of θ produce a maximum horizontal velocity?

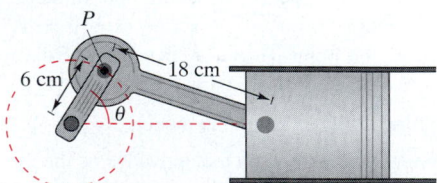

66. Electric Field Intensity The equation

$$E = \frac{kqx}{(x^2 + a^2)^{3/2}}$$

gives the electric field intensity on the axis of a uniformly charged ring, where q is the total charge, k is a constant, and a is the radius of the ring. At what value of x is E maximum?

Linear and Quadratic Approximations In Exercises 67–70, use a graphing utility to graph the function. Then graph the linear and quadratic approximations

$$P_1(x) = f(a) + f'(a)(x - a)$$

and

$$P_2(x) = f(a) + f'(a)(x - a) + \tfrac{1}{2}f''(a)(x - a)^2$$

in the same viewing rectangle. Compare the values of f, P_1, and P_2 and their first derivatives at $x = a$. How do the approximations change as you move away from $x = a$?

Function	Value of a
67. $f(x) = 2(\sin x + \cos x)$	$a = 0$
68. $f(x) = \sqrt{1 - x}$	$a = 0$
69. $f(x) = 2(\sin x + \cos x)$	$a = \dfrac{\pi}{4}$
70. $f(x) = \dfrac{\sqrt{x}}{x - 1}$	$a = 2$

71. Use a graphing utility to graph $y = x\sin(1/x)$. Show that the graph is concave downward to the right of $x = 1/\pi$.

72. Show that the point of inflection of $f(x) = x(x - 6)^2$ lies midway between the relative extrema of f.

73. Prove that every cubic function with three distinct real zeros has a point of inflection whose x-coordinate is the average of the three zeros.

74. Show that the cubic polynomial $p(x) = ax^3 + bx^2 + cx + d$ has exactly one point of inflection (x_0, y_0), where

$$x_0 = \frac{-b}{3a} \quad \text{and} \quad y_0 = \frac{2b^3}{27a^2} - \frac{bc}{3a} + d.$$

Use this formula to find the point of inflection of $p(x) = x^3 - 3x^2 + 2$.

75. Darboux's Theorem Prove Darboux's Theorem: Let $f(x)$ be differentiable on $[a, b]$, $f'(a) = y_1$, and $f'(b) = y_2$. If d lies between y_1 and y_2, then there exists c in (a, b) such that $f'(c) = d$.

76. Writing Discuss the advantages and disadvantages of the First and Second Derivative Tests. Illustrate your discussion with examples.

True or False? In Exercises 77–80, determine whether the statement is true or false. If it is false, explain why or give an example that shows it is false.

77. The graph of every cubic polynomial has precisely one point of inflection.

78. The graph of $f(x) = 1/x$ is concave downward for $x < 0$ and concave upward for $x > 0$, and thus it has a point of inflection at $x = 0$.

79. The maximum value of $y = 3\sin x + 2\cos x$ is 5.

80. The maximum slope of the graph of $y = \sin(bx)$ is b.

Limits at Infinity • Horizontal Asymptotes

Limits at Infinity

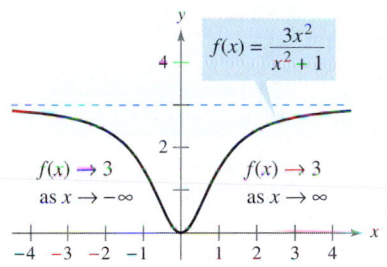

The limit of $f(x)$ as x approaches $-\infty$ or ∞ is 3.

Figure 3.33

This section discusses the "end behavior" of a function on an *infinite* interval. Consider the graph of

$$f(x) = \frac{3x^2}{x^2 + 1}$$

as shown in Figure 3.33. Graphically, you can see that the values of $f(x)$ appear to approach 3 as x increases without bound or decreases without bound. You can come to the same conclusions numerically, as shown in the table.

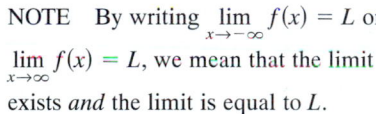

x	$-\infty \leftarrow$	-100	-10	-1	0	1	10	100	$\rightarrow \infty$
$f(x)$	$3 \leftarrow$	2.9997	2.97	1.5	0	1.5	2.97	2.9997	$\rightarrow 3$

$f(x)$ approaches 3. $f(x)$ approaches 3.

The table suggests that the value of $f(x)$ approaches 3 as x increases without bound ($x \rightarrow \infty$). Similarly, $f(x)$ approaches 3 as $x \rightarrow -\infty$. These **limits at infinity** are denoted by

NOTE   By writing $\lim\limits_{x \to -\infty} f(x) = L$ or $\lim\limits_{x \to \infty} f(x) = L$, we mean that the limit exists *and* the limit is equal to L.

$$\lim_{x \to -\infty} f(x) = 3 \qquad \text{\color{red} Limit at negative infinity}$$

and

$$\lim_{x \to \infty} f(x) = 3. \qquad \text{\color{red} Limit at positive infinity}$$

To say that a statement is true as x increases *without bound* means that for some (large) real number M, the statement is true for *all* x in the interval $\{x: x > M\}$. The following definition uses this concept.

Definition of Limits at Infinity

Let L be a real number.

1. The statement $\lim\limits_{x \to \infty} f(x) = L$ means that for each $\varepsilon > 0$ there exists an $M > 0$ such that $|f(x) - L| < \varepsilon$ whenever $x > M$.

2. The statement $\lim\limits_{x \to -\infty} f(x) = L$ means that for each $\varepsilon > 0$ there exists an $N < 0$ such that $|f(x) - L| < \varepsilon$ whenever $x < N$.

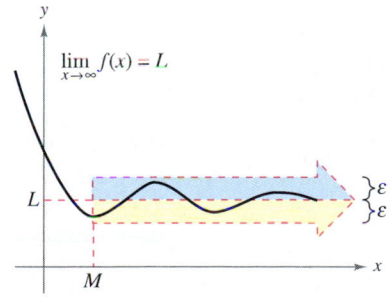

$f(x)$ is within ε units of L as $x \rightarrow \infty$.

Figure 3.34

The definition of a limit at infinity is illustrated in Figure 3.34. In this figure, note that for a given positive number ε there exists a positive number M such that, for $x > M$, the graph of f will lie between the horizontal lines given by $y = L + \varepsilon$ and $y = L - \varepsilon$.

Horizontal Asymptotes

In Figure 3.34, the graph of f approaches the line $y = L$ as x increases without bound. We call the line $y = L$ a **horizontal asymptote** of the graph of f.

Definition of Horizontal Asymptote

The line $y = L$ is a **horizontal asymptote** of the graph of f if

$$\lim_{x \to -\infty} f(x) = L \quad \text{or} \quad \lim_{x \to \infty} f(x) = L.$$

NOTE   From this definition, it follows that the graph of a *function* of x can have at most two horizontal asymptotes—one to the right and one to the left.

Limits at infinity have many of the same properties of limits discussed in Section 1.3. For example, if $\lim_{x \to \infty} f(x)$ and $\lim_{x \to \infty} g(x)$ both exist, then

$$\lim_{x \to \infty} [f(x) + g(x)] = \lim_{x \to \infty} f(x) + \lim_{x \to \infty} g(x)$$

and

$$\lim_{x \to \infty} [f(x)g(x)] = \left[\lim_{x \to \infty} f(x) \right]\left[\lim_{x \to \infty} g(x) \right].$$

Similar properties hold for limits at $-\infty$.

When evaluating limits at infinity, the following theorem is helpful. (A proof of this theorem is given in the appendix.)

THEOREM 3.10   Limits at Infinity

If r is a positive rational number and c is any real number, then

$$\lim_{x \to \infty} \frac{c}{x^r} = 0.$$

Furthermore, if x^r is defined when $x < 0$, then $\displaystyle \lim_{x \to -\infty} \frac{c}{x^r} = 0.$

EXAMPLE 1   Evaluating a Limit at Infinity

Find the limit: $\displaystyle \lim_{x \to \infty} \left(5 - \frac{2}{x^2} \right).$

Solution   Using Theorem 3.10, you can write the following.

$$\lim_{x \to \infty} \left(5 - \frac{2}{x^2} \right) = \lim_{x \to \infty} 5 - \lim_{x \to \infty} \frac{2}{x^2}$$

$$= 5 - 0$$

$$= 5$$

EXAMPLE 2   Evaluating a Limit at Infinity

Find the limit: $\lim\limits_{x \to \infty} \dfrac{2x - 1}{x + 1}$.

Solution   Note that both the numerator and the denominator approach infinity as x approaches infinity.

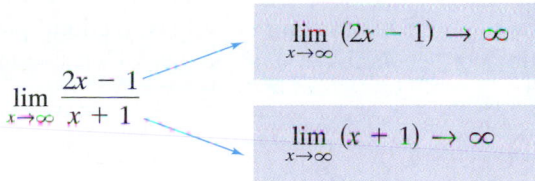

This results in $\dfrac{\infty}{\infty}$, an **indeterminate form**. To resolve this problem, you can divide both the numerator and the denominator by x. After dividing, the limit may be evaluated as follows.

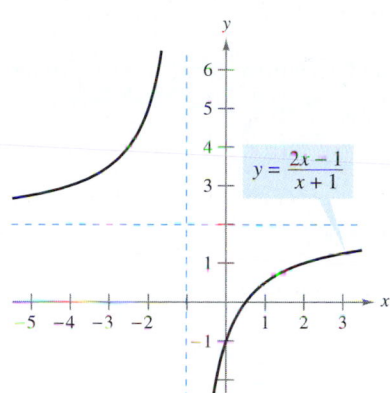

$y = 2$ is a horizontal asymptote.
Figure 3.35

$$\lim_{x \to \infty} \frac{2x - 1}{x + 1} = \lim_{x \to \infty} \frac{\dfrac{2x - 1}{x}}{\dfrac{x + 1}{x}} \qquad \text{Divide numerator and denominator by } x.$$

$$= \lim_{x \to \infty} \frac{2 - (1/x)}{1 + (1/x)}$$

$$= \frac{\lim\limits_{x \to \infty} 2 - \lim\limits_{x \to \infty} \dfrac{1}{x}}{\lim\limits_{x \to \infty} 1 + \lim\limits_{x \to \infty} \dfrac{1}{x}} \qquad \text{Take limits of numerator and denominator.}$$

NOTE   When you encounter an indeterminate form such as the one in Example 2, we suggest dividing by the highest power of x in the *denominator*.

$$= \frac{2 - 0}{1 + 0} \qquad \text{Apply Theorem 3.10.}$$

$$= 2$$

Thus, the line $y = 2$ is a horizontal asymptote to the right. By taking the limit as $x \to -\infty$, you can see that $y = 2$ is also a horizontal asymptote to the left. The graph of the function is shown in Figure 3.35.

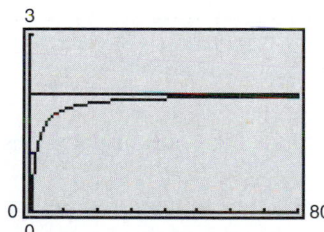

As x increases, the graph of f moves closer and closer to the line $y = 2$.
Figure 3.36

TECHNOLOGY   You can test the reasonableness of the limit found in Example 2 by evaluating $f(x)$ for a few large positive values of x. For instance,

$$f(100) \approx 1.9703, \quad f(1000) \approx 1.9970, \quad \text{and} \quad f(10{,}000) \approx 1.9997.$$

Another way to test the reasonableness of the limit is to use a graphing utility. For instance, in Figure 3.36, the graph of

$$f(x) = \frac{2x - 1}{x + 1}$$

is shown with the horizontal line $y = 2$. Note that as x increases, the graph of f moves closer and closer to its horizontal asymptote.

EXAMPLE 3   A Comparison of Three Rational Functions

Find each of the limits.

a. $\displaystyle\lim_{x\to\infty}\frac{2x+5}{3x^2+1}$ **b.** $\displaystyle\lim_{x\to\infty}\frac{2x^2+5}{3x^2+1}$ **c.** $\displaystyle\lim_{x\to\infty}\frac{2x^3+5}{3x^2+1}$

Solution

a. Attempting to evaluate the limit produces the indeterminate form ∞/∞. Divide both the numerator and the denominator by x^2.

$$\lim_{x\to\infty}\frac{2x+5}{3x^2+1}=\lim_{x\to\infty}\frac{(2/x)+(5/x^2)}{3+(1/x^2)}=\frac{0+0}{3+0}=\frac{0}{3}=0$$

b. Divide both the numerator and the denominator by x^2.

$$\lim_{x\to\infty}\frac{2x^2+5}{3x^2+1}=\lim_{x\to\infty}\frac{2+(5/x^2)}{3+(1/x^2)}=\frac{2+0}{3+0}=\frac{2}{3}$$

c. Divide both the numerator and the denominator by x^2.

$$\lim_{x\to\infty}\frac{2x^3+5}{3x^2+1}=\lim_{x\to\infty}\frac{2x+(5/x^2)}{3+(1/x^2)}=\frac{\infty}{3}$$

You can conclude that the limit *does not exist* because the numerator increases without bound while the denominator approaches 3.

Guidelines for Finding Limits of Rational Functions

1. If the degree of the numerator is *less than* the degree of the denominator, then the limit of the rational function is 0.

2. If the degree of the numerator is *equal to* the degree of the denominator, then the limit of the rational function is the ratio of the leading coefficients.

3. If the degree of the numerator is *greater than* the degree of the denominator, then the limit of the rational function does not exist.

Use these guidelines to check the results in Example 3. These limits seem reasonable when you consider that for large values of x, the highest-power term of the rational function is the most "influential" in determining the limit. For instance, the limit as x approaches infinity of the function

$$f(x)=\frac{1}{x^2+1}$$

is 0 because the denominator overpowers the numerator as x increases or decreases without bound, as shown in Figure 3.37.

The function shown in Figure 3.37 is a special case of a type of curve studied by the Italian mathematician Maria Gaetana Agnesi. The general form of this function is

$$f(x)=\frac{8a^3}{x^2+4a^2}\qquad\text{Witch of Agnesi}$$

and, through a mistranslation of the Italian word *vertéré*, the curve has come to be known as the witch of Agnesi. Agnesi's work with this curve first appeared in a comprehensive text on calculus that was published in 1748.

MARIA AGNESI (1718–1799)

Agnesi was one of a handful of women to receive credit for significant contributions to mathematics before the twentieth century. In her early twenties, she wrote the first text that included both differential and integral calculus. By age 30, she was an honorary member of the faculty at the University of Bologna.

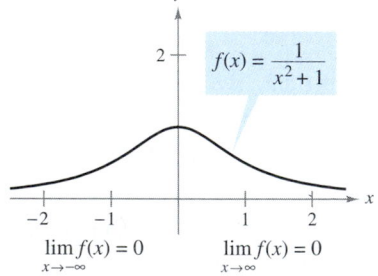

$f(x)=\dfrac{1}{x^2+1}$

$\displaystyle\lim_{x\to-\infty}f(x)=0\qquad\lim_{x\to\infty}f(x)=0$

f has a horizontal asymptote at $y=0$.

Figure 3.37

FOR FURTHER INFORMATION For more information on the contributions of women to mathematics, see the article "Why Women Succeed in Mathematics" by Mona Fabricant, Sylvia Svitak, and Patricia Clark Kenschaft in the February 1990 issue of *Mathematics Teacher*.

In Figure 3.37, you can see that the function $f(x) = 1/(x^2 + 1)$ approaches the same horizontal asymptote to the right and to the left. This is always true of rational functions. Functions that are not rational, however, may approach different horizontal asymptotes to the right and to the left. This is demonstrated in Example 4.

EXAMPLE 4   A Function with Two Horizontal Asymptotes

Determine each of the limits.

a. $\displaystyle\lim_{x\to\infty} \frac{3x - 2}{\sqrt{2x^2 + 1}}$ **b.** $\displaystyle\lim_{x\to-\infty} \frac{3x - 2}{\sqrt{2x^2 + 1}}$

Solution

a. For $x > 0$, you can write $x = \sqrt{x^2}$. Thus, dividing both the numerator and the denominator by x produces

$$\frac{3x - 2}{\sqrt{2x^2 + 1}} = \frac{\dfrac{3x - 2}{x}}{\dfrac{\sqrt{2x^2 + 1}}{\sqrt{x^2}}} = \frac{3 - \dfrac{2}{x}}{\sqrt{\dfrac{2x^2 + 1}{x^2}}} = \frac{3 - \dfrac{2}{x}}{\sqrt{2 + \dfrac{1}{x^2}}}$$

and you can take the limit as follows.

$$\lim_{x\to\infty} \frac{3x - 2}{\sqrt{2x^2 + 1}} = \lim_{x\to\infty} \frac{3 - \dfrac{2}{x}}{\sqrt{2 + \dfrac{1}{x^2}}} = \frac{3 - 0}{\sqrt{2 + 0}} = \frac{3}{\sqrt{2}}$$

b. For $x < 0$, you can write $x = -\sqrt{x^2}$. Thus, dividing both the numerator and the denominator by x produces

$$\frac{3x - 2}{\sqrt{2x^2 + 1}} = \frac{\dfrac{3x - 2}{x}}{\dfrac{\sqrt{2x^2 + 1}}{-\sqrt{x^2}}} = \frac{3 - \dfrac{2}{x}}{-\sqrt{\dfrac{2x^2 + 1}{x^2}}} = \frac{3 - \dfrac{2}{x}}{-\sqrt{2 + \dfrac{1}{x^2}}}$$

and you can take the limit as follows.

$$\lim_{x\to-\infty} \frac{3x - 2}{\sqrt{2x^2 + 1}} = \lim_{x\to-\infty} \frac{3 - \dfrac{2}{x}}{-\sqrt{2 + \dfrac{1}{x^2}}} = \frac{3 - 0}{-\sqrt{2 + 0}} = -\frac{3}{\sqrt{2}}$$

The graph of $f(x) = (3x - 2)/\sqrt{2x^2 + 1}$ is shown in Figure 3.38.

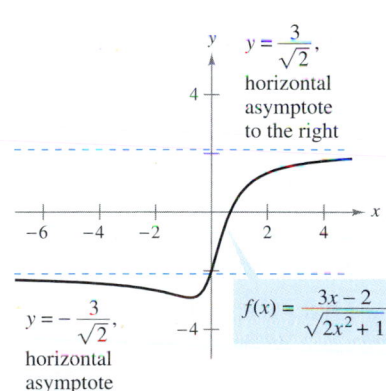

Functions that are not rational may have different right and left horizontal asymptotes.
Figure 3.38

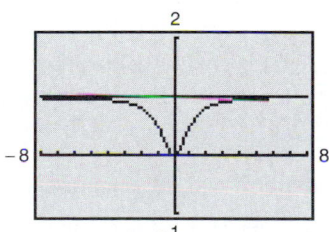

The horizontal asymptote appears to be the line $y = 1$ but is actually $y = 2$.
Figure 3.39

TECHNOLOGY   If you use a graphing utility to help estimate a limit, be sure that you also confirm the estimate analytically—the pictures shown by a graphing utility can be misleading. For instance, Figure 3.39 shows one view of the graph of

$$y = \frac{2x^3 + 1000x^2 + x}{x^3 + 1000x^2 + x + 1000}.$$

From this view, one could be convinced that the graph has $y = 1$ as a horizontal asymptote. An analytical approach shows that the horizontal asymptote is actually $y = 2$. Confirm this by enlarging the viewing rectangle on the graphing utility.

In Section 1.3 (Example 9), you saw how the Squeeze Theorem can be used to evaluate limits involving trigonometric functions. This theorem is also valid for limits at infinity.

EXAMPLE 5   Limits Involving Trigonometric Functions

Determine each of the limits.

a. $\lim\limits_{x \to \infty} \sin x$ **b.** $\lim\limits_{x \to \infty} \dfrac{\sin x}{x}$

Solution

a. As x approaches infinity, the sine function oscillates between 1 and -1. Thus, this limit does not exist.

b. Because $-1 \le \sin x \le 1$, it follows that for $x > 0$,

$$-\frac{1}{x} \le \frac{\sin x}{x} \le \frac{1}{x}$$

where $\lim\limits_{x \to \infty} (-1/x) = 0$ and $\lim\limits_{x \to \infty} (1/x) = 0$. Therefore, by the Squeeze Theorem, you can obtain

$$\lim_{x \to \infty} \frac{\sin x}{x} = 0$$

as indicated in Figure 3.40.

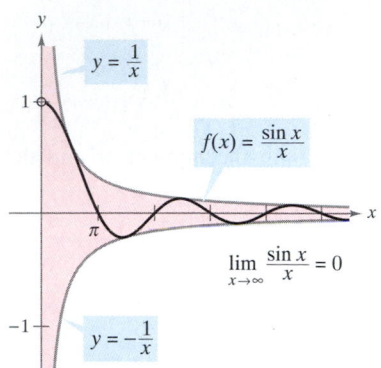

As x increases without bound, $f(x)$ approaches 0.
Figure 3.40

EXAMPLE 6   Oxygen Level in a Pond

Suppose that $f(t)$ measures the level of oxygen in a pond, where $f(t) = 1$ is the normal (unpolluted) level and the time t is measured in weeks. When $t = 0$, organic waste is dumped into the pond, and as the waste material oxidizes, the level of oxygen in the pond is

$$f(t) = \frac{t^2 - t + 1}{t^2 + 1}.$$

What percent of the normal level of oxygen exists in the pond after 1 week? After 2 weeks? After 10 weeks? What is the limit as t approaches infinity?

Solution   When $t = 1, 2,$ and 10, the levels of oxygen are as follows.

$$f(1) = \frac{1^2 - 1 + 1}{1^2 + 1} = \frac{1}{2} = 50\% \qquad \text{1 week}$$

$$f(2) = \frac{2^2 - 2 + 1}{2^2 + 1} = \frac{3}{5} = 60\% \qquad \text{2 weeks}$$

$$f(10) = \frac{10^2 - 10 + 1}{10^2 + 1} = \frac{91}{101} \approx 90.1\% \qquad \text{10 weeks}$$

To take the limit as t approaches infinity, divide the numerator and the denominator by t^2 to obtain

$$\lim_{x \to \infty} \frac{t^2 - t + 1}{t^2 + 1} = \lim_{x \to \infty} \frac{1 - (1/t) + (1/t^2)}{1 + (1/t^2)} = \frac{1 - 0 + 0}{1 + 0} = 1 = 100\%.$$

(See Figure 3.41.)

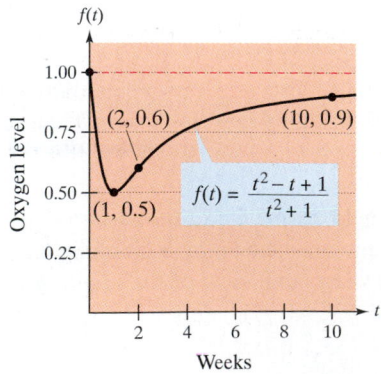

The level of oxygen in a pond approaches the normal level of 1 as t approaches ∞.
Figure 3.41

EXERCISES FOR SECTION 3.5

In Exercises 1–6, match the function with one of the graphs [(a), (b), (c), (d), (e), or (f)] using horizontal asymptotes as an aid.

(a)
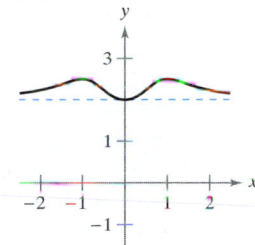

(b)
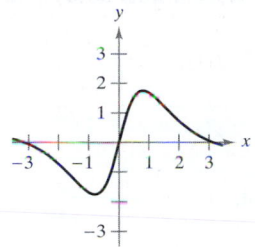

(c)
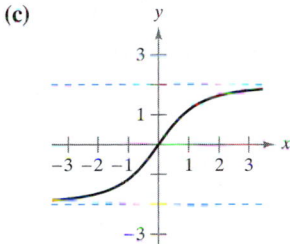

(d)
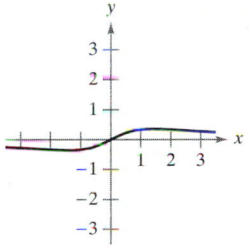

(e)
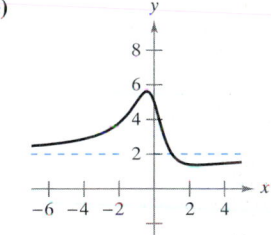

(f)
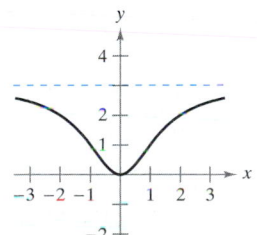

1. $f(x) = \dfrac{3x^2}{x^2 + 2}$

2. $f(x) = \dfrac{2x}{\sqrt{x^2 + 2}}$

3. $f(x) = \dfrac{x}{x^2 + 2}$

4. $f(x) = 2 + \dfrac{x^2}{x^4 + 1}$

5. $f(x) = \dfrac{4 \sin x}{x^2 + 1}$

6. $f(x) = \dfrac{2x^2 - 3x + 5}{x^2 + 1}$

Numerical and Graphical Analysis   In Exercises 7–10, use a graphing utility to complete the table and estimate the limit as x approaches infinity. Then use a graphing utility to graph the function and estimate the limit graphically.

x	10^0	10^1	10^2	10^3	10^4	10^5	10^6
$f(x)$							

7. $f(x) = \dfrac{4x + 3}{2x - 1}$

8. $f(x) = \dfrac{2x^2}{x + 1}$

9. $f(x) = \dfrac{-6x}{\sqrt{4x^2 + 5}}$

10. $f(x) = 5 - \dfrac{1}{x^2 + 1}$

In Exercises 11–24, find the limit.

11. $\displaystyle \lim_{x \to \infty} \dfrac{2x - 1}{3x + 2}$

12. $\displaystyle \lim_{x \to \infty} \dfrac{5x^3 + 1}{10x^3 - 3x^2 + 7}$

13. $\displaystyle \lim_{x \to \infty} \dfrac{x}{x^2 - 1}$

14. $\displaystyle \lim_{x \to \infty} \dfrac{2x^{10} - 1}{10x^{11} - 3}$

15. $\displaystyle \lim_{x \to -\infty} \dfrac{5x^2}{x + 3}$

16. $\displaystyle \lim_{x \to \infty} \left(2x - \dfrac{1}{x^2} \right)$

17. $\displaystyle \lim_{x \to -\infty} \dfrac{x}{\sqrt{x^2 - x}}$

18. $\displaystyle \lim_{x \to \infty} \dfrac{x}{\sqrt{x^2 + 1}}$

19. $\displaystyle \lim_{x \to \infty} \dfrac{2x + 1}{\sqrt{x^2 - x}}$

20. $\displaystyle \lim_{x \to -\infty} \dfrac{-3x + 1}{\sqrt{x^2 + x}}$

21. $\displaystyle \lim_{x \to \infty} \dfrac{\sin 2x}{x}$

22. $\displaystyle \lim_{x \to \infty} \dfrac{x - \cos x}{x}$

23. $\displaystyle \lim_{x \to \infty} \dfrac{1}{2x + \sin x}$

24. $\displaystyle \lim_{x \to \infty} \sin \dfrac{1}{x}$

In Exercises 25 and 26, find the limit. (*Hint:* Let $x = 1/t$ and find the limit as $t \to 0^+$.)

25. $\displaystyle \lim_{x \to \infty} x \sin \dfrac{1}{x}$

26. $\displaystyle \lim_{x \to \infty} x \tan \dfrac{1}{x}$

 In Exercises 27–30, find the limit (*Hint:* Treat the expression as a fraction whose denominator is 1, and rationalize the numerator.) Use a graphing utility to verify your result.

27. $\displaystyle \lim_{x \to -\infty} \left(x + \sqrt{x^2 + 3} \right)$

28. $\displaystyle \lim_{x \to \infty} \left(2x - \sqrt{4x^2 + 1} \right)$

29. $\displaystyle \lim_{x \to \infty} \left(x - \sqrt{x^2 + x} \right)$

30. $\displaystyle \lim_{x \to -\infty} \left(3x + \sqrt{9x^2 - x} \right)$

Numerical, Graphical, and Analytic Analysis   In Exercises 31–34, use a graphing utility to complete the table and estimate the limit as x approaches infinity. Then use a graphing utility to graph the function and estimate the limit graphically. Finally, find the limit analytically and compare your results with the estimates.

x	10^0	10^1	10^2	10^3	10^4	10^5	10^6
$f(x)$							

31. $f(x) = x - \sqrt{x(x - 1)}$

32. $f(x) = x^2 - x\sqrt{x(x - 1)}$

33. $f(x) = x \sin \dfrac{1}{2x}$

34. $f(x) = \dfrac{x + 1}{x\sqrt{x}}$

In Exercises 35–52, sketch the graph of the equation. Look for extrema, intercepts, symmetry, and asymptotes as necessary. Use a graphing utility to verify your result.

35. $y = \dfrac{2 + x}{1 - x}$

36. $y = \dfrac{x - 3}{x - 2}$

37. $y = \dfrac{x}{x^2 - 4}$

38. $y = \dfrac{2x}{9 - x^2}$

39. $y = \dfrac{x^2}{x^2 + 9}$

40. $y = \dfrac{x^2}{x^2 - 9}$

41. $y = \dfrac{2x^2}{x^2 - 4}$

42. $y = \dfrac{2x^2}{x^2 + 4}$

43. $xy^2 = 4$

44. $x^2 y = 4$

45. $y = \dfrac{2x}{1 - x}$

46. $y = \dfrac{2x}{1 - x^2}$

47. $y = 2 - \dfrac{3}{x^2}$

48. $y = 1 + \dfrac{1}{x}$

49. $y = 3 + \dfrac{2}{x}$

50. $y = 4\left(1 - \dfrac{1}{x^2}\right)$

51. $y = \dfrac{x^3}{\sqrt{x^2 - 4}}$

52. $y = \dfrac{x}{\sqrt{x^2 - 4}}$

In Exercises 53–60, use a symbolic differentiation utility to analyze the graph of the function. Label any extrema and/or asymptotes that exist.

53. $f(x) = 5 - \dfrac{1}{x^2}$

54. $f(x) = \dfrac{x^2}{x^2 - 1}$

55. $f(x) = \dfrac{x}{x^2 - 4}$

56. $f(x) = \dfrac{1}{x^2 - x - 2}$

57. $f(x) = \dfrac{x - 2}{x^2 - 4x + 3}$

58. $f(x) = \dfrac{x + 1}{x^2 + x + 1}$

59. $f(x) = \dfrac{3x}{\sqrt{4x^2 + 1}}$

60. $f(x) = \dfrac{2 \sin 2x}{x}$

61. Use a graphing utility to graph each function and verify that each has two horizontal asymptotes.

(a) $f(x) = \dfrac{|x|}{x + 1}$

(b) $f(x) = \dfrac{2x}{\sqrt{x^2 + 1}}$

62. Given the function $f(x) = 5x^3 - 3x^2 + 10$, find $\lim\limits_{x \to \infty} h(x)$, if possible.

(a) $h(x) = \dfrac{f(x)}{x^2}$

(b) $h(x) = \dfrac{f(x)}{x^3}$

(c) $h(x) = \dfrac{f(x)}{x^4}$

In Exercises 63 and 64, (a) use a graphing utility to graph f and g in the same viewing rectangle, (b) verify algebraically that f and g represent the same function, and (c) zoom out sufficiently far so that the graph appears as a line. What equation does this line appear to have? (Note that the points at which the function is not continuous are not readily seen when you zoom out.)

63. $f(x) = \dfrac{x^3 - 3x^2 + 2}{x(x - 3)}$

$g(x) = x + \dfrac{2}{x(x - 3)}$

64. $f(x) = -\dfrac{x^3 - 2x^2 + 2}{2x^2}$

$g(x) = -\dfrac{1}{2}x + 1 - \dfrac{1}{x^2}$

65. **Think About It** The graph of a function f is shown below.

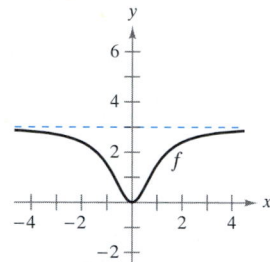

(a) Sketch f'.

(b) Use the graphs to estimate $\lim\limits_{x \to \infty} f(x)$ and $\lim\limits_{x \to \infty} f'(x)$.

(c) Explain the answers you gave in part (b).

66. **Engine Efficiency** The efficiency of an internal combustion engine is

$$\text{Efficiency (\%)} = 100\left[1 - \dfrac{1}{(v_1/v_2)^c}\right]$$

where v_1/v_2 is the ratio of the uncompressed gas to the compressed gas and c is a positive constant dependent on the engine design. Find the limit of the efficiency as the compression ratio approaches infinity.

67. **Average Cost** A business has a cost of $C = 0.5x + 500$ for producing x units. The average cost per unit is

$$\overline{C} = \dfrac{C}{x}.$$

Find the limit of $\overline{C}$ as x approaches infinity.

68. A line with slope m passes through the point $(0, 4)$.

(a) Write the distance d between the line and the point $(3, 1)$ as a function of m.

(b) Use a graphing utility to graph the equation in part (a).

(c) Find $\lim\limits_{m \to \infty} d(m)$ and $\lim\limits_{m \to -\infty} d(m)$. Interpret the results geometrically.

69. *Modeling Data* A heat probe is attached to the heat exchanger of a heating system. The temperature T (degrees Celsius) is recorded t seconds after the furnace is started. The results for the first 2 minutes are recorded in the table.

t	0	15	30	45	60
T	25.2°	36.9°	45.5°	51.4°	56.0°

t	75	90	105	120
T	59.6°	62.0°	64.0°	65.2°

(a) Use the regression capabilities of a graphing utility to find a model of the form $T_1 = at^2 + bt + c$ for the data.

(b) Use a graphing utility to graph T_1.

(c) A rational model for the data is

$$T_2 = \frac{2468 + 155t}{2(50 + t)}.$$

Use a graphing utility to graph the model.

(d) Find $T_1(0)$ and $T_2(0)$.

(e) Find $\lim\limits_{t \to \infty} T_2$.

(f) Interpret the result in part (e) in the context of the problem. Is it possible to do this type of analysis using T_1? Explain.

70. *Modeling Data* The data in the table give the number N (in thousands) of high school graduates at the end of each decade for the years 1900 through 1990. *(Source: U.S. Department of Education)*

Year	1900	1910	1920	1930	1940
N	62	111	231	592	1140

Year	1950	1960	1970	1980	1990
N	1063	1627	2589	2748	2505

A model for these data is

$$N = \frac{68{,}436.82 + 4731.82t}{1000 - 23.37t + 0.16t^2}, \quad 0 \le t \le 90$$

where t is the time in years, with $t = 0$ corresponding to 1990.

(a) Use a graphing utility to plot the data and graph the model.

(b) Use the model to estimate the number of high school graduates in 1975.

(c) Approximate the year when the number of graduates was greatest.

(d) Use a symbolic differentiation utility to determine the time when the rate of increase in the number of graduates was greatest.

(e) Why should this model *not* be used to predict the number of graduates in future years?

True or False? In Exercises 71 and 72, determine whether the statement is true or false. If it is false, explain why or give an example that shows it is false.

71. If $f'(x) > 0$ for all real numbers x, then f increases without bound.

72. If $f''(x) < 0$ for all real numbers x, then f decreases without bound.

73. *Think About It* Sketch a graph of a differentiable function f that satisfies the following conditions and has $x = 2$ as its only critical number.

$f'(x) < 0$ for $x < 2$

$f'(x) > 0$ for $x > 2$

$\lim\limits_{x \to -\infty} f(x) = \lim\limits_{x \to \infty} f(x) = 6$

74. *Think About It* Is it possible to sketch a graph of a function that satisfies the conditions of Exercise 73 and has *no* points of inflection? Explain.

75. Prove that if $p(x) = a_n x^n + \cdots + a_1 x + a_0$ and $q(x) = b_m x^m + \cdots + b_1 x + b_0 \, (a_n \neq 0, b_m \neq 0)$, then

$$\lim_{x \to \infty} \frac{p(x)}{q(x)} = \begin{cases} 0, & n < m \\ \dfrac{a_n}{b_m}, & n = m \\ \pm\infty, & n > m. \end{cases}$$

SECTION PROJECT

Graph the functions

$$f(x) = \frac{2x^2 - x}{x - 1} \quad \text{and} \quad g(x) = 2x + 1$$

in the same viewing rectangle, and zoom out a few times.

(a) Describe the behavior of the graphs of f and g as $x \to \infty$ and $x \to -\infty$.

(b) Use long division of polynomials to show that

$$f(x) = 2x + 1 + \frac{1}{x - 1}.$$

What does this mean about the shapes of the graphs of f and g as $x \to \infty$ and $x \to -\infty$?

(c) Show that f can also be written as follows.

$$f(x) = \frac{2x - 1}{1 - 1/x}.$$

Does this mean that the graph of f should resemble that of $h(x) = 2x - 1$ as $x \to \infty$ and $x \to -\infty$? Explain.

(d) Apply the techniques above to write a short paragraph about the shape of the graph of

$$y = \frac{1 + 2x - 2x^2}{2x}$$

as $x \to \infty$ and $x \to -\infty$.

SECTION 3.6   A Summary of Curve Sketching

Summary of Curve-Sketching Techniques

Summary of Curve-Sketching Techniques

It would be difficult to overstate the importance of using graphs in mathematics. Descartes's introduction of analytic geometry contributed significantly to the rapid advances in calculus that began during the mid-seventeenth century. In the words of Lagrange, "As long as algebra and geometry traveled separate paths their advance was slow and their applications limited. But when these two sciences joined company, they drew from each other fresh vitality and thenceforth marched on at a rapid pace toward perfection."

So far, you have studied several concepts that are useful in analyzing the graph of a function.

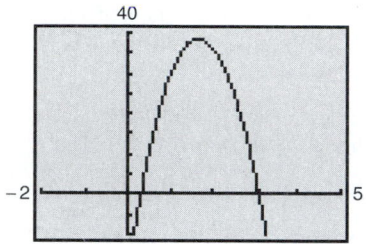

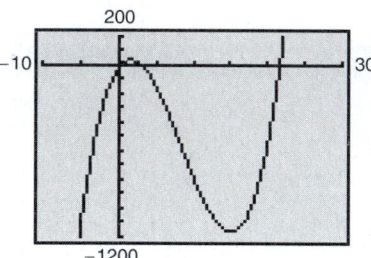

Different viewing rectangles for the same graph
Figure 3.42

- x-intercepts and y-intercepts   (Section P.1)
- Symmetry   (Section P.1)
- Domain and range   (Section P.3)
- Continuity   (Section 1.4)
- Vertical asymptotes   (Section 1.5)
- Differentiability   (Section 2.1)
- Relative extrema   (Section 3.1)
- Concavity   (Section 3.4)
- Points of inflection   (Section 3.4)
- Horizontal asymptotes   (Section 3.5)

When you are sketching the graph of a function, either by hand or with a graphing utility, remember that normally you cannot show the *entire* graph. The decision as to which part of the graph you choose to show is often crucial. For instance, which of the viewing rectangles in Figure 3.42 better represents the graph of

$$f(x) = x^3 - 25x^2 + 74x - 20?$$

By seeing both views, it is clear that the second viewing rectangle gives a more complete representation of the graph. But would a third viewing rectangle reveal other interesting portions of the graph? To answer this, you need to use calculus to interpret the first and second derivatives. Here are some guidelines for determining a good viewing rectangle for the graph of a function.

Guidelines for Analyzing the Graph of a Function

1. Determine the domain and range of the function.

2. Determine the intercepts and asymptotes of the graph.

3. Locate the x-values for which $f'(x)$ and $f''(x)$ are either zero or undefined. Use the results to determine relative extrema and points of inflection.

NOTE   In these guidelines, note the importance of *algebra* (as well as calculus) for solving the equations $f(x) = 0$, $f'(x) = 0$, and $f''(x) = 0$.

EXAMPLE 1   Analyzing the Graph of a Rational Function

Analyze the graph of $f(x) = \dfrac{2(x^2 - 9)}{x^2 - 4}$.

Solution

$$\textit{First derivative:}\quad f'(x) = \frac{20x}{(x^2 - 4)^2}$$

$$\textit{Second derivative:}\quad f''(x) = \frac{-20(3x^2 + 4)}{(x^2 - 4)^3}$$

$$\textit{x-intercepts:}\quad (-3, 0), (3, 0)$$

$$\textit{y-intercept:}\quad \left(0, \tfrac{9}{2}\right)$$

$$\textit{Vertical asymptotes:}\quad x = -2, x = 2$$

$$\textit{Horizontal asymptote:}\quad y = 2$$

$$\textit{Critical number:}\quad x = 0$$

$$\textit{Possible points of inflection:}\quad \text{None}$$

$$\textit{Domain:}\quad \text{All real numbers except } x = \pm 2$$

$$\textit{Symmetry:}\quad \text{With respect to } y\text{-axis}$$

$$\textit{Test intervals:}\quad (-\infty, -2), (-2, 0), (0, 2), (2, \infty)$$

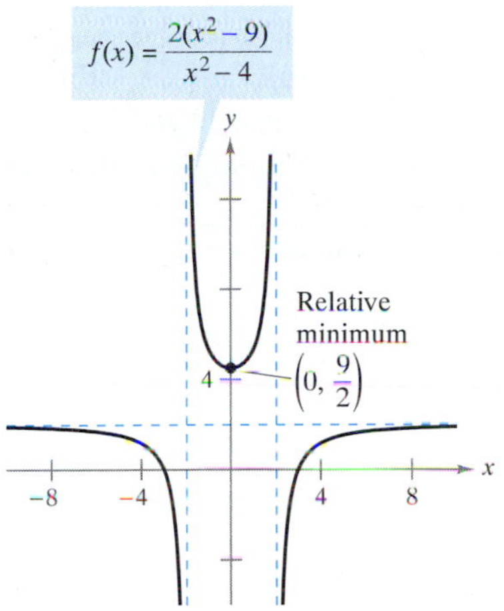

The table shows how the test intervals are used to determine several characteristics of the graph. The graph of f is shown in Figure 3.43.

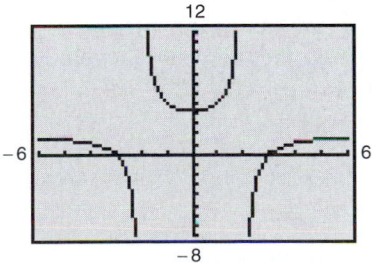

$f(x) = \dfrac{2(x^2 - 9)}{x^2 - 4}$

Relative minimum $\left(0, \tfrac{9}{2}\right)$

Using calculus, you can be certain that you have determined all characteristics of the graph of f.

Figure 3.43

	$f(x)$	$f'(x)$	$f''(x)$	**Characteristic of Graph**
$-\infty < x < -2$		$-$	$-$	Decreasing, concave down
$x = -2$	Undefined	Undefined	Undefined	Vertical asymptote
$-2 < x < 0$		$-$	$+$	Decreasing, concave up
$x = 0$	$\tfrac{9}{2}$	0	$+$	Relative minimum
$0 < x < 2$		$+$	$+$	Increasing, concave up
$x = 2$	Undefined	Undefined	Undefined	Vertical asymptote
$2 < x < \infty$		$+$	$-$	Increasing, concave down

FOR FURTHER INFORMATION For more information on the use of technology to graph rational functions, see the article "Graphs of Rational Functions for Computer Assisted Calculus" by Stan Byrd and Terry Walters in the September 1991 issue of *The College Mathematics Journal*.

Be sure you understand all of the implications of creating a table such as that shown in Example 1. Because of the use of calculus, you can *be sure* that the graph has no relative extrema or points of inflection other than those indicated in Figure 3.43.

Without using the type of analysis outlined in Example 1, it is easy to obtain an incomplete view of a graph's basic characteristics. For instance, Figure 3.44 shows a view of the graph of

$$g(x) = \frac{2(x^2 - 9)(x - 20)}{(x^2 - 4)(x - 21)}.$$

From this view, it appears that the graph of g is about the same as the graph of f shown in Figure 3.43. The graphs of these two functions, however, differ significantly. Try enlarging the viewing rectangle to see the differences.

By not using calculus you may overlook important characteristics of the graph of g.

Figure 3.44

EXAMPLE 2   Analyzing the Graph of a Rational Function

Analyze the graph of $f(x) = \dfrac{x^2 - 2x + 4}{x - 2}$.

Solution

$$\text{\textit{First derivative:}} \quad f'(x) = \frac{x(x - 4)}{(x - 2)^2}$$

$$\text{\textit{Second derivative:}} \quad f''(x) = \frac{8}{(x - 2)^3}$$

$$\text{\textit{x-intercepts:}} \quad \text{None}$$

$$\text{\textit{y-intercept:}} \quad (0, -2)$$

$$\text{\textit{Vertical asymptote:}} \quad x = 2$$

$$\text{\textit{Horizontal asymptotes:}} \quad \text{None}$$

$$\text{\textit{Critical numbers:}} \quad x = 0, x = 4$$

$$\text{\textit{Possible points of inflection:}} \quad \text{None}$$

$$\text{\textit{Domain:}} \quad \text{All real numbers except } x = 2$$

$$\text{\textit{Test intervals:}} \quad (-\infty, 0), (0, 2), (2, 4), (4, \infty)$$

The analysis of the graph of f is shown in the table, and the graph is shown in Figure 3.45.

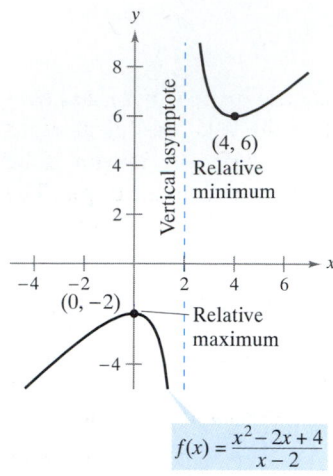

$f(x) = \dfrac{x^2 - 2x + 4}{x - 2}$

Figure 3.45

	$f(x)$	$f'(x)$	$f''(x)$	**Characteristic of Graph**
$-\infty < x < 0$		$+$	$-$	Increasing, concave down
$x = 0$	-2	0	$-$	Relative maximum
$0 < x < 2$		$-$	$-$	Decreasing, concave down
$x = 2$	Undefined	Undefined	Undefined	Vertical asymptote
$2 < x < 4$		$-$	$+$	Decreasing, concave up
$x = 4$	6	0	$+$	Relative minimum
$4 < x < \infty$		$+$	$+$	Increasing, concave up

Although the graph of the function in Example 2 has no horizontal asymptote, it does have a slant asymptote. The graph of a rational function (having no common factors) has a **slant asymptote** if the degree of the numerator exceeds the degree of the denominator by 1. To find the slant asymptote, use division to rewrite the rational function as the sum of a first-degree polynomial and another rational function.

$$f(x) = \frac{x^2 - 2x + 4}{x - 2} \qquad \color{red}{\text{Rewrite using long division.}}$$

$$= x + \frac{4}{x - 2} \qquad \color{red}{y = x \text{ is a slant asymptote.}}$$

In Figure 3.46, note that the graph of f approaches the slant asymptote $y = x$ as x approaches $-\infty$ or ∞. Try using a graphing utility to graph $f(x)$ and the line $y = x$ in the same viewing rectangle.

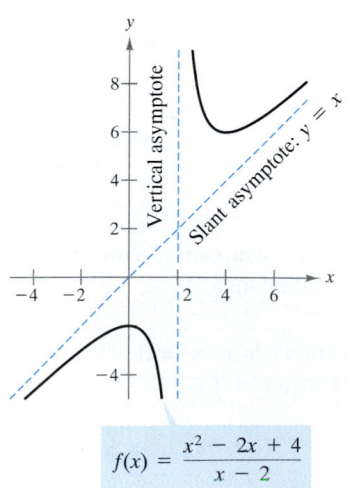

$f(x) = \dfrac{x^2 - 2x + 4}{x - 2}$

A slant asymptote
Figure 3.46

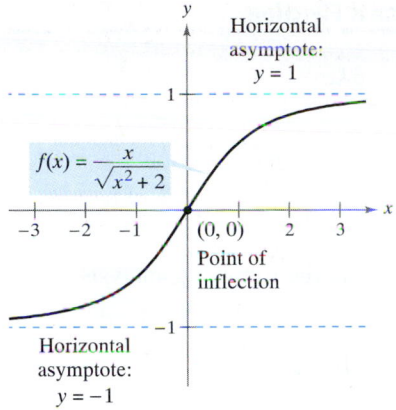

Horizontal asymptote: $y = 1$

$f(x) = \dfrac{x}{\sqrt{x^2 + 2}}$

$(0, 0)$
Point of inflection

Horizontal asymptote: $y = -1$

Figure 3.47

EXAMPLE 3   Analyzing the Graph of a Radical Function

Analyze the graph of $f(x) = \dfrac{x}{\sqrt{x^2 + 2}}$.

Solution

$$f'(x) = \frac{2}{(x^2 + 2)^{3/2}} \qquad f''(x) = -\frac{6x}{(x^2 + 2)^{5/2}}$$

The graph has only one intercept, $(0, 0)$. It has no vertical asymptotes, but it has two horizontal asymptotes: $y = 1$ (to the right) and $y = -1$ (to the left). The function has no critical numbers and one possible point of inflection (at $x = 0$). The domain of the function is all real numbers, and the graph is symmetric with respect to the origin. The analysis of the graph of f is shown in the table, and the graph is shown in Figure 3.47.

	$f(x)$	$f'(x)$	$f''(x)$	**Characteristic of Graph**
$-\infty < x < 0$		$+$	$+$	Increasing, concave up
$x = 0$	0	$\dfrac{1}{\sqrt{2}}$	0	Point of inflection
$0 < x < \infty$		$+$	$-$	Increasing, concave down

EXAMPLE 4   Analyzing the Graph of a Radical Function

Analyze the graph of $f(x) = 2x^{5/3} - 5x^{4/3}$.

Solution

$$f'(x) = \frac{10}{3}x^{1/3}(x^{1/3} - 2) \qquad f''(x) = \frac{20(x^{1/3} - 1)}{9x^{2/3}}$$

The function has two intercepts: $(0, 0)$ and $\left(\frac{125}{8}, 0\right)$. There are no horizontal or vertical asymptotes. The function has two critical numbers ($x = 0$ and $x = 8$) and two possible points of inflection ($x = 0$ and $x = 1$). The domain is all real numbers. The analysis of the graph of f is shown in the table, and the graph is shown in Figure 3.48.

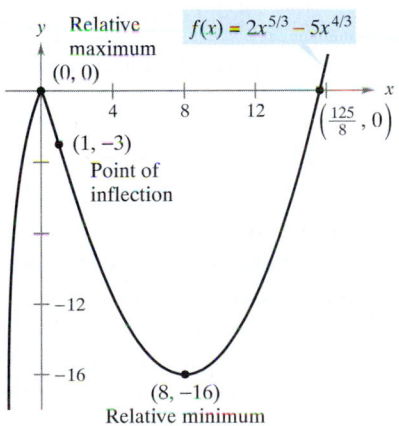

Relative maximum $(0, 0)$

$f(x) = 2x^{5/3} - 5x^{4/3}$

$\left(\frac{125}{8}, 0\right)$

$(1, -3)$
Point of inflection

$(8, -16)$
Relative minimum

Figure 3.48

	$f(x)$	$f'(x)$	$f''(x)$	**Characteristic of Graph**
$-\infty < x < 0$		$+$	$-$	Increasing, concave down
$x = 0$	0	0	Undefined	Relative maximum
$0 < x < 1$		$-$	$-$	Decreasing, concave down
$x = 1$	-3	$-$	0	Point of inflection
$1 < x < 8$		$-$	$+$	Decreasing, concave up
$x = 8$	-16	0	$+$	Relative minimum
$8 < x < \infty$		$+$	$+$	Increasing, concave up

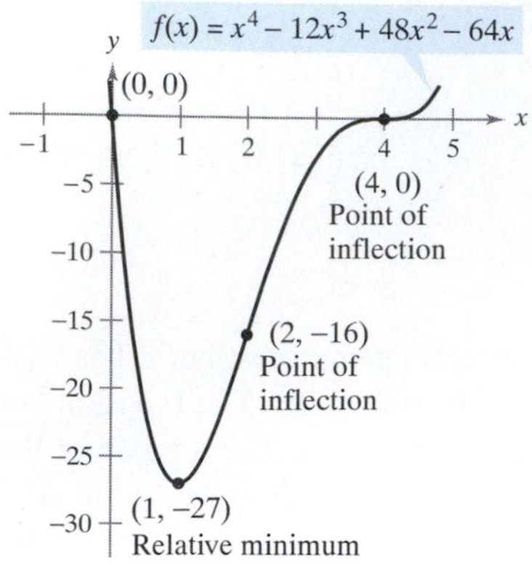

$f(x) = x^4 - 12x^3 + 48x^2 - 64x$

(0, 0)

(4, 0)
Point of
inflection

(2, −16)
Point of
inflection

(1, −27)
Relative minimum

(a)

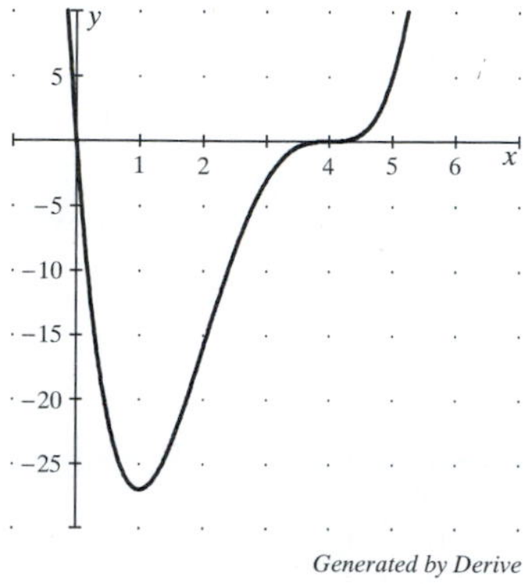

Generated by Derive

(b)

A polynomial function of even degree must have at least one relative extremum.
Figure 3.49

> **EXAMPLE 5** **Analyzing the Graph of a Polynomial Function**

Analyze the graph of $f(x) = x^4 - 12x^3 + 48x^2 - 64x$.

Solution Begin by factoring to obtain

$$f(x) = x^4 - 12x^3 + 48x^2 - 64x$$
$$= x(x - 4)^3.$$

Then, using the factored form of $f(x)$, you can perform the following analysis.

First derivative: $f'(x) = 4(x - 1)(x - 4)^2$
Second derivative: $f''(x) = 12(x - 4)(x - 2)$
x-intercepts: $(0, 0), (4, 0)$
y-intercept: $(0, 0)$
Vertical asymptotes: None
Horizontal asymptotes: None
Critical numbers: $x = 1, x = 4$
Possible points of inflection: $x = 2, x = 4$
Domain: All real numbers
Test intervals: $(-\infty, 1), (1, 2), (2, 4), (4, \infty)$

The analysis of the graph of f is shown in the table, and the graph is shown in Figure 3.49(a). Using a computer algebra system such as Derive (see Figure 3.49b) can help you verify your analysis.

	$f(x)$	$f'(x)$	$f''(x)$	**Characteristic of Graph**
$-\infty < x < 1$		−	+	Decreasing, concave up
$x = 1$	−27	0	+	Relative minimum
$1 < x < 2$		+	+	Increasing, concave up
$x = 2$	−16	+	0	Point of inflection
$2 < x < 4$		+	−	Increasing, concave down
$x = 4$	0	0	0	Point of inflection
$4 < x < \infty$		+	+	Increasing, concave up

The fourth-degree polynomial function in Example 5 has one relative minimum and no relative maxima. In general, a polynomial function of degree n can have *at most* $n - 1$ relative extrema, and *at most* $n - 2$ points of inflection. Moreover, polynomial functions of even degree must have *at least* one relative extremum.

Remember from the leading coefficient test described in Section P.3 that the "end behavior" of the graph of a polynomial function is determined by its leading coefficient and its degree. For instance, because the polynomial in Example 5 has a positive leading coefficient, the graph moves up to the right. Moreover, because the degree is even, the graph also moves up to the left.

EXAMPLE 6   Analyzing the Graph of a Trigonometric Function

Analyze the graph of $f(x) = \dfrac{\cos x}{1 + \sin x}$.

Solution   Because the function has a period of 2π, you can restrict the analysis of the graph to the interval $(-\pi/2, 3\pi/2)$.

$$\text{First derivative:}\quad f'(x) = -\frac{1}{1 + \sin x}$$

$$\text{Second derivative:}\quad f''(x) = \frac{\cos x}{(1 + \sin x)^2}$$

$$\text{Period:}\quad 2\pi$$

$$\text{x-intercept:}\quad \left(\frac{\pi}{2}, 0\right)$$

$$\text{y-intercept:}\quad (0, 1)$$

$$\text{Vertical asymptotes:}\quad x = -\frac{\pi}{2}, x = \frac{3\pi}{2}$$

$$\text{Horizontal asymptotes:}\quad \text{None}$$

$$\text{Critical numbers:}\quad \text{None}$$

$$\text{Possible points of inflection:}\quad x = \frac{\pi}{2}$$

$$\text{Domain:}\quad \text{All real numbers except } x = \frac{3 + 4n}{2}\pi$$

$$\text{Test intervals:}\quad \left(-\frac{\pi}{2}, \frac{\pi}{2}\right), \left(\frac{\pi}{2}, \frac{3\pi}{2}\right)$$

The analysis of the graph of f on the interval $(-\pi/2, 3\pi/2)$ is shown in the table, and the graph is shown in Figure 3.50(a). Compare this with the graph generated by the computer algebra system Derive in Figure 3.50(b).

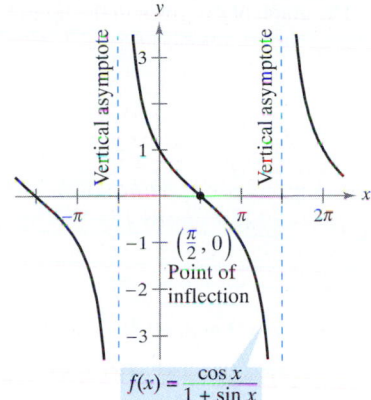

$$f(x) = \frac{\cos x}{1 + \sin x}$$

(a)

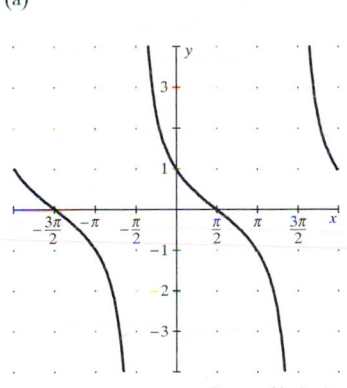

Generated by Derive

(b)

Figure 3.50

	$f(x)$	$f'(x)$	$f''(x)$	Characteristic of Graph
$x = -\dfrac{\pi}{2}$	Undefined	Undefined	Undefined	Vertical asymptote
$-\dfrac{\pi}{2} < x < \dfrac{\pi}{2}$		$-$	$+$	Decreasing, concave up
$x = \dfrac{\pi}{2}$	0	$-\dfrac{1}{2}$	0	Point of inflection
$\dfrac{\pi}{2} < x < \dfrac{3\pi}{2}$		$-$	$-$	Decreasing, concave down
$x = \dfrac{3\pi}{2}$	Undefined	Undefined	Undefined	Vertical asymptote

NOTE   The work involved in sketching the graph of a trigonometric function can be lessened sometimes by using trigonometric identities. For instance, the function in Example 6 can be rewritten as

$$f(x) = \frac{\cos x}{1 + \sin x} = \cot\left(\frac{x}{2} + \frac{\pi}{4}\right).$$

In this form, you can recognize the familiar cotangent graph shown in Figure 3.50.

1. Match the graph of f in the left column with that of its derivative in the right column.

f

(a)
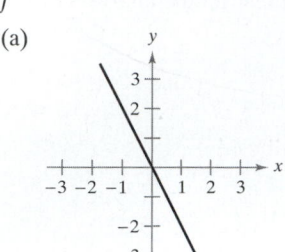

(b)
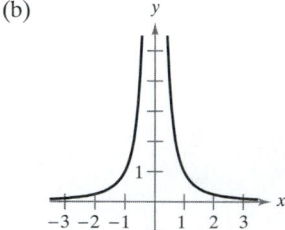

(c)
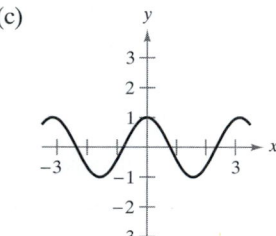

(d)
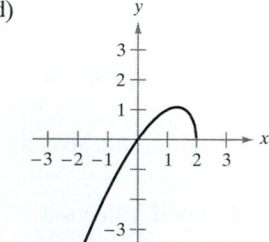

f'

(A)
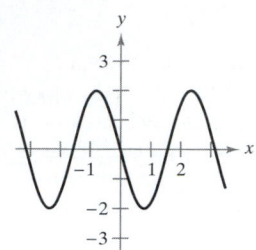

(B)
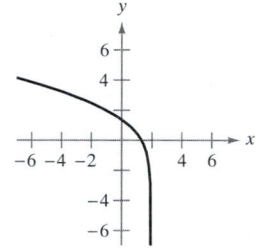

(C)
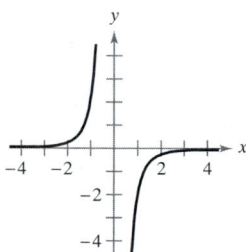

(D)
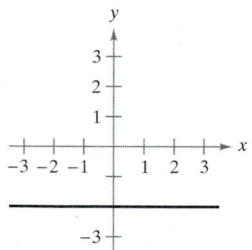

2. *Think About It* Suppose $f'(t) < 0$ for all t in the interval $(2, 8)$. Explain why $f(3) > f(5)$.

3. *Think About It* Suppose $f'(x) = \frac{2}{3}$ for all x and $f(0) = 1$. Find $f(6)$.

4. *Think About It* Suppose $f(0) = 3$ and $2 \le f'(x) \le 4$ for all x in the interval $[-5, 5]$. Determine the greatest and least possible values of $f(2)$.

5. *Graphical Reasoning* The graph of f is given in the figure.

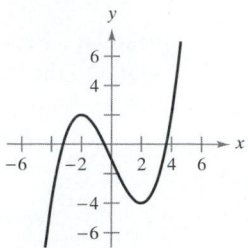

(a) For which values of x is $f'(x)$ zero? Positive? Negative?

(b) For which values of x is $f''(x)$ zero? Positive? Negative?

(c) On what interval is f' an increasing function?

(d) For which value of x is $f'(x)$ minimum? For this value of x, how does the rate of change of f compare with the rate of change of f for other values of x? Explain.

6. *Investigation* Consider the function

$$f(x) = \frac{3x^n}{x^4 + 1}$$

for nonnegative integer values of n.

(a) Discuss the relationship between the value of n and the symmetry of the graph.

(b) For which values of n will the x-axis be the horizontal asymptote?

(c) For which value of n will $y = 3$ be the horizontal asymptote?

(d) What is the asymptote of the graph when $n = 5$?

(e) Use a graphing utility to graph f for the indicated values of n in the table. Use the graph to determine the number of extrema M and the number of inflection points N of the graph.

n	0	1	2	3	4	5
M						
N						

In Exercises 7–24, make use of domain, range, symmetry, asymptotes, intercepts, relative extrema, and/or points of inflection to sketch a graph of the function. You can use a graphing utility to verify your results.

7. $y = x^3 - 3x^2 + 3$

8. $y = -\frac{1}{3}(x^3 - 3x + 2)$

9. $y = 2 - x - x^3$

10. $f(x) = \frac{1}{3}(x - 1)^3 + 2$

11. $f(x) = 3x^3 - 9x + 1$

12. $f(x) = (x + 1)(x - 2)(x - 5)$

13. $y = 3x^4 + 4x^3$

14. $y = 3x^4 - 6x^2$

15. $f(x) = x^4 - 4x^3 + 16x$

16. $f(x) = x^4 - 8x^3 + 18x^2 - 16x + 5$

17. $y = x^5 - 5x$ **18.** $y = (x - 1)^5$

19. $y = |2x - 3|$ **20.** $y = |x^2 - 6x + 5|$

21. $y = x\sqrt{4 - x}$ **22.** $y = x\sqrt{4 - x^2}$

23. $y = 3x^{2/3} - 2x$ **24.** $y = 3x^{2/3} - x^2$

In Exercises 25–28, sketch a graph of the function over the indicated interval. Use a graphing utility to verify your graph.

Function	Interval
25. $y = \sin x - \frac{1}{18}\sin 3x$	$0 \le x \le 2\pi$
26. $y = \cos x - \frac{1}{2}\cos 2x$	$0 \le x \le 2\pi$
27. $y = 2x - \tan x$	$-\dfrac{\pi}{2} < x < \dfrac{\pi}{2}$
28. $y = 2x + \cot x$	$0 < x < \pi$

In Exercises 29–40, sketch a graph of the function. Label any intercepts, relative extrema, points of inflection, or asymptotes. Use a graphing utility to verify your results.

29. $y = \dfrac{x^2}{x^2 + 3}$ **30.** $y = \dfrac{x}{x^2 + 1}$

31. $y = \dfrac{1}{x - 2} - 3$ **32.** $y = \dfrac{x^2 + 1}{x^2 - 2}$

33. $y = \dfrac{2x}{x^2 - 1}$ **34.** $f(x) = \dfrac{x + 2}{x}$

35. $g(x) = x + \dfrac{4}{x^2 + 1}$ **36.** $f(x) = x + \dfrac{32}{x^2}$

37. $f(x) = \dfrac{x^2 + 1}{x}$ **38.** $f(x) = \dfrac{x^3}{x^2 - 1}$

39. $y = \dfrac{x^2 - 6x + 12}{x - 4}$ **40.** $y = \dfrac{2x^2 - 5x + 5}{x - 2}$

In Exercises 41–44, use a symbolic differentiation utility to analyze and graph the function. Identify any relative extrema, points of inflection, and asymptotes.

41. $f(x) = \dfrac{20x}{x^2 + 1} - \dfrac{1}{x}$

42. $f(x) = 5\left(\dfrac{1}{x - 4} - \dfrac{1}{x + 2}\right)$

43. $f(x) = \dfrac{x}{\sqrt{x^2 + 7}}$

44. $f(x) = \dfrac{4x}{\sqrt{x^2 + 15}}$

In Exercises 45 and 46, use a graphing utility to graph the function. Use the graph to determine whether or not it is possible for the graph of a function to cross its horizontal asymptote. Do you think it is possible for the graph of a function to cross its vertical asymptote? Why or why not?

45. $f(x) = \dfrac{4(x - 1)^2}{x^2 - 4x + 5}$ **46.** $g(x) = \dfrac{3x^4 - 5x + 3}{x^4 + 1}$

Writing **In Exercises 47 and 48, use a graphing utility to graph the function. Explain why there is no vertical asymptote when a superficial examination of the function may indicate that there should be one.**

47. $h(x) = \dfrac{6 - 2x}{3 - x}$ **48.** $g(x) = \dfrac{x^2 + x - 2}{x - 1}$

Writing **In Exercises 49 and 50, use a graphing utility to graph the function and determine the slant asymptote of the graph. Zoom out repeatedly and describe how the graph on the display appears to change. Why does this occur?**

49. $f(x) = -\dfrac{x^2 - 3x - 1}{x - 2}$ **50.** $g(x) = \dfrac{2x^2 - 8x - 15}{x - 5}$

Think About It **In Exercises 51–54, create a function whose graph has the indicated characteristics. (The answer is not unique.)**

51. Vertical asymptote: $x = 5$

Horizontal asymptote: $y = 0$

52. Vertical asymptote: $x = -3$

Horizontal asymptote: None

53. Vertical asymptote: $x = 5$

Slant asymptote: $y = 3x + 2$

54. Vertical asymptote: $x = 0$

Slant asymptote: $y = -x$

55. *Graphical Reasoning*   Consider the function

$$f(x) = \dfrac{ax}{(x - b)^2}.$$

(a) Determine the effect on the graph of f if $b \ne 0$ and a is varied. Consider cases where a is positive and a is negative.

(b) Determine the effect on the graph of f if $a \ne 0$ and b is varied.

56. Consider the function

$$f(x) = \tfrac{1}{2}(ax)^2 - (ax), \quad a \ne 0.$$

(a) Determine the changes (if any) in the intercepts, extrema, and concavity of the graph of f when a is varied.

(b) In the same viewing rectangle, use a graphing utility to graph the function for four different values of a.

Think About It In Exercises 57–62, use the graph (of f' or f'') to sketch a graph of the function f. (*Hint:* There is more than one correct answer.)

57.
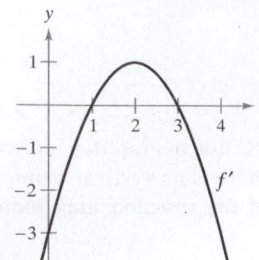

58.
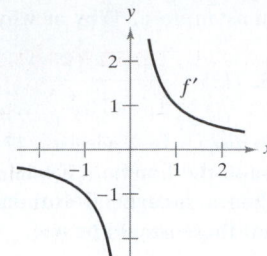

59.
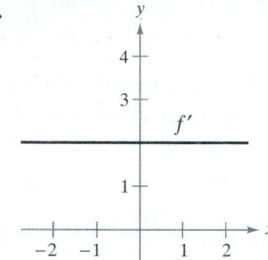

60.
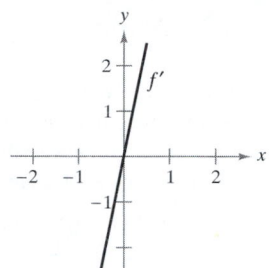

61.
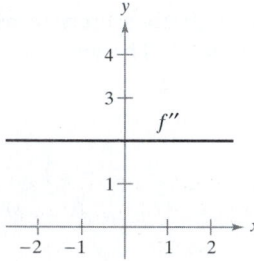

62.
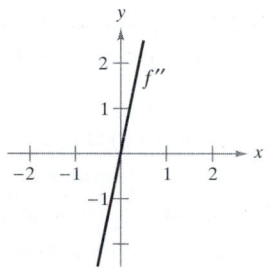

Think About It In Exercises 63 and 64, the graphs of f, f', and f'' are shown on the same set of coordinate axes. Which is which?

63.
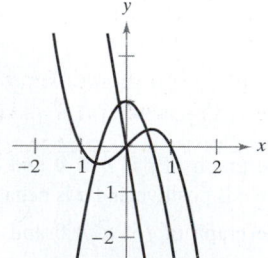

64.
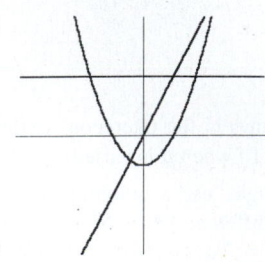

Generated by Maple

Think About It In Exercises 65–70, determine conditions for the coefficients of

$$f(x) = ax^3 + bx^2 + cx + d$$

such that the graph of f resembles the given graph.

65.
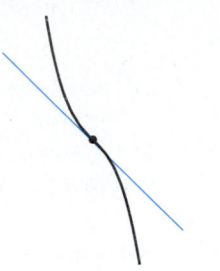

66.
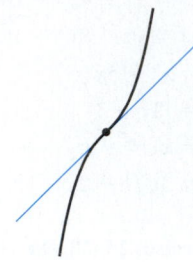

67.
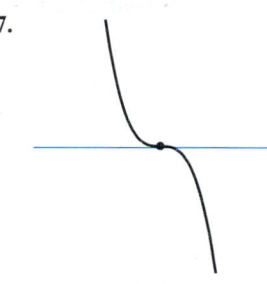

68.
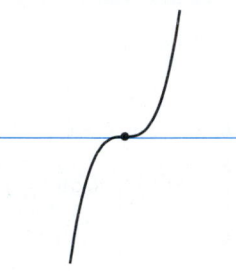

69.
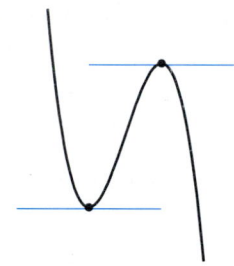

70.
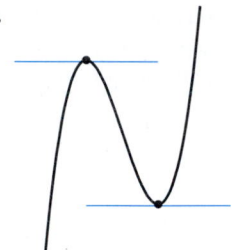

In Exercises 71–74, use the graph of f' to sketch a graph of f and the graph of f''.

71.
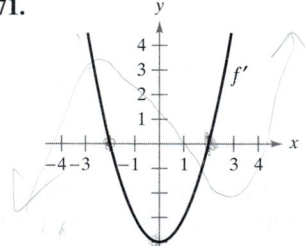

72.
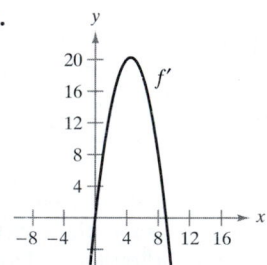

73.
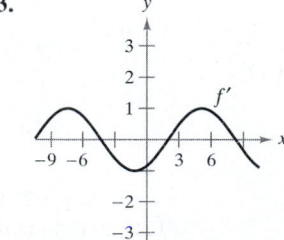

74.
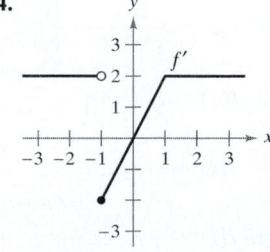

(Submitted by Bill Fox, Moberly Area Community College, Moberly, MO.)

Applied Minimum and Maximum Problems

Applied Minimum and Maximum Problems

One of the most common applications of calculus involves the determination of minimum and maximum values. Consider how frequently you hear or read terms such as greatest profit, least cost, least time, greatest voltage, optimum size, least size, greatest strength, and greatest distance. Before outlining a general problem-solving strategy for such problems, let's look at an example.

EXAMPLE 1   Finding Maximum Volume

A manufacturer wants to design an open box having a square base and a surface area of 108 square inches, as shown in Figure 3.51. What dimensions will produce a box with maximum volume?

Solution   Because the box has a square base, its volume is

$$V = x^2h. \qquad \qquad \text{Primary equation}$$

(This equation is called the **primary equation** because it gives a formula for the quantity to be optimized.) The surface area of the box is

$$S = (\text{area of base}) + (\text{area of four sides})$$
$$S = x^2 + 4xh = 108. \qquad \text{Secondary equation}$$

Because V is to be maximized, you want to express V as a function of just one variable. To do this, you can solve the equation $x^2 + 4xh = 108$ for h in terms of x to obtain $h = (108 - x^2)/(4x)$. Substituting into the primary equation produces

$$V = x^2h \qquad \qquad \text{Function of two variables}$$

$$= x^2\left(\frac{108 - x^2}{4x}\right) \qquad \text{Substitute for } h.$$

$$= 27x - \frac{x^3}{4}. \qquad \qquad \text{Function of one variable}$$

Before finding which x-value will yield a maximum value of V, you should determine the *feasible domain*. That is, what values of x make sense in this problem? You know that $V \geq 0$. You also know that x must be nonnegative and that the area of the base ($A = x^2$) is at most 108. Thus, the feasible domain is

$$0 \leq x \leq \sqrt{108}. \qquad \qquad \text{Feasible domain}$$

To maximize V, you can find the critical numbers of the volume function.

$$\frac{dV}{dx} = 27 - \frac{3x^2}{4} = 0 \qquad \text{Set derivative equal to 0.}$$

$$3x^2 = 108$$

$$x = \pm 6 \qquad \qquad \text{Critical numbers}$$

Evaluating V at the critical number ($x = 6$) in the domain and at the endpoints of the domain produces $V(0) = 0$, $V(6) = 108$, and $V(\sqrt{108}) = 0$. Thus, V is maximum when $x = 6$ and the dimensions of the box are $6 \times 6 \times 3$ inches.

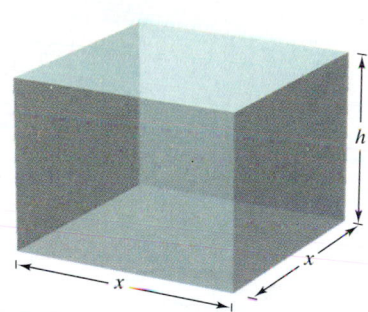

Open box with square base:
$S = x^2 + 4xh = 108$
Figure 3.51

TECHNOLOGY   You can verify your answer by using a graphing utility to graph the volume

$$V = 27x - \frac{x^3}{4}.$$

Use a viewing rectangle in which $0 \leq x \leq \sqrt{108} \approx 10.4$ and $0 \leq y \leq 120$, and the trace feature to determine the maximum value of V.

In Example 1, you should realize that there are infinitely many open boxes having 108 square inches of surface area. To begin solving the problem, you might ask yourself which basic shape would seem to yield a maximum volume. Should the box be tall, squat, or more nearly cubical?

You might even try calculating a few volumes, as shown in Figure 3.52, to see if you can get a better feeling for what the optimum dimensions should be. Remember that you are not ready to begin solving a problem until you have clearly identified what the problem is.

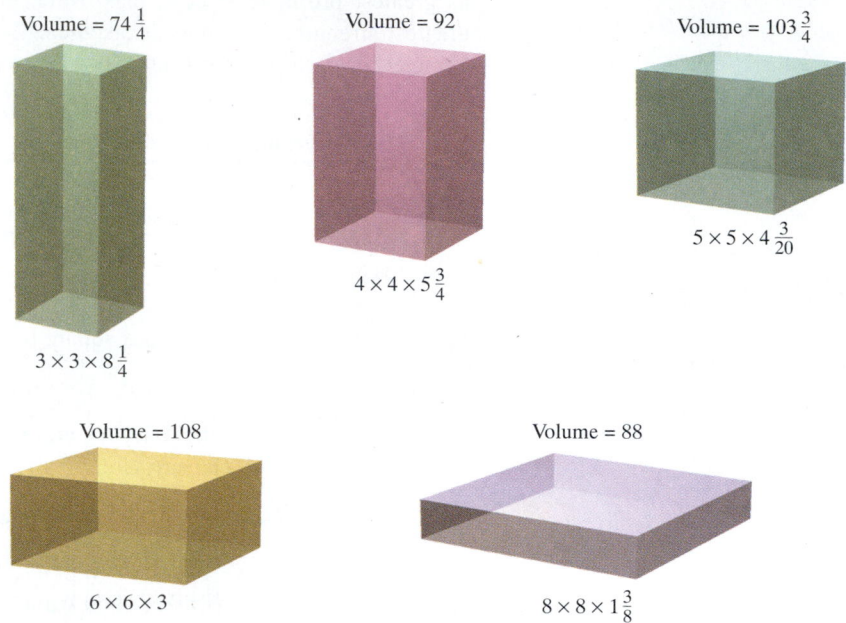

Volume $= 74\frac{1}{4}$

$3 \times 3 \times 8\frac{1}{4}$

Volume $= 92$

$4 \times 4 \times 5\frac{3}{4}$

Volume $= 103\frac{3}{4}$

$5 \times 5 \times 4\frac{3}{20}$

Volume $= 108$

$6 \times 6 \times 3$

Volume $= 88$

$8 \times 8 \times 1\frac{3}{8}$

Which box has the greatest volume?
Figure 3.52

Example 1 illustrates the following strategy for solving applied minimum and maximum problems.

NOTE   When performing Step 5, recall that to determine the maximum or minimum value of a continuous function f on a closed interval, you should compare the values of f at its critical numbers with the values of f at the endpoints of the interval.

Problem-Solving Strategy for Applied Minimum and Maximum Problems

1. Assign symbols to all *given* quantities and quantities *to be determined*. When feasible, make a sketch.

2. Write a **primary equation** for the quantity that is to be maximized (or minimized). (A review of several useful formulas from geometry is presented inside the front cover.)

3. Reduce the primary equation to one having a *single independent variable*. This may involve the use of **secondary equations** relating the independent variables of the primary equation.

4. Determine the domain of the primary equation. That is, determine the values for which the stated problem makes sense.

5. Determine the desired maximum or minimum value by the calculus techniques discussed in Sections 3.1 through 3.4.

EXAMPLE 2   Finding Minimum Distance

Which points on the graph of $y = 4 - x^2$ are closest to the point $(0, 2)$?

Solution   Figure 3.53 indicates that there are two points at a minimum distance from the point $(0, 2)$. The distance between the point $(0, 2)$ and a point (x, y) on the graph of $y = 4 - x^2$ is given by

$$d = \sqrt{(x - 0)^2 + (y - 2)^2}. \qquad \text{\color{red}Primary equation}$$

Using the secondary equation $y = 4 - x^2$, you can rewrite the primary equation as

$$d = \sqrt{x^2 + (4 - x^2 - 2)^2} = \sqrt{x^4 - 3x^2 + 4}.$$

Because d is smallest when the expression inside the radical is smallest, you need only find the critical numbers of $f(x) = x^4 - 3x^2 + 4$. Note that the domain of f is the entire real line. Moreover, differentiation yields

$$f'(x) = 4x^3 - 6x = 2x(2x^2 - 3) = 0$$

$$x = 0, \ \sqrt{\frac{3}{2}}, \ -\sqrt{\frac{3}{2}}.$$

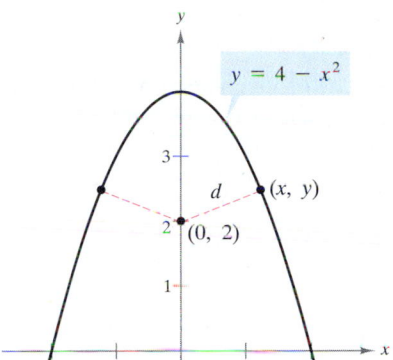

The quantity to be minimized is distance:
$d = \sqrt{(x - 0)^2 + (y - 2)^2}$.
Figure 3.53

The First Derivative Test verifies that $x = 0$ yields a relative maximum, whereas both $x = \sqrt{3/2}$ and $x = -\sqrt{3/2}$ yield a minimum distance. Hence, the closest points are $\left(\sqrt{3/2}, 5/2\right)$ and $\left(-\sqrt{3/2}, 5/2\right)$.

EXAMPLE 3   Finding Minimum Area

A rectangular page is to contain 24 square inches of print. The margins at the top and bottom of the page are to be $1\frac{1}{2}$ inches, and the margins on the left and right are to be 1 inch (see Figure 3.54). What should the dimensions of the page be so that the least amount of paper is used?

Solution   Let A be the area to be minimized.

$$A = (x + 3)(y + 2) \qquad \text{\color{red}Primary equation}$$

The printed area inside the margins is given by

$$24 = xy. \qquad \text{\color{red}Secondary equation}$$

Solving this equation for y produces $y = 24/x$. Substitution into the primary equation produces

$$A = (x + 3)\left(\frac{24}{x} + 2\right) = 30 + 2x + \frac{72}{x}. \qquad \text{\color{red}Function of one variable}$$

Because x must be positive, you are interested only in values of A for $x > 0$. To find the critical numbers, differentiate with respect to x.

$$\frac{dA}{dx} = 2 - \frac{72}{x^2} = 0 \qquad \Longrightarrow \qquad x^2 = 36$$

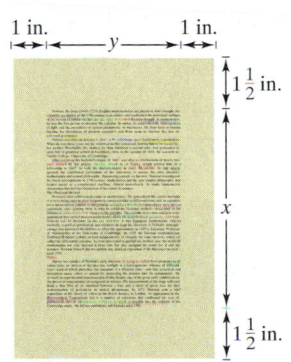

1 in.   1 in.
$1\frac{1}{2}$ in.
$1\frac{1}{2}$ in.

The quantity to be minimized is area:
$A = (x + 3)(y + 2)$.
Figure 3.54

Thus, the critical numbers are $x = \pm 6$. You do not have to consider -6 because it is outside the domain. The First Derivative Test confirms that A is a minimum when $x = 6$. Therefore, $y = \frac{24}{6} = 4$ and the dimensions of the page should be $x + 3 = 9$ inches by $y + 2 = 6$ inches.

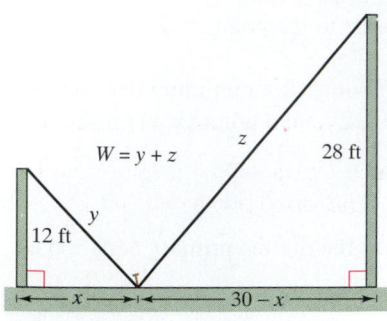

The quantity to be minimized is length. From the diagram, you can see that x varies between 0 and 30.

Figure 3.55

EXAMPLE 4 Finding Minimum Length

Two posts, one 12 feet high and the other 28 feet high, stand 30 feet apart. They are to be stayed by two wires, attached to a single stake, running from ground level to the top of each post. Where should the stake be placed to use the least wire?

Solution   Let W be the wire length to be minimized. Using Figure 3.55, you can write

$$W = y + z. \qquad \text{Primary equation}$$

In this problem, rather than solving for y in terms of z (or vice versa), you can solve for both y and z in terms of a third variable x, as shown in Figure 3.55. From the Pythagorean Theorem, you obtain

$$x^2 + 12^2 = y^2 \qquad \Longrightarrow \qquad y = \sqrt{x^2 + 144}$$
$$(30 - x)^2 + 28^2 = z^2 \qquad \Longrightarrow \qquad z = \sqrt{x^2 - 60x + 1684}.$$

Thus, W is given by

$$W = y + z$$
$$= \sqrt{x^2 + 144} + \sqrt{x^2 - 60x + 1684}, \qquad 0 \le x \le 30.$$

Differentiating W with respect to x yields

$$\frac{dW}{dx} = \frac{x}{\sqrt{x^2 + 144}} + \frac{x - 30}{\sqrt{x^2 - 60x + 1684}}.$$

By letting $dW/dx = 0$, you obtain

$$\frac{x}{\sqrt{x^2 + 144}} + \frac{x - 30}{\sqrt{x^2 - 60x + 1684}} = 0$$
$$x\sqrt{x^2 - 60x + 1684} = (30 - x)\sqrt{x^2 + 144}$$
$$x^2(x^2 - 60x + 1684) = (30 - x)^2(x^2 + 144)$$
$$x^4 - 60x^3 + 1684x^2 = x^4 - 60x^3 + 1044x^2 - 8640x + 129{,}600$$
$$640x^2 + 8640x - 129{,}600 = 0$$
$$320(x - 9)(2x + 45) = 0$$
$$x = 9, \ -22.5.$$

Because $x = -22.5$ is not in the interval $[0, 30]$ and

$$W(0) \approx 53.04, \qquad W(9) = 50, \qquad \text{and} \qquad W(30) \approx 60.31$$

you can conclude that the wire should be staked at 9 feet from the 12-foot pole.

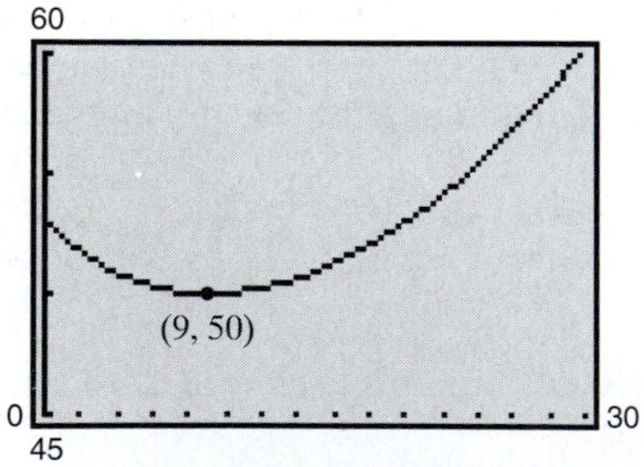

TECHNOLOGY   From Example 4, you can see that applied optimization problems can involve a lot of algebra. If you have access to a graphing utility, you can confirm that $x = 9$ yields a minimum value of W by sketching the graph of

$$W = \sqrt{x^2 + 144} + \sqrt{x^2 - 60x + 1684}$$

as shown in Figure 3.56.

60

(9, 50)

0
45
30

You can confirm the minimum value of W with a graphing utility.

Figure 3.56

In each of the first four examples, the extreme value occurs at a critical number. Although this happens often, remember that an extreme value can also occur at an endpoint of an interval, as shown in Example 5.

EXAMPLE 5 An Endpoint Maximum

Four feet of wire is to be used to form a square and a circle. How much of the wire should be used for the square and how much should be used for the circle to enclose the maximum total area?

Solution The total area (see Figure 3.57) is given by

$$A = (\text{area of square}) + (\text{area of circle})$$

$$A = x^2 + \pi r^2. \qquad \color{red}{\textit{Primary equation}}$$

Because the total amount of wire is 4 feet, you obtain

$$4 = (\text{perimeter of square}) + (\text{circumference of circle})$$

$$4 = 4x + 2\pi r.$$

Thus, $r = 2(1 - x)/\pi$, and by substituting into the primary equation you have

$$A = x^2 + \pi\left[\frac{2(1 - x)}{\pi}\right]^2$$

$$= x^2 + \frac{4(1 - x)^2}{\pi}$$

$$= \frac{1}{\pi}[(\pi + 4)x^2 - 8x + 4].$$

The feasible domain is $0 \le x \le 1$ restricted by the square's perimeter. Because

$$\frac{dA}{dx} = \frac{2(\pi + 4)x - 8}{\pi}$$

the only critical number in $(0, 1)$ is $x = 4/(\pi + 4) \approx 0.56$. Therefore, using

$$A(0) \approx 1.273, \quad A(0.56) \approx 0.56, \quad \text{and} \quad A(1) = 1$$

you can conclude that the maximum area occurs when $x = 0$. That is, *all* the wire is used for the circle.

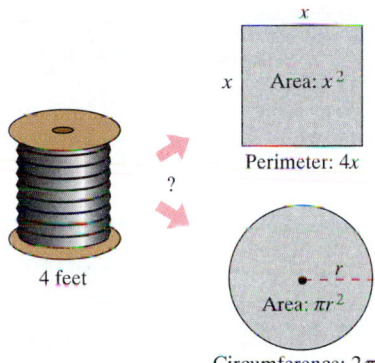

The quantity to be maximized is area:
$A = x^2 + \pi r^2.$
Figure 3.57

NOTE What would the answer be if Example 5 asked for the dimensions needed to enclose the *minimum* total area?

Let's review the primary equations developed in the first five examples. As applications go, these five examples are fairly simple, and yet the resulting primary equations are quite complicated.

$$V = 27x - \frac{x^3}{4} \qquad\qquad W = \sqrt{x^2 + 144} + \sqrt{x^2 - 60x + 1684}$$

$$d = \sqrt{x^4 - 3x^2 + 4} \qquad A = \frac{1}{\pi}[(\pi + 4)x^2 - 8x + 4]$$

$$A = 30 + 2x + \frac{72}{x}$$

You must expect that real-life applications often involve equations that are *at least as complicated* as these five. Remember that one of the main goals of this course is to learn to use calculus to analyze equations that initially seem formidable.

EXERCISES FOR SECTION 3.7

1. **Numerical, Graphical, and Analytic Analysis**   Find two positive numbers whose sum is 110 and whose product is a maximum.

   (a) Analytically complete six rows of a table such as the one below. (The first two rows are shown.)

First Number x	Second Number	Product P
10	$110 - 10$	$10(110 - 10) = 1000$
20	$110 - 20$	$20(110 - 20) = 1800$

   (b) Use a graphing utility to generate additional rows of the table. Use the table to estimate the solution. (*Hint:* Use the table feature of the graphing utility.)

   (c) Write the product P as a function of x.

   (d) Use a graphing utility to graph the function in part (c) and estimate the solution from the graph.

   (e) Use calculus to find the critical number of the function in part (c). Then find the two numbers.

In Exercises 2–6, find two positive numbers that satisfy the given requirements.

2. The sum is S and the product is a maximum.

3. The product is 192 and the sum is a minimum.

4. The product is 192 and the sum of the first plus three times the second is a minimum.

5. The second number is the reciprocal of the first and the sum is a minimum.

6. The sum of the first and twice the second is 100 and the product is a maximum.

In Exercises 7 and 8, find the length and width of a rectangle that has the given perimeter and a maximum area.

7. Perimeter: 100 meters

8. Perimeter: P units

In Exercises 9 and 10, find the length and width of a rectangle that has the given area and a minimum perimeter.

9. Area: 64 square feet

10. Area: A square centimeters

In Exercises 11 and 12, find the point on the graph of the function that is closest to the given point.

Function	Point
11. $f(x) = \sqrt{x}$	$(4, 0)$
12. $f(x) = x^2$	$\left(2, \frac{1}{2}\right)$

13. **Chemical Reaction**   In an autocatalytic chemical reaction, the product formed is a catalyst for the reaction. If Q_0 is the amount of the original substance and x is the amount of catalyst formed, the rate of chemical reaction is

$$\frac{dQ}{dx} = kx(Q_0 - x).$$

For what value of x will the rate of chemical reaction be greatest?

14. **Traffic Control**   On a given day, the flow rate F (cars per hour) on a congested roadway is

$$F = \frac{v}{22 + 0.02v^2}$$

where v is the speed of the traffic in miles per hour. What speed will maximize the flow rate on the road?

15. **Area**   A farmer plans to fence a rectangular pasture adjacent to a river. The pasture must contain 180,000 square meters in order to provide enough grass for the herd. What dimensions would require the least amount of fencing if no fencing is needed along the river?

16. **Area**   A rancher has 200 feet of fencing with which to enclose two adjacent rectangular corrals (see figure). What dimensions should be used so that the enclosed area will be a maximum?

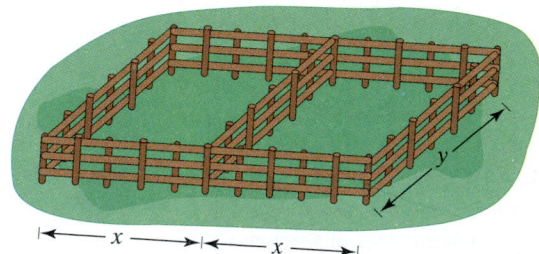

17. **Volume**

   (a) Verify that each of the rectangular solids shown in the figure has a surface area of 150 square inches.

   (b) Find the volume of each.

   (c) Determine the dimensions of a rectangular solid (with a square base) of maximum volume if its surface area is 150 square inches.

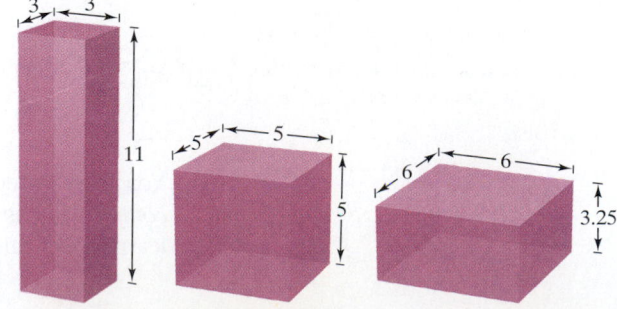

18. *Numerical, Graphical, and Analytic Analysis* An open box of maximum volume is to be made from a square piece of material, 24 inches on a side, by cutting equal squares from the corners and turning up the sides (see figure).

(a) Analytically complete six rows of a table such as the one below. (The first two rows are shown.) Use the table to guess the maximum volume.

Height	Length and Width	Volume
1	$24 - 2(1)$	$1[24 - 2(1)]^2 = 484$
2	$24 - 2(2)$	$2[24 - 2(2)]^2 = 800$

(b) Write the volume V as a function of x.

(c) Use calculus to find the critical number of the function in part (b) and find the maximum value.

(d) Use a graphing utility to graph the function in part (b) and verify the maximum volume from the graph.

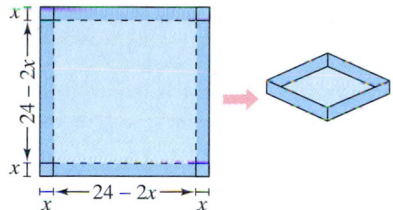

19. (a) Solve Exercise 18 given that the square piece of material is s meters on a side.

(b) If the dimensions of the square piece of material are doubled, how does the volume change?

20. *Numerical, Graphical, and Analytic Analysis* A physical fitness room consists of a rectangle with a semicircle on each end. A 200-meter running track runs around the outside of the room.

(a) Draw a figure to represent the problem. Let x and y represent the length and width of the rectangle.

(b) Analytically complete six rows of a table such as the one below. (The first two rows are shown.) Use the table to guess the maximum area of the rectangular region.

Length x	Width y	Area
10	$\frac{2}{\pi}(100 - 10)$	$(10)\frac{2}{\pi}(100 - 10) \approx 573$
20	$\frac{2}{\pi}(100 - 20)$	$(20)\frac{2}{\pi}(100 - 20) \approx 1019$

(c) Write the area A as a function of x.

(d) Use calculus to find the critical number of the function in part (c) and find the maximum value.

(e) Use a graphing utility to graph the function in part (c) and verify the maximum area from the graph.

21. *Area* A Norman window is constructed by adjoining a semicircle to the top of an ordinary rectangular window (see figure). Find the dimensions of a Norman window of maximum area if the total perimeter is 16 feet.

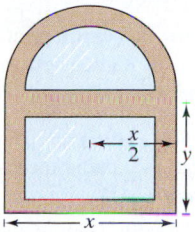

22. *Area* A rectangle is bounded by the x- and y-axes and the graph of $y = (6 - x)/2$ (see figure). What length and width should the rectangle have so that its area is a maximum?

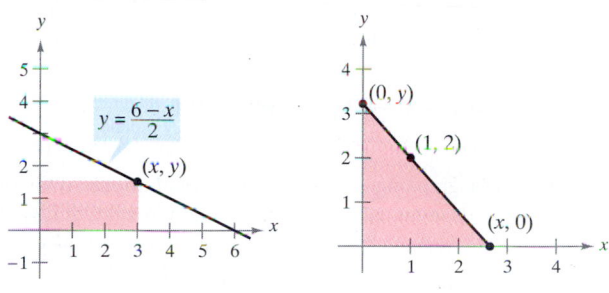

Figure for 22   **Figure for 23**

23. *Length* A right triangle is formed in the first quadrant by the x- and y-axes and a line through the point $(1, 2)$ (see figure).

(a) Write the length L of the hypotenuse as a function of x.

(b) Use a graphing utility to graphically approximate x such that the length of the hypotenuse is a minimum.

(c) Find the vertices of the triangle such that its area is a minimum.

24. *Area* Find the dimensions of the largest isosceles triangle that can be inscribed in a circle of radius 4 (see figure).

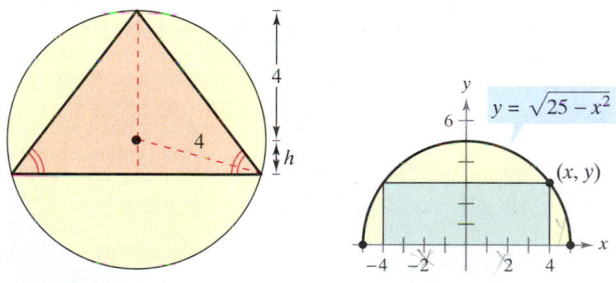

Figure for 24   **Figure for 25**

25. *Area* A rectangle is bounded by the x-axis and the semicircle $y = \sqrt{25 - x^2}$ (see figure). What length and width should the rectangle have so that its area is a maximum?

26. Area   Find the dimensions of the largest rectangle that can be inscribed in a semicircle of radius r (see Exercise 25).

27. Area   Find the dimensions of the trapezoid of greatest area that can be inscribed in a semicircle of radius r.

28. Area   A page is to contain 30 square inches of print. The margins on each side are 1 inch. Find the dimensions of the page such that the least amount of paper is used.

29. Numerical, Graphical, and Analytic Analysis   A right circular cylinder is to be designed to hold 22 cubic inches of a soft drink (approximately 12 fluid ounces).

(a) Analytically complete six rows of a table such as the one below. (The first two rows are shown.)

Radius r	Height	Surface Area
0.2	$\dfrac{22}{\pi(0.2)^2}$	$2\pi(0.2)\left[0.2 + \dfrac{22}{\pi(0.2)^2}\right] \approx 220.3$
0.4	$\dfrac{22}{\pi(0.4)^2}$	$2\pi(0.4)\left[0.4 + \dfrac{22}{\pi(0.4)^2}\right] \approx 111.1$

(b) Use a graphing utility to generate additional rows of the table. Use the table to estimate the minimum surface area. (*Hint:* Use the table feature of the graphing utility.)

(c) Write the surface area S as a function of r.

(d) Use a graphing utility to graph the function in part (c) and estimate the minimum surface area from the graph.

(e) Use calculus to find the critical number of the function in part (c) and find dimensions that will yield the minimum surface area.

30. Surface Area   Use calculus to find the required dimensions for the cylinder in Exercise 29 if its volume is V_0 cubic units.

31. Volume   A rectangular package to be sent by a postal service can have a maximum combined length and girth (perimeter of a cross section) of 108 inches (see figure). Find the dimensions of the package of maximum volume that can be sent. (Assume the cross section is square.)

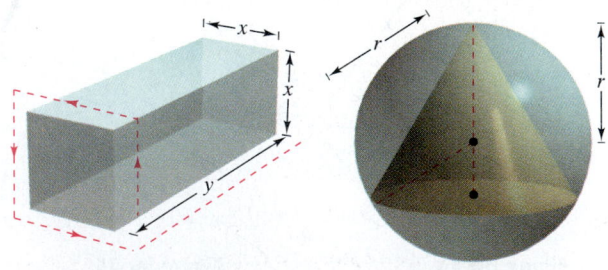

Figure for 31   **Figure for 33**

32. Volume   Rework Exercise 31 for a cylindrical package. (The cross section is circular.)

33. Volume   Find the volume of the largest right circular cone that can be inscribed in a sphere of radius r (see figure).

34. Volume   Find the volume of the largest right circular cylinder that can be inscribed in a sphere of radius r.

35. Surface Area   A solid is formed by adjoining two hemispheres to the ends of a right circular cylinder. The total volume of the solid is 12 cubic centimeters. Find the radius of the cylinder that produces the minimum surface area.

36. Cost   An industrial tank of the shape described in Exercise 35 must have a volume of 3000 cubic feet. The hemispherical ends cost twice as much per square foot of surface area as the sides. Find the dimensions that will minimize cost.

37. Area   The sum of the perimeters of an equilateral triangle and a square is 10. Find the dimensions of the triangle and the square that produce a minimum total area.

38. Area   Twenty feet of wire is to be used to form two figures. In each of the following cases, how much should be used for each figure so that the total enclosed area is maximum?

(a) Equilateral triangle and square

(b) Square and regular pentagon

(c) Regular pentagon and regular hexagon

(d) Regular hexagon and circle

What can you conclude from this pattern? {*Hint:* The area of a regular polygon with n sides of length x is $A = (n/4)[\cot(\pi/n)]x^2$.}

39. Beam Strength   A wooden beam has a rectangular cross section of height h and width w (see figure). The strength S of the beam is directly proportional to the width and the square of the height. What are the dimensions of the strongest beam that can be cut from a round log of diameter 24 inches? (*Hint:* $S = kh^2w$, where k is the proportionality constant.)

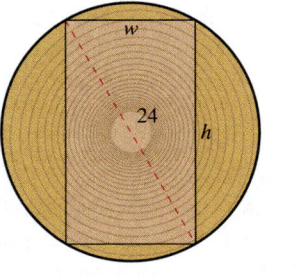

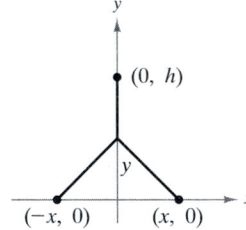

Figure for 39   **Figure for 40**

40. Minimum Length   Two factories are located at the coordinates $(-x, 0)$ and $(x, 0)$ with their power supply located at the point $(0, h)$ (see figure). Find y such that the total amount of power line from the power supply to the factories is a minimum.

41. Projectile Range   The range R of a projectile fired with an initial velocity v_0 at an angle θ with the horizontal is

$$R = \frac{v_0^2 \sin 2\theta}{g}$$

where g is the acceleration due to gravity. Find the angle θ such that the range is a maximum.

SECTION 3.7 Optimization Problems **213**

 42. *Traffic Flow* The police department must determine the speed limit on a bridge such that the flow rate of cars is maximum per unit time. The greater the speed limit, the farther apart the cars must be in order to keep a safe stopping distance. Experimental data on the stopping distance d (in meters) for various velocities v (in kilometers per hour) are given in the table.

v	20	40	60	80	100
d	5.1	13.7	27.2	44.2	66.4

(a) Convert the speeds v in the table to the speeds s in meters per second. Use the regression capabilities of a graphing utility to find a model of the form $d(s) = as^2 + bs + c$ for the data.

(b) Consider two consecutive vehicles of average length 5.5 meters, traveling at a safe speed on the bridge. Let T be the difference between the times (in seconds) when the front bumpers of the vehicles pass a given point on the bridge. Verify that this difference in times is given by

$$T = \frac{d(s)}{s} + \frac{5.5}{s}.$$

(c) Use a graphing utility to graph the function T and estimate the speed s that minimizes the time between vehicles.

(d) Use calculus to determine the speed that minimizes T. What is the minimum value of T? Convert the required speed to kilometers per hour.

(e) Find the optimal distance between vehicles for the posted speed limit determined in part (d).

43. *Conjecture* Consider the functions $f(x) = \frac{1}{2}x^2$ and $g(x) = \frac{1}{16}x^4 - \frac{1}{2}x^2$ on the domain $[0, 4]$.

(a) Use a graphing utility to graph the functions on the specified domain.

(b) Write the vertical distance d between the functions as a function of x and use calculus to find the value of x for which d is maximum.

(c) Find the equations of the tangent lines to the graphs of f and g at the critical number found in part (b). Graph the tangent lines. What is the relationship between the lines?

(d) Make a conjecture about the relationship between tangent lines to the graphs of two functions at the value of x at which the vertical distance between the functions is greatest, and prove your conjecture.

44. *Illumination* A light source is located over the center of a circular table of diameter 4 feet (see figure). Find the height h of the light source such that the illumination I at the perimeter of the table is maximum if $I = k(\sin \alpha)/s^2$, where s is the slant height, α is the angle at which the light strikes the table, and k is a constant.

45. *Illumination* The illumination from a light source is directly proportional to the strength of the source and inversely proportional to the square of the distance from the source. Two light sources of intensities I_1 and I_2 are d units apart. What point on the line segment joining the two sources has the least illumination?

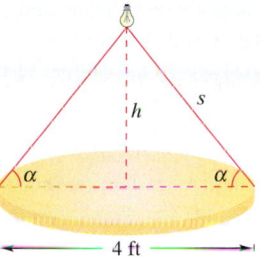

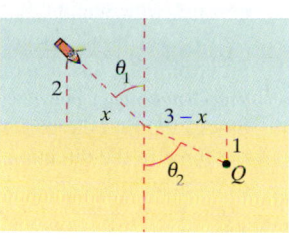

Figure for 44   **Figure for 46**

46. *Minimum Time* A man is in a boat 2 miles from the nearest point on the coast. He is to go to a point Q, 3 miles down the coast and 1 mile inland (see figure). If he can row at 2 miles per hour and walk at 4 miles per hour, toward what point on the coast should he row in order to reach point Q in the least time?

47. *Minimum Time* The conditions are the same as in Exercise 46 except that the man can row at v_1 miles per hour and walk at v_2 miles per hour. If θ_1 and θ_2 are the magnitudes of the angles, show that the man will reach point Q in the least time when

$$\frac{\sin \theta_1}{v_1} = \frac{\sin \theta_2}{v_2}.$$

48. *Minimum Time* When light waves, traveling in a transparent medium, strike the surface of a second transparent medium, they change directions. This change of direction is called refraction and is defined by **Snell's Law of Refraction,**

$$\frac{\sin \theta_1}{v_1} = \frac{\sin \theta_2}{v_2}$$

where θ_1 and θ_2 are the magnitudes of the angles shown in the figure and v_1 and v_2 are the velocities of light in the two media. Show that this problem is equivalent to that of Exercise 47, and that light waves traveling from P to Q follow the path of minimum time.

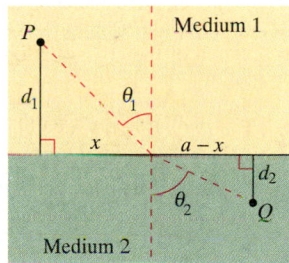

49. Sketch the graph of $f(x) = 2 - 2 \sin x$ on the interval $[0, \pi/2]$.

(a) Find the distance from the origin to the y-intercept and the distance from the origin to the x-intercept.

(b) Express the distance d from the origin to a point on the graph of f as a function of x. Use your graphing utility to graph d and find the minimum distance.

(c) Use calculus and the root feature of a graphing utility to find the value of x that minimizes the function d on the interval $[0, \pi/2]$. What is the minimum distance? *(Submitted by Tim Chapell, Penn Valley Community College, Kansas City, MO.)*

50. Minimum Force A component is designed to slide a block of steel with weight W across a table and into a chute (see figure.) The motion of the block is resisted by a frictional force proportional to its apparent weight. (Let k be the constant of proportionality.) Find the minimum force F needed to slide the block, and find the corresponding value of θ. (*Hint:* $F \cos \theta$ is the force in the direction of motion, and $F \sin \theta$ is the amount of force tending to lift the block. Therefore, the apparent weight of the block is $W - F \sin \theta$.)

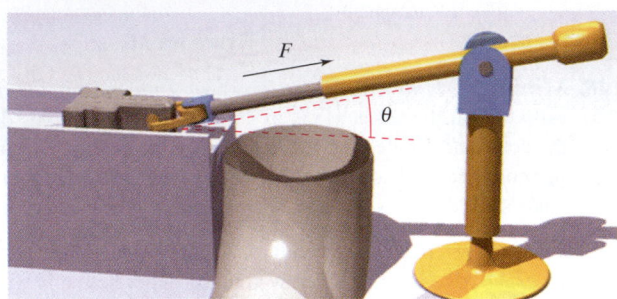

51. Volume A sector with central angle θ is cut from a circle of radius 12 inches (see figure), and the edges of the sector are brought together to form a cone. Find the magnitude of θ so that the volume of the cone is a maximum.

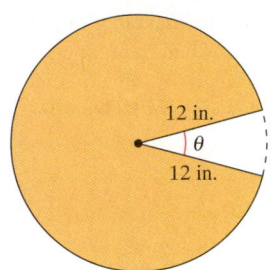

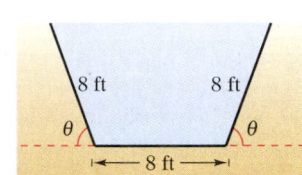

Figure for 51   **Figure for 52**

52. Numerical, Graphical, and Analytic Analysis The cross sections of an irrigation canal are isosceles trapezoids of which three sides are 8 feet long (see figure). Determine the angle of elevation θ of the sides so that the area of the cross section is a maximum by completing the following.

(a) Analytically complete six rows of a table such as the one below. (The first two rows are shown.)

Base 1	Base 2	Altitude	Area
8	$8 + 16 \cos 10°$	$8 \sin 10°$	≈ 22.21
8	$8 + 16 \cos 20°$	$8 \sin 20°$	≈ 42.5

(b) Use a graphing utility to generate additional rows of the table and estimate the maximum cross-sectional area. (*Hint:* Use the table feature of the graphing utility.)

(c) Write the cross-sectional area A as a function of θ.

(d) Use calculus to find the critical number of the function in part (c) and find the angle that will yield the maximum cross-sectional area.

(e) Use a graphing utility to graph the function in part (c) and verify the maximum cross-sectional area.

53. Numerical, Graphical, and Analytic Analysis A 2-meter-high fence is 7 meters from the side of a grain storage bin. A grain elevator must reach from ground level outside the fence to the storage bin (see figure). Complete the following to find the length of the shortest grain elevator.

(a) Analytically complete six rows of a table such as the one below. (The first two rows are shown.)

θ	L_1	L_2	$L_1 + L_2$
0.1	$\dfrac{2}{\sin(0.1)}$	$\dfrac{7}{\cos(0.1)}$	≈ 27.1
0.2	$\dfrac{2}{\sin(0.2)}$	$\dfrac{7}{\cos(0.2)}$	≈ 17.2

(b) Use a graphing utility to generate additional rows of the table. Use the table to estimate the minimum length. (*Hint:* Use the table feature of the graphing utility.)

(c) Write the length L as a function of θ.

(d) Use a graphing utility to graph the function in part (c). Use the graph to estimate the minimum length. How does your estimate compare with that in part (b)?

(e) Use calculus to find the critical number of the function and the angle that will yield the minimum length.

(f) Determine the height of the grain storage bin for the elevator of minimum length.

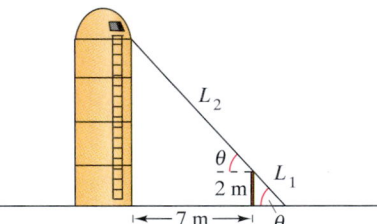

54. Friction The efficiency E of a screw with square threads is

$$E = \frac{\tan \phi (1 - \mu \tan \phi)}{\mu + \tan \phi}$$

where μ is the coefficient of sliding friction and ϕ is the angle of inclination of the threads to a plane perpendicular to the axis of the screw. Find the angle ϕ that yields maximum efficiency when $\mu = 0.1$.

55. Writing The figures show a rectangle, a circle, and a semi-circle inscribed in a triangle bounded by the coordinate axes and the first quadrant portion of the line with intercepts $(3, 0)$ and $(0, 4)$. Find the dimensions of each inscribed figure such that its area is maximum. State whether calculus was helpful in finding the required dimensions. Explain your reasoning.

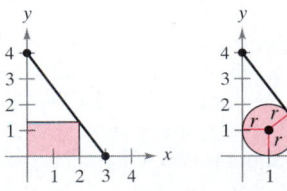
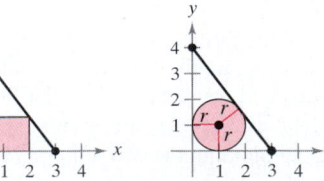
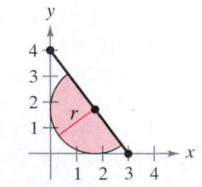

SECTION **3.8** **Newton's Method**

Newton's Method • Algebraic Solutions of Polynomial Equations

Newton's Method

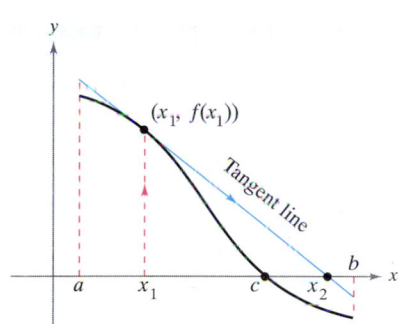

(a)

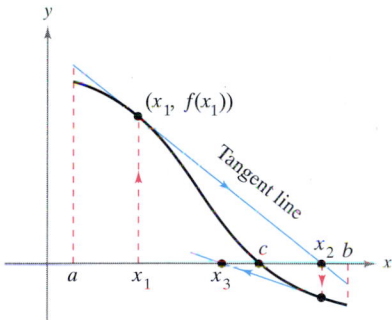

(b)

The *x*-intercept of the tangent line approximates the zero of *f*.
Figure 3.58

In this section you will study a technique for approximating the real zeros of a function. The technique is called **Newton's Method,** and it uses tangent lines to approximate the graph of the function near its *x*-intercepts.

To see how Newton's Method works, consider a function *f* that is continuous on the interval $[a, b]$ and differentiable on the interval (a, b). If $f(a)$ and $f(b)$ differ in sign, then, by the Intermediate Value Theorem, *f* must have at least one zero in the interval (a, b). Suppose you estimate this zero to occur at

$$x = x_1 \qquad \text{First estimate}$$

as shown in Figure 3.58(a). Newton's Method is based on the assumption that the graph of *f* and the tangent line at $(x_1, f(x_1))$ both cross the *x*-axis at *about* the same point. Because you can easily calculate the *x*-intercept for this tangent line, you can use it as a second (and, usually, better) estimate for the zero of *f*. The tangent line passes through the point $(x_1, f(x_1))$ with a slope of $f'(x_1)$. In point-slope form, the equation of the tangent line is therefore

$$y - f(x_1) = f'(x_1)(x - x_1)$$
$$y = f'(x_1)(x - x_1) + f(x_1).$$

Letting $y = 0$ and solving for *x* produces

$$x = x_1 - \frac{f(x_1)}{f'(x_1)}.$$

Thus, from the initial estimate x_1 you obtain a new estimate

$$x_2 = x_1 - \frac{f(x_1)}{f'(x_1)}. \qquad \text{Second estimate (see Figure 3.58b)}$$

You can improve on x_2 and calculate yet a third estimate

$$x_3 = x_2 - \frac{f(x_2)}{f'(x_2)}. \qquad \text{Third estimate}$$

Repeated application of this process is called Newton's Method.

Newton's Method for Approximating the Zeros of a Function

Let $f(c) = 0$, where *f* is differentiable on an open interval containing *c*. Then, to approximate *c*, use the following steps.

1. Make an initial estimate x_1 that is "close to" *c*. (A graph is helpful.)
2. Determine a new approximation

$$x_{n+1} = x_n - \frac{f(x_n)}{f'(x_n)}.$$

3. If $|x_n - x_{n+1}|$ is within the desired accuracy, let x_{n+1} serve as the final approximation. Otherwise, return to Step 2 and calculate a new approximation.

Each successive application of this procedure is called an **iteration.**

NOTE For many functions, just a few iterations of Newton's Method will produce approximations having very small errors, as shown in Example 1.

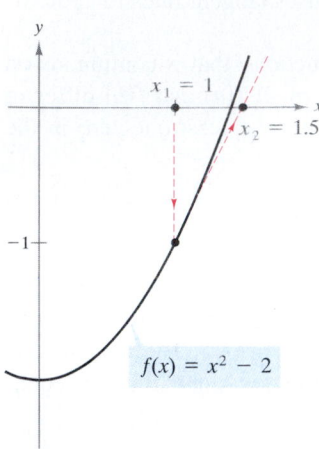

The first iteration of Newton's Method
Figure 3.59

EXAMPLE 1 Using Newton's Method

Calculate three iterations of Newton's Method to approximate a zero of $f(x) = x^2 - 2$. Use $x_1 = 1$ as the initial guess.

Solution Because $f(x) = x^2 - 2$, you have $f'(x) = 2x$, and the iterative process is given by the formula

$$x_{n+1} = x_n - \frac{f(x_n)}{f'(x_n)} = x_n - \frac{x_n^2 - 2}{2x_n}.$$

The calculations for three iterations are shown in the table.

n	x_n	$f(x_n)$	$f'(x_n)$	$\dfrac{f(x_n)}{f'(x_n)}$	$x_n - \dfrac{f(x_n)}{f'(x_n)}$
1	1.000000	-1.000000	2.000000	-0.500000	1.500000
2	1.500000	0.250000	3.000000	0.083333	1.416667
3	1.416667	0.006945	2.833334	0.002451	1.414216
4	1.414216				

Of course, in this case you know that the two zeros of the function are $\pm\sqrt{2}$. To six decimal places, $\sqrt{2} = 1.414214$. Thus, after only three iterations of Newton's Method, you have obtained an approximation that is within 0.000002 of an actual root. The first iteration of this process is shown in Figure 3.59.

EXAMPLE 2 Using Newton's Method

Use Newton's Method to approximate the zeros of

$$f(x) = 2x^3 + x^2 - x + 1.$$

Continue the iterations until two successive approximations differ by less than 0.0001.

Solution Begin by sketching a graph of f, as shown in Figure 3.60. From the graph, you can observe that the function has only one zero, which occurs near $x = -1.2$. Next, differentiate f and form the iterative formula

$$x_{n+1} = x_n - \frac{f(x_n)}{f'(x_n)} = x_n - \frac{2x_n^3 + x_n^2 - x_n + 1}{6x_n^2 + 2x_n - 1}.$$

The calculations are shown in the table.

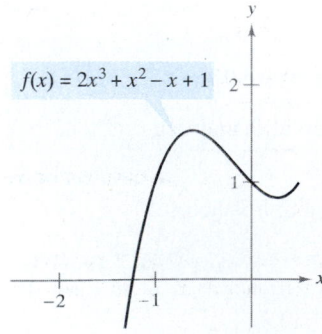

$f(x) = 2x^3 + x^2 - x + 1$

After three iterations of Newton's Method, the zero of f is approximated to the desired accuracy.
Figure 3.60

n	x_n	$f(x_n)$	$f'(x_n)$	$\dfrac{f(x_n)}{f'(x_n)}$	$x_n - \dfrac{f(x_n)}{f'(x_n)}$
1	-1.20000	0.18400	5.24000	0.03511	-1.23511
2	-1.23511	-0.00771	5.68276	-0.00136	-1.23375
3	-1.23375	0.00001	5.66533	0.00000	-1.23375
4	-1.23375				

Because two successive approximations differ by less than the required 0.0001, you can estimate the zero of f to be -1.23375.

TECHNOLOGY Newton's Method is especially useful in conjunction with a computer or programmable calculator. The following program is for a Texas Instruments *TI-83* graphing calculator. Programs for other computers and calculators would involve similar steps. (To run the program, enter the function as Y_1 and enter the initial guess as X.)

```
PROGRAM:NEWTON
:(Xmax–Xmin)/100→D
:1→I
:Disp "INITIAL GUESS"
:Input X
:Lbl 1
:X–Y1/nDeriv(Y1,X,X,D)→R
:If abs(X–R)≤abs(X/1ᴇ10)
:Goto 2
:R→X
:I+1→I
:Goto 1
:Lbl 2
:Disp "ROOT=",R
:Disp "ITER=",I
```

When, as in Examples 1 and 2, the approximations approach a limit, the sequence $x_1, x_2, x_3, \ldots, x_n, \ldots$ is said to **converge.** Moreover, if the limit is c, it can be shown that c must be a zero of f.

Newton's Method does not always yield a convergent sequence. One way this can happen is shown in Figure 3.61. Because Newton's Method involves division by $f'(x_n)$, it is clear that the method will fail if the derivative is zero for any x_n in the sequence. When you encounter this problem, you can usually overcome it by choosing a different value for x_1. Another way Newton's Method can fail is illustrated in the next example.

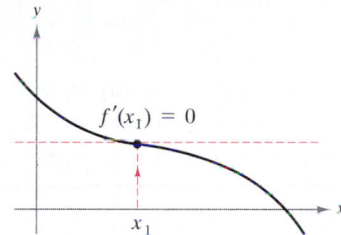

Newton's Method fails to converge if $f'(x_n) = 0$.
Figure 3.61

EXAMPLE 3 An Example in Which Newton's Method Fails

Using $x_1 = 0.1$, show that Newton's Method fails to converge for $f(x) = x^{1/3}$.

Solution Because $f'(x) = \frac{1}{3} x^{-2/3}$, the iterative formula is

$$x_{n+1} = x_n - \frac{f(x_n)}{f'(x_n)} = x_n - \frac{x_n^{1/3}}{\frac{1}{3}x_n^{-2/3}} = x_n - 3x_n = -2x_n.$$

The calculations are shown in the table. This table and Figure 3.62 indicate that x_n continues to increase in magnitude as $n \to \infty$, and thus the limit of the sequence does not exist.

n	x_n	$f(x_n)$	$f'(x_n)$	$\dfrac{f(x_n)}{f'(x_n)}$	$x_n - \dfrac{f(x_n)}{f'(x_n)}$
1	0.10000	0.46416	1.54720	0.30000	−0.20000
2	−0.20000	−0.58480	0.97467	−0.60000	0.40000
3	0.40000	0.73681	0.61401	1.20000	−0.80000
4	−0.80000	−0.92832	0.38680	−2.40000	1.60000

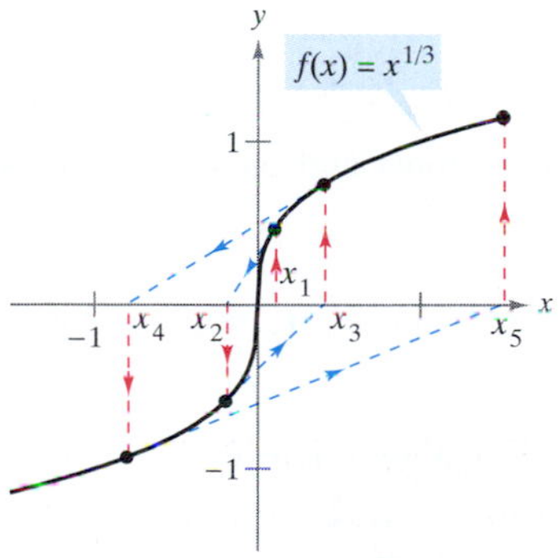

Newton's Method fails to converge for every *x*-value other than the actual zero of *f*.
Figure 3.62

NOTE In Example 3, the initial estimate $x_1 = 0.1$ fails to produce a convergent sequence. Try showing that Newton's Method also fails for every other choice of x_1 (other than the actual zero).

NIELS HENRIK ABEL (1802–1829)

EVARISTE GALOIS (1811–1832)

Although the lives of both Abel and Galois were brief, their work in the fields of analysis and abstract algebra was far-reaching.

It can be shown that a condition sufficient to produce convergence of Newton's Method to a zero of f is that

$$\left| \frac{f(x)\, f''(x)}{[f'(x)]^2} \right| < 1 \qquad \text{Condition for convergence}$$

on an open interval containing the zero. For instance, in Example 1 this test would yield $f(x) = x^2 - 2,\ f'(x) = 2x,\ f''(x) = 2$, and

$$\left| \frac{f(x)\, f''(x)}{[f'(x)]^2} \right| = \left| \frac{(x^2 - 2)(2)}{4x^2} \right| = \left| \frac{1}{2} - \frac{1}{x^2} \right|. \qquad \text{Example 1}$$

On the interval $(1, 3)$, this quantity is less than 1 and therefore the convergence of Newton's Method is guaranteed. On the other hand, in Example 3, you have $f(x) = x^{1/3},\ f'(x) = \frac{1}{3}x^{-2/3},\ f''(x) = -\frac{2}{9}x^{-5/3}$, and

$$\left| \frac{f(x)\, f''(x)}{[f'(x)]^2} \right| = \left| \frac{x^{1/3}(-2/9)(x^{-5/3})}{(1/9)(x^{-4/3})} \right| = 2 \qquad \text{Example 3}$$

which is not less than 1 for any value of x, so Newton's Method fails to converge.

Algebraic Solutions of Polynomial Equations

The zeros of some functions, such as

$$f(x) = x^3 - 2x^2 - x + 2$$

can be found by simple algebraic techniques, such as factoring. The zeros of other functions, such as

$$f(x) = x^3 - x + 1$$

cannot be found by *elementary* algebraic methods. This particular function has only one real zero, and by using more advanced algebraic techniques you can determine the zero to be

$$x = -\sqrt[3]{\frac{3 - \sqrt{23/3}}{6}} - \sqrt[3]{\frac{3 + \sqrt{23/3}}{6}}.$$

Because the *exact* solution is written in terms of square roots and cube roots, it is called a **solution by radicals.**

NOTE   Try approximating the real zero of $f(x) = x^3 - x + 1$ and compare your result with the exact solution shown above.

The determination of radical solutions of a polynomial equation is one of the fundamental problems of algebra. The earliest such result is the Quadratic Formula, which dates back at least to Babylonian times. The general formula for the zeros of a cubic function was developed much later. In the sixteenth century an Italian mathematician, Jerome Cardan, published a method for finding radical solutions to cubic and quartic equations. Then, for 300 years, the problem of finding a general quintic formula remained open. Finally, in the nineteenth century, the problem was answered independently by two young mathematicians. Niels Henrik Abel, a Norwegian mathematician, and Evariste Galois, a French mathematician, proved that it is not possible to solve a *general* fifth- (or higher-) degree polynomial equation by radicals. Of course, you can solve particular fifth-degree equations such as $x^5 - 1 = 0$, but Abel and Galois were able to show that no general *radical* solution exists.

EXERCISES FOR SECTION 3.8

In Exercises 1–4, complete two iterations of Newton's Method for the function using the indicated initial guess.

1. $f(x) = x^2 - 3$, $x_1 = 1.7$ **2.** $f(x) = 3x^2 - 2$, $x_1 = 1$

3. $f(x) = \sin x$, $x_1 = 3$ **4.** $f(x) = \tan x$, $x_1 = 0.1$

In Exercises 5–12, approximate the zero(s) of the function. Use Newton's Method and continue the process until two successive approximations differ by less than 0.001. Then find the zero(s) using a graphing utility and compare the results.

5. $f(x) = x^3 + x - 1$ **6.** $f(x) = x^5 + x - 1$

7. $f(x) = 3\sqrt{x - 1} - x$ **8.** $f(x) = x^3 + 3$

9. $f(x) = x^3 - 3.9x^2 + 4.79x - 1.881$

10. $f(x) = x^4 - 10x^2 - 11$

11. $f(x) = x + \sin(x + 1)$

12. $f(x) = x^3 - \cos x$

In Exercises 13–16, apply Newton's Method to approximate the *x*-value of the indicated point(s) of intersection of the two graphs. Continue the process until two successive approximations differ by less than 0.001. [*Hint:* Let $h(x) = f(x) - g(x)$.]

13. $f(x) = 2x + 1$ **14.** $f(x) = 3 - x$

$g(x) = \sqrt{x + 4}$ $g(x) = 1/(x^2 + 1)$

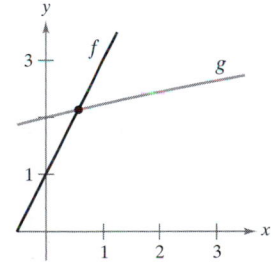 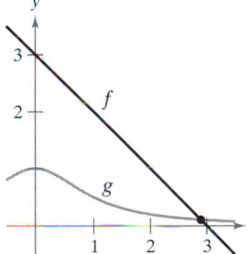

15. $f(x) = x$ **16.** $f(x) = x^2$

$g(x) = \tan x$ $g(x) = \cos x$

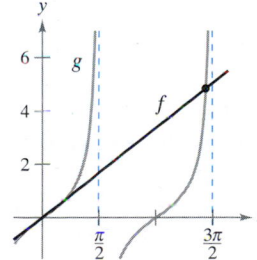 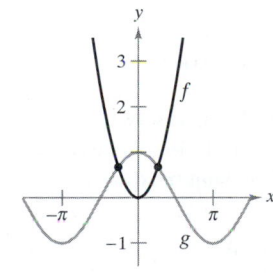

Fixed Point In Exercises 17 and 18, approximate the fixed point of the function to two decimal places. [A *fixed point* x_0 of a function *f* is a value of *x* such that $f(x_0) = x_0$.]

17. $f(x) = \cos x$ **18.** $f(x) = \cot x$, $0 < x < \pi$

19. *Writing* Consider the function $f(x) = x^3 - 3x^2 + 3$.

(a) Use a graphing utility to obtain the graph of *f*.

(b) Use Newton's Method with $x_1 = 1$ as an initial guess.

(c) Repeat part (b) using $x_1 = \frac{1}{4}$ as an initial guess and observe that the result is different.

(d) To understand why the results in parts (b) and (c) are different, sketch the tangent lines to the graph of *f* at the points $(1, f(1))$ and $\left(\frac{1}{4}, f\left(\frac{1}{4}\right)\right)$. Find the *x*-intercept of each tangent line and compare the intercepts with the first iteration of Newton's Method using the respective initial guesses.

(e) Write a short paragraph summarizing how Newton's Method works. Use the results of this exercise to describe why it is important to select the initial guess carefully.

20. *Writing* Repeat the steps in Exercise 19 for the function $f(x) = \sin x$ with initial guesses of $x_1 = 1.8$ and $x_1 = 3$.

In Exercises 21–24, apply Newton's Method using the indicated initial guess, and explain why the method fails.

21. $y = 2x^3 - 6x^2 + 6x - 1$, $x_1 = 1$

22. $y = 4x^3 - 12x^2 + 12x - 3$, $x_1 = \frac{3}{2}$

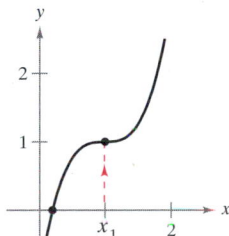 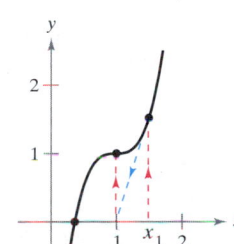

Figure for 21 **Figure for 22**

23. $y = -x^3 + 3x^2 - x + 1$, $x_1 = 1$

24. $f(x) = 2\sin x + \cos 2x$, $x_1 = \frac{3\pi}{2}$

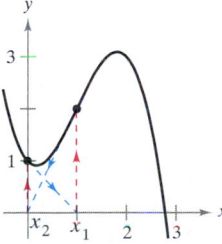 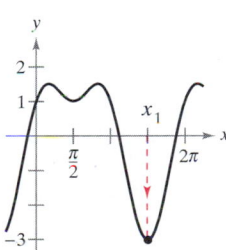

Figure for 23 **Figure for 24**

In Exercises 25 and 26, use Newton's Method to obtain a general rule for approximating the required radical.

25. $x = \sqrt{a}$ [*Hint:* Consider $f(x) = x^2 - a$.]

26. $x = \sqrt[n]{a}$ [*Hint:* Consider $f(x) = x^n - a$.]

In Exercises 27–30, use the results of Exercises 25 and 26 to approximate the indicated radical to three decimal places.

27. $\sqrt{7}$ 28. $\sqrt{5}$

29. $\sqrt[4]{6}$ 30. $\sqrt[3]{15}$

31. Use Newton's Method to show that the equation

$$x_{n+1} = x_n(2 - ax_n)$$

can be used to approximate $1/a$ if x_1 is an initial guess of the reciprocal of a. Note that this method of approximating reciprocals uses only the operations of multiplication and subtraction. [*Hint:* Consider $f(x) = (1/x) - a$.]

32. Use the result of Exercise 31 to approximate the indicated reciprocal to three decimal places.

(a) $\frac{1}{3}$ (b) $\frac{1}{11}$

In Exercises 33 and 34, approximate π to three decimal places using Newton's Method and the given function.

33. $f(x) = 1 + \cos x$ 34. $f(x) = \tan x$

In Exercises 35 and 36, approximate the critical number of f on the interval $[0, \pi]$. Sketch the graph of f, labeling any extrema.

35. $f(x) = x \cos x$ 36. $f(x) = x \sin x$

In Exercises 37–40, we review some typical problems from the previous sections of this chapter. In each case, use Newton's Method to approximate the solution.

37. *Minimum Distance* Find the point on the graph of $f(x) = 4 - x^2$ that is closest to the point $(1, 0)$.

38. *Minimum Distance* Find the point on the graph of $f(x) = x^2$ that is closest to the point $(4, -3)$.

39. *Minimum Time* You are in a boat 2 miles from the nearest point on the coast (see figure). You are to go to a point Q, which is 3 miles down the coast and 1 mile inland. You can row at 3 miles per hour and walk at 4 miles per hour. Toward what point on the coast should you row in order to reach Q in the least time?

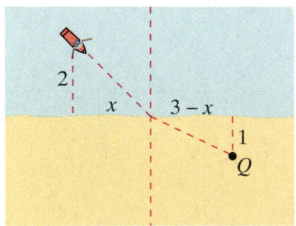

40. *Medicine* The concentration C of a certain chemical in the bloodstream t hours after injection into muscle tissue is given by $C = (3t^2 + t)/(50 + t^3)$. When is the concentration greatest?

41. *Advertising Costs* A company that produces portable cassette players estimates that the profit for selling a particular model is given by

$$P = -76x^3 + 4830x^2 - 320,000, \quad 0 \le x \le 60$$

where P is the profit in dollars and x is the advertising expense in 10,000s of dollars (see figure). According to this model, find the smaller of two advertising amounts that yield a profit P of \$2,500,000.

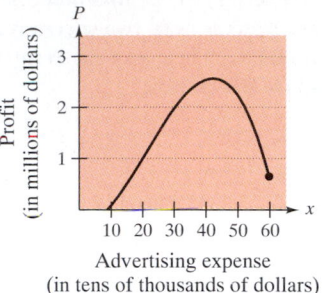

Advertising expense
(in tens of thousands of dollars)

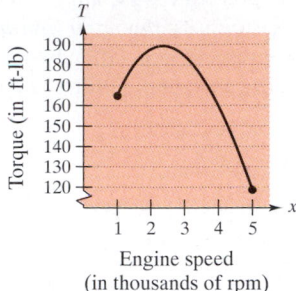

Engine speed
(in thousands of rpm)

Figure for 41   **Figure for 42**

42. *Engine Power* The torque produced by a compact automobile engine is approximated by the model

$$T = 0.808x^3 - 17.974x^2 + 71.248x + 110.843, \quad 1 \le x \le 5$$

where T is the torque in foot-pounds and x is the engine speed in thousands of revolutions per minute (see figure). Approximate the two engine speeds that yield a torque T of 170 foot-pounds.

True or False? **In Exercises 43–46, determine whether the statement is true or false. If it is false, explain why or give an example that shows it is false.**

43. The zeros of $f(x) = p(x)/q(x)$ coincide with the zeros of $p(x)$.

44. If the coefficients of a polynomial function are all positive, then the polynomial has no positive zeros.

45. If $f(x)$ is a cubic polynomial such that $f'(x)$ is never zero, then any initial guess will force Newton's Method to converge to the zero of f.

46. The roots of $\sqrt{f(x)} = 0$ coincide with the roots of $f(x) = 0$.

In Exercises 47 and 48, *write a computer program or use a spreadsheet* to find the zeros of a function using Newton's Method. Approximate the zeros of the function accurate to three decimal places. The output should be a table with the following headings.

$$n, \quad x_n, \quad f(x_n), \quad f'(x_n), \quad \frac{f(x_n)}{f'(x_n)}, \quad x_n - \frac{f(x_n)}{f'(x_n)}$$

47. $f(x) = \frac{1}{4}x^3 - 3x^2 + \frac{3}{4}x - 2$

48. $f(x) = \sqrt{4 - x^2} \sin(x - 2)$

SECTION 3.9 Differentials

EXPLORATION

Tangent Line Approximation Use a graphing utility to sketch the graph of

$$f(x) = x^2.$$

In the same viewing rectangle, sketch the graph of the tangent line to the graph of f at the point $(1, 1)$. Zoom in twice on the point of tangency. Does your graphing utility distinguish between the two graphs? Use the trace feature to compare the two graphs. As the x-values get closer to 1, what can you say about the y-values?

Linear Approximations • Differentials • Error Propagation • Calculating Differentials

Linear Approximations

Newton's Method (Section 3.8) is an example of the use of a tangent line to a graph to approximate the graph. In this section, you will study other situations in which the graph of a function can be approximated by a straight line.

To begin, consider a function f that is differentiable at c. The equation for the tangent line at the point $(c, f(c))$ is given by

$$y - f(c) = f'(c)(x - c)$$
$$y = f(c) + f'(c)(x - c).$$

Because c is a constant, y is a linear function of x. Moreover, by restricting the values of x to be sufficiently close to c, the values of y can be used as approximations (to any desired accuracy) of the values of the function f. In other words, as $x \to c$, the limit of y is $f(c)$.

EXAMPLE 1 Using a Tangent Line Approximation

Find the tangent line approximation of

$$f(x) = 1 + \sin x \qquad \text{Original function}$$

at the point $(0, 1)$. Then use a table to compare the y-values of the linear function with those of $f(x)$ in an open interval containing $x = 0$.

Solution The derivative of f is

$$f'(x) = \cos x. \qquad \text{First derivative}$$

Thus, the equation of the tangent line to the graph of f at the point $(0, 1)$ is

$$y - f(0) = f'(0)(x - 0)$$
$$y - 1 = (1)(x - 0)$$
$$y = 1 + x. \qquad \text{Tangent line approximation}$$

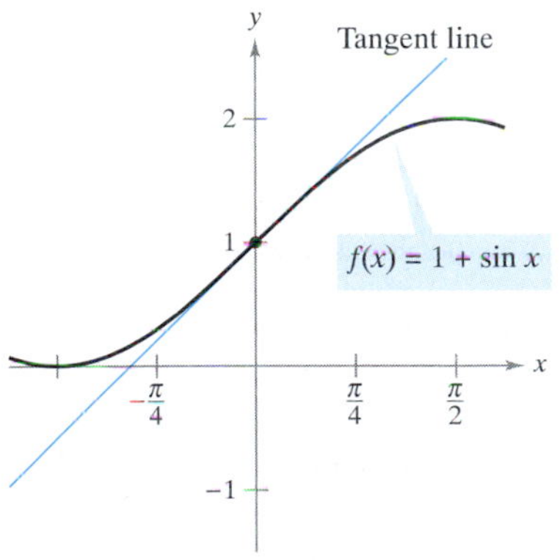

The tangent line approximation of f at the point (0, 1)
Figure 3.63

The table compares the values of y given by this linear approximation with the values of $f(x)$ near $x = 0$. Notice that the closer x is to 0, the better the approximation is. This conclusion is reinforced by the graph shown in Figure 3.63.

x	-0.5	-0.1	-0.01	0	0.01	0.1	0.5
$f(x) = 1 + \sin x$	0.521	0.9002	0.9900002	1	1.0099998	1.0998	1.479
$y = 1 + x$	0.5	0.9	0.99	1	1.01	1.1	1.5

NOTE Be sure you see that this linear approximation of $f(x) = 1 + \sin x$ depends on the point of tangency. At a different point on the graph of f, you would obtain a different tangent line approximation.

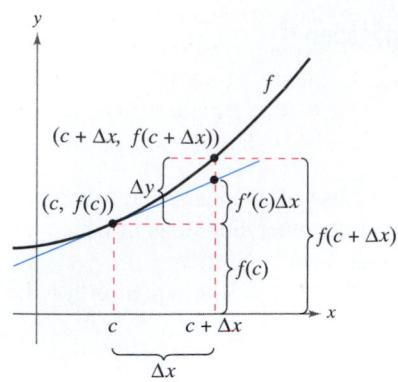

When Δx is small, $\Delta y = f(c + \Delta x) - f(c)$ is approximated by $f'(c)\Delta x$.

Figure 3.64

Differentials

When the tangent line to the graph of f at the point $(c, f(c))$

$$y = f(c) + f'(c)(x - c) \qquad \text{Tangent line at } (c, f(c))$$

is used as an approximation to the graph of f, the quantity $x - c$ is called the *change in x*, and is denoted by Δx, as shown in Figure 3.64. When Δx is small, the change in y (denoted by Δy) can be approximated as follows.

$$\Delta y = f(c + \Delta x) - f(c) \qquad \text{Actual change in } y$$
$$\approx f'(c)\Delta x \qquad \text{Approximate change in } y$$

For such an approximation, the quantity Δx is traditionally denoted by dx, and is called the **differential of x.** The expression $f'(x)\,dx$ is denoted by dy, and is called the **differential of y.**

> ### Definition of Differentials
>
> Let $y = f(x)$ represent a function that is differentiable in an open interval containing x. The **differential of x** (denoted by dx) is any nonzero real number. The **differential of y** (denoted by dy) is
>
> $$dy = f'(x)dx.$$

In many types of applications, the differential of y can be used as an approximation of the change in y. That is,

$$\Delta y \approx dy \qquad \text{or} \qquad \Delta y \approx f'(x)dx.$$

EXAMPLE 2   Comparing Δy and dy

Let $y = x^2$. Find dy when $x = 1$ and $dx = 0.01$. Compare this value with Δy for $x = 1$ and $\Delta x = 0.01$.

Solution   Because $y = f(x) = x^2$, you have $f'(x) = 2x$, and the differential dy is given by

$$dy = f'(x)dx \qquad \text{Differential of } y$$
$$= f'(1)(0.01)$$
$$= 2(0.01)$$
$$= 0.02.$$

Now, using $\Delta x = 0.01$, the change in y is

$$\Delta y = f(x + \Delta x) - f(x) \qquad \text{Change in } y$$
$$= f(1.01) - f(1)$$
$$= (1.01)^2 - 1^2$$
$$= 0.0201.$$

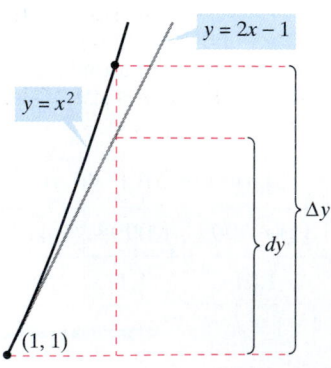

The change in y, Δy, is approximated by the differential of y, dy.

Figure 3.65

Figure 3.65 shows the geometric comparison of dy and Δy. Try comparing other values of dy and Δy. You will see that the values become closer to each other as dx (or Δx) approaches 0.

Error Propagation

Physicists and engineers tend to make liberal use of the approximation of Δy by dy. One way this occurs in practice is in the estimation of errors propagated by physical measuring devices. For example, if you let x represent the measured value of a variable and let $x + \Delta x$ represent the exact value, then Δx is the *error in measurement*. Finally, if the measured value x is used to compute another value $f(x)$, the difference between $f(x + \Delta x)$ and $f(x)$ is the **propagated error.**

$$\underbrace{f(\overbrace{x + \Delta x}^{\text{Measurement error}}) - \overbrace{f(x)}^{\text{Propagated error}} = \Delta y}_{}$$

Exact value   Measured value

EXAMPLE 3   Estimation of Error

The radius of a ball bearing is measured to be 0.7 inch, as shown in Figure 3.66. If the measurement is correct to within 0.01 inch, estimate the propagated error in the volume V of the ball bearing.

Solution   The formula for the volume of a sphere is $V = \frac{4}{3}\pi r^3$, where r is the radius of the sphere. Thus, you can write

$$r = 0.7 \qquad \text{Measured radius}$$

and

$$-0.01 \leq \Delta r \leq 0.01. \qquad \text{Possible error}$$

To approximate the propagated error in the volume, differentiate V to obtain $dV/dr = 4\pi r^2$ and write

$$\Delta V \approx dV \qquad \text{Approximate } \Delta V \text{ by } dV.$$
$$= 4\pi r^2 dr$$
$$= 4\pi (0.7)^2(\pm 0.01) \qquad \text{Substitute for } r \text{ and } dr.$$
$$\approx \pm 0.06158 \text{ in}^3.$$

So the volume has a propagated error of about 0.06 cubic inches.

Would you say that the propagated error in Example 3 is large or small? The answer is best given in *relative* terms by comparing dV with V. The ratio

$$\frac{dV}{V} = \frac{4\pi r^2 dr}{\frac{4}{3}\pi r^3} \qquad \text{Ratio of } dV \text{ to } V$$

$$= \frac{3 dr}{r} \qquad \text{Simplify.}$$

$$\approx \frac{3}{0.7}(\pm 0.01) \qquad \text{Substitute for } dr \text{ and } r.$$

$$\approx \pm 0.0429$$

is called the **relative error.** The corresponding **percent error** is approximately 4.29%.

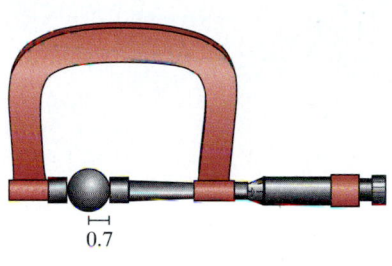

Ball bearing with measured radius that is correct to within 0.01 inch
Figure 3.66

0.7

Calculating Differentials

Each of the differentiation rules that you studied in Chapter 2 can be written in **differential form.** For example, suppose u and v are differentiable functions of x. By the definition of differentials, you have $du = u'dx$ and $dv = v'dx$. Therefore, you can write the differential form of the Product Rule as follows.

$$d[uv] = \frac{d}{dx}[uv]dx \qquad \text{Differential of } uv$$
$$= [uv' + vu']dx \qquad \text{Product Rule}$$
$$= uv'dx + vu'dx$$
$$= u\,dv + v\,du$$

Differential Formulas

Let u and v be differentiable functions of x.

Constant multiple:	$d[cu] = c\,du$
Sum or difference:	$d[u \pm v] = du \pm dv$
Product:	$d[uv] = u\,dv + v\,du$
Quotient:	$d\left[\dfrac{u}{v}\right] = \dfrac{v\,du - u\,dv}{v^2}$

EXAMPLE 4   Finding Differentials

Function	Derivative	Differential
a. $y = x^2$	$\dfrac{dy}{dx} = 2x$	$dy = 2x\,dx$
b. $y = 2\sin x$	$\dfrac{dy}{dx} = 2\cos x$	$dy = 2\cos x\,dx$
c. $y = x\cos x$	$\dfrac{dy}{dx} = -x\sin x + \cos x$	$dy = (-x\sin x + \cos x)dx$
d. $y = \dfrac{1}{x}$	$\dfrac{dy}{dx} = -\dfrac{1}{x^2}$	$dy = -\dfrac{dx}{x^2}$

GOTTFRIED WILHELM LEIBNIZ (1646–1716)

Both Leibniz and Newton are credited with creating calculus. It was Leibniz, however, who tried to broaden calculus by developing rules and formal notation. He often spent days choosing an appropriate notation for a new concept.

The notation in Example 4 is called the **Leibniz notation** for derivatives and differentials, named after the German mathematician Gottfried Wilhelm Leibniz. The beauty of this notation is that it provides an easy way to remember several important calculus formulas by making it seem as though the formulas were derived from algebraic manipulations of differentials. For instance, in Leibniz notation, the *Chain Rule*

$$\frac{dy}{dx} = \frac{dy}{du}\frac{du}{dx}$$

would appear to be true because the du's cancel. Even though this reasoning is *incorrect*, the notation does help one remember the Chain Rule.

EXAMPLE 5   Finding the Differential of a Composite Function

$$y = f(x) = \sin 3x \qquad\qquad \text{Original function}$$
$$f'(x) = 3 \cos 3x \qquad\qquad \text{Apply Chain Rule.}$$
$$dy = f'(x)dx = 3 \cos 3x \, dx \qquad\qquad \text{Differential form}$$

EXAMPLE 6   Finding the Differential of a Composite Function

$$y = f(x) = (x^2 + 1)^{1/2} \qquad\qquad \text{Original function}$$
$$f'(x) = \frac{1}{2}(x^2 + 1)^{-1/2}(2x) = \frac{x}{\sqrt{x^2 + 1}} \qquad \text{Apply Chain Rule.}$$
$$dy = f'(x)dx = \frac{x}{\sqrt{x^2 + 1}} \, dx \qquad\qquad \text{Differential form}$$

Differentials can be used to approximate function values. To do this for the function given by $y = f(x)$, you use the formula

$$f(x + \Delta x) \approx f(x) + dy = f(x) + f'(x)dx$$

which is derived from the approximation $\Delta y = f(x + \Delta x) - f(x) \approx dy$. The key to using this formula is to choose a value for x that makes the calculations easier, as shown in Example 7.

EXAMPLE 7   Approximating Function Values

Use differentials to approximate $\sqrt{16.5}$.

Solution   Using $f(x) = \sqrt{x}$, you can write

$$f(x + \Delta x) \approx f(x) + f'(x)dx = \sqrt{x} + \frac{1}{2\sqrt{x}} \, dx.$$

Now, choosing $x = 16$ and $dx = 0.5$, you obtain the following approximation.

$$f(x + \Delta x) = \sqrt{16.5}$$
$$\approx \sqrt{16} + \frac{1}{2\sqrt{16}} (0.5)$$
$$= 4 + \left(\frac{1}{8}\right)\left(\frac{1}{2}\right)$$
$$= 4 + \frac{1}{16}$$
$$= 4.0625$$

NOTE   The use of differentials to approximate function values has diminished with the availability of calculators and computers. Using a calculator, however, you can determine that

$$\sqrt{16.5} \approx 4.0620$$

which indicates the accuracy of the differential method in Example 7.

In Exercises 1–6, find the equation of the tangent line T to the function f at the indicated point. Use this linear approximation to complete the table.

x	1.9	1.99	2	2.01	2.1
$f(x)$					
$T(x)$					

	Function	Point
1.	$f(x) = x^2$	$(2, 4)$
2.	$f(x) = 1/x^2$	$(2, 1/4)$
3.	$f(x) = x^5$	$(2, 32)$
4.	$f(x) = \sqrt{x}$	$\left(2, \sqrt{2}\right)$
5.	$f(x) = \sin x$	$(2, \sin 2)$
6.	$f(x) = \csc x$	$(2, \csc 2)$

In Exercises 7–10, use the information to evaluate and compare Δy and dy.

7. $y = x^3$ $x = 1$ $\Delta x = dx = 0.1$

8. $y = 1 - 2x^2$ $x = 0$ $\Delta x = dx = -0.1$

9. $y = x^4 + 1$ $x = -1$ $\Delta x = dx = 0.01$

10. $y = 2x + 1$ $x = 2$ $\Delta x = dx = 0.01$

In Exercises 11–20, find the differential dy of the given function.

11. $y = 3x^2 - 4$

12. $y = 2x^{3/2}$

13. $y = \dfrac{x + 1}{2x - 1}$

14. $y = \sqrt{x^2 - 4}$

15. $y = x\sqrt{1 - x^2}$

16. $y = \sqrt{x} + \dfrac{1}{\sqrt{x}}$

17. $y = \dfrac{\sec^2 x}{x^2 + 1}$

18. $y = x \sin x$

19. $y = \dfrac{1}{3}\cos\left(\dfrac{6\pi x - 1}{2}\right)$

20. $y = x - \tan^2 x$

21. *Area*   The measurement of the side of a square is found to be 12 inches, with a possible error of $\frac{1}{64}$ inch. Use differentials to approximate the possible propagated error in computing the area of the square.

22. *Area*   The measurements of the base and altitude of a triangle are found to be 36 and 50 centimeters. The possible error in each measurement is 0.25 centimeter. Use differentials to approximate the possible propagated error in computing the area of the triangle.

23. *Area*   The measurement of the radius of the end of a log is found to be 14 inches, with a possible error of $\frac{1}{4}$ inch. Use differentials to approximate the possible propagated error in computing the area of the end of the log.

24. *Volume and Surface Area*   The measurement of the edge of a cube is found to be 12 inches, with a possible error of 0.03 inch. Use differentials to approximate the maximum possible propagated error in computing

(a) the volume of the cube.

(b) the surface area of the cube.

25. *Area*   The measurement of a side of a square is found to be 15 centimeters. The possible error in measuring the side is 0.05 centimeter.

(a) Approximate the percent error in computing the area of the square.

(b) Estimate the maximum allowable percent error in measuring the side if the error in computing the area cannot exceed 2.5%.

26. *Circumference*   The measurement of the circumference of a circle is found to be 56 centimeters. The possible error in measuring the circumference is 1.2 centimeter.

(a) Approximate the percent error in computing the area of the circle.

(b) Estimate the maximum allowable percent error in measuring the circumference if the error in computing the area cannot exceed 3%.

27. *Volume and Surface Area*   The radius of a sphere is claimed to be 6 inches, with a possible error of 0.02 inch. Use differentials to approximate the maximum possible error in calculating (a) the volume of the sphere, (b) the surface area of the sphere, and (c) the relative errors in parts (a) and (b).

28. *Profit*   The profit P for a company is given by

$$P = (500x - x^2) - \left(\tfrac{1}{2}x^2 - 77x + 3000\right).$$

Approximate the change and percent change in profit as production changes from $x = 115$ to $x = 120$ units.

In Exercises 29 and 30, the thickness of the shell is 0.2 centimeter. Use differentials to approximate the volume of the shell.

29. A cylindrical shell with height 40 centimeters and radius 5 centimeters

30. A spherical shell of radius 100 centimeters

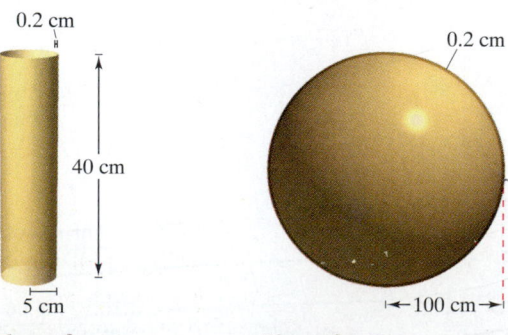

Figure for 29   **Figure for 30**

31. *Pendulum* The period of a pendulum is given by

$$T = 2\pi \sqrt{\frac{L}{g}}$$

where L is the length of the pendulum in feet, g is the acceleration due to gravity, and T is the time in seconds. Suppose that the pendulum has been subjected to an increase in temperature such that the length has increased by $\frac{1}{2}\%$.

(a) Find the approximate percent change in the period.

(b) Using the result in part (a), find the approximate error in this pendulum clock in one day.

32. *Pendulum* Suppose a pendulum of length L is to be used to find the acceleration of gravity at a given point on the earth's surface. Use the formula

$$T = 2\pi \sqrt{\frac{L}{g}}$$

to approximate the percent error in the value of g if you can determine the period T to within 0.1 percent of its true value.

33. *Ohm's Law* A current of I amperes passes through a resistor of R ohms. **Ohm's Law** states that the voltage E applied to the resistor is

$$E = IR.$$

If the voltage is constant, show that the magnitude of the relative error in R caused by a change in I is equal in magnitude to the relative error in I.

34. *Environment* The cost in dollars of removing $p\%$ of the air pollutants in the stack emission of a utility company that burns coal to generate electricity is

$$C = \frac{80{,}000p}{100 - p}, \quad 0 \le p < 100.$$

Use differentials to approximate the increase in cost if the government requires the utility company to remove 2% more of the pollutants and p is currently (a) 40% and (b) 75%.

35. *Triangle Measurements* The measurement of one side of a right triangle is found to be 9.5 inches, and the angle opposite that side is 26°45′ with a possible error of 15′.

(a) Approximate the percent error in computing the length of the hypotenuse.

(b) Estimate the maximum allowable percent error in measuring the angle if the error in computing the length of the hypotenuse cannot exceed 2%.

36. *Area* Approximate the percent error in computing the area of the triangle in Exercise 35.

37. *Projectile Motion* The range R of a projectile is

$$R = \frac{v_0^2}{32}(\sin 2\theta)$$

where v_0 is the initial velocity in feet per second and θ is the angle of elevation. If $v_0 = 2200$ feet per second and θ is changed from 10° to 11°, use differentials to approximate the change in the range.

38. *Surveying* A surveyor standing 50 feet from the base of a large tree measures the angle of elevation to the top of the tree as 71.5°. How accurately must the angle be measured if the percent error in estimating the height of the tree is to be less than 6%?

In Exercises 39–42, use differentials to approximate the value of the function. Compare your answer with that of a calculator.

39. $\sqrt{99.4}$ **40.** $\sqrt[3]{28}$

41. $\sqrt[4]{624}$ **42.** $(2.99)^3$

Writing **In Exercises 43–46, give a short explanation of why the approximation is valid.**

43. $0.99^4 \approx 1 - 4(0.01)$ **44.** $\sqrt{4.02} \approx 2 + \dfrac{1}{4}(0.02)$

45. $\sec 0.03 \approx 1 + 0(0.03)$ **46.** $\tan 0.05 \approx 0 + 1(0.05)$

True or False? **In Exercises 47–50, determine whether the statement is true or false. If it is false, explain why or give an example that shows it is false.**

47. If $y = x + c$, then $dy = dx$.

48. If $y = ax + b$, then $\Delta y / \Delta x = dy/dx$.

49. If y is differentiable, then $\lim_{\Delta x \to 0} (\Delta y - dy) = 0$.

50. If $y = f(x)$, f is increasing and differentiable, and $\Delta x > 0$, then $\Delta y \ge dy$.

51. *Graphical Reasoning* The area of a square of side x is given by $A(x) = x^2$.

(a) Compute dA and ΔA in terms of x and Δx.

(b) Use the figure to identify the region whose area is dA.

(c) Use the figure to identify the region whose area is $\Delta A - dA$.

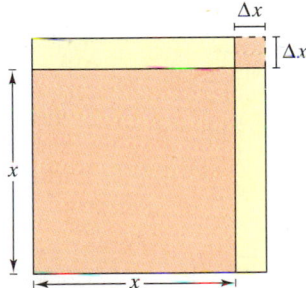

52. Show that if $y = f(x)$ is a differentiable function, then

$$\Delta y - dy = \varepsilon \Delta x$$

where $\varepsilon \to 0$ as $\Delta x \to 0$.

| SECTION | 3.10 | Business and Economics Applications |

Business and Economics Applications

Business and Economics Applications

In Section 2.6, you learned that one of the most common ways to measure change is with respect to time. In this section, you will study some important rates of change in economics that are not measured with respect to time. For example, economists refer to **marginal profit, marginal revenue,** and **marginal cost** as the rates of change of the profit, revenue, and cost with respect to the number of units produced or sold.

Summary of Business Terms and Formulas

Basic Terms	*Basic Formulas*
x is the number of units produced (or sold)	
p is the price per unit	
R is the total revenue from selling x units	$R = xp$
C is the total cost of producing x units	
$\overline{C}$ is the average cost per unit	$\overline{C} = \dfrac{C}{x}$
P is the total profit from selling x units	$P = R - C$

The **break-even point** is the number of units for which $R = C$.

Marginals

$\dfrac{dR}{dx} = $ (Marginal revenue) $\approx$ (*extra* revenue from selling one additional unit)

$\dfrac{dC}{dx} = $ (Marginal cost) $\approx$ (*extra* cost of producing one additional unit)

$\dfrac{dP}{dx} = $ (Marginal profit) $\approx$ (*extra* profit from selling one additional unit)

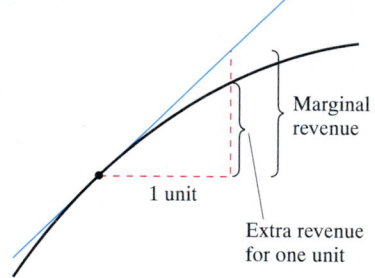

Marginal revenue

1 unit

Extra revenue for one unit

A revenue function
Figure 3.67

In this summary, note that marginals can be used to approximate the *extra* revenue, cost, or profit associated with selling or producing one additional unit. This is illustrated graphically for marginal revenue in Figure 3.67.

EXPLORATION

Graphs of Revenue Functions The graph of the revenue function shown in Figure 3.67 is concave down. How can that be? Shouldn't a revenue function be linear, of the form

$R = xp$

where x is the number of units and p is the price? Discuss this question within your group or class. Consider that the number of units sold x may itself be a function of the price per unit p. For instance, by lowering the price per unit, you may be able to increase the number of units sold.

EXAMPLE 1   Using Marginals as Approximations

A manufacturer determines that the profit derived from selling x units of a certain item is given by $P = 0.0002x^3 + 10x$.

a. Find the marginal profit for a production level of 50 units.

b. Compare this with the actual gain in profit obtained by increasing the production from 50 to 51 units. (See Figure 3.68.)

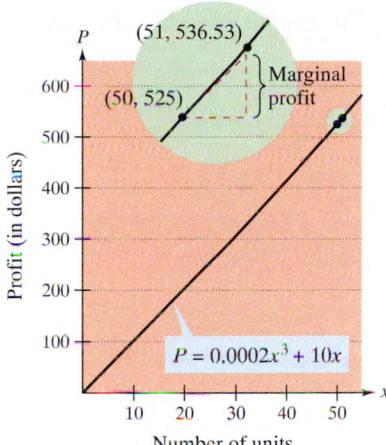

Marginal profit is the extra profit from selling one additional unit.
Figure 3.68

Solution

a. The marginal profit is given by

$$\frac{dP}{dx} = 0.0006x^2 + 10.$$

When $x = 50$, the marginal profit is

$$\frac{dP}{dx} = (0.0006)(50)^2 + 10 = \$11.50. \qquad \text{Marginal profit}$$

b. For $x = 50$ and 51, the actual profits are

$$P = (0.0002)(50)^3 + 10(50) = 25 + 500 = \$525.00$$
$$P = (0.0002)(51)^3 + 10(51) = 26.53 + 510 = \$536.53.$$

Thus, the additional profit obtained by increasing the production level from 50 to 51 units is

$$536.53 - 525.00 = \$11.53. \qquad \text{Extra profit for one unit}$$

The profit function in Example 1 is unusual in that the profit continues to increase as long as the number of units sold increases. In practice, it is more common to encounter situations in which sales can be increased only by lowering the price per item. Such reductions in price ultimately cause the profit to decline. The number of units x that consumers are willing to purchase at a given price p per unit is defined as the **demand function**

$$p = f(x). \qquad \text{Demand function}$$

EXAMPLE 2   Finding the Demand Function

A business sells 2000 items per month at a price of \$10 each. It is predicted that monthly sales will increase by 250 items for each \$0.25 reduction in price. Find the demand function corresponding to this prediction.

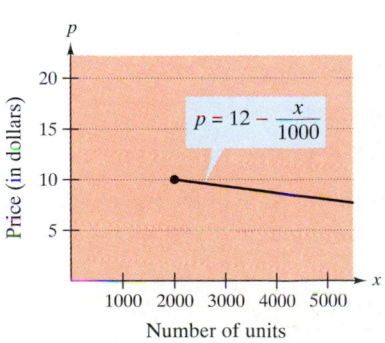

A demand function p
Figure 3.69

Solution   From the given prediction, x increases 250 units each time p drops \$0.25 from the original cost of \$10. This is described by the equation

$$x = 2000 + 250\left(\frac{10 - p}{0.25}\right) = 12{,}000 - 1000p$$

or

$$p = 12 - \frac{x}{1000}, \qquad x \geq 2000. \qquad \text{Demand function}$$

The graph of the demand function is shown in Figure 3.69.

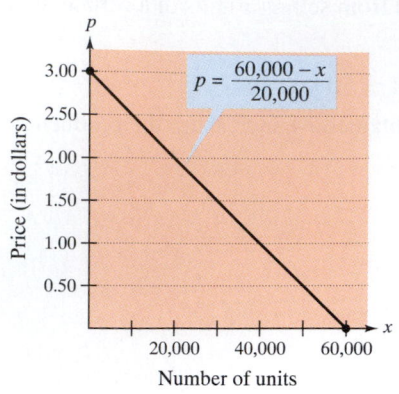

As the price decreases, more hamburgers are sold.
Figure 3.70

EXAMPLE 3   Finding the Marginal Revenue

A fast-food restaurant has determined that the monthly demand for its hamburgers is

$$p = \frac{60,000 - x}{20,000}.$$

Find the increase in revenue per hamburger (marginal revenue) for monthly sales of 20,000 hamburgers. (See Figure 3.70.)

Solution   Because the total revenue is given by $R = xp$, you have

$$R = xp = x\left(\frac{60,000 - x}{20,000}\right) = \frac{1}{20,000}(60,000x - x^2)$$

and the marginal revenue is

$$\frac{dR}{dx} = \frac{1}{20,000}(60,000 - 2x).$$

When $x = 20,000$, the marginal revenue is

$$\frac{dR}{dx} = \frac{1}{20,000}[60,000 - 2(20,000)] = \frac{20,000}{20,000} = \$1/\text{unit}.$$

NOTE   The demand function in Example 3 is typical in that a high demand corresponds to a low price, as shown in Figure 3.70.

EXAMPLE 4   Finding the Marginal Profit

Suppose that in Example 3 the cost of producing x hamburgers is

$$C = 5000 + 0.56x.$$

Find the total profit and the marginal profit for 20,000, for 24,400, and for 30,000 units.

Solution   Because $P = R - C$, you can use the revenue function in Example 3 to obtain

$$P = \frac{1}{20,000}(60,000x - x^2) - 5000 - 0.56x$$

$$= 2.44x - \frac{x^2}{20,000} - 5000.$$

Thus, the marginal profit is

$$\frac{dP}{dx} = 2.44 - \frac{x}{10,000}.$$

The table shows the total profit and the marginal profit for each of the three indicated demands. Figure 3.71 shows the graph of the profit function.

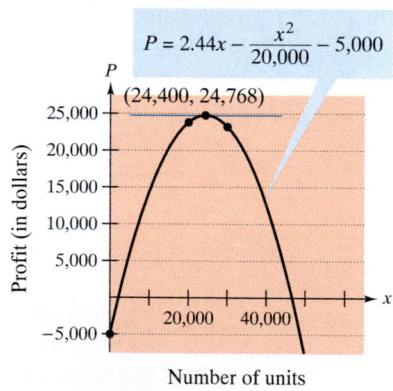

The maximum profit corresponds to the point where the marginal profit is 0. When more than 24,400 hamburgers are sold, the marginal profit is negative–increasing production beyond this point will *reduce* rather than increase profit.
Figure 3.71

Demand	20,000	24,400	30,000
Profit	\$23,800	\$24,768	\$23,200
Marginal profit	\$0.44	\$0.00	−\$0.56

EXAMPLE 5   Finding the Maximum Profit

In marketing a certain item, a business has discovered that the demand for the item is represented by

$$p = \frac{50}{\sqrt{x}}.$$ Demand function

The cost of producing x items is given by $C = 0.5x + 500$. Find the price per unit that yields a maximum profit (see Figure 3.72).

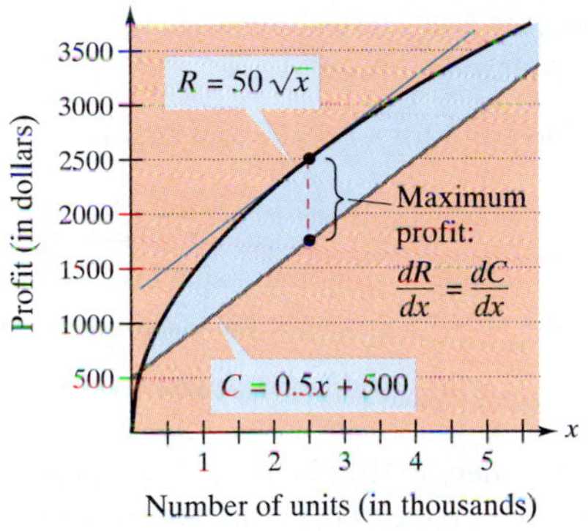

Solution   From the given cost function, you obtain

$$P = R - C = xp - (0.5x + 500).$$ Primary equation

Substituting for p (from the demand function) produces

$$P = x\left(\frac{50}{\sqrt{x}}\right) - (0.5x + 500) = 50\sqrt{x} - 0.5x - 500.$$

Setting the marginal profit equal to 0

$$\frac{dP}{dx} = \frac{25}{\sqrt{x}} - 0.5 = 0$$

yields $x = 2500$. From this, you can conclude that the maximum profit occurs when the price is

$$p = \frac{50}{\sqrt{2500}} = \frac{50}{50} = \$1.00.$$

NOTE   To find the maximum profit in Example 5, we differentiated the profit function, $P = R - C$, and set dP/dx equal to 0. From the equation

$$\frac{dP}{dx} = \frac{dR}{dx} - \frac{dC}{dx} = 0$$

it follows that the maximum profit occurs when the marginal revenue is equal to the marginal cost, as shown in Figure 3.72.

Maximum profit occurs when $\frac{dR}{dx} = \frac{dC}{dx}$.

Figure 3.72

EXAMPLE 6   Minimizing the Average Cost

A company estimates that the cost (in dollars) of producing x units of a certain product is given by $C = 800 + 0.04x + 0.0002x^2$. Find the production level that minimizes the average cost per unit.

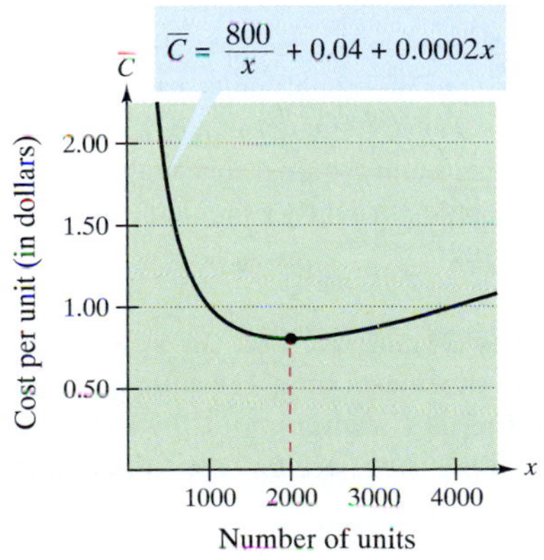

Solution   Substituting from the given equation for C produces

$$\overline{C} = \frac{C}{x} = \frac{800 + 0.04x + 0.0002x^2}{x} = \frac{800}{x} + 0.04 + 0.0002x.$$

Setting the derivative $d\overline{C}/dx$ equal to 0 yields

$$\frac{d\overline{C}}{dx} = -\frac{800}{x^2} + 0.0002 = 0$$

$$x^2 = \frac{800}{0.0002} = 4{,}000{,}000 \Rightarrow x = 2000 \text{ units.}$$

(See Figure 3.73.)

Minimum average cost occurs when $\frac{d\overline{C}}{dx} = 0$.

Figure 3.73

EXERCISES FOR SECTION 3.10

1. **Think About It** The figure shows the cost C of producing x units of a product.

   (a) What is $C(0)$ called?

   (b) Sketch a graph of the marginal cost function.

   (c) Does the marginal cost function have an extremum? If so, describe what it means in economic terms.

2. **Think About It** The figure shows the cost C and revenue R for producing and selling x units of a product.

   (a) Sketch a graph of the marginal revenue function.

   (b) Sketch a graph of the profit function. Approximate the position of the value of x for which profit is maximum.

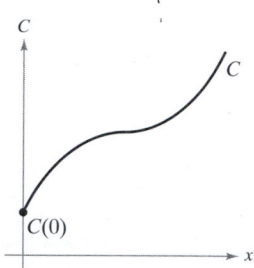

Figure for 1

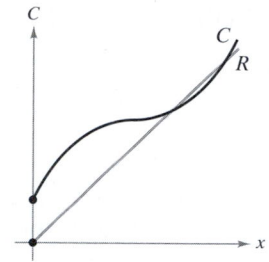

Figure for 2

In Exercises 3–6, find the number of units x that produces a maximum revenue R.

3. $R = 900x - 0.1x^2$

4. $R = 600x^2 - 0.02x^3$

5. $R = \dfrac{1,000,000x}{0.02x^2 + 1800}$

6. $R = 30x^{2/3} - 2x$

In Exercises 7–10, find the number of units x that produces the minimum average cost per unit $\overline{C}$.

7. $C = 0.125x^2 + 20x + 5000$

8. $C = 0.001x^3 - 5x + 250$

9. $C = 3000x - x^2\sqrt{300 - x}$

10. $C = \dfrac{2x^3 - x^2 + 5000x}{x^2 + 2500}$

In Exercises 11–14, find the price per unit p that produces the maximum profit P.

Cost Function	Demand Function
11. $C = 100 + 30x$	$p = 90 - x$
12. $C = 2400x + 5200$	$p = 6000 - 0.4x^2$
13. $C = 4000 - 40x + 0.02x^2$	$p = 50 - \dfrac{x}{100}$
14. $C = 35x + 2\sqrt{x - 1}$	$p = 40 - \sqrt{x - 1}$

Average Cost In Exercises 15 and 16, use the cost function to find the value of x at which the average cost is a minimum. For that value of x, show that the marginal cost and average cost are equal.

15. $C = 2x^2 + 5x + 18$

16. $C = x^3 - 6x^2 + 13x$

17. Prove that the average cost is a minimum at the value of x where the average cost equals the marginal cost.

18. **Maximum Profit** The profit for a certain company is

$$P = 230 + 20s - \tfrac{1}{2}s^2$$

where s is the amount (in hundreds of dollars) spent on advertising. What amount of advertising gives the maximum profit?

19. **Numerical, Graphical, and Analytic Analysis** The cost per unit for the production of a radio is $60. The manufacturer charges $90 per unit for orders of 100 or less. To encourage large orders, the manufacturer reduces the charge by $0.15 per radio for each unit ordered in excess of 100 (for example, there would be a charge of $87 per radio for an order size of 120).

   (a) Analytically complete six rows of a table such as the one below. (The first two rows are shown.)

x	Price	Profit
102	$90 - 2(0.15)$	$102[90 - 2(0.15)] - 102(60) = 3029.40$
104	$90 - 4(0.15)$	$104[90 - 4(0.15)] - 104(60) = 3057.60$

   (b) Use a graphing utility to generate additional rows of the table. Use the table to estimate the maximum profit. (*Hint:* Use the table feature of the graphing utility.)

   (c) Write the profit P as a function of x.

   (d) Use calculus to find the critical number of the function in part (c) and find the required order size.

   (e) Use a graphing utility to graph the function in part (c) and verify the maximum profit from the graph.

20. **Maximum Profit** A real estate office handles 50 apartment units. When the rent is $720 per month, all units are occupied. However, on the average, for each $40 increase in rent, one unit becomes vacant. Each occupied unit requires an average of $48 per month for service and repairs. What rent should be charged to obtain the maximum profit?

Minimum Cost In Exercises 21 and 22, find the speed v, in miles per hour, that will minimize costs on a 110-mile delivery trip. The cost per hour for fuel is C dollars, and the driver is paid W dollars per hour. (Assume there are no costs other than wages and fuel.)

21. Fuel cost: $C = \dfrac{v^2}{600}$

Driver: $W = \$5$

22. Fuel costs: $C = \dfrac{v^2}{500}$

Driver: $W = \$7.50$

23. *Minimum Cost* A power station is on one side of a river that is $\frac{1}{2}$ mile wide, and a factory is 6 miles downstream on the other side. It costs $12 per foot to run power lines over land and $16 per foot to run them underwater. Find the most economical path for the transmission line from the power station to the factory.

24. *Minimum Cost* An offshore oil well is 2 kilometers off the coast. The refinery is 4 kilometers down the coast. If laying pipe in the ocean is twice as expensive as on land, what path should the pipe follow in order to minimize the cost?

Minimum Cost **In Exercises 25–28, consider a fuel distribution center located at the origin of the rectangular coordinate system (units in miles; see figures). The center supplies three factories with coordinates (4, 1), (5, 6), and (10, 3). A trunk line will run from the distribution center along the line $y = mx$, and feeder lines will run to the three factories. The objective is to find m such that the lengths of the feeder lines are minimized.**

25. Minimize the sum of the squares of the lengths of the vertical feeder lines

$$S_1 = (4m - 1)^2 + (5m - 6)^2 + (10m - 3)^2.$$

Find the equation for the trunk line by this method and then determine the sum of the lengths of the feeder lines.

26. Minimize the sum of the absolute values of the lengths of the vertical feeder lines

$$S_2 = |4m - 1| + |5m - 6| + |10m - 3|.$$

Find the equation for the trunk line by this method and then determine the sum of the lengths of the feeder lines. (*Hint:* Use a graphing utility to graph the function S_2 and approximate the required critical number.)

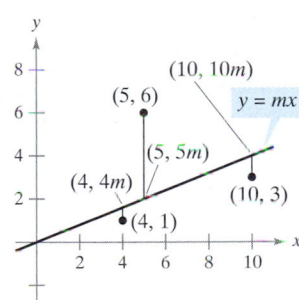

Figure for 25 and 26

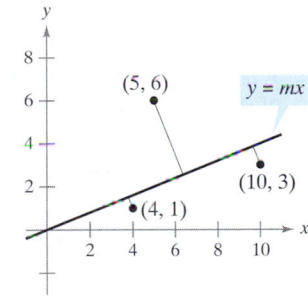

Figure for 27

27. Minimize the sum of the perpendicular distances (see Exercises 75–81 in Section P.2) from the trunk line to the factories given by

$$S_3 = \frac{|4m - 1|}{\sqrt{m^2 + 1}} + \frac{|5m - 6|}{\sqrt{m^2 + 1}} + \frac{|10m - 3|}{\sqrt{m^2 + 1}}.$$

Find the equation for the trunk line by this method and then determine the sum of the lengths of the feeder lines. (*Hint:* Use a graphing utility to graph the function S_3 and approximate the required critical number.)

28. *Writing* Write a paragraph comparing the methods of Exercises 25–27. Which method is easier to use? Which method best meets the stated objective of minimizing the amount of feeder line required? Explain.

29. *Maximum Profit* Assume that the amount of money deposited in a bank is proportional to the square of the interest rate the bank pays on this money. Furthermore, the bank can reinvest this money at 12%. Find the interest rate the bank should pay to maximize profit. (Use the simple interest formula.)

30. *Maximum Revenue* When a wholesaler sold a certain product at $25 per unit, sales were 800 units per week. After a price increase of $5, the average number of units sold dropped to 775 per week. Assume that the demand function is linear, and find the price that will maximize the total revenue.

31. *Minimum Cost* The ordering and transportation cost C of the components used in manufacturing a certain product is

$$C = 100\left(\frac{200}{x^2} + \frac{x}{x + 30}\right), \quad 1 \le x$$

where C is measured in thousands of dollars and x is the order size in hundreds. Find the order size that minimizes the cost. (*Hint:* Use Newton's Method or the root feature of a graphing utility.)

32. *Average Cost* A company estimates that the cost in dollars of producing x units of a certain product is

$$C = 800 + 0.4x + 0.02x^2 + 0.0001x^3.$$

Find the production level that minimizes the average cost per unit. (*Hint:* Use Newton's Method or the root feature of a graphing utility.)

33. *Revenue* The revenue R for a company selling x units is

$$R = 900x - 0.1x^2.$$

Use differentials to approximate the change in revenue if sales increase from $x = 3000$ to $x = 3100$ units.

34. *Profit* The profit P for a company selling x units is

$$P = (500x - x^2) - \left(\tfrac{1}{2}x^2 - 77x + 3000\right).$$

Use differentials to approximate the change and percent change in profit as production changes from $x = 175$ to $x = 180$ units.

35. *Analytic and Graphical Analysis* A manufacturer of fertilizer finds that the national sales of fertilizer roughly follow the seasonal pattern

$$F = 100{,}000\left\{1 + \sin\left[\frac{2\pi(t - 60)}{365}\right]\right\}$$

where F is measured in pounds. Time t is measured in days, with $t = 1$ corresponding to January 1.

(a) Use calculus to determine the day of the year when the maximum amount of fertilizer is sold.

(b) Use a graphing utility to graph the function and approximate the day of the year when sales are minimum.

36. Modeling Data The table shows the monthly wholesale distribution of gasoline G in billions of gallons for the year 1994 in the United States. The time in months is represented by t, with $t = 1$ corresponding to January. (*Source: Federal Highway Electronic Bulletin Board*)

t	1	2	3	4	5	6
G	8.91	9.18	9.79	9.83	10.37	10.16

t	7	8	9	10	11	12
G	10.37	10.81	10.03	9.97	9.85	9.51

A model for these data is

$$G = 9.90 - 0.64 \cos\left(\frac{\pi t}{6} - 0.62\right).$$

(a) Use a graphing utility to plot the data and graph the model.

(b) Use the model to approximate the month when wholesale gasoline sales were greatest.

(c) What factor in the model causes the seasonal variation in sales of gasoline? What part of the model gives the average monthly sales of gasoline?

(d) Suppose the Department of Energy added the term $0.02t$ to the model. What does the inclusion of this term mean? Use this model to estimate the maximum monthly consumption in the year 2000.

37. Railroad Revenues The annual revenue (in millions of dollars) for Union Pacific for the years 1985–1994 can be modeled by

$$R = 4.7t^4 - 193.5t^3 + 2941.7t^2 - 19,294.7t + 52,012$$

where $t = 0$ corresponds to 1980. (*Source: Union Pacific Corporation*)

(a) During which year (between 1985 and 1994) was Union Pacific's revenue the least?

(b) During which year was the revenue the greatest?

(c) Find the revenues for the years in which the revenue was the least and greatest.

(d) Use a graphing utility to confirm the results in parts (a) and (b).

38. Modeling Data The manager of a department store recorded the quarterly sales S (in thousands of dollars) of a new seasonal product over a period of 2 years, as shown in the table, where t is the time in quarters, with $t = 1$ corresponding to the winter quarter of 1996.

t	1	2	3	4	5	6	7	8
S	7.5	6.2	5.3	7.0	9.1	7.8	6.9	8.6

(a) Use a graphing utility to plot the data.

(b) Find a model of the form $S = a + bt + c \sin \beta t$ for the data. (*Hint:* Start by finding β. Next, use a graphing utility to find $a + bt$. Finally, approximate c.)

(c) Use a graphing utility to graph the model and make any adjustments necessary to obtain a better fit.

(d) Use the model to predict the maximum quarterly sales in the year 2000.

39. Think About It Match each graph with the function it best represents—a demand function, a revenue function, a cost function, or a profit function. Explain your reasoning. [The graphs are labeled (a), (b), (c), and (d).]

(a)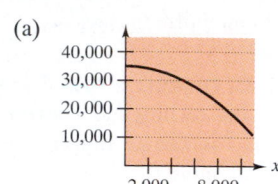

(b)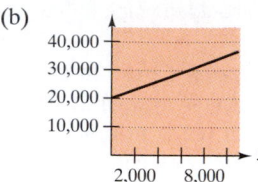

(c)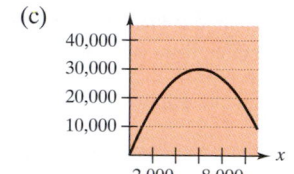

(d)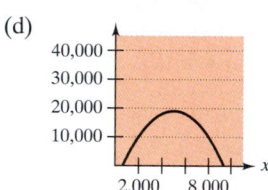

40. Diminishing Returns The profit P (in thousands of dollars) for a company spending an amount s (in thousands of dollars) on advertising is

$$P = -\tfrac{1}{10}s^3 + 6s^2 + 400.$$

(a) Find the amount of money the company should spend on advertising in order to yield a maximum profit.

(b) The *point of diminishing returns* is the point at which the rate of growth of the profit function begins to decline. Find the point of diminishing returns.

The relative responsiveness of consumers to a change in the price of an item is called the *price elasticity of demand*. If $p = f(x)$ is a differentiable demand function, the price elasticity of demand is

$$\eta = \frac{p/x}{dp/dx}.$$

For a given price, if $|\eta| < 1$, the demand is *inelastic*, and if $|\eta| > 1$, the demand is *elastic*.

Elasticity In Exercises 41–44, find η for the demand function at the indicated x-value. Is the demand elastic, inelastic, or neither at the indicated x-value?

41. $p = 400 - 3x$
$x = 20$

42. $p = 5 - 0.03x$
$x = 100$

43. $p = 400 - 0.5x^2$
$x = 20$

44. $p = \dfrac{500}{x + 2}$
$x = 23$

REVIEW EXERCISES FOR CHAPTER 3

1. *Think About It* Give the definition of a critical number, and graph an arbitrary function f showing the different types of critical numbers.

2. Consider the odd function f that is continuous, is differentiable, and has the functional values shown in the table.

x	-5	-4	-1	0	2	3	6
$f(x)$	1	3	2	0	-1	-4	0

(a) Determine $f(4)$.

(b) Determine $f(-3)$.

(c) Plot the points and make a possible sketch of the graph of f on the interval $[-6, 6]$. What is the smallest number of critical points in the interval? Explain.

(d) Does there exist at least one real number c in the interval $(-6, 6)$ where $f'(c) = -1$? Explain.

(e) Is it possible that $\lim_{x \to 0} f(x)$ does not exist? Explain.

(f) Is it necessary that $f'(x)$ exists at $x = 2$? Explain.

In Exercises 3 and 4, locate the absolute extrema of the function on the closed interval. Use a graphing utility to graph the function over the indicated interval to confirm your results.

3. $g(x) = 2x + 5 \cos x, \quad [0, 2\pi]$

4. $f(x) = \dfrac{x}{\sqrt{x^2 + 1}}, \quad [0, 2]$

5. *Think About It* Consider the function $f(x) = 3 - |x - 4|$.

(a) Graph the function and verify that $f(1) = f(7)$.

(b) Note that $f'(x)$ is not equal to zero for any x in $[1, 7]$. Explain why this does not contradict Rolle's Theorem.

6. *Think About It* Can the Mean Value Theorem be applied to the function $f(x) = 1/x^2$ on the interval $[-2, 1]$? Explain.

In Exercises 7–10, find the point(s) guaranteed by the Mean Value Theorem for the indicated interval.

7. $f(x) = x^{2/3}$

$1 \le x \le 8$

8. $f(x) = \dfrac{1}{x}$

$1 \le x \le 4$

9. $f(x) = x - \cos x$

$-\dfrac{\pi}{2} \le x \le \dfrac{\pi}{2}$

10. $f(x) = \sqrt{x} - 2x$

$0 \le x \le 4$

11. For the function $f(x) = Ax^2 + Bx + C$, determine the value of c guaranteed by the Mean Value Theorem on the interval $[x_1, x_2]$.

12. Demonstrate the result of Exercise 11 for $f(x) = 2x^2 - 3x + 1$ on the interval $[0, 4]$.

In Exercises 13–16, find the critical numbers (if any) and the open intervals on which the function is increasing or decreasing.

13. $f(x) = (x - 1)^2(x - 3)$

14. $g(x) = (x + 1)^3$

15. $h(x) = \sqrt{x}(x - 3)$

16. $f(x) = \sin x + \cos x, \quad 0 \le x \le 2\pi$

In Exercises 17 and 18, use the First Derivative Test to find any relative extrema of the function. Use a graphing utility to verify your results.

17. $h(t) = \dfrac{1}{4}t^4 - 8t$

18. $g(x) = \dfrac{3}{2} \sin\left(\dfrac{\pi x}{2} - 1\right), \quad [0, 4]$

In Exercises 19 and 20, find any points of inflection of the function.

19. $f(x) = x + \cos x$

$0 \le x \le 2\pi$

20. $f(x) = (x + 2)^2(x - 4)$

In Exercises 21–24, find the limit.

21. $\lim_{x \to \infty} \dfrac{2x^2}{3x^2 + 5}$

22. $\lim_{x \to \infty} \dfrac{2x}{3x^2 + 5}$

23. $\lim_{x \to \infty} \dfrac{5 \cos x}{x}$

24. $\lim_{x \to \infty} \dfrac{3x}{\sqrt{x^2 + 4}}$

In Exercises 25–28, find any vertical and horizontal asymptotes of the graph of the function. Use a graphing utility to verify your results.

25. $h(x) = \dfrac{2x + 3}{x - 4}$

26. $g(x) = \dfrac{5x^2}{x^2 + 2}$

27. $f(x) = \dfrac{3}{x} - 2$

28. $f(x) = \dfrac{3x}{\sqrt{x^2 + 2}}$

In Exercises 29–32, use a graphing utility to graph the function. Use the graph to approximate any relative extrema or asymptotes.

29. $f(x) = x^3 + \dfrac{243}{x}$

30. $f(x) = |x^3 - 3x^2 + 2x|$

31. $f(x) = \dfrac{x - 1}{1 + 3x^2}$

32. $g(x) = \dfrac{\pi^2}{3} - 4 \cos x + \cos 2x$

In Exercises 33–50, make use of domain, range, symmetry, asymptotes, intercepts, relative extrema, and/or points of inflection to obtain an accurate graph of the function.

33. $f(x) = 4x - x^2$

34. $f(x) = 4x^3 - x^4$

35. $f(x) = x\sqrt{16 - x^2}$

36. $f(x) = (x^2 - 4)^2$

37. $f(x) = (x - 1)^3(x - 3)^2$

38. $f(x) = (x - 3)(x + 2)^3$

39. $f(x) = x^{1/3}(x + 3)^{2/3}$

40. $f(x) = (x - 2)^{1/3}(x + 1)^{2/3}$

41. $f(x) = \dfrac{x + 1}{x - 1}$

42. $f(x) = \dfrac{2x}{1 + x^2}$

43. $f(x) = \dfrac{4}{1 + x^2}$

44. $f(x) = \dfrac{x^2}{1 + x^4}$

45. $f(x) = x^3 + x + \dfrac{4}{x}$

46. $f(x) = x^2 + \dfrac{1}{x}$

47. $f(x) = |x^2 - 9|$

48. $f(x) = |x - 1| + |x - 3|$

49. $f(x) = x + \cos x, \qquad 0 \le x \le 2\pi$

50. $f(x) = \dfrac{1}{\pi}(2 \sin \pi x - \sin 2\pi x), \qquad -1 \le x \le 1$

51. *Minimum Distance* At noon, ship A is 100 kilometers due east of ship B. Ship A is sailing west at 12 kilometers per hour, and ship B is sailing south at 10 kilometers per hour. At what time will the ships be nearest to each other, and what will this distance be?

52. *Maximum Area* Find the dimensions of the rectangle of maximum area, with sides parallel to the coordinate axes, that can be inscribed in the ellipse given by

$$\frac{x^2}{144} + \frac{y^2}{16} = 1.$$

53. *Minimum Length* A right triangle in the first quadrant has the coordinate axes as sides, and the hypotenuse passes through the point $(1, 8)$. Find the vertices of the triangle such that the length of the hypotenuse is minimum.

54. *Minimum Length* The wall of a building is to be braced by a beam that must pass over a parallel fence 5 feet high and 4 feet from the building. Find the length of the shortest beam that can be used.

55. *Maximum Area* Three sides of a trapezoid have the same length s. Of all such possible trapezoids, show that the one of maximum area has a fourth side of length $2s$.

56. *Maximum Area* Show that the greatest area of any rectangle inscribed in a triangle is one-half that of the triangle.

57. *Minimum Distance* Find the length of the longest pipe that can be carried level around a right-angle corner at the intersection of two corridors of widths 4 feet and 6 feet. (Do not use trigonometry.)

58. *Minimum Distance* Rework Exercise 57, given corridors of widths a meters and b meters.

59. *Minimum Distance* A hallway of width 6 feet meets a hallway of width 9 feet at right angles. Find the length of the longest pipe that can be carried level around this corner. [*Hint:* If L is the length of the pipe, show that

$$L = 6 \csc \theta + 9 \csc\left(\frac{\pi}{2} - \theta\right)$$

where θ is the angle between the pipe and the wall of the narrower hallway.]

60. *Minimum Distance* Rework Exercise 59, given that one hallway is of width a meters and the other is of width b meters. Show that the result is the same as in Exercise 58.

61. *Harmonic Motion* The height of an object attached to a spring is given by the harmonic equation

$$y = \tfrac{1}{3} \cos 12t - \tfrac{1}{4} \sin 12t$$

where y is measured in inches and t is measured in seconds.

(a) Calculate the height and velocity of the object when $t = \pi/8$ second.

(b) Show that the maximum displacement of the object is $\frac{5}{12}$ inch.

(c) Find the period P of y. Also, find the frequency f (number of oscillations per second) if $f = 1/P$.

62. *Writing* The general equation giving the height of an oscillating object attached to a spring is

$$y = A \sin \sqrt{\frac{k}{m}}\, t + B \cos \sqrt{\frac{k}{m}}\, t$$

where k is the spring constant and m is the mass of the object.

(a) Show that the maximum displacement of the object is $\sqrt{A^2 + B^2}$.

(b) Show that the object oscillates with a frequency of

$$f = \frac{1}{2\pi} \sqrt{\frac{k}{m}}.$$

(c) How is the frequency changed if the stiffness k of the spring is increased?

(d) How is the frequency changed if the mass m of the object is increased?

In Exercises 63 and 64, use Newton's Method to approximate any real zeros of the function accurate to three decimal places. Use the root-finding capabilities of a graphing utility to verify your results.

63. $f(x) = x^3 - 3x - 1$

64. $f(x) = x^3 + 2x + 1$

In Exercises 65 and 66, use Newton's Method to approximate, to three decimal places, the x-value of the points of intersection of the equations. Use a graphing utility to verify your results.

65. $y = x^4$

$\quad\; y = x + 3$

66. $y = \sin \pi x$

$\quad\; y = 1 - x$

In Exercises 67 and 68, find the differential dy.

67. $y = x(1 - \cos x)$ **68.** $y = \sqrt{36 - x^2}$

69. *Modeling Data* Outlays for national defense S (in billions of dollars) for the years 1970 through 1995 are given in the table, where t is the time in years, with $t = 0$ corresponding to 1970.   *(Source: U.S. Office of Management and Budget)*

t	0	1	2	3	4	5	6
S	81.7	78.9	79.2	76.7	79.3	86.5	89.6

t	7	8	9	10	11	12	13
S	97.2	104.5	116.3	134.0	157.5	185.3	209.9

t	14	15	16	17	18	19	20
S	227.4	252.7	273.4	282.0	290.4	303.6	299.3

t	21	22	23	24	25
S	273.3	298.4	291.1	281.6	271.6

(a) Use the regression capabilities of a graphing utility to find a model of the form $S = at^3 + bt^2 + ct + d$ for the data.

(b) Use a graphing utility to plot the data and graph the model.

(c) For the years given in the table, when does the model indicate that the outlay for national defense is at a maximum?

(d) For the years given in the table, when does the model indicate that the outlay for national defense is increasing at the greatest rate?

70. *Modeling Data* Petroleum imports I (in billions of dollars) for the years 1980 through 1994 are given in the table, where t is the time in years, with $t = 0$ corresponding to 1980. *(Source: "Highlights of U.S. Export and Import Trade" and the U.S. Bureau of Census)*

t	0	1	2	3	4
I	77.6	75.6	59.4	52.3	55.9

t	5	6	7	8	9
I	49.6	34.1	41.5	38.8	49.1

t	10	11	12	13	14
I	60.5	50.1	50.4	49.7	49.6

(a) Use the regression capabilities of a graphing utility to find a model of the form

$I = at^4 + bt^3 + ct^2 + dt + e$

for the data.

(b) Use a graphing utility to plot the data and graph the model.

(c) For the years given in the table, when does the model indicate that the cost of petroleum imports is at a minimum?

(d) For the years given in the table, when does the model indicate the cost of petroleum imports is decreasing at the greatest rate?

71. *Surface Area and Volume* The diameter of a sphere is measured to be 18 centimeters with a maximum possible error of 0.05 centimeter. Use the differential to approximate the possible propagated error and percent error in calculating the surface area and the volume of the sphere.

72. *Demand Function* A company finds that the demand for its commodity is

$p = 75 - \frac{1}{4}x.$

If x changes from 7 to 8, find and compare the values of Δp and dp.

73. *Profit* Find the maximum profit if the demand equation is $p = 36 - 4x$ and the total cost is $C = 2x^2 + 6$.

74. *Profit* The cost C of producing x units per day is

$C = \frac{1}{4}x^2 + 62x + 125$

and the price p per unit is

$p = 75 - \frac{1}{3}x.$

(a) What daily output produces maximum profit?

(b) What daily output produces minimum average cost?

(c) Find the elasticity of demand.

75. *Revenue* For groups of 80 or more, a charter bus company determines the rate per person according to the following formula.

Rate $= \$8.00 - \$0.05(n - 80)$, $n \geq 80$

What number of passengers will give the bus company maximum revenue?

76. *Cost* The cost of fuel for running a locomotive is proportional to the $\frac{3}{2}$ power of the speed $(s^{3/2})$ and is \$50 per hour for a speed of 25 miles per hour. Other fixed costs amount to an average of \$100 per hour. Find the speed that will minimize the cost per mile.

77. *Inventory Cost* The cost of inventory depends on the ordering and storage costs according to the inventory model

$C = \left(\frac{Q}{x}\right)s + \left(\frac{x}{2}\right)r.$

Determine the order size that will minimize the cost, assuming that sales occur at a constant rate, Q is the number of units sold per year, r is the cost of storing one unit for 1 year, s is the cost of placing an order, and x is the number of units per order.

78. *Taxes* The demand and cost equations for a certain product are $p = 600 - 3x$ and $C = 0.3x^2 + 6x + 600$, where p is the price per unit, x is the number of units, and C is the cost of producing x units. If t is the excise tax per unit, the profit for producing x units is $P = xp - C - xt$. Find the maximum profit for

(a) $t = 5$,   (b) $t = 10$,   and   (c) $t = 20$.

79. Find the maximum and minimum points on the graph of

$$x^2 + 4y^2 - 2x - 16y + 13 = 0$$

(a) without using calculus.

(b) using calculus.

80. Consider the function $f(x) = x^n$ for positive integer values of n.

(a) For what values of n does the function have a relative minimum at the origin?

(b) For what values of n does the function have a point of inflection at the origin?

True or False? **In Exercises 81 and 82, determine whether the statement is true or false. If it is false, explain why or give an example that shows it is false.**

81. If c is a critical number of f, then f has a relative extremum at $x = c$.

82. The function $y = f(x)$ can have at most one horizontal asymptote.

83. *Writing* A newspaper headline states that "The rate of growth of the national deficit is decreasing." What does this mean? What does it imply about the graph of the deficit as a function of time?

84. *Modeling Data* The manager of a store recorded the annual sales S (in thousands of dollars) of a product over a period of 7 years, as shown in the table, where t is the time in years, with $t = 1$ corresponding to 1991.

t	1	2	3	4	5	6	7
S	5.4	6.9	11.5	15.5	19.0	22.0	23.6

(a) Use the regression capabilities of a graphing utility to find a model of the form $S = at^3 + bt^2 + ct + d$ for the data.

(b) Use a graphing utility to plot the data and graph the model.

(c) Use calculus to find the time t when sales were increasing at the greatest rate.

(d) Do you think the model would be accurate for predicting future sales? Explain.

SECTION PROJECT

Whenever the Connecticut River reaches a level of 105 feet above sea level, two Northampton, Massachusetts flood control station operators begin a round-the-clock river watch. Every two hours, they check the height of the river, using a scale marked off in tenths of a foot, and record the data in a log book. In the spring of 1996, the flood watch lasted from April 4, when the river reached 105 feet and was rising at 0.2 foot per hour, until April 25, when the level subsided again to 105 feet. Between those dates, their log shows that the river rose and fell several times, at one point coming close to the 115-foot mark. If the river had reached 115 feet, the city would have closed down Mount Tom Road (Route 5, south of Northampton).

The graph below shows the *rate of change* of the level of the river during one portion of the flood watch. Use the graph to answer the following questions.

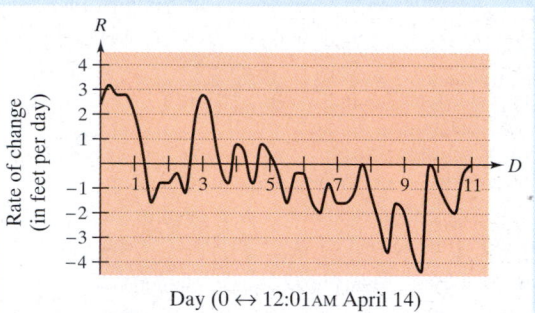

Day ($0 \leftrightarrow$ 12:01AM April 14)

(a) On what date was the river rising most rapidly? How do you know?

(b) On what date was the river falling most rapidly? How do you know?

(c) There were two dates in a row on which the river rose, then fell, then rose again during the course of the day. On which days did this occur, and how do you know?

(d) At one minute past midnight, April 14, the river level was 111.0 feet. Estimate its height 24 hours later and 48 hours later. Explain how you made your estimates.

(e) The river crested at 114.4 feet. On what date do you think this occurred?

(Submitted by Mary Murphy, Smith College, Northampton, MA)

4

Integration

The Four-Stroke Rotary Cycle

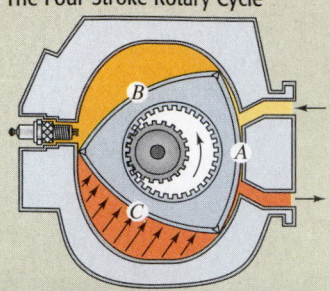

A sweeps out exhaust, B begins compression, and C is nearly finished with expansion.

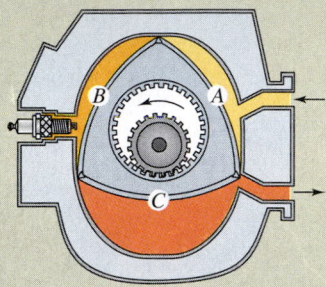

A moves back to allow intake while B continues compression. C begins to push out exhaust.

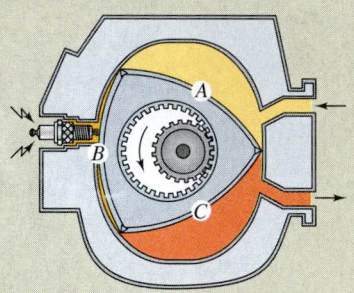

Ignition takes place in B, A continues intake, and C continues exhaust.

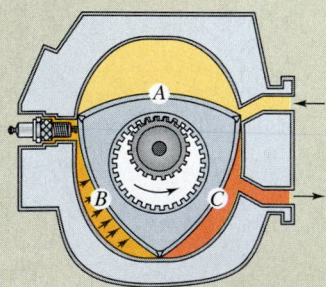

Intake is nearly complete in A, B expands following ignition, and C is nearly finished with exhaust.

Named for Felix Wankel, who developed its basic principles in the 1950s, the Wankel rotary engine presents an alternative to the piston engine commonly used in automobiles. Many auto makers, including Mercedes-Benz, Citroen, and Ford, have experimented with rotary engines. By far the greatest number of Wankel-powered vehicles have been put on the road by Mazda, whose current rotary engine design is the RX-7.

The Wankel rotary engine has several advantages over the piston engine. A rotary engine is approximately half the size and weight of a piston engine of equivalent power. Compared with about 97 major moving parts in a V-8 engine, the typical two-rotor rotary engine has only three major moving parts. As a result, the Wankel engine has lower labor and material costs and less internal energy waste.

Although many different designs are possible for the rotary engine, the most common configuration is a two-lobed housing containing a three-sided rotor. The size of the rotor in comparison with the size of the housing cavity is critical in determining the compression ratio and thus the combustion efficiency.

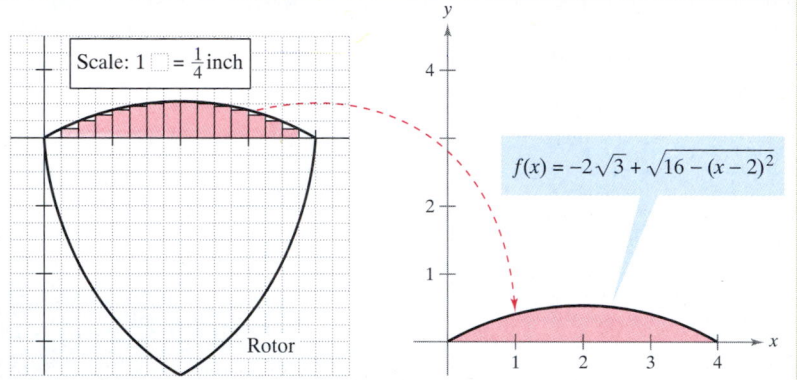

The rotor in a Wankel engine is shaped like a slightly bulged equilateral triangle. The area of this "triangle" is used by engineers to optimize the size of the combustion chamber.

QUESTIONS

1. The region shown in the figure on the right is bounded above by the graph of

$$f(x) = -2\sqrt{3} + \sqrt{16 - (x - 2)^2}$$

and below by the x-axis. Describe different ways in which you might approximate the area of the region. Then choose one of the ways and use it to obtain an approximation. What type of accuracy do you think your approximation has?

2. Now that you have found one approximation for the area of the region, describe a way that you can improve your approximation. Does your strategy allow you to obtain an approximation that is arbitrarily close to the actual area? Explain.

3. Use your approximation to estimate the area of the "bulged triangle" shown in the figure on the left.

The concepts presented here will be explored further in this chapter. For an extension of this application, see the lab series that accompanies this text.

Antiderivatives • Notation for Antiderivatives •
Basic Integration Rules • Initial Conditions and Particular Solutions

Antiderivatives

Suppose you were asked to find a function F whose derivative is

$$f(x) = 3x^2.$$

From your knowledge of derivatives, you would probably say that

$$F(x) = x^3 \text{ because } \frac{d}{dx}[x^3] = 3x^2.$$

The function F is an *antiderivative* of f. In general, a function F is an **antiderivative** of f on an interval I if $F'(x) = f(x)$ for all x in I.

Note that F is called *an* antiderivative of f, rather than *the* antiderivative of f. To see why, observe that

$$F_1(x) = x^3, \ F_2(x) = x^3 - 5, \text{ and } F_3(x) = x^3 + 97$$

are all antiderivatives of $f(x) = 3x^2$. In fact, for any constant C, the function given by $F(x) = x^3 + C$ is an antiderivative of f.

> **THEOREM 4.1   Representation of Antiderivatives**
>
> If F is an antiderivative of f on an interval I, then G is an antiderivative of f on the interval I if and only if G is of the form
>
> $$G(x) = F(x) + C, \text{ for all } x \text{ in } I$$
>
> where C is a constant.

Proof   The proof of one direction is straightforward. That is, if $G(x) = F(x) + C$, $F'(x) = f(x)$, and C is a constant, then

$$G'(x) = \frac{d}{dx}[F(x) + C] = F'(x) + 0 = f(x).$$

To prove the other direction, you can define a function H such that

$$H(x) = G(x) - F(x).$$

If H is not constant on the interval I, there must exist a and b ($a < b$) in the interval such that $H(a) \neq H(b)$. Moreover, because H is differentiable on (a, b), you can apply the Mean Value Theorem to conclude that there exists some c in (a, b) such that

$$H'(c) = \frac{H(b) - H(a)}{b - a}.$$

Because $H(b) \neq H(a)$, it follows that $H'(c) \neq 0$. However, because $G'(c) = F'(c)$, you know that $H'(c) = G'(c) - F'(c) = 0$, which contradicts the fact that $H'(c) \neq 0$. Consequently, you can conclude that $H(x)$ is a constant, C. Therefore, $G(x) - F(x) = C$ and it follows that $G(x) = F(x) + C$.

Using Theorem 4.1, you can represent the entire family of antiderivatives of a function by adding a constant to a *known* antiderivative. For example, knowing that $D_x[x^2] = 2x$, you can represent the family of *all* antiderivatives of $f(x) = 2x$ by

$$G(x) = x^2 + C \qquad \text{Family of all antiderivatives of } f(x) = 2x$$

where C is a constant. The constant C is called the **constant of integration**. The family of functions represented by G is the **general antiderivative** of f, and $G(x) = x^2 + C$ is the **general solution** of the *differential equation*

$$G'(x) = 2x. \qquad \text{Differential equation}$$

A **differential equation** in x and y is an equation that involves x, y, and derivatives of y. For instance, $y' = 3x$ and $y' = x^2 + 1$ are examples of differential equations.

EXAMPLE 1   Solving a Differential Equation

Find the general solution of the differential equation $y' = 2$.

Solution   To begin, you need to find a function whose derivative is 2. One such function is

$$y = 2x.$$

Now, you can use Theorem 4.1 to conclude that the general solution of the differential equation is

$$y = 2x + C.$$

The graphs of several functions of the form $y = 2x + C$ are shown in Figure 4.1.

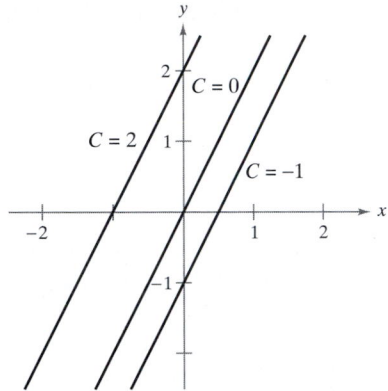

Functions of the form $y = 2x + C$

Figure 4.1

Notation for Antiderivatives

When solving a differential equation of the form

$$\frac{dy}{dx} = f(x)$$

it is convenient to write it in the equivalent differential form

$$dy = f(x)\,dx.$$

The operation of finding all solutions of this equation is called **antidifferentiation** (or **indefinite integration**) and is denoted by an integral sign $\int$. The general solution is denoted by

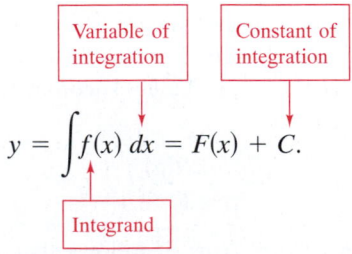

$$y = \int f(x)\,dx = F(x) + C.$$

The expression $\int f(x)\,dx$ is read as the *antiderivative of f with respect to x*. Thus, the differential dx serves to identify x as the variable of integration. The term **indefinite integral** is a synonym for antiderivative.

In this text, whenever we write $\int f(x)\,dx = F(x) + C$, we mean that F is an antiderivative of f *on an interval*.

Basic Integration Rules

The inverse nature of integration and differentiation can be verified by substituting $F'(x)$ for $f(x)$ in the indefinite integration definition to obtain

$$\int F'(x)\,dx = F(x) + C.$$

Integration is the "inverse" of differentiation.

Moreover, if $\int f(x)\,dx = F(x) + C$, then

$$\frac{d}{dx}\left[\int f(x)\,dx\right] = f(x).$$

Differentiation is the "inverse" of integration.

These two equations allow you to obtain integration formulas directly from differentiation formulas, as shown in the following summary.

Basic Integration Rules

Differentiation Formula	*Integration Formula*
$\dfrac{d}{dx}[C] = 0$	$\displaystyle\int 0\,dx = C$
$\dfrac{d}{dx}[kx] = k$	$\displaystyle\int k\,dx = kx + C$
$\dfrac{d}{dx}[kf(x)] = kf'(x)$	$\displaystyle\int kf(x)\,dx = k\int f(x)\,dx$
$\dfrac{d}{dx}[f(x) \pm g(x)] = f'(x) \pm g'(x)$	$\displaystyle\int [f(x) \pm g(x)]\,dx = \int f(x)\,dx \pm \int g(x)\,dx$
$\dfrac{d}{dx}[x^n] = nx^{n-1}$	$\displaystyle\int x^n\,dx = \frac{x^{n+1}}{n+1} + C, \quad n \neq -1$ **Power Rule**
$\dfrac{d}{dx}[\sin x] = \cos x$	$\displaystyle\int \cos x\,dx = \sin x + C$
$\dfrac{d}{dx}[\cos x] = -\sin x$	$\displaystyle\int \sin x\,dx = -\cos x + C$
$\dfrac{d}{dx}[\tan x] = \sec^2 x$	$\displaystyle\int \sec^2 x\,dx = \tan x + C$
$\dfrac{d}{dx}[\sec x] = \sec x \tan x$	$\displaystyle\int \sec x \tan x\,dx = \sec x + C$
$\dfrac{d}{dx}[\cot x] = -\csc^2 x$	$\displaystyle\int \csc^2 x\,dx = -\cot x + C$
$\dfrac{d}{dx}[\csc x] = -\csc x \cot x$	

NOTE   Note that the Power Rule for integration has the restriction that $n \neq -1$. The evaluation of $\int 1/x\,dx$ must wait until the introduction of the natural logarithm function in Chapter 5.

EXAMPLE 2 **Applying the Basic Integration Rules**

Describe the antiderivatives of $3x$.

Solution

$$\int 3x \, dx = 3 \int x \, dx \qquad \text{Constant Multiple Rule}$$

$$= 3 \int x^1 \, dx \qquad \text{Rewrite } (x = x^1).$$

$$= 3 \left(\frac{x^2}{2} \right) + C \qquad \text{Power Rule } (n = 1)$$

$$= \frac{3}{2} x^2 + C \qquad \text{Simplify.}$$

When indefinite integrals are evaluated, a strict application of the basic integration rules tends to produce complicated constants of integration. For instance, in Example 2, we could have written

$$\int 3x \, dx = 3 \int x \, dx$$

$$= 3 \left(\frac{x^2}{2} + C \right)$$

$$= \frac{3}{2} x^2 + 3C.$$

However, because C represents *any* constant, it is both cumbersome and unnecessary to write $3C$ as the constant of integration, and we choose the simpler form, $\frac{3}{2}x^2 + C$.

In Example 2, note that the general pattern of integration is similar to that of differentiation.

Original integral ⇒ Rewrite ⇒ Integrate ⇒ Simplify

EXAMPLE 3 **Rewriting Before Integrating**

Original Integral	*Rewrite*	*Integrate*	*Simplify*
a. $\displaystyle\int \frac{1}{x^3} \, dx$	$\displaystyle\int x^{-3} \, dx$	$\dfrac{x^{-2}}{-2} + C$	$-\dfrac{1}{2x^2} + C$
b. $\displaystyle\int \sqrt{x} \, dx$	$\displaystyle\int x^{1/2} \, dx$	$\dfrac{x^{3/2}}{3/2} + C$	$\dfrac{2}{3} x^{3/2} + C$
c. $\displaystyle\int 2 \sin x \, dx$	$2 \displaystyle\int \sin x \, dx$	$2(-\cos x) + C$	$-2 \cos x + C$

Remember that you can check your answer to an antidifferentiation problem by differentiating. For instance, in Example 3b, you can check that $\frac{2}{3} x^{3/2} + C$ is the correct antiderivative by differentiating the answer to obtain

$$D_x \left[\frac{2}{3} x^{3/2} + C \right] = \left(\frac{2}{3} \right) \left(\frac{3}{2} \right) x^{1/2} = \sqrt{x}.$$

The basic integration rules listed earlier in this section allow you to integrate *any* polynomial function, as demonstrated in Example 4.

EXAMPLE 4   Integrating Polynomial Functions

a. $\int dx = \int 1 \, dx$

$ = x + C$

b. $\int (x + 2) \, dx = \int x \, dx + \int 2 \, dx$

$ = \dfrac{x^2}{2} + C_1 + 2x + C_2$

$ = \dfrac{x^2}{2} + 2x + C \qquad\qquad C = C_1 + C_2$

The second line in the solution is usually omitted.

c. $\int (3x^4 - 5x^2 + x) \, dx = 3\left(\dfrac{x^5}{5}\right) - 5\left(\dfrac{x^3}{3}\right) + \dfrac{x^2}{2} + C$

$ = \dfrac{3}{5}x^5 - \dfrac{5}{3}x^3 + \dfrac{1}{2}x^2 + C$

EXAMPLE 5   Rewriting Before Integrating

$\int \dfrac{x + 1}{\sqrt{x}} \, dx = \int \left(\dfrac{x}{\sqrt{x}} + \dfrac{1}{\sqrt{x}}\right) dx$ Rewrite as two fractions.

$ = \int (x^{1/2} + x^{-1/2}) \, dx$ Rewrite with fractional exponents.

$ = \dfrac{x^{3/2}}{3/2} + \dfrac{x^{1/2}}{1/2} + C$ Integrate.

$ = \dfrac{2}{3}x^{3/2} + 2x^{1/2} + C$ Simplify.

NOTE   When integrating quotients, do not integrate the numerator and denominator separately. This is no more valid in integration than it is in differentiation. For instance, in Example 5, be sure you understand that

$\int \dfrac{x + 1}{\sqrt{x}} \, dx \neq \dfrac{\int (x + 1) \, dx}{\int \sqrt{x} \, dx}.$

EXAMPLE 6   Rewriting Before Integrating

$\int \dfrac{\sin x}{\cos^2 x} \, dx = \int \left(\dfrac{1}{\cos x}\right)\left(\dfrac{\sin x}{\cos x}\right) dx$ Rewrite as a product.

$ = \int \sec x \tan x \, dx$ Rewrite using trigonometric identities.

$ = \sec x + C$ Integrate.

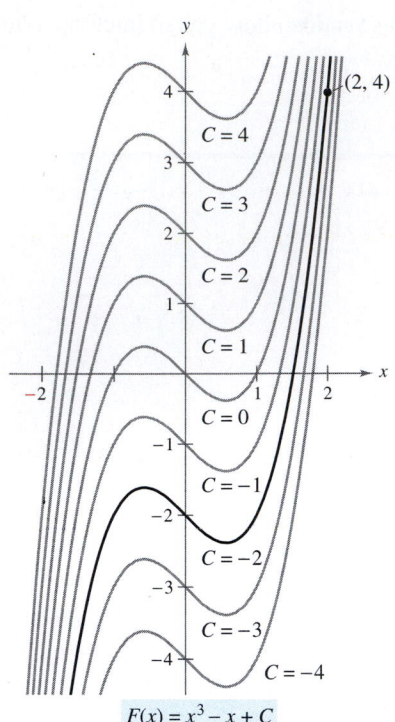

The particular solution that satisfies
the initial condition $F(2) = 4$ is
$F(x) = x^3 - x - 2$.
Figure 4.2

Initial Conditions and Particular Solutions

You have already seen that the equation $y = \int f(x)\,dx$ has many solutions (each differing from the others by a constant). This means that the graphs of any two anti-derivatives of f are vertical translations of each other. For example, Figure 4.2 shows the graphs of several antiderivatives of the form

$$y = \int (3x^2 - 1)\,dx = x^3 - x + C \qquad \text{General solution}$$

for various integer values of C. Each of these antiderivatives is a solution of the differential equation

$$\frac{dy}{dx} = 3x^2 - 1.$$

In many applications of integration, you are given enough information to determine a **particular solution**. To do this you need only know the value of $y = F(x)$ for one value of x. (This information is called an **initial condition**.) For example, in Figure 4.2, only one curve passes through the point $(2, 4)$. To find this curve, you can use the following information.

$$F(x) = x^3 - x + C \qquad \text{General solution}$$
$$F(2) = 4 \qquad \text{Initial condition}$$

By using the initial condition in the general solution, you can determine that $F(2) = 8 - 2 + C = 4$, which implies that $C = -2$. Thus, you obtain

$$F(x) = x^3 - x - 2. \qquad \text{Particular solution}$$

EXAMPLE 7   Finding a Particular Solution

Find the general solution of

$$F'(x) = \frac{1}{x^2}, \quad x > 0$$

and find the particular solution that satisfies the initial condition $F(1) = 0$.

Solution   To find the general solution, integrate to obtain

$$F(x) = \int \frac{1}{x^2}\,dx$$

$$= \int x^{-2}\,dx$$

$$= \frac{x^{-1}}{-1} + C$$

$$= -\frac{1}{x} + C. \qquad \text{General solution}$$

Using the initial condition $F(1) = 0$, you can solve for C as follows.

$$F(1) = -\frac{1}{1} + C = 0 \quad \Longrightarrow \quad C = 1$$

Thus, the particular solution, as shown in Figure 4.3, is

$$F(x) = -\frac{1}{x} + 1, \quad x > 0. \qquad \text{Particular solution}$$

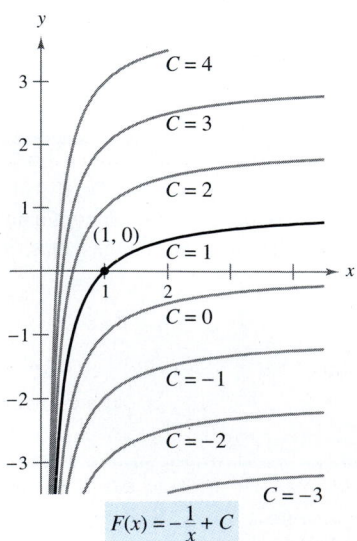

The particular solution that satisfies
the initial condition $F(1) = 0$ is
$F(x) = -(1/x) + 1, x > 0$.
Figure 4.3

So far in this section we have been using x as the variable of integration. In applications, it is often convenient to use a different variable. For instance, in the following example involving *time*, the variable of integration is t.

EXAMPLE 8   Solving a Vertical Motion Problem

A ball is thrown upward with an initial velocity of 64 feet per second from an initial height of 80 feet, as shown in Figure 4.4.

a. Find the position function giving the height s as a function of the time t.

b. When does the ball hit the ground?

Solution

a. Let $t = 0$ represent the initial time. The two given initial conditions can be written as follows.

$$s(0) = 80 \qquad \text{Initial height is 80 feet.}$$
$$s'(0) = 64 \qquad \text{Initial velocity is 64 feet per second.}$$

Using -32 feet per second per second as the acceleration due to gravity, you can write

$$s''(t) = -32$$
$$s'(t) = \int s''(t)\, dt = \int -32\, dt = -32t + C_1.$$

Using the initial velocity, you obtain $s'(0) = 64 = -32(0) + C_1$, which implies that $C_1 = 64$. Next, by integrating $s'(t)$, you obtain

$$s(t) = \int s'(t)\, dt = \int (-32t + 64)\, dt = -16t^2 + 64t + C_2$$

and using the initial height, you obtain

$$s(0) = 80 = -16(0^2) + 64(0) + C_2$$

which implies that $C_2 = 80$. Therefore, the position function is

$$s(t) = -16t^2 + 64t + 80.$$

b. Using the position function found in part (a), you can find the time that the ball hits the ground by solving the equation $s(t) = 0$.

$$s(t) = -16t^2 + 64t + 80 = 0$$
$$-16(t + 1)(t - 5) = 0$$
$$t = -1, 5$$

Because t must be positive, you can conclude that the ball hits the ground 5 seconds after it was thrown.

Example 8 shows how to use calculus to analyze vertical motion problems in which the acceleration is determined by a gravitational force. You can use a similar strategy to analyze other linear motion problems (vertical or horizontal) in which the acceleration (or deceleration) is the result of some other force, as you can see in Exercises 67–76.

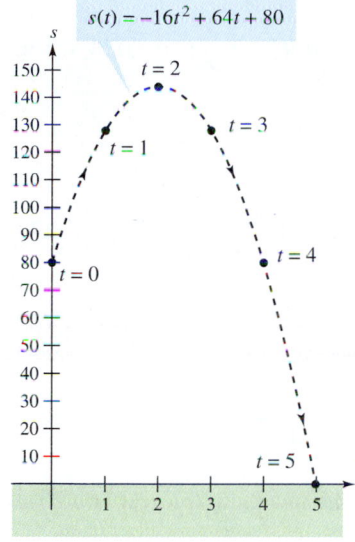

$s(t) = -16t^2 + 64t + 80$

Position of a ball at time t
Figure 4.4

NOTE   In Example 8, note that the position function has the form

$$s(t) = \tfrac{1}{2} g t^2 + v_0 t + s_0,$$

where $g = -32$, v_0 is the initial velocity, and s_0 is the initial height, as presented in Section 2.2.

Before you begin the exercise set, be sure you realize that one of the most important steps in integration is *rewriting the integrand* in a form that fits the basic integration rules. To further illustrate this point, here are some additional examples.

Original Integral	Rewrite	Integrate	Simplify
$\displaystyle\int \frac{2}{\sqrt{x}}\,dx$	$\displaystyle 2\int x^{-1/2}\,dx$	$\displaystyle 2\left(\frac{x^{1/2}}{1/2}\right)+C$	$4x^{1/2}+C$
$\displaystyle\int (t^2+1)^2\,dt$	$\displaystyle\int (t^4+2t^2+1)\,dt$	$\displaystyle\frac{t^5}{5}+2\left(\frac{t^3}{3}\right)+t+C$	$\displaystyle\frac{1}{5}t^5+\frac{2}{3}t^3+t+C$
$\displaystyle\int \frac{x^3+3}{x^2}\,dx$	$\displaystyle\int (x+3x^{-2})\,dx$	$\displaystyle\frac{x^2}{2}+3\left(\frac{x^{-1}}{-1}\right)+C$	$\displaystyle\frac{1}{2}x^2-\frac{3}{x}+C$
$\displaystyle\int \sqrt[3]{x}(x-4)\,dx$	$\displaystyle\int (x^{4/3}-4x^{1/3})\,dx$	$\displaystyle\frac{x^{7/3}}{7/3}-4\left(\frac{x^{4/3}}{4/3}\right)+C$	$\displaystyle\frac{3}{7}x^{4/3}(x-7)+C$

EXERCISES FOR SECTION 4.1

In Exercises 1–4, verify the statement by showing that the derivative of the right side is equal to the integrand of the left side.

1. $\displaystyle\int\left(-\frac{9}{x^4}\right)dx = \frac{3}{x^3}+C$

2. $\displaystyle\int\left(4x^3-\frac{1}{x^2}\right)dx = x^4+\frac{1}{x}+C$

3. $\displaystyle\int (x-2)(x+2)\,dx = \frac{1}{3}x^3-4x+C$

4. $\displaystyle\int\frac{x^2-1}{x^{3/2}}\,dx = \frac{2(x^2+3)}{3\sqrt{x}}+C$

In Exercises 5–10, complete the table using the examples at the top of this page as a model.

Original Integral	Rewrite	Integrate	Simplify
5. $\displaystyle\int \sqrt[3]{x}\,dx$			
6. $\displaystyle\int \frac{1}{x^2}\,dx$			
7. $\displaystyle\int \frac{1}{x\sqrt{x}}\,dx$			
8. $\displaystyle\int x(x^2+3)\,dx$			
9. $\displaystyle\int \frac{1}{2x^3}\,dx$			
10. $\displaystyle\int \frac{1}{(2x)^3}\,dx$			

In Exercises 11–14, find the general solution of the differential equation and check the result by differentiation.

11. $\displaystyle\frac{dy}{dt}=3t^2$

12. $\displaystyle\frac{dr}{d\theta}=\pi$

13. $\displaystyle\frac{dy}{dx}=x^{3/2}$

14. $\displaystyle\frac{dy}{dx}=3x^{-4}$

In Exercises 15–30, evaluate the indefinite integral and check the result by differentiation.

15. $\displaystyle\int (x^3+2)\,dx$

16. $\displaystyle\int (x^2-2x+3)\,dx$

17. $\displaystyle\int (x^{3/2}+2x+1)\,dx$

18. $\displaystyle\int\left(\sqrt{x}+\frac{1}{2\sqrt{x}}\right)dx$

19. $\displaystyle\int \sqrt[3]{x^2}\,dx$

20. $\displaystyle\int \left(\sqrt[4]{x^3}+1\right)dx$

21. $\displaystyle\int \frac{1}{x^3}\,dx$

22. $\displaystyle\int \frac{1}{x^4}\,dx$

23. $\displaystyle\int \frac{x^2+x+1}{\sqrt{x}}\,dx$

24. $\displaystyle\int \frac{x^2+1}{x^2}\,dx$

25. $\displaystyle\int (x+1)(3x-2)\,dx$

26. $\displaystyle\int (2t^2-1)^2\,dt$

27. $\displaystyle\int y^2\sqrt{y}\,dy$

28. $\displaystyle\int (1+3t)t^2\,dt$

29. $\displaystyle\int dx$

30. $\displaystyle\int 3\,dt$

In Exercises 31–38, evaluate the trigonometric integral and check the result by differentiation.

31. $\displaystyle\int (2\sin x+3\cos x)\,dx$

32. $\displaystyle\int (t^2-\sin t)\,dt$

33. $\displaystyle\int (1-\csc t\cot t)\,dt$

34. $\displaystyle\int (\theta^2+\sec^2\theta)\,d\theta$

35. $\displaystyle\int (\sec^2\theta-\sin\theta)\,d\theta$

36. $\displaystyle\int \sec y(\tan y-\sec y)\,dy$

37. $\displaystyle\int (\tan^2 y+1)\,dy$

38. $\displaystyle\int \frac{\sin x}{1-\sin^2 x}\,dx$

In Exercises 39 and 40, sketch the graphs of the function $g(x) = f(x) + C$ for $C = -2$, $C = 0$, and $C = 3$ on the same set of coordinate axes.

39. $f(x) = \cos x$

40. $f(x) = \sqrt{x}$

In Exercises 41–44, the graph of the derivative of a function is given. Sketch the graphs of *two* functions that have the given derivative. (There is more than one correct answer.)

41.
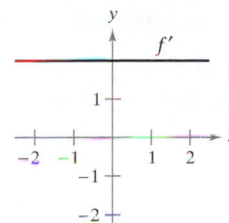

42.
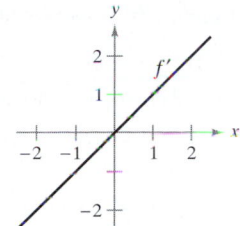

43.
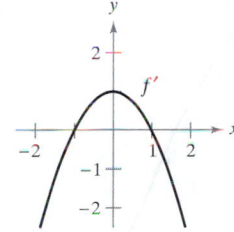

44.
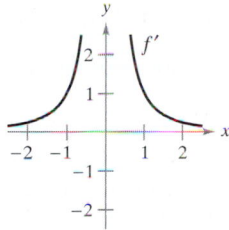

In Exercises 45–48, find the equation for y, given the derivative and the indicated point on the curve.

45. $\dfrac{dy}{dx} = 2x - 1$

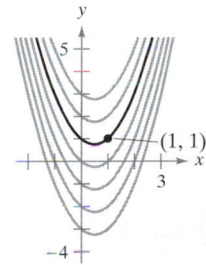

46. $\dfrac{dy}{dx} = 2(x - 1)$

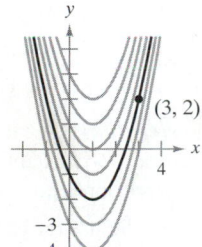

47. $\dfrac{dy}{dx} = \cos x$

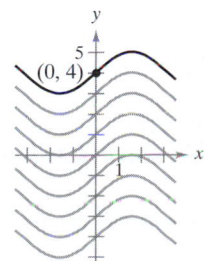

48. $\dfrac{dy}{dx} = -\dfrac{1}{x^2}$, $x > 0$

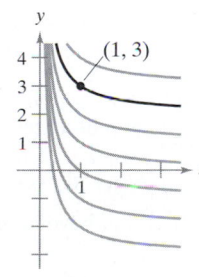

Direction Fields   In Exercises 49 and 50, a differential equation, a point, and a direction field are given. A *direction field* consists of line segments with slopes given by the differential equation. These line segments give a visual perspective of the directions of the solutions of the differential equation. (a) Sketch two approximate solutions of the differential equation on the direction field, one of which passes through the indicated point. (b) Use integration to find the particular solution of the differential equation and use a graphing utility to graph the solution. Compare the result with the sketches in part (a).

49. $\dfrac{dy}{dx} = \dfrac{1}{2}x - 1$, $(4, 2)$

50. $\dfrac{dy}{dx} = x^2 - 1$, $(-1, 3)$

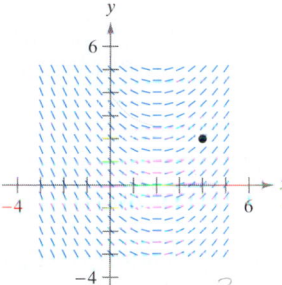

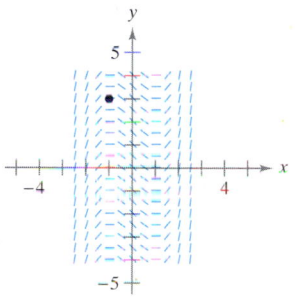

In Exercises 51–54, solve the differential equation.

51. $f''(x) = 2$, $f'(2) = 5$, $f(2) = 10$

52. $f''(x) = x^2$, $f'(0) = 6$, $f(0) = 3$

53. $f''(x) = x^{-3/2}$, $f'(4) = 2$, $f(0) = 0$

54. $f''(x) = \sin x$, $f'(0) = 1$, $f(0) = 6$

55. *Tree Growth*   An evergreen nursery usually sells a certain shrub after 6 years of growth and shaping. The growth rate during those 6 years is approximated by

$$\frac{dh}{dt} = 1.5t + 5$$

where t is the time in years and h is the height in centimeters. The seedlings are 12 centimeters tall when planted ($t = 0$).

(a) Find the height after t years.

(b) How tall are the shrubs when they are sold?

56. *Population Growth*   The rate of growth dP/dt of a population of bacteria is proportional to the square root of t, where P is the population size and t is the time in days ($0 \le t \le 10$). That is,

$$\frac{dP}{dt} = k\sqrt{t}.$$

The initial size of the population is 500. After 1 day, the population has grown to 600. Estimate the population after 7 days.

Vertical Motion In Exercises 57–60, use $a(t) = -32$ feet per second per second as the acceleration due to gravity. (Neglect air resistance.)

57. A ball is thrown vertically upward from the ground with an initial velocity of 60 feet per second. How high will the ball go?

58. Show that the height above the ground of an object thrown upward from a point s_0 feet above the ground with an initial velocity of v_0 feet per second is given by the function

    $$f(t) = -16t^2 + v_0 t + s_0.$$

59. With what initial velocity must an object be thrown upward (from ground level) to reach the top of the Washington Monument (approximately 550 feet)?

60. A balloon, rising vertically with a velocity of 16 feet per second, releases a sandbag at the instant it is 64 feet above the ground.

    (a) How many seconds after its release will the bag strike the ground?

    (b) At what velocity will it hit the ground?

Vertical Motion In Exercises 61–64, use $a(t) = -9.8$ meters per second per second as the acceleration due to gravity. (Neglect air resistance.)

61. Show that the height above the ground of an object thrown upward from a point s_0 meters above the ground with an initial velocity of v_0 meters per second is given by the function

    $$f(t) = -4.9t^2 + v_0 t + s_0.$$

62. *Grand Canyon* The Grand Canyon is 1600 meters deep at its deepest point. A rock is dropped from the rim above this point. Express the height of the rock as a function of the time t in seconds. How long will it take the rock to hit the canyon floor?

63. A baseball is thrown upward from ground level with a velocity of 10 meters per second. Determine its maximum height.

64. With what initial velocity must an object be thrown upward (from ground level) to reach a maximum height of 200 meters?

65. *Lunar Gravity* On the moon, the acceleration due to gravity is -1.6 meters per second per second. A stone is dropped from a cliff on the moon and hits the surface of the moon 20 seconds later. How far did it fall? What was its velocity at impact?

66. *Escape Velocity* The minimum velocity required for an object to escape earth's gravitational pull is obtained from the solution of the equation

    $$\int v \, dv = -GM \int \frac{1}{y^2} \, dy$$

    where v is the velocity of the object projected from the earth, y is the distance from the center of the earth, G is the gravitational constant, and M is the mass of the earth.

Show that v and y are related by the equation

$$v^2 = v_0^2 + 2GM\left(\frac{1}{y} - \frac{1}{R}\right)$$

where v_0 is the initial velocity of the object and R is the radius of the earth.

Rectilinear Motion In Exercises 67–70, consider a particle moving along the x-axis where $x(t)$ is the position of the particle at time t, $x'(t)$ is its velocity, and $x''(t)$ is its acceleration.

67. $x(t) = t^3 - 6t^2 + 9t - 2,$ $0 \le t \le 5$

    (a) Find the velocity and acceleration of the particle.

    (b) Find the open t-intervals on which the particle is moving to the right.

    (c) Find the velocity of the particle when the acceleration is 0.

68. Repeat Exercise 67 for the position function

    $$x(t) = (t - 1)(t - 3)^2, \quad 0 \le t \le 5.$$

69. A particle moves along the x-axis at a velocity of $v(t) = 1/\sqrt{t}$, $t > 0$. At time $t = 1$, its position is $x = 4$. Find the acceleration and position functions for the particle.

70. A particle, initially at rest, moves along the x-axis such that its acceleration at time $t > 0$ is given by $a(t) = \cos t$. At the time $t = 0$, its position is $x = 3$.

    (a) Find the velocity and position functions for the particle.

    (b) Find the values of t for which the particle is at rest.

71. *Acceleration* The maker of a certain automobile advertises that it takes 13 seconds to accelerate from 25 kilometers per hour to 80 kilometers per hour. Assuming constant acceleration, compute the following.

    (a) The acceleration in meters per second per second

    (b) The distance the car travels during the 13 seconds

72. *Deceleration* A car traveling at 45 miles per hour is brought to a stop, at constant deceleration, 132 feet from where the brakes are applied.

    (a) How far has the car moved when its speed has been reduced to 30 miles per hour?

    (b) How far has the car moved when its speed has been reduced to 15 miles per hour?

    (c) Draw the real number line from 0 to 132, and plot the points found in parts (a) and (b). What can you conclude?

73. *Acceleration* At the instant the traffic light turns green, a car that has been waiting at an intersection starts with a constant acceleration of 6 feet per second per second. At the same instant, a truck traveling with a constant velocity of 30 feet per second passes the car.

    (a) How far beyond its starting point will the car pass the truck?

    (b) How fast will the car be traveling when it passes the truck?

74. *Think About It* Two cars starting from rest accelerate to 65 miles per hour in 30 seconds. The velocity of each car is shown in the figure. Are the cars side by side at the end of the 30-second time interval? Explain.

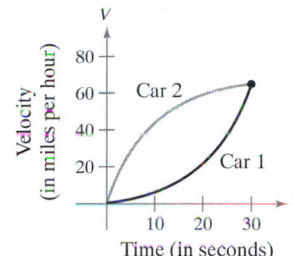

75. *Acceleration* Assume that a fully loaded plane starting from rest has a constant acceleration while moving down a runway. The plane requires 0.7 mile of runway and a speed of 160 miles per hour in order to lift off. What is the plane's acceleration?

76. *Airplane Separation* Two airplanes are in a straight-line landing pattern and, according to FAA regulations, must keep at least a 3-mile separation. Airplane A is 10 miles from touchdown and is gradually slowing its speed from 150 miles per hour to a landing speed of 100 miles per hour. Airplane B is 17 miles from touchdown and is gradually slowing its speed from 250 miles per hour to a landing speed of 115 miles per hour.

(a) Assuming the deceleration of each airplane is constant, find the position functions s_1 and s_2 for airplane A and airplane B. Let $t = 0$ represent the times when the airplanes are 10 and 17 miles from the airport.

(b) Use a graphing utility to graph the position functions.

(c) Find a formula for the magnitude of the distance d between the two airplanes as a function of t. Use a graphing utility to graph d. Is $d < 3$ for some time prior to the landing of airplane A? If so, find that time.

(d) If the airplanes do not keep the required separation, determine how much the 250 miles per hour speed for airplane B should be reduced in order to meet FAA requirements.

Marginal Cost **In Exercises 77 and 78, find the cost function and average cost for the given marginal cost and fixed cost (cost when $x = 0$).**

	Marginal Cost	Fixed Cost
77.	$\dfrac{dC}{dx} = 2x - 12$	$50
78.	$\dfrac{dC}{dx} = \dfrac{\sqrt[4]{x}}{10} + 10$	$2300

In Exercises 79 and 80, find the revenue and demand functions for the given marginal revenue.

79. $\dfrac{dR}{dx} = 100 - 5x$ **80.** $\dfrac{dR}{dx} = 100 - 6x - 2x^2$

True or False? **In Exercises 81–84, determine whether the statement is true or false. If it is false, explain why or give an example that shows it is false.**

81. Each antiderivative of an nth-degree polynomial function is an $(n + 1)$st-degree polynomial function.

82. If $p(x)$ is a polynomial function, then p has exactly one antiderivative whose graph contains the origin.

83. If $F(x)$ and $G(x)$ are antiderivatives of $f(x)$, then $F(x) = G(x) + C$.

84. If $f'(x) = g(x)$, then $\int g(x)\,dx = f(x) + C$.

85. *Think About It* Use the graph of f' in the figure to answer the following, given that $f(0) = -4$.

(a) Approximate the slope of f at $x = 4$. Explain.

(b) Is it possible that $f(2) = -1$? Explain.

(c) Is $f(5) - f(4) > 0$? Explain.

(d) Approximate the value of x where f is maximum. Explain.

(e) Approximate any intervals in which the graph of f is concave upward and any intervals in which it is concave downward. Approximate the x-coordinates of any points of inflection.

(f) Approximate the x-coordinate of the minimum of $f''(x)$.

(g) Sketch an approximate graph of f.

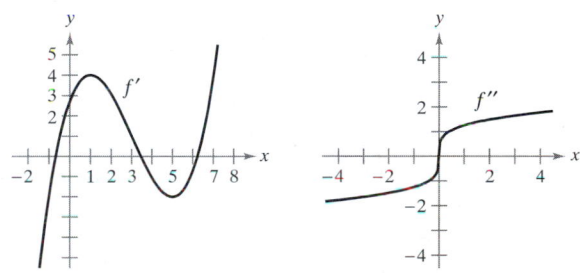

Figure for 85   **Figure for 86**

86. *Think About It* The graphs of f and f' each pass through the origin. Use the graph of f'' shown in the figure to sketch the graphs of f and f'.

87. *Acceleration* Galileo Galilei (1564–1642) stated the following proposition concerning falling objects: The time in which any space is traversed by a uniformly accelerating body is equal to the time in which that same space would be traversed by the same body moving at a uniform speed whose value is the mean of the highest speed of the accelerating body and the speed just before acceleration began. Use the techniques of this section to verify this proposition.

88. Let $s(x)$ and $c(x)$ be two functions satisfying $s'(x) = c(x)$ and $c'(x) = -s(x)$ for all x. If $s(0) = 0$ and $c(0) = 1$, prove that $[s(x)]^2 + [c(x)]^2 = 1$.

SECTION 4.2 Area

Sigma Notation • Area • The Area of a Plane Region • Upper and Lower Sums

Sigma Notation

In the preceding section, you studied antidifferentiation. In this section, you will look further into a problem introduced in Section 1.1—that of finding the area of a region in the plane. At first glance, these two ideas may seem unrelated, but you will discover in Section 4.4 that they are closely related by an extremely important theorem called the Fundamental Theorem of Calculus.

We begin this section by introducing a concise notation for sums. This notation is called **sigma notation** because it uses the uppercase Greek letter sigma, written as Σ.

Sigma Notation

The sum of n terms $a_1, a_2, a_3, \ldots, a_n$ is written as

$$\sum_{i=1}^{n} a_i = a_1 + a_2 + a_3 + \cdots + a_n$$

where i is the **index of summation,** a_i is the i**th term** of the sum, and the **upper and lower bounds of summation** are n and 1.

NOTE   The upper and lower bounds must be constant *with respect to the index of summation.* However, the lower bound doesn't have to be 1. Any integer less than or equal to the upper bound is legitimate.

EXAMPLE 1   Examples of Sigma Notation

a. $\displaystyle\sum_{i=1}^{6} i = 1 + 2 + 3 + 4 + 5 + 6$

b. $\displaystyle\sum_{i=0}^{5} (i + 1) = 1 + 2 + 3 + 4 + 5 + 6$

c. $\displaystyle\sum_{j=3}^{7} j^2 = 3^2 + 4^2 + 5^2 + 6^2 + 7^2$

d. $\displaystyle\sum_{k=1}^{n} \frac{1}{n}(k^2 + 1) = \frac{1}{n}(1^2 + 1) + \frac{1}{n}(2^2 + 1) + \cdots + \frac{1}{n}(n^2 + 1)$

e. $\displaystyle\sum_{i=1}^{n} f(x_i)\,\Delta x = f(x_1)\,\Delta x + f(x_2)\,\Delta x + \cdots + f(x_n)\,\Delta x$

From parts (a) and (b), notice that the same sum can be represented in different ways using sigma notation.

Although any variable can be used as the index of summation i, j, and k are often used. Notice in Example 1 that the index of summation does not appear in the terms of the expanded sum.

FOR FURTHER INFORMATION For a geometric interpretation of summation formulas, see the article, "Looking at $\displaystyle\sum_{k=1}^{n} k$ and $\displaystyle\sum_{k=1}^{n} k^2$ Geometrically" by Eric Hegblom in the October 1993 issue of *Mathematics Teacher.*

The following properties of summation can be derived using the associative and commutative properties of addition and the distributive property of addition over multiplication. (In the first property, k is a constant.)

1. $\displaystyle\sum_{i=1}^{n} ka_i = k\sum_{i=1}^{n} a_i$

2. $\displaystyle\sum_{i=1}^{n} (a_i \pm b_i) = \sum_{i=1}^{n} a_i \pm \sum_{i=1}^{n} b_i$

The next theorem lists some useful formulas for sums of powers. A proof of this theorem is given in the appendix.

THEOREM 4.2   Summation Formulas

1. $\displaystyle\sum_{i=1}^{n} c = cn$ 2. $\displaystyle\sum_{i=1}^{n} i = \frac{n(n+1)}{2}$

3. $\displaystyle\sum_{i=1}^{n} i^2 = \frac{n(n+1)(2n+1)}{6}$ 4. $\displaystyle\sum_{i=1}^{n} i^3 = \frac{n^2(n+1)^2}{4}$

EXAMPLE 2   Evaluating a Sum

Evaluate $\displaystyle\sum_{i=1}^{n} \frac{i+1}{n^2}$ for $n = 10, 100, 1000,$ and $10,000$.

Solution   Applying Theorem 4.2, you can write

$$\sum_{i=1}^{n} \frac{i+1}{n^2} = \frac{1}{n^2}\sum_{i=1}^{n}(i+1) \qquad \text{Factor constant } 1/n^2 \text{ out of sum.}$$

$$= \frac{1}{n^2}\left(\sum_{i=1}^{n} i + \sum_{i=1}^{n} 1\right) \qquad \text{Write as two sums.}$$

$$= \frac{1}{n^2}\left[\frac{n(n+1)}{2} + n\right] \qquad \text{Apply Theorem 4.2.}$$

$$= \frac{1}{n^2}\left[\frac{n^2 + 3n}{2}\right]$$

$$= \frac{n+3}{2n}. \qquad \text{Simplify.}$$

Now you can evaluate the sum by substituting the appropriate values of n, as shown in the table at the left.

n	$\displaystyle\sum_{i=1}^{n}\frac{i+1}{n^2} = \frac{n+3}{2n}$
10	0.65000
100	0.51500
1000	0.50150
10,000	0.50015

In the table, note that the sum appears to approach a limit as n increases. Although the discussion of limits at infinity in Section 3.5 applies to a variable x, where x can be any real number, many of the same results hold true for limits involving the variable n, where n is restricted to positive integer values. Thus, to find the limit of $(n+3)/2n$ as n approaches infinity, you can write

$$\lim_{n\to\infty} \frac{n+3}{2n} = \frac{1}{2}.$$

THE SUM OF THE FIRST 100 INTEGERS

Carl Friedrich Gauss's (1777–1855) teacher asked him to add all the integers from 1 to 100. When Gauss returned with the correct answer only after a few moments, the teacher could only look at him in astounded silence. This is what Gauss did:

$$\begin{array}{r} 1 + 2 + 3 + \cdots + 100 \\ 100 + 99 + 98 + \cdots + 1 \\ \hline 101 + 101 + 101 + \cdots + 101 \end{array}$$

$$\frac{100 \times 101}{2} = 5050$$

This is generalized by Theorem 4.2, where

$$\sum_{i=1}^{100} i = \frac{100(101)}{2} = 5050.$$

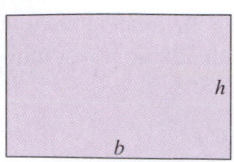

Rectangle: $A = bh$
Figure 4.5

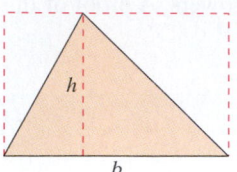

Triangle: $A = \frac{1}{2}bh$
Figure 4.6

ARCHIMEDES (287–212 B.C.)

Archimedes used the method of exhaustion to derive formulas for the areas of ellipses, parabolic segments, and sectors of a spiral. He is considered to have been the greatest applied mathematician of antiquity.

FOR FURTHER INFORMATION For an alternative development of the formula for the area of a circle, see the article "Proof Without Words: Area of a Disk is πR^2" by Russell Jay Hendel in the June 1990 issue of the *Mathematics Magazine.*

Area

In Euclidean geometry, the simplest type of plane region is a rectangle. Although people often say that the *formula* for the area of a rectangle is $A = bh$, as shown in Figure 4.5, it is actually more proper to say that this is the *definition* of the **area of a rectangle.**

From this definition, you can develop formulas for the areas of many other plane regions. For example, to determine the area of a triangle, you can form a rectangle whose area is twice that of the triangle, as shown in Figure 4.6. Once you know how to find the area of a triangle, you can determine the area of any polygon by subdividing the polygon into triangular regions, as shown in Figure 4.7.

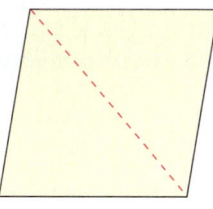
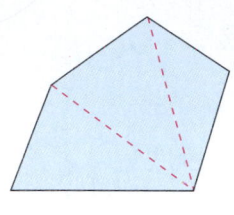

Parallelogram   Hexagon   Polygon
Figure 4.7

Finding the areas of regions other than polygons is more difficult. The ancient Greeks were able to determine formulas for the areas of some general regions (principally those bounded by conics) by the *exhaustion* method. The clearest description of this method was given by Archimedes. Essentially, the method is a limiting process in which the area is squeezed between two polygons—one inscribed in the region and one circumscribed about the region.

For instance, in Figure 4.8 the area of a circular region is approximated by an n-sided inscribed polygon and an n-sided circumscribed polygon. For each value of n the area of the inscribed polygon is less than the area of the circle, and the area of the circumscribed polygon is greater than the area of the circle. Moreover, as n increases, the areas of both polygons become better and better approximations of the area of the circle.

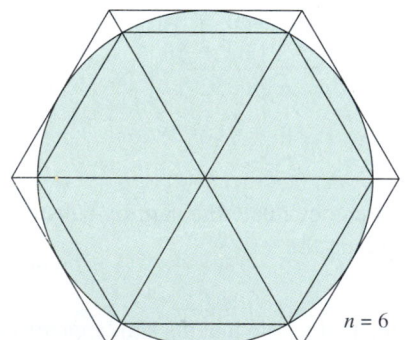
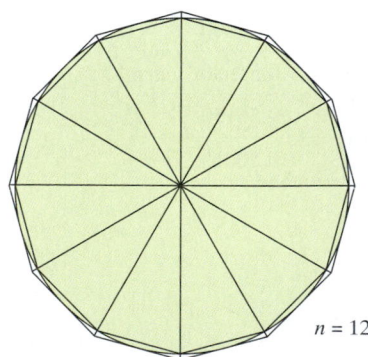

$n = 6$ $n = 12$

The exhaustion method for finding the area of a circular region
Figure 4.8

In the remaining examples in this section, we use a process that is similar to that used by Archimedes to determine the area of a plane region.

The Area of a Plane Region

We investigate the general problem of finding the area of a region in the plane with an example.

EXAMPLE 3   Approximating the Area of a Plane Region

Use the five rectangles in Figure 4.9(a) and (b) to find *two* approximations of the area of the region lying between the graph of

$$f(x) = -x^2 + 5$$

and the *x*-axis between $x = 0$ and $x = 2$.

Solution

a. The right endpoints of the five intervals are $\frac{2}{5}i$, where $i = 1, 2, 3, 4, 5$. The width of each rectangle is $\frac{2}{5}$, and the height of each rectangle can be obtained by evaluating f at the right endpoint of each interval.

$$\left[0, \frac{2}{5}\right], \left[\frac{2}{5}, \frac{4}{5}\right], \left[\frac{4}{5}, \frac{6}{5}\right], \left[\frac{6}{5}, \frac{8}{5}\right], \left[\frac{8}{5}, \frac{10}{5}\right]$$

Evaluate f at the right endpoints of these intervals.

The sum of the areas of the five rectangles is

$$\sum_{i=1}^{5} f\left(\frac{2i}{5}\right)\left(\frac{2}{5}\right) = \sum_{i=1}^{5}\left[-\left(\frac{2i}{5}\right)^2 + 5\right]\left(\frac{2}{5}\right) = \frac{162}{25} = 6.48.$$

Because each of the five rectangles lies *inside* the parabolic region, you can conclude that the area of the parabolic region is *greater* than 6.48.

b. The left endpoints of the five intervals are $\frac{2}{5}(i-1)$, where $i = 1, 2, 3, 4, 5$. The width of each rectangle is $\frac{2}{5}$, and the height of each rectangle can be obtained by evaluating f at the left endpoint of each interval.

$$\sum_{i=1}^{5} f\left(\frac{2i-2}{5}\right)\left(\frac{2}{5}\right) = \sum_{i=1}^{5}\left[-\left(\frac{2i-2}{5}\right)^2 + 5\right]\left(\frac{2}{5}\right) = \frac{202}{25} = 8.08.$$

Because the parabolic region lies within the union of the five rectangular regions, you can conclude that the area of the parabolic region is *less* than 8.08.

By combining the results in parts (a) and (b), you can conclude that

$$6.48 < (\text{Area of region}) < 8.08.$$

NOTE   By increasing the number of rectangles used in Example 3, you can obtain closer and closer approximations of the area of the region. For instance, using 25 rectangles of width $\frac{2}{25}$ each, you can conclude that

$$7.17 < (\text{Area of region}) < 7.49.$$

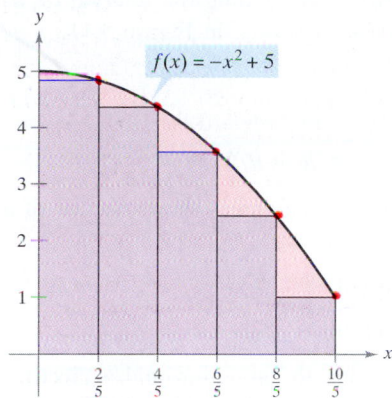

(a) The area of a parabolic region is greater than the area of the rectangles.

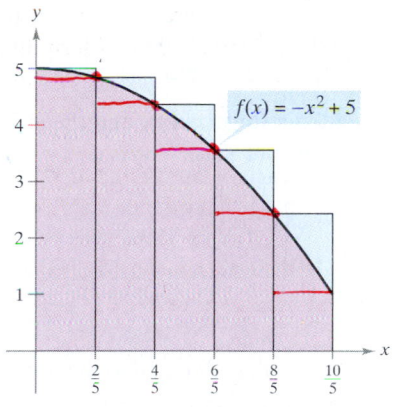

(b) The area of the parabolic region is less than the area of the rectangles.

Figure 4.9

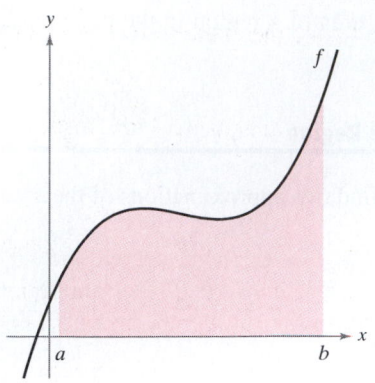

The region under a curve
Figure 4.10

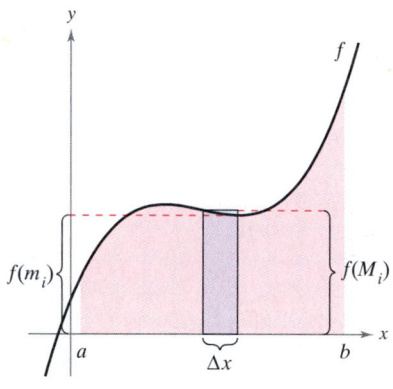

The interval $[a, b]$ is divided into n subintervals of length $\Delta x = \dfrac{b - a}{n}$.

Figure 4.11

Upper and Lower Sums

The procedure used in Example 3 can be generalized as follows. Consider a plane region bounded above by the graph of a nonnegative, continuous function $y = f(x)$, as shown in Figure 4.10. The region is bounded below by the x-axis, and the left and right boundaries of the region are the vertical lines $x = a$ and $x = b$.

To approximate the area of the region, begin by subdividing the interval $[a, b]$ into n subintervals, each of length $\Delta x = (b - a)/n$, as shown in Figure 4.11. The endpoints of the intervals are as follows.

$$\underbrace{a + 0(\Delta x)}_{a\,=\,x_0} < \underbrace{a + 1(\Delta x)}_{x_1} < \underbrace{a + 2(\Delta x)}_{x_2} < \cdots < \underbrace{a + n(\Delta x)}_{x_n\,=\,b}$$

Because f is continuous, the Extreme Value Theorem guarantees the existence of a minimum and a maximum value of $f(x)$ in *each* subinterval.

$f(m_i) =$ Minimum value of $f(x)$ in ith subinterval

$f(M_i) =$ Maximum value of $f(x)$ in ith subinterval

Next, define an **inscribed rectangle** lying *inside* the ith subregion and a **circumscribed rectangle** extending *outside* the ith subregion. The height of the ith inscribed rectangle is $f(m_i)$ and the height of the ith circumscribed rectangle is $f(M_i)$. For *each* i, the area of the inscribed rectangle is less than or equal to the area of the circumscribed rectangle.

$$\left(\begin{array}{c}\text{Area of inscribed}\\ \text{rectangle}\end{array}\right) = f(m_i)\,\Delta x \le f(M_i)\,\Delta x = \left(\begin{array}{c}\text{Area of circumscribed}\\ \text{rectangle}\end{array}\right)$$

The sum of the areas of the inscribed rectangles is called a **lower sum,** and the sum of the areas of the circumscribed rectangles is called an **upper sum.**

$$\text{Lower sum} = s(n) = \sum_{i=1}^{n} f(m_i)\,\Delta x \qquad \textcolor{red}{\text{Area of inscribed rectangles}}$$

$$\text{Upper sum} = S(n) = \sum_{i=1}^{n} f(M_i)\,\Delta x \qquad \textcolor{red}{\text{Area of circumscribed rectangles}}$$

From Figure 4.12, you can see that the lower sum $s(n)$ is less than or equal to the upper sum $S(n)$. Moreover, the actual area of the region lies between these two sums.

$$s(n) \le (\text{Area of region}) \le S(n)$$

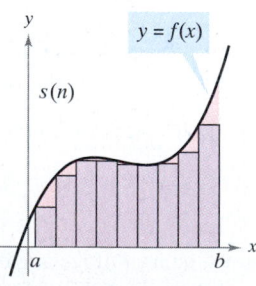

Area of inscribed rectangles is less than area of region.

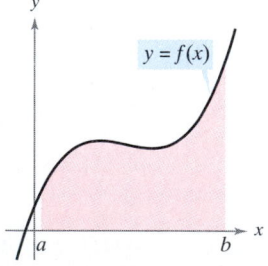

Area of region

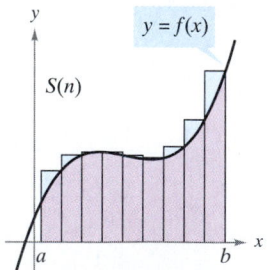

Area of circumscribed rectangles is greater than area of region.

Figure 4.12

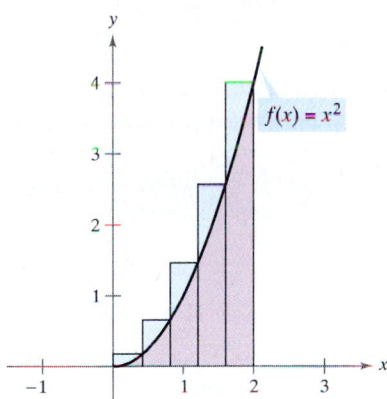

Inscribed rectangles

Circumscribed rectangles
Figure 4.13

EXAMPLE 4   Finding Upper and Lower Sums for a Region

Find the upper and lower sums for the region bounded by the graph of $f(x) = x^2$ and the x-axis between $x = 0$ and $x = 2$.

Solution   To begin, partition the interval $[0, 2]$ into n subintervals, each of length

$$\Delta x = \frac{b - a}{n} = \frac{2 - 0}{n} = \frac{2}{n}.$$

Figure 4.13 shows the endpoints of the subintervals and several inscribed and circumscribed rectangles. Because f is increasing in the interval $[0, 2]$, the minimum value on each subinterval occurs at the left endpoint, and the maximum value occurs at the right endpoint.

Left Endpoints

$$m_i = 0 + (i - 1)\left(\frac{2}{n}\right) = \frac{2(i - 1)}{n}$$

Right Endpoints

$$M_i = 0 + i\left(\frac{2}{n}\right) = \frac{2i}{n}$$

Using the left endpoints, the lower sum is

$$s(n) = \sum_{i=1}^{n} f(m_i)\, \Delta x = \sum_{i=1}^{n} f\left[\frac{2(i - 1)}{n}\right]\left(\frac{2}{n}\right)$$

$$= \sum_{i=1}^{n} \left[\frac{2(i - 1)}{n}\right]^2\left(\frac{2}{n}\right)$$

$$= \sum_{i=1}^{n} \left(\frac{8}{n^3}\right)(i^2 - 2i + 1)$$

$$= \frac{8}{n^3}\left(\sum_{i=1}^{n} i^2 - 2\sum_{i=1}^{n} i + \sum_{i=1}^{n} 1\right)$$

$$= \frac{8}{n^3}\left\{\frac{n(n + 1)(2n + 1)}{6} - 2\left[\frac{n(n + 1)}{2}\right] + n\right\}$$

$$= \frac{4}{3n^3}(2n^3 - 3n^2 + n)$$

$$= \frac{8}{3} - \frac{4}{n} + \frac{4}{3n^2}.$$

Using the right endpoints, the upper sum is

$$S(n) = \sum_{i=1}^{n} f(M_i)\, \Delta x = \sum_{i=1}^{n} f\left(\frac{2i}{n}\right)\left(\frac{2}{n}\right)$$

$$= \sum_{i=1}^{n} \left(\frac{2i}{n}\right)^2\left(\frac{2}{n}\right)$$

$$= \sum_{i=1}^{n} \left(\frac{8}{n^3}\right)i^2$$

$$= \frac{8}{n^3}\left[\frac{n(n + 1)(2n + 1)}{6}\right]$$

$$= \frac{4}{3n^3}(2n^3 + 3n^2 + n)$$

$$= \frac{8}{3} + \frac{4}{n} + \frac{4}{3n^2}.$$

EXPLORATION

For the region given in Example 4, evaluate the lower sum

$$s(n) = \frac{8}{3} - \frac{4}{n} + \frac{4}{3n^2}$$

and the upper sum

$$S(n) = \frac{8}{3} + \frac{4}{n} + \frac{4}{3n^2}$$

for $n = 10$, 100, and 1000. Use your results to determine the area of the region.

Example 4 illustrates some important things about lower and upper sums. First, notice that for any value of n, the lower sum is less than (or equal to) the upper sum.

$$s(n) = \frac{8}{3} - \frac{4}{n} + \frac{4}{3n^2} < \frac{8}{3} + \frac{4}{n} + \frac{4}{3n^2} = S(n)$$

Second, the difference between these two sums lessens as n increases. In fact, if you take the limit as $n \to \infty$, both the upper sum and the lower sum approach $\frac{8}{3}$.

$$\lim_{n \to \infty} s(n) = \lim_{n \to \infty} \left(\frac{8}{3} - \frac{4}{n} + \frac{4}{3n^2} \right) = \frac{8}{3} \qquad \text{Lower sum limit}$$

$$\lim_{n \to \infty} S(n) = \lim_{n \to \infty} \left(\frac{8}{3} + \frac{4}{n} + \frac{4}{3n^2} \right) = \frac{8}{3} \qquad \text{Upper sum limit}$$

The next theorem shows that the equivalence of the limits (as $n \to \infty$) of the upper and lower sums is not mere coincidence. It is true for all functions that are continuous and nonnegative on the closed interval $[a, b]$. The proof of this theorem is best left to a course in advanced calculus.

THEOREM 4.3   Limit of the Lower and Upper Sums

Let f be continuous and nonnegative on the interval $[a, b]$. The limits as $n \to \infty$ of both the lower and upper sums exist and are equal to each other. That is,

$$\lim_{n \to \infty} s(n) = \lim_{n \to \infty} \sum_{i=1}^{n} f(m_i)\, \Delta x$$

$$= \lim_{n \to \infty} \sum_{i=1}^{n} f(M_i)\, \Delta x$$

$$= \lim_{n \to \infty} S(n)$$

where $\Delta x = (b - a)/n$ and $f(m_i)$ and $f(M_i)$ are the minimum and maximum values of f on the ith subinterval.

Because the same limit is attained for both the minimum value $f(m_i)$ and the maximum value $f(M_i)$, it follows from the Squeeze Theorem (Theorem 1.8) that the choice of x in the ith subinterval does not affect the limit. This means that you are free to choose an *arbitrary* x-value in the ith subinterval, as in the following *definition of the area of a region in the plane.*

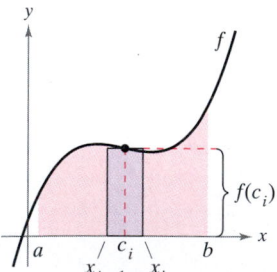

The length of the ith subinterval is $\Delta x = x_i - x_{i-1}$.
Figure 4.14

Definition of the Area of a Region in the Plane

Let f be continuous and nonnegative on the interval $[a, b]$. The **area** of the region bounded by the graph of f, the x-axis, and the vertical lines $x = a$ and $x = b$ is

$$\text{Area} = \lim_{n \to \infty} \sum_{i=1}^{n} f(c_i)\, \Delta x, \qquad x_{i-1} \le c_i \le x_i$$

where $\Delta x = (b - a)/n$ (see Figure 4.14).

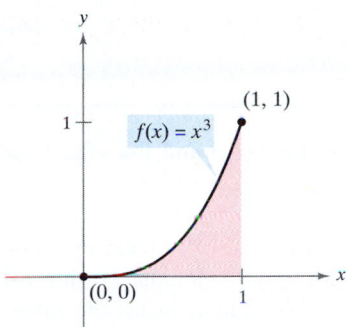

The area of the region bounded by the graph of f, the x-axis, $x = 0$, and $x = 1$ is $\frac{1}{4}$.

Figure 4.15

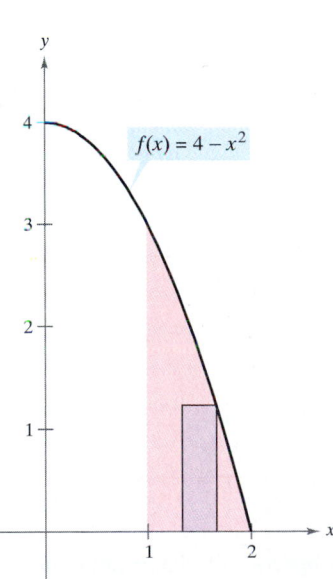

The area of the region bounded by the graph of f, the x-axis, $x = 1$, and $x = 2$ is $\frac{5}{3}$.

Figure 4.16

EXAMPLE 5   Finding Area by the Limit Definition

Find the area of the region bounded by the graph $f(x) = x^3$, the x-axis, and the vertical lines $x = 0$ and $x = 1$, as shown in Figure 4.15.

Solution   Begin by noting that f is continuous and nonnegative on the interval $[0, 1]$. Next, partition the interval $[0, 1]$ into n equal subintervals, each of length $\Delta x = 1/n$. According to the definition of area, you can choose any x-value in the ith subinterval. For this example, the right endpoints $(c_i = i/n)$ are convenient.

$$\text{Area} = \lim_{n \to \infty} \sum_{i=1}^{n} f(c_i)\, \Delta x = \lim_{n \to \infty} \sum_{i=1}^{n} \left(\frac{i}{n}\right)^3 \left(\frac{1}{n}\right)$$

$$= \lim_{n \to \infty} \frac{1}{n^4} \sum_{i=1}^{n} i^3$$

$$= \lim_{n \to \infty} \frac{1}{n^4} \left[\frac{n^2(n+1)^2}{4}\right]$$

$$= \lim_{n \to \infty} \left(\frac{1}{4} + \frac{1}{2n} + \frac{1}{4n^2}\right)$$

$$= \frac{1}{4}$$

Thus, the area of the region is $\frac{1}{4}$.

EXAMPLE 6   Finding Area by the Limit Definition

Find the area of the region bounded by the graph of $f(x) = 4 - x^2$, the x-axis, and the vertical lines $x = 1$ and $x = 2$, as shown in Figure 4.16.

Solution   The function f is continuous and nonnegative on the interval $[1, 2]$, and so you begin by partitioning the interval into n equal subintervals, each of length $\Delta x = 1/n$. Choosing the right endpoint, $c_i = 1 + (i/n)$, of each subinterval, you obtain the following.

$$\text{Area} = \lim_{n \to \infty} \sum_{i=1}^{n} f(c_i)\, \Delta x = \lim_{n \to \infty} \sum_{i=1}^{n} \left[4 - \left(1 + \frac{i}{n}\right)^2\right]\left(\frac{1}{n}\right)$$

$$= \lim_{n \to \infty} \sum_{i=1}^{n} \left(3 - \frac{2i}{n} - \frac{i^2}{n^2}\right)\left(\frac{1}{n}\right)$$

$$= \lim_{n \to \infty} \left(\frac{1}{n} \sum_{i=1}^{n} 3 - \frac{2}{n^2} \sum_{i=1}^{n} i - \frac{1}{n^3} \sum_{i=1}^{n} i^2\right)$$

$$= \lim_{n \to \infty} \left[3 - \left(1 + \frac{1}{n}\right) - \left(\frac{1}{3} + \frac{1}{2n} + \frac{1}{6n^2}\right)\right]$$

$$= 3 - 1 - \frac{1}{3}$$

$$= \frac{5}{3}$$

Thus, the area of the region is $\frac{5}{3}$.

The last example in this section looks at a region that is bounded by the y-axis (rather than the x-axis).

EXAMPLE 7   A Region Bounded by the y-Axis

Find the area of the region bounded by the graph of $f(y) = y^2$ and the y-axis for $0 \le y \le 1$, as shown in Figure 4.17.

Solution   When f is a continuous, nonnegative function of y, you still can use the same basic procedure illustrated in Examples 5 and 6. Begin by partitioning the interval $[0, 1]$ into n equal subintervals, each of length $\Delta y = 1/n$. Then, using the upper endpoints $c_i = i/n$, you obtain the following.

$$\text{Area} = \lim_{n \to \infty} \sum_{i=1}^{n} f(c_i)\, \Delta y = \lim_{n \to \infty} \sum_{i=1}^{n} \left(\frac{i}{n}\right)^2 \left(\frac{1}{n}\right)$$

$$= \lim_{n \to \infty} \frac{1}{n^3} \sum_{i=1}^{n} i^2$$

$$= \lim_{n \to \infty} \frac{n(n+1)(2n+1)}{6n^3}$$

$$= \lim_{n \to \infty} \left(\frac{1}{3} + \frac{1}{2n} + \frac{1}{6n^2}\right)$$

$$= \frac{1}{3}$$

Thus, the area of the region is $\frac{1}{3}$.

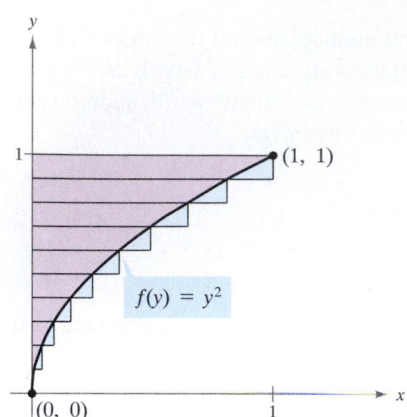

The area of the region bounded by the graph of f and the y-axis for $0 \le y \le 1$ is $\frac{1}{3}$.

Figure 4.17

EXERCISES FOR SECTION 4.2

In Exercises 1–6, find the sum. Use the summation capabilities of a graphing utility to verify your result.

1. $\displaystyle\sum_{i=1}^{5} (2i + 1)$ **2.** $\displaystyle\sum_{k=2}^{5} (k+1)(k-3)$

3. $\displaystyle\sum_{k=0}^{4} \frac{1}{k^2 + 1}$ **4.** $\displaystyle\sum_{j=3}^{5} \frac{1}{j}$

5. $\displaystyle\sum_{k=1}^{4} c$ **6.** $\displaystyle\sum_{i=1}^{4} [(i-1)^2 + (i+1)^3]$

In Exercises 7–14, use sigma notation to write the sum.

7. $\dfrac{1}{3(1)} + \dfrac{1}{3(2)} + \dfrac{1}{3(3)} + \cdots + \dfrac{1}{3(9)}$

8. $\dfrac{5}{1+1} + \dfrac{5}{1+2} + \dfrac{5}{1+3} + \cdots + \dfrac{5}{1+15}$

9. $\left[2\left(\dfrac{1}{8}\right) + 3\right] + \left[2\left(\dfrac{2}{8}\right) + 3\right] + \cdots + \left[2\left(\dfrac{8}{8}\right) + 3\right]$

10. $\left[1 - \left(\dfrac{1}{4}\right)^2\right] + \left[1 - \left(\dfrac{2}{4}\right)^2\right] + \cdots + \left[1 - \left(\dfrac{4}{4}\right)^2\right]$

11. $\left[\left(\dfrac{2}{n}\right)^3 - \dfrac{2}{n}\right]\left(\dfrac{2}{n}\right) + \cdots + \left[\left(\dfrac{2n}{n}\right)^3 - \dfrac{2n}{n}\right]\left(\dfrac{2}{n}\right)$

12. $\left[1 - \left(\dfrac{2}{n} - 1\right)^2\right]\left(\dfrac{2}{n}\right) + \cdots + \left[1 - \left(\dfrac{2n}{n} - 1\right)^2\right]\left(\dfrac{2}{n}\right)$

13. $\left[2\left(1 + \dfrac{3}{n}\right)^2\right]\left(\dfrac{3}{n}\right) + \cdots + \left[2\left(1 + \dfrac{3n}{n}\right)^2\right]\left(\dfrac{3}{n}\right)$

14. $\left(\dfrac{1}{n}\right)\sqrt{1 + \left(\dfrac{0}{n}\right)^2} + \cdots + \left(\dfrac{1}{n}\right)\sqrt{1 - \left(\dfrac{n-1}{n}\right)^2}$

In Exercises 15–20, use the properties of sigma notation and summation formulas to evaluate the sum. Use the summation capabilities of a graphing utility to verify your result.

15. $\displaystyle\sum_{i=1}^{20} 2i$ **16.** $\displaystyle\sum_{i=1}^{15} (2i - 3)$

17. $\displaystyle\sum_{i=1}^{20} (i - 1)^2$

18. $\displaystyle\sum_{i=1}^{10} (i^2 - 1)$

19. $\displaystyle\sum_{i=1}^{15} i(i - 1)^2$

20. $\displaystyle\sum_{i=1}^{10} i(i^2 + 1)$

In Exercises 21 and 22, use the summation capabilities of a graphing utility to evaluate the sum. Then use the properties of sigma notation and summation formulas to verify the sum.

21. $\displaystyle\sum_{i=1}^{20} (i^2 + 3)$

22. $\displaystyle\sum_{i=1}^{15} (i^3 - 2i)$

In Exercises 23–28, find the limit of $s(n)$ as $n \to \infty$.

23. $s(n) = \left(\dfrac{4}{3n^3}\right)(2n^3 + 3n^2 + n)$

24. $s(n) = \left(\dfrac{8}{3} + \dfrac{4}{n} + \dfrac{4}{3n^2}\right)$

25. $s(n) = \dfrac{81}{n^4}\left[\dfrac{n^2(n + 1)^2}{4}\right]$

26. $s(n) = \dfrac{64}{n^3}\left[\dfrac{n(n + 1)(2n + 1)}{6}\right]$

27. $s(n) = \dfrac{18}{n^2}\left[\dfrac{n(n + 1)}{2}\right]$

28. $s(n) = \dfrac{1}{n^2}\left[\dfrac{n(n + 1)}{2}\right]$

In Exercises 29–34, find a formula for the sum of n terms. Use the formula to find the limit as $n \to \infty$.

29. $\displaystyle\lim_{n\to\infty} \sum_{i=1}^{n} \dfrac{16i}{n^2}$

30. $\displaystyle\lim_{n\to\infty} \sum_{i=1}^{n} \left(\dfrac{2i}{n}\right)\left(\dfrac{2}{n}\right)$

31. $\displaystyle\lim_{n\to\infty} \sum_{i=1}^{n} \dfrac{1}{n^3}(i - 1)^2$

32. $\displaystyle\lim_{n\to\infty} \sum_{i=1}^{n} \left(1 + \dfrac{2i}{n}\right)^2\left(\dfrac{2}{n}\right)$

33. $\displaystyle\lim_{n\to\infty} \sum_{i=1}^{n} \left(1 + \dfrac{i}{n}\right)\left(\dfrac{2}{n}\right)$

34. $\displaystyle\lim_{n\to\infty} \sum_{i=1}^{n} \left(1 + \dfrac{2i}{n}\right)^3\left(\dfrac{2}{n}\right)$

In Exercises 35–38, use upper and lower sums to approximate the area of the region using the indicated number of subintervals (of equal length).

35. $y = \sqrt{x}$

36. $y = \sqrt{x} + 1$

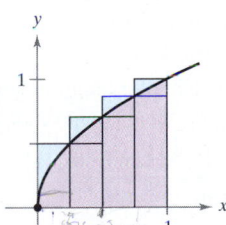

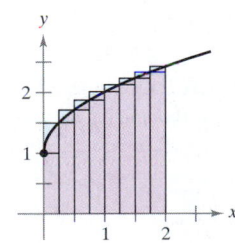

37. $y = \dfrac{1}{x}$

38. $y = \sqrt{1 - x^2}$

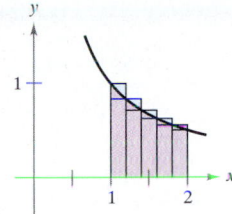

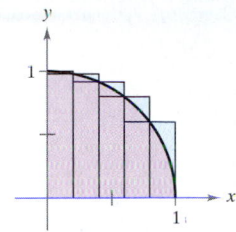

39. *Numerical Reasoning*   Consider a triangle of area 2 bounded by the graphs of $y = x$, $y = 0$, and $x = 2$.

(a) Sketch the graph of the region.

(b) Divide the interval $[0, 2]$ into n equal subintervals and show that the endpoints are

$$0 < 1\left(\dfrac{2}{n}\right) < \cdots < (n - 1)\left(\dfrac{2}{n}\right) < n\left(\dfrac{2}{n}\right).$$

(c) Show that $s(n) = \displaystyle\sum_{i=1}^{n} \left[(i - 1)\left(\dfrac{2}{n}\right)\right]\left(\dfrac{2}{n}\right)$.

(d) Show that $S(n) = \displaystyle\sum_{i=1}^{n} \left[i\left(\dfrac{2}{n}\right)\right]\left(\dfrac{2}{n}\right)$.

(e) Complete the table.

n	5	10	50	100
$s(n)$				
$S(n)$				

(f) Show that $\displaystyle\lim_{n\to\infty} s(n) = \lim_{n\to\infty} S(n) = 2$.

40. *Numerical Reasoning*   Consider a trapezoid of area 4 bounded by the graphs of $y = x$, $y = 0$, $x = 1$, and $x = 3$.

(a) Sketch the graph of the region.

(b) Divide the interval $[1, 3]$ into n equal subintervals and show that the endpoints are

$$1 < 1 + 1\left(\dfrac{2}{n}\right) < \cdots < 1 + (n - 1)\left(\dfrac{2}{n}\right) < 1 + n\left(\dfrac{2}{n}\right).$$

(c) Show that $s(n) = \displaystyle\sum_{i=1}^{n} \left[1 + (i - 1)\left(\dfrac{2}{n}\right)\right]\left(\dfrac{2}{n}\right)$.

(d) Show that $S(n) = \displaystyle\sum_{i=1}^{n} \left[1 + i\left(\dfrac{2}{n}\right)\right]\left(\dfrac{2}{n}\right)$.

(e) Complete the table.

n	5	10	50	100
$s(n)$				
$S(n)$				

(f) Show that $\displaystyle\lim_{n\to\infty} s(n) = \lim_{n\to\infty} S(n) = 4$.

In Exercises 41–48, use the limit process to find the area of the region between the graph of the function and the *x*-axis over the indicated interval. Sketch the region.

Function	Interval
41. $y = -2x + 3$	$[0, 1]$
42. $y = 3x - 4$	$[2, 5]$
43. $y = x^2 + 2$	$[0, 1]$
44. $y = 1 - x^2$	$[-1, 1]$
45. $y = 27 - x^3$	$[1, 3]$
46. $y = 2x - x^3$	$[0, 1]$
47. $y = x^2 - x^3$	$[-1, 1]$
48. $y = x^2 - x^3$	$[-1, 0]$

In Exercises 49 and 50, use the limit process to find the area of the region between the graph of the function and the *y*-axis over the indicated *y*-interval. Sketch the region.

49. $f(y) = 3y, 0 \leq y \leq 2$ **50.** $f(y) = y^2, 0 \leq y \leq 3$

In Exercises 51–54, use the *Midpoint Rule*

$$\text{Area} \approx \sum_{i=1}^{n} f\left(\frac{x_i + x_{i-1}}{2}\right)\Delta x$$

with $n = 4$ to approximate the area of the region bounded by the graph of the function and the *x*-axis over the indicated interval.

Function	Interval
51. $f(x) = x^2 + 3$	$[0, 2]$
52. $f(x) = x^2 + 4x$	$[0, 4]$
53. $f(x) = \tan x$	$\left[0, \dfrac{\pi}{4}\right]$
54. $f(x) = \sin x$	$\left[0, \dfrac{\pi}{2}\right]$

 Write a program for a graphing utility to approximate areas by using the Midpoint Rule. Assume the function is positive over the indicated interval and the subintervals are of equal width. In Exercises 55–58, use the program to approximate the area between the function and the *x*-axis over the indicated interval, and complete the table.

n	4	8	12	16	20
Approximate area					

Function	Interval
55. $f(x) = \sqrt{x}$	$[0, 4]$
56. $f(x) = \dfrac{8}{x^2 + 1}$	$[2, 6]$
57. $f(x) = \tan\left(\dfrac{\pi x}{8}\right)$	$[1, 3]$
58. $f(x) = \cos \sqrt{x}$	$[0, 2]$

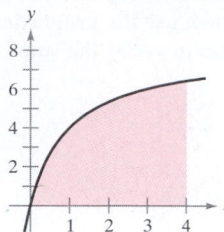 **59. *Graphical Reasoning*** Consider the region bounded by the graphs of

$$f(x) = \frac{8x}{x + 1},$$

$x = 0, x = 4$, and $y = 0$, as shown in the figure.

(a) Redraw the figure, and complete and shade the rectangles representing the lower sum when $n = 4$. Find this lower sum.

(b) Redraw the figure, and complete and shade the rectangles representing the upper sum when $n = 4$. Find this upper sum.

(c) Redraw the figure, and complete and shade the rectangles whose heights are determined by the functional values at the midpoint of each subinterval when $n = 4$. Find this sum using the Midpoint Rule.

(d) Verify the following formulas for approximating the area of the region using n subintervals of equal width.

$$\text{Lower sum: } s(n) = \sum_{i=1}^{n} f\left[(i - 1)\frac{4}{n}\right]\left(\frac{4}{n}\right)$$

$$\text{Upper sum: } S(n) = \sum_{i=1}^{n} f\left[(i)\frac{4}{n}\right]\left(\frac{4}{n}\right)$$

$$\text{Midpoint Rule: } M(n) = \sum_{i=1}^{n} f\left[\left(i - \frac{1}{2}\right)\frac{4}{n}\right]\left(\frac{4}{n}\right)$$

(e) Use a graphing utility and the formulas in part (d) to complete the table.

n	4	8	20	100	200
$s(n)$					
$S(n)$					
$M(n)$					

(f) Explain why $s(n)$ increases and $S(n)$ decreases for increasing n, as shown in the table in part (e).

60. Use a graphing utility to complete the table for approximations of the area of the region bounded by the graphs of $f(x) = \sqrt[3]{x}$, $x = 0, x = 8$, and $y = 0$.

n	10	20	50	100	200
$s(n)$					
$S(n)$					
$M(n)$					

Approximation In Exercises 61 and 62, determine which value best approximates the area of the region between the *x*-axis and the graph of the function over the indicated interval. (Make your selection on the basis of a sketch of the region and *not* by performing calculations.)

61. $f(x) = 4 - x^2$, $[0, 2]$

(a) -2 (b) 6   (c) 10   (d) 3   (e) 8

62. $f(x) = \sin \dfrac{\pi x}{4}$, $[0, 4]$

(a) 3   (b) 1   (c) -2 (d) 8   (e) 6

True or False? In Exercises 63 and 64, determine whether the statement is true or false. If it is false, explain why or give an example that shows it is false.

63. The sum of the first *n* positive integers is $n(n+1)/2$.

64. If *f* is continuous and nonnegative on $[a, b]$, then the limits as $n \to \infty$ of its lower sum $s(n)$ and upper sum $S(n)$ both exist and are equal.

65. *Monte Carlo Method* The following program approximates the area of the region under the graph of a monotonic function and above the *x*-axis between $x = a$ and $x = b$. Run the program for $a = 0$ and $b = \pi/2$ for several values of N2. Explain why the Monte Carlo Method works. *(Adaptation of Monte Carlo Method program, from James M. Sconyers, "Approximation of Area Under a Curve," MATHEMATICS TEACHER 77, no. 2 (February 1984). Copyright © 1984 by the National Council of Teachers of Mathematics. Reprinted with permission.)*

```
10   DEF FNF(X)=SIN(X)
20   A=0
30   B=1.570796
40   PRINT "Input Number of Random Points"
50   INPUT N2
60   N1=0
70   IF FNF(A)>FNF(B) THEN YMAX=FNF(A) ELSE
     YMAX=FNF(B)
80   FOR I=1 TO N2
90   X=A+(B-A)*RND(1)
100  Y=YMAX*RND(1)
110  IF Y>=FNF(X) THEN GOTO 130
120  N1=N1+1
130  NEXT I
140  AREA=(N1/N2)*(B-A)*YMAX
150  PRINT "Approximate Area:"; AREA
160  END
```

66. *Graphical Reasoning* Consider an *n*-sided regular polygon inscribed in a circle of radius *r*. Join the vertices of the polygon to the center of the circle, forming *n* congruent triangles (see figure).

(a) Determine the central angle θ in terms of *n*.

(b) Show that the area of each triangle is $\frac{1}{2}r^2 \sin\theta$.

(c) Let A_n be the sum of the areas of the *n* triangles. Find $\displaystyle\lim_{n \to \infty} A_n$.

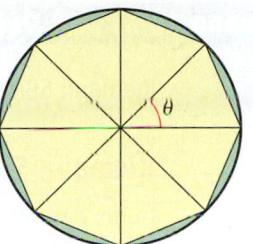

Figure for 66

Figure for 67

67. *Writing* Use the figure to write a short paragraph explaining why the formula

$$1 + 2 + \cdots + n = \tfrac{1}{2}n(n+1)$$

is valid for all positive integers *n*.

68. Verify the formula

$$\sum_{i=1}^{n} i^2 = \frac{n(n+1)(2n+1)}{6}$$

by showing the following.

(a) $(1 + i)^3 - i^3 = 3i^2 + 3i + 1$

(b) $-1 + (n + 1)^3 = \displaystyle\sum_{i=1}^{n}(3i^2 + 3i + 1)$

(c) $\displaystyle\sum_{i=1}^{n} i^2 = \frac{n(n+1)(2n+1)}{6}$

69. *Modeling Data* The table lists the measurements of a lot bounded by a stream and two straight roads that meet at right angles, where *x* and *y* are measured in feet (see figure).

x	0	50	100	150	200	250	300
y	450	362	305	268	245	156	0

(a) Use the regression capabilities of a graphing utility to find a model of the form

$$y = ax^3 + bx^2 + cx + d.$$

(b) Use a graphing utility to plot the data and graph the model.

(c) Use the model in part (a) to estimate the area of the lot.

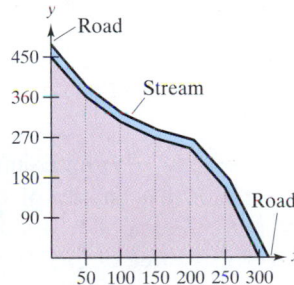

Riemann Sums • Definite Integrals • Properties of Definite Integrals

Riemann Sums

In the definition of area given in Section 4.2, the partitions have subintervals of *equal length*. This was done only for computational convenience. We begin this section with an example that shows that it is not necessary to have subintervals of equal length.

EXAMPLE 1   **A Partition with Subintervals of Unequal Lengths**

Consider the region bounded by the graph of $f(x) = \sqrt{x}$ and the x-axis for $0 \le x \le 1$, as shown in Figure 4.18. Evaluate the limit

$$\lim_{n \to \infty} \sum_{i=1}^{n} f(c_i)\, \Delta x_i$$

where c_i is the right endpoint of the partition given by $x_i = i^2/n^2$ and Δx_i is the width of the ith interval.

Solution   The length of the ith interval is given by

$$\Delta x_i = \frac{i^2}{n^2} - \frac{(i-1)^2}{n^2}$$

$$= \frac{i^2 - i^2 + 2i - 1}{n^2}$$

$$= \frac{2i - 1}{n^2}.$$

Thus, the limit is

$$\lim_{n \to \infty} \sum_{i=1}^{n} f(c_i)\, \Delta x_i = \lim_{n \to \infty} \sum_{i=1}^{n} \sqrt{\frac{i^2}{n^2}}\left(\frac{2i-1}{n^2}\right)$$

$$= \lim_{n \to \infty} \frac{1}{n^3} \sum_{i=1}^{n} (2i^2 - i)$$

$$= \lim_{n \to \infty} \frac{1}{n^3}\left[2\left(\frac{n(n+1)(2n+1)}{6}\right) - \frac{n(n+1)}{2}\right]$$

$$= \lim_{n \to \infty} \frac{4n^3 + 3n^2 - n}{6n^3}$$

$$= \frac{2}{3}.$$

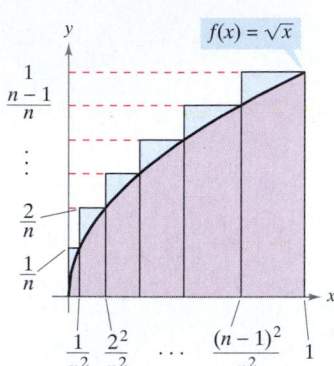

The subintervals do not have equal lengths.
Figure 4.18

From Example 7 in Section 4.2, you know that the region shown in Figure 4.19 has an area of $\frac{1}{3}$. Because the square bounded by $0 \le x \le 1$ and $0 \le y \le 1$ has an area of 1, you can conclude that the area of the region shown in Figure 4.18 has an area of $\frac{2}{3}$. This agrees with the limit found in Example 1, even though that example used a partition having subintervals of unequal lengths. The reason this particular partition gave the proper area is that as *n increases, the length of the largest subinterval approaches zero.* This is a key feature of the development of definite integrals.

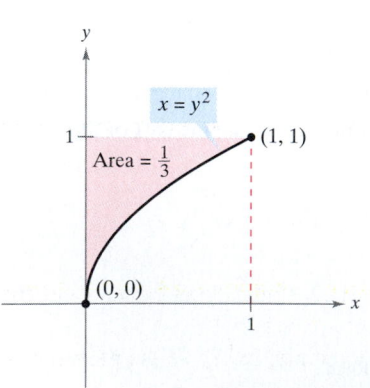

The area of the region bounded by the graph of $x = y^2$ and the y-axis for $0 \le y \le 1$ is $\frac{1}{3}$.
Figure 4.19

GEORG FRIEDRICH BERNHARD RIEMANN
(1826–1866)

German mathematician Riemann did his most famous work in the areas of non-Euclidean geometry, differential equations, and number theory. It was Riemann's results in physics and mathematics that formed the structure on which Einstein's theory of general relativity was based.

The Granger Collection

In the preceding section, the limit of a sum was used to define the area of a region in the plane. Finding area by this means is only one of *many* applications involving the limit of a sum. A similar approach can be used to determine quantities as diverse as arc length, average value, centroids, volumes, work, and surface areas. The following development is named after Georg Friedrich Bernhard Riemann. Although the definite integral had been defined and used long before the time of Riemann, he generalized the concept to cover a broader category of functions.

In the following definition of a **Riemann sum**, note that the function f has no restrictions other than being on the interval $[a, b]$. (In the preceding section, the function f was assumed to be continuous and nonnegative because we were dealing with the area under a curve.)

Definition of a Riemann Sum

Let f be defined on the closed interval $[a, b]$, and let Δ be a partition of $[a, b]$ given by

$$a = x_0 < x_1 < x_2 < \cdots < x_{n-1} < x_n = b,$$

where Δx_i is the length of the ith subinterval. If c_i is *any* point in the ith subinterval, then the sum

$$\sum_{i=1}^{n} f(c_i)\,\Delta x_i, \qquad x_{i-1} \le c_i \le x_i$$

is called a **Riemann sum** of f for the partition Δ.

NOTE The sums in Section 4.2 are examples of Riemann sums, but there are more general Riemann sums than the ones covered there.

The length of the largest subinterval of a partition Δ is the **norm** of the partition and is denoted by $\|\Delta\|$. If every subinterval is of equal length, the partition is **regular** and the norm is denoted by

$$\|\Delta\| = \Delta x = \frac{b - a}{n}. \qquad \text{Regular partition}$$

For a general partition, the norm is related to the number of subintervals of $[a, b]$ in the following way.

$$\frac{b - a}{\|\Delta\|} \le n \qquad \text{General partition}$$

Thus, the number of subintervals in a partition approaches infinity as the norm of the partition approaches 0. That is, $\|\Delta\| \to 0$ implies that $n \to \infty$.

The converse of this statement is not true. For example, let Δ_n be the partition of the interval $[0, 1]$ given by

$$0 < \frac{1}{2^n} < \frac{1}{2^{n-1}} < \cdots < \frac{1}{8} < \frac{1}{4} < \frac{1}{2} < 1.$$

As shown in Figure 4.20, for any positive value of n, the norm of the partition Δ_n is $\frac{1}{2}$. Thus, letting n approach infinity does not force $\|\Delta\|$ to approach 0. In a regular partition, however, the statements $\|\Delta\| \to 0$ and $n \to \infty$ are equivalent.

$\|\Delta\| = \frac{1}{2}$

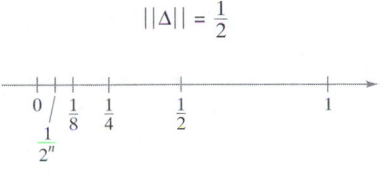

$n \to \infty$ does not imply that $\|\Delta\| \to 0$.
Figure 4.20

Definite Integrals

To define the definite integral, consider the following limit.

$$\lim_{\|\Delta\|\to 0} \sum_{i=1}^{n} f(c_i)\,\Delta x_i = L$$

To say that this limit exists means that for $\varepsilon > 0$ there exists a $\delta > 0$ such that for every partition with $\|\Delta\| < \delta$ it follows that

$$\left| L - \sum_{i=1}^{n} f(c_i)\,\Delta x_i \right| < \varepsilon.$$

(This must be true for any choice of c_i in the ith subinterval of Δ.)

FOR FURTHER INFORMATION For insight into the history of the definite integral, see the article "The Evolution of Integration" by A. Shenitzer and J. Steprāns in the January 1994 issue of *The American Mathematical Monthly*.

Definition of a Definite Integral

If f is defined on the closed interval $[a, b]$ and the limit

$$\lim_{\|\Delta\|\to 0} \sum_{i=1}^{n} f(c_i)\,\Delta x_i$$

exists, then f is **integrable** on $[a, b]$ and the limit is denoted by

$$\lim_{\|\Delta\|\to 0} \sum_{i=1}^{n} f(c_i)\,\Delta x_i = \int_{a}^{b} f(x)\,dx.$$

The limit is called the **definite integral** of f from a to b. The number a is the **lower limit** of integration, and the number b is the **upper limit** of integration.

It is not a coincidence that the notation for definite integrals is similar to that used for indefinite integrals. You will see why in the next section when we discuss the Fundamental Theorem of Calculus. For now it is important to see that definite integrals and indefinite integrals are different identities. A definite integral is a *number*, whereas an indefinite integral is a *family of functions*.

A sufficient condition for a function f to be integrable on $[a, b]$ is that it is continuous on $[a, b]$. A proof of this theorem is beyond the scope of this text.

THEOREM 4.4   Continuity Implies Integrability

If a function f is continuous on the closed interval $[a, b]$, then f is integrable on $[a, b]$.

EXPLORATION

The Converse of Theorem 4.4   Is the converse of Theorem 4.4 true? That is, if a function is integrable, does it have to be continuous? Explain your reasoning and give examples.

Describe the relationship among continuity, differentiability, and integrability. Which is the strongest condition? Which is the weakest? Which conditions imply other conditions?

EXAMPLE 2   Evaluating a Definite Integral as a Limit

Evaluate the definite integral $\displaystyle\int_{-2}^{1} 2x\, dx$.

Solution   The function $f(x) = 2x$ is integrable on the interval $[-2, 1]$, because it is continuous on $[-2, 1]$. Moreover, the definition of integrability implies that any partition whose norm approaches 0 can be used to determine the limit. For computational convenience, define Δ by subdividing $[-2, 1]$ into n subintervals of equal length

$$\Delta x_i = \Delta x = \frac{b - a}{n} = \frac{3}{n}.$$

Choosing c_i as the right endpoint of each subinterval produces

$$c_i = a + i(\Delta x) = -2 + \frac{3i}{n}.$$

Thus, the definite integral is given by

$$\int_{-2}^{1} 2x\, dx = \lim_{\|\Delta\| \to 0} \sum_{i=1}^{n} f(c_i)\, \Delta x_i = \lim_{n \to \infty} \sum_{i=1}^{n} f(c_i)\, \Delta x$$

$$= \lim_{n \to \infty} \sum_{i=1}^{n} 2\left(-2 + \frac{3i}{n}\right)\left(\frac{3}{n}\right)$$

$$= \lim_{n \to \infty} \frac{6}{n} \sum_{i=1}^{n} \left(-2 + \frac{3i}{n}\right)$$

$$= \lim_{n \to \infty} \frac{6}{n} \left\{-2n + \frac{3}{n}\left[\frac{n(n + 1)}{2}\right]\right\}$$

$$= \lim_{n \to \infty} \left(-12 + 9 + \frac{9}{n}\right)$$

$$= -3.$$

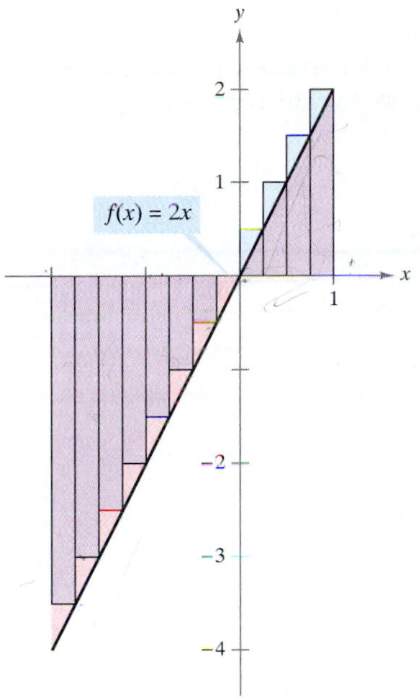

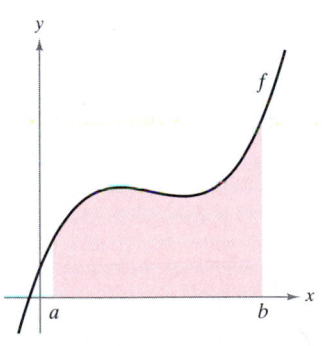

Because the definite integral is negative, it does not represent the area of the region.
Figure 4.21

Because the definite integral in Example 2 is negative, it *does not* represent the area of the region shown in Figure 4.21. Definite integrals can be positive, negative, or zero. For a definite integral to be interpreted as an area (as defined in Section 4.2), the function f must be continuous and nonnegative on $[a, b]$, as stated in the following theorem. (Proof of this theorem is straightforward—you simply use the definition of area given in Section 4.2.)

You can use a definite integral to find the area of the region bounded by the graph of f, the x-axis, $x = a$, and $x = b$.
Figure 4.22

THEOREM 4.5   The Definite Integral as the Area of a Region

If f is continuous and nonnegative on the closed interval $[a, b]$, then the area of the region bounded by the graph of f, the x-axis, and the vertical lines $x = a$ and $x = b$ is given by

$$\text{Area} = \int_{a}^{b} f(x)\, dx.$$

(See Figure 4.22.)

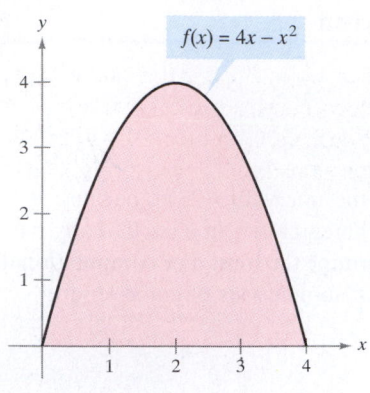

Area $= \displaystyle\int_0^4 (4x - x^2)\, dx$

Figure 4.23

As an example of Theorem 4.5, consider the region bounded by the graph of

$$f(x) = 4x - x^2$$

and the x-axis, as shown in Figure 4.23. Because f is continuous and nonnegative on the closed interval $[0, 4]$, the area of the region is

$$\text{Area} = \int_0^4 (4x - x^2)\, dx.$$

A straightforward technique for evaluating a definite integral such as this will be discussed in Section 4.4. For now, however, you can evaluate a definite integral in two ways—you can use the limit definition *or* you can check to see whether the definite integral represents the area of a common geometric region such as a rectangle, triangle, or semicircle.

EXAMPLE 3 Areas of Common Geometric Figures

Sketch the region corresponding to each definite integral. Then evaluate each integral using a geometric formula.

a. $\displaystyle\int_1^3 4\, dx$ **b.** $\displaystyle\int_0^3 (x + 2)\, dx$ **c.** $\displaystyle\int_{-2}^2 \sqrt{4 - x^2}\, dx$

Solution A sketch of each region is shown in Figure 4.24.

a. This region is a rectangle of height 4 and width 2.

$$\int_1^3 4\, dx = (\text{Area of rectangle}) = 4(2) = 8$$

b. This region is a trapezoid with an altitude of 3 and parallel bases of lengths 2 and 5. The formula for the area of a trapezoid is $\frac{1}{2}h(b_1 + b_2)$.

$$\int_0^3 (x + 2)\, dx = (\text{Area of trapezoid}) = \frac{1}{2}(3)(2 + 5) = \frac{21}{2}$$

c. This region is a semicircle of radius 2. The formula for the area of a semicircle is $\frac{1}{2}\pi r^2$.

$$\int_{-2}^2 \sqrt{4 - x^2}\, dx = (\text{Area of semicircle}) = \frac{1}{2}\pi(2^2) = 2\pi$$

NOTE The variable of integration in a definite integral is sometimes called a *dummy variable* because it can be replaced by any other variable without changing the value of the integral. For instance, the definite integrals.

$$\int_0^3 (x + 2)\, dx$$

and

$$\int_0^3 (t + 2)\, dt$$

have the same value.

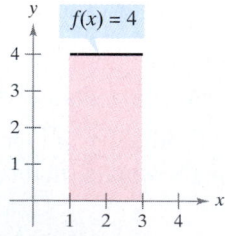 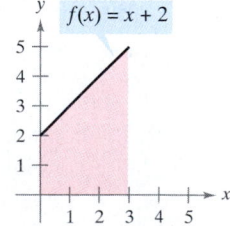 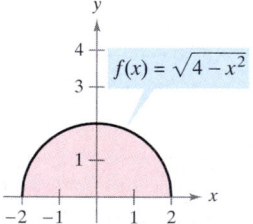

Figure 4.24

Properties of Definite Integrals

The definition of the definite integral of f on the interval $[a, b]$ specifies that $a < b$. Now, however, it is convenient to extend the definition to cover cases in which $a = b$ or $a > b$. Geometrically, the following two special definitions seem reasonable. For instance, it makes sense to define the area of a region of zero width and finite height to be 0.

Definition of Two Special Definite Integrals

1. If f is defined at $x = a$, then $\displaystyle\int_a^a f(x)\,dx = 0$.

2. If f is integrable on $[a, b]$, then $\displaystyle\int_b^a f(x)\,dx = -\int_a^b f(x)\,dx$.

 EXAMPLE 4   Evaluating Definite Integrals

Evaluate each of the definite integrals.

a. $\displaystyle\int_\pi^\pi \sin x\,dx$ **b.** $\displaystyle\int_3^0 (x + 2)\,dx$

Solution

a. Because the sine function is defined at $x = \pi$, and the upper and lower limits of integration are equal, you can write

$$\int_\pi^\pi \sin x\,dx = 0.$$

b. This integral is the same as that given in Example 3b except that the upper and lower limits are interchanged. Because the integral in Example 3b has a value of $\frac{21}{2}$, you can write

$$\int_3^0 (x + 2)\,dx = -\int_0^3 (x + 2)\,dx = -\frac{21}{2}.$$

THEOREM 4.6   Additive Interval Property

If f is integrable on the three closed intervals determined by a, b, and c, then

$$\int_a^b f(x)\,dx = \int_a^c f(x)\,dx + \int_c^b f(x)\,dx.$$

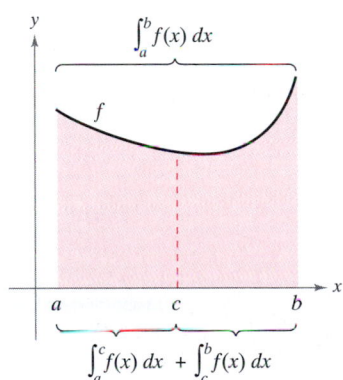

The larger region can be divided at $x = c$ into two subregions whose intersection is a line segment. Because the line segment has zero area, it follows that the area of the larger region is equal to the sum of the areas of the two smaller regions.

Figure 4.25

This theorem is valid for any integrable function and for any three numbers a, b, and c. Rather than give a formal proof for the general case, however, it seems more instructive to give a geometric argument for the case in which $a < c < b$ and f is continuous and nonnegative, as indicated in Figure 4.25.

Because the definite integral is defined as the limit of a sum, it inherits the properties of summation given at the top of page 253.

THEOREM 4.7   Properties of Definite Integrals

If f and g are integrable on $[a, b]$ and k is a constant, then the functions of kf and $f \pm g$ are integrable on $[a, b]$, and

1. $\displaystyle \int_a^b kf(x)\, dx = k \int_a^b f(x)\, dx$

2. $\displaystyle \int_a^b [f(x) \pm g(x)]\, dx = \int_a^b f(x)\, dx \pm \int_a^b g(x)\, dx.$

NOTE   Property 2 of Theorem 4.7 can be extended to cover any finite number of functions. For example,

$$\int_a^b [f(x) + g(x) + h(x)]\, dx = \int_a^b f(x)\, dx + \int_a^b g(x)\, dx + \int_a^b h(x)\, dx.$$

EXAMPLE 5   Evaluation of a Definite Integral

Evaluate $\displaystyle \int_1^3 (-x^2 + 4x - 3)\, dx$ using each of the following values.

$$\int_1^3 x^2\, dx = \frac{26}{3}, \qquad \int_1^3 x\, dx = 4, \qquad \int_1^3 dx = 2$$

Solution

$$\int_1^3 (-x^2 + 4x - 3)\, dx = \int_1^3 (-x^2)\, dx + \int_1^3 4x\, dx + \int_1^3 (-3)\, dx$$

$$= -\int_1^3 x^2\, dx + 4\int_1^3 x\, dx - 3\int_1^3 dx$$

$$= -\left(\frac{26}{3}\right) + 4(4) - 3(2)$$

$$= \frac{4}{3}$$

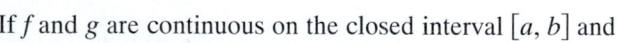

If f and g are continuous on the closed interval $[a, b]$ and

$$0 \le f(x) \le g(x)$$

for $a \le x \le b$, the following properties are true. First, the area of the region bounded by the graph of f and the x-axis (between a and b) must be nonnegative. Second, this area must be less than or equal to the area of the region bounded by the graph of g and the x-axis (between a and b), as shown in Figure 4.26. These two results are generalized in the following theorem. (A proof of this theorem is given in the appendix.)

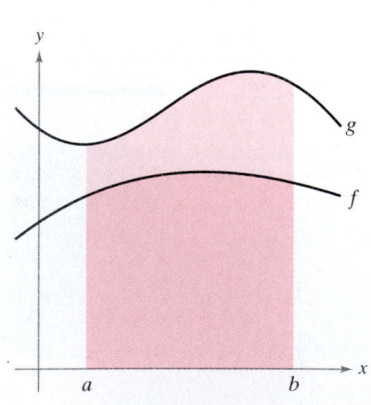

$$\int_a^b f(x)\, dx \le \int_a^b g(x)\, dx$$

Figure 4.26

THEOREM 4.8   Preservation of Inequality

1. If f is integrable and nonnegative on the closed interval $[a, b]$, then

$$0 \le \int_a^b f(x)\, dx.$$

2. If f and g are integrable on the closed interval $[a, b]$ and $f(x) \le g(x)$ for every x in $[a, b]$, then

$$\int_a^b f(x)\, dx \le \int_a^b g(x)\, dx.$$

EXERCISES FOR SECTION 4.3

In Exercises 1–10, set up a definite integral that yields the area of the given region. (Do not evaluate the integral.)

1. $f(x) = 3$

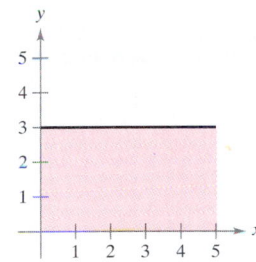

2. $f(x) = 4 - 2x$

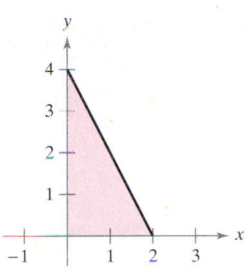

3. $f(x) = 4 - |x|$

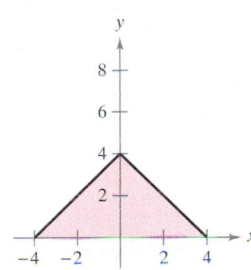

4. $f(x) = x^2$

5. $f(x) = 4 - x^2$

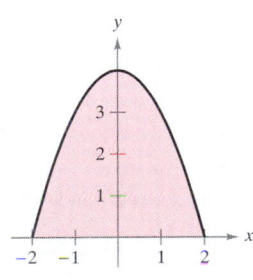

6. $f(x) = \dfrac{1}{x^2 + 1}$

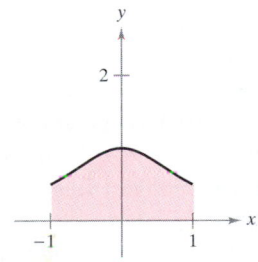

7. $f(x) = \sin x$

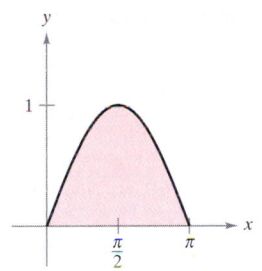

8. $f(x) = \tan x$

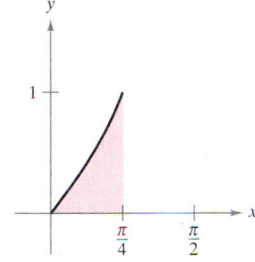

9. $g(y) = y^3$

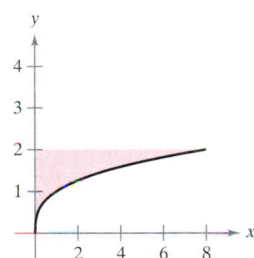

10. $f(y) = (y - 2)^2$

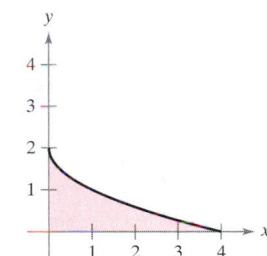

In Exercises 11–20, sketch the region whose area is given by the definite integral. Then use a geometric formula to evaluate the integral $(a > 0, r > 0)$.

11. $\displaystyle\int_0^3 4\, dx$

12. $\displaystyle\int_{-a}^a 4\, dx$

13. $\displaystyle\int_0^4 x\, dx$

14. $\displaystyle\int_0^4 \frac{x}{2}\, dx$

15. $\displaystyle\int_0^2 (2x + 5)\, dx$

16. $\displaystyle\int_0^5 (5 - x)\, dx$

17. $\displaystyle\int_{-1}^1 (1 - |x|)\, dx$

18. $\displaystyle\int_{-a}^a (a - |x|)\, dx$

19. $\displaystyle\int_{-3}^3 \sqrt{9 - x^2}\, dx$

20. $\displaystyle\int_{-r}^r \sqrt{r^2 - x^2}\, dx$

21. Given $\int_0^5 f(x)\,dx = 10$ and $\int_5^7 f(x)\,dx = 3$, find

(a) $\displaystyle\int_0^7 f(x)\,dx.$

(b) $\displaystyle\int_5^0 f(x)\,dx.$

(c) $\displaystyle\int_5^5 f(x)\,dx.$

(d) $\displaystyle\int_0^5 3f(x)\,dx.$

22. Given $\int_0^3 f(x)\,dx = 4$ and $\int_3^6 f(x)\,dx = -1$, find

(a) $\displaystyle\int_0^6 f(x)\,dx.$

(b) $\displaystyle\int_6^3 f(x)\,dx.$

(c) $\displaystyle\int_3^3 f(x)\,dx.$

(d) $\displaystyle\int_3^6 -5f(x)\,dx.$

23. Given $\int_2^6 f(x)\,dx = 10$ and $\int_2^6 g(x)\,dx = -2$, find

(a) $\displaystyle\int_2^6 [f(x) + g(x)]\,dx.$

(b) $\displaystyle\int_2^6 [g(x) - f(x)]\,dx.$

(c) $\displaystyle\int_2^6 2g(x)\,dx.$

(d) $\displaystyle\int_2^6 3f(x)\,dx.$

24. Given $\int_{-1}^1 f(x)\,dx = 0$ and $\int_0^1 f(x)\,dx = 5$, find

(a) $\displaystyle\int_{-1}^0 f(x)\,dx.$

(b) $\displaystyle\int_0^1 f(x)\,dx - \int_{-1}^0 f(x)\,dx.$

(c) $\displaystyle\int_{-1}^1 3f(x)\,dx.$

(d) $\displaystyle\int_0^1 3f(x)\,dx.$

In Exercises 25–30, evaluate the definite integral by the limit definition.

25. $\displaystyle\int_4^{10} 6\,dx$

26. $\displaystyle\int_{-2}^3 x\,dx$

27. $\displaystyle\int_{-1}^1 x^3\,dx$

28. $\displaystyle\int_0^1 x^3\,dx$

29. $\displaystyle\int_1^2 (x^2 + 1)\,dx$

30. $\displaystyle\int_1^2 4x^2\,dx$

In Exercises 31–34, express the limit as a definite integral on the interval $[a, b]$, where c_i is any point in the ith subinterval.

Limit	*Interval*
31. $\lim\limits_{\|\Delta\|\to 0} \sum\limits_{i=1}^n (3c_i + 10)\,\Delta x_i$	$[-1, 5]$
32. $\lim\limits_{\|\Delta\|\to 0} \sum\limits_{i=1}^n 6c_i(4 - c_i)^2\,\Delta x_i$	$[0, 4]$
33. $\lim\limits_{\|\Delta\|\to 0} \sum\limits_{i=1}^n \sqrt{c_i^2 + 4}\,\Delta x_i$	$[0, 3]$
34. $\lim\limits_{\|\Delta\|\to 0} \sum\limits_{i=1}^n \left(\dfrac{3}{c_i^2}\right)\Delta x_i$	$[1, 3]$

Write a program for your graphing utility to approximate a definite integral using the Riemann sum

$$\sum_{i=1}^n f(c_i)\,\Delta x_i$$

where the subintervals are of equal width. The output should give three approximations of the integral where c_i is the left-hand endpoint $L(n)$, midpoint $M(n)$, and right-hand endpoint $R(n)$ of each subinterval. In Exercises 35–38, use the program to approximate the definite integral and complete the table.

n	4	8	12	16	20
$L(n)$					
$M(n)$					
$R(n)$					

35. $\displaystyle\int_0^3 x\sqrt{3 - x}\,dx$

36. $\displaystyle\int_0^3 \dfrac{5}{x^2 + 1}\,dx$

37. $\displaystyle\int_0^{\pi/2} \sin^2 x\,dx$

38. $\displaystyle\int_0^3 x \sin x\,dx$

Think About It   **In Exercises 39–42, use the figure to fill in the blank with the symbol <, >, or =.**

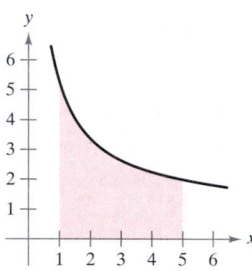

39. The interval $[1, 5]$ is partitioned into n subintervals of equal width Δx, and x_i is the left endpoint of the ith subinterval.

$$\sum_{i=1}^n f(x_i)\,\Delta x \quad \boxed{} \quad \int_1^5 f(x)\,dx$$

40. The interval $[1, 5]$ is partitioned into n subintervals of equal width Δx, and x_i is the right endpoint of the ith subinterval.

$$\sum_{i=1}^n f(x_i)\,\Delta x \quad \boxed{} \quad \int_1^5 f(x)\,dx$$

41. The interval $[1, 5]$ is partitioned into n subintervals of equal width Δx, and x_i is the midpoint of the ith subinterval.

$$\sum_{i=1}^n f(x_i)\,\Delta x \quad \boxed{} \quad \int_1^5 f(x)\,dx$$

42. Let T be the average of the results of Exercises 39 and 40.

$$T \quad \boxed{} \quad \int_1^5 f(x)\,dx$$

43. *Think About It* The graph of f below consists of line segments and a semicircle. Evaluate each definite integral by using geometric formulas.

(a) $\displaystyle\int_0^2 f(x)\,dx$ (b) $\displaystyle\int_2^6 f(x)\,dx$

(c) $\displaystyle\int_{-4}^2 f(x)\,dx$ (d) $\displaystyle\int_{-4}^6 f(x)\,dx$

(e) $\displaystyle\int_{-4}^6 |f(x)|\,dx$ (f) $\displaystyle\int_{-4}^6 [f(x)+2]\,dx$

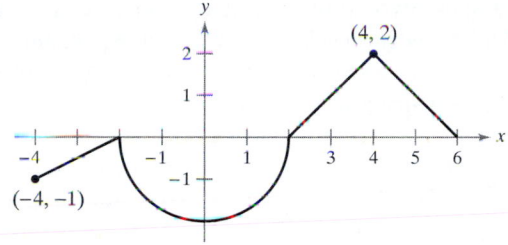

44. *Think About It* Consider the function f that is continuous on the interval $[-5,5]$ and for which $\int_0^5 f(x)\,dx = 4$. Evaluate each integral.

(a) $\displaystyle\int_0^5 [f(x)+2]\,dx$ (b) $\displaystyle\int_{-2}^3 f(x+2)\,dx$

(c) $\displaystyle\int_{-5}^5 f(x)\,dx$ (f is even.)   (d) $\displaystyle\int_{-5}^5 f(x)\,dx$ (f is odd.)

In Exercises 45 and 46, determine which value best approximates the definite integral. Make your selection on the basis of a sketch.

45. $\displaystyle\int_0^4 \sqrt{x}\,dx$

(a) 5   (b) -3 (c) 10   (d) 2   (e) 8

46. $\displaystyle\int_0^{1/2} 4\cos \pi x\,dx$

(a) 4   (b) $\frac{4}{3}$ (c) 16   (d) 2π (e) -6

True or False? **In Exercises 47–52, determine whether the statement is true or false. If it is false, explain why or give an example that shows it is false.**

47. $\displaystyle\int_a^b [f(x)+g(x)]\,dx = \int_a^b f(x)\,dx + \int_a^b g(x)\,dx$

48. $\displaystyle\int_a^b f(x)g(x)\,dx = \left[\int_a^b f(x)\,dx\right]\left[\int_a^b g(x)\,dx\right]$

49. If the norm of a partition approaches zero, then the number of subintervals approaches infinity.

50. If f is increasing on $[a,b]$, then the minimum value of $f(x)$ on $[a,b]$ is $f(a)$.

51. The value of $\displaystyle\int_a^b f(x)\,dx$ must be positive.

52. If $\displaystyle\int_a^b f(x)\,dx > 0$, then f is nonnegative for all x in $[a,b]$.

53. Find the Riemann sum for $f(x) = x^2 + 3x$ over the interval $[0,8]$, where $x_0 = 0$, $x_1 = 1$, $x_2 = 3$, $x_3 = 7$, and $x_4 = 8$, and where $c_1 = 1$, $c_2 = 2$, $c_3 = 5$, and $c_4 = 8$.

54. Find the Riemann sum for $f(x) = \sin x$ over the interval $[0,2\pi]$, where $x_0 = 0$, $x_1 = \pi/4$, $x_2 = \pi/3$, $x_3 = \pi$, and $x_4 = 2\pi$, and where $c_1 = \pi/6$, $c_2 = \pi/3$, $c_3 = 2\pi/3$, and $c_4 = 3\pi/2$.

In Exercises 55 and 56, use Example 1 as a model to evaluate the limit

$$\lim_{n\to\infty}\sum_{i=1}^n f(c_i)\,\Delta x_i$$

over the region bounded by the graphs of the equations.

55. $f(x) = \sqrt{x}$, $y = 0$, $x = 0$, $x = 2$

    (*Hint:* Let $c_i = 2i^2/n^2$.)

56. $f(x) = \sqrt[3]{x}$, $y = 0$, $x = 0$, $x = 1$

    (*Hint:* Let $c_i = i^3/n^3$.)

57. *Think About It* Determine whether the function

$$f(x) = \frac{1}{x-4}$$

is integrable on the interval $[3,5]$. Explain.

58. *Think About It* Determine whether the function

$$f(x) = \begin{cases} 1, & x \text{ is rational} \\ 0, & x \text{ is irrational} \end{cases}$$

is integrable on the interval $[0,1]$. Explain.

59. *Think About It* Give an example of a function that is integrable on the interval $[-1,1]$, but not continuous on $[-1,1]$.

60. Evaluate, if possible, the integral $\displaystyle\int_0^2 [\![x]\!]\,dx$.

61. Determine $\displaystyle\lim_{x\to\infty}\frac{1}{n^3}[1^2 + 2^2 + 3^2 + \cdots + n^2]$ by using an appropriate Riemann sum.

62. Suppose f is integrable on $[a,b]$ and $m \le f(x) \le M$ for all x in $[a,b]$, $m > 0$, and $M > 0$. Prove that

$$m(b-a) \le \int_a^b f(x)\,dx \le M(b-a).$$

Use the result to estimate $\displaystyle\int_0^1 \sqrt{1+x^4}\,dx$.

63. Prove that if f is a continuous function on a closed interval $[a,b]$, then

$$\left|\int_a^b f(x)\,dx\right| \le \int_a^b |f(x)|\,dx.$$

SECTION 4.4 **The Fundamental Theorem of Calculus**

The Fundamental Theorem of Calculus •
The Mean Value Theorem for Integrals • Average Value of a Function •
The Second Fundamental Theorem of Calculus

The Fundamental Theorem of Calculus

You have now been introduced to the two major branches of calculus: differential calculus (introduced with the tangent line problem) and integral calculus (introduced with the area problem). At this point, these two problems might seem unrelated—but there is a very close connection. The connection was discovered independently by Isaac Newton and Gottfried Leibniz. The connection is stated in a theorem that is appropriately called the **Fundamental Theorem of Calculus.**

Informally, the theorem states that differentiation and (definite) integration are inverse operations, in the same sense that division and multiplication are inverse operations. To see how Newton and Leibniz might have anticipated this relationship, consider the approximations shown in Figure 4.27. When we define the slope of the tangent line, we used the *quotient* $\Delta y/\Delta x$ (the slope of the secant line). Similarly, when we defined the area of a region under a curve, we used the *product* $\Delta y \Delta x$ (the area of a rectangle). Thus, at least in the primitive approximation stage, the operations of differentiation and definite integration appear to have an inverse relationship in the same sense that division and multiplication are inverse operations. The Fundamental Theorem of Calculus states that the limit processes (used to define the derivative and definite integral) preserve this inverse relationship.

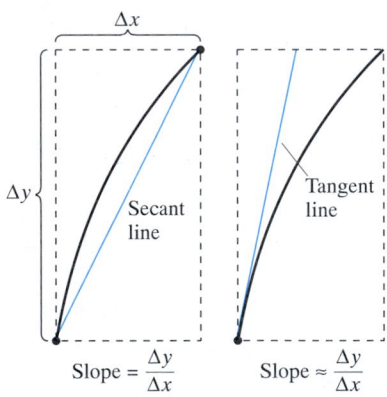
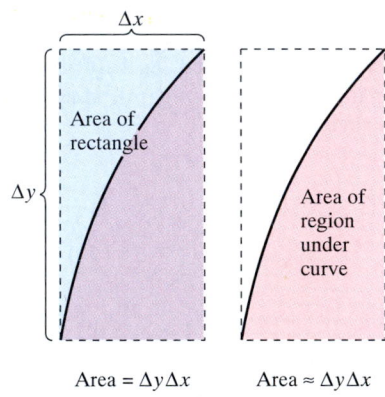

(a) Differentiation (b) Definite integration

Differentiation and definite integration have an "inverse" relationship.
Figure 4.27

THEOREM 4.9   The Fundamental Theorem of Calculus

If a function f is continuous on the closed interval $[a, b]$ and F is an antiderivative of f on the interval $[a, b]$, then

$$\int_a^b f(x)\,dx = F(b) - F(a).$$

Proof   The key to the proof is in writing the difference $F(b) - F(a)$ in a convenient form. Let Δ be the following partition of $[a, b]$.

$$a = x_0 < x_1 < x_2 < \cdots < x_{n-1} < x_n = b$$

By pairwise subtraction and addition of like terms, you can write

$$F(b) - F(a) = F(x_n) - F(x_{n-1}) + F(x_{n-1}) - \cdots - F(x_1) + F(x_1) - F(x_0)$$

$$= \sum_{i=1}^{n} [F(x_i) - F(x_{i-1})].$$

By the Mean Value Theorem, you know that there exists a number c_i in the ith subinterval such that

$$F'(c_i) = \frac{F(x_i) - F(x_{i-1})}{x_i - x_{i-1}}.$$

Because $F'(c_i) = f(c_i)$, you can let $\Delta x_i = x_i - x_{i-1}$ and obtain

$$F(b) - F(a) = \sum_{i=1}^{n} f(c_i)\Delta x_i.$$

This important equation tells you that by applying the Mean Value Theorem you can always find a collection of c_i's such that the *constant* $F(b) - F(a)$ is a Riemann sum of f on $[a, b]$. Taking the limit (as $\|\Delta\| \to 0$) produces

$$F(b) - F(a) = \int_a^b f(x)\, dx.$$

The following guidelines can help you understand the use of the Fundamental Theorem of Calculus.

Guidelines for Using the Fundamental Theorem of Calculus

1. *Provided you can find* an antiderivative of f, you now have a way to evaluate a definite integral without having to use the limit of a sum.

2. When applying the Fundamental Theorem of Calculus, the following notation is convenient.

$$\int_a^b f(x)\, dx = F(x) \Big]_a^b$$

$$= F(b) - F(a)$$

   For instance, to evaluate $\int_1^3 x^3\, dx$, you can write

$$\int_1^3 x^3\, dx = \frac{x^4}{4}\Big]_1^3 = \frac{3^4}{4} - \frac{1^4}{4} = \frac{81}{4} - \frac{1}{4} = 20.$$

3. It is not necessary to include a constant of integration C in the antiderivative because

$$\int_a^b f(x)\, dx = \Big[F(x) + C \Big]_a^b$$

$$= [F(b) + C] - [F(a) + C]$$

$$= F(b) - F(a).$$

EXAMPLE 1 Evaluating a Definite Integral

a. $\displaystyle\int_{1}^{2} (x^2 - 3)\, dx = \left[\frac{x^3}{3} - 3x \right]_{1}^{2} = \left(\frac{8}{3} - 6 \right) - \left(\frac{1}{3} - 3 \right) = -\frac{2}{3}$

b. $\displaystyle\int_{1}^{4} 3\sqrt{x}\, dx = 3 \int_{1}^{4} x^{1/2}\, dx = 3 \left[\frac{x^{3/2}}{3/2} \right]_{1}^{4} = 2(4)^{3/2} - 2(1)^{3/2} = 14$

c. $\displaystyle\int_{0}^{\pi/4} \sec^2 x\, dx = \tan x \Big]_{0}^{\pi/4} = 1 - 0 = 1$

EXAMPLE 2 A Definite Integral Involving Absolute Value

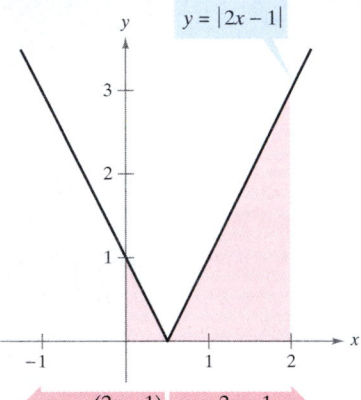

The definite integral of y on $[0, 2]$ is $\frac{5}{2}$.

Figure 4.28

Evaluate $\displaystyle\int_{0}^{2} |2x - 1|\, dx$.

Solution Using Figure 4.28 and the definition of absolute value, you can rewrite the integrand as follows.

$$|2x - 1| = \begin{cases} -(2x - 1), & x < \frac{1}{2} \\ 2x - 1, & x \geq \frac{1}{2} \end{cases}$$

From this, you can rewrite the integral in two parts.

$$\int_{0}^{2} |2x - 1|\, dx = \int_{0}^{1/2} -(2x - 1)\, dx + \int_{1/2}^{2} (2x - 1)\, dx$$

$$= \left[-x^2 + x \right]_{0}^{1/2} + \left[x^2 - x \right]_{1/2}^{2}$$

$$= \left(-\frac{1}{4} + \frac{1}{2} \right) - (0 + 0) + (4 - 2) - \left(\frac{1}{4} - \frac{1}{2} \right)$$

$$= \frac{5}{2}$$

EXAMPLE 3 Using the Fundamental Theorem to Find Area

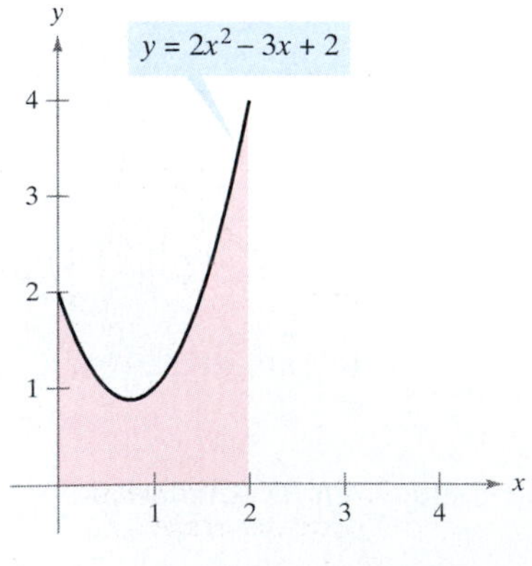

Find the area of the region bounded by the graph of $y = 2x^2 - 3x + 2$, the x-axis, and the vertical lines $x = 0$ and $x = 2$, as shown in Figure 4.29.

Solution Note that $y > 0$ on the interval $[0, 2]$.

$$\text{Area} = \int_{0}^{2} (2x^2 - 3x + 2)\, dx \qquad \textcolor{red}{\text{Integrate between } x = 0 \text{ and } x = 2.}$$

$$= \left[\frac{2x^3}{3} - \frac{3x^2}{2} + 2x \right]_{0}^{2} \qquad \textcolor{red}{\text{Find antiderivative, } F(x).}$$

$$= \left(\frac{16}{3} - 6 + 4 \right) - (0 - 0 + 0) \qquad \textcolor{red}{\text{Evaluate } F(2) - F(0).}$$

$$= \frac{10}{3} \qquad \textcolor{red}{\text{Simplify.}}$$

The area of the region bounded by the graph of y, the x-axis, $x = 0$, and $x = 2$ is $\frac{10}{3}$.

Figure 4.29

The Mean Value Theorem for Integrals

In Section 4.2, you saw that the area of a region under a curve is greater than the area of an inscribed rectangle and less than the area of a circumscribed rectangle. The Mean Value Theorem for Integrals states that somewhere "between" the inscribed and circumscribed rectangles there is a rectangle whose area is precisely equal to the area of the region under the curve, as shown in Figure 4.30.

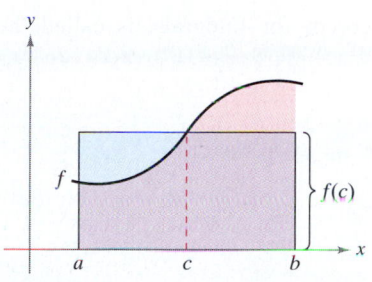

Mean value rectangle:

$$f(c)(b - a) = \int_a^b f(x)\,dx$$

Figure 4.30

> **THEOREM 4.10   Mean Value Theorem for Integrals**
>
> If f is continuous on the closed interval $[a, b]$, then there exists a number c in the closed interval $[a, b]$ such that
>
> $$\int_a^b f(x)\,dx = f(c)(b - a).$$

Proof

Case 1: If f is constant on the interval $[a, b]$, the theorem is clearly valid because c can be any point in $[a, b]$.

Case 2: If f is not constant on $[a, b]$, then, by the Extreme Value Theorem, you can choose $f(m)$ and $f(M)$ to be the minimum and maximum values of f on $[a, b]$. Because $f(m) \le f(x) \le f(M)$ for all x in $[a, b]$, you can apply Theorem 4.8 to write the following.

$$\int_a^b f(m)\,dx \le \int_a^b f(x)\,dx \le \int_a^b f(M)\,dx \qquad \text{See Figure 4.31.}$$

$$f(m)(b - a) \le \int_a^b f(x)\,dx \le f(M)(b - a)$$

$$f(m) \le \frac{1}{b - a}\int_a^b f(x)\,dx \le f(M)$$

From the third inequality, you can apply the Intermediate Value Theorem to conclude that there exists some c in $[a, b]$ such that

$$f(c) = \frac{1}{b - a}\int_a^b f(x)\,dx \qquad \text{or} \qquad f(c)(b - a) = \int_a^b f(x)\,dx.$$

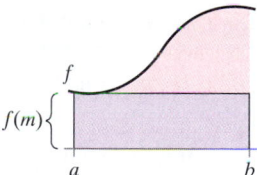
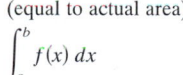
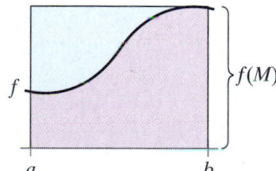

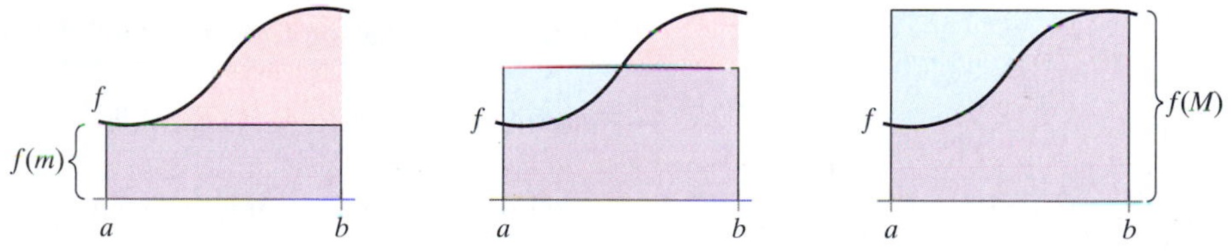

Inscribed rectangle
(less than actual area)

$$\int_a^b f(m)\,dx = f(m)(b - a)$$

Mean value rectangle
(equal to actual area)

$$\int_a^b f(x)\,dx$$

Circumscribed rectangle
(greater than actual area)

$$\int_a^b f(M)\,dx = f(M)(b - a)$$

Figure 4.31

NOTE Notice that Theorem 4.10 does not specify how to determine c. It merely guarantees the existence of at least one number c in the interval.

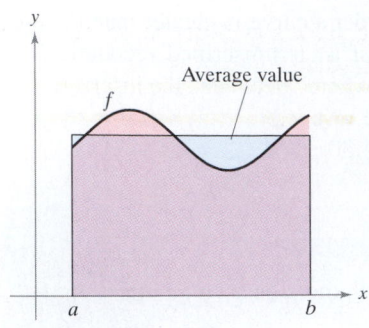

Average value $= \dfrac{1}{b - a} \displaystyle\int_a^b f(x)\, dx$

Figure 4.32

Average Value of a Function

The value of $f(c)$, given in the Mean Value Theorem for Integrals, is called the **average value** of f on the interval $[a, b]$.

> **Definition of the Average Value of a Function on an Interval**
>
> If f is integrable on the closed interval $[a, b]$, then the **average value** of f on the interval is
> $$\frac{1}{b - a}\int_a^b f(x)\, dx.$$

NOTE Notice in Figure 4.32 that the area of the region under the graph of f is equal to the area of the rectangle whose height is the average value.

To see why the average value of f is defined in this way, suppose that you partition $[a, b]$ into n subintervals of equal width $\Delta x = (b - a)/n$. If c_i is any point in the ith subinterval, the arithmetic average (or mean) of the function values at the c_i's is given by

$$a_n = \frac{1}{n}\big[f(c_1) + f(c_2) + \cdots + f(c_n)\big]. \qquad \text{Average of } f(c_1), \ldots, f(c_n)$$

By multiplying and dividing by $(b - a)$, you can write the average as

$$a_n = \frac{1}{n}\sum_{i=1}^{n} f(c_i)\left(\frac{b - a}{b - a}\right) = \frac{1}{b - a}\sum_{i=1}^{n} f(c_i)\left(\frac{b - a}{n}\right)$$

$$= \frac{1}{b - a}\sum_{i=1}^{n} f(c_i)\,\Delta x.$$

Finally, taking the limit as $n \to \infty$ produces the average value of f on the interval $[a, b]$, as given in the definition above.

This development of the average value of a function on an interval is only one of many practical uses of definite integrals to represent summation processes. In Chapter 6, you will study other applications, such as volume, arc length, centers of mass, and work.

EXAMPLE 4   Finding the Average Value of a Function

Find the average value of $f(x) = 3x^2 - 2x$ on the interval $[1, 4]$.

Solution   The average value is given by

$$\frac{1}{b - a}\int_a^b f(x)\, dx = \frac{1}{3}\int_1^4 (3x^2 - 2x)\, dx$$

$$= \frac{1}{3}\Big[x^3 - x^2\Big]_1^4$$

$$= \frac{1}{3}\big[64 - 16 - (1 - 1)\big] = \frac{48}{3} = 16.$$

(See Figure 4.33.)

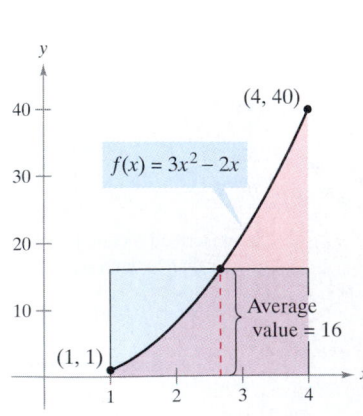

Figure 4.33

The first person to fly at a speed greater than the speed of sound was Charles Yeager. On October 14, 1947, flying in an *X-1* rocket plane at an altitude of 12.8 kilometers, Yeager was clocked at 299.5 meters per second. If Yeager had been flying at an altitude under 10.375 kilometers, his speed of 299.5 meters per second would not have "broken the sound barrier." The illustration above shows the *X-1* and its *B-29* mother plane.

EXAMPLE 5   The Speed of Sound

At different altitudes in earth's atmosphere, sound travels at different speeds. The speed of sound $s(x)$ (in meters per second) can be modeled by

$$s(x) = \begin{cases} -4x + 341, & 0 \le x < 11.5 \\ 295, & 11.5 \le x < 22 \\ \frac{3}{4}x + 278.5, & 22 \le x < 32 \\ \frac{3}{2}x + 254.5, & 32 \le x < 50 \\ -\frac{3}{2}x + 404.5, & 50 \le x < 80 \end{cases}$$

where x is the altitude in kilometers (see Figure 4.34). What is the average speed of sound over the interval $[0, 80]$?

Solution   Begin by integrating $s(x)$ over the interval $[0, 80]$. To do this, you can break the integral into five parts.

$$\int_0^{11.5} s(x)\, dx = \int_0^{11.5} (-4x + 341)\, dx = 3657$$

$$\int_{11.5}^{22} s(x)\, dx = \int_{11.5}^{22} (295)\, dx = 3097.5$$

$$\int_{22}^{32} s(x)\, dx = \int_{22}^{32} \left(\tfrac{3}{4}x + 278.5\right) dx = 2987.5$$

$$\int_{32}^{50} s(x)\, dx = \int_{32}^{50} \left(\tfrac{3}{2}x + 254.5\right) dx = 5688$$

$$\int_{50}^{80} s(x)\, dx = \int_{50}^{80} \left(-\tfrac{3}{2}x + 404.5\right) dx = 9210$$

By adding the values of the five integrals, you have

$$\int_0^{80} s(x)\, dx = 24{,}640.$$

Therefore, the average speed of sound from an altitude of 0 kilometers to an altitude of 80 kilometers is

$$\text{Average speed} = \frac{1}{80} \int_0^{80} s(x)\, dx = \frac{24{,}640}{80} = 308 \text{ meters per second.}$$

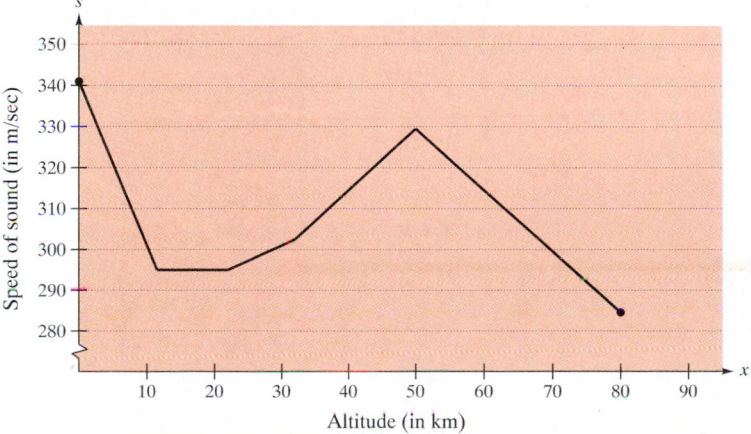

Speed of sound depends on altitude.
Figure 4.34

The Second Fundamental Theorem of Calculus

When we defined the definite integral of f on the interval $[a, b]$, we used the constant b as the upper limit of integration and x as the variable of integration. We now look at a slightly different situation in which the variable x is used as the upper limit of integration. To avoid the confusion of using x in two different ways, we temporarily switch to using t as the variable of integration. (Remember that the definite integral is *not* a function of its variable of integration.)

The Definite Integral as a Number

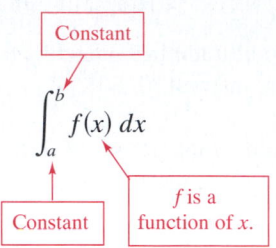

The Definite Integral as a Function of x

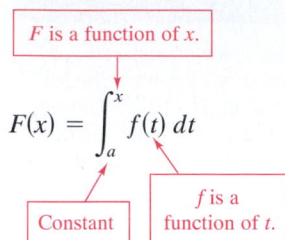

$$\int_a^b f(x)\, dx$$

Constant; f is a function of x.

$$F(x) = \int_a^x f(t)\, dt$$

F is a function of x. Constant; f is a function of t.

EXAMPLE 6   The Definite Integral as a Function

Evaluate the function

$$F(x) = \int_0^x \cos t\, dt$$

at $x = 0,\ \pi/6,\ \pi/4,\ \pi/3,$ and $\pi/2$.

Solution   You could evaluate five different definite integrals, one for each of the given upper limits. However, it is much simpler to fix x (as a constant) temporarily and apply the Fundamental Theorem once, to obtain

$$\int_0^x \cos t\, dt = \sin t\ \Big]_0^x$$

$$= \sin x - \sin 0$$

$$= \sin x.$$

Now, using $F(x) = \sin x$, you can obtain the results shown in Figure 4.35.

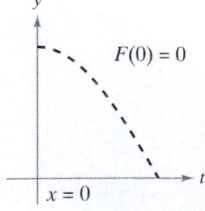
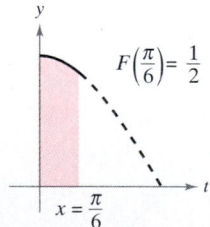
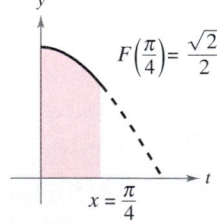
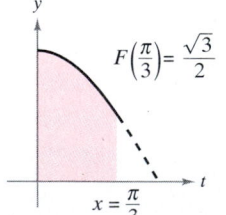
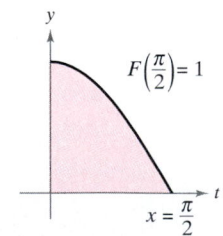

$F(x) = \displaystyle\int_0^x \cos t\, dt$ is the area under the curve $f(t) = \cos t$ from 0 to x.

Figure 4.35

In Example 6, note that the derivative of F is the original integrand (with only the variable changed). That is

$$\frac{d}{dx}[F(x)] = \frac{d}{dx}[\sin x] = \frac{d}{dx}\left[\int_0^x \cos t \, dt\right] = \cos x.$$

We generalize this result in the following theorem, called the **Second Fundamental Theorem of Calculus.**

THEOREM 4.11   The Second Fundamental Theorem of Calculus

If f is continuous on an open interval I containing a, then, for every x in the interval,

$$\frac{d}{dx}\left[\int_a^x f(t) \, dt\right] = f(x).$$

Proof   Begin by defining F as

$$F(x) = \int_a^x f(t) \, dt.$$

Then, by the definition of the derivative, you can write

$$F'(x) = \lim_{\Delta x \to 0} \frac{F(x + \Delta x) - F(x)}{\Delta x}$$

$$= \lim_{\Delta x \to 0} \frac{1}{\Delta x}\left[\int_a^{x + \Delta x} f(t) \, dt - \int_a^x f(t) \, dt\right]$$

$$= \lim_{\Delta x \to 0} \frac{1}{\Delta x}\left[\int_a^{x + \Delta x} f(t) \, dt + \int_x^a f(t) \, dt\right]$$

$$= \lim_{\Delta x \to 0} \frac{1}{\Delta x}\left[\int_x^{x + \Delta x} f(t) \, dt\right].$$

From the Mean Value Theorem for Integrals (assuming $\Delta x > 0$), you know there exists a number c in the interval $[x, x + \Delta x]$ such that the integral in the expression above is equal to $f(c) \Delta x$. Moreover, because $x \leq c \leq x + \Delta x$, it follows that $c \to x$ as $\Delta x \to 0$. Thus, you obtain

$$F'(x) = \lim_{\Delta x \to 0}\left[\frac{1}{\Delta x} f(c) \Delta x\right]$$

$$= \lim_{\Delta x \to 0} f(c)$$

$$= f(x).$$

A similar argument can be made for $\Delta x < 0$.   ■

NOTE   Using the area model for definite integrals, you can view the approximation

$$f(x) \Delta x \approx \int_x^{x + \Delta x} f(t) \, dt$$

as saying that the area of the rectangle of height $f(x)$ and width Δx is approximately equal to the area of the region lying between the graph of f and the x-axis on the interval $[x, x + \Delta x]$, as shown in Figure 4.36.

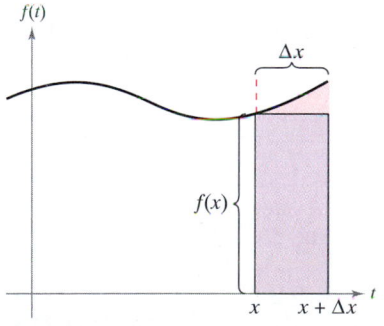

$$f(x) \, \Delta x \approx \int_x^{x + \Delta x} f(t) \, dt$$

Figure 4.36

Note that the Second Fundamental Theorem of Calculus tells you that if a function is continuous, you can be sure that it has an antiderivative. This antiderivative need not, however, be an elementary function. (Recall the discussion of elementary functions in Section P.3.)

EXAMPLE 7   Using the Second Fundamental Theorem of Calculus

Evaluate $\dfrac{d}{dx}\left[\displaystyle\int_0^x \sqrt{t^2+1}\,dt\right]$.

Solution Note that $f(t)=\sqrt{t^2+1}$ is continuous on the entire real line. Thus, using the Second Fundamental Theorem of Calculus, you can write

$$\frac{d}{dx}\left[\int_0^x \sqrt{t^2+1}\,dt\right]=\sqrt{x^2+1}.$$

The differentiation shown in Example 7 is a straightforward application of the Second Fundamental Theorem of Calculus. The next example shows how this theorem can be combined with the Chain Rule to find the derivative of a function.

EXAMPLE 8   Using the Second Fundamental Theorem of Calculus

Find the derivative of $F(x)=\displaystyle\int_{\pi/2}^{x^3}\cos t\,dt$

Solution Using $u=x^3$, you can apply the Second Fundamental Theorem of Calculus with the Chain Rule as follows.

$$F'(x)=\frac{dF}{du}\frac{du}{dx}$$
$$=\frac{d}{du}\left[\int_{\pi/2}^{u}\cos t\,dt\right]\frac{du}{dx}$$
$$=(\cos u)(3x^2)$$
$$=(\cos x^3)(3x^2)$$

Because the integrand in Example 8 is easily integrated, you can verify the derivative as follows.

$$F(x)=\int_{\pi/2}^{x^3}\cos t\,dt$$
$$=\sin t\,\Big]_{\pi/2}^{x^3}$$
$$=\sin x^3-\sin\frac{\pi}{2}$$
$$=(\sin x^3)-1$$

In this form, you can apply the Chain Rule to verify that the derivative is the same as that obtained in Example 8.

$$F'(x)=(\cos x^3)(3x^2)$$

EXERCISES FOR SECTION 4.4

Think About It   In Exercises 1–4, use a graphing utility to graph the integrand. Use the graph to determine whether the definite integral is positive, negative, or zero.

1. $\int_0^{\pi} \dfrac{4}{x^2 + 1}\, dx$

2. $\int_0^{\pi} \cos x\, dx$

3. $\int_{-2}^{2} x\sqrt{x^2 + 1}\, dx$

4. $\int_{-2}^{2} x\sqrt{2 - x}\, dx$

In Exercises 5–24, evaluate the definite integral of the algebraic function. Use a graphing utility to verify your result.

5. $\int_0^{1} 2x\, dx$

6. $\int_2^{7} 3\, dv$

7. $\int_{-1}^{0} (x - 2)\, dx$

8. $\int_2^{5} (-3v + 4)\, dv$

9. $\int_{-1}^{1} (t^2 - 2)\, dt$

10. $\int_0^{3} (3x^2 + x - 2)\, dx$

11. $\int_0^{1} (2t - 1)^2\, dt$

12. $\int_{-1}^{1} (t^3 - 9t)\, dt$

13. $\int_1^{2} \left(\dfrac{3}{x^2} - 1\right) dx$

14. $\int_{-2}^{-1} \left(u - \dfrac{1}{u^2}\right) du$

15. $\int_1^{4} \dfrac{u - 2}{\sqrt{u}}\, du$

16. $\int_{-3}^{3} v^{1/3}\, dv$

17. $\int_{-1}^{1} (\sqrt[3]{t} - 2)\, dt$

18. $\int_1^{8} \sqrt{\dfrac{2}{x}}\, dx$

19. $\int_0^{1} \dfrac{x - \sqrt{x}}{3}\, dx$

20. $\int_0^{2} (2 - t)\sqrt{t}\, dt$

21. $\int_{-1}^{0} (t^{1/3} - t^{2/3})\, dt$

22. $\int_{-8}^{-1} \dfrac{x - x^2}{2\sqrt[3]{x}}\, dx$

23. $\int_0^{3} |2x - 3|\, dx$

24. $\int_0^{4} |x^2 - 4x + 3|\, dx$

In Exercises 25–30, evaluate the definite integral of the trigonometric function. Use a graphing utility to verify your result.

25. $\int_0^{\pi} (1 + \sin x)\, dx$

26. $\int_0^{\pi/4} \dfrac{1 - \sin^2 \theta}{\cos^2 \theta}\, d\theta$

27. $\int_{-\pi/6}^{\pi/6} \sec^2 x\, dx$

28. $\int_{\pi/4}^{\pi/2} (2 - \csc^2 x)\, dx$

29. $\int_{-\pi/3}^{\pi/3} 4 \sec \theta \tan \theta\, d\theta$

30. $\int_{-\pi/2}^{\pi/2} (2t + \cos t)\, dt$

31. ***Depreciation***   A company purchases a new machine for which the rate of depreciation is $dV/dt = 10{,}000(t - 6)$, $0 \le t \le 5$, where V is the value of the machine after t years. Set up and evaluate the definite integral that yields the total loss of value of the machine over the first 3 years.

32. ***Buffon's Needle Experiment***   A horizontal plane is ruled with parallel lines 2 inches apart. If a 2-inch needle is tossed randomly onto the plane, the probability that the needle will touch a line is

$$P = \dfrac{2}{\pi} \int_0^{\pi/2} \sin \theta\, d\theta$$

where θ is the acute angle between the needle and any one of the parallel lines. Find this probability.

In Exercises 33–38, determine the area of the indicated region.

33. $y = x - x^2$

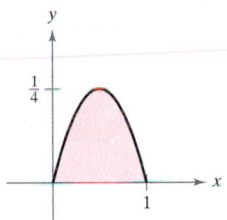

34. $y = 1 - x^4$

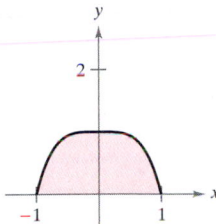

35. $y = (3 - x)\sqrt{x}$

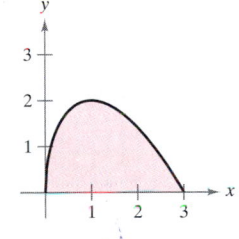

36. $y = \dfrac{1}{x^2}$

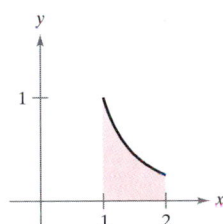

37. $y = \cos x$

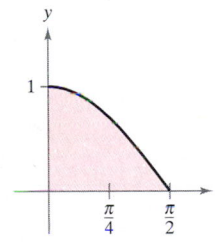

38. $y = x + \sin x$

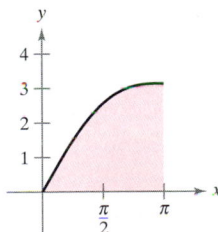

In Exercises 39–42, find the area of the region bounded by the graphs of the equations.

39. $y = 3x^2 + 1$, $x = 0$, $x = 2$, $y = 0$

40. $y = 1 + \sqrt{x}$, $x = 0$, $x = 4$, $y = 0$

41. $y = x^3 + x$, $x = 2$, $y = 0$

42. $y = -x^2 + 3x$, $y = 0$

In Exercises 43–46, find the value of c guaranteed by the *Mean Value Theorem for Integrals* for the function over the indicated interval.

Function	Interval
43. $f(x) = x - 2\sqrt{x}$	$[0, 2]$
44. $f(x) = \dfrac{9}{x^3}$	$[1, 3]$
45. $f(x) = 2\sec^2 x$	$[-\pi/4, \pi/4]$
46. $f(x) = \cos x$	$[-\pi/3, \pi/3]$

In Exercises 47–50, use a graphing utility to graph the function over the indicated interval. Find the average value of the function over the interval and all values of x in the interval for which the function equals its average value.

Function	Interval
47. $f(x) = 4 - x^2$	$[-2, 2]$
48. $f(x) = \dfrac{x^2 + 1}{x^2}$	$\left[\dfrac{1}{2}, 2\right]$
49. $f(x) = \sin x$	$[0, \pi]$
50. $f(x) = \cos x$	$[0, \pi/2]$

Think About It   In Exercises 51–56, use the graph of f shown in the figure. The shaded region A has an area of 1.5, and $\int_0^6 f(x)\,dx = 3.5$. Use this information to fill in the blanks.

51. $\displaystyle\int_0^2 f(x)\,dx = $ ▨

52. $\displaystyle\int_2^6 f(x)\,dx = $ ▨

53. $\displaystyle\int_0^6 |f(x)|\,dx = $ ▨

54. $\displaystyle\int_0^2 -2f(x)\,dx = $ ▨

55. $\displaystyle\int_0^6 [2 + f(x)]\,dx = $ ▨

56. The average value of f over the interval $[0, 6]$ is ▨.

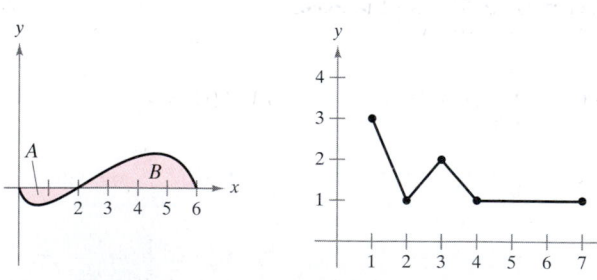

Figure for 51–56   **Figure for 57**

57. **Think About It**   The graph of f is given in the figure.

(a) Evaluate $\displaystyle\int_1^7 f(x)\,dx$.

(b) Determine the average value of f on the interval $[1, 7]$.

(c) Determine the answers to parts (a) and (b) if the graph is translated two units upward.

58. *Average Profit*   A company introduces a new product, and the profit in thousands of dollars over the first 6 months is approximated by the model

$$P = 5(\sqrt{t} + 30), \qquad t = 1, 2, 3, 4, 5, 6.$$

(a) Use the model to complete the table and use the entries to calculate (arithmetically) the average profit over the first 6 months.

t	1	2	3	4	5	6
P						

(b) Find the average value of the profit function by integration and compare the result with that in part (a). (Integrate over the interval $[0.5, 6.5]$.)

(c) What, if any, is the advantage of using the approximation of the average given by the definite integral? (Note that the integral approximation utilizes all real values of t in the interval rather than just integers.)

59. *Modeling Data*   A life insurance company needs a model to approximate the death rate of citizens during the years they are in the work force. The table gives the death rate R per 1000 for individuals of age x.   *(Source: Department of Health and Human Services)*

x	20	30	40	50	60
R	1.0	1.4	2.3	4.7	11.7

A model for these data is

$$R = -91.1 - 6.313x + 0.035x^2 + 45.794\sqrt{x},$$

$20 \le x \le 60$.

(a) Use a graphing utility to plot the data and graph the model.

(b) Find the rate of increase of the death rate when $x = 40$ and $x = 50$.

(c) Find the average death rate for people between the ages of 30 and 40 and for people between the ages of 50 and 60.

60. *Blood Flow*   The velocity v of the flow of blood at a distance r from the central axis of an artery of radius R is $v = k(R^2 - r^2)$, where k is the constant of proportionality. Find the average rate of flow of blood along a radius of the artery. (Use 0 and R as the limits of integration.)

61. *Force*   The force F (in newtons) of a hydraulic cylinder in a press is proportional to the square of $\sec x$, where x is the distance (in meters) that the cylinder is extended in its cycle. The domain of F is $[0, \pi/3]$, and $F(0) = 500$.

(a) Find F as a function of x.

(b) Find the average force exerted by the press over the interval $[0, \pi/3]$.

62. *Respiratory Cycle*   The volume V in liters of air in the lungs during a 5-second respiratory cycle is approximated by the model $V = 0.1729t + 0.1522t^2 - 0.0374t^3$ where t is the time in seconds. Approximate the average volume of air in the lungs during one cycle.

63. Modeling Data A department store manager wants to estimate the number of customers that enter the store from noon until closing at 9 P.M. The table shows the number of customers N entering the store during a randomly selected minute each hour from $t - 1$ to t, with $t = 0$ corresponding to noon.

t	1	2	3	4	5	6	7	8	9
N	6	7	9	12	15	14	11	7	2

(a) Draw a histogram of the data.

(b) Estimate the total number of customers entering the store between noon and 9 P.M.

(c) Use the regression capabilities of a graphing utility to find a model of the form

$$N(t) = at^3 + bt^2 + ct + d$$

for the data.

(d) Use a graphing utility to plot the data and graph the model.

(e) Use a graphing utility to evaluate $\int_0^9 N(t)\, dt$, and use the result to estimate the number of customers entering the store between noon and 9 P.M. Compare this with your answer in part (b).

(f) Estimate the average number of customers entering the store per minute between 3 P.M. and 7 P.M.

64. Modeling Data In the manufacturing process of a product, there is a repetitive heating cycle of 4 minutes. During a review of the process, the flow R (cubic feet per minute) of natural gas was measured in 1-minute intervals and the results were recorded in the table.

t	0	1	2	3	4
R	0	62	76	38	0

(a) Use a graphing utility to find a model of the form $R = at^4 + bt^3 + ct^2 + dt + e$ for the data.

(b) Use a graphing utility to plot the data and graph the model.

(c) Use the Fundamental Theorem of Calculus to approximate the number of cubic feet of natural gas used in one heating cycle.

65. Modeling Data A radio-controlled experimental vehicle is tested on a straight track. It starts from rest, and its velocity v (meters per second) is recorded in the table every 10 seconds for 1 minute.

t	0	10	20	30	40	50	60
v	0	5	21	40	62	78	83

(a) Use a graphing utility to find a model of the form $v = at^3 + bt^2 + ct + d$ for the data.

(b) Use a graphing utility to plot the data and graph the model.

(c) Use the Fundamental Theorem of Calculus to approximate the distance traveled by the vehicle during the test.

66. Use the function f in the figure and the function g defined by

$$g(x) = \int_0^x f(t)\, dt.$$

(a) Complete the table.

x	0	1	2	3	4	5	6	7	8	9	10
$g(x)$											

(b) Plot the points from the table in part (a).

(c) Where does g have its minimum? Explain.

(d) Which four consecutive points are collinear? Explain.

(e) Between which two consecutive points does g increase at the greatest rate? Explain.

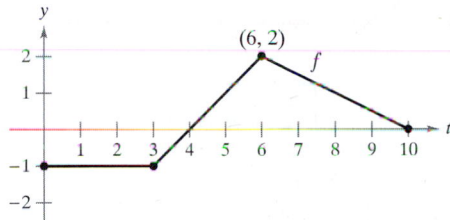

In Exercises 67–72, (a) integrate to find F as a function of x and (b) demonstrate the Second Fundamental Theorem of Calculus by differentiating the result in part (a).

67. $F(x) = \displaystyle\int_0^x (t + 2)\, dt$

68. $F(x) = \displaystyle\int_0^x t(t^2 + 1)\, dt$

69. $F(x) = \displaystyle\int_8^x \sqrt[3]{t}\, dt$

70. $F(x) = \displaystyle\int_4^x \sqrt{t}\, dt$

71. $F(x) = \displaystyle\int_{\pi/4}^x \sec^2 t\, dt$

72. $F(x) = \displaystyle\int_{\pi/3}^x \sec t \tan t\, dt$

In Exercises 73–78, use the Second Fundamental Theorem of Calculus to find $F'(x)$.

73. $F(x) = \displaystyle\int_{-2}^x (t^2 - 2t)\, dt$

74. $F(x) = \displaystyle\int_1^x \sqrt[4]{t}\, dt$

75. $F(x) = \displaystyle\int_{-1}^x \sqrt{t^4 + 1}\, dt$

76. $F(x) = \displaystyle\int_0^x \tan^4 t\, dt$

77. $F(x) = \displaystyle\int_0^x t \cos t\, dt$

78. $F(x) = \displaystyle\int_1^x \dfrac{t^2}{t^2 + 1}\, dt$

In Exercises 79–84, find $F'(x)$.

79. $F(x) = \displaystyle\int_x^{x+2} (4t + 1)\, dt$

80. $F(x) = \displaystyle\int_{-x}^x t^3\, dt$

81. $F(x) = \displaystyle\int_0^{\sin x} \sqrt{t}\, dt$

82. $F(x) = \displaystyle\int_2^{x^2} \dfrac{1}{t^2}\, dt$

83. $F(x) = \displaystyle\int_0^{x^3} \sin t^2\, dt$

84. $F(x) = \displaystyle\int_0^{3x} \sqrt{1 + t^3}\, dt$

In Exercises 85 and 86, sketch a graph of the function

$$F(x) = \int_0^x f(t)\, dt.$$

State any relationship that may exist between the extrema and inflection points on the graphs of f and F.

85.

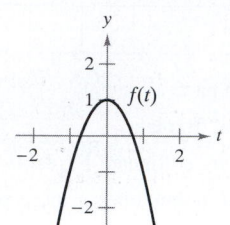

86.

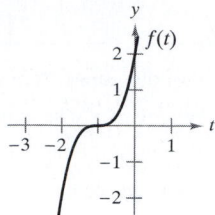

87. *Cost*   The total cost of purchasing and maintaining a piece of equipment for x years is

$$C(x) = 5000\left(25 + 3\int_0^x t^{1/4}\, dt\right).$$

(a) Perform the integration to write C as a function of x.

(b) Find $C(1)$, $C(5)$, and $C(10)$.

88. *Area*   The area A between the graph of the function $g(t) = 4 - 4/t^2$ and the t-axis over the interval $[1, x]$ is

$$A(x) = \int_1^x \left(4 - \frac{4}{t^2}\right) dt.$$

(a) Find the horizontal asymptote of the graph of g.

(b) Integrate to find A as a function of x. Does the graph of A have a horizontal asymptote? Explain.

True or False?   **In Exercises 89–91, determine whether the statement is true or false. If it is false, explain why or give an example that shows it is false.**

89. If $F'(x) = G'(x)$ on the interval $[a, b]$, then $F(b) - F(a) = G(b) - G(a)$.

90. If f is continuous on $[a, b]$, then f is integrable on $[a, b]$.

91. $\displaystyle\int_{-1}^1 x^{-2}\, dx = \left[-x^{-1}\right]_{-1}^1 = (-1) - 1 = -2$

92. Prove: $\dfrac{d}{dx}\left[\displaystyle\int_{u(x)}^{v(x)} f(t)\, dt\right] = f(v(x))v'(x) - f(u(x))u'(x).$

93. Show that the function

$$f(x) = \int_0^{1/x} \frac{1}{t^2 + 1}\, dt + \int_0^x \frac{1}{t^2 + 1}\, dt$$

is constant for $x > 0$.

94. Let $G(x) = \displaystyle\int_0^x \left[s\int_0^s f(t)\,dt\right] ds$, where f is continuous for all real t. Find

(a) $G(0)$.   (b) $G'(0)$.   (c) $G''(x)$.   (d) $G''(0)$.

95. *Area*   Archimedes showed that the area of a parabolic arch is equal to $\frac{2}{3}$ the product of the base and the height (see figure).

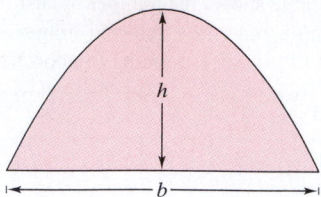

(a) Graph the parabolic arch bounded by $y = 9 - x^2$ and the x-axis. Use an appropriate definite integral to find the area A.

(b) Find the base and height of the arch in part (a) and verify that $A = \frac{2}{3}bh$.

(c) Verify Archimedes's formula for the parabolic arch bounded by $y = 5x - x^2$ and the x-axis.

Rectilinear Motion   **In Exercises 96–98, consider a particle moving along the x-axis where $x(t)$ is the position of the particle at time t, $x'(t)$ is its velocity, and $\int_a^b |x'(t)|\, dt$ is the distance the particle travels in the interval of time.**

96. The position function is $x(t) = t^3 - 6t^2 + 9t - 2$, $0 \le t \le 5$. Find the total distance the particle travels in 5 units of time.

97. Repeat Exercise 96 for the position function given by $x(t) = (t - 1)(t - 3)^2$, $0 \le t \le 5$.

98. A particle moves along the x-axis with velocity $v(t) = 1/\sqrt{t}$, $t > 0$. At time $t = 1$, its position is $x = 4$. Find the total distance traveled by the particle on the interval $1 \le t \le 4$.

SECTION PROJECT

Use a graphing utility to graph the function $y_1 = \sin^2 t$ on the interval $0 \le t \le \pi$. Let $F(x)$ be the following function of x.

$$F(x) = \int_0^x \sin^2 t\, dt$$

(a) Complete the table and explain why the values of F are increasing.

x	0	$\pi/6$	$\pi/3$	$\pi/2$	$2\pi/3$	$5\pi/6$	π
$F(x)$							

(b) Use the integration capabilities of a graphing utility to graph F.

(c) Use the differentiation capabilities of a graphing utility to graph $F'(x)$. How is this graph related to the graph in part (b)?

(d) Verify that the derivative of $y = (1/2)t - (\sin 2t)/4$ is $\sin^2 t$. Graph y and write a short paragraph about how this graph is related to those in parts (b) and (c).

Pattern Recognition • Change of Variables • The General Power Rule
for Integration • Change of Variables for Definite Integrals •
Integration of Even and Odd Functions

Pattern Recognition

In this section you will study techniques for integrating composite functions. The discussion is split into two parts—*pattern recognition* and *change of variables*. Both techniques involve a *u*-substitution. With pattern recognition you perform the substitution mentally, and with change of variables you write the substitution steps.

The role of substitution in integration is comparable to the role of the Chain Rule in differentiation. Recall that for differentiable functions given by $y = F(u)$ and $u = g(x)$, the Chain Rule states that

$$\frac{d}{dx}[F(g(x))] = F'(g(x))g'(x).$$

From the definition of an antiderivative, it follows that

$$\int F'(g(x))g'(x)\, dx = F(g(x)) + C$$

$$= F(u) + C.$$

These results are summarized in the following theorem.

NOTE   The statement of Theorem 4.12 doesn't tell how to distinguish between $f(g(x))$ and $g'(x)$ in the integrand. As you become more experienced at integration, your skill in doing this will increase. Of course, part of the key is familiarity with derivatives.

> **THEOREM 4.12   Antidifferentiation of a Composite Function**
>
> Let g be a function whose range is an interval I, and let f be a function that is continuous on I. If g is differentiable on its domain and F is an antiderivative of f on I, then
>
> $$\int f(g(x))g'(x)\, dx = F(g(x)) + C.$$
>
> If $u = g(x)$, then $du = g'(x)\, dx$ and
>
> $$\int f(u)\, du = F(u) + C.$$

STUDY TIP   There are several techniques for applying substitution, each differing slightly from the others. However, you should remember that the goal is the same with every technique— *you are trying to find an antiderivative of the integrand.*

> ## EXPLORATION
>
> ***Recognizing Patterns***   The integrand in each of the following integrals fits the pattern $f(g(x))g'(x)$. Identify the pattern and use the result to evaluate the integral.
>
> **(a)** $\displaystyle\int 2x(x^2 + 1)^4\, dx$ **(b)** $\displaystyle\int 3x^2\sqrt{x^3 + 1}\, dx$ **(c)** $\displaystyle\int \sec^2 x(\tan x + 3)\, dx$
>
> The next three integrals are similar to the first three. Show how you can multiply and divide by a constant to evaluate these integrals.
>
> **(d)** $\displaystyle\int x(x^2 + 1)^4\, dx$ **(e)** $\displaystyle\int x^2\sqrt{x^3 + 1}\, dx$ **(f)** $\displaystyle\int 2\sec^2 x(\tan x + 3)\, dx$

Examples 1 and 2 show how to apply the theorem *directly*, by recognizing the presence of $f(g(x))$ and $g'(x)$. Note that the composite function in the integrand has an *outside function f* and an *inside function g*. Moreover, the derivative $g'(x)$ is present as a factor of the integrand.

$$\int \underset{\text{Inside function}}{\underline{f(\underbrace{g(x)}}})\underbrace{g'(x)}_{\substack{\text{Derivative of}\\\text{inside function}}}\, dx = F(g(x)) + C$$

Outside function

EXAMPLE 1   Recognizing the $f(g(x))g'(x)$ Pattern

Evaluate $\displaystyle\int (x^2 + 1)^2(2x)\, dx.$

Solution   Letting $g(x) = x^2 + 1$, you obtain

$$g'(x) = 2x$$

and

$$f(g(x)) = [g(x)]^2.$$

From this, you can recognize that the integrand follows the $f(g(x))g'(x)$ pattern. Using the Power Rule for integration and Theorem 4.12, you can write

$$\int \overbrace{(x^2 + 1)^2}^{[g(x)]^2}\overbrace{(2x)}^{g'(x)}\, dx = \frac{1}{3}(x^2 + 1)^3 + C.$$

Try using the Chain Rule to check that the derivative of $\frac{1}{3}(x^2 + 1)^3 + C$ is the integrand of the original integral.

TECHNOLOGY   Try using a symbolic integration utility, such as *Maple, Derive, Mathematica, Mathcad,* or the *TI-92*, to solve the integrals given in Example 1 and 2. Do you obtain the same antiderivatives that are listed in the examples?

EXAMPLE 2   Recognizing the $f(g(x))g'(x)$ Pattern

Evaluate $\displaystyle\int 5\cos 5x\, dx.$

Solution   Letting $g(x) = 5x$, you obtain

$$g'(x) = 5$$

and

$$f(g(x)) = \cos[g(x)].$$

From this, you can recognize that the integrand follows the $f(g(x))g'(x)$ pattern. Using the Cosine Rule for integration and Theorem 4.12, you can write

$$\int \overbrace{(\cos 5x)}^{\cos[g(x)]}\overbrace{(5)}^{g'(x)}\, dx = \sin 5x + C.$$

You can check this by differentiating $\sin 5x + C$ to obtain the original integrand.

The integrands in Examples 1 and 2 fit the $f(g(x))g'(x)$ pattern exactly—you only had to recognize the pattern. You can extend this technique considerably with the Constant Multiple Rule

$$\int kf(x)\,dx = k\int f(x)\,dx.$$

Many integrands contain the essential part (the variable part) of $g'(x)$ but are missing a constant multiple. In such cases, you can multiply and divide by the necessary constant multiple.

EXAMPLE 3   Multiplying and Dividing by a Constant

Evaluate $\int x(x^2 + 1)^2\,dx.$

Solution   This is similar to the integral given in Example 1, except that the integrand is missing a factor of 2. Recognizing that $2x$ is the derivative of $x^2 + 1$, you can let $g(x) = x^2 + 1$ and supply the $2x$ as follows.

$$\int x(x^2 + 1)^2\,dx = \int (x^2 + 1)^2 \left(\frac{1}{2}\right)(2x)\,dx \qquad \text{Multiply and divide by 2.}$$

$$= \frac{1}{2}\int (x^2 + 1)^2\,(2x)\,dx \qquad \text{Constant Multiple Rule}$$

$$= \frac{1}{2}\int [g(x)]^2\,g'(x)\,dx$$

$$= \frac{1}{2}\frac{[g(x)]^3}{3} + C \qquad \text{Integrate.}$$

$$= \frac{1}{6}(x^2 + 1)^3 + C$$

In practice, most people would not write as many steps as are shown in Example 3. For instance, you could evaluate the integral by simply writing

$$\int x(x^2 + 1)^2\,dx = \frac{1}{2}\int (x^2 + 1)^2\,2x\,dx$$

$$= \frac{1}{2}\frac{(x^2 + 1)^3}{3} + C$$

$$= \frac{1}{6}(x^2 + 1)^3 + C.$$

NOTE   Be sure you see that the *Constant* Multiple Rule applies only to *constants*. You cannot multiply and divide by a variable and then move the variable outside the integral sign. For instance,

$$\int (x^2 + 1)^2\,dx \neq \frac{1}{2x}\int (x^2 + 1)^2\,(2x)\,dx.$$

After all, if it were legitimate to move variable quantities outside the integral sign, you could move the entire integrand out and simplify the whole process. But the result would be incorrect.

Change of Variables

With a formal **change of variables,** you completely rewrite the integral in terms of u and du (or any other convenient variable). Although this procedure can involve more written steps than the pattern recognition illustrated in Examples 1 to 3, it is useful for complicated integrands. The change of variable technique uses the Leibniz notation for the differential. That is, if $u = g(x)$, then $du = g'(x)\,dx$, and the integral in Theorem 4.12 takes the form

$$\int f(g(x))g'(x)\,dx = \int f(u)\,du = F(u) + C.$$

EXAMPLE 4   Change of Variables

Evaluate $\displaystyle\int \sqrt{2x - 1}\,dx.$

Solution   First, let u be the inner function, $u = 2x - 1$. Then calculate the differential du to be $du = 2\,dx$. Now, using $\sqrt{2x - 1} = \sqrt{u}$ and $dx = du/2$, substitute to obtain the following.

$$\int \sqrt{2x - 1}\,dx = \int \sqrt{u}\left(\frac{du}{2}\right) \qquad \text{\color{red}Integral in terms of } u$$

$$= \frac{1}{2}\int u^{1/2}\,du$$

$$= \frac{1}{2}\left(\frac{u^{3/2}}{3/2}\right) + C \qquad \text{\color{red}Antiderivative in terms of } u$$

$$= \frac{1}{3}u^{3/2} + C$$

$$= \frac{1}{3}(2x - 1)^{3/2} + C \qquad \text{\color{red}Antiderivative in terms of } x$$

> **STUDY TIP**   Because integration is usually more difficult than differentiation, you should always check your answer to an integration problem by differentiating. For instance, in Example 4 you should differentiate $\frac{1}{3}(2x - 1)^{3/2} + C$ to verify that you obtain the original integrand.

EXAMPLE 5   Change of Variables

Evaluate $\displaystyle\int x\sqrt{2x - 1}\,dx.$

Solution   As in the previous example, let $u = 2x - 1$ and obtain $dx = du/2$. Because the integrand contains a factor of x, you must also solve for x in terms of u, as follows.

$$u = 2x - 1 \quad \Longrightarrow \quad x = (u + 1)/2 \qquad \text{\color{red}Solve for } x \text{ in terms of } u.$$

Now, using substitution, you obtain the following.

$$\int x\sqrt{2x - 1}\,dx = \int \left(\frac{u + 1}{2}\right) u^{1/2}\left(\frac{du}{2}\right)$$

$$= \frac{1}{4}\int (u^{3/2} + u^{1/2})\,du$$

$$= \frac{1}{4}\left(\frac{u^{5/2}}{5/2} + \frac{u^{3/2}}{3/2}\right) + C$$

$$= \frac{1}{10}(2x - 1)^{5/2} + \frac{1}{6}(2x - 1)^{3/2} + C$$

To complete the change of variables in Example 5, we solved for x in terms of u. Sometimes this is very difficult. Fortunately it is not always necessary, as shown in the next example.

EXAMPLE 6 Change of Variables

Evaluate $\displaystyle\int \sin^2 3x \cos 3x \, dx$.

Solution Because $\sin^2 3x = (\sin 3x)^2$, you can let $u = \sin 3x$. Then

$$du = (\cos 3x)(3) \, dx.$$

Now, because $\cos 3x \, dx$ is part of the given integral, you can write

$$\frac{du}{3} = \cos 3x \, dx.$$

Substituting u and $du/3$ in the given integral yields the following.

$$\int \sin^2 3x \cos 3x \, dx = \int u^2 \frac{du}{3}$$

$$= \frac{1}{3}\int u^2 \, du$$

$$= \frac{1}{3}\left(\frac{u^3}{3}\right) + C$$

$$= \frac{1}{9}\sin^3 3x + C$$

STUDY TIP When making a change of variables, be sure that your answer is written using the same variables as in the original integrand. For instance, in Example 6, you should not leave your answer as

$$\frac{1}{9}u^3 + C,$$

but rather, replace u by $\sin 3x$.

You can check this by differentiating.

$$\frac{d}{dx}\left[\frac{1}{9}\sin^3 3x\right] = \left(\frac{1}{9}\right)(3)(\sin 3x)^2(\cos 3x)(3)$$

$$= \sin^2 3x \cos 3x$$

Because differentiation produces the original integrand, you know that you have obtained the correct antiderivative.

We summarize the steps used for integration by substitution in the following guidelines.

Guidelines for Making a Change of Variables

1. Choose a substitution $u = g(x)$. Usually, it is best to choose the *inner* part of a composite function, such as a quantity raised to a power.
2. Compute $du = g'(x) \, dx$.
3. Rewrite the integral in terms of the variable u.
4. Evaluate the resulting integral in terms of u.
5. Replace u by $g(x)$ to obtain an antiderivative in terms of x.
6. Check your answer by differentiating.

The General Power Rule for Integration

One of the most common u-substitutions involves quantities in the integrand that are raised to a power. Because of the importance of this type of substitution, it is given a special name—the **General Power Rule.** A proof of this rule follows directly from the (simple) Power Rule for integration, together with Theorem 4.12.

> **THEOREM 4.13   The General Power Rule for Integration**
>
> If g is a differentiable function of x, then
>
> $$\int [g(x)]^n \, g'(x) \, dx = \frac{[g(x)]^{n+1}}{n+1} + C, \quad n \neq -1.$$
>
> Equivalently, if $u = g(x)$, then
>
> $$\int u^n \, du = \frac{u^{n+1}}{n+1} + C, \quad n \neq -1.$$

EXAMPLE 7   Substitution and the General Power Rule

a. $\displaystyle \int 3(3x-1)^4 \, dx = \int \overbrace{(3x-1)^4}^{u^4}\overbrace{(3)dx}^{du} = \overbrace{\frac{(3x-1)^5}{5}}^{u^5/5} + C$

b. $\displaystyle \int (2x+1)(x^2+x)dx = \int \overbrace{(x^2+x)^1}^{u^1}\overbrace{(2x+1)dx}^{du} = \overbrace{\frac{(x^2+x)^2}{2}}^{u^2/2} + C$

c. $\displaystyle \int 3x^2\sqrt{x^3-2}\, dx = \int \overbrace{(x^3-2)^{1/2}}^{u^{1/2}}\overbrace{(3x^2)dx}^{du} = \overbrace{\frac{(x^3-2)^{3/2}}{3/2}}^{u^{3/2}/(3/2)} + C$

d. $\displaystyle \int \frac{-4x}{(1-2x^2)^2}\, dx = \int \overbrace{(1-2x^2)^{-2}}^{u^{-2}}\overbrace{(-4x)\, dx}^{du} = \overbrace{\frac{(1-2x^2)^{-1}}{-1}}^{u^{-1}/(-1)} + C$

e. $\displaystyle \int \cos^2 x \sin x \, dx = -\int \overbrace{(\cos x)^2}^{u^2}\overbrace{(-\sin x)\, dx}^{du} = -\overbrace{\frac{(\cos x)^3}{3}}^{u^3/3} + C$

Some integrals whose integrands involve a quantity raised to a power cannot be evaluated by the General Power Rule. Consider the two integrals

$$\int x(x^2+1)^2 \, dx \quad \text{and} \quad \int (x^2+1)^2 \, dx.$$

The substitution $u = x^2 + 1$ works in the first integral but not in the second. (In the second, the substitution fails because the integrand lacks the factor x needed for du.) Fortunately, *for this particular integral*, you can expand the integrand as $(x^2+1)^2 = x^4 + 2x^2 + 1$ and use the (simple) Power Rule to integrate each term.

EXPLORATION

Suppose you were asked to evaluate one of the following integrals. Which one would you choose? Explain your reasoning.

(a) $\displaystyle \int \sqrt{x^3+1}\, dx$ or

$\displaystyle \int x^2\sqrt{x^3+1}\, dx$

(b) $\displaystyle \int x[\tan(x^2)\sec(x^2)]\, dx$ or

$\displaystyle \int x\tan(x^2)\, dx$

Change of Variables for Definite Integrals

When using u-substitution with a definite integral, it is often convenient to determine the limits of integration for the variable u rather than to convert the antiderivative back to the variable x and evaluate at the original limits. This change of variables is stated explicitly in the next theorem. The proof follows from Theorem 4.12 combined with the Fundamental Theorem of Calculus.

> **THEOREM 4.14   Change of Variables for Definite Integrals**
>
> If the function $u = g(x)$ has a continuous derivative on the closed interval $[a, b]$ and f is continuous on the range of g, then
>
> $$\int_a^b f(g(x))g'(x)\,dx = \int_{g(a)}^{g(b)} f(u)\,du.$$

EXAMPLE 8   Change of Variables

Evaluate $\displaystyle\int_0^1 x(x^2 + 1)^3 dx.$

Solution   To evaluate this integral, let $u = x^2 + 1$. Then, you obtain

$$u = x^2 + 1 \Rightarrow du = 2x\,dx.$$

Before substituting, determine the new upper and lower limits of integration.

Lower Limit	Upper Limit
When $x = 0$, $u = 0^2 + 1 = 1$.	When $x = 1$, $u = 1^2 + 1 = 2$.

Now, you can substitute to obtain

$$\int_0^1 x(x^2 + 1)^3\,dx = \frac{1}{2}\int_0^1 (x^2 + 1)^3 (2x)\,dx \qquad \text{Integration limits for } x$$

$$= \frac{1}{2}\int_1^2 u^3\,du \qquad \text{Integration limits for } u$$

$$= \frac{1}{2}\left[\frac{u^4}{4}\right]_1^2$$

$$= \frac{1}{2}\left(4 - \frac{1}{4}\right)$$

$$= \frac{15}{8}.$$

Try converting the antiderivative $\frac{1}{2}(u^4/4)$ back to the variable x and evaluate the definite integral at the original limits of integration. You should obtain the same result.

EXAMPLE 9   **Change of Variables**

Evaluate $A = \displaystyle\int_1^5 \frac{x}{\sqrt{2x - 1}}\, dx$.

Solution   To evaluate this integral, let $u = \sqrt{2x - 1}$. Then, you obtain

$$u^2 = 2x - 1$$
$$u^2 + 1 = 2x$$
$$\frac{u^2 + 1}{2} = x$$
$$u\, du = dx. \qquad \textcolor{red}{\text{Differentiate both sides.}}$$

Before substituting, determine the new upper and lower limits of integration.

Lower Limit	*Upper Limit*
When $x = 1$, $u = \sqrt{2 - 1} = 1$.	When $x = 5$, $u = \sqrt{10 - 1} = 3$.

Now, substitute to obtain

$$\int_1^5 \frac{x}{\sqrt{2x - 1}}\, dx = \int_1^3 \frac{1}{u}\left(\frac{u^2 + 1}{2}\right) u\, du$$

$$= \frac{1}{2}\int_1^3 (u^2 + 1)\, du$$

$$= \frac{1}{2}\left[\frac{u^3}{3} + u\right]_1^3$$

$$= \frac{1}{2}\left(9 + 3 - \frac{1}{3} - 1\right)$$

$$= \frac{16}{3}.$$

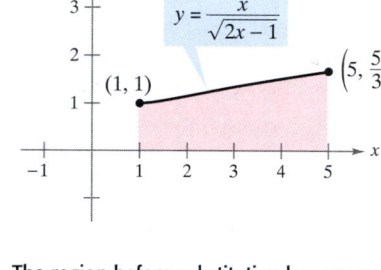

The region before substitution has an area of $\frac{16}{3}$.
Figure 4.37

Geometrically, you can interpret the equation

$$\int_1^5 \frac{x}{\sqrt{2x - 1}}\, dx = \int_1^3 \frac{u^2 + 1}{2}\, du$$

to mean that the two *different* regions shown in Figures 4.37 and 4.38 have the *same* area.

When evaluating definite integrals by substitution, it is possible for the upper limit of integration of the u-variable form to be smaller than the lower limit. If this happens, don't rearrange the limits. Simply evaluate as usual. For example, after substituting $u = \sqrt{1 - x}$ in the integral

$$\int_0^1 x^2(1 - x)^{1/2}\, dx$$

you obtain $u = \sqrt{1 - 1} = 0$ when $x = 1$, and $u = \sqrt{1 - 0} = 1$ when $x = 0$. Thus, the correct u-variable form of this integral is

$$-2\int_1^0 (1 - u^2)^2 u^2\, du.$$

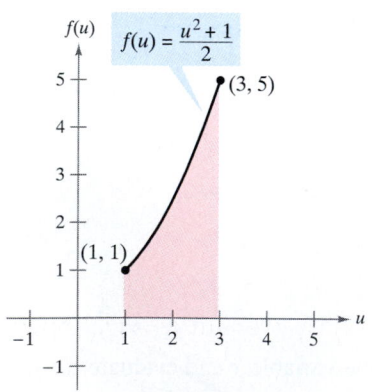

The region after substitution has an area of $\frac{16}{3}$.
Figure 4.38

Integration of Even and Odd Functions

Even with a change of variables, integration can be difficult. Occasionally, you can simplify the evaluation of a definite integral (over an interval that is symmetric about the y-axis or about the origin) by recognizing the integrand to be an even or odd function (see Figure 4.39).

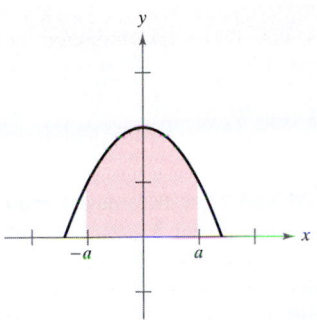

Even function

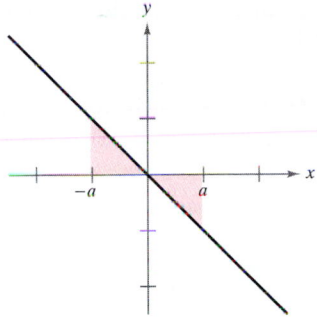

Odd function
Figure 4.39

> ### THEOREM 4.15   Integration of Even and Odd Functions
>
> Let f be integrable on the closed interval $[-a, a]$.
>
> 1. If f is an *even* function, then $\displaystyle\int_{-a}^{a} f(x)\, dx = 2\int_{0}^{a} f(x)\, dx$.
>
> 2. If f is an *odd* function, then $\displaystyle\int_{-a}^{a} f(x)\, dx = 0$.

Proof   Because f is even, you know that $f(x) = f(-x)$. Using Theorem 4.12 with the substitution $u = -x$ produces

$$\int_{-a}^{0} f(x)\, dx = \int_{a}^{0} f(-u)(-du) = -\int_{a}^{0} f(u)\, du = \int_{0}^{a} f(u)\, du = \int_{0}^{a} f(x)\, dx.$$

Finally, using Theorem 4.6, you obtain

$$\int_{-a}^{a} f(x)\, dx = \int_{-a}^{0} f(x)\, dx + \int_{0}^{a} f(x)\, dx$$

$$= \int_{0}^{a} f(x)\, dx + \int_{0}^{a} f(x)\, dx = 2\int_{0}^{a} f(x)\, dx.$$

This proves the first property. The proof of the second property is left to you.

EXAMPLE 10   Integration of an Odd Function

Evaluate $\displaystyle\int_{-\pi/2}^{\pi/2} (\sin^3 x \cos x + \sin x \cos x)\, dx$.

Solution   Letting $f(x) = \sin^3 x \cos x + \sin x \cos x$ produces

$$f(-x) = \sin^3(-x)\cos(-x) + \sin(-x)\cos(-x)$$

$$= -\sin^3 x \cos x - \sin x \cos x = -f(x).$$

Thus, f is an odd function, and because $[-\pi/2, \pi/2]$ is symmetric about the origin, you can apply Theorem 4.15 to conclude that

$$\int_{-\pi/2}^{\pi/2} (\sin^3 x \cos x + \sin x \cos x)\, dx = 0.$$

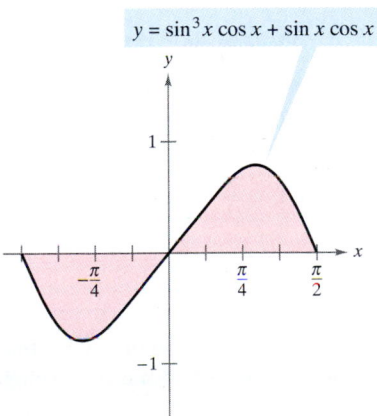

$y = \sin^3 x \cos x + \sin x \cos x$

Because f is an odd function,
$\displaystyle\int_{-\pi/2}^{\pi/2} f(x)\, dx = 0.$

Figure 4.40

NOTE   From Figure 4.40 you can see that the two regions on either side of the y-axis have the same area. However, because one lies below the x-axis and one lies above it, integration produces a cancellation effect. (We will say more about this in Section 6.1.)

EXERCISES FOR SECTION 4.5

In Exercises 1–6, complete the table by identifying u and du for the integral.

$$\int f(g(x))g'(x)\,dx \qquad u = g(x) \qquad du = g'(x)\,dx$$

1. $\int (5x^2 + 1)^2 (10x)\,dx$

2. $\int x^2 \sqrt{x^3 + 1}\,dx$

3. $\int \dfrac{x}{\sqrt{x^2 + 1}}\,dx$

4. $\int \sec 2x \tan 2x\,dx$

5. $\int \tan^2 x \sec^2 x\,dx$

6. $\int \dfrac{\cos x}{\sin^2 x}\,dx$

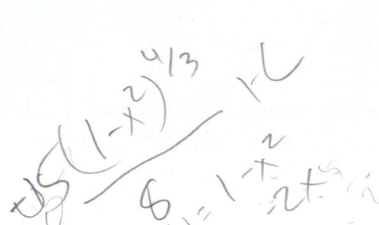

In Exercises 7–26, evaluate the indefinite integral and check the result by differentiation.

7. $\int (1 + 2x)^4 (2)\,dx$

8. $\int (x^2 - 1)^3 (2x)\,dx$

9. $\int \sqrt{9 - x^2}\,(-2x)\,dx$

10. $\int (1 - 2x^2)^3(-4x)\,dx$

11. $\int x^2(x^3 - 1)^4\,dx$

12. $\int x(4x^2 + 3)^3\,dx$

13. $\int 5x \sqrt[3]{1 - x^2}\,dx$

14. $\int u^3 \sqrt{u^4 + 2}\,du$

15. $\int \dfrac{x^2}{(1 + x^3)^2}\,dx$

16. $\int \dfrac{x^2}{(16 - x^3)^2}\,dx$

17. $\int \left(1 + \dfrac{1}{t}\right)^3 \left(\dfrac{1}{t^2}\right) dt$

18. $\int \left[x^2 + \dfrac{1}{(3x)^2}\right] dx$

19. $\int \dfrac{1}{\sqrt{2x}}\,dx$

20. $\int \dfrac{1}{2\sqrt{x}}\,dx$

21. $\int \dfrac{x^2 + 3x + 7}{\sqrt{x}}\,dx$

22. $\int \dfrac{t + 2t^2}{\sqrt{t}}\,dt$

23. $\int t^2 \left(t - \dfrac{2}{t}\right) dt$

24. $\int \left(\dfrac{t^3}{3} + \dfrac{1}{4t^2}\right) dt$

25. $\int (9 - y)\sqrt{y}\,dy$

26. $\int 2\pi y(8 - y^{3/2})\,dy$

In Exercises 27–30, solve the differential equation.

27. $\dfrac{dy}{dx} = 4x + \dfrac{4x}{\sqrt{16 - x^2}}$

28. $\dfrac{dy}{dx} = \dfrac{10x^2}{\sqrt{1 + x^3}}$

29. $\dfrac{dy}{dx} = \dfrac{x + 1}{(x^2 + 2x - 3)^2}$

30. $\dfrac{dy}{dx} = \dfrac{x - 4}{\sqrt{x^2 - 8x + 1}}$

Direction Fields In Exercises 31 and 32, a differential equation, a point, and a direction field are given. A *direction field* consists of line segments with slopes given by the differential equation. These line segments give a visual perspective of the directions of the solutions of the differential equation. (a) Sketch two approximate solutions of the differential equation on the direction field, one of which passes through the indicated point. (b) Use integration to find the particular solution of the differential equation and use a graphing utility to graph the solution. Compare the result with the sketches in part (a).

31. $\dfrac{dy}{dx} = x\sqrt{4 - x^2}$, $(2, 2)$

32. $\dfrac{dy}{dx} = x \cos x^2$, $(0, 1)$

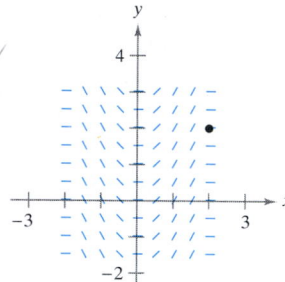
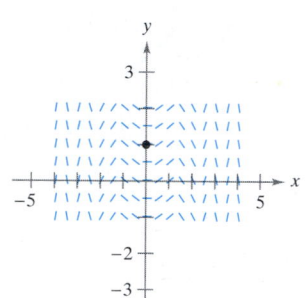

In Exercises 33–44, evaluate the indefinite integral.

33. $\int \sin 2x\,dx$

34. $\int x \sin x^2\,dx$

35. $\int \dfrac{1}{\theta^2} \cos \dfrac{1}{\theta}\,d\theta$

36. $\int \cos 6x\,dx$

37. $\int \sin 2x \cos 2x\,dx$

38. $\int \sec(1 - x) \tan(1 - x)\,dx$

39. $\int \tan^4 x \sec^2 x\,dx$

40. $\int \sqrt{\cot x}\,\csc^2 x\,dx$

41. $\int \dfrac{\csc^2 x}{\cot^3 x}\,dx$

42. $\int \dfrac{\sin x}{\cos^2 x}\,dx$

43. $\int \cot^2 x\,dx$

44. $\int \csc^2 \left(\dfrac{x}{2}\right) dx$

In Exercises 45 and 46, find an equation for the function f that has the indicated derivative and whose graph passes through the given point.

Derivative	Point
45. $f'(x) = \cos \dfrac{x}{2}$	$(0, 3)$
46. $f'(x) = \pi \sec \pi x \tan \pi x$	$\left(\frac{1}{3}, 1\right)$

In Exercises **47–54**, evaluate the indefinite integral by the method shown in Example 5.

47. $\displaystyle\int x\sqrt{x+2}\,dx$

48. $\displaystyle\int x\sqrt{2x+1}\,dx$

49. $\displaystyle\int x^2\sqrt{1-x}\,dx$

50. $\displaystyle\int (x+1)\sqrt{2-x}\,dx$

51. $\displaystyle\int \frac{x^2-1}{\sqrt{2x-1}}\,dx$

52. $\displaystyle\int \frac{2x-1}{\sqrt{x+3}}\,dx$

53. $\displaystyle\int \frac{-x}{(x+1)-\sqrt{x+1}}\,dx$

54. $\displaystyle\int t\sqrt[3]{t-4}\,dt$

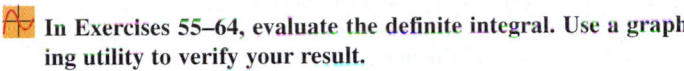 In Exercises **55–64**, evaluate the definite integral. Use a graphing utility to verify your result.

55. $\displaystyle\int_{-1}^{1} x(x^2+1)^3\,dx$

56. $\displaystyle\int_{0}^{1} x\sqrt{1-x^2}\,dx$

57. $\displaystyle\int_{0}^{4} \frac{1}{\sqrt{2x+1}}\,dx$

58. $\displaystyle\int_{0}^{2} \frac{x}{\sqrt{1+2x^2}}\,dx$

59. $\displaystyle\int_{1}^{9} \frac{1}{\sqrt{x}\left(1+\sqrt{x}\right)^2}\,dx$

60. $\displaystyle\int_{0}^{2} x\sqrt[3]{4+x^2}\,dx$

61. $\displaystyle\int_{1}^{2} (x-1)\sqrt{2-x}\,dx$

62. $\displaystyle\int_{0}^{4} \frac{x}{\sqrt{2x+1}}\,dx$

63. $\displaystyle\int_{0}^{\pi/2} \cos\left(\frac{2x}{3}\right)dx$

64. $\displaystyle\int_{\pi/3}^{\pi/2} (x+\cos x)\,dx$

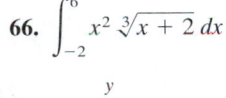 In Exercises **65–70**, find the area of the region. Use a graphing utility to verify your result.

65. $\displaystyle\int_{0}^{7} x\sqrt[3]{x+1}\,dx$

66. $\displaystyle\int_{-2}^{6} x^2\sqrt[3]{x+2}\,dx$

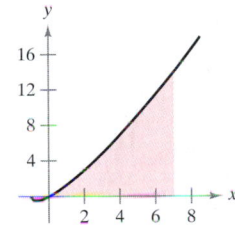

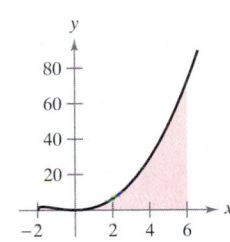

67. $y = 2\sin x + \sin 2x$

68. $y = \sin x + \cos 2x$

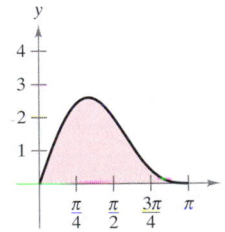

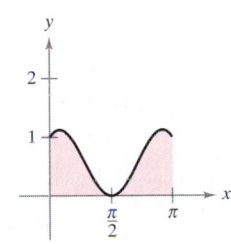

69. $\displaystyle\int_{\pi/2}^{2\pi/3} \sec^2\left(\frac{x}{2}\right)dx$

70. $\displaystyle\int_{\pi/12}^{\pi/4} \csc 2x \cot 2x\,dx$

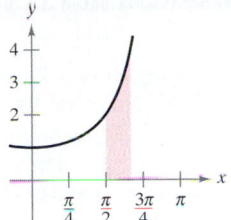

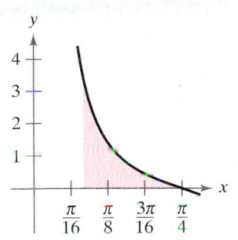

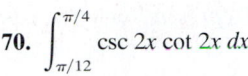 In Exercises **71–76**, use a graphing utility to evaluate the integral. Graph the region whose area is given by the definite integral.

71. $\displaystyle\int_{0}^{4} \frac{x}{\sqrt{2x+1}}\,dx$

72. $\displaystyle\int_{0}^{2} x^3\sqrt{x+2}\,dx$

73. $\displaystyle\int_{3}^{7} x\sqrt{x-3}\,dx$

74. $\displaystyle\int_{1}^{5} x^2\sqrt{x-1}\,dx$

75. $\displaystyle\int_{0}^{3} \left(\theta+\cos\frac{\theta}{6}\right)d\theta$

76. $\displaystyle\int_{0}^{\pi/2} \sin 2x\,dx$

Writing In Exercises **77** and **78**, perform the integration in two ways. Explain any difference in the forms of the answers.

77. $\displaystyle\int (2x-1)^2\,dx$

78. $\displaystyle\int \sin x \cos x\,dx$

79. Use $\displaystyle\int_{0}^{2} x^2\,dx = \frac{8}{3}$ to evaluate the definite integrals without using the Fundamental Theorem of Calculus.

(a) $\displaystyle\int_{-2}^{0} x^2\,dx$

(b) $\displaystyle\int_{-2}^{2} x^2\,dx$

(c) $\displaystyle\int_{0}^{2} -x^2\,dx$

(d) $\displaystyle\int_{-2}^{0} 3x^2\,dx$

80. Use the symmetry of the graphs of the sine and cosine functions as an aid in evaluating each of the integrals.

(a) $\displaystyle\int_{-\pi/4}^{\pi/4} \sin x\,dx$

(b) $\displaystyle\int_{-\pi/4}^{\pi/4} \cos x\,dx$

(c) $\displaystyle\int_{-\pi/2}^{\pi/2} \cos x\,dx$

(d) $\displaystyle\int_{-\pi/2}^{\pi/2} \sin x \cos x\,dx$

In Exercises **81** and **82**, write the integral as the sum of the integral of an odd function and the integral of an even function. Use this simplification to evaluate the integral.

81. $\displaystyle\int_{-4}^{4} (x^3+6x^2-2x-3)\,dx$

82. $\displaystyle\int_{-\pi}^{\pi} (\sin 3x + \cos 3x)\,dx$

83. Depreciation The rate of depreciation dV/dt of a machine is inversely proportional to the square of $t + 1$, where V is the value of the machine t years after it was purchased. If the initial value of the machine was \$500,000, and its value decreased \$100,000 in the first year, estimate its value after 4 years.

84. Cash Flow The rate of disbursement dQ/dt of a 2 million dollar federal grant is proportional to the square of $100 - t$. Time t is measured in days $(0 \le t \le 100)$, and Q is the amount that remains to be disbursed. Find the amount that remains to be disbursed after 50 days. Assume that all the money will be disbursed in 100 days.

85. Marginal Cost The marginal cost for a certain commodity has been determined to be

$$\frac{dC}{dx} = \frac{12}{\sqrt[3]{12x + 1}}.$$

(a) Find the cost function if $C = 100$ when $x = 13$.

(b) Use a graphing utility to graph the marginal cost function and the cost function in the same viewing rectangle.

86. Gasoline Reserves The minimum stockpile level of gasoline in the United States can be approximated by the model

$$Q = 217 + 13 \cos \frac{\pi(t - 3)}{6}$$

where Q is measured in millions of barrels of gasoline and t is the time in months, with $t = 1$ corresponding to January. Find the average minimum level given by this model during the following periods.

(a) The first quarter $(0 \le t \le 3)$

(b) The second quarter $(3 \le t \le 6)$

(c) The entire year $(0 \le t \le 12)$

87. Sales The sales of a seasonal product are given by the model

$$S = 74.50 + 43.75 \sin \frac{\pi t}{6}$$

where S is measured in thousands of units and t is the time in months, with $t = 1$ corresponding to January. Find the average sales for the following periods.

(a) The first quarter $(0 \le t \le 3)$

(b) The second quarter $(3 \le t \le 6)$

(c) The entire year $(0 \le t \le 12)$

88. Water Supply A model for the flow rate of water at a pumping station on a given day is

$$R(t) = 53 + 7 \sin \left(\frac{\pi t}{6} + 3.6 \right) + 9 \cos \left(\frac{\pi t}{12} + 8.9 \right)$$

where $0 \le t \le 24$. R is the flow rate in thousands of gallons per hour, and t is the time in hours.

(a) Use a graphing utility to graph the rate function and approximate the maximum flow rate at the pumping station.

(b) Approximate the total volume of water pumped in 1 day.

89. Electricity The oscillating current in an electrical circuit is

$$I = 2 \sin(60\pi t) + \cos(120\pi t)$$

where I is measured in amperes and t is measured in seconds. Find the average current for each of the following time intervals.

(a) $0 \le t \le \frac{1}{60}$ (b) $0 \le t \le \frac{1}{240}$ (c) $0 \le t \le \frac{1}{30}$

90. Graphical Analysis Consider the functions f and g, where

$$f(x) = 6 \sin x \cos^2 x \quad \text{and} \quad g(t) = \int_0^t f(x) \, dx.$$

(a) Use a graphing utility to graph f and g in the same viewing rectangle.

(b) Explain why g is nonnegative.

(c) Identify the points on the graph of g that correspond to the extrema of f.

(d) Does each of the zeros of f correspond to an extrema of g? Explain.

(e) Consider the function $h(t) = \int_{\pi/2}^t f(x) \, dx$. Use a graphing utility to graph h. What is the relationship between g and h? Verify your conjecture.

True or False? In Exercises 91–96, determine whether the statement is true or false. If it is false, explain why or give an example that shows it is false.

91. $\int (2x + 1)^2 \, dx = \frac{1}{3}(2x + 1)^3 + C$

92. $\int x(x^2 + 1) \, dx = \frac{1}{2}x^2 \left(\frac{1}{3}x^3 + x \right) + C$

93. $\int_{-10}^{10} (ax^3 + bx^2 + cx + d) \, dx = 2 \int_0^{10} (bx^2 + d) \, dx$

94. $\int_a^b \sin x \, dx = \int_a^{b + 2\pi} \sin x \, dx$

95. $4 \int \sin x \cos x \, dx = -\cos 2x + C$

96. $\int \sin^2 2x \cos 2x \, dx = \frac{1}{3} \sin^3 2x + C$

97. Show that if f is continuous on the entire real line, then

$$\int_a^b f(x + h) \, dx = \int_{a+h}^{b+h} f(x) \, dx.$$

98. Let f be continuous on the interval $[0, b]$. Show that

$$\int_0^b \frac{f(x)}{f(x) + f(b - x)} \, dx = \frac{b}{2}.$$

Use this result to evaluate

$$\int_0^1 \frac{\sin x}{\sin(1 - x) + \sin x} \, dx.$$

SECTION 4.6 Numerical Integration

The Trapezoidal Rule • Simpson's Rule • Error Analysis

The Trapezoidal Rule

Some elementary functions simply do not have antiderivatives that are elementary functions. For example, there is no elementary function that has any of the following functions as its derivative.

$$\sqrt[3]{x}\sqrt{1-x}, \quad \sqrt{x}\cos x, \quad \frac{\cos x}{x}, \quad \sqrt{1-x^3}, \quad \sin x^2$$

If you need to evaluate a definite integral involving a function whose antiderivative cannot be found, the Fundamental Theorem of Calculus cannot be applied, and you must resort to an approximation technique. Two such techniques are described in this section.

One way to approximate a definite integral is to use n trapezoids, as shown in Figure 4.41. In the development of this method, assume that f is continuous and positive on the interval $[a, b]$. Thus, the definite integral

$$\int_a^b f(x)\, dx$$

represents the area of the region bounded by the graph of f and the x-axis, from $x = a$ to $x = b$. First, partition the interval $[a, b]$ into n equal subintervals, each of width $\Delta x = (b - a)/n$, such that

$$a = x_0 < x_1 < x_2 < \cdots < x_n = b.$$

Then form a trapezoid for each subinterval, as shown in Figure 4.42. The area of the ith trapezoid is

$$\text{Area of } i\text{th trapezoid} = \left[\frac{f(x_{i-1}) + f(x_i)}{2}\right]\left(\frac{b-a}{n}\right).$$

This implies that the sum of the areas of the n trapezoids is

$$\text{Area} = \left(\frac{b-a}{n}\right)\left[\frac{f(x_0) + f(x_1)}{2} + \cdots + \frac{f(x_{n-1}) + f(x_n)}{2}\right]$$

$$= \left(\frac{b-a}{2n}\right)[f(x_0) + f(x_1) + f(x_1) + f(x_2) + \cdots + f(x_{n-1}) + f(x_n)]$$

$$= \left(\frac{b-a}{2n}\right)[f(x_0) + 2f(x_1) + 2f(x_2) + \cdots + 2f(x_{n-1}) + f(x_n)].$$

Letting $\Delta x = (b - a)/n$, you can take the limit as $n \to \infty$, to obtain

$$\lim_{n\to\infty}\left(\frac{b-a}{2n}\right)[f(x_0) + 2f(x_1) + \cdots + 2f(x_{n-1}) + f(x_n)]$$

$$= \lim_{n\to\infty}\left[\frac{[f(a) - f(b)]\,\Delta x}{2} + \sum_{i=1}^{n} f(x_i)\,\Delta x\right]$$

$$= \lim_{n\to\infty}\frac{[f(a) - f(b)](b - a)}{2n} + \lim_{n\to\infty}\sum_{i=1}^{n} f(x_i)\,\Delta x$$

$$= 0 + \int_a^b f(x)\, dx.$$

The result is summarized in the following theorem.

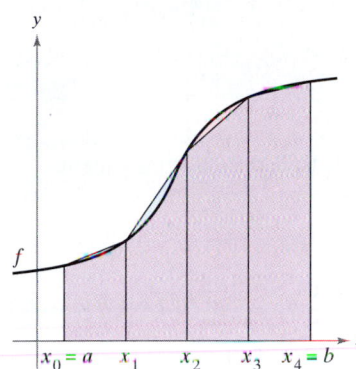

The area of the region can be approximated using four trapezoids.

Figure 4.41

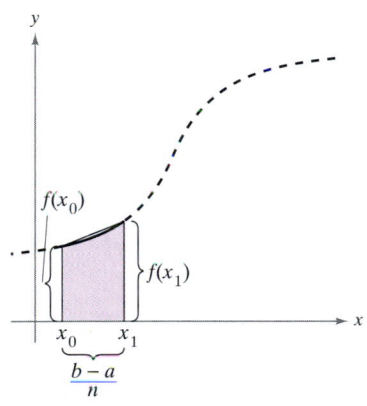

The area of the first trapezoid is

$$\left[\frac{f(x_0) + f(x_1)}{2}\right]\left(\frac{b-a}{n}\right).$$

Figure 4.42

> **THEOREM 4.16   The Trapezoidal Rule**
>
> Let f be continuous on $[a, b]$. The Trapezoidal Rule for approximating $\int_a^b f(x)\, dx$ is given by
>
> $$\int_a^b f(x)\, dx \approx \frac{b-a}{2n}\left[f(x_0) + 2f(x_1) + 2f(x_2) + \cdots + 2f(x_{n-1}) + f(x_n)\right].$$
>
> Moreover, as $n \to \infty$, the right-hand side approaches $\int_a^b f(x)\, dx$.

NOTE   Observe that the coefficients in the Trapezoidal Rule have the following pattern.

$$1 \quad 2 \quad 2 \quad 2 \quad \cdots \quad 2 \quad 2 \quad 1$$

EXAMPLE 1   Approximation with the Trapezoidal Rule

Use the Trapezoidal Rule to approximate

$$\int_0^\pi \sin x\, dx.$$

Compare the results for $n = 4$ and $n = 8$, as shown in Figure 4.43.

Solution   When $n = 4$, $\Delta x = \pi/4$, and you obtain

$$\int_0^\pi \sin x\, dx \approx \frac{\pi}{8}\left(\sin 0 + 2\sin\frac{\pi}{4} + 2\sin\frac{\pi}{2} + 2\sin\frac{3\pi}{4} + \sin\pi\right)$$

$$= \frac{\pi}{8}\left(0 + \sqrt{2} + 2 + \sqrt{2} + 0\right) = \frac{\pi\left(1 + \sqrt{2}\right)}{4} \approx 1.896.$$

When $n = 8$, $\Delta x = \pi/8$, and you obtain

$$\int_0^\pi \sin x\, dx \approx \frac{\pi}{16}\left(\sin 0 + 2\sin\frac{\pi}{8} + 2\sin\frac{\pi}{4} + 2\sin\frac{3\pi}{8} + 2\sin\frac{\pi}{2}\right.$$

$$\left. + 2\sin\frac{5\pi}{8} + 2\sin\frac{3\pi}{4} + 2\sin\frac{7\pi}{8} + \sin\pi\right)$$

$$= \frac{\pi}{16}\left(2 + 2\sqrt{2} + 4\sin\frac{\pi}{8} + 4\sin\frac{3\pi}{8}\right) \approx 1.974.$$

For this particular integral, you could have found an antiderivative and determined that the exact area of the region is 2.

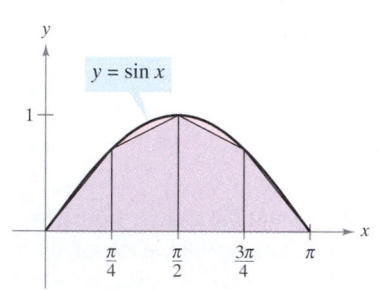

$y = \sin x$

Four subintervals

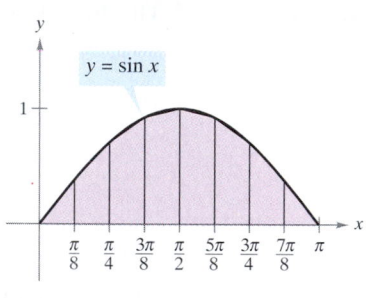

$y = \sin x$

Eight subintervals

Trapezoidal approximations
Figure 4.43

TECHNOLOGY   Most graphing utilities and symbolic integration utilities have built-in programs that can be used to approximate the value of a definite integral. Try using such a program to approximate the integral in Example 1. How close is your approximation?

When you use such a program, you need to be aware of its limitations. Often, you are given no indication of the degree of accuracy of the approximation. Other times, you may be given an approximation that is completely wrong. For instance, try using a built-in numerical integration program to evaluate

$$\int_{-1}^2 \frac{1}{x}\, dx.$$

Your calculator should give an error message. Does yours?

NOTE　There are two important points that should be made concerning the Trapezoidal Rule (or the Midpoint Rule). First, the approximation tends to become more accurate as n increases. For instance, in Example 1, if $n = 16$, the Trapezoidal Rule yields an approximation of 1.994. Second, although you could have used the Fundamental Theorem to evaluate the integral in Example 1, this theorem cannot be used to evaluate an integral as simple as

$$\int_0^\pi \sin x^2 \, dx$$

because $\sin x^2$ has no elementary antiderivative. Yet, the Trapezoidal Rule can be applied easily to this integral.

It is interesting to compare the Trapezoidal Rule with the Midpoint Rule given in Section 4.2 (Exercises 51–54). For the Trapezoidal Rule, you average the function values at the endpoints of the subintervals, but for the Midpoint Rule you take the function values of the subinterval midpoints.

$$\int_a^b f(x) \, dx \approx \sum_{i=1}^n f\left(\frac{x_i + x_{i-1}}{2}\right) \Delta x \qquad \text{Midpoint Rule}$$

$$\int_a^b f(x) \, dx \approx \sum_{i=1}^n \left(\frac{f(x_i) + f(x_{i-1})}{2}\right) \Delta x \qquad \text{Trapezoidal Rule}$$

Simpson's Rule

One way to view the trapezoidal approximation of a definite integral is to say that on each subinterval you approximate f by a *first*-degree polynomial. In Simpson's Rule, named after the English mathematician Thomas Simpson (1710–1761), you take this procedure one step further and approximate f by *second*-degree polynomials.

Before presenting Simpson's Rule, we list a theorem for evaluating integrals of polynomials of degree 2 (or less).

THEOREM 4.17　Integral of $p(x) = Ax^2 + Bx + C$

If $p(x) = Ax^2 + Bx + C$, then

$$\int_a^b p(x) \, dx = \left(\frac{b - a}{6}\right)\left[p(a) + 4p\left(\frac{a + b}{2}\right) + p(b)\right].$$

Proof

$$\int_a^b p(x) \, dx = \int_a^b (Ax^2 + Bx + C) \, dx$$

$$= \left[\frac{Ax^3}{3} + \frac{Bx^2}{2} + Cx\right]_a^b$$

$$= \frac{A(b^3 - a^3)}{3} + \frac{B(b^2 - a^2)}{2} + C(b - a)$$

$$= \left(\frac{b - a}{6}\right)\left[2A(a^2 + ab + b^2) + 3B(b + a) + 6C\right]$$

By expansion and collection of terms, the expression inside the brackets becomes

$$\underbrace{(Aa^2 + Ba + C)}_{p(a)} + \underbrace{4\left[A\left(\frac{b + a}{2}\right)^2 + B\left(\frac{b + a}{2}\right) + C\right]}_{4p\left(\frac{a + b}{2}\right)} + \underbrace{(Ab^2 + Bb + C)}_{p(b)}$$

and you can write

$$\int_a^b p(x) \, dx = \left(\frac{b - a}{6}\right)\left[p(a) + 4p\left(\frac{a + b}{2}\right) + p(b)\right].$$

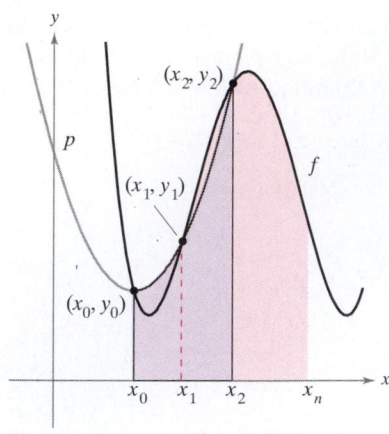

$$\int_{x_0}^{x_2} p(x)\, dx \approx \int_{x_0}^{x_2} f(x)\, dx$$

Figure 4.44

To develop Simpson's Rule for approximating a definite integral, you again partition the interval $[a, b]$ into n equal subintervals, each of width $\Delta x = (b - a)/n$. This time, however, n is required to be even, and the subintervals are grouped in pairs such that

$$a = \underbrace{x_0 < x_1 < x_2}_{[x_0,\, x_2]} < \underbrace{x_3 < x_4}_{[x_2,\, x_4]} < \cdots < \underbrace{x_{n-2} < x_{n-1} < x_n}_{[x_{n-2},\, x_n]} = b.$$

On each (double) subinterval $[x_{i-2}, x_i]$, you can approximate f by a polynomial p of degree less than or equal to 2. For example, on the subinterval $[x_0, x_2]$, choose the polynomial of least degree passing through the points (x_0, y_0), (x_1, y_1), and (x_2, y_2), as shown in Figure 4.44. Now, using p as an approximation of f on this subinterval, you have, by Theorem 4.17,

$$\int_{x_0}^{x_2} f(x)\, dx \approx \int_{x_0}^{x_2} p(x)\, dx = \frac{x_2 - x_0}{6}\left[p(x_0) + 4p\left(\frac{x_0 + x_2}{2}\right) + p(x_2)\right]$$

$$= \frac{2[(b - a)/n]}{6}\left[p(x_0) + 4p(x_1) + p(x_2)\right]$$

$$= \frac{b - a}{3n}\left[f(x_0) + 4f(x_1) + f(x_2)\right].$$

Repeating this procedure on the entire interval $[a, b]$ produces the following theorem.

> **THEOREM 4.18   Simpson's Rule (n is even)**
>
> Let f be continuous on $[a, b]$. Simpson's Rule for approximating $\int_a^b f(x)\, dx$ is
>
> $$\int_a^b f(x)\, dx \approx \frac{b - a}{3n}[f(x_0) + 4f(x_1) + 2f(x_2) + 4f(x_3) + \cdots$$
> $$+ 4f(x_{n-1}) + f(x_n)].$$
>
> Moreover, as $n \to \infty$, the right-hand side approaches $\int_a^b f(x)\, dx$.

NOTE   Observe that the coefficients in Simpson's Rule have the following pattern.

1 4 2 4 2 4 . . . 4 2 4 1

In Example 1, the Trapezoidal Rule was used to estimate $\int_0^\pi \sin x\, dx$. In the next example, Simpson's Rule is applied to the same integral.

EXAMPLE 2   Approximation with Simpson's Rule

NOTE   In Example 1, the Trapezoidal Rule with $n = 8$ approximated $\int_0^\pi \sin x\, dx$ as 1.974. In Example 2, Simpson's Rule with $n = 8$ gave an approximation of 2.0003. The antiderivative would produce the true value of 2.

Use Simpson's Rule to approximate

$$\int_0^\pi \sin x\, dx.$$

Compare the results for $n = 4$ and $n = 8$.

Solution   When $n = 4$, you have

$$\int_0^\pi \sin x\, dx \approx \frac{\pi}{12}\left(\sin 0 + 4\sin\frac{\pi}{4} + 2\sin\frac{\pi}{2} + 4\sin\frac{3\pi}{4} + \sin \pi\right)$$

$$\approx 2.005.$$

When $n = 8$, you have $\displaystyle\int_0^\pi \sin x\, dx \approx 2.0003.$

Error Analysis

If you must use an approximation technique, it is important to know how accurate you can expect the approximation to be. The following theorem, which we list without proof, gives the formulas for estimating the errors involved in the use of Simpson's Rule and the Trapezoidal Rule.

THEOREM 4.19   Errors in the Trapezoidal Rule and Simpson's Rule

If f has a continuous second derivative on $[a, b]$, then the error E in approximating $\int_a^b f(x)\,dx$ by the Trapezoidal Rule is

$$E \le \frac{(b-a)^3}{12n^2}\left[\max |f''(x)|\right], \quad a \le x \le b. \qquad \text{Trapezoidal Rule}$$

Moreover, if f has a continuous fourth derivative on $[a, b]$, then the error E in approximating $\int_a^b f(x)\,dx$ by Simpson's Rule is

$$E \le \frac{(b-a)^5}{180n^4}\left[\max |f^{(4)}(x)|\right], \quad a \le x \le b. \qquad \text{Simpson's Rule}$$

TECHNOLOGY If you have access to a symbolic integration utility, try using it to evaluate the definite integral in Example 3. You should obtain a value of

$$\int_0^1 \sqrt{1 + x^2}\,dx = \tfrac{1}{2}\left[\sqrt{2} + \ln\left(1 + \sqrt{2}\right)\right]$$

$$\approx 1.14779.$$

("ln" represents the natural logarithmic function, which you will study in Section 5.1.)

Theorem 4.19 states that the errors generated by the Trapezoidal Rule and Simpson's Rule have upper bounds dependent on the extreme values of $f''(x)$ and $f^{(4)}(x)$ in the interval $[a, b]$. Furthermore, these errors can be made arbitrarily small by *increasing n*, provided that f'' and $f^{(4)}$ are continuous and therefore bounded in $[a, b]$.

EXAMPLE 3   The Approximate Error in the Trapezoidal Rule

Determine a value of n such that the Trapezoidal Rule will approximate the value of $\int_0^1 \sqrt{1 + x^2}\,dx$ with an error that is less than 0.01.

Solution Begin by letting $f(x) = \sqrt{1 + x^2}$ and finding the second derivative of f.

$$f'(x) = x(1 + x^2)^{-1/2} \text{ and } f''(x) = (1 + x^2)^{-3/2}$$

The maximum value of $|f''(x)|$ on the interval $[0, 1]$ is $|f''(0)| = 1$. Thus, by Theorem 4.19, you can write

$$E \le \frac{(b-a)^3}{12n^2}|f''(0)| \le \frac{1}{12n^2}(1) = \frac{1}{12n^2}.$$

To obtain an error E that is less than 0.01, you must choose n such that $1/(12n^2) \le 1/100$.

$$100 \le 12n^2 \implies n \ge \sqrt{\tfrac{100}{12}} \approx 2.89$$

Therefore, you can choose $n = 3$ (because n must be greater than or equal to 2.89) and apply the Trapezoidal Rule, as shown in Figure 4.45, to obtain

$$\int_0^1 \sqrt{1 + x^2}\,dx \approx \frac{1}{6}\left[\sqrt{1 + 0^2} + 2\sqrt{1 + \left(\tfrac{1}{3}\right)^2} + 2\sqrt{1 + \left(\tfrac{2}{3}\right)^2} + \sqrt{1 + 1^2}\right]$$

$$\approx 1.154.$$

Thus, with an error no larger than 0.01, you know that

$$1.144 \le \int_0^1 \sqrt{1 + x^2}\,dx \le 1.164.$$

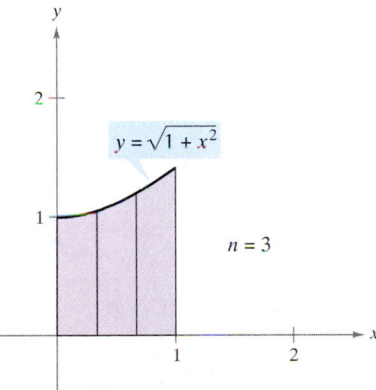

$$1.144 \le \int_0^1 \sqrt{1 + x^2}\,dx \le 1.164$$

Figure 4.45

EXERCISES FOR SECTION 4.6

In Exercises 1–10, use the Trapezoidal Rule and Simpson's Rule to approximate the value of the definite integral for the indicated value of n. Round your answer to four decimal places and compare the results with the exact value of the definite integral.

1. $\int_0^2 x^2\,dx, \quad n = 4$

2. $\int_0^1 \left(\dfrac{x^2}{2} + 1\right) dx, \quad n = 4$

3. $\int_0^2 x^3\,dx, \quad n = 4$

4. $\int_1^2 \dfrac{1}{x^2}\,dx, \quad n = 4$

5. $\int_0^2 x^3\,dx, \quad n = 8$

6. $\int_0^8 \sqrt[3]{x}\,dx, \quad n = 8$

7. $\int_4^9 \sqrt{x}\,dx, \quad n = 8$

8. $\int_1^3 (4 - x^2)\,dx, \quad n = 4$

9. $\int_1^2 \dfrac{1}{(x + 1)^2}\,dx, \quad n = 4$

10. $\int_0^2 x\sqrt{x^2 + 1}\,dx, \quad n = 4$

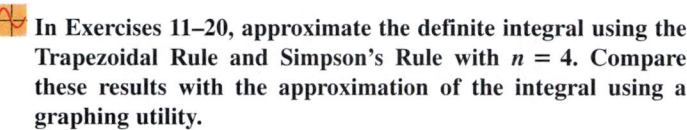

In Exercises 11–20, approximate the definite integral using the Trapezoidal Rule and Simpson's Rule with $n = 4$. Compare these results with the approximation of the integral using a graphing utility.

11. $\int_0^2 \sqrt{1 + x^3}\,dx$

12. $\int_0^2 \dfrac{1}{\sqrt{1 + x^3}}\,dx$

13. $\int_0^1 \sqrt{x}\,\sqrt{1 - x}\,dx$

14. $\int_{\pi/2}^{\pi} \sqrt{x}\sin x\,dx$

15. $\int_0^{\sqrt{\pi/2}} \cos x^2\,dx$

16. $\int_0^{\sqrt{\pi/4}} \tan x^2\,dx$

17. $\int_1^{1.1} \sin x^2\,dx$

18. $\int_0^{\pi/2} \sqrt{1 + \cos^2 x}\,dx$

19. $\int_0^{\pi/4} x\tan x\,dx$

20. $\int_0^{\pi} f(x)\,dx, \quad f(x) = \begin{cases} \dfrac{\sin x}{x}, & x > 0 \\ 1, & x = 0 \end{cases}$

In Exercises 21 and 22, use the error formulas in Theorem 4.19 to find the maximum possible error in approximating the integral, with $n = 4$, using (a) the Trapezoidal Rule and (b) Simpson's Rule.

21. $\int_0^2 x^3\,dx$

22. $\int_0^1 \dfrac{1}{x + 1}\,dx$

In Exercises 23 and 24, use the error formulas in Theorem 4.19 to find n such that the error in the approximation of the definite integral is less than 0.00001 using (a) the Trapezoidal Rule and (b) Simpson's Rule.

23. $\int_1^3 \dfrac{1}{x}\,dx$

24. $\int_0^1 \dfrac{1}{1 + x}\,dx$

In Exercises 25–28, use a symbolic differentiation utility and the error formulas to find n such that the error in the approximation of the definite integral is less than 0.00001 using (a) the Trapezoidal Rule and (b) Simpson's Rule.

25. $\int_0^2 \sqrt{1 + x}\,dx$

26. $\int_0^2 (x + 1)^{2/3}\,dx$

27. $\int_0^1 \tan x^2\,dx$

28. $\int_0^1 \sin x^2\,dx$

29. Prove that Simpson's Rule is exact when approximating the integral of a cubic polynomial function, and demonstrate the result for
$$\int_0^1 x^3\,dx, \quad n = 2.$$

30. Write a program for a graphing utility to approximate a definite integral using the Trapezoidal Rule and Simpson's Rule. Start with the program written in Section 4.3, Exercises 35–38, and note that the Trapezoidal Rule can be written as
$$T(n) = \tfrac{1}{2}\left[L(n) + R(n)\right]$$
and Simpson's Rule can be written as
$$S(n) = \tfrac{1}{3}\left[T(n/2) + 2M(n/2)\right].$$
[Recall that $L(n)$, $M(n)$, and $R(n)$ represent the Riemann sums using the left-hand endpoint, midpoint, and right-hand endpoint of subintervals of equal width.]

In Exercises 31–34, use the program in Exercise 30 to approximate the definite integral and complete the table.

n	$L(n)$	$M(n)$	$R(n)$	$T(n)$	$S(n)$
4					
8					
10					
12					
16					
20					

31. $\int_0^4 \sqrt{2 + 3x^2}\,dx$

32. $\int_0^1 \sqrt{1 - x^2}\,dx$

33. $\int_0^4 \sin \sqrt{x}\,dx$

34. $\int_1^2 \dfrac{\sin x}{x}\,dx$

35. **Area** Use Simpson's Rule with $n = 14$ to approximate the area of the region bounded by the graphs of $y = \sqrt{x}\cos x$, $y = 0$, $x = 0$, and $x = \pi/2$.

36. Circumference The **elliptic integral**

$$8\sqrt{3}\int_0^{\pi/2}\sqrt{1-\tfrac{2}{3}\sin^2\theta}\,d\theta$$

gives the circumference of an ellipse. Use Simpson's Rule with $n=8$ to approximate the circumference.

37. Work To determine the size of the motor required to operate a press, a company must know the amount of work done when the press moves an object linearly 5 feet. The variable force to move the object is

$$F(x)=100x\sqrt{125-x^3}$$

where F is given in pounds and x gives the position of the unit in feet. Use Simpson's Rule with $n=12$ to approximate the work W (in foot-pounds) done through one cycle if

$$W=\int_0^5 F(x)\,dx.$$

38. Approximation of Pi Use Simpson's Rule with $n=6$ to approximate π using the equation

$$\pi=\int_0^1\frac{4}{1+x^2}\,dx.$$

(In Section 5.9, you will be able to evaluate this integral using the inverse tangent function.)

39. Area To estimate the surface area of a pond, a surveyor takes several measurements, as shown in the figure. Estimate the surface area of the pond using (a) the Trapezoidal Rule and (b) Simpson's Rule.

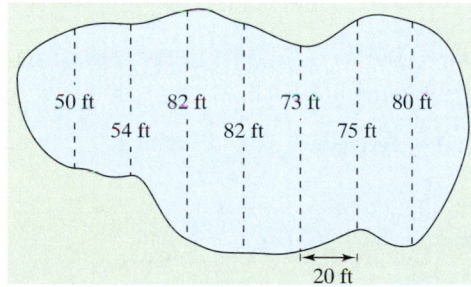

40. The table lists several measurements gathered in an experiment to approximate an unknown continuous function $y=f(x)$. Approximate the integral $\int_0^2 f(x)\,dx$ using (a) the Trapezoidal Rule and (b) Simpson's Rule.

x	0.00	0.25	0.50	0.75	1.00
y	4.32	4.36	4.58	5.79	6.14

x	1.25	1.50	1.75	2.00
y	7.25	7.64	8.08	8.14

Area In Exercises 41 and 42, use the Trapezoidal Rule to estimate the number of square meters of land in a lot where x and y are measured in meters, as shown in the figures. The land is bounded by a stream and two straight roads that meet at right angles.

41.

x	y
0	125
100	125
200	120
300	112
400	90
500	90
600	95
700	88
800	75
900	35
1000	0

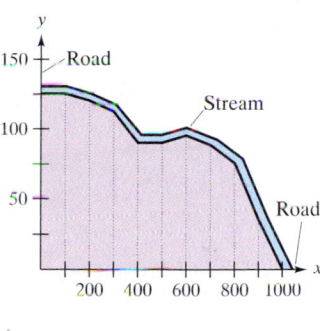

42.

x	y
0	75
10	81
20	84
30	76
40	67
50	68
60	69
70	72
80	68
90	56
100	42
110	23
120	0

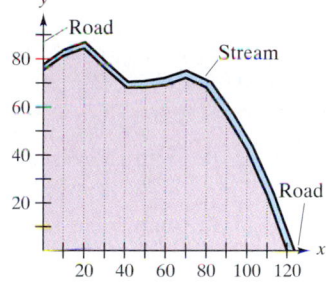

43. Use Simpson's Rule with $n=10$ and a symbolic integration utility to approximate t in the integral equation

$$\int_0^t \sin\sqrt{x}\,dx=2.$$

44. Determine whether the Trapezoidal Rule overestimates or underestimates a definite integral if the graph of the integrand is (a) concave up and (b) concave down.

45. Let $L(x)=\int_1^x \frac{1}{t}\,dt$ for all $x>0$.

(a) Find $L(1)$.

(b) Find $L'(x)$ and $L'(1)$.

(c) Use the Trapezoidal Rule to approximate the value of x (to three decimal places) for which $L(x)=1$.

(d) Prove that $L(x_1x_2)=L(x_1)+L(x_2)$, for $x_1>0$ and $x_2>0$.

REVIEW EXERCISES FOR CHAPTER 4

In Exercises 1 and 2, use the graph of f' to sketch a graph of f.

1.

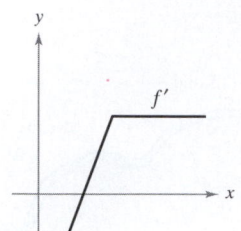

2.

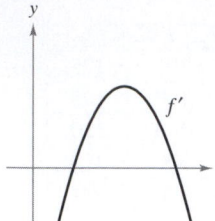

In Exercises 3–8, find the indefinite integral.

3. $\int (2x^2 + x - 1)\, dx$

4. $\int \dfrac{2}{\sqrt[3]{3x}}\, dx$

5. $\int \dfrac{x^3 + 1}{x^2}\, dx$

6. $\int \dfrac{x^3 - 2x^2 + 1}{x^2}\, dx$

7. $\int (4x - 3\sin x)\, dx$

8. $\int (5\cos x - 2\sec^2 x)\, dx$

9. Find the particular solution of the differential equation $f'(x) = -2x$ whose graph passes through the point $(-1, 1)$.

10. Find the particular solution of the differential equation $f''(x) = 6(x - 1)$ whose graph passes through the point $(2, 1)$ and at that point is tangent to the line $3x - y - 5 = 0$.

11. **Velocity and Acceleration** An airplane taking off from a runway travels 3600 feet before lifting off. If it starts from rest, moves with constant acceleration, and makes the run in 30 seconds, with what speed does it lift off?

12. **Velocity and Acceleration** The speed of a car traveling in a straight line is reduced from 45 to 30 miles per hour in a distance of 264 feet. Find the distance in which the car can be brought to rest from 30 miles per hour, assuming the same constant deceleration.

13. **Velocity and Acceleration** A ball is thrown vertically upward from ground level with an initial velocity of 96 feet per second.

    (a) How long will it take the ball to rise to its maximum height?

    (b) What is the maximum height?

    (c) When is the velocity of the ball one-half the initial velocity?

    (d) What is the height of the ball when its velocity is one-half the initial velocity?

14. **Velocity and Acceleration** Repeat Exercise 13 for an initial velocity of 40 meters per second.

15. Write in the sigma notation (a) the sum of the first ten positive odd integers, (b) the sum of the cubes of the first n positive integers, and (c) $6 + 10 + 14 + 18 + \cdots + 42$.

16. Evaluate each of the following sums for $x_1 = 2$, $x_2 = -1$, $x_3 = 5$, $x_4 = 3$, and $x_5 = 7$.

    (a) $\dfrac{1}{5}\displaystyle\sum_{i=1}^{5} x_i$

    (b) $\displaystyle\sum_{i=1}^{5} \dfrac{1}{x_i}$

    (c) $\displaystyle\sum_{i=1}^{5} (2x_i - x_i^2)$

    (d) $\displaystyle\sum_{i=2}^{5} (x_i - x_{i-1})$

17. Consider the region bounded by $y = mx$, $y = 0$, $x = 0$, and $x = b$.

    (a) Find the upper and lower sums to approximate the area of the region when $\Delta x = b/4$.

    (b) Find the upper and lower sums to approximate the area of the region when $\Delta x = b/n$.

    (c) Find the area of the region by letting n approach infinity in both sums in part (b). Show that in each case you obtain the formula for the area of a triangle.

    (d) Find the area of the region using the Fundamental Theorem of Calculus.

18. (a) Find the area of the region bounded by the graphs of $y = x^3$, $y = 0$, $x = 1$, and $x = 3$ by the limit definition.

    (b) Find the area of the region using the Fundamental Theorem of Calculus.

In Exercises 19 and 20, use the given values to find the value of each definite integral.

19. If $\displaystyle\int_{2}^{6} f(x)\, dx = 10$ and $\displaystyle\int_{2}^{6} g(x)\, dx = 3$, find

    (a) $\displaystyle\int_{2}^{6} [f(x) + g(x)]\, dx$.

    (b) $\displaystyle\int_{2}^{6} [f(x) - g(x)]\, dx$.

    (c) $\displaystyle\int_{2}^{6} [2f(x) - 3g(x)]\, dx$.

    (d) $\displaystyle\int_{2}^{6} 5f(x)\, dx$.

20. If $\displaystyle\int_{0}^{3} f(x)\, dx = 4$ and $\displaystyle\int_{3}^{6} f(x)\, dx = -1$, find

    (a) $\displaystyle\int_{0}^{6} f(x)\, dx$.

    (b) $\displaystyle\int_{6}^{3} f(x)\, dx$.

    (c) $\displaystyle\int_{4}^{4} f(x)\, dx$.

    (d) $\displaystyle\int_{3}^{6} -10 f(x)\, dx$.

In Exercises 21–34, find the indefinite integral.

21. $\int (x^2 + 1)^3\, dx$

22. $\int \left(x + \dfrac{1}{x}\right)^2 dx$

23. $\int \dfrac{x^2}{\sqrt{x^3 + 3}}\, dx$

24. $\int x^2 \sqrt{x^3 + 3}\, dx$

25. $\int x(1 - 3x^2)^4\, dx$

26. $\int \dfrac{x + 3}{(x^2 + 6x - 5)^2}\, dx$

27. $\displaystyle\int \sin^3 x \cos x \, dx$

28. $\displaystyle\int x \sin 3x^2 \, dx$

29. $\displaystyle\int \frac{\sin \theta}{\sqrt{1 - \cos \theta}} \, d\theta$

30. $\displaystyle\int \frac{\cos x}{\sqrt{\sin x}} \, dx$

31. $\displaystyle\int \tan^n x \sec^2 x \, dx, \quad n \neq -1$

32. $\displaystyle\int \sec 2x \tan 2x \, dx$

33. $\displaystyle\int (1 + \sec \pi x)^2 \sec \pi x \, \tan \pi x \, dx$

34. $\displaystyle\int \cot^4 \alpha \csc^2 \alpha \, d\alpha$

In Exercises 35–46, use the Fundamental Theorem of Calculus to evaluate the definite integral. Use a graphing utility to verify your result.

35. $\displaystyle\int_0^4 (2 + x) \, dx$

36. $\displaystyle\int_{-1}^1 (t^2 + 2) \, dt$

37. $\displaystyle\int_{-1}^1 (4t^3 - 2t) \, dt$

38. $\displaystyle\int_3^6 \frac{x}{3\sqrt{x^2 - 8}} \, dx$

39. $\displaystyle\int_0^3 \frac{1}{\sqrt{1 + x}} \, dx$

40. $\displaystyle\int_0^1 x^2(x^3 + 1)^3 \, dx$

41. $\displaystyle\int_4^9 x\sqrt{x} \, dx$

42. $\displaystyle\int_1^2 \left(\frac{1}{x^2} - \frac{1}{x^3}\right) dx$

43. $\displaystyle 2\pi \int_0^1 (y + 1)\sqrt{1 - y} \, dy$

44. $\displaystyle 2\pi \int_{-1}^0 x^2 \sqrt{x + 1} \, dx$

45. $\displaystyle\int_0^\pi \cos \frac{x}{2} \, dx$

46. $\displaystyle\int_{-\pi/4}^{\pi/4} \sin 2x \, dx$

In Exercises 47–52, sketch the graph of the region whose area is given by the integral and find the area.

47. $\displaystyle\int_1^3 (2x - 1) \, dx$

48. $\displaystyle\int_0^2 (x + 4) \, dx$

49. $\displaystyle\int_3^4 (x^2 - 9) \, dx$

50. $\displaystyle\int_{-1}^2 (-x^2 + x + 2) \, dx$

51. $\displaystyle\int_0^1 (x - x^3) \, dx$

52. $\displaystyle\int_0^1 \sqrt{x} \, (1 - x) \, dx$

In Exercises 53–56, sketch the region bounded by the graphs of the equations and determine its area.

53. $y = \dfrac{4}{\sqrt{x + 1}}, \; y = 0, \; x = 0, \; x = 8$

54. $y = x, \; y = x^5$

55. $y = \sec^2 x, \; y = 0, \; x = 0, \; x = \dfrac{\pi}{3}$

56. $y = \cos x, \; y = 0, \; x = -\dfrac{\pi}{4}, \; x = \dfrac{\pi}{4}$

In Exercises 57–60, find the average value of the function over the interval. Find the values of x where the function assumes its average value and graph the function.

Function	Interval
57. $f(x) = \dfrac{1}{\sqrt{x - 1}}$	$[5, 10]$
58. $f(x) = x^3$	$[0, 2]$
59. $f(x) = x$	$[0, 4]$
60. $f(x) = x^2 - \dfrac{1}{x^2}$	$[1, 2]$

In Exercises 61 and 62, use the Trapezoidal Rule and Simpson's Rule with $n = 4$, and use the integration capabilities of a graphing utility, to approximate the definite integral. Compare the results.

61. $\displaystyle\int_1^2 \frac{1}{1 + x^3} \, dx$

62. $\displaystyle\int_0^1 \frac{x^{3/2}}{3 - x^2} \, dx$

63. *Air Conditioning Costs* The temperature in degrees Fahrenheit is $T = 72 + 12 \sin[\pi(t - 8)/12]$ where t is time in hours, with $t = 0$ representing midnight. Suppose the hourly cost of cooling a house is $0.10 per degree.

(a) Find the cost C of cooling the house if its thermostat is set at 72°F by evaluating the integral

$$C = 0.1 \int_8^{20} \left[72 + 12 \sin \frac{\pi(t - 8)}{12} - 72\right] dt.$$

(See figure.)

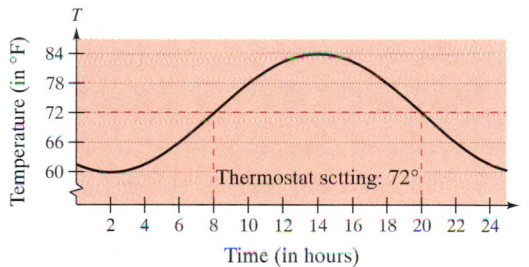

(b) Find the savings from resetting the thermostat to 78°F by evaluating the integral

$$C = 0.1 \int_{10}^{18} \left[72 + 12 \sin \frac{\pi(t - 8)}{12} - 78\right] dt.$$

(See figure.)

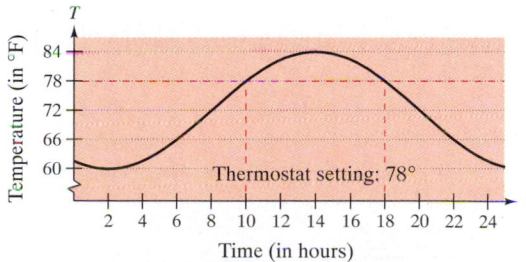

64. *Production Scheduling* A manufacturer of fertilizer finds that national sales of fertilizer follow the seasonal pattern

$$F = 100,000 \left[1 + \sin \frac{2\pi(t - 60)}{365} \right]$$

where F is measured in pounds and t is time in days, with $t = 1$ representing January 1. The manufacturer wants to set up a schedule to produce a uniform amount each day. What should this amount be?

65. *Respiratory Cycle* For a person at rest, the rate of air intake v, in liters per second, during a respiratory cycle is

$$v = 0.85 \sin \frac{\pi t}{3}$$

where t is time in seconds. Find the volume, in liters, of air inhaled during one cycle by integrating the function over the interval $[0, 3]$.

66. *Respiratory Cycle* After exercising a few minutes, a person has a respiratory cycle for which the rate of air intake is

$$v = 1.75 \sin \frac{\pi t}{2}.$$

Find the volume, in liters, of air inhaled during one cycle by integrating the function over the interval $[0, 2]$. How much does the person's lung capacity increase as a result of exercising? (Compare your answer with that found in Exercise 65.)

67. Suppose that gasoline is increasing in price according to the equation

$$p = 1.20 + 0.04t$$

where p is the dollar price per gallon and $t = 0$ represents the year 1990. If an automobile is driven 15,000 miles a year and gets M miles per gallon, the annual fuel cost is

$$C = \frac{15,000}{M} \int_{t}^{t+1} p \, ds.$$

Estimate the annual fuel cost for the years (a) 2000 and (b) 2005.

Probability In Exercises 68 and 69, the function

$$f(x) = kx^n (1 - x)^m, \qquad 0 \le x \le 1$$

where $n > 0$, $m > 0$, and k is a constant, can be used to represent various probability distributions. If k is chosen such that

$$\int_0^1 f(x) \, dx = 1$$

the probability that x will fall between a and b $(0 \le a \le b \le 1)$ is

$$P_{a,b} = \int_a^b f(x) \, dx.$$

68. The probability of recall in a certain experiment is

$$P_{a,b} = \int_a^b \frac{15}{4} x \sqrt{1 - x} \, dx$$

where x represents the percent of recall. (See figure.)

(a) For a randomly chosen individual, what is the probability that he or she will recall between 50% and 75% of the material?

(b) What is the median percent recall? That is, for what value of b is it true that the probability from 0 to b is 0.5?

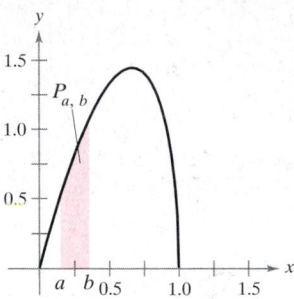
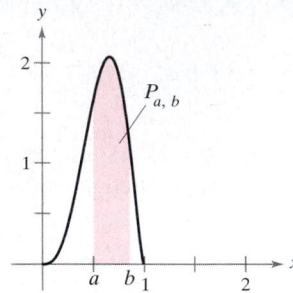

Figure for 68   **Figure for 69**

69. The probability of finding iron in ore samples taken from a certain region is

$$P_{a,b} = \int_a^b \frac{1155}{32} x^3 (1 - x)^{3/2} \, dx.$$

(See figure.) What is the probability that a sample will contain between

(a) 0% and 25%?

(b) 50% and 100%?

True or False? In Exercises 70–74, determine whether the statement is true or false. If it is false, explain why or give an example that shows it is false.

70. $\int x f(x) \, dx = x \int f(x) \, dx$

71. $\int (1/x) \, dx = -\left(\dfrac{1}{x^2} \right) + C$

72. If $f(x) = -f(-x)$ on the interval $[-a, a]$, then $\int_{-a}^{a} f(x) \, dx = 0$.

73. The average value of the sine function over an interval of length 2π is 0.

74. $\int \tan x \, dx = \sec^2 x + C$

Logarithmic, Exponential, and Other Transcendental Functions

5

In 1953, Chevrolet introduced the Corvette, the first mass-produced automobile with a plastic body.

What do Corvette fenders, panty hose, and garbage bags have in common? They are all made of plastic. The Greek word *plastikos*, meaning "able to be shaped," was modified to name the most versatile family of materials ever created. Since Bakelite was introduced in 1909, the plastics industry has steadily expanded to the point where, today, plastics are used in nearly every aspect of our daily lives.

Several methods are used to shape plastic products, one of the most common being to pour hot, syrupy *plastic resin* into a mold or cast. The temperature of the molten resin is over 300°F. The mold is then cooled in a chiller system that is kept at 58°F before the part is ejected from the mold. To minimize the cost, it helps to eject the parts quickly, allowing the mold to be reused as soon as possible. Yet ejecting the part when it is too hot can cause warping or punctures. The rate at which objects cool is therefore of great interest.

To illustrate the rate of cooling, the *Texas Instruments Calculator-Based Laboratory (CBL) System* was used to measure the temperature of a cup of water over a 40-second period. The room temperature was measured at 69.55°F, and the water temperature at time $t = 0$ was measured at 165.58°F. The results are shown in the following scatter plot.

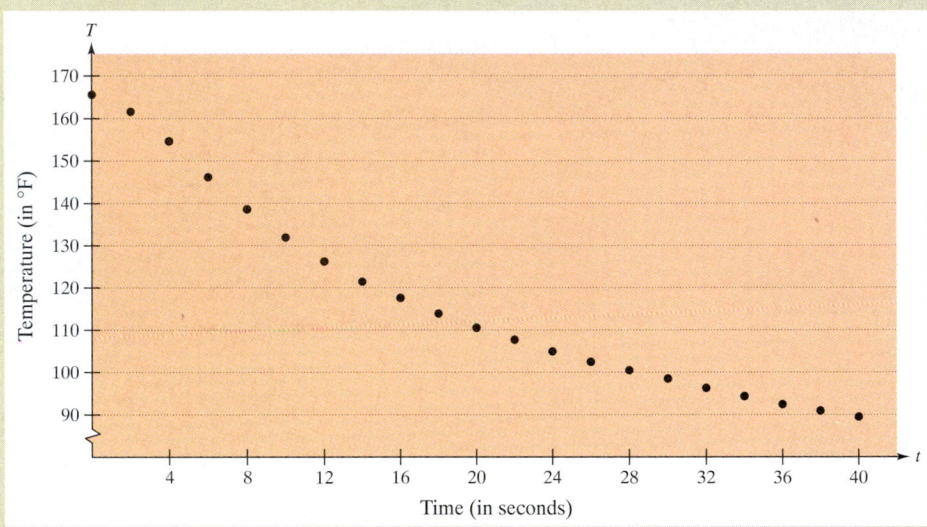

Time (in seconds)

QUESTIONS

1. Describe the pattern of the temperature points over time. Does the *rate* at which the temperature changes seem to increase, decrease, or remain constant?

2. Imagine a curve running through the data points. How would you expect the curve to behave as the value of t increases? Would you expect the curve to intersect the line $T = 69.55$? Explain your reasoning.

3. Would the derivative of a function modeling the data points be increasing, decreasing, or constant? Explain your reasoning.

4. The data in the scatter plot can be modeled using a function of the form

$$T = a \cdot b^t + c.$$

Find values of a, b, and c that produce a reasonable model.

The concepts presented here will be explored further in this chapter. For an extension of this application, see the lab series that accompanies this text.

Leo Hendrik Baekeland attempted to create a shellac by combining phenol and formaldehyde. The experiment "failed" in that it did not result in shellac, but it did form the first completely synthetic plastic resin. Bakelite is still used today in the automotive and electronics industry.

The Natural Logarithmic Function • The Number *e* •
The Derivative of the Natural Logarithmic Function

The Natural Logarithmic Function

Recall that the Power Rule

$$\int x^n \, dx = \frac{x^{n+1}}{n+1} + C, \qquad n \neq -1 \qquad \text{Power Rule}$$

has an important disclaimer—it doesn't apply when $n = -1$. Consequently, we have not yet found an antiderivative for the function $f(x) = 1/x$. In this section, we will use the Second Fundamental Theorem of Calculus to *define* such a function. This antiderivative is a function that we have not encountered previously in the text. It is neither algebraic nor trigonometric, but falls into a new class of functions called *logarithmic functions*. This particular function is the **natural logarithmic function.**

JOHN NAPIER (1550–1617)

Logarithms were invented by the Scottish mathematician John Napier. Although he did not introduce the *natural* logarithmic function, it is sometimes called the *Napierian* logarithm.

> **Definition of the Natural Logarithmic Function**
>
> The **natural logarithmic function** is defined by
>
> $$\ln x = \int_1^x \frac{1}{t} \, dt, \qquad x > 0.$$
>
> The domain of the natural logarithmic function is the set of all positive real numbers.

From the definition, you can see that $\ln x$ is positive for $x > 1$ and negative for $0 < x < 1$, as shown in Figure 5.1. Moreover, $\ln(1) = 0$, because the upper and lower limits of integration are equal when $x = 1$.

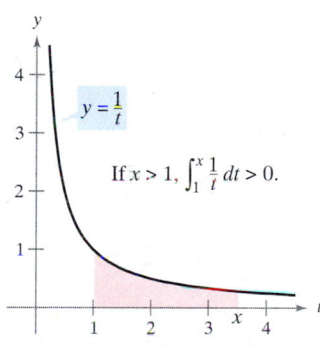

If $x > 1$, then $\ln x > 0$.

Figure 5.1

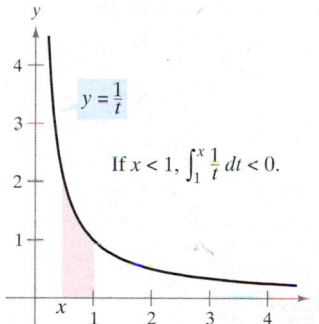

If $0 < x < 1$, then $\ln x < 0$.

EXPLORATION

Graphing the Natural Logarithmic Function   Using *only* the definition of the natural logarithmic function, sketch a graph of the function. Explain your reasoning.

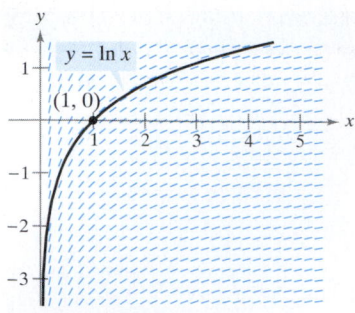

Each small line segment has a slope of $\dfrac{1}{x}$.

Figure 5.2

NOTE   Direction fields can be helpful in getting a visual perspective of the directions of the solutions of a differential equation.

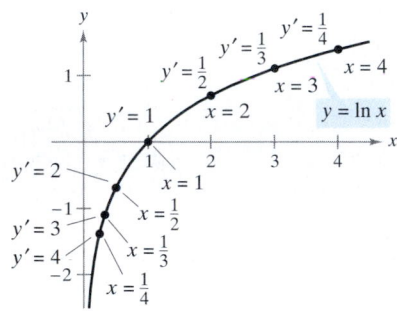

The natural logarithmic function is increasing, and its graph is concave downward.

Figure 5.3

To sketch the graph of $y = \ln x$, you can think of the natural logarithmic function as an *antiderivative* given by the differential equation

$$\frac{dy}{dx} = \frac{1}{x}.$$

Figure 5.2 is a computer-generated graph, called a *direction field*, showing small line segments of slope $1/x$. The graph of $y = \ln x$ is the solution that passes through the point $(1, 0)$.

The following theorem lists some basic properties of the natural logarithmic function.

THEOREM 5.1   Properties of the Natural Logarithmic Function

The natural logarithmic function has the following properties.

1. The domain is $(0, \infty)$ and the range is $(-\infty, \infty)$.
2. The function is continuous, increasing, and one-to-one.
3. The graph is concave downward.

Proof   The domain of $f(x) = \ln x$ is $(0, \infty)$ by definition. Moreover, the function is continuous because it is differentiable. It is increasing because its derivative

$$f'(x) = \frac{1}{x} \qquad \text{First derivative}$$

is positive for $x > 0$, as shown in Figure 5.3. It is concave downward because

$$f''(x) = -\frac{1}{x^2} \qquad \text{Second derivative}$$

is negative for $x > 0$. We leave the proof that f is one-to-one as an exercise (see Exercise 96). The following limits imply that its range is the entire real line.

$$\lim_{x \to 0^+} \ln x = -\infty \qquad \text{and} \qquad \lim_{x \to \infty} \ln x = \infty$$

Verification of these two limits is given in the appendix.   ▬▬▬

Using the definition of the natural logarithmic function, you can prove several important properties involving operations with natural logarithms. If you are already familiar with logarithms, you will recognize that these properties are characteristic of all logarithms.

THEOREM 5.2   Logarithmic Properties

If a and b are positive numbers and n is rational, then the following properties are true.

1. $\ln(1) = 0$
2. $\ln(ab) = \ln a + \ln b$
3. $\ln(a^n) = n \ln a$
4. $\ln\left(\dfrac{a}{b}\right) = \ln a - \ln b$

Proof We have already discussed the first property. The proof of the second property follows from the fact that two antiderivatives of the same function differ at most by a constant. From the Second Fundamental Theorem of Calculus and the definition of the natural logarithmic function, you know that

$$\frac{d}{dx}[\ln x] = \frac{1}{x}.$$

Thus, consider the two derivatives

$$\frac{d}{dx}[\ln(ax)] = \frac{a}{ax} = \frac{1}{x}$$

and

$$\frac{d}{dx}[\ln a + \ln x] = 0 + \frac{1}{x} = \frac{1}{x}.$$

Because $\ln(ax)$ and $(\ln a + \ln x)$ are both antiderivatives of $1/x$, they must differ at most by a constant

$$\ln(ax) = \ln a + \ln x + C.$$

By letting $x = 1$, you can see that $C = 0$. The third property can be proved similarly by comparing the derivatives of $\ln(x^n)$ and $n \ln x$. Finally, using the second and third properties, you can prove the fourth property.

$$\ln\left(\frac{a}{b}\right) = \ln[a(b^{-1})]$$

$$= \ln a + \ln(b^{-1})$$

$$= \ln a - \ln b$$

EXAMPLE 1   Expanding Logarithmic Expressions

a. $\ln\dfrac{10}{9} = \ln 10 - \ln 9$ Property 4

b. $\ln\sqrt{3x + 2} = \ln(3x + 2)^{1/2}$ Rewrite with rational exponent.

$\qquad\qquad = \dfrac{1}{2}\ln(3x + 2)$ Property 3

c. $\ln\dfrac{6x}{5} = \ln(6x) - \ln 5$ Property 4

$\qquad\quad = \ln 6 + \ln x - \ln 5$ Property 2

d. $\ln\dfrac{(x^2 + 3)^2}{x\sqrt[3]{x^2 + 1}} = \ln(x^2 + 3)^2 - \ln\left(x\sqrt[3]{x^2 + 1}\right)$

$\qquad\qquad = 2\ln(x^2 + 3) - \left[\ln x + \ln(x^2 + 1)^{1/3}\right]$

$\qquad\qquad = 2\ln(x^2 + 3) - \ln x - \ln(x^2 + 1)^{1/3}$

$\qquad\qquad = 2\ln(x^2 + 3) - \ln x - \dfrac{1}{3}\ln(x^2 + 1)$

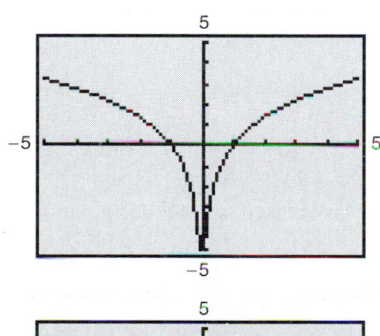

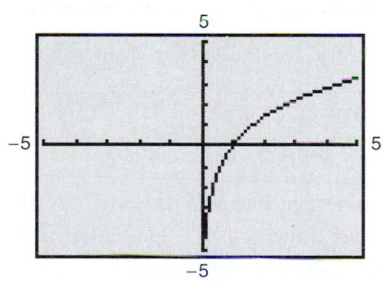

Try using a graphing utility to compare the graphs of

$f(x) = \ln x^2$

and

$g(x) = 2\ln x.$

Which of the above is the graph of f? Which is the graph of g?

Figure 5.4

NOTE   When using the properties of logarithms to rewrite logarithmic functions, you must check to see whether the domain of the rewritten function is the same as the domain of the original. For instance, the domain of $f(x) = \ln x^2$ is all real numbers except $x = 0$, and the domain of $g(x) = 2\ln x$ is all positive real numbers. (See Figure 5.4.)

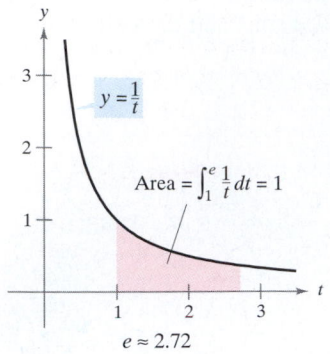

e is the base for the natural logarithm because ln *e* = 1.
Figure 5.5

The Number *e*

It is likely that you have previously studied logarithms in an algebra course. There, without the benefit of calculus, logarithms would have been defined in terms of a **base** number. For example, common logarithms have a base of 10 because $\log_{10} 10 = 1$. (We will say more about this in Section 5.5.)

To define the **base for the natural logarithm,** we use the fact that the natural logarithmic function is continuous, is one-to-one, and has a range of $(-\infty, \infty)$. Hence, there must be a unique real number x such that $\ln x = 1$, as shown in Figure 5.5. This number is denoted by the letter e. It can be shown that e is irrational and has the following decimal approximation.

$$e \approx 2.71828182846$$

Definition of *e*

The letter e denotes the positive real number such that

$$\ln e = \int_1^e \frac{1}{t} \, dt = 1.$$

FOR FURTHER INFORMATION For more information about the number e, see the article "Unexpected Occurrences of the Number e" by Harris S. Shultz and Bill Leonard in the October 1989 issue of *Mathematics Magazine.*

Once you know that $\ln e = 1$, you can use logarithmic properties to evaluate the natural logarithms of several other numbers. For example, by using the property

$$\ln(e^n) = n \ln e$$
$$= n(1)$$
$$= n$$

you can evaluate $\ln(e^n)$ for various powers of n, as shown in the table and in Figure 5.6.

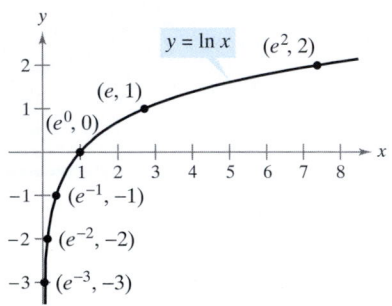

If $x = e^n$, then ln $x = n$.
Figure 5.6

x	$\frac{1}{e^3} \approx 0.050$	$\frac{1}{e^2} \approx 0.135$	$\frac{1}{e} \approx 0.368$	$e^0 = 1$	$e \approx 2.718$	$e^2 \approx 7.389$
ln x	-3	-2	-1	0	1	2

The logarithms given in the table above are convenient because the x-values are integer powers of e. Most logarithmic expressions are, however, best evaluated with a calculator.

EXAMPLE 2   Evaluating Natural Logarithmic Expressions

a. $\ln 2 \approx 0.693$

b. $\ln 32 \approx 3.466$

c. $\ln 0.1 \approx -2.303$

The Derivative of the Natural Logarithmic Function

The derivative of the natural logarithmic function is given in Theorem 5.3. The first part of the theorem follows from the definition of the natural logarithmic function as an antiderivative. The second part of the theorem is simply the Chain Rule version of the first part.

> **THEOREM 5.3 Derivative of the Natural Logarithmic Function**
>
> Let u be a differentiable function of x.
>
> **1.** $\dfrac{d}{dx}[\ln x] = \dfrac{1}{x}, x > 0$ **2.** $\dfrac{d}{dx}[\ln u] = \dfrac{1}{u}\dfrac{du}{dx} = \dfrac{u'}{u}, u > 0$

EXAMPLE 3 Differentiation of Logarithmic Functions

a. $\dfrac{d}{dx}[\ln(2x)] = \dfrac{u'}{u} = \dfrac{2}{2x} = \dfrac{1}{x}$ $u = 2x$

b. $\dfrac{d}{dx}[\ln(x^2 + 1)] = \dfrac{u'}{u} = \dfrac{2x}{x^2 + 1}$ $u = x^2 + 1$

c. $\dfrac{d}{dx}[x \ln x] = x\left(\dfrac{d}{dx}[\ln x]\right) + (\ln x)\left(\dfrac{d}{dx}[x]\right)$ Product Rule

$\qquad\qquad = x\left(\dfrac{1}{x}\right) + (\ln x)(1)$

$\qquad\qquad = 1 + \ln x$

d. $\dfrac{d}{dx}[(\ln x)^3] = 3(\ln x)^2 \dfrac{d}{dx}[\ln x]$ Chain Rule

$\qquad\qquad = 3(\ln x)^2 \dfrac{1}{x}$

Napier used logarithmic properties to simplify *calculations* involving products, quotients, and powers. Of course, given the availability of calculators, there is now little need for this particular application of logarithms. However, there is great value in using logarithmic properties to simplify *differentiation* involving products, quotients, and powers.

EXAMPLE 4 Logarithmic Properties as Aids to Differentiation

Differentiate $f(x) = \ln\sqrt{x + 1}$.

Solution Because

$$f(x) = \ln\sqrt{x + 1} = \ln(x + 1)^{1/2} = \frac{1}{2}\ln(x + 1)$$

you can write the following.

$$f'(x) = \frac{1}{2}\left(\frac{1}{x + 1}\right) = \frac{1}{2(x + 1)}$$

EXPLORATION

Use a graphing utility to graph

$$y_1 = \frac{1}{x}$$

and

$$y_2 = \frac{d}{dx}[\ln x]$$

in the same viewing rectangle in which $0.1 \le x \le 5$ and $-2 \le y \le 8$. Explain why the graphs appear to be identical.

EXAMPLE 5 **Logarithmic Properties as Aids to Differentiation**

Differentiate $f(x) = \ln \dfrac{x(x^2 + 1)^2}{\sqrt{2x^3 - 1}}$.

Solution

$$f(x) = \ln \frac{x(x^2 + 1)^2}{\sqrt{2x^3 - 1}} \qquad \text{Original function}$$

$$= \ln x + 2 \ln(x^2 + 1) - \frac{1}{2} \ln(2x^3 - 1) \qquad \text{Rewrite before differentiating.}$$

$$f'(x) = \frac{1}{x} + 2\left(\frac{2x}{x^2 + 1}\right) - \frac{1}{2}\left(\frac{6x^2}{2x^3 - 1}\right) \qquad \text{Differentiate.}$$

$$= \frac{1}{x} + \frac{4x}{x^2 + 1} - \frac{3x^2}{2x^3 - 1} \qquad \text{Simplify.}$$

NOTE In Examples 4 and 5, be sure that you see the benefit of applying logarithmic properties *before* differentiating. Consider, for instance, the difficulty of direct differentiation of the function given in Example 5.

On occasion, it is convenient to use logarithms as aids in differentiating *non*logarithmic functions. This procedure is called **logarithmic differentiation.**

EXAMPLE 6 **Logarithmic Differentiation**

Find the derivative of $y = \dfrac{(x - 2)^2}{\sqrt{x^2 + 1}}, x \neq 2$.

Solution Note that $y > 0$ and hence that $\ln y$ is defined. Begin by taking the natural logarithms of both sides of the equation. Then apply logarithmic properties and differentiate implicitly. Finally, solve for y'.

$$\ln y = \ln \frac{(x - 2)^2}{\sqrt{x^2 + 1}} \qquad \text{Take ln of both sides.}$$

$$\ln y = 2 \ln(x - 2) - \frac{1}{2} \ln(x^2 + 1) \qquad \text{Logarithmic properties}$$

$$\frac{y'}{y} = 2\left(\frac{1}{x - 2}\right) - \frac{1}{2}\left(\frac{2x}{x^2 + 1}\right) \qquad \text{Differentiate.}$$

$$= \frac{2}{x - 2} - \frac{x}{x^2 + 1}$$

$$y' = y\left(\frac{2}{x - 2} - \frac{x}{x^2 + 1}\right) \qquad \text{Solve for } y'.$$

$$= \frac{(x - 2)^2}{\sqrt{x^2 + 1}}\left[\frac{x^2 + 2x + 2}{(x - 2)(x^2 + 1)}\right] \qquad \text{Substitute for } y.$$

$$= \frac{(x - 2)(x^2 + 2x + 2)}{(x^2 + 1)^{3/2}} \qquad \text{Simplify.}$$

Because the natural logarithm is undefined for negative numbers, you will often encounter expressions of the form $\ln |u|$. The following theorem states that you can differentiate functions of the form $y = \ln |u|$ as if the absolute value sign were not present.

THEOREM 5.4   Derivative Involving Absolute Value

If u is a differentiable function of x such that $u \neq 0$, then

$$\frac{d}{dx}[\ln |u|] = \frac{u'}{u}.$$

Proof   If $u > 0$, then $|u| = u$, and the result follows from Theorem 5.3. If $u < 0$, then $|u| = -u$, and you have

$$\frac{d}{dx}[\ln |u|] = \frac{d}{dx}[\ln(-u)]$$

$$= \frac{-u'}{-u}$$

$$= \frac{u'}{u}.$$

EXAMPLE 7   Derivative Involving Absolute Value

Find the derivative of $f(x) = \ln |\cos x|$.

Solution   Using Theorem 5.4, let $u = \cos x$ and write

$$\frac{d}{dx}[\ln |\cos x|] = \frac{u'}{u}$$

$$= \frac{-\sin x}{\cos x}$$

$$= -\tan x.$$

EXAMPLE 8   Finding Relative Extrema

Locate the relative extrema of $y = \ln(x^2 + 2x + 3)$.

Solution   Differentiating y, you obtain

$$\frac{dy}{dx} = \frac{2x + 2}{x^2 + 2x + 3}.$$

Because $dy/dx = 0$ when $x = -1$, you can apply the First Derivative Test and conclude that the point $(-1, \ln 2)$ is a relative minimum. Because there are no other critical points, it follows that this is the only relative extremum (see Figure 5.7).

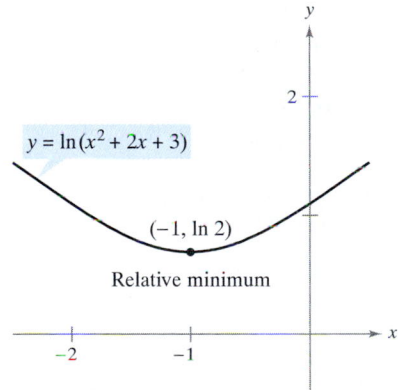

$y = \ln(x^2 + 2x + 3)$

$(-1, \ln 2)$

Relative minimum

The derivative of y changes from negative to positive at $x = -1$.

Figure 5.7

EXERCISES FOR SECTION 5.1

1. Complete the table below. Use a graphing utility and Simpson's Rule with $n = 10$ to approximate the integral

$$\int_1^x \frac{1}{t}\, dt.$$

x	0.5	1.5	2	2.5	3	3.5	4
$\int_1^x (1/t)\, dt$							

2. (a) Plot the points generated in Exercise 1 and connect them with a smooth curve. Compare the result with the graph of $y = \ln x$.

   (b) Use a graphing utility to graph $y = \displaystyle\int_1^x (1/t)\, dt$ for $0.2 \le x \le 4$. Compare the result with the graph of $y = \ln x$.

In Exercises 3–6, use the graph of $y = \ln x$ to match the function with its graph. [The graphs are labeled (a), (b), (c), and (d).]

(a)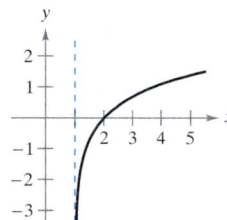

(b)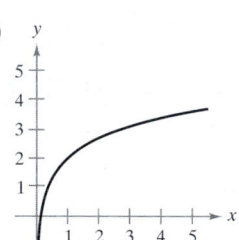

(c)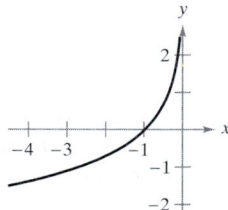

(d)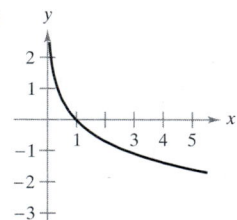

3. $f(x) = \ln x + 2$

4. $f(x) = -\ln x$

5. $f(x) = \ln(x - 1)$

6. $f(x) = -\ln(-x)$

In Exercises 7–12, sketch the graph of the function and state its domain.

7. $f(x) = 3 \ln x$

8. $f(x) = -2 \ln x$

9. $f(x) = \ln 2x$

10. $f(x) = \ln|x|$

11. $f(x) = \ln(x - 1)$

12. $g(x) = 2 + \ln x$

In Exercises 13 and 14, use the properties of logarithms to approximate the indicated logarithm, given that $\ln 2 \approx 0.6931$ and $\ln 3 \approx 1.0986$.

13. (a) $\ln 6$ (b) $\ln \frac{2}{3}$ (c) $\ln 81$ (d) $\ln \sqrt{3}$

14. (a) $\ln 0.25$ (b) $\ln 24$ (c) $\ln \sqrt[3]{12}$ (d) $\ln \frac{1}{72}$

In Exercises 15–24, use the properties of logarithms to write the expression as a sum, difference, and/or multiple of logarithms.

15. $\ln \frac{2}{3}$

16. $\ln \frac{1}{5}$

17. $\ln \dfrac{xy}{z}$

18. $\ln(xyz)$

19. $\ln \sqrt{2^3}$

20. $\ln \sqrt{a - 1}$

21. $\ln\left(\dfrac{x^2 - 1}{x^3}\right)^3$

22. $\ln 3e^2$

23. $\ln z(z - 1)^2$

24. $\ln \dfrac{1}{e}$

In Exercises 25–30, write the expression as a logarithm of a single quantity.

25. $\ln(x - 2) - \ln(x + 2)$

26. $3 \ln x + 2 \ln y - 4 \ln z$

27. $\frac{1}{3}[2 \ln(x + 3) + \ln x - \ln(x^2 - 1)]$

28. $2[\ln x - \ln(x + 1) - \ln(x - 1)]$

29. $2 \ln 3 - \frac{1}{2} \ln(x^2 + 1)$

30. $\frac{3}{2}[\ln(x^2 + 1) - \ln(x + 1) - \ln(x - 1)]$

In Exercises 31 and 32, show that $f = g$ by using a graphing utility to graph f and g in the same viewing rectangle.

31. $f(x) = \ln \dfrac{x^2}{4}, x > 0, \quad g(x) = 2 \ln x - \ln 4$

32. $f(x) = \ln \sqrt{x(x^2 + 1)}, \quad g(x) = \frac{1}{2}[\ln x + \ln(x^2 + 1)]$

In Exercises 33–36, find the limit.

33. $\displaystyle\lim_{x \to 3^+} \ln(x - 3)$

34. $\displaystyle\lim_{x \to 6^-} \ln(6 - x)$

35. $\displaystyle\lim_{x \to 2^-} \ln[x^2(3 - x)]$

36. $\displaystyle\lim_{x \to 5^+} \ln \dfrac{x}{\sqrt{x - 4}}$

In Exercises 37–40, find the slope of the tangent line to the logarithmic function at the point $(1, 0)$.

37. $y = \ln x^3$

38. $y = \ln x^{3/2}$

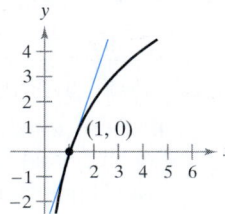

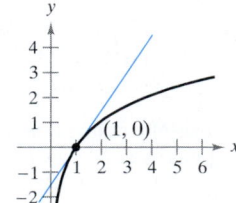

39. $y = \ln x^2$

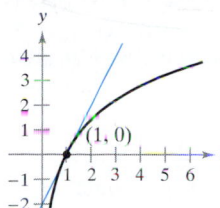

40. $y = \ln x^{1/2}$

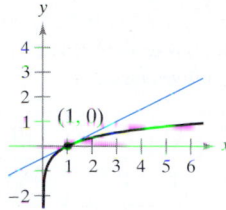

In Exercises 41–66, find the derivative of the function.

41. $g(x) = \ln x^2$

42. $h(x) = \ln(x^2 + 3)$

43. $y = (\ln x)^4$

44. $y = x \ln x$

45. $y = \ln\left(x\sqrt{x^2 - 1}\right)$

46. $y = \ln\sqrt{x^2 - 4}$

47. $f(x) = \ln\left(\dfrac{x}{x^2 + 1}\right)$

48. $f(x) = \ln\left(\dfrac{x}{x + 1}\right)$

49. $g(t) = \dfrac{\ln t}{t^2}$

50. $h(t) = \dfrac{\ln t}{t}$

51. $y = \ln(\ln x^2)$

52. $y = \ln(\ln x)$

53. $y = \ln\sqrt{\dfrac{x + 1}{x - 1}}$

54. $y = \ln\sqrt{\dfrac{x - 1}{x + 1}}$

55. $f(x) = \ln\left(\dfrac{\sqrt{4 + x^2}}{x}\right)$

56. $f(x) = \ln\left(x + \sqrt{4 + x^2}\right)$

57. $y = \dfrac{-\sqrt{x^2 + 1}}{x} + \ln\left(x + \sqrt{x^2 + 1}\right)$

58. $y = \dfrac{-\sqrt{x^2 + 4}}{2x^2} - \dfrac{1}{4}\ln\left(\dfrac{2 + \sqrt{x^2 + 4}}{x}\right)$

59. $y = \ln|\sin x|$

60. $y = \ln|\sec x|$

61. $y = \ln\left|\dfrac{\cos x}{\cos x - 1}\right|$

62. $y = \ln|\sec x + \tan x|$

63. $y = \ln\left|\dfrac{-1 + \sin x}{2 + \sin x}\right|$

64. $y = \ln\sqrt{1 + \sin^2 x}$

65. $f(x) = \sin 2x \ln x^2$

66. $g(x) = \displaystyle\int_{1}^{\ln x} (t^2 + 3)\,dt$

In Exercises 67 and 68, (a) find an equation of the tangent line to the graph of f at the indicated point, (b) use a graphing utility to graph the function and its tangent line at the point, and (c) use the derivative feature of a graphing utility to confirm your results.

Function	Point
67. $y = 3x^2 - \ln x$	$(1, 3)$
68. $y = 4 - x^2 - \ln\left(\frac{1}{2}x + 1\right)$	$(0, 4)$

In Exercises 69 and 70, use implicit differentiation to find dy/dx.

69. $x^2 - 3\ln y + y^2 = 10$

70. $\ln xy + 5x = 30$

In Exercises 71 and 72, show that the function is a solution of the differential equation.

Function	Differential Equation
71. $y = 2\ln x + 3$	$xy'' + y' = 0$
72. $y = x\ln x - 4x$	$x + y - xy' = 0$

In Exercises 73–78, find any relative extrema and inflection points of the function. Use a graphing utility to confirm your results.

73. $y = \dfrac{x^2}{2} - \ln x$

74. $y = x - \ln x$

75. $y = x\ln x$

76. $y = \dfrac{\ln x}{x}$

77. $y = \dfrac{x}{\ln x}$

78. $y = x^2 \ln x$

Linear and Quadratic Approximations   **In Exercises 79 and 80, use a graphing utility to graph the function. Then graph**

$$P_1(x) = f(1) + f'(1)(x - 1)$$

and

$$P_2(x) = f(1) + f'(1)(x - 1) + \tfrac{1}{2}f''(1)(x - 1)^2$$

in the same viewing rectangle. Compare the values of f, P_1, and P_2 and their first derivatives at $x = 1$.

79. $f(x) = \ln x$

80. $f(x) = x\ln x$

In Exercises 81 and 82, use Newton's Method to approximate, to three decimal places, the x-coordinate of the point of intersection of the graphs of the two equations. Use a graphing utility to verify your result.

81. $y = \ln x$
$y = -x$

82. $y = \ln x$
$y = 3 - x$

In Exercises 83–88, find dy/dx using logarithmic differentiation.

83. $y = x\sqrt{x^2 - 1}$

84. $y = \sqrt{(x - 1)(x - 2)(x - 3)}$

85. $y = \dfrac{x^2\sqrt{3x - 2}}{(x - 1)^2}$

86. $y = \sqrt[3]{\dfrac{x^2 + 1}{x^2 - 1}}$

87. $y = \dfrac{x(x - 1)^{3/2}}{\sqrt{x + 1}}$

88. $y = \dfrac{(x + 1)(x + 2)}{(x - 1)(x - 2)}$

89. *Sound Intensity* The relationship between the number of decibels β and the intensity of a sound I in watts per centimeter squared is

$$\beta = 10 \log_{10}\left(\frac{I}{10^{-16}}\right).$$

Use the properties of logarithms to write the formula in simpler form, and determine the number of decibels of a sound with an intensity of 10^{-10} watts per square centimeter.

90. *Home Mortgage* The term t (in years) of a $120,000 home mortgage at 10% interest can be approximated by

$$t = \frac{5.315}{-6.7968 + \ln x}, \quad x > 1000$$

where x is the monthly payment in dollars.

(a) Use a graphing utility to graph the model.

(b) Use the model to approximate the term of a home mortgage for which the monthly payment is $1167.41. What is the total amount paid?

(c) Use the model to approximate the term of a home mortgage for which the monthly payment is $1068.45. What is the total amount paid?

(d) Find the instantaneous rate of change of t with respect to x when $x = 1167.41$ and $x = 1068.45$.

(e) Write a short paragraph describing the benefit of the higher monthly payment.

91. *Modeling Data* The table gives the temperature T (°F) at which water boils at selected pressures p (pounds per square inch). *(Source: Standard Handbook of Mechanical Engineers)*

p	5	10	14.696 (1 atm)	20
T	162.24°	193.21°	212.00°	227.96°

p	30	40	60	80	100
T	250.33°	267.25°	292.71°	312.03°	327.81°

A model that approximates these data is

$$T = 87.97 + 34.96 \ln p + 7.91 \sqrt{p}.$$

(a) Use a graphing utility to plot the data and graph the model.

(b) Find the rate of change of T with respect to p when $p = 10$ and $p = 70$.

(c) Use a graphing utility to graph T'. Find

$$\lim_{p \to \infty} T'(p)$$

and interpret the result in the context of the problem.

92. *Think About It* Let f be a function that is positive and differentiable on the entire real line. Let $g(x) = \ln f(x)$.

(a) If the graph of g is increasing, must the graph of f be increasing? Explain.

(b) If the graph of f is concave upward, must the graph of g be concave upward? Explain.

 93. *Modeling Data* The atmospheric pressure decreases with increasing altitude. At sea level, the average air pressure is *one atmosphere* (1.033227 kilograms per square centimeter). The table gives the pressure p (in atmospheres) at a given altitude h (in kilometers).

h	0	5	10	15	20	25
p	1	0.55	0.25	0.12	0.06	0.02

(a) Use a graphing utility to find a model of the form $p = a + b \ln h$ for the data. Explain why the result is an error message.

(b) Use a graphing utility to find the logarithmic model $h = a + b \ln p$ for the data.

(c) Use a graphing utility to plot the data and graph the logarithmic model.

(d) Use the model to estimate the altitude at which the pressure is 0.75 atmosphere.

(e) Use the model to estimate the pressure at an altitude of 13 kilometers.

(f) Find the rate of change of pressure when $h = 5$ and $h = 20$. Interpret the results in the context of the problem.

94. *Tractrix* A person walking along a dock drags a boat by a 10-meter rope. The boat travels along a path known as a *tractrix* (see figure). The equation of this path is

$$y = 10 \ln\left(\frac{10 + \sqrt{100 - x^2}}{x}\right) - \sqrt{100 - x^2}.$$

(a) Use a graphing utility to graph the function.

(b) What is the slope of this path when $x = 5$ and $x = 9$?

(c) What does the slope of the path approach as $x \to 10$?

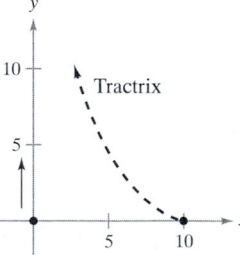

95. *Conjecture* Use a graphing utility to graph f and g in the same viewing rectangle and determine which is increasing at the faster rate for "large" values of x. What can you conclude about the rate of growth of the natural logarithmic function?

(a) $f(x) = \ln x, \qquad g(x) = \sqrt{x}$

(b) $f(x) = \ln x, \qquad g(x) = \sqrt[4]{x}$

96. Prove that the natural logarithmic function is one-to-one.

True or False? **In Exercises 97 and 98, determine whether the statement is true or false. If it is false, explain why or give an example that shows it is false.**

97. $\ln(x + 25) = \ln x + \ln 25$

98. If $y = \ln \pi$, then $y' = 1/\pi$.

SECTION **5.2** **The Natural Logarithmic Function and Integration**

Log Rule for Integration • Integrals of Trigonometric Functions

Log Rule for Integration

The differentiation rules

$$\frac{d}{dx}[\ln |x|] = \frac{1}{x} \qquad \text{and} \qquad \frac{d}{dx}[\ln |u|] = \frac{u'}{u}$$

that you studied in the previous section produce the following integration rules.

THEOREM 5.5   Log Rule for Integration

Let u be a differentiable function of x.

1. $\displaystyle\int \frac{1}{x}\, dx = \ln |x| + C$ **2.** $\displaystyle\int \frac{1}{u}\, du = \ln |u| + C$

Because $du = u'\, dx$, the second formula can also be written as

$$\int \frac{u'}{u}\, dx = \ln |u| + C. \qquad\qquad \text{Alternative form of Log Rule}$$

EXAMPLE 1   Using the Log Rule for Integration

$$\int \frac{2}{x}\, dx = 2\int \frac{1}{x}\, dx$$
$$= 2\ln|x| + C \qquad\qquad \text{Log Rule for Integration}$$
$$= \ln(x^2) + C \qquad\qquad \text{Property of logarithms}$$

Because x^2 cannot be negative, the absolute value is unnecessary in the final form of the antiderivative.

EXAMPLE 2   Using the Log Rule with a Change of Variable

Evaluate $\displaystyle\int \frac{1}{4x - 1}\, dx$.

Solution   If you let $u = 4x - 1$, then $du = 4\, dx$.

$$\int \frac{1}{4x - 1}\, dx = \frac{1}{4}\int \left(\frac{1}{4x - 1}\right)4\, dx \qquad \text{Multiply and divide by 4.}$$
$$= \frac{1}{4}\int \frac{1}{u}\, du \qquad\qquad \text{Substitute: } u = 4x - 1.$$
$$= \frac{1}{4}\ln |u| + C \qquad\qquad \text{Apply Log Rule.}$$
$$= \frac{1}{4}\ln |4x - 1| + C \qquad\qquad \text{Back-substitute.}$$

EXPLORATION

Integrating Rational Functions
Early in Chapter 4, you learned rules that allowed you to integrate *any* polynomial function. The Log Rule presented in this section goes a long way toward enabling you to integrate rational functions. For instance, each of the following functions can be integrated with the Log Rule.

$\dfrac{2}{x}$ Example 1

$\dfrac{1}{4x - 1}$ Example 2

$\dfrac{x}{x^2 + 1}$ Example 3

$\dfrac{3x^2 + 1}{x^3 + x}$ Example 4a

$\dfrac{x + 1}{x^2 + 2x}$ Example 4c

$\dfrac{1}{3x + 2}$ Example 4d

$\dfrac{x^2 + x + 1}{x^2 + 1}$ Example 5

$\dfrac{2x}{(x + 1)^2}$ Example 6

There are still some rational functions that cannot be integrated using the Log Rule. Give examples of these functions, and explain your reasoning.

Example 3 uses the alternative form of the Log Rule. To apply this rule, look for quotients in which the numerator is the derivative of the denominator.

EXAMPLE 3   Finding Area with the Log Rule

Find the area of the region bounded by the graph of

$$y = \frac{x}{x^2 + 1}$$

the x-axis, and the line $x = 3$.

Solution   From Figure 5.8, you can see that the area of the region is given by the definite integral

$$\int_0^3 \frac{x}{x^2 + 1} \, dx.$$

If you let $u = x^2 + 1$, then $u' = 2x$. To apply the Log Rule, multiply and divide by 2 as follows.

$$\int_0^3 \frac{x}{x^2 + 1} \, dx = \frac{1}{2} \int_0^3 \frac{2x}{x^2 + 1} \, dx$$

$$= \frac{1}{2} \left[\ln(x^2 + 1) \right]_0^3 \qquad \int \frac{u'}{u} \, dx = \ln |u| + C$$

$$= \frac{1}{2} (\ln 10 - \ln 1)$$

$$= \frac{1}{2} \ln 10 \qquad\qquad \ln 1 = 0$$

$$\approx 1.151$$

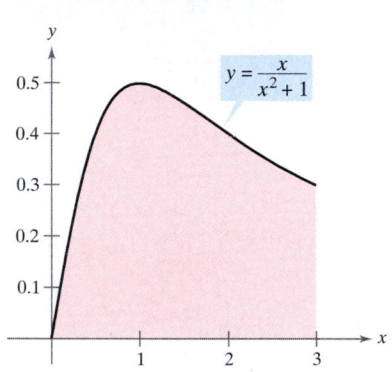

Area $= \displaystyle\int_0^3 \frac{x}{x^2 + 1} \, dx$

The area of the region bounded by the graph of y, the x-axis, and $x = 3$ is $\frac{1}{2}$ ln 10.
Figure 5.8

EXAMPLE 4   Recognizing Quotient Forms of the Log Rule

a. $\displaystyle\int \frac{3x^2 + 1}{x^3 + x} \, dx = \ln |x^3 + x| + C \qquad u = x^3 + x$

b. $\displaystyle\int \frac{\sec^2 x}{\tan x} \, dx = \ln |\tan x| + C \qquad u = \tan x$

c. $\displaystyle\int \frac{x + 1}{x^2 + 2x} \, dx = \frac{1}{2} \int \frac{2x + 2}{x^2 + 2x} \, dx \qquad u = x^2 + 2x$

$$= \frac{1}{2} \ln |x^2 + 2x| + C$$

d. $\displaystyle\int \frac{1}{3x + 2} \, dx = \frac{1}{3} \int \frac{3}{3x + 2} \, dx \qquad u = 3x + 2$

$$= \frac{1}{3} \ln |3x + 2| + C$$

With antiderivatives involving logarithms, it is easy to obtain forms that look quite different but are still equivalent. For instance, which of the following are equivalent to the antiderivative listed in Example 4d?

$$\ln |(3x + 2)^{1/3}| + C, \quad \frac{1}{3} \ln \left| x + \tfrac{2}{3} \right| + C, \quad \ln |3x + 2|^{1/3} + C$$

Integrals to which the Log Rule can be applied often appear in disguised form. For instance, if a rational function has a *numerator of degree greater than or equal to that of the denominator,* division may reveal a form to which you can apply the Log Rule. This is illustrated in Example 5.

EXAMPLE 5   Using Long Division Before Integrating

Evaluate $\int \dfrac{x^2 + x + 1}{x^2 + 1}\, dx$.

Solution   Begin by using long division to rewrite the integrand.

$$\dfrac{x^2 + x + 1}{x^2 + 1} \quad \Longrightarrow \quad x^2 + 1\overline{\smash{\big)}\,x^2 + x + 1} \quad \Longrightarrow \quad 1 + \dfrac{x}{x^2 + 1}$$

with the long division showing quotient 1, $\dfrac{x^2 \qquad + 1}{x}$

Now, you can integrate to obtain

$$
\begin{aligned}
\int \frac{x^2 + x + 1}{x^2 + 1}\, dx &= \int \left(1 + \frac{x}{x^2 + 1}\right) dx \\
&= \int dx + \frac{1}{2}\int \frac{2x}{x^2 + 1}\, dx \\
&= x + \frac{1}{2}\ln(x^2 + 1) + C.
\end{aligned}
$$

Check this result by differentiating to obtain the original integrand.   ▬▬▬

The next example gives another instance in which the use of the Log Rule is disguised. In this case, a change of variables helps you recognize the Log Rule.

EXAMPLE 6   Change of Variables with the Log Rule

Evaluate $\int \dfrac{2x}{(x + 1)^2}\, dx$.

Solution   If you let $u = x + 1$, then $du = dx$ and $x = u - 1$.

$$
\begin{aligned}
\int \frac{2x}{(x + 1)^2}\, dx &= \int \frac{2(u - 1)}{u^2}\, du && \text{Substitute: } u = x + 1. \\
&= 2\int \left(\frac{u}{u^2} - \frac{1}{u^2}\right) du && \text{Rewrite as two fractions.} \\
&= 2\int \frac{du}{u} - 2\int u^{-2}\, du && \text{Rewrite as two integrals.} \\
&= 2\ln|u| - 2\left(\frac{u^{-1}}{-1}\right) + C && \text{Integrate.} \\
&= 2\ln|u| + \frac{2}{u} + C && \text{Simplify.} \\
&= 2\ln|x + 1| + \frac{2}{x + 1} + C && \text{Back-substitute.}
\end{aligned}
$$

Check this result by differentiating to obtain the original integrand.   ▬▬▬

TECHNOLOGY   If you have access to a symbolic integration utility, try using it to evaluate the indefinite integrals in Examples 5 and 6. How does the form of the antiderivative that it gives you compare with that given in Examples 5 and 6?

As you study the methods shown in Examples 5 and 6, be aware that both methods involve rewriting a disguised integrand so that it fits one or more of the basic integration formulas. In the remaining sections of Chapter 5 and in Chapter 7 we will devote much time to integration techniques. To master these techniques, you must recognize the "form-fitting" nature of integration. In this sense, integration is not nearly as straightforward as differentiation. Differentiation takes the form

"Here is the question; what is the answer?"

Integration is more like

"Here is the answer; what is the question?"

We suggest the following guidelines for integration.

STUDY TIP Because integration is more difficult than differentiation, keep in mind that you can check your answer to an integration problem by differentiating the answer. For instance, in Example 7, the derivative of $y = \ln |\ln x| + C$ is $y' = 1/(x \ln x)$.

Guidelines for Integration

1. Memorize a basic list of integration formulas. (Including those given in this section, you now have 12 formulas: the Power Rule, the Log Rule, and ten trigonometric rules. By the end of Section 5.9, this list will have expanded to 20 basic rules.

2. Find an integration formula that resembles all or part of the integrand, and, by trial and error, find a choice of u that will make the integrand conform to the formula.

3. If you cannot find a u-substitution that works, try altering the integrand. You might try a trigonometric identity, multiplication and division by the same quantity, or addition and subtraction of the same quantity. Be creative.

4. If you have access to computer software that will find antiderivatives symbolically, use it.

EXAMPLE 7   *u*-Substitution and the Log Rule

Solve the differential equation $\dfrac{dy}{dx} = \dfrac{1}{x \ln x}$.

Solution The solution can be written as an indefinite integral.

$$y = \int \frac{1}{x \ln x}\, dx$$

Because the integrand is a quotient whose denominator is raised to the first power, you should try the Log Rule. There are three basic choices for u. The choices $u = x$ and $u = x \ln x$ fail to fit the u'/u form of the Log Rule. However, the third choice does fit. Letting $u = \ln x$ produces $u' = 1/x$, and you obtain the following.

$$\int \frac{1}{x \ln x}\, dx = \int \frac{1/x}{\ln x}\, dx \qquad \text{Divide numerator and denominator by } x.$$

$$= \int \frac{u'}{u}\, dx \qquad \text{Substitute: } u = \ln x.$$

$$= \ln |u| + C \qquad \text{Apply Log Rule.}$$

$$= \ln |\ln x| + C \qquad \text{Back-substitute.}$$

So, the solution is $y = \ln |\ln x| + C$.

Integrals of Trigonometric Functions

In Section 4.1, you looked at six trigonometric integration rules—the six that correspond directly to differentiation rules. With the Log Rule, you can now complete the set of basic trigonometric integration formulas.

EXAMPLE 8   Using a Trigonometric Identity

Evaluate $\int \tan x \, dx$.

Solution   This integral does not seem to fit any formulas on our basic list. However, by using a trigonometric identity, you obtain the following.

$$\int \tan x \, dx = \int \frac{\sin x}{\cos x} \, dx$$

Knowing that $D_x[\cos x] = -\sin x$, you can let $u = \cos x$ and write

$$\int \tan x \, dx = -\int \frac{-\sin x}{\cos x} \, dx \qquad \text{Trigonometric identity}$$

$$= -\int \frac{u'}{u} \, dx \qquad \text{Substitute: } u = \cos x.$$

$$= -\ln |u| + C \qquad \text{Apply Log Rule.}$$

$$= -\ln |\cos x| + C. \qquad \text{Back-substitute.}$$

Example 8 uses a trigonometric identity to derive an integration rule for the tangent function. In the next example, we take a rather unusual step (multiplying and dividing by the same quantity) to derive an integration rule for the secant function.

EXAMPLE 9   Derivation of the Secant Formula

Evaluate $\int \sec x \, dx$.

Solution   Consider the following procedure.

$$\int \sec x \, dx = \int \sec x \left(\frac{\sec x + \tan x}{\sec x + \tan x} \right) dx$$

$$= \int \frac{\sec^2 x + \sec x \tan x}{\sec x + \tan x} \, dx$$

Letting u be the denominator of this quotient produces

$$u = \sec x + \tan x \quad \Longrightarrow \quad u' = \sec x \tan x + \sec^2 x.$$

Therefore, you can conclude that

$$\int \sec x \, dx = \int \frac{\sec^2 x + \sec x \tan x}{\sec x + \tan x} \, dx \qquad \text{Rewrite integrand.}$$

$$= \int \frac{u'}{u} \, dx \qquad \text{Substitute: } u = \sec x + \tan x.$$

$$= \ln |u| + C \qquad \text{Apply Log Rule.}$$

$$= \ln |\sec x + \tan x| + C. \qquad \text{Back-substitute.}$$

With the results of Examples 8 and 9, you now have integration formulas for $\sin x$, $\cos x$, $\tan x$, and $\sec x$. All six trigonometric rules are summarized below.

NOTE Using trigonometric identities and properties of logarithms, you could rewrite these six integration rules in other forms. For instance, you could write

$$\int \csc u \, du = \ln |\csc u - \cot u| + C.$$

(See Exercises 47–50.)

Integrals of the Six Basic Trigonometric Functions

$$\int \sin u \, du = -\cos u + C \qquad \int \cos u \, du = \sin u + C$$

$$\int \tan u \, du = -\ln |\cos u| + C \qquad \int \cot u \, du = \ln |\sin u| + C$$

$$\int \sec u \, du = \ln |\sec u + \tan u| + C \qquad \int \csc u \, du = -\ln |\csc u + \cot u| + C$$

EXAMPLE 10 Integrating Trigonometric Functions

Evaluate $\displaystyle\int_0^{\pi/4} \sqrt{1 + \tan^2 x} \, dx$.

Solution Using $1 + \tan^2 x = \sec^2 x$, you can write

$$\int_0^{\pi/4} \sqrt{1 + \tan^2 x} \, dx = \int_0^{\pi/4} \sqrt{\sec^2 x} \, dx$$

$$= \int_0^{\pi/4} \sec x \, dx$$

$$= \ln |\sec x + \tan x| \Big]_0^{\pi/4}$$

$$= \ln(\sqrt{2} + 1) - \ln 1$$

$$\approx 0.8814.$$

NOTE Although it is not always true that $\sqrt{a^2} = a$ for all real values of a, $\sqrt{\sec^2 x} = |\sec x| = \sec x$ for $0 \le x \le \pi/4$ because $\sec x \ge 0$ on that interval. This is useful in Example 10.

EXAMPLE 11 Electromotive Force

The electromotive force E of a particular electrical circuit is given by $E = 3 \sin 2t$, where E is measured in volts and t is measured in seconds. Find the average value of E as t ranges from 0 to 0.5 second.

Solution The average value of E is given by

$$\text{Average value} = \frac{1}{0.5 - 0} \int_0^{0.5} 3 \sin 2t \, dt$$

$$= 6 \int_0^{0.5} \sin 2t \, dt$$

$$= 6\left(\frac{1}{2}\right) \int_0^{0.5} (\sin 2t)(2) dt \qquad \color{red}{u = 2t, \, du = 2dt}$$

$$= 3\left[-\cos 2t \right]_0^{0.5}$$

$$\approx 1.379 \text{ volts.}$$

EXERCISES FOR SECTION 5.2

In Exercises 1–18, find the indefinite integral.

1. $\displaystyle\int \frac{1}{x+1}\,dx$

2. $\displaystyle\int \frac{1}{x-5}\,dx$

3. $\displaystyle\int \frac{1}{3-2x}\,dx$

4. $\displaystyle\int \frac{1}{6x+1}\,dx$

5. $\displaystyle\int \frac{x}{x^2+1}\,dx$

6. $\displaystyle\int \frac{x^2}{3-x^3}\,dx$

7. $\displaystyle\int \frac{x^2-4}{x}\,dx$

8. $\displaystyle\int \frac{x}{\sqrt{9-x^2}}\,dx$

9. $\displaystyle\int \frac{x^2+2x+3}{x^3+3x^2+9x}\,dx$

10. $\displaystyle\int \frac{x+3}{x^2+6x+7}\,dx$

11. $\displaystyle\int \frac{(\ln x)^2}{x}\,dx$

12. $\displaystyle\int \frac{1}{x\ln(x^2)}\,dx$

13. $\displaystyle\int \frac{1}{\sqrt{x}+1}\,dx$

14. $\displaystyle\int \frac{1}{x^{2/3}(1+x^{1/3})}\,dx$

15. $\displaystyle\int \frac{\sqrt{x}}{\sqrt{x}-3}\,dx$

16. $\displaystyle\int \frac{1}{1+\sqrt{2x}}\,dx$

17. $\displaystyle\int \frac{2x}{(x-1)^2}\,dx$

18. $\displaystyle\int \frac{x(x-2)}{(x-1)^3}\,dx$

In Exercises 19–26, find the indefinite integral of the trigonometric function.

19. $\displaystyle\int \frac{\cos\theta}{\sin\theta}\,d\theta$

20. $\displaystyle\int \tan 5\theta\,d\theta$

21. $\displaystyle\int \csc 2x\,dx$

22. $\displaystyle\int \sec\frac{x}{2}\,dx$

23. $\displaystyle\int \frac{\cos t}{1+\sin t}\,dt$

24. $\displaystyle\int \frac{\sin x}{1+\cos x}\,dx$

25. $\displaystyle\int \frac{\sec x\tan x}{\sec x-1}\,dx$

26. $\displaystyle\int (\sec t+\tan t)\,dt$

In Exercises 27–30, solve the differential equation. Use a graphing utility to graph three solutions, one of which passes through the indicated point.

27. $\dfrac{dy}{dx}=\dfrac{3}{2-x}$, $(1,0)$

28. $\dfrac{dy}{dx}=\dfrac{2x}{x^2-9}$, $(0,4)$

29. $\dfrac{ds}{d\theta}=\tan 2\theta$, $(0,2)$

30. $\dfrac{dr}{dt}=\dfrac{\sec^2 t}{\tan t+1}$, $(\pi,4)$

Direction Fields   **In Exercises 31 and 32, a differential equation, a point, and a direction field are given. (a) Sketch two approximate solutions of the differential equation on the direction field, one of which passes through the indicated point. (b) Use integration to find the particular solution of the differential equation and use a graphing utility to graph the solution. Compare the result with the sketches in part (a).**

31. $\dfrac{dy}{dx}=\dfrac{1}{x+2}$, $(0,1)$

32. $\dfrac{dy}{dx}=\dfrac{\ln x}{x}$, $(1,-2)$

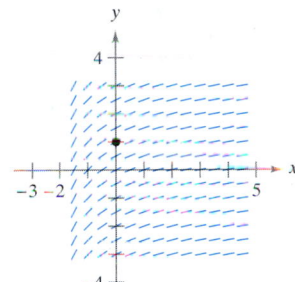
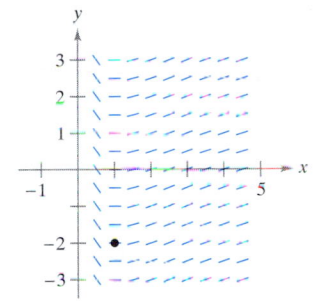

In Exercises 33–40, evaluate the definite integral. Use a graphing utility to verify your result.

33. $\displaystyle\int_0^4 \frac{5}{3x+1}\,dx$

34. $\displaystyle\int_{-1}^1 \frac{1}{x+2}\,dx$

35. $\displaystyle\int_1^e \frac{(1+\ln x)^2}{x}\,dx$

36. $\displaystyle\int_e^{e^2} \frac{1}{x\ln x}\,dx$

37. $\displaystyle\int_0^2 \frac{x^2-2}{x+1}\,dx$

38. $\displaystyle\int_0^1 \frac{x-1}{x+1}\,dx$

39. $\displaystyle\int_1^2 \frac{1-\cos\theta}{\theta-\sin\theta}\,d\theta$

40. $\displaystyle\int_{0.1}^{0.2} (\csc 2\theta-\cot 2\theta)^2\,d\theta$

In Exercises 41–46, use a symbolic integration utility to evaluate the integral. Graph the integrand.

41. $\displaystyle\int \frac{1}{1+\sqrt{x}}\,dx$

42. $\displaystyle\int \frac{1-\sqrt{x}}{1+\sqrt{x}}\,dx$

43. $\displaystyle\int \cos(1-x)\,dx$

44. $\displaystyle\int \frac{\tan^2 2x}{\sec 2x}\,dx$

45. $\displaystyle\int_{\pi/4}^{\pi/2} (\csc x - \sin x)\,dx$

46. $\displaystyle\int_{-\pi/4}^{\pi/4} \frac{\sin^2 x-\cos^2 x}{\cos x}\,dx$

In Exercises 47–50, show that the two formulas are equivalent.

47. $\displaystyle\int \tan x \, dx = -\ln |\cos x| + C$

$\displaystyle\int \tan x \, dx = \ln |\sec x| + C$

48. $\displaystyle\int \cot x \, dx = \ln |\sin x| + C$

$\displaystyle\int \cot x \, dx = -\ln |\csc x| + C$

49. $\displaystyle\int \sec x \, dx = \ln |\sec x + \tan x| + C$

$\displaystyle\int \sec x \, dx = -\ln |\sec x - \tan x| + C$

50. $\displaystyle\int \csc x \, dx = -\ln |\csc x + \cot x| + C$

$\displaystyle\int \csc x \, dx = \ln |\csc x - \cot x| + C$

In Exercises 51–54, find $F'(x)$.

51. $F(x) = \displaystyle\int_1^x \frac{1}{t} \, dt$

52. $F(x) = \displaystyle\int_0^x \tan t \, dt$

53. $F(x) = \displaystyle\int_x^{3x} \frac{1}{t} \, dt$

54. $F(x) = \displaystyle\int_1^{x^2} \frac{1}{t} \, dt$

Approximation **In Exercises 55 and 56, determine which value best approximates the area of the region between the x-axis and the function over the given interval. (Make your selection on the basis of a sketch of the region and *not* by performing any calculations.)**

55. $f(x) = \sec x$, $[0, 1]$

    (a) 6   (b) -6 (c) $\frac{1}{2}$ (d) 1.25   (e) 3

56. $f(x) = \dfrac{2x}{x^2 + 1}$, $[0, 4]$

    (a) 3   (b) 7   (c) -2 (d) 5   (e) 1

Area **In Exercises 57–60, find the area of the region bounded by the graphs of the equations. Use a graphing utility to graph the region and verify your result.**

57. $y = \dfrac{x^2 + 4}{x}$, $x = 1$, $x = 4$, $y = 0$

58. $y = \dfrac{x + 5}{x}$, $x = 1$, $x = 5$, $y = 0$

59. $y = 2 \sec \dfrac{\pi x}{6}$, $x = 0$, $x = 2$, $y = 0$

60. $y = 2x - \tan(0.3x)$, $x = 1$, $x = 4$, $y = 0$

61. *Population Growth* A population of bacteria is changing at a rate of

$$\frac{dP}{dt} = \frac{3000}{1 + 0.25t}$$

where t is the time in days. The initial population (when $t = 0$) is 1000. Write an equation that gives the population at any time t, and find the population when $t = 3$ days.

62. *Heat Transfer* Find the time required for an object to cool from 300°F to 250°F by evaluating

$$t = \frac{10}{\ln 2} \int_{250}^{300} \frac{1}{T - 100} \, dT$$

where t is time in minutes.

63. *Average Price* The demand equation for a product is

$$p = \frac{90{,}000}{400 + 3x}.$$

Find the *average* price p on the interval $40 \le x \le 50$.

64. *Sales* The rate of change in sales S is inversely proportional to time t ($t > 1$) measured in weeks. Find S as a function of t if sales after 2 and 4 weeks are 200 units and 300 units.

65. *Orthogonal Trajectory*

  (a) Use a graphing utility to graph the equation $2x^2 - y^2 = 8$.

  (b) Evaluate the integral to find y^2 in terms of x.

$$y^2 = e^{-\int (1/x) \, dx}$$

    For a particular value of the constant of integration, graph the result on the same screen used in part (a).

  (c) Verify that the tangents to the graphs of parts (a) and (b) are perpendicular at the points of intersection.

66. Graph the function

$$f_k(x) = \frac{x^k - 1}{k}$$

for $k = 1$, 0.5, and 0.1 on $[0, 10]$. Find $\displaystyle\lim_{k \to 0} f_k(x)$.

True or False? **In Exercises 67–70, determine whether the statement is true or false. If it is false, explain why or give an example that shows it is false.**

67. $(\ln x)^{1/2} = \frac{1}{2}(\ln x)$

68. $\int \ln x \, dx = (1/x) + C$

69. $\displaystyle\int \frac{1}{x} \, dx = \ln |cx|$, $c \ne 0$

70. $\displaystyle\int_{-1}^2 \frac{1}{x} \, dx = \Big[\ln |x| \Big]_{-1}^2 = \ln 2 - \ln 1 = \ln 2$

Inverse Functions • Existence of an Inverse Function •
Derivative of an Inverse Function

Inverse Functions

Recall from Section P.3 that a function can be represented by a set of ordered pairs. For instance, the function $f(x) = x + 3$ and $A = \{1, 2, 3, 4\}$ to $B = \{4, 5, 6, 7\}$ can be written as

$$f: \{(1, 4), (2, 5), (3, 6), (4, 7)\}.$$

By interchanging the first and second coordinates of each ordered pair, you can form the **inverse function** of f. This function is denoted by f^{-1}. It is a function from B to A, and can be written as

$$f^{-1}: \{(4, 1), (5, 2), (6, 3), (7, 4)\}.$$

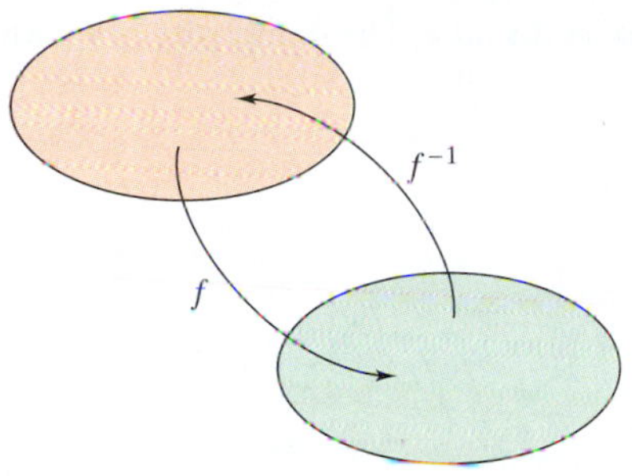

Note that the domain of f is equal to the range of f^{-1}, and vice versa, as shown in Figure 5.9. The functions f and f^{-1} have the effect of "undoing" each other. That is, when you form the composition of f with f^{-1} or the composition of f^{-1} with f, you obtain the identity function.

$$f(f^{-1}(x)) = x \quad \text{and} \quad f^{-1}(f(x)) = x$$

Domain of f = range of f^{-1}
Domain of f^{-1} = range of f
Figure 5.9

EXPLORATION

Finding Inverse Functions
Explain how to "undo" each of the following functions. Then use your explanation to write the inverse function of f.

(a) $f(x) = x - 5$

(b) $f(x) = 6x$

(c) $f(x) = \dfrac{x}{2}$

(d) $f(x) = 3x + 2$

(e) $f(x) = x^3$

(f) $f(x) = 4(x - 2)$

Use a graphing utility to graph each function and its inverse in the same "square" viewing rectangle. What observation can you make about each pair of graphs?

Definition of Inverse Function

A function g is the **inverse** of the function f if

$$f(g(x)) = x \quad \text{for each } x \text{ in the domain of } g$$

and

$$g(f(x)) = x \quad \text{for each } x \text{ in the domain of } f.$$

The function g is denoted by f^{-1} (read "f inverse").

NOTE Although the notation used to denote an inverse function resembles *exponential notation*, it is a different use of -1 as a superscript. That is, in general, $f^{-1}(x) \neq 1/f(x)$.

Here are some important observations about inverse functions.

1. If g is the inverse of f, then f is the inverse of g.

2. The domain of f^{-1} is equal to the range of f, and the range of f^{-1} is equal to the domain of f.

3. A function need not have an inverse, but if it does, the inverse is unique (see Exercise 95).

You can think of f^{-1} as undoing what has been done by f. For example, subtraction can be used to undo addition, and division can be used to undo multiplication. Use the definition of an inverse function to check the following inverses.

$$f(x) = x + c \quad \text{and} \quad f^{-1}(x) = x - c \quad \text{are inverses of each other.}$$

$$f(x) = cx \quad \text{and} \quad f^{-1}(x) = \frac{x}{c}, c \neq 0, \quad \text{are inverses of each other.}$$

EXAMPLE 1 Verifying Inverse Functions

Show that the functions are inverses of each other.

$$f(x) = 2x^3 - 1 \quad \text{and} \quad g(x) = \sqrt[3]{\frac{x+1}{2}}$$

Solution Because the domains and ranges of both f and g consist of all real numbers, you can conclude that both composite functions exist for all x. The composite of f with g is given by

$$\begin{aligned} f(g(x)) &= 2\left(\sqrt[3]{\frac{x+1}{2}}\right)^3 - 1 \\ &= 2\left(\frac{x+1}{2}\right) - 1 \\ &= x + 1 - 1 \\ &= x. \end{aligned}$$

The composite of g with f is given by

$$\begin{aligned} g(f(x)) &= \sqrt[3]{\frac{(2x^3 - 1) + 1}{2}} \\ &= \sqrt[3]{\frac{2x^3}{2}} \\ &= \sqrt[3]{x^3} \\ &= x. \end{aligned}$$

Because $f(g(x)) = x$ and $g(f(x)) = x$, you can conclude that f and g are inverses of each other (see Figure 5.10).

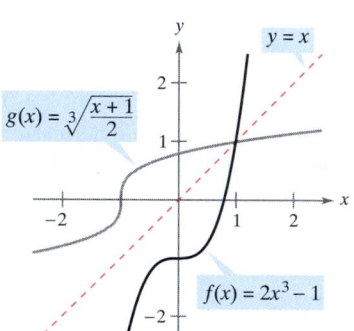

f and *g* are inverses of each other.
Figure 5.10

STUDY TIP In Example 1, try comparing the functions f and g verbally.

For f: First cube x, then multiply by 2, then subtract 1.

For g: First add 1, then divide by 2, then take the cube root.

Do you see the "undoing pattern"?

In Figure 5.10, the graphs of f and $g = f^{-1}$ appear to be mirror images of each other with respect to the line $y = x$. The graph of f^{-1} is a **reflection** of the graph of f in the line $y = x$. This idea is generalized in the following theorem.

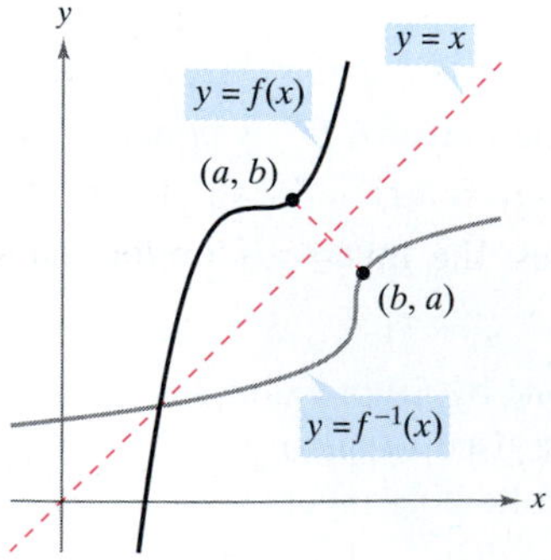

The graph of f^{-1} is a reflection of the graph of f in the line $y = x$.
Figure 5.11

THEOREM 5.6 Reflective Property of Inverse Functions

The graph of f contains the point (a, b) if and only if the graph of f^{-1} contains the point (b, a).

Proof If (a, b) is on the graph of f, then $f(a) = b$ and you can write

$$f^{-1}(b) = f^{-1}(f(a)) = a.$$

Thus, (b, a) is on the graph of f^{-1}, as shown in Figure 5.11. A similar argument will prove the theorem in the other direction.

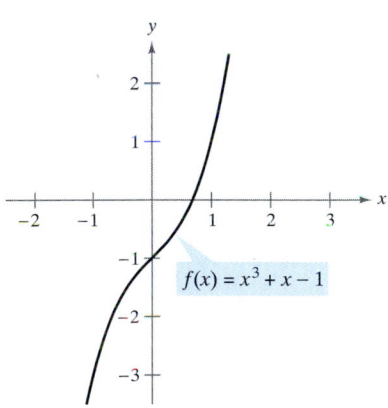

If a horizontal line intersects the graph of f twice, then f is not one-to-one.

Figure 5.12

Existence of an Inverse Function

Not every function has an inverse, and Theorem 5.6 suggests a graphical test for those that do—the **horizontal line test** for an inverse function. This test states that a function f has an inverse if and only if every horizontal line intersects the graph of f at most once (see Figure 5.12). The following theorem formally states why the horizontal line test is valid. (Recall from Section 3.3 that a function is *strictly monotonic* if it is either increasing on its entire domain or decreasing on its entire domain.)

THEOREM 5.7   The Existence of an Inverse Function

1. A function has an inverse if and only if it is one-to-one.

2. If f is strictly monotonic on its entire domain, then it is one-to-one and therefore has an inverse.

Proof   To prove the second part of the theorem, recall from Section P.3 that f is one-to-one if for x_1 and x_2 in its domain

$$f(x_1) = f(x_2) \quad \Longrightarrow \quad x_1 = x_2.$$

The *contrapositive* of this implication is logically equivalent and states that

$$x_1 \neq x_2 \quad \Longrightarrow \quad f(x_1) \neq f(x_2).$$

Now, choose x_1 and x_2 in the domain of f. If $x_1 \neq x_2$, then, because f is strictly monotonic, it follows that either

$$f(x_1) < f(x_2) \quad \text{or} \quad f(x_1) > f(x_2).$$

In either case, $f(x_1) \neq f(x_2)$. Thus, f is one-to-one on the interval. The proof of the first part is left as an exercise (see Exercise 96).

EXAMPLE 2   The Existence of an Inverse

Which of the functions has an inverse?

a. $f(x) = x^3 + x - 1$ **b.** $f(x) = x^3 - x + 1$

Solution

a. From the graph of f given in Figure 5.13(a), it appears that f is increasing over its entire domain. To verify this, note that the derivative, $f'(x) = 3x^2 + 1$, is positive for all real values of x. Therefore, f is strictly monotonic and it must have an inverse.

b. From the graph in Figure 5.13(b) you can see that the function does not pass the horizontal line test. In other words, it is not one-to-one. For instance, f has the same value when $x = -1$, 0, and 1.

$$f(-1) = f(1) = f(0) = 1 \qquad \text{Not one-to-one}$$

Therefore, by Theorem 5.7, f does not have an inverse.

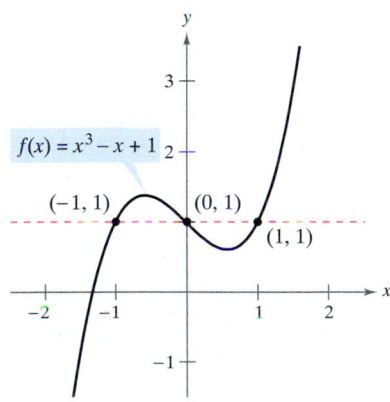

(a) Because f is increasing over its entire domain, it has an inverse.

(b) Because f is not one-to-one, it does not have an inverse.

Figure 5.13

NOTE   Often it is easier to prove that a function has an inverse than to find the inverse. For instance, it would be difficult algebraically to find the inverse of the function in Example 2a.

The following guidelines suggest a procedure for finding the inverse of a function.

> **Guidelines for Finding the Inverse of a Function**
>
> 1. Use Theorem 5.7 to determine whether the function given by $y = f(x)$ has an inverse.
> 2. Solve for x as a function of y: $x = g(y) = f^{-1}(y)$.
> 3. Interchange x and y. The resulting equation is $y = f^{-1}(x)$.
> 4. Define the domain of f^{-1} to be the range of f.
> 5. Verify that $f(f^{-1}(x)) = x$ and $f^{-1}(f(x)) = x$.

EXAMPLE 3 Finding the Inverse of a Function

Find the inverse of $f(x) = \sqrt{2x - 3}$.

Solution The function has an inverse because it is increasing on its entire domain (see Figure 5.14). To find an equation for the inverse, let $y = f(x)$ and solve for x in terms of y.

$$\sqrt{2x - 3} = y \qquad\qquad \text{Let } y = f(x).$$

$$2x - 3 = y^2 \qquad\qquad \text{Square both sides.}$$

$$x = \frac{y^2 + 3}{2} \qquad\qquad \text{Solve for } x.$$

$$y = \frac{x^2 + 3}{2} \qquad\qquad \text{Interchange } x \text{ and } y.$$

$$f^{-1}(x) = \frac{x^2 + 3}{2} \qquad\qquad \text{Replace } y \text{ by } f^{-1}(x).$$

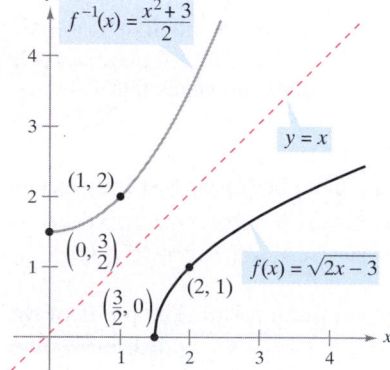

The domain of f^{-1}, $[0, \infty)$, is the range of f.
Figure 5.14

The domain of f^{-1} is the range of f, which is $[0, \infty)$. You can verify this result as follows.

$$f(f^{-1}(x)) = \sqrt{2\left(\frac{x^2 + 3}{2}\right) - 3} = \sqrt{x^2} = x, \qquad x \geq 0$$

$$f^{-1}(f(x)) = \frac{\left(\sqrt{2x - 3}\right)^2 + 3}{2} = \frac{2x - 3 + 3}{2} = x, \qquad x \geq \frac{3}{2}$$

NOTE Remember that any letter can be used to represent the independent variable. Thus,

$$f^{-1}(y) = \frac{y^2 + 3}{2}$$

$$f^{-1}(x) = \frac{x^2 + 3}{2}$$

$$f^{-1}(s) = \frac{s^2 + 3}{2}$$

all represent the same function.

Theorem 5.7 is useful in the following type of problem. Suppose you are given a function that is *not* one-to-one on its domain. By restricting the domain to an interval on which the function is strictly monotonic, you can conclude that the new function *is* one-to-one on the restricted domain.

EXAMPLE 4   Testing Whether a Function is One-to-One

Show that the sine function $f(x) = \sin x$ is not one-to-one on the entire real line. Then show that $[-\pi/2, \pi/2]$ is the largest interval, centered at the origin, for which f is strictly monotonic.

Solution   It is clear that f is not one-to-one, because many different x-values yield the same y-value. For instance, $\sin(0) = 0 = \sin(\pi)$. Moreover, f is increasing on the open interval $(-\pi/2, \pi/2)$, because its derivative.

$$f'(x) = \cos x$$

is positive there. Finally, because the left and right endpoints correspond to relative extrema of the sine function, you can conclude that f is increasing on the closed interval $[-\pi/2, \pi/2]$ *and* that in any larger interval the function is not strictly monotonic (see Figure 5.15).

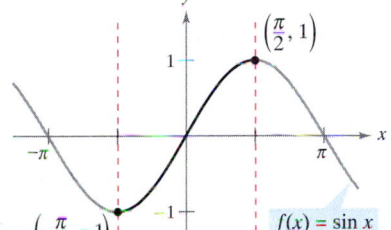

f is one-to-one on the interval
$[-\pi/2, \pi/2]$.
Figure 5.15

Derivative of an Inverse Function

The next two theorems discuss the derivative of an inverse function. The reasonableness of Theorem 5.8 follows from the reflective property of inverse functions as shown in Figure 5.11. Proofs of the two theorems are given in the appendix.

THEOREM 5.8   Continuity and Differentiability of Inverse Functions

Let f be a function whose domain is an interval I. If f has an inverse, then the following statements are true.

1. If f is continuous on its domain, then f^{-1} is continuous on its domain.
2. If f is increasing on its domain, then f^{-1} is increasing on its domain.
3. If f is decreasing on its domain, then f^{-1} is decreasing on its domain.
4. If f is differentiable at c and $f'(c) \neq 0$, then f^{-1} is differentiable at $f(c)$.

EXPLORATION

Graph the inverse functions

$$f(x) = x^3$$

and

$$g(x) = x^{1/3}.$$

Calculate the slope of f at $(1, 1)$, $(2, 8)$, and $(3, 27)$, and the slope of g at $(1, 1)$, $(8, 2)$, and $(27, 3)$. What do you observe? What happens at $(0, 0)$?

THEOREM 5.9   The Derivative of an Inverse Function

Let f be a function that is differentiable on an interval I. If f has an inverse function g, then g is differentiable at any x for which $f'(g(x)) \neq 0$. Moreover,

$$g'(x) = \frac{1}{f'(g(x))}, \qquad f'(g(x)) \neq 0.$$

EXAMPLE 5 Evaluating the Derivative of an Inverse Function

Let $f(x) = \frac{1}{4}x^3 + x - 1$.

a. What is the value of $f^{-1}(x)$ when $x = 3$?

b. What is the value of $(f^{-1})'(x)$ when $x = 3$?

Solution Notice that f is one-to-one and hence has an inverse.

a. Because $f(x) = 3$ when $x = 2$, you know that $f^{-1}(3) = 2$.

b. Because the function f is differentiable and has an inverse, you can apply Theorem 5.9 (with $g = f^{-1}$) to write

$$(f^{-1})'(3) = \frac{1}{f'(f^{-1}(3))} = \frac{1}{f'(2)}.$$

Moreover, using $f'(x) = \frac{3}{4}x^2 + 1$, you can conclude that

$$(f^{-1})'(3) = \frac{1}{f'(2)} = \frac{1}{\frac{3}{4}(2^2) + 1} = \frac{1}{4}.$$

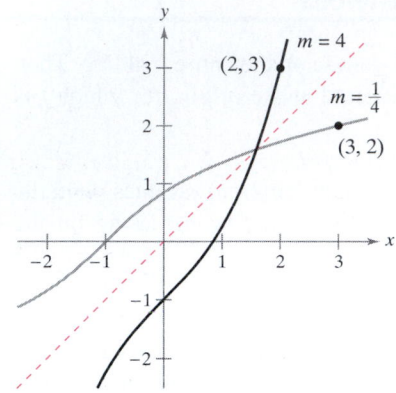

The graphs of the inverse functions f and f^{-1} have reciprocal slopes at points (a, b) and (b, a).
Figure 5.16

   In Example 5, note that at the point $(2, 3)$ the slope of the graph of f is 4 and at the point $(3, 2)$ the slope of the graph of f^{-1} is $\frac{1}{4}$ (see Figure 5.16). This reciprocal relationship (which follows from Theorem 5.9) is sometimes written as

$$\frac{dy}{dx} = \frac{1}{dx/dy}.$$

EXAMPLE 6 Graphs of Inverse Functions Have Reciprocal Slopes

Let $f(x) = x^2$ (for $x \geq 0$) and let $f^{-1}(x) = \sqrt{x}$. Show that the slopes of the graphs of f and f^{-1} are reciprocals at each of the following points.

a. $(2, 4)$ and $(4, 2)$

b. $(3, 9)$ and $(9, 3)$

Solution The derivatives of f and f^{-1} are given by

$$f'(x) = 2x \qquad \text{and} \qquad (f^{-1})'(x) = \frac{1}{2\sqrt{x}}.$$

a. At $(2, 4)$, the slope of the graph of f is $f'(2) = 2(2) = 4$. At $(4, 2)$, the slope of the graph of f^{-1} is

$$(f^{-1})'(4) = \frac{1}{2\sqrt{4}} = \frac{1}{2(2)} = \frac{1}{4}.$$

b. At $(3, 9)$, the slope of the graph of f is $f'(3) = 2(3) = 6$. At $(9, 3)$, the slope of the graph of f^{-1} is

$$(f^{-1})'(9) = \frac{1}{2\sqrt{9}} = \frac{1}{2(3)} = \frac{1}{6}.$$

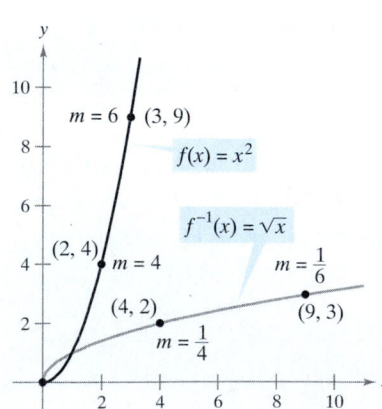

At $(0, 0)$, the derivative of f is 0, and the derivative of f^{-1} does not exist.
Figure 5.17

Thus, in both cases, the slopes are reciprocals, as shown in Figure 5.17.

EXERCISES FOR SECTION 5.3

In Exercises 1–8, show that f and g are inverse functions (a) algebraically and (b) graphically.

1. $f(x) = 5x + 1,$ $g(x) = (x - 1)/5$
2. $f(x) = 3 - 4x,$ $g(x) = (3 - x)/4$
3. $f(x) = x^3,$ $g(x) = \sqrt[3]{x}$
4. $f(x) = 1 - x^3,$ $g(x) = \sqrt[3]{1 - x}$
5. $f(x) = \sqrt{x - 4},$ $g(x) = x^2 + 4, \ \ x \geq 0$
6. $f(x) = 9 - x^2, \ \ x \geq 0,$ $g(x) = \sqrt{9 - x}$
7. $f(x) = 1/x,$ $g(x) = 1/x$
8. $f(x) = \dfrac{1}{1 + x}, \ \ x \geq 0,$ $g(x) = \dfrac{1 - x}{x}, \ \ 0 < x \leq 1$

In Exercises 9–12, match the graph of the function with the graph of its inverse. [The graphs of the inverse functions are labeled (a), (b), (c), and (d).]

(a)
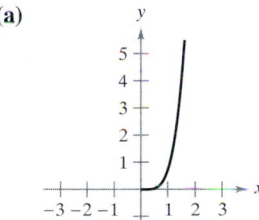

(b)
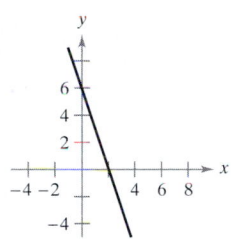

(c)
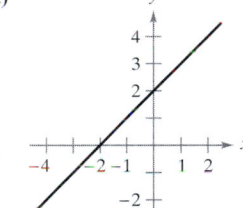

(d)
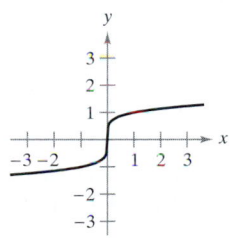

9.
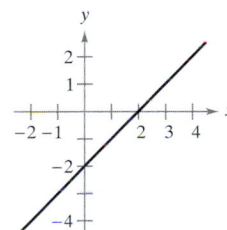

10.
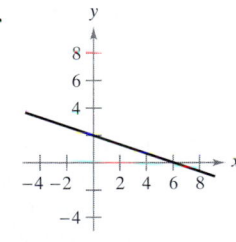

11.
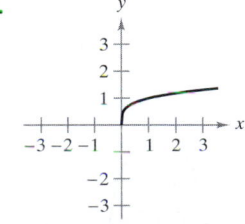

12.
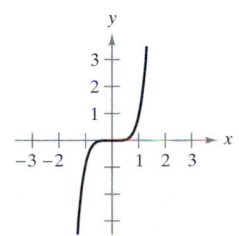

In Exercises 13–20, find the inverse of f. Graph (by hand) f and f^{-1}. Describe the relationship between the graphs.

13. $f(x) = 2x - 3$
14. $f(x) = 3x$
15. $f(x) = x^5$
16. $f(x) = x^3 + 1$
17. $f(x) = \sqrt{x}$
18. $f(x) = x^2, \ \ x \geq 0$
19. $f(x) = \sqrt{4 - x^2}, \ \ x \geq 0$
20. $f(x) = \sqrt{x^2 - 4}, \ \ x \geq 2$

 In Exercises 21–26, find the inverse of f. Use a graphing utility to graph f and f^{-1} in the same viewing rectangle. Describe the relationship between the graphs.

21. $f(x) = \sqrt[3]{x - 1}$
22. $f(x) = 3\sqrt[5]{2x - 1}$
23. $f(x) = x^{2/3}, \ \ x \geq 0$
24. $f(x) = x^{3/5}$
25. $f(x) = \dfrac{x}{\sqrt{x^2 + 7}}$
26. $f(x) = \dfrac{x + 2}{x}$

 In Exercises 27 and 28, find the inverse function of f over the indicated interval. Use a graphing utility to graph f and f^{-1} in the same viewing rectangle. Describe the relationship between the graphs.

Function	Interval
27. $f(x) = \dfrac{x}{x^2 - 4}$	$-2 < x < 2$
28. $f(x) = 2 - \dfrac{3}{x^2}$	$0 < x < 10$

Graphical Reasoning In Exercises 29–32, (a) use a graphing utility to graph the function, (b) use the drawing feature of a graphing utility to draw the inverse of the function, and (c) determine whether the graph of the inverse relation is an inverse function. Explain your reasoning.

29. $f(x) = x^3 + x + 4$
30. $h(x) = x\sqrt{4 - x^2}$
31. $g(x) = \dfrac{3x^2}{x^2 + 1}$
32. $f(x) = \dfrac{4x}{\sqrt{x^2 + 15}}$

In Exercises 33 and 34, use the graph of the function f to complete the table and sketch the graph of f^{-1}.

33.
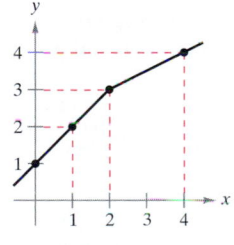

34.
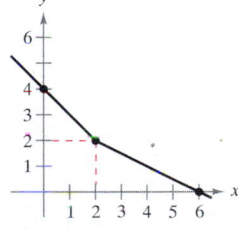

x	1	2	3	4
$f^{-1}(x)$				

x	0	2	4
$f^{-1}(x)$			

35. ***Cost*** Suppose you need 50 pounds of two commodities costing $1.25 and $1.60 per pound.

(a) Verify that the total cost is

$$y = 1.25x + 1.60(50 - x)$$

where x is the number of pounds of the less expensive commodity.

(b) Find the inverse of the cost function. What does each variable represent in the inverse function?

(c) Use the context of the problem to determine the domain of the inverse function.

(d) Determine the number of pounds of the less expensive commodity purchased if the total cost is $73.

36. ***Think About It*** The function

$$f(x) = k(2 - x - x^3)$$

is one-to-one and $f^{-1}(3) = -2$. Find k.

In Exercises 37–40, use the horizontal line test to determine whether the function is one-to-one on its entire domain and therefore has an inverse.

37. $f(x) = \frac{3}{4}x + 6$

38. $f(x) = 5x - 3$

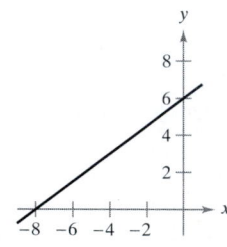

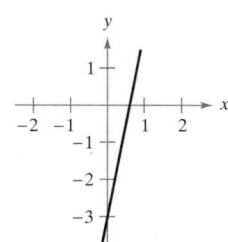

39. $f(\theta) = \sin\theta$

40. $f(x) = \dfrac{x^2}{x^2 + 4}$

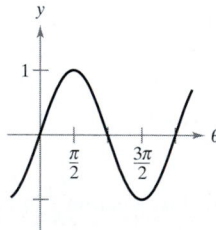

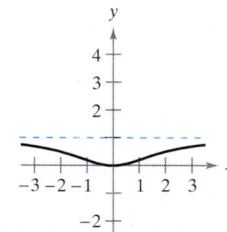

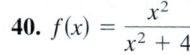 In Exercises 41–46, use a graphing utility to graph the function. Determine whether the function is one-to-one on its entire domain.

41. $h(s) = \dfrac{1}{s - 2} - 3$

42. $g(t) = \dfrac{1}{\sqrt{t^2 + 1}}$

43. $f(x) = \ln x$

44. $f(x) = 3x\sqrt{x + 1}$

45. $g(x) = (x + 5)^3$

46. $h(x) = |x + 4| - |x - 4|$

In Exercises 47–52, use the derivative to determine whether the function is strictly monotonic on its entire domain and therefore has an inverse.

47. $f(x) = (x + a)^3 + b$

48. $f(x) = \cos\dfrac{3x}{2}$

49. $f(x) = \dfrac{x^4}{4} - 2x^2$

50. $f(x) = x^3 - 6x^2 + 12x$

51. $f(x) = 2 - x - x^3$

52. $f(x) = \ln(x - 3)$

In Exercises 53–58, show that f is strictly monotonic on the indicated interval and therefore has an inverse on that interval.

Function	Interval
53. $f(x) = (x - 4)^2$	$[4, \infty)$
54. $f(x) = |x + 2|$	$[-2, \infty)$
55. $f(x) = \dfrac{4}{x^2}$	$(0, \infty)$
56. $f(x) = \tan x$	$\left(-\dfrac{\pi}{2}, \dfrac{\pi}{2}\right)$
57. $f(x) = \cos x$	$[0, \pi]$
58. $f(x) = \sec x$	$\left[0, \dfrac{\pi}{2}\right)$

Think About It In Exercises 59 and 60, the derivative of the function has the same sign for all x in its domain, but the function is not one-to-one. Explain.

59. $f(x) = \tan x$

60. $f(x) = \dfrac{x}{x^2 - 4}$

In Exercises 61–64, determine whether the function is one-to-one. If it is, find its inverse.

61. $f(x) = \sqrt{x - 2}$

62. $f(x) = -3$

63. $f(x) = |x - 2|, \quad x \le 2$

64. $f(x) = ax + b, \quad a \ne 0$

In Exercises 65–68, delete part of the graph of the function so that the part that remains is one-to-one. Find the inverse of the remaining part and give the domain of the inverse. (*Note:* There is more than one correct answer.)

65. $f(x) = (x - 3)^2$

66. $f(x) = 16 - x^4$

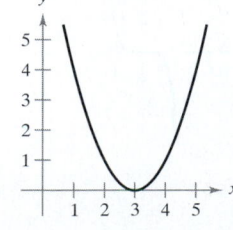

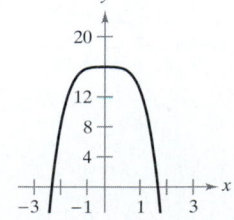

67. $f(x) = |x + 3|$

68. $f(x) = |x - 3|$

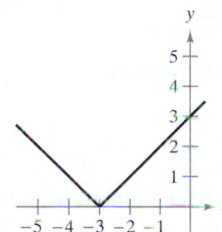

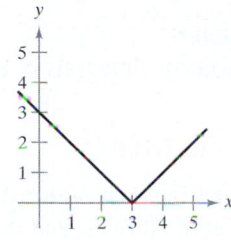

Think About It   In Exercises 69–72, decide whether the function has an inverse. If so, what is the inverse?

69. $g(t)$ is the volume of water that has passed through a water line t minutes after a control valve is opened.

70. $h(t)$ is the height of the tide t hours after midnight, where $0 \le t < 24$.

71. $C(t)$ is the cost of a long distance call lasting t minutes.

72. $A(r)$ is the area of a circle of radius r.

In Exercises 73–78, find $(f^{-1})'(a)$ for the function f and real number a.

Function	Real Number
73. $f(x) = x^3 + 2x - 1$	$a = 2$
74. $f(x) = 2x^5 + x^3 + 1$	$a = -2$
75. $f(x) = \sin x, \quad -\dfrac{\pi}{2} \le x \le \dfrac{\pi}{2}$	$a = \dfrac{1}{2}$
76. $f(x) = \cos 2x, \quad 0 \le x \le \dfrac{\pi}{2}$	$a = 1$
77. $f(x) = x^3 - \dfrac{4}{x}$	$a = 6$
78. $f(x) = \sqrt{x - 4}$	$a = 2$

In Exercises 79–82, (a) find the domains of f and f^{-1}, (b) find the ranges of f and f^{-1}, (c) graph f and f^{-1}, and (d) show that the slopes of the graphs of f and f^{-1} are reciprocals at the indicated points.

Functions	Point
79. $f(x) = x^3$	$\left(\dfrac{1}{2}, \dfrac{1}{8}\right)$
$\quad f^{-1}(x) = \sqrt[3]{x}$	$\left(\dfrac{1}{8}, \dfrac{1}{2}\right)$
80. $f(x) = 3 - 4x$	$(1, -1)$
$\quad f^{-1}(x) = \dfrac{3 - x}{4}$	$(-1, 1)$
81. $f(x) = \sqrt{x - 4}$	$(5, 1)$
$\quad f^{-1}(x) = x^2 + 4$	$(1, 5)$
82. $f(x) = \dfrac{1}{1 + x^2}, \quad x \ge 0$	$\left(1, \dfrac{1}{2}\right)$
$\quad f^{-1}(x) = \sqrt{\dfrac{1 - x}{x}}$	$\left(\dfrac{1}{2}, 1\right)$

In Exercises 83 and 84, find dy/dx at the indicated point for the equation.

83. $x = y^3 - 7y^2 + 2$
$\quad (-4, 1)$

84. $x = 2\ln(y^2 - 3)$
$\quad (0, 4)$

In Exercises 85–88, use the functions

$$f(x) = \tfrac{1}{8}x - 3 \quad \text{and} \quad g(x) = x^3$$

to find the indicated value.

85. $(f^{-1} \circ g^{-1})(1)$

86. $(g^{-1} \circ f^{-1})(-3)$

87. $(f^{-1} \circ f^{-1})(6)$

88. $(g^{-1} \circ g^{-1})(-4)$

In Exercises 89–92, use the functions

$$f(x) = x + 4 \quad \text{and} \quad g(x) = 2x - 5$$

to find the indicated function.

89. $g^{-1} \circ f^{-1}$

90. $f^{-1} \circ g^{-1}$

91. $(f \circ g)^{-1}$

92. $(g \circ f)^{-1}$

93. Prove that if f and g are one-to-one functions, then $(f \circ g)^{-1}(x) = (g^{-1} \circ f^{-1})(x)$.

94. Prove that if f has an inverse, then $(f^{-1})^{-1} = f$.

95. Prove that if a function has an inverse, then the inverse is unique.

96. Prove that a function has an inverse if and only if it is one-to-one.

True or False?   In Exercises 97–100, determine whether the statement is true or false. If it is false, explain why or give an example that shows it is false.

97. If f is an even function, then f^{-1} exists.

98. If the inverse of f exists, then the y-intercept of f is an x-intercept of f^{-1}.

99. If $f(x) = x^n$ where n is odd, then f^{-1} exists.

100. There exists no function f such that $f = f^{-1}$.

101. Is the converse of the second part of Theorem 5.7 true? That is, if a function is one-to-one (and hence has an inverse), then must the function be strictly monotonic? If so, prove it. If not, give a counterexample.

102. Let f be twice-differentiable and one-to-one on an open interval I. Show that its inverse g satisfies

$$g''(x) = -\frac{f''(g(x))}{[f'(g(x))]^3}.$$

If f is increasing and concave downward, what is the concavity of $f^{-1} = g$?

103. If $f(x) = \displaystyle\int_2^x \frac{dt}{\sqrt{1 + t^4}}$, find $(f^{-1})'(0)$.

The Natural Exponential Function •
Derivatives of Exponential Functions • Integrals of Exponential Functions

The Natural Exponential Function

The function $f(x) = \ln x$ is increasing on its entire domain, and hence it has an inverse f^{-1}. The domain of f^{-1} is the set of all reals, and the range is the set of positive reals, as shown in Figure 5.18. Thus, for any real number x,

$$f(f^{-1}(x)) = \ln\left[f^{-1}(x)\right] = x. \qquad \text{\color{red}\textit{x} is any real number.}$$

If x happens to be rational, then

$$\ln(e^x) = x \ln e = x(1) = x. \qquad \text{\color{red}\textit{x} is a rational number.}$$

Because the natural logarithmic function is one-to-one, you can conclude that $f^{-1}(x)$ and e^x agree for *rational* values of x. The following definition extends the meaning of e^x to include *all* real values of x.

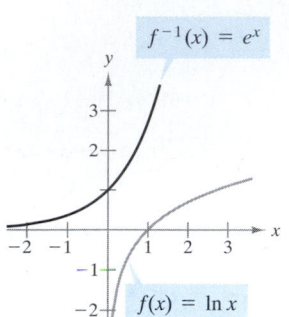

The inverse of the natural logarithmic function is the natural exponential function.
Figure 5.18

Definition of the Natural Exponential Function

The inverse of the natural logarithmic function $f(x) = \ln x$ is called the **natural exponential function** and is denoted by

$$f^{-1}(x) = e^x.$$

That is,

$$y = e^x \qquad \text{if and only if} \qquad x = \ln y.$$

The inverse relationship between the natural logarithmic function and the natural exponential function can be summarized as follows.

$$\ln(e^x) = x \qquad \text{and} \qquad e^{\ln x} = x \qquad\qquad \text{\color{red}Inverse relationships}$$

EXAMPLE 1   Solving Exponential Equations

Solve $7 = e^{x+1}$.

Solution   You can convert from exponential form to logarithmic form by *taking the natural log of both sides* of the equation.

$$7 = e^{x+1} \qquad\qquad \text{\color{red}Original equation}$$
$$\ln 7 = \ln(e^{x+1}) \qquad\qquad \text{\color{red}Take natural log of both sides.}$$
$$\ln 7 = x + 1 \qquad\qquad \text{\color{red}Apply inverse property.}$$
$$-1 + \ln 7 = x \qquad\qquad \text{\color{red}Solve for \textit{x}.}$$
$$0.946 \approx x \qquad\qquad \text{\color{red}Use a calculator.}$$

Check this solution in the original equation.

EXAMPLE 2   Solving a Logarithmic Equation

Solve $\ln(2x - 3) = 5$.

Solution   To convert from logarithmic form to exponential form, you can *exponentiate both sides* of the logarithmic equation.

$$\ln(2x - 3) = 5 \qquad \text{Original equation}$$
$$e^{\ln(2x-3)} = e^5 \qquad \text{Exponentiate both sides.}$$
$$2x - 3 = e^5 \qquad \text{Apply inverse property.}$$
$$x = \tfrac{1}{2}(e^5 + 3) \qquad \text{Solve for } x.$$
$$x \approx 75.707 \qquad \text{Use a calculator.}$$

The familiar rules for operating with rational exponents can be extended to the natural exponential function, as indicated in the following theorem.

THEOREM 5.10   Operations with Exponential Functions

Let a and b be any real numbers.

1. $e^a e^b = e^{a+b}$ 2. $\dfrac{e^a}{e^b} = e^{a-b}$

Proof   To prove Property 1, you can write

$$\ln(e^a e^b) = \ln(e^a) + \ln(e^b)$$
$$= a + b$$
$$= \ln(e^{a+b}).$$

Because the natural log function is one-to-one, you can conclude that

$$e^a e^b = e^{a+b}.$$

The proof of the second property is left to you (see Exercise 108).

In Section 5.3, you learned that an inverse function f^{-1} shares many properties with f. Thus, the natural exponential function inherits the following properties from the natural logarithmic function (see Figure 5.19).

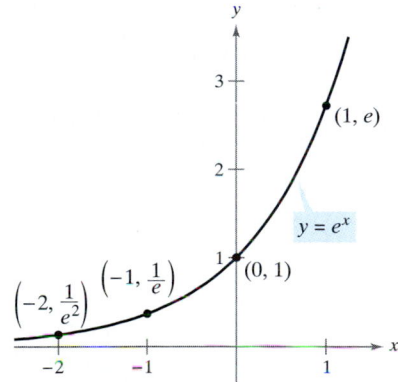

The natural exponential function is increasing, and its graph is concave upward.

Figure 5.19

Properties of the Natural Exponential Function

1. The domain of $f(x) = e^x$ is $(-\infty, \infty)$, and the range is $(0, \infty)$.
2. The function $f(x) = e^x$ is continuous, increasing, and one-to-one on its entire domain.
3. The graph of $f(x) = e^x$ is concave upward on its entire domain.
4. $\displaystyle\lim_{x \to -\infty} e^x = 0$ and $\displaystyle\lim_{x \to \infty} e^x = \infty$

Derivatives of Exponential Functions

One of the most intriguing (and useful) characteristics of the natural exponential function is that *it is its own derivative*. In other words, it is a solution to the differential equation $y' = y$. This result is stated in the next theorem.

THEOREM 5.11   The Derivative of the Natural Exponential Function

Let u be a differentiable function of x.

1. $\dfrac{d}{dx}[e^x] = e^x$ **2.** $\dfrac{d}{dx}[e^u] = e^u \dfrac{du}{dx}$

Proof   To prove Property 1, use the fact that $\ln e^x = x$, and differentiate both sides of the equation.

$$\ln e^x = x \qquad \text{Definition of exponential function}$$

$$\frac{d}{dx}[\ln e^x] = \frac{d}{dx}[x] \qquad \text{Differentiate both sides.}$$

$$\frac{1}{e^x}\frac{d}{dx}[e^x] = 1$$

$$\frac{d}{dx}[e^x] = e^x$$

The derivative of e^u follows from the Chain Rule. ■

NOTE   You can interpret this theorem geometrically by saying that the slope of the graph of $f(x) = e^x$ at any point (x, e^x) is equal to the y-coordinate of the point.

EXAMPLE 3   Differentiating Exponential Functions

a. $\dfrac{d}{dx}[e^{2x-1}] = e^u \dfrac{du}{dx} = 2e^{2x-1}$ $u = 2x - 1$

b. $\dfrac{d}{dx}[e^{-3/x}] = e^u \dfrac{du}{dx} = \left(\dfrac{3}{x^2}\right)e^{-3/x} = \dfrac{3e^{-3/x}}{x^2}$ $u = -\dfrac{3}{x}$

EXAMPLE 4   Locating Relative Extrema

Find the relative extrema of $f(x) = xe^x$.

Solution   The derivative of f is given by

$$f'(x) = x(e^x) + e^x(1) \qquad \text{Product Rule}$$

$$= e^x(x + 1).$$

Because e^x is never 0, the derivative is 0 only when $x = -1$. Moreover, by the First Derivative Test, you can determine that this corresponds to a relative minimum, as shown in Figure 5.20. Because the derivative $f'(x) = e^x(x + 1)$ is defined for all x, there are no other critical points. ■

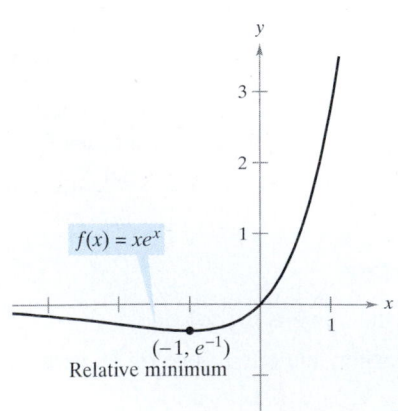

$f(x) = xe^x$

$(-1, e^{-1})$
Relative minimum

The derivative of f changes from negative to positive at $x = -1$.
Figure 5.20

EXAMPLE 5   The Normal Probability Density Function

Show that the *normal probability density function*

$$f(x) = \frac{1}{\sqrt{2\pi}} e^{-x^2/2}$$

has points of inflection when $x = \pm 1$.

Solution   To locate possible points of inflection, find the x-values for which the second derivative is 0.

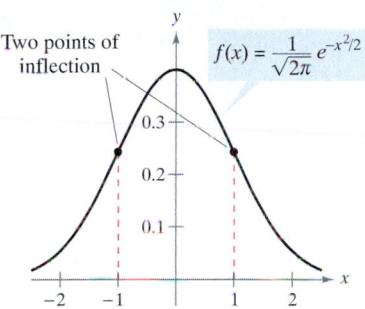

Two points of inflection

$f(x) = \dfrac{1}{\sqrt{2\pi}} e^{-x^2/2}$

The bell-shaped curve given by a normal probability density function

Figure 5.21

$f(x) = \dfrac{1}{\sqrt{2\pi}} e^{-x^2/2}$	Original function
$f'(x) = \dfrac{1}{\sqrt{2\pi}} (-x)e^{-x^2/2}$	First derivative
$f''(x) = \dfrac{1}{\sqrt{2\pi}} [(-x)(-x)e^{-x^2/2} + (-1)e^{-x^2/2}]$	Product Rule
$\qquad = \dfrac{1}{\sqrt{2\pi}} (e^{-x^2/2})(x^2 - 1)$	Second derivative

Therefore, $f''(x) = 0$ when $x = \pm 1$, and you can apply the techniques of Chapter 3 to conclude that these values yield the two points of inflection shown in Figure 5.21.

NOTE   The general form of a normal probability density function (whose mean is 0) is given by

$$f(x) = \frac{1}{\sigma\sqrt{2\pi}} e^{-x^2/2\sigma^2}$$

where σ is the standard deviation (σ is the lowercase Greek letter sigma). This "bell-shaped curve" has points of inflection when $x = \pm\sigma$.

EXAMPLE 6   M.D.s in the United States

For 1980 through 1993, the number y of medical doctors in the United States can be modeled by

$$y = 476{,}260 e^{0.026663t}$$

where $t = 0$ represents 1980. At what rate was the number of M.D.'s changing in 1988?   *(Source: American Medical Association)*

Solution   The derivative of the given model is

$$y' = (0.026663)(476{,}260)e^{0.026663t}$$

$$\approx 12{,}699 e^{0.026663t}.$$

By evaluating the derivative when $t = 8$, you can conclude that the rate of change in 1988 was about

15,718 doctors per year.

The graph of this model is shown in Figure 5.22.

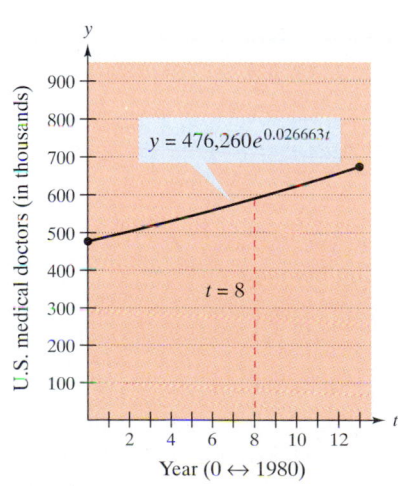

$y = 476{,}260 e^{0.026663t}$

$t = 8$

U.S. medical doctors (in thousands)

Year ($0 \leftrightarrow 1980$)

Figure 5.22

Integrals of Exponential Functions

Each differentiation formula in Theorem 5.11 has a corresponding integration formula.

> **THEOREM 5.12 Integration Rules for Exponential Functions**
>
> Let u be a differentiable function of x.
>
> **1.** $\displaystyle\int e^x \, dx = e^x + C$ **2.** $\displaystyle\int e^u \, du = e^u + C$

EXAMPLE 7 Integrating Exponential Functions

Evaluate $\int e^{3x+1} \, dx$.

Solution If you let $u = 3x + 1$, then $du = 3 \, dx$.

$$\int e^{3x+1} dx = \frac{1}{3}\int e^{3x+1}(3) \, dx \qquad \text{Multiply and divide by 3.}$$

$$= \frac{1}{3}\int e^u \, du \qquad \text{Substitute.}$$

$$= \frac{1}{3}e^u + C \qquad \text{Apply Exponential Rule.}$$

$$= \frac{e^{3x+1}}{3} + C \qquad \text{Back-substitute.}$$

NOTE In Example 7, the missing *constant* factor 3 was introduced to create $du = 3 \, dx$. However, remember that you cannot introduce a missing *variable* factor in the integrand. For instance,

$$\int e^{-x^2} dx \neq \frac{1}{x}\int e^{-x^2}(x \, dx).$$

EXAMPLE 8 Integrating Exponential Functions

Evaluate $\int 5xe^{-x^2} \, dx$.

Solution If you let $u = -x^2$, then $du = -2x \, dx$ or $x \, dx = -du/2$.

$$\int 5xe^{-x^2} dx = \int 5e^{-x^2}(x \, dx) \qquad \text{Regroup integrand.}$$

$$= \int 5e^u\left(-\frac{du}{2}\right) \qquad \text{Substitute.}$$

$$= -\frac{5}{2}\int e^u \, du \qquad \text{Factor } -\tfrac{5}{2} \text{ out of integral.}$$

$$= -\frac{5}{2}e^u + C \qquad \text{Apply Exponential Rule.}$$

$$= -\frac{5}{2}e^{-x^2} + C \qquad \text{Back-substitute.}$$

EXAMPLE 9   **Integrating Exponential Functions**

a. $\displaystyle \int \frac{e^{1/x}}{x^2}\,dx = -\int \overbrace{e^{1/x}}^{e^u}\overbrace{\left(-\frac{1}{x^2}\right)dx}^{du}$ $u = \dfrac{1}{x}$

$\qquad\qquad\quad = -e^{1/x} + C$

b. $\displaystyle \int \sin x\, e^{\cos x}\,dx = -\int \overbrace{e^{\cos x}}^{e^u}\overbrace{(-\sin x\,dx)}^{du}$ $u = \cos x$

$\qquad\qquad\qquad\quad = -e^{\cos x} + C$

EXAMPLE 10   **Finding Areas Bounded by Exponential Functions**

a. $\displaystyle \int_0^1 e^{-x}\,dx = -e^{-x}\Big]_0^1$ See Figure 5.23(a).

$\qquad\qquad\quad = -e^{-1} - (-1)$

$\qquad\qquad\quad = 1 - \dfrac{1}{e}$

$\qquad\qquad\quad \approx 0.632$

b. $\displaystyle \int_0^1 \frac{e^x}{1 + e^x}\,dx = \ln(1 + e^x)\Big]_0^1$ See Figure 5.23(b).

$\qquad\qquad\qquad = \ln(1 + e) - \ln 2$

$\qquad\qquad\qquad \approx 0.620$

c. $\displaystyle \int_{-1}^0 \left[e^x \cos(e^x)\right]dx = \sin(e^x)\Big]_{-1}^0$ See Figure 5.23(c).

$\qquad\qquad\qquad\quad = \sin 1 - \sin(e^{-1})$

$\qquad\qquad\qquad\quad \approx 0.482$

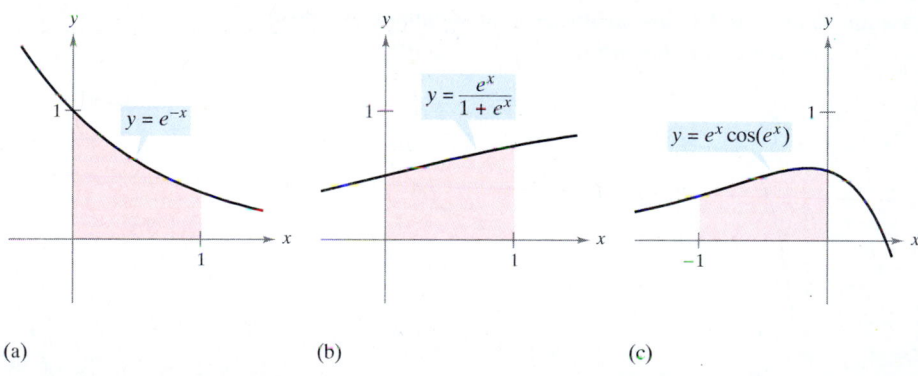

(a)   (b)   (c)

Areas bounded by exponential functions
Figure 5.23

EXERCISES FOR SECTION 5.4

In Exercises 1–4, write the exponential equation as a logarithmic equation or vice versa.

1. $e^0 = 1$

2. $e^{-2} = 0.1353 \ldots$

3. $\ln 2 = 0.6931 \ldots$

4. $\ln 0.5 = -0.6931 \ldots$

In Exercises 5–8, solve for x.

5. $e^{\ln x} = 4$

6. $e^{\ln 2x} = 12$

7. $\ln x = 2$

8. $\ln x^2 = 10$

In Exercises 9–12, sketch the graph of the function.

9. $y = e^{-x}$

10. $y = \frac{1}{2}e^x$

11. $y = e^{-x^2}$

12. $y = e^{-x/2}$

  13. Use a graphing utility to graph $f(x) = e^x$ and the given function in the same viewing rectangle. How are the two graphs related?
(a) $g(x) = e^{x-2}$ (b) $h(x) = -\frac{1}{2}e^x$ (c) $q(x) = e^{-x} + 3$

14. Use a graphing utility to graph the function. Use the graph to determine any asymptotes of the function.

(a) $f(x) = \dfrac{8}{1 + e^{-0.5x}}$ (b) $g(x) = \dfrac{8}{1 + e^{-0.5/x}}$

In Exercises 15–18, illustrate that the functions are inverses of each other by graphing both functions on the same set of coordinate axes.

15. $f(x) = e^{2x}$
   $g(x) = \ln \sqrt{x}$

16. $f(x) = e^{x/3}$
   $g(x) = \ln x^3$

17. $f(x) = e^x - 1$
   $g(x) = \ln(x + 1)$

18. $f(x) = e^{x-1}$
   $g(x) = 1 + \ln x$

In Exercises 19–22, match the equation with the correct graph. Assume that a and C are arbitrary real numbers such that $a > 0$. [The graphs are labeled (a), (b), (c), and (d).]

(a)
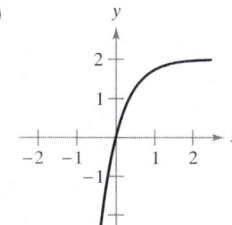

(b)
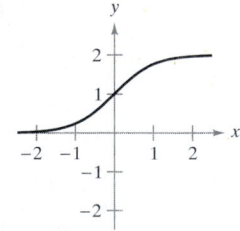

(c)
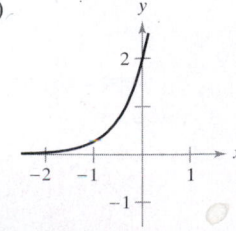

(d)
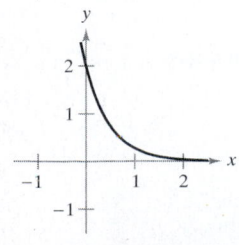

19. $y = Ce^{ax}$

20. $y = Ce^{-ax}$

21. $y = C(1 - e^{-ax})$

22. $y = \dfrac{C}{1 + e^{-ax}}$

23. *Graphical Analysis*   Use a graphing utility to graph

$$f(x) = \left(1 + \frac{0.5}{x}\right)^x \quad \text{and} \quad g(x) = e^{0.5}$$

in the same viewing rectangle. What is the relationship between f and g as $x \to \infty$?

24. *Conjecture*   Use the result of Exercise 23 to make a conjecture about the value of

$$\left(1 + \frac{r}{x}\right)^x$$

as $x \to \infty$.

In Exercises 25 and 26, compare the given number with the number e. Is the number less than or greater than e?

25. $\left(1 + \dfrac{1}{1{,}000{,}000}\right)^{1{,}000{,}000}$ (See Exercise 24.)

26. $1 + 1 + \dfrac{1}{2} + \dfrac{1}{6} + \dfrac{1}{24} + \dfrac{1}{120} + \dfrac{1}{720} + \dfrac{1}{5040}$

In Exercises 27 and 28, find the slope of the tangent line to the function at the point $(0, 1)$.

27. (a) $y = e^{3x}$

(b) $y = e^{-3x}$

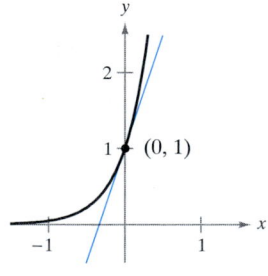

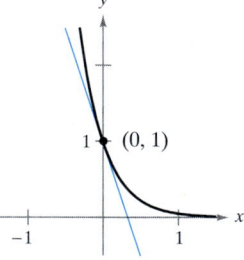

28. (a) $y = e^{2x}$

(b) $y = e^{-2x}$

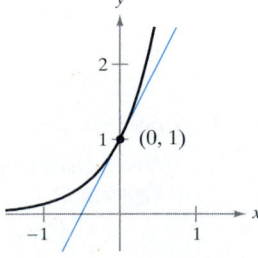

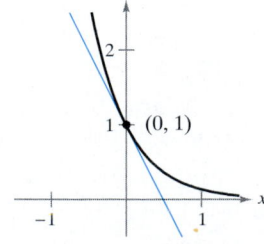

In Exercises 29–48, find the derivative of the function.

29. $f(x) = e^{2x}$

30. $f(x) = e^{1-x}$

31. $y = e^{-2x+x^2}$

32. $y = e^{-x^2}$

33. $y = e^{\sqrt{x}}$

34. $y = x^2 e^{-x}$

35. $g(t) = (e^{-t} + e^t)^3$

36. $g(t) = e^{-1/t^2}$

37. $y = \ln(e^{x^2})$

38. $y = \ln\left(\dfrac{1 + e^x}{1 - e^x}\right)$

39. $y = \ln(1 + e^{2x})$

40. $y = \ln\dfrac{e^x + e^{-x}}{2}$

41. $y = \dfrac{2}{e^x + e^{-x}}$

42. $y = \dfrac{e^x - e^{-x}}{2}$

43. $y = x^2 e^x - 2xe^x + 2e^x$

44. $y = xe^x - e^x$

45. $f(x) = e^{-x} \ln x$

46. $f(x) = e^3 \ln x$

47. $y = e^x(\sin x + \cos x)$

48. $y = \ln e^x$

In Exercises 49 and 50, use implicit differentiation to find dy/dx.

49. $xe^y - 10x + 3y = 0$

50. $e^{xy} + x^2 - y^2 = 10$

In Exercises 51 and 52, find the second derivative of the function.

51. $f(x) = (3 + 2x)e^{-3x}$

52. $g(x) = \sqrt{x} + e^x \ln x$

In Exercises 53 and 54, show that the function $y = f(x)$ is a solution of the differential equation.

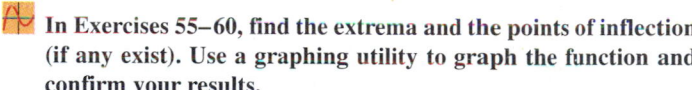

53. $y = e^x(\cos \sqrt{2}x + \sin \sqrt{2}x)$

$y'' - 2y' + 3y = 0$

54. $y = e^x(3 \cos 2x - 4 \sin 2x)$

$y'' - 2y' + 5y = 0$

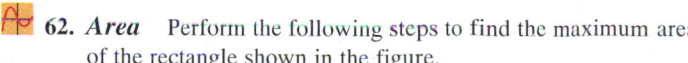

In Exercises 55–60, find the extrema and the points of inflection (if any exist). Use a graphing utility to graph the function and confirm your results.

55. $f(x) = \dfrac{1}{\sqrt{2\pi}} e^{-(x^2/2)}$

56. $f(x) = \dfrac{e^x - e^{-x}}{2}$

57. $f(x) = \dfrac{e^x + e^{-x}}{2}$

58. $f(x) = xe^{-x}$

59. $f(x) = x^2 e^{-x}$

60. $f(x) = -2 + e^{3x}(4 - 2x)$

61. *Area* Find the area of the largest rectangle that can be inscribed under the curve $y = e^{-x^2}$ in the first and second quadrants.

62. *Area* Perform the following steps to find the maximum area of the rectangle shown in the figure.

(a) Solve for c in the equation $f(c) = f(c + x)$.

(b) Use the result in part (a) to write the area A as a function of x. [*Hint:* $A = xf(c)$]

(c) Use a graphing utility to graph the area function. Use the graph to approximate the dimensions of the rectangle of maximum area. Determine the required area.

(d) Use a graphing utility to graph the expression for c found in part (a). Use the graph to approximate

$$\lim_{x \to 0^+} c \quad \text{and} \quad \lim_{x \to \infty} c.$$

Use this result to describe the changes in dimensions and position of the rectangle for $0 < x < \infty$.

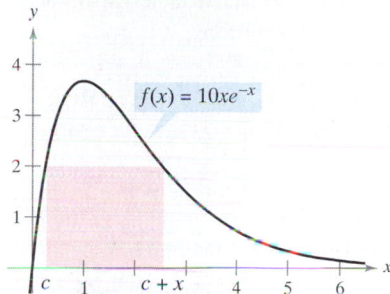

$f(x) = 10xe^{-x}$

63. Verify that the function

$$y = \dfrac{L}{1 + ae^{-x/b}}, \qquad a > 0, b > 0, L > 0$$

increases at a maximum rate when $y = L/2$.

64. Find the point on the graph of $y = e^{-x}$ where the normal line to the curve passes through the origin. (Use Newton's Method or the root-finding capabilities of a graphing utility.)

65. Find, to three decimal places, the value of x such that

$$e^{-x} = x.$$

(Use Newton's Method or the root-finding capabilities of a graphing utility.)

66. *Depreciation* The value V of an item t years after it is purchased is

$$V = 15{,}000e^{-0.6286t}, \qquad 0 \le t \le 10.$$

(a) Use a graphing utility to graph the function.

(b) Find the rate of change of V with respect to t when $t = 1$ and $t = 5$.

(c) Use a graphing utility to graph the tangent line to the function when $t = 1$ and $t = 5$.

67. *Writing* Consider the function

$$f(x) = \dfrac{2}{1 + e^{1/x}}.$$

(a) Use a graphing utility to graph f.

(b) Write a short paragraph explaining why the graph has a horizontal asymptote at $y = 1$ and why the function has a nonremovable discontinuity at $x = 0$.

68. *Harmonic Motion* The displacement from equilibrium of a mass oscillating on the end of a spring suspended from a ceiling is

$$y = 1.56e^{-0.22t} \cos 4.9t$$

where y is the displacement in feet and t is the time in seconds. Use a graphing utility to graph the displacement function on the interval $[0, 10]$. Find a value of t past which the displacement is less than 3 inches from equilibrium.

69. *Modeling Data* A meteorologist measures the atmospheric pressure P (in kilograms per square meter) at altitude h (in kilometers). The data are shown below.

h	0	5	10	15	20
P	10,332	5583	2376	1240	517

(a) Use a graphing utility to plot the points $(h, \ln P)$. Use the regression capabilities of the graphing utility to find a linear model for the revised data points.

(b) The line in part (a) has the form $\ln P = ah + b$. Write the equation in exponential form.

(c) Use a graphing utility to plot the original data and graph the exponential model in part (b).

(d) Find the rate of change of the pressure when $h = 5$ and $h = 18$.

70. *Modeling Data* A 1990 Chevrolet Beretta with a six-cylinder engine, automatic transmission, and air conditioning had a retail price of $11,500. A local dealership had the following guide for the approximate value of the car for the years 1990 through 1995. (*Source: National Automobile Dealer's Association*)

Year	1990	1991	1992
Value	$11,500	$9315	$9200

Year	1993	1994	1995
Value	$7935	$7130	$6095

In each of the following, let V represent the value of the automobile in the year t, with $t = 0$ corresponding to 1990.

(a) Use the regression capabilities of a graphing utility to find linear and quadratic models for the data. Use the graphing utility to plot the data and graph the models.

(b) What does the slope represent in the linear model in part (a)?

(c) Use the regression capabilities of a graphing utility to find a linear model for the points $(t, \ln V)$. Write the resulting equation of the form $\ln V = at + b$ in exponential form.

(d) Determine the horizontal asymptote of the exponential model in part (c). Interpret its meaning in the context of the problem.

(e) Find the rate of decrease in the value of the car when $t = 1$ and $t = 5$.

Linear and Quadratic Approximations In Exercises 71 and 72, use a graphing utility to graph the function. Then graph

$$P_1(x) = f(0) + f'(0)(x - 0)$$

and

$$P_2(x) = f(0) + f'(0)(x - 0) + \tfrac{1}{2}f''(0)(x - 0)^2$$

in the same viewing rectangle. Compare the values of f, P_1, and P_2 and their first derivatives at $x = 0$.

71. $f(x) = e^{x/2}$ **72.** $f(x) = e^{-x^2/2}$

73. *Finding a Pattern* Use a graphing utility to compare the graph of the function $y = e^x$ with the graphs of each of the following functions.

(a) $y_1 = 1 + \dfrac{x}{1!}$

(b) $y_2 = 1 + \dfrac{x}{1!} + \dfrac{x^2}{2!}$

(c) $y_3 = 1 + \dfrac{x}{1!} + \dfrac{x^2}{2!} + \dfrac{x^3}{3!}$

74. Identify the pattern of successive polynomials in Exercise 73. Extend the pattern one more term and compare the graph of the resulting polynomial function with the graph of $y = e^x$. What do you think this pattern implies?

In Exercises 75–92, evaluate the integral.

75. $\displaystyle\int e^{5x}(5)\,dx$

76. $\displaystyle\int e^{-x^4}(-4x^3)\,dx$

77. $\displaystyle\int_0^1 e^{-2x}\,dx$

78. $\displaystyle\int_1^2 e^{1-x}\,dx$

79. $\displaystyle\int \dfrac{e^{-x}}{1 + e^{-x}}\,dx$

80. $\displaystyle\int \dfrac{e^{2x}}{1 + e^{2x}}\,dx$

81. $\displaystyle\int_1^3 \dfrac{e^{3/x}}{x^2}\,dx$

82. $\displaystyle\int_0^{\sqrt{2}} xe^{-(x^2/2)}\,dx$

83. $\displaystyle\int e^x\sqrt{1 - e^x}\,dx$

84. $\displaystyle\int \dfrac{e^x - e^{-x}}{e^x + e^{-x}}\,dx$

85. $\displaystyle\int \dfrac{e^x + e^{-x}}{e^x - e^{-x}}\,dx$

86. $\displaystyle\int \dfrac{2e^x - 2e^{-x}}{(e^x + e^{-x})^2}\,dx$

87. $\displaystyle\int \dfrac{5 - e^x}{e^{2x}}\,dx$

88. $\displaystyle\int \dfrac{e^{2x} + 2e^x + 1}{e^x}\,dx$

89. $\displaystyle\int e^{\sin \pi x} \cos \pi x\,dx$

90. $\displaystyle\int e^{\tan 2x} \sec^2 2x\,dx$

91. $\displaystyle\int e^{-x} \tan(e^{-x})\,dx$

92. $\displaystyle\int \ln(e^{2x-1})\,dx$

In Exercises 93 and 94, solve the differential equation.

93. $\dfrac{dy}{dx} = xe^{ax^2}$

94. $\dfrac{dy}{dx} = (e^x - e^{-x})^2$

In Exercises 95 and 96, find the particular solution of the differential equation that satisfies the initial conditions.

95. $f''(x) = \frac{1}{2}(e^x + e^{-x})$,
$f(0) = 1, f'(0) = 0$

96. $f''(x) = \sin x + e^{2x}$,
$f(0) = \frac{1}{4}, f'(0) = \frac{1}{2}$

Direction Fields   **In Exercises 97 and 98, a differential equation, a point, and a direction field are given. (a) Sketch two approximate solutions of the differential equation on the direction field, one of which passes through the indicated point. (b) Use integration to find the particular solution of the differential equation and use a graphing utility to graph the solution. Compare the result with the sketches in part (a).**

97. $\dfrac{dy}{dx} = 2e^{-x/2}$, $(0, 1)$

98. $\dfrac{dy}{dx} = xe^{-0.2x^2}$, $\left(0, -\dfrac{3}{2}\right)$

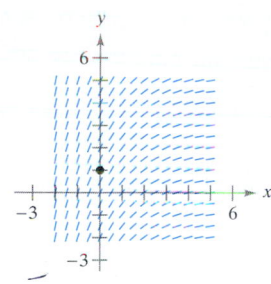

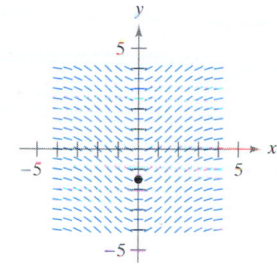

Area   **In Exercises 99–102, find the area of the region bounded by the graphs of the equations. Use a graphing utility to graph the region and verify your result.**

99. $y = e^x, y = 0, x = 0, x = 5$

100. $y = e^{-x}, y = 0, x = a, x = b$

101. $y = xe^{-(x^2/2)}, y = 0, x = 0, x = \sqrt{2}$

102. $y = e^{-2x} + 2, y = 0, x = 0, x = 2$

103. Given the exponential function $f(x) = e^x$, show that

(a) $f(u - v) = \dfrac{f(u)}{f(v)}$.

(b) $f(kx) = [f(x)]^k$.

104. Approximate each integral using the Midpoint Rule, the Trapezoidal Rule, and Simpson's Rule with $n = 12$. Then use the integration capabilities of a graphing utility to approximate the integrals and compare the results.

(a) $\displaystyle\int_0^4 \sqrt{x}\, e^x\, dx$

(b) $\displaystyle\int_0^2 2xe^{-x}\, dx$

105. *Probability*   A car battery has an average lifetime of 48 months with a standard deviation of 6 months. The battery lives are normally distributed. The probability that a given battery will last between 48 months and 60 months is

$$0.0665 \int_{48}^{60} e^{-0.0139(t - 48)^2}\, dt.$$

Use the integration capabilities of a graphing utility to approximate the integral. Interpret the resulting probability.

106. Given $e^x \geq 1$ for $x \geq 0$, it follows that

$$\int_0^x e^t\, dt \geq \int_0^x 1\, dt.$$

Perform this integration to derive the inequality $e^x \geq 1 + x$ for $x \geq 0$.

107. *Modeling Data*   A valve on a storage tank is opened for 4 hours to release a chemical in a manufacturing process. The flow rate R (in liters per hour) at time t (in hours) is given in the table.

t	0	1	2	3	4
R	425	240	118	71	36

(a) Use the regression capabilities of a graphing utility to find a linear model for the points $(t, \ln R)$. Write the resulting equation of the form $\ln R = at + b$ in exponential form.

(b) Use a graphing utility to plot the data and graph the exponential model.

(c) Use the definite integral to approximate the number of liters of chemical released during the 4 hours.

108. Prove that $\dfrac{e^a}{e^b} = e^{a-b}$.

109. Let $f(x) = \dfrac{\ln x}{x}$.

(a) Graph f on $(0, \infty)$ and show that f is strictly decreasing on $[e, \infty)$.

(b) Show that if $e \leq A < B$, then $A^B > B^A$.

(c) Use part (b) to show that $e^\pi > \pi^e$.

SECTION 5.5   Bases Other than e and Applications

Bases Other than e • Differentiation and Integration • Applications of Exponential Functions

Bases Other than e

The **base** of the natural exponential function is e. This "natural" base can be used to assign a meaning to a general base a.

> #### Definition of Exponential Function to Base a
>
> If a is a positive real number ($a \neq 1$) and x is any real number, then the **exponential function to the base a** is denoted by a^x and is defined by
>
> $$a^x = e^{(\ln a)x}.$$
>
> If $a = 1$, then $y = 1^x = 1$ is a constant function.

These functions obey the usual laws of exponents. For instance, here are some familiar properties.

1. $a^0 = 1$ 2. $a^x a^y = a^{x+y}$

3. $\dfrac{a^x}{a^y} = a^{x-y}$ 4. $(a^x)^y = a^{xy}$

EXAMPLE 1   Radioactive Half-Life Model

The half-life of carbon-14 is about 5730 years. If 1 gram of carbon-14 is present in a sample, how much will be present in 10,000 years?

Solution Let $t = 0$ represent the present time and let y represent the amount (in grams) of carbon-14 in the sample. Using a base of $\frac{1}{2}$, you can model y by the equation

$$y = \left(\frac{1}{2}\right)^{t/5730}.$$

Notice that when $t = 5730$, the amount is reduced to half of the original amount.

$$y = \left(\frac{1}{2}\right)^{5730/5730} = \frac{1}{2} \text{ gram}$$

When $t = 11,460$, the amount is reduced to a quarter of the original amount, and so on. To find the amount of carbon-14 after 10,000 years, substitute 10,000 for t.

$$y = \left(\frac{1}{2}\right)^{10,000/5730}$$

$$\approx 0.30 \text{ gram}$$

The graph of y is shown in Figure 5.24.

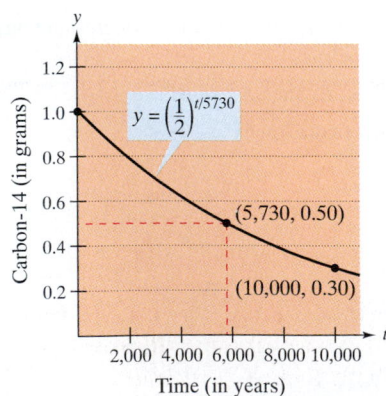

The half-life of carbon-14 is about 5730 years.
Figure 5.24

Logarithmic functions to bases other than *e* can be defined in much the same way as exponential functions to other bases are defined.

Definition of Logarithmic Function to Base *a*

If *a* is a positive real number $(a \neq 1)$ and *x* is any positive real number, then the **logarithmic function to the base *a*** is denoted by $\log_a x$ and is defined as

$$\log_a x = \frac{1}{\ln a} \ln x.$$

Logarithmic functions to the base *a* have properties similar to those of the natural logarithmic function given in Theorem 5.2.

1. $\log_a 1 = 0$ Log of 1
2. $\log_a xy = \log_a x + \log_a y$ Log of a product
3. $\log_a x^n = n \log_a x$ Log of a power
4. $\log_a \dfrac{x}{y} = \log_a x - \log_a y$ Log of a quotient

From the definitions of the exponential and logarithmic functions to the base *a*, it follows that $f(x) = a^x$ and $g(x) = \log_a x$ are inverse functions of each other.

Properties of Inverse Functions

1. $y = a^x$ if and only if $x = \log_a y$
2. $a^{\log_a x} = x, \quad$ for $x > 0$
3. $\log_a a^x = x, \quad$ for all x

The logarithmic function to the base 10 is called the **common logarithmic function.** Thus, for common logarithms, $y = 10^x$ if and only if $x = \log_{10} y$.

EXAMPLE 2   Bases Other than *e*

Solve for *x* in each of the following equations.

a. $3^x = \dfrac{1}{81}$ **b.** $\log_2 x = -4$

Solution

a. To solve this equation, you can apply the logarithmic function to the base 3 to both sides of the equation.

$$3^x = \frac{1}{81}$$

$$\log_3 3^x = \log_3 \frac{1}{81}$$

$$x = \log_3 3^{-4}$$

$$x = -4$$

b. To solve this equation, you can apply the exponential function to the base 2 to both sides of the equation.

$$\log_2 x = -4$$

$$2^{\log_2 x} = 2^{-4}$$

$$x = \frac{1}{2^4}$$

$$x = \frac{1}{16}$$

Differentiation and Integration

To differentiate exponential and logarithmic functions to other bases, you have three options: (1) use the definitions of a^x and $\log_a x$ and differentiate using the rules for the natural exponential and logarithmic functions, (2) use logarithmic differentiation, or (3) use the following differentiation rules for bases other than e.

THEOREM 5.13 Derivatives for Bases Other than e

Let a be a positive real number $(a \neq 1)$ and let u be a differentiable function of x.

1. $\dfrac{d}{dx}[a^x] = (\ln a)a^x$ **2.** $\dfrac{d}{dx}[a^u] = (\ln a)a^u \dfrac{du}{dx}$

3. $\dfrac{d}{dx}[\log_a x] = \dfrac{1}{(\ln a)x}$ **4.** $\dfrac{d}{dx}[\log_a u] = \dfrac{1}{(\ln a)u}\dfrac{du}{dx}$

Proof By definition, $a^x = e^{(\ln a)x}$. Therefore, you can prove the first rule by letting $u = (\ln a)x$ and differentiating with base e to obtain

$$\frac{d}{dx}[a^x] = \frac{d}{dx}[e^{(\ln a)x}] = e^u \frac{du}{dx} = e^{(\ln a)x}(\ln a) = (\ln a)a^x.$$

To prove the third rule, you can write

$$\frac{d}{dx}[\log_a x] = \frac{d}{dx}\left[\frac{1}{\ln a}\ln x\right] = \frac{1}{\ln a}\left(\frac{1}{x}\right) = \frac{1}{(\ln a)x}.$$

The second and fourth rules are simply the Chain Rule versions of the first and third rules.

NOTE These differentiation rules are similar to those for the natural exponential function and natural logarithmic function. In fact, they differ only by the constant factors $\ln a$ and $1/\ln a$. This points out one reason why, for calculus, e is the most convenient base.

EXAMPLE 3 Differentiating Functions to Other Bases

Find the derivative of each of the following.

a. $y = 2^x$ **b.** $y = 2^{3x}$ **c.** $y = \log_{10} \cos x$

Solution

a. $y' = \dfrac{d}{dx}[2^x] = (\ln 2)2^x$

b. $y' = \dfrac{d}{dx}[2^{3x}] = (\ln 2)2^{3x}(3) = (3 \ln 2)2^{3x}$

Try writing 2^{3x} as 8^x and differentiating to see that you obtain the same result.

c. $y' = \dfrac{d}{dx}[\log_{10} \cos x] = \dfrac{-\sin x}{(\ln 10)\cos x} = -\dfrac{1}{\ln 10}\tan x$

Occasionally, an integrand involves an exponential function to a base other than *e*. When this occurs, there are two options: (1) convert to base *e* using the formula $a^x = e^{(\ln a)x}$ and then integrate, or (2) integrate directly, using the integration formula (which follows from Theorem 5.13).

$$\int a^x \, dx = \left(\frac{1}{\ln a}\right)a^x + C$$

EXAMPLE 4   Integrating an Exponential Function to Another Base

Evaluate $\int 2^x \, dx$.

Solution

$$\int 2^x \, dx = \frac{1}{\ln 2} 2^x + C$$

When we introduced the Power Rule, $D_x[x^n] = nx^{n-1}$, in Chapter 2, we required the exponent *n* to be a rational number. We now extend the rule to cover any real value of *n*. Try to prove this theorem using logarithmic differentiation.

THEOREM 5.14   The Power Rule for Real Exponents

Let *n* be any real number and let *u* be a differentiable function of *x*.

1. $\frac{d}{dx}[x^n] = nx^{n-1}$ **2.** $\frac{d}{dx}[u^n] = nu^{n-1}\frac{du}{dx}$

The next example compares the derivatives of four types of functions. Each function uses a different differentiation formula, depending on whether the base and exponent are constants or variables.

EXAMPLE 5   Comparing Variables and Constants

a. $\frac{d}{dx}[e^e] = 0$ Constant Rule

b. $\frac{d}{dx}[e^x] = e^x$ Exponential Rule

c. $\frac{d}{dx}[x^e] = ex^{e-1}$ Power Rule

d. $y = x^x$ Logarithmic differentiation

$\ln y = x \ln x$

$\frac{y'}{y} = x\left(\frac{1}{x}\right) + (\ln x)(1) = 1 + \ln x$

$y' = y(1 + \ln x) = x^x(1 + \ln x)$

NOTE   Be sure you see that there is no simple differentiation rule for calculating the derivative of $y = x^x$. In general, if $y = u(x)^{v(x)}$, you need to use logarithmic differentiation.

Applications of Exponential Functions

n	A
1	$1080.00
2	$1081.60
4	$1082.43
12	$1083.00
365	$1083.28

Suppose P dollars is deposited at an annual interest rate r (in decimal form). If interest accumulates in the account, what is the balance in the account at the end of 1 year? The answer depends on the number of times n the interest is compounded according to the formula

$$A = P\left(1 + \frac{r}{n}\right)^n.$$

For instance, the result for a deposit of $1000 at 8% interest compounded n times a year is shown in the upper table at the left.

As n increases, the balance A approaches a limit. To develop this limit, we use the following theorem. To test the reasonableness of this theorem, try evaluating $[(x + 1)/x]^x$ for several values of x, as shown in the lower table at the left. (A proof of this theorem is given in the appendix.)

x	$\left(\dfrac{x+1}{x}\right)^x$
10	2.59374
100	2.70481
1000	2.71692
10,000	2.71815
100,000	2.71827
1,000,000	2.71828

THEOREM 5.15   A Limit Involving e

$$\lim_{x \to \infty}\left(1 + \frac{1}{x}\right)^x = \lim_{x \to \infty}\left(\frac{x+1}{x}\right)^x = e$$

Now, let's take another look at the formula for the balance of A in an account in which the interest is compounded n times per year. By taking the limit as n approaches infinity, you obtain the following.

$$A = \lim_{n \to \infty} P\left(1 + \frac{r}{n}\right)^n \qquad \text{Take limit as } n \to \infty.$$

$$= P \lim_{n \to \infty}\left[\left(1 + \frac{1}{n/r}\right)^{n/r}\right]^r \qquad \text{Rewrite.}$$

$$= P\left[\lim_{x \to \infty}\left(1 + \frac{1}{x}\right)^x\right]^r \qquad \text{Let } x = n/r.\ \text{Then } x \to \infty \text{ as } n \to \infty.$$

$$= Pe^r \qquad \text{Apply Theorem 5.15.}$$

This limit produces the balance after 1 year of **continuous compounding.** Thus, for a deposit of $1000 at 8% interest compounded continuously, the balance at the end of 1 year would be

$$A = 1000e^{0.08} \approx \$1083.29.$$

These results are summarized as follows.

Summary of Compound Interest Formulas

Let P = amount of deposit, t = number of years, A = balance after t years, r = annual interest rate (decimal form), and n = the number of compoundings per year.

1. Compounding n times per year: $A = P\left(1 + \dfrac{r}{n}\right)^{nt}$

2. Compounded continuously: $A = Pe^{rt}$

EXAMPLE 6   Comparing Continuous and Quarterly Compounding

A deposit of $2500 is made in an account that pays an annual interest rate of 5%. Find the balance in the account at the end of 5 years if the interest is compounding (a) quarterly, (b) monthly, and (c) continuously.

Solution

a. $A = P\left(1 + \dfrac{r}{n}\right)^{nt} = 2500\left(1 + \dfrac{0.05}{4}\right)^{4(5)}$ Quarterly compounding

$\qquad = 2500(1.0125)^{20}$

$\qquad \approx \$3205.09$

b. $A = P\left(1 + \dfrac{r}{n}\right)^{nt} = 2500\left(1 + \dfrac{0.05}{12}\right)^{12(5)}$ Monthly compounding

$\qquad \approx 2500(1.0041667)^{60}$

$\qquad \approx \$3208.40$

c. $A = Pe^{rt} = 2500\left[e^{0.05(5)}\right]$ Continuous compounding

$\qquad = 2500e^{0.25}$

$\qquad \approx \$3210.06$

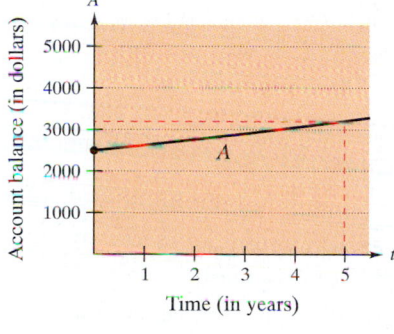

The balance in a savings account grows exponentially.
Figure 5.25

Figure 5.25 shows how the balance increases over the 5-year period. Notice that the scale used in the figure does not graphically distinguish among the three types of exponential growth in (a), (b), and (c).

EXAMPLE 7   Bacterial Culture Growth

A bacterial culture is growing according to the *logistics growth function*

$$y = \frac{1.25}{1 + 0.25e^{-0.4t}}, \qquad t \ge 0$$

where *y* is the weight of the culture in grams and *t* is the time in hours. Find the weight of the culture after (a) 0 hours, (b) 1 hour, and (c) 10 hours. (d) What is the limit as *t* approaches infinity?

Solution

a. When $t = 0$, $y = \dfrac{1.25}{1 + 0.25e^{-0.4(0)}} = \dfrac{1.25}{1.25} = 1$ gram.

b. When $t = 1$, $y = \dfrac{1.25}{1 + 0.25e^{-0.4(1)}} \approx 1.071$ grams.

c. When $t = 10$, $y = \dfrac{1.25}{1 + 0.25e^{-0.4(10)}} \approx 1.244$ grams.

d. Finally, taking the limit as *t* approaches infinity, you obtain

$$\lim_{t \to \infty} \frac{1.25}{1 + 0.25e^{-0.4t}} = \frac{1.25}{1 + 0} = 1.25 \text{ grams.}$$

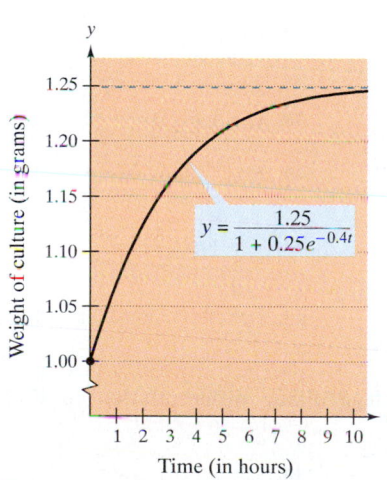

The limit of the weight of the culture as $t \to \infty$ is 1.25 grams.
Figure 5.26

The graph of the function is shown in Figure 5.26.

EXERCISES FOR SECTION 5.5

In Exercises 1–4, evaluate the expression without using a calculator.

1. $\log_2\left(\frac{1}{8}\right)$

2. $\log_{27} 9$

3. $\log_7 1$

4. $\log_a \frac{1}{a}$

In Exercises 5–8, write the exponential equation as a logarithmic equation or vice versa.

5. (a) $2^3 = 8$
(b) $3^{-1} = \frac{1}{3}$

6. (a) $27^{2/3} = 9$
(b) $16^{3/4} = 8$

7. (a) $\log_{10} 0.01 = -2$
(b) $\log_{0.5} 8 = -3$

8. (a) $\log_3 \frac{1}{9} = -2$
(b) $49^{1/2} = 7$

In Exercises 9–14, solve for x or b.

9. (a) $\log_{10} 1000 = x$
(b) $\log_{10} 0.1 = x$

10. (a) $\log_4 \frac{1}{64} = x$
(b) $\log_5 25 = x$

11. (a) $\log_3 x = -1$
(b) $\log_2 x = -4$

12. (a) $\log_b 27 = 3$
(b) $\log_b 125 = 3$

13. (a) $x^2 - x = \log_5 25$
(b) $3x + 5 = \log_2 64$

14. (a) $\log_3 x + \log_3(x - 2) = 1$
(b) $\log_{10}(x + 3) - \log_{10} x = 1$

In Exercises 15–20, sketch the graph of the function by hand.

15. $y = 3^x$

16. $y = 3^{x-1}$

17. $y = \left(\frac{1}{3}\right)^x$

18. $y = 2^{x^2}$

19. $h(x) = 5^{x-2}$

20. $y = 3^{-|x|}$

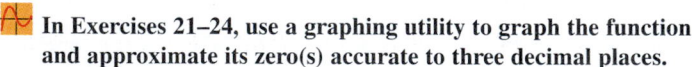 In Exercises 21–24, use a graphing utility to graph the function and approximate its zero(s) accurate to three decimal places.

21. $g(x) = 6(2^{1-x}) - 25$

22. $f(t) = 300(1.0075^{12t}) - 735.41$

23. $h(s) = 32 \log_{10}(s - 2) + 15$

24. $g(x) = 1 - 2 \log_{10}[x(x - 3)]$

In Exercises 25 and 26, illustrate that the functions are inverses of each other by sketching their graphs on the same set of coordinate axes.

25. $f(x) = 4^x$
$g(x) = \log_4 x$

26. $f(x) = 3^x$
$g(x) = \log_3 x$

27. *Think About It*   The table of values was obtained from evaluating a function. Determine which of the statements may be true and which must be false, and explain why.

(a) y is an exponential function of x.

(b) y is a logarithmic function of x.

(c) x is an exponential function of y.

(d) y is a linear function of x.

x	1	2	8
y	0	1	3

28. *Think About It*   Consider the function $f(x) = \log_{10} x$.

(a) What is the domain of f?

(b) Find f^{-1}.

(c) If x is a real number between 1000 and 10,000, determine the interval in which $f(x)$ will be found.

(d) Determine the interval in which x will be found if $f(x)$ is negative.

(e) If $f(x)$ is increased by one unit, x must have been increased by what factor?

(f) Find the ratio of x_1 to x_2 given that $f(x_1) = 3n$ and $f(x_2) = n$.

In Exercises 29–44, find the derivative of the function.

29. $f(x) = 4^x$

30. $g(x) = 2^{-x}$

31. $y = 5^{x-2}$

32. $y = x(7^{-3x})$

33. $g(t) = t^2 2^t$

34. $f(t) = \dfrac{3^{2t}}{t}$

35. $h(\theta) = 2^{-\theta} \cos \pi\theta$

36. $g(\alpha) = 5^{-\alpha/2} \sin 2\alpha$

37. $y = \log_3 x$

38. $y = \log_{10} 2x$

39. $f(x) = \log_2 \dfrac{x^2}{x - 1}$

40. $h(x) = \log_3 \dfrac{x\sqrt{x - 1}}{2}$

41. $y = \log_5 \sqrt{x^2 - 1}$

42. $y = \log_{10} \dfrac{x^2 - 1}{x}$

43. $g(t) = \dfrac{10 \log_4 t}{t}$

44. $f(t) = t^{3/2} \log_2 \sqrt{t + 1}$

In Exercises 45–48, use logarithmic differentiation to find dy/dx.

45. $y = x^{2/x}$

46. $y = x^{x-1}$

47. $y = (x - 2)^{x+1}$

48. $y = (1 + x)^{1/x}$

49. *Ordering Functions*   Order the functions $f(x) = \log_2 x$, $g(x) = x^x$, $h(x) = x^2$, and $k(x) = 2^x$ from the one with the greatest rate of growth to the one with the smallest rate of growth for "large" values of x.

50. Given the exponential function $f(x) = a^x$, show that

(a) $f(u + v) = f(u) \cdot f(v)$.

(b) $f(2x) = [f(x)]^2$.

51. *Inflation*   If the annual rate of inflation averages 5% over the next 10 years, the approximate cost C of goods or services during any year in that decade is

$$C(t) = P(1.05)^t$$

where t is the time in years and P is the present cost.

(a) If the price of an oil change for your car is presently $24.95, estimate the price 10 years from now.

(b) Find the rate of change of C with respect to t when $t = 1$ and $t = 8$.

(c) Verify that the rate of change of C is proportional to C. What is the constant of proportionality?

52. *Depreciation*   After t years, the value of a car purchased for $20,000 is

$$V(t) = 20{,}000\left(\tfrac{3}{4}\right)^t.$$

(a) Use a graphing utility to graph the function and determine the value of the car 2 years after it was purchased.

(b) Find the rate of change of V with respect to t when $t = 1$ and $t = 4$.

(c) Use a graphing utility to graph $V'(t)$ and determine the horizontal asymptote of $V'(t)$. Interpret its meaning in the context of the problem.

Compound Interest   **In Exercises 53–56, complete the table to determine the balance A for P dollars invested at rate r for t years and compounded n times per year.**

n	1	2	4	12	365	Continuous compounding
A						

53. $P = \$1000$

$r = 3\tfrac{1}{2}\%$

$t = 10$ years

54. $P = \$2500$

$r = 6\%$

$t = 20$ years

55. $P = \$1000$

$r = 5\%$

$t = 30$ years

56. $P = \$2500$

$r = 5\%$

$t = 40$ years

Compound Interest   **In Exercises 57–60, complete the table to determine the amount of money P (present value) that should be invested at rate r to produce a balance of $100,000 in t years.**

t	1	10	20	30	40	50
P						

57. $r = 5\%$

Compounded continuously

58. $r = 6\%$

Compounded continuously

59. $r = 5\%$

Compounded monthly

60. $r = 7\%$

Compounded daily

61. *Compound Interest*   Assume that you can earn 6% on an investment, compounded daily. Which of the following options would yield the greatest balance in 8 years?

(a) $20,000 now

(b) $30,000 in 8 years

(c) $8000 now and $20,000 in 4 years

(d) $9000 now, $9000 in 4 years, and $9000 in 8 years

62. *Compound Interest*   Consider a deposit of $100 placed in an account for 20 years at $r\%$ compounded continuously. Use a graphing utility to graph the exponential functions giving the growth of the investment over the 20 years for each of the following interest rates. Compare the ending balances for each of the rates.

(a) $r = 3\%$ (b) $r = 5\%$ (c) $r = 6\%$

63. *Timber Yield*   The yield V (in millions of cubic feet per acre) for a stand of timber at age t is

$$V = 6.7e^{(-48.1)/t}$$

where t is measured in years.

(a) Find the limiting volume of wood per acre as t approaches infinity.

(b) Find the rate at which the yield is changing when $t = 20$ years and $t = 60$ years.

64. *Learning Theory*   In a group project in learning theory, a mathematical model for the proportion P of correct responses after n trials was found to be

$$P = \frac{0.83}{1 + e^{-0.2n}}.$$

(a) Find the limiting proportion of correct responses as n approaches infinity.

(b) Find the rate at which P is changing after $n = 3$ trials and $n = 10$ trials.

65. *Forest Defoliation*   To estimate the amount of defoliation caused by the gypsy moth during a year, a forester counts the number of egg masses on $\tfrac{1}{40}$ of an acre the preceding fall. The percent of defoliation y is approximated by

$$y = \frac{300}{3 + 17e^{-0.0625x}}$$

where x is the number of egg masses in thousands.   (*Source: USDA Forest Service*)

(a) Use a graphing utility to graph the function.

(b) Estimate the percent of defoliation if 2000 egg masses are counted.

(c) Estimate the number of egg masses that existed if you observe that approximately $\tfrac{2}{3}$ of a forest is defoliated.

(d) Use calculus to estimate the value of x for which y is increasing most rapidly.

66. *Population Growth* A lake is stocked with 500 fish, and their population increases according to the logistics curve

$$p(t) = \frac{10,000}{1 + 19e^{-t/5}}$$

where t is measured in months.

(a) Use a graphing utility to graph the function.

(b) What is the limiting size of the fish population?

(c) At what rates is the fish population changing at the end of 1 month and at the end of 10 months?

(d) After how many months is the population increasing most rapidly?

67. *Modeling Data* The table gives the national health expenditures y (in billions of dollars) for the years 1984 through 1993, with $x = 4$ corresponding to 1984. *(Source: U.S. Health Care Financing Administration)*

x	4	5	6	7	8
y	396.0	434.5	466.0	506.2	562.3

x	9	10	11	12	13
y	623.9	696.6	755.6	820.3	884.2

(a) Use a graphing utility to find an exponential model for the data. Use the graphing utility to plot the data and graph the exponential model.

(b) Use a graphing utility to find a logarithmic model for the data. Use the graphing utility to plot the data and graph the logarithmic model.

(c) Which model better approximates the data?

(d) Find the rate of growth of each model for the year 2000. If the rate of growth of health care expenditures could be slowed, which may be the better model for the future? Explain.

68. *Comparing Models* The amount y (in billions of dollars) given to philanthropy (from individuals, foundations, corporations, and charitable bequests) for the years 1983 through 1993 in the United States is given in the table, with $x = 3$ corresponding to 1983. *(Source: AAFRC Trust for Philanthropy)*

x	3	4	5	6	7	8
y	63.2	68.8	73.2	83.9	90.3	98.4

x	9	10	11	12	13
y	107.0	111.9	117.1	121.9	126.2

(a) Use the regression capabilities of a graphing utility to find the following models for the data.

$$y_1 = ax + b \qquad y_2 = a + b \ln x$$
$$y_3 = ab^x \qquad y_4 = ax^b$$

(b) Use a graphing utility to plot the data and graph each model. Which model do you think best fits the data?

(c) Interpret the slope of the linear model in the context of the problem.

(d) Find the rate of change of each model for the year 1992. Which model is increasing at the greatest rate in 1992?

In Exercises 69–76, evaluate the integral.

69. $\displaystyle\int 3^x \, dx$

70. $\displaystyle\int 4^{-x} \, dx$

71. $\displaystyle\int_{-1}^{2} 2^x \, dx$

72. $\displaystyle\int_{-2}^{0} (3^3 - 5^2) \, dx$

73. $\displaystyle\int x(5^{-x^2}) \, dx$

74. $\displaystyle\int (3 - x)7^{(3-x)^2} \, dx$

75. $\displaystyle\int \frac{3^{2x}}{1 + 3^{2x}} \, dx$

76. $\displaystyle\int 2^{\sin x} \cos x \, dx$

Direction Fields **In Exercises 77 and 78, a differential equation, a point, and a direction field are given. (a) Sketch two approximate solutions of the differential equation on the direction field, one of which passes through the indicated point. (b) Use integration to find the particular solution of the differential equation and use a graphing utility to graph the solution. Compare the result with the sketches in part (a).**

77. $\dfrac{dy}{dx} = 0.4^{x/3}, \quad \left(0, \frac{1}{2}\right)$

78. $\dfrac{dy}{dx} = e^{\sin x} \cos x, \quad (\pi, 2)$

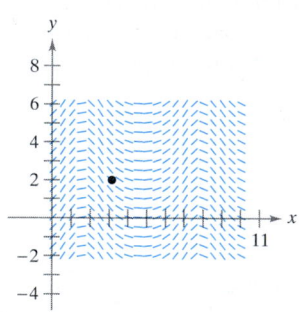

79. *Conjecture*

(a) Use a graphing utility to approximate the integrals of the functions

$$f(t) = 4\left(\frac{3}{8}\right)^{2t/3}, \quad g(t) = 4\left(\frac{\sqrt[3]{9}}{4}\right)^{t},$$

and $h(t) = 4e^{-0.653886t}$

on the interval $[0, 4]$.

(b) Use a graphing utility to graph the three functions.

(c) Use the results in parts (a) and (b) to make a conjecture about the three functions. Could you make the conjecture using only part (a)? Explain. Prove your conjecture analytically.

80. *Area* Find the area of the region bounded by the graphs of $y = 3^x, y = 0, x = 0,$ and $x = 3$.

81. *Continuous Cash Flow* The present value P of a continuous cash flow of $2000 per year earning 6% compounded continuously over 10 years is

$$P = \int_0^{10} 2000 e^{-0.06t}\, dt.$$

Find P.

82. Complete the table to demonstrate that e can also be defined as $\lim_{x \to 0^+} (1 + x)^{1/x}$.

x	1	10^{-1}	10^{-2}	10^{-4}	10^{-6}
$(1 + x)^{1/x}$					

In Exercises 83 and 84, find an exponential function that fits the experimental data collected over time *t*.

83.

t	0	1	2	3	4
y	1200.00	720.00	432.00	259.20	155.52

84.

t	0	1	2	3	4
y	600.00	630.00	661.50	694.58	729.30

True or False? **In Exercises 85–90, determine whether the statement is true or false. If it is false, explain why or give an example that shows it is false.**

85. $e = 271,801/99,990$.

86. If $f(x) = \ln x$, then $f(e^{n+1}) - f(e^n) = 1$ for any value of n.

87. The functions $f(x) = 2 + e^x$ and $g(x) = \ln(x - 2)$ are inverses of each other.

88. The exponential function $y = Ce^x$ is a solution of the differential equation $d^n y/dx^n = y, n = 1, 2, 3, \ldots$.

89. The graphs of $f(x) = e^x$ and $g(x) = e^{-x}$ meet at right angles.

90. If $f(x) = g(x)e^x$, then the only zeros of f are the zeros of g.

91. Solve the logistics differential equation

$$\frac{dy}{dt} = \frac{8}{25} y \left(\frac{5}{4} - y \right), \quad y(0) = 1$$

and obtain the logistics growth function of Example 7.

$$\left[Hint: \frac{1}{y\left(\frac{5}{4} - y\right)} = \frac{4}{5}\left(\frac{1}{y} + \frac{1}{\frac{5}{4} - y}\right) \right]$$

92. Find an equation of the tangent line to $y = x^{\sin x}$ at $\left(\dfrac{\pi}{2}, \dfrac{\pi}{2} \right)$.

SECTION PROJECT

Let $f(x) = \begin{cases} |x|^x, & x \neq 0 \\ 1, & x = 0. \end{cases}$

(a) Use a graphing utility to graph f in the viewing rectangle $-3 \le x \le 3, -2 \le y \le 2$. What is the domain of f?

(b) Use the zoom and trace features of a graphing utility to estimate

$$\lim_{x \to 0} f(x).$$

(c) Write a short paragraph explaining why the function f is continuous for all real numbers.

(d) Visually estimate the slope of f at the point $(0, 1)$.

(e) Explain why the derivative of a function can be approximated by the formula

$$\frac{f(x + h) - f(x - h)}{2h}$$

for small values of h. Use this formula to approximate the slope of f at the point $(0, 1)$.

$$f'(0) \approx \frac{f(0 + h) - f(0 - h)}{2h} = \frac{f(h) - f(-h)}{2h}$$

What do you think the slope of the graph of f is at $(0, 1)$?

(f) Find a formula for the derivative of f and determine $f'(0)$. Write a short paragraph explaining how a graphing utility might lead you to approximate the slope of a graph incorrectly.

(g) Use your formula for the derivative of f to find the relative extrema of f. Verify your answer with a graphing utility.

FOR FURTHER INFORMATION For more information on using graphing utilities to estimate slope, see the article "Computer-Aided Delusions" by Richard L. Hall in the September 1993 issue of *The College Mathematics Journal*.

SECTION $\boxed{5.6}$ **Differential Equations: Growth and Decay**

Differential Equations • Growth and Decay Models

Differential Equations

Up to now in the text, you have learned to solve only two types of differential equations—those of the forms

$$y' = f(x) \quad \text{and} \quad y'' = f(x).$$

In this section, you will learn how to solve a more general type of differential equation. The strategy is to rewrite the equation so that each variable occurs on only one side of the equation. This strategy is called **separation of variables.**

EXAMPLE 1 Using Separation of Variables

NOTE When you integrate both sides of the equation in Example 1, you don't need to add a constant of integration to both sides of the equation. If you did, you would obtain the same results as in Example 1.

$$\int y \, dy = \int 2x \, dx$$

$$\tfrac{1}{2} y^2 + C_2 = x^2 + C_3$$

$$\tfrac{1}{2} y^2 = x^2 + (C_3 - C_2)$$

Solve the differential equation $y' = 2x/y$.

Solution

$y' = \dfrac{2x}{y}$	Original equation
$yy' = 2x$	Multiply both sides by y.
$\displaystyle\int yy' \, dx = \int 2x \, dx$	Integrate with respect to x.
$\displaystyle\int y \, dy = \int 2x \, dx$	$dy = y' \, dx$
$\dfrac{1}{2} y^2 = x^2 + C_1$	Apply Power Rule.
$y^2 - 2x^2 = C$	Rewrite, letting $C = 2C_1$.

Thus, the general solution is given by

$$y^2 - 2x^2 = C.$$

You can use implicit differentiation to check this result.

In practice, most people prefer to use Leibniz notation and differentials when applying separation of variables. With this notation, the solution of Example 1 is as follows.

$$\frac{dy}{dx} = \frac{2x}{y}$$

$$y \, dy = 2x \, dx$$

$$\int y \, dy = \int 2x \, dx$$

$$\frac{1}{2} y^2 = x^2 + C_1$$

$$y^2 - 2x^2 = C$$

Growth and Decay Models

In many applications, the rate of change of a variable y is proportional to the value of y. If y is a function of time t, the proportionality can be written as follows.

Rate of change of y is   proportional to y.

$$\frac{dy}{dt} = ky$$

The general solution of this differential equation is given in the following theorem.

> **THEOREM 5.16   Exponential Growth and Decay Model**
>
> If y is a differentiable function of t such that $y > 0$ and $y' = ky$, for some constant k, then
>
> $$y = Ce^{kt}.$$
>
> C is the **initial value** of y, and k is the **proportionality constant.** Exponential growth occurs when $k > 0$, and **exponential decay** occurs when $k < 0$.

NOTE   Differentiate the function $y = Ce^{kt}$ with respect to t, and verify that $y' = ky$.

Proof

$$y' = ky \qquad \text{Original equation}$$

$$\frac{y'}{y} = k \qquad \text{Separate variables.}$$

$$\int \frac{y'}{y}\, dt = \int k\, dt \qquad \text{Integrate with respect to } t.$$

$$\int \frac{1}{y}\, dy = \int k\, dt \qquad dy = y'\, dt$$

$$\ln y = kt + C_1 \qquad \text{Find antiderivative of each side.}$$

$$y = e^{C_1}e^{kt} \qquad \text{Solve for } y.$$

$$y = Ce^{kt} \qquad \text{Let } C = e^{C_1}.$$

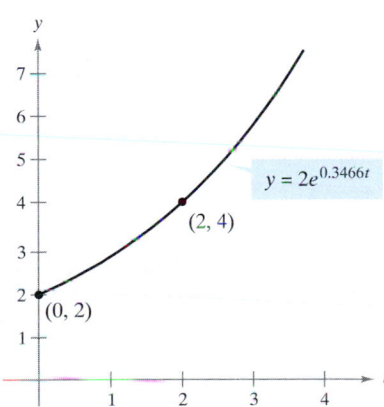

$y = 2e^{0.3466t}$

(2, 4)

(0, 2)

If the rate of change of y is proportional to y, then y follows an exponential model.
Figure 5.27

EXAMPLE 2   Using an Exponential Growth Model

The rate of change of y is proportional to y. When $t = 0$, $y = 2$. When $t = 2$, $y = 4$. What is the value of y when $t = 3$?

Solution   Because $y' = ky$, you know that y and t are related by the equation $y = Ce^{kt}$. You can find the values of the constants C and k by applying the initial conditions.

$$2 = Ce^0 \quad \Longrightarrow \quad C = 2 \qquad \text{When } t = 0, y = 2.$$

$$4 = 2e^{2k} \quad \Longrightarrow \quad k = \frac{1}{2}\ln 2 \approx 0.3466 \qquad \text{When } t = 2, y = 4.$$

Therefore, the model is $y \approx 2e^{0.3466t}$, and when $t = 3$, the value of y is $2e^{0.3466(3)} \approx 5.657$ (see Figure 5.27).

Radioactive decay is measured in terms of *half-life*—the number of years required for half of the atoms in a sample of radioactive material to decay. The half-lives of some common radioactive isotopes are as follows.

Uranium (U^{238})   4,510,000,000 years
Plutonium (Pu^{239})   24,360 years
Carbon (C^{14})   5730 years
Radium (Ra^{226})   1620 years
Einsteinium (Es^{254})   270 days
Nobelium (No^{257})   23 seconds

EXAMPLE 3   Radioactive Decay

Suppose that 10 grams of the plutonium isotope Pu-239 was released in the Chernobyl nuclear accident. How long will it take for the 10 grams to decay to 1 gram?

Solution   Let y represent the mass (in grams) of the plutonium. Because the rate of decay is proportional to y, you know that

$$y = Ce^{kt}$$

where t is the time in years. To find the values of the constants C and k, apply the initial conditions. Using the fact that $y = 10$ when $t = 0$, you can write

$$10 = Ce^{k(0)} = Ce^0$$

which implies that $C = 10$. Next, using the fact that $y = 5$ when $t = 24{,}360$, you can write

$$5 = 10e^{k(24,360)}$$

$$\frac{1}{2} = e^{24,360k}$$

$$\frac{1}{24,360} \ln \frac{1}{2} = k$$

$$-2.8454 \times 10^{-5} \approx k.$$

Therefore, the model is

$$y = 10e^{-0.000028454t}. \qquad \color{red}{\text{Half-life model}}$$

To find the time it would take for 10 grams to decay to 1 gram, you can solve for t in the equation

$$1 = 10e^{-0.000028454t}.$$

The solution is approximately 80,923 years.

The worst nuclear accident in history happened in 1986 at the Chernobyl nuclear plant near Kiev in the Ukraine. An explosion destroyed one of the plant's four reactors, releasing large amounts of radioactive isotopes into the atmosphere. Two of the four reactors are still in operation.

NOTE   The exponential decay model in Example 3 could also be written as $y = 10\left(\frac{1}{2}\right)^{t/24,360}$. This model is much easier to derive, but for some applications, it is not as convenient to use.

From Example 3, notice that in an exponential growth or decay problem, it is easy to solve for C when you are given the value of y at $t = 0$. The next example demonstrates a procedure for solving for C and k when you do not know the value of y at $t = 0$.

EXAMPLE 4   Population Growth

Suppose an experimental population of fruit flies increases according to the law of exponential growth. There were 100 flies after the second day of the experiment and 300 flies after the fourth day. Approximately how many flies were in the original population?

Solution   Let $y = Ce^{kt}$ be the number of flies at time t, where t is measured in days. Because $y = 100$ when $t = 2$ and $y = 300$ when $t = 4$, you can write

$$100 = Ce^{2k} \quad \text{and} \quad 300 = Ce^{4k}.$$

From the first equation, you know that $C = 100e^{-2k}$. Substituting this value into the second equation produces the following.

$$300 = 100e^{-2k}e^{4k}$$
$$300 = 100e^{2k}$$
$$\ln 3 = 2k$$
$$\frac{1}{2}\ln 3 = k$$
$$0.5493 \approx k$$

Therefore, the exponential growth model is $y = Ce^{0.5493t}$. To solve for C, reapply the condition $y = 100$ when $t = 2$ and obtain

$$100 = Ce^{0.5493(2)}$$
$$C = 100e^{-1.0986} \approx 33.$$

Thus, the original population (when $t = 0$) consisted of approximately $y = C = 33$ flies, as indicated in Figure 5.28.

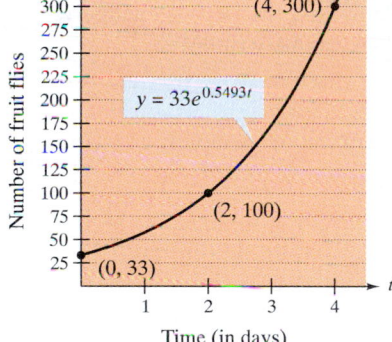

Figure 5.28

EXAMPLE 5   Declining Sales

Four months after it stops advertising, a manufacturing company notices that its sales have dropped from 100,000 units per month to 80,000 units per month. If the sales follow an exponential pattern of decline, what will they be after another 2 months?

Solution   Use the exponential decay model $y = Ce^{kt}$, where t is measured in months, as shown in Figure 5.29. From the initial condition ($t = 0$), you know that $C = 100,000$. Moreover, because $y = 80,000$ when $t = 4$, you have

$$80,000 = 100,000e^{4k}$$
$$0.8 = e^{4k}$$
$$\ln(0.8) = 4k$$
$$-0.0558 \approx k.$$

Thus, after 2 more months ($t = 6$), you can expect the monthly sales rate to be

$$y \approx 100,000e^{-0.0558(6)}$$
$$\approx 71,500 \text{ units.}$$

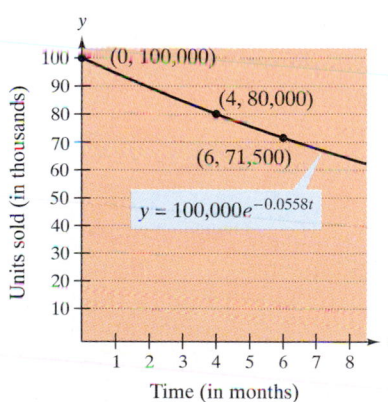

Figure 5.29

In Examples 2 to 5, you did not actually have to solve the differential equation

$$y' = ky.$$

(This was done once in the proof of Theorem 5.16.) The next example demonstrates a problem whose solution involves the separation of variables technique. The example concerns **Newton's Law of Cooling,** which states that the rate of change in the temperature of an object is proportional to the difference between the object's temperature and the temperature of the surrounding medium.

EXAMPLE 6 Newton's Law of Cooling

Let y represent the temperature (in °F) of an object in a room whose temperature is kept at a constant 60°. If the object cools from 100° to 90° in 10 minutes, how much longer will it take for its temperature to decrease to 80°?

Solution From Newton's Law of Cooling, you know that the rate of change in y is proportional to the difference between y and 60. This can be written as

$$y' = k(y - 60), \qquad 80 \le y \le 100.$$

To solve this differential equation, you can use separation of variables, as follows.

$$\frac{dy}{dt} = k(y - 60) \qquad \text{Differential equation}$$

$$\left(\frac{1}{y - 60}\right) dy = k \, dt \qquad \text{Separate variables.}$$

$$\int \frac{1}{y - 60} \, dy = \int k \, dt \qquad \text{Integrate both sides.}$$

$$\ln |y - 60| = kt + C_1 \qquad \text{Find antiderivative of each side.}$$

Because $y > 60$, $|y - 60| = y - 60$, and you can omit the absolute value signs. Using exponential notation, you have

$$y - 60 = e^{kt + C_1} \quad \Longrightarrow \quad y = 60 + Ce^{kt}. \qquad C = e^{C_1}$$

Using $y = 100$ when $t = 0$, you obtain $100 = 60 + Ce^{k(0)} = 60 + C$, which implies that $C = 40$. Because $y = 90$ when $t = 10$, it follows that

$$90 = 60 + 40e^{k(10)}$$

$$30 = 40e^{10k}$$

$$k = \tfrac{1}{10} \ln \tfrac{3}{4} \approx -0.02877.$$

Thus, the model is

$$y = 60 + 40e^{-0.02877t} \qquad \text{Cooling model}$$

and finally, when $y = 80$, you obtain

$$80 = 60 + 40e^{-0.02877t}$$

$$20 = 40e^{-0.02877t}$$

$$\tfrac{1}{2} = e^{-0.02877t}$$

$$\ln \tfrac{1}{2} = -0.02877t$$

$$t \approx 24.09 \text{ minutes.}$$

Therefore, it will require about 14.09 *more* minutes for the object to cool to a temperature of 80° (see Figure 5.30).

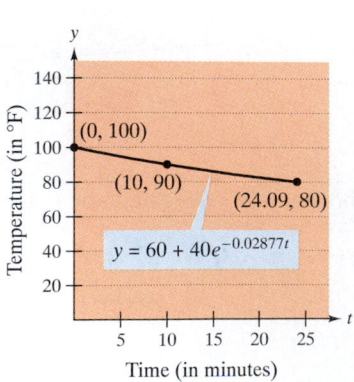

Figure 5.30

EXERCISES FOR SECTION 5.6

In Exercises 1–6, solve the differential equation.

1. $y' = \dfrac{5x}{y}$ **2.** $y' = \dfrac{\sqrt{x}}{2y}$

3. $y' = \sqrt{x}\,y$ **4.** $y' = x(1 + y)$

5. $(1 + x^2)y' - 2xy = 0$ **6.** $xy + y' = 100x$

Direction Fields   In Exercises 7 and 8, a differential equation, a point, and a direction field are given. (a) Sketch two approximate solutions of the differential equation on the direction field, one of which passes through the indicated point. (b) Use integration to find the particular solution of the differential equation and use a graphing utility to graph the solution. Compare the result with the sketch in part (a).

7. $\dfrac{dy}{dx} = x(6 - y)$, $(0, 0)$ **8.** $\dfrac{dy}{dx} = xy$, $\left(0, \tfrac{1}{2}\right)$

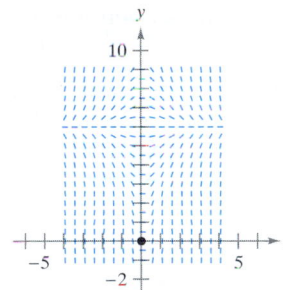 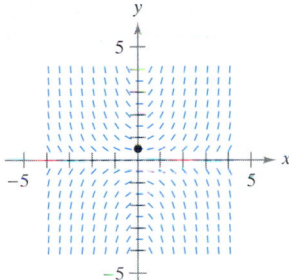

Think About It   In Exercises 9–12, write and solve the differential equation that models the verbal statement.

9. The rate of change of Q with respect to t is inversely proportional to the square of t.

10. The rate of change of P with respect to t is proportional to $10 - t$.

11. The rate of change of N with respect to s in proportional to $250 - s$.

12. The rate of change of y with respect to x varies jointly as x and $L - y$.

In Exercises 13–16, find the function $y = f(t)$ passing through the point $(0, 10)$ with the given first derivative. Use a graphing utility to graph the solution.

13. $\dfrac{dy}{dt} = \dfrac{1}{2}t$ **14.** $\dfrac{dy}{dt} = -\dfrac{3}{4}\sqrt{t}$

15. $\dfrac{dy}{dt} = -\dfrac{1}{2}y$ **16.** $\dfrac{dy}{dt} = \dfrac{3}{4}y$

In Exercises 17–20, find the exponential function $y = Ce^{kt}$ that passes through the two given points.

17. 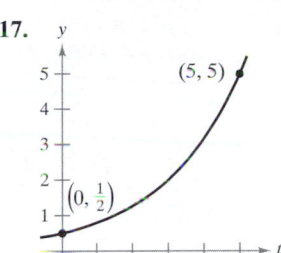   **18.**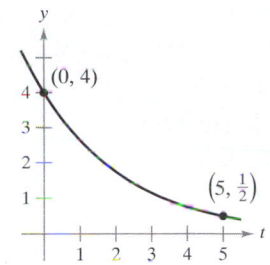

19. 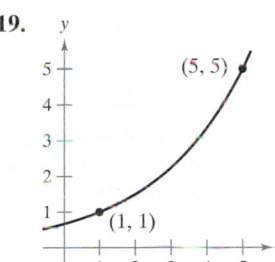   **20.**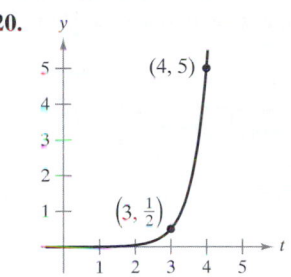

Radioactive Decay   In Exercises 21–26, complete the table for the radioactive isotope.

Isotope	Half-Life (in years)	Initial Quantity	Amount After 1000 years	Amount After 10,000 years
21. Ra^{226}	1620	10g		
22. Ra^{226}	1620		1.5g	
23. C^{14}	5730			2g
24. C^{14}	5730	3g		
25. Pu^{239}	24,360		2.1g	
26. Pu^{239}	24,360			0.4g

27. ***Radioactive Decay***   Radioactive radium has a half-life of approximately 1620 years. What percent of a given amount remains after 100 years?

28. ***Carbon Dating***   Carbon-14 dating assumes that the carbon dioxide on earth today has the same radioactive content as it did centuries ago. If this is true, the amount of C^{14} absorbed by a tree that grew several centuries ago should be the same as the amount of C^{14} absorbed by a tree growing today. A piece of ancient charcoal contains only 15% as much of the radioactive carbon as a piece of modern charcoal. How long ago was the tree burned to make the ancient charcoal? (The half-life of C^{14} is 5730 years.)

Compound Interest **In Exercises 29–34, complete the table for a savings account in which interest is compounded continuously.**

Initial Investment	Annual Rate	Time to Double	Amount After 10 Years
29. $1000	6%		
30. $20,000	$5\frac{1}{2}$%		
31. $750		$7\frac{3}{4}$ yr	
32. $10,000		5 yr	
33. $500			$1292.85
34. $2000			$5436.56

Compound Interest **In Exercises 35 and 36, find the principal P that must be invested at rate r, compounded monthly, so that $500,000 will be available for retirement in t years.**

35. $r = 7\frac{1}{2}\%, \quad t = 20$ **36.** $r = 6\%, \quad t = 40$

Compound Interest **In Exercises 37 and 38, find the time necessary for $1000 to double if it is invested at a rate of r compounded (a) annually, (b) monthly, (c) daily, and (d) continuously.**

37. $r = 7\%$ **38.** $r = 6\frac{1}{2}\%$

Population **In Exercises 39–42, the table gives the population (in millions) of a city in 1990 and the projected population (in millions) for the year 2000. Find the exponential growth model $y = Ce^{kt}$ for the population by letting $t = 0$ correspond to 1990. Use the model to predict the population of the city in 2010.**

City	1900	2000
39. Dhaka, Bangladesh	4.22	6.49
40. Houston, Texas	2.30	2.65
41. Detroit, Michigan	3.00	2.74
42. London, United Kingdom	9.17	8.57

43. *Think About It*

(a) In Exercises 39–42, what constant in the equation $y = Ce^{kt}$ represents the growth rate? Write a sentence that describes the growth rate in common English.

(b) In Exercises 39 and 41, you can see that one population is increasing whereas the other is decreasing. What constant in the equation $y = Ce^{kt}$ reflects this difference? Explain.

44. *Modeling Data* One hundred bacteria are started in a culture and the number N of bacteria is counted each hour for 5 hours. The results are given in the table, where t is the time in hours.

t	0	1	2	3	4	5
N	100	126	151	198	243	297

(a) Use the regression capabilities of a graphing utility to find an exponential model for the data.

(b) Use the model to estimate the time required for the population to quadruple in size.

45. *Atmospheric Pressure* Atmospheric pressure P (measured in millimeters of mercury) decreases exponentially with increasing altitude x (measured in meters). The pressure is 760 millimeters of mercury at sea level ($x = 0$) and 672.71 millimeters of mercury at an altitude of 1000 meters. Find the pressure at an altitude of 3000 meters.

46. *Revenue* Because of a slump in the economy, a company finds that its annual revenues have dropped from $742,000 in 1996 to $632,000 in 1998. If the revenue is following an exponential pattern of decline, what is the expected revenue for 1999? (Let $t = 0$ represent 1996.)

47. *Learning Curve* The management at a certain factory has found that a worker can produce at most 30 units in a day. The learning curve for the number of units N produced per day after a new employee has worked t days is

$$N = 30(1 - e^{kt}).$$

After 20 days on the job, a particular worker produces 19 units.

(a) Find the learning curve for this worker.

(b) How many days should pass before this worker is producing 25 units per day?

48. *Learning Curve* If in Exercise 47 management requires a new employee to produce at least 20 units per day after 30 days on the job, find (a) the learning curve that describes this minimum requirement and (b) the number of days before a minimal achiever is producing 25 units per day.

49. *Sales* The sales S (in thousands of units) of a new product after it has been on the market for t years is

$$S = Ce^{k/t}.$$

(a) Find S as a function of t if 5000 units have been sold after 1 year and the saturation point for the market is 30,000 (that is, $\lim_{t \to \infty} S = 30$).

(b) How many units will have been sold after 5 years?

(c) Use a graphing utility to graph this sales function.

50. *Sales* The sales S (in thousands of units) of a new product after it has been on the market for t years is

$$S = 30(1 - e^{kt}).$$

(a) Find S as a function of t if 5000 units have been sold after 1 year.

(b) How many units will saturate this market?

(c) How many units will have been sold after 5 years?

(d) Use a graphing utility to graph this sales function.

51. *Forestry* The value of a tract of timber is

$$V(t) = 100,000e^{0.8\sqrt{t}}$$

where t is the time in years, with $t = 0$ corresponding to 1998. If money earns interest continuously at 10%, the present value of the timber at any time t is

$$A(t) = V(t)e^{-0.10t}.$$

Find the year in which the timber should be harvested to maximize the present value function.

52. Modeling Data   The table gives the net receipts and the amounts required to service the national debt of the United States from 1986 through 1995. The monetary amounts are given in billions of dollars.   *(Source: U.S. Office of Management and Budget)*

Year	1986	1987	1988	1989	1990
Receipts	769.1	854.1	909.0	990.7	1031.3
Interest	136.0	138.7	151.8	169.3	184.2

Year	1991	1992	1993	1994	1995
Receipts	1054.3	1090.5	1153.5	1257.7	1346.4
Interest	194.5	199.4	198.8	203.0	234.2

(a) Use the regression capabilities of a graphing utility to find an exponential model R for the receipts and a logarithmic model I for the amount required to service the debt. Let t represent the time in years, with $t = 0$ corresponding to 1980.

(b) Use a graphing utility to plot the receipt data and graph the exponential model. Based on the model, what is the continuous rate of growth of the receipts?

(c) Use a graphing utility to plot the interest data and graph the logarithmic model.

(d) Find a function $P(t)$ that approximates the percent of the receipts that is required to service the national debt. Use a graphing utility to graph this function.

53. Sound Intensity   The level of sound β, in decibels, with an intensity of I is

$$\beta(I) = 10 \log_{10} \frac{I}{I_0}$$

where I_0 is an intensity of 10^{-16} watts per square centimeter, corresponding roughly to the faintest sound that can be heard. Determine $\beta(I)$ for the following.

(a) $I = 10^{-14}$ watts per square centimeter (whisper)

(b) $I = 10^{-9}$ watts per square centimeter (busy street corner)

(c) $I = 10^{-6.5}$ watts per square centimeter (air hammer)

(d) $I = 10^{-4}$ watts per square centimeter (threshold of pain)

54. Noise Level   With the installation of noise suppression materials, the noise level in an auditorium was reduced from 93 to 80 decibels. Use the function in Exercise 53 to find the percent decrease in the intensity level of the noise as a result of the installation of these materials.

55. Earthquake Intensity   On the Richter scale, the magnitude R of an earthquake of intensity I is

$$R = \frac{\ln I - \ln I_0}{\ln 10}$$

where I_0 is the minimum intensity used for comparison. Assume that $I_0 = 1$.

(a) Find the intensity of the 1906 San Francisco earthquake $(R = 8.3)$.

(b) Find the factor by which the intensity is increased if the Richter scale measurement is doubled.

(c) Find dR/dI.

56. Newton's Law of Cooling   When an object is removed from a furnace and placed in an environment with a constant temperature of 90°F, its core temperature is 1500°F. One hour after it is removed, the core temperature is 1120°F. Find the core temperature 5 hours after the object is removed from the furnace.

57. Newton's Law of Cooling   A thermometer is taken from a room at 72°F to the outdoors where the temperature is 20°F. Determine the reading on the thermometer after 5 minutes, if the reading drops to 48°F after 1 minute.

58. Comparing Models   The time t (in seconds) required to attain a speed of s miles per hour from a standing start for a 1995 Dodge Avenger is given in the table.   *(Source: Road & Track, March 1995)*

s	30	40	50	60	70	80	90
t	3.4	5.0	7.0	9.3	12.0	15.8	20.0

(a) Use a graphing utility to find an exponential model for the data.

(b) Use a graphing utility to plot the data and graph the model.

(c) Find t and dt/ds for $s = 55$.

59. Home Mortgage   A $120,000 home mortgage for 35 years at $9\frac{1}{2}\%$ has a monthly payment of $985.93. Part of the monthly payment goes for the interest charge on the unpaid balance and the remainder of the payment is used to reduce the principal. The amount that goes for interest is

$$u = M - \left(M - \frac{Pr}{12}\right)\left(1 + \frac{r}{12}\right)^{12t}$$

and the amount that goes toward reduction of the principal is

$$v = \left(M - \frac{Pr}{12}\right)\left(1 + \frac{r}{12}\right)^{12t}.$$

In these formulas, P is the size of the mortgage, r is the interest rate, M is the monthly payment, and t is the time in years.

(a) Use a graphing utility to graph each function in the same viewing rectangle. (The viewing rectangle should show all 35 years of mortgage payments.)

(b) In the early years of the mortgage, the larger part of the monthly payment goes for what purpose? Approximate the time when the monthly payment is evenly divided between interest and principal reduction.

(c) Use the graphs to make a conjecture about the relationship between the slopes of the tangent lines to the two curves for a specified value of t. Give an analytical argument to verify your conjecture. Find $u'(15)$ and $v'(15)$.

(d) Repeat parts (a) and (b) for a repayment period of 20 years $(M = \$1118.56)$. What can you conclude?

SECTION **5.7** **Differential Equations: Separation of Variables**

General and Particular Solutions • Separation of Variables •
Homogeneous Differential Equations • Applications

General and Particular Solutions

Several times in the text, we have identified physical phenomena that can be described by differential equations. For example, in Section 5.6, you saw that problems involving radioactive decay, population growth, and Newton's Law of Cooling can be formulated in terms of differential equations.

A function $y = f(x)$ is called a **solution** of a differential equation if the equation is satisfied when y and its derivatives are replaced by $f(x)$ and its derivatives. For example, differentiation and substitution would show that $y = e^{-2x}$ is a solution of the differential equation $y' + 2y = 0$. It can be shown that every solution of this differential equation is of the form

$$y = Ce^{-2x} \qquad \text{General solution of } y' + 2y = 0$$

where C is any real number. This solution is called the **general solution.** Some differential equations have **singular solutions** that cannot be written as special cases of the general solution. However, we will not consider such solutions in this text. The **order** of a differential equation is determined by the highest-order derivative in the equation. For instance, $y' = 4y$ is a first-order differential equation.

In Section 4.1, Example 8, you saw that the second-order differential equation $s''(t) = -32$ has the general solution

$$s(t) = -16t^2 + C_1 t + C_2 \qquad \text{General solution of } s''(t) = -32$$

which contains two arbitrary constants. It can be shown that a differential equation of order n has a general solution with n arbitrary constants.

EXAMPLE 1 Verifying Solutions

Determine whether or not each of the given functions is a solution of the differential equation $y'' - y = 0$.

a. $y = \sin x$ **b.** $y = 4e^{-x}$ **c.** $y = Ce^x$

Solution

a. Because $y = \sin x$, $y' = \cos x$, and $y'' = -\sin x$, it follows that

$$y'' - y = -\sin x - \sin x = -2 \sin x \neq 0.$$

Hence, $y = \sin x$ is *not* a solution.

b. Because $y = 4e^{-x}$, $y' = -4e^{-x}$, and $y'' = 4e^{-x}$, it follows that

$$y'' - y = 4e^{-x} - 4e^{-x} = 0.$$

Hence, $y = 4e^{-x}$ *is* a solution.

c. Because $y = Ce^x$, $y' = Ce^x$, and $y'' = Ce^x$, it follows that

$$y'' - y = Ce^x - Ce^x = 0.$$

Hence, $y = Ce^x$ *is* a solution for any value of C.

Geometrically, the general solution of a first-order differential equation represents a family of curves known as **solution curves,** one for each value assigned to the arbitrary constant. For instance, you can verify that every function of the form

$$y = \frac{C}{x}$$ General solution of $xy' + y = 0$

is a solution of the differential equation $xy' + y = 0$. Figure 5.31 shows some of the solution curves corresponding to different values of C.

Particular solutions of a differential equation are obtained from **initial conditions** that give the value of the dependent variable or one of its derivatives for a particular value of the independent variable. The term "initial condition" stems from the fact that, often in problems involving time, the value of the dependent variable or one of its derivatives is known at the *initial* time $t = 0$. For instance, the second-order differential equation $s''(t) = -32$ having the general solution

$$s(t) = -16t^2 + C_1 t + C_2$$ General solution $s''(t) = -32$

might have the following initial conditions.

$$s(0) = 80, \qquad s'(0) = 64$$ Initial conditions

In this case, the initial conditions yield the particular solution

$$s(t) = -16t^2 + 64t + 80.$$ Particular solution

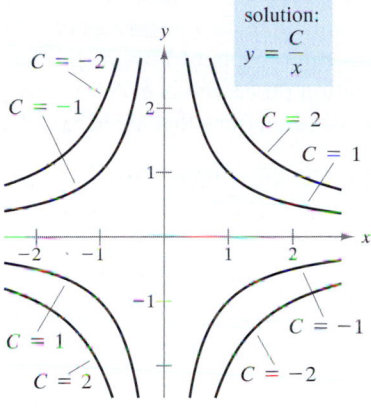

General
solution:
$$y = \frac{C}{x}$$

Solution curves for $xy' + y = 0$
Figure 5.31

 EXAMPLE 2   Finding a Particular Solution

For the differential equation

$$xy' - 3y = 0$$

verify that $y = Cx^3$ is a solution, and find the particular solution determined by the initial condition $y = 2$ when $x = -3$.

Solution   You know that $y = Cx^3$ is a solution because $y' = 3Cx^2$ and

$$xy' - 3y = x(3Cx^2) - 3(Cx^3)$$
$$= 0.$$

Furthermore, the initial condition $y = 2$ when $x = -3$ yields

$$y = Cx^3$$ General solution
$$2 = C(-3)^3$$ Substitute initial condition.
$$-\frac{2}{27} = C$$ Solve for C.

and you can conclude that the particular solution is

$$y = -\frac{2x^3}{27}.$$ Particular solution

Try checking this solution by substituting for y and y' in the original differential equation.

NOTE   To determine a particular solution, the number of initial conditions must match the number of constants in the general solution.

Separation of Variables

Consider a differential equation that can be written in the form

$$M(x) + N(y)\frac{dy}{dx} = 0$$

FOR FURTHER INFORMATION For an example from engineering of a differential equation that is separable, see the article "Designing a Rose Cutter" by J. S. Hartzler in the January 1995 issue of *The College Mathematics Journal*.

where M is a continuous function of x alone and N is a continuous function of y alone. As you saw in the preceding section, for this type of equation, all x terms can be collected with dx and all y terms with dy, and a solution can be obtained by integration. Such equations are said to be **separable,** and the solution procedure is called *separation of variables*. Here are some examples of differential equations that are separable.

Original Differential Equation	*Rewritten with Variables Separated*
$x^2 + 3y\dfrac{dy}{dx} = 0$	$3y\,dy = -x^2\,dx$
$\sin x\,y' = \cos x$	$dy = \cot x\,dx$
$\dfrac{xy'}{e^y + 1} = 2$	$\dfrac{1}{e^y + 1}\,dy = \dfrac{2}{x}\,dx$

EXAMPLE 3 Separation of Variables

Find the general solution of

$$(x^2 + 4)\frac{dy}{dx} = xy.$$

Solution To begin, note that $y = 0$ is a solution. To find other solutions, assume that $y \neq 0$ and separate variables as follows.

$$(x^2 + 4)dy = xy\,dx \qquad \text{Differential form}$$

$$\frac{dy}{y} = \frac{x}{x^2 + 4}\,dx \qquad \text{Separate variables.}$$

Now, integrate to obtain

$$\int \frac{dy}{y} = \int \frac{x}{x^2 + 4}\,dx \qquad \text{Integrate.}$$

$$\ln|y| = \frac{1}{2}\ln(x^2 + 4) + C_1$$

$$= \ln\sqrt{x^2 + 4} + C_1$$

$$|y| = e^{C_1}\sqrt{x^2 + 4}$$

$$y = \pm e^{C_1}\sqrt{x^2 + 4}.$$

Because $y = 0$ is also a solution, you can write the general solution as

$$y = C\sqrt{x^2 + 4}. \qquad \text{General solution}$$

NOTE We encourage you to *check your solutions* throughout this chapter. In Example 3 you can check the solution $y = C\sqrt{x^2 + 4}$ by differentiating and substituting into the original equation.

In some cases it is not feasible to write the general solution in the explicit form $y = f(x)$. The next example illustrates such a solution. Implicit differentiation can be used to verify this solution.

EXAMPLE 4   Finding a Particular Solution

Given the initial condition $y(0) = 1$, find the particular solution of the equation

$$xy\,dx + e^{-x^2}(y^2 - 1)\,dy = 0.$$

Solution   Note that $y = 0$ is a solution of the differential equation—but this solution does not satisfy the initial condition. Hence, you can assume that $y \neq 0$. To separate variables, you must rid the first term of y and the second term of e^{-x^2}. Thus, you should multiply by e^{x^2}/y and obtain the following.

$$xy\,dx + e^{-x^2}(y^2 - 1)\,dy = 0$$

$$e^{-x^2}(y^2 - 1)\,dy = -xy\,dx$$

$$\int\left(y - \frac{1}{y}\right)dy = \int -xe^{x^2}\,dx$$

$$\frac{y^2}{2} - \ln y = -\frac{1}{2}e^{x^2} + C$$

From the initial condition $y(0) = 1$, you have $\frac{1}{2} - 0 = -\frac{1}{2} + C$, which implies that $C = 1$. Thus, the particular solution has the implicit form

$$\frac{y^2}{2} - \ln y = -\frac{1}{2}e^{x^2} + 1$$

$$y^2 - \ln y^2 + e^{x^2} = 2.$$

EXAMPLE 5   Finding a Particular Solution Curve

Find the equation of the curve that passes through the point $(1, 3)$ and has a slope of y/x^2 at the point (x, y), as shown in Figure 5.32.

Solution   Because the slope of the curve is given by y/x^2, you have

$$\frac{dy}{dx} = \frac{y}{x^2}$$

with the initial condition $y(1) = 3$. Separating variables and integrating produces

$$\int \frac{dy}{y} = \int \frac{dx}{x^2}, \qquad y \neq 0$$

$$\ln|y| = -\frac{1}{x} + C_1$$

$$y = e^{-(1/x)+C_1} = Ce^{-1/x}.$$

Because $y = 3$ when $x = 1$, it follows that $3 = Ce^{-1}$ and $C = 3e$. Therefore, the equation of the specified curve is

$$y = (3e)e^{-1/x} = 3e^{(x-1)/x}, \qquad x > 0.$$

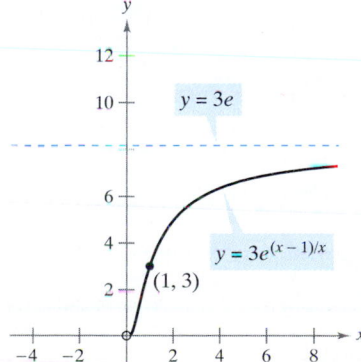

$y = 3e$

$y = 3e^{(x-1)/x}$

$(1, 3)$

Figure 5.32

Homogeneous Differential Equations

NOTE The notation $f(x, y)$ is used to denote a function of two variables in much the same way as $f(x)$ denotes a function of one variable. You will study functions of two variables in detail in Chapter 12.

Some differential equations that are not separable in x and y can be made separable by a change of variables. This is true for differential equations of the form $y' = f(x, y)$, where f is a **homogeneous function.** The function given by $f(x, y)$ is **homogeneous of degree n** if

$$f(tx, ty) = t^n f(x, y)$$ Homogeneous function of degree n

where n is a real number.

EXAMPLE 6 Verifying Homogeneous Functions

a. $f(x, y) = x^2 y - 4x^3 + 3xy^2$ is a homogeneous function of degree 3 because

$$\begin{aligned} f(tx, ty) &= (tx)^2(ty) - 4(tx)^3 + 3(tx)(ty)^2 \\ &= t^3(x^2 y) - t^3(4x^3) + t^3(3xy^2) \\ &= t^3(x^2 y - 4x^3 + 3xy^2) \\ &= t^3 f(x, y). \end{aligned}$$

b. $f(x, y) = xe^{x/y} + y\sin(y/x)$ is a homogeneous function of degree 1 because

$$\begin{aligned} f(tx, ty) &= txe^{tx/ty} + ty\sin\frac{ty}{tx} \\ &= t\left(xe^{x/y} + y\sin\frac{y}{x}\right) \\ &= tf(x, y). \end{aligned}$$

c. $f(x, y) = x + y^2$ is not a homogeneous function because

$$f(tx, ty) = tx + t^2 y^2 = t(x + ty^2) \neq t^n(x + y^2).$$

d. $f(x, y) = x/y$ is a homogeneous function of degree 0 because

$$f(tx, ty) = \frac{tx}{ty} = t^0\frac{x}{y}.$$

Definition of Homogeneous Differential Equation

A **homogeneous differential equation** is an equation of the form

$$M(x, y)\,dx + N(x, y)dy = 0$$

where M and N are homogeneous functions of the same degree.

EXAMPLE 7 Homogeneous Differential Equations

a. $(x^2 + xy)\,dx + y^2\,dy = 0$ is homogeneous of degree 2.

b. $(x^2 + 1)\,dx + y^2\,dy = 0$ is *not* a homogeneous differential equation.

To solve a homogeneous differential equation by the method of separation of variables, we use the following change of variables theorem.

THEOREM 5.17   Change of Variables for Homogeneous Equations

If $M(x, y)\, dx + N(x, y)\, dy = 0$ is homogeneous, then it can be transformed into a differential equation whose variables are separable by the substitution

$$y = vx$$

where v is a differentiable function of x.

EXAMPLE 8   Solving a Homogeneous Differential Equation

Find the general solution of $(x^2 - y^2)\, dx + 3xy\, dy = 0$.

STUDY TIP   The substitution $y = vx$ will yield a differential equation that is separable with respect to the variables x and v. You must express your final solution, however, in terms of x and y.

Solution   Because $(x^2 - y^2)$ and $3xy$ are both homogeneous of degree 2, let $y = vx$ to obtain $dy = x\, dv + v\, dx$. Then, by substitution, you have

$$(x^2 - v^2x^2)\, dx + 3x(vx)\overbrace{(x\, dv + v\, dx)}^{dy} = 0$$
$$(x^2 + 2v^2x^2)\, dx + 3x^3v\, dv = 0$$
$$x^2(1 + 2v^2)\, dx + x^2(3vx)\, dv = 0.$$

Dividing by x^2 and separating variables produces

$$(1 + 2v^2)\, dx = -3vx\, dv$$
$$\int \frac{dx}{x} = \int \frac{-3v}{1 + 2v^2}\, dv$$
$$\ln |x| = -\frac{3}{4} \ln(1 + 2v^2) + C_1$$
$$4 \ln |x| = -3 \ln(1 + 2v^2) + \ln |C|$$
$$\ln x^4 = \ln |C(1 + 2v^2)^{-3}|$$
$$x^4 = C(1 + 2v^2)^{-3}.$$

Substituting for v produces the following general solution.

$$x^4 = C\left[1 + 2\left(\frac{y}{x}\right)^2\right]^{-3}$$
$$\left(1 + 2\frac{y^2}{x^2}\right)^3 x^4 = C$$
$$(x^2 + 2y^2)^3 = Cx^2 \qquad \text{General solution}$$

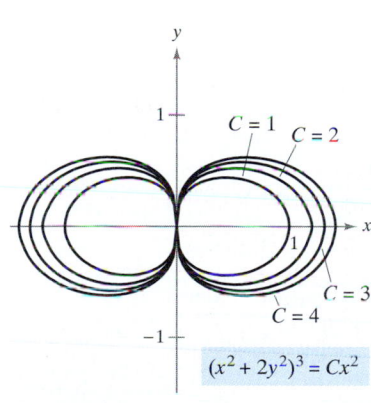

$(x^2 + 2y^2)^3 = Cx^2$

General solutions of
$(x^2 - y^2)\, dx + 3xy\, dy = 0$
Figure 5.33

TECHNOLOGY   If you have access to a graphing utility, try using it to graph several of the solutions in Example 8. For instance, Figure 5.33 shows the graphs of $(x^2 + 2y^2)^3 = Cx^2$ for $C = 1, 2, 3,$ and 4.

Applications

EXAMPLE 9 Wildlife Population

The rate of change of the number of coyotes $N(t)$ in a population is directly proportional to $650 - N(t)$, where t is the time in years. When $t = 0$, the population is 300, and when $t = 2$, the population has increased to 500. Find the population when $t = 3$.

Solution Because the rate of change of the population is proportional to $650 - N(t)$, you can write the following differential equation.

$$\frac{dN}{dt} = k(650 - N)$$

You can solve this equation using separation of variables.

$$dN = k(650 - N)\, dt \qquad \text{Differential form}$$

$$\frac{dN}{650 - N} = k\, dt \qquad \text{Separate variables.}$$

$$-\ln|650 - N| = kt + C_1 \qquad \text{Integrate.}$$

$$\ln|650 - N| = -kt - C_1$$

$$650 - N = e^{-kt - C_1} \qquad \text{Assume } N < 650.$$

$$N = 650 - Ce^{-kt} \qquad \text{General solution}$$

Using $N = 300$ when $t = 0$, you can conclude that $C = 350$, which produces

$$N = 650 - 350e^{-kt}.$$

Then, using $N = 500$ when $t = 2$, it follows that

$$500 = 650 - 350e^{-2k} \quad \Longrightarrow \quad e^{-2k} = \tfrac{3}{7} \quad \Longrightarrow \quad k \approx 0.4236.$$

Thus, the model for the coyote population is

$$N = 650 - 350e^{-0.4236t}. \qquad \text{Model for population}$$

When $t = 3$, you can approximate the population to be

$$N = 650 - 350e^{-0.4236(3)} \approx 552 \text{ coyotes.}$$

The model for the population is shown in Figure 5.34.

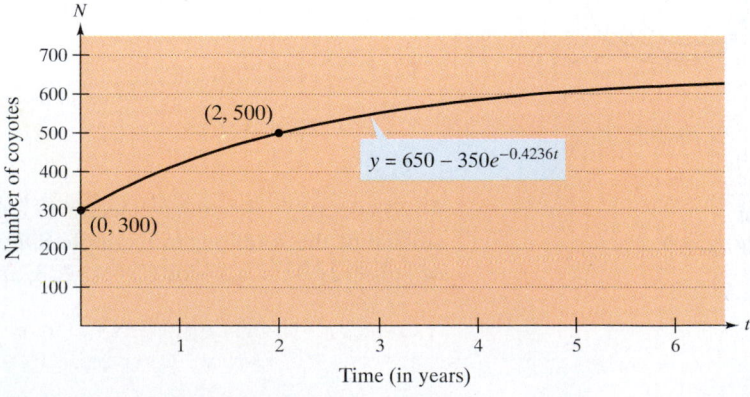

Model for coyote population
Figure 5.34

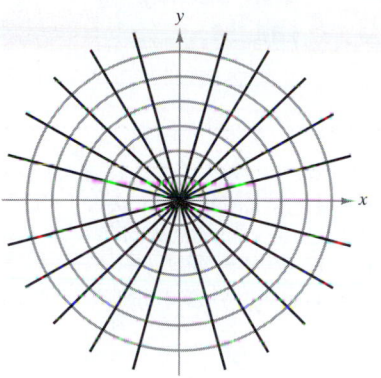

Each line $y = Kx$ is an orthogonal trajectory to the family of circles.

Figure 5.35

A common problem in electrostatics, thermodynamics, and hydrodynamics involves finding a family of curves, each of which is orthogonal to all members of a given family of curves. For example, Figure 5.35 shows a family of circles

$$x^2 + y^2 = C \qquad \text{Family of circles}$$

each of which intersects the lines in the family

$$y = Kx \qquad \text{Family of lines}$$

at right angles. Two such families of curves are said to be **mutually orthogonal,** and each curve in one of the families is called an **orthogonal trajectory** of the other family. In electrostatics, lines of force are orthogonal to the *equipotential curves.* In thermodynamics, the flow of heat across a plane surface is orthogonal to the *isothermal curves.* In hydrodynamics, the flow (stream) lines are orthogonal trajectories of the *velocity potential curves.*

EXAMPLE 10   Finding Orthogonal Trajectories

Describe the orthogonal trajectories for the family of curves given by

$$y = \frac{C}{x}$$

for $C \neq 0$. Sketch several members of each family.

Solution   First, solve the given equation for C and write $xy = C$. Then, by differentiating implicitly with respect to x, you obtain the differential equation

$$xy' + y = 0 \qquad \text{Differential equation}$$

$$x\frac{dy}{dx} = -y$$

$$\frac{dy}{dx} = -\frac{y}{x}. \qquad \text{Slope of given family}$$

Because y' represents the slope of the given family of curves at (x, y), it follows that the orthogonal family has the negative reciprocal slope x/y, and we write

$$\frac{dy}{dx} = \frac{x}{y}. \qquad \text{Slope of orthogonal family}$$

Now you can find the orthogonal family by separating variables and integrating.

$$\int y \, dy = \int x \, dx$$

$$\frac{y^2}{2} = \frac{x^2}{2} + C_1$$

Therefore, each orthogonal trajectory is a hyperbola given by

$$\frac{y^2}{K} - \frac{x^2}{K} = 1. \qquad 2C_1 = K \neq 0$$

The centers are at the origin, and the transverse axes are vertical for $K > 0$ and horizontal for $K < 0$. Several trajectories are shown in Figure 5.36.

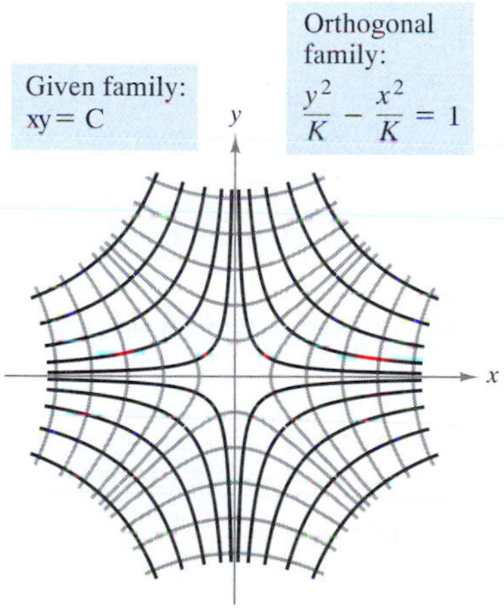

Given family:
$xy = C$

Orthogonal family:
$\dfrac{y^2}{K} - \dfrac{x^2}{K} = 1$

Orthogonal trajectories

Figure 5.36

EXERCISES FOR SECTION 5.7

In Exercises 1–6, verify the solution of the differential equation.

Solution	*Differential Equation*
1. $y = Ce^{4x}$	$\dfrac{dy}{dx} = 4y$
2. $x^2 + y^2 = Cy$	$y' = \dfrac{2xy}{x^2 - y^2}$
3. $y = C_1 \cos x + C_2 \sin x$	$y'' + y = 0$
4. $y = C_1 e^{-x} \cos x + C_2 e^{-x} \sin x$	$y'' + 2y' + 2y = 0$
5. $y = -\cos x \ln |\sec x + \tan x|$	$y'' + y = \tan x$
6. $y = \frac{2}{3}(e^{-2x} + e^x)$	$y'' + 2y' = 2e^x$

In Exercises 7–12, determine whether the function is a solution of the differential equation $y^{(4)} - 16y = 0$.

7. $y = 3 \cos x$

8. $y = 3 \cos 2x$

9. $y = e^{-2x}$

10. $y = 5 \ln x$

11. $y = C_1 e^{2x} + C_2 e^{-2x} + C_3 \sin 2x + C_4 \cos 2x$

12. $y = 5e^{-2x} + 3 \cos 2x$

In Exercises 13–18, determine whether the function is a solution of the differential equation $xy' - 2y = x^3 e^x$.

13. $y = x^2$

14. $y = x^2 e^x$

15. $y = x^2(2 + e^x)$

16. $y = \sin x$

17. $y = \ln x$

18. $y = x^2 e^x - 5x^2$

19. *Think About It* It is known that $y = Ce^{kx}$ is a solution of the differential equation

$$\frac{dy}{dx} = 0.07y.$$

Is it possible to determine C or k from the information given? If so, find its value.

20. *Think About It* It is known that $y = A \sin \omega t$ is a solution of the differential equation

$$\frac{d^2y}{dt^2} + 16y = 0.$$

Find the value of ω.

In Exercises 21 and 22, some of the curves corresponding to different values of C in the general solution of the differential equation are given. Find the particular solution that passes through the point indicated on the graph.

Solution	*Differential Equation*
21. $y^2 = Cx^3$	$2xy' - 3y = 0$
22. $2x^2 - y^2 = C$	$yy' - 2x = 0$

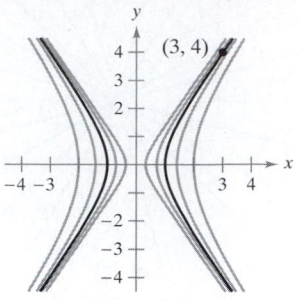

Figure for 21   **Figure for 22**

In Exercises 23 and 24, the general solution of the differential equation is given. Use a graphing utility to graph the particular solutions for the given values of C.

23. $4yy' - x = 0$
$\quad 4y^2 - x^2 = C$
$\quad C = 0, C = \pm 1, C = \pm 4$

24. $yy' + x = 0$
$\quad x^2 + y^2 = C$
$\quad C = 0, C = 1, C = 4$

In Exercises 25–30, verify that the general solutions satisfy the differential equation. Then find the particular solution that satisfies the initial condition.

25. $y = Ce^{-2x}$
$\quad y' + 2y = 0$
$\quad y = 3$ when $x = 0$

26. $2x^2 + 3y^2 = C$
$\quad 2x + 3yy' = 0$
$\quad y = 2$ when $x = 1$

27. $y = C_1 \sin 3x + C_2 \cos 3x$
$\quad y'' + 9y = 0$
$\quad y = 2$ when $x = \pi/6$
$\quad y' = 1$ when $x = \pi/6$

28. $y = C_1 + C_2 \ln x$
$\quad xy'' + y' = 0$
$\quad y = 0$ when $x = 2$
$\quad y' = \frac{1}{2}$ when $x = 2$

29. $y = C_1 x + C_2 x^3$
$\quad x^2 y'' - 3xy' + 3y = 0$
$\quad y = 0$ when $x = 2$
$\quad y' = 4$ when $x = 2$

30. $y = e^{2x/3}(C_1 + C_2 x)$
$\quad 9y'' - 12y' + 4y = 0$
$\quad y = 4$ when $x = 0$
$\quad y = 0$ when $x = 3$

In Exercises 31–38, use integration to find a general solution of the differential equation.

31. $\dfrac{dy}{dx} = 3x^2$

32. $\dfrac{dy}{dx} = \dfrac{x}{1 + x^2}$

33. $\dfrac{dy}{dx} = \dfrac{x - 2}{x}$

34. $\dfrac{dy}{dx} = x \cos x^2$

35. $\dfrac{dy}{dx} = \sin 2x$

36. $\dfrac{dy}{dx} = \tan^2 x$

37. $\dfrac{dy}{dx} = x\sqrt{x - 3}$

38. $\dfrac{dy}{dx} = xe^{x^2}$

In Exercises 39–48, find the general solution of the differential equation.

39. $\dfrac{dy}{dx} = \dfrac{x}{y}$

40. $\dfrac{dy}{dx} = \dfrac{x^2 + 2}{3y^2}$

41. $\dfrac{dr}{ds} = 0.05r$

42. $\dfrac{dr}{ds} = 0.05s$

43. $(2 + x)y' = 3y$

44. $xy' = y$

45. $yy' = \sin x$

46. $\sqrt{1 - 4x^2}\, y' = x$

47. $y \ln x - xy' = 0$

48. $yy' - 2e^x = 0$

In Exercises 49–58, find the particular solution of the differential equation that satisfies the initial condition.

Differential Equation	Initial Condition
49. $yy' - e^x = 0$	$y(0) = 4$
50. $\sqrt{x} + \sqrt{y}\, y' = 0$	$y(1) = 4$
51. $y(x + 1) + y' = 0$	$y(-2) = 1$
52. $xyy' - \ln x = 0$	$y(1) = 0$
53. $y(1 + x^2)y' - x(1 + y^2) = 0$	$y(0) = \sqrt{3}$
54. $y\sqrt{1 - x^2}\, y' - x\sqrt{1 - y^2} = 0$	$y(0) = 1$
55. $\dfrac{du}{dv} = uv \sin v^2$	$u(0) = 1$
56. $\dfrac{dr}{ds} = e^{r+s}$	$r(1) = 0$
57. $dP - kP\, dt = 0$	$P(0) = P_0$
58. $dT + k(T - 70)\, dt = 0$	$T(0) = 140$

In Exercises 59 and 60, find an equation for the curve that passes through the point and has the indicated slope.

Point	Slope
59. $(1, 1)$	$y' = -\dfrac{9x}{16y}$
60. $(8, 2)$	$y' = \dfrac{2y}{3x}$

In Exercises 61 and 62, find all functions f having the indicated property.

61. The tangent to the graph of f at the point (x, y) intersects the x-axis at $(x + 2, 0)$.

62. All tangents to the graph of f pass through the origin.

In Exercises 63–68, determine whether the function is homogeneous, and if it is, determine its degree.

63. $f(x, y) = x^3 - 4xy^2 + y^3$

64. $f(x, y) = \dfrac{xy}{\sqrt{x^2 + y^2}}$

65. $f(x, y) = 2 \ln xy$

66. $f(x, y) = \tan(x + y)$

67. $f(x, y) = 2 \ln \dfrac{x}{y}$

68. $f(x, y) = \tan \dfrac{y}{x}$

In Exercises 69–74, solve the homogeneous differential equation.

69. $y' = \dfrac{x + y}{2x}$

70. $y' = \dfrac{x^3 + y^3}{xy^2}$

71. $y' = \dfrac{x - y}{x + y}$

72. $y' = \dfrac{x^2 + y^2}{2xy}$

73. $y' = \dfrac{xy}{x^2 - y^2}$

74. $y' = \dfrac{3x + 2y}{x}$

In Exercises 75–78, find the particular solution that satisfies the initial condition.

Differential Equation	Initial Condition
75. $x\, dy - (2xe^{-y/x} + y)dx = 0$	$y(1) = 0$
76. $-y^2\, dx + x(x + y)dy = 0$	$y(1) = 1$
77. $\left(x \sec \dfrac{y}{x} + y\right)dx - x\, dy = 0$	$y(1) = 0$
78. $(2x^2 + y^2)dx + xy\, dy = 0$	$y(1) = 0$

Direction Fields   **Consider the first-order differential equation**

$$\frac{dy}{dx} = F(x, y).$$

The function F assigns to each point (x_0, y_0) in its domain a direction (slope) of the solution of the differential equation. To visualize this, draw a short mark at each of several points to indicate the direction associated with each point and obtain a direction field. This field can be used to approximate graphically the solutions of the differential equation. In Exercises 79–82, sketch a few solutions of the differential equation on the direction field and then find the general solution analytically.

79. $\dfrac{dy}{dx} = x$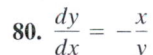

80. $\dfrac{dy}{dx} = -\dfrac{x}{y}$

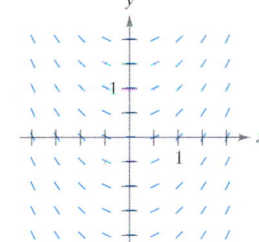

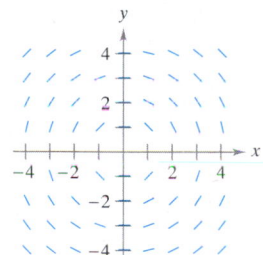

81. $\dfrac{dy}{dx} = 4 - y$

82. $\dfrac{dy}{dx} = 0.25x(4 - y)$

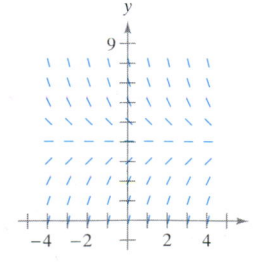

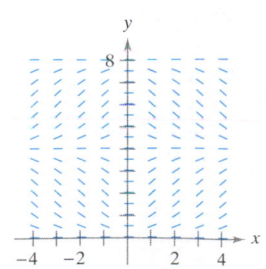

Direction Fields In Exercises 83–86, (a) write a differential equation for the statement, (b) match the differential equation with a possible direction field, and (c) verify your result by using a graphing utility to graph a direction field for the differential equation. [The direction fields are labeled (a), (b), (c), and (d).]

(a)
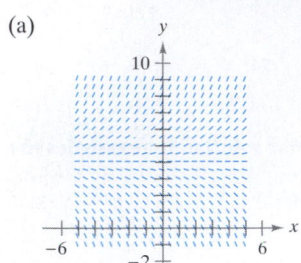

(b)
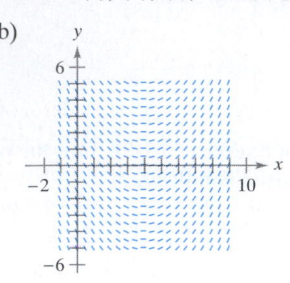

(c)
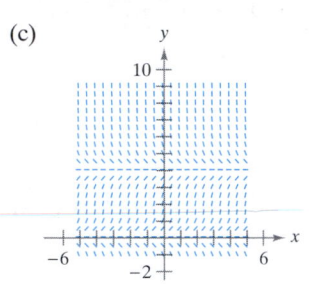

(d)
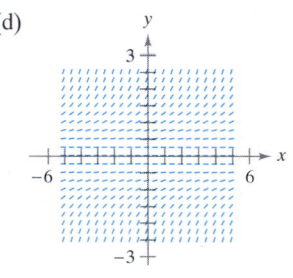

83. The rate of change of y with respect to x is proportional to the difference between y and 4.

84. The rate of change of y with respect to x is proportional to the difference between x and 4.

85. The rate of change of y with respect to x is proportional to the product of y and the difference between y and 4.

86. The rate of change of y with respect to x is proportional to y^2.

87. **Sales Increase** Let S represent sales of a new product (in thousands of units), let L represent the maximum level of sales (in thousands of units), and let t represent time (in months). The rate of change of S with respect to t varies jointly as the product of S and $L - S$ (see figure).

(a) Write the differential equation for the sales model if $L = 100$, $S = 10$ when $t = 0$, and $S = 20$ when $t = 1$. Verify that $S = L/(1 + Ce^{-kt})$.

(b) At what time is the growth in sales increasing most rapidly?

(c) Use a graphing utility to graph the sales function.

(d) Sketch the solution in part (a) on the direction field shown in the figure.

(e) If the estimated maximum level of sales is correct, use the direction field to describe the shape of the solution curves for sales if, at some period of time, sales exceed L.

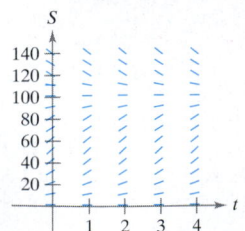

88. **Population Growth** The rate of growth of a population of fruit flies is proportional to the size of the population at any instant. If there were 180 flies after the second day of the experiment and 300 after the fourth day, how many flies were in the original population?

89. **Radioactive Decay** The rate of decomposition of radio-active radium is proportional to the amount present at any time. The half-life of radioactive radium is 1620 years. What percent of a present amount will remain after 25 years?

90. **Chemical Reaction** In a chemical reaction, a certain compound changes into another compound at a rate proportional to the unchanged amount. If initially there is 20 grams of the original compound, and there is 16 grams after 1 hour, when will 75 percent of the compound be changed?

91. **Sailing** Ignoring resistance, a sailboat starting from rest accelerates (dv/dt) at a rate proportional to the difference between the velocities of the wind and the boat.

(a) Write the velocity as a function of time if the wind is blowing at 20 knots, and after 1 minute the boat is moving at 5 knots.

(b) Use the result in part (a) to write the distance traveled by the boat as a function of time.

92. **Radio Reception** In hilly areas, radio reception may be poor. Consider a situation where an FM transmitter is located at the point $(-1, 1)$ behind a hill modeled by the graph of $y = x - x^2$, and a radio receiver is on the opposite side of the hill. (Assume that the x-axis represents ground level at the base of the hill.)

(a) What is the closest the radio can be to the hill so that reception is unobstructed?

(b) Write the closest position x of the radio as a function of h if the transmitter is located at $(-1, h)$.

(c) Use a graphing utility to graph the function from part (b). Determine the vertical asymptote of the function and interpret the result.

In Exercises 93–98, find the orthogonal trajectories of the family. Use a graphing utility to graph several members of each family.

93. $x^2 + y^2 = C$

94. $2x^2 - y^2 = C$

95. $x^2 = Cy$

96. $y^2 = 2Cx$

97. $y^2 = Cx^3$

98. $y = Ce^x$

True or False? In Exercises 99–102, determine whether the statement is true or false. If it is false, explain why or give an example that shows it is false.

99. If $y = f(x)$ is a solution of a first-order differential equation, then $y = f(x) + C$ is also a solution.

100. The differential equation $y' = xy - 2y + x - 2$ can be written in separated variables form.

101. The function $f(x, y) = x^2 + xy + 2$ is homogeneous.

102. The families $x^2 + y^2 = 2Cy$ and $x^2 + y^2 = 2Kx$ are mutually orthogonal.

SECTION 5.8   **Inverse Trigonometric Functions and Differentiation**

Inverse Trigonometric Functions • Derivatives of Inverse Trigonometric Functions • Review of Basic Differentiation Rules

Inverse Trigonometric Functions

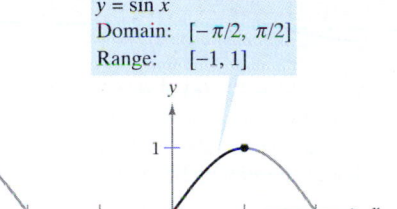

$y = \sin x$
Domain: $[-\pi/2, \pi/2]$
Range: $[-1, 1]$

The sine function is one-to-one on $[-\pi/2, \pi/2]$.
Figure 5.37

This section begins with a rather startling statement: *None of the six basic trigonometric functions has an inverse.* This statement is true because all six trigonometric functions are periodic and hence not one-to-one. In this section you will examine these six functions to see whether their domains can be redefined in such a way that they will have inverses on the *restricted domains*.

In Example 4 of Section 5.3, you saw that the sine function is increasing (and therefore is one-to-one) on the interval $[-\pi/2, \pi/2]$ (see Figure 5.37). On this interval you can define the inverse of the *restricted* sine function to be

$$y = \arcsin x \qquad \text{if and only if} \qquad \sin y = x$$

where $-1 \le x \le 1$ and $-\pi/2 \le \arcsin x \le \pi/2$.

Under suitable restrictions, each of the six trigonometric functions is one-to-one and so has an inverse, as indicated in the following definition. (The term "iff" is used to represent the phrase "if and only if.")

Definition of Inverse Trigonometric Functions

Function	Domain	Range
$y = \arcsin x$ iff $\sin y = x$	$-1 \le x \le 1$	$-\dfrac{\pi}{2} \le y \le \dfrac{\pi}{2}$
$y = \arccos x$ iff $\cos y = x$	$-1 \le x \le 1$	$0 \le y \le \pi$
$y = \arctan x$ iff $\tan y = x$	$-\infty < x < \infty$	$-\dfrac{\pi}{2} < y < \dfrac{\pi}{2}$
$y = \operatorname{arccot} x$ iff $\cot y = x$	$-\infty < x < \infty$	$0 < y < \pi$
$y = \operatorname{arcsec} x$ iff $\sec y = x$	$\lvert x \rvert \ge 1$	$0 \le y \le \pi, \; y \ne \dfrac{\pi}{2}$
$y = \operatorname{arccsc} x$ iff $\csc y = x$	$\lvert x \rvert \ge 1$	$-\dfrac{\pi}{2} \le y \le \dfrac{\pi}{2}, \; y \ne 0$

NOTE   The term "arcsin x" is read as the "arcsine of x" or sometimes the "angle whose sine is x." An alternative notation for the inverse sine function is "$\sin^{-1} x$."

EXPLORATION

The Inverse Secant Function   In the definition above, the inverse secant function is defined by restricting the domain of the secant function to the intervals $\left[0, \dfrac{\pi}{2}\right) \cup \left(\dfrac{\pi}{2}, \pi\right]$. Most other texts and reference books agree with this, but some disagree. What other domains might make sense? Explain your reasoning graphically. Most calculators do not have a key for the inverse secant function. How can you use a calculator to evaluate the inverse secant function?

The graphs of the six inverse trigonometric functions are shown in Figure 5.38.

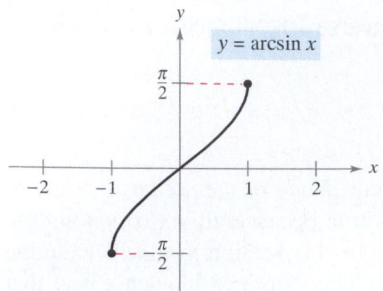

Domain: $[-1, 1]$
Range: $[-\pi/2, \pi/2]$

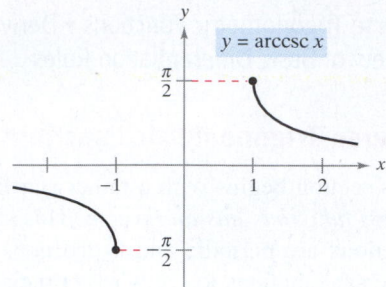

Domain: $(-\infty, -1] \cup [1, \infty)$
Range: $[-\pi/2, 0) \cup (0, \pi/2]$

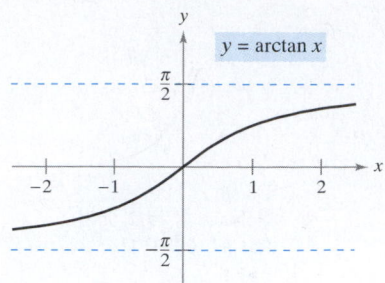

Domain: $(-\infty, \infty)$
Range: $(-\pi/2, \pi/2)$

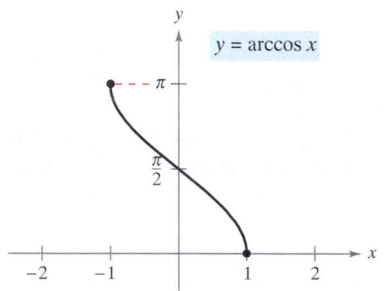

Domain: $[-1, 1]$
Range: $[0, \pi]$

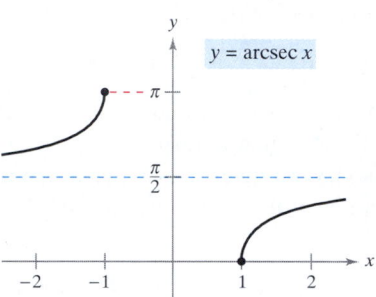

Domain: $(-\infty, -1] \cup [1, \infty)$
Range: $[0, \pi/2) \cup (\pi/2, \pi]$

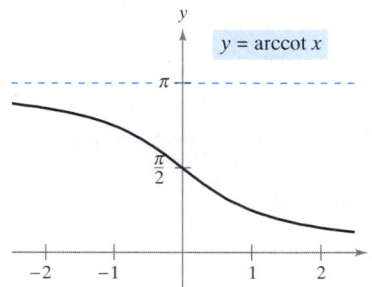

Domain: $(-\infty, \infty)$
Range: $(0, \pi)$

Graphs of the six
inverse trigonometric functions
Figure 5.38

NOTE   When evaluating inverse
trigonometric functions, remember that
they denote *angles in radian measure.*

EXAMPLE 1   Evaluating Inverse Trigonometric Functions

Evaluate each of the following.

a. $\arcsin\left(-\dfrac{1}{2}\right)$ **b.** $\arccos 0$ **c.** $\arctan \sqrt{3}$ **d.** $\arcsin (0.3)$

Solution

a. By definition, $y = \arcsin\left(-\dfrac{1}{2}\right)$ implies that $\sin y = -\dfrac{1}{2}.$ In the interval $[-\pi/2, \pi/2]$, the correct value of y is $-\pi/6$.

$$\arcsin\left(-\frac{1}{2}\right) = -\frac{\pi}{6}$$

b. By definition, $y = \arccos 0$ implies that $\cos y = 0$. In the interval $[0, \pi]$, you have $y = \pi/2$.

$$\arccos 0 = \frac{\pi}{2}$$

c. By definition, $y = \arctan \sqrt{3}$ implies that $\tan y = \sqrt{3}$. In the interval $(-\pi/2, \pi/2)$, you have $y = \pi/3$.

$$\arctan \sqrt{3} = \frac{\pi}{3}$$

d. Using a calculator set in *radian mode* produces

$$\arcsin (0.3) \approx 0.3047.$$

EXPLORATION

Graph $y = \arccos(\cos x)$ for $-4\pi \le x \le 4\pi$. Why isn't the graph the same as the graph of $y = x$?

Inverse functions have the properties

$$f(f^{-1}(x)) = x \quad \text{and} \quad f^{-1}(f(x)) = x.$$

When applying these properties to inverse trigonometric functions, remember that the trigonometric functions have inverses only in restricted domains. For x-values outside these domains, these two properties do not hold. For example, $\arcsin(\sin \pi)$ is equal to 0, not π.

Inverse Properties

If $-1 \le x \le 1$ and $-\pi/2 \le y \le \pi/2$, then

$$\sin(\arcsin x) = x \quad \text{and} \quad \arcsin(\sin y) = y.$$

If $-\pi/2 < y < \pi/2$, then

$$\tan(\arctan x) = x \quad \text{and} \quad \arctan(\tan y) = y.$$

If $|x| \ge 1$ and $0 \le y < \pi/2$ or $\pi/2 < y \le \pi$, then

$$\sec(\text{arcsec } x) = x \quad \text{and} \quad \text{arcsec}(\sec y) = y.$$

Similar properties hold for the other inverse trigonometric functions.

EXAMPLE 2   Using Inverse Properties to Solve an Equation

$$\arctan(2x - 3) = \frac{\pi}{4} \qquad \text{Original equation}$$

$$\tan[\arctan(2x - 3)] = \tan\frac{\pi}{4} \qquad \text{Take tangent of both sides.}$$

$$2x - 3 = 1 \qquad \text{Apply inverse property.}$$

$$x = 2 \qquad \text{Solve for } x.$$

Some problems in calculus require you to evaluate expressions such as $\cos(\arcsin x)$, as illustrated in Example 3.

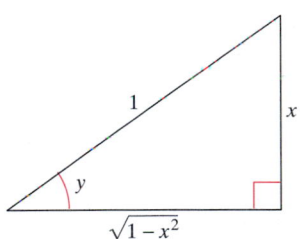

$y = \arcsin x$
Figure 5.39

EXAMPLE 3   Using Right Triangles

a. Given $y = \arcsin x$, where $0 < y < \pi/2$, find $\cos y$.
b. Given $y = \text{arcsec}\left(\sqrt{5}/2\right)$, find $\tan y$.

Solution
a. Because $y = \arcsin x$, you know that $\sin y = x$. This relationship between x and y can be represented by a right triangle, as shown in Figure 5.39.

$$\cos y = \cos(\arcsin x) = \frac{\text{adj.}}{\text{hyp.}} = \sqrt{1 - x^2}$$

(This result is also valid for $-\pi/2 < y < 0$.)

b. Use the right triangle shown in Figure 5.40.

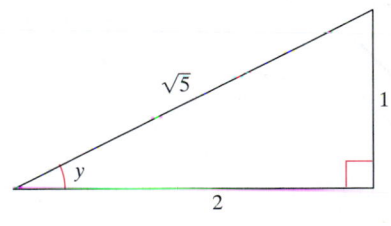

$y = \text{arcsec}\dfrac{\sqrt{5}}{2}$
Figure 5.40

$$\tan y = \tan\left[\text{arcsec}\left(\frac{\sqrt{5}}{2}\right)\right] = \frac{\text{opp.}}{\text{adj.}} = \frac{1}{2}$$

Derivatives of Inverse Trigonometric Functions

NOTE There is no common agreement on the definition of arcsec x (or arccsc x) for negative values of x. When we defined the range of the arcsecant, we chose to preserve the reciprocal identity

$$\operatorname{arcsec} x = \arccos \frac{1}{x}.$$

For example, to evaluate $\operatorname{arcsec}(-2)$, you can write

$$\operatorname{arcsec}(-2) = \arccos(-0.5) \approx 2.09.$$

One of the consequences of the definition of the inverse secant function given in this text is that its graph has a positive slope at every x-value in its domain. (See Figure 5.38.) This accounts for the absolute value sign in the formula for the deriva- tive of arcsec x.

In Section 5.1 you saw that the derivative of the *transcendental* function $f(x) = \ln x$ is the *algebraic* function $f'(x) = 1/x$. You will now see that the derivatives of the inverse trigonometric functions also are algebraic (even though the inverse trigonometric functions are themselves transcendental).

The following theorem lists the derivatives of the six inverse trigonometric functions. Note that the derivatives of arccos u, arccot u, and arccsc u are the *negatives* of the derivatives of arcsin u, arctan u, and arcsec u, respectively.

THEOREM 5.18   Derivatives of Inverse Trigonometric Functions

Let u be a differentiable function of x.

$$\frac{d}{dx}[\arcsin u] = \frac{u'}{\sqrt{1-u^2}} \qquad \frac{d}{dx}[\arccos u] = \frac{-u'}{\sqrt{1-u^2}}$$

$$\frac{d}{dx}[\arctan u] = \frac{u'}{1+u^2} \qquad \frac{d}{dx}[\operatorname{arccot} u] = \frac{-u'}{1+u^2}$$

$$\frac{d}{dx}[\operatorname{arcsec} u] = \frac{u'}{|u|\sqrt{u^2-1}} \qquad \frac{d}{dx}[\operatorname{arccsc} u] = \frac{-u'}{|u|\sqrt{u^2-1}}$$

To derive these formulas, you can use implicit differentiation. For instance, if $y = \arcsin x$, then $\sin y = x$ and $(\cos y) y' = 1$. (See Exercise 64.)

EXAMPLE 4   Differentiating Inverse Trigonometric Functions

a. $\dfrac{d}{dx}[\arcsin(2x)] = \dfrac{2}{\sqrt{1-(2x)^2}} = \dfrac{2}{\sqrt{1-4x^2}}$

b. $\dfrac{d}{dx}[\arctan(3x)] = \dfrac{3}{1+(3x)^2} = \dfrac{3}{1+9x^2}$

c. $\dfrac{d}{dx}[\arcsin \sqrt{x}] = \dfrac{(1/2)x^{-1/2}}{\sqrt{1-x}} = \dfrac{1}{2\sqrt{x}\sqrt{1-x}} = \dfrac{1}{2\sqrt{x-x^2}}$

d. $\dfrac{d}{dx}[\operatorname{arcsec} e^{2x}] = \dfrac{2e^{2x}}{e^{2x}\sqrt{(e^{2x})^2-1}} = \dfrac{2e^{2x}}{e^{2x}\sqrt{e^{4x}-1}} = \dfrac{2}{\sqrt{e^{4x}-1}}$

The absolute value sign is not necessary because $e^{2x} > 0$.

TECHNOLOGY If your graphing utility does not have the arcsecant function, you can obtain its graph using

$$f(x) = \operatorname{arcsec} x = \arccos \frac{1}{x}.$$

EXAMPLE 5   A Derivative That Can Be Simplified

Differentiate $y = \arcsin x + x\sqrt{1-x^2}$.

Solution

$$y' = \frac{1}{\sqrt{1-x^2}} + x\left(\frac{1}{2}\right)(-2x)(1-x^2)^{-1/2} + \sqrt{1-x^2}$$

$$= \frac{1}{\sqrt{1-x^2}} - \frac{x^2}{\sqrt{1-x^2}} + \sqrt{1-x^2}$$

$$= \sqrt{1-x^2} + \sqrt{1-x^2}$$

$$= 2\sqrt{1-x^2}$$

NOTE From Example 5, you can see one of the benefits of inverse trigonometric functions—they can be used to integrate common algebraic functions. For instance, from the result shown in the example, it follows that,

$$\int \sqrt{1-x^2}\, dx$$

$$= \frac{1}{2}\left(\arcsin x + x\sqrt{1-x^2}\right).$$

EXAMPLE 6   Analyzing an Inverse Trigonometric Graph

Analyze the graph of $y = (\arctan x)^2$.

Solution   From the derivative

$$y' = 2(\arctan x)\left(\frac{1}{1 + x^2}\right)$$

$$= \frac{2\arctan x}{1 + x^2}$$

you can see that the only critical number is $x = 0$. By the First Derivative Test, this value corresponds to a relative minimum. From the second derivative

$$y'' = \frac{(1 + x^2)\left(\dfrac{2}{1 + x^2}\right) - (2\arctan x)(2x)}{(1 + x^2)^2}$$

$$= \frac{2(1 - 2x\arctan x)}{(1 + x^2)^2}$$

it follows that points of inflection occur when $2x\arctan x = 1$. Using Newton's Method, these points occur when $x \approx \pm 0.765$. Finally, because

$$\lim_{x \to \pm \infty} (\arctan x)^2 = \frac{\pi^2}{4}$$

it follows that the graph has a horizontal asymptote at $y = \pi^2/4$. The graph is shown in Figure 5.41.

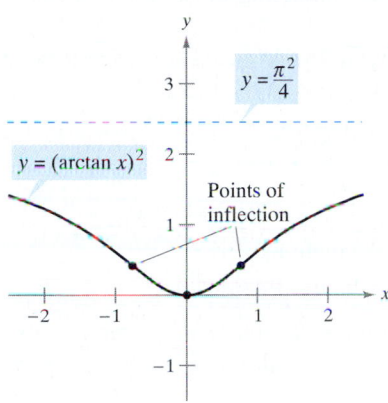

The graph of $y = (\arctan x)^2$ has a horizontal asymptote at $y = \pi^2/4$.
Figure 5.41

EXAMPLE 7   Maximizing an Angle

A photographer is taking a picture of a 4-foot painting hung in an art gallery. The camera lens is 1 foot below the lower edge of the painting, as shown in Figure 5.42. How far should the camera be from the painting to maximize the angle subtended by the camera lens?

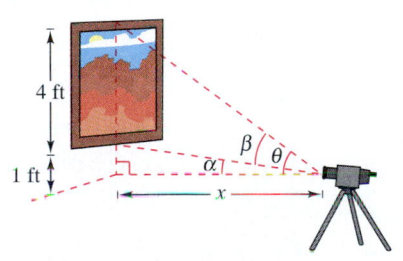

The camera should be 2.236 feet from the painting to maximize the angle β.
Figure 5.42

Solution   In Figure 5.42, let β be the angle to be maximized.

$$\beta = \theta - \alpha$$

$$= \operatorname{arccot} \frac{x}{5} - \operatorname{arccot} x$$

Differentiating produces

$$\frac{d\beta}{dx} = \frac{-1/5}{1 + (x^2/25)} - \frac{-1}{1 + x^2}$$

$$= \frac{-5}{25 + x^2} + \frac{1}{1 + x^2}$$

$$= \frac{4(5 - x^2)}{(25 + x^2)(1 + x^2)}.$$

Because $d\beta/dx = 0$ when $x = \sqrt{5}$, you can conclude from the First Derivative Test that this distance yields a maximum value of β. Thus, the distance is $x \approx 2.236$ feet and the angle is $\beta \approx 0.7297$ radians $\approx 41.81°$.

GALILEO GALILEI (1564–1642)

Galileo's approach to science departed from the accepted Aristotelian view that nature had describable *qualities*, such as "fluidity" and "potentiality." He chose to describe the physical world in terms of measurable *quantities*, such as time, distance, force, and mass.

Review of Basic Differentiation Rules

In the 1600s, Europe was ushered into the scientific age by such great thinkers as Descartes, Galileo, Huygens, Newton, and Kepler. These men believed that nature is governed by basic laws—laws that can, for the most part, be written in terms of mathematical equations. One of the most influential publications of this period—*Dialogue on the Great World Systems*, by Galileo Galilei—has become a classic description of modern scientific thought.

As mathematics has developed during the past few hundred years, a small number of elementary functions has proven sufficient for modeling most* phenomena in physics, chemistry, biology, engineering, economics, and a variety of other fields. An **elementary function** is a function from the following list or one that can be formed as the sum, product, quotient, or composition of functions in the list.

Algebraic Functions	*Transcendental Functions*
Polynomial functions	Logarithmic functions
Rational functions	Exponential functions
Functions involving radicals	Trigonometric functions
	Inverse trigonometric functions

With the differentiation rules introduced so far in the text, you can differentiate *any* elementary function. For convenience, we summarize these differentiation rules here.

Basic Differentiation Rules for Elementary Functions

1. $\dfrac{d}{dx}[cu] = cu'$

2. $\dfrac{d}{dx}[u \pm v] = u' \pm v'$

3. $\dfrac{d}{dx}[uv] = uv' + vu'$

4. $\dfrac{d}{dx}\left[\dfrac{u}{v}\right] = \dfrac{vu' - uv'}{v^2}$

5. $\dfrac{d}{dx}[c] = 0$

6. $\dfrac{d}{dx}[u^n] = nu^{n-1}u'$

7. $\dfrac{d}{dx}[x] = 1$

8. $\dfrac{d}{dx}[|u|] = \dfrac{u}{|u|}(u'),\ u \neq 0$

9. $\dfrac{d}{dx}[\ln u] = \dfrac{u'}{u}$

10. $\dfrac{d}{dx}[e^u] = e^u u'$

11. $\dfrac{d}{dx}[\log_a u] = \dfrac{u'}{(\ln a)u}$

12. $\dfrac{d}{dx}[a^u] = (\ln a)a^u u'$

13. $\dfrac{d}{dx}[\sin u] = (\cos u)u'$

14. $\dfrac{d}{dx}[\cos u] = -(\sin u)u'$

15. $\dfrac{d}{dx}[\tan u] = (\sec^2 u)u'$

16. $\dfrac{d}{dx}[\cot u] = -(\csc^2 u)u'$

17. $\dfrac{d}{dx}[\sec u] = (\sec u \tan u)u'$

18. $\dfrac{d}{dx}[\csc u] = -(\csc u \cot u)u'$

19. $\dfrac{d}{dx}[\arcsin u] = \dfrac{u'}{\sqrt{1 - u^2}}$

20. $\dfrac{d}{dx}[\arccos u] = \dfrac{-u'}{\sqrt{1 - u^2}}$

21. $\dfrac{d}{dx}[\arctan u] = \dfrac{u'}{1 + u^2}$

22. $\dfrac{d}{dx}[\operatorname{arccot} u] = \dfrac{-u'}{1 + u^2}$

23. $\dfrac{d}{dx}[\operatorname{arcsec} u] = \dfrac{u'}{|u|\sqrt{u^2 - 1}}$

24. $\dfrac{d}{dx}[\operatorname{arccsc} u] = \dfrac{-u'}{|u|\sqrt{u^2 - 1}}$

Some important functions used in engineering and science (such as Bessel functions and gamma functions) are not elementary functions.

EXERCISES FOR SECTION 5.8

Numerical and Graphical Analysis **In Exercises 1 and 2, (a)** use a graphing utility to complete the table, **(b)** plot the points in the table and graph the function by hand, **(c)** use a graphing utility to graph the function and compare the result with your hand-drawn graph in part (b), and **(d)** determine any intercepts and symmetry of the graph.

x	-1	-0.8	-0.6	-0.4	-0.2
y					

x	0	0.2	0.4	0.6	0.8	1
y						

1. $y = \arcsin x$

2. $y = \arccos x$

3. *True or False?* Decide whether the following statement is true or false, and explain: Because $\cos(-\pi/3) = \frac{1}{2}$, it follows that $\arccos \frac{1}{2} = -\pi/3$.

4. Determine the missing coordinates of the points on the graph of the function.

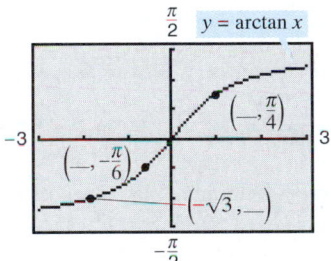

In Exercises 5–12, evaluate the expression without using a calculator.

5. $\arcsin \frac{1}{2}$

6. $\arcsin 0$

7. $\arccos \frac{1}{2}$

8. $\arccos 0$

9. $\arctan \dfrac{\sqrt{3}}{3}$

10. $\operatorname{arccot}(-1)$

11. $\operatorname{arccsc}(-\sqrt{2})$

12. $\arccos\left(-\dfrac{\sqrt{3}}{2}\right)$

In Exercises 13–16, use a calculator to approximate the inverse trigonometric function. Round your answer to two decimal places.

13. $\arccos(-0.8)$

14. $\arcsin(-0.39)$

15. $\operatorname{arcsec} 1.269$

16. $\arctan(-3)$

17. *Think About It* Explain why $\tan \pi = 0$ does not imply that $\arctan 0 = \pi$.

18. Use a graphing utility to confirm that $f(x) = \sin x$ and $g(x) = \arcsin x$ are inverse functions. (Remember to restrict the domain of f properly.)

In Exercises 19–22, evaluate the expression without using a calculator. (*Hint:* See Example 3.)

19. (a) $\sin\left(\arctan \dfrac{3}{4}\right)$

   (b) $\sec\left(\arcsin \dfrac{4}{5}\right)$

20. (a) $\tan\left(\arccos \dfrac{\sqrt{2}}{2}\right)$

   (b) $\cos\left(\arcsin \dfrac{5}{13}\right)$

21. (a) $\cot\left[\arcsin\left(-\dfrac{1}{2}\right)\right]$

   (b) $\csc\left[\arctan\left(-\dfrac{5}{12}\right)\right]$

22. (a) $\sec\left[\arctan\left(-\dfrac{3}{5}\right)\right]$

   (b) $\tan\left[\arcsin\left(-\dfrac{5}{6}\right)\right]$

In Exercises 23–30, write the expression in algebraic form.

23. $\cos(\arcsin 2x)$

24. $\sec(\arctan 3x)$

25. $\sin(\operatorname{arcsec} x)$

26. $\cos(\operatorname{arccot} x)$

27. $\tan\left(\operatorname{arcsec} \dfrac{x}{3}\right)$

28. $\sec[\arcsin(x-1)]$

29. $\csc\left(\arctan \dfrac{x}{\sqrt{2}}\right)$

30. $\cos\left(\arcsin \dfrac{x-h}{r}\right)$

In Exercises 31 and 32, use a graphing utility to graph f and g in the same viewing rectangle to verify that they are equal. Explain why they are equal. Identify any asymptotes of the graphs.

31. $f(x) = \sin(\arctan 2x), \quad g(x) = \dfrac{2x}{\sqrt{1+4x^2}}$

32. $f(x) = \tan\left(\arccos \dfrac{x}{2}\right), \quad g(x) = \dfrac{\sqrt{4-x^2}}{x}$

In Exercises 33 and 34, verify each identity.

33. (a) $\operatorname{arccsc} x = \arcsin \dfrac{1}{x}, \quad |x| \geq 1$

   (b) $\arctan x + \arctan \dfrac{1}{x} = \dfrac{\pi}{2}, \quad x > 0$

34. (a) $\arcsin(-x) = -\arcsin x, \quad |x| \leq 1$

   (b) $\arccos(-x) = \pi - \arccos x, \quad |x| \leq 1$

In Exercises 35–38, sketch the graph of the function. Use a graphing utility to verify your graph.

35. $f(x) = \arcsin(x-1)$

36. $f(x) = \arctan x + \dfrac{\pi}{2}$

37. $f(x) = \operatorname{arcsec} 2x$

38. $f(x) = \arccos \dfrac{x}{4}$

In Exercises 39–42, solve the equation for x.

39. $\arcsin(3x - \pi) = \frac{1}{2}$ **40.** $\arctan 2x = -1$

41. $\arcsin\sqrt{2x} = \arccos\sqrt{x}$ **42.** $\arccos x = \text{arcsec } x$

In Exercises 43–56, find the derivative of the function.

43. $f(x) = 2\arcsin(x - 1)$ **44.** $f(t) = \arcsin t^2$

45. $g(x) = 3\arccos\dfrac{x}{2}$ **46.** $f(x) = \text{arcsec } 2x$

47. $f(x) = \arctan\dfrac{x}{a}$ **48.** $f(x) = \arctan\sqrt{x}$

49. $g(x) = \dfrac{\arcsin 3x}{x}$ **50.** $h(x) = x\arctan x$

51. $h(t) = \sin(\arccos t)$

52. $f(x) = \arcsin x + \arccos x$

53. $y = \dfrac{1}{2}\left(\dfrac{1}{2}\ln\dfrac{x+1}{x-1} + \arctan x\right)$

54. $y = \frac{1}{2}\left(x\sqrt{1 - x^2} + \arcsin x\right)$

55. $y = x\arcsin x + \sqrt{1 - x^2}$

56. $y = x\arctan 2x - \frac{1}{4}\ln(1 + 4x^2)$

Linear and Quadratic Approximations   **In Exercises 57 and 58, use a symbolic differentiation utility to find the linear approximation**

$$P_1(x) = f(a) + f'(a)(x - a)$$

and the quadratic approximation

$$P_2(x) = f(a) + f'(a)(x - a) + \frac{1}{2}f''(a)(x - a)^2$$

to the function f at x = a. Sketch the graph of the function and its linear and quadratic approximations.

57. $f(x) = \arcsin x$ **58.** $f(x) = \arctan x$
     $a = \frac{1}{2}$ $a = 1$

In Exercises 59 and 60, find any relative extrema of the function.

59. $f(x) = \text{arcsec } x - x$ **60.** $f(x) = \arcsin x - 2x$

61. *Angular Rate of Change*   In a free-fall experiment, an object is dropped from a height of 256 feet. A camera on the ground 500 feet from the point of impact records the fall of the object.

(a) Find the position function giving the height of the object at time t assuming the object is released at time $t = 0$. At what time will the object reach ground level?

(b) Find the rate of change of the angle of elevation of the camera when $t = 1$ and $t = 2$.

62. *Angular Rate of Change*   A television camera at ground level is filming the lift-off of a space shuttle at a point 750 meters from the launch pad. Let θ be the angle of elevation of the shuttle and let s be the distance between the camera and the shuttle. Write θ as a function of s for the period of time when the shuttle is moving vertically. Differentiate the result to find $d\theta/dt$ in terms of s and ds/dt.

63. Prove that

$$\arctan x + \arctan y = \arctan\frac{x + y}{1 - xy}, \quad xy \neq 1.$$

Use this formula to show that

$$\arctan\frac{1}{2} + \arctan\frac{1}{3} = \frac{\pi}{4}.$$

64. Verify each of the following differentiation formulas.

(a) $\dfrac{d}{dx}[\arcsin u] = \dfrac{u'}{\sqrt{1 - u^2}}$

(b) $\dfrac{d}{dx}[\arctan u] = \dfrac{u'}{1 + u^2}$

(c) $\dfrac{d}{dx}[\text{arcsec } u] = \dfrac{u'}{|u|\sqrt{u^2 - 1}}$

(d) $\dfrac{d}{dx}[\arccos u] = \dfrac{-u'}{\sqrt{1 - u^2}}$

(e) $\dfrac{d}{dx}[\text{arccot } u] = \dfrac{-u'}{1 + u^2}$

(f) $\dfrac{d}{dx}[\text{arccsc } u] = \dfrac{-u'}{|u|\sqrt{u^2 - 1}}$

65. *Existence of an Inverse*   Determine the values of k such that the function

$$f(x) = kx + \sin x$$

has an inverse.

66. *Think About It*   Use a graphing utility to graph

$$f(x) = \sin x \quad \text{and} \quad g(x) = \arcsin(\sin x).$$

(a) Why isn't the graph of g the line $y = x$?

(b) Determine the extrema of g.

True or False?   **In Exercises 67–70, determine whether the statement is true or false. If it is false, explain why or give an example that shows it is false.**

67. The slope of the graph of the inverse tangent is positive for all x.

68. The range of $y = \arcsin x$ is $[0, \pi]$.

69. $\dfrac{d}{dx}[\arctan(\tan x)] = 1$ for all x in the domain.

70. $\arcsin^2 x + \arccos^2 x = 1$

Integrals Involving Inverse Trigonometric Functions • Completing the Square •
Review of Basic Integration Rules

Integrals Involving Inverse Trigonometric Functions

The derivatives of the six inverse trigonometric functions fall into three pairs. In each
pair, the derivative of one function is the negative of the other. For example,

$$\frac{d}{dx}[\arcsin x] = \frac{1}{\sqrt{1-x^2}}$$

and

$$\frac{d}{dx}[\arccos x] = -\frac{1}{\sqrt{1-x^2}}.$$

When listing the *antiderivative* that corresponds to each of the inverse trigonometric
functions, you need to use only one member from each pair. It is conventional to use
arcsin x as the antiderivative of $1/\sqrt{1-x^2}$, rather than $-\arccos x$. The next theorem
gives one antiderivative formula for each of the three pairs. The proofs of these integration rules are left to you (Exercise 48).

NOTE   For a proof of part 2 of
Theorem 5.19, see the article "A Direct
Proof of the Integral Formula for
Arctangent" by Arnold J. Insel in the
May 1989 issue of *The College
Mathematics Journal*.

THEOREM 5.19   Integrals Involving Inverse Trigonometric Functions

Let u be a differentiable function of x, and let $a > 0$.

1. $\displaystyle\int \frac{du}{\sqrt{a^2-u^2}} = \arcsin \frac{u}{a} + C$

2. $\displaystyle\int \frac{du}{a^2+u^2} = \frac{1}{a}\arctan \frac{u}{a} + C$

3. $\displaystyle\int \frac{du}{u\sqrt{u^2-a^2}} = \frac{1}{a}\operatorname{arcsec} \frac{|u|}{a} + C$

EXAMPLE 1   Integration with Inverse Trigonometric Functions

a. $\displaystyle\int \frac{dx}{\sqrt{4-x^2}} = \arcsin \frac{x}{2} + C$

b. $\displaystyle\int \frac{dx}{2+9x^2} = \frac{1}{3}\int \frac{3\,dx}{\left(\sqrt{2}\right)^2 + (3x)^2}$ $u = 3x,\ a = \sqrt{2}$

$$= \frac{1}{3\sqrt{2}}\arctan \frac{3x}{\sqrt{2}} + C$$

c. $\displaystyle\int \frac{dx}{x\sqrt{4x^2-9}} = \int \frac{2\,dx}{2x\sqrt{(2x)^2-3^2}}$ $u = 2x,\ a = 3$

$$= \frac{1}{3}\operatorname{arcsec} \frac{|2x|}{3} + C$$

The integrals in Example 1 are fairly straightforward applications of integration
formulas. Unfortunately, this is not typical. The inverse trigonometric integration
formulas can be disguised in many ways.

TECHNOLOGY Computer software that can perform symbolic integration is useful for integrating functions such as the one in Example 2. When using such software, however, you must remember that it can fail to find an antiderivative for two reasons. First, some elementary functions simply do not have antiderivatives that are elementary functions. Second, every symbolic integration utility has limitations—you might have entered a function that the software was not programmed to handle. You should also remember that antiderivatives involving trigonometric functions or logarithmic functions can be written in many different forms. For instance, when we use a symbolic integration utility to evaluate the integral in Example 2, we obtain

$$\int \frac{dx}{\sqrt{e^{2x} - 1}} = \arctan \sqrt{e^{2x} - 1} + C.$$

Try showing that this antiderivative is equivalent to that obtained in Example 2.

EXAMPLE 2   Integration by Substitution

Evaluate $\displaystyle\int \frac{dx}{\sqrt{e^{2x} - 1}}$.

Solution As it stands, this integral doesn't fit any of the three inverse trigonometric formulas. Using the substitution $u = e^x$, however, produces the following.

$$u = e^x \quad\Longrightarrow\quad du = e^x\,dx \quad\Longrightarrow\quad dx = \frac{du}{e^x} = \frac{du}{u}$$

With this substitution, you can integrate as follows.

$$\int \frac{dx}{\sqrt{e^{2x} - 1}} = \int \frac{dx}{\sqrt{(e^x)^2 - 1}} \qquad \text{Write } e^{2x} \text{ as } (e^x)^2.$$

$$= \int \frac{du/u}{\sqrt{u^2 - 1}} \qquad \text{Substitute.}$$

$$= \int \frac{du}{u\sqrt{u^2 - 1}} \qquad \text{Rewrite to fit Arcsecant Rule.}$$

$$= \operatorname{arcsec} \frac{|u|}{1} + C \qquad \text{Apply Arcsecant Rule.}$$

$$= \operatorname{arcsec} e^x + C \qquad \text{Back-substitute.}$$

EXAMPLE 3   Rewriting as the Sum of Two Quotients

Evaluate $\displaystyle\int \frac{x + 2}{\sqrt{4 - x^2}}\,dx$.

Solution This integral does not appear to fit any of the basic integration formulas. By splitting the integrand into two parts, however, you can see that the first part can be evaluated with the Power Rule and the second part yields an inverse sine function.

$$\int \frac{x + 2}{\sqrt{4 - x^2}}\,dx = \int \frac{x}{\sqrt{4 - x^2}}\,dx + \int \frac{2}{\sqrt{4 - x^2}}\,dx$$

$$= -\frac{1}{2}\int (4 - x^2)^{-1/2}(-2x)\,dx + 2\int \frac{1}{\sqrt{4 - x^2}}\,dx$$

$$= -\frac{1}{2}\left[\frac{(4 - x^2)^{1/2}}{1/2}\right] + 2\arcsin \frac{x}{2} + C$$

$$= -\sqrt{4 - x^2} + 2\arcsin \frac{x}{2} + C$$

Completing the Square

Completing the square helps when quadratic functions are involved in the integrand. For example, the quadratic $x^2 + bx + c$ can be written as the difference of two squares by adding and subtracting $(b/2)^2$.

$$x^2 + bx + c = x^2 + bx + \left(\frac{b}{2}\right)^2 - \left(\frac{b}{2}\right)^2 + c = \left(x + \frac{b}{2}\right)^2 - \left(\frac{b}{2}\right)^2 + c$$

EXAMPLE 4   Completing the Square

Evaluate $\displaystyle\int \frac{dx}{x^2 - 4x + 7}$.

Solution You can write the denominator as the sum of two squares as follows.

$$x^2 - 4x + 7 = (x^2 - 4x + 4) - 4 + 7 = (x - 2)^2 + 3 = u^2 + a^2$$

Now, in this completed square form, let $u = x - 2$ and $a = \sqrt{3}$.

$$\int \frac{dx}{x^2 - 4x + 7} = \int \frac{dx}{(x - 2)^2 + 3} = \frac{1}{\sqrt{3}} \arctan \frac{x - 2}{\sqrt{3}} + C$$

If the leading coefficient is not 1, it helps to factor before completing the square. For instance, you can complete the square of $2x^2 - 8x + 10$ as follows.

$$2x^2 - 8x + 10 = 2(x^2 - 4x + 5)$$
$$= 2(x^2 - 4x + 4 - 4 + 5)$$
$$= 2[(x - 2)^2 + 1]$$

To complete the square when the coefficient of x^2 is negative, use the same "factoring process" illustrated above. For instance, you can complete the square for $3x - x^2$ as follows.

$$3x - x^2 = -(x^2 - 3x)$$
$$= -\left[x^2 - 3x + \left(\tfrac{3}{2}\right)^2 - \left(\tfrac{3}{2}\right)^2\right]$$
$$= \left(\tfrac{3}{2}\right)^2 - \left(x - \tfrac{3}{2}\right)^2$$

EXAMPLE 5   Completing the Square (Negative Leading Coefficient)

Find the area of the region bounded by the graph of

$$f(x) = \frac{1}{\sqrt{3x - x^2}},$$

the x-axis, and the lines $x = \tfrac{3}{2}$ and $x = \tfrac{9}{4}$.

Solution From Figure 5.43, you can see that the area is given by

$$\text{Area} = \int_{3/2}^{9/4} \frac{1}{\sqrt{3x - x^2}} \, dx.$$

Using the completed square form derived above, you can integrate as follows.

$$\int_{3/2}^{9/4} \frac{dx}{\sqrt{3x - x^2}} = \int_{3/2}^{9/4} \frac{dx}{\sqrt{(3/2)^2 - [x - (3/2)]^2}}$$
$$= \arcsin \frac{x - (3/2)}{3/2} \Bigg]_{3/2}^{9/4}$$
$$= \arcsin \frac{1}{2} - \arcsin 0$$
$$= \frac{\pi}{6}$$
$$\approx 0.524$$

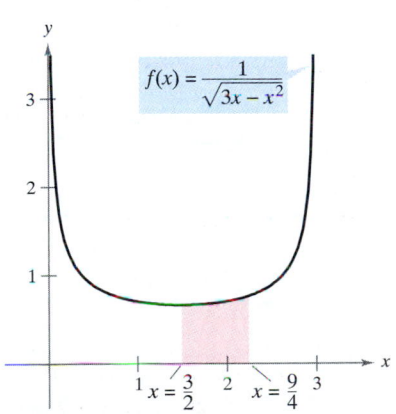

$f(x) = \dfrac{1}{\sqrt{3x - x^2}}$

$x = \dfrac{3}{2}$ $x = \dfrac{9}{4}$

The area of the region bounded by the graph of f, the x-axis, $x = \tfrac{3}{2}$, and $x = \tfrac{9}{4}$ is $\pi/6$.
Figure 5.43

TECHNOLOGY With definite integrals such as the one given in Example 5, remember that you can resort to a numerical solution. For instance, applying Simpson's Rule (with $n = 6$) to the integral in the example, you obtain

$$\int_{3/2}^{9/4} \frac{1}{\sqrt{3x - x^2}} = dx \approx 0.523599.$$

This differs from the exact value of the integral ($\pi/6 \approx 0.5235988$) by less than one millionth.

Review of Basic Integration Rules

You have now completed the introduction of the **basic integration rules.** To be efficient at applying these rules, you should have practiced enough so that each rule is committed to memory.

Basic Integration Rules ($a > 0$)

1. $\displaystyle\int kf(u)\,du = k\int f(u)\,du$

2. $\displaystyle\int [f(u) \pm g(u)]\,du = \int f(u)\,du \pm \int g(u)\,du$

3. $\displaystyle\int du = u + C$

4. $\displaystyle\int u^n du = \frac{u^{n+1}}{n+1} + C, \quad n \neq -1$

5. $\displaystyle\int \frac{du}{u} = \ln |u| + C$

6. $\displaystyle\int e^u\,du = e^u + C$

7. $\displaystyle\int a^u\,du = \left(\frac{1}{\ln a}\right)a^u + C$

8. $\displaystyle\int \sin u\,du = -\cos u + C$

9. $\displaystyle\int \cos u\,du = \sin u + C$

10. $\displaystyle\int \tan u\,du = -\ln |\cos u| + C$

11. $\displaystyle\int \cot u\,du = \ln |\sin u| + C$

12. $\displaystyle\int \sec u\,du = \ln |\sec u + \tan u| + C$

13. $\displaystyle\int \csc u\,du = -\ln |\csc u + \cot u| + C$

14. $\displaystyle\int \sec^2 u\,du = \tan u + C$

15. $\displaystyle\int \csc^2 u\,du = -\cot u + C$

16. $\displaystyle\int \sec u \tan u\,du = \sec u + C$

17. $\displaystyle\int \csc u \cot u\,du = -\csc u + C$

18. $\displaystyle\int \frac{du}{\sqrt{a^2 - u^2}} = \arcsin \frac{u}{a} + C$

19. $\displaystyle\int \frac{du}{a^2 + u^2} = \frac{1}{a} \arctan \frac{u}{a} + C$

20. $\displaystyle\int \frac{du}{u\sqrt{u^2 - a^2}} = \frac{1}{a} \operatorname{arcsec} \frac{|u|}{a} + C$

You can learn a lot about the nature of integration by comparing this list with the summary of differentiation rules given in the previous section. For differentiation, you now have rules that allow you to differentiate *any* elementary function. For integration, this is far from true.

The integration rules listed above are primarily those that we happened on when developing differentiation rules. We do not find integration rules for the antiderivative of a general product or quotient, the natural logarithmic function, or the inverse trigonometric functions. More importantly, you cannot apply any of the rules in this list unless you can create the proper *du* corresponding to the *u* in the formula. The point is that we need to work more on integration techniques, which we will do in Chapter 7. The next two examples should give you a better feeling for the integration problems that you *can* and *cannot* do with the techniques and rules you now know.

EXAMPLE 6   Comparing Integration Problems

Evaluate as many of the following integrals as you can using the formulas and techniques you have studied so far in the text.

a. $\displaystyle\int \frac{dx}{x\sqrt{x^2 - 1}}$ **b.** $\displaystyle\int \frac{x\,dx}{\sqrt{x^2 - 1}}$ **c.** $\displaystyle\int \frac{dx}{\sqrt{x^2 - 1}}$

Solution

a. You *can* evaluate this integral (it fits the Arcsecant Rule).

$$\int \frac{dx}{x\sqrt{x^2 - 1}} = \operatorname{arcsec} |x| + C$$

b. You *can* evaluate this integral (it fits the Power Rule).

$$\int \frac{x\,dx}{\sqrt{x^2 - 1}} = \frac{1}{2}\int (x^2 - 1)^{-1/2}(2x)\,dx$$

$$= \frac{1}{2}\left[\frac{(x^2 - 1)^{1/2}}{1/2}\right] + C$$

$$= \sqrt{x^2 - 1} + C$$

c. You *cannot* evaluate this integral using present techniques. (You should scan the list of basic integration rules to verify this conclusion.)

EXAMPLE 7   Comparing Integration Problems

Evaluate as many of the following integrals as you can using the formulas and techniques you have studied so far in the text.

a. $\displaystyle\int \frac{dx}{x \ln x}$ **b.** $\displaystyle\int \frac{\ln x\,dx}{x}$ **c.** $\displaystyle\int \ln x\,dx$

Solution

a. You *can* evaluate this integral (it fits the Log Rule).

$$\int \frac{dx}{x \ln x} = \int \frac{1/x}{\ln x}\,dx$$

$$= \ln |\ln x| + C$$

b. You *can* evaluate this integral (it fits the Power Rule).

$$\int \frac{\ln x\,dx}{x} = \int \left(\frac{1}{x}\right)(\ln x)^1\,dx$$

$$= \frac{(\ln x)^2}{2} + C$$

c. You *cannot* evaluate this integral using present techniques.

NOTE   Note in Examples 6 and 7 that the *simplest* functions are the ones that you cannot yet integrate.

EXERCISES FOR SECTION 5.9

In Exercises 1–20, evaluate the integral.

1. $\displaystyle\int_0^{1/6} \frac{1}{\sqrt{1-9x^2}}\,dx$

2. $\displaystyle\int_0^1 \frac{dx}{\sqrt{4-x^2}}$

3. $\displaystyle\int_0^{\sqrt{3}/2} \frac{1}{1+4x^2}\,dx$

4. $\displaystyle\int_{\sqrt{3}}^3 \frac{1}{9+x^2}\,dx$

5. $\displaystyle\int \frac{1}{x\sqrt{4x^2-1}}\,dx$

6. $\displaystyle\int \frac{1}{4+(x-1)^2}\,dx$

7. $\displaystyle\int \frac{x^3}{x^2+1}\,dx$

8. $\displaystyle\int \frac{x^4-1}{x^2+1}\,dx$

9. $\displaystyle\int \frac{1}{\sqrt{1-(x+1)^2}}\,dx$

10. $\displaystyle\int \frac{t}{t^4+16}\,dt$

11. $\displaystyle\int \frac{t}{\sqrt{1-t^4}}\,dt$

12. $\displaystyle\int \frac{1}{x\sqrt{x^4-4}}\,dx$

13. $\displaystyle\int_0^{1/\sqrt{2}} \frac{\arcsin x}{\sqrt{1-x^2}}\,dx$

14. $\displaystyle\int_0^{1/\sqrt{2}} \frac{\arccos x}{\sqrt{1-x^2}}\,dx$

15. $\displaystyle\int_{-1/2}^0 \frac{x}{\sqrt{1-x^2}}\,dx$

16. $\displaystyle\int_{-\sqrt{3}}^0 \frac{x}{1+x^2}\,dx$

17. $\displaystyle\int \frac{e^{2x}}{4+e^{4x}}\,dx$

18. $\displaystyle\int_1^2 \frac{1}{3+(x-2)^2}\,dx$

19. $\displaystyle\int_{\pi/2}^{\pi} \frac{\sin x}{1+\cos^2 x}\,dx$

20. $\displaystyle\int \frac{1}{\sqrt{x}(1+x)}\,dx$

In Exercises 21–32, evaluate the integral. (Complete the square, if necessary.)

21. $\displaystyle\int_0^2 \frac{dx}{x^2-2x+2}$

22. $\displaystyle\int_{-3}^{-1} \frac{dx}{x^2+6x+13}$

23. $\displaystyle\int \frac{2x}{x^2+6x+13}\,dx$

24. $\displaystyle\int \frac{2x-5}{x^2+2x+2}\,dx$

25. $\displaystyle\int \frac{1}{\sqrt{-x^2-4x}}\,dx$

26. $\displaystyle\int \frac{1}{\sqrt{-x^2+2x}}\,dx$

27. $\displaystyle\int \frac{x+2}{\sqrt{-x^2-4x}}\,dx$

28. $\displaystyle\int \frac{x-1}{\sqrt{x^2-2x}}\,dx$

29. $\displaystyle\int_2^3 \frac{2x-3}{\sqrt{4x-x^2}}\,dx$

30. $\displaystyle\int \frac{1}{(x-1)\sqrt{x^2-2x}}\,dx$

31. $\displaystyle\int \frac{x}{x^4+2x^2+2}\,dx$

32. $\displaystyle\int \frac{x}{\sqrt{9+8x^2-x^4}}\,dx$

Think About It **In Exercises 33–36, determine which of the given integrals can be evaluated using the basic integration formulas you have studied so far in the text.**

33. (a) $\displaystyle\int \frac{1}{\sqrt{1-x^2}}\,dx$ (b) $\displaystyle\int \frac{x}{\sqrt{1-x^2}}\,dx$ (c) $\displaystyle\int \frac{1}{x\sqrt{1-x^2}}\,dx$

34. (a) $\displaystyle\int e^{x^2}\,dx$ (b) $\displaystyle\int xe^{x^2}\,dx$ (c) $\displaystyle\int \frac{1}{x^2}e^{1/x}\,dx$

35. (a) $\displaystyle\int \sqrt{x-1}\,dx$ (b) $\displaystyle\int x\sqrt{x-1}\,dx$ (c) $\displaystyle\int \frac{x}{\sqrt{x-1}}\,dx$

36. (a) $\displaystyle\int \frac{1}{1+x^4}\,dx$ (b) $\displaystyle\int \frac{x}{1+x^4}\,dx$ (c) $\displaystyle\int \frac{x^3}{1+x^4}\,dx$

In Exercises 37 and 38, use substitution to evaluate the integral.

37. $\displaystyle\int \sqrt{e^t-3}\,dt$

38. $\displaystyle\int \frac{\sqrt{x-2}}{x+1}\,dx$

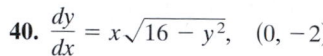 **Direction Fields** **In Exercises 39 and 40, a differential equation, a point, and a direction field are given. (a) Sketch two approximate solutions of the differential equation on the direction field, one of which passes through the indicated point. (b) Use integration to find the particular solution of the differential equation and use a graphing utility to graph the solution. Compare the result with the sketches in part (a).**

39. $\dfrac{dy}{dx} = \dfrac{3}{1+x^2}$, $(0,0)$

40. $\dfrac{dy}{dx} = x\sqrt{16-y^2}$, $(0,-2)$

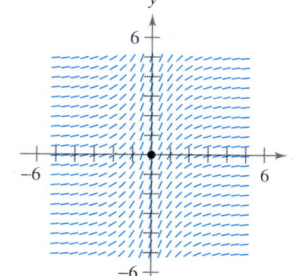

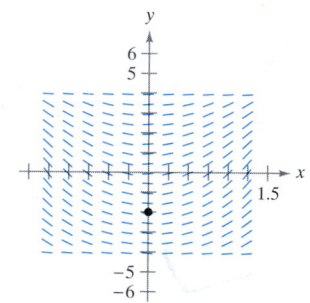

In Exercises 41 and 42, find the area of the region bounded by the graphs of the equations.

41. $y = \dfrac{1}{x^2-2x+5}$, $y = 0$, $x = 1$, $x = 3$

42. $y = \dfrac{1}{\sqrt{4-x^2}}$, $y = 0$, $x = 0$, $x = 1$

43. **Approximation** Determine which value best approximates the area of the region between the x-axis and the function

$$f(x) = \frac{1}{\sqrt{1-x^2}}$$

over the interval $[-0.5, 0.5]$. (Make your selection on the basis of a sketch of the region and *not* by performing any calculations.)

(a) 4   (b) -3 (c) 1   (d) 2   (e) 3

44. Approximation   Sketch the region whose area is represented by the integral

$$\int_0^1 \arcsin x \, dx$$

and use the integration capabilities of a graphing utility to approximate the area.

 45. (a) Show that

$$\int_0^1 \frac{4}{1 + x^2} \, dx = \pi.$$

(b) Approximate the number π using Simpson's Rule (with $n = 6$) and the integral in part (a).

(c) Approximate the number π by using the integration capabilities of a graphing utility.

46. Investigation   Consider the function

$$F(x) = \frac{1}{2} \int_x^{x+2} \frac{2}{t^2 + 1} \, dt.$$

(a) Write a short paragraph giving a geometric interpretation of the function $F(x)$ relative to the function

$$f(x) = \frac{2}{x^2 + 1}.$$

Use what you have written to guess the value of x that will make F maximum.

(b) Perform the specified integration to find an alternative form of $F(x)$. Use calculus to locate the value of x that will make F maximum and compare the result with your guess in part (a).

 47. Consider the integral

$$\int \frac{1}{\sqrt{6x - x^2}} \, dx.$$

(a) Evaluate the integral by completing the square of the radicand.

(b) Evaluate the integral by making the substitution $u = \sqrt{x}$.

(c) The antiderivatives in parts (a) and (b) appear significantly different. Use a graphing utility to graph each in the same viewing rectangle and determine the relationship between the two antiderivatives. Find the domain of each.

48. Verify the following rules by differentiating $(a > 0)$.

(a) $\displaystyle\int \frac{du}{\sqrt{a^2 - u^2}} = \arcsin \frac{u}{a} + C$

(b) $\displaystyle\int \frac{du}{a^2 + u^2} = \frac{1}{a} \arctan \frac{u}{a} + C$

(c) $\displaystyle\int \frac{du}{u\sqrt{u^2 - a^2}} = \frac{1}{a} \operatorname{arcsec} \frac{|u|}{a} + C$

 49. Vertical Motion   An object is projected upward from ground level with an initial velocity of 500 feet per second. In this exercise, the goal is to analyze the motion of the object during its upward flight.

(a) If air resistance is neglected, find the velocity of the object as a function of time. Use a graphing utility to graph this function.

(b) Use the result in part (a) to find the position function and determine the maximum height attained by the object.

(c) If the air resistance is proportional to the square of the velocity, you obtain the equation

$$\frac{dv}{dt} = -(32 + kv^2)$$

where -32 feet per second per second is the acceleration due to gravity and k is a constant. Find the velocity as a function of time by solving the equation

$$\int \frac{dv}{32 + kv^2} = -\int dt.$$

(d) Use a graphing utility to graph the velocity function $v(t)$ in part (c) if $k = 0.001$. Use the graph to approximate the time t_0 at which the object reaches its maximum height.

(e) Use the integration capabilities of a graphing utility to approximate the integral

$$\int_0^{t_0} v(t) \, dt$$

where $v(t)$ and t_0 are those found in part (d). This is the approximation to the maximum height of the object.

(f) Explain the difference between the results in part (b) and part (e).

FOR FURTHER INFORMATION   For more information on this topic, see "What Goes Up Must Come Down; Will Air Resistance Make It Return Sooner, or Later?" by John Lekner in the January 1982 issue of *Mathematics Magazine*.

50. Graph $y_1 = \dfrac{x}{1 + x^2}$, $y_2 = \arctan x$, and $y_3 = x$ on $[0, 10]$. Prove that $\dfrac{x}{1 + x^2} < \arctan x < x$ for $x > 0$.

SECTION **5.10** **Hyperbolic Functions**

Hyperbolic Functions • Differentiation and Integration of Hyperbolic Functions • Inverse Hyperbolic Functions • Differentiation and Integration of Inverse Hyperbolic Functions

Hyperbolic Functions

In this section you will look briefly at a special class of exponential functions called **hyperbolic functions.** The name *hyperbolic function* arose from comparison of the area of a semicircular region, as shown in Figure 5.44, with the area of a region under a hyperbola, as shown in Figure 5.45. The integral for the semicircular region involves an *inverse trigonometric (circular) function:*

$$\int_{-1}^{1} \sqrt{1 - x^2} \, dx = \frac{1}{2}\left[x\sqrt{1 - x^2} + \arcsin x \right]_{-1}^{1} = \frac{\pi}{2} \approx 1.571.$$

The integral for the hyperbolic region involves an inverse hyperbolic function:

$$\int_{-1}^{1} \sqrt{1 + x^2} \, dx = \frac{1}{2}\left[x\sqrt{1 + x^2} + \sinh^{-1} x \right]_{-1}^{1} \approx 2.296.$$

This is only one of many ways in which the hyperbolic functions are similar to the trigonometric functions.

JOHANN HEINRICH LAMBERT (1728–1777)

The first person to publish a comprehensive study on hyperbolic functions was Johann Heinrich Lambert, a Swiss-German mathematician and colleague of Euler.

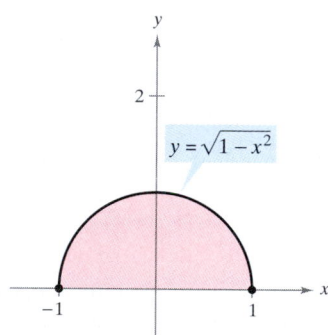

Circle: $x^2 + y^2 = 1$
Figure 5.44

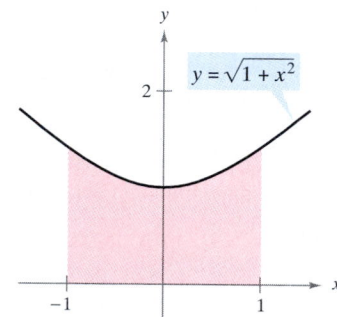

Hyperbola: $-x^2 + y^2 = 1$
Figure 5.45

FOR FURTHER INFORMATION For more information on the development of hyperbolic functions, see the article "An Introduction to Hyperbolic Functions in Elementary Calculus" by Jerome Rosenthal in the April 1986 issue of *Mathematics Teacher.*

Definition of the Hyperbolic Functions

$$\sinh x = \frac{e^x - e^{-x}}{2} \qquad \operatorname{csch} x = \frac{1}{\sinh x}, \quad x \neq 0$$

$$\cosh x = \frac{e^x + e^{-x}}{2} \qquad \operatorname{sech} x = \frac{1}{\cosh x}$$

$$\tanh x = \frac{\sinh x}{\cosh x} \qquad \coth x = \frac{1}{\tanh x}, \quad x \neq 0$$

NOTE $\sinh x$ is read as "the hyperbolic sine of x," $\cosh x$ as "the hyperbolic cosine of x," and so on.

The graphs of the six hyperbolic functions and their domains and ranges are shown in Figure 5.46. Note that the graphs of $\sinh x$ and $\cosh x$ can be obtained by *addition of ordinates* using the exponential functions $f(x) = \frac{1}{2}e^x$ and $g(x) = \frac{1}{2}e^{-x}$.

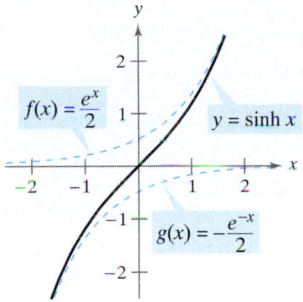

Domain: $(-\infty, \infty)$
Range: $(-\infty, \infty)$

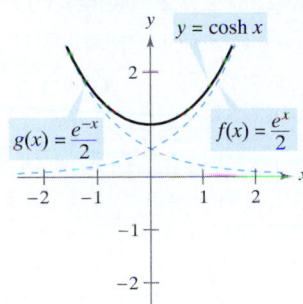

Domain: $(-\infty, \infty)$
Range: $[1, \infty)$

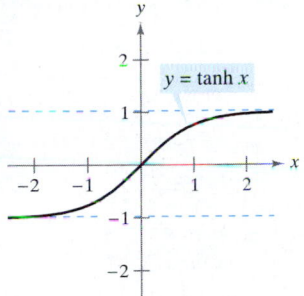

Domain: $(-\infty, \infty)$
Range: $(-1, 1)$

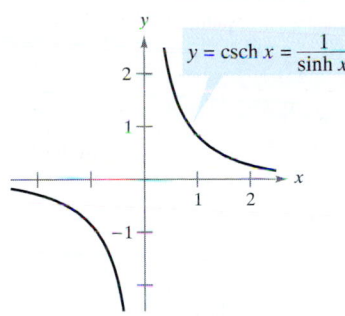

Domain: $(-\infty, 0) \cup (0, \infty)$
Range: $(-\infty, 0) \cup (0, \infty)$

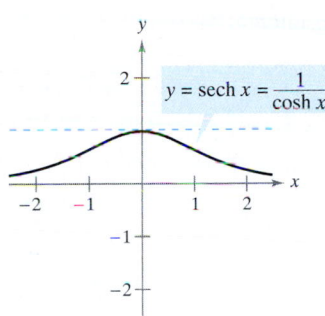

Domain: $(-\infty, \infty)$
Range: $(0, 1]$

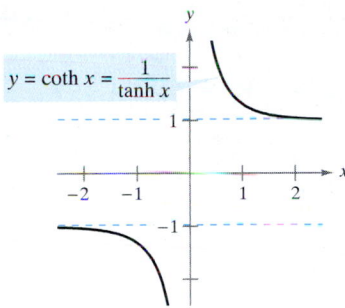

Domain: $(-\infty, 0) \cup (0, \infty)$
Range: $(-\infty, -1) \cup (1, \infty)$

Graphs of the six hyperbolic functions
Figure 5.46

Many of the trigonometric identities have corresponding *hyperbolic identities.* For instance,

$$\cosh^2 x - \sinh^2 x = \left(\frac{e^x + e^{-x}}{2}\right)^2 - \left(\frac{e^x - e^{-x}}{2}\right)^2$$

$$= \frac{e^{2x} + 2 + e^{-2x}}{4} - \frac{e^{2x} - 2 + e^{-2x}}{4}$$

$$= \frac{4}{4}$$

$$= 1$$

FOR FURTHER INFORMATION To understand geometrically the relationship between the hyperbolic and exponential functions, see the article "A Short Proof Linking the Hyperbolic and Geometric Functions" by Michael J. Seery in the *AMATYC Review,* Volume 15 (Number 2).

and

$$2 \sinh x \cosh x = 2\left(\frac{e^x - e^{-x}}{2}\right)\left(\frac{e^x + e^{-x}}{2}\right)$$

$$= \frac{e^{2x} - e^{-2x}}{2}$$

$$= \sinh 2x.$$

Hyperbolic Identities

$$\cosh^2 x - \sinh^2 x = 1 \qquad \sinh(x + y) = \sinh x \cosh y + \cosh x \sinh y$$

$$\tanh^2 x + \operatorname{sech}^2 x = 1 \qquad \sinh(x - y) = \sinh x \cosh y - \cosh x \sinh y$$

$$\coth^2 x - \operatorname{csch}^2 x = 1 \qquad \cosh(x + y) = \cosh x \cosh y + \sinh x \sinh y$$

$$\cosh(x - y) = \cosh x \cosh y - \sinh x \sinh y$$

$$\sinh^2 x = \frac{-1 + \cosh 2x}{2} \qquad \cosh^2 x = \frac{1 + \cosh 2x}{2}$$

$$\sinh 2x = 2 \sinh x \cosh x \qquad \cosh 2x = \cosh^2 x + \sinh^2 x$$

Differentiation and Integration of Hyperbolic Functions

Because the hyperbolic functions are written in terms of e^x and e^{-x}, you can easily derive rules for their derivatives. The following theorem lists these derivatives with the corresponding integration rules.

THEOREM 5.20   Derivatives and Integrals of Hyperbolic Functions

Let u be a differentiable function of x.

$$\frac{d}{dx}[\sinh u] = (\cosh u)u' \qquad\qquad \int \cosh u \, du = \sinh u + C$$

$$\frac{d}{dx}[\cosh u] = (\sinh u)u' \qquad\qquad \int \sinh u \, du = \cosh u + C$$

$$\frac{d}{dx}[\tanh u] = (\operatorname{sech}^2 u)u' \qquad\qquad \int \operatorname{sech}^2 u \, du = \tanh u + C$$

$$\frac{d}{dx}[\coth u] = -(\operatorname{csch}^2 u)u' \qquad\qquad \int \operatorname{csch}^2 u \, du = -\coth u + C$$

$$\frac{d}{dx}[\operatorname{sech} u] = -(\operatorname{sech} u \tanh u)u' \qquad\qquad \int \operatorname{sech} u \tanh u \, du = -\operatorname{sech} u + C$$

$$\frac{d}{dx}[\operatorname{csch} u] = -(\operatorname{csch} u \coth u)u' \qquad\qquad \int \operatorname{csch} u \coth u \, du = -\operatorname{csch} u + C$$

Proof

$$\frac{d}{dx}[\sinh x] = \frac{d}{dx}\left[\frac{e^x - e^{-x}}{2}\right] = \frac{e^x + e^{-x}}{2} = \cosh x$$

$$\frac{d}{dx}[\tanh x] = \frac{d}{dx}\left[\frac{\sinh x}{\cosh x}\right] = \frac{\cosh x(\cosh x) - \sinh x(\sinh x)}{\cosh^2 x}$$

$$= \frac{1}{\cosh^2 x} = \operatorname{sech}^2 x$$

In Exercises 83 and 87, you are asked to prove some of the other differentiation rules.

EXAMPLE 1   **Differentiation of Hyperbolic Functions**

a. $\dfrac{d}{dx}\left[\sinh(x^2 - 3)\right] = 2x\cosh(x^2 - 3)$

b. $\dfrac{d}{dx}\left[\ln(\cosh x)\right] = \dfrac{\sinh x}{\cosh x} = \tanh x$

c. $\dfrac{d}{dx}\left[x\sinh x - \cosh x\right] = x\cosh x + \sinh x - \sinh x = x\cosh x$

EXAMPLE 2   **Finding Relative Extrema**

Find the relative extrema of $f(x) = (x - 1)\cosh x - \sinh x$.

Solution   Begin by setting the first derivative of f equal to 0.

$$f'(x) = (x - 1)\sinh x + \cosh x - \cosh x = 0$$
$$(x - 1)\sinh x = 0$$

Thus, the critical numbers are $x = 1$ and $x = 0$. Using the Second Derivative Test, you can verify that the point $(0, -1)$ yields a relative maximum and the point $(1, -\sinh 1)$ yields a relative minimum, as shown in Figure 5.47. Try using a graphing utility to confirm this result. If your graphing utility does not have hyperbolic functions, you can use exponential functions as follows.

$$f(x) = (x - 1)\left(\tfrac{1}{2}\right)(e^x + e^{-x}) - \tfrac{1}{2}(e^x - e^{-x})$$
$$= \tfrac{1}{2}(xe^x + xe^{-x} - e^x - e^{-x} - e^x + e^{-x})$$
$$= \tfrac{1}{2}(xe^x + xe^{-x} - 2e^x)$$

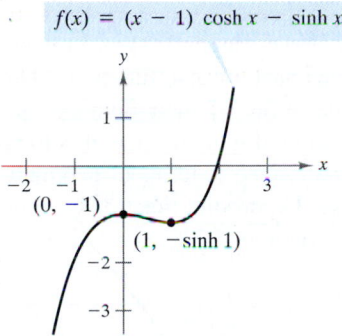

$f(x) = (x - 1)\cosh x - \sinh x$

$f''(0) < 0$, so $(0, -1)$ is a relative maximum. $f''(1) > 0$, so $(1, -\sinh 1)$ is a relative minimum.
Figure 5.47

When a uniform flexible cable, such as a telephone wire, is suspended from two points, it takes the shape of a **catenary,** as discussed in Example 3.

EXAMPLE 3   **Hanging Power Cables**

Power cables are suspended between two towers, forming the catenary shown in Figure 5.48. The equation for this catenary is

$$y = a\cosh\frac{x}{a}.$$

The distance between the two towers is $2b$. Find the slope of the catenary at the point where the cable meets the right-hand tower.

Solution   Differentiating produces

$$y' = a\left(\frac{1}{a}\right)\sinh\frac{x}{a} = \sinh\frac{x}{a}.$$

At the point $(b, a\cosh(b/a))$, the slope (from the left) is given by

$$m = \sinh\frac{b}{a}.$$

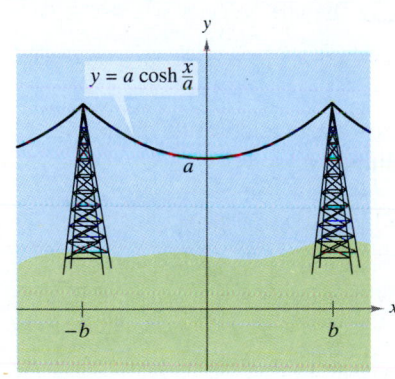

$y = a\cosh\dfrac{x}{a}$

Catenary
Figure 5.48

EXAMPLE 4 Integrating a Hyperbolic Function

$$\int \cosh 2x \sinh^2 2x \, dx = \frac{1}{2} \int (\sinh 2x)^2 (2 \cosh 2x) \, dx \qquad u = \sinh 2x$$

$$= \frac{1}{2} \left[\frac{(\sinh 2x)^3}{3} \right] + C$$

$$= \frac{\sinh^3 2x}{6} + C$$

Inverse Hyperbolic Functions

Unlike trigonometric functions, hyperbolic functions are *not* periodic. In fact, by looking back at Figure 5.46, you can see that four of the six hyperbolic functions are actually one-to-one (the hyperbolic sine, tangent, cosecant, and cotangent). Thus, you can apply Theorem 5.7 to conclude that these four functions have inverse functions. The other two (the hyperbolic cosine and secant) are one-to-one if their domains are restricted to the positive real numbers, and for this restricted domain they also have inverse functions. Because the hyperbolic functions are defined in terms of exponential functions, it is not surprising to find that the inverse hyperbolic functions can be written in terms of logarithmic functions, as shown in Theorem 5.21.

THEOREM 5.21   Inverse Hyperbolic Functions

Function	Domain
$\sinh^{-1} x = \ln\left(x + \sqrt{x^2 + 1}\right)$	$(-\infty, \infty)$
$\cosh^{-1} x = \ln\left(x + \sqrt{x^2 - 1}\right)$	$[1, \infty)$
$\tanh^{-1} x = \dfrac{1}{2} \ln \dfrac{1 + x}{1 - x}$	$(-1, 1)$
$\coth^{-1} x = \dfrac{1}{2} \ln \dfrac{x + 1}{x - 1}$	$(-\infty, -1) \cup (1, \infty)$
$\operatorname{sech}^{-1} x = \ln \dfrac{1 + \sqrt{1 - x^2}}{x}$	$(0, 1]$
$\operatorname{csch}^{-1} x = \ln \left(\dfrac{1}{x} + \dfrac{\sqrt{1 + x^2}}{|x|} \right)$	$(-\infty, 0) \cup (0, \infty)$

Proof The proof of this theorem is a straightforward application of the properties of the exponential and logarithmic functions. For example, if

$$f(x) = \sinh x = \frac{e^x - e^{-x}}{2} \qquad \text{and} \qquad g(x) = \ln\left(x + \sqrt{x^2 + 1}\right)$$

you can show that $f(g(x)) = x$ and $g(f(x)) = x$, which implies that g is the inverse of f.

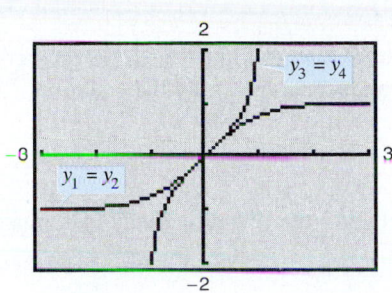

Graphs of hyperbolic tangent function and inverse hyperbolic tangent function
Figure 5.49

TECHNOLOGY   You can use a graphing utility to confirm graphically the results of Theorem 5.21. For instance, try sketching the graphs of the following functions.

$y_1 = \tanh x$ Hyperbolic tangent

$y_2 = \dfrac{e^x - e^{-x}}{e^x + e^{-x}}$ Definition of hyperbolic tangent

$y_3 = \tanh^{-1} x$ Inverse hyperbolic tangent

$y_4 = \dfrac{1}{2} \ln \dfrac{1 + x}{1 - x}$ Definition of inverse hyperbolic tangent

The resulting display is shown in Figure 5.49. As you watch the graphs being traced out, notice that $y_1 = y_2$ and $y_3 = y_4$. Also notice that the graph of y_1 is the reflection of the graph of y_3 in the line $y = x$.

The graphs of the inverse hyperbolic functions are shown in Figure 5.50.

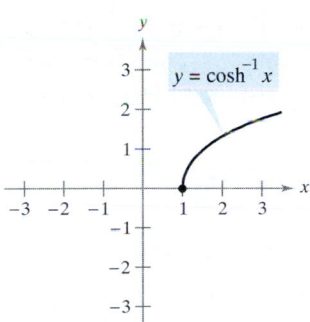

Domain: $[1, \infty)$
Range: $[0, \infty)$

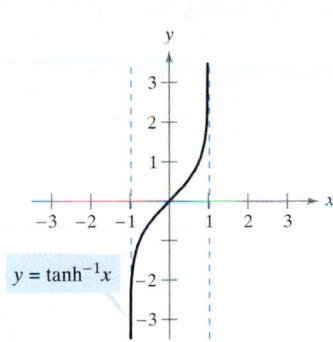

Domain: $(-1, 1)$
Range: $(-\infty, \infty)$

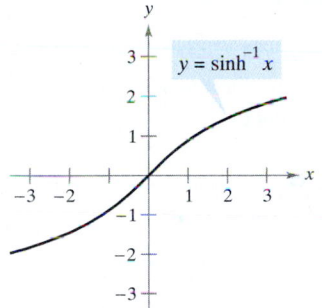

Domain: $(-\infty, \infty)$
Range: $(-\infty, \infty)$

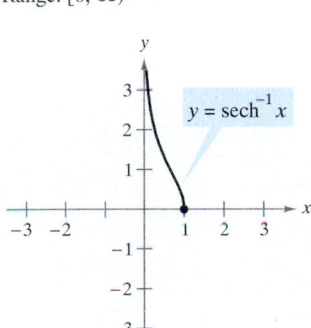

Domain: $(0, 1]$
Range: $[0, \infty)$

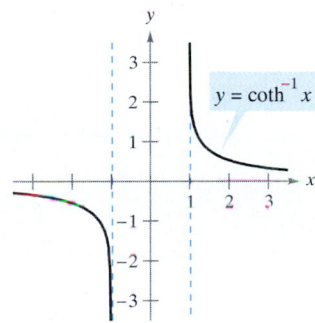

Domain: $(-\infty, -1) \cup (1, \infty)$
Range: $(-\infty, 0) \cup (0, \infty)$

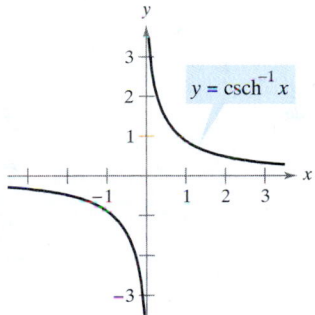

Domain: $(-\infty, 0) \cup (0, \infty)$
Range: $(-\infty, 0) \cup (0, \infty)$

Graphs of the six inverse hyperbolic functions
Figure 5.50

The inverse hyperbolic secant can be used to define a curve called a *tractrix* or *pursuit curve,* as discussed in Example 5.

EXAMPLE 5 A Tractrix

A person is holding a rope that is tied to a boat, as shown in Figure 5.51. As the person walks along the dock, the boat travels along a **tractrix,** given by the equation

$$y = a \operatorname{sech}^{-1} \frac{x}{a} - \sqrt{a^2 - x^2}$$

where a is the length of the rope. If $a = 20$ feet, find the distance the person must walk to bring the boat 5 feet from the dock.

Solution   In figure 5.51, notice that the distance the person has walked is given by

$$y_1 = y + \sqrt{20^2 - x^2} = \left(20 \operatorname{sech}^{-1} \frac{x}{20} - \sqrt{20^2 - x^2}\right) + \sqrt{20^2 - x^2}$$

$$= 20 \operatorname{sech}^{-1} \frac{x}{20}.$$

When $x = 5$, this distance is

$$y_1 = 20 \operatorname{sech}^{-1} \frac{5}{20} = 20 \ln \frac{1 + \sqrt{1 - (1/4)^2}}{1/4}$$

$$= 20 \ln\left(4 + \sqrt{15}\right)$$

$$\approx 41.27 \text{ feet.}$$

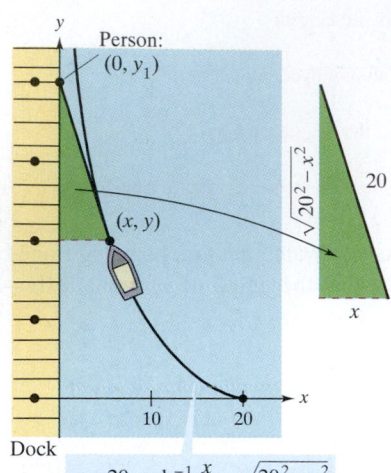

Person: $(0, y_1)$

(x, y)

$\sqrt{20^2 - x^2}$

20

Dock

10   20

$$y = 20 \operatorname{sech}^{-1} \frac{x}{20} - \sqrt{20^2 - x^2}$$

A person must walk 41.27 feet to bring the boat 5 feet from the dock.
Figure 5.51

Differentiation and Integration of Inverse Hyperbolic Functions

The derivatives of the inverse hyperbolic functions, which resemble the derivatives of the inverse trigonometric functions, are listed in Theorem 5.22 with the corresponding integration formulas (in logarithmic form). You can verify each of these formulas by applying the logarithmic definitions of the inverse hyperbolic functions. (See Exercises 84–86.)

THEOREM 5.22   Differentiation and Integration Involving Inverse Hyperbolic Functions

Let u be a differentiable function of x.

$$\frac{d}{dx}[\sinh^{-1} u] = \frac{u'}{\sqrt{u^2 + 1}} \qquad \frac{d}{dx}[\cosh^{-1} u] = \frac{u'}{\sqrt{u^2 - 1}}$$

$$\frac{d}{dx}[\tanh^{-1} u] = \frac{u'}{1 - u^2} \qquad \frac{d}{dx}[\coth^{-1} u] = \frac{u'}{1 - u^2}$$

$$\frac{d}{dx}[\operatorname{sech}^{-1} u] = \frac{-u'}{u\sqrt{1 - u^2}} \qquad \frac{d}{dx}[\operatorname{csch}^{-1} u] = \frac{-u'}{|u|\sqrt{1 + u^2}}$$

$$\int \frac{du}{\sqrt{u^2 \pm a^2}} = \ln\left(u + \sqrt{u^2 \pm a^2}\right) + C$$

$$\int \frac{du}{a^2 - u^2} = \frac{1}{2a} \ln \left|\frac{a + u}{a - u}\right| + C$$

$$\int \frac{du}{u\sqrt{a^2 \pm u^2}} = -\frac{1}{a} \ln \frac{a + \sqrt{a^2 \pm u^2}}{|u|} + C$$

EXAMPLE 6   **More About a Tractrix**

For the tractrix given in Example 5, show that the boat is always pointing toward the person.

Solution   For a point (x, y) on a tractrix, the slope of the graph gives the direction of the boat, as shown in Figure 5.51.

$$y' = \frac{d}{dx}\left[20\,\text{sech}^{-1}\frac{x}{20} - \sqrt{20^2 - x^2}\right]$$

$$= -20\left(\frac{1}{20}\right)\left[\frac{1}{(x/20)\sqrt{1 - (x/20)^2}}\right] - \left(\frac{1}{2}\right)\left(\frac{-2x}{\sqrt{20^2 - x^2}}\right)$$

$$= \frac{-20^2}{x\sqrt{20^2 - x^2}} + \frac{x}{\sqrt{20^2 - x^2}}$$

$$= -\frac{\sqrt{20^2 - x^2}}{x}$$

However, from Figure 5.51, you can see that the slope of the line segment connecting the point $(0, y_1)$ with the point (x, y) is also $m = \left(-\sqrt{20^2 - x^2}\right)/x$. Thus, the boat is always pointing toward the person. (It is because of this property that a tractrix is called a *pursuit curve*.

EXAMPLE 7   **Integration Using Inverse Hyperbolic Functions**

Evaluate $\displaystyle\int \frac{dx}{x\sqrt{4 - 9x^2}}$.

Solution   Let $a = 2$ and $u = 3x$.

$$\int \frac{dx}{x\sqrt{4 - 9x^2}} = \int \frac{3\,dx}{(3x)\sqrt{4 - 9x^2}} \qquad\qquad \int \frac{du}{u\sqrt{a^2 - u^2}}$$

$$= -\frac{1}{2}\ln\frac{2 + \sqrt{4 - 9x^2}}{|3x|} + C \qquad\qquad -\frac{1}{a}\ln\frac{a + \sqrt{a^2 - u^2}}{|u|} + C$$

EXAMPLE 8   **Integration Using Inverse Hyperbolic Functions**

Evaluate $\displaystyle\int \frac{dx}{5 - 4x^2}$.

Solution   Let $a = \sqrt{5}$ and $u = 2x$.

$$\int \frac{dx}{5 - 4x^2} = \frac{1}{2}\int \frac{2\,dx}{\left(\sqrt{5}\right)^2 - (2x)^2} \qquad\qquad \int \frac{du}{a^2 - u^2}$$

$$= \frac{1}{2}\frac{1}{2\sqrt{5}}\ln\left|\frac{\sqrt{5} + 2x}{\sqrt{5} - 2x}\right| + C \qquad\qquad \frac{1}{2a}\ln\left|\frac{a + u}{a - u}\right| + C$$

$$= \frac{1}{4\sqrt{5}}\ln\left|\frac{\sqrt{5} + 2x}{\sqrt{5} - 2x}\right| + C$$

EXERCISES FOR SECTION 5.10

In Exercises 1–6, evaluate the function. If the value is not a rational number, give the answer to three-decimal-place accuracy.

1. (a) $\sinh 3$
   (b) $\tanh(-2)$
2. (a) $\cosh 0$
   (b) $\operatorname{sech} 1$
3. (a) $\operatorname{csch}(\ln 2)$
   (b) $\coth(\ln 5)$
4. (a) $\sinh^{-1} 0$
   (b) $\tanh^{-1} 0$
5. (a) $\cosh^{-1} 2$
   (b) $\operatorname{sech}^{-1} \frac{2}{3}$
6. (a) $\operatorname{csch}^{-1} 2$
   (b) $\coth^{-1} 3$

In Exercises 7–12, verify the identity.

7. $\tanh^2 x + \operatorname{sech}^2 x = 1$
8. $\cosh^2 x = \dfrac{1 + \cosh 2x}{2}$
9. $\sinh(x + y) = \sinh x \cosh y + \cosh x \sinh y$
10. $\sinh 2x = 2 \sinh x \cosh x$
11. $\sinh 3x = 3 \sinh x + 4 \sinh^3 x$
12. $\cosh x + \cosh y = 2 \cosh \dfrac{x + y}{2} \cosh \dfrac{x - y}{2}$

In Exercises 13 and 14, use the value of the given hyperbolic function to find the other hyperbolic functions.

13. $\sinh x = \frac{3}{2}$, $\cosh x = $ ▨ , $\tanh x = $ ▨
    $\operatorname{csch} x = $ ▨ , $\operatorname{sech} x = $ ▨ , $\coth x = $ ▨
14. $\sinh x = $ ▨ , $\cosh x = $ ▨ , $\tanh x = \frac{1}{2}$
    $\operatorname{csch} x = $ ▨ , $\operatorname{sech} x = $ ▨ , $\coth x = $ ▨

In Exercises 15–28, find the derivative of the function.

15. $y = \sinh(1 - x^2)$
16. $y = \coth 3x$
17. $f(x) = \ln(\sinh x)$
18. $g(x) = \ln(\cosh x)$
19. $y = \ln\left(\tanh \dfrac{x}{2}\right)$
20. $y = x \sinh x - \cosh x$
21. $h(x) = \dfrac{1}{4} \sinh 2x - \dfrac{x}{2}$
22. $h(t) = t - \coth t$
23. $f(t) = \arctan(\sinh t)$
24. $f(x) = e^{\sinh x}$
25. $g(x) = x^{\cosh x}$
26. $g(x) = \operatorname{sech}^2 3x$
27. $y = (\cosh x - \sinh x)^2$
28. $y = \operatorname{sech}(x + 1)$

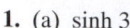 **In Exercises 29 and 30, find any relative extrema of the function and use a graphing utility to confirm your result.**

29. $f(x) = \sin x \sinh x - \cos x \cosh x$, $-4 \le x \le 4$
30. $f(x) = x \cosh(x - 1) - \sinh(x - 1)$

In Exercises 31 and 32, use a graphing utility to graph the function and approximate any relative extrema of the function.

31. $g(x) = x \operatorname{sech} x$
32. $h(x) = 2 \tanh x - x$

In Exercises 33 and 34, show that the function satisfies the differential equation.

Function	Differential Equation
33. $y = a \sinh x$	$y''' - y' = 0$
34. $y = a \cosh x$	$y'' - y = 0$

Linear and Quadratic Approximations   **In Exercises 35 and 36, use a symbolic differentiation utility to find the linear approximation**

$$P_1(x) = f(a) + f'(a)(x - a)$$

and the quadratic approximation

$$P_2(x) = f(a) + f'(a)(x - a) + \tfrac{1}{2} f''(a)(x - a)^2$$

to the function f at $x = a$. Use a graphing utility to graph the function and its linear and quadratic approximations.

35. $f(x) = \tanh x$
    $a = 1$
36. $f(x) = \cosh x$
    $a = 0$

In Exercises 37–52, evaluate the integral.

37. $\displaystyle\int \sinh(1 - 2x)\, dx$
38. $\displaystyle\int \dfrac{\cosh \sqrt{x}}{\sqrt{x}}\, dx$
39. $\displaystyle\int \cosh^2(x - 1) \sinh(x - 1)\, dx$
40. $\displaystyle\int \dfrac{\sinh x}{1 + \sinh^2 x}\, dx$
41. $\displaystyle\int \dfrac{\cosh x}{\sinh x}\, dx$
42. $\displaystyle\int \operatorname{sech}^2(2x - 1)\, dx$
43. $\displaystyle\int x \operatorname{csch}^2 \dfrac{x^2}{2}\, dx$
44. $\displaystyle\int \operatorname{sech}^3 x \tanh x\, dx$
45. $\displaystyle\int \dfrac{\operatorname{csch}(1/x) \coth(1/x)}{x^2}\, dx$
46. $\displaystyle\int \sinh^2 x\, dx$
47. $\displaystyle\int_0^4 \dfrac{1}{25 - x^2}\, dx$
48. $\displaystyle\int_0^4 \dfrac{1}{\sqrt{25 - x^2}}\, dx$
49. $\displaystyle\int_0^{\sqrt{2}/4} \dfrac{2}{\sqrt{1 - 4x^2}}\, dx$
50. $\displaystyle\int \dfrac{2}{x\sqrt{1 + 4x^2}}\, dx$
51. $\displaystyle\int \dfrac{x}{x^4 + 1}\, dx$
52. $\displaystyle\int \dfrac{\cosh x}{\sqrt{9 - \sinh^2 x}}\, dx$

In Exercises 53–60, find the derivative of the function.

53. $y = \cosh^{-1}(3x)$

54. $y = \tanh^{-1}\dfrac{x}{2}$

55. $y = \sinh^{-1}(\tan x)$

56. $y = \text{sech}^{-1}(\cos 2x), \quad 0 < x < \pi/4$

57. $y = \coth^{-1}(\sin 2x)$

58. $y = (\text{csch}^{-1}\, x)^2$

59. $y = 2x \sinh^{-1}(2x) - \sqrt{1 + 4x^2}$

60. $y = x \tanh^{-1} x + \ln\sqrt{1 - x^2}$

Tractrix **In Exercises 61 and 62, use the equation of the tractrix** $y = a\,\text{sech}^{-1}\dfrac{x}{a} - \sqrt{a^2 - x^2}, \quad a > 0.$

61. Find dy/dx.

62. Let L be the tangent line at the point P to the tractrix. If L intersects the y-axis at the point Q, show that the distance between P and Q is a.

In Exercises 63–70, find the indefinite integral using the formulas of Theorem 5.22.

63. $\displaystyle\int \dfrac{1}{\sqrt{1 + e^{2x}}}\,dx$

64. $\displaystyle\int \dfrac{e^x}{1 - e^{2x}}\,dx$

65. $\displaystyle\int \dfrac{1}{\sqrt{x}\sqrt{1 + x}}\,dx$

66. $\displaystyle\int \dfrac{\sqrt{x}}{\sqrt{1 + x^3}}\,dx$

67. $\displaystyle\int \dfrac{-1}{4x - x^2}\,dx$

68. $\displaystyle\int \dfrac{dx}{(x - 1)\sqrt{x^2 - 2x + 2}}$

69. $\displaystyle\int \dfrac{1}{1 - 4x - 2x^2}\,dx$

70. $\displaystyle\int \dfrac{dx}{(x + 1)\sqrt{2x^2 + 4x + 8}}$

In Exercises 71–74, solve the differential equation.

71. $\dfrac{dy}{dx} = \dfrac{1}{\sqrt{80 + 8x - 16x^2}}$

72. $\dfrac{dy}{dx} = \dfrac{1}{(x - 1)\sqrt{-4x^2 + 8x - 1}}$

73. $\dfrac{dy}{dx} = \dfrac{x^3 - 21x}{5 + 4x - x^2}$

74. $\dfrac{dy}{dx} = \dfrac{1 - 2x}{4x - x^2}$

In Exercises 75–78, find the area of the region bounded by the graphs of the equations.

75. $y = \text{sech}\dfrac{x}{2}, \quad y = 0, \quad x = -4, \quad x = 4$

76. $y = \tanh 2x, \quad y = 0, \quad x = 2$

77. $y = \dfrac{5x}{\sqrt{x^4 + 1}}, \quad y = 0, \quad x = 2$

78. $y = \dfrac{6}{\sqrt{x^2 - 4}}, \quad y = 0, \quad x = 3, \quad x = 5$

79. ***Chemical Reactions*** Suppose that chemicals A and B combine in a 3-to-1 ratio to form a compound. The amount of compound x being produced at any time t is proportional to the unchanged amounts of A and B remaining in the solution. Thus, if 3 kilograms of A is mixed with 2 kilograms of B, you have

$$\frac{dx}{dt} = k\left(3 - \frac{3x}{4}\right)\left(2 - \frac{x}{4}\right) = \frac{3k}{16}(x^2 - 12x + 32).$$

If 1 kilogram of the compound is formed after 10 minutes, find the amount formed after 20 minutes by solving the integral

$$\int \frac{3k}{16}\,dt = \int \frac{dx}{x^2 - 12x + 32}.$$

80. ***Vertical Motion*** An object is dropped from a height of 400 feet.

(a) Find the velocity of the object as a function of time (neglect air resistance on the object).

(b) Use the result in part (a) to find the position function.

(c) If the air resistance is proportional to the square of the velocity, then

$$\frac{dv}{dt} = -32 + kv^2$$

where -32 feet per second per second is the acceleration due to gravity and k is a constant. Show that the velocity v as a function of time is

$$v(t) = -\sqrt{\frac{32}{k}}\,\tanh\!\left(\sqrt{32k}\,t\right)$$

by performing the following integration and simplifying the result.

$$\int \frac{dv}{32 - kv^2} = -\int dt$$

(d) Use the result in part (c) to find $\displaystyle\lim_{t \to \infty} v(t)$ and give its interpretation.

(e) Integrate the velocity function in part (c) and find the position s of the object as a function of t. Use a graphing utility to graph the position function when $k = 0.01$ and the position function in part (b) in the same viewing rectangle. Estimate the additional time required for the object to reach ground level when air resistance is not neglected.

81. ***Writing*** Give a written description of what you believe would happen if k were increased in Exercise 80. Then test your assertion with a particular value of k.

82. Show that $\arctan(\sinh x) = \arcsin(\tanh x)$.

In Exercises 83–87, verify the differentiation formula.

83. $\dfrac{d}{dx}[\cosh x] = \sinh x$

84. $\dfrac{d}{dx}[\text{sech}^{-1} x] = \dfrac{-1}{x\sqrt{1 - x^2}}$

85. $\dfrac{d}{dx}[\cosh^{-1} x] = \dfrac{1}{\sqrt{x^2 - 1}}$

86. $\dfrac{d}{dx}[\sinh^{-1} x] = \dfrac{1}{\sqrt{x^2 + 1}}$

87. $\dfrac{d}{dx}[\text{sech } x] = -\text{sech } x \tanh x$

Superstock, Inc.

St. Louis Arch   The Gateway Arch in St. Louis, Missouri was constructed using the hyperbolic cosine function. The equation used to construct the arch was

$$y = 693.8597 - 68.7672 \cosh 0.0100333x,$$
$$-299.2239 \le x \le 299.2239$$

where x and y are measured in feet. Cross sections of the arch are equilateral triangles and (x, y) traces the path of the centers of mass of the cross-sectional triangles. For each value of x, the area of the cross-sectional triangle is

$$A = 125.1406 \cosh 0.0100333x.$$

(*Source:* Owner's Manual for the Gateway Arch, *Saint Louis, MO, by William Thayer*)

(a) How high above the ground is the center of the highest triangle? (At ground level, $y = 0$.)

(b) What is the height of the arch? (*Hint:* For an equilateral triangle, $A = \sqrt{3}c^2$, where c is one-half the base of the triangle, and the center of mass of the triangle is located at two-thirds the height of the triangle.)

(c) How wide is the arch at ground level?

REVIEW EXERCISES FOR CHAPTER 5

In Exercises 1 and 2, sketch the graph of the function by hand. Identify any asymptotes of the graph.

1. $f(x) = \ln x + 3$

2. $f(x) = \ln(x - 3)$

In Exercises 3 and 4, use the properties of logarithms to write the expression as a sum, difference, and/or multiple of logarithms.

3. $\ln \sqrt[5]{\dfrac{4x^2 - 1}{4x^2 + 1}}$

4. $\ln[(x^2 + 1)(x - 1)]$

In Exercises 5 and 6, write the expression as the logarithm of a single quantity.

5. $\ln 3 + \frac{1}{3}\ln(4 - x^2) - \ln x$

6. $3[\ln x - 2\ln(x^2 + 1)] + 2\ln 5$

True or False?   **In Exercises 7 and 8, determine whether the statement is true or false.**

7. The domain of the function $f(x) = \ln x$ is the set of all real numbers.

8. $\ln(x + y) = \ln x + \ln y$

In Exercises 9 and 10, solve the equation for x.

9. $\ln \sqrt{x + 1} = 2$

10. $\ln x + \ln(x - 3) = 0$

In Exercises 11–18, find the derivative of the function.

11. $g(x) = \ln \sqrt{x}$

12. $h(x) = \ln \dfrac{x(x - 1)}{x - 2}$

13. $f(x) = x\sqrt{\ln x}$

14. $f(x) = \ln[x(x^2 - 2)^{2/3}]$

15. $y = \dfrac{1}{b^2}\left[\ln(a + bx) + \dfrac{a}{a + bx}\right]$

16. $y = \dfrac{1}{b^2}[a + bx - a\ln(a + bx)]$

17. $y = -\dfrac{1}{a}\ln \dfrac{a + bx}{x}$

18. $y = -\dfrac{1}{ax} + \dfrac{b}{a^2}\ln \dfrac{a + bx}{x}$

In Exercises 19–26, evaluate the integral.

19. $\displaystyle\int \dfrac{1}{7x - 2}\,dx$

20. $\displaystyle\int \dfrac{x}{x^2 - 1}\,dx$

21. $\displaystyle\int \dfrac{\sin x}{1 + \cos x}\,dx$

22. $\displaystyle\int \dfrac{\ln \sqrt{x}}{x}\,dx$

23. $\displaystyle\int_1^4 \dfrac{x + 1}{x}\,dx$

24. $\displaystyle\int_1^e \dfrac{\ln x}{x}\,dx$

25. $\displaystyle\int_0^{\pi/3} \sec \theta\,d\theta$

26. $\displaystyle\int_0^{\pi/4} \tan\left(\dfrac{\pi}{4} - x\right)dx$

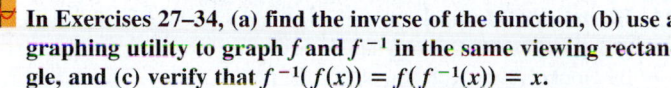

In Exercises 27–34, (a) find the inverse of the function, (b) use a graphing utility to graph f and f^{-1} in the same viewing rectangle, and (c) verify that $f^{-1}(f(x)) = f(f^{-1}(x)) = x$.

27. $f(x) = \frac{1}{2}x - 3$ **28.** $f(x) = 5x - 7$

29. $f(x) = \sqrt{x + 1}$ **30.** $f(x) = x^3 + 2$

31. $f(x) = \sqrt[3]{x + 1}$ **32.** $f(x) = x^2 - 5, \quad x \geq 0$

33. $f(x) = \ln \sqrt{x}$ **34.** $f(x) = e^{1-x}$

In Exercises 35–38, sketch the graph of the function by hand.

35. $y = e^{-x/2}$ **36.** $g(x) = 6(2^{-x^2})$

37. $h(x) = -3 \arcsin 2x$ **38.** $f(x) = 2 \arctan(x + 3)$

In Exercises 39 and 40, evaluate the expression without using a calculator. (*Hint:* Make a sketch of a right triangle.)

39. (a) $\sin\left(\arcsin \frac{1}{2}\right)$ **40.** (a) $\tan(\text{arccot } 2)$

   (b) $\cos\left(\arcsin \frac{1}{2}\right)$ (b) $\cos\left(\text{arcsec } \sqrt{5}\right)$

In Exercises 41–58, find the derivative of the function.

41. $f(x) = \ln(e^{-x^2})$ **42.** $g(x) = \ln \dfrac{e^x}{1 + e^x}$

43. $g(t) = t^2 e^t$ **44.** $h(z) = e^{-z^2/2}$

45. $y = \sqrt{e^{2x} + e^{-2x}}$ **46.** $y = x^{2x+1}$

47. $f(x) = 3^{x-1}$ **48.** $f(x) = (4e)^x$

49. $g(x) = \dfrac{x^2}{e^x}$ **50.** $f(\theta) = \frac{1}{2}e^{\sin 2\theta}$

51. $y = \tan(\arcsin x)$

52. $y = \arctan(x^2 - 1)$

53. $y = x \text{ arcsec } x$

54. $y = \frac{1}{2} \arctan e^{2x}$

55. $y = x(\arcsin x)^2 - 2x + 2\sqrt{1 - x^2} \arcsin x$

56. $y = \sqrt{x^2 - 4} - 2 \text{ arcsec}(x/2), \quad 2 < x < 4$

57. $y = 2x - \cosh \sqrt{x}$

58. $y = x \tanh^{-1} 2x$

In Exercises 59 and 60, use implicit differentiation to find dy/dx.

59. $y \ln x + y^2 = 0$

60. $\cos x^2 = xe^y$

61. *Think About It*   Find the derivative of each function, given that a is constant.

   (a) $y = x^a$,   (b) $y = a^x$,   (c) $y = x^x$,   (d) $y = a^a$

62. *Compound Interest*   How large a deposit, at 7 percent interest compounded continuously, must be made to obtain a balance of $10,000 in 15 years?

63. *Compound Interest*   A deposit earns interest at a rate of r percent compounded continuously and doubles in value in 10 years. Find r.

64. *Climb Rate*   The time t (in minutes) for a small plane to climb to an altitude of h feet is

$$t = 50 \log_{10} \frac{18,000}{18,000 - h}$$

where 18,000 feet is the plane's absolute ceiling.

(a) Determine the domain of the function appropriate for the context of the problem.

(b) Use a graphing utility to graph the time function and identify any asymptotes.

(c) As the plane approaches its absolute ceiling, what can be concluded about the time required to further increase its altitude?

(d) Find the time when the altitude is increasing at the greatest rate.

In Exercises 65–80, evaluate the integral.

65. $\displaystyle\int xe^{-3x^2}\,dx$ **66.** $\displaystyle\int \frac{e^{1/x}}{x^2}\,dx$

67. $\displaystyle\int \frac{e^{4x} - e^{2x} + 1}{e^x}\,dx$ **68.** $\displaystyle\int \frac{e^{2x} - e^{-2x}}{e^{2x} + e^{-2x}}\,dx$

69. $\displaystyle\int \frac{e^x}{e^x - 1}\,dx$ **70.** $\displaystyle\int x^2 e^{x^3+1}\,dx$

71. $\displaystyle\int \frac{1}{e^{2x} + e^{-2x}}\,dx$ **72.** $\displaystyle\int \frac{1}{3 + 25x^2}\,dx$

73. $\displaystyle\int \frac{x}{\sqrt{1 - x^4}}\,dx$ **74.** $\displaystyle\int \frac{1}{16 + x^2}\,dx$

75. $\displaystyle\int \frac{x}{16 + x^2}\,dx$ **76.** $\displaystyle\int \frac{4 - x}{\sqrt{4 - x^2}}\,dx$

77. $\displaystyle\int \frac{\arctan(x/2)}{4 + x^2}\,dx$ **78.** $\displaystyle\int \frac{\arcsin x}{\sqrt{1 - x^2}}\,dx$

79. $\displaystyle\int \frac{x}{\sqrt{x^4 - 1}}\,dx$ **80.** $\displaystyle\int x^2 \text{ sech}^2 x^3\,dx$

In Exercises 81 and 82, find the area of the region bounded by the graphs of the equations.

81. $y = xe^{-x^2}, \quad y = 0, \quad x = 0, \quad x = 4$

82. $y = \dfrac{1}{x^2 + 1}, \quad y = 0, \quad x = 0, \quad x = 1$

In Exercises 83–88, solve the differential equation.

83. $\dfrac{dy}{dx} = \dfrac{x^2 + 3}{x}$ **84.** $\dfrac{dy}{dx} = \dfrac{e^{-2x}}{1 + e^{-2x}}$

85. $y' - 2xy = 0$ **86.** $y' - e^y \sin x = 0$

87. $\dfrac{dy}{dx} = \dfrac{x^2 + y^2}{2xy}$ **88.** $\dfrac{dy}{dx} = \dfrac{3(x + y)}{x}$

89. Use the differential equation $y' = \sqrt{1 - y^2}$ and the direction field shown in the figure to answer each of the following.

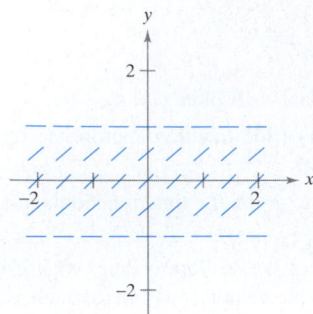

(a) Sketch several solution curves to the differential equation on the direction field.

(b) When is the rate of change of the solution the greatest? When is it the least?

(c) Find the general solution of the differential equation. Compare the result with the sketches in part (a).

90. Show that $y = e^x(a \cos 3x + b \sin 3x)$ satisfies the differential equation $y'' - 2y' + 10y = 0$.

 91. Find the orthogonal trajectories of the family $(x - C)^2 + y^2 = C^2$ and use a graphing utility to sketch several members of the family.

92. *Air Pressure* Under ideal conditions, air pressure decreases continuously with height above sea level at a rate proportional to the pressure at that height. If the barometer reads 30 inches at sea level and 15 inches at 18,000 feet, find the barometric pressure at 35,000 feet.

93. *Probability* Two numbers between 0 and 10 are chosen at random. The probability that their product is less than $n(0 < n < 100)$ is

$$P = \frac{1}{100}\left(n + \int_{n/10}^{10} \frac{n}{x}\, dx\right).$$

(a) What is the probability that the product is less than 25?

(b) What is the probability that the product is less than 50?

94. *Vertical Motion* A falling object encounters air resistance that is proportional to its velocity. If the acceleration due to gravity is considered to be -9.8 meters per second per second, the net change in velocity is

$$\frac{dv}{dt} = kv - 9.8.$$

(a) Find the velocity of the object as a function of time if the initial velocity is v_0.

(b) Use the result in part (a) to find the limit of the velocity as t approaches infinity.

(c) Integrate the velocity function found in part (a) to find the position function s.

95. *Harmonic Motion* A weight of mass m is attached to a spring and oscillates with simple harmonic motion (see figure). By Hooke's Law, you can determine that

$$\int \frac{dy}{\sqrt{A^2 - y^2}} = \int \sqrt{\frac{k}{m}}\, dt$$

where A is the maximum displacement, t is the time, and k is a constant. Find y as a function of t, given that $y = 0$ when $t = 0$.

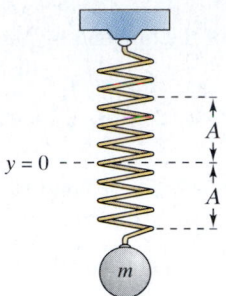

96. *Modeling Data* For the years 1850 through 1990, the ordered pairs (t, P) give the population of the United States P (in millions), where t is the time in years, with $t = 0$ corresponding to 1900. *(Source: U.S. Bureau of Census)*

$(-50, 23.2), (-40, 31.4), (-30, 39.8), (-20, 50.2),$

$(-10, 62.9), (0, 76.0), (10, 92.0), (20, 105.7),$

$(30, 122.8), (40, 131.7), (50, 150.7), (60, 178.5),$

$(70, 203.3), (80, 226.5), (90, 248.7)$

(a) Use the regression capabilities of a graphing utility to find the following models for the data.

Linear: $y = at + b$

Quadratic: $y = at^2 + bt + c$

Exponential: $y = ab^t$

(b) Use a graphing utility to plot the data and graph each of the models.

(c) Evaluate each model and the rate of growth of each model for the year 1998. Which model seems to be the best?

97. *Fuel Economy* A certain automobile gets 28 miles per gallon of gasoline for speeds up to 50 miles per hour. Over 50 miles per hour, the number of miles per gallon drops at the rate of 12 percent for each 10 miles per hour.

(a) If s is the speed and y is the number of miles per gallon, find y as a function of s by solving the differential equation

$$\frac{dy}{ds} = -0.012y, \quad s > 50.$$

(b) Use the function in part (a) to complete the table.

Speed	50	55	60	65	70
Miles per gallon					

Applications of Integration

6

HOOVER DAM

Hoover Dam, one of the highest concrete dams in the world, uses a gravity-arch construction. It relies on both the walls of the Black Canyon and its own mass to hold back the waters of the Colorado River.

Dams were originally built to ensure water supplies during dry seasons. As technical knowledge has increased, they have begun serving other functions. Today, dams may be built to create recreational lakes, to power generators, and to prevent flooding. Every new dam creates concerns. Along with its benefits, the dam may upset an area's ecology and force the relocation of people and wildlife. Also, a poorly constructed dam endangers the entire surrounding region, creating the possibility of a massive disaster.

There are several designs used in dam construction, one of which is the arch dam. This design curves toward the water it contains, and is usually built in narrow canyons. The force of the water presses the edges of the dam against the walls of the canyon, so the natural rock helps support the structure. This added support means that the arch dam can be built with less construction materials than its gravity-supported counterpart.

A cross section of a typical arch dam can be modeled as shown in the figure at the lower left. The model for this cross section is as follows.

$$f(x) = \begin{cases} 0.03x^2 + 7.1x + 350, & -70 \le x \le -16 \\ 389, & -16 < x < 0 \\ -6.593x + 389, & 0 \le x \le 59 \end{cases}$$

To form the arch dam, this cross section is swung through an arc, rotating it about the y-axis. The number of degrees through which it is rotated and the length of the axis of rotation vary, depending primarily on how much the water level varies. A possible configuration shows a rotation of 150° and an axis of rotation of 150 feet.

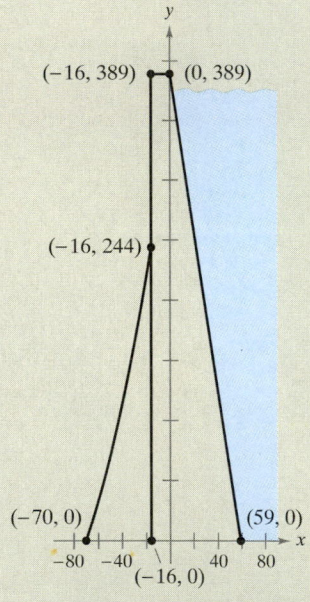

Cross section of an arch dam

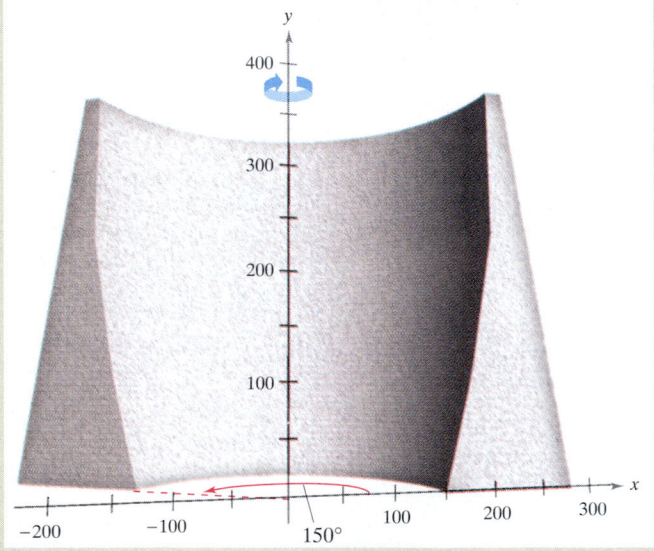

QUESTIONS

1. Find the area of a cross section of the dam.

2. Describe a strategy for estimating the volume of concrete that would be needed to build this dam.

3. Use the strategy to estimate the volume of concrete needed to build the dam described on this page.

The concepts presented here will be explored further in this chapter. For an extension of this application, see the lab series that accompanies this text.

FOR FURTHER INFORMATION For more information on the calculus of dam design. see *Calculus, Understanding Change,* a three-part, half-hour video produced by COMAP and funded by the National Science Foundation.

Area of a Region Between Two Curves •
Area of a Region Between Intersecting Curves

Area of a Region Between Two Curves

With a few modifications you can extend the application of definite integrals from the area of a region *under* a curve to the area of a region *between* two curves. Consider two functions f and g that are continuous on the interval $[a, b]$. If, as in Figure 6.1, the graphs of both f and g lie above the x-axis, and the graph of g lies below the graph of f, you can geometrically interpret the area of the region between the graphs as the area of the region under the graph of g subtracted from the area of the region under the graph of f, as shown in Figure 6.2.

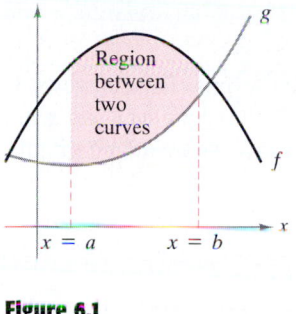

Figure 6.1

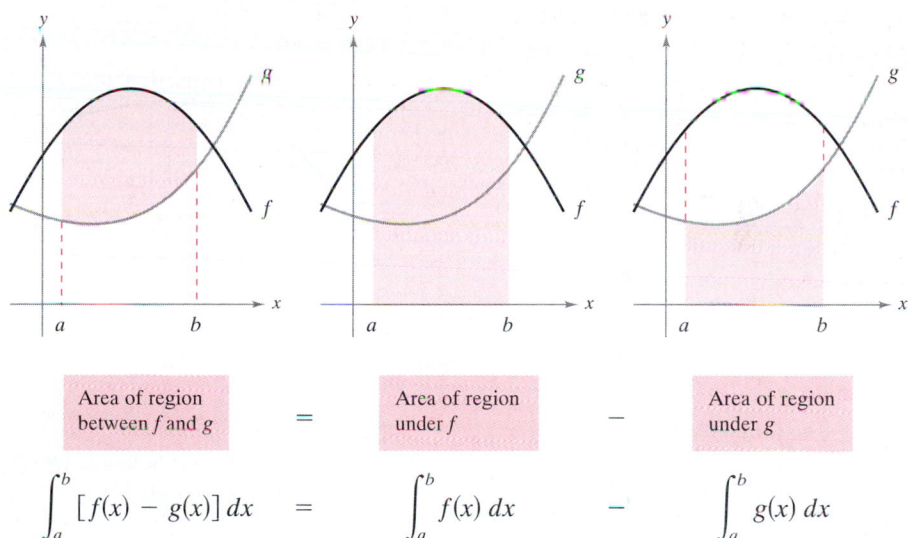

Area of region between f and g	$=$	Area of region under f	$-$	Area of region under g

$$\int_a^b \left[f(x) - g(x) \right] dx \quad = \quad \int_a^b f(x)\, dx \quad - \quad \int_a^b g(x)\, dx$$

Figure 6.2

To verify the reasonableness of the result shown in Figure 6.2, you can partition the interval $[a, b]$ into n subintervals, each of width Δx. Then, as shown in Figure 6.3, sketch a **representative rectangle** of width Δx and height $f(x_i) - g(x_i)$, where x_i is in the ith interval. The area of this representative rectangle is

$$\Delta A_i = (\text{height})(\text{width}) = \left[f(x_i) - g(x_i) \right] \Delta x.$$

By adding the areas of the n rectangles and taking the limit as $\|\Delta\| \to 0$ $(n \to \infty)$, you obtain

$$\lim_{n \to \infty} \sum_{i=1}^{n} \left[f(x_i) - g(x_i) \right] \Delta x.$$

Because f and g are continuous on $[a, b]$, $f - g$ is also continuous on $[a, b]$ and the limit exists. Therefore, the area of the given region is

$$\text{Area} = \lim_{n \to \infty} \sum_{i=1}^{n} \left[f(x_i) - g(x_i) \right] \Delta x$$

$$= \int_a^b \left[f(x) - g(x) \right] dx.$$

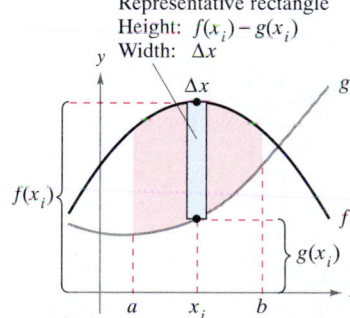

Representative rectangle
Height: $f(x_i) - g(x_i)$
Width: Δx

Figure 6.3

Area of a Region Between Two Curves

If f and g are continuous on $[a, b]$ and $g(x) \leq f(x)$ for all x in $[a, b]$, then the area of the region bounded by the graphs of f and g and the vertical lines $x = a$ and $x = b$ is

$$A = \int_a^b [f(x) - g(x)] \, dx.$$

In Figure 6.1, the graphs of f and g are shown above the x-axis. This, however is not necessary. The same integrand $[f(x) - g(x)]$ can be used as long as f and g are continuous and $g(x) \leq f(x)$ on the interval $[a, b]$. This result is summarized graphically in Figure 6.4.

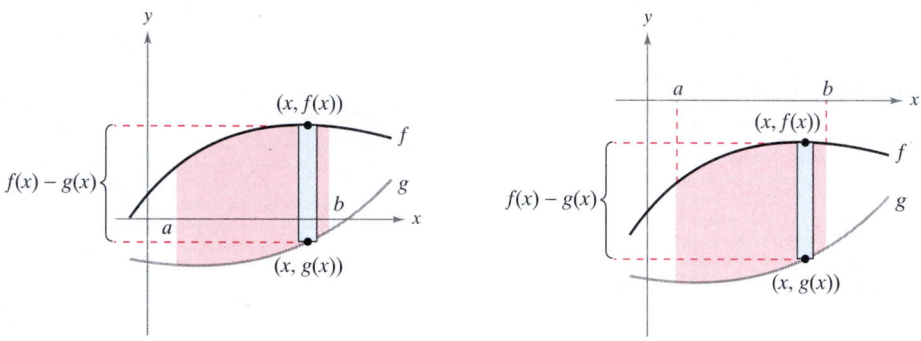

Figure 6.4

NOTE   The height of a representative rectangle is $f(x) - g(x)$ regardless of the relative position of the x-axis, as shown in Figure 6.4.

Representative rectangles are used throughout this chapter in various applications of integration. A vertical rectangle (of width Δx) implies integration with respect to x, whereas a horizontal rectangle (of width Δy) implies integration with respect to y.

EXAMPLE 1   Finding the Area of a Region Between Two Curves

Find the area of the region bounded by the graphs of $y = x^2 + 2$, $y = -x$, $x = 0$, and $x = 1$.

Solution   Let $g(x) = -x$ and $f(x) = x^2 + 2$. Then $g(x) \leq f(x)$ for all x in $[0, 1]$, as shown in Figure 6.5. Thus, the area of the representative rectangle is

$$\Delta A = [f(x) - g(x)] \, \Delta x = [(x^2 + 2) - (-x)] \, \Delta x$$

and the area of the region is

$$A = \int_a^b [f(x) - g(x)] \, dx = \int_0^1 [(x^2 + 2) - (-x)] \, dx$$

$$= \left[\frac{x^3}{3} + \frac{x^2}{2} + 2x \right]_0^1$$

$$= \frac{1}{3} + \frac{1}{2} + 2$$

$$= \frac{17}{6}.$$

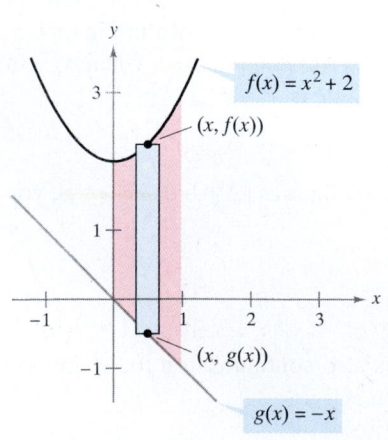

Region bounded by the graph of f, the graph of g, $x = 0$, and $x = 1$

Figure 6.5

Area of a Region Between Intersecting Curves

In Example 1, the graphs of $f(x) = x^2 + 2$ and $g(x) = -x$ do not intersect, and the values of a and b are given explicitly. A more common problem involves the area of a region bounded by two *intersecting* graphs, where the values of a and b must be calculated.

EXAMPLE 2   A Region Lying Between Two Intersecting Graphs

Find the area of the region bounded by the graphs of $f(x) = 2 - x^2$ and $g(x) = x$.

Solution   In Figure 6.6, notice that the graphs of f and g have two points of intersection. To find the x-coordinates of these points, set $f(x)$ and $g(x)$ equal to each other and solve for x.

$$2 - x^2 = x \qquad \text{Set } f(x) \text{ equal to } g(x).$$
$$-x^2 - x + 2 = 0 \qquad \text{Write in standard form.}$$
$$-(x + 2)(x - 1) = 0 \qquad \text{Factor.}$$
$$x = -2 \text{ or } 1 \qquad \text{Solve for } x.$$

Thus, $a = -2$ and $b = 1$. Because $g(x) \le f(x)$ on the interval $[-2, 1]$, the representative rectangle has an area of

$$\Delta A = [f(x) - g(x)]\,\Delta x = [(2 - x^2) - x]\,\Delta x$$

and the area of the region is

$$A = \int_{-2}^{1} [(2 - x^2) - x]\,dx = \left[-\frac{x^3}{3} - \frac{x^2}{2} + 2x \right]_{-2}^{1}$$
$$= \frac{9}{2}.$$

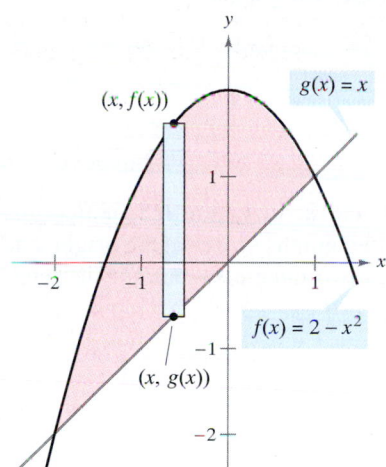

Region bounded by the graph of f and the graph of g
Figure 6.6

EXAMPLE 3   A Region Lying Between Two Intersecting Graphs

The sine and cosine curves intersect infinitely many times, bounding regions of equal areas, as shown in Figure 6.7. Find the area of one of these regions.

Solution

$$\sin x = \cos x \qquad \text{Set } f(x) \text{ equal to } g(x).$$
$$\frac{\sin x}{\cos x} = 1 \qquad \text{Divide both sides by } \cos x.$$
$$\tan x = 1 \qquad \text{Trigonometric identity}$$
$$x = \frac{\pi}{4} \text{ or } \frac{5\pi}{4}, \qquad 0 \le x \le 2\pi \qquad \text{Solve for } x.$$

Thus, $a = \pi/4$ and $b = 5\pi/4$. Because $\sin x \ge \cos x$ on the interval $[\pi/4, 5\pi/4]$, the area of the region is

$$A = \int_{\pi/4}^{5\pi/4} [\sin x - \cos x]\,dx = \left[-\cos x - \sin x \right]_{\pi/4}^{5\pi/4}$$
$$= 2\sqrt{2}.$$

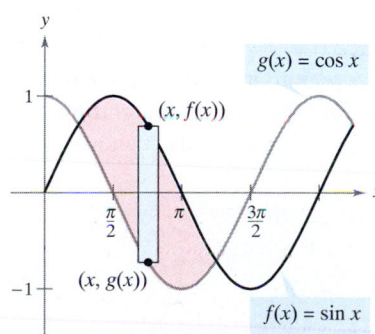

One of the regions bounded by the graphs of the sine and cosine functions
Figure 6.7

If two curves intersect at *more* than two points, then to find the area of the region between the curves, you must find all points of intersection and check to see which curve is above the other in each interval determined by these points.

EXAMPLE 4   Curves That Intersect at More than Two Points

Find the area of the region between the graphs of $f(x) = 3x^3 - x^2 - 10x$ and $g(x) = -x^2 + 2x$.

Solution   Begin by setting $f(x)$ and $g(x)$ equal to each other and solving for x. This yields the x-values of each point of intersection of the two graphs.

$3x^3 - x^2 - 10x = -x^2 + 2x$	Set $f(x)$ equal to $g(x)$.
$3x^3 - 12x = 0$	Write in standard form.
$3x(x^2 - 4) = 0$	Factor.
$x = -2, 0, 2$	Solve for x.

Thus, the two graphs intersect when $x = -2, 0$, and 2. In Figure 6.8, notice that $g(x) \leq f(x)$ on the interval $[-2, 0]$. However, the two graphs switch at the origin, and $f(x) \leq g(x)$ on the interval $[0, 2]$. Hence, you need two integrals—one for the interval $[-2, 0]$ and one for $[0, 2]$.

$$A = \int_{-2}^{0} [f(x) - g(x)]\, dx + \int_{0}^{2} [g(x) - f(x)]\, dx$$

$$= \int_{-2}^{0} (3x^3 - 12x)\, dx + \int_{0}^{2} (-3x^3 + 12x)\, dx$$

$$= \left[\frac{3x^4}{4} - 6x^2 \right]_{-2}^{0} + \left[\frac{-3x^4}{4} + 6x^2 \right]_{0}^{2}$$

$$= -(12 - 24) + (-12 + 24)$$

$$= 24$$

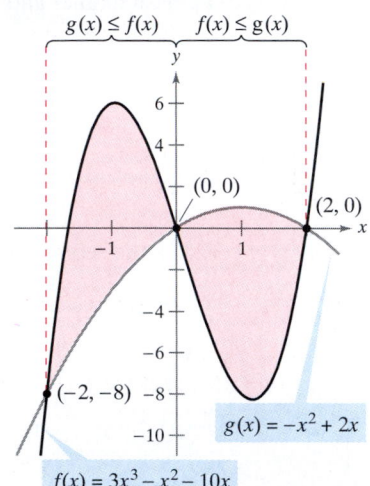

In $[-2, 0]$, $g(x) \leq f(x)$, and in $[0, 2]$, $f(x) \leq g(x)$.
Figure 6.8

NOTE   In Example 4, notice that you get an incorrect result if you integrate from -2 to 2. Such integration produces

$$\int_{-2}^{2} [f(x) - g(x)]\, dx = \int_{-2}^{2} (3x^3 - 12x)\, dx = 0.$$

If the graph of a function of y is a boundary of a region, it is often convenient to use representative rectangles that are *horizontal* and find the area by integrating with respect to y. In general, to determine the area between two curves, you can use

$$A = \int_{x_1}^{x_2} \underbrace{[(\text{top curve}) - (\text{bottom curve})]}_{\text{in variable } x}\, dx \qquad \text{Vertical rectangles}$$

$$A = \int_{y_1}^{y_2} \underbrace{[(\text{right curve}) - (\text{left curve})]}_{\text{in variable } y}\, dy \qquad \text{Horizontal rectangles}$$

where (x_1, y_1) and (x_2, y_2) are either adjacent points of intersection of the two curves involved or points on the specified boundary lines.

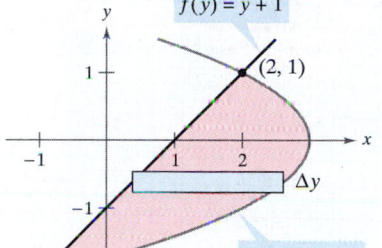

Horizontal rectangles (integration with respect to y)
Figure 6.9

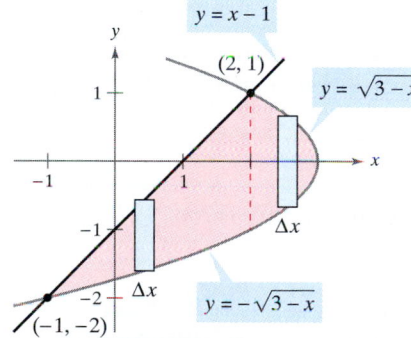

Vertical rectangles (integration with respect to x)
Figure 6.10

EXAMPLE 5   Horizontal Representative Rectangles

Find the area of the region bounded by the graphs of $x = 3 - y^2$ and $x = y + 1$.

Solution   Consider $g(y) = 3 - y^2$ and $f(y) = y + 1$. These two curves intersect when $y = -2$ and $y = 1$, as shown in Figure 6.9. Because $f(y) \leq g(y)$ on this interval, you have

$$\Delta A = [g(y) - f(y)]\,\Delta y = [(3 - y^2) - (y + 1)]\,\Delta y.$$

Hence, the area is

$$A = \int_{-2}^{1} [(3 - y^2) - (y + 1)]\,dy$$

$$= \int_{-2}^{1} (-y^2 - y + 2)\,dy$$

$$= \left[\frac{-y^3}{3} - \frac{y^2}{2} + 2y \right]_{-2}^{1}$$

$$= \left(-\frac{1}{3} - \frac{1}{2} + 2 \right) - \left(\frac{8}{3} - 2 - 4 \right)$$

$$= \frac{9}{2}.$$

NOTE   In Example 5, notice that by integrating with respect to y we need only one integral. If we had integrated with respect to x, we would have needed two integrals

$$A = \int_{-1}^{2} \left[(x - 1) + \sqrt{3 - x} \right] dx + \int_{2}^{3} \left(\sqrt{3 - x} + \sqrt{3 - x} \right) dx$$

because the upper boundary changes at $x = 2$, as shown in Figure 6.10.

In this section, we developed the integration formula for the area between two curves by using a rectangle as the *representative element*. For each new application in the remaining sections of this chapter, we will construct an appropriate representative element using precalculus formulas you already know. Each integration formula then will be obtained by summing or accumulating these representative elements.

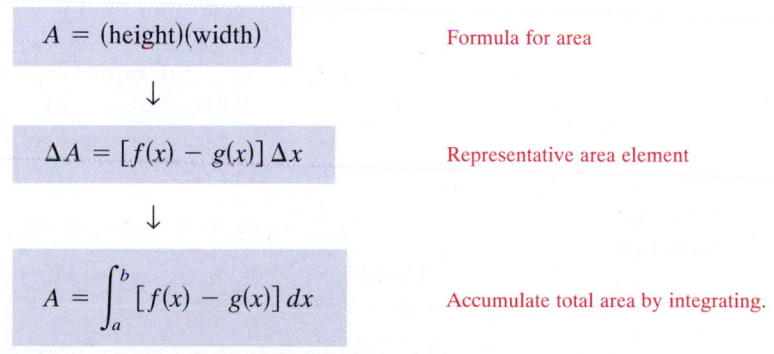

For example, in this section we developed the area formula as follows.

$$A = (\text{height})(\text{width})$$ Formula for area

$$\downarrow$$

$$\Delta A = [f(x) - g(x)]\,\Delta x$$ Representative area element

$$\downarrow$$

$$A = \int_{a}^{b} [f(x) - g(x)]\,dx$$ Accumulate total area by integrating.

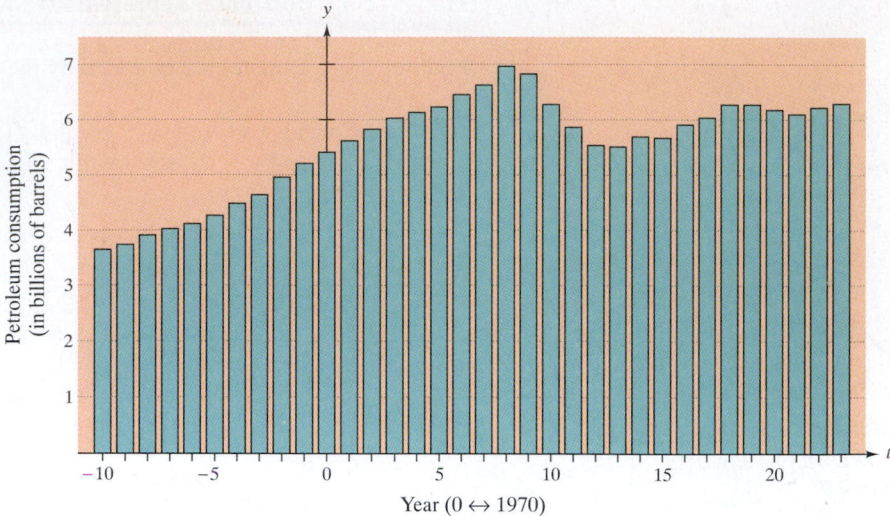

From 1960 to 1979, the consumption of petroleum in the United States followed a pattern that was approximately linear. In the late 1970s, however, when crude oil prices increased dramatically, the consumption pattern changed, as shown in the bar graph. *(Source: U.S. Energy Information Administration)*
Figure 6.11

EXAMPLE 6   Consumption of Petroleum

Petroleum consumption patterns have changed in recent years (see Figure 6.11). The rate of consumption (in billions of barrels per year) of petroleum in the United States from 1960 to 1979 can be modeled by

$$f(t) = 0.18t + 5.38, \qquad -10 \leq t \leq 9$$

where $t = 0$ corresponds to the year starting January 1, 1970. From 1979 through 1993, the rate of consumption can be modeled by

$$g(t) = -0.0029t^3 + 0.149t^2 - 2.42t + 18.38, \qquad 9 \leq t \leq 23$$

as shown in Figure 6.12. Find the total amount of fuel saved from 1979 through 1993 as a result of fuel being consumed at the post-1979 rate rather than at the pre-1979 rate.

Solution   Because the graph of the pre-1979 model lies above the post-1979 graph on the interval $[9, 23]$, the amount of petroleum saved is given by

$$\text{Fuel saved} = \int_9^{23} [f(t) - g(t)] \, dt.$$

Using the given models for f and g, you can find this amount to be

$$\text{Fuel saved} = \int_9^{23} (0.0029t^3 - 0.149t^2 + 2.6t - 13) \, dt$$

$$= \left[0.000725t^4 - 0.049667t^3 + 1.3t^2 - 13t \right]_9^{23}$$

$$\approx 30.44 \text{ billion barrels.}$$

Therefore, approximately 30.44 billion barrels of petroleum were saved.

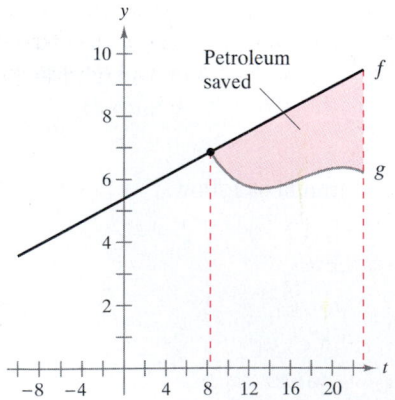

f: Pre-1979 consumption rate
g: Post-1979 consumption rate
Figure 6.12

EXERCISES FOR SECTION 6.1

In Exercises 1–6, set up the definite integral that gives the area of the region.

1. $f(x) = x^2 - 6x$
 $g(x) = 0$

2. $f(x) = x^2 + 2x + 1$
 $g(x) = 2x + 5$

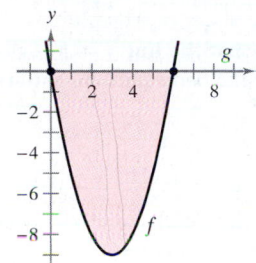 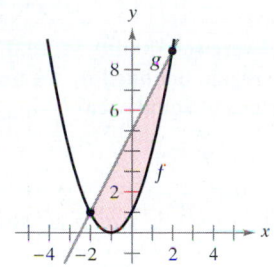

3. $f(x) = x^2 - 4x + 3$
 $g(x) = -x^2 + 2x + 3$

4. $f(x) = x^2$
 $g(x) = x^3$

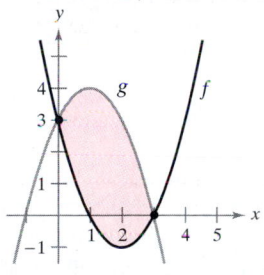 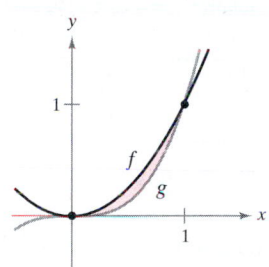

5. $f(x) = 3(x^3 - x)$
 $g(x) = 0$

6. $f(x) = (x - 1)^3$
 $g(x) = x - 1$

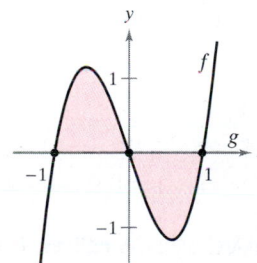 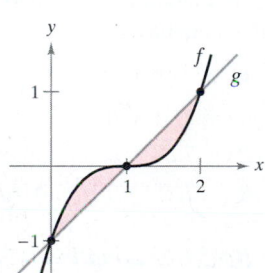

In Exercises 7–10, the integrand of the definite integral is a difference of two functions. Sketch the graph of each function and shade the region whose area is represented by the integral.

7. $\displaystyle\int_0^4 \left[(x + 1) - \frac{x}{2} \right] dx$

8. $\displaystyle\int_{-1}^1 \left[(1 - x^2) - (x^2 - 1) \right] dx$

9. $\displaystyle\int_0^6 \left[4(2^{-x/3}) - \frac{x}{6} \right] dx$

10. $\displaystyle\int_{-\pi/3}^{\pi/3} (2 - \sec x)\, dx$

Approximation In Exercises 11 and 12, determine which value best approximates the area of the region bounded by the graphs of f and g. (Make your selection on the basis of a sketch of the region and *not* by performing any calculations.)

11. $f(x) = x + 1$, $g(x) = (x - 1)^2$
 (a) -2 (b) 2 (c) 10 (d) 4 (e) 8

12. $f(x) = 2 - \frac{1}{2}x$, $g(x) = 2 - \sqrt{x}$
 (a) 1 (b) 6 (c) -3 (d) 3 (e) 4

In Exercises 13–26, sketch the region bounded by the graphs of the algebraic functions and find the area of the region.

13. $f(x) = x^2 - 4x$, $g(x) = 0$

14. $f(x) = 3 - 2x - x^2$, $g(x) = 0$

15. $f(x) = x^2 + 2x + 1$, $g(x) = 3x + 3$

16. $f(x) = -x^2 + 4x + 2$, $g(x) = x + 2$

17. $y = x$, $y = 2 - x$, $y = 0$

18. $y = \dfrac{1}{x^2}$, $y = 0$, $x = 1$, $x = 5$

19. $f(x) = \sqrt{3x} + 1$, $g(x) = x + 1$

20. $f(x) = \sqrt[3]{x}$, $g(x) = x$

21. $f(y) = y^2$, $g(y) = y + 2$

22. $f(y) = y(2 - y)$, $g(y) = -y$

23. $f(y) = y^2 + 1$, $g(y) = 0$, $y = -1$, $y = 2$

24. $f(y) = \dfrac{y}{\sqrt{16 - y^2}}$, $g(y) = 0$, $y = 3$

25. $f(x) = \dfrac{4}{x}$, $x = 0$, $y = 1$, $y = 4$

26. $g(x) = \dfrac{4}{2 - x}$, $y = 4$, $x = 0$

In Exercises 27–36, use a graphing utility to graph the region bounded by the graphs of the functions and use the integration capabilities of the graphing utility to find the area of the region.

27. $f(x) = x(x^2 - 3x + 3)$, $g(x) = x^2$

28. $f(x) = x^3 - 2x + 1$, $g(x) = -2x$, $x = 1$

29. $y = x^2 - 4x + 3$, $y = 3 + 4x - x^2$

30. $y = x^4 - 2x^2$, $y = 2x^2$

31. $f(x) = x^4 - 4x^2$, $g(x) = x^2 - 4$

32. $f(x) = x^4 - 4x^2$, $g(x) = x^3 - 4x$

33. $f(x) = 1/(1 + x^2)$, $g(x) = \frac{1}{2}x^2$

34. $f(x) = 6x/(x^2 + 1)$, $y = 0$, $0 \le x \le 3$

35. $y = \sqrt{1 + x^3}, \quad y = \frac{1}{2}x + 2, \quad x = 0$

36. $y = x\sqrt{\dfrac{4 - x}{4 + x}}, \quad y = 0, \quad x = 4$

In Exercises 37–40, sketch the region bounded by the graphs of the transcendental functions and find the area of the region.

37. $f(x) = 2 \sin x, \quad g(x) = \tan x, \quad -\dfrac{\pi}{3} \le x \le \dfrac{\pi}{3}$

38. $f(x) = \sin 2x, \quad g(x) = \cos x, \quad \dfrac{\pi}{6} \le x \le \dfrac{5\pi}{6}$

39. $f(x) = xe^{-x^2}, \quad y = 0, \quad 0 \le x \le 1$

40. $f(x) = 3^x, \quad g(x) = 2x + 1$

In Exercises 41–44, use a graphing utility to graph the region bounded by the graphs of the functions and use the integration capabilities of the graphing utility to find the area of the region.

41. $f(x) = 2 \sin x + \sin 2x, \quad y = 0, \quad 0 \le x \le \pi$

42. $f(x) = 2 \sin x + \cos 2x, \quad y = 0, \quad 0 \le x \le \pi$

43. $f(x) = \dfrac{1}{x^2}e^{1/x}, \quad y = 0, \quad 1 \le x \le 3$

44. $g(x) = \dfrac{4 \ln x}{x}, \quad y = 0, \quad x = 5$

In Exercises 45 and 46, (a) use a graphing utility to graph the region bounded by the graphs of the equations. (b) Set up the integral giving the area of the region. Can you evaluate the integral by hand? (c) Use the integration capabilities of a graphing utility to approximate the area.

45. $y = \sqrt{\dfrac{x^3}{4 - x}}, \quad y = 0, \quad x = 3$

46. $y = \sqrt{x}\, e^x, \quad y = 0, \quad x = 0, \quad x = 1$

In Exercises 47 and 48, use integration to find the area of the triangle having the given vertices.

47. $(0, 0), (a, 0), (b, c)$ **48.** $(2, -3), (4, 6), (6, 1)$

In Exercises 49 and 50, set up and evaluate the definite integral that gives the area of the region bounded by the graph of the function and the tangent line to the graph at the indicated point.

49. $f(x) = x^3, (1, 1)$ **50.** $f(x) = \dfrac{1}{x^2 + 1}, \left(1, \dfrac{1}{2}\right)$

51. *Think About It* The graphs of $y = x^4 - 2x^2 + 1$ and $y = 1 - x^2$ intersect at three points. However, the area between the curves *can* be found by a single integral. Explain why this is so, and write an integral for this area.

52. *Think About It* The area of the region bounded by the graphs of $y = x^3$ and $y = x$ *cannot* be found by the single integral

$$\int_{-1}^{1} (x^3 - x)\, dx.$$

Explain why this is so. Use symmetry to write a single integral that does represent the area.

In Exercises 53 and 54, find b such that the line $y = b$ divides the region bounded by the graphs of the two equations into two regions of equal area.

53. $y = 9 - x^2, \quad y = 0$ **54.** $y = 9 - |x|, \quad y = 0$

In Exercises 55 and 56, evaluate the limit and sketch the graph of the region whose area is represented by the limit.

55. $\displaystyle\lim_{\|\Delta\| \to 0} \sum_{i=1}^{n} (x_i - x_i^2)\, \Delta x$

where $x_i = i/n$ and $\Delta x = 1/n$

56. $\displaystyle\lim_{\|\Delta\| \to 0} \sum_{i=1}^{n} (4 - x_i^2)\, \Delta x$

where $x_i = -2 + (4i/n)$ and $\Delta x = 4/n$

Revenue In Exercises 57 and 58, two models R_1 and R_2 are given for revenue (in billions of dollars per year) for a large corporation. The model R_1 gives projected annual revenues from 2000 to 2005, with $t = 0$ corresponding to 2000, and R_2 gives projected revenues if there is a decrease in the rate of growth of corporate sales over the period. Approximate the total reduction in revenue if corporate sales are actually closer to the model R_2.

57. $R_1 = 7.21 + 0.58t$
$R_2 = 7.21 + 0.45t$

58. $R_1 = 7.21 + 0.26t + 0.02t^2$
$R_2 = 7.21 + 0.1t + 0.01t^2$

59. *Beef Consumption* For the years 1985 through 1994, the rate of consumption of beef (in billions of pounds per year) in the United States can be modeled by

$$f(t) = \begin{cases} 27.77 - 0.36t, & 5 \le t \le 10 \\ 21.00 + 0.27t, & 10 \le t \le 14 \end{cases}$$

where t is the time in years, with $t = 5$ corresponding to 1985. (*Source: U.S. Department of Agriculture*)

(a) Use a graphing utility to graph the model.

(b) Suppose the rate of beef consumption for 1990 through 1994 had continued to follow the model for the years 1985 through 1990. How much less beef would have been consumed from 1990 through 1994?

60. *Fuel Cost* For the years 2000 to 2010, the projected rate of fuel cost C (in millions of dollars per year) for a corporation is

$$C_1 = 568.50 + 7.15t$$

where t is the time in years, with $t = 0$ corresponding to 2000. Because of the installation of fuel-saving equipment, a more accurate model for the rate of fuel costs for the period is

$$C_2 = 525.60 + 6.43t.$$

Approximate the savings for the 10-year period owing to the installation of the new equipment.

61. *Profit* The chief financial officer of a company reports that profits for the past fiscal year were $893,000. The officer predicts that profits for the next 5 years will grow at a continuous annual rate somewhere between $3\frac{1}{2}\%$ and 5%. Estimate the cumulative difference in total profit over the 5 years based on the predicted range of growth rates.

62. *Think About It* Congress is debating two proposals for eliminating the annual budget deficits by the year 2004. The rate of decrease for each proposal is shown in the figure. From the viewpoint of minimizing the cumulative federal deficit, which is the better proposal? Explain.

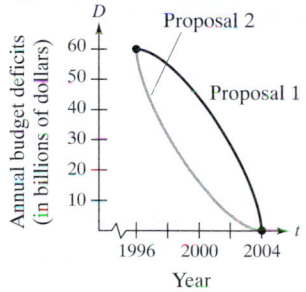

Figure for 62

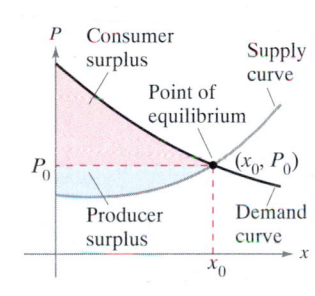

Figure for 67 and 68

63. *Area* The shaded region in the figure consists of the set of all points whose distances to the center of the square are less than the distances to the edges of the square. Find the area of the region.

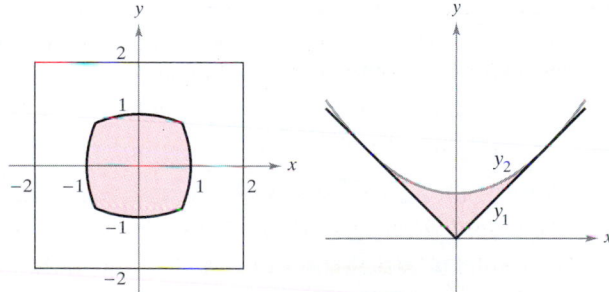

Figure for 63   Figure for 64

64. *Mechanical Design* The surface of a machine part is the region between the graphs of $y_1 = |x|$ and $y_2 = 0.08x^2 + k$ (see figure).

(a) Find k if the parabola is tangent to the graph of y_1.

(b) Find the surface area of the machine part.

65. *Building Design* Concrete sections for a new building have the dimensions (in meters) and shape shown in the figure.

(a) Find the area of the face of the section superimposed on the rectangular coordinate system.

(b) Find the volume of concrete in one of the sections by multiplying the area in part (a) by 2 meters.

(c) One cubic meter of concrete weighs 5000 pounds. Find the weight of the section.

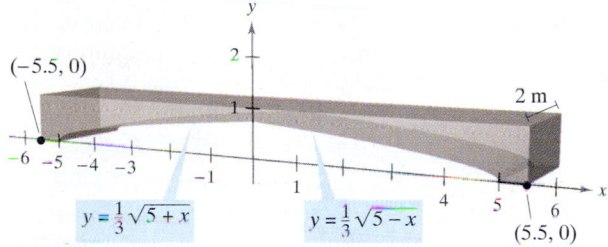

66. *Building Design* To decrease the weight and to aid in the hardening process, the concrete sections in Exercise 65 often are not solid. Rework Exercise 65 to allow for cylindrical openings such as those shown in the figure.

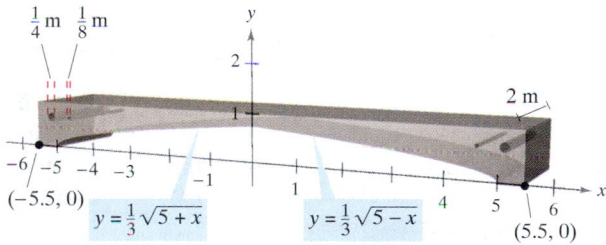

Consumer and Producer Surplus **In Exercises 67 and 68, find the consumer surplus and producer surplus for the supply and demand curves. The consumer surplus and producer surplus are represented by the areas shown in the figure.**

Demand Function	Supply Function
67. $p_1(x) = 50 - 0.5x$	$p_2(x) = 0.125x$
68. $p_1(x) = 1000 - 0.4x^2$	$p_2(x) = 42x$

True or False? **In Exercises 69 and 70, determine whether the statement is true or false. If it is false, explain why or give an example that shows it is false.**

69. If the area of the region bounded by the graphs of f and g is 1, then the area of the region bounded by the graphs of $h(x) = f(x) + C$ and $k(x) = g(x) + C$ is also 1.

70. If $\displaystyle\int_a^b [f(x) - g(x)]\,dx = A$, then $\displaystyle\int_a^b [g(x) - f(x)]\,dx = -A$.

The Disc Method • The Washer Method • Solids with Known Cross Sections

The Disc Method

In Chapter 4, we mentioned that area is only *one* of the many applications of the definite integral. Another important application is its use in finding the volume of a three-dimensional solid. In this section you will study a particular type of three-dimensional solid—one whose cross sections are similar. We begin with solids of revolution. Such solids are used commonly in engineering and manufacturing. Some examples are axles, funnels, pills, bottles, and pistons, as indicated in Figure 6.13.

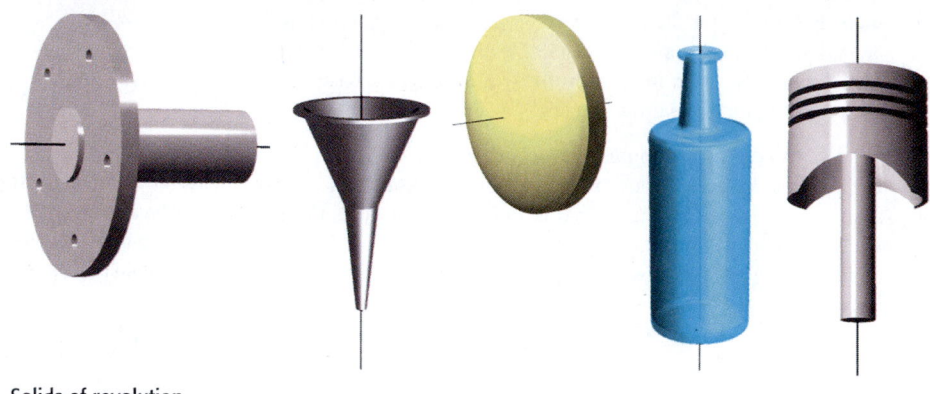

Solids of revolution
Figure 6.13

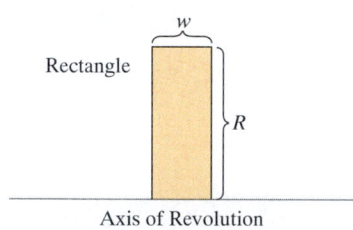

Rectangle

w

R

Axis of Revolution

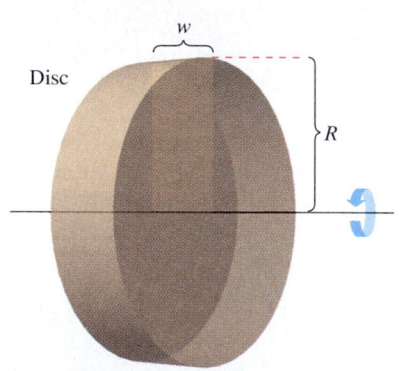

Disc

w

R

Volume of a disc: $\pi R^2 w$
Figure 6.14

If a region in the plane is revolved about a line, the resulting solid is a **solid of revolution,** and the line is called the **axis of revolution.** The simplest such solid is a right circular cylinder or **disc,** which is formed by revolving a rectangle about an axis adjacent to one side of the rectangle, as shown in Figure 6.14. The volume of such a disc is

Volume of disc = (area of disc)(width of disc)
= $\pi R^2 w$

where R is the radius of the disc and w is the width.

To see how to use the volume of a disc to find the volume of a general solid of revolution, consider a solid of revolution formed by revolving the plane region in Figure 6.15 about the indicated axis. To determine the volume of this solid, consider a representative rectangle in the plane region. When this rectangle is revolved about the axis of revolution, it generates a representative disc whose volume is

$$\Delta V = \pi R^2 \Delta x.$$

Approximating the volume of the solid by n such discs of width Δx and radius $R(x_i)$ produces

$$\text{Volume of solid} \approx \sum_{i=1}^{n} \pi [R(x_i)]^2 \Delta x$$

$$= \pi \sum_{i=1}^{n} [R(x_i)]^2 \Delta x.$$

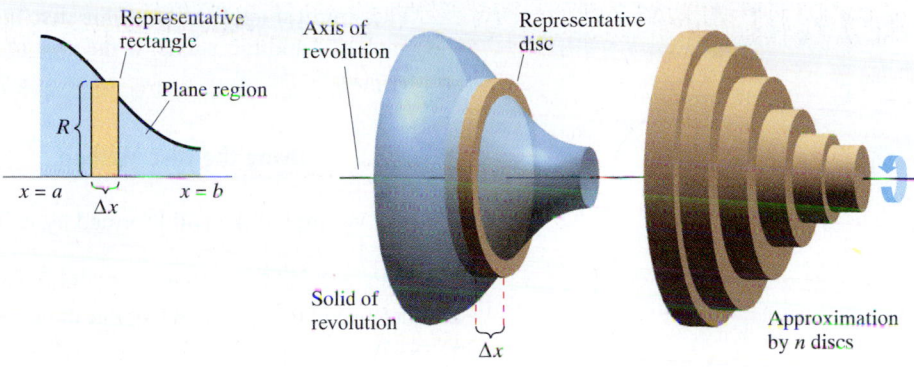

Disc method
Figure 6.15

This approximation appears to become better and better as $\|\Delta\| \to 0$ $(n \to \infty)$. Therefore, you can define the volume of the solid as

$$\text{Volume of solid} = \lim_{\|\Delta\| \to 0} \pi \sum_{i=1}^{n} [R(x_i)]^2 \, \Delta x$$

$$= \pi \int_a^b [R(x)]^2 \, dx.$$

Schematically, the disc method looks like this.

Known Precalculus Formula		*Representative Element*		*New Integration Formula*
Volume of disc $V = \pi R^2 w$	⇨	$\Delta V = \pi [R(x_i)]^2 \, \Delta x$	⇨	Solid of revolution $V = \pi \int_a^b [R(x)]^2 \, dx$

A similar formula can be derived if the axis of revolution is vertical.

(handwritten annotations:)
$Y = X + 3$
$X + 3$ use this because plugging in x will get a y value that is radius
$V = \pi \int_a^b [R(x)]^2 \, dx$

$R(x)$
Horizontal axis of revolution

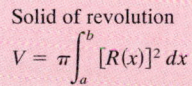

(handwritten annotations:)
use y + 3 because plugging in y will get an x that is radius
$V = \pi \int_c^d [R(y)]^2 \, dy$

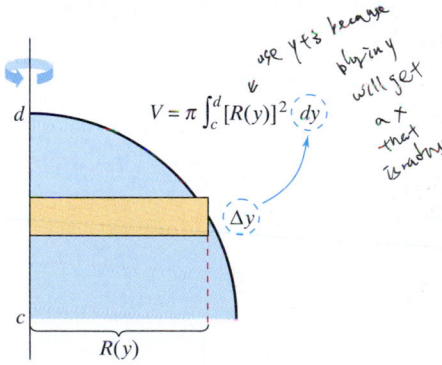

Vertical axis of revolution
Figure 6.16

The Disc Method

To find the volume of a solid of revolution with the **disc method**, use one of the following, as indicated in Figure 6.16.

Horizontal Axis of Revolution	*Vertical Axis of Revolution*
Volume $= V = \pi \int_a^b [R(x)]^2 \, dx$	Volume $= V = \pi \int_c^d [R(y)]^2 \, dy$

NOTE   In Figure 6.16, note that you can determine the variable of integration by placing a representative rectangle in the *plane* region "perpendicular" to the axis of revolution. If the width of the rectangle is Δx, integrate with respect to x, and if the width of the rectangle is Δy, integrate with respect to y.

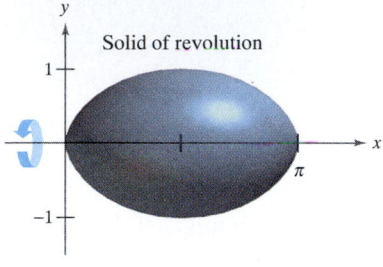

$f(x) = \sqrt{\sin x}$

$R(x)$

$\dfrac{\pi}{2}$ Δx π

Plane region

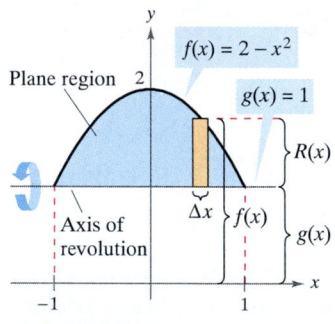

Solid of revolution

π

Figure 6.17

The simplest application of the disc method involves a plane region bounded by the graph of f and the x-axis. If the axis of revolution is the x-axis, the radius $R(x)$ is simply $f(x)$.

EXAMPLE 1   Using the Disc Method

Find the volume of the solid formed by revolving the region bounded by the graph of

$$f(x) = \sqrt{\sin x}$$

and the x-axis $(0 \le x \le \pi)$ about the x-axis.

Solution   From the representative rectangle in the upper graph in Figure 6.17, you can see that the radius of this solid is

$$R(x) = f(x) = \sqrt{\sin x}.$$

Thus, the volume of the solid of revolution is

$$V = \pi \int_a^b [R(x)]^2 \, dx = \pi \int_0^\pi \left(\sqrt{\sin x}\right)^2 dx$$

$$= \pi \int_0^\pi \sin x \, dx$$

$$= \pi \Big[-\cos x \Big]_0^\pi$$

$$= \pi(1 + 1)$$

$$= 2\pi.$$

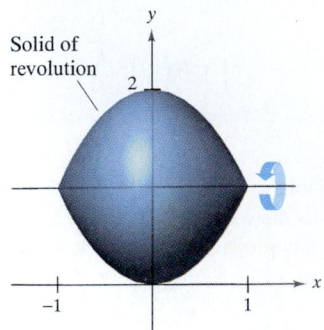

$f(x) = 2 - x^2$

Plane region   2

$g(x) = 1$

$R(x)$

Axis of
revolution

Δx $f(x)$

$g(x)$

-1 1

Plane region

Solid of
revolution

2

-1 1

Figure 6.18

EXAMPLE 2   Revolving About a Line That Is Not a Coordinate Axis

Find the volume of the solid formed by revolving the region bounded by $f(x) = 2 - x^2$ and $g(x) = 1$ about the line $y = 1$, as shown in Figure 6.18.

Solution   By equating $f(x)$ and $g(x)$, you can determine that the two graphs intersect when $x = \pm 1$. To find the radius, subtract $g(x)$ from $f(x)$.

$$R(x) = f(x) - g(x)$$

$$= (2 - x^2) - 1$$

$$= 1 - x^2$$

Finally, integrate between -1 and 1 to find the volume.

$$V = \pi \int_a^b [R(x)]^2 \, dx = \pi \int_{-1}^1 (1 - x^2)^2 \, dx$$

$$= \pi \int_{-1}^1 (1 - 2x^2 + x^4) \, dx$$

$$= \pi \left[x - \frac{2x^3}{3} + \frac{x^5}{5} \right]_{-1}^1$$

$$= \frac{16\pi}{15}$$

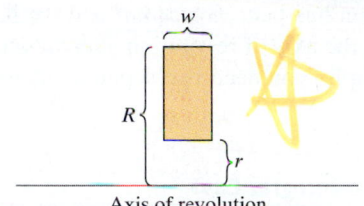

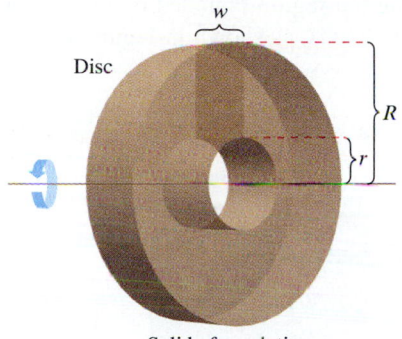

Figure 6.19

The Washer Method

The disc method can be extended to cover solids of revolution with holes by replacing the representative disc with a representative **washer.** The washer is formed by revolving a rectangle about an axis, as shown in Figure 6.19. If r and R are the inner and outer radii of the washer and w is the width of the washer, the volume is given by

Volume of washer $= \pi(R^2 - r^2)w$.

To see how this concept can be used to find the volume of a solid of revolution, consider a region bounded by an **outer radius** $R(x)$ and an **inner radius** $r(x)$, as shown in Figure 6.20. If the region is revolved about its axis of revolution, the volume of the resulting solid is given by

$$V = \pi \int_a^b \left([R(x)]^2 - [r(x)]^2 \right) dx.$$ Washer method

Note that the integral involving the inner radius represents the volume of the hole and is *subtracted* from the integral involving the outer radius.

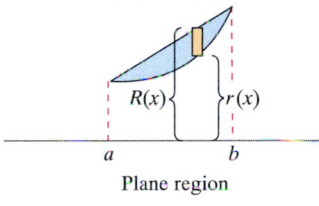

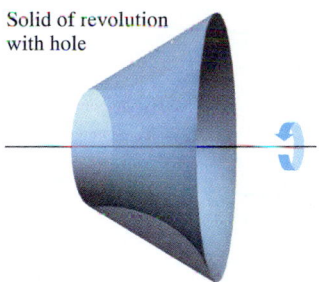

Figure 6.20

EXAMPLE 3   Using the Washer Method

Find the volume of the solid formed by revolving the region bounded by the graphs of $y = \sqrt{x}$ and $y = x^2$ about the x-axis, as shown in Figure 6.21.

Solution   In Figure 6.21, you can see that the outer and inner radii are as follows.

$R(x) = \sqrt{x}$ Outer radius

$r(x) = x^2$ Inner radius

Integrating between 0 and 1 produces

$$V = \pi \int_a^b \left\{ [R(x)]^2 - [r(x)]^2 \right\} dx$$

$$= \pi \int_0^1 (x - x^4) \, dx$$

$$= \pi \left[\frac{x^2}{2} - \frac{x^5}{5} \right]_0^1$$

$$= \frac{3\pi}{10}.$$

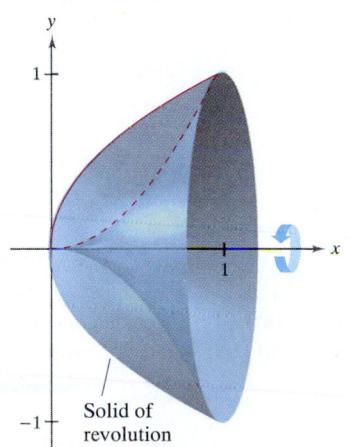

Solid of revolution
Figure 6.21

In each example so far, the axis of revolution has been *horizontal* and we have integrated with respect to x. In the next example, the axis of revolution is *vertical* and you must integrate with respect to y. In this example, you need two separate integrals to compute the volume.

EXAMPLE 4   Integrating with Respect to y, Two-Integral Case

Find the volume of the solid formed by revolving the region bounded by the graphs of $y = x^2 + 1$, $y = 0$, $x = 0$, and $x = 1$ about the y-axis, as shown in Figure 6.22.

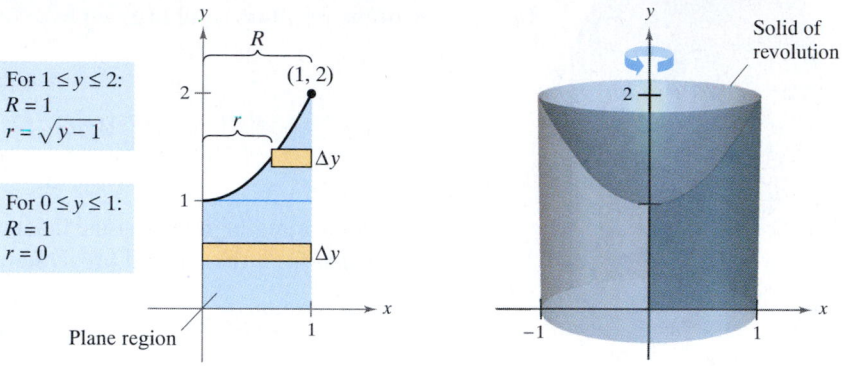

For $1 \leq y \leq 2$:
$R = 1$
$r = \sqrt{y - 1}$

For $0 \leq y \leq 1$:
$R = 1$
$r = 0$

Plane region

Solid of revolution

Figure 6.22

Solution   For the region shown in Figure 6.22, the outer radius is simply $R = 1$. There is, however, no convenient formula that represents the inner radius. When $0 \leq y \leq 1$, $r = 0$, but when $1 \leq y \leq 2$, r is determined by the equation $y = x^2 + 1$, which implies that $r = \sqrt{y - 1}$.

$$r(y) = \begin{cases} 0, & 0 \leq y \leq 1 \\ \sqrt{y - 1}, & 1 \leq y \leq 2 \end{cases}$$

Using this definition of the inner radius, you can use two integrals to find the volume.

$$V = \pi \int_0^1 (1^2 - 0^2)\, dy + \pi \int_1^2 \left[1^2 - \left(\sqrt{y - 1} \right)^2 \right] dy$$

$$= \pi \int_0^1 1\, dy + \pi \int_1^2 (2 - y)\, dy$$

$$= \pi \left[y \right]_0^1 + \pi \left[2y - \frac{y^2}{2} \right]_1^2$$

$$= \pi + \pi \left(4 - 2 - 2 + \frac{1}{2} \right) = \frac{3\pi}{2}$$

Note that the first integral $\pi \int_0^1 1\, dy$ represents the volume of a right circular cylinder of radius 1 and height 1. This portion of the volume could have been determined without using calculus.

TECHNOLOGY   Some graphing utilities allow you to view a solid of revolution. If you have access to such software, try using it to sketch some of the solids of revolution described in this section. For instance, the solid in Example 4 might appear like that shown in Figure 6.23.

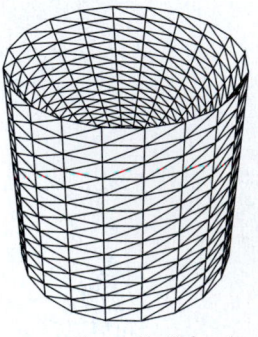

Generated by Mathematica

Figure 6.23

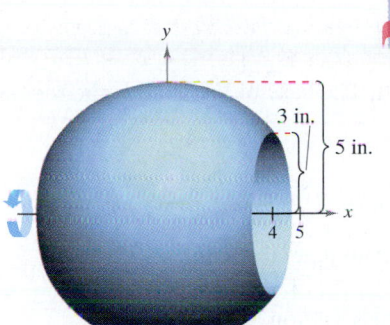

Solid of revolution

(a)

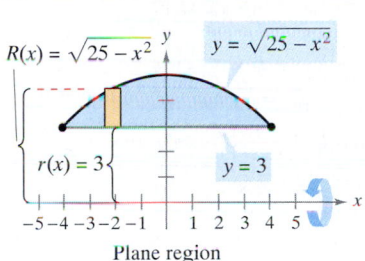

Plane region

(b)

Figure 6.24

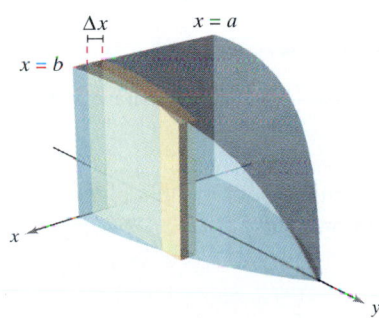

(a) Cross sections perpendicular to x-axis

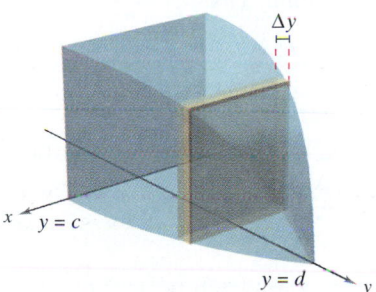

(b) Cross sections perpendicular to y-axis

Figure 6.25

⬤ **EXAMPLE 5   Manufacturing**

A manufacturer drills a hole through the center of a metal sphere of radius 5 inches, as shown in Figure 6.24(a). The hole has a radius of 3 inches. What is the volume of the resulting metal ring?

Solution   You can imagine the ring to be generated by a segment of the circle whose equation is $x^2 + y^2 = 25$, as shown in Figure 6.24(b). Because the radius of the hole is 3 inches, you can let $y = 3$ and solve the equation $x^2 + y^2 = 25$ to determine that the limits of integration are $x = \pm 4$. Thus, the inner and outer radii are

$$r(x) = 3 \quad \text{and} \quad R(x) = \sqrt{25 - x^2}$$

and the volume is given by

$$V = \pi \int_a^b \{ [R(x)]^2 - [r(x)]^2 \} \, dx$$

$$= \pi \int_{-4}^4 \left[\left(\sqrt{25 - x^2} \right)^2 - (3)^2 \right] dx$$

$$= \pi \int_{-4}^4 (16 - x^2) \, dx$$

$$= \pi \left[16x - \frac{x^3}{3} \right]_{-4}^4$$

$$= \frac{256\pi}{3} \text{ cubic inches.}$$

Solids with Known Cross Sections

With the disc method, you can find the volume of a solid having a circular cross section whose area is

$$A = \pi R^2.$$

This method can be generalized to solids of any shape, as long as you know a formula for the area of an arbitrary cross section. Some common cross sections are squares, rectangles, triangles, semicircles, and trapezoids.

> **Volumes of Solids with Known Cross Sections**
>
> 1. For cross sections of area $A(x)$ taken perpendicular to the x-axis,
>
> $$\text{Volume} = \int_a^b A(x) \, dx$$
>
> as shown in Figure 6.25(a).
>
> 2. For cross sections of area $A(y)$ taken perpendicular to the y-axis,
>
> $$\text{Volume} = \int_c^d A(y) \, dy$$
>
> as shown in Figure 6.25(b).

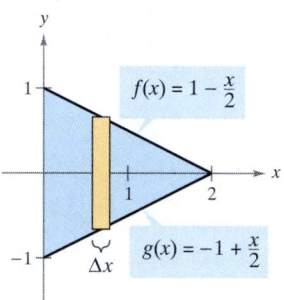

Cross sections are equilateral triangles.

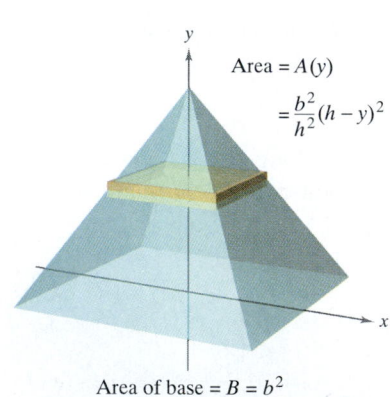

Triangular base in *xy*-plane

Figure 6.26

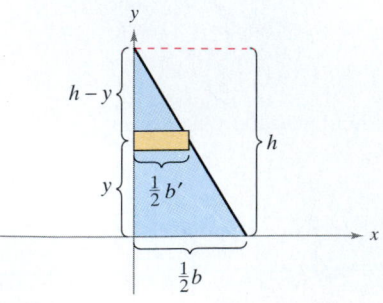

Figure 6.27

EXAMPLE 6   Triangular Cross Sections

Find the volume of the solid shown in Figure 6.26. The base of the solid is the region bounded by the lines

$$f(x) = 1 - \frac{x}{2}, \qquad g(x) = -1 + \frac{x}{2}, \qquad \text{and} \qquad x = 0.$$

The cross sections perpendicular to the *x*-axis are equilateral triangles.

Solution   The base and area of each triangular cross section are as follows.

$$\text{Base} = \left(1 - \frac{x}{2}\right) - \left(-1 + \frac{x}{2}\right) = 2 - x \qquad \text{Length of base}$$

$$\text{Area} = \frac{\sqrt{3}}{4}(\text{base})^2 \qquad \text{Area of equilateral triangle}$$

$$A(x) = \frac{\sqrt{3}}{4}(2 - x)^2 \qquad \text{Area of cross section}$$

Because *x* ranges from 0 to 2, the volume of the solid is

$$V = \int_a^b A(x)\,dx = \int_0^2 \frac{\sqrt{3}}{4}(2 - x)^2\,dx$$

$$= -\frac{\sqrt{3}}{4}\left[\frac{(2 - x)^3}{3}\right]_0^2$$

$$= \frac{2\sqrt{3}}{3}.$$

EXAMPLE 7   An Application to Geometry

Prove that the volume of a pyramid with a square base is $V = \frac{1}{3}hB$, where *h* is the height of the pyramid and *B* is the area of the base.

Solution   As shown in Figure 6.27, you can intersect the pyramid with a plane parallel to the base at height *y* to form a square cross section whose sides are of length *b′*. Using similar triangles, you can show that

$$\frac{b'}{b} = \frac{h - y}{h} \qquad \text{or} \qquad b' = \frac{b}{h}(h - y)$$

where *b* is the length of the sides of the base of the pyramid. Thus,

$$A(y) = (b')^2 = \frac{b^2}{h^2}(h - y)^2.$$

Integrating between 0 and *h* produces

$$V = \int_0^h \frac{b^2}{h^2}(h - y)^2\,dy = \frac{b^2}{h^2}\int_0^h (h - y)^2\,dy$$

$$= -\left(\frac{b^2}{h^2}\right)\left[\frac{(h - y)^3}{3}\right]_0^h$$

$$= \frac{b^2}{h^2}\left(\frac{h^3}{3}\right)$$

$$= \frac{1}{3}hB. \qquad B = b^2$$

EXERCISES FOR SECTION 6.2

In Exercises 1–6, set up and evaluate the integral that gives the volume of the solid formed by revolving the region about the *x*-axis.

1. $y = -x + 1$

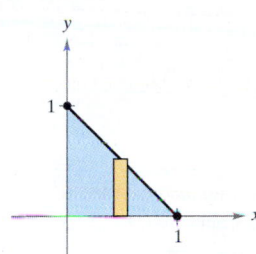

2. $y = 4 - x^2$

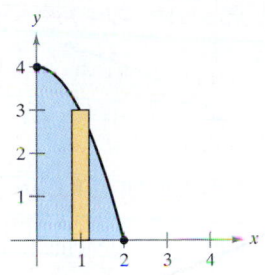

3. $y = \sqrt{x}$

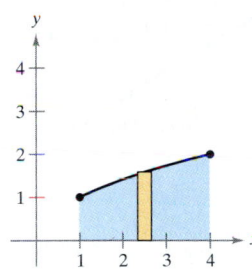

4. $y = \sqrt{4 - x^2}$

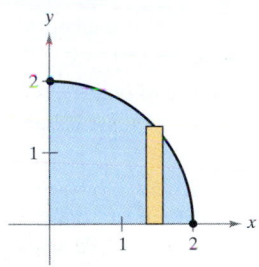

5. $y = x^2$, $y = x^3$

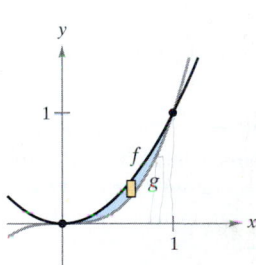

6. $y = 2$, $y = 4 - \dfrac{x^2}{4}$

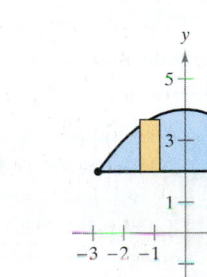

In Exercises 7–10, set up and evaluate the integral that gives the volume of the solid formed by revolving the region about the *y*-axis.

7. $y = x^2$

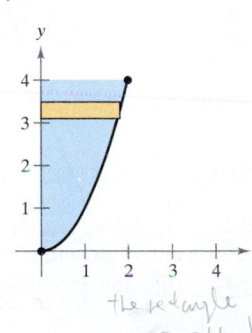

the rectangle
is attached to y
No space.

8. $y = \sqrt{16 - x^2}$

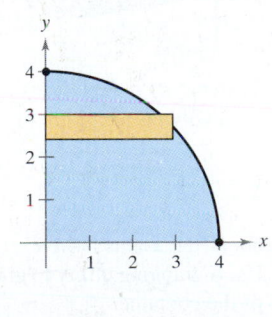

9. $y = x^{2/3}$

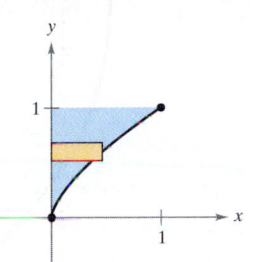

10. $x = -y^2 + 4y$

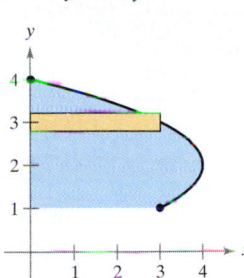

In Exercises 11–14, find the volume of the solid generated by revolving the region bounded by the graphs of the equations about the indicated lines.

11. $y = \sqrt{x}$, $y = 0$, $x = 4$
  (a) the *x*-axis   (b) the *y*-axis
  (c) the line $x = 4$ (d) the line $x = 6$

12. $y = 2x^2$, $y = 0$, $x = 2$
  (a) the *y*-axis   (b) the *x*-axis
  (c) the line $y = 8$ (d) the line $x = 2$

13. $y = x^2$, $y = 4x - x^2$
  (a) the *x*-axis   (b) the line $y = 6$

14. $y = 6 - 2x - x^2$, $y = x + 6$
  (a) the *x*-axis   (b) the line $y = 3$

In Exercises 15–18, find the volume of the solid generated by revolving the region bounded by the graphs of the equations about the line $y = 4$.

15. $y = x$, $y = 3$, $x = 0$

16. $y = x^2$, $y = 4$

17. $y = \dfrac{1}{x}$, $y = 0$, $x = 1$, $x = 4$

18. $y = \sec x$, $y = 0$, $0 \le x \le \dfrac{\pi}{3}$

In Exercises 19–22, find the volume of the solid generated by revolving the region bounded by the graphs of the equations about the line $x = 6$.

19. $y = x$, $y = 0$, $y = 4$, $x = 6$

20. $y = 6 - x$, $y = 0$, $y = 4$, $x = 0$

21. $x = y^2$, $x = 4$

22. $xy = 6$, $y = 2$, $y = 6$, $x = 6$

In Exercises 23–28, find the volume of the solid generated by revolving the region bounded by the graphs of the equations about the x-axis.

23. $y = \dfrac{1}{\sqrt{x+1}}, \quad y = 0, \quad x = 0, \quad x = 3$

24. $y = x\sqrt{4 - x^2}, \quad y = 0$

25. $y = \dfrac{1}{x}, \quad y = 0, \quad x = 1, \quad x = 4$

26. $y = \dfrac{3}{x+1}, \quad y = 0, \quad x = 0, \quad x = 8$

27. $y = e^{-x}, \quad y = 0, \quad x = 0, \quad x = 1$

28. $y = e^{x/2}, y = 0, \quad x = 0, \quad x = 4$

In Exercises 29 and 30, find the volume of the solid generated by revolving the region bounded by the graphs of the equations about the y-axis.

29. $y = 3(2 - x), \quad y = 0, \quad x = 0$

30. $y = 9 - x^2, \quad y = 0, \quad x = 2, \quad x = 3$

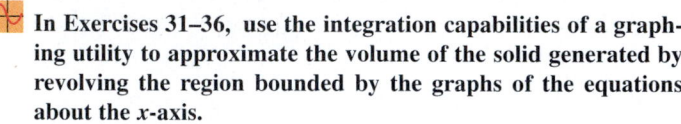 **In Exercises 31–36, use the integration capabilities of a graphing utility to approximate the volume of the solid generated by revolving the region bounded by the graphs of the equations about the x-axis.**

31. $y = \sin x, \quad y = 0, \quad x = 0, \quad x = \pi$

32. $y = \cos x, \quad y = 0, \quad x = 0, \quad x = \dfrac{\pi}{2}$

33. $y = e^{-x^2}, \quad y = 0, \quad x = 0, \quad x = 2$

34. $y = \ln x, \quad y = 0, \quad x = 1, \quad x = 3$

35. $y = e^{x/2} + e^{-x/2}, \quad y = 0, \quad x = -1, \quad x = 2$

36. $y = 2\arctan(0.2x), \quad y = 0, \quad x = 0, \quad x = 5$

Approximation **In Exercises 37 and 38, determine which value best approximates the volume of the solid generated by revolving the region bounded by the graphs of the equations about the x-axis. (Make your selection on the basis of a sketch of the solid and *not* by performing any calculations.)**

37. $y = e^{-x^2/2}, \quad y = 0, \quad x = 0, \quad x = 2$

  (a) 3   (b) −5   (c) 10   (d) 7   (e) 20

38. $y = \arctan x, \quad y = 0, \quad x = 0, \quad x = 1$

  (a) 10   (b) $\frac{3}{4}$ (c) 5   (d) −6   (e) 15

39. *Think About It*

  (a) The region bounded by the parabola $y = 4x - x^2$ and the x-axis is revolved about the x-axis. Find the volume of the resulting solid.

  (b) If the equation of the parabola in part (a) were changed to $y = 4 - x^2$, would the volume of the solid generated be different? Explain.

40. *Think About It* The region in the figure is revolved about the indicated axes and line. Order the volumes of the resulting solids from least to greatest. Explain your reasoning.

  (a) x-axis   (b) y-axis   (c) $x = 8$

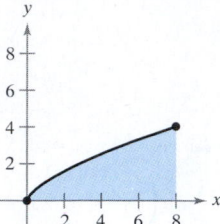

41. If the portion of the line $y = \frac{1}{2}x$ lying in the first quadrant is revolved about the x-axis, a cone is generated. Find the volume of the cone extending from $x = 0$ to $x = 6$.

42. Use the disc method to verify that the volume of a right circular cone is $\frac{1}{3}\pi r^2 h$, where r is the radius of the base and h is the height.

43. Use the disc method to verify that the volume of a sphere is $\frac{4}{3}\pi r^3$.

44. A sphere of radius r is cut by a plane h ($h < r$) units above the equator. Find the volume of the solid (spherical segment) above the plane.

45. A cone with a base of radius r and height H is cut by a plane parallel to and h units above the base. Find the volume of the solid (frustum of a cone) below the plane.

46. The region bounded by $y = \sqrt{x}$, $y = 0$, $x = 0$, and $x = 4$ is revolved about the x-axis.

  (a) Find the value of x in the interval $[0, 4]$ that divides the solid into two parts of equal volume.

  (b) Find the values of x in the interval $[0, 4]$ that divide the solid into three parts of equal volume.

47. *Volume of a Fuel Tank* A tank on the wing of a jet aircraft is formed by revolving the region bounded by the graph of $y = \frac{1}{8}x^2\sqrt{2 - x}$ and the x-axis about the x-axis (see figure), where x and y are measured in meters. Find the tank's volume.

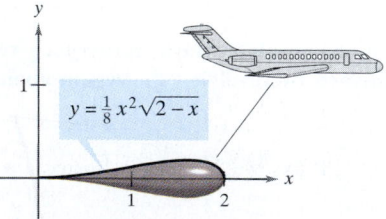

48. *Volume of a Lab Glass* A glass container can be modeled by revolving the graph of

$$y = \begin{cases} \sqrt{0.1x^3 - 2.2x^2 + 10.9x + 22.2}, & 0 \le x \le 11.5 \\ 2.95, & 11.5 < x \le 15 \end{cases}$$

about the x-axis, where x and y are measured in centimeters. Use a graphing utility to graph the function and find the volume of the container.

49. Find the volume of the solid generated if the upper half of the ellipse $9x^2 + 25y^2 = 225$ is revolved about

    (a) the x-axis to form a prolate spheroid (shaped like a football).

    (b) the y-axis to form an oblate spheroid (shaped like half of an M&M candy).

50. *Minimum Volume*   The arc of $y = 4 - (x^2/4)$ on the interval $[0, 4]$ is revolved about the line $y = b$ (see figure).

    (a) Find the volume of the resulting solid as a function of b.

    (b) Use a graphing utility to graph the function in part (a), and use the graph to approximate the value of b that minimizes the volume of the solid.

    (c) Use calculus to find the value of b that minimizes the volume of the solid, and compare the result with the answer to part (b).

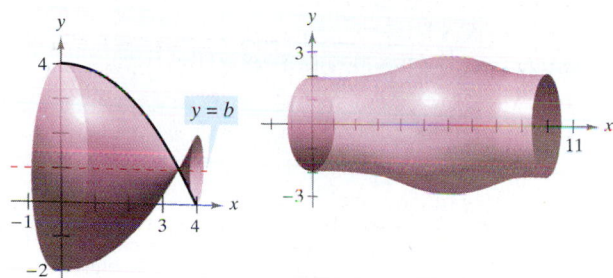

Figure for 50   **Figure for 52**

51. *Water Depth in a Tank*   A tank on a water tower is a sphere of radius 50 feet. Determine the depths of the water when the tank is filled to one-fourth and three-fourths of its total capacity. (*Note:* Use the root-finding capabilities of a graphing utility after evaluating the definite integral.)

52. *Modeling Data*   A draftsman is asked to determine the amount of material required to produce a machine part (see figure). The diameters d of the part at equally spaced points x are listed in the table. The measurements are listed in centimeters.

x	0	1	2	3	4	5
d	4.2	3.8	4.2	4.7	5.2	5.7

x	6	7	8	9	10
d	5.8	5.4	4.9	4.4	4.6

    (a) Use these data with Simpson's Rule to approximate the volume of the part.

    (b) Use the regression capabilities of a graphing utility to find a fourth-degree polynomial through the points representing the radius of the solid. Plot the data and graph the model.

    (c) Use a graphing utility to approximate the definite integral yielding the volume of the part. Compare the result with the answer to part (a).

53. *Think About It*   Match each integral with the solid whose volume it represents, and give the dimensions of each solid.

    (a) Right circular cylinder   (b) Ellipsoid

    (c) Sphere   (d) Right circular cone   (e) Torus

    (i) $\pi \displaystyle\int_0^h \left(\frac{rx}{h}\right)^2 dx$

    (ii) $\pi \displaystyle\int_0^h r^2\, dx$

    (iii) $\pi \displaystyle\int_{-r}^r \left(\sqrt{r^2 - x^2}\right)^2 dx$

    (iv) $\pi \displaystyle\int_{-b}^b \left(\sqrt{1 - \frac{x^2}{b^2}}\right)^2 dx$

    (v) $\pi \displaystyle\int_{-r}^r \left[\left(R + \sqrt{r^2 - x^2}\right)^2 - \left(R - \sqrt{r^2 - x^2}\right)^2\right] dx$

54. Find the volume of concrete in a ramp that is 2 meters wide and whose cross sections are right triangles with base 8 meters and height 1 meter (see figure).

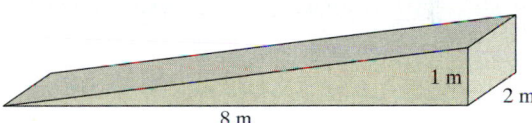

55. Find the volume of the solid whose base is bounded by the circle $x^2 + y^2 = 4$, with the indicated cross sections taken perpendicular to the x-axis.

    (a) Squares   (b) Equilateral triangles

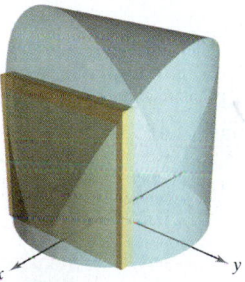 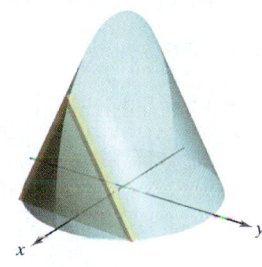

    (c) Semicircles   (d) Isosceles right triangles

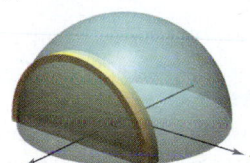 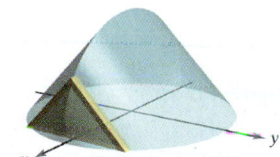

56. Find the volume of the solid whose base is bounded by the graphs of $y = x + 1$ and $y = x^2 - 1$, with the indicated cross sections taken perpendicular to the x-axis.

(a) Squares

(b) Rectangles of height 1

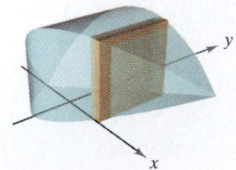
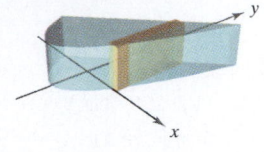

57. The base of a solid is bounded by $y = x^3$, $y = 0$, and $x = 1$. Find the volume of the solid for each of the following cross sections (taken perpendicular to the y-axis): (a) squares, (b) semicircles, (c) equilateral triangles, and (d) semiellipses whose heights are twice the lengths of their bases.

58. Find the volume of the solid of intersection (the solid common to both) of the two right circular cylinders of radius r whose axes meet at right angles (see figure).

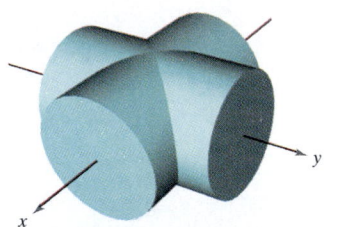

Two intersecting cylinders   Solid of intersection

FOR FURTHER INFORMATION For more information on this problem, see the article "Estimating the Volumes of Solid Figures with Curved Surfaces" by Donald Cohen in the May 1991 issue of *Mathematics Teacher*.

59. *Volume of Oil* A barrel of diameter d lies on a stand so that its axis makes an angle of 20° with its horizontal (see figure). The amount of motor oil in the barrel just covers the bottom of the barrel. Approximate the volume of oil.

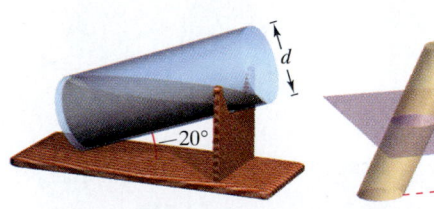
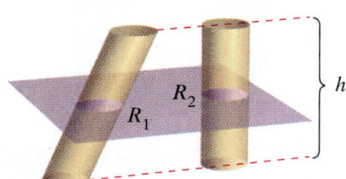

Area of R_1 = area of R_2

Figure for 59   **Figure for 60**

60. *Cavalieri's Theorem* Prove that if two solids have equal altitudes and all plane sections parallel to their bases and at equal distances from their bases have equal areas, then the solids have the same volume (see figure).

61. Two planes cut a right circular cylinder to form a wedge. One plane is perpendicular to the axis of the cylinder and the second makes an angle of θ degrees with the first (see figure).

(a) Find the volume of the wedge if $\theta = 45°$.

(b) Find the volume of the wedge for an arbitrary angle θ. Assuming that the cylinder has sufficient length, how does the volume of the wedge change as θ increases from 0° to 90°?

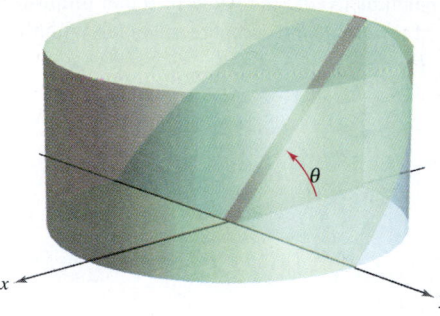

62. A manufacturer drills a hole through the center of a metal sphere of radius R. The hole has a radius r. Find the volume of the resulting ring.

63. For the metal sphere in Exercises 62, let $R = 5$. What value of r will produce a ring whose volume is exactly half the volume of the sphere.

64. The solid shown in the figure has cross sections bounded by the graph of

$$|x|^a + |y|^a = 1$$

where $1 \le a \le 2$.

(a) Describe the cross section when $a = 1$ and $a = 2$.

(b) Describe a procedure for approximating the volume of the solid.

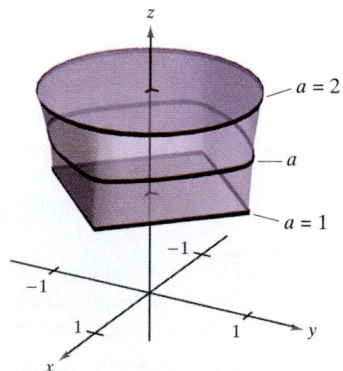

SECTION 6.3 Volume: The Shell Method

The Shell Method • Comparison of Disc and Shell Methods

The Shell Method

In this section, you will study an alternative method for finding the volume of a solid of revolution. This method is called the **shell method** because it uses cylindrical shells. We will compare the advantages of the disc and shell methods later in this section.

To begin, consider a representative rectangle as shown in Figure 6.28, where w is the width of the rectangle, h is the height of the rectangle, and p is the distance between the axis of revolution and the *center* of the rectangle. When this rectangle is revolved about its axis of revolution, it forms a cylindrical shell (or tube) of thickness w. To find the volume of this shell, consider two cylinders. The radius of the larger cylinder corresponds to the outer radius of the shell, and the radius of the smaller cylinder corresponds to the inner radius of the shell. Because p is the average radius of the shell, you know the outer radius is $p + (w/2)$ and the inner radius is $p - (w/2)$.

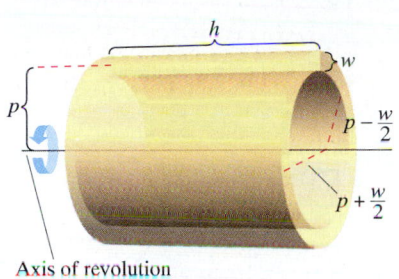

Figure 6.28

$$p + \frac{w}{2} \qquad \text{Outer radius}$$

$$p - \frac{w}{2} \qquad \text{Inner radius}$$

Thus, the volume of the shell is

Volume of shell = (volume of cylinder) − (volume of hole)

$$= \pi \left(p + \frac{w}{2}\right)^2 h - \pi \left(p - \frac{w}{2}\right)^2 h$$

$$= 2\pi p h w$$

$$= 2\pi (\text{average radius})(\text{height})(\text{thickness}).$$

You can use this formula to find the volume of a solid of revolution. Assume that the plane region in Figure 6.29 is revolved about a line to form the indicated solid. If you consider a horizontal rectangle of width Δy, then, as the plane region is revolved about a line parallel to the x-axis, the rectangle generates a representative shell whose volume is

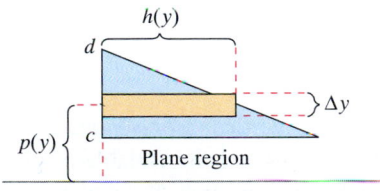

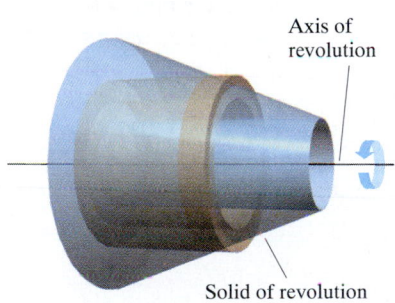

Figure 6.29

$$\Delta V = 2\pi [p(y)h(y)]\Delta y.$$

You can approximate the volume of the solid by n such shells of thickness Δy, height $h(y_i)$, and average radius $p(y_i)$.

$$\text{Volume of solid} \approx \sum_{i=1}^{n} 2\pi [p(y_i)h(y_i)]\Delta y = 2\pi \sum_{i=1}^{n} [p(y_i)h(y_i)]\Delta y$$

This approximation appears to become better and better as $\|\Delta\| \to 0 \ (n \to \infty)$. Therefore, we define the volume of the solid to be

$$\text{Volume of solid} = \lim_{\|\Delta\| \to 0} 2\pi \sum_{i=1}^{n} [p(y_i)h(y_i)]\Delta y$$

$$= 2\pi \int_{c}^{d} [p(y)h(y)] \, dy.$$

The Shell Method

To find the volume of a solid of revolution with the **shell method,** use one of the following, as shown in Figure 6.30.

Horizontal Axis of Revolution	*Vertical Axis of Revolution*
Volume = $V = 2\pi \displaystyle\int_c^d p(y)h(y)\,dy$	Volume = $V = 2\pi \displaystyle\int_a^b p(x)h(x)\,dx$

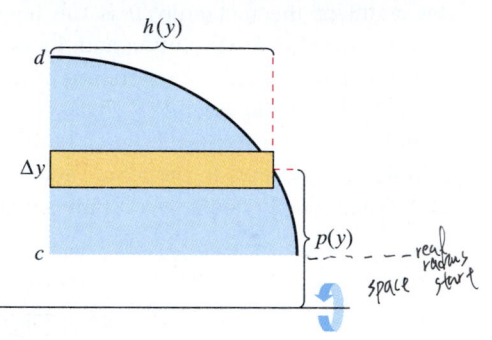
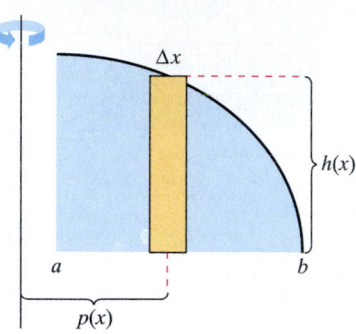

Horizontal axis of revolution   Vertical axis of revolution
Figure 6.30

EXAMPLE 1   Using the Shell Method to Find Volume

Find the volume of the solid of revolution formed by revolving the region bounded by

$$y = x - x^3$$

and the x-axis $(0 \le x \le 1)$ about the y-axis.

Solution   Because the axis of revolution is vertical, use a vertical representative rectangle, as shown in Figure 6.31. The width Δx indicates that x is the variable of integration. The distance from the center of the rectangle to the axis of revolution is $p(x) = x$, and the height of the rectangle is $h(x) = x - x^3$. Because x ranges from 0 to 1, the volume of the solid is

$$V = 2\pi \int_a^b p(x)h(x)\,dx = 2\pi \int_0^1 x(x - x^3)\,dx$$

$$= 2\pi \int_0^1 (-x^4 + x^2)\,dx$$

$$= 2\pi \left[-\frac{x^5}{5} + \frac{x^3}{3} \right]_0^1$$

$$= 2\pi \left(-\frac{1}{5} + \frac{1}{3} + 0 - 0 \right)$$

$$= \frac{4\pi}{15}.$$

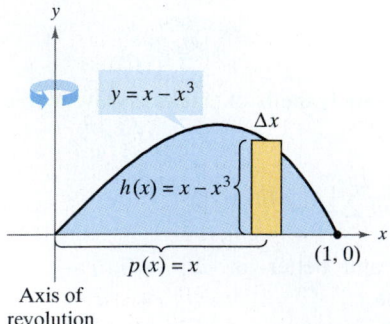

Axis of revolution

Figure 6.31

EXAMPLE 2   Using the Shell Method to Find Volume

Find the volume of the solid of revolution formed by revolving the region bounded by the graph of $x = e^{-y^2}$ and the y-axis $(0 \le y \le 1)$ about the x-axis.

Solution   Because the axis of revolution is horizontal, use a horizontal representative rectangle, as shown in Figure 6.32. The width Δy indicates that y is the variable of integration. The distance from the center of the rectangle to the axis of revolution is $p(y) = y$, and the height of the rectangle is $h(y) = e^{-y^2}$. Because y ranges from 0 to 1, the volume of the solid is

$$V = 2\pi \int_c^d p(y)h(y)\, dy = 2\pi \int_0^1 ye^{-y^2}\, dy$$

$$= -\pi \left[e^{-y^2} \right]_0^1$$

$$= \pi \left(1 - \frac{1}{e} \right)$$

$$\approx 1.986.$$

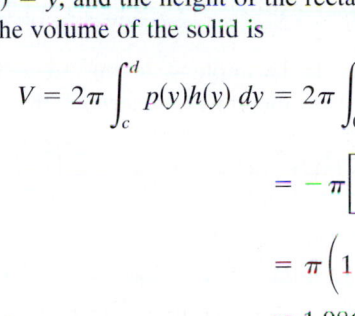

Figure 6.32

NOTE   To see the advantage of using the shell method in Example 2, solve the equation $x = e^{-y^2}$ for y.

$$y = \begin{cases} 1, & 0 \le x \le 1/e \\ \sqrt{-\ln x}, & 1/e < x \le 1 \end{cases}$$

Then use this equation to find the volume using the disc method.

Comparison of Disc and Shell Methods

The disc and shell methods can be distinguished as follows. For the disc method, the representative rectangle is always *perpendicular* to the axis of revolution, whereas for the shell method, the representative rectangle is always *parallel* to the axis of revolution as shown in Figure 6.33.

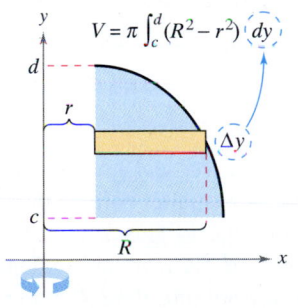

Vertical axis
of revolution

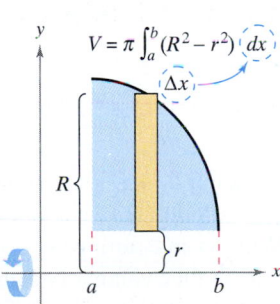

Horizontal axis
of revolution

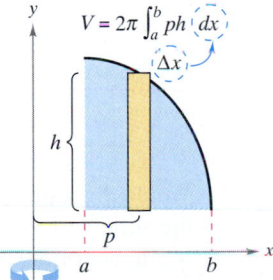

Vertical axis
of revolution

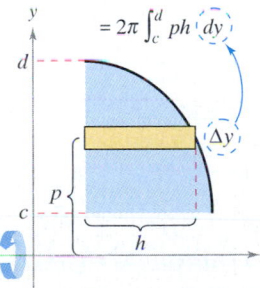

Horizontal axis
of revolution

Disc method: Representative rectangle is perpendicular to the axis of revolution.

Shell method: Representative rectangle is parallel to the axis of revolution.

Figure 6.33

Often, one method is more convenient to use than the other. The following example illustrates a case in which the shell method is preferable.

 EXAMPLE 3   Shell Method Preferable

Find the volume of the solid formed by revolving the region bounded by the graphs of

$$y = x^2 + 1, \quad y = 0, \quad x = 0, \quad \text{and} \quad x = 1$$

about the y-axis.

Solution   In Example 4 in the preceding section, you saw that the disc method requires two integrals to determine the volume of this solid. See Figure 6.34(a).

$$V = \pi \int_0^1 (1^2 - 0^2)\, dy + \pi \int_1^2 \left[1^2 - \left(\sqrt{y-1}\right)^2\right] dy \qquad \text{Disc method}$$

$$= \pi \int_0^1 1\, dy + \pi \int_1^2 (2 - y)\, dy$$

$$= \pi \Big[y\Big]_0^1 + \pi \left[2y - \frac{y^2}{2}\right]_1^2$$

$$= \pi \left(1 + 4 - 2 - 2 + \frac{1}{2}\right)$$

$$= \frac{3\pi}{2}$$

In Figure 6.34(b), you can see that the shell method requires only one integral to find the volume.

$$V = 2\pi \int_a^b p(x)h(x)\, dx \qquad \text{Shell method}$$

$$= 2\pi \int_0^1 x(x^2 + 1)\, dx$$

$$= 2\pi \left[\frac{x^4}{4} + \frac{x^2}{2}\right]_0^1$$

$$= 2\pi \left(\frac{3}{4}\right)$$

$$= \frac{3\pi}{2}$$

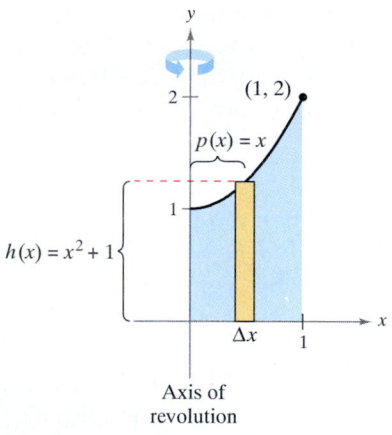

For $1 \le y \le 2$:
$R = 1$
$r = \sqrt{y-1}$

For $0 \le y \le 1$:
$R = 1$
$r = 0$

Axis of revolution

(a) Disc method

$p(x) = x$

$h(x) = x^2 + 1$

Axis of revolution

(b) Shell method

Figure 6.34

FOR FURTHER INFORMATION To learn more about the disc and shell methods see the article "The Disc and Shell Method" by Charles A. Cable in the February 1984 issue of *The American Mathematical Monthly.*

Suppose the region in Example 3 were revolved about the vertical line $x = 1$. Would the resulting solid of revolution have a greater volume or a smaller volume than the solid in Example 3? Without integrating, you should be able to reason that the resulting solid would have a smaller volume because "more" of the revolved region would be closer to the axis of revolution. To confirm this, try solving the following integral, which gives the volume of the solid.

$$V = 2\pi \int_0^1 (1 - x)(x^2 + 1)\, dx \qquad p(x) = 1 - x$$

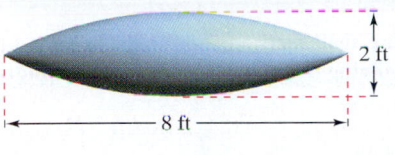

Pontoon

Figure 6.35

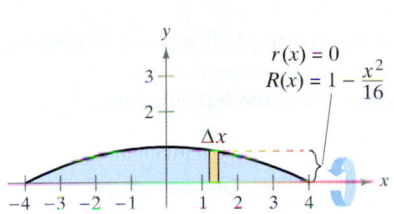

(a) Disc method

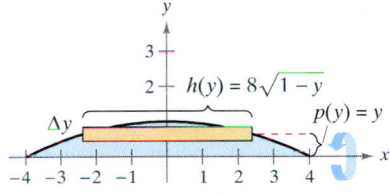

(b) Shell method

Figure 6.36

EXAMPLE 4   Volume of a Pontoon

A pontoon is to be made in the shape shown in Figure 6.35. The pontoon is designed by rotating the graph of

$$y = 1 - \frac{x^2}{16}, \qquad -4 < x \le 4$$

about the x-axis, where x and y are measured in feet. Find the volume of the pontoon.

Solution   Refer to Figure 6.36(a) and use the disc method as follows.

$$V = \pi \int_{-4}^{4} \left(1 - \frac{x^2}{16}\right)^2 dx \qquad \text{Disc method}$$

$$= \pi \int_{-4}^{4} \left(1 - \frac{x^2}{8} + \frac{x^4}{256}\right) dx$$

$$= \pi \left[x - \frac{x^3}{24} + \frac{x^5}{1280}\right]_{-4}^{4}$$

$$= \frac{64\pi}{15} \approx 13.4 \text{ cubic feet}$$

Try using Figure 6.36(b) to set up the integral for the volume using the shell method. Does the integral seem more complicated?

For the shell method in Example 4, you would have to solve for x in terms of y in the equation $y = 1 - (x^2/16)$. Sometimes, solving for x is very difficult (or even impossible). In such cases you must use a vertical rectangle (of width Δx), thus making x the variable of integration. The position (horizontal or vertical) of the axis of revolution then determines the method to be used. This is illustrated in Example 5.

EXAMPLE 5   Shell Method Necessary

Find the volume of the solid formed by revolving the region bounded by the graphs of $y = x^3 + x + 1$, $y = 1$, and $x = 1$ about the line $x = 2$, as shown in Figure 6.37.

Solution   In the equation $y = x^3 + x + 1$, you cannot easily solve for x in terms of y. (See the discussion at the end of Section 3.8.) Therefore, the variable of integration must be x, and you should choose a vertical representative rectangle. Because the rectangle is parallel to the axis of revolution, use the shell method and obtain

$$V = 2\pi \int_{a}^{b} p(x)h(x)\, dx = 2\pi \int_{0}^{1} (2 - x)(x^3 + x + 1 - 1)\, dx$$

$$= 2\pi \int_{0}^{1} (-x^4 + 2x^3 - x^2 + 2x)\, dx$$

$$= 2\pi \left[-\frac{x^5}{5} + \frac{x^4}{2} - \frac{x^3}{3} + x^2\right]_{0}^{1}$$

$$= 2\pi \left(-\frac{1}{5} + \frac{1}{2} - \frac{1}{3} + 1\right)$$

$$= \frac{29\pi}{15}.$$

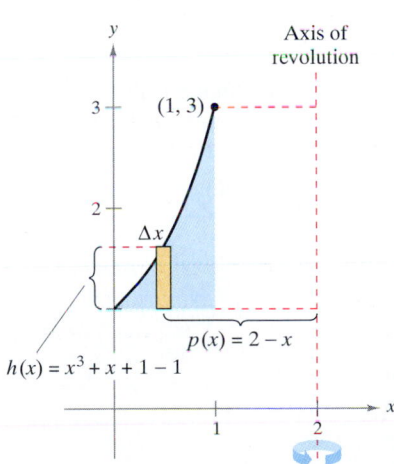

Figure 6.37

In Exercises 1–12, use the shell method to set up and evaluate the integral that gives the volume of the solid generated by revolving the plane region about the *y*-axis.

1. $y = x$

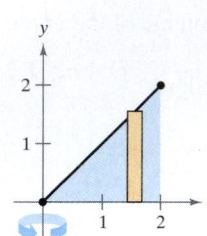

2. $y = 1 - x$

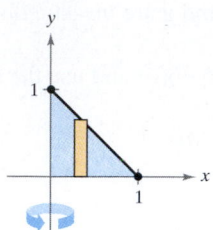

3. $y = \sqrt{x}$

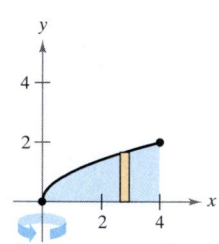

4. $y = x^2 + 4$

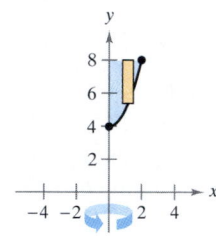

5. $y = x^2, y = 0, x = 2$

6. $y = x^2, y = 0, x = 4$

7. $y = x^2, y = 4x - x^2$

8. $y = 4 - x^2, y = 0$

9. $y = 4x - x^2, x = 0, y = 4$

10. $y = 2x, y = 4, x = 0$

11. $y = \dfrac{1}{\sqrt{2\pi}} e^{-x^2/2}, y = 0, x = 0, x = 1$

12. $y = \begin{cases} \dfrac{\sin x}{x}, & x > 0 \\ 1, & x = 0 \end{cases}, y = 0, x = 0, x = \pi$

In Exercises 13–16, use the shell method to set up and evaluate the integral that gives the volume of the solid generated by revolving the plane region about the *x*-axis.

13. $y = x$

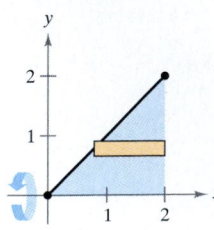

14. $y = 2 - x$

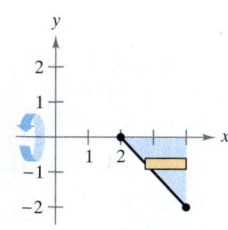

15. $y = \dfrac{1}{x}, x = 1, x = 2, y = 0$

16. $x + y^2 = 9, x = 0$

In Exercises 17–20, use the shell method to find the volume of the solid generated by revolving the plane region about the indicated line.

17. $y = x^2, y = 4x - x^2$, about the line $x = 4$

18. $y = x^2, y = 4x - x^2$, about the line $x = 2$

19. $y = 4x - x^2, y = 0$, about the line $x = 5$

20. $y = \sqrt{x}, y = 0, x = 4$, about the line $x = 6$

In Exercises 21–24, use the disc *or* the shell method to find the volume of the solid generated by revolving the region bounded by the graphs of the equations about the indicated line.

21. $y = x^3, y = 0, x = 2$

  (a) the *x*-axis   (b) the *y*-axis   (c) the line $x = 4$

22. $y = \dfrac{1}{x^2}, y = 0, x = 1, x = 4$

  (a) the *x*-axis   (b) the *y*-axis   (c) the line $y = 1$

23. $x^{1/2} + y^{1/2} = a^{1/2}, x = 0, y = 0$

  (a) the *x*-axis   (b) the *y*-axis   (c) the line $x = a$

24. $x^{2/3} + y^{2/3} = a^{2/3}, \quad a > 0$ (hypocycloid)

  (a) the *x*-axis   (b) the *y*-axis

In Exercises 25–28, (a) use a graphing utility to graph the plane region bounded by the graphs of the equations and (b) use the integration capabilities of the graphing utility to approximate the volume of the solid generated by revolving the region about the *y*-axis.

25. $x^{4/3} + y^{4/3} = 1, x = 0, y = 0$, first quadrant

26. $y = \sqrt{1 - x^3}, y = 0, x = 0$

27. $y = \sqrt[3]{(x - 2)^2(x - 6)^2}, y = 0, x = 2, x = 6$

28. $y = \dfrac{2}{1 + e^{1/x}}, y = 0, x = 1, x = 3$

29. *Think About It*   The region in the figure is revolved about the indicated axes and line. Order the volumes of the resulting solids from least to greatest. Explain your reasoning.

  (a) *x*-axis   (b) *y*-axis   (c) $x = 5$

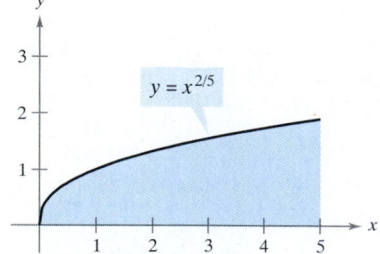

30. Use integration to confirm your results in Exercise 29, where the region is bounded by the graphs of $y = x^{2/5}$, $y = 0$, and $x = 5$.

Think About It In Exercises 31 and 32, determine which value best approximates the volume of the solid generated by revolving the region bounded by the graphs of the equations about the y-axis. (Make your selection on the basis of a sketch of the solid and *not* by performing any calculations.)

31. $y = 2e^{-x}$, $y = 0$, $x = 0$, $x = 2$

(a) $\frac{3}{2}$ (b) -2 (c) 4   (d) 7.5   (e) 15

32. $y = \tan x$, $y = 0$, $x = 0$, $x = \dfrac{\pi}{4}$

(a) 3.5   (b) $-\frac{9}{4}$ (c) 8   (d) 10   (e) 1

33. *Machine Part* A solid is generated by revolving the region bounded by $y = \frac{1}{2}x^2$ and $y = 2$ about the y-axis. A hole, centered along the axis of revolution, is drilled through this solid so that one-fourth of the volume is removed. Find the diameter of the hole.

34. *Machine Part* A solid is generated by revolving the region bounded by $y = \sqrt{9 - x^2}$ and $y = 0$ about the y-axis. A hole, centered along the axis of revolution, is drilled through this solid so that one-third of the volume is removed. Find the diameter of the hole.

35. A hole is cut through the center of a sphere of radius r. The height of the remaining spherical ring is h, as shown in the figure. Show that the volume of the ring is $V = \pi h^3/6$. (*Note:* The volume is independent of r.)

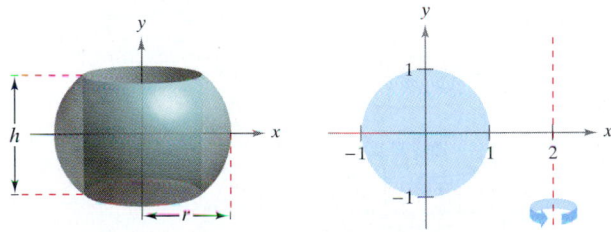

Figure for 35   **Figure for 36**

36. *Volume of a Torus* A torus is formed by revolving the region bounded by the circle $x^2 + y^2 = 1$ about the line $x = 2$, as shown in the figure. Find the volume of this "doughnut-shaped" solid. (*Hint:* The integral $\int_{-1}^{1} \sqrt{1 - x^2}\, dx$ represents the area of a semicircle.)

37. *Volume of a Torus* Repeat Exercise 36 for a torus formed by revolving the region bounded by the circle $x^2 + y^2 = r^2$ about the line $x = R$, where $r < R$.

38. *Volume of a Segment of a Sphere* Let a sphere of radius r be cut by a plane, thus forming a segment of height h. Show that the volume of this segment is $\frac{1}{3}\pi h^2(3r - h)$.

Think About It In Exercises 39 and 40, give a geometric argument that explains why the integrals have equal values.

39. $\pi \displaystyle\int_{1}^{5} (x - 1)\, dx$, $2\pi \displaystyle\int_{0}^{2} y[5 - (y^2 + 1)]\, dy$

40. $\pi \displaystyle\int_{0}^{2} [16 - (2y)^2]\, dy$, $2\pi \displaystyle\int_{0}^{4} x\left(\dfrac{x}{2}\right) dx$

41. ***Think About It***   Match each of the integrals with the solid whose volume it represents, and give the dimensions of each solid.

(a) Right circular cone   (b) Torus   (c) Sphere

(d) Right circular cylinder   (e) Ellipsoid

(i) $2\pi \displaystyle\int_{0}^{r} hx\, dx$

(ii) $2\pi \displaystyle\int_{0}^{r} hx \left(1 - \dfrac{x}{r}\right) dx$

(iii) $2\pi \displaystyle\int_{0}^{r} 2x\sqrt{r^2 - x^2}\, dx$

(iv) $2\pi \displaystyle\int_{0}^{b} 2ax \sqrt{1 - \dfrac{x^2}{b^2}}\, dx$

(v) $2\pi \displaystyle\int_{-r}^{r} (R - x)(2\sqrt{r^2 - x^2})\, dx$

42. *Volume of a Storage Shed* A storage shed has a circular base of diameter 80 feet (see figure). Starting at the center, the interior height is measured every 10 feet and recorded in the table.

x	0	10	20	30	40
Height	50	45	40	20	0

(a) Use Simpson's Rule to approximate the volume of the building.

(b) Note that the roof line consists of two line segments. Find the equations of the line segments and use integration to find the volume of the shed.

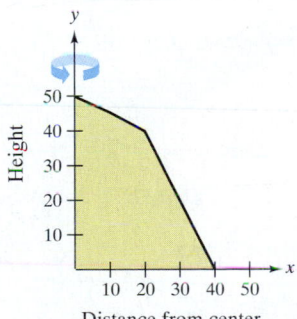

Distance from center

43. Modeling Data A pond is approximately circular, with a diameter of 400 feet (see figure). Starting at the center, the depth of the water is measured every 25 feet and recorded in the table.

x	0	25	50	75	100
Depth	20	19	19	17	15

x	125	150	175	200
Depth	14	10	6	0

(a) Use Simpson's Rule to approximate the volume of water in the pond.

(b) Use a graphing utility to find a quadratic model for the depths recorded in the table. Use the graphing utility to plot the depths and graph the model.

(c) Use the integration capabilities of a graphing utility and the model in part (b) to approximate the volume of water in the pond.

(d) Use the result in part (c) to approximate the number of gallons of water in the pond if 1 cubic foot of water is approximately 7.48 gallons.

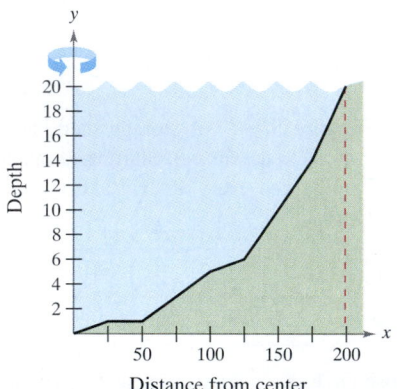

The Oblateness of Saturn Saturn is the most oblate of the nine planets in our solar system. Its equatorial radius is 60,268 kilometers and its polar radius is 54,364 kilometers.

(a) Find the ratio of the volumes of the sphere and the oblate ellipsoid shown below.

(b) If a planet were spherical and had the same volume as Saturn, what would its radius be?

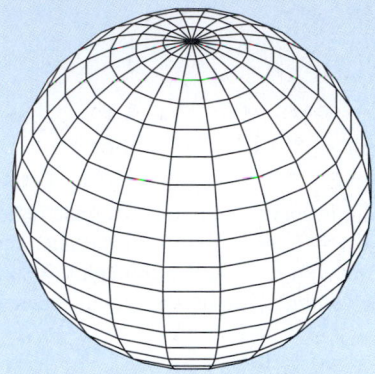

Computer model of "spherical Saturn," whose equatorial radius is equal to its polar radius. The equation of the cross section passing through the pole is

$$x^2 + y^2 = 60{,}268^2.$$

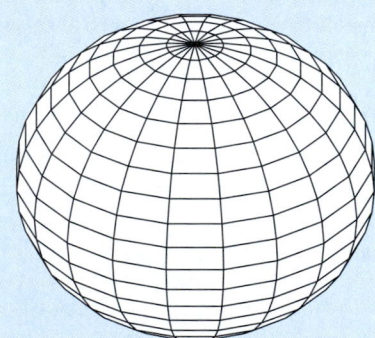

Computer model of "oblate Saturn," whose equatorial radius is greater than its polar radius. The equation of the cross section passing through the pole is

$$\frac{x^2}{60{,}268^2} + \frac{y^2}{54{,}364^2} = 1.$$

The color-enchanced photo of Saturn at the left was taken by Voyager 1. In the photograph, the oblateness of Saturn is clearly visible.

Arc Length • Area of a Surface of Revolution

Arc Length

In this section, definite integrals are used to find the arc length of a plane curve and the area of a surface of revolution. In both cases, we approximate an arc (a segment of a curve) by straight line segments whose lengths are given by the familiar distance formula

$$d = \sqrt{(x_2 - x_1)^2 + (y_2 - y_1)^2}.$$

A **rectifiable** curve is one that has a finite arc length. You will see that a sufficient condition for the graph of a function f to be rectifiable between $(a, f(a))$ and $(b, f(b))$ is that f' be continuous on $[a, b]$. Such a function is **continuously differentiable** on $[a, b]$, and its graph on the interval $[a, b]$ is a **smooth curve**.

Consider a function $y = f(x)$ that is continuously differentiable on the interval $[a, b]$. You can approximate the graph of f by n line segments whose endpoints are determined by the partition

$$a = x_0 < x_1 < x_2 < \cdots < x_n = b$$

as shown in Figure 6.38. By letting $\Delta x_i = x_i - x_{i-1}$ and $\Delta y_i = y_i - y_{i-1}$, you can approximate the length of the graph by

$$s \approx \sum_{i=1}^{n} \sqrt{(\Delta x_i)^2 + (\Delta y_i)^2}$$

$$= \sum_{i=1}^{n} \sqrt{1 + \left(\frac{\Delta y_i}{\Delta x_i}\right)^2} \, (\Delta x_i).$$

This approximation appears to become better and better as $\|\Delta\| \to 0 \, (n \to \infty)$. Therefore, we define the length of the graph to be

$$s = \lim_{\|\Delta\| \to 0} \sum_{i=1}^{n} \sqrt{1 + \left(\frac{\Delta y_i}{\Delta x_i}\right)^2} \, (\Delta x_i).$$

Because $f'(x)$ exists for each x in (x_{i-1}, x_i), the Mean Value Theorem guarantees the existence of c_i in (x_{i-1}, x_i) such that

$$f(x_i) - f(x_{i-1}) = f'(c_i)(x_i - x_{i-1})$$

$$\frac{\Delta y_i}{\Delta x_i} = f'(c_i).$$

Because f' is continuous on $[a, b]$, it follows that $\sqrt{1 + [f'(x)]^2}$ is also continuous (and hence integrable) on $[a, b]$, which implies that

$$s = \lim_{\|\Delta\| \to 0} \sum_{i=1}^{n} \sqrt{1 + [f'(c_i)]^2} \, (\Delta x_i)$$

$$= \int_a^b \sqrt{1 + [f'(x)]^2} \, dx.$$

We call s the **arc length** of f between a and b.

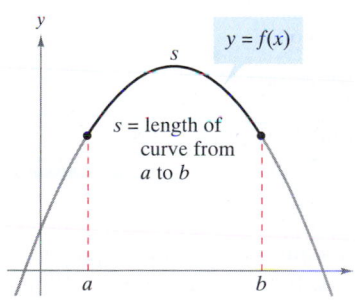

CHRISTIAN HUYGENS (1629–1695)

The Dutch mathematician Christian Huygens, who invented the pendulum clock, and James Gregory (1638–1675), a Scottish mathematician, both made early contributions to the problem of finding the length of a rectifiable curve.

Corbis-Bettmann

Figure 6.38

Definition of Arc Length

Let the function given by $y = f(x)$ represent a smooth curve on the interval $[a, b]$. The **arc length** of f between a and b is

$$s = \int_a^b \sqrt{1 + [f'(x)]^2}\,dx.$$

Similarly, for a smooth curve given by $x = g(y)$, the **arc length** of g between c and d is

$$s = \int_c^d \sqrt{1 + [g'(y)]^2}\,dy.$$

Because the definition of arc length can be applied to a linear function, you can check to see that this new definition agrees with the standard distance formula for the length of a line segment. This is done in Example 1.

EXAMPLE 1   The Length of a Line Segment

Find the arc length from (x_1, y_1) to (x_2, y_2) on the graph of $f(x) = mx + b$ as shown in Figure 6.39.

Solution   Because

$$m = f'(x) = \frac{y_2 - y_1}{x_2 - x_1}$$

it follows that

$$
\begin{aligned}
s &= \int_{x_1}^{x_2} \sqrt{1 + [f'(x)]^2}\,dx \\
&= \int_{x_1}^{x_2} \sqrt{1 + \left(\frac{y_2 - y_1}{x_2 - x_1}\right)^2}\,dx \\
&= \left[\sqrt{\frac{(x_2 - x_1)^2 + (y_2 - y_1)^2}{(x_2 - x_1)^2}}\,(x)\right]_{x_1}^{x_2} \\
&= \sqrt{\frac{(x_2 - x_1)^2 + (y_2 - y_1)^2}{(x_2 - x_1)^2}}\,(x_2 - x_1) \\
&= \sqrt{(x_2 - x_1)^2 + (y_2 - y_1)^2}
\end{aligned}
$$

which is the formula for the distance between two points in the plane.

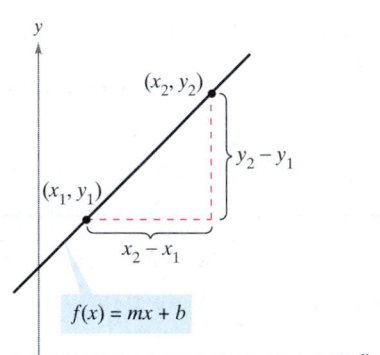

The arc length of the graph of f from (x_1, y_1) and (x_2, y_2) is the same as the standard distance formula.
Figure 6.39

TECHNOLOGY   Definite integrals representing arc length often are very difficult to evaluate. In this section we present a few examples. In the next chapter, with more advanced integration techniques, you will be able to tackle more difficult arc length problems. In the meantime, remember that you can always use a numerical integration program to approximate an arc length. For instance, try using the numerical integration feature of a graphing utility to approximate the arc lengths in Examples 2 and 3.

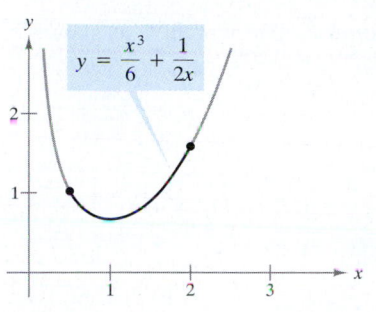

The arc length of the graph of y on $\left[\frac{1}{2}, 2\right]$ is $\frac{33}{16}$.

Figure 6.40

EXAMPLE 2   Finding Arc Length

Find the arc length of the graph of $y = x^3/6 + 1/(2x)$ on the interval $\left[\frac{1}{2}, 2\right]$, as shown in Figure 6.40.

Solution   Using

$$\frac{dy}{dx} = \frac{3x^2}{6} - \frac{1}{2x^2} = \frac{1}{2}\left(x^2 - \frac{1}{x^2}\right)$$

yields an arc length of

$$s = \int_a^b \sqrt{1 + \left(\frac{dy}{dx}\right)^2}\, dx = \int_{1/2}^2 \sqrt{1 + \left[\frac{1}{2}\left(x^2 - \frac{1}{x^2}\right)\right]^2}\, dx$$

$$= \int_{1/2}^2 \sqrt{\frac{1}{4}\left(x^4 + 2 + \frac{1}{x^4}\right)}\, dx$$

$$= \int_{1/2}^2 \frac{1}{2}\left(x^2 + \frac{1}{x^2}\right)\, dx$$

$$= \frac{1}{2}\left[\frac{x^3}{3} - \frac{1}{x}\right]_{1/2}^2$$

$$= \frac{1}{2}\left(\frac{13}{6} + \frac{47}{24}\right) = \frac{33}{16}.$$

EXAMPLE 3   Finding Arc Length

Find the arc length of the graph of $(y - 1)^3 = x^2$ on the interval $[0, 8]$, as shown in Figure 6.41.

Solution   Begin by solving for x in terms of y: $x = \pm(y - 1)^{3/2}$. Choosing the positive value of x produces

$$\frac{dx}{dy} = \frac{3}{2}(y - 1)^{1/2}.$$

The x-interval $[0, 8]$ corresponds to the y-interval $[1, 5]$, and the arc length is

$$s = \int_c^d \sqrt{1 + \left(\frac{dx}{dy}\right)^2}\, dy = \int_1^5 \sqrt{1 + \left[\frac{3}{2}(y - 1)^{1/2}\right]^2}\, dy$$

$$= \int_1^5 \sqrt{\frac{9}{4}y - \frac{5}{4}}\, dy$$

$$= \frac{1}{2}\int_1^5 \sqrt{9y - 5}\, dy$$

$$= \frac{1}{18}\left[\frac{(9y - 5)^{3/2}}{3/2}\right]_1^5$$

$$= \frac{1}{27}(40^{3/2} - 4^{3/2})$$

$$\approx 9.0734.$$

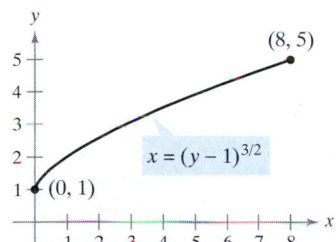

The arc length of the graph of y on $[0, 8]$ is approximately 9.0734.

Figure 6.41

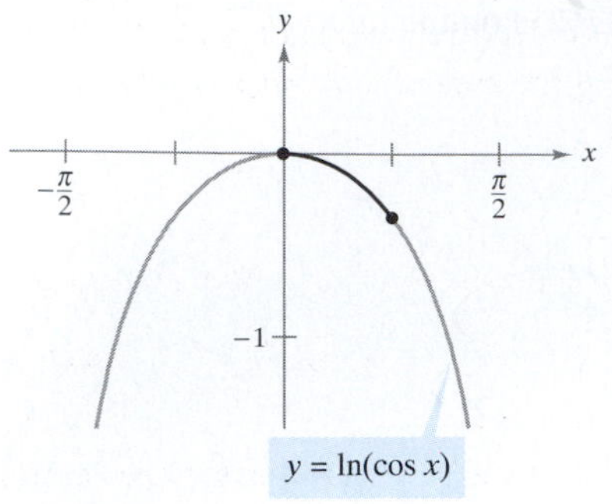

EXAMPLE 4 **Finding Arc Length**

Find the arc length of the graph of $y = \ln(\cos x)$ from $x = 0$ to $x = \pi/4$, as shown in Figure 6.42.

Solution Using

$$\frac{dy}{dx} = -\frac{\sin x}{\cos x} = -\tan x$$

yields an arc length of

$$s = \int_a^b \sqrt{1 + \left(\frac{dy}{dx}\right)^2}\, dx = \int_0^{\pi/4} \sqrt{1 + \tan^2 x}\, dx$$

$$= \int_0^{\pi/4} \sqrt{\sec^2 x}\, dx \qquad \text{\color{red}Trigonometric identity}$$

$$= \int_0^{\pi/4} \sec x\, dx$$

$$= \left[\ln|\sec x + \tan x|\right]_0^{\pi/4}$$

$$= \ln\left(\sqrt{2} + 1\right) - \ln 1$$

$$\approx 0.8814.$$

The arc length of the graph of y on $\left[0, \frac{\pi}{4}\right]$ is approximately 0.8814.
Figure 6.42

EXAMPLE 5 **Length of a Cable**

An electric cable is hung between two towers that are 200 feet apart, as shown in Figure 6.43. The cable takes the shape of a catenary whose equation is

$$y = 75(e^{x/150} + e^{-x/150}) = 150 \cosh \frac{x}{150}.$$

Find the arc length of the cable between the two towers.

Solution Because $y' = \frac{1}{2}(e^{x/150} - e^{-x/150})$, you can write

$$(y')^2 = \frac{1}{4}(e^{x/75} - 2 + e^{-x/75})$$

and

$$1 + (y')^2 = \frac{1}{4}(e^{x/75} + 2 + e^{-x/75}) = \left[\frac{1}{2}(e^{x/150} + e^{-x/150})\right]^2.$$

Therefore, the arc length of the cable is

$$s = \int_a^b \sqrt{1 + (y')^2}\, dx = \frac{1}{2}\int_{-100}^{100}(e^{x/150} + e^{-x/150})\, dx$$

$$= 75\left[e^{x/150} - e^{-x/150}\right]_{-100}^{100}$$

$$= 150(e^{2/3} - e^{-2/3})$$

$$\approx 215 \text{ feet.}$$

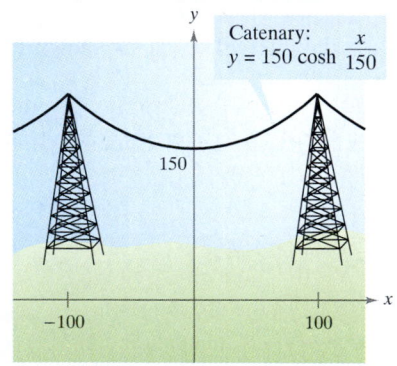

The arc length of the cable is approximately 215 feet.
Figure 6.43

Area of a Surface of Revolution

In Sections 6.2 and 6.3, integration was used to calculate the volume of a solid of revolution. We now look at a procedure for finding the area of a surface of revolution.

> **Definition of a Surface of Revolution**
>
> If the graph of a continuous function is revolved about a line, the resulting surface is a **surface of revolution.**

The area of a surface of revolution is derived from the formula for the lateral surface area of the frustum of a right circular cone. Consider the line segment in Figure 6.44, where L is the length of the line segment, r_1 is the radius at the left end of the line segment, and r_2 is the radius at the right end of the line segment. When the line segment is revolved about its axis of revolution, it forms a frustum of a right circular cone, with

$$S = 2\pi r L \qquad \text{Lateral surface area of frustum}$$

where

$$r = \frac{1}{2}(r_1 + r_2). \qquad \text{Average radius of frustum}$$

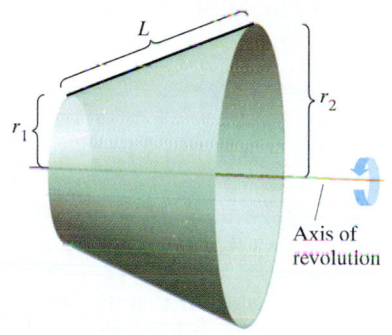

Figure 6.44

(In Exercise 49, you are asked to verify the formula for S.)

Suppose the graph of a function f, having a continuous derivative on the interval $[a, b]$, is revolved about the x-axis to form a surface of revolution, as shown in Figure 6.45. Let Δ be a partition of $[a, b]$, with subintervals of width Δx_i. Then the line segment of length

$$\Delta L_i = \sqrt{\Delta x_i^2 + \Delta y_i^2}$$

generates a frustum of a cone. Let r_i be the average radius of this frustum. By the Intermediate Value Theorem, a point d_i exists (in the ith subinterval) such that $r_i = f(d_i)$. The lateral surface area ΔS_i of the frustum is

$$\Delta S_i = 2\pi r_i \Delta L_i$$
$$= 2\pi f(d_i)\sqrt{\Delta x_i^2 + \Delta y_i^2}$$
$$= 2\pi f(d_i)\sqrt{1 + \left(\frac{\Delta y_i}{\Delta x_i}\right)^2}\,\Delta x_i.$$

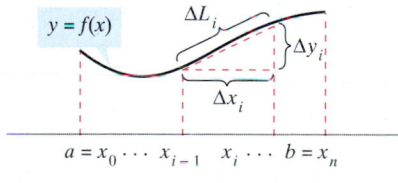

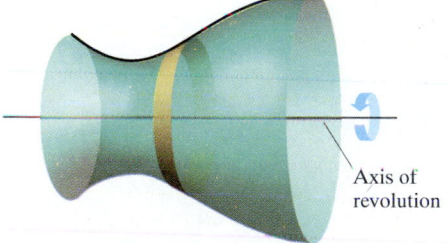

Figure 6.45

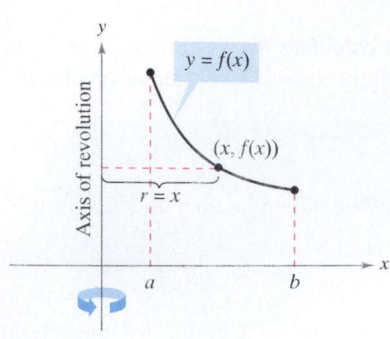

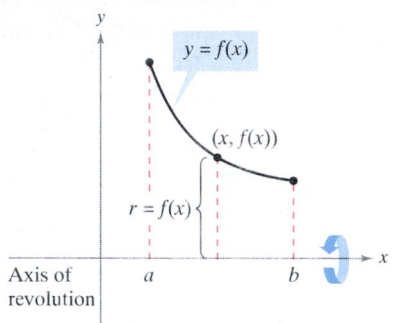

Figure 6.46

By the Mean Value Theorem, a point c_i exists in (x_{i-1}, x_i) such that

$$f'(c_i) = \frac{f(x_i) - f(x_{i-1})}{x_i - x_{i-1}} = \frac{\Delta y_i}{\Delta x_i}.$$

Therefore, $\Delta S_i = 2\pi f(d_i)\sqrt{1 + [f'(c_i)]^2}\,\Delta x_i$, and the total surface area can be approximated by

$$S \approx 2\pi \sum_{i=1}^{n} f(d_i)\sqrt{1 + [f'(c_i)]^2}\,\Delta x_i.$$

It can be shown that the limit of the right side as $\|\Delta\| \to 0$ $(n \to \infty)$ is

$$S = 2\pi \int_a^b f(x)\sqrt{1 + [f'(x)]^2}\,dx.$$

In a similar manner, if the graph of f is revolved about the y-axis then S is

$$S = 2\pi \int_a^b x\sqrt{1 + [f'(x)]^2}\,dx.$$

In both formulas for S, you can regard the products $2\pi f(x)$ and $2\pi x$ as the circumference of the circle traced by a point (x, y) on the graph of f as it is revolved about the x- or y-axis (Figure 6.46). In one case the radius is $r = f(x)$, and in the other case the radius is $r = x$. Moreover, by appropriately adjusting r, you can generalize the formula for surface area to cover *any* horizontal or vertical axis of revolution, as indicated in the following definition.

Definition of the Area of a Surface of Revolution

Let $y = f(x)$ have a continuous derivative on the interval $[a, b]$. The area S of the surface of revolution formed by revolving the graph of f about a horizontal or vertical axis is

$$S = 2\pi \int_a^b r(x)\sqrt{1 + [f'(x)]^2}\,dx \qquad \text{\textcolor{red}{y is a function of x.}}$$

where $r(x)$ is the distance between the graph of f and the axis of revolution. If $x = g(y)$ on the interval $[c, d]$, then the surface area is

$$S = 2\pi \int_c^d r(y)\sqrt{1 + [g'(y)]^2}\,dy \qquad \text{\textcolor{red}{x is a function of y.}}$$

where $r(y)$ is the distance between the graph of g and the axis of revolution.

The formulas in this definition are sometimes written as

$$S = 2\pi \int_a^b r(x)\,ds \qquad \text{\textcolor{red}{y is a function of x.}}$$

and

$$S = 2\pi \int_c^d r(y)\,ds \qquad \text{\textcolor{red}{x is a function of y.}}$$

where $ds = \sqrt{1 + [f'(x)]^2}\,dx$ and $ds = \sqrt{1 + [g'(y)]^2}\,dy$, respectively.

EXAMPLE 6   The Area of a Surface of Revolution

Find the area of the surface formed by revolving the graph of

$$f(x) = x^3$$

on the interval $[0, 1]$ about the x-axis, as shown in Figure 6.47.

Solution   The distance between the x-axis and the graph of f is $r(x) = f(x)$, and because $f'(x) = 3x^2$, the surface area is

$$S = 2\pi \int_a^b r(x) \sqrt{1 + [f'(x)]^2}\, dx$$

$$= 2\pi \int_0^1 x^3 \sqrt{1 + (3x^2)^2}\, dx$$

$$= \frac{2\pi}{36} \int_0^1 (36x^3)(1 + 9x^4)^{1/2}\, dx$$

$$= \frac{\pi}{18} \left[\frac{(1 + 9x^4)^{3/2}}{3/2} \right]_0^1$$

$$= \frac{\pi}{27} (10^{3/2} - 1)$$

$$\approx 3.563.$$

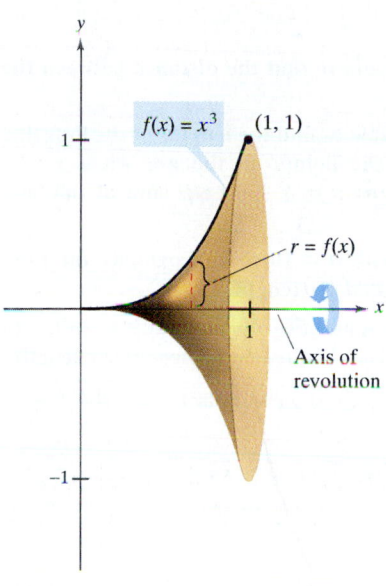

Figure 6.47

EXAMPLE 7   The Area of a Surface of Revolution

Find the area of the surface formed by revolving the graph of

$$f(x) = x^2$$

on the interval $\left[0, \sqrt{2}\right]$ about the y-axis, as shown in Figure 6.48.

Solution   In this case, the distance between the graph of f and the y-axis is $r(x) = x$. Using $f'(x) = 2x$, you can determine that the surface area is

$$S = 2\pi \int_a^b r(x) \sqrt{1 + [f'(x)]^2}\, dx$$

$$= 2\pi \int_0^{\sqrt{2}} x \sqrt{1 + (2x)^2}\, dx$$

$$= \frac{2\pi}{8} \int_0^{\sqrt{2}} (1 + 4x^2)^{1/2}(8x)\, dx$$

$$= \frac{\pi}{4} \left[\frac{(1 + 4x^2)^{3/2}}{3/2} \right]_0^{\sqrt{2}}$$

$$= \frac{\pi}{6} \left[(1 + 8)^{3/2} - 1 \right]$$

$$= \frac{13\pi}{3}.$$

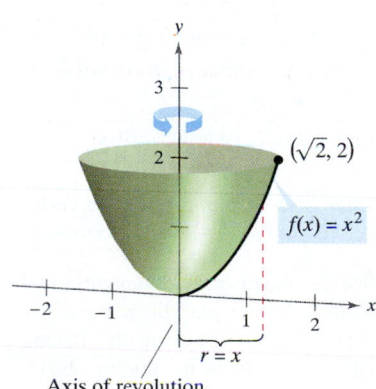

Figure 6.48

EXERCISES FOR SECTION 6.4

In Exercises 1 and 2, find the distance between the points by using (a) the Distance Formula and (b) integration.

1. $(0, 0)$, $(5, 12)$ **2.** $(1, 2)$, $(7, 10)$

In Exercises 3–8, find the arc length of the graph of the function over the indicated interval.

3. $y = \frac{2}{3}x^{3/2} + 1$, $[0, 1]$ **4.** $y = x^{3/2} - 1$, $[0, 4]$

5. $y = \frac{3}{2}x^{2/3}$, $[1, 8]$ **6.** $y = \frac{x^5}{10} + \frac{1}{6x^3}$, $[1, 2]$

7. $y = \frac{x^4}{8} + \frac{1}{4x^2}$, $[1, 2]$ **8.** $y = \frac{1}{2}(e^x + e^{-x})$, $[0, 2]$

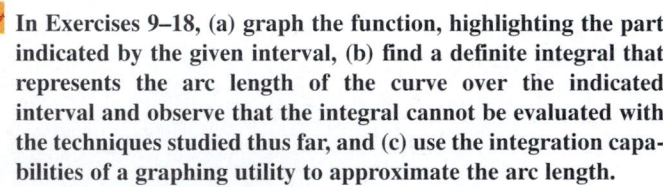 In Exercises 9–18, (a) graph the function, highlighting the part indicated by the given interval, (b) find a definite integral that represents the arc length of the curve over the indicated interval and observe that the integral cannot be evaluated with the techniques studied thus far, and (c) use the integration capabilities of a graphing utility to approximate the arc length.

Function	Interval
9. $y = 4 - x^2$	$0 \le x \le 2$
10. $y = x^2 + x - 2$	$-2 \le x \le 1$
11. $y = \dfrac{1}{x}$	$1 \le x \le 3$
12. $y = \dfrac{1}{x + 1}$	$0 \le x \le 1$
13. $y = \sin x$	$0 \le x \le \pi$
14. $y = \cos x$	$-\dfrac{\pi}{2} \le x \le \dfrac{\pi}{2}$
15. $x = e^{-y}$	$0 \le y \le 2$
16. $y = \ln x$	$1 \le x \le 5$
17. $y = 2 \arctan x$	$0 \le x \le 1$
18. $x = \sqrt{36 - y^2}$	$0 \le y \le 3$

Approximation In Exercises 19 and 20, determine which value best approximates the length of the arc represented by the integral. (Make your selection on the basis of a sketch of the arc and *not* by performing any calculations.)

19. $\displaystyle\int_0^2 \sqrt{1 + \left[\frac{d}{dx}\left(\frac{5}{x^2 + 1}\right)\right]^2}\, dx$

   (a) 25   (b) 5   (c) 2   (d) -4 (e) 3

20. $\displaystyle\int_0^{\pi/4} \sqrt{1 + \left[\frac{d}{dx}(\tan x)\right]^2}\, dx$

   (a) 3   (b) -2 (c) 4   (d) $\dfrac{4\pi}{3}$ (e) 1

Approximation In Exercises 21 and 22, approximate the arc length of the graph of the function over the interval $[0, 4]$ in four ways.

(a) Use the Distance Formula to find the distance between the endpoints of the arc.

(b) Use the Distance Formula to find the lengths of the four line segments connecting the points on the arc when $x = 0$, $x = 1$, $x = 2$, $x = 3$, and $x = 4$. Find the sum of the four lengths.

(c) Use Simpson's Rule with $n = 10$ to approximate the integral yielding the indicated arc length.

(d) Use the integration capabilities of a graphing utility to approximate the integral yielding the indicated arc length.

21. $f(x) = x^3$ **22.** $f(x) = (x^2 - 4)^2$

23. *Think About It* The figure shows the graphs of the functions $y_1 = x$, $y_2 = \frac{1}{2}x^{3/2}$, $y_3 = \frac{1}{4}x^2$, and $y_4 = \frac{1}{8}x^{5/2}$ on the interval $[0, 4]$.

(a) Label the functions.

(b) List the functions in order of increasing arc length.

(c) Verify your answer in part (b) by approximating each arc length accurate to three decimal places.

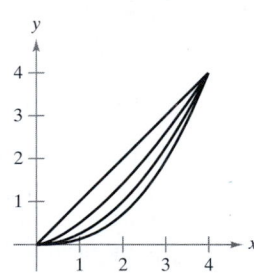   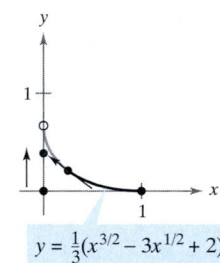

$y = \frac{1}{3}(x^{3/2} - 3x^{1/2} + 2)$

Figure for 23   **Figure for 25**

24. *Think About It* Explain why the two integrals are equal.

$$\int_1^e \sqrt{1 + \frac{1}{x^2}}\, dx = \int_0^1 \sqrt{1 + e^{2x}}\, dx$$

Use the integration capabilities of a graphing utility to verify that the integrals are equal.

25. *Length of Pursuit* A fleeing object leaves the origin and moves up the y-axis (see figure). At the same time, a pursuer leaves the point $(1, 0)$ and always moves toward the fleeing object. If the pursuer's speed is twice that of the fleeing object, the equation of the path is

$$y = \frac{1}{3}(x^{3/2} - 3x^{1/2} + 2).$$

How far has the fleeing object traveled when it is caught? Show that the pursuer has traveled twice as far.

26. Roof Area A barn is 100 feet long and 40 feet wide (see figure). A cross section of the roof is the inverted catenary

$$y = 31 - 10(e^{x/20} + e^{-x/20}).$$

Find the number of square feet of roofing on the barn.

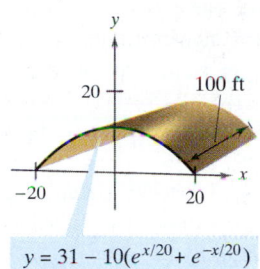

$$y = 31 - 10(e^{x/20} + e^{-x/20})$$

Figure for 26

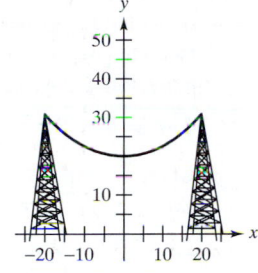

Figure for 27

27. Length of a Catenary Electric wires suspended between two towers form a catenary (see figure) modeled by the equation

$$y = 20 \cosh \frac{x}{20}, \quad -20 \le x \le 20$$

where x and y are measured in meters. Find the length of the suspended cable if the towers are 40 meters apart.

28. Length of Gateway Arch The Gateway Arch in St. Louis, Missouri is modeled by

$$y = 693.8597 - 68.7672 \cosh 0.0100333x,$$

$$-299.2239 \le x \le 299.2239.$$

(See the Section Project in Section 5.10.) Find the length of this curve (see figure).

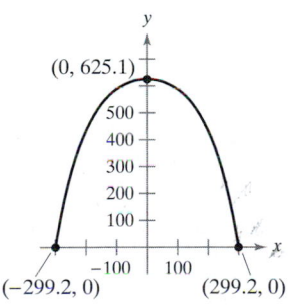

Figure for 28

29. Find the arc length from $(0, 3)$ clockwise to $\left(2, \sqrt{5}\right)$ along the circle $x^2 + y^2 = 9$.

30. Find the arc length from $(-3, 4)$ clockwise to $(4, 3)$ along the circle $x^2 + y^2 = 25$. Show that the result is one-fourth the circumference of the circle.

In Exercises 31–34, set up and evaluate the definite integral for the area of the surface generated by revolving the curve about the x-axis.

31. $y = \frac{1}{3}x^3, \quad [0, 3]$

32. $y = \sqrt{x}, \quad [1, 4]$

33. $y = \frac{x^3}{6} + \frac{1}{2x}, \quad [1, 2]$

34. $y = \frac{x}{2}, \quad [0, 6]$

In Exercises 35 and 36, set up and evaluate the definite integral that gives the area of the surface of revolution generated by revolving the curve about the y-axis.

Function	Interval
35. $y = \sqrt[3]{x} + 2$	$[1, 8]$
36. $y = 4 - x^2$	$[0, 2]$

In Exercises 37 and 38, use the integration capabilities of a graphing utility to approximate the surface area of the solid of revolution.

Function	Interval
37. $y = \sin x$	$[0, \pi]$
revolved about the x-axis	
38. $y = \ln x$	$[1, e]$
revolved about the y-axis	

39. A right circular cone is generated by revolving the region bounded by $y = hx/r$, $y = h$, and $x = 0$ about the y-axis. Verify that the lateral surface area of the cone is

$$S = \pi r \sqrt{r^2 + h^2}.$$

40. A sphere of radius r is generated by revolving the graph of $y = \sqrt{r^2 - x^2}$ about the x-axis. Verify that the surface area of the sphere is $4\pi r^2$.

41. Find the area of the zone of a sphere formed by revolving the graph of $y = \sqrt{9 - x^2}$, $0 \le x \le 2$, about the y-axis.

42. Find the area of the zone of a sphere formed by revolving the graph of $y = \sqrt{r^2 - x^2}$, $0 \le x \le a$, about the y-axis. Assume that $a < r$.

43. Bulb Design An ornamental light bulb is designed by revolving the graph of

$$y = \tfrac{1}{3}x^{1/2} - x^{3/2}, \quad 0 \le x \le \tfrac{1}{3}$$

about the x-axis, where x and y are measured in feet (see figure). Find the surface area of the bulb and use the result to approximate the amount of glass needed to make the bulb. (Assume that the glass is 0.015 inch thick.)

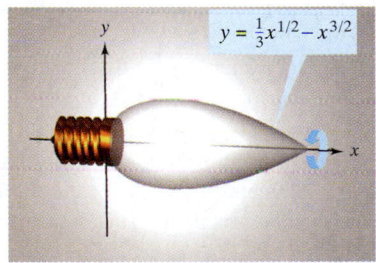

$$y = \tfrac{1}{3}x^{1/2} - x^{3/2}$$

44. Modeling Data   The circumference C (in inches) of a vase is measured at 3-inch intervals starting at its base. The measurements are given in the table, where y is the vertical distance in inches from the base.

y	0	3	6	9	12	15	18
C	50	65.5	70	66	58	51	48

(a) Use the data to approximate the volume of the vase by summing the volumes of approximating discs.

(b) Use the data to approximate the outside surface area (excluding the base) of the vase by summing the outside surface areas of approximating frustums of right circular cones.

(c) Use a graphing utility to find a cubic model for the points (y, r) where $r = C/(2\pi)$. Use the graphing utility to plot the points and graph the model.

(d) Use the model in part (c) and the integration capabilities of a graphing utility to approximate the volume and outside surface area of the vase. Compare the results with your answers in parts (a) and (b).

45. Individual Project   Select a solid of revolution from everyday life. Measure the radius of the solid at a minimum of seven points along its axis. Use the data to approximate the volume of the solid and the surface area of the lateral sides of the solid.

46. Let R be the region bounded by $y = 1/x$, the x-axis, $x = 1$, and $x = b$, where $b > 1$. Let D be the solid formed when R is revolved about the x-axis.

(a) Find the volume V of D.

(b) Express the surface area S as an integral.

(c) Show that V approaches a finite limit as $b \to \infty$.

(d) Show that $S \to \infty$ as $b \to \infty$.

47. Let f be rectifiable on the interval $[a, b]$, and let

$$s(x) = \int_a^x \sqrt{1 + [f'(t)]^2}\, dt.$$

(a) Find $\dfrac{ds}{dx}$.

(b) Find ds and $(ds)^2$.

(c) If $f(t) = t^{3/2}$, find $s(x)$ on $[1, 3]$. What is $s(2)$?

 48. Think About It   Consider the equation $\dfrac{x^2}{9} + \dfrac{y^2}{4} = 1$.

(a) Use a graphing utility to graph the equation.

(b) Set up the definite integral for finding the first quadrant arc length of the graph in part (a).

(c) Compare the interval of integration in part (b) and the domain of the integrand. Is it possible to evaluate the definite integral? Is it possible to use Simpson's Rule to evaluate the definite integral? Explain. (This problem will be addressed in Section 7.8.)

49. (a) Given a circular sector with radius L and central angle θ (see figure), show that the area of the sector is given by

$$S = \frac{1}{2}L^2\theta.$$

(b) By joining the straight line edges of the sector in part (a), a right circular cone is formed (see figure) and the lateral surface area of the cone is the same as the area of the sector. Show that the area is

$$S = \pi r L$$

where r is the radius of the base of the cone. (*Hint:* The arc length of the sector equals the circumference of the base of the cone.)

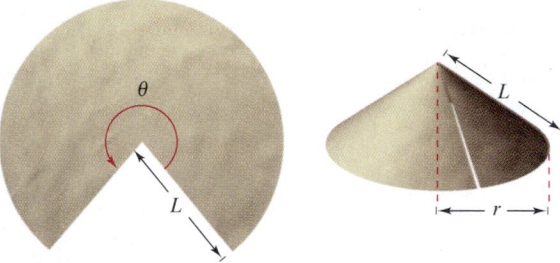

Figure for 49(a)   **Figure for 49(b)**

(c) Use the result in part (b) to verify that the formula for the lateral surface area of the frustum of a cone with slant height L and radii r_1 and r_2 (see figure) is

$$S = \pi (r_1 + r_2)L.$$

(*Note:* This formula was used to develop the integral for finding the surface area of a surface of revolution.)

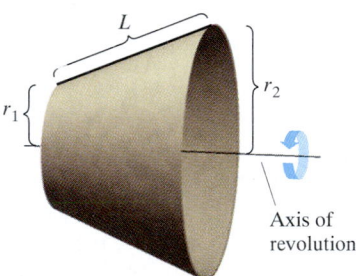

Figure for 49(c)

50. Writing   Read the article "Arc Length, Area and the Arcsine Function" by Andrew M. Rockett in the March 1983 issue of *Mathematics Magazine*. Then write a paragraph explaining how the arcsine function can be defined in terms of an arc length.

Work Done by a Constant Force • Work Done by a Variable Force

Work Done by a Constant Force

The concept of work is important to scientists and engineers for determining the energy needed to perform various jobs. For instance, it is useful to know the amount of work done when a crane lifts a steel girder, when a spring is compressed, when a rocket is propelled into the air, or when a truck pulls a load along a highway.

In general, we say that **work** is done by a force when it moves an object. If the force applied to the object is *constant*, we have the following definition of work.

> **Definition of Work Done by a Constant Force**
>
> If an object is moved a distance D in the direction of an applied constant force F, then the **work** W done by the force is defined as $W = FD$.

There are many types of forces—centrifugal, electromotive, and gravitational, to name a few. A **force** can be thought of as a *push* or a *pull*; a force changes the state of rest or state of motion of a body. For gravitational forces on the earth, it is common to use units of measure corresponding to the weight of an object.

EXAMPLE 1   Lifting an Object

Determine the work done in lifting a 50-pound object 4 feet.

Solution   The magnitude of the required force F is the weight of the object, as shown in Figure 6.49. Thus, the work done in lifting the object 4 feet is

$$W = FD \qquad \text{Work = (force)(distance)}$$
$$= 50(4) \qquad \text{Force = 50 pounds, distance = 4 feet}$$
$$= 200 \text{ foot-pounds.}$$

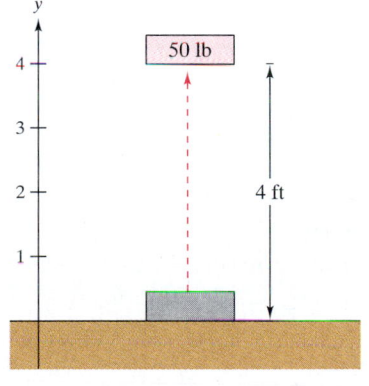

The work done in lifting a 50-pound object 4 feet is 200 foot-pounds.
Figure 6.49

In the U.S. measurement system, work is typically expressed in foot-pounds (ft · lb), inch-pounds, or foot-tons. In the centimeter-gram-second (C-G-S) system, the basic unit of force is the **dyne**—the force required to produce an acceleration of 1 centimeter per second per second on a mass of 1 gram. In this system, work is typically expressed in dyne-centimeters (ergs) or newton-meters (joules), where 1 joule = 10^7 ergs.

> ### EXPLORATION
>
> *How Much Work?*   In Example 1, it required 200 foot-pounds of work to lift the 50-pound object 4 feet, vertically off the ground. Suppose that once you lifted the object, you held it and walked a horizontal distance of 4 feet. Would this require an additional 200 foot-pounds of work? Explain your reasoning.

Work Done by a Variable Force

In Example 1, the force involved is *constant*. If a *variable* force is applied to an object, calculus is needed to determine the work done, because the amount of force changes as the object changes positions. For instance, the force required to compress a spring increases as the spring is compressed.

Suppose that an object is moved along a straight line from $x = a$ to $x = b$ by a continuously varying force $F(x)$. Let Δ be a partition that divides the interval $[a, b]$ into n subintervals determined by

$$a = x_0 < x_1 < x_2 < \cdots < x_n = b$$

and let $\Delta x_i = x_i - x_{i-1}$. For each i, choose c_i such that $x_{i-1} \leq c_i \leq x_i$. Then at c_i the force is given by $F(c_i)$. Because F is continuous, you can approximate the work done in moving the object through the ith subinterval by the increment

$$\Delta W_i = F(c_i)\Delta x_i$$

as shown in Figure 6.50. Thus, the total work done as the object moves from a to b is approximated by

$$W \approx \sum_{i=1}^{n} \Delta W_i$$

$$= \sum_{i=1}^{n} F(c_i)\Delta x_i.$$

This approximation appears to become better and better as $\|\Delta\| \to 0 \ (n \to \infty)$. Therefore, we define the work to be

$$W = \lim_{\|\Delta\| \to 0} \sum_{i=1}^{n} F(c_i)\Delta x_i$$

$$= \int_a^b F(x)\, dx.$$

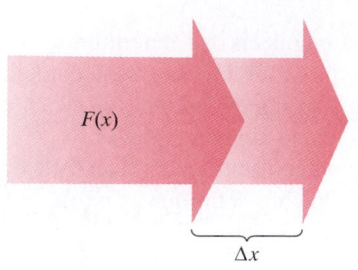

$F(x)$

Δx

The amount of force changes as an object changes position (Δx).
Figure 6.50

Definition of Work Done by a Variable Force

If an object is moved along a straight line by a continuously varying force $F(x)$, then the **work** W done by the force as the object is moved from $x = a$ to $x = b$ is

$$W = \lim_{\|\Delta\| \to 0} \sum_{i=1}^{n} \Delta W_i$$

$$= \int_a^b F(x)\, dx.$$

The remaining examples in this section use some well-known physical laws. The discoveries of many of these laws occurred during the same period in which calculus was being developed. In fact, during the seventeenth and eighteenth centuries, there was little difference between physicists and mathematicians. One such physicist-mathematician was Emilie de Breteuil. Breteuil was instrumental in synthesizing the work of many other scientists, including Newton, Leibniz, Huygens, Kepler, and Descartes. Her physics text *Institutions* was widely used for many years.

Corbis-Bettmann

EMILIE DE BRETEUIL (1706–1749)

Another major work by de Breteuil was the translation of Newton's "Philosophiae Naturalis Principia Mathematica" into French. Her translation and commentaries greatly contributed to the acceptance of Newtonian science in Europe.

The following three laws of physics were developed by Robert Hooke (1635–1703), Isaac Newton (1642–1727), and Charles Coulomb (1736–1806).

1. **Hooke's Law:** The force F required to compress or stretch a spring (within its elastic limits) is proportional to the distance d that the spring is compressed or stretched from its original length. That is,

$$F = kd$$

where the constant of proportionality k (the spring constant) depends on the specific nature of the spring.

2. **Law of Universal Gravitation:** The force F of attraction between two particles of masses m_1 and m_2 is proportional to the product of the masses and inversely proportional to the square of the distance d between the two particles. That is,

$$F = k\frac{m_1 m_2}{d^2}.$$

If m_1 and m_2 are given in grams and d in centimeters, F will be in dynes for a value of $k = 6.670 \times 10^{-8}$ cubic centimeter per gram-second squared.

3. **Coulomb's Law:** The force between two charges q_1 and q_2 in a vacuum is proportional to the product of the charges and inversely proportional to the square of the distance d between the two charges. That is,

$$F = k\frac{q_1 q_2}{d^2}.$$

If q_1 and q_2 are given in electrostatic units and d in centimeters, F will be in dynes for a value of $k = 1$.

EXPLORATION

The work done in compressing the spring in Example 2 from $x = 3$ inches to $x = 6$ inches is 3375 inch-pounds. Should the work done in compressing the spring from $x = 0$ inches to $x = 3$ inches be more than, the same as, or less than this? Explain.

EXAMPLE 2   Compressing a Spring

A force of 750 pounds compresses a spring 3 inches from its natural length of 15 inches. Find the work done in compressing the spring an additional 3 inches.

Solution   By Hooke's Law, the force $F(x)$ required to compress the spring x units (from its natural length) is $F(x) = kx$. Using the given data, it follows that $F(3) = 750 = (k)(3)$ and thus $k = 250$ and $F(x) = 250x$, as shown in Figure 6.51. To find the increment of work, assume that the force required to compress the spring over a small increment Δx is nearly constant. Thus, the increment of work is

$$\Delta W = (\text{force})(\text{distance increment}) = (250x)\,\Delta x.$$

Because the spring is compressed from $x = 3$ to $x = 6$ inches less than its natural length, the work required is

$$W = \int_3^6 250x \, dx$$

$$= 125x^2 \Big]_3^6 = 4500 - 1125 = 3375 \text{ inch-pounds.}$$

Note that you do not integrate from $x = 0$ to $x = 6$ because you were asked to determine the work done in compressing the spring an *additional* 3 inches (not including the first 3 inches).

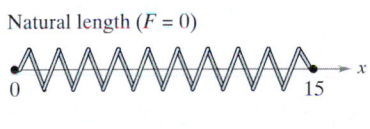

Natural length ($F = 0$)

Compressed 3 inches ($F = 750$)

Compressed x inches ($F = 250x$)

Figure 6.51

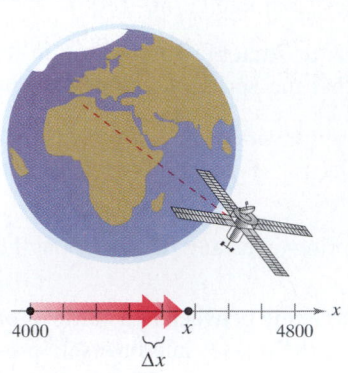

The work required to move a space module 800 miles above the earth is approximately 1.056×10^{11} foot-pounds.

Figure 6.52

EXAMPLE 3   Moving a Space Module into Orbit

A space module weighs 15 tons on the surface of the earth. How much work is done in propelling the module to a height of 800 miles above the earth, as shown in Figure 6.52? (Use 4000 miles as the radius of the earth. Do not consider the effect of air resistance or the weight of the propellant.)

Solution   Because the weight of a body varies inversely as the square of its distance from the center of the earth, the force $F(x)$ exerted by gravity is

$$F(x) = \frac{C}{x^2}. \qquad \text{\textcolor{red}{\textit{C} is the constant of proportionality.}}$$

Because the module weighs 15 tons on the surface of the earth and the radius of the earth is approximately 4000 miles, you have

$$15 = \frac{C}{(4000)^2}$$
$$240{,}000{,}000 = C.$$

Thus, the increment of work is

$$\Delta W = (\text{force})(\text{distance increment})$$
$$= \frac{240{,}000{,}000}{x^2} \Delta x.$$

Finally, because the module is propelled from $x = 4000$ to $x = 4800$ miles, the total work done is

$$W = \int_{4000}^{4800} \frac{240{,}000{,}000}{x^2}\, dx$$
$$= \frac{-240{,}000{,}000}{x} \Bigg]_{4000}^{4800}$$
$$= -50{,}000 + 60{,}000$$
$$= 10{,}000 \text{ mile-tons}$$
$$\approx 1.056 \times 10^{11} \text{ foot-pounds.}$$

In the C-G-S system, using a conversion factor of 1 foot-pound ≈ 1.35582 joules, the work done is

$$W \approx 1.432 \times 10^{11} \text{ joules.}$$

The solutions to Examples 2 and 3 conform to our development of work as the summation of increments in the form

$$\Delta W = (\text{force})(\text{distance increment}) = (F)(\Delta x).$$

Another way to formulate the increment of work is

$$\Delta W = (\text{force increment})(\text{distance}) = (\Delta F)(x).$$

This second interpretation of ΔW is useful in problems involving the movement of nonrigid substances such as fluids and chains.

EXAMPLE 4   Emptying a Tank of Oil

A spherical tank of radius 8 feet is half full of oil that weighs 50 pounds per cubic foot. Find the work required to pump oil out through a hole in the top of the tank.

Solution   Consider the oil to be subdivided into discs of thickness Δy and radius x, as shown in Figure 6.53. Because the increment of force for each disc is given by its weight, you have

$$\Delta F = \text{weight}$$
$$= \left(\frac{50 \text{ pounds}}{\text{cubic foot}}\right)(\text{volume})$$
$$= 50(\pi x^2 \Delta y) \text{ pounds}.$$

For a circle of radius 8 and center at $(0, 8)$, you have

$$x^2 + (y - 8)^2 = 8^2$$
$$x^2 = 16y - y^2$$

and you can write the force increment as

$$\Delta F = 50(\pi x^2 \Delta y)$$
$$= 50\pi(16y - y^2)\,\Delta y.$$

In Figure 6.53, note that a disc y feet from the bottom of the tank must be moved a distance of $(16 - y)$ feet. Therefore, the increment of work is

$$\Delta W = \Delta F(16 - y)$$
$$= 50\pi(16y - y^2)\,\Delta y(16 - y)$$
$$= 50\pi(256y - 32y^2 + y^3)\,\Delta y.$$

Because the tank is half full, y ranges from 0 to 8, and the work required to empty the tank is

$$W = \int_0^8 50\pi(256y - 32y^2 + y^3)\,dy$$
$$= 50\pi\left[128y^2 - \frac{32}{3}y^3 + \frac{y^4}{4}\right]_0^8$$
$$= 50\pi\left(\frac{11{,}264}{3}\right)$$
$$\approx 589{,}782 \text{ foot-pounds}.$$

To estimate the reasonableness of the result in Example 4, consider that the weight of the oil in the tank is

$$\left(\frac{1}{2}\right)(\text{volume})(\text{density}) = \frac{1}{2}\left(\frac{4}{3}\pi 8^3\right)(50)$$
$$\approx 53{,}616.5 \text{ pounds}.$$

Lifting the entire half-tank of oil 8 feet would involve work of $8(53{,}616.5) \approx 428{,}932$ foot-pounds. Because the oil is actually lifted between 8 and 16 feet, it seems reasonable that the work done is 589,782 foot-pounds.

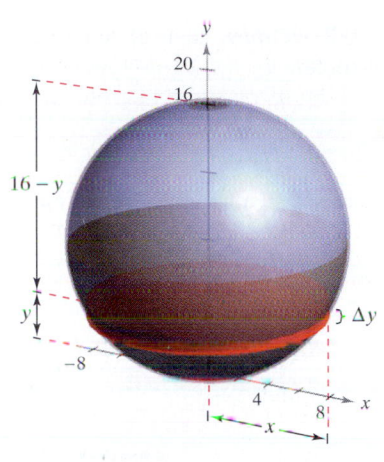

The work required to pump oil out through a hole in the top of the tank is approximately 589,782 foot-pounds.

Figure 6.53

The work required to raise one end of the chain 20 feet is 1000 foot-pounds.
Figure 6.54

EXAMPLE 5 Lifting a Chain

A 20-foot chain weighing 5 pounds per foot is lying coiled on the ground. How much work is required to raise one end of the chain to a height of 20 feet so that it is fully extended, as shown in Figure 6.54?

Solution Imagine that the chain is divided into small sections, each of length Δy. Then the weight of each section is the increment of force

$$\Delta F = (\text{weight}) = \left(\frac{5 \text{ pounds}}{\text{foot}}\right)(\text{length}) = 5\Delta y.$$

Because a typical section (initially on the ground) is raised to a height of y, the increment of work is

$$\Delta W = (\text{force increment})(\text{distance}) = (5 \Delta y)y = 5y \Delta y.$$

Because y ranges from 0 to 20, the total work is

$$W = \int_0^{20} 5y \, dy = \frac{5y^2}{2}\bigg]_0^{20} = \frac{5(400)}{2} = 1000 \text{ foot-pounds.}$$

In the next example we consider a piston of radius r in a cylindrical casing, as shown in Figure 6.55. As the gas in the cylinder expands, the piston moves and work is done. If p represents the pressure of the gas (in pounds per square foot) against the piston head and V represents the volume of the gas (in cubic feet), the work increment involved in moving the piston Δx feet is

$$\Delta W = (\text{force})(\text{distance increment}) = F(\Delta x) = p(\pi r^2) \Delta x = p \, \Delta V.$$

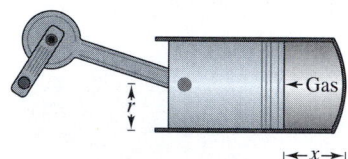

Work done by expanding gas
Figure 6.55

Thus, as the volume of the gas expands from V_0 to V_1, the work done in moving the piston is

$$W = \int_{V_0}^{V_1} p \, dV.$$

Assuming the pressure of the gas to be inversely proportional to its volume, we have $p = k/V$ and the integral for work becomes

$$W = \int_{V_0}^{V_1} \frac{k}{V} \, dV.$$

EXAMPLE 6 Work Done by an Expanding Gas

A quantity of gas with an initial volume of 1 cubic foot and pressure of 500 pounds per square foot expands to a volume of 2 cubic feet. Find the work done by the gas. (Assume that the pressure is inversely proportional to the volume).

Solution Because $p = k/V$ and $p = 500$ when $V = 1$, we have $k = 500$. Thus, the work is

$$W = \int_{V_0}^{V_1} \frac{k}{V} \, dv = \int_1^2 \frac{500}{V} \, dv = 500 \ln |V| \bigg]_1^2 \approx 346.6 \text{ foot-pounds.}$$

EXERCISES FOR SECTION 6.5

Constant Force **In Exercises 1–4, determine the work done by the constant force.**

1. A 100-pound bag of sugar is lifted 10 feet.

2. An electric hoist lifts a 2400-pound car 6 feet.

3. A force of 112 newtons is required to slide a cement block 4 meters in a construction project.

4. The locomotive of a freight train pulls its cars with a constant force of 9 tons a distance of one-half mile.

5. ***Think About It*** The graphs show the force F_i (in pounds) required to move an object 9 feet along the x-axis. Order the force functions from the one that yields the least work to the one that yields the most work without doing any calculations.

(a)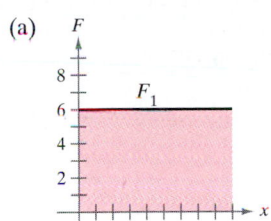

(b)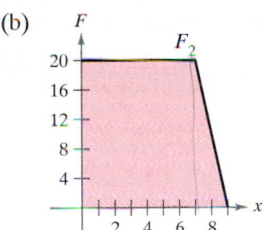

(c)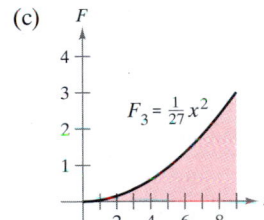

(d)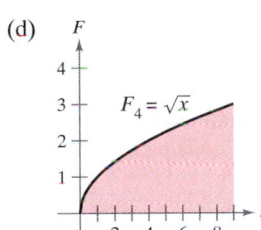

6. Verify your answer to Exercise 5 by calculating the work for each force function.

Hooke's Law **In Exercises 7–14, use Hooke's Law to determine the variable force in the spring problem.**

7. A force of 5 pounds compresses a 15-inch spring a total of 4 inches. How much work is done in compressing the spring 7 inches?

8. How much work is done in compressing the spring in Exercise 7 from a length of 10 inches to a length of 6 inches?

9. A force of 250 newtons stretches a spring 30 centimeters. How much work is done in stretching the spring from 20 centimeters to 50 centimeters?

10. A force of 800 newtons stretches a spring 70 centimeters on a mechanical device for driving fence posts. Find the work done in stretching the spring the required 70 centimeters.

11. A force of 15 pounds stretches a spring 6 inches in an exercise machine. Find the work done in stretching the spring 1 foot from its natural position.

12. An overhead garage door has two springs, one on each side of the door. A force of 15 pounds is required to stretch each spring 1 foot. Because of the pulley system, the springs stretch only one-half the distance the door travels. Find the work done by the pair of springs if the door moves a total of 8 feet and the springs are at their natural length when the door is open.

13. Eighteen foot-pounds of work is required to stretch a spring 4 inches from its natural length. Find the work required to stretch the spring an additional 3 inches.

14. Seven and one-half foot-pounds of work is required to compress a spring 2 inches from its natural length. Find the work required to compress the spring an additional one-half inch.

15. ***Propulsion*** Neglecting air resistance and the weight of the propellant, determine the work done in propelling a 4-ton satellite to a height of

    (a) 200 miles above the earth.

    (b) 400 miles above the earth.

16. ***Propulsion*** Use the information in Exercise 15 to write the work W of the propulsion system as a function of the height h of the satellite above the earth. Find the limit (if it exists) of W as h approaches infinity.

17. ***Propulsion*** Neglecting air resistance and the weight of the propellant, determine the work done in propelling a 10-ton satellite to the height of

    (a) 11,000 miles above the earth.

    (b) 22,000 miles above the earth.

18. ***Propulsion*** A lunar module weighs 12 tons on the surface of the earth. How much work is done in propelling the module from the surface of the moon to a height of 50 miles? Consider the radius of the moon to be 1100 miles and its force of gravity to be one-sixth that of the earth's.

19. ***Pumping Water*** A rectangular tank with a base 4 feet by 5 feet and a height of 4 feet is full of water (see figure). (The water weighs 62.4 pounds per cubic foot.) How much work is done in pumping water out over the top edge in order to empty (a) half of the tank? (b) all of the tank?

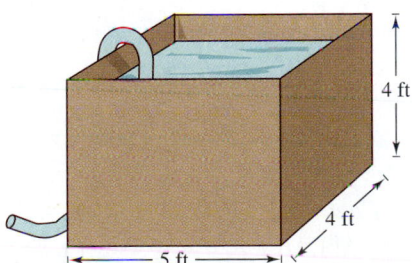

20. ***Think About It*** Explain why the answer in part (b) of Exercise 19 is not twice the answer in part (a).

21. *Pumping Water* A cylindrical water tank 4 meters high with a radius of 2 meters is buried so that the top of the tank is 1 meter below ground level (see figure). (The weight of water is 1000 kilograms per cubic meter.) How much work is done in pumping a full tank of water up to ground level?

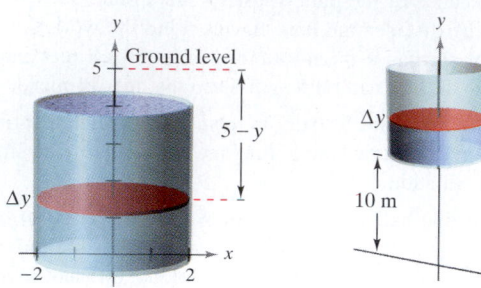

Figure for 21   **Figure for 22**

22. *Pumping Water* Suppose the tank in Exercise 21 is located on the tower so that the bottom of the tank is 10 meters above the level of a stream (see figure). How much work is done in filling the tank half full of water through a hole in the bottom, using water from the stream?

23. *Pumping Water* An open tank has the shape of a right circular cone (see figure). The tank is 8 feet across the top and 6 feet high. How much work is done in emptying the tank by pumping the water over the top edge?

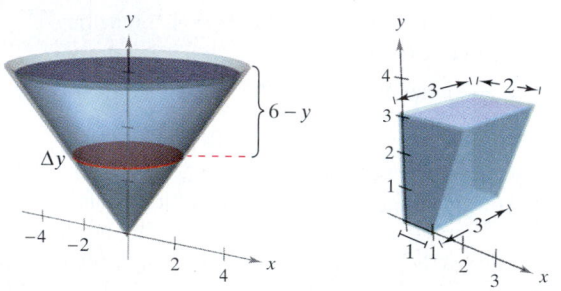

Figure for 23   **Figure for 26**

24. *Pumping Water* If water is pumped in through the bottom of the tank in Exercise 23, how much work is done to fill the tank

(a) to a depth of 2 feet?

(b) from a depth of 4 feet to a depth of 6 feet?

25. *Pumping Water* A hemispherical tank of radius 6 feet is positioned so that its base is circular. How much work is required to fill the tank with water through a hole in the base if the water source is at the base?

26. *Pumping Diesel Fuel* The fuel tank on a large truck has trapezoidal cross sections with dimensions (in feet) shown in the figure. Assume that the engine is approximately 2 feet above the top of the fuel tank and that diesel fuel weighs approximately 55.6 pounds per cubic foot. Find the work done by the fuel pump in raising a full tank of fuel to the level of the engine.

Pumping Gasoline In Exercises 27 and 28, find the work done in pumping gasoline that weighs 42 pounds per cubic foot. (*Hint:* Evaluate one integral by a geometric formula and the other by observing that the integrand is an odd function.)

27. A cylindrical gasoline tank 3 feet in diameter and 4 feet long is carried on the back of a truck and is used to fuel tractors. The axis of the tank is horizontal. Find the work done in pumping the entire contents of the fuel tank into a tractor if the opening on the tractor tank is 5 feet above the top of the tank in the truck.

28. The top of a cylindrical storage tank for gasoline at a service station is 4 feet below ground level. The axis of the tank is horizontal and its diameter and length are 5 feet and 12 feet. Find the work done in pumping the entire contents of the full tank to a height of 3 feet above ground level.

Lifting a Chain In Exercises 29–32, consider a 15-foot chain hanging from a winch 15 feet above ground level. Find the work done by the winch in winding up the specified amount of chain, if the chain weighs 3 pounds per foot.

29. Wind up the entire chain.

30. Wind up one-third of the chain.

31. Run the winch until the bottom of the chain is at the 10-foot level.

32. Wind up the entire chain with a 100-pound load attached to it.

Lifting a Chain In Exercises 33 and 34, consider a 15-foot hanging chain that weighs 3 pounds per foot. Find the work done in lifting the chain vertically to the indicated position.

33. Take the bottom of the chain and raise it to the 15-foot level, leaving the chain doubled and still hanging vertically (see figure).

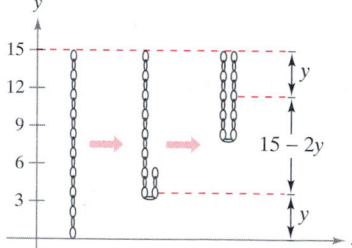

34. Repeat Exercise 33 raising the bottom of the chain to the 12-foot level.

Demolition Crane In Exercises 35 and 36, consider a demolition crane with a 500-pound ball suspended from a 40-foot cable that weighs 1 pound per foot.

35. Find the work required to wind up 15 feet of the apparatus.

36. Find the work required to wind up all 40 feet of the apparatus.

Boyle's Law   In Exercises 37 and 38, find the work done by the gas for the given volume and pressure. Assume that the pressure is inversely proportional to the volume. (See Example 6.)

37. A quantity of gas with an initial volume of 2 cubic feet and pressure of 1000 pounds per square foot expands to a volume of 3 cubic feet.

38. A quantity of gas with an initial volume of 1 cubic foot and pressure of 2000 pounds per square foot expands to a volume of 4 cubic feet.

39. *Electric Force*   Two electrons repel each other with a force that varies inversely as the square of the distance between them. If one electron is fixed at the point $(2, 4)$, find the work done in moving the second electron from $(-2, 4)$ to $(1, 4)$.

40. *Modeling Data*   The hydraulic cylinder on a woodsplitter has a 4-inch bore (diameter) and a stroke of 2 feet. The hydraulic pump creates a maximum pressure of 2000 pounds per square inch. Therefore, the maximum force created by the cylinder is $2000(\pi 2^2) = 8000\pi$ pounds.

(a) Find the work done through one extension of the cylinder given that the maximum force is required.

(b) The force exerted in splitting a piece of wood is variable. Measurements of the force obtained when a piece of wood was split are given in the table. The variable x measures the extension of the cylinder in feet, and F is the force in pounds. Use Simpson's Rule to approximate the work done in splitting the piece of wood.

x	0	$\frac{1}{3}$	$\frac{2}{3}$	1
$F(x)$	0	20,000	22,000	15,000

x	$\frac{4}{3}$	$\frac{5}{3}$	2
$F(x)$	10,000	5000	0

(c) Use the regression capabilities of a graphing utility to find a fourth-degree polynomial model for the data. Plot the data and graph the model.

(d) Use the model in part (c) to approximate the extension of the cylinder when the force is maximum.

(e) Use the model in part (c) to approximate the work done in splitting the piece of wood.

Hydraulic Press   In Exercises 41–44, use the integration capabilities of a graphing utility to approximate the work done by a press in a manufacturing process. A model for the variable force F (pounds) and the distance x (feet) the press moves are given.

Force	Interval
41. $F(x) = 1000[1.8 - \ln(x + 1)]$	$0 \le x \le 5$
42. $F(x) = \dfrac{e^{x^2} - 1}{100}$	$0 \le x \le 4$
43. $F(x) = 100x\sqrt{125 - x^3}$	$0 \le x \le 5$
44. $F(x) = 1000 \sinh x$	$0 \le x \le 2$

Tidal Energy   Tidal power plants use "tidal energy" to produce electrical energy. To construct a tidal power plant, a dam is built to separate a bay from the sea. Electrical energy is produced as the water flows back and forth between the bay and the sea. The amount of "natural energy" produced depends on the volume of the basin and the tidal range—the vertical distance between high and low tides. (Throughout the world, several natural bays have tidal ranges in excess of 15 feet; the Bay of Fundy in Nova Scotia has a tidal range of 47.5 feet.)

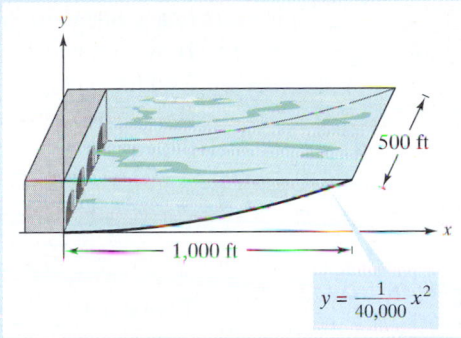

$$y = \frac{1}{40,000} x^2$$

(a) Consider a bay with a rectangular base, as shown in the figure. The bay has a tidal range of 25 feet, with low tide corresponding to $y = 0$. How much water does the bay hold at high tide?

(b) The amount of energy produced during the filling (or the emptying) of the bay is proportional to the amount of work required to fill (or empty) the bay. How much work is required to fill the bay with seawater? (Use a seawater density of 64 pounds per cubic foot.)

The Bay of Fundy in Nova Scotia has an extreme tidal range, as displayed in the greatly contrasting photos above.

FOR FURTHER INFORMATION   For more information on tidal power, see the article "LaRance: Six Years of Operating a Tidal Power Plant in France" by J. Cotillon in the October 1984 issue of *Water Power Magazine*.

Mass • Center of Mass in a One-Dimensional System •
Center of Mass in a Two-Dimensional System •
Center of Mass of a Planar Lamina • Theorem of Pappus

Mass

In this section you will study several important applications of integration that are related to **mass**. Mass is a measure of a body's resistance to changes in motion, and is independent of the particular gravitational system in which the body is located. However, because so many applications involving mass occur on earth's surface, we tend to equate an object's mass with its *weight*. This is not technically correct. Weight is a type of force and as such is dependent on gravity. Force and mass are related by the equation

$$\text{Force} = (\text{mass})(\text{acceleration}).$$

The table below lists some commonly used measures of mass and force, together with their conversion factors.

System of Measurement	Measure of Mass	Measure of Force
U.S.	Slug	Pound = (slug)(ft/sec²)
International	Kilogram	Newton = (kilogram)(m/sec²)
C-G-S	Gram	Dyne = (gram)(cm/sec²)

Conversions:

1 pound = 4.448 newtons   1 slug = 14.59 kilograms
1 newton = 0.2248 pound   1 kilogram = 0.06854 slug
1 dyne = 0.000002247 pound   1 gram = 0.00006854 slug
1 dyne = 0.00001 newton   1 meter = 0.3048 foot

EXAMPLE 1   Mass on the Surface of the Earth

Find the mass (in slugs) of an object whose weight at sea level is 1 pound.

Solution   Using 32 feet per second per second as the acceleration due to gravity produces

$$\text{Mass} = \frac{\text{force}}{\text{acceleration}} \qquad \color{red}{\text{Force} = (\text{mass})(\text{acceleration})}$$

$$= \frac{1 \text{ pound}}{32 \text{ feet per second per second}}$$

$$= 0.03125 \frac{\text{pound}}{\text{foot per second per second}}$$

$$= 0.03125 \text{ slug}.$$

Because many applications occur on the earth's surface, this amount of mass is called a **pound mass.**

Center of Mass in a One-Dimensional System

We will consider two types of moments of a mass–the **moment about a point** and the **moment about a line.** To define these two moments, consider an idealized situation in which a mass m is concentrated at a point. If x is the distance between this point mass and another point P, the **moment of m about the point P** is

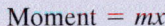

$$\text{Moment} = mx$$

and x is the **length of the moment arm.**

The concept of moment can be demonstrated simply by a seesaw, as illustrated in Figure 6.56. Suppose a child of mass 20 kilograms sits 2 meters to the left of fulcrum P, and an older child of mass 30 kilograms sits 2 meters to the right of P. From experience, you know that the seesaw will begin to rotate clockwise, moving the larger child down. This rotation occurs because the moment produced by the child on the left is less than the moment produced by the child on the right.

$$\text{Left moment} = (20)(2) = 40 \text{ kilogram-meters}$$

$$\text{Right moment} = (30)(2) = 60 \text{ kilogram-meters}$$

To balance the seesaw, the two moments must be equal. For example, if the larger child moved to a position $\frac{4}{3}$ meters from the fulcrum, the seesaw would balance, because each child would produce a moment of 40 kilogram-meters.

To generalize this, you can introduce a coordinate line on which the origin corresponds to the fulcrum, as shown in Figure 6.57. Suppose several point masses are located on the x-axis. The measure of the tendency of this system to rotate about the origin is the **moment about the origin,** and it is defined as the sum of the n products $m_i x_i$.

$$M_0 = m_1 x_1 + m_2 x_2 + \cdots + m_n x_n$$

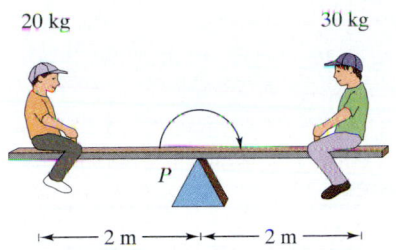

20 kg   30 kg

P

|← 2 m →|← 2 m →|

The seesaw will balance when the left and the right moments are equal.
Figure 6.56

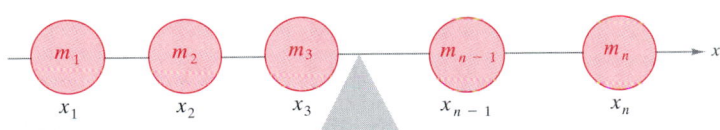

If $m_1 x_1 + m_2 x_2 + \cdots + m_n x_n = 0$, the system is in equilibrium.
Figure 6.57

If M_0 is 0, the system is said to be in **equilibrium.**

For a system that is not in equilibrium, the **center of mass** is defined as the point $\bar{x}$ at which the fulcrum could be relocated to attain equilibrium. If the system were translated $\bar{x}$ units, each coordinate x_i would become $(x_i - \bar{x})$, and because the moment of the translated system is 0, you have

$$\sum_{i=1}^{n} m_i(x_i - \bar{x}) = \sum_{i=1}^{n} m_i x_i - \sum_{i=1}^{n} m_i \bar{x} = 0.$$

Solving for $\bar{x}$ produces

$$\bar{x} = \frac{\displaystyle\sum_{i=1}^{n} m_i x_i}{\displaystyle\sum_{i=1}^{n} m_i} = \frac{\text{moment of system about origin}}{\text{total mass of system}}.$$

If $m_1 x_1 + m_2 x_2 + \cdots + m_n x_n = 0$, the system is in equilibrium.

> **Moments and Center of Mass: One-Dimensional System**
>
> Let the point masses $m_1, m_2, \ldots, m_n$ be located at $x_1, x_2, \ldots, x_n$.
>
> 1. The **moment about the origin** is $M_0 = m_1 x_1 + m_2 x_2 + \cdots + m_n x_n$.
>
> 2. The **center of mass** is $\bar{x} = \dfrac{M_0}{m}$, where $m = m_1 + m_2 + \cdots + m_n$ is the **total mass** of the system.

EXAMPLE 2   The Center of Mass of a Linear System

Find the center of mass of the linear system shown in Figure 6.58.

NOTE   In Example 2, where should you locate the fulcrum so that the point masses will be in equilibrium?

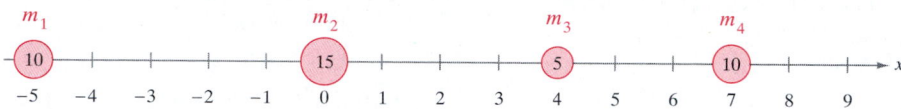

Figure 6.58

Solution   The moment about the origin is

$$
\begin{aligned}
M_0 &= m_1 x_1 + m_2 x_2 + m_3 x_3 + m_4 x_4 \\
&= 10(-5) + 15(0) + 5(4) + 10(7) \\
&= -50 + 0 + 20 + 70 \\
&= 40.
\end{aligned}
$$

Because the total mass of the system is $m = 10 + 15 + 5 + 10 = 40$, the center of mass is

$$
\bar{x} = \frac{M_0}{m} = \frac{40}{40} = 1.
$$

Rather than define the moment of a *mass*, you could define the moment of a *force*. In this context, the center of mass is called the **center of gravity.** Suppose that a system of point masses $m_1, m_2, \ldots, m_n$ is located at $x_1, x_2, \ldots, x_n$. Then, because force $=$ (mass)(acceleration), the total force of the system is

$$
\begin{aligned}
F &= m_1 a + m_2 a + \cdots + m_n a \\
&= ma.
\end{aligned}
$$

The **torque** (moment) about the origin is

$$
\begin{aligned}
T_0 &= (m_1 a)x_1 + (m_2 a)x_2 + \cdots + (m_n a)x_n \\
&= M_0 a
\end{aligned}
$$

and the **center of gravity** is

$$
\frac{T_0}{F} = \frac{M_0 a}{ma} = \frac{M_0}{m} = \bar{x}.
$$

Therefore, the center of gravity and the center of mass have the same location.

Center of Mass in a Two-Dimensional System

You can extend the concept of moment to two dimensions by considering a system of masses located in the xy-plane at the points (x_1, y_1), (x_2, y_2), . . . , (x_n, y_n) as shown in Figure 6.59. Rather than defining a single moment (with respect to the origin), we define two moments—one with respect to the x-axis and one with respect to the y-axis.

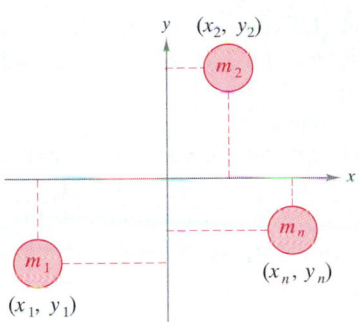

In a two-dimensional system, there is a moment about the y-axis, M_y, and a moment about the x-axis, M_x.

Figure 6.59

Moments and Center of Mass: Two-Dimensional System

Let the point masses $m_1, m_2, . . . , m_n$ be located at (x_1, y_1), (x_2, y_2), . . . , (x_n, y_n).

1. The **moment about the y-axis** is $M_y = m_1 x_1 + m_2 x_2 + \cdots + m_n x_n$.
2. The **moment about the x-axis** is $M_x = m_1 y_1 + m_2 y_2 + \cdots + m_n y_n$.
3. The **center of mass** $(\bar{x}, \bar{y})$ (or **center of gravity**) is

$$\bar{x} = \frac{M_y}{m} \qquad \text{and} \qquad \bar{y} = \frac{M_x}{m}$$

where $m = m_1 + m_2 + \cdots + m_n$ is the **total mass** of the system.

The moment of a system of masses in the plane can be taken about any horizontal or vertical line. In general, the moment about a line is the sum of the product of the masses and the *directed distances* from the points to the line.

$$\text{Moment} = m_1(y_1 - b) + m_2(y_2 - b) + \cdots + m_n(y_n - b) \qquad \text{Horizontal line } y = b$$

$$\text{Moment} = m_1(x_1 - a) + m_2(x_2 - a) + \cdots + m_n(x_n - a) \qquad \text{Vertical line } x = a$$

EXAMPLE 3   The Center of Mass of a Two-Dimensional System

Find the center of mass of a system of point masses $m_1 = 6$, $m_2 = 3$, $m_3 = 2$, and $m_4 = 9$, located at

$$(3, -2), (0, 0), (-5, 3), \text{ and } (4, 2)$$

as shown in Figure 6.60.

Solution

$$
\begin{aligned}
m &= 6 &&+ 3 &&+ 2 &&+ 9 &&= 20 &&\text{Mass}\\
M_y &= 6(3) &&+ 3(0) &&+ 2(-5) &&+ 9(4) &&= 44 &&\text{Moment about } y\text{-axis}\\
M_x &= 6(-2) &&+ 3(0) &&+ 2(3) &&+ 9(2) &&= 12 &&\text{Moment about } x\text{-axis}
\end{aligned}
$$

Therefore,

$$\bar{x} = \frac{M_y}{m} = \frac{44}{20} = \frac{11}{5}$$

and

$$\bar{y} = \frac{M_x}{m} = \frac{12}{20} = \frac{3}{5}$$

and thus the center of mass is $\left(\frac{11}{5}, \frac{3}{5}\right)$.

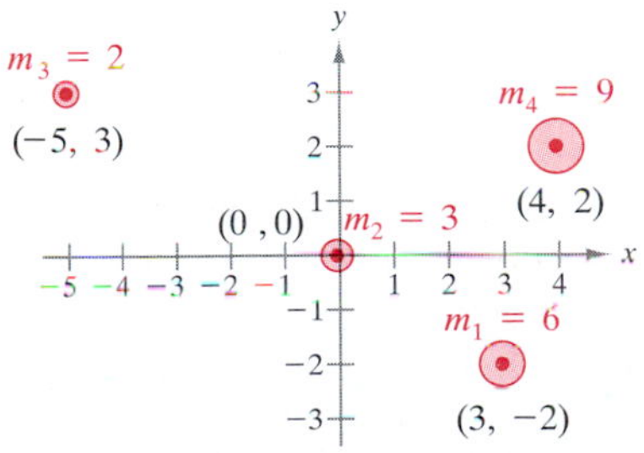

The center of mass of the system is $\left(\frac{11}{5}, \frac{3}{5}\right)$.

Figure 6.60

Center of Mass of a Planar Lamina

So far in this section we have assumed the total mass of a system to be distributed at discrete points in a plane or on a line. We now consider a thin, flat plate of material of constant density called a **planar lamina** (see Figure 6.61). **Density** is a measure of mass per unit of volume, such as grams per cubic centimeter. For planar laminas, however, density is considered to be a measure of mass per unit of area. Density is denoted by ρ, the lowercase Greek letter rho.

Consider an irregularly shaped planar lamina of uniform density ρ, bounded by the graphs of $y = f(x)$, $y = g(x)$, and $a \leq x \leq b$, as shown in Figure 6.62. The mass of this region is given by

$$m = \text{(density)(area)}$$
$$= \rho \int_a^b [f(x) - g(x)]\, dx$$
$$= \rho A$$

where A is the area of the region. To find the center of mass of this lamina, partition the interval $[a, b]$ into n equal subintervals of width Δx. Let x_i be the center of the ith subinterval. You can approximate the portion of the lamina lying in the ith subinterval by a rectangle whose height is $h = f(x_i) - g(x_i)$. Because the density of the rectangle is ρ, its mass is

$$m_i = \text{(density)(area)}$$
$$= \rho \underbrace{[f(x_i) - g(x_i)]}_{\text{Height}} \underbrace{\Delta x}_{\text{Width}}.$$
$$\quad\;\; \text{Density}$$

Now, considering this mass to be located at the center (x_i, y_i) of the rectangle, the directed distance from the x-axis to (x_i, y_i) is $y_i = [f(x_i) + g(x_i)]/2$. Thus, the moment of m_i about the x-axis is

$$\text{Moment} = \text{(mass)(distance)}$$
$$= m_i y_i = \rho[f(x_i) - g(x_i)]\Delta x \left[\frac{f(x_i) + g(x_i)}{2}\right].$$

Summing the moments and taking the limit as $n \to \infty$ suggest the definitions below.

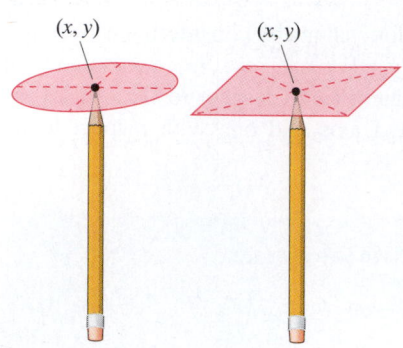

You can think of the center of mass $(\bar{x}, \bar{y})$ of a lamina as its balancing point. For a circular lamina, the center of mass is the center of the circle. For a rectangular lamina, the center of mass is the center of the rectangle.

Figure 6.61

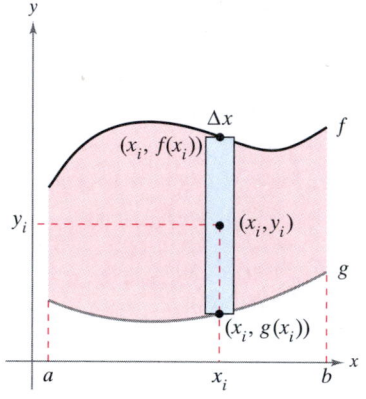

Planar lamina of uniform density ρ
Figure 6.62

Moments and Center of Mass of a Planar Lamina

Let f and g be continuous functions such that $f(x) \geq g(x)$ on $[a, b]$ and consider the planar lamina of uniform density ρ bounded by the graphs of $y = f(x)$, $y = g(x)$, and $a \leq x \leq b$.

1. The **moments about the x- and y-axes** are

$$M_x = \rho \int_a^b \left[\frac{f(x) + g(x)}{2}\right][f(x) - g(x)]\, dx$$

$$M_y = \rho \int_a^b x[f(x) - g(x)]\, dx.$$

2. The **center of mass** $(\bar{x}, \bar{y})$ is given by $\bar{x} = \dfrac{M_y}{m}$ and $\bar{y} = \dfrac{M_x}{m}$, where $m = \rho \int_a^b [f(x) - g(x)]\, dx$ is the mass of the lamina.

> **EXAMPLE 4**   **The Center of Mass of a Planar Lamina**

Find the center of mass of the lamina of uniform density ρ bounded by the graph of $f(x) = 4 - x^2$ and the x-axis.

Solution Because the center of mass lies on the axis of symmetry, you know that $\bar{x} = 0$. Moreover, the mass of the lamina is

$$m = \rho \int_{-2}^{2} (4 - x^2)\, dx$$

$$= \rho \left[4x - \frac{x^3}{3} \right]_{-2}^{2}$$

$$= \frac{32\rho}{3}.$$

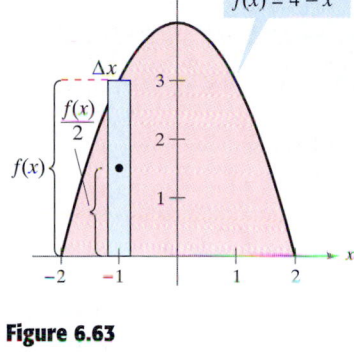

Figure 6.63

To find the moment about the x-axis, place a representative rectangle in the region, as shown in Figure 6.63. The distance from the x-axis to the center of this rectangle is

$$y_i = \frac{f(x)}{2} = \frac{4 - x^2}{2}.$$

Because the mass of the representative rectangle is

$$\rho f(x)\, \Delta x = \rho (4 - x^2)\, \Delta x$$

you have

$$M_x = \rho \int_{-2}^{2} \frac{4 - x^2}{2} (4 - x^2)\, dx$$

$$= \frac{\rho}{2} \int_{-2}^{2} (16 - 8x^2 + x^4)\, dx$$

$$= \frac{\rho}{2} \left[16x - \frac{8x^3}{3} + \frac{x^5}{5} \right]_{-2}^{2}$$

$$= \frac{256\rho}{15}$$

and $\bar{y}$ is given by

$$\bar{y} = \frac{M_x}{m} = \frac{256\rho/15}{32\rho/3} = \frac{8}{5}.$$

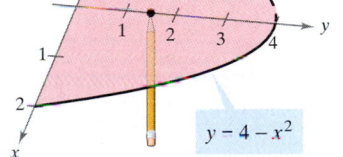

The center of mass is the balancing point.
Figure 6.64

Thus, the center of mass (the balancing point) of the lamina is $\left(0, \frac{8}{5}\right)$, as shown in Figure 6.64.

The density ρ in Example 4 is a common factor of both the moments and the mass, and as such cancels out of the quotients representing the coordinates of the center of mass. Thus, the center of mass of a lamina of *uniform* density depends only on the shape of the lamina and not on its density. For this reason, the point

$$(\bar{x}, \bar{y}) \qquad \text{Center of mass or centroid}$$

is sometimes called the center of mass of a *region* in the plane, or the **centroid** of the region. In other words, to find the centroid of a region in the plane, you simply assume that the region has a constant density of $\rho = 1$ and compute the corresponding center of mass.

EXAMPLE 5 The Centroid of a Plane Region

Find the centroid of the region bounded by the graphs of $f(x) = 4 - x^2$ and $g(x) = x + 2$.

Solution The two graphs intersect at the points $(-2, 0)$ and $(1, 3)$, as shown in Figure 6.65. Thus, the area of the region is

$$A = \int_{-2}^{1} [f(x) - g(x)]\, dx = \int_{-2}^{1} (2 - x - x^2)\, dx = \frac{9}{2}.$$

The centroid $(\bar{x}, \bar{y})$ of the region has the following coordinates.

$$\bar{x} = \frac{1}{A} \int_{-2}^{1} x[(4 - x^2) - (x + 2)]\, dx = \frac{2}{9} \int_{-2}^{1} (-x^3 - x^2 + 2x)\, dx$$

$$= \frac{2}{9} \left[-\frac{x^4}{4} - \frac{x^3}{3} + x^2 \right]_{-2}^{1} = -\frac{1}{2}$$

$$\bar{y} = \frac{1}{A} \int_{-2}^{1} \left[\frac{(4 - x^2) + (x + 2)}{2} \right] [(4 - x^2) - (x + 2)]\, dx$$

$$= \frac{2}{9} \left(\frac{1}{2} \right) \int_{-2}^{1} (-x^2 + x + 6)(-x^2 - x + 2)\, dx$$

$$= \frac{1}{9} \int_{-2}^{1} (x^4 - 9x^2 - 4x + 12)\, dx$$

$$= \frac{1}{9} \left[\frac{x^5}{5} - 3x^3 - 2x^2 + 12x \right]_{-2}^{1} = \frac{12}{5}.$$

Thus, the centroid of the region is $(\bar{x}, \bar{y}) = \left(-\frac{1}{2}, \frac{12}{5} \right)$.

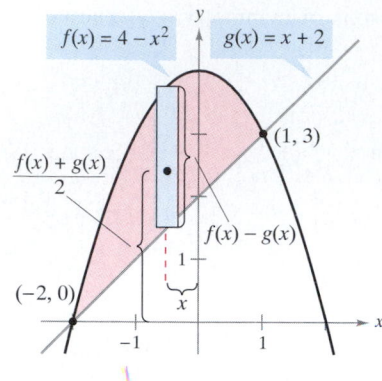

$f(x) = 4 - x^2$ $g(x) = x + 2$

$(1, 3)$

$\dfrac{f(x) + g(x)}{2}$

$f(x) - g(x)$

$(-2, 0)$

The centroid of the region is $\left(-\frac{1}{2}, \frac{12}{5} \right)$.

Figure 6.65

For simple plane regions, you may be able to find the centroid without resorting to integration.

EXAMPLE 6 The Centroid of a Simple Plane Region

Find the centroid of the region shown in Figure 6.66(a).

Solution By superimposing a coordinate system on the region, as shown in Figure 6.66(b), you can locate the centroids of the three rectangles at

$$\left(\frac{1}{2}, \frac{3}{2} \right), \quad \left(\frac{5}{2}, \frac{1}{2} \right), \quad \text{and} \quad (5, 1).$$

Using these three points, you can find the centroid of the region.

$$A = \text{area of region} = 3 + 3 + 4 = 10$$

$$\bar{x} = \frac{(1/2)(3) + (5/2)(3) + (5)(4)}{10} = \frac{29}{10} = 2.9$$

$$\bar{y} = \frac{(3/2)(3) + (1/2)(3) + (1)(4)}{10} = \frac{10}{10} = 1$$

Thus, the centroid of the region is $(2.9, 1)$.

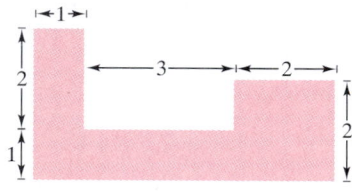

(a) Original region

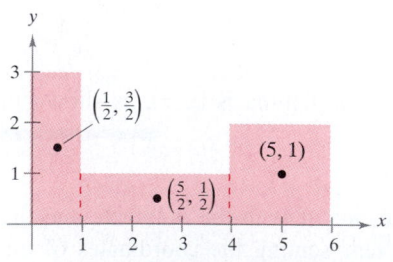

$\left(\frac{1}{2}, \frac{3}{2} \right)$

$(5, 1)$

$\left(\frac{5}{2}, \frac{1}{2} \right)$

(b) The centroid of the region is $(2.9, 1)$.

Figure 6.66

NOTE In Example 6, notice that $(2.9, 1)$ is not the "average" of $\left(\frac{1}{2}, \frac{3}{2} \right)$, $\left(\frac{5}{2}, \frac{1}{2} \right)$, and $(5, 1)$.

Theorem of Pappus

The final topic in this section is a useful theorem credited to Pappus of Alexandria (ca. 300 A.D.), a Greek mathematician whose eight-volume *Mathematical Collection* is a record of much of classical Greek mathematics. We delay the proof of this theorem until Section 13.4 (Exercise 47).

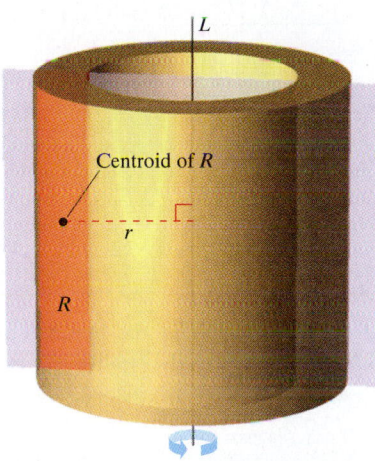

The volume V is $2\pi rA$ where A is the area of region R.
Figure 6.67

> **THEOREM 6.1   The Theorem of Pappus**
>
> Let R be a region in a plane and let L be a line in the same plane such that L does not intersect the interior of R, as shown in Figure 6.67. If r is the distance between the centroid of R and the line, then the volume V of the solid of revolution formed by revolving R about the line is
>
> $$V = 2\pi rA$$
>
> where A is the area of R. (Note that $2\pi r$ is the distance traveled by the centroid as the region is revolved about the line).

The Theorem of Pappus can be used to find the volume of a torus, as shown in the following example. Recall that a **torus** is a doughnut-shaped solid formed by revolving a circular region about a line that lies in the same plane as the circle (but does not intersect the circle).

EXAMPLE 7   Finding Volume by the Theorem of Pappus

Find the volume of the torus formed by revolving the circular region bounded by $(x - 2)^2 + y^2 = 1$ about the y-axis, as shown in Figure 6.68(a).

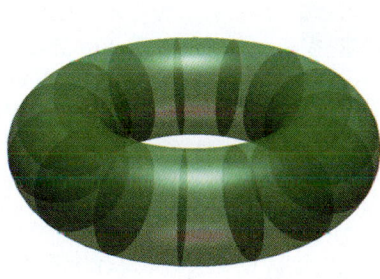

Torus

(a)

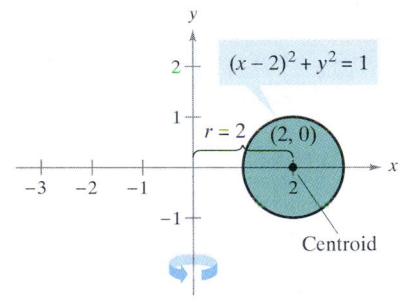

(b)

Figure 6.68

Solution   In Figure 6.68(b), you can see that the centroid of the circular region is $(2, 0)$. Thus, the distance between the centroid and the axis of revolution is $r = 2$. Because the area of the circular region is $A = \pi$, the volume of the torus is

$$\begin{aligned}
V &= 2\pi rA \\
&= 2\pi(2)(\pi) \\
&= 4\pi^2 \\
&\approx 39.5.
\end{aligned}$$

EXERCISES FOR SECTION 6.6

In Exercises 1–4, find the center of mass of the point masses lying on the *x*-axis.

1. $m_1 = 6, m_2 = 3, m_3 = 5$

   $x_1 = -5, x_2 = 1, x_3 = 3$

2. $m_1 = 7, m_2 = 4, m_3 = 3, m_4 = 8$

   $x_1 = -3, x_2 = -2, x_3 = 5, x_4 = 6$

3. $m_1 = 1, m_2 = 1, m_3 = 1, m_4 = 1, m_5 = 1$

   $x_1 = 7, x_2 = 8, x_3 = 12, x_4 = 15, x_5 = 18$

4. $m_1 = 12, m_2 = 1, m_3 = 6, m_4 = 3, m_5 = 11$

   $x_1 = -3, x_2 = -2, x_3 = -1, x_4 = 0, x_5 = 4$

5. *Graphical Reasoning*

   (a) Translate each point mass in Exercise 3 to the right five units and determine the resulting center of mass.

   (b) Translate each point mass in Exercise 4 to the left three units and determine the resulting center of mass.

6. *Conjecture* Use the result of Exercise 5 to make a conjecture about the change in the center of mass that results when each point mass is translated *k* units horizontally.

Statics Problems In Exercises 7 and 8, consider a beam of length *L* with a fulcrum *x* feet from one end (see figure). If there are objects with weights W_1 and W_2 placed on opposite ends of the beam, find *x* such that the system is in equilibrium.

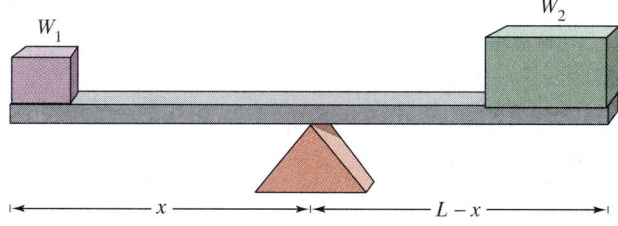

7. Two children weighing 50 pounds and 75 pounds are going to play on a seesaw that is 10 feet long.

8. In order to move a 550-pound rock, a person weighing 200 pounds wants to balance it on a beam that is 5 feet long.

In Exercise 9–12, find the center of mass of the given system of point masses.

9.

m_i	5	1	3
(x_i, y_i)	(2, 2)	(−3, 1)	(1, −4)

10.

m_i	10	2	5
(x_i, y_i)	(1, −1)	(5, 5)	(−4, 0)

11.

m_i	3	4
(x_i, y_i)	(−2, −3)	(−1, 0)

m_i	2	1	6
(x_i, y_i)	(7, 1)	(0, 0)	(−3, 0)

12.

m_i	4	2	$\frac{5}{2}$	5
(x_i, y_i)	(2, 3)	(−1, 5)	(6, 8)	(2, −2)

In Exercises 13–24, find M_x, M_y, and $(\bar{x}, \bar{y})$ for the laminas of uniform density ρ bounded by the graphs of the equations.

13. $y = \sqrt{x}, y = 0, x = 4$

14. $y = x^2, y = 0, x = 4$

15. $y = x^2, y = x^3$

16. $y = \sqrt{x}, y = x$

17. $y = -x^2 + 4x + 2, y = x + 2$

18. $y = \sqrt{3x} + 1, y = x + 1$

19. $y = x^{2/3}, y = 0, x = 8$

20. $y = x^{2/3}, y = 4$

21. $x = 4 - y^2, x = 0$

22. $x = 2y - y^2, x = 0$

23. $x = -y, x = 2y - y^2$

24. $x = y + 2, x = y^2$

In Exercises 25–28, set up and evaluate the integrals for finding the area and moments about the *x*- and *y*-axes for the region bounded by the graphs of the equations. (Assume $\rho = 1$.)

25. $y = x^2, y = x$

26. $y = \frac{1}{x}, y = 0, 1 \le x \le 4$

27. $y = 2x + 4, y = 0, 0 \le x \le 3$

28. $y = x^2 - 4, y = 0$

In Exercises 29–32, use a graphing utility to graph the region bounded by the graphs of the equations. Use the integration capabilities of the graphing utility to approximate the centroid of the region.

29. $y = 10x\sqrt{125 - x^3}, y = 0$

30. $y = xe^{-x/2}, y = 0, x = 0, \ x = 4$

31. *Prefabricated End Section of a Building*

   $y = 5\sqrt[3]{400 - x^2}, y = 0$

32. *Witch of Agnesi*

   $y = 8/(x^2 + 4), y = 0, x = -2, x = 2$

In Exercises 33–38, find and/or verify the centroid of the common region used in engineering.

33. Triangle Show that the centroid of the triangle with vertices $(-a, 0)$, $(a, 0)$, and (b, c) is the point of intersection of the medians (see figure).

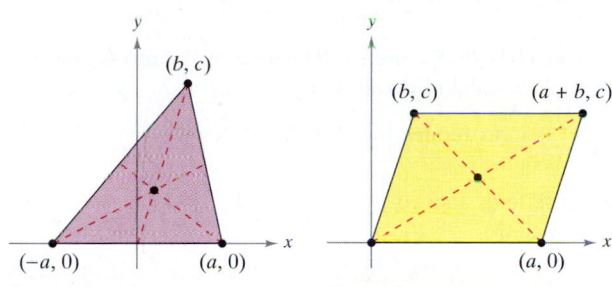

Figure for 33   **Figure for 34**

34. Parallelogram Show that the centroid of the parallelogram with vertices $(0, 0)$, $(a, 0)$, (b, c), and $(a + b, c)$ is the point of intersection of the diagonals (see figure).

35. Trapezoid Find the centroid of the trapezoid with vertices $(0, 0)$, $(0, a)$, (c, b), and $(c, 0)$. Show that it is the intersection of the line connecting the midpoints of the parallel sides and the line connecting the extended parallel sides, as shown in the figure.

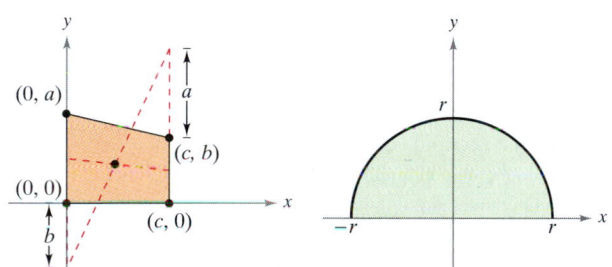

Figure for 35   **Figure for 36**

36. Semicircle Find the centroid of the region bounded by the graphs of $y = \sqrt{r^2 - x^2}$ and $y = 0$ (see figure).

37. Semiellipse Find the centroid of the region bounded by the graphs of $y = \dfrac{b}{a}\sqrt{a^2 - x^2}$ and $y = 0$ (see figure).

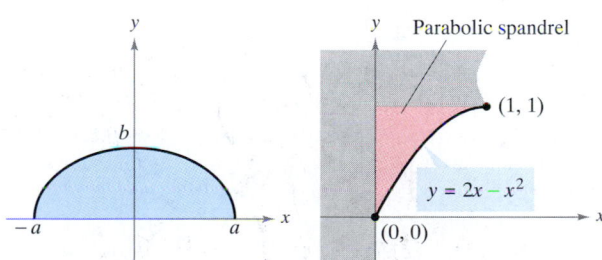

Figure for 37   **Figure for 38**

38. Parabolic Spandrel Find the centroid of the **parabolic spandrel** shown in the figure.

39. Graphical Reasoning Consider the region bounded by the graphs of $y = x^2$ and $y = b$, where $b > 0$.

(a) Sketch a graph of the region.

(b) Use the graph in part (a) to determine $\bar{x}$. Explain.

(c) Set up the integral for finding M_y. Because of the form of the integrand, the value of the integral can be obtained without integrating. What is the form of the integrand and what is the value of the integral? Compare with the result in part (b).

(d) Use the graph in part (a) to determine whether $\bar{y} > \dfrac{b}{2}$ or $\bar{y} < \dfrac{b}{2}$. Explain.

(e) Use integration to verify your answer in part (b).

40. Graphical and Numerical Reasoning Consider the region bounded by the graphs of $y = x^{2n}$ and $y = b$, where $b > 0$ and n is a positive integer.

(a) Set up the integral for finding M_y. Because of the form of the integrand, the value of the integral can be obtained without integrating. What is the form of the integrand and what is the value of the integral? Compare with the result in part (b).

(b) Is $\bar{y} > \dfrac{b}{2}$ or $\bar{y} < \dfrac{b}{2}$? Explain.

(c) Use integration to find $\bar{y}$ as a function of n.

(d) Use the result in part (c) to complete the table.

n	1	2	3	4
$\bar{y}$				

(e) Find $\displaystyle \lim_{n \to \infty} \bar{y}$.

(f) Give a geometric explanation of the result in part (e).

41. Modeling Data The manufacturer of glass for a window in a conversion van needs to approximate its center of mass. A coordinate system is superimposed on a prototype of the glass (see figure). The measurements (in centimeters) for the right half of the symmetric piece of glass are given in the table.

x	0	10	20	30	40
y	30	29	26	20	0

(a) Use Simpson's Rule to approximate the center of mass of the glass.

(b) Use a graphing utility to find a fourth-degree polynomial model for the data.

(c) Use the integration capabilities of a graphing utility and the model to approximate the center of mass of the glass. Compare with the result in part (a).

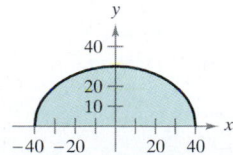

42. *Modeling Data*   The manufacturer of a boat needs to approximate the center of mass of a section of the hull. A coordinate system is superimposed on a prototype (see figure). The measurements (in feet) for the right half of the symmetric prototype are listed in the table.

x	0	0.5	1.0	1.5	2
l	1.50	1.45	1.30	0.99	0
d	0.50	0.48	0.43	0.33	0

(a) Use Simpson's Rule to approximate the center of mass of the hull section.

(b) Use a graphing utility to find fourth-degree polynomial models for both curves shown in the figure. Plot the data and graph the models.

(c) Use the integration capabilities of a graphing utility and the model to approximate the center of mass of the hull section. Compare with the result in part (a).

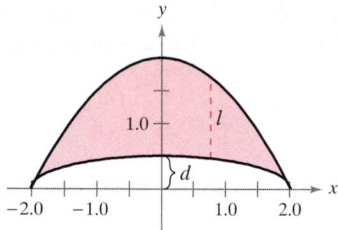

In Exercises 43–46, introduce an appropriate coordinate system and find the coordinates of the center of mass of the planar lamina. (The answer depends on the position of the coordinate system.)

43.

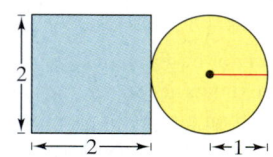

44.

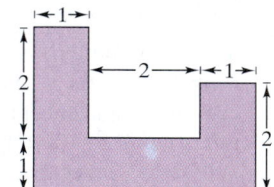

45.

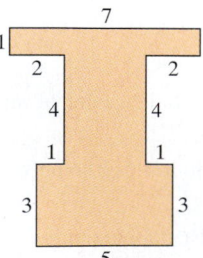

46.

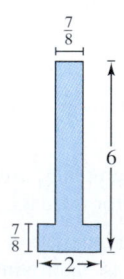

47. Find the center of mass of the lamina in Exercise 43 if the circular portion of the lamina has twice the density of the square portion of the lamina.

48. Find the center of mass of the lamina in Exercise 43 if the square portion of the lamina has twice the density of the circular portion of the lamina.

In Exercises 49–52, use the Theorem of Pappus to find the volume of the solid of revolution.

49. The torus formed by revolving the circle $(x - 5)^2 + y^2 = 16$ about the y-axis

50. The torus formed by revolving the circle $x^2 + (y - 3)^2 = 4$ about the x-axis

51. The solid formed by revolving the region bounded by the graphs of $y = x$, $y = 4$, and $x = 0$ about the x-axis

52. The solid formed by revolving the region bounded by the graphs of $y = \sqrt{x - 1}$, $y = 0$, and $x = 5$ about the y-axis

In Exercises 53 and 54, use the *Second Theorem of Pappus*, which is stated as follows. If a segment of a plane curve C is revolved about an axis that does not intersect the curve (except possibly at its endpoints), the area S of the resulting surface of revolution is given by the product of the length of C times the distance d traveled by the centroid of C.

53. A sphere is formed by revolving the graph of

$$y = \sqrt{r^2 - x^2}$$

about the x-axis. Use the formula for surface area, $S = 4\pi r^2$, to find the centroid of the semicircle $y = \sqrt{r^2 - x^2}$.

54. A torus is formed by revolving the graph of

$$(x - 1)^2 + y^2 = 1$$

about the y-axis. Find the surface area of the torus.

55. Let $n \geq 1$ be constant, and consider the region bounded by $f(x) = x^n$, the x-axis, and $x = 1$. Find the centroid of this region. As $n \to \infty$, what does the region look like, and where is its centroid?

Fluid Pressure and Fluid Force

Fluid Pressure and Fluid Force

Swimmers know that the deeper an object is submerged in a fluid, the greater the pressure on the object. **Pressure** is defined as the force per unit of area over the surface of a body. For example, because a column of water that is 10 feet in height and 1 inch square weighs 4.3 pounds, the *fluid* pressure at a depth of 10 feet of water is 4.3 pounds per square inch.* At 20 feet, this would increase to 8.6 pounds per square inch, and in general the pressure is proportional to the depth of the object in the fluid.

BLAISE PASCAL (1623–1662)

Pascal is well known for his work in many areas of mathematics and physics and also for his influence on Leibniz. Although much of Pascal's work in calculus was intuitive and lacked the rigor of modern mathematics, he nevertheless anticipated many important results.

The Granger Collection

> ### Definition of Fluid Pressure
>
> The **pressure** on an object at depth h in a liquid is
>
> $$\text{Pressure} = P = wh$$
>
> where w is the weight-density of the liquid per unit of volume.

Below are some common weight-densities of fluids in pounds per cubic foot.

Ethyl alcohol	49.4
Gasoline	41.0–43.0
Glycerin	78.6
Kerosene	51.2
Mercury	849.0
Seawater	64.0
Water	62.4

When calculating fluid pressure, you can use an important (and rather suprising) physical law called **Pascal's Principle**, named after the French mathematician Blaise Pascal. Pascal's Principle states that the pressure exerted by a fluid at a depth h is transmitted equally *in all directions*. For example, in Figure 6.69, the pressure at the indicated depth is the same for all three objects. Because fluid pressure is given in terms of force per unit area $(P = F/A)$, the fluid force on a *submerged horizontal surface* of area A is

$$\text{Fluid force} = F = PA = (\text{pressure})(\text{area}).$$

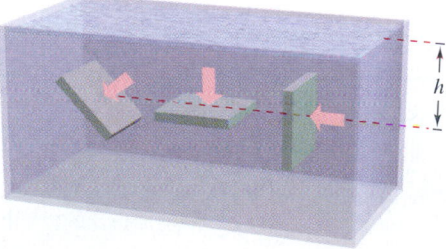

The pressure at h is the same for all three objects.
Figure 6.69

The total pressure on an object in 10 feet of water would also include the pressure due to the earth's atmosphere. At sea level, atmospheric pressure is approximately 14.7 pounds per square inch.

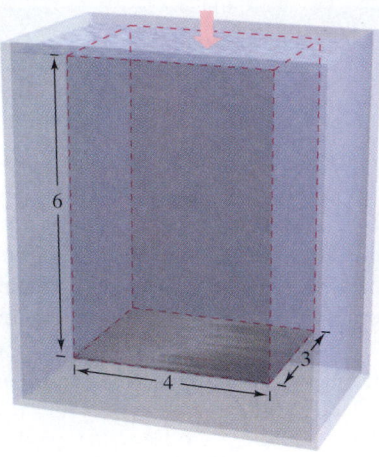

The fluid force on a horizontal metal sheet is equal to the fluid pressure times the area.
Figure 6.70

EXAMPLE 1 Fluid Force on a Submerged Sheet

Find the fluid force on a rectangular metal sheet measuring 3 feet by 4 feet that is submerged in 6 feet of water, as shown in Figure 6.70.

Solution Because the weight-density of water is 62.4 pounds per cubic foot and the sheet is submerged in 6 feet of water, the fluid pressure is

$$P = (62.4)(6) \qquad P = wh$$

$$= 374.4 \text{ pounds per square foot.}$$

Because the total area of the sheet is $A = (3)(4) = 12$ square feet, the fluid force is

$$F = PA = \left(374.4 \ \frac{\text{pounds}}{\text{square foot}}\right) (12 \text{ square feet})$$

$$= 4492.8 \text{ pounds.}$$

This result is independent of the size of the body of water. The fluid force would be the same in a swimming pool or lake.

In Example 1, the fact that the sheet is rectangular and horizontal means that you do not need the methods of calculus to solve the problem. We now look at a surface that is submerged vertically in a fluid. The problem is more difficult because the pressure is not constant over the surface.

Suppose a vertical plate is submerged in a fluid of weight-density w (per unit of volume), as shown in Figure 6.71. To determine the total force against *one side* of the region from depth c to depth d, you can subdivide the interval $[c, d]$ into n subintervals, each of width Δy. Next, consider the representative rectangle of width Δy and length $L(y_i)$, where y_i is in the ith subinterval. The force against this representative rectangle is

$$\Delta F_i = w(\text{depth})(\text{area})$$

$$= wh(y_i)L(y_i)\,\Delta y.$$

The force against n such rectangles is

$$\sum_{i=1}^{n} \Delta F_i = w \sum_{i=1}^{n} h(y_i)L(y_i)\,\Delta y.$$

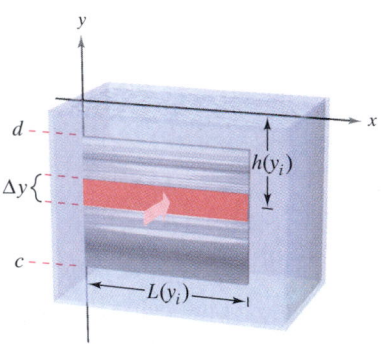

Calculus methods must be used to find the fluid force on a vertical metal plate.
Figure 6.71

Note that w is considered to be constant and is factored out of the summation. Therefore, taking the limit as $\|\Delta\| \to 0$ ($n \to \infty$) suggests the following definition.

> ### Definition of Force Exerted by a Fluid
>
> The **force F exerted by a fluid** of constant weight-density w (per unit of volume) against a submerged vertical plane region from $y = c$ to $y = d$ is
>
> $$F = w \lim_{\|\Delta\| \to 0} \sum_{i=1}^{n} h(y_i)L(y_i)\,\Delta y = w \int_{c}^{d} h(y)L(y)\,dy$$
>
> where $h(y)$ is the depth of the fluid at y and $L(y)$ is the horizontal length of the region at y.

EXAMPLE 2   Fluid Force on a Vertical Surface

A vertical gate in a dam has the shape of an isosceles trapezoid 8 feet across the top and 6 feet across the bottom, with a height of 5 feet, as shown in Figure 6.72 (a). What is the fluid force on the gate if the top of the gate is 4 feet below the surface of the water?

Solution   In setting up a mathematical model for this problem, you are at liberty to locate the x- and y-axis in several different ways. A convenient approach is to let the y-axis bisect the gate and place the x-axis at the surface of the water, as shown in Figure 6.72 (b). Thus, the depth of the water at y in feet is

Depth $= h(y) = -y$.

To find the length $L(y)$ of the region at y, we find the equation of the line forming the right side of the gate. Because this line passes through the points $(3, -9)$ and $(4, -4)$, its equation is

$$y - (-9) = \frac{-4 - (-9)}{4 - 3}(x - 3)$$

$$y + 9 = 5(x - 3)$$

$$y = 5x - 24$$

$$x = \frac{y + 24}{5}.$$

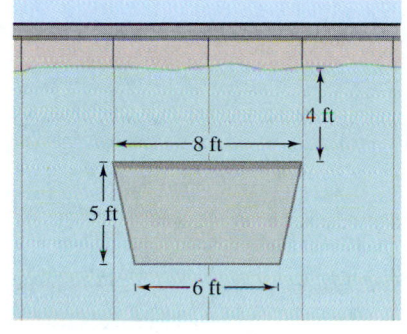

(a) Water gate in a dam

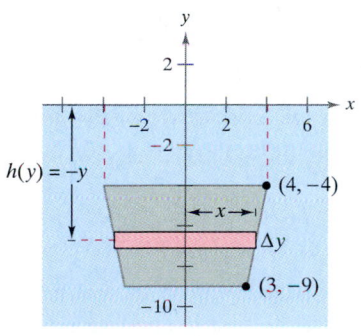

(b) The fluid force against the gate is 13,936 pounds.

Figure 6.72

In Figure 6.72 (b) you can see that the length of the region at y is

Length $= 2x$

$$= \frac{2}{5}(y + 24)$$

$$= L(y).$$

Finally, by integrating from $y = -9$ to $y = -4$, you can calculate the fluid force to be

$$F = w \int_c^d h(y)L(y)\, dy$$

$$= 62.4 \int_{-9}^{-4} (-y)\left(\frac{2}{5}\right)(y + 24)\, dy$$

$$= -62.4 \left(\frac{2}{5}\right)\int_{-9}^{-4} (y^2 + 24y)\, dy$$

$$= -62.4 \left(\frac{2}{5}\right)\left[\frac{y^3}{3} + 12y^2\right]_{-9}^{-4}$$

$$= -62.4 \left(\frac{2}{5}\right)\left(\frac{-1675}{3}\right)$$

$$= 13{,}936 \text{ pounds.}$$

NOTE   In Example 2, we let the x-axis coincide with the surface of the water. This was convenient, but arbitrary. In choosing a coordinate system to represent a physical situation, you should consider various possibilities. Often you can simplify the calculations in a problem by locating the coordinate system to take advantage of special characteristics of the problem, such as symmetry.

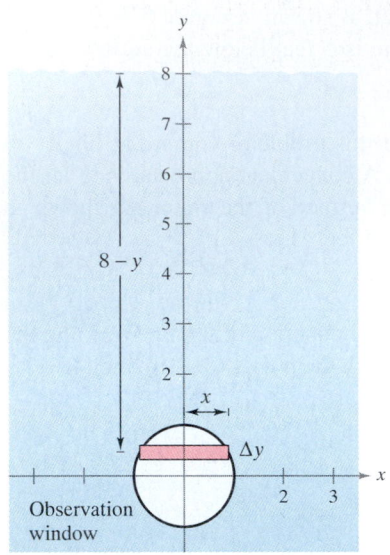

The fluid force on the window is 1608.5 pounds.
Figure 6.73

EXAMPLE 3   Fluid Force on a Vertical Surface

A circular observation window on a marine science ship has a radius of 1 foot, and the center of the window is 8 feet below water level, as shown in Figure 6.73. What is the fluid force on the window?

Solution   To take advantage of symmetry, locate a coordinate system such that the origin coincides with the center of the window, as shown in Figure 6.73. The depth at y is then

$$\text{Depth } = h(y) = 8 - y.$$

The horizontal length of the window is $2x$, and you can use the equation for the circle, $x^2 + y^2 = 1$, to solve for x as follows.

$$\text{Length } = 2x = 2\sqrt{1 - y^2} = L(y)$$

Finally, because y ranges from -1 to 1, and using 64 pounds per cubic foot as the weight-density of seawater, you have

$$F = w \int_c^d h(y)L(y)\, dy$$

$$= 64 \int_{-1}^1 (8 - y)(2)\sqrt{1 - y^2}\, dy.$$

Initially it looks as if this integral would be difficult to solve. However, if you break the integral into two parts and apply symmetry, the solution is simple.

$$F = 64(16) \int_{-1}^1 \sqrt{1 - y^2}\, dy - 64(2) \int_{-1}^1 y\sqrt{1 - y^2}\, dy$$

The second integral is 0 (because the integrand is odd and the limits of integration are symmetric to the origin). Moreover, by recognizing that the first integral represents the area of a semicircle of radius 1, you obtain

$$F = 64(16)\left(\frac{\pi}{2}\right) - 64(2)(0)$$

$$= 512\pi$$

$$\approx 1608.5 \text{ pounds.}$$

So, the fluid force on the window is 1608.5 pounds.

TECHNOLOGY   To confirm the result obtained in Example 3, you might have considered using Simpson's Rule to approximate the value of

$$128 \int_{-1}^1 (8 - x)\sqrt{1 - x^2}\, dx.$$

From the graph of

$$f(x) = (8 - x)\sqrt{1 - x^2}$$

however, you can see that f is not differentiable when $x = \pm 1$ (see Figure 6.74). This means that you cannot apply Theorem 4.19 in Section 4.6 to determine the potential error in Simpson's Rule. Without knowing the potential error, the approximation is of little value. Try using a graphing utility to approximate the integral.

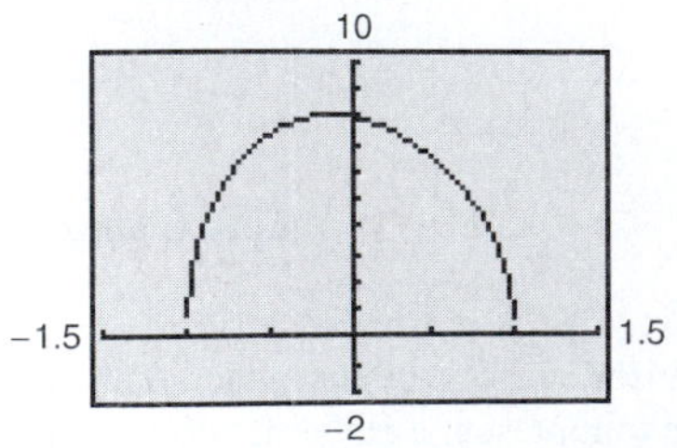

f is not differentiable at $x = \pm 1$.
Figure 6.74

EXERCISES FOR SECTION 6.7

Force on a Submerged Sheet   In Exercises 1 and 2, the area of the top side of a piece of sheet metal is given. The sheet metal is submerged horizontally in 5 feet of water. Find the fluid force on the top side.

1. 3 square feet   **2.** 18 square feet

Buoyant Force   In Exercises 3 and 4, find the buoyant force of a rectangular solid of the given dimensions submerged in water so that the top side is parallel to the surface of the water. The buoyant force is the difference between the fluid forces on the top and bottom sides of the solid.

3.
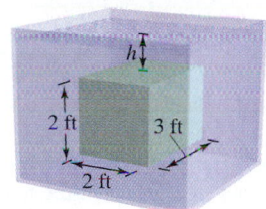

4.
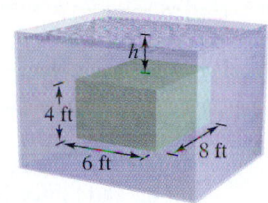

Fluid Force on a Tank Wall   In Exercises 5–10, find the fluid force on the vertical side of the tank, where the dimensions are given in feet. Assume that the tank is full of water.

5. Rectangle
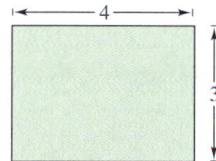

6. Triangle
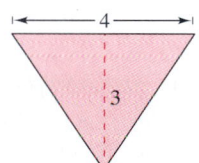

7. Trapezoid
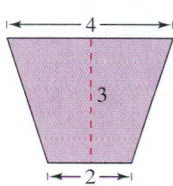

8. Semicircle
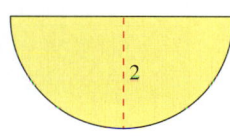

9. Parabola, $y = x^2$
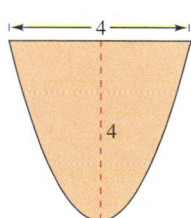

10. Semiellipse,
$y = -\frac{1}{2}\sqrt{36 - 9x^2}$
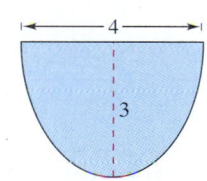

Fluid Force of Water   In Exercises 11–14, find the fluid force on the vertical plate submerged in water, where the dimensions are given in meters and the density of water is 1000 kilograms per cubic meter.

11. Square
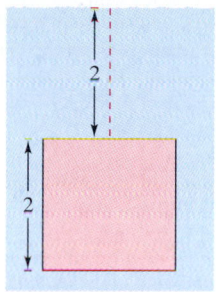

12. Square
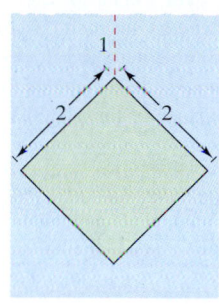

13. Triangle
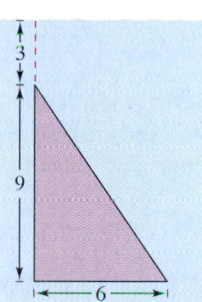

14. Rectangle
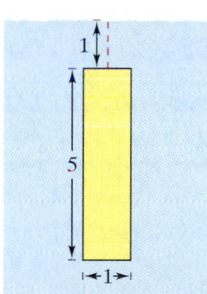

Force on a Concrete Form   In Exercises 15–18, the figure is the vertical side of a form for poured concrete that weighs 140.7 pounds per cubic foot. Determine the force on this part of the concrete form.

15. Rectangle
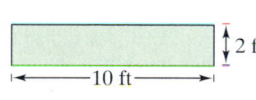

16. Semiellipse,
$y = -\frac{3}{4}\sqrt{16 - x^2}$
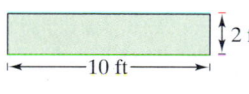

17. Rectangle
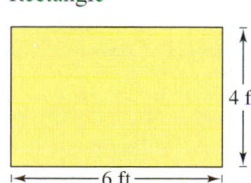

18. Triangle
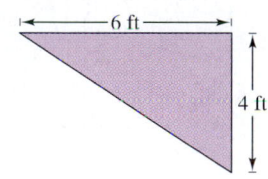

19. Fluid Force of Gasoline A cylindrical gasoline tank is placed so that the axis of the cylinder is horizontal. Find the fluid force on a circular end of the tank if the tank is half full, assuming that the diameter is 3 feet and the gasoline weighs 42 pounds per cubic foot.

20. Fluid Force of Gasoline Repeat Exercise 19 for a tank that is full. (Evaluate one integral by a geometric formula and the other by observing that the integrand is an odd function.)

21. Fluid Force on a Circular Plate A circular plate of radius r feet is submerged vertically in a tank of fluid that weighs w pounds per cubic foot. The center of the circle is k ($k > r$) feet below the surface of the fluid. Show that the fluid force on the surface of the plate is

$$F = wk(\pi r^2).$$

(Evaluate one integral by a geometric formula and the other by observing that the integrand is an odd function.)

22. Fluid Force on a Rectangular Plate A rectangular plate of height h feet and base b feet is submerged vertically in a tank of fluid that weighs w pounds per cubic foot. The center is k feet below the surface of the fluid, where $h \leq k/2$. Show that the fluid force on the surface of the plate is

$$F = wkhb.$$

23. Submarine Porthole A porthole on a vertical side of a submarine (submerged in seawater) is 1 foot square. Find the fluid force on the porthole, assuming that the center of the square is 15 feet below the surface.

24. Submarine Porthole Repeat Exercise 23 for a circular porthole that has a diameter of 1 foot. The center is 15 feet below the surface.

25. Boat Design The vertical stern of a boat with a superimposed coordinate system is shown in the figure. The table gives the width w of the stern at indicated values of y. Find the fluid force against the stern if the measurements are given in feet.

y	0	$\frac{1}{2}$	1	$\frac{3}{2}$	2	$\frac{5}{2}$	3	$\frac{7}{2}$	4
w	0	3	5	8	9	10	10.25	10.5	10.5

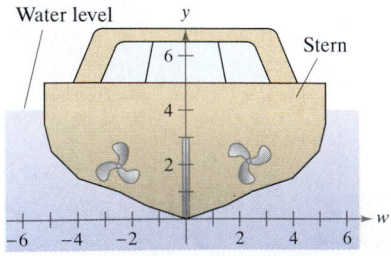

 26. Irrigation Canal Gate The vertical cross section of an irrigation canal is modeled by

$$f(x) = \frac{5x^2}{x^2 + 4}$$

where x is measured in feet and $x = 0$ corresponds to the center of the canal. Use the integration capabilities of a graphing utility to approximate the fluid force against a vertical gate used to stop the flow of water if the water is 3 feet deep.

 In Exercises 27 and 28, use the integration capabilities of a graphing utility to approximate the fluid force on the vertical plate bounded by the x-axis and the top half of the graph of the equation. Assume that the base of the plate is 12 feet beneath the surface of the water.

27. $x^{2/3} + y^{2/3} = 4^{2/3}$ **28.** $\dfrac{x^2}{28} + \dfrac{y^2}{16} = 1$

29. Think About It

(a) Approximate the depth of the water in the tank in Exercise 5 if the fluid force is one-half as great as when the tank is full.

(b) Explain why the answer in part (a) is not $\frac{3}{2}$.

30. Swimming Pool Design A swimming pool is 20 feet wide, 40 feet long, 4 feet deep at one end, and 8 feet deep at the other end (see figure). The bottom is an inclined plane. Find the fluid force on each of the vertical walls.

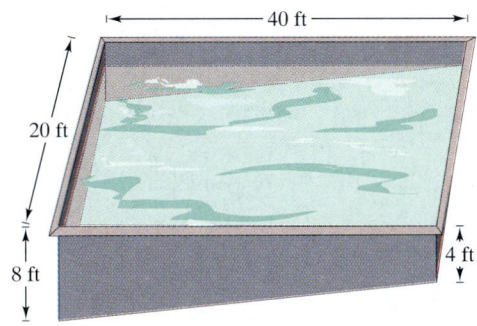

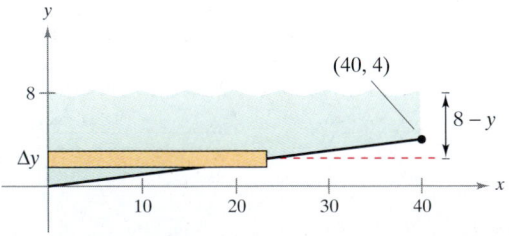

(*Hint:* You must set up two integrals, one for $0 \leq y \leq 4$ and one for $4 \leq y \leq 8$.)

REVIEW EXERCISES FOR CHAPTER 6

Area **In Exercises 1–10, sketch the region bounded by the graphs of the equations and determine the area of the region.**

1. $y = \dfrac{1}{x^2}$, $y = 0$, $x = 1$, $x = 5$ 2. $y = \dfrac{1}{x^2}$, $y = 4$, $x = 5$

3. $y = \dfrac{1}{x^2 + 1}$, $y = 0$, $x = -1$, $x = 1$

4. $x = y^2 - 2y$, $x = -1$, $y = 0$

5. $y = x$, $y = x^3$

6. $x = y^2 + 1$, $x = y + 3$

7. $y = e^x$, $y = e^2$, $x = 0$

8. $y = \csc x$, $y = 2$ (one region)

9. $y = \sin x$, $y = \cos x$, $\dfrac{\pi}{4} \le x \le \dfrac{5\pi}{4}$

10. $x = \cos y$, $x = \dfrac{1}{2}$, $\dfrac{\pi}{3} \le y \le \dfrac{7\pi}{3}$

In Exercises 11–14, use a graphing utility to graph the region bounded by the graphs of the functions and use the integration capabilities of the graphing utility to find the area of the region.

11. $y = x^2 - 8x + 3$, $y = 3 + 8x - x^2$

12. $y = x^2 - 4x + 3$, $y = x^3$, $x = 0$

13. $\sqrt{x} + \sqrt{y} = 1$, $y = 0$, $x = 0$

14. $y = x^4 - 2x^2$, $y = 2x^2$

Area **In Exercises 15–18, set up integrals for finding the area of the region bounded by the graphs of the equations using vertical and horizontal representative rectangles. Find the area of the region by evaluating the easier of the two integrals.**

15. $x = y^2 - 2y$, $x = 0$ 16. $y = \sqrt{x-1}$, $y = \dfrac{x-1}{2}$

17. $y = 1 - \dfrac{x}{2}$, $y = x - 2$, $y = 1$

18. $y = \sqrt{x-1}$, $y = 2$, $y = 0$, $x = 0$

19. *Think About It* A person has two job offers. The starting salary for each is $30,000 and after 10 years of service each will pay $56,000. The salary increases for each offer are shown in the figure. From a strictly monetary viewpoint, which is the better offer? Explain.

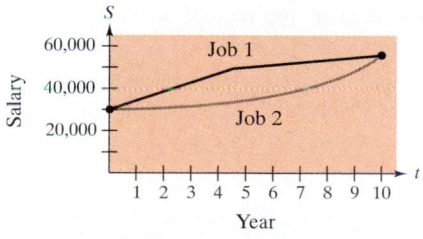

20. *Credit Card Sales* The annual sales s (in millions of dollars) for VISA, Mastercard, and American Express from 1983 through 1992 can be modeled by

$s = 15.9696t - 6.318$	VISA
$s = 8.581t + 6.965$	Mastercard
$s = 6.214t + 10.345$	American Express

where $3 \le t \le 12$ represents the 10-year period from 1983 through 1992. *(Source: Credit Card News)*

(a) From 1983 through 1992, how much more were VISA's sales than Mastercard's sales?

(b) From 1983 through 1992, how much more were VISA's sales than American Express' sales?

Volume **In Exercises 21–28, find the volume of the solid generated by revolving the plane region bounded by the equations about the indicated lines.**

21. $y = x$, $y = 0$, $x = 4$
   (a) the x-axis   (b) the y-axis
   (c) the line $x = 4$ (d) the line $x = 6$

22. $y = \sqrt{x}$, $y = 2$, $x = 0$
   (a) the x-axis   (b) the line $y = 2$
   (c) the y-axis   (d) the line $x = -1$

23. $\dfrac{x^2}{16} + \dfrac{y^2}{9} = 1$ (a) the y-axis (oblate spheroid)
                            (b) the x-axis (prolate spheroid)

24. $\dfrac{x^2}{a^2} + \dfrac{y^2}{b^2} = 1$ (a) the y-axis (oblate spheroid)
                            (b) the x-axis (prolate spheroid)

25. $y = \dfrac{1}{x^4 + 1}$, $y = 0$, $x = 0$, $x = 1$

   revolved about the y-axis

26. $y = \dfrac{1}{\sqrt{1 + x^2}}$, $y = 0$, $x = -1$, $x = 1$

   revolved about the x-axis

27. $y = 1/(1 + \sqrt{x - 2})$, $y = 0$, $x = 2$, $x = 6$

   revolved about the y-axis

28. $y = e^{-x}$, $y = 0$, $x = 0$, $x = 1$

   revolved about the x-axis

In Exercises 29 and 30, consider the region bounded by the graphs of the equations $y = x\sqrt{x + 1}$ and $y = 0$.

29. *Area* Find the area of the region.

30. *Volume* Find the volume of the solid generated by revolving the region about (a) the x-axis and (b) the y-axis.

31. Depth of Gasoline in a Tank A gasoline tank is an oblate spheroid generated by revolving the region bounded by the graph of $(x^2/16) + (y^2/9) = 1$ about the y-axis, where x and y are measured in feet. Find the depth of the gasoline in the tank when it is filled to one-fourth its capacity.

32. Magnitude of the Base The base of a solid is a circle of radius a, and its vertical cross sections are equilateral triangles. Find the radius of the circle if the volume of the solid is 10 cubic meters.

Arc Length In Exercises 33 and 34, find the arc length of the graph of the function over the indicated interval.

33. $f(x) = \dfrac{4}{5}x^{5/4}$, $[0, 4]$ **34.** $y = \dfrac{1}{6}x^3 + \dfrac{1}{2x}$, $[1, 3]$

35. Length of a Catenary A cable of a suspension bridge forms a catenary modeled by the equation

$$y = 300 \cosh\left(\frac{x}{2000}\right) - 280, \quad -2000 \le x \le 2000$$

where x and y are measured in feet. Use a graphing utility to approximate the length of the cable.

36. Approximation Determine which value best approximates the length of the arc represented by the integral

$$\int_0^{\pi/4} \sqrt{1 + (\sec^2 x)^2}\, dx.$$

(Make your selection on the basis of a sketch of the arc and *not* by performing any calculations.)

(a) -2 (b) 1 (c) π (d) 4 (e) 3

37. Surface Area Use integration to find the lateral surface area of a right circular cone of height 4 and radius 3.

38. Surface Area The region bounded by the graphs of $y = 2\sqrt{x}$, $y = 0$, and $x = 3$ is revolved about the x-axis. Find the surface area of the solid generated.

39. Work Find the work done in stretching a spring from its natural length of 10 inches to a length of 15 inches, if a force of 4 pounds is needed to stretch it 1 inch from its natural position.

40. Work Find the work done in stretching a spring from its natural length of 9 inches to double that length. The force required to stretch the spring is 50 pounds.

41. Work A water well has an 8-inch casing (diameter) and is 175 feet deep. If the water is 25 feet from the top of the well, determine the amount of work done in pumping it dry, assuming that no water enters the well while it is being pumped.

42. Work Repeat Exercise 41, assuming that water enters the well at a rate of 4 gallons per minute and the pump works at a rate of 12 gallons per minute. How many gallons are pumped in this case?

43. Work A chain 10 feet long weighs 5 pounds per foot and is hung from a platform 20 feet above the ground. How much work is required to raise the entire chain to the 20-foot level?

44. Work A windlass, 200 feet above ground level on the top of a building, uses a cable weighing 4 pounds per foot. Find the work done in winding up the cable if

(a) one end is at ground level.

(b) there is a 300-pound load attached to the end of the cable.

45. Work The work done by a variable force in a press is 80 foot-pounds. The press moves a distance of 4 feet and the force is a quadratic of the form $F = ax^2$. Find a.

46. Work Find the work done by the force F shown in the figure.

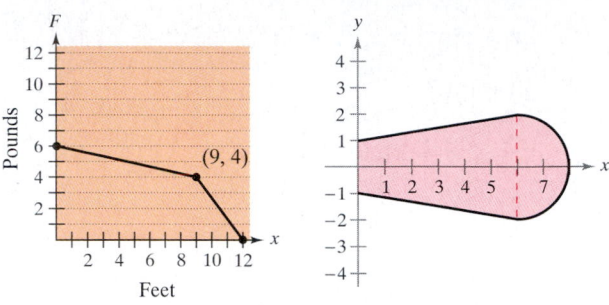

Figure for 46   **Figure for 51**

Centroids In Exercises 47–50, find the centroid of the region bounded by the graphs of the equations.

47. $\sqrt{x} + \sqrt{y} = \sqrt{a}$, $x = 0$, $y = 0$

48. $y = x^2$, $y = 2x + 3$

49. $y = a^2 - x^2$, $y = 0$

50. $y = x^{2/3}$, $y = \frac{1}{2}x$

51. Centroid A blade on an industrial fan has the configuration of a semicircle attached to a trapezoid (see figure). Find the centroid of the blade.

52. Centroid Cut an irregular shape from a piece of cardboard.

(a) Hold a pencil vertically and move the object on the pencil point until the centroid is located.

(b) Divide the object into representative elements. Make the necessary measurements and numerically approximate the centroid. Compare with the result in part (a).

53. Fluid Force A swimming pool is 5 feet deep at one end and 10 feet deep at the other, and the bottom is an inclined plane. The length and width of the pool are 40 feet and 20 feet. If the pool is full of water, what is the fluid force on each of the vertical walls?

54. Fluid Force Show that the fluid force against any vertical region in a liquid is the product of the weight per cubic volume of the liquid, the area of the region, and the depth of the centroid of the region.

55. Fluid Force Using the result of Exercise 54, find the fluid force on one side of a vertical circular plate of radius 4 feet that is submerged in water so that its center is 5 feet below the surface.

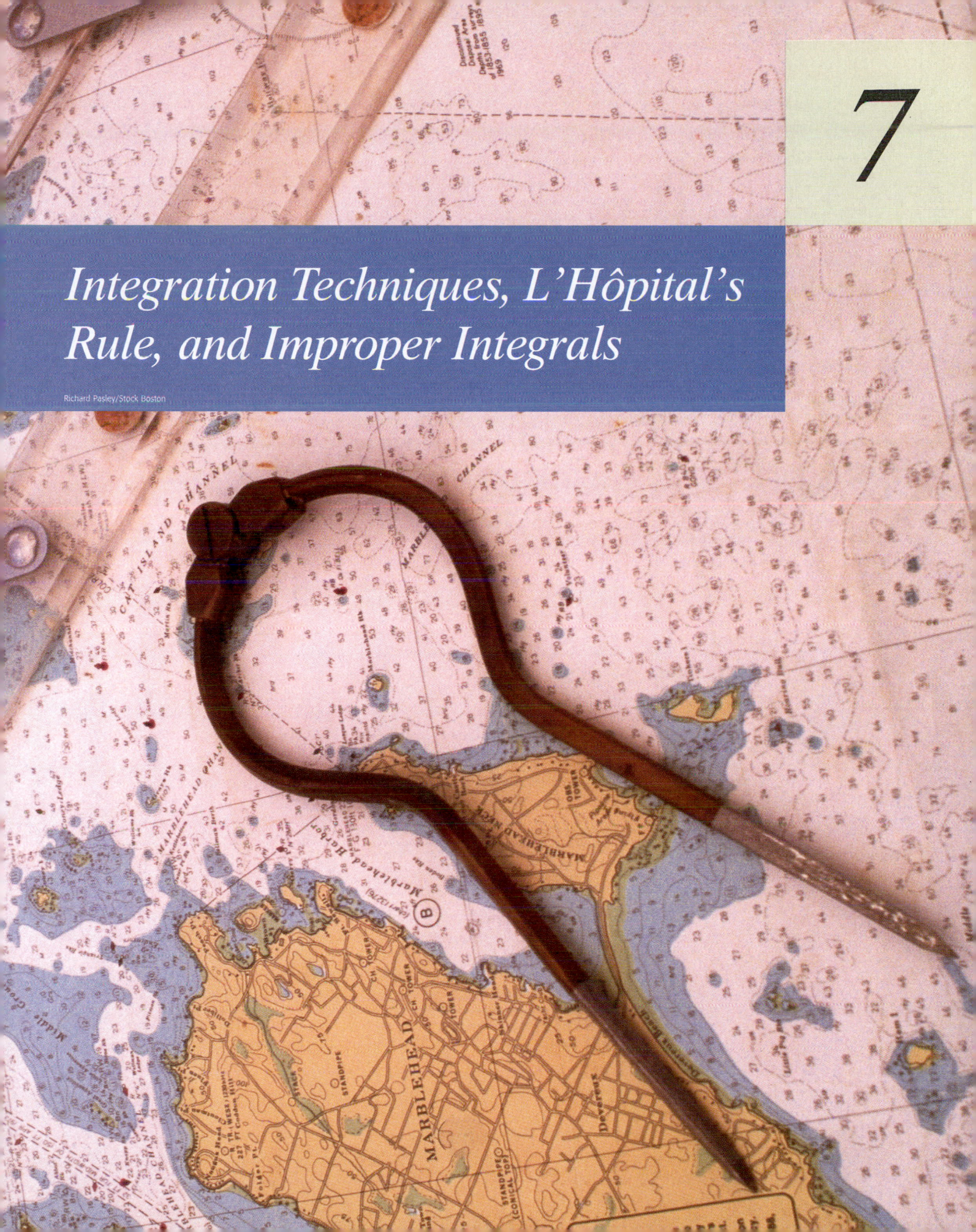

Integration Techniques, L'Hôpital's Rule, and Improper Integrals

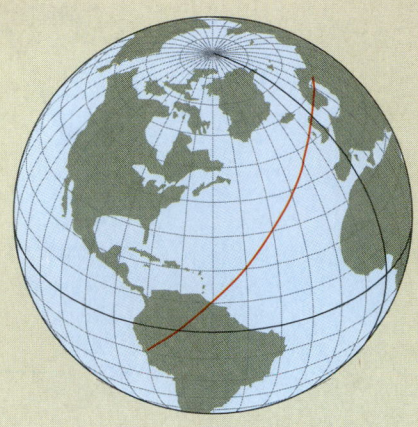

Globe: flight with constant 45° bearing

When flying or sailing, pilots expect to be given a steady compass course to follow. On a standard flat map, this is difficult because a steady compass course results in a curved line, as shown in the top left and middle left figures.

For curved lines to appear as straight lines on a flat map, Flemish geographer Gerardus Mercator (1512–1594) realized that latitude lines must be stretched horizontally by a scaling factor of $\sec \phi$, where ϕ is the angle of the latitude line. For the map to preserve the angles between latitude and longitude lines, the lengths of longitude lines are also stretched by a scaling factor of $\sec \phi$ at latitude ϕ. The Mercator map has latitude lines that are not equidistant, as shown in the bottom left figure.

To calculate these vertical lengths, imagine a globe with latitude lines marked at angles of every $\Delta\phi$ radians, with $\Delta\phi = \phi_i - \phi_{i-1}$. The arc length of consecutive latitude lines is $R\Delta\phi$. On the Mercator map, the vertical distance between the equator and the first latitude line is $R\Delta\phi \sec \phi_1$. The vertical distance between the first and second latitude lines is $R\Delta\phi \sec \phi_2$. The vertical distance between the second and third latitude lines is $R\Delta\phi \sec \phi_3$, and so on, as shown in the figure below.

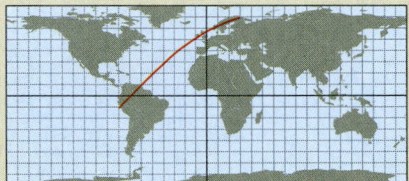

Standard flat map: flight with constant 45° bearing

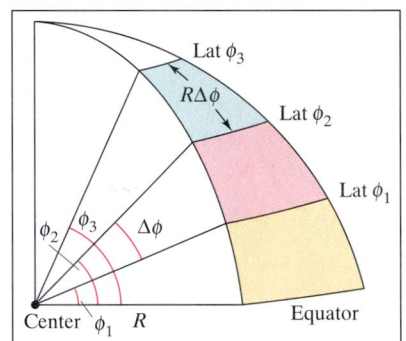

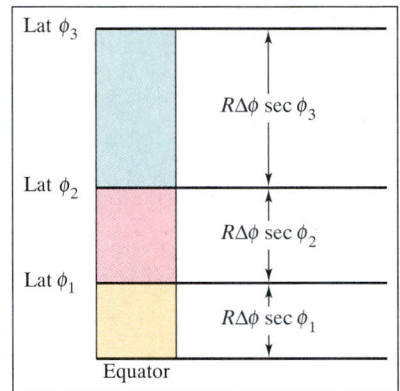

On a globe, the angle between consecutive latitude lines is $\Delta\phi$, and the arc length between them is $R\Delta\phi$ (see the above left figure). On a Mercator map, the vertical distance between the ith and $(i-1)$st latitude lines is $R\Delta\phi \sec \phi_i$, and the distance from the equator to the ith latitude line is approximately

$$R\Delta\phi \sec \phi_1 + R\Delta\phi \sec \phi_2 + \ldots + R\Delta\phi \sec \phi_i \quad \text{(see figure above)}.$$

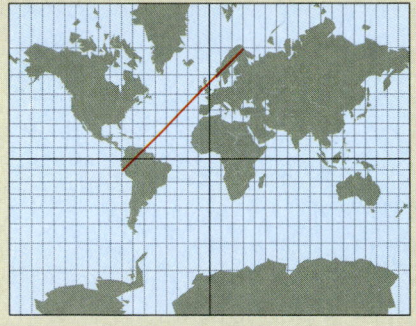

Mercator map: flight with constant 45° bearing

QUESTIONS

1. Use summation notation to write an expression to calculate how far from the equator to draw the line representing latitude ϕ_n.

2. In the calculations above, Mercator realized that the smaller the value used for $\Delta\phi$, the better the map became (better in the sense that straight lines could be used to plot steady compass courses). From your knowledge of calculus, how could you use Mercator's observation to calculate the total vertical distance of a latitude line from the equator?

3. Use the result of Question 2 to find how far from the equator to place latitude lines whose angles are 10°, 20°, 30°, 40°, and 50°. (Use a globe radius of $R = 6$ inches.)

4. What problem do you encounter when you attempt to calculate how far from the equator to place the North Pole?

The concepts presented here will be explored further in this chapter. For an extension of this application, see the lab series that accompanies this text.

Fitting Integrands to Basic Rules

Fitting Integrands to Basic Rules

In this chapter, you will study several integration techniques that greatly expand the set of integrals to which the basic integration rules (see page 388) can be applied. A major step in solving any integration problem is recognizing the proper basic integration rule to be used. This is not easy. As demonstrated in Example 1, slight differences in the integrand can lead to very different solution techniques.

EXAMPLE 1   A Comparison of Three Similar Integrals

Evaluate each of the following integrals.

a. $\displaystyle\int \frac{4}{x^2 + 9}\, dx$ **b.** $\displaystyle\int \frac{4x}{x^2 + 9}\, dx$ **c.** $\displaystyle\int \frac{4x^2}{x^2 + 9}\, dx$

Solution

a. Use the Arctangent Rule and let $u = x$ and $a = 3$.

$$\int \frac{4}{x^2 + 9}\, dx = 4\int \frac{1}{x^2 + 3^2}\, dx$$

$$= 4\left(\frac{1}{3}\arctan\frac{x}{3}\right) + C$$

$$= \frac{4}{3}\arctan\frac{x}{3} + C$$

b. Here the Arctangent Rule does not apply because the numerator contains a factor of x. Consider the Log Rule and let $u = x^2 + 9$. Then $du = 2x\, dx$, and you have

$$\int \frac{4x}{x^2 + 9}\, dx = 2\int \frac{2x\, dx}{x^2 + 9}$$

$$= 2\int \frac{du}{u}$$

$$= 2\ln|u| + C$$

$$= 2\ln(x^2 + 9) + C.$$

c. Because the degree of the numerator is equal to the degree of the denominator, you should first use division to rewrite the improper rational function as the sum of a polynomial and a proper rational function.

$$\int \frac{4x^2}{x^2 + 9}\, dx = \int \left(4 - \frac{36}{x^2 + 9}\right) dx$$

$$= \int 4\, dx - 36\int \frac{1}{x^2 + 9}\, dx$$

$$= 4x - 36\left(\frac{1}{3}\arctan\frac{x}{3}\right) + C$$

$$= 4x - 12\arctan\frac{x}{3} + C$$

NOTE   Notice in Example 1c that some preliminary algebra was required before applying the rules for integration, and that subsequently more than one rule was needed to evaluate the resulting integral.

EXAMPLE 2 **Using Two Basic Rules to Solve a Single Integral**

Evaluate $\int_0^1 \dfrac{x + 3}{\sqrt{4 - x^2}}\, dx$.

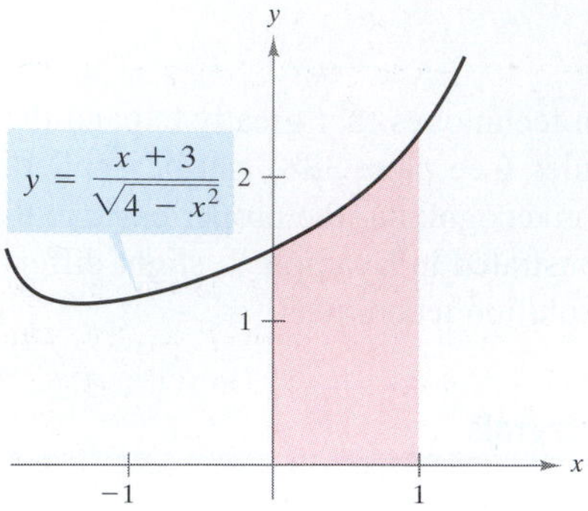

Solution Begin by writing the integral as the sum of two integrals. Then apply the Power Rule and the Arcsine Rule as follows.

$$\int_0^1 \frac{x + 3}{\sqrt{4 - x^2}}\, dx = \int_0^1 \frac{x}{\sqrt{4 - x^2}}\, dx + \int_0^1 \frac{3}{\sqrt{4 - x^2}}\, dx$$

$$= -\frac{1}{2}\int_0^1 (4 - x^2)^{-1/2}(-2x)\, dx + 3\int_0^1 \frac{1}{\sqrt{2^2 - x^2}}\, dx$$

$$= \left[-(4 - x^2)^{1/2} + 3\arcsin\frac{x}{2} \right]_0^1$$

$$= \left(-\sqrt{3} + \tfrac{1}{2}\pi \right) - (-2 + 0)$$

$$\approx 1.839$$

(See Figure 7.1.)

$y = \dfrac{x + 3}{\sqrt{4 - x^2}}$

The area of the region is approximately 1.839.

Figure 7.1

TECHNOLOGY Simpson's Rule can be used to give a good approximation of the value of the integral in Example 2 (for $n = 10$, the approximation is 1.839). When using numerical integration, however, you should be aware that Simpson's Rule does not give good approximations when one or both of the limits of integration are near a vertical asymptote. For instance, using the Fundamental Theorem of Calculus, you can obtain

$$\int_0^{1.99} \frac{x + 3}{\sqrt{4 - x^2}}\, dx \approx 6.213.$$

Applying Simpson's Rule (with $n = 10$) to this integral produces an approximation of 6.889.

EXAMPLE 3 **A Substitution Involving $a^2 - u^2$**

Evaluate $\int \dfrac{x^2}{\sqrt{16 - x^6}}\, dx$.

Solution Because the radical in the denominator can be written in the form

$$\sqrt{a^2 - u^2} = \sqrt{4^2 - (x^3)^2}$$

you can try the substitution $u = x^3$. Then $du = 3x^2\, dx$, and you have

$$\int \frac{x^2}{\sqrt{16 - x^6}}\, dx = \frac{1}{3}\int \frac{3x^2\, dx}{\sqrt{16 - (x^3)^2}}$$

$$= \frac{1}{3}\int \frac{du}{\sqrt{4^2 - u^2}}$$

$$= \frac{1}{3}\arcsin\frac{u}{4} + C$$

$$= \frac{1}{3}\arcsin\frac{x^3}{4} + C.$$

STUDY TIP Rules 18, 19, and 20 on page 388 all have expressions involving the sum or difference of two squares:

$a^2 - [f(x)]^2$

$a^2 + [f(x)]^2$

$[f(x)]^2 - a^2$

With such an expression, consider the substitution $u = f(x)$, as in Example 3.

Surprisingly, two of the most commonly overlooked integration rules are the Log Rule and the Power Rule. Notice in the next two examples how these two integration rules can be disguised.

EXAMPLE 4   A Disguised Form of the Log Rule

Evaluate $\displaystyle\int \frac{1}{1 + e^x}\, dx$.

Solution   The integral does not appear to fit any of the basic rules. However, the quotient form suggests the Log Rule. If you let $u = 1 + e^x$, then $du = e^x\, dx$. You can obtain the required du by adding and subtracting e^x in the numerator, as follows.

$$\int \frac{1}{1 + e^x}\, dx = \int \frac{1 + e^x - e^x}{1 + e^x}\, dx \qquad \text{Add and subtract } e^x \text{ in numerator.}$$

$$= \int \left(\frac{1 + e^x}{1 + e^x} - \frac{e^x}{1 + e^x} \right) dx$$

$$= \int dx - \int \frac{e^x\, dx}{1 + e^x}$$

$$= x - \ln(1 + e^x) + C$$

NOTE   There is usually more than one way to solve an integration problem. For instance, in Example 4, try integrating by multiplying the numerator and denominator by e^{-x} to obtain an integral of the form $-\int du/u$. See if you can get the same answer by this procedure. (Be careful: the answer will appear in a different form.)

EXAMPLE 5   A Disguised Form of the Power Rule

Evaluate $\int (\cot x)[\ln(\sin x)]\, dx$.

Solution   Again, the integral does not appear to fit any of the basic rules. However, considering the two primary choices for u ($u = \cot x$ and $u = \ln \sin x$), you can see that the second choice is the appropriate one because

$$u = \ln \sin x$$

$$du = \frac{\cos x}{\sin x}\, dx$$

$$du = \cot x\, dx.$$

Thus, you have

$$\int (\cot x)[\ln(\sin x)]\, dx = \int u\, du = \frac{u^2}{2} + C$$

$$= \frac{1}{2}[\ln(\sin x)]^2 + C.$$

NOTE   In Example 5, try *checking* that the derivative of

$$\frac{1}{2}[\ln(\sin x)]^2 + C$$

is the integrand of the original integral.

Trigonometric identities can often be used to fit integrals to one of the basic integration rules.

EXAMPLE 6 Using Trigonometric Identities

Evaluate $\int \tan^2 2x \, dx$.

TECHNOLOGY If you have access to symbolic integration software, try using it to evaluate the integrals in this section. Compare the *form* of the antiderivative given by the software with the form obtained by hand. Sometimes the forms will be the same, but often they will differ. For instance, why is the antiderivative $\ln 2x + C$ equivalent to the antiderivative $\ln x + C$?

Solution Note that $\tan^2 u$ is not in the list of basic integration rules. However, $\sec^2 u$ is in the list. This suggests the trigonometric identity $\tan^2 u = \sec^2 u - 1$. If you let $u = 2x$, then $du = 2 \, dx$ and

$$\int \tan^2 2x \, dx = \frac{1}{2} \int \tan^2 u \, du$$

$$= \frac{1}{2} \int (\sec^2 u - 1) \, du$$

$$= \frac{1}{2} \int \sec^2 u \, du - \frac{1}{2} \int du$$

$$= \frac{1}{2} \tan u - \frac{u}{2} + C$$

$$= \frac{1}{2} \tan 2x - x + C.$$

We conclude this section with a summary of the common procedures for fitting integrands to the basic integration rules.

Procedures for Fitting Integrands to Basic Rules

Technique	*Example*
Expand (numerator)	$(1 + e^x)^2 = 1 + 2e^x + e^{2x}$
Separate numerator	$\dfrac{1 + x}{x^2 + 1} = \dfrac{1}{x^2 + 1} + \dfrac{x}{x^2 + 1}$
Complete the square	$\dfrac{1}{\sqrt{2x - x^2}} = \dfrac{1}{\sqrt{1 - (x - 1)^2}}$
Divide improper rational function	$\dfrac{x^2}{x^2 + 1} = 1 - \dfrac{1}{x^2 + 1}$
Add and subtract terms in numerator	$\dfrac{2x}{x^2 + 2x + 1} = \dfrac{2x + 2 - 2}{x^2 + 2x + 1} = \dfrac{2x + 2}{x^2 + 2x + 1} - \dfrac{2}{(x + 1)^2}$
Use trigonometric identities	$\cot^2 x = \csc^2 x - 1$
Multiply and divide by Pythagorean conjugate	$\dfrac{1}{1 + \sin x} = \left(\dfrac{1}{1 + \sin x}\right)\left(\dfrac{1 - \sin x}{1 - \sin x}\right) = \dfrac{1 - \sin x}{1 - \sin^2 x}$
	$= \dfrac{1 - \sin x}{\cos^2 x} = \sec^2 x - \dfrac{\sin x}{\cos^2 x}$

NOTE Remember that you can separate numerators but not denominators. Watch out for this common error when fitting integrands to basic rules.

$$\frac{1}{x^2 + 1} \neq \frac{1}{x^2} + \frac{1}{1} \qquad \text{Do not separate denominators.}$$

EXERCISES FOR SECTION 7.1

In Exercises 1–4, select the correct antiderivative.

1. $\dfrac{dy}{dx} = \dfrac{x}{\sqrt{x^2 + 1}}$

   (a) $2\sqrt{x^2 + 1} + C$ (b) $\sqrt{x^2 + 1} + C$

   (c) $\frac{1}{2}\sqrt{x^2 + 1} + C$ (d) $\ln(x^2 + 1) + C$

2. $\dfrac{dy}{dx} = \dfrac{x}{x^2 + 1}$

   (a) $\ln\sqrt{x^2 + 1} + C$ (b) $\dfrac{2x}{(x^2 + 1)^2} + C$

   (c) $\arctan x + C$ (d) $\ln(x^2 + 1) + C$

3. $\dfrac{dy}{dx} = \dfrac{1}{x^2 + 1}$

   (a) $\ln\sqrt{x^2 + 1} + C$ (b) $\dfrac{2x}{(x^2 + 1)^2} + C$

   (c) $\arctan x + C$ (d) $\ln(x^2 + 1) + C$

4. $\dfrac{dy}{dx} = x\cos(x^2 + 1)$

   (a) $2x\sin(x^2 + 1) + C$ (b) $-\frac{1}{2}\sin(x^2 + 1) + C$

   (c) $\frac{1}{2}\sin(x^2 + 1) + C$ (d) $-2x\sin(x^2 + 1) + C$

In Exercises 5–14, select the basic integration formula you would use to evaluate the integral, and identify u and a when appropriate.

5. $\displaystyle\int (3x - 2)^4\, dx$

6. $\displaystyle\int \dfrac{2t - 1}{t^2 - t + 2}\, dt$

7. $\displaystyle\int \dfrac{1}{\sqrt{x}\left(1 - 2\sqrt{x}\right)}\, dx$

8. $\displaystyle\int \dfrac{2}{(2t - 1)^2 + 4}\, dt$

9. $\displaystyle\int \dfrac{3}{\sqrt{1 - t^2}}\, dt$

10. $\displaystyle\int \dfrac{-2x}{\sqrt{x^2 - 4}}\, dx$

11. $\displaystyle\int t\sin t^2\, dt$

12. $\displaystyle\int \sec 3x \tan 3x\, dx$

13. $\displaystyle\int \cos x e^{\sin x}\, dx$

14. $\displaystyle\int \dfrac{1}{x\sqrt{x^2 - 4}}\, dx$

In Exercises 15–46, evaluate the indefinite integral.

15. $\displaystyle\int (-2x + 5)^{3/2}\, dx$

16. $\displaystyle\int \dfrac{2}{(t - 9)^2}\, dt$

17. $\displaystyle\int \left[v + \dfrac{1}{(3v - 1)^3}\right] dv$

18. $\displaystyle\int x\sqrt{4 - 2x^2}\, dx$

19. $\displaystyle\int \dfrac{t^2 - 3}{-t^3 + 9t + 1}\, dt$

20. $\displaystyle\int \dfrac{2x}{x - 4}\, dx$

21. $\displaystyle\int \dfrac{x^2}{x - 1}\, dx$

22. $\displaystyle\int \dfrac{x + 1}{\sqrt{x^2 + 2x - 4}}\, dx$

23. $\displaystyle\int \dfrac{e^x}{1 + e^x}\, dx$

24. $\displaystyle\int \left(\dfrac{1}{3x - 1} - \dfrac{1}{3x + 1}\right) dx$

25. $\displaystyle\int (1 + 2x^2)^2\, dx$

26. $\displaystyle\int x\left(1 + \dfrac{1}{x}\right)^3 dx$

27. $\displaystyle\int x\cos 2\pi x^2\, dx$

28. $\displaystyle\int \sec 4u\, du$

29. $\displaystyle\int \csc \pi x \cot \pi x\, dx$

30. $\displaystyle\int \dfrac{\sin x}{\sqrt{\cos x}}\, dx$

31. $\displaystyle\int e^{5x}\, dx$

32. $\displaystyle\int \csc^2 x e^{\cot x}\, dx$

33. $\displaystyle\int \dfrac{2}{e^{-x} + 1}\, dx$

34. $\displaystyle\int \dfrac{1}{2e^x - 3}\, dx$

35. $\displaystyle\int \dfrac{1 + \sin x}{\cos x}\, dx$

36. $\displaystyle\int \dfrac{1}{\sec x - 1}\, dx$

37. $\displaystyle\int \dfrac{2t - 1}{t^2 + 4}\, dt$

38. $\displaystyle\int \dfrac{3}{t^2 + 1}\, dt$

39. $\displaystyle\int \dfrac{-1}{\sqrt{1 - (2t - 1)^2}}\, dt$

40. $\displaystyle\int \dfrac{1}{4 + 3x^2}\, dx$

41. $\displaystyle\int \dfrac{\tan(2/t)}{t^2}\, dt$

42. $\displaystyle\int \dfrac{e^{1/t}}{t^2}\, dt$

43. $\displaystyle\int \dfrac{3}{\sqrt{6x - x^2}}\, dx$

44. $\displaystyle\int \dfrac{1}{(x - 1)\sqrt{4x^2 - 8x + 3}}\, dx$

45. $\displaystyle\int \dfrac{4}{4x^2 + 4x + 65}\, dx$

46. $\displaystyle\int \dfrac{1}{\sqrt{2 - 2x - x^2}}\, dx$

Direction Fields **In Exercises 47 and 48, a differential equation, a point, and a direction field are given. (a) Sketch two approximate solutions of the differential equation on the direction field, one of which passes through the indicated point. (b) Use integration to find the particular solution of the differential equation and use a graphing utility to graph the solution. Compare the result with the sketches in part (a).**

47. $\dfrac{ds}{dt} = \dfrac{t}{\sqrt{1 - t^4}}$, $\left(0, -\frac{1}{2}\right)$

48. $\dfrac{dy}{dx} = \tan^2(2x)$, $(0, 0)$

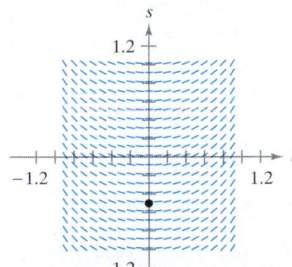

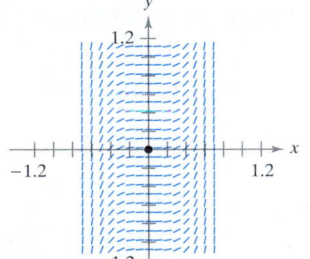

In Exercises 49–52, solve the differential equation.

49. $\dfrac{dy}{dx} = (1 + e^x)^2$

50. $\dfrac{dr}{dt} = \dfrac{(1 + e^t)^2}{e^t}$

51. $(4 + \tan^2 x)y' = \sec^2 x$

52. $y' = \dfrac{1}{x\sqrt{4x^2 - 1}}$

 In Exercises 53–60, evaluate the definite integral. Use the integration capabilities of a graphing utility to verify your result.

53. $\displaystyle\int_0^{\pi/4} \cos 2x \, dx$

54. $\displaystyle\int_0^{\pi} \sin^2 t \cos t \, dt$

55. $\displaystyle\int_0^1 xe^{-x^2} \, dx$

56. $\displaystyle\int_1^e \dfrac{1 - \ln x}{x} \, dx$

57. $\displaystyle\int_0^4 \dfrac{2x}{\sqrt{x^2 + 9}} \, dx$

58. $\displaystyle\int_1^2 \dfrac{x - 2}{x} \, dx$

59. $\displaystyle\int_0^{2/\sqrt{3}} \dfrac{1}{4 + 9x^2} \, dx$

60. $\displaystyle\int_0^4 \dfrac{1}{\sqrt{25 - x^2}} \, dx$

 In Exercises 61–64, use symbolic integration software to evaluate the integral. Use the symbolic integration software to graph two antiderivatives. Describe the relationship between the two graphs of the antiderivatives.

61. $\displaystyle\int \dfrac{1}{x^2 + 4x + 13} \, dx$

62. $\displaystyle\int \dfrac{x - 2}{x^2 + 4x + 13} \, dx$

63. $\displaystyle\int \dfrac{1}{1 + \sin \theta} \, d\theta$

64. $\displaystyle\int \left(\dfrac{e^x + e^{-x}}{2} \right)^3 \, dx$

65. Determine the constants a and b such that

$$\sin x + \cos x = a \sin(x + b).$$

Use this result to integrate $\displaystyle\int \dfrac{dx}{\sin x + \cos x}$.

 66. Think About It Use a graphing utility to graph the function $f(x) = \frac{1}{5}(x^3 - 7x^2 + 10x)$. Use the graph to determine whether

$$\int_0^5 f(x) \, dx$$

is positive or negative. Explain.

Approximation **In Exercises 67 and 68, determine which value best approximates the area of the region between the x-axis and the function over the given interval. (Make your selection on the basis of a sketch of the region and *not* by performing any calculations.)**

67. $f(x) = \dfrac{4x}{x^2 + 1}$, $[0, 2]$

(a) 3   (b) 1   (c) −8   (d) 8   (e) 10

68. $f(x) = \dfrac{4}{x^2 + 1}$, $[0, 2]$

(a) 3   (b) 1   (c) −4   (d) 4   (e) 10

Area **In Exercises 69 and 70, find the area of the region bounded by the graph(s) of the equation(s).**

69. $y^2 = x^2(1 - x^2)$

70. $y = \sin 2x$, $y = 0$, $x = 0$, $x = \pi/2$

71. *Area* The graphs of $f(x) = x$ and $g(x) = ax^2$ intersect at the points $(0, 0)$ and $(1/a, 1/a)$. Find a $(a > 0)$ such that the area of the region bounded by the graphs of these two functions is $\frac{2}{3}$.

72. *Interpreting an Integral* You are given the integral

$$\int_0^2 2\pi x^2 \, dx$$

but are not told what it represents. (There is more than one correct answer for each part.)

(a) Sketch the region whose area is given by the integral.

(b) Sketch the solid whose volume is given by the integral if the disc method is used.

(c) Sketch the solid whose volume is given by the integral if the shell method is used.

73. *Volume* The region bounded by $y = e^{-x^2}, y = 0, x = 0$, and $x = b$ $(b > 0)$ is revolved about the y-axis.

(a) Find the volume of the solid generated if $b = 1$.

(b) Find b such that the volume of the generated solid is $\frac{4}{3}$ cubic units.

74. *Average Value* Compute the average value of each of the functions over the indicated interval.

(a) $f(x) = \sin nx$, $0 \le x \le \pi/n$, n is a positive integer

(b) $f(x) = \dfrac{1}{1 + x^2}$, $-3 \le x \le 3$

75. *Centroid* Find the x-coordinate of the centroid of the region bounded by the graphs of

$$y = \dfrac{5}{\sqrt{25 - x^2}}, \quad y = 0, \quad x = 0, \quad \text{and} \quad x = 4.$$

76. *Surface Area* Find the area of the surface formed by revolving the graph of $y = 2\sqrt{x}$ on the interval $[0, 9]$ about the x-axis.

Arc Length **In Exercises 77 and 78, use the integration capabilities of a graphing utility to approximate the arc length of the curve over the indicated interval.**

77. $y = \tan \pi x$, $\left[0, \frac{1}{4}\right]$

78. $y = x^{2/3}$, $[1, 8]$

True or False? **In Exercises 79 and 80, determine whether the statement is true or false. If it is false, explain why or give an example that shows it is false.**

79. $\displaystyle\int \dfrac{x^3}{x^4 - 1} \, dx = \dfrac{1}{4} \int \dfrac{du}{u}$, $u = x^4 - 1$

80. $\displaystyle\int \dfrac{dx}{1 + \sin^2 x} = \int \dfrac{du}{a^2 + u^2}$, $u = \sin x, a = 1$

Integration by Parts • Tabular Method

Integration by Parts

In this section you will study an important integration technique called **integration by parts.** This technique can be applied to a wide variety of functions and is particularly useful for integrands involving *products* of algebraic and transcendental functions. For instance, integration by parts works well with integrals such as

$$\int x \ln x \, dx, \quad \int x^2 e^x \, dx, \quad \text{and} \quad \int e^x \sin x \, dx.$$

Integration by parts is based on the formula for the derivative of a product

$$\frac{d}{dx}[uv] = u\frac{dv}{dx} + v\frac{du}{dx}$$

$$= uv' + vu'$$

where both u and v are differentiable functions of x. If u' and v' are continuous, you can integrate both sides of this equation to obtain

$$uv = \int uv' \, dx + \int vu' \, dx$$

$$= \int u \, dv + \int v \, du.$$

By rewriting this equation, you obtain the following theorem.

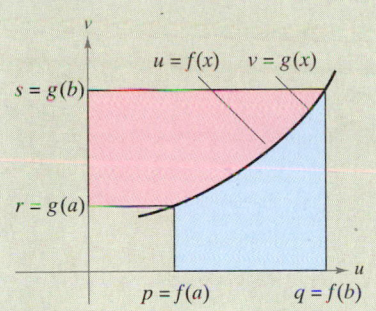
> **THEOREM 7.1   Integration by Parts**
>
> If u and v are functions of x and have continuous derivatives, then
>
> $$\int u \, dv = uv - \int v \, du.$$

This formula expresses the original integral in terms of another integral. Depending on the choices of u and dv, it may be easier to evaluate the second integral than the original one. Because the choices of u and dv are critical in the integration by parts process, we provide the following guidelines.

> **Guidelines for Integration by Parts**
>
> 1. Try letting dv be the most complicated portion of the integrand that fits a basic integration rule. Then u will be the remaining factor(s) of the integrand.
>
> 2. Try letting u be the portion of the integrand whose derivative is a function simpler than u. Then dv will be the remaining factor(s) of the integrand.

EXAMPLE 1 Integration by Parts

Evaluate $\int xe^x \, dx$.

Solution To apply integration by parts, you need to write the integral in the form $\int u \, dv$. There are several ways to do this.

$$\int \underbrace{(x)}_{u} \underbrace{(e^x dx)}_{dv}, \quad \int \underbrace{(e^x)}_{u} \underbrace{(x\, dx)}_{dv}, \quad \int \underbrace{(1)}_{u} \underbrace{(xe^x\, dx)}_{dv}, \quad \int \underbrace{(xe^x)}_{u} \underbrace{(dx)}_{dv}$$

The guidelines on page 481 suggest choosing the first option because the derivative of $u = x$ is simpler than x, and $dv = e^x \, dx$ is the most complicated portion of the integrand that fits a basic integration formula.

$$dv = e^x \, dx \quad \Longrightarrow \quad v = \int dv = \int e^x \, dx = e^x$$

$$u = x \quad \Longrightarrow \quad du = dx$$

NOTE In Example 1, note that it is not necessary to include a constant of integration when solving

$$v = \int e^x \, dx = e^x + C_1.$$

To illustrate this, replace $v = e^x$ by $v = e^x + C_1$ and apply integration by parts to see that you obtain the same result.

Now, integration by parts produces the following.

$$\int u \, dv = uv - \int v \, du$$

$$\int xe^x \, dx = xe^x - \int e^x \, dx$$

$$= xe^x - e^x + C$$

To check this result, try differentiating $xe^x - e^x + C$ to see that you obtain the original integrand.

EXAMPLE 2 Integration by Parts

Evaluate $\int x^2 \ln x \, dx$.

Solution In this case, x^2 is more easily integrated than $\ln x$. Furthermore, the derivative of $\ln x$ is simpler than $\ln x$. Therefore, you should let $dv = x^2 \, dx$.

$$dv = x^2 \, dx \quad \Longrightarrow \quad v = \int x^2 \, dx = \frac{x^3}{3}$$

$$u = \ln x \quad \Longrightarrow \quad du = \frac{1}{x} \, dx$$

Integration by parts produces the following.

$$\int x^2 \ln x \, dx = \frac{x^3}{3} \ln x - \int \left(\frac{x^3}{3}\right)\left(\frac{1}{x}\right) dx$$

$$= \frac{x^3}{3} \ln x - \frac{1}{3} \int x^2 \, dx$$

$$= \frac{x^3}{3} \ln x - \frac{x^3}{9} + C$$

TECHNOLOGY Try graphing

$$\int x^2 \ln x \, dx \text{ and } \frac{x^3}{3} \ln x - \frac{x^3}{9}$$

on your graphing utility. Do you get the same graph? (This will take a while, so be patient.)

You can check this result by differentiating.

$$\frac{d}{dx}\left[\frac{x^3}{3} \ln x - \frac{x^3}{9}\right] = \frac{x^3}{3}\left(\frac{1}{x}\right) + (\ln x)(x^2) - \frac{x^2}{3} = x^2 \ln x$$

One surprising application of integration by parts involves integrands consisting of a single factor, such as $\int \ln x \, dx$ or $\int \arcsin x \, dx$. In such cases, you should let $dv = dx$, as illustrated in the next example.

EXAMPLE 3   An Integrand with a Single Term

Evaluate $\displaystyle\int_0^1 \arcsin x \, dx$.

Solution   Let $dv = dx$.

$$dv = dx \quad\Longrightarrow\quad v = \int dx = x$$

$$u = \arcsin x \quad\Longrightarrow\quad du = \frac{1}{\sqrt{1-x^2}} \, dx$$

Integration by parts now produces the following.

$$\int \arcsin x \, dx = x \arcsin x - \int \frac{x}{\sqrt{1-x^2}} \, dx$$

$$= x \arcsin x + \frac{1}{2} \int (1-x^2)^{-1/2} \, (-2x) \, dx$$

$$= x \arcsin x + \sqrt{1-x^2} + C$$

Using this antiderivative, you can evaluate the definite integral as follows.

$$\int_0^1 \arcsin x \, dx = \left[x \arcsin x + \sqrt{1-x^2} \right]_0^1$$

$$= \frac{\pi}{2} - 1$$

$$\approx 0.571$$

The area represented by this definite integral is shown in Figure 7.2.

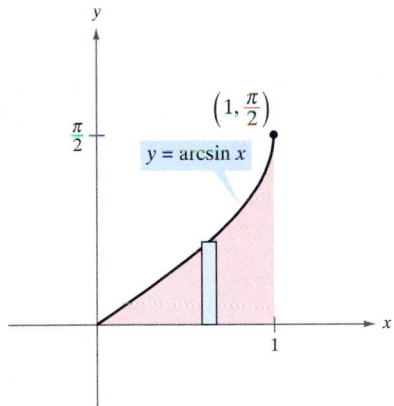

The area of the region is approximately 0.571.

Figure 7.2

TECHNOLOGY   Remember that there are two ways to use technology to evaluate a definite integral: (1) you can use a numerical approximation such as the Trapezoidal Rule or Simpson's Rule, or (2) you can use a symbolic integration utility to find the antiderivative and then apply the Fundamental Theorem of Calculus. Both methods have shortcomings. To find the possible error in a numerical method, the integrand must have a second derivative (Trapezoid Rule) or a fourth derivative (Simpson's Rule) in the interval of integration: the integrand in Example 3 fails to meet this requirement. To apply the Fundamental Theorem of Calculus, the symbolic integration utility must be able to find the antiderivative.

Which method would you use to evaluate

$$\int_0^1 \arctan x \, dx?$$

Which method would you use to evaluate

$$\int_0^1 \arctan x^2 \, dx?$$

Some integrals require repeated use of the integration by parts formula.

EXAMPLE 4   Repeated Use of Integration by Parts

Evaluate $\int x^2 \sin x \, dx$.

Solution   The factors x^2 and $\sin x$ are equally easy to integrate. However, the derivative of x^2 becomes simpler, whereas the derivative of $\sin x$ does not. Therefore, you should let $u = x^2$.

$$dv = \sin x \, dx \quad \Longrightarrow \quad v = \int \sin x \, dx = -\cos x$$

$$u = x^2 \quad \Longrightarrow \quad du = 2x \, dx$$

Now, integration by parts produces the following.

$$\int x^2 \sin x \, dx = -x^2 \cos x + \int 2x \cos x \, dx \qquad \text{First use of parts}$$

This first use of integration by parts has succeeded in simplifying the original integral, but the integral on the right still doesn't fit a basic integration rule. To evaluate that integral, you can apply integration by parts again. This time, let $u = 2x$.

$$dv = \cos x \, dx \quad \Longrightarrow \quad v = \int \cos x \, dx = \sin x$$

$$u = 2x \quad \Longrightarrow \quad du = 2 \, dx$$

Now, integration by parts produces

$$\int 2x \cos x \, dx = 2x \sin x - \int 2 \sin x \, dx \qquad \text{Second use of parts}$$

$$= 2x \sin x + 2 \cos x + C.$$

Combining these two results, you can write

$$\int x^2 \sin x \, dx = -x^2 \cos x + 2x \sin x + 2 \cos x + C.$$

When making repeated applications of integration by parts, you need to be careful not to interchange the substitutions in successive applications. For instance, in Example 4, the first substitution was $u = x^2$ and $dv = \sin x \, dx$. If, in the second application, you had switched the substitution to $u = \cos x$ and $dv = 2x$, you would have obtained

$$\int x^2 \sin x \, dx = -x^2 \cos x + \int 2x \cos x \, dx$$

$$= -x^2 \cos x + x^2 \cos x + \int x^2 \sin x \, dx$$

$$= \int x^2 \sin x \, dx$$

thus undoing the previous integration and returning to the *original* integral. When making repeated applications of integration by parts, you should also watch for the appearance of a *constant multiple* of the original integral. For instance, this occurs when you use integration by parts to evaluate $\int e^x \cos 2x \, dx$, and also occurs in the next example.

EXPLORATION

Try to evaluate

$$\int e^x \cos 2x \, dx$$

by letting $u = \cos 2x$ and $dv = e^x \, dx$ in the first substitution. For the second substitution, let $u = \sin 2x$ and $dv = e^x \, dx$.

EXAMPLE 5   Integration by Parts

NOTE   The integral in Example 5 is an important one. In Section 7.4 (Example 5), you will see that it is used to find the arc length of a parabolic segment.

Evaluate $\int \sec^3 x \, dx$.

Solution   The most complicated portion of the integrand that can be easily integrated is $\sec^2 x$, so you should let $dv = \sec^2 x \, dx$ and $u = \sec x$.

$$dv = \sec^2 x \, dx \quad \Longrightarrow \quad v = \int \sec^2 x \, dx = \tan x$$

$$u = \sec x \quad \Longrightarrow \quad du = \sec x \tan x \, dx$$

Integration by parts produces the following.

$$\int \sec^3 x \, dx = \sec x \tan x - \int \sec x \tan^2 x \, dx$$

$$= \sec x \tan x - \int \sec x (\sec^2 x - 1) \, dx$$

$$= \sec x \tan x - \int \sec^3 x \, dx + \int \sec x \, dx$$

$$2 \int \sec^3 x \, dx = \sec x \tan x + \int \sec x \, dx \qquad \text{Collect like integrals.}$$

$$\int \sec^3 x \, dx = \frac{1}{2} \sec x \tan x + \frac{1}{2} \ln |\sec x + \tan x| + C$$

STUDY TIP   The trigonometric identities

$$\sin^2 x = \frac{1 - \cos 2x}{2}$$

$$\cos^2 x = \frac{1 + \cos 2x}{2}$$

play an important role in this chapter.

EXAMPLE 6   Finding a Centroid

A machine part is modeled by the region bounded by the graph of $y = \sin x$ and the x-axis, $0 \le x \le \pi/2$, as shown in Figure 7.3. Find the centroid of this region.

Solution   Begin by finding the area of the region.

$$A = \int_0^{\pi/2} \sin x \, dx = \left[-\cos x \right]_0^{\pi/2} = 1$$

Now, you can find the coordinates of the centroid as follows.

$$\bar{y} = \frac{1}{A} \int_0^{\pi/2} \frac{\sin x}{2} (\sin x) \, dx = \frac{1}{4} \int_0^{\pi/2} (1 - \cos 2x) \, dx = \frac{1}{4} \left[x - \frac{\sin 2x}{2} \right]_0^{\pi/2} = \frac{\pi}{8}$$

You can evaluate the integral for $\bar{x}$, $(1/A) \int_0^{\pi/2} x \sin x \, dx$, with integration by parts. To do this, let $dv = \sin x \, dx$ and $u = x$. This produces $v = -\cos x$ and $du = dx$, and you can write

$$\int x \sin x \, dx = -x \cos x + \int \cos x \, dx$$

$$= -x \cos x + \sin x + C.$$

Finally, you can determine $\bar{x}$ to be

$$\bar{x} = \frac{1}{A} \int_0^{\pi/2} x \sin x \, dx = \left[-x \cos x + \sin x \right]_0^{\pi/2} = 1.$$

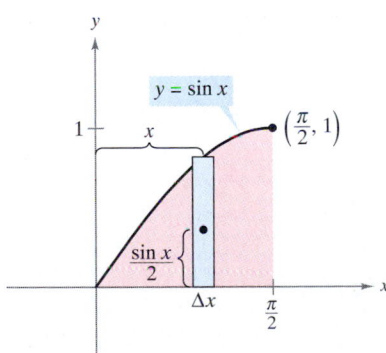

The centroid of the region, $\left(1, \dfrac{\pi}{8}\right)$, is its balancing point.

Figure 7.3

Therefore, the centroid of the region is $(1, \pi/8)$.

As you gain experience in using integration by parts, your skill in determining u and dv will increase. The following summary lists several common integrals with suggestions for the choice of u and dv.

Summary of Common Integrals Using Integration by Parts

1. For integrals of the form

$$\int x^n e^{ax}\, dx, \qquad \int x^n \sin ax\, dx, \qquad \text{or} \qquad \int x^n \cos ax\, dx$$

let $u = x^n$ and let $dv = e^{ax}\, dx$, $\sin ax\, dx$, or $\cos ax\, dx$.

2. For integrals of the form

$$\int x^n \ln x\, dx, \qquad \int x^n \arcsin ax\, dx, \qquad \text{or} \qquad \int x^n \arctan ax\, dx$$

let $u = \ln x$, $\arcsin ax$, or $\arctan ax$ and let $dv = x^n\, dx$.

3. For integrals of the form

$$\int e^{ax} \sin bx\, dx \qquad \text{or} \qquad \int e^{ax} \cos bx\, dx$$

let $u = \sin bx$ or $\cos bx$ and let $dv = e^{ax}\, dx$.

Tabular Method

In problems involving repeated applications of integration by parts, a tabular method, illustrated in Example 7, can help to organize the work. This method works well for integrals of the form $\int x^n \sin ax\, dx$, $\int x^n \cos ax\, dx$, and $\int x^n e^{ax}\, dx$.

EXAMPLE 7   Using the Tabular Method

Evaluate $\int x^2 \sin 4x\, dx$.

Solution   Begin as usual by letting $u = x^2$ and $dv = v'\, dx = \sin 4x\, dx$. Next, create a table consisting of three columns, as follows.

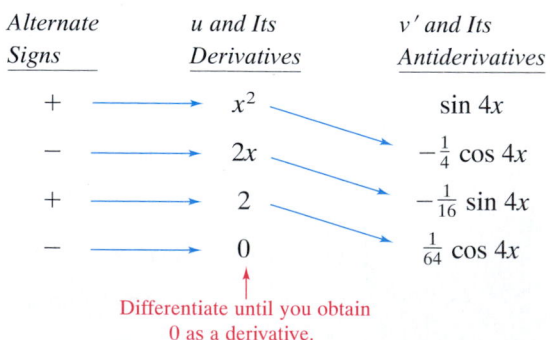

Alternate Signs	u and Its Derivatives	v' and Its Antiderivatives
$+$	x^2	$\sin 4x$
$-$	$2x$	$-\frac{1}{4}\cos 4x$
$+$	2	$-\frac{1}{16}\sin 4x$
$-$	0	$\frac{1}{64}\cos 4x$

Differentiate until you obtain 0 as a derivative.

FOR FURTHER INFORMATION For more information on the tabular method, see the article "More on Tabular Integration by Parts" by Leonard Gillman in the November 1991 issue of *The College Mathematics Journal.*

The solution is obtained by adding the *signed* products of the diagonal entries:

$$\int x^2 \sin 4x\, dx = -\frac{1}{4}x^2 \cos 4x + \frac{1}{8}x \sin 4x + \frac{1}{32}\cos 4x + C.$$

EXERCISES FOR SECTION 7.2

1. Match the antiderivative in the left column with the correct integral in the right column.

(a) $y = \sin x - x \cos x$ (i) $\int \ln x \, dx$

(b) $y = x^2 \sin x + 2x \cos x - 2 \sin x$ (ii) $\int x \sin x \, dx$

(c) $y = x^2 e^x - 2xe^x + 2e^x$ (iii) $\int x^2 e^x \, dx$

(d) $y = -x + x \ln x$ (iv) $\int x^2 \cos x \, dx$

2. *Think About It*   Explain in your own words the guidelines for integration by parts.

In Exercises 3–8, identify u and dv for evaluating the integral by integration by parts. (Do not evaluate the integral.)

3. $\displaystyle\int xe^{2x} \, dx$ **4.** $\displaystyle\int x^2 e^{2x} \, dx$

5. $\displaystyle\int (\ln x)^2 \, dx$ **6.** $\displaystyle\int \ln 3x \, dx$

7. $\displaystyle\int x \sec^2 x \, dx$ **8.** $\displaystyle\int x^2 \cos x \, dx$

In Exercises 9–30, evaluate the integral. (*Note:* Solve by the simplest method—not all require integration by parts.)

9. $\displaystyle\int xe^{-2x} \, dx$ **10.** $\displaystyle\int \frac{x}{e^x} \, dx$

11. $\displaystyle\int x^3 e^x \, dx$ **12.** $\displaystyle\int \frac{e^{1/t}}{t^2} \, dt$

13. $\displaystyle\int x^2 e^{x^3} \, dx$ **14.** $\displaystyle\int x^3 \ln x \, dx$

15. $\displaystyle\int t \ln(t + 1) \, dt$ **16.** $\displaystyle\int \frac{1}{x(\ln x)^3} \, dx$

17. $\displaystyle\int \frac{(\ln x)^2}{x} \, dx$ **18.** $\displaystyle\int \frac{\ln x}{x^2} \, dx$

19. $\displaystyle\int \frac{xe^{2x}}{(2x + 1)^2} \, dx$ **20.** $\displaystyle\int \frac{x^3 e^{x^2}}{(x^2 + 1)^2} \, dx$

21. $\displaystyle\int (x^2 - 1)e^x \, dx$ **22.** $\displaystyle\int \frac{\ln 2x}{x^2} \, dx$

23. $\displaystyle\int x\sqrt{x - 1} \, dx$ **24.** $\displaystyle\int \frac{x}{\sqrt{2 + 3x}} \, dx$

25. $\displaystyle\int x \cos x \, dx$ **26.** $\displaystyle\int \theta \sec \theta \tan \theta \, d\theta$

27. $\displaystyle\int \arctan x \, dx$ **28.** $\displaystyle\int \arccos x \, dx$

29. $\displaystyle\int e^{2x} \sin x \, dx$ **30.** $\displaystyle\int e^x \cos 2x \, dx$

In Exercises 31–36, solve the differential equation.

31. $y' = xe^{x^2}$ **32.** $y' = \ln x$

33. $\dfrac{dy}{dt} = \dfrac{t^2}{\sqrt{2 + 3t}}$ **34.** $\dfrac{dy}{dt} = x^2\sqrt{x - 1}$

35. $(\cos y)y' = 2x$ **36.** $y' = \arctan \dfrac{x}{2}$

Direction Fields   In Exercises 37 and 38, a differential equation, a point, and a direction field are given. (a) Sketch two approximate solutions of the differential equation on the direction field, one of which passes through the indicated point. (b) Use integration to find the particular solution of the differential equation and use a graphing utility to graph the solution. Compare the result with the sketches in part (a).

37. $\dfrac{dy}{dx} = x\sqrt{y} \cos x, \ (0, 4)$ **38.** $\dfrac{dy}{dx} = e^{-x/3} \sin 2x, \ \left(0, -\frac{18}{37}\right)$

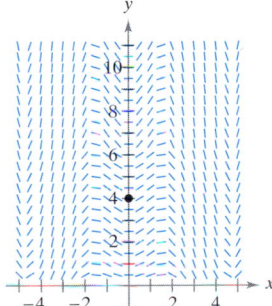
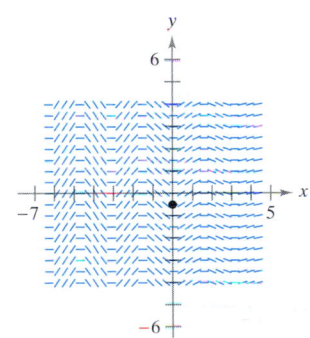

In Exercises 39–44, evaluate the definite integral. Use a graphing utility to confirm your result.

39. $\displaystyle\int_0^\pi x \sin 2x \, dx$ **40.** $\displaystyle\int_0^1 x \arcsin x^2 \, dx$

41. $\displaystyle\int_0^1 e^x \sin x \, dx$ **42.** $\displaystyle\int_0^1 x^2 e^x \, dx$

43. $\displaystyle\int_0^{\pi/2} x \cos x \, dx$ **44.** $\displaystyle\int_0^1 \ln(1 + x^2) \, dx$

In Exercises 45–50, use the tabular method to evaluate the integral.

45. $\displaystyle\int x^2 e^{2x} \, dx$ **46.** $\displaystyle\int x^4 e^{-x} \, dx$

47. $\displaystyle\int x^3 \sin x \, dx$ **48.** $\displaystyle\int x^3 \cos 2x \, dx$

49. $\displaystyle\int x \sec^2 x \, dx$ **50.** $\displaystyle\int x^2(x - 2)^{3/2} \, dx$

In Exercises 51–54, use a symbolic integration utility to evaluate the integral.

51. $\int t^3 e^{-4t} \, dt$

52. $\int \alpha^4 \sin \pi \alpha \, d\alpha$

53. $\int_0^{\pi/2} e^{-2x} \sin 3x \, dx$

54. $\int_0^5 x^4 \, (25 - x^2)^{3/2} \, dx$

55. Integrate $\int 2x\sqrt{2x - 3} \, dx$

   (a) by parts, letting $dv = \sqrt{2x - 3} \, dx$.

   (b) by substitution, letting $u = \sqrt{2x - 3}$.

56. Integrate $\int x\sqrt{4 + x} \, dx$

   (a) by parts, letting $dv = \sqrt{4 + x} \, dx$.

   (b) by substitution, letting $u = 4 + x$.

57. Integrate $\int \dfrac{x^3}{\sqrt{4 + x^2}} \, dx$

   (a) by parts, letting $dv = \left(x/\sqrt{4 + x^2}\right) dx$.

   (b) by substitution, letting $u = 4 + x^2$.

58. Integrate $\int x\sqrt{4 - x} \, dx$

   (a) by parts, letting $dv = \sqrt{4 - x} \, dx$.

   (b) by substitution, letting $u = 4 - x$.

In Exercises 59 and 60, use a symbolic integration utility to evaluate the integral for $n = 0, 1, 2,$ and 3. Use the result to obtain a general rule for the integral for any positive integer n and test your results for $n = 4$.

59. $\int x^n \ln x \, dx$

60. $\int x^n e^x \, dx$

In Exercises 61–66, use integration by parts to verify the formula. (For Exercises 61–64, assume that n is a positive integer.)

61. $\int x^n \sin x \, dx = -x^n \cos x + n \int x^{n-1} \cos x \, dx$

62. $\int x^n \cos x \, dx = x^n \sin x - n \int x^{n-1} \sin x \, dx$

63. $\int x^n \ln x \, dx = \dfrac{x^{n+1}}{(n+1)^2}\left[-1 + (n+1) \ln x\right] + C$

64. $\int x^n e^{ax} \, dx = \dfrac{x^n e^{ax}}{a} - \dfrac{n}{a} \int x^{n-1} e^{ax} \, dx$

65. $\int e^{ax} \sin bx \, dx = \dfrac{e^{ax}(a \sin bx - b \cos bx)}{a^2 + b^2} + C$

66. $\int e^{ax} \cos bx \, dx = \dfrac{e^{ax}(a \cos bx + b \sin bx)}{a^2 + b^2} + C$

In Exercises 67–70, evaluate the integral by using the appropriate formula from Exercises 61–66.

67. $\int x^3 \ln x \, dx$

68. $\int x^2 \cos x \, dx$

69. $\int e^{2x} \cos 3x \, dx$

70. $\int x^3 e^{2x} \, dx$

Area **In Exercises 71–74, use a graphing utility to sketch the region bounded by the graphs of equations, and find the area of the region.**

71. $y = xe^{-x}, y = 0, x = 4$

72. $y = \frac{1}{9} xe^{-x/3}, y = 0, x = 0, x = 3$

73. $y = e^{-x} \sin \pi x, y = 0, x = 0, x = 1$

74. $y = x \sin x, y = 0, x = 0, x = \pi$

75. ***Area, Volume, and Centroid***   Given the region bounded by the graphs of $y = \ln x, y = 0,$ and $x = e$, find

   (a) the area of the region.

   (b) the volume of the solid generated by revolving the region about the x-axis.

   (c) the volume of the solid generated by revolving the region about the y-axis.

   (d) the centroid of the region.

76. ***Centroid***   Find the centroid of the region bounded by the graphs of $y = \arcsin x, x = 0,$ and $y = \pi/2$. How is this problem related to Example 6 in this section?

77. ***Average Displacement***   A damping force affects the vibration of a spring so that the displacement of the spring is

$$y = e^{-4t} \, (\cos 2t + 5 \sin 2t).$$

Find the average value of y on the interval from $t = 0$ to $t = \pi$.

78. ***Memory Model***   A model for the ability M of a child to memorize, measured on a scale from 0 to 10, is

$$M = 1 + 1.6t \ln t, \quad 0 < t \le 4$$

where t is the child's age in years. Find the average value of this function

   (a) between the child's first and second birthdays.

   (b) between the child's third and fourth birthdays.

Present Value **In Exercises 79 and 80, find the present value P of a continuous income flow of $c(t)$ dollars per year if**

$$P = \int_0^{t_1} c(t)e^{-rt} \, dt$$

where t_1 is the time in years and r is the annual interest rate compounded continuously.

79. $c(t) = 100{,}000 + 4000t, r = 5\%, t_1 = 10$

80. $c(t) = 30{,}000 + 500t, r = 7\%, t_1 = 5$

Integrals Used to Find Fourier Coefficients   In Exercises 81 and 82, verify the value of the definite integral, where n is a positive integer.

81. $\displaystyle\int_{-\pi}^{\pi} x \sin nx \, dx = \begin{cases} \dfrac{2\pi}{n}, & n \text{ is odd} \\[2mm] -\dfrac{2\pi}{n}, & n \text{ is even} \end{cases}$

82. $\displaystyle\int_{-\pi}^{\pi} x^2 \cos nx \, dx = \dfrac{(-1)^n \, 4\pi}{n^2}$

83. *Vibrating String*   A string stretched between the two points $(0, 0)$ and $(2, 0)$ is plucked by displacing the string h units at its midpoint. The motion of the string is modeled by a **Fourier Sine Series** whose coefficients are given by

$$b_n = h \int_0^1 x \sin \frac{n\pi x}{2} \, dx + h \int_1^2 (-x + 2) \sin \frac{n\pi x}{2} \, dx.$$

Find b_n.

84. Find the fallacy in the following argument that $0 = 1$.

$dv = dx$ ➡ $v = \int dx = x$

$u = \dfrac{1}{x}$ ➡ $du = -\dfrac{1}{x^2} \, dx$

$0 + \displaystyle\int \frac{dx}{x} = \left(\frac{1}{x}\right)(x) - \int\left(-\frac{1}{x^2}\right)(x) \, dx = 1 + \int \frac{dx}{x}$

Hence, $0 = 1$.

85. Let $y = f(x)$ be positive and strictly increasing on the interval $0 < a \le x \le b$. Consider the region R bounded by the graphs of $y = f(x)$, $y = 0$, $x = a$, and $x = b$. If R is revolved about the y-axis, show that the disc method and shell method yield the same volume.

86. Consider the differential equation $f'(x) = xe^{-x}$ with the initial condition $f(0) = 0$.

(a) Use integration to solve the differential equation.

(b) Use a graphing utility to graph the solution of the differential equation.

(c) *Euler's Method*   From the definition of the derivative it follows that for "small" Δx

$$f'(x) \approx \frac{f(x + \Delta x) - f(x)}{\Delta x}$$

$$f(x + \Delta x) \approx f(x) + [f'(x)] \, \Delta x.$$

Consider points of the form

$$(x_n, y_n) = (n \, \Delta x, y_{n-1} + f'(x_{n-1}) \, \Delta x)$$

where $(x_0, y_0) = (0, 0)$. Starting with $n = 0$, use the recursive capabilities of a graphing utility to generate 80 points of this form when $\Delta x = 0.05$. Use the graphing utility to plot the points and compare the result with the graph in part (b).

(d) Repeat part (c) by generating 40 points when $\Delta x = 0.1$.

(e) Give a geometric explanation of the process described in part (c). Why do you think the result in part (c) is a better approximation of the solution than the result in part (d)?

87. *Euler's Method*   Consider the differential equation

$$f'(x) = \cos \sqrt{x}$$

with the initial condition $f(0) = 2$.

(a) Try solving the differential equation by integration. Can you perform the integration?

(b) Starting with $n = 0$, use the recursive capabilities of a graphing utility to generate 80 points of the form shown in part (c) of Exercise 86 when $\Delta x = 0.05$. Plot the points for an approximation of the graph of the solution of the differential equation.

SECTION PROJECT

Rocket Velocity   The velocity (in feet per second) of a rocket whose initial mass is m (including fuel) is

$$v = gt + u \ln \frac{m}{m - rt}, \quad t < \frac{m}{r}$$

where u is the expulsion speed of the fuel, r is the rate at which the fuel is consumed, and $g = -32$ feet per second per second is the acceleration due to gravity. Find the position equation for a rocket for which $m = 50{,}000$ pounds, $u = 12{,}000$ feet per second, and $r = 400$ pounds per second. What is the height of the rocket when $t = 100$ seconds? (Assume that the rocket was fired from ground level and is moving straight up.)

NASA

The space shuttle *Columbia,* after four test flights, made its first operational flight on November 16, 1982. It stayed in orbit 5 days. The second space shuttle, the *Challenger,* made its maiden flight in April 1983. The *Atlantis* was launched on October 3, 1985. On September 29, 1988, the *Discovery* was launched, returning 4 days later to Edwards Air Force Base in California. On May 7, 1992, the fifth space shuttle, the *Endeavor,* was launched.

SECTION **7.3** **Trigonometric Integrals**

Integrals Involving Powers of Sine and Cosine •
Integrals Involving Powers of Secant and Tangent •
Integrals Involving Sine-Cosine Products with Different Angles

Integrals Involving Powers of Sine and Cosine

SHEILA SCOTT MACINTYRE (1910–1960)

Sheila Scott Macintyre published her first paper on the asymptotic periods of integral functions in 1935. She completed her doctorate work at Aberdeen University where she taught. In 1958 she accepted a visiting research fellowship at the University of Cincinnati.

In this section you will study techniques for evaluating integrals of the form

$$\int \sin^m x \cos^n x \, dx \qquad \text{and} \qquad \int \sec^m x \tan^n x \, dx$$

where either m or n is a positive integer. To find antiderivatives for these forms, try to break them into combinations of trigonometric integrals to which you can apply the Power Rule.

For instance, you can evaluate $\int \sin^5 x \cos x \, dx$ with the Power Rule by letting $u = \sin x$. Then, $du = \cos x \, dx$ and you have

$$\int \sin^5 x \cos x \, dx = \int u^5 \, du = \frac{u^6}{6} + C = \frac{\sin^6 x}{6} + C.$$

To break up $\int \sin^m x \cos^n x \, dx$ into forms to which you can apply the Power Rule, use the following identities.

$$\sin^2 x + \cos^2 x = 1 \qquad\qquad \text{Pythagorean identity}$$

$$\sin^2 x = \frac{1 - \cos 2x}{2} \qquad\qquad \text{Half-angle identity for } \sin^2 x$$

$$\cos^2 x = \frac{1 + \cos 2x}{2} \qquad\qquad \text{Half-angle identity for } \cos^2 x$$

Guidelines for Evaluating Integrals Involving Sine and Cosine

1. If the power of the sine is odd and positive, save one sine factor and convert the remaining factors to cosine. Then, expand and integrate.

$$\int \underbrace{\sin^{2k+1} x}_{\text{Odd}} \cos^n x \, dx = \int \underbrace{(\sin^2 x)^k}_{\text{Convert to cosine}} \cos^n x \underbrace{\sin x \, dx}_{\text{Save for } du} = \int (1 - \cos^2 x)^k \cos^n x \sin x \, dx$$

2. If the power of the cosine is odd and positive, save one cosine factor and convert the remaining factors to sine. Then, expand and integrate.

$$\int \sin^m x \underbrace{\cos^{2k+1} x}_{\text{Odd}} \, dx = \int \sin^m x \underbrace{(\cos^2 x)^k}_{\text{Convert to sine}} \underbrace{\cos x \, dx}_{\text{Save for } du} = \int \sin^m x \, (1 - \sin^2 x)^k \cos x \, dx$$

3. If the powers of *both* the sine and cosine are even and nonnegative, make repeated use of the identities

$$\sin^2 x = \frac{1 - \cos 2x}{2} \qquad \text{and} \qquad \cos^2 x = \frac{1 + \cos 2x}{2}$$

to convert the integrand to odd powers of the cosine. Then proceed as in guideline 2.

EXAMPLE 1   Power of Sine is Odd and Positive

Evaluate $\int \sin^3 x \cos^4 x \, dx$.

TECHNOLOGY Try using a symbolic integration utility to evaluate the integral in Example 1. When we did this, we obtained

$$\int \sin^3 x \cos^4 x \, dx =$$

$$-\cos^5 x \left(\frac{1}{7} \sin^2 x + \frac{2}{35} \right) + C.$$

Is this equivalent to the result obtained in Example 1?

Solution Because you expect to use the Power Rule with $u = \cos x$, *save one sine factor to form du* and convert the remaining sine factors to cosines.

$$\int \sin^3 x \cos^4 x \, dx = \int \sin^2 x \cos^4 x \, (\sin x) \, dx$$

$$= \int (1 - \cos^2 x)\cos^4 x \sin x \, dx$$

$$= \int (\cos^4 x - \cos^6 x)\sin x \, dx$$

$$= \int \cos^4 x \sin x \, dx - \int \cos^6 x \sin x \, dx$$

$$= -\int \cos^4 x(-\sin x) \, dx + \int \cos^6 x(-\sin x) \, dx$$

$$= -\frac{\cos^5 x}{5} + \frac{\cos^7 x}{7} + C$$

In Example 1, *both* of the powers m and n happened to be positive integers. However, the same strategy will work as long as either m or n is odd and positive. For instance, in the next example the power of the cosine is 3, but the power of the sine is $-\frac{1}{2}$.

EXAMPLE 2   Power of Cosine is Odd and Positive

Evaluate $\displaystyle\int_{\pi/6}^{\pi/3} \frac{\cos^3 x}{\sqrt{\sin x}} \, dx$.

Solution

$$\int_{\pi/6}^{\pi/3} \frac{\cos^3 x}{\sqrt{\sin x}} \, dx = \int_{\pi/6}^{\pi/3} \frac{(1 - \sin^2 x)(\cos x)}{\sqrt{\sin x}} \, dx$$

$$= \int_{\pi/6}^{\pi/3} [(\sin x)^{-1/2} \cos x - (\sin x)^{3/2} \cos x] \, dx$$

$$= \left[\frac{(\sin x)^{1/2}}{1/2} - \frac{(\sin x)^{5/2}}{5/2} \right]_{\pi/6}^{\pi/3}$$

$$= 2\left(\frac{\sqrt{3}}{2} \right)^{1/2} - \frac{2}{5}\left(\frac{\sqrt{3}}{2} \right)^{5/2} - \sqrt{2} + \frac{\sqrt{32}}{80}$$

$$\approx 0.239$$

Figure 7.4 shows the region whose area is represented by this integral.

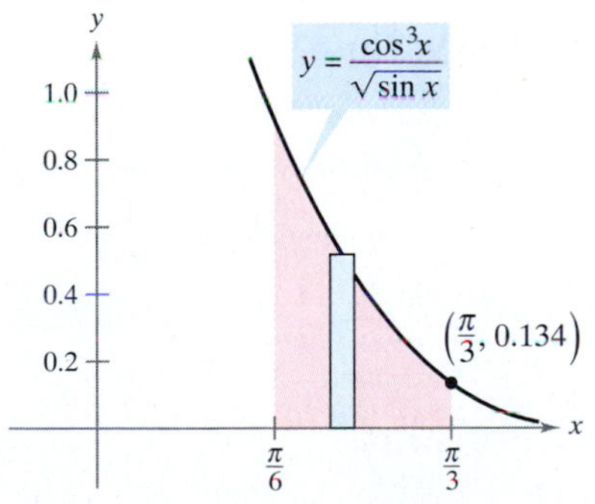

The area of the region is approximately 0.239.

Figure 7.4

JOHN WALLIS (1616–1703)

Wallis did much of his work in calculus prior to Newton and Leibniz, and he influenced the thinking of both of these men. Wallis is also credited with introducing the present symbol (∞) for infinity.

EXAMPLE 3   Power of Cosine is Even and Nonnegative

Evaluate $\displaystyle\int \cos^4 x \, dx$.

Solution   Because m and n are both even and nonnegative ($m = 0$), you can replace $\cos^4 x$ by $[(1 + \cos 2x)/2]^2$.

$$\int \cos^4 x \, dx = \int \left(\frac{1 + \cos 2x}{2}\right)^2 dx$$

$$= \int \left(\frac{1}{4} + \frac{\cos 2x}{2} + \frac{\cos^2 2x}{4}\right) dx$$

$$= \int \left[\frac{1}{4} + \frac{\cos 2x}{2} + \frac{1}{4}\left(\frac{1 + \cos 4x}{2}\right)\right] dx$$

$$= \frac{3}{8}\int dx + \frac{1}{4}\int 2 \cos 2x \, dx + \frac{1}{32}\int 4 \cos 4x \, dx$$

$$= \frac{3x}{8} + \frac{\sin 2x}{4} + \frac{\sin 4x}{32} + C$$

Try using a symbolic differentiation utility to verify this. Can you simplify the derivative to obtain the original integrand?

In Example 3, suppose that you were to evaluate the *definite* integral from 0 to $\pi/2$. You would obtain

$$\int_0^{\pi/2} \cos^4 x \, dx = \left[\frac{3x}{8} + \frac{\sin 2x}{4} + \frac{\sin 4x}{32}\right]_0^{\pi/2}$$

$$= \left(\frac{3\pi}{16} + 0 + 0\right) - (0 + 0 + 0)$$

$$= \frac{3\pi}{16}.$$

Note that the only term that contributes to the solution is $3x/8$. This observation is generalized in the following formulas developed by John Wallis.

Wallis's Formulas

1. If n is odd ($n \geq 3$), then

$$\int_0^{\pi/2} \cos^n x \, dx = \left(\frac{2}{3}\right)\left(\frac{4}{5}\right)\left(\frac{6}{7}\right) \cdots \left(\frac{n-1}{n}\right).$$

2. If n is even ($n \geq 2$), then

$$\int_0^{\pi/2} \cos^n x \, dx = \left(\frac{1}{2}\right)\left(\frac{3}{4}\right)\left(\frac{5}{6}\right) \cdots \left(\frac{n-1}{n}\right)\left(\frac{\pi}{2}\right).$$

These formulas are also valid if $\cos^n x$ is replaced by $\sin^n x$. (You are asked to prove both formulas in Exercise 85.)

Integrals Involving Powers of Secant and Tangent

The following guidelines can help you evaluate integrals of the form $\int \sec^m x \tan^n x\, dx$.

Guidelines for Evaluating Integrals Involving Secant and Tangent

1. If the power of the secant is even and positive, save a secant-squared factor and convert the remaining factors to tangents. Then expand and integrate.

$$\underbrace{\int \sec^{2k} x}_{\text{Even}} \tan^n x\, dx = \int (\sec^2 x)^{k-1} \tan^n x \underbrace{\sec^2 x\, dx}_{\substack{\text{Convert to tangents} \quad \text{Save for } du}} = \int (1 + \tan^2 x)^{k-1} \tan^n x \sec^2 x\, dx$$

2. If the power of the tangent is odd and positive, save a secant-tangent factor and convert the remaining factors to secants. Then expand and integrate.

$$\int \sec^m x \underbrace{\tan^{2k+1} x}_{\text{Odd}}\, dx = \int \sec^{m-1} x \underbrace{(\tan^2 x)^k}_{\text{Convert to secants}} \underbrace{\sec x \tan x\, dx}_{\text{Save for } du} = \int \sec^{m-1} x (\sec^2 x - 1)^k \sec x \tan x\, dx$$

3. If there are no secant factors and the power of the tangent is even and positive, convert a tangent-squared factor to a secant-squared factor, then expand and repeat if necessary.

$$\int \tan^n x\, dx = \int \tan^{n-2} x \underbrace{(\tan^2 x)}_{\text{Convert to secants}}\, dx = \int \tan^{n-2} x (\sec^2 x - 1)\, dx$$

4. If the integral is of the form $\int \sec^m x\, dx$, where m is odd and positive, use integration by parts, as illustrated in Example 5 in the previous section.

5. If none of the first four guidelines applies, try converting to sines and cosines.

EXAMPLE 4 Power of Tangent is Odd and Positive

Evaluate $\displaystyle\int \frac{\tan^3 x}{\sqrt{\sec x}}\, dx$.

Solution

$$\int \frac{\tan^3 x}{\sqrt{\sec x}}\, dx = \int (\sec x)^{-1/2} \tan^3 x\, dx$$

$$= \int (\sec x)^{-3/2} (\tan^2 x)(\sec x \tan x)\, dx$$

$$= \int (\sec x)^{-3/2} (\sec^2 x - 1)(\sec x \tan x)\, dx$$

$$= \int [(\sec x)^{1/2} - (\sec x)^{-3/2}](\sec x \tan x)\, dx$$

$$= \frac{2}{3}(\sec x)^{3/2} + 2(\sec x)^{-1/2} + C$$

NOTE   In Example 5, the power of the tangent is odd and positive. Thus, you could also evaluate the integral with the procedure described in guideline 2 on page 493. In Exercise 71, you are asked to show that the results obtained by these two procedures differ only by a constant.

EXAMPLE 5   Power of Secant is Even and Positive

Evaluate $\int \sec^4 3x \tan^3 3x \, dx$.

Solution   Let $u = \tan 3x$, then $du = 3 \sec^2 3x \, dx$ and you can write

$$\int \sec^4 3x \tan^3 3x \, dx = \int \sec^2 3x \tan^3 3x (\sec^2 3x) \, dx$$

$$= \int (1 + \tan^2 3x) \tan^3 3x (\sec^2 3x) \, dx$$

$$= \frac{1}{3} \int (\tan^3 3x + \tan^5 3x)(3 \sec^2 3x) \, dx$$

$$= \frac{1}{3}\left(\frac{\tan^4 3x}{4} + \frac{\tan^6 3x}{6} \right) + C$$

$$= \frac{\tan^4 3x}{12} + \frac{\tan^6 3x}{18} + C.$$

EXAMPLE 6   Power of Tangent is Even

Evaluate $\int_0^{\pi/4} \tan^4 x \, dx$.

Solution   Because there are no secant factors, you can begin by converting a tangent-squared factor to a secant-squared factor.

$$\int \tan^4 x \, dx = \int \tan^2 x (\tan^2 x) \, dx$$

$$= \int \tan^2 x (\sec^2 x - 1) \, dx$$

$$= \int \tan^2 x \sec^2 x \, dx - \int \tan^2 x \, dx$$

$$= \int \tan^2 x \sec^2 x \, dx - \int (\sec^2 x - 1) \, dx$$

$$= \frac{\tan^3 x}{3} - \tan x + x + C$$

You can evaluate the definite integral as follows.

$$\int_0^{\pi/4} \tan^4 x \, dx = \left[\frac{\tan^3 x}{3} - \tan x + x \right]_0^{\pi/4}$$

$$= \frac{\pi}{4} - \frac{2}{3}$$

$$\approx 0.119$$

The area represented by the definite integral is shown in Figure 7.5. Try using Simpson's Rule to approximate this integral. With $n = 10$, you should obtain an approximation that is within 0.00001 of the actual value.

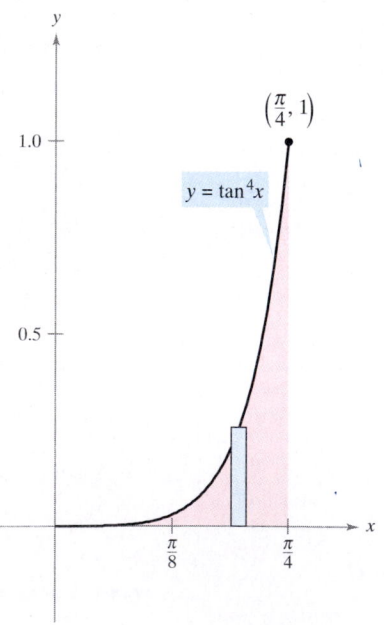

$y = \tan^4 x$

$\left(\frac{\pi}{4}, 1 \right)$

The area of the region is approximately 0.119.

Figure 7.5

For integrals involving powers of cotangents and cosecants, you can follow a strategy similar to that used for powers of tangents and secants. Also, when integrating trigonometric functions, remember that it sometimes helps to convert the entire integrand to powers of sines and cosines.

EXAMPLE 7   Converting to Sines and Cosines

Evaluate $\displaystyle\int \frac{\sec x}{\tan^2 x}\, dx$.

Solution Because the first four guidelines on page 493 do not apply, try converting the integrand to sines and cosines. In this case, you are able to integrate the resulting powers of sine and cosine as follows.

$$\int \frac{\sec x}{\tan^2 x}\, dx = \int \left(\frac{1}{\cos x}\right)\left(\frac{\cos x}{\sin x}\right)^2 dx$$

$$= \int (\sin x)^{-2}(\cos x)\, dx$$

$$= -(\sin x)^{-1} + C$$

$$= -\csc x + C$$

Integrals Involving Sine-Cosine Products with Different Angles

FOR FURTHER INFORMATION For more information on integrals involving sine-cosine products with different angles, see the article "Integrals of Products of Sine and Cosine with Different Arguments" by Sherrie J. Nicol in the March 1993 issue of *The College Mathematics Journal*.

Integrals involving the products of sines and cosines of two *different* angles occur in many applications. In such instances you can use the following product-to-sum identities.

$$\sin mx \sin nx = \frac{1}{2}(\cos[(m-n)x] - \cos[(m+n)x])$$

$$\sin mx \cos nx = \frac{1}{2}(\sin[(m-n)x] + \sin[(m+n)x])$$

$$\cos mx \cos nx = \frac{1}{2}(\cos[m-n)x] + \cos[(m+n)x])$$

EXAMPLE 8   Using Product-to-Sum Identities

Evaluate $\int \sin 5x \cos 4x\, dx$.

Solution Considering the second product-to-sum identity above, you can write the following.

$$\int \sin 5x \cos 4x\, dx = \frac{1}{2}\int (\sin x + \sin 9x)\, dx$$

$$= \frac{1}{2}\left(-\cos x - \frac{\cos 9x}{9}\right) + C$$

$$= -\frac{\cos x}{2} - \frac{\cos 9x}{18} + C$$

EXERCISES FOR SECTION 7.3

1. Consider the function $f(x) = \sin^4 x + \cos^4 x$.

    (a) Use the power-reducing formulas to write the function in terms of the first power of the cosine.

    (b) Determine another way of rewriting the function. Use a graphing utility to verify your result.

    (c) Determine a trigonometric expression to add to the function so that it becomes a perfect square trinomial. Rewrite the function as a perfect square trinomial minus the term that you added. Use a graphing utility to verify your result.

    (d) Rewrite the result in part (c) in terms of the sine of a double angle. Use a graphing utility to verify your result.

    (e) In how many ways have you rewritten the trigonometric function? When rewriting a trigonometric expression, your result may not be the same as another person's result. Does this mean that one of you is wrong? Explain.

2. Match the antiderivative in the left column with the correct integral in the right column.

    (a) $y = \sec x$
    (i) $\displaystyle\int \sin x \tan^2 x \, dx$

    (b) $y = \cos x + \sec x$
    (ii) $8\displaystyle\int \cos^4 x \, dx$

    (c) $y = x - \tan x + \frac{1}{3}\tan^3 x$
    (iii) $\displaystyle\int \sin x \sec^2 x \, dx$

    (d) $y = 3x + 2\sin x \cos^3 x + 3\sin x \cos x$
    (iv) $\displaystyle\int \tan^4 x \, dx$

In Exercises 3–12, evaluate the integral involving sine and cosine.

3. $\displaystyle\int \cos^3 x \sin x \, dx$ 4. $\displaystyle\int \cos^3 x \sin^2 x \, dx$

5. $\displaystyle\int \sin^5 2x \cos 2x \, dx$ 6. $\displaystyle\int \sin^3 x \, dx$

7. $\displaystyle\int \sin^5 x \cos^2 x \, dx$ 8. $\displaystyle\int \cos^3 \frac{x}{3} \, dx$

9. $\displaystyle\int \cos^2 3x \, dx$ 10. $\displaystyle\int \sin^2 2x \, dx$

11. $\displaystyle\int x \sin^2 x \, dx$ 12. $\displaystyle\int x^2 \sin^2 x \, dx$

In Exercises 13–16, verify Wallis's Formulas by evaluating the integral.

13. $\displaystyle\int_0^{\pi/2} \cos^3 x \, dx = \frac{2}{3}$ 14. $\displaystyle\int_0^{\pi/2} \cos^5 x \, dx = \frac{8}{15}$

15. $\displaystyle\int_0^{\pi/2} \cos^7 x \, dx = \frac{16}{35}$ 16. $\displaystyle\int_0^{\pi/2} \sin^2 x \, dx = \frac{\pi}{4}$

In Exercises 17–32, evaluate the integral involving secant and tangent.

17. $\displaystyle\int \sec 3x \, dx$ 18. $\displaystyle\int \sec^2 (2x - 1) \, dx$

19. $\displaystyle\int \sec^4 5x \, dx$ 20. $\displaystyle\int \sec^6 \frac{x}{2} \, dx$

21. $\displaystyle\int \sec^3 \pi x \, dx$ 22. $\displaystyle\int \tan^2 x \, dx$

23. $\displaystyle\int \tan^5 \frac{x}{4} \, dx$ 24. $\displaystyle\int \tan^3 \frac{\pi x}{2} \sec^2 \frac{\pi x}{2} \, dx$

25. $\displaystyle\int \sec^2 x \tan x \, dx$ 26. $\displaystyle\int \tan^3 t \sec^3 t \, dt$

27. $\displaystyle\int \tan^2 x \sec^2 x \, dx$ 28. $\displaystyle\int \tan^5 2x \sec^2 2x \, dx$

29. $\displaystyle\int \sec^6 4x \tan 4x \, dx$ 30. $\displaystyle\int \sec^2 \frac{x}{2} \tan \frac{x}{2} \, dx$

31. $\displaystyle\int \sec^3 x \tan x \, dx$ 32. $\displaystyle\int \tan^3 3x \, dx$

In Exercises 33–36, solve the differential equation.

33. $\dfrac{dr}{d\theta} = \sin^4 \pi\theta$ 34. $\dfrac{ds}{d\alpha} = \sin^2 \dfrac{\alpha}{2} \cos^2 \dfrac{\alpha}{2}$

35. $y' = \tan^3 3x \sec 3x$ 36. $y' = \sqrt{\tan x} \sec^4 x$

Direction Fields **In Exercises 37 and 38, a differential equation, a point, and a direction field are given. (a) Sketch two approximate solutions of the differential equation on the direction field, one of which passes through the indicated point. (b) Use integration to find the particular solution of the differential equation and use a graphing utility to graph the solution. Compare the result with the sketches in part (a).**

37. $\dfrac{dy}{dx} = \sin^2 x$, $(0, 0)$ 38. $\dfrac{dy}{dx} = \sec^2 x \tan^2 x$, $\left(0, -\dfrac{1}{4}\right)$

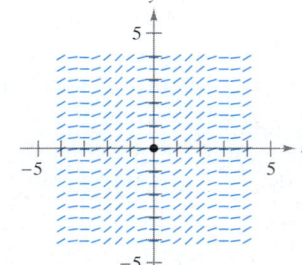

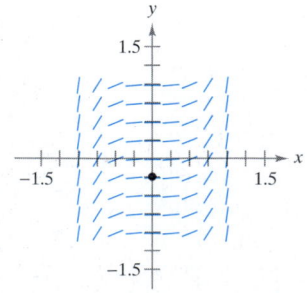

In Exercises 39–42, evaluate the integral.

39. $\displaystyle\int \sin 3x \cos 2x \, dx$

40. $\displaystyle\int \cos 3\theta \cos(-2\theta) \, d\theta$

41. $\displaystyle\int \sin\theta \sin 3\theta \, d\theta$

42. $\displaystyle\int \sin(-4x) \cos 3x \, dx$

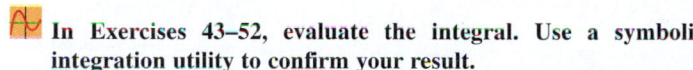 **In Exercises 43–52, evaluate the integral. Use a symbolic integration utility to confirm your result.**

43. $\displaystyle\int \cot^3 2x \, dx$

44. $\displaystyle\int \tan^4 \frac{x}{2} \sec^4 \frac{x}{2} \, dx$

45. $\displaystyle\int \csc^4 \theta \, d\theta$

46. $\displaystyle\int \csc^2 3x \cot 3x \, dx$

47. $\displaystyle\int \frac{\cot^2 t}{\csc t} \, dt$

48. $\displaystyle\int \frac{\cot^3 t}{\csc t} \, dt$

49. $\displaystyle\int \frac{1}{\sec x \tan x} \, dx$

50. $\displaystyle\int \frac{\sin^2 x - \cos^2 x}{\cos x} \, dx$

51. $\displaystyle\int (\tan^4 t - \sec^4 t) \, dt$

52. $\displaystyle\int \frac{1 - \sec t}{\cos t - 1} \, dt$

In Exercises 53–60, evaluate the definite integral. Use a graphing utility to confirm your result.

53. $\displaystyle\int_{-\pi}^{\pi} \sin^2 x \, dx$

54. $\displaystyle\int_0^{\pi/4} \tan^2 x \, dx$

55. $\displaystyle\int_0^{\pi/4} \tan^3 x \, dx$

56. $\displaystyle\int_0^{\pi/4} \sec^2 t \sqrt{\tan t} \, dt$

57. $\displaystyle\int_0^{\pi/2} \frac{\cos t}{1 + \sin t} \, dt$

58. $\displaystyle\int_{-\pi}^{\pi} \sin 3\theta \cos\theta \, d\theta$

59. $\displaystyle\int_{-\pi/2}^{\pi/2} \cos^3 x \, dx$

60. $\displaystyle\int_{-\pi/2}^{\pi/2} (\sin^2 x + 1) \, dx$

In Exercises 61–66, use a symbolic integration utility to evaluate the integral. Graph the antiderivatives for two different values of the constant of integration.

61. $\displaystyle\int \cos^4 \frac{x}{2} \, dx$

62. $\displaystyle\int \sin^2 x \cos^2 x \, dx$

63. $\displaystyle\int \sec^5 \pi x \, dx$

64. $\displaystyle\int \tan^3(1 - x) \, dx$

65. $\displaystyle\int \sec^5 \pi x \tan \pi x \, dx$

66. $\displaystyle\int \sec^4(1 - x) \tan (1 - x) \, dx$

In Exercises 67–70, use a symbolic integration utility to evaluate the integral.

67. $\displaystyle\int_0^{\pi/4} \sin 2\theta \sin 3\theta \, d\theta$

68. $\displaystyle\int_0^{\pi/2} (1 - \cos \theta)^2 \, d\theta$

69. $\displaystyle\int_0^{\pi/2} \sin^4 x \, dx$

70. $\displaystyle\int_0^{\pi/2} \sin^6 x \, dx$

In Exercises 71 and 72, (a) find the indefinite integral in two different ways. (b) Use a graphing utility to graph the antiderivative (without the constant of integration) obtained by each method to show that the results differ only by a constant. (c) Verify analytically that the results differ only by a constant.

71. $\displaystyle\int \sec^4 3x \tan^3 3x \, dx$

72. $\displaystyle\int \sec^2 x \tan x \, dx$

73. *Area*   Find the area of the region bounded by the graphs of the equations $y = \sin^2 \pi x$, $y = 0$, $x = 0$, and $x = 1$.

74. *Volume*   Find the volume of the solid generated by revolving the region bounded by the graphs of the equations $y = \tan x$, $y = 0$, $x = -\pi/4$, and $x = \pi/4$ about the x-axis.

Volume and Centroid   In Exercises 75 and 76, for the region bounded by the graphs of the equations, find (a) the volume of the solid formed by revolving the region about the x-axis and (b) the centroid of the region.

75. $y = \sin x$, $y = 0$, $x = 0$, $x = \pi$

76. $y = \cos x$, $y = 0$, $x = 0$, $x = \dfrac{\pi}{2}$

In Exercises 77–80, use integration by parts to verify the reduction formula.

77. $\displaystyle\int \sin^n x \, dx = -\frac{\sin^{n-1} x \cos x}{n} + \frac{n-1}{n} \int \sin^{n-2} x \, dx$

78. $\displaystyle\int \cos^n x \, dx = \frac{\cos^{n-1} x \sin x}{n} + \frac{n-1}{n} \int \cos^{n-2} x \, dx$

79. $\displaystyle\int \cos^m x \sin^n x \, dx = -\frac{\cos^{m+1} x \sin^{n-1} x}{m+n} +$
$\dfrac{n-1}{m+n} \displaystyle\int \cos^m x \sin^{n-2} x \, dx$

80. $\displaystyle\int \sec^n x \, dx = \frac{1}{n-1} \sec^{n-2} x \tan x + \frac{n-2}{n-1} \int \sec^{n-2} x \, dx$

In Exercises 81–84, use the results of Exercises 77–80, to evaluate the integral.

81. $\displaystyle\int \sin^5 x \, dx$

82. $\displaystyle\int \cos^4 x \, dx$

83. $\displaystyle\int \sec^4 \frac{2\pi x}{5} \, dx$

84. $\displaystyle\int \sin^4 x \cos^2 x \, dx$

85. *Wallis's Formulas*   Use the result of Exercise 78 to prove the following versions of Wallis's Formulas.

(a) If n is odd $(n \geq 3)$, then

$$\int_0^{\pi/2} \cos^n x \, dx = \left(\frac{2}{3}\right)\left(\frac{4}{5}\right)\left(\frac{6}{7}\right) \cdots \left(\frac{n-1}{n}\right).$$

(b) If n is even $(n \geq 2)$, then

$$\int_0^{\pi/2} \cos^n x \, dx = \left(\frac{1}{2}\right)\left(\frac{3}{4}\right)\left(\frac{5}{6}\right) \cdots \left(\frac{n-1}{n}\right)\left(\frac{\pi}{2}\right).$$

86. The **inner product** of two functions f and g on $[a, b]$ is given by $\langle f, g \rangle = \int_a^b f(x)g(x)\,dx$. Two distinct functions f and g are said to be **orthogonal** if $\langle f, g \rangle = 0$. Show that the following set of functions is orthogonal on $[-\pi, \pi]$.

$$\{\sin x, \sin 2x, \sin 3x, \ldots, \cos x, \cos 2x, \cos 3x, \ldots\}$$

87. *Modeling Data* The table gives the normal maximum (high) and minimum (low) temperatures for Erie, Pennsylvania for each month of a year. *(Source: NOAA)*

Month	Jan	Feb	Mar	Apr	May	Jun
Max	30.9	32.2	41.1	53.7	64.6	74.0
Min	18.0	17.7	25.8	36.1	45.4	55.2

Month	Jul	Aug	Sep	Oct	Nov	Dec
Max	78.2	77.0	71.0	60.1	47.1	35.7
Min	59.9	59.4	53.1	43.2	34.3	24.2

The maximum and minimum temperatures can be modeled by

$$f(t) = a_0 + a_1 \cos \frac{\pi t}{6} + b_1 \sin \frac{\pi t}{6}$$

where a_0, a_1, and b_1 are as follows.

$$a_0 = \frac{1}{12} \int_0^{12} f(t)\,dt$$

$$a_1 = \frac{1}{6} \int_0^{12} f(t) \cos \frac{\pi t}{6}\,dt$$

$$b_1 = \frac{1}{6} \int_0^{12} f(t) \sin \frac{\pi t}{6}\,dt$$

(a) Approximate the model $H(t)$ for the maximum temperatures. Let $t = 0$ correspond to January. (*Hint:* Use Simpson's Rule to approximate the integrals and use the January data twice.)

(b) Repeat part (a) for a model $L(t)$ for the minimum temperature data.

(c) Use a graphing utility to compare each model with the actual data. During what part of the year is the difference between the maximum and minimum temperatures greatest?

88. Use a trigonometric substitution and Wallis's Formulas to show that

$$\int_{-1}^1 (1 - x^2)^n\,dx = \frac{2^{2n+1}(n!)^2}{(2n + 1)!}$$

where n is a positive integer.

Power Lines Power lines are constructed by stringing wire between supports and adjusting the tension on each span. The wire hangs between supports in the shape of a catenary, as shown in the figure.

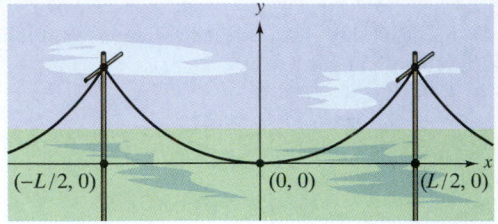

Let T be the tension (in pounds) on a span of wire, let u be the density (in pounds per foot), let $g \approx 32.2$ be the acceleration due to gravity (in feet per second per second), and let L be the distance (in feet) between the supports. Then the equation of the catenary is

$$y = \frac{T}{ug}\left(\cosh \frac{ugx}{T} - 1\right)$$

where x and y are measured in feet.

(a) Find the length of the wire between two spans.

(b) To measure the tension in a span, power line workers use the *return wave method.* The wire is struck at one support, creating a wave in the line, and the time t (in seconds) it takes for the wave to make a round trip is measured. The velocity v (in feet per second) is given by $v = \sqrt{T/u}$. How long does it take the wave to make a round trip between supports?

(c) The sag s (in inches) can be obtained by evaluating y when $x = L/2$ in the equation for the catenary (and multiplying by 12). In practice, however, power line workers use the "lineman's equation" given by $s \approx 12.075t^2$. Use the fact that $[\cosh(ugL/2T) + 1] \approx 2$ to derive this equation.

FOR FURTHER INFORMATION For more information on the mathematics of power lines, see the article "Constructing Power Lines" by Thomas O'Neil in the *UMAP Journal,* Volume 4, Number 3, 1983.

Trigonometric Substitution • Applications

Trigonometric Substitution

Now that you can evaluate integrals involving powers of trigonometric functions, you can use **trigonometric substitution** to evaluate integrals involving the radicals

$$\sqrt{a^2 - u^2}, \qquad \sqrt{a^2 + u^2}, \qquad \text{and} \qquad \sqrt{u^2 - a^2}.$$

The objective with trigonometric substitution is to eliminate the radical in the integrand. You do this with the Pythagorean identities

$$\cos^2 \theta = 1 - \sin^2 \theta, \quad \sec^2 \theta = 1 + \tan^2 \theta, \quad \text{and} \quad \tan^2 \theta = \sec^2 \theta - 1.$$

For example, if $a > 0$, let $u = a \sin \theta$, where $-\pi/2 \leq \theta \leq \pi/2$. Then

$$\begin{aligned}
\sqrt{a^2 - u^2} &= \sqrt{a^2 - a^2 \sin^2 \theta} \\
&= \sqrt{a^2(1 - \sin^2 \theta)} \\
&= \sqrt{a^2 \cos^2 \theta} \\
&= a \cos \theta.
\end{aligned}$$

Note that $\cos \theta \geq 0$, because $-\pi/2 \leq \theta \leq \pi/2$.

EXPLORATION

Integrating a Radical Function
Up to this point in the text, you have not evaluated the following integral.

$$\int_{-1}^{1} \sqrt{1 - x^2} \, dx$$

From geometry, you should be able to find the exact value of this integral— what is it? Using numerical integration, with Simpson's Rule or the Trapezoidal Rule, you can't be sure of the accuracy of the approximation. Why?

Try finding the exact value using the substitution

$$x = \sin \theta \quad \text{and} \quad dx = \cos \theta \, d\theta.$$

Does your answer agree with the value you obtained using geometry?

Trigonometric Substitution ($a > 0$)

1. For integrals involving $\sqrt{a^2 - u^2}$, let

    $$u = a \sin \theta.$$

    Then $\sqrt{a^2 - u^2} = a \cos \theta$, where $-\pi/2 \leq \theta \leq \pi/2$.

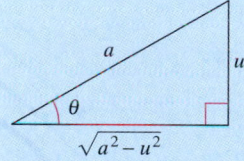

2. For integrals involving $\sqrt{a^2 + u^2}$, let

    $$u = a \tan \theta.$$

    Then $\sqrt{a^2 + u^2} = a \sec \theta$, where $-\pi/2 < \theta < \pi/2$.

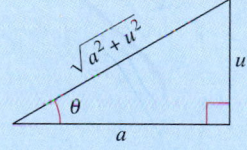

3. For integrals involving $\sqrt{u^2 - a^2}$, let

    $$u = a \sec \theta.$$

    Then $\sqrt{u^2 - a^2} = \pm a \tan \theta$, where $0 \leq \theta < \pi/2$ or $\pi/2 < \theta \leq \pi$. Use the positive value if $u > a$ and the negative value if $u < -a$.

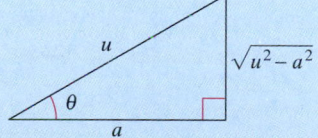

NOTE   The restrictions on θ ensure that the function that defines the substitution is one-to-one. In fact, these are the same intervals over which the arcsine, arctangent, and arcsecant are defined.

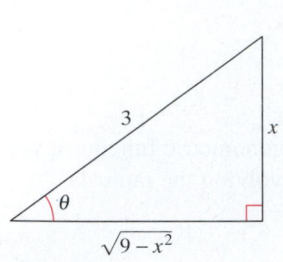

$\sin \theta = \dfrac{x}{3},\ \cot \theta = \dfrac{\sqrt{9-x^2}}{x}$

Figure 7.6

EXAMPLE 1 Trigonometric Substitution: $u = a \sin \theta$

Evaluate $\displaystyle\int \frac{dx}{x^2\sqrt{9-x^2}}$.

Solution First, note that none of the basic integration rules in Section 5.9 applies. To use trigonometric substitution, you should observe that $\sqrt{9-x^2}$ is of the form $\sqrt{a^2-u^2}$. Hence, you can use the substitution

$$x = a \sin \theta = 3 \sin \theta.$$

Using differentiation and the triangle shown in Figure 7.6, you obtain

$$dx = 3 \cos \theta\, d\theta, \qquad \sqrt{9-x^2} = 3 \cos \theta, \qquad \text{and} \qquad x^2 = 9 \sin^2 \theta.$$

Therefore, trigonometric substitution yields the following.

$$\int \frac{dx}{x^2\sqrt{9-x^2}} = \int \frac{3 \cos \theta\, d\theta}{(9 \sin^2 \theta)(3 \cos \theta)} \qquad \text{Substitute.}$$

$$= \frac{1}{9}\int \frac{d\theta}{\sin^2 \theta} \qquad \text{Simplify.}$$

$$= \frac{1}{9}\int \csc^2 \theta\, d\theta \qquad \text{Trigonometric identity}$$

$$= -\frac{1}{9}\cot \theta + C \qquad \text{Apply Cosecant Rule.}$$

$$= -\frac{1}{9}\left(\frac{\sqrt{9-x^2}}{x}\right) + C \qquad \text{Substitute for } \cot \theta.$$

$$= -\frac{\sqrt{9-x^2}}{9x} + C$$

Note that the triangle in Figure 7.6 can be used to convert the θ's back to x's as follows.

$$\cot \theta = \frac{\text{adj.}}{\text{opp.}} = \frac{\sqrt{9-x^2}}{x}$$

TECHNOLOGY Try using a symbolic integration utility to integrate each of the following.

$$\int \frac{dx}{\sqrt{9-x^2}} \qquad \int \frac{dx}{x\sqrt{9-x^2}} \qquad \int \frac{dx}{x^2\sqrt{9-x^2}} \qquad \int \frac{dx}{x^3\sqrt{9-x^2}}$$

Then try using trigonometric substitution to duplicate the results obtained with the symbolic integration utility.

In Section 5.10, you saw how the inverse hyperbolic functions can be used to evaluate the integrals

$$\int \frac{du}{\sqrt{u^2 \pm a^2}}, \qquad \int \frac{du}{a^2 - u^2}, \qquad \text{and} \qquad \int \frac{du}{u\sqrt{a^2 \pm u^2}}.$$

You can also evaluate these integrals using trigonometric substitution. This is illustrated in the next example.

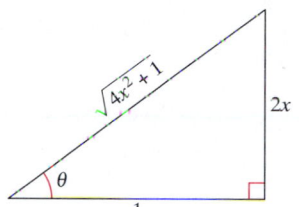

$\tan \theta = 2x, \sec \theta = \sqrt{4x^2 + 1}$
Figure 7.7

EXAMPLE 2   Trigonometric Substitution: $u = a \tan \theta$

Evaluate $\displaystyle\int \frac{dx}{\sqrt{4x^2 + 1}}$.

Solution   Let $u = 2x$, $a = 1$, and $2x = \tan \theta$, as shown in Figure 7.7. Then,

$$dx = \frac{1}{2} \sec^2 \theta \, d\theta \qquad \text{and} \qquad \sqrt{4x^2 + 1} = \sec \theta.$$

Trigonometric substitution produces the following.

$$\int \frac{1}{\sqrt{4x^2 + 1}} \, dx = \frac{1}{2} \int \frac{\sec^2 \theta \, d\theta}{\sec \theta} \qquad\qquad \text{Substitute.}$$

$$= \frac{1}{2} \int \sec \theta \, d\theta \qquad\qquad \text{Simplify.}$$

$$= \frac{1}{2} \ln |\sec \theta + \tan \theta| + C \qquad\qquad \text{Apply Secant Rule.}$$

$$= \frac{1}{2} \ln \left| \sqrt{4x^2 + 1} + 2x \right| + C \qquad\qquad \text{Back-substitute.}$$

Try checking this result with a symbolic integration utility. Is the result given in this form or in the form of an inverse hyperbolic function?

You can extend the use of trigonometric substitution to cover integrals involving expressions such as $(a^2 - u^2)^{n/2}$ by writing the expression as

$$(a^2 - u^2)^{n/2} = \left(\sqrt{a^2 - u^2} \right)^n.$$

EXAMPLE 3   Trigonometric Substitution: Rational Powers

Evaluate $\displaystyle\int \frac{dx}{(x^2 + 1)^{3/2}}$.

Solution   Begin by writing $(x^2 + 1)^{3/2}$ as $\left(\sqrt{x^2 + 1} \right)^3$. Then, let $a = 1$ and $u = x = \tan \theta$, as shown in Figure 7.8. Using

$$dx = \sec^2 \theta \, d\theta \qquad \text{and} \qquad \sqrt{x^2 + 1} = \sec \theta$$

you can apply trigonometric substitution as follows.

$$\int \frac{dx}{(x^2 + 1)^{3/2}} = \int \frac{dx}{\left(\sqrt{x^2 + 1} \right)^3} \qquad\qquad \text{Rewrite denominator.}$$

$$= \int \frac{\sec^2 \theta \, d\theta}{\sec^3 \theta} \qquad\qquad \text{Substitute.}$$

$$= \int \frac{d\theta}{\sec \theta} \qquad\qquad \text{Simplify.}$$

$$= \int \cos \theta \, d\theta \qquad\qquad \text{Trigonometric identity}$$

$$= \sin \theta + C \qquad\qquad \text{Apply Cosine Rule.}$$

$$= \frac{x}{\sqrt{x^2 + 1}} + C \qquad\qquad \text{Back-substitute.}$$

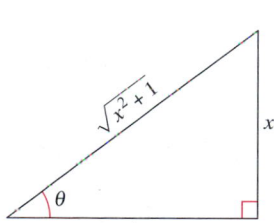

$\tan \theta = x, \sin \theta = \dfrac{x}{\sqrt{x^2 + 1}}$
Figure 7.8

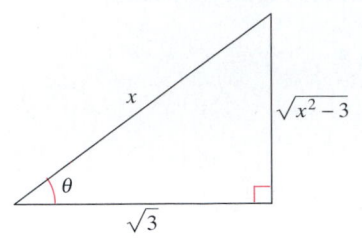

$$\sec \theta = \frac{x}{\sqrt{3}}, \tan \theta = \frac{\sqrt{x^2 - 3}}{\sqrt{3}}$$

Figure 7.9

For definite integrals, it is often convenient to determine the integration limits for θ that avoid converting back to x. You might want to review this procedure in Section 4.5, Examples 8 and 9.

EXAMPLE 4   Converting the Limits of Integration

Evaluate $\displaystyle\int_{\sqrt{3}}^{2} \frac{\sqrt{x^2 - 3}}{x}\, dx$.

Solution   Because $\sqrt{x^2 - 3}$ has the form $\sqrt{u^2 - a^2}$, you can consider $u = x$, $a = \sqrt{3}$, and let $x = \sqrt{3}\sec\theta$, as shown in Figure 7.9. Then,

$$dx = \sqrt{3}\sec\theta\tan\theta\, d\theta \qquad \text{and} \qquad \sqrt{x^2 - 3} = \sqrt{3}\tan\theta.$$

To determine the upper and lower limits of integration, use the substitution $x = \sqrt{3}\sec\theta$, as follows.

Lower Limit	Upper Limit
When $x = \sqrt{3}$, $\sec\theta = 1$ and $\theta = 0$.	When $x = 2$, $\sec\theta = \dfrac{2}{\sqrt{3}}$ and $\theta = \dfrac{\pi}{6}$.

Therefore, you have

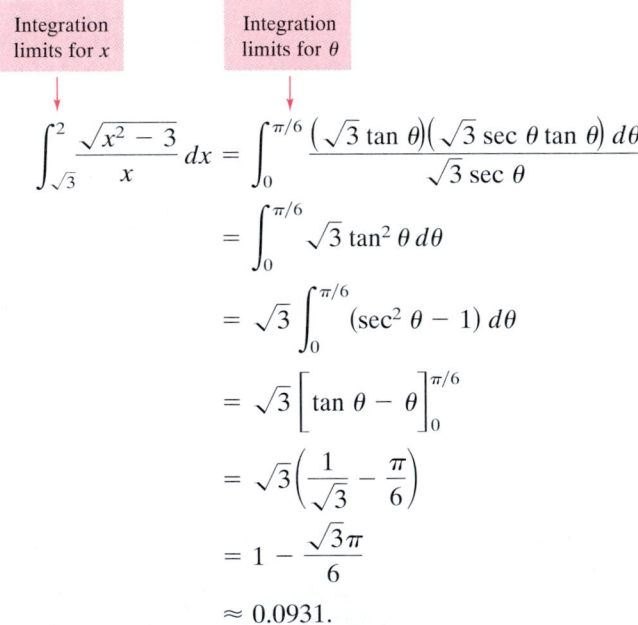

$$\int_{\sqrt{3}}^{2} \frac{\sqrt{x^2 - 3}}{x}\, dx = \int_{0}^{\pi/6} \frac{\left(\sqrt{3}\tan\theta\right)\left(\sqrt{3}\sec\theta\tan\theta\right)\, d\theta}{\sqrt{3}\sec\theta}$$

$$= \int_{0}^{\pi/6} \sqrt{3}\tan^2\theta\, d\theta$$

$$= \sqrt{3}\int_{0}^{\pi/6} (\sec^2\theta - 1)\, d\theta$$

$$= \sqrt{3}\left[\tan\theta - \theta\right]_{0}^{\pi/6}$$

$$= \sqrt{3}\left(\frac{1}{\sqrt{3}} - \frac{\pi}{6}\right)$$

$$= 1 - \frac{\sqrt{3}\pi}{6}$$

$$\approx 0.0931.$$

NOTE   In Example 4, try converting back to variable x and evaluating the antiderivative at the original limits of integration. You should obtain

$$\int_{\sqrt{3}}^{2} \frac{\sqrt{x^2 - 3}}{x}\, dx = \sqrt{3}\left[\frac{\sqrt{x^2 - 3}}{\sqrt{3}} - \text{arcsec}\frac{x}{\sqrt{3}}\right]_{\sqrt{3}}^{2}.$$

When using trigonometric substitution to evaluate definite integrals, you must be careful to check that the values of θ lie in the intervals discussed at the beginning of this section. For instance, if in Example 4 you had been asked to evaluate the definite integral

$$\int_{-2}^{-\sqrt{3}} \frac{\sqrt{x^2 - 3}}{x}\, dx$$

then using $u = x$ and $a = \sqrt{3}$ in the interval $\left[-2, -\sqrt{3}\right]$ would imply that $u < -a$. Thus, when determining the upper and lower limits of integration, you would have to choose θ such that $\pi/2 < \theta \le \pi$. In this case the integral would be evaluated as follows.

$$\int_{-2}^{-\sqrt{3}} \frac{\sqrt{x^2 - 3}}{x}\, dx = \int_{5\pi/6}^{\pi} \frac{\left(-\sqrt{3}\,\tan\theta\right)\left(\sqrt{3}\,\sec\theta\,\tan\theta\right) d\theta}{\sqrt{3}\,\sec\theta}$$

$$= \int_{5\pi/6}^{\pi} -\sqrt{3}\,\tan^2\theta\, d\theta$$

$$= -\sqrt{3} \int_{5\pi/6}^{\pi} (\sec^2\theta - 1)\, d\theta$$

$$= -\sqrt{3} \left[\tan\theta - \theta\right]_{5\pi/6}^{\pi}$$

$$= -\sqrt{3}\left[(0 - \pi) - \left(-\frac{1}{\sqrt{3}} - \frac{5\pi}{6}\right)\right]$$

$$= -1 + \frac{\sqrt{3}\,\pi}{6}$$

$$\approx -0.0931$$

Trigonometric substitution can be used with *completing the square* (see Section 5.9). For instance, try evaluating the following integral.

$$\int \sqrt{x^2 - 2x}\, dx$$

To begin, you could complete the square and write the integral as

$$\int \sqrt{(x - 1)^2 - 1^2}\, dx.$$

Trigonometric substitution can be used to evaluate the three integrals listed in the following theorem. These integrals will be encountered several times in the remainder of the text. When this happens, we will simply refer to this theorem. (In Exercise 73, you are asked to verify the formulas given in the theorem.)

THEOREM 7.2   Special Integration Formulas ($a > 0$)

1. $\displaystyle \int \sqrt{a^2 - u^2}\, du = \frac{1}{2}\left(a^2 \arcsin\frac{u}{a} + u\sqrt{a^2 - u^2}\right) + C$

2. $\displaystyle \int \sqrt{u^2 - a^2}\, du = \frac{1}{2}\left(u\sqrt{u^2 - a^2} - a^2 \ln\left|u + \sqrt{u^2 + a^2}\right|\right) + C,\, u > a$

3. $\displaystyle \int \sqrt{u^2 + a^2}\, du = \frac{1}{2}\left(u\sqrt{u^2 + a^2} + a^2 \ln\left|u + \sqrt{u^2 + a^2}\right|\right) + C$

Applications

EXAMPLE 5 Finding Arc Length

Find the arc length of the graph of $f(x) = \frac{1}{2}x^2$ from $x = 0$ to $x = 1$ (see Figure 7.10).

Solution   (Refer to the arc length formula in Section 6.4.)

$$s = \int_0^1 \sqrt{1 + [f'(x)]^2}\, dx \qquad \text{Formula for arc length}$$

$$= \int_0^1 \sqrt{1 + x^2}\, dx \qquad f'(x) = x$$

$$= \int_0^{\pi/4} \sec^3 \theta\, d\theta \qquad \text{Let } a = 1 \text{ and } x = \tan \theta.$$

$$= \frac{1}{2}\left[\sec \theta \tan \theta + \ln \left| \sec \theta + \tan \theta \right| \right]_0^{\pi/4} \qquad \text{Example 5, Section 7.2}$$

$$= \frac{1}{2}\left[\sqrt{2} + \ln\left(\sqrt{2} + 1 \right) \right] \approx 1.148$$

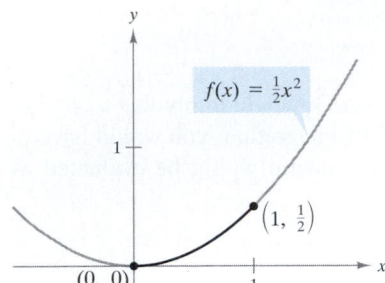

The arc length of the curve from $(0, 0)$ to $\left(1, \frac{1}{2}\right)$ is approximately 1.148.

Figure 7.10

EXAMPLE 6 Comparing Two Fluid Forces

A sealed barrel of oil (weighing 48 pounds per cubic foot) is floating in seawater (weighing 64 pounds per cubic foot), as shown in Figures 7.11 and 7.12. (The barrel is not completely full of oil—on its side, the top 0.2 foot of the barrel is empty.) Compare the fluid forces against one end of the barrel from the inside and from the outside.

Solution   In Figure 7.12, locate the coordinate system with the origin at the center of the circle given by $x^2 + y^2 = 1$. To find the fluid force against an end of the barrel *from the inside*, integrate between -1 and 0.8 (using a weight of $w = 48$).

$$F = w \int_c^d h(y)L(y)\, dy \qquad \text{General equation}$$

$$F_{\text{inside}} = 48 \int_{-1}^{0.8} (0.8 - y)(2)\sqrt{1 - y^2}\, dy$$

$$= 76.8 \int_{-1}^{0.8} \sqrt{1 - y^2}\, dy - 96 \int_{-1}^{0.8} y\sqrt{1 - y^2}\, dy$$

To find the fluid force *from the outside*, integrate between -1 and 0.4 (using a weight of $w = 64$).

$$F_{\text{outside}} = 64 \int_{-1}^{0.4} (0.4 - y)(2)\sqrt{1 - y^2}\, dy$$

$$= 51.2 \int_{-1}^{0.4} \sqrt{1 - y^2}\, dy - 128 \int_{-1}^{0.4} y\sqrt{1 - y^2}\, dy$$

The barrel is not quite full of oil—the top 0.2 foot of the barrel is empty.

Figure 7.11

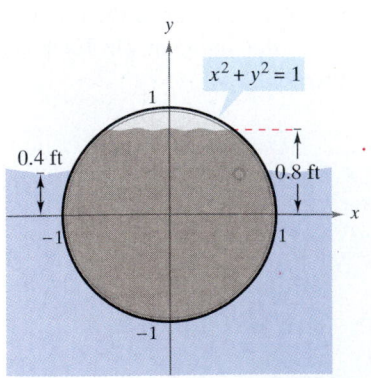

Figure 7.12

We leave the details of integration for you to complete in Exercise 66. Intuitively, would you say that the force from the oil (the inside) or the force from the seawater (the outside) is greater? By evaluating these two integrals, you can determine that

$$F_{\text{inside}} \approx 121.3 \text{ pounds} \qquad \text{and} \qquad F_{\text{outside}} \approx 93.0 \text{ pounds}.$$

In Exercises 1–4, match the antiderivative with the correct integral. [Integrals are labeled (a), (b), (c), and (d).]

(a) $\displaystyle\int \frac{x^2}{\sqrt{16 - x^2}}\,dx$

(b) $\displaystyle\int \frac{\sqrt{x^2 + 16}}{x}\,dx$

(c) $\displaystyle\int \sqrt{7 + 6x - x^2}\,dx$

(d) $\displaystyle\int \frac{x^2}{\sqrt{x^2 - 16}}\,dx$

1. $4\ln\left|\dfrac{\sqrt{x^2 + 16} - 4}{x}\right| + \sqrt{x^2 + 16} + C$

2. $8\ln\left|\sqrt{x^2 - 16} + x\right| + \dfrac{x\sqrt{x^2 - 16}}{2} + C$

3. $8\arcsin\dfrac{x}{4} - \dfrac{x\sqrt{16 - x^2}}{2} + C$

4. $8\arcsin\dfrac{x - 3}{4} + \dfrac{(x - 3)\sqrt{7 + 6x - x^2}}{2} + C$

In Exercises 5–8, evaluate the indefinite integral using the substitution $x = 5\sin\theta$.

5. $\displaystyle\int \frac{1}{(25 - x^2)^{3/2}}\,dx$

6. $\displaystyle\int \frac{1}{x^2\sqrt{25 - x^2}}\,dx$

7. $\displaystyle\int \frac{\sqrt{25 - x^2}}{x}\,dx$

8. $\displaystyle\int \frac{x^2}{\sqrt{25 - x^2}}\,dx$

In Exercises 9–12, evaluate the indefinite integral using the substitution $x = 2\sec\theta$.

9. $\displaystyle\int \frac{1}{\sqrt{x^2 - 4}}\,dx$

10. $\displaystyle\int \frac{\sqrt{x^2 - 4}}{x}\,dx$

11. $\displaystyle\int x^3\sqrt{x^2 - 4}\,dx$

12. $\displaystyle\int \frac{x^3}{\sqrt{x^2 - 4}}\,dx$

In Exercises 13–16, evaluate the indefinite integral using the substitution $x = \tan\theta$.

13. $\displaystyle\int x\sqrt{1 + x^2}\,dx$

14. $\displaystyle\int \frac{x^3}{\sqrt{1 + x^2}}\,dx$

15. $\displaystyle\int \frac{1}{(1 + x^2)^2}\,dx$

16. $\displaystyle\int \frac{x^2}{(1 + x^2)^2}\,dx$

In Exercises 17 and 18, use Theorem 7.2 to evaluate the integral.

17. $\displaystyle\int \sqrt{4 + 9x^2}\,dx$

18. $\displaystyle\int \sqrt{1 + x^2}\,dx$

In Exercises 19–38, evaluate the integral.

19. $\displaystyle\int \frac{x}{\sqrt{x^2 + 9}}\,dx$

20. $\displaystyle\int \frac{1}{\sqrt{25 - x^2}}\,dx$

21. $\displaystyle\int_0^2 \sqrt{16 - 4x^2}\,dx$

22. $\displaystyle\int_0^2 x\sqrt{16 - 4x^2}\,dx$

23. $\displaystyle\int \frac{1}{\sqrt{x^2 - 9}}\,dx$

24. $\displaystyle\int \frac{t}{(1 - t^2)^{3/2}}\,dt$

25. $\displaystyle\int \frac{\sqrt{1 - x^2}}{x^4}\,dx$

26. $\displaystyle\int \frac{\sqrt{4x^2 + 9}}{x^4}\,dx$

27. $\displaystyle\int \frac{1}{x\sqrt{4x^2 + 9}}\,dx$

28. $\displaystyle\int \frac{1}{(x^2 + 3)^{3/2}}\,dx$

29. $\displaystyle\int \frac{x}{(x^2 + 3)^{3/2}}\,dx$

30. $\displaystyle\int \frac{1}{x\sqrt{4x^2 + 16}}\,dx$

31. $\displaystyle\int e^{2x}\sqrt{1 + e^{2x}}\,dx$

32. $\displaystyle\int (x + 1)\sqrt{x^2 + 2x + 2}\,dx$

33. $\displaystyle\int e^x\sqrt{1 - e^{2x}}\,dx$

34. $\displaystyle\int \frac{\sqrt{1 - x}}{\sqrt{x}}\,dx$

35. $\displaystyle\int \frac{1}{4 + 4x^2 + x^4}\,dx$

36. $\displaystyle\int \frac{x^3 + x + 1}{x^4 + 2x^2 + 1}\,dx$

37. $\displaystyle\int \operatorname{arcsec} 2x\,dx$

38. $\displaystyle\int x\arcsin x\,dx$

In Exercises 39–42, complete the square and evaluate the integral.

39. $\displaystyle\int \frac{1}{\sqrt{4x - x^2}}\,dx$

40. $\displaystyle\int \frac{x^2}{\sqrt{2x - x^2}}\,dx$

41. $\displaystyle\int \frac{x}{\sqrt{x^2 + 4x + 8}}\,dx$

42. $\displaystyle\int \frac{x}{\sqrt{x^2 - 6x + 5}}\,dx$

In Exercises 43–46, evaluate the integral using (a) the given integration limits and (b) the limits obtained by trigonometric substitution.

43. $\displaystyle\int_0^{\sqrt{3}/2} \frac{t^2}{(1 - t^2)^{3/2}}\,dt$

44. $\displaystyle\int_0^{\sqrt{3}/2} \frac{1}{(1 - t^2)^{5/2}}\,dt$

45. $\displaystyle\int_0^3 \frac{x^3}{\sqrt{x^2 + 9}}\,dx$

46. $\displaystyle\int_0^{5/3} \sqrt{25 - 9x^2}\,dx$

In Exercises 47–50, use a symbolic integration utility to evaluate the integral. Verify the result by differentiation.

47. $\displaystyle\int \frac{x^2}{\sqrt{x^2 + 10x + 9}}\,dx$

48. $\displaystyle\int (x^2 + 2x + 11)^{3/2}\,dx$

49. $\displaystyle\int \frac{x^2}{\sqrt{x^2 - 1}}\,dx$

50. $\displaystyle\int x^2\sqrt{x^2 - 4}\,dx$

51. Area   Find the area enclosed by the ellipse shown in the figure.

$$\frac{x^2}{a^2} + \frac{y^2}{b^2} = 1$$

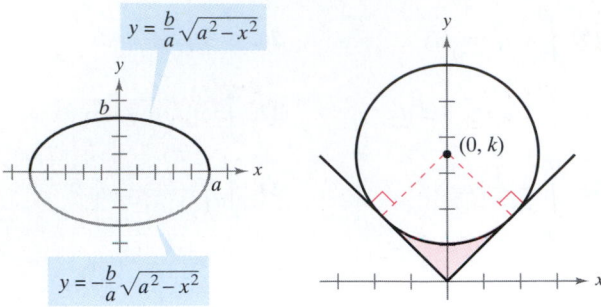

$y = \frac{b}{a}\sqrt{a^2 - x^2}$

$y = -\frac{b}{a}\sqrt{a^2 - x^2}$

Figure for 51   **Figure for 52**

52. Mechanical Design   The surface of a machine part is the region between the graphs of $y = |x|$ and $x^2 + (y - k)^2 = 25$ (see figure).

(a) Find k if the circle is tangent to the graph of $y = |x|$.

(b) Find the area of the surface of the machine part.

(c) Find the area of the surface of the machine part as a function of the radius r of the circle.

Volume of a Torus   In Exercises 53 and 54, find the volume of the torus generated by revolving the region bounded by the graph of the circle about the y-axis.

53. $(x - 3)^2 + y^2 = 1$ (see figure)

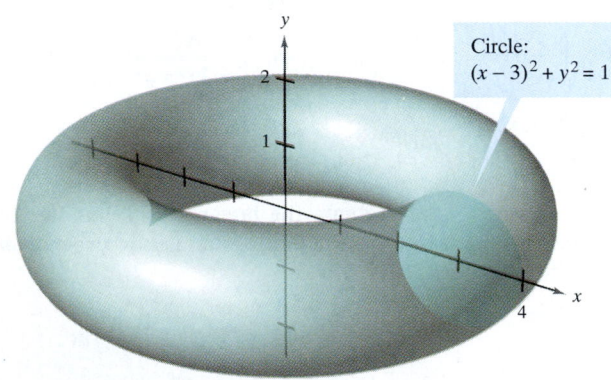

Circle:
$(x - 3)^2 + y^2 = 1$

54. $(x - h)^2 + y^2 = r^2$, $h > r$

Arc Length   In Exercises 55 and 56, find the arc length of the plane curve over the indicated interval.

Function	Interval
55. $y = \ln x$	$[1, 5]$
56. $y = x^2$	$[0, 3]$

57. Arc Length   Show that the length of one arch of the sine curve is equal to the length of one arch of the cosine curve.

58. Volume   The axis of a storage tank in the form of a right circular cylinder is horizontal (see figure). The radius and length of the tank are 1 meter and 3 meters.

(a) Determine the volume of fluid in the tank as a function of its depth d.

(b) Use a graphing utility to graph the function in part (a).

(c) Design a dip stick for the tank with markings of $\frac{1}{4}$, $\frac{1}{2}$, and $\frac{3}{4}$.

(d) If fluid is entering the tank at a rate of $\frac{1}{4}$ cubic meters per minute, determine the rate of change of depth of the fluid as a function of its depth d.

(e) Use a graphing utility to graph the function in part (d). When will the rate of change of depth be minimum? Does this agree with your intuition? Explain.

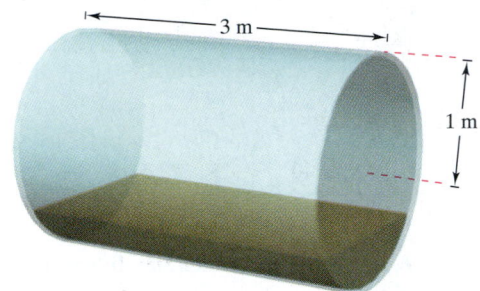

3 m

1 m

Projectile Motion   In Exercises 59 and 60, (a) use a graphing utility to graph the path of a projectile that follows the path given by the graph of the equation, (b) determine the range of the projectile, and (c) use the integration capabilities of a graphing utility to determine the distance the projectile travels.

59. $y = x - 0.005x^2$ **60.** $y = x - \dfrac{x^2}{72}$

61. Surface Area   Find the surface area of the solid generated by revolving the region bounded by the graphs of $y = x^2$, $y = 0$, $x = 0$, and $x = \sqrt{2}$ about the x-axis.

Centroid   In Exercises 62 and 63, find the centroid of the region determined by the graphs of the inequalities.

62. $y \leq 3/\sqrt{x^2 + 9}$, $y \geq 0$, $x \geq -4$, $x \leq 4$

63. $y \leq \frac{1}{4}x^2$, $(x - 4)^2 + y^2 \leq 16$, $y \geq 0$

64. Average Field Strength   The field strength H of a magnet of length $2L$ on a particle r units from the center of the magnet is

$$H = \frac{2mL}{(r^2 + L^2)^{3/2}}$$

where $\pm m$ are the poles of the magnet (see figure). Find the average field strength as the particle moves from 0 to R units from the center by evaluating the integral

$$\frac{1}{R} \int_0^R \frac{2mL}{(r^2 + L^2)^{3/2}}\, dr.$$

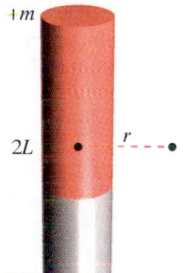

Figure for 64   **Figure for 65**

65. Fluid Force   Find the fluid force on a circular observation window of radius 1 foot in a vertical wall of a large water-filled tank at a fish hatchery for each of the indicated depths (see figure). Use trigonometric substitution to evaluate the one integral. (Recall that in Section 6.7 in a similar problem, you evaluated one integral by a geometric formula and the other by observing that the integrand was odd.)

(a) The center of the window is 3 feet below the water's surface.

(b) The center of the window is d feet below the water's surface ($d > 1$).

66. Fluid Force   Evaluate the following two integrals, which yield the fluid forces given in Example 6.

(a) $F_{inside} = 48 \int_{-1}^{0.8} (0.8 - y)(2)\sqrt{1 - y^2}\, dy$

(b) $F_{outside} = 64 \int_{-1}^{0.4} (0.4 - y)(2)\sqrt{1 - y^2}\, dy$

67. Tractrix   A person moves from the origin along the positive y-axis pulling a weight at the end of a 12-meter rope (see figure). Initially, the weight is located at the point $(12, 0)$.

(a) Show that the slope of the tangent line of the path of the weight is

$$\frac{dy}{dx} = -\frac{\sqrt{144 - x^2}}{x}.$$

(b) Use the result in part (a) to find the equation of the path of the weight. Use a graphing utility to graph the path and compare it with the figure.

(c) Find any vertical asymptotes in the graph in part (b).

(d) When the person has reached the point $(0, 12)$, how far has the weight moved?

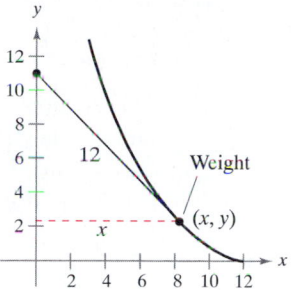

Figure for 67

68. Modeling Data   For the years 1988 through 1993, the average size S (in thousands of dollars) of ordinary life insurance policies in force in the United States is given in the table. (Source: American Council of Life Insurance)

Year	1988	1989	1990	1991	1992	1993
S	31.4	34.4	37.9	41.5	43.0	45.8

A model for these data is

$$S = \sqrt{1445.6 + 228.5t - 4.4t^2}$$

where t is the time in years, with $t = 0$ corresponding to 1990.

(a) Use a graphing utility to graph the model for $-2 \le t \le 3$.

(b) Find the rate of increase of S when $t = 1$.

(c) Use the model and integration to predict the average value of S for the years 1998 through 2000.

True or False?   In Exercises 69–72, determine whether the statement is true or false. If it is false, explain why or give an example that shows it is false.

69. If $x = \sin \theta$, then

$$\int \frac{dx}{\sqrt{1 - x^2}} = \int d\theta.$$

70. If $x = \sec \theta$, then

$$\int \frac{\sqrt{x^2 - 1}}{x}\, dx = \int \sec \theta \tan \theta\, d\theta.$$

71. If $x = \tan \theta$, then

$$\int_0^{\sqrt{3}} \frac{dx}{(1 + x^2)^{3/2}} = \int_0^{4\pi/3} \cos \theta\, d\theta.$$

72. If $x = \sin \theta$, then

$$\int_{-1}^{1} x^2 \sqrt{1 - x^2}\, dx = 2 \int_0^{\pi/2} \sin^2 \theta \cos^2 \theta\, d\theta.$$

73. Use trigonometric substitution to verify the integration formulas given in Theorem 7.2.

SECTION **7.5** **Partial Fractions**

Partial Fractions • Linear Factors • Quadratic Factors

Partial Fractions

This section examines a procedure for decomposing a rational function into simpler rational functions to which you can apply the basic integration formulas. This procedure is called the **method of partial fractions.** To see the benefit of the method of partial fractions, consider the integral

$$\int \frac{1}{x^2 - 5x + 6} \, dx.$$

To evaluate this integral *without* partial fractions, you can complete the square and use trigonometric substitution (see Figure 7.13) to obtain the following.

$$\int \frac{1}{x^2 - 5x + 6} \, dx = \int \frac{dx}{(x - 5/2)^2 - (1/2)^2} \qquad {\color{red} a = \tfrac{1}{2}, x - \tfrac{5}{2} = \tfrac{1}{2} \sec \theta}$$

$$= \int \frac{(1/2) \sec \theta \tan \theta \, d\theta}{(1/4) \tan^2 \theta} \qquad {\color{red} dx = \tfrac{1}{2} \sec \theta \tan \theta \, d\theta}$$

$$= 2 \int \csc \theta \, d\theta$$

$$= 2 \ln |\csc \theta - \cot \theta| + C$$

$$= 2 \ln \left| \frac{2x - 5}{2\sqrt{x^2 - 5x + 6}} - \frac{1}{2\sqrt{x^2 - 5x + 6}} \right| + C$$

$$= 2 \ln \left| \frac{x - 3}{\sqrt{x^2 - 5x + 6}} \right| + C$$

$$= 2 \ln \left| \frac{\sqrt{x - 3}}{\sqrt{x - 2}} \right| + C$$

$$= \ln \left| \frac{x - 3}{x - 2} \right| + C$$

$$= \ln |x - 3| - \ln |x - 2| + C$$

Now, suppose you had observed that

$$\frac{1}{x^2 - 5x + 6} = \frac{1}{x - 3} - \frac{1}{x - 2} \qquad {\color{red} \text{Partial fraction decomposition}}$$

Then you could evaluate the integral easily, as follows.

$$\int \frac{1}{x^2 - 5x + 6} \, dx = \int \left(\frac{1}{x - 3} - \frac{1}{x - 2} \right) dx$$

$$= \ln |x - 3| - \ln |x - 2| + C$$

This method is clearly preferable to trigonometric substitution. However, its use depends on the ability to factor the denominator, $x^2 - 5x + 6$, and to find the **partial fractions**

$$\frac{1}{x - 3} \qquad \text{and} \qquad -\frac{1}{x - 2}.$$

In this section, you will study techniques for finding partial fraction decompositions.

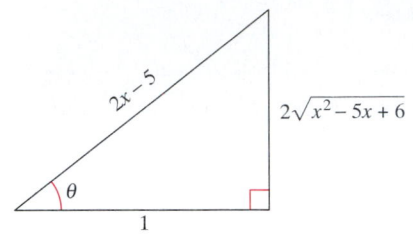

$$\sec \theta = 2x - 5$$
Figure 7.13

JOHN BERNOULLI (1667–1748)

The method of partial fractions was introduced by John Bernoulli, a Swiss mathematician who was instrumental in the early development of calculus. John Bernoulli was a professor at the University of Basel and taught many outstanding students, the most famous of whom was Leonhard Euler.

STUDY TIP In precalculus you learned how to combine functions such as

$$\frac{1}{x-2} + \frac{-1}{x+3} = \frac{5}{(x-2)(x+3)}.$$

The method of partial fractions shows you how to reverse this process.

$$\frac{5}{(x-2)(x+3)} = \frac{?}{x-2} + \frac{?}{x+3}$$

Recall from algebra that every polynomial with real coefficients can be factored into linear and irreducible quadratic factors.* For instance, the polynomial $x^5 + x^4 - x - 1$ can be written as

$$x^5 + x^4 - x - 1 = x^4(x+1) - (x+1)$$
$$= (x^4 - 1)(x+1)$$
$$= (x^2 + 1)(x^2 - 1)(x+1)$$
$$= (x^2 + 1)(x+1)(x-1)(x+1)$$
$$= (x-1)(x+1)^2(x^2+1)$$

where $(x-1)$ is a linear factor, $(x+1)^2$ is a repeated linear factor, and (x^2+1) is an irreducible quadratic factor. Using this factorization, you can write the partial fraction decomposition of the rational expression

$$\frac{N(x)}{x^5 + x^4 - x - 1}$$

where $N(x)$ is a polynomial of degree less than 5, as follows.

$$\frac{N(x)}{(x-1)(x+1)^2(x^2+1)} = \frac{A}{x-1} + \frac{B}{x+1} + \frac{C}{(x+1)^2} + \frac{Dx+E}{x^2+1}$$

Decomposition of $N(x)/D(x)$ into Partial Fractions

1. **Divide if improper:** If $N(x)/D(x)$ is an improper fraction (that is, if the degree of the numerator is greater than or equal to the degree of the denominator), divide the denominator into the numerator to obtain

$$\frac{N(x)}{D(x)} = (\text{a polynomial}) + \frac{N_1(x)}{D(x)}$$

where the degree of $N_1(x)$ is less than the degree of $D(x)$. Then apply steps 2, 3, and 4 to the proper rational expression $N_1(x)/D(x)$.

2. **Factor denominator:** Completely factor the denominator into factors of the form

$$(px+q)^m \quad \text{and} \quad (ax^2+bx+c)^n$$

where ax^2+bx+c is irreducible.

3. **Linear factors:** For *each* factor of the form $(px+q)^m$, the partial fraction decomposition must include the following sum of m fractions.

$$\frac{A_1}{(px+q)} + \frac{A_2}{(px+q)^2} + \cdots + \frac{A_m}{(px+q)^m}$$

4. **Quadratic factors:** For *each* factor of the form $(ax^2+bx+c)^n$, the partial fraction decomposition must include the following sum of n fractions.

$$\frac{B_1x+C_1}{ax^2+bx+c} + \frac{B_2x+C_2}{(ax^2+bx+c)^2} + \cdots + \frac{B_nx+C_n}{(ax^2+bx+c)^n}$$

*For a review of factorization techniques, see Precalculus, 4th edition, by Larson and Hostetler or Precalculus: A Graphing Approach, 2nd edition, by Larson, Hostetler, and Edwards (Boston, Massachusetts: Houghton Mifflin, 1997).

Linear Factors

Algebraic techniques for determining the constants in the numerators of a partial decomposition with linear or repeated linear factors are demonstrated in Examples 1 and 2.

EXAMPLE 1 Distinct Linear Factors

Write the partial fraction decomposition for $\dfrac{1}{x^2 - 5x + 6}$.

Solution Because $x^2 - 5x + 6 = (x - 3)(x - 2)$, you should include one partial fraction for each factor and write

$$\frac{1}{x^2 - 5x + 6} = \frac{A}{x - 3} + \frac{B}{x - 2}$$

where A and B are to be determined. Multiplying this equation by the least common denominator $(x - 3)(x - 2)$ yields the **basic equation**

$$1 = A(x - 2) + B(x - 3). \qquad \text{Basic equation}$$

Because this equation is to be true for all x, you can substitute any *convenient* values for x to obtain equations in A and B. The most convenient values are the ones that make particular factors equal to 0.

To solve for A, let $x = 3$ and obtain

$$1 = A(3 - 2) + B(3 - 3) \qquad \text{Let } x = 3 \text{ in basic equation.}$$
$$1 = A(1) + B(0)$$
$$A = 1.$$

To solve for B, let $x = 2$ and obtain

$$1 = A(2 - 2) + B(2 - 3) \qquad \text{Let } x = 2 \text{ in basic equation.}$$
$$1 = A(0) + B(-1)$$
$$B = -1.$$

Therefore, the decomposition is

$$\frac{1}{x^2 - 5x + 6} = \frac{1}{x - 3} - \frac{1}{x - 2}$$

as indicated at the beginning of this section.

NOTE Note that the substitutions for x in Example 1 are chosen for their convenience in determining values for A and B; $x = 2$ is chosen to eliminate the term $A(x - 2)$, and $x = 3$ is chosen to eliminate the term $B(x - 3)$. The goal is to make *convenient* substitutions whenever possible.

FOR FURTHER INFORMATION To learn a different method for finding the partial fraction decomposition, called the Heavyside Method, see the article "Calculus to Algebra Connections in Partial Fraction Decomposition" by Joseph Wiener and Will Watkins in *The AMATYC Review*, Volume 15, Number 1.

Be sure you see that the method of partial fractions is practical only for integrals of rational functions whose denominators factor "nicely." For instance, if the denominator in Example 1 were changed to $x^2 - 5x + 5$, its factorization as

$$x^2 - 5x + 5 = \left[x + \frac{5 + \sqrt{5}}{2}\right]\left[x - \frac{5 - \sqrt{5}}{2}\right]$$

would be too cumbersome to use with partial fractions. In such cases, you should use completing the square or a symbolic integration utility to perform the integration. If you do this, you should obtain

$$\int \frac{1}{x^2 - 5x + 5}\, dx = \frac{\sqrt{5}}{5}\ln\left|2x - \sqrt{5} - 5\right| - \frac{\sqrt{5}}{5}\ln\left|2x + \sqrt{5} - 5\right| + C.$$

EXAMPLE 2   Repeated Linear Factors

Evaluate $\displaystyle\int \frac{5x^2 + 20x + 6}{x^3 + 2x^2 + x}\, dx.$

Solution   Because

$$x^3 + 2x^2 + x = x(x^2 + 2x + 1)$$
$$= x(x + 1)^2$$

FOR FURTHER INFORMATION For an alternative approach to using partial fractions, see the article " A Shortcut in Partial Fractions" by Xun-Cheng Huang in the November 1991 issue of *The College Mathematics Journal.*

you should include one fraction for *each power* of x and $(x + 1)$ and write

$$\frac{5x^2 + 20x + 6}{x(x + 1)^2} = \frac{A}{x} + \frac{B}{x + 1} + \frac{C}{(x + 1)^2}.$$

Multiplying by the least common denominator $x(x + 1)^2$ yields the *basic equation*

$$5x^2 + 20x + 6 = A(x + 1)^2 + Bx(x + 1) + Cx. \qquad \text{Basic equation}$$

To solve for A, let $x = 0$. This eliminates the B and C terms and yields

$$6 = A(1) + 0 + 0$$
$$A = 6.$$

To solve for C, let $x = -1$. This eliminates the A and B terms and yields

$$5 - 20 + 6 = 0 + 0 - C$$
$$C = 9.$$

The most convenient choices for x have been used, so to find the value of B, you can use *any other value* of x along with the calculated values of A and C. Using $x = 1$, $A = 6$, and $C = 9$ produces

$$5 + 20 + 6 = A(4) + B(2) + C$$
$$31 = 6(4) + 2B + 9$$
$$-2 = 2B$$
$$B = -1.$$

Therefore, it follows that

$$\int \frac{5x^2 + 20x + 6}{x(x + 1)^2}\, dx = \int \left(\frac{6}{x} - \frac{1}{x + 1} + \frac{9}{(x + 1)^2} \right) dx$$

TECHNOLOGY   Most computer algebra systems, such as *Derive*, *Maple, Mathcad, Mathematica,* and the *TI-92,* can be used to convert a rational function to its partial fraction decomposition. For instance, using *Maple,* you obtain the following.

> convert$\left(\dfrac{5x^2 + 20x + 6}{x^3 + 2x^2 + x}, parfrac, x \right)$

$\dfrac{6}{x} + \dfrac{9}{(x + 1)^2} - \dfrac{1}{x + 1}$

$$= 6 \ln |x| - \ln |x + 1| + 9\frac{(x + 1)^{-1}}{-1} + C$$

$$= \ln \left| \frac{x^6}{x + 1} \right| - \frac{9}{x + 1} + C.$$

Try checking this result by differentiating. Include algebra in your check, simplifying the derivative until you have obtained the original integrand.

NOTE   It is necessary to make as many substitutions for x as there are unknowns $(A, B, C, \ldots)$ to be determined. For instance, in Example 2, we made three substitutions $(x = -1, x = 0,$ and $x = 1)$ to solve for $C, A,$ and B.

Quadratic Factors

When using the method of partial fractions with *linear* factors, a convenient choice of x immediately yields a value for one of the coefficients. With *quadratic* factors, a system of linear equations usually has to be solved, regardless of the choice of x.

EXAMPLE 3   Distinct Linear and Quadratic Factors

Evaluate $\displaystyle\int \frac{2x^3 - 4x - 8}{(x^2 - x)(x^2 + 4)}\, dx$.

Solution   Because

$$(x^2 - x)(x^2 + 4) = x(x - 1)(x^2 + 4)$$

you should include one partial fraction for each factor and write

$$\frac{2x^3 - 4x - 8}{x(x - 1)(x^2 + 4)} = \frac{A}{x} + \frac{B}{x - 1} + \frac{Cx + D}{x^2 + 4}.$$

Multiplying by the least common denominator $x(x - 1)(x^2 + 4)$ yields the *basic equation*

$$2x^3 - 4x - 8 = A(x - 1)(x^2 + 4) + Bx(x^2 + 4) + (Cx + D)(x)(x - 1).$$

To solve for A, let $x = 0$ and obtain

$$-8 = A(-1)(4) + 0 + 0 \quad \Longrightarrow \quad A = 2.$$

To solve for B, let $x = 1$ and obtain

$$-10 = 0 + B(5) + 0 \quad \Longrightarrow \quad B = -2.$$

At this point, C and D are yet to be determined. You can find these remaining constants by choosing two other values for x and solving the resulting system of linear equations. If $x = -1$, then, using $A = 2$ and $B = -2$, you can write

$$-6 = (2)(-2)(5) + (-2)(-1)(5) + (-C + D)(-1)(-2)$$
$$2 = -C + D.$$

If $x = 2$, you have

$$0 = (2)(1)(8) + (-2)(2)(8) + (2C + D)(2)(1)$$
$$8 = 2C + D.$$

Solving the linear system by subtracting the first equation from the second

$$-C + D = 2$$
$$2C + D = 8$$

yields $C = 2$. Consequently, $D = 4$, and it follows that

$$\int \frac{2x^3 - 4x - 8}{x(x - 1)(x^2 + 4)}\, dx =$$

$$\int \left(\frac{2}{x} - \frac{2}{x - 1} + \frac{2x}{x^2 + 4} + \frac{4}{x^2 + 4} \right) dx =$$

$$2 \ln |x| - 2 \ln |x - 1| + \ln(x^2 + 4) + 2 \arctan \frac{x}{2} + C.$$

In Examples 1, 2, and 3, we began the solution of the basic equation by substituting values of x that made the linear factors equal to 0. This method works well when the partial fraction decomposition involves linear factors. However, if the decomposition involves only quadratic factors, an alternative procedure is often more convenient.

EXAMPLE 4 Repeated Quadratic Factors

Evaluate $\displaystyle\int \frac{8x^3 + 13x}{(x^2 + 2)^2}\, dx$.

Solution Include one partial fraction for each power of $(x^2 + 2)$ and write

$$\frac{8x^3 + 13x}{(x^2 + 2)^2} = \frac{Ax + B}{x^2 + 2} + \frac{Cx + D}{(x^2 + 2)^2}.$$

Multiplying by the least common denominator $(x^2 + 2)^2$ yields the *basic equation*

$$8x^3 + 13x = (Ax + B)(x^2 + 2) + Cx + D.$$

Expanding the basic equation and collecting like terms produces

$$8x^3 + 13x = Ax^3 + 2Ax + Bx^2 + 2B + Cx + D$$
$$8x^3 + 13x = Ax^3 + Bx^2 + (2A + C)x + (2B + D).$$

Now, you can equate the coefficients of like terms on opposite sides of the equation.

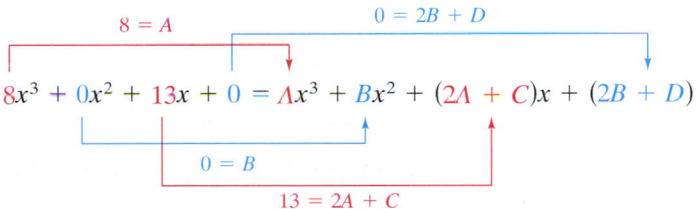

Using the known values $A = 8$ and $B = 0$, you can write the following.

$$13 = 2A + C = 2(8) + C \quad\Longrightarrow\quad C = -3$$
$$0 = 2B + D = 2(0) + D \quad\Longrightarrow\quad D = 0$$

Finally, you can conclude that

$$\int \frac{8x^3 + 13x}{(x^2 + 2)^2}\, dx = \int \left(\frac{8x}{x^2 + 2} + \frac{-3x}{(x^2 + 2)^2} \right) dx$$

$$= 4\ln(x^2 + 2) + \frac{3}{2(x^2 + 2)} + C.$$

TECHNOLOGY Try using a symbolic integration utility to evaluate the integral in Example 4—you might find that the form of the antiderivative is different. For instance, when we used a symbolic integration utility to work Example 4, we obtained

$$\int \frac{8x^3 + 13x}{(x^2 + 2)^2}\, dx = \ln(x^8 + 8x^6 + 24x^4 + 32x^2 + 16) + \frac{3}{2(x^2 + 2)} + C.$$

Is this result equivalent to that obtained in Example 4?

When integrating rational expressions, keep in mind that for *improper* rational expressions such as

$$\frac{N(x)}{D(x)} = \frac{2x^3 + x^2 - 7x + 7}{x^2 + x - 2}$$

you must first divide to obtain

$$\frac{N(x)}{D(x)} = 2x - 1 + \frac{-2x + 5}{x^2 + x - 2}.$$

The proper rational expression is then decomposed into its partial fractions by the usual methods. Here are some guidelines for solving the basic equation that is obtained in a partial fraction decomposition.

Guidelines for Solving the Basic Equation

Linear Factors

1. Substitute the *roots* of the distinct linear factors into the basic equation.
2. For repeated linear factors, use the coefficients determined in guideline 1 to rewrite the basic equation. Then substitute *other* convenient values of x and solve for the remaining coefficients.

Quadratic Factors

1. Expand the basic equation.
2. Collect terms according to powers of x.
3. Equate the coefficients of like powers to obtain a system of linear equations involving, A, B, C, and so on.
4. Solve the system of linear equations.

Before concluding this section, here are a few things you should remember. First, it is not necessary to use the partial fractions technique on all rational functions. For instance, the following integral is evaluated more easily by the Log Rule.

$$\int \frac{x^2 + 1}{x^3 + 3x - 4}\,dx = \frac{1}{3}\int \frac{3x^2 + 3}{x^3 + 3x - 4}\,dx = \frac{1}{3}\ln|x^3 + 3x - 4| + C$$

Second, if the integrand is not in reduced form, reducing it may eliminate the need for partial fractions, as shown in the following integral.

$$\int \frac{x^2 - x - 2}{x^3 - 2x - 4}\,dx = \int \frac{(x+1)(x-2)}{(x-2)(x^2+2x+2)}\,dx$$

$$= \int \frac{x+1}{x^2+2x+2}\,dx$$

$$= \frac{1}{2}\ln|x^2 + 2x + 2| + C$$

Finally, partial fractions can be used with some quotients involving transcendental functions. For instance, the substitution $u = \sin x$ allows you to write

$$\int \frac{\cos x}{\sin x(\sin x - 1)}\,dx = \int \frac{du}{u(u-1)}. \qquad u = \sin x,\, du = \cos x\,dx$$

EXERCISES FOR SECTION 7.5

In Exercises 1–6, give the form of the partial fraction decomposition of the rational expression. Do not solve for the constants.

1. $\dfrac{5}{x^2 - 10x}$

2. $\dfrac{4x^2 + 3}{(x - 5)^3}$

3. $\dfrac{2x - 3}{x^3 + 10x}$

4. $\dfrac{x - 2}{x^2 + 4x + 3}$

5. $\dfrac{16x}{x^3 - 10x^2}$

6. $\dfrac{2x - 1}{x(x^2 + 1)^2}$

In Exercises 7–18, use partial fractions (linear factors) to evaluate the integral.

7. $\displaystyle\int \dfrac{1}{x^2 - 1}\, dx$

8. $\displaystyle\int \dfrac{1}{4x^2 - 9}\, dx$

9. $\displaystyle\int \dfrac{3}{x^2 + x - 2}\, dx$

10. $\displaystyle\int \dfrac{x + 1}{x^2 + 4x + 3}\, dx$

11. $\displaystyle\int \dfrac{5 - x}{2x^2 + x - 1}\, dx$

12. $\displaystyle\int \dfrac{3x^2 - 7x - 2}{x^3 - x}\, dx$

13. $\displaystyle\int \dfrac{x^2 + 12x + 12}{x^3 - 4x}\, dx$

14. $\displaystyle\int \dfrac{x^3 - x + 3}{x^2 + x - 2}\, dx$

15. $\displaystyle\int \dfrac{2x^3 - 4x^2 - 15x + 5}{x^2 - 2x - 8}\, dx$

16. $\displaystyle\int \dfrac{x + 2}{x^2 - 4x}\, dx$

17. $\displaystyle\int \dfrac{4x^2 + 2x - 1}{x^3 + x^2}\, dx$

18. $\displaystyle\int \dfrac{2x - 3}{(x - 1)^2}\, dx$

In Exercises 19–26, use partial fractions (linear and quadratic factors) to evaluate the integral.

19. $\displaystyle\int \dfrac{x^2 - 1}{x^3 + x}\, dx$

20. $\displaystyle\int \dfrac{x}{x^3 - 1}\, dx$

21. $\displaystyle\int \dfrac{x^2}{x^4 - 2x^2 - 8}\, dx$

22. $\displaystyle\int \dfrac{2x^2 + x + 8}{(x^2 + 4)^2}\, dx$

23. $\displaystyle\int \dfrac{x}{16x^4 - 1}\, dx$

24. $\displaystyle\int \dfrac{x^2 - 4x + 7}{x^3 - x^2 + x + 3}\, dx$

25. $\displaystyle\int \dfrac{x^2 + 5}{x^3 - x^2 + x + 3}\, dx$

26. $\displaystyle\int \dfrac{x^2 + x + 3}{x^4 + 6x^2 + 9}\, dx$

In Exercises 27–30, evaluate the definite integral. Use a graphing utility to verify your result.

27. $\displaystyle\int_0^1 \dfrac{3}{2x^2 + 5x + 2}\, dx$

28. $\displaystyle\int_1^5 \dfrac{x - 1}{x^2(x + 1)}\, dx$

29. $\displaystyle\int_1^2 \dfrac{x + 1}{x(x^2 + 1)}\, dx$

30. $\displaystyle\int_0^1 \dfrac{x^2 - x}{x^2 + x + 1}\, dx$

In Exercises 31–38, use a symbolic integration utility to determine the antiderivative that passes through the indicated point. Use the utility to graph the resulting antiderivative.

Integral	*Point*
31. $\displaystyle\int \dfrac{3x}{x^2 - 6x + 9}\, dx$	$(4, 0)$
32. $\displaystyle\int \dfrac{6x^2 + 1}{x^2(x - 1)^3}\, dx$	$(2, 1)$
33. $\displaystyle\int \dfrac{x^2 + x + 2}{(x^2 + 2)^2}\, dx$	$(0, 1)$
34. $\displaystyle\int \dfrac{x^3}{(x^2 - 4)^2}\, dx$	$(3, 4)$
35. $\displaystyle\int \dfrac{2x^2 - 2x + 3}{x^3 - x^2 - x - 2}\, dx$	$(3, 10)$
36. $\displaystyle\int \dfrac{x(2x - 9)}{x^3 - 6x^2 + 12x - 8}\, dx$	$(3, 2)$
37. $\displaystyle\int \dfrac{x^2 - x + 2}{x^3 - x^2 + x - 1}\, dx$	$(2, 6)$
38. $\displaystyle\int \dfrac{1}{x^2 - 4}\, dx$	$(6, 4)$

In Exercises 39–44, use substitution to evaluate the integral.

39. $\displaystyle\int \dfrac{\sin x}{\cos x(\cos x - 1)}\, dx$

40. $\displaystyle\int \dfrac{\sin x}{\cos x + \cos^2 x}\, dx$

41. $\displaystyle\int \dfrac{3 \cos x}{\sin^2 x + \sin x - 2}\, dx$

42. $\displaystyle\int \dfrac{\sec^2 x}{\tan x(\tan x + 1)}\, dx$

43. $\displaystyle\int \dfrac{e^x}{(e^x - 1)(e^x + 4)}\, dx$

44. $\displaystyle\int \dfrac{e^x}{(e^{2x} + 1)(e^x - 1)}\, dx$

In Exercises 45–48, use the method of partial fractions to verify the integration formula.

45. $\displaystyle\int \dfrac{1}{x(a + bx)}\, dx = \dfrac{1}{a} \ln \left| \dfrac{x}{a + bx} \right| + C$

46. $\displaystyle\int \dfrac{1}{a^2 - x^2}\, dx = \dfrac{1}{2a} \ln \left| \dfrac{a + x}{a - x} \right| + C$

47. $\displaystyle\int \dfrac{x}{(a + bx)^2}\, dx = \dfrac{1}{b^2} \left(\dfrac{a}{a + bx} + \ln |a + bx| \right) + C$

48. $\displaystyle\int \dfrac{1}{x^2(a + bx)}\, dx = -\dfrac{1}{ax} - \dfrac{b}{a^2} \ln \left| \dfrac{x}{a + bx} \right| + C$

49. Approximation Determine which value best approximates the area of the region between the x-axis and the graph of the function $10/[x(x^2 + 1)]$ over the interval $[1, 3]$. (Make your selection on the basis of a sketch of the region and not by performing any calculations.)

(a) -6 (b) 6 (c) 3 (d) 5 (e) 8

50. Area Find the area of the region bounded by the graphs of $y = 7/(16 - x^2)$ and $y = 1$.

51. Volume and Centroid Consider the region bounded by the graphs of

$$y = 2x/(x^2 + 1), \quad y = 0, x = 0, \text{ and } x = 3.$$

(a) Find the volume of the solid generated by revolving the region about the x-axis.

(b) Find the centroid of the region.

52. Logistics Growth In Chapter 5, the exponential growth equation was derived from the assumption that the rate of growth was proportional to the existing quantity. In practice, there often exists some upper limit L past which growth cannot occur. In such cases, we assume the rate of growth to be proportional not only to the existing quantity, but also to the difference between the existing quantity y and the upper limit L. That is,

$$\frac{dy}{dt} = ky(L - y).$$

In integral form, we can express this relationship as

$$\int \frac{dy}{y(L - y)} = \int k \, dt.$$

(a) A direction field for the differential equation $dy/dt = y(3 - y)$ is shown. Draw a possible solution to the differential equation if $y(0) = 5$, and another if $y(0) = \frac{1}{2}$.

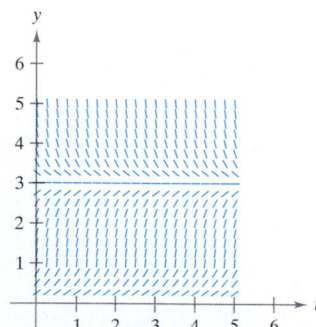

(b) Where $y(0)$ is greater than 3, what is the sign of the slope of the solution?

(c) For $y > 0$, find $\lim_{t \to \infty} y(t)$.

(d) Evaluate the two integrals above and solve for y as a function of t, where y_0 is the initial quantity.

(e) Use the result in part (d) to find and graph the solutions in part (a). Use a graphing utility to graph the solutions and compare the results with the solutions in part (a).

(f) The graph of the function y is called a **logistics curve**. Show that the rate of growth is maximum at the point of inflection, and that this occurs when $y = L/2$.

53. Epidemic Model A single infected individual enters a community of n susceptible individuals. Let x be the number of newly infected individuals at time t. The common epidemic model assumes that the disease spreads at a rate proportional to the product of the total number infected and the number not yet infected. Thus

$$\frac{dx}{dt} = k(x + 1)(n - x)$$

and you obtain

$$\int \frac{1}{(x + 1)(n - x)} \, dx = \int k \, dt.$$

Solve for x as a function of t.

54. Chemical Reactions In a chemical reaction, one unit of compound Y and one unit of compound Z are converted into a single unit of compound X. If x is the amount of compound X formed, and the rate of formation of X is proportional to the product of the amounts of unconverted compounds Y and Z, then

$$\frac{dx}{dt} = k(y_0 - x)(z_0 - x)$$

where y_0 and z_0 are the initial amounts of compounds Y and Z. From the above equation you obtain

$$\int \frac{1}{(y_0 - x)(z_0 - x)} \, dx = \int k \, dt.$$

(a) Perform the two integrations and solve for x in terms of t.

(b) Use the result in part (a) to find x as $t \to \infty$ if (1) $y_0 < z_0$, (2) $y_0 > z_0$, and (3) $y_0 = z_0$.

55. Use partial fractions to evaluate

$$\int_0^1 \frac{x}{1 + x^4} \, dx.$$

56. Suppose the denominator of a rational function can be factored into distinct linear factors

$$D(x) = (x - c_1)(x - c_2) \cdots (x - c_n),$$

for a positive integer n and distinct real numbers $c_1, c_2, \ldots, c_n$. If N is a polynomial of degree less than n, show that

$$\frac{N(x)}{D(x)} = \frac{P_1}{x - c_1} + \frac{P_2}{x - c_2} + \cdots + \frac{P_n}{x - c_n},$$

where $P_k = N(c_k)/D'(c_k)$ for $k = 1, 2, \ldots, n$. Note that this is the partial fraction decomposition of $N(x)/D(x)$.

57. Use the results of Exercise 56 to find the partial fraction decomposition of

$$\frac{x^3 - 3x^2 + 1}{x^4 - 13x^2 + 12x}.$$

Integration by Tables • Reduction Formulas • Rational Functions of Sine and Cosine

Integration by Tables

So far in this chapter you have studied several integration techniques that can be used with the basic integration rules. But merely knowing *how* to use the various techniques is not enough. You also need to know *when* to use them. Integration is first and foremost a problem of recognition. That is, you must recognize which rule or technique to apply to obtain an antiderivative. Frequently, a slight alteration of an integrand will require a different integration technique (or produce a function whose antiderivative is not an elementary function), as shown below.

$$\int x \ln x \, dx = \frac{x^2}{2} \ln x - \frac{x^2}{4} + C \qquad \text{Integration by parts}$$

$$\int \frac{\ln x}{x} \, dx = \frac{(\ln x)^2}{2} + C \qquad \text{Power Rule}$$

$$\int \frac{1}{x \ln x} \, dx = \ln |\ln x| + C \qquad \text{Log Rule}$$

$$\int \frac{x}{\ln x} \, dx = ? \qquad \text{Not an elementary function}$$

TECHNOLOGY   A symbolic integration utility consists, in part, of a database of integration formulas. The primary difference between using a symbolic integration utility and using tables of integrals is that with a symbolic integration utility the computer searches through the database to find a fit. With integration tables, *you* must do the searching.

Many people find tables of integrals to be a valuable supplement to the integration techniques discussed in this chapter. Tables of common integrals can be found in the appendix. **Integration by tables** is not a "cure-all" for all of the difficulties that can accompany integration—using tables of integrals requires considerable thought and insight and often involves substitution.

Each integration formula in the appendix can be developed using one or more of the techniques in this chapter. You should try to verify several of the formulas. For instance, Formula 4.

$$\int \frac{u}{(a + bu)^2} \, du = \frac{1}{b^2} \left(\frac{a}{a + bu} + \ln |a + bu| \right) + C \qquad \text{Formula 4}$$

can be verified using the method of partial fractions, and Formula 19

$$\int \frac{\sqrt{a + bu}}{u} \, du = 2\sqrt{a + bu} + a \int \frac{du}{u\sqrt{a + bu}} \qquad \text{Formula 19}$$

can be verified using integration by parts. Note that the integrals in the appendix are classified according to forms involving the following.

u^n	$(a + bu)$
$(a + bu + cu^2)$	$\sqrt{a + bu}$
$(a^2 \pm u^2)$	$\sqrt{u^2 \pm a^2}$
$\sqrt{a^2 - u^2}$	Trigonometric functions
Inverse trigonometric functions	Exponential functions
Logarithmic functions	

EXAMPLE 1   Integration by Tables

Evaluate $\int \dfrac{dx}{x\sqrt{x - 1}}$.

Solution   Because the expression inside the radical is linear, you should consider forms involving $\sqrt{a + bu}$.

$$\int \frac{du}{u\sqrt{a + bu}} = \frac{2}{\sqrt{-a}} \arctan \sqrt{\frac{a + bu}{-a}} + C \qquad \text{Formula 17 } (a < 0)$$

Let $a = -1$, $b = 1$, and $u = x$. Then $du = dx$, and you can write

$$\int \frac{dx}{x\sqrt{x - 1}} = 2 \arctan \sqrt{x - 1} + C.$$

EXAMPLE 2   Integration by Tables

Evaluate $\int x\sqrt{x^4 - 9}\,dx$.

Solution   Because the radical has the form $\sqrt{u^2 - a^2}$, you should consider Formula 26.

$$\int \sqrt{u^2 - a^2}\,du = \frac{1}{2}\left(u\sqrt{u^2 - a^2} - a^2 \ln \left|u + \sqrt{u^2 - a^2}\right|\right) + C$$

Let $u = x^2$ and $a = 3$. Then $du = 2x\,dx$, and you have

$$\int x\sqrt{x^4 - 9}\,dx = \frac{1}{2}\int \sqrt{(x^2)^2 - 3^2}\,(2x)\,dx$$

$$= \frac{1}{4}\left(x^2\sqrt{x^4 - 9} - 9 \ln\left|x^2 + \sqrt{x^4 - 9}\right|\right) + C.$$

EXAMPLE 3   Integration by Tables

Evaluate $\int \dfrac{x}{1 + e^{-x^2}}\,dx$.

Solution   Of the forms involving e^u, consider the following.

$$\int \frac{du}{1 + e^u} = u - \ln(1 + e^u) + C \qquad \text{Formula 84}$$

Let $u = -x^2$. Then $du = -2x\,dx$, and you have

$$\int \frac{x}{1 + e^{-x^2}}\,dx = -\frac{1}{2}\int \frac{-2x\,dx}{1 + e^{-x^2}}$$

$$= -\frac{1}{2}\left[-x^2 - \ln(1 + e^{-x^2})\right] + C$$

$$= \frac{1}{2}\left[x^2 + \ln(1 + e^{-x^2})\right] + C.$$

TECHNOLOGY   Example 3 shows the importance of having several solution techniques at your disposal. This integral is not difficult to solve with a table, but when we entered it into a well-known symbolic integration utility, the utility was unable to find the antiderivative.

Reduction Formulas

Several of the integrals in the integration tables have the form $\int f(x)\,dx = g(x) + \int h(x)\,dx$. Such integration formulas are called **reduction formulas** because they reduce a given integral to the sum of a function and a simpler integral.

EXAMPLE 4   Using a Reduction Formula

Evaluate $\displaystyle\int x^3 \sin x\,dx$.

Solution   Consider the following three formulas.

$$\int u \sin u\,du = \sin u - u \cos u + C \qquad\qquad \text{Formula 52}$$

$$\int u^n \sin u\,du = -u^n \cos u + n \int u^{n-1} \cos u\,du \qquad \text{Formula 54}$$

$$\int u^n \cos u\,du = u^n \sin u - n \int u^{n-1} \sin u\,du \qquad \text{Formula 55}$$

Using Formula 54, Formula 55, and then Formula 52 produces

$$\int x^3 \sin x\,dx = -x^3 \cos x + 3 \int x^2 \cos x\,dx$$

$$= -x^3 \cos x + 3\left(x^2 \sin x - 2 \int x \sin x\,dx \right)$$

$$= -x^3 \cos x + 3x^2 \sin x + 6x \cos x - 6 \sin x + C.$$

EXAMPLE 5   Using a Reduction Formula

Evaluate $\displaystyle\int \frac{\sqrt{3-5x}}{2x}\,dx$.

Solution   Consider the following two formulas.

$$\int \frac{du}{u\sqrt{a+bu}} = \frac{1}{\sqrt{a}} \ln \left| \frac{\sqrt{a+bu} - \sqrt{a}}{\sqrt{a+bu} + \sqrt{a}} \right| + C \qquad \text{Formula 17 } (a > 0)$$

$$\int \frac{\sqrt{a+bu}}{u}\,du = 2\sqrt{a+bu} + a \int \frac{du}{u\sqrt{a+bu}} \qquad \text{Formula 19}$$

Using Formula 19, with $a = 3$, $b = -5$, and $u = x$, produces

$$\frac{1}{2} \int \frac{\sqrt{3-5x}}{x}\,dx = \frac{1}{2}\left(2\sqrt{3-5x} + 3 \int \frac{dx}{x\sqrt{3-5x}} \right)$$

$$= \sqrt{3-5x} + \frac{3}{2} \int \frac{dx}{x\sqrt{3-5x}}.$$

Using Formula 17, with $a = 3$, $b = -5$, and $u = x$, produces

$$\int \frac{\sqrt{3-5x}}{2x}\,dx = \sqrt{3-5x} + \frac{3}{2}\left(\frac{1}{\sqrt{3}} \ln \left| \frac{\sqrt{3-5x} - \sqrt{3}}{\sqrt{3-5x} + \sqrt{3}} \right| \right) + C$$

$$= \sqrt{3-5x} + \frac{\sqrt{3}}{2} \ln \left| \frac{\sqrt{3-5x} - \sqrt{3}}{\sqrt{3-5x} + \sqrt{3}} \right| + C.$$

NOTE   Sometimes when you use symbolic integration utilities you obtain results that look very different, but are actually equivalent. We used several to evaluate the integral in Example 5, as follows.

Maple

$$\int \frac{\sqrt{3-5x}}{2x}\,dx =$$

$$\sqrt{3-5x} -$$

$$\sqrt{3} \operatorname{arctanh}\left(\tfrac{1}{3}\sqrt{3-5x}\sqrt{3} \right)$$

Derive

$$\int \frac{\sqrt{3-5x}}{2x}\,dx =$$

$$\sqrt{3} \ln \left[\frac{\sqrt{(3-5x)} - \sqrt{3}}{\sqrt{x}} \right] +$$

$$\sqrt{(3-5x)}$$

Mathematica

$$\int \frac{\sqrt{3-5x}}{2x}\,dx =$$

$$\text{Sqrt}[3-5x] - \text{Sqrt}[3]\,\text{ArcTanh}$$

$$\left[\frac{\text{Sqrt}[3-5x]}{\text{Sqrt}[3]} \right]$$

Mathcad

$$\int \frac{\sqrt{3-5x}}{2x}\,dx =$$

$$\sqrt{3-5x} +$$

$$\tfrac{1}{2}\sqrt{3} \ln\left[-\tfrac{1}{5} \frac{(-6 + 5x + 2\sqrt{3}\sqrt{3-5x})}{x} \right]$$

Notice that symbolic integration utilities do not include a constant of integration.

Rational Functions of Sine and Cosine

EXAMPLE 6   Integration by Tables

Evaluate $\displaystyle\int \frac{\sin 2x}{2 + \cos x}\, dx$.

Solution   Substituting $2 \sin x \cos x$ for $\sin 2x$ produces

$$\int \frac{\sin 2x}{2 + \cos x}\, dx = 2 \int \frac{\sin x \cos x}{2 + \cos x}\, dx.$$

A check of the forms involving $\sin u$ or $\cos u$ in the appendix shows that none of those listed applies. Therefore, you can consider forms involving $a + bu$. For example,

$$\int \frac{u\, du}{a + bu} = \frac{1}{b^2}(bu - a \ln |a + bu|) + C. \qquad \text{Formula 3}$$

Let $a = 2$, $b = 1$, and $u = \cos x$. Then $du = -\sin x\, dx$, and you have

$$\begin{aligned}
2 \int \frac{\sin x \cos x}{2 + \cos x}\, dx &= -2 \int \frac{\cos x (-\sin x\, dx)}{2 + \cos x} \\
&= -2(\cos x - 2 \ln |2 + \cos x|) + C \\
&= -2 \cos x + 4 \ln |2 + \cos x| + C.
\end{aligned}$$

Example 6 involves a rational expression of $\sin x$ and $\cos x$. If you are unable to find an integral of this form in the integration tables, try using the following special substitution to convert the trigonometric expression to a standard rational expression.

Substitution for Rational Functions of Sine and Cosine

For integrals involving rational functions of sine and cosine, the substitution

$$u = \frac{\sin x}{1 + \cos x} = \tan \frac{x}{2}$$

yields

$$\cos x = \frac{1 - u^2}{1 + u^2}, \quad \sin x = \frac{2u}{1 + u^2}, \quad \text{and} \quad dx = \frac{2\, du}{1 + u^2}.$$

Proof   From the substitution for u, it follows that

$$u^2 = \frac{\sin^2 x}{(1 + \cos x)^2} = \frac{1 - \cos^2 x}{(1 + \cos x)^2} = \frac{1 - \cos x}{1 + \cos x}.$$

Solving for $\cos x$ produces $\cos x = (1 - u^2)/(1 + u^2)$. To find $\sin x$, write $u = \sin x/(1 + \cos x)$ as

$$\sin x = u(1 + \cos x) = u \left(1 + \frac{1 - u^2}{1 + u^2} \right) = \frac{2u}{1 + u^2}.$$

Finally, to find dx, consider $u = \tan(x/2)$. Then you have $\arctan u = x/2$ and $dx = (2\, du)/(1 + u^2)$.

EXERCISES FOR SECTION 7.6

In Exercises 1 and 2, use a table of integrals with forms involving $a + bu$ to evaluate the integral.

1. $\displaystyle\int \frac{x^2}{1 + x}\, dx$

2. $\displaystyle\int \frac{1}{2x^2(2x - 1)^2}\, dx$

In Exercises 3 and 4, use a table of integrals with forms involving $\sqrt{u^2 \pm a^2}$ to evaluate the integral.

3. $\displaystyle\int e^x \sqrt{1 + e^{2x}}\, dx$

4. $\displaystyle\int \frac{\sqrt{x^2 - 4}}{x}\, dx$

In Exercises 5 and 6, use a table of integrals with forms involving $\sqrt{a^2 - u^2}$ to evaluate the integral.

5. $\displaystyle\int \frac{1}{x^2\sqrt{1 - x^2}}\, dx$

6. $\displaystyle\int \frac{x}{\sqrt{9 - x^4}}\, dx$

In Exercises 7–10, use a table of integrals with forms involving the trigonometric functions to evaluate the integral.

7. $\displaystyle\int \sin^4 2x\, dx$

8. $\displaystyle\int \frac{\cos^3 \sqrt{x}}{\sqrt{x}}\, dx$

9. $\displaystyle\int \frac{1}{\sqrt{x}\,(1 - \cos \sqrt{x})}\, dx$

10. $\displaystyle\int \frac{1}{1 - \tan 5x}\, dx$

In Exercises 11 and 12, use a table of integrals with forms involving e^u to evaluate the integral.

11. $\displaystyle\int \frac{1}{1 + e^{2x}}\, dx$

12. $\displaystyle\int e^{-2x} \cos 3x\, dx$

In Exercises 13 and 14, use a table of integrals with forms involving $\ln u$ to evaluate the integral.

13. $\displaystyle\int x^3 \ln x\, dx$

14. $\displaystyle\int (\ln x)^3\, dx$

In Exercises 15–18, find the indefinite integral (a) using integration tables and (b) using the indicated method.

Integral	Method
15. $\displaystyle\int x^2 e^x\, dx$	Integration by parts
16. $\displaystyle\int x^4 \ln x\, dx$	Integration by parts
17. $\displaystyle\int \frac{1}{x^2(x + 1)}\, dx$	Partial fractions
18. $\displaystyle\int \frac{1}{x^2 - 75}\, dx$	Partial fractions

In Exercises 19–50, use the integration tables to evaluate the integral.

19. $\displaystyle\int xe^{x^2}\, dx$

20. $\displaystyle\int \frac{x}{\sqrt{1 + x}}\, dx$

21. $\displaystyle\int x \operatorname{arcsec}(x^2 + 1)\, dx$

22. $\displaystyle\int \operatorname{arcsec} 2x\, dx$

23. $\displaystyle\int x^2 \ln x\, dx$

24. $\displaystyle\int x \sin x\, dx$

25. $\displaystyle\int \frac{1}{x^2\sqrt{x^2 - 4}}\, dx$

26. $\displaystyle\int \frac{x^2}{(3x - 5)^2}\, dx$

27. $\displaystyle\int \frac{2x}{(1 - 3x)^2}\, dx$

28. $\displaystyle\int \frac{1}{x^2 + 2x + 2}\, dx$

29. $\displaystyle\int e^x \arccos e^x\, dx$

30. $\displaystyle\int \frac{\theta^2}{1 - \sin \theta^3}\, d\theta$

31. $\displaystyle\int \frac{x}{1 - \sec x^2}\, dx$

32. $\displaystyle\int \frac{e^x}{1 - \tan e^x}\, dx$

33. $\displaystyle\int \frac{\cos x}{1 + \sin^2 x}\, dx$

34. $\displaystyle\int \frac{1}{t[1 + (\ln t)^2]}\, dt$

35. $\displaystyle\int \frac{\cos \theta}{3 + 2\sin \theta + \sin^2 \theta}\, d\theta$

36. $\displaystyle\int \sqrt{3 + x^2}\, dx$

37. $\displaystyle\int \frac{1}{x^2\sqrt{2 + 9x^2}}\, dx$

38. $\displaystyle\int x^2\sqrt{2 + 9x^2}\, dx$

39. $\displaystyle\int t^4 \cos t\, dt$

40. $\displaystyle\int \sqrt{x} \arctan x^{3/2}\, dx$

41. $\displaystyle\int \frac{\ln x}{x(3 + 2\ln x)}\, dx$

42. $\displaystyle\int \frac{e^x}{(1 - e^{2x})^{3/2}}\, dx$

43. $\displaystyle\int \frac{x}{(x^2 - 6x + 10)^2}\, dx$

44. $\displaystyle\int (2x - 3)^2\sqrt{(2x - 3)^2 + 4}\, dx$

45. $\displaystyle\int \frac{x}{\sqrt{x^4 - 6x^2 + 5}}\, dx$

46. $\displaystyle\int \frac{\cos x}{\sqrt{\sin^2 x + 1}}\, dx$

47. $\displaystyle\int \frac{x^3}{\sqrt{4 - x^2}}\, dx$

48. $\displaystyle\int \sqrt{\frac{3 - x}{3 + x}}\, dx$

49. $\displaystyle\int \frac{e^{3x}}{(1 + e^x)^3}\, dx$

50. $\displaystyle\int \sec^5 \theta\, d\theta$

In Exercises 51–56, verify the integration formula.

51. $\displaystyle\int \frac{u^2}{(a + bu)^2}\, du =$
$$\frac{1}{b^3}\left(bu - \frac{a^2}{a + bu} - 2a \ln|a + bu|\right) + C$$

52. $\displaystyle\int \frac{u^n}{\sqrt{a + bu}}\, du =$
$$\frac{2}{(2n + 1)b}\left(u^n \sqrt{a + bu} - na \int \frac{u^{n-1}}{\sqrt{a + bu}}\, du\right)$$

53. $\displaystyle\int \frac{1}{(u^2 \pm a^2)^{3/2}} \, du = \frac{\pm u}{a^2 \sqrt{u^2 \pm a^2}} + C$

54. $\displaystyle\int u^n \cos u \, du = u^n \sin u - n \int u^{n-1} \sin u \, du$

55. $\displaystyle\int \arctan u \, du = u \arctan u - \ln\sqrt{1 + u^2} + C$

56. $\displaystyle\int (\ln u)^n \, du = u(\ln u)^n - n \int (\ln u)^{n-1} \, du$

In Exercises 57–62, use a symbolic integration utility to determine the antiderivative that passes through the indicated point. Use the utility to graph the resulting antiderivative.

Integral	Point
57. $\displaystyle\int \frac{1}{x^{3/2}\sqrt{1-x}} \, dx$	$\left(\frac{1}{2}, 5\right)$
58. $\displaystyle\int x\sqrt{x^2 + 2x} \, dx$	$(0, 0)$
59. $\displaystyle\int \frac{\sqrt{2 - 2x - x^2}}{x + 1} \, dx$	$(0, \sqrt{2})$
60. $\displaystyle\int \frac{1}{(x^2 - 6 + 10)^2} \, dx$	$(3, 0)$
61. $\displaystyle\int \frac{1}{\sin \theta \tan \theta} \, d\theta$	$\left(\frac{\pi}{4}, 2\right)$
62. $\displaystyle\int \frac{\sin \theta}{(\cos \theta)(1 + \sin \theta)} \, d\theta$	$(0, 1)$

In Exercises 63–70, evaluate the integral.

63. $\displaystyle\int \frac{1}{2 - 3\sin \theta} \, d\theta$

64. $\displaystyle\int \frac{\sin \theta}{1 + \cos^2 \theta} \, d\theta$

65. $\displaystyle\int_0^{\pi/2} \frac{1}{1 + \sin \theta + \cos \theta} \, d\theta$

66. $\displaystyle\int_0^{\pi/2} \frac{1}{3 - 2\cos \theta} \, d\theta$

67. $\displaystyle\int \frac{\sin \theta}{3 - 2\cos \theta} \, d\theta$

68. $\displaystyle\int \frac{\sin \theta}{1 + \sin \theta} \, d\theta$

69. $\displaystyle\int \frac{\cos \sqrt{\theta}}{\sqrt{\theta}} \, d\theta$

70. $\displaystyle\int \frac{1}{\sec \theta - \tan \theta} \, d\theta$

Area   **In Exercises 71 and 72, find the area of the region bounded by the graphs of the equations.**

71. $y = \dfrac{x}{\sqrt{x + 1}}, y = 0, x = 8$

72. $y = \dfrac{x}{1 + e^{x^2}}, y = 0, x = 2$

73. Work   A hydraulic cylinder on an industrial machine pushes a steel block a distance of x feet $(0 \le x \le 5)$, where the variable force required is

$$F(x) = 2000xe^{-x} \text{ pounds.}$$

Find the work done in pushing the block the full 5 feet through the machine.

74. Work   Repeat Exercise 73, using a force of

$$F(x) = \frac{500x}{\sqrt{26 - x^2}} \text{ pounds.}$$

75. Building Design   The cross section of a precast concrete beam for a building is bounded by the graphs of the equations

$$x = \frac{2}{\sqrt{1 + y^2}}, x = \frac{-2}{\sqrt{1 + y^2}}, y = 0 \text{ and } y = 3$$

where x and y are measured in feet. The length of the beam is 20 feet (see figure).

(a) Find the volume V and the weight W of the concrete in the beam. Assume the concrete weighs 148 pounds per cubic foot.

(b) Find the centroid of a cross section of the beam.

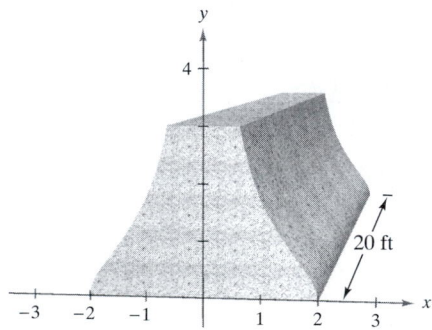

76. Average Population Size   A population is growing according to the logistics model

$$N = \frac{5000}{1 + e^{4.8 - 1.9t}}$$

where t is the time in days. Find the average population over the interval $[0, 2]$.

In Exercises 77 and 78, use a graphing utility to (a) solve the integral equation for the constant k and (b) graph the region whose area is given by the integral.

77. $\displaystyle\int_0^4 \frac{k}{2 + 3x} \, dx = 10$

78. $\displaystyle\int_0^k 6x^2 e^{-x/2} \, dx = 50$

Indeterminate Forms • L'Hôpital's Rule

Indeterminate Forms

Recall from Chapters 1 and 3 that the forms $0/0$ and ∞/∞ are called *indeterminate* because they do not guarantee that a limit exists, nor do they indicate what the limit is, if one does exist. When you encountered one of these indeterminate forms earlier in the text, you attempted to rewrite the expression by using various algebraic techniques.

Indeterminate Form	Limit	Algebraic Technique
$\dfrac{0}{0}$	$\displaystyle\lim_{x\to-1}\frac{2x^2-2}{x+1} = \lim_{x\to-1} 2(x-1)$ $= -4$	Divide numerator and denominator by $(x+1)$.
$\dfrac{\infty}{\infty}$	$\displaystyle\lim_{x\to\infty}\frac{3x^2-1}{2x^2+1} = \lim_{x\to\infty}\frac{3-(1/x^2)}{2+(1/x^2)}$ $= \dfrac{3}{2}$	Divide numerator and denominator by x^2.

Occasionally, you can extend these algebraic techniques to find limits of transcendental functions. For instance, the limit

$$\lim_{x\to0}\frac{e^{2x}-1}{e^x-1}$$

produces the indeterminate form $0/0$. Factoring and then dividing produces

$$\lim_{x\to0}\frac{e^{2x}-1}{e^x-1} = \lim_{x\to0}\frac{(e^x+1)(e^x-1)}{e^x-1} = \lim_{x\to0}(e^x+1) = 2.$$

However, not all indeterminate forms can be evaluated by algebraic manipulation. This is particularly true when *both* algebraic and transcendental functions are involved. For instance, the limit

$$\lim_{x\to0}\frac{e^{2x}-1}{x}$$

produces the indeterminate form $0/0$. Rewriting the expression to obtain

$$\lim_{x\to0}\left(\frac{e^{2x}}{x}-\frac{1}{x}\right)$$

merely produces another indeterminate form, $\infty - \infty$. Of course, you could use technology to estimate the limit, as shown in the table and in Figure 7.14. From the table and the graph, the limit appears to be 2. (This limit will be verified in Example 1.)

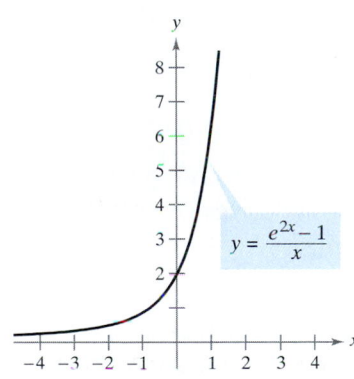

The limit as x approaches 0 appears to be 2.
Figure 7.14

x	-1	-0.1	-0.01	-0.001	0	0.001	0.01	0.1	1
$\dfrac{e^{2x}-1}{x}$	0.865	1.813	1.980	1.998	?	2.002	2.020	2.214	6.389

GUILLAUME L'HÔPITAL (1661–1704)

L'Hôpital's Rule is named after the French mathematician Guillaume Francois Antoine de L'Hôpital. L'Hôpital is credited with writing the first text on differential calculus (in 1696) in which the rule publicly appeared. It was recently discovered that the rule and its proof were written in a letter from John Bernoulli to L'Hôpital. ". . . I acknowledge that I owe very much to the bright minds of the Bernoulli brothers. . . . I have made free use of their discoveries . . . ," said L'Hôpital.

L'Hôpital's Rule

To find the limit illustrated in Figure 7.14, you can use a theorem called **L'Hôpital's Rule.** This theorem states that under certain conditions the limit of the quotient $f(x)/g(x)$ is determined by the limit of the quotient of the derivatives

$$\frac{f'(x)}{g'(x)}.$$

To prove this theorem, you can use a more general result called the **Extended Mean Value Theorem.**

THEOREM 7.3   The Extended Mean Value Theorem

If f and g are differentiable on an open interval (a, b) and continuous on $[a, b]$ such that $g'(x) \neq 0$ for any x in (a, b), then there exists a point c in (a, b) such that

$$\frac{f'(c)}{g'(c)} = \frac{f(b) - f(a)}{g(b) - g(a)}.$$

NOTE   To see why this is called the Extended Mean Value Theorem, consider the special case in which $g(x) = x$. For this case, you obtain the "standard" Mean Value Theorem as presented in Section 3.2.

The Extended Mean Value Theorem and L'Hôpital's Rule are both proved in the appendix.

THEOREM 7.4   L'Hôpital's Rule

Let f and g be functions that are differentiable on an open interval (a, b) containing c, except possibly at c itself. Assume that $g'(x) \neq 0$ for all x in (a, b), except possibly at c itself. If the limit of $f(x)/g(x)$ as x approaches c produces the indeterminate form $0/0$, then

$$\lim_{x \to c} \frac{f(x)}{g(x)} = \lim_{x \to c} \frac{f'(x)}{g'(x)}$$

provided the limit on the right exists (or is infinite). This result also applies if the limit of $f(x)/g(x)$ as x approaches c produces any one of the indeterminate forms ∞/∞, $(-\infty)/\infty$, $\infty/(-\infty)$, or $(-\infty)/(-\infty)$.

NOTE   People occasionally use L'Hôpital's Rule incorrectly by applying the Quotient Rule to $f(x)/g(x)$. Be sure you see that the rule involves $f'(x)/g'(x)$, not the derivative of $f(x)/g(x)$.

FOR FURTHER INFORMATION For more information on the necessity of the restriction that $g'(x)$ be nonzero for all x in (a, b), except possibly at c, see the article "Counterexamples to L'Hôpital's Rule" by R. P. Boas in the October 1986 issue of *The American Mathematical Monthly.*

L'Hôpital's Rule can also be applied to one-sided limits. For instance, if the limit of $f(x)/g(x)$ as x approaches c *from the right* produces the indeterminate form $0/0$, then

$$\lim_{x \to c+} \frac{f(x)}{g(x)} = \lim_{x \to c+} \frac{f'(x)}{g'(x)}$$

provided the limit exists (or is infinite).

EXPLORATION

Numerical and Graphical Approaches   Use a numerical or a graphical approach to approximate each of the following limits.

(a) $\lim\limits_{x \to 0} \dfrac{2^{2x} - 1}{x}$

(b) $\lim\limits_{x \to 0} \dfrac{3^{2x} - 1}{x}$

(c) $\lim\limits_{x \to 0} \dfrac{4^{2x} - 1}{x}$

(d) $\lim\limits_{x \to 0} \dfrac{5^{2x} - 1}{x}$

What pattern do you observe? Does an analytic approach have an advantage for these limits? If so, explain your reasoning.

EXAMPLE 1   Indeterminate Form 0/0

Evaluate $\lim\limits_{x \to 0} \dfrac{e^{2x} - 1}{x}$.

Solution   Because direct substitution results in the indeterminate form 0/0

$$\lim_{x \to 0} \frac{e^{2x} - 1}{x}$$

$$\lim_{x \to 0} (e^{2x} - 1) = 0$$

$$\lim_{x \to 0} x = 0$$

you can apply L'Hôpital's Rule as follows.

$$\lim_{x \to 0} \frac{e^{2x} - 1}{x} = \lim_{x \to 0} \frac{\dfrac{d}{dx}[e^{2x} - 1]}{\dfrac{d}{dx}[x]} \qquad \text{Apply L'Hôpital's Rule.}$$

$$= \lim_{x \to 0} \frac{2e^{2x}}{1} \qquad \text{Differentiate numerator and denominator.}$$

$$= 2 \qquad \text{Evaluate the limit.}$$

NOTE   In writing the string of equations in Example 1, you actually do not know that the first limit is equal to the second until you have shown that the second limit exists. In other words, if the second limit had not existed, it would not have been permissible to apply L'Hôpital's Rule.

Another form of L'Hôpital's Rule states that if the limit of $f(x)/g(x)$ as x approaches ∞ (or $-\infty$) produces the indeterminate form 0/0 or ∞/∞, then

$$\lim_{x \to \infty} \frac{f(x)}{g(x)} = \lim_{x \to \infty} \frac{f'(x)}{g'(x)}$$

provided the limit on the right exists.

EXAMPLE 2   Indeterminate Form ∞/∞

Evaluate $\lim\limits_{x \to \infty} \dfrac{\ln x}{x}$.

Solution   Because direct substitution results in the indeterminate form ∞/∞, you can apply L'Hôpital's Rule to obtain

$$\lim_{x \to \infty} \frac{\ln x}{x} = \lim_{x \to \infty} \frac{\dfrac{d}{dx}[\ln x]}{\dfrac{d}{dx}[x]} \qquad \text{Apply L'Hôpital's Rule.}$$

$$= \lim_{x \to \infty} \frac{1}{x} \qquad \text{Differentiate numerator and denominator.}$$

$$= 0. \qquad \text{Evaluate the limit.}$$

NOTE   Try graphing $y_1 = \ln x$ and $y_2 = x$ in the same viewing rectangle. Which function grows faster as x approaches ∞? How is this observation related to Example 2?

Occasionally it is necessary to apply L'Hôpital's Rule more than once to remove an indeterminate form, as illustrated in Example 3.

EXAMPLE 3   Applying L'Hôpital's Rule More than Once

Evaluate $\lim\limits_{x \to -\infty} \dfrac{x^2}{e^{-x}}$.

Solution   Because direct substitution results in the indeterminate form ∞/∞, you can apply L'Hôpital's Rule.

$$\lim_{x \to -\infty} \frac{x^2}{e^{-x}} = \lim_{x \to -\infty} \frac{\dfrac{d}{dx}[x^2]}{\dfrac{d}{dx}[e^{-x}]} = \lim_{x \to -\infty} \frac{2x}{-e^{-x}}$$

This limit yields the indeterminate form $(-\infty)/(-\infty)$, so you can apply L'Hôpital's Rule again to obtain

$$\lim_{x \to -\infty} \frac{2x}{-e^{-x}} = \lim_{x \to -\infty} \frac{\dfrac{d}{dx}[2x]}{\dfrac{d}{dx}[-e^{-x}]} = \lim_{x \to -\infty} \frac{2}{e^{-x}} = 0.$$

In addition to the forms $0/0$ and ∞/∞, there are other indeterminate forms such as $0 \cdot \infty$, 1^{∞}, ∞^0, 0^0, and $\infty - \infty$. For example, consider the following four limits that lead to the indeterminate form $0 \cdot \infty$.

$$\underbrace{\lim_{x \to 0} (x)\left(\frac{1}{x}\right)}_{\text{Limit is 1.}}, \qquad \underbrace{\lim_{x \to 0} (x)\left(\frac{2}{x}\right)}_{\text{Limit is 2.}}, \qquad \underbrace{\lim_{x \to \infty} (x)\left(\frac{1}{e^x}\right)}_{\text{Limit is 0.}}, \qquad \underbrace{\lim_{x \to \infty} (e^x)\left(\frac{1}{x}\right)}_{\text{Limit is } \infty.}$$

Because each limit is different, it is clear that the form $0 \cdot \infty$ is indeterminate in the sense that it does not determine the value (or even the existence) of the limit. The following examples indicate methods for evaluating these forms. Basically, you attempt to convert each of these forms to $0/0$ or ∞/∞ so that L'Hôpital's Rule can be applied.

EXAMPLE 4   Indeterminate Form $0 \cdot \infty$

Evaluate $\lim\limits_{x \to \infty} e^{-x}\sqrt{x}$.

Solution   Because direct substitution produces the indeterminate form $0 \cdot \infty$, you should try to rewrite the limit to fit the form $0/0$ or ∞/∞. In this case, you can rewrite the limit to fit the second form.

$$\lim_{x \to \infty} e^{-x}\sqrt{x} = \lim_{x \to \infty} \frac{\sqrt{x}}{e^x}$$

Now, by L'Hôpital's Rule, you have

$$\lim_{x \to \infty} \frac{\sqrt{x}}{e^x} = \lim_{x \to \infty} \frac{1/(2\sqrt{x})}{e^x} = \lim_{x \to \infty} \frac{1}{2\sqrt{x}\,e^x} = 0.$$

If rewriting a limit in one of the forms $0/0$ or ∞/∞ does not seem to work, try the other form. For instance, in Example 4 you can write the limit as

$$\lim_{x\to\infty} e^{-x}\sqrt{x} = \lim_{x\to\infty} \frac{e^{-x}}{x^{-1/2}}$$

which yields the indeterminate form $0/0$. As it happens, applying L'Hôpital's Rule to this limit produces

$$\lim_{x\to\infty} \frac{e^{-x}}{x^{-1/2}} = \lim_{x\to\infty} \frac{-e^{-x}}{-1/(2x^{3/2})}$$

which also yields the indeterminate form $0/0$.

The indeterminate forms 1^{∞}, ∞^0, and 0^0 arise from limits of functions that have variable bases and variable exponents. When we encountered this type of function in Section 5.5 we used logarithmic differentiation to find the derivative. You can use a similar procedure when taking limits, as indicated in the next example. (Note that this example serves as an alternative proof of Theorem 5.15.)

EXAMPLE 5   Indeterminate Form 1^{∞}

Evaluate $\lim_{x\to\infty} \left(1 + \dfrac{1}{x}\right)^x$.

Solution Because direct substitution yields the indeterminate form 1^{∞}, you can proceed as follows. To begin, assume that the limit exists and is equal to y.

$$y = \lim_{x\to\infty} \left(1 + \frac{1}{x}\right)^x$$

Taking the natural logarithm of both sides produces

$$\ln y = \ln\left[\lim_{x\to\infty} \left(1 + \frac{1}{x}\right)^x\right].$$

Because the natural logarithmic function is continuous, you can write the following.

$$\ln y = \lim_{x\to\infty}\left[x \ln\left(1 + \frac{1}{x}\right)\right] \qquad \text{Indeterminate form } \infty \cdot 0$$

$$= \lim_{x\to\infty}\left(\frac{\ln[1 + (1/x)]}{1/x}\right) \qquad \text{Indeterminate form } 0/0$$

$$= \lim_{x\to\infty}\left(\frac{(-1/x^2)\{1/[1 + (1/x)]\}}{-1/x^2}\right) \qquad \text{L'Hôpital's Rule}$$

$$= \lim_{x\to\infty} \frac{1}{1 + (1/x)}$$

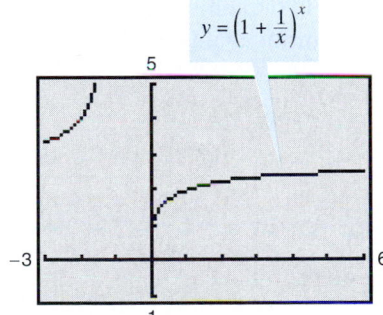

$y = \left(1 + \dfrac{1}{x}\right)^x$

The limit of $[1 + (1/x)]^x$ as x approaches infinity is e.
Figure 7.15

Now, because you have shown that $\ln y = 1$, you can conclude that $y = e$ and obtain

$$\lim_{x\to\infty} \left(1 + \frac{1}{x}\right)^x = e.$$

You can use a graphing utililty to confirm this result, as shown in Figure 7.15.

L'Hôpital's Rule can also be applied to one-sided limits, as demonstrated in Examples 6 and 7.

EXAMPLE 6   Indeterminate Form 0^0

Evaluate $\lim\limits_{x \to 0^+} (\sin x)^x$.

Solution   Because direct substitution produces the indeterminate form 0^0, you can proceed as follows. To begin, assume that the limit exists and is equal to y.

$$y = \lim_{x \to 0^+} (\sin x)^x \qquad\qquad\qquad \text{Indeterminate form } 0^0$$

$$\ln y = \ln\left[\lim_{x \to 0^+} (\sin x)^x \right] \qquad\qquad \text{Take log of both sides.}$$

$$= \lim_{x \to 0^+} \left[\ln(\sin x)^x \right] \qquad\qquad \text{Continuity}$$

$$= \lim_{x \to 0^+} \left[x \ln(\sin x) \right] \qquad\qquad \text{Indeterminate form } 0 \cdot (-\infty)$$

$$= \lim_{x \to 0^+} \frac{\ln(\sin x)}{1/x} \qquad\qquad \text{Indeterminate form } -\infty/\infty$$

$$= \lim_{x \to 0^+} \frac{\cot x}{-1/x^2} \qquad\qquad \text{L'Hôpital's Rule}$$

$$= \lim_{x \to 0^+} \frac{-x^2}{\tan x} \qquad\qquad \text{Indeterminate form } 0/0$$

$$= \lim_{x \to 0^+} \frac{-2x}{\sec^2 x} = 0 \qquad\qquad \text{L'Hôpital's Rule}$$

Now, because $\ln y = 0$, you can conclude that $y = e^0 = 1$, and it follows that

$$\lim_{x \to 0^+} (\sin x)^x = 1.$$

TECHNOLOGY   When evaluating complicated limits such as the one in Example 6, it is helpful to check the reasonableness of the solution with a calculator or with a graphing utility. For instance, the calculations in the following table and the graph in Figure 7.16 are consistent with the conclusion that $(\sin x)^x$ approaches 1 as x approaches 0 from the right.

x	1.0	0.1	0.01	0.001	0.0001	0.00001
$(\sin x)^x$	0.8415	0.7942	0.9550	0.9931	0.9991	0.9999

Try using a computer or graphing utility to estimate the following limits.

$$\lim_{x \to 0} (1 - \cos x)^x$$

and

$$\lim_{x \to 0^+} (\tan x)^x$$

Then see if you can verify your estimates analytically.

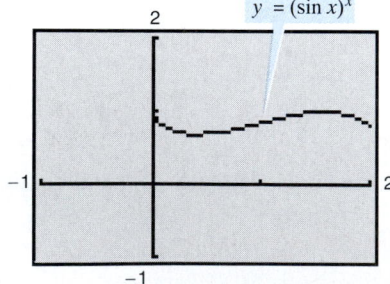

$y = (\sin x)^x$

The limit of $(\sin x)^x$ is 1 as x approaches 0 from the right.
Figure 7.16

EXAMPLE 7   Indeterminate Form $\infty - \infty$

Evaluate $\displaystyle\lim_{x\to1^+}\left(\dfrac{1}{\ln x}-\dfrac{1}{x-1}\right)$.

Solution Because direct substitution yields the indeterminate form $\infty - \infty$, you should try to rewrite the expression to produce a form to which you can apply L'Hôpital's Rule. In this case, you can combine the two fractions to obtain

$$\lim_{x\to1^+}\left(\frac{1}{\ln x}-\frac{1}{x-1}\right)=\lim_{x\to1^+}\left[\frac{x-1-\ln x}{(x-1)\ln x}\right].$$

Now, because direct substitution produces the indeterminate form $0/0$, you can apply L'Hôpital's Rule to obtain

$$\lim_{x\to1^+}\left(\frac{1}{\ln x}-\frac{1}{x-1}\right)=\lim_{x\to1^+}\frac{\dfrac{d}{dx}[x-1-\ln x]}{\dfrac{d}{dx}[(x-1)\ln x]}$$

$$=\lim_{x\to1^+}\left[\frac{1-(1/x)}{(x-1)(1/x)+\ln x}\right]$$

$$=\lim_{x\to1^+}\left(\frac{x-1}{x-1+x\ln x}\right).$$

This limit also yields the indeterminate form $0/0$, so you can apply L'Hôpital's Rule again to obtain

$$\lim_{x\to1^+}\left(\frac{1}{\ln x}-\frac{1}{x-1}\right)=\lim_{x\to1^+}\left[\frac{1}{1+x(1/x)+\ln x}\right]$$

$$=\frac{1}{2}.$$

We have identified the forms $0/0,\ \infty/\infty,\ \infty-\infty,\ 0\cdot\infty,\ 0^0,\ 1^\infty$, and ∞^0 as *indeterminate*. There are similar forms that you should recognize as "determinate."

$$\infty+\infty\to\infty \qquad\qquad \text{Limit is positive infinity.}$$
$$-\infty-\infty\to-\infty \qquad\qquad \text{Limit is negative infinity.}$$
$$0^\infty\to0 \qquad\qquad\qquad \text{Limit is zero.}$$
$$0^{-\infty}\to\infty \qquad\qquad\quad \text{Limit is positive infinity.}$$

(You are asked to verify two of these in Exercises 83 and 84.)

As a final comment, we remind you that L'Hôpital's Rule can be applied only to quotients leading to the indeterminate forms $0/0$ and ∞/∞. For instance, the following application of L'Hôpital's Rule is *incorrect*.

$$\lim_{x\to0}\frac{e^x}{x}=\lim_{x\to0}\frac{e^x}{1}=1 \qquad\qquad \text{Incorrect use of L'Hôpital's Rule}$$

The reason this application is incorrect is that, even though the limit of the denominator is 0, the limit of the numerator is 1, which means that the hypotheses of L'Hôpital's Rule have not been satisfied.

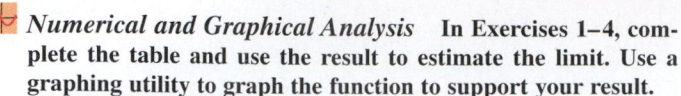 *Numerical and Graphical Analysis* In Exercises 1–4, complete the table and use the result to estimate the limit. Use a graphing utility to graph the function to support your result.

1. $\displaystyle \lim_{x \to 0} \frac{\sin 5x}{\sin 2x}$

x	-0.1	-0.01	-0.001	0.001	0.01	0.1
$f(x)$						

2. $\displaystyle \lim_{x \to 0} \frac{1 - e^x}{x}$

x	-0.1	-0.01	-0.001	0.001	0.01	0.1
$f(x)$						

3. $\displaystyle \lim_{x \to \infty} x^5 e^{-x/100}$

x	1	10	10^2	10^3	10^4	10^5
$f(x)$						

4. $\displaystyle \lim_{x \to \infty} \frac{6x}{\sqrt{3x^2 - 2x}}$

x	1	10	10^2	10^3	10^4	10^5
$f(x)$						

In Exercises 5–10, evaluate the limit (a) using techniques from Chapters 1 and 3 and (b) using L'Hôpital's Rule.

5. $\displaystyle \lim_{x \to 3} \frac{2(x - 3)}{x^2 - 9}$

6. $\displaystyle \lim_{x \to -1} \frac{2x^2 - x - 3}{x + 1}$

7. $\displaystyle \lim_{x \to 3} \frac{\sqrt{x + 1} - 2}{x - 3}$

8. $\displaystyle \lim_{x \to 0} \frac{\sin 4x}{2x}$

9. $\displaystyle \lim_{x \to \infty} \frac{5x^2 - 3x + 1}{3x^2 - 5}$

10. $\displaystyle \lim_{x \to \infty} \frac{2x + 1}{4x^2 + x}$

In Exercises 11–30, evaluate the limit using L'Hôpital's Rule if necessary. (In Exercise 17, n is a positive integer.)

11. $\displaystyle \lim_{x \to 2} \frac{x^2 - x - 2}{x - 2}$

12. $\displaystyle \lim_{x \to -1} \frac{x^2 - x - 2}{x + 1}$

13. $\displaystyle \lim_{x \to 0} \frac{\sqrt{4 - x^2} - 2}{x}$

14. $\displaystyle \lim_{x \to 2^-} \frac{\sqrt{4 - x^2}}{x - 2}$

15. $\displaystyle \lim_{x \to 0} \frac{e^x - (1 - x)}{x}$

16. $\displaystyle \lim_{x \to 0^+} \frac{e^x - (1 + x)}{x^3}$

17. $\displaystyle \lim_{x \to 0^+} \frac{e^x - (1 + x)}{x^n}$

18. $\displaystyle \lim_{x \to 1} \frac{\ln x}{x^2 - 1}$

19. $\displaystyle \lim_{x \to 0} \frac{\sin 2x}{\sin 3x}$

20. $\displaystyle \lim_{x \to 0} \frac{\sin ax}{\sin bx}$

21. $\displaystyle \lim_{x \to 0} \frac{\arcsin x}{x}$

22. $\displaystyle \lim_{x \to 1} \frac{\arctan x - (\pi/4)}{x - 1}$

23. $\displaystyle \lim_{x \to \infty} \frac{3x^2 - 2x + 1}{2x^2 + 3}$

24. $\displaystyle \lim_{x \to \infty} \frac{x - 1}{x^2 + 2x + 3}$

25. $\displaystyle \lim_{x \to \infty} \frac{x^2 + 2x + 3}{x - 1}$

26. $\displaystyle \lim_{x \to \infty} \frac{x^2}{e^x}$

27. $\displaystyle \lim_{x \to \infty} \frac{x}{\sqrt{x^2 + 1}}$

28. $\displaystyle \lim_{x \to \infty} \frac{\sin x}{x - \pi}$

29. $\displaystyle \lim_{x \to \infty} \frac{\ln x}{x}$

30. $\displaystyle \lim_{x \to \infty} \frac{e^x}{x}$

 In Exercises 31–44, (a) describe the type of indeterminate form (if any) that is obtained by direct substitution. (b) Evaluate the limit using L'Hôpital's Rule if necessary. (c) Use a graphing utility to graph the function and verify the result in part (b). (For a geometric approach to Exercise 31, see the article by John H. Mathews in the May 1992 issue of *The College Mathematics Journal*.)

31. $\displaystyle \lim_{x \to 0^+} (-x \ln x)$

32. $\displaystyle \lim_{x \to 0^+} x^2 \cot x$

33. $\displaystyle \lim_{x \to \infty} \left(x \sin \frac{1}{x} \right)$

34. $\displaystyle \lim_{x \to \infty} x \tan \frac{1}{x}$

35. $\displaystyle \lim_{x \to 0^+} x^{1/x}$

36. $\displaystyle \lim_{x \to 0^+} (e^x + x)^{1/x}$

37. $\displaystyle \lim_{x \to \infty} x^{1/x}$

38. $\displaystyle \lim_{x \to \infty} \left(1 + \frac{1}{x} \right)^x$

39. $\displaystyle \lim_{x \to 0^+} (1 + x)^{1/x}$

40. $\displaystyle \lim_{x \to \infty} (1 + x)^{1/x}$

41. $\displaystyle \lim_{x \to 2^+} \left(\frac{8}{x^2 - 4} - \frac{x}{x - 2} \right)$

42. $\displaystyle \lim_{x \to 2^+} \left(\frac{1}{x^2 - 4} - \frac{\sqrt{x - 1}}{x^2 - 4} \right)$

43. $\displaystyle \lim_{x \to 1^+} \left(\frac{3}{\ln x} - \frac{2}{x - 1} \right)$

44. $\displaystyle \lim_{x \to 0^+} \left(\frac{1}{x} - \frac{1}{x^2} \right)$

 In Exercises 45–48, use a graphing utility to (a) graph the function and (b) find the required limit (if it exists).

45. $\displaystyle \lim_{x \to 3} \frac{x - 3}{\ln(2x - 5)}$

46. $\displaystyle \lim_{x \to 0^+} (\sin x)^x$

47. $\displaystyle \lim_{x \to \infty} \left(\sqrt{x^2 + 5x + 2} - x \right)$

48. $\displaystyle \lim_{x \to \infty} \frac{x^3}{e^{2x}}$

49. *Think About It* Find the differentiable functions f and g that satisfy the specified condition such that $\displaystyle \lim_{x \to 5} f(x) = 0$ and $\displaystyle \lim_{x \to 5} g(x) = 0$.

(a) $\displaystyle \lim_{x \to 5} \frac{f(x)}{g(x)} = 10$ (b) $\displaystyle \lim_{x \to 5} \frac{f(x)}{g(x)} = 0$ (c) $\displaystyle \lim_{x \to 5} \frac{f(x)}{g(x)} = \infty$

(*Note:* There are many correct answers.)

50. *Think About It*   Find differentiable functions f and g such that

$$\lim_{x \to \infty} f(x) = \lim_{x \to \infty} g(x) = \infty$$

$$\lim_{x \to \infty} [f(x) - g(x)] = 25.$$

(*Note:* There are many correct answers.)

Comparing Functions   In Exercises 51–56, use L'Hôpital's Rule to determine the comparative rates of increase of the functions

$$f(x) = x^m, \quad g(x) = e^{nx}, \quad \text{and} \quad h(x) = (\ln x)^n$$

where $n > 0, m > 0$, and $x \to \infty$.

51. $\displaystyle \lim_{x \to \infty} \frac{x^2}{e^{5x}}$ **52.** $\displaystyle \lim_{x \to \infty} \frac{x^3}{e^{2x}}$

53. $\displaystyle \lim_{x \to \infty} \frac{(\ln x)^3}{x}$ **54.** $\displaystyle \lim_{x \to \infty} \frac{(\ln x)^2}{x^3}$

55. $\displaystyle \lim_{x \to \infty} \frac{(\ln x)^n}{x^m}$ **56.** $\displaystyle \lim_{x \to \infty} \frac{x^m}{e^{nx}}$

57. *Numerical Approach*   Complete the table to show that x eventually "overpowers" $(\ln x)^4$.

x	10	10^2	10^4	10^6	10^8	10^{10}
$\dfrac{(\ln x)^4}{x}$						

58. *Numerical Approach*   Complete the table to show that e^x eventually "overpowers" x^5.

x	1	5	10	20	30	40	50	100
$\dfrac{e^x}{x^5}$								

In Exercises 59–62, find any asymptotes and relative extrema that may exist and use a graphing utility to graph the function. (*Hint:* Some of the limits required in finding asymptotes have been found in preceding exercises.)

59. $y = x^{1/x}, \quad x > 0$ **60.** $y = x^x, \quad x > 0$

61. $y = 2xe^{-x}$ **62.** $y = \dfrac{\ln x}{x}$

Think About It   In Exercises 63–66, L'Hôpital's Rule is used *incorrectly.* Describe the error.

63. $\displaystyle \lim_{x \to 0} \frac{e^{2x} - 1}{e^x} = \lim_{x \to 0} \frac{2e^{2x}}{e^x} = \lim_{x \to 0} 2e^x = 2$

64. $\displaystyle \lim_{x \to 0} \frac{\sin \pi x - 1}{x} = \lim_{x \to 0} \frac{\pi \cos \pi x}{1} = \pi$

65. $\displaystyle \lim_{x \to \infty} x \cos \frac{1}{x} = \lim_{x \to \infty} \frac{\cos(1/x)}{1/x}$

$$= \lim_{x \to \infty} \frac{[-\sin(1/x)](1/x^2)}{-1/x^2}$$

$$= 0$$

66. $\displaystyle \lim_{x \to \infty} \frac{e^{-x}}{1 + e^{-x}} = \lim_{x \to \infty} \frac{-e^{-x}}{-e^{-x}} = \lim_{x \to \infty} 1 = 1$

67. *Analytical Approach*   Consider $\displaystyle \lim_{x \to \infty} \frac{x}{\sqrt{x^2 + 1}}$.

(a) Find the limit analytically without trying to use L'Hôpital's Rule.

(b) Show that L'Hôpital's Rule fails.

(c) Use a graphing utility to graph the function and approximate the limit from the graph. Compare the result with that in part (a).

68. *Compound Interest*   The formula for the amount A in a savings account compounded n times per year for t years at an interest rate r and an initial deposit of P is

$$A = P\left(1 + \frac{r}{n}\right)^{nt}.$$

Use L'Hôpital's Rule to show that the limiting formula as the number of compoundings per year becomes infinite is

$$A = Pe^{rt}.$$

69. *Velocity in a Resisting Medium*   The velocity of an object falling through a resisting medium such as air or water is

$$v = \frac{32}{k}\left(1 - e^{-kt} + \frac{v_0 k e^{-kt}}{32}\right)$$

where v_0 is the initial velocity, t is the time in seconds, and k is the resistence constant of the medium. Use L'Hôpital's Rule to find the formula for the velocity of a falling body in a vacuum by fixing v_0 and t and letting k approach zero. (Assume that the downward direction is positive.)

70. *The Gamma Function*   The Gamma Function $\Gamma(n)$ is defined in terms of the integral of the function

$$f(x) = x^{n-1}e^{-x}, \quad n > 0.$$

Show that for any fixed value of n, the limit of $f(x)$ as x approaches infinity is zero.

71. *Area*   Find the limit, as x approaches 0, of the ratio of the area of the triangle to the total shaded area in the figure.

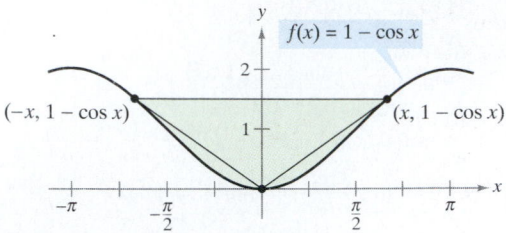

72. Use a graphing utility to graph

$$f(x) = \frac{x^k - 1}{k}$$

for $k = 1, 0.1$, and 0.01. Then evaluate the limit

$$\lim_{k \to 0^+} \frac{x^k - 1}{k}.$$

In Exercises 73–76, apply the Extended Mean Value Theorem to the functions f and g on the indicated interval. Find all values c in the interval (a, b) such that

$$\frac{f'(c)}{g'(c)} = \frac{f(b) - f(a)}{g(b) - g(a)}.$$

Functions	*Interval*
73. $f(x) = x^3$	$[0, 1]$
$g(x) = x^2 + 1$	
74. $f(x) = \dfrac{1}{x}$	$[1, 2]$
$g(x) = x^2 - 4$	
75. $f(x) = \sin x$	$\left[0, \dfrac{\pi}{2}\right]$
$g(x) = \cos x$	
76. $f(x) = \ln x$	$[1, 4]$
$g(x) = x^3$	

True or False?   **In Exercises 77–80, determine whether the statement is true or false. If it is false, explain why or give an example that shows it is false.**

77. $\lim\limits_{x \to 0} \left[\dfrac{x^2 + x + 1}{x} \right] = \lim\limits_{x \to 0} \left[\dfrac{2x + 1}{1} \right] = 1$

78. If $y = e^x/x^2$, then $y' = e^x/2x$.

79. If $p(x)$ is a polynomial, then $\lim\limits_{x \to \infty} \left[p(x)/e^x \right] = 0$.

80. If $\lim\limits_{x \to \infty} \dfrac{f(x)}{g(x)} = 1$, then $\lim\limits_{x \to \infty} \left[f(x) - g(x) \right] = 0$.

81. In Chapter 1 we used a geometrical argument (see figure) to prove that

$$\lim_{\theta \to 0} \frac{\sin \theta}{\theta} = 1.$$

   (a) Express the area of the triangle $\triangle ABD$ in terms of θ.

   (b) Express the area of the shaded region in terms of θ.

   (c) Express the ratio R of the area of $\triangle ABD$ to that of the shaded region.

   (d) Find $\lim\limits_{\theta \to 0} R$.

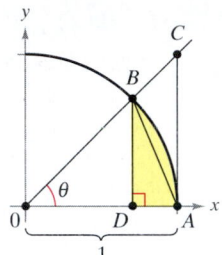

82. Sketch the graph of

$$g(x) = \begin{cases} e^{-1/x^2}, & x \neq 0 \\ 0, & x = 0 \end{cases}$$

and determine $g'(0)$.

83. Prove that if $f(x) \geq 0$, $\lim\limits_{x \to a} f(x) = 0$, and $\lim\limits_{x \to a} g(x) = \infty$, then

$$\lim_{x \to a} f(x)^{g(x)} = 0.$$

84. Prove that if $f(x) \geq 0$, $\lim\limits_{x \to a} f(x) = 0$, and $\lim\limits_{x \to a} g(x) = -\infty$, then

$$\lim_{x \to a} f(x)^{g(x)} = \infty.$$

85. Prove the following generalization of the Mean Value Theorem. If f is twice differentiable on the closed interval $[a, b]$, then

$$f(b) - f(a) = f'(a)(b - a) - \int_a^b f''(t)(t - b)\, dt.$$

86. Show that the indeterminate form 0^0 is not always equal to 1 by evaluating

$$\lim_{x \to 0^+} x^{\ln 2/(1 + \ln x)}.$$

Improper Integrals with Infinite Limits of Integration • Improper Integrals with Infinite Discontinuities

Improper Integrals with Infinite Limits of Integration

The definition of a definite integral

$$\int_a^b f(x)\,dx$$

requires that the interval $[a, b]$ be finite. Furthermore, the Fundamental Theorem of Calculus, by which you have been evaluating definite integrals, requires that f be continuous on $[a, b]$. In this section you will study a procedure for evaluating integrals that do not satisfy these requirements—usually because either one or both of the limits of integration are infinite, or f has a finite number of infinite discontinuities in the interval $[a, b]$. Integrals that possess either property are **improper integrals.** Note that a function f is said to have an **infinite discontinuity** at c if, *from the right or left,*

$$\lim_{x \to c} f(x) = \infty \qquad \text{or} \qquad \lim_{x \to c} f(x) = -\infty.$$

To get an idea of how to evaluate an improper integral, consider the integral

$$\int_1^b \frac{dx}{x^2} = -\frac{1}{x}\bigg]_1^b = -\frac{1}{b} + 1 = 1 - \frac{1}{b}$$

which can be interpreted as the area of the shaded region shown in Figure 7.17. Taking the limit as $b \to \infty$ produces

$$\int_1^\infty \frac{dx}{x^2} = \lim_{b \to \infty}\left(\int_1^b \frac{dx}{x^2}\right) = \lim_{b \to \infty}\left(1 - \frac{1}{b}\right) = 1.$$

This improper integral can be interpreted as the area of the *unbounded* region between the graph of $f(x) = 1/x^2$ and the *x*-axis (to the right of $x = 1$).

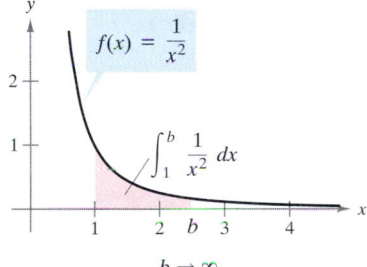

The unbounded region has an area of 1.
Figure 7.17

Definition of Improper Integrals with Infinite Integration Limits

1. If f is continuous on the interval $[a, \infty)$, then

$$\int_a^\infty f(x)\,dx = \lim_{b \to \infty}\int_a^b f(x)\,dx.$$

2. If f is continuous on the interval $(-\infty, b]$, then

$$\int_{-\infty}^b f(x)\,dx = \lim_{a \to -\infty}\int_a^b f(x)\,dx.$$

3. If f is continuous on the interval $(-\infty, \infty)$, then

$$\int_{-\infty}^\infty f(x)\,dx = \int_{-\infty}^c f(x)\,dx + \int_c^\infty f(x)\,dx$$

where c is any real number.

In the first two cases, the improper integral **converges** if the limit exists—otherwise, the improper integral **diverges.** In the third case, the improper integral on the left diverges if either of the improper integrals on the right diverges.

EXAMPLE 1 An Improper Integral That Diverges

Evaluate $\displaystyle\int_1^\infty \frac{dx}{x}$.

Solution

$$\int_1^\infty \frac{dx}{x} = \lim_{b\to\infty} \int_1^b \frac{dx}{x} \qquad \text{Take limit as } b \to \infty.$$

$$= \lim_{b\to\infty} \left[\ln x \right]_1^b \qquad \text{Apply Log Rule.}$$

$$= \lim_{b\to\infty} (\ln b - 0) \qquad \text{Apply Fundamental Theorem of Calculus.}$$

$$= \infty \qquad \text{Evaluate limit.}$$

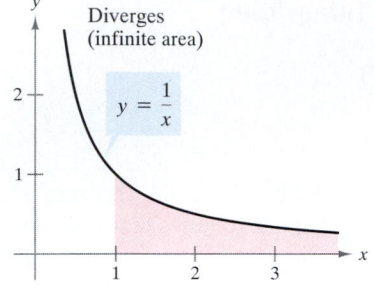

y Diverges (infinite area)

$y = \dfrac{1}{x}$

This unbounded region has an infinite area.
Figure 7.18

NOTE Try comparing the regions shown in Figures 7.17 and 7.18. They look similar, yet the region in Figure 7.17 has a finite area of 1 and the region in Figure 7.18 has an infinite area.

EXAMPLE 2 Improper Integrals That Converge

Evaluate each of the improper integrals.

a. $\displaystyle\int_0^\infty e^{-x}\, dx$ **b.** $\displaystyle\int_0^\infty \frac{1}{x^2 + 1}\, dx$

Solution

a. $\displaystyle\int_0^\infty e^{-x}\, dx = \lim_{b\to\infty} \int_0^b e^{-x}\, dx$

$$= \lim_{b\to\infty} \left[-e^{-x} \right]_0^b$$

$$= \lim_{b\to\infty} (-e^{-b} + 1)$$

$$= 1$$

(See Figure 7.19.)

b. $\displaystyle\int_0^\infty \frac{1}{x^2 + 1}\, dx = \lim_{b\to\infty} \int_0^b \frac{1}{x^2 + 1}\, dx$

$$= \lim_{b\to\infty} \left[\arctan x \right]_0^b$$

$$= \lim_{b\to\infty} \arctan b$$

$$= \frac{\pi}{2}$$

(See Figure 7.20.)

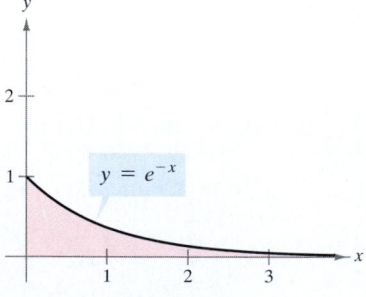

$y = e^{-x}$

The area of the unbounded region is 1.
Figure 7.19

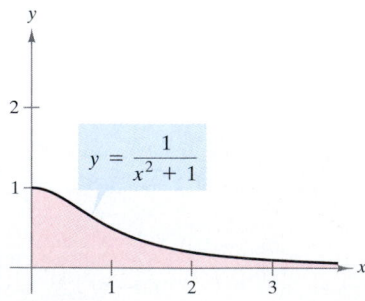

$y = \dfrac{1}{x^2 + 1}$

The area of the unbounded region is $\pi/2$.
Figure 7.20

In the following example, note how L'Hôpital's Rule can be used to evaluate an improper integral.

EXAMPLE 3   Using L'Hôpital's Rule with an Improper Integral

Evaluate $\displaystyle\int_{1}^{\infty} (1 - x)e^{-x}\, dx$.

Solution   Use integration by parts, with $dv = e^{-x}\, dx$ and $u = (1 - x)$.

$$\int (1 - x)e^{-x}\, dx = -e^{-x}(1 - x) - \int e^{-x}\, dx$$
$$= -e^{-x} + xe^{-x} + e^{-x} + C$$
$$= xe^{-x} + C$$

Now, apply the definition of an improper integral.

$$\int_{1}^{\infty} (1 - x)e^{-x}\, dx = \lim_{b \to \infty}\left[xe^{-x} \right]_{1}^{b} = \left(\lim_{b \to \infty} \frac{b}{e^{b}} \right) - \frac{1}{e}$$

Finally, using L'Hôpital's Rule on the right-hand limit produces

$$\lim_{b \to \infty} \frac{b}{e^{b}} = \lim_{b \to \infty} \frac{1}{e^{b}} = 0$$

from which you can conclude that

$$\int_{1}^{\infty} (1 - x)e^{-x}\, dx = -\frac{1}{e}.$$

(See Figure 7.21.)

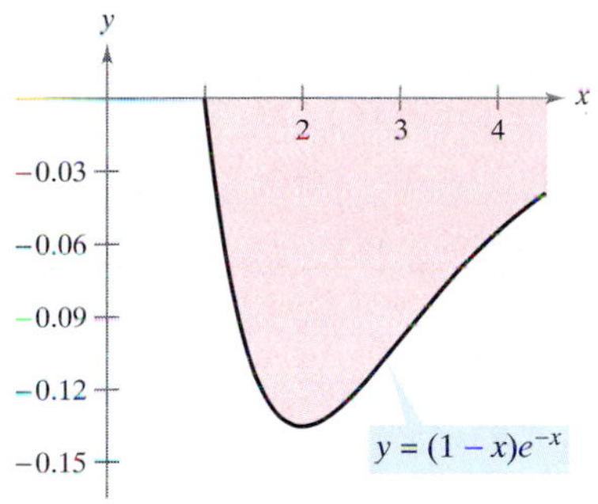

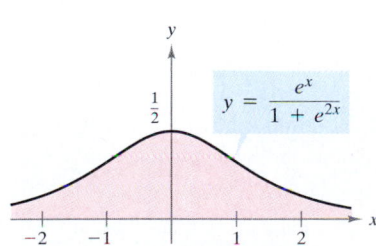

The area of the unbounded region is $|-1/e|$.
Figure 7.21

EXAMPLE 4   Infinite Upper and Lower Limits of Integration

Evaluate $\displaystyle\int_{-\infty}^{\infty} \frac{e^{x}}{1 + e^{2x}}\, dx$.

Solution   Note that the integrand is continuous on $(-\infty, \infty)$. To evaluate the integral, you can break it into two parts, choosing $c = 0$ as a convenient value.

$$\int_{-\infty}^{\infty} \frac{e^{x}}{1 + e^{2x}}\, dx = \int_{-\infty}^{0} \frac{e^{x}}{1 + e^{2x}}\, dx + \int_{0}^{\infty} \frac{e^{x}}{1 + e^{2x}}\, dx$$
$$= \lim_{b \to -\infty}\left[\arctan e^{x} \right]_{b}^{0} + \lim_{b \to \infty}\left[\arctan e^{x} \right]_{0}^{b}$$
$$= \lim_{b \to -\infty}\left(\frac{\pi}{4} - \arctan e^{b} \right) + \lim_{b \to \infty}\left(\arctan e^{b} - \frac{\pi}{4} \right)$$
$$= \frac{\pi}{4} - 0 + \frac{\pi}{2} - \frac{\pi}{4}$$
$$= \frac{\pi}{2}$$

The area of the unbounded region is $\pi/2$.
Figure 7.22

$y = \dfrac{e^{x}}{1 + e^{2x}}$

(See Figure 7.22.)

EXAMPLE 5   Sending a Space Module into Orbit

In Example 3 of Section 6.5, you found that it would require 10,000 mile-tons of work to propel a 15-ton space module to a height of 800 miles above the earth. How much work is required to propel the module an unlimited distance away from the earth's surface?

Solution   At first you might think that an infinite amount of work would be required. But if this were the case, it would be impossible to send rockets into outer space. Because this has been done, the work required must be finite. You can determine the work in the following manner. Using the integral of Example 3, Section 6.5, replace the upper bound of 4800 miles by ∞ and write

$$
\begin{aligned}
W &= \int_{4000}^{\infty} \frac{240{,}000{,}000}{x^2}\, dx \\
&= \lim_{b \to \infty} \left[-\frac{240{,}000{,}000}{x} \right]_{4000}^{b} \\
&= \lim_{b \to \infty} \left(-\frac{240{,}000{,}000}{b} + \frac{240{,}000{,}000}{4000} \right) \\
&= 60{,}000 \text{ mile-tons} \\
&= 6.336 \times 10^{11} \text{ foot-pounds.}
\end{aligned}
$$

(See Figure 7.23.)

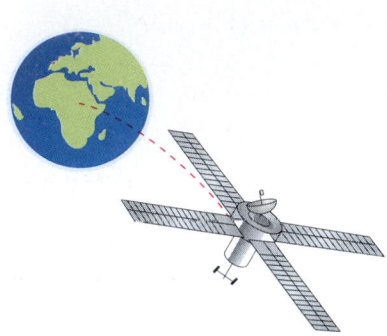

The work required to move a space module an unlimited distance away from the earth is approximately 6.336×10^{11} foot-pounds.
Figure 7.23

Improper Integrals with Infinite Discontinuities

The second basic type of improper integral is one that has an infinite discontinuity *at or between* the limits of integration.

Definition of Improper Integrals with Infinite Discontinuities

1. If f is continuous on the interval $[a, b)$ and has an infinite discontinuity at b, then

$$
\int_{a}^{b} f(x)\, dx = \lim_{c \to b^-} \int_{a}^{c} f(x)\, dx.
$$

2. If f is continuous on the interval $(a, b]$ and has an infinite discontinuity at a, then

$$
\int_{a}^{b} f(x)\, dx = \lim_{c \to a^+} \int_{c}^{b} f(x)\, dx.
$$

3. If f is continuous on the interval $[a, b]$, except for some c in (a, b) at which f has an infinite discontinuity, then

$$
\int_{a}^{b} f(x)\, dx = \int_{a}^{c} f(x)\, dx + \int_{c}^{b} f(x)\, dx.
$$

In the first two cases, the improper integral **converges** if the limit exists—otherwise, the improper integral **diverges**. In the third case, the improper integral on the left diverges if either of the improper integrals on the right diverges.

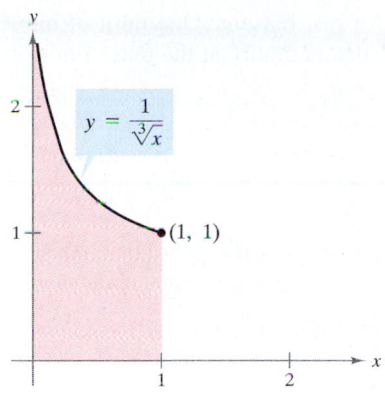

The area of the unbounded region is 3/2.
Figure 7.24

EXAMPLE 6   An Improper Integral with an Infinite Discontinuity

Evaluate $\displaystyle\int_0^1 \frac{dx}{\sqrt[3]{x}}$.

Solution   The integrand has an infinite discontinuity at $x = 0$, as shown in Figure 7.24. You can evaluate this integral as follows.

$$\int_0^1 x^{-1/3}\,dx = \lim_{b \to 0^+} \left[\frac{x^{2/3}}{2/3} \right]_b^1$$

$$= \lim_{b \to 0^+} \frac{3}{2}(1 - b^{2/3})$$

$$= \frac{3}{2}$$

EXAMPLE 7   An Improper Integral That Diverges

Evaluate $\displaystyle\int_0^2 \frac{dx}{x^3}$.

Solution   Because the integrand has an infinite discontinuity at $x = 0$, you can write the following.

$$\int_0^2 \frac{dx}{x^3} = \lim_{b \to 0^+} \left[-\frac{1}{2x^2} \right]_b^2 = \lim_{b \to 0^+} \left(-\frac{1}{8} + \frac{1}{2b^2} \right) = \infty$$

Thus, you can conclude that the improper integral diverges.

EXAMPLE 8   An Improper Integral with an Interior Discontinuity

Evaluate $\displaystyle\int_{-1}^2 \frac{dx}{x^3}$.

Solution   This integral is improper because the integrand has an infinite discontinuity at the interior point $x = 0$, as shown in Figure 7.25. Thus, you can write the following.

$$\int_{-1}^2 \frac{dx}{x^3} = \int_{-1}^0 \frac{dx}{x^3} + \int_0^2 \frac{dx}{x^3}$$

From Example 7 you know that the second integral diverges. Therefore, the original improper integral also diverges.

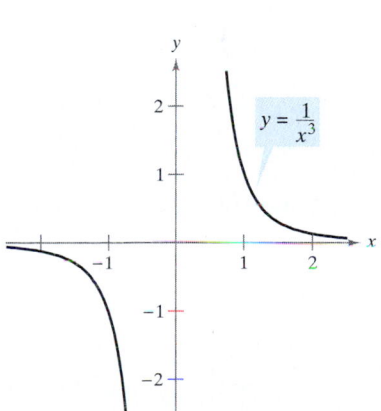

The improper integral $\displaystyle\int_{-1}^2 1/x^3\,dx$ diverges.
Figure 7.25

NOTE   Remember to check for infinite discontinuities at interior points as well as endpoints when determining whether an integral is improper. For instance, if you had not recognized that the integral in Example 8 was improper, you would have obtained the *incorrect* result

$$\int_{-1}^2 \frac{dx}{x^3} = \left[\frac{-1}{2x^2} \right]_{-1}^2 = -\frac{1}{8} + \frac{1}{2} = \frac{3}{8}.$$ Incorrect evaluation

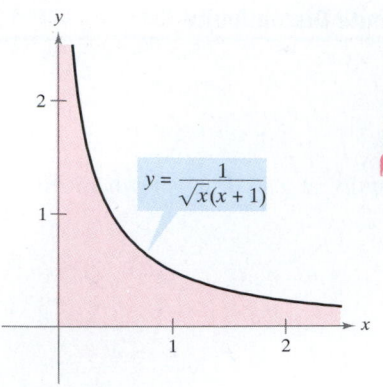

The area of the unbounded region is π.
Figure 7.26

The integral in the next example is improper for *two* reasons. One limit of integration is infinite, and the integrand has an infinite discontinuity at the outer limit of integration, as shown in Figure 7.26.

EXAMPLE 9   A Doubly Improper Integral

Evaluate $\displaystyle\int_0^\infty \frac{dx}{\sqrt{x}(x+1)}$.

Solution   To evaluate this integral, split it at a convenient point (say, $x = 1$) and write

$$\int_0^\infty \frac{dx}{\sqrt{x}(x+1)} = \int_0^1 \frac{dx}{\sqrt{x}(x+1)} + \int_1^\infty \frac{dx}{\sqrt{x}(x+1)}$$

$$= \lim_{b \to 0^+} \left[2 \arctan \sqrt{x} \right]_b^1 + \lim_{c \to \infty} \left[2 \arctan \sqrt{x} \right]_1^c$$

$$= 2\left(\frac{\pi}{4}\right) - 0 + 2\left(\frac{\pi}{2}\right) - 2\left(\frac{\pi}{4}\right)$$

$$= \pi.$$

EXAMPLE 10   An Application Involving Arc Length

Use the formula for arc length to show that the circumference of the circle $x^2 + y^2 = 1$ is 2π.

Solution   To simplify the work, consider the quarter circle given by $y = \sqrt{1 - x^2}$, where $0 \le x \le 1$. The function y is differentiable for any x in this interval except $x = 1$. Therefore, the arc length of the quarter circle is given by the improper integral

$$s = \int_0^1 \sqrt{1 + (y')^2}\, dx$$

$$= \int_0^1 \sqrt{1 + \left(\frac{-x}{\sqrt{1 - x^2}}\right)^2}\, dx$$

$$= \int_0^1 \frac{dx}{\sqrt{1 - x^2}}.$$

This integral is improper, because it has an infinite discontinuity at $x = 1$. Thus, you can write

$$s = \int_0^1 \frac{dx}{\sqrt{1 - x^2}}$$

$$= \lim_{b \to 1^-} \left[\arcsin x \right]_0^b$$

$$= \frac{\pi}{2} - 0 = \frac{\pi}{2}.$$

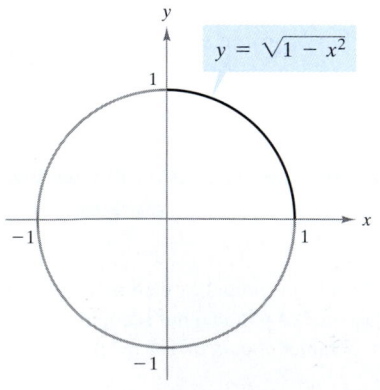

The circumference of the circle is 2π.
Figure 7.27

Finally, multiplying by 4, you can conclude that the circumference of the circle is $4s = 2\pi$, as shown in Figure 7.27.

We conclude this section with a useful theorem describing the convergence or divergence of a common type of improper integral. The proof of this theorem is left as an exercise (see Exercise 35).

> **THEOREM 7.5   A Special Type of Improper Integral**
>
> $$\int_1^\infty \frac{dx}{x^p} = \begin{cases} \dfrac{1}{p-1}, & \text{if } p > 1 \\ \text{diverges}, & \text{if } p \leq 1 \end{cases}$$

EXAMPLE 11   An Application Involving A Solid of Revolution

The solid formed by revolving (about the x-axis) the *unbounded* region lying between the graph of $f(x) = 1/x$ and the x-axis ($x \geq 1$) is called **Gabriel's Horn.** (See Figure 7.28.) Show that this solid has a finite volume and an infinite surface area.

FOR FURTHER INFORMATION For more information on solids that have finite volumes and infinite surface areas, see the article "Supersolids: Solids Having Finite Volume and Infinite Surfaces" by William P. Love in the January 1989 issue of *Mathematics Teacher*.

Solution   Using the disc method and Theorem 7.5, you can determine the volume to be

$$V = \pi \int_1^\infty \left(\frac{1}{x}\right)^2 dx \qquad \text{Theorem 7.5, } p = 2 > 1$$

$$= \pi \left(\frac{1}{2-1}\right)$$

$$= \pi.$$

The surface area is given by

$$S = 2\pi \int_1^\infty f(x) \sqrt{1 + [f'(x)]^2}\, dx = 2\pi \int_1^\infty \frac{1}{x} \sqrt{1 + \frac{1}{x^4}}\, dx.$$

Because

$$\sqrt{1 + \frac{1}{x^4}} > 1$$

on the interval $[1, \infty)$, and the improper integral

$$\int_1^\infty \frac{1}{x}\, dx$$

diverges, you can conclude that the improper integral

$$\int_1^\infty \frac{1}{x} \sqrt{1 + \frac{1}{x^4}}\, dx$$

also diverges. (See Exercise 38.) Thus, the surface area is infinite.

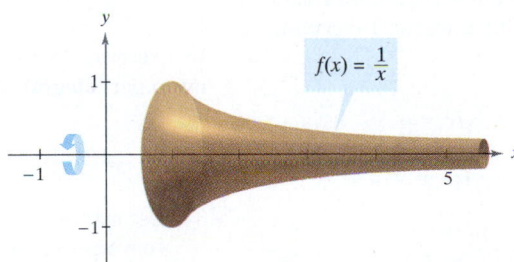

Gabriel's Horn has a finite volume and an infinite surface area.
Figure 7.28

EXERCISES FOR SECTION 7.8

In Exercises 1–6, state why the integral is improper and determine its divergence or convergence. Evaluate the integral if it converges.

1. $\int_0^4 \frac{1}{\sqrt{x}}\, dx$

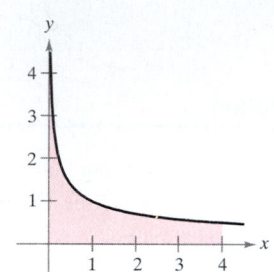

2. $\int_3^4 \frac{1}{(x-3)^{3/2}}\, dx$

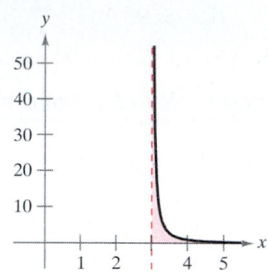

3. $\int_0^2 \frac{1}{(x-1)^2}\, dx$

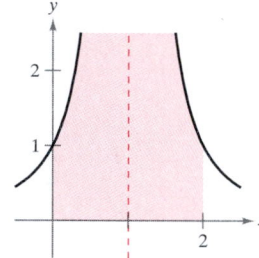

4. $\int_0^2 \frac{1}{(x-1)^{2/3}}\, dx$

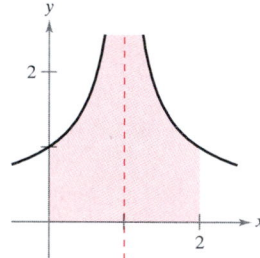

5. $\int_0^\infty e^{-x}\, dx$

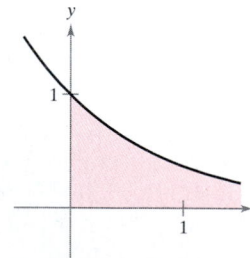

6. $\int_{-\infty}^0 e^{2x}\, dx$

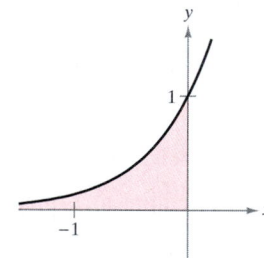

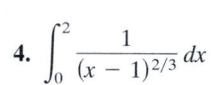 **Writing** In Exercises 7 and 8, explain why the evaluation of the integral is *incorrect*. Use the integration capabilities of a graphing utility to attempt to evaluate the integral. Determine whether it gives the correct answer.

7. $\int_{-1}^1 \frac{1}{x^2}\, dx = -2$

8. $\int_0^\infty e^{-x}\, dx = 0$

In Exercises 9–22, determine the divergence or convergence of the integral with infinite limits. Evaluate the integral if it converges.

9. $\int_{-\infty}^0 xe^{-2x}\, dx$

10. $\int_0^\infty xe^{-x}\, dx$

11. $\int_0^\infty x^2 e^{-x}\, dx$

12. $\int_0^\infty (x-1)e^{-x}\, dx$

13. $\int_1^\infty \frac{1}{x^2}\, dx$

14. $\int_1^\infty \frac{1}{\sqrt{x}}\, dx$

15. $\int_0^\infty e^{-x}\cos x\, dx$

16. $\int_0^\infty e^{-ax}\sin bx\, dx, \quad a > 0$

17. $\int_{-\infty}^\infty \frac{1}{1+x^2}\, dx$

18. $\int_0^\infty \frac{x^3}{(x^2+1)^2}\, dx$

19. $\int_0^\infty \frac{1}{e^x + e^{-x}}\, dx$

20. $\int_0^\infty \frac{e^x}{1+e^x}\, dx$

21. $\int_0^\infty \cos \pi x\, dx$

22. $\int_0^\infty \sin \frac{x}{2}\, dx$

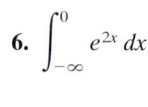 In Exercises 23–34, determine the divergence or convergence of the improper integral. Evaluate the integral if it converges, and check your results with the integration capabilities of a graphing utility.

23. $\int_0^1 \frac{1}{x^2}\, dx$

24. $\int_0^1 \frac{1}{x}\, dx$

25. $\int_0^8 \frac{1}{\sqrt[3]{8-x}}\, dx$

26. $\int_0^e \ln x\, dx$

27. $\int_0^1 x \ln x\, dx$

28. $\int_0^{\pi/2} \sec \theta\, d\theta$

29. $\int_0^{\pi/2} \tan \theta\, d\theta$

30. $\int_0^2 \frac{1}{\sqrt{4-x^2}}\, dx$

31. $\int_2^4 \frac{1}{\sqrt{x^2-4}}\, dx$

32. $\int_0^2 \frac{1}{4-x^2}\, dx$

33. $\int_0^2 \frac{1}{\sqrt[3]{x-1}}\, dx$

34. $\int_0^2 \frac{1}{(x-1)^{4/3}}\, dx$

In Exercises 35 and 36, determine all values of p for which the improper integral converges.

35. $\int_1^\infty \frac{1}{x^p}\, dx$

36. $\int_0^1 \frac{1}{x^p}\, dx$

37. Use mathematical induction to verify that the following integral converges for any positive integer n.

$$\int_0^\infty x^n e^{-x}\, dx$$

38. Given continuous functions f and g such that $0 \le f(x) \le g(x)$ on the interval $[a, \infty)$, prove the following.

(a) If $\int_a^\infty g(x)\,dx$ converges, then $\int_a^\infty f(x)\,dx$ converges.

(b) If $\int_a^\infty f(x)\,dx$ diverges, then $\int_a^\infty g(x)\,dx$ diverges.

In Exercises 39–48, use the results of Exercises 35–38 to determine whether the improper integral converges or diverges.

39. $\int_0^1 \frac{1}{x^3}\,dx$

40. $\int_0^1 \frac{1}{\sqrt[3]{x}}\,dx$

41. $\int_1^\infty \frac{1}{x^3}\,dx$

42. $\int_0^\infty x^4 e^{-x}\,dx$

43. $\int_1^\infty \frac{1}{x^2 + 5}\,dx$

44. $\int_2^\infty \frac{1}{\sqrt{x-1}}\,dx$

45. $\int_2^\infty \frac{1}{\sqrt[3]{x(x-1)}}\,dx$

46. $\int_1^\infty \frac{1}{\sqrt{x}(x+1)}\,dx$

47. $\int_0^\infty e^{-x^2}\,dx$

48. $\int_2^\infty \frac{1}{\sqrt{x}\,\ln x}\,dx$

Laplace Transforms Let $f(t)$ be a function defined for all positive values of t. The Laplace Transform of $f(t)$ is defined by

$$F(s) = \int_0^\infty e^{-st} f(t)\,dt$$

if the improper integral exists. Laplace Transforms are used to solve differential equations. In Exercises 49–56, find the Laplace Transform of the function.

49. $f(t) = 1$

50. $f(t) = t$

51. $f(t) = t^2$

52. $f(t) = e^{at}$

53. $f(t) = \cos at$

54. $f(t) = \sin at$

55. $f(t) = \cosh at$

56. $f(t) = \sinh at$

Area and Volume In Exercises 57 and 58, consider the region satisfying the inequalities. (a) Find the area of the region. (b) Find the volume of the solid generated by revolving the region about the x-axis. (c) Find the volume of the solid generated by revolving the region about the y-axis.

57. $y \le e^{-x},\ y \ge 0,\ x \ge 0$

58. $y \le 1/x^2,\ y \ge 0,\ x \ge 1$

59. *Arc Length* Sketch the graph of the hypocycloid of four cusps

$$x^{2/3} + y^{2/3} = 1$$

and find its perimeter.

60. *Surface Area* The region bounded by

$$(x - 2)^2 + y^2 = 1$$

is revolved about the y-axis to form a torus. Find the surface area of the torus.

61. *The Gamma Function* The Gamma Function $\Gamma(n)$ is defined by

$$\Gamma(n) = \int_0^\infty x^{n-1} e^{-x}\,dx, \quad n > 0.$$

(a) Find $\Gamma(1)$, $\Gamma(2)$, and $\Gamma(3)$.

(b) Use integration by parts to show that $\Gamma(n + 1) = n\Gamma(n)$.

(c) Express $\Gamma(n)$ in terms of factorial notation where n is a positive integer.

62. *Work* A 5-ton rocket is fired from the surface of the earth into outer space.

(a) How much work is required to overcome the earth's gravitational force?

(b) How far has the rocket traveled when half the total work has occurred?

Probability A nonnegative function f is called a *probability density function* if

$$\int_{-\infty}^\infty f(t)\,dt = 1.$$

The *probability* that x lies between a and b is given by

$$P(a \le x \le b) = \int_a^b f(t)\,dt.$$

The *expected value* of x is given by

$$E(x) = \int_{-\infty}^\infty t f(t)\,dt.$$

In Exercises 63 and 64, (a) show that the nonnegative function is a probability density function, (b) find $P(0 \le x \le 4)$, and (c) find $E(x)$.

63. $f(t) = \begin{cases} \frac{1}{7} e^{-t/7}, & t \ge 0 \\ 0, & t < 0 \end{cases}$

64. $f(t) = \begin{cases} \frac{2}{5} e^{-2t/5}, & t \ge 0 \\ 0, & t < 0 \end{cases}$

Capitalized Cost In Exercises 65 and 66, find the capitalized cost C of an asset (a) for $n = 5$ years, (b) for $n = 10$ years, and (c) forever. The capitalized cost is given by

$$C = C_0 + \int_0^n c(t) e^{-rt}\,dt$$

where C_0 is the original investment, t is the time in years, r is the annual rate compounded continuously, and $c(t)$ is the annual cost of maintenance.

65. $C_0 = \$650{,}000$
$c(t) = \$25{,}000$
$r = 0.06$

66. $C_0 = \$650{,}000$
$c(t) = \$25{,}000(1 + 0.08t)$
$r = 0.06$

67. *Electromagnetic Theory* Find the value of the following integral used in electromagnetic theory.

$$P = k \int_1^\infty \frac{1}{(a^2 + x^2)^{3/2}}\,dx$$

68. Writing

(a) The improper integrals

$$\int_1^\infty \frac{1}{x}\,dx \quad \text{and} \quad \int_1^\infty \frac{1}{x^2}\,dx$$

diverge and converge, respectively. Describe the essential differences between the integrands that cause one integral to converge and the other to diverge.

(b) Sketch a graph of the function $y = \sin x/x$ over the interval $(1, \infty)$. Use your knowledge of the definite integral to make an inference as to whether or not the integral

$$\int_1^\infty \frac{\sin x}{x}\,dx$$

converges. Give reasons for your answer.

(c) Use one iteration of integration by parts on the integral in part (b) to determine its divergence or convergence.

69. Let $I_n = \int_0^\infty \frac{x^{2n-1}}{(x^2 + 1)^{n+3}}\,dx, \quad n \geq 1.$

Prove that $I_n = \left(\dfrac{n-1}{n+2}\right) I_{n-1}$

and then evaluate each of the following.

(a) $\int_0^\infty \dfrac{x}{(x^2 + 1)^4}\,dx$

(b) $\int_0^\infty \dfrac{x^3}{(x^2 + 1)^5}\,dx$

(c) $\int_0^\infty \dfrac{x^5}{(x^2 + 1)^6}\,dx$

70. Normal Probability In 1992, the mean height of American men between 18 and 24 years old was 70 inches, and the standard deviation was 3 inches. If an 18- to 24-year-old man was chosen at random from the population, the probability that he was 6 feet tall or taller is

$$P(72 \leq x < \infty) = \int_{72}^\infty \frac{1}{3\sqrt{2\pi}} e^{-(x-70)^2/18}\,dx.$$

(Source: National Center for Health Statistics)

(a) Use a graphing utility to graph the integrand. Use the graphing utility to convince yourself that the area between the x-axis and the integrand is 1.

(b) Use a graphing utility to approximate $P(72 \leq x < \infty)$.

(c) Approximate $0.5 - P(70 \leq x \leq 72)$ using a graphing utility. Use the graph in part (a) to explain why this result is the same as the answer in part (b).

True or False? **In Exercises 71–74, determine whether the statement is true or false. If it is false, explain why or give an example that shows it is false.**

71. If f is continuous on $[0, \infty)$ and $\lim\limits_{x\to\infty} f(x) = 0$, then $\int_0^\infty f(x)\,dx$ converges.

72. If f is continuous on $[0, \infty)$ and $\int_0^\infty f(x)\,dx$ diverges, then $\lim\limits_{x\to\infty} f(x) \neq 0$.

73. If f' is continuous on $[0, \infty)$ and $\lim\limits_{x\to\infty} f(x) = 0$, then $\int_0^\infty f'(x)\,dx = -f(0)$.

74. If the graph of f is symmetric with respect to the origin or the y-axis, then $\int_0^\infty f(x)\,dx$ converges if and only if $\int_{-\infty}^\infty f(x)\,dx$ converges.

REVIEW EXERCISES FOR CHAPTER 7

In Exercises 1 and 2, use integration by parts to evaluate the integral.

1. $\displaystyle\int e^{2x} \sin 3x\,dx$

2. $\displaystyle\int (x^2 - 1)e^x\,dx$

In Exercises 3–6, evaluate the trigonometric integral.

3. $\displaystyle\int \cos^3(\pi x - 1)\,dx$

4. $\displaystyle\int \sin^2 \frac{\pi x}{2}\,dx$

5. $\displaystyle\int \sec^4 \frac{x}{2}\,dx$

6. $\displaystyle\int \tan\theta \sec^4\theta\,d\theta$

In Exercises 7 and 8, use trigonometric substitution to evaluate the integral.

7. $\displaystyle\int \frac{-12}{x^2\sqrt{4 - x^2}}\,dx$

8. $\displaystyle\int \frac{\sqrt{x^2 - 9}}{x}\,dx, \quad x > 3$

In Exercises 9 and 10, use partial fractions to evaluate the integral.

9. $\displaystyle\int \frac{x^2 + 2x}{x^3 - x^2 + x - 1}\,dx$

10. $\displaystyle\int \frac{4x - 2}{3(x - 1)^2}\,dx$

In Exercises 11–36, evaluate the integral.

11. $\displaystyle\int \frac{x^2}{x^2 + 2x - 15}\,dx$

12. $\displaystyle\int \frac{3}{2x\sqrt{9x^2 - 1}}\,dx, \quad x > \frac{1}{3}$

13. $\displaystyle\int \frac{1}{1 - \sin\theta}\,d\theta$

14. $\displaystyle\int x^2 \sin 2x\,dx$

15. $\displaystyle\int \frac{\ln(2x)}{x^2}\,dx$

16. $\displaystyle\int 2x\sqrt{2x - 3}\,dx$

17. $\displaystyle\int \sqrt{4 - x^2}\,dx$

18. $\displaystyle\int \frac{\sec^2\theta}{\tan\theta(\tan\theta - 1)}\,d\theta$

19. $\displaystyle\int \frac{3x^3 + 4x}{(x^2 + 1)^2}\, dx$

20. $\displaystyle\int \sqrt{\frac{x - 2}{x + 2}}\, dx$

21. $\displaystyle\int \frac{16}{\sqrt{16 - x^2}}\, dx$

22. $\displaystyle\int \frac{\sin\theta}{1 + 2\cos^2\theta}\, d\theta$

23. $\displaystyle\int \frac{x}{x^2 + 4x + 8}\, dx$

24. $\displaystyle\int \frac{x}{x^2 - 4x + 8}\, dx$

25. $\displaystyle\int \theta \sin\theta \cos\theta\, d\theta$

26. $\displaystyle\int \frac{\csc\sqrt{2x}}{\sqrt{x}}\, dx$

27. $\displaystyle\int (\sin\theta + \cos\theta)^2\, d\theta$

28. $\displaystyle\int \cos 2\theta(\sin\theta + \cos\theta)^2\, d\theta$

29. $\displaystyle\int \sqrt{1 + \cos x}\, dx$

30. $\displaystyle\int \ln\sqrt{x^2 - 1}\, dx$

31. $\displaystyle\int \cos x \ln(\sin x)\, dx$

32. $\displaystyle\int \frac{x^4 + 2x^2 + x + 1}{(x^2 + 1)^2}\, dx$

33. $\displaystyle\int x \arcsin 2x\, dx$

34. $\displaystyle\int e^x \arctan e^x\, dx$

35. $\displaystyle\int \frac{x^{1/4}}{1 + x^{1/2}}\, dx$

36. $\displaystyle\int \sqrt{1 + \sqrt{x}}\, dx$

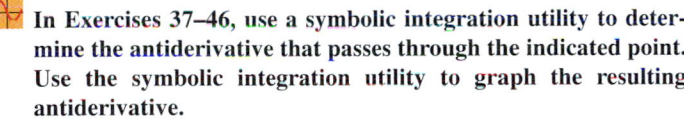 **In Exercises 37–46, use a symbolic integration utility to determine the antiderivative that passes through the indicated point. Use the symbolic integration utility to graph the resulting antiderivative.**

Integral	*Point*
37. $\displaystyle\int \frac{x^4}{(x - 1)^3}\, dx$	$(2, 4)$
38. $\displaystyle\int \frac{4x^2 - 1}{(2x)(x^2 + 2x + 1)}\, dx$	$(1, 0)$
39. $\displaystyle\int \frac{6x^2 - 3x + 14}{x^3 - 2x^2 + 4x - 8}\, dx$	$(3, 2)$
40. $\displaystyle\int \frac{x^3 - 6x^2 + x - 4}{1 - x^4}\, dx$	$(4, 1)$
41. $\displaystyle\int \frac{1}{2 - 3\sin\theta}\, d\theta$	$(1, 1)$
42. $\displaystyle\int \frac{1}{\sec\theta - \tan\theta}\, d\theta$	$(\pi, 2)$
43. $\displaystyle\int \frac{1}{1 + \sin\theta + \cos\theta}\, d\theta$	$(0, 0)$
44. $\displaystyle\int \frac{1}{3 - 2\cos\theta}\, d\theta$	$(0, 0)$
45. $\displaystyle\int \frac{\sin\theta}{3 - 2\cos\theta}\, d\theta$	$(0, 0)$
46. $\displaystyle\int \frac{\sin\theta}{1 + \sin\theta}\, d\theta$	$(0, 1)$

In Exercises 47–50, solve the differential equation.

47. $\displaystyle\frac{dy}{dx} = \frac{9}{x^2 - 9}$

48. $\displaystyle\frac{dy}{dx} = \frac{\sqrt{4 - x^2}}{2x}$

49. $y' = \ln(x^2 + x)$

50. $y' = \sqrt{1 - \cos\theta}$

In Exercises 51–54, evaluate the integral using the indicated methods.

51. $\displaystyle\int \frac{1}{x^2\sqrt{4 + x^2}}\, dx$

   (a) Trigonometric substitution

   (b) Substitution: $x = 2/u$

52. $\displaystyle\int \frac{1}{x\sqrt{4 + x^2}}\, dx$

   (a) Trigonometric substitution

   (b) Substitution: $u^2 = 4 + x^2$

53. $\displaystyle\int \frac{x^3}{\sqrt{4 + x^2}}\, dx$

   (a) Trigonometric substitution

   (b) Substitution: $u^2 = 4 + x^2$

   (c) Integration by parts: $dv = \left(x/\sqrt{4 + x^2}\right) dx$

54. $\displaystyle\int x\sqrt{4 + x}\, dx$

   (a) Trigonometric substitution

   (b) Substitution: $u^2 = 4 + x$

   (c) Substitution: $u = 4 + x$

   (d) Integration by parts: $dv = \sqrt{4 + x}\, dx$

In Exercises 55–60, evaluate the definite integral. Use a graphing utility to verify your result.

55. $\displaystyle\int_2^{\sqrt{5}} x(x^2 - 4)^{3/2}\, dx$

56. $\displaystyle\int_0^1 \frac{x}{(x - 2)(x - 4)}\, dx$

57. $\displaystyle\int_1^4 \frac{\ln x}{x}\, dx$

58. $\displaystyle\int_0^2 xe^{3x}\, dx$

59. $\displaystyle\int_0^{\pi} x \sin x\, dx$

60. $\displaystyle\int_0^3 \frac{x}{\sqrt{1 + x}}\, dx$

Area **In Exercises 61 and 62, find the area of the region bounded by the graphs of the equations.**

61. $y = x\sqrt{4 - x}, \quad y = 0$

62. $y = \dfrac{1}{25 - x^2}, \quad y = 0, x = 0, x = 4$

In Exercises 63 and 64, find the centroid of the region bounded by the graphs of the equations.

63. $y = \sqrt{1 - x^2}, \quad y = 0$

64. $(x - 1)^2 + y^2 = 1, \quad (x - 4)^2 + y^2 = 4$

In Exercises 65 and 66, evaluate each definite integral to two-decimal-place accuracy. Which, if any, requires a numerical method or a graphing utility to evaluate the integral?

65. (a) $\int_0^1 e^x \, dx$ (b) $\int_0^1 xe^x \, dx$

   (c) $\int_0^1 xe^{x^2} \, dx$ (d) $\int_0^1 e^{x^2} \, dx$

66. (a) $\int_0^{\pi/2} \cos x \, dx$ (b) $\int_0^{\pi/2} \cos^2 x \, dx$

   (c) $\int_0^{\pi/2} \cos x^2 \, dx$ (d) $\int_0^{\pi/2} \cos \sqrt{x} \, dx$

Arc Length In Exercises 67 and 68, approximate to two decimal places the arc length of the curve over the given interval.

Function	Interval
67. $y = \sin x$	$[0, \pi]$
68. $y = \sin^2 x$	$[0, \pi]$

In Exercises 69–76, use L'Hôpital's Rule to evaluate the limit.

69. $\lim_{x \to 1} \dfrac{(\ln x)^2}{x - 1}$ 70. $\lim_{x \to 0} \dfrac{\sin \pi x}{\sin 2 \pi x}$

71. $\lim_{x \to \infty} \dfrac{e^{2x}}{x^2}$ 72. $\lim_{x \to \infty} xe^{-x^2}$

73. $\lim_{x \to \infty} (\ln x)^{2/x}$ 74. $\lim_{x \to 1^+} (x - 1)^{\ln x}$

75. $\lim_{x \to \infty} 1000 \left(1 + \dfrac{0.09}{n} \right)^n$ 76. $\lim_{x \to 1^+} \left(\dfrac{2}{\ln x} - \dfrac{2}{x - 1} \right)$

In Exercises 77–80, determine the convergence or divergence of the improper integral. Evaluate the integral if it converges.

77. $\int_0^{16} \dfrac{1}{\sqrt[4]{x}} \, dx$ 78. $\int_0^1 \dfrac{6}{x - 1} \, dx$

79. $\int_1^{\infty} x^2 \ln x \, dx$ 80. $\int_0^{\infty} \dfrac{e^{-1/x}}{x^2} \, dx$

81. **Present Value** The board of directors of a corporation is calculating the price to pay for a business that is forecast to yield a continuous flow of profit of $500,000 per year. If money will earn a nominal rate of 5% per year compounded continuously, what is the present value of the business

(a) for 20 years?

(b) forever (in perpetuity)?

(*Note:* The present value for t_0 years is $\int_0^{t_0} 500{,}000e^{-0.05t} \, dt$.)

82. **Volume** Find the volume of the solid generated by revolving the region bounded by the graphs of $y = xe^{-x}$, $y = 0$, and $x = 0$ about the x-axis.

83. **Probability** The average length (from beak to tail) of different species of warblers in the eastern United States is approximately normally distributed with a mean of 12.9 centimeters and a standard deviation of 0.95 centimeter (see figure). The probability that a randomly selected warbler has a length between a and b centimeters is

$$P(a \le x \le b) = \dfrac{1}{0.95\sqrt{2\pi}} \int_a^b e^{-(x - 12.9)^2/2(0.95)^2} \, dx.$$

Use a graphing utility to approximate the probability that a randomly selected warbler has a length of (a) 13 centimeters or greater and (b) 15 centimeters or greater. (*Source: Peterson's Field Guide: Eastern Birds*)

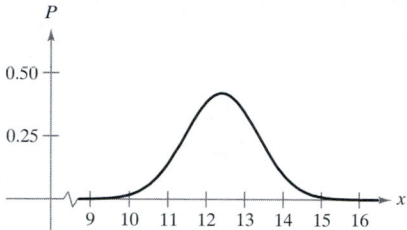

84. Using the inequality

$$\frac{1}{x^5} + \frac{1}{x^{10}} + \frac{1}{x^{15}} < \frac{1}{x^5 - 1} < \frac{1}{x^5} + \frac{1}{x^{10}} + \frac{2}{x^{15}}$$

for $x \ge 2$, approximate $\int_2^{\infty} \dfrac{1}{x^5 - 1} \, dx$.

85. Verify the reduction formula

$$\int (\ln x)^n \, dx = x(\ln x)^n - n \int (\ln x)^{n-1} \, dx.$$

86. Verify the reduction formula

$$\int \tan^n x \, dx = \frac{1}{n - 1} \tan^{n-1} x - \int \tan^{n-2} x \, dx.$$

True or False? In Exercises 87–90, determine whether the statement is true or false. If it is false, explain why or give an example that shows it is false.

87. $\int \dfrac{\ln x^2}{x} \, dx = \int u \, du, \quad u = \ln x^2$

88. If $u = \sqrt[3]{3x + 1}$, then $dx = u^2 \, du$.

89. $\int_{-1}^1 \sqrt{x^2 - x^3} \, dx = \int_{-1}^1 x\sqrt{1 - x} \, dx$

90. $\int_1^{\infty} \dfrac{1}{x^p} \, dx$ converges if $p > 1$.

91. Show that

$$\int_x^1 \frac{1}{1 + t^2} \, dt = \int_1^{1/x} \frac{1}{1 + t^2} \, dt$$

by evaluating each integral and then using the identity

$$\arctan x + \arctan \frac{1}{x} = \frac{\pi}{2}, \quad x > 0.$$

Infinite Series

Eric Haines

8

From *The Desktop Fractal Design System* by Michael Barnsley

A FERN FRACTAL

Fractals are *self-similar*, as seen in the fern fractal. When magnifying a small portion of a fractal image, you see an image similar to the original fractal.

Why is geometry often described as "cold" and "dry"? One reason lies in its inability to describe the shape of a cloud, a mountain, a coastline, or a tree. Clouds are not spheres, mountains are not cones, coastlines are not circles, and bark is not smooth, nor does lightning travel in a straight line. . . . Nature exhibits not simply a higher degree but an altogether different level of complexity.

Benoit Mandelbrot (1924–)

To meet the challenge of creating a geometry capable of describing nature, Mandelbrot developed *fractal* geometry. Fractal sets come in diverse forms. Some are curves, others are disconnected "dust," and still others are such odd forms that there are no existing geometric terms to describe them.

One of the "classic" fractals is the Koch snowflake, named after the Swedish mathematician Helge von Koch (1870–1924). It is sometimes classified as a "coast-line curve" because of the way a coastline appears increasingly more complex with magnification. To describe the Koch snowflake, Mandelbrot coined the term *teragon*, which translates literally from the Greek words for "monster curve."

The construction of the Koch snowflake begins with an equilateral triangle whose sides are one unit long. In the first iteration, a triangle with sides one-third unit long is added in the center of each side of the original. In the second iteration, a triangle with sides one-ninth unit long is added in the center of each side. Successive iterations continue this process—without stopping.

Eric Haines

Eric Haines generated the sphereflake fractal shown on the previous page. It is a three-dimensional version of the Koch snowflake. You are asked to prove that its surface area is infinite in Exercise 74 on page 566.

QUESTIONS

1. Write a formula that describes the side length of the triangles that will be added in the *n*th iteration.

2. Make a table of the perimeter of the original triangle and of the teragon in the first three iterations, as shown above. Write an expression describing the perimeter of the teragon after the *n*th iteration. What do you expect will happen to the perimeter as *n* approaches infinity?

3. Make a table of the area of the teragon in the first four iterations. Write an expression describing the area after the *n*th iteration. What do you expect will happen to the area as *n* approaches infinity?

4. Is it possible for a closed and bounded region in the plane to have a finite area and an infinite perimeter? Explain your reasoning.

The concepts presented here will be explored further in this chapter. For an extension of this application, see the lab series that accompanies this text.

Sequences • Limit of a Sequence • Pattern Recognition for Sequences •
Monotonic Sequences and Bounded Sequences

Sequences

In mathematics, the word "sequence" is used in much the same way as in ordinary English. To say that a collection of objects or events is *in sequence* usually means that the collection is ordered so that it has an identified first member, second member, third member, and so on.

Mathematically, a **sequence** is defined as a function whose domain is the set of positive integers. Although a sequence is a function, it is common to represent sequences by subscript notation rather than by the standard function notation. For instance, in the sequence

$$1, \quad 2, \quad 3, \quad 4, \quad \ldots, \quad n, \quad \ldots$$
$$\downarrow \quad \downarrow \quad \downarrow \quad \downarrow \qquad\qquad \downarrow \qquad\qquad \text{Sequence}$$
$$a_1, \quad a_2, \quad a_3, \quad a_4, \quad \ldots, \quad a_n, \quad \ldots$$

1 is mapped onto a_1, 2 is mapped onto a_2, and so on. The numbers $a_1, a_2, a_3, \ldots, a_n,$ $\ldots$ are the **terms** of the sequence. The number a_n is the **nth term** of the sequence, and the entire sequence is denoted by $\{a_n\}$.

NOTE   Occasionally, it is convenient to begin a sequence with a_0, so that the terms of the sequence become

$$a_0, a_1, a_2, a_3, \ldots, a_n, \ldots$$

EXAMPLE 1   Listing the Terms of a Sequence

a. The terms of the sequence $\{a_n\} = \{3 + (-1)^n\}$ are

$$3 + (-1)^1, 3 + (-1)^2, 3 + (-1)^3, 3 + (-1)^4, \ldots$$
$$2, 4, 2, 4, \ldots.$$

b. The terms of the sequence $\{b_n\} = \left\{\dfrac{2n}{1+n}\right\}$ are

$$\frac{2\cdot 1}{1+1}, \frac{2\cdot 2}{1+2}, \frac{2\cdot 3}{1+3}, \frac{2\cdot 4}{1+4}, \ldots$$

$$\frac{2}{2}, \frac{4}{3}, \frac{6}{4}, \frac{8}{5}, \ldots.$$

c. The terms of the sequence $\{c_n\} = \left\{\dfrac{n^2}{2^n - 1}\right\}$ are

$$\frac{1^2}{2^1 - 1}, \frac{2^2}{2^2 - 1}, \frac{3^2}{2^3 - 1}, \frac{4^2}{2^4 - 1}, \ldots$$

$$\frac{1}{1}, \frac{4}{3}, \frac{9}{7}, \frac{16}{15}, \ldots.$$

Limit of a Sequence

The primary focus of this chapter concerns sequences whose terms approach limiting values. Such sequences are said to **converge.** For instance, the sequence $\{1/2^n\}$

$$\frac{1}{2}, \frac{1}{4}, \frac{1}{8}, \frac{1}{16}, \frac{1}{32}, \cdots$$

converges to 0, as indicated in the following definition.

> ### Definition of the Limit of a Sequence
>
> Let L be a real number. The **limit** of a sequence $\{a_n\}$ is L, written as
>
> $$\lim_{n \to \infty} a_n = L$$
>
> if for each $\varepsilon > 0$, there exists $M > 0$ such that $|a_n - L| < \varepsilon$ whenever $n > M$. Sequences that have limits **converge,** whereas sequences that do not have limits **diverge.**

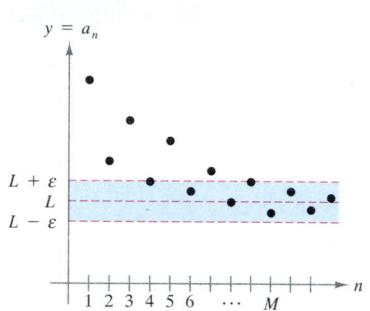

For $n > M$, the terms of the sequence all lie within ε units of L.
Figure 8.1

Graphically, this definition says that eventually (for $n > M$) the terms of a sequence that converges to L will lie within the band between the lines $y = L + \varepsilon$ and $y = L - \varepsilon$, as illustrated in Figure 8.1.

If a sequence $\{a_n\}$ agrees with a function f at every positive integer, and if $f(x)$ approaches a limit L as $x \to \infty$, the sequence must converge to the same limit L.

> ### THEOREM 8.1   Limit of a Sequence
>
> Let f be a function of a real variable such that
>
> $$\lim_{x \to \infty} f(x) = L.$$
>
> If $\{a_n\}$ is a sequence such that $f(n) = a_n$ for every positive integer n, then
>
> $$\lim_{n \to \infty} a_n = L.$$

EXAMPLE 2   Finding the Limit of a Sequence

Find the limit of the sequence whose nth term is $a_n = \left(1 + \dfrac{1}{n}\right)^n$.

Solution   You know from Theorem 5.15 that

$$\lim_{x \to \infty} \left(1 + \frac{1}{x}\right)^x = e.$$

Therefore, you can apply Theorem 8.1 to conclude that

$$\lim_{n \to \infty} a_n = \lim_{n \to \infty} \left(1 + \frac{1}{n}\right)^n = e.$$

The following properties of limits of sequences parallel those given for limits of functions of a real variable on page 188.

> **THEOREM 8.2   Properties of Limits of Sequences**
>
> Let $\lim\limits_{n\to\infty} a_n = L$ and $\lim\limits_{n\to\infty} b_n = K$.
>
> 1. $\lim\limits_{n\to\infty} (a_n \pm b_n) = L \pm K$ 2. $\lim\limits_{n\to\infty} ca_n = cL,$ c is any real number
>
> 3. $\lim\limits_{n\to\infty} (a_n b_n) = LK$ 4. $\lim\limits_{n\to\infty} \dfrac{a_n}{b_n} = \dfrac{L}{K},$ $b_n \neq 0$ and $K \neq 0$

EXAMPLE 3   Determining Convergence or Divergence

a. Because the sequence $\{a_n\} = \{3 + (-1)^n\}$ has terms

$$2, 4, 2, 4, \ldots$$

that alternate between 2 and 4, the limit

$$\lim_{n\to\infty} a_n$$

does not exist. Thus, the sequence diverges.

b. For $\{b_n\} = \left\{\dfrac{n}{1 - 2n}\right\}$, you can divide the numerator and denominator by n to obtain

$$\lim_{n\to\infty} \frac{n}{1 - 2n} = \lim_{n\to\infty} \frac{1}{(1/n) - 2} = -\frac{1}{2}$$

which implies that the sequence converges to $-\frac{1}{2}$.

EXAMPLE 4   Using L'Hôpital's Rule to Determine Convergence

Show that the sequence whose nth term is $a_n = \dfrac{n^2}{2^n - 1}$ converges.

Solution   Consider the function of a real variable

$$f(x) = \frac{x^2}{2^x - 1}.$$

Applying L'Hôpital's Rule twice produces

$$\lim_{x\to\infty} \frac{x^2}{2^x - 1} = \lim_{x\to\infty} \frac{2x}{(\ln 2)2^x} = \lim_{x\to\infty} \frac{2}{(\ln 2)^2 2^x} = 0.$$

Because $f(n) = a_n$ for every positive integer, you can apply Theorem 8.1 to conclude that

$$\lim_{n\to\infty} \frac{n^2}{2^n - 1} = 0.$$

Thus, the sequence converges to 0.

TECHNOLOGY   Use a graphing utility to graph the function in Example 4. Notice that as x approaches infinity, the value of the function gets closer and closer to 0. If you have access to a graphing utility that can generate terms of a sequence, try using it to calculate the first 20 terms of the sequence in Example 4. Then view the terms to observe numerically that the sequence converges to 0.

To simplify some of the formulas developed in this chapter, we use the symbol $n!$ (read "n factorial"). Let n be a positive integer; then **n factorial** is given by

$$n! = 1 \cdot 2 \cdot 3 \cdot 4 \cdots (n-1) \cdot n.$$

Zero factorial is given by $0! = 1$. From this definition, you can see that $1! = 1$, $2! = 1 \cdot 2 = 2$, $3! = 1 \cdot 2 \cdot 3 = 6$, and so on. Factorials follow the same conventions for order of operations as exponents. That is, just as $2x^3$ and $(2x)^3$ imply different orders of operations, $2n!$ and $(2n)!$ imply the following orders.

$$2n! = 2(n!) = 2(1 \cdot 2 \cdot 3 \cdot 4 \cdots n)$$

and

$$(2n)! = 1 \cdot 2 \cdot 3 \cdot 4 \cdots n \cdot (n+1) \cdots 2n$$

Another useful limit theorem that can be rewritten for sequences is the Squeeze Theorem from Section 1.3.

> **THEOREM 8.3   Squeeze Theorem for Sequences**
>
> If
> $$\lim_{n\to\infty} a_n = L = \lim_{n\to\infty} b_n$$
> and there exists an integer N such that $a_n \le c_n \le b_n$ for all $n > N$, then
> $$\lim_{n\to\infty} c_n = L.$$

EXAMPLE 5   Using the Squeeze Theorem

Show that the sequence $\{c_n\} = \left\{ (-1)^n \dfrac{1}{n!} \right\}$ converges, and find its limit.

Solution   To apply the Squeeze Theorem, you must find two convergent sequences that can be related to the given sequence. Two possibilities are $a_n = -1/2^n$ and $b_n = 1/2^n$, both of which converge to 0. By comparing the term $n!$ with 2^n, you can see that

$$n! = 1 \cdot 2 \cdot 3 \cdot 4 \cdot 5 \cdot 6 \cdots n = 24 \cdot \underbrace{5 \cdot 6 \cdots n}_{n - 4 \text{ factors}} \qquad (n \ge 4)$$

and

$$2^n = 2 \cdot 2 \cdot 2 \cdot 2 \cdot 2 \cdot 2 \cdots 2 = 16 \cdot \underbrace{2 \cdot 2 \cdots 2}_{n - 4 \text{ factors}}. \qquad (n \ge 4)$$

This implies that for $n \ge 4$, $2^n < n!$, and you have

$$\frac{-1}{2^n} \le (-1)^n \frac{1}{n!} \le \frac{1}{2^n}, \qquad n \ge 4$$

as illustrated in Figure 8.2. Therefore, by Squeeze Theorem it follows that

$$\lim_{n\to\infty} (-1)^n \frac{1}{n!} = 0.$$

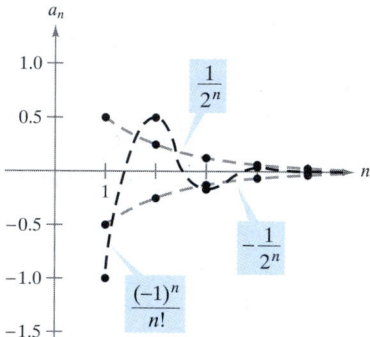

For $n \ge 4$, $(-1)^n/n!$ is squeezed between $-1/2^n$ and $1/2^n$.

Figure 8.2

NOTE   Example 5 suggests something about the rate at which $n!$ increases as $n \to \infty$. From Figure 8.2 you can see that both $1/2^n$ and $1/n!$ approach 0 as $n \to \infty$. Yet $1/n!$ approaches 0 so much faster than $1/2^n$ does that

$$\lim_{n\to\infty} \frac{1/n!}{1/2^n} = \lim_{n\to\infty} \frac{2^n}{n!} = 0.$$

In fact, it can be shown that for any fixed number k,

$$\lim_{n\to\infty} \frac{k^n}{n!} = 0.$$

This means that *the factorial function grows faster than any exponential function.*

In Example 5, the sequence $\{c_n\}$ has both positive and negative terms. For this sequence, it happens that the sequence of absolute values, $\{|c_n|\}$, also converges to 0. You can show this by the Squeeze Theorem using the inequality

$$0 \le \frac{1}{n!} \le \frac{1}{2^n}, \qquad n \ge 4.$$

In such cases, it is often convenient to consider the sequence of absolute values—and then apply Theorem 8.4, which states that if the absolute value sequence converges to 0, the original signed sequence also converges to 0.

THEOREM 8.4   Absolute Value Theorem

For the sequence $\{a_n\}$, if

$$\lim_{n \to \infty} |a_n| = 0 \qquad \text{then} \qquad \lim_{n \to \infty} a_n = 0.$$

Proof   Consider the two sequences $\{|a_n|\}$ and $\{-|a_n|\}$. Because both of these sequences converge to 0 and

$$-|a_n| \le a_n \le |a_n|$$

you can use the Squeeze Theorem to conclude that $\{a_n\}$ converges to 0.

Pattern Recognition for Sequences

Sometimes the terms of a sequence are generated by some rule that does not explicitly identify the nth term of the sequence. In such cases, you may be required to discover a *pattern* in the sequence and to describe the nth term. Once the nth term has been specified, you can investigate the convergence or divergence of the sequence.

EXAMPLE 6   Finding the nth Term of a Sequence

Find a sequence $\{a_n\}$ whose first five terms are

$$\frac{2}{1}, \frac{4}{3}, \frac{8}{5}, \frac{16}{7}, \frac{32}{9}, \ \cdots$$

and then determine whether the particular sequence you have chosen converges or diverges.

Solution   First, note that the numerators are successive powers of 2, and the denominators form the sequence of positive odd integers. By comparing a_n with n, you have the following pattern.

$$\frac{2^1}{1}, \frac{2^2}{3}, \frac{2^3}{5}, \frac{2^4}{7}, \frac{2^5}{9}, \ \cdots, \frac{2^n}{2n - 1}$$

Using L'Hôpital's Rule to evaluate the limit of $f(x) = 2^x/(2x - 1)$, you obtain

$$\lim_{x \to \infty} \frac{2^x}{2x - 1} = \lim_{x \to \infty} \frac{2^x (\ln 2)}{2} = \infty \qquad \Longrightarrow \qquad \lim_{n \to \infty} \frac{2^n}{2n - 1} = \infty.$$

Hence, the sequence *diverges*.

Without a specific rule for generating the terms of a sequence or some knowledge of the context in which the terms of the sequence are obtained, it is not possible to determine the convergence or divergence of the sequence merely from its first several terms. For instance, although the first three terms of the four sequences given next are identical, the first two sequences converge to 0, the third sequence converges to $\frac{1}{9}$, and the fourth sequence diverges.

$$\{a_n\} : \frac{1}{2}, \frac{1}{4}, \frac{1}{8}, \frac{1}{16}, \cdots , \frac{1}{2^n}, \cdots$$

$$\{b_n\} : \frac{1}{2}, \frac{1}{4}, \frac{1}{8}, \frac{1}{15}, \cdots , \frac{6}{(n+1)(n^2-n+6)}, \cdots$$

$$\{c_n\} : \frac{1}{2}, \frac{1}{4}, \frac{1}{8}, \frac{7}{62}, \cdots , \frac{n^2-3n+3}{9n^2-25n+18}, \cdots$$

$$\{d_n\} : \frac{1}{2}, \frac{1}{4}, \frac{1}{8}, 0, \cdots , \frac{-n(n+1)(n-4)}{6(n^2+3n-2)}, \cdots$$

The process of determining an nth term from the pattern observed in the first several terms of a sequence is an example of *inductive reasoning*.

EXAMPLE 7   Finding the nth Term of a Sequence

Determine an nth term for a sequence whose first five terms are

$$-\frac{2}{1}, \frac{8}{2}, -\frac{26}{6}, \frac{80}{24}, -\frac{242}{120}, \cdots$$

and then decide whether the sequence converges or diverges.

Solution   Note that the numerators are 1 less than 3^n. Hence, you can reason that the numerators are given by the rule $3^n - 1$. Factoring the denominators produces

$$1 = 1$$
$$2 = 1 \cdot 2$$
$$6 = 1 \cdot 2 \cdot 3$$
$$24 = 1 \cdot 2 \cdot 3 \cdot 4$$
$$120 = 1 \cdot 2 \cdot 3 \cdot 4 \cdot 5 \cdots.$$

This suggests that the denominators are represented by $n!$. Finally, because the signs alternate, you can write the nth term as

$$a_n = (-1)^n \left(\frac{3^n - 1}{n!} \right).$$

From the discussion about the growth of $n!$, it follows that

$$\lim_{n \to \infty} |a_n| = \lim_{n \to \infty} \frac{3^n - 1}{n!} = 0.$$

Applying Theorem 8.4, you can conclude that

$$\lim_{n \to \infty} a_n = 0.$$

Thus, the sequence $\{a_n\}$ converges to 0.

Monotonic Sequences and Bounded Sequences

So far you have determined the convergence of a sequence by finding its limit. Even if you cannot determine the limit of a particular sequence, it still may be useful to know whether the sequence converges. Theorem 8.5 identifies a test for convergence of sequences without determining the limit. First, we give some preliminary definitions.

Definition of a Monotonic Sequence

A sequence $\{a_n\}$ is **monotonic** if its terms are nondecreasing

$$a_1 \leq a_2 \leq a_3 \leq \cdots \leq a_n \leq \cdots$$

or if its terms are nonincreasing

$$a_1 \geq a_2 \geq a_3 \geq \cdots \geq a_n \geq \cdots.$$

EXAMPLE 8   Determining Whether a Sequence is Monotonic

Determine whether each sequence having the given nth term is monotonic.

a. $a_n = 3 + (-1)^n$ **b.** $b_n = \dfrac{2n}{1+n}$ **c.** $c_n = \dfrac{n^2}{2^n - 1}$

Solution

a. This sequence alternates between 2 and 4. Therefore, it is not monotonic.

b. This sequence is monotonic because each successive term is larger than its predecessor. To see this, compare the terms b_n and b_{n+1}. [Note that, because n is positive, you can multiply both sides of the inequality by $(1 + n)$ and $(2 + n)$ without reversing the inequality sign.]

$$b_n = \frac{2n}{1+n} \overset{?}{<} \frac{2(n+1)}{1+(n+1)} = b_{n+1}$$

$$2n(2+n) \overset{?}{<} (1+n)(2n+2)$$

$$4n + 2n^2 \overset{?}{<} 2 + 4n + 2n^2$$

$$0 < 2$$

Starting with the final inequality, which is valid, you can reverse the steps to conclude that the original inequality it also valid.

c. This sequence is not monotonic, because the second term is larger than the first term, and larger than the third. (Note that if we drop the first term, the remaining sequence $c_2, c_3, c_4, \ldots$ is monotonic.)

Figure 8.3 graphically illustrates these three sequences.

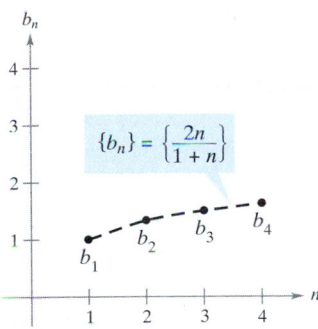

(a) Not monotonic

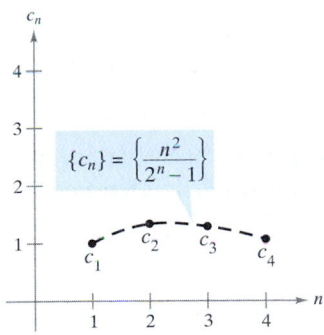

(b) Monotonic

(c) Not monotonic

Figure 8.3

NOTE   In Example 8b, another way to see that the sequence is monotonic is to argue that the derivative of the corresponding differentiable function $f(x) = 2x/(1 + x)$ is positive for all x. This implies that f is increasing, which in turn implies that $\{a_n\}$ is increasing.

NOTE   All three sequences shown in Figure 8.3 are bounded. To see this, consider the following

$$2 \leq a_n \leq 4$$
$$1 \leq b_n \leq 2$$
$$0 \leq c_n \leq \frac{4}{3}$$

Definition of a Bounded Sequence

1. A sequence $\{a_n\}$ is **bounded above** if there is a real number M such that $a_n \leq M$ for all n. The number M is called an **upper bound** of the sequence.

2. A sequence $\{a_n\}$ is **bounded below** if there is a real number N such that $N \leq a_n$ for all n. The number N is called a **lower bound** of the sequence.

3. A sequence $\{a_n\}$ is **bounded** if it is bounded above and bounded below.

One important property of the real numbers is that they are **complete.** Informally, this means that there are no holes or gaps on the real number line. (The set of rational numbers does not have the completeness property.) The completeness axiom for real numbers can be used to conclude that if a sequence has an upper bound, it must have a **least upper bound** (an upper bound that is smaller than all other upper bounds for the sequence). For example, the least upper bound of the sequence $\{a_n\} = \{n/(n + 1)\}$,

$$\frac{1}{2}, \frac{2}{3}, \frac{3}{4}, \frac{4}{5}, \dots, \frac{n}{n + 1}, \dots$$

is 1. We use the completeness axiom in the proof of Theorem 8.5.

THEOREM 8.5   Bounded Monotonic Sequences

If a sequence $\{a_n\}$ is bounded and monotonic, then it converges.

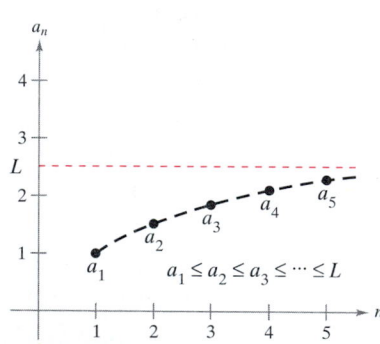

Every bounded nondecreasing sequence converges.

Figure 8.4

Proof   Assume that the sequence is nondecreasing, as shown in Figure 8.4. For the sake of simplicity, also assume that each term in the sequence is positive. Because the sequence is bounded, there must exist an upper bound M such that

$$a_1 \leq a_2 \leq a_3 \leq \dots \leq a_n \leq \dots \leq M.$$

From the completeness axiom, it follows that there is a least upper bound L such that

$$a_1 \leq a_2 \leq a_3 \leq \dots \leq a_n \leq \dots \leq L.$$

For $\varepsilon > 0$, if follows that $L - \varepsilon < L$, and therefore $L - \varepsilon$ cannot be an upper bound for the sequence. Consequently, at least one term of $\{a_n\}$ is greater than $L - \varepsilon$. That is, $L - \varepsilon < a_N$ for some positive integer N. Because the terms of $\{a_n\}$ are nondecreasing, it follows that $a_N \leq a_n$ for $n > N$. You now know that $L - \varepsilon < a_N \leq a_n \leq L < L + \varepsilon$, for every $n > N$. It follows that $|a_n - L| < \varepsilon$ for $n > N$, which by definition means that $\{a_n\}$ converges to L. The proof for a nonincreasing sequence is similar.

EXAMPLE 9   Bounded and Monotonic Sequences

a. The sequence $\{a_n\} = \{1/n\}$ is both bounded and monotonic and thus, by Theorem 8.5, must converge.

b. The divergent sequence $\{b_n\} = \{n^2/(n + 1)\}$ is monotonic, but not bounded. (It *is* bounded below.)

c. The divergent sequence $\{c_n\} = \{(-1)^n\}$ is bounded, but not monotonic.

EXERCISES FOR SECTION 8.1

In Exercises 1–8, write the first five terms of the sequence.

1. $a_n = 2^n$

2. $a_n = \dfrac{n}{n+1}$

3. $a_n = \left(-\dfrac{1}{2}\right)^n$

4. $a_n = \sin \dfrac{n\pi}{2}$

5. $a_n = \dfrac{(-1)^{n(n+1)/2}}{n^2}$

6. $a_n = 5 - \dfrac{1}{n} + \dfrac{1}{n^2}$

7. $a_n = \dfrac{3^n}{n!}$

8. $a_n = \dfrac{3n!}{(n-1)!}$

In Exercises 9–12, write the first five terms of the recursively defined sequence.

9. $a_1 = 3,\ a_{k+1} = 2(a_k - 1)$

10. $a_1 = 4,\ a_{k+1} = \left(\dfrac{k+1}{2}\right)a_k$

11. $a_1 = 32,\ a_{k+1} = \tfrac{1}{2}a_k$

12. $a_1 = 6,\ a_{k+1} = \tfrac{1}{3}a_k^2$

In Exercises 13–16, match the sequence with its graph. [The graphs are labeled (a), (b), (c), and (d).]

(a)
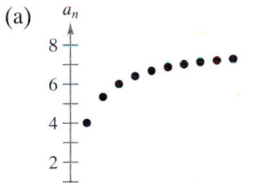

(b)
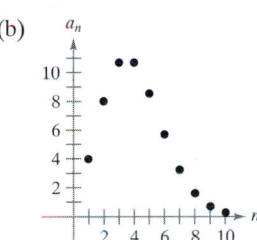

(c)
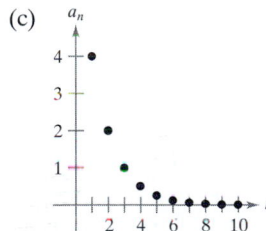

(d)
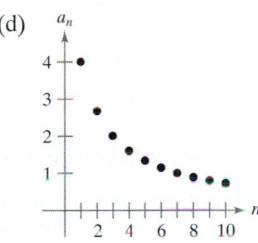

13. $a_n = \dfrac{8}{n+1}$

14. $a_n = \dfrac{8n}{n+1}$

15. $a_n = 4(0.5)^{n-1}$

16. $a_n = \dfrac{4^n}{n!}$

In Exercises 17–22, use a graphing utility to graph the first ten terms of the sequence.

17. $a_n = \dfrac{2}{3}n$

18. $a_n = 2 - \dfrac{4}{n}$

19. $a_n = 16(-0.5)^{n-1}$

20. $a_n = 8(0.75)^{n-1}$

21. $a_n = \dfrac{2n}{n+1}$

22. $a_n = \dfrac{3n^2}{n^2+1}$

In Exercises 23–26, write the next two *apparent* terms of the sequence. Describe the pattern you used to find these terms.

23. $2, 5, 8, 11, \ldots$

24. $\tfrac{7}{2}, 4, \tfrac{9}{2}, 5, \ldots$

25. $3, -\tfrac{3}{2}, \tfrac{3}{4}, -\tfrac{3}{8}, \ldots$

26. $5, 10, 20, 40, \ldots$

In Exercises 27–32, simplify the ratio of factorials.

27. $\dfrac{10!}{8!}$

28. $\dfrac{25!}{23!}$

29. $\dfrac{(n+1)!}{n!}$

30. $\dfrac{(n+2)!}{n!}$

31. $\dfrac{(2n-1)!}{(2n+1)!}$

32. $\dfrac{(2n+2)!}{(2n)!}$

In Exercises 33–46, write an expression for the nth term of the sequence. (There is more than one correct answer.)

33. $1, 4, 7, 10, \ldots$

34. $3, 7, 11, 15, \ldots$

35. $-1, 2, 7, 14, 23, \ldots$

36. $1, -\tfrac{1}{4}, \tfrac{1}{9}, -\tfrac{1}{16}, \ldots$

37. $\tfrac{2}{3}, \tfrac{3}{4}, \tfrac{4}{5}, \tfrac{5}{6}, \ldots$

38. $2, \tfrac{3}{3}, \tfrac{4}{5}, \tfrac{5}{7}, \tfrac{6}{9}, \ldots$

39. $2, -1, \tfrac{1}{2}, -\tfrac{1}{4}, \tfrac{1}{8}, \ldots$

40. $-\tfrac{1}{2}, \tfrac{1}{3}, -\tfrac{2}{9}, \tfrac{4}{27}, -\tfrac{8}{81}, \ldots$

41. $2, 1 + \tfrac{1}{2}, 1 + \tfrac{1}{3}, 1 + \tfrac{1}{4}, 1 + \tfrac{1}{5}, \ldots$

42. $1 + \tfrac{1}{2}, 1 + \tfrac{3}{4}, 1 + \tfrac{7}{8}, 1 + \tfrac{15}{16}, 1 + \tfrac{31}{32}, \ldots$

43. $\dfrac{1}{2\cdot 3}, \dfrac{2}{3\cdot 4}, \dfrac{3}{4\cdot 5}, \dfrac{4}{5\cdot 6}, \ldots$

44. $1, \tfrac{1}{2}, \tfrac{1}{6}, \tfrac{1}{24}, \tfrac{1}{120}, \ldots$

45. $1, -\dfrac{1}{1\cdot 3}, \dfrac{1}{1\cdot 3\cdot 5}, -\dfrac{1}{1\cdot 3\cdot 5\cdot 7}, \ldots$

46. $1, x, \dfrac{x^2}{2}, \dfrac{x^3}{6}, \dfrac{x^4}{24}, \dfrac{x^5}{120}, \ldots$

In Exercises 47–50, use a graphing utility to graph the first ten terms of the sequence. Use the graph to make an inference about the convergence or divergence of the sequence. Verify your inference analytically and, if the sequence converges, find its limit.

47. $a_n = \dfrac{n+1}{n}$

48. $a_n = \dfrac{1}{n^{3/2}}$

49. $a_n = \cos \dfrac{n\pi}{2}$

50. $a_n = 3 - \dfrac{1}{2^n}$

In Exercises 51–66, determine the convergence or divergence of the sequence with the given nth term. If the sequence converges, find its limit.

51. $a_n = (-1)^n \left(\dfrac{n}{n+1}\right)$

52. $a_n = 1 + (-1)^n$

53. $a_n = \dfrac{3n^2 - n + 4}{2n^2 + 1}$

54. $a_n = \dfrac{\sqrt{n}}{\sqrt{n} + 1}$

55. $a_n = \dfrac{1 + (-1)^n}{n}$

56. $a_n = \dfrac{\ln(n^2)}{n}$

57. $a_n = \dfrac{3^n}{4^n}$

58. $a_n = (0.5)^n$

59. $a_n = \dfrac{(n + 1)!}{n!}$

60. $a_n = \dfrac{(n - 2)!}{n!}$

61. $a_n = \dfrac{n - 1}{n} - \dfrac{n}{n - 1}$

62. $a_n = \dfrac{n^2}{2n + 1} - \dfrac{n^2}{2n - 1}$

63. $a_n = \dfrac{n^p}{e^n} \ (p > 0, n \geq 2)$

64. $a_n = n \sin \dfrac{1}{n}$

65. $a_n = \left(1 + \dfrac{k}{n}\right)^n$

66. $a_n = 2^{1/n}$

In Exercises 67–76, determine whether the sequence with the given nth term is monotonic. Discuss the boundedness of the sequence. Use a graphing utility to confirm your results.

67. $a_n = 4 - \dfrac{1}{n}$

68. $a_n = \dfrac{4n}{n + 1}$

69. $a_n = \dfrac{\cos n}{n}$

70. $a_n = ne^{-n/2}$

71. $a_n = (-1)^n \left(\dfrac{1}{n}\right)$

72. $a_n = \left(-\dfrac{2}{3}\right)^n$

73. $a_n = \left(\dfrac{2}{3}\right)^n$

74. $a_n = \left(\dfrac{3}{2}\right)^n$

75. $a_n = \sin \dfrac{n\pi}{6}$

76. $a_n = \dfrac{n}{2^{n+2}}$

In Exercises 77–80, (a) use Theorem 8.5 to show that the sequence with the given nth term converges and (b) use a graphing utility to graph the first ten terms of the sequence and find its limit.

77. $a_n = 5 + \dfrac{1}{n}$

78. $a_n = 3 - \dfrac{4}{n}$

79. $a_n = \dfrac{1}{3}\left(1 - \dfrac{1}{3^n}\right)$

80. $a_n = 4 + \dfrac{1}{2^n}$

81. **Think About It**   Give an example of a sequence satisfying the given condition or explain why no such sequence exists. (There is more than one correct answer.)

   (a) A monotonically increasing sequence that converges to 10.

   (b) A monotonically increasing bounded sequence that does not converge.

   (c) A sequence that converges to $\frac{3}{4}$.

   (d) An unbounded sequence that converges to 100.

82. Find a sequence that converges to $\frac{3}{8}$. (There are many such sequences.) What is the first term in the sequence that is within 0.001 of the limit?

83. **Compound Interest**   Consider the sequence $\{A_n\}$ whose nth term is given by

$$A_n = P\left(1 + \dfrac{r}{12}\right)^n$$

where P is the principal, A_n is the amount at compound interest after n months, and r is the interest rate compounded annually.

   (a) Is $\{A_n\}$ a convergent sequence? Explain.

   (b) Find the first ten terms of the sequence if $P = \$9000$ and $r = 0.115$.

84. **Investment**   A deposit of $100 is made at the beginning of each month in an account at an annual interest rate of 12% compounded monthly. The balance in the account after n months is

$$A_n = 100(101)[(1.01)^n - 1].$$

   (a) Compute the first six terms of the sequence $\{A_n\}$.

   (b) Find the balance after 5 years by computing the 60th term of the sequence.

   (c) Find the balance after 20 years by computing the 240th term of the sequence.

85. **Government Expenditures**   A government program that currently costs taxpayers $2.5 billion per year is cut back by 20 percent per year.

   (a) Write an expression for the amount budgeted for this program after n years.

   (b) Compute the budgets for the first 4 years.

   (c) Determine the convergence or divergence of the sequence of reduced budgets. If the sequence converges, find its limit.

86. **Inflation**   If the rate of inflation is $4\frac{1}{2}\%$ per year and the average price of a car is currently $16,000, the average price after n years is

$$P_n = \$16,000(1.045)^n.$$

Compute the average price for the next 5 years.

87. **Modeling Data**   The average cost per day for a hospital room from 1988 to 1993 is given in the table, where a_n is the average cost in dollars and n is the year, with $n = 0$ corresponding to 1990.   (Source: American Hospital Association)

n	-2	-1	0	1	2	3
a_n	586	637	687	752	820	881

   (a) Use the regression capabilities of a graphing utility to find a model of the form

$$a_n = kn + b, \qquad n = -2, -1, 0, 1, 2, 3$$

   for the data. Use the graphing utility to plot the points and graph the model.

   (b) Use the model to predict the cost in the year 2000.

88. *Modeling Data*   The net income a_n (in millions of dollars) of Wal-Mart from 1987 to 1996 is given below as ordered pairs of the form (n, a_n), where n is the year, with $n = 0$ corresponding to 1990.   *(Source: 1996 Wal-Mart Annual Report)*

$(-3, 451), (-2, 628), (-1, 838), (0, 1076), (1, 1291)$

$(2, 1609), (3, 1995), (4, 2333), (5, 2681), (6, 2740)$

(a) Use the regression capabilities of a graphing utility to find a model of the form

$a_n = bn^2 + cn + d, \quad n = -3, -1, \ldots, 6$

for the data. Graphically compare the points and the model.

(b) Use the model to predict the net income in the year 2000.

89. *Population Growth*   Consider an idealized population with the characteristic that each member of the population produces one offspring at the end of every time period. If each member has a life span of three time periods and the population begins with ten newborn members, then the following table gives the population during the first five time periods.

	Time Period				
Age Bracket	1	2	3	4	5
0–1	10	10	20	40	70
1–2		10	10	20	40
2–3			10	10	20
Total	10	20	40	70	130

The sequence for the total population has the property that

$S_n = S_{n-1} + S_{n-2} + S_{n-3}, \quad n > 3.$

Find the total population during the next five time periods.

90. *Federal Debt*   It took more than 200 years for the United States to accumulate a \$1 trillion debt. Then it took just 8 years to get to \$3 trillion. The federal debt during the decade of the 1980s is approximated by the model

$a_n - 0.905e^{0.134n}, \quad n = 0, 1, 2, \ldots, 10$

where a_n is the debt in trillions and n is the year, with $n = 0$ corresponding to 1980. Find the terms of this finite sequence and construct a bar graph that represents the sequence. *(Source: U.S. Treasury Department)*

91. *Comparing Exponential and Factorial Growth*   Consider the sequence $a_n = 10^n/n!$.

(a) Find two consecutive terms that are equal in magnitude.

(b) Are the terms following those found in part (a) increasing or decreasing?

(c) In Section 7.7, Exercises 51–56, it was shown that for "large" values of the independent variable an exponential function increased more rapidly than a polynomial function. From the result in part (b), what inference can you make about the rate of growth of an exponential function versus a factorial function for "large" integer values of n?

92. Compute the first six terms of the sequence

$\{a_n\} = \{(1 + 1/n)^n\}.$

If the sequence converges, find its limit.

93. Compute the first six terms of the sequence $\{a_n\} = \{\sqrt[n]{n}\}$. If the sequence converges, find its limit.

94. Prove that if $\{s_n\}$ converges to L and $L > 0$, then there exists a number N such that $s_n > 0$ for $n > N$.

95. *Fibonacci Sequence*   In the study of the progeny of rabbits, Fibonacci (ca. 1175–ca. 1250) encountered the sequence now bearing his name. It is defined recursively by

$a_{n+2} = a_n + a_{n+1}, \quad$ where $a_1 = 1$ and $a_2 = 1.$

(a) Write the first 12 terms of the sequence.

(b) Write the first ten terms of the sequence defined by

$b_n = \dfrac{a_{n+1}}{a_n}, \quad$ for $n \geq 1.$

(c) Using the definition in part (b), show that

$b_n = 1 + \dfrac{1}{b_{n-1}}.$

(d) The **golden ratio** ρ can be defined by $\lim\limits_{n \to \infty} b_n = \rho$. Show that

$\rho = 1 + 1/\rho$

and solve this equation for ρ.

96. Complete the proof of Theorem 8.5.

True or False?   In Exercises 97–100, determine whether the statement is true or false. If it is false, explain why or give an example that shows it is false.

97. If $\{a_n\}$ converges to 3 and $\{b_n\}$ converges to 2, then $\{a_n + b_n\}$ converges to 5.

98. If $\{a_n\}$ converges, then $\lim\limits_{n \to \infty} (a_n - a_{n+1}) = 0.$

99. If $n > 1$, then $n! = n(n-1)!$.

100. If $\{a_n\}$ converges, then $\{a_n/n\}$ converges to 0.

101. Consider the sequence

$\sqrt{2}, \sqrt{2 + \sqrt{2}}, \sqrt{2 + \sqrt{2 + \sqrt{2}}}, \ldots$

where $a_n = \sqrt{2 + a_{n-1}}$ for $n \geq 2$. Compute the first five terms of this sequence. Find $\lim\limits_{n \to \infty} a_n$.

102. *Conjecture*   Let $x_0 = 1$ and consider the sequence x_n given by the formula

$x_n = \dfrac{1}{2}x_{n-1} + \dfrac{1}{x_{n-1}}, \quad n = 1, 2, \ldots.$

Use a graphing utility to compute the first ten terms of the sequence and make a conjecture about the limit of the sequence.

Infinite Series • Geometric Series • *n*th-Term Test for Divergence

Infinite Series

One important application of infinite sequences is in representing "infinite summations." Informally, if $\{a_n\}$ is an infinite sequence, then

$$\sum_{n=1}^{\infty} a_n = a_1 + a_2 + a_3 + \cdots + a_n + \cdots \qquad \text{Infinite series}$$

is an **infinite series** (or simply a **series**). The numbers $a_1, a_2, a_3, \ldots$ are the **terms** of the series. For some series it is convenient to begin the index at $n = 0$ (or some other integer). As a typesetting convention, it is common to represent an infinite series as simply $\Sigma\, a_n$. In such cases, the starting value for the index must be taken from the context of the statement.

To find the sum of an infinite series, consider the following **sequence of partial sums.**

$$S_1 = a_1$$
$$S_2 = a_1 + a_2$$
$$S_3 = a_1 + a_2 + a_3$$
$$\vdots$$
$$S_n = a_1 + a_2 + a_3 + \cdots + a_n$$

If this sequence of partial sums converges, the series is said to converge and has the sum indicated in the following definition.

Definition of Convergent and Divergent Series

For the infinite series $\Sigma\, a_n$, the **nth partial sum** is given by

$$S_n = a_1 + a_2 + \cdots + a_n.$$

If the sequence of partial sums $\{S_n\}$ converges to S, then the series $\Sigma\, a_n$ **converges.** The limit S is called the **sum of the series.**

$$S = a_1 + a_2 + \cdots + a_n + \cdots$$

If $\{S_n\}$ diverges, then the series **diverges.**

STUDY TIP As you study this chapter, you will see that there are two basic questions involving infinite series. Does a series converge or does it diverge? If a series converges, what is its sum? These questions are not always easy to answer, especially the second one.

EXPLORATION

Finding the Sum of an Infinite Series Find the sum of each infinite series. Explain your reasoning.

(a) $0.1 + 0.01 + 0.001 + 0.0001 + \cdots$ 　(b) $\frac{3}{10} + \frac{3}{100} + \frac{3}{1000} + \frac{3}{10,000} + \cdots$

(c) $1 + \frac{1}{2} + \frac{1}{4} + \frac{1}{8} + \frac{1}{16} + \cdots$ 　(d) $\frac{15}{100} + \frac{15}{10,000} + \frac{15}{1,000,000} + \cdots$

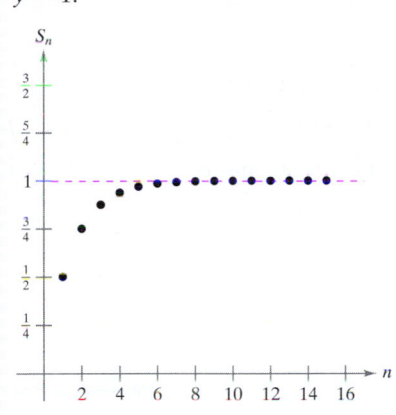

Figure 8.5

NOTE You can geometrically determine the partial sums of the series in Example 1a using Figure 8.6.

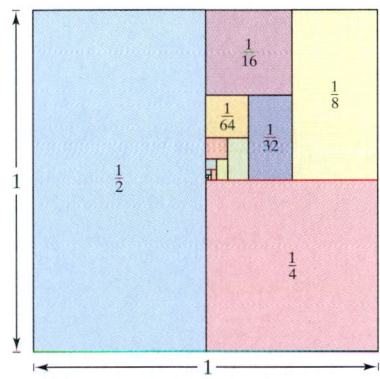

Figure 8.6

FOR FURTHER INFORMATION To learn more about the partial sums of infinite series, see the article "Six Ways to Sum a Series" by Dan Kalmon in the November 1993 issue of *The College Mathematics Journal.*

EXAMPLE 1   Convergent and Divergent Series

a. The series

$$\sum_{n=1}^{\infty} \frac{1}{2^n} = \frac{1}{2} + \frac{1}{4} + \frac{1}{8} + \frac{1}{16} + \cdots$$

has the following partial sums.

$$S_1 = \frac{1}{2}$$

$$S_2 = \frac{1}{2} + \frac{1}{4} = \frac{3}{4}$$

$$S_3 = \frac{1}{2} + \frac{1}{4} + \frac{1}{8} = \frac{7}{8}$$

$$\vdots$$

$$S_n = \frac{1}{2} + \frac{1}{4} + \frac{1}{8} + \cdots + \frac{1}{2^n} = \frac{2^n - 1}{2^n}$$

Because

$$\lim_{n \to \infty} \frac{2^n - 1}{2^n} = 1$$

it follows that the series converges and its sum is 1.

b. The nth partial sum of the series

$$\sum_{n=1}^{\infty} \left(\frac{1}{n} - \frac{1}{n+1} \right) = \left(1 - \frac{1}{2} \right) + \left(\frac{1}{2} - \frac{1}{3} \right) + \left(\frac{1}{3} - \frac{1}{4} \right) + \cdots$$

is given by

$$S_n = 1 - \frac{1}{n+1}.$$

Because the limit of S_n is 1, the series converges and its sum is 1.

c. The series

$$\sum_{n=1}^{\infty} 1 = 1 + 1 + 1 + 1 + \cdots$$

diverges because $S_n = n$ and the sequence of partial sums diverges.

The series in Example 1b is a **telescoping series.** That is, it is of the form

$$(b_1 - b_2) + (b_2 - b_3) + (b_3 - b_4) + (b_4 - b_5) + \cdots. \qquad \text{Telescoping series}$$

Note that b_2 is canceled by the second term, b_3 is canceled by the third term, and so on. Because the nth partial sum of this series is

$$S_n = b_1 - b_{n+1}$$

it follows that a telescoping series will converge if and only if b_n approaches a finite number as $n \to \infty$. Moreover, if the series converges, its sum is

$$S = b_1 - \lim_{n \to \infty} b_{n+1}.$$

EXAMPLE 2   Writing a Series in Telescoping Form

Find the sum of the series $\displaystyle\sum_{n=1}^{\infty} \frac{2}{4n^2 - 1}$.

Solution   Using partial fractions, you can write

$$a_n = \frac{2}{4n^2 - 1} = \frac{2}{(2n - 1)(2n + 1)} = \frac{1}{2n - 1} - \frac{1}{2n + 1}.$$

From this telescoping form, you can see that the nth partial sum is

$$S_n = \left(\frac{1}{1} - \frac{1}{3}\right) + \left(\frac{1}{3} - \frac{1}{5}\right) + \cdots + \left(\frac{1}{2n - 1} - \frac{1}{2n + 1}\right) = 1 - \frac{1}{2n + 1}.$$

Thus, the series converges and its sum is 1. That is,

$$\sum_{n=1}^{\infty} \frac{2}{4n^2 - 1} = \lim_{n \to \infty} S_n = \lim_{n \to \infty}\left(1 - \frac{1}{2n + 1}\right) = 1.$$

EXPLORATION

In "Proof Without Words," by Benjamin G. Klein and Irl C. Bivens, the authors present the following diagram. Explain why the final statement below the diagram is valid. How is this result related to Theorem 8.6?

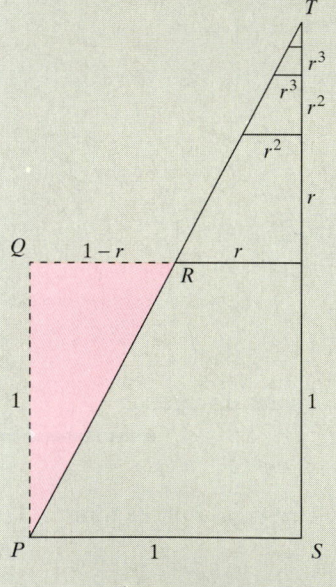

$\Delta PQR \approx \Delta TSP$

$$1 + r + r^2 + r^3 + \cdots = \frac{1}{1 - r}$$

Exercise taken from "Proof Without Words" by Benjamin G. Klein and Irl C. Bivens, MATHEMATICS MAGAZINE, October 1988, by permission of the authors.

Geometric Series

The series given in Example 1a is a **geometric series.** In general, the series given by

$$\sum_{n=0}^{\infty} ar^n = a + ar + ar^2 + \cdots + ar^n + \cdots, \quad a \neq 0 \qquad \text{Geometric series}$$

is a **geometric series** with ratio r.

> ### THEOREM 8.6   Convergence of a Geometric Series
>
> A geometric series with ratio r diverges if $|r| \geq 1$. If $0 < |r| < 1$, then the series converges to the sum
>
> $$\sum_{n=0}^{\infty} ar^n = \frac{a}{1 - r}, \qquad 0 < |r| < 1.$$

Proof   It is easy to see that the series diverges if $r = \pm 1$. If $r \neq \pm 1$, then $S_n = a + ar + ar^2 + \cdots + ar^{n-1}$. Multiplication by r yields

$$rS_n = ar + ar^2 + ar^3 + \cdots + ar^n.$$

Subtracting the second equation from the first produces $S_n - rS_n = a - ar^n$. Therefore, $S_n(1 - r) = a(1 - r^n)$, and the nth partial sum is

$$S_n = \frac{a}{1 - r}(1 - r^n).$$

If $0 < |r| < 1$, it follows that $r^n \to 0$ as $n \to \infty$, and you obtain

$$\lim_{n \to \infty} S_n = \lim_{n \to \infty}\left[\frac{a}{1 - r}(1 - r^n)\right] = \frac{a}{1 - r}\left[\lim_{n \to \infty}(1 - r^n)\right] = \frac{a}{1 - r}$$

which means that the series *converges* and its sum is $a/(1 - r)$. We leave it to you to show that the series diverges if $|r| > 1$.

EXAMPLE 3   Convergent and Divergent Geometric Series

a. The geometric series

$$\sum_{n=0}^{\infty} \frac{3}{2^n} = \sum_{n=0}^{\infty} 3\left(\frac{1}{2}\right)^n = 3(1) + 3\left(\frac{1}{2}\right) + 3\left(\frac{1}{2}\right)^2 + \cdots$$

has a ratio of $r = \frac{1}{2}$ with $a = 3$. Because $0 < |r| < 1$, the series converges and its sum is

$$S = \frac{a}{1-r} = \frac{3}{1-(1/2)} = 6.$$

b. The geometric series

$$\sum_{n=0}^{\infty} \left(\frac{3}{2}\right)^n = 1 + \frac{3}{2} + \frac{9}{4} + \frac{27}{8} + \cdots$$

has a ratio of $r = \frac{3}{2}$. Because $|r| \geq 1$, the series diverges.

The formula for the sum of a geometric series can be used to write a repeating decimal as the ratio of two integers, as demonstrated in the next example.

EXAMPLE 4   A Geometric Series for a Repeating Decimal

Use a geometric series to express $0.080\overline{808}$ as the ratio of two integers.

Solution   For the repeating decimal $0.080\overline{808}$, you can write

$$0.080808\ldots = \frac{8}{10^2} + \frac{8}{10^4} + \frac{8}{10^6} + \frac{8}{10^8} + \cdots = \sum_{n=0}^{\infty} \left(\frac{8}{10^2}\right)\left(\frac{1}{10^2}\right)^n.$$

For this series, you have $a = 8/10^2$ and $r = 1/10^2$. Thus,

$$0.080808\ldots = \frac{a}{1-r} = \frac{8/10^2}{1-(1/10^2)} = \frac{8}{99}.$$

Try dividing 8 by 99 on a calculator to see that it produces $0.080\overline{808}$.

The convergence of a series is not affected by removal of a finite number of terms from the beginning of the series. For instance, the geometric series

$$\sum_{n=4}^{\infty} \left(\frac{1}{2}\right)^n \quad \text{and} \quad \sum_{n=0}^{\infty} \left(\frac{1}{2}\right)^n$$

both converge. Furthermore, because the sum of the second series is $a/(1-r) = 2$, you can conclude that the sum of the first series is

$$S = 2 - \left[\left(\frac{1}{2}\right)^0 + \left(\frac{1}{2}\right)^1 + \left(\frac{1}{2}\right)^2 + \left(\frac{1}{2}\right)^3\right]$$

$$= 2 - \frac{15}{8}$$

$$= \frac{1}{8}.$$

TECHNOLOGY   Try using a graphing utility or writing a computer program that computes the sum of the first 20 terms of the sequence in Example 3a. You should obtain a sum of about 5.999997.

The following properties are direct consequences of the corresponding properties
of limits of sequences.

THEOREM 8.7   Properties of Infinite Series

If $\Sigma\, a_n = A$, $\Sigma\, b_n = B$, and c is a real number, then the following series converge
to the indicated sums.

1. $\displaystyle\sum_{n=1}^{\infty} ca_n = cA$ **2.** $\displaystyle\sum_{n=1}^{\infty} (a_n + b_n) = A + B$ **3.** $\displaystyle\sum_{n=1}^{\infty} (a_n - b_n) = A - B$

---

*n*th-Term Test for Divergence

The following theorem states that if a series converges, the limit of its nth term must
be 0.

THEOREM 8.8   Limit of *n*th Term of a Convergent Series

If the series $\Sigma\, a_n$ converges, then the sequence $\{a_n\}$ converges to 0.

Proof   Assume that

$$\sum_{n=1}^{\infty} a_n = \lim_{n\to\infty} S_n = L.$$

Then, because $S_n = S_{n-1} + a_n$ and

$$\lim_{n\to\infty} S_n = \lim_{n\to\infty} S_{n-1} = L$$

it follows that

$$L = \lim_{n\to\infty} S_n = \lim_{n\to\infty} (S_{n-1} + a_n) = \lim_{n\to\infty} S_{n-1} + \lim_{n\to\infty} a_n$$

$$= L + \lim_{n\to\infty} a_n$$

which implies that $\{a_n\}$ converges to 0.   ▬▬▬

The contrapositive of Theorem 8.8 provides a useful test for *divergence*. This
nth-Term Test for Divergence states that if the limit of the nth term of a series does
not converge to 0, the series must diverge.

THEOREM 8.9   *n*th-Term Test for Divergence

If the sequence $\{a_n\}$ does not converge to 0, then the series $\Sigma\, a_n$ diverges.

NOTE   Be sure you see that this theorem does *not* state that the series $\Sigma\, a_n$ converges if $\{a_n\}$
converges to 0.

EXAMPLE 5   Using the *n*th-Term Test for Divergence

a. For the series $\displaystyle\sum_{n=0}^{\infty} 2^n$, you have

$$\lim_{n\to\infty} 2^n = \infty.$$

Thus, the limit of the *n*th term is not 0, and the series *diverges*.

b. For the series $\displaystyle\sum_{n=1}^{\infty} \frac{n!}{2n! + 1}$, you have

$$\lim_{n\to\infty} \frac{n!}{2n! + 1} = \frac{1}{2}.$$

Thus, the limit of the *n*th term is not 0, and the series *diverges*.

c. For the series $\displaystyle\sum_{n=1}^{\infty} \frac{1}{n}$, you have

$$\lim_{n\to\infty} \frac{1}{n} = 0.$$

Because the limit of the *n*th term is 0, the *n*th-Term Test for Divergence does *not* apply and you can draw no conclusions about convergence or divergence. (In the next section, you will see that this particular series diverges.)

STUDY TIP   The series in Example 5c will play an important role in this chapter.

$$\sum_{n=1}^{\infty} \frac{1}{n} = 1 + \frac{1}{2} + \frac{1}{3} + \frac{1}{4} + \cdots$$

You will see that this series diverges even though the *n*th term approaches 0 as *n* approaches ∞.

EXAMPLE 6   Bouncing Ball Problem

A ball is dropped from a height of 6 feet and begins bouncing, as shown in Figure 8.7. The height of each bounce is three-fourths the height of the previous bounce. Find the total vertical distance traveled by the ball.

Solution   When the ball hits the ground for the first time, it has traveled a distance of $D_1 = 6$. For subsequent bounces, let D_n be the distance traveled up *and* down. For example, D_2 and D_3 are as follows.

$$D_2 = \underbrace{6\left(\tfrac{3}{4}\right)}_{\text{Up}} + \underbrace{6\left(\tfrac{3}{4}\right)}_{\text{Down}} = 12\left(\tfrac{3}{4}\right) \qquad D_3 = \underbrace{6\left(\tfrac{3}{4}\right)\left(\tfrac{3}{4}\right)}_{\text{Up}} + \underbrace{6\left(\tfrac{3}{4}\right)\left(\tfrac{3}{4}\right)}_{\text{Down}} = 12\left(\tfrac{3}{4}\right)^2$$

By continuing this process, it can be determined that the total vertical distance is

$$D = 6 + 12\left(\tfrac{3}{4}\right) + 12\left(\tfrac{3}{4}\right)^2 + 12\left(\tfrac{3}{4}\right)^3 + \cdots$$

$$= 6 + 12 \sum_{n=0}^{\infty} \left(\tfrac{3}{4}\right)^{n+1}$$

$$= 6 + 12\left(\tfrac{3}{4}\right) \sum_{n=0}^{\infty} \left(\tfrac{3}{4}\right)^n$$

$$= 6 + 9\left(\frac{1}{1 - \tfrac{3}{4}}\right) = 6 + 9(4)$$

$$= 42 \text{ feet.}$$

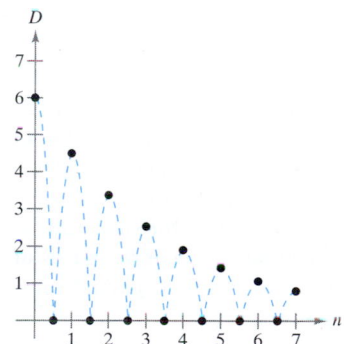

The height of each bounce is three-fourths the height of the previous bounce.
Figure 8.7

EXERCISES FOR SECTION 8.2

In Exercises 1–6, find the first five terms of the sequence of partial sums.

1. $1 + \frac{1}{4} + \frac{1}{9} + \frac{1}{16} + \frac{1}{25} + \cdots$

2. $\frac{1}{2 \cdot 3} + \frac{2}{3 \cdot 4} + \frac{3}{4 \cdot 5} + \frac{4}{5 \cdot 6} + \frac{5}{6 \cdot 7} + \cdots$

3. $3 - \frac{9}{2} + \frac{27}{4} - \frac{81}{8} + \frac{243}{16} - \cdots$

4. $\frac{1}{1} + \frac{1}{3} + \frac{1}{5} + \frac{1}{7} + \frac{1}{9} + \frac{1}{11} + \cdots$

5. $\sum_{n=1}^{\infty} \frac{3}{2^{n-1}}$

6. $\sum_{n=1}^{\infty} \frac{(-1)^{n+1}}{n!}$

In Exercises 7–16, verify that the infinite series diverges.

7. $\sum_{n=1}^{\infty} \frac{n}{n+1} = \frac{1}{2} + \frac{2}{3} + \frac{3}{4} + \frac{4}{5} + \cdots$

8. $\sum_{n=1}^{\infty} \frac{n}{2n+3} = \frac{1}{5} + \frac{2}{7} + \frac{3}{9} + \frac{4}{11} + \cdots$

9. $\sum_{n=1}^{\infty} \frac{n^2}{n^2 + 1}$

10. $\sum_{n=1}^{\infty} \frac{n}{\sqrt{n^2 + 1}}$

11. $\sum_{n=0}^{\infty} 3\left(\frac{3}{2}\right)^n$

12. $\sum_{n=0}^{\infty} \left(\frac{4}{3}\right)^n$

13. $\sum_{n=0}^{\infty} 1000(1.055)^n$

14. $\sum_{n=0}^{\infty} 2(-1.03)^n$

15. $\sum_{n=1}^{\infty} \frac{2^n + 1}{2^{n+1}}$

16. $\sum_{n=1}^{\infty} \frac{n!}{2^n}$

In Exercises 17–22, verify that the infinite series converges.

17. $\sum_{n=0}^{\infty} 2\left(\frac{3}{4}\right)^n = 2 + \frac{3}{2} + \frac{9}{8} + \frac{27}{32} + \frac{81}{128} + \cdots$

18. $\sum_{n=1}^{\infty} 2\left(-\frac{1}{2}\right)^n = 2 - 1 + \frac{1}{2} - \frac{1}{4} + \frac{1}{8} - \cdots$

19. $\sum_{n=0}^{\infty} (0.9)^n = 1 + 0.9 + 0.81 + 0.729 + \cdots$

20. $\sum_{n=0}^{\infty} (-0.6)^n = 1 - 0.6 + 0.36 - 0.216 + \cdots$

21. $\sum_{n=1}^{\infty} \frac{1}{n(n+1)}$ (Use partial fractions.)

22. $\sum_{n=1}^{\infty} \frac{1}{n(n+2)}$ (Use partial fractions.)

In Exercises 23–26, match the series with the graph of its sequence of partial sums. [The graphs are labeled (a), (b), (c), and (d).] Use the graph to estimate the sum of the series. Confirm your answer analytically.

(a)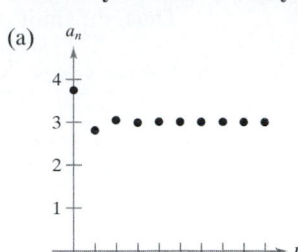

(b)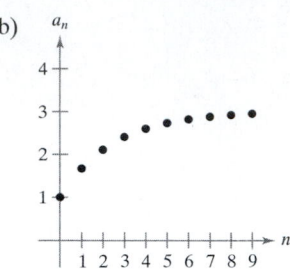

(c)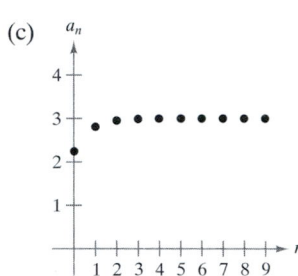

(d)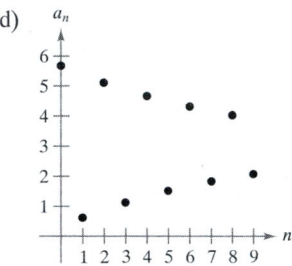

23. $\sum_{n=0}^{\infty} \frac{9}{4}\left(\frac{1}{4}\right)^n$

24. $\sum_{n=0}^{\infty} \left(\frac{2}{3}\right)^n$

25. $\sum_{n=0}^{\infty} \frac{15}{4}\left(-\frac{1}{4}\right)^n$

26. $\sum_{n=0}^{\infty} \frac{17}{3}\left(-\frac{8}{9}\right)^n$

Numerical, Graphical, and Analytic Analysis   In Exercises 27–30, (a) find the sum of the series, (b) use a graphing utility to find the indicated partial sum S_n and complete the table, (c) use a graphing utility to graph the first ten terms of the sequence of partial sums and a horizontal line representing the sum, and (d) explain the relationship between the magnitude of the terms of the series and the rate at which the sequence of partial sums approaches the sum of the series.

n	5	10	20	50	100
S_n					

27. $\sum_{n=1}^{\infty} 2(0.9)^{n-1}$

28. $\sum_{n=1}^{\infty} \frac{4}{n(n+4)}$

29. $\sum_{n=1}^{\infty} 10(0.25)^{n-1}$

30. $\sum_{n=1}^{\infty} 5\left(-\frac{1}{3}\right)^{n-1}$

In Exercises 31–44, find the sum of the convergent series.

31. $\sum_{n=0}^{\infty} \left(\frac{1}{2}\right)^n$

32. $\sum_{n=0}^{\infty} 2\left(\frac{2}{3}\right)^n$

33. $\displaystyle\sum_{n=0}^{\infty} \left(-\frac{1}{2}\right)^n$

34. $\displaystyle\sum_{n=0}^{\infty} 2\left(-\frac{2}{3}\right)^n$

35. $1 + 0.1 + 0.01 + 0.001 + \cdots$

36. $8 + 6 + \frac{9}{2} + \frac{27}{8} + \cdots$

37. $3 - 1 + \frac{1}{3} - \frac{1}{9} + \cdots$

38. $4 - 2 + 1 - \frac{1}{2} + \cdots$

39. $\displaystyle\sum_{n=2}^{\infty} \frac{1}{n^2 - 1}$

40. $\displaystyle\sum_{n=1}^{\infty} \frac{1}{n(n+1)}$

41. $\displaystyle\sum_{n=1}^{\infty} \frac{4}{n(n+2)}$

42. $\displaystyle\sum_{n=1}^{\infty} \frac{1}{(2n+1)(2n+3)}$

43. $\displaystyle\sum_{n=0}^{\infty} \left(\frac{1}{2^n} - \frac{1}{3^n}\right)$

44. $\displaystyle\sum_{n=1}^{\infty} [(0.7)^n + (0.9)^n]$

In Exercises 45–48, express the repeating decimal as a geometric series, and write its sum as the ratio of two integers.

45. $0.\overline{4}$

46. $0.23\overline{23}$

47. $0.075\overline{75}$

48. $0.215\overline{15}$

In Exercises 49–60, determine the convergence or divergence of the series.

49. $\displaystyle\sum_{n=1}^{\infty} \frac{n+10}{10n+1}$

50. $\displaystyle\sum_{n=1}^{\infty} \frac{n+1}{2n-1}$

51. $\displaystyle\sum_{n=1}^{\infty} \left(\frac{1}{n} - \frac{1}{n+2}\right)$

52. $\displaystyle\sum_{n=1}^{\infty} \frac{1}{n(n+3)}$

53. $\displaystyle\sum_{n=1}^{\infty} \frac{3n-1}{2n+1}$

54. $\displaystyle\sum_{n=1}^{\infty} \frac{2^n}{n^2}$

55. $\displaystyle\sum_{n=0}^{\infty} \frac{4}{2^n}$

56. $\displaystyle\sum_{n=0}^{\infty} \frac{1}{4^n}$

57. $\displaystyle\sum_{n=0}^{\infty} (1.075)^n$

58. $\displaystyle\sum_{n=1}^{\infty} \frac{2^n}{100}$

59. $\displaystyle\sum_{n=2}^{\infty} \frac{n}{\ln n}$

60. $\displaystyle\sum_{n=1}^{\infty} \left(1 + \frac{k}{n}\right)^n$

In Exercises 61 and 62, (a) find the common ratio of the geometric series, (b) write the function that gives the sum of the series, and (c) use a graphing utility to graph the function and the partial sum S_2.

61. $1 + x + x^2 + x^3 + \cdots$

62. $1 - \frac{x}{2} + \frac{x^2}{4} - \frac{x^3}{8} + \cdots$

In Exercises 63 and 64, use a graphing utility to graph the function. Identify the horizontal asymptote of the graph and determine its relationship to the sum of the series.

Function	Series
63. $f(x) = 3\left[\dfrac{1-(0.5)^x}{1-0.5}\right]$	$\displaystyle\sum_{n=0}^{\infty} 3\left(\frac{1}{2}\right)^n$
64. $f(x) = 2\left[\dfrac{1-(0.8)^x}{1-0.8}\right]$	$\displaystyle\sum_{n=0}^{\infty} 2\left(\frac{4}{5}\right)^n$

Writing   **In Exercises 65 and 66, use a graphing utility to determine the first term that is less than 0.0001 in each of the convergent series. Note that the answers are very different. Explain how this will affect the rate at which the series converges.**

65. $\displaystyle\sum_{n=1}^{\infty} \frac{1}{n(n+1)}, \qquad \sum_{n=1}^{\infty} \left(\frac{1}{8}\right)^n$

66. $\displaystyle\sum_{n=1}^{\infty} \frac{1}{2^n}, \qquad \sum_{n=1}^{\infty} (0.01)^n$

67. ***Marketing***   A company producing a new product estimates the annual sales to be 8000 units. Each year 10% of the units that have been sold will become inoperative. Thus, 8000 units will be in use after 1 year, $[8000 + 0.9(8000)]$ units will be in use after 2 years, and so on. How many units will be in use after n years?

68. ***Multiplier Effect***   The annual spending by tourists in a resort city is $100 million. Approximately 75% of that revenue is again spent in the resort city, and of that amount approximately 75% is again spent in the same city, and so on. Write the geometric series that gives the total amount of spending generated by the $100 million and find the sum of the series.

69. ***Distance***   A ball is dropped from a height of 16 feet. Each time it drops h feet, it rebounds $0.81h$ feet. Find the total distance traveled by the ball.

70. ***Time***   The ball in Exercise 69 takes the following times for each fall.

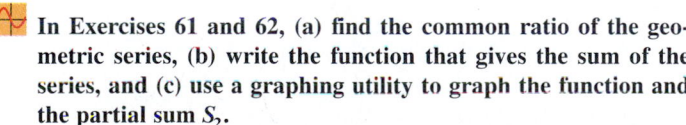

$s_1 = -16t^2 + 16,$	$s_1 = 0$ if $t = 1$
$s_2 = -16t^2 + 16(0.81),$	$s_2 = 0$ if $t = 0.9$
$s_3 = -16t^2 + 16(0.81)^2,$	$s_3 = 0$ if $t = (0.9)^2$
$s_4 = -16t^2 + 16(0.81)^3,$	$s_4 = 0$ if $t = (0.9)^3$
$\vdots$	$\vdots$
$s_n = -16t^2 + 16(0.81)^{n-1},$	$s_n = 0$ if $t = (0.9)^{n-1}$

Beginning with s_2, the ball takes the same amount of time to bounce up as it does to fall, and thus the total time elapsed before it comes to rest is

$$t = 1 + 2\sum_{n=1}^{\infty} (0.9)^n.$$

Find this total.

Probability **In Exercises 71 and 72, the random variable *n* represents the number of units of a certain product sold per day in a store. The probability distribution of *n* is given by *P(n)*. Find the probability that two units are sold in a given day [*P(2)*] and show that**

$$\sum_{n=0}^{\infty} P(n) = 1.$$

71. $P(n) = \frac{1}{2}\left(\frac{1}{2}\right)^n$ **72.** $P(n) = \frac{1}{3}\left(\frac{2}{3}\right)^n$

73. *Area* The sides of a square are 16 inches in length. A new square is formed by connecting the midpoints of the sides of the original square, and two of the triangles outside the second square are shaded (see figure). Determine the area of the shaded region (a) if this process is continued five more times and (b) if this pattern of shading is continued infinitely.

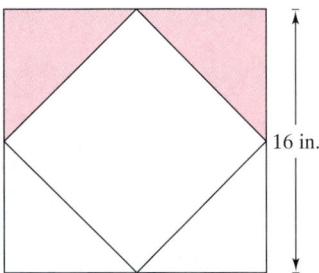

16 in.

In Exercises 74–76, use the formula for the *n*th partial sum of a geometric series

$$\sum_{i=0}^{n-1} ar^i = \frac{a(1 - r^n)}{1 - r}.$$

74. *Sphereflake* The sphereflake shown below is a computer-generated fractal that was created by Eric Haines, 3D/Eye Inc. The radius of the large sphere is 1. To the large sphere, nine spheres of radius $\frac{1}{3}$ are attached. To each of these, nine spheres of radius $\frac{1}{9}$ are attached. This process is continued infinitely. Prove that the sphereflake has an infinite surface area.

Eric Haines, 1987

This sphereflake was created by Eric Haines, using a recursive program.

75. *Income* Suppose you go to work at a company that pays $0.01 for the first day, $0.02 for the second day, $0.04 for the third day, and so on. If the daily wage keeps doubling, what would your total income be for working (a) 29 days, (b) 30 days, and (c) 31 days?

76. *Annuities* When an employee receives a paycheck at the end of each month, *P* dollars is invested in a retirement account. These deposits are made each month for *t* years and the account earns interest at the annual percentage rate *r*. If the interest is compounded monthly, the amount *A* in the account at the end of *t* years is

$$A = P + P\left(1 + \frac{r}{12}\right) + \cdots + P\left(1 + \frac{r}{12}\right)^{12t-1}$$

$$= P\left(\frac{12}{r}\right)\left[\left(1 + \frac{r}{12}\right)^{12t} - 1\right].$$

If the interest is compounded continuously, the amount *A* in the account after *t* years is

$$A = P + Pe^{r/12} + Pe^{2r/12} + Pe^{(12t-1)r/12}$$

$$= \frac{P(e^{rt} - 1)}{e^{r/12} - 1}.$$

Verify the formulas for the sums given above.

Annuities **In Exercises 77–80, consider making monthly deposits of *P* dollars in a savings account at an annual interest rate *r*. Use the results of Exercise 76 to find the balance *A* after *t* years if the interest is compounded (a) monthly and (b) continuously.**

77. $P = \$50$ $r = 3\%$ $t = 20$ years
78. $P = \$75$ $r = 5\%$ $t = 25$ years
79. $P = \$100$ $r = 4\%$ $t = 40$ years
80. $P = \$20$ $r = 6\%$ $t = 50$ years

81. Prove that $0.75 = 0.749999\ldots$.

82. Prove that every decimal with a repeating pattern of digits is a rational number.

83. Show that the series

$$\sum_{n=1}^{\infty} a_n$$

can be written in the telescoping form

$$\sum_{n=1}^{\infty} [(c - S_{n-1}) - (c - S_n)]$$

where $S_0 = 0$ and S_n is the *n*th partial sum.

84. Let $\Sigma\, a_n$ be a convergent series, and let

$$R_N = a_{N+1} + a_{N+2} + \cdots$$

be the remainder of the series after the first *N* terms. Prove that

$$\lim_{N \to \infty} R_N = 0.$$

85. *Salary*   You accept a job with a salary of $30,000 for the first year. Suppose, during the next 39 years, you receive a 5% raise each year. What would be your total compensation over the 40-year period?

86. *Profit*   The annual profit for the H. J. Heinz Company from 1980 through 1989 can be approximated by the model

$$a_n = 167.5e^{0.12n}, \qquad n = 0, 1, 2, \ldots, 9$$

where a_n is the annual profit in millions of dollars and n represents the year, with $n = 0$ corresponding to 1980. Use the formula for the sum of a geometric sequence to approximate the total profit earned during this 10-year period.

87. Find two divergent series $\Sigma\, a_n$ and $\Sigma\, b_n$ such that $\Sigma(a_n + b_n)$ converges.

88. Given two infinite series $\Sigma\, a_n$ and $\Sigma\, b_n$ such that $\Sigma\, a_n$ converges and $\Sigma\, b_n$ diverges, prove that $\Sigma(a_n + b_n)$ diverges.

True or False?   **In Exercises 89–92, determine whether the statement is true or false. If it is false, explain why or give an example that shows it is false.**

89. If $\displaystyle\lim_{n\to\infty} a_n = 0$, then $\displaystyle\sum_{n=1}^{\infty} a_n$ converges.

90. If $\displaystyle\sum_{n=1}^{\infty} a_n = L$, then $\displaystyle\sum_{n=0}^{\infty} a_n = L + a_0$.

91. If $|r| < 1$, then $\displaystyle\sum_{n=1}^{\infty} ar^n = a/(1 - r)$.

92. The series $\displaystyle\sum_{n=1}^{\infty} \frac{n}{1000(n + 1)}$ diverges.

93. *Writing*   Read the article "The Exponential-Decay Law Applied to Medical Dosages" by Gerald M. Armstrong and Calvin P. Midgley in the February 1987 issue of *Mathematics Teacher*. Then write a paragraph on how a geometric sequence can be used to find the total amount of a drug that remains in a patient's system after n equal dosages have been administered (at equal time intervals).

94. Prove that

$$\frac{1}{r} + \frac{1}{r^2} + \frac{1}{r^3} + \cdots = \frac{1}{r - 1}$$

for $|r| > 1$.

SECTION PROJECT

Cantor's Disappearing Table   The following procedure shows how to make a table disappear by removing only half of the table!

(a) Original table has a length of L.

(b) Remove $\frac{1}{4}$ of the table centered at the midpoint. Each remaining piece has a length that is less than $\frac{1}{2}L$.

(c) Remove $\frac{1}{8}$ of the table by taking sections of length $\frac{1}{16}L$ from the centers of each of the two remaining pieces. Now, you have removed $\frac{1}{4} + \frac{1}{8}$ of the table. Each remaining piece has a length that is less than $\frac{1}{4}L$.

(d) Remove $\frac{1}{16}$ of the table by taking sections of length $\frac{1}{64}L$ from the centers of each of the four remaining pieces. Now, you have removed $\frac{1}{4} + \frac{1}{8} + \frac{1}{16}$ of the table. Each remaining piece has a length that is less than $\frac{1}{8}L$.

Will continuing this process cause the table to disappear, even though you have only removed half of the table? Why?

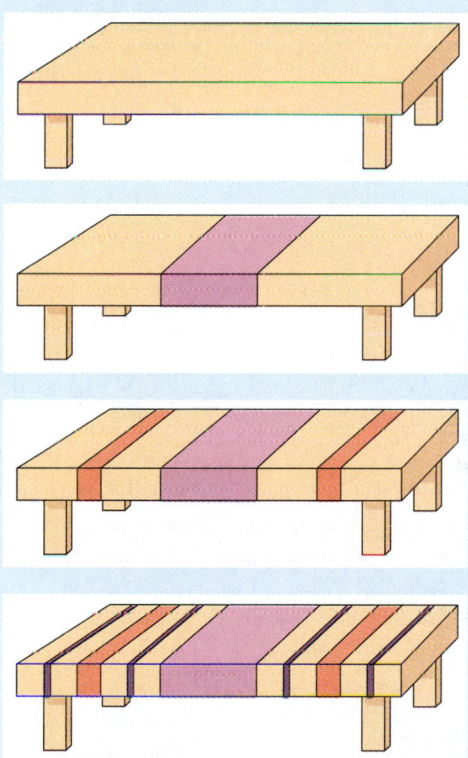

FOR FURTHER INFORMATION   Read the article "Cantor's Disappearing Table" by Larry E. Knop in the November 1985 issue of the *College Mathematics Journal*.

The Integral Test • *p*-Series and Harmonic Series

The Integral Test

In this and the following section, you will study several convergence tests that apply to series with *positive* terms.

> **THEOREM 8.10   The Integral Test**
>
> If f is positive, continuous, and decreasing for $x \geq 1$ and $a_n = f(n)$, then
>
> $$\sum_{n=1}^{\infty} a_n \quad \text{and} \quad \int_1^{\infty} f(x)\, dx$$
>
> either both converge or both diverge.

Proof   Begin by partitioning the interval $[1, n]$ into $n - 1$ unit intervals, as shown in Figure 8.8. The total areas of the inscribed rectangles and the circumscribed rectangles are as follows.

$$\sum_{i=2}^{n} f(i) = f(2) + f(3) + \cdots + f(n) \qquad \text{Inscribed area}$$

$$\sum_{i=1}^{n-1} f(i) = f(1) + f(2) + \cdots + f(n-1) \qquad \text{Circumscribed area}$$

The exact area under the graph of f from $x = 1$ to $x = n$ lies between the inscribed and circumscribed areas.

$$\sum_{i=2}^{n} f(i) \leq \int_1^{n} f(x)\, dx \leq \sum_{i=1}^{n-1} f(i)$$

Using the nth partial sum, $S_n = f(1) + f(2) + \cdots + f(n)$, you can write this inequality as

$$S_n - f(1) \leq \int_1^{n} f(x)\, dx \leq S_{n-1}.$$

Now, assuming that $\int_1^{\infty} f(x)\, dx$ converges to L, it follows that for $n \geq 1$

$$S_n - f(1) \leq L \qquad \Longrightarrow \qquad S_n \leq L + f(1).$$

Consequently, $\{S_n\}$ is bounded and monotonic, and by Theorem 8.5 it converges. Thus, $\Sigma\, a_n$ converges. For the other direction of the proof, assume that the improper integral diverges. Then $\int_1^{n} f(x)\, dx$ approaches infinity as $n \to \infty$, and the inequality $S_{n-1} \geq \int_1^{n} f(x)\, dx$ implies that $\{S_n\}$ diverges. Thus, $\Sigma\, a_n$ diverges.

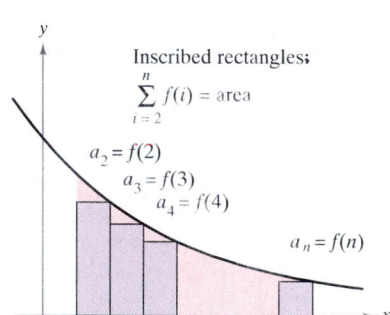

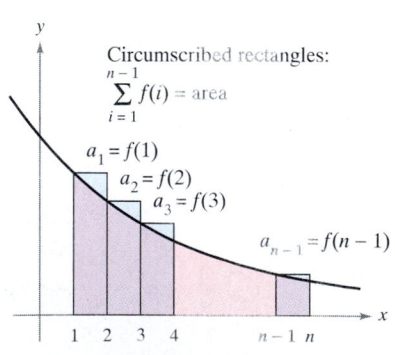

Figure 8.8

NOTE   Remember that the convergence or divergence of $\Sigma\, a_n$ is not affected by deleting the first N terms. Similarly, if the conditions for the Integral Test are satisfied for all $x \geq N > 1$, you can simply use the integral $\int_N^{\infty} f(x)\, dx$ to test for convergence or divergence. (This is illustrated in Example 4.)

EXAMPLE 1   Using the Integral Test

Apply the Integral Test to the series $\displaystyle\sum_{n=1}^{\infty} \frac{n}{n^2 + 1}$.

Solution   Because $f(x) = x/(x^2 + 1)$ satisfies the conditions for the Integral Test (check this), you can integrate to obtain

$$
\begin{aligned}
\int_1^\infty \frac{x}{x^2 + 1}\,dx &= \frac{1}{2}\int_1^\infty \frac{2x}{x^2 + 1}\,dx \\[4pt]
&= \frac{1}{2}\lim_{b\to\infty}\int_1^b \frac{2x}{x^2 + 1}\,dx \\[4pt]
&= \frac{1}{2}\lim_{b\to\infty}\Big[\ln(x^2 + 1)\Big]_1^b \\[4pt]
&= \frac{1}{2}\lim_{b\to\infty}\big[\ln(b^2 + 1) - \ln 2\big] \\[4pt]
&= \infty.
\end{aligned}
$$

Thus, the series *diverges*.

EXAMPLE 2   Using the Integral Test

Apply the Integral Test to the series $\displaystyle\sum_{n=1}^{\infty} \frac{1}{n^2 + 1}$.

Solution   Because $f(x) = 1/(x^2 + 1)$ satisfies the conditions for the Integral Test, you can integrate to obtain

$$
\begin{aligned}
\int_1^\infty \frac{1}{x^2 + 1}\,dx &= \lim_{b\to\infty}\int_1^b \frac{1}{x^2 + 1}\,dx \\[4pt]
&= \lim_{b\to\infty}\Big[\arctan x\Big]_1^b \\[4pt]
&= \lim_{b\to\infty}(\arctan b - \arctan 1) \\[4pt]
&= \frac{\pi}{2} - \frac{\pi}{4} \\[4pt]
&= \frac{\pi}{4}.
\end{aligned}
$$

Thus, the series *converges* (see Figure 8.9).

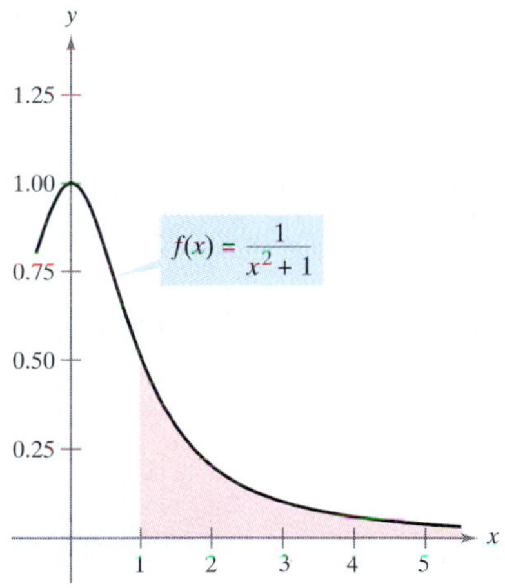

$f(x) = \dfrac{1}{x^2 + 1}$

Because the improper integral converges, the infinite series also converges.
Figure 8.9

TECHNOLOGY   In Example 2, the fact that the improper integral converges to $\pi/4$ does not imply that the infinite series converges to $\pi/4$. To approximate the sum of the series, you can use the inequality

$$
\sum_{n=1}^{N} \frac{1}{n^2 + 1} \le \sum_{n=1}^{\infty} \frac{1}{n^2 + 1} \le \sum_{n=1}^{N} \frac{1}{n^2 + 1} + \int_N^\infty \frac{1}{x^2 + 1}\,dx.
$$

(See Exercise 32.) The larger the value of N, the better the approximation. For instance, using $N = 200$ produces $1.072 \le \Sigma 1/(n^2 + 1) \le 1.077$.

p-Series and Harmonic Series

In the remainder of this section, we investigate a second type of series that has a simple arithmetic test for convergence or divergence. A series of the form

$$\sum_{n=1}^{\infty} \frac{1}{n^p} = \frac{1}{1^p} + \frac{1}{2^p} + \frac{1}{3^p} + \cdots \qquad \text{\textcolor{red}{\textit{p}-series}}$$

is a ***p*-series,** where *p* is a positive constant. For $p = 1$, the series

$$\sum_{n=1}^{\infty} \frac{1}{n} = 1 + \frac{1}{2} + \frac{1}{3} + \cdots \qquad \text{\textcolor{red}{Harmonic series}}$$

is the **harmonic series.** A **general harmonic series** is of the form $\Sigma 1/(an + b)$. In music, strings of the same material, diameter, and tension, whose lengths form a harmonic series, produce harmonic tones.

The Integral Test is convenient for establishing the convergence or divergence of *p*-series. This is shown in the proof of Theorem 8.11.

THEOREM 8.11   Convergence of *p*-Series

The *p*-series

$$\sum_{n=1}^{\infty} \frac{1}{n^p} = \frac{1}{1^p} + \frac{1}{2^p} + \frac{1}{3^p} + \frac{1}{4^p} + \cdots$$

1. converges if $p > 1$,   and   **2.** diverges if $0 < p \le 1$.

Proof   The proof follows from the Integral Test and from Theorem 7.5, which states that

$$\int_{1}^{\infty} \frac{1}{x^p} \, dx$$

converges if $p > 1$ and diverges if $0 < p \le 1$.

EXAMPLE 3   Convergent and Divergent *p*-Series

a. From Theorem 8.11, it follows that the *harmonic series*

$$\sum_{n=1}^{\infty} \frac{1}{n} = \frac{1}{1} + \frac{1}{2} + \frac{1}{3} + \cdots \qquad \text{\textcolor{red}{$p = 1$}}$$

diverges.

b. From Theorem 8.11, it follows that the *p*-series

$$\sum_{n=1}^{\infty} \frac{1}{n^2} = \frac{1}{1^2} + \frac{1}{2^2} + \frac{1}{3^2} + \cdots \qquad \text{\textcolor{red}{$p = 2$}}$$

converges.

NOTE   The sum of the series in Example 3b can be shown to be $\pi^2/6$. (This was shown by Leonhard Euler, but the proof is too difficult to present here.) Be sure you see that the Integral Test does not tell you that the sum of the series is equal to the value of the integral. For instance, the sum of the series in Example 3b is

$$\sum_{n=1}^{\infty} \frac{1}{n^2} = \frac{\pi^2}{6} \approx 1.645$$

but the value of the corresponding improper integral is

$$\int_{1}^{\infty} \frac{1}{x^2} \, dx = 1.$$

EXAMPLE 4   Testing a Series for Convergence

Determine whether the following series converges or diverges.

$$\sum_{n=2}^{\infty} \frac{1}{n \ln n}$$

Solution   This series is similar to the divergent harmonic series. If its terms were larger than those of the harmonic series, you would expect it to diverge. However, because its terms are smaller, you are not sure what to expect. Using the Integral Test, with $f(x) = 1/(x \ln x)$, you can see that the series diverges.

$$\int_2^{\infty} \frac{1}{x \ln x} \, dx = \int_2^{\infty} \frac{1/x}{\ln x} \, dx$$

$$= \lim_{b \to \infty} \left[\ln(\ln x) \right]_2^b$$

$$= \lim_{b \to \infty} \left[\ln(\ln b) - \ln(\ln 2) \right]$$

$$= \infty$$

EXERCISES FOR SECTION 8.3

In Exercises 1–10, use the Integral Test to determine the convergence or divergence of the series.

1. $\displaystyle\sum_{n=1}^{\infty} \frac{1}{n+1}$

2. $\displaystyle\sum_{n=1}^{\infty} n e^{-n}$

3. $\displaystyle\sum_{n=1}^{\infty} e^{-n}$

4. $\displaystyle\sum_{n=1}^{\infty} \frac{1}{4n+1}$

5. $\dfrac{1}{2} + \dfrac{1}{5} + \dfrac{1}{10} + \dfrac{1}{17} + \dfrac{1}{26} + \cdots$

6. $\dfrac{1}{3} + \dfrac{1}{5} + \dfrac{1}{7} + \dfrac{1}{9} + \dfrac{1}{11} + \cdots$

7. $\dfrac{\ln 2}{2} + \dfrac{\ln 3}{3} + \dfrac{\ln 4}{4} + \dfrac{\ln 5}{5} + \dfrac{\ln 6}{6} + \cdots$

8. $\dfrac{1}{4} + \dfrac{2}{7} + \dfrac{3}{12} + \cdots + \dfrac{n}{n^2 + 3} + \cdots$

9. $\displaystyle\sum_{n=1}^{\infty} \frac{n^{k-1}}{n^k + c}$, k is a positive integer

10. $\displaystyle\sum_{n=1}^{\infty} n^k e^{-n}$, k is a positive integer

In Exercises 11 and 12, use the Integral Test to determine the convergence or divergence of the *p*-series.

11. $\displaystyle\sum_{n=1}^{\infty} \frac{1}{n^3}$

12. $\displaystyle\sum_{n=1}^{\infty} \frac{1}{n^{1/3}}$

In Exercises 13 and 14, find the positive values of *p* for which the series converges.

13. $\displaystyle\sum_{n=2}^{\infty} \frac{1}{n(\ln n)^p}$

14. $\displaystyle\sum_{n=2}^{\infty} \frac{\ln n}{n^p}$

In Exercises 15–22, use Theorem 8.11 to determine the convergence or divergence of the *p*-series.

15. $\displaystyle\sum_{n=1}^{\infty} \frac{1}{\sqrt[5]{n}}$

16. $\displaystyle\sum_{n=1}^{\infty} \frac{1}{n^{4/3}}$

17. $1 + \dfrac{1}{\sqrt{2}} + \dfrac{1}{\sqrt{3}} + \dfrac{1}{\sqrt{4}} + \cdots$

18. $1 + \dfrac{1}{4} + \dfrac{1}{9} + \dfrac{1}{16} + \dfrac{1}{25} + \cdots$

19. $1 + \dfrac{1}{2\sqrt{2}} + \dfrac{1}{3\sqrt{3}} + \dfrac{1}{4\sqrt{4}} + \dfrac{1}{5\sqrt{5}} + \cdots$

20. $1 + \dfrac{1}{\sqrt[3]{4}} + \dfrac{1}{\sqrt[3]{9}} + \dfrac{1}{\sqrt[3]{16}} + \dfrac{1}{\sqrt[3]{25}} + \cdots$

21. $\displaystyle\sum_{n=1}^{\infty} \frac{1}{n^{1.04}}$

22. $\displaystyle\sum_{n=1}^{\infty} \frac{1}{n^{\pi}}$

In Exercises 23–26, match the series with the graph of its sequence of partial sums. [The graphs are labeled (a), (b), (c), and (d).] Determine the convergence or divergence of the series.

(a)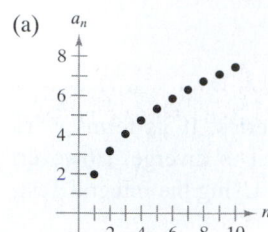

(b)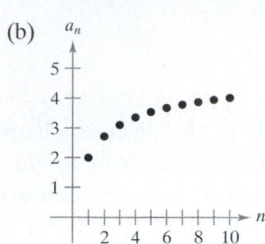

(c)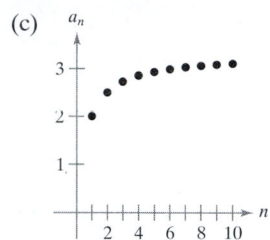

(d)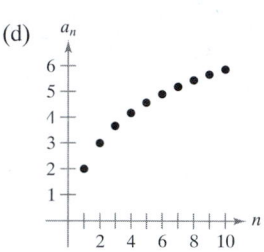

23. $\displaystyle\sum_{n=1}^{\infty} \frac{2}{\sqrt[4]{n^3}}$

24. $\displaystyle\sum_{n=1}^{\infty} \frac{2}{n}$

25. $\displaystyle\sum_{n=1}^{\infty} \frac{2}{n\sqrt{n}}$

26. $\displaystyle\sum_{n=1}^{\infty} \frac{2}{n^2}$

27. **Writing** In Exercises 23–26, $\displaystyle\lim_{n\to\infty} a_n = 0$ for each series but they do not all converge. Is this a contradiction of Theorem 8.9? Why do you think some converge and others diverge?

28. **Numerical and Graphical Analysis** (a) Use a graphing utility to find the indicated partial sum S_n and complete the table. (b) Use a graphing utility to graph the first ten terms of the sequence of partial sums. (c) Compare the rate at which the sequence of partial sums approaches the sum of the series for each series.

n	5	10	20	50	100
S_n					

(a) $\displaystyle\sum_{n=1}^{\infty} 3\left(\frac{1}{5}\right)^{n-1} = \frac{15}{4}$ (b) $\displaystyle\sum_{n=1}^{\infty} \frac{1}{n^2} = \frac{\pi^2}{6}$

29. **Numerical Reasoning** Because the harmonic series diverges, it follows that for any positive real number M there exists a positive integer N such that the partial sum

$$\sum_{n=1}^{N} \frac{1}{n} > M.$$

(a) Use a graphing utility to complete the table.

M	2	4	6	8
N				

(b) As the real number M increases in equal increments, does the number N increase in equal increments? Explain.

30. The **Riemann zeta function** for real numbers is defined for all x for which the series

$$\zeta(x) = \sum_{n=1}^{\infty} n^{-x}$$

converges. Find the domain of the function.

31. Let f be a positive, continuous, and decreasing function for $x \geq 1$, such that $a_n = f(n)$. Prove that if the series

$$\sum_{n=1}^{\infty} a_n$$

converges to S, then the remainder $R_N = S - S_N$ is bounded by

$$0 \leq R_N \leq \int_{N}^{\infty} f(x)\, dx.$$

32. Show that the result of Exercise 31 can be written as

$$\sum_{n=1}^{N} a_n \leq \sum_{n=1}^{\infty} a_n \leq \sum_{n=1}^{N} a_n + \int_{N}^{\infty} f(x)\, dx.$$

In Exercises 33–38, use the result of Exercise 31 to approximate the sum of the convergent series using the indicated number of terms. Include an estimate of the maximum error for your approximation.

33. $\displaystyle\sum_{n=1}^{\infty} \frac{1}{n^4}$

Six terms

34. $\displaystyle\sum_{n=1}^{\infty} \frac{1}{n^5}$

Four terms

35. $\displaystyle\sum_{n=1}^{\infty} \frac{1}{n^2 + 1}$

Ten terms

36. $\displaystyle\sum_{n=1}^{\infty} \frac{1}{(n+1)[\ln(n+1)]^3}$

Ten terms

37. $\displaystyle\sum_{n=1}^{\infty} ne^{-n^2}$

Four terms

38. $\displaystyle\sum_{n=1}^{\infty} e^{-n}$

Four terms

In Exercises 39–42, use the result of Exercises 31 to find N such that $R_N \leq 0.001$ for the convergent series.

39. $\displaystyle\sum_{n=1}^{\infty} \frac{1}{n^4}$

40. $\displaystyle\sum_{n=1}^{\infty} \frac{1}{n^{3/2}}$

41. $\displaystyle\sum_{n=1}^{\infty} e^{-5n}$

42. $\displaystyle\sum_{n=1}^{\infty} \frac{1}{n^2 + 1}$

43. **Think About It** A friend in your calculus class tells you that the following series converges because the terms are very small and approach 0 rapidly. Is your friend correct? Explain.

$$\frac{1}{10,000} + \frac{1}{10,001} + \frac{1}{10,002} + \cdots$$

44. Ten terms are used to approximate a convergent *p*-series. Therefore, the remainder is a function of *p* and is

$$0 \le R_{10}(p) \le \int_{10}^{\infty} \frac{1}{x^p}\,dx, \qquad p > 1.$$

(a) Perform the integration in the inequality.

(b) Use a graphing utility to represent the inequality graphically.

(c) Identify any asymptotes of the error function and interpret their meaning.

45. (a) Show that

$$\sum_{n=2}^{\infty} \frac{1}{n^{1.1}}$$

converges and

$$\sum_{n=2}^{\infty} \frac{1}{n \ln n}$$

diverges.

(b) Compare the first five terms of each series.

(c) Find $n > 3$ such that

$$\frac{1}{n^{1.1}} < \frac{1}{n \ln n}.$$

46. *Euler's Constant*   Let

$$S_n = \sum_{k=1}^{n} \frac{1}{k} = 1 + \frac{1}{2} + \cdots + \frac{1}{n}.$$

(a) Show that $\ln(n + 1) \le S_n \le 1 + \ln n$.

(b) Show that the sequence $\{a_n\} = \{S_n - \ln n\}$ is bounded.

(c) Show that the sequence $\{a_n\}$ is decreasing.

(d) Show that a_n converges to a limit γ (called Euler's constant).

(e) Approximate γ using a_{100}.

47. *Think About It*   Find a series such that the *n*th term goes to 0, but the series diverges.

48. Find the sum of the series

$$\sum_{n=2}^{\infty} \ln\left(1 - \frac{1}{n^2}\right).$$

Review   In Exercises 49–60, determine the convergence or divergence of the series.

49. $\displaystyle\sum_{n=1}^{\infty} \frac{1}{2n-1}$

50. $\displaystyle\sum_{n=2}^{\infty} \frac{1}{n\sqrt{n^2-1}}$

51. $\displaystyle\sum_{n=1}^{\infty} \frac{1}{n\sqrt[4]{n}}$

52. $\displaystyle 3\sum_{n=1}^{\infty} \frac{1}{n^{0.95}}$

53. $\displaystyle\sum_{n=0}^{\infty} \left(\frac{2}{3}\right)^n$

54. $\displaystyle\sum_{n=0}^{\infty} (1.075)^n$

55. $\displaystyle\sum_{n=1}^{\infty} \frac{n}{\sqrt{n^2+1}}$

56. $\displaystyle\sum_{n=1}^{\infty} \left(\frac{1}{n^2} - \frac{1}{n^3}\right)$

57. $\displaystyle\sum_{n=1}^{\infty} \left(1 + \frac{1}{n}\right)^n$

58. $\displaystyle\sum_{n=2}^{\infty} \ln n$

59. $\displaystyle\sum_{n=2}^{\infty} \frac{1}{n(\ln n)^3}$

60. $\displaystyle\sum_{n=2}^{\infty} \frac{\ln n}{n^3}$

SECTION PROJECT

The harmonic series

$$\sum_{n=1}^{\infty} \frac{1}{n} = 1 + \frac{1}{2} + \frac{1}{3} + \frac{1}{4} + \cdots + \frac{1}{n} + \cdots$$

is one of the most important series in this chapter. Even though its terms tend to zero as *n* increases,

$$\lim_{n\to\infty} \frac{1}{n} = 0,$$

the harmonic series diverges. In other words, even though the terms are getting smaller and smaller, the sum "adds up to infinity."

(a) One way to show that the harmonic series diverges is due to J. Bernoulli. He grouped the terms of the harmonic series as follows:

$$1 + \frac{1}{2} + \underbrace{\frac{1}{3} + \frac{1}{4}}_{> \frac{1}{2}} + \underbrace{\frac{1}{5} + \cdots + \frac{1}{8}}_{> \frac{1}{2}} + \underbrace{\frac{1}{9} + \cdots + \frac{1}{16}}_{> \frac{1}{2}} +$$

$$\underbrace{\frac{1}{17} + \cdots + \frac{1}{32}}_{> \frac{1}{2}} + \cdots$$

Write a short paragraph explaining how you can use this grouping to show that the harmonic series diverges.

(b) Use the proof of the Integral Test, Theorem 8.10, to show that

$$\ln(n + 1) \le 1 + \frac{1}{2} + \frac{1}{3} + \frac{1}{4} + \cdots + \frac{1}{n} \le 1 + \ln n.$$

(c) Use part (b) to determine how many terms *M* you would need so that

$$\sum_{n=1}^{M} \frac{1}{n} > 50.$$

(d) Show that the sum of the first million terms of the harmonic series is less than 15.

(e) Show that the following inequalities are valid.

$$\ln \tfrac{21}{10} \le \tfrac{1}{10} + \tfrac{1}{11} + \cdots + \tfrac{1}{20} \le \ln \tfrac{20}{9}$$

$$\ln \tfrac{201}{100} \le \tfrac{1}{100} + \tfrac{1}{101} + \cdots + \tfrac{1}{200} \le \ln \tfrac{200}{99}$$

(f) Use the ideas in part (e) to find the limit

$$\lim_{m\to\infty} \sum_{n=m}^{2m} \frac{1}{n}.$$

SECTION   **8.4**   **Comparisons of Series**

Direct Comparison Test • Limit Comparison Test

Direct Comparison Test

For the convergence tests developed so far, the terms of the series had to be fairly simple and the series had to have special characteristics in order for the convergence tests to be applied. A slight deviation from these special characteristics can make a test nonapplicable. For example, in the following pairs, the second series cannot be tested by the same convergence test as the first series even though it is similar to the first.

1. $\displaystyle\sum_{n=0}^{\infty} \frac{1}{2^n}$ is geometric, but $\displaystyle\sum_{n=0}^{\infty} \frac{n}{2^n}$ is not.

2. $\displaystyle\sum_{n=1}^{\infty} \frac{1}{n^3}$ is a p-series, but $\displaystyle\sum_{n=1}^{\infty} \frac{1}{n^3 + 1}$ is not.

3. $a_n = \dfrac{n}{(n^2 + 3)^2}$ is easily integrated, but $b_n = \dfrac{n^2}{(n^2 + 3)^2}$ is not.

In this section you will study two additional tests for positive-term series. These two tests greatly expand the variety of series you are able to test for convergence or divergence. They allow you to *compare* a series having complicated terms with a simpler series whose convergence or divergence is known.

THEOREM 8.12   **Direct Comparison Test**

Let $0 < a_n \le b_n$ for all n.

1. If $\displaystyle\sum_{n=1}^{\infty} b_n$ converges, then $\displaystyle\sum_{n=1}^{\infty} a_n$ converges.

2. If $\displaystyle\sum_{n=1}^{\infty} a_n$ diverges, then $\displaystyle\sum_{n=1}^{\infty} b_n$ diverges.

Proof   To prove the first property, let $L = \displaystyle\sum_{n=1}^{\infty} b_n$ and let

$$S_n = a_1 + a_2 + \cdots + a_n.$$

Because $0 < a_n \le b_n$, the sequence $S_1, S_2, S_3, \ldots$ is nondecreasing and bounded above by L; hence, it must converge. Because

$$\lim_{n\to\infty} S_n = \sum_{n=1}^{\infty} a_n$$

it follows that $\Sigma\, a_n$ converges. The second property is logically equivalent to the first.

NOTE   As stated, the Direct Comparison Test requires that $0 < a_n \le b_n$ for all n. Because the convergence of a series is not dependent on its first several terms, you could modify the test to require only that $0 < a_n \le b_n$ for all n greater than some integer N.

EXAMPLE 1   Using the Direct Comparison Test

Determine the convergence or divergence of $\displaystyle\sum_{n=1}^{\infty} \frac{1}{2 + 3^n}$.

Solution   This series resembles

$$\sum_{n=1}^{\infty} \frac{1}{3^n}.$$ Convergent geometric series

Term-by-term comparison yields

$$a_n = \frac{1}{2 + 3^n} < \frac{1}{3^n} = b_n, \quad n \ge 1.$$

Thus, by the Direct Comparison Test, the series converges.

EXAMPLE 2   Using the Direct Comparison Test

Determine the convergence or divergence of $\displaystyle\sum_{n=1}^{\infty} \frac{1}{2 + \sqrt{n}}$.

Solution   This series resembles

$$\sum_{n=1}^{\infty} \frac{1}{n^{1/2}}.$$ Divergent p-series

Term-by-term comparison yields

$$\frac{1}{2 + \sqrt{n}} \le \frac{1}{\sqrt{n}}, \quad n \ge 1$$

which *does not* meet the requirements for divergence. (Remember that if term-by-term comparison reveals a series that is *smaller* than a divergent series, the Direct Comparison Test tells you nothing.) Still expecting the series to diverge, you can compare the given series with

$$\sum_{n=1}^{\infty} \frac{1}{n}.$$ Divergent harmonic series

In this case, term-by-term comparison yields

$$a_n = \frac{1}{n} \le \frac{1}{2 + \sqrt{n}} = b_n, \quad n \ge 4$$

and, by the Direct Comparison Test, the given series diverges.

NOTE   To verify the last inequality in Example 2, try showing that $2 + \sqrt{n} \le n$ whenever $n \ge 4$.

Remember that both parts of the Direct Comparison Test require that $0 < a_n \le b_n$. Informally, the test says the following about the two series with nonnegative terms.

1. If the "larger" series converges, the "smaller" series must also converge.
2. If the "smaller" series diverges, the "larger" series must also diverge.

Limit Comparison Test

Often a given series closely resembles a *p*-series or a geometric series, yet you cannot establish the term-by-term comparison necessary to apply the Direct Comparison Test. Under these circumstances you may be able to apply a second comparison test, called the **Limit Comparison Test.**

THEOREM 8.13   Limit Comparison Test

Suppose that $a_n > 0$, $b_n > 0$, and

$$\lim_{n \to \infty} \left(\frac{a_n}{b_n} \right) = L$$

where *L* is *finite and positive*. Then the two series $\Sigma\, a_n$ and $\Sigma\, b_n$ either both converge or both diverge.

NOTE As with the Direct Comparison Test, the Limit Comparison Test could be modified to require only that a_n and b_n be positive for all *n* greater than some integer *N*.

Proof Because $a_n > 0$, $b_n > 0$, and $(a_n/b_n) \to L$ as $n \to \infty$, there exists $N > 0$ such that

$$0 < (a_n/b_n) < (L + 1), \quad \text{for } n \geq N.$$

This implies that

$$0 < a_n < (L + 1)b_n.$$

Hence, by the Direct Comparison Test, the convergence of $\Sigma\, b_n$ implies the convergence of $\Sigma\, a_n$. Similarly, the fact that

$$\lim_{n \to \infty} \left(\frac{b_n}{a_n} \right) = \frac{1}{L}$$

can be used to show that the convergence of $\Sigma\, a_n$ implies the convergence of $\Sigma\, b_n$.

EXAMPLE 3   Using the Limit Comparison Test

Show that the following general harmonic series diverges.

$$\sum_{n=1}^{\infty} \frac{1}{an + b}, \quad a > 0, \quad b > 0$$

Solution By comparison with

$$\sum_{n=1}^{\infty} \frac{1}{n} \qquad\qquad \text{\color{red}Divergent harmonic series}$$

we have

$$\lim_{n \to \infty} \frac{1/(an + b)}{1/n} = \lim_{n \to \infty} \frac{n}{an + b} = \frac{1}{a}.$$

Because this limit is greater than 0, you can conclude from the Limit Comparison Test that the given series diverges.

The Limit Comparison Test works well for comparing a "messy" algebraic series with a *p*-series. In choosing an appropriate *p*-series, you must choose one with an *n*th term of the same magnitude as the *n*th term of the given series.

Given Series	*Comparison Series*	*Conclusion*
$\displaystyle\sum_{n=1}^{\infty} \frac{1}{3n^2 - 4n + 5}$	$\displaystyle\sum_{n=1}^{\infty} \frac{1}{n^2}$	Both series converge.
$\displaystyle\sum_{n=1}^{\infty} \frac{1}{\sqrt{3n - 2}}$	$\displaystyle\sum_{n=1}^{\infty} \frac{1}{\sqrt{n}}$	Both series diverge.
$\displaystyle\sum_{n=1}^{\infty} \frac{n^2 - 10}{4n^5 + n^3}$	$\displaystyle\sum_{n=1}^{\infty} \frac{n^2}{n^5} = \sum_{n=1}^{\infty} \frac{1}{n^3}$	Both series converge.

In other words, when choosing a series for comparison, you can disregard all but the *highest powers of n* in both the numerator and the denominator.

EXAMPLE 4 Using the Limit Comparison Test

Determine the convergence or divergence of $\displaystyle\sum_{n=1}^{\infty} \frac{\sqrt{n}}{n^2 + 1}$.

Solution Disregarding all but the highest powers of *n* in the numerator and the denominator, you can compare the series with

$$\sum_{n=1}^{\infty} \frac{\sqrt{n}}{n^2} = \sum_{n=1}^{\infty} \frac{1}{n^{3/2}}. \qquad \text{Convergent } p\text{-series}$$

Because

$$\lim_{n\to\infty} \frac{a_n}{b_n} = \lim_{n\to\infty} \left(\frac{\sqrt{n}}{n^2 + 1}\right)\left(\frac{n^{3/2}}{1}\right) = \lim_{n\to\infty} \frac{n^2}{n^2 + 1} = 1$$

you can conclude by the Limit Comparison Test that the given series converges.

EXAMPLE 5 Using the Limit Comparison Test

Determine the convergence or divergence of $\displaystyle\sum_{n=1}^{\infty} \frac{n2^n}{4n^3 + 1}$.

Solution A reasonable comparison would be with the series

$$\sum_{n=1}^{\infty} \frac{2^n}{n^2}.$$

Note that this series diverges by the *n*th-Term Test. From the limit

$$\lim_{n\to\infty} \frac{a_n}{b_n} = \lim_{n\to\infty} \left(\frac{n2^n}{4n^3 + 1}\right)\left(\frac{n^2}{2^n}\right) = \lim_{n\to\infty} \frac{1}{4 + (1/n^3)} = \frac{1}{4}$$

you can conclude that the given series diverges.

EXERCISES FOR SECTION 8.4

1. *Graphical Analysis* The figures show the graphs of the first ten terms, and the graphs of the first ten terms of the sequence of partial sums, of each series.

$$\sum_{n=1}^{\infty} \frac{6}{n^{3/2}}, \quad \sum_{n=1}^{\infty} \frac{6}{n^{3/2} + 3}, \quad \text{and} \quad \sum_{n=1}^{\infty} \frac{6}{n\sqrt{n^2 + 0.5}}$$

(a) Identify the series in each figure.

(b) Which series is a *p*-series? Does it converge or diverge?

(c) For the series that are not *p*-series, how do the magnitudes of the terms compare with the magnitudes of the terms of the *p*-series? What conclusion can you draw about the convergence or divergence of the series?

(d) Explain the relationship between the magnitudes of the terms of the series and the magnitudes of the terms of the partial sums.

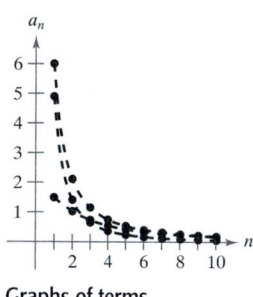

Graphs of terms

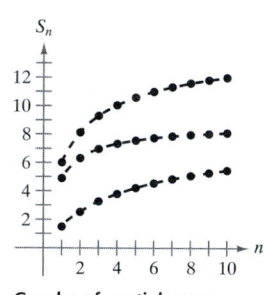

Graphs of partial sums

2. *Graphical Analysis* The figures show the graphs of the first ten terms, and the graphs of the first ten terms of the sequence of partial sums, of each series.

$$\sum_{n=1}^{\infty} \frac{2}{\sqrt{n}}, \quad \sum_{n=1}^{\infty} \frac{2}{\sqrt{n} - 0.5}, \quad \text{and} \quad \sum_{n=1}^{\infty} \frac{4}{\sqrt{n + 0.5}}$$

(a) Identify the series in each figure.

(b) Which series is a *p*-series? Does it converge or diverge?

(c) For the series that are not *p*-series, how do the magnitudes of the terms compare with the magnitudes of the terms of the *p*-series? What conclusion can you draw about the convergence or divergence of the series?

(d) Explain the relationship between the magnitudes of the terms of the series and the magnitudes of the terms of the partial sums.

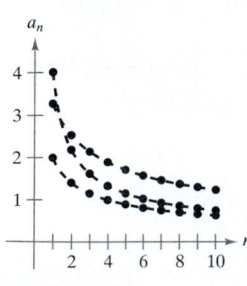

Graphs of terms

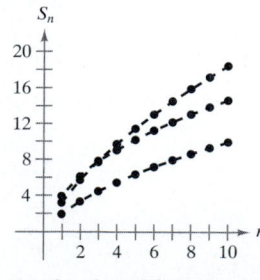

Graphs of partial sums

In Exercises 3–14, use the Direct Comparison Test to determine the convergence or divergence of the series.

3. $\displaystyle\sum_{n=1}^{\infty} \frac{1}{n^2 + 1}$

4. $\displaystyle\sum_{n=1}^{\infty} \frac{1}{3n^2 + 2}$

5. $\displaystyle\sum_{n=2}^{\infty} \frac{1}{n - 1}$

6. $\displaystyle\sum_{n=2}^{\infty} \frac{1}{\sqrt{n} - 1}$

7. $\displaystyle\sum_{n=0}^{\infty} \frac{1}{3^n + 1}$

8. $\displaystyle\sum_{n=0}^{\infty} \frac{2^n}{3^n + 5}$

9. $\displaystyle\sum_{n=2}^{\infty} \frac{\ln n}{n + 1}$

10. $\displaystyle\sum_{n=1}^{\infty} \frac{1}{\sqrt{n^3 + 1}}$

11. $\displaystyle\sum_{n=0}^{\infty} \frac{1}{n!}$

12. $\displaystyle\sum_{n=1}^{\infty} \frac{1}{3\sqrt[4]{n} - 1}$

13. $\displaystyle\sum_{n=0}^{\infty} e^{-n^2}$

14. $\displaystyle\sum_{n=1}^{\infty} \frac{4^n}{3^n - 1}$

In Exercises 15–28, use the Limit Comparison Test to determine the convergence or divergence of the series.

15. $\displaystyle\sum_{n=1}^{\infty} \frac{n}{n^2 + 1}$

16. $\displaystyle\sum_{n=2}^{\infty} \frac{1}{\sqrt{n^2 - 1}}$

17. $\displaystyle\sum_{n=0}^{\infty} \frac{1}{\sqrt{n^2 + 1}}$

18. $\displaystyle\sum_{n=1}^{\infty} \frac{1}{2^n - 5}$

19. $\displaystyle\sum_{n=1}^{\infty} \frac{2n^2 - 1}{3n^5 + 2n + 1}$

20. $\displaystyle\sum_{n=1}^{\infty} \frac{5n - 3}{n^2 - 2n + 5}$

21. $\displaystyle\sum_{n=1}^{\infty} \frac{n + 3}{n(n + 2)}$

22. $\displaystyle\sum_{n=1}^{\infty} \frac{1}{n(n^2 + 1)}$

23. $\displaystyle\sum_{n=1}^{\infty} \frac{1}{n\sqrt{n^2 + 1}}$

24. $\displaystyle\sum_{n=1}^{\infty} \frac{n}{(n + 1)2^{n-1}}$

25. $\displaystyle\sum_{n=1}^{\infty} \frac{n^{k-1}}{n^k + 1}, \quad k > 2$

26. $\displaystyle\sum_{n=1}^{\infty} \frac{1}{n + \sqrt{n^2 + 1}}$

27. $\displaystyle\sum_{n=1}^{\infty} \sin \frac{1}{n}$

28. $\displaystyle\sum_{n=1}^{\infty} \tan \frac{1}{n}$

In Exercises 29–36, test for convergence or divergence, using each test at least once. Identify the test used.

(a) **nth-Term Test** (b) **Geometric Series Test**

(c) **p-Series Test** (d) **Telescoping Series Test**

(e) **Integral Test** (f) **Direct Comparison Test**

(g) **Limit Comparison Test**

29. $\displaystyle\sum_{n=1}^{\infty} \frac{\sqrt{n}}{n}$

30. $\displaystyle\sum_{n=0}^{\infty} 5\left(-\frac{1}{5}\right)^n$

31. $\displaystyle\sum_{n=1}^{\infty} \frac{1}{3^n + 2}$

32. $\displaystyle\sum_{n=4}^{\infty} \frac{1}{3n^2 - 2n - 15}$

33. $\displaystyle\sum_{n=1}^{\infty} \frac{n}{2n + 3}$

34. $\displaystyle\sum_{n=1}^{\infty} \left(\frac{1}{n+1} - \frac{1}{n+2} \right)$

35. $\displaystyle\sum_{n=1}^{\infty} \frac{n}{(n^2+1)^2}$

36. $\displaystyle\sum_{n=1}^{\infty} \frac{3}{n(n+3)}$

37. Use the Limit Comparison Test with the harmonic series to show that the series $\Sigma\, a_n$ (where $0 < a_n < a_{n-1}$) diverges if

$$\lim_{n\to\infty} n a_n \neq 0.$$

38. Prove that, if $P(n)$ and $Q(n)$ are polynomials of degree j and k, then the series

$$\sum_{n=1}^{\infty} \frac{P(n)}{Q(n)}$$

converges if $j < k - 1$ and diverges if $j \geq k - 1$.

In Exercises 39–42, use the polynomial test given in Exercise 38 to determine whether the series converges or diverges.

39. $\frac{1}{2} + \frac{2}{5} + \frac{3}{10} + \frac{4}{17} + \frac{5}{26} + \cdots$

40. $\frac{1}{3} + \frac{1}{8} + \frac{1}{15} + \frac{1}{24} + \frac{1}{35} + \cdots$

41. $\displaystyle\sum_{n=1}^{\infty} \frac{1}{n^3 + 1}$

42. $\displaystyle\sum_{n=1}^{\infty} \frac{n^2}{n^3 + 1}$

In Exercises 43 and 44, use the divergence test given in Exercise 37 to show that the series diverges.

43. $\displaystyle\sum_{n=1}^{\infty} \frac{n^3}{5n^4 + 3}$

44. $\displaystyle\sum_{n=2}^{\infty} \frac{1}{\ln n}$

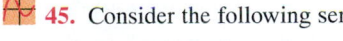

45. Consider the following series and its sum.

$$\sum_{n=1}^{\infty} \frac{1}{(2n-1)^2} = \frac{\pi^2}{8}$$

(a) Verify that the series converges.

(b) Use a graphing utility to complete the table.

n	5	10	20	50	100
S_n					

(c) Find the sum of the series

$$\sum_{n=3}^{\infty} \frac{1}{(2n-1)^2}$$

by hand. Describe how you found the sum.

(d) Use a graphing utility to find the sum of the series

$$\sum_{n=10}^{\infty} \frac{1}{(2n-1)^2}.$$

46. *Think About It*   It appears that the terms of the series

$$\frac{1}{1000} + \frac{1}{1001} + \frac{1}{1002} + \frac{1}{1003} + \cdots$$

are less than the corresponding terms of the convergent series

$$1 + \frac{1}{4} + \frac{1}{9} + \frac{1}{16} + \cdots.$$

If the statement above is correct, the first series converges. Is this correct? Why or why not? Make a statement about how the divergence or convergence of a series is affected by inclusion or exclusion of the first finite number of terms.

True or False?   **In Exercises 47–50, determine whether the statement is true or false. If it is false, explain why or give an example that shows it is false.**

47. If $0 < a_n \leq b_n$ and $\displaystyle\sum_{n=1}^{\infty} a_n$ converges, then $\displaystyle\sum_{n=1}^{\infty} b_n$ diverges.

48. If $0 < a_{n+10} \leq b_n$ and $\displaystyle\sum_{n=1}^{\infty} b_n$ converges, then $\displaystyle\sum_{n=1}^{\infty} a_n$ converges.

49. If $a_n + b_n \leq c_n$ and $\displaystyle\sum_{n=1}^{\infty} c_n$ converges, then the series $\displaystyle\sum_{n=1}^{\infty} a_n$ and $\displaystyle\sum_{n=1}^{\infty} b_n$ both converge. (Assume that the terms of all three series are positive.)

50. If $a_n \leq b_n + c_n$ and $\displaystyle\sum_{n=1}^{\infty} a_n$ diverges, then the series $\displaystyle\sum_{n=1}^{\infty} b_n$ and $\displaystyle\sum_{n=1}^{\infty} c_n$ both diverge. (Assume that the terms of all three series are positive.)

51. Prove that if the nonnegative series

$$\sum_{n=1}^{\infty} a_n \quad \text{and} \quad \sum_{n=1}^{\infty} b_n$$

converge, then so does the series

$$\sum_{n=1}^{\infty} a_n b_n.$$

52. Use the result of Exercise 51 to prove that if the nonnegative series

$$\sum_{n=1}^{\infty} a_n$$

converges, then so does the series

$$\sum_{n=1}^{\infty} a_n^2.$$

53. Find two series that demonstrate the result of Exercise 51.

54. Find two series that demonstrate the result of Exercise 52.

55. Suppose that $\Sigma\, a_n$ and $\Sigma\, b_n$ are series with positive terms. Prove the following.

  (a) If $\displaystyle\lim_{n\to\infty}\frac{a_n}{b_n}=0$ and $\Sigma\, b_n$ converges, then $\Sigma\, a_n$ also converges.

  (b) If $\displaystyle\lim_{n\to\infty}\frac{a_n}{b_n}=\infty$ and $\Sigma\, b_n$ diverges, then $\Sigma\, a_n$ also diverges.

56. Find two series that demonstrate the result of Exercise 55.

57. ***Think About It*** The figure shows the first 20 terms of the convergent series

$$\sum_{n=1}^{\infty} a_n$$

and the first 20 terms of the series

$$\sum_{n=1}^{\infty} a_n^2.$$

Identify the two series and explain your reasoning in making the selection.

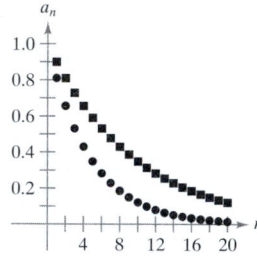

58. ***Investigation*** Consider an equilateral triangle with sides of length 9. Center equilateral triangles with sides of length 3 on each side of the first triangle. Center equilateral triangles with sides of length 1 on each side of the second set of triangles. Continue this process of centering equilateral triangles on the previous set of triangles where the length of the sides of each set is $\frac{1}{3}$ that of the previous set. This forms the **Koch snowflake** as described in Motivating the Chapter on page 546. Use infinite series to find (if possible) the area and the perimeter of the figure below.

SECTION PROJECT

Solera Method Most wines are produced entirely from grapes grown in a single year. Sherry, however, is a complex mixture of older wines with new wines. This is done with a sequence of barrels (called a solera) stacked on top of each other, as shown in the photo.

Everton/The Image Works

The oldest wine is in the bottom tier of barrels, and the newest is in the top tier. Each year, half of each barrel in the bottom tier is bottled as sherry. The bottom barrels are then refilled with the wine from the barrels above. This process is repeated throughout the solera, with new wine being added to the top barrels. A mathematical model for the amount of n-year-old wine that is removed from a solera (with k tiers) each year is

$$f(n,k)=\left(\frac{n-1}{k-1}\right)\left(\frac{1}{2}\right)^{n+1}, \qquad k \le n.$$

(a) Consider a solera that has five tiers, numbered $k = 1, 2, 3, 4,$ and 5. In 1980 ($n = 0$), half of each barrel in the top tier (tier 1) was refilled with new wine. How much of this wine was removed from the solera in 1981? In 1982? In 1983? . . . In 1995? During which year(s) was the greatest amount of the 1980 wine removed from the solera?

(b) In part (a), let a_n be the amount of 1980 wine that is removed from the solera in year n. Evaluate

$$\sum_{n=0}^{\infty} a_n.$$

FOR FURTHER INFORMATION See the article "Finding Vintage Concentrations in a Sherry Solera" by Rhodes Peele and John T. MacQueen in the 1990 *UMAP Modules.*

SECTION | 8.5 | Alternating Series

Alternating Series • Alternating Series Remainder •
Absolute and Conditional Convergence • Rearrangement of Series

Alternating Series

So far, most series we have dealt with have had positive terms. In this section and the following section, you will study series that contain both positive and negative terms. The simplest such series is an **alternating series,** whose terms alternate in sign. For example, the geometric series

$$\sum_{n=0}^{\infty} \left(-\frac{1}{2}\right)^n = \sum_{n=0}^{\infty} (-1)^n \frac{1}{2^n}$$

$$= 1 - \frac{1}{2} + \frac{1}{4} - \frac{1}{8} + \frac{1}{16} - \cdots$$

is an *alternating geometric series* with $r = -\frac{1}{2}$. Alternating series occur in two ways: either the odd terms are negative or the even terms are negative.

THEOREM 8.14   Alternating Series Test

Let $a_n > 0$. The alternating series

$$\sum_{n=1}^{\infty} (-1)^n a_n \quad \text{and} \quad \sum_{n=1}^{\infty} (-1)^{n+1} a_n$$

converge if the following two conditions are met.

1. $\lim\limits_{x \to \infty} a_n = 0$ **2.** $a_{n+1} \le a_n$, for all n

---

NOTE   The second condition in the Alternating Series Test can be modified to require only that $0 < a_{n+1} \le a_n$ for all n greater than some integer N.

Proof   Consider the alternating series $\sum (-1)^{n+1} a_n$. For this series, the partial sum (where $2n$ is even)

$$S_{2n} = (a_1 - a_2) + (a_3 - a_4) + (a_5 - a_6) + \cdots + (a_{2n-1} - a_{2n})$$

has all nonnegative terms, and therefore $\{S_{2n}\}$ is a nondecreasing sequence. But you can also write

$$S_{2n} = a_1 - (a_2 - a_3) - (a_4 - a_5) - \cdots - (a_{2n-2} - a_{2n-1}) - a_{2n}$$

which implies that $S_{2n} \le a_1$ for every integer n. Thus $\{S_{2n}\}$ is a bounded, nondecreasing sequence that converges to some value L. Because $S_{2n-1} - a_{2n} = S_{2n}$ and $a_{2n} \to 0$, you have

$$\lim_{n \to \infty} S_{2n-1} = \lim_{n \to \infty} S_{2n} + \lim_{n \to \infty} a_{2n}$$

$$= L + \lim_{n \to \infty} a_{2n}$$

$$= L.$$

Because both S_{2n} and S_{2n-1} converge to the same limit L, it follows that $\{S_n\}$ also converges to L. Consequently, the given alternating series converges.

EXAMPLE 1 Using the Alternating Series Test

NOTE The series in Example 1 is called the *alternating harmonic series*—more is said about this series in Example 6.

Determine the convergence or divergence of $\displaystyle\sum_{n=1}^{\infty} (-1)^{n+1}\frac{1}{n}$.

Solution Because

$$\frac{1}{n+1} \le \frac{1}{n}$$

for all n and the limit (as $n \to \infty$) of $1/n$ is 0, you can apply the Alternating Series Test to conclude that the series converges.

EXAMPLE 2 Using the Alternating Series Test

Determine the convergence or divergence of $\displaystyle\sum_{n=1}^{\infty} \frac{n}{(-2)^{n-1}}$.

Solution To apply the Alternating Series Test, note that, for $n \ge 1$,

$$\frac{1}{2} \le \frac{n}{n+1}$$

$$\frac{2^{n-1}}{2^n} \le \frac{n}{n+1}$$

$$(n+1)2^{n-1} \le n2^n$$

$$\frac{n+1}{2^n} \le \frac{n}{2^{n-1}}.$$

Hence, $a_{n+1} = (n+1)/2^n \le n/2^{n-1} = a_n$ for all n. Furthermore, by L'Hôpital's Rule,

$$\lim_{x \to \infty} \frac{x}{2^{x-1}} = \lim_{x \to \infty} \frac{1}{2^{x-1}(\ln 2)} = 0 \quad \Longrightarrow \quad \lim_{n \to \infty} \frac{n}{2^{n-1}} = 0.$$

Therefore, by the Alternating Series Test, the series converges.

EXAMPLE 3 Cases for Which the Alternating Series Test Fails

NOTE In Example 3a, remember that whenever a series does not pass the first condition of the Alternating Series Test, you can use the nth-Term Test for Divergence to conclude that the series diverges.

a. The alternating series

$$\sum_{n=1}^{\infty} \frac{(-1)^{n+1}(n+1)}{n} = \frac{2}{1} - \frac{3}{2} + \frac{4}{3} - \frac{5}{4} + \frac{6}{5} - \cdots$$

passes the second condition of the Alternating Series Test because $a_{n+1} \le a_n$ for all n. You cannot apply the Alternating Series Test, however, because the series does not pass the first condition. In fact, the series diverges.

b. The alternating series

$$\frac{2}{1} - \frac{1}{1} + \frac{2}{2} - \frac{1}{2} + \frac{2}{3} - \frac{1}{3} + \frac{2}{4} - \frac{1}{4} + \cdots$$

passes the first condition because a_n approaches 0 as $n \to \infty$. You cannot apply the Alternating Series Test, however, because the series does not pass the second condition. To conclude that the series diverges, you can argue that S_{2N} equals the Nth partial sum of the divergent harmonic series. This implies that the sequence of partial sums diverges. Hence, the series diverges.

Alternating Series Remainder

For a convergent alternating series, the partial sum S_N can be a useful approximation for the sum S of the series. Just how close S_N is to S is stated in the following theorem.

> **THEOREM 8.15   Alternating Series Remainder**
>
> If a convergent alternating series satisfies the condition $a_{n+1} \leq a_n$, then the absolute value of the remainder R_N involved in approximating the sum S by S_N is less than (or equal to) the first neglected term. That is,
>
> $$|S - S_N| = |R_N| \leq a_{N+1}.$$

Proof   The series obtained by deleting the first N terms of the given series satisfies the conditions of the Alternating Series Test and has a sum of R_N.

$$R_N = S - S_N = \sum_{n=1}^{\infty} (-1)^{n+1} a_n - \sum_{n=1}^{N} (-1)^{n+1} a_n$$

$$= (-1)^N a_{N+1} + (-1)^{N+1} a_{N+2} + (-1)^{N+2} a_{N+3} + \cdots$$

$$= (-1)^N (a_{N+1} - a_{N+2} + a_{N+3} - \cdots)$$

$$|R_N| = a_{N+1} - a_{N+2} + a_{N+3} - a_{N+4} + a_{N+5} - \cdots$$

$$= a_{N+1} - (a_{N+2} - a_{N+3}) - (a_{N+4} - a_{N+5}) - \cdots \leq a_{N+1}$$

Consequently, $|S - S_N| = |R_N| \leq a_{N+1}$, which establishes the theorem.

EXAMPLE 4   Approximating the Sum of an Alternating Series

Approximate the sum of the following series by its first six terms.

$$\sum_{n=1}^{\infty} (-1)^{n+1} \left(\frac{1}{n!} \right) = \frac{1}{1!} - \frac{1}{2!} + \frac{1}{3!} - \frac{1}{4!} + \frac{1}{5!} - \frac{1}{6!} + \cdots$$

Solution   The series converges by the Alternating Series Test because

$$\frac{1}{(n+1)!} \leq \frac{1}{n!} \quad \text{and} \quad \lim_{n \to \infty} \frac{1}{n!} = 0.$$

The sum of the first six terms is

$$S_6 = 1 - \frac{1}{2} + \frac{1}{6} - \frac{1}{24} + \frac{1}{120} - \frac{1}{720} \approx 0.63194$$

and, by the Alternating Series Remainder, you have

$$|S - S_6| = |R_6| \leq a_7 = \frac{1}{5040} \approx 0.0002.$$

Therefore, the sum S lies between $0.63194 - 0.0002$ and $0.63194 + 0.0002$, and you have

$$0.63174 \leq S \leq 0.63214.$$

TECHNOLOGY   Later, in Section 8.10, you will be able to show that the series in Example 4 converges to

$$\frac{e - 1}{e} \approx 0.63212.$$

For now, try using a computer to obtain an approximation of the sum of the series. How many terms do you need to obtain an approximation that is within 0.00001 unit of the actual sum?

Absolute and Conditional Convergence

Occasionally, a series may have both positive and negative terms and not be an alternating series. For instance, the series

$$\sum_{n=1}^{\infty} \frac{\sin n}{n^2} = \frac{\sin 1}{1} + \frac{\sin 2}{4} + \frac{\sin 3}{9} + \cdots$$

has both positive and negative terms, yet it is not an alternating series. One way to obtain some information about the convergence of this series is to investigate the convergence of the series

$$\sum_{n=1}^{\infty} \left| \frac{\sin n}{n^2} \right|.$$

By direct comparison, you have $|\sin n| \leq 1$ for all n, so

$$\left| \frac{\sin n}{n^2} \right| \leq \frac{1}{n^2}, \quad n \geq 1.$$

Thus, by the Direct Comparison Test, the series $\sum |(\sin n)/n^2|$ converges. But the question still is "Does the original series converge?" The next theorem tells us that the answer is yes.

THEOREM 8.16   Absolute Convergence

If the series $\sum |a_n|$ converges, then the series $\sum a_n$ also converges.

Proof Because $0 \leq a_n + |a_n| \leq 2|a_n|$ for all n, the series

$$\sum_{n=1}^{\infty} (a_n + |a_n|)$$

converges by comparison with the convergent series

$$\sum_{n=1}^{\infty} 2|a_n|.$$

Furthermore, because $a_n = (a_n + |a_n|) - |a_n|$, you can write

$$\sum_{n=1}^{\infty} a_n = \sum_{n=1}^{\infty} (a_n + |a_n|) - \sum_{n=1}^{\infty} |a_n|$$

where both series on the right converge. Hence it follows that $\sum a_n$ converges.

The converse of Theorem 8.16 is not true. For instance, the **alternating harmonic series**

$$\sum_{n=1}^{\infty} \frac{(-1)^{n+1}}{n} = \frac{1}{1} - \frac{1}{2} + \frac{1}{3} - \frac{1}{4} + \cdots$$

converges by the Alternating Series Test. Yet the harmonic series diverges. This type of convergence is called **conditional.**

Definition of Absolute and Conditional Convergence

1. $\Sigma\, a_n$ is **absolutely convergent** if $\Sigma\, |a_n|$ converges.
2. $\Sigma\, a_n$ is **conditionally convergent** if $\Sigma\, a_n$ converges but $\Sigma\, |a_n|$ diverges.

EXAMPLE 5   Absolute and Conditional Convergence

Determine whether each of the following series is convergent or divergent. Classify any convergent series as absolutely or conditionally convergent.

a. $\displaystyle\sum_{n=1}^{\infty} \frac{(-1)^{n(n+1)/2}}{3^n} = -\frac{1}{3} - \frac{1}{9} + \frac{1}{27} + \frac{1}{81} - \cdots$

b. $\displaystyle\sum_{n=1}^{\infty} \frac{(-1)^n}{\ln(n+1)} = -\frac{1}{\ln 2} + \frac{1}{\ln 3} - \frac{1}{\ln 4} + \frac{1}{\ln 5} - \cdots$

c. $\displaystyle\sum_{n=0}^{\infty} \frac{(-1)^n\, n!}{2^n} = \frac{0!}{2^0} - \frac{1!}{2^1} + \frac{2!}{2^2} - \frac{3!}{2^3} + \cdots$

d. $\displaystyle\sum_{n=1}^{\infty} \frac{(-1)^n}{\sqrt{n}} = -\frac{1}{\sqrt{1}} + \frac{1}{\sqrt{2}} - \frac{1}{\sqrt{3}} + \frac{1}{\sqrt{4}} - \cdots$

Solution

a. This is *not* an alternating series. However, because

$$\sum_{n=1}^{\infty} \left| \frac{(-1)^{n(n+1)/2}}{3^n} \right| = \sum_{n=1}^{\infty} \frac{1}{3^n}$$

is a convergent geometric series, you can apply Theorem 8.16 to conclude that the given series is *absolutely* convergent (and hence convergent).

b. In this case, the Alternating Series Test indicates that the given series converges. However, the series

$$\sum_{n=1}^{\infty} \left| \frac{(-1)^n}{\ln(n+1)} \right| = \frac{1}{\ln 2} + \frac{1}{\ln 3} + \frac{1}{\ln 4} + \cdots$$

diverges by direct comparison with the terms of the harmonic series. Therefore, the given series is *conditionally* convergent.

c. By the nth-Term Test for Divergence, you can conclude that this series diverges.

d. The given series can be shown to be convergent by the Alternating Series Test. Moreover, because the p-series

$$\sum_{n=1}^{\infty} \left| \frac{(-1)^n}{\sqrt{n}} \right| = \frac{1}{\sqrt{1}} + \frac{1}{\sqrt{2}} + \frac{1}{\sqrt{3}} + \frac{1}{\sqrt{4}} + \cdots$$

diverges, the given series is *conditionally* convergent.

Rearrangement of Series

A finite sum as $(1 + 3 - 2 + 5 - 4)$ can be rearranged without changing the value of the sum. This is not necessarily true of an infinite series—it depends on whether the series is absolutely convergent (every rearrangement has the same sum) or conditionally convergent.

EXAMPLE 6   Rearrangement of a Series

FOR FURTHER INFORMATION Georg Friedrich Riemann (1826-1866) proved that if $\Sigma\, a_n$ is conditionally convergent and S is any real number, the terms of the series can be rearranged to converge to S. For more on this topic, see the article "Riemann's Rearrangement Theorem" by Stewart Galanor in the November 1987 issue of *Mathematics Teacher*.

The alternating harmonic series converges to $\ln 2$. That is,

$$\sum_{n=1}^{\infty} (-1)^{n+1} \frac{1}{n} = \frac{1}{1} - \frac{1}{2} + \frac{1}{3} - \frac{1}{4} + \cdots = \ln 2.$$

(See Exercise 43, Section 8.10.) Rearrange the series to produce a different sum.

Solution   Consider the following rearrangement.

$$1 - \frac{1}{2} - \frac{1}{4} + \frac{1}{3} - \frac{1}{6} - \frac{1}{8} + \frac{1}{5} - \frac{1}{10} - \frac{1}{12} + \frac{1}{7} - \frac{1}{14} - \cdots$$

$$= \left(1 - \frac{1}{2}\right) - \frac{1}{4} + \left(\frac{1}{3} - \frac{1}{6}\right) - \frac{1}{8} + \left(\frac{1}{5} - \frac{1}{10}\right) - \frac{1}{12} + \left(\frac{1}{7} - \frac{1}{14}\right) - \cdots$$

$$= \frac{1}{2} - \frac{1}{4} + \frac{1}{6} - \frac{1}{8} + \frac{1}{10} - \frac{1}{12} + \frac{1}{14} - \cdots$$

$$= \frac{1}{2}\left(1 - \frac{1}{2} + \frac{1}{3} - \frac{1}{4} + \frac{1}{5} - \frac{1}{6} + \frac{1}{7} - \cdots\right) = \frac{1}{2}(\ln 2)$$

By rearranging the terms, you obtain a sum that is half the original sum.

EXERCISES FOR SECTION 8.5

In Exercises 1–4, match the series with the graph of its sequence of partial sums. [The graphs are labeled (a), (b), (c), and (d).]

(a)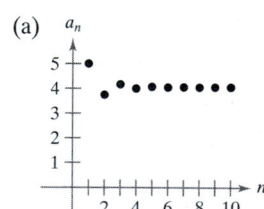

(b)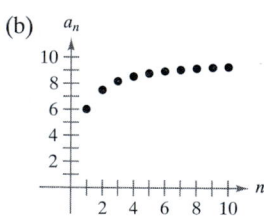

(c)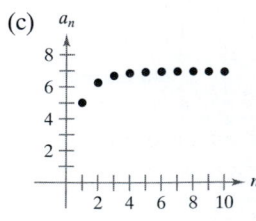

(d)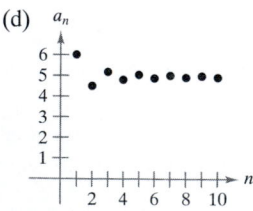

1. $\displaystyle\sum_{n=1}^{\infty} \frac{6}{n^2}$

2. $\displaystyle\sum_{n=1}^{\infty} \frac{(-1)^{n-1}\,6}{n^2}$

3. $\displaystyle\sum_{n=1}^{\infty} \frac{10}{n2^n}$

4. $\displaystyle\sum_{n=1}^{\infty} \frac{(-1)^{n-1}\,10}{n2^n}$

Numerical and Graphical Analysis   In Exercises 5–8, (a) use a graphing utility to find the indicated partial sum S_n and complete the table. (b) Use a graphing utility to graph the first ten terms of the sequence of partial sums and a horizontal line representing the sum. (c) What pattern exists between the plot of the successive points in part (b) relative to the horizontal line representing the sum of the series? Do the distances between the successive points and the horizontal line increase or decrease? (d) Discuss the relationship between the answers in part (c) and the Alternating Series Remainder as given in Theorem 8.15.

n	1	2	3	4	5	6	7	8	9	10
S_n										

5. $\displaystyle\sum_{n=1}^{\infty} \frac{(-1)^{n-1}}{2n-1} = \frac{\pi}{4}$

6. $\displaystyle\sum_{n=1}^{\infty} \frac{(-1)^{n-1}}{(n-1)!} = \frac{1}{e}$

7. $\displaystyle\sum_{n=1}^{\infty} \frac{(-1)^{n-1}}{n^2} = \frac{\pi^2}{12}$

8. $\displaystyle\sum_{n=1}^{\infty} \frac{(-1)^{n-1}}{(2n-1)!} = \sin 1$

In Exercises 9–28, determine the convergence or divergence of the series.

9. $\displaystyle\sum_{n=1}^{\infty} \frac{(-1)^{n+1}}{n}$

10. $\displaystyle\sum_{n=1}^{\infty} \frac{(-1)^{n+1}n}{2n-1}$

11. $\displaystyle\sum_{n=1}^{\infty} \frac{(-1)^{n+1}}{2n-1}$

12. $\displaystyle\sum_{n=2}^{\infty} \frac{(-1)^n}{\ln n}$

13. $\displaystyle\sum_{n=1}^{\infty} \frac{(-1)^n n^2}{n^2+1}$

14. $\displaystyle\sum_{n=1}^{\infty} \frac{(-1)^{n+1}n}{n^2+1}$

15. $\displaystyle\sum_{n=1}^{\infty} \frac{(-1)^n}{\sqrt{n}}$

16. $\displaystyle\sum_{n=1}^{\infty} \frac{(-1)^{n+1}n^3}{n^3+6}$

17. $\displaystyle\sum_{n=1}^{\infty} \frac{(-1)^{n+1}(n+1)}{\ln(n+1)}$

18. $\displaystyle\sum_{n=1}^{\infty} \frac{(-1)^{n+1}\ln(n+1)}{n+1}$

19. $\displaystyle\sum_{n=1}^{\infty} \sin\frac{(2n-1)\pi}{2}$

20. $\displaystyle\sum_{n=1}^{\infty} \cos n\pi$

21. $\displaystyle\sum_{n=1}^{\infty} \frac{1}{n}\sin\frac{(2n-1)\pi}{2}$

22. $\displaystyle\sum_{n=1}^{\infty} \frac{1}{n}\cos n\pi$

23. $\displaystyle\sum_{n=0}^{\infty} \frac{(-1)^n}{n!}$

24. $\displaystyle\sum_{n=0}^{\infty} \frac{(-1)^n}{(2n)!}$

25. $\displaystyle\sum_{n=1}^{\infty} \frac{(-1)^{n+1}\sqrt{n}}{n+2}$

26. $\displaystyle\sum_{n=1}^{\infty} \frac{(-1)^{n+1}\sqrt{n}}{\sqrt[3]{n}}$

27. $\displaystyle\sum_{n=1}^{\infty} \frac{2(-1)^{n+1}}{e^n-e^{-n}} = \sum_{n=1}^{\infty} (-1)^{n+1}\operatorname{csch} n$

28. $\displaystyle\sum_{n=1}^{\infty} \frac{2(-1)^{n+1}}{e^n+e^{-n}} = \sum_{n=1}^{\infty} (-1)^{n+1}\operatorname{sech} n$

In Exercises 29–34, (a) use Theorem 8.15 to determine the number of terms required to approximate the sum of the convergent series with an error of less than 0.001, and (b) use a graphing utility to approximate the sum of the series with an error of less than 0.001.

29. $\displaystyle\sum_{n=0}^{\infty} \frac{(-1)^n}{n!} = \frac{1}{e}$

30. $\displaystyle\sum_{n=0}^{\infty} \frac{(-1)^n}{2^n n!} = \frac{1}{\sqrt{e}}$

31. $\displaystyle\sum_{n=0}^{\infty} \frac{(-1)^n}{(2n+1)!} = \sin 1$

32. $\displaystyle\sum_{n=0}^{\infty} \frac{(-1)^n}{(2n)!} = \cos 1$

33. $\displaystyle\sum_{n=1}^{\infty} \frac{(-1)^{n+1}}{n} = \ln 2$

34. $\displaystyle\sum_{n=1}^{\infty} \frac{(-1)^{n+1}}{n4^n} = \ln\frac{5}{4}$

In Exercises 35 and 36, use Theorem 8.15 to determine the number of terms required to approximate the sum of the series with an error of less than 0.001.

35. $\displaystyle\sum_{n=1}^{\infty} \frac{(-1)^{n+1}}{2n^3-1}$

36. $\displaystyle\sum_{n=1}^{\infty} \frac{(-1)^{n+1}}{n^4}$

In Exercises 37–52, determine whether the series converges conditionally or absolutely, or diverges.

37. $\displaystyle\sum_{n=1}^{\infty} \frac{(-1)^{n+1}}{(n+1)^2}$

38. $\displaystyle\sum_{n=1}^{\infty} \frac{(-1)^{n+1}}{n+1}$

39. $\displaystyle\sum_{n=1}^{\infty} \frac{(-1)^{n+1}}{\sqrt{n}}$

40. $\displaystyle\sum_{n=1}^{\infty} \frac{(-1)^{n+1}}{n\sqrt{n}}$

41. $\displaystyle\sum_{n=1}^{\infty} \frac{(-1)^{n+1}n^2}{(n+1)^2}$

42. $\displaystyle\sum_{n=1}^{\infty} \frac{(-1)^{n+1}(2n+3)}{n+10}$

43. $\displaystyle\sum_{n=2}^{\infty} \frac{(-1)^n}{\ln n}$

44. $\displaystyle\sum_{n=0}^{\infty} (-1)^n e^{-n^2}$

45. $\displaystyle\sum_{n=2}^{\infty} \frac{(-1)^n n}{n^3-1}$

46. $\displaystyle\sum_{n=1}^{\infty} \frac{(-1)^{n+1}}{n^{1.5}}$

47. $\displaystyle\sum_{n=0}^{\infty} \frac{(-1)^n}{(2n+1)!}$

48. $\displaystyle\sum_{n=0}^{\infty} \frac{(-1)^n}{\sqrt[3]{n+1}}$

49. $\displaystyle\sum_{n=0}^{\infty} \frac{\cos n\pi}{n+1}$

50. $\displaystyle\sum_{n=1}^{\infty} (-1)^{n+1}\arctan n$

51. $\displaystyle\sum_{n=1}^{\infty} \frac{\cos n\pi}{n^2}$

52. $\displaystyle\sum_{n=1}^{\infty} \frac{\sin[(2n-1)\pi/2]}{n}$

53. Prove that the alternating p-series

$$\sum_{n=1}^{\infty} (-1)^n \left(\frac{1}{n^p}\right)$$

converges if $p > 0$.

54. Prove that if $\Sigma\,|a_n|$ converges, then $\Sigma\,a_n^2$ converges. Is the converse true? If not, give an example that shows it is false.

55. Use the result of Exercise 53 to give an example of an alternating p-series that converges, but whose corresponding p-series diverges.

56. Give an example of a series that demonstrates the statement you proved in Exercise 54.

57. Find all values of x for which the series $\Sigma\,(x^n/n)$ (a) converges absolutely and (b) converges conditionally.

58. The following argument, that $0 = 1$, is *incorrect*. Describe the error.

$$0 = 0 + 0 + 0 + \cdots$$
$$= (1-1) + (1-1) + (1-1) + \cdots$$
$$= 1 + (-1+1) + (-1+1) + \cdots$$
$$= 1 + 0 + 0 + \cdots$$
$$= 1$$

True or False?   **In Exercises 59 and 60, determine whether the statement is true or false. If false, explain why.**

59. If both $\Sigma\,a_n$ and $\Sigma\,(-a_n)$ converge, then $\Sigma\,|a_n|$ converges.

60. If $\Sigma\,a_n$ does not converge, then $\Sigma\,|a_n|$ does not converge.

The Ratio Test • The Root Test • Strategies for Testing Series

The Ratio Test

This section begins with a test for absolute convergence—the **Ratio Test.**

> **THEOREM 8.17 Ratio Test**
>
> Let $\Sigma\, a_n$ be a series with nonzero terms.
>
> 1. $\Sigma\, a_n$ converges absolutely if $\lim\limits_{n\to\infty} \left| \dfrac{a_{n+1}}{a_n} \right| < 1$.
>
> 2. $\Sigma\, a_n$ diverges if $\lim\limits_{n\to\infty} \left| \dfrac{a_{n+1}}{a_n} \right| > 1$ or $\lim\limits_{n\to\infty} \left| \dfrac{a_{n+1}}{a_n} \right| = \infty$.
>
> 3. The Ratio Test is inconclusive if $\lim\limits_{n\to\infty} \left| \dfrac{a_{n+1}}{a_n} \right| = 1$.

Proof To prove Property 1, assume that

$$\lim_{n\to\infty} \left| \frac{a_{n+1}}{a_n} \right| = r < 1$$

and choose R such that $0 \le r < R < 1$. By the definition of the limit of a sequence, there exists some $N > 0$ such that $|a_{n+1}/a_n| < R$ for all $n > N$. Therefore, you can write the following inequalities.

$$|a_{N+1}| < |a_N|R$$
$$|a_{N+2}| < |a_{N+1}|R < |a_N|R^2$$
$$|a_{N+3}| < |a_{N+2}|R < |a_{N+1}|R^2 < |a_N|R^3$$
$$\vdots$$

The geometric series $\Sigma\,|a_N|R^n = |a_N|R + |a_N|R^2 + \cdots + |a_N|R^n + \cdots$ converges, and so, by the Direct Comparison Test, the series

$$\sum_{n=1}^{\infty} |a_{N+n}| = |a_{N+1}| + |a_{N+2}| + \cdots + |a_{N+n}| + \cdots$$

also converges. This in turn implies that the series $\Sigma\,|a_n|$ converges, because discarding a finite number of terms ($n = N - 1$) does not affect convergence. Consequently, by Theorem 8.16, the series $\Sigma\, a_n$ converges absolutely. The proof of Property 2 is similar and is left as an exercise (see Exercise 68).

NOTE The fact that the Ratio Test is inconclusive when $|a_{n+1}/a_n| \to 1$ can be seen by comparing the two series $\Sigma\,(1/n)$ and $\Sigma\,(1/n^2)$. The first series diverges and the second one converges, but in both cases

$$\lim_{n\to\infty} \left| \frac{a_{n+1}}{a_n} \right| = 1.$$

Although the Ratio Test is not a cure for all ills related to tests for convergence, it is particularly useful for series that *converge rapidly*. Series involving factorials or exponentials are frequently of this type.

EXAMPLE 1   Using the Ratio Test

Determine the convergence or divergence of $\displaystyle\sum_{n=0}^{\infty} \frac{2^n}{n!}$.

Solution   Because $a_n = 2^n/n!$, you can write the following.

$$\lim_{n\to\infty} \left|\frac{a_{n+1}}{a_n}\right| = \lim_{n\to\infty} \left[\frac{2^{n+1}}{(n+1)!} \div \frac{2^n}{n!}\right]$$

$$= \lim_{n\to\infty} \left[\frac{2^{n+1}}{(n+1)!} \cdot \frac{n!}{2^n}\right]$$

$$= \lim_{n\to\infty} \frac{2}{n+1}$$

$$= 0$$

Therefore, the series converges.

STUDY TIP   A step frequently used in applications of the Ratio Test involves simplifying quotients of factorials. In Example 1, for instance, notice that

$$\frac{n!}{(n+1)!} = \frac{n!}{(n+1)n!} = \frac{1}{n+1}.$$

EXAMPLE 2   Using the Ratio Test

Determine whether each of the following series converges or diverges.

a. $\displaystyle\sum_{n=0}^{\infty} \frac{n^2 2^{n+1}}{3^n}$ **b.** $\displaystyle\sum_{n=1}^{\infty} \frac{n^n}{n!}$

Solution

a. This series converges because the limit of $|a_{n+1}/a_n|$ is less than 1.

$$\lim_{n\to\infty} \left|\frac{a_{n+1}}{a_n}\right| = \lim_{n\to\infty} \left[(n+1)^2\left(\frac{2^{n+2}}{3^{n+1}}\right)\left(\frac{3^n}{n^2 2^{n+1}}\right)\right]$$

$$= \lim_{n\to\infty} \frac{2(n+1)^2}{3n^2}$$

$$= \frac{2}{3} < 1$$

b. This series diverges because the limit of $|a_{n+1}/a_n|$ is greater than 1.

$$\lim_{n\to\infty} \left|\frac{a_{n+1}}{a_n}\right| = \lim_{n\to\infty} \left[\frac{(n+1)^{n+1}}{(n+1)!}\left(\frac{n!}{n^n}\right)\right]$$

$$= \lim_{n\to\infty} \left[\frac{(n+1)^{n+1}}{(n+1)}\left(\frac{1}{n^n}\right)\right]$$

$$= \lim_{n\to\infty} \frac{(n+1)^n}{n^n}$$

$$= \lim_{n\to\infty} \left(1 + \frac{1}{n}\right)^n$$

$$= e > 1$$

EXAMPLE 3   A Failure of the Ratio Test

Determine the convergence or divergence of $\displaystyle\sum_{n=1}^{\infty} (-1)^n \frac{\sqrt{n}}{n+1}$.

Solution   The limit of $\left|a_{n+1}/a_n\right|$ is equal to 1.

$$
\begin{aligned}
\lim_{n\to\infty} \left|\frac{a_{n+1}}{a_n}\right| &= \lim_{n\to\infty} \left[\left(\frac{\sqrt{n+1}}{n+2}\right)\left(\frac{n+1}{\sqrt{n}}\right)\right] \\
&= \lim_{n\to\infty}\left[\sqrt{\frac{n+1}{n}}\left(\frac{n+1}{n+2}\right)\right] \\
&= \sqrt{1}\,(1) \\
&= 1
\end{aligned}
$$

Thus, the Ratio Test is inconclusive. To determine whether the series converges, you need to try a different test. In this case, you can apply the Alternating Series Test. To show that $a_{n+1} \leq a_n$, let

$$
f(x) = \frac{\sqrt{x}}{x+1}.
$$

Then the derivative is

$$
f'(x) = \frac{-x+1}{2\sqrt{x}(x+1)^2}.
$$

Because the derivative is negative for $x > 1$, you know that f is a decreasing function. Also, by L'Hôpital's Rule,

$$
\begin{aligned}
\lim_{x\to\infty} \frac{\sqrt{x}}{x+1} &= \lim_{x\to\infty} \frac{1/(2\sqrt{x})}{1} \\
&= \lim_{x\to\infty} \frac{1}{2\sqrt{x}} \\
&= 0.
\end{aligned}
$$

Therefore, by the Alternating Series Test, the series converges.

The series in Example 3 is *conditionally convergent*. This follows from the fact that the series

$$
\sum_{n=1}^{\infty} |a_n|
$$

diverges $\left(\text{by the Limit Comparison Test with } \Sigma\, 1/\sqrt{n}\right)$, but the series

$$
\sum_{n=1}^{\infty} a_n
$$

converges.

TECHNOLOGY   A computer or programmable calculator can reinforce the conclusion that the series in Example 3 converges *conditionally*. By adding the first 100 terms of the series, you obtain a sum of about -0.2. (The sum of the first 100 terms of the series $\Sigma\, |a_n|$ is about 17.)

The Root Test

The next test for convergence or divergence of series works especially well for series involving nth powers. The proof of this theorem is similar to that given for the Ratio Test, and we leave it as an exercise (see Exercise 69).

THEOREM 8.18   Root Test

Let $\Sigma\, a_n$ be a series with nonzero terms.

1. $\Sigma\, a_n$ converges absolutely if $\displaystyle\lim_{n\to\infty} \sqrt[n]{|a_n|} < 1$.

2. $\Sigma\, a_n$ diverges if $\displaystyle\lim_{n\to\infty} \sqrt[n]{|a_n|} > 1$ or $\displaystyle\lim_{n\to\infty} \sqrt[n]{|a_n|} = \infty$.

3. The ~~Ratio~~ Root Test is inconclusive if $\displaystyle\lim_{n\to\infty} \sqrt[n]{|a_n|} = 1$.

EXAMPLE 4   Using the Root Test

Determine the convergence or divergence of $\displaystyle\sum_{n=1}^{\infty} \frac{e^{2n}}{n^n}$.

Solution   You can apply the Root Test as follows.

$$
\begin{aligned}
\lim_{n\to\infty} \sqrt[n]{|a_n|} &= \lim_{n\to\infty} \sqrt[n]{\frac{e^{2n}}{n^n}} \\
&= \lim_{n\to\infty} \frac{e^{2n/n}}{n^{n/n}} \\
&= \lim_{n\to\infty} \frac{e^2}{n} \\
&= 0 < 1
\end{aligned}
$$

Because this limit is less than 1, you can conclude that the series converges absolutely (and hence converges).

FOR FURTHER INFORMATION For more information on the usefulness of the Root Test, see the article "N! and the Root Test" by Charles C. Mumma II in the August-September 1986 issue of *The American Mathematical Monthly*.

To see the usefulness of the Root Test for the series in Example 4, try applying the Ratio Test to that series. When you do this, you obtain the following.

$$
\begin{aligned}
\lim_{n\to\infty} \left| \frac{a_{n+1}}{a_n} \right| &= \lim_{n\to\infty} \left[\frac{e^{2(n+1)}}{(n+1)^{n+1}} \div \frac{e^{2n}}{n^n} \right] \\
&= \lim_{n\to\infty} e^2 \frac{n^n}{(n+1)^{n+1}} \\
&= \lim_{n\to\infty} e^2 \left(\frac{n}{n+1} \right)^n \left(\frac{1}{n+1} \right) \\
&= 0
\end{aligned}
$$

Note that this limit is not as easily evaluated as the limit obtained by the Root Test in Example 4.

Strategies for Testing Series

You have now studied ten tests for determining the convergence or divergence of an infinite series. (See the summary in the table on page 593.) Skill in choosing and applying the various tests will come only with practice. Below is a useful checklist for choosing an appropriate test.

Guidelines for Testing a Series for Convergence or Divergence

1. Does the nth term approach 0? If not, the series diverges.
2. Is the series one of the special types—geometric, p-series, telescoping, or alternating?
3. Can the Integral Test, the Root Test, or the Ratio Test be applied?
4. Can the series be compared favorably to one of the special types?

In some instances, more than one test is applicable. However, your objective should be to learn to choose the most efficient test.

EXAMPLE 5 Applying the Strategies for Testing Series

Determine the convergence or divergence of each series.

a. $\displaystyle\sum_{n=1}^{\infty} \frac{n+1}{3n+1}$ **b.** $\displaystyle\sum_{n=1}^{\infty} \left(\frac{\pi}{6}\right)^n$ **c.** $\displaystyle\sum_{n=1}^{\infty} ne^{-n^2}$ **d.** $\displaystyle\sum_{n=1}^{\infty} \frac{1}{3n+1}$

e. $\displaystyle\sum_{n=1}^{\infty} (-1)^n \frac{3}{4n+1}$ **f.** $\displaystyle\sum_{n=1}^{\infty} \frac{n!}{10^n}$ **g.** $\displaystyle\sum_{n=1}^{\infty} \left(\frac{n+1}{2n+1}\right)^n$

Solution

a. For this series, the limit of the nth term is not 0 $\left(a_n \to \frac{1}{3} \text{ as } n \to \infty\right)$. Thus, by the nth-Term Test, the series diverges.

b. This series is geometric. Moreover, because the common ratio of the terms is less than 1 in absolute value $(r = \pi/6)$, you can conclude that the series converges.

c. Because the function $f(x) = xe^{-x^2}$ is easily integrated, you can use the Integral Test to conclude that the series converges.

d. The nth term of this series can be compared to the nth term of the harmonic series. After using the Limit Comparison Test, you can conclude that the series diverges.

e. This is an alternating series whose nth term approaches 0. Because $a_{n+1} \leq a_n$, you can use the Alternating Series Test to conclude that the series converges.

f. The nth term of this series involves a factorial, which indicates that the Ratio Test may work well. After applying the Ratio Test, you can conclude that the series diverges.

g. The nth term of this series involves a variable that is raised to the nth power, which indicates that the Root Test may work well. After applying the Root Test, you can conclude that the series converges.

Summary of Tests for Series

Test	Series	Converges	Diverges	Comment
nth-Term	$\displaystyle\sum_{n=1}^{\infty} a_n$		$\displaystyle\lim_{n\to\infty} a_n \neq 0$	This test cannot be used to show convergence.
Geometric Series	$\displaystyle\sum_{n=0}^{\infty} ar^n$	$\|r\| < 1$	$\|r\| \geq 1$	Sum: $S = \dfrac{a}{1-r}$
Telescoping Series	$\displaystyle\sum_{n=1}^{\infty} (b_n - b_{n+1})$	$\displaystyle\lim_{n\to\infty} b_n = L$		Sum: $S = b_1 - L$
p-Series	$\displaystyle\sum_{n=1}^{\infty} \dfrac{1}{n^p}$	$p > 1$	$p \leq 1$	
Alternating Series	$\displaystyle\sum_{n=1}^{\infty} (-1)^{n-1} a_n$	$0 < a_{n+1} \leq a_n$ and $\displaystyle\lim_{n\to\infty} a_n = 0$		Remainder: $\|R_N\| \leq a_{N+1}$
Integral (f is continuous, positive, and decreasing)	$\displaystyle\sum_{n=1}^{\infty} a_n,$ $a_n = f(n) \geq 0$	$\displaystyle\int_1^{\infty} f(x)\, dx$ converges	$\displaystyle\int_1^{\infty} f(x)\, dx$ diverges	Remainder: $0 < R_N < \displaystyle\int_N^{\infty} f(x)\, dx$
Root	$\displaystyle\sum_{n=1}^{\infty} a_n$	$\displaystyle\lim_{n\to\infty} \sqrt[n]{\|a_n\|} < 1$	$\displaystyle\lim_{n\to\infty} \sqrt[n]{\|a_n\|} > 1$	Test is inconclusive if $\displaystyle\lim_{n\to\infty} \sqrt[n]{\|a_n\|} = 1.$
Ratio	$\displaystyle\sum_{n=1}^{\infty} a_n$	$\displaystyle\lim_{n\to\infty} \left\|\dfrac{a_{n+1}}{a_n}\right\| < 1$	$\displaystyle\lim_{n\to\infty} \left\|\dfrac{a_{n+1}}{a_n}\right\| > 1$	Test is inconclusive if $\displaystyle\lim_{n\to\infty} \left\|\dfrac{a_{n+1}}{a_n}\right\| = 1.$
Direct Comparison ($a_n, b_n > 0$)	$\displaystyle\sum_{n=1}^{\infty} a_n$	$0 < a_n \leq b_n$ and $\displaystyle\sum_{n=1}^{\infty} b_n$ converges	$0 < b_n \leq a_n$ and $\displaystyle\sum_{n=1}^{\infty} b_n$ diverges	
Limit Comparison ($a_n, b_n > 0$)	$\displaystyle\sum_{n=1}^{\infty} a_n$	$\displaystyle\lim_{n\to\infty} \dfrac{a_n}{b_n} = L > 0$ and $\displaystyle\sum_{n=1}^{\infty} b_n$ converges	$\displaystyle\lim_{n\to\infty} \dfrac{a_n}{b_n} = L > 0$ and $\displaystyle\sum_{n=1}^{\infty} b_n$ diverges	

EXERCISES FOR SECTION 8.6

In Exercises 1–4, verify the formula.

1. $\dfrac{(n+1)!}{(n-2)!} = (n+1)(n)(n-1)$

2. $\dfrac{(2k-2)!}{(2k)!} = \dfrac{1}{(2k)(2k-1)}$

3. $1 \cdot 3 \cdot 5 \cdots (2k-1) = \dfrac{(2k)!}{2^k k!}$

4. $\dfrac{1}{1 \cdot 3 \cdot 5 \cdots (2k-5)} = \dfrac{2^k k!(2k-3)(2k-1)}{(2k)!}, \quad k \geq 3$

In Exercises 5–8, match the series with the graph of its sequence of partial sums. [The graphs are labeled (a), (b), (c), and (d).]

(a)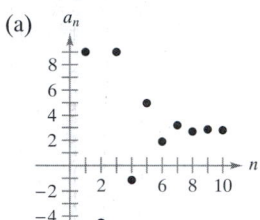

(b)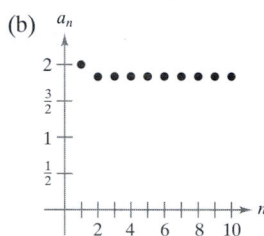

(c)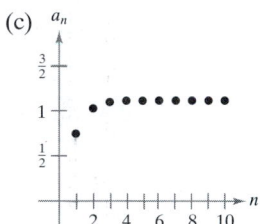

(d)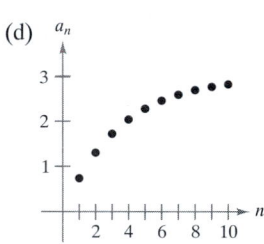

5. $\displaystyle\sum_{n=1}^{\infty} n\left(\dfrac{3}{4}\right)^n$

6. $\displaystyle\sum_{n=1}^{\infty} \left(\dfrac{3}{4}\right)^n\left(\dfrac{1}{n!}\right)$

7. $\displaystyle\sum_{n=1}^{\infty} \dfrac{(-3)^{n+1}}{n!}$

8. $\displaystyle\sum_{n=1}^{\infty} \dfrac{(-1)^{n-1}4}{(2n)!}$

Numerical, Graphical, and Analytic Analysis **In Exercises 9 and 10, (a) verify that the series converges. (b) Use a graphing utility to find the indicated partial sum S_n and complete the table. (c) Use a graphing utility to graph the first ten terms of the sequence of partial sums. (d) Use the table to estimate the sum of the series. (e) Explain the relationship between the magnitude of the terms of the series and the rate at which the sequence of partial sums approaches the sum of the series.**

n	5	10	15	20	25
S_n					

9. $\displaystyle\sum_{n=1}^{\infty} n^2\left(\dfrac{5}{8}\right)^n$

10. $\displaystyle\sum_{n=1}^{\infty} \dfrac{n^2+1}{n!}$

In Exercises 11–30, use the Ratio Test to test for convergence or divergence of the series.

11. $\displaystyle\sum_{n=0}^{\infty} \dfrac{n!}{3^n}$

12. $\displaystyle\sum_{n=1}^{\infty} n\left(\dfrac{2}{3}\right)^n$

13. $\displaystyle\sum_{n=0}^{\infty} \dfrac{3^n}{n!}$

14. $\displaystyle\sum_{n=1}^{\infty} n\left(\dfrac{3}{2}\right)^n$

15. $\displaystyle\sum_{n=1}^{\infty} \dfrac{n}{2^n}$

16. $\displaystyle\sum_{n=1}^{\infty} \dfrac{n^2}{2^n}$

17. $\displaystyle\sum_{n=1}^{\infty} \dfrac{2^n}{n^2}$

18. $\displaystyle\sum_{n=1}^{\infty} \dfrac{(-1)^{n+1}(n+2)}{n(n+1)}$

19. $\displaystyle\sum_{n=0}^{\infty} \dfrac{(-1)^n \, 2^n}{n!}$

20. $\displaystyle\sum_{n=1}^{\infty} \dfrac{(-1)^{n-1}(3/2)^n}{n^2}$

21. $\displaystyle\sum_{n=1}^{\infty} \dfrac{n!}{n3^n}$

22. $\displaystyle\sum_{n=1}^{\infty} \dfrac{(2n)!}{n^5}$

23. $\displaystyle\sum_{n=0}^{\infty} \dfrac{4^n}{n!}$

24. $\displaystyle\sum_{n=1}^{\infty} \dfrac{n^n}{n!}$

25. $\displaystyle\sum_{n=0}^{\infty} \dfrac{3^n}{(n+1)^n}$

26. $\displaystyle\sum_{n=0}^{\infty} \dfrac{(n!)^2}{(3n)!}$

27. $\displaystyle\sum_{n=0}^{\infty} \dfrac{4^n}{3^n+1}$

28. $\displaystyle\sum_{n=0}^{\infty} \dfrac{(-1)^n 2^{4n}}{(2n+1)!}$

29. $\displaystyle\sum_{n=0}^{\infty} \dfrac{(-1)^{n+1}n!}{1 \cdot 3 \cdot 5 \cdots (2n+1)}$

30. $\displaystyle\sum_{n=1}^{\infty} \dfrac{(-1)^n 2 \cdot 4 \cdot 6 \cdots (2n)}{2 \cdot 5 \cdot 8 \cdots (3n-1)}$

In Exercises 31 and 32, verify that the Ratio Test is inconclusive for the p-series.

31. (a) $\displaystyle\sum_{n=1}^{\infty} \dfrac{1}{n^{3/2}}$

(b) $\displaystyle\sum_{n=1}^{\infty} \dfrac{1}{n^{1/2}}$

32. (a) $\displaystyle\sum_{n=1}^{\infty} \dfrac{1}{n^3}$

(b) $\displaystyle\sum_{n=1}^{\infty} \dfrac{1}{n^p}$

In Exercises 33–40, use the Root Test to test for convergence or divergence of the series.

33. $\displaystyle\sum_{n=1}^{\infty} \left(\dfrac{n}{2n+1}\right)^n$

34. $\displaystyle\sum_{n=1}^{\infty} \left(\dfrac{2n}{n+1}\right)^n$

35. $\displaystyle\sum_{n=2}^{\infty} \dfrac{(-1)^n}{(\ln n)^n}$

36. $\displaystyle\sum_{n=1}^{\infty} \left(\dfrac{-2n}{3n+1}\right)^{3n}$

37. $\displaystyle\sum_{n=1}^{\infty} \left(2\sqrt[n]{n}+1\right)^n$

38. $\displaystyle\sum_{n=0}^{\infty} e^{-n}$

39. $\dfrac{1}{(\ln 3)^3} + \dfrac{1}{(\ln 4)^4} + \dfrac{1}{(\ln 5)^5} + \dfrac{1}{(\ln 6)^6} + \cdots$

40. $1 + \dfrac{2}{3} + \dfrac{3}{3^2} + \dfrac{4}{3^3} + \dfrac{5}{3^4} + \dfrac{6}{3^5} + \cdots$

In Exercises 41–58, test for convergence or divergence using any appropriate test from this chapter. Identify the test used.

41. $\displaystyle\sum_{n=1}^{\infty} \dfrac{(-1)^{n+1}5}{n}$

42. $\displaystyle\sum_{n=1}^{\infty} \dfrac{5}{n}$

43. $\displaystyle\sum_{n=1}^{\infty} \dfrac{3}{n\sqrt{n}}$

44. $\displaystyle\sum_{n=1}^{\infty} \left(\dfrac{\pi}{4}\right)^n$

45. $\displaystyle\sum_{n=1}^{\infty} \dfrac{2n}{n+1}$

46. $\displaystyle\sum_{n=1}^{\infty} \dfrac{n}{2n^2+1}$

47. $\displaystyle\sum_{n=1}^{\infty} \dfrac{(-1)^n 3^{n-2}}{2^n}$

48. $\displaystyle\sum_{n=1}^{\infty} \dfrac{10}{3\sqrt{n^3}}$

49. $\displaystyle\sum_{n=1}^{\infty} \dfrac{10n+3}{n2^n}$

50. $\displaystyle\sum_{n=1}^{\infty} \dfrac{2^n}{4n^2-1}$

51. $\displaystyle\sum_{n=1}^{\infty} \dfrac{\cos n}{2^n}$

52. $\displaystyle\sum_{n=2}^{\infty} \dfrac{(-1)^n}{n \ln n}$

53. $\displaystyle\sum_{n=1}^{\infty} \dfrac{n7^n}{n!}$

54. $\displaystyle\sum_{n=1}^{\infty} \dfrac{\ln n}{n^2}$

55. $\displaystyle\sum_{n=1}^{\infty} \dfrac{(-1)^n 3^{n-1}}{n!}$

56. $\displaystyle\sum_{n=1}^{\infty} \dfrac{(-1)^n 3^n}{n2^n}$

57. $\displaystyle\sum_{n=1}^{\infty} \dfrac{(-3)^n}{3 \cdot 5 \cdot 7 \cdots (2n+1)}$

58. $\displaystyle\sum_{n=1}^{\infty} \dfrac{3 \cdot 5 \cdot 7 \cdots (2n+1)}{18^n(2n-1)n!}$

In Exercises 59–62, identify the two series that are the same.

59. (a) $\displaystyle\sum_{n=1}^{\infty} \dfrac{n5^n}{n!}$

(b) $\displaystyle\sum_{n=0}^{\infty} \dfrac{n5^n}{n!}$

(c) $\displaystyle\sum_{n=0}^{\infty} \dfrac{(n+1)5^{n+1}}{(n+1)!}$

60. (a) $\displaystyle\sum_{n=4}^{\infty} n\left(\dfrac{3}{4}\right)^n$

(b) $\displaystyle\sum_{n=0}^{\infty} (n+1)\left(\dfrac{3}{4}\right)^n$

(c) $\displaystyle\sum_{n=1}^{\infty} n\left(\dfrac{3}{4}\right)^{n-1}$

61. (a) $\displaystyle\sum_{n=0}^{\infty} \dfrac{(-1)^n}{(2n+1)!}$

(b) $\displaystyle\sum_{n=1}^{\infty} \dfrac{(-1)^{n-1}}{(2n-1)!}$

(c) $\displaystyle\sum_{n=1}^{\infty} \dfrac{(-1)^{n-1}}{(2n+1)!}$

62. (a) $\displaystyle\sum_{n=2}^{\infty} \dfrac{(-1)^n}{(n-1)2^{n-1}}$

(b) $\displaystyle\sum_{n=1}^{\infty} \dfrac{(-1)^{n+1}}{n2^n}$

(c) $\displaystyle\sum_{n=0}^{\infty} \dfrac{(-1)^{n+1}}{(n+1)2^n}$

In Exercises 63 and 64, write an equivalent series with the index of summation beginning at $n = 0$.

63. $\displaystyle\sum_{n=1}^{\infty} \dfrac{n}{4^n}$

64. $\displaystyle\sum_{n=2}^{\infty} \dfrac{2^n}{(n-2)!}$

In Exercises 65 and 66, (a) determine the number of terms required to approximate the sum of the series with an error less than 0.0001, and (b) use a graphing utility to approximate the sum of the series with an error less than 0.0001.

65. $\displaystyle\sum_{k=1}^{\infty} \dfrac{(-3)^k}{2^k k!}$

66. $\displaystyle\sum_{k=0}^{\infty} \dfrac{(-3)^k}{1 \cdot 3 \cdot 5 \cdots (2k+1)}$

67. *Writing* You are told that the terms of a positive series appear to approach zero rapidly as n approaches infinity. In fact, $a_7 \le 0.0001$. Given no other information, does this imply that the series converges? Support your conclusion with examples.

68. Prove Property 2 of Theorem 8.17.

69. Prove Theorem 8.18. (*Hint for Property 1:* If the limit equals $r < 1$, choose a real number R such that $r < R < 1$. By the definition of the limit, there exists some $N > 0$ such that $\sqrt[n]{|a_n|} < R$ for $n > N$.)

70. *Writing* Read the article "A Differentiation Test for Absolute Convergence" by Yaser S. Abu-Mostafa in the September 1984 issue of *Mathematics Magazine*. Then write a paragraph that describes the test. Support your answer with examples of series that converge and examples of series that diverge.

Polynomial Approximations of Elementary Functions •
Taylor and Maclaurin Polynomials • Remainder of a Taylor Polynomial

Polynomial Approximations of Elementary Functions

The goal of this section is to show how polynomial functions can be used as approximations for other elementary functions. To find a polynomial function P that approximates another function f, begin by choosing a number c in the domain of f at which f and P have the same value. That is,

$$P(c) = f(c). \qquad \text{Graphs of } f \text{ and } P \text{ pass through } (c, f(c)).$$

The approximating polynomial is said to be **expanded about** c or **centered at** c. Geometrically, the requirement that $P(c) = f(c)$ means that the graph of P passes through the point $(c, f(c))$. Of course, there are many polynomials whose graphs pass through the point $(c, f(c))$. Your task is to find a polynomial whose graph resembles the graph of f near this point. One way to do this is to impose the additional requirement that the slope of the polynomial function be the same as the slope of the graph of f at the point $(c, f(c))$.

$$P'(c) = f'(c) \qquad \text{Graphs of } f \text{ and } P \text{ have the same slope at } (c, f(c)).$$

With these two requirements, you can obtain a simple linear approximation of f, as shown in Figure 8.10.

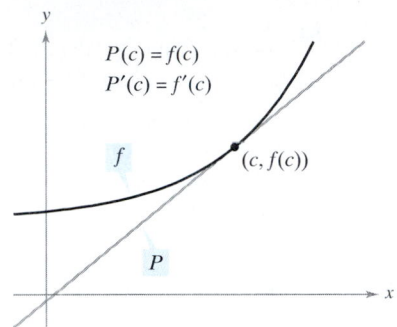

$P(c) = f(c)$
$P'(c) = f'(c)$

Near $(c, f(c))$, the graph of P can be used to approximate the graph of f.
Figure 8.10

EXAMPLE 1 First-Degree Polynomial Approximation of $f(x) = e^x$

For the function $f(x) = e^x$, find a first-degree polynomial function

$$P_1(x) = a_0 + a_1 x$$

whose value and slope agree with the value and slope of f at $x = 0$.

Solution Because $f(x) = e^x$ and $f'(x) = e^x$, the value of the slope of f, at $x = 0$, are given by

$$f(0) = e^0 = 1$$

and

$$f'(0) = e^0 = 1.$$

Because $P_1(x) = a_0 + a_1 x$, you can use the condition that $P_1(0) = f(0)$ to conclude that $a_0 = 1$. Moreover, because $P_1{}'(x) = a_1$, you can use the condition that $P_1{}'(0) = f'(0)$ to conclude that $a_1 = 1$. Therefore,

$$P_1(x) = 1 + x.$$

Figure 8.11 shows the graphs of $P_1(x) = 1 + x$ and $f(x) = e^x$. ▬▬▬▬

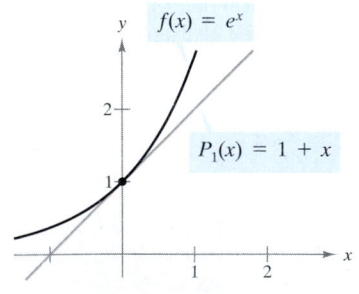

$f(x) = e^x$

$P_1(x) = 1 + x$

P_1 is the first-degree polynomial approximation of $f(x) = e^x$.
Figure 8.11

NOTE Example 1 isn't the first time you have used a linear function to approximate another function. The same procedure was used as the basis for Newton's Method in Section 3.8.

In Figure 8.11 you can see that, at points near $(0, 1)$, the graph of

$$P_1(x) = 1 + x \qquad \text{1st-degree approximation}$$

is reasonably close to the graph of $f(x) = e^x$. However, as you move away from $(0, 1)$, the graphs move farther from each other and the accuracy of the approximation decreases. To improve the approximation, you can impose yet another requirement— that the values of the second derivatives of P and f agree when $x = 0$. The polynomial, P_2, of least degree that satisfies all three requirements $P_2(0) = f(0)$, $P_2'(0) = f'(0)$, and $P_2''(0) = f''(0)$ can be shown to be

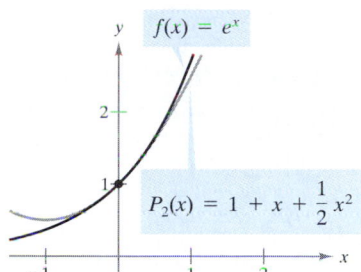

$$P_2(x) = 1 + x + \frac{1}{2}x^2. \qquad \text{2nd-degree approximation}$$

Moreover, in Figure 8.12, you can see that P_2 is a better approximation of f than P. If you continue this pattern, requiring that the values of $P_n(x)$ and its first n derivatives match those of $f(x) = e^x$ at $x = 0$, you obtain the following.

P_2 is the second-degree polynomial approximation of $f(x) = e^x$.

Figure 8.12

$$P_n(x) = 1 + x + \frac{1}{2}x^2 + \frac{1}{3!}x^3 + \cdots + \frac{1}{n!}x^n \qquad \text{nth-degree approximation}$$

$$\approx e^x$$

EXAMPLE 2   Third-Degree Polynomial Approximation of $f(x) = e^x$

Construct a table comparing the values of the polynomial

$$P_3(x) = 1 + x + \frac{1}{2}x^2 + \frac{1}{3!}x^3 \qquad \text{3rd-degree approximation}$$

with $f(x) = e^x$ for several values of x near 0.

Solution   Using a calculator or a computer, you can obtain the results shown in the table below. Note that for $x = 0$, the two functions have the same value, but that as x moves farther away from 0, the accuracy of the approximating polynomial $P_3(x)$ decreases.

x	-1.0	-0.2	-0.1	0.0	0.1	0.2	1.0
e^x	0.3679	0.81873	0.904837	1	1.105171	1.22140	2.7183
$P_3(x)$	0.3333	0.81867	0.904833	1	1.105167	1.22133	2.6667

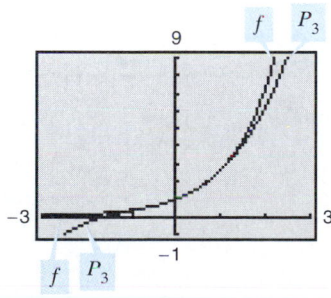

P_3 is the third-degree polynomial approximation of $f(x) = e^x$.

Figure 8.13

TECHNOLOGY   A graphing utility can be used to compare the graph of the approximating polynomial with the graph of the function f. For instance, in Figure 8.13, the graph of

$$P_3(x) = 1 + x + \tfrac{1}{2}x^2 + \tfrac{1}{6}x^3 \qquad \text{3rd-degree approximation}$$

is compared with the graph of $f(x) = e^x$. If you have access to a graphing utility, try comparing the graphs of

$$P_4(x) = 1 + x + \tfrac{1}{2}x^2 + \tfrac{1}{6}x^3 + \tfrac{1}{24}x^4 \qquad \text{4th-degree approximation}$$

$$P_5(x) = 1 + x + \tfrac{1}{2}x^2 + \tfrac{1}{6}x^3 + \tfrac{1}{24}x^4 + \tfrac{1}{120}x^5 \qquad \text{5th-degree approximation}$$

$$P_6(x) = 1 + x + \tfrac{1}{2}x^2 + \tfrac{1}{6}x^3 + \tfrac{1}{24}x^4 + \tfrac{1}{120}x^5 + \tfrac{1}{720}x^6 \qquad \text{6th-degree approximation}$$

with the graph of f. What do you notice?

The Granger Collection

BROOK TAYLOR (1685–1731)

Although Taylor was not the first to seek polynomial approximations of transcendental functions, his account published in 1715 was one of the first comprehensive works on the subject.

Taylor and Maclaurin Polynomials

The polynomial approximation of $f(x) = e^x$ given in Example 2 is expanded about $c = 0$. For expansions about an arbitrary value of c, it is convenient to write the polynomial in the form

$$P_n(x) = a_0 + a_1(x - c) + a_2(x - c)^2 + a_3(x - c)^3 + \cdots + a_n(x - c)^n.$$

In this form, repeated differentiation produces

$$P_n'(x) = a_1 + 2a_2(x - c) + 3a_3(x - c)^2 + \cdots + na_n(x - c)^{n-1}$$
$$P_n''(x) = 2a_2 + 2(3a_3)(x - c) + \cdots + n(n - 1)a_n(x - c)^{n-2}$$
$$P_n'''(x) = 2(3a_3) + \cdots + n(n - 1)(n - 2)a_n(x - c)^{n-3}$$
$$\vdots$$
$$P_n^{(n)}(x) = n(n - 1)(n - 2) \cdots (2)(1)a_n.$$

Letting $x = c$, you then obtain

$$P_n(c) = a_0, \quad P_n'(c) = a_1, \quad P_n''(c) = 2a_2, \quad \ldots, \quad P_n^{(n)}(c) = n!a_n$$

and because the value of f and its first n derivatives must agree with the value of P_n and its first n derivatives at $x = c$, it follows that

$$f(c) = a_0, \quad f'(c) = a_1, \quad \frac{f''(c)}{2!} = a_2, \quad \ldots, \quad \frac{f^{(n)}(c)}{n!} = a_n.$$

With these coefficients, you can obtain the following definition of **Taylor polynomials,** named after the English mathematician Brook Taylor, and **Maclaurin polynomials,** named after the English mathematician Colin Maclaurin (1698–1746).

Definition of nth Taylor Polynomial and Maclaurin Polynomial

If f has n derivatives at c, then the polynomial

$$P_n(x) = f(c) + f'(c)(x - c) + \frac{f''(c)}{2!}(x - c)^2 + \cdots + \frac{f^{(n)}(c)}{n!}(x - c)^n$$

is called the **nth Taylor polynomial for f at c.** If $c = 0$, then

$$P_n(x) = f(0) + f'(0)x + \frac{f''(0)}{2!}x^2 + \frac{f'''(0)}{3!}x^3 + \cdots + \frac{f^{(n)}(0)}{n!}x^n$$

is called the **nth Maclaurin polynomial for f.**

EXAMPLE 3 A Maclaurin Polynomial for $f(x) = e^x$

From the discussion on page 597, the nth Maclaurin polynomial for $f(x) = e^x$ is given by

$$P_n(x) = 1 + x + \frac{1}{2}x^2 + \frac{1}{3!}x^3 + \cdots + \frac{1}{n!}x^n.$$

EXAMPLE 4   **Finding Taylor Polynomials for ln x**

Find the Taylor polynomials P_0, P_1, P_2, P_3, and P_4 for $f(x) = \ln x$ centered at $c = 1$.

Solution   Expanding about $c = 1$ yields the following.

$$f(x) = \ln x \qquad\qquad f(1) = 0$$

$$f'(x) = \frac{1}{x} \qquad\qquad f'(1) = 1$$

$$f''(x) = -\frac{1}{x^2} \qquad\qquad f''(1) = -1$$

$$f'''(x) = \frac{2!}{x^3} \qquad\qquad f'''(1) = 2$$

$$f^{(4)}(x) = -\frac{3!}{x^4} \qquad\qquad f^{(4)}(1) = -6$$

Therefore, the Taylor polynomials are as follows.

$$P_0(x) = f(1) = 0$$
$$P_1(x) = f(1) + f'(1)(x - 1) = (x - 1)$$
$$P_2(x) = f(1) + f'(1)(x - 1) + \frac{f''(1)}{2!}(x - 1)^2$$

$$= (x - 1) - \frac{1}{2}(x - 1)^2$$

$$P_3(x) = f(1) + f'(1)(x - 1) + \frac{f''(1)}{2!}(x - 1)^2 + \frac{f'''(1)}{3!}(x - 1)^3$$

$$= (x - 1) - \frac{1}{2}(x - 1)^2 + \frac{1}{3}(x - 1)^3$$

$$P_4(x) = f(1) + f'(1)(x - 1) + \frac{f''(1)}{2!}(x - 1)^2 + \frac{f'''(1)}{3!}(x - 1)^3$$

$$\qquad + \frac{f^{(4)}(1)}{4!}(x - 1)^4$$

$$= (x - 1) - \frac{1}{2}(x - 1)^2 + \frac{1}{3}(x - 1)^3 - \frac{1}{4}(x - 1)^4$$

Figure 8.14 compares the graphs of P_1, P_2, P_3, and P_4 with the graph of $f(x) = \ln x$. Note that near $x = 1$ the graphs are nearly indistinguishable. For instance, $P_4(0.9) \approx -0.105358$ and $\ln(0.9) \approx -0.105361$.

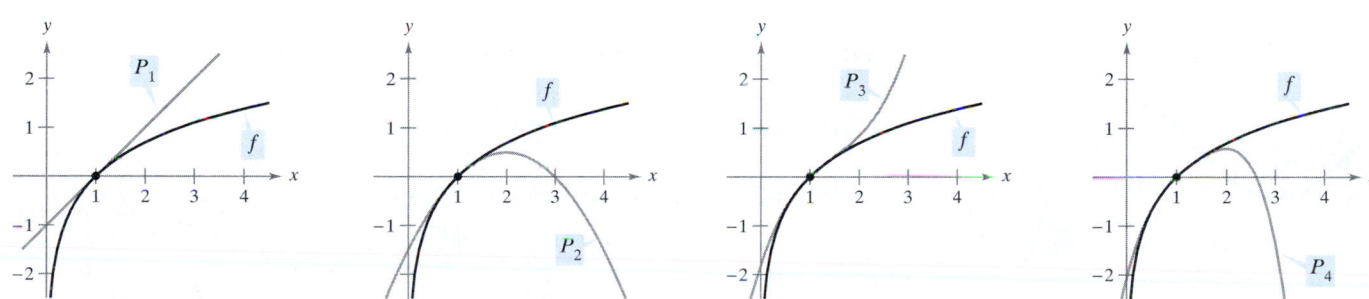

As n increases, the graph of P_n becomes a better and better approximation of the graph of $f(x) = \ln x$ near $x = 1$.
Figure 8.14

EXAMPLE 5   Finding Maclaurin Polynomials for cos x

Find the Maclaurin polynomials P_0, P_2, P_4, and P_6 for $f(x) = \cos x$. Use $P_6(x)$ to approximate the value of $\cos(0.1)$.

Solution   Expanding about $c = 0$ yields the following.

$$f(x) = \cos x \qquad\qquad f(0) = 1$$
$$f'(x) = -\sin x \qquad\qquad f'(0) = 0$$
$$f''(x) = -\cos x \qquad\qquad f''(0) = -1$$
$$f'''(x) = \sin x \qquad\qquad f'''(0) = 0$$

Through repeated differentiation, you can see that the pattern $1, 0, -1, 0$ continues, and you obtain the following Maclaurin polynomials.

$$P_0(x) = 1$$

$$P_2(x) = 1 - \frac{1}{2!}x^2$$

$$P_4(x) = 1 - \frac{1}{2!}x^2 + \frac{1}{4!}x^4$$

$$P_6(x) = 1 - \frac{1}{2!}x^2 + \frac{1}{4!}x^4 - \frac{1}{6!}x^6$$

Using $P_6(x)$, you obtain the approximation $\cos(0.1) \approx 0.995004165$, which coincides with the calculator value to nine decimal places. Figure 8.15 compares the graphs of $f(x) = \cos x$ and P_6.

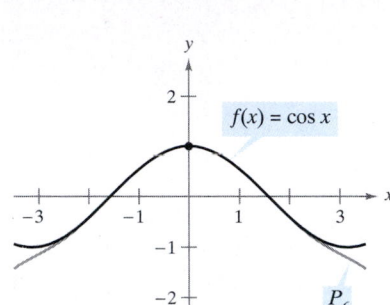

Near $(0, 1)$, the graph of P_6 can be used to approximate the graph of $f(x) = \cos x$.
Figure 8.15

Note in Example 5 that the Maclaurin polynomials for $\cos x$ have only even powers of x. Similarly, the Maclaurin polynomials for $\sin x$ have only odd powers of x (see Exercise 11). This is not generally true of the Taylor polynomials for $\sin x$ and $\cos x$ expanded about $c \neq 0$, as you can see in the next example.

EXAMPLE 6   Finding a Taylor Polynomial for sin x

Find the third Taylor polynomial for $f(x) = \sin x$, expanded about $c = \pi/6$.

Solution   Expanding about $c = \pi/6$ yields the following.

$$f(x) = \sin x \qquad\qquad f\left(\frac{\pi}{6}\right) = \frac{1}{2}$$

$$f'(x) = \cos x \qquad\qquad f'\left(\frac{\pi}{6}\right) = \frac{\sqrt{3}}{2}$$

$$f''(x) = -\sin x \qquad\qquad f''\left(\frac{\pi}{6}\right) = -\frac{1}{2}$$

$$f'''(x) = -\cos x \qquad\qquad f'''\left(\frac{\pi}{6}\right) = -\frac{\sqrt{3}}{2}$$

Thus, the third Taylor polynomial for $f(x) = \sin x$, expanded about $c = \pi/6$, is

$$P_3(x) = f\left(\frac{\pi}{6}\right) + f'\left(\frac{\pi}{6}\right)\left(x - \frac{\pi}{6}\right) + \frac{f''\left(\frac{\pi}{6}\right)}{2!}\left(x - \frac{\pi}{6}\right)^2 + \frac{f'''\left(\frac{\pi}{6}\right)}{3!}\left(x - \frac{\pi}{6}\right)^3$$

$$= \frac{1}{2} + \frac{\sqrt{3}}{2}\left(x - \frac{\pi}{6}\right) - \frac{1}{2(2!)}\left(x - \frac{\pi}{6}\right)^2 - \frac{\sqrt{3}}{2(3!)}\left(x - \frac{\pi}{6}\right)^3.$$

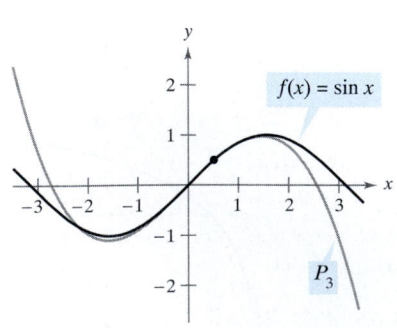

Near $(\pi/6, 1/2)$, the graph of P_3 can be used to approximate the graph of $f(x) = \sin x$.
Figure 8.16

Figure 8.16 compares the graphs of $f(x) = \sin x$ and P_3.

Taylor polynomials and Maclaurin polynomials can be used to approximate the value of a function at a specific point. For instance, to approximate the value of $\ln(1.1)$, you can use Taylor polynomials for $f(x) = \ln x$ expanded about $c = 1$, as shown in Example 4, or you can use Maclaurin polynomials, as shown in Example 7.

EXAMPLE 7   Approximation Using Maclaurin Polynomials

Use a fourth Maclaurin polynomial to approximate the value of $\ln(1.1)$.

Solution   Because 1.1 is closer to 1 than to 0, you should consider Maclaurin polynomials for the function $g(x) = \ln(1 + x)$.

$$
\begin{aligned}
g(x) &= \ln(1 + x) & g(0) &= 0 \\
g'(x) &= (1 + x)^{-1} & g'(0) &= 1 \\
g''(x) &= -(1 + x)^{-2} & g''(0) &= -1 \\
g'''(x) &= 2(1 + x)^{-3} & g'''(0) &= 2 \\
g^{(4)}(x) &= -6(1 + x)^{-4} & g^{(4)}(0) &= -6
\end{aligned}
$$

Note that you obtain the same coefficients as in Example 4. Therefore, the fourth Maclaurin polynomial for $g(x) = \ln(1 + x)$ is

$$P_4(x) = g(0) + g'(0)x + \frac{g''(0)}{2!}x^2 + \frac{g'''(0)}{3!}x^3 + \frac{g^{(4)}(0)}{4!}x^4$$

$$= x - \frac{1}{2}x^2 + \frac{1}{3}x^3 - \frac{1}{4}x^4.$$

Consequently,

$$\ln(1.1) = \ln(1 + 0.1) \approx P_4(0.1) \approx 0.0953083.$$

Check to see that the fourth Taylor polynomial (from Example 4), evaluated at $x = 1.1$, yields the same result.

n	$P_n(0.1)$
1	0.1000000
2	0.0950000
3	0.0953333
4	0.0953083

As n increases, the value of $P_n(0.1)$ gets closer and closer to the value of $\ln(1.1)$.

The table at the left illustrates the accuracy of the Taylor polynomial approximation of the calculator value of $\ln(1.1)$. You can see that as n becomes larger, $P_n(0.1)$ approaches the calculator value of 0.0953102.

On the other hand, the table below illustrates that as you move away from the expansion point $c = 1$, the accuracy of the approximation decreases.

Fourth Taylor Polynomial Approximation of $\ln(1 + x)$

x	0.0	0.1	0.5	0.75	1.0
$\ln(1 + x)$	0.0000000	0.0953102	0.4054651	0.5596158	0.6931472
$P_4(x)$	0.0000000	0.0953083	0.4010417	0.5302734	0.5833333

These two tables illustrate two very important points about the accuracy of Taylor (or Maclaurin) polynomials for use in approximations.

1. The approximation is usually better at x-values close to c than at x-values far from c.

2. The approximation is usually better for higher-degree Taylor (or Maclaurin) polynomials than for those of lower degree.

Remainder of a Taylor Polynomial

An approximation technique is of little value without some idea of its accuracy. To measure the accuracy of approximating a function value $f(x)$ by the Taylor polynomial $P_n(x)$, you can use the concept of a **remainder** $R_n(x)$, defined as follows.

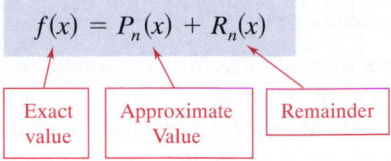

$$f(x) = P_n(x) + R_n(x)$$

Exact value   Approximate Value   Remainder

Thus, $R_n(x) = f(x) - P_n(x)$. The absolute value of $R_n(x)$ is called the **error** associated with the approximation. That is,

$$\text{Error} = |R_n(x)| = |f(x) - P_n(x)|.$$

The next theorem gives a general procedure for estimating the remainder associated with a Taylor polynomial. This important theorem is called **Taylor's Theorem,** and the remainder given in the theorem is called the **Lagrange form of the remainder.** (The proof of the theorem is lengthy, and is given in the appendix.)

THEOREM 8.19   Taylor's Theorem

If a function f is differentiable through order $n + 1$ in an interval I containing c, then, for each x in I, there exists z between x and c such that

$$f(x) = f(c) + f'(c)(x - c) + \frac{f''(c)}{2!}(x - c)^2 + \cdots + \frac{f^{(n)}(c)}{n!}(x - c)^n + R_n(x)$$

where

$$R_n(x) = \frac{f^{(n+1)}(z)}{(n+1)!}(x - c)^{n+1}.$$

NOTE   For $n = 0$, Taylor's Theorem states that if f is differentiable in an interval I containing c, then, for each x in I, there exists z between x and c such that

$$f(x) = f(c) + f'(z)(x - c)$$

or

$$f'(z) = \frac{f(x) - f(c)}{x - c}.$$

Do you recognize this special case of Taylor's Theorem? (It is the Mean Value Theorem.)

When applying Taylor's Theorem, you should not expect to be able to find the exact value of z. (If you could do this, an approximation would not be necessary.) Rather, you try to find bounds for $f^{(n+1)}(z)$ from which you are able to tell how large the remainder $R_n(x)$ is.

EXAMPLE 8   Determining the Accuracy of an Approximation

The third Maclaurin polynomial for $\sin x$ is given by

$$P_3(x) = x - \frac{x^3}{3!}.$$

Use Taylor's Theorem to approximate $\sin(0.1)$ by $P_3(0.1)$ and determine the accuracy of the approximation.

Solution   Using Taylor's Theorem, you have

$$\sin x = x - \frac{x^3}{3!} + R_3(x) = x - \frac{x^3}{3!} + \frac{f^{(4)}(z)}{4!} x^4$$

where $0 < z < 0.1$. Therefore,

$$\sin(0.1) \approx 0.1 - \frac{(0.1)^3}{3!} \approx 0.1 - 0.000167 = 0.099833.$$

Because $f^{(4)}(z) = \sin z$, it follows that the error $|R_3(0.1)|$ can be bounded as follows.

$$0 < R_3(0.1) = \frac{\sin z}{4!} (0.1)^4 < \frac{0.0001}{4!} \approx 0.000004$$

This implies that

$$0.099833 < \sin(0.1) = 0.099833 + R_3(x) < 0.099833 + 0.000004$$
$$0.099833 < \sin(0.1) < 0.099837.$$

NOTE   Try using a calculator to verify the results obtained in Examples 8 and 9. For Example 8, you obtain

$$\sin(0.1) \approx 0.0998334.$$

For Example 9, you obtain

$$P_3(1.2) \approx 0.1827$$

and

$$\ln(1.2) \approx 0.1823.$$

EXAMPLE 9   Approximating a Value to a Desired Accuracy

Determine the degree of the Taylor polynomial $P_n(x)$ expanded about $c = 1$ that should be used to approximate $\ln(1.2)$ so that the error is less than 0.001.

Solution   Following the pattern of Example 4, you can see that the $(n + 1)$st derivative of $f(x) = \ln x$ is given by

$$f^{(n+1)}(x) = (-1)^n \frac{n!}{x^{n+1}}.$$

Using Taylor's Theorem, you know that the error $|R_n(1.2)|$ is given by

$$|R_n(1.2)| = \left| \frac{f^{(n+1)}(z)}{(n+1)!} (1.2 - 1)^{n+1} \right| = \frac{n!}{z^{n+1}} \left[\frac{1}{(n+1)!} \right] (0.2)^{n+1}$$

$$= \frac{(0.2)^{n+1}}{z^{n+1}(n+1)}$$

where $1 < z < 1.2$. In this interval, $|R_n(1.2)|$ is largest when $z = 1$. Thus, you are seeking a value of n such that

$$\frac{(0.2)^{n+1}}{(1)^{n+1}(n+1)} < 0.001 \quad \Longrightarrow \quad 1000 < (n+1)5^{n+1}.$$

By trial and error, you can determine that the smallest value of n that satisfies this inequality is $n = 3$. Thus, you would need the third Taylor polynomial to achieve the desired accuracy in approximating $\ln(1.2)$.

EXERCISES FOR SECTION 8.7

In Exercises 1–4, match the Taylor polynomial approximation of the function $f(x) = e^{-x^2/2}$ with the correct graph. [The graphs are labeled (a), (b), (c), and (d).]

(a)
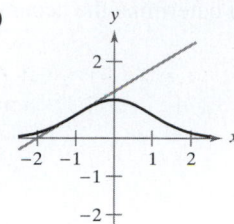

(b)
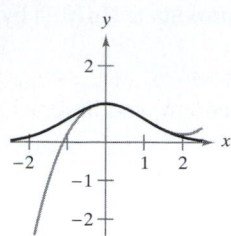

(c)
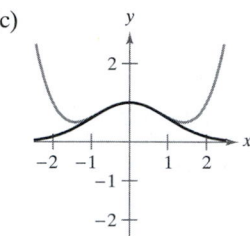

(d)
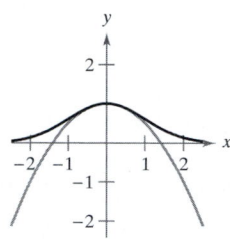

1. $g(x) = -\frac{1}{2}x^2 + 1$

2. $g(x) = \frac{1}{8}x^4 - \frac{1}{2}x^2 + 1$

3. $g(x) = e^{-1/2}[(x + 1) + 1]$

4. $g(x) = e^{-1/2}\left[\frac{1}{3}(x - 1)^3 - (x - 1) + 1\right]$

 5. **Conjecture** Consider the function $f(x) = \cos x$ and its Maclaurin polynomials $P_2, P_4,$ and P_6 (see Example 5).

(a) Use a graphing utility to graph f and the indicated polynomial approximations.

(b) Evaluate and compare the values of $f^{(n)}(0)$ and $P_n^{(n)}(0)$ for $n = 2, 4,$ and 6.

(c) Use the results in part (b) to make a conjecture about $f^{(n)}(0)$ and $P_n^{(n)}(0)$.

6. **Conjecture** Consider the function $f(x) = x^2e^x$.

(a) Find the Maclaurin polynomials $P_2, P_3,$ and P_4 for f.

(b) Use a graphing utility to graph $f, P_2, P_3,$ and P_4.

(c) Evaluate and compare the values of $f^{(n)}(0)$ and $P_n^{(n)}(0)$ for $n = 2, 3,$ and 4.

(d) Use the results in part (c) to make a conjecture about $f^{(n)}(0)$ and $P_n^{(n)}(0)$.

In Exercises 7–16, find the Maclaurin polynomial of degree n for the function.

7. $f(x) = e^{-x}, \quad n = 3$

8. $f(x) = e^{-x}, \quad n = 5$

9. $f(x) = e^{2x}, \quad n = 4$

10. $f(x) = e^{3x}, \quad n = 4$

11. $f(x) = \sin x, \quad n = 5$

12. $f(x) = \sin \pi x, \quad n = 3$

13. $f(x) = xe^x, \quad n = 4$

14. $f(x) = x^2e^{-x}, \quad n = 4$

15. $f(x) = \dfrac{1}{x + 1}, \quad n = 4$

16. $f(x) = \sec x, \quad n = 2$

In Exercises 17–20, find the nth Taylor polynomial centered at c.

17. $f(x) = \dfrac{1}{x}, \quad n = 4, \ c = 1$

18. $f(x) = \sqrt{x}, \quad n = 4, \ c = 4$

19. $f(x) = \ln x, \quad n = 4, \ c = 1$

20. $f(x) = x^2 \cos x, \quad n = 2, \ c = \pi$

 In Exercises 21 and 22, use a symbolic differentiation utility to find the indicated Taylor polynomials for the function f. Graph the function and the Taylor polynomials.

21. $f(x) = \tan x$

(a) $n = 3, \ c = 0$

(b) $n = 5, \ c = 0$

(c) $n = 3, \ c = \pi/4$

22. $f(x) = \dfrac{1}{x^2 + 1}$

(a) $n = 2, \ c = 0$

(b) $n = 4, \ c = 0$

(c) $n = 4, \ c = 1$

23. **Numerical and Graphical Approximations**

(a) Use the Maclaurin polynomials, $P_1(x), P_3(x), P_5(x),$ and $P_7(x)$ for $f(x) = \sin x$ to complete the table.

x	0	0.25	0.50	0.75	1.00
$\sin x$	0	0.2474	0.4794	0.6816	0.8415
$P_1(x)$					
$P_3(x)$					
$P_5(x)$					
$P_7(x)$					

(b) Use a graphing utility to graph $f(x) = \sin x$ and the Maclaurin polynomials in part (a).

(c) Describe the change in accuracy of a polynomial approximation as the distance from the point where the polynomial is centered increases.

24. **Numerical and Graphical Approximations**

(a) Use the Taylor polynomials $P_1(x)$ and $P_4(x)$ for $f(x) = \ln x$ centered at $c = 1$ to complete the table.

x	1.00	1.25	1.50	1.75	2.00
$\ln x$	0	0.2231	0.4055	0.5596	0.6931
$P_1(x)$					
$P_4(x)$					

(b) Use a graphing utility to graph $f(x) = \ln x$ and the Taylor polynomials in part (a).

(c) Describe the change in accuracy of polynomial approximations as the degree increases.

Numerical and Graphical Approximations **In Exercises 25 and 26, (a) find the Maclaurin polynomial $P_3(x)$ for $f(x)$, (b) complete the table for $f(x)$ and $P_3(x)$, and (c) sketch the graphs of $f(x)$ and $P_3(x)$ on the same set of coordinate axes.**

x	-0.75	-0.50	-0.25	0	0.25	0.50	0.75
$f(x)$							
$P_3(x)$							

25. $f(x) = \arcsin x$ **26.** $f(x) = \arctan x$

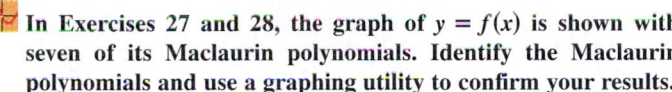 **In Exercises 27 and 28, the graph of $y = f(x)$ is shown with seven of its Maclaurin polynomials. Identify the Maclaurin polynomials and use a graphing utility to confirm your results.**

27.
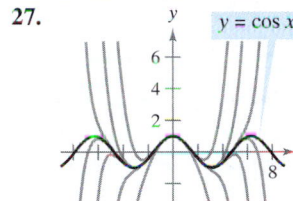
$y = \cos x$

28.
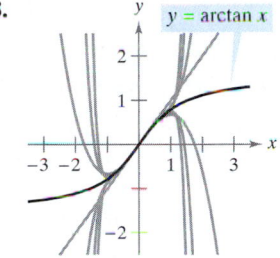
$y = \arctan x$

In Exercises 29–32, approximate the function at the given value of x, using the polynomial found in the indicated exercise.

29. $f(x) = e^{-x}$, $f\left(\frac{1}{2}\right)$, Exercise 7

30. $f(x) = x^2 e^{-x}$, $f\left(\frac{1}{4}\right)$, Exercise 14

31. $f(x) = \ln x$, $f(1.2)$, Exercise 19

32. $f(x) = x^2 \cos x$, $f\left(\frac{7\pi}{8}\right)$, Exercise 20

In Exercises 33–36, use Taylor's Theorem to determine the accuracy of the approximation.

33. $\cos(0.3) \approx 1 - \dfrac{(0.3)^2}{2!} + \dfrac{(0.3)^4}{4!}$

34. $e \approx 1 + 1 + \dfrac{1^2}{2!} + \dfrac{1^3}{3!} + \dfrac{1^4}{4!} + \dfrac{1^5}{5!}$

35. $\arcsin(0.4) \approx 0.4 + \dfrac{(0.4)^3}{2 \cdot 3}$

36. $\arctan(0.5) \approx 0.5 - \dfrac{(0.5)^3}{3}$

In Exercises 37 and 38, determine the degree of the Maclaurin polynomial required for the error in the approximation of the function at the indicated value of x to be less than 0.001.

37. $\sin(0.3)$

38. $e^{0.75}$

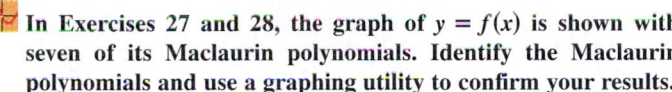 **In Exercises 39 and 40, determine the degree of the Maclaurin polynomial required for the error in the approximation of the function at the indicated value of x to be less than 0.0001. Use a symbolic differentiation utility to obtain and evaluate the required derivatives.**

39. $f(x) = \ln(x + 1)$,   approximate $f(1.5)$.

40. $f(x) = \cos(\pi x^2)$,   approximate $f(0.6)$.

In Exercises 41 and 42, determine the values of x for which the function can be replaced by the Taylor Polynomial if the error cannot exceed 0.001.

41. $f(x) = e^x \approx 1 + x + \dfrac{x^2}{2!} + \dfrac{x^3}{3!}$, $x < 0$

42. $f(x) = \sin x \approx x - \dfrac{x^3}{3!}$

43. *Comparing Maclaurin Polynomials*

(a) Compare the Maclaurin polynomials of degree 4 and degree 5, respectively, for the functions
$$f(x) = e^x \quad \text{and} \quad g(x) = xe^x.$$
What is the relationship between them?

(b) Use the result in part (a) and the Maclaurin polynomial of degree 5 for $f(x) = \sin x$ to find a Maclaurin polynomial of degree 6 for the function $g(x) = x \sin x$.

(c) Use the result in part (a) and the Maclaurin polynomial of degree 5 for $f(x) = \sin x$ to find a Maclaurin polynomial of degree 4 for the function $g(x) = (\sin x)/x$.

44. *Differentiating Maclaurin Polynomials*

(a) Differentiate the Maclaurin polynomial of degree 5 for $f(x) = \sin x$ and compare the result with the Maclaurin polynomial of degree 4 for $g(x) = \cos x$.

(b) Differentiate the Maclaurin polynomial of degree 6 for $f(x) = \cos x$ and compare the result with the Maclaurin polynomial of degree 5 for $g(x) = \sin x$.

(c) Differentiate the Maclaurin polynomial of degree 4 for $f(x) = e^x$. Describe the relationship between the two series.

45. Prove that if f is an odd function then, its nth Maclaurin polynomial contains only terms with odd powers of x.

46. Prove that if f is an even function then, its nth Maclaurin polynomial contains only terms with even powers of x.

47. Let $P_n(x)$ be the nth Taylor polynomial for f at c. Prove that $P_n(c) = f(c)$ and $P^{(k)}(c) = f^{(k)}(c)$ for $1 \le k \le n$. (See Exercises 5 and 6.)

48. *Writing*   The proof in Exercise 47 guarantees that the Taylor polynomial and its derivatives agree with the function and its derivatives at $x = c$. Use the graphs and tables in Exercises 23–26 to discuss what happens to the accuracy of the Taylor polynomial as you move away from $x = c$.

SECTION 8.8 Power Series

Power Series • Radius and Interval of Convergence • Endpoint Convergence •
Differentiation and Integration of Power Series

Power Series

In Section 8.7, we introduced the concept of approximating functions by Taylor polynomials. For instance, the function $f(x) = e^x$ can be *approximated* by its Maclaurin polynomials as follows.

$e^x \approx 1 + x$	1st-degree polynomial
$e^x \approx 1 + x + \dfrac{x^2}{2}$	2nd-degree polynomial
$e^x \approx 1 + x + \dfrac{x^2}{2} + \dfrac{x^3}{3!}$	3rd-degree polynomial
$e^x \approx 1 + x + \dfrac{x^2}{2} + \dfrac{x^3}{3!} + \dfrac{x^4}{4!}$	4th-degree polynomial
$e^x \approx 1 + x + \dfrac{x^2}{2} + \dfrac{x^3}{3!} + \dfrac{x^4}{4!} + \dfrac{x^5}{5!}$	5th-degree polynomial

In that section, you saw that the higher the degree of the approximating polynomial, the better the approximation becomes.

In this and the next two sections, you will see that several important types of functions, including

$$f(x) = e^x$$

can be represented *exactly* by an infinite series called a **power series.** For example, the power series representation for e^x is

$$e^x = 1 + x + \frac{x^2}{2} + \frac{x^3}{3!} + \cdots + \frac{x^n}{n!} + \cdots.$$

For each real number x, it can be shown that the infinite series on the right converges to the number e^x. Before doing this, however, we will discuss some preliminary results dealing with power series—beginning with the following definition.

EXPLORATION

Graphical Reasoning Use a graphing utility to approximate the graphs of the following power series near $x = 0$. (Use the first several terms of each series.) Each series represents a well-known function. What is the function?

(a) $\displaystyle\sum_{n=0}^{\infty} \frac{(-1)^n x^n}{n!}$

(b) $\displaystyle\sum_{n=0}^{\infty} \frac{(-1)^n x^{2n}}{(2n)!}$

(c) $\displaystyle\sum_{n=0}^{\infty} \frac{(-1)^n x^{2n+1}}{(2n+1)!}$

(d) $\displaystyle\sum_{n=0}^{\infty} \frac{(-1)^n x^{2n+1}}{2n+1}$

(e) $\displaystyle\sum_{n=0}^{\infty} \frac{2^n x^n}{n!}$

Definition of Power Series

If x is a variable, then an infinite series of the form

$$\sum_{n=0}^{\infty} a_n x^n = a_0 + a_1 x + a_2 x^2 + a_3 x^3 + \cdots + a_n x^n + \cdots$$

is called a **power series.** More generally, series of the form

$$\sum_{n=0}^{\infty} a_n (x - c)^n = a_0 + a_1(x - c) + a_2(x - c)^2 + \cdots + a_n(x - c)^n + \cdots$$

is called a **power series centered at c,** where c is a constant.

NOTE To simplify the notation for power series, we agree that $(x - c)^0 = 1$, even if $x = c$.

EXAMPLE 1   Power Series

a. The following power series is centered at 0.

$$\sum_{n=0}^{\infty} \frac{x^n}{n!} = 1 + x + \frac{x^2}{2} + \frac{x^3}{3!} + \cdots$$

b. The following power series is centered at -1.

$$\sum_{n=0}^{\infty} (-1)^n (x + 1)^n = 1 - (x + 1) + (x + 1)^2 - (x + 1)^3 + \cdots$$

c. The following power series is centered at 1.

$$\sum_{n=1}^{\infty} \frac{1}{n} (x - 1)^n = (x - 1) + \frac{1}{2} (x - 1)^2 + \frac{1}{3} (x - 1)^3 + \cdots$$

Radius and Interval of Convergence

A power series in x can be viewed as a function of x

$$f(x) = \sum_{n=0}^{\infty} a_n(x - c)^n$$

where the *domain of f* is the set of all x for which the power series converges. Determination of the domain of a power series is the primary concern in this section. Of course, every power series converges at its center c because

$$f(c) = \sum_{n=0}^{\infty} a_n(c - c)^n$$

$$= a_0(1) + 0 + 0 + \cdots + 0 + \cdots$$

$$= a_0.$$

Thus, c always lies in the domain of f. The following important theorem (which we present without proof) states that the domain of a power series can take three basic forms: a single point, an interval centered at c, or the entire real line, as shown in Figure 8.17.

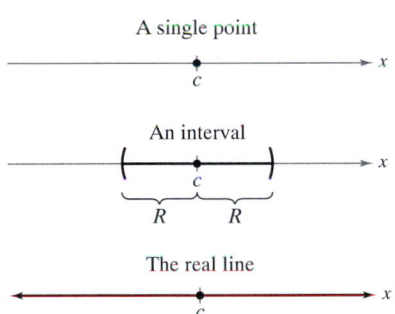

A single point

An interval

The real line

The domain of a power series has only three basic forms: a single point, an interval centered at c, or the entire real line.

Figure 8.17

STUDY TIP To determine the radius of convergence of a power series, use the Ratio Test, as demonstrated in Examples 2, 3, and 4.

THEOREM 8.20   Convergence of a Power Series

For a power series centered at c, precisely one of the following is true.

1. The series converges only at c.

2. There exists a real number $R > 0$ such that the series converges absolutely for $|x - c| < R$, and diverges for $|x - c| > R$.

3. The series converges absolutely for all x.

The number R is the **radius of convergence** of the power series. If the series converges only at c, the radius of convergence is $R = 0$, and if the series converges for all x, the radius of convergence is $R = \infty$. The set of all values of x for which the power series converges is the **interval of convergence** of the power series.

EXAMPLE 2   Finding the Radius of Convergence

Find the radius of convergence of $\displaystyle\sum_{n=0}^{\infty} n!x^n$.

Solution   For $x = 0$, you obtain

$$f(0) = \sum_{n=0}^{\infty} n!0^n = 1 + 0 + 0 + \cdots = 1.$$

For any fixed value of x such that $|x| > 0$, let $u_n = n!x^n$. Then

$$\lim_{n\to\infty} \left|\frac{u_{n+1}}{u_n}\right| = \lim_{n\to\infty} \left|\frac{(n+1)!x^{n+1}}{n!x^n}\right|$$

$$= |x| \lim_{n\to\infty} (n+1)$$

$$= \infty.$$

Therefore, by the Ratio Test, the series diverges for $|x| > 0$ and converges only at its center, 0. Hence, the radius of convergence is $R = 0$.

EXAMPLE 3   Finding the Radius of Convergence

Find the radius of convergence of $\displaystyle\sum_{n=0}^{\infty} 3(x-2)^n$.

Solution   For $x \neq 2$, let $u_n = 3(x-2)^n$. Then

$$\lim_{n\to\infty} \left|\frac{u_{n+1}}{u_n}\right| = \lim_{n\to\infty} \left|\frac{3(x-2)^{n+1}}{3(x-2)^n}\right|$$

$$= \lim_{n\to\infty} |x-2|$$

$$= |x-2|.$$

By the Ratio Test, the series converges if $|x-2| < 1$ and diverges if $|x-2| > 1$. Therefore, the radius of convergence of the series is $R = 1$.

EXAMPLE 4   Finding the Radius of Convergence

Find the radius of convergence of $\displaystyle\sum_{n=0}^{\infty} \frac{(-1)^n x^{2n+1}}{(2n+1)!}$.

Solution   Let $u_n = (-1)^n x^{2n+1}/(2n+1)!$. Then

$$\lim_{n\to\infty} \left|\frac{u_{n+1}}{u_n}\right| = \lim_{n\to\infty} \left|\frac{[(-1)^{n+1} x^{2n+3}]/(2n+3)!}{[(-1)^n x^{2n+1}]/(2n+1)!}\right|$$

$$= \lim_{n\to\infty} \frac{x^2}{(2n+3)(2n+2)}.$$

For any *fixed* value of x, this limit is 0. Thus, by the Ratio Test, the series converges for all x. Therefore, the radius of convergence is $R = \infty$.

Endpoint Convergence

Note that for a power series whose radius of convergence is a finite number R, Theorem 8.20 says nothing about the convergence at the *endpoints* of the interval of convergence. Each endpoint must be tested separately for convergence or divergence. As a result, the interval of convergence of a power series can take any one of the six forms shown in Figure 8.18.

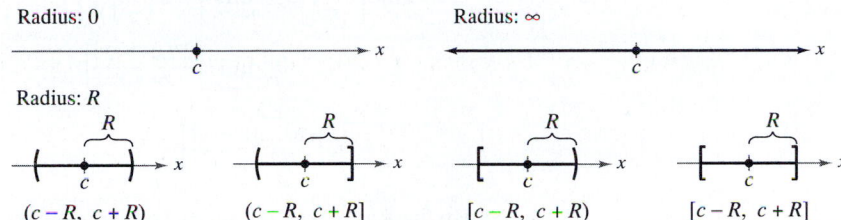

Radius: 0   Radius: ∞

Radius: R

$(c - R,\ c + R)$ $(c - R,\ c + R]$ $[c - R,\ c + R)$ $[c - R,\ c + R]$

Intervals of convergence
Figure 8.18

EXAMPLE 5   Finding the Interval of Convergence

Find the interval of convergence of $\displaystyle\sum_{n=1}^{\infty} \frac{x^n}{n}$.

Solution   Letting $u_n = x^n/n$ produces

$$\lim_{n \to \infty} \left| \frac{u_{n+1}}{u_n} \right| = \lim_{n \to \infty} \left| \frac{x^{n+1}/(n+1)}{x^n/n} \right|$$

$$= \lim_{n \to \infty} \left| \frac{nx}{n+1} \right|$$

$$= |x|.$$

Therefore, by the Ratio Test, the radius of convergence is $R = 1$. Moreover, because the series is centered at 0, it converges in the interval $(-1, 1)$. This interval, however, is not necessarily the *interval of convergence*. To determine this, you must test for convergence at each endpoint. When $x = 1$, you obtain the *divergent* harmonic series

$$\sum_{n=1}^{\infty} \frac{1}{n} = \frac{1}{1} + \frac{1}{2} + \frac{1}{3} + \cdots. \qquad \text{Diverges when } x = 1$$

When $x = -1$, you obtain the *convergent* alternating harmonic series

$$\sum_{n=1}^{\infty} \frac{(-1)^n}{n} = -1 + \frac{1}{2} - \frac{1}{3} + \frac{1}{4} - \cdots. \qquad \text{Converges when } x = -1$$

Therefore, the interval of convergence for the series is $[-1, 1)$, as shown in Figure 8.19.

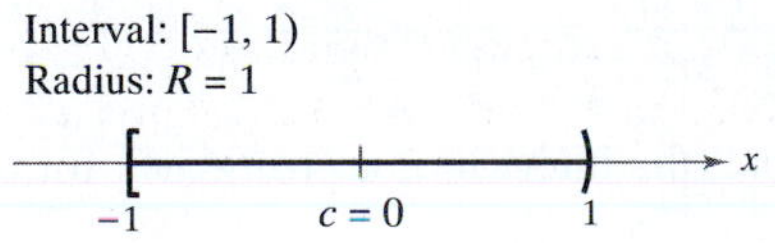

Interval: $[-1, 1)$
Radius: $R = 1$

$$-1 \qquad c = 0 \qquad 1$$

Figure 8.19

EXAMPLE 6 **Finding the Interval of Convergence**

Find the interval of convergence of $\displaystyle\sum_{n=0}^{\infty} \frac{(-1)^n(x+1)^n}{2^n}$.

Solution Letting $u_n = (-1)^n(x+1)^n/2^n$ produces

$$\lim_{n\to\infty} \left| \frac{u_{n+1}}{n} \right| = \lim_{n\to\infty} \left| \frac{(-1)^{n+1}(x+1)^{n+1}/2^{n+1}}{(-1)^n(x+1)^n/2^n} \right|$$

$$= \lim_{n\to\infty} \left| \frac{2^n(x+1)}{2^{n+1}} \right|$$

$$= \left| \frac{x+1}{2} \right|.$$

By the Ratio Test, the series converges if $|(x+1)/2| < 1$ or $|x+1| < 2$. Hence, the radius of convergence is $R = 2$. Because the series is centered at $x = -1$, it will converge in the interval $(-3, 1)$. Furthermore, at the endpoints you have

$$\sum_{n=0}^{\infty} \frac{(-1)^n(-2)^n}{2^n} = \sum_{n=0}^{\infty} \frac{2^n}{2^n} = \sum_{n=0}^{\infty} 1 \qquad \text{\color{red}Diverges when } x = -3$$

and

$$\sum_{n=0}^{\infty} \frac{(-1)^n(2)^n}{2^n} = \sum_{n=0}^{\infty} (-1)^n \qquad \text{\color{red}Diverges when } x = 1$$

both of which diverge. Thus, the interval of convergence is $(-3, 1)$, as shown in Figure 8.20.

Interval: $(-3, 1)$
Radius: $R = 2$

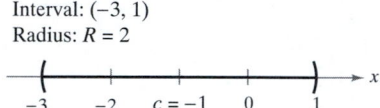

Figure 8.20

EXAMPLE 7 **Finding the Interval of Convergence**

Find the interval of convergence of $\displaystyle\sum_{n=1}^{\infty} \frac{x^n}{n^2}$.

Solution Letting $u_n = x^n/n^2$ produces

$$\lim_{n\to\infty} \left| \frac{u_{n+1}}{u_n} \right| = \lim_{n\to\infty} \left| \frac{x^{n+1}/(n+1)^2}{x^n/n^2} \right|$$

$$= \lim_{n\to\infty} \left| \frac{n^2 x}{(n+1)^2} \right| = |x|.$$

Thus, the radius of convergence is $R = 1$. Because the series is centered at $x = 0$, it converges in the interval $(-1, 1)$. When $x = 1$, you obtain the *convergent p*-series

$$\sum_{n=1}^{\infty} \frac{1}{n^2} = \frac{1}{1^2} + \frac{1}{2^2} + \frac{1}{3^2} + \frac{1}{4^2} + \cdots. \qquad \text{\color{red}Converges when } x = 1$$

When $x = -1$, you obtain the *convergent* alternating series

$$\sum_{n=1}^{\infty} \frac{(-1)^n}{n^2} = -\frac{1}{1^2} + \frac{1}{2^2} - \frac{1}{3^2} + \frac{1}{4^2} - \cdots. \qquad \text{\color{red}Converges when } x = -1$$

Therefore, the interval of convergence for the given series is $[-1, 1]$.

Differentiation and Integration of Power Series

Power series representation of functions has played an important role in the development of calculus. In fact, much of Newton's work with differentiation and integration was done in the context of power series—especially his work with complicated algebraic functions and transcendental functions. Euler, Lagrange, Leibniz, and the Bernoullis all used power series extensively in calculus.

Once you have defined a function with a power series, it is natural to wonder how you can determine the characteristics of the function. Is it continuous? Differentiable? Integrable? Theorem 8.21, which we state without proof, answers these questions.

JAMES GREGORY (1638–1675)

One of the earliest mathematicians to work with power series was a Scotsman, James Gregory. He developed a power series method for interpolating table values—a method that was later used by Brook Taylor in the development of Taylor polynomials and Taylor series, covered in Section 8.10.

THEOREM 8.21   Properties of Functions Defined by Power Series

If the function given by

$$f(x) = \sum_{n=0}^{\infty} a_n(x-c)^n = a_0 + a_1(x-c) + a_2(x-c)^2 + a_3(x-c)^3 + \cdots$$

has a radius of convergence of $R > 0$, then, on the interval $(c - R, c + R)$, f is continuous, differentiable, and integrable. Moreover, the derivative and antiderivative of f are as follows.

1. $f'(x) = \sum_{n=1}^{\infty} na_n(x-c)^{n-1} = a_1 + 2a_2(x-c) + 3a_3(x-c)^2 + \cdots$

2. $\int f(x)\, dx = C + \sum_{n=0}^{\infty} a_n \frac{(x-c)^{n+1}}{n+1}$

$$= C + a_0(x-c) + a_1 \frac{(x-c)^2}{2} + a_2 \frac{(x-c)^3}{3} + \cdots$$

The *radius of convergence* of the series obtained by differentiating or integrating a power series is the same as that of the original power series. The *interval of convergence*, however, may differ as a result of the behavior at the endpoints.

Theorem 8.21 states that, in many ways, a function defined by a power series behaves like a polynomial. It is continuous in its interval of convergence, and both its derivative and its antiderivative can be determined by differentiating and integrating each term of the given power series. For instance, the derivative of the power series

$$f(x) = \sum_{n=0}^{\infty} \frac{x^n}{n!} = 1 + x + \frac{x^2}{2} + \frac{x^3}{3!} + \frac{x^4}{4!} + \cdots$$

is

$$f'(x) = 1 + (2)\frac{x}{2} + (3)\frac{x^2}{3!} + (4)\frac{x^3}{4!} + \cdots$$

$$= 1 + x + \frac{x^2}{2} + \frac{x^3}{3!} + \frac{x^4}{4!} + \cdots$$

$$= f(x).$$

Notice that $f'(x) = f(x)$. Do you recognize this function?

EXAMPLE 8 **Intervals of Convergence for $f(x)$, $f'(x)$, and $\int f(x)\, dx$**

Consider the function given by

$$f(x) = \sum_{n=1}^{\infty} \frac{x^n}{n} = x + \frac{x^2}{2} + \frac{x^3}{3} + \cdots.$$

Find the intervals of convergence for each of the following.

a. $\int f(x)\, dx$ **b.** $f(x)$ **c.** $f'(x)$

Solution By Theorem 8.21, you have

$$f'(x) = \sum_{n=1}^{\infty} x^{n-1}$$

$$= 1 + x + x^2 + x^3 + \cdots$$

and

$$\int f(x)\, dx = C + \sum_{n=1}^{\infty} \frac{x^{n+1}}{n(n+1)}$$

$$= C + \frac{x^2}{1 \cdot 2} + \frac{x^3}{2 \cdot 3} + \frac{x^4}{3 \cdot 4} + \cdots.$$

By the Ratio Test, you can show that each series has a radius of convergence of $R = 1$. Considering the interval $(-1, 1)$, you have the following.

a. For $\int f(x)\, dx$, the series

$$\sum_{n=1}^{\infty} \frac{x^{n+1}}{n(n+1)} \qquad \text{\color{red} Interval of convergence: } [-1, 1]$$

converges for $x = \pm 1$, and its interval of convergence is $[-1, 1]$.

b. For $f(x)$, the series

$$\sum_{n=1}^{\infty} \frac{x^n}{n} \qquad \text{\color{red} Interval of convergence: } [-1, 1)$$

converges for $x = -1$ and diverges for $x = 1$. Hence, its interval of convergence is $[-1, 1)$.

c. For $f'(x)$, the series

$$\sum_{n=1}^{\infty} x^{n-1} \qquad \text{\color{red} Interval of convergence: } (-1, 1)$$

diverges for $x = \pm 1$, and its interval of convergence is $(-1, 1)$.

From Example 8, it appears that of the three series, the one for the derivative, $f'(x)$, is the least likely to converge at the endpoints. In fact, it can be shown that if the series for $f'(x)$ converges at the endpoints $x = c \pm R$, the series for $f(x)$ will also converge there.

EXERCISES FOR SECTION 8.8

In Exercises 1–6, find the radius of convergence of the power series.

1. $\displaystyle\sum_{n=0}^{\infty} (-1)^n \frac{x^n}{n+1}$

2. $\displaystyle\sum_{n=0}^{\infty} (4x)^n$

3. $\displaystyle\sum_{n=1}^{\infty} \frac{(2x)^n}{n^2}$

4. $\displaystyle\sum_{n=0}^{\infty} \frac{(-1)^n x^n}{2^n}$

5. $\displaystyle\sum_{n=0}^{\infty} \frac{(2x)^n}{n!}$

6. $\displaystyle\sum_{n=0}^{\infty} \frac{(2n)! x^n}{n!}$

In Exercises 7–30, find the interval of convergence of the power series. (Be sure to include a check for convergence at the endpoints of the interval.)

7. $\displaystyle\sum_{n=0}^{\infty} \left(\frac{x}{2}\right)^n$

8. $\displaystyle\sum_{n=0}^{\infty} \left(\frac{x}{k}\right)^n, \quad k > 0$

9. $\displaystyle\sum_{n=1}^{\infty} \frac{(-1)^n x^n}{n}$

10. $\displaystyle\sum_{n=0}^{\infty} (-1)^{n+1} n x^n$

11. $\displaystyle\sum_{n=0}^{\infty} \frac{x^n}{n!}$

12. $\displaystyle\sum_{n=0}^{\infty} \frac{(3x)^n}{(2n)!}$

13. $\displaystyle\sum_{n=0}^{\infty} (2n)! \left(\frac{x}{2}\right)^n$

14. $\displaystyle\sum_{n=0}^{\infty} \frac{(-1)^n x^n}{(n+1)(n+2)}$

15. $\displaystyle\sum_{n=1}^{\infty} \frac{(-1)^{n+1} x^n}{4^n}$

16. $\displaystyle\sum_{n=0}^{\infty} \frac{(-1)^n n! (x-4)^n}{3^n}$

17. $\displaystyle\sum_{n=1}^{\infty} \frac{(-1)^{n+1} (x-5)^n}{n 5^n}$

18. $\displaystyle\sum_{n=0}^{\infty} \frac{(x-2)^{n+1}}{(n+1) 3^{n+1}}$

19. $\displaystyle\sum_{n=0}^{\infty} \frac{(-1)^{n+1} (x-1)^{n+1}}{n+1}$

20. $\displaystyle\sum_{n=1}^{\infty} \frac{(-1)^{n+1} (x-c)^n}{n c^n}$

21. $\displaystyle\sum_{n=1}^{\infty} \frac{(x-c)^{n-1}}{c^{n-1}}, \quad c > 0$

22. $\displaystyle\sum_{n=1}^{\infty} \frac{(-1)^{n+1} x^{2n-1}}{2n-1}$

23. $\displaystyle\sum_{n=1}^{\infty} \frac{n}{n+1} (-2x)^{n-1}$

24. $\displaystyle\sum_{n=0}^{\infty} \frac{(-1)^n x^{2n}}{n!}$

25. $\displaystyle\sum_{n=0}^{\infty} \frac{x^{2n+1}}{(2n+1)!}$

26. $\displaystyle\sum_{n=1}^{\infty} \frac{n! x^n}{(2n)!}$

27. $\displaystyle\sum_{n=1}^{\infty} \frac{k(k+1)(k+2)\cdots(k+n-1)x^n}{n!}, \quad k \geq 1$

28. $\displaystyle\sum_{n=1}^{\infty} \left[\frac{2 \cdot 4 \cdot 6 \cdots 2n}{3 \cdot 5 \cdot 7 \cdots (2n+1)}\right] x^{2n+1}$

29. $\displaystyle\sum_{n=1}^{\infty} \frac{(-1)^{n+1} 3 \cdot 7 \cdot 11 \cdots (4n-1)(x-3)^n}{4^n}$

30. $\displaystyle\sum_{n=1}^{\infty} \frac{n! (x-c)^n}{1 \cdot 3 \cdot 5 \cdots (2n-1)}$

In Exercises 31–34, find the intervals of convergence of (a) $f(x)$, (b) $f'(x)$, (c) $f''(x)$, and (d) $\int f(x)\,dx$. Include a check for convergence at the endpoints.

31. $f(x) = \displaystyle\sum_{n=0}^{\infty} \left(\frac{x}{2}\right)^n$

32. $f(x) = \displaystyle\sum_{n=1}^{\infty} \frac{(-1)^{n+1} (x-5)^n}{n 5^n}$

33. $f(x) = \displaystyle\sum_{n=0}^{\infty} \frac{(-1)^{n+1} (x-1)^{n+1}}{n+1}$

34. $f(x) = \displaystyle\sum_{n=1}^{\infty} \frac{(-1)^{n+1} (x-1)^n}{n}$

In Exercises 35–38, match the graph of the first ten terms of the sequence of partial sums of the series

$$g(x) = \sum_{n=0}^{\infty} \left(\frac{x}{3}\right)^n$$

with the indicated value of the function. [The graphs are labeled (a), (b), (c), and (d).] Explain how you made your choice.

(a)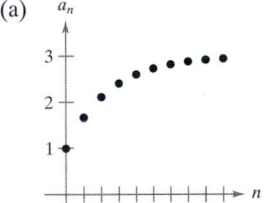

(b)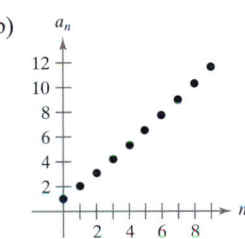

(c)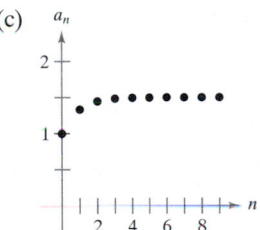

(d)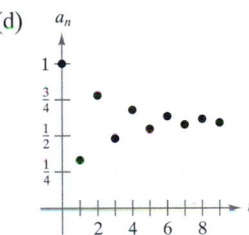

35. $g(1)$

36. $g(2)$

37. $g(3.1)$

38. $g(-2)$

39. Let

$$f(x) = \sum_{n=0}^{\infty} \frac{(-1)^n x^{2n+1}}{(2n+1)!} \quad \text{and} \quad g(x) = \sum_{n=0}^{\infty} \frac{(-1)^n x^{2n}}{(2n)!}.$$

(a) Find the intervals of convergence of f and g.

(b) Show that $f'(x) = g(x)$.

(c) Show that $g'(x) = -f(x)$.

(d) Identify the functions f and g.

40. Let

$$f(x) = \sum_{n=0}^{\infty} \frac{x^n}{n!}.$$

(a) Find the interval of convergence of f.

(b) Show that $f'(x) = f(x)$.

(c) Show that $f(0) = 1$.

(d) Identify the function f.

In Exercises 41 and 42, show that the function represented by the power series is a solution of the differential equation.

41. $y = \sum_{n=0}^{\infty} \frac{x^{2n}}{2^n n!}, \quad y'' - xy' - y = 0$

42. $y = 1 + \sum_{n=1}^{\infty} \frac{(-1)^n x^{4n}}{2^{2n} n! \cdot 3 \cdot 7 \cdot 11 \cdots (4n-1)}, \quad y'' + x^2 y = 0$

43. Bessel Function   The Bessel function of order 0 is

$$J_0(x) = \sum_{k=0}^{\infty} \frac{(-1)^k x^{2k}}{2^{2k} (k!)^2}.$$

(a) Show that the series converges for all x.

(b) Show that the series is a solution of the differential equation $x^2 J_0'' + x J_0' + x^2 J_0 = 0$.

(c) Use a graphing utility to graph the polynomial composed of the first four terms of J_0.

(d) Approximate $\int_0^1 J_0 \, dx$ accurate to two decimal places.

44. Bessel Function   The Bessel function of order 1 is

$$J_1(x) = x \sum_{k=0}^{\infty} \frac{(-1)^k x^{2k}}{2^{2k+1} k!(k+1)!}.$$

(a) Show that the series converges for all x.

(b) Show that the series is a solution of the differential equation

$$x^2 J_1'' + x J_1' + (x^2 - 1) J_1 = 0.$$

(c) Use a graphing utility to graph the polynomial composed of the first four terms of J_1.

(d) Show that $J_0'(x) = -J_1(x)$.

In Exercises 45–48, use a graphing utility to graph the power series. The series represents a well-known function. Identify the function from the graph of its power series.

45. $f(x) = \sum_{n=0}^{\infty} (-1)^n \frac{x^{2n}}{(2n)!}$

46. $f(x) = \sum_{n=0}^{\infty} (-1)^n \frac{x^{2n+1}}{(2n+1)!}$

47. $f(x) = \sum_{n=0}^{\infty} (-1)^n x^n, \quad -1 < x < 1$

48. $f(x) = \sum_{n=0}^{\infty} (-1)^n \frac{x^{2n+1}}{2n+1}, \quad -1 \le x \le 1$

49. Investigation   In Exercise 7 you found that the interval of convergence of the geometric series

$$\sum_{n=0}^{\infty} \left(\frac{x}{2}\right)^n$$

is $(-2, 2)$.

(a) Find the sum of the series when $x = \frac{3}{4}$. Use a graphing utility to graph the first six terms of the sequence of partial sums and the horizontal line representing the sum of the series.

(b) Repeat part (a) for $x = -\frac{3}{4}$.

(c) Write a short paragraph comparing the rate of convergence of the partial sums with the sum of the series in parts (a) and (b). How do the plots of the partial sums differ as they converge toward the sum of the series?

(d) Given any positive real number M, there exists a positive integer N such that the partial sum

$$\sum_{n=0}^{N} \left(\frac{3}{2}\right)^n > M.$$

Use a graphing utility to complete the table.

M	10	100	1000	10,000
N				

50. Write a series equivalent to

$$\sum_{n=0}^{\infty} \frac{x^{2n+1}}{(2n+1)!}$$

where the index of summation has been adjusted to begin at $n = 1$.

True or False?   **In Exercises 51–54, determine whether the statement is true or false. If it is false, explain why or give an example that shows it is false.**

51. If the power series $\sum_{n=0}^{\infty} a_n x^n$ converges for $x = 2$, then it also converges for $x = -2$.

52. If the power series $\sum_{n=0}^{\infty} a_n x^n$ converges for $x = 2$, then it also converges for $x = -1$.

53. If the interval of convergence for $\sum_{a=0}^{\infty} a_n x^n$ is $(-1, 1)$, then the interval of convergence for $\sum_{n=0}^{\infty} a_n (x-1)^2$ is $(0, 2)$.

54. If $f(x) = \sum_{n=0}^{\infty} a_n x^n$ converges for $|x| < 2$, then $\int_0^1 f(x) \, dx = \sum_{n=0}^{\infty} \frac{a_n}{n+1}$.

Geometric Power Series • Operations with Power Series

Geometric Power Series

In this section and the next, you will study several techniques for finding a power series that represents a given function.

Consider the function given by $f(x) = 1/(1 - x)$. The form of f closely resembles the sum of a geometric series

$$\sum_{n=0}^{\infty} ar^n = \frac{a}{1 - r}, \quad |r| < 1.$$

In other words, if you let $a = 1$ and $r = x$, a power series representation for $1/(1 - x)$, centered at 0, is

$$\frac{1}{1 - x} = \sum_{n=0}^{\infty} x^n$$

$$= 1 + x + x^2 + x^3 + \cdots, \quad |x| < 1.$$

Of course, this series represents $f(x) = 1/(1 - x)$ only on the interval $(-1, 1)$, whereas f is defined for all $x \neq 1$, as shown in Figure 8.21. To represent f in another interval, you must develop a different series. For instance, to obtain the power series centered at -1, you could write

$$\frac{1}{1 - x} = \frac{1}{2 - (x + 1)} = \frac{1/2}{1 - [(x + 1)/2]} = \frac{a}{1 - r}$$

which implies that $a = \frac{1}{2}$ and $r = (x + 1)/2$. Thus, for $|x + 1| < 2$, you have

$$\frac{1}{1 - x} = \sum_{n=0}^{\infty} \left(\frac{1}{2}\right)\left(\frac{x + 1}{2}\right)^n$$

$$= \frac{1}{2}\left[1 + \frac{(x + 1)}{2} + \frac{(x + 1)^2}{4} + \frac{(x + 1)^3}{8} + \cdots\right], \quad |x + 1| < 2$$

which converges on the interval $(-3, 1)$.

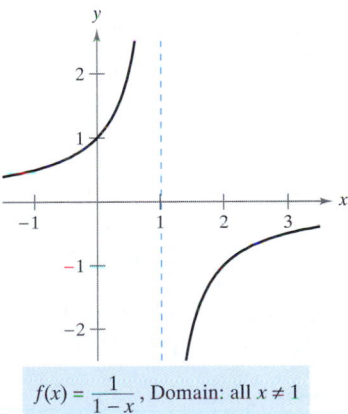

$f(x) = \dfrac{1}{1 - x}$, Domain: all $x \neq 1$ $f(x) = \displaystyle\sum_{n=0}^{\infty} x^n$, Domain: all $-1 < x < 1$

Figure 8.21

JOSEPH FOURIER (1768–1830)

Some of the early work in representing functions by power series was done by the French mathematician Joseph Fourier. Fourier's work is important in the history of calculus, partly because it forced eighteenth century mathematicians to question the then-prevailing narrow concept of a function. Both Cauchy and Dirichlet were motivated by Fourier's work with series, and in 1837 Dirichlet published the general definition of a function that is used today.

EXAMPLE 1   Finding a Geometric Power Series Centered at 0

Find a power series for $f(x) = \dfrac{4}{x + 2}$, centered at 0.

Solution   Writing $f(x)$ in the form $a/(1 - r)$ produces

$$\frac{4}{2 + x} = \frac{2}{1 - (-x/2)} = \frac{a}{1 - r}$$

which implies that $a = 2$ and $r = -x/2$. Thus, the power series for $f(x)$ is

$$\frac{4}{x + 2} = \sum_{n=0}^{\infty} ar^n$$

$$= \sum_{n=0}^{\infty} 2\left(-\frac{x}{2}\right)^n$$

$$= 2\left(1 - \frac{x}{2} + \frac{x^2}{4} - \frac{x^3}{8} + \cdots\right).$$

This power series converges when

$$\left|-\frac{x}{2}\right| < 1$$

which implies that the interval of convergence is $(-2, 2)$.

Long Division

$$\begin{array}{r}
2 - \ x + \frac{1}{2}x^2 - \frac{1}{4}x^3 + \cdots \\
2 + x \overline{)4} \\
\underline{4 + 2x} \\
- 2x \\
\underline{- 2x - \ x^2} \\
x^2 \\
\underline{x^2 + \frac{1}{2}x^3} \\
-\frac{1}{2}x^3 \\
\underline{-\frac{1}{2}x^3 - \frac{1}{4}x^4}
\end{array}$$

Another way to determine a power series for a rational function such as the one in Example 1 is to use long division. For instance, by dividing $2 + x$ into 4, you obtain the result shown at the left.

EXAMPLE 2   Finding a Geometric Power Series Centered at 1

Find a power series for $f(x) = \dfrac{1}{x}$, centered at 1.

Solution   Writing $f(x)$ in the form $a/(1 - r)$ produces

$$\frac{1}{x} = \frac{1}{1 - (-x + 1)} = \frac{a}{1 - r}$$

which implies that $a = 1$ and $r = 1 - x = -(x - 1)$. Thus, the power series for $f(x)$ is

$$\frac{1}{x} = \sum_{n=0}^{\infty} ar^n$$

$$= \sum_{n=0}^{\infty} [-(x - 1)]^n$$

$$= 1 - (x - 1) + (x - 1)^2 - (x - 1)^3 + \cdots.$$

This power series converges when

$$|x - 1| < 1$$

which implies that the interval of convergence is $(0, 2)$.

Operations with Power Series

The versatility of geometric power series will be shown later in this section, following a discussion of power series operations. These operations, used with differentiation and integration, provide a means of developing power series for a variety of elementary functions. (For simplicity, the following properties are stated for a series centered at 0.)

> **THEOREM 8.22   Operations with Power Series**
>
> Let $f(x) = \Sigma\, a_n x^n$ and $g(x) = \Sigma\, b_n x^n$.
>
> **1.** $f(kx) = \displaystyle\sum_{n=0}^{\infty} a_n k^n x^n$
>
> **2.** $f(x^N) = \displaystyle\sum_{n=0}^{\infty} a_n x^{nN}$
>
> **3.** $f(x) \pm g(x) = \displaystyle\sum_{n=0}^{\infty} (a_n \pm b_n) x^n$

The operations described in Theorem 8.22 can change the interval of convergence for the resulting series. For example, in the following addition, the interval of convergence for the sum is the *intersection* of the intervals of convergence of the two original series.

$$\underbrace{\sum_{n=0}^{\infty} x^n}_{(-1,\,1)} + \underbrace{\sum_{n=0}^{\infty} \left(\frac{x}{2}\right)^n}_{(-2,\,2)} = \underbrace{\sum_{n=0}^{\infty} \left(1 + \frac{1}{2^n}\right) x^n}_{(-1,\,1)}$$

$$(-1,\,1) \;\cap\; (-2,\,2) \;=\; (-1,\,1)$$

EXAMPLE 3   Adding Two Power Series

Find a power series, centered at 0, for $f(x) = \dfrac{3x - 1}{x^2 - 1}$.

Solution   Using partial fractions, you can write $f(x)$ as

$$\frac{3x - 1}{x^2 - 1} = \frac{2}{x + 1} + \frac{1}{x - 1}.$$

By adding the two geometric power series

$$\frac{2}{x + 1} = \frac{2}{1 - (-x)} = \sum_{n=0}^{\infty} 2(-1)^n x^n, \quad |x| < 1$$

and

$$\frac{1}{x - 1} = \frac{-1}{1 - x} = -\sum_{n=0}^{\infty} x^n, \quad |x| < 1$$

you obtain the following power series.

$$\frac{3x - 1}{x^2 - 1} = \sum_{n=0}^{\infty} [2(-1)^n - 1] x^n = 1 - 3x + x^2 - 3x^3 + x^4 - \cdots$$

The interval of convergence for this power series is $(-1, 1)$.

EXAMPLE 4 Finding a Power Series by Integration

Find a power series for $f(x) = \ln x$, centered at 1.

Solution From Example 2, you know that

$$\frac{1}{x} = \sum_{n=0}^{\infty} (-1)^n (x-1)^n. \qquad \text{Interval of convergence, } (0, 2)$$

Integrating this series produces

$$\ln x = \int \frac{1}{x}\, dx + C$$

$$= C + \sum_{n=0}^{\infty} (-1)^n \frac{(x-1)^{n+1}}{n+1}.$$

By letting $x = 1$, you can conclude that $C = 0$. Therefore,

$$\ln x = \sum_{n=0}^{\infty} (-1)^n \frac{(x-1)^{n+1}}{n+1}$$

$$= \frac{(x-1)}{1} - \frac{(x-1)^2}{2} + \frac{(x-1)^3}{3} - \frac{(x-1)^4}{4} + \cdots. \qquad \text{Interval of convergence, } (0, 2]$$

Note that the series converges at $x = 2$. This is consistent with the observation in the preceding section that integration of a power series may alter the convergence at the endpoints of the interval of convergence.

TECHNOLOGY In Section 8.7, we used the fourth-degree Taylor polynomial for the natural logarithmic function

$$\ln x \approx (x-1) - \frac{(x-1)^2}{2} + \frac{(x-1)^3}{3} - \frac{(x-1)^4}{4}$$

to approximate $\ln(1.1)$.

$$\ln(1.1) \approx (0.1) - \frac{1}{2}(0.1)^2 + \frac{1}{3}(0.1)^3 - \frac{1}{4}(0.1)^4$$

$$\approx 0.0953083$$

You now know from Example 4 that this polynomial represents the first four terms of the power series for $\ln x$. Moreover, using the Alternating Series Remainder, you can determine that the error in this approximation is less than

$$|R_4| \le |a_5|$$

$$= \frac{1}{5}(0.1)^5$$

$$= 0.000002.$$

During the seventeenth and eighteenth centuries, mathematical tables for logarithms and values of other transcendental functions were computed in this manner. The use for such numerical techniques is far from outdated, because it is precisely by such means that many modern calculating devices are programmed to evaluate transcendental functions.

EXAMPLE 5   Finding a Power Series by Integration

Find a power series for $g(x) = \arctan x$, centered at 0.

Solution   Because $D_x[\arctan x] = 1/(1 + x^2)$, you can use the series

$$f(x) = \frac{1}{1 + x} = \sum_{n=0}^{\infty} (-1)^n x^n. \qquad \text{Interval of convergence, } (-1, 1)$$

Substituting x^2 for x produces

$$f(x^2) = \frac{1}{1 + x^2} = \sum_{n=0}^{\infty} (-1)^n x^{2n}.$$

Finally, by integrating, you obtain

$$\arctan x = \int \frac{1}{1 + x^2}\, dx + C$$

$$= C + \sum_{n=0}^{\infty} (-1)^n \frac{x^{2n+1}}{2n + 1}$$

$$= \sum_{n=0}^{\infty} (-1)^n \frac{x^{2n+1}}{2n + 1} \qquad \text{Let } x = 0, \text{ then } C = 0.$$

$$= x - \frac{x^3}{3} + \frac{x^5}{5} - \frac{x^7}{7} + \cdots. \qquad \text{Interval of convergence, } (-1, 1)$$

SRINIVASA RAMANUJAN (1887–1920)

Series that can be used to approximate π have interested mathematicians for the past 300 years. An amazing series for approximating $1/\pi$ was discovered by the Indian mathematician Srinivasa Ramanujan in 1914. Each successive term of Ramanujan's series adds roughly eight more correct digits to the value of $1/\pi$. (For more information about Ramanujan's work, see the article "Ramanujan and Pi" by Jonathan M. Borwein and Peter B. Borwein in the February 1988 issue of *Scientific American*.)

It can be shown that the power series developed for $\arctan x$ in Example 5 also converges (to $\arctan x$) for $x = \pm 1$. For instance, when $x = 1$, you can write

$$\arctan 1 = 1 - \frac{1}{3} + \frac{1}{5} - \frac{1}{7} + \cdots = \frac{\pi}{4}.$$

However, this series (developed by James Gregory in 1671) does not give us a practical way of approximating π because it converges so slowly that hundreds of terms would have to be used to obtain reasonable accuracy. Example 6 shows how to use *two* different arctangent series to obtain a very good approximation of π using only a few terms. This approximation was developed by John Machin in 1706.

EXAMPLE 6   Approximating π with a Series

Use the trigonometric identity

$$4 \arctan \frac{1}{5} - \arctan \frac{1}{239} = \frac{\pi}{4}$$

to approximate the number π [see Exercise 44(b)].

Solution   By using only five terms from each of the series for $\arctan(1/5)$ and $\arctan(1/239)$, you obtain

$$4\left(4 \arctan \frac{1}{5} - \arctan \frac{1}{239}\right) \approx 3.1415926$$

which agrees with the decimal representation of π with an error of less than 0.0000001.

EXERCISES FOR SECTION 8.9

In Exercises 1–4, find a geometric power series for the function, centered at 0, (a) by the technique shown in Examples 1 and 2 and (b) by long division.

1. $f(x) = \dfrac{1}{2 - x}$

2. $f(x) = \dfrac{3}{4 - x}$

3. $f(x) = \dfrac{1}{2 + x}$

4. $f(x) = \dfrac{1}{1 + x}$

In Exercises 5–16, find a power series for the function, centered at c, and determine the interval of convergence.

5. $f(x) = \dfrac{1}{2 - x}, \quad c = 5$

6. $f(x) = \dfrac{3}{4 - x}, \quad c = -2$

7. $f(x) = \dfrac{3}{2x - 1}, \quad c = 0$

8. $f(x) = \dfrac{3}{2x - 1}, \quad c = 2$

9. $g(x) = \dfrac{1}{2x - 5}, \quad c = -3$

10. $h(x) = \dfrac{1}{2x - 5}, \quad c = 0$

11. $f(x) = \dfrac{3}{x + 2}, \quad c = 0$

12. $f(x) = \dfrac{4}{3x + 2}, \quad c = 2$

13. $g(x) = \dfrac{3x}{x^2 + x - 2}, \quad c = 0$

14. $g(x) = \dfrac{4x - 7}{2x^2 + 3x - 2}, \quad c = 0$

15. $f(x) = \dfrac{2}{1 - x^2}, \quad c = 0$

16. $f(x) = \dfrac{4}{4 + x^2}, \quad c = 0$

In Exercises 17–26, use the power series

$$\frac{1}{1 + x} = \sum_{n=0}^{\infty} (-1)^n x^n$$

to determine a power series, centered at 0, for the function. Identify the interval of convergence.

17. $h(x) = \dfrac{-2}{x^2 - 1} = \dfrac{1}{1 + x} + \dfrac{1}{1 - x}$

18. $h(x) = \dfrac{2x}{x^2 - 1} = \dfrac{1}{1 + x} - \dfrac{1}{1 - x}$

19. $f(x) = -\dfrac{1}{(x + 1)^2} = \dfrac{d}{dx}\left[\dfrac{1}{x + 1}\right]$

20. $f(x) = \dfrac{2}{(x + 1)^3} = \dfrac{d^2}{dx^2}\left[\dfrac{1}{x + 1}\right]$

21. $f(x) = \ln(x + 1) = \displaystyle\int \dfrac{1}{x + 1}\, dx$

22. $f(x) = \ln(1 - x^2) = \displaystyle\int \dfrac{1}{1 + x}\, dx - \int \dfrac{1}{1 - x}\, dx$

23. $g(x) = \dfrac{1}{x^2 + 1}$

24. $f(x) = \ln(x^2 + 1)$

25. $h(x) = \dfrac{1}{4x^2 + 1}$

26. $f(x) = \arctan 2x$

Graphical and Numerical Analysis In Exercises 27 and 28, let

$$S_n = x - \frac{x^2}{2} + \frac{x^3}{3} - \frac{x^4}{4} + \cdots \pm \frac{x^n}{n}.$$

Use a graphing utility to confirm the inequality graphically. Then complete the table to confirm the inequality numerically.

x	0.0	0.2	0.4	0.6	0.8	1.0
S_n						
$\ln(x + 1)$						
S_{n+1}						

27. $S_2 \le \ln(x + 1) \le S_3$

28. $S_4 \le \ln(x + 1) \le S_5$

In Exercises 29–32, match the polynomial approximation of the function $f(x) = \arctan x$ with the correct graph. [The graphs are labeled (a), (b), (c), and (d).]

(a)
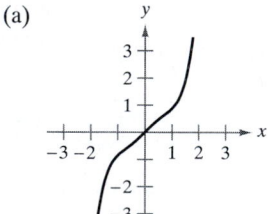

(b)
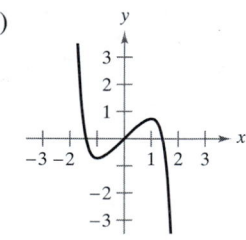

(c)
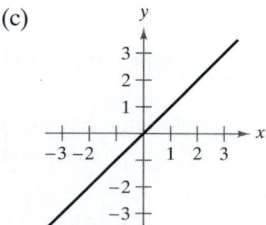

(d)
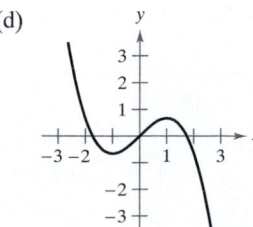

29. $g(x) = x$

30. $g(x) = x - \dfrac{x^3}{3}$

31. $g(x) = x - \dfrac{x^3}{3} + \dfrac{x^5}{5}$

32. $g(x) = x - \dfrac{x^3}{3} + \dfrac{x^5}{5} - \dfrac{x^7}{7}$

33. *Think About It* Use the results of Exercises 29–32 to make a geometric argument for why the series approximations of $f(x) = \arctan x$ have only odd powers of x.

34. *Conjecture* Use the results of Exercises 29–32 to make a conjecture about the degree of series approximations of $f(x) = \arctan x$ that have relative extrema.

In Exercises 35–38, use the series for $f(x) = \arctan x$ to approximate the value, using $R_N \le 0.001$.

35. $\arctan \dfrac{1}{4}$

36. $\displaystyle\int_0^{3/4} \arctan x^2 \, dx$

37. $\displaystyle\int_0^{1/2} \dfrac{\arctan x^2}{x} \, dx$

38. $\displaystyle\int_0^{1/2} x^2 \arctan x \, dx$

In Exercises 39–42, use the power series

$$\frac{1}{1-x} = \sum_{n=0}^{\infty} x^n.$$

39. Find the series representation of $\dfrac{1}{(1-x)^2}$ and determine its interval of convergence.

40. Adjust the index of summation for the series found in Exercise 39 to begin with $n = 0$.

41. *Probability* If a fair coin is tossed repeatedly, the probability that the first head occurs on the nth toss is

$$P(n) = \left(\frac{1}{2}\right)^n.$$

When this game is repeated many times, the average number of tosses required until the first head occurs is

$$E(n) = \sum_{n=1}^{\infty} nP(n).$$

(This value is called the *expected value* of n.) Use the results of Exercises 39 and 40 to find $E(n)$. Is the answer what you expected? Why or why not?

42. Use the results of Exercises 39 and 40 to find the sum of each of the following series.

(a) $\dfrac{1}{3} \displaystyle\sum_{n=1}^{\infty} n\left(\frac{2}{3}\right)^n$

(b) $\dfrac{1}{10} \displaystyle\sum_{n=1}^{\infty} n\left(\frac{9}{10}\right)^n$

43. Prove that

$$\arctan x + \arctan y = \arctan \frac{x+y}{1-xy} \quad \text{for } xy \ne 1$$

provided the value of the left side of the equation is between $-\pi/2$ and $\pi/2$.

44. Use the result of Exercise 43 to verify the identity.

(a) $\arctan \dfrac{120}{119} - \arctan \dfrac{1}{239} = \dfrac{\pi}{4}$

(b) $4 \arctan \dfrac{1}{5} - \arctan \dfrac{1}{239} = \dfrac{\pi}{4}$

[*Hint:* Use Exercise 43 twice to find $4 \arctan \frac{1}{5}$. Then use part (a).]

In Exercises 45 and 46, (a) verify the given equation and (b) use the equation and the series for the arctangent to approximate π to two-decimal-place accuracy.

45. $2 \arctan \dfrac{1}{2} - \arctan \dfrac{1}{7} = \dfrac{\pi}{4}$

46. $2 \arctan \dfrac{2}{3} - \arctan \dfrac{7}{17} = \dfrac{\pi}{4}$

In Exercises 47–52, find the sum of the convergent series. Explain how you obtained the sum.

47. $\displaystyle\sum_{n=1}^{\infty} (-1)^{n+1} \frac{1}{2^n n}$

48. $\displaystyle\sum_{n=1}^{\infty} (-1)^{n+1} \frac{1}{3^n n}$

49. $\displaystyle\sum_{n=1}^{\infty} (-1)^{n+1} \frac{2^n}{5^n n}$

50. $\displaystyle\sum_{n=0}^{\infty} (-1)^n \frac{1}{2n+1}$

51. $\displaystyle\sum_{n=0}^{\infty} (-1)^n \frac{1}{2^{2n+1}(2n+1)}$

52. $\displaystyle\sum_{n=1}^{\infty} (-1)^{n+1} \frac{1}{3^{2n-1}(2n-1)}$

53. *Writing* One of the series in Exercises 47–52 converges to its sum at a much slower rate than the other five series. Which is it? Explain why this series converges so slowly. Use a graphing utility to illustrate the rate of convergence.

54. Prove that $\displaystyle\sum_{n=0}^{\infty} \frac{(-1)^n}{3^n(2n+1)} = \frac{\pi}{2\sqrt{3}}$.

55. Use a graphing utility and 50 terms of the series

$$f(x) = \sum_{n=1}^{\infty} \frac{(-1)^{n+1}(x-1)^n}{n}, \quad 0 < x \le 2$$

to approximate $f(0.5)$. (The actual sum is $\ln 0.5$.)

Taylor Series and Maclaurin Series • Binomial Series •
Deriving Taylor Series from a Basic List

Taylor Series and Maclaurin Series

In Section 8.9, you derived power series for several functions using geometric series with term-by-term differentiation or integration. In this section you will study a *general* procedure for deriving the power series for a function that has derivatives of all orders. The following theorem gives the form that *every* (convergent) power series must take.

Colin Maclaurin (1698–1746)

The development of power series to represent functions is credited to the combined work of many seventeenth and eighteenth century mathematicians. Gregory, Newton, John and James Bernoulli, Leibniz, Euler, Lagrange, Wallis, and Fourier all contributed to this work. However, the two names that are most commonly associated with power series are Brook Taylor (1685-1731) and Colin Maclaurin.

> **THEOREM 8.23   The Form of a Convergent Power Series**
>
> If f is represented by a power series $f(x) = \Sigma\, a_n(x - c)^n$ for all x in an open interval I containing c, then $a_n = f^{(n)}(c)/n!$ and
>
> $$f(x) = f(c) + f'(c)(x - c) + \frac{f''(c)}{2!}(x - c)^2 + \cdots + \frac{f^{(n)}(c)}{n!}(x - c)^n + \cdots.$$

Proof   Suppose the power series $\Sigma\, a_n(x - c)^n$ has a radius of convergence R. Then, by Theorem 8.21, you know that the nth derivative of f exists for $|x - c| < R$, and by successive differentiation you obtain the following.

$$f^{(0)}(x) = a_0 + a_1(x - c) + a_2(x - c)^2 + a_3(x - c)^3 + a_4(x - c)^4 + \cdots$$
$$f^{(1)}(x) = a_1 + 2a_2(x - c) + 3a_3(x - c)^2 + 4a_4(x - c)^3 + \cdots$$
$$f^{(2)}(x) = 2a_2 + 3!a_3(x - c) + 4 \cdot 3a_4(x - c)^2 + \cdots$$
$$f^{(3)}(x) = 3!a_3 + 4!a_4(x - c) + \cdots$$
$$\vdots$$
$$f^{(n)}(x) = n!a_n + (n + 1)!a_{n+1}(x - c) + \cdots$$

Evaluating each of these derivatives at $x = c$ yields

$$f^{(0)}(c) = 0!a_0$$
$$f^{(1)}(c) = 1!a_1$$
$$f^{(2)}(c) = 2!a_2$$
$$f^{(3)}(c) = 3!a_3$$

and, in general, $f^{(n)}(c) = n!a_n$. By solving for a_n, you find that the coefficients of the power series representation of $f(x)$ are

$$a_n = \frac{f^{(n)}(c)}{n!}.$$

Notice that the coefficients of the power series in Theorem 8.23 are precisely the coefficients of the Taylor polynomials for $f(x)$ at c as defined in Section 8.7. For this reason, the series is called the **Taylor series** for $f(x)$ at c.

Definition of Taylor and Maclaurin Series

If a function f has derivatives of all orders at $x = c$, then the series

$$\sum_{n=0}^{\infty} \frac{f^{(n)}(c)}{n!}(x - c)^n = f(c) + f'(c)(x - c) + \cdots + \frac{f^{(n)}(c)}{n!}(x - c)^n + \cdots$$

is called the **Taylor series for $f(x)$ at c.** Moreover, if $c = 0$, then the series is the **Maclaurin series for f.**

If you know the pattern for the coefficients of the Taylor polynomials for a function, you can extend the pattern easily to form the corresponding Taylor series. For instance, in Example 4 of Section 8.7, you found the fourth Taylor polynomial for $\ln x$, centered at 1, to be

$$P_4(x) = (x - 1) - \frac{1}{2}(x - 1)^2 + \frac{1}{3}(x - 1)^3 - \frac{1}{4}(x - 1)^4.$$

From this pattern, you can obtain the Taylor series for $\ln x$ centered at $c = 1$,

$$(x - 1) - \frac{1}{2}(x - 1)^2 + \cdots + \frac{(-1)^{n+1}}{n}(x - 1)^n + \cdots.$$

EXAMPLE 1   Forming a Power Series

Use the function $f(x) = \sin x$ to form the Maclaurin series

$$\sum_{n=0}^{\infty} \frac{f^{(n)}(0)}{n!}x^n = f(0) + f'(0)x + \frac{f''(0)}{2!}x^2 + \frac{f^{(3)}(0)}{3!}x^3 + \frac{f^{(4)}(0)}{4!}x^4 + \cdots$$

and determine the interval of convergence.

Solution   Successive differentiation of $f(x)$ yields

$$f(x) = \sin x \qquad\qquad f(0) = 0$$
$$f'(x) = \cos x \qquad\qquad f'(0) = 1$$
$$f''(x) = -\sin x \qquad\qquad f''(0) = 0$$
$$f^{(3)}(x) = -\cos x \qquad\qquad f^{(3)}(0) = -1$$
$$f^{(4)}(x) = \sin x \qquad\qquad f^{(4)}(0) = 0$$
$$f^{(5)}(x) = \cos x \qquad\qquad f^{(5)}(0) = 1$$

and so on. The pattern repeats after the third derivative. Hence, the power series is as follows.

$$\sum_{n=0}^{\infty} \frac{f^{(n)}(0)}{n!}x^n = f(0) + f'(0)x + \frac{f''(0)}{2!}x^2 + \frac{f^{(3)}(0)}{3!}x^3 + \frac{f^{(4)}(0)}{4!}x^4 + \cdots$$

$$\sum_{n=0}^{\infty} \frac{(-1)^n x^{2n+1}}{(2n + 1)!} = x - \frac{x^3}{3!} + \frac{x^5}{5!} - \frac{x^7}{7!} + \cdots$$

By the Ratio Test, you can conclude that this series converges for all x.

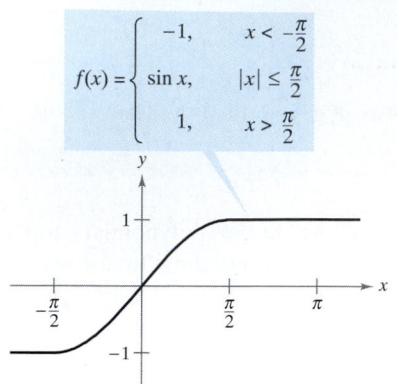

$$f(x) = \begin{cases} -1, & x < -\dfrac{\pi}{2} \\ \sin x, & |x| \le \dfrac{\pi}{2} \\ 1, & x > \dfrac{\pi}{2} \end{cases}$$

Figure 8.22

Notice that in Example 1 we do not conclude that the power series converges to $\sin x$ for all x. We simply conclude that the power series converges to some function, but we are not sure what function it is. This is a subtle, but important, point in dealing with Taylor or Maclaurin series. To persuade yourself that the series

$$f(c) + f'(c)(x - c) + \frac{f''(c)}{2!}(x - c)^2 + \cdots + \frac{f^{(n)}(c)}{n!}(x - c)^n + \cdots$$

might converge to a function other than f, remember that the derivatives are being evaluated at a single point. It can easily happen that another function will agree with the values of $f^{(n)}(x)$ when $x = c$ and disagree at other x-values. For instance, if you formed the power series (centered at 0) for the function shown in Figure 8.22, you would obtain the same series as in Example 1. You know that the series converges for all x, and yet it obviously cannot converge to both $f(x)$ and $\sin x$ for all x.

NOTE   Don't confuse this observation with the result given in Theorem 8.23. That theorem says that *if a power series converges to* $f(x)$, the series must be a Taylor series. The theorem does not say that every series formed with the Taylor coefficients $a_n = f^{(n)}(c)/n!$ will converge to $f(x)$.

THEOREM 8.24   Convergence of Taylor Series

If a function f has derivatives of all orders in an open interval I centered at c, then the equality

$$f(x) = \sum_{n=0}^{\infty} \frac{f^{(n)}(c)}{n!}(x - c)^n$$

holds if and only if there exists a z between x and c such that

$$\lim_{n \to \infty} R_n(x) = \lim_{n \to \infty} \frac{f^{(n+1)}(z)}{(n + 1)!}(x - c)^{n+1} = 0$$

for every x in I.

Proof   For a Taylor series, the nth partial sum coincides with the nth Taylor polynomial. That is, $S_n(x) = P_n(x)$. Moreover, because

$$P_n(x) = f(x) - R_n(x)$$

it follows that

$$\lim_{n \to \infty} S_n(x) = \lim_{n \to \infty} P_n(x)$$
$$= \lim_{n \to \infty} [f(x) - R_n(x)]$$
$$= f(x) - \lim_{n \to \infty} R_n(x).$$

Hence, for a given x, the Taylor series (the sequence of partial sums) converges to $f(x)$ if and only if $R_n(x) \to 0$ as $n \to \infty$.

NOTE   Stated another way, Theorem 8.24 says that a power series formed with Taylor coefficients $a_n = f^{(n)}(c)/n!$ converges to the function from which it was derived at precisely those values for which the remainder approaches 0 as $n \to \infty$.

In Example 1, we derived the power series from the sine function and we also concluded that the series converges to some function on the entire real line. In Example 2, you will see that the series actually converges to sin x.

EXAMPLE 2   A Convergent Maclaurin Series

Show that the Maclaurin series for $f(x) = \sin x$ converges to sin x for all x.

Solution   Using the result in Example 1, you need to show that

$$\sin x = x - \frac{x^3}{3!} + \frac{x^5}{5!} - \frac{x^7}{7!} + \cdots + \frac{(-1)^n x^{2n+1}}{(2n+1)!} + \cdots$$

is true for all x. Because

$$f^{(n+1)}(x) = \pm\sin x$$

or

$$f^{(n+1)}(x) = \pm\cos x$$

you know that $\left|f^{(n+1)}(z)\right| \le 1$ for every real number z. Therefore, for any fixed x, you can apply Taylor's Theorem (Theorem 8.19) to conclude that

$$0 \le |R_n(x)| = \left|\frac{f^{(n+1)}(z)}{(n+1)!} x^{n+1}\right| \le \frac{|x|^{n+1}}{(n+1)!}.$$

From the discussion in Section 8.1 regarding the relative rates of convergence of exponential and factorial sequences, it follows that for a fixed x

$$\lim_{n\to\infty} \frac{|x|^{n+1}}{(n+1)!} = 0.$$

Finally, by the Squeeze Theorem, it follows that for all x, $R_n(x) \to 0$ as $n \to \infty$. Hence, by Theorem 8.24, the Maclaurin series for sin x converges to sin x for all x.

Figure 8.23 visually illustrates the convergence of the Maclaurin series for sin x by comparing the graphs of the Maclaurin polynomials $P_1(x)$, $P_3(x)$, $P_5(x)$, and $P_7(x)$ with the graph of the sine function. Notice that as the degree of the polynomial increases, its graph more closely resembles that of the sine function.

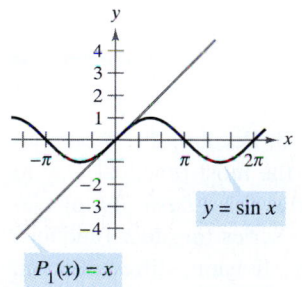

$P_1(x) = x$

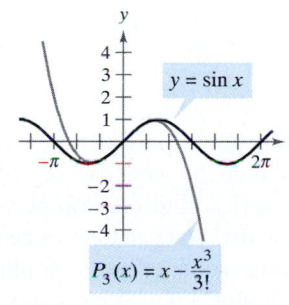

$P_3(x) = x - \dfrac{x^3}{3!}$

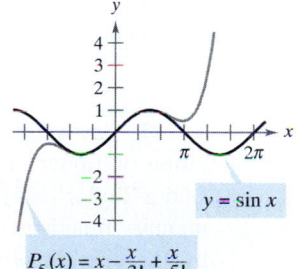

$P_5(x) = x - \dfrac{x^3}{3!} + \dfrac{x^5}{5!}$

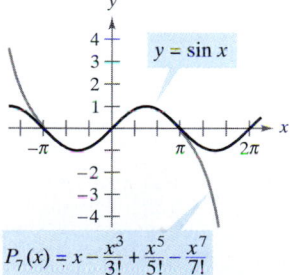

$P_7(x) = x - \dfrac{x^3}{3!} + \dfrac{x^5}{5!} - \dfrac{x^7}{7!}$

As n increases, the graph of P_n more closely resembles the sine function.
Figure 8.23

The steps for finding a Taylor series for $f(x)$ at c are summarized in the following list.

> **Guidelines for Finding a Taylor Series**
>
> 1. Differentiate $f(x)$ several times and evaluate each derivative at c.
>
> $$f(c), f'(c), f''(c), f'''(c), \cdots, f^{(n)}(c), \cdots$$
>
>   Try to recognize a pattern in these numbers.
> 2. Use the sequence developed in the first step to form the Taylor coefficients $a_n = f^{(n)}(c)/n!$, and determine the interval of convergence for the resulting power series
>
> $$f(c) + f'(c)(x - c) + \frac{f''(c)}{2!}(x - c)^2 + \cdots + \frac{f^{(n)}(c)}{n!}(x - c)^n + \cdots.$$
> 3. Within this interval of convergence, determine whether or not the series converges to $f(x)$.

The direct determination of Taylor or Maclaurin coefficients using successive differentiation can be difficult, and the next example illustrates a shortcut for finding the coefficients indirectly—using the coefficients of a known Taylor or Maclaurin series.

EXAMPLE 3 Maclaurin Series for a Composite Function

Find the Maclaurin series for $f(x) = \sin x^2$.

Solution To find the coefficients for this Maclaurin series *directly*, you must calculate successive derivatives of $f(x) = \sin x^2$. By calculating just the first two,

$$f'(x) = 2x \cos x^2 \quad \text{and} \quad f''(x) = -4x^2 \sin x^2 + 2 \cos x^2$$

you can see that this task would be quite cumbersome. Fortunately, there is an alternative. Suppose you first consider the Maclaurin series for $\sin x$ found in Example 1.

$$g(x) = \sin x = x - \frac{x^3}{3!} + \frac{x^5}{5!} - \frac{x^7}{7!} + \cdots$$

Now, because $\sin x^2 = g(x^2)$, you can substitute x^2 for x in the series for $\sin x$ to obtain

$$\sin x^2 = g(x^2) = x^2 - \frac{x^6}{3!} + \frac{x^{10}}{5!} - \frac{x^{14}}{7!} + \cdots.$$

Be sure to understand the point illustrated in Example 3. Because direct computation of Taylor or Maclaurin coefficients can be tedious, the most practical way to find a Taylor or Maclaurin series is to develop power series for a *basic list* of elementary functions. From this list, you can determine power series for other functions by the operations of addition, subtraction, multiplication, division, differentiation, integration, or composition with known power series.

Binomial Series

Before presenting the basic list for elementary functions, we develop one more series—for a function of the form $f(x) = (1 + x)^k$. This produces the **binomial series.**

EXAMPLE 4   Binomial Series

Find the Maclaurin series for $f(x) = (1 + x)^k$ and determine its radius of convergence.

Solution   By successive differentiation, you have

$$f(x) = (1 + x)^k \qquad\qquad f(0) = 1$$
$$f'(x) = k(1 + x)^{k-1} \qquad\qquad f'(0) = k$$
$$f''(x) = k(k - 1)(1 + x)^{k-2} \qquad\qquad f''(0) = k(k - 1)$$
$$f'''(x) = k(k - 1)(k - 2)(1 + x)^{k-3} \qquad f'''(0) = k(k - 1)(k - 2)$$
$$\vdots \qquad\qquad\qquad\qquad \vdots$$
$$f^{(n)}(x) = k \cdots (k - n + 1)(1 + x)^{k-n} \quad f^{(n)}(0) = k(k - 1) \cdots (k - n + 1)$$

which produces the series

$$1 + kx + \frac{k(k - 1)x^2}{2} + \cdots + \frac{k(k - 1) \cdots (k - n + 1)x^n}{n!} + \cdots.$$

Because $a_{n+1}/a_n \to 1$, you can apply the Ratio Test to conclude that the radius of convergence is $R = 1$. Thus, the series converges to *some* function in the interval $(-1, 1)$.

Note that in Example 4 we showed that the Taylor series for $(1 + x)^k$ converges to *some* function in the interval $(-1, 1)$. However, we did not show that the series actually converges to $(1 + x)^k$. To do this, you could show that the remainder $R_n(x)$ converges to 0, as in Example 2.

EXAMPLE 5   Finding a Binomial Series

Find the power series for $f(x) = \sqrt[3]{1 + x}$.

Solution   Using the binomial series

$$(1 + x)^k = 1 + kx + \frac{k(k - 1)x^2}{2!} + \frac{k(k - 1)(k - 2)x^3}{3!} + \cdots$$

let $k = \frac{1}{3}$ and write

$$(1 + x)^{1/3} = 1 + \frac{x}{3} - \frac{2x^2}{3^2 2!} + \frac{2 \cdot 5x^3}{3^3 3!} - \frac{2 \cdot 5 \cdot 8x^4}{3^4 4!} + \cdots$$

which converges for $-1 \le x \le 1$.

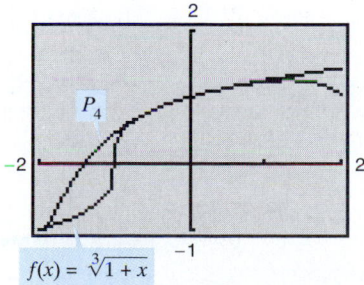

$f(x) = \sqrt[3]{1 + x}$

Figure 8.24

⋮ **TECHNOLOGY**   Try using a graphing utility to confirm the result in Example 5. When you graph the functions

$$f(x) = (1 + x)^{1/3} \quad \text{and} \quad P_4(x) = 1 + \frac{x}{3} - \frac{x^2}{9} + \frac{5x^3}{81} - \frac{10x^4}{243}$$

in the same viewing rectangle, you should obtain the result shown in Figure 8.24.

Deriving Taylor Series from a Basic List

In the following list, we provide the power series for several elementary functions with the corresponding intervals of convergence.

Power Series for Elementary Functions

Function	Interval of Convergence
$\dfrac{1}{x} = 1 - (x - 1) + (x - 1)^2 - (x - 1)^3 + (x - 1)^4 - \cdots + (-1)^n (x - 1)^n + \cdots$	$0 < x < 2$
$\dfrac{1}{1 + x} = 1 - x + x^2 - x^3 + x^4 - x^5 + \cdots + (-1)^n x^n + \cdots$	$-1 < x < 1$
$\ln x = (x - 1) - \dfrac{(x - 1)^2}{2} + \dfrac{(x - 1)^3}{3} - \dfrac{(x - 1)^4}{4} + \cdots + \dfrac{(-1)^{n-1}(x - 1)^n}{n} + \cdots$	$0 < x \leq 2$
$e^x = 1 + x + \dfrac{x^2}{2!} + \dfrac{x^3}{3!} + \dfrac{x^4}{4!} + \dfrac{x^5}{5!} + \cdots + \dfrac{x^n}{n!} + \cdots$	$-\infty < x < \infty$
$\sin x = x - \dfrac{x^3}{3!} + \dfrac{x^5}{5!} - \dfrac{x^7}{7!} + \dfrac{x^9}{9!} - \cdots + \dfrac{(-1)^n x^{2n+1}}{(2n + 1)!} + \cdots$	$-\infty < x < \infty$
$\cos x = 1 - \dfrac{x^2}{2!} + \dfrac{x^4}{4!} - \dfrac{x^6}{6!} + \dfrac{x^8}{8!} - \cdots + \dfrac{(-1)^n x^{2n}}{(2n)!} + \cdots$	$-\infty < x < \infty$
$\arctan x = x - \dfrac{x^3}{3} + \dfrac{x^5}{5} - \dfrac{x^7}{7} + \dfrac{x^9}{9} - \cdots + \dfrac{(-1)^n x^{2n+1}}{2n + 1} + \cdots$	$-1 \leq x \leq 1$
$\arcsin x = x + \dfrac{x^3}{2 \cdot 3} + \dfrac{1 \cdot 3x^5}{2 \cdot 4 \cdot 5} + \dfrac{1 \cdot 3 \cdot 5x^7}{2 \cdot 4 \cdot 6 \cdot 7} + \cdots + \dfrac{(2n)!x^{2n+1}}{(2^n n!)^2(2n + 1)} + \cdots$	$-1 \leq x \leq 1$
$(1 + x)^k = 1 + kx + \dfrac{k(k - 1)x^2}{2!} + \dfrac{k(k - 1)(k - 2)x^3}{3!} + \dfrac{k(k - 1)(k - 2)(k - 3)x^4}{4!} + \cdots$	$-1 < x < 1^*$

The convergence at $x = \pm 1$ depends on the value k.

NOTE   The binomial series is valid for noninteger values of k. Moreover, if k happens to be a positive integer, the binomial series reduces to a simple binomial expansion.

EXAMPLE 6   Deriving a Power Series from a Basic List

Find the power series for $f(x) = \cos \sqrt{x}$.

Solution   Using the power series

$$\cos x = 1 - \frac{x^2}{2!} + \frac{x^4}{4!} - \frac{x^6}{6!} + \frac{x^8}{8!} - \cdots$$

you can replace x by $\sqrt{x}$ to obtain the series

$$\cos \sqrt{x} = 1 - \frac{x}{2!} + \frac{x^2}{4!} - \frac{x^3}{6!} + \frac{x^4}{8!} - \cdots.$$

This series converges for all x in the domain of $\cos \sqrt{x}$—that is, for $x \geq 0$.

EXAMPLE 7   A Power Series for sin² x

Find the power series for $f(x) = \sin^2 x$.

Solution   Consider rewriting $\sin^2 x$ as follows.

$$\sin^2 x = \frac{1 - \cos 2x}{2} = \frac{1}{2} - \frac{\cos 2x}{2}$$

Now, use the series for $\cos x$.

$$\cos x = 1 - \frac{x^2}{2!} + \frac{x^4}{4!} - \frac{x^6}{6!} + \frac{x^8}{8!} - \cdots$$

$$\cos 2x = 1 - \frac{2^2}{2!}x^2 + \frac{2^4}{4!}x^4 - \frac{2^6}{6!}x^6 + \frac{2^8}{8!}x^8 - \cdots$$

$$-\frac{1}{2}\cos 2x = -\frac{1}{2} + \frac{2}{2!}x^2 - \frac{2^3}{4!}x^4 + \frac{2^5}{6!}x^6 - \frac{2^7}{8!}x^8 + \cdots$$

$$\sin^2 x = \frac{1}{2} - \frac{1}{2}\cos 2x = \frac{1}{2} - \frac{1}{2} + \frac{2}{2!}x^2 - \frac{2^3}{4!}x^4 + \frac{2^5}{6!}x^6 - \frac{2^7}{8!}x^8 + \cdots$$

$$= \frac{2}{2!}x^2 - \frac{2^3}{4!}x^4 + \frac{2^5}{6!}x^6 - \frac{2^7}{8!}x^8 + \cdots$$

This series converges for $-\infty < x < \infty$.

As mentioned in the previous section, power series can be used to obtain tables of values of transcendental functions. They are also useful for estimating the values of definite integrals for which antiderivatives cannot be found. The next example demonstrates this use.

EXAMPLE 8   *Power Series Approximation of a Definite Integral*

Use a power series to approximate

$$\int_0^1 e^{-x^2}\, dx$$

with an error of less than 0.01.

Solution   Replacing x with $-x^2$ in the series for e^x produces the following.

$$e^{-x^2} = 1 - x^2 + \frac{x^4}{2!} - \frac{x^6}{3!} + \frac{x^8}{4!} - \cdots$$

$$\int_0^1 e^{-x^2}\, dx = \left[x - \frac{x^3}{3} + \frac{x^5}{5 \cdot 2!} - \frac{x^7}{7 \cdot 3!} + \frac{x^9}{9 \cdot 4!} - \cdots \right]_0^1$$

$$= 1 - \frac{1}{3} + \frac{1}{10} - \frac{1}{42} + \frac{1}{216} - \cdots$$

Summing the first *four* terms, you have

$$\int_0^1 e^{-x^2}\, dx \approx 0.74$$

which, by the Alternating Series Test, has an error of less than $\frac{1}{216} \approx 0.005$.

EXERCISES FOR SECTION 8.10

In Exercises 1–10, use the definition to find the Taylor series (centered at *c*) for the function.

1. $f(x) = e^{2x}, \quad c = 0$

2. $f(x) = e^{-2x}, \quad c = 0$

3. $f(x) = \cos x, \quad c = \dfrac{\pi}{4}$

4. $f(x) = \sin x, \quad c = \dfrac{\pi}{4}$

5. $f(x) = \ln x, \quad c = 1$

6. $f(x) = e^x, \quad c = 1$

7. $f(x) = \sin 2x \quad c = 0$

8. $f(x) = \ln(x^2 + 1), \quad c = 0$

9. $f(x) = \sec x, \quad c = 0$ (first three nonzero terms)

10. $f(x) = \tan x, \quad c = 0$ (first three nonzero terms)

In Exercises 11–16, use the binomial series to find the Maclaurin series of the function.

11. $f(x) = \dfrac{1}{(1 + x)^2}$

12. $f(x) = \dfrac{1}{\sqrt{1 - x}}$

13. $f(x) = \dfrac{1}{\sqrt{4 + x^2}}$

14. $f(x) = \sqrt{1 + x}$

15. $f(x) = \sqrt{1 + x^2}$

16. $f(x) = \sqrt{1 + x^3}$

In Exercises 17–28, find the Maclaurin series for the function. (Use the table of power series for elementary functions.)

17. $f(x) = e^{x^2/2}$

18. $g(x) = e^{-3x}$

19. $g(x) = \sin 2x$

20. $h(x) = x \cos x$

21. $f(x) = \cos x^{3/2}$

22. $f(x) = \cos 3x$

23. $g(x) = \dfrac{\sin x}{x}$

24. $f(x) = \dfrac{\arcsin x}{x}$

25. $f(x) = \tfrac{1}{2}(e^x - e^{-x}) = \sinh x$

26. $f(x) = \tfrac{1}{2}(e^x + e^{-x}) = \cosh x$

27. $f(x) = \cos^2 x$

[*Hint:* $\cos^2 x = \tfrac{1}{2}(1 + \cos 2x)$]

28. $f(x) = \sinh^{-1} x = \ln\left(x + \sqrt{x^2 + 1}\right)$

$\left(\text{\textit{Hint:} Integrate the series for } \dfrac{1}{\sqrt{x^2 + 1}}.\right)$

In Exercises 29 and 30, use a power series and the fact that $i^2 = -1$ to verify the formula.

29. $g(x) = \dfrac{1}{2i}(e^{ix} - e^{-ix}) = \sin x$

30. $g(x) = \tfrac{1}{2}(e^{ix} + e^{-ix}) = \cos x$

In Exercises 31–36, find the first four nonzero terms of the Maclaurin series of the function by multiplying or dividing the appropriate power series in the table of power series for elementary functions on page 628. Use a graphing utility to obtain a graph of the function and its corresponding polynomial approximation.

31. $f(x) = e^x \sin x$

32. $g(x) = e^x \cos x$

33. $h(x) = \cos x \ln(1 + x)$

34. $f(x) = e^x \ln(1 + x)$

35. $g(x) = \dfrac{\sin x}{1 + x}$

36. $f(x) = \dfrac{e^x}{1 + x}$

In Exercises 37–40, match the polynomial and its graph. [The graphs are labeled (a), (b), (c), and (d).] Factor a common factor from each polynomial and identify the function approximated by the remaining Taylor polynomial.

(a)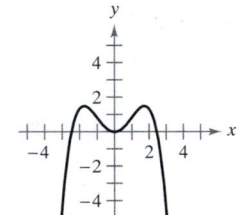

(b)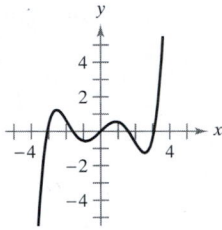

(c)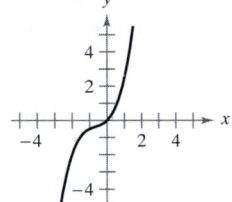

(d)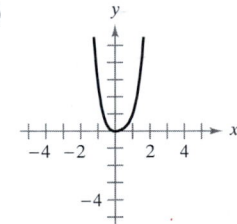

37. $y = x^2 - \dfrac{x^4}{3!}$

38. $y = x - \dfrac{x^3}{2!} + \dfrac{x^5}{4!}$

39. $y = x + x^2 + \dfrac{x^3}{2!}$

40. $y = x^2 - x^3 + x^4$

In Exercises 41 and 42, find a Maclaurin series for $f(x)$.

41. $f(x) = \displaystyle\int_0^x (e^{-t^2} - 1)\, dt$

42. $f(x) = \displaystyle\int_0^x \sqrt{1 + t^3}\, dt$

In Exercises 43–46, verify the sum. Then use a graphing utility to approximate the sum with an error of less than 0.0001.

43. $\displaystyle\sum_{n=1}^{\infty} (-1)^{n+1} \dfrac{1}{n} = \ln 2$

44. $\displaystyle\sum_{n=0}^{\infty} (-1)^n \left[\dfrac{1}{(2n + 1)!}\right] = \sin 1$

45. $\displaystyle\sum_{n=0}^{\infty} \dfrac{2^n}{n!} = e^2$

46. $\displaystyle\sum_{n=1}^{\infty} (-1)^{n-1} \left(\dfrac{1}{n!}\right) = \dfrac{e - 1}{e}$

In Exercises 47 and 48, use the series representation of the function f to find $\lim\limits_{x \to 0} f(x)$ (if it exists.)

47. $f(x) = \dfrac{1 - \cos x}{x}$

48. $f(x) = \dfrac{\sin x}{x}$

In Exercises 49–54, use power series to approximate the integral with an error of less than 0.0001. (In Exercises 49, 50, and 54, assume that the integrand is defined as 1 when $x = 0$.)

49. $\displaystyle\int_0^1 \dfrac{\sin x}{x}\, dx$

50. $\displaystyle\int_0^{1/2} \dfrac{\arctan x}{x}\, dx$

51. $\displaystyle\int_0^{\pi/2} \sqrt{x}\cos x\, dx$

52. $\displaystyle\int_{0.5}^1 \cos \sqrt{x}\, dx$

53. $\displaystyle\int_{0.1}^{0.3} \sqrt{1 + x^3}\, dx$

54. $\displaystyle\int_0^{1/2} \dfrac{\ln(x + 1)}{x}\, dx$

Probability **In Exercises 55 and 56, approximate the normal probability with an error of less than 0.0001 where the probability is given by**

$$P(a < x < b) = \dfrac{1}{\sqrt{2\pi}}\int_a^b e^{-x^2/2}\, dx.$$

55. $P(0 < x < 1)$

56. $P(1 < x < 2)$

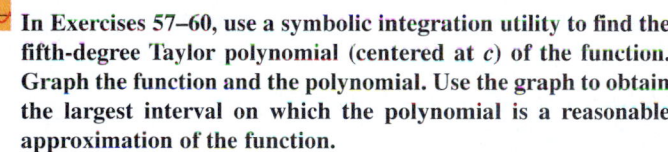

In Exercises 57–60, use a symbolic integration utility to find the fifth-degree Taylor polynomial (centered at c) of the function. Graph the function and the polynomial. Use the graph to obtain the largest interval on which the polynomial is a reasonable approximation of the function.

57. $f(x) = x \cos 2x, \quad c = 0$

58. $f(x) = \sin \dfrac{x}{2} \ln(1 + x), \quad c = 0$

59. $g(x) = \sqrt{x}\ln x, \quad c = 1$

60. $h(x) = \sqrt[3]{x}\arctan x, \quad c = 1$

61. *Projectile Motion* A projectile fired from the ground follows the trajectory given by

$$y = \left(\tan\theta - \dfrac{g}{kv_0\cos\theta}\right)x - \dfrac{g}{k^2}\ln\left(1 - \dfrac{kx}{v_0\cos\theta}\right)$$

where v_0 is the initial speed, θ is the angle of projection, g is the acceleration due to gravity, and k is the drag factor caused by air resistance. Using the power series representation

$$\ln(1 + x) = x - \dfrac{x^2}{2} + \dfrac{x^3}{3} - \dfrac{x^4}{4} + \cdots, \quad -1 < x < 1$$

verify that the trajectory can be rewritten as

$$y = (\tan\theta)x + \dfrac{gx^2}{2v_0{}^2\cos^2\theta} + \dfrac{kgx^3}{3v_0\cos^3\theta} + \dfrac{k^2 gx^4}{4v_0\cos^4\theta} + \cdots.$$

62. *Projectile Motion* Use the result of Exercise 61 to determine the series for the path of a projectile projected from ground level at an angle of $\theta = 60°$ with an initial speed of $v_0 = 64$ feet per second and a drag factor of $k = \frac{1}{16}$.

63. *Investigation* Consider the function f defined by

$$f(x) = \begin{cases} e^{-1/x^2}, & x \neq 0 \\ 0, & x = 0. \end{cases}$$

(a) Sketch a graph of the function.

(b) Use the alternative form of the definition of the derivative (Section 2.1) and L'Hôpital's Rule to show that $f'(0) = 0$. [By continuing this process, it can be shown that $f^{(n)}(0) = 0$ for $n > 1$.]

(c) Using the result in part (b), find the Maclaurin series for f. Does the series converge to f?

64. *Investigation*

(a) Find the power series centered at 0 for the function

$$f(x) = \dfrac{\ln(x^2 + 1)}{x^2}.$$

(b) Use a graphing utility to graph f and the eighth-degree Taylor polynomial $P_8(x)$ for f.

(c) Complete the following table, where

$$F(x) = \int_0^x \dfrac{\ln(t^2 + 1)}{t^2}\, dt \quad \text{and}$$

$$G(x) = \int_0^x P_8(t)\, dt.$$

x	0.25	0.50	0.75	1.00	1.50	2.00
$F(x)$						
$G(x)$						

(d) Describe the relationship between the graphs of f and P_8 and the results given in the table in part (c).

65. Prove that $\lim\limits_{n \to 0} \dfrac{x^n}{n!} = 0$ for any real x.

66. Prove that e is irrational. $\Big[$ *Hint:* Assume that $e = p/q$ is rational (p, q integers) and consider

$$e = 1 + 1 + \dfrac{1}{2!} + \cdots + \dfrac{1}{n!} + \cdots.\Big]$$

REVIEW EXERCISES FOR CHAPTER 8

In Exercises 1 and 2, find the general term of the sequence.

1. $1, \dfrac{1}{2}, \dfrac{1}{6}, \dfrac{1}{24}, \dfrac{1}{120}, \cdots$

2. $\dfrac{1}{2}, \dfrac{2}{5}, \dfrac{3}{10}, \dfrac{4}{17}, \cdots$

In Exercises 3–6, match the sequence with its graph. [The graphs are labeled (a), (b), (c), and (d).]

(a)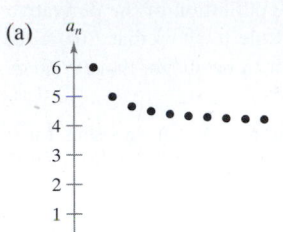

(b)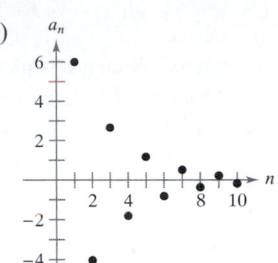

(c)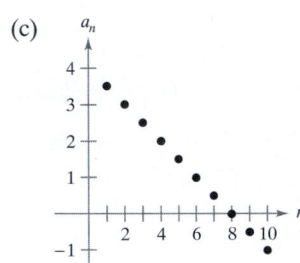

(d)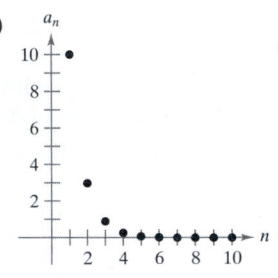

3. $a_n = 4 + \dfrac{2}{n}$

4. $a_n = 4 - \dfrac{1}{2}n$

5. $a_n = 10(0.3)^{n-1}$

6. $a_n = 6\left(-\dfrac{2}{3}\right)^{n-1}$

In Exercises 7 and 8, use a graphing utility to graph the first ten terms of the sequence. Use the graph to make an inference about the convergence or divergence of the sequence. Verify your inference analytically, and if the sequence converges, find its limit.

7. $a_n = \dfrac{5n + 2}{n}$

8. $a_n = \sin \dfrac{n\pi}{2}$

In Exercises 9–16, determine the convergence or divergence of the sequence with the given general term. (b and c are positive real numbers.)

9. $a_n = \dfrac{n + 1}{n^2}$

10. $a_n = \dfrac{1}{\sqrt{n}}$

11. $a_n = \dfrac{n^3}{n^2 + 1}$

12. $a_n = \dfrac{n}{\ln n}$

13. $a_n = \sqrt{n + 1} - \sqrt{n}$

14. $a_n = \left(1 + \dfrac{1}{2n}\right)^n$

15. $a_n = \dfrac{\sin \sqrt{n}}{\sqrt{n}}$

16. $a_n = (b^n + c^n)^{1/n}$

17. *Compound Interest*   A deposit of $5000 is made in an account that earns 5% interest compounded quarterly. The balance in the account after n quarters is

$$A_n = 5000\left(1 + \dfrac{0.05}{4}\right)^n, \quad n = 1, 2, 3, \cdots.$$

(a) Compute the first eight terms of the sequence.

(b) Find the balance in the account after 10 years by computing the 40th term of the sequence.

18. *Depreciation*   A company buys a machine for $120,000. During the next 5 years it will depreciate at a rate of 30% per year. (That is, at the end of each year the depreciated valued is 70% of what it was at the beginning of the year.)

(a) Find the formula for the nth term of a sequence that gives the value V of the machine t full years after it was purchased.

(b) Find the depreciated value of the machine at the end of 5 full years.

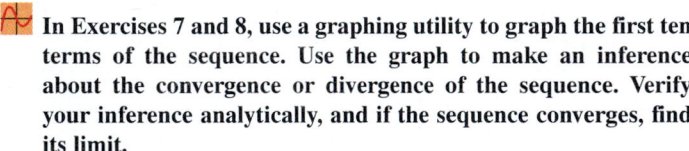 *Numerical, Graphical, and Analytic Analysis*   **In Exercises 19–22, (a) use a graphing utility to find the indicated partial sum S_k and complete the table, (b) use a graphing utility to graph the first ten terms of the sequence of partial sums, and (c) determine the convergence or divergence of the series.**

k	5	10	15	20	25
S_k					

19. $\displaystyle\sum_{n=1}^{\infty} \left(\dfrac{3}{2}\right)^{n-1}$

20. $\displaystyle\sum_{n=1}^{\infty} \dfrac{(-1)^{n+1}}{2n}$

21. $\displaystyle\sum_{n=1}^{\infty} \dfrac{(-1)^{n+1}}{(2n)!}$

22. $\displaystyle\sum_{n=1}^{\infty} \dfrac{1}{n(n + 1)}$

In Exercises 23–26, find the sum of the series.

23. $\displaystyle\sum_{n=0}^{\infty} \left(\dfrac{2}{3}\right)^n$

24. $\displaystyle\sum_{n=0}^{\infty} \dfrac{2^{n+2}}{3^n}$

25. $\displaystyle\sum_{n=0}^{\infty} \left(\dfrac{1}{2^n} - \dfrac{1}{3^n}\right)$

26. $\displaystyle\sum_{n=0}^{\infty} \left[\left(\dfrac{2}{3}\right)^n - \dfrac{1}{(n + 1)(n + 2)}\right]$

In Exercises 27 and 28, express the repeating decimal as the ratio of two integers.

27. $0.\overline{09}$

28. $0.\overline{923076}$

29. *Bouncing Ball* A ball is dropped from a height of 8 meters. Each time it drops h meters, it rebounds $0.7h$ meters. Find the total distance traveled by the ball.

30. *Total Compensation* Suppose you accept a job that pays a salary of $32,000 the first year. During the next 39 years, you will receive a 5.5% raise each year. Find your total salary over the 40-year period.

31. *Compound Interest* A deposit of $200 is made at the end of each month for 2 years in an account that pays 6%, compounded continuously. Determine the balance in the account at the end of 2 years.

32. *Compound Interest* A deposit of $100 is made at the end of each month for 10 years in an account that pays 6.5%, compounded monthly. Determine the balance in the account at the end of 10 years.

Numerical, Graphical, and Analytic Analysis In Exercises 33 and 34, (a) verify that the series converges, (b) use a graphing utility to find the indicated partial sum S_n and complete the table, (c) use a graphing utility to graph the first ten terms of the sequence of partial sums, and (d) use the table to estimate the sum of the series.

n	5	10	15	20	25
S_n					

33. $\displaystyle\sum_{n=1}^{\infty} n\left(\frac{3}{5}\right)^n$

34. $\displaystyle\sum_{n=1}^{\infty} \frac{(-1)^{n-1}n}{n^3 + 5}$

In Exercises 35–46, determine the convergence or divergence of the series.

35. $\displaystyle\sum_{n=1}^{\infty} \frac{2^n}{n^3}$

36. $\displaystyle\sum_{n=1}^{\infty} \frac{1}{\sqrt[4]{n^3}}$

37. $\displaystyle\sum_{n=1}^{\infty} \frac{1}{\sqrt{n^3 + 2n}}$

38. $\displaystyle\sum_{n=1}^{\infty} \frac{n + 1}{n(n + 2)}$

39. $\displaystyle\sum_{n=1}^{\infty} \frac{n}{e^{n^2}}$

40. $\displaystyle\sum_{n=1}^{\infty} \frac{n!}{e^n}$

41. $\displaystyle\sum_{n=1}^{\infty} \frac{(-1)^n n}{\ln n}$

42. $\displaystyle\sum_{n=1}^{\infty} \frac{(-1)^n \sqrt{n}}{n + 1}$

43. $\displaystyle\sum_{n=1}^{\infty} \left(\frac{1}{n^2} - \frac{1}{n}\right)$

44. $\displaystyle\sum_{n=1}^{\infty} \left(\frac{1}{n^2} - \frac{1}{2^n}\right)$

45. $\displaystyle\sum_{n=1}^{\infty} \frac{1 \cdot 3 \cdot 5 \cdots (2n - 1)}{2 \cdot 4 \cdot 6 \cdots (2n)}$

46. $\displaystyle\sum_{n=1}^{\infty} \frac{1 \cdot 3 \cdot 5 \cdots (2n - 1)}{2 \cdot 5 \cdot 8 \cdots (3n - 1)}$

47. *Writing* Use a graphing utility to complete the table for (a) $p = 2$ and (b) $p = 5$. Write a short paragraph describing and comparing the entries in the table.

N	5	10	20	30	40
$\displaystyle\sum_{n=1}^{N} \frac{1}{n^P}$					
$\displaystyle\int_{N}^{\infty} \frac{1}{x^P} \, dx$					

48. *Writing* You are told that the terms of a positive series appear to approach zero very slowly as n approaches infinity. (In fact, $a_{75} = 0.7$.) If you are given no other information, can you conclude that the series diverges? Support your answer with an example.

In Exercises 49–54, find the interval of convergence of the power series.

49. $\displaystyle\sum_{n=0}^{\infty} \left(\frac{x}{10}\right)^n$

50. $\displaystyle\sum_{n=0}^{\infty} (2x)^n$

51. $\displaystyle\sum_{n=0}^{\infty} \frac{(-1)^n (x - 2)^n}{(n + 1)^2}$

52. $\displaystyle\sum_{n=1}^{\infty} \frac{3^n (x - 2)^n}{n}$

53. $\displaystyle\sum_{n=0}^{\infty} n!(x - 2)^n$

54. $\displaystyle\sum_{n=0}^{\infty} \frac{(x - 2)^n}{2^n}$

In Exercises 55–62, find the power series for the function centered at c.

55. $f(x) = \sin x, \quad c = \dfrac{3\pi}{4}$

56. $f(x) = \cos x, \quad c = -\dfrac{\pi}{4}$

57. $f(x) = 3^x, \quad c = 0$

58. $f(x) = \csc x, \quad c = \dfrac{\pi}{2}$ (first three terms)

59. $f(x) = \dfrac{1}{x}, \quad c = -1$

60. $f(x) = \sqrt{x}, \quad c = 4$

61. $g(x) = \dfrac{2}{3 - x}, \quad c = 0$

62. $h(x) = \dfrac{1}{(1 + x)^3}, \quad c = 0$

In Exercises 63–68, find the sum of the convergent series. Explain how you obtained the sum. (*Hint:* Use the power series for elementary functions.)

63. $\displaystyle\sum_{n=1}^{\infty} (-1)^{n+1} \frac{1}{4^n n}$

64. $\displaystyle\sum_{n=1}^{\infty} (-1)^{n+1} \frac{1}{5^n n}$

65. $\displaystyle\sum_{n=0}^{\infty} \frac{1}{2^n n!}$

66. $\displaystyle\sum_{n=0}^{\infty} \frac{2^n}{3^n n!}$

67. $\displaystyle\sum_{n=0}^{\infty} (-1)^n \frac{2^{2n}}{3^{2n} (2n)!}$

68. $\displaystyle\sum_{n=0}^{\infty} (-1)^n \frac{1}{3^{2n+1} (2n + 1)!}$

69. *Writing* One of the series in Exercises 37 and 39 converges to its sum at much slower rate than the other series. Which is it? Explain why this series converges so slowly. Use a graphing utility to illustrate the rate of convergence.

70. Find the Maclaurin series for $f(x) = xe^x$. Integrate the series term-by-term over the closed interval $[0, 1]$, and show that

$$\sum_{n=0}^{\infty} \frac{1}{(n + 2)n!} = 1.$$

71. *Forming Maclaurin Series* Determine the first four terms of the Maclaurin series for e^{2x}

(a) by using the definition of the Maclaurin series and the formula for the coefficient of the nth term, $a_n = f^{(n)}(0)/n!$.

(b) by replacing x by $2x$ in the series for e^x.

(c) by multiplying the series for e^x by itself, because $e^{2x} = e^x \cdot e^x$.

72. *Forming Maclaurin Series* Follow the pattern of Exercise 71 to find the first four terms of the series for $\sin 2x$. (*Hint:* $\sin 2x = 2 \sin x \cos x$.)

In Exercises 73 and 74, find a function represented by the series and give the domain of the function.

73. $1 + \dfrac{2}{3}x + \dfrac{4}{9}x^2 + \dfrac{8}{27}x^3 + \cdots$

74. $8 - 2(x - 3) + \dfrac{1}{2}(x - 3)^2 - \dfrac{1}{8}(x - 3)^3 + \cdots$

In Exercises 75–78, find the series representation of the function defined by the integral.

75. $\displaystyle\int_0^x \frac{\sin t}{t}\, dt$

76. $\displaystyle\int_0^x \cos \frac{\sqrt{t}}{2}\, dt$

77. $\displaystyle\int_0^x \frac{\ln(t + 1)}{t}\, dt$

78. $\displaystyle\int_0^x \frac{e^t - 1}{t}\, dt$

In Exercises 79 and 80, use power series to find the limit (if it exists). Verify the result by using L'Hôpital's Rule.

79. $\displaystyle\lim_{x \to 0} \frac{\arctan x}{\sqrt{x}}$

80. $\displaystyle\lim_{x \to 0} \frac{\arcsin x}{x}$

In Exercises 81 and 82, show that the function defined by the series is a solution of the differential equation.

81. $y = \displaystyle\sum_{n=0}^{\infty} (-1)^n \frac{x^{2n}}{4^n (n!)^2}$

$x^2 y'' + x y' + x^2 y = 0$

82. $y = \displaystyle\sum_{n=0}^{\infty} \frac{(-3)^n x^{2n}}{2^n n!}$

$y'' + 3xy' + 3y = 0$

In Exercises 83–86, use a Taylor polynomial to approximate the function with an error of less than 0.001.

83. $\sin 95°$

84. $\cos(0.75)$

85. $\ln(1.75)$

86. $e^{-0.25}$

87. A Taylor polynomial centered at 0 will be used to approximate the cosine function. Find the degree of the polynomial required to obtain the desired accuracy over the indicated interval.

Maximum Error	Interval
(a) 0.001	$[-0.5, 0.5]$
(b) 0.001	$[-1, 1]$
(c) 0.0001	$[-0.5, 0.5]$
(d) 0.0001	$[-2, 2]$

88. Use a graphing utility to graph the cosine function and the Taylor polynomials in Exercise 87.

Anglo-Australian Observatory 1986

9

Conics, Parametric Equations, and Polar Coordinates

Planets outside our own solar system are difficult to find because they are so dim compared with their parent stars. To discover these planets, astronomers rely on the influence that the planet may have on the star. An orbiting planet's gravitational pull drags the star back and forth as the planet rotates around it. This wobbling results in a subtle red-blue shift in the color of the star's light, known as the Doppler effect. Using a spectrometer, astronomers can monitor a star's Doppler variations, and use the results to calculate details pertaining to the orbiting body.

It was this technique that allowed Geoffrey Marcy and Paul Butler, of San Francisco State University, to identify a body rotating around the star 70 Virginis. They theorize that it is a large planet, 6.6 times as massive as Jupiter, although there is a small probability that it is a brown dwarf star. Marcy and Butler have calculated that the planet, named 70 Vir B, completes an orbit once every 116.6 days.

According to the astronomers, the planet's orbit is an ellipse with an eccentricity of 0.4, and a major axis length of 0.86 AU. (An astronomical unit, or AU, is the mean distance from the earth to the sun, about 93 million miles.) Placed on a rectangular coordinate system and centered at the origin, the equation for this ellipse is

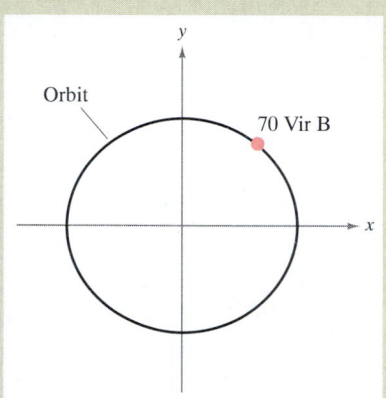

$$\frac{x^2}{0.1849} + \frac{y^2}{0.1553} = 1$$

as shown in the graph at the left.

Rather than using Cartesian coordinates and centering the orbit at the origin, however, astronomers find it convenient to use polar coordinates. Using the sun as the main reference point, or the pole, each point is defined by its distance r from the sun and its angle θ from the horizontal. With the star 70 Virginis as the pole, the new planet's orbit is

$$r = \frac{0.3612}{1 - 0.4 \cos \theta}.$$

Polar equation for orbit of 70 Vir B

Kepler's second law of planetary motion allows you to set up the proportion

$$\frac{t}{\text{Period}} = \frac{\text{area of segment}}{\text{area of ellipse}} = \frac{\frac{1}{2} \int_\alpha^\beta \left(\frac{0.3612}{1 - 0.4 \cos \theta} \right)^2 d\theta}{0.5324}$$

which you can solve to find the time t (in days) that it takes this particular planet to move in its orbit from $\theta = \alpha$ to $\theta = \beta$.

FOR FURTHER INFORMATION For more information on the discovery of the new planet 70 Vir B, see "Searching for Other Worlds" in the Feb. 5, 1996 edition of *Time*.

QUESTIONS

1. Set your graphing utility to polar mode and enter the polar equation for the orbit of 70 Vir B. Graph the equation using a window with θ varying from 0 to π. Then graph the equation again with θ varying from 0 to 2π, and again with θ varying from 0 to 4π. What do you observe?

2. When θ varies from 0 to π, the planet moves through half of its orbit. Starting with $\theta = 0$, what value of θ corresponds to one-fourth of the orbit? Explain.

3. Use the result of Question 2 to estimate the time it takes the planet to travel from $\theta = 0$ through one-quarter of its orbit. Then estimate the time it takes to travel through the second quarter of its orbit. Are these times the same? Describe the motion of this planet. When does it have a maximum speed? When does it have a minimum speed?

The concepts presented here will be explored further in this chapter. For an extension of this application, see the lab series that accompanies this text.

SECTION 9.1 Conics and Calculus

Conic Sections • Parabolas • Ellipses • Hyperbolas

Conic Sections

Each **conic section** (or simply **conic**) can be described as the intersection of a plane and a double-napped cone. Notice in Figure 9.1 that for the four basic conics, the intersecting plane does not pass through the vertex of the cone. When the plane passes through the vertex, the resulting figure is a **degenerate conic,** as shown in Figure 9.2.

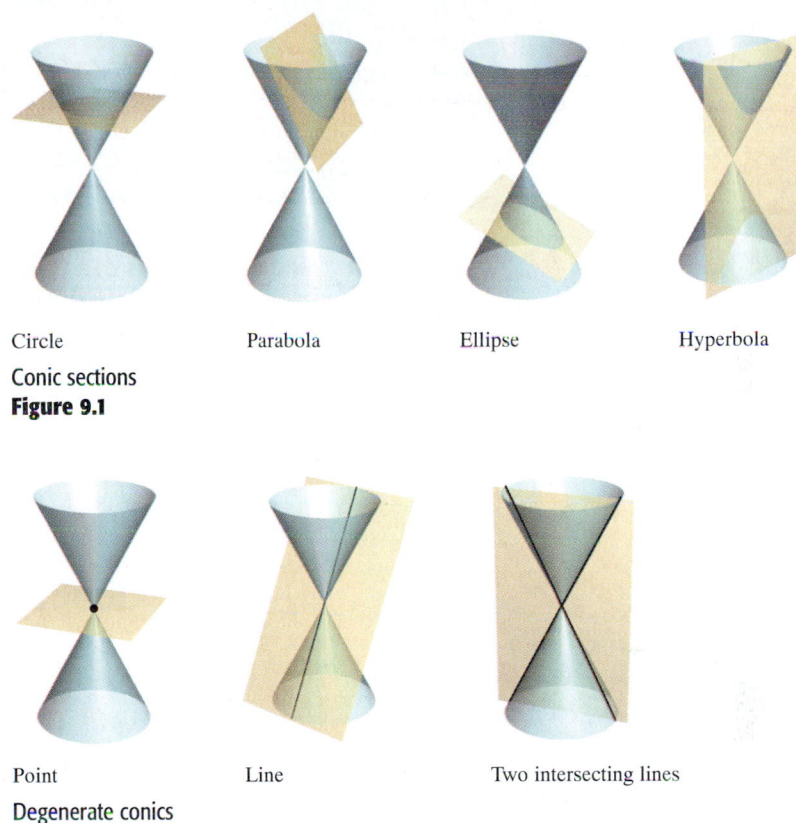

Circle   Parabola   Ellipse   Hyperbola

Conic sections
Figure 9.1

Point   Line   Two intersecting lines

Degenerate conics
Figure 9.2

HYPATIA (370–415 A.D.)

The Greeks discovered conic sections sometime between 600 and 300 B.C. By the beginning of the Alexandrian period, enough was known about conics for Apollonius (262–190 B.C.) to produce an eight-volume work on the subject. Later, toward the end of the Alexandrian period, Hypatia wrote a textbook entitled *On the Conics of Apollonius.* Her death marked the end of major mathematical discoveries in Europe for several hundred years.

The early Greeks were largely concerned with the geometric properties of conics. It was not until 1900 years later, in the early seventeenth century, that the broader applicability of conics became apparent. Conics then played a prominent role in the development of calculus.

There are several ways to study conics. You could begin as the Greeks did by defining the conics in terms of the intersections of planes and cones, or you could define them algebraically in terms of the general second-degree equation

$$Ax^2 + Bxy + Cy^2 + Dx + Ey + F = 0. \qquad \text{General second-degree equation}$$

However, a third approach, in which each of the conics is defined as a **locus** (collection) of points satisfying a certain geometric property, suits our needs best. For example, a circle can be defined as the collection of all points (x, y) that are equidistant from a fixed point (h, k). This locus definition easily produces the standard equation of a circle,

$$(x - h)^2 + (y - k)^2 = r^2. \qquad \text{Standard equation of a circle}$$

FOR FURTHER INFORMATION To learn more about the mathematical activities of Hypatia, see the article "Hypatia and Her Mathematics" by Michael A. B. Deakin in the March 1994 issue of *The American Mathematical Monthly.*

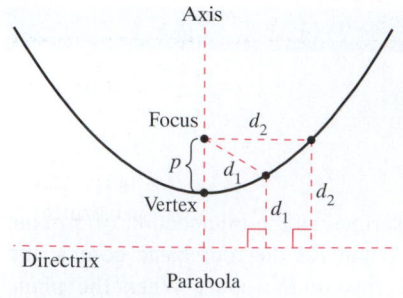

Figure 9.3

Parabolas

A **parabola** is the set of all points (x, y) that are equidistant from a fixed line (**directrix**) and a fixed point (**focus**) not on the line. The midpoint between the focus and the directrix is the **vertex,** and the line passing through the focus and the vertex is the **axis** of the parabola. Note in Figure 9.3 that a parabola is symmetric with respect to its axis.

THEOREM 9.1   Standard Equation of a Parabola

The **standard form** of the equation of a parabola with vertex (h, k) and the directrix $y = k - p$ is

$$(x - h)^2 = 4p(y - k). \qquad \text{Vertical axis}$$

For directrix $x = h - p$, the equation is

$$(y - k)^2 = 4p(x - h). \qquad \text{Horizontal axis}$$

The focus lies on the axis p units (*directed distance*) from the vertex.

EXAMPLE 1   Finding the Focus of a Parabola

Find the focus of the parabola given by $y = -\frac{1}{2}x^2 - x + \frac{1}{2}$.

Solution   To find the focus, convert to standard form by completing the square.

$$y = \frac{1}{2} - x - \frac{1}{2}x^2 \qquad \text{Original equation}$$

$$y = \frac{1}{2}(1 - 2x - x^2) \qquad \text{Factor out } \tfrac{1}{2}.$$

$$2y = 1 - (x^2 + 2x) \qquad \text{Group terms.}$$

$$2y = 2 - (x^2 + 2x + 1) \qquad \text{Add and subtract 1 on right side.}$$

$$x^2 + 2x + 1 = -2y + 2$$

$$(x + 1)^2 = -2(y - 1) \qquad \text{Standard form}$$

Comparing this equation with $(x - h)^2 = 4p(y - k)$, you can conclude that

$$h = -1, \qquad k = 1, \qquad \text{and} \qquad p = -\frac{1}{2}.$$

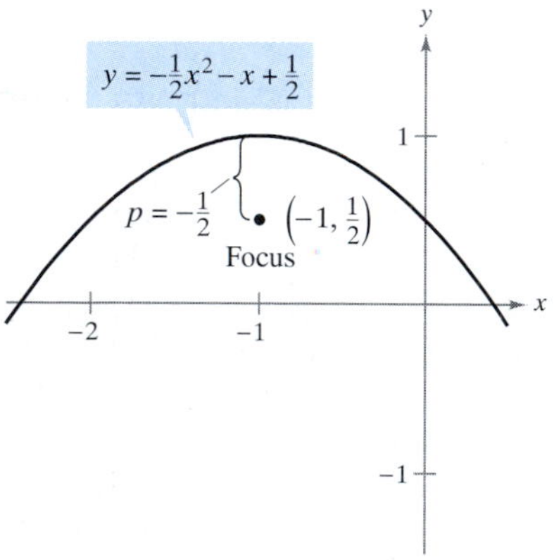

Because p is negative, the parabola opens downward, as shown in Figure 9.4. Therefore, the focus of the parabola is p units from the vertex, or

$$(h, k + p) = \left(-1, \frac{1}{2}\right). \qquad \text{Focus}$$

$y = -\frac{1}{2}x^2 - x + \frac{1}{2}$

$p = -\frac{1}{2}$ Focus $\left(-1, \frac{1}{2}\right)$

Parabola with a vertical axis, $p < 0$
Figure 9.4

A line segment that passes through the focus of a parabola and has endpoints on the parabola is called a **focal chord.** The specific focal chord perpendicular to the axis of the parabola is the **latus rectum.** In the next example, we determine the length of the latus rectum and the length of the corresponding intercepted arc.

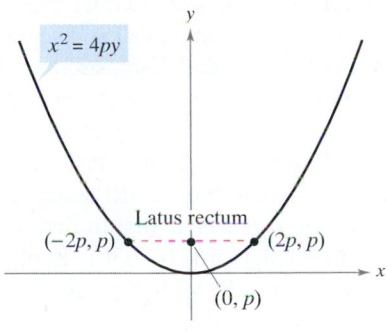

Length of latus rectum: $4p$
Arc length: $4.59p$

The length of the latus rectum is less than the length of the intercepted arc.
Figure 9.5

EXAMPLE 2   **Focal Chord Length and Arc Length**

Find the length of the latus rectum of the parabola given by $x^2 = 4py$. Then find the length of the parabolic arc intercepted by the latus rectum.

Solution   Because the latus rectum passes through the focus $(0, p)$ and is perpendicular to the y-axis, the coordinates of its endpoints are $(-x, p)$ and (x, p). Substituting p for y in the equation of the parabola produces

$$x^2 = 4p(p) \quad \Longrightarrow \quad x = \pm 2p.$$

Thus, the endpoints of the latus rectum are $(-2p, p)$ and $(2p, p)$, and you can conclude that its length is $4p$, as shown in Figure 9.5. In contrast, the length of the intercepted arc is given by the following.

$$
\begin{aligned}
s &= \int_{-2p}^{2p} \sqrt{1 + (y')^2}\, dx \\
&= 2\int_{0}^{2p} \sqrt{1 + \left(\frac{x}{2p}\right)^2}\, dx \\
&= \frac{1}{p}\int_{0}^{2p} \sqrt{4p^2 + x^2}\, dx \\
&= \frac{1}{2p}\left[x\sqrt{4p^2 + x^2} + 4p^2 \ln\left|x + \sqrt{4p^2 + x^2}\right| \right]_{0}^{2p} \qquad \text{Theorem 7.2} \\
&= \frac{1}{2p}\left[2p\sqrt{8p^2} + 4p^2 \ln\left(2p + \sqrt{8p^2}\right) - 4p^2 \ln(2p) \right] \\
&= 2p\left[\sqrt{2} + \ln\left(1 + \sqrt{2}\right) \right] \\
&\approx 4.59p
\end{aligned}
$$

One widely used property of a parabola is its reflective property. In physics, a surface is called **reflective** if the tangent line at any point on the surface makes equal angles with an incoming ray and the resulting outgoing ray. The angle corresponding to the incoming ray is the **angle of incidence,** and the angle corresponding to the outgoing ray is the **angle of reflection.** One example of a reflective surface is a flat mirror.

Another type of reflective surface is that formed by revolving a parabola about its axis. A special property of parabolic reflectors is that they allow us to direct all incoming rays parallel to the axis through the focus of the parabola—this is the principle behind the design of the parabolic mirrors used in reflecting telescopes. Conversely, all light rays emanating from the focus of a parabolic reflector used in a flashlight are parallel, as shown in Figure 9.6.

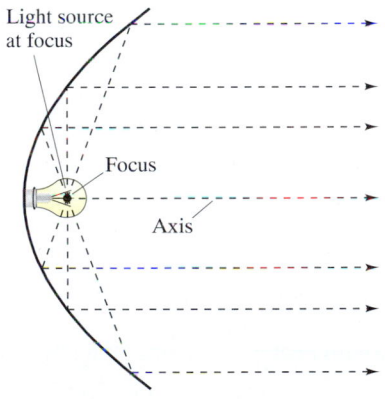

Parabolic reflector: light is reflected in parallel rays.
Figure 9.6

THEOREM 9.2   Reflective Property of a Parabola

Let P be a point on a parabola. The tangent line to the parabola at the point P makes equal angles with the following two lines.

1. The line passing through P and the focus.
2. The line passing through P parallel to the axis of the parabola.

NICOLAUS COPERNICUS (1473–1543)

Copernicus began to study planetary motion when asked to revise the calendar. At that time, the exact length of the year could not be accurately predicted using the theory that the earth was the center of the universe.

Ellipses

More than a thousand years after the close of the Alexandrian period of Greek mathematics, Western civilization finally began a Renaissance of mathematical and scientific discovery. One of the principal figures in this rebirth was the Polish astronomer Nicolaus Copernicus. In his work *On the Revolutions of the Heavenly Spheres,* Copernicus claimed that all of the planets, including the earth, revolved about the sun in circular orbits. Although some of Copernicus's claims were invalid, the controversy set off by his heliocentric theory motivated astronomers to search for a mathematical model to explain the observed movements of the sun and planets. The first to find the correct model was the German astronomer Johannes Kepler (1571–1630). Kepler discovered that the planets move about the sun in elliptical orbits, with the sun not as the center but as a focal point of the orbit.

The use of ellipses to explain the movement of the planets is only one of many practical and aesthetic uses. As with parabolas, we begin our study of this second type of conic by defining it as a locus of points. Now, however, we use *two* focal points rather than one.

An **ellipse** is the set of all points (x, y) the sum of whose distances from two distinct fixed points (**foci**) is constant. (See Figure 9.7.) The line through the foci intersects the ellipse at two points, called the **vertices.** The chord joining the vertices is the **major axis,** and its midpoint is the **center** of the ellipse. The chord perpendicular to the major axis at the center is the **minor axis** of the ellipse.

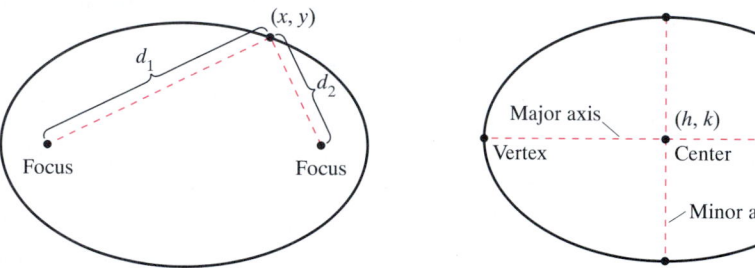

Figure 9.7

THEOREM 9.3   Standard Equation of an Ellipse

The standard form of the equation of an ellipse, with center (h, k) and major and minor axes of lengths $2a$ and $2b$, where $a > b$, is

$$\frac{(x - h)^2}{a^2} + \frac{(y - k)^2}{b^2} = 1 \qquad \text{Major axis is horizontal.}$$

$$\frac{(x - h)^2}{b^2} + \frac{(y - k)^2}{a^2} = 1. \qquad \text{Major axis is vertical.}$$

The foci lie on the major axis, c units from the center, with $c^2 = a^2 - b^2$.

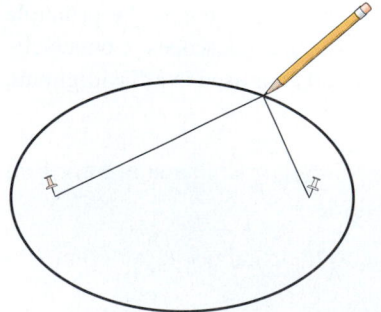

Figure 9.8

NOTE   You can visualize the definition of an ellipse by imagining two thumbtacks placed at the foci, as shown in Figure 9.8. If the ends of a fixed length of string are fastened to the thumbtacks and the string is drawn taut with a pencil, the path traced by the pencil will be an ellipse.

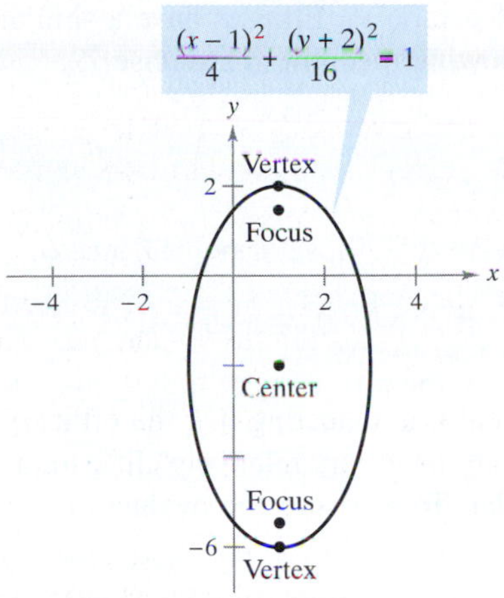

$$\frac{(x-1)^2}{4} + \frac{(y+2)^2}{16} = 1$$

Ellipse with a vertical major axis
Figure 9.9

NOTE   If the constant term $F = -8$ in the equation in Example 3 had been greater than or equal to 8, we would have obtained one of the following degenerate cases.

1. $F = 8$, single point, $(1, -2)$:

$$\frac{(x-1)^2}{4} + \frac{(y+2)^2}{16} = 0$$

2. $F > 8$, no solution points:

$$\frac{(x-1)^2}{4} + \frac{(y+2)^2}{16} < 0$$

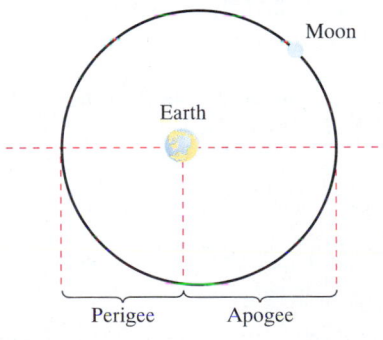

Figure 9.10

EXAMPLE 3   Completing the Square

Find the center, vertices, and foci of the ellipse given by

$$4x^2 + y^2 - 8x + 4y - 8 = 0.$$

Solution   By completing the square, you can write the given equation in standard form.

$$4x^2 + y^2 - 8x + 4y - 8 = 0 \qquad \text{Original equation}$$
$$4x^2 - 8x + y^2 + 4y = 8$$
$$4(x^2 - 2x + 1) + (y^2 + 4y + 4) = 8 + 4 + 4$$
$$4(x-1)^2 + (y+2)^2 = 16$$
$$\frac{(x-1)^2}{4} + \frac{(y+2)^2}{16} = 1 \qquad \text{Standard form}$$

Thus, the major axis is parallel to the y-axis, where $h = 1$, $k = -2$, $a = 4$, $b = 2$, and $c = \sqrt{16 - 4} = 2\sqrt{3}$. Therefore, you obtain the following.

Center: $(1, -2)$ (h, k)
Vertices: $(1, -6)$ and $(1, 2)$ $(h, k \pm a)$
Foci: $\left(1, -2 - 2\sqrt{3}\right)$ and $\left(1, -2 + 2\sqrt{3}\right)$ $(h, k \pm c)$

The graph of the ellipse is shown in Figure 9.9.

EXAMPLE 4   The Orbit of the Moon

The moon orbits the earth in an elliptical path with the center of the earth at one focus, as shown in Figure 9.10. The major and minor axes of the orbit have lengths of 768,806 kilometers and 767,746 kilometers. Find the greatest and least distances (the apogee and perigee) from the earth's center to the moon's center.

Solution   Begin by solving for a and b.

$$2a = 768{,}806 \qquad \text{Length of major axis}$$
$$a = 384{,}403 \qquad \text{Solve for } a.$$
$$2b = 767{,}746 \qquad \text{Length of minor axis}$$
$$b = 383{,}873 \qquad \text{Solve for } b.$$

Now, using these values, you can solve for c as follows.

$$c = \sqrt{a^2 - b^2}$$
$$\approx 20{,}179$$

Therefore, the greatest distance between the center of the earth and the center of the moon is

$$a + c \approx 404{,}582 \text{ km}$$

and the least distance is

$$a - c \approx 364{,}224 \text{ km}.$$

FOR FURTHER INFORMATION For more information on some uses of the reflective properties of conics, see the article "Parabolic Mirrors, Elliptic and Hyperbolic Lenses" by Mohsen Maesumi in the June-July 1992 issue of *The American Mathematical Monthly*. Also see the article "The Geometry of Microwave Antennas" by William R. Parzynski in the April 1984 issue of *Mathematics Teacher*.

Theorem 9.2 presents a reflective property of parabolas. Ellipses have a similar reflective property. You are asked to prove the following theorem in Exercise 72.

THEOREM 9.4   Reflective Property of an Ellipse

Let P be a point on an ellipse. The tangent line to the ellipse at point P makes equal angles with the lines through P and the foci.

One of the reasons that astronomers had difficulty in detecting that the orbits of the planets are ellipses is that the foci of the planetary orbits are relatively close to the center of the sun, making the orbits nearly circular. To measure the ovalness of an ellipse, we use the concept of **eccentricity.**

Definition of Eccentricity of an Ellipse

The **eccentricity** e of an ellipse is given by the ratio

$$e = \frac{c}{a}.$$

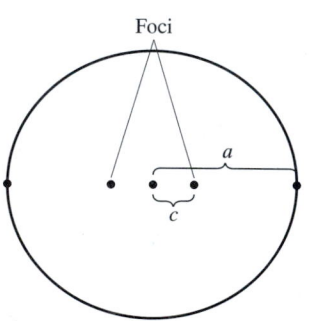

(a) $\frac{c}{a}$ is small.

To see how this ratio is used to describe the shape of an ellipse, note that because the foci of an ellipse are located along the major axis between the vertices and the center, it follows that

$$0 < c < a.$$

For an ellipse that is nearly circular, the foci are close to the center and the ratio c/a is small, and for an elongated ellipse, the foci are close to the vertices and the ratio is close to 1, as shown in Figure 9.11. Note that $0 < e < 1$ for every ellipse.

The orbit of the moon has an eccentricity of $e = 0.0549$, and the eccentricities of the nine planetary orbits are as follows.

Mercury:	$e = 0.2056$	Saturn:	$e = 0.0543$
Venus:	$e = 0.0068$	Uranus:	$e = 0.0460$
Earth:	$e = 0.0167$	Neptune:	$e = 0.0082$
Mars:	$e = 0.0934$	Pluto:	$e = 0.2481$
Jupiter:	$e = 0.0484$		

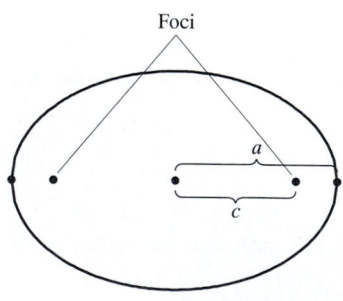

(b) $\frac{c}{a}$ is close to 1.

Eccentricity is the ratio $\frac{c}{a}$.

Figure 9.11

You can use integration to show that the area of an ellipse is $A = \pi ab$. For instance, the area of the ellipse $(x^2/a^2) + (y^2/b^2) = 1$ is given by

$$A = 4 \int_0^a \frac{b}{a} \sqrt{a^2 - x^2} \, dx$$

$$= \frac{4b}{a} \int_0^{\pi/2} a^2 \cos^2 \theta \, d\theta. \qquad \text{Trigonometric substitution } x = a \sin \theta.$$

However, it is not so simple to find the *circumference* of an ellipse. The next example shows how to use eccentricity to set up an "elliptic integral" for the circumference of an ellipse.

EXAMPLE 5   Finding the Circumference of an Ellipse

Show that the circumference of the ellipse $(x^2/a^2) + (y^2/b^2) = 1$ is

$$4a \int_0^{\pi/2} \sqrt{1 - e^2 \sin^2 \theta} \, d\theta. \qquad e = \frac{c}{a}$$

Solution   Because the given ellipse is symmetric with respect to both the x-axis and the y-axis, you know that its circumference C is four times the arc length of $y = (b/a)\sqrt{a^2 - x^2}$ in the first quadrant. The function y is differentiable for all x in the interval $[0, a]$ except at $x = a$. Thus, the circumference is given by the following improper integral.

$$C = \lim_{d \to a} 4 \int_0^d \sqrt{1 + (y')^2} \, dx = 4 \int_0^a \sqrt{1 + (y')^2} \, dx = 4 \int_0^a \sqrt{1 + \frac{b^2 x^2}{a^2(a^2 - x^2)}} \, dx$$

Using the trigonometric substitution $x = a \sin \theta$, you obtain

$$C = 4 \int_0^{\pi/2} \sqrt{1 + \frac{b^2 \sin^2 \theta}{a^2 \cos^2 \theta}} (a \cos \theta) \, d\theta$$

$$= 4 \int_0^{\pi/2} \sqrt{a^2 \cos^2 \theta + b^2 \sin^2 \theta} \, d\theta$$

$$= 4 \int_0^{\pi/2} \sqrt{a^2(1 - \sin^2 \theta) + b^2 \sin^2 \theta} \, d\theta$$

$$= 4 \int_0^{\pi/2} \sqrt{a^2 - (a^2 - b^2)\sin^2 \theta} \, d\theta.$$

Because $e^2 = c^2/a^2 = (a^2 - b^2)/a^2$, you can rewrite this integral as

$$C = 4a \int_0^{\pi/2} \sqrt{1 - e^2 \sin^2 \theta} \, d\theta.$$

> **AREA AND CIRCUMFERENCE OF AN ELLIPSE**
>
> In his work with elliptic orbits in the early 1600's, Johannes Kepler successfully developed a formula for the area of an ellipse, $A = \pi ab$. He was less successful in developing a formula for the circumference of an ellipse, however, the best he could do was to give the approximate formula $C = \pi(a + b)$.

A great deal of time has been devoted to the study of elliptic integrals. Such integrals generally do not have elementary antiderivatives. To find the circumference of an ellipse, you must usually resort to an approximation technique.

EXAMPLE 6   Approximating the Value of an Elliptic Integral

Use the elliptic integral in Example 5 to approximate the circumference of the ellipse $(x^2/25) + (y^2/16) = 1.$

Solution   Because $e^2 = c^2/a^2 = (a^2 - b^2)/a^2 = 9/25$, you have

$$C = (4)(5) \int_0^{\pi/2} \sqrt{1 - \frac{9 \sin^2 \theta}{25}} \, d\theta.$$

Applying Simpson's Rule with $n = 4$ produces

$$C \approx 20 \left(\frac{\pi}{6} \right) \left(\frac{1}{4} \right) [1 + 4(0.9733) + 2(0.9055) + 4(0.8323) + 0.8]$$

$$\approx 28.36.$$

Hyperbolas

The definition of a hyperbola is similar to that of an ellipse. For an ellipse, the *sum* of the distances between the foci and a point on the ellipse is fixed, whereas for a hyperbola, the absolute value of the *difference* between these distances is fixed.

A **hyperbola** is the set of all points (x, y) for which the absolute value of the difference between the distances from two distinct fixed points (foci) is constant. (See Figure 9.12.) The line through the two foci intersects a hyperbola at two points called the **vertices**. The line segment connecting the vertices is the **transverse axis,** and the midpoint of the transverse axis is the **center** of the hyperbola. One distinguishing feature of a hyperbola is that its graph has two separate *branches*.

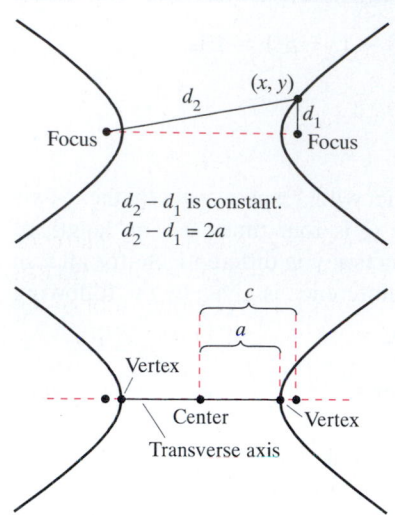

Figure 9.12

THEOREM 9.5   Standard Equation of a Hyperbola

The standard form of the equation of a hyperbola with center at (h, k) is

$$\frac{(x - h)^2}{a^2} - \frac{(y - k)^2}{b^2} = 1 \qquad \text{Transverse axis is horizontal.}$$

or

$$\frac{(y - k)^2}{a^2} - \frac{(x - h)^2}{b^2} = 1. \qquad \text{Transverse axis is vertical.}$$

The vertices are a units from the center, and the foci are c units from the center. Moreover, $b^2 = c^2 - a^2$.

An important aid in sketching the graph of a hyperbola is the determination of its **asymptotes,** as shown in Figure 9.13. Each hyperbola has two asymptotes that intersect at the center of the hyperbola. The asymptotes pass through the vertices of a rectangle of dimensions $2a$ by $2b$, with its center at (h, k). The line segment of length $2b$ joining $(h, k + b)$ and $(h, k - b)$ is referred to as the **conjugate axis** of the hyperbola.

THEOREM 9.6   Asymptotes of a Hyperbola

For a *horizontal* transverse axis, the equations of the asymptotes are

$$y = k + \frac{b}{a}(x - h) \qquad \text{and} \qquad y = k - \frac{b}{a}(x - h).$$

For a *vertical* transverse axis, the equations of the asymptotes are

$$y = k + \frac{a}{b}(x - h) \qquad \text{and} \qquad y = k - \frac{a}{b}(x - h).$$

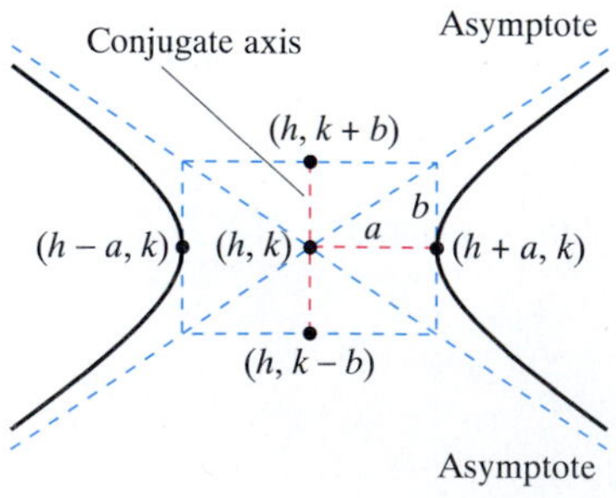

In Figure 9.13 you can see that the asymptotes coincide with the diagonals of the rectangle with dimensions $2a$ and $2b$, centered at (h, k). This provides you with a quick means of sketching the asymptotes, which in turn aids in sketching the hyperbola.

Figure 9.13

> **EXAMPLE 7   Using Asymptotes to Sketch a Hyperbola**

Sketch the graph of the hyperbola whose equation is $4x^2 - y^2 = 16$.

> **TECHNOLOGY**   You can use a graphing utility to verify the graph obtained in Example 7 by solving the original equation for y and graphing the following.
>
> $$y_1 = \sqrt{4x^2 - 16}$$
>
> $$y_2 = -\sqrt{4x^2 - 16}$$

Solution   Begin by rewriting the equation in standard form.

$$\frac{x^2}{2^2} - \frac{y^2}{4^2} = 1$$

The transverse axis is horizontal and the vertices occur at $(-2, 0)$ and $(2, 0)$. The ends of the conjugate axis occur at $(0, -4)$ and $(0, 4)$. Using these four points, you can sketch the rectangle shown in Figure 9.14(a). By drawing the asymptotes through the corners of this rectangle, you can complete the sketch shown in Figure 9.14(b).

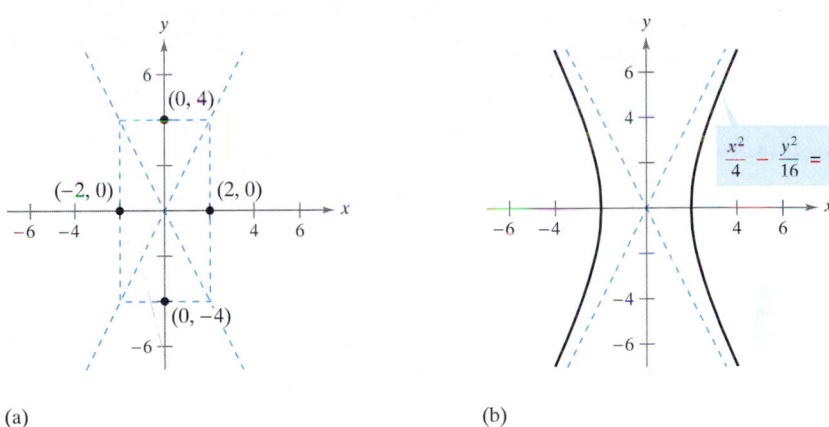

(a)   (b)

Figure 9.14

As with an ellipse, the **eccentricity** of a hyperbola is $e = c/a$. Because $c > a$ for hyperbolas, it follows that $e > 1$ for hyperbolas. If the eccentricity is large, the branches of the hyperbola are nearly flat. If the eccentricity is close to 1, the branches of the hyperbola are more pointed, as shown in Figure 9.15.

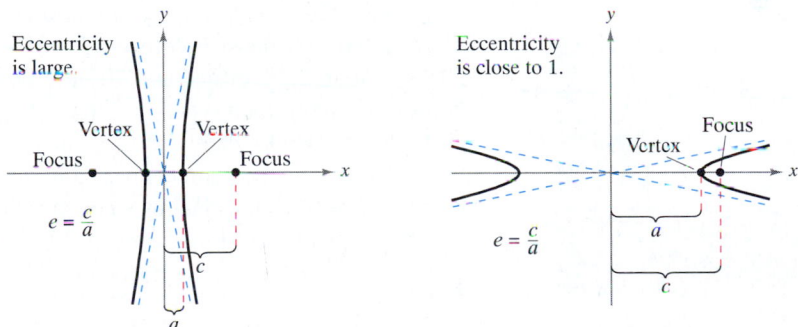

Figure 9.15

> **Definition of Eccentricity of a Hyperbola**
>
> The eccentricity e of a hyperbola is given by the ratio
>
> $$e = \frac{c}{a}.$$

The following application was developed during World War II. It shows how the properties of hyperbolas can be used in radar and other detection systems.

EXAMPLE 8   A Hyperbolic Detection System

Two microphones, 1 mile apart, record an explosion. Microphone A receives the sound 2 seconds before microphone B. Where was the explosion?

Solution   Assuming that sound travels at 1100 feet per second, you know that the explosion took place 2200 feet farther from B than from A, as shown in Figure 9.16. The locus of all points that are 2200 feet closer to A than to B is one branch of the hyperbola $(x^2/a^2) - (y^2/b^2) = 1$, where

$$c = \frac{1 \text{ mile}}{2} = \frac{5280 \text{ ft}}{2} = 2640 \text{ ft}$$

and

$$a = \frac{2200 \text{ ft}}{2} = 1100 \text{ ft.}$$

Thus, it follows that

$$b^2 = c^2 - a^2 = 5,759,600$$

and you can conclude that the explosion occurred somewhere on the right branch of the hyperbola given by

$$\frac{x^2}{1,210,000} - \frac{y^2}{5,759,600} = 1.$$

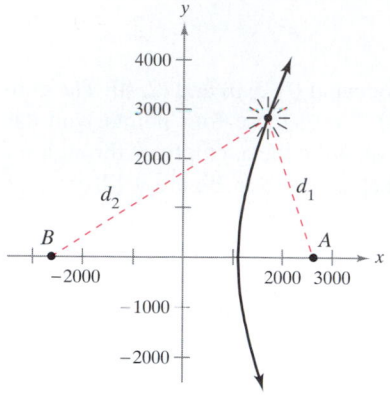

$2c = 5280$
$d_2 - d_1 = 2a = 2200$

The locus of all points that are 2200 feet closer to A than to B is one branch of the hyperbola with $c = 2640$ and $a = 1100$.
Figure 9.16

In Example 8, you were able to determine only the hyperbola on which the explosion occurred, but not the exact location of the explosion. If, however, you had received the sound at a third position, C, then two other hyperbolas would be determined. The exact location of the explosion would be the point at which these three hyperbolas intersect.

Another interesting application of conics involves the orbits of comets in our solar system. Of the 610 comets identified prior to 1970, 245 have elliptical orbits, 295 have parabolic orbits, and 70 have hyperbolic orbits. The center of the sun is a focus of each orbit, and each orbit has a vertex at the point at which the comet is closest to the sun. Undoubtedly, many comets with parabolic or hyperbolic orbits have not been identified—such comets pass through our solar system once. Only comets with elliptical orbits such as Halley's comet remain in our solar system.

The type of orbit for a comet can be determined as follows.

1. Ellipse: $v < \sqrt{2GM/p}$
2. Parabola: $v = \sqrt{2GM/p}$
3. Hyperbola: $v > \sqrt{2GM/p}$

In these three formulas, p is the distance between one vertex and one focus of the comet's orbit (in meters), v is the velocity of the comet at the vertex (in meters per second), $M \approx 1.991 \times 10^{30}$ kilograms is the mass of the sun, and $G \approx 6.67 \times 10^{-11}$ cubic meters per kilogram-second squared is the gravitational constant.

CAROLINE HERSCHEL (1750–1848)

The first woman to be credited with detecting a new comet was the English astronomer Caroline Herschel. During her long life, Caroline Herschel discovered a total of eight new comets.

EXERCISES FOR SECTION 9.1

In Exercises 1–8, match the equation with its graph. [The graphs are labeled (a), (b), (c), (d), (e), (f), (g), and (h).]

(a)
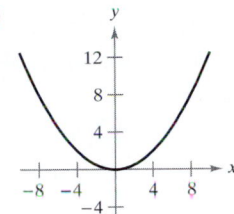

(b)
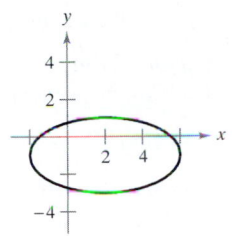

(c)
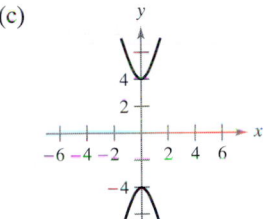

(d)
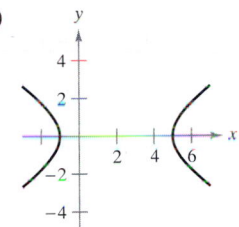

(e)
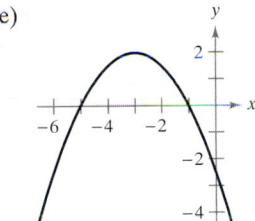

(f)
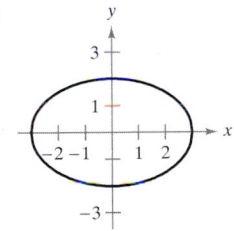

(g)
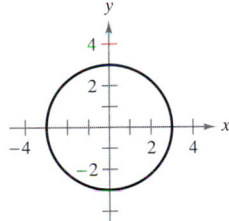

(h)
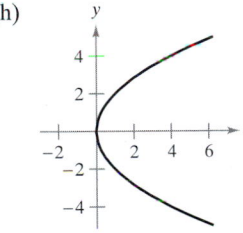

1. $y^2 = 4x$

2. $x^2 = 8y$

3. $(x + 3)^2 = -2(y - 2)$

4. $\dfrac{(x - 2)^2}{16} + \dfrac{(y + 1)^2}{4} = 1$

5. $\dfrac{x^2}{9} + \dfrac{y^2}{4} = 1$

6. $\dfrac{x^2}{9} + \dfrac{y^2}{9} = 1$

7. $\dfrac{y^2}{16} - \dfrac{x^2}{1} = 1$

8. $\dfrac{(x - 2)^2}{9} - \dfrac{y^2}{4} = 1$

In Exercises 9–16, find the vertex, focus, and directrix of the parabola, and sketch its graph.

9. $y^2 = -6x$

10. $x^2 + 8y = 0$

11. $(x + 3) + (y - 2)^2 = 0$

12. $(x - 1)^2 + 8(y + 2) = 0$

13. $y^2 - 4y - 4x = 0$

14. $y^2 + 6y + 8x + 25 = 0$

15. $x^2 + 4x + 4y - 4 = 0$

16. $y^2 + 4y + 8x - 12 = 0$

In Exercises 17–20, find the vertex, focus, and directrix of the parabola. Then use a graphing utility to graph the parabola.

17. $y^2 + x + y = 0$

18. $y = -\dfrac{1}{6}(x^2 + 4x - 2)$

19. $y^2 - 4x - 4 = 0$

20. $x^2 - 2x + 8y + 9 = 0$

In Exercises 21–28, find an equation of the parabola.

21. Vertex: $(3, 2)$
Focus: $(1, 2)$

22. Vertex: $(-1, 2)$
Focus: $(-1, 0)$

23. Vertex: $(0, 4)$
Directrix: $y = -2$

24. Focus: $(2, 2)$
Directrix: $x = -2$

25.
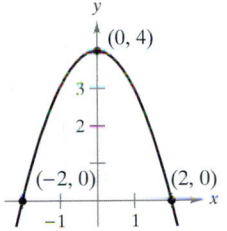

26.
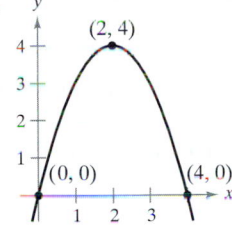

27. Axis is parallel to y-axis; graph passes through $(0, 3)$, $(3, 4)$, and $(4, 11)$.

28. Directrix: $y = -2$; endpoints of latus rectum are $(0, 2)$ and $(8, 2)$.

29. *Solar Collector* A solar collector for heating water is constructed with a sheet of stainless steel that is formed into the shape of a parabola (see figure). The water will flow through a pipe that is located at the focus of the parabola. At what distance is the pipe from the vertex?

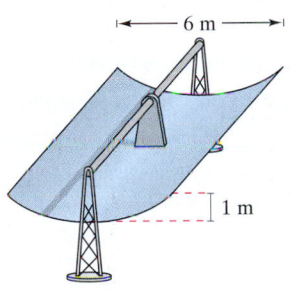

Figure for 29   **Figure for 30**

30. *Beam Deflection* A simply supported beam that is 16 meters long has a load concentrated at the center (see figure). The deflection of the beam at its center is 3 centimeters. Assume that the shape of the deflected beam is parabolic.

(a) Find an equation of the parabola. (Assume that the origin is at the center of the beam.)

(b) How far from the center of the beam is the deflection 1 centimeter?

31. Find an equation of the tangent line to the parabola $y = ax^2$ at $x = x_0$. Prove that the x-intercept of this tangent line is $(x_0/2, 0)$.

32. (a) Prove that any two distinct tangent lines to a parabola intersect.

(b) Demonstrate the result in part (a) by finding the point of intersection of the tangent lines to the parabola $x^2 - 4x - 4y = 0$ at the points $(0, 0)$ and $(6, 3)$.

33. (a) Prove that if any two tangent lines to a parabola intersect at right angles, their point of intersection must lie on the directrix.

(b) Demonstrate the result in part (a) by showing that the tangent lines to the parabola $x^2 - 4x - 4y + 8 = 0$ at the points $(-2, 5)$ and $(3, \frac{5}{4})$ intersect at right angles, and that the point of intersection lies on the directrix.

34. Find the point on the graph of $x^2 = 8y$ that is closest to the focus of the parabola.

35. *Radio and Television Reception*   In mountainous areas, reception of radio and television is sometimes poor. Consider an idealized case where a hill is represented by the graph of the parabola $y = x - x^2$, a transmitter is located at the point $(-1, 1)$, and a receiver is located on the other side of the hill at the point $(x_0, 0)$. What is the closest the receiver can be to the hill so that the reception is unobstructed?

36. *Modeling Data*   Per capita whole milk consumption C (in pounds) in the United States for selected years is given in the table.   *(Source: U.S. Department of Agriculture)*

Year	1970	1975	1980	1985	1990	1993
C	213.5	174.9	141.7	119.7	87.6	77.8

(a) Use the regression capabilities of a graphing utility to find a quadratic model for the data where t is the time in years, with $t = 0$ corresponding to 1970.

(b) Use a graphing utility to plot the data and graph the model.

(c) Use the model to estimate the year when consumption was a minimum.

37. *Architecture*   A church window is bounded on top by a parabola and below by the arc of a circle (see figure). Find the surface area of the window.

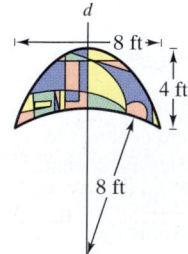

Figure for 37   **Figure for 39**

38. *Arc Length*   Find the arc length of the parabola $4x - y^2 = 0$ over the interval $0 \le y \le 4$.

39. *Bridge Design*   A cable of a suspension bridge is suspended (in the shape of a parabola) between two towers that are 120 meters apart and 20 meters above the roadway (see figure). The cables touch the roadway midway between the towers.

(a) Find an equation for the parabolic shape of each cable.

(b) Find the length of the parabolic supporting cable.

40. *Surface Area*   A satellite-signal receiving dish is formed by revolving the parabola given by the graph of

$$x^2 = 20y$$

about the y-axis. If the radius of the dish is r feet, verify that the surface area of the dish is given by

$$2\pi \int_0^r x \sqrt{1 + \left(\frac{x}{10}\right)^2}\, dx = \frac{\pi}{15}[(100 + r^2)^{3/2} - 1000].$$

In Exercises 41–46, find the center, foci, vertices, and eccentricity of the ellipse, and sketch its graph.

41. $x^2 + 4y^2 = 4$

42. $5x^2 + 7y^2 = 70$

43. $\dfrac{(x - 1)^2}{9} + \dfrac{(y - 5)^2}{25} = 1$

44. $(x + 2)^2 + \dfrac{(y + 4)^2}{1/4} = 1$

45. $9x^2 + 4y^2 + 36x - 24y + 36 = 0$

46. $16x^2 + 25y^2 - 32x + 50y + 31 = 0$

In Exercises 47–50, find the center, foci, and vertices of the ellipse. Use a graphing utility to graph the ellipse. (Explain how you used the graphing utility to obtain the graph.)

47. $12x^2 + 20y^2 - 12x + 40y - 37 = 0$

48. $36x^2 + 9y^2 + 48x - 36y + 43 = 0$

49. $x^2 + 2y^2 - 3x + 4y + 0.25 = 0$

50. $2x^2 + y^2 + 4.8x - 6.4y + 3.12 = 0$

In Exercises 51–56, find an equation of the ellipse.

51. Center: $(0, 0)$
Focus: $(2, 0)$
Vertex: $(3, 0)$

52. Vertices: $(0, 2), (4, 2)$
Eccentricity: $\frac{1}{2}$

53. Vertices: $(3, 1), (3, 9)$
Minor axis length: 6

54. Foci: $(0, \pm 5)$
Major axis length: 14

55. Center: $(0, 0)$
Major axis: horizontal
Points on the ellipse:
$(3, 1), (4, 0)$

56. Center: $(1, 2)$
Major axis: vertical
Points on the ellipse:
$(1, 6), (3, 2)$

57. Sketch a graph of an ellipse that consists of all points (x, y) such that the sum of the distances between (x, y) and two fixed points is 16 units, and the foci are located at the centers of the two sets of concentric circles in the figure.

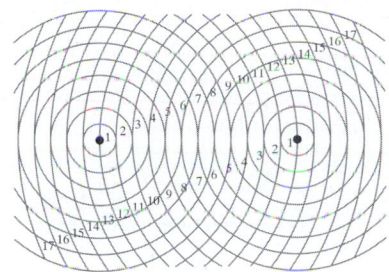

58. *Writing*   On page 640, it was noted that an ellipse can be drawn using two thumbtacks, a string of fixed length (greater than the distance between the tacks), and a pencil. If the ends of the string are fastened at the tacks and the string is drawn taut with a pencil, the path traced by the pencil will be an ellipse.

(a) What is the length of the string in terms of a?

(b) Explain why the path is an ellipse.

59. *Construction of a Semielliptical Arch*   A fireplace arch is to be constructed in the shape of a semiellipse. The opening is to have a height of 2 feet at the center and a width of 5 feet along the base (see figure). The contractor draws the outline of the ellipse by the method shown in Exercise 58. Where should the tacks be placed and what should be the length of the piece of string?

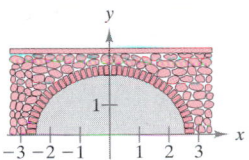

60. *Orbit of the Earth*   The earth moves in an elliptical orbit with the sun at one of the foci. The length of half of the major axis is 14,957,000 kilometers, and the eccentricity is 0.0167. Find the smallest distance (*perihelion*) and the greatest distance (*aphelion*) of the earth from the sun.

61. *Satellite Orbit*   If the apogee and the perigee of an elliptical orbit of an earth satellite are given by A and P, show that the eccentricity of the orbit is

$$e = \frac{A - P}{A + P}.$$

62. *Explorer 18*   On November 26, 1963, the United States launched Explorer 18. Its low and high points over the surface of the earth were 119 miles and 122,000 miles. Find the eccentricity of its elliptical orbit.

63. *Halley's Comet*   Probably the most famous of all comets, Halley's comet, has an elliptical orbit with the sun at the focus. Its maximum distance from the sun is approximately 35.34 AU (astronomical unit $\approx 92.956 \times 10^6$ miles), and its minimum distance is approximately 0.59 AU. Find the eccentricity of the orbit.

64. The equation of an ellipse with its center at the origin can be written as

$$\frac{x^2}{a^2} + \frac{y^2}{a^2(1 - e^2)} = 1.$$

Show that as $e \to 0$, with a remaining fixed, the ellipse approaches a circle.

In Exercises 65 and 66, determine the points at which dy/dx is zero or undefined to locate the endpoints of the major and minor axes of the ellipse.

65. $16x^2 + 9y^2 + 96x + 36y + 36 = 0$

66. $9x^2 + 4y^2 + 36x - 24y + 36 = 0$

67. Consider a particle traveling clockwise on the elliptical path $x^2/100 + y^2/25 = 1$. The particle leaves the orbit at the point $(-8, 3)$ and travels in a straight line tangent to the ellipse. At what point will the particle cross the y-axis.

68. *Volume*   The water tank on a fire truck is 16 feet long, and its cross sections are ellipses. Find the volume of water in the partially filled tank as shown in the figure.

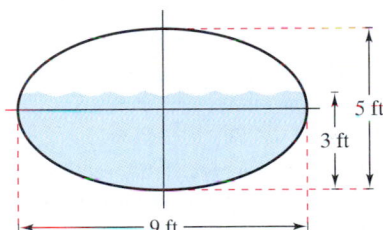

Area and Volume   In Exercises 69 and 70, find (a) the area of the region bounded by the ellipse, (b) the volume and surface area of the solid generated by revolving the region about its major axis (prolate spheroid), and (c) the volume and surface area of the solid generated by revolving the region about its minor axis (oblate spheroid).

69. $\dfrac{x^2}{4} + \dfrac{y^2}{1} = 1$ **70.** $\dfrac{x^2}{16} + \dfrac{y^2}{9} = 1$

 71. *Arc Length*   Use the integration capabilities of a graphing utility to approximate to two-decimal-place accuracy the elliptical integral representing the circumference of the ellipse.

$$\frac{x^2}{9} + \frac{y^2}{16} = 1$$

72. Prove that the tangent line to an ellipse at a point P makes equal angles with lines through P and the foci (see figure). [*Hint:* (1) Find the slope of the tangent line at P, (2) find the slopes of the lines through P and each focus, and (3) use the formula for the tangent of the angle between two lines.]

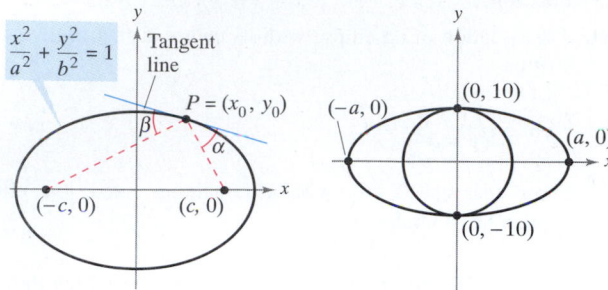

Figure for 72   **Figure for 73**

73. *Geometry* The area of the ellipse in the figure is twice the area of the circle. What is the length of the major axis?

 74. *Conjecture*

(a) Show that the equation of an ellipse can be written as

$$\frac{(x - h)^2}{a^2} + \frac{(y - k)^2}{a^2(1 - e^2)} = 1.$$

(b) Use a graphing utility to graph the ellipse

$$\frac{(x - 2)^2}{4} + \frac{(y - 3)^2}{4(1 - e^2)} = 1$$

for $e = 0.95$, $e = 0.75$, $e = 0.5$, $e = 0.25$, and $e = 0$.

(c) Use the results in part (b) to make a conjecture about the change in the shape of the ellipse as e approaches 0.

In Exercises 75–82, find the center, foci, and vertices of the hyperbola, and sketch its graph using asymptotes as an aid.

75. $y^2 - \dfrac{x^2}{4} = 1$

76. $\dfrac{x^2}{36} - \dfrac{y^2}{4} = 1$

77. $\dfrac{(x - 1)^2}{4} - \dfrac{(y + 2)^2}{1} = 1$

78. $\dfrac{(y + 1)^2}{144} - \dfrac{(x - 4)^2}{25} = 1$

79. $9x^2 - y^2 - 36x - 6y + 18 = 0$

80. $y^2 - 9x^2 + 36x - 72 = 0$

81. $x^2 - 9y^2 + 2x - 54y - 80 = 0$

82. $9x^2 - 4y^2 + 54x + 8y + 78 = 0$

 In Exercises 83–86, find the center, foci, and vertices of the hyperbola. Use a graphing utility to graph the hyperbola and its asymptotes.

83. $9y^2 - x^2 + 2x + 54y + 62 = 0$

84. $9x^2 - y^2 + 54x + 10y + 55 = 0$

85. $3x^2 - 2y^2 - 6x - 12y - 27 = 0$

86. $3y^2 - x^2 + 6x - 12y = 0$

In Exercises 87–94, find an equation of the hyperbola.

87. Vertices: $(\pm 1, 0)$
Asymptotes: $y = \pm 3x$

88. Vertices: $(0, \pm 3)$
Asymptotes: $y = \pm 3x$

89. Vertices: $(2, \pm 3)$
Point on graph: $(0, 5)$

90. Vertices: $(2, \pm 3)$
Foci: $(2, \pm 5)$

91. Center: $(0, 0)$
Vertex: $(0, 2)$
Focus: $(0, 4)$

92. Center: $(0, 0)$
Vertex: $(3, 0)$
Focus: $(5, 0)$

93. Vertices: $(0, 2), (6, 2)$
Asymptotes: $y = \frac{2}{3}x$
$y = 4 - \frac{2}{3}x$

94. Focus: $(10, 0)$
Asymptotes: $y = \pm \frac{3}{4}x$

95. Find an equation of the hyperbola such that for any point on the hyperbola, the difference between its distance from the points $(2, 2)$ and $(10, 2)$ is 6.

96. Sketch a graph of a hyperbola that consists of all points (x, y) such that the difference of the distances between (x, y) and two fixed points is 10 units, and the foci are located at the centers of the two sets of concentric circles in the figure.

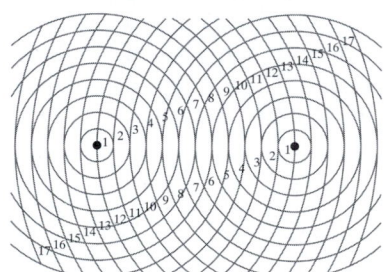

97. Find an equation of the hyperbola such that for any point on the hyperbola, the difference between its distances from the points $(-3, 0)$ and $(-3, 3)$ is 2.

98. Consider a hyperbola centered at the origin with a horizontal transverse axis. Use the definition of a hyperbola to derive its standard form.

99. *Sound Location* A rifle positioned at point $(-c, 0)$ is fired at a target positioned at point $(c, 0)$. A person hears the sound of the rifle and the sound of the bullet hitting the target at the same time. Show that the person is positioned on one branch of the hyperbola given by

$$\frac{x^2}{c^2 v_s^2 / v_m^2} - \frac{y^2}{c^2(v_m^2 - v_s^2)/v_m^2} = 1$$

where v_m is the muzzle velocity of the rifle and v_s is the speed of sound $= 1100$ feet per second.

100. *Navigation*   LORAN (long distance radio navigation) for aircraft and ships uses synchronized pulses transmitted by widely separated transmitting stations. These pulses travel at the speed of light (186,000 miles per second). The difference in the times of arrival of these pulses at an aircraft or ship is constant on a hyperbola having the transmitting stations as foci. Assume that two stations, 300 miles apart, are positioned on the rectangular coordinate system at $(-150, 0)$ and $(150, 0)$ and that a ship is traveling on a path with coordinates $(x, 75)$ (see figure). Find the x-coordinate of the position of the ship if the time difference between the pulses from the transmitting stations is 1000 microseconds (0.001 second).

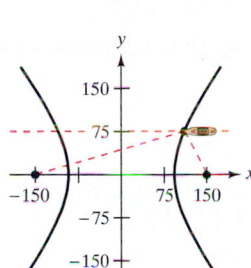

Figure for 100

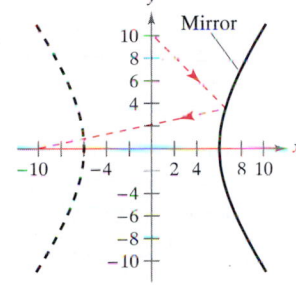

Figure for 101

101. *Hyberbolic Mirror*   A hyperbolic mirror (used in some telescopes) has the property that a light ray directed at the focus will be reflected to the other focus. The mirror in the figure has the equation $(x^2/36) - (y^2/64) = 1$. At which point on the mirror will light from the point $(0, 10)$ be reflected to the other focus?

In Exercises 102 and 103, find equations for **(a)** the tangent lines and **(b)** the normal lines to the hyperbola for the given value of x.

102. $\dfrac{x^2}{9} - y^2 = 1, \quad x = 6$ **103.** $\dfrac{y^2}{4} - \dfrac{x^2}{2} = 1, \quad x = 4$

104. Show that the equation of the tangent line to

$$\frac{x^2}{a^2} - \frac{y^2}{b^2} = 1$$

at the point (x_0, y_0) is

$$\frac{x_0 x}{a^2} - \frac{y_0 y}{b^2} = 1.$$

105. Show that the ellipse

$$\frac{x^2}{a^2} + \frac{2y^2}{b^2} = 1$$

and the hyperbola

$$\frac{x^2}{a^2 - b^2} - \frac{2y^2}{b^2} = 1$$

intersect at right angles.

True or False?   **In Exercises 106–112, determine whether the statement is true or false. If it is false, explain why or give an example that shows it is false.**

106. It is possible for a parabola to intersect its directrix.

107. The point on a parabola closest to its focus is its vertex.

108. If C is the circumference of the ellipse

$$\frac{x^2}{a^2} + \frac{y^2}{b^2} = 1, \quad b < a,$$

then $2\pi b \leq C \leq 2\pi a$.

109. The graph of

$$\frac{x^2}{4} + y^4 = 1$$

is an ellipse.

110. If $D \neq 0$ or $E \neq 0$, then the graph of

$$y^2 - x^2 + Dx + Ey = 0$$

is a hyperbola.

111. If the asymptotes of the hyperbola

$$\frac{x^2}{a^2} - \frac{y^2}{b^2} = 1$$

intersect at right angles, then $a = b$.

112. Every tangent line to a hyperbola intersects the hyperbola only at the point of tangency.

113. Prove that the graph of the equation

$$Ax^2 + Cy^2 + Dx + Ey + F = 0$$

is one of the following (except in degenerate cases).

Conic	Condition
(a) Circle	$A = C$
(b) Parabola	$A = 0$ or $C = 0$ (but not both)
(c) Ellipse	$AC > 0$
(d) Hyperbola	$AC < 0$

In Exercises 114–123, classify the graph of the equation as a circle, a parabola, an ellipse, or a hyperbola.

114. $x^2 + 4y^2 - 6x + 16y + 21 = 0$

115. $4x^2 - y^2 - 4x - 3 = 0$

116. $y^2 - 4y - 4x = 0$

117. $25x^2 - 10x - 200y - 119 = 0$

118. $4x^2 + 4y^2 - 16y + 15 = 0$

119. $y^2 - 4y = x + 5$

120. $9x^2 + 9y^2 - 36x + 6y + 34 = 0$

121. $2x(x - y) = y(3 - y - 2x)$

122. $3(x - 1)^2 = 6 + 2(y + 1)^2$

123. $9(x + 3)^2 = 36 - 4(y - 2)^2$

Plane Curves and Parametric Equations • Eliminating the Parameter •
Finding Parametric Equations • The Tautochrone and Brachistochrone Problems

Plane Curves and Parametric Equations

Until now, we have been representing a graph by a single equation involving *two* variables. In this section you will study situations in which *three* variables are used to represent a curve in the plane.

Consider the path followed by an object that is propelled into the air at an angle of 45°. If the initial velocity of the object is 48 feet per second, the object travels the parabolic path given by

$$y = -\frac{x^2}{72} + x \qquad \text{Rectangular equation}$$

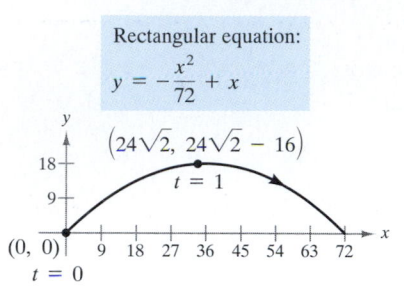

Rectangular equation:
$$y = -\frac{x^2}{72} + x$$

$\left(24\sqrt{2},\ 24\sqrt{2} - 16\right)$

$t = 1$

$(0, 0)$ 9   18   27   36   45   54   63   72

$t = 0$

Parametric equations:
$x = 24\sqrt{2}\,t$
$y = -16t^2 + 24\sqrt{2}\,t$

Curvilinear motion: two variables for position, one variable for time
Figure 9.17

as shown in Figure 9.17. However, this equation does not tell the whole story. Although it does tell you *where* the object has been, it doesn't tell you *when* the object was at a given point (x, y). To determine this time, you can introduce a third variable t, called a **parameter**. By writing both x and y as functions of t, you obtain the **parametric equations**

$$x = 24\sqrt{2}\,t \qquad \text{Parametric equation for } x$$

and

$$y = -16t^2 + 24\sqrt{2}\,t. \qquad \text{Parametric equation for } y$$

From this set of equations, you can determine that at the time $t = 0$, the object is at the point $(0, 0)$. Similarly, at time $t = 1$, the object is at the point $(24\sqrt{2}, 24\sqrt{2} - 16)$, and so on. (We will discuss a method for determining this particular set of parametric equations—the equations of motion—later, in Section 11.3.)

For this particular motion problem, x and y are continuous functions of t, and the resulting path is called a **plane curve**.

Definition of a Plane Curve

If f and g are continuous functions of t on an interval I, then the equations

$$x = f(t) \quad \text{and} \quad y = g(t)$$

are called **parametric equations** and t is called the **parameter.** The set of points (x, y) obtained as t varies over the interval I is called the **graph** of the parametric equations. Taken together, the parametric equations and the graph are called a **plane curve,** denoted by C.

NOTE   At times it is important to distinguish between a graph (the set of points) and a curve (the points together with their defining parametric equations). When it is important, we will make the distinction explicit. When it is not important, we will use C to represent the graph or the curve.

When sketching (by hand) a curve represented by a pair of parametric equations, you can plot points in the *xy*-plane. Each set of coordinates (x, y) is determined from a value chosen for the parameter *t*. By plotting the resulting points in order of *increasing* values of *t*, the curve is traced out in a specific direction. This is called the **orientation** of the curve.

EXAMPLE 1   Sketching a Curve

Sketch the curve described by the parametric equations

$$x = t^2 - 4 \quad \text{and} \quad y = \frac{t}{2}, \quad -2 \le t \le 3.$$

Solution   For values of *t* on the given interval, the parametric equations yield the points (x, y) shown in the table.

t	-2	-1	0	1	2	3
x	0	-3	-4	-3	0	5
y	-1	$-\frac{1}{2}$	0	$\frac{1}{2}$	1	$\frac{3}{2}$

By plotting these points in order of increasing *t* and using the continuity of *f* and *g*, you obtain the curve *C* shown in Figure 9.18. Note that the arrows on the curve indicate its orientation as *t* increases from -2 to 3.

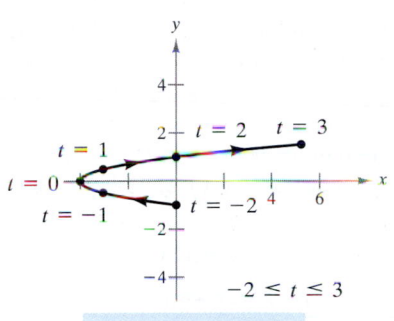

Parametric equations:
$$x = t^2 - 4 \text{ and } y = \frac{t}{2}$$

Figure 9.18

NOTE   From the vertical line test, you can see that the graph shown in Figure 9.18 does not define *y* as a function of *x*. This points out one benefit of parametric equations—they can be used to represent graphs that are more general than graphs of functions.

It often happens that two different sets of parametric equations have the same graph. For example, the set of parametric equations

$$x = 4t^2 - 4 \quad \text{and} \quad y = t, \quad -1 \le t \le \frac{3}{2}$$

has the same graph as the set given in Example 1. However, comparing the values of *t* in Figures 9.18 and 9.19, you can see that the second graph is traced out more *rapidly* (considering *t* as time) than the first graph. Thus, in applications, different parametric representations can be used to represent various *speeds* at which objects travel along a given path.

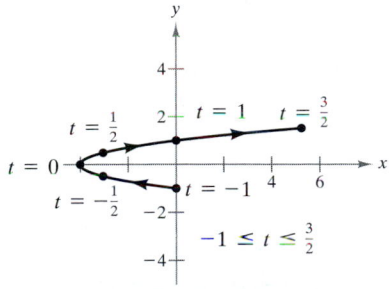

Parametric equations:
$$x = 4t^2 - 4 \text{ and } y = t$$

Figure 9.19

⋮ **TECHNOLOGY**   Most graphing utilities have a parametric graphing mode. If you
⋮ have access to such a utility (either a graphing calculator or graphing software),
⋮ try using it to confirm the graphs shown in Figures 9.18 and 9.19. Does the curve
⋮ given by
⋮
⋮ $$x = 4t^2 - 8t \quad \text{and} \quad y = 1 - t, \quad -\tfrac{1}{2} \le t \le 2$$
⋮
⋮ represent the same graph as that shown in Figures 9.18 and 9.19? What do you
⋮ notice about the *orientation* of this curve?

Eliminating the Parameter

Finding a rectangular equation that represents the graph of a set of parametric equations is called **eliminating the parameter**. For instance, you can eliminate the parameter from the set of parametric equations in Example 1 as follows.

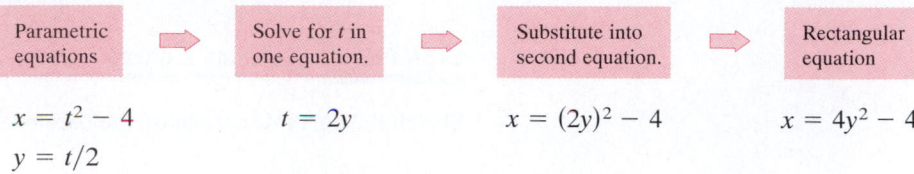

Parametric equations	Solve for t in one equation.	Substitute into second equation.	Rectangular equation
$x = t^2 - 4$ $y = t/2$	$t = 2y$	$x = (2y)^2 - 4$	$x = 4y^2 - 4$

Once you have eliminated the parameter, you can recognize that the equation $x = 4y^2 - 4$ represents a parabola with a horizontal axis and vertex at $(-4, 0)$, as shown in Figure 9.18.

The range of x and y implied by the parametric equations may be altered by the change to rectangular form. In such instances the domain of the rectangular equation must be adjusted so that its graph matches the graph of the parametric equations. Such a situation is demonstrated in the next example.

EXAMPLE 2 Adjusting the Domain After Eliminating the Parameter

Sketch the curve represented by the equations

$$x = \frac{1}{\sqrt{t + 1}} \quad \text{and} \quad y = \frac{t}{t + 1}, \quad t > -1$$

by eliminating the parameter and adjusting the domain of the resulting rectangular equation.

Solution Begin by solving one of the parametric equations for t. For instance, you can solve the first equation for t as follows.

$$x = \frac{1}{\sqrt{t + 1}} \qquad \text{Parametric equation for } x$$

$$x^2 = \frac{1}{t + 1} \qquad \text{Square both sides.}$$

$$t + 1 = \frac{1}{x^2}$$

$$t = \frac{1}{x^2} - 1 = \frac{1 - x^2}{x^2} \qquad \text{Solve for } t.$$

Now, substituting into the parametric equation for y produces the following.

$$y = \frac{t}{t + 1} \qquad \text{Parametric equation for } y$$

$$y = \frac{(1 - x^2)/x^2}{[(1 - x^2)/x^2] + 1} \qquad \text{Substitute } (1 - x^2)/x^2 \text{ for } t.$$

$$y = 1 - x^2 \qquad \text{Simplify.}$$

The rectangular equation, $y = 1 - x^2$, is defined for all values of x, but from the parametric equation for x you can see that the curve is defined only when $t > -1$. This implies that you should restrict the domain of x to positive values, as shown in Figure 9.20.

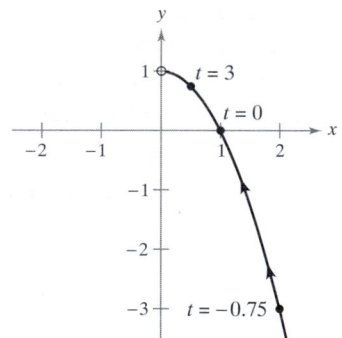

Parametric equations:
$x = \dfrac{1}{\sqrt{t + 1}}, \; y = \dfrac{t}{t + 1}, \; t > -1$

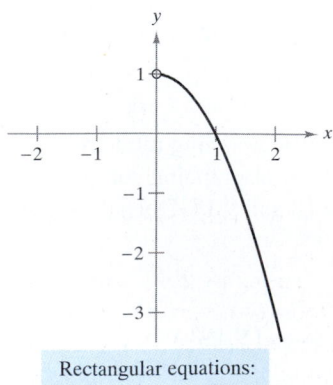

Rectangular equations:
$y = 1 - x^2, \; x > 0$

Figure 9.20

It is not necessary for the parameter in a set of parametric equations to represent time. The next example uses an *angle* as the parameter.

EXAMPLE 3   Using Trigonometry to Eliminate a Parameter

Sketch the curve represented by

$$x = 3 \cos \theta \quad \text{and} \quad y = 4 \sin \theta, \quad 0 \leq \theta \leq 2\pi$$

by eliminating the parameter and finding the corresponding rectangular equation.

Solution   Begin by solving for $\cos \theta$ and $\sin \theta$ in the given equations.

$$\cos \theta = \frac{x}{3} \quad \text{and} \quad \sin \theta = \frac{y}{4} \qquad \text{Solve for } \cos \theta \text{ and } \sin \theta.$$

Next, make use of the identity $\sin^2 \theta + \cos^2 \theta = 1$ to form an equation involving only x and y.

$$\cos^2 \theta + \sin^2 \theta = 1 \qquad \text{Trigonometric identity}$$

$$\left(\frac{x}{3}\right)^2 + \left(\frac{y}{4}\right)^2 = 1 \qquad \text{Substitute.}$$

$$\frac{x^2}{9} + \frac{y^2}{16} = 1 \qquad \text{Rectangular equation}$$

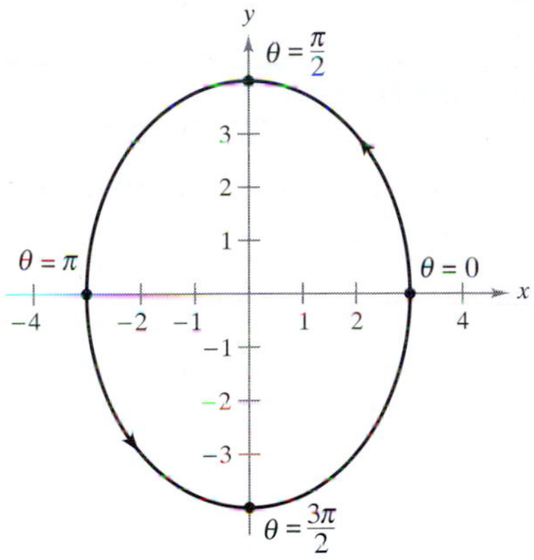

From this rectangular equation you can see that the graph is an ellipse centered at $(0, 0)$, with vertices at $(0, 4)$ and $(0, -4)$ and minor axis of length $2b = 6$, as shown in Figure 9.21. Note that the elliptic curve is traced out *counterclockwise* as θ varies from 0 to 2π.

Parametric equations:
$x = 3 \cos \theta, \; y = 4 \sin \theta$
Rectangular equation:
$$\frac{x^2}{9} + \frac{y^2}{16} = 1$$

Figure 9.21

Using the technique shown in Example 3, you can conclude that the graph of the parametric equations

$$x = h + a \cos \theta \quad \text{and} \quad y = k + b \sin \theta, \quad 0 \leq \theta \leq 2\pi$$

is the ellipse (traced counterclockwise) given by

$$\frac{(x - h)^2}{a^2} + \frac{(y - k)^2}{b^2} = 1.$$

The graph of the parametric equations

$$x = h + a \sin \theta \quad \text{and} \quad y = k + b \cos \theta, \quad 0 \leq \theta \leq 2\pi$$

is also the ellipse (traced clockwise) given by

$$\frac{(x - h)^2}{a^2} + \frac{(y - k)^2}{b^2} = 1.$$

If you have access to a graphing utility, try using it in parametric mode to sketch several ellipses.

In Examples 2 and 3, it is important to realize that eliminating the parameter is primarily an *aid to curve sketching*. If the parametric equations represent the path of a moving object, the graph alone is not sufficient to describe the object's motion. You still need the parametric equations to tell you the *position*, *direction*, and *speed* at a given time.

Finding Parametric Equations

The first three examples in this section illustrated techniques for sketching the graph represented by a set of parametric equations. We now look at the reverse problem. How can you determine a set of parametric equations for a given graph or a given physical description? From the discussion following Example 1, you know that such a representation is not unique. This is demonstrated further in the following example, which finds two different parametric representations for a given graph.

EXAMPLE 4   Finding Parametric Equations for a Given Graph

Find a set of parametric equations to represent the graph of $y = 1 - x^2$, using each of the following parameters.

a. $t = x$ **b.** the slope $m = \dfrac{dy}{dx}$ at the point (x, y)

Solution

a. Letting $x = t$ produces the parametric equations

$$x = t \quad \text{and} \quad y = 1 - x^2 = 1 - t^2.$$

b. To express x and y in terms of the parameter m, you can proceed as follows.

$$m = \frac{dy}{dx} = -2x \qquad \text{Differentiate } y = 1 - x^2.$$

$$x = -\frac{m}{2} \qquad \text{Solve for x.}$$

This produces a parametric equation for x. To obtain a parametric equation for y, substitute $-m/2$ for x in the original equation.

$$y = 1 - x^2 \qquad \text{Original rectangular equation}$$

$$y = 1 - \left(-\frac{m}{2}\right)^2 \qquad \text{Substitute } -m/2 \text{ for x.}$$

$$y = 1 - \frac{m^2}{4} \qquad \text{Simplify.}$$

Thus, the parametric equations are

$$x = -\frac{m}{2} \quad \text{and} \quad y = 1 - \frac{m^2}{4}.$$

In Figure 9.22, note that the resulting curve has a right-to-left orientation as determined by the direction of increasing values of slope m. For part (a), the curve would have the opposite orientation.

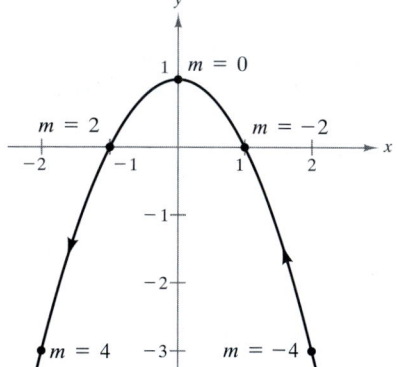

Rectangular equation: $y = 1 - x^2$
Parametric equations:
$x = -\dfrac{m}{2} \quad y = 1 - \dfrac{m^2}{4}$

Figure 9.22

TECHNOLOGY   To be efficient at using a graphing utility, it is important that you develop skill in representing a graph by a set of parametric equations. The reason for this is that many graphing utilities have only three graphing modes—(1) functions, (2) parametric equations, and (3) polar equations. Most graphing utilities are not programmed to sketch the graph of a general equation. For instance, suppose you want to sketch the graph of the hyperbola $x^2 - y^2 = 1$. To sketch the graph in function mode, you need two equations: $y = \sqrt{x^2 - 1}$ and $y = -\sqrt{x^2 - 1}$. In parametric mode, you can represent the graph by $x = \sec t$ and $y = \tan t$.

FOR FURTHER INFORMATION For more information on cycloids, see the article "The Geometry of Rolling Curves" by John Bloom and Lee Whitt in the June-July 1981 issue of *The American Mathematical Monthly*.

EXAMPLE 5   Parametric Equations for a Cycloid

Determine the curve traced by a point P on the circumference of a circle of radius a rolling along a straight line in a plane. Such a curve is called a **cycloid**.

Solution   Let the parameter θ be the measure of the circle's rotation, and let the point $P = (x, y)$ begin at the origin. When $\theta = 0$, P is at the origin. When $\theta = \pi$, P is at a maximum point $(\pi a, 2a)$. When $\theta = 2\pi$, P is back on the x-axis at $(2\pi a, 0)$. From Figure 9.23, you can see that $\angle APC = 180° - \theta$. Hence,

$$\sin \theta = \sin(180° - \theta) = \sin(\angle APC) = \frac{AC}{a} = \frac{BD}{a}$$

$$\cos \theta = -\cos(180° - \theta) = -\cos(\angle APC) = \frac{AP}{-a}$$

which implies that

$$AP = -a \cos \theta \quad \text{and} \quad BD = a \sin \theta.$$

Because the circle rolls along the x-axis, you know that $OD = \overset{\frown}{PD} = a\theta$. Furthermore, because $BA = DC = a$, you have

$$x = OD - BD = a\theta - a \sin \theta$$
$$y = BA + AP = a - a \cos \theta.$$

Therefore, the parametric equations are

$$x = a(\theta - \sin \theta) \quad \text{and} \quad y = a(1 - \cos \theta).$$

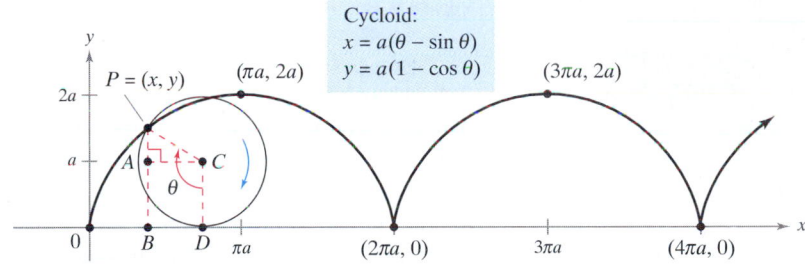

Figure 9.23

TECHNOLOGY   Some graphing utilities allow you to simulate the motion of an object that is moving in the plane or in space. If you have access to such a utility, try using it to trace out the path of the cycloid shown in Figure 9.23.

The cycloid in Figure 9.23 has sharp corners at the values $x = 2n\pi a$. Notice that the derivatives $x'(\theta)$ and $y'(\theta)$ are both zero at the points for which $\theta = 2n\pi$.

$$x(\theta) = a(\theta - \sin \theta) \qquad y(\theta) = a(1 - \cos \theta)$$
$$x'(\theta) = a - a \cos \theta \qquad y'(\theta) = a \sin \theta$$
$$x'(2n\pi) = 0 \qquad y'(2n\pi) = 0$$

Between these points, the cycloid is called **smooth**.

Definition of a Smooth Curve

A curve C represented by $x = f(t)$ and $y = g(t)$ on an interval I is called **smooth** if f' and g' are continuous on I and not simultaneously 0, except possibly at the endpoints of I. The curve C is called **piecewise smooth** if it is smooth on each subinterval of some partition of I.

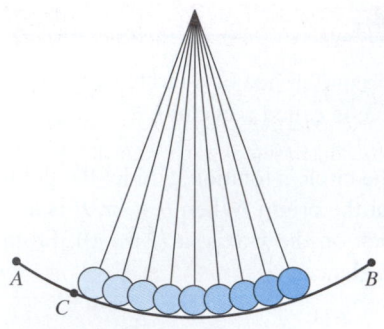

The time required to complete a full swing of the pendulum when starting from point *C* is only approximately the same as when starting from point *A*.
Figure 9.24

JAMES BERNOULLI (1654–1705)

James Bernoulli, also called Jacques, was the older brother of John. He was one of several accomplished mathematicians of the Swiss Bernoulli family. James's mathematical accomplishments have given him a prominent place in the early development of calculus.

FOR FURTHER INFORMATION To see a proof of the famous brachistochrone problem, see the article "A New Minimization Proof for the Brachistochrone" by Gary Lawlor in March 1996 issue of *The American Mathematical Monthly*.

The Tautochrone and Brachistochrone Problems

The type of curve described in Example 5 is related to one of the most famous pairs of problems in the history of calculus. The first problem (called the **tautochrone problem**) began with Galileo's discovery that the time required to complete a full swing of a given pendulum is *approximately* the same whether it makes a large movement at high speeds or a small movement at lower speeds (see Figure 9.24). Late in his life, Galileo (1564–1642) realized that he could use this principle to construct a clock. However, he was not able to conquer the mechanics of actual construction. Christian Huygens (1629–1695) was the first to design and construct a working model. In his work with pendulums, Huygens realized that a pendulum does not take *exactly* the same time to complete swings of varying lengths. (This doesn't affect a pendulum clock, because the length of the circular arc is kept constant by giving the pendulum a slight boost each time it passes its lowest point.) But, in studying the problem, Huygens discovered that a ball rolling back and forth on an inverted cycloid does complete each cycle in exactly the same time.

An inverted cycloid is the path down which a ball will roll in the shortest time.
Figure 9.25

The second problem, posed by John Bernoulli in 1696, is called the **brachistochrone problem**—in Greek, *brachys* means short and *chronos* means time. The problem was to determine the path down which a particle will slide from point *A* to point *B* in the *shortest time*. Several mathematicians took up the challenge, and the following year the problem was solved by Newton, Leibniz, L'Hôpital, John Bernoulli, and James Bernoulli. As it turns out, the solution is not a straight line from *A* to *B*, but an inverted cycloid passing through the points *A* and *B*, as shown in Figure 9.25. The amazing part of the solution is that a particle starting at rest at *any* other point *C* of the cycloid between *A* and *B* will take exactly the same time to reach *B*, as indicated in Figure 9.26.

A ball starting at point *C* takes the same time to reach point *B* as one that starts at point *A*.
Figure 9.26

EXERCISES FOR SECTION 9.2

 1. Consider the parametric equations $x = \sqrt{t}$ and $y = 1 - t$.

(a) Complete the table.

t	0	1	2	3	4
x					
y					

(b) Plot the points (x, y) generated in the table, and sketch a graph of the parametric equations. Indicate the orientation of the graph.

(c) Use a graphing utility to confirm your graph in part (b).

(d) Find the rectangular equation by eliminating the parameter. Compare the graph in part (b) with the graph of the rectangular equation.

2. Consider the parametric equations $x = 4\cos^2\theta$ and $y = 2\sin\theta$.

(a) Complete the table.

θ	$-\dfrac{\pi}{2}$	$-\dfrac{\pi}{4}$	0	$\dfrac{\pi}{4}$	$\dfrac{\pi}{2}$
x					
y					

(b) Plot the points (x, y) generated in the table, and sketch a graph of the parametric equations. Indicate the orientation of the graph.

(c) Use a graphing utility to confirm your graph in part (b).

(d) Find the rectangular equation by eliminating the parameter. Compare the graph in part (b) with the graph of the rectangular equation.

(e) If values of θ were selected from the interval $[\pi/2, 3\pi/2]$ for the table in part (a), would the graph in part (b) be different? Explain.

In Exercises 3–16, sketch the curve represented by the parametric equations (indicate the direction of the curve), and write the corresponding rectangular equation by eliminating the parameter.

3. $x = 3t - 1,\ y = 2t + 1$ **4.** $x = 3 - 2t,\ y = 2 + 3t$

5. $x = t + 1,\ y = t^2$ **6.** $x = \sqrt[3]{t},\ y = 1 - t$

7. $x = t^3,\ y = \dfrac{t^2}{2}$ **8.** $x = t^2 + t,\ y = t^2 - t$

9. $x = t - 1,\ y = \dfrac{t}{t - 1}$ **10.** $x = 1 + \dfrac{1}{t},\ y = t - 1$

11. $x = 2t,\ y = |t - 2|$

12. $x = |t - 1|,\ y = t + 2$

13. $x = \sec\theta,\ y = \cos\theta,\ 0 \le \theta < \pi/2,\ \pi/2 < \theta \le \pi$

14. $x = \tan^2\theta,\ y = \sec^2\theta$

15. $x = 3\cos\theta,\ y = 3\sin\theta$

16. $x = \cos\theta,\ y = 3\sin\theta$

In Exercises 17–28, use a graphing utility to sketch the curve represented by the parametric equations (indicate the direction of the curve). Eliminate the parameter and write the corresponding rectangular equation.

17. $x = 4\sin 2\theta,\ y = 2\cos 2\theta$ **18.** $x = \cos\theta,\ y = 2\sin 2\theta$

19. $x = 4 + 2\cos\theta$ **20.** $x = 4 + 2\cos\theta$
$\quad\ y = -1 + \sin\theta$ $\quad\ y = -1 + 2\sin\theta$

21. $x = 4 + 2\cos\theta$ **22.** $x = \sec\theta$
$\quad\ y = -1 + 4\sin\theta$ $\quad\ y = \tan\theta$

23. $x = 4\sec\theta,\ y = 3\tan\theta$ **24.** $x = \cos^3\theta,\ y = \sin^3\theta$

25. $x = t^3,\ y = 3\ln t$ **26.** $x = \ln 2t,\ y = t^2$

27. $x = e^{-t},\ y = e^{3t}$ **28.** $x = e^{2t},\ y = e^t$

Comparing Plane Curves   In Exercises 29–32, determine any differences between the curves of the parametric equations. Are the graphs the same? Are the orientations the same? Are the curves smooth?

29. (a) $x = t$ (b) $x = \cos\theta$
$\qquad y = 2t + 1$ $\qquad y = 2\cos\theta + 1$

$\quad$ (c) $x = e^{-t}$ (d) $x = e^t$
$\qquad y = 2e^{-t} + 1$ $\qquad y = 2e^t + 1$

30. (a) $x = 2\cos\theta$ (b) $x = \sqrt{4t^2 - 1}/|t|$
$\qquad y = 2\sin\theta$ $\qquad y = 1/t$

$\quad$ (c) $x = \sqrt{t}$ (d) $x = -\sqrt{4 - e^{2t}}$
$\qquad y = \sqrt{4 - t}$ $\qquad y = e^t$

31. (a) $x = \cos\theta$ (b) $x = \cos(-\theta)$
$\qquad y = 2\sin^2\theta$ $\qquad y = 2\sin^2(-\theta)$
$\qquad 0 < \theta < \pi$ $\qquad 0 < \theta < \pi$

32. (a) $x = t + 1,\ y = t^3$ (b) $x = -t + 1,\ y = (-t)^3$

33. *Conjecture*

(a) Use a graphing utility to sketch the curves represented by the two sets of parametric equations.

$\quad x = 4\cos t \qquad x = 4\cos(-t)$

$\quad y = 3\sin t \qquad y = 3\sin(-t)$

(b) Describe the change in the graph when the sign of the parameter is changed.

(c) Make a conjecture about the change in the graph of parametric equations when the sign of the parameter is changed.

(d) Test your conjecture with another set of parametric equations.

34. *Writing*   Review Exercises 29–33 and write a short paragraph describing how the graphs of curves represented by different sets of parametric equations can differ even though eliminating the parameter from each yields the same rectangular equation.

In Exercises 35–38, eliminate the parameter and obtain the standard form of the rectangular equation.

35. Line through (x_1, y_1) and (x_2, y_2):

$x = x_1 + t(x_2 - x_1)$, $y = y_1 + t(y_2 - y_1)$

36. Circle: $x = h + r \cos \theta$, $y = k + r \sin \theta$

37. Ellipse: $x = h + a \cos \theta$

$y = k + b \sin \theta$

38. Hyperbola: $x = h + a \sec \theta$

$y = k + b \tan \theta$

In Exercises 39–46, use the results of Exercises 35–38 to find a set of parametric equations for the line or conic.

39. Line: Passes through $(0, 0)$ and $(5, -2)$

40. Line: Passes through $(1, 4)$ and $(5, -2)$

41. Circle: Center: $(2, 1)$; Radius: 4

42. Circle: Center: $(-3, 1)$; Radius: 3

43. Ellipse: Vertices: $(\pm 5, 0)$; Foci: $(\pm 4, 0)$

44. Ellipse: Vertices: $(4, 7)$, $(4, -3)$; Foci: $(4, 5)$, $(4, -1)$

45. Hyperbola: Vertices: $(\pm 4, 0)$; Foci: $(\pm 5, 0)$

46. Hyperbola: Vertices: $(0, \pm 1)$; Foci: $(0, \pm 2)$

In Exercises 47–50, find two different sets of parametric equations for the given rectangular equation.

47. $y = 3x - 2$ **48.** $y = \dfrac{1}{x}$

49. $y = x^3$ **50.** $y = x^2$

In Exercises 51–58, use a graphing utility to graph the curve represented by the parametric equations. Indicate the direction of the curve. Identify any points at which the curve is not smooth.

51. Cycloid: $x = 2(\theta - \sin \theta)$, $y = 2(1 - \cos \theta)$

52. Cycloid: $x = \theta + \sin \theta$, $y = 1 - \cos \theta$

53. Prolate cycloid: $x = \theta - \frac{3}{2} \sin \theta$, $y = 1 - \frac{3}{2} \cos \theta$

54. Prolate cycloid: $x = 2\theta - 4 \sin \theta$, $y = 2 - 4 \cos \theta$

55. Hypocycloid: $x = 3 \cos^3 \theta$, $y = 3 \sin^3 \theta$

56. Curtate cycloid: $x = 2\theta - \sin \theta$, $y = 2 - \cos \theta$

57. Witch of Agnesi: $x = 2 \cot \theta$, $y = 2 \sin^2 \theta$

58. Folium of Descartes: $x = 3t/(1 + t^3)$, $y = 3t^2/(1 + t^3)$

In Exercises 59–62, match the set of parametric equations with the correct graph. [The graphs are labeled (a), (b), (c), and (d).]

(a)

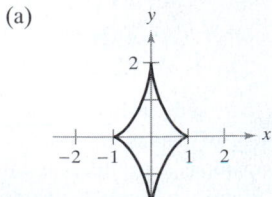

(b)

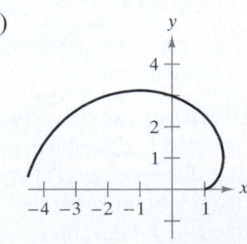

(c)

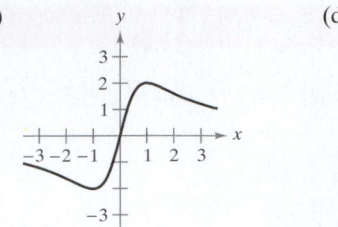

(d)

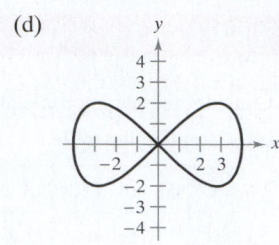

59. Lissajous curve: $x = 4 \cos \theta$, $y = 2 \sin 2\theta$

60. Evolute of ellipse: $x = \cos^3 \theta$, $y = 2 \sin^3 \theta$

61. Involute of circle: $x = \cos \theta + \theta \sin \theta$, $y = \sin \theta - \theta \cos \theta$

62. Serpentine curve; $x = \cot \theta$, $y = 4 \sin \theta \cos \theta$

63. *Curtate Cycloid* A wheel of radius a rolls along a straight line without slipping. The curve traced by a point P that is b units from the center $(b < a)$ is called a **curtate cycloid** (see figure). Use the angle θ to find a set of parametric equations for this curve.

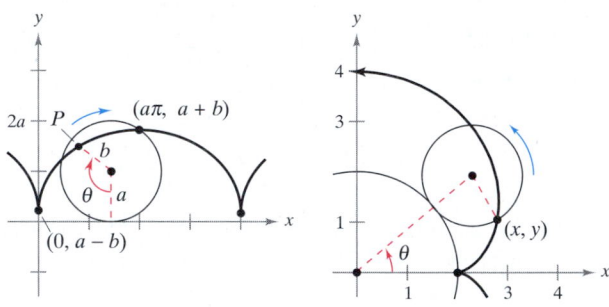

Figure for 63   **Figure for 64**

64. *Epicycloid* A circle of radius 1 rolls around the outside of a circle of radius 2 without slipping. The curve traced by a point on the circumference of the smaller circle is called an **epicycloid** (see figure). Use the angle θ to find a set of parametric equations for this curve.

True or False? **In Exercises 65–68, determine whether the statement is true or false. If it is false, explain why or give an example that shows it is false.**

65. The two sets of parametric equations $x = t$, $y = t^2 + 1$ and $x = 3t$, $y = 9t^2 + 1$ correspond to the same rectangular equation.

66. The graph of the parametric equations $x = t^2$ and $y = t^2$ is the line $y = x$.

67. If y is a function of t and x is a function of t, then y is a function of x.

68. If $f(t_1) = 0$ and $g(t_2) = 0$, then the curve represented by the parametric equations $x = f(t)$ and $y = g(t)$ passes through the origin.

Projectile Motion   In Exercises 69 and 70, consider a projectile launched at a height *h* feet above the ground and at an angle *θ* with the horizontal. If the initial velocity is v_0 feet per second, the path of the projectile is modeled by the parametric equations

$$x = (v_0 \cos \theta)t \quad \text{and} \quad y = h + (v_0 \sin \theta)t - 16t^2.$$

69. *Baseball* The center field fence in a ballpark is 10 feet high and 400 feet from home plate. The ball is hit 3 feet above the ground. It leaves the bat at an angle of *θ* degrees with the horizontal at a speed of 100 miles per hour (see figure).

(a) Write a set of parametric equations for the path of the ball.

(b) Use a graphing utility to graph the path of the ball if *θ* = 15°. Is the hit a home run?

(c) Use a graphing utility to graph the path of the ball if *θ* = 23°. Is the hit a home run?

(d) Find the minimum angle for the ball to leave the bat in order for the hit to be a home run.

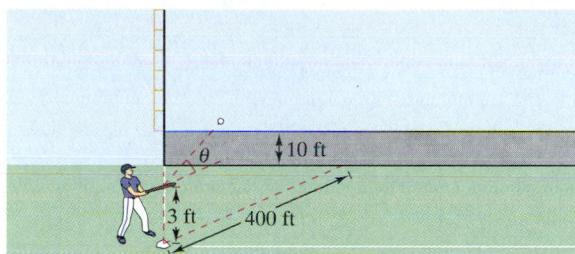

70. A rectangular equation for the path of a projectile is

$$y = 5 + x - 0.005x^2.$$

(a) Eliminate the parameter *t* from the position function for the motion of a projectile to show that the rectangular equation is

$$y = -\frac{16 \sec^2\theta}{v_0{}^2}x^2 + (\tan \theta)\, x + h.$$

(b) Use the result in part (a) to find *h*, v_0, and *θ*. Find the parametric equations of the path.

(c) Use a graphing utility to graph the rectangular equation for the path of the projectile. Confirm your answer in part (b) by sketching the curve represented by the parametric equations.

(d) Use a graphing utility to approximate the maximum height of the projectile and its range.

71. Consider the curve given by the parametric equations

$$x(t) = \frac{1 - t^2}{1 + t^2} \quad \text{and} \quad y(t) = \frac{2t}{1 + t^2}.$$

Graph this curve in the *xy*-plane. Use the graph to show that $(t^2 - 1)^2 + (2t)^2 = (t^2 + 1)^2$. Illustrate the formula for *t* = 2 and *t* = 3.

SECTION PROJECT

In Greek, the word *cycloid* means *wheel*, the word *hypocycloid* means *under the wheel*, and the word *epicycloid* means *upon the wheel*. Match the hypocycloid or epicycloid with its graph. [The graphs are labeled (a), (b), (c), (d), (e), and (f).]

Hypocycloid, H(A, B)

Path traced by a fixed point on a circle of radius *B* as it rolls around the *inside* of a circle of radius *A*.

$$x = (A - B)\cos t + B \cos\!\left(\frac{A - B}{B}\right)t$$

$$y = (A - B)\sin t - B \sin\!\left(\frac{A - B}{B}\right)t$$

Epicycloid, E(A, B)

Path traced by a fixed point on a circle of radius *B* as it rolls around the *outside* of a circle of radius *A*.

$$x = (A + B)\cos t - B \cos\!\left(\frac{A + B}{B}\right)t$$

$$y = (A + B)\sin t - B \sin\!\left(\frac{A + B}{B}\right)t$$

I. *H*(8, 3)   II. *E*(8,3)

III. *H*(8, 7)   IV. *E*(24, 3)

V. *H*(24, 7)   VI. *E*(24, 7)

(a) 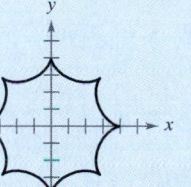   (b)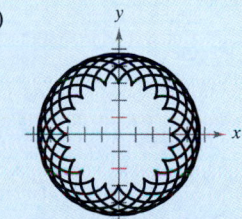

(c) 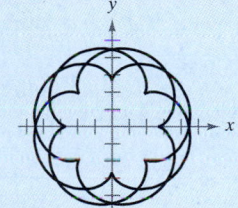   (d)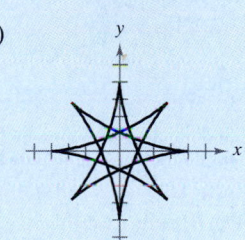

(e) 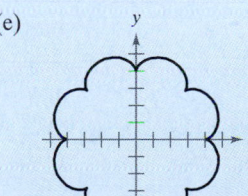   (f)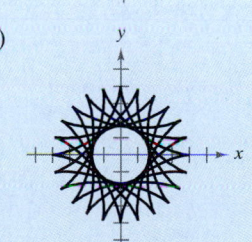

Exercises based on "Mathematical Discovery via Computer Graphics: Hypocycloids and Epicycloids" by Florence S. Gordon and Sheldon P. Gordon, COLLEGE MATHEMATICS JOURNAL, November 1984, p. 441. Used by permission of the authors.

Slope and Tangent Lines • Arc Length • Area of a Surface of Revolution

Slope and Tangent Lines

Now that you can represent a graph in the plane by a set of parametric equations, it is natural to ask how to use calculus to study plane curves. To begin, let's take another look at the projectile represented by the parametric equations

$$x = 24\sqrt{2}\,t \quad \text{and} \quad y = -16t^2 + 24\sqrt{2}\,t$$

as shown in Figure 9.27. From Section 9.2, you know that these equations enable you to locate the position of the projectile at a given time. You also know that the object is initially projected at an angle of 45°. But how can you find the angle θ representing the object's direction at some other time t? The following theorem answers this question by giving a formula for the slope of the tangent line as a function of t.

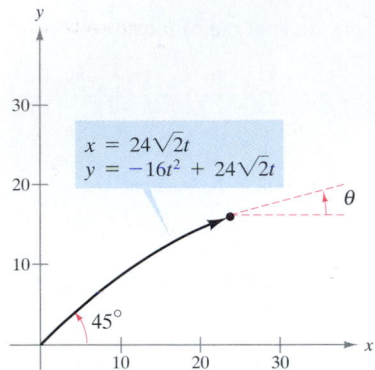

$x = 24\sqrt{2}\,t$
$y = -16t^2 + 24\sqrt{2}\,t$

At time t, the angle of elevation of the projectile is θ, the slope of the tangent line at that point.
Figure 9.27

> **THEOREM 9.7   Parametric Form of the Derivative**
>
> If a smooth curve C is given by the equations $x = f(t)$ and $y = g(t)$, then the slope of C at (x, y) is
>
> $$\frac{dy}{dx} = \frac{dy/dt}{dx/dt}, \qquad \frac{dx}{dt} \neq 0.$$

Proof   In Figure 9.28, consider $\Delta t > 0$ and let

$$\Delta y = g(t + \Delta t) - g(t) \qquad \text{and} \qquad \Delta x = f(t + \Delta t) - f(t).$$

Because $\Delta x \to 0$ as $\Delta t \to 0$, you can write

$$\frac{dy}{dx} = \lim_{\Delta x \to 0} \frac{\Delta y}{\Delta x}$$

$$= \lim_{\Delta t \to 0} \frac{g(t + \Delta t) - g(t)}{f(t + \Delta t) - f(t)}.$$

Dividing both the numerator and denominator by Δt, you can use the differentiability of f and g to conclude that

$$\frac{dy}{dx} = \lim_{\Delta t \to 0} \frac{[g(t + \Delta t) - g(t)]/\Delta t}{[f(t + \Delta t) - f(t)]/\Delta t}$$

$$= \frac{\displaystyle\lim_{\Delta t \to 0} \frac{g(t + \Delta t) - g(t)}{\Delta t}}{\displaystyle\lim_{\Delta t \to 0} \frac{f(t + \Delta t) - f(t)}{\Delta t}}$$

$$= \frac{g'(t)}{f'(t)}$$

$$= \frac{dy/dt}{dx/dt}.$$

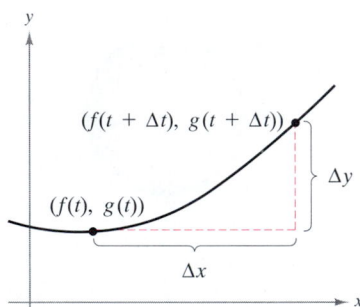

$(f(t + \Delta t),\ g(t + \Delta t))$

Δy

$(f(t),\ g(t))$

Δx

The slope of the secant line through the points $(f(t), g(t))$ and $(f(t + \Delta t),\ g(t + \Delta t))$ is $\Delta y/\Delta x$.
Figure 9.28

EXAMPLE 1 Differentiation and Parametric Form

Find dy/dx for the curve given by $x = \sin t$ and $y = \cos t$.

Solution

$$\frac{dy}{dx} = \frac{dy/dt}{dx/dt} = \frac{-\sin t}{\cos t} = -\tan t$$

Because dy/dx is a function of t, you can use Theorem 9.7 repeatedly to find *higher-order* derivatives. For instance,

$$\frac{d^2y}{dx^2} = \frac{d}{dx}\left[\frac{dy}{dx}\right] = \frac{\frac{d}{dt}\left[\frac{dy}{dx}\right]}{dx/dt} \qquad \text{Second derivative}$$

$$\frac{d^3y}{dx^3} = \frac{d}{dx}\left[\frac{d^2y}{dx^2}\right] = \frac{\frac{d}{dt}\left[\frac{d^2y}{dx^2}\right]}{dx/dt}. \qquad \text{Third derivative}$$

EXAMPLE 2 Finding Slope and Concavity

For the curve given by

$$x = \sqrt{t} \qquad \text{and} \qquad y = \frac{1}{4}(t^2 - 4), \qquad t \geq 0$$

find the slope and concavity at the point $(2, 3)$.

Solution Because

$$\frac{dy}{dx} = \frac{dy/dt}{dx/dt} = \frac{(1/2)t}{(1/2)t^{-1/2}} = t^{3/2}$$

you can find the second derivative to be

$$\frac{d^2y}{dx^2} = \frac{\frac{d}{dt}[dy/dx]}{dx/dt} = \frac{\frac{d}{dt}[t^{3/2}]}{dx/dt} = \frac{(3/2)t^{1/2}}{(1/2)t^{-1/2}} = 3t.$$

At $(x, y) = (2, 3)$, it follows that $t = 4$, and the slope is

$$\frac{dy}{dx} = (4)^{3/2} = 8.$$

Moreover, when $t = 4$, the second derivative is

$$\frac{d^2y}{dx^2} = 3(4) = 12 > 0$$

and you can conclude that the graph is concave upward at $(2, 3)$, as shown in Figure 9.29.

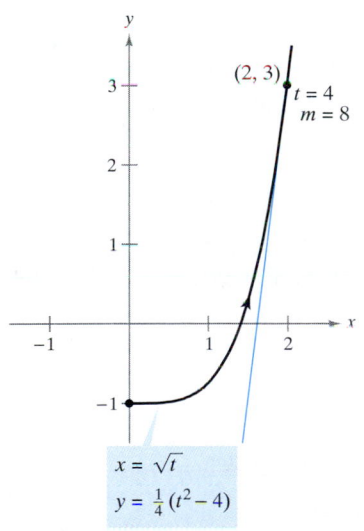

The graph is concave upward at $(2, 3)$, when $t = 4$.
Figure 9.29

Because the parametric equations $x = f(t)$ and $y = g(t)$ need not define y as a function of x, it follows that a plane curve can loop and cross itself. At such points the curve may have more than one tangent line, as shown in the next example.

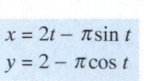

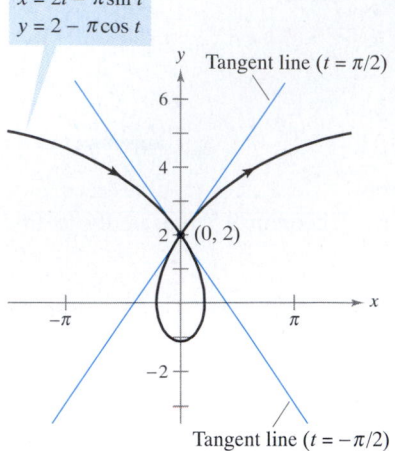

This prolate cycloid has two tangent lines at the point (0, 2).

Figure 9.30

 EXAMPLE 3   A Curve with Two Tangent Lines at a Point

The **prolate cycloid** given by

$$x = 2t - \pi \sin t \quad \text{and} \quad y = 2 - \pi \cos t$$

crosses itself at the point (0, 2), as shown in Figure 9.30. Find the equations of both tangent lines at this point.

Solution   Because $x = 0$ and $y = 2$ when $t = \pm \pi/2$, and

$$\frac{dy}{dx} = \frac{dy/dt}{dx/dt} = \frac{\pi \sin t}{2 - \pi \cos t}$$

you have $dy/dx = -\pi/2$ when $t = -\pi/2$ and $dy/dx = \pi/2$ when $t = \pi/2$. Therefore, the two tangent lines at (0, 2) are

$$y - 2 = -\left(\frac{\pi}{2}\right)x \qquad \text{Tangent line when } t = -\frac{\pi}{2}$$

$$y - 2 = \left(\frac{\pi}{2}\right)x. \qquad \text{Tangent line when } t = \frac{\pi}{2}$$

If $dy/dt = 0$ and $dx/dt \neq 0$ when $t = t_0$, the curve represented by $x = f(t)$ and $y = g(t)$ has a *horizontal* tangent at $(f(t_0), g(t_0))$. For instance, in Example 3, the given curve has a horizontal tangent at the point $(0, 2 - \pi)$ (when $t = 0$). Similarly, if $dx/dt = 0$ and $dy/dt \neq 0$ when $t = t_0$, the curve represented by $x = f(t)$ and $y = g(t)$ has a *vertical* tangent at $(f(t_0), g(t_0))$.

Arc Length

You have seen how parametric equations can be used to describe the path of a particle moving in the plane. We now develop a formula for determining the *distance* traveled by the particle along its path.

Recall from Section 6.4 that the formula for the arc length of a curve C given by $y = h(x)$ over the interval $[x_0, x_1]$ is

$$s = \int_{x_0}^{x_1} \sqrt{1 + [h'(x)]^2} \, dx = \int_{x_0}^{x_1} \sqrt{1 + \left(\frac{dy}{dx}\right)^2} \, dx.$$

If C is represented by the parametric equations $x = f(t)$ and $y = g(t)$, $a \leq t \leq b$, and if $dx/dt = f'(t) > 0$, you can write

$$s = \int_{x_0}^{x_1} \sqrt{1 + \left(\frac{dy}{dx}\right)^2} \, dx = \int_{x_0}^{x_1} \sqrt{1 + \left(\frac{dy/dt}{dx/dt}\right)^2} \, dx$$

$$= \int_a^b \sqrt{\frac{(dx/dt)^2 + (dy/dt)^2}{(dx/dt)^2}} \, \frac{dx}{dt} \, dt$$

$$= \int_a^b \sqrt{\left(\frac{dx}{dt}\right)^2 + \left(\frac{dy}{dt}\right)^2} \, dt$$

$$= \int_a^b \sqrt{[f'(t)]^2 + [g'(t)]^2} \, dt.$$

NOTE   When applying the arc length formula to a curve, be sure that the curve is traced out only once on the interval of integration. For instance, the circle given by $x = \cos t$ and $y = \sin t$ is traced out once on the interval $0 \le t \le 2\pi$, but is traced out twice on the interval $0 \le t \le 4\pi$.

> **THEOREM 9.8   Arc Length in Parametric Form**
>
> If a smooth curve C is given by $x = f(t)$ and $y = g(t)$ such that C does not intersect itself on the interval $a \le t \le b$ (except possibly at the endpoints), then the arc length of C over the interval is given by
>
> $$s = \int_a^b \sqrt{\left(\frac{dx}{dt}\right)^2 + \left(\frac{dy}{dt}\right)^2}\, dt = \int_a^b \sqrt{[f'(t)]^2 + [g'(t)]^2}\, dt.$$

In the preceding section you saw that if a circle rolls along a line, a point on its circumference will trace a path called a cycloid. If the circle rolls around the circumference of another circle, the path of the point is an **epicycloid.** The next example shows how to find the arc length of an epicycloid.

> ARCH OF A CYCLOID
>
> The arc length of an arch of a cycloid was first calculated in 1658 by British architect and mathematician Christopher Wren, famous for rebuilding many buildings and churches in London, including St. Paul's Cathedral.

EXAMPLE 4   Finding Arc Length

A circle of radius 1 rolls around the circumference of a larger circle of radius 4, as shown in Figure 9.31. The epicycloid traced by a point on the circumference of the smaller circle is given by

$$x = 5 \cos t - \cos 5t \qquad \text{and} \qquad y = 5 \sin t - \sin 5t.$$

Find the distance traveled by the point in one complete trip about the larger circle.

Solution   Before applying Theorem 9.8, note in Figure 9.31 that the curve has sharp points when $t = 0$ and $t = \pi/2$. Between these two points, dx/dt and dy/dt are not simultaneously 0. Thus, the portion of the curve generated from $t = 0$ to $t = \pi/2$ is smooth. To find the total distance traveled by the point, you can find the arc length of that portion lying in the first quadrant and multiply by 4.

$$s = 4\int_0^{\pi/2} \sqrt{\left(\frac{dx}{dt}\right)^2 + \left(\frac{dy}{dt}\right)^2}\, dt$$

$$= 4\int_0^{\pi/2} \sqrt{(-5\sin t + 5\sin 5t)^2 + (5\cos t - 5\cos 5t)^2}\, dt$$

$$= 20\int_0^{\pi/2} \sqrt{2 - 2\sin t \sin 5t - 2\cos t \cos 5t}\, dt$$

$$= 20\int_0^{\pi/2} \sqrt{2 - 2\cos 4t}\, dt$$

$$= 20\int_0^{\pi/2} \sqrt{4\sin^2 2t}\, dt$$

$$= 40\int_0^{\pi/2} \sin 2t\, dt$$

$$= -20\left[\cos 2t\right]_0^{\pi/2}$$

$$= 40$$

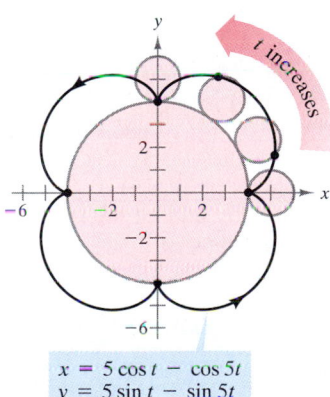

$$x = 5 \cos t - \cos 5t$$
$$y = 5 \sin t - \sin 5t$$

An epicycloid is traced by a point on the smaller circle as it rolls around the larger circle.

Figure 9.31

For the epicycloid shown in Figure 9.31, an arc length of 40 seems about right because the circumference of a circle of radius 6 is $2\pi r = 12\pi \approx 37.7$.

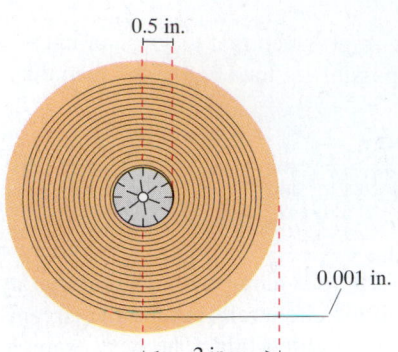

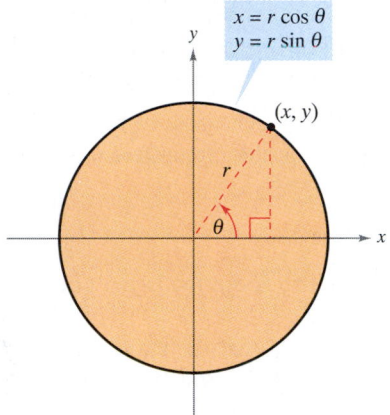

It takes approximately 982 feet of tape to fill the reel.

Figure 9.32

NOTE The graph of $r = a\theta$ is called the **spiral of Archimedes.** The graph of $r = \theta/2000\pi$ (in Example 5) is of this form.

EXAMPLE 5 Length of a Recording Tape

A recording tape 0.001 inch thick is wound around a reel whose inner radius is 0.5 inch and outer radius is 2 inches, as shown in Figure 9.32. How much tape is required to fill the reel?

Solution To create a model for this problem, assume that as the tape is wound around the reel its distance r from the center increases linearly at a rate of 0.001 inch per revolution, or

$$r = (0.001)\frac{\theta}{2\pi} = \frac{\theta}{2000\pi}, \qquad 1000\pi \le \theta \le 4000\pi$$

where θ is measured in radians. You can determine the coordinates of the point (x, y) corresponding to a given radius to be

$$x = r \cos \theta$$

and

$$y = r \sin \theta.$$

Substituting for r, you obtain the parametric equations

$$x = \left(\frac{\theta}{2000\pi}\right) \cos \theta \qquad \text{and} \qquad y = \left(\frac{\theta}{2000\pi}\right) \sin \theta.$$

You can use the arc length formula to determine the total length of the tape to be

$$s = \int_{1000\pi}^{4000\pi} \sqrt{\left(\frac{dx}{d\theta}\right)^2 + \left(\frac{dy}{d\theta}\right)^2} \, d\theta$$

$$= \frac{1}{2000\pi} \int_{1000\pi}^{4000\pi} \sqrt{(-\theta \sin \theta + \cos \theta)^2 + (\theta \cos \theta + \sin \theta)^2} \, d\theta$$

$$= \frac{1}{2000\pi} \int_{1000\pi}^{4000\pi} \sqrt{\theta^2 + 1} \, d\theta \qquad \color{red}{\text{Integration tables (the appendix), Formula 26}}$$

$$= \frac{1}{2000\pi} \left(\frac{1}{2}\right) \left[\theta\sqrt{\theta^2 + 1} + \ln \left| \theta + \sqrt{\theta^2 + 1} \right| \right]_{1000\pi}^{4000\pi}$$

$$\approx 11{,}781 \text{ in.} \approx 982 \text{ ft.}$$

FOR FURTHER INFORMATION For more information on the mathematics of recording tape, see "Tape Counters" by Richard L. Roth in the August-September 1992 issue of *The American Mathematical Monthly.*

The length of the tape in Example 5 can be approximated by adding the circumferences of circular pieces of tape. The smallest circle has a radius of 0.501 and the largest has a radius of 2.

$$s \approx 2\pi(0.501) + 2\pi(0.502) + 2\pi(0.503) + \cdots + 2\pi(2.000)$$

$$= \sum_{i=1}^{1500} 2\pi(0.5 + 0.001i)$$

$$= 2\pi[1500(0.5) + 0.001(1500)(1501)/2]$$

$$\approx 11{,}786 \text{ in.}$$

Area of a Surface of Revolution

You can use the formula for the area of a surface of revolution in rectangular form to develop a formula for surface area in parametric form.

STUDY TIP These formulas are easy to remember if you think of the differential of arc length as

$$ds = \sqrt{\left(\frac{dx}{dt}\right)^2 + \left(\frac{dy}{dt}\right)^2}\, dt.$$

Then the formulas are written as follows.

1. $S = 2\pi \displaystyle\int_a^b g(t)\, ds$

2. $S = 2\pi \displaystyle\int_a^b f(t)\, ds$

> **THEOREM 9.9   Area of a Surface of Revolution**
>
> If a smooth curve C given by $x = f(t)$ and $y = g(t)$ does not cross itself on an interval $a \le t \le b$, then the area S of the surface of revolution formed by revolving C about the coordinate axes is given by the following.
>
> 1. $S = 2\pi \displaystyle\int_a^b g(t) \sqrt{\left(\frac{dx}{dt}\right)^2 + \left(\frac{dy}{dt}\right)^2}\, dt$ Revolution about x-axis: $g(t) \ge 0$
>
> 2. $S = 2\pi \displaystyle\int_a^b f(t) \sqrt{\left(\frac{dx}{dt}\right)^2 + \left(\frac{dy}{dt}\right)^2}\, dt$ Revolution about y-axis: $f(t) > 0$

EXAMPLE 6   Finding the Area of a Surface of Revolution

Let C be the arc of the circle

$$x^2 + y^2 = 9$$

from $(3, 0)$ to $\left(3/2, 3\sqrt{3}/2\right)$, as shown in Figure 9.33. Find the area of the surface formed by revolving C about the x-axis.

Solution You can represent C parametrically by the equations

$$x = 3 \cos t \quad \text{and} \quad y = 3 \sin t, \quad 0 \le t \le \pi/3.$$

(Note that you can determine the interval for t by observing that $t = 0$ when $x = 3$ and $t = \pi/3$ when $x = 3/2$.) On this interval, C is smooth and y is nonnegative, and you can apply Theorem 9.9 to obtain a surface area of

$$S = 2\pi \int_0^{\pi/3} (3 \sin t) \sqrt{(-3 \sin t)^2 + (3 \cos t)^2}\, dt$$

$$= 6\pi \int_0^{\pi/3} \sin t \sqrt{9(\sin^2 t + \cos^2 t)}\, dt$$

$$= 6\pi \int_0^{\pi/3} 3 \sin t\, dt$$

$$= -18\pi \left[\cos t\right]_0^{\pi/3}$$

$$= -18\pi \left(\frac{1}{2} - 1\right)$$

$$= 9\pi.$$

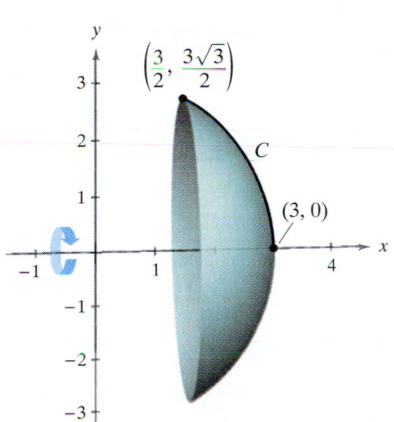

This surface of revolution has a surface area of 9π.

Figure 9.33

EXERCISES FOR SECTION 9.3

In Exercises 1–10, find dy/dx and d^2y/dx^2, and evaluate each at the indicated value of the parameter.

Parametric Equations	Point
1. $x = 2t, y = 3t - 1$	$t = 3$
2. $x = \sqrt{t}, y = 3t - 1$	$t = 1$
3. $x = t + 1, y = t^2 + 3t$	$t = -1$
4. $x = t^2 + 3t, y = t + 1$	$t = 0$
5. $x = 2 \cos \theta, y = 2 \sin \theta$	$\theta = \dfrac{\pi}{4}$
6. $x = \cos \theta, y = 3 \sin \theta$	$\theta = 0$
7. $x = 2 + \sec \theta, y = 1 + 2 \tan \theta$	$\theta = \dfrac{\pi}{6}$
8. $x = \sqrt{t}, y = \sqrt{t - 1}$	$t = 2$
9. $x = \cos^3 \theta, y = \sin^3 \theta$	$\theta = \dfrac{\pi}{4}$
10. $x = \theta - \sin \theta, y = 1 - \cos \theta$	$\theta = \pi$

In Exercises 11 and 12, find an equation of the tangent line at the indicated points on the curve.

11. $x = 2 \cot \theta$
$y = 2 \sin^2 \theta$

12. $x = 2 - 3 \cos \theta$
$y = 3 + 2 \sin \theta$

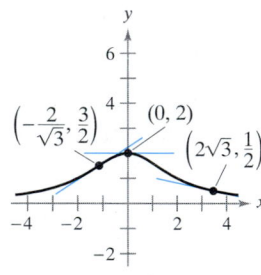

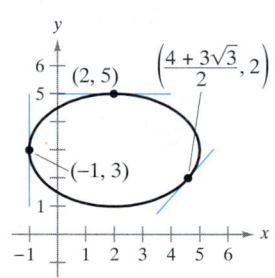

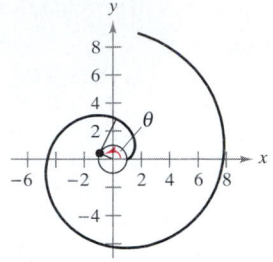 **In Exercises 13–16, (a) use a graphing utility to graph the curve represented by the parametric equations, (b) use a graphing utility to find dx/dt, dy/dt, and dy/dx at the indicated value of the parameter, (c) find an equation of the tangent line to the curve at the indicated value of the parameter, and (d) confirm the result in part (c) by using a graphing utility to graph the tangent line.**

Parametric Equations	Parameter
13. $x = 2t, y = t^2 - 1$	$t = 2$
14. $x = t - 1, y = \dfrac{1}{t} + 1$	$t = 1$
15. $x = t^2 - t + 2, y = t^3 - 3t$	$t = -1$
16. $x = 4 \cos \theta, y = 3 \sin \theta$	$\theta = \dfrac{3\pi}{4}$

In Exercises 17 and 18, find all points (if any) of horizontal and vertical tangency to the portion of the curve shown.

17. Involute of a circle:
$x = \cos \theta + \theta \sin \theta$
$y = \sin \theta - \theta \cos \theta$

18. $x = 2\theta$
$y = 2(1 - \cos \theta)$

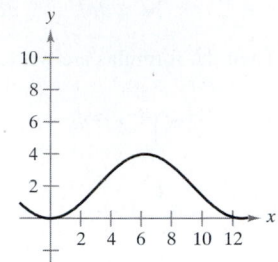

In Exercises 19–28, find all points (if any) of horizontal and vertical tangency to the curve. Use a graphing utility to sketch the curve and confirm your results.

19. $x = 1 - t, y = t^2$

20. $x = t + 1, y = t^2 + 3t$

21. $x = 1 - t, y = t^3 - 3t$

22. $x = t^2 - t + 2, y = t^3 - 3t$

23. $x = 3 \cos \theta, y = 3 \sin \theta$

24. $x = \cos \theta, y = 2 \sin 2\theta$

25. $x = 4 + 2 \cos \theta$
$y = -1 + \sin \theta$

26. $x = 4 \cos^2 \theta$
$y = 2 \sin \theta$

27. $x = \sec \theta, y = \tan \theta$

28. $x = \cos^2 \theta, y = \cos \theta$

29. *Think About It* Sketch a graph of a curve defined by the parametric equations $x = g(t)$ and $y = f(t)$ such that $dx/dt > 0$ and $dy/dt < 0$ for all real numbers t.

30. *Think About It* Sketch a graph of a curve defined by the parametric equations $x = g(t)$ and $y = f(t)$ such that $dx/dt < 0$ and $dy/dt < 0$ for all real numbers t.

Arc Length **In Exercises 31–36, find the arc length of the given curve on the indicated interval.**

Parametric Equations	Interval
31. $x = e^{-t} \cos t, y = e^{-t} \sin t$	$0 \le t \le \dfrac{\pi}{2}$
32. $x = t^2, y = 4t^3 - 1$	$-1 \le t \le 1$
33. $x = t^2, y = 2t$	$0 \le t \le 2$
34. $x = \arcsin t, y = \ln \sqrt{1 - t^2}$	$0 \le t \le \frac{1}{2}$
35. $x = \sqrt{t}, y = 3t - 1$	$0 \le t \le 1$
36. $x = t, y = \dfrac{t^5}{10} + \dfrac{1}{6t^3}$	$1 \le t \le 2$

Arc Length In Exercises 37–40, find the arc length of the curve on the interval $[0, 2\pi]$.

37. Hypocycloid perimeter: $x = a \cos^3 \theta, y = a \sin^3 \theta$

38. Circle circumference: $x = a \cos \theta, y = a \sin \theta$

39. Cycloid arch: $x = a(\theta - \sin \theta), y = a(1 - \cos \theta)$

40. Involute of a circle: $x = \cos \theta + \theta \sin \theta, y = \sin \theta - \theta \cos \theta$

41. ***Path of a Projectile*** The path of a projectile is modeled by the parametric equations

$$x = (90 \cos 30°)t \qquad \text{and} \qquad y = (90 \sin 30°)t - 16t^2$$

where x and y are measured in feet. Use a graphing utility to perform the following.

(a) Graph the path of the projectile.

(b) Approximate the range of the projectile.

(c) Use the integration capabilities of the graphing utility to approximate the arc length of the path. Compare this result with the range of the projectile.

42. ***Circumference of an Ellipse*** Use the integration capabilities of a graphing utility to approximate the circumference of the ellipse given by the parametric equations $x = 3 \cos \theta$ and $y = 4 \sin \theta$.

43. ***Writing***

(a) Use a graphing utility to graph each of the following sets of parametric equations.

$$\begin{array}{ll} x = t - \sin t & x = 2t - \sin(2t) \\ y = 1 - \cos t & y = 1 - \cos(2t) \\ 0 \le t \le 2\pi & 0 \le t \le \pi \end{array}$$

(b) Compare the graphs of the two sets of parametric equations in part (a). If the curve represents the motion of a particle and t is time, what can you infer about the average speed of the particle on the paths represented by the two sets of parametric equations?

(c) Without graphing the curve, determine the time required for a particle to traverse the same path as in parts (a) and (b) if the path is modeled by

$$x = \tfrac{1}{2}t - \sin\left(\tfrac{1}{2}t\right) \qquad \text{and} \qquad y = 1 - \cos\left(\tfrac{1}{2}t\right).$$

44. ***Folium of Descartes*** Given the parametric equations

$$x = \frac{4t}{1 + t^3} \qquad \text{and} \qquad y = \frac{4t^2}{1 + t^3}$$

use a graphing utility to perform the following.

(a) Sketch the curve described by the parametric equations.

(b) Find the points of horizontal tangency to the curve.

(c) Use the integration capabilities of the graphing utility to approximate the arc length of the closed loop. (*Hint:* Use symmetry and integrate over the interval $0 \le t \le 1$.)

Surface Area In Exercises 45–50, find the area of the surface generated by revolving the curve about the given axis.

45. $x = t, y = 2t, \quad 0 \le t \le 4, \quad$ (a) x-axis (b) y-axis

46. $x = t, y = 4 - 2t, \quad 0 \le t \le 2, \quad$ (a) x-axis (b) y-axis

47. $x = 4 \cos \theta, y = 4 \sin \theta, \quad 0 \le \theta \le \dfrac{\pi}{2}, \quad y$-axis

48. $x = t^3, y = t + 2, \quad 1 \le t \le 2, \quad y$-axis

49. $x = a \cos^3 \theta, y = a \sin^3 \theta, \quad 0 \le \theta \le \pi, \quad x$-axis

50. $x = a \cos \theta, y = b \sin \theta, \quad 0 \le \theta \le 2\pi,$

(a) x-axis (b) y-axis

51. ***Surface Area*** A portion of a sphere of radius r is removed by cutting out a circular cone with its vertex at the center of the sphere. Find the surface area removed from the sphere if the vertex of the cone forms an angle of 2θ.

52. Use integration by substitution to show that if y is a continuous function of x on the interval $a \le x \le b$, where $x = f(t)$ and $y = g(t)$, then

$$\int_a^b y \, dx = \int_{t_1}^{t_2} g(t) f'(t) \, dt,$$

where $f(t_1) = a$, $f(t_2) = b$, and both g and f' are continuous on $[t_1, t_2]$.

Centroid In Exercises 53 and 54, find the centroid of the region bounded by the graph of the parametric equations and the coordinate axes. (Use the result in Exercise 52.)

53. $x = \sqrt{t}, y = 4 - t$ **54.** $x = \sqrt{4 - t}, y = \sqrt{t}$

Volume In Exercises 55 and 56, find the volume of the solid formed by revolving the region bounded by the graphs of the given equations about the x-axis. (Use the result in Exercise 52.)

55. $x = 3 \cos \theta, y = 3 \sin \theta$ **56.** $x = \cos \theta, y = 3 \sin \theta$

Area In Exercises 57 and 58, find the area of the region. (Use the result in Exercise 52.)

57. $x = 2 \sin^2 \theta$
$y = 2 \sin^2 \theta \tan \theta$
$0 \le \theta < \dfrac{\pi}{2}$

58. $x = 2 \cot \theta$
$y = 2 \sin^2 \theta$
$0 < \theta < \pi$

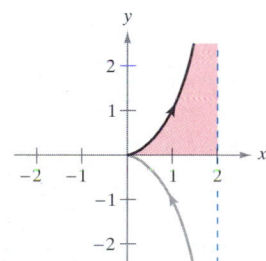

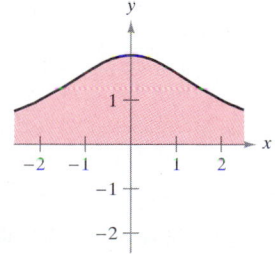

Areas of Simple Closed Curves   In Exercises 59–64, use a symbolic integration utility and the result in Exercise 52 to match the closed curve with its area. (These exercises were adapted from the article "The Surveyor's Area Formula" by Bart Braden in the September 1986 issue of the *College Mathematics Journal* by permission of the author.)

(a) $\frac{8}{3}ab$ (b) $\frac{3}{8}\pi a^2$ (c) $2\pi a^2$

(d) πab (e) $2\pi ab$ (f) $6\pi a^2$

59. Ellipse: $(0 \le t \le 2\pi)$

$x = b \cos t$

$y = a \sin t$

60. Asteroid: $(0 \le t \le 2\pi)$

$x = a \cos^3 t$

$y = a \sin^3 t$

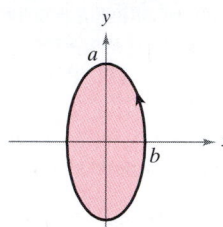

61. Cardioid: $(0 \le t \le 2\pi)$

$x = 2a \cos t - a \cos 2t$

$y = 2a \sin t - a \sin 2t$

62. Deltoid: $(0 \le t \le 2\pi)$

$x = 2a \cos t + a \cos 2t$

$y = 2a \sin t - a \sin 2t$

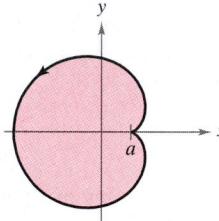
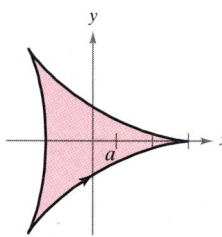

63. Hourglass: $(0 \le t \le 2\pi)$

$x = a \sin 2t$

$y = b \sin t$

64. Teardrop: $(0 \le t \le 2\pi)$

$x = 2a \cos t - a \sin 2t$

$y = b \sin t$

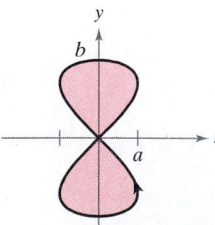
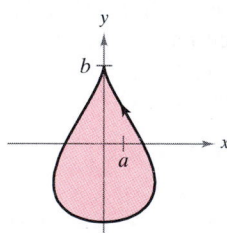

65. Use a graphing utility to graph the curve given by

$$x = \frac{1 - t^2}{1 + t^2}, \quad y = \frac{2t}{1 + t^2}, \quad -20 \le t \le 20.$$

(a) Describe the graph and confirm your result analytically.

(b) Discuss the speed at which the curve is traced as t increases from -20 to 20.

66. *Tractrix*   A person moves from the origin along the positive y-axis pulling a weight at the end of a 12-meter rope. Initially, the weight is located at the point $(12, 0)$.

(a) In Exercise 67 of Section 7.4, it was shown that the path of the weight is modeled by the rectangular equation

$$y = -12 \ln\left(\frac{12 - \sqrt{144 - x^2}}{x}\right) - \sqrt{144 - x^2}$$

where $0 < x \le 12$. Use a graphing utility to graph the rectangular equation.

(b) Use a graphing utility to graph the parametric equations

$$x = 12 \operatorname{sech} \frac{t}{12} \quad \text{and} \quad y = t - 12 \tanh \frac{t}{12}$$

where $t \ge 0$. How does this graph compare with the graph in part (a)? Which graph (if either) do you think is a better representation of the path?

(c) Use the parametric equations for the tractrix to verify that the distance from the y-intercept of the tangent line to the point of tangency is independent of the location of the point of tangency.

67. *Air Traffic Control*   A controller spots two planes at the same altitude flying toward each other (see figure). Their flight paths are S 20° W and S 45° E. One plane is 150 miles from point P with a speed of 375 miles per hour. The other is 190 miles from point P with a speed of 450 miles per hour.

(a) Find parametric equations for the path of each plane where t is the time in hours, with $t = 0$ corresponding to the time at which the air traffic controller spots the planes.

(b) Use the result in part (a) to write the distance between the planes as a function of t.

(c) Use a graphing utility to graph the function in part (b). When will the distance between the planes be minimum? If the planes must keep a separation of at least 3 miles, is the requirement met?

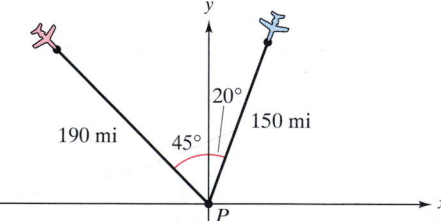

68. Let $x = g(t)$ and $y = f(t)$ be the parametric equations for a planar curve C. Interpret the Extended Mean Value Theorem in this case.

True or False?   In Exercises 69 and 70, determine whether the statement is true or false. If it is false, explain why or give an example that shows it is false.

69. If $x = f(t)$ and $y = g(t)$, then $d^2y/dx^2 = g''(t)/f''(t)$.

70. The curve given by $x = t^3$, $y = t^2$ has a horizontal tangent at the origin because $dy/dt = 0$ when $t = 0$.

Polar Coordinates • Coordinate Conversion • Polar Graphs •
Slope and Tangent Lines • Special Polar Graphs

Polar Coordinates

So far, we have been representing graphs as collections of points (x, y) on the rectangular coordinate system. The corresponding equations for these graphs have been in either rectangular or parametric form. In this section we introduce a coordinate system called the **polar coordinate system.**

To form the polar coordinate system in the plane, we fix a point O, called the **pole** (or **origin**), and construct from O an initial ray called the **polar axis,** as shown in Figure 9.34. Then each point P in the plane can be assigned **polar coordinates** (r, θ), as follows.

$r = $ *directed distance* from O to P

$\theta = $ *directed angle*, counterclockwise from polar axis to segment $\overline{OP}$

Figure 9.35 shows three points on the polar coordinate system. Notice that in this system, it is convenient to locate points with respect to a grid of concentric circles intersected by **radial lines** through the pole.

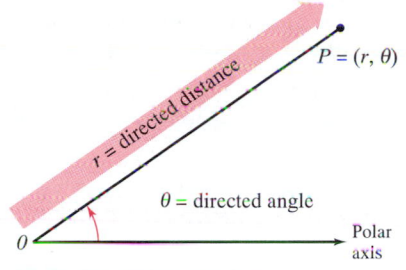

Polar coordinates
Figure 9.34

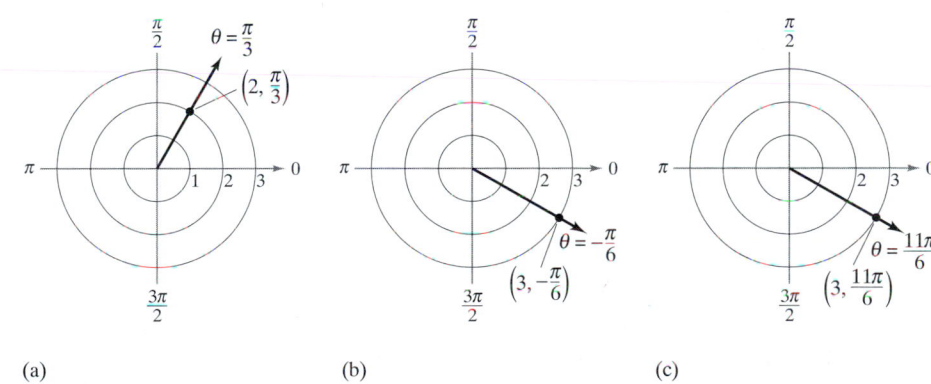

(a)   (b)   (c)

Figure 9.35

With rectangular coordinates, each point (x, y) has a unique representation. This is not true with polar coordinates. For instance, the coordinates (r, θ) and $(r, 2\pi + \theta)$ represent the same point [see parts (b) and (c) in Figure 9.35]. Also, because r is a *directed distance*, the coordinates (r, θ) and $(-r, \theta + \pi)$ represent the same point. In general, the point (r, θ) can be written as

$$(r, \theta) = (r, \theta + 2n\pi)$$

or

$$(r, \theta) = (-r, \theta + (2n + 1)\pi)$$

where n is any integer. Moreover, the pole is represented by $(0, \theta)$, where θ is any angle.

POLAR COORDINATES

The mathematician credited with first using polar coordinates was James Bernoulli, who introduced them in 1691. However, there is some evidence that it may have been Isaac Newton who first used them.

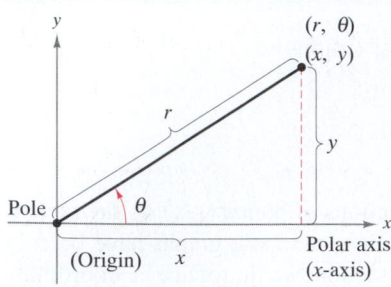

Relating polar and rectangular coordinates
Figure 9.36

Coordinate Conversion

To establish the relationship between polar and rectangular coordinates, let the polar axis coincide with the positive x-axis and the pole with the origin, as shown in Figure 9.36. Because (x, y) lies on a circle of radius r, it follows that $r^2 = x^2 + y^2$. Moreover, for $r > 0$, the definition of the trigonometric functions implies that

$$\tan \theta = \frac{y}{x}, \qquad \cos \theta = \frac{x}{r}, \qquad \text{and} \qquad \sin \theta = \frac{y}{r}.$$

If $r < 0$, you can show that the same relationships hold.

THEOREM 9.10 Coordinate Conversion

The polar coordinates (r, θ) of a point are related to the rectangular coordinates (x, y) of the point as follows.

1. $x = r \cos \theta$
$\quad y = r \sin \theta$

2. $\tan \theta = \dfrac{y}{x}$
$\quad r^2 = x^2 + y^2$

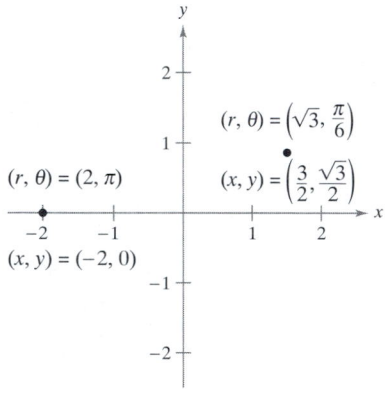

To convert from polar to rectangular coordinates, let $x = r \cos \theta$ and $y = r \sin \theta$.
Figure 9.37

EXAMPLE 1 Polar-to-Rectangular Conversion

a. For the point $(r, \theta) = (2, \pi)$,

$$x = r \cos \theta = 2 \cos \pi = -2 \qquad \text{and} \qquad y = r \sin \theta = 2 \sin \pi = 0.$$

Thus, the rectangular coordinates are $(x, y) = (-2, 0)$.

b. For the point $(r, \theta) = \left(\sqrt{3}, \pi/6 \right)$,

$$x = \sqrt{3} \cos \frac{\pi}{6} = \frac{3}{2} \qquad \text{and} \qquad y = \sqrt{3} \sin \frac{\pi}{6} = \frac{\sqrt{3}}{2}.$$

Thus, the rectangular coordinates are $(x, y) = \left(3/2, \sqrt{3}/2 \right)$.

(See Figure 9.37.)

EXAMPLE 2 Rectangular-to-Polar Conversion

a. For the second quadrant point $(x, y) = (-1, 1)$,

$$\tan \theta = \frac{y}{x} = -1 \qquad \Longrightarrow \qquad \theta = \frac{3\pi}{4}.$$

Because θ was chosen to be in the same quadrant as (x, y), you should use a positive value of r.

$$\begin{aligned} r &= \sqrt{x^2 + y^2} \\ &= \sqrt{(-1)^2 + (1)^2} \\ &= \sqrt{2} \end{aligned}$$

This implies that *one* set of polar coordinates is $(r, \theta) = \left(\sqrt{2}, 3\pi/4 \right)$.

b. Because the point $(x, y) = (0, 2)$ lies on the positive y-axis, we choose $\theta = \pi/2$ and $r = 2$, and one set of polar coordinates is $(r, \theta) = (2, \pi/2)$.

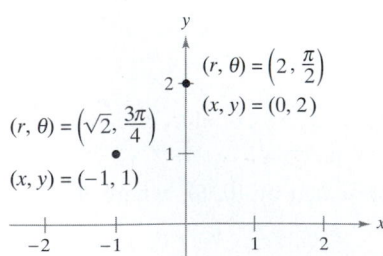

To convert from rectangular to polar coordinates, let $\tan \theta = y/x$ and $r = \sqrt{x^2 + y^2}$.
Figure 9.38

(See Figure 9.38.)

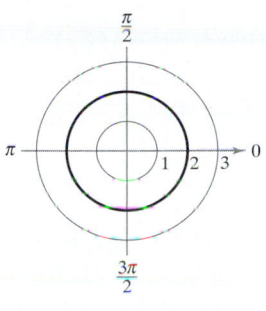

(a) Circle: $r = 2$

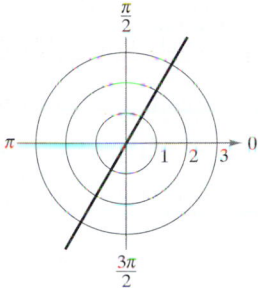

(b) Radial line: $\theta = \dfrac{\pi}{3}$

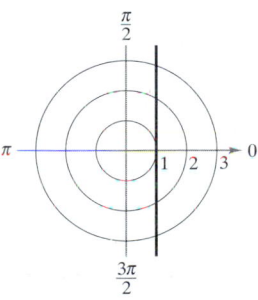

(c) Vertical line: $r = \sec \theta$

Figure 9.39

Polar Graphs

One way to sketch the graph of a polar equation is to convert to rectangular coordinates and then sketch the graph of the rectangular equation.

EXAMPLE 3   Graphing Polar Equations

Describe the graph of each of the following polar equations. Confirm each description by converting to a rectangular equation.

a. $r = 2$ **b.** $\theta = \dfrac{\pi}{3}$ **c.** $r = \sec \theta$

Solution   (See Figure 9.39.)

a. The graph of the polar equation $r = 2$ consists of all points that are two units from the pole. In other words, this graph is a circle centered at the origin with a radius of 2. You can confirm this by using the relationship $r^2 = x^2 + y^2$ to obtain the rectangular equation

$$x^2 + y^2 = 2^2. \qquad \text{Rectangular equation}$$

b. The graph of the polar equation $\theta = \pi/3$ consists of all points on the line that makes an angle of $\pi/3$ with the positive x-axis. You can confirm this by using the relationship $\tan \theta = y/x$ to obtain the rectangular equation

$$y = \sqrt{3}\,x. \qquad \text{Rectangular equation}$$

c. The graph of the polar equation $r = \sec \theta$ is not evident by simple inspection, so you can begin by converting to rectangular form using the relationship $r \cos \theta = x$.

$$r = \sec \theta \qquad \text{Polar equation}$$
$$r \cos \theta = 1$$
$$x = 1 \qquad \text{Rectangular equation}$$

From the rectangular equation, you can see that the graph is a vertical line.

TECHNOLOGY   Sketching the graphs of complicated polar equations *by hand* can be tedious. With technology, however, the task is not difficult. If you have access to a graphing utility that has a polar mode, try using it to sketch the graphs in the exercise set. If your graphing utility doesn't have a polar mode, but does have a parametric mode, you can sketch the graph of $r = f(\theta)$ by writing the equation as

$$x = f(\theta) \cos \theta$$
$$y = f(\theta) \sin \theta.$$

For instance, the graph of $r = \frac{1}{2}\theta$ shown in Figure 9.40 was produced with a graphing calculator in parametric mode. To sketch the graph, we entered the parametric equations

$$x = \frac{1}{2}\theta \cos \theta$$

$$y = \frac{1}{2}\theta \sin \theta$$

and let the values of θ vary from -4π to 4π. This curve is of the form $r = a\theta$ and is called a **spiral of Archimedes.**

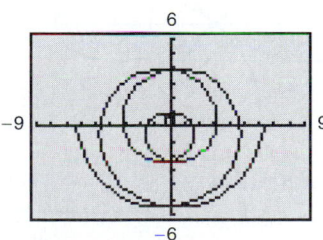

Spiral of Archimedes
Figure 9.40

NOTE One way to sketch the graph of $r = 2 \cos 3\theta$ *by hand* is to make a table of values.

θ	0	$\dfrac{\pi}{6}$	$\dfrac{\pi}{3}$	$\dfrac{\pi}{2}$	$\dfrac{2\pi}{3}$
r	2	0	-2	0	2

By extending the table and plotting the points, you will obtain the curve shown in Example 4.

EXAMPLE 4 **Sketching a Polar Graph**

Use a graphing utility to sketch the graph of $r = 2 \cos 3\theta$.

Solution Begin by writing the polar equation in parametric form.

$$x = 2 \cos 3\theta \cos \theta \qquad \text{and} \qquad y = 2 \cos 3\theta \sin \theta$$

After some experimentation, you will find that the entire curve, which is called a **rose curve,** can be sketched by letting θ vary from 0 to π, as shown in Figure 9.41. If you try duplicating this graph with a graphing utility, you will find that by letting θ vary from 0 to 2π, you will actually trace the entire curve *twice*.

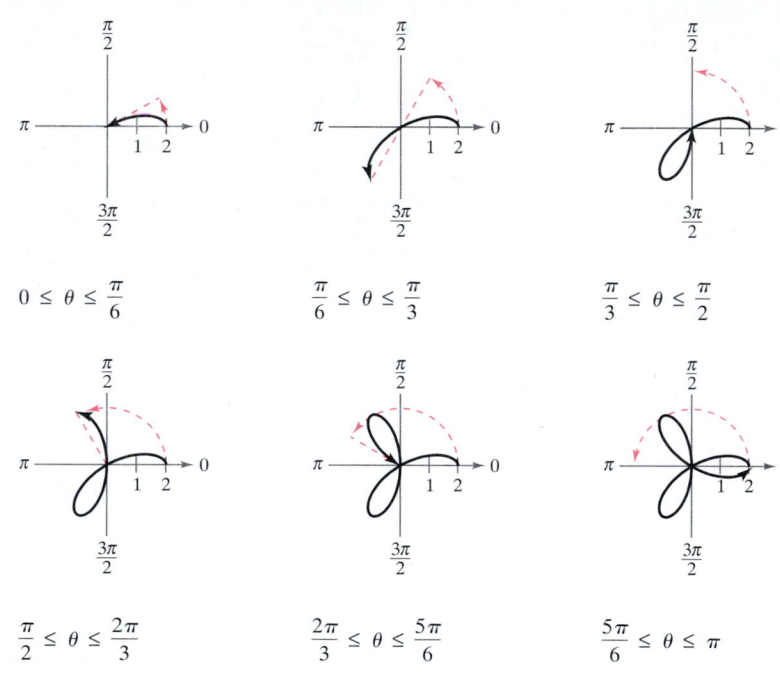

Figure 9.41

If you have access to a graphing utility, try using it to experiment with other rose curves (they are of the form $r = a \cos n\theta$ or $r = a \sin n\theta$). For instance, Figure 9.42 shows the graphs of two other rose curves.

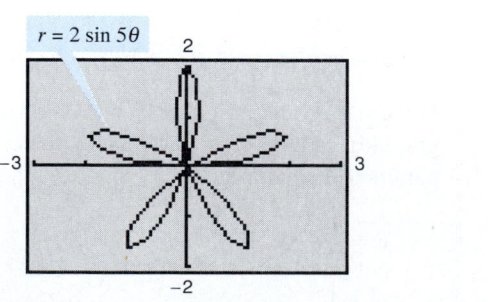

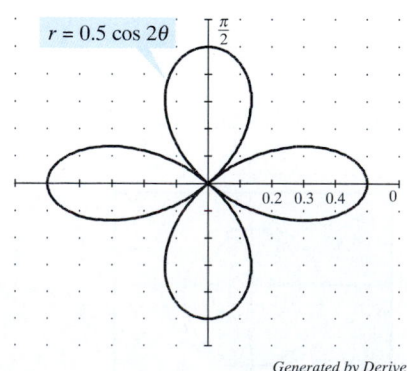

$r = 0.5 \cos 2\theta$

Generated by Derive

Rose curves
Figure 9.42

Slope and Tangent Lines

To find the slope of a tangent line to a polar graph, consider a differentiable function given by $r = f(\theta)$. To convert to polar form, use the parametric equations

$$x = r\cos\theta = f(\theta)\cos\theta \quad\text{and}\quad y = r\sin\theta = f(\theta)\sin\theta.$$

Using the parametric form of dy/dx given in Theorem 9.7, you have

$$\frac{dy}{dx} = \frac{dy/d\theta}{dx/d\theta} = \frac{f(\theta)\cos\theta + f'(\theta)\sin\theta}{-f(\theta)\sin\theta + f'(\theta)\cos\theta}$$

which establishes the following theorem.

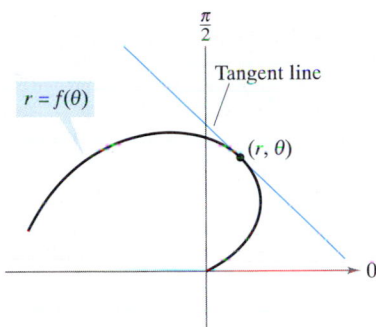

$r = f(\theta)$

Tangent line

(r, θ)

Tangent line to polar curve
Figure 9.43

THEOREM 9.11   Slope in Polar Form

If f is a differentiable function of θ, then the *slope* of the tangent line to the graph of $r = f(\theta)$ at the point (r, θ) is

$$\frac{dy}{dx} = \frac{dy/d\theta}{dx/d\theta} = \frac{f(\theta)\cos\theta + f'(\theta)\sin\theta}{-f(\theta)\sin\theta + f'(\theta)\cos\theta}$$

provided that $dx/d\theta \neq 0$ at (r, θ). (See Figure 9.43.)

From Theorem 9.11, you can make the following observations.

1. Solutions to $\dfrac{dy}{d\theta} = 0$ yield horizontal tangents, provided that $\dfrac{dx}{d\theta} \neq 0$.

2. Solutions to $\dfrac{dx}{d\theta} = 0$ yield vertical tangents, provided that $\dfrac{dy}{d\theta} \neq 0$.

If $dy/d\theta$ and $dx/d\theta$ are *simultaneously* 0, no conclusion can be drawn about tangent lines.

EXAMPLE 5   Finding Horizontal and Vertical Tangent Lines

Find the horizontal and vertical tangent lines of $r = \sin\theta$, $0 \leq \theta \leq \pi$.

Solution   Begin by writing the equation in parametric form.

$$x = r\cos\theta = \sin\theta\cos\theta$$

and

$$y = r\sin\theta = \sin\theta\sin\theta = \sin^2\theta$$

Next, differentiate x and y with respect to θ and set each derivative equal to 0.

$$\frac{dx}{d\theta} = \cos^2\theta - \sin^2\theta = \cos 2\theta = 0 \quad\Longrightarrow\quad \theta = \frac{\pi}{4}, \frac{3\pi}{4}$$

$$\frac{dy}{d\theta} = 2\sin\theta\cos\theta = \sin 2\theta = 0 \quad\Longrightarrow\quad \theta = 0, \frac{\pi}{2}$$

Thus, the graph has vertical tangent lines at $\left(\sqrt{2}/2, \pi/4\right)$ and $\left(\sqrt{2}/2, 3\pi/4\right)$, and it has horizontal tangent lines at $(0, 0)$ and $(1, \pi/2)$, as shown in Figure 9.44.

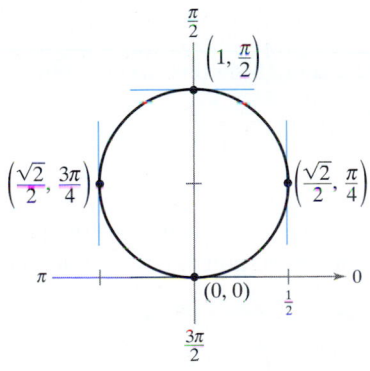

Horizontal and vertical tangent lines of
$r = \sin\theta$
Figure 9.44

EXAMPLE 6   Finding Horizontal and Vertical Tangent Lines

Find the horizontal and vertical tangents to the graph of $r = 2(1 - \cos \theta)$.

Solution   Using $y = r \sin \theta$, differentiate and set $dy/d\theta$ equal to 0.

$$y = r \sin \theta = 2(1 - \cos \theta) \sin \theta$$

$$\frac{dy}{d\theta} = 2[(1 - \cos \theta)(\cos \theta) + \sin \theta(\sin \theta)]$$

$$= -2(2 \cos \theta + 1)(\cos \theta - 1) = 0.$$

Thus, $\cos \theta = -\frac{1}{2}$ and $\cos \theta = 1$, and you can conclude that $dy/d\theta = 0$ when $\theta = 2\pi/3$, $4\pi/3$, and 0. Similarly, using $x = r \cos \theta$, you have

$$x = r \cos \theta = 2 \cos \theta - 2 \cos^2 \theta$$

$$\frac{dx}{d\theta} = -2 \sin \theta + 4 \cos \theta \sin \theta = 2 \sin \theta(2 \cos \theta - 1) = 0.$$

Thus, $\sin \theta = 0$ or $\cos \theta = \frac{1}{2}$, and you can conclude that $dx/d\theta = 0$ when $\theta = 0$, π, $\pi/3$, and $5\pi/3$. From these results, and from the graph shown in Figure 9.45, you can conclude that the graph has horizontal tangents at $(3, 2\pi/3)$ and $(3, 4\pi/3)$, and has vertical tangents at $(1, \pi/3)$, $(1, 5\pi/3)$, and $(4, \pi)$. This graph is called a **cardioid.** Note that both derivatives ($dy/d\theta$ and $dx/d\theta$) are 0 when $\theta = 0$. Using this information alone, you don't know whether the graph has a horizontal or vertical tangent line at the pole. From Figure 9.45, however, you can see that the graph has a cusp at the pole.

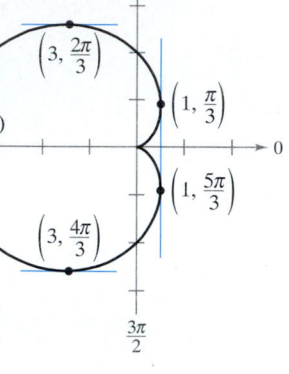

Horizontal and vertical tangent lines of
$r = 2(1 - \cos \theta)$
Figure 9.45

Theorem 9.11 has an important consequence. Suppose the graph of $r = f(\theta)$ passes through the pole when $\theta = \alpha$ and $f'(\alpha) \neq 0$. Then the formula for dy/dx simplifies as follows.

$$\frac{dy}{dx} = \frac{f'(\alpha) \sin \alpha + f(\alpha) \cos \alpha}{f'(\alpha) \cos \alpha - f(\alpha) \sin \alpha} = \frac{f'(\alpha) \sin \alpha + 0}{f'(\alpha) \cos \alpha - 0} = \frac{\sin \alpha}{\cos \alpha} = \tan \alpha$$

Thus, the line $\theta = \alpha$ is tangent to the graph at the pole, $(0, \alpha)$.

THEOREM 9.12   Tangent Lines at the Pole

If $f(\alpha) = 0$ and $f'(\alpha) \neq 0$, then the line $\theta = \alpha$ is tangent at the pole to the graph of $r = f(\theta)$.

Theorem 9.12 is useful because it states that the zeros of $r = f(\theta)$ can be used to find the tangent lines at the pole. Note that because a polar curve can cross the pole more than once, it can have more than one tangent line at the pole. For example, the rose curve

$$f(\theta) = 2 \cos 3\theta$$

has three tangent lines at the pole, as shown in Figure 9.46. For this curve, $f(\theta) = 2 \cos 3\theta$ is 0 when θ is $\pi/6$, $\pi/2$, and $5\pi/6$. Moreover, the derivative $f'(\theta) = -6 \sin 3\theta$ is not 0 for these values of θ.

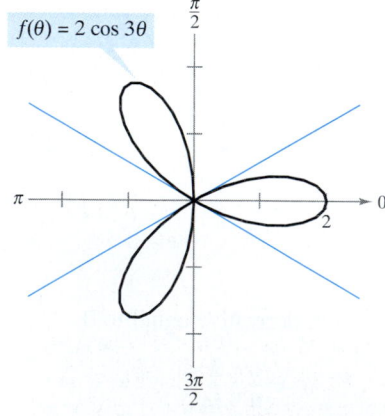

$f(\theta) = 2 \cos 3\theta$

This rose curve has three tangent lines
($\theta = \pi/6$, $\theta = \pi/2$, and $\theta = 5\pi/6$) at the pole.
Figure 9.46

Special Polar Graphs

Several important types of graphs have equations that are simpler in polar form than in rectangular form. For example, the polar equation of a circle having a radius of a and centered at the origin is simply $r = a$. Later in the text you will come to appreciate this benefit. For now, we summarize some other types of graphs that have simpler equations in polar form. (Conics are considered in Section 9.6.)

Limaçons

$r = a \pm b \cos \theta$
$r = a \pm b \sin \theta$
$(a > 0, b > 0)$

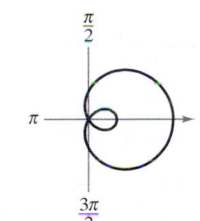

$\dfrac{a}{b} < 1$

Limaçon with inner loop

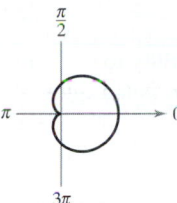

$\dfrac{a}{b} = 1$

Cardioid (heart-shaped)

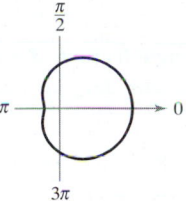

$1 < \dfrac{a}{b} < 2$

Dimpled limaçon

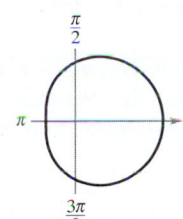

$\dfrac{a}{b} \geq 2$

Convex limaçon

Rose Curves

n petals if n is odd
$2n$ petals if n is even
$(n \geq 2)$

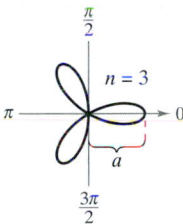

$r = a \cos n\theta$
Rose curve

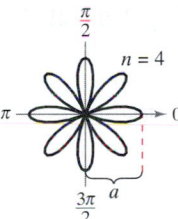

$r = a \cos n\theta$
Rose curve

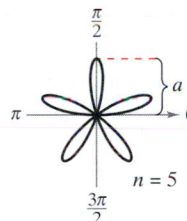

$r = a \sin n\theta$
Rose curve

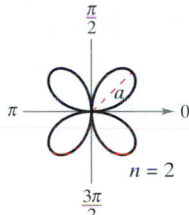

$r = a \sin n\theta$
Rose curve

Circles and Lemniscates

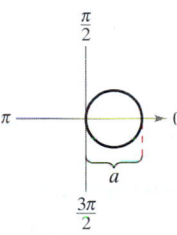

$r = a \cos \theta$
Circle

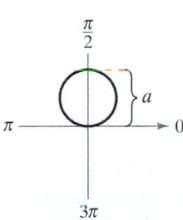

$r = a \sin \theta$
Circle

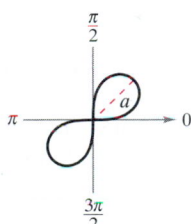

$r^2 = a^2 \sin 2\theta$
Lemniscate

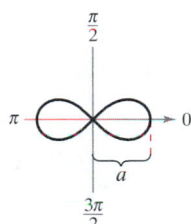

$r^2 = a^2 \cos 2\theta$
Lemniscate

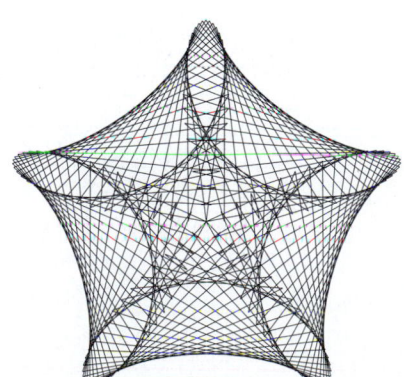

TECHNOLOGY   The rose curves described above are of the form $r = a \cos n\theta$ or $r = a \sin n\theta$, where n is a positive integer that is greater than or equal to 2. Try using a graphing utility to sketch the graph of $r = a \cos n\theta$ or $r = a \sin n\theta$ for some noninteger values of n. Are these graphs also rose curves? For example, try sketching the graph of $r = \cos \frac{2}{3}\theta$, $0 \leq \theta \leq 6\pi$.

FOR FURTHER INFORMATION   For more information on rose curves and related curves, see the article "A Rose is a Rose . . ." by Peter M. Maurer in the August-September 1987 issue of *The American Mathematical Monthly*. (The computer-generated graph at the left is the result of an algorithm that Maurer calls "The Rose.")

EXERCISES FOR SECTION 9.4

In Exercises 1–6, plot the point in polar coordinates and find the corresponding rectangular coordinates for the point.

1. $(4, 3\pi/6)$
2. $(-1, 5\pi/4)$
3. $(-4, -\pi/3)$
4. $(0, -7\pi/6)$
5. $(\sqrt{2}, 2.36)$
6. $(-3, -1.57)$

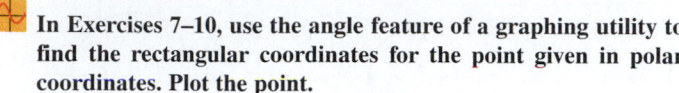 **In Exercises 7–10, use the angle feature of a graphing utility to find the rectangular coordinates for the point given in polar coordinates. Plot the point.**

7. $(5, 3\pi/4)$
8. $(-2, 11\pi/6)$
9. $(-3.5, 2.5)$
10. $(8.25, 1.3)$

In Exercises 11–14, the rectangular coordinates of a point are given. Plot the point and find *two* sets of polar coordinates for the point for $0 \le \theta < 2\pi$.

11. $(1, 1)$
12. $(0, -5)$
13. $(-3, 4)$
14. $(3, -1)$

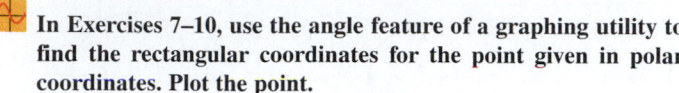 **In Exercises 15–18, use the angle feature of a graphing utility to find one set of polar coordinates for the point given in rectangular coordinates.**

15. $(3, -2)$
16. $(3\sqrt{2}, 3\sqrt{2})$
17. $\left(\frac{5}{2}, \frac{4}{3}\right)$
18. $(0, -5)$

19. Plot the point $(4, 3.5)$ if the point is given (a) in rectangular coordinates and (b) in polar coordinates.

20. *Graphical Reasoning*

   (a) Set the window format of a graphing utility on rectangular coordinates and locate the cursor at any position off the coordinate axes. Move the cursor horizontally and describe any changes in the displayed coordinates of the points. Repeat the process moving the cursor vertically.

   (b) Set the window format of a graphing utility on polar coordinates and locate the cursor at any position off the coordinate axes. Move the cursor horizontally and describe any changes in the displayed coordinates of the points. Repeat the process moving the cursor vertically.

   (c) Why are the results in parts (a) and (b) different?

In Exercises 21–28, convert the rectangular equation to polar form and sketch its graph.

21. $x^2 + y^2 = a^2$
22. $x^2 + y^2 - 2ax = 0$
23. $y = 4$
24. $x = 10$
25. $3x - y + 2 = 0$
26. $xy = 4$
27. $y^2 = 9x$
28. $(x^2 + y^2)^2 - 9(x^2 - y^2) = 0$

In Exercises 29–36, convert the polar equation to rectangular form and sketch its graph.

29. $r = 3$
30. $r = -2$
31. $r = \sin \theta$
32. $r = 3 \cos \theta$
33. $r = \theta$
34. $\theta = \dfrac{5\pi}{6}$
35. $r = 3 \sec \theta$
36. $r = 2 \csc \theta$

37. Convert the equation

$$r = 2(h \cos \theta + k \sin \theta)$$

to rectangular form and verify that it is the equation of a circle. Find the radius and the rectangular coordinates of the center of the circle.

38. *Distance Formula*

   (a) Verify that the Distance Formula for the distance between the two points (r_1, θ_1) and (r_2, θ_2) in polar coordinates is

   $$d = \sqrt{r_1^2 + r_2^2 - 2r_1 r_2 \cos(\theta_1 - \theta_2)}.$$

   (b) Describe the position of the points relative to each other if $\theta_1 = \theta_2$. Simplify the Distance Formula for this case. Is the simplification what you expected? Explain.

   (c) Simplify the Distance Formula if $\theta_1 - \theta_2 = 90°$. Is the simplification what you expected? Explain.

   (d) Choose two points on the polar coordinate system and find the distance between them. Then choose different polar representations of the same two points and apply the Distance Formula again. Discuss the result.

In Exercises 39–42, use the result of Exercise 38 to approximate the distance between the two points in polar coordinates.

39. $\left(4, \dfrac{2\pi}{3}\right), \left(2, \dfrac{\pi}{6}\right)$
40. $\left(10, \dfrac{7\pi}{6}\right), (3, \pi)$
41. $(2, 0.5), (7, 1.2)$
42. $(4, 2.5), (12, 1)$

In Exercises 43 and 44, find dy/dx and the slope of the tangent lines shown on the graph of the polar equation.

43. $r = 2 + 3 \sin \theta$
44. $r = 2(1 - \sin \theta)$

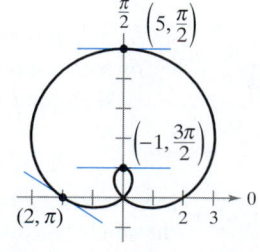

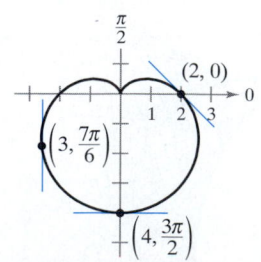

In Exercises 45–48, use a graphing utility (a) to graph the polar equation, (b) draw the tangent line at the given value of θ, and (c) find dy/dx at the given value of θ. (*Hint:* Let the increment between the values of θ equal $\pi/24$.)

45. $r = 3(1 - \cos \theta)$, $\theta = \dfrac{\pi}{2}$ **46.** $r = 3 - 2 \cos \theta$, $\theta = 0$

47. $r = 3 \sin \theta$, $\theta = \dfrac{\pi}{3}$ **48.** $r = 4$, $\theta = \dfrac{\pi}{4}$

 In Exercises 49 and 50, find the points of horizontal and vertical tangency (if any) to the polar curve.

49. $r = 1 + \sin \theta$ **50.** $r = a \sin \theta$

In Exercises 51 and 52, find the points of horizontal tangency (if any) to the polar curve.

51. $r = 2 \csc \theta + 3$ **52.** $r = a \sin \theta \cos^2 \theta$

 In Exercises 53–56, use a graphing utility to graph the polar equation and find all points of horizontal tangency.

53. $r = 4 \sin \theta \cos^2 \theta$ **54.** $r = 3 \cos 2\theta \sec \theta$

55. $r = 2 \csc \theta + 5$ **56.** $r = 2 \cos(3\theta - 2)$

In Exercises 57–62, sketch the graph of the polar equation and find the tangents at the pole.

57. $r = 3 \sin \theta$ **58.** $r = 3(1 - \cos \theta)$

59. $r = 2 \cos 3\theta$ **60.** $r = -\sin 5\theta$

61. $r = 3 \sin 2\theta$ **62.** $r = 3 \cos 2\theta$

In Exercises 63–70, sketch the graph of the polar equation.

63. $r = 3 - 2 \cos \theta$ **64.** $r = 5 - 4 \sin \theta$

65. $r = 3 \csc \theta$ **66.** $r = \dfrac{6}{2 \sin \theta - 3 \cos \theta}$

67. $r = 2\theta$ **68.** $r = \dfrac{1}{\theta}$

69. $r^2 = 4 \cos 2\theta$ **70.** $r^2 = 4 \sin \theta$

 In Exercises 71–80, use a graphing utility to graph the polar equation. Find an interval for θ over which the graph is traced *only once.*

71. $r = 3 - 4 \cos \theta$ **72.** $r = 2(1 - 2 \sin \theta)$

73. $r = 2 + \sin \theta$ **74.** $r = 4 + 3 \cos \theta$

75. $r = \dfrac{2}{1 + \cos \theta}$ **76.** $r = \dfrac{2}{4 - 3 \sin \theta}$

77. $r = 2 \cos\left(\dfrac{3\theta}{2}\right)$ **78.** $r = 3 \sin\left(\dfrac{5\theta}{2}\right)$

79. $r^2 = 4 \sin 2\theta$ **80.** $r^2 = \dfrac{1}{\theta}$

 In Exercises 81–84, use a graphing utility to graph the equation and show that the indicated line is an asymptote of the graph.

Name of Graph	Polar Equation	Asymptote
81. Conchoid	$r = 2 - \sec \theta$	$x = -1$
82. Conchoid	$r = 2 + \csc \theta$	$y = 1$
83. Hyperbolic spiral	$r = 2/\theta$	$y = 2$
84. Strophoid	$r = 2 \cos 2\theta \sec \theta$	$x = -2$

85. Sketch the graph of $r = 4 \sin \theta$ over each interval.

(a) $0 \le \theta \le \dfrac{\pi}{2}$ (b) $\dfrac{\pi}{2} \le \theta \le \pi$ (c) $-\dfrac{\pi}{2} \le \theta \le \dfrac{\pi}{2}$

 86. *Think About It*   Use a graphing utility to graph the polar equation $r = 6[1 + \cos(\theta - \phi)]$ for (a) $\phi = 0$, (b) $\phi = \pi/4$, and (c) $\phi = \pi/2$. Use the graphs to describe the effect of the angle ϕ. Write the equation as a function of $\sin \theta$ for part (c).

87. Verify that if the curve whose polar equation is $r = f(\theta)$ is rotated about the pole through an angle ϕ, then an equation for the rotated curve is $r = f(\theta - \phi)$.

88. The polar form of an equation for a curve is $r = f(\sin \theta)$. Show that the form becomes

(a) $r = f(-\cos \theta)$ if the curve is rotated counterclockwise $\pi/2$ radians about the pole.

(b) $r = f(-\sin \theta)$ if the curve is rotated counterclockwise π radians about the pole.

(c) $r = f(\cos \theta)$ if the curve is rotated counterclockwise $3\pi/2$ radians about the pole.

In Exercises 89–92, use the results of Exercises 87 and 88.

 89. Write an equation for the limaçon $r = 2 - \sin \theta$ after it has been rotated by the given amount. Verify the results by using a graphing utility to graph the rotated limaçon.

(a) $\pi/4$ (b) $\pi/2$ (c) π (d) $3\pi/2$

 90. Write an equation for the rose curve $r = 2 \sin 2\theta$ after it has been rotated by the given amount. Verify the results by using a graphing utility to graph the rotated rose curve.

(a) $\dfrac{\pi}{6}$ (b) $\dfrac{\pi}{2}$ (c) $\dfrac{2\pi}{3}$ (d) π

91. Sketch the graph of each equation.

(a) $r = 1 - \sin \theta$ (b) $r = 1 - \sin\left(\theta - \dfrac{\pi}{4}\right)$

92. Prove that the tangent of the angle ψ $(0 \le \psi \le \pi/2)$ between the radial line and the tangent line at the point (r, θ) on the graph of $r = f(\theta)$ (see figure) is given by $\tan \psi = |r/(dr/d\theta)|$.

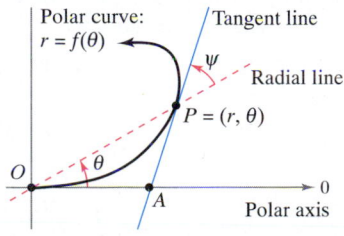

In Exercises 93–98, use the result of Exercise 92 to find the angle ψ between the radial and tangent lines to the graph for the indicated value of θ. Use a graphing utility to graph the polar equation, the radial line, and the tangent line for the indicated value of θ. Identify the angle ψ.

Polar Equation	Value of θ
93. $r = 2(1 - \cos\theta)$	$\theta = \pi$
94. $r = 3(1 - \cos\theta)$	$\theta = 3\pi/4$
95. $r = 2\cos 3\theta$	$\theta = \pi/6$
96. $r = 4\sin 2\theta$	$\theta = \pi/6$
97. $r = \dfrac{6}{1 - \cos\theta}$	$\theta = 2\pi/3$
98. $r = 5$	$\theta = \pi/6$

99. ***The Butterfly Curve*** Use a graphing utility to produce the curve shown below. The curve is given by

$$r = e^{\cos\theta} - 2\cos 4\theta + \sin^5 \frac{\theta}{12}.$$

Over what interval must θ vary to produce the curve?

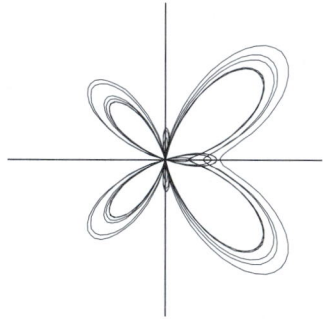

FOR FURTHER INFORMATION For more information on this curve, see the article "A Study in Step Size" by Temple H. Fay in the April 1997 issue of the *Mathematics Magazine*.

True or False? In Exercises 100–103, determine whether the statement is true or false. If it is false, explain why or give an example that shows it is false.

100. If (r_1, θ_1) and (r_2, θ_2) represent the same point on the polar coordinate system, then $|r_1| = |r_2|$.

101. If (r, θ_1) and (r, θ_2) represent the same point on the polar coordinate system, then $\theta_1 = \theta_2 + 2\pi n$ for some integer n.

102. If $x \geq 0$, then the point (x, y) on the rectangular coordinate system can be represented by (r, θ) on the polar coordinate system, where $r = \sqrt{x^2 + y^2}$ and $\theta = \arctan(y/x)$.

103. The polar equations $r = \sin 2\theta$ and $r = -\sin 2\theta$ have the same graph.

104. ***Heart to Bell*** Use a graphing utility to graph the polar equation

$$r = \cos 5\theta + n\cos\theta$$

for $0 \leq \theta < \pi$ for the integers $n = -5$ to $n = 5$. What values of n produce the "heart" portion of the curve? What values of n produce the "bell"? This curve, created by Michael W. Chamberlin, appeared in the January 1994 issue of *The College Mathematics Journal*.

SECTION PROJECT

Anamorphic Art Use the anamorphic transformations

$$r = y + 16 \quad \text{and} \quad \theta = -\frac{\pi}{8}x, \quad -\frac{3\pi}{4} \leq \theta \leq \frac{3\pi}{4}$$

to sketch the transformed polar image of the rectangular graph. When the reflection (in a cylindrical mirror centered at the pole) of each polar image is viewed from the polar axis, the viewer will see the original rectangular image.

(a) $y = 3$　　　(b) $x = 2$

(c) $y = x + 5$　　(d) $x^2 + (y - 5)^2 = 5^2$

Museum of Science and Industry in Manchester, England

This example of anamorphic art is from the Museum of Science and Industry in Manchester, England. When the reflection of the transformed "polar painting" is viewed in the mirror, the viewer sees faces.

FOR FURTHER INFORMATION For more information on anamorphic art, see the article "Anamorphisms" by Philip Hickin in the July 1992 issue of the *Mathematical Gazette*.

Area of a Polar Region • Points of Intersection of Polar Graphs •
Arc Length in Polar Form • Area of a Surface of Revolution

Area of a Polar Region

The development of a formula for the area of a polar region parallels that for the area of a region on the rectangular coordinate system, but uses *sectors* of a circle instead of rectangles as the basic element of area. In Figure 9.47, note that the area of a circular sector of radius r is given by $\frac{1}{2}\theta r^2$, provided θ is measured in radians.

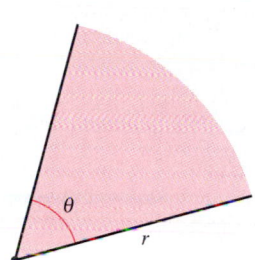

The area of a sector of a circle is $A = \frac{1}{2}\theta r^2$.
Figure 9.47

Consider the function given by $r = f(\theta)$, where f is continuous and nonnegative in the interval given by $\alpha \leq \theta \leq \beta$. The region bounded by the graph of f and the radial lines $\theta = \alpha$ and $\theta = \beta$ is shown in Figure 9.48. To find the area of this region, partition the interval $[a, \beta]$ into n equal subintervals,

$$\alpha = \theta_0 < \theta_1 < \theta_2 < \cdots < \theta_{n-1} < \theta_n = \beta.$$

Then, approximate the area of the region by the sum of the areas of the n sectors.

$$\text{Radius of } i\text{th sector} = f(\theta_i)$$

$$\text{Central angle of } i\text{th sector} = \frac{\beta - \alpha}{n} = \Delta\theta$$

$$A \approx \sum_{i=1}^{n} \left(\frac{1}{2}\right) \Delta\theta [f(\theta_i)]^2$$

Taking the limit as $n \to \infty$ produces

$$A = \lim_{n \to \infty} \frac{1}{2} \sum_{i=1}^{n} [f(\theta_i)]^2 \Delta\theta$$

$$= \frac{1}{2} \int_{\alpha}^{\beta} [f(\theta)]^2 \, d\theta$$

which leads to the following theorem.

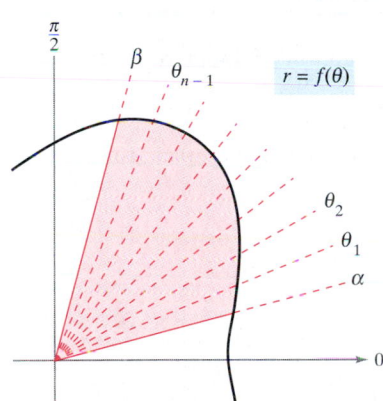

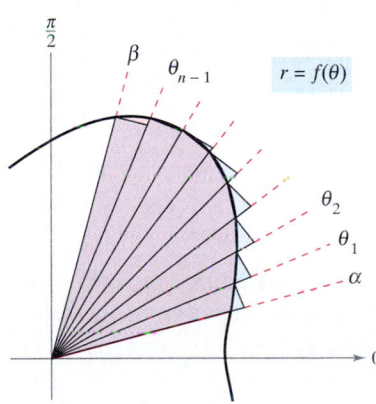

Figure 9.48

THEOREM 9.13   Area in Polar Coordinates

If f is continuous and nonnegative on the interval $[\alpha, \beta]$, then the area of the region bounded by the graph of $r = f(\theta)$ between the radial lines $\theta = \alpha$ and $\theta = \beta$ is given by

$$A = \frac{1}{2} \int_{\alpha}^{\beta} [f(\theta)]^2 \, d\theta$$

$$= \frac{1}{2} \int_{\alpha}^{\beta} r^2 \, d\theta.$$

NOTE   You can use the same formula to find the area of a region bounded by the graph of a continuous *nonpositive* function. However, the formula is not necessarily valid if f takes on both positive *and* negative values in the interval $[\alpha, \beta]$.

$r = 3 \cos 3\theta$

The area of one petal of the rose curve that lies between the radial lines $\theta = -\pi/6$ and $\theta = \pi/6$ is $3\pi/4$.
Figure 9.49

EXAMPLE 1   Finding the Area of a Polar Region

Find the area of *one petal* of the rose curve given by $r = 3 \cos 3\theta$.

Solution   In Figure 9.49, you can see that the right petal is traced as θ increases from $-\pi/6$ to $\pi/6$. Thus, the area is

$$A = \frac{1}{2} \int_{\alpha}^{\beta} r^2 \, d\theta = \frac{1}{2} \int_{-\pi/6}^{\pi/6} (3 \cos 3\theta)^2 \, d\theta$$

$$= \frac{9}{2} \int_{-\pi/6}^{\pi/6} \frac{1 + \cos 6\theta}{2} \, d\theta$$

$$= \frac{9}{4} \left[\theta + \frac{\sin 6\theta}{6} \right]_{-\pi/6}^{\pi/6} = \frac{9}{4} \left(\frac{\pi}{6} + \frac{\pi}{6} \right) = \frac{3\pi}{4}.$$

To find the area of the region lying inside all three petals of the rose curve in Example 1, you could not simply integrate between 0 and 2π. In doing this you would obtain $9\pi/2$, which is twice the area of the three petals—the duplication occurs because the rose curve is traced *twice* as θ increases from 0 to 2π.

EXAMPLE 2   Finding the Area Bounded by a Single Curve

Find the area of the region lying between the inner and outer loops of the limaçon $r = 1 - 2 \sin \theta$.

Solution   In Figure 9.50, note that the inner loop is traced as θ increases from $\pi/6$ to $5\pi/6$. Thus, the area inside the *inner loop* is

$$A_1 = \frac{1}{2} \int_{\pi/6}^{5\pi/6} (1 - 2 \sin \theta)^2 \, d\theta$$

$$= \frac{1}{2} \int_{\pi/6}^{5\pi/6} (1 - 4 \sin \theta + 4 \sin^2 \theta) \, d\theta$$

$$= \frac{1}{2} \int_{\pi/6}^{5\pi/6} \left[1 - 4 \sin \theta + 4 \left(\frac{1 - \cos 2\theta}{2} \right) \right] d\theta$$

$$= \frac{1}{2} \int_{\pi/6}^{5\pi/6} (3 - 4 \sin \theta - 2 \cos 2\theta) \, d\theta$$

$$= \frac{1}{2} \left[3\theta + 4 \cos \theta - \sin 2\theta \right]_{\pi/6}^{5\pi/6}$$

$$= \frac{1}{2} \left(2\pi - 3\sqrt{3} \right) = \pi - \frac{3\sqrt{3}}{2}.$$

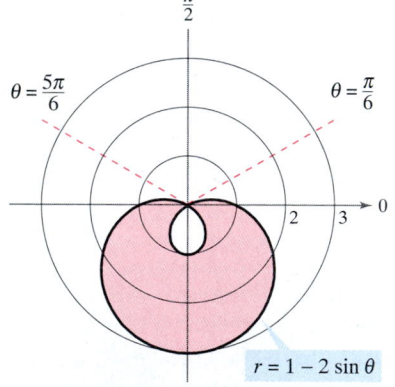

$\theta = \frac{5\pi}{6}$ $\theta = \frac{\pi}{6}$

$r = 1 - 2 \sin \theta$

The area between the inner and outer loops is approximately 8.34.
Figure 9.50

In a similar way, you can integrate from $5\pi/6$ to $13\pi/6$ to find that the area of the region lying inside the *outer loop* is $A_2 = 2\pi + (3\sqrt{3}/2)$. The area of the region lying between the two loops is the difference of A_2 and A_1.

$$A = A_2 - A_1 = \left(2\pi + \frac{3\sqrt{3}}{2} \right) - \left(\pi - \frac{3\sqrt{3}}{2} \right) = \pi + 3\sqrt{3} \approx 8.34$$

Points of Intersection of Polar Graphs

Because a point may be represented in different ways in polar coordinates, care must be taken in determining the points of intersection of two polar graphs. For example, consider the points of intersection of the graphs of

$$r = 1 - 2 \cos \theta \quad \text{and} \quad r = 1$$

as shown in Figure 9.51. If, as with rectangular equations, you attempted to find the points of intersection by solving the two equations simultaneously, you would obtain the following.

$r = 1 - 2 \cos \theta$	First equation
$1 = 1 - 2 \cos \theta$	Substitute $r = 1$ from 2nd equation into 1st equation.
$\cos \theta = 0$	Simplify.
$\theta = \dfrac{\pi}{2}, \dfrac{3\pi}{2}$	Solve for θ.

FOR FURTHER INFORMATION For more information on using technology to find points of intersection, see the article "Finding Points of Intersection of Polar-Coordinate Graphs" by Warren W. Esty in the September 1991 issue of *Mathematics Teacher.*

The corresponding points of intersection are $(1, \pi/2)$ and $(1, 3\pi/2)$. However, from Figure 9.51 you can see that there is a *third* point of intersection that did not show up when the two polar equations were solved simultaneously. (This is one reason we stress sketching a graph when finding the area of a polar region.) The reason the third point was not found is that it does not occur with the same coordinates in the two graphs. On the graph of $r = 1$, the point occurs with coordinates $(1, \pi)$, but on the graph of $r = 1 - 2 \cos \theta$, the point occurs with coordinates $(-1, 0)$.

You can compare the problem of finding points of intersection of two polar graphs with that of finding collision points of two satellites in intersecting orbits about the earth, as shown in Figure 9.52. The satellites will not collide as long as they reach the points of intersection at different times (θ-values). A collision will occur only at the points of intersection that are "simultaneous points"—those reached at the same time (θ-value).

NOTE   Because the pole can be represented by $(0, \theta)$, where θ is *any* angle, you should check separately for the pole when hunting for points of intersection.

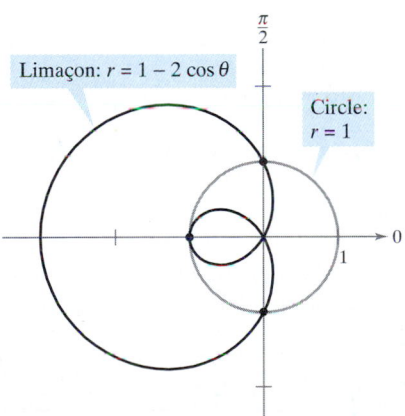

Three points of intersection: $(1, \pi/2)$, $(-1, 0)$, $(1, 3\pi/2)$
Figure 9.51

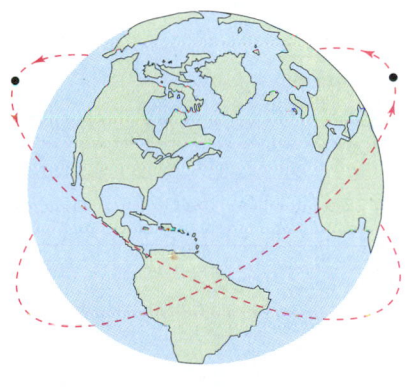

The paths of satellites can cross without causing a collision.
Figure 9.52

EXAMPLE 3   Finding the Area of a Region Between Two Curves

Find the area of the region common to the two regions bounded by the following curves.

$$r = -6 \cos \theta \qquad \text{Circle}$$
$$r = 2 - 2 \cos \theta \qquad \text{Cardioid}$$

Solution   Because both curves are symmetric with respect to the x-axis, you can work with the upper half-plane, as shown in Figure 9.53. The gray shaded region lies between the circle and the radial line $\theta = 2\pi/3$. Because the circle has coordinates $(0, \pi/2)$ at the pole, you can integrate between $\pi/2$ and $2\pi/3$ to obtain the area of this region. The region that is shaded red is bounded by the radial lines $\theta = 2\pi/3$ and $\theta = \pi$ and the cardioid. Thus, you can find the area of this second region by integrating between $2\pi/3$ and π. The sum of these two integrals gives the area of the common region lying *above* the radial line $\theta = \pi$.

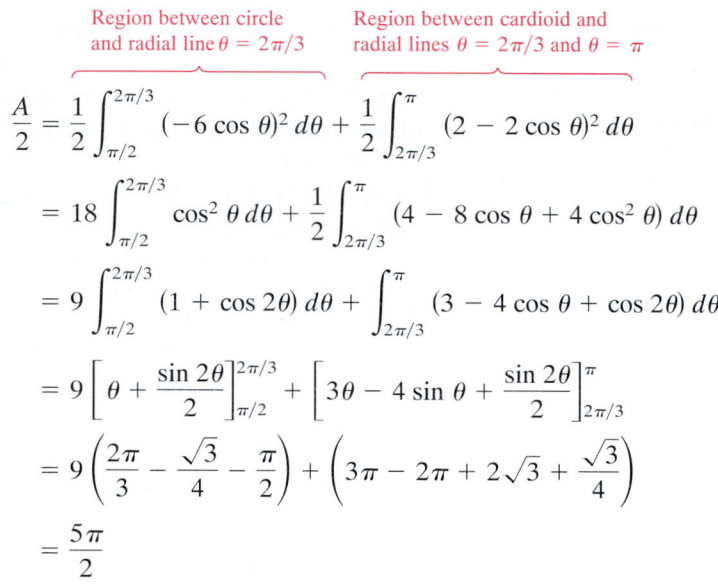

Region between circle and radial line $\theta = 2\pi/3$ ⎴   Region between cardioid and radial lines $\theta = 2\pi/3$ and $\theta = \pi$ ⎴

$$\frac{A}{2} = \frac{1}{2} \int_{\pi/2}^{2\pi/3} (-6 \cos \theta)^2 \, d\theta + \frac{1}{2} \int_{2\pi/3}^{\pi} (2 - 2 \cos \theta)^2 \, d\theta$$

$$= 18 \int_{\pi/2}^{2\pi/3} \cos^2 \theta \, d\theta + \frac{1}{2} \int_{2\pi/3}^{\pi} (4 - 8 \cos \theta + 4 \cos^2 \theta) \, d\theta$$

$$= 9 \int_{\pi/2}^{2\pi/3} (1 + \cos 2\theta) \, d\theta + \int_{2\pi/3}^{\pi} (3 - 4 \cos \theta + \cos 2\theta) \, d\theta$$

$$= 9 \left[\theta + \frac{\sin 2\theta}{2} \right]_{\pi/2}^{2\pi/3} + \left[3\theta - 4 \sin \theta + \frac{\sin 2\theta}{2} \right]_{2\pi/3}^{\pi}$$

$$= 9 \left(\frac{2\pi}{3} - \frac{\sqrt{3}}{4} - \frac{\pi}{2} \right) + \left(3\pi - 2\pi + 2\sqrt{3} + \frac{\sqrt{3}}{4} \right)$$

$$= \frac{5\pi}{2}$$

Finally, multiplying by 2, you can conclude that the total area is 5π.

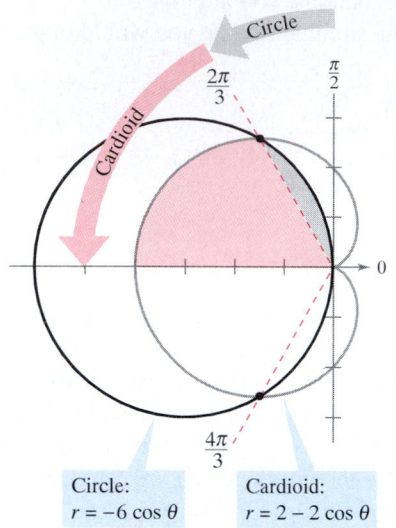

Circle:
$r = -6 \cos \theta$ Cardioid:
$r = 2 - 2 \cos \theta$

Figure 9.53

NOTE   To check the reasonableness of the result obtained in Example 3, note that the area of the circular region is $\pi r^2 = 9\pi$. Thus, it seems reasonable that the area of the region lying inside the circle and the cardioid is 5π.

To see the benefit of polar coordinates for finding the area in Example 3, consider the following integral, which gives the comparable area in rectangular coordinates.

$$\frac{A}{2} = \int_{-4}^{-3/2} \sqrt{2\sqrt{1 - 2x} - x^2 - 2x + 2} \, dx + \int_{-3/2}^{0} \sqrt{-x^2 - 6x} \, dx$$

Try using the integration capabilities of a graphing utility to show that you obtain the same area as that found in Example 3.

Arc Length in Polar Form

NOTE   When applying the arc length formula to a polar curve, be sure that the curve is traced out only once on the interval of integration. For instance, the rose curve given by $r = \cos 3\theta$ is traced out once on the interval $0 \leq \theta \leq \pi$, but is traced out twice on the interval $0 \leq \theta \leq 2\pi$.

The formula for the length of a polar arc can be obtained from the arc length formula for a curve described by parametric equations. (See Exercise 58.)

> **THEOREM 9.14   Arc Length of a Polar Curve**
>
> Let f be a function whose derivative is continuous on an interval $\alpha \leq \theta \leq \beta$. The length of the graph of $r = f(\theta)$ from $\theta = \alpha$ to $\theta = \beta$ is
>
> $$s = \int_{\alpha}^{\beta} \sqrt{[f(\theta)]^2 + [f'(\theta)]^2}\, d\theta = \int_{\alpha}^{\beta} \sqrt{r^2 + \left(\frac{dr}{d\theta}\right)^2}\, d\theta.$$

EXAMPLE 4   Finding the Length of a Polar Curve

Find the length of the arc from $\theta = 0$ to $\theta = 2\pi$ for the cardioid

$$r = f(\theta) = 2 - 2\cos\theta$$

as shown in Figure 9.54.

Solution   Because $f'(\theta) = 2\sin\theta$, you can find the arc length as follows.

$$
\begin{aligned}
s &= \int_{\alpha}^{\beta} \sqrt{[f(\theta)]^2 + [f'(\theta)]^2}\, d\theta && \text{\color{red}Formula for arc length} \\[4pt]
&= \int_{0}^{2\pi} \sqrt{(2 - 2\cos\theta)^2 + (2\sin\theta)^2}\, d\theta \\[4pt]
&= 2\sqrt{2} \int_{0}^{2\pi} \sqrt{1 - \cos\theta}\, d\theta \\[4pt]
&= 2\sqrt{2} \int_{0}^{2\pi} \sqrt{2\sin^2\frac{\theta}{2}}\, d\theta \\[4pt]
&= 4 \int_{0}^{2\pi} \sin\frac{\theta}{2}\, d\theta && \text{\color{red}$\sin\frac{\theta}{2} \geq 0$ for $0 \leq \theta \leq 2\pi$} \\[4pt]
&= 8 \left[-\cos\frac{\theta}{2} \right]_{0}^{2\pi} \\[4pt]
&= 8(1 + 1) \\[4pt]
&= 16
\end{aligned}
$$

In the fifth step of the solution, it is legitimate to write

$$\sqrt{2\sin^2(\theta/2)} = \sqrt{2}\sin(\theta/2)$$

rather than $\sqrt{2\sin^2(\theta/2)} = \sqrt{2}\,|\sin(\theta/2)|$ because $\sin(\theta/2) \geq 0$ for $0 \leq \theta \leq 2\pi$.

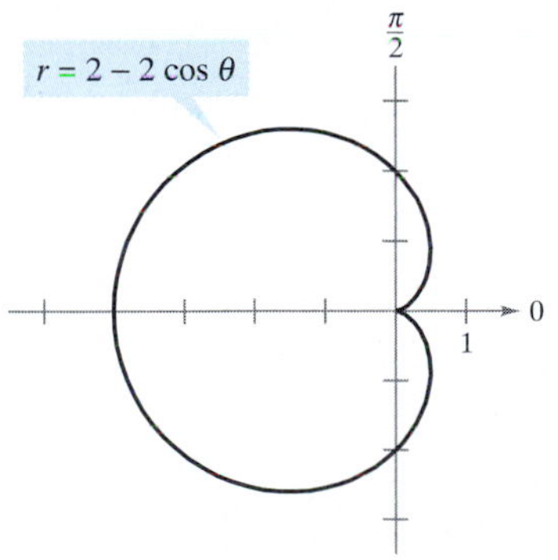

NOTE   Using Figure 9.54, you can determine the reasonableness of this answer by comparing it with the circumference of a circle. For example, a circle of radius $\frac{5}{2}$ has a circumference of $5\pi \approx 15.7$.

$r = 2 - 2\cos\theta$

The arc length of this cardioid is 16.
Figure 9.54

Area of a Surface of Revolution

The polar coordinate version of the formulas for the area of a surface of revolution can be obtained from the parametric versions given in Theorem 9.9, using the equations $x = r \cos \theta$ and $y = r \sin \theta$.

THEOREM 9.15   Area of a Surface of Revolution

Let f be a function whose derivative is continuous on an interval $\alpha \leq \theta \leq \beta$. The area of the surface formed by revolving the graph of $r = f(\theta)$ from $\theta = \alpha$ to $\theta = \beta$ about the indicated line is as follows.

1. $S = 2\pi \displaystyle\int_{\alpha}^{\beta} f(\theta) \sin \theta \sqrt{[f(\theta)]^2 + [f'(\theta)]^2}\, d\theta$ About polar axis

2. $S = 2\pi \displaystyle\int_{\alpha}^{\beta} f(\theta) \cos \theta \sqrt{[f(\theta)]^2 + [f'(\theta)]^2}\, d\theta$ About line $\theta = \dfrac{\pi}{2}$

NOTE   When using Theorem 9.15, check to see that the graph of $r = f(\theta)$ is traced only once on the interval $\alpha \leq \theta \leq \beta$. For example, the circle given by $r = \cos \theta$ is traced once on the interval $0 \leq \theta \leq \pi$.

EXAMPLE 5   Finding the Area of a Surface of Revolution

Find the area of the surface formed by revolving the circle $r = f(\theta) = \cos \theta$ about the line $\theta = \pi/2$, as shown in Figure 9.55.

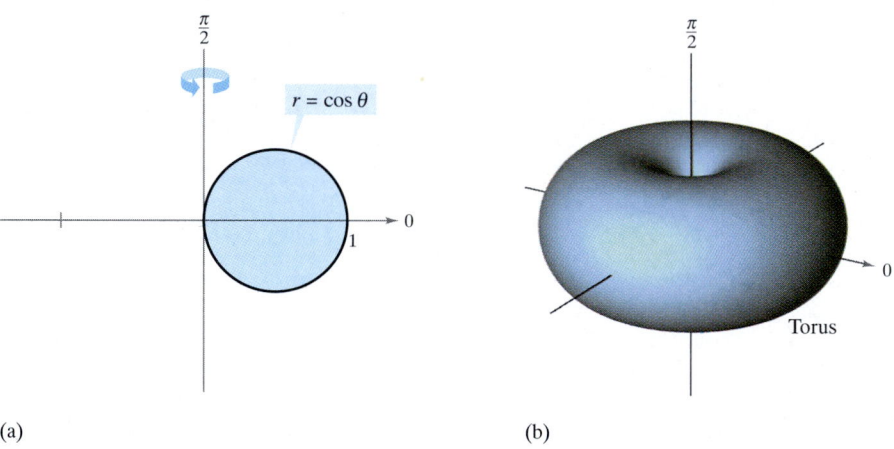

$r = \cos \theta$

(a)   (b)

Torus

The surface area of this torus is π^2.
Figure 9.55

Solution   You can use the second formula given in Theorem 9.15 with $f'(\theta) = -\sin \theta$. Because the circle is traced once as θ increases from 0 to π, we have

$$S = 2\pi \int_{\alpha}^{\beta} f(\theta) \cos \theta \sqrt{[f(\theta)]^2 + [f'(\theta)]^2}\, d\theta$$

$$= 2\pi \int_{0}^{\pi} \cos \theta (\cos \theta) \sqrt{\cos^2 \theta + \sin^2 \theta}\, d\theta$$

$$= 2\pi \int_{0}^{\pi} \cos^2 \theta\, d\theta$$

$$= \pi \int_{0}^{\pi} (1 + \cos 2\theta)\, d\theta$$

$$= \pi \left[\theta + \frac{\sin 2\theta}{2} \right]_{0}^{\pi} = \pi^2.$$

EXERCISES FOR SECTION 9.5

In Exercises 1 and 2, find the area of the region bounded by the graph of the polar equation by (a) a geometric formula, and (b) integration.

1. $r = 8 \sin \theta$

2. $r = 3 \cos \theta$

In Exercises 3–8, find the area of the region.

3. One petal of $r = 2 \cos 3\theta$

4. One petal of $r = 4 \sin 2\theta$

5. One petal of $r = \cos 2\theta$

6. One petal of $r = \cos 5\theta$

7. Interior of $r = 1 - \sin \theta$

8. Interior of $r = 1 - \sin \theta$ (above the polar axis)

In Exercises 9–12, use a graphing utility to graph the polar equation and find the area of the indicated region.

9. Inner loop of $r = 1 + 2 \cos \theta$

10. Inner loop of $r = 3 + 4 \sin \theta$

11. Between the loops of $r = 1 + 2 \cos \theta$

12. Between the loops of $r = 2(1 + 2 \sin \theta)$

In Exercises 13–22, find the points of intersection of the graphs of the equation.

13. $r = 1 + \cos \theta$
$r = 1 - \cos \theta$

14. $r = 3(1 + \sin \theta)$
$r = 3(1 - \sin \theta)$

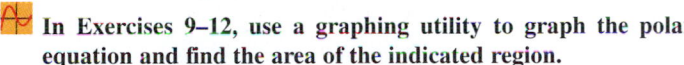

15. $r = 1 + \cos \theta$
$r = 1 - \sin \theta$

16. $r = 2 - 3 \cos \theta$
$r = \cos \theta$

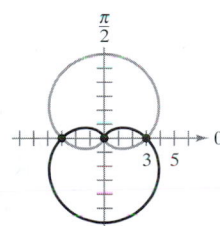

17. $r = 4 - 5 \sin \theta$
$r = 3 \sin \theta$

18. $r = 1 + \cos \theta$
$r = 3 \cos \theta$

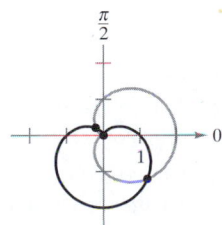
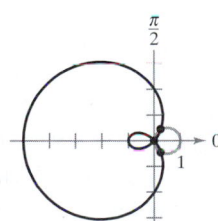

19. $r = \dfrac{\theta}{2}$
$r = 2$

20. $\theta = \dfrac{\pi}{4}$
$r = 2$

21. $r = 4 \sin 2\theta$
$r = 2$

22. $r = 3 + \sin \theta$
$r = 2 \csc \theta$

In Exercises 23 and 24, use a graphing utility to approximate the points of intersection of the graphs of the polar equations. Confirm your results analytically.

23. $r = 2 + 3 \cos \theta$
$r = \dfrac{\sec \theta}{2}$

24. $r = 3(1 - \cos \theta)$
$r = \dfrac{6}{1 - \cos \theta}$

Writing   **In Exercises 25 and 26, use a graphing utility to find the point of intersection of the graphs of the polar equations. Watch the graphs as they are traced in the viewing rectangle. Explain why the pole is not a point of intersection obtained by solving the equations simultaneously.**

25. $r = \cos \theta$
$r = 2 - 3 \sin \theta$

26. $r = 4 \sin \theta$
$r = 2(1 + \sin \theta)$

In Exercises 27–32, use a graphing utility to graph the polar equation and find the area of the indicated region.

27. Common interior of $r = 4 \sin 2\theta$ and $r = 2$

28. Common interior of $r = 3(1 + \sin \theta)$ and $r = 3(1 - \sin \theta)$

29. Common interior of $r = 3 - 2 \sin \theta$ and $r = -3 + 2 \sin \theta$

30. Common interior of $r = 3 - 2 \sin \theta$ and $r = 3 - 2 \cos \theta$

31. Common interior of $r = 4 \sin \theta$ and $r = 2$

32. Inside $r = 3 \sin \theta$ and outside $r = 2 - \sin \theta$

In Exercises 33–36, find the area of the region.

33. Inside $r = a(1 + \cos \theta)$ and outside $r = a \cos \theta$

34. Inside $r = 2a \cos \theta$ and outside $r = a$

35. Common interior of $r = a(1 + \cos \theta)$ and $r = a \sin \theta$

36. Region bounded by the graphs of

$$r = \dfrac{ab}{a \sin \theta + b \cos \theta}, \quad \theta = 0, \quad \text{and} \quad \theta = \dfrac{\pi}{2}$$

37. *Antenna Radiation*   The radiation from a transmitting antenna is not uniform in all directions. The intensity from a particular antenna is modeled by

$$r = a \cos^2 \theta.$$

(a) Convert the polar equation to rectangular form.

(b) Use a graphing utility to graph the model for $a = 4$ and $a = 6$.

(c) Find the area of the geographical region between the two curves in part (b).

38. *Area* The area inside one or more of the three interlocking circles

$$r = 2a \cos \theta, \quad r = 2a \sin \theta, \quad \text{and} \quad r = a$$

is divided into seven regions. Find the area of each region.

39. *Conjecture* Find the area of the region enclosed by $r = a \cos(n\theta)$ for $n = 1, 2, 3, \ldots$. Use the results to make a conjecture about the area enclosed by the function if n is even and if n is odd.

40. *Area* Sketch the strophoid

$$r = \sec \theta - 2 \cos \theta, \quad -\frac{\pi}{2} < \theta < \frac{\pi}{2}.$$

Convert this equation to Cartesian coordinates. Find the area enclosed by the loop.

In Exercises 41–44, find the length of the graph over the indicated interval.

Polar Equation	Interval
41. $r = a$	$0 \le \theta \le 2\pi$
42. $r = 2a \cos \theta$	$-\frac{\pi}{2} \le \theta \le \frac{\pi}{2}$
43. $r = 1 + \sin \theta$	$0 \le \theta \le 2\pi$
44. $r = 5(1 + \cos \theta)$	$0 \le \theta \le 2\pi$

In Exercises 45–50, use a graphing utility to graph the polar equation over the indicated interval. Use the integration capabilities of the graphing utility to approximate the length of the graph accurate to two decimal places.

Polar Equation	Interval
45. $r = 2\theta$	$0 \le \theta \le \frac{\pi}{2}$
46. $r = \sec \theta$	$0 \le \theta \le \frac{\pi}{3}$
47. $r = \dfrac{1}{\theta}$	$\pi \le \theta \le 2\pi$
48. $r = e^\theta$	$0 \le \theta \le \pi$
49. $r = \sin(3 \cos \theta)$	$0 \le \theta \le \pi$
50. $r = 2 \sin(2 \cos \theta)$	$0 \le \theta \le \pi$

In Exercises 51–54, find the area of the surface formed by revolving the curve about the given line.

Polar Equation	Interval	Axis of Revolution
51. $r = 2 \cos \theta$	$0 \le \theta \le \frac{\pi}{2}$	Polar axis
52. $r = a \cos \theta$	$0 \le \theta \le \frac{\pi}{2}$	$\theta = \frac{\pi}{2}$
53. $r = e^{a\theta}$	$0 \le \theta \le \frac{\pi}{2}$	$\theta = \frac{\pi}{2}$
54. $r = a(1 + \cos \theta)$	$0 \le \theta \le \pi$	Polar axis

In Exercises 55 and 56, use the integration capabilities of a graphing utility to approximate to two decimal places the area of the surface formed by revolving the curve about the polar axis.

Polar Equation	Interval
55. $r = 4 \cos 2\theta$	$0 \le \theta \le \frac{\pi}{4}$
56. $r = \theta$	$0 \le \theta \le \pi$

57. *Surface Area of a Torus* Find the surface area of the torus generated by revolving the circle given by $r = a$ about the line $r = b \sec \theta$ where $0 < a < b$. [*Hint:* Derive the integral that yields the surface area for revolving the graph of $r = f(\theta)$ about the line $r = b \sec \theta$.]

58. Use the formula for the arc length of a curve in parametric form to derive the formula for the arc length of a polar curve.

59. *Approximating Area* Consider the circle $r = 8 \cos \theta$.

(a) Find the area of the circle.

(b) Complete the table giving the areas A of the sector of the circle between $\theta = 0$ and the values of θ in the table.

θ	0.2	0.4	0.6	0.8	1.0	1.2	1.4
A							

(c) Use the table in part (b) to approximate the values of θ for which the sector of the circle composes $\frac{1}{4}$, $\frac{1}{2}$, and $\frac{3}{4}$ of the total area of the circle.

(d) Use the graphing utility to approximate to two-decimal-place accuracy the angles θ for which the sector of the circle composes $\frac{1}{4}$, $\frac{1}{2}$, and $\frac{3}{4}$ of the total area of the circle.

(e) Do the results in part (d) depend on the radius of the circle? Explain.

True or False? **In Exercises 60–63, determine whether the statement is true or false. If it is false, explain why or give an example that shows it is false.**

60. The area of the region enclosed by the circle $r = \sin \theta$ is given by the definite integral

$$\frac{1}{2} \int_0^{2\pi} \sin^2 \theta \, d\theta.$$

61. If $f(\theta) > 0$ for all θ and $g(\theta) < 0$ for all θ, then the graphs of $r = f(\theta)$ and $r = g(\theta)$ do not intersect.

62. If $f(\theta) = g(\theta)$ for $\theta = 0$, $\pi/2$, and $3\pi/2$, then the graphs of $r = f(\theta)$ and $r = g(\theta)$ have at least four points of intersection.

63. If n is an even integer, then the area of the region enclosed by $r = \sin(n\theta)$ is twice the area of the region enclosed by $r = \sin[(n + 1)\theta]$.

SECTION | **9.6** | **Polar Equations of Conics and Kepler's Laws**

Polar Equations of Conics • Kepler's Laws

Polar Equations of Conics

In this chapter you have seen that the rectangular equations of ellipses and hyperbolas take simple forms when the origin lies at their *centers*. As it happens, there are many important applications of conics in which it is more convenient to use one of the *foci* as the reference point (the origin) for the coordinate system. For example, the sun lies at a focus of the earth's orbit. Similarly, the light source of a parabolic reflector lies at its focus. In this section you will see that polar equations of conics take simple forms if one of the foci lies at the pole.

The following theorem uses the concept of *eccentricity*, as defined in Section 9.1, to classify the three basic types of conics. A proof of this theorem is given in the appendix.

THEOREM 9.16   Classification of Conics by Eccentricity

The locus of a point in the plane whose distance from a fixed point (*focus*) has a constant ratio to its distance from a fixed line (*directrix*) is a conic. The constant ratio e is the *eccentricity* of the conic.

1. The conic is an ellipse if $0 < e < 1$.
2. The conic is a parabola if $e = 1$.
3. The conic is a hyperbola if $e > 1$.

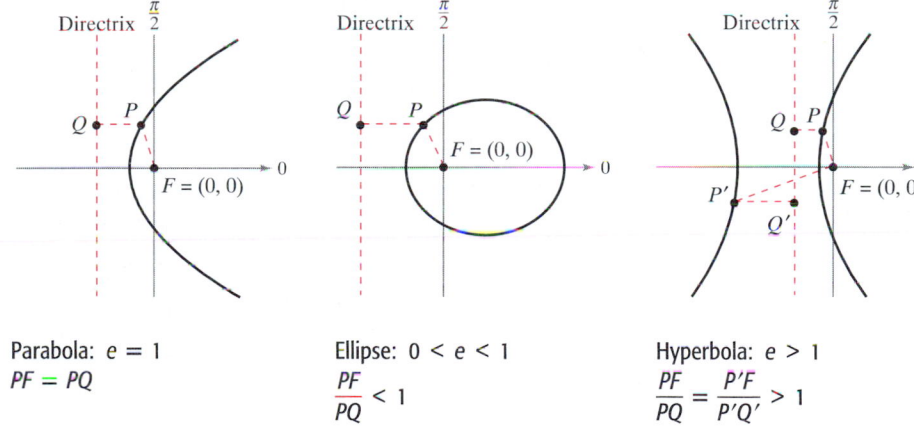

Parabola: $e = 1$
$PF = PQ$

Ellipse: $0 < e < 1$
$\dfrac{PF}{PQ} < 1$

Hyperbola: $e > 1$
$\dfrac{PF}{PQ} = \dfrac{P'F}{P'Q'} > 1$

Figure 9.56

In Figure 9.56, note that for each type of conic the pole corresponds to the fixed point (focus) given in the definition. The benefit of this location can be seen in the proof of the following theorem.

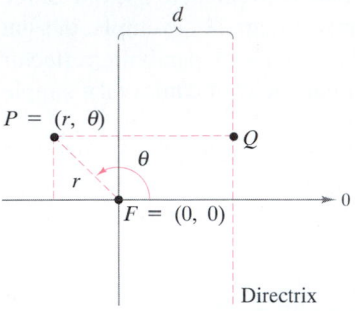

Figure 9.57

> **THEOREM 9.17   Polar Equations of Conics**
>
> The graph of a polar equation of the form
>
> $$r = \frac{ed}{1 \pm e \cos \theta} \qquad \text{or} \qquad r = \frac{ed}{1 \pm e \sin \theta}$$
>
> is a conic, where $e > 0$ is the eccentricity and $|d|$ is the distance between the focus at the pole and its corresponding directrix.

Proof We give a proof for $r = ed/(1 + e \cos \theta)$ with $d > 0$. In Figure 9.57, consider a vertical directrix d units to the right of the focus $F = (0, 0)$. If $P = (r, \theta)$ is a point on the graph of $r = ed/(1 + e \cos \theta)$, the distance between P and the directrix can be shown to be

$$PQ = |d - x| = |d - r \cos \theta| = \left| \frac{r(1 + e \cos \theta)}{e} - r \cos \theta \right| = \left| \frac{r}{e} \right|.$$

Because the distance between P and the pole is simply $PF = |r|$, the ratio of PF to PQ is $PF/PQ = |r|/|r/e| = |e| = e$ and, by Theorem 9.16, the graph of the equation must be a conic. The proofs of the other cases are similar.

The four types of equations indicated in Theorem 9.17 can be classified as follows, where $d > 0$.

a. Horizontal directrix above the pole: $r = \dfrac{ed}{1 + e \sin \theta}$

b. Horizontal directrix below the pole: $r = \dfrac{ed}{1 - e \sin \theta}$

c. Vertical directrix to the right of the pole: $r = \dfrac{ed}{1 + e \cos \theta}$

d. Vertical directrix to the left of the pole: $r = \dfrac{ed}{1 - e \cos \theta}$

Figure 9.58 illustrates these four possibilities for a parabola.

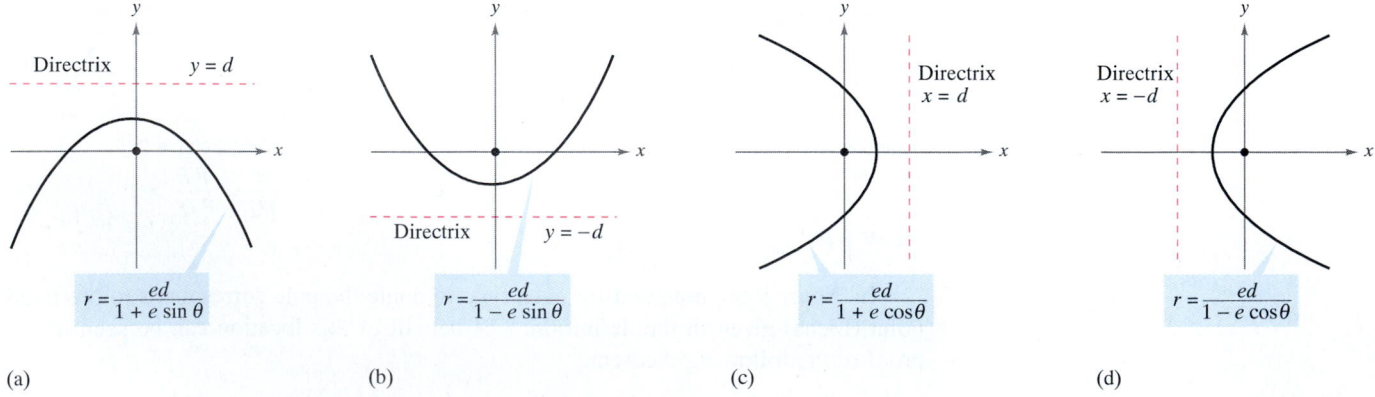

(a)   (b)   (c)   (d)

The four types of polar equations for a parabola
Figure 9.58

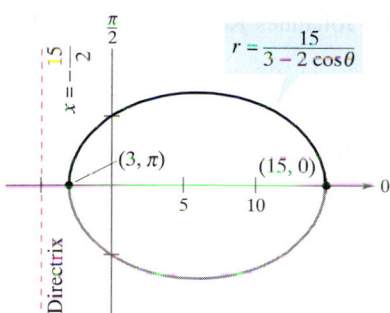

$$r = \frac{15}{3 - 2\cos\theta}$$

The graph of the conic is an ellipse with $e = \frac{2}{3}$.

Figure 9.59

EXAMPLE 1   Determining a Conic from Its Equation

Sketch the graph of the conic given by $r = \dfrac{15}{3 - 2\cos\theta}$.

Solution   To determine the type of conic, rewrite the equation as

$$r = \frac{15}{3 - 2\cos\theta}$$

$$= \frac{5}{1 - (2/3)\cos\theta}.$$

Thus the graph is an ellipse with $e = \frac{2}{3}$. You can sketch the upper half of the ellipse by plotting points from $\theta = 0$ to $\theta = \pi$, as shown in Figure 9.59. Then, using symmetry with respect to the polar axis, you can sketch the lower half.

For the ellipse in Figure 9.59, the major axis is horizontal and the vertices lie at $(15, 0)$ and $(3, \pi)$. Thus, the length of the *major* axis is $2a = 18$. To find the length of the *minor* axis, you can use the equations $e = c/a$ and $b^2 = a^2 - c^2$ to conclude

$$b^2 = a^2 - c^2 = a^2 - (ea)^2 = a^2(1 - e^2). \qquad \text{Ellipse}$$

Because $e = \frac{2}{3}$, you have

$$b^2 = 9^2\left[1 - \left(\tfrac{2}{3}\right)^2\right] = 45$$

which implies that $b = \sqrt{45} = 3\sqrt{5}$. Thus, the length of the minor axis is $2b = 6\sqrt{5}$. A similar analysis for hyperbolas yields

$$b^2 = c^2 - a^2 = (ea)^2 - a^2 = a^2(e^2 - 1). \qquad \text{Hyperbola}$$

EXAMPLE 2   Sketching a Conic from Its Polar Equation

Sketch the graph of the polar equation $r = \dfrac{32}{3 + 5\sin\theta}$.

Solution   Dividing the numerator and denominator by 3 produces

$$r = \frac{32/3}{1 + (5/3)\sin\theta}.$$

Because $e = \frac{5}{3} > 1$, the graph is a hyperbola. Because $d = \frac{32}{5}$, the directrix is the line $y = \frac{32}{5}$. The transverse axis of the hyperbola lies on the line $\theta = \pi/2$, and the vertices occur at

$$(r, \theta) = \left(4, \frac{\pi}{2}\right) \qquad \text{and} \qquad (r, \theta) = \left(-16, \frac{3\pi}{2}\right).$$

Because the length of the transverse axis is 12, you can see that $a = 6$. To find b, write

$$b^2 = a^2(e^2 - 1) = 6^2\left[\left(\frac{5}{3}\right)^2 - 1\right] = 64.$$

Therefore, $b = 8$. Finally, you can use a and b to determine the asymptotes of the hyperbola and obtain the sketch shown in Figure 9.60.

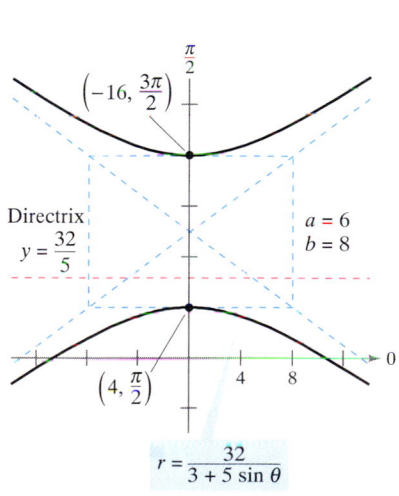

$$r = \frac{32}{3 + 5\sin\theta}$$

The graph of the conic is a hyperbola with $e = \frac{5}{3}$.

Figure 9.60

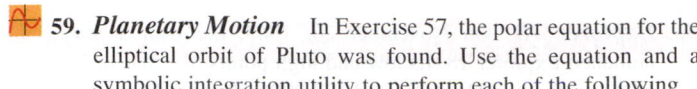

59. *Planetary Motion* In Exercise 57, the polar equation for the elliptical orbit of Pluto was found. Use the equation and a symbolic integration utility to perform each of the following.

(a) Approximate the area swept out by a ray from the sun to the planet as θ increases from 0 to $\pi/9$. Use this result to determine the number of years for the planet to move through this arc if the period of one revolution around the sun is 248 years.

(b) By trial and error, approximate the angle α so that the area swept out by a ray from the sun to the planet as θ increases from π to α equals the area found in part (a) (see figure). Does the ray sweep through a larger or smaller angle than in part (a) to generate the same area? Why is this the case?

(c) Approximate the distances the planet traveled in parts (a) and (b). Use these distances to approximate the average number of kilometers per year the planet traveled in the two cases.

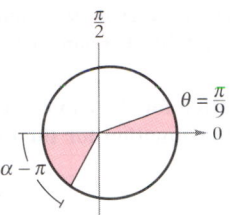

Figure for 59

60. What conic section does the following polar equation represent?

$$r = a \sin \theta + b \cos \theta$$

61. Show that the graphs of the following equations intersect at right angles.

$$r = \frac{ed}{1 + \sin \theta} \quad \text{and} \quad r = \frac{ed}{1 - \sin \theta}$$

REVIEW EXERCISES FOR CHAPTER 9

In Exercises 1–4, match the equation with the correct graph. [The graphs are labeled (a), (b), (c), and (d).]

(a)
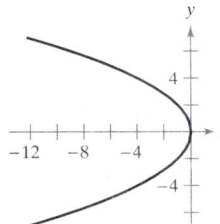

(b)
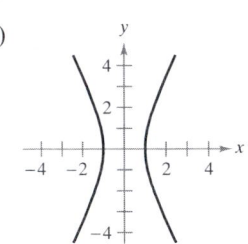

(c)
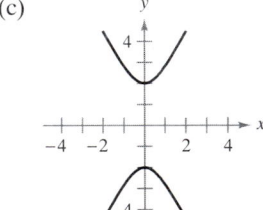

(d)
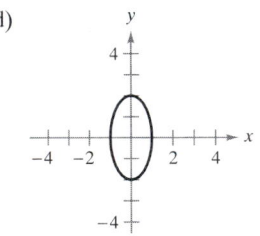

1. $4x^2 + y^2 = 4$

2. $4x^2 - y^2 = 4$

3. $y^2 = -4x$

4. $y^2 - 4x^2 = 4$

In Exercises 5–10, analyze each equation and sketch its graph. Use a graphing utility to confirm your results.

5. $16x^2 + 16y^2 - 16x + 24y - 3 = 0$

6. $y^2 - 12y - 8x + 20 = 0$

7. $3x^2 - 2y^2 + 24x + 12y + 24 = 0$

8. $4x^2 + y^2 - 16x + 15 = 0$

9. $3x^2 + 2y^2 - 12x + 12y + 29 = 0$

10. $4x^2 - 4y^2 - 4x + 8y - 11 = 0$

In Exercises 11 and 12, find an equation of the parabola.

11. Vertex: $(0, 2)$
Directrix: $x = -3$

12. Vertex: $(4, 2)$
Focus: $(4, 0)$

In Exercises 13 and 14, find an equation of the ellipse.

13. Vertices: $(-3, 0)$, $(7, 0)$; Foci: $(0, 0)$, $(4, 0)$

14. Center: $(0, 0)$; Solution points: $(1, 2)$, $(2, 0)$

In Exercises 15 and 16, use a graphing utility to approximate the perimeter of the ellipse.

15. $\dfrac{x^2}{9} + \dfrac{y^2}{4} = 1$

16. $\dfrac{x^2}{4} + \dfrac{y^2}{25} = 1$

17. A line is tangent to the parabola $y = x^2 - 2x + 2$ and perpendicular to the line $y = x - 2$. Find the equation of the line.

18. *Satellite Antenna* A cross section of a large parabolic antenna is modeled by the graph of $y = x^2/200$, $-100 \le x \le 100$. The receiving and transmitting equipment is positioned at the focus. (a) Find the coordinates of the focus. (b) Find the surface area of the antenna.

19. Consider a fire truck with a water tank 16 feet long whose vertical cross sections are ellipses modeled by the equation $x^2/16 + y^2/9 = 1$.

(a) Find the volume of the tank.

(b) Find the force on the end of the tank when it is full of water. (The density of water is 62.4 pounds per cubic foot.)

(c) Find the depth of the water in the tank if it is $\frac{3}{4}$ full (by volume) and the truck is on level ground.

(d) Approximate the tank's surface area.

20. Consider the region bounded by the ellipse

$$\frac{x^2}{a^2} + \frac{y^2}{b^2} = 1, \quad \text{eccentricity } e = c/a.$$

(a) Show that the area of the region is πab.

(b) Show that the solid (oblate spheroid) generated by revolving the region about the minor axis of the ellipse has a volume of $V = 4\pi a^2 b/3$ and a surface area of

$$S = 2\pi a^2 + \pi\left(\frac{b^2}{e}\right) \ln\left(\frac{1 + e}{1 - e}\right).$$

(c) Show that the solid (prolate spheroid) generated by revolving the region about the major axis of the ellipse has a volume of $V = 4\pi ab^2/3$ and a surface area of

$$S = 2\pi b^2 + 2\pi\left(\frac{ab}{e}\right) \arcsin e.$$

(d) Use the results in parts (b) and (c) to find the volumes and surface areas of the prolate and oblate spheroids generated by revolving the region bounded by the graph of

$$\frac{x^2}{9} + \frac{y^2}{4} = 1$$

about the appropriate axis.

In Exercises 21–30, (a) find dy/dx and all points of horizontal tangency, (b) eliminate the parameter where possible, and (c) sketch the curve represented by the parametric equations.

21. $x = 1 + 4t, \; y = 2 - 3t$

22. $x = t + 4, \; y = t^2$

23. $x = \dfrac{1}{t}, \; y = 2t + 3$

24. $x = \dfrac{1}{t}, \; y = t^2$

25. $x = \dfrac{1}{2t + 1}$
$y = \dfrac{1}{t^2 - 2t}$

26. $x = 2t - 1$
$y = \dfrac{1}{t^2 - 2t}$

27. $x = 3 + 2\cos\theta$
$y = 2 + 5\sin\theta$

28. $x = 6\cos\theta$
$y = 6\sin\theta$

29. $x = \cos^3\theta$
$y = 4\sin^3\theta$

30. $x = e^t$
$y = e^{-t}$

In Exercises 31 and 32, (a) use a graphing utility to sketch the curve represented by the parametric equations, (b) use the graphing utility to find $dx/d\theta$, $dy/d\theta$, and dy/dx for $\theta = \pi/6$, and (c) use the graphing utility to graph the tangent line to the curve when $\theta = \pi/6$.

31. $x = \cot\theta$
$y = \sin 2\theta$

32. $x = 2\theta - \sin\theta$
$y = 2 - \cos\theta$

In Exercises 33–36, find a parametric representation of the line or conic.

33. Line: Passes through $(-2, 6)$ and $(3, 2)$

34. Circle: Center at $(5, 3)$; radius: 2

35. Ellipse: Center at $(-3, 4)$; horizontal major axis of length 8 and minor axis of length 6

36. Hyperbola: Vertices at $(0, \pm 4)$; foci at $(0, \pm 5)$

37. Rotary Engine The rotary engine was developed by Felix Wankel in the 1950's (see page 240). It features a rotor, which is a modified equilateral triangle. The rotor moves in a chamber that, in two dimensions, is an epitrochoid. Use a graphing utility to graph the chamber modeled by the parametric equations.

$$x = \cos 3\theta + 5\cos\theta \quad \text{and} \quad y = \sin 3\theta + 5\sin\theta.$$

38. Hypocycloids A hypocycloid has the parametric equations

$$x = (a - b)\cos t + b\cos\left(\frac{a - b}{b}t\right)$$

and

$$y = (a - b)\sin t - b\sin\left(\frac{a - b}{b}t\right).$$

Use a graphing utility to graph the hypocycloid for each of the following values of a and b.

(a) $a = 2, b = 1$ (b) $a = 3, b = 1$ (c) $a = 4, b = 1$

(d) $a = 10, b = 1$ (e) $a = 3, b = 2$ (f) $a = 4, b = 3$

39. Cycloids Eliminate the parameter from

$$x = a(\theta - \sin\theta) \quad \text{and} \quad y = a(1 - \cos\theta)$$

to show that the rectangular equation of a cycloid is

$$x = a\arccos\left(\frac{a - y}{a}\right) \pm \sqrt{2ay - y^2}.$$

40. Involute of a Circle The involute of a circle is described by the endpoint P of a string that is held taut as it is unwound from a spool that does not turn (see figure). Show that a parametric representation of the involute is

$$x = r(\cos\theta + \theta\sin\theta) \quad \text{and} \quad y = r(\sin\theta - \theta\cos\theta).$$

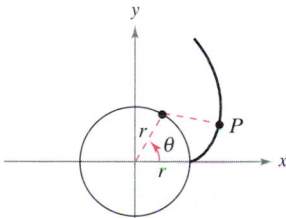

In Exercises 41 and 42, find the length of the curve represented by the parametric equations over the given interval.

41. $x = r(\cos\theta + \theta\sin\theta)$
$y = r(\sin\theta - \theta\cos\theta)$
$0 \leq \theta \leq \pi$

42. $x = 6\cos\theta$
$y = 6\sin\theta$
$0 \leq \theta \leq \pi$

In Exercises 43 and 44, the rectangular coordinates of a point are given. Plot the point and find *two* sets of polar coordinates for the point for $0 \le \theta \le 2\pi$.

43. $(4, -4)$　　　　　　　　**44.** $(-1, 3)$

In Exercises 45–52, convert the polar equation to rectangular form.

45. $r = 3 \cos \theta$　　　　　**46.** $r = 10$

47. $r = -2(1 + \cos \theta)$　**48.** $r = \dfrac{1}{2 - \cos \theta}$

49. $r^2 = \cos 2\theta$　　　　**50.** $r = 4 \sec\left(\theta - \dfrac{\pi}{3}\right)$

51. $r = 4 \cos 2\theta \sec \theta$　**52.** $\theta = \dfrac{3\pi}{4}$

In Exercises 53–56, convert the rectangular equation to polar form.

53. $(x^2 + y^2)^2 = ax^2 y$

54. $x^2 + y^2 - 4x = 0$

55. $x^2 + y^2 = a^2 \left(\arctan \dfrac{y}{x}\right)^2$

56. $(x^2 + y^2)\left(\arctan \dfrac{y}{x}\right)^2 = a^2$

In Exercises 57–70, sketch a graph of the polar equation.

57. $r = 4$　　　　　　　　**58.** $\theta = \dfrac{\pi}{12}$

59. $r = -\sec \theta$　　　　**60.** $r = 3 \csc \theta$

61. $r = -2(1 + \cos \theta)$　**62.** $r = 3 - 4 \cos \theta$

63. $r = 4 - 3 \cos \theta$　　**64.** $r = 2\theta$

65. $r = -3 \cos 2\theta$　　　**66.** $r = \cos 5\theta$

67. $r^2 = 4 \sin^2 2\theta$　　**68.** $r^2 = \cos 2\theta$

69. $r = \dfrac{2}{1 - \sin \theta}$　　**70.** $r = \dfrac{4}{5 - 3 \cos \theta}$

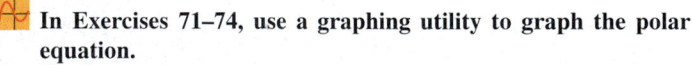 **In Exercises 71–74, use a graphing utility to graph the polar equation.**

71. $r = \dfrac{3}{\cos(\theta - \pi/4)}$　**72.** $r = 2 \sin \theta \cos^2 \theta$

73. $r = 4 \cos 2\theta \sec \theta$　**74.** $r = 4(\sec \theta - \cos \theta)$

In Exercises 75 and 76, (a) find the tangents at the pole, (b) find all points of horizontal and vertical tangency, and (c) use a graphing utility to graph the polar equation and draw a tangent line to the graph for $\theta = \pi/6$.

75. $r = 1 - 2 \cos \theta$　　**76.** $r^2 = 4 \sin 2\theta$

In Exercises 77 and 78, show that the graphs of the polar equations are orthogonal at the points of intersection. Use a graphing utility to confirm your results graphically.

77. $r = 1 + \cos \theta$　　　**78.** $r = a \sin \theta$
　$r = 1 - \cos \theta$　　　　　$r = a \cos \theta$

79. Find the angle between the circle $r = 3 \sin \theta$ and the limaçon $r = 4 - 5 \sin \theta$ at the point of intersection $(3/2, \pi/6)$.

80. *True or False?*　There is a unique polar coordinate representation of each point in the plane. Explain.

In Exercises 81–88, use a graphing utility to graph the polar equation. Set up an integral for finding the area of the indicated region and use the integration capabilities of a graphing utility to approximate the integral accurate to two decimal places.

81. Interior of $r = 2 + \cos \theta$

82. Interior of $r = 5(1 - \sin \theta)$

83. Interior of $r = \sin \theta \cos^2 \theta$

84. Interior of $r = 4 \sin 3\theta$

85. Interior of $r^2 = 4 \sin 2\theta$

86. Common interior of $r = 3$ and $r^2 = 18 \sin 2\theta$

87. Common interior of $r = 4 \cos \theta$ and $r = 2$

88. Region bounded by the polar axis and $r = e^{\theta}$ for $0 \le \theta \le \pi$

In Exercises 89 and 90, find the perimeter of the curve represented by the polar equation.

89. $r = a(1 - \cos \theta)$　　**90.** $r = a \cos 2\theta$

In Exercises 91–96, find a polar equation for the line or conic.

91. Circle　　　　Center: $(5, \pi/2)$
　　　　　　　　Solution point: $(0, 0)$

92. Line　　　　　Solution point: $(0, 0)$
　　　　　　　　Slope: $\sqrt{3}$

93. Parabola　　　Vertex: $(2, \pi)$
　　　　　　　　Focus: $(0, 0)$

94. Parabola　　　Vertex: $(2, \pi/2)$
　　　　　　　　Focus: $(0, 0)$

95. Ellipse　　　　Vertices: $(5, 0), (1, \pi)$
　　　　　　　　One focus: $(0, 0)$

96. Hyperbola　　Vertices: $(1, 0), (7, 0)$
　　　　　　　　One focus: $(0, 0)$

10

Vectors and the Geometry of Space

The oldest bridges in existence are similar in design to the Clapper Bridge in Devon, England, which is thought to date back to the thirteenth century.

Bridges have been around since primitive people first threw a tree trunk across a stream. The oldest known bridge was constructed of stone slabs. Since primitive times, engineers have strived to construct bridges that are longer, sturdier, and more aesthetically pleasing than their predecessors.

One of the most commonly used bridge designs is the suspension bridge. This type of bridge is used in situations that require long single spans. The roadway is hung from cables that are supported by stationary towers. Well-designed suspension bridges, such as the Golden Gate Bridge (shown on the preceding page) and the Brooklyn Bridge, can be functional for many years. On the other hand, a poor design can result in tragedy. For a suspension bridge to be stable, the forces acting on its main cables must be in equilibrium. This equilibrium occurs when the main cables are in the shape of parabolas.

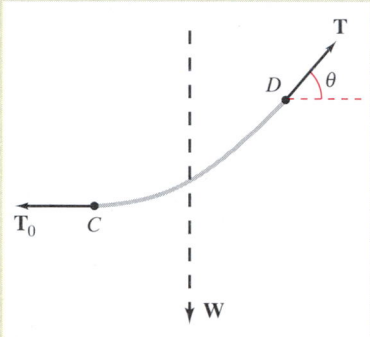 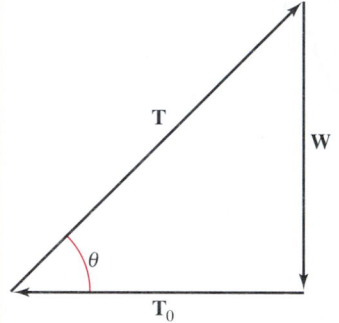

The forces acting on the main cable are shown in the diagram above as directed line segments, indicating both the magnitude and the direction of the forces. At the center of the parabolic cable, the tension force T_0 is horizontal. T is the tension force at point D, and is directed along the tangent at point D. The uniformly distributed load supported by the section CD of the cable is represented by W.

QUESTIONS

1. The forces acting on the cable are related by a "force triangle," as shown above. Use trigonometric functions to relate the magnitudes $\|T_0\|$, $\|T\|$, and $\|W\|$ of the vectors T_0, T, and W.

2. The main cable of the Golden Gate Bridge is suspended from towers 520 feet above the roadway at either end of a 4200-foot span. The low point in the center of the cable is 6 feet above the roadway. Given that the cable hangs in the shape of a parabola, find an equation describing the shape of the cable.

3. Use the equation you found in Question 2 to find the acute angle θ of the force T at a point located 400 feet horizontally from the center of the span. Express the magnitude of T in terms of T_0 and W.

4. Find the angle and magnitude of T at a point located 200 feet horizontally from the center of the span.

5. In general, where in a suspension cable would you expect the magnitude of T to be the greatest? Where would you expect the magnitude of T to be the least? Explain.

In 1930 Joseph Strauss designed the Golden Gate Bridge in San Francisco, still considered to be one of the world's greatest civil engineering masterpieces.

The concepts presented here will be explored further in this chapter. For an extension of this application, see the lab series that accompanies this text.

SECTION | *10.1* | **Vectors in the Plane**

Component Form of a Vector • Vector Operations • Standard Unit Vectors •
Applications of Vectors

Component Form of a Vector

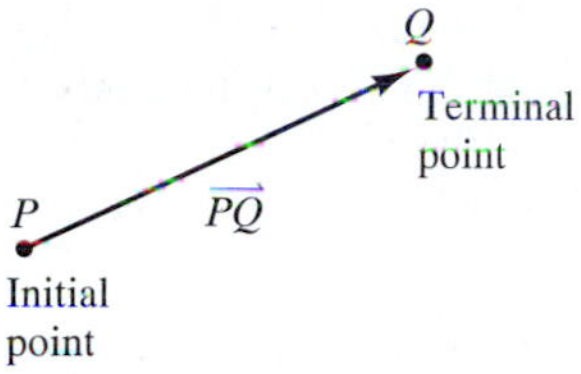

Many quantities in geometry and physics, such as area, volume, temperature, mass,
and time, can be characterized by single real numbers scaled to appropriate units of
measure. We call these **scalar quantities,** and the real number associated with each is
called a **scalar**

Other quantities, such as force, velocity, and acceleration, involve both magnitude and direction and cannot be characterized completely by single real numbers. A
directed line segment is used to represent such a quantity, as shown in Figure 10.1.
The directed line segment $\overrightarrow{PQ}$ has **initial point** P and **terminal point** Q, and its
length is denoted by $\|\overrightarrow{PQ}\|$. Directed line segments that have the same length and
direction are **equivalent,** as shown in Figure 10.2. The set of all directed line segments that are equivalent to a given directed line segment $\overrightarrow{PQ}$ is a **vector in the plane**
and is denoted by $\mathbf{v} = \overrightarrow{PQ}$. In typeset material, vectors are usually denoted by lowercase, boldface letters such as $\mathbf{u}$, $\mathbf{v}$, and $\mathbf{w}$. When written by hand, however, vectors are
often denoted by letters with arrows above them, such as $\overrightarrow{u}$, $\overrightarrow{v}$, and $\overrightarrow{w}$.

Be sure you see that a vector in the plane can be represented by many different
directed line segments—all pointing in the same direction and all of the same length.

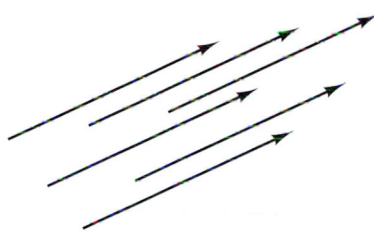

A directed line segment
Figure 10.1

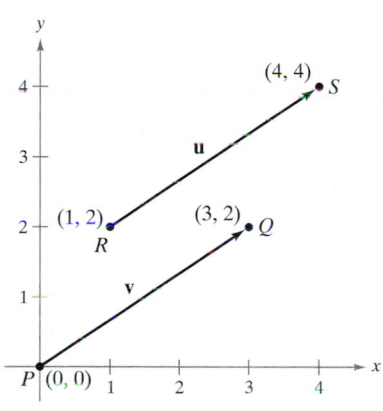

Equivalent directed line segments
Figure 10.2

EXAMPLE 1   Vector Representation by Directed Line Segments

Let $\mathbf{v}$ be represented by the directed line segment from $(0, 0)$ to $(3, 2)$, and let $\mathbf{u}$ be
represented by the directed line segment from $(1, 2)$ to $(4, 4)$. Show that $\mathbf{v} = \mathbf{u}$.

Solution   Let $P(0, 0)$ and $Q(3, 2)$ be the initial and terminal points of $\mathbf{v}$, and let $R(1, 2)$
and $S(4, 4)$ be the initial and terminal points of $\mathbf{u}$, as shown in Figure 10.3. You can
use the Distance Formula to show that $\overrightarrow{PQ}$ and $\overrightarrow{RS}$ have the *same length*.

$$\|\overrightarrow{PQ}\| = \sqrt{(3 - 0)^2 + (2 - 0)^2} = \sqrt{13} \qquad \text{Length of } \overrightarrow{PQ}$$
$$\|\overrightarrow{RS}\| = \sqrt{(4 - 1)^2 + (4 - 2)^2} = \sqrt{13} \qquad \text{Length of } \overrightarrow{RS}$$

Both line segments have the *same direction*, because they both are directed toward the
upper right on lines having the same slope.

$$\text{Slope of } \overrightarrow{PQ} = \frac{2 - 0}{3 - 0} = \frac{2}{3}$$

and

$$\text{Slope of } \overrightarrow{RS} = \frac{4 - 2}{4 - 1} = \frac{2}{3}$$

Because $\overrightarrow{PQ}$ and $\overrightarrow{RS}$ have the same length and direction, you can conclude that the
two vectors are equal. That is,

$$\mathbf{v} = \mathbf{u}.$$

The vectors $\mathbf{u}$ and $\mathbf{v}$ are equal.
Figure 10.3

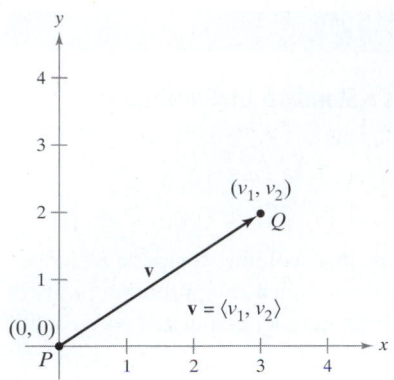

The standard position of a vector
Figure 10.4

The directed line segment whose initial point is the origin is often the most convenient representative of a set of equivalent directed line segments such as those shown in Figure 10.3. This representation of **v** is said to be in **standard position.** A directed line segment whose initial point is at the origin can be uniquely represented by the coordinates of its terminal point $Q(v_1, v_2)$, as shown in Figure 10.4.

Definition of Component Form of a Vector in the Plane

If **v** is a vector in the plane whose initial point is the origin and whose terminal point is (v_1, v_2), then the **component form of v** is given by

$$\mathbf{v} = \langle v_1, v_2 \rangle.$$

The coordinates v_1 and v_2 are called the **components of v.** If both the initial point and the terminal point lie at the origin, then **v** is called the **zero vector** and is denoted by $\mathbf{0} = \langle 0, 0 \rangle.$

This definition implies that two vectors $\mathbf{u} = \langle u_1, u_2 \rangle$ and $\mathbf{v} = \langle v_1, v_2 \rangle$ are **equal** if and only if $u_1 = v_1$ and $u_2 = v_2$.

The following procedures can be used to convert directed line segments to component form or vice versa.

NOTE It is important to understand that a vector represents a *set* of directed line segments (each having the same length and direction). In practice, however, it is common not to distinguish between a vector and one of its representatives.

1. If $P(p_1, p_2)$ and $Q(q_1, q_2)$ are the initial and terminal points of a directed line segment, the component form of the vector **v** represented by $\overrightarrow{PQ}$ is $\langle v_1, v_2 \rangle = \langle q_1 - p_1, q_2 - p_2 \rangle$. Moreover, the **length of v** is

$$\|\mathbf{v}\| = \sqrt{(q_1 - p_1)^2 + (q_2 - p_2)^2} \qquad \text{Length of a vector}$$
$$= \sqrt{v_1^2 + v_2^2}.$$

2. If $\mathbf{v} = \langle v_1, v_2 \rangle$, **v** can be represented by the directed line segment, in standard position, from $P(0, 0)$ to $Q(v_1, v_2)$.

The length of **v** is also called the **norm of v.** If $\|\mathbf{v}\| = 1$, **v** is a **unit vector.** Moreover, $\|\mathbf{v}\| = 0$ if and only if **v** is the zero vector **0.**

EXAMPLE 2 **Finding the Component Form and Length of a Vector**

Find the component form and length of the vector **v** that has initial point $(3, -7)$ and terminal point $(-2, 5)$.

Solution Let $P(3, -7) = (p_1, p_2)$ and $Q(-2, 5) = (q_1, q_2)$. Then the components of $\mathbf{v} = \langle v_1, v_2 \rangle$ are

$$v_1 = q_1 - p_1 = -2 - 3 = -5$$
$$v_2 = q_2 - p_2 = 5 - (-7) = 12.$$

Thus, as shown in Figure 10.5, $\mathbf{v} = \langle -5, 12 \rangle$, and the length of **v** is

$$\|\mathbf{v}\| = \sqrt{(-5)^2 + 12^2}$$
$$= \sqrt{169}$$
$$= 13.$$

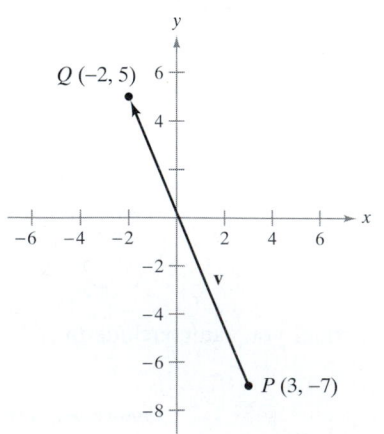

Component form of **v**: $\mathbf{v} = \langle -5, 12 \rangle$
Figure 10.5

Vector Operations

> ### Definitions of Vector Addition and Scalar Multiplication
>
> Let $\mathbf{u} = \langle u_1, u_2 \rangle$ and $\mathbf{v} = \langle v_1, v_2 \rangle$ be vectors and let k be a scalar.
>
> 1. The **vector sum** of $\mathbf{u}$ and $\mathbf{v}$ is the vector $\mathbf{u} + \mathbf{v} = \langle u_1 + v_1, u_2 + v_2 \rangle$.
> 2. The **scalar multiple** of k and $\mathbf{u}$ is the vector $k\mathbf{u} = \langle ku_1, ku_2 \rangle$.
> 3. The **negative** of $\mathbf{v}$ is the vector $-\mathbf{v} = (-1)\mathbf{v} = \langle -v_1, -v_2 \rangle$.
> 4. The **difference** of $\mathbf{u}$ and $\mathbf{v}$ is $\mathbf{u} - \mathbf{v} = \mathbf{u} + (-\mathbf{v}) = \langle u_1 - v_1, u_2 - v_2 \rangle$.

Geometrically, the product of a vector $\mathbf{v}$ and a scalar k is the vector that is k times as long as $\mathbf{v}$, as shown in Figure 10.6. If k is positive, $k\mathbf{v}$ has the same direction as $\mathbf{v}$. If k is negative, $k\mathbf{v}$ has the opposite direction.

To add two vectors geometrically, position them (without changing their magnitudes or directions) so that the initial point of one coincides with the terminal point of the other, as shown in Figure 10.7. The vector $\mathbf{u} + \mathbf{v}$ (called the **resultant vector**) is the diagonal of a parallelogram having $\mathbf{u}$ and $\mathbf{v}$ as its adjacent sides.

The scalar multiplication of $\mathbf{v}$
Figure 10.6

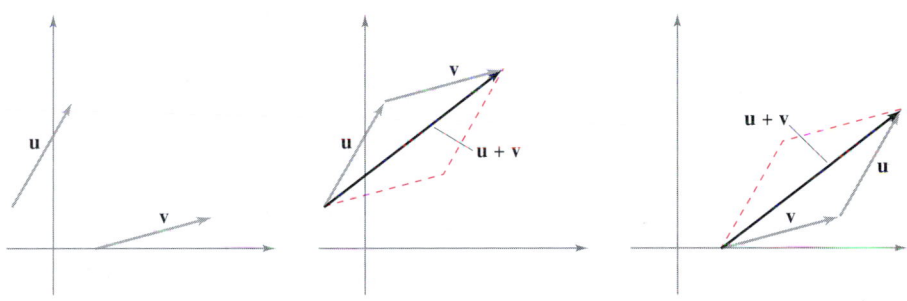

To find $\mathbf{u} + \mathbf{v}$,   (1) move the initial point of $\mathbf{v}$ to the terminal point of $\mathbf{u}$, or   (2) move the initial point of $\mathbf{u}$ to the terminal point of $\mathbf{v}$.

Figure 10.7

Figure 10.8 shows the equivalence of the geometric and algebraic definitions of vector addition and scalar multiplication, and presents (at far right) a geometric interpretation of $\mathbf{u} - \mathbf{v}$.

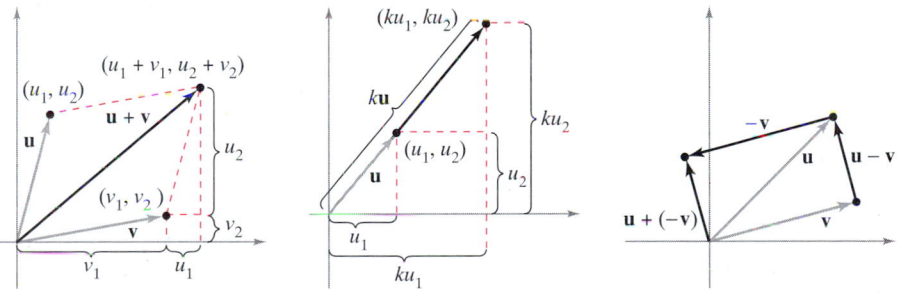

Vector addition   Scalar multiplication   Vector subtraction
Figure 10.8

WILLIAM ROWAN HAMILTON (1805–1865)

Some of the earliest work with vectors was done by the Irish mathematician William Rowan Hamilton. Hamilton spent many years developing a system of vector like quantities called quaternions. Although Hamilton was convinced of the benefits of quaternions, the operations he defined did not produce good models for physical phenomena. It wasn't until the latter half of the nineteenth century that the Scottish physicist James Maxwell (1831–1879) restructured Hamilton's quaternions in a form useful for representing physical quantities such as force, velocity, and acceleration.

EXAMPLE 3   **Vector Operations**

Given $\mathbf{v} = \langle -2, 5 \rangle$ and $\mathbf{w} = \langle 3, 4 \rangle$, find each of the following vectors.

a. $\frac{1}{2}\mathbf{v}$ **b.** $\mathbf{w} - \mathbf{v}$ **c.** $\mathbf{v} + 2\mathbf{w}$

Solution

a. $\frac{1}{2}\mathbf{v} = \left\langle \frac{1}{2}(-2), \frac{1}{2}(5) \right\rangle = \left\langle -1, \frac{5}{2} \right\rangle$

b. $\mathbf{w} - \mathbf{v} = \langle w_1 - v_1, w_2 - v_2 \rangle = \langle 3 - (-2), 4 - 5 \rangle = \langle 5, -1 \rangle$

c. Using $2\mathbf{w} = \langle 6, 8 \rangle$, you have

$$\mathbf{v} + 2\mathbf{w} = \langle -2, 5 \rangle + \langle 6, 8 \rangle = \langle -2 + 6, 5 + 8 \rangle = \langle 4, 13 \rangle.$$

Vector addition and scalar multiplication share many properties of ordinary arithmetic, as shown in the following theorem.

THEOREM 10.1   Properties of Vector Operations

Let $\mathbf{u}$, $\mathbf{v}$, and $\mathbf{w}$ be vectors in the plane, and let c and d be scalars.

1. $\mathbf{u} + \mathbf{v} = \mathbf{v} + \mathbf{u}$ Commutative property
2. $(\mathbf{u} + \mathbf{v}) + \mathbf{w} = \mathbf{u} + (\mathbf{v} + \mathbf{w})$ Associative property
3. $\mathbf{u} + \mathbf{0} = \mathbf{u}$ Additive identity property
4. $\mathbf{u} + (-\mathbf{u}) = \mathbf{0}$ Additive inverse property
5. $c(d\mathbf{u}) = (cd)\mathbf{u}$
6. $(c + d)\mathbf{u} = c\mathbf{u} + d\mathbf{u}$ Distributive property
7. $c(\mathbf{u} + \mathbf{v}) = c\mathbf{u} + c\mathbf{v}$ Distributive property
8. $1(\mathbf{u}) = \mathbf{u}$, $0(\mathbf{u}) = \mathbf{0}$

Proof   The proof of the *associative property* of vector addition uses the associative property of addition of real numbers.

$$
\begin{aligned}
(\mathbf{u} + \mathbf{v}) + \mathbf{w} &= [\langle u_1, u_2 \rangle + \langle v_1, v_2 \rangle] + \langle w_1, w_2 \rangle \\
&= \langle u_1 + v_1, u_2 + v_2 \rangle + \langle w_1, w_2 \rangle \\
&= \langle (u_1 + v_1) + w_1, (u_2 + v_2) + w_2 \rangle \\
&= \langle u_1 + (v_1 + w_1), u_2 + (v_2 + w_2) \rangle \\
&= \langle u_1, u_2 \rangle + \langle v_1 + w_1, v_2 + w_2 \rangle = \mathbf{u} + (\mathbf{v} + \mathbf{w})
\end{aligned}
$$

Similarly, the proof of the following *distributive property* depends on the distributive property of real numbers.

$$
\begin{aligned}
(c + d)\mathbf{u} &= (c + d)\langle u_1, u_2 \rangle \\
&= \langle (c + d)u_1, (c + d)u_2 \rangle \\
&= \langle cu_1 + du_1, cu_2 + du_2 \rangle \\
&= \langle cu_1, cu_2 \rangle + \langle du_1, du_2 \rangle = c\mathbf{u} + d\mathbf{u}
\end{aligned}
$$

The other properties can be proved in a similar manner.

Any set of vectors (with an accompanying set of scalars) that satisfies the eight properties given in Theorem 10.1 is a **vector space**. The eight properties are the *vector space axioms*. Thus, this theorem states that the set of vectors in the plane (with the set of real numbers) forms a vector space. For more information about vector spaces, see *Elementary Linear Algebra*, 3rd edition, by Larson and Edwards (Boston: Houghton Mifflin Company, 1996).

EMMY NOETHER (1882–1935)

One person who contributed to our knowledge of axiomatic systems was the German mathematician Emmy Noether. Noether is generally recognized as the leading woman mathematician in recent history.

FOR FURTHER INFORMATION For more information on Emmy Noether, see the article "Emmy Noether, Greatest Woman Mathematician" by Clark Kimberling in the March 1982 issue of *The Mathematics Teacher*.

THEOREM 10.2　Length of a Scalar Multiple

Let $\mathbf{v}$ be a vector and c be a scalar. Then

$$\|c\mathbf{v}\| = |c|\,\|\mathbf{v}\|.$$ $|c|$ is the absolute value of c.

Proof　Because $c\mathbf{v} = \langle cv_1, cv_2 \rangle$, it follows that

$$
\begin{aligned}
\|c\mathbf{v}\| = \|\langle cv_1, cv_2 \rangle\| &= \sqrt{(cv_1)^2 + (cv_2)^2} \\
&= \sqrt{c^2 v_1^2 + c^2 v_2^2} \\
&= \sqrt{c^2(v_1^2 + v_2^2)} \\
&= |c|\sqrt{v_1^2 + v_2^2} \\
&= |c|\,\|\mathbf{v}\|.
\end{aligned}
$$

In many applications of vectors, it is useful to find a unit vector that has the same direction as a given vector. The following theorem gives a procedure for doing this.

THEOREM 10.3　Unit Vector in the Direction of v

If $\mathbf{v}$ is a nonzero vector in the plane, then the vector

$$\mathbf{u} = \frac{\mathbf{v}}{\|\mathbf{v}\|} = \frac{1}{\|\mathbf{v}\|}\mathbf{v}$$

has length 1 and the same direction as $\mathbf{v}$.

Proof　Because $1/\|\mathbf{v}\|$ is positive and $\mathbf{u} = (1/\|\mathbf{v}\|)\mathbf{v}$, you can conclude that $\mathbf{u}$ has the same direction as $\mathbf{v}$. To see that $\|\mathbf{u}\| = 1$, note that

$$
\begin{aligned}
\|\mathbf{u}\| &= \left\| \left(\frac{1}{\|\mathbf{v}\|} \right) \mathbf{v} \right\| \\
&= \left| \frac{1}{\|\mathbf{v}\|} \right| \|\mathbf{v}\| \\
&= \frac{1}{\|\mathbf{v}\|} \|\mathbf{v}\| \\
&= 1.
\end{aligned}
$$

Thus, $\mathbf{u}$ has length 1 and the same direction as $\mathbf{v}$.

In Theorem 10.3, $\mathbf{u}$ is called a **unit vector in the direction of v.** The process of multiplying $\mathbf{v}$ by $1/\|\mathbf{v}\|$ to get a unit vector is called **normalization of v.**

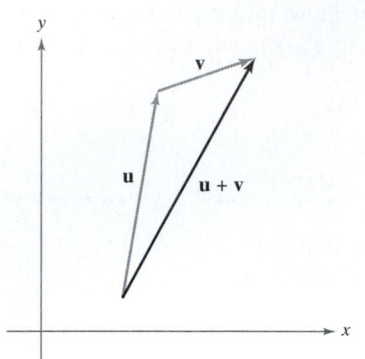

Triangle inequality
Figure 10.9

EXAMPLE 4 Finding a Unit Vector

Find a unit vector in the direction of $\mathbf{v} = \langle -2, 5 \rangle$ and verify that it has length 1.

Solution   From Theorem 10.3, the unit vector in the direction of $\mathbf{v}$ is

$$\frac{\mathbf{v}}{\|\mathbf{v}\|} = \frac{\langle -2, 5 \rangle}{\sqrt{(-2)^2 + (5)^2}} = \frac{1}{\sqrt{29}} \langle -2, 5 \rangle = \left\langle \frac{-2}{\sqrt{29}}, \frac{5}{\sqrt{29}} \right\rangle.$$

This vector has length 1, because

$$\sqrt{\left(\frac{-2}{\sqrt{29}}\right)^2 + \left(\frac{5}{\sqrt{29}}\right)^2} = \sqrt{\frac{4}{29} + \frac{25}{29}} = \sqrt{\frac{29}{29}} = 1.$$

Generally, the length of the sum of two vectors is not equal to the sum of their lengths. To see this, consider the vectors $\mathbf{u}$ and $\mathbf{v}$ as shown in Figure 10.9. By considering $\mathbf{u}$ and $\mathbf{v}$ as two sides of a triangle, you can see that the length of the third side is $\|\mathbf{u} + \mathbf{v}\|$, and you have

$$\|\mathbf{u} + \mathbf{v}\| \le \|\mathbf{u}\| + \|\mathbf{v}\|.$$

Equality occurs only if the vectors $\mathbf{u}$ and $\mathbf{v}$ have the *same* direction. This result is called the **triangle inequality** for vectors. (You are asked to prove this in Exercise 67, Section 10.3.)

Standard Unit Vectors

The unit vectors $\langle 1, 0 \rangle$ and $\langle 0, 1 \rangle$ are called the **standard unit vectors** in the plane and are denoted by

$$\mathbf{i} = \langle 1, 0 \rangle \qquad \text{and} \qquad \mathbf{j} = \langle 0, 1 \rangle \qquad \text{\color{red}Standard unit vectors}$$

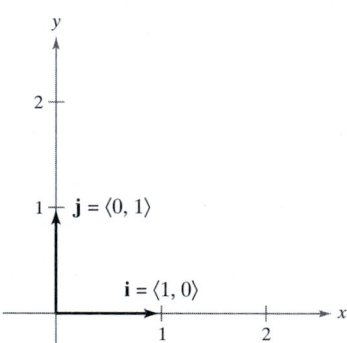

Standard unit vectors **i** and **j**
Figure 10.10

as shown in Figure 10.10. These vectors can be used to represent any vector, as follows.

$$\mathbf{v} = \langle v_1, v_2 \rangle = \langle v_1, 0 \rangle + \langle 0, v_2 \rangle = v_1 \langle 1, 0 \rangle + v_2 \langle 0, 1 \rangle = v_1 \mathbf{i} + v_2 \mathbf{j}$$

The vector $\mathbf{v} = v_1 \mathbf{i} + v_2 \mathbf{j}$ is called a **linear combination** of **i** and **j**. The scalars v_1 and v_2 are called the **horizontal** and **vertical components of v.**

EXAMPLE 5 Writing a Vector as a Linear Combination of Unit Vectors

Let $\mathbf{u}$ be the vector with initial point $(2, -5)$ and terminal point $(-1, 3)$, and let $\mathbf{v} = 2\mathbf{i} - \mathbf{j}$. Write each of the following vectors as a linear combination of **i** and **j**.

a. $\mathbf{u}$ **b.** $\mathbf{w} = 2\mathbf{u} - 3\mathbf{v}$

Solution

a. $\mathbf{u} = \langle q_1 - p_1, q_2 - p_2 \rangle = \langle -1 - 2, 3 - (-5) \rangle = \langle -3, 8 \rangle = -3\mathbf{i} + 8\mathbf{j}$

b. $\mathbf{w} = 2\mathbf{u} - 3\mathbf{v} = 2(-3\mathbf{i} + 8\mathbf{j}) - 3(2\mathbf{i} - \mathbf{j})$

$$= -6\mathbf{i} + 16\mathbf{j} - 6\mathbf{i} + 3\mathbf{j}$$

$$= -12\mathbf{i} + 19\mathbf{j}$$

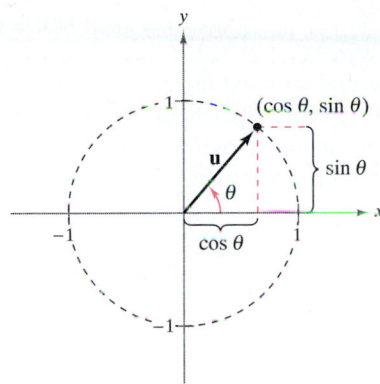

The angle θ from the positive x-axis to the vector **u**

Figure 10.11

If **u** is a unit vector such that θ is the angle (measured counterclockwise) from the positive x-axis to **u**, then the terminal point of **u** lies on the unit circle, and you have

$$\mathbf{u} = \langle \cos\theta, \sin\theta \rangle = \cos\theta\mathbf{i} + \sin\theta\mathbf{j} \qquad \text{Unit vector}$$

as shown in Figure 10.11. Moreover, it follows that any other nonzero vector **v** making an angle θ with the positive x-axis has the same direction as **u**, and you can write

$$\mathbf{v} = \|\mathbf{v}\|\langle \cos\theta, \sin\theta \rangle = \|\mathbf{v}\|\cos\theta\mathbf{i} + \|\mathbf{v}\|\sin\theta\mathbf{j}.$$

EXAMPLE 6   Writing a Vector of Given Length and Direction

The vector **v** has a length of 3 and makes an angle of $30° = \pi/6$ with the positive x-axis. Write **v** as a linear combination of the unit vectors **i** and **j**.

Solution   Because the angle between **v** and the positive x-axis is $\theta = \pi/6$, you can write the following.

$$\mathbf{v} = \|\mathbf{v}\|\cos\theta\mathbf{i} + \|\mathbf{v}\|\sin\theta\mathbf{j} = 3\cos\frac{\pi}{6}\mathbf{i} + 3\sin\frac{\pi}{6}\mathbf{j}$$

$$= \frac{3\sqrt{3}}{2}\mathbf{i} + \frac{3}{2}\mathbf{j}$$

Applications of Vectors

There are many applications of vectors in physics and engineering. We conclude this section by considering two examples.

EXAMPLE 7   Finding the Resultant Force

Two tugboats are pushing an ocean liner, as shown in Figure 10.12. Each boat is exerting a force of 400 pounds. What is the resultant force on the ocean liner?

Solution   Using Figure 10.12, you can represent the forces exerted by the first and second tugboats as

$$\mathbf{F}_1 = 400\langle \cos 20°, \sin 20° \rangle$$
$$= 400\cos(20°)\mathbf{i} + 400\sin(20°)\mathbf{j}$$
$$\mathbf{F}_2 = 400\langle \cos(-20°), \sin(-20°) \rangle$$
$$= 400\cos(20°)\mathbf{i} - 400\sin(20°)\mathbf{j}.$$

To obtain the resultant force on the ocean liner, add these two forces.

$$\mathbf{F} = \mathbf{F}_1 + \mathbf{F}_2$$
$$= 400\cos(20°)\mathbf{i} + 400\sin(20°)\mathbf{j} + 400\cos(20°)\mathbf{i} - 400\sin(20°)\mathbf{j}$$
$$= 800\cos(20°)\mathbf{i}$$
$$\approx 752\mathbf{i}$$

Thus, the resultant force on the ocean liner is approximately 752 pounds in the direction of the positive x-axis.

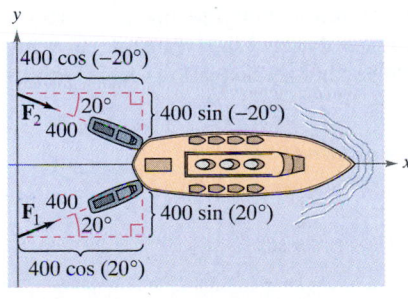

The resultant force on the ocean liner is approximately 752 pounds in the direction of the positive x-axis.

Figure 10.12

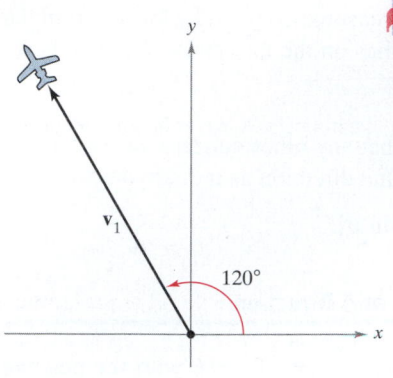

Direction without wind

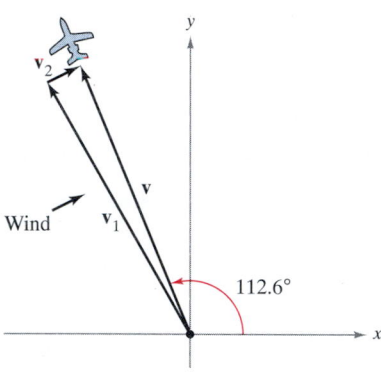

Direction with wind
Figure 10.13

EXAMPLE 8   Finding a Velocity

An airplane is traveling at a fixed altitude with a negligible wind factor. The plane is headed N 30° W (30° west of north) at a speed of 500 miles per hour, as shown in Figure 10.13. As the plane reaches a certain point, it encounters wind with a velocity of 70 miles per hour in the direction E 45° N. What are the resultant speed and direction of the plane?

Solution   Using Figure 10.13, you can represent the velocity of the plane as

$$\mathbf{v}_1 = 500\cos(120°)\mathbf{i} + 500\sin(120°)\mathbf{j}.$$

The velocity of the wind is represented by the vector

$$\mathbf{v}_2 = 70\cos(45°)\mathbf{i} + 70\sin(45°)\mathbf{j}.$$

The resultant velocity of the plane is

$$\begin{aligned}
\mathbf{v} &= \mathbf{v}_1 + \mathbf{v}_2 \\
&= 500\cos(120°)\mathbf{i} + 500\sin(120°)\mathbf{j} + 70\cos(45°)\mathbf{i} + 70\sin(45°)\mathbf{j} \\
&\approx -200.5\mathbf{i} + 482.5\mathbf{j}
\end{aligned}$$

To find the speed and direction, write $\mathbf{v} = \|\mathbf{v}\|(\cos\theta\,\mathbf{i} + \sin\theta\,\mathbf{j})$. Because

$$\|\mathbf{v}\| \approx \sqrt{(-200.5)^2 + (482.5)^2} \approx 522.5$$

you can write

$$\mathbf{v} \approx 522.5\left(\frac{-200.5}{522.5}\mathbf{i} + \frac{482.5}{522.5}\mathbf{j}\right) \approx 522.5\left[\cos(112.6°)\mathbf{i} + \sin(112.6°)\mathbf{j}\right].$$

The new speed of the plane, as altered by the wind, is approximately 522.5 miles per hour in a path that makes an angle of 112.6° with the positive x-axis.

LAB SERIES
Lab 10.1

EXERCISES FOR SECTION 10.1

In Exercises 1–4, (a) find the component form of the vector v and (b) sketch the vector with its initial point at the origin.

1.
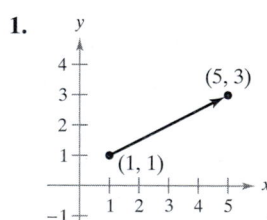

2.
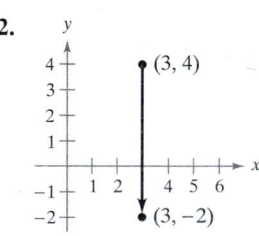

3.
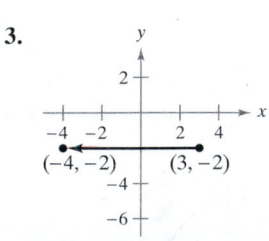

4.
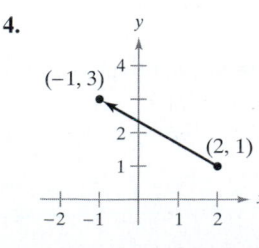

In Exercises 5–12, the initial and terminal points of a vector v are given. (a) Sketch the given directed line segment, (b) write the vector in component form, and (c) sketch the vector with its initial point at the origin.

	Initial Point	*Terminal Point*
5.	$(1, 2)$	$(5, 5)$
6.	$(3, -5)$	$(4, 7)$
7.	$(10, 2)$	$(6, -1)$
8.	$(0, -4)$	$(-5, -1)$
9.	$(6, 2)$	$(6, 6)$
10.	$(7, -1)$	$(-3, -1)$
11.	$\left(\frac{3}{2}, \frac{4}{3}\right)$	$\left(\frac{1}{2}, 3\right)$
12.	$(0.12, 0.60)$	$(0.84, 1.25)$

A red number indicates that a detailed solution can be found in the Study and Solutions Guide.

In Exercises 13 and 14, sketch each scalar multiple of v.

13. $v = \langle 2, 3 \rangle$ (a) $2v$ (b) $-3v$ (c) $\frac{7}{2}v$ (d) $\frac{2}{3}v$

14. $v = \langle -1, 5 \rangle$ (a) $4v$ (b) $-\frac{1}{2}v$ (c) $0v$ (d) $-6v$

In Exercises 15–18, use the figure to sketch a graph of the indicated vector.

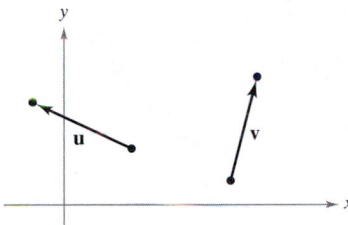

15. $-u$ **16.** $2u$

17. $u - v$ **18.** $u + 2v$

In Exercises 19–22, find the vector v where $u = \langle 2, -1 \rangle$ and $w = \langle 1, 2 \rangle$. Illustrate the vector operations geometrically.

19. $v = \frac{3}{2}u$ **20.** $v = u + w$

21. $v = u + 2w$ **22.** $v = u - 2w$

In Exercises 23–28, find a and b such that $v = au + bw$, where $u = \langle 1, 2 \rangle$ and $w = \langle 1, -1 \rangle$.

23. $v = \langle 2, 1 \rangle$ **24.** $v = \langle 0, 3 \rangle$

25. $v = \langle 3, 0 \rangle$ **26.** $v = \langle 3, 3 \rangle$

27. $v = \langle 1, 1 \rangle$ **28.** $v = \langle -1, 7 \rangle$

In Exercises 29 and 30, the vector v and its initial point are given. Find the terminal point.

29. $v = \langle -1, 3 \rangle$, initial point $(4, 2)$

30. $v = \langle 4, -9 \rangle$, initial point $(3, 2)$

In Exercises 31–36, find the magnitude of v.

31. $v = \langle 4, 3 \rangle$ **32.** $v = \langle 12, -5 \rangle$

33. $v = 6i - 5j$ **34.** $v = -10i + 3j$

35. $v = 4j$ **36.** $v = i - j$

In Exercises 37–40, find the following.

(a) $\|u\|$ (b) $\|v\|$ (c) $\|u + v\|$

(d) $\left\|\dfrac{u}{\|u\|}\right\|$ (e) $\left\|\dfrac{v}{\|v\|}\right\|$ (f) $\left\|\dfrac{u + v}{\|u + v\|}\right\|$

37. $u = \langle 1, -1 \rangle$ **38.** $u = \langle 0, 1 \rangle$
  $v = \langle -1, 2 \rangle$ $v = \langle 3, -3 \rangle$

39. $u = \langle 1, \frac{1}{2} \rangle$ **40.** $u = \langle 2, -4 \rangle$
  $v = \langle 2, 3 \rangle$ $v = \langle 5, 5 \rangle$

In Exercises 41 and 42, demonstrate the triangle inequality using the vectors u and v.

41. $u = \langle 2, 1 \rangle$ **42.** $u = \langle -3, 2 \rangle$
  $v = \langle 5, 4 \rangle$ $v = \langle 1, -2 \rangle$

In Exercises 43–46, find the vector v with the given magnitude and the same direction as u.

Magnitude	Direction
43. $\|v\| = 4$	$u = \langle 1, 1 \rangle$
44. $\|v\| = 4$	$u = \langle -1, 1 \rangle$
45. $\|v\| = 2$	$u = \langle \sqrt{3}, 3 \rangle$
46. $\|v\| = 3$	$u = \langle 0, 3 \rangle$

In Exercises 47–50, find a unit vector (a) parallel to and (b) normal to the graph of $f(x)$ at the indicated point.

Function	Point
47. $f(x) = x^3$	$(1, 1)$
48. $f(x) = x^3$	$(-2, -8)$
49. $f(x) = \sqrt{25 - x^2}$	$(3, 4)$
50. $f(x) = \tan x$	$\left(\dfrac{\pi}{4}, 1\right)$

In Exercises 51–54, find the component form of v given its magnitude and the angle it makes with the positive x-axis.

Magnitude	Angle
51. $\|v\| = 3$	$\theta = 0°$
52. $\|v\| = 1$	$\theta = 45°$
53. $\|v\| = 2$	$\theta = 150°$
54. $\|v\| = 1$	$\theta = 3.5°$

In Exercises 55–58, find the component form of $u + v$ given the magnitudes of u and v and the angles that u and v make with the positive x-axis.

55. $\|u\| = 1$, $\theta_u = 0°$ **56.** $\|u\| = 4$, $\theta_u = 0°$
  $\|v\| = 3$, $\theta_v = 45°$ $\|v\| = 2$, $\theta_v = 60°$

57. $\|u\| = 2$, $\theta_u = 4$ **58.** $\|u\| = 5$, $\theta_u = -0.5$
  $\|v\| = 1$, $\theta_v = 2$ $\|v\| = 5$, $\theta_v = 0.5$

In Exercises 59 and 60, find the component form of v given the magnitudes of u and $u + v$ and the angles that u and $u + v$ make with the positive x-axis.

59. $\|u\| = 1$, $\theta = 45°$ **60.** $\|u\| = 4$, $\theta = 30°$
  $\|u + v\| = \sqrt{2}$, $\theta = 90°$ $\|u + v\| = 6$, $\theta = 120°$

61. *Programming* You are given the magnitudes of **u** and **v** and the angles **u** and **v** make with the positive *x*-axis. Write a program for a graphing utility in which the output is the following.

(a) **u** + **v**

(b) $\|\mathbf{u} + \mathbf{v}\|$

(c) The angle **u** + **v** makes with the positive *x*-axis

62. Use the program of Exercise 61 to find the magnitude and direction of the resultant of the vectors.

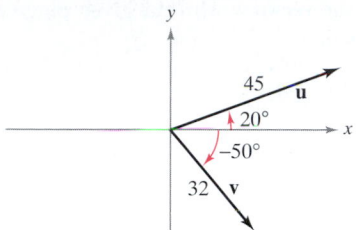

In Exercises 63 and 64, use a graphing utility to find the magnitude and direction of the resultant of the vectors.

63.

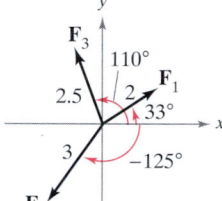

64.

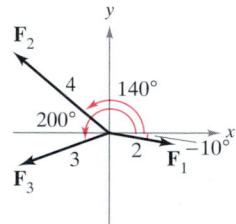

65. *Think About It* Consider two forces of equal magnitude acting on a point.

(a) If the magnitude of the resultant is the sum of the magnitudes of the two forces, make a conjecture about the angle between the forces.

(b) If the resultant of the forces is **0**, make a conjecture about the angle between the forces.

(c) Can the magnitude of the resultant be greater than the sum of the magnitudes of the two forces? Explain.

66. *Graphical Reasoning* Consider two forces $\mathbf{F}_1 = \langle 20, 0 \rangle$ and $\mathbf{F}_2 = 10\langle \cos \theta, \sin \theta \rangle$.

(a) Find $\|\mathbf{F}_1 + \mathbf{F}_2\|$.

(b) Determine the magnitude of the resultant as a function of θ. Use a graphing utility to graph the function for $0 \le \theta < 2\pi$.

(c) Use the graph in part (b) to determine the range of the function. What is its maximum and for what value of θ does it occur? What is its minimum and for what value of θ does it occur?

(d) Explain why the magnitude of the resultant is never 0.

67. *Numerical and Graphical Analysis* Forces with magnitudes of 180 newtons and 275 newtons act on a hook (see figure). The angle between the two forces is θ degrees.

(a) If $\theta = 30°$, find the direction and magnitude of the resultant (vector sum) of these two forces.

(b) Express the magnitude M of the resultant and direction α of the resultant as functions of θ where $0° \le \theta \le 180°$.

(c) Use a graphing utility to complete the table.

θ	0°	30°	60°	90°	120°	150°	180°
M							
α							

(d) Use a graphing utility to graph the two functions M and α.

(e) Explain why one of the functions decreases for increasing θ whereas the other does not.

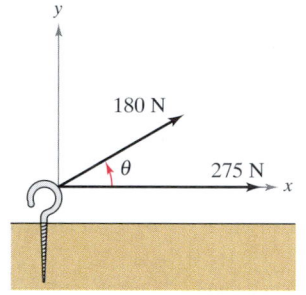

Figure for 67

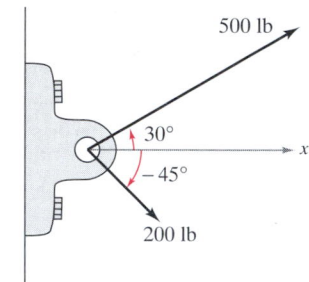

Figure for 68

68. *Resultant Force* Forces with magnitudes of 500 pounds and 200 pounds act on a machine part at angles of 30° and −45° with the *x*-axis (see figure). Find the direction and magnitude of the resultant of these forces.

69. *Resultant Force* Three forces with magnitudes of 75 pounds, 100 pounds, and 125 pounds act on an object at angles of 30°, 45°, and 120° with the positive *x*-axis. Find the direction and magnitude of the resultant of these forces.

70. *Resultant Force* Three forces with magnitudes of 300 newtons, 180 newtons, and 250 newtons act on an object at angles of −30°, 45°, and 135° with the positive *x*-axis. Find the direction and magnitude of the resultant of these forces.

71. Three vertices of a parallelogram are (1, 2), (3, 1), and (8, 4). Find the three possible fourth vertices (see figure).

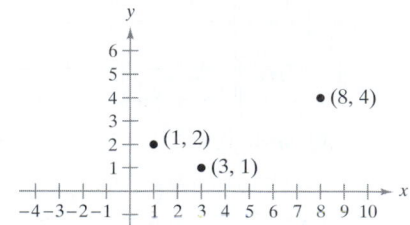

The symbol indicates an exercise in which you are instructed to use graphing technology or a symbolic computer algebra system. The solutions of other exercises may also be facilitated by use of appropriate technology.

72. Use vectors to find the points of trisection of the line segment with endpoints $(1, 2)$ and $(7, 5)$.

73. Numerical and Graphical Analysis   A tetherball weighing 1 pound is pulled outward from the pole by a horizontal force $\mathbf{u}$ until the rope makes an angle of θ degrees with the pole (see figure).

(a) Determine the resulting tension in the rope and the magnitude of $\mathbf{u}$ when $\theta = 30°$.

(b) Write the tension T in the rope and the magnitude of $\mathbf{u}$ as functions of θ. Determine the domains of the functions.

(c) Use a graphing utility to complete the table.

θ	0°	10°	20°	30°	40°	50°	60°
T							
$\|\mathbf{u}\|$							

(d) Use a graphing utility to graph the two functions for $0° \leq \theta \leq 60°$.

(e) Compare T and $\|\mathbf{u}\|$ as θ increases.

(f) Find (if possible) $\lim\limits_{\theta \to \pi/2^-} T$ and $\lim\limits_{\theta \to \pi/2^-} \|\mathbf{u}\|$. Are the results what you expected? Explain.

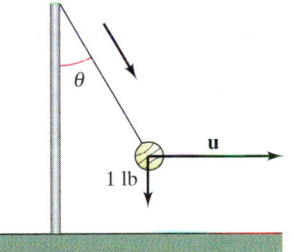

Figure for 73

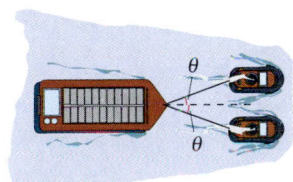

Figure for 74

74. Numerical and Graphical Analysis   A loaded barge is being towed by two tugboats, and the magnitude of the resultant is 6000 pounds directed along the axis of the barge (see figure). Each towline makes an angle of θ degrees with the axis of the barge.

(a) Find the tension in the towlines if $\theta = 20°$.

(b) Write the tension T of each line as a function of θ. Determine the domain of the function.

(c) Use a graphing utility to complete the table.

θ	10°	20°	30°	40°	50°	60°
T						

(d) Use a graphing utility to graph the tension function.

(e) Explain why the tension increases as θ increases.

75. Projectile Motion   A gun with a muzzle velocity of 1200 feet per second is fired at an angle of 6° above the horizontal. Find the vertical and horizontal components of the velocity.

76. Shared Load   To carry a 100-pound cylindrical weight, two workers lift on the ends of short ropes tied to an eyelet on the top center of the cylinder. One rope makes a 20° angle away from the vertical and the other a 30° angle (see figure).

(a) Find each rope's tension if the resultant force is vertical.

(b) Find the vertical component of each worker's force.

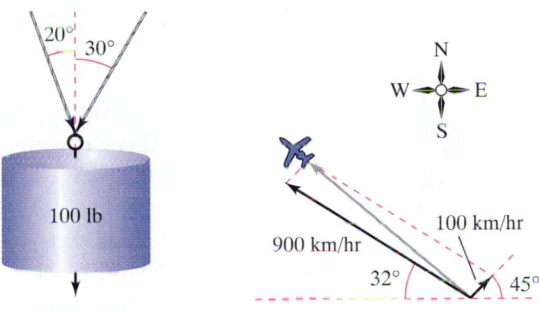

Figure for 76   **Figure for 77**

77. Navigation   An airplane is headed 32° north of west. Its speed with respect to the air is 900 kilometers per hour. The wind at the plane's altitude is from the southwest at 100 kilometers per hour (see figure). What is the true direction of the plane, and what is its speed with respect to the ground?

78. Navigation   A plane flies at a constant groundspeed of 450 miles per hour due east and encounters a 50 mile per hour wind from the northwest. Find the airspeed and compass direction that will allow the plane to maintain its groundspeed and eastward direction.

79. If $\mathbf{F}_1 + \mathbf{F}_2 + \mathbf{F}_3 = \mathbf{0}$, find T_2 and T_3.

$$\mathbf{F}_1 = -3600\mathbf{j}, \qquad \mathbf{F}_2 = T_2(\cos 35°\mathbf{i} - \sin 35°\mathbf{j}),$$
$$\mathbf{F}_3 = T_3(\cos 92°\mathbf{i} + \sin 92°\mathbf{j})$$

80. Prove that $\mathbf{u} = (\cos \theta)\mathbf{i} - (\sin \theta)\mathbf{j}$ and $\mathbf{v} = (\sin \theta)\mathbf{i} + (\cos \theta)\mathbf{j}$ are unit vectors for any angle θ.

81. Geometry   Using vectors, prove that the line segment joining the midpoints of two sides of a triangle is parallel to, and one half the length of, the third side.

82. Geometry   Using vectors, prove that the diagonals of a parallelogram bisect each other.

83. Prove that the vector $\mathbf{w} = \|\mathbf{u}\|\mathbf{v} + \|\mathbf{v}\|\mathbf{u}$ bisects the angle between $\mathbf{u}$ and $\mathbf{v}$.

True or False?   In Exercises 84–89, determine whether the statement is true or false. If it is false, explain why or give an example that shows it is false.

84. If $\mathbf{u}$ and $\mathbf{v}$ have the same magnitude and direction, then $\mathbf{u} = \mathbf{v}$.

85. If $\mathbf{u}$ is a unit vector in the direction of $\mathbf{v}$, then $\mathbf{v} = \|\mathbf{v}\|\mathbf{u}$.

86. If $\mathbf{u} = a\mathbf{i} + b\mathbf{j}$ is a unit vector, then $a^2 + b^2 = 1$.

87. If $\mathbf{v} = a\mathbf{i} + b\mathbf{j} = \mathbf{0}$, then $a = -b$.

88. If $a = b$, then $\|a\mathbf{i} + b\mathbf{j}\| = \sqrt{2}a$.

89. If $\mathbf{u}$ and $\mathbf{v}$ have the same magnitude but opposite directions, then $\mathbf{u} + \mathbf{v} = \mathbf{0}$.

Coordinates in Space • Vectors in Space • Application

Coordinates in Space

Up to this point in the text, you have been primarily concerned with the two-dimensional coordinate system. Much of the remaining part of your study of calculus will involve the three-dimensional coordinate system.

Before extending the concept of a vector to three dimensions, we introduce the **three-dimensional coordinate system.** You can construct this system by passing a *z*-axis perpendicular to both the *x*- and *y*-axes at the origin. Figure 10.14 shows the positive portion of each coordinate axis. Taken as pairs, the axes determined three **coordinate planes:** the *xy*-plane, the *xz*-plane, and the *yz*-plane. These three coordinate planes separate three-space into eight **octants.** The first octant is the one for which all three coordinates are positive. In this three-dimensional system, a point *P* in space is determined by an ordered triple (*x, y, z*), where *x, y,* and *z* are as follows.

$$x = \text{directed distance from } yz\text{-plane to } P$$
$$y = \text{directed distance from } xz\text{-plane to } P$$
$$z = \text{directed distance from } xy\text{-plane to } P$$

Several points are shown in Figure 10.15.

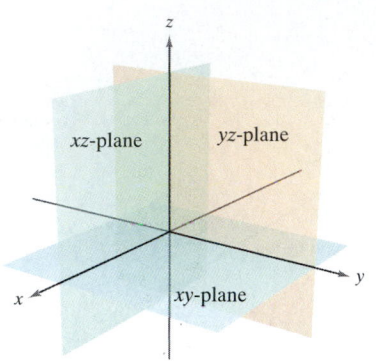

The three-dimensional coordinate system
Figure 10.14

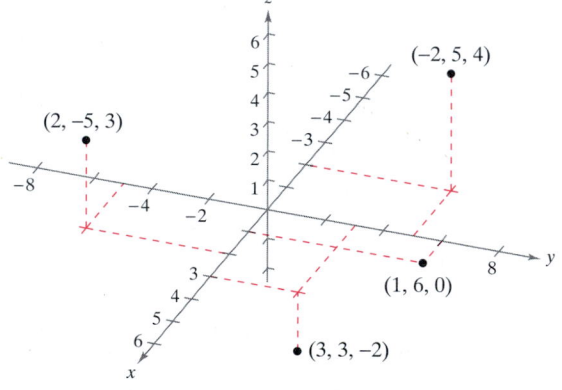

Points in the three-dimensional coordinate system are represented by ordered triples.
Figure 10.15

A three-dimensional coordinate system can have either a **left-handed** or a **right-handed** orientation. To determine the orientation of a system, imagine that you are standing at the origin, with your arms pointing in the direction of the positive *x*- and *y*-axes, and the *z*-axis pointing up, as shown in Figure 10.16. The system is right-handed or left-handed depending on which hand points along the *x*-axis. In this text we work exclusively with the right-handed system.

NOTE   To help visualize points or objects in a three-dimensional system, try using the CD-ROM three-dimensional software that accompanies this text.

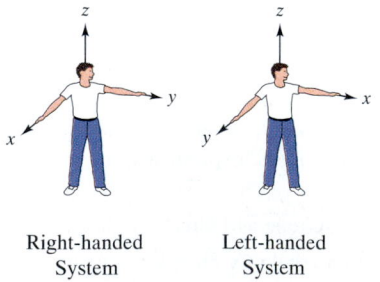

Right-handed   Left-handed
System   System

Figure 10.16

Many of the formulas established for the two-dimensional coordinate system can be extended to three dimensions. For example, to find the **distance between two points in space,** you can use the Pythagorean Theorem twice, as shown in Figure 10.17. By doing this, you will obtain the formula for the distance between the points (x_1, y_1, z_1) and (x_2, y_2, z_2).

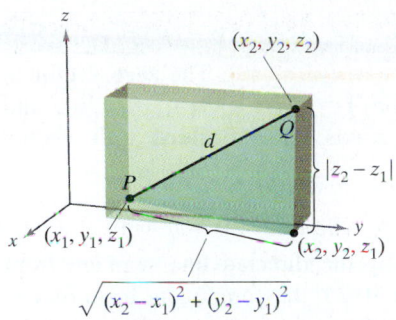

$$d = \sqrt{(x_2 - x_1)^2 + (y_2 - y_1)^2 + (z_2 - z_1)^2} \qquad \text{Distance Formula}$$

The distance between two points in space
Figure 10.17

EXAMPLE 1   Finding the Distance Between Two Points in Space

The distance between the points $(2, -1, 3)$ and $(1, 0, -2)$ is

$$d = \sqrt{(1 - 2)^2 + (0 + 1)^2 + (-2 - 3)^2}$$
$$= \sqrt{1 + 1 + 25}$$
$$= \sqrt{27}$$
$$= 3\sqrt{3}.$$

A **sphere** with center at (x_0, y_0, z_0) and radius r is defined to be the set of all points (x, y, z) such that the distance between (x, y, z) and (x_0, y_0, z_0) is r. You can use the Distance Formula to find the **standard equation of a sphere** of radius r, centered at (x_0, y_0, z_0). If (x, y, z) is an arbitrary point on the sphere, the equation of the sphere is

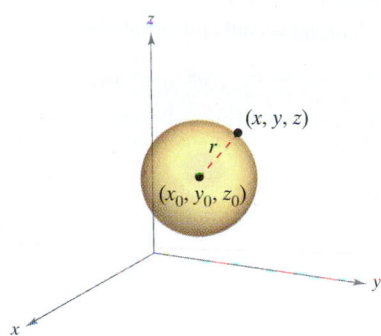

$$(x - x_0)^2 + (y - y_0)^2 + (z - z_0)^2 = r^2 \qquad \text{Equation of sphere}$$

Figure 10.18

as shown in Figure 10.18. Moreover, the midpoint of the line segment joining the points (x_1, y_1, z_1) and (x_2, y_2, z_2) has coordinates

$$\left(\frac{x_1 + x_2}{2}, \frac{y_1 + y_2}{2}, \frac{z_1 + z_2}{2} \right). \qquad \text{Midpoint Rule}$$

EXAMPLE 2   Finding the Equation of a Sphere

Find the standard equation for the sphere that has the points $(5, -2, 3)$ and $(0, 4, -3)$ as endpoints of a diameter.

Solution   By the Midpoint Rule, the center of the sphere is

$$\left(\frac{5 + 0}{2}, \frac{-2 + 4}{2}, \frac{3 - 3}{2} \right) = \left(\frac{5}{2}, 1, 0 \right).$$

By the Distance Formula, the radius is

$$r = \sqrt{\left(0 - \frac{5}{2} \right)^2 + (4 - 1)^2 + (-3 - 0)^2} = \sqrt{\frac{97}{4}} = \frac{\sqrt{97}}{2}.$$

Therefore, the standard equation of the sphere is

$$\left(x - \frac{5}{2} \right)^2 + (y - 1)^2 + (z - 0)^2 = \frac{97}{4}.$$

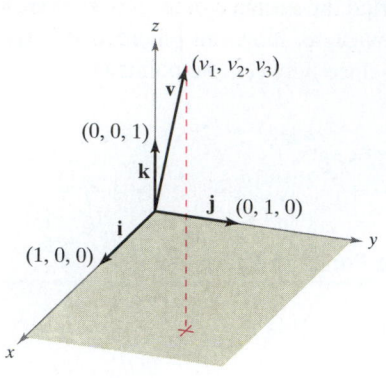

The standard unit vectors in space
Figure 10.19

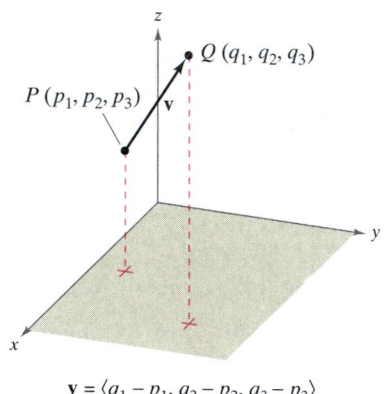

$$v = \langle q_1 - p_1, q_2 - p_2, q_3 - p_3 \rangle$$

Figure 10.20

Vectors in Space

In space, vectors are denoted by ordered triples $\mathbf{v} = \langle v_1, v_2, v_3 \rangle$. The **zero vector** is denoted by $\mathbf{0} = \langle 0, 0, 0 \rangle$. Using the unit vectors $\mathbf{i} = \langle 1, 0, 0 \rangle$, $\mathbf{j} = \langle 0, 1, 0 \rangle$, and $\mathbf{k} = \langle 0, 0, 1 \rangle$ in the direction of the positive z-axis, the **standard unit vector notation** for $\mathbf{v}$ is

$$\mathbf{v} = v_1\mathbf{i} + v_2\mathbf{j} + v_3\mathbf{k}$$

as shown in Figure 10.19. If $\mathbf{v}$ is represented by the directed line segment from $P(p_1, p_2, p_3)$ to $Q(q_1, q_2, q_3)$, as shown in Figure 10.20, the component form of $\mathbf{v}$ is given by subtracting the coordinates of the initial point from the coordinates of the terminal point, as follows.

$$\mathbf{v} = \langle v_1, v_2, v_3 \rangle = \langle q_1 - p_1, q_2 - p_2, q_3 - p_3 \rangle$$

Vectors in Space

Let $\mathbf{u} = \langle u_1, u_2, u_3 \rangle$ and $\mathbf{v} = \langle v_1, v_2, v_3 \rangle$ be vectors in space and let c be a scalar.

1. *Equality of Vectors:* $\mathbf{u} = \mathbf{v}$ if and only if $u_1 = v_1$, $u_2 = v_2$, and $u_3 = v_3$.
2. *Component Form:* If $\mathbf{v}$ is represented by the directed line segment from $P(p_1, p_2, p_3)$ to $Q(q_1, q_2, q_3)$, then

$$\mathbf{v} = \langle v_1, v_2, v_3 \rangle = \langle q_1 - p_1, q_2 - p_2, q_3 - p_3 \rangle.$$

3. *Length:* $\|\mathbf{v}\| = \sqrt{v_1^2 + v_2^2 + v_3^2}$
4. *Unit Vector in the Direction of* $\mathbf{v}$: $\dfrac{\mathbf{v}}{\|\mathbf{v}\|} = \left(\dfrac{1}{\|\mathbf{v}\|}\right) \langle v_1, v_2, v_3 \rangle, \quad \mathbf{v} \neq \mathbf{0}$
5. *Vector Addition:* $\mathbf{v} + \mathbf{u} = \langle v_1 + u_1, v_2 + u_2, v_3 + u_3 \rangle$
6. *Scalar Multiplication:* $c\mathbf{v} = \langle cv_1, cv_2, cv_3 \rangle$

NOTE The properties of vector addition and scalar multiplication given in Theorem 10.1 are also valid for vectors in space.

 EXAMPLE 3 **Finding the Component Form of a Vector in Space**

Find the component form and length of the vector $\mathbf{v}$ having initial point $(-2, 3, 1)$ and terminal point $(0, -4, 4)$. Then find a unit vector in the direction of $\mathbf{v}$.

Solution The component form of $\mathbf{v}$ is

$$\mathbf{v} = \langle q_1 - p_1, q_2 - p_2, q_3 - p_3 \rangle = \langle 0 - (-2), -4 - 3, 4 - 1 \rangle$$
$$= \langle 2, -7, 3 \rangle$$

which implies that its length is

$$\|\mathbf{v}\| = \sqrt{(2)^2 + (-7)^2 + (3)^2} = \sqrt{62}.$$

The unit vector in the direction of $\mathbf{v}$ is

$$\mathbf{u} = \frac{\mathbf{v}}{\|\mathbf{v}\|} = \frac{1}{\sqrt{62}} \langle 2, -7, 3 \rangle.$$

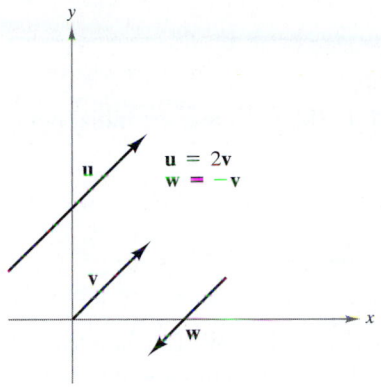

Parallel vectors
Figure 10.21

Recall from the definition of scalar multiplication that positive scalar multiples of a nonzero vector **v** have the same direction as **v**, whereas negative multiples have the direction opposite of **v**. In general, two nonzero vectors **u** and **v** are **parallel** if there is some scalar c such that $\mathbf{u} = c\mathbf{v}$.

Definition of Parallel Vectors

Two nonzero vectors **u** and **v** are **parallel** if there is some scalar c such that $\mathbf{u} = c\mathbf{v}$.

For example, in Figure 10.21, the vectors **u, v,** and **w** are parallel because $\mathbf{u} = 2\mathbf{v}$ and $\mathbf{w} = -\mathbf{v}$.

EXAMPLE 4   Parallel Vectors

Vector **w** has initial point $(2, -1, 3)$ and terminal point $(-4, 7, 5)$. Which of the following vectors is parallel to **w**?

a. $\mathbf{u} = \langle 3, -4, -1 \rangle$ **b.** $\mathbf{v} = \langle 12, -16, 4 \rangle$

Solution   Begin by writing **w** in component form.

$$\mathbf{w} = \langle -4 - 2, 7 - (-1), 5 - 3 \rangle = \langle -6, 8, 2 \rangle$$

a. Because $\mathbf{u} = \langle 3, -4, -1 \rangle = -\frac{1}{2}\langle -6, 8, 2 \rangle = -\frac{1}{2}\mathbf{w}$, you can conclude that **u** *is* parallel to **w**.

b. In this case, you want to find a scalar c such that

$$\langle 12, -16, 4 \rangle = c\langle -6, 8, 2 \rangle.$$
$$12 = -6c \rightarrow c = -2$$
$$-16 = 8c \rightarrow c = -2$$
$$4 = 2c \rightarrow c = 2$$

Because there is no c for which the equation has a solution, the vectors are *not* parallel.

EXAMPLE 5   Using Vectors to Determine Collinear Points

Determine whether the points $P(1, -2, 3)$, $Q(2, 1, 0)$, and $R(4, 7, -6)$ lie on the same line.

Solution   The component forms of $\overrightarrow{PQ}$ and $\overrightarrow{PR}$ are

$$\overrightarrow{PQ} = \langle 2 - 1, 1 - (-2), 0 - 3 \rangle = \langle 1, 3, -3 \rangle$$

and

$$\overrightarrow{PR} = \langle 4 - 1, 7 - (-2), -6 - 3 \rangle = \langle 3, 9, -9 \rangle.$$

These two vectors have a common initial point. Hence, P, Q, and R lie on the same line if and only if $\overrightarrow{PQ}$ and $\overrightarrow{PR}$ are parallel—which they are because $\overrightarrow{PR} = 3\,\overrightarrow{PQ}$, as shown in Figure 10.22.

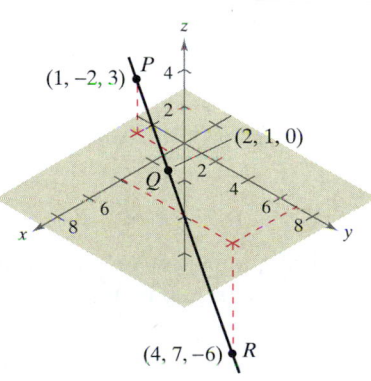

The points P, Q, and R lie on the same line.
Figure 10.22

EXAMPLE 6   Standard Unit Vector Notation

a. Write the vector $\mathbf{v} = 4\mathbf{i} - 5\mathbf{k}$ in component form.

b. Find the terminal point of the vector $\mathbf{v} = 7\mathbf{i} - \mathbf{j} + 3\mathbf{k}$, given that the initial point is $P(-2, 3, 5)$.

Solution

a. Because $\mathbf{j}$ is missing, its component is 0 and

$$\mathbf{v} = 4\mathbf{i} - 5\mathbf{k} = \langle 4, 0, -5 \rangle.$$

b. You need to find $Q(q_1, q_2, q_3)$ such that $\mathbf{v} = \overrightarrow{PQ} = 7\mathbf{i} - \mathbf{j} + 3\mathbf{k}$. This implies that $q_1 - (-2) = 7$, $q_2 - 3 = -1$, and $q_3 - 5 = 3$. The solution of these three equations is $q_1 = 5$, $q_2 = 2$, and $q_3 = 8$. Therefore, Q is $(5, 2, 8)$.

Application

EXAMPLE 7   Measuring Force

A television camera weighing 120 pounds is supported by a tripod, as shown in Figure 10.23. Represent the force exerted on each leg of the tripod as a vector. (Assume that the weight is distributed equally among the three legs.)

Solution   Let the vectors $\mathbf{F}$, $\mathbf{F}$, and $\mathbf{F}$ represent the forces exerted on the three legs. From Figure 10.23, you can determine the directions of $\mathbf{F}_1$, $\mathbf{F}_2$, and $\mathbf{F}_3$ to be as follows.

$$\overrightarrow{PQ}_1 = \langle 0 - 0, -1 - 0, 0 - 4 \rangle = \langle 0, -1, -4 \rangle$$

$$\overrightarrow{PQ}_2 = \left\langle \frac{\sqrt{3}}{2} - 0, \frac{1}{2} - 0, 0 - 4 \right\rangle = \left\langle \frac{\sqrt{3}}{2}, \frac{1}{2}, -4 \right\rangle$$

$$\overrightarrow{PQ}_3 = \left\langle -\frac{\sqrt{3}}{2} - 0, \frac{1}{2} - 0, 0 - 4 \right\rangle = \left\langle -\frac{\sqrt{3}}{2}, \frac{1}{2}, -4 \right\rangle$$

Because each leg has the same length, and the total force is distributed equally among the three legs, you know that $\|\mathbf{F}_1\| = \|\mathbf{F}_2\| = \|\mathbf{F}_3\|$. Hence, there exists a constant c such that

$$\mathbf{F}_1 = c\langle 0, -1, -4 \rangle, \quad \mathbf{F}_2 = c\left\langle \frac{\sqrt{3}}{2}, \frac{1}{2}, -4 \right\rangle, \quad \text{and} \quad \mathbf{F}_3 = c\left\langle -\frac{\sqrt{3}}{2}, \frac{1}{2}, -4 \right\rangle.$$

Let the total force exerted by the object be given by $\mathbf{F} = -120\mathbf{k}$. Then, using the fact that

$$\mathbf{F} = \mathbf{F}_1 + \mathbf{F}_2 + \mathbf{F}_3$$

you can conclude that $\mathbf{F}_1$, $\mathbf{F}_2$, and $\mathbf{F}_3$ each has a vertical component of -40. This implies that $c(-4) = -40$ and $c = 10$. Therefore, the forces exerted on the legs can be represented by

$$\mathbf{F}_1 = \langle 0, -10, -40 \rangle$$
$$\mathbf{F}_2 = \langle 5\sqrt{3}, 5, -40 \rangle$$
$$\mathbf{F}_3 = \langle -5\sqrt{3}, 5, -40 \rangle.$$

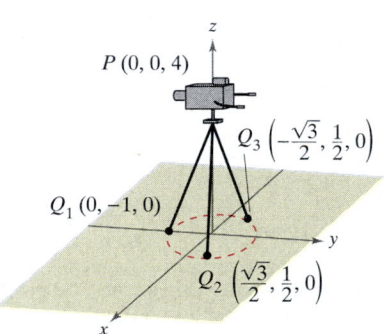

Figure 10.23

EXERCISES FOR SECTION 10.2

In Exercises 1–4, plot the points on the same three-dimensional coordinate system.

1. (a) $(2, 1, 3)$ (b) $(-1, 2, 1)$
2. (a) $(3, -2, 5)$ (b) $\left(\frac{3}{2}, 4, -2\right)$
3. (a) $(5, -2, 2)$ (b) $(5, -2, -2)$
4. (a) $(0, 4, -5)$ (b) $(4, 0, 5)$

In Exercises 5 and 6, approximate the coordinates of the points.

5.
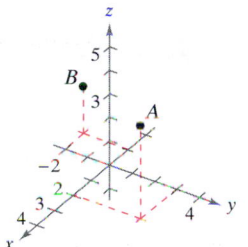

6.
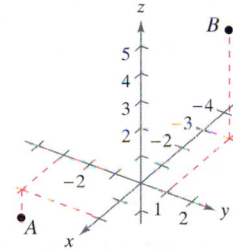

In Exercises 7–10, find the coordinates of the point.

7. The point is located three units behind the yz-plane, four units to the right of the xz-plane, and five units above the xy-plane.

8. The point is located seven units in front of the yz-plane, two units to the left of the xz-plane, and one unit below the xy-plane.

9. The point is located on the x-axis, ten units in front of the yz-plane.

10. The point is located in the yz-plane, three units to the right of the xz-plane, and two units above the xy-plane.

11. **Think About It** What is the z-coordinate of any point in the xy-plane?

12. **Think About It** What is the x-coordinate of any point in the yz-plane?

In Exercises 13–16, determine the location of the point (x, y, z) so that the condition(s) is (are) satisfied.

13. $xy > 0, \quad z = -3$
14. $xy < 0, \quad z = 4$
15. $xyz < 0$
16. $xyz > 0$

In Exercises 17–20, find the lengths of the sides of the triangle with the indicated vertices, and determine whether the triangle is a right triangle, an isosceles triangle, or neither.

17. $(0, 0, 0), (2, 2, 1), (2, -4, 4)$
18. $(5, 3, 4), (7, 1, 3), (3, 5, 3)$
19. $(1, -3, -2), (5, -1, 2), (-1, 1, 2)$
20. $(5, 0, 0), (0, 2, 0), (0, 0, -3)$

21. **Think About It** The triangle in Exercise 17 is translated five units upward along the z-axis. Determine the coordinates of the translated triangle.

22. **Think About It** The triangle in Exercise 18 is translated three units to the right along the y-axis. Determine the coordinates of the translated triangle.

In Exercises 23 and 24, find the coordinates of the midpoint of the line segment joining the points.

23. $(5, -9, 7), (-2, 3, 3)$ 24. $(4, 0, -6), (8, 8, 20)$

In Exercises 25–28, find the standard form of the equation of the sphere.

25. Center: $(0, 2, 5)$
    Radius: 2

26. Center: $(4, -1, 1)$
    Radius: 5

27. Endpoints of a diameter: $(2, 0, 0), (0, 6, 0)$

28. Center: $(-2, 1, 1)$, tangent to the xy-plane

In Exercises 29–32, complete the square to write the equation of the sphere in standard form, and find the center and radius of the sphere.

29. $x^2 + y^2 + z^2 - 2x + 6y + 8z + 1 = 0$
30. $x^2 + y^2 + z^2 + 9x - 2y + 10z + 19 = 0$
31. $9x^2 + 9y^2 + 9z^2 - 6x + 18y + 1 = 0$
32. $4x^2 + 4y^2 + 4z^2 - 4x - 32y + 8z + 33 = 0$

In Exercises 33–36, (a) find the component form of the vector v and (b) sketch the vector with its initial point at the origin.

33.
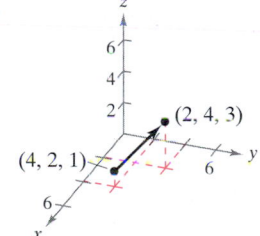

34.
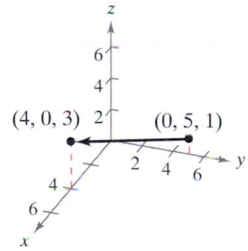

35.
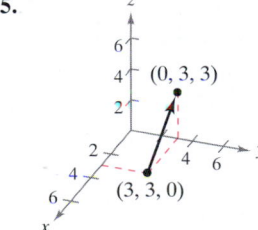

36.
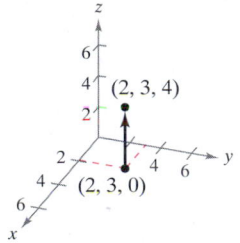

In Exercises 37 and 38, the initial and terminal points of a vector v are given. (a) Sketch the directed line segment, (b) find the component form of the vector, and (c) sketch the vector with its initial point at the origin.

37. Initial point: $(-1, 2, 3)$

    Terminal point: $(3, 3, 4)$

38. Initial point: $(2, -1, -2)$

    Terminal point: $(-4, 3, 7)$

In Exercises 39 and 40, the vector v and its initial point are given. Find the terminal point.

39. $\mathbf{v} = \langle 3, -5, 6 \rangle$

    Initial point: $(0, 6, 2)$

40. $\mathbf{v} = \left\langle 0, \frac{1}{2}, -\frac{1}{3} \right\rangle$

    Initial point: $\left(3, 0, -\frac{2}{3} \right)$

In Exercises 41 and 42, sketch each scalar multiple of v.

41. $\mathbf{v} = \langle 1, 2, 2 \rangle$

    (a) $2\mathbf{v}$ (b) $-\mathbf{v}$ (c) $\frac{3}{2}\mathbf{v}$ (d) $0\mathbf{v}$

42. $\mathbf{v} = \langle 2, -2, 1 \rangle$

    (a) $-\mathbf{v}$ (b) $2\mathbf{v}$ (c) $\frac{1}{2}\mathbf{v}$ (d) $\frac{5}{2}\mathbf{v}$

In Exercises 43–48, find the vector z, given $\mathbf{u} = \langle 1, 2, 3 \rangle$, $\mathbf{v} = \langle 2, 2, -1 \rangle$, and $\mathbf{w} = \langle 4, 0, -4 \rangle$.

43. $\mathbf{z} = \mathbf{u} - \mathbf{v}$

44. $\mathbf{z} = \mathbf{u} - \mathbf{v} + 2\mathbf{w}$

45. $\mathbf{z} = 2\mathbf{u} + 4\mathbf{v} - \mathbf{w}$

46. $\mathbf{z} = 5\mathbf{u} - 3\mathbf{v} - \frac{1}{2}\mathbf{w}$

47. $2\mathbf{z} - 3\mathbf{u} = \mathbf{w}$

48. $2\mathbf{u} + \mathbf{v} - \mathbf{w} + 3\mathbf{z} = \mathbf{0}$

In Exercises 49–52, determine which of the vectors are parallel to z. Use a graphing utility to confirm your results.

49. $\mathbf{z} = \langle 3, 2, -5 \rangle$

    (a) $\langle -6, -4, 10 \rangle$ (b) $\left\langle 2, \frac{4}{3}, -\frac{10}{3} \right\rangle$

    (c) $\langle 6, 4, 10 \rangle$ (d) $\langle 1, -4, 2 \rangle$

50. $\mathbf{z} = \frac{1}{2}\mathbf{i} - \frac{2}{3}\mathbf{j} + \frac{3}{4}\mathbf{k}$

    (a) $6\mathbf{i} - 4\mathbf{j} + 9\mathbf{k}$ (b) $-\mathbf{i} + \frac{4}{3}\mathbf{j} - \frac{3}{2}\mathbf{k}$

    (c) $12\mathbf{i} + 9\mathbf{k}$ (d) $\frac{3}{4}\mathbf{i} - \mathbf{j} + \frac{9}{8}\mathbf{k}$

51. z has initial point $(1, -1, 3)$ and terminal point $(-2, 3, 5)$.

    (a) $-6\mathbf{i} + 8\mathbf{j} + 4\mathbf{k}$ (b) $4\mathbf{j} + 2\mathbf{k}$

52. z has initial point $(3, 2, -1)$ and terminal point $(-1, -3, 5)$.

    (a) $\langle 0, 5, -6 \rangle$ (b) $\langle 8, 10, -12 \rangle$

In Exercises 53–56, use vectors to determine whether the points lie in a straight line.

53. $(0, -2, -5), (3, 4, 4), (2, 2, 1)$

54. $(1, -1, 5), (0, -1, 6), (3, -1, 3)$

55. $(1, 2, 4), (2, 5, 0), (0, 1, 5)$

56. $(0, 0, 0), (1, 3, -2), (2, -6, 4)$

In Exercises 57 and 58, use vectors to show that the points form the vertices of a parallelogram.

57. $(2, 9, 1), (3, 11, 4), (0, 10, 2), (1, 12, 5)$

58. $(1, 1, 3), (9, -1, -2), (11, 2, -9), (3, 4, -4)$

In Exercises 59–64, find the magnitude of v.

59. $\mathbf{v} = \langle 0, 0, 0 \rangle$

60. $\mathbf{v} = \langle 1, 0, 3 \rangle$

61. $\mathbf{v} = \mathbf{i} - 2\mathbf{j} - 3\mathbf{k}$

62. $\mathbf{v} = -4\mathbf{i} + 3\mathbf{j} + 7\mathbf{k}$

63. Initial point of $\mathbf{v}$: $(1, -3, 4)$

    Terminal point of $\mathbf{v}$: $(1, 0, -1)$

64. Initial point of $\mathbf{v}$: $(0, -1, 0)$

    Terminal point of $\mathbf{v}$: $(1, 2, -2)$

In Exercises 65–68, find a unit vector (a) in the direction of u and (b) in the direction opposite of u.

65. $\mathbf{u} = \langle 2, -1, 2 \rangle$

66. $\mathbf{u} = \langle 6, 0, 8 \rangle$

67. $\mathbf{u} = \langle 3, 2, -5 \rangle$

68. $\mathbf{u} = \langle 8, 0, 0 \rangle$

69. ***Programming*** You are given the component forms of the vectors $\mathbf{u}$ and $\mathbf{v}$. Write a program for a graphing utility in which the output is (a) the component form of $\mathbf{u} + \mathbf{v}$, (b) $\|\mathbf{u} + \mathbf{v}\|$, (c) $\|\mathbf{u}\|$, and (d) $\|\mathbf{v}\|$.

70. Run the program you wrote in Exercise 69 for the vectors $\mathbf{u} = \langle -1, 3, 4 \rangle$ and $\mathbf{v} = \langle 5, 4.5, -6 \rangle$.

In Exercises 71 and 72, determine the values of c that satisfy the equation. Let $\mathbf{u} = \mathbf{i} + 2\mathbf{j} + 3\mathbf{k}$ and $\mathbf{v} = 2\mathbf{i} + 2\mathbf{j} - \mathbf{k}$.

71. $\|c\mathbf{v}\| = 5$

72. $\|c\mathbf{u}\| = 3$

In Exercises 73–76, find the vector v with the given magnitude and direction u.

	Magnitude	Direction
73.	10	$\mathbf{u} = \langle 0, 3, 3 \rangle$
74.	3	$\mathbf{u} = \langle 1, 1, 1 \rangle$
75.	$\frac{3}{2}$	$\mathbf{u} = \langle 2, -2, 1 \rangle$
76.	$\sqrt{5}$	$\mathbf{u} = \langle -4, 6, 2 \rangle$

In Exercises 77 and 78, sketch the vector v and write its component form.

77. **v** lies in the *yz*-plane, has magnitude 2, and makes an angle of 30° with the positive *y*-axis.

78. **v** lies in the *xz*-plane, has magnitude 5, and makes an angle of 45° with the positive *z*-axis.

In Exercises 79 and 80, use vectors to find the point that lies two-thirds of the way from P to Q.

79. $P(4, 3, 0),\quad Q(1, -3, 3)$

80. $P(1, 2, 5),\quad Q(6, 8, 2)$

81. Let $\mathbf{u} = \mathbf{i} + \mathbf{j}$, $\mathbf{v} = \mathbf{j} + \mathbf{k}$, and $\mathbf{w} = a\mathbf{u} + b\mathbf{v}$.

(a) Sketch **u** and **v**.

(b) If $\mathbf{w} = \mathbf{0}$, show that a and b must both be zero.

(c) Find a and b such that $\mathbf{w} = \mathbf{i} + 2\mathbf{j} + \mathbf{k}$.

(d) Show that no choice of a and b yields $\mathbf{w} = \mathbf{i} + 2\mathbf{j} + 3\mathbf{k}$.

82. *Think About It* The initial and terminal points of the vector **v** are (x_1, y_1, z_1) and (x, y, z). Describe the set of all points (x, y, z) such that $\|\mathbf{v}\| = 4$.

83. *Numerical, Graphical, and Analytic Analysis* The lights in an auditorium are 24-pound discs of radius 18 inches. Each disc is supported by three equally spaced wires that are *L* inches long (see figure).

(a) Write the tension *T* in each wire as a function of *L*. Determine the domain of the function.

(b) Use a graphing utility and the model in part (a) to complete the table.

L	20	25	30	35	40	45	50
T							

(c) Use a graphing utility to graph the model in part (a). Determine the asymptotes of the graph.

(d) Confirm the asymptotes of the graph in part (c) analytically.

(e) Determine the minimum length of each cable if a cable is designed to carry a maximum load of 10 pounds.

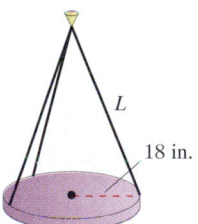

84. *Think About It* Suppose the length of each cable in Exercise 83 has a fixed length $L = a$, and the radius of each disc is r_0 inches. Make a conjecture about the limit.

$$\lim_{r_0 \to a} T$$

and give a reason for your answer.

85. *Diagonal of a Cube* Find the component form of the unit vector **v** in the direction of the diagonal of the cube shown in the figure.

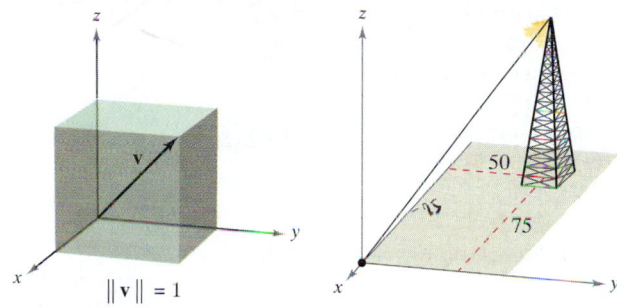

Figure for 85   **Figure for 86**

86. *Tower Guy Wire* The guy wire to a 100-foot tower has a tension of 550 pounds. Using the distances shown in the figure, write the component form of the vector **F** representing the tension in the wire.

87. *Load Supports* Find the tension in each of the supporting cables in the figure if the weight of the crate is 500 newtons.

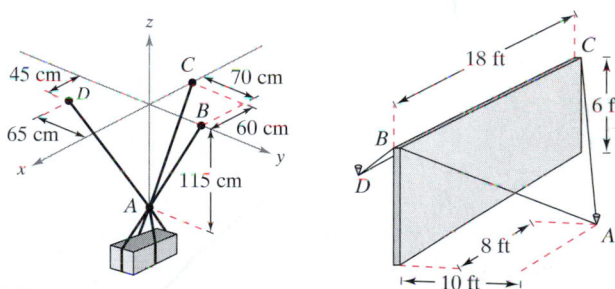

Figure for 87   **Figure for 88**

88. *Building Construction* A precast concrete wall is temporarily kept in its vertical position by ropes (see figure). Find the total force exerted on the pin at position *A* if the tensions in *AB* and *AC* are 420 pounds and 650 pounds.

89. Write an equation whose graph consists of the set of points $P(x, y, z)$ that are twice as far from $A(0, -1, 1)$ as from $B(1, 2, 0)$.

SECTION | *10.3* | **The Dot Product of Two Vectors**

The Dot Product • Angle Between Two Vectors • Direction Cosines •
Projections and Vector Components • Work

The Dot Product

So far you have studied two operations with vectors—vector addition and multiplication by a scalar—each of which yields another vector. In this section you will study a third vector operation, called the **dot product.** This product yields a scalar, rather than a vector.

Definition of Dot Product

The **dot product** of $\mathbf{u} = \langle u_1, u_2 \rangle$ and $\mathbf{v} = \langle v_1, v_2 \rangle$ is

$$\mathbf{u} \cdot \mathbf{v} = u_1 v_1 + u_2 v_2.$$

The **dot product** of $\mathbf{u} = \langle u_1, u_2, u_3 \rangle$ and $\mathbf{v} = \langle v_1, v_2, v_3 \rangle$ is

$$\mathbf{u} \cdot \mathbf{v} = u_1 v_1 + u_2 v_2 + u_3 v_3.$$

NOTE   The dot product of two vectors is also called the **inner product** (or scalar product) of two vectors.

THEOREM 10.4   Properties of the Dot Product

Let $\mathbf{u}$, $\mathbf{v}$, and $\mathbf{w}$ be vectors in the plane or in space and let c be a scalar.

1. $\mathbf{u} \cdot \mathbf{v} = \mathbf{v} \cdot \mathbf{u}$ Commutative property
2. $\mathbf{u} \cdot (\mathbf{v} + \mathbf{w}) = \mathbf{u} \cdot \mathbf{v} + \mathbf{u} \cdot \mathbf{w}$ Distributive property
3. $c(\mathbf{u} \cdot \mathbf{v}) = c\mathbf{u} \cdot \mathbf{v} = \mathbf{u} \cdot c\mathbf{v}$
4. $\mathbf{0} \cdot \mathbf{v} = 0$
5. $\mathbf{v} \cdot \mathbf{v} = \|\mathbf{v}\|^2$

Proof   To prove the first property, let $\mathbf{u} = \langle u_1, u_2, u_3 \rangle$ and $\mathbf{v} = \langle v_1, v_2, v_3 \rangle$. Then

$$\begin{aligned}
\mathbf{u} \cdot \mathbf{v} &= u_1 v_1 + u_2 v_2 + u_3 v_3 \\
&= v_1 u_1 + v_2 u_2 + v_3 u_3 \\
&= \mathbf{v} \cdot \mathbf{u}.
\end{aligned}$$

For the fifth property, let $\mathbf{v} = \langle v_1, v_2, v_3 \rangle$. Then

$$\begin{aligned}
\mathbf{v} \cdot \mathbf{v} &= v_1^2 + v_2^2 + v_3^2 \\
&= \left(\sqrt{v_1^2 + v_2^2 + v_3^2} \right)^2 \\
&= \|\mathbf{v}\|^2.
\end{aligned}$$

Proofs of the other properties are left to you.

EXPLORATION

Interpreting a Dot Product   Several vectors are shown below on the unit circle. Find the dot products of several pairs of vectors. Then find the angle between each pair that you used. Make a conjecture about the relationship between the dot product of two vectors and the angle between the vectors.

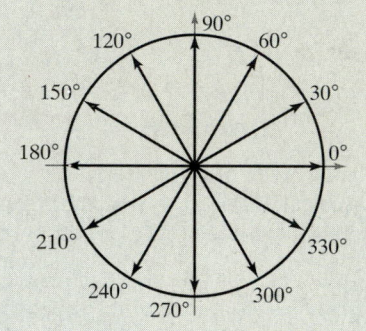

EXAMPLE 1   Finding Dot Products

Given $\mathbf{u} = \langle 2, -2 \rangle$, $\mathbf{v} = \langle 5, 8 \rangle$, and $\mathbf{w} = \langle -4, 3 \rangle$, find each of the following.

a. $\mathbf{u} \cdot \mathbf{v}$ **b.** $(\mathbf{u} \cdot \mathbf{v})\mathbf{w}$ **c.** $\mathbf{u} \cdot (2\mathbf{v})$ **d.** $\|\mathbf{w}\|^2$

Solution

a. $\mathbf{u} \cdot \mathbf{v} = \langle 2, -2 \rangle \cdot \langle 5, 8 \rangle = 2(5) + (-2)(8) = -6$
b. $(\mathbf{u} \cdot \mathbf{v})\mathbf{w} = -6\langle -4, 3 \rangle = \langle 24, -18 \rangle$
c. $\mathbf{u} \cdot (2\mathbf{v}) = 2(\mathbf{u} \cdot \mathbf{v}) = 2(-6) = -12$
d. $\|\mathbf{w}\|^2 = \mathbf{w} \cdot \mathbf{w} = \langle -4, 3 \rangle \cdot \langle -4, 3 \rangle = (-4)(-4) + (3)(3) = 25$

Notice that the result of part (b) is a *vector* quantity, whereas the results of the other three parts are *scalar* quantities.

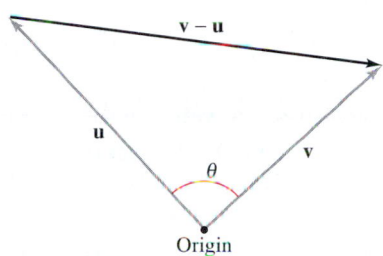

The angle between two vectors
Figure 10.24

Angle Between Two Vectors

The **angle between two nonzero vectors** is the angle θ, $0 \le \theta \le \pi$, between their respective standard position vectors, as shown in Figure 10.24. The next theorem shows how to find this angle using the dot product. (Note that we do not define the angle between the zero vector and another vector.)

> **THEOREM 10.5   Angle Between Two Vectors**
>
> If θ is the angle between two nonzero vectors $\mathbf{u}$ and $\mathbf{v}$, then
>
> $$\cos \theta = \frac{\mathbf{u} \cdot \mathbf{v}}{\|\mathbf{u}\| \|\mathbf{v}\|}.$$

Proof   Consider the triangle determined by vectors, $\mathbf{u}$, $\mathbf{v}$, and $\mathbf{v} - \mathbf{u}$, as shown in Figure 10.24. By the Law of Cosines, you can write

$$\|\mathbf{v} - \mathbf{u}\|^2 = \|\mathbf{u}\|^2 + \|\mathbf{v}\|^2 - 2\|\mathbf{u}\| \|\mathbf{v}\| \cos \theta.$$

Using the properties of the dot product, the left side can be rewritten as

$$\begin{aligned}
\|\mathbf{v} - \mathbf{u}\|^2 &= (\mathbf{v} - \mathbf{u}) \cdot (\mathbf{v} - \mathbf{u}) \\
&= (\mathbf{v} - \mathbf{u}) \cdot \mathbf{v} - (\mathbf{v} - \mathbf{u}) \cdot \mathbf{u} \\
&= \mathbf{v} \cdot \mathbf{v} - \mathbf{u} \cdot \mathbf{v} - \mathbf{v} \cdot \mathbf{u} + \mathbf{u} \cdot \mathbf{u} \\
&= \|\mathbf{v}\|^2 - 2\mathbf{u} \cdot \mathbf{v} + \|\mathbf{u}\|^2
\end{aligned}$$

and substitution back into the Law of Cosines yields

$$\begin{aligned}
\|\mathbf{v}\|^2 - 2\mathbf{u} \cdot \mathbf{v} + \|\mathbf{u}\|^2 &= \|\mathbf{u}\|^2 + \|\mathbf{v}\|^2 - 2\|\mathbf{u}\| \|\mathbf{v}\| \cos \theta \\
-2\mathbf{u} \cdot \mathbf{v} &= -2\|\mathbf{u}\| \|\mathbf{v}\| \cos \theta \\
\cos \theta &= \frac{\mathbf{u} \cdot \mathbf{v}}{\|\mathbf{u}\| \|\mathbf{v}\|}.
\end{aligned}$$

If the angle between two vectors is known, rewriting Theorem 10.5 in the form

$$\mathbf{u} \cdot \mathbf{v} = \|\mathbf{u}\| \|\mathbf{v}\| \cos \theta \qquad \text{Alternative form of dot product}$$

produces an alternative way to calculate the dot product. From this form, you can see that because $\|\mathbf{u}\|$ and $\|\mathbf{v}\|$ are always positive, $\mathbf{u} \cdot \mathbf{v}$ and $\cos \theta$ will always have the same sign. Figure 10.25 shows the possible orientations of two vectors.

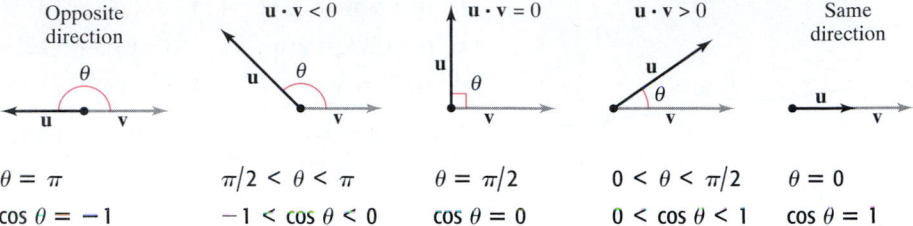

Opposite direction	$\mathbf{u} \cdot \mathbf{v} < 0$	$\mathbf{u} \cdot \mathbf{v} = 0$	$\mathbf{u} \cdot \mathbf{v} > 0$	Same direction
$\theta = \pi$	$\pi/2 < \theta < \pi$	$\theta = \pi/2$	$0 < \theta < \pi/2$	$\theta = 0$
$\cos \theta = -1$	$-1 < \cos \theta < 0$	$\cos \theta = 0$	$0 < \cos \theta < 1$	$\cos \theta = 1$

Figure 10.25

> **Definition of Orthogonal Vectors**
>
> The vectors $\mathbf{u}$ and $\mathbf{v}$ are **orthogonal** if $\mathbf{u} \cdot \mathbf{v} = 0$.

NOTE   The terms "perpendicular," "orthogonal," and "normal" all mean essentially the same thing—meeting at right angles. However, we usually say that two vectors are *orthogonal*, two lines or planes are *perpendicular*, and a vector is *normal* to a given line or plane.

From this definition, it follows that the zero vector is orthogonal to every vector $\mathbf{u}$, because $\mathbf{0} \cdot \mathbf{u} = 0$. Moreover, for $0 \leq \theta \leq \pi$, you know that $\cos \theta = 0$ if and only if $\theta = \pi/2$. Thus, you can use Theorem 10.5 to conclude that two *nonzero* vectors are orthogonal if and only if the angle between them is $\pi/2$.

 EXAMPLE 2   Finding the Angle Between Two Vectors

For $\mathbf{u} = \langle 3, -1, 2 \rangle$, $\mathbf{v} = \langle -4, 0, 2 \rangle$, $\mathbf{w} = \langle 1, -1, -2 \rangle$, and $\mathbf{z} = \langle 2, 0, -1 \rangle$, find the angle between each of the following pairs of vectors.

a. $\mathbf{u}$ and $\mathbf{v}$ **b.** $\mathbf{u}$ and $\mathbf{w}$ **c.** $\mathbf{v}$ and $\mathbf{z}$

Solution

a. $\cos \theta = \dfrac{\mathbf{u} \cdot \mathbf{v}}{\|\mathbf{u}\| \|\mathbf{v}\|} = \dfrac{-12 + 4}{\sqrt{14}\sqrt{20}} = \dfrac{-8}{2\sqrt{14}\sqrt{5}} = \dfrac{-4}{\sqrt{70}}$

   Because $\mathbf{u} \cdot \mathbf{v} < 0$, $\theta = \arccos \dfrac{-4}{\sqrt{70}} \approx 2.069$ radians.

b. $\cos \theta = \dfrac{\mathbf{u} \cdot \mathbf{w}}{\|\mathbf{u}\| \|\mathbf{w}\|} = \dfrac{3 + 1 - 4}{\sqrt{14}\sqrt{6}} = \dfrac{0}{\sqrt{84}} = 0$

   Because $\mathbf{u} \cdot \mathbf{w} = 0$, $\mathbf{u}$ and $\mathbf{w}$ are *orthogonal*. Furthermore, $\theta = \pi/2$.

c. $\cos \theta = \dfrac{\mathbf{v} \cdot \mathbf{z}}{\|\mathbf{v}\| \|\mathbf{z}\|} = \dfrac{-8 + 0 - 2}{\sqrt{20}\sqrt{5}} = \dfrac{-10}{\sqrt{100}} = -1$

   Consequently, $\theta = \pi$. Note that $\mathbf{v}$ and $\mathbf{z}$ are parallel, with $\mathbf{v} = -2\mathbf{z}$.

Direction Cosines

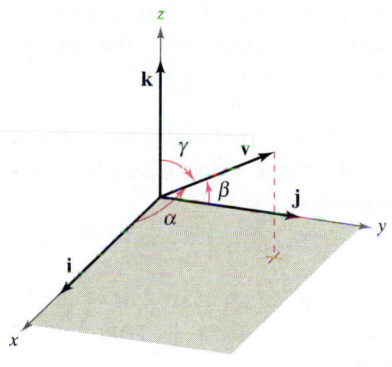

Direction angles
Figure 10.26

For a vector in the plane, you have seen that it is convenient to measure direction in terms of the angle, measured counterclockwise, *from* the positive *x*-axis *to* the vector. In space it is more convenient to measure direction in terms of the angles *between* the nonzero vector **v** and the three unit vectors **i**, **j**, and **k**, as shown in Figure 10.26. The angles α, β, and γ are the **direction angles of v**, and $\cos \alpha$, $\cos \beta$, and $\cos \gamma$ are the **direction cosines of v**. Because

$$\mathbf{v} \cdot \mathbf{i} = \|\mathbf{v}\| \|\mathbf{i}\| \cos \alpha = \|\mathbf{v}\| \cos \alpha$$

and

$$\mathbf{v} \cdot \mathbf{i} = \langle v_1, v_2, v_3 \rangle \cdot \langle 1, 0, 0 \rangle = v_1$$

it follows that $\cos \alpha = v_1/\|\mathbf{v}\|$. By similar reasoning with the unit vectors **j** and **k**, you have

$$\cos \alpha = \frac{v_1}{\|\mathbf{v}\|} \qquad \text{α is angle between $\mathbf{v}$ and $\mathbf{i}$.}$$

$$\cos \beta = \frac{v_2}{\|\mathbf{v}\|} \qquad \text{β is angle between $\mathbf{v}$ and $\mathbf{j}$.}$$

$$\cos \gamma = \frac{v_3}{\|\mathbf{v}\|}. \qquad \text{γ is angle between $\mathbf{v}$ and $\mathbf{k}$.}$$

Consequently, any nonzero vector **v** in space has the normalized form

$$\frac{\mathbf{v}}{\|\mathbf{v}\|} = \frac{v_1}{\|\mathbf{v}\|}\mathbf{i} + \frac{v_2}{\|\mathbf{v}\|}\mathbf{j} + \frac{v_3}{\|\mathbf{v}\|}\mathbf{k} = \cos \alpha\, \mathbf{i} + \cos \beta\, \mathbf{j} + \cos \gamma\, \mathbf{k}$$

and because $\mathbf{v}/\|\mathbf{v}\|$ is a unit vector, it follows that

$$\cos^2 \alpha + \cos^2 \beta + \cos^2 \gamma = 1.$$

EXAMPLE 3   Finding Direction Angles

Find the direction cosines and angles for the vector $\mathbf{v} = 2\mathbf{i} + 3\mathbf{j} + 4\mathbf{k}$, and show that $\cos^2 \alpha + \cos^2 \beta + \cos^2 \gamma = 1$.

α = angle between **v** and **i**
β = angle between **v** and **j**
γ = angle between **v** and **k**

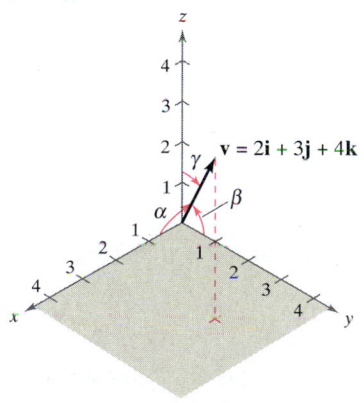

The direction angles of **v**
Figure 10.27

Solution   Because $\|\mathbf{v}\| = \sqrt{2^2 + 3^2 + 4^2} = \sqrt{29}$, you can write the following.

$$\cos \alpha = \frac{v_1}{\|\mathbf{v}\|} = \frac{2}{\sqrt{29}} \quad \Longrightarrow \quad \alpha \approx 68.2°$$

$$\cos \beta = \frac{v_2}{\|\mathbf{v}\|} = \frac{3}{\sqrt{29}} \quad \Longrightarrow \quad \beta \approx 56.1°$$

$$\cos \gamma = \frac{v_3}{\|\mathbf{v}\|} = \frac{4}{\sqrt{29}} \quad \Longrightarrow \quad \gamma \approx 42.0°$$

Furthermore, the sum of the squares of the direction cosines is

$$\cos^2 \alpha + \cos^2 \beta + \cos^2 \gamma = \frac{4}{29} + \frac{9}{29} + \frac{16}{29}$$

$$= \frac{29}{29}$$

$$= 1.$$

(See Figure 10.27.)

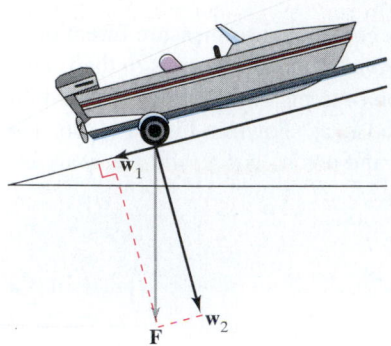

The force due to gravity pulls the boat against the ramp and down the ramp.
Figure 10.28

Projections and Vector Components

You have already seen applications in which two vectors are added to produce a resultant vector. Many applications in physics and engineering pose the reverse problem—decomposing a given vector into the sum of two **vector components.** To see the usefulness of this procedure, we look at a physical example.

Consider a boat on an inclined ramp, as shown in Figure 10.28. The force **F** due to gravity pulls the boat *down* the ramp and *against* the ramp. These two forces, $\mathbf{w}_1$ and $\mathbf{w}_2$, are orthogonal—they are called the **vector components** of **F.**

$$\mathbf{F} = \mathbf{w}_1 + \mathbf{w}_2 \qquad \text{Vector components of } \mathbf{F}$$

The forces $\mathbf{w}_1$ and $\mathbf{w}_2$ help you analyze the effect of gravity on the boat. For example, $\mathbf{w}_1$ indicates the force necessary to keep the boat from rolling down the ramp whereas $\mathbf{w}_2$ indicates the force that the tires must withstand.

Definition of Projection and Vector Components

Let **u** and **v** be nonzero vectors. Moreover, let $\mathbf{u} = \mathbf{w}_1 + \mathbf{w}_2$, where $\mathbf{w}_1$ is parallel to **v** and $\mathbf{w}_2$ is orthogonal to **v**, as shown in Figure 10.29.

1. $\mathbf{w}_1$ is called the **projection of u onto v** or the **vector component of u along v**, and is denoted by $\mathbf{w}_1 = \text{proj}_\mathbf{v}\mathbf{u}$.

2. $\mathbf{w}_2 = \mathbf{u} - \mathbf{w}_1$ is called the **vector component of u orthogonal to v.**

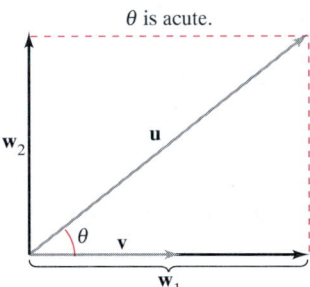
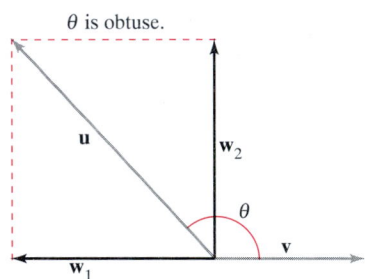

$\mathbf{w}_1 = \text{proj}_\mathbf{v}\mathbf{u} = $ projection of **u** onto **v** = vector component of **u** along **v**
$\mathbf{w}_2 = $ vector component of **u** orthogonal to **v**
Figure 10.29

EXAMPLE 4   **Finding a Vector Component of u Orthogonal to v**

Find the vector component of $\mathbf{u} = \langle 7, 4\rangle$ that is orthogonal to $\mathbf{v} = \langle 2, 3\rangle$, given that $\mathbf{w}_1 = \text{proj}_\mathbf{v}\mathbf{u} = \langle 4, 6\rangle$ and

$$\mathbf{u} = \langle 7, 4\rangle = \mathbf{w}_1 + \mathbf{w}_2.$$

Solution   Because $\mathbf{u} = \mathbf{w}_1 + \mathbf{w}_2$, where $\mathbf{w}_1$ is parallel to **v**, it follows that $\mathbf{w}_2$ is the vector component of **u** orthogonal to **v**. Thus, you have

$$\mathbf{w}_2 = \mathbf{u} - \mathbf{w}_1 = \langle 7, 4\rangle - \langle 4, 6\rangle = \langle 3, -2\rangle.$$

Check to see that $\mathbf{w}_2$ is orthogonal to **v**, as shown in Figure 10.30.

$\mathbf{u} = \mathbf{w}_1 + \mathbf{w}_2$
Figure 10.30

From Example 4, you can see that it is easy to find the vector component $\mathbf{w}_2$ once you have found the projection, $\mathbf{w}_1$, of $\mathbf{u}$ onto $\mathbf{v}$. To find this projection, use the dot product, shown in the theorem below and proven in Exercise 68.

NOTE Note the distinction between the terms "component" and "vector component." For example, using the standard unit vectors with $\mathbf{u} = u_1\mathbf{i} + u_2\mathbf{j}$, u_1 is the *component* of $\mathbf{u}$ in the direction of $\mathbf{i}$ and $u_1\mathbf{i}$ is the *vector component* in the direction of $\mathbf{i}$.

> **THEOREM 10.6   Projection Using the Dot Product**
>
> If $\mathbf{u}$ and $\mathbf{v}$ are nonzero vectors, then the projection of $\mathbf{u}$ onto $\mathbf{v}$ is given by
>
> $$\text{proj}_{\mathbf{v}}\mathbf{u} = \left(\frac{\mathbf{u} \cdot \mathbf{v}}{\|\mathbf{v}\|^2}\right)\mathbf{v}.$$

The projection of $\mathbf{u}$ onto $\mathbf{v}$ can be written as a scalar multiple of a unit vector in the direction of $\mathbf{v}$. That is,

$$\left(\frac{\mathbf{u} \cdot \mathbf{v}}{\|\mathbf{v}\|^2}\right)\mathbf{v} = \left(\frac{\mathbf{u} \cdot \mathbf{v}}{\|\mathbf{v}\|}\right)\frac{\mathbf{v}}{\|\mathbf{v}\|} = (k)\frac{\mathbf{v}}{\|\mathbf{v}\|} \quad\Longrightarrow\quad k = \frac{\mathbf{u} \cdot \mathbf{v}}{\|\mathbf{v}\|} = \|\mathbf{u}\|\cos\theta.$$

The scalar k is called the **component of u in the direction of v.**

EXAMPLE 5   Decomposing a Vector into Vector Components

Find the projection of $\mathbf{u}$ onto $\mathbf{v}$ and the vector component of $\mathbf{u}$ orthogonal to $\mathbf{v}$ for the vectors $\mathbf{u} = 3\mathbf{i} - 5\mathbf{j} + 2\mathbf{k}$ and $\mathbf{v} = 7\mathbf{i} + \mathbf{j} - 2\mathbf{k}$. (See Figure 10.31.)

Solution   The projection of $\mathbf{u}$ onto $\mathbf{v}$ is

$$\mathbf{w}_1 = \left(\frac{\mathbf{u} \cdot \mathbf{v}}{\|\mathbf{v}\|^2}\right)\mathbf{v} = \left(\frac{12}{54}\right)(7\mathbf{i} + \mathbf{j} - 2\mathbf{k}) = \frac{14}{9}\mathbf{i} + \frac{2}{9}\mathbf{j} - \frac{4}{9}\mathbf{k}.$$

The vector component of $\mathbf{u}$ orthogonal to $\mathbf{v}$ is the vector

$$\mathbf{w}_2 = \mathbf{u} - \mathbf{w}_1 = (3\mathbf{i} - 5\mathbf{j} + 2\mathbf{k}) - \left(\frac{14}{9}\mathbf{i} + \frac{2}{9}\mathbf{j} - \frac{4}{9}\mathbf{k}\right) = \frac{13}{9}\mathbf{i} - \frac{47}{9}\mathbf{j} + \frac{22}{9}\mathbf{k}.$$

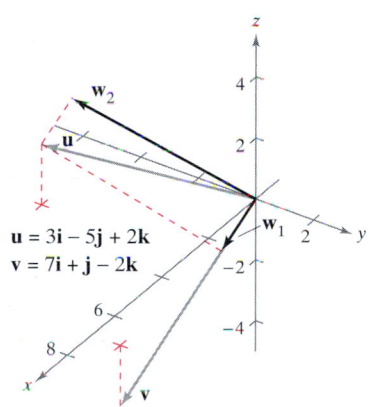

$\mathbf{u} = 3\mathbf{i} - 5\mathbf{j} + 2\mathbf{k}$
$\mathbf{v} = 7\mathbf{i} + \mathbf{j} - 2\mathbf{k}$

$\mathbf{u} = \mathbf{w}_1 + \mathbf{w}_2$
Figure 10.31

EXAMPLE 6   Finding a Force

A 600-pound boat sits on a ramp inclined at $30°$, as shown in Figure 10.32. What force is required to keep the boat from rolling down the ramp?

Solution   Because the force due to gravity is vertical and downward, you can represent the gravitational force by the vector $\mathbf{F} = -600\mathbf{j}$. To find the force required to keep the boat from rolling down the ramp, we project $\mathbf{F}$ onto a unit vector $\mathbf{v}$ in the direction of the ramp, as follows.

$$\mathbf{v} = \cos 30°\mathbf{i} + \sin 30°\mathbf{j} = \frac{\sqrt{3}}{2}\mathbf{i} + \frac{1}{2}\mathbf{j} \qquad \textcolor{red}{\text{Unit vector along ramp}}$$

Therefore, the projection of $\mathbf{F}$ onto $\mathbf{v}$ is given by

$$\mathbf{w}_1 = \text{proj}_{\mathbf{v}}\mathbf{F} = \left(\frac{\mathbf{F} \cdot \mathbf{v}}{\|\mathbf{v}\|^2}\right)\mathbf{v} = (\mathbf{F} \cdot \mathbf{v})\mathbf{v} = (-600)\left(\frac{1}{2}\right)\mathbf{v} = -300\left(\frac{\sqrt{3}}{2}\mathbf{i} + \frac{1}{2}\mathbf{j}\right).$$

The magnitude of this force is 300, and therefore a force of 300 pounds is required to keep the boat from rolling down the ramp.

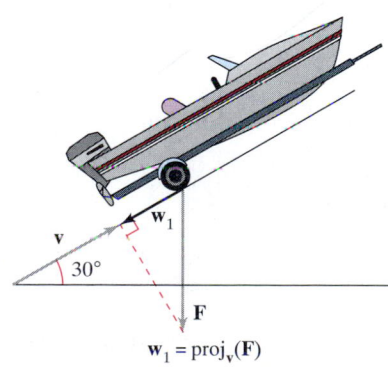

$\mathbf{w}_1 = \text{proj}_{\mathbf{v}}(\mathbf{F})$

Figure 10.32

Work

The work W done by the constant force $\mathbf{F}$ acting along the line of motion of an object is given by

$$W = (\text{magnitude of force})(\text{distance}) = \|\mathbf{F}\| \, \|\overrightarrow{PQ}\|$$

as shown in Figure 10.33(a). If the constant force $\mathbf{F}$ is not directed along the line of motion, you can see from Figure 10.33(b) that the work W done by the force is

$$W = \|\text{proj}_{\overrightarrow{PQ}} \mathbf{F}\| \, \|\overrightarrow{PQ}\| = (\cos\theta)\|\mathbf{F}\| \, \|\overrightarrow{PQ}\| = \mathbf{F} \cdot \overrightarrow{PQ}.$$

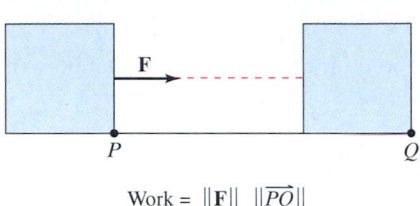

Work $= \|\mathbf{F}\| \; \|\overrightarrow{PQ}\|$

(a) Force acts along the line of motion.

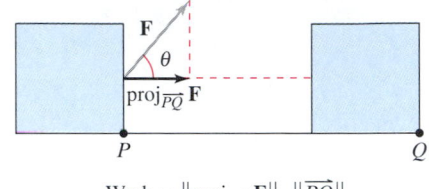

Work $= \|\text{proj}_{\overrightarrow{PQ}} \mathbf{F}\| \; \|\overrightarrow{PQ}\|$

(b) Force acts at angle θ with the line of motion.

Figure 10.33

We summarize this notion of work in the following definition.

Definition of Work

The work W done by a constant force $\mathbf{F}$ as its point of application moves along the vector $\overrightarrow{PQ}$ is given by either of the following.

1. $W = \|\text{proj}_{\overrightarrow{PQ}} \mathbf{F}\| \, \|\overrightarrow{PQ}\|$ Projection form

2. $W = \mathbf{F} \cdot \overrightarrow{PQ}$ Dot product form

---

EXAMPLE 7 Finding Work

To close a sliding door, a person pulls on a rope with a constant force of 50 pounds at a constant angle of 60°, as shown in Figure 10.34. Find the work done in moving the door 12 feet to its closed position.

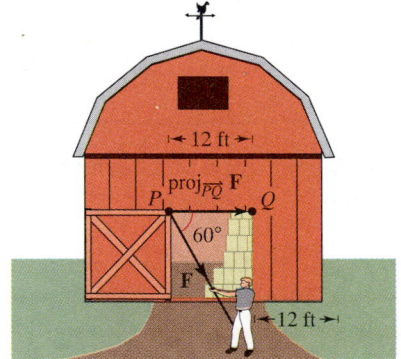

Figure 10.34

Solution Using a projection, you can calculate the work as follows.

$$\begin{aligned}
W &= \|\text{proj}_{\overrightarrow{PQ}} \mathbf{F}\| \, \|\overrightarrow{PQ}\| \\
&= \cos(60°)\|\mathbf{F}\| \, \|\overrightarrow{PQ}\| \\
&= \frac{1}{2}(50)(12) \\
&= 300 \text{ foot-pounds}
\end{aligned}$$

EXERCISES FOR SECTION 10.3

In Exercises 1–6, find (a) u · v, (b) u · u, (c) ‖u‖², (d) (u · v)v, and (e) u · (2v).

1. $u = \langle 3, 4 \rangle$
  $v = \langle 2, -3 \rangle$

2. $u = \langle 5, 12 \rangle$
  $v = \langle -3, 2 \rangle$

3. $u = \langle 2, -3, 4 \rangle$
  $v = \langle 0, 6, 5 \rangle$

4. $u = i$
  $v = i$

5. $u = 2i - j + k$
  $v = i - k$

6. $u = 2i + j - 2k$
  $v = i - 3j + 2k$

7. *Revenue*   The vector $u = \langle 3240, 1450, 2235 \rangle$ gives the number of units for products X, Y, and Z. The vector $v = \langle 2.22, 1.85, 3.25 \rangle$ gives the price (in dollars) per unit for each product. Find the dot product $u \cdot v$, and explain what information it gives.

8. *True or False?*   If $u \cdot v = u \cdot w$ and $u \neq 0$, then is it true that $v = w$? If it is false, explain why or give an example that shows it is false.

In Exercises 9 and 10, find u · v.

9. $\|u\| = 8$, $\|v\| = 5$, and the angle between u and v is $\pi/3$.

10. $\|u\| = 40$, $\|v\| = 25$, and the angle between u and v is $5\pi/6$.

In Exercises 11–18, find the angle θ between the vectors.

11. $u = \langle 1, 1 \rangle$, $v = \langle 2, -2 \rangle$

12. $u = \langle 3, 1 \rangle$, $v = \langle 2, -1 \rangle$

13. $u = 3i + j$
  $v = -2i + 4j$

14. $u = \cos\left(\dfrac{\pi}{6}\right)i + \sin\left(\dfrac{\pi}{6}\right)j$
  $v = \cos\left(\dfrac{3\pi}{4}\right)i + \sin\left(\dfrac{3\pi}{4}\right)j$

15. $u = \langle 1, 1, 1 \rangle$
  $v = \langle 2, 1, -1 \rangle$

16. $u = 2i + 3j + k$
  $v = -3i + 2j$

17. $u = 3i + 4j$
  $v = -2j + 3k$

18. $u = 2i - 3j + k$
  $v = i - 2j + k$

19. *Programming*   Given vectors u and v in component form, write a program for a graphing utility in which the output is (a) $\|u\|$, (b) $\|v\|$, and (c) the angle between u and v.

20.   Use Exercise 19 to find the angle between the vectors.
  (a) $u = \langle 3, 4 \rangle$, $v = \langle -7, 5 \rangle$
  (b) $u = \langle 8, -4, 2 \rangle$, $v = \langle 2, 5, 2 \rangle$

In Exercises 21–28, determine whether u and v are orthogonal, parallel, or neither.

21. $u = \langle 4, 0 \rangle$, $v = \langle 1, 1 \rangle$

22. $u = \langle 2, 18 \rangle$, $v = \left\langle \frac{3}{2}, -\frac{1}{6} \right\rangle$

23. $u = \langle 4, 3 \rangle$, $v = \left\langle \frac{1}{2}, -\frac{2}{3} \right\rangle$

24. $u = -\frac{1}{3}(i - 2j)$, $v = 2i - 4j$

25. $u = j + 6k$
  $v = i - 2j - k$

26. $u = -2i + 3j - k$
  $v = 2i + j - k$

27. $u = \langle 2, -3, 1 \rangle$
  $v = \langle -1, -1, -1 \rangle$

28. $u = \langle \cos\theta, \sin\theta, -1 \rangle$
  $v = \langle \sin\theta, -\cos\theta, 0 \rangle$

29.   Use vectors to prove that the diagonals of a rhombus are perpendicular.

30. *Bond Angle*   Consider a regular tetrahedron with vertices $(0, 0, 0)$, $(k, k, 0)$, $(k, 0, k)$, and $(0, k, k)$ where k is a positive real number.
  (a) Sketch the graph of the tetrahedron.
  (b) Find the length of each edge.
  (c) Use the dot product to find the angle between any two edges.
  (d) Find the angle between the line segments from the centroid $(k/2, k/2, k/2)$ to two vertices. This is the *bond angle* for a molecule such as CH_4, or $PbCl_4$ where the structure of the molecule is a tetrahedron.

31. *Think About It*   What is known about θ, the angle between two nonzero vectors u and v, if
  (a) $u \cdot v = 0$?   (b) $u \cdot v > 0$?   (c) $u \cdot v < 0$?

32. *True or False?*   If u and v are orthogonal to w, then is it true that $u + v$ is orthogonal to w? If it is true, prove it. Otherwise, explain why it is false or give an example that shows it is false.

33.   Consider the vectors $u = \langle \cos\alpha, \sin\alpha, 0 \rangle$ and $v = \langle \cos\beta, \sin\beta, 0 \rangle$, where $\alpha > \beta$. Find the dot product of the vectors and use the result to prove the identity
$$\cos(\alpha - \beta) = \cos\alpha \cos\beta + \sin\alpha \sin\beta.$$

34.   Consider the two curves $y_1 = x^2$ and $y_2 = x^{1/3}$. Find the unit tangent vectors to each curve at their points of intersection. Find the angles between the curves at their points of intersection.

In Exercises 35–38, find the direction cosines of u and demonstrate that the sum of the squares of the direction cosines is 1.

35. $u = i + 2j + 2k$

36. $u = 3i - j + 5k$

37. $u = \langle 0, 6, -4 \rangle$

38. $u = \langle a, b, c \rangle$

In Exercises 39 and 40, use a graphing utility to find the magnitude and direction angles of the resultant of forces F_1 and F_2 with initial points at the origin. The magnitude and terminal point of each vector are given.

	Vector	Magnitude	Terminal Point
39.	F_1	50 lb	$(10, 5, 3)$
	F_2	80 lb	$(12, 7, -5)$
40.	F_1	300 N	$(-20, -10, 5)$
	F_2	100 N	$(5, 15, 0)$

41. Find the angle between a cube's diagonal and one of its edges.

42. Find the angle between the diagonal of a cube and the diagonal of one of its sides.

43. *Load-Supporting Cables* A load is supported by three cables, as shown in the figure. Find the direction angles of the load-supporting cable *OA*.

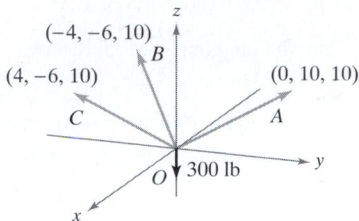

44. *Load-Supporting Cables* Determine the weight of the load if the tension in the cable *OA* in Exercise 43 is 200 newtons.

In Exercises 45–48, (a) find the projection of u onto v, and (b) find the vector component of u orthogonal to v.

45. $\mathbf{u} = \langle 2, 3 \rangle$, $\mathbf{v} = \langle 5, 1 \rangle$

46. $\mathbf{u} = \langle 2, -3 \rangle$, $\mathbf{v} = \langle 3, 2 \rangle$

47. $\mathbf{u} = \langle 2, 1, 2 \rangle$
$\mathbf{v} = \langle 0, 3, 4 \rangle$

48. $\mathbf{u} = \langle 0, 4, 1 \rangle$
$\mathbf{v} = \langle 0, 2, 3 \rangle$

49. *Programming* Given vectors **u** and **v** in component form, write a program for a graphing utility in which the output is the component form of the projection of **u** onto **v**.

50. Use the program you wrote in Exercise 49 to find the projection of **u** onto **v**.

(a) $\mathbf{u} = \langle 3, 4 \rangle$, $\mathbf{v} = \langle 8, 2 \rangle$

(b) $\mathbf{u} = \langle 5, 6, 2 \rangle$, $\mathbf{v} = \langle -1, 3, 4 \rangle$

Think About It **In Exercises 51 and 52, use the figure to mentally determine the projection of u onto v. (The coordinates of the terminal points of the vectors in standard position are given.) Verify your results analytically.**

51.
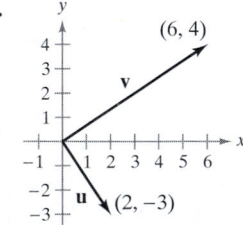

52.
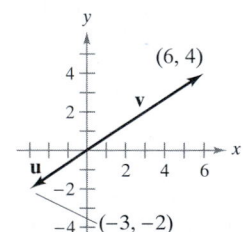

53. *Think About It* What can be said about the vectors **u** and **v** if (a) the projection of **u** onto **v** equals **u** and (b) the projection of **u** onto **v** equals **0**?

54. *Think About It* If the projection of **u** onto **v** has the same magnitude as the projection of **v** onto **u**, is it true that $\|\mathbf{u}\| = \|\mathbf{v}\|$?

In Exercises 55–58, find two vectors in opposite directions that are orthogonal to the vector u. (The answers are not unique.)

55. $\mathbf{u} = \frac{1}{2}\mathbf{i} - \frac{2}{3}\mathbf{j}$

56. $\mathbf{u} = -8\mathbf{i} + 3\mathbf{j}$

57. $\mathbf{u} = \langle 3, 1, -2 \rangle$

58. $\mathbf{u} = \langle 0, -3, 6 \rangle$

59. *Braking Load* A 32,000-pound truck is parked on a 15° slope (see figure). Assume the only force to overcome is that due to gravity. Find (a) the force required to keep the truck from rolling down the hill, and (b) the force perpendicular to the hill.

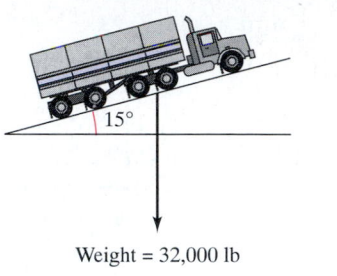

Weight = 32,000 lb

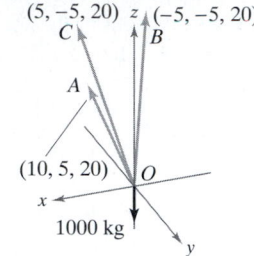

Figure for 59   **Figure for 60**

60. *Load-Supporting Cables* Find the magnitude of the projection of the load-supporting cable *OA* onto the positive *z*-axis as shown in the figure.

61. *Work* An object is pulled 10 feet across a floor, using a force of 85 pounds. Find the work done if the direction of the force is 60° above the horizontal (see figure).

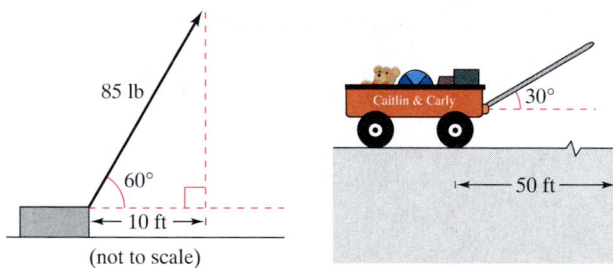

Figure for 61   **Figure for 62**

62. *Work* A toy wagon is pulled by exerting a force of 15 pounds on a handle that makes a 30° angle with the horizontal (see figure). Find the work done in pulling the wagon 50 feet.

Work **In Exercises 63 and 64, find the work done in moving a particle from *P* to *Q* if the magnitude and direction of the force are given by v.**

63. $P(0, 0, 0)$, $Q(4, 7, 5)$, $\mathbf{v} = \langle 1, 4, 8 \rangle$

64. $P(1, 3, 0)$, $Q(-3, 5, 10)$, $\mathbf{v} = -2\mathbf{i} + 3\mathbf{j} + 6\mathbf{k}$

65. Prove that $\|\mathbf{u} - \mathbf{v}\|^2 = \|\mathbf{u}\|^2 + \|\mathbf{v}\|^2 - 2\mathbf{u} \cdot \mathbf{v}$.

66. Prove the **Cauchy-Schwarz Inequality** $|\mathbf{u} \cdot \mathbf{v}| \leq \|\mathbf{u}\| \|\mathbf{v}\|$.

67. Prove the triangle inequality $\|\mathbf{u} + \mathbf{v}\| \leq \|\mathbf{u}\| + \|\mathbf{v}\|$.

68. Prove Theorem 10.6.

The Cross Product • The Triple Scalar Product

The Cross Product

Many applications in physics, engineering, and geometry involve finding a vector in space that is orthogonal to two given vectors. In this section you will study a product that will yield such a vector. It is called the **cross product,** and it is most conveniently defined and calculated using the standard unit vector form.

Definition of Cross Product of Two Vectors in Space

Let $\mathbf{u} = u_1\mathbf{i} + u_2\mathbf{j} + u_3\mathbf{k}$ and $\mathbf{v} = v_1\mathbf{i} + v_2\mathbf{j} + v_3\mathbf{k}$ be vectors in space. The **cross product** of $\mathbf{u}$ and $\mathbf{v}$ is the vector

$$\mathbf{u} \times \mathbf{v} = (u_2v_3 - u_3v_2)\mathbf{i} - (u_1v_3 - u_3v_1)\mathbf{j} + (u_1v_2 - u_2v_1)\mathbf{k}.$$

NOTE   Be sure you see that this definition applies only to three-dimensional vectors. The cross product is not defined for two-dimensional vectors.

A convenient way to calculate $\mathbf{u} \times \mathbf{v}$ is to use the following *determinant form* with cofactor expansion. (This 3×3 determinant form is used simply to help remember the formula for the cross product—it is technically not a determinant because the entries of the corresponding matrix are not all real numbers.)

$$\mathbf{u} \times \mathbf{v} = \begin{vmatrix} \mathbf{i} & \mathbf{j} & \mathbf{k} \\ u_1 & u_2 & u_3 \\ v_1 & v_2 & v_3 \end{vmatrix}$$

← Put "**u**" in Row 2.
← Put "**v**" in Row 3.

$$= \begin{vmatrix} \mathbf{i} & \mathbf{j} & \mathbf{k} \\ u_1 & u_2 & u_3 \\ v_1 & v_2 & v_3 \end{vmatrix}\mathbf{i} - \begin{vmatrix} \mathbf{i} & \mathbf{j} & \mathbf{k} \\ u_1 & u_2 & u_3 \\ v_1 & v_2 & v_3 \end{vmatrix}\mathbf{j} + \begin{vmatrix} \mathbf{i} & \mathbf{j} & \mathbf{k} \\ u_1 & u_2 & u_3 \\ v_1 & v_2 & v_3 \end{vmatrix}\mathbf{k}$$

$$= \begin{vmatrix} u_2 & u_3 \\ v_2 & v_3 \end{vmatrix}\mathbf{i} - \begin{vmatrix} u_1 & u_3 \\ v_1 & v_3 \end{vmatrix}\mathbf{j} + \begin{vmatrix} u_1 & u_2 \\ v_1 & v_2 \end{vmatrix}\mathbf{k}$$

$$= (u_2v_3 - u_3v_2)\mathbf{i} - (u_1v_3 - u_3v_1)\mathbf{j} + (u_1v_2 - u_2v_1)\mathbf{k}$$

Note the minus sign in front of the **j**-component. Each of the three 2×2 determinants can be evaluated by using the following diagonal pattern.

$$\begin{vmatrix} a & b \\ e & d \end{vmatrix} = ad - bc$$

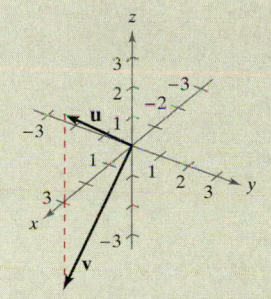
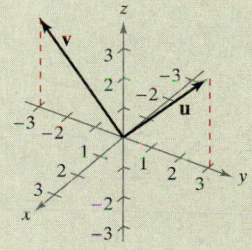
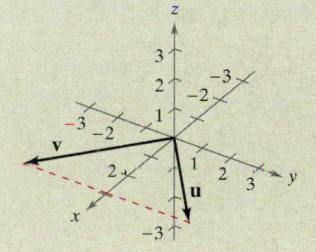

EXAMPLE 1   Finding the Cross Product

Given $\mathbf{u} = \mathbf{i} - 2\mathbf{j} + \mathbf{k}$ and $\mathbf{v} = 3\mathbf{i} + \mathbf{j} - 2\mathbf{k}$, find each of the following.

a. $\mathbf{u} \times \mathbf{v}$ **b.** $\mathbf{v} \times \mathbf{u}$ **c.** $\mathbf{v} \times \mathbf{v}$

Solution

$$\mathbf{a.}\ \mathbf{u} \times \mathbf{v} = \begin{vmatrix} \mathbf{i} & \mathbf{j} & \mathbf{k} \\ 1 & -2 & 1 \\ 3 & 1 & -2 \end{vmatrix} = \begin{vmatrix} -2 & 1 \\ 1 & -2 \end{vmatrix}\mathbf{i} - \begin{vmatrix} 1 & 1 \\ 3 & -2 \end{vmatrix}\mathbf{j} + \begin{vmatrix} 1 & -2 \\ 3 & 1 \end{vmatrix}\mathbf{k}$$

$$= (4 - 1)\mathbf{i} - (-2 - 3)\mathbf{j} + (1 + 6)\mathbf{k}$$

$$= 3\mathbf{i} + 5\mathbf{j} + 7\mathbf{k}$$

$$\mathbf{b.}\ \mathbf{v} \times \mathbf{u} = \begin{vmatrix} \mathbf{i} & \mathbf{j} & \mathbf{k} \\ 3 & 1 & -2 \\ 1 & -2 & 1 \end{vmatrix} = \begin{vmatrix} 1 & -2 \\ -2 & 1 \end{vmatrix}\mathbf{i} - \begin{vmatrix} 3 & -2 \\ 1 & 1 \end{vmatrix}\mathbf{j} + \begin{vmatrix} 3 & 1 \\ 1 & -2 \end{vmatrix}\mathbf{k}$$

$$= (1 - 4)\mathbf{i} - (3 + 2)\mathbf{j} + (-6 - 1)\mathbf{k}$$

$$= -3\mathbf{i} - 5\mathbf{j} - 7\mathbf{k}$$

Note that this result is the negative of that in part (a).

$$\mathbf{c.}\ \mathbf{v} \times \mathbf{v} = \begin{vmatrix} \mathbf{i} & \mathbf{j} & \mathbf{k} \\ 3 & 1 & -2 \\ 3 & 1 & -2 \end{vmatrix} = \mathbf{0}$$

The results obtained in Example 1 suggest some interesting *algebraic* properties of the cross product. For instance, $\mathbf{u} \times \mathbf{v} = -(\mathbf{v} \times \mathbf{u})$, and $\mathbf{v} \times \mathbf{v} = \mathbf{0}$. These properties, and several others, are summarized in the following theorem.

THEOREM 10.7   Algebraic Properties of the Cross Product

Let $\mathbf{u}$, $\mathbf{v}$, and $\mathbf{w}$ be vectors in space, and let c be a scalar.

1. $\mathbf{u} \times \mathbf{v} = -(\mathbf{v} \times \mathbf{u})$
2. $\mathbf{u} \times (\mathbf{v} + \mathbf{w}) = (\mathbf{u} \times \mathbf{v}) + (\mathbf{u} \times \mathbf{w})$
3. $c(\mathbf{u} \times \mathbf{v}) = (c\mathbf{u}) \times \mathbf{v} = \mathbf{u} \times (c\mathbf{v})$
4. $\mathbf{u} \times \mathbf{0} = \mathbf{0} \times \mathbf{u} = \mathbf{0}$
5. $\mathbf{u} \times \mathbf{u} = \mathbf{0}$
6. $\mathbf{u} \cdot (\mathbf{v} \times \mathbf{w}) = (\mathbf{u} \times \mathbf{v}) \cdot \mathbf{w}$

Proof   To prove Property 1, let $\mathbf{u} = u_1\mathbf{i} + u_2\mathbf{j} + u_3\mathbf{k}$ and $\mathbf{v} = v_1\mathbf{i} + v_2\mathbf{j} + v_3\mathbf{k}$. Then,

$$\mathbf{u} \times \mathbf{v} = (u_2v_3 - u_3v_2)\mathbf{i} - (u_1v_3 - u_3v_1)\mathbf{j} + (u_1v_2 - u_2v_1)\mathbf{k}$$

and

$$\mathbf{v} \times \mathbf{u} = (v_2u_3 - v_3u_2)\mathbf{i} - (v_1u_3 - v_3u_1)\mathbf{j} + (v_1u_2 - v_2u_1)\mathbf{k}$$

which implies that $\mathbf{u} \times \mathbf{v} = -(\mathbf{v} \times \mathbf{u})$. Proofs of Properties 2, 3, 5, and 6 are left as exercises (see Exercises 45–48).

Note that Property 1 in Theorem 10.7 indicates that the cross product is *not commutative.* In particular, this property indicates that the vectors $\mathbf{u} \times \mathbf{v}$ and $\mathbf{v} \times \mathbf{u}$ have equal lengths but opposite directions. The following theorem lists some other *geometric* properties of the cross product of two vectors.

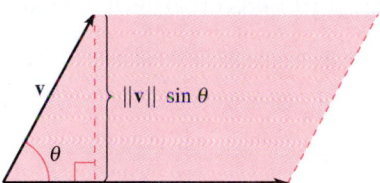

The vectors **u** and **v** form adjacent sides of a parallelogram.
Figure 10.35

> **THEOREM 10.8   Geometric Properties of the Cross Product**
>
> Let $\mathbf{u}$ and $\mathbf{v}$ be nonzero vectors in space, and let θ be the angle between $\mathbf{u}$ and $\mathbf{v}$.
>
> 1. $\mathbf{u} \times \mathbf{v}$ is orthogonal to both $\mathbf{u}$ and $\mathbf{v}$.
> 2. $\|\mathbf{u} \times \mathbf{v}\| = \|\mathbf{u}\|\|\mathbf{v}\| \sin \theta$.
> 3. $\mathbf{u} \times \mathbf{v} = \mathbf{0}$ if and only if $\mathbf{u}$ and $\mathbf{v}$ are scalar multiples of each other.
> 4. $\|\mathbf{u} \times \mathbf{v}\| =$ area of parallelogram having $\mathbf{u}$ and $\mathbf{v}$ as adjacent sides.

Proof   To prove Property 2, note that because $\cos \theta = (\mathbf{u} \cdot \mathbf{v})/(\|\mathbf{u}\| \|\mathbf{v}\|)$, it follows that

$$
\begin{aligned}
\|\mathbf{u}\| \|\mathbf{v}\| \sin \theta &= \|\mathbf{u}\| \|\mathbf{v}\| \sqrt{1 - \cos^2 \theta} \\
&= \|\mathbf{u}\| \|\mathbf{v}\| \sqrt{1 - \frac{(\mathbf{u} \cdot \mathbf{v})^2}{\|\mathbf{u}\|^2 \|\mathbf{v}\|^2}} \\
&= \sqrt{\|\mathbf{u}\|^2 \|\mathbf{v}\|^2 - (\mathbf{u} \cdot \mathbf{v})^2} \\
&= \sqrt{(u_1^2 + u_2^2 + u_3^2)(v_1^2 + v_2^2 + v_3^2) - (u_1v_1 + u_2v_2 + u_3v_3)^2} \\
&= \sqrt{(u_2v_3 - u_3v_2)^2 + (u_1v_3 - u_3v_1)^2 + (u_1v_2 - u_2v_1)^2} \\
&= \|\mathbf{u} \times \mathbf{v}\|.
\end{aligned}
$$

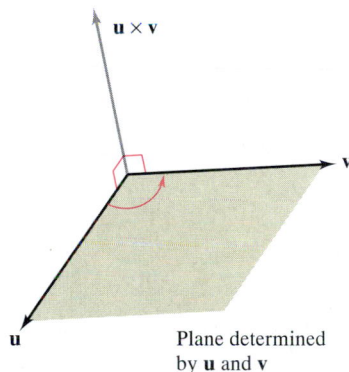

Plane determined by **u** and **v**

To prove Property 4, refer to Figure 10.35, which is a parallelogram having $\mathbf{v}$ and $\mathbf{u}$ as adjacent sides. Because the height of the parallelogram is $\|\mathbf{v}\| \sin \theta$, the area is

$$
\begin{aligned}
\text{Area} &= (\text{base})(\text{height}) \\
&= \|\mathbf{u}\| \|\mathbf{v}\| \sin \theta \\
&= \|\mathbf{u} \times \mathbf{v}\|.
\end{aligned}
$$

Proofs of Properties 1 and 3 are left as exercises (see Exercises 49 and 50).

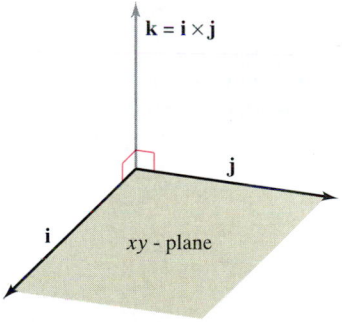

Right-handed systems
Figure 10.36

NOTE   It follows from Properties 1 and 2 in Theorem 10.8 that if $\mathbf{n}$ is a unit vector orthogonal to both $\mathbf{u}$ and $\mathbf{v}$, then

$$
\mathbf{u} \times \mathbf{v} = \pm(\|\mathbf{u}\| \|\mathbf{v}\| \sin \theta)\mathbf{n}.
$$

Both $\mathbf{u} \times \mathbf{v}$ and $\mathbf{v} \times \mathbf{u}$ are perpendicular to the plane determined by $\mathbf{u}$ and $\mathbf{v}$. One way to remember the orientation of the vectors $\mathbf{u}$, $\mathbf{v}$, and $\mathbf{u} \times \mathbf{v}$ is to compare them with the unit vectors $\mathbf{i}$, $\mathbf{j}$, and $\mathbf{k} = \mathbf{i} \times \mathbf{j}$, as shown in Figure 10.36. The three vectors $\mathbf{u}$, $\mathbf{v}$, and $\mathbf{u} \times \mathbf{v}$ form a *right-handed system,* whereas the three vectors $\mathbf{u}$, $\mathbf{v}$, and $\mathbf{v} \times \mathbf{u}$ form a *left-handed system.*

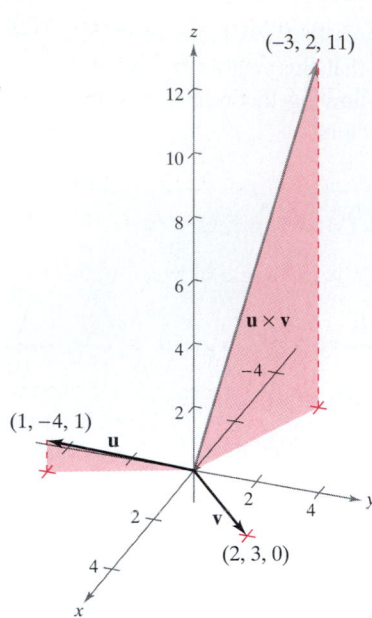

The vector $\mathbf{u} \times \mathbf{v}$ is orthogonal to both $\mathbf{u}$ and $\mathbf{v}$.

Figure 10.37

EXAMPLE 2 Using the Cross Product

Find a unit vector that is orthogonal to both

$$\mathbf{u} = \mathbf{i} - 4\mathbf{j} + \mathbf{k} \quad \text{and} \quad \mathbf{v} = 2\mathbf{i} + 3\mathbf{j}.$$

Solution The cross product $\mathbf{u} \times \mathbf{v}$, as shown in Figure 10.37, is orthogonal to both $\mathbf{u}$ and $\mathbf{v}$.

$$\mathbf{u} \times \mathbf{v} = \begin{vmatrix} \mathbf{i} & \mathbf{j} & \mathbf{k} \\ 1 & -4 & 1 \\ 2 & 3 & 0 \end{vmatrix}$$

$$= -3\mathbf{i} + 2\mathbf{j} + 11\mathbf{k}$$

Because $\|\mathbf{u} \times \mathbf{v}\| = \sqrt{(-3)^2 + 2^2 + 11^2} = \sqrt{134}$, a unit vector orthogonal to both $\mathbf{u}$ and $\mathbf{v}$ is

$$\frac{\mathbf{u} \times \mathbf{v}}{\|\mathbf{u} \times \mathbf{v}\|} = -\frac{3}{\sqrt{134}}\mathbf{i} + \frac{2}{\sqrt{134}}\mathbf{j} + \frac{11}{\sqrt{134}}\mathbf{k}.$$

NOTE In Example 2, note that you could have used the cross product $\mathbf{v} \times \mathbf{u}$ to form a unit vector that is orthogonal to both $\mathbf{u}$ and $\mathbf{v}$. With that choice, you would have obtained the negative of the unit vector found in the example.

EXAMPLE 3 Geometric Application of the Cross Product

Show that the quadrilateral with vertices at the following points is a parallelogram, and find its area.

$$A = (5, 2, 0) \qquad B = (2, 6, 1)$$
$$C = (2, 4, 7) \qquad D = (5, 0, 6)$$

Solution From Figure 10.38 you can see that the sides of the quadrilateral correspond to the following four vectors.

$$\overrightarrow{AB} = -3\mathbf{i} + 4\mathbf{j} + \mathbf{k} \qquad \overrightarrow{CD} = 3\mathbf{i} - 4\mathbf{j} - \mathbf{k} = -\overrightarrow{AB}$$
$$\overrightarrow{AD} = 0\mathbf{i} - 2\mathbf{j} + 6\mathbf{k} \qquad \overrightarrow{CB} = 0\mathbf{i} + 2\mathbf{j} - 6\mathbf{k} = -\overrightarrow{AD}$$

Thus, $\overrightarrow{AB}$ is parallel to $\overrightarrow{CD}$ and $\overrightarrow{AD}$ is parallel to $\overrightarrow{CB}$, and you can conclude that the quadrilateral is a parallelogram with $\overrightarrow{AB}$ and $\overrightarrow{AD}$ as adjacent sides. Moreover, because

$$\overrightarrow{AB} \times \overrightarrow{AD} = \begin{vmatrix} \mathbf{i} & \mathbf{j} & \mathbf{k} \\ -3 & 4 & 1 \\ 0 & -2 & 6 \end{vmatrix}$$

$$= 26\mathbf{i} + 18\mathbf{j} + 6\mathbf{k}$$

the area of the parallelogram is

$$\|\overrightarrow{AB} \times \overrightarrow{AD}\| = \sqrt{1036} \approx 32.19.$$

Is the parallelogram a rectangle? You can tell whether it is by finding the angle between the vectors $\overrightarrow{AB}$ and $\overrightarrow{AD}$.

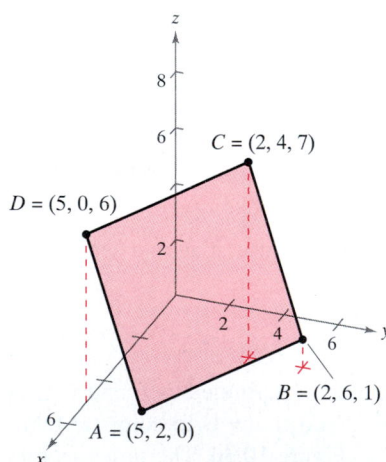

The area of the parallelogram is approximately 32.19.

Figure 10.38

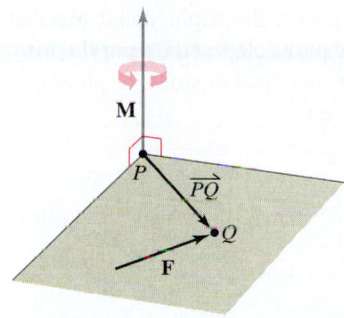

The moment of **F** about P
Figure 10.39

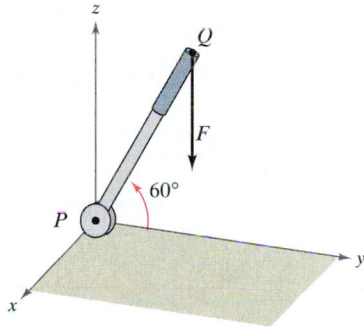

A vertical force of 50 pounds is applied at point Q.
Figure 10.40

In physics, the cross product can be used to measure **torque**—the **moment M of a force F about a point P**, as shown in Figure 10.39. If the point of application of the force is Q, the moment of **F** about P is given by

$$\mathbf{M} = \overrightarrow{PQ} \times \mathbf{F}. \qquad \text{Moment of \textbf{F} about } P.$$

The magnitude of the moment **M** measures the tendency of the vector $\overrightarrow{PQ}$ to rotate counterclockwise (using the right-hand rule) about an axis directed along the vector **M**.

EXAMPLE 4   An Application of the Cross Product

A vertical force of 50 pounds is applied to the end of a 1-foot lever that is attached to an axle at point P, as shown in Figure 10.40. Find the moment of this force about the point P when $\theta = 60°$.

Solution   If you represent the 50-pound force as $\mathbf{F} = -50\mathbf{k}$ and the lever as

$$\overrightarrow{PQ} = \cos(60°)\mathbf{j} + \sin(60°)\mathbf{k} = \frac{1}{2}\mathbf{j} + \frac{\sqrt{3}}{2}\mathbf{k}$$

the moment of **F** about P is given by

$$\mathbf{M} = \overrightarrow{PQ} \times \mathbf{F} = \begin{vmatrix} \mathbf{i} & \mathbf{j} & \mathbf{k} \\ 0 & \dfrac{1}{2} & \dfrac{\sqrt{3}}{2} \\ 0 & 0 & -50 \end{vmatrix} = -25\mathbf{i}.$$

The magnitude of this moment is 25 foot-pounds.

NOTE   In Example 4, note that the moment (the tendency of the lever to rotate about its axle) is dependent on the angle θ. When $\theta = \pi/2$, the moment is **0**. The moment is greatest when $\theta = 0$.

The Triple Scalar Product

For vectors **u**, **v**, and **w** in space, the dot product of **u** and $\mathbf{v} \times \mathbf{w}$

$$\mathbf{u} \cdot (\mathbf{v} \times \mathbf{w})$$

is called the **triple scalar product**. The proof of this theorem is left as an exercise (see Exercise 53).

NOTE   The value of a determinant is multiplied by -1 if two rows are interchanged. After two such interchanges, the value of the determinant will be unchanged. Thus, the following triple scalar products are equivalent.

$$\mathbf{u} \cdot (\mathbf{v} \times \mathbf{w}) =$$
$$\mathbf{v} \cdot (\mathbf{w} \times \mathbf{u}) =$$
$$\mathbf{w} \cdot (\mathbf{u} \times \mathbf{v})$$

> ### THEOREM 10.9   The Triple Scalar Product
>
> For $\mathbf{u} = u_1\mathbf{i} + u_2\mathbf{j} + u_3\mathbf{k}$, $\mathbf{v} = v_1\mathbf{i} + v_2\mathbf{j} + v_3\mathbf{k}$, and $\mathbf{w} = w_1\mathbf{i} + w_2\mathbf{j} + w_3\mathbf{k}$, the triple scalar product is given by
>
> $$\mathbf{u} \cdot (\mathbf{v} \times \mathbf{w}) = \begin{vmatrix} u_1 & u_2 & u_3 \\ v_1 & v_2 & v_3 \\ w_1 & w_2 & w_3 \end{vmatrix}.$$

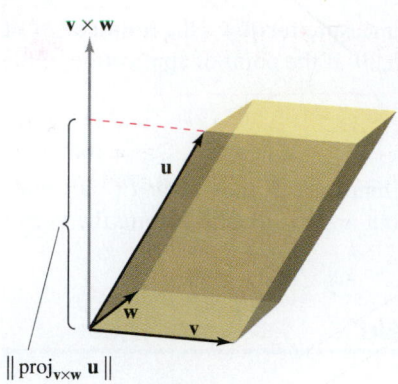

$\mathbf{v} \times \mathbf{w}$

u

w   v

$\|\text{proj}_{\mathbf{v} \times \mathbf{w}} \mathbf{u}\|$

Area of base $= \|\mathbf{v} \times \mathbf{w}\|$
Volume of parallelepiped $= |\mathbf{u} \cdot (\mathbf{v} \times \mathbf{w})|$
Figure 10.41

If the vectors **u**, **v**, and **w** do not lie in the same plane, the triple scalar product **u** · (**v** × **w**) can be used to determine the volume of the parallelepiped (a polyhedron, all of whose faces are parallelograms) with **u**, **v**, and **w** as adjacent sides, as shown in Figure 10.41. This is established in the following theorem.

> **THEOREM 10.10   Geometric Property of Triple Scalar Product**
>
> The volume V of a parallelepiped with vectors **u**, **v**, and **w** as adjacent sides is given by
>
> $$V = |\mathbf{u} \cdot (\mathbf{v} \times \mathbf{w})|.$$

Proof   In Figure 10.41, note that

$$\|\mathbf{v} \times \mathbf{w}\| = \text{area of base} \quad \text{and} \quad \|\text{proj}_{\mathbf{v} \times \mathbf{w}} \mathbf{u}\| = \text{height of parallelepiped.}$$

Therefore, the volume is

$$
\begin{aligned}
V &= (\text{height})(\text{area of base}) = \|\text{proj}_{\mathbf{v} \times \mathbf{w}} \mathbf{u}\| \, \|\mathbf{v} \times \mathbf{w}\| \\
&= \frac{|\mathbf{u} \cdot (\mathbf{v} \times \mathbf{w})|}{\|\mathbf{v} \times \mathbf{w}\|} \|\mathbf{v} \times \mathbf{w}\| \\
&= |\mathbf{u} \cdot (\mathbf{v} \times \mathbf{w})|.
\end{aligned}
$$

EXAMPLE 5   Volume by the Triple Scalar Product

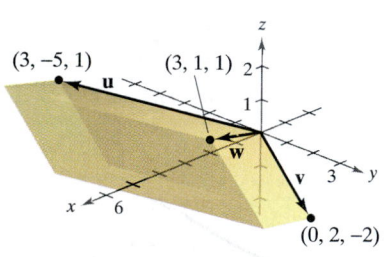

(3, −5, 1)   (3, 1, 1)
u
w
v
x   6
(0, 2, −2)

The parallelepiped has a volume of 36.
Figure 10.42

Find the volume of the parallelepiped having $\mathbf{u} = 3\mathbf{i} - 5\mathbf{j} + \mathbf{k}$, $\mathbf{v} = 2\mathbf{j} - 2\mathbf{k}$, and $\mathbf{w} = 3\mathbf{i} + \mathbf{j} + \mathbf{k}$ as adjacent edges (see Figure 10.42).

Solution   By Theorem 10.10, you have

$$
\begin{aligned}
V &= |\mathbf{u} \cdot (\mathbf{v} \times \mathbf{w})| \\
&= \begin{vmatrix} 3 & -5 & 1 \\ 0 & 2 & -2 \\ 3 & 1 & 1 \end{vmatrix} \\
&= 3 \begin{vmatrix} 2 & -2 \\ 1 & 1 \end{vmatrix} - (-5) \begin{vmatrix} 0 & -2 \\ 3 & 1 \end{vmatrix} + (1) \begin{vmatrix} 0 & 2 \\ 3 & 1 \end{vmatrix} \\
&= 3(4) + 5(6) = 1(-6) \\
&= 36.
\end{aligned}
$$

A natural consequence of Theorem 10.10 is that the volume of the parallelepiped is 0 if and only if the three vectors are coplanar. That is, if the vectors $\mathbf{u} = \langle u_1, u_2, u_3 \rangle$, $\mathbf{v} = \langle v_1, v_2, v_3 \rangle$, and $\mathbf{w} = \langle w_1, w_2, w_3 \rangle$ have the same initial point, they lie in the same plane if and only if

$$\mathbf{u} \cdot (\mathbf{v} \times \mathbf{w}) = \begin{vmatrix} u_1 & u_2 & u_3 \\ v_1 & v_2 & v_3 \\ w_1 & w_2 & w_3 \end{vmatrix} = 0.$$

EXERCISES FOR SECTION 10.4

In Exercises 1–6, find the cross product of the unit vectors and sketch your result.

1. $j \times i$

2. $i \times j$

3. $j \times k$

4. $k \times j$

5. $i \times k$

6. $k \times i$

In Exercises 7–12, find $u \times v$ and show that it is orthogonal to both u and v.

7. $u = \langle 2, -3, 1 \rangle$
   $v = \langle 1, -2, 1 \rangle$

8. $u = \langle -1, 1, 2 \rangle$
   $v = \langle 0, 1, 0 \rangle$

9. $u = \langle 12, -3, 0 \rangle$
   $v = \langle -2, 5, 0 \rangle$

10. $u = \langle -10, 0, 6 \rangle$
    $v = \langle 7, 0, 0 \rangle$

11. $u = i + j + k$
    $v = 2i + j - k$

12. $u = j + 6k$
    $v = i - 2j + k$

Think About It   In Exercises 13–16, use the vectors u and v shown in the figure to sketch a vector in the direction of the indicated cross product in a right-handed system.

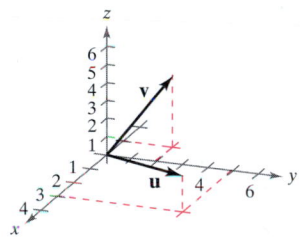

13. $u \times v$

14. $v \times u$

15. $(-v) \times u$

16. $u \times (u \times v)$

In Exercises 17–20, use a computer algebra system to find $u \times v$ and a unit vector orthogonal to u and v.

17. $u = \langle 4, -3.5, 7 \rangle$
    $v = \langle -1, 8, 4 \rangle$

18. $u = \langle -8, -6, 4 \rangle$
    $v = \langle 10, -12, -2 \rangle$

19. $u = -3i + 2j - 5k$
    $v = \frac{1}{2}i - \frac{3}{4}j + \frac{1}{10}k$

20. $u = \frac{2}{3}k$
    $v = \frac{1}{2}i + 6k$

21. *Programming*   Given the vectors u and v in component form, write a program for a graphing utility in which the output is $u \times v$ and $\|u \times v\|$.

22. Use the program you wrote in Exercise 21 to find $u \times v$ and $\|u \times v\|$.

(a) $u = \langle 8, -4, 2 \rangle$
    $v = \langle 2, 5, 2 \rangle$

(b) $u = \langle -2, 6, 10 \rangle$
    $v = \langle 3, 8, 5 \rangle$

Area   In Exercises 23–26, find the area of the parallelogram that has the given vectors as adjacent sides. Use a computer algebra system or a graphing utility to verify your result.

23. $u = j$
    $v = j + k$

24. $u = i + j + k$
    $v = j + k$

25. $u = \langle 3, 2, -1 \rangle$
    $v = \langle 1, 2, 3 \rangle$

26. $u = \langle 2, -1, 0 \rangle$
    $v = \langle -1, 2, 0 \rangle$

Area   In Exercises 27 and 28, verify that the points are the vertices of a parallelogram, and find its area.

27. $(1, 1, 1)$, $(2, 3, 4)$, $(6, 5, 2)$, $(7, 7, 5)$

28. $(2, -1, 1)$, $(5, 1, 4)$, $(0, 1, 1)$, $(3, 3, 4)$

Area   In Exercises 29–32, find the area of the triangle with the given vertices. (*Hint:* $\frac{1}{2}\|u \times v\|$ is the area of the triangle having u and v as adjacent sides.)

29. $(0, 0, 0)$, $(1, 2, 3)$, $(-3, 0, 0)$

30. $(2, -3, 4)$, $(0, 1, 2)$, $(-1, 2, 0)$

31. $(1, 3, 5)$, $(3, 3, 0)$, $(-2, 0, 5)$

32. $(1, 2, 0)$, $(-2, 1, 0)$, $(0, 0, 0)$

In Exercises 33–36, find $u \cdot (v \times w)$.

33. $u = i$
    $v = j$
    $w = k$

34. $u = \langle 1, 1, 1 \rangle$
    $v = \langle 2, 1, 0 \rangle$
    $w = \langle 0, 0, 1 \rangle$

35. $u = \langle 2, 0, 1 \rangle$
    $v = \langle 0, 3, 0 \rangle$
    $w = \langle 0, 0, 1 \rangle$

36. $u = \langle 2, 0, 0 \rangle$
    $v = \langle 1, 1, 1 \rangle$
    $w = \langle 0, 2, 2 \rangle$

Volume   In Exercises 37 and 38, use the triple scalar product to find the volume of the parallelepiped having adjacent edges u, v, and w.

37. $u = i + j$
    $v = j + k$
    $w = i + k$

38. $u = \langle 1, 3, 1 \rangle$
    $v = \langle 0, 5, 5 \rangle$
    $w = \langle 4, 0, 4 \rangle$

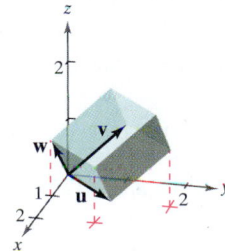

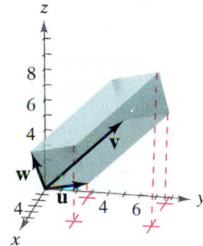

Volume **In Exercises 39 and 40, find the volume of the parallelepiped with the given vertices (see figures).**

39. $(0, 0, 0)$, $(3, 0, 0)$, $(0, 5, 1)$, $(3, 5, 1)$
$(2, 0, 5)$, $(5, 0, 5)$, $(2, 5, 6)$, $(5, 5, 6)$

40. $(0, 0, 0)$, $(1, 1, 0)$, $(1, 0, 2)$, $(0, 1, 1)$
$(2, 1, 2)$, $(1, 1, 3)$, $(1, 2, 1)$, $(2, 2, 3)$

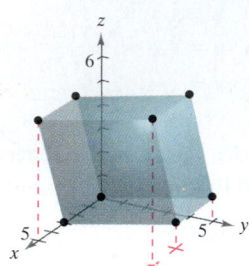

Figure for 39   **Figure for 40**

41. ***Torque***   A child applies the brakes on a bicycle by applying a downward force of 20 pounds on the pedal when the crank makes a 40° angle with the horizontal (see figure). Find the torque at P if the crank is 6 inches in length.

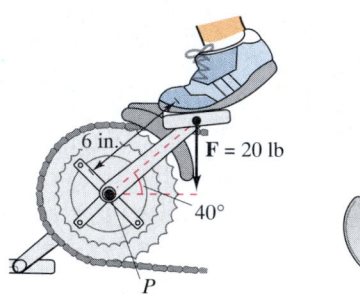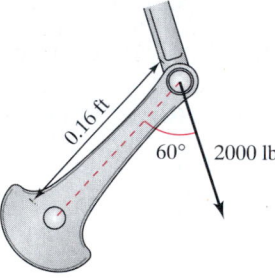

Figure for 41   **Figure for 42**

42. ***Torque***   Both the magnitude and the direction of the force on a crankshaft change as the crankshaft rotates. Find the torque on the crankshaft using the position and data shown in the figure.

43. ***Optimization***   A force of 200 pounds acts on the bracket shown in the figure.

(a) Determine the vector $\overrightarrow{AB}$ and the vector $\mathbf{F}$ representing the force. ($\mathbf{F}$ will be in terms of θ.)

(b) Find the magnitude of the moment about A by evaluating $\|\overrightarrow{AB} \times \mathbf{F}\|$.

(c) Use the result of part (b) to determine the magnitude of the moment when $\theta = 30°$.

(d) Use the result of part (b) to determine the angle θ when the magnitude of the moment is maximum. At that angle, what is the relationship between the vectors $\mathbf{F}$ and $\overrightarrow{AB}$? Is it what you expected? Why or why not?

(e) Use a graphing utility to graph the function for the magnitude of the moment about A for $0° \le \theta \le 180°$. Find the zero of the function in the given domain. Interpret the meaning of the zero in the context of the problem.

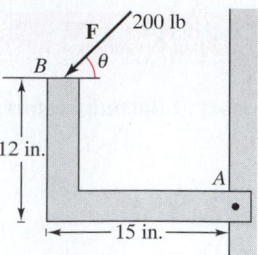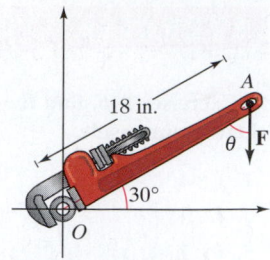

Figure for 43   **Figure for 44**

44. ***Optimization***   A force of 60 pounds acts on the pipe wrench shown in the figure.

(a) Find the magnitude of the moment about O by evaluating $\|\overrightarrow{OA} \times \mathbf{F}\|$. Use a graphing utility to graph the resulting function of θ.

(b) Use the result of part (a) to determine the magnitude of the moment when $\theta = 45°$.

(c) Use the result of part (a) to determine the angle θ when the magnitude of the moment is maximum. Is the answer what you expected? Why or why not?

In Exercises 45–52, prove the property of the cross product.

45. $\mathbf{u} \times (\mathbf{v} + \mathbf{w}) = (\mathbf{u} \times \mathbf{v}) + (\mathbf{u} \times \mathbf{w})$

46. $c(\mathbf{u} \times \mathbf{v}) = (c\mathbf{u}) \times \mathbf{v} = \mathbf{u} \times (c\mathbf{v})$

47. $\mathbf{u} \times \mathbf{u} = \mathbf{0}$

48. $\mathbf{u} \cdot (\mathbf{v} \times \mathbf{w}) = (\mathbf{u} \times \mathbf{v}) \cdot \mathbf{w}$

49. $\mathbf{u} \times \mathbf{v}$ is orthogonal to both $\mathbf{u}$ and $\mathbf{v}$.

50. $\mathbf{u} \times \mathbf{v} = \mathbf{0}$ if and only if $\mathbf{u}$ and $\mathbf{v}$ are scalar multiples of each other.

51. $\|\mathbf{u} \times \mathbf{v}\| = \|\mathbf{u}\| \|\mathbf{v}\|$ if $\mathbf{u}$ and $\mathbf{v}$ are orthogonal.

52. $\mathbf{u} \times (\mathbf{v} \times \mathbf{w}) = (\mathbf{u} \cdot \mathbf{w})\mathbf{v} - (\mathbf{u} \cdot \mathbf{v})\mathbf{w}$

53. Prove Theorem 10.9.

54. ***Think About It***   If the magnitudes of two vectors are doubled, how will the magnitude of the cross product of the vectors change? Explain.

55. ***Think About It***   The vertices of a triangle in space are (x_1, y_1, z_1), (x_2, y_2, z_2), and (x_3, y_3, z_3). Explain how to find a vector perpendicular to the triangle.

56. ***True or False?***   It is possible to find the cross product of two vectors in the plane. Explain.

57. Consider the vectors $\mathbf{u} = \langle \cos \alpha, \sin \alpha, 0 \rangle$ and $\mathbf{v} = \langle \cos \beta, \sin \beta, 0 \rangle$, where $\alpha > \beta$. Find the cross product of the vectors and use the result to prove the identity

$$\sin(\alpha - \beta) = \sin \alpha \cos \beta - \cos \alpha \sin \beta.$$

58. ***Writing***   Read the article "Tooth Tables: Solution of a Dental Problem by Vector Algebra" by Gary Hosler Meisters in the November 1982 issue of *Mathematics Magazine*. Then write a paragraph explaining how vectors and vector algebra can be used in the construction of dental inlays.

Lines in Space • Planes in Space • Sketching Planes in Space •
Distances Between Points, Planes, and Lines

Lines in Space

In the plane, *slope* is used to determine an equation of a line. In space, it is more convenient to use *vectors* to determine the equation of a line.

In Figure 10.43, consider the line L through the point $P(x_1, y_1, z_1)$ and parallel to the vector $\mathbf{v} = \langle a, b, c \rangle$. The vector $\mathbf{v}$ is the **direction vector** for the line L, and a, b, and c are the **direction numbers**. One way of describing the line L is to say that it consists of all points $Q(x, y, z)$ for which the vector $\overrightarrow{PQ}$ is parallel to $\mathbf{v}$. This means that $\overrightarrow{PQ}$ is a scalar multiple of $\mathbf{v}$, and you can write $\overrightarrow{PQ} = t\mathbf{v}$, where t is scalar (a real number).

direction # not unit vector

$$\overrightarrow{PQ} = \langle x - x_1, y - y_1, z - z_1 \rangle = \langle at, bt, ct \rangle = t\mathbf{v}$$

By equating corresponding components, you can obtain the **parametric equations** of a line in space.

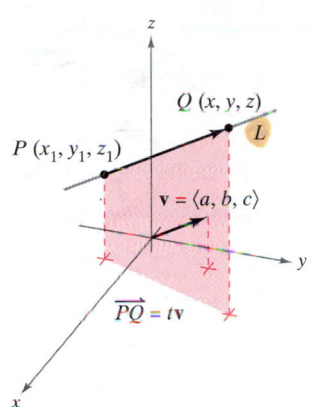

Line L and its direction vector $\mathbf{v}$
Figure 10.43

THEOREM 10.11   Parametric Equations of a Line in Space

A line L parallel to the vector $\mathbf{v} = \langle a, b, c \rangle$ and passing through the point $P(x_1, y_1, z_1)$ is represented by the **parametric equations**

$$x = x_1 + at, \quad y = y_1 + bt, \quad \text{and} \quad z = z_1 + ct.$$

If the direction numbers a, b, and c are all nonzero, you can eliminate the parameter t to obtain the **symmetric equations** of a line.

$$\frac{x - x_1}{a} = \frac{y - y_1}{b} = \frac{z - z_1}{c}$$ Symmetric equations

EXAMPLE 1   Finding Parametric and Symmetric Equations

Find parametric and symmetric equations of the line L that passes through the point $(1, -2, 4)$ and is parallel to $\mathbf{v} = \langle 2, 4, -4 \rangle$.

Solution To find a set of parametric equations of the line, use the coordinates $x_1 = 1$, $y_1 = -2$, and $z_1 = 4$ and direction numbers $a = 2$, $b = 4$, and $c = -4$ (see Figure 10.44).

$$x = 1 + 2t, \quad y = -2 + 4t, \quad z = 4 - 4t$$ Parametric equations

Because a, b, and c are all nonzero, a set of symmetric equations is

$$\frac{x - 1}{2} = \frac{y + 2}{4} = \frac{z - 4}{-4}.$$ Symmetric equations

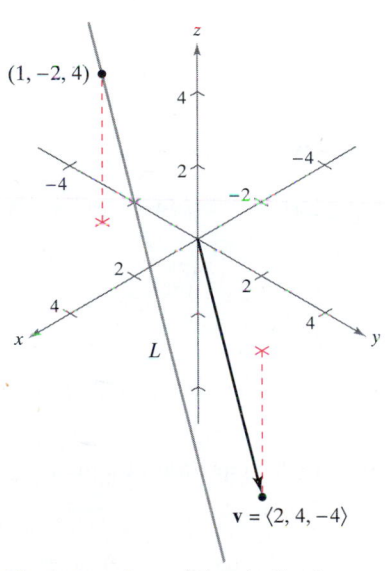

The vector $\mathbf{v}$ is parallel to the line L.
Figure 10.44

Neither the parametric equations nor the symmetric equations of a given line are unique. For instance, in Example 1, by letting $t = 1$ in the parametric equations you would obtain the point $(3, 2, 0)$. Using this point with the direction numbers $a = 2$, $b = 4$, and $c = -4$ would produce the different parametric equations

$$x = 3 + 2t, \quad y = 2 + 4t, \quad \text{and} \quad z = -4t.$$

EXAMPLE 2   Parametric Equations of a Line Through Two Points

Find a set of parametric equations of the line that passes through the points $(-2, 1, 0)$ and $(1, 3, 5)$.

Solution   Begin by using the points $P(-2, 1, 0)$ and $Q(1, 3, 5)$ to find a direction vector for the line passing through P and Q, given by

$$\mathbf{v} = \overrightarrow{PQ} = \langle 1 - (-2), 3 - 1, 5 - 0 \rangle = \langle 3, 2, 5 \rangle = \langle a, b, c \rangle.$$

Using the direction numbers $a = 3$, $b = 2$, and $c = 5$ with the point $P(-2, 1, 0)$, you can obtain the parametric equations

$$x = -2 + 3t, \quad y = 1 + 2t, \quad \text{and} \quad z = 5t.$$

NOTE   As t varies over all real numbers, the parametric equations in Example 2 determine the points (x, y, z) on the line. In particular, note that $t = 0$ and $t = 1$ give the original points $(-2, 1, 0)$ and $(1, 3, 5)$.

Planes in Space

You have seen how an equation of a line in space can be obtained from a point on the line and a vector *parallel* to it. You will now see that an equation of a plane in space can be obtained from a point in the plane and a vector *normal* (perpendicular) to it.

Consider the plane containing the point $P(x_1, y_1, z_1)$ having a nonzero normal vector $\mathbf{n} = \langle a, b, c \rangle$, as shown in Figure 10.45. This plane consists of all points $Q(x, y, z)$ for which vector $\overrightarrow{PQ}$ is orthogonal to $\mathbf{n}$. Using the dot product, you can write the following.

$$\mathbf{n} \cdot \overrightarrow{PQ} = 0$$
$$\langle a, b, c \rangle \cdot \langle x - x_1, y - y_1, z - z_1 \rangle = 0$$
$$a(x - x_1) + b(y - y_1) + c(z - z_1) = 0$$

The third equation of the plane is said to be in **standard form.**

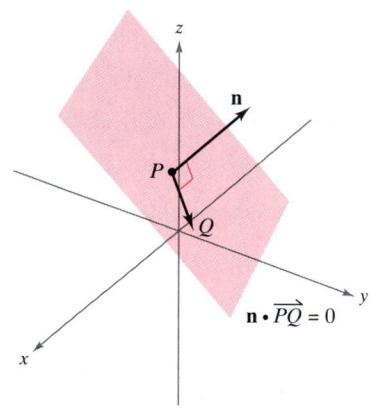

The normal vector $\mathbf{n}$ is orthogonal to each vector $\overrightarrow{PQ}$ in the plane.
Figure 10.45

THEOREM 10.12   Standard Equation of a Plane in Space

The plane containing the point (x_1, y_1, z_1) and having a normal vector $\mathbf{n} = \langle a, b, c \rangle$ can be represented, in **standard form,** by the equation

$$a(x - x_1) + b(y - y_1) + c(z - z_1) = 0.$$

By regrouping terms, you obtain the **general form** of the equation of a plane in space,

$$ax + by + cz + d = 0. \qquad \text{General form of equation of plane}$$

Given the general form of the equation of a plane, it is easy to find a normal vector to the plane. Simply use the coefficients of x, y, and z and write $\mathbf{n} = \langle a, b, c \rangle$.

EXAMPLE 3   Finding an Equation of a Plane in Three-Space

Find the general equation of the plane containing the points $(2, 1, 1)$, $(0, 4, 1)$, and $(-2, 1, 4)$.

Solution   To apply Theorem 10.12 you need a point in the plane and a vector that is normal to the plane. There are three choices for the point, but no normal vector is given. To obtain a normal vector, use the cross product of vectors $\mathbf{u}$ and $\mathbf{v}$ extending from the point $(2, 1, 1)$ to the points $(0, 4, 1)$ and $(-2, 1, 4)$, as shown in Figure 10.46. The component forms of $\mathbf{u}$ and $\mathbf{v}$ are

$$\mathbf{u} = \langle 0 - 2, 4 - 1, 1 - 1 \rangle = \langle -2, 3, 0 \rangle$$
$$\mathbf{v} = \langle -2 - 2, 1 - 1, 4 - 1 \rangle = \langle -4, 0, 3 \rangle$$

and it follows that

$$\mathbf{n} = \mathbf{u} \times \mathbf{v}$$
$$= \begin{vmatrix} \mathbf{i} & \mathbf{j} & \mathbf{k} \\ -2 & 3 & 0 \\ -4 & 0 & 3 \end{vmatrix}$$
$$= 9\mathbf{i} + 6\mathbf{j} + 12\mathbf{k}$$
$$= \langle a, b, c \rangle$$

is normal to the given plane. Using the direction numbers for $\mathbf{n}$ and the point $(x_1, y_1, z_1) = (2, 1, 1)$, you can determine an equation of the plane to be

$$a(x - x_1) + b(y - y_1) + c(z - z_1) = 0$$
$$9(x - 2) + 6(y - 1) + 12(z - 1) = 0 \qquad \textcolor{red}{\text{Standard form}}$$
$$9x + 6y + 12z - 36 = 0$$
$$3x + 2y + 4z - 12 = 0. \qquad \textcolor{red}{\text{General form}}$$

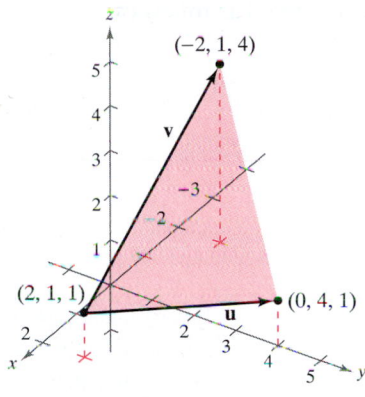

A plane determined by $\mathbf{u}$ and $\mathbf{v}$
Figure 10.46

NOTE   In Example 3, check to see that each of the three original points satisfies the equation $3x + 2y + 4z - 12 = 0$.

Two distinct planes in three-space either are parallel or intersect in a line. If they intersect, you can determine the angle between them from the angle between their normal vectors, as shown in Figure 10.47. Specifically, if vectors $\mathbf{n}_1$ and $\mathbf{n}_2$ are normal to two intersecting planes, the angle θ between the normal vectors is equal to the angle between the two planes and is given by

$$\cos\theta = \frac{|\mathbf{n}_1 \cdot \mathbf{n}_2|}{\|\mathbf{n}_1\|\,\|\mathbf{n}_2\|}. \qquad \textcolor{red}{\text{Angle between two planes}}$$

Consequently, two planes with normal vectors $\mathbf{n}_1$ and $\mathbf{n}_2$ are

1. *perpendicular* if $\mathbf{n}_1 \cdot \mathbf{n}_2 = 0$.

2. *parallel* if $\mathbf{n}_1$ is a scalar multiple of $\mathbf{n}_2$.

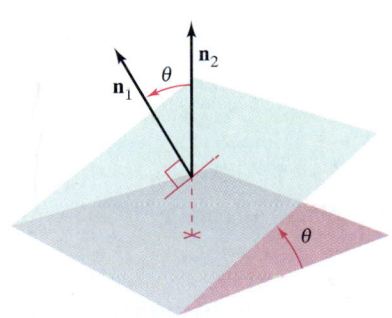

The angle θ between two planes
Figure 10.47

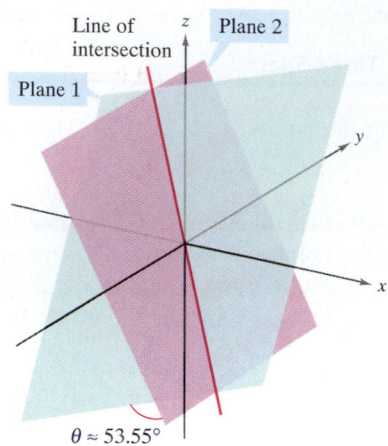

Line of intersection

The angle between the planes is approximately 53.55°.
Figure 10.48

EXAMPLE 4 Finding the Line of Intersection of Two Planes

Find the angle between the two planes given by

$$x - 2y + z = 0 \qquad \text{Equation for plane 1}$$
$$2x + 3y - 2z = 0 \qquad \text{Equation for plane 2}$$

and find parametric equations of their line of intersection (see Figure 10.48).

Solution The normal vectors for the planes are $\mathbf{n}_1 = \langle 1, -2, 1 \rangle$ and $\mathbf{n}_2 = \langle 2, 3, -2 \rangle$. Consequently, the angle between the two planes is determined as follows.

$$\cos \theta = \frac{|\mathbf{n}_1 \cdot \mathbf{n}_2|}{\|\mathbf{n}_1\| \|\mathbf{n}_2\|} \qquad \text{Cosine of angle between } \mathbf{n}_1 \text{ and } \mathbf{n}_2$$

$$= \frac{|-6|}{\sqrt{6}\sqrt{17}}$$

$$= \frac{6}{\sqrt{102}}$$

$$\approx 0.59409 \qquad \theta \approx \arccos 0.59409$$

This implies that the angle between the two planes is $\theta \approx 53.55°$. You can find the line of intersection of the two planes by simultaneously solving the two linear equations representing the planes. One way to do this is to multiply the first equation by -2 and add the result to the second equation.

$$
\begin{array}{lll}
x - 2y + z = 0 & \Longrightarrow & -2x + 4y - 2z = 0 \\
2x + 3y - 2z = 0 & & \underline{2x + 3y - 2z = 0} \\
& & 7y - 4z = 0 \quad \Longrightarrow \quad y = \dfrac{4z}{7}
\end{array}
$$

Substituting $y = 4z/7$ back into one of the original equations, you can determine that $x = z/7$. Finally, by letting $t = z/7$, you obtain the parametric equations

$$x = t, \quad y = 4t, \quad \text{and} \quad z = 7t \qquad \text{Line of intersection}$$

which indicate that 1, 4, and 7 are direction numbers for the line of intersection.

Note that the direction numbers in Example 4 can be obtained from the cross product of the two normal vectors as follows.

$$
\mathbf{n}_1 \times \mathbf{n}_2 = \begin{vmatrix} \mathbf{i} & \mathbf{j} & \mathbf{k} \\ 1 & -2 & 1 \\ 2 & 3 & -2 \end{vmatrix}
$$

$$
= \begin{vmatrix} -2 & 1 \\ 3 & -2 \end{vmatrix} \mathbf{i} - \begin{vmatrix} 1 & 1 \\ 2 & -2 \end{vmatrix} \mathbf{j} + \begin{vmatrix} 1 & -2 \\ 2 & 3 \end{vmatrix} \mathbf{k}
$$

$$
= \mathbf{i} + 4\mathbf{j} + 7\mathbf{k}
$$

This means that the line of intersection of the two planes is parallel to the cross product of their normal vectors.

NOTE The three-dimensional rotating software that accompanies this text can help you visualize surfaces as those shown in Figure 10.48. If you have access to this software, we suggest that you use it to help your spatial intuition in this section and in all other sections in the text that deal with vectors, curves, or surfaces in space.

Sketching Planes in Space

If a plane in space intersects one of the coordinate planes, we call the line of intersection the **trace** of the given plane in the coordinate plane. To sketch a plane in space, it is helpful to find its points of intersection with the coordinate axes and its traces in the coordinate planes. For example, consider the plane given by

$$3x + 2y + 4z = 12. \qquad \text{Equation of plane}$$

We find the xy-trace by letting $z = 0$ and sketching the line

$$3x + 2y = 12 \qquad \text{xy-trace}$$

in the xy-plane. This line intersects the x-axis at $(4, 0, 0)$ and the y-axis at $(0, 6, 0)$. In Figure 10.49, we continue this process by finding the yz-trace and the xz-trace, and then shading in the triangular region lying in the first octant.

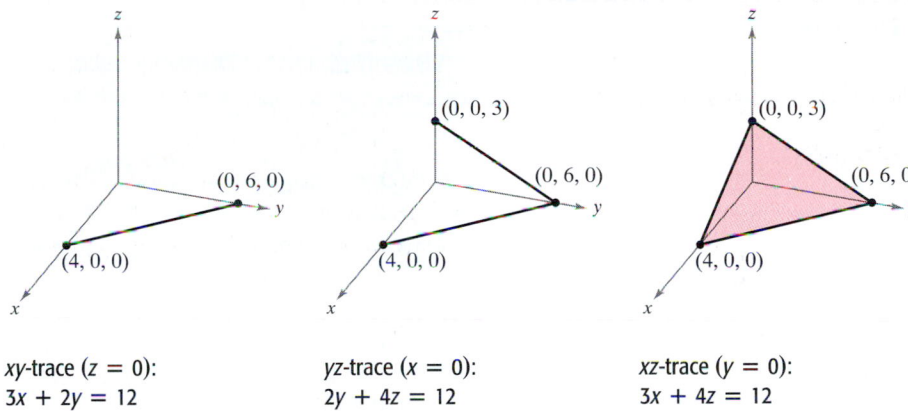

xy-trace ($z = 0$):
$3x + 2y = 12$

yz-trace ($x = 0$):
$2y + 4z = 12$

xz-trace ($y = 0$):
$3x + 4z = 12$

Traces of the plane $3x + 2y + 4z = 12$
Figure 10.49

If the equation of a plane has a missing variable such as $2x + z = 1$, the plane must be *parallel to the axis* represented by the missing variable, as shown in Figure 10.50. If two variables are missing from the equation of a plane, it is *parallel to the coordinate plane* represented by the missing variables, as shown in Figure 10.51.

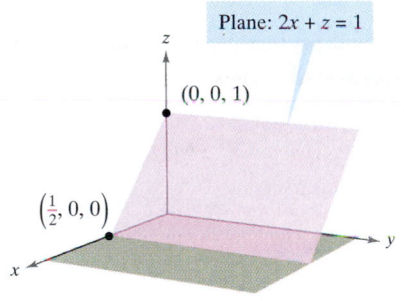

Figure 10.50

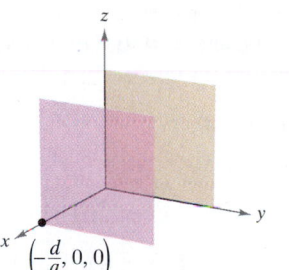

Plane $ax + d = 0$ is
parallel to yz-plane.
Figure 10.51

Plane $by + d = 0$ is
parallel to xz-plane.

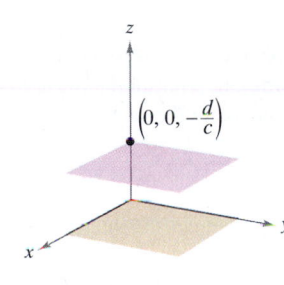

Plane $cz + d = 0$ is
parallel to xy-plane.

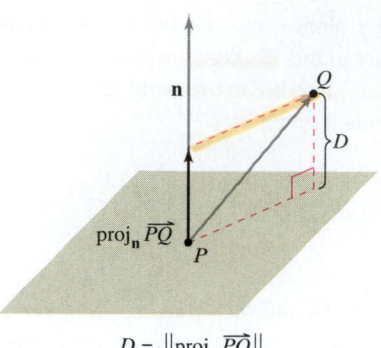

$$D = \|\text{proj}_\mathbf{n} \overrightarrow{PQ}\|$$

The distance between a point and a plane
Figure 10.52

Distances Between Points, Planes, and Lines

We conclude this section with a discussion of two basic types of problems involving distance in space.

1. Finding the distance between a point and a plane
2. Finding the distance between a point and a line

The solutions of these problems illustrate the versatility and usefulness of vectors in coordinate geometry: the first problem uses the *dot product* of two vectors, and the second problem uses the *cross product*.

The distance D between a point Q and a plane is the length of the shortest line segment connecting Q to the plane, as shown in Figure 10.52. If P is *any* point in the plane, you can find this distance by projecting the vector $\overrightarrow{PQ}$ onto the normal vector $\mathbf{n}$. The length of this projection is the desired distance.

> ### THEOREM 10.13   Distance Between a Point and a Plane
>
> The distance between a plane and a point Q (not in the plane) is
>
> $$D = \|\text{proj}_\mathbf{n} \overrightarrow{PQ}\| = \frac{|\overrightarrow{PQ} \cdot \mathbf{n}|}{\|\mathbf{n}\|}$$
>
> where P is a point in the plane and $\mathbf{n}$ is normal to the plane.

To find a point in the plane given by $ax + by + cz + d = 0$ $(a \neq 0)$, let $y = 0$ and $z = 0$. Then, from the equation $ax + d = 0$, you can conclude that the point $(-d/a, 0, 0)$ lies in the plane.

EXAMPLE 5   Finding the Distance Between a Point and a Plane

Find the distance between the point $Q(1, 5, -4)$ and the plane given by

$$3x - y + 2z = 6.$$

Solution   You know that $\mathbf{n} = \langle 3, -1, 2 \rangle$ is normal to the given plane. To find a point in the plane, let $y = 0$ and $z = 0$, and obtain the point $P(2, 0, 0)$. The vector from P to Q is given by

$$\overrightarrow{PQ} = \langle 1 - 2, 5 - 0, -4 - 0 \rangle$$
$$= \langle -1, 5, -4 \rangle.$$

Using the distance formula given in Theorem 10.13 produces

$$D = \frac{|\overrightarrow{PQ} \cdot \mathbf{n}|}{\|\mathbf{n}\|} = \frac{|\langle -1, 5, -4 \rangle \cdot \langle 3, -1, 2 \rangle|}{\sqrt{9 + 1 + 4}}$$
$$= \frac{|-3 - 5 - 8|}{\sqrt{14}}$$
$$= \frac{16}{\sqrt{14}}.$$

NOTE   The choice of the point P in Example 5 is arbitrary. Try choosing a different point in the plane to verify that you obtain the same distance.

From Theorem 10.13, you can determine that the distance between the point $Q(x_0, y_0, z_0)$ and the plane given by $ax + by + cz + d = 0$ is

$$D = \frac{|a(x_0 - x_1) + b(y_0 - y_1) + c(z_0 - z_1)|}{\sqrt{a^2 + b^2 + c^2}}$$

or

$$D = \frac{|ax_0 + by_0 + cz_0 + d|}{\sqrt{a^2 + b^2 + c^2}}$$ Distance between point and plane

where $P(x_1, y_1, z_1)$ is a point in the plane and $d = -(ax_1 + by_1 + cz_1)$.

EXAMPLE 6   Finding the Distance Between Two Parallel Planes

Find the distance between the two parallel planes given by

$$3x - y + 2z - 6 = 0 \quad \text{and} \quad 6x - 2y + 4z + 4 = 0.$$

Solution   The two planes are shown in Figure 10.53. To find the distance between the planes, choose a point in the first plane, say $(x_0, y_0, z_0) = (2, 0, 0)$. Then, from the second plane, you can determine that $a = 6$, $b = -2$, $c = 4$, and $d = 4$, and conclude that the distance is

$$D = \frac{|ax_0 + by_0 + cz_0 + d|}{\sqrt{a^2 + b^2 + c^2}}$$

$$= \frac{|6(2) + (-2)(0) + (4)(0) + 4|}{\sqrt{6^2 + (-2)^2 + 4^2}} = \frac{16}{\sqrt{56}} = \frac{8}{\sqrt{14}} \approx 2.14.$$

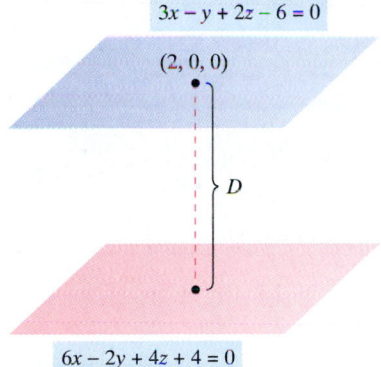

$3x - y + 2z - 6 = 0$

$(2, 0, 0)$

D

$6x - 2y + 4z + 4 = 0$

The distance between the parallel planes is approximately 2.14.
Figure 10.53

The formula for the distance between a point and a line in space resembles that for the distance between a point and a plane—except that you replace the dot product with the cross product and the normal vector **n** with a direction vector for the line.

THEOREM 10.14   Distance Between a Point and a Line in Space

The distance between a point Q and a line in space is given by

$$D = \frac{\|\overrightarrow{PQ} \times \mathbf{u}\|}{\|\mathbf{u}\|}$$

where **u** is the direction vector for the line and P is a point on the line.

Proof   In Figure 10.54, let D be the distance between the point Q and the given line. Then $D = \|\overrightarrow{PQ}\| \sin \theta$, where θ is the angle between **u** and $\overrightarrow{PQ}$. By Theorem 10.8, you have

$$\|\mathbf{u}\| \|\overrightarrow{PQ}\| \sin \theta = \|\mathbf{u} \times \overrightarrow{PQ}\| = \|\overrightarrow{PQ} \times \mathbf{u}\|.$$

Consequently,

$$D = \|\overrightarrow{PQ}\| \sin \theta = \frac{\|\overrightarrow{PQ} \times \mathbf{u}\|}{\|\mathbf{u}\|}.$$

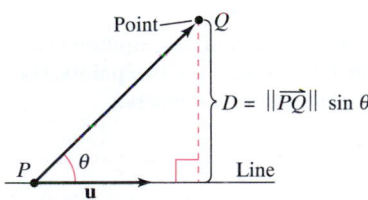

Point Q

$D = \|\overrightarrow{PQ}\| \sin \theta$

P θ Line

u

The distance between a point and a line
Figure 10.54

EXAMPLE 7   Finding the Distance Between a Point and a Line

Find the distance between the point $Q(3, -1, 4)$ and the line given by

$$x = -2 + 3t, \quad y = -2t, \quad \text{and} \quad z = 1 + 4t.$$

Solution   Using the direction numbers 3, -2, and 4, you know that the direction vector for the line is

$$\mathbf{u} = \langle 3, -2, 4 \rangle. \qquad \qquad \text{Direction vector for line}$$

To find a point on the line, let $t = 0$ and obtain

$$P = (-2, 0, 1). \qquad \qquad \text{Point on the line}$$

Thus,

$$\overrightarrow{PQ} = \langle 3 - (-2), -1 - 0, 4 - 1 \rangle = \langle 5, -1, 3 \rangle$$

and you can form the cross product

$$\overrightarrow{PQ} \times \mathbf{u} = \begin{vmatrix} \mathbf{i} & \mathbf{j} & \mathbf{k} \\ 5 & -1 & 3 \\ 3 & -2 & 4 \end{vmatrix} = 2\mathbf{i} - 11\mathbf{j} - 7\mathbf{k} = \langle 2, -11, -7 \rangle.$$

Finally, using Theorem 10.14, you can find the distance to be

$$D = \frac{\| \overrightarrow{PQ} \times \mathbf{u} \|}{\| \mathbf{u} \|} = \frac{\sqrt{174}}{\sqrt{29}} = \sqrt{6} \approx 2.45. \qquad \text{(See Figure 10.55.)}$$

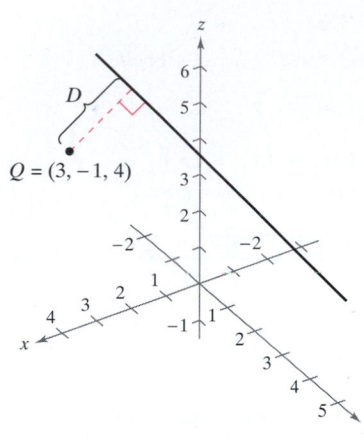

The distance between the point Q and the line is $\sqrt{6} \approx 2.45$.
Figure 10.55

EXERCISES FOR SECTION 10.5

In Exercises 1 and 2, the figure shows the graph of a line given by the parametric equations. (a) Draw an arrow on the line to indicate its orientation. (b) Find the coordinates of two points, P and Q, on the line. Determine the vector $\overrightarrow{PQ}$. What is the relationship between the components of the vector and the coefficients of t in the parametric equations? Why is this true? (c) Determine the coordinates of any points of intersection with the coordinate planes. If the line does not intersect a coordinate plane, explain why.

1. $x = 1 + 3t$
   $y = 2 - t$
   $z = 2 + 5t$

2. $x = 2 - 3t$
   $y = 2$
   $z = 1 - t$

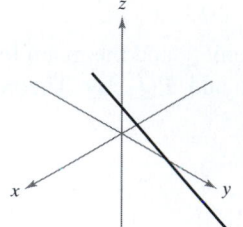

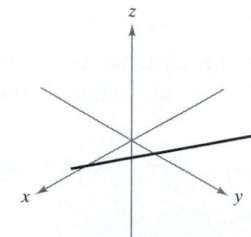

In Exercises 3–8, find sets of (a) parametric equations and (b) symmetric equations of the line through the point parallel to the indicated vector or line. (For each line, express the direction numbers as integers.)

Point	Parallel to
3. $(0, 0, 0)$	$\mathbf{v} = \langle 1, 2, 3 \rangle$
4. $(0, 0, 0)$	$\mathbf{v} = \langle -2, \frac{5}{2}, 1 \rangle$
5. $(-2, 0, 3)$	$\mathbf{v} = 2\mathbf{i} + 4\mathbf{j} - 2\mathbf{k}$
6. $(-2, 0, 3)$	$\mathbf{v} = 6\mathbf{i} + 3\mathbf{j}$
7. $(1, 0, 1)$	$x = 3 + 3t, y = 5 - 2t, z = -7 + t$
8. $(-3, 5, 4)$	$\dfrac{x - 1}{3} = \dfrac{y + 1}{-2} = z - 3$

In Exercises 9 and 10, find sets of (a) parametric equations and (b) symmetric equations of the line through the two points. (For each line, express the direction numbers as integers.)

9. $(5, -3, -2), \left(-\frac{2}{3}, \frac{2}{3}, 1\right)$

10. $(1, 0, 1), (1, 3, -2)$

In Exercises 11 and 12, find a set of parametric equations of the line.

11. The line passes through the point $(2, 3, 4)$ and is parallel to the xz-plane and the yz-plane.

12. The line passes through the point $(2, 3, 4)$ and is perpendicular to the plane given by $3x + 2y - z = 6$.

In Exercises 13 and 14, determine which points lie on the line L.

13. The line L passes through the point $(-2, 3, 1)$ and is parallel to the vector $\mathbf{v} = 4\mathbf{i} - \mathbf{k}$.

  (a) $(2, 3, 0)$ (b) $(-6, 3, 2)$ (c) $(2, 1, 0)$ (d) $(6, 3, -2)$

14. The line L passes through the points $(2, 0, -3)$ and $(4, 2, -2)$.

  (a) $(4, 1, -2)$ (b) $\left(\frac{5}{2}, \frac{1}{2}, -\frac{11}{4}\right)$ (c) $(-1, -3, -4)$

In Exercises 15–18, determine whether the lines intersect, and if so, find the point of intersection and the cosine of the angle of intersection.

15. $x = 4t + 2, y = 3, z = -t + 1$

  $x = 2s + 2, y = 2s + 3, z = s + 1$

16. $x = -3t + 1, y = 4t + 1, z = 2t + 4$

  $x = 3s + 1, y = 2s + 4, z = -s + 1$

17. $\dfrac{x}{3} = \dfrac{y - 2}{-1} = z + 1, \quad \dfrac{x - 1}{4} = y + 2 = \dfrac{z + 3}{-3}$

18. $\dfrac{x - 2}{-3} = \dfrac{y - 2}{6} = z - 3, \quad \dfrac{x - 3}{2} = y + 5 = \dfrac{z + 2}{4}$

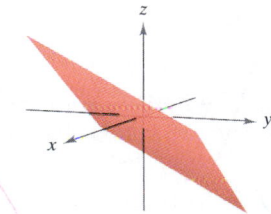 **In Exercises 19 and 20, use a computer algebra system to graph the pair of intersecting lines and find the point of intersection.**

19. $x = 2t + 3, y = 5t - 2, z = -t + 1$

  $x = -2s + 7, y = s + 8, z = 2s - 1$

20. $x = 2t - 1, y = -4t + 10, z = t$

  $x = -5s - 12, y = 3s + 11, z = -2s - 4$

Cross Product **In Exercises 21 and 22, (a) find the coordinates of three points P, Q, and R in the plane, and determine the vectors $\overrightarrow{PQ}$ and $\overrightarrow{PR}$. (b) Find $\overrightarrow{PQ} \times \overrightarrow{PR}$. What is the relationship between the components of the cross product and the coefficients in the equation of the plane? Why is this true?**

21. $4x - 3y - 6z = 6$ **22.** $2x + 3y + 4z = 4$

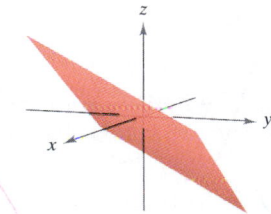   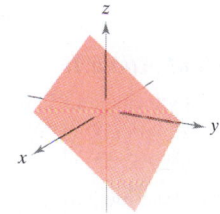

In Exercises 23–28, find an equation of the plane passing through the point perpendicular to the indicated vector or line.

Point	Perpendicular to
23. $(2, 1, 2)$	$\mathbf{n} = \mathbf{i}$
24. $(1, 0, -3)$	$\mathbf{n} = \mathbf{k}$
25. $(3, 2, 2)$	$\mathbf{n} = 2\mathbf{i} + 3\mathbf{j} - \mathbf{k}$
26. $(0, 0, 0)$	$\mathbf{n} = -3\mathbf{i} + 2\mathbf{k}$
27. $(0, 0, 6)$	$x = 1 - t, y = 2 + t, z = 4 - 2t$
28. $(3, 2, 2)$	$\dfrac{x - 1}{4} = y + 2 = \dfrac{z + 3}{-3}$

In Exercises 29–40, find an equation of the plane.

29. The plane passes through $(0, 0, 0)$, $(1, 2, 3)$, and $(-2, 3, 3)$.

30. The plane passes through $(1, 2, -3)$, $(2, 3, 1)$, and $(0, -2, -1)$.

31. The plane passes through $(1, 2, 3)$, $(3, 2, 1)$, and $(-1, -2, 2)$.

32. The plane passes through the point $(1, 2, 3)$ and is parallel to the yz-plane.

33. The plane passes through the point $(1, 2, 3)$ and is parallel to the xy-plane.

34. The plane contains the y-axis and makes an angle of $\pi/6$ with the positive x-axis.

35. The plane contains the lines given by

  $\dfrac{x - 1}{-2} = y - 4 = z$ and $\dfrac{x - 2}{-3} = \dfrac{y - 1}{4} = \dfrac{z - 2}{-1}$.

36. The plane passes through the point $(2, 2, 1)$ and contains the line given by

  $\dfrac{x}{2} = \dfrac{y - 4}{-1} = z$.

37. The plane passes through the points $(2, 2, 1)$ and $(-1, 1, -1)$ and is perpendicular to the plane $2x - 3y + z = 3$.

38. The plane passes through the points $(3, 2, 1)$ and $(3, 1, -5)$ and is perpendicular to the plane $6x + 7y + 2z = 10$.

39. The plane passes through the points $(1, -2, -1)$ and $(2, 5, 6)$ and is parallel to the x-axis.

40. The plane passes through the points $(4, 2, 1)$ and $(-3, 5, 7)$ and is parallel to the z-axis.

In Exercises 41–46, determine whether the planes are parallel, orthogonal, or neither. If they are neither parallel nor orthogonal, find the angle of intersection.

41. $5x - 3y + z = 4$ **42.** $3x + y - 4z = 3$

  $x + 4y + 7z = 1$ $-9x - 3y + 12z = 4$

43. $x - 3y + 6z = 4$ **44.** $3x + 2y - z = 7$

  $5x + y - z = 4$ $x - 4y + 2z = 0$

45. $x - 5y - z = 1$ **46.** $2x - z = 1$

  $5x - 25y - 5z = -3$ $4x + y + 8z = 10$

In Exercises 47–52, mark the intercepts and sketch a graph of the plane.

47. $4x + 2y + 6z = 12$ **48.** $3x + 6y + 2z = 6$

49. $2x - y + 3z = 4$ **50.** $2x - y + z = 4$

51. $y + z = 5$ **52.** $x + 2y = 4$

In Exercises 53–56, use a computer algebra system to graph the plane.

53. $2x + y - z = 6$ **54.** $x - 3z = 3$

55. $-5x + 4y - 6z = -8$ **56.** $2.1x - 4.7y - z = -3$

In Exercises 57 and 58, find a set of parametric equations for the line of intersection of the planes.

57. $3x + 2y - z = 7$ **58.** $x - 3y + 6z = 4$
$\quad\ \ x - 4y + 2z = 0$ $\qquad\ 5x + y - z = 4$

In Exercises 59–62, find the point of intersection (if any) of the plane and the line. Also determine whether the line lies in the plane.

59. $2x - 2y + z = 12, \quad x - \dfrac{1}{2} = \dfrac{y + (3/2)}{-1} = \dfrac{z + 1}{2}$

60. $2x + 3y = -5, \quad \dfrac{x - 1}{4} = \dfrac{y}{2} = \dfrac{z - 3}{6}$

61. $2x + 3y = 10, \quad \dfrac{x - 1}{3} = \dfrac{y + 1}{-2} = z - 3$

62. $5x + 3y = 17, \quad \dfrac{x - 4}{2} = \dfrac{y + 1}{-3} = \dfrac{z + 2}{5}$

In Exercises 63 and 64, find the distance between the point and the plane.

63. $(0, 0, 0)$
$\quad 2x + 3y + z = 12$

64. $(1, 2, 3)$
$\quad 2x - y + z = 4$

In Exercises 65 and 66, find the distance between the planes.

65. $x - 3y + 4z = 10$
$\quad x - 3y + 4z = 6$

66. $2x - 4z = 4$
$\quad 2x - 4z = 10$

In Exercises 67 and 68, find the distance between the point and the line given by the set of parametric equations.

67. $(1, 5, -2);\quad x = 4t - 2, y = 3, z = -t + 1$

68. $(4, 1, -2);\quad x = 2t + 2, y = 2t, z = t - 3$

69. *Modeling Data*   Per capita consumption (in pounds) of different types of milk in United States for selected years is given in the table. Consumptions of skim milk, lowfat milk, and whole milk are represented by the variables x, y, and z. (*Source: U.S. Department of Agriculture*)

Year	1970	1975	1980	1985
x	11.6	11.5	11.6	12.6
y	29.8	53.2	70.1	83.3
z	213.5	174.9	141.7	119.7

Year	1990	1991	1992	1993
x	22.9	23.9	25.0	26.7
y	98.3	99.7	99.4	97.1
z	87.6	84.7	81.5	77.8

A model for the data is given by

$0.987x + 1.71y + z = 276.$

(a) Complete a fourth row in the table using the model to approximate z for the given values of x and y. Compare the approximations with the actual values of z.

(b) According to this model, any increases in consumption of two types of milk will have what effect on the consumption of the third type?

(c) Because x, y, and z must be nonnegative, sketch the traces of the plane and the first octant portion of the plane.

70. *Optimization*   Consider the line given by the parametric equations

$x = -t + 3, \quad y = \tfrac{1}{2}t + 1, \quad z = 2t - 1$

and the point $(4, 3, s)$ for any real number s.

(a) Write the distance between the point and the line as a function of s.

(b) Use a graphing utility to graph the function in part (a). Use the graph to find the value of s such that the distance between the point and the line is minimum.

(c) Use the zoom feature of a graphing utility to zoom out several times on the graph in part (b). Does it appear that the graph has slant asymptotes? Explain. If it appears to have slant asymptotes, find them.

71. *Think About It*

(a) Describe and find an equation for the surface generated by all points (x, y, z) that are four units from the point $(3, -2, 5)$.

(b) Describe and find an equation for the surface generated by all points (x, y, z) that are four units from the plane

$4x - 3y + z = 10.$

72. *Think About It* Consider the two nonzero vectors **u** and **v**. Describe the geometric figure generated by the terminal points of the following vectors where s and t represent all real numbers.

(a) $t\mathbf{v}$ (b) $\mathbf{u} + t\mathbf{v}$ (c) $s\mathbf{u} + t\mathbf{v}$

73. *Mechanical Design* A chute at the top of a grain elevator of a combine funnels the grain into a bin (see figure). Find the angle between two adjacent sides.

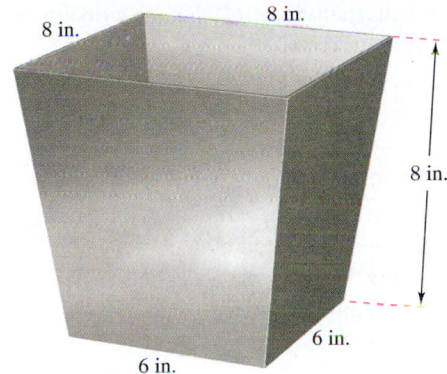

8 in.   8 in.

8 in.

6 in.

6 in.

74. If $a_1, b_1, c_1,$ and a_2, b_2, c_2 are two sets of direction numbers for the same line, show that there exists a scalar d such that $a_1 = a_2 d,\ b_1 = b_2 d,$ and $c_1 = c_2 d.$

True or False? **In Exercises 75 and 76, determine whether the statement is true or false. If it is false, explain why or give an example that shows it is false.**

75. If $\mathbf{v} = a_1\mathbf{i} + b_1\mathbf{j} + c_1\mathbf{k}$ is any vector in the plane given by $a_2 x + b_2 y + c_2 z + d_2 = 0,$ then $a_1 a_2 + b_1 b_2 + c_1 c_2 = 0.$

76. Every pair of lines in space are either intersecting or parallel.

77. Consider the plane that passes through the points P, R, and S. Show that the distance from a point Q to this plane is

$$\text{Distance} = \frac{|\mathbf{u} \cdot (\mathbf{v} \times \mathbf{w})|}{\|\mathbf{u} \times \mathbf{v}\|}$$

where $\mathbf{u} = \vec{PR},\ \mathbf{v} = \vec{PS},$ and $\mathbf{w} = \vec{PQ}.$

78. Show that the distance between the parallel planes $ax + by + cz + d_1 = 0$ and $ax + by + cz + d_2 = 0$ is

$$\text{Distance} = \frac{|d_1 - d_2|}{\sqrt{a^2 + b^2 + c^2}}.$$

We have developed two distance formulas in this section—the distance between a point and a plane, and the distance between a point and a line. In this project you will study a third distance problem—the distance between two skew lines. Two lines in space are *skew* if they are neither parallel nor intersecting (see figure below).

(a) Consider the following two lines in space.

$$L_1: x = 4 + 5t,\ y = 5 + 5t,\ z = 1 - 4t$$
$$L_2: x = 4 + s,\ y = -6 + 8s,\ z = 7 - 3s$$

(i) Show that these lines are not parallel.

(ii) Show that these lines do not intersect, and hence are skew lines.

(iii) Show that the two lines lie in parallel planes.

(iv) Find the distance between the parallel planes from part (iii). This is the distance between the original skew lines.

(b) Use the procedure in part (a) to find the distance between the lines

$$L_1: x = 2t,\ y = 4t,\ z = 6t$$
$$L_2: x = 1 - s,\ y = 4 + s,\ z = -1 + s.$$

(c) Use the procedure in part (a) to find the distance between the lines

$$L_1: x = 3t,\ y = 2 - t,\ z = -1 + t$$
$$L_2: x = 1 + 4s,\ y = -2 + s,\ z = -3 - 3s$$

(d) Develop a formula for finding the distance between the skew lines

$$L_1: x = x_1 + a_1 t,\ y = y_1 + b_1 t,\ z = z_1 + c_1 t$$
$$L_2: x = x_2 + a_2 s,\ y = y_2 + b_2 s,\ z = z_2 + c_2 s$$

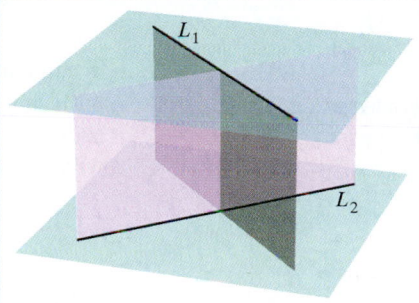

Cylindrical Surfaces • Quadric Surfaces • Surfaces of Revolution

Cylindrical Surfaces

The first five sections of this chapter contained the vector portion of the preliminary work necessary to study vector calculus and the calculus of space. In this and the next section, you will study surfaces in space and alternative coordinate systems for space. You have already studied two special types of surfaces.

1. Spheres: $(x - x_0)^2 + (y - y_0)^2 + (z - z_0)^2 = r^2$ Section 10.2
2. Planes: $ax + by + cz + d = 0$ Section 10.5

A third type of surface in space is called a **cylindrical surface,** or simply a **cylinder.** To define a cylinder, consider the familiar right circular cylinder shown in Figure 10.56. You can imagine that this cylinder is generated by a vertical line moving around the circle $x^2 + y^2 = a^2$ in the xy-plane. This circle is called a **generating curve** for the cylinder, as indicated in the following definition.

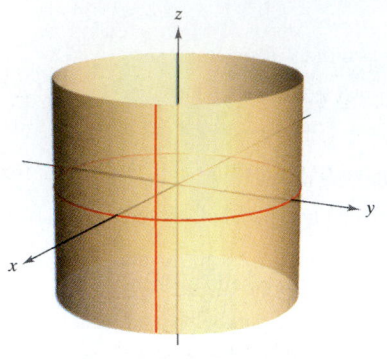

Right circular cylinder:
$x^2 + y^2 = a^2$

Rulings are parallel to z-axis.
Figure 10.56

Definition of a Cylinder

Let C be a curve in a plane and let L be a line not in a parallel plane. The set of all lines parallel to L and intersecting C is called a **cylinder.** C is called the **generating curve** (or **directrix**) of the cylinder, and the parallel lines are called **rulings.**

NOTE   Without loss of generality, you can assume that C lies in one of the three coordinate planes. Moreover, in this text we restrict the discussion to *right* cylinders—cylinders whose rulings are perpendicular to the coordinate plane containing C, as shown in Figure 10.57.

For the right circular cylinder shown in Figure 10.56, the equation of the generating curve is

$$x^2 + y^2 = a^2$$ Equation of generating curve in xy-plane

To find an equation for the cylinder, note that you can generate any one of the rulings by fixing the values of x and y and then allowing z to take on all real values. In this sense the value of z is arbitrary and is, therefore, not included in the equation. In other words, the equation of this cylinder is simply the equation of its generating curve.

$$x^2 + y^2 = a^2$$ Equation of cylinder in space

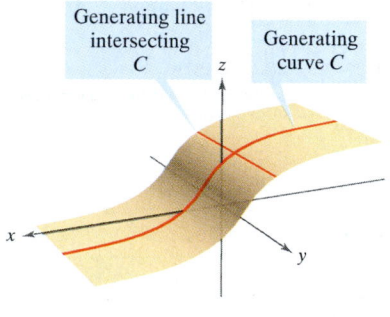

Generating line intersecting C

Generating curve C

Cylinder: Rulings intersect C and are parallel to the given line.
Figure 10.57

Equations of Cylinders

The equation of a cylinder whose rulings are parallel to one of the coordinate axes contains only the variables corresponding to the other two axes.

EXAMPLE 1   Sketching a Cylinder

Sketch the surface represented by each of the following equations.

a. $z = y^2$ **b.** $z = \sin x, \quad 0 \le x \le 2\pi$

Solution

a. The graph is a cylinder whose generating curve, $z = y^2$, is a parabola in the yz-plane. The rulings of the cylinder are parallel to the x-axis, as shown in Figure 10.58(a).

b. The graph is a cylinder generated by the sine curve in the xz-plane. The rulings are parallel to the y-axis, as shown in Figure 10.58(b).

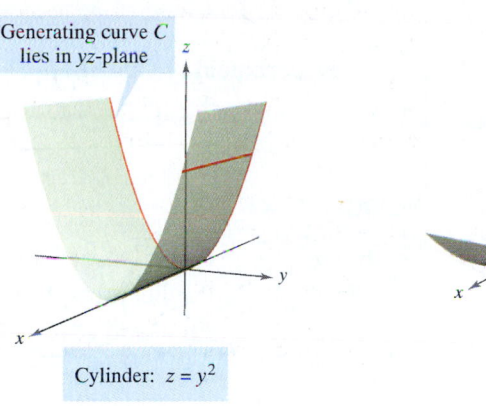

Generating curve C lies in yz-plane

Cylinder: $z = y^2$

(a) Rulings are parallel to x-axis.

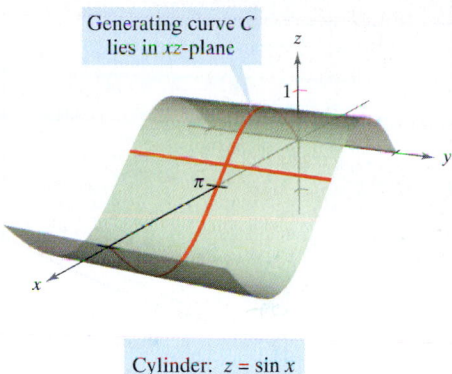

Generating curve C lies in xz-plane

Cylinder: $z = \sin x$

(b) Rulings are parallel to y-axis.

Figure 10.58

Quadric Surfaces

STUDY TIP   In the table on pages 750 and 751, only one of several orientations of each quadric surface is shown. If the surface is oriented along a different axis, its standard equation will change accordingly, as illustrated in Examples 2 and 3. The fact that the two types of paraboloids have one variable raised to the first power can be helpful in classifying quadric surfaces. The other four types of basic quadric surfaces have equations that are of *second degree* in all three variables.

The fourth basic type of surface in space is a **quadric surface.** Quadric surfaces are the three-dimensional analogs of conic sections.

Quadric Surface

The equation of a **quadric surface** in space is a second-degree equation of the form

$$Ax^2 + By^2 + Cz^2 + Dxy + Exz + Fyz + Gx + Hy + Iz + J = 0.$$

There are six basic types of quadric surfaces: **ellipsoid, hyperboloid of one sheet, hyperboloid of two sheets, elliptic cone, elliptic paraboloid,** and **hyperbolic paraboloid.**

The intersection of a surface with a plane is called the **trace of the surface** in the plane. To visualize a surface in space, it is helpful to determine its traces in some well-chosen planes. The traces of quadric surfaces are conics. These traces, together with the **standard form** of the equation of each quadric surface, are shown in the table on pages 750 and 751.

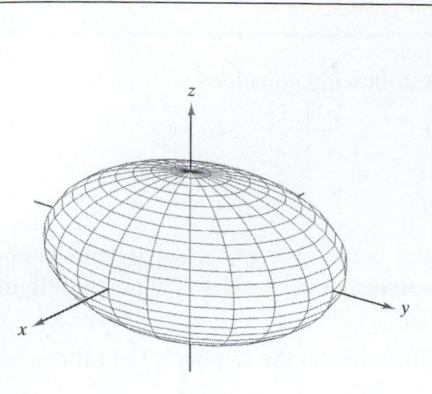

Ellipsoid

$$\frac{x^2}{a^2} + \frac{y^2}{b^2} + \frac{z^2}{c^2} = 1$$

Trace	*Plane*
Ellipse	Parallel to xy-plane
Ellipse	Parallel to xz-plane
Ellipse	Parallel to yz-plane

The surface is a sphere if
$a = b = c \neq 0$.

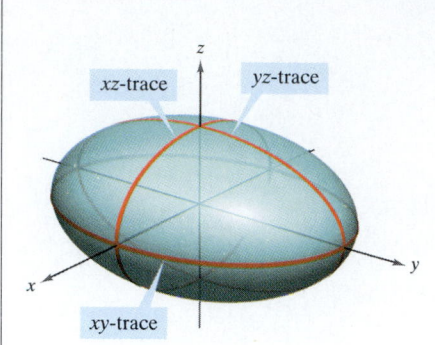

Hyperboloid of One Sheet

$$\frac{x^2}{a^2} + \frac{y^2}{b^2} - \frac{z^2}{c^2} = 1$$

Trace	*Plane*
Ellipse	Parallel to xy-plane
Hyperbola	Parallel to xz-plane
Hyperbola	Parallel to yz-plane

The axis of the hyperboloid
corresponds to the variable whose
coefficient is negative.

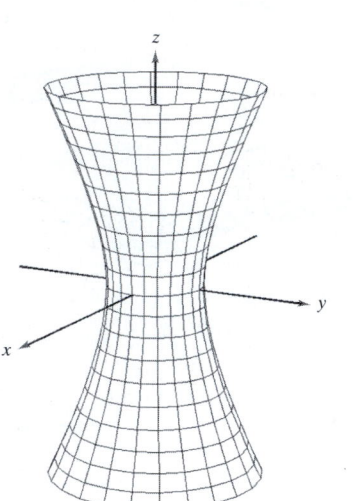

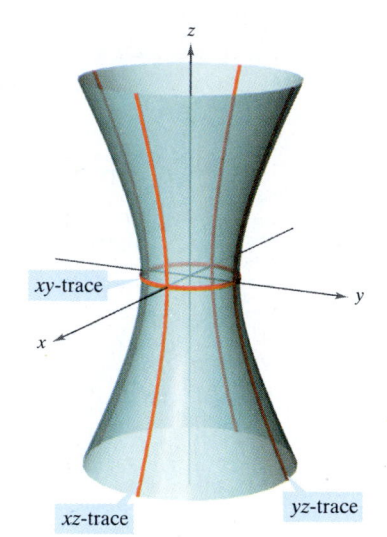

Hyperboloid of Two Sheets

$$\frac{z^2}{c^2} - \frac{x^2}{a^2} - \frac{y^2}{b^2} = 1$$

Trace	*Plane*
Ellipse	Parallel to xy-plane
Hyperbola	Parallel to xz-plane
Hyperbola	Parallel to yz-plane

The axis of the hyperboloid
corresponds to the variable whose
coefficient is positive. There is
no trace in the coordinate plane
perpendicular to this axis.

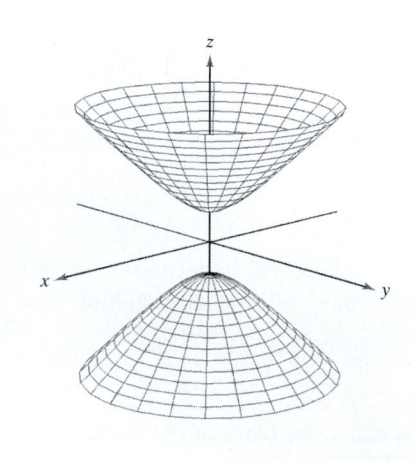

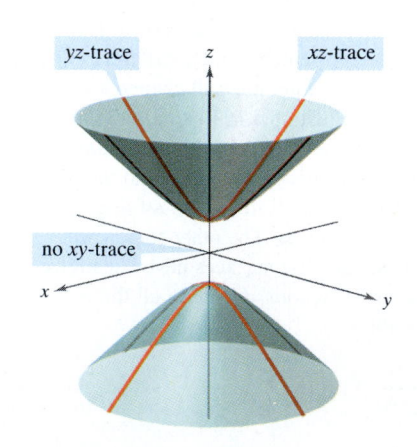

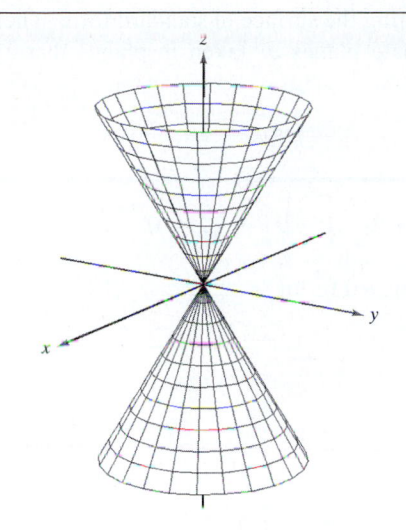

Elliptic Cone

$$\frac{x^2}{a^2} + \frac{y^2}{b^2} - \frac{z^2}{c^2} = 0$$

Trace	*Plane*
Ellipse	Parallel to xy-plane
Hyperbola	Parallel to xz-plane
Hyperbola	Parallel to yz-plane

The axis of the cone corresponds to the variable whose coefficient is negative. The traces in the coordinate planes parallel to this axis are intersecting lines.

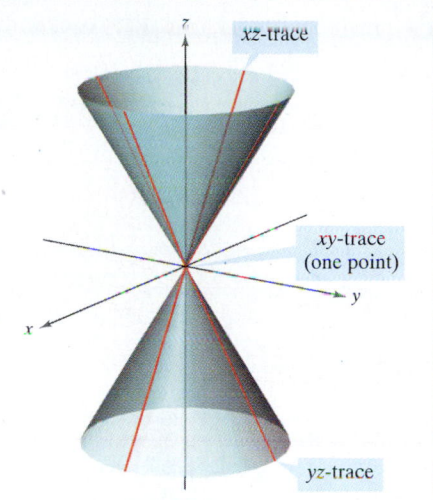

Elliptic Paraboloid

$$z = \frac{x^2}{a^2} + \frac{y^2}{b^2}$$

Trace	*Plane*
Ellipse	Parallel to xy-plane
Parabola	Parallel to xz-plane
Parabola	Parallel to yz-plane

The axis of the paraboloid corresponds to the variable raised to the first power.

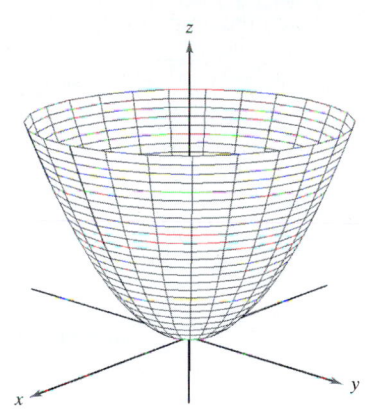

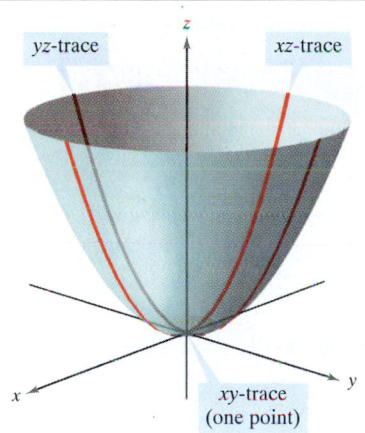

Hyperbolic Paraboloid

$$z = \frac{y^2}{b^2} - \frac{x^2}{a^2}$$

Trace	*Plane*
Hyperbola	Parallel to xy-plane
Parabola	Parallel to xz-plane
Parabola	Parallel to yz-plane

The axis of the paraboloid corresponds to the variable raised to the first power.

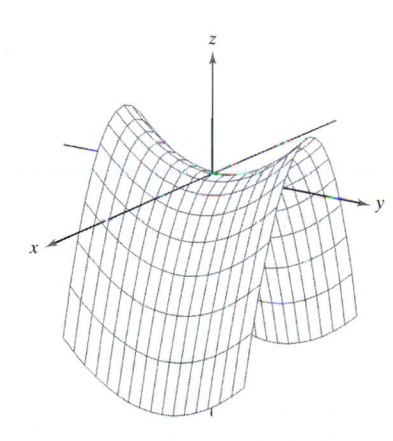

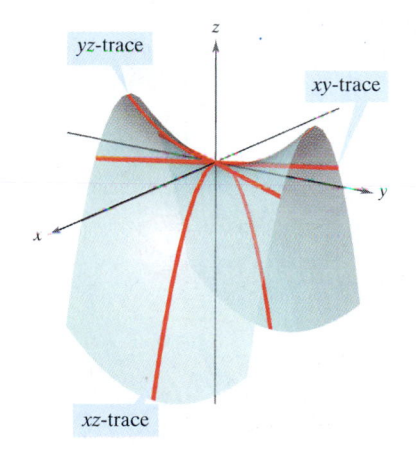

To classify a quadric surface, begin by writing the surface in standard form. Then, determine several traces taken in the coordinate planes *or* taken in planes that are parallel to the coordinate planes.

EXAMPLE 2   Sketching a Quadric Surface

Classify and sketch the surface given by $4x^2 - 3y^2 + 12z^2 + 12 = 0$.

Solution   Begin by writing the equation in standard form.

$$4x^2 - 3y^2 + 12z^2 + 12 = 0 \qquad \text{Original equation}$$

$$\frac{x^2}{-3} + \frac{y^2}{4} - z^2 - 1 = 0 \qquad \text{Divide by } -12.$$

$$\frac{y^2}{4} - \frac{x^2}{3} - \frac{z^2}{1} = 1 \qquad \text{Standard form}$$

From the table on pages 750 and 751, you can conclude that the surface is a hyperboloid of two sheets with the *y*-axis as its axis. To sketch the graph of this surface, it helps to find the traces in the coordinate planes.

xy-trace $(z = 0)$: $\dfrac{y^2}{4} - \dfrac{x^2}{3} = 1$ Hyperbola

xz-trace $(y = 0)$: $\dfrac{x^2}{3} + \dfrac{z^2}{1} = -1$ No trace

yz-trace $(x = 0)$: $\dfrac{y^2}{4} - \dfrac{z^2}{1} = 1$ Hyperbola

The graph is shown in Figure 10.59.

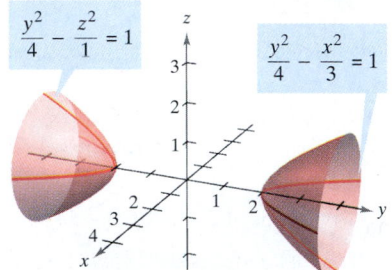

$\dfrac{y^2}{4} - \dfrac{z^2}{1} = 1$

$\dfrac{y^2}{4} - \dfrac{x^2}{3} = 1$

Hyperboloid of two sheets:

$$\frac{y^2}{4} - \frac{x^2}{3} - z^2 = 1$$

Figure 10.59

EXAMPLE 3   Sketching a Quadric Surface

Classify and sketch the surface given by $x - y^2 - 4z^2 = 0$.

Solution   Because *x* is raised only to the first power, the surface is a paraboloid. The axis of the paraboloid is the *x*-axis. In the standard form, the equation is

$$x = y^2 + 4z^2. \qquad \text{Standard form}$$

Some convenient traces are as follows.

xy-trace $(z = 0)$: $x = y^2$ Parabola
xz-trace $(y = 0)$: $x = 4z^2$ Parabola

parallel to yz-plane $(x = 4)$: $\dfrac{y^2}{4} + \dfrac{z^2}{1} = 1$ Ellipse

The surface is an *elliptic* paraboloid, as shown in Figure 10.60.

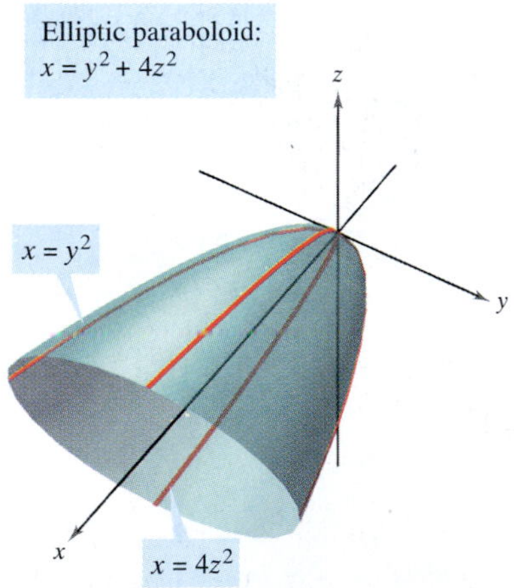

Elliptic paraboloid:
$x = y^2 + 4z^2$

$x = y^2$

$x = 4z^2$

Figure 10.60

Some second-degree equations in *x*, *y*, and *z* do not represent one of the basic types of quadric surfaces. Here are two examples.

$$x^2 + y^2 + z^2 = 0 \qquad \text{Single point}$$

$$x^2 + y^2 = 1 \qquad \text{Right circular cylinder}$$

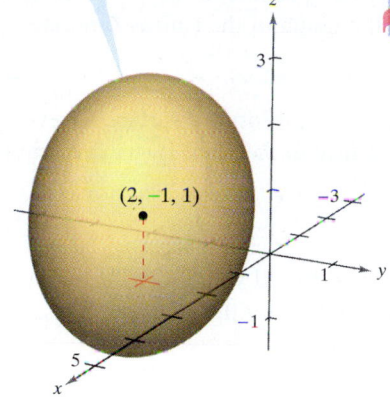

$$\frac{(x-2)^2}{4} + \frac{(y+1)^2}{2} + \frac{(z-1)^2}{4} = 1$$

$(2, -1, 1)$

An ellipsoid centered at $(2, -1, 1)$
Figure 10.61

For a quadric surface not centered at the origin, you can form the standard equation by completing the square, as demonstrated in Example 4.

⯈ **EXAMPLE 4   A Quadric Surface Not Centered at the Origin**

Classify and sketch the surface given by

$$x^2 + 2y^2 + z^2 - 4x + 4y - 2z + 3 = 0.$$

Solution   Completing the square for each variable produces the following.

$$(x^2 - 4x +\ \ ) + 2(y^2 + 2y +\ \ ) + (z^2 - 2z +\ \ ) = -3$$
$$(x^2 - 4x + 4) + 2(y^2 + 2y + 1) + (z^2 - 2z + 1) = -3 + 4 + 2 + 1$$
$$(x - 2)^2 + 2(y + 1)^2 + (z - 1)^2 = 4$$
$$\frac{(x - 2)^2}{4} + \frac{(y + 1)^2}{2} + \frac{(z - 1)^2}{4} = 1$$

From this equation, you can see that the quadric surface is an ellipsoid that is centered at $(2, -1, 1)$. Its graph is shown in Figure 10.61.

TECHNOLOGY   A computer graphing utility can help you visualize a surface in space.* Most such utilities create three-dimensional illusions by sketching several traces of the surface and then applying a "hidden-line" routine that blocks out portions of the surface that lie behind other portions of the surface. Two examples of figures that were generated by *Mathematica* are shown below.

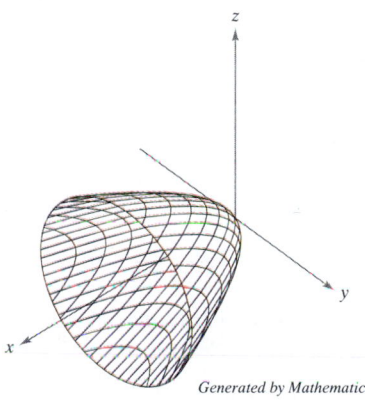

Generated by Mathematica   *Generated by Mathematica*

Elliptic Paraboloid   Hyperbolic Paraboloid

$$x = \frac{y^2}{2} + \frac{z^2}{2}$$ $$z = \frac{y^2}{16} - \frac{x^2}{16}$$

Using a graphing utility to sketch the graph of a surface in space requires practice. For one thing, you must know enough about the surface to be able to specify a *viewing box* that gives a representative view of the surface. Also, you can often improve the view of a surface by rotating the axes. For instance, note that the elliptic paraboloid in the figure is seen from a line of sight that is "higher" than the line of sight used to view the hyperbolic paraboloid.

---

*Some 3-D graphing utilities require surfaces to be entered with parametric equations. For a discussion of this technique, see Section 14.5.

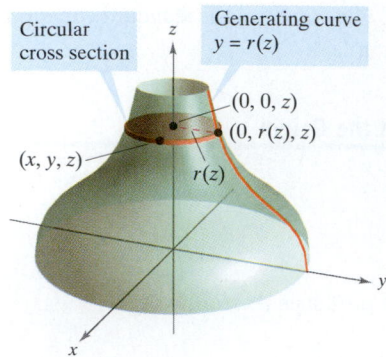

Circular cross section

Generating curve $y = r(z)$

$(0, 0, z)$

$(0, r(z), z)$

(x, y, z)

$r(z)$

A surface of revolution
Figure 10.62

Surfaces of Revolution

The fifth special type of surface we will study is called a **surface of revolution.** In Section 6.4, you studied a method for finding the *area* of such a surface. We now look at a procedure for finding its *equation*. Consider the graph of the **radius function**

$$y = r(z) \qquad \text{Generating curve}$$

in the yz-plane. If this graph is revolved about the z-axis, it forms a surface of revolution, as shown in Figure 10.62. The trace of the surface in the plane $z = z_0$ is a circle whose radius is $r(z_0)$ and whose equation is

$$x^2 + y^2 = [r(z_0)]^2. \qquad \text{Circular trace in plane: } z = z_0$$

Replacing z_0 with z produces an equation that is valid for all values of z. In a similar manner, we can obtain equations for surfaces of revolution for the other two axes, and we summarize the results as follows.

Surface of Revolution

If the graph of a radius function r is revolved about one of the coordinate axes, the equation of the resulting surface of revolution has one of the following forms.

1. Revolved about the x-axis: $y^2 + z^2 = [r(x)]^2$
2. Revolved about the y-axis: $x^2 + z^2 = [r(y)]^2$
3. Revolved about the z-axis: $x^2 + y^2 = [r(z)]^2$

EXAMPLE 5   Finding an Equation for a Surface of Revolution

a. An equation for the surface of revolution formed by revolving the graph of

$$y = \frac{1}{z} \qquad \text{Radius function}$$

about the z-axis is

$$x^2 + y^2 = [r(z)]^2 \qquad \text{Revolved about the } z\text{-axis}$$

$$x^2 + y^2 = \left(\frac{1}{z}\right)^2. \qquad \text{Substitute } 1/z \text{ for } r(z).$$

b. To find an equation for the surface formed by revolving the graph of $9x^2 = y^3$ about the y-axis, solve for x in terms of y to obtain

$$x = \tfrac{1}{3}y^{3/2} = r(y). \qquad \text{Radius function}$$

Thus, the equation for this surface is

$$x^2 + z^2 = [r(y)]^2 \qquad \text{Revolved about the } y\text{-axis}$$

$$x^2 + z^2 = \left(\tfrac{1}{3}y^{3/2}\right)^2 \qquad \text{Substitute } \tfrac{1}{3}y^{3/2} \text{ for } r(y).$$

$$x^2 + z^2 = \tfrac{1}{9}y^3. \qquad \text{Equation of surface}$$

The graph is shown in Figure 10.63.

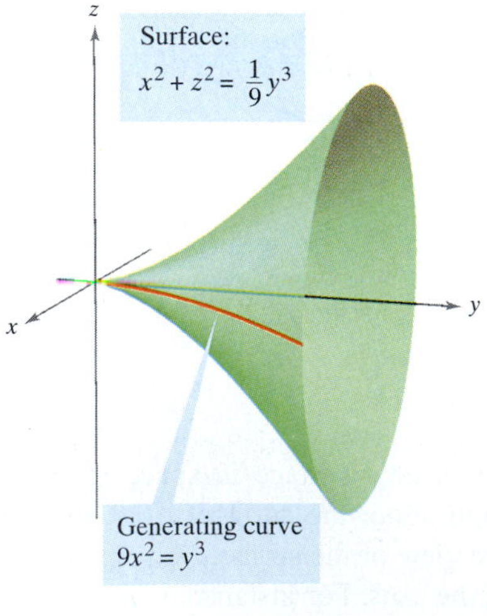

Surface:
$x^2 + z^2 = \frac{1}{9}y^3$

Generating curve
$9x^2 = y^3$

Figure 10.63

The generating curve for a surface of revolution is not unique. For instance, the surface

$$x^2 + z^2 = e^{-2y}$$

can be formed by revolving either the graph of $x = e^{-y}$ about the y-axis or the graph of $z = e^{-y}$ about the y-axis, as shown in Figure 10.64.

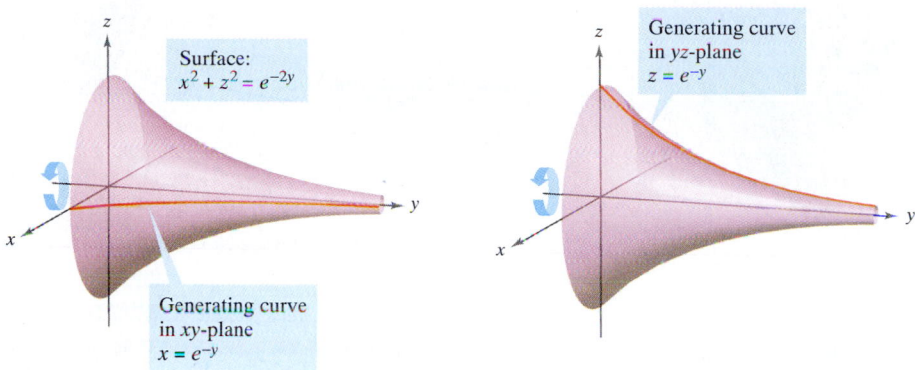

Surface:
$x^2 + z^2 = e^{-2y}$

Generating curve
in xy-plane
$x = e^{-y}$

Generating curve
in yz-plane
$z = e^{-y}$

Figure 10.64

EXAMPLE 6   Finding a Generating Curve for a Surface of Revolution

Find a generating curve and the axis of revolution for the surface given by

$$x^2 + 3y^2 + z^2 = 9.$$

Solution   You now know that the equation has one of the following forms.

$$x^2 + y^2 = [r(z)]^2 \qquad \text{Revolved about } z\text{-axis}$$
$$y^2 + z^2 = [r(x)]^2 \qquad \text{Revolved about } x\text{-axis}$$
$$x^2 + z^2 = [r(y)]^2 \qquad \text{Revolved about } y\text{-axis}$$

Because the coefficients of x^2 and z^2 are equal, you should choose the third form and write

$$x^2 + z^2 = 9 - 3y^2.$$

The y-axis is the axis of revolution. You can choose a generating curve from either of the following traces.

$$x^2 = 9 - 3y^2 \qquad \text{Trace in } xy\text{-plane}$$
$$z^2 = 9 - 3y^2 \qquad \text{Trace in } yz\text{-plane}$$

For example, using the first trace, the generating curve is the semiellipse given by

$$x = \sqrt{9 - 3y^2}. \qquad \text{Generating curve}$$

The graph of this surface is shown in Figure 10.65.

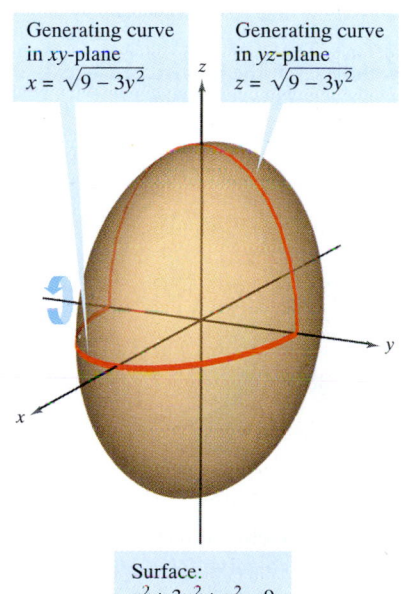

Generating curve
in xy-plane
$x = \sqrt{9 - 3y^2}$

Generating curve
in yz-plane
$z = \sqrt{9 - 3y^2}$

Surface:
$x^2 + 3y^2 + z^2 = 9$

Figure 10.65

EXERCISES FOR SECTION 10.6

In Exercises 1–6, match the equation with its graph. [The graphs are labeled (a), (b), (c), (d), (e), and (f).]

(a)
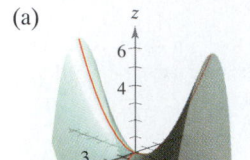

(b)
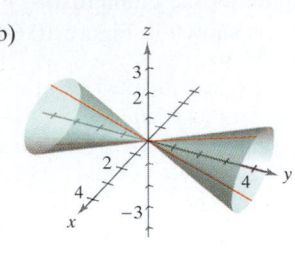

(c)
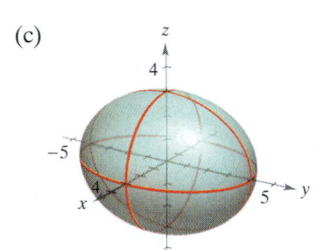

(d)
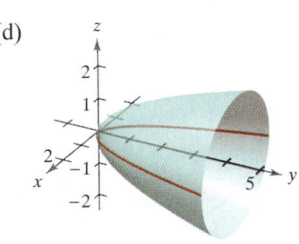

(e)
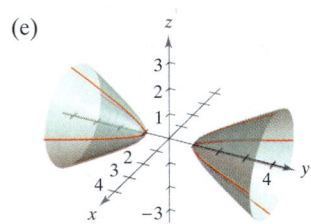

(f)
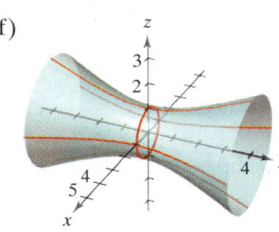

1. $\dfrac{x^2}{9} + \dfrac{y^2}{16} + \dfrac{z^2}{9} = 1$

2. $15x^2 - 4y^2 + 15z^2 = -4$

3. $4x^2 - y^2 + 4z^2 = 4$

4. $y^2 = 4x^2 + 9z^2$

5. $4x^2 - 4y + z^2 = 0$

6. $4x^2 - y^2 + 4z = 0$

In Exercises 7–16, describe and sketch the surface.

7. $z = 3$

8. $x = 4$

9. $y^2 + z^2 = 9$

10. $x^2 + z^2 = 16$

11. $x^2 - y = 0$

12. $y^2 + z = 4$

13. $4x^2 + y^2 = 4$

14. $y^2 - z^2 = 4$

15. $z - \sin y = 0$

16. $z - e^y = 0$

17. *Think About It* The four figures are graphs of the quadric surface $z = x^2 + y^2$. Match each of the four graphs with the point in space from which the paraboloid is viewed. The four points are $(0, 0, 20)$, $(0, 20, 0)$, $(20, 0, 0)$, and $(10, 10, 20)$.

(a)
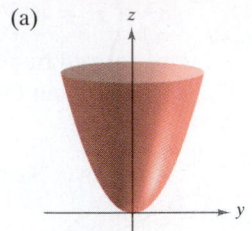

(b)
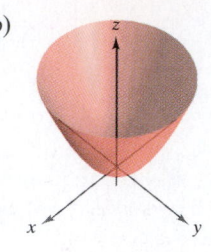

(c)
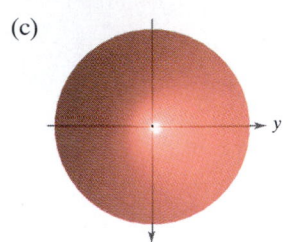

(d)
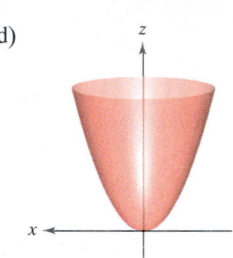

Figures for 17

18. Use a graphing utility to sketch a view of the cylinder $y^2 + z^2 = 4$ from each of the following points.

(a) $(10, 0, 0)$ (b) $(0, 10, 0)$ (c) $(10, 10, 10)$

In Exercises 19–30, identify and sketch the quadric surface. Use a computer algebra system to confirm your sketch.

19. $x^2 + \dfrac{y^2}{4} + z^2 = 1$

20. $\dfrac{x^2}{9} + \dfrac{y^2}{16} + \dfrac{z^2}{16} = 1$

21. $16x^2 - y^2 + 16z^2 = 4$

22. $z^2 - x^2 - \dfrac{y^2}{4} = 1$

23. $x^2 - y + z^2 = 0$

24. $z = 4x^2 + y^2$

25. $x^2 - y^2 + z = 0$

26. $3z = -y^2 + x^2$

27. $z^2 = x^2 + \dfrac{y^2}{4}$

28. $x^2 = 2y^2 + 2z^2$

29. $16x^2 + 9y^2 + 16z^2 - 32x - 36y + 36 = 0$

30. $4x^2 + y^2 - 4z^2 - 16x - 6y - 16z + 9 = 0$

In Exercises 31–40, use a computer algebra system to graph the surface. (*Hint:* It may be necessary to solve for z and acquire two equations to graph the surface.)

31. $z = 2\sin x$

32. $z = x^2 + 0.5y^2$

33. $z^2 = x^2 + 4y^2$

34. $4y = x^2 + z^2$

35. $x^2 + y^2 = \left(\dfrac{2}{z}\right)^2$

36. $x^2 + y^2 = e^{-z}$

37. $z = 4 - \sqrt{|xy|}$

38. $z = \dfrac{-x}{8 + x^2 + y^2}$

39. $4x^2 - y^2 + 4z^2 = -16$

40. $9x^2 + 4y^2 - 8z^2 = 72$

In Exercises 41–44, sketch the region bounded by the graphs of the equations.

41. $z = 2\sqrt{x^2 + y^2}$, $z = 2$

42. $z = \sqrt{4 - x^2}$, $y = \sqrt{4 - x^2}$, $x = 0$, $y = 0$, $z = 0$

43. $x^2 + y^2 = 1$, $x + z = 2$, $z = 0$

44. $z = \sqrt{4 - x^2 - y^2}$, $y = 2z$, $z = 0$

In Exercises 45–50, find an equation for the surface of revolution generated by revolving the curve in the indicated coordinate plane about the given axis.

Equation of Curve	Coordinate Plane	Axis of Revolution
45. $z^2 = 4y$	yz-plane	y-axis
46. $z = 2y$	yz-plane	y-axis
47. $z = 2y$	yz-plane	z-axis
48. $2z = \sqrt{4 - x^2}$	xz-plane	x-axis
49. $xy = 2$	xy-plane	x-axis
50. $z = \ln y$	yz-plane	z-axis

In Exercises 51 and 52, find an equation of a generating curve given the equation of its surface of revolution.

51. $x^2 + y^2 - 2z = 0$

52. $x^2 + z^2 = \sin^2 y$

In Exercises 53 and 54, use the shell method to find the volume of the solid below the surface of revolution and above the xy-plane.

53. The curve $z = 4x - x^2$ in the xz-plane is revolved about the z-axis.

54. The curve $z = \sin y$ $(0 \le y \le \pi)$ in the yz-plane is revolved about the z-axis.

In Exercises 55 and 56, analyze the trace when the surface

$$z = \tfrac{1}{2}x^2 + \tfrac{1}{4}y^2$$

is intersected by the indicated planes.

55. Find the length of the major and minor axes and the coordinates of the foci of the ellipse generated when the surface is intersected by the planes given by

(a) $z = 2$ and   (b) $z = 8$.

56. Find the coordinates of the focus of the parabola formed when the surface is intersected by the planes given by

(a) $y = 4$ and   (b) $x = 2$.

57. *Shape of the Earth* Because of the forces caused by its rotation, the earth is an oblate ellipsoid rather than a sphere. The equatorial radius is 3963 miles and the polar radius is 3942 miles. Find an equation of the ellipsoid. (Assume the center of the earth is at the origin and the trace formed by the plane $z = 0$ corresponds to the equator.)

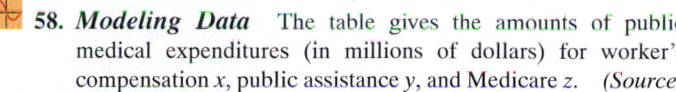 **58. *Modeling Data*** The table gives the amounts of public medical expenditures (in millions of dollars) for worker's compensation x, public assistance y, and Medicare z. (*Source: U.S. Health Care Financing Administration*)

Year	1980	1985	1990	1991	1992	1993
x	5.1	8.0	16.1	17.2	19.0	21.1
y	28.0	44.5	80.5	99.1	113.5	122.9
z	37.5	72.2	112.1	123.3	138.3	154.2

A mathematical model for the data is

$$0.551x^2 - 0.014y^2 - 19.868x + 1.856y + z + 3.512 = 0.$$

(a) Use the model to approximate z for the given values of x and y.

(b) Use a computer algebra system to graph the model for $5 \le x \le 22$ and $25 \le y \le 125$.

(c) Determine the concavity of the traces parallel to the xz-plane. Interpret the result in the context of the problem.

(d) Determine the concavity of the traces parallel to the yz-plane. Interpret the result in the context of the problem.

59. Determine the intersection of the hyperbolic paraboloid $z = y^2/b^2 - x^2/a^2$ with the plane $bx + ay - z = 0$. (Assume $a, b > 0$.)

60. *Think About It* Three types of classic "topological" surfaces are shown below. The sphere and torus have both an "inside" and an "outside." Does the Klein bottle have both an inside and an outside? Explain.

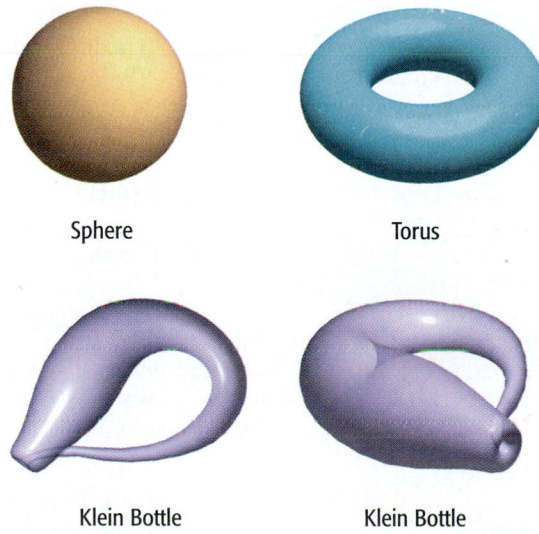

Sphere   Torus

Klein Bottle   Klein Bottle

Cylindrical Coordinates • Spherical Coordinates

Cylindrical Coordinates

You have already seen that some two-dimensional graphs are easier to represent in polar coordinates than in rectangular coordinates. A similar situation exists for surfaces in space. In this section, you will study two alternative space-coordinate systems. The first, the **cylindrical coordinate system,** is an extension of polar coordinates to space.

> ### The Cylindrical Coordinate System
>
> In a **cylindrical coordinate system,** a point P in space is represented by an ordered triple (r, θ, z).
>
> 1. (r, θ) is a polar representation of the projection of P in the xy-plane.
> 2. z is the directed distance from (r, θ) to P.

Rectangular coordinates:
$x = r \cos \theta$
$y = r \sin \theta$
$z = z$

Cylindrical coordinates:
$r^2 = x^2 + y^2$
$\tan \theta = \dfrac{y}{x}$
$z = z$

Figure 10.66

To convert from rectangular to cylindrical coordinates (or vice versa), use the following conversion guidelines for polar coordinates, as illustrated in Figure 10.66.

Cylindrical to rectangular:

$$x = r \cos \theta, \qquad y = r \sin \theta, \qquad z = z$$

Rectangular to cylindrical:

$$r^2 = x^2 + y^2, \qquad \tan \theta = \frac{y}{x}, \qquad z = z$$

The point $(0, 0, 0)$ is called the **pole.** Moreover, because the representation of a point in the polar coordinate system in not unique, it follows that the representation in the cylindrical coordinate system is also not unique.

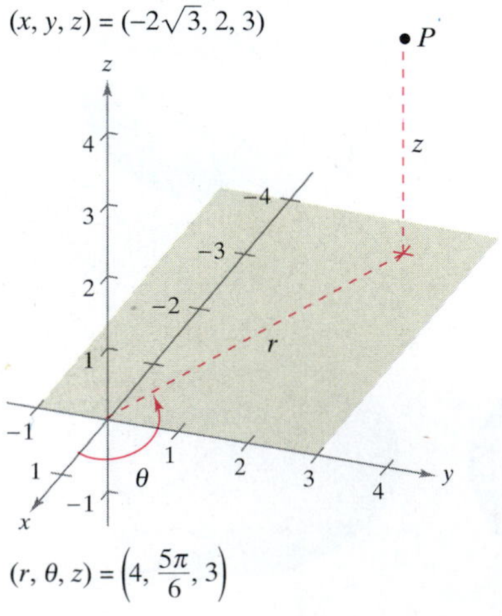

EXAMPLE 1   Changing from Cylindrical to Rectangular Coordinates

Express the point $(r, \theta, z) = (4, 5\pi/6, 3)$ in rectangular coordinates.

Solution   Using the *cylindrical-to-rectangular* conversion equations produces

$$x = 4 \cos \frac{5\pi}{6} = 4\left(-\frac{\sqrt{3}}{2}\right) = -2\sqrt{3}$$

$$y = 4 \sin \frac{5\pi}{6} = 4\left(\frac{1}{2}\right) = 2$$

$$z = 3.$$

Thus, in rectangular coordinates, the point is $(x, y, z) = \left(-2\sqrt{3}, 2, 3\right)$, as shown in Figure 10.67.

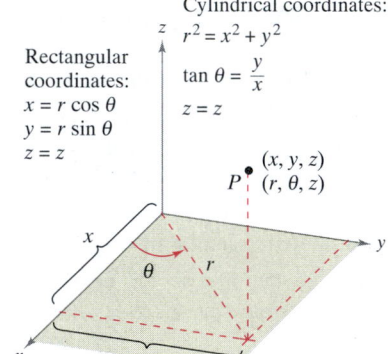

$(x, y, z) = (-2\sqrt{3}, 2, 3)$

$(r, \theta, z) = \left(4, \dfrac{5\pi}{6}, 3\right)$

Figure 10.67

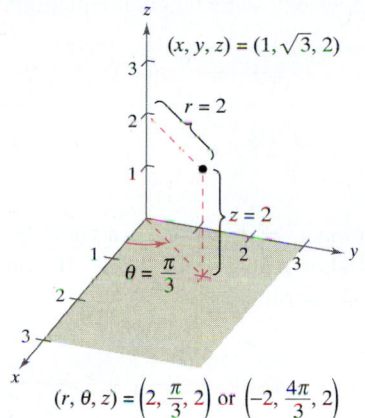

$(x, y, z) = (1, \sqrt{3}, 2)$

$r = 2$

$z = 2$

$\theta = \dfrac{\pi}{3}$

$(r, \theta, z) = \left(2, \dfrac{\pi}{3}, 2\right)$ or $\left(-2, \dfrac{4\pi}{3}, 2\right)$

Figure 10.68

EXAMPLE 2   Changing from Rectangular to Cylindrical Coordinates

Express the point $(x, y, z) = \left(1, \sqrt{3}, 2\right)$ in cylindrical coordinates.

Solution   Use the *rectangular-to-cylindrical* conversion equations.

$$r = \pm\sqrt{1 + 3} = \pm 2$$

$$\tan \theta = \sqrt{3} \qquad \Longrightarrow \qquad \theta = \arctan\left(\sqrt{3}\right) + n\pi = \frac{\pi}{3} + n\pi$$

$$z = 2$$

You have two choices for r and infinitely many choices for θ. As shown in Figure 10.68, two convenient representations of the point are

$$\left(2, \frac{\pi}{3}, 2\right) \qquad\qquad r > 0 \text{ and } \theta \text{ in Quadrant I}$$

$$\left(-2, \frac{4\pi}{3}, 2\right). \qquad r < 0 \text{ and } \theta \text{ in Quadrant III}$$

Cylindrical coordinates are especially convenient for representing cylindrical surfaces and surfaces of revolution with the z-axis as the axis of symmetry, as shown in Figure 10.69.

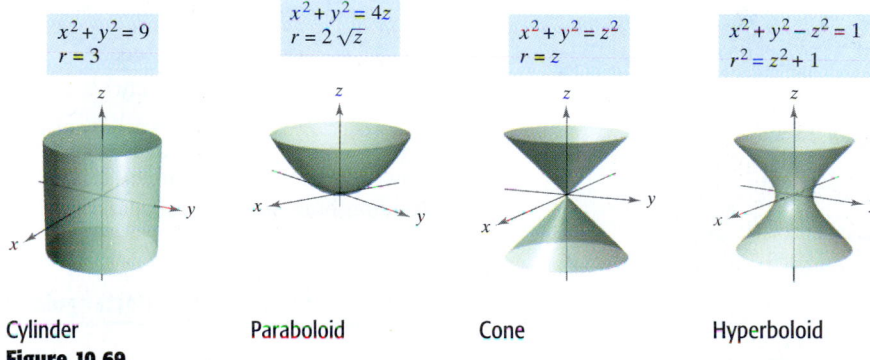

$x^2 + y^2 = 9$
$r = 3$

$x^2 + y^2 = 4z$
$r = 2\sqrt{z}$

$x^2 + y^2 = z^2$
$r = z$

$x^2 + y^2 - z^2 = 1$
$r^2 = z^2 + 1$

Cylinder   Paraboloid   Cone   Hyperboloid
Figure 10.69

Vertical planes containing the z-axis and horizontal planes also have simple cylindrical coordinate equations, as shown in Figure 10.70.

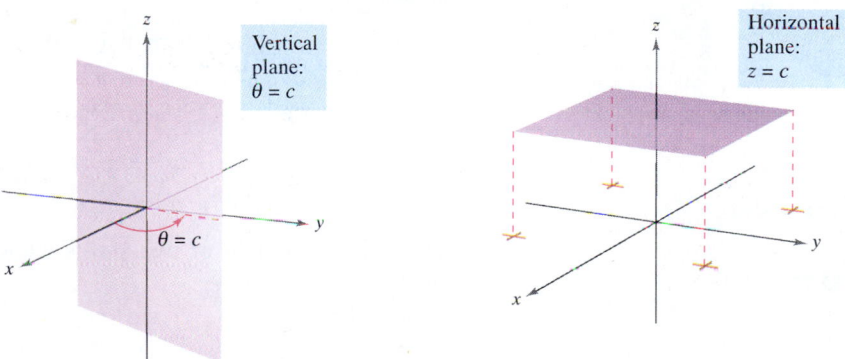

Vertical plane:
$\theta = c$

$\theta = c$

Horizontal plane:
$z = c$

Figure 10.70

Rectangular
coordinates:
$x^2 + y^2 = 4z^2$

Cylindrical
coordinates:
$r^2 = 4z^2$

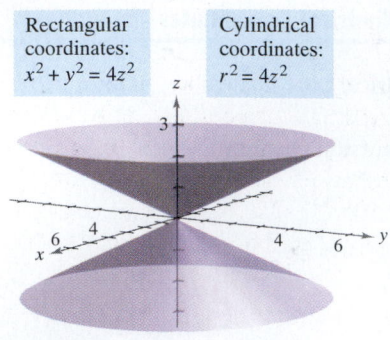

Figure 10.71

Rectangular
coordinates:
$y^2 = x$

Cylindrical
coordinates:
$r = \csc \theta \cot \theta$

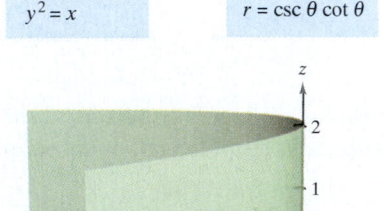

Figure 10.72

Cylindrical coordinates:
$r^2 \cos 2\theta + z^2 + 1 = 0$

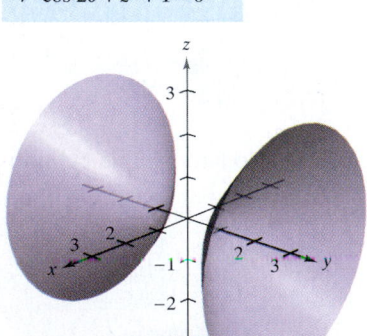

Rectangular coordinates:
$y^2 - x^2 - z^2 = 1$

Figure 10.73

EXAMPLE 3 Rectangular-to-Cylindrical Conversion

Find equations in cylindrical coordinates for the surfaces whose rectangular equations are as follows.

a. $x^2 + y^2 = 4z^2$

b. $y^2 = x$

Solution

a. From the preceding section, you know that the graph $x^2 + y^2 = 4z^2$ is a "double-napped" cone with its axis along the z-axis, as shown in Figure 10.71. If you replace $x^2 + y^2$ with r^2, the equation in cylindrical coordinates is

$$x^2 + y^2 = 4z^2 \qquad \text{Rectangular coordinate equation}$$
$$r^2 = 4z^2. \qquad \text{Cylindrical coordinate equation}$$

b. The graph of the surface $y^2 = x$ is a parabolic cylinder with rulings parallel to the z-axis, as shown in Figure 10.72. By replacing y^2 with $r^2 \sin^2 \theta$ and x with $r \cos \theta$, you obtain the following equation in cylindrical coordinates.

$$y^2 = x \qquad \text{Rectangular equation}$$
$$r^2 \sin^2 \theta = r \cos \theta \qquad \text{Substitute } r \sin \theta \text{ for } y \text{ and } r \cos \theta \text{ for } x.$$
$$r(r \sin^2 \theta - \cos \theta) = 0 \qquad \text{Collect terms and factor.}$$
$$r \sin^2 \theta - \cos \theta = 0 \qquad \text{Divide both sides by } r.$$
$$r = \frac{\cos \theta}{\sin^2 \theta} \qquad \text{Solve for } r.$$
$$r = \csc \theta \cot \theta \qquad \text{Cylindrical equation}$$

Note that this equation includes a point for which $r = 0$, so nothing was lost by dividing both sides by the factor r.

EXAMPLE 4 Cylindrical-to-Rectangular Conversion

Find a rectangular equation for the graph represented by the cylindrical equation

$$r^2 \cos 2\theta + z^2 + 1 = 0.$$

Solution

$$r^2 \cos 2\theta + z^2 + 1 = 0 \qquad \text{Cylindrical equation}$$
$$r^2(\cos^2 \theta - \sin^2 \theta) + z^2 + 1 = 0 \qquad \text{Trigonometric identity}$$
$$r^2 \cos^2 \theta - r^2 \sin^2 \theta + z^2 = -1$$
$$x^2 - y^2 + z^2 = -1 \qquad \text{Replace } r \cos \theta \text{ with } x \text{ and } r \sin \theta \text{ with } y.$$
$$y^2 - x^2 - z^2 = 1 \qquad \text{Rectangular equation}$$

This is a hyperboloid of two sheets whose axis lies along the y-axis, as shown in Figure 10.73.

Spherical Coordinates

In the **spherical coordinate system,** each point is represented by an ordered triple: the first coordinate is a distance, and the second and third coordinates are angles. This system is similar to the latitude-longitude system used to identify points on the surface of the earth. For example, the point on the surface of the earth whose latitude is 40° North (of the equator) and whose longitude is 80° West (of the prime meridian) is shown in Figure 10.74. Assuming that the earth is spherical and has a radius of 4000 miles, you would label this point as

$$(4000, -80°, 50°).$$

Radius   80° clockwise from   50° down from
prime meridian   North Pole

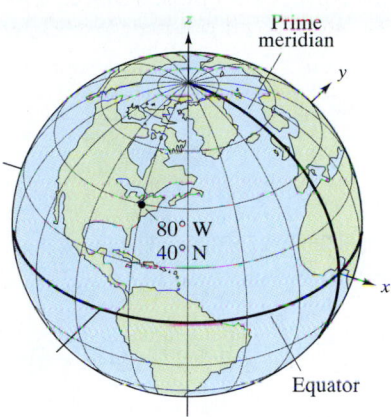

Figure 10.74

The Spherical Coordinate System

In a **spherical coordinate system,** a point P in space is represented by an ordered triple (ρ, θ, ϕ).

1. ρ is the distance between P and the origin, $\rho \geq 0$.

2. θ is the same angle used in cylindrical coordinates for $r \geq 0$.

3. ϕ is the angle *between* the positive z-axis and the line segment $\overrightarrow{OP}$, $0 \leq \phi \leq \pi$.

Note that the first and third coordinates, ρ and ϕ, are nonnegative. ρ is the lowercase Greek letter rho, and ϕ is the lowercase Greek letter phi.

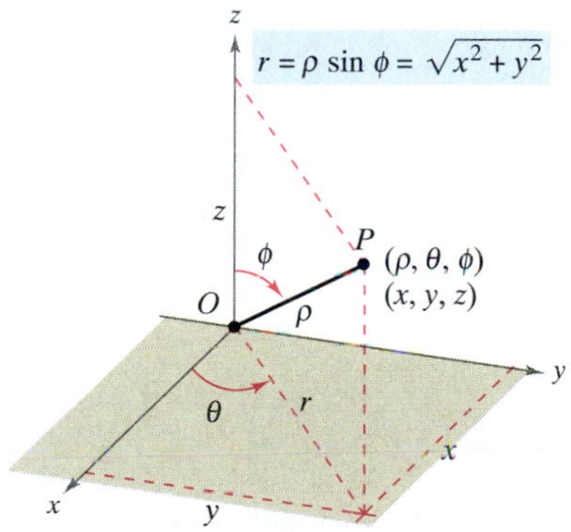

The relationship between rectangular and spherical coordinates is illustrated in Figure 10.75. To convert from one system to the other, use the following.

Spherical to rectangular:

$$x = \rho \sin \phi \cos \theta, \qquad y = \rho \sin \phi \sin \theta, \qquad z = \rho \cos \phi$$

Rectangular to spherical:

$$\rho^2 = x^2 + y^2 + z^2, \qquad \tan \theta = \frac{y}{x}, \qquad \phi = \arccos\left(\frac{z}{\sqrt{x^2 + y^2 + z^2}}\right)$$

To change coordinates between the cylindrical and spherical systems, use the following.

Spherical to cylindrical ($r \geq 0$):

$$r^2 = \rho^2 \sin^2 \phi, \qquad \theta = \theta, \qquad z = \rho \cos \phi$$

Cylindrical to spherical ($r \geq 0$):

$$\rho = \sqrt{r^2 + z^2}, \qquad \theta = \theta, \qquad \phi = \arccos\left(\frac{z}{\sqrt{r^2 + z^2}}\right)$$

Spherical coordinates
Figure 10.75

The spherical coordinate system is useful primarily for surfaces in space that have a *point* or *center* of symmetry. For example, Figure 10.76 shows three surfaces with simple spherical equations.

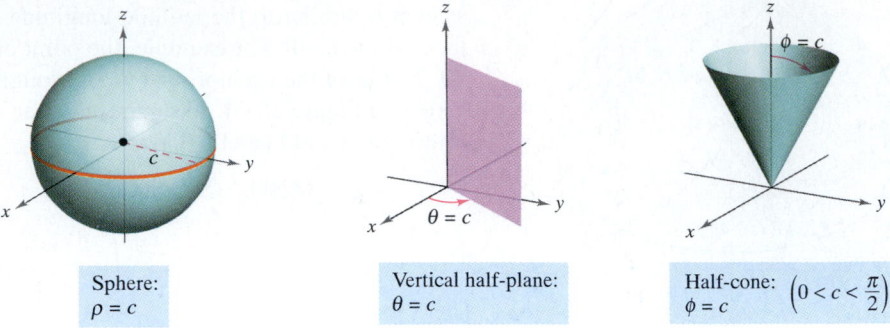

Sphere:
$\rho = c$

Vertical half-plane:
$\theta = c$

Half-cone: $\left(0 < c < \dfrac{\pi}{2}\right)$
$\phi = c$

Figure 10.76

EXAMPLE 5   Rectangular-to-Spherical Conversion

Find an equation in spherical coordinates for the surface represented by each of the following rectangular coordinate equations.

a. Cone: $x^2 + y^2 = z^2$

b. Sphere: $x^2 + y^2 + z^2 - 4z = 0$

Solution

a. Making the appropriate replacements for x, y, and z in the given equation yields the following.

$$x^2 + y^2 = z^2$$
$$\rho^2 \sin^2 \phi \cos^2 \theta + \rho^2 \sin^2 \phi \sin^2 \theta = \rho^2 \cos^2 \phi$$
$$\rho^2 \sin^2 \phi (\cos^2 \theta + \sin^2 \theta) = \rho^2 \cos^2 \phi$$
$$\rho^2 \sin^2 \phi = \rho^2 \cos^2 \phi$$
$$\frac{\sin^2 \phi}{\cos^2 \phi} = 1 \qquad {\color{red}\rho \geq 0}$$
$$\tan^2 \phi = 1 \qquad {\color{red}\phi = \pi/4 \text{ or } \phi = 3\pi/4}$$

The equation $\phi = \pi/4$ represents the *upper* half-cone, and the equation $\phi = 3\pi/4$ represents the *lower* half-cone.

b. Because $\rho^2 = x^2 + y^2 + z^2$ and $z = \rho \cos \phi$, the given equation has the following spherical form.

$$\rho^2 - 4\rho \cos \phi = 0 \qquad \Longrightarrow \qquad \rho(\rho - 4 \cos \phi) = 0$$

Temporarily discarding the possibility that $\rho = 0$, you have the spherical equation

$$\rho - 4 \cos \phi = 0 \qquad \text{or} \qquad \rho = 4 \cos \phi.$$

Note that the solution set for this equation includes a point for which $\rho = 0$, so nothing is lost by discarding the factor ρ. The sphere represented by the equation $\rho = 4 \cos \phi$ is shown in Figure 10.77.

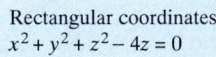

Rectangular coordinates:
$x^2 + y^2 + z^2 - 4z = 0$

Spherical:
$\rho = 4 \cos \phi$

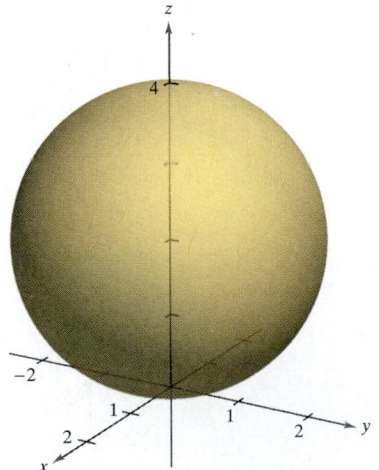

Figure 10.77

EXERCISES FOR SECTION 10.7

In Exercises 1–6, convert the point from rectangular coordinates to cylindrical coordinates.

1. $(0, 5, 1)$ **2.** $(2\sqrt{2}, -2\sqrt{2}, 4)$

3. $(1, \sqrt{3}, 4)$ **4.** $(\sqrt{3}, -1, 2)$

5. $(2, -2, -4)$ **6.** $(-3, 2 -1)$

In Exercises 7–12, convert the point from cylindrical coordinates to rectangular coordinates.

7. $(5, 0, 2)$ **8.** $(4, \pi/2, -2)$

9. $(2, \pi/3, 2)$ **10.** $(3, -\pi/4, 1)$

11. $(4, 7\pi/6, 3)$ **12.** $(1, 3\pi/2, 1)$

In Exercises 13–20, find an equation in rectangular coordinates for the equation in cylindrical coordinates, and sketch its graph.

13. $r = 2$ **14.** $z = 2$

15. $\theta = \pi/6$ **16.** $r = \frac{1}{2}z$

17. $r = 2\sin\theta$ **18.** $r = 2\cos\theta$

19. $r^2 + z^2 = 4$ **20.** $z = r^2\sin^2\theta$

In Exercises 21–26, convert the point from rectangular coordinates to spherical coordinates.

21. $(4, 0, 0)$ **22.** $(1, 1, 1)$

23. $(-2, 2\sqrt{3}, 4)$ **24.** $(2, 2, 4\sqrt{2})$

25. $(\sqrt{3}, 1, 2\sqrt{3})$ **26.** $(-4, 0, 0)$

In Exercises 27–32, convert the point from spherical coordinates to rectangular coordinates.

27. $(4, \pi/6, \pi/4)$ **28.** $(12, 3\pi/4, \pi/9)$

29. $(12, -\pi/4, 0)$ **30.** $(9, \pi/4, \pi)$

31. $(5, \pi/4, 3\pi/4)$ **32.** $(6, \pi, \pi/2)$

33. *Programming*

(a) Write a program for a graphing utility that converts a point from rectangular coordinates to spherical coordinates.

(b) Use the program in part (a) to convert the point $(3, -4, 2)$ from rectangular coordinates to spherical coordinates.

34. *Programming*

(a) Write a program for a graphing utility that converts a point from spherical coordinates to rectangular coordinates.

(b) Use the program in part (a) to convert the point $(5, 1, 0.5)$ from spherical coordinates to rectangular coordinates.

In Exercises 35–42, find an equation in rectangular coordinates for the equation in spherical coordinates, and sketch its graph.

35. $\rho = 2$ **36.** $\theta = \dfrac{3\pi}{4}$

37. $\phi = \dfrac{\pi}{6}$ **38.** $\phi = \dfrac{\pi}{2}$

39. $\rho = 4\cos\phi$ **40.** $\rho = 2\sec\phi$

41. $\rho = \csc\phi$ **42.** $\rho = 4\csc\phi\sec\theta$

In Exercises 43–48, convert the point from cylindrical coordinates to spherical coordinates.

43. $(4, \pi/4, 0)$ **44.** $(2, 2\pi/3, -2)$

45. $(4, -\pi/6, 6)$ **46.** $(-4, \pi/3, 4)$

47. $(12, \pi, 5)$ **48.** $(4, \pi/2, 3)$

In Exercises 49–54, convert the point from spherical coordinates to cylindrical coordinates.

49. $(10, \pi/6, \pi/2)$ **50.** $(4, \pi/18, \pi/2)$

51. $(6, -\pi/6, \pi/3)$ **52.** $(5, -5\pi/6, \pi)$

53. $(8, 7\pi/6, \pi/6)$ **54.** $(7, \pi/4, 3\pi/4)$

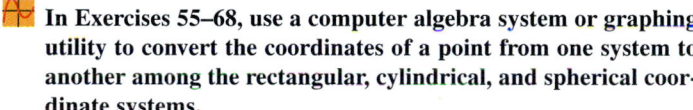 In Exercises 55–68, use a computer algebra system or graphing utility to convert the coordinates of a point from one system to another among the rectangular, cylindrical, and spherical coordinate systems.

	Rectangular	Cylindrical	Spherical
55.	$(4, 6, 3)$		
56.	$(6, -2, -3)$		
57.		$(5, \pi/9, 8)$	
58.		$(10, -0.75, 6)$	
59.			$(20, 2\pi/3, \pi/4)$
60.			$(7.5, 0.25, 1)$
61.	$(3, -2, 2)$		
62.	$(3\sqrt{2}, 3\sqrt{2}, -3)$		
63.	$(5/2, 4/3, -3/2)$		
64.	$(0, -5, 4)$		
65.		$(5, 3\pi/4, -5)$	
66.		$(-2, 11\pi/6, 3)$	
67.		$(-3.5, 2.5, 6)$	
68.		$(8.25, 1.3, -4)$	

In Exercises 69–74, match the equation (expressed in terms of cylindrical or spherical coordinates) with its graph. [The graphs are labeled (a), (b), (c), (d), (e), and (f).]

(a)

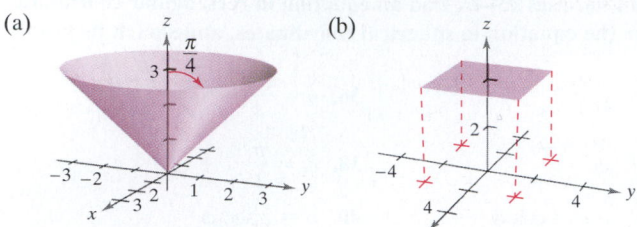

(b)

(c)

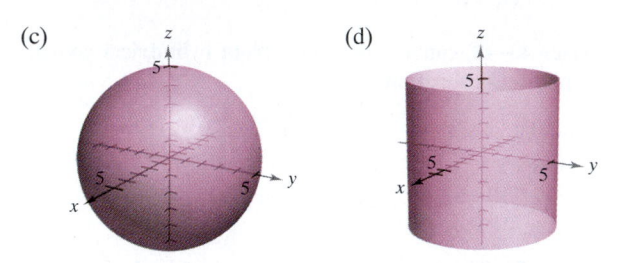

(d)

(e)

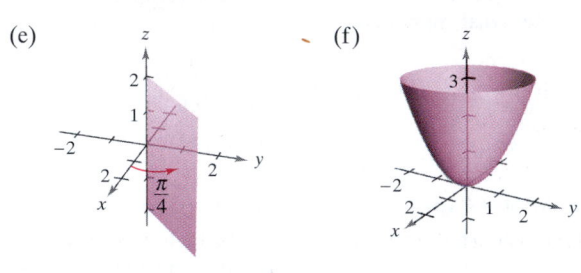

(f)

69. $r = 5$

70. $\theta = \dfrac{\pi}{4}$

71. $\rho = 5$

72. $\phi = \dfrac{\pi}{4}$

73. $r^2 = z$

74. $\rho = 4 \sec \phi$

In Exercises 75–82, convert the rectangular equation to (a) cylindrical coordinates and (b) spherical coordinates.

75. $x^2 + y^2 + z^2 = 16$

76. $4(x^2 + y^2) = z^2$

77. $x^2 + y^2 + z^2 - 2z = 0$

78. $x^2 + y^2 = z$

79. $x^2 + y^2 = 4y$

80. $x^2 + y^2 = 16$

81. $x^2 - y^2 = 9$

82. $y = 4$

In Exercises 83–86, sketch the solid that has the given description in cylindrical coordinates.

83. $0 \le \theta \le \pi/2, 0 \le r \le 2, 0 \le z \le 4$

84. $-\pi/2 \le \theta \le \pi/2, 0 \le r \le 3, 0 \le z \le r \cos \theta$

85. $0 \le \theta \le 2\pi, 0 \le r \le a, r \le z \le a$

86. $0 \le \theta \le 2\pi, 2 \le r \le 4, z^2 \le -r^2 + 6r - 8$

In Exercises 87 and 88, sketch the solid that has the given description in spherical coordinates.

87. $0 \le \theta \le 2\pi, 0 \le \phi \le \pi/6, 0 \le \rho \le a \sec \phi$

88. $0 \le \theta \le 2\pi, \pi/4 \le \phi \le \pi/2, 0 \le \rho \le 1$

Think About It **In Exercises 89–92, find inequalities that describe the solid, and state the coordinate system used. Position the solid on the coordinate system such that the inequalities are as simple as possible.**

89. A cube with each edge 10 centimeters long

90. A cylindrical shell 8 meters long with an inside diameter of 0.75 meter and an outside diameter of 1.25 meters.

91. A spherical shell with inside and outside radii of 4 inches and 6 inches

92. The solid that remains after a hole 1 inch in diameter is drilled through the center of a sphere 6 inches in diameter

93. Identify the curve of intersection of the surfaces (in cylindrical coordinates) $z = \sin \theta$ and $r = 1$.

94. Identify the curve of intersection of the surfaces (in spherical coordinates) $\rho = 2 \sec \phi$ and $\rho = 4$.

SECTION PROJECT

Los Angeles is located at 34.05° North latitude and 118.24° West longitude, and Rio de Janeiro, Brazil is located at 22.90° South latitude and 43.22° West longitude (see figure). Assume that the earth is spherical and has a radius of 4000 miles.

(a) Find the spherical coordinates for the location of each city.

(b) Find the rectangular coordinates for the location of each city.

(c) Find the angle (in radians) between the vectors from the center of the earth to each city.

(d) Find the great-circle distance s between the cities. (*Hint:* $s = r\theta$)

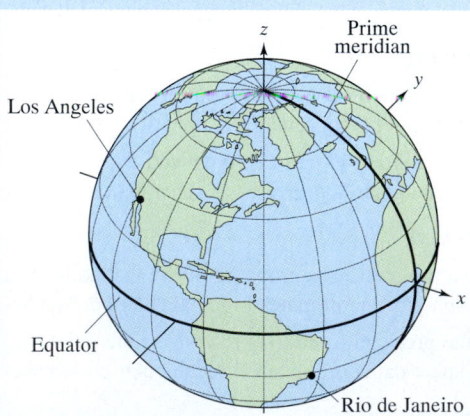

(e) Repeat parts (a)–(d) for the cities of Boston, located at 42.36° North latitude and 71.06° West longitude, and Honolulu, located at 21.31° North latitude and 157.86° West longitude.

REVIEW EXERCISES FOR CHAPTER 10

In Exercises 1 and 2, let $u = \overrightarrow{PQ}$ and $v = \overrightarrow{PR}$, and find (a) the component forms of u and v, (b) the magnitude of v, (c) u · v, (d) 2u + v, (e) the vector component of u in the direction of v, and (f) the vector component of u orthogonal to v.

1. $P(1, 2), Q(4, 1), R(5, 4)$
2. $P(-2, -1), Q(5, -1), R(2, 4)$

3. Find the coordinates of the point in the xy-plane four units to the right of the xz-plane and five units behind the yz-plane.

4. Find the coordinates of the point located on the y-axis and seven units to the left of the xz-plane.

In Exercises 5 and 6, determine the location of the point (x, y, z) so that the given condition is satisfied.

5. $yz > 0$
6. $xy < 0$

In Exercises 7 and 8, find the standard form of the equation of the sphere.

7. Center: $(3, -2, 6)$; Diameter: 15
8. Endpoints of a diameter: $(0, 0, 4), (4, 6, 0)$

In Exercises 9 and 10, find the center and radius of the sphere and sketch its graph.

9. $x^2 + y^2 + z^2 - 4x - 6y + 4 = 0$
10. $x^2 + y^2 + z^2 - 10x + 6y - 4z + 34 = 0$

In Exercises 11 and 12, let $u = \overrightarrow{PQ}$ and $v = \overrightarrow{PR}$, and find (a) the component forms of u and v, (b) u · v, (c) u × v, (d) an equation of the plane containing P, Q, and R, and (e) a set of parametric equations of the line through P and Q.

11. $P(5, 0, 0), Q(4, 4, 0), R(2, 0, 6)$
12. $P(2, -1, 3), Q(0, 5, 1), R(5, 5, 0)$

In Exercises 13 and 14, determine whether the vectors are orthogonal, parallel, or neither.

13. $\langle 7, -2, 3 \rangle, \langle -1, 4, 5 \rangle$
14. $\langle -4, 3, -6 \rangle, \langle 16, -12, 24 \rangle$

In Exercises 15–18, find the angle θ between the vectors u and v.

15. $u = 5[\cos(3\pi/4)\mathbf{i} + \sin(3\pi/4)\mathbf{j}]$
    $v = 2[\cos(2\pi/3)\mathbf{i} + \sin(2\pi/3)\mathbf{j}]$
16. $u = \langle 4, -1, 5 \rangle$, $\quad v = \langle 3, 2, -2 \rangle$
17. $u = \langle 10, -5, 15 \rangle$, $\quad v = \langle -2, 1, -3 \rangle$
18. $u = \langle 1, 0, -3 \rangle$, $\quad v = \langle 2, -2, 1 \rangle$

In Exercises 19–22, find the component form of u.

19. The angle, measured counterclockwise, from the positive x-axis to u is 135°, and $\|u\| = 4$. (Assume that u is in the plane.)

20. The angle between u and the positive x-axis is 180°, and $\|u\| = 8$. (Assume that u is in the plane.)

21. u is perpendicular to the plane $x - 3y + 4z = 0$, and $\|u\| = 3$.

22. u is a unit vector perpendicular to the lines

$$x = 4 - t, y = 3 + 2t, z = 1 + 5t$$
$$x = -3 + 7s, y = -2 + s, z = 1 + 2s.$$

In Exercises 23–32, let $u = \langle 3, -2, 1 \rangle$, $v = \langle 2, -4, -3 \rangle$, and $w = \langle -1, 2, 2 \rangle$.

23. Find $\|u\|$.
24. Find the angle between u and v.
25. Show that $u \cdot u = \|u\|^2$.
26. Determine a unit vector perpendicular to the plane containing v and w.
27. Determine the projection of w onto u.
28. Show that $u \times v = -(v \times u)$.
29. Show that $u \cdot (v + w) = u \cdot v + u \cdot w$.
30. Show that $u \times (v + w) = (u \times v) + (u \times w)$.
31. Find the volume of the solid whose edges are u, v, and w.
32. Find the work done in moving an object along the vector u, if the applied force is w.

33. **Load Supports** Find the tension in each of the supporting cables in the figure.

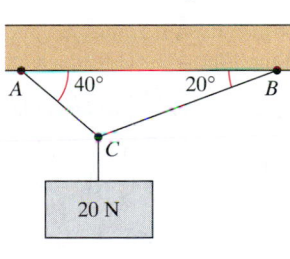

34. **Equilibrium** A 100-pound collar slides on a frictionless vertical rod (see figure). Find the distance y for which the system is in equilibrium if the counterweight weighs 120 pounds.

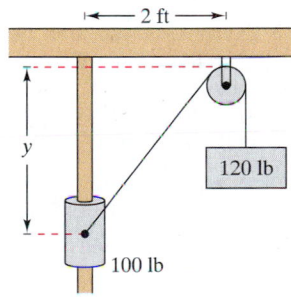

Figure for 33   **Figure for 34**

35. *Torque* The specifications for a tractor state that the torque on a bolt with head size 7/8 inch cannot exceed 200 foot-pounds. Determine the maximum force $\|\mathbf{F}\|$ that can be applied to the wrench in the figure.

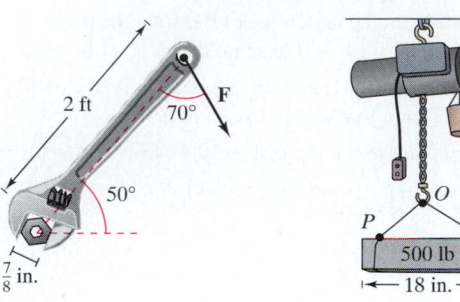

Figure for 35   **Figure for 36**

36. *Minimum Length* In a manufacturing process, an electric hoist lifts 500-pound ingots (see figure). The length of the cable connecting points P, O, and Q is L inches. (Assume that O is at the midpoint of the cable.)

(a) Write the tension T in the cable as a function of L. What is the domain of the function?

(b) Use the function in part (a) to complete the table.

L	19	20	21	22	23	24	25
T							

(c) Use a graphing utility to graph the tension function.

(d) Find the shortest cable connecting points P, O, and Q that can be used if the tension in the cable cannot exceed 400 pounds.

(e) Find (if possible) $\lim\limits_{L \to \infty} T$. Interpret the result in the context of the problem.

In Exercises 37–40, find (a) a set of parametric equations and (b) a set of symmetric equations for the line.

37. The line passes through the point $(1, 2, 3)$ and is perpendicular to the xz-plane.

38. The line passes through the point $(1, 2, 3)$ and is parallel to the line given by $x = y = z$.

39. The intersection of the planes $3x - 3y - 7z = -4$ and $x - y + 2z = 3$.

40. The line passes through the point $(0, 1, 4)$ and is perpendicular to $\mathbf{u} = \langle 2, -5, 1 \rangle$ and $\mathbf{v} = \langle -3, 1, 4 \rangle$.

In Exercises 41 and 42, find an equation of the plane.

41. The plane contains the lines $(x - 1)/(-2) = y = z + 1$ and $(x + 1)/(-2) = y - 1 = z - 2$.

42. The plane passes through the points $(-3, -4, 2)$, $(-3, 4, 1)$, and $(1, 1, -2)$.

43. Find the distance between $(1, 0, 2)$ and the plane $2x - 3y + 6z = 6$.

44. Find the distance between $5x - 3y + z = 2$ and $5x - 3y + z = -3$.

In Exercises 45 and 46, find the distance between the lines.

45. $\dfrac{x}{1} = \dfrac{y}{2} = \dfrac{z}{3}$

$\dfrac{x + 1}{-1} = \dfrac{y}{3} = \dfrac{z + 2}{2}$

46. $x - 4 = \dfrac{y - 3}{-1} = \dfrac{z - 7}{3}$

$\dfrac{x + 3}{-5} = \dfrac{y - 7}{2} = \dfrac{z + 5}{-6}$

In Exercises 47–54, sketch the graph of the surface.

47. $x + 2y + 3z = 6$

48. $y = z^2$

49. $y = \frac{1}{2}z$

50. $y = \cos z$

51. $\dfrac{x^2}{16} + \dfrac{y^2}{9} + z^2 = 1$

52. $16x^2 + 16y^2 - 9z^2 = 0$

53. $\dfrac{x^2}{16} - \dfrac{y^2}{9} + z^2 = -1$

54. $\dfrac{x^2}{25} + \dfrac{y^2}{4} - \dfrac{z^2}{100} = 1$

55. *Machine Design* The top of a rubber bushing designed to absorb vibrations in an automobile is the surface of revolution generated by revolving the curve $z = \frac{1}{2}y^2 + 1(0 \le y \le 2)$ in the yz-plane about the z-axis (see figure).

(a) Find an equation for the surface of revolution.

(b) If all measurements are in centimeters and the bushing is set on the xy-plane, use the shell method to find its volume.

(c) Suppose the bushing has a hole of diameter 1 cm through its center and parallel to the axis of revolution. Find the volume of the rubber bushing.

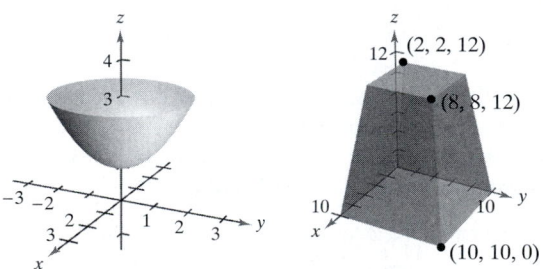

Figure for 55   **Figure for 56**

56. *Machine Design* A tractor fuel tank has the shape and dimensions shown in the figure. In fabricating the tank, it is necessary to know the angle between two adjacent sides. Find the angle.

In Exercises 57 and 58, convert the point from rectangular coordinates to (a) cylindrical coordinates and (b) spherical coordinates.

57. $\left(-2\sqrt{2}, 2\sqrt{2}, 2\right)$

58. $\left(\sqrt{3}/4, 3/4, 3\sqrt{3}/2\right)$

In Exercises 59 and 60, find an equation of the surface in (a) cylindrical coordinates and (b) spherical coordinates.

59. $x^2 - y^2 = 2z$

60. $x^2 + y^2 + z^2 = 16$

Vector-Valued Functions

11

Duomo

Lyn St. James of Daytona Beach, Florida has qualified six times for the Indianapolis 500. In 1996, her qualifying speed was more than 224 miles per hour.

Just northwest of Indianapolis, Indiana sits the Indianapolis Motor Speedway, host to the internationally famous Indianapolis 500 Mile Race. The $2\frac{1}{2}$-mile track consists of front and back stretches that measure 3300 feet each, north and south straightaways that are each 660 feet long, and four 1320-foot-long turns.

A race car's position on a turn (see figure) could be modeled by the vector function

$$\mathbf{r}(t) = x(t)\mathbf{i} + y(t)\mathbf{j} \qquad \text{Position function}$$

$$= 840(\cos 0.349t)\mathbf{i} + 840(\sin 0.349t)\mathbf{j}, \qquad 0 \le t \le 4.5.$$

The velocity (comprising both speed and direction) can also be represented by a vector function. Any change in velocity, whether in speed or direction, is called acceleration. If the velocity is constant, you do not "feel" the motion and the acceleration is zero. You might say that you cannot feel motion, you can only feel *changes* in motion.

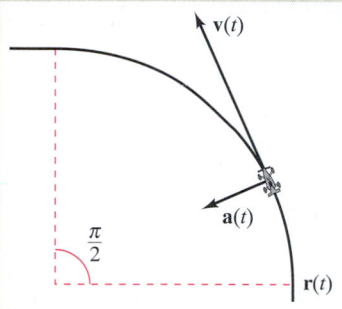

QUESTIONS

1. Use the function **r** to make a table of the car's positions for several values of t between $t = 0$ and $t = 4.5$.

t	0	0.5	1	1.5	2	2.5	3	3.5	4	4.5
$x(t)$										
$y(t)$										

Plot your results. Describe the car's path during this interval of time.

2. Use a graphing utility to graph the car's path. Do the results agree with your conclusion in Question 1?

3. As the graphing utility plots the car's position, does the car appear to be taking the turn at a constant speed? Does it appear to be taking the turn at a constant velocity? Explain the distinction between these two questions.

4. In this chapter you will learn that the velocity for an object whose position function is **r** is given by

$$\mathbf{v}(t) = x'(t)\mathbf{i} + y'(t)\mathbf{j}.$$

Find the velocity vector for the car described above. Use your result to find a function that describes the speed of the car at any time. Is the speed of the car constant? Explain your reasoning.

5. The acceleration vector for the car is given by $\mathbf{a}(t) = x''(t)\mathbf{i} + y''(t)\mathbf{j}$. Describe the physical relationship between the car's velocity and acceleration. Is the driver "feeling" any acceleration? Explain your reasoning.

As of 1996, Indianapolis Speedway had been the scene of 82 international automobile races, beginning in 1911. The winner of the first race was Ray Harroun with a speed of 74.6 miles per hour.

The concepts presented here will be explored further in this chapter. For an extension of this application, see the lab series that accompanies this text.

Space Curves and Vector-Valued Functions • Limits and Continuity

Space Curves and Vector-Valued Functions

In Section 9.2, a *plane curve* was defined as the set of ordered pairs $(f(t), g(t))$ together with their defining parametric equations

$$x = f(t) \quad \text{and} \quad y = g(t)$$

where f and g are continuous functions of t on an interval I. This definition can be extended naturally to three-dimensional space as follows. A **space curve** C is the set of all ordered triples $(f(t), g(t), h(t))$ together with their defining parametric equations

$$x = f(t), \quad y = g(t), \quad \text{and} \quad z = h(t)$$

where f, g, and h are continuous functions of t on an interval I.

Before looking at examples of space curves, we introduce a new type of function, called a **vector-valued function**, that maps real numbers onto vectors.

Definition of a Vector-Valued Function

A function of the form

$$\mathbf{r}(t) = f(t)\mathbf{i} + g(t)\mathbf{j} \qquad \text{Plane}$$

or

$$\mathbf{r}(t) = f(t)\mathbf{i} + g(t)\mathbf{j} + h(t)\mathbf{k} \qquad \text{Space}$$

is a **vector-valued function**, where the **component functions** f, g, and h are real-valued functions of the parameter t. Vector-valued functions are sometimes denoted as $\mathbf{r}(t) = \langle f(t), g(t) \rangle$ or $\mathbf{r}(t) = \langle f(t), g(t), h(t) \rangle$.

NOTE Technically, a curve in the plane or in space consists of a collection of points *and* the defining parametric equations. Two different curves can have the same graph. For instance, each of the curves given by $\mathbf{r} = \sin t\, \mathbf{i} + \cos t\, \mathbf{j}$ and $\mathbf{r} = \sin t^2\, \mathbf{i} + \cos t^2\, \mathbf{j}$ has the unit circle as its graph, but these equations do not represent the same curve—because the circle is traced out in different ways on the graphs.

Be sure you see the distinction between the vector-valued function $\mathbf{r}$ and the real-valued functions f, g, and h. All are functions of the real variable t, but $\mathbf{r}(t)$ is a vector, whereas $f(t)$, $g(t)$, and $h(t)$ are real numbers (for each specific value of t).

Vector-valued functions serve dual roles in the representation of curves. By letting the parameter t represent time, you can use a vector-valued function to represent *motion* along a curve. Or, in the more general case, you can use a vector-valued function to *trace the graph* of a curve. In either case, the terminal point of the position vector $\mathbf{r}(t)$ coincides with the point (x, y) or (x, y, z) on the curve given by the parametric equations, as shown in Figure 11.1. The arrowhead on the curve indicates the curve's *orientation* by pointing in the direction of increasing values of t.

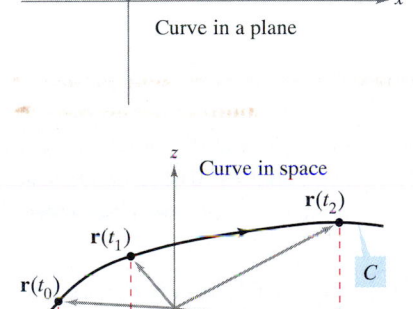

Curve in a plane

Curve in space

Curve C is traced out by the terminal point of position vector $\mathbf{r}(t)$.
Figure 11.1

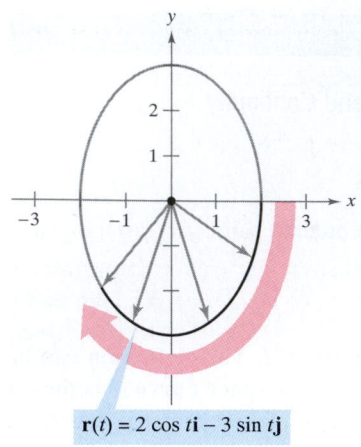

$\mathbf{r}(t) = 2\cos t\,\mathbf{i} - 3\sin t\,\mathbf{j}$

The ellipse is tracked clockwise as t increases from 0 to 2π.
Figure 11.2

Unless stated otherwise, the **domain** of a vector-valued function $\mathbf{r}$ is considered to be the intersection of the domains of the component functions f, g, and h. For instance, the domain of

$$\mathbf{r}(t) = (\ln t)\mathbf{i} + \sqrt{1 - t}\,\mathbf{j} + t\mathbf{k}$$

is the interval $(0, 1]$.

EXAMPLE 1   Sketching a Plane Curve

Sketch the plane curve represented by the vector-valued function

$$\mathbf{r}(t) = 2\cos t\,\mathbf{i} - 3\sin t\,\mathbf{j}, \qquad 0 \le t \le 2\pi.$$

Solution   From the position vector $\mathbf{r}(t)$, you can write the parametric equations $x = 2\cos t$ and $y = -3\sin t$. Solving for $\cos t$ and $\sin t$ and using the identity $\cos^2 t + \sin^2 t = 1$ produces the rectangular equation

$$\frac{x^2}{2^2} + \frac{y^2}{3^2} = 1.$$

The graph of this rectangular equation is the ellipse shown in Figure 11.2. The curve has a *clockwise* orientation. That is, as t increases from 0 to 2π, the position vector $\mathbf{r}(t)$ moves clockwise, and its terminal point traces the ellipse.

EXAMPLE 2   Sketching a Space Curve

Sketch the space curve represented by the vector-valued function

$$\mathbf{r}(t) = 4\cos t\,\mathbf{i} + 4\sin t\,\mathbf{j} + t\mathbf{k}, \qquad 0 \le t \le 4\pi.$$

Solution   From the first two parametric equations $x = 4\cos t$ and $y = 4\sin t$, you can obtain

$$x^2 + y^2 = 16.$$

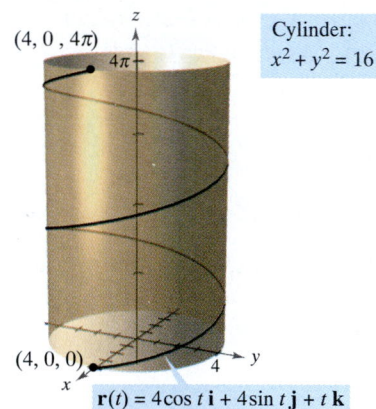

As t increases from 0 to 4π, two spirals on the helix are traced out.
Figure 11.3

This means that the curve lies on a right circular cylinder of radius 4, centered about the z-axis. To locate the curve on this cylinder, you can use the third parametric equation $z = t$. In Figure 11.3, note that as t increases from 0 to 4π the point (x, y, z) spirals up the cylinder to produce a **helix**. A real-life example of a helix is shown in the drawing at lower left.

In Examples 1 and 2, you were given a vector-valued function and asked to sketch the corresponding curve. The next two examples concern the reverse problem—finding a vector-valued function to represent a given graph. Of course, if the graph is described parametrically, representation by a vector-valued function is straightforward. For instance, to represent the line in space given by

$$x = 2 + t, \qquad y = 3t, \qquad \text{and} \qquad z = 4 - t$$

you can simply use the vector-valued function given by

$$\mathbf{r}(t) = (2 + t)\mathbf{i} + 3t\mathbf{j} + (4 - t)\mathbf{k}.$$

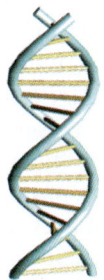

The DNA molecule, discovered in 1962 by Francis Crick and James D. Watson, is in the shape of a double helix.

If a set of parametric equations for the graph is not given, the problem of representing the graph by a vector-valued function boils down to finding a set of parametric equations.

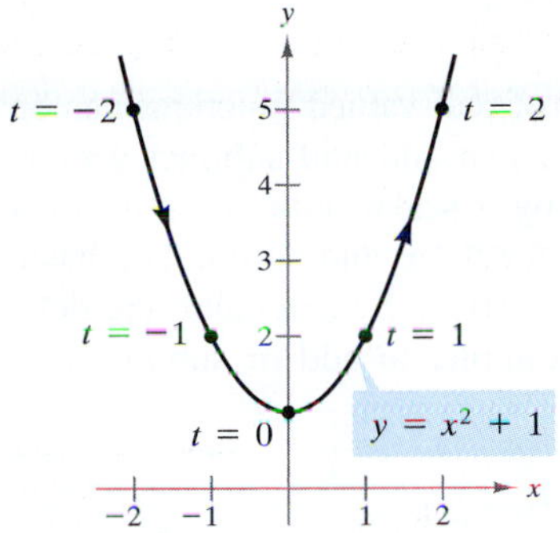

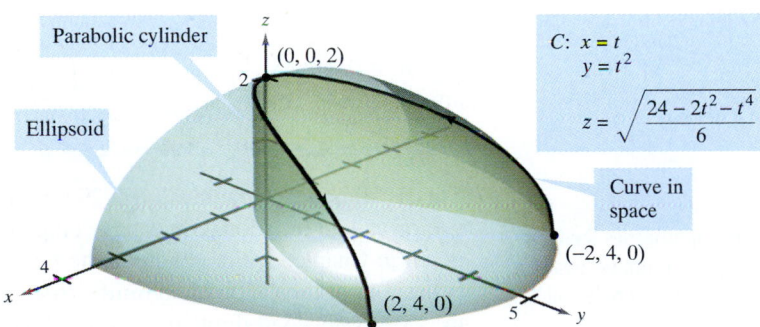

There are many ways to parametrize this graph. One way is to let $x = t$.
Figure 11.4

EXAMPLE 3   Representing a Graph by a Vector-Valued Function

Represent the parabola given by $y = x^2 + 1$ by a vector-valued function.

Solution Although there are many ways to choose the parameter t, a natural choice is to let $x = t$. Then $y = t^2 + 1$ and you have

$$\mathbf{r}(t) = t\mathbf{i} + (t^2 + 1)\mathbf{j}.$$

Note in Figure 11.4 the orientation produced by this particular choice of parameter. Had you chosen $x = -t$ as the parameter, the curve would have been oriented in the opposite direction.

EXAMPLE 4   Representing a Graph by a Vector-Valued Function

Sketch the graph C represented by the intersection of the semiellipsoid

$$\frac{x^2}{12} + \frac{y^2}{24} + \frac{z^2}{4} = 1, \qquad z \geq 0$$

and the parabolic cylinder $y = x^2$. Then, find a vector-valued function to represent the graph.

Solution The intersection of the two surfaces is shown in Figure 11.5. As in Example 3, a natural choice of parameter is $x = t$. For this choice, you can use the given equation $y = x^2$ to obtain $y = t^2$. Then, it follows that

$$\frac{z^2}{4} = 1 - \frac{x^2}{12} - \frac{y^2}{24} = 1 - \frac{t^2}{12} - \frac{t^4}{24} = \frac{24 - 2t^2 - t^4}{24}.$$

NOTE   Curves in space can be specified in various ways. For instance, the curve in Example 4 is described as the intersection of two surfaces in space.

Because the curve lies above the xy-plane, you should choose the positive square root for z and obtain the following parametric equations.

$$x = t, \qquad y = t^2, \qquad \text{and} \qquad z = \sqrt{\frac{24 - 2t^2 - t^4}{6}}$$

The resulting vector-valued function is

$$\mathbf{r}(t) = t\mathbf{i} + t^2\mathbf{j} + \sqrt{\frac{24 - 2t^2 - t^4}{6}}\,\mathbf{k}, \qquad -2 \leq t \leq 2.$$

From the points $(-2, 4, 0)$ and $(2, 4, 0)$ shown in Figure 11.5, you can see that the curve is traced as t increases from -2 to 2.

The curve C is the intersection of the semiellipsoid and the parabolic cylinder.
Figure 11.5

Limits and Continuity

Many techniques and definitions used in the calculus of real-valued functions can be applied to vector-valued functions. For instance, you can add and subtract vector-valued functions, multiply a vector-valued function by a scalar, take the limit of a vector-valued function, differentiate a vector-valued function, and so on. The basic approach is to capitalize on the linearity of vector operations by extending the definitions on a component-by-component basis. For example, to add or subtract two vector-valued functions (in the plane), you can write

$$\mathbf{r}_1(t) + \mathbf{r}_2(t) = [f_1(t)\mathbf{i} + g_1(t)\mathbf{j}] + [f_2(t)\mathbf{i} + g_2(t)\mathbf{j}] \qquad \text{Sum}$$
$$= [f_1(t) + f_2(t)]\mathbf{i} + [g_1(t) + g_2(t)]\mathbf{j}$$
$$\mathbf{r}_1(t) - \mathbf{r}_2(t) = [f_1(t)\mathbf{i} + g_1(t)\mathbf{j}] - [f_2(t)\mathbf{i} + g_2(t)\mathbf{j}] \qquad \text{Difference}$$
$$= [f_1(t) - f_2(t)]\mathbf{i} + [g_1(t) - g_2(t)]\mathbf{j}.$$

Similarly, to multiply and divide a vector-valued function by a scalar, you can write

$$c\mathbf{r}(t) = c[f_1(t)\mathbf{i} + g_1(t)\mathbf{j}] \qquad \text{Scalar multiplication}$$
$$= cf_1(t)\mathbf{i} + cg_1(t)\mathbf{j}$$
$$\frac{\mathbf{r}(t)}{c} = \frac{[f_1(t)\mathbf{i} + g_1(t)\mathbf{j}]}{c}, \qquad c \neq 0 \qquad \text{Scalar division}$$
$$= \frac{f_1(t)}{c}\mathbf{i} + \frac{g_1(t)}{c}\mathbf{j}.$$

This component-by-component extension of operations with real-valued functions to vector-valued functions is further illustrated in the following definition of the limit of a vector-valued function.

Definition of the Limit of a Vector-Valued Function

1. If **r** is a vector-valued function such that $\mathbf{r}(t) = f(t)\mathbf{i} + g(t)\mathbf{j}$, then

$$\lim_{t \to a} \mathbf{r}(t) = \left[\lim_{t \to a} f(t)\right]\mathbf{i} + \left[\lim_{t \to a} g(t)\right]\mathbf{j} \qquad \text{Plane}$$

   provided f and g have limits as $t \to a$.

2. If **r** is a vector-valued function such that $\mathbf{r}(t) = f(t)\mathbf{i} + g(t)\mathbf{j} + h(t)\mathbf{k}$, then

$$\lim_{t \to a} \mathbf{r}(t) = \left[\lim_{t \to a} f(t)\right]\mathbf{i} + \left[\lim_{t \to a} g(t)\right]\mathbf{j} + \left[\lim_{t \to a} h(t)\right]\mathbf{k} \qquad \text{Space}$$

   provided f, g, and h have limits as $t \to a$.

If $\mathbf{r}(t)$ approaches the vector **L** as $t \to a$, the length of the vector $\mathbf{r}(t) - \mathbf{L}$ approaches 0. That is,

$$\|\mathbf{r}(t) - \mathbf{L}\| \to 0 \qquad \text{as} \qquad t \to a.$$

This is illustrated graphically in Figure 11.6. With this definition of the limit of a vector-valued function, you can develop vector versions of most of the limit theorems given in Chapter 1. For example, the limit of the sum of two vector-valued functions is the sum of their individual limits. Also, you can use the orientation of the curve $\mathbf{r}(t)$ to define one-sided limits of vector-valued functions. The next definition extends the notion of continuity to vector-valued functions.

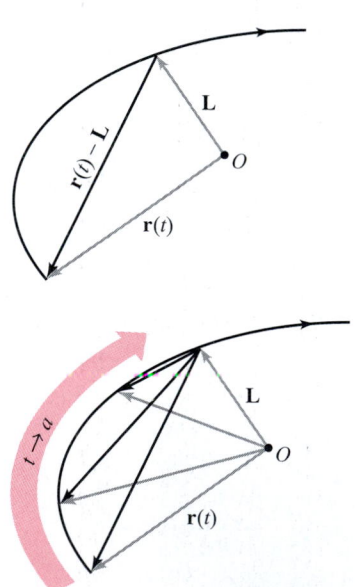

As t approaches a, $\mathbf{r}(t)$ approaches the limit **L**. For the limit **L** to exist, it is not necessary that $\mathbf{r}(a)$ be defined or that $\mathbf{r}(a)$ be equal to **L**.

Figure 11.6

Definition of Continuity of a Vector-Valued Function

A vector-valued function **r** is **continuous at the point** given by $t = a$ if the limit of **r**(t) exists as $t \rightarrow a$ and

$$\lim_{t \to a} \mathbf{r}(t) = \mathbf{r}(a).$$

A vector-valued function **r** is **continuous on an interval** I if it is continuous at every point in the interval.

From this definition, it follows that a vector-valued function is continuous at $t = a$ if and only if each of its component functions is continuous at $t = a$.

EXAMPLE 5   Continuity of Vector-Valued Functions

Discuss the continuity of the vector-valued function given by

$$\mathbf{r}(t) = t\mathbf{i} + a\mathbf{j} + (a^2 - t^2)\mathbf{k}, \qquad \textcolor{red}{a \text{ is a constant.}}$$

at $t = 0$.

Solution   As t approaches 0, the limit is

$$\lim_{t \to 0} \mathbf{r}(t) = \left[\lim_{t \to 0} t \right]\mathbf{i} + \left[\lim_{t \to 0} a \right]\mathbf{j} + \left[\lim_{t \to 0} (a^2 - t^2) \right]\mathbf{k}$$

$$= 0\mathbf{i} + a\mathbf{j} + a^2\mathbf{k}$$

$$= a\mathbf{j} + a^2\mathbf{k}.$$

Because

$$\mathbf{r}(0) = (0)\mathbf{i} + (a)\mathbf{j} + (a^2)\mathbf{k} = a\mathbf{j} + a^2\mathbf{k}$$

you can conclude that **r** is continuous at $t = 0$. By similar reasoning, you can conclude that the vector-valued function **r** is continuous at all real-number values of t.

For each value of a, the curve represented by the vector-valued function in Example 5,

$$\mathbf{r}(t) = t\mathbf{i} + a\mathbf{j} + (a^2 - t^2)\mathbf{k}, \qquad \textcolor{red}{a \text{ is a constant.}}$$

is a parabola. You can think of each parabola as the intersection of the vertical plane $y = a$ and the hyperbolic paraboloid

$$y^2 - x^2 = z$$

as shown in Figure 11.7.

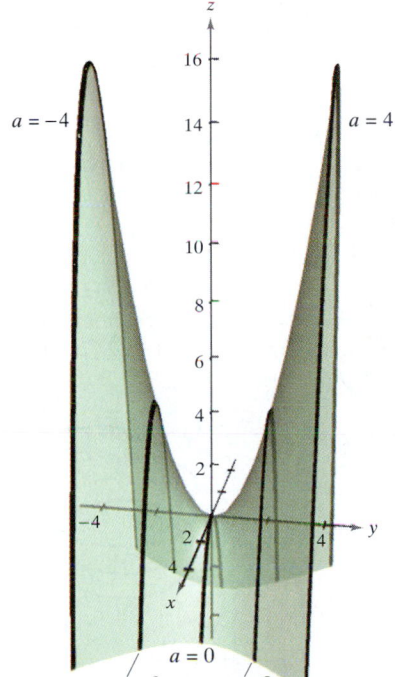

For each value of a, the curve represented by the vector-valued function
$$\mathbf{r}(t) = t\mathbf{i} + a\mathbf{j} + (a^2 - t^2)\mathbf{k} \text{ is a}$$
parabola.
Figure 11.7

TECHNOLOGY   Almost any type of three-dimensional sketch is difficult to do by hand, but sketching curves in space is especially difficult. The problem is in trying to create the illusion of three dimensions. Graphing utilities use a variety of techniques to add "three-dimensionality" to sketches of space curves: one way is to show the curve on a surface, as in Figure 11.7.

EXERCISES FOR SECTION 11.1

In Exercises 1–8, find the domain of the vector-valued function.

1. $\mathbf{r}(t) = 5t\mathbf{i} - 4t\mathbf{j} - \dfrac{1}{t}\mathbf{k}$

2. $\mathbf{r}(t) = \sqrt{4 - t^2}\,\mathbf{i} + t^2\mathbf{j} - 6t\mathbf{k}$

3. $\mathbf{r}(t) = \ln t\,\mathbf{i} - e^t\mathbf{j} - t\mathbf{k}$

4. $\mathbf{r}(t) = \sin t\,\mathbf{i} + 4\cos t\,\mathbf{j} + t\mathbf{k}$

5. $\mathbf{r}(t) = \mathbf{F}(t) + \mathbf{G}(t)$ where

$\mathbf{F}(t) = \cos t\,\mathbf{i} - \sin t\,\mathbf{j} + \sqrt{t}\,\mathbf{k}$, $\mathbf{G}(t) = \cos t\,\mathbf{i} + \sin t\,\mathbf{j}$

6. $\mathbf{r}(t) = \mathbf{F}(t) - \mathbf{G}(t)$ where

$\mathbf{F}(t) = \ln t\,\mathbf{i} + 5t\mathbf{j} - 3t^2\mathbf{k}$, $\mathbf{G}(t) = \mathbf{i} + 4t\mathbf{j} - 3t^2\mathbf{k}$

7. $\mathbf{r}(t) = \mathbf{F}(t) \times \mathbf{G}(t)$ where

$\mathbf{F}(t) = \sin t\,\mathbf{i} + \cos t\,\mathbf{j}$, $\mathbf{G}(t) = \sin t\,\mathbf{j} + \cos t\,\mathbf{k}$

8. $\mathbf{r}(t) = \mathbf{F}(t) \times \mathbf{G}(t)$ where

$\mathbf{F}(t) = t^3\mathbf{i} - t\mathbf{j} + t\mathbf{k}$, $\mathbf{G}(t) = \sqrt[3]{t}\,\mathbf{i} + \dfrac{1}{t+1}\mathbf{j} + (t+2)\mathbf{k}$

In Exercises 9–12, evaluate (if possible) the vector-valued function at the indicated value of t.

9. $\mathbf{r}(t) = \frac{1}{2}t^2\mathbf{i} - (t-1)\mathbf{j}$

(a) $\mathbf{r}(1)$ (b) $\mathbf{r}(0)$ (c) $\mathbf{r}(s+1)$ (d) $\mathbf{r}(2 + \Delta t) - \mathbf{r}(2)$

10. $\mathbf{r}(t) = \cos t\,\mathbf{i} + 2\sin t\,\mathbf{j}$

(a) $\mathbf{r}(0)$ (b) $\mathbf{r}(\pi/4)$ (c) $\mathbf{r}(\theta - \pi)$

(d) $\mathbf{r}(\pi/6 + \Delta t) - \mathbf{r}(\pi/6)$

11. $\mathbf{r}(t) = \ln t\,\mathbf{i} + \dfrac{1}{t}\mathbf{j} + 3t\mathbf{k}$

(a) $\mathbf{r}(2)$ (b) $\mathbf{r}(-3)$ (c) $\mathbf{r}(t-4)$

(d) $\mathbf{r}(1 + \Delta t) - \mathbf{r}(1)$

12. $\mathbf{r}(t) = \sqrt{t}\,\mathbf{i} + t^{3/2}\mathbf{j} + e^{-t/4}\mathbf{k}$

(a) $\mathbf{r}(0)$ (b) $\mathbf{r}(4)$ (c) $\mathbf{r}(c+2)$

(d) $\mathbf{r}(9 + \Delta t) - \mathbf{r}(9)$

In Exercises 13 and 14, find $\|\mathbf{r}(t)\|$.

13. $\mathbf{r}(t) = \sin 3t\,\mathbf{i} + \cos 3t\,\mathbf{j} + t\mathbf{k}$

14. $\mathbf{r}(t) = \sqrt{t}\,\mathbf{i} + 3t\mathbf{j} - 4t\mathbf{k}$

Think About It **In Exercises 15 and 16, find $\mathbf{r}(t) \cdot \mathbf{u}(t)$. Is the result a vector-valued function? Explain.**

15. $\mathbf{r}(t) = (3t - 1)\mathbf{i} + \frac{1}{4}t^3\mathbf{j} + 4\mathbf{k}$

$\mathbf{u}(t) = t^2\mathbf{i} - 8\mathbf{j} + t^3\mathbf{k}$

16. $\mathbf{r}(t) = \langle 3\cos t, 2\sin t, t - 2 \rangle$

$\mathbf{u}(t) = \langle 4\sin t, -6\cos t, t^2 \rangle$

In Exercises 17–20, match the equation with its graph. [The graphs are labeled (a), (b), (c), and (d).]

(a) 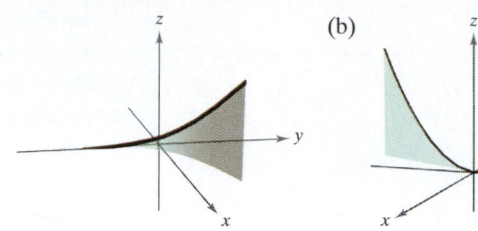   (b)

(c) 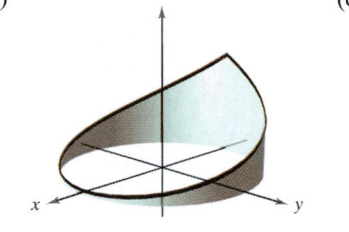   (d)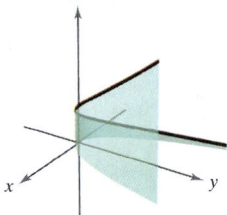

17. $\mathbf{r}(t) = t\mathbf{i} + 2t\mathbf{j} + t^2\mathbf{k}$, $-2 \le t \le 2$

18. $\mathbf{r}(t) = \cos(\pi t)\mathbf{i} + \sin(\pi t)\mathbf{j} + t^2\mathbf{k}$, $-1 \le t \le 1$

19. $\mathbf{r}(t) = t\mathbf{i} + t^2\mathbf{j} + e^{0.75t}\mathbf{k}$, $-2 \le t \le 2$

20. $\mathbf{r}(t) = t\mathbf{i} + \ln t\,\mathbf{j} + \dfrac{2t}{3}\mathbf{k}$, $0.1 \le t \le 5$

21. Think About It The four figures below are graphs of the vector-valued function

$$\mathbf{r}(t) = 4\cos t\,\mathbf{i} + 4\sin t\,\mathbf{j} + \dfrac{t}{4}\mathbf{k}.$$

Match the four graphs with the point in space from which the helix is viewed. The four points are $(0, 0, 20)$, $(20, 0, 0)$, $(-20, 0, 0)$, and $(10, 20, 10)$.

(a)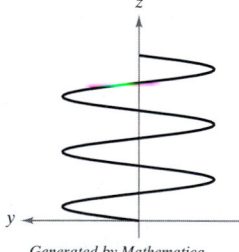

Generated by Mathematica

(b)

Generated by Mathematica

(c)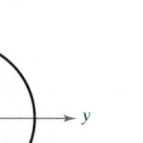

Generated by Mathematica

(d)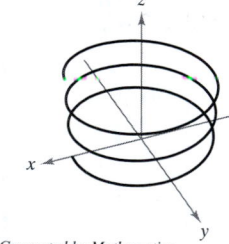

Generated by Mathematica

22. Sketch three graphs of the vector-valued function

$$\mathbf{r}(t) = t\mathbf{i} + t\mathbf{j} + 2\mathbf{k}$$

as viewed from the points

(a) $(0, 0, 20)$ (b) $(10, 0, 0)$ (c) $(5, 5, 5)$.

In Exercises 23–32, sketch the curve represented by the vector-valued function and give the orientation of the curve.

23. $\mathbf{r}(t) = 3t\mathbf{i} + (t - 1)\mathbf{j}$

24. $\mathbf{r}(t) = 2 \cos t\mathbf{i} + 2 \sin t\mathbf{j}$

25. $\mathbf{r}(t) = (-t + 1)\mathbf{i} + (4t + 2)\mathbf{j} + (2t + 3)\mathbf{k}$

26. $\mathbf{r}(t) = t\mathbf{i} + (2t - 5)\mathbf{j} + 3t\mathbf{k}$

27. $\mathbf{r}(t) = 2 \cos t\mathbf{i} + 2 \sin t\mathbf{j} + t\mathbf{k}$

28. $\mathbf{r}(t) = 3 \cos t\mathbf{i} + 4 \sin t\mathbf{j} + \dfrac{t}{2}\mathbf{k}$

29. $\mathbf{r}(t) = 2 \sin t\mathbf{i} + 2 \cos t\mathbf{j} + e^{-t}\mathbf{k}$

30. $\mathbf{r}(t) = t\mathbf{i} + t^2\mathbf{j} + \dfrac{3t}{2}\mathbf{k}$

31. $\mathbf{r}(t) = \left\langle t, t^2, \frac{2}{3}t^3 \right\rangle$

32. $\mathbf{r}(t) = \left\langle \cos t + t \sin t, \sin t - t \cos t, t \right\rangle$

In Exercises 33–36, use a computer algebra system to graph the vector-valued function and identify the common curve.

33. $\mathbf{r}(t) = -\dfrac{1}{2}t^2\mathbf{i} + t\mathbf{j} - \dfrac{\sqrt{3}}{2}t^2\mathbf{k}$

34. $\mathbf{r}(t) = t\mathbf{i} - \dfrac{\sqrt{3}}{2}t^2\mathbf{j} + \dfrac{1}{2}t^2\mathbf{k}$

35. $\mathbf{r}(t) = \sin t\mathbf{i} + \left(\dfrac{\sqrt{3}}{2} \cos t - \dfrac{1}{2}t \right)\mathbf{j} + \left(\dfrac{1}{2} \cos t + \dfrac{\sqrt{3}}{2} \right)\mathbf{k}$

36. $\mathbf{r}(t) = -\sqrt{2} \sin t\mathbf{i} + 2 \cos t\mathbf{j} + \sqrt{2} \sin t\mathbf{k}$

37. *Think About It* Use a computer algebra system to graph the vector-valued function

$$\mathbf{r}(t) = 2 \cos t\mathbf{i} + 2 \sin t\mathbf{j} + \tfrac{1}{2}t\mathbf{k}.$$

For each of the following, make a conjecture about the transformation (if any) of the graph. Use the computer algebra system to check your conjectures.

(a) $\mathbf{u}(t) = 2(\cos t - 1)\mathbf{i} + 2 \sin t\mathbf{j} + \tfrac{1}{2}t\mathbf{k}$

(b) $\mathbf{u}(t) = 2 \cos t\mathbf{i} + 2 \sin t\mathbf{j} + 2t\mathbf{k}$

(c) $\mathbf{u}(t) = 2 \cos(-t)\mathbf{i} + 2 \sin(-t)\mathbf{j} + \tfrac{1}{2}(-t)\mathbf{k}$

(d) $\mathbf{u}(t) = \tfrac{1}{2}t\mathbf{i} + 2 \sin t\mathbf{j} + 2 \cos t\mathbf{k}$

(e) $\mathbf{u}(t) = 6 \cos t\mathbf{i} + 6 \sin t\mathbf{j} + \tfrac{1}{2}t\mathbf{k}$

38. *Think About It* Given a vector-valued function $\mathbf{r}(t)$, is the graph of the vector-valued function $\mathbf{u}(t) = \mathbf{r}(t - 2)$ a horizontal translation of the graph of $\mathbf{r}(t)$? Explain.

In Exercises 39–42, represent the plane curve by a vector-valued function. (There are many correct answers.)

39. $y = 4 - x$

40. $y = 4 - x^2$

41. $x^2 + y^2 = 25$

42. $\dfrac{x^2}{25} + \dfrac{y^2}{16} = 1$

43. A particle moves on a straight line path that passes through the points $(2, 3, 0)$ and $(0, 8, 8)$. Find a vector-valued function for the path. Use a computer algebra system to graph your function. (There are many correct answers.)

44. The outer edge of a playground slide is in the shape of a helix of radius 1.5 meters. The slide has a height of 2 meters and makes one complete revolution from top to bottom. Find a vector-valued function for the helix. Use a computer algebra system to graph your function. (There are many correct answers.)

In Exercises 45–48, find vector-valued functions forming the boundaries of the region in the figure. State the interval for the parameter of each function.

45.

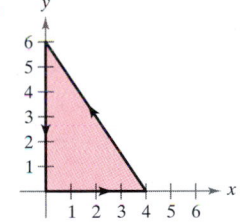

46.

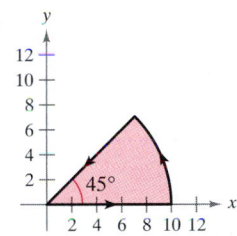

47.

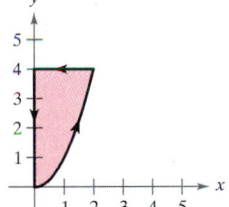

48.

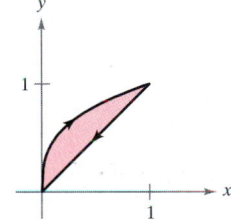

In Exercises 49–56, sketch the space curve represented by the intersection of the surfaces. Then represent the curve by a vector-valued function using the given parameter.

Surfaces	Parameter
49. $z = x^2 + y^2, \; x + y = 0$	$x = t$
50. $z = x^2 + y^2, \; z = 4$	$x = 2 \cos t$
51. $x^2 + y^2 = 4, \; z = x^2$	$x = 2 \sin t$
52. $4x^2 + y^2 + 4z^2 = 16, \; x = y^2$	$y = t$
53. $x^2 + y^2 + z^2 = 4, \; x + z = 2$	$x = 1 + \sin t$
54. $x^2 + y^2 + z^2 = 10, \; x + y = 4$	$x = 2 + \sin t$
55. $x^2 + z^2 = 4, \; y^2 + z^2 = 4$	$x = t$ (first octant)
56. $x^2 + y^2 + z^2 = 16, \; xy = 4$	$x = t$ (first octant)

In Exercises 57–62, evaluate the limit.

57. $\lim\limits_{t \to 2} \left(t\mathbf{i} + \dfrac{t^2 - 4}{t^2 - 2t}\mathbf{j} + \dfrac{1}{t}\mathbf{k} \right)$

58. $\lim\limits_{t \to 0} \left(e^t\mathbf{i} + \dfrac{\sin t}{t}\mathbf{j} + e^{-t}\mathbf{k} \right)$

59. $\lim\limits_{t \to 0} \left(t^2\mathbf{i} + 3t\mathbf{j} + \dfrac{1 - \cos t}{t}\mathbf{k} \right)$

60. $\lim\limits_{t \to 1} \left(\sqrt{t}\,\mathbf{i} + \dfrac{\ln t}{t^2 - 1}\mathbf{j} + 2t^2\mathbf{k} \right)$

61. $\lim\limits_{t \to 0} \left(\dfrac{1}{t}\mathbf{i} + \cos t\mathbf{j} + \sin t\mathbf{k} \right)$

62. $\lim\limits_{t \to \infty} \left(e^{-t}\mathbf{i} + \dfrac{1}{t}\mathbf{j} + \dfrac{t}{t^2 + 1}\mathbf{k} \right)$

In Exercises 63–68, determine the interval(s) on which the vector-valued function is continuous.

63. $\mathbf{r}(t) = t\mathbf{i} + \dfrac{1}{t}\mathbf{j}$

64. $\mathbf{r}(t) = \sqrt{t}\,\mathbf{i} + \sqrt{t - 1}\,\mathbf{j}$

65. $\mathbf{r}(t) = t\mathbf{i} + \arcsin t\mathbf{j} + (t - 1)\mathbf{k}$

66. $\mathbf{r}(t) = \sin t\mathbf{i} + \cos t\mathbf{j} + \ln t\mathbf{k}$

67. $\mathbf{r}(t) = \langle e^{-t}, t^2, \tan t \rangle$

68. $\mathbf{r}(t) = \langle 8, \sqrt{t}, \sqrt[3]{t} \rangle$

69. Let $\mathbf{r}(t)$ and $\mathbf{u}(t)$ be vector-valued functions whose limits exist as $t \to c$. Prove that

$$\lim_{t \to c} [\mathbf{r}(t) \times \mathbf{u}(t)] = \lim_{t \to c} \mathbf{r}(t) \times \lim_{t \to c} \mathbf{u}(t).$$

70. Let $\mathbf{r}(t)$ and $\mathbf{u}(t)$ be vector-valued functions whose limits exist as $t \to c$. Prove that

$$\lim_{t \to c} [\mathbf{r}(t) \cdot \mathbf{u}(t)] = \lim_{t \to c} \mathbf{r}(t) \cdot \lim_{t \to c} \mathbf{u}(t).$$

71. Prove that if $\mathbf{r}$ is a vector-valued function that is continuous at c, then $\|\mathbf{r}\|$ is continuous at c.

72. Verify that the converse of Exercise 71 is not true by finding a vector-valued function $\mathbf{r}$ such that $\|\mathbf{r}\|$ is continuous at c but $\mathbf{r}$ is not continuous at c.

True or False? **In Exercises 73 and 74, determine whether the statement is true or false. If it is false, explain why or give an example that shows it is false.**

73. If f, g, and h are first-degree polynomial functions, then the curve given by $x = f(t)$, $y = g(t)$, and $z = h(t)$ is a line.

74. If the curve given by $x = f(t)$, $y = g(t)$, and $z = h(t)$ is a line, then f, g, and h are first-degree polynomial functions of t.

SECTION PROJECT

On page 190 in Section 3.5, you studied a famous curve called the **witch of Agnesi.** In this project you will take a closer look at this function.

Consider a circle of radius a centered on the y-axis at $(0, a)$, as shown in the figure. Let A be a point on the horizontal line $y = 2a$, and let B be the point where the segment OA intersects the circle. A point P is on the witch of Agnesi if P lies on the horizontal line through B and on the vertical line through A.

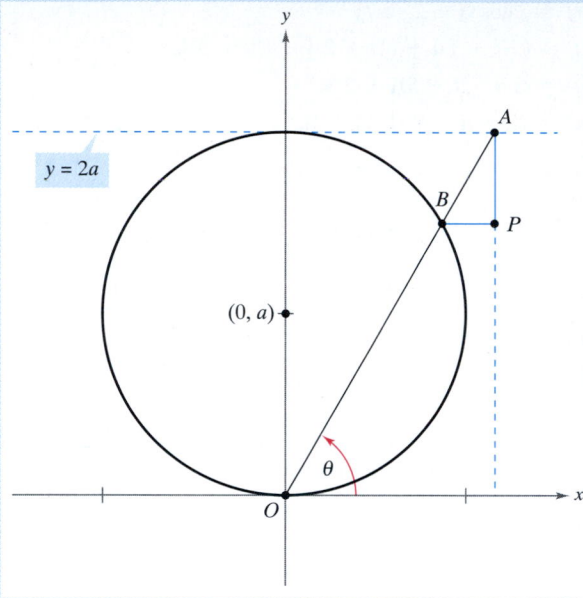

(a) Show that the point A is traced out by the vector-valued function

$$\mathbf{r}_A(\theta) = 2a \cot \theta\mathbf{i} + 2a\mathbf{j}\qquad 0 < \theta < \pi$$

where θ is the angle OA makes with the positive x-axis.

(b) Show that the point B is traced out by the vector-valued function

$$\mathbf{r}_B(\theta) = a \sin 2\theta\mathbf{i} + a(1 - \cos 2\theta)\mathbf{j}\qquad 0 < \theta < \pi.$$

(c) Combine these results to find the vector-valued function $\mathbf{r}(\theta)$ for the witch of Agnesi. Use your graphing utility to graph this curve for $a = 1$.

(d) Describe the following limits.

$$\lim_{\theta \to 0^+} \mathbf{r}(\theta)\qquad \lim_{\theta \to \pi^-} \mathbf{r}(\theta)$$

(e) Eliminate the parameter θ and determine the rectangular equation of the witch of Agnesi. Use a graphing utility to graph this function for $a = 1$ and compare your graph with that obtained in part (c).

Differentiation of Vector-Valued Functions • Integration of Vector-Valued Functions

Differentiation of Vector-Valued Functions

In Sections 11.3–11.5, you will study several important applications involving the calculus of vector-valued functions. In preparation for that study, this section is devoted to the mechanics of differentiation and integration of vector-valued functions.

The definition of the derivative of a vector-valued function parallels that given for real-valued functions.

Definition of the Derivative of a Vector-Valued Function

The **derivative of a vector-valued function r** is defined by

$$\mathbf{r}'(t) = \lim_{\Delta t \to 0} \frac{\mathbf{r}(t + \Delta t) - \mathbf{r}(t)}{\Delta t}$$

for all t for which the limit exists. If $\mathbf{r}'(c)$ exists, then **r** is **differentiable** at c. If $\mathbf{r}'(c)$ exists for all c in an open interval I, then **r** is **differentiable** on the interval I. Differentiability of vector-valued functions can be extended to closed intervals by considering one-sided limits.

NOTE   In addition to $\mathbf{r}'(t)$, other notations for the derivative of a vector-valued function are

$$D_t[\mathbf{r}(t)], \quad \frac{d}{dt}[\mathbf{r}(t)], \quad \text{and} \quad \frac{d\mathbf{r}}{dt}.$$

Differentiation of vector-valued functions can be done on a *component-by-component basis*. To see why this is true, consider the function given by $\mathbf{r}(t) = f(t)\mathbf{i} + g(t)\mathbf{j}$. Applying the definition of the derivative produces the following.

$$\mathbf{r}'(t) = \lim_{\Delta t \to 0} \frac{\mathbf{r}(t + \Delta t) - \mathbf{r}(t)}{\Delta t}$$

$$= \lim_{\Delta t \to 0} \frac{f(t + \Delta t)\mathbf{i} + g(t + \Delta t)\mathbf{j} - f(t)\mathbf{i} - g(t)\mathbf{j}}{\Delta t}$$

$$= \lim_{\Delta t \to 0} \left\{ \left[\frac{f(t + \Delta t) - f(t)}{\Delta t} \right]\mathbf{i} + \left[\frac{g(t + \Delta t) - g(t)}{\Delta t} \right]\mathbf{j} \right\}$$

$$= \left\{ \lim_{\Delta t \to 0} \left[\frac{f(t + \Delta t) - f(t)}{\Delta t} \right] \right\}\mathbf{i} + \left\{ \lim_{\Delta t \to 0} \left[\frac{g(t + \Delta t) - g(t)}{\Delta t} \right] \right\}\mathbf{j}$$

$$= f'(t)\mathbf{i} + g'(t)\mathbf{j}$$

This important result is listed in the following theorem. Note that the derivative of the vector-valued function **r** is itself a vector-valued function. You can see from Figure 11.8 that $\mathbf{r}'(t)$ is a vector tangent to the curve given by $\mathbf{r}(t)$ and pointing in the direction of increasing t-values.

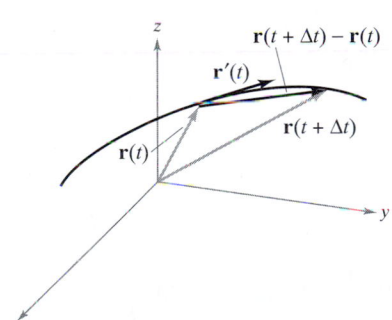

Figure 11.8

> **THEOREM 11.1   Differentiation of Vector-Valued Functions**
>
> **1.** If $\mathbf{r}(t) = f(t)\mathbf{i} + g(t)\mathbf{j}$, where f and g are differentiable functions of t, then
>
> $$\mathbf{r}'(t) = f'(t)\mathbf{i} + g'(t)\mathbf{j}. \qquad \text{Plane}$$
>
> **2.** If $\mathbf{r}(t) = f(t)\mathbf{i} + g(t)\mathbf{j} + h(t)\mathbf{k}$, where f, g, and h are differentiable functions of t, then
>
> $$\mathbf{r}'(t) = f'(t)\mathbf{i} + g'(t)\mathbf{j} + h'(t)\mathbf{k}. \qquad \text{Space}$$

EXAMPLE 1   Differentiation of Vector-Valued Functions

Find the derivative of each of the following vector-valued functions.

a. $\mathbf{r}(t) = t^2\mathbf{i} - 4\mathbf{j}$ **b.** $\mathbf{r}(t) = \dfrac{1}{t}\mathbf{i} + \ln t\mathbf{j} + e^{2t}\mathbf{k}$

Solution Differentiating on a component-by-component basis produces the following.

a. $\mathbf{r}'(t) = 2t\mathbf{i} - 0\mathbf{j} = 2t\mathbf{i}$

b. $\mathbf{r}'(t) = -\dfrac{1}{t^2}\mathbf{i} + \dfrac{1}{t}\mathbf{j} + 2e^{2t}\mathbf{k}$

Higher-order derivatives of vector-valued functions are obtained by successive differentiation of each component function.

EXAMPLE 2   Higher-Order Differentiation

For the vector-valued function given by $\mathbf{r}(t) = \cos t\mathbf{i} + \sin t\mathbf{j} + 2t\mathbf{k}$, find each of the following.

a. $\mathbf{r}'(t)$ **b.** $\mathbf{r}''(t)$ **c.** $\mathbf{r}'(t) \cdot \mathbf{r}''(t)$ **d.** $\mathbf{r}'(t) \times \mathbf{r}''(t)$

Solution

a. $\mathbf{r}'(t) = -\sin t\mathbf{i} + \cos t\mathbf{j} + 2\mathbf{k}$

b. $\mathbf{r}''(t) = -\cos t\mathbf{i} - \sin t\mathbf{j} + 0\mathbf{k} = -\cos t\mathbf{i} - \sin t\mathbf{j}$

c. $\mathbf{r}'(t) \cdot \mathbf{r}''(t) = \sin t\cos t - \sin t\cos t = 0$

d. $\mathbf{r}'(t) \times \mathbf{r}''(t) = \begin{vmatrix} \mathbf{i} & \mathbf{j} & \mathbf{k} \\ -\sin t & \cos t & 2 \\ -\cos t & -\sin t & 0 \end{vmatrix}$

$$= \begin{vmatrix} \cos t & 2 \\ -\sin t & 0 \end{vmatrix}\mathbf{i} - \begin{vmatrix} -\sin t & 2 \\ -\cos t & 0 \end{vmatrix}\mathbf{j} + \begin{vmatrix} -\sin t & \cos t \\ -\cos t & -\sin t \end{vmatrix}\mathbf{k}$$

$$= 2\sin t\mathbf{i} - 2\cos t\mathbf{j} + \mathbf{k}$$

Note that the dot product in part (c) is a *real-valued* function, not a vector-valued function.

The parametrization of the curve represented by the vector-valued function

$$\mathbf{r}(t) = f(t)\mathbf{i} + g(t)\mathbf{j} + h(t)\mathbf{k}$$

is **smooth on an open interval** I if f', g', and h' are continuous on I and $\mathbf{r}'(t) \neq \mathbf{0}$ for any value of t in the interval I.

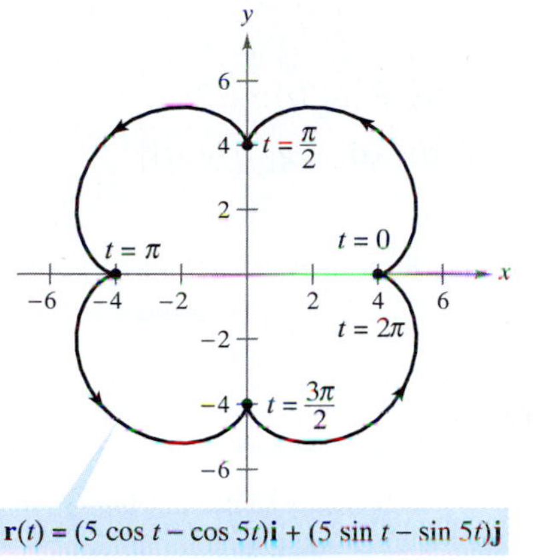

EXAMPLE 3   Finding Intervals on Which a Curve Is Smooth

Find the intervals on which the epicycloid C given by

$$\mathbf{r}(t) = (5 \cos t - \cos 5t)\mathbf{i} + (5 \sin t - \sin 5t)\mathbf{j}, \quad 0 \leq t \leq 2\pi$$

is smooth.

Solution   The derivative of $\mathbf{r}$ is

$$\mathbf{r}'(t) = (-5 \sin t + 5 \sin 5t)\mathbf{i} + (5 \cos t - 5 \cos 5t)\mathbf{j}.$$

In the interval $[0, 2\pi]$, the only values of t for which $\mathbf{r}'(t) = 0\mathbf{i} + 0\mathbf{j}$ are $t = 0, \pi/2$, $\pi, 3\pi/2$, and 2π. Therefore, you can conclude that C is smooth in the intervals

$$\left(0, \frac{\pi}{2}\right), \left(\frac{\pi}{2}, \pi\right), \left(\pi, \frac{3\pi}{2}\right), \quad \text{and} \quad \left(\frac{3\pi}{2}, 2\pi\right)$$

as shown in Figure 11.9.

$\mathbf{r}(t) = (5 \cos t - \cos 5t)\mathbf{i} + (5 \sin t - \sin 5t)\mathbf{j}$

The epicycloid is not smooth at the points where it intersects the axes.
Figure 11.9

NOTE   In Figure 11.9, note that the curve is not smooth at points at which the curve makes an abrupt change in direction. Such points are called **cusps** or **nodes**.

Most of the differentiation rules in Chapter 2 have counterparts for vector-valued functions, and several are listed in the following theorem. Note that the theorem contains three versions of "product rules." Property 3 gives the derivative of the product of a real-valued function f and a vector-valued function $\mathbf{r}$, Property 4 gives the derivative of the dot product of two vector-valued functions, and Property 5 gives the derivative of the cross product of two vector-valued functions (in space). (Property 5 applies only to three-dimensional vector-valued functions, because the cross product is not defined for two-dimensional vectors.)

THEOREM 11.2   Properties of the Derivative

Let $\mathbf{r}$ and $\mathbf{u}$ be differentiable vector-valued functions of t, let f be a differentiable real-valued function of t, and let c be a scalar.

1. $D_t[c\mathbf{r}(t)] = c\mathbf{r}'(t)$
2. $D_t[\mathbf{r}(t) \pm \mathbf{u}(t)] = \mathbf{r}'(t) \pm \mathbf{u}'(t)$
3. $D_t[f(t)\mathbf{r}(t)] = f(t)\mathbf{r}'(t) + f'(t)\mathbf{r}(t)$
4. $D_t[\mathbf{r}(t) \cdot \mathbf{u}(t)] = \mathbf{r}(t) \cdot \mathbf{u}'(t) + \mathbf{r}'(t) \cdot \mathbf{u}(t)$
5. $D_t[\mathbf{r}(t) \times \mathbf{u}(t)] = \mathbf{r}(t) \times \mathbf{u}'(t) + \mathbf{r}'(t) \times \mathbf{u}(t)$
6. $D_t[\mathbf{r}(f(t))] = \mathbf{r}'(f(t))f'(t)$
7. If $\mathbf{r}(t) \cdot \mathbf{r}(t) = c$, then $\mathbf{r}(t) \cdot \mathbf{r}'(t) = 0$.

Proof   To prove Property 4, let

$$\mathbf{r}(t) = f_1(t)\mathbf{i} + g_1(t)\mathbf{j} \quad \text{and} \quad \mathbf{u}(t) = f_2(t)\mathbf{i} + g_2(t)\mathbf{j}$$

where f_1 and g_1 are differentiable functions of t. Then,

$$\mathbf{r}(t) \cdot \mathbf{u}(t) = f_1(t)f_2(t) + g_1(t)g_2(t)$$

and it follows that

$$\begin{aligned} D_t[\mathbf{r}(t) \cdot \mathbf{u}(t)] &= f_1(t)f_2'(t) + f_1'(t)f_2(t) + g_1(t)g_2'(t) + g_1'(t)g_2(t) \\ &= [f_1(t)f_2'(t) + g_1(t)g_2'(t)] + [f_1'(t)f_2(t) + g_1'(t)g_2(t)] \\ &= \mathbf{r}(t) \cdot \mathbf{u}'(t) + \mathbf{r}'(t) \cdot \mathbf{u}(t). \end{aligned}$$

Proofs of the other properties are left as exercises (see Exercises 55–59 and Exercise 62).

EXPLORATION

Let $\mathbf{r}(t) = \cos t\mathbf{i} + \sin t\mathbf{j}$. Sketch the graph of $\mathbf{r}(t)$. Explain why the graph is a circle of radius 1 centered at the origin. Calculate $\mathbf{r}(\pi/4)$ and $\mathbf{r}'(\pi/4)$. Position the vector $\mathbf{r}'(\pi/4)$ so that its initial point is at the terminal point of $\mathbf{r}(\pi/4)$. What do you observe? Show that $\mathbf{r}(t) \cdot \mathbf{r}(t)$ is constant and that $\mathbf{r}(t) \cdot \mathbf{r}'(t) = 0$ for all t. How does this example relate to Property 7 of Theorem 11.2?

EXAMPLE 4   Using Properties of the Derivative

For the vector-valued functions given by

$$\mathbf{r}(t) = \frac{1}{t}\mathbf{i} - \mathbf{j} + (\ln t)\mathbf{k} \quad \text{and} \quad \mathbf{u}(t) = t^2\mathbf{i} - 2t\mathbf{j} + \mathbf{k}$$

find

a. $D_t[\mathbf{r}(t) \cdot \mathbf{u}(t)]$ and   **b.** $D_t[\mathbf{u}(t) \times \mathbf{u}'(t)]$.

Solution

a. Because $\mathbf{r}'(t) = -\dfrac{1}{t^2}\mathbf{i} + \dfrac{1}{t}\mathbf{k}$ and $\mathbf{u}'(t) = 2t\mathbf{i} - 2\mathbf{j}$, you have

$$\begin{aligned} D_t[\mathbf{r}(t) \cdot \mathbf{u}(t)] &= \mathbf{r}(t) \cdot \mathbf{u}'(t) + \mathbf{r}'(t) \cdot \mathbf{u}(t) \\ &= \left(\frac{1}{t}\mathbf{i} - \mathbf{j} + \ln t\mathbf{k} \right) \cdot (2t\mathbf{i} - 2\mathbf{j}) \\ &\quad + \left(-\frac{1}{t^2}\mathbf{i} + \frac{1}{t}\mathbf{k} \right) \cdot (t^2\mathbf{i} - 2t\mathbf{j} + \mathbf{k}) \\ &= 2 + 2 + (-1) + \frac{1}{t} \\ &= 3 + \frac{1}{t}. \end{aligned}$$

b. Because $\mathbf{u}'(t) = 2t\mathbf{i} - 2\mathbf{j}$ and $\mathbf{u}''(t) = 2\mathbf{i}$, you have

$$D_t[\mathbf{u}(t) \times \mathbf{u}'(t)] = [\mathbf{u}(t) \times \mathbf{u}''(t)] + [\mathbf{u}'(t) \times \mathbf{u}'(t)]$$

$$= \begin{vmatrix} \mathbf{i} & \mathbf{j} & \mathbf{k} \\ t^2 & -2t & 1 \\ 2 & 0 & 0 \end{vmatrix} + \mathbf{0}$$

$$= \begin{vmatrix} -2t & 1 \\ 0 & 0 \end{vmatrix}\mathbf{i} - \begin{vmatrix} t^2 & 1 \\ 2 & 0 \end{vmatrix}\mathbf{j} + \begin{vmatrix} t^2 & -2t \\ 2 & 0 \end{vmatrix}\mathbf{k}$$

$$= 0\mathbf{i} - (-2)\mathbf{j} + 4t\mathbf{k}$$

$$= 2\mathbf{j} + 4t\mathbf{k}.$$

NOTE   Try reworking parts (a) and (b) in Example 4 by first forming the dot and cross products and then differentiating to see that you obtain the same results.

Integration of Vector-Valued Functions

The following definition is a rational consequence of the definition of the derivative of a vector-valued function.

Definition of Integration of Vector-Valued Functions

1. If $\mathbf{r}(t) = f(t)\mathbf{i} + g(t)\mathbf{j}$, where f and g are continuous on $[a, b]$, then the **indefinite integral (antiderivative)** of $\mathbf{r}$ is

$$\int \mathbf{r}(t)\,dt = \left[\int f(t)\,dt \right]\mathbf{i} + \left[\int g(t)\,dt \right]\mathbf{j} \qquad \text{Plane}$$

and its **definite integral** over the interval $a \le t \le b$ is

$$\int_a^b \mathbf{r}(t)\,dt = \left[\int_a^b f(t)\,dt \right]\mathbf{i} + \left[\int_a^b g(t)\,dt \right]\mathbf{j}.$$

2. If $\mathbf{r}(t) = f(t)\mathbf{i} + g(t)\mathbf{j} + h(t)\mathbf{k}$, where f, g, and h are continuous on $[a, b]$, then the **indefinite integral (antiderivative)** of $\mathbf{r}$ is

$$\int \mathbf{r}(t)\,dt = \left[\int f(t)\,dt \right]\mathbf{i} + \left[\int g(t)\,dt \right]\mathbf{j} + \left[\int h(t)\,dt \right]\mathbf{k} \qquad \text{Space}$$

and its **definite integral** over the interval $a \le t \le b$ is

$$\int_a^b \mathbf{r}(t)\,dt = \left[\int_a^b f(t)\,dt \right]\mathbf{i} + \left[\int_a^b g(t)\,dt \right]\mathbf{j} + \left[\int_a^b h(t)\,dt \right]\mathbf{k}.$$

The antiderivative of a vector-valued function is a family of vector-valued functions all differing by a constant vector $\mathbf{C}$. For instance, if $\mathbf{r}(t)$ is a three-dimensional vector-valued function, then for the indefinite integral $\int \mathbf{r}(t)\,dt$, you obtain three constants of integration

$$\int f(t)\,dt = F(t) + C_1, \qquad \int g(t)\,dt = G(t) + C_2, \qquad \int h(t)\,dt = H(t) + C_3$$

where $F'(t) = f(t)$, $G'(t) = g(t)$, and $H'(t) = h(t)$. These three *scalar* constants produce one *vector* constant of integration,

$$\int \mathbf{r}(t)\,dt = [F(t) + C_1]\mathbf{i} + [G(t) + C_2]\mathbf{j} + [H(t) + C_3]\mathbf{k}$$

$$= [F(t)\mathbf{i} + G(t)\mathbf{j} + H(t)\mathbf{k}] + [C_1\mathbf{i} + C_2\mathbf{j} + C_3\mathbf{k}]$$

$$= \mathbf{R}(t) + \mathbf{C}$$

where $\mathbf{R}'(t) = \mathbf{r}(t)$.

EXAMPLE 5   Integrating a Vector-Valued Function

Evaluate the indefinite integral

$$\int (t\,\mathbf{i} + 3\mathbf{j})\,dt.$$

Solution   Integrating on a component-by-component basis produces

$$\int (t\,\mathbf{i} + 3\mathbf{j})\,dt = \frac{t^2}{2}\mathbf{i} + 3t\mathbf{j} + \mathbf{C}.$$

Example 6 shows how to evaluate the definite integral of a vector-valued function.

EXAMPLE 6   Definite Integral of a Vector-Valued Function

Evaluate the integral

$$\int_0^1 \mathbf{r}(t)\, dt = \int_0^1 \left(\sqrt[3]{t}\, \mathbf{i} + \frac{1}{t+1}\mathbf{j} + e^{-t}\, \mathbf{k} \right) dt.$$

Solution

$$\int_0^1 \mathbf{r}(t)\, dt = \left(\int_0^1 t^{1/3}\, dt \right)\mathbf{i} + \left(\int_0^1 \frac{1}{t+1}\, dt \right)\mathbf{j} + \left(\int_0^1 e^{-t}\, dt \right)\mathbf{k}$$

$$= \left[\left(\frac{3}{4} \right)t^{4/3} \right]_0^1 \mathbf{i} + \left[\ln|t+1| \right]_0^1 \mathbf{j} + \left[-e^{-t} \right]_0^1 \mathbf{k}$$

$$= \frac{3}{4}\mathbf{i} + (\ln 2)\mathbf{j} + \left(1 - \frac{1}{e} \right)\mathbf{k}$$

As with real-valued functions, you can narrow the family of antiderivatives of a vector-valued function $\mathbf{r}'$ down to a single antiderivative by imposing an initial condition on the vector-valued function $\mathbf{r}$. This is demonstrated in the next example.

EXAMPLE 7   The Antiderivative of a Vector-Valued Function

Find the antiderivative of

$$\mathbf{r}'(t) = \cos 2t\mathbf{i} - 2 \sin t\mathbf{j} + \frac{1}{1+t^2}\mathbf{k}$$

that satisfies the initial condition $\mathbf{r}(0) = 3\mathbf{i} - 2\mathbf{j} + \mathbf{k}$.

Solution

$$\mathbf{r}(t) = \int \mathbf{r}'(t)\, dt$$

$$= \left(\int \cos 2t\, dt \right)\mathbf{i} + \left(\int -2 \sin t\, dt \right)\mathbf{j} + \left(\int \frac{1}{1+t^2}\, dt \right)\mathbf{k}$$

$$= \left(\frac{1}{2} \sin 2t + C_1 \right)\mathbf{i} + (2 \cos t + C_2)\mathbf{j} + (\arctan t + C_3)\mathbf{k}$$

Letting $t = 0$ and using the fact that $\mathbf{r}(0) = 3\mathbf{i} - 2\mathbf{j} + \mathbf{k}$, you have

$$\mathbf{r}(0) = (0 + C_1)\mathbf{i} + (2 + C_2)\mathbf{j} + (0 + C_3)\mathbf{k}$$

$$= 3\mathbf{i} + (-2)\mathbf{j} + \mathbf{k}.$$

Equating corresponding components produces

$$C_1 = 3, \quad 2 + C_2 = -2, \quad \text{and} \quad C_3 = 1.$$

Thus, the antiderivative that satisfies the given initial condition is

$$\mathbf{r}(t) = \left(\frac{1}{2} \sin 2t + 3 \right)\mathbf{i} + (2 \cos t - 4)\mathbf{j} + (\arctan t + 1)\mathbf{k}.$$

EXERCISES FOR SECTION 11.2

In Exercises 1–4, (a) sketch the plane curve represented by the vector-valued function, and (b) sketch the vectors $\mathbf{r}(t_0)$ and $\mathbf{r}'(t_0)$ for the indicated value of t_0. Position the vectors such that the initial point of $\mathbf{r}(t_0)$ is at the origin and the initial point of $\mathbf{r}'(t_0)$ is at the terminal point of $\mathbf{r}(t_0)$. What is the relationship between $\mathbf{r}'(t_0)$ and the curve?

1. $\mathbf{r}(t) = t^2\mathbf{i} + t\mathbf{j}$ $t_0 = 2$

2. $\mathbf{r}(t) = t\mathbf{i} + t^3\mathbf{j}$ $t_0 = 1$

3. $\mathbf{r}(t) = \cos t\mathbf{i} + \sin t\mathbf{j}$ $t_0 = \dfrac{\pi}{2}$

4. $\mathbf{r}(t) = t^2\mathbf{i} + \dfrac{1}{t}\mathbf{j}$ $t_0 = 2$

 5. *Investigation*   Consider the vector-valued function

$\mathbf{r}(t) = t\mathbf{i} + t^2\mathbf{j}$.

(a) Sketch the graph of $\mathbf{r}(t)$. Use a graphing utility to verify your graph.

(b) Sketch the vectors $\mathbf{r}(1/4)$, $\mathbf{r}(1/2)$, and $\mathbf{r}(1/2) - \mathbf{r}(1/4)$ on the graph in part (a).

(c) Compare the vector $\mathbf{r}'(1/4)$ with the vector

$\dfrac{\mathbf{r}(1/2) - \mathbf{r}(1/4)}{1/2 - 1/4}$.

 6. *Investigation*   Consider the vector-valued function

$\mathbf{r}(t) = t\mathbf{i} + (4 - t^2)\mathbf{j}$.

(a) Sketch the graph of $\mathbf{r}(t)$. Use a graphing utility to verify your graph.

(b) Sketch the vectors $\mathbf{r}(1)$, $\mathbf{r}(1.25)$, and $\mathbf{r}(1.25) - \mathbf{r}(1)$ on the graph in part (a).

(c) Compare the vector $\mathbf{r}'(1)$ with the vector

$\dfrac{\mathbf{r}(1.25) - \mathbf{r}(1)}{1.25 - 1}$.

In Exercises 7 and 8, (a) sketch the space curve represented by the vector-valued function, and (b) sketch the vectors $\mathbf{r}(t_0)$ and $\mathbf{r}'(t_0)$ for the indicated value of t_0.

7. $\mathbf{r}(t) = 2\cos t\mathbf{i} + 2\sin t\mathbf{j} + t\mathbf{k}$ $t_0 = \dfrac{3\pi}{2}$

8. $\mathbf{r}(t) = t\mathbf{i} + t^2\mathbf{j} + \dfrac{3}{2}\mathbf{k}$ $t_0 = 2$

In Exercises 9 and 10, a vector-valued function and its graph are given. The graph also shows the unit vectors $\mathbf{r}'(t_0)/\|\mathbf{r}'(t_0)\|$ and $\mathbf{r}''(t_0)/\|\mathbf{r}''(t_0)\|$. Find these two unit vectors and identify them on the graph.

9. $\mathbf{r}(t) = \cos(\pi t)\mathbf{i} + \sin(\pi t)\mathbf{j} + t^2\mathbf{k}$ $t_0 = -\dfrac{1}{4}$

10. $\mathbf{r}(t) = t\mathbf{i} + t^2\mathbf{j} + e^{0.75t}\mathbf{k}$ $t_0 = \dfrac{1}{4}$

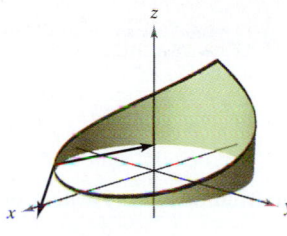
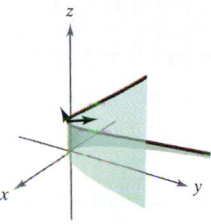

Figure for 9   **Figure for 10**

In Exercises 11–18, find $\mathbf{r}'(t)$.

11. $\mathbf{r}(t) = 6t\mathbf{i} - 7t^2\mathbf{j} + t^3\mathbf{k}$

12. $\mathbf{r}(t) = \dfrac{1}{t}\mathbf{i} + 16t\mathbf{j} + \dfrac{t^2}{2}\mathbf{k}$

13. $\mathbf{r}(t) = a\cos^3 t\mathbf{i} + a\sin^3 t\mathbf{j} + \mathbf{k}$

14. $\mathbf{r}(t) = \sqrt{t}\mathbf{i} + t\sqrt{t}\mathbf{j} + \ln t\mathbf{k}$

15. $\mathbf{r}(t) = e^{-t}\mathbf{i} + 4\mathbf{j}$

16. $\mathbf{r}(t) = \langle \sin t - t\cos t, \cos t + t\sin t, t^2 \rangle$

17. $\mathbf{r}(t) = \langle t\sin t, t\cos t, t \rangle$

18. $\mathbf{r}(t) = \langle \arcsin t, \arccos t, 0 \rangle$

In Exercises 19 and 20, find the following.

(a) $\mathbf{r}'(t)$ (b) $\mathbf{r}''(t)$

(c) $D_t[\mathbf{r}(t) \cdot \mathbf{u}(t)]$ (d) $D_t[3\mathbf{r}(t) - \mathbf{u}(t)]$

(e) $D_t[\mathbf{r}(t) \times \mathbf{u}(t)]$ (f) $D_t[\|\mathbf{r}(t)\|]$, $t > 0$

19. $\mathbf{r}(t) = t\mathbf{i} + 3t\mathbf{j} + t^2\mathbf{k}$, $\mathbf{u}(t) = 4t\mathbf{i} + t^2\mathbf{j} + t^3\mathbf{k}$

20. $\mathbf{r}(t) = t^2\mathbf{i} + \sin t\mathbf{j} + \cos t\mathbf{k}$, $\mathbf{u}(t) = \dfrac{1}{t^2}\mathbf{i} + \sin t\mathbf{j} + \cos t\mathbf{k}$

 In Exercises 21 and 22, find the angle θ as a function of t between $\mathbf{r}(t)$ and $\mathbf{r}'(t)$. Use a graphing utility to graph $\theta(t)$. Use the graph to find any extrema of the function. Find any values of t at which the vectors are orthogonal.

21. $\mathbf{r}(t) = 3\sin t\mathbf{i} + 4\cos t\mathbf{j}$

22. $\mathbf{r}(t) = t^2\mathbf{i} + t\mathbf{j}$

In Exercises 23–32, find the open interval(s) on which the curve given by the vector-valued function is smooth.

23. $\mathbf{r}(t) = t^2\mathbf{i} + t^3\mathbf{j}$

24. $\mathbf{r}(t) = \dfrac{1}{t-1}\mathbf{i} + 3t\mathbf{j}$

25. $\mathbf{r}(\theta) = 2\cos^3 \theta\mathbf{i} + 3\sin^3 \theta\mathbf{j}$

26. $\mathbf{r}(\theta) = (\theta + \sin \theta)\mathbf{i} + (1 - \cos \theta)\mathbf{j}$

27. $\mathbf{r}(\theta) = (\theta - 2\sin\theta)\mathbf{i} + (1 - 2\cos \theta)\mathbf{j}$

28. $\mathbf{r}(t) = \dfrac{3t}{1 + t^3}\mathbf{i} + \dfrac{3t^2}{1 + t^3}\mathbf{j}$

29. $\mathbf{r}(t) = (t - 1)\mathbf{i} + \dfrac{1}{t}\mathbf{j} - t^2\mathbf{k}$

30. $\mathbf{r}(t) = e^t\mathbf{i} - e^{-t}\mathbf{j} + 3t\mathbf{k}$

31. $\mathbf{r}(t) = t\mathbf{i} - 3t\mathbf{j} + \tan t\mathbf{k}$

32. $\mathbf{r}(t) = \sqrt{t}\,\mathbf{i} + (t^2 - 1)\mathbf{j} + \tfrac{1}{4}t\mathbf{k}$

In Exercises 33 and 34, use the definition of the derivative to find r′(t).

33. $\mathbf{r}(t) = (3t + 2)\mathbf{i} + (1 - t^2)\mathbf{j}$

34. $\mathbf{r}(t) = \sqrt{t}\,\mathbf{i} + \dfrac{3}{t}\mathbf{j} - 2t\mathbf{k}$

35. *Writing* The three components of the derivative of the vector-valued function $\mathbf{u}$ are positive at $t = t_0$. Describe the behavior of $\mathbf{u}$ at $t = t_0$.

36. *Writing* The z-component of the derivative of the vector-valued function $\mathbf{u}$ is 0 for t in the domain of the function. What does this information imply about the graph of $\mathbf{u}$?

In Exercises 37–44, evaluate the indefinite integral.

37. $\displaystyle\int (2t\mathbf{i} + \mathbf{j} + \mathbf{k})\,dt$

38. $\displaystyle\int (3t^2\,\mathbf{i} + 4t\mathbf{j} - 8t^3\mathbf{k})\,dt$

39. $\displaystyle\int \left(\dfrac{1}{t}\mathbf{i} + \mathbf{j} - t^{3/2}\,\mathbf{k}\right) dt$

40. $\displaystyle\int \left(\ln t\mathbf{i} + \dfrac{1}{t}\mathbf{j} + \mathbf{k}\right) dt$

41. $\displaystyle\int \left[(2t - 1)\mathbf{i} + 4t^3\mathbf{j} + 3\sqrt{t}\,\mathbf{k}\right] dt$

42. $\displaystyle\int (e^t\,\mathbf{i} + \sin t\mathbf{j} + \cos t\mathbf{k})\,dt$

43. $\displaystyle\int \left(\sec^2 t\mathbf{i} + \dfrac{1}{1 + t^2}\mathbf{j}\right) dt$

44. $\displaystyle\int (e^{-t} \sin t\mathbf{i} + e^{-t} \cos t\mathbf{j})\,dt$

In Exercises 45–50, find r(t) for the given conditions.

45. $\mathbf{r}'(t) = 4e^{2t}\mathbf{i} + 3e^t\mathbf{j}, \quad \mathbf{r}(0) = 2\mathbf{i}$

46. $\mathbf{r}'(t) = 2t\mathbf{j} + \sqrt{t}\,\mathbf{k}, \quad \mathbf{r}(0) = \mathbf{i} + \mathbf{j}$

47. $\mathbf{r}''(t) = -32\mathbf{j}$
$\mathbf{r}'(0) = 600\sqrt{3}\mathbf{i} + 600\mathbf{j}, \quad \mathbf{r}(0) = \mathbf{0}$

48. $\mathbf{r}''(t) = -4 \cos t\mathbf{j} - 3 \sin t\mathbf{k}$
$\mathbf{r}'(0) = 3\mathbf{k}, \quad \mathbf{r}(0) = 4\mathbf{j}$

49. $\mathbf{r}'(t) = te^{-t^2}\mathbf{i} - e^{-t}\mathbf{j} + \mathbf{k}, \quad \mathbf{r}(0) = \tfrac{1}{2}\mathbf{i} - \mathbf{j} + \mathbf{k}$

50. $\mathbf{r}'(t) = \dfrac{1}{1 + t^2}\mathbf{i} + \dfrac{1}{t^2}\mathbf{j} + \dfrac{1}{t}\mathbf{k}, \quad \mathbf{r}(1) = 2\mathbf{i}$

In Exercises 51–54, evaluate the definite integral.

51. $\displaystyle\int_0^1 (8t\mathbf{i} + t\mathbf{j} - \mathbf{k})\,dt$

52. $\displaystyle\int_{-1}^1 (t\mathbf{i} + t^3\mathbf{j} + \sqrt[3]{t}\,\mathbf{k})\,dt$

53. $\displaystyle\int_0^{\pi/2} [(a \cos t)\mathbf{i} + (a \sin t)\mathbf{j} + \mathbf{k}]\,dt$

54. $\displaystyle\int_0^3 (e^t\mathbf{i} + te^t\,\mathbf{k})\,dt$

In Exercises 55–62, prove the given property. In each case, assume that r, u, and v are differentiable vector-valued functions of t, f is a differentiable real-valued function of t, and c is a scalar.

55. $D_t[c\mathbf{r}(t)] = c\mathbf{r}'(t)$

56. $D_t[\mathbf{r}(t) \pm \mathbf{u}(t)] = \mathbf{r}'(t) \pm \mathbf{u}'(t)$

57. $D_t[f(t)\mathbf{r}(t)] = f(t)\mathbf{r}'(t) + f'(t)\mathbf{r}(t)$

58. $D_t[\mathbf{r}(t) \times \mathbf{u}(t)] = \mathbf{r}(t) \times \mathbf{u}'(t) + \mathbf{r}'(t) \times \mathbf{u}(t)$

59. $D_t[\mathbf{r}(f(t))] = \mathbf{r}'(f(t))f'(t)$

60. $D_t[\mathbf{r}(t) \times \mathbf{r}'(t)] = \mathbf{r}(t) \times \mathbf{r}''(t)$

61. $D_t\{\mathbf{r}(t) \cdot [\mathbf{u}(t) \times \mathbf{v}(t)]\} = \mathbf{r}'(t) \cdot [\mathbf{u}(t) \times \mathbf{v}(t)] +$
$\mathbf{r}(t) \cdot [\mathbf{u}'(t) \times \mathbf{v}(t)] + \mathbf{r}(t) \cdot [\mathbf{u}(t) \times \mathbf{v}'(t)]$

62. If $\mathbf{r}(t) \cdot \mathbf{r}(t)$ is a constant, then $\mathbf{r}(t) \cdot \mathbf{r}'(t) = 0$.

True or False? **In Exercises 63 and 64, determine whether the statement is true or false. If it is false, explain why or give an example that shows it is false.**

63. $\dfrac{d}{dt}\left[\|\mathbf{r}(t)\|\right] = \|\mathbf{r}'(t)\|$

64. If $\mathbf{r}$ and $\mathbf{u}$ are differentiable vector-valued functions of t, then $D_t[\mathbf{r}(t) \cdot \mathbf{u}(t)] = \mathbf{r}'(t) \cdot \mathbf{u}'(t)$.

Velocity and Acceleration • Projectile Motion

Velocity and Acceleration

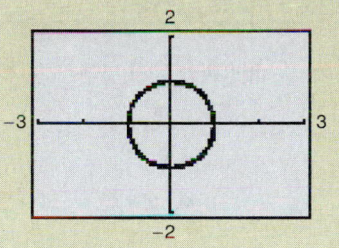
You are now ready to combine your study of parametric equations, curves, vectors, and vector-valued functions to form a model for motion along a curve. We begin by looking at the motion of an object in the plane. (The motion of an object in space can be developed similarly.)

As an object moves along a curve in the plane, the coordinates x and y of its center of mass are each functions of time t. Rather than using f and g to represent these two functions, it is convenient to write $x = x(t)$ and $y = y(t)$. Thus, the position vector $\mathbf{r}(t)$ takes the form

$$\mathbf{r}(t) = x(t)\mathbf{i} + y(t)\mathbf{j}. \qquad \text{Position vector}$$

The beauty of this vector model for representing motion is that you can use the first and second derivatives of the vector-valued function $\mathbf{r}$ to find the object's velocity and acceleration. (Recall from the preceding chapter that velocity and acceleration are both vector quantities having magnitude and direction.) To find the velocity and acceleration vectors at a given time t, consider a point $Q(x(t + \Delta t), y(t + \Delta t))$ that is approaching the point $P(x(t), y(t))$ along the curve C given by $\mathbf{r}(t) = x(t)\mathbf{i} + y(t)\mathbf{j}$, as shown in Figure 11.10. As $\Delta t \to 0$, the direction of the vector $\overrightarrow{PQ}$ (denoted by $\Delta \mathbf{r}$) approaches the *direction of motion* at time t.

$$\Delta \mathbf{r} = \mathbf{r}(t + \Delta t) - \mathbf{r}(t)$$

$$\frac{\Delta \mathbf{r}}{\Delta t} = \frac{\mathbf{r}(t + \Delta t) - \mathbf{r}(t)}{\Delta t}$$

$$\lim_{\Delta t \to 0} \frac{\Delta \mathbf{r}}{\Delta t} = \lim_{\Delta t \to 0} \frac{\mathbf{r}(t + \Delta t) - \mathbf{r}(t)}{\Delta t}$$

If this limit exists (and is not equal to the zero vector), it is defined to be the **velocity vector** or **tangent vector** to the curve at point P. Note that this is the same limit used to define $\mathbf{r}'(t)$. Thus, the direction of $\mathbf{r}'(t)$ gives the direction of motion at time t. Moreover, the magnitude of the vector $\mathbf{r}'(t)$

$$\|\mathbf{r}'(t)\| = \|x'(t)\mathbf{i} + y'(t)\mathbf{j}\| = \sqrt{[x'(t)]^2 + [y'(t)]^2}$$

gives the **speed** of the object at time t. Similarly, you can use $\mathbf{r}''(t)$ to represent acceleration, as indicated in the definition at the top of page 786.

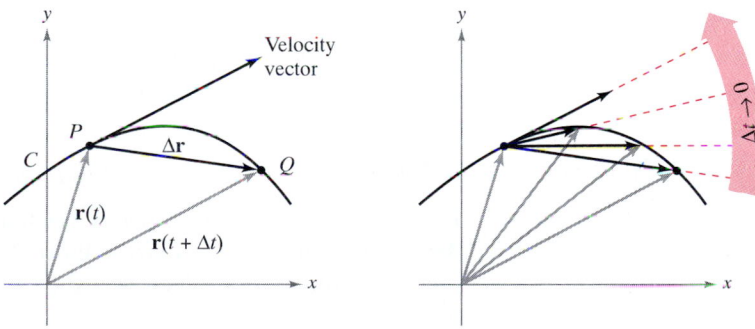

As $\Delta t \to 0$, $\dfrac{\Delta \mathbf{r}}{\Delta t}$ approaches the velocity vector.

Figure 11.10

Definition of Velocity and Acceleration

If x and y are twice-differentiable functions of t, and $\mathbf{r}$ is a vector-valued function given by $\mathbf{r}(t) = x(t)\mathbf{i} + y(t)\mathbf{j}$, then the velocity vector, acceleration vector, and speed at time t are as follows.

$$\text{Velocity} = \mathbf{v}(t) \quad = \mathbf{r}'(t) \quad = x'(t)\mathbf{i} + y'(t)\mathbf{j}$$
$$\text{Acceleration} = \mathbf{a}(t) \quad = \mathbf{r}''(t) \quad = x''(t)\mathbf{i} + y''(t)\mathbf{j}$$
$$\text{Speed} = \|\mathbf{v}(t)\| = \|\mathbf{r}'(t)\| = \sqrt{[x'(t)]^2 + [y'(t)]^2}$$

For motion along a space curve, the definitions are similar. That is, if $\mathbf{r}(t) = x(t)\mathbf{i} + y(t)\mathbf{j} + z(t)\mathbf{k}$, you have

$$\text{Velocity} = \mathbf{v}(t) \quad = \mathbf{r}'(t) \quad = x'(t)\mathbf{i} + y'(t)\mathbf{j} + z'(t)\mathbf{k}$$
$$\text{Acceleration} = \mathbf{a}(t) \quad = \mathbf{r}''(t) \quad = x''(t)\mathbf{i} + y''(t)\mathbf{j} + z''(t)\mathbf{k}$$
$$\text{Speed} = \|\mathbf{v}(t)\| = \|\mathbf{r}'(t)\| = \sqrt{[x'(t)]^2 + [y'(t)]^2 + [z'(t)]^2}.$$

EXAMPLE 1   Finding Velocity and Acceleration Along a Plane Curve

NOTE   In Example 1, note that the velocity and acceleration vectors are orthogonal at any point in time. This is characteristic of motion at a constant speed. (See Exercise 49.)

Find the velocity vector, speed, and acceleration vector of a particle that moves along the plane curve C described by

$$\mathbf{r}(t) = 2 \sin \frac{t}{2} \mathbf{i} + 2 \cos \frac{t}{2} \mathbf{j}. \qquad \text{\color{red}Position vector}$$

Solution   The velocity vector is

$$\mathbf{v}(t) = \mathbf{r}'(t) = \cos \frac{t}{2} \mathbf{i} - \sin \frac{t}{2} \mathbf{j}. \qquad \text{\color{red}Velocity vector}$$

The speed (at any time) is

$$\|\mathbf{r}'(t)\| = \sqrt{\cos^2 \frac{t}{2} + \sin^2 \frac{t}{2}} = 1. \qquad \text{\color{red}Speed}$$

The acceleration vector is

$$\mathbf{a}(t) = \mathbf{r}''(t) = -\frac{1}{2} \sin \frac{t}{2} \mathbf{i} - \frac{1}{2} \cos \frac{t}{2} \mathbf{j}. \qquad \text{\color{red}Acceleration vector}$$

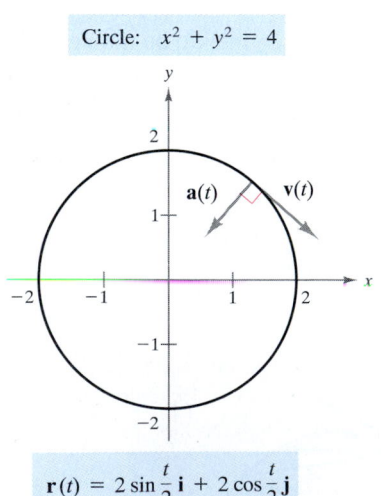

Circle: $x^2 + y^2 = 4$

$$\mathbf{r}(t) = 2 \sin \frac{t}{2} \mathbf{i} + 2 \cos \frac{t}{2} \mathbf{j}$$

The particle moves around the circle at constant speed.
Figure 11.11

The parametric equations for the curve in Example 1 are $x = 2 \sin(t/2)$ and $y = 2 \cos(t/2)$. By eliminating the parameter t, you can obtain the rectangular equation

$$x^2 + y^2 = 4. \qquad \text{\color{red}Rectangular equation}$$

Thus, the curve is a circle of radius 2 centered at the origin, as shown in Figure 11.11. Because the velocity vector $\mathbf{v}(t) = \cos(t/2)\mathbf{i} - \sin(t/2)\mathbf{j}$ has a constant magnitude but a changing direction as t increases, the particle moves around the circle at a constant speed.

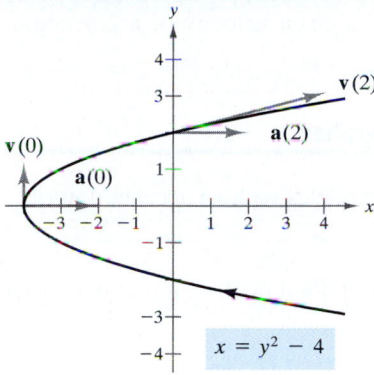

$$r(t) = (t^2 - 4)\,\mathbf{i} + t\mathbf{j}$$

At each point on the curve, the acceleration vector points to the right.
Figure 11.12

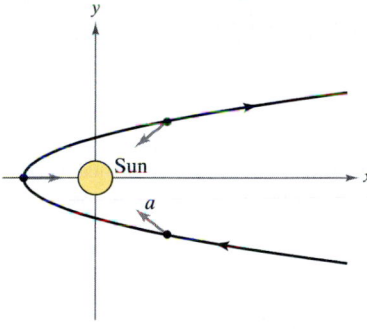

At each point in the comet's orbit, the acceleration vector points toward the sun.
Figure 11.13

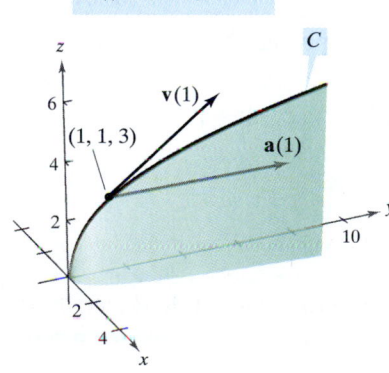

Curve:
$$r(t) = t\mathbf{i} + t^3\mathbf{j} + 3t\mathbf{k}$$

Figure 11.14

EXAMPLE 2   Sketching Velocity and Acceleration Vectors in the Plane

Sketch the path of an object moving along the plane curve given by

$$\mathbf{r}(t) = (t^2 - 4)\mathbf{i} + t\mathbf{j} \qquad\qquad \text{Position vector}$$

and find the velocity and acceleration vectors when $t = 0$ and $t = 2$.

Solution   Using the parametric equations $x = t^2 - 4$ and $y = t$, you can determine that the curve is a parabola given by $x = y^2 - 4$, as shown in Figure 11.12. The velocity vector (at any time) is

$$\mathbf{v}(t) = \mathbf{r}'(t) = 2t\mathbf{i} + \mathbf{j} \qquad\qquad \text{Velocity vector}$$

and the acceleration vector (at any time) is

$$\mathbf{a}(t) = \mathbf{r}''(t) = 2\mathbf{i}. \qquad\qquad \text{Acceleration vector}$$

When $t = 0$, the velocity and acceleration vectors are given by

$$\mathbf{v}(0) = 2(0)\mathbf{i} + \mathbf{j} = \mathbf{j} \quad \text{and} \quad \mathbf{a}(0) = 2\mathbf{i}.$$

When $t = 2$, the velocity and acceleration vectors are given by

$$\mathbf{v}(2) = 2(2)\mathbf{i} + \mathbf{j} = 4\mathbf{i} + \mathbf{j} \quad \text{and} \quad \mathbf{a}(2) = 2\mathbf{i}.$$

For the object moving along the path shown in Figure 11.12, note that the acceleration vector is constant (it has a magnitude of 2 and points to the right). This implies that the speed of the object is decreasing as the object moves toward the vertex of the parabola, and the speed is increasing as the object moves away from the vertex of the parabola.

This type of motion is *not* characteristic of comets that travel on parabolic paths through our solar system. For such comets, the acceleration vector always points to the origin (the sun), which implies that the comet's speed increases as it approaches the vertex of the path and decreases as it moves away from the vertex. (See Figure 11.13.)

EXAMPLE 3   Sketching Velocity and Acceleration Vectors in Space

Sketch the path of an object moving along the space curve C given by

$$\mathbf{r}(t) = t\mathbf{i} + t^3\mathbf{j} + 3t\mathbf{k}, \quad t \geq 0 \qquad\qquad \text{Position vector}$$

and find the velocity and acceleration vectors when $t = 1$.

Solution   Using the parametric equations $x = t$ and $y = t^3$, you can determine that the path of the object lies on the cubic cylinder given by $y = x^3$. Moreover, because $z = 3t$, the object starts as $(0, 0, 0)$ and moves upward as t increases, as shown in Figure 11.14. Because $\mathbf{r}(t) = t\mathbf{i} + t^3\mathbf{j} + 3t\mathbf{k}$, you have

$$\mathbf{v}(t) = \mathbf{r}'(t) = \mathbf{i} + 3t^2\mathbf{j} + 3\mathbf{k} \qquad\qquad \text{Velocity vector}$$

and

$$\mathbf{a}(t) = \mathbf{r}''(t) = 6t\mathbf{j}. \qquad\qquad \text{Acceleration vector}$$

When $t = 1$, the velocity and acceleration vectors are given by

$$\mathbf{v}(1) = \mathbf{r}'(1) = \mathbf{i} + 3\mathbf{j} + 3\mathbf{k} \quad \text{and} \quad \mathbf{a}(1) = \mathbf{r}''(1) = 6\mathbf{j}.$$

So far in this section, we have concentrated on finding the velocity and acceleration by differentiating the position function. Many practical applications involve the reverse problem—finding the position function for a given velocity or acceleration. This is demonstrated in the next example.

EXAMPLE 4 Finding a Position Function by Integration

An object starts from rest at the point $P(1, 2, 0)$ and moves with an acceleration of

$$\mathbf{a}(t) = \mathbf{j} + 2\mathbf{k} \qquad \text{\color{red}Acceleration vector}$$

where $\|\mathbf{a}(t)\|$ is measured in feet per second per second. Find the location of the object after $t = 2$ seconds.

Solution From the description of the object's motion, you can deduce the following *initial conditions*. Because the object starts from rest, you have

$$\mathbf{v}(0) = \mathbf{0}.$$

Moreover, because the object starts at the point $(x, y, z) = (1, 2, 0)$, you have

$$\begin{aligned} \mathbf{r}(0) &= x(0)\mathbf{i} + y(0)\mathbf{j} + z(0)\mathbf{k} \\ &= 1\mathbf{i} + 2\mathbf{j} + 0\mathbf{k} \\ &= \mathbf{i} + 2\mathbf{j}. \end{aligned}$$

To find the position function, you should integrate twice, each time using one of the initial conditions to solve for the constant of integration. The velocity vector is

$$\mathbf{v}(t) = \int \mathbf{a}(t)\, dt = \int (\mathbf{j} + 2\mathbf{k})\, dt = t\mathbf{j} + 2t\mathbf{k} + \mathbf{C}$$

where $\mathbf{C} = C_1\mathbf{i} + C_2\mathbf{j} + C_3\mathbf{k}$. Letting $t = 0$ and applying the initial condition $\mathbf{v}(0) = \mathbf{0}$, you obtain

$$\mathbf{v}(0) = C_1\mathbf{i} + C_2\mathbf{j} + C_3\mathbf{k} = \mathbf{0} \quad \color{red}\Longrightarrow \quad \color{black} C_1 = C_2 = C_3 = 0.$$

Thus, the *velocity* at any time t is

$$\mathbf{v}(t) = t\mathbf{j} + 2t\mathbf{k}. \qquad \text{\color{red}Velocity vector}$$

Integrating once more produces

$$\mathbf{r}(t) = \int \mathbf{v}(t)\, dt = \int (t\mathbf{j} + 2t\mathbf{k})\, dt = \frac{t^2}{2}\mathbf{j} + t^2\mathbf{k} + \mathbf{C}$$

where $\mathbf{C} = C_4\mathbf{i} + C_5\mathbf{j} + C_6\mathbf{k}$. Letting $t = 0$ and applying the initial condition $\mathbf{r}(0) = \mathbf{i} + 2\mathbf{j}$, you have

$$\mathbf{r}(0) = C_4\mathbf{i} + C_5\mathbf{j} + C_6\mathbf{k} = \mathbf{i} + 2\mathbf{j} \quad \color{red}\Longrightarrow \quad \color{black} C_4 = 1, C_5 = 2, C_6 = 0.$$

Thus, the *position* vector is

$$\mathbf{r}(t) = \mathbf{i} + \left(\frac{t^2}{2} + 2 \right)\mathbf{j} + t^2\mathbf{k}. \qquad \text{\color{red}Position vector}$$

The location of the object after 2 seconds is given by $\mathbf{r}(2) = \mathbf{i} + 4\mathbf{j} + 4\mathbf{k}$, as shown in Figure 11.15.

Curve:

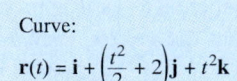

$$\mathbf{r}(t) = \mathbf{i} + \left(\frac{t^2}{2} + 2 \right)\mathbf{j} + t^2\mathbf{k}$$

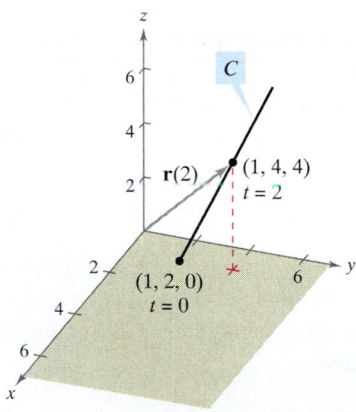

The object takes 2 seconds to move from point (1, 2, 0) to point (1, 4, 4) along C.
Figure 11.15

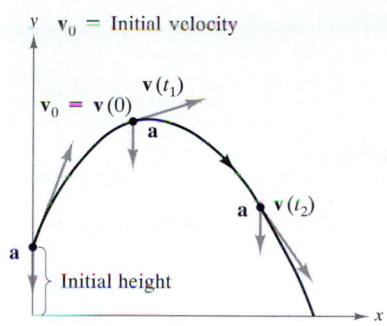

y $\mathbf{v}_0$ = Initial velocity

$\mathbf{v}(t_1)$

$\mathbf{v}_0 = \mathbf{v}(0)$
  $\mathbf{a}$

  $\mathbf{a}$ $\mathbf{v}(t_2)$

$\mathbf{a}$

Initial height

                                    x

The parabolic path of a projectile
Figure 11.16

Projectile Motion

We now have the machinery to derive the parametric equations for the path of a projectile. We assume that gravity is the only force acting on the projectile after it is launched. Hence, the motion occurs in a vertical plane, which we represent by the xy-coordinate system with the origin as a point on the earth's surface, as shown in Figure 11.16. For a projectile of mass m, the force due to gravity is

$$\mathbf{F} = -mg\mathbf{j} \qquad \text{Force due to gravity}$$

where the gravitational constant is $g = 32$ feet per second per second, or 9.81 meters per second per second. By **Newton's Second Law of Motion,** this same force produces an acceleration $\mathbf{a} = \mathbf{a}(t)$, and satisfies the equation $\mathbf{F} = m\mathbf{a}$. Consequently, the acceleration of the projectile is given by $m\mathbf{a} = -mg\mathbf{j}$, which implies that

$$\mathbf{a} = -g\mathbf{j}. \qquad \text{Acceleration of projectile}$$

EXAMPLE 5   Derivation of the Position Function for a Projectile

A projectile of mass m is launched from an initial position $\mathbf{r}_0$ with an initial velocity $\mathbf{v}_0$. Find its position vector as a function of time.

Solution   Begin with the acceleration $\mathbf{a}(t) = -g\mathbf{j}$ and integrate twice.

$$\mathbf{v}(t) = \int \mathbf{a}(t)\, dt = \int -g\mathbf{j}\, dt = -gt\mathbf{j} + \mathbf{C}_1$$

$$\mathbf{r}(t) = \int \mathbf{v}(t)\, dt = \int (-gt\mathbf{j} + \mathbf{C}_1)\, dt = -\frac{1}{2}gt^2\mathbf{j} + \mathbf{C}_1 t + \mathbf{C}_2$$

You can use the facts that $\mathbf{v}(0) = \mathbf{v}_0$ and $\mathbf{r}(0) = \mathbf{r}_0$ to solve for the constant vectors $\mathbf{C}_1$ and $\mathbf{C}_2$. Doing this produces $\mathbf{C}_1 = \mathbf{v}_0$ and $\mathbf{C}_2 = \mathbf{r}_0$. Therefore, the position vector is

$$\mathbf{r}(t) = -\frac{1}{2}gt^2\mathbf{j} + t\mathbf{v}_0 + \mathbf{r}_0. \qquad \text{Position vector}$$

In many projectile problems, the constant vectors $\mathbf{r}_0$ and $\mathbf{v}_0$ are not given explicitly. Often you are given initial height h, the initial speed v_0, and the angle θ at which the projectile is launched, as shown in Figure 11.17. From the given height, you can deduce that $\mathbf{r}_0 = h\mathbf{j}$. Because the speed gives the magnitude of the initial velocity, it follows that $v_0 = \|\mathbf{v}_0\|$ and you can write

$$\begin{aligned}
\mathbf{v}_0 &= x\mathbf{i} + y\mathbf{j} \\
&= (\|\mathbf{v}_0\| \cos \theta)\mathbf{i} + (\|\mathbf{v}_0\| \sin \theta)\mathbf{j} \\
&= v_0 \cos \theta \mathbf{i} + v_0 \sin \theta \mathbf{j}.
\end{aligned}$$

Thus, the position vector can be written in the form

$$\mathbf{r}(t) = -\frac{1}{2}gt^2\mathbf{j} + t\mathbf{v}_0 + \mathbf{r}_0 \qquad \text{Position vector}$$

$$= -\frac{1}{2}gt^2\mathbf{j} + tv_0 \cos \theta \mathbf{i} + tv_0 \sin \theta \mathbf{j} + h\mathbf{j}$$

$$= (v_0 \cos \theta)t\mathbf{i} + \left[h + (v_0 \sin \theta)t - \frac{1}{2}gt^2 \right]\mathbf{j}.$$

$\|\mathbf{v}_0\| = v_0$ = initial speed
$\|\mathbf{r}_0\| = h$ = initial height

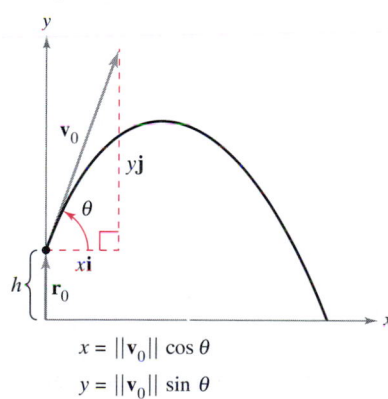

y

$\mathbf{v}_0$

  $y\mathbf{j}$

θ

  $x\mathbf{i}$

h $\mathbf{r}_0$

                      x

$x = \|\mathbf{v}_0\| \cos \theta$
$y = \|\mathbf{v}_0\| \sin \theta$

The initial conditions for launching a projectile
Figure 11.17

THEOREM 11.3   Position Function for a Projectile

Neglecting air resistance, the path of a projectile launched from an initial height h with initial speed v_0 and angle of elevation θ is described by the vector function

$$\mathbf{r}(t) = (v_0 \cos \theta)t\mathbf{i} + \left[h + (v_0 \sin \theta)t - \frac{1}{2}gt^2 \right]\mathbf{j}$$

where g is the gravitational constant.

EXAMPLE 6   Describing the Path of a Baseball

A baseball is hit 3 feet above ground at 100 feet per second and at an angle of $\pi/4$ with respect to the ground, as shown in Figure 11.18. Find the maximum height reached by the baseball. Will it clear a 10-foot high fence located 300 feet from home plate?

Solution   You are given $h = 3$, $v_0 = 100$, and $\theta = \pi/4$. Thus, using $g = 32$ feet per second per second produces

$$\mathbf{r}(t) = \left(100 \cos \frac{\pi}{4} \right)t\mathbf{i} + \left[3 + \left(100 \sin \frac{\pi}{4} \right)t - 16t^2 \right]\mathbf{j}$$

$$= \left(50\sqrt{2}\,t \right)\mathbf{i} + (3 + 50\sqrt{2}\,t - 16t^2)\mathbf{j}$$

$$\mathbf{v}(t) = \mathbf{r}'(t) = 50\sqrt{2}\,\mathbf{i} + (50\sqrt{2} - 32t)\mathbf{j}.$$

The maximum height occurs when

$$y'(t) = 50\sqrt{2} - 32t = 0$$

which implies that

$$t = \frac{25\sqrt{2}}{16} \approx 2.21 \text{ seconds.}$$

Hence, the maximum height reached by the ball is

$$y = 3 + 50\sqrt{2}\left(\frac{25\sqrt{2}}{16} \right) - 16\left(\frac{25\sqrt{2}}{16} \right)^2$$

$$= \frac{649}{8}$$

$$\approx 81 \text{ feet.}\qquad\qquad \textit{Maximum height when } t \approx 2.21$$

The ball is 300 feet from where it was hit when

$$300 = x(t) = 50\sqrt{2}\,t.$$

Solving this equation for t produces $t = 3\sqrt{2} \approx 4.24$ seconds. At this time, the height of the ball is

$$y = 3 + 50\sqrt{2}\left(3\sqrt{2} \right) - 16\left(3\sqrt{2} \right)^2$$

$$= 303 - 288$$

$$= 15 \text{ feet.}\qquad\qquad \textit{Height when } t \approx 4.24$$

Therefore, the ball clears the 10-foot fence for a home run.

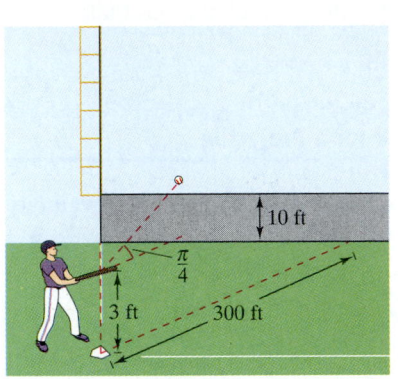

The batter hits the ball with an initial speed of 100 feet per second.

Figure 11.18

EXERCISES FOR SECTION 11.3

In Exercises 1–8, the position function **r** describes the path of an object moving in the *xy*-plane. Sketch a graph of the path and sketch the velocity and acceleration vectors at the given point.

Position Function	Point
1. $\mathbf{r}(t) = 3t\mathbf{i} + (t - 1)\mathbf{j}$	$(3, 0)$
2. $\mathbf{r}(t) = (6 - t)\mathbf{i} + t\mathbf{j}$	$(3, 3)$
3. $\mathbf{r}(t) = t^2\mathbf{i} + t\mathbf{j}$	$(4, 2)$
4. $\mathbf{r}(t) = t^3\mathbf{i} + t^2\mathbf{j}$	$(1, 1)$
5. $\mathbf{r}(t) = 2\cos t\,\mathbf{i} + 2\sin t\,\mathbf{j}$	$(\sqrt{2}, \sqrt{2})$
6. $\mathbf{r}(t) = 2\cos t\,\mathbf{i} + 3\sin t\,\mathbf{j}$	$(2, 0)$
7. $\mathbf{r}(t) = \langle t - \sin t, 1 - \cos t \rangle$	$(\pi, 2)$
8. $\mathbf{r}(t) = \langle e^{-t}, e^{t} \rangle$	$(1, 1)$

In Exercises 9–16, the position function **r** describes the path of an object moving in space. Find the velocity, speed, and acceleration of the object.

9. $\mathbf{r}(t) = t\mathbf{i} + (2t - 5)\mathbf{j} + 3t\mathbf{k}$

10. $\mathbf{r}(t) = 4t\mathbf{i} + 4t\mathbf{j} + 2t\mathbf{k}$

11. $\mathbf{r}(t) = t\mathbf{i} + t^2\mathbf{j} + \dfrac{t^2}{2}\mathbf{k}$

12. $\mathbf{r}(t) = t\mathbf{i} + 3t\mathbf{j} + \dfrac{t^2}{2}\mathbf{k}$

13. $\mathbf{r}(t) = t\mathbf{i} + t\mathbf{j} + \sqrt{9 - t^2}\,\mathbf{k}$

14. $\mathbf{r}(t) = t^2\mathbf{i} + t\mathbf{j} + 2t^{3/2}\mathbf{k}$

15. $\mathbf{r}(t) = \langle 4t, 3\cos t, 3\sin t \rangle$

16. $\mathbf{r}(t) = \langle e^t \cos t, e^t \sin t, e^t \rangle$

Linear Approximation In Exercises 17 and 18, the graph of the vector-valued function **r**(*t*) and a tangent vector to the graph at $t = t_0$ are given.

(a) Find a set of parametric equations for the tangent line to the graph at $t = t_0$.

(b) Use the equations for the line to approximate $\mathbf{r}(t_0 + 0.1)$.

17. $\mathbf{r}(t) = \langle t, -t^2, \frac{1}{4}t^3 \rangle$, $t_0 = 1$

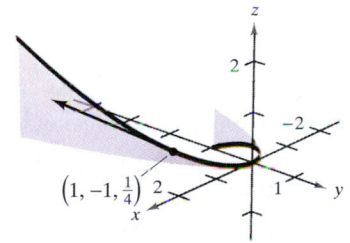

18. $\mathbf{r}(t) = \langle t, \sqrt{25 - t^2}, \sqrt{25 - t^2} \rangle$, $t_0 = 3$

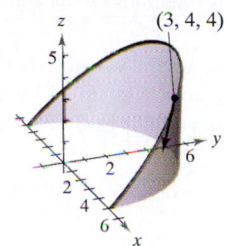

In Exercises 19–22, use the given acceleration function to find the velocity and position functions. Then find the position at time $t = 2$.

19. $\mathbf{a}(t) = \mathbf{i} + \mathbf{j} + \mathbf{k}$
  $\mathbf{v}(0) = 0$, $\mathbf{r}(0) = 0$

20. $\mathbf{a}(t) = \mathbf{i} + \mathbf{k}$
  $\mathbf{v}(0) = 5\mathbf{j}$, $\mathbf{r}(0) = 0$

21. $\mathbf{a}(t) + t\mathbf{j} + t\mathbf{k}$
  $\mathbf{v}(1) = 5\mathbf{j}$, $\mathbf{r}(1) = 0$

22. $\mathbf{a}(t) = -\cos t\,\mathbf{i} - \sin t\,\mathbf{j}$
  $\mathbf{v}(0) = \mathbf{j} + \mathbf{k}$, $\mathbf{r}(0) = \mathbf{i}$

Projectile Motion In Exercises 23–36, use the model for projectile motion, assuming there is no air resistance.

23. Find the vector-valued function for the path of a projectile launched at a height of 10 feet above the ground with an initial velocity of 88 feet per second and at an angle of 30° above the horizontal. Use a graphing utility to sketch the path of the projectile.

24. Determine the maximum height and range of a projectile fired at a height of 3 feet above the ground with an initial velocity of 900 feet per second and at an angle of 45° above the horizontal.

25. A baseball, hit 3 feet above the ground, leaves the bat at an angle of 45° and is caught by an outfielder 300 feet from home plate. What is the initial speed of the ball, and how high does it rise if it is caught 3 feet above the ground?

26. A baseball player at second base throws a ball 90 feet to the player at first base. The ball is thrown at 50 miles per hour at an angle of 15° above the horizontal. At what height does the player at first base catch the ball if the ball is thrown from a height of 5 feet?

27. The quarterback of a football team releases a pass at a height of 7 feet above the playing field, and the football is caught by a receiver 30 yards directly downfield at a height of 4 feet. The pass is released at an angle of 35° with the horizontal.

  (a) Find the speed of the football when it is released.

  (b) Find the maximum height of the football.

  (c) Find the time the receiver has to reach the proper position after the quarterback releases the football.

28. The center field fence in a ballpark is 10 feet high and 400 feet from home plate. A ball is hit 3 feet above the ground and leaves the bat at a speed of 100 miles per hour.

(a) If the ball leaves the bat at an angle of $\theta = \theta_0$ with the horizontal, write the vector-valued function for the path of the ball.

(b) Use a graphing utility to graph the vector-valued function for $\theta_0 = 10°$, $\theta_0 = 15°$, $\theta_0 = {}_2 0°$, and $\theta_0 = 25°$. Use the graphs to approximate the minimum angle required for the hit to be a home run.

(c) Determine analytically the minimum angle required for the hit to be a home run.

29. A bale ejector consists of two variable-speed belts at the end of a baler. Its purpose is to toss bales into a trailing wagon. In loading the back of a wagon, a bale must be thrown to a position 8 feet above and 16 feet behind the ejector. Find the minimum initial speed of the bale and the corresponding angle at which it must be ejected from the baler.

30. A bomber is flying at an altitude of 30,000 feet at a speed of 540 miles per hour (792 feet per second) (see figure). When should the bomb be released for it to hit the target? (Give your answer in terms of the angle of depression from the plane to the target.) What is the speed of the bomb at the time of impact?

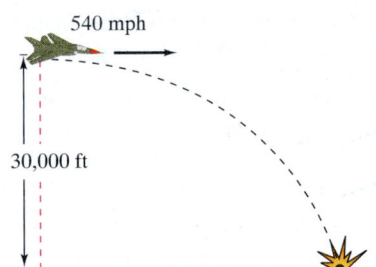

31. A shot fired from a gun with a muzzle velocity of 1200 feet per second is to hit a target 3000 feet away. Determine the minimum angle of elevation of the gun.

32. A projectile is fired from ground level at an angle of 10° with the horizontal. Find the minimum initial velocity necessary if the projectile is to have a range of 100 feet.

33. Use a graphing utility to graph the paths of a projectile for the indicated values of θ and v_0. For each case, use the graph to approximate the maximum height and range of the projectile. (Assume that the projectile is launched from ground level.)

(a) $\theta = 10°$, $v_0 = 66$ ft/sec

(b) $\theta = 10°$, $v_0 = 146$ ft/sec

(c) $\theta = 45°$, $v_0 = 66$ ft/sec

(d) $\theta = 45°$, $v_0 = 146$ ft/sec

(e) $\theta = 60°$, $v_0 = 66$ ft/sec

(f) $\theta = 60°$, $v_0 = 146$ ft/sec

34. Eliminate the parameter t from the position function for the motion of a projectile to show that the rectangular equation is

$$y = -\frac{16 \sec^2 \theta}{v_0{}^2} x^2 + (\tan \theta)x + h.$$

35. The path of a ball is given by the rectangular equation

$$y = x - 0.005x^2.$$

Use the result of Exercise 34 to find the position function. Then find the speed and direction of the ball at the point when it has traveled 60 feet horizontally.

36. Find the angle at which an object must be thrown to obtain

(a) the maximum range and

(b) the maximum height.

Projectile Motion   **In Exercises 37 and 38, use the model for projectile motion, assuming there is no resistance. [$a(t) = -9.8$ meters per second per second]**

37. Determine the maximum height and range of a projectile fired at a height of 1.5 meters above the ground with an initial velocity of 100 meters per second and at an angle of 30° above the horizontal.

38. A projectile is fired from ground level at an angle of 8° with the horizontal. Find the minimum velocity necessary if the projectile is to have a range of 50 meters.

Cycloidal Motion   **In Exercises 39 and 40, consider the motion of a point (or particle) on the circumference of a rolling circle. As the circle rolls, it generates the cycloid**

$$\mathbf{r}(t) = b(\omega t - \sin \omega t)\mathbf{i} + b(1 - \cos \omega t)\mathbf{j}$$

where ω is the constant angular velocity of the circle.

39. Find the velocity and acceleration vectors of the particle. Use the results to determine the times at which the speed of the particle will be (a) zero and (b) maximized.

40. Find the maximum speed of a point on the circumference of an automobile tire of radius 1 foot when the automobile is traveling at 55 miles per hour. Compare this speed with the speed of the automobile.

Circular Motion   **In Exercises 41–44, consider a particle moving on a circular path of radius b described by**

$$\mathbf{r}(t) = b \cos \omega t\, \mathbf{i} + b \sin \omega t\, \mathbf{j}$$

where $\omega = d\theta/dt$ is the constant angular velocity.

41. Find the velocity vector and show that it is orthogonal to $\mathbf{r}(t)$.

42. (a) Show that the speed of the particle is $b\omega$.

(b) Use a graphing utility in parametric mode to sketch the circle for $b = 6$. Try different values of ω. Does the graphing utility draw the circle faster for greater values of ω?

43. Find the acceleration vector and show that its direction is always toward the center of the circle.

44. Show that the magnitude of the acceleration vector is $\omega^2 b$.

Circular Motion   **In Exercises 45 and 46, use the results of Exercises 41–44.**

45. A stone weighing 1 pound is attached to a 2-foot string and is whirled horizontally (see figure). The string will break under a force of 10 pounds. Find the maximum speed the stone can attain without breaking the string. $\left(\text{Use } \mathbf{F} = m\mathbf{a}, \text{ where } m = \frac{1}{32}.\right)$

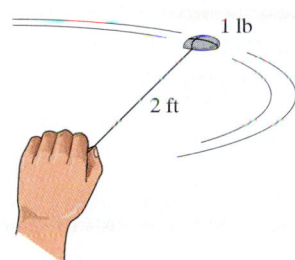

2 ft

1 lb

Figure for 45

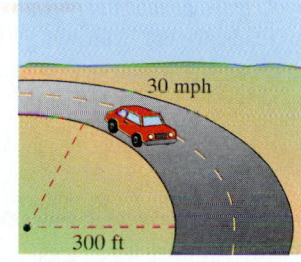

30 mph

300 ft

Figure for 46

46. A 3000-pound automobile is negotiating a circular interchange of radius 300 feet at 30 miles per hour (see figure). Assuming the roadway to be level, find the force between the tires and the road so that the car stays on the circular path and does not skid. (Use $\mathbf{F} = m\mathbf{a}$, where $m = 3000/32$.) Find the angle at which the roadway should be banked so that no lateral frictional force is exerted on the tires of the automobile.

47. *Shot-Put Throw*   The path of a shot thrown at an angle θ is

$$\mathbf{r}(t) = (v_0 \cos\theta)t\,\mathbf{i} + \left[h + (v_0 \sin\theta)t - \frac{1}{2}gt^2\right]\mathbf{j}$$

where v_0 is the initial speed, h is the initial height, t is the time in seconds, and g is the acceleration due to gravity. Verify that the shot will remain in the air for a total of

$$t = \frac{v_0 \sin\theta + \sqrt{v_0{}^2 \sin^2\theta + 2gh}}{g} \quad \text{seconds}$$

and will travel a horizontal distance of

$$\frac{v_0{}^2 \cos\theta}{g}\left(\sin\theta + \sqrt{\sin^2\theta + \frac{2gh}{v_0{}^2}}\right) \quad \text{feet.}$$

48. *Shot-Put Throw*   A shot is thrown from a height of $h = 6$ feet with an initial speed of $v_0 = 45$ feet per second. Find the total time of travel and the total horizontal distance traveled if the shot is thrown at an angle of $\theta = 42.5°$ with the horizontal.

49. Prove that if an object is traveling at a constant speed, its velocity and acceleration vectors are orthogonal.

50. Prove that an object moving in a straight line at a constant speed has an acceleration of 0.

51. *Investigation*   An object moves on an elliptical path given by the vector-valued function $\mathbf{r}(t) = 6 \cos t\,\mathbf{i} + 3 \sin t\,\mathbf{j}$.

(a) Find $\mathbf{v}(t)$, $\|\mathbf{v}(t)\|$, and $\mathbf{a}(t)$.

(b) Use a graphing utility to complete the table.

t	0	$\frac{\pi}{4}$	$\frac{\pi}{2}$	$\frac{2\pi}{3}$	π
Speed					

(c) Graph the elliptical path and the velocity and acceleration vectors at the values of t given in the table in part (b).

(d) Use the results in parts (b) and (c) to describe the geometric relationship between the velocity and acceleration vectors when the speed of the particle is increasing, and when it is decreasing.

52. *Writing*   Consider a particle moving on the path

$$\mathbf{r}_1(t) = x(t)\,\mathbf{i} + y(t)\,\mathbf{j} + z(t)\,\mathbf{k}.$$

Discuss any changes in the position, velocity, or acceleration of the particle if its position is given by the vector-valued function $\mathbf{r}_2(t) = \mathbf{r}_1(2t)$. Generalize the results for the position function $\mathbf{r}_3(t) = \mathbf{r}_1(\omega t)$.

SECTION PROJECT

You want to toss an object to a friend who is riding a Ferris wheel (see figure). The following parametric equations give the path of the friend $\mathbf{r}_1(t)$ and the path of the object $\mathbf{r}_2(t)$. Distance is measured in meters and time is measured in seconds.

$$\mathbf{r}_1(t) = 15\left(\sin\frac{\pi t}{10}\right)\mathbf{i} + \left(16 - 15\cos\frac{\pi t}{10}\right)\mathbf{j}$$

$$\mathbf{r}_2(t) = [22 - 8.03(t - t_0)]\mathbf{i} +$$
$$[1 + 11.47(t - t_0) - 4.9(t - t_0)^2]\mathbf{j}$$

(a) Locate your friend's position on the Ferris wheel at time $t = 0$.

(b) Determine the number of revolutions per minute of the Ferris wheel.

(c) What are the speed and angle of inclination (in degrees) at which the object is thrown at time $t = t_0$?

(d) Use a graphing utility to graph the vector-valued functions using a value of t_0 that allows your friend to be within reach of the object. (Do this by trial and error.) Explain the significance of t_0.

(e) Find the approximate time your friend should be able to catch the object. Approximate the speeds of your friend and the object at that time.

Tangent Vectors and Normal Vectors •
Tangential and Normal Components of Acceleration

Tangent Vectors and Normal Vectors

In the preceding section, you learned that the velocity vector points in the direction of motion. This observation leads to the following definition, which applies to any smooth curve—not just to those for which the parameter represents time.

> **Definition of Unit Tangent Vector**
>
> Let C be a smooth curve represented by $\mathbf{r}$ on an open interval I. The **unit tangent vector** $\mathbf{T}(t)$ at t is defined to be
>
> $$\mathbf{T}(t) = \frac{\mathbf{r}'(t)}{\|\mathbf{r}'(t)\|}, \qquad \mathbf{r}'(t) \neq \mathbf{0}.$$

NOTE   Recall that a curve is *smooth* on an interval if $\mathbf{r}'$ is continuous and nonzero on the interval. Thus, "smoothness" is sufficient to guarantee that a curve has a unit tangent vector.

EXAMPLE 1   Finding the Unit Tangent Vector

Find the unit tangent vector to the curve given by

$$\mathbf{r}(t) = t\mathbf{i} + t^2\mathbf{j}$$

when $t = 1$.

Solution   The derivative of $\mathbf{r}(t)$ is

$$\mathbf{r}'(t) = \mathbf{i} + 2t\mathbf{j}. \qquad \text{Derivative of } \mathbf{r}(t)$$

Thus, the unit tangent vector is

$$\mathbf{T}(t) = \frac{\mathbf{r}'(t)}{\|\mathbf{r}'(t)\|} \qquad \text{Definition of } \mathbf{T}(t)$$

$$= \frac{1}{\sqrt{1 + 4t^2}}(\mathbf{i} + 2t\mathbf{j}). \qquad \text{Substitute for } \mathbf{r}'(t).$$

When $t = 1$, the unit tangent vector is

$$\mathbf{T}(1) = \frac{1}{\sqrt{5}}(\mathbf{i} + 2\mathbf{j})$$

as shown in Figure 11.19.

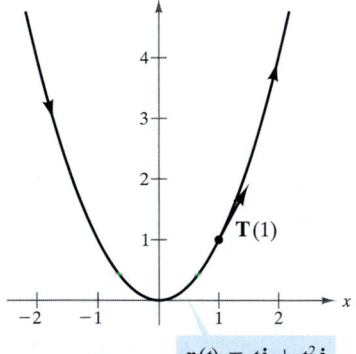

$$\mathbf{r}(t) = t\mathbf{i} + t^2\mathbf{j}$$

The direction of the unit tangent vector depends on the orientation of the curve.
Figure 11.19

NOTE   In Example 1, note that the direction of the unit tangent vector depends on the orientation of the curve. For instance, if the parabola in Figure 11.19 were given by

$$\mathbf{r}(t) = -(t - 2)\mathbf{i} + (t - 2)^2\mathbf{j},$$

$\mathbf{T}(1)$ would still represent the unit tangent vector at the point $(1, 1)$, but it would point in the opposite direction. (Try verifying this.)

The **tangent line to a curve** at a point is the line passing through the point and parallel to the unit tangent vector. In Example 2, the unit tangent vector is used to find the tangent line at a point on a helix.

EXAMPLE 2   Finding the Tangent Line at a Point on a Curve

Find $\mathbf{T}(t)$ and then find a set of parametric equations for the tangent line to the helix given by

$$\mathbf{r}(t) = 2 \cos t \mathbf{i} + 2 \sin t \mathbf{j} + t \mathbf{k}$$

at the point corresponding to $t = \pi/4$.

Solution   The derivative of $\mathbf{r}(t)$ is $\mathbf{r}'(t) = -2 \sin t \mathbf{i} + 2 \cos t \mathbf{j} + \mathbf{k}$, which implies that $\|\mathbf{r}'(t)\| = \sqrt{4 \sin^2 t + 4 \cos^2 t + 1} = \sqrt{5}$. Therefore, the unit tangent vector is

$$\mathbf{T}(t) = \frac{\mathbf{r}'(t)}{\|\mathbf{r}'(t)\|} = \frac{1}{\sqrt{5}}(-2 \sin t \mathbf{i} + 2\cos t \mathbf{j} + \mathbf{k}).$$

When $t = \pi/4$, the unit tangent vector is

$$\mathbf{T}\left(\frac{\pi}{4}\right) = \frac{1}{\sqrt{5}}\left(-2\frac{\sqrt{2}}{2}\mathbf{i} + 2\frac{\sqrt{2}}{2}\mathbf{j} + \mathbf{k}\right)$$

$$= \frac{1}{\sqrt{5}}(-\sqrt{2}\,\mathbf{i} + \sqrt{2}\,\mathbf{j} + \mathbf{k}).$$

Using the direction numbers $a = -\sqrt{2}$, $b = \sqrt{2}$, and $c = 1$, and the point $(x_1, y_1, z_1) = (\sqrt{2}, \sqrt{2}, \pi/4)$, you can obtain the following parametric equations (given with parameter s).

$$x = x_1 + as = \sqrt{2} - \sqrt{2}s$$
$$y = y_1 + bs = \sqrt{2} + \sqrt{2}s$$
$$z = z_1 + cs = \frac{\pi}{4} + s$$

This tangent line is shown in Figure 11.20.

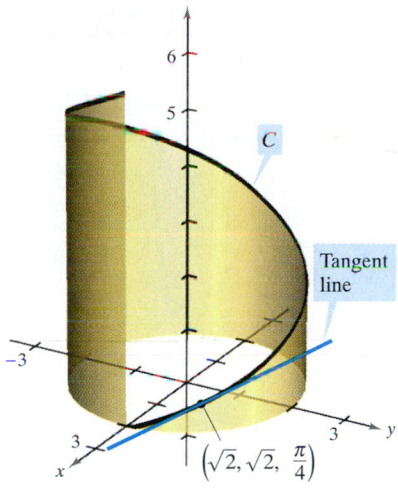

Curve:
$\mathbf{r}(t) = 2 \cos t\mathbf{i} + 2 \sin t\mathbf{j} + t\mathbf{k}$

C

Tangent line

$\left(\sqrt{2}, \sqrt{2}, \frac{\pi}{4}\right)$

The tangent line to a curve at a point is determined by the unit tangent vector at the point.
Figure 11.20

In Example 2, there are infinitely many vectors that are orthogonal to the tangent vector $\mathbf{T}(t)$. One of these is the vector $\mathbf{T}'(t)$. This follows from Property 7 of Theorem 11.2. That is,

$$\mathbf{T}(t) \cdot \mathbf{T}(t) = \|\mathbf{T}(t)\|^2 = 1 \quad \Longrightarrow \quad \mathbf{T}(t) \cdot \mathbf{T}'(t) = 0.$$

By normalizing the vector $\mathbf{T}'(t)$, you obtain a special vector called the **principal unit normal vector**, as indicated in the following definition.

Definition of Principal Unit Normal Vector

Let C be a smooth curve represented by $\mathbf{r}$ on an open interval I. If $\mathbf{T}'(t) \neq \mathbf{0}$, then the **principal unit normal vector** at t is defined to be

$$\mathbf{N}(t) = \frac{\mathbf{T}'(t)}{\|\mathbf{T}'(t)\|}.$$

EXAMPLE 3   Finding the Principal Unit Normal Vector

Find $\mathbf{N}(t)$ and $\mathbf{N}(1)$ for the curve represented by

$$\mathbf{r}(t) = 3t\mathbf{i} + 2t^2\mathbf{j}.$$

Solution By differentiating, you obtain

$$\mathbf{r}'(t) = 3\mathbf{i} + 4t\mathbf{j} \qquad \text{and} \qquad \|\mathbf{r}'(t)\| = \sqrt{9 + 16t^2}$$

which implies that the unit tangent vector is

$$\begin{aligned}
\mathbf{T}(t) &= \frac{\mathbf{r}'(t)}{\|\mathbf{r}'(t)\|} \\[6pt]
&= \frac{1}{\sqrt{9 + 16t^2}}(3\mathbf{i} + 4t\mathbf{j}). \qquad \text{\color{red}Unit tangent vector}
\end{aligned}$$

Using Theorem 11.2, differentiate $\mathbf{T}(t)$ with respect to t to obtain

$$\begin{aligned}
\mathbf{T}'(t) &= \frac{1}{\sqrt{9 + 16t^2}}(4\mathbf{j}) - \frac{16t}{(9 + 16t^2)^{3/2}}(3\mathbf{i} + 4t\mathbf{j}) \\[6pt]
&= \frac{12}{(9 + 16t^2)^{3/2}}(-4t\mathbf{i} + 3\mathbf{j}) \\[6pt]
\|\mathbf{T}'(t)\| &= 12\sqrt{\frac{9 + 16t^2}{(9 + 16t^2)^3}} = \frac{12}{9 + 16t^2}.
\end{aligned}$$

Therefore, the principal unit normal vector is

$$\begin{aligned}
\mathbf{N}(t) &= \frac{\mathbf{T}'(t)}{\|\mathbf{T}'(t)\|} \\[6pt]
&= \frac{1}{\sqrt{9 + 16t^2}}(-4t\mathbf{i} + 3\mathbf{j}). \qquad \text{\color{red}Principal unit normal vector}
\end{aligned}$$

When $t = 1$, the principal unit normal vector is

$$\mathbf{N}(1) = \frac{1}{5}(-4\mathbf{i} + 3\mathbf{j})$$

as shown in Figure 11.21.

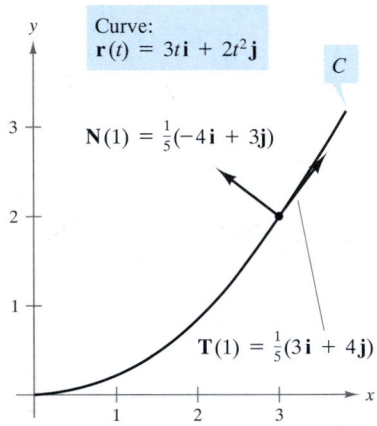

Curve:
$\mathbf{r}(t) = 3t\mathbf{i} + 2t^2\mathbf{j}$

$\mathbf{N}(1) = \frac{1}{5}(-4\mathbf{i} + 3\mathbf{j})$

$\mathbf{T}(1) = \frac{1}{5}(3\mathbf{i} + 4\mathbf{j})$

The principal unit normal vector points toward the concave side of the curve.
Figure 11.21

The principal unit normal vector can be difficult to evaluate algebraically. For plane curves, you can simplify the algebra by finding

$$\mathbf{T}(t) = x(t)\mathbf{i} + y(t)\mathbf{j} \qquad \text{\color{red}Unit tangent vector}$$

and observing that $\mathbf{N}(t)$ must be either

$$\mathbf{N}_1(t) = y(t)\mathbf{i} - x(t)\mathbf{j} \qquad \text{or} \qquad \mathbf{N}_2(t) = -y(t)\mathbf{i} + x(t)\mathbf{j}.$$

Because $\sqrt{[x(t)]^2 + [y(t)]^2} = 1$, it follows that both $\mathbf{N}_1(t)$ and $\mathbf{N}_2(t)$ are unit normal vectors. The *principal* unit normal vector $\mathbf{N}$ is the one that points toward the concave side of the curve, as indicated in Figure 11.21 (see Exercise 49). This also holds for curves in space. That is, for an object moving along a curve C in space, the vector $\mathbf{T}(t)$ points in the direction the object is moving, whereas the vector $\mathbf{N}(t)$ is orthogonal to $\mathbf{T}(t)$ and points in the direction the object is turning, as shown in Figure 11.22.

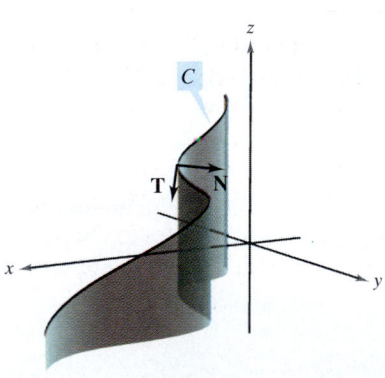

At any point on a curve, a unit normal vector is orthogonal to the unit tangent vector. The *principal* unit normal vector points in the direction the curve is turning.
Figure 11.22

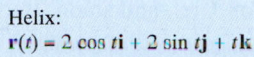

Helix:
$\mathbf{r}(t) = 2 \cos t\mathbf{i} + 2 \sin t\mathbf{j} + t\mathbf{k}$

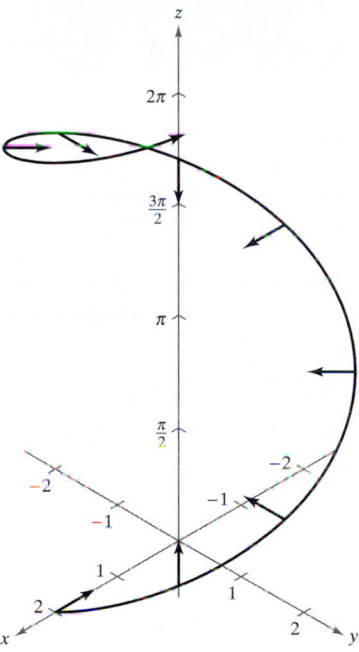

$\mathbf{N}(t)$ is horizontal and points toward the z-axis.

Figure 11.23

EXAMPLE 4   Finding the Principal Unit Normal Vector

Find the principal unit normal vector for the helix given by

$$\mathbf{r}(t) = 2 \cos t\mathbf{i} + 2 \sin t\mathbf{j} + t\mathbf{k}.$$

Solution   From Example 2, you know that the unit tangent vector is

$$\mathbf{T}(t) = \frac{1}{\sqrt{5}}(-2 \sin t\mathbf{i} + 2 \cos t\mathbf{j} + \mathbf{k}).$$

Thus, $\mathbf{T}'(t)$ is given by

$$\mathbf{T}'(t) = \frac{1}{\sqrt{5}}(-2 \cos t\mathbf{i} - 2 \sin t\mathbf{j}).$$

Because $\|\mathbf{T}'(t)\| = 2/\sqrt{5}$, it follows that the principal unit normal vector is

$$
\begin{aligned}
\mathbf{N}(t) &= \frac{\mathbf{T}'(t)}{\|\mathbf{T}'(t)\|} \\
&= \frac{1}{2}(-2 \cos t\mathbf{i} - 2 \sin t\mathbf{j}) \\
&= -\cos t\mathbf{i} - \sin t\mathbf{j}.
\end{aligned}
$$

Note that this vector is horizontal and points toward the z-axis, as shown in Figure 11.23.

Tangential and Normal Components of Acceleration

We now return to the problem of describing the motion of an object along a curve. In the preceding section, you saw that for an object traveling at a *constant speed*, the velocity and acceleration vectors are perpendicular. This seems reasonable, because the speed would not be constant if any acceleration were acting in the direction of motion. You can verify this observation by noting that

$$\mathbf{r}''(t) \cdot \mathbf{r}'(t) = 0$$

if $\|\mathbf{r}'(t)\|$ is a constant. (See Property 7 of Theorem 11.2.)

However, for an object traveling at a *variable speed*, the velocity and acceleration vectors are not necessarily perpendicular. For instance, you saw that the acceleration vector for a projectile always points down, regardless of the direction of motion.

In general, part of the acceleration (the tangential component) acts in the line of motion, and part (the normal component) acts perpendicular to the line of motion. In order to determine these two components, you can use the unit vectors $\mathbf{T}(t)$ and $\mathbf{N}(t)$, which serve in much the same way as do $\mathbf{i}$ and $\mathbf{j}$ in representing vectors in the plane. The following theorem states that the acceleration vector lies in the plane determined by $\mathbf{T}(t)$ and $\mathbf{N}(t)$.

THEOREM 11.4   Acceleration Vector

If $\mathbf{r}(t)$ is the position vector for a smooth curve C and $\mathbf{N}(t)$ exists, then the acceleration vector $\mathbf{a}(t)$ lies in the plane determined by $\mathbf{T}(t)$ and $\mathbf{N}(t)$.

Proof To simplify the notation, we write $\mathbf{T}$ for $\mathbf{T}(t)$, $\mathbf{T}'$ for $\mathbf{T}'(t)$, and so on. Because $\mathbf{T} = \mathbf{r}'/\|\mathbf{r}'\| = \mathbf{v}/\|\mathbf{v}\|$, it follows that

$$\mathbf{v} = \|\mathbf{v}\|\mathbf{T}.$$

By differentiating, you obtain

$$\mathbf{a} = \mathbf{v}' = D_t[\|\mathbf{v}\|]\mathbf{T} + \|\mathbf{v}\|\mathbf{T}' \qquad \text{Product Rule}$$

$$= D_t[\|\mathbf{v}\|]\mathbf{T} + \|\mathbf{v}\|\mathbf{T}'\left(\frac{\|\mathbf{T}'\|}{\|\mathbf{T}'\|}\right)$$

$$= D_t[\|\mathbf{v}\|]\mathbf{T} + \|\mathbf{v}\|\,\|\mathbf{T}'\|\,\mathbf{N}. \qquad \mathbf{N} = \mathbf{T}'/\|\mathbf{T}'\|$$

Because $\mathbf{a}$ is written as a linear combination of $\mathbf{T}$ and $\mathbf{N}$, it must lie in the plane determined by $\mathbf{T}$ and $\mathbf{N}$.

The coefficients of $\mathbf{T}$ and $\mathbf{N}$ in the proof of Theorem 11.4 are called the **tangential** and **normal components of acceleration** and are denoted by $a_{\mathbf{T}} = D_t[\|\mathbf{v}\|]$ and $a_{\mathbf{N}} = \|\mathbf{v}\|\,\|\mathbf{T}'\|$. Thus, you can write

$$\mathbf{a}(t) = a_{\mathbf{T}}\mathbf{T}(t) + a_{\mathbf{N}}\mathbf{N}(t).$$

The following theorem gives some convenient formulas for $a_{\mathbf{N}}$ and $a_{\mathbf{T}}$.

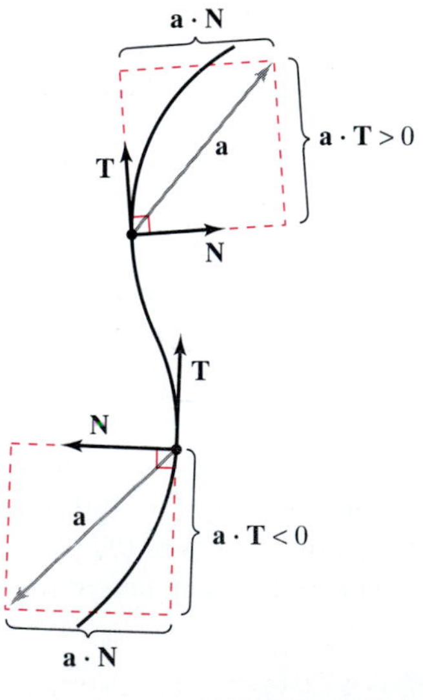

> **THEOREM 11.5 Tangential and Normal Components of Acceleration**
>
> If $\mathbf{r}(t)$ is the position vector for a smooth curve C [for which $\mathbf{N}(t)$ exists], then the tangential and normal components of acceleration are as follows.
>
> $$a_{\mathbf{T}} = D_t[\|\mathbf{v}\|] = \mathbf{a} \cdot \mathbf{T} = \frac{\mathbf{v} \cdot \mathbf{a}}{\|\mathbf{v}\|}$$
>
> $$a_{\mathbf{N}} = \|\mathbf{v}\|\,\|\mathbf{T}'\| = \mathbf{a} \cdot \mathbf{N} = \frac{\|\mathbf{v} \times \mathbf{a}\|}{\|\mathbf{v}\|} = \sqrt{\|\mathbf{a}\|^2 - a_{\mathbf{T}}^2}$$
>
> Note that $a_{\mathbf{N}} \geq 0$. The normal component of acceleration is also called the **centripetal component of acceleration.**

Proof Note that $\mathbf{a}$ lies in the plane of $\mathbf{T}$ and $\mathbf{N}$. Thus, you can use Figure 11.24 to conclude that, for any time t, the component of the projection of the acceleration vector onto $\mathbf{T}$ is given by $a_{\mathbf{T}} = \mathbf{a} \cdot \mathbf{T}$, and onto $\mathbf{N}$ is given by $a_{\mathbf{N}} = \mathbf{a} \cdot \mathbf{N}$. Moreover, because $\mathbf{a} = \mathbf{v}'$ and $\mathbf{T} = \mathbf{v}/\|\mathbf{v}\|$, you have

$$a_{\mathbf{T}} = \mathbf{a} \cdot \mathbf{T} = \mathbf{T} \cdot \mathbf{a} = \frac{\mathbf{v}}{\|\mathbf{v}\|} \cdot \mathbf{a} = \frac{\mathbf{v} \cdot \mathbf{a}}{\|\mathbf{v}\|}.$$

In Exercises 51 and 52, you are asked to prove the other parts of the theorem.

The tangential and normal components of acceleration are obtained by projecting $\mathbf{a}$ onto $\mathbf{T}$ and $\mathbf{N}$.
Figure 11.24

NOTE The formulas from Theorem 11.5, together with several other formulas from this chapter, are summarized on page 811.

EXAMPLE 5   Tangential and Normal Components of Acceleration

Find the tangential and normal components of acceleration for the position function given by $\mathbf{r}(t) = 3t\mathbf{i} - t\mathbf{j} + t^2\mathbf{k}$.

Solution   Begin by finding the velocity, speed, and acceleration.

$$\mathbf{v}(t) = \mathbf{r}'(t) = 3\mathbf{i} - \mathbf{j} + 2t\mathbf{k}$$
$$\|\mathbf{v}(t)\| = \sqrt{9 + 1 + 4t^2} = \sqrt{10 + 4t^2}$$
$$\mathbf{a}(t) = \mathbf{r}''(t) = 2\mathbf{k}$$

By Theorem 11.5, the tangential component of acceleration is

$$a_{\mathbf{T}} = \frac{\mathbf{v} \cdot \mathbf{a}}{\|\mathbf{v}\|} = \frac{4t}{\sqrt{10 + 4t^2}}$$

and because

$$\mathbf{v} \times \mathbf{a} = \begin{vmatrix} \mathbf{i} & \mathbf{j} & \mathbf{k} \\ 3 & -1 & 2t \\ 0 & 0 & 2 \end{vmatrix} = -2\mathbf{i} - 6\mathbf{j}$$

the normal component of acceleration is

$$a_{\mathbf{N}} = \frac{\|\mathbf{v} \times \mathbf{a}\|}{\|\mathbf{v}\|} = \frac{\sqrt{4 + 36}}{\sqrt{10 + 4t^2}} = \frac{2\sqrt{10}}{\sqrt{10 + 4t^2}}.$$

NOTE   In Example 5, you could have used the alternative formula for $a_{\mathbf{N}}$ as follows.

$$a_{\mathbf{N}} = \sqrt{\|\mathbf{a}\|^2 - a_{\mathbf{T}}^2} = \sqrt{(2)^2 - \frac{16t^2}{10 + 4t^2}} = \frac{2\sqrt{10}}{\sqrt{10 + 4t^2}}$$

EXAMPLE 6   Finding $a_{\mathbf{T}}$ and $a_{\mathbf{N}}$ for a Circular Helix

Find the tangential and normal components of acceleration for the helix given by $\mathbf{r}(t) = b\cos t\mathbf{i} + b\sin t\mathbf{j} + ct\mathbf{k}, b > 0$.

Solution

$$\mathbf{v}(t) = \mathbf{r}'(t) = -b\sin t\mathbf{i} + b\cos t\mathbf{j} + c\mathbf{k}$$
$$\|\mathbf{v}(t)\| = \sqrt{b^2\sin^2 t + b^2\cos^2 t + c^2} = \sqrt{b^2 + c^2}$$
$$\mathbf{a}(t) = \mathbf{r}''(t) = -b\cos t\mathbf{i} - b\sin t\mathbf{j}$$

By Theorem 11.5, the tangential component of acceleration is

$$a_{\mathbf{T}} = \frac{\mathbf{v} \cdot \mathbf{a}}{\|\mathbf{v}\|} = \frac{b^2\sin t\cos t - b^2\sin t\cos t + 0}{\sqrt{b^2 + c^2}} = 0.$$

Moreover, because $\|\mathbf{a}(t)\| = \sqrt{b^2\cos^2 t + b^2\sin^2 t} = b$, you can use the alternative formula for the normal component of acceleration to obtain

$$a_{\mathbf{N}}\sqrt{\|\mathbf{a}(t)\|^2 - a_{\mathbf{T}}^2} = \sqrt{b^2 - 0^2} = b.$$

Note that the normal component of acceleration is equal to the magnitude of the acceleration. In other words, because the speed is constant, the acceleration is perpendicular to the velocity. (See Figure 11.25.)

NOTE   The normal component of acceleration is equal to the radius of the cylinder around which the helix is spiraling.

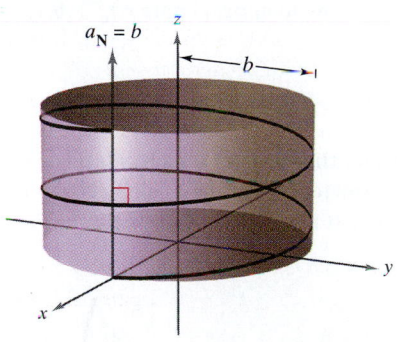

For this parametrization, the normal component of acceleration is equal to the magnitude of the acceleration.
Figure 11.25

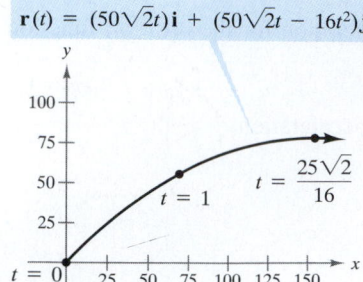

$\mathbf{r}(t) = (50\sqrt{2}t)\mathbf{i} + (50\sqrt{2}t - 16t^2)\mathbf{j}$

The path of a projectile
Figure 11.26

EXAMPLE 7   Projectile Motion

The position function for the projectile shown in Figure 11.26 is given by

$$\mathbf{r}(t) = (50\sqrt{2}\,t)\mathbf{i} + (50\sqrt{2}\,t - 16t^2)\mathbf{j}. \qquad \text{\color{red}Position vector}$$

Find the tangential component of acceleration when $t = 0$, 1, and $25\sqrt{2}/16$.

Solution

$$\mathbf{v}(t) = 50\sqrt{2}\,\mathbf{i} + (50\sqrt{2} - 32t)\mathbf{j} \qquad \text{\color{red}Velocity vector}$$

$$\|\mathbf{v}(t)\| = 2\sqrt{50^2 - 16(50)\sqrt{2}t + 16^2 t^2} \qquad \text{\color{red}Speed}$$

$$\mathbf{a}(t) = -32\mathbf{j} \qquad \text{\color{red}Acceleration vector}$$

The tangential component of acceleration is

$$a_{\mathbf{T}}(t) = \frac{\mathbf{v}(t) \cdot \mathbf{a}(t)}{\|\mathbf{v}(t)\|} = \frac{-32(50\sqrt{2} - 32t)}{2\sqrt{50^2 - 16(50)\sqrt{2}t + 16^2 t^2}}.$$

At the specified times, you have

$$a_{\mathbf{T}}(0) = \frac{-32(50\sqrt{2})}{100} = -16\sqrt{2} \approx -22.6$$

$$a_{\mathbf{T}}(1) = \frac{-32(50\sqrt{2} - 32)}{2\sqrt{50^2 - 16(50)\sqrt{2} + 16^2}} \approx -15.4$$

$$a_{\mathbf{T}}\!\left(\frac{25\sqrt{2}}{16}\right) = \frac{-32(50\sqrt{2} - 50\sqrt{2})}{50\sqrt{2}} = 0.$$

You can see from Figure 11.26 that, at the maximum height, when $t = 25\sqrt{2}/16$, the tangential component is 0. This is reasonable because the direction of motion is horizontal at the point and the tangential component of the acceleration is equal to the horizontal component of the acceleration.

EXERCISES FOR SECTION 11.4

In Exercises 1–6, find the unit tangent vector T(t) and find a set of parametric equations for the line tangent to the space curve at point P.

1. $\mathbf{r}(t) = t\mathbf{i} + t^2\mathbf{j} + t\mathbf{k}, \quad P(0, 0, 0)$

2. $\mathbf{r}(t) = t\mathbf{i} + t^2\mathbf{j} + \frac{2}{3}\mathbf{k}, \quad P\!\left(1, 1, \frac{2}{3}\right)$

3. $\mathbf{r}(t) = 2\cos t\,\mathbf{i} + 2\sin t\,\mathbf{j} + t\mathbf{k}, \quad P(2, 0, 0)$

4. $\mathbf{r}(t) = \langle t, t, \sqrt{4 - t^2} \rangle, \quad P\!\left(1, 1, \sqrt{3}\right)$

5. $\mathbf{r}(t) = \langle 2\cos t, 2\sin t, 4 \rangle \quad P(\sqrt{2}, \sqrt{2}, 4)$

6. $\mathbf{r}(t) = \langle 2\sin t, 2\cos t, 4\sin^2 t \rangle, \quad P(1, \sqrt{3}, 1)$

In Exercises 7 and 8, use a graphing utility to graph the space curve. Then find T(t) and find a set of parametric equations for the line tangent to the space curve at point P. Graph the tangent line.

7. $\mathbf{r}(t) = \langle t, t^2, 2t^3/3 \rangle, \quad P(3, 9, 18)$

8. $\mathbf{r}(t) = 3\cos t\,\mathbf{i} + 4\sin t\,\mathbf{j} + \frac{1}{2}t\mathbf{k}, \quad P(0, 4, \pi/4)$

Linear Approximation   **In Exercises 9 and 10, find a set of parametric equations for the tangent line to the graph at $t = t_0$ and use the equations for the line to approximate $\mathbf{r}(t_0 + 0.1)$.**

9. $\mathbf{r}(t) = \langle t, \ln t, \sqrt{t} \rangle, \quad t_0 = 1$

10. $\mathbf{r}(t) = \langle e^{-t}, 2\cos t, 2\sin t \rangle, \quad t_0 = 0$

In Exercises 11 and 12, verify that the space curves intersect at the given values of the parameters. Find the angle between the tangent vectors to the curves at the point of intersection.

11. $\mathbf{r}(t) = \langle t - 2, t^2, \frac{1}{2}t \rangle, \quad t = 4$

$\mathbf{u}(s) = \langle \frac{1}{4}s, 2s, \sqrt[3]{s} \rangle, \quad s = 8$

12. $\mathbf{r}(t) = \langle t, \cos t, \sin t \rangle, \quad t = 0$

$\mathbf{u}(s) = \langle -\frac{1}{2}\sin^2 s - \sin s, 1 - \frac{1}{2}\sin^2 s - \sin s,$
$\frac{1}{2}\sin s \cos s + \frac{1}{2}s \rangle, \; s = 0$

In Exercises 13–16, find $\mathbf{v}(t)$, $\mathbf{a}(t)$, $\mathbf{T}(t)$ and $\mathbf{N}(t)$ (if it exists) for an object moving along the path given by the vector-valued function $\mathbf{r}(t)$. Use the results to determine the form of the path. Is the speed of the object constant or changing?

13. $\mathbf{r}(t) = 4t\mathbf{i}$ **14.** $\mathbf{r}(t) = 4t\mathbf{i} - 2t\mathbf{j}$

15. $\mathbf{r}(t) = 4t^2\mathbf{i}$ **16.** $\mathbf{r}(t) = t^2\mathbf{j} + \mathbf{k}$

In Exercises 17–22, find $\mathbf{T}(t)$, $\mathbf{N}(t)$, $a_{\mathbf{T}}$, and $a_{\mathbf{N}}$ at the given time t for the plane curve $\mathbf{r}(t)$.

17. $\mathbf{r}(t) = t\mathbf{i} + \dfrac{1}{t}\mathbf{j}$, $t = 1$

18. $\mathbf{r}(t) = t\mathbf{i} + t^2\mathbf{j}$, $t = 1$

19. $\mathbf{r}(t) = e^t \cos t\mathbf{i} + e^t \sin t\mathbf{j}$, $t = \dfrac{\pi}{2}$

20. $\mathbf{r}(t) = a \cos \omega t\mathbf{i} + b \sin \omega t\mathbf{j}$, $t = 0$

21. $\mathbf{r}(t) = \langle \cos \omega t + \omega t \sin \omega t, \sin \omega t - \omega t \cos \omega t \rangle$, $t = t_0$

22. $\mathbf{r}(t) = \langle \omega t - \sin \omega t, 1 - \cos \omega t \rangle$, $t = t_0$

Circular Motion   In Exercises 23–26, consider an object moving according to the position function

$\mathbf{r}(t) = a \cos \omega t\, \mathbf{i} + a \sin \omega t\, \mathbf{j}$.

23. Find $\mathbf{T}(t)$, $\mathbf{N}(t)$, $a_{\mathbf{T}}$, and $a_{\mathbf{N}}$.

24. Determine the directions of $\mathbf{T}$ and $\mathbf{N}$ relative to the position function $\mathbf{r}$.

25. Determine the speed of the object at any time t and explain its value relative to the value of $a_{\mathbf{T}}$.

26. If the angular velocity ω is halved, by what factor is $a_{\mathbf{N}}$ changed?

In Exercises 27 and 28, sketch the graph of the plane curve given by the vector-valued function, and, at the point on the curve determined by $\mathbf{r}(t_0)$, sketch the vectors $\mathbf{T}$ and $\mathbf{N}$. Note that $\mathbf{N}$ points toward the concave side of the curve.

Function	Time
27. $\mathbf{r}(t) = t\mathbf{i} + \dfrac{1}{t}\mathbf{j}$	$t_0 = 2$
28. $\mathbf{r}(t) = 2 \cos t\mathbf{i} + 2\sin t\mathbf{j}$	$t_0 = \dfrac{\pi}{4}$

29. ***Cycloidal Motion***   The figure shows the path of a particle modeled by the vector-valued function

$\mathbf{r}(t) = \langle \pi t - \sin \pi t, 1 - \cos \pi t \rangle$.

The figure also shows the vectors $\mathbf{v}(t)/\|\mathbf{v}(t)\|$ and $\mathbf{a}(t)/\|\mathbf{a}(t)\|$ at the indicated values of t.

(a) Find $a_{\mathbf{T}}$ and $a_{\mathbf{N}}$ at $t = \frac{1}{2}$, $t = 1$, and $t = \frac{3}{2}$.

(b) Determine whether the speed of the particle is increasing or decreasing at each of the indicated values of t. Give reasons for your answers.

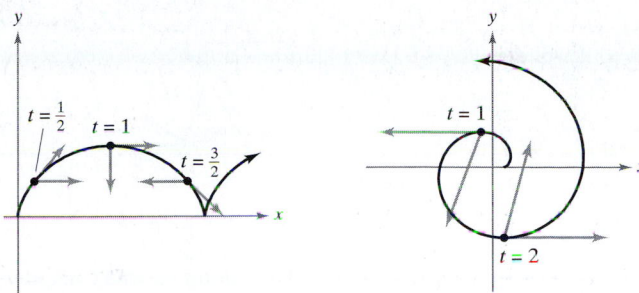

Figure for 29   **Figure for 30**

30. ***Motion Along an Involute of a Circle***   The figure shows a particle moving along a path modeled by

$\mathbf{r}(t) = \langle \cos \pi t + \pi t \sin \pi t, \sin \pi t - \pi t \cos \pi t \rangle$.

The figure also shows the vectors $\mathbf{v}(t)$ and $\mathbf{a}(t)$ for $t = 1$ and $t = 2$.

(a) Find $a_{\mathbf{T}}$ and $a_{\mathbf{N}}$ at $t = 1$ and $t = 2$.

(b) Determine whether the speed of the particle is increasing or decreasing at each of the indicated values of t. Give reasons for your answers.

In Exercises 31–34, find $\mathbf{T}(t)$, $\mathbf{N}(t)$, $a_{\mathbf{T}}$, and $a_{\mathbf{N}}$ at the given time t for the space curve $\mathbf{r}(t)$.

Function	Time
31. $\mathbf{r}(t) = t\mathbf{i} + 2t\mathbf{j} - 3t\mathbf{k}$	$t = 1$
32. $\mathbf{r}(t) = 4t\mathbf{i} - 4t\mathbf{j} + 2t\mathbf{k}$	$t = 2$
33. $\mathbf{r}(t) = t\mathbf{i} + t^2\mathbf{j} + \dfrac{t^2}{2}\mathbf{k}$	$t = 1$
34. $\mathbf{r}(t) = e^t \cos t\mathbf{i} + e^t \sin t\mathbf{j} + e^t\mathbf{k}$	$t = 0$

In Exercises 35 and 36, use a graphing utility to graph the space curve. Then find $\mathbf{T}(t)$, $\mathbf{N}(t)$, $a_{\mathbf{T}}$, $a_{\mathbf{N}}$ at the given time t. Sketch $\mathbf{T}(t)$ and $\mathbf{N}(t)$ on the space curve.

Function	Time
35. $\mathbf{r}(t) = 4t\mathbf{i} + 3 \cos t\mathbf{j} + 3 \sin t\mathbf{k}$	$t = \dfrac{\pi}{2}$
36. $\mathbf{r}(t) = t\mathbf{i} + 3t^2\mathbf{j} + \dfrac{t^2}{2}\mathbf{k}$	$t = 2$

In Exercises 37 and 38, find the vectors $\mathbf{T}$ and $\mathbf{N}$, and the unit binormal vector $\mathbf{B} = \mathbf{T} \times \mathbf{N}$, for the vector-valued function $\mathbf{r}(t)$ at the indicated value of t.

37. $\mathbf{r}(t) = 2 \cos t\mathbf{i} + 2 \sin t\mathbf{j} + \dfrac{t}{2}\mathbf{k}$

$t_0 = \dfrac{\pi}{2}$

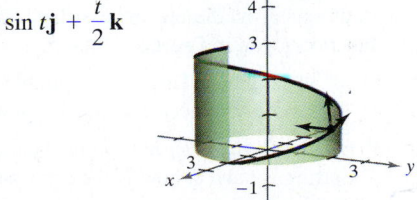

38. $\mathbf{r}(t) = t\mathbf{i} + t^2\mathbf{j} + \dfrac{t^3}{3}\mathbf{k}$

$t_0 = 1$

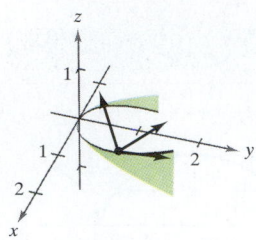

39. *Air Traffic Control* Because of a storm, ground controllers instruct a pilot of a plane flying at an altitude of 4 miles to make a 90° turn and climb to an altitude of 4.2 miles. The model for the path of the plane during this maneuver is

$\mathbf{r}(t) = \langle 10 \cos 10\pi t, 10 \sin 10\pi t, 4 + 4t \rangle, \quad 0 \le t \le \frac{1}{20}$

where t is the time in hours and $\mathbf{r}$ is the distance in miles.

(a) Determine the speed of the plane.

(b) Use a computer algebra system to calculate $a_{\mathbf{T}}$ and $a_{\mathbf{N}}$. Why is one of these equal to 0?

40. *Projectile Motion* A projectile is launched with an initial velocity of 100 feet per second at a height of 5 feet and at an angle of 30° with the horizontal.

(a) Determine the vector-valued function for the path of the projectile.

(b) Use a graphing utility to graph the path and approximate the maximum height and range of the projectile.

(c) Find $\mathbf{v}(t)$, $\|\mathbf{v}(t)\|$, and $\mathbf{a}(t)$.

(d) Use a graphing utility to complete the table.

t	0.5	1.0	1.5	2.0	2.5	3.0
Speed						

(e) Use a computer algebra system to find $\mathbf{T}(t)$, $\mathbf{N}(t)$, $a_{\mathbf{T}}$, and $a_{\mathbf{N}}$, and verify that $\mathbf{a} = a_{\mathbf{T}}\mathbf{T} + a_{\mathbf{N}}\mathbf{N}$.

(f) Use a graphing utility to graph the scalar functions $a_{\mathbf{T}}$ and $a_{\mathbf{N}}$. How is the speed of the projectile changing when $a_{\mathbf{T}}$ and $a_{\mathbf{N}}$ have opposite signs?

41. *Projectile Motion* Find the tangential and normal components of acceleration for a projectile fired at an angle θ with the horizontal at an initial speed of v_0. What are the components when the projectile is at its maximum height?

42. *Projectile Motion* A plane flying at an altitude of 30,000 feet at a speed of 540 miles per hour (792 feet per second) releases a bomb. Find the tangential and normal components of acceleration acting on the bomb.

43. *Centripetal Acceleration* An object is spinning at a constant speed on the end of a string, according to the position function given in Exercises 23–26.

(a) If the angular velocity ω is doubled, how is the centripetal component of acceleration changed?

(b) If the angular velocity is unchanged but the length of the string is halved, how is the centripetal component of acceleration changed?

44. *Centripetal Force* An object of mass m moves at a constant speed v in a circular path of radius r. The force required to produce the centripetal component of acceleration is called the centripetal force and is given by $F = mv^2/r$. Newton's Law of Universal Gravitation is given by $F = GMm/d^2$, where d is the distance between the centers of the two bodies of masses M and m, and G is a gravitational constant. Use this law to show that the speed required for circular motion is $v = \sqrt{GM/r}$.

Orbital Speed In Exercises 45–48, use the result of Exercise 44 to find the speed necessary for the given circular orbit around the earth. Let $GM = 9.56 \times 10^4$ cubic miles per second per second, and assume the radius of the earth is 4000 miles.

45. The orbit of a space shuttle 100 miles above the surface of the earth

46. The orbit of a space shuttle 200 miles above the surface of the earth

47. The orbit of a heat capacity mapping satellite 385 miles above the surface of the earth (see figure).

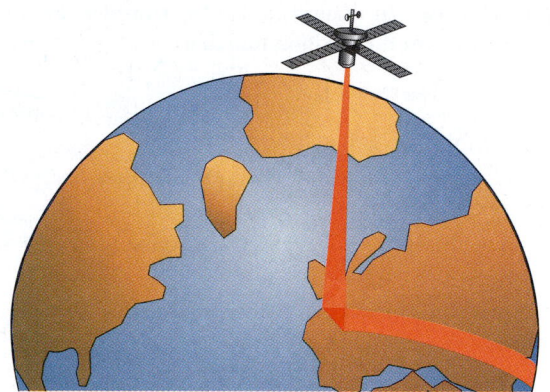

48. The orbit of a SYNCOM satellite r miles above the surface of the earth that is in geosynchronous orbit. [The satellite completes one orbit per sidereal day (23 hours, 56 minutes), and thus appears to remain stationary above a point on the earth.]

49. Prove that the principal unit normal vector $\mathbf{N}$ points toward the concave side of a plane curve.

50. Prove that the vector $\mathbf{T}'(t)$ is $\mathbf{0}$ for an object moving in a straight line.

51. Prove that $a_{\mathbf{N}} = \dfrac{\|\mathbf{v} \times \mathbf{a}\|}{\|\mathbf{v}\|}$.

52. Prove that $a_{\mathbf{N}} = \sqrt{\|\mathbf{a}\|^2 - a_{\mathbf{T}}^2}$.

Arc Length • Arc Length Parameter • Curvature • Applications

Arc Length

In Section 9.3, you saw that the arc length of a smooth *plane* curve C given by the parametric equations $x = x(t)$ and $y = y(t)$, $a \leq t \leq b$, is

$$s = \int_a^b \sqrt{[x'(t)]^2 + [y'(t)]^2}\, dt.$$

In vector form, where C is given by $\mathbf{r}(t) = x(t)\mathbf{i} + y(t)\mathbf{j}$, you can rewrite this equation for arc length as

$$s = \int_a^b \|\mathbf{r}'(t)\|\, dt.$$

The formula for the arc length of a plane curve has a natural extension to a smooth curve in *space*, as stated in the following theorem.

> **THEOREM 11.6   Arc Length of a Space Curve**
>
> If C is a smooth curve given by $\mathbf{r}(t) = x(t)\mathbf{i} + y(t)\mathbf{j} + z(t)\mathbf{k}$, on an interval $[a, b]$, then the arc length of C on the interval is
>
> $$s = \int_a^b \sqrt{[x'(t)]^2 + [y'(t)]^2 + [z'(t)]^2}\, dt = \int_a^b \|\mathbf{r}'(t)\|\, dt.$$

EXAMPLE 1   Finding the Arc Length of a Curve in Space

Find the arc length of the curve given by

$$\mathbf{r}(t) = t\mathbf{i} + \frac{4}{3}t^{3/2}\mathbf{j} + \frac{1}{2}t^2\mathbf{k}$$

from $t = 0$ to $t = 2$, as shown in Figure 11.27.

Solution   Using $x(t) = t$, $y(t) = \frac{4}{3}t^{3/2}$, and $z(t) = \frac{1}{2}t^2$, you obtain $x'(t) = 1$, $y'(t) = 2t^{1/2}$, and $z'(t) = t$. Thus, the arc length from $t = 0$ to $t = 2$ is given by

$$
\begin{aligned}
s &= \int_0^2 \sqrt{[x'(t)]^2 + [y'(t)]^2 + [z'(t)]^2}\, dt \\
&= \int_0^2 \sqrt{1 + 4t + t^2}\, dt \\
&= \int_0^2 \sqrt{(t + 2)^2 - 3}\, dt \qquad \text{\textcolor{red}{Integration tables (the appendix), Formula 26}} \\
&= \left[\frac{t + 2}{2}\sqrt{(t + 2)^2 - 3} - \frac{3}{2}\ln\left| (t + 2) + \sqrt{(t + 2)^2 - 3} \right| \right]_0^2 \\
&= 2\sqrt{13} - \frac{3}{2}\ln\left(4 + \sqrt{13}\right) - 1 + \frac{3}{2}\ln 3 \\
&\approx 4.816.
\end{aligned}
$$

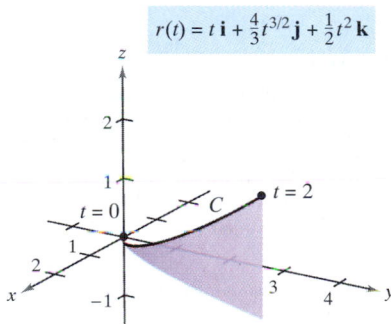

$$r(t) = t\,\mathbf{i} + \tfrac{4}{3}t^{3/2}\mathbf{j} + \tfrac{1}{2}t^2\,\mathbf{k}$$

As t increases from 0 to 2, the vector $\mathbf{r}(t)$ traces out a curve whose length is approximately 4.816.

Figure 11.27

Curve:
$\mathbf{r}(t) = b \cos t\mathbf{i} + b \sin t\mathbf{j} + \sqrt{1 - b^2}\, t\mathbf{k}$

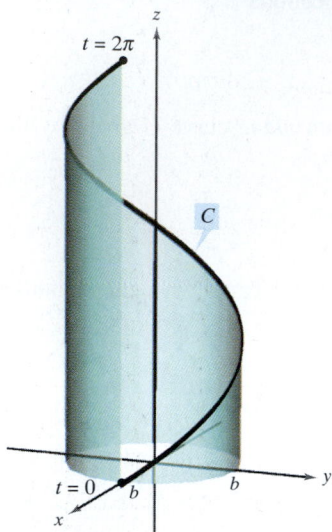

One turn of a helix
Figure 11.28

$s(t) = \int_a^t \sqrt{[x'(u)]^2 + [y'(u)]^2 + [z'(u)]^2}\, du$

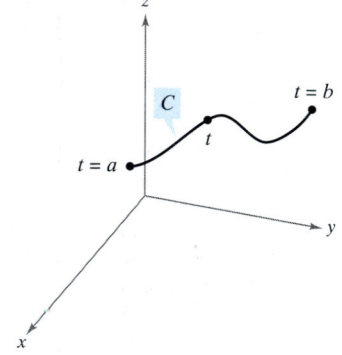

Figure 11.29

EXAMPLE 2 **Finding the Arc Length of a Helix**

Find the length of one turn of the helix given by

$$\mathbf{r}(t) = b \cos t\mathbf{i} + b \sin t\mathbf{j} + \sqrt{1 - b^2}\, t\mathbf{k}$$

as shown in Figure 11.28.

Solution Because $\mathbf{r}'(t) = -b \sin t\mathbf{i} + b \cos t\mathbf{j} + \sqrt{1 - b^2}\, \mathbf{k}$, the arc length of one turn is

$$s = \int_0^{2\pi} \|\mathbf{r}'(t)\|\, dt$$

$$= \int_0^{2\pi} \sqrt{b^2(\sin^2 t + \cos^2 t) + (1 - b^2)}\, dt$$

$$= \int_0^{2\pi} dt$$

$$= 2\pi.$$

Arc Length Parameter

You have seen that curves can be represented by vector-valued functions in different ways, depending on the choice of parameter. For *motion* along a curve, the convenient parameter is time t. However, for studying the *geometric properties* of a curve, the convenient parameter is often arc length s.

Definition of Arc Length Function

Let C be a smooth curve given by $\mathbf{r}(t)$ defined on the closed interval $[a, b]$. For $a \leq t \leq b$, the **arc length function** is given by

$$s(t) = \int_a^t \|\mathbf{r}'(u)\|\, du = \int_a^t \sqrt{[x'(u)]^2 + [y'(u)]^2 + [z'(u)]^2}\, du.$$

The arc length s is called the **arc length parameter.** (See Figure 11.29.)

NOTE The arc length function s is *nonnegative*. It measures the distance along C from the initial point $(x(a), y(a), z(a))$ to the point $(x(t), y(t), z(t))$.

Using the definition of the arc length function and the Second Fundamental Theorem of Calculus, you can conclude that

$$\frac{ds}{dt} = \|\mathbf{r}'(t)\|. \qquad \text{\color{red}Derivative of arc length function}$$

In differential form, you can write

$$ds = \|\mathbf{r}'(t)\|\, dt.$$

EXAMPLE 3   **Finding the Arc Length Function for a Line**

Find the arc length function $s(t)$ for the line segment given by

$$\mathbf{r}(t) = (3 - 3t)\mathbf{i} + 4t\mathbf{j}, \qquad 0 \le t \le 1$$

and express $\mathbf{r}$ as a function of the parameter s. (See Figure 11.30.)

Solution   Because $\mathbf{r}'(t) = -3\mathbf{i} + 4\mathbf{j}$ and

$$\|\mathbf{r}'(t)\| = \sqrt{3^2 + 4^2} = 5$$

you have

$$s(t) = \int_0^t \|\mathbf{r}'(u)\| \, du$$

$$= \int_0^t 5 \, du$$

$$= 5t.$$

Using $s = 5t$ (or $t = s/5$), you can rewrite $\mathbf{r}$ using the arc length parameter as follows.

$$\mathbf{r}(s) = \left(3 - \tfrac{3}{5}s\right)\mathbf{i} + \tfrac{4}{5}s\mathbf{j}, \qquad 0 \le s \le 5.$$

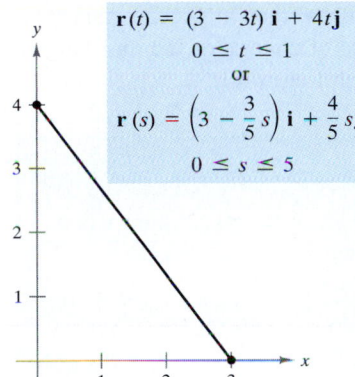

$$\mathbf{r}(t) = (3 - 3t)\,\mathbf{i} + 4t\mathbf{j}$$
$$0 \le t \le 1$$
or
$$\mathbf{r}(s) = \left(3 - \frac{3}{5}s\right)\mathbf{i} + \frac{4}{5}s\mathbf{j}$$
$$0 \le s \le 5$$

The line segment from $(3, 0)$ to $(0, 4)$ can be parametrized using the arc length parameter s.
Figure 11.30

One of the advantages of writing a vector-valued function in terms of the arc length parameter is that $\|\mathbf{r}'(s)\| = 1$. For instance, in Example 3, you have

$$\|\mathbf{r}'(s)\| = \sqrt{\left(-\frac{3}{5}\right)^2 + \left(\frac{4}{5}\right)^2} = 1.$$

Thus, for a smooth curve C, represented by $\mathbf{r}(s)$, where s is the arc length parameter, the arc length between a and b is

$$\text{Length of arc} = \int_a^b \|\mathbf{r}'(s)\| \, ds$$

$$= \int_a^b ds = b - a$$

$$= \text{length of interval.}$$

Furthermore, if t is *any* parameter such that $\|\mathbf{r}'(t)\| = 1$, then t must be the arc length parameter. These results are summarized in the following theorem, which we state without proof.

THEOREM 11.7   Arc Length Parameter

If C is a smooth curve given by

$$\mathbf{r}(s) = x(s)\mathbf{i} + y(s)\mathbf{j} \quad \text{or} \quad \mathbf{r}(s) = x(s)\mathbf{i} + y(s)\mathbf{j} + z(s)\mathbf{k}$$

where s is the arc length parameter, then

$$\|\mathbf{r}'(s)\| = 1.$$

Moreover, if t is *any* parameter for the vector-valued function $\mathbf{r}$ such that $\|\mathbf{r}'(t)\| = 1$, then t must be the arc length parameter.

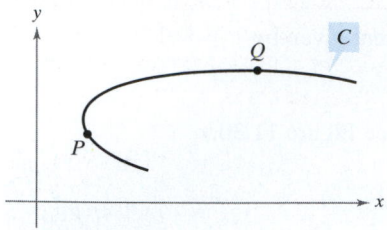

Curvature at P is greater than at Q.
Figure 11.31

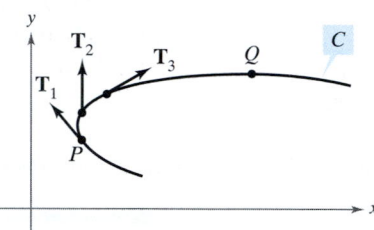

The magnitude of the rate of change of **T** with respect to the arc length is the curvature of a curve.
Figure 11.32

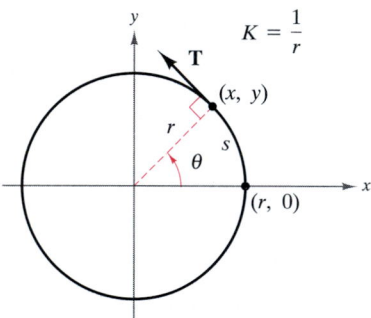

The curvature of a circle is constant.
Figure 11.33

Curvature

An important use of the arc length parameter is to find **curvature**—the measure of how sharply a curve bends. For instance, in Figure 11.31 the curve bends more sharply at P than at Q, and we say that the curvature is greater at P than at Q. You can calculate curvature by calculating the magnitude of the rate of change of the unit tangent vector **T** with respect to the arc length s, as indicated in Figure 11.32.

> ### Definition of Curvature
>
> Let C be a smooth curve (in the plane *or* in space) given by $\mathbf{r}(s)$, where s is the arc length parameter. The **curvature** at s is given by
>
> $$K = \left\| \frac{d\mathbf{T}}{ds} \right\| = \|\mathbf{T}'(s)\|.$$

A circle has the same curvature at any point. Moreover, the curvature and the radius of the circle are inversely related. That is, a circle with a large radius has a small curvature, and a circle with a small radius has a large curvature. This inverse relationship is made explicit in the following example.

EXAMPLE 4   Finding the Curvature of a Circle

Show that the curvature of a circle of radius r is $K = 1/r$.

Solution   Without loss of generality you can consider the circle to be centered at the origin. Let (x, y) be any point on the circle and let s be the length of the arc from $(r, 0)$ to (x, y), as shown in Figure 11.33. By letting θ be the central angle of the circle, you can represent the circle by

$$\mathbf{r}(\theta) = r \cos \theta \mathbf{i} + r \sin \theta \mathbf{j}. \qquad \text{\textcolor{red}{θ is the parameter.}}$$

Using the formula for the length of a circular arc $s = r\theta$, you can rewrite $\mathbf{r}(\theta)$ in terms of the arc length parameter as follows.

$$\mathbf{r}(s) = r \cos \frac{s}{r} \mathbf{i} + r \sin \frac{s}{r} \mathbf{j} \qquad \text{\textcolor{red}{Arc length s is the parameter.}}$$

Thus, $\mathbf{r}'(s) = -\sin \frac{s}{r} \mathbf{i} + \cos \frac{s}{r} \mathbf{j}$, and it follows that $\|\mathbf{r}'(s)\| = 1$, which implies that the unit tangent vector is

$$\mathbf{T}(s) = \frac{\mathbf{r}'(s)}{\|\mathbf{r}'(s)\|} = -\sin \frac{s}{r} \mathbf{i} + \cos \frac{s}{r} \mathbf{j}$$

and the curvature is given by

$$K = \|\mathbf{T}'(s)\| = \left\| -\frac{1}{r} \cos \frac{s}{r} \mathbf{i} - \frac{1}{r} \sin \frac{s}{r} \mathbf{j} \right\| = \frac{1}{r}$$

at every point on the circle.

NOTE   Because a straight line doesn't "curve," its curvature should be 0. Try checking this by finding the curvature of the line given by

$$\mathbf{r}(s) = \left(3 - \frac{3}{5} s \right) \mathbf{i} + \frac{4}{5} s \mathbf{j}.$$

In Example 4, the curvature was found by applying the definition directly. This requires that the curve be written in terms of the arc length parameter s. The following theorem gives two other formulas for finding the curvature of a curve written in terms of an arbitrary parameter t. We leave the proof of this theorem as an exercise [see Exercise 76, parts (a) and (b)].

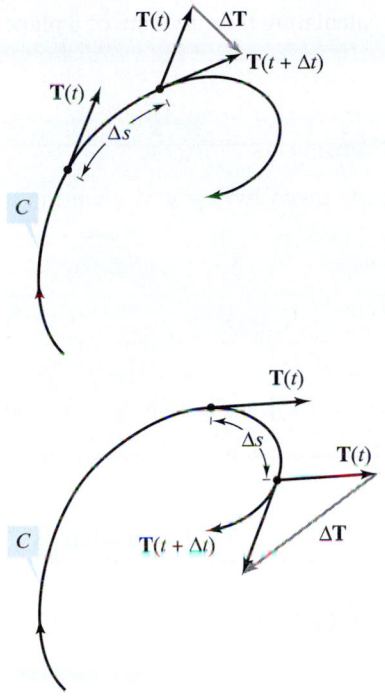

Figure 11.34

> ### THEOREM 11.8   Formulas for Curvature
>
> If C is a smooth curve given by $\mathbf{r}(t)$, then the curvature of C at t is given by
>
> $$K = \frac{\|\mathbf{T}'(t)\|}{\|\mathbf{r}'(t)\|} = \frac{\|\mathbf{r}'(t) \times \mathbf{r}''(t)\|}{\|\mathbf{r}'(t)\|^3}.$$

Because $\|\mathbf{r}'(t)\| = ds/dt$, the first formula implies that curvature is the ratio of the rate of change in the tangent vector $\mathbf{T}$ to the rate of change in arc length. To see the reasonableness of this, let Δt be a "small number." Then,

$$\frac{\mathbf{T}'(t)}{ds/dt} \approx \frac{[\mathbf{T}(t + \Delta t) - \mathbf{T}(t)]/\Delta t}{[s(t + \Delta t) - s(t)]/\Delta t} = \frac{\mathbf{T}(t + \Delta t) - \mathbf{T}(t)}{s(t + \Delta t) - s(t)} = \frac{\Delta \mathbf{T}}{\Delta s}.$$

In other words, for a given Δs, the greater the length of $\Delta \mathbf{T}$, the more the curve bends at t, as shown in Figure 11.34.

EXAMPLE 5   Finding the Curvature of a Space Curve

Find the curvature of the curve given by $\mathbf{r}(t) = 2t\mathbf{i} + t^2\mathbf{j} - \frac{1}{3}t^3\mathbf{k}$.

Solution   It is not apparent whether or not this parameter is arc length, so you should use the formula $K = \|\mathbf{T}'(t)\|/\|\mathbf{r}'(t)\|$.

$$\mathbf{r}'(t) = 2\mathbf{i} + 2t\mathbf{j} - t^2\mathbf{k}$$

$$\|\mathbf{r}'(t)\| = \sqrt{4 + 4t^2 + t^4} = t^2 + 2 \qquad \text{Length of } \mathbf{r}'(t)$$

$$\mathbf{T}(t) = \frac{\mathbf{r}'(t)}{\|\mathbf{r}'(t)\|} = \frac{2\mathbf{i} + 2t\mathbf{j} - t^2\mathbf{k}}{t^2 + 2}$$

$$\mathbf{T}'(t) = \frac{(t^2 + 2)(2\mathbf{j} - 2t\mathbf{k}) - (2t)(2\mathbf{i} - 2t\mathbf{j} - t^2\mathbf{k})}{(t^2 + 2)^2}$$

$$= \frac{-4t\mathbf{i} + (4 - 2t^2)\mathbf{j} - 4t\mathbf{k}}{(t^2 + 2)^2}$$

$$\|\mathbf{T}'(t)\| = \frac{\sqrt{16t^2 + 16 - 16t^2 + 4t^4 + 16t^2}}{(t^2 + 2)^2}$$

$$= \frac{2(t^2 + 2)}{(t^2 + 2)^2}$$

$$= \frac{2}{t^2 + 2} \qquad \text{Length of } \mathbf{T}'(t)$$

Therefore,

$$K = \frac{\|\mathbf{T}'(t)\|}{\|\mathbf{r}'(t)\|} = \frac{2}{(t^2 + 2)^2}. \qquad \text{Curvature}$$

The following theorem presents a formula for calculating the curvature of a plane curve given by $y = f(x)$.

THEOREM 11.9   Curvature in Rectangular Coordinates

If C is the graph of a twice-differentiable function given by $y = f(x)$, then the curvature at the point (x, y) is given by

$$K = \frac{|y''|}{[1 + (y')^2]^{3/2}}.$$

Proof By representing the curve C by $\mathbf{r}(x) = x\mathbf{i} + f(x)\mathbf{j} + 0\mathbf{k}$ (where x is the parameter), you obtain $\mathbf{r}'(x) = \mathbf{i} + f'(x)\mathbf{j}$

$$\|\mathbf{r}'(x)\| = \sqrt{1 + [f'(x)]^2}$$

and $\mathbf{r}''(x) = f''(x)\mathbf{j}$. Because $\mathbf{r}'(x) \times \mathbf{r}''(x) = f''(x)\mathbf{k}$, it follows that the curvature is

$$K = \frac{\|\mathbf{r}'(x) \times \mathbf{r}''(x)\|}{\|\mathbf{r}'(x)\|^3} = \frac{|f''(x)|}{\{1 + [f'(x)]^2\}^{3/2}} = \frac{|y''|}{[1 + (y')^2]^{3/2}}.$$

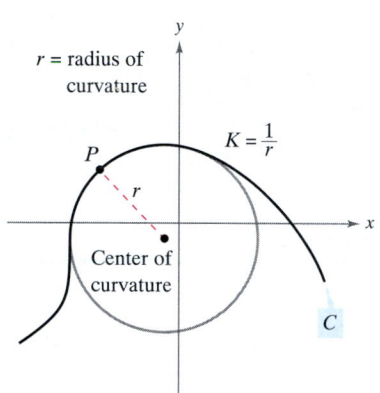

$r = $ radius of curvature

$K = \dfrac{1}{r}$

The circle of curvature
Figure 11.35

Let C be a curve with curvature K at point P. The circle passing through point P with radius $r = 1/K$ is called the **circle of curvature** if the circle lies on the concave side of the curve and shares a common tangent line with the curve at point P. The radius is called the **radius of curvature** at P, and the center of the circle is called the **center of curvature.**

The circle of curvature gives us a nice way to estimate graphically the curvature K at a point P on a curve. Using a compass, you can sketch a circle that snuggles up against the concave side of the curve at point P, as shown in Figure 11.35. If the circle has a radius of r, you can estimate the curvature to be $K = 1/r$.

EXAMPLE 6   Finding Curvature in Rectangular Coordinates

Find the curvature of the parabola given by $y = x - \frac{1}{4}x^2$ at $x = 2$. Sketch the circle of curvature at $(2, 1)$.

Solution The curvature at $x = 2$ is as follows.

$$y' = 1 - \frac{x}{2} \qquad\qquad y' = 0$$

$$y'' = -\frac{1}{2} \qquad\qquad y'' = -\frac{1}{2}$$

$$K = \frac{|y''|}{[1 + (y')^2]^{3/2}} \qquad K = \frac{1}{2}$$

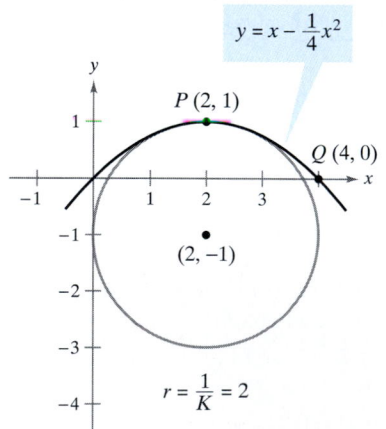

$y = x - \frac{1}{4}x^2$

$P(2, 1)$

$Q(4, 0)$

$(2, -1)$

$r = \dfrac{1}{K} = 2$

The circle of curvature
Figure 11.36

Because the curvature at $P(2, 1)$ is $\frac{1}{2}$, it follows that the radius of the circle of curvature at that point is 2. Thus, the center of curvature is $(2, -1)$, as shown in Figure 11.36. [In the figure, note that the curve has the greatest curvature at P. Try showing that the curvature at $Q(4, 0)$ is $1/2^{5/2} \approx 0.177$.]

Arc length and curvature are closely related to the tangential and normal components of acceleration. The tangential component of acceleration is the rate of change of the speed, which in turn is the rate of change of the arc length. This component is negative as a moving object slows down and positive as it speeds up—regardless of whether the object is turning or traveling in a straight line. Thus, the tangential component is solely a function of the arc length and is independent of the curvature.

On the other hand, the normal component of acceleration is a function of **both** speed and curvature. This component measures the acceleration acting perpendicular to the direction of motion. To see why the normal component is affected by both speed and curvature, imagine that you are driving a car around a turn, as shown in Figure 11.37. If your speed is high and the turn is sharp, you feel yourself thrown against the car door. By lowering your speed *or* taking a more gentle turn, you are able to lessen this sideways thrust.

The next theorem explicitly states the relationship among speed, curvature, and the components of acceleration.

The amount of thrust felt by passengers in a car that is turning depends on two things—the speed of the car and the sharpness of the turn.

Figure 11.37

THEOREM 11.10   Acceleration, Speed, and Curvature

If $\mathbf{r}(t)$ is the position vector for a smooth curve C, then the acceleration vector is given by

$$\mathbf{a}(t) = \frac{d^2s}{dt^2}\mathbf{T} + K\left(\frac{ds}{dt}\right)^2\mathbf{N}$$

where K is the curvature of C and ds/dt is the speed.

NOTE   Note that Theorem 11.10 gives additional formulas for $a_{\mathbf{T}}$ and $a_{\mathbf{N}}$.

Proof   For the position vector $\mathbf{r}(t)$, you have

$$\mathbf{a}(t) = a_{\mathbf{T}}\mathbf{T} + a_{\mathbf{N}}\mathbf{N} = D_t[\|\mathbf{v}\|]\mathbf{T} + \|\mathbf{v}\|\,\|\mathbf{T}'\|\mathbf{N}$$

$$= \frac{d^2s}{dt^2}\mathbf{T} + \frac{ds}{dt}(\|\mathbf{v}\|K)\mathbf{N} = \frac{d^2s}{dt^2}\mathbf{T} + K\left(\frac{ds}{dt}\right)^2\mathbf{N}.$$

EXAMPLE 7   Tangential and Normal Components of Acceleration

Find $a_{\mathbf{T}}$ and $a_{\mathbf{N}}$ for the curve given by $\mathbf{r}(t) = 2t\mathbf{i} + t^2\mathbf{j} - \frac{1}{3}t^3\mathbf{k}$.

Solution   From Example 5, you know that

$$\frac{ds}{dt} = \|\mathbf{r}'(t)\| = t^2 + 2 \quad \text{and} \quad K = \frac{2}{(t^2 + 2)^2}.$$

Therefore,

$$a_{\mathbf{T}} = \frac{d^2s}{dt^2} = 2t \qquad\qquad \text{Tangential component}$$

and

$$a_{\mathbf{N}} = K\left(\frac{ds}{dt}\right)^2 = \frac{2}{(t^2 + 2)^2}(t^2 + 2)^2 = 2. \qquad \text{Normal component}$$

Applications

There are many applications in physics and engineering dynamics that involve the relationships among speed, arc length, curvature, and acceleration. One such application concerns frictional force.

Suppose a moving object with mass m is in contact with a stationary object. The total force required to produce an acceleration **a** along a given path is

$$\mathbf{F} = m\mathbf{a} = m\left(\frac{d^2s}{dt^2}\right)\mathbf{T} + mK\left(\frac{ds}{dt}\right)^2\mathbf{N}$$

$$= ma_{\mathbf{T}}\mathbf{T} + ma_{\mathbf{N}}\mathbf{N}.$$

The portion of this total force that is supplied by the stationary object is called the **force of friction.** For example, if a car moving with constant speed is rounding a turn, the roadway exerts a frictional force that keeps the car from sliding off the road. If the car is not sliding, the frictional force is perpendicular to the direction of motion and has magnitude equal to the normal component of acceleration, as shown in Figure 11.38. The potential frictional force of a road around a turn can be increased by banking the roadway.

Force of friction

The force of friction is perpendicular to the direction of motion.
Figure 11.38

EXAMPLE 8   Frictional Force

A 360-kilogram go-cart is driven at a speed of 60 kilometers per hour around a circular racetrack of radius 12 meters. To keep the cart from skidding off course, what frictional force must the track surface exert on the tires? (See Figure 11.39.)

Solution   The frictional force must equal the mass times the normal component of acceleration. For this circular path, you know that the curvature is

$$K = \frac{1}{12}.$$
Curvature of circular racetrack

Therefore, the frictional force is

$$ma_{\mathbf{N}} = mK\left(\frac{ds}{dt}\right)^2$$

$$= (360 \text{ kg})\left(\frac{1}{12 \text{ m}}\right)\left(\frac{60{,}000 \text{ m}}{3600 \text{ sec}}\right)^2$$

$$\approx 8333 \text{ (kg)(m)/sec}^2.$$

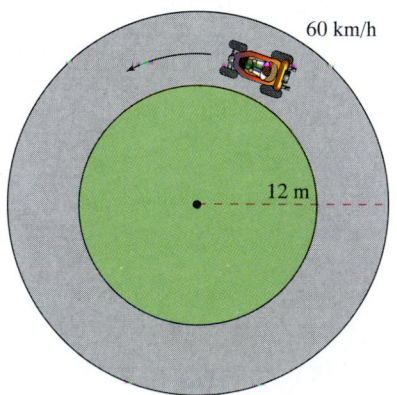

60 km/h

12 m

Figure 11.39

Summary of Velocity, Acceleration, and Curvature

Let C be a curve (in the plane or in space) given by the position function

$\mathbf{r}(t) = x(t)\mathbf{i} + y(t)\mathbf{j}$ Curve in the plane

$\mathbf{r}(t) = x(t)\mathbf{i} + y(t)\mathbf{j} + z(t)\mathbf{k}.$ Curve in space

Velocity vector, speed, and acceleration vector:

$\mathbf{v}(t) = \mathbf{r}'(t)$

$\|\mathbf{v}(t)\| = \dfrac{ds}{dt} = \|\mathbf{r}'(t)\|$

$\mathbf{a}(t) = \mathbf{r}''(t) = a_{\mathbf{T}}\mathbf{T}(t) + a_{\mathbf{N}}\mathbf{N}(t)$

Unit tangent vector and principal unit normal vector:

$\mathbf{T}(t) = \dfrac{\mathbf{r}'(t)}{\|\mathbf{r}'(t)\|}$ and $\mathbf{N}(t) = \dfrac{\mathbf{T}'(t)}{\|\mathbf{T}'(t)\|}$

Components of acceleration:

$a_{\mathbf{T}} = \mathbf{a} \cdot \mathbf{T} = \dfrac{\mathbf{v} \cdot \mathbf{a}}{\|\mathbf{v}\|} = \dfrac{d^2s}{dt^2}$

$a_{\mathbf{N}} = \mathbf{a} \cdot \mathbf{N} = \sqrt{\|\mathbf{a}\|^2 - a_{\mathbf{T}}^2} = \dfrac{\|\mathbf{v} \times \mathbf{a}\|}{\|\mathbf{v}\|} = K\left(\dfrac{ds}{dt}\right)^2$

Formulas for curvature in the plane:

$K = \dfrac{|y''|}{[1 + (y')^2]^{3/2}}$ C given by $y = f(x)$

$K = \dfrac{|x'y'' - y'x''|}{[(x')^2 + (y')^2]^{3/2}}$ C given by $x = x(t), y = y(t)$

Formulas for curvature in the plane or in space:

$K = \|\mathbf{T}'(s)\| = \|\mathbf{r}''(s)\|$ s is arc length parameter.

$K = \dfrac{\|\mathbf{T}'(t)\|}{\|\mathbf{r}'(t)\|} = \dfrac{\|\mathbf{r}'(t) \times \mathbf{r}''(t)\|}{\|\mathbf{r}'(t)\|^3}$ t is general parameter.

$K = \dfrac{\mathbf{a}(t) \cdot \mathbf{N}(t)}{\|\mathbf{v}(t)\|^2}$

Cross product formulas apply only to curves in space.

LAB SERIES
Lab 11.1

EXERCISES FOR SECTION 11.5

In Exercises 1–4, sketch the plane curve and find its length over the indicated interval.

Function	Interval
1. $\mathbf{r}(t) = t\mathbf{i} + 3t\mathbf{j}$	$[0, 4]$
2. $\mathbf{r}(t) = t\mathbf{i} + t^2\mathbf{k}$	$[0, 4]$
3. $\mathbf{r}(t) = a\cos^3 t\,\mathbf{i} + a\sin^3 t\,\mathbf{j}$	$[0, 2\pi]$
4. $\mathbf{r}(t) = a\cos t\,\mathbf{i} + a\sin t\,\mathbf{j}$	$[0, 2\pi]$

In Exercises 5–8, sketch the space curve and find its length over the indicated interval.

Function	Interval
5. $\mathbf{r}(t) = 2t\mathbf{i} - 3t\mathbf{j} + t\mathbf{k}$	$[0, 2]$
6. $\mathbf{r}(t) = \langle 4t, 3\cos t, 3\sin t\rangle$	$\left[0, \dfrac{\pi}{2}\right]$
7. $\mathbf{r}(t) = a\cos t\,\mathbf{i} + a\sin t\,\mathbf{j} + bt\,\mathbf{k}$	$[0, 2\pi]$
8. $\mathbf{r}(t) = \langle \sin t - t\cos t, \cos t + t\sin t, t^2\rangle$	$\left[0, \dfrac{\pi}{2}\right]$

In Exercises 9 and 10, use the integration capabilities of a graphing utility to approximate the length of the space curve over the indicated interval.

Function	Interval
9. $\mathbf{r}(t) = t^2\mathbf{i} + t\mathbf{j} + \ln t\mathbf{k}$	$1 \le t \le 3$
10. $\mathbf{r}(t) = \sin \pi t\,\mathbf{i} + \cos \pi t\,\mathbf{j} + t^3\mathbf{k}$	$0 \le t \le 2$

11. *Investigation*   Consider the graph of the vector-valued function $\mathbf{r}(t) = t\mathbf{i} + (4 - t^2)\mathbf{j} + t^3\mathbf{k}$ on the interval $[0, 2]$.

(a) Approximate the length of the curve by finding the length of the line segment connecting its endpoints.

(b) Approximate the length of the curve by summing the lengths of the line segments connecting the terminal points of the vectors $\mathbf{r}(0)$, $\mathbf{r}(0.5)$, $\mathbf{r}(1)$, $\mathbf{r}(1.5)$, and $\mathbf{r}(2)$.

(c) Describe how you could obtain a more accurate approximation by continuing the processes in parts (a) and (b).

(d) Use the integration capabilities of a graphing utility to approximate the length of the curve. Compare this result with the answers in parts (a) and (b).

12. *Investigation*   Repeat Exercise 11 for the vector-valued function $\mathbf{r}(t) = 6 \cos(\pi t/4)\mathbf{i} + 2 \sin(\pi t/4)\mathbf{j} + t\mathbf{k}$.

13. *Investigation*   Consider the helix represented by the vector-valued function $\mathbf{r}(t) = \langle 2 \cos t, 2 \sin t, t \rangle$.

(a) Express the length of the arc s on the helix as a function of t by evaluating the integral

$$s = \int_0^t \sqrt{[x'(u)]^2 + [y'(u)]^2 + [z'(u)]^2}\, du.$$

(b) Solve for t in the relationship derived in part (a), and substitute the result into the original set of parametric equations. This yields a parametrization of the curve in terms of the arc length parameter s.

(c) Find the coordinates of the point on the helix when the length of the arc is $s = \sqrt{5}$ and $s = 4$.

(d) Verify that $\|\mathbf{r}'(s)\| = 1$.

14. *Investigation*   Repeat Exercise 13 for the curve represented by the vector-valued function

$$\mathbf{r}(t) = \left\langle 4(\sin t - t \cos t), 4(\cos t + t \sin t), \tfrac{3}{2} t^2 \right\rangle.$$

In Exercises 15–18, find the curvature K of the curve where s is the arc length parameter.

15. $\mathbf{r}(s) = \left(1 + \dfrac{\sqrt{2}}{2} s\right)\mathbf{i} + \left(1 - \dfrac{\sqrt{2}}{2} s\right)\mathbf{j}$

16. $\mathbf{r}(s) = (3 + s)\mathbf{i} + \mathbf{j}$

17. Helix in Exercise 13

18. Curve in Exercise 14

In Exercises 19–22, find the curvature K of the plane curve at the indicated value of the parameter.

19. $\mathbf{r}(t) = 4t\mathbf{i} - 2t\mathbf{j},\quad t = 1$

20. $\mathbf{r}(t) = t^2\mathbf{j} + \mathbf{k},\quad t = 0$

21. $\mathbf{r}(t) = t\mathbf{i} + \dfrac{1}{t}\mathbf{j},\quad t = 1$

22. $\mathbf{r}(t) = t\mathbf{i} + t^2\mathbf{j},\quad t = 1$

In Exercises 23–34, find the curvature K of the curve.

23. $\mathbf{r}(t) = 4 \cos 2\pi t\,\mathbf{i} + 4 \sin 2\pi t\,\mathbf{j}$

24. $\mathbf{r}(t) = 2 \cos \pi t\,\mathbf{i} + \sin \pi t\,\mathbf{j}$

25. $\mathbf{r}(t) = a \cos \omega t\,\mathbf{i} + a \sin \omega t\,\mathbf{j}$

26. $\mathbf{r}(t) = a \cos \omega t\,\mathbf{i} + b \sin \omega t\,\mathbf{j}$

27. $\mathbf{r}(t) = e^t \cos t\,\mathbf{i} + e^t \sin t\,\mathbf{j}$

28. $\mathbf{r}(t) = \langle a(\omega t - \sin \omega t), a(1 - \cos \omega t) \rangle$

29. $\mathbf{r}(t) = \langle \cos \omega t + \omega t \sin \omega t, \sin \omega t - \omega t \cos \omega t \rangle$

30. $\mathbf{r}(t) = 4t\mathbf{i} - 4t\mathbf{j} + 2t\mathbf{k}$

31. $\mathbf{r}(t) = t\mathbf{i} + t^2\mathbf{j} + \dfrac{t^2}{2}\mathbf{k}$

32. $\mathbf{r}(t) = t\mathbf{i} + 3t^2\mathbf{j} + \dfrac{t^2}{2}\mathbf{k}$

33. $\mathbf{r}(t) = 4t\mathbf{i} + 3 \cos t\,\mathbf{j} + 3 \sin t\,\mathbf{k}$

34. $\mathbf{r}(t) = e^t \cos t\,\mathbf{i} + e^t \sin t\,\mathbf{j} + e^t\,\mathbf{k}$

In Exercises 35–40, find the curvature and radius of curvature of the plane curve at the indicated value of x.

Function	Point
35. $y = 3x - 2$	$x = a$
36. $y = mx + b$	$x = a$
37. $y = 2x^2 + 3$	$x = -1$
38. $y = x + \dfrac{1}{x}$	$x = 1$
39. $y = \sqrt{a^2 - x^2}$	$x = 0$
40. $y = \tfrac{3}{4}\sqrt{16 - x^2}$	$x = 0$

Writing   In Exercises 41 and 42, two circles of curvature to the graph of the function are given. (a) Find the equation of the smaller circle, and (b) write a short paragraph explaining why the circles have different radii.

41. $f(x) = \sin x$

42. $f(x) = 4x^2/(x^2 + 3)$

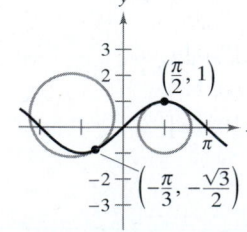

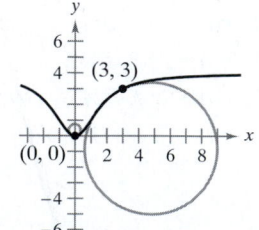

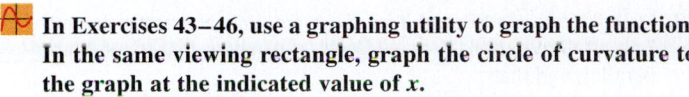

In Exercises 43–46, use a graphing utility to graph the function. In the same viewing rectangle, graph the circle of curvature to the graph at the indicated value of x.

43. $y = x + \dfrac{1}{x}, \quad x = 1$ **44.** $y = \ln x, \quad x = 1$

45. $y = e^x, \quad x = 0$ **46.** $y = \frac{1}{3} x^3, \quad x = 1$

In Exercises 47–50, (a) find the point on the curve at which the curvature K is a maximum and (b) find the limit of K as $x \to \infty$.

47. $y = (x - 1)^2 + 3$ **48.** $y = x^3$

49. $y = x^{2/3}$ **50.** $y = \ln x$

51. Find all points on the graph of

$$y = (x - 1)^3 + 3$$

at which the curvature is zero.

52. *Writing*   Use the result of Exercise 51 to write a paragraph describing the points on the graph of $y = f(x)$ at which the curvature is 0.

53. Show that the curvature is greatest at the endpoint of the major axis and is least at the endpoints of the minor axis for the ellipse given by $x^2 + 4y^2 = 4$.

54. Verify that the curvature at any point (x, y) on the graph of $y = \cosh x$ is y^2.

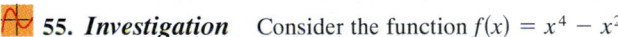

55. *Investigation*   Consider the function $f(x) = x^4 - x^2$.

(a) Use a computer algebra system to find the curvature K of the curve as a function of x.

(b) Use the result of part (a) to find the circle of curvature to the graph of f when $x = 0$ and $x = 1$. Use a computer algebra system to graph the function and the two circles of curvature.

(c) Graph the function $K(x)$ and compare it with the graph of $f(x)$. For example, do the extrema of f and K occur at the same critical numbers? Explain.

56. *Investigation*   The surface of a goblet is formed by revolving the graph of the function

$$y = \frac{1}{4} x^{8/5}, \quad 0 \le x \le 5$$

about the y-axis. The measurements are given in centimeters.

(a) Use a graphing utility to graph the surface.

(b) Find the volume of the goblet.

(c) Find the curvature K of the generating curve as a function of x. Use a graphing utility to graph K.

(d) If a spherical object is dropped into the goblet, is it possible for it to touch the bottom? Explain.

57. A sphere of radius 4 is dropped into the paraboloid given by $z = x^2 + y^2$.

(a) How close will the sphere come to the vertex of the paraboloid?

(b) What is the radius of the largest sphere that will touch the vertex?

58. *Investigation*   Find all a and b such that the two curves given by

$$y_1 = ax(b - x) \quad \text{and} \quad y_2 = \dfrac{x}{x + 2}$$

intersect at only one point and have a common tangent line and equal curvature at the point. Sketch a graph for each set of values for a and b.

59. *Think About It*   Given a twice-differentiable function $y = f(x)$, determine its curvature at a relative extremum. Can the curvature ever be greater than it is at a relative extremum? Why or why not?

60. *Speed*   The smaller the curvature in a bend of a road, the faster a car can travel. Assume that the maximum speed around a turn is inversely proportional to the square root of the curvature. A car moving on the path

$$y = \frac{1}{3} x^3$$

(x and y are measured in miles) can safely go 30 miles per hour at $\left(1, \frac{1}{3}\right)$. How fast can it go at $\left(\frac{3}{2}, \frac{9}{8}\right)$?

61. Let C be given by $y = f(x)$. Show that the center of curvature for (x, y) on C is $(x_0, y_0) = (x - y'z, y + z)$, where

$$z = \dfrac{1 + (y')^2}{y''}.$$

Find the center of curvature for $y = e^x$ at the point $(0, 1)$ on the curve.

62. A curve C is given by the polar equation $r = f(\theta)$. Show that the curvature K at the point (r, θ) is

$$K = \dfrac{\left|2(r')^2 - rr'' + r^2\right|}{[(r')^2 + r^2]^{3/2}}.$$

[*Hint:* Represent the curve by $\mathbf{r}(\theta) = r \cos \theta \mathbf{i} + r \sin \theta \mathbf{j}$.]

In Exercises 63–66, use the result of Exercise 62 to find the curvature of the polar curve.

63. $r = 1 + \sin \theta$ **64.** $r = \theta$

65. $r = a \sin \theta$ **66.** $r = e^\theta$

67. Given the polar curve $r = e^{a\theta}, a > 0$, find the curvature K and determine the limit of K as (a) $\theta \to \infty$ and (b) $a \to \infty$.

68. Show that the formula for the curvature of a polar curve $r = f(\theta)$ given in Exercise 62 reduces to

$$K = 2/|r'|$$

for the curvature *at the pole.*

In Exercises 69 and 70, use the result of Exercise 68 to find the curvature of the rose curve at the pole.

69. $r = 4 \sin 2\theta$

70. $r = 6 \cos 3\theta$

71. For a smooth curve given by the parametric equations $x = f(t)$ and $y = g(t)$, prove that the curvature is given by

$$K = \frac{|f'(t)g''(t) - g'(t)f''(t)|}{\{[f'(t)]^2 + [g'(t)]^2\}^{3/2}}.$$

72. Use the result of Exercise 71 to find the curvature K of the curve represented by the parametric equations $x(t) = t^3$ and $y(t) = \frac{1}{2}t^2$. Use a graphing utility to graph K and determine any horizontal asymptotes. Interpret the asymptotes in the context of the problem.

73. Use the result of Exercise 71 to find the curvature K of the cycloid represented by the parametric equations

$$x(\theta) = a(\theta - \sin \theta) \quad \text{and} \quad y(\theta) = a(1 - \cos \theta).$$

What are the minimum and maximum values of K?

74. Use Theorem 11.10 to find $a_{\mathbf{T}}$ and $a_{\mathbf{N}}$ for each curve given by the vector-valued function.

(a) $\mathbf{r}(t) = 3t^2\mathbf{i} + (3t - t^3)\mathbf{j}$

(b) $\mathbf{r}(t) = t^2\mathbf{i} + 2t\mathbf{j} + t^2\mathbf{k}$

75. *Frictional Force* A 4000-pound vehicle is driven at a speed of 30 miles per hour on a circular interchange of radius 100 feet. To keep the vehicle from skidding off course, what frictional force must the road surface exert on the tires?

76. Use the definition of curvature in space, $K = \|\mathbf{T}'(s)\| = \|\mathbf{r}''(s)\|$, to verify the following three formulas.

(a) $K = \dfrac{\|\mathbf{T}'(t)\|}{\|\mathbf{r}'(t)\|}$

(b) $K = \dfrac{\|\mathbf{r}'(t) \times \mathbf{r}''(t)\|}{\|\mathbf{r}'(t)\|^3}$

(c) $K = \dfrac{\mathbf{a}(t) \cdot \mathbf{N}(t)}{\|\mathbf{v}(t)\|^2}$

Kepler's Laws **In Exercises 77–84, you are asked to verify Kepler's Law of Planetary Motion. For these exercises, assume that each planet moves in an orbit given by the vector-valued function r. Let $r = \|\mathbf{r}\|$, let G represent the universal gravitational constant, let M represent the mass of the sun, and let m represent the mass of the planet.**

77. Prove that $\mathbf{r} \cdot \mathbf{r}' = r\dfrac{dr}{dt}$.

78. Using Newton's Second Law, $\mathbf{F} = m\mathbf{a}$, and Newton's Second Law of Gravitation, $\mathbf{F} = -(GmM/r^3)\mathbf{r}$, show that $\mathbf{a}$ and $\mathbf{r}$ are parallel, and that

$$\mathbf{r}(t) \times \mathbf{r}'(t) = \mathbf{L}$$

is a constant vector. Hence, $\mathbf{r}(t)$ moves in a *fixed plane*, orthogonal to $\mathbf{L}$.

79. Prove that $\dfrac{d}{dt}\left[\dfrac{\mathbf{r}}{r}\right] = \dfrac{1}{r^3}\{[\mathbf{r} \times \mathbf{r}'] \times \mathbf{r}\}$.

80. Show that

$$\frac{\mathbf{r}'}{GM} \times \mathbf{L} - \frac{\mathbf{r}}{r} = \mathbf{e}$$

is a constant vector.

81. Prove Kepler's First Law: Each planet moves in an elliptical orbit with the sun as a focus.

82. Assume that the elliptical orbit

$$r = \frac{ed}{1 + e \cos \theta}$$

is in the *xy*-plane, with $\mathbf{L}$ along the *z*-axis. Prove that

$$\|\mathbf{L}\| = r^2\frac{d\theta}{dt}.$$

83. Prove Kepler's Second Law: Each ray from the sun to a planet sweeps out equal areas of the ellipse in equal times.

84. Prove Kepler's Third Law: The square of the period of a planet's orbit is proportional to the cube of the mean distance between the planet and the sun.

85. *Cornu Spiral* The cornu spiral is given by

$$x = \int_0^s \cos \frac{\pi u^2}{2} \, du \quad \text{and} \quad y = \int_0^s \sin \frac{\pi u^2}{2} \, du.$$

The spiral shown in the figure was plotted over the interval $-\pi \le s \le \pi$.

(a) Find the arc length of this curve from $s = 0$ to $s = a$.

(b) Find the curvature of the graph when $s = a$.

(c) The cornu spiral was discovered by James Bernoulli. He found that the spiral has an amazing relationship between curvature and arc length. What is this relationship?

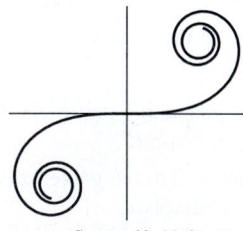

Generated by Mathematica

REVIEW EXERCISES FOR CHAPTER 11

In Exercises 1–4, (a) find the domain of r and (b) determine the values (if any) of t for which the function is continuous.

1. $\mathbf{r}(t) = t\mathbf{i} + \csc t\,\mathbf{k}$

2. $\mathbf{r}(t) = \sqrt{t}\,\mathbf{i} + \dfrac{1}{t-4}\mathbf{j} + \mathbf{k}$

3. $\mathbf{r}(t) = \ln t\,\mathbf{i} + t\mathbf{j} + t\mathbf{k}$

4. $\mathbf{r}(t) = (2t+1)\mathbf{i} + t^2\mathbf{j} + t\mathbf{k}$

In Exercises 5 and 6, evaluate (if possible) the vector-valued function at the indicated value of t.

5. $\mathbf{r}(t) = (2t+1)\mathbf{i} + t^2\mathbf{j} - \frac{1}{3}t^3\mathbf{k}$

    (a) $\mathbf{r}(0)$ (b) $\mathbf{r}(-2)$ (c) $\mathbf{r}(c-1)$ (d) $\mathbf{r}(1 + \Delta t) - \mathbf{r}(1)$

6. $\mathbf{r}(t) = 3\cos t\,\mathbf{i} + (1 - \sin t)\mathbf{j} - t\mathbf{k}$

    (a) $\mathbf{r}(0)$ (b) $\mathbf{r}\!\left(\frac{\pi}{2}\right)$ (c) $\mathbf{r}(s - \pi)$ (d) $\mathbf{r}(\pi + \Delta t) - \mathbf{r}(\pi)$

Think About It In Exercises 7 and 8, find $\mathbf{r}(t) \cdot \mathbf{u}(t)$. Is the result a vector-valued function? Explain.

7. $\mathbf{r}(t) = \tan t\,\mathbf{i} - \sec t\,\mathbf{j} + (1 - t)\mathbf{k}$

   $\mathbf{u}(t) = 2\cos t\,\mathbf{i} + 3\sin t\cos t\,\mathbf{j} + 1\mathbf{k}$

8. $\mathbf{r}(t) = \langle e^t, e^{-t}, 2\rangle$, $\mathbf{u}(t) = \langle 4e^{-t}, 2, t\rangle$

In Exercises 9 and 10, sketch the plane curve represented by the vector-valued function.

9. $\mathbf{r}(t) = \langle \cos t, 2\sin^2 t\rangle$ 10. $\mathbf{r}(t) = \langle t, t/(t-1)\rangle$

In Exercises 11–14, use a computer algebra system to graph the space curve represented by the vector-valued function.

11. $\mathbf{r}(t) = \mathbf{i} + t\mathbf{j} + t^2\mathbf{k}$ 12. $\mathbf{r}(t) = 2t\mathbf{i} + t\mathbf{j} + t^2\mathbf{k}$

13. $\mathbf{r}(t) = \langle 1, \sin t, 1\rangle$ 14. $\mathbf{r}(t) = \langle 2\cos t, t, 2\sin t\rangle$

In Exercises 15 and 16, use a graphing utility to graph the space curve represented by the vector-valued function.

15. $\mathbf{r}(t) = \left\langle t, \ln t, \frac{1}{2}t^2\right\rangle$ 16. $\mathbf{r}(t) = \left\langle \frac{1}{2}t, \sqrt{t}, \frac{1}{4}t^3\right\rangle$

In Exercises 17 and 18, find vector-valued functions forming the boundaries of the region in the figure.

17.
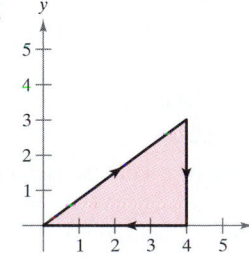

18.
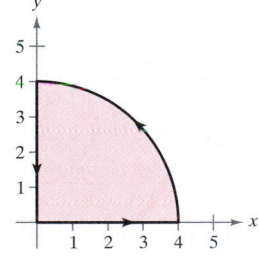

19. A particle moves on a straight-line path that passes through the points $(-2, -3, 8)$ and $(5, 1, -2)$. Find a vector-valued function for the path. (There are many correct answers.)

20. The outer edge of a spiral staircase is in the shape of a helix of radius 2 meters. The staircase has a height of 2 meters and is three-fourths of one complete revolution from bottom to top. Find a vector-valued function for the helix. (There are many correct answers.)

In Exercises 21 and 22, sketch the space curve represented by the intersection of the surfaces. Use the parameter $x = t$ to find a vector-valued function for the space curve.

21. $z = x^2 + y^2$, $x + y = 0$

22. $x^2 + z^2 = 4$, $x - y = 0$

In Exercises 23 and 24, find the limit.

23. $\displaystyle \lim_{t\to 2^-}\left(t^2\mathbf{i} + \sqrt{4 - t^2}\,\mathbf{j} + \mathbf{k}\right)$

24. $\displaystyle \lim_{t\to 0}\left(\dfrac{\sin 2t}{t}\mathbf{i} + e^{-t}\mathbf{j} + e^t\mathbf{k}\right)$

In Exercises 25 and 26, find the following.

(a) $\mathbf{r}'(t)$ (b) $\mathbf{r}''(t)$ (c) $D_t\,[\mathbf{r}(t) \cdot \mathbf{u}(t)]$

(d) $D_t\,[\mathbf{u}(t) - 2\mathbf{r}(t)]$ (e) $D_t\,[\|\mathbf{r}(t)\|]$, $t > 0$ (f) $D_t\,[\mathbf{r}(t) \times \mathbf{u}(t)]$

25. $\mathbf{r}(t) = 3t\mathbf{i} + (t-1)\mathbf{j}$, $\mathbf{u}(t) = t\mathbf{i} + t^2\mathbf{j} + \frac{2}{3}t^3\mathbf{k}$

26. $\mathbf{r}(t) = \sin t\,\mathbf{i} + \cos t\,\mathbf{j} + t\mathbf{k}$, $\mathbf{u}(t) = \sin t\,\mathbf{i} + \cos t\,\mathbf{j} + \dfrac{1}{t}\mathbf{k}$

In Exercises 27 and 28, find a set of parametric equations for the line tangent to the space curve at the indicated point.

27. $\mathbf{r}(t) = 2\cos t\,\mathbf{i} + 2\sin t\,\mathbf{j} + t\mathbf{k}$, $t = \dfrac{3\pi}{4}$

28. $\mathbf{r}(t) = t\mathbf{i} + t^2\mathbf{j} + \frac{2}{3}t^3\mathbf{k}$, $t = 2$

29. *Writing* The x- and y-components of the derivative of the vector-valued function $\mathbf{u}$ are positive at $t = t_0$, and the z-component is negative. Describe the behavior of $\mathbf{u}$ at $t = t_0$.

30. *Writing* The x-component of the derivative of the vector-valued function $\mathbf{u}$ is 0 for t in the domain of the function. What does this information imply about the graph of $\mathbf{u}$?

Linear Approximation In Exercises 31 and 32, find a set of parametric equations for the tangent line to the graph of the vector-valued function at $t = t_0$. Use the equations for the line to approximate $\mathbf{r}(t_0 + 0.1)$.

31. $\mathbf{r}(t) = \ln(t - 3)\mathbf{i} + t^2\mathbf{j} + \frac{1}{2}t\mathbf{k}$, $t_0 = 4$

32. $\mathbf{r}(t) = 3\cosh t\,\mathbf{i} + \sinh t\,\mathbf{j} - 2t\mathbf{k}$, $t_0 = 0$

In Exercises 33–36, find the indefinite integral.

33. $\displaystyle\int (\cos t\,\mathbf{i} + t\cos t\,\mathbf{j})\,dt$

34. $\displaystyle\int (\ln t\,\mathbf{i} + t\ln t\,\mathbf{j} + \mathbf{k})\,dt$

35. $\displaystyle\int \|\cos t\,\mathbf{i} + \sin t\,\mathbf{j} + t\,\mathbf{k}\|\,dt$

36. $\displaystyle\int (t\,\mathbf{j} + t^2\,\mathbf{k}) \times (\mathbf{i} + t\,\mathbf{j} + t\,\mathbf{k})\,dt$

In Exercises 37 and 38, find r(t) for the given condition.

37. $\mathbf{r}'(t) = 2t\,\mathbf{i} + e^t\,\mathbf{j} + e^{-t}\,\mathbf{k}, \quad \mathbf{r}(0) = \mathbf{i} + 3\mathbf{j} - 5\mathbf{k}$
38. $\mathbf{r}'(t) = \sec t\,\mathbf{i} + \tan t\,\mathbf{j} + t^2\,\mathbf{k}, \quad \mathbf{r}(0) = 3\mathbf{k}$

In Exercises 39 and 40, evaluate the definite integral.

39. $\displaystyle\int_{-2}^{2} (3t\,\mathbf{i} + 2t^2\,\mathbf{j} - t^3\,\mathbf{k})\,dt$

40. $\displaystyle\int_{0}^{1} \left(\sqrt{t}\,\mathbf{j} + t\sin t\,\mathbf{k}\right)dt$

In Exercises 41 and 42, the position function r describes the path of an object moving in space. Find the velocity, speed, and acceleration of the object.

41. $\mathbf{r}(t) = \langle \cos^3 t, \sin^3 t, 3t\rangle$
42. $\mathbf{r}(t) = \langle t, -\tan t, e^t\rangle$

Projectile Motion **In Exercises 43–46, use the model for projectile motion, assuming there is no air resistance. [$a(t) = -32$ ft/sec² or $a(t) = -9.8\text{m}/\text{sec}^2$]**

43. A projectile is fired from ground level at an angle of elevation of 30°. Find the range of the projectile if the initial velocity is 75 feet per second.

44. The center of a truckbed is 6 feet below and 4 feet horizontally from the end of a horizontal conveyor that is discharging gravel (see figure). Determine the speed ds/dt at which the conveyor belt should be moving so that the gravel falls onto the center of the truck bed.

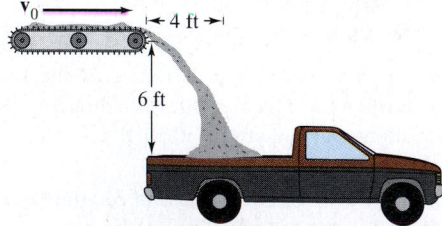

45. A projectile is fired from ground level at an angle of 20° with the horizontal. Find the minimum initial velocity if the projectile has a range of 80 meters.

46. Use a graphing utility to graph the paths of a projectile if $v_0 = 20$ meters per second and

   (a) $\theta = 30°$, (b) $\theta = 45°$, and (c) $\theta = 60°$.

   Use the graphs to approximate the maximum height and range of the projectile for each case.

In Exercises 47–54, find the velocity, speed, and acceleration at time t. Then find a · T, a · N, and the curvature at time t.

47. $\mathbf{r}(t) = 5t\,\mathbf{i}$
48. $\mathbf{r}(t) = (1 + 4t)\mathbf{i} + (2 - 3t)\mathbf{j}$

49. $\mathbf{r}(t) = t\,\mathbf{i} + \sqrt{t}\,\mathbf{j}$
50. $\mathbf{r}(t) = 2(t + 1)\mathbf{i} + \dfrac{2}{t+1}\,\mathbf{j}$

51. $\mathbf{r}(t) = e^t\,\mathbf{i} + e^{-t}\,\mathbf{j}$
52. $\mathbf{r}(t) = t\cos t\,\mathbf{i} + t\sin t\,\mathbf{j}$

53. $\mathbf{r}(t) = t\,\mathbf{i} + t^2\,\mathbf{j} + \dfrac{1}{2}t^2\,\mathbf{k}$
54. $\mathbf{r}(t) = (t - 1)\mathbf{i} + t\,\mathbf{j} + \dfrac{1}{t}\,\mathbf{k}$

In Exercises 55 and 56, use a symbolic integration utility to find the length of the space curve over the indicated interval.

55. $\mathbf{r}(t) = \frac{1}{2}t\,\mathbf{i} + \sin t\,\mathbf{j} + \cos t\,\mathbf{k}, \quad 0 \le t \le \pi$
56. $\mathbf{r}(t) = e^t\sin t\,\mathbf{i} + e^t\cos t\,\mathbf{k}, \quad 0 \le t \le \pi$

57. ***Satellite Orbit*** Find the speed necessary for a satellite to maintain a circular orbit 600 miles above the surface of the earth.

58. ***Centripetal Force*** An automobile in a circular traffic exchange is traveling at twice the posted speed. By what factor is the centripetal force increased over that which would occur at the posted speed?

59. ***Writing*** A civil engineer designs a highway as indicated in the figure. BC is an arc of the circle. AB and CD are straight lines tangent to the circular arc. Criticize the design.

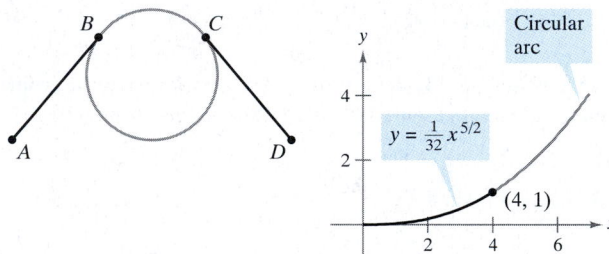

Figure for 59   **Figure for 60**

60. ***Writing*** A highway has an exit ramp that begins at the origin of a coordinate system and follows the curve $y = \frac{1}{32}x^{5/2}$ to the point $(4, 1)$ (see figure). Then it follows a circular path whose curvature is that given by the curve at $(4, 1)$. What is the radius of the circular arc? Explain why the curve and the circular arc should have the same curvature at $(4, 1)$.

61. ***Investigation*** Consider the vector-valued function $\mathbf{r}(t) = \langle t\cos \pi t, t\sin \pi t\rangle, \quad 0 \le t \le 2$.

   (a) Use a graphing utility to graph the function.
   (b) Find the length of the arc in part (a).
   (c) Find the curvature K as a function of t. Find the curvature when t is 0, 1, and 2.
   (d) Use a graphing utility to graph the function K.
   (e) Find (if possible) $\lim_{t \to \infty} K$.
   (f) Using the result in part (e), make a conjecture about the graph of $\mathbf{r}$ as $t \to \infty$.

12

Functions of Several Variables

From 1970 through today, home satellite dishes have evolved from large, unwieldy things sitting in yards to small units that can be attached to roofs or other stationary objects. The shape of these dishes, however, has remained the same. Each dish, whether large or small, has the shape of a circular paraboloid. This shape allows the dish to receive signals whose strength is only a few billionths of a watt.

To model a circular paraboloid, you can modify the elliptic paraboloid formula described in Section 10.6.

$$z = \frac{x^2}{a^2} + \frac{y^2}{a^2} \qquad \text{\color{red}{Circular paraboloid}}$$

Older home satellite dishes have diameters of 10 feet. Today's consumers, subscribing to a direct-broadcast satellite system, may have satellite dishes with diameters as small as $1\frac{1}{2}$ feet.

For this model, the paraboloid opens up and has its vertex at the origin. For a satellite dish, the axis of the paraboloid should pass through the satellite so that all incoming signals are parallel to the paraboloid's axis. When incoming rays strike the surface of the paraboloid at a point P, they reflect at the same angle as they would if they were reflecting off a plane that was tangent to the surface at point P, as shown in the left-hand figure below.

All parabolas have a special reflective property. One way to describe the property is to say that *all incoming rays that are parallel to the axis of the parabola reflect directly through the focus of the parabola*, as shown in the right-hand figure below.

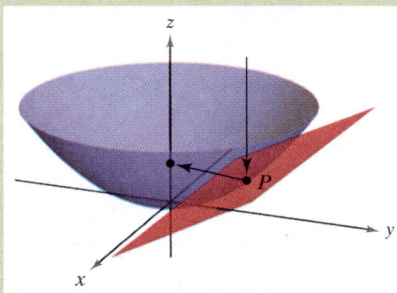
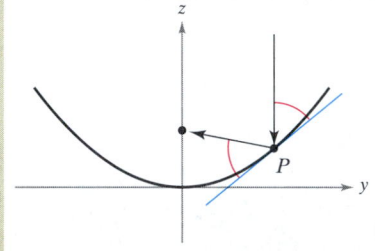

QUESTIONS

1. You are designing a satellite receiving dish. At what point should you place the receiver? Explain your reasoning.

2. Other than satellite receiving dishes, what other common objects use the reflective property of parabolas in their design?

3. Consider the circular paraboloid given by

   $$z = x^2 + y^2$$

   and consider the point $P(1, 1, 2)$ on this surface. You can characterize the tangent plane to the surface at this point by saying it is the only plane whose intersection with the paraboloid consists of the single point P. Find an equation for this tangent plane. Explain your strategy.

4. Does it make sense to talk about the "slope" of the tangent plane in Question 3? If so, what is the slope of this plane? How might you use differentiation to discover the slope of this plane?

The concepts presented here will be explored further in this chapter. For an extension of this application, see the lab series that accompanies this text.

SECTION | 12.1 | **Introduction to Functions of Several Variables**

Functions of Several Variables • The Graph of a Function of Two Variables •
Level Curves • Level Surfaces • Computer Graphics • Functions of Several Variables

Functions of Several Variables

So far in this text, we have dealt only with functions of single (independent) variables. Many familiar quantities, however, are functions of two or more variables. For instance, the work done by a force $(W = FD)$ and the volume of a right circular cylinder $(V = \pi r^2 h)$ are both functions of two variables. The volume of a rectangular solid $(V = lwh)$ is a function of three variables. The notation for a function of two or more variables is similar to that for a function of a single variable. Here are two examples.

$$z = f\underbrace{(x, y)}_{\text{2 variables}} = x^2 + xy$$

and

$$w = f\underbrace{(x, y, z)}_{\text{3 variables}} = x + 2y - 3z$$

Definition of a Function of Two Variables

Let D be a set of ordered pairs of real numbers. If to each ordered pair (x, y) in D there corresponds a unique real number $f(x, y)$, then f is called a **function of x and y**. The set D is the **domain** of f, and the corresponding set of values for $f(x, y)$ is the **range** of f.

For the function given by $z = f(x, y)$, we call x and y the **independent variables** and z the **dependent variable**.

Similar definitions can be given to functions of three, four, or n variables, where the domains consist of ordered triples (x_1, x_2, x_3), quadruples (x_1, x_2, x_3, x_4), and n-tuples $(x_1, x_2, \ldots , x_n)$. In all cases, the range is a set of real numbers. In this chapter, we limit the discussion to functions of two or three variables.

As with functions of one variable, the most common way to describe a function of several variables is with an *equation*, and unless otherwise restricted, you can assume that the domain is the set of all points for which the equation is defined. For instance, the domain of the function given by

$$f(x, y) = x^2 + y^2$$

is assumed to be the entire xy-plane. Similarly, the domain of

$$f(x, y) = \ln xy$$

is the set of all points (x, y) in the plane for which $xy > 0$. This consists of all points in the first and third quadrants.

MARY FAIRFAX SOMERVILLE (1780–1872)

Somerville was interested in the problem of creating geometric models for functions of several variables. Her most well-known book, *The Mechanics of the Heavens*, was published in 1831.

Archive Photos

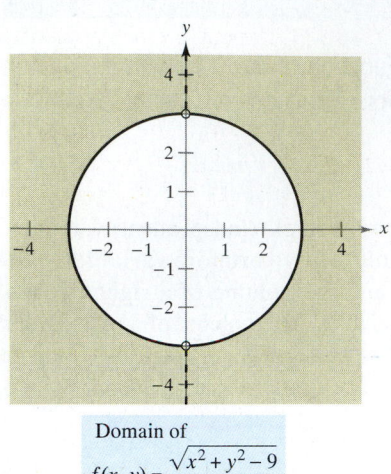

Domain of
$$f(x, y) = \frac{\sqrt{x^2 + y^2 - 9}}{x}$$

Figure 12.1

EXAMPLE 1 Domains of Functions of Several Variables

Find the domain of each of the following functions.

a. $f(x, y) = \dfrac{\sqrt{x^2 + y^2 - 9}}{x}$ **b.** $g(x, y, z) = \dfrac{x}{\sqrt{9 - x^2 - y^2 - z^2}}$

Solution

a. The function f is defined for all points (x, y) such that $x \neq 0$ and

$$x^2 + y^2 \geq 9.$$

Thus, the domain is the set of all points lying on or outside the circle $x^2 + y^2 = 9$, *except* those on the y-axis, as shown in Figure 12.1.

b. The function g is defined for all points (x, y, z) such that

$$x^2 + y^2 + z^2 < 9.$$

Consequently, the domain is the set of all points (x, y, z) lying inside a sphere of radius 3 that is centered at the origin.

Functions of several variables can be combined in the same ways as functions of single variables. For instance, you can form the sum, difference, product, and quotient of two functions of two variables as follows.

$$(f \pm g)(x, y) = f(x, y) \pm g(x, y) \qquad \text{Sum or difference}$$
$$(fg)(x, y) = f(x, y)g(x, y) \qquad \text{Product}$$
$$\frac{f}{g}(x, y) = \frac{f(x, y)}{g(x, y)}, \quad g(x, y) \neq 0 \qquad \text{Quotient}$$

You cannot form the composite of two functions of several variables. However, if h is a function of several variables and g is a function of a single variable, you can form the **composite** function $(g \circ h)(x, y)$ as follows.

$$(g \circ h)(x, y) = g(h(x, y)) \qquad \text{Composition}$$

The domain of this composite function consists of all (x, y) in the domain of h such that $h(x, y)$ is in the domain of g. For example, the function given by

$$f(x, y) = \sqrt{16 - 4x^2 - y^2}$$

can be viewed as the composite of the function of two variables given by $h(x, y) = 16 - 4x^2 - y^2$ and the function of a single variable given by $g(u) = \sqrt{u}$. The domain of this function is the set of all points lying on or inside the ellipse given by $4x^2 + y^2 = 16$.

A function that can be expressed as a sum of functions of the form $cx^m y^n$ (where c is a real number and m and n are nonnegative integers) is called a **polynomial function** of two variables. For instance, the functions given by

$$f(x, y) = x^2 + y^2 - 2xy + x + 2 \quad \text{and} \quad g(x, y) = 3xy^2 + x - 2$$

are polynomial functions of two variables. A **rational function** is the quotient of two polynomial functions. Similar terminology is used for functions of more than two variables.

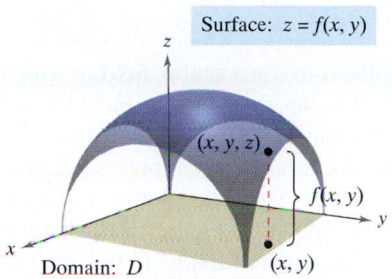

Surface: $z = f(x, y)$

(x, y, z)

$f(x, y)$

(x, y)

Domain: D

Figure 12.2

The Graph of a Function of Two Variables

As with functions of a single variable, you can learn a lot about the behavior of a function of two variables by sketching its graph. The **graph** of a function f of two variables is the set of all points (x, y, z) for which $z = f(x, y)$ and (x, y) is in the domain of f. This graph can be interpreted geometrically as a *surface in space*, as discussed in Sections 10.5 and 10.6. In Figure 12.2, note that the graph of $z = f(x, y)$ is a surface whose projection onto the xy-plane is D, the domain of f. To each point (x, y) in D there corresponds a point (x, y, z) on the surface, and, conversely, to each point (x, y, z) on the surface there corresponds a point (x, y) in D.

EXAMPLE 2   Describing the Graph of a Function of Two Variables

What is the range of $f(x, y) = \sqrt{16 - 4x^2 - y^2}$? Describe the graph of f.

Solution   The domain D implied by the equation for f is the set of all points (x, y) such that $16 - 4x^2 - y^2 \geq 0$. Thus, D is the set of all points lying on or inside the ellipse given by

$$\frac{x^2}{4} + \frac{y^2}{16} = 1.$$

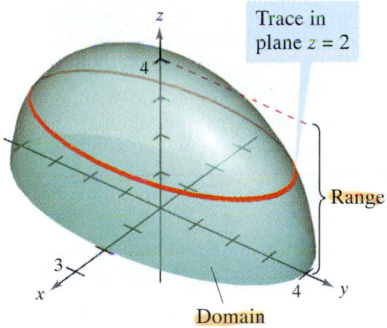

Surface: $z = \sqrt{16 - 4x^2 - y^2}$

Trace in plane $z = 2$

Range

Domain

The graph of $f(x, y) = \sqrt{16 - 4x^2 - y^2}$ is the upper half of an ellipsoid.

Figure 12.3

The range of f is all values $z = f(x, y)$ such that $0 \leq z \leq \sqrt{16}$ or

$$0 \leq z \leq 4. \qquad \text{Range of } f$$

A point (x, y, z) is on the graph of f if and only if

$$z = \sqrt{16 - 4x^2 - y^2}$$
$$z^2 = 16 - 4x^2 - y^2$$
$$4x^2 + y^2 + z^2 = 16$$
$$\frac{x^2}{4} + \frac{y^2}{16} + \frac{z^2}{16} = 1, \quad 0 \leq z \leq 4.$$

From Section 10.6, you know that the graph of f is the upper half of an ellipsoid, as shown in Figure 12.3.

To sketch a surface in space *by hand*, it helps to use traces in planes parallel to the coordinate planes, as shown in Figure 12.3. For example, to find the trace of the surface in the plane $z = 2$, substitute $z = 2$ in the equation $z = \sqrt{16 - 4x^2 - y^2}$ and obtain

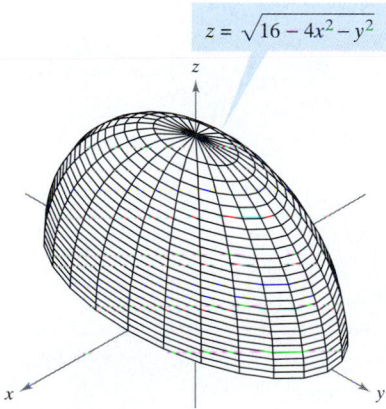

$z = \sqrt{16 - 4x^2 - y^2}$

Figure 12.4

$$2 = \sqrt{16 - 4x^2 - y^2} \qquad \Longrightarrow \qquad \frac{x^2}{3} + \frac{y^2}{12} = 1.$$

Thus, the trace is an ellipse centered at the point $(0, 0, 2)$ with major and minor axes of lengths $4\sqrt{3}$ and $2\sqrt{3}$.

Traces are also used with most three-dimensional graphing utilities. For instance, Figure 12.4 shows a computer-generated version of the surface given in Example 2. For this sketch, the computer took 25 traces parallel to the xy-plane and 12 traces in vertical planes.

If you have access to a three-dimensional graphing utility, try using it to sketch the graphs of several surfaces.

Level Curves

A second way to visualize a function of two variables is to use a **scalar field** in which the scalar $z = f(x, y)$ is assigned to the point (x, y). A scalar field can be characterized by **level curves** (or **contour lines**) along which the value of $f(x, y)$ is constant. For instance, the weather map in Figure 12.5 shows level curves of equal pressure called **isobars.** In weather maps for which the level curves represent points of equal temperature, the level curves are called **isotherms,** as shown in Figure 12.6. Another common use of level curves is in representing electric potential fields. In this type of map, the level curves are called **equipotential lines.**

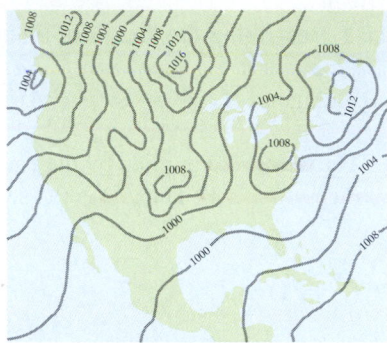

Level curves show the lines of equal pressure (isobars) measured in millibars.
Figure 12.5

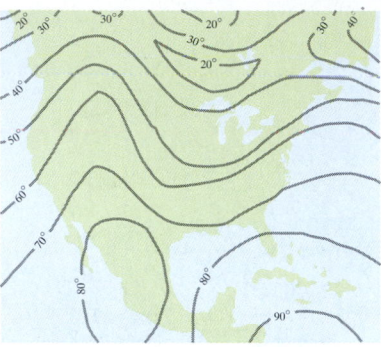

Level curves show the lines of equal temperature (isotherms) measured in degrees Fahrenheit.
Figure 12.6

Contour maps are commonly used to show regions of the earth's surface, with the level curves representing the height above sea level. This type of map is called a **topographic map.** For example, the mountain shown in Figure 12.7 is represented by the topographic map in Figure 12.8.

A contour map depicts the variation of z with respect to x and y by the spacing between level curves. Much space between level curves indicates that z is changing slowly, whereas little space indicates a rapid change in z. Furthermore, to give a good three-dimensional illusion in a contour map, it is important to choose c-values that are *evenly spaced.*

Figure 12.7

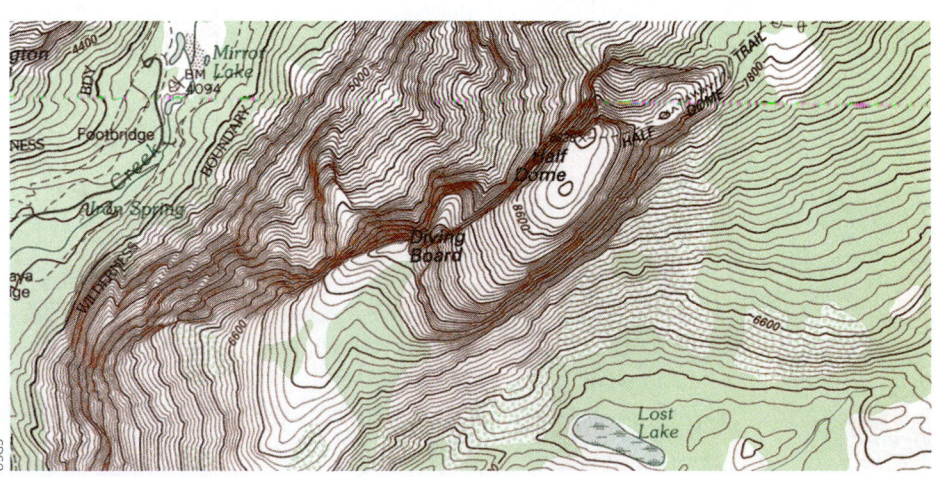

Figure 12.8

EXAMPLE 3 Sketching a Contour Map

The hemisphere given by $f(x, y) = \sqrt{64 - x^2 - y^2}$ is shown in Figure 12.9. Sketch a contour map for this surface using level curves corresponding to $c = 0, 1, 2, \ldots, 8$.

Solution For each value of c , the equation given by $f(x, y) = c$ is a circle (or point) in the xy-plane. For example, when $c_1 = 0$ the level curve is

$$x^2 + y^2 = 64 \qquad \text{Circle of radius 8}$$

which is a circle of radius 8. Figure 12.10 shows the nine level curves for the hemisphere.

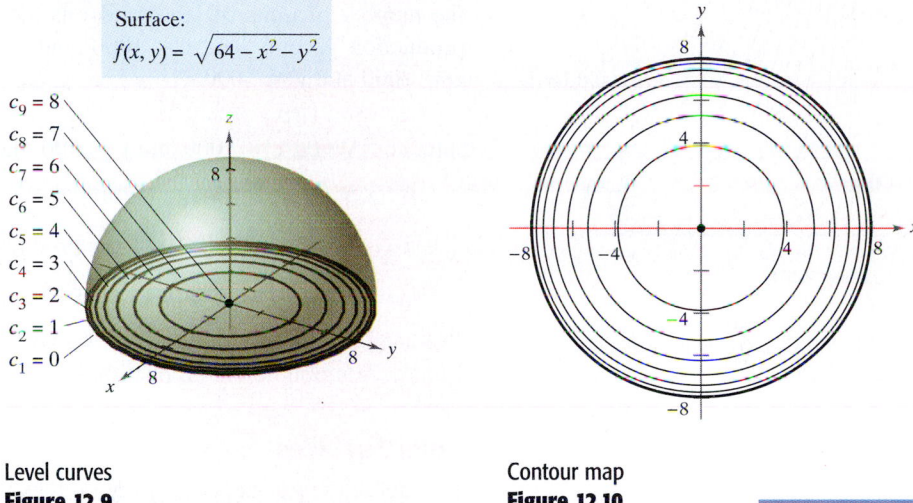

Surface:
$f(x, y) = \sqrt{64 - x^2 - y^2}$
$c_9 = 8$
$c_8 = 7$
$c_7 = 6$
$c_6 = 5$
$c_5 = 4$
$c_4 = 3$
$c_3 = 2$
$c_2 = 1$
$c_1 = 0$
z
8
x
8
8
y

Level curves
Figure 12.9

y
8
4
−8
−4
4
8
x
−4
−8

Contour map
Figure 12.10

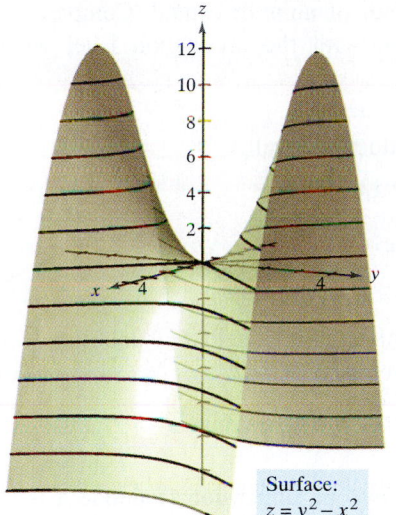

z
12
10
8
6
4
2
x
4
4
y
Surface:
$z = y^2 - x^2$

Hyperbolic paraboloid
Figure 12.11

EXAMPLE 4 Sketching a Contour Map

The hyperbolic paraboloid given by

$$z = y^2 - x^2$$

is shown in Figure 12.11. Sketch a contour map for this surface.

Solution For each value of c, we let $f(x, y) = c$ and sketch the resulting level curve in the xy-plane. For this function, each of the level curves $(c \neq 0)$ is a hyperbola whose asymptotes are the lines $y = \pm x$. If $c < 0$, the transverse axis is horizontal. For instance, the level curve for $c = -4$ is given by

$$\frac{x^2}{2^2} - \frac{y^2}{2^2} = 1. \qquad \text{Hyperbola with horizontal transverse axis}$$

If $c > 0$, the transverse axis is vertical. For instance, the level curve for $c = 4$ is given by

$$\frac{y^2}{2^2} - \frac{x^2}{2^2} = 1. \qquad \text{Hyperbola with vertical transverse axis}$$

If $c = 0$, the level curve is the degenerate conic representing the intersecting asymptotes, as shown in Figure 12.12.

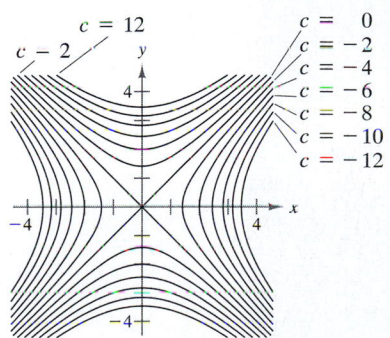

c = 12
c = 2
y
4
c = 0
c = −2
c = −4
c = −6
c = −8
c = −10
c = −12
x
−4
4
−4

Hyperbolic level curves (at increments of 2)
Figure 12.12

One example of a function of two variables used in economics is the **Cobb-Douglas production function.** This function is used as a model to represent the number of units produced by varying amounts of labor and capital. If x measures the units of labor and y measures the units of capital, the number of units produced is given by

$$f(x, y) = Cx^a y^{1-a}$$

where C is constant and $0 < a < 1$.

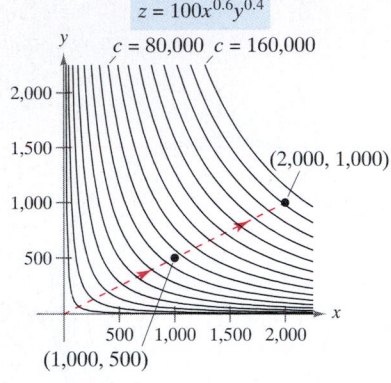

$z = 100x^{0.6}y^{0.4}$

$c = 80,000$ $c = 160,000$

$(2,000, 1,000)$

$(1,000, 500)$

Level curves (at increments of 10,000)
Figure 12.13

EXAMPLE 5   The Cobb-Douglas Production Function

A manufacturer estimates a production function to be $f(x, y) = 100x^{0.6}y^{0.4}$, where x is the number of units of labor and y is the number of units of capital. Compare the production level when $x = 1000$ and $y = 500$ with the production level when $x = 2000$ and $y = 1000$.

Solution   When $x = 1000$ and $y = 500$, the production level is

$$f(1000, 500) = 100(1000^{0.6})(500^{0.4}) \approx 75,786.$$

When $x = 2000$ and $y = 1000$, the production level is

$$f(2000, 1000) = 100(2000^{0.6})(1000^{0.4}) \approx 151,572.$$

The level curves of $z = f(x, y)$ are shown in Figure 12.13. Note that by doubling *both* x and y, you double the production level (see Exercise 72).

Level Surfaces

The concept of a level curve can be extended by one dimension to define a **level surface.** If f is a function of three variables and c is a constant, the graph of the equation $f(x, y, z) = c$ is a **level surface** of the function f, as shown in Figure 12.14.

With computers, engineers and scientists have developed other ways to view functions of three variables. For instance, Figure 12.15 shows a computer simulation that uses color to represent the optimal strain distribution of a car door.

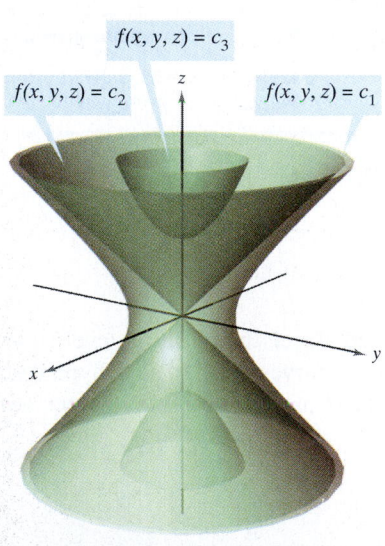

$f(x, y, z) = c_3$

$f(x, y, z) = c_2$ $f(x, y, z) = c_1$

Level surfaces of f
Figure 12.14

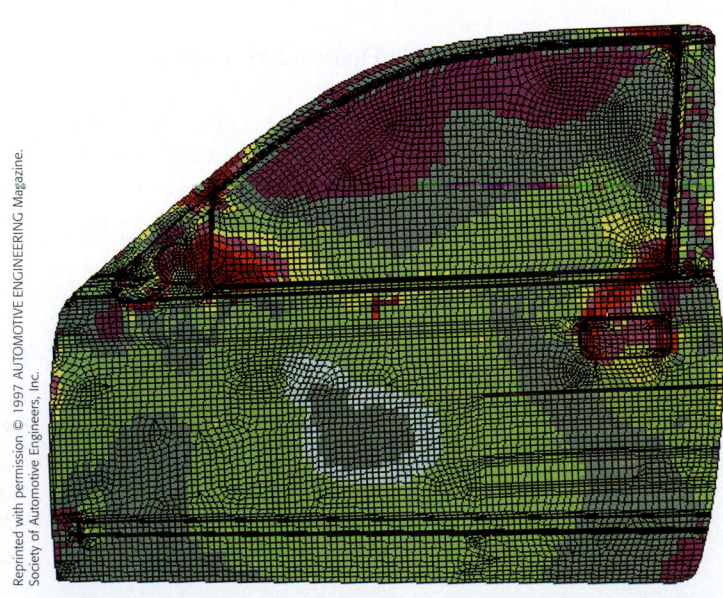

Reprinted with permission © 1997 AUTOMOTIVE ENGINEERING Magazine. Society of Automotive Engineers, Inc.

Figure 12.15

EXAMPLE 6   **Level Surfaces**

Describe the level surfaces of the function

$$f(x, y, z) = 4x^2 + y^2 + z^2.$$

Solution   Each level surface has an equation of the form

$$4x^2 + y^2 + z^2 = c. \qquad \text{Equation of level surface}$$

Therefore, the level surfaces are ellipsoids (whose cross sections parallel to the yz-plane are circles). As c increases, the radii of the circular cross sections increase according to the square root of c. For example, the level surfaces corresponding to the values $c = 0$, $c = 4$, and $c = 16$ are as follows.

$$4x^2 + y^2 + z^2 = 0 \qquad \text{Level surface for } c = 0 \text{ (Single point)}$$

$$\frac{x^2}{1} + \frac{y^2}{4} + \frac{z^2}{4} = 1 \qquad \text{Level surface for } c = 4 \text{ (Ellipsoid)}$$

$$\frac{x^2}{4} + \frac{y^2}{16} + \frac{z^2}{16} = 1 \qquad \text{Level surface for } c = 16 \text{ (Ellipsoid)}$$

These level surfaces are shown in Figure 12.16.

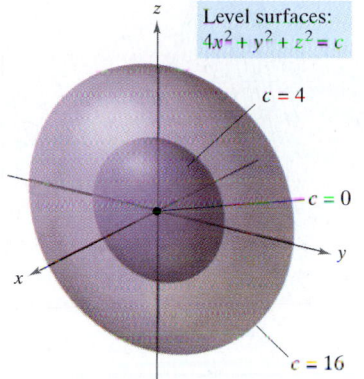

Level surfaces:
$4x^2 + y^2 + z^2 = c$

$c = 4$

$c = 0$

$c = 16$

Figure 12.16

NOTE   If the function in Example 6 represented the *temperature* at the point (x, y, z), the level surfaces shown in Figure 12.16 would be called **isothermal surfaces.**

Computer Graphics

The problem of sketching the graph of a surface in space can be simplified by using a computer. Although there are several types of three-dimensional graphing utilities, most use some form of trace analysis to give the illusion of three-dimensions. To use such a graphing utility, you usually need to enter the equation of the surface, the region in the xy-plane over which the surface is to be plotted, and the number of traces to be taken. For instance, to sketch the surface given by

$$f(x, y) = (x^2 + y^2)e^{1-x^2-y^2}$$

you might choose the following bounds for x, y, and z.

$$-3 \le x \le 3 \qquad \text{Bounds for } x$$

$$-3 \le y \le 3 \qquad \text{Bounds for } y$$

$$0 \le z \le 3 \qquad \text{Bounds for } z$$

Figure 12.17 shows a computer-generated sketch of this surface using 26 traces taken parallel to the yz-plane. To heighten the three-dimensional effect, the program uses a "hidden line" routine. That is, it begins by plotting the traces in the foreground (those corresponding to the largest x-values), and then, as each new trace is plotted, the program determines whether all or only part of the next trace should be shown.

The graphs on page 826 show a variety of surfaces that were plotted by computer. If you have access to a computer drawing program, try using it to reproduce these surfaces. Remember also that the three-dimensional graphics in this text can be viewed and rotated using the three-dimensional software that accompanies the text and is also incorporated into *Interactive Calculus*, the interactive version of the text.

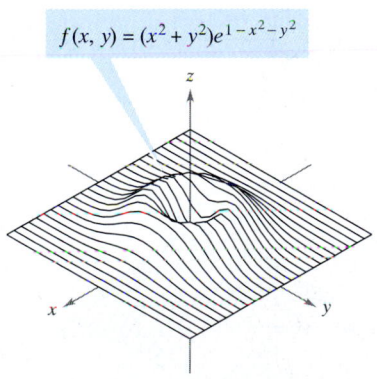

$f(x, y) = (x^2 + y^2)e^{1-x^2-y^2}$

Figure 12.17

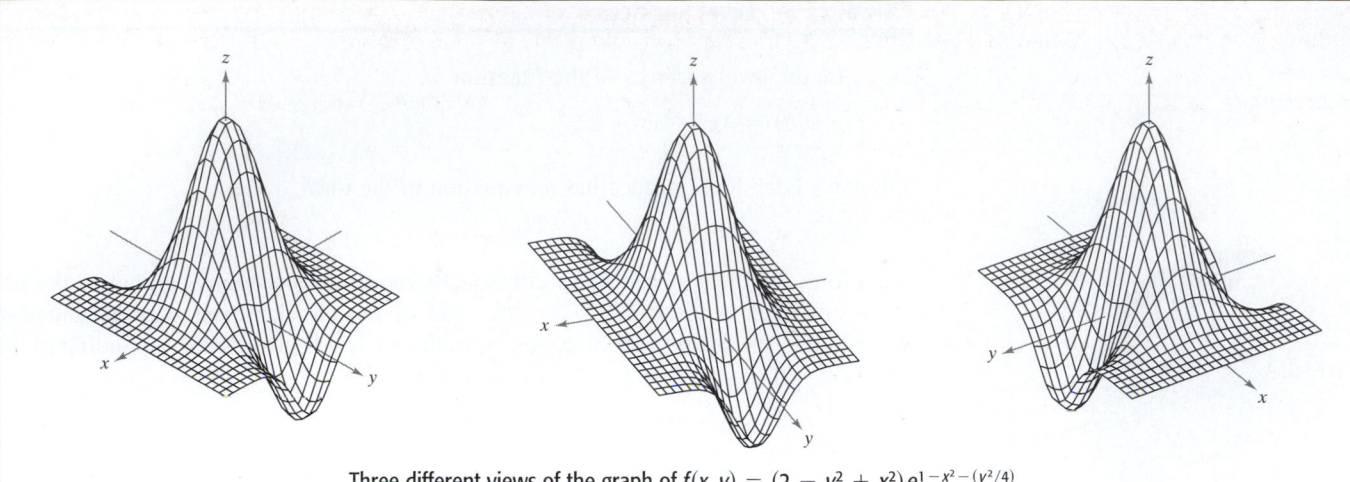

Three different views of the graph of $f(x, y) = (2 - y^2 + x^2)e^{1 - x^2 - (y^2/4)}$

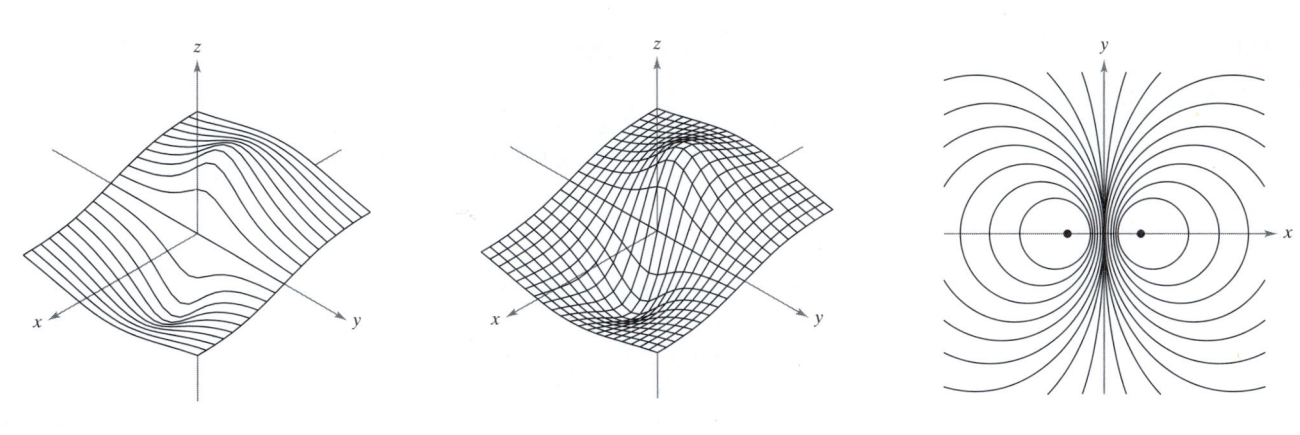

Single traces   Double traces   Level curves

Traces and level curves of the graph of $f(x, y) = \dfrac{-4x}{x^2 + y^2 + 1}$

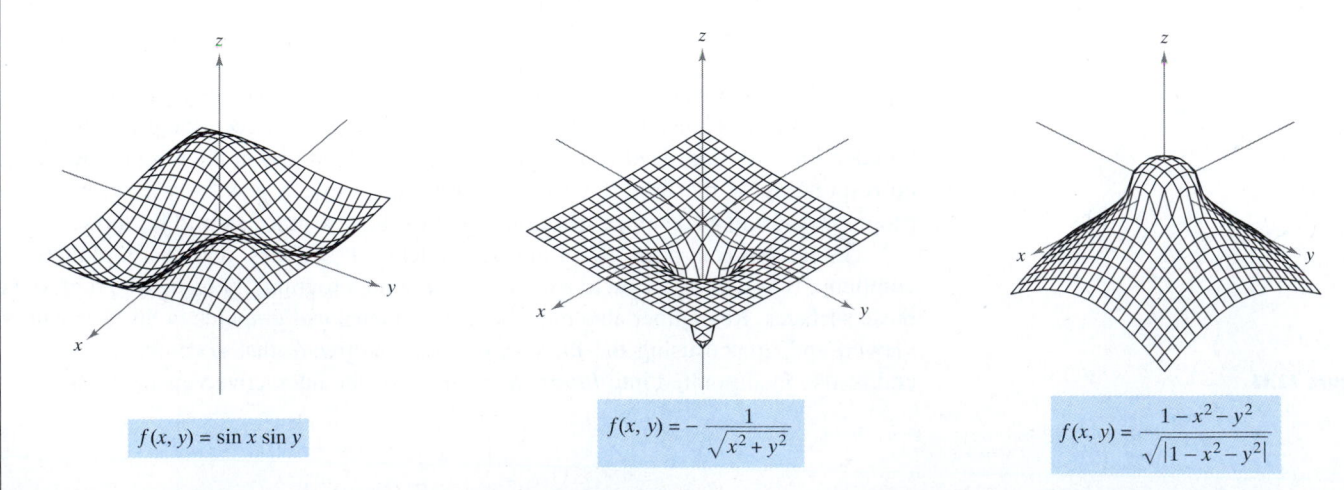

$f(x, y) = \sin x \sin y$ $f(x, y) = -\dfrac{1}{\sqrt{x^2 + y^2}}$ $f(x, y) = \dfrac{1 - x^2 - y^2}{\sqrt{|1 - x^2 - y^2|}}$

EXERCISES FOR SECTION 12.1

In Exercises 1–4, determine whether z is a function of x and y.

1. $x^2z + yz - xy = 10$ **2.** $xz^2 + 2xy - y^2 = 4$

3. $\dfrac{x^2}{4} + \dfrac{y^2}{9} + z^2 = 1$ **4.** $z + x \ln y - 8 = 0$

In Exercises 5–16, find and simplify the function values.

5. $f(x, y) = x/y$

    (a) $(3, 2)$ (b) $(-1, 4)$ (c) $(30, 5)$

    (d) $(5, y)$ (e) $(x, 2)$ (f) $(5, t)$

6. $f(x, y) = 4 - x^2 - 4y^2$

    (a) $(0, 0)$ (b) $(0, 1)$ (c) $(2, 3)$

    (d) $(1, y)$ (e) $(x, 0)$ (f) $(t, 1)$

7. $f(x, y) = xe^y$

    (a) $(5, 0)$ (b) $(3, 2)$ (c) $(2, -1)$

    (d) $(5, y)$ (e) $(x, 2)$ (f) (t, t)

8. $g(x, y) = \ln|x + y|$

    (a) $(2, 3)$ (b) $(5, 6)$ (c) $(e, 0)$

    (d) $(0, 1)$ (e) $(2, -3)$ (f) (e, e)

9. $h(x, y, z) = \dfrac{xy}{z}$

    (a) $(2, 3, 9)$ (b) $(1, 0, 1)$

10. $f(x, y, z) = \sqrt{x + y + z}$

    (a) $(0, 5, 4)$ (b) $(6, 8, -3)$

11. $f(x, y) = x \sin y$

    (a) $\left(2, \dfrac{\pi}{4}\right)$ (b) $(3, 1)$

12. $V(r, h) = \pi r^2 h$

    (a) $(3, 10)$ (b) $(5, 2)$

13. $g(x, y) = \displaystyle\int_x^y (2t - 3)\, dt$

    (a) $(0, 4)$ (b) $(1, 4)$

14. $g(x, y) = \displaystyle\int_x^y \dfrac{1}{t}\, dt$

    (a) $(4, 1)$ (b) $(6, 3)$

15. $f(x, y) = x^2 - 2y$

    (a) $\dfrac{f(x + \Delta x, y) - f(x, y)}{\Delta x}$

    (b) $\dfrac{f(x, y + \Delta y) - f(x, y)}{\Delta y}$

16. $f(x, y) = 3xy + y^2$

    (a) $\dfrac{f(x + \Delta x, y) - f(x, y)}{\Delta x}$

    (b) $\dfrac{f(x, y + \Delta y) - f(x, y)}{\Delta y}$

In Exercises 17–28, describe the domain and range of the function.

17. $f(x, y) = \sqrt{4 - x^2 - y^2}$ **18.** $f(x, y) = \sqrt{4 - x^2 - 4y^2}$

19. $f(x, y) = \arcsin(x + y)$ **20.** $f(x, y) = \arccos(y/x)$

21. $f(x, y) = \ln(4 - x - y)$ **22.** $f(x, y) = \ln(4 - xy)$

23. $z = \dfrac{x + y}{xy}$ **24.** $z = \dfrac{xy}{x - y}$

25. $f(x, y) = e^{x/y}$ **26.** $f(x, y) = x^2 + y^2$

27. $g(x, y) = \dfrac{1}{xy}$ **28.** $g(x, y) = x\sqrt{y}$

29. ***Think About It***   The graphs labeled (a), (b), (c), and (d) are graphs of the function $f(x, y) = -4x/(x^2 + y^2 + 1)$. Match the four graphs with the points in space from which the surface is viewed. The four points are $(20, 15, 25)$, $(-15, 10, 20)$, $(20, 20, 0)$, and $(20, 0, 0)$.

(a)   (b)

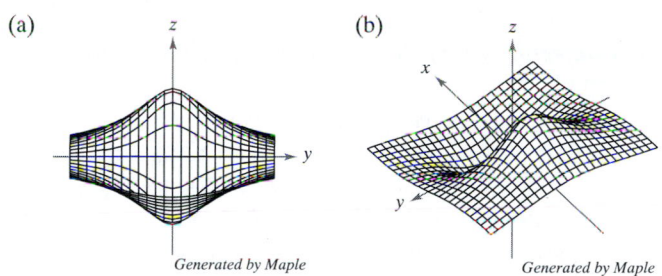

Generated by Maple   *Generated by Maple*

(c)   (d)

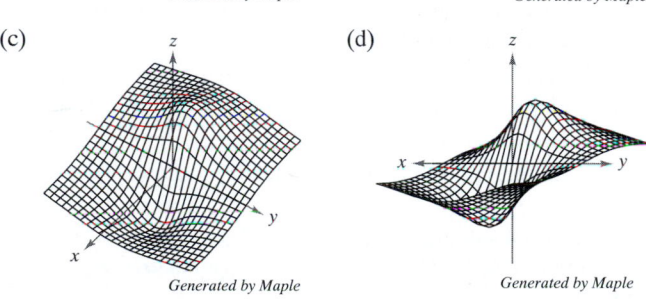

Generated by Maple   *Generated by Maple*

30. ***Think About It***   Use the function given in Exercise 29.

    (a) Find the domain and range of the function.

    (b) Identify the points in the xy-plane where the function value is 0.

    (c) Does the surface pass through all the octants of the rectangular coordinate system? Give reasons for your answer.

In Exercises 31–38, sketch the graph of the surface given by the function.

31. $f(x, y) = 5$ **32.** $f(x, y) = 6 - 2x - 3y$

33. $f(x, y) = y^2$ **34.** $g(x, y) = \frac{1}{2}x$

35. $z = 4 - x^2 - y^2$ **36.** $z = \sqrt{x^2 + y^2}$

37. $f(x, y) = e^{-x}$

38. $f(x, y) = \begin{cases} xy, & x \ge 0, y \ge 0 \\ 0, & x < 0 \text{ or } y < 0 \end{cases}$

In Exercises 39–42, use a computer algebra system to graph the function.

39. $z = y^2 - x^2 + 1$

40. $z = \frac{1}{12}\sqrt{144 - 16x^2 - 9y^2}$

41. $f(x, y) = x^2 e^{(-xy/2)}$

42. $f(x, y) = x \sin y$

43. ***Conjecture*** Consider the function $f(x, y) = x^2 + y^2$.

  (a) Sketch the graph of the surface given by f.

  (b) Make a conjecture about the relationship between the graphs of f and $g(x, y) = f(x, y) + 2$. Use a computer algebra system to confirm your answer.

  (c) Make a conjecture about the relationship between the graphs of f and $g(x, y) = f(x, y - 2)$. Use a computer algebra system to confirm your answer.

  (d) Make a conjecture about the relationship between the graphs of f and $g(x, y) = 4 - f(x, y)$. Use a computer algebra system to confirm your answer.

  (e) On the surface in part (a), sketch the graphs of $z = f(1, y)$ and $z = f(x, 1)$.

44. ***Conjecture*** Consider the function $f(x, y) = xy$ for $x \geq 0$ and $y \geq 0$.

  (a) Sketch the graph of the surface given by f.

  (b) Make a conjecture about the relationship between the graphs of f and $g(x, y) = f(x, y) - 3$. Use a computer algebra system to confirm your answer.

  (c) Make a conjecture about the relationship between the graphs of f and $g(x, y) = -f(x, y)$. Use a computer algebra system to confirm your answer.

  (d) Make a conjecture about the relationship between the graphs of f and $g(x, y) = \frac{1}{2}f(x, y)$. Use a computer algebra system to confirm your answer.

  (e) On the surface in part (a), sketch the graph of $z = f(x, x)$.

In Exercises 45–48, match the graph of the surface with one of the contour maps. [The contour maps are labeled (a), (b), (c), and (d).]

(a)

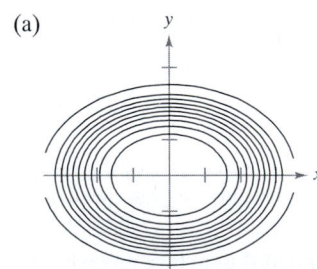

(b)

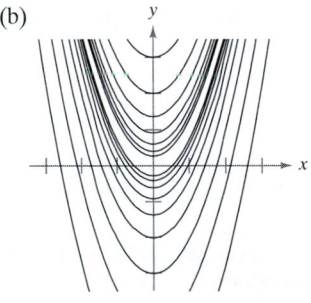

(c)

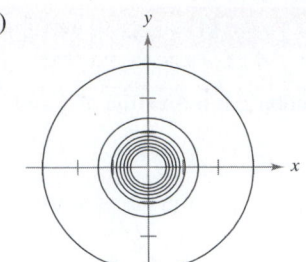

(d)

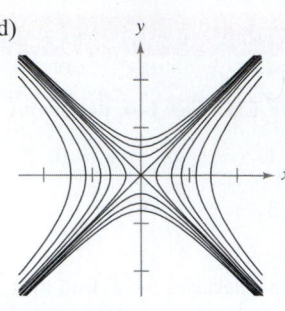

45. $f(x, y) = e^{1 - x^2 - y^2}$

46. $f(x, y) = e^{1 - x^2 + y^2}$

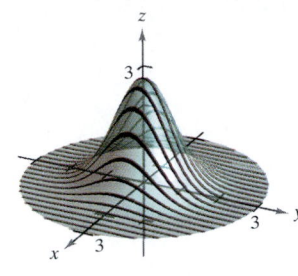

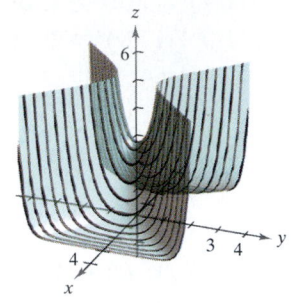

47. $f(x, y) = \ln|y - x^2|$

48. $f(x, y) = \cos\left(\dfrac{x^2 + 2y^2}{4}\right)$

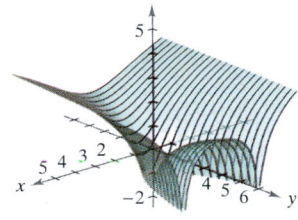

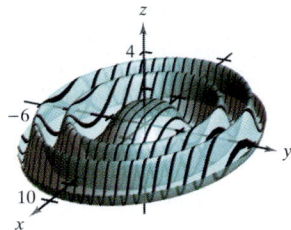

In Exercises 49–56, describe the level curves of the function. Sketch the level curves for the given c-values.

49. $z = x + y$ $c = -1, 0, 2, 4$

50. $z = 6 - 2x - 3y$ $c = 0, 2, 4, 6, 8, 10$

51. $z = \sqrt{25 - x^2 - y^2}$ $c = 0, 1, 2, 3, 4, 5$

52. $f(x, y) = x^2 + y^2$ $c = 0, 2, 4, 6, 8$

53. $f(x, y) = xy$ $c = \pm 1, \pm 2, \ldots, \pm 6$

54. $f(x, y) = e^{xy}$ $c = 1, 2, 3, 4, \frac{1}{2}, \frac{1}{3}, \frac{1}{4}$

55. $f(x, y) = \dfrac{x}{x^2 + y^2}$ $c = \pm\frac{1}{2}, \pm 1, \pm\frac{3}{2}, \pm 2$

56. $f(x, y) = \ln(x - y)$ $c = 0, \pm\frac{1}{2}, \pm 1, \pm\frac{3}{2}, \pm 2$

57. ***Think About It*** All of the level curves of the surface given by $z = f(x, y)$ are concentric circles. Does this imply that the graph of f is a hemisphere? Illustrate your answer with an example.

58. ***Think About It*** Construct a function whose level curves are lines passing through the origin.

Writing In Exercises 59 and 60, use the graphs of the level curves (*c*-values evenly spaced) of the function *f* to write a description of a possible graph of *f*. Is the graph of *f* unique? Explain.

59.

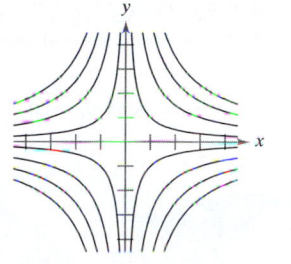

60.

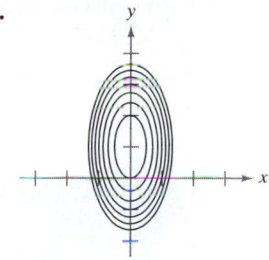

61. *Investment* In 1998, an investment of $1000 was made in a bond earning 10% compounded annually. Assume that the buyer pays tax at rate *R* and the annual rate of inflation is *I*. In the year 2008, the value *V* of the investment in constant 1998 dollars is

$$V(I, R) = 1000\left[\frac{1 + 0.10(1 - R)}{1 + I}\right]^{10}.$$

Use this function of two variables to complete the table.

	\multicolumn{3}{c}{Inflation Rate}	
Tax rate	**0**	**0.03**	**0.05**
0			
0.28			
0.35			

62. *Investment* A principal of $1000 is deposited in a savings account that earns an interest rate of *r* (expressed as a decimal), compounded continuously. The amount *A*(*r*, *t*) after *t* years is

$$A(r, t) = 1000e^{rt}.$$

Use this function of two variables to complete the table.

	\multicolumn{4}{c}{Number of Years}			
Rate	**5**	**10**	**15**	**20**
0.08				
0.10				
0.12				
0.14				

In Exercises 63–68, sketch the graph of the level surface $f(x, y, z) = c$ at the indicated value of *c*.

63. $f(x, y, z) = x - 2y + 3z$ $c = 6$
64. $f(x, y, z) = 4x + y + 2z$ $c = 4$
65. $f(x, y, z) = x^2 + y^2 + z^2$ $c = 9$
66. $f(x, y, z) = x^2 + y^2 - z$ $c = 1$
67. $f(x, y, z) = 4x^2 + 4y^2 - z^2$ $c = 0$
68. $f(x, y, z) = \sin x - z$ $c = 0$

69. *Forestry* The **Doyle Log Rule** is one of several methods used to determine the lumber yield of a log (in board-feet) in terms of its diameter *d* (in inches) and its length *L* (in feet). The number of board-feet is

$$N(d, L) = \left(\frac{d - 4}{4}\right)^2 L.$$

(a) Find the number of board-feet of lumber in a log 22 inches in diameter and 12 feet in length.

(b) Find $N(30, 12)$.

70. *Queuing Model* The average length of time that a customer waits in line for service is

$$W(x, y) = \frac{1}{x - y}, \quad x > y$$

where *y* is the average arrival rate, expressed as the number of customers per unit of time, and *x* is the average service rate, expressed in the same units. Evaluate each of the following.

(a) $W(15, 10)$ (b) $W(12, 9)$ (c) $W(12, 6)$ (d) $W(4, 2)$

71. *Temperature Distribution* The temperature *T* (in degrees Celsius) at any point (*x*, *y*) in a circular steel plate of radius 10 meters is

$$T = 600 - 0.75x^2 - 0.75y^2$$

where *x* and *y* are measured in meters. Sketch some of the isothermal curves.

72. *Cobb-Douglas Production Function* Use the Cobb-Douglas production function (see Example 5) to show that if the number of units of labor and the number of units of capital are doubled, the production level is also doubled.

73. *Construction Cost* A rectangular box with an open top has a length of *x* feet, a width of *y* feet, and a height of *z* feet. Express the cost *C* of constructing the box as a function of *x*, *y*, and *z* if it costs $0.75 per square foot to build the base and $0.40 per square foot to build the sides.

74. *Volume* A propane tank is constructed by welding hemispheres to the ends of a right circular cylinder. Write the volume *V* of the tank as a function of *r* and *l*, where *r* is the radius of the cylinder and hemispheres and *l* is the length of the cylinder.

75. Ideal Gas Law According to the Ideal Gas Law, $PV = kT$, where P is pressure, V is volume, T is temperature (in Kelvins), and k is a constant of proportionality. A tank contains 2600 cubic inches of nitrogen at a pressure of 20 pounds per square inch and a temperature of 300 K.

(a) Determine k.

(b) Express P as a function of V and T and describe the level curves.

76. Modeling Data The table gives the net sales x (in billions of dollars), the total assets y (in billions of dollars), and the shareholder's equity z (in billions of dollars) for Wal-Mart for the years 1991 through 1996. *(Source: 1996 Annual Report)*

Year	1991	1992	1993	1994	1995	1996
x	32.6	43.9	55.5	67.3	82.5	93.6
y	11.4	15.4	20.6	26.4	32.8	37.5
z	5.4	7.0	8.8	10.8	12.7	14.8

A model for these data is

$$z = f(x, y) = 0.083x + 0.160y + 0.890.$$

(a) Use a graphing utility and the model to approximate z for the given values of x and y.

(b) Which of the two variables in this model has the greater influence on shareholder's equity?

(c) Simplify the expression for $f(x, 25)$ and interpret its meaning in the context of the problem.

77. Meteorology Meteorologists measure the atmospheric pressure in millibars. From these observations they create weather maps on which the curves of equal atmospheric pressure (isobars) are drawn (see figure). If the closer the isobars the higher the wind speed, match points *A, B,* and *C* with (a) highest pressure, (b) lowest pressure, and (c) highest wind velocity.

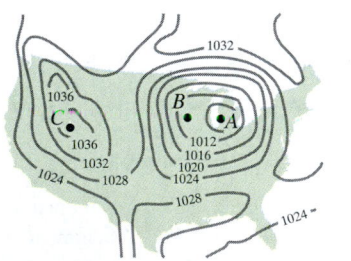

Figure for 77

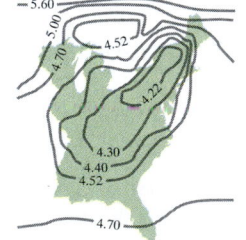

Figure for 78

78. Acid Rain The acidity of rainwater is measured in units called pH. A pH of 7 is neutral, smaller values are increasingly acidic, and larger values are increasingly alkaline. The map shows the curves of equal pH and gives evidence that downwind of heavily industrialized areas the acidity has been increasing. Using the level curves on the map, determine the direction of the prevailing winds in the northeastern United States.

79. Air Conditioner Use The contour map in the figure represents the estimated annual hours of air conditioner use for an average household in 1991. *(Source: Association of Home Appliance Manufacturers)*

(a) Discuss the use of color to represent the level curves.

(b) Do the level curves correspond to equally spaced annual usage hours? Explain.

(c) Describe how to obtain a more detailed contour map.

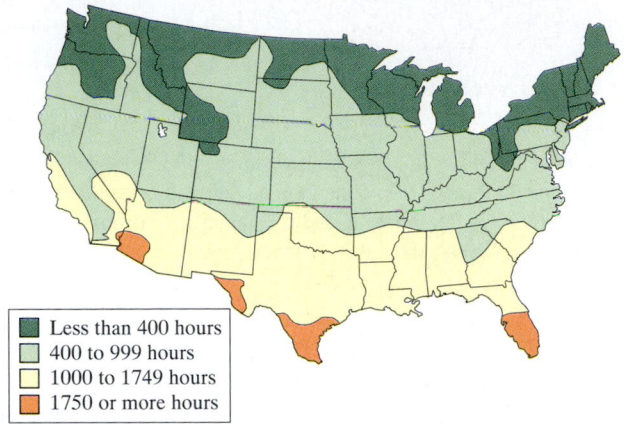

■ Less than 400 hours
□ 400 to 999 hours
□ 1000 to 1749 hours
■ 1750 or more hours

80. Annual Temperature Range The colored regions shown on the weather map of the world depict the normal annual temperature range of each region on earth. Which two geographic features have the greatest effect on temperature range?

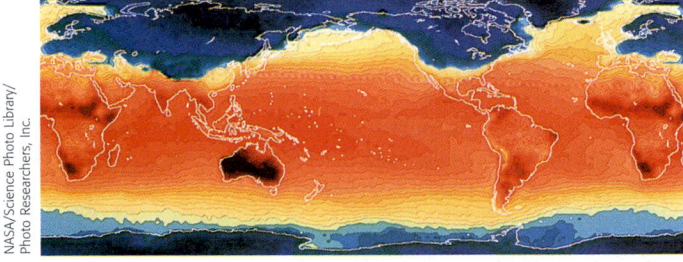

NASA/Science Photo Library/
Photo Researchers, Inc.

True or False? **In Exercises 81–84, determine whether the statement is true or false. If it is false, explain why or give an example that shows it is false.**

81. If $f(x_0, y_0) = f(x_1, y_1)$, then $x_0 = x_1$ and $y_0 = y_1$.

82. A vertical line can intersect the graph of $z = f(x, y)$ at most once.

83. If f is a function, then $f(ax, ay) = a^2 f(x, y)$.

84. The graph of $f(x, y) = x^2 - y^2$ is a hyperbolic paraboloid.

Neighborhoods in the Plane • Limit of a Function of Two Variables •
Continuity of a Function of Two Variables •
Continuity of a Function of Three Variables

Neighborhoods in the Plane

In this section, you will study limits and continuity involving functions of two or three variables. The section begins with functions of two variables. At the end of the section, the concepts are extended to functions of three variables.

We begin our discussion of the limit of a function of two variables by defining a two-dimensional analog to an interval on the real line. Using the formula for the distance between two points (x, y) and (x_0, y_0) in the plane, you can define the **δ-neighborhood** about (x_0, y_0) to be the **disc** centered at (x_0, y_0) with radius $\delta > 0$

$$\left\{(x, y): \sqrt{(x - x_0)^2 + (y - y_0)^2} < \delta\right\} \qquad \text{Open disc}$$

as shown in Figure 12.18. When this formula contains the *less than* inequality, $<$, the disc is called **open,** and when it contains the *less than or equal to* inequality, $\leq$, the disc is called **closed.** This corresponds to the use of $<$ and $\leq$ to define open and closed intervals.

A point (x_0, y_0) in a plane region R is an **interior point** of R if there exists a δ-neighborhood about (x_0, y_0) that lies entirely in R, as shown in Figure 12.19. If every point in R is an interior point, then R is an **open region.** A point (x_0, y_0) is a **boundary point** of R if every open disc centered at (x_0, y_0) contains points inside R *and* points outside R. By definition, a region must contain its interior points, but it need not contain its boundary points. If a region contains all its boundary points, the region is **closed.** A region that contains some but not all of its boundary points is neither open nor closed.

SONYA KOVALEVSKY (1850–1891)

Much of the terminology used to define limits and continuity of a function of two or three variables was introduced by the German mathematician Karl Weierstrass (1815–1897). Weierstrass's rigorous approach to limits and other topics in calculus gained him the reputation as the "father of modern analysis." Weierstrass was a gifted teacher. One of his best-known students was the Russian mathematician Sonya Kovalevsky, who applied many of Weierstrass's techniques to problems in mathematical physics and became one of the first women to gain acceptance as a research mathematician.

FOR FURTHER INFORMATION For more information on Sonya Kovalevsky, see the article "S. Kovalevsky: A Mathematical Lesson" by Karen D. Rappaport in the October 1981 issue of *The American Mathematical Monthly.*

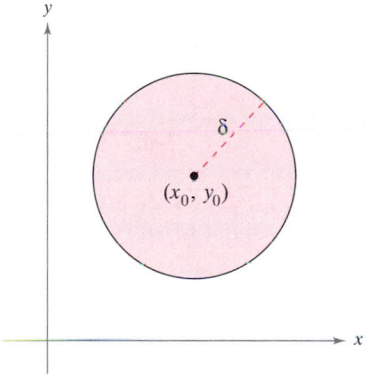

An open disc
Figure 12.18

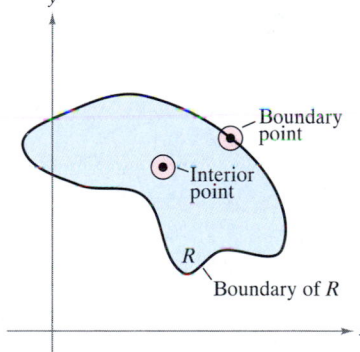

The boundary and interior points of a region R
Figure 12.19

Limit of a Function of Two Variables

Definition of the Limit of a Function of Two Variables

Let f be a function of two variables defined, except possibly at (x_0, y_0), on an open disc centered at (x_0, y_0), and let L be a real number. Then

$$\lim_{(x, y) \to (x_0, y_0)} f(x, y) = L$$

if for each $\varepsilon > 0$ there corresponds a $\delta > 0$ such that

$$|f(x, y) - L| < \varepsilon \quad \text{whenever} \quad 0 < \sqrt{(x - x_0)^2 + (y - y_0)^2} < \delta.$$

NOTE Graphically, this definition of a limit implies that for any point $(x, y) \neq (x_0, y_0)$ in the disc of radius δ, the value $f(x, y)$ lies between $L + \varepsilon$ and $L - \varepsilon$, as shown in Figure 12.20.

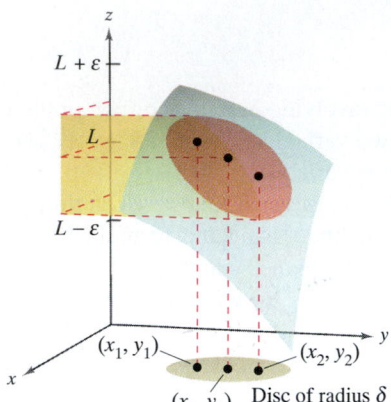

For any (x, y) in the circle of radius δ, the value $f(x, y)$ lies between $L + \varepsilon$ and $L - \varepsilon$.

Figure 12.20

The definition of the limit of a function of two variables is similar to the definition of the limit of a function of a single variable, yet there is a critical difference. To determine whether a function of a single variable possesses a limit, you need only test the approach from two directions—from the left and from the right. If the function approaches the same limit from the right and from the left, you can conclude that the limit exists. However, for a function of two variables, the statement

$$(x, y) \to (x_0, y_0)$$

means that the point (x, y) is allowed to approach (x_0, y_0) from any "direction." If the value of

$$\lim_{(x, y) \to (x_0, y_0)} f(x, y)$$

is not the same for all possible approaches, or **paths,** to (x_0, y_0), the limit does not exist.

EXAMPLE 1   Verifying a Limit by the Definition

Show that

$$\lim_{(x, y) \to (a, b)} x = a.$$

Solution   Let $(x, y) = x$ and $L = a$. You need to show that for each $\varepsilon > 0$, there exists a δ-neighborhood about (a, b) such that

$$|f(x, y) - L| = |x - a| < \varepsilon$$

whenever $(x, y) \neq (a, b)$ lies in the neighborhood. You can first observe that from

$$0 < \sqrt{(x - a)^2 + (y - b)^2} < \delta$$

it follows that

$$|f(x, y) - a| = |x - a| = \sqrt{(x - a)^2} \leq \sqrt{(x - a)^2 + (y - b)^2} < \delta.$$

Thus, you can choose $\delta = \varepsilon$, and the limit is verified.

Limits of functions of several variables have the same properties regarding sums, differences, products, and quotients as do limits of functions of single variables. (See Theorem 1.2 in Section 1.3.) Some of these properties are used in the next example.

EXAMPLE 2   Verifying a Limit

Evaluate each of the following limits.

a. $\displaystyle\lim_{(x,\,y)\to(1,\,2)} \frac{5x^2y}{x^2 + y^2}$ **b.** $\displaystyle\lim_{(x,\,y)\to(0,\,0)} \frac{5x^2y}{x^2 + y^2}$

Solution

a. By using the properties of limits of products and sums, you obtain

$$\lim_{(x,\,y)\to(1,\,2)} 5x^2y = 5(1^2)(2) = 10$$

and

$$\lim_{(x,\,y)\to(1,\,2)} (x^2 + y^2) = (1^2 + 2^2) = 5.$$

Because the limit of a quotient is equal to the quotient of the limits (and the denominator is not 0), you have

$$\lim_{(x,\,y)\to(1,\,2)} \frac{5x^2y}{x^2 + y^2} = \frac{10}{5}$$
$$= 2.$$

b. In this case, the limits of the numerator and of the denominator are both 0, and so you cannot determine the existence (or nonexistence) of a limit by taking the limits of the numerator and denominator separately and then dividing. However, from the graph of f in Figure 12.21, it seems reasonable that the limit might be 0. Thus, you try applying the definition to $L = 0$. First, note that

$$|y| \le \sqrt{x^2 + y^2} \qquad \text{and} \qquad \frac{x^2}{x^2 + y^2} \le 1.$$

Then, in a δ-neighborhood about $(0, 0)$, you have $0 < \sqrt{x^2 + y^2} < \delta$, and it follows that, for $(x, y) \ne (0, 0)$,

$$|f(x, y) - 0| = \left| \frac{5x^2y}{x^2 + y^2} \right|$$
$$= 5|y|\left(\frac{x^2}{x^2 + y^2} \right)$$
$$\le 5|y|$$
$$\le 5\sqrt{x^2 + y^2}$$
$$< 5\delta.$$

Thus, you can choose $\delta = \varepsilon/5$ and conclude that

$$\lim_{(x,\,y)\to(0,\,0)} \frac{5x^2y}{x^2 + y^2} = 0.$$

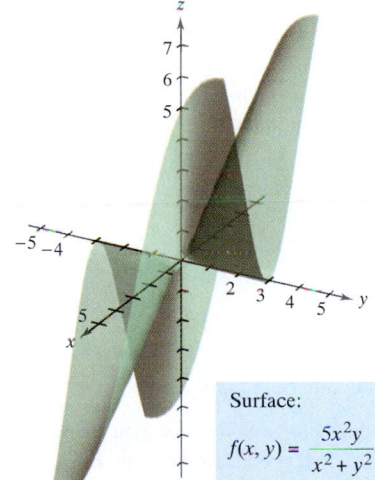

Surface:

$$f(x, y) = \frac{5x^2y}{x^2 + y^2}$$

Figure 12.21

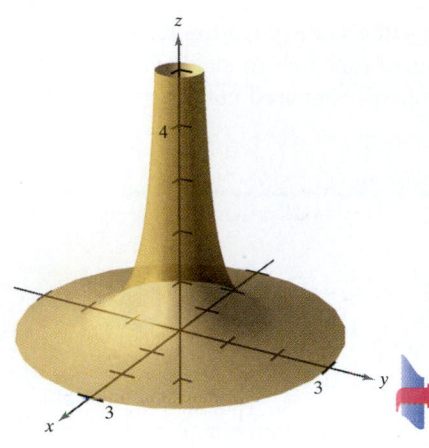

$$\lim_{(x, y) \to (0, 0)} \frac{1}{x^2 + y^2} \text{ does not exist.}$$
Figure 12.22

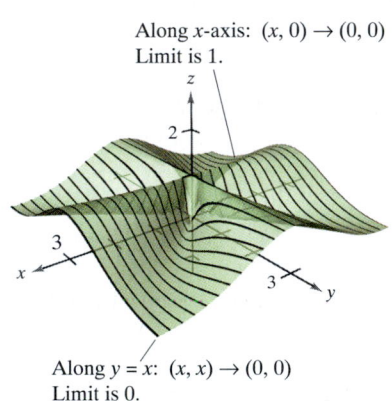

Along x-axis: $(x, 0) \to (0, 0)$
Limit is 1.

Along $y = x$: $(x, x) \to (0, 0)$
Limit is 0.

$$\lim_{(x, y) \to (0, 0)} \left(\frac{x^2 - y^2}{x^2 + y^2}\right)^2 \text{ does not exist.}$$
Figure 12.23

For some functions, it is easy to recognize that a limit does not exist. For instance, it is clear that the limit

$$\lim_{(x, y) \to (0, 0)} \frac{1}{x^2 + y^2}$$

does not exist because the values of $f(x, y)$ increase without bound as (x, y) approaches $(0, 0)$ along *any* path (see Figure 12.22).

For other functions, it is not so easy to recognize that a limit does not exist. For instance, the next example describes a limit that does not exist because the function approaches different values along different paths.

➤ **EXAMPLE 3   A Limit That Does Not Exist**

Show that the following limit does not exist.

$$\lim_{(x, y) \to (0, 0)} \left(\frac{x^2 - y^2}{x^2 + y^2}\right)^2$$

Solution   The domain of the function given by

$$f(x, y) = \left(\frac{x^2 - y^2}{x^2 + y^2}\right)^2$$

consists of all points in the xy-plane except for the point $(0, 0)$. To show that the limit as (x, y) approaches $(0, 0)$ does not exist, consider approaching $(0, 0)$ along two different "paths," as shown in Figure 12.23. Along the x-axis, every point is of the form $(x, 0)$, and the limit along this approach is

$$\lim_{(x, 0) \to (0, 0)} \left(\frac{x^2 - 0^2}{x^2 + 0^2}\right)^2 = \lim_{(x, 0) \to (0, 0)} (1)^2 = 1. \qquad \text{\textcolor{red}{Limit along } } x \text{\textcolor{red}{-axis}}$$

However, if (x, y) approaches $(0, 0)$ along the line $y = x$, you obtain

$$\lim_{(x, x) \to (0, 0)} \left(\frac{x^2 - x^2}{x^2 + x^2}\right)^2 = \lim_{(x, x) \to (0, 0)} \left(\frac{0}{2x^2}\right)^2 = 0. \qquad \text{\textcolor{red}{Limit along line } } y = x$$

This means that in any open disc centered at $(0, 0)$ there are points (x, y) at which f takes on the value 1, and other points at which f takes on the value 0. For instance, $f(x, y) = 1$ at the points

$$(1, 0), (0.1, 0), (0.01, 0), \text{ and } (0.001, 0)$$

and $f(x, y) = 0$ at the points

$$(1, 1), (0.1, 0.1), (0.01, 0.01), \text{ and } (0.001, 0.001).$$

Hence, f does not have a limit as $(x, y) \to (0, 0)$.

NOTE   In Example 3, you could conclude that the limit does not exist because you found two approaches that produced different limits. If two approaches had produced the same limit, you still could not have concluded that the limit exists. To form such a conclusion, you must show that the limit is the same along *all* possible approaches.

Continuity of a Function of Two Variables

NOTE   This definition of continuity can be extended to *boundary points* of the open region R by considering a special type of limit in which (x, y) is allowed to approach (x_0, y_0) along paths lying in the region R. This notion is similar to that of one-sided limits, as discussed in Chapter 1.

Notice in Example 2(a) that the limit of

$$f(x, y) = 5x^2y/(x^2 + y^2)$$

as $(x, y) \to (1, 2)$ can be evaluated by direct substitution. That is, the limit is $f(1, 2) = 2$. In such cases the function f is said to be **continuous** at the point $(1, 2)$.

Definition of Continuity of a Function of Two Variables

A function f of two variables is **continuous at a point** (x_0, y_0) in an open region R if $f(x_0, y_0)$ is equal to the limit of $f(x, y)$ as (x, y) approaches (x_0, y_0). That is,

$$\lim_{(x, y) \to (x_0, y_0)} f(x, y) = f(x_0, y_0).$$

The function f is **continuous in the open region** R if it is continuous at every point in R.

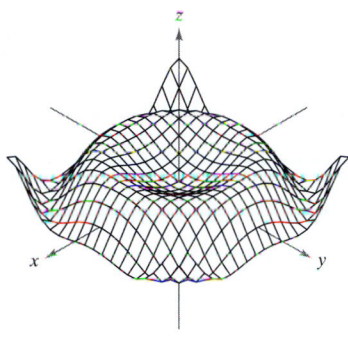

Surface: $f(x, y) = \frac{1}{2}\sin(x^2 + y^2)$

The function f is continuous at every point in the plane.
Figure 12.24

In Example 2(b), it is shown that the function

$$f(x, y) = \frac{5x^2y}{x^2 + y^2}$$

is not continuous at $(0, 0)$. However, because the limit at this point exists, you can remove the discontinuity by defining f at $(0, 0)$ as being equal to its limit there. Such a discontinuity is called **removable**. In Example 3, the function

$$f(x, y) = \left(\frac{x^2 - y^2}{x^2 + y^2}\right)^2$$

is also shown not to be continuous at $(0, 0)$, but this discontinuity is **nonremovable.**

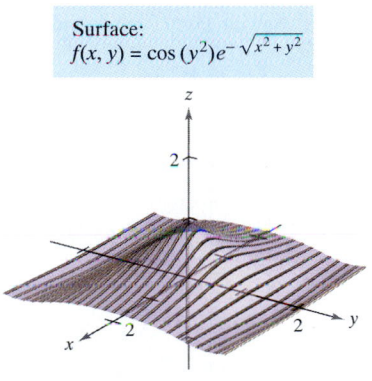

Surface:
$f(x, y) = \cos(y^2)e^{-\sqrt{x^2 + y^2}}$

The function f is continuous at every point in the plane.
Figure 12.25

THEOREM 12.1   Continuous Functions of Two Variables

If k is a real number and f and g are continuous at (x_0, y_0), then the following functions are continuous at (x_0, y_0).

1. Scalar multiple: kf
2. Sum and difference: $f \pm g$
3. Product: fg
4. Quotient: f/g, if $g(x_0, y_0) \neq 0$

Theorem 12.1 establishes the continuity of *polynomial* and *rational* functions at every point in their domains. Furthermore, the continuity of other types of functions can be extended naturally from one to two variables. For instance, the functions whose graphs are shown in Figures 12.24 and 12.25 are continuous at every point in the plane.

The next theorem states conditions under which a composite function is continuous.

THEOREM 12.2   Continuity of a Composite Function

If h is continuous at (x_0, y_0) and g is continuous at $h(x_0, y_0)$, then the composite function given by $(g \circ h)(x, y) = g(h(x, y))$ is continuous at (x_0, y_0). That is,

$$\lim_{(x, y) \to (x_0, y_0)} g(h(x, y)) = g(h(x_0, y_0)).$$

NOTE   Note in Theorem 12.2 that h is a function of two variables and g is a function of one variable.

EXAMPLE 4   Testing for Continuity

Discuss the continuity of each of the following functions.

a. $f(x, y) = \dfrac{x - 2y}{x^2 + y^2}$ **b.** $g(x, y) = \dfrac{2}{y - x^2}$

Solution

a. Because a rational function is continuous at every point in its domain, you can conclude that f is continuous at each point in the xy-plane except at $(0, 0)$, as shown in Figure 12.26.

b. The function given by $g(x, y) = 2/(y - x^2)$ is continuous except at the points at which the denominator is 0, $y - x^2 = 0$. Thus, you can conclude that the function is continuous at all points except those lying on the parabola $y = x^2$. Inside this parabola, you have $y > x^2$, and the surface represented by the function lies above the xy-plane, as shown in Figure 12.27. Outside the parabola, $y < x^2$, and the surface lies below the xy-plane.

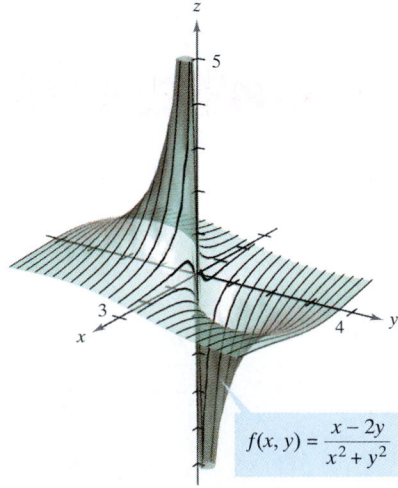

The function f is not continuous at $(0, 0)$.
Figure 12.26

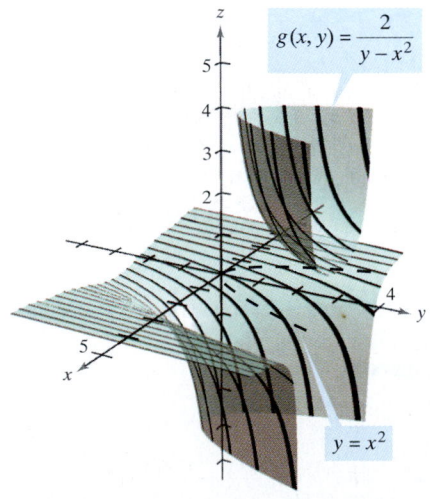

The function g is not continuous on the parabola $y = x^2$.
Figure 12.27

Continuity of a Function of Three Variables

The preceding definitions of limits and continuity can be extended to functions of three variables by considering points (x, y, z) within the *open sphere*

$$(x - x_0)^2 + (y - y_0)^2 + (z - z_0)^2 < \delta^2. \qquad \text{Open sphere}$$

The radius of this sphere is δ, and the sphere is centered at (x_0, y_0, z_0), as shown in Figure 12.28. A point (x_0, y_0, z_0) in a region R in space is an **interior point** of R if there exists a δ-sphere about (x_0, y_0, z_0) that lies entirely in R. If every point in R is an interior point, then R is called **open.**

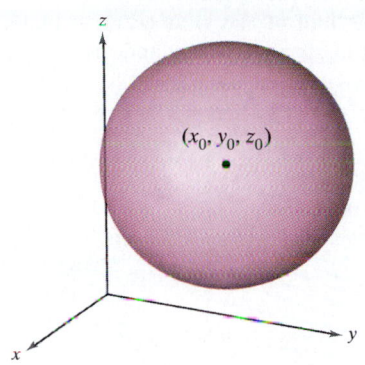

(x_0, y_0, z_0)

Open sphere in space
Figure 12.28

> ### Definition of Continuity of a Function of Three Variables
>
> A function f of three variables is **continuous at a point** (x_0, y_0, z_0) in an open region R if $f(x_0, y_0, z_0)$ is defined and is equal to the limit of $f(x, y, z)$ as (x, y, z) approaches (x_0, y_0, z_0). That is,
>
> $$\lim_{(x, y, z) \to (x_0, y_0, z_0)} f(x, y, z) = f(x_0, y_0, z_0).$$
>
> The function f is **continuous in the open region** R if it is continuous at every point in R.

EXAMPLE 5   Testing Continuity of a Function of Three Variables

The function

$$f(x, y, z) = \frac{1}{x^2 + y^2 - z}$$

is continuous at each point in space except at the points on the paraboloid given by $z = x^2 + y^2$.

EXERCISES FOR SECTION 12.2

In Exercises 1–4, find the indicated limit by using the limits

$$\lim_{(x, y) \to (a, b)} f(x, y) = 5 \text{ and } \lim_{(x, y) \to (a, b)} g(x, y) = 3.$$

1. $\displaystyle\lim_{(x, y) \to (a, b)} [f(x, y) - g(x, y)]$

2. $\displaystyle\lim_{(x, y) \to (a, b)} \left[\frac{4f(x, y)}{g(x, y)} \right]$

3. $\displaystyle\lim_{(x, y) \to (a, b)} [f(x, y)g(x, y)]$

4. $\displaystyle\lim_{(x, y) \to (a, b)} \left[\frac{f(x, y) - g(x, y)}{f(x, y)} \right]$

In Exercises 5–14, find the limit and discuss the continuity of the function.

5. $\displaystyle\lim_{(x, y) \to (2, 1)} (x + 3y^2)$

6. $\displaystyle\lim_{(x, y) \to (0, 0)} (5x + y + 1)$

7. $\displaystyle\lim_{(x, y) \to (2, 4)} \frac{x + y}{x - y}$

8. $\displaystyle\lim_{(x, y) \to (1, 1)} \frac{x}{\sqrt{x + y}}$

9. $\displaystyle\lim_{(x, y) \to (0, 1)} \frac{\arcsin(x/y)}{1 + xy}$

10. $\displaystyle\lim_{(x, y) \to (\pi/4, 2)} y \sin xy$

11. $\displaystyle\lim_{(x, y) \to (0, 0)} e^{xy}$

12. $\displaystyle\lim_{(x, y) \to (1, 1)} \frac{xy}{x^2 + y^2}$

13. $\displaystyle\lim_{(x, y, z) \to (1, 2, 5)} \sqrt{x + y + z}$

14. $\displaystyle\lim_{(x, y, z) \to (2, 0, 1)} xe^{yz}$

In Exercises 15–18, discuss the continuity of the function and evaluate the limit of $f(x, y)$ (if it exists) as $(x, y) \to (0, 0)$.

15. $f(x, y) = e^{xy}$

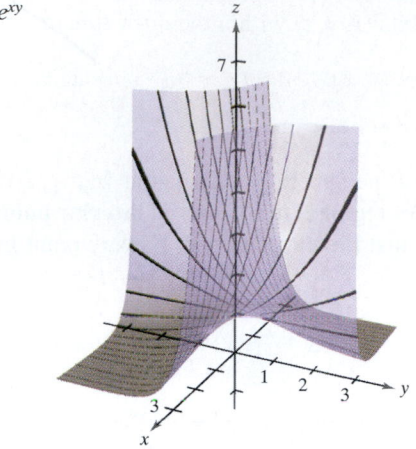

16. $f(x, y) = \dfrac{x^2}{(x^2 + 1)(y^2 + 1)}$

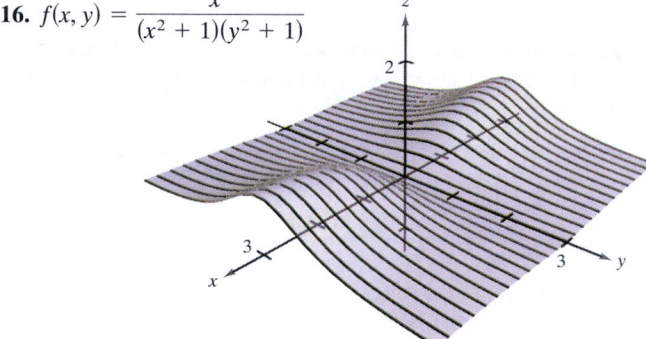

17. $f(x, y) = \ln(x^2 + y^2)$

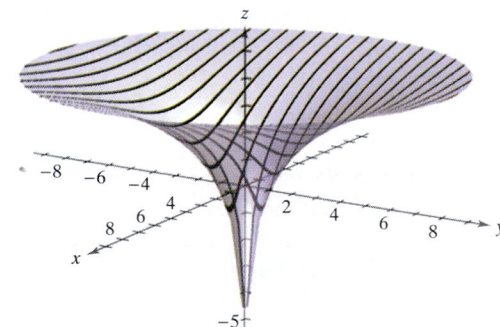

18. $f(x, y) = 1 - \dfrac{\cos(x^2 + y^2)}{x^2 + y^2}$

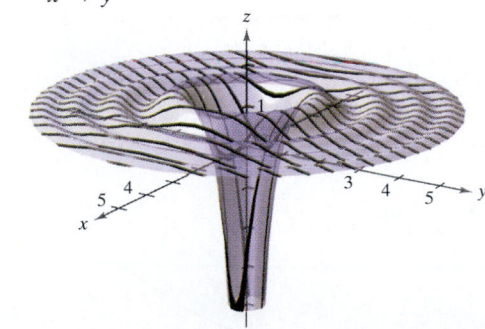

In Exercises 19–22, use a graphing utility to make a table showing the values of $f(x, y)$ at the indicated points. Use the result to make a conjecture about the limit of $f(x, y)$ as $(x, y) \to (0, 0)$. Determine whether the limit exists analytically and discuss the continuity of the function.

19. $f(x, y) = \dfrac{xy}{x^2 + y^2}$

Path: $y = 0$

Points: $(1, 0)$,

$(0.5, 0)$, $(0.1, 0)$,

$(0.01, 0)$, $(0.001, 0)$

Path: $y = x$

Points: $(1, 1)$,

$(0.5, 0.5)$,

$(0.1, 0.1)$,

$(0.01, 0.01)$,

$(0.001, 0.001)$

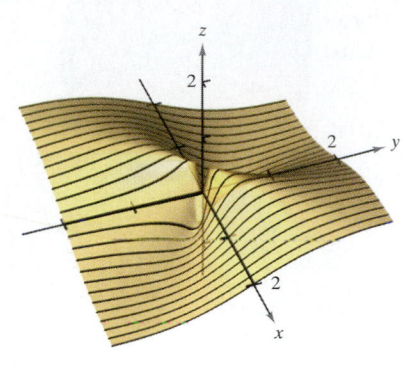

20. $f(x, y) = \dfrac{y}{x^2 + y^2}$

Path: $y = 0$

Points: $(1, 0)$,

$(0.5, 0)$,

$(0.1, 0)$,

$(0.01, 0)$,

$(0.001, 0)$

Path: $y = x$

Points: $(1, 1)$,

$(0.5, 0.5)$,

$(0.1, 0.1)$,

$(0.01, 0.01)$,

$(0.001, 0.001)$

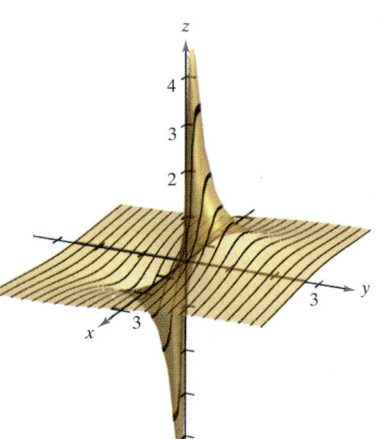

21. $f(x, y) = -\dfrac{xy^2}{x^2 + y^4}$

Path: $x = y^2$

Points: $(1, 1)$,

$(0.25, 0.5)$,

$(0.01, 0.1)$,

$(0.0001, 0.01)$,

$(0.000001, 0.001)$

Path: $x = -y^2$

Points: $(-1, 1)$,

$(-0.25, 0.5)$,

$(-0.01, 0.1)$,

$(-0.0001, 0.01)$,

$(-0.000001, 0.001)$

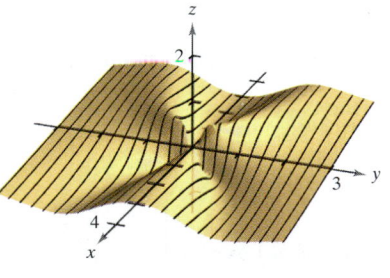

22. $f(x, y) = \dfrac{2x - y^2}{2x^2 + y}$

Path: $y = 0$

Points: $(1, 0)$,

$(0.25, 0)$,

$(0.01, 0)$,

$(0.001, 0)$,

$(0.000001, 0)$

Path: $y = x$

Points: $(1, 1)$,

$(0.25, 0.25)$,

$(0.01, 0.01)$,

$(0.001, 0.001)$,

$(0.0001, 0.0001)$

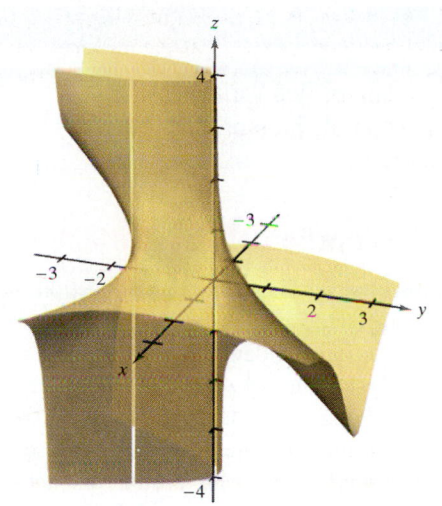

In Exercises 23–28, use a computer algebra system to graph the function and find $\lim\limits_{(x, y)\to(0, 0)} f(x, y)$ **(if it exists).**

23. $f(x, y) = \sin x + \sin y$

24. $f(x, y) = \sin\dfrac{1}{x} + \cos\dfrac{1}{x}$

25. $f(x, y) = \dfrac{x^2 y}{x^4 + 4y^2}$

26. $f(x, y) = \dfrac{x^2 + y^2}{x^2 y}$

27. $f(x, y) = \dfrac{xy^3}{x^2 + 2y^6}$

28. $f(x, y) = \dfrac{2xy}{x^2 + y^2 + 1}$

In Exercises 29–32, use polar coordinates to find the limit. [Hint: Let $x = r\cos\theta$ and $y = r\sin\theta$, and note that $(x, y)\to(0, 0)$ is equivalent to $r \to 0$.]

29. $\lim\limits_{(x, y)\to(0, 0)} \dfrac{\sin(x^2 + y^2)}{x^2 + y^2}$

30. $\lim\limits_{(x, y)\to(0, 0)} \dfrac{xy^2}{x^2 + y^2}$

31. $\lim\limits_{(x, y)\to(0, 0)} \dfrac{x^3 + y^3}{x^2 + y^2}$

32. $\lim\limits_{(x, y)\to(0, 0)} \dfrac{x^2 y^2}{x^2 + y^2}$

In Exercises 33–36, discuss the continuity of the function.

33. $f(x, y, z) = \dfrac{1}{\sqrt{x^2 + y^2 + z^2}}$

34. $f(x, y, z) = \dfrac{z}{x^2 + y^2 - 4}$

35. $f(x, y, z) = \dfrac{\sin z}{e^x + e^y}$

36. $f(x, y, z) = xy\sin z$

In Exercises 37–40, discuss the continuity of the composite function $f \circ g$.

37. $f(t) = t^2$

$g(x, y) = 3x - 2y$

38. $f(t) = \dfrac{1}{t}$

$g(x, y) = x^2 + y^2$

39. $f(t) = \dfrac{1}{t}$

$g(x, y) = 3x - 2y$

40. $f(t) = \dfrac{1}{4 - t}$

$g(x, y) = x^2 + y^2$

In Exercises 41–44, find each of the following limits.

(a) $\lim\limits_{\Delta x \to 0} \dfrac{f(x + \Delta x, y) - f(x, y)}{\Delta x}$

(b) $\lim\limits_{\Delta y \to 0} \dfrac{f(x, y + \Delta y) - f(x, y)}{\Delta y}$

41. $f(x, y) = x^2 - 4y$

42. $f(x, y) = x^2 + y^2$

43. $f(x, y) = 2x + xy - 3y$

44. $f(x, y) = \sqrt{y}\,(y + 1)$

45. Prove that

$$\lim\limits_{(x, y)\to(a, b)} \left[f(x, y) + g(x, y) \right] = L_1 + L_2$$

where $f(x, y)$ approaches L_1 and $g(x, y)$ approaches L_2 as $(x, y) \to (a, b)$.

46. Prove that if f is continuous and $f(a, b) < 0$, there exists a δ-neighborhood about (a, b) such that $f(x, y) < 0$ for every point (x, y) in the neighborhood.

47. *Think About It*   If $f(2, 3) = 4$, can you conclude anything about

$$\lim\limits_{(x, y)\to(2, 3)} f(x, y)?$$

Give reasons for your answer.

48. *Think About It*   If

$$\lim\limits_{(x, y)\to(2, 3)} f(x, y) = 4$$

can you conclude anything about $f(2, 3)$? Give reasons for your answer.

True or False?   **In Exercises 49–52, determine whether the statement is true or false. If it is false, explain why or give an example that shows it is false.**

49. If $\lim\limits_{(x, y)\to(0, 0)} f(x, y) = 0$, then $\lim\limits_{x\to0} f(x, 0) = 0$.

50. If $\lim\limits_{(x, y)\to(0, 0)} f(0, y) = 0$, then $\lim\limits_{(x, y)\to(0, 0)} f(x, y) = 0$.

51. If f is continuous for all nonzero x and y, and $f(0, 0) = 0$, then $\lim\limits_{(x, y)\to(0, 0)} f(x, y) = 0$.

52. If g and h are continuous functions of x and y, and $f(x, y) = g(x) + h(y)$, then f is continuous.

Partial Derivatives of a Function of Two Variables •
Partial Derivatives of a Function of Three or More Variables •
Higher-Order Partial Derivatives

Partial Derivatives of a Function of Two Variables

In applications of functions of several variables, the question often arises, "How will a function be affected by a change in one of its independent variables?" You can answer this by considering the independent variables one at a time. For example, to determine the effect of a catalyst in an experiment, a chemist could conduct the experiment several times using varying amounts of the catalyst, while keeping constant other variables such as temperature and pressure. You can use a similar procedure to determine the rate of change of a function f with respect to one of its several independent variables. This process is called **partial differentiation,** and the result is referred to as the **partial derivative** of f with respect to the chosen independent variable.

JEAN LE ROND D'ALEMBERT (1717–1783)

The introduction of partial derivatives followed Newton's and Leibniz's work in calculus by several years. Between 1730 and 1760, Leonhard Euler and Jean Le Rond d'Alembert separately published several papers on dynamics, in which they established much of the theory of partial derivatives. These papers used functions of two or more variables to study problems involving equilibrium, fluid motion, and vibrating strings.

Definition of Partial Derivatives of a Function of Two Variables

If $z = f(x, y)$, then the **first partial derivatives** of f with respect to x and y are the functions f_x and f_y defined by

$$f_x(x, y) = \lim_{\Delta x \to 0} \frac{f(x + \Delta x, y) - f(x, y)}{\Delta x}$$

$$f_y(x, y) = \lim_{\Delta y \to 0} \frac{f(x, y + \Delta y) - f(x, y)}{\Delta y}$$

provided the limits exist.

This definition indicates that if $z = f(x, y)$, then to find f_x you *consider y constant* and differentiate with respect to x. Similarly, to find f_y, you *consider x constant* and differentiate with respect to y.

EXAMPLE 1 Finding Partial Derivatives

Find the partial derivatives f_x and f_y for the function

$$f(x, y) = 3x - x^2 y^2 + 2x^3 y.$$

Solution Considering y to be constant and differentiating with respect to x produces

$$f_x(x, y) = 3 - 2xy^2 + 6x^2 y.$$

Considering x to be constant and differentiating with respect to y produces

$$f_y(x, y) = -2x^2 y + 2x^3.$$

Notation for First Partial Derivatives

For $z = f(x, y)$, the partial derivatives f_x and f_y are denoted by

$$\frac{\partial}{\partial x} f(x, y) = f_x(x, y) = z_x = \frac{\partial z}{\partial x}$$

and

$$\frac{\partial}{\partial y} f(x, y) = f_y(x, y) = z_y = \frac{\partial z}{\partial y}.$$

The first partials evaluated at the point (a, b) are denoted by

$$\left.\frac{\partial z}{\partial x}\right|_{(a, b)} = f_x(a, b) \qquad \text{and} \qquad \left.\frac{\partial z}{\partial y}\right|_{(a, b)} = f_y(a, b).$$

EXAMPLE 2   Finding and Evaluating Partial Derivatives

For $f(x, y) = xe^{x^2y}$, find f_x and f_y, and evaluate each at the point $(1, \ln 2)$.

Solution Because

$$f_x(x, y) = xe^{x^2y}(2xy) + e^{x^2y}$$

the partial derivative of f with respect to x at $(1, \ln 2)$ is

$$f_x(1, \ln 2) = e^{\ln 2}(2 \ln 2) + e^{\ln 2} = 4 \ln 2 + 2.$$

Because

$$f_y(x, y) = xe^{x^2y}(x^2) = x^3 e^{x^2y}$$

the partial derivative of f with respect to y at $(1, \ln 2)$ is

$$f_y(1, \ln 2) = e^{\ln 2} = 2.$$

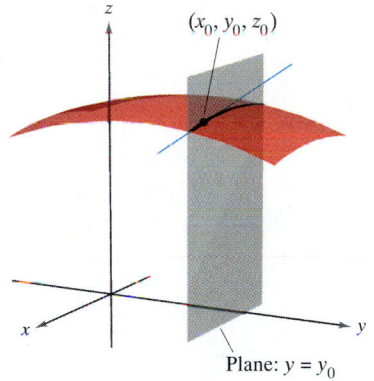

$\dfrac{\partial f}{\partial x} =$ (slope in x-direction)

Figure 12.29

The partial derivatives of a function of two variables, $z = f(x, y)$, have a useful geometric interpretation. If $y = y_0$, then $z = f(x, y_0)$ represents the curve formed by intersecting the surface $z = f(x, y)$ with the plane $y = y_0$, as shown in Figure 12.29. Therefore,

$$f_x(x_0, y_0) = \lim_{\Delta x \to 0} \frac{f(x_0 + \Delta x, y_0) - f(x_0, y_0)}{\Delta x}$$

represents the slope of this curve at the point $(x_0, y_0, f(x_0, y_0))$. Note that both the curve and the tangent line lie in the plane $y = y_0$. Similarly,

$$f_y(x_0, y_0) = \lim_{\Delta y \to 0} \frac{f(x_0, y_0 + \Delta y) - f(x_0, y_0)}{\Delta y}$$

represents the slope of the curve given by the intersection of $z = f(x, y)$ and the plane $x = x_0$ at $(x_0, y_0, f(x_0, y_0))$, as shown in Figure 12.30.

Informally, we say that the values of $\partial f/\partial x$ and $\partial f/\partial y$ at the point (x_0, y_0, z_0) denote the **slopes of the surface in the x- and y-directions**.

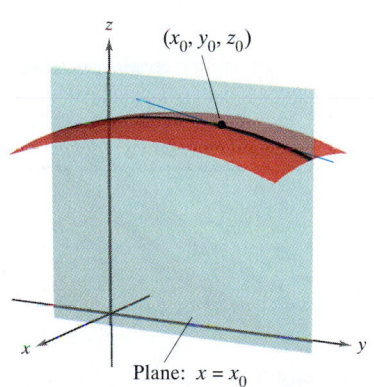

$\dfrac{\partial f}{\partial y} =$ (slope in y-direction)

Figure 12.30

EXAMPLE 3 **Finding the Slopes of a Surface in the x- and y-Directions**

Find the slopes of the surface given by

$$f(x, y) = -\frac{x^2}{2} - y^2 + \frac{25}{8}$$

at the point $\left(\frac{1}{2}, 1, 2\right)$ in the x-direction and in the y-direction.

Solution The partial derivatives of f with respect to x and y are

$$f_x(x, y) = -x \qquad \text{and} \qquad f_y(x, y) = -2y.$$

Thus, in the x-direction, the slope is

$$f_x\left(\frac{1}{2}, 1\right) = -\frac{1}{2} \qquad \text{Figure 12.31(a)}$$

and in the y-direction, the slope is

$$f_y\left(\frac{1}{2}, 1\right) = -2. \qquad \text{Figure 12.31(b)}$$

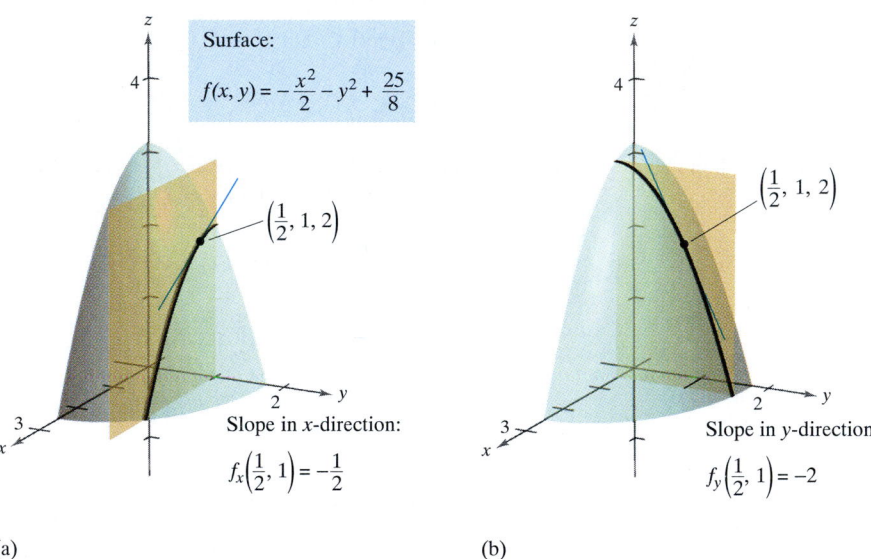

Surface:
$$f(x, y) = -\frac{x^2}{2} - y^2 + \frac{25}{8}$$

$\left(\frac{1}{2}, 1, 2\right)$

Slope in x-direction:
$$f_x\left(\frac{1}{2}, 1\right) = -\frac{1}{2}$$

$\left(\frac{1}{2}, 1, 2\right)$

Slope in y-direction:
$$f_y\left(\frac{1}{2}, 1\right) = -2$$

(a)   (b)

Figure 12.31

EXAMPLE 4 **Finding the Slopes of a Surface in the x- and y-Directions**

Find the slopes of the surface given by

$$f(x, y) = 1 - (x - 1)^2 - (y - 2)^2$$

at the point $(1, 2, 1)$ in the x-direction and in the y-direction.

Solution The partial derivatives of f with respect to x and y are

$$f_x(x, y) = -2(x - 1) \qquad \text{and} \qquad f_y(x, y) = -2(y - 2).$$

Thus, at the point $(1, 2, 1)$, the slopes in the x- and y-directions are

$$f_x(1, 2) = -2(1 - 1) = 0 \qquad \text{and} \qquad f_y(1, 2) = -2(2 - 2) = 0$$

as indicated in Figure 12.32.

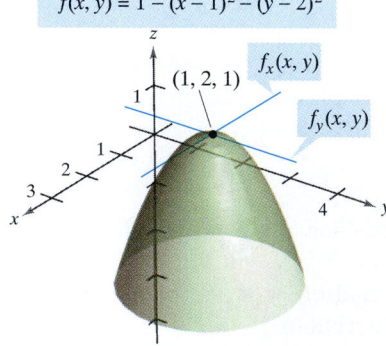

Surface:
$$f(x, y) = 1 - (x - 1)^2 - (y - 2)^2$$

$f_x(x, y)$

$(1, 2, 1)$

$f_y(x, y)$

Figure 12.32

No matter how many variables are involved, partial derivatives of several variables can be interpreted as *rates of change*.

EXAMPLE 5   Using Partial Derivatives to Find Rates of Change

The area of a parallelogram with adjacent sides a and b and included angle θ is given by $A = ab \sin \theta$, as shown in Figure 12.33.

a. Find the rate of change of A with respect to a for $a = 10$, $b = 20$, and $\theta = \dfrac{\pi}{6}$.

b. Find the rate of change of A with respect to θ for $a = 10$, $b = 20$, and $\theta = \dfrac{\pi}{6}$.

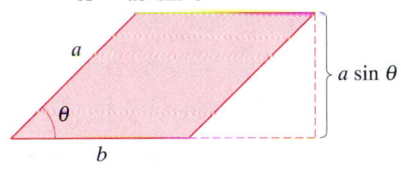

$A = ab \sin \theta$

The area of the parallelogram is $ab \sin \theta$.

Figure 12.33

Solution

a. To find the rate of change of the area with respect to a, hold b and θ constant and differentiate with respect to a to obtain

$$\frac{\partial A}{\partial a} = b \sin \theta \qquad\qquad \text{Find partial with respect to } a.$$

$$\frac{\partial A}{\partial a} = 20 \sin \frac{\pi}{6} = 10. \qquad\qquad \text{Substitute for } b, \text{ and } \theta.$$

b. To find the rate of change of the area with respect to θ, hold a and b constant and differentiate with respect to θ to obtain

$$\frac{\partial A}{\partial \theta} = ab \cos \theta \qquad\qquad \text{Find partial with respect to } \theta.$$

$$\frac{\partial A}{\partial \theta} = 200 \cos(\pi/6) = 100\sqrt{3}. \qquad\qquad \text{Substitute for } a, b, \text{ and } \theta.$$

Partial Derivatives of a Function of Three or More Variables

The concept of a partial derivative can be extended naturally to functions of three or more variables. For instance, if $w = f(x, y, z)$, there are three partial derivatives, each of which is formed by holding two of the variables constant. That is, to define the partial derivative of w with respect to x, consider y and z to be constant and differentiate with respect to x. A similar process is used to find the derivatives of w with respect to y and with respect to z.

$$\frac{\partial w}{\partial x} = f_x(x, y, z) = \lim_{\Delta x \to 0} \frac{f(x + \Delta x, y, z) - f(x, y, z)}{\Delta x}$$

$$\frac{\partial w}{\partial y} = f_y(x, y, z) = \lim_{\Delta y \to 0} \frac{f(x, y + \Delta y, z) - f(x, y, z)}{\Delta y}$$

$$\frac{\partial w}{\partial z} = f_z(x, y, z) = \lim_{\Delta z \to 0} \frac{f(x, y, z + \Delta z) - f(x, y, z)}{\Delta z}$$

In general, if $w = f(x_1, x_2, \ldots, x_n)$, there are n partial derivatives denoted by

$$\frac{\partial w}{\partial x_k} = f_{x_k}(x_1, x_2, \ldots, x_n), \quad k = 1, 2, \ldots, n.$$

To find the partial derivative with respect to one of the variables, hold the other variables constant and differentiate with respect to the given variable.

EXAMPLE 6 **Finding Partial Derivatives**

a. To find the partial derivative of $f(x, y, z) = xy + yz^2 + xz$ with respect to z, consider x and y to be constant and obtain

$$\frac{\partial}{\partial z}[xy + yz^2 + xz] = 2yz + x.$$

b. To find the partial derivative of $f(x, y, z) = z\sin(xy^2 + 2z)$ with respect to z, consider x and y to be constant. Then, using the Product Rule, you obtain

$$\frac{\partial}{\partial z}[z\sin(xy^2 + 2z)] = (z)\frac{\partial}{\partial z}[\sin(xy^2 + 2z)] + \sin(xy^2 + 2z)\frac{\partial}{\partial z}[z]$$

$$= (z)[\cos(xy^2 + 2z)](2) + \sin(xy^2 + 2z)$$

$$= 2z\cos(xy^2 + 2z) + \sin(xy^2 + 2z).$$

c. To find the partial derivative of $f(x, y, z, w) = (x + y + z)/w$ with respect to w, consider x, y, and z to be constant and obtain

$$\frac{\partial}{\partial w}\left[\frac{x + y + z}{w}\right] = -\frac{x + y + z}{w^2}.$$

Higher-Order Partial Derivatives

As is true for ordinary derivatives, it is possible to take second, third, and higher partial derivatives of a function of several variables, provided such derivatives exist. Higher-order derivatives are denoted by the order in which the differentiation occurs. For instance, the function $z = f(x, y)$ has the following second partial derivatives.

1. Differentiate twice with respect to x:

$$\frac{\partial}{\partial x}\left(\frac{\partial f}{\partial x}\right) = \frac{\partial^2 f}{\partial x^2} = f_{xx}.$$

2. Differentiate twice with respect to y:

$$\frac{\partial}{\partial y}\left(\frac{\partial f}{\partial y}\right) = \frac{\partial^2 f}{\partial y^2} = f_{yy}.$$

3. Differentiate first with respect to x and then with respect to y:

$$\frac{\partial}{\partial y}\left(\frac{\partial f}{\partial x}\right) = \frac{\partial^2 f}{\partial y\partial x} = f_{xy}.$$

4. Differentiate first with respect to y and then with respect to x:

$$\frac{\partial}{\partial x}\left(\frac{\partial f}{\partial y}\right) = \frac{\partial^2 f}{\partial x\partial y} = f_{yx}.$$

The third and fourth cases are called **mixed partial derivatives.**

NOTE Note that the two types of notation for mixed partials have different conventions for indicating the order of differentiation.

$$\frac{\partial}{\partial y}\left(\frac{\partial f}{\partial x}\right) = \frac{\partial^2 f}{\partial y\partial x} \qquad \text{Right-to-left order}$$

$$(f_x)_y = f_{xy} \qquad \text{Left-to-right order}$$

You can remember the order by observing that in both notations, you differentiate first with respect to the variable "nearest" f.

EXAMPLE 7 Finding Second Partial Derivatives

Find the second partial derivatives of $f(x, y) = 3xy^2 - 2y + 5x^2y^2$, and determine the value of $f_{xy}(-1, 2)$.

Solution Begin by finding the first partial derivatives with respect to x and y.

$$f_x(x, y) = 3y^2 + 10xy^2 \qquad \text{and} \qquad f_y(x, y) = 6xy - 2 + 10x^2y$$

Then, differentiate each of these with respect to x and y.

$$f_{xx}(x, y) = 10y^2 \qquad \text{and} \qquad f_{yy}(x, y) = 6x + 10x^2$$
$$f_{xy}(x, y) = 6y + 20xy \qquad \text{and} \qquad f_{yx}(x, y) = 6y + 20xy$$

At $(-1, 2)$, the value of f_{xy} is $f_{xy}(-1, 2) = 12 - 40 = -28$.

NOTE   Notice in Example 7 that the two mixed partials are equal. Sufficient conditions for this occurrence are given in Theorem 12.3.

THEOREM 12.3 Equality of Mixed Partial Derivatives

If f is a function of x and y such that f_{xy} and f_{yx} are continuous on an open disc R, then, for every (x, y) in R,

$$f_{xy}(x, y) = f_{yx}(x, y).$$

Theorem 12.3 also applies to a function f of *three or more variables* so long as all second partial derivatives are continuous. For example, if $w = f(x, y, z)$ and all the second partial derivatives are continuous in an open region R, then at each point in R the order of differentiation in the mixed second partial derivatives is irrelevant. If the third partial derivatives of f are also continuous, the order of differentiation of the mixed third partial derivatives is irrelevant.

EXAMPLE 8 Finding Higher-Order Partial Derivatives

Show that $f_{xz} = f_{zx}$ and $f_{xzz} = f_{zxz} = f_{zzx}$ for the function given by

$$f(x, y, z) = ye^x + x \ln z.$$

Solution

First partials:

$$f_x(x, y, z) = ye^x + \ln z, \qquad f_z(x, y, z) = \frac{x}{z}$$

Second partials (note that the first two are equal):

$$f_{xz}(x, y, z) = \frac{1}{z}, \qquad f_{zx}(x, y, z) = \frac{1}{z}, \qquad f_{zz}(x, y, z) = -\frac{x}{z^2}$$

Third partials (note that all three are equal):

$$f_{xzz}(x, y, z) = -\frac{1}{z^2}, \qquad f_{zxz}(x, y, z) = -\frac{1}{z^2}, \qquad f_{zzx}(x, y, z) = -\frac{1}{z^2}$$

Think About It   In Exercises 1–4, use the graph of the surface to determine the sign of the indicated partial derivative.

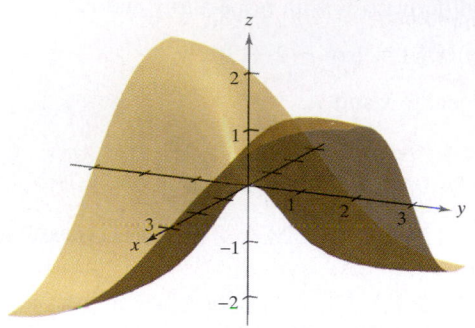

1. $f_x(4, 1)$ **2.** $f_y(-1, -2)$

3. $f_y(4, 1)$ **4.** $f_x(-1, -1)$

In Exercises 5–24, find both first partial derivatives.

5. $f(x, y) = 2x - 3y + 5$ **6.** $f(x, y) = x^2 - 3y^2 + 7$

7. $z = x\sqrt{y}$ **8.** $z = x^2 - 3xy + y^2$

9. $z = x^2 e^{2y}$ **10.** $z = xe^{x/y}$

11. $z = \ln(x^2 + y^2)$ **12.** $z = \ln\sqrt{xy}$

13. $z = \ln\dfrac{x + y}{x - y}$ **14.** $z = \dfrac{x^2}{2y} + \dfrac{4y^2}{x}$

15. $h(x, y) = e^{-(x^2+y^2)}$ **16.** $g(x, y) = \ln\sqrt{x^2 + y^2}$

17. $f(x, y) = \sqrt{x^2 + y^2}$ **18.** $f(x, y) = \dfrac{xy}{x^2 + y^2}$

19. $z = \tan(2x - y)$ **20.** $z = \sin 3x \cos 3y$

21. $z = e^y \sin xy$ **22.** $z = \cos(x^2 + y^2)$

23. $f(x, y) = \displaystyle\int_x^y (t^2 - 1)\, dt$

24. $f(x, y) = \displaystyle\int_x^y (2t + 1)\, dt + \int_y^x (2t - 1)\, dt$

In Exercises 25–28, use the limit definition of partial derivatives to find $f_x(x, y)$ and $f_y(x, y)$.

25. $f(x, y) = 2x + 3y$ **26.** $f(x, y) = \dfrac{1}{x + y}$

27. $f(x, y) = \sqrt{x + y}$ **28.** $f(x, y) = x^2 - 2xy + y^2$

In Exercises 29–32, find the slopes of the surface in the x- and y-directions at the indicated point.

29. $g(x, y) = 4 - x^2 - y^2$, $(1, 1, 2)$

30. $h(x, y) = x^2 - y^2$, $(-2, 1, 3)$

31. $z = e^{-x} \cos y$, $(0, 0, 1)$

32. $z = \sin(2x - y)$, $(\pi/4, \pi/3, 1/2)$

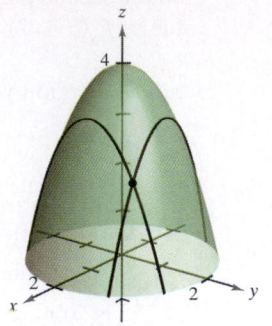

Figure for 29

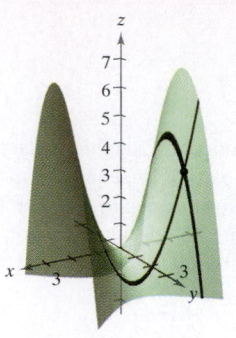

Figure for 30

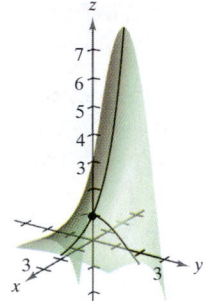

Figure for 31

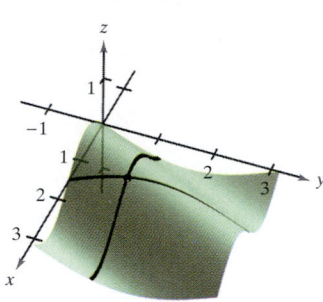

Figure for 32

In Exercises 33–36, evaluate f_x and f_y at the indicated point.

Function	Point
33. $f(x, y) = \arctan\dfrac{y}{x}$	$\left(2, -2, -\dfrac{\pi}{4}\right)$
34. $f(x, y) = \arcsin xy$	$\left(1, 1, \dfrac{\pi}{2}\right)$
35. $f(x, y) = xy/(x - y)$	$(2, -2, -1)$
36. $f(x, y) = \dfrac{4xy}{\sqrt{x^2 + y^2}}$	$(1, 0, 0)$

37. *Think About It*   Sketch the graph of a function $z = f(x, y)$ whose derivatives f_x and f_y are always positive.

38. *Think About It*   Sketch the graph of a function $z = f(x, y)$ whose derivative f_x is always negative and whose derivative f_y is always positive.

In Exercises 39–42, use a computer algebra system to graph the curve formed by the intersection of the surface and the plane. Find the slope of the curve at the given point.

Surface	Plane	Point
39. $z = \sqrt{49 - x^2 - y^2}$	$x = 2$	$(2, 3, 6)$
40. $z = x^2 + 4y^2$	$y = 1$	$(2, 1, 8)$
41. $z = 9x^2 - y^2$	$y = 3$	$(1, 3, 0)$
42. $z = 9x^2 - y^2$	$x = 1$	$(1, 3, 0)$

In Exercises 43–46, for $f(x, y)$, find all values of x and y such that $f_x(x, y) = 0$ and $f_y(x, y) = 0$ simultaneously.

43. $f(x, y) = x^2 + 4xy + y^2 - 4x + 16y + 3$

44. $f(x, y) = 3x^3 - 12xy + y^3$

45. $f(x, y) = \dfrac{1}{x} + \dfrac{1}{y} + xy$

46. $f(x, y) = \ln(x^2 + y^2 + 1)$

In Exercises 47–54, find the four second partial derivatives. Observe that the second mixed partials are equal.

47. $z = x^2 - 2xy + 3y^2$

48. $z = x^4 - 3x^2y^2 + y^4$

49. $z = \sqrt{x^2 + y^2}$

50. $z = \ln(x - y)$

51. $z = e^x \tan y$

52. $z = xe^y + ye^x$

53. $z = \arctan \dfrac{y}{x}$

54. $z = \sin(x - 2y)$

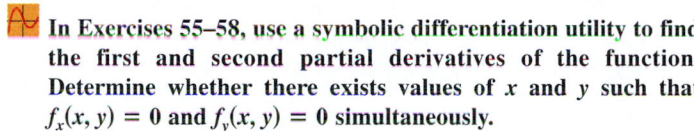 In Exercises 55–58, use a symbolic differentiation utility to find the first and second partial derivatives of the function. Determine whether there exists values of x and y such that $f_x(x, y) = 0$ and $f_y(x, y) = 0$ simultaneously.

55. $f(x, y) = x \sec y$

56. $f(x, y) = \sqrt{9 - x^2 - y^2}$

57. $f(x, y) = \ln \dfrac{x}{x^2 + y^2}$

58. $f(x, y) = \dfrac{xy}{x - y}$

In Exercises 59–64, find the first partial derivatives with respect to x, y, and z.

59. $w = \sqrt{x^2 + y^2 + z^2}$

60. $w = \dfrac{xy}{x + y + z}$

61. $F(x, y, z) = \ln\sqrt{x^2 + y^2 + z^2}$

62. $G(x, y, z) = \dfrac{1}{\sqrt{1 - x^2 - y^2 - z^2}}$

63. $H(x, y, z) = \sin(x + 2y + 3z)$

64. $f(x, y, z) = 3x^2y - 5xyz + 10yz^2$

In Exercises 65–68, show that the mixed partials f_{xyy}, f_{yxy}, and f_{yyx} are equal.

65. $f(x, y, z) = xyz$

66. $f(x, y, z) = x^2 - 3xy + 4yz + z^3$

67. $f(x, y, z) = e^{-x} \sin yz$

68. $f(x, y, z) = \dfrac{x}{y + z}$

Laplace's Equation   In Exercises 69–72, show that the function satisfies Laplace's equation $\partial^2 z/\partial x^2 + \partial^2 z/\partial y^2 = 0$.

69. $z = 5xy$

70. $z = \frac{1}{2}(e^y - e^{-y})\sin x$

71. $z = e^x \sin y$

72. $z = \arctan \dfrac{y}{x}$

Wave Equation   In Exercises 73 and 74, show that the function satisfies the wave equation $\partial^2 z/\partial t^2 = c^2(\partial^2 z/\partial x^2)$.

73. $z = \sin(x - ct)$

74. $z = \sin \omega ct \sin \omega x$

Heat Equation   In Exercises 75 and 76, show that the function satisfies the heat equation $\partial z/\partial t = c^2(\partial^2 z/\partial x^2)$.

75. $z = e^{-t} \cos \dfrac{x}{c}$

76. $z = e^{-t} \sin \dfrac{x}{c}$

77. *Marginal Costs*   A company manufactures two types of wood-burning stoves: a free-standing model and a fireplace-insert model. The cost function for producing x freestanding and y fireplace-insert stoves is

$$C = 32\sqrt{xy} + 175x + 205y + 1050.$$

(a) Find the marginal costs ($\partial C/\partial x$ and $\partial C/\partial y$) when $x = 80$ and $y = 20$.

(b) When additional production is required, which model stove results in the cost increasing at a faster rate? How can this be determined from the cost model?

78. *Marginal Productivity*   Consider the Cobb-Douglas production function

$$f(x, y) = 100x^{0.6}y^{0.4}.$$

When $x = 1000$ and $y = 500$, find

(a) the marginal productivity of labor, $\partial f/\partial x$.

(b) the marginal productivity of capital, $\partial f/\partial y$.

79. *Think About It*   Let N be the number of applicants to a university, p the charge for food and housing at the university, and t the tuition. Suppose that N is a function of p and t such that $\partial N/\partial p < 0$ and $\partial N/\partial t < 0$. What information is gained by noticing that both partials are negative?

80. *Investment*   The value of an investment of \$1000 earning 10% compounded annually is

$$V(I, R) = 1000\left[\dfrac{1 + 0.10(1 - R)}{1 + I}\right]^{10}$$

where I is the annual rate of inflation and R is the tax rate for the person making the investment. Calculate $V_I(0.03, 0.28)$ and $V_R(0.03, 0.28)$. Determine whether the tax rate or the rate of inflation is the greater "negative" factor on the growth of the investment.

81. *Temperature Distribution*   The temperature at any point (x, y) in a steel plate is

$$T = 500 - 0.6x^2 - 1.5y^2$$

where x and y are measured in meters. At the point $(2, 3)$, find the rate of change of the temperature with respect to the distance moved along the plate in the directions of the x- and y-axes.

82. *Apparent Temperature* A measure of what hot weather feels like to two average persons is the Apparent Temperature Index. A model for this index is

$$A = 0.885t - 22.4h + 1.20th - 0.544,$$

where A is the apparent temperature in °C, t is the air temperature, and h is the relative humidity in decimal form. (*Source: The UMAP Journal, Fall 1984*)

(a) Find $\partial A/\partial t$ and $\partial A/\partial h$ when $t = 30°$ and $h = 0.80$.

(b) Which has a greater effect on A, air temperature or humidity? Explain.

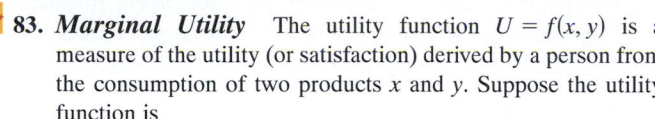 **83. *Marginal Utility*** The utility function $U = f(x, y)$ is a measure of the utility (or satisfaction) derived by a person from the consumption of two products x and y. Suppose the utility function is

$$U = -5x^2 + xy - 3y^2.$$

(a) Determine the marginal utility of product x.

(b) Determine the marginal utility of product y.

(c) When $x = 2$ and $y = 3$, should a person consume one more unit of product x or one more unit of product y? Explain your reasoning.

(d) Use a computer algebra system to graph the function. Interpret the marginal utilities of products x and y graphically.

84. *Ideal Gas Law* The Ideal Gas Law states $PV = nRT$, where P is pressure, V is volume, n is the number of moles of gas, R is a fixed constant (the gas constant), and T is absolute temperature. Show that

$$\frac{\partial T}{\partial P}\frac{\partial P}{\partial V}\frac{\partial V}{\partial T} = -1.$$

85. Consider the function defined by

$$f(x, y) = \begin{cases} \dfrac{xy(x^2 - y^2)}{x^2 + y^2}, & (x, y) \neq (0, 0) \\ 0, & (x, y) = (0, 0). \end{cases}$$

(a) Find $f_x(x, y)$ and $f_y(x, y)$ for $(x, y) \neq (0, 0)$.

(b) Use the definition of partial derivatives to find $f_x(0, 0)$ and $f_y(0, 0)$.

$$\left[Hint: f_x(0, 0) = \lim_{\Delta x \to 0} \frac{f(\Delta x, 0) - f(0, 0)}{\Delta x}. \right]$$

(c) Use the definition of partial derivatives to find $f_{xy}(0, 0)$ and $f_{yx}(0, 0)$.

(d) Using Theorem 12.3 and the result in part (c), what can be said about f_{xy} or f_{yx}?

True or False? **In Exercises 86–89, determine whether the statement is true or false. If it is false, explain why or give an example that shows it is false.**

86. If $z = f(x, y)$ and $\partial z/\partial x = \partial z/\partial y$, then $z = c(x + y)$.

87. If $z = f(x)g(y)$, then $(\partial z/\partial x) + (\partial z/\partial y) = f'(x)g(y) + f(x)g'(y)$.

88. If $z = e^{xy}$, then $\dfrac{\partial^2 z}{\partial y \partial x} = (xy + 1)e^{xy}$.

89. If a cylindrical surface $z = f(x, y)$ has rulings parallel to the y-axis, then $\partial z/\partial y = 0$.

90. Let $f(x, y) = \displaystyle\int_x^y \sqrt{1 + t^3} \, dt$. Find $f_x(x, y)$ and $f_y(x, y)$.

SECTION PROJECT

Read the article "Moiré Fringes and the Conic Sections" by Mike Cullen in the November 1990 issue of *The College Mathematics Journal*. The article describes how two families of level curves given by

$$f(x, y) = a \quad \text{and} \quad g(x, y) = b$$

can form Moiré patterns. After reading the article, write a paper explaining how the expression

$$\frac{\partial f}{\partial x} \cdot \frac{\partial g}{\partial x} + \frac{\partial f}{\partial y} \cdot \frac{\partial g}{\partial y}$$

is related to the Moiré patterns formed by intersecting the two families of level curves. Use one of the following patterns as an example in your paper.

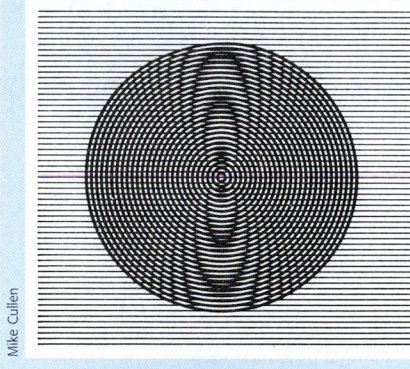

Mike Cullen

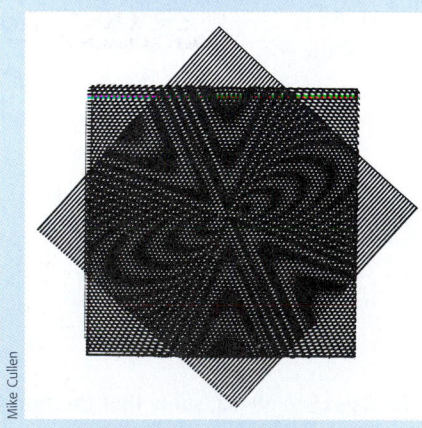

Mike Cullen

Increments and Differentials • Differentiability • Approximation by Differentials

Increments and Differentials

In this section, we generalize the concepts of increments and differentials to functions of two or more variables. Recall from Section 3.9 that for $y = f(x)$, the differential of y was defined as

$$dy = f'(x)\, dx.$$

Similar terminology is used for a function of two variables, $z = f(x, y)$. That is, Δx and Δy are the **increments of x and y,** and the **increment of z** is given by

$$\Delta z = f(x + \Delta x, y + \Delta y) - f(x, y). \qquad \text{Increment of } z$$

Definition of Total Differential

If $z = f(x, y)$ and Δx and Δy are increments of x and y, then the **differentials** of the independent variables x and y are

$$dx = \Delta x \qquad \text{and} \qquad dy = \Delta y$$

and the **total differential** of the dependent variable z is

$$dz = \frac{\partial z}{\partial x}\, dx + \frac{\partial z}{\partial y}\, dy = f_x(x, y)\, dx + f_y(x, y)\, dy.$$

This definition can be extended to a function of three or more variables. For instance, if $w = f(x, y, z, u)$, then $dx = \Delta x$, $dy = \Delta y$, $dz = \Delta z$, $du = \Delta u$, and the total differential of w is

$$dw = \frac{\partial w}{\partial x}\, dx + \frac{\partial w}{\partial y}\, dy + \frac{\partial w}{\partial z}\, dz + \frac{\partial w}{\partial u}\, du.$$

EXAMPLE 1   Finding the Total Differential

a. The total differential dz for $z = 2x \sin y - 3x^2 y^2$ is

$$dz = \frac{\partial z}{\partial x}\, dx + \frac{\partial z}{\partial y}\, dy$$
$$= (2 \sin y - 6xy^2)\, dx + (2x \cos y - 6x^2 y)\, dy.$$

b. The total differential dw for $w = x^2 + y^2 + z^2$ is

$$dw = \frac{\partial w}{\partial x}\, dx + \frac{\partial w}{\partial y}\, dy + \frac{\partial w}{\partial z}\, dz$$
$$= 2x\, dx + 2y\, dy + 2z\, dz.$$

Differentiability

In Section 3.9, you learned that for a *differentiable* function given by $y = f(x)$, you can use the differential $dy = f'(x)\,dx$ as an approximation (for small Δx) to the value $\Delta y = f(x + \Delta x) - f(x)$. When a similar approximation is possible for a function of two variables, the function is said to be **differentiable.** This is stated explicitly in the following definition.

Definition of Differentiability

A function f given by $z = f(x, y)$ is **differentiable** at (x_0, y_0) if Δz can be expressed in the form

$$\Delta z = f_x(x_0, y_0)\Delta x + f_y(x_0, y_0)\Delta y + \varepsilon_1 \Delta x + \varepsilon_2 \Delta y$$

where both ε_1 and $\varepsilon_2 \to 0$ as $(\Delta x, \Delta y) \to (0, 0)$. The function f is **differentiable in a region R** if it is differentiable at each point in R.

EXAMPLE 2   Showing That a Function Is Differentiable

Show that the function given by $f(x, y) = x^2 + 3y$ is differentiable at every point in the plane.

Solution   Letting $z = f(x, y)$, the increment of z at an arbitrary point (x, y) in the plane is

$$\begin{aligned}
\Delta z &= f(x + \Delta x, y + \Delta y) - f(x, y) \\
&= (x^2 + 2x\Delta x + \Delta x^2) + 3(y + \Delta y) - (x^2 + 3y) \\
&= 2x\Delta x + \Delta x^2 + 3\Delta y \\
&= 2x(\Delta x) + 3(\Delta y) + \Delta x(\Delta x) + 0(\Delta y) \\
&= f_x(x, y)\Delta x + f_y(x, y)\Delta y + \varepsilon_1 \Delta x + \varepsilon_2 \Delta y
\end{aligned}$$

where $\varepsilon_1 = \Delta x$ and $\varepsilon_2 = 0$. Because $\varepsilon_1 \to 0$ and $\varepsilon_2 \to 0$ as $(\Delta x, \Delta y) \to (0, 0)$, it follows that f is differentiable at every point in the plane.

Be sure you see that the term "differentiable" is used differently for functions of two variables than for functions of one variable. A function of one variable is differentiable at a point if its derivative exists at the point. However, for a function of two variables, the existence of the partial derivatives f_x and f_y does not guarantee that the function is differentiable (see Example 5). The following theorem gives a *sufficient* condition for differentiability of a function of two variables.

THEOREM 12.4   Sufficient Condition for Differentiability

If f is a function of x and y, where f_x and f_y are continuous in an open region R, then f is differentiable on R.

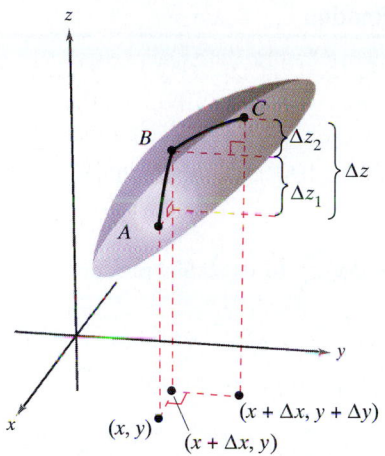

$$\Delta z = f(x + \Delta x, y + \Delta y) - f(x, y)$$

Figure 12.34

Proof   Let S be the surface defined by $z = f(x, y)$, where f, f_x, and f_y are continuous at (x, y). Let A, B, and C be points on surface S, as shown in Figure 12.34. From this figure, you can see that the change in f from point A to point C is given by

$$\Delta z = f(x + \Delta x, y + \Delta y) - f(x, y)$$
$$= [f(x + \Delta x, y) - f(x, y)] + [f(x + \Delta x, y + \Delta y) - f(x + \Delta x, y)]$$
$$= \Delta z_1 + \Delta z_2.$$

Between A and B, y is fixed and x changes. Hence, by the Mean Value Theorem, there is a value x_1 between x and $x + \Delta x$ such that

$$\Delta z_1 = f(x + \Delta x, y) - f(x, y) = f_x(x_1, y)\Delta x.$$

Similarly, between B and C, x is fixed and y changes, and there is a value y_1 between y and $y + \Delta y$ such that

$$\Delta z_2 = f(x + \Delta x, y + \Delta y) - f(x + \Delta x, y) = f_y(x + \Delta x, y_1)\Delta y.$$

By combining these two results, you can write

$$\Delta z = \Delta z_1 + \Delta z_2 = f_x(x_1, y)\Delta x + f_y(x + \Delta x, y_1)\Delta y.$$

If you define ε_1 and ε_2 as

$$\varepsilon_1 = f_x(x_1, y) - f_x(x, y) \qquad \text{and} \qquad \varepsilon_2 = f_y(x + \Delta x, y_1) - f_y(x, y)$$

it follows that

$$\Delta z = \Delta z_1 + \Delta z_2 = [\varepsilon_1 + f_x(x, y)]\Delta x + [\varepsilon_2 + f_y(x, y)]\Delta y$$
$$= [f_x(x, y)\Delta x + f_y(x, y)\Delta y] + \varepsilon_1 \Delta x + \varepsilon_2 \Delta y.$$

By the continuity of f_x and f_y and the fact that $x \le x_1 \le x + \Delta x$ and $y \le y_1 \le y + \Delta y$, it follows that $\varepsilon_1 \to 0$ and $\varepsilon_2 \to 0$ as $\Delta x \to 0$ and $\Delta y \to 0$. Therefore, by definition, f is differentiable.

Approximation by Differentials

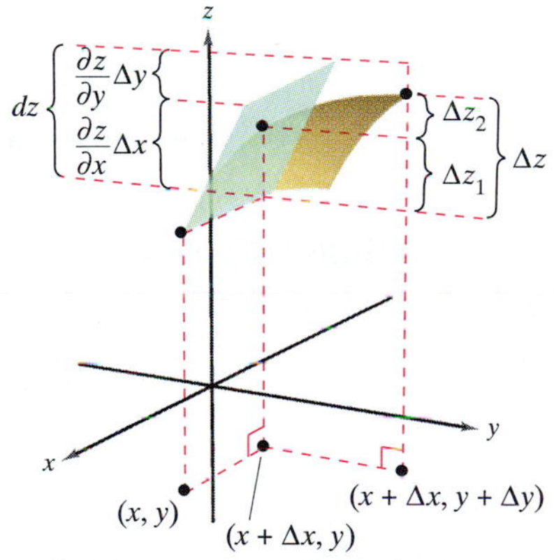

Theorem 12.4 tells you that you can choose $(x + \Delta x, y + \Delta y)$ close enough to (x, y) to make $\varepsilon_1 \Delta x$ and $\varepsilon_2 \Delta y$ insignificant. In other words, for small Δx and Δy, you can use the approximation

$$\Delta z \approx dz.$$

This approximation is illustrated graphically in Figure 12.35. Recall that the partial derivatives $\partial z / \partial x$ and $\partial z / \partial y$ can be interpreted as the slopes of the surface in the x- and y-directions. This means that

$$dz = \frac{\partial z}{\partial x}\Delta x + \frac{\partial z}{\partial y}\Delta y$$

represents the change in height of a plane that is tangent to the surface at the point $(x, y, f(x, y))$. Because a plane in space is represented by a linear equation in the variables x, y, and z, the approximation of Δz by dz is called a **linear approximation.** You will learn more about this geometric interpretation in Section 12.7.

The exact change in z is Δz. This change can be approximated by the differential dz.

Figure 12.35

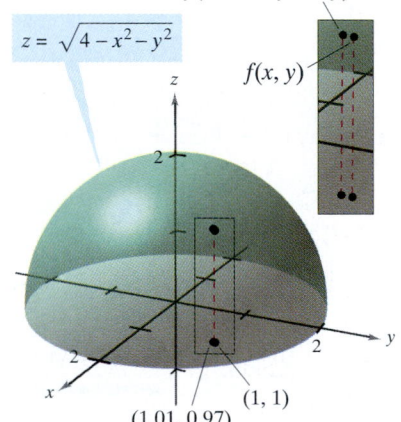

$z = \sqrt{4 - x^2 - y^2}$

$f(x + \Delta x, y + \Delta y)$

$f(x, y)$

$(1, 1)$

$(1.01, 0.97)$

As (x, y) moves from $(1, 1)$ to the point $(1.01, 0.97)$, the value of $f(x, y)$ changes by about 0.0137.

Figure 12.36

> **EXAMPLE 3** **Using a Differential as an Approximation**

Use the differential dz to approximate the change in

$$z = \sqrt{4 - x^2 - y^2}$$

as (x, y) moves from the point $(1, 1)$ to the point $(1.01, 0.97)$. Compare this approximation with the exact change in z.

Solution Letting $(x, y) = (1, 1)$ and $(x + \Delta x, y + \Delta y) = (1.01, 0.97)$ produces

$$dx = \Delta x = 0.01 \quad \text{and} \quad dy = \Delta y = -0.03.$$

Thus, the change in z can be approximated by

$$\Delta z \approx dz$$

$$= \frac{\partial z}{\partial x} dx + \frac{\partial z}{\partial y} dy$$

$$= \frac{-x}{\sqrt{4 - x^2 - y^2}} \Delta x + \frac{-y}{\sqrt{4 - x^2 - y^2}} \Delta y.$$

When $x = 1$ and $y = 1$, you have

$$\Delta z \approx -\frac{1}{\sqrt{2}} (0.01) - \frac{1}{\sqrt{2}} (-0.03)$$

$$= \frac{0.02}{\sqrt{2}}$$

$$= \sqrt{2} (0.01)$$

$$\approx 0.0141. \qquad \text{Approximation of } \Delta z \text{ by } dz$$

In Figure 12.36 you can see that the exact change corresponds to the difference in the heights of two points on the surface of a hemisphere. This difference is given by

$$\Delta z = f(1.01, 0.97) - f(1, 1)$$

$$= \sqrt{4 - (1.01)^2 - (0.97)^2} - \sqrt{4 - 1^2 - 1^2}$$

$$\approx 0.0137. \qquad \text{Actual value of } \Delta z$$

A function of three variables $w = f(x, y, z)$ is called **differentiable** at (x, y, z) provided that

$$\Delta w = f(x + \Delta x, y + \Delta y, z + \Delta z) - f(x, y, z)$$

can be expressed in the form

$$\Delta w = f_x \Delta x + f_y \Delta y + f_z \Delta z + \varepsilon_1 \Delta x + \varepsilon_2 \Delta y + \varepsilon_3 \Delta z$$

where ε_1, ε_2, and $\varepsilon_3 \to 0$ as $(\Delta x, \Delta y, \Delta z) \to (0, 0, 0)$. With this definition of differentiability, Theorem 12.4 has the following extension for functions of three variables: If f is a function of x, y, and z, where f, f_x, f_y, and f_z are continuous in an open region R, then f is differentiable on R.

In Section 3.9, you used differentials to approximate the propagated error introduced by an error in measurement. This application of differentials is further illustrated in Example 4.

EXAMPLE 4   Error Analysis

The possible error involved in measuring each dimension of a rectangular box is ±0.1 millimeter. The dimensions of the box are $x = 50$ centimeters, $y = 20$ centimeters, and $z = 15$ centimeters, as shown in Figure 12.37. Use dV to estimate the propagated error and the relative error in the calculated volume of the box.

Solution   The volume of the box is given by $V = xyz$, and thus

$$dV = \frac{\partial V}{\partial x}\,dx + \frac{\partial V}{\partial y}\,dy + \frac{\partial V}{\partial z}\,dz$$

$$= yz\,dx + xz\,dy + xy\,dz.$$

Using 0.1 millimeter = 0.01 centimeter, you have $dx = dy = dz = \pm0.01$, and the propagated error is approximately

$$dV = (20)(15)(\pm0.01) + (50)(15)(\pm0.01) + (50)(20)(\pm0.01)$$

$$= 300(\pm0.01) + 750(\pm0.01) + 1000(\pm0.01)$$

$$= 2050(\pm0.01) = \pm20.5 \text{ cm}^3.$$

Because the measured volume is

$$V = (50)(20)(15) = 15,000 \text{ cm}^3$$

the relative error, $\Delta V/V$, is approximately

$$\frac{\Delta V}{V} \approx \frac{dV}{V} = \frac{20.5}{15,000} \approx 0.14\%.$$

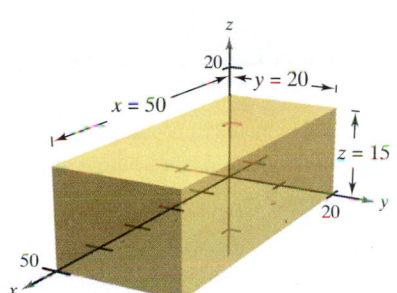

Volume = xyz
Figure 12.37

As is true for a function of a single variable, if a function in two or more variables is differentiable at a point, it is also continuous there.

THEOREM 12.5   Differentiability Implies Continuity

If a function of x and y is differentiable at (x_0, y_0), then it is continuous at (x_0, y_0).

Proof   Let f be differentiable at (x_0, y_0), where $z = f(x, y)$. Then

$$\Delta z = [f_x(x_0, y_0) + \varepsilon_1]\Delta x + [f_y(x_0, y_0) + \varepsilon_2]\Delta y$$

where both ε_1 and $\varepsilon_2 \to 0$ as $(\Delta x, \Delta y) \to (0, 0)$. However, by definition, you know that Δz is given by

$$\Delta z = f(x_0 + \Delta x, y_0 + \Delta y) - f(x_0, y_0).$$

Letting $x = x_0 + \Delta x$ and $y = y_0 + \Delta y$ produces

$$f(x, y) - f(x_0, y_0) = [f_x(x_0, y_0) + \varepsilon_1]\Delta x + [f_y(x_0, y_0) + \varepsilon_2]\Delta y$$

$$= [f_x(x_0, y_0) + \varepsilon_1](x - x_0) + [f_y(x_0, y_0) + \varepsilon_2](y - y_0).$$

Taking the limit as $(x, y) \to (x_0, y_0)$, you have

$$\lim_{(x, y)\to(x_0, y_0)} f(x, y) = f(x_0, y_0)$$

which means that f is continuous at (x_0, y_0).

Remember that the existence of f_x and f_y is not sufficient to guarantee differentiability, as illustrated in the next example.

EXAMPLE 5   A Function That Is Not Differentiable

Show that $f_x(0, 0)$ and $f_y(0, 0)$ both exist, but that f is not differentiable at $(0, 0)$ where f is defined as

$$f(x, y) = \begin{cases} \dfrac{-3xy}{x^2 + y^2}, & \text{if } (x, y) \neq (0, 0) \\ 0, & \text{if } (x, y) = (0, 0). \end{cases}$$

TECHNOLOGY   Try using a graphing utility to graph the function given in Example 5. For instance, the graph shown below was generated by *Mathematica*.

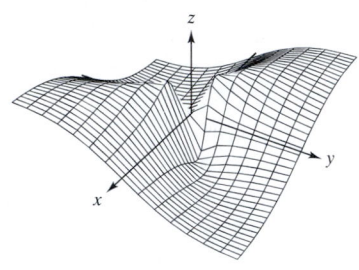

Solution   You can show that f is not differentiable at $(0, 0)$ by showing that it is not continuous at this point. To see that f is not continuous at $(0, 0)$, look at the values of $f(x, y)$ along two different approaches to $(0, 0)$, as shown in Figure 12.38. Along the line $y = x$, the limit is

$$\lim_{(x, x) \to (0, 0)} f(x, y) = \lim_{(x, x) \to (0, 0)} \frac{-3x^2}{2x^2} = -\frac{3}{2}$$

whereas along $y = -x$ you have

$$\lim_{(x, -x) \to (0, 0)} f(x, y) = \lim_{(x, -x) \to (0, 0)} \frac{3x^2}{2x^2} = \frac{3}{2}.$$

Thus, the limit of $f(x, y)$ as $(x, y) \to (0, 0)$ does not exist, and you can conclude that f is not continuous at $(0, 0)$. Therefore, by Theorem 12.5, you know that f is not differentiable at $(0, 0)$. On the other hand, by the definition of the partial derivatives f_x and f_y, you have

$$f_x(0, 0) = \lim_{\Delta x \to 0} \frac{f(\Delta x, 0) - f(0, 0)}{\Delta x} = \lim_{\Delta x \to 0} \frac{0 - 0}{\Delta x} = 0$$

and

$$f_y(0, 0) = \lim_{\Delta y \to 0} \frac{f(0, \Delta y) - f(0, 0)}{\Delta y} = \lim_{\Delta y \to 0} \frac{0 - 0}{\Delta y} = 0.$$

Thus, the partial derivatives at $(0, 0)$ exist.

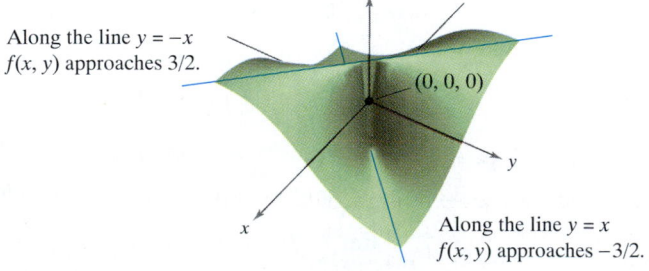

$$f(x, y) = \begin{cases} \dfrac{-3xy}{x^2 + y^2}, & (x, y) \neq (0, 0) \\ 0, & (x, y) = (0, 0) \end{cases}$$

Along the line $y = -x$
$f(x, y)$ approaches 3/2.

$(0, 0, 0)$

Along the line $y = x$
$f(x, y)$ approaches $-3/2$.

Figure 12.38

EXERCISES FOR SECTION 12.4

In Exercises 1–10, find the total differential.

1. $z = 3x^2y^3$

2. $z = \dfrac{x^2}{y}$

3. $z = \dfrac{-1}{x^2 + y^2}$

4. $z = e^x \sin y$

5. $z = x \cos y - y \cos x$

6. $z = \frac{1}{2}(e^{x^2+y^2} - e^{-x^2-y^2})$

7. $w = 2z^3y \sin x$

8. $w = e^x \cos y + z$

9. $w = \dfrac{x + y}{z - 2y}$

10. $w = x^2yz^2 + \sin yz$

In Exercises 11–16, (a) evaluate $f(1, 2)$ and $f(1.05, 2.1)$ and calculate Δz, and (b) use the total differential dz to approximate Δz.

11. $f(x, y) = 9 - x^2 - y^2$

12. $f(x, y) = \sqrt{x^2 + y^2}$

13. $f(x, y) = x \sin y$

14. $f(x, y) = xy$

15. $f(x, y) = 3x - 4y$

16. $f(x, y) = \dfrac{x}{y}$

In Exercises 17–20, find $z = f(x, y)$ and use the total differential to approximate the quantity.

17. $\sqrt{(5.05)^2 + (3.1)^2} - \sqrt{5^2 + 3^2}$

18. $(2.03)^2(1 + 8.9)^3 - 2^2(1 + 9)^3$

19. $\dfrac{1 - (3.05)^2}{(5.95)^2} - \dfrac{1 - 3^2}{6^2}$

20. $\sin[(1.05)^2 + (0.95)^2] - \sin(1^2 + 1^2)$

21. *Area*   The area of the rectangle in the figure is $A = lh$. Find dA and identify the regions in the figure whose areas are given by the terms of dA. What region represents the difference between ΔA and dA?

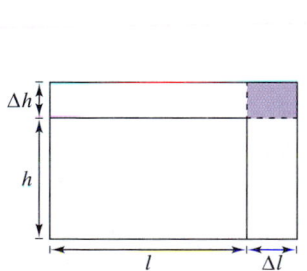
Figure for 21

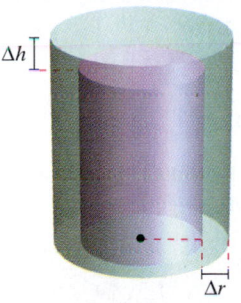
Figure for 22

22. *Volume*   The volume of the right circular cylinder in the figure is $V = \pi r^2 h$. Find dV and identify the solids in the figure whose volumes are given by the terms of dV. What solid represents the difference between ΔV and dV?

23. *Numerical Analysis*   A right circular cone of height $h = 6$ and radius $r = 3$ is constructed, and in the process errors Δr and Δh are made in the radius and height. Complete the table to show the relationship between ΔV and dV for the indicated errors.

Δr	Δh	dV or dS	ΔV or ΔS	$\Delta V - dV$ or $\Delta S - dS$
0.1	0.1			
0.1	−0.1			
0.001	0.002			
−0.0001	0.0002			

24. *Numerical Analysis*   The height and radius of a right circular cone are measured as $h = 20$ meters and $r = 8$ meters. In the process of measuring, errors Δr and Δh are made. If S is the lateral surface area of a cone, complete the table above to show the relationship between ΔS and dS for the indicated errors.

25. *Volume*   The radius r and height h of a right circular cylinder are measured with possible errors of 4% and 2%. Approximate the maximum possible percent error in measuring the volume.

26. *Rectangular to Polar Coordinates*   A rectangular coordinate system is placed over a map and the coordinates of a point of interest are $(8.5, 3.2)$. There is a possible error of 0.05 in each coordinate. Approximate the maximum possible error in measuring the polar coordinates for the point.

27. *Area*   A triangle is measured and two adjacent sides are found to be 3 and 4 inches long, with an included angle of $\pi/4$. The possible errors in measurements are $\frac{1}{16}$ inch in the sides and 0.02 radian in the angle. Approximate the maximum possible error in the computation of the area.

28. *Acceleration*   The centripetal acceleration of a particle moving in a circle is

$$a = \frac{v^2}{r}$$

where v is the velocity and r is the radius of the circle. Approximate the maximum percent error in measuring the acceleration due to errors of 2% in v and 1% in r.

29. *Power*   Electrical power P is given by

$$P = \frac{E^2}{R}$$

where E is voltage and R is resistance. Approximate the maximum percent error in calculating power if 200 volts is applied to a 4000-ohm resistor and the possible percent errors in measuring E and R are 2% and 3%.

30. Volume A trough is 16 feet long (see figure). Its cross sections are isosceles triangles; each of the two equal sides is 18 inches long. The angle between the two equal sides is θ.

(a) Express the volume of the trough as a function of θ and determine the value of θ such that the volume is a maximum.

(b) Approximate the change from the maximum volume if the maximum error in the linear measurements is one-half inch and the maximum error in the angle measure is $2°$.

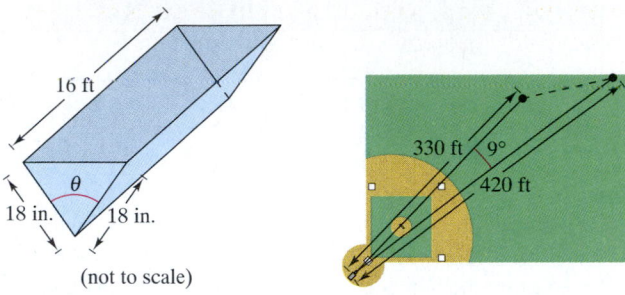

(not to scale)

Figure for 30   **Figure for 31**

31. Baseball A baseball player in center field is playing approximately 330 feet from a television camera that is behind home plate. A batter hits a fly ball that goes to a wall 420 feet from the camera (see figure).

(a) Approximate the number of feet that the center fielder had to run to make the catch if the camera turned $9°$ to follow the play.

(b) Approximate the maximum possible error in the result in part (a) if the position of the center fielder could be in error by as much as 6 feet and the maximum error in measuring the rotation of the camera is $1°$.

32. Resistance The total resistance R of two resistors connected in parallel is

$$\frac{1}{R} = \frac{1}{R_1} + \frac{1}{R_2}.$$

Approximate the change in R as R_1 is increased from 10 ohms to 10.5 ohms and R_2 is decreased from 15 ohms to 13 ohms.

33. Inductance The inductance L (in microhenrys) of a straight nonmagnetic wire in free space is

$$L = 0.00021\left(\ln\frac{2h}{r} - 0.75\right)$$

where h is the length of the wire in millimeters and r is the radius of a circular cross section. Approximate L when $r = 2 \pm \frac{1}{16}$ millimeters and $h = 100 \pm \frac{1}{100}$ millimeters.

34. Pendulum The period T of a pendulum of length L is $T = 2\pi\sqrt{L/g}$, where g is the acceleration due to gravity. A pendulum is moved from the Canal Zone, where $g = 32.09$ feet per second per second, to Greenland, where $g = 32.24$ feet per second per second. Because of the change in temperature, the length of the pendulum changes from 2.5 feet to 2.48 feet. Approximate the change in the period of the pendulum.

35. Think About It The figure shows a rectangle that is approximately $l = 6$ centimeters long and $h = 1$ centimeter high.

(a) Draw a rectangular strip along the rectangular region showing a small increase in length.

(b) Draw a rectangular strip along the rectangular region showing a small increase in height.

(c) Use the results in parts (a) and (b) to identify the measurement that has more effect on the area A of the rectangle.

(d) Verify your answer in part (c) analytically by finding dA if $dl = 0.01$ and $dh = 0.01$.

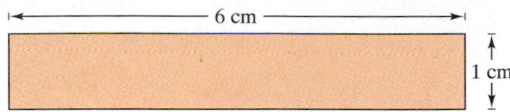

36. Think About It Consider converting a point $(5 \pm 0.05, \pi/18 \pm 0.05)$ in polar coordinates to rectangular coordinates (x, y).

(a) Use a geometric argument to determine whether the accuracy in x is more dependent on the accuracy in r or on the accuracy in θ. Explain. Verify your answer analytically.

(b) Use a geometric argument to determine whether the accuracy in y is more dependent on the accuracy in r or on the accuracy in θ. Explain. Verify your answer analytically.

37. Writing Use the results in Exercises 35 and 36 to write a short paragraph discussing the importance of making accurate measurements and identifying variables that have the greater effects in a formula being applied.

38. Interdisciplinary Problem Consider measurements and formulas you are using, or have used, in other science or engineering courses. Show how to apply differentials to these measurements and formulas to estimate possible propagated errors.

In Exercises 39–42, show that the function is differentiable by finding values for ε_1 and ε_2 as designated in the definition of differentiability, and verify that both ε_1 and $\varepsilon_2 \to 0$ as $(\Delta x, \Delta y) \to (0, 0)$.

39. $f(x, y) = x^2 - 2x + y$ **40.** $f(x, y) = x^2 + y^2$

41. $f(x, y) = x^2 y$ **42.** $f(x, y) = 5x - 10y + y^3$

In Exercises 43 and 44, use the function to prove that (a) $f_x(0, 0)$ and $f_y(0, 0)$ exist, and (b) f is not differentiable at $(0, 0)$.

43. $f(x, y) = \begin{cases} \dfrac{3x^2 y}{x^4 + y^2}, & (x, y) \neq (0, 0) \\ 0, & (x, y) = (0, 0) \end{cases}$

44. $f(x, y) = \begin{cases} \dfrac{2x^2 y^2}{x^4 + y^4}, & (x, y) \neq (0, 0) \\ 0, & (x, y) = (0, 0) \end{cases}$

SECTION 12.5 Chain Rules for Functions of Several Variables

Chain Rules for Functions of Several Variables •
Implicit Partial Differentiation

Chain Rules for Functions of Several Variables

Your work with differentials in the preceding section provides the basis for the extension of the Chain Rule to functions of two variables. There are two cases–the first case involves w as a function of x and y, where x and y are functions of a single independent variable t. (A proof of this theorem is given in the appendix.)

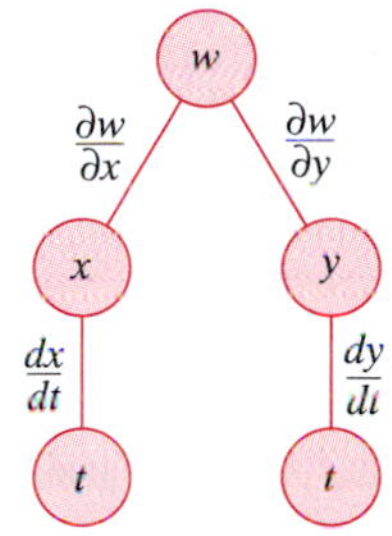

Chain Rule: one independent variable w is a function of x and y, which are each functions of t. This diagram represents the derivative of w with respect to t.
Figure 12.39

THEOREM 12.6 Chain Rule: One Independent Variable

Let $w = f(x, y)$, where f is a differentiable function of x and y. If $x = g(t)$ and $y = h(t)$, where g and h are differentiable functions of t, then w is a differentiable function of t, and

$$\frac{dw}{dt} = \frac{\partial w}{\partial x}\frac{dx}{dt} + \frac{\partial w}{\partial y}\frac{dy}{dt}.$$

NOTE The Chain Rule presented in this theorem can be represented schematically as shown in Figure 12.39.

EXAMPLE 1 Using the Chain Rule with One Independent Variable

Let $w = x^2y - y^2$, where $x = \sin t$ and $y = e^t$. Find dw/dt when $t = 0$.

Solution By the Chain Rule for one independent variable, you have

$$\begin{aligned}\frac{dw}{dt} &= \frac{\partial w}{\partial x}\frac{dx}{dt} + \frac{\partial w}{\partial y}\frac{dy}{dt}\\ &= 2xy(\cos t) + (x^2 - 2y)e^t.\end{aligned}$$

When $t = 0$, $x = 0$, and $y = 1$, it follows that

$$\frac{dw}{dt} = 0 - 2 = -2.$$

The Chain Rules presented in this section provide alternative techniques for solving many problems in single-variable calculus. For instance, in Example 1, you could have used single-variable techniques to find dw/dt by first writing w as a function of t,

$$\begin{aligned}w &= x^2y - y^2\\ &= (\sin t)^2(e^t) - (e^t)^2\\ &= e^t\sin^2 t - e^{2t}\end{aligned}$$

and then differentiating as usual. (Try doing this.)

The Chain Rule in Theorem 12.6 can be extended to any number of variables. For example, if each x_i is a differentiable function of a single variable t, then for

$$w = f(x_1, x_2, \ldots, x_n)$$

you have

$$\frac{dw}{dt} = \frac{\partial w}{\partial x_1}\frac{dx_1}{dt} + \frac{\partial w}{\partial x_2}\frac{dx_2}{dt} + \cdots + \frac{\partial w}{\partial x_n}\frac{dx_n}{dt}.$$

EXAMPLE 2   An Application of a Chain Rule to Related Rates

Two objects are traveling in elliptical paths given by the following parametric equations.

$$x_1 = 4\cos t \quad \text{and} \quad y_1 = 2\sin t \qquad \text{First object}$$
$$x_2 = 2\sin 2t \quad \text{and} \quad y_2 = 3\cos 2t \qquad \text{Second object}$$

At what rate is the distance between the two objects changing when $t = \pi$?

Solution   From Figure 12.40, you can see that the distance s between the two objects is given by

$$s = \sqrt{(x_2 - x_1)^2 + (y_2 - y_1)^2}$$

and when $t = \pi$, you have $x_1 = -4$, $y_1 = 0$, $x_2 = 0$, $y_2 = 3$, and

$$s = \sqrt{(0 + 4)^2 + (3 - 0)^2} = 5.$$

When $t = \pi$, the partial derivatives of s are as follows.

$$\frac{\partial s}{\partial x_1} = \frac{-(x_2 - x_1)}{\sqrt{(x_2 - x_1)^2 + (y_2 - y_1)^2}} = -\frac{1}{5}(0 + 4) = -\frac{4}{5}$$

$$\frac{\partial s}{\partial y_1} = \frac{-(y_2 - y_1)}{\sqrt{(x_2 - x_1)^2 + (y_2 - y_1)^2}} = -\frac{1}{5}(3 - 0) = -\frac{3}{5}$$

$$\frac{\partial s}{\partial x_2} = \frac{(x_2 - x_1)}{\sqrt{(x_2 - x_1)^2 + (y_2 - y_1)^2}} = \frac{1}{5}(0 + 4) = \frac{4}{5}$$

$$\frac{\partial s}{\partial y_2} = \frac{(y_2 - y_1)}{\sqrt{(x_2 - x_1)^2 + (y_2 - y_1)^2}} = \frac{1}{5}(3 - 0) = \frac{3}{5}$$

When $t = \pi$, the derivatives of x_1, y_1, x_2, and y_2 are

$$\frac{dx_1}{dt} = -4\sin t = 0 \qquad \frac{dy_1}{dt} = 2\cos t = -2$$

$$\frac{dx_2}{dt} = 4\cos 2t = 4 \qquad \frac{dy_2}{dt} = -6\sin 2t = 0.$$

Therefore, using the appropriate Chain Rule, you know that the distance is changing at the rate of

$$\frac{ds}{dt} = \frac{\partial s}{\partial x_1}\frac{dx_1}{dt} + \frac{\partial s}{\partial y_1}\frac{dy_1}{dt} + \frac{\partial s}{\partial x_2}\frac{dx_2}{dt} + \frac{\partial s}{\partial y_2}\frac{dy_2}{dt}$$

$$= \left(-\frac{4}{5}\right)(0) + \left(-\frac{3}{5}\right)(-2) + \left(\frac{4}{5}\right)(4) + \left(\frac{3}{5}\right)(0)$$

$$= \frac{22}{5}.$$

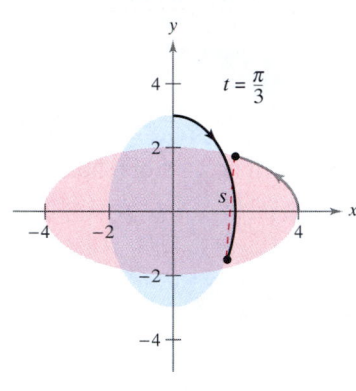
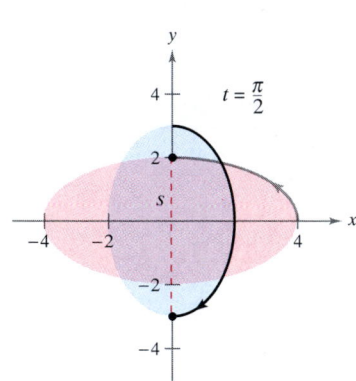
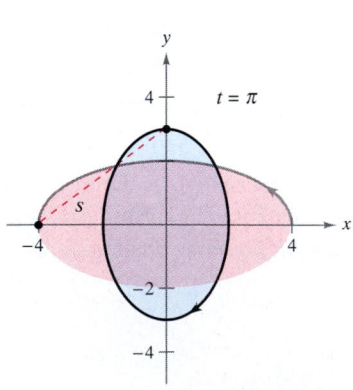

Paths of two objects traveling in elliptical orbits
Figure 12.40

In Example 2, note that s is the function of four *intermediate* variables, $x_1, y_1, x_2,$ and $y_2,$ which in turn are each functions of a single variable $t.$ Another type of composite function is one in which the intermediate variables are themselves functions of more than one variable. For instance, if $w = f(x, y),$ where $x = g(s, t)$ and $y = h(s, t),$ it follows that w is a function of s and $t,$ and you can consider the partial derivatives of w with respect to s and $t.$ One way to find these partial derivatives is to write w as a function of s and t explicitly by substituting the equations $x = g(s, t)$ and $y = h(s, t)$ into the equation $w = f(x, y).$ Then you can find the partial derivatives in the usual way, as demonstrated in the next example.

EXAMPLE 3   Finding Partial Derivatives by Substitution

Find $\partial w/\partial s$ and $\partial w/\partial t$ for $w = 2xy,$ where $x = s^2 + t^2$ and $y = s/t.$

Solution   Begin by substituting $x = s^2 + t^2$ and $y = s/t$ into the equation $w = 2xy$ to obtain

$$w = 2xy$$
$$= 2(s^2 + t^2)\left(\frac{s}{t}\right)$$
$$= 2\left(\frac{s^3}{t} + st\right).$$

Then, to find $\partial w/\partial s,$ hold t constant and differentiate with respect to $s.$

$$\frac{\partial w}{\partial s} = 2\left(\frac{3s^2}{t} + t\right) = \frac{6s^2 + 2t^2}{t}$$

Similarly, to find $\partial w/\partial t,$ hold s constant and differentiate with respect to t to obtain

$$\frac{\partial w}{\partial t} = 2\left(-\frac{s^3}{t^2} + s\right) = 2\left(\frac{-s^3 + st^2}{t^2}\right) = \frac{2st^2 - 2s^3}{t^2}.$$

Theorem 12.7 gives an alternative method for finding the partial derivatives in Example 3—without explicitly writing w as a function of s and $t.$

THEOREM 12.7   Chain Rule: Two Independent Variables

Let $w = f(x, y),$ where f is a differentiable function of x and $y.$ If $x = g(s, t)$ and $y = h(s, t)$ such that the first partials $\partial x/\partial s,$ $\partial x/\partial t,$ $\partial y/\partial s,$ and $\partial y/\partial t$ all exist, then $\partial w/\partial s$ and $\partial w/\partial t$ exist and are given by

$$\frac{\partial w}{\partial s} = \frac{\partial w}{\partial x}\frac{\partial x}{\partial s} + \frac{\partial w}{\partial y}\frac{\partial y}{\partial s} \qquad \text{and} \qquad \frac{\partial w}{\partial t} = \frac{\partial w}{\partial x}\frac{\partial x}{\partial t} + \frac{\partial w}{\partial y}\frac{\partial y}{\partial t}.$$

Proof   To obtain $\partial w/\partial s,$ hold t constant and apply Theorem 12.6 to obtain the desired result. Similarly, for $\partial w/\partial t$ hold s constant and apply Theorem 12.6.

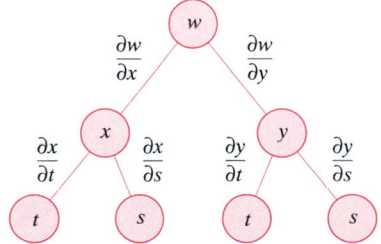

Chain Rule: two independent variables
Figure 12.41

NOTE   The Chain Rule in this theorem is shown schematically in Figure 12.41.

EXAMPLE 4 **The Chain Rule with Two Independent Variables**

Use the Chain Rule to find $\partial w/\partial s$ and $\partial w/\partial t$ for

$$w = 2xy$$

where $x = s^2 + t^2$ and $y = s/t$.

Solution (These same partials were found in Example 3.) Using Theorem 12.7, you can hold t constant and differentiate with respect to s to obtain

$$\frac{\partial w}{\partial s} = \frac{\partial w}{\partial x}\frac{\partial x}{\partial s} + \frac{\partial w}{\partial y}\frac{\partial y}{\partial s}$$

$$= 2y(2s) + 2x\left(\frac{1}{t}\right)$$

$$= 4\left(\frac{s^2}{t}\right) + \frac{2s^2 + 2t^2}{t} \qquad \text{Substitute } (s/t) \text{ for } y \text{ and } s^2 + t^2 \text{ for } x.$$

$$= \frac{6s^2 + 2t^2}{t}.$$

Similarly, holding s constant gives

$$\frac{\partial w}{\partial t} = \frac{\partial w}{\partial x}\frac{\partial x}{\partial t} + \frac{\partial w}{\partial y}\frac{\partial y}{\partial t}$$

$$= 2y(2t) + 2x\left(\frac{-s}{t^2}\right)$$

$$= 2\left(\frac{s}{t}\right)(2t) + 2(s^2 + t^2)\left(\frac{-s}{t^2}\right) \qquad \text{Substitute } (s/t) \text{ for } y \text{ and } s^2 + t^2 \text{ for } x.$$

$$= 4s - \frac{2s^3 + 2st^2}{t^2}$$

$$= \frac{4st^2 - 2s^3 - 2st^2}{t^2}$$

$$= \frac{2st^2 - 2s^3}{t^2}.$$

The Chain Rule in Theorem 12.7 can also be extended to any number of variables. For example, if w is a differentiable function of the n variables $x_1, x_2, \ldots, x_n$ where each x_i is a differentiable function of the m variables $t_1, t_2, \ldots, t_m$, then for

$$w = f(x_1, x_2, \ldots, x_n)$$

you obtain the following.

$$\frac{\partial w}{\partial t_1} = \frac{\partial w}{\partial x_1}\frac{\partial x_1}{\partial t_1} + \frac{\partial w}{\partial x_2}\frac{\partial x_2}{\partial t_1} + \cdots + \frac{\partial w}{\partial x_n}\frac{\partial x_n}{\partial t_1}$$

$$\frac{\partial w}{\partial t_2} = \frac{\partial w}{\partial x_1}\frac{\partial x_1}{\partial t_2} + \frac{\partial w}{\partial x_2}\frac{\partial x_2}{\partial t_2} + \cdots + \frac{\partial w}{\partial x_n}\frac{\partial x_n}{\partial t_2}$$

$$\vdots$$

$$\frac{\partial w}{\partial t_m} = \frac{\partial w}{\partial x_1}\frac{\partial x_1}{\partial t_m} + \frac{\partial w}{\partial x_2}\frac{\partial x_2}{\partial t_m} + \cdots + \frac{\partial w}{\partial x_n}\frac{\partial x_n}{\partial t_m}$$

EXAMPLE 5   The Chain Rule for a Function of Three Variables

Find $\partial w / \partial s$ and $\partial w / \partial t$ when $s = 1$ and $t = 2\pi$ for the function given by

$$w = xy + yz + xz$$

where $x = s \cos t$, $y = s \sin t$, and $z = t$.

Solution   By extending the result of Theorem 12.7, you have

$$\frac{\partial w}{\partial s} = \frac{\partial w}{\partial x}\frac{\partial x}{\partial s} + \frac{\partial w}{\partial y}\frac{\partial y}{\partial s} + \frac{\partial w}{\partial z}\frac{\partial z}{\partial s}$$

$$= (y + z)(\cos t) + (x + z)(\sin t) + (y + x)(0)$$

$$= (y + z)(\cos t) + (x + z)(\sin t).$$

When $s = 1$ and $t = 2\pi$, you have $x = 1$, $y = 0$, and $z = 2\pi$. Therefore, $\partial w / \partial s = 2\pi(1) + (1 + 2\pi)(0) + 0 = 2\pi$. Furthermore,

$$\frac{\partial w}{\partial t} = \frac{\partial w}{\partial x}\frac{\partial x}{\partial t} + \frac{\partial w}{\partial y}\frac{\partial y}{\partial t} + \frac{\partial w}{\partial z}\frac{\partial z}{\partial t}$$

$$= (y + z)(-s \sin t) + (x + z)(s \cos t) + (y + x)(1)$$

and for $s = 1$ and $t = 2\pi$ it follows that

$$\frac{\partial w}{\partial t} = (0 + 2\pi)(0) + (1 + 2\pi)(1) + (0 + 1)(1)$$

$$= 2 + 2\pi.$$

Implicit Partial Differentiation

We conclude this section with an application of the Chain Rule to determine the derivative of a function defined *implicitly*. Suppose that x and y are related by the equation $F(x, y) = 0$, where it is assumed that $y = f(x)$ is a differentiable function of x. To find dy/dx, you could use the techniques discussed in Section 2.5. However, you will see that the Chain Rule provides a convenient alternative. If you consider the function given by

$$w = F(x, y) = F(x, f(x))$$

you can apply Theorem 12.6 to obtain

$$\frac{dw}{dx} = F_x(x, y)\frac{dx}{dx} + F_y(x, y)\frac{dy}{dx}.$$

Because $w = F(x, y) = 0$ for all x in the domain of f, you know that $dw/dx = 0$ and you have

$$F_x(x, y)\frac{dx}{dx} + F_y(x, y)\frac{dy}{dx} = 0.$$

Now, if $F_y(x, y) \neq 0$, you can use the fact that $dx/dx = 1$ to conclude that

$$\frac{dy}{dx} = -\frac{F_x(x, y)}{F_y(x, y)}.$$

A similar procedure can be used to find the partial derivatives of functions of several variables that are defined implicitly.

THEOREM 12.8   Chain Rule: Implicit Differentiation

If the equation $F(x, y) = 0$ defines y implicitly as a differentiable function of x, then

$$\frac{dy}{dx} = -\frac{F_x(x, y)}{F_y(x, y)}, \qquad F_y(x, y) \neq 0.$$

If the equation $F(x, y, z) = 0$ defines z implicitly as a differentiable function of x and y, then

$$\frac{\partial z}{\partial x} = -\frac{F_x(x, y, z)}{F_z(x, y, z)} \quad \text{and} \quad \frac{\partial z}{\partial y} = -\frac{F_y(x, y, z)}{F_z(x, y, z)}, \qquad F_z(x, y, z) \neq 0.$$

This theorem can be extended to differentiable functions defined implicitly with any number of variables.

EXAMPLE 6   Finding a Derivative Implicitly

Find dy/dx, given $y^3 + y^2 - 5y - x^2 + 4 = 0$.

NOTE   Compare the solution of Example 6 with the solution of Example 2 in Section 2.5.

Solution   Begin by defining a function F as

$$F(x, y) = y^3 + y^2 - 5y - x^2 + 4.$$

Then, using Theorem 12.8, you have

$$F_x(x, y) = -2x \quad \text{and} \quad F_y(x, y) = 3y^2 + 2y - 5$$

and it follows that

$$\frac{dy}{dx} = -\frac{F_x(x, y)}{F_y(x, y)} = \frac{-(-2x)}{3y^2 + 2y - 5} = \frac{2x}{3y^2 + 2y - 5}.$$

EXAMPLE 7   Finding Partial Derivatives Implicitly

Find $\partial z/\partial x$ and $\partial z/\partial y$, given $3x^2z - x^2y^2 + 2z^3 + 3yz - 5 = 0$.

Solution   To apply Theorem 12.8, let

$$F(x, y, z) = 3x^2z - x^2y^2 + 2z^3 + 3yz - 5.$$

Then

$$F_x(x, y, z) = 6xz - 2xy^2$$
$$F_y(x, y, z) = -2x^2y + 3z$$
$$F_z(x, y, z) = 3x^2 + 6z^2 + 3y$$

and you obtain

$$\frac{\partial z}{\partial x} = -\frac{F_x}{F_z} = \frac{2xy^2 - 6xz}{3x^2 + 6z^2 + 3y}$$

$$\frac{\partial z}{\partial y} = -\frac{F_y}{F_z} = \frac{2x^2y - 3z}{3x^2 + 6z^2 + 3y}.$$

EXERCISES FOR SECTION 12.5

In Exercises 1–4, find dw/dt using the appropriate Chain Rule.

1. $w = x^2 + y^2$

$x = e^t, \; y = e^{-t}$

2. $w = \sqrt{x^2 + y^2}$

$x = \sin t, \; y = e^t$

3. $w = x \sec y$

$x = e^t, \; y = \pi - t$

4. $w = \ln \dfrac{y}{x}$

$x = \cos t, \; y = \sin t$

In Exercises 5–10, find dw/dt (a) using the appropriate Chain Rule and (b) by converting w to a function of t before differentiating.

5. $w = xy, \; x = 2 \sin t, \; y = \cos t$

6. $w = \cos(x - y), \; x = t^2, \; y = 1$

7. $w = x^2 + y^2 + z^2, \; x = e^t \cos t, \; y = e^t \sin t, \; z = e^t$

8. $w = xy \cos z, \; x = t, \; y = t^2, \; z = \arccos t$

9. $w = xy + xz + yz, \; x = t - 1, \; y = t^2 - 1, \; z = t$

10. $w = xyz, \; x = t^2, \; y = 2t, \; z = e^{-t}$

In Exercises 11 and 12, find d^2w/dt^2 using the appropriate Chain Rule. Evaluate d^2w/dt^2 at the given value of t.

11. $w = \arctan(2xy), \; x = \cos t, \; y = \sin t, \; t = 0$

12. $w = \dfrac{x^2}{y}, \; x = t^2, \; y = t + 1, \; t = 1$

In Exercises 13–16, find $\partial w/\partial s$ and $\partial w/\partial t$ using the appropriate Chain Rule, and evaluate each partial derivative at the indicated values of s and t.

Function	Point
13. $w = x^2 + y^2$	$s = 2, \; t = -1$
$x = s + t, \; y = s - t$	
14. $w = y^3 - 3x^2 y$	$s = 0, \; t = 1$
$x = e^s, \; y = e^t$	
15. $w = x^2 - y^2$	$s = 3, \; t = \dfrac{\pi}{4}$
$x = s \cos t, \; y = s \sin t$	
16. $w = \sin(2x + 3y)$	$s = 0, \; t = \dfrac{\pi}{2}$
$x = s + t, \; y = s - t$	

In Exercises 17–20, find $\partial w/\partial r$ and $\partial w/\partial \theta$ (a) using the appropriate Chain Rule and (b) by converting w to a function of r and θ before differentiating.

17. $w = x^2 - 2xy + y^2, \; x = r + \theta, \; y = r - \theta$

18. $w = \sqrt{4 - 2x^2 - 2y^2}, \; x = r \cos \theta, \; y = r \sin \theta$

19. $w = \arctan \dfrac{y}{x}, \; x = r \cos \theta, \; y = r \sin \theta$

20. $w = \dfrac{xy}{z}, \; x = r + \theta, \; y = r - \theta, \; z = \theta^2$

In Exercises 21–24, differentiate implicitly to find dy/dx.

21. $x^2 - 3xy + y^2 - 2x + y - 5 = 0$

22. $\sin x + \sec xy - 3 = 0$

23. $\ln \sqrt{x^2 + y^2} + xy = 4$

24. $\dfrac{x}{x^2 + y^2} - y^2 = 6$

In Exercises 25–32, differentiate implicitly to find the first partial derivatives of z.

25. $x^2 + y^2 + z^2 = 25$

26. $xz + yz + xy = 0$

27. $\tan(x + y) + \tan(y + z) = 1$

28. $z = e^x \sin(y + z)$

29. $x^2 + 2yz + z^2 = 1$

30. $x + \sin(y + z) = 0$

31. $e^{xz} + xy = 0$

32. $x \ln y + y^2 z + z^2 = 8$

In Exercises 33–36, differentiate implicitly to find the first partial derivatives of w.

33. $xyz + xzw - yzw + w^2 = 5$

34. $x^2 + y^2 + z^2 + 6xw - 8w^2 = 5$

35. $\cos xy + \sin yz + wz = 20$

36. $w - \sqrt{x - y} - \sqrt{y - z} = 0$

Homogeneous Functions **In Exercises 37–40, the function f is homogeneous of degree n if $f(tx, ty) = t^n f(x, y)$. Determine the degree of the homogeneous function, and show that**

$$xf_x(x, y) + yf_y(x, y) = nf(x, y).$$

37. $f(x, y) = \dfrac{xy}{\sqrt{x^2 + y^2}}$

38. $f(x, y) = x^3 - 3xy^2 + y^3$

39. $f(x, y) = e^{x/y}$

40. $f(x, y) = \dfrac{x^2}{\sqrt{x^2 + y^2}}$

41. ***Area*** Let θ be the angle between equal sides of an isosceles triangle and let x be the length of these sides. If x is increasing at $\frac{1}{2}$ meter per hour and θ is increasing at $\pi/90$ radian per hour, find the rate of increase of the area when $x = 6$ and $\theta = \pi/4$.

42. ***Volume and Surface Area*** The radius of a right circular cylinder is increasing at a rate of 6 inches per minute, and the height is decreasing at a rate of 4 inches per minute. What is the rate of change of the volume and surface area when the radius is 12 inches and the height is 36 inches?

43. ***Volume and Surface Area*** Repeat Exercise 42 for a right circular cone.

44. Volume and Surface Area   The two radii of the frustum of a right circular cone are increasing at a rate of 4 centimeters per minute, and the height is increasing at a rate of 12 centimeters per minute (see figure). Find the rate at which the volume and surface area are changing when the two radii are 15 centimeters and 25 centimeters, and the height is 10 centimeters.

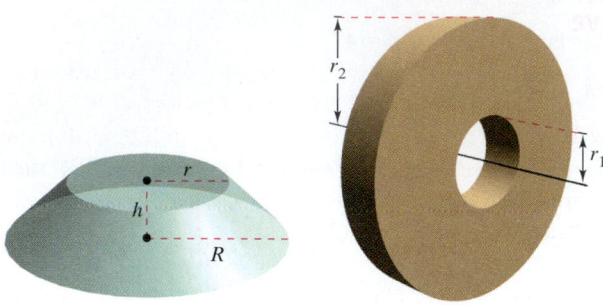

Figure for 44   **Figure for 45**

45. Moment of Inertia   An annular cylinder has an inside radius of r_1 and an outside radius of r_2 (see figure). Its moment of inertia is

$$I = \tfrac{1}{2}m(r_1^2 + r_2^2)$$

where m is the mass. Find the rate at which I is changing at the instant the radii are 6 centimeters and 8 centimeters if the two radii are increasing at a rate of 2 centimeters per second.

46. Ideal Gas Law   The Ideal Gas Law is $pV = mRT$, where R is a constant and m is a constant mass. If p and V are functions of time, find dT/dt, the rate at which the temperature changes with respect to time.

47. Projectile Motion   A projectile is launched at an angle of $45°$ with the horizontal and with an initial velocity of 64 feet per second. A television camera is located in the plane of the path of the projectile 50 feet behind the launch site (see figure).

(a) Find parametric equations for the path of the projectile in terms of the parameter t representing time.

(b) Express the angle α that the camera makes with the horizontal in terms of x and y and in terms of t.

(c) Use the results in part (b) to find $d\alpha/dt$.

(d) Use a graphing utility to obtain a graph of α in terms of t. Is the graph symmetric to the axis of the parabolic arch of the projectile? At what time is the rate of change of α greatest?

(e) At what time is the angle α maximum? Does this occur when the projectile is at its greatest height?

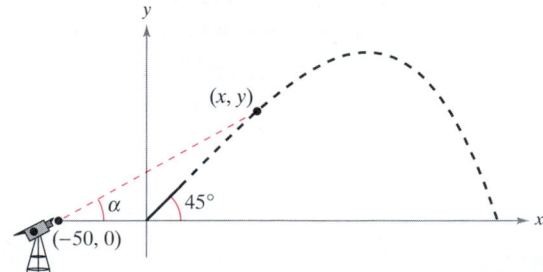

48. Projectile Motion   Consider the distance d between the launch site and the projectile in Exercise 47.

(a) Express the distance d in terms of x and y and in terms of the parameter t.

(b) Use the results in part (a) to find the rate of change of d.

(c) Find the rate of change of the distance when $t = 2$.

(d) When is the rate of change of d minimum during the flight of the projectile? Does it occur at the time the projectile reaches its maximum height?

49. Show that if $f(x, y)$ is homogeneous of degree n, then

$$xf_x(x, y) + yf_y(x, y) = nf(x, y).$$

[*Hint*: Let $g(t) = f(tx, ty) = t^n f(x, y)$. Find $g'(t)$ and then let $t = 1$.]

50. Show that

$$\frac{\partial w}{\partial u} + \frac{\partial w}{\partial v} = 0$$

for $w = f(x, y)$, $x = u - v$, and $y = v - u$.

51. Demonstrate the result in Exercise 50 for $w = (x - y) \sin(y - x)$.

52. Consider the function $w = f(x, y)$ where $x = r \cos \theta$ and $y = r \sin \theta$. Prove each of the following.

(a) $\dfrac{\partial w}{\partial x} = \dfrac{\partial w}{\partial r} \cos \theta - \dfrac{\partial w}{\partial \theta} \dfrac{\sin \theta}{r}$

$\dfrac{\partial w}{\partial y} = \dfrac{\partial w}{\partial r} \sin \theta + \dfrac{\partial w}{\partial \theta} \dfrac{\cos \theta}{r}$

(b) $\left(\dfrac{\partial w}{\partial x}\right)^2 + \left(\dfrac{\partial w}{\partial y}\right)^2 = \left(\dfrac{\partial w}{\partial r}\right)^2 + \left(\dfrac{1}{r^2}\right)\left(\dfrac{\partial w}{\partial \theta}\right)^2$

53. Demonstrate the result in Exercise 52(b) for $w = \arctan(y/x)$.

54. Cauchy-Riemann Equations   Given the functions $u(x, y)$ and $v(x, y)$, show that the **Cauchy-Riemann differential equations**

$$\frac{\partial u}{\partial x} = \frac{\partial v}{\partial y} \quad \text{and} \quad \frac{\partial u}{\partial y} = -\frac{\partial v}{\partial x}$$

can be written in polar coordinate form as

$$\frac{\partial u}{\partial r} = \frac{1}{r}\frac{\partial v}{\partial \theta} \quad \text{and} \quad \frac{\partial v}{\partial r} = -\frac{1}{r}\frac{\partial u}{\partial \theta}.$$

55. Demonstrate the result in Exercise 54 for the functions $u = \ln\sqrt{x^2 + y^2}$ and $v = \arctan\dfrac{y}{x}$.

56. Wave Equation   Show that

$$u(x, t) = \tfrac{1}{2}[f(x - ct) + f(x + ct)]$$

is a solution to the one-dimensional wave equation

$$\frac{\partial^2 u}{\partial t^2} = c^2 \frac{\partial^2 u}{\partial x^2}.$$

(This equation describes the small transverse vibration of an elastic string such as those on certain musical instruments.)

Directional Derivative • The Gradient of a Function of Two Variables •
Applications of the Gradient • Functions of Three Variables

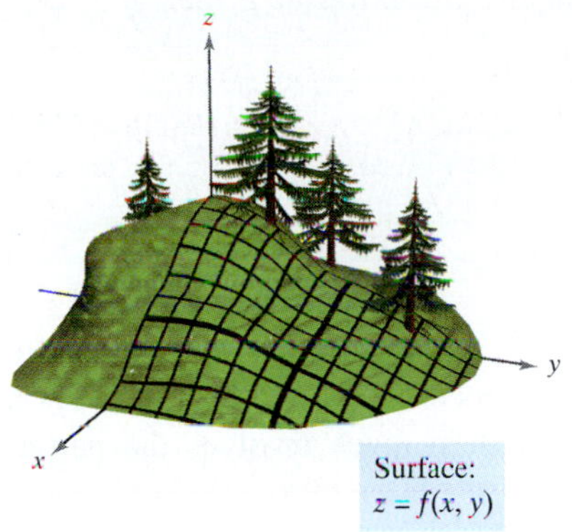

Directional Derivative

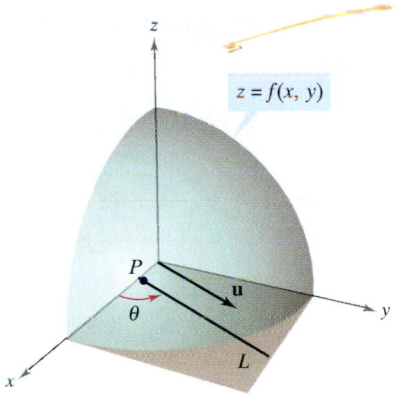

Surface:
$z = f(x, y)$

Figure 12.42

Suppose you are standing on the hillside pictured in Figure 12.42 and want to determine the hill's incline toward the z-axis. If the hill were represented by $z = f(x, y)$, you would already know how to determine the slopes in two different directions—the slope in the y-direction would be given by the partial derivative $f_y(x, y)$, and the slope in the x-direction would be given by the partial derivative $f_x(x, y)$. In this section, you will see that these two partial derivatives can be used to find the slope in *any* direction.

To determine the slope at a point on a surface, we define a new type of derivative called a **directional derivative.** We begin by letting $z = f(x, y)$ be a *surface* and $P(x_0, y_0)$ a *point* in the domain of f, as shown in Figure 12.43. The "direction" of the directional derivative is given by a unit vector

$$\mathbf{u} = \cos\theta\,\mathbf{i} + \sin\theta\,\mathbf{j}$$

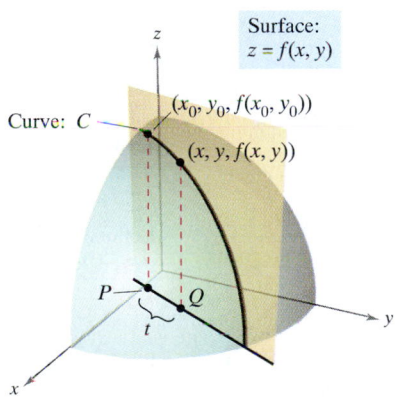

$z = f(x, y)$

P
θ
$\mathbf{u}$
L

Figure 12.43

where θ is the angle the vector makes with the positive x-axis. To find the desired slope, reduce the problem to two dimensions by intersecting the surface with a vertical plane passing through the point P and parallel to $\mathbf{u}$, as shown in Figure 12.44. This vertical plane intersects the surface to form a curve C. The slope of the surface at $(x_0, y_0, f(x_0, y_0))$ in the direction of $\mathbf{u}$ is defined as the slope of the curve C at that point.

Informally, you can write the slope of the curve C as a limit that looks much like those used in single-variable calculus. The vertical plane used to form C intersects the xy-plane in a line L, represented by the parametric equations

line in space.

$$x = x_0 + t\cos\theta$$

and

$$y = y_0 + t\sin\theta$$

so that for any value of t, the point $Q(x, y)$ lies on the line L. For each of the points P and Q, there is a corresponding point on the surface.

$(x_0, y_0, f(x_0, y_0))$ Point above P
$(x, y, f(x, y))$ Point above Q

Moreover, because the distance between P and Q is

$$\sqrt{(x - x_0)^2 + (y - y_0)^2} = \sqrt{(t\cos\theta)^2 + (t\sin\theta)^2} = |t|$$

you can write the slope of the secant line through $(x_0, y_0, f(x_0, y_0))$ and $(x, y, f(x, y))$ as

$$\frac{f(x, y) - f(x_0, y_0)}{t} = \frac{f(x_0 + t\cos\theta, y_0 + t\sin\theta) - f(x_0, y_0)}{t}.$$

Finally, by letting t approach 0, you arrive at the following definition.

Surface:
$z = f(x, y)$

Curve: C
$(x_0, y_0, f(x_0, y_0))$
$(x, y, f(x, y))$

P
Q
t

Figure 12.44

Definition of Directional Derivative

Let f be a function of two variables x and y and let $\mathbf{u} = \cos\theta\,\mathbf{i} + \sin\theta\,\mathbf{j}$ be a unit vector. Then the **directional derivative of f in the direction of u**, denoted by $D_{\mathbf{u}}f$, is

$$D_{\mathbf{u}}f(x, y) = \lim_{t \to 0} \frac{f(x + t\cos\theta, y + t\sin\theta) - f(x, y)}{t}$$

provided this limit exists.

Calculating directional derivatives by this definition is similar to finding the derivative of a function of one variable by the limit process (given in Section 2.1). A simpler "working" formula for finding directional derivatives involves the partial derivatives f_x and f_y.

THEOREM 12.9   Directional Derivative

If f is a differentiable function of x and y, then the directional derivatives of f in the direction of the unit vector $\mathbf{u} = \cos\theta\,\mathbf{i} + \sin\theta\,\mathbf{j}$ is

$$D_{\mathbf{u}}f(x, y) = f_x(x, y)\cos\theta + f_y(x, y)\sin\theta.$$

Proof   For a fixed point (x_0, y_0), let $x = x_0 + t\cos\theta$ and let $y = y_0 + t\sin\theta$. Then, let $g(t) = f(x, y)$. Because f is differentiable, you can apply the Chain Rule given in Theorem 12.7 to obtain

$$g'(t) = f_x(x, y)x'(t) + f_y(x, y)y'(t) = f_x(x, y)\cos\theta + f_y(x, y)\sin\theta.$$

If $t = 0$, then $x = x_0$ and $y = y_0$, so

$$g'(0) = f_x(x_0, y_0)\cos\theta + f_y(x_0, y_0)\sin\theta.$$

By the definition of $g'(t)$, it is also true that

$$g'(0) = \lim_{t \to 0} \frac{g(t) - g(0)}{t} = \lim_{t \to 0} \frac{f(x_0 + t\cos\theta, y_0 + t\sin\theta) - f(x_0, y_0)}{t}.$$

Consequently, $D_{\mathbf{u}}f(x_0, y_0) = f_x(x_0, y_0)\cos\theta + f_y(x_0, y_0)\sin\theta.$

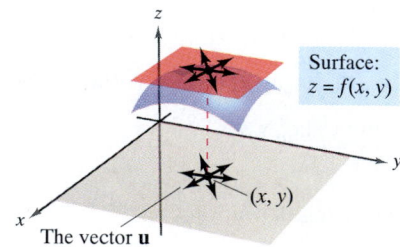

Surface:
$z = f(x, y)$

(x, y)

The vector **u**

Figure 12.45

There are infinitely many directional derivatives to a surface at a given point—one for each direction specified by $\mathbf{u}$, as indicated in Figure 12.45. Two of these are the partial derivatives f_x and f_y.

1. Direction of positive x-axis $(\theta = 0)$: $\mathbf{u} = \cos 0\,\mathbf{i} + \sin 0\,\mathbf{j} = \mathbf{i}$

$$D_{\mathbf{i}}f(x, y) = f_x(x, y)\cos 0 + f_y(x, y)\sin 0 = f_x(x, y)$$

2. Direction of positive y-axis $(\theta = \pi/2)$: $\mathbf{u} = \cos\frac{\pi}{2}\,\mathbf{i} + \sin\frac{\pi}{2}\,\mathbf{j} = \mathbf{j}$

$$D_{\mathbf{j}}f(x, y) = f_x(x, y)\cos\frac{\pi}{2} + f_y(x, y)\sin\frac{\pi}{2} = f_y(x, y)$$

EXAMPLE 1   Finding a Directional Derivative

Find the directional derivative of

$$f(x, y) = 4 - x^2 - \tfrac{1}{4}y^2 \qquad \text{Surface}$$

at $(1, 2)$ in the direction of

$$\mathbf{u} = \left(\cos\frac{\pi}{3}\right)\mathbf{i} + \left(\sin\frac{\pi}{3}\right)\mathbf{j}. \qquad \text{Direction}$$

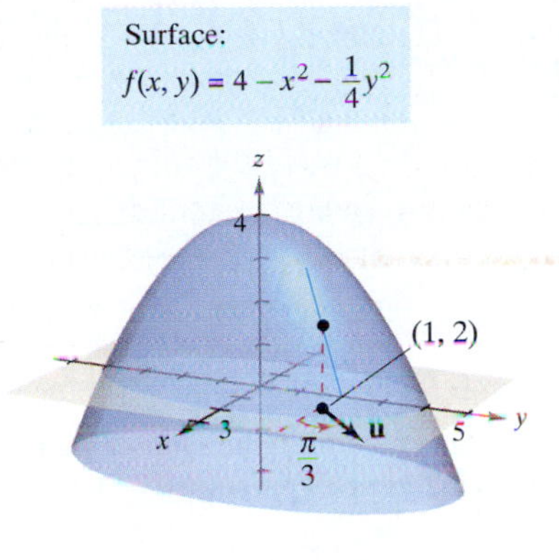

Solution Because f_x and f_y are continuous, f is differentiable, and you can apply Theorem 12.9.

$$D_{\mathbf{u}} f(x, y) = f_x(x, y) \cos\theta + f_y(x, y) \sin\theta$$

$$= (-2x)\cos\theta + \left(-\frac{y}{2}\right)\sin\theta$$

Evaluating at $\theta = \pi/3$, $x = 1$, and $y = 2$ produces

$$D_{\mathbf{u}} f(1, 2) = (-2)\left(\frac{1}{2}\right) + (-1)\left(\frac{\sqrt{3}}{2}\right)$$

$$= -1 - \frac{\sqrt{3}}{2}$$

$$\approx -1.866.$$

(See Figure 12.46.)

Surface:
$f(x, y) = 4 - x^2 - \tfrac{1}{4}y^2$

Figure 12.46

NOTE Note in Figure 12.46 that you can interpret the directional derivative as giving the slope of the surface at the point $(1, 2, 2)$ in the direction of the unit vector $\mathbf{u}$.

We have been specifying direction by a unit vector $\mathbf{u}$. If the direction is given by a vector whose length is not 1, we must normalize the vector before applying the formula in Theorem 12.9.

EXAMPLE 2   Finding a Directional Derivative

Find the directional derivative of

$$f(x, y) = x^2 \sin 2y \qquad \text{Surface}$$

at $(1, \pi/2)$ in the direction of

$$\mathbf{v} = 3\mathbf{i} - 4\mathbf{j}. \qquad \text{Direction}$$

Solution Because f_x and f_y are continuous, f is differentiable, and you can apply Theorem 12.9. Begin by finding a unit vector in the direction of $\mathbf{v}$.

$$\mathbf{u} = \frac{\mathbf{v}}{\|\mathbf{v}\|} = \frac{3}{5}\mathbf{i} - \frac{4}{5}\mathbf{j} = \cos\theta\,\mathbf{i} + \sin\theta\,\mathbf{j}$$

Using this unit vector, you have

$$D_{\mathbf{u}} f(x, y) = (2x \sin 2y)(\cos\theta) + (2x^2 \cos 2y)(\sin\theta)$$

$$D_{\mathbf{u}} f\left(1, \frac{\pi}{2}\right) = (2 \sin \pi)\left(\frac{3}{5}\right) + (2 \cos \pi)\left(-\frac{4}{5}\right)$$

$$= (0)\left(\frac{3}{5}\right) + (-2)\left(-\frac{4}{5}\right)$$

$$= \frac{8}{5}.$$

(See Figure 12.47.)

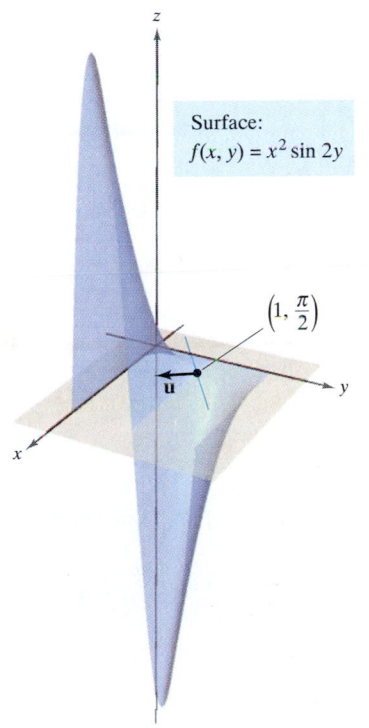

Surface:
$f(x, y) = x^2 \sin 2y$

Figure 12.47

The Gradient of a Function of Two Variables

The **gradient** of a function of two variables is a vector-valued function of two variables. This function has many important uses, some of which are described later in this section.

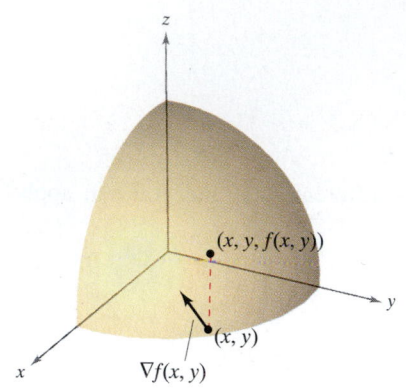

The gradient of f is a vector in the xy-plane.
Figure 12.48

Definition of Gradient of a Function of Two Variables

Let $z = f(x, y)$ be a function of x and y such that f_x and f_y exist. Then the **gradient of f,** denoted by $\nabla f(x, y)$ is the vector

$$\nabla f(x, y) = f_x(x, y)\mathbf{i} + f_y(x, y)\mathbf{j}.$$

We read ∇f as "del f." Another notation for the gradient is **grad** $f(x, y)$. In Figure 12.48, note that for each (x, y), the gradient $\nabla f(x, y)$ is a vector in the plane (not a vector in space).

NOTE No value is assigned to the symbol ∇ by itself. It is an operator in the same sense that d/dx is an operator. When ∇ operates on $f(x, y)$, it produces the vector $\nabla f(x, y)$.

EXAMPLE 3 Finding the Gradient of a Function

Find the gradient of $f(x, y) = y \ln x + xy^2$ at the point $(1, 2)$.

Solution Using $f_x(x, y) = \dfrac{y}{x} + y^2$ and $f_y(x, y) = \ln x + 2xy$, you have

$$\nabla f(x, y) = \left(\frac{y}{x} + y^2\right)\mathbf{i} + (\ln x + 2xy)\mathbf{j}.$$

At the point $(1, 2)$, the gradient is

$$\nabla f(1, 2) = \left(\frac{2}{1} + 2^2\right)\mathbf{i} + [\ln 1 + 2(1)(2)]\mathbf{j}$$
$$= 6\mathbf{i} + 4\mathbf{j}.$$

Because the gradient of f is a vector, you can write the directional derivative of f in the direction of $\mathbf{u}$ as

$$D_{\mathbf{u}} f(x, y) = [f_x(x, y)\mathbf{i} + f_y(x, y)\mathbf{j}] \cdot [\cos \theta \mathbf{i} + \sin \theta \mathbf{j}].$$

In other words, the directional derivative is the dot product of the gradient and the direction vector. This useful result is summarized in the following theorem.

THEOREM 12.10 Alternative Form of the Directional Derivative

If f is a differentiable function of x and y, then the directional derivative of f in the direction of the unit vector $\mathbf{u}$ is

$$D_{\mathbf{u}} f(x, y) = \nabla f(x, y) \cdot \mathbf{u}.$$

EXAMPLE 4   Using $\nabla f(x, y)$ to Find a Directional Derivative

Find the directional derivative of

$$f(x, y) = 3x^2 - 2y^2$$

at $\left(-\frac{3}{4}, 0\right)$ in the direction from $P\left(-\frac{3}{4}, 0\right)$ to $Q(0, 1)$.

Solution   Because the partials of f are continuous, f is differentiable and you can apply Theorem 12.10. A vector in the specified direction is

$$\overrightarrow{PQ} = \mathbf{v} = \left(0 + \frac{3}{4}\right)\mathbf{i} + (1 - 0)\mathbf{j}$$

$$= \frac{3}{4}\mathbf{i} + \mathbf{j}$$

and a unit vector in this direction is

$$\mathbf{u} = \frac{\mathbf{v}}{\|\mathbf{v}\|} = \frac{3}{5}\mathbf{i} + \frac{4}{5}\mathbf{j}. \qquad \text{Unit vector in direction of } \overrightarrow{PQ}$$

Because $\nabla f(x, y) = f_x(x, y)\mathbf{i} + f_y(x, y)\mathbf{j} = 6x\mathbf{i} - 4y\mathbf{j}$, the gradient at $\left(-\frac{3}{4}, 0\right)$ is

$$\nabla f\left(-\frac{3}{4}, 0\right) = -\frac{9}{2}\mathbf{i} + 0\mathbf{j}. \qquad \text{Gradient at } \left(-\frac{3}{4}, 0\right)$$

Consequently, at $\left(-\frac{3}{4}, 0\right)$ the directional derivative is

$$D_{\mathbf{u}} f\left(-\frac{3}{4}, 0\right) = \nabla f\left(-\frac{3}{4}, 0\right) \cdot \mathbf{u}$$

$$= \left(-\frac{9}{2}\mathbf{i} + 0\mathbf{j}\right) \cdot \left(\frac{3}{5}\mathbf{i} + \frac{4}{5}\mathbf{j}\right)$$

$$= -\frac{27}{10}. \qquad \text{Directional derivative at } \left(-\frac{3}{4}, 0\right)$$

(See Figure 12.49.)

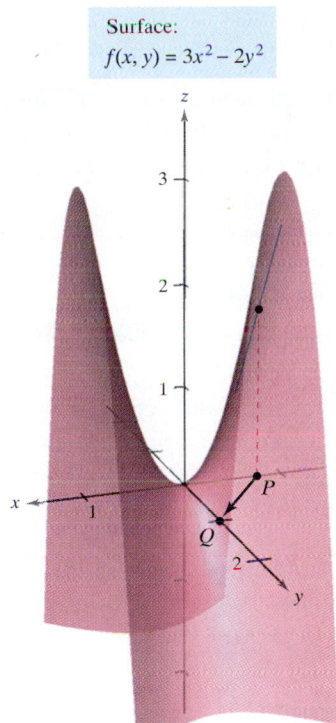

Surface:
$f(x, y) = 3x^2 - 2y^2$

Figure 12.49

Applications of the Gradient

You have already seen that there are many directional derivatives at the point (x, y) on a surface. In many applications we would like to know in which direction to move so that $f(x, y)$ increases most rapidly. This direction is called the direction of steepest ascent, and it is given by the gradient, as stated in the following theorem.

NOTE   Part 2 of Theorem 12.11 says that at the point (x, y), f increases most rapidly in the direction of the gradient, $\nabla f(x, y)$.

THEOREM 12.11   Properties of the Gradient

Let f be differentiable at the point (x, y).

1. If $\nabla f(x, y) = \mathbf{0}$, then $D_{\mathbf{u}} f(x, y) = 0$ for all $\mathbf{u}$.
2. The direction of *maximum* increase of f is given by $\nabla f(x, y)$. The maximum value of $D_{\mathbf{u}} f(x, y)$ is $\|\nabla f(x, y)\|$.
3. The direction of *minimum* increase of f is given by $-\nabla f(x, y)$. The minimum value of $D_{\mathbf{u}} f(x, y)$ is $-\|\nabla f(x, y)\|$.

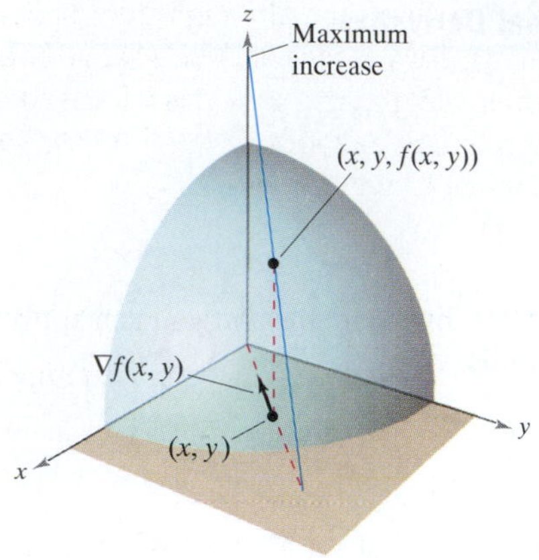

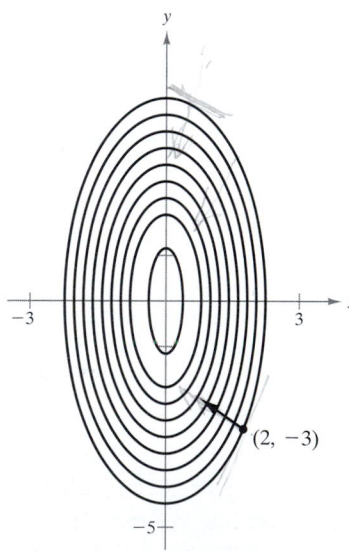

The gradient of f is a vector in the xy-plane that points in the direction of maximum increase on the surface given by $z = f(x, y)$.
Figure 12.50

Proof   If $\nabla f(x, y) = \mathbf{0}$, then for any direction (any $\mathbf{u}$), you have

$$D_{\mathbf{u}} f(x, y) = \nabla f(x, y) \cdot \mathbf{u} = (0\mathbf{i} + 0\mathbf{j}) \cdot (\cos \theta \mathbf{i} + \sin \theta \mathbf{j}) = 0.$$

If $\nabla f(x, y) \neq \mathbf{0}$, then let ϕ be the angle between $\nabla f(x, y)$ and a unit vector $\mathbf{u}$. Using the dot product, you can apply Theorem 10.5 to conclude that

$$\begin{aligned} D_{\mathbf{u}} f(x, y) &= \nabla f(x, y) \cdot \mathbf{u} \\ &= \|\nabla f(x, y)\| \, \|\mathbf{u}\| \cos \phi \\ &= \|\nabla f(x, y)\| \cos \phi \end{aligned}$$

and it follows that the maximum value of $D_{\mathbf{u}} f(x, y)$ will occur when $\cos \phi = 1$. Thus, $\phi = 0$, and the maximum value for the directional derivative occurs when $\mathbf{u}$ has the same direction as $\nabla f(x, y)$. Moreover, this largest value for $D_{\mathbf{u}} f(x, y)$ is precisely

$$\|\nabla f(x, y)\| \cos \phi = \|\nabla f(x, y)\|.$$

Similarly, the minimum value of $D_{\mathbf{u}} f(x, y)$ can be obtained by letting $\phi = \pi$ so that $\mathbf{u}$ points in the direction opposite that of $\nabla f(x, y)$, as indicated in Figure 12.50.

To visualize one of the properties of the gradient, imagine a skier coming down a mountainside. If fx, y denotes the altitude of the skier, then $-\nabla f(x, y)$ indicates the *compass direction* the skier should take to ski the path of steepest descent. (Remember that the gradient indicates direction in the xy-plane and does not itself point up or down the mountainside.)

As another illustration of the gradient, consider the temperature $T(x, y)$ at any point (x, y) on a flat metal plate. In this case, $\nabla T(x, y)$ gives the direction of greatest temperature increase at the point (x, y), as illustrated in the next example.

EXAMPLE 5   Finding the Direction of Maximum Increase

The temperature in degrees Celsius on the surface of a metal plate is

$$T(x, y) = 20 - 4x^2 - y^2$$

where x and y are measured in centimeters. In what direction from $(2, -3)$ does the temperature increase most rapidly? What is this rate of increase?

Solution   The gradient is

$$\begin{aligned} \nabla T(x, y) &= T_x(x, y)\mathbf{i} + T_y(x, y)\mathbf{j} \\ &= -8x\mathbf{i} - 2y\mathbf{j}. \end{aligned}$$

It follows that the direction of maximum increase is given by

$$\nabla T(2, -3) = -16\mathbf{i} + 6\mathbf{j}$$

as shown in Figure 12.51, and the rate of increase is

$$\begin{aligned} \|\nabla T(2, -3)\| &= \sqrt{256 + 36} \\ &= \sqrt{292} \\ &\approx 17.09° \text{ per centimeter.} \end{aligned}$$

Level curves:
$T(x, y) = 20 - 4x^2 - y^2$

The direction of most rapid increase in temperature at $(2, -3)$ is given by $-16\mathbf{i} + 6\mathbf{j}$.
Figure 12.51

The solution presented in Example 5 can be misleading. Although the gradient points in the direction of maximum temperature increase, it does not necessarily point toward the hottest spot on the plate. In other words, the gradient provides a local solution to finding an increase relative to the temperature at the point $(2, -3)$. *Once you leave that position, the direction of maximum increase may change.*

EXAMPLE 6   Finding the Path of a Heat-Seeking Particle

A heat-seeking particle is located at the point $(2, -3)$ on a metal plate whose temperature at (x, y) is

$$T(x, y) = 20 - 4x^2 - y^2.$$

Find the path of the particle as it continuously moves in the direction of maximum temperature increase.

Solution   Let the path be represented by the position function

$$\mathbf{r}(t) = x(t)\mathbf{i} + y(t)\mathbf{j}.$$

A tangent vector at each point $(x(t), y(t))$ is given by

$$\mathbf{r}'(t) = \frac{dx}{dt}\mathbf{i} + \frac{dy}{dt}\mathbf{j}.$$

Because the particle seeks maximum temperature increase, the directions of $\mathbf{r}'(t)$ and $\nabla T(x, y) = -8x\mathbf{i} - 2y\mathbf{j}$ are the same at each point on the path. Thus,

$$-8x = k\frac{dx}{dt} \quad \text{and} \quad -2y = k\frac{dy}{dt}$$

where k depends on t. By solving each equation for dt/k and equating the results, you obtain

$$\frac{dx}{-8x} = \frac{dy}{-2y}.$$

The solution of this differential equation is $x = Cy^4$. Because the particle starts at the point $(2, -3)$, you can determine that $C = 2/81$. Thus, the path of the heat-seeking particle is

$$x = \frac{2}{81}y^4.$$

The path is shown in Figure 12.52.

Level curves:
$T(x, y) = 20 - 4x^2 - y^2$

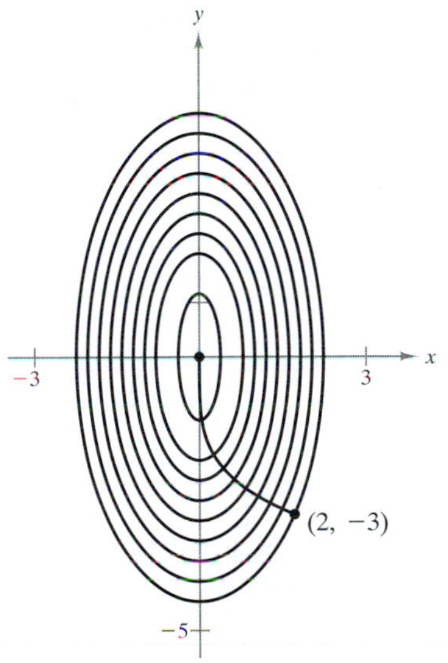

$(2, -3)$

Path followed by a heat-seeking particle
Figure 12.52

In Figure 12.52, the path of the particle (determined by the gradient at each point) appears to be orthogonal to each of the level curves. This becomes clear when you consider that the temperature $T(x, y)$ is constant along a given level curve. Hence, at any point (x, y) on the curve, the rate of change of T in the direction of a unit tangent vector $\mathbf{u}$ is 0, and you can write

$$\nabla f(x, y) \cdot \mathbf{u} = D_{\mathbf{u}}T(x, y) = 0. \qquad \text{u is a unit tangent vector.}$$

Because the dot product of $\nabla f(x, y)$ and $\mathbf{u}$ is 0, you can conclude that they must be orthogonal. This result is stated in the following theorem.

THEOREM 12.12   Gradient Is Normal to Level Curves

If f is differentiable at (x_0, y_0) and $\nabla f(x_0, y_0) \neq \mathbf{0}$, then $\nabla f(x_0, y_0)$ is normal to the level curve through (x_0, y_0).

EXAMPLE 7   Finding a Normal Vector to a Level Curve

Sketch the level curve corresponding to $c = 0$ for the function given by

$$f(x, y) = y - \sin x$$

and find a normal vector at several points on the curve.

Solution   The level curve for $c = 0$ is given by

$$0 = y - \sin x$$
$$y = \sin x$$

as shown in Figure 12.53(a). Because the gradient vector of f at (x, y) is

$$\nabla f(x, y) = f_x(x, y)\mathbf{i} + f_y(x, y)\mathbf{j}$$
$$= -\cos x\mathbf{i} + \mathbf{j}$$

you can use Theorem 12.12 to conclude that $\nabla f(x, y)$ is normal to the level curve at the point (x, y). Some gradient vectors are

$$\nabla f(-\pi, 0) = \mathbf{i} + \mathbf{j}$$
$$\nabla f\left(-\frac{2\pi}{3}, -\frac{\sqrt{3}}{2}\right) = \frac{1}{2}\mathbf{i} + \mathbf{j}$$
$$\nabla f\left(-\frac{\pi}{2}, -1\right) = \mathbf{j}.$$

Several others are shown in Figure 12.53(b).

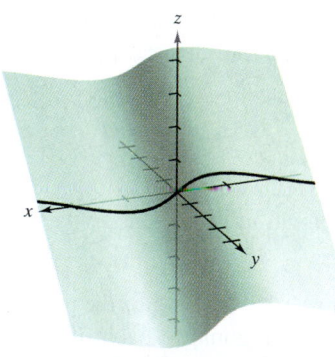

(a) The surface is given by $f(x, y) = y - \sin x$.

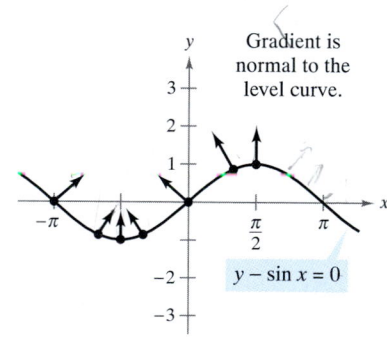

Gradient is normal to the level curve.

$$y - \sin x = 0$$

(b) The level curve is given by $f(x, y) = 0$.

Figure 12.53

Functions of Three Variables

The definitions of the directional derivative and the gradient can be extended naturally to functions of three or more variables. As often happens, some of the geometric interpretation is lost in the generalization from functions of two variables to those of three variables. For example, you cannot interpret the directional derivative of a function of three variables to represent slope.

The definitions and properties of the directional derivative and the gradient of a function of three variables are given in the following summary.

NOTE   You can generalize Theorem 12.12 to functions of three variables. Under suitable hypotheses,

$$\nabla f(x_0, y_0, z_0)$$

is normal to the *level surface* through (x_0, y_0, z_0).

Directional Derivative and Gradient for Three Variables

Let f be a function of x, y, and z, with continuous first partial derivatives. The **directional derivative of f** in the direction of a unit vector $\mathbf{u} = a\mathbf{i} + b\mathbf{j} + c\mathbf{k}$ is given by

$$D_{\mathbf{u}} f(x, y, z) = af_x(x, y, z) + bf_y(x, y, z) + cf_z(x, y, z).$$

The **gradient of f** is defined to be

$$\nabla f(x, y, z) = f_x(x, y, z)\mathbf{i} + f_y(x, y, z)\mathbf{j} + f_z(x, y, z)\mathbf{k}.$$

Properties of the gradient are as follows.

1. $D_{\mathbf{u}} f(x, y, z) = \nabla f(x, y, z) \cdot \mathbf{u}$
2. If $\nabla f(x, y, z) = \mathbf{0}$, then $D_{\mathbf{u}} f(x, y, z) = 0$ for all $\mathbf{u}$.
3. The direction of *maximum* increase of f is given by $\nabla f(x, y, z)$. The maximum value of $D_{\mathbf{u}} f(x, y, z)$ is

    $\|\nabla f(x, y, z)\|.$ Maximum value of $D_{\mathbf{u}} f(x, y, z)$

4. The direction of *minimum* increase of f is given by $-\nabla f(x, y, z)$. The minimum value of $D_{\mathbf{u}} f(x, y, z)$ is

    $-\|\nabla f(x, y, z)\|.$ Minimum value of $D_{\mathbf{u}} f(x, y, z)$

EXAMPLE 8   Finding the Gradient for a Function of Three Variables

Find $\nabla f(x, y, z)$ for the function given by

$$f(x, y, z) = x^2 + y^2 - 4z$$

and find the direction of maximum increase of f at the point $(2, -1, 1)$.

Solution   The gradient vector is given by

$$\begin{aligned}
\nabla f(x, y, z) &= f_x(x, y, z)\mathbf{i} + f_y(x, y, z)\mathbf{j} + f_z(x, y, z)\mathbf{k} \\
&= 2x\mathbf{i} + 2y\mathbf{j} - 4\mathbf{k}.
\end{aligned}$$

Hence, it follows that the direction of maximum increase at $(2, -1, 1)$ is

$$\nabla f(2, -1, 1) = 4\mathbf{i} - 2\mathbf{j} - 4\mathbf{k}.$$

In Exercises 1–12, find the directional derivative of the function at *P* in the direction of **v**.

1. $f(x, y) = 3x - 4xy + 5y$,
$P(1, 2)$, $\mathbf{v} = \frac{1}{2}(\mathbf{i} + \sqrt{3}\mathbf{j})$

2. $f(x, y) = x^2 - y^2$, $P(4, 3)$, $\mathbf{v} = \frac{\sqrt{2}}{2}(\mathbf{i} + \mathbf{j})$

3. $f(x, y) = xy$, $P(2, 3)$, $\mathbf{v} = \mathbf{i} + \mathbf{j}$

4. $f(x, y) = \frac{x}{y}$, $P(1, 1)$, $\mathbf{v} = -\mathbf{j}$

5. $g(x, y) = \sqrt{x^2 + y^2}$, $P(3, 4)$, $\mathbf{v} = 3\mathbf{i} - 4\mathbf{j}$

6. $g(x, y) = \arcsin xy$, $P(1, 0)$, $\mathbf{v} = \mathbf{i} + 5\mathbf{j}$

7. $h(x, y) = e^x \sin y$, $P\left(1, \frac{\pi}{2}\right)$, $\mathbf{v} = -\mathbf{i}$

8. $h(x, y) = e^{-(x^2 + y^2)}$, $P(0, 0)$, $\mathbf{v} = \mathbf{i} + \mathbf{j}$

9. $f(x, y, z) = xy + yz + xz$,
$P(1, 1, 1)$, $\mathbf{v} = 2\mathbf{i} + \mathbf{j} - \mathbf{k}$

10. $f(x, y, z) = x^2 + y^2 + z^2$,
$P(1, 2, -1)$, $\mathbf{v} = \mathbf{i} - 2\mathbf{j} + 3\mathbf{k}$

11. $h(x, y, z) = x \arctan yz$,
$P(4, 1, 1)$, $\mathbf{v} = \langle 1, 2, -1 \rangle$

12. $h(x, y, z) = xyz$, $P(2, 1, 1)$, $\mathbf{v} = \langle 2, 1, 2 \rangle$

In Exercises 13–16, find the directional derivative of the function in the direction of $\mathbf{u} = \cos \theta \mathbf{i} + \sin \theta \mathbf{j}$.

13. $f(x, y) = x^2 + y^2$, $\theta = \frac{\pi}{4}$

14. $f(x, y) = \frac{y}{x + y}$, $\theta = -\frac{\pi}{6}$

15. $f(x, y) = \sin(2x - y)$, $\theta = -\frac{\pi}{3}$

16. $g(x, y) = xe^y$, $\theta = \frac{2\pi}{3}$

In Exercises 17–20, find the directional derivative of the function at *P* in the direction of *Q*.

17. $f(x, y) = x^2 + 4y^2$
$P(3, 1)$, $Q(1, -1)$

18. $f(x, y) = \cos(x + y)$
$P(0, \pi)$, $Q\left(\frac{\pi}{2}, 0\right)$

19. $h(x, y, z) = \ln(x + y + z)$
$P(1, 0, 0)$, $Q(4, 3, 1)$

20. $g(x, y, z) = xye^z$
$P(2, 4, 0)$, $Q(0, 0, 0)$

Investigation   In Exercises 21 and 22, (a) use the graph to estimate the components of the vector in the direction of the maximum rate of increase in the function at the indicated point. (b) Find the gradient at the point and compare it with your estimate in part (a). (c) In what direction would the function be decreasing at the greatest rate? Explain.

21. $f(x, y) = \frac{1}{10}(x^2 - 3xy + y^2)$, $(1, 2)$

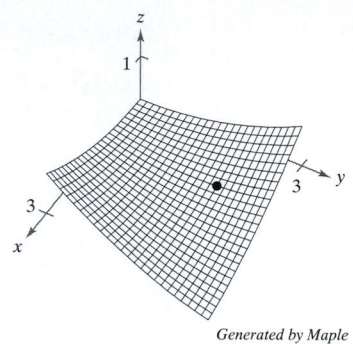

Generated by Maple

22. $f(x, y) = \frac{1}{2}y\sqrt{x}$, $(1, 2)$

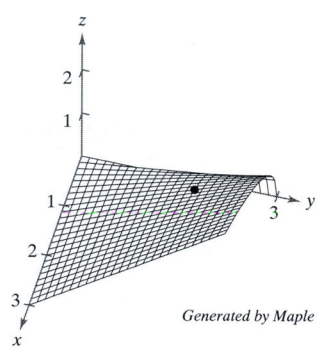

Generated by Maple

23. *Investigation*   Consider the function
$$f(x, y) = x^2 - y^2$$
at the point $(4, -3, 7)$.

(a) Use a computer algebra system to graph the surface represented by the function.

(b) Determine the directional derivative $D_{\mathbf{u}} f(4, -3)$ as a function of θ where $\mathbf{u} = \cos \theta \mathbf{i} + \sin \theta \mathbf{j}$. Use a computer algebra system to graph the function on the interval $[0, 2\pi]$.

(c) Approximate the zeros of the function in part (b) and interpret each in the context of the problem.

(d) Approximate the critical numbers of the function in part (b) and interpret each in the context of the problem.

(e) Find $\|\nabla f(4, -3)\|$ and explain its relationship to the answers in part (d).

(f) Use a computer algebra system to graph the level curve of the function f at the level $c = 7$. On this curve, graph the vector in the direction of $\nabla f(4, -3)$, and state its relationship to the level curve.

24. *Investigation* The figure below gives the level curve of the function

$$f(x, y) = \frac{8y}{1 + x^2 + y^2}$$

at the level $c = 2$.

(a) Analytically verify that the curve is a circle.

(b) At the point $(\sqrt{3}, 2)$ on the level curve, sketch the vector showing the direction of the greatest rate of increase of the function.

(c) At the point $(\sqrt{3}, 2)$ on the level curve, sketch the vector such that the directional derivative is 0.

(d) Use a computer algebra system to graph the surface to verify your answer in parts (a) to (c).

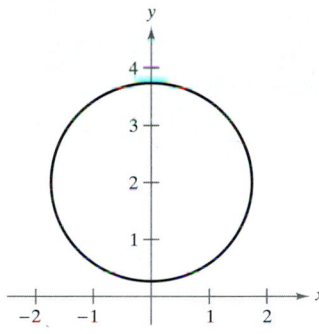

In Exercises 25–32, find the gradient of the function and the maximum value of the directional derivative at the indicated point.

Function	Point
25. $h(x, y) = x \tan y$	$\left(2, \dfrac{\pi}{4}\right)$
26. $h(x, y) = y \cos(x - y)$	$\left(0, \dfrac{\pi}{3}\right)$
27. $g(x, y) = \ln \sqrt[3]{x^2 + y^2}$	$(1, 2)$
28. $g(x, y) = ye^{-x^2}$	$(0, 5)$
29. $f(x, y, z) = \sqrt{x^2 + y^2 + z^2}$	$(1, 4, 2)$
30. $f(x, y, z) = xe^{yz}$	$(2, 0, -4)$
31. $w = \dfrac{1}{\sqrt{1 - x^2 - y^2 - z^2}}$	$(0, 0, 0)$
32. $w = xy^2z^2$	$(2, 1, 1)$

In Exercises 33–40, use the function

$$f(x, y) = 3 - \frac{x}{3} - \frac{y}{2}.$$

33. Sketch the graph of f in the first octant and plot the point $(3, 2, 1)$ on the surface.

34. Find $D_{\mathbf{u}} f(3, 2)$ where $\mathbf{u} = \cos\theta\mathbf{i} + \sin\theta\mathbf{j}$.

(a) $\theta = \dfrac{\pi}{4}$ (b) $\theta = \dfrac{2\pi}{3}$

35. Find $D_{\mathbf{u}} f(3, 2)$ where $\mathbf{u} = \cos\theta\mathbf{i} + \sin\theta\mathbf{j}$.

(a) $\theta = \dfrac{4\pi}{3}$ (b) $\theta = -\dfrac{\pi}{6}$

36. Find $D_{\mathbf{u}} f(3, 2)$ where $\mathbf{u} = \mathbf{v}/\|\mathbf{v}\|$.

(a) $\mathbf{v} = \mathbf{i} + \mathbf{j}$ (b) $\mathbf{v} = -3\mathbf{i} - 4\mathbf{j}$

37. Find $D_{\mathbf{u}} f(3, 2)$ where $\mathbf{u} = \mathbf{v}/\|\mathbf{v}\|$.

(a) $\mathbf{v}$ is the vector from $(1, 2)$ to $(-2, 6)$.

(b) $\mathbf{v}$ is the vector from $(3, 2)$ to $(4, 5)$.

38. Find $\nabla f(x, y)$.

39. Find the maximum value of the directional derivative at $(3, 2)$.

40. Find a unit vector $\mathbf{u}$ orthogonal to $\nabla f(3, 2)$ and calculate $D_{\mathbf{u}} f(3, 2)$. Discuss the geometric meaning of the result.

In Exercises 41–44, use the function

$$f(x, y) = 9 - x^2 - y^2.$$

41. Sketch the graph of f in the first octant and plot the point $(1, 2, 4)$ on the surface.

42. Find $D_{\mathbf{u}} f(1, 2)$ where $\mathbf{u} = \cos\theta\mathbf{i} + \sin\theta\mathbf{j}$.

(a) $\theta = -\dfrac{\pi}{4}$ (b) $\theta = \dfrac{\pi}{3}$

43. Find $\nabla f(1, 2)$ and $\|\nabla f(1, 2)\|$.

44. Find a unit vector $\mathbf{u}$ orthogonal to $\nabla f(1, 2)$ and calculate $D_{\mathbf{u}} f(1, 2)$. Discuss the geometric meaning of the result.

In Exercises 45–48, find a normal vector to the level curve $f(x, y) = c$ at P.

45. $f(x, y) = x^2 + y^2$
$c = 25, \quad P(3, 4)$

46. $f(x, y) = 6 - 2x - 3y$
$c = 6, \quad P(0, 0)$

47. $f(x, y) = \dfrac{x}{x^2 + y^2}$
$c = \frac{1}{2}, \quad P(1, 1)$

48. $f(x, y) = xy$
$c = -3, \quad P(-1, 3)$

In Exercises 49–52, use the gradient to find a unit normal vector to the graph of the equation at the indicated point. Sketch your results.

Equation	Point
49. $4x^2 - y = 6$	$(2, 10)$
50. $3x^2 - 2y^2 = 1$	$(1, 1)$
51. $9x^2 + 4y^2 = 40$	$(2, -1)$
52. $xe^y - y = 5$	$(5, 0)$

53. *Temperature Distribution* The temperature at the point (x, y) on a metal plate is

$$T = \frac{x}{x^2 + y^2}.$$

Find the direction of greatest increase in heat from the point $(3, 4)$.

54. *Topography* The surface of a mountain is modeled by the equation

$$h(x, y) = 4000 - 0.001x^2 - 0.004y^2.$$

Suppose that a mountain climber is at the point (500, 300, 3390). In what direction should the climber move in order to ascend at the greatest rate?

Heat-Seeking Path In Exercises 55 and 56, find the path followed by a heat-seeking particle placed at point P on a metal plate with a temperature field $T(x, y)$.

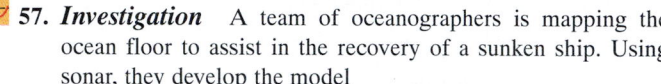

Temperature Field	Point
55. $T(x, y) = 400 - 2x^2 - y^2$	$P(10, 10)$
56. $T(x, y) = 50 - x^2 - 2y^2$	$P(4, 3)$

57. *Investigation* A team of oceanographers is mapping the ocean floor to assist in the recovery of a sunken ship. Using sonar, they develop the model

$$D = 250 + 30x^2 + 50 \sin \frac{\pi y}{2}, \quad 0 \le x \le 2, 0 \le y \le 2$$

where D is the depth in meters, and x and y are the distances in kilometers.

(a) Use a computer algebra system to graph the surface.

(b) Because the graph in part (a) is showing depth, it is not a map of the ocean floor. How could the model be changed so that the graph of the ocean floor could be obtained?

(c) What is the depth of the ship if it is located at the coordinates $x = 1$ and $y = 0.5$?

(d) Determine the steepness of the ocean floor in the positive x-direction from the position of the ship.

(e) Determine the steepness of the ocean floor in the positive y-direction from the position of the ship.

(f) Determine the direction of the greatest rate of change of depth from the position of the ship.

58. *Temperature* The temperature at the point (x, y) on a metal plate is modeled by

$$T(x, y) = 400e^{-(x^2+y)/2}, \quad x \ge 0, y \ge 0.$$

(a) Use a computer algebra system to graph the temperature distribution function.

(b) Find the directions of no change in heat on the plate from the point (3, 5).

(c) Find the direction of greatest increase in heat from the point (3, 5).

59. *Meteorology* Meteorologists measure the atmospheric pressure in units called millibars. From these observations they create weather maps on which the curves of equal atmospheric pressure (isobars) are drawn (see figure). These are level curves to the function $P(x, y)$ yielding the pressure at any point. Sketch the gradients to the isobars at the points A, B, and C. Although the magnitudes of the gradients are unknown, their lengths relative to each other can be estimated. At which of the three points is the wind speed greatest if the speed increases as the pressure gradient increases?

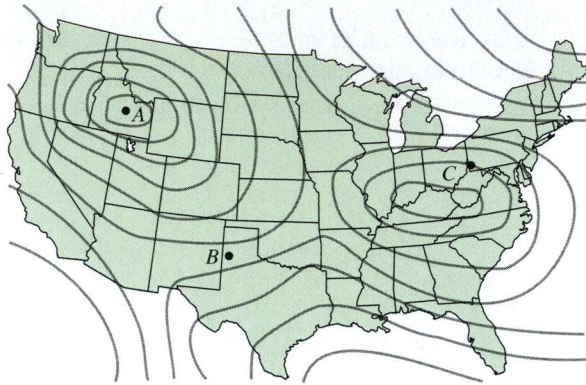

60. *Writing* Write a paragraph describing the directional derivative of the function f in the direction of $\mathbf{u} = \cos \theta \mathbf{i} + \sin \theta \mathbf{j}$ if (a) $\theta = 0°$ and (b) $\theta = 90°$.

True or False? In Exercises 61–64, determine whether the statement is true or false. If it is false, explain why or give an example that shows it is false.

61. If $f(x, y) = \sqrt{1 - x^2 - y^2}$, then $D_{\mathbf{u}} f(0, 0) = 0$ for any unit vector $\mathbf{u}$.

62. If $f(x, y) = x + y$, then $-1 \le D_{\mathbf{u}} f(x, y) \le 1$.

63. If $D_{\mathbf{u}} f(x, y)$ exists, then $D_{\mathbf{u}} f(x, y) = -D_{-\mathbf{u}} f(x, y)$.

64. If $D_{\mathbf{u}} f(x_0, y_0) = c$ for any unit vector $\mathbf{u}$, then $c = 0$.

65. Find a function f such that

$$\nabla f = e^x \cos y \mathbf{i} - e^x \sin y \mathbf{j} + z \mathbf{k}.$$

66. Let $f(x, y) = \begin{cases} \dfrac{4xy}{x^2 + y^2}, & (x, y) \ne (0, 0) \\ 0, & (x, y) = (0, 0). \end{cases}$

Let $\mathbf{u} = (1/\sqrt{2})(\mathbf{i} + \mathbf{j})$. Does the directional derivative of f at $P(0, 0)$ in the direction of $\mathbf{u}$ exist? If $f(0, 0)$ were defined as 2 instead of 0, would $\mathbf{u}$ exist?

SECTION **12.7**   **Tangent Planes and Normal Lines**

Tangent Plane and Normal Line to a Surface • The Angle of Inclination of a Plane • A Comparison of the Gradients $\nabla f(x, y)$ and $\nabla F(x, y, z)$

Tangent Plane and Normal Line to a Surface

So far we have represented surfaces in space primarily by equations of the form

$$z = f(x, y). \qquad \text{Equation of a surface } S$$

In the development to follow, however, it is convenient to use the more general representation $F(x, y, z) = 0$. For a surface S given by $z = f(x, y)$, you can convert to the general form by defining F as

$$F(x, y, z) = f(x, y) - z.$$

Because $f(x, y) - z = 0$, you can consider S to be the level surface of F given by

$$F(x, y, z) = 0. \qquad \text{Alternative equation of surface } S$$

EXAMPLE 1   Writing an Equation of a Surface

For the function given by $F(x, y, z) = x^2 + y^2 + z^2 - 4$, describe the level surface given by $F(x, y, z) = 0$.

Solution   The level surface given by $F(x, y, z) = 0$ can be written as

$$x^2 + y^2 + z^2 = 4$$

which is a sphere of radius 2 whose center is at the origin.

You have seen many examples of the usefulness of normal lines in applications involving curves. Normal lines are equally important in analyzing surfaces and solids. For example, consider the collision of two billiard balls. When a stationary ball is struck at a point P on its surface, it moves along the **line of impact** determined by P and the center of the ball. The impact can occur in *two* ways. If the cue ball is moving along the line of impact, it stops dead and imparts all of its momentum to the stationary ball, as shown in Figure 12.54. If the cue ball is not moving along the line of impact, it is deflected to one side or the other and retains part of its momentum. That part of the momentum that is transferred to the stationary ball occurs along the line of impact, *regardless* of the direction of the cue ball, as shown in Figure 12.55. This line of impact is called the **normal line** to the surface of the ball at the point P.

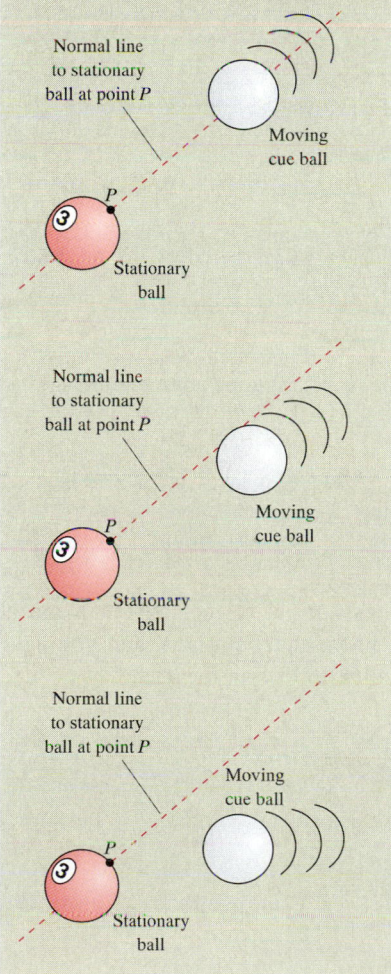
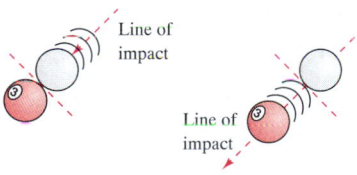

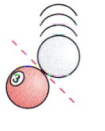

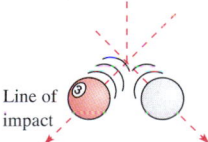

Figure 12.54

Figure 12.55

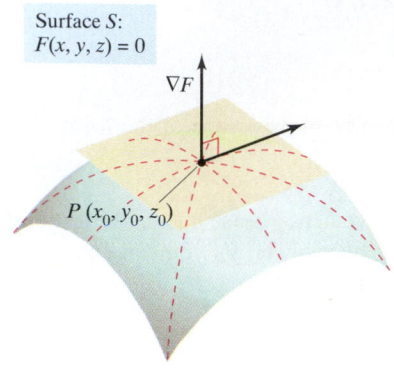

Surface S:
$F(x, y, z) = 0$

∇F

$P(x_0, y_0, z_0)$

Tangent plane to surface S at P
Figure 12.56

In the process of finding a normal line to a surface, you are also able to solve the problem of finding a **tangent plane** to the surface. Let S be a surface given by $F(x, y, z) = 0$, and let $P(x_0, y_0, z_0)$ be a point on S. Let C be a curve on S through P that is defined by the vector-valued function $\mathbf{r}(t) = x(t)\mathbf{i} + y(t)\mathbf{j} + z(t)\mathbf{k}$. Then, for all t

$$F(x(t), y(t), z(t)) = 0.$$

If F is differentiable and $x'(t)$, $y'(t)$, and $z'(t)$ all exist, it follows from the Chain Rule that

$$0 = F'(t) = F_x(x, y, z)x'(t) + F_y(x, y, z)y'(t) + F_z(x, y, z)z'(t).$$

At (x_0, y_0, z_0), the equivalent vector form is

$$0 = \underbrace{\nabla F(x_0, y_0, z_0)}_{\text{Gradient}} \cdot \underbrace{\mathbf{r}'(t_0)}_{\substack{\text{Tangent} \\ \text{Vector}}}.$$

This result means that the gradient at P is orthogonal to the tangent vector of every curve on S through P. Thus, all tangent lines at P lie in a plane that is normal to $\nabla F(x_0, y_0, z_0)$ and contains P, as shown in Figure 12.56.

Definition of Tangent Plane and Normal Line

Let F be differentiable at the point $P(x_0, y_0, z_0)$ on the surface S given by $F(x, y, z) = 0$ such that $\nabla F(x_0, y_0, z_0) \neq \mathbf{0}$.

1. The plane through P that is normal to $\nabla F(x_0, y_0, z_0)$ is called the **tangent plane to S at P**.

2. The line through P having the direction of $\nabla F(x_0, y_0, z_0)$ is called the **normal line to S at P**.

NOTE   In the remainder of this section, we assume $\nabla F(x_0, y_0, z_0)$ to be nonzero unless stated otherwise.

To find an equation for the tangent plane to S at (x_0, y_0, z_0), let (x, y, z) be an arbitrary point in the tangent plane. Then the vector

$$\mathbf{v} = (x - x_0)\mathbf{i} + (y - y_0)\mathbf{j} + (z - z_0)\mathbf{k}$$

lies in the tangent plane. Because $\nabla F(x_0, y_0, z_0)$ is normal to the tangent plane at (x_0, y_0, z_0), it must be orthogonal to every vector in the tangent plane, and you have $\nabla F(x_0, y_0, z_0) \cdot \mathbf{v} = 0$, which leads to the result in the following theorem.

THEOREM 12.13   Equation of Tangent Plane

If F is differentiable at (x_0, y_0, z_0), then an equation of the tangent plane to the surface given by $F(x, y, z) = 0$ at (x_0, y_0, z_0) is

$$F_x(x_0, y_0, z_0)(x - x_0) + F_y(x_0, y_0, z_0)(y - y_0) + F_z(x_0, y_0, z_0)(z - z_0) = 0.$$

EXAMPLE 2   Finding an Equation of a Tangent Plane

Find an equation of the tangent plane to the hyperboloid given by

$$z^2 - 2x^2 - 2y^2 = 12$$

at the point $(1, -1, 4)$.

Solution   Begin by writing the equation of the surface as

$$z^2 - 2x^2 - 2y^2 - 12 = 0.$$

Then, considering

$$F(x, y, z) = z^2 - 2x^2 - 2y^2 - 12$$

you have

$$F_x(x, y, z) = -4x, \quad F_y(x, y, z) = -4y, \quad \text{and} \quad F_z(x, y, z) = 2z.$$

At the point $(1, -1, 4)$ the partial derivatives are

$$F_x(1, -1, 4) = -4, \quad F_y(1, -1, 4) = 4, \quad \text{and} \quad F_z(1, -1, 4) = 8.$$

Therefore, an equation of the tangent plane at $(1, -1, 4)$ is

$$-4(x - 1) + 4(y + 1) + 8(z - 4) = 0$$
$$-4x + 4 + 4y + 4 + 8z - 32 = 0$$
$$-4x + 4y + 8z - 24 = 0$$
$$x - y - 2z + 6 = 0.$$

Figure 12.57 shows a portion of the hyperboloid and tangent plane.

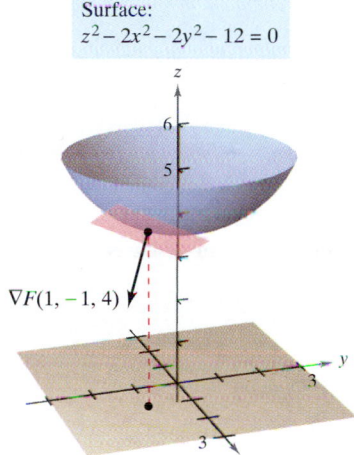

Surface:
$z^2 - 2x^2 - 2y^2 - 12 = 0$

$\nabla F(1, -1, 4)$

Tangent plane to surface
Figure 12.57

TECHNOLOGY   Some three-dimensional graphing utilities are capable of sketching tangent planes to surfaces. Two examples are shown below.

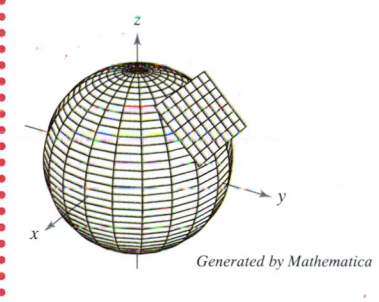

Generated by Mathematica

Sphere: $x^2 + y^2 + z^2 = 1$

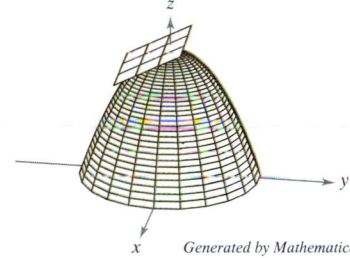

Generated by Mathematica

Paraboloid: $z = 2 - x^2 - y^2$

To find the equation of the tangent plane at a point on a surface given by $z = f(x, y)$, you can define the function F by

$$F(x, y, z) = f(x, y) - z.$$

Then S is given by the level surface $F(x, y, z) = 0$, and by Theorem 12.13 an equation of the tangent plane to S at the point (x_0, y_0, z_0) is

$$f_x(x_0, y_0)(x - x_0) + f_y(x_0, y_0)(y - y_0) - (z - z_0) = 0.$$

EXAMPLE 3   **Finding an Equation of the Tangent Plane**

Find the equation of the tangent plane to the paraboloid

$$z = 1 - \frac{1}{10}(x^2 + 4y^2)$$

at the point $\left(1, 1, \frac{1}{2}\right)$.

Solution   From $z = f(x, y) = 1 - \frac{1}{10}(x^2 + 4y^2)$, you obtain

$$f_x(x, y) = -\frac{x}{5} \quad \Rightarrow \quad f_x(1, 1) = -\frac{1}{5}$$

and

$$f_y(x, y) = -\frac{4y}{5} \quad \Rightarrow \quad f_y(1, 1) = -\frac{4}{5}.$$

Therefore, an equation of the tangent plane at $\left(1, 1, \frac{1}{2}\right)$ is

$$f_x(1, 1)(x - 1) + f_y(1, 1)(y - 1) - (z - \tfrac{1}{2}) = 0$$
$$-\frac{1}{5}(x - 1) - \frac{4}{5}(y - 1) - (z - \tfrac{1}{2}) = 0$$
$$-\frac{1}{5}x - \frac{4}{5}y - z + \frac{3}{2} = 0.$$

This tangent plane is shown in Figure 12.58.

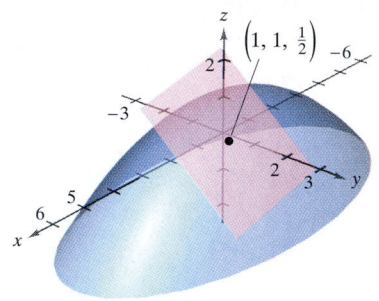

Surface:
$f(x, y) = 1 - \frac{1}{10}(x^2 + 4y^2)$

Tangent plane to surface
Figure 12.58

The gradient $\nabla F(x, y, z)$ gives a convenient way to find equations of normal lines, as shown in Example 4.

EXAMPLE 4   **Finding an Equation of a Normal Line to a Surface**

Find a set of symmetric equations for the normal line to the surface given by

$$xyz = 12$$

at the point $(2, -2, -3)$.

Solution   Begin by letting

$$F(x, y, z) = xyz - 12.$$

Then, the gradient is given by

$$\nabla F(x, y, z) = F_x(x, y, z)\mathbf{i} + F_y(x, y, z)\mathbf{j} + F_z(x, y, z)\mathbf{k}$$
$$= yz\mathbf{i} + xz\mathbf{j} + xy\mathbf{k}$$

and at the point $(2, -2, -3)$ you have

$$\nabla F(2, -2, -3) = (-2)(-3)\mathbf{i} + (2)(-3)\mathbf{j} + (2)(-2)\mathbf{k}$$
$$= 6\mathbf{i} - 6\mathbf{j} - 4\mathbf{k}.$$

The normal line at $(2, -2, -3)$ has direction numbers $6, -6,$ and -4, and the corresponding set of symmetric equations is

$$\frac{x - 2}{6} = \frac{y + 2}{-6} = \frac{z + 3}{-4}.$$

(See Figure 12.59.)

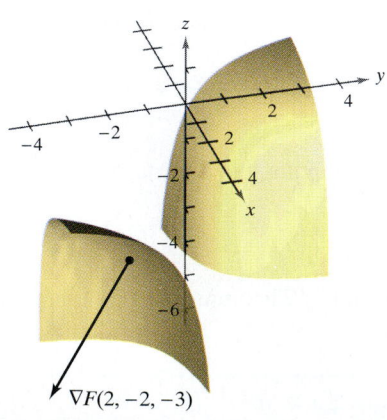

Surface: $xyz = 12$

$\nabla F(2, -2, -3)$

Figure 12.59

Knowing that the gradient $\nabla F(x, y, z)$ is normal to the surface given by $F(x, y, z) = 0$ allows you to solve a variety of problems dealing with surfaces and curves in space.

EXAMPLE 5   Finding the Equation of a Tangent Line to a Curve

Describe the tangent line to the curve of intersection of the surfaces

$$x^2 + 2y^2 + 2z^2 = 20 \qquad \text{Ellipsoid}$$
$$x^2 + y^2 + z = 4 \qquad \text{Paraboloid}$$

at the point $(0, 1, 3)$, as shown in Figure 12.60.

Solution   Begin by finding the gradients to both surfaces at the point $(0, 1, 3)$.

Ellipsoid	Paraboloid
$F(x, y, z) = x^2 + 2y^2 + 2z^2 - 20$	$G(x, y, z) = x^2 + y^2 + z - 4$
$\nabla F(x, y, z) = 2x\mathbf{i} + 4y\mathbf{j} + 4z\mathbf{k}$	$\nabla G(x, y, z) = 2x\mathbf{i} + 2y\mathbf{j} + \mathbf{k}$
$\nabla F(0, 1, 3) = 4\mathbf{j} + 12\mathbf{k}$	$\nabla G(0, 1, 3) = 2\mathbf{j} + \mathbf{k}$

The cross product of these two gradients is a vector that is tangent to both surfaces at the point $(0, 1, 3)$.

$$\nabla F(0, 1, 3) \times \nabla G(0, 1, 3) = \begin{vmatrix} \mathbf{i} & \mathbf{j} & \mathbf{k} \\ 0 & 4 & 12 \\ 0 & 2 & 1 \end{vmatrix} = -20\mathbf{i}.$$

Thus, the tangent line to the curve of intersection of the two surfaces at the point $(0, 1, 3)$ is a line that is parallel to the x-axis and passes through the point $(0, 1, 3)$.

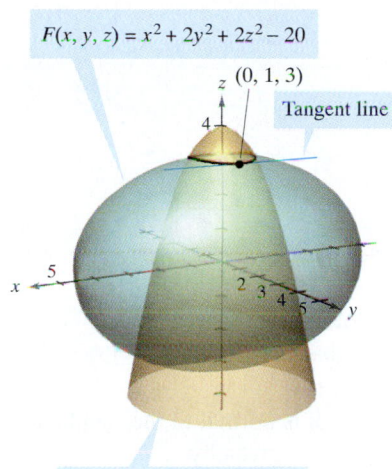

$F(x, y, z) = x^2 + 2y^2 + 2z^2 - 20$

$G(x, y, z) = x^2 + y^2 + z - 4$

Figure 12.60

The Angle of Inclination of a Plane

Another use of the gradient $\nabla F(x, y, z)$ is to determine the angle of inclination of the tangent plane to a surface. The **angle of inclination** of a plane is defined to be the angle θ, $0 \le \theta \le \pi/2$, between the given plane and the xy-plane, as shown in Figure 12.61. (The angle of inclination of a horizontal plane is defined to be zero.) Because the vector $\mathbf{k}$ is normal to the xy-plane, you can use the formula for the cosine of the angle between two planes (given in Section 10.5) to conclude that the angle of inclination of a plane with normal vector $\mathbf{n}$ is given by

$$\cos \theta = \frac{|\mathbf{n} \cdot \mathbf{k}|}{\|\mathbf{n}\| \, \|\mathbf{k}\|} = \frac{|\mathbf{n} \cdot \mathbf{k}|}{\|\mathbf{n}\|}. \qquad \text{Angle of inclination of a plane}$$

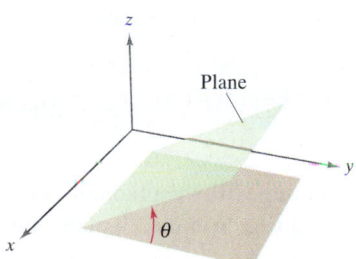

The angle of inclination
Figure 12.61

EXAMPLE 6 **Finding the Angle of Inclination of a Tangent Plane**

Find the angle of inclination of the tangent plane to the ellipsoid given by

$$\frac{x^2}{12} + \frac{y^2}{12} + \frac{z^2}{3} = 1$$

at the point $(2, 2, 1)$.

Solution If you let

$$F(x, y, z) = \frac{x^2}{12} + \frac{y^2}{12} + \frac{z^2}{3} - 1$$

the gradient of F at the point $(2, 2, 1)$ is given by

$$\nabla F(x, y, z) = \frac{x}{6}\mathbf{i} + \frac{y}{6}\mathbf{j} + \frac{2z}{3}\mathbf{k}$$

$$\nabla F(2, 2, 1) = \frac{1}{3}\mathbf{i} + \frac{1}{3}\mathbf{j} + \frac{2}{3}\mathbf{k}.$$

Because $\nabla F(2, 2, 1)$ is normal to the tangent plane and $\mathbf{k}$ is normal to the xy-plane, it follows that the angle of inclination of the tangent plane is given by

$$\cos\theta = \frac{|\nabla F(2, 2, 1) \cdot \mathbf{k}|}{\|\nabla F(2, 2, 1)\|} = \frac{2/3}{\sqrt{(1/3)^2 + (1/3)^2 + (2/3)^2}} = \sqrt{\frac{2}{3}}$$

which implies that $\theta = \arccos\sqrt{2/3} \approx 35.3°$, as shown in Figure 12.62.

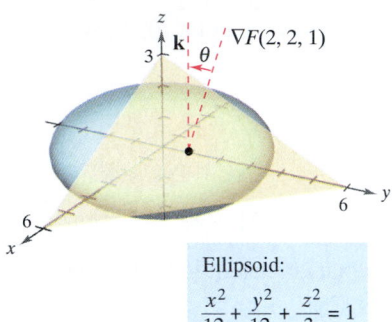

Ellipsoid:
$$\frac{x^2}{12} + \frac{y^2}{12} + \frac{z^2}{3} = 1$$

Figure 12.62

NOTE A special case of the procedure shown in Example 6 is worth noting. The angle of inclination θ of the tangent plane to the surface $z = f(x, y)$ at (x_0, y_0, z_0) is given by

$$\cos\theta = \frac{1}{\sqrt{[f_x(x_0, y_0)]^2 + [f_y(x_0, y_0)]^2 + 1}}.$$

Alternative formula for angle of inclination (See Exercise 56.)

A Comparison of the Gradients $\nabla f(x, y)$ and $\nabla F(x, y, z)$

We conclude this section with a comparison of the gradients $\nabla f(x, y)$ and $\nabla F(x, y, z)$. In the previous section, you saw that the gradient of a function f of two variables is normal to the *level curves* of f. Specifically, Theorem 12.12 stated that if f is differentiable at (x_0, y_0) and $\nabla f(x_0, y_0) \neq \mathbf{0}$, then $\nabla f(x_0, y_0)$ is normal to the level curve through (x_0, y_0). Having developed normal lines to surfaces, you can now extend this result to a function of three variables.

THEOREM 12.14 Gradient Is Normal to Level Surfaces

If F is differentiable at (x_0, y_0, z_0) and $\nabla F(x_0, y_0, z_0) \neq \mathbf{0}$, then $\nabla F(x_0, y_0, z_0)$ is normal to the level surface through (x_0, y_0, z_0).

When working with the gradients $\nabla f(x, y)$ and $\nabla F(x, y, z)$, be sure you remember that $\nabla f(x, y)$ is a vector in the xy-plane and $\nabla F(x, y, z)$ is a vector in space.

EXERCISES FOR SECTION 12.7

In Exercises 1–10, find a unit normal vector to the surface at the indicated point. [*Hint*: Normalize the gradient vector $\nabla F(x, y, z)$.]

Surface	Point
1. $x + y + z = 4$	$(2, 0, 2)$
2. $x^2 + y^2 + z^2 = 11$	$(3, 1, 1)$
3. $z = \sqrt{x^2 + y^2}$	$(3, 4, 5)$
4. $z = x^3$	$(2, 1, 8)$
5. $x^2 y^4 - z = 0$	$(1, 2, 16)$
6. $x^2 + 3y + z^3 = 9$	$(2, -1, 2)$
7. $z - x \sin y = 4$	$\left(6, \dfrac{\pi}{6}, 7\right)$
8. $ze^{x^2 - y^2} - 3 = 0$	$(2, 2, 3)$
9. $\ln\left(\dfrac{x}{y - z}\right) = 0$	$(1, 4, 3)$
10. $\sin(x - y) - z = 2$	$\left(\dfrac{\pi}{3}, \dfrac{\pi}{6}, -\dfrac{3}{2}\right)$

In Exercises 11–14, find an equation of the tangent plane to the surface at the indicated point.

11. $z = 25 - x^2 - y^2$, $(3, 1, 15)$

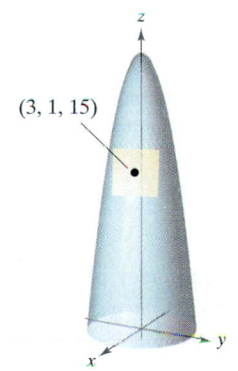

$(3, 1, 15)$

12. $z = \sqrt{x^2 + y^2}$, $(3, 4, 5)$

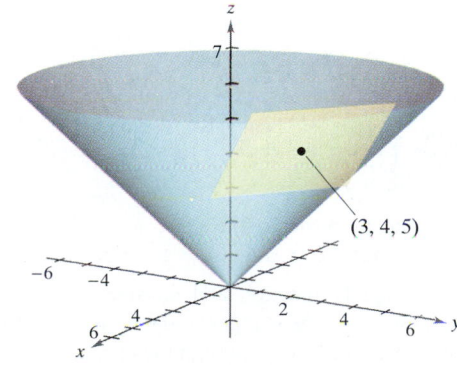

$(3, 4, 5)$

13. $f(x, y) = \dfrac{y}{x}$, $(1, 2, 2)$

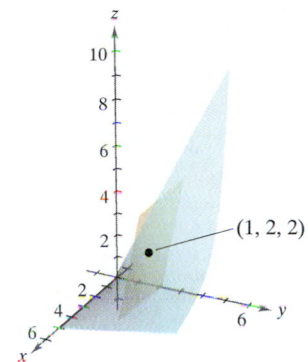

$(1, 2, 2)$

14. $g(x, y) = \arctan \dfrac{y}{x}$, $(1, 0, 0)$

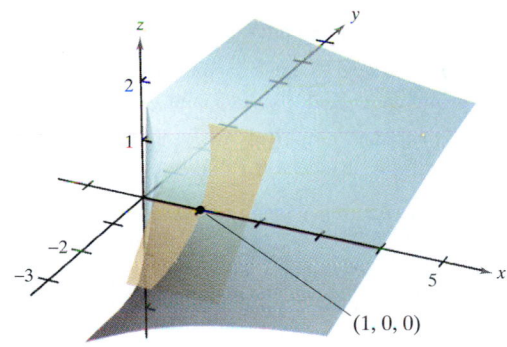

$(1, 0, 0)$

In Exercises 15–24, find an equation of the tangent plane to the surface at the indicated point.

Surface	Point
15. $g(x, y) = x^2 - y^2$	$(5, 4, 9)$
16. $f(x, y) = 2 - \frac{2}{3}x - y$	$(3, -1, 1)$
17. $z = e^x(\sin y + 1)$	$\left(0, \dfrac{\pi}{2}, 2\right)$
18. $z = x^3 - 3xy + y^3$	$(1, 2, 3)$
19. $h(x, y) = \ln \sqrt{x^2 + y^2}$	$(3, 4, \ln 5)$
20. $h(x, y) = \cos y$	$\left(5, \dfrac{\pi}{4}, \dfrac{\sqrt{2}}{2}\right)$
21. $x^2 + 4y^2 + z^2 = 36$	$(2, -2, 4)$
22. $x^2 + 2z^2 = y^2$	$(1, 3, -2)$
23. $xy^2 + 3x - z^2 = 4$	$(2, 1, -2)$
24. $y = x(2z - 1)$	$(4, 4, 1)$

In Exercises 25–30, find an equation of the tangent plane and find symmetric equations of the normal line to the surface at the indicated point.

Surface	Point
25. $x^2 + y^2 + z = 9$	$(1, 2, 4)$
26. $x^2 + y^2 + z^2 = 9$	$(1, 2, 2)$
27. $xy - z = 0$	$(-2, -3, 6)$
28. $x^2 + y^2 - z^2 = 0$	$(5, 12, 13)$
29. $z = \arctan(y/x)$	$(1, 1, \pi/4)$
30. $xyz = 10$	$(1, 2, 5)$

 31. ***Investigation*** Consider the function

$$f(x, y) = \frac{4xy}{(x^2 + 1)(y^2 + 1)}$$

on the intervals $-2 \le x \le 2$ and $0 \le y \le 3$.

(a) Find a set of parametric equations of the normal line and an equation of the tangent plane to the surface at the point $(1, 1, 1)$.

(b) Repeat part (a) for the point $(-1, 2, -\frac{4}{5})$.

(c) Use a computer algebra system to graph the surface, the normal lines, and the tangent planes found in parts (a) and (b).

(d) Use analytical and graphical analyses to write a brief description of the surface at the two indicated points.

32. ***Think About It*** For some surfaces, the normal lines at any point pass through the same geometrical object. What is the common geometrical object for a sphere? What is the common geometrical object for a right circular cylinder? Explain.

In Exercises 33–38, (a) find symmetric equations of the tangent line to the curve of intersection of the surfaces at the indicated point, and (b) find the cosine of the angle between the gradient vectors at this point. State whether or not the surfaces are orthogonal at the point of intersection.

Surfaces	Point
33. $x^2 + y^2 = 5$, $z = x$	$(2, 1, 2)$
34. $z = x^2 + y^2$, $z = 4 - y$	$(2, -1, 5)$
35. $x^2 + z^2 = 25$, $y^2 + z^2 = 25$	$(3, 3, 4)$
36. $z = \sqrt{x^2 + y^2}$, $2x + y + 2z = 20$	$(3, 4, 5)$
37. $x^2 + y^2 + z^2 = 6$, $x - y - z = 0$	$(2, 1, 1)$
38. $z = x^2 + y^2$, $x + y + 6z = 33$	$(1, 2, 5)$

 39. Consider the functions

$$f(x, y) = 6 - x^2 - y^2/4 \quad \text{and} \quad g(x, y) = 2x + y.$$

(a) Find a set of parametric equations of the tangent line to the curve of intersection of the surfaces at the point $(1, 2, 4)$, and find the angle between the gradient vectors.

(b) Use a computer algebra system to graph the surfaces. Graph the tangent line found in part (a).

 40. Consider the functions

$$f(x, y) = \sqrt{16 - x^2 - y^2 + 2x - 4y} \quad \text{and}$$

$$g(x, y) = \frac{\sqrt{2}}{2}\sqrt{1 - 3x^2 + y^2 + 6x + 4y}.$$

(a) Use a computer algebra system to graph the first-octant portion of the surfaces represented by f and g.

(b) Find two first-octant points on the curve of the intersection and show that the surfaces are orthogonal at these points.

(c) These surfaces are orthogonal along the curve of intersection. Does part (b) prove this fact? Explain.

In Exercises 41–44, find the angle of inclination θ of the tangent plane to the given surface at the indicated point.

Surface	Point
41. $3x^2 + 2y^2 - z = 15$	$(2, 2, 5)$
42. $xy - z^2 = 0$	$(2, 2, 2)$
43. $x^2 - y^2 + z = 0$	$(1, 2, 3)$
44. $x^2 + y^2 = 5$	$(2, 1, 3)$

 In Exercises 45 and 46, find the point on the surface where the tangent plane is horizontal. Use a computer algebra system to graph the surface and the horizontal tangent plane. Describe the surface where the tangent plane is horizontal.

45. $z = 3 - x^2 - y^2 + 6y$

46. $z = 3x^2 + 2y^2 - 3x + 4y - 5$

In Exercises 47 and 48, find the path of a heat-seeking particle in the temperature field T, starting at the indicated point.

47. $T(x, y, z) = 400 - 2x^2 - y^2 - 4z^2$, $(4, 3, 10)$

48. $T(x, y, z) = 100 - 3x - y - z^2$, $(2, 2, 5)$

In Exercises 49 and 50, show that the tangent plane to the quadric surface at the point (x_0, y_0, z_0) can be written in the given form.

49. Ellipsoid: $\dfrac{x^2}{a^2} + \dfrac{y^2}{b^2} + \dfrac{z^2}{c^2} = 1$

   Plane: $\dfrac{x_0 x}{a^2} + \dfrac{y_0 y}{b^2} + \dfrac{z_0 z}{c^2} = 1$

50. Hyperboloid: $\dfrac{x^2}{a^2} + \dfrac{y^2}{b^2} - \dfrac{z^2}{c^2} = 1$

   Plane: $\dfrac{x_0 x}{a^2} + \dfrac{y_0 y}{b^2} - \dfrac{z_0 z}{c^2} = 1$

51. Show that any tangent plane to the cone $z^2 = a^2 x^2 + b^2 y^2$ passes through the origin.

52. Let f be a differentiable function and consider the surface $z = xf(y/x)$. Show that the tangent plane at any point $P(x_0, y_0, z_0)$ on the surface passes through the origin.

53. *Approximation*   Consider the following approximations centered at $(0, 0)$ for a function $f(x, y)$.

Linear approximation:

$$P_1(x, y) = f(0, 0) + f_x(0, 0)x + f_y(0, 0)y$$

Quadratic approximation:

$$P_2(x, y) = f(0, 0) + f_x(0, 0)x + f_y(0, 0)y +$$
$$\tfrac{1}{2} f_{xx}(0, 0)x^2 + f_{xy}(0, 0)xy + \tfrac{1}{2} f_{yy}(0, 0)y^2$$

[Note that the linear approximation is the tangent plane to the surface at $(0, 0, f(0, 0))$.]

(a) Find the linear approximations of $f(x, y) = e^{(x-y)}$ centered at $(0, 0)$.

(b) Find the quadratic approximation of $f(x, y) = e^{(x-y)}$ centered at $(0, 0)$.

(c) If $x = 0$ in the quadratic approximation, you obtain the second-degree Taylor polynomial for what function? Answer the same question for $y = 0$.

(d) Complete the table.

x	y	$f(x, y)$	$P_1(x, y)$	$P_2(x, y)$
0	0			
0	0.1			
0.2	0.1			
0.2	0.5			
1	0.5			

(e) Use a graphing utility to graph the surfaces $z = f(x, y)$, $z = P_1(x, y)$, and $z = P_2(x, y)$.

54. *Approximation*   Repeat Exercise 53 for the function $f(x, y) = \cos(x + y)$.

55. Prove Theorem 12.14.

56. Prove that the angle of inclination θ of the tangent plane to the surface $z = f(x, y)$ at the point (x_0, y_0, z_0) is given by

$$\cos \theta = \frac{1}{\sqrt{[f_x(x_0, y_0)]^2 + [f_y(x_0, y_0)]^2 + 1}}.$$

Wildflower Diversity   The diversity of wildflowers in a meadow can be measured by counting the number of daisies, buttercups, shooting stars, and so on. If there are n types of wildflowers, each with a proportion p_i of the total population, it follows that $p_1 + p_2 + \cdots + p_n = 1$. The measure of diversity of the population is defined as

$$H = -\sum_{i=1}^{n} p_i \log_2 p_i.$$

In this definition, it is understood that $p_i \log_2 p_i = 0$ when $p_i = 0$. The tables show proportions of wildflowers in a meadow in May, June, August, and September.

May

Flower type	1	2	3	4
Proportion	$\frac{5}{16}$	$\frac{5}{16}$	$\frac{5}{16}$	$\frac{1}{16}$

June

Flower type	1	2	3	4
Proportion	$\frac{1}{4}$	$\frac{1}{4}$	$\frac{1}{4}$	$\frac{1}{4}$

August

Flower type	1	2	3	4
Proportion	$\frac{1}{4}$	0	$\frac{1}{4}$	$\frac{1}{2}$

September

Flower type	1	2	3	4
Proportion	0	0	0	1

(a) Determine the wildflower diversity for each month. How would you interpret September's diversity? Which month had the greatest diversity?

(b) If the meadow contains ten types of wildflowers in roughly equal proportions, is the diversity of the population greater than or less than the diversity of a similar distribution of four types of flowers? What type of distribution (of ten types of wildflowers) would produce maximum diversity?

(c) Let H_n represent the maximum diversity of n types of wildflowers. Does H_n approach a limit as $n \to \infty$?

FOR FURTHER INFORMATION   Biologists use the concept of diversity to measure the proportions of different types of organisms within an environment. For more information on this technique, see the article "Information Theory and Biological Diversity" by Steven Kolmes and Kevin Mitchell in the 1990 *UMAP Modules*.

Absolute Extrema and Relative Extrema • The Second Partials Test

Absolute Extrema and Relative Extrema

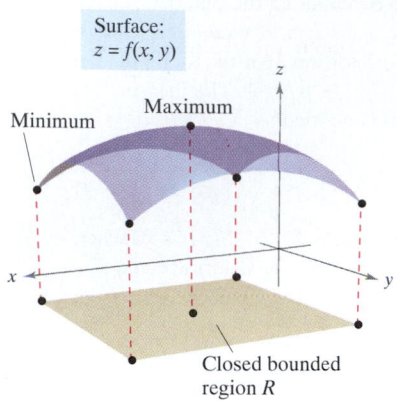

Surface:
$z = f(x, y)$

Minimum   Maximum

Closed bounded
region R

R contains point(s) at which $f(x, y)$ is a minimum and point(s) at which $f(x, y)$ is a maximum.

Figure 12.63

In Chapter 3, you studied techniques for finding the extreme values of a function of a single variable. In this section, we extend these techniques to functions of two variables. For example, in Theorem 12.15 the Extreme Value Theorem for a function of a single variable is extended to a function of two variables.

Consider the continuous function f of two variables, defined on a closed bounded region R. The values $f(a, b)$ and $f(c, d)$ such that

$$f(a, b) \leq f(x, y) \leq f(c, d) \qquad \text{(a, b) and (c, d) are in R.}$$

for all (x, y) in R are called the **minimum** and **maximum** of f in the region R, as shown in Figure 12.63. Recall from Section 12.2 that a region in the plane is *closed* if it contains all of its boundary points. The Extreme Value Theorem deals with a region in the plane that is both closed and *bounded*. A region in the plane is called **bounded** if it is a subregion of a closed disc in the plane.

> **THEOREM 12.15 Extreme Value Theorem**
>
> Let f be a continuous function of two variables x and y defined on a closed bounded region R in the xy-plane.
>
> 1. There is at least one point in R where f takes on a minimum value.
> 2. There is at least one point in R where f takes on a maximum value.

The Granger Collection

KARL WEIERSTRASS (1815–1897)

Although the Extreme Value Theorem had been used by earlier mathematicians, the first to provide a rigorous proof was the German mathematician Karl Weierstrass. Weierstrass also provided rigorous justifications for many other mathematical results already in common use. We are indebted to him for much of the logical foundation on which modern calculus is built.

A minimum is also called an **absolute minimum** and a maximum is also called an **absolute maximum.** As in single-variable calculus, we distinguish between absolute extrema and **relative extrema.**

> **Definition of Relative Extrema**
>
> Let f be a function defined on a region R containing (x_0, y_0).
>
> 1. The function f has a **relative minimum** at (x_0, y_0) if
>
> $$f(x, y) \geq f(x_0, y_0)$$
>
>   for all (x, y) in an *open* disc containing (x_0, y_0).
> 2. The function f has a **relative maximum** at (x_0, y_0) if
>
> $$f(x, y) \leq f(x_0, y_0)$$
>
>   for all (x, y) in an *open* disc containing (x_0, y_0).

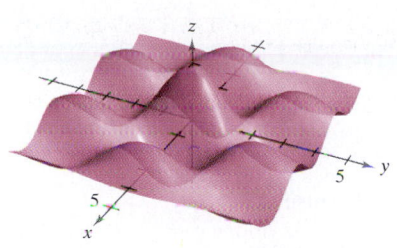

Relative extrema
Figure 12.64

To say that f has a relative maximum at (x_0, y_0) means that the point (x_0, y_0, z_0) is at least as high as all nearby points on the graph of $z = f(x, y)$. Similarly, f has a relative minimum at (x_0, y_0) if (x_0, y_0, z_0) is at least as low as all nearby points on the graph. (See Figure 12.64.)

To locate relative extrema of f, you can investigate the points at which the gradient of f is **0** or undefined. Such points are called **critical points** of f.

> **Definition of Critical Point**
>
> Let f be defined on an open region R containing (x_0, y_0). The point (x_0, y_0) is a **critical point** of f if one of the following is true.
>
> 1. $f_x(x_0, y_0) = 0$ and $f_y(x_0, y_0) = 0$
> 2. $f_x(x_0, y_0)$ or $f_y(x_0, y_0)$ does not exist.

Recall from Theorem 12.11 that if f is differentiable and

$$\nabla f(x_0, y_0) = f_x(x_0, y_0)\mathbf{i} + f_y(x_0, y_0)\mathbf{j}$$
$$= 0\mathbf{i} + 0\mathbf{j}$$

then every directional derivative at (x_0, y_0) must be 0. In Exercise 56 you are asked to show that this implies that the function has a horizontal tangent plane at the point (x_0, y_0), as shown in Figure 12.65. It appears that such a point is a likely location of a relative extremum. This is confirmed by Theorem 12.16.

EXPLORATION

Try using a graphing utility to sketch the graph of $z = x^3 - 3xy + y^3$ using the bounds $0 \le x \le 3$, $0 \le y \le 3$, and $-3 \le z \le 3$. This view makes it appear as though the surface has an absolute minimum. But does it?

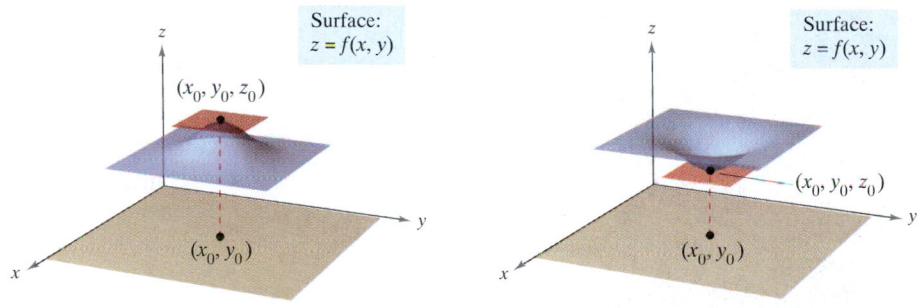

Relative maximum   Relative minimum
Figure 12.65

> **THEOREM 12.16   Relative Extrema Occur Only at Critical Points**
>
> If f has a relative extremum at (x_0, y_0) on an open region R, then (x_0, y_0) is a critical point of f.

EXAMPLE 1   Finding a Relative Extremum

Determine the relative extrema of

$$f(x, y) = 2x^2 + y^2 + 8x - 6y + 20.$$

Solution   Begin by finding the critical points of f. Because

$$f_x(x, y) = 4x + 8 \qquad \text{Partial with respect to } x$$

and

$$f_y(x, y) = 2y - 6 \qquad \text{Partial with respect to } y$$

are defined for all x and y, the only critical points are those for which both first partial derivatives are 0. To locate these points, let $f_x(x, y)$ and $f_y(x, y)$ be 0, and solve the system of equations

$$4x + 8 = 0 \quad \text{and} \quad 2y - 6 = 0$$

to obtain the critical point $(-2, 3)$. By completing the square, you can conclude that for all $(x, y) \neq (-2, 3)$,

$$f(x, y) = 2(x + 2)^2 + (y - 3)^2 + 3 > 3.$$

Therefore, a relative *minimum* of f occurs at $(-2, 3)$. The value of the relative minimum is $f(-2, 3) = 3$, as shown in Figure 12.66.

Example 1 shows a relative minimum occurring at one type of critical point—the type for which both $f_x(x, y)$ and $f_y(x, y)$ are 0. The next example concerns a relative maximum that occurs at the other type of critical point—the type for which either $f_x(x, y)$ or $f_y(x, y)$ is undefined.

EXAMPLE 2   Finding a Relative Extremum

Determine the relative extrema of $f(x, y) = 1 - (x^2 + y^2)^{1/3}$.

Solution   Because

$$f_x(x, y) = -\frac{2x}{3(x^2 + y^2)^{2/3}} \qquad \text{Partial with respect to } x$$

and

$$f_y(x, y) = -\frac{2y}{3(x^2 + y^2)^{2/3}} \qquad \text{Partial with respect to } y$$

it follows that both partial derivatives are defined for all points in the xy-plane except for $(0, 0)$. Moreover, because the partial derivatives cannot both be 0 unless both x and y are 0, you can conclude that $(0, 0)$ is the only critical point. In Figure 12.67, note that $f(0, 0)$ is 1. For all other (x, y) it is clear that

$$f(x, y) = 1 - (x^2 + y^2)^{1/3} < 1.$$

Therefore, f has a relative *maximum* at $(0, 0)$.

NOTE   In Example 2, $f_x(x, y) = 0$ for every point on the y-axis other than $(0, 0)$. However, because $f_y(x, y)$ is nonzero, these are not critical points. Remember that *one* of the partials must be undefined or *both* must be 0 in order to yield a critical point.

Surface:
$f(x, y) = 2x^2 + y^2 + 8x - 6y + 20$

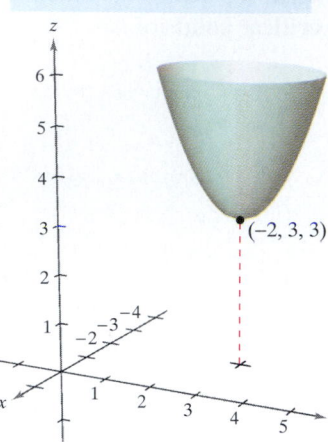

The function $z = f(x, y)$ has a relative minimum at $(-2, 3)$.
Figure 12.66

Surface:
$f(x, y) = 1 - (x^2 + y^2)^{1/3}$

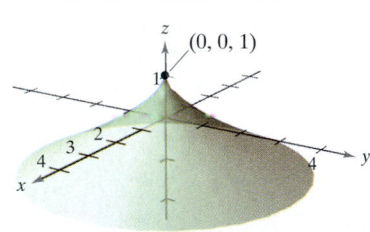

$f_x(x, y)$ and $f_y(x, y)$ are undefined at $(0, 0)$.
Figure 12.67

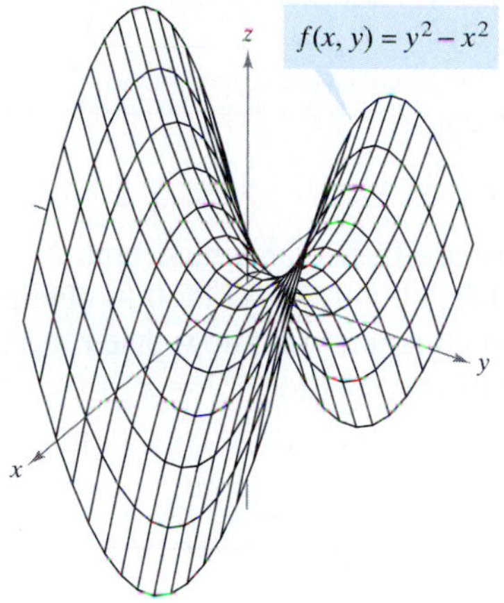

Saddle point at $(0, 0, 0)$:
$f_x(0, 0) = f_y(0, 0) = 0$
Figure 12.68

The Second Partials Test

Theorem 12.16 tells you that to find relative extrema you need only examine values of $f(x, y)$ at critical points. However, as is true for a function of one variable, the critical points of a function of two variables do not always yield relative maxima or minima. Some critical points yield **saddle points,** which are neither relative maxima nor relative minima.

As an example of a critical point that does not yield a relative extremum, consider the surface given by

$$f(x, y) = y^2 - x^2 \qquad \textcolor{red}{\text{Hyperbolic paraboloid}}$$

as shown in Figure 12.68. At the point $(0, 0)$, both partial derivatives are 0. The function f does not, however, have a relative extremum at this point because in any open disc centered at $(0, 0)$ the function takes on both negative values (along the x-axis) *and* positive values (along the y-axis). Thus, the point $(0, 0, 0)$ is a saddle point of the surface. (The name "saddle point" comes from the fact that the surface shown in Figure 12.68 resembles a saddle.)

For the functions in Examples 1 and 2, it was relatively easy to determine the relative extrema, because each function was either given, or able to be written, in completed square form. For more complicated functions, algebraic arguments are less convenient and it is better to rely on the analytical means presented in the following Second Partials Test. This is the two-variable counterpart of the Second Derivative Test for functions of one variable. The proof of this theorem is best left to a course in advanced calculus.

> ### THEOREM 12.17   Second Partials Test
>
> Let f have continuous second partial derivatives on an open region containing a point (a, b) for which
>
> $$f_x(a, b) = 0 \quad \text{and} \quad f_y(a, b) = 0.$$
>
> To test for relative extrema of f, consider the quantity
>
> $$d = f_{xx}(a, b)f_{yy}(a, b) - [f_{xy}(a, b)]^2.$$
>
> 1. If $d > 0$ and $f_{xx}(a, b) > 0$, then f has a **relative minimum** at (a, b).
> 2. If $d > 0$ and $f_{xx}(a, b) < 0$, then f has a **relative maximum** at (a, b).
> 3. If $d < 0$, then $(a, b, f(a, b))$ is a **saddle point**.
> 4. The test is inconclusive if $d = 0$.

NOTE   If $d > 0$, then $f_{xx}(a, b)$ and $f_{yy}(a, b)$ must have the same sign. This means that $f_{xx}(a, b)$ can be replaced by $f_{yy}(a, b)$ in the first two parts of the test.

A convenient device for remembering the formula for d in the Second Partials Test is given by the 2×2 determinant

$$d = \begin{vmatrix} f_{xx}(a, b) & f_{xy}(a, b) \\ f_{yx}(a, b) & f_{yy}(a, b) \end{vmatrix}$$

where $f_{xy}(a, b) = f_{yx}(a, b)$ by Theorem 12.3.

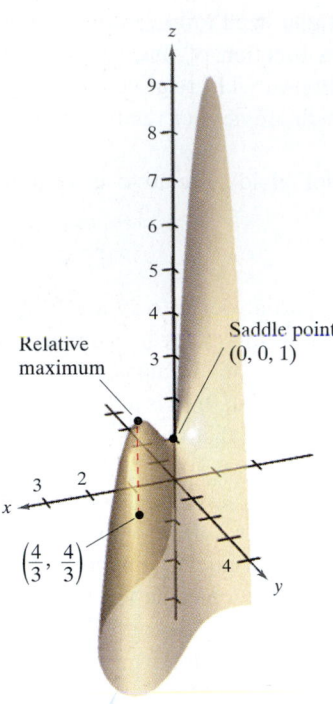

$$f(x, y) = -x^3 + 4xy - 2y^2 + 1$$

Figure 12.69

EXAMPLE 3   Using the Second Partials Test

Find the relative extrema of $f(x, y) = -x^3 + 4xy - 2y^2 + 1$.

Solution   Begin by finding the critical points of f. Because

$$f_x(x, y) = -3x^2 + 4y \quad \text{and} \quad f_y(x, y) = 4x - 4y$$

are defined for all x and y, the only critical points are those for which both first partial derivatives are 0. To locate these points, let $f_x(x, y)$ and $f_y(x, y)$ be 0 to obtain $-3x^2 + 4y = 0$ and $4x - 4y = 0$. From the second equation you know that $x = y$, and, by substitution into the first equation, you obtain two solutions: $y = x = 0$ and $y = x = \frac{4}{3}$. Because

$$f_{xx}(x, y) = -6x, \quad f_{yy}(x, y) = -4, \quad \text{and} \quad f_{xy}(x, y) = 4$$

it follows that, for the critical point $(0, 0)$,

$$d = f_{xx}(0, 0)f_{yy}(0, 0) - [f_{xy}(0, 0)]^2 = 0 - 16 < 0$$

and, by the Second Partials Test, you can conclude that $(0, 0, 1)$ is a saddle point of f. Furthermore, for the critical point $\left(\frac{4}{3}, \frac{4}{3}\right)$,

$$d = f_{xx}\left(\tfrac{4}{3}, \tfrac{4}{3}\right) f_{yy}\left(\tfrac{4}{3}, \tfrac{4}{3}\right) - \left[f_{xy}\left(\tfrac{4}{3}, \tfrac{4}{3}\right)\right]^2$$
$$= -8(-4) - 16 = 16 > 0$$

and because $f_{xx}\left(\frac{4}{3}, \frac{4}{3}\right) = -8 < 0$ you can conclude that f has a relative maximum at $\left(\frac{4}{3}, \frac{4}{3}\right)$, as shown in Figure 12.69.

The Second Partials Test can fail to find relative extrema in two ways. If either of the first partial derivatives is undefined, you cannot use the test. Also, if

$$d = f_{xx}(a, b)f_{yy}(a, b) - [f_{xy}(a, b)]^2 = 0$$

the test fails. In such cases, you can try a sketch or some other approach, as demonstrated in the next example.

EXAMPLE 4   Failure of the Second Partials Test

Find the relative extrema of $f(x, y) = x^2y^2$.

Solution   Because $f_x(x, y) = 2xy^2$ and $f_y(x, y) = 2x^2y$, you know that both partial derivatives are 0 if $x = 0$ or $y = 0$. That is, every point along the x- or y-axis is a critical point. Moreover, because

$$f_{xx}(x, y) = 2y^2, \quad f_{yy}(x, y) = 2x^2, \quad \text{and} \quad f_{xy}(x, y) = 4xy$$

you know that if either $x = 0$ or $y = 0$, then

$$d = f_{xx}(x, y)f_{yy}(x, y) - [f_{xy}(x, y)]^2$$
$$= 4x^2y^2 - 16x^2y^2 = -12x^2y^2 = 0.$$

Thus, the Second Partials Test fails. However, because $f(x, y) = 0$ for every point along the x- or y-axis and $f(x, y) = x^2y^2 > 0$ for all other points, you can conclude that each of these critical points yields an absolute minimum, as shown in Figure 12.70.

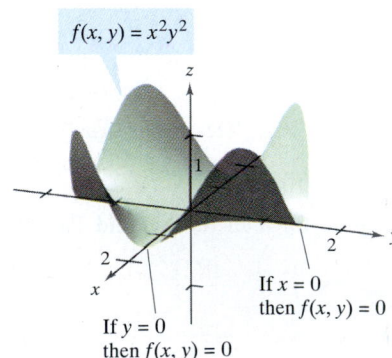

$$f(x, y) = x^2y^2$$

Figure 12.70

Absolute extrema of a function can occur in two ways. First, some relative extrema also happen to be absolute extrema. For instance, in Example 1, $f(-2, 3)$ is an absolute minimum of the function. (On the other hand, the relative maximum found in Example 3 is not an absolute maximum of the function.) Second, absolute extrema can occur at a boundary point of the domain. This is illustrated in Example 5.

EXAMPLE 5   Finding Absolute Extrema

Find the absolute extrema of the function

$$f(x, y) = \sin xy$$

on the closed region given by $0 \le x \le \pi$ and $0 \le y \le 1$.

Solution   From the partial derivatives

$$f_x(x, y) = y \cos xy \quad \text{and} \quad f_y(x, y) = x \cos xy$$

you can see that each point lying on the hyperbola given by $xy = \pi/2$ is a critical point. These points each yield the value

$$f(x, y) = \sin\left(\frac{\pi}{2}\right) = 1$$

which you know is the absolute maximum, as shown in Figure 12.71. The only other critical point of f *lying in the given region* is $(0, 0)$. It yields an absolute minimum of 0, because

$$0 \le xy \le \pi$$

implies that

$$0 \le \sin xy \le 1.$$

To hunt for other absolute extrema, you should consider the four boundaries of the region formed by taking traces with the vertical planes $x = 0$, $x = \pi$, $y = 0$, and $y = 1$. In doing this, you will find that $\sin xy = 0$ at all points on the x-axis, at all points on the y-axis, and at the point $(\pi, 1)$. Each of these points yields an absolute minimum for the surface, as shown in Figure 12.71.

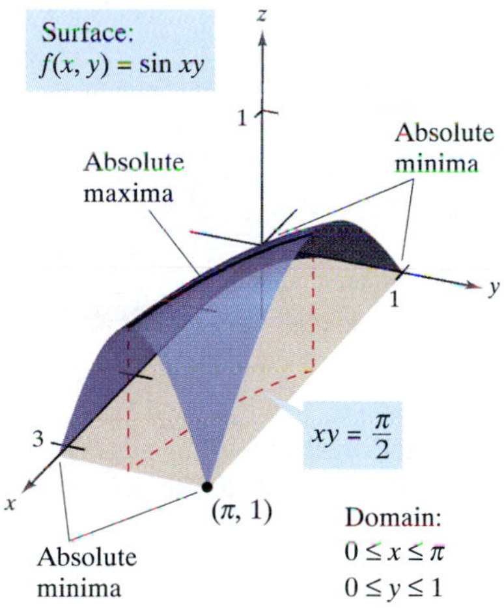

Surface:
$f(x, y) = \sin xy$

Absolute maxima

Absolute minima

$xy = \dfrac{\pi}{2}$

$(\pi, 1)$

Absolute minima

Domain:
$0 \le x \le \pi$
$0 \le y \le 1$

Figure 12.71

The concepts of relative extrema and critical points can be extended to functions of three or more variables. If all first partial derivatives of

$$w = f(x_1, x_2, x_3, \ldots, x_n)$$

exist, it can be shown that a relative maximum or minimum can occur at $(x_1, x_2, x_3, \ldots, x_n)$ only if every first partial derivative is 0 at that point. This means that the critical points are obtained by solving the following system of equations.

$$f_{x_1}(x_1, x_2, x_3, \ldots, x_n) = 0$$
$$f_{x_2}(x_1, x_2, x_3, \ldots, x_n) = 0$$
$$\vdots$$
$$f_{x_n}(x_1, x_2, x_3, \ldots, x_n) = 0$$

The extension of Theorem 12.17 to three or more variables is also possible, although we will not consider such an extension in this text.

In Exercises 1–6, identify any extrema of the function by recognizing its given form or its form after completing the square. Verify your results by using the partial derivatives to locate any critical points and test for relative extrema. Use a computer algebra system to graph the function and label any extrema.

1. $g(x, y) = (x - 1)^2 + (y - 3)^2$

2. $g(x, y) = 9 - (x - 3)^2 - (y + 2)^2$

3. $f(x, y) = \sqrt{x^2 + y^2 + 1}$

4. $f(x, y) = \sqrt{25 - (x - 2)^2 - y^2}$

5. $f(x, y) = x^2 + y^2 + 2x - 6y + 6$

6. $f(x, y) = -x^2 - y^2 + 4x + 8y - 11$

In Exercises 7–20, examine each function for relative extrema and saddle points.

7. $f(x, y) = 2x^2 + 2xy + y^2 + 2x - 3$

8. $f(x, y) = -x^2 - 5y^2 + 8x - 10y - 13$

9. $f(x, y) = -5x^2 + 4xy - y^2 + 16x + 10$

10. $f(x, y) = x^2 + 6xy + 10y^2 - 4y + 4$

11. $z = 2x^2 + 3y^2 - 4x - 12y + 13$

12. $z = -3x^2 - 2y^2 + 3x - 4y + 5$

13. $h(x, y) = x^2 - y^2 - 2x - 4y - 4$

14. $g(x, y) = 120x + 120y - xy - x^2 - y^2$

15. $h(x, y) = x^2 - 3xy - y^2$

16. $g(x, y) = xy$

17. $f(x, y) = x^3 - 3xy + y^3$

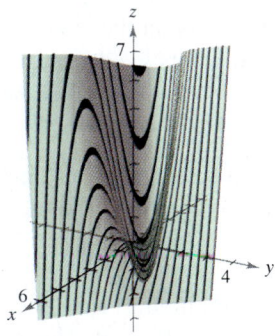

18. $f(x, y) = 4xy - x^4 - y^4$

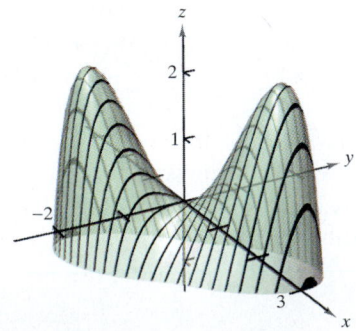

19. $z = e^{-x} \sin y$

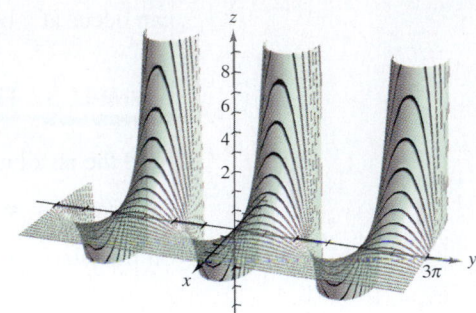

20. $z = \left(\dfrac{1}{2} - x^2 + y^2\right)e^{1 - x^2 - y^2}$

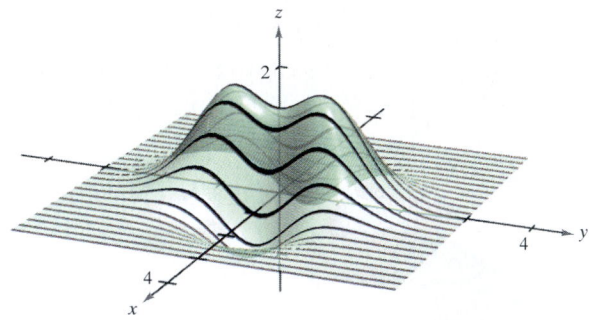

 In Exercises 21–24, use a computer algebra system to graph the surface and locate any relative extrema and saddle points.

21. $z = \dfrac{-4x}{x^2 + y^2 + 1}$

22. $z = (x^2 + 4y^2)e^{1 - x^2 - y^2}$

23. $f(x, y) = y^3 - 3yx^2 - 3y^2 - 3x^2 + 1$

24. $z = e^{xy}$

Think About It In Exercises 25–28, determine whether there is a relative maximum, a relative minimum, a saddle point, or insufficient information to determine the nature of the function $f(x, y)$ at the critical point (x_0, y_0).

25. $f_{xx}(x_0, y_0) = 9$, $f_{yy}(x_0, y_0) = 4$, $f_{xy}(x_0, y_0) = 6$

26. $f_{xx}(x_0, y_0) = -3$, $f_{yy}(x_0, y_0) = -8$, $f_{xy}(x_0, y_0) = 2$

27. $f_{xx}(x_0, y_0) = -9$, $f_{yy}(x_0, y_0) = 6$, $f_{xy}(x_0, y_0) = 10$

28. $f_{xx}(x_0, y_0) = 25$, $f_{yy}(x_0, y_0) = 8$, $f_{xy}(x_0, y_0) = 10$

Think About It In Exercises 29–32, sketch the graph of an arbitrary function f satisfying the given conditions. State whether the function has any extrema or saddle points. (There are many correct answers.)

29. $f_x(x, y) > 0$ and $f_y(x, y) < 0$ for all (x, y)

30. All of the first and second partial derivatives of f are 0.

31. $f_x(0, 0) = 0$, $f_y(0, 0) = 0$

$$f_x(x, y)\begin{cases} < 0, & x < 0 \\ > 0, & x > 0 \end{cases}, \quad f_y(x, y)\begin{cases} > 0, & y < 0 \\ < 0, & y > 0 \end{cases}$$

$f_{xx}(x, y) > 0$, $f_{yy}(x, y) < 0$, and $f_{xy}(x, y) = 0$ for all (x, y)

32. $f_x(2, 1) = 0$, $f_y(2, 1) = 0$

$$f_x(x, y)\begin{cases} > 0, & x < 2 \\ < 0, & x > 2 \end{cases}, \quad f_y(x, y)\begin{cases} > 0, & y < 1 \\ < 0, & y > 1 \end{cases}$$

$f_{xx}(x, y) < 0$, $f_{yy}(x, y) < 0$, and $f_{xy}(x, y) = 0$ for all (x, y)

33. A function f has continuous second partial derivatives on an open region containing the critical point $(3, 7)$. The function has a minimum at $(3, 7)$. Determine the interval for $f_{xy}(3, 7)$ if $f_{xx}(3, 7) = 2$ and $f_{yy}(3, 7) = 8$.

34. A function f has continuous second partial derivatives on an open region containing the critical point (a, b). If $f_x(a, b)$ and $f_y(a, b)$ have opposite signs, what is implied? Explain.

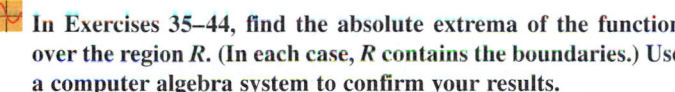 **In Exercises 35–44, find the absolute extrema of the function over the region R. (In each case, R contains the boundaries.) Use a computer algebra system to confirm your results.**

35. $f(x, y) = 12 - 3x - 2y$

    R: The triangular region in the xy-plane with vertices $(2, 0)$, $(0, 1)$, and $(1, 2)$.

36. $f(x, y) = (2x - y)^2$

    R: The triangular region in the xy-plane with vertices $(2, 0)$, $(0, 1)$, and $(1, 2)$.

37. $f(x, y) = 3x^2 + 2y^2 - 4y$

    R: The region in the xy-plane bounded by the graphs of $y = x^2$ and $y = 4$.

38. $f(x, y) = 2x - 2xy + y^2$

    R: The region in the xy-plane bounded by the graphs of $y = x^2$ and $y = 1$.

39. $f(x, y) = x^2 + xy$, $R = \{(x, y) : |x| \le 2, |y| \le 1\}$

40. $f(x, y) = x^2 + 2xy + y^2$, $R = \{(x, y) : |x| \le 2, |y| \le 1\}$

41. $f(x, y) = x^2 + 2xy + y^2$, $R = \{(x, y) : x^2 + y^2 \le 8\}$

42. $f(x, y) = x^2 - 4xy$

    $R = \{(x, y) : 0 \le x \le 4, 0 \le y \le \sqrt{x}\}$

43. $f(x, y) = \dfrac{4xy}{(x^2 + 1)(y^2 + 1)}$

    $R = \{(x, y) : 0 \le x \le 1, 0 \le y \le 1\}$

44. $f(x, y) = \dfrac{4xy}{(x^2 + 1)(y^2 + 1)}$

    $R = \{(x, y) : x \ge 0, y \ge 0, x^2 + y^2 \le 1\}$

In Exercises 45–50, find the critical points and test for relative extrema. List the critical points for which the Second Partials Test fails.

45. $f(x, y) = x^3 + y^3$

46. $f(x, y) = x^3 + y^3 - 3x^2 + 6y^2 + 3x + 12y + 7$

47. $f(x, y) = (x - 1)^2(y + 4)^2$

48. $f(x, y) = \sqrt{(x - 1)^2 + (y + 2)^2}$

49. $f(x, y) = x^{2/3} + y^{2/3}$

50. $f(x, y) = (x^2 + y^2)^{2/3}$

In Exercises 51 and 52, find the critical points of the function and, from the form of the function, determine whether each point is a relative maximum or relative minimum.

51. $f(x, y, z) = x^2 + (y - 3)^2 + (z + 1)^2$

52. $f(x, y, z) = 4 - [x(y - 1)(z + 2)]^2$

53. _Think About It_   The figure shows the level curves for an unknown function $f(x, y)$. What, if any, information can be given about f at the point A?

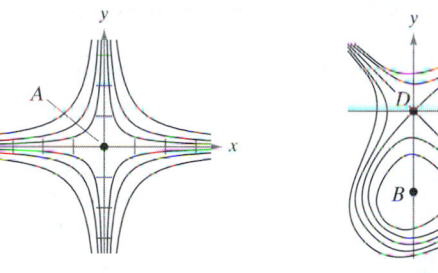

Figure for 53   **Figure for 54**

54. _Think About It_   The figure shows the level curves for an unknown function $f(x, y)$. What, if any, information can be given about the points A, B, C, and D?

 **55. _Investigation_**   Consider the function

$$f(x, y) = (\alpha x^2 + \beta y^2)e^{-(x^2 + y^2)}, \quad 0 < |\alpha| < \beta.$$

    (a) Use a graphing utility to graph the function for $\alpha = 1$ and $\beta = 2$, and identify any extrema or saddle points.

    (b) Use a graphing utility to graph the function for $\alpha = -1$ and $\beta = 2$, and identify any extrema or saddle points.

    (c) Generalize the results in parts (a) and (b) for the function f.

56. Prove that if f is a differentiable function such that $\nabla f(x_0, y_0) = \mathbf{0}$, then the tangent plane at (x_0, y_0) is horizontal.

True or False?   **In Exercises 57–60, determine whether the statement is true or false. If it is false, explain why or give an example that shows it is false.**

57. If f has a relative maximum at (x_0, y_0, z_0), then $f_x(x_0, y_0) = f_y(x_0, y_0) = 0$.

58. The function given by $f(x, y) = \sqrt[3]{x^2 + y^2}$ has a relative minimum at the origin.

59. If f is continuous for all x and y and has two relative minima, then f must have at least one relative maximum.

60. If $f_x(x_0, y_0) = 0$ and $f_y(x_0, y_0) = 0$, then at the point (x_0, y_0, z_0), the tangent plane to the surface given by $z = f(x, y)$ is horizontal.

Applied Optimization Problems • The Method of Least Squares

Applied Optimization Problems

In this section, we survey a few of the many applications of extrema of functions of two (or more) variables.

EXAMPLE 1   Finding Maximum Volume

A rectangular box is resting on the xy-plane with one vertex at the origin. The opposite vertex lies in the plane

$$6x + 4y + 3z = 24$$

as shown in Figure 12.72. Find the maximum volume of such a box.

Solution   Let x, y, and z represent the length, width, and height of the box. Because one vertex of the box lies in the plane $6x + 4y + 3z = 24$, you know that $z = \frac{1}{3}(24 - 6x - 4y)$, and you can write the volume xyz of the box as a function of two variables.

$$V(x, y) = (x)(y)\left[\tfrac{1}{3}(24 - 6x - 4y)\right]$$
$$= \tfrac{1}{3}(24xy - 6x^2y - 4xy^2)$$

By setting the first partial derivatives equal to 0

$$V_x(x, y) = \tfrac{1}{3}(24y - 12xy - 4y^2) = \frac{y}{3}(24 - 12x - 4y) = 0$$

$$V_y(x, y) = \tfrac{1}{3}(24x - 6x^2 - 8xy) = \frac{x}{3}(24 - 6x - 8y) = 0$$

you obtain the critical points $(0, 0)$ and $\left(\tfrac{4}{3}, 2\right)$. At $(0, 0)$ the volume is 0, so that point does not yield a maximum volume. At the point $\left(\tfrac{4}{3}, 2\right)$, you can apply the Second Partials Test.

$$V_{xx}(x, y) = -4y$$
$$V_{yy}(x, y) = \frac{-8x}{3}$$
$$V_{xy}(x, y) = \tfrac{1}{3}(24 - 12x - 8y)$$

Because

$$V_{xx}\left(\tfrac{4}{3}, 2\right)V_{yy}\left(\tfrac{4}{3}, 2\right) - \left[V_{xy}\left(\tfrac{4}{3}, 2\right)\right]^2 = (-8)\left(-\tfrac{32}{9}\right) - \left(-\tfrac{8}{3}\right)^2 = \tfrac{64}{3} > 0$$

and

$$V_{xx}\left(\tfrac{4}{3}, 2\right) = -8 < 0$$

you can conclude from the Second Partials Test that the maximum volume is

$$V\left(\tfrac{4}{3}, 2\right) = \tfrac{1}{3}\left[24\left(\tfrac{4}{3}\right)(2) - 6\left(\tfrac{4}{3}\right)^2(2) - 4\left(\tfrac{4}{3}\right)(2^2)\right]$$
$$= \tfrac{64}{9} \text{ cubic units.}$$

(Note that the volume is 0 at the boundary points of the triangular domain of V.)

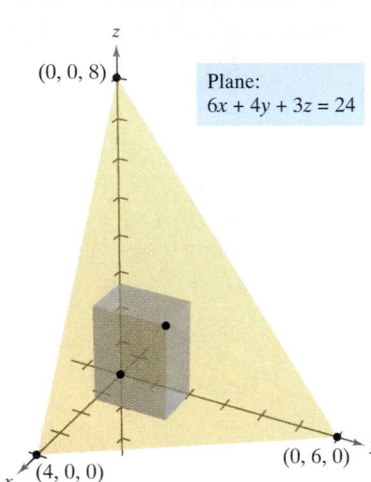

$(0, 0, 8)$

Plane:
$6x + 4y + 3z = 24$

$(0, 6, 0)$

$(4, 0, 0)$

The maximum volume of the box is $\tfrac{64}{9}$ cubic units.
Figure 12.72

NOTE   In many applied problems, the domain of the function to be optimized is a closed bounded region. To find minimum or maximum points, you must not only test critical points, but also consider the values of the function at points on the boundary.

In Section 3.10 you studied several applications of extrema in economics and business. In practice, such applications often involve more than one independent variable. For instance, a company may produce several models of one type of product. The price per unit and profit per unit are usually different for each model. Moreover, the demand for each model is often a function of the prices of the other models (as well as its own price). The next example illustrates an application involving two products.

EXAMPLE 2   Finding the Maximum Profit

The profit obtained by producing x units of product A and y units of product B is approximated by the model

$$P(x, y) = 8x + 10y - (0.001)(x^2 + xy + y^2) - 10,000.$$

Find the production level that produces a maximum profit.

Solution   The partial derivatives of the profit function are

$$P_x(x, y) = 8 - (0.001)(2x + y)$$

and

$$P_y(x, y) = 10 - (0.001)(x + 2y).$$

FOR FURTHER INFORMATION For more information on the use of mathematics in economics, see the article "Mathematical Methods of Economics" by Joel Franklin in the April 1983 issue of *The American Mathematical Monthly.*

By setting these partial derivatives equal to 0, you obtain the following system of equations.

$$8 - (0.001)(2x + y) = 0$$
$$10 - (0.001)(x + 2y) = 0$$

After simplifying, this system of linear equations can be written as

$$2x + \ y = \ \ 8,000$$
$$x + 2y = 10,000.$$

Solving this system produces $x = 2000$ and $y = 4000$. The second partial derivatives of P are

$$P_{xx}(2000, 4000) = -0.002$$
$$P_{yy}(2000, 4000) = -0.002$$
$$P_{xy}(2000, 4000) = -0.001.$$

Moreover, because $P_{xx} < 0$ and

$$P_{xx}(2000, 4000)P_{yy}(2000, 4000) - [P_{xy}(2000, 4000)]^2 =$$
$$(-0.002)^2 - (-0.001)^2 > 0$$

you can conclude that the production level of $x = 2000$ units and $y = 4000$ units yields a *maximum* profit.

NOTE   In Example 2, we assumed that the manufacturing plant is able to produce the required number of units to yield a maximum profit. In actual practice, the production would be bounded by physical constraints. You will study such constrained optimization problems in the next section.

The Method of Least Squares

Many of the examples in this text have involved **mathematical models.** For instance, Example 2 involves a quadratic model for profit. There are several ways to develop such models; one is called the **method of least squares.**

In constructing a model to represent a particular phenomenon, the goals are simplicity and accuracy. Of course, these goals often conflict. For instance, a simple linear model for the points in Figure 12.73 is

$$y = 1.8566x - 5.0246.$$

However, Figure 12.74 shows that by choosing the slightly more complicated quadratic model*

$$y = 0.1996x^2 - 0.7281x + 1.3749$$

you can achieve greater accuracy.

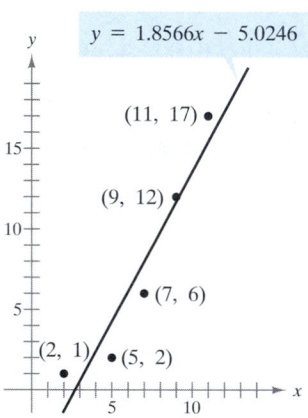

Figure 12.73

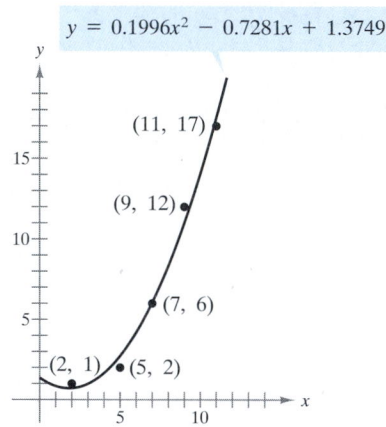

Figure 12.74

As a measure of how well the model $y = f(x)$ fits the collection of points

$$\{(x_1, y_1), (x_2, y_2), (x_3, y_3), \ldots, (x_n, y_n)\}$$

you can add the squares of the differences between the actual y-values and the values given by the model to obtain the **sum of the squared errors**

$$S = \sum_{i=1}^{n} [f(x_i) - y_i]^2. \qquad \text{Sum of the squared errors}$$

Graphically, S can be interpreted as the sum of the squares of the vertical distances between the graph of f and the given points in the plane, as shown in Figure 12.75. If the model is perfect, then $S = 0$. However, when perfection is not feasible, we settle for a model that minimizes S. Statisticians call the *linear model* that minimizes S the **least squares regression line.** The proof that this line actually minimizes S involves the minimizing of a function of two variables.

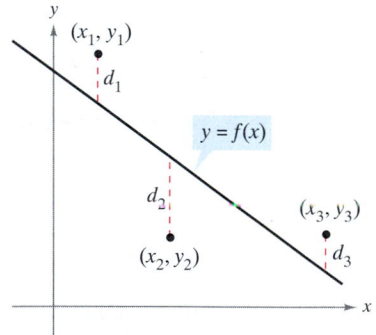

Sum of the squared errors:
$S = d_1^2 + d_2^2 + d_3^2$
Figure 12.75

---

*A method for finding the least squares quadratic model for a collection of data is described in Exercise 37.

The Granger Collection

ADRIEN-MARIE LEGENDRE (1752–1833)

The method of least squares was introduced by the French mathematician Adrien-Marie Legendre. Legendre is best known for his work in geometry. In fact, his text *Elements of Geometry* was so popular in the United States that it continued to be used for 33 editions, spanning a period of more than 100 years.

THEOREM 12.18   Least Squares Regression Line

The **least squares regression line** for $\{(x_1, y_1), (x_2, y_2), \ldots, (x_n, y_n)\}$ is given by $f(x) = ax + b$, where

$$a = \frac{n\sum_{i=1}^{n} x_i y_i - \sum_{i=1}^{n} x_i \sum_{i=1}^{n} y_i}{n\sum_{i=1}^{n} x_i^2 - \left(\sum_{i=1}^{n} x_i\right)^2} \qquad \text{and} \qquad b = \frac{1}{n}\left(\sum_{i=1}^{n} y_i - a\sum_{i=1}^{n} x_i\right).$$

Proof   Let $S(a, b)$ represent the sum of the squared errors for the model $f(x) = ax + b$ and the given set of points. That is,

$$S(a, b) = \sum_{i=1}^{n} [f(x_i) - y_i]^2$$

$$= \sum_{i=1}^{n} (ax_i + b - y_i)^2$$

where the points (x_i, y_i) represent constants. Because S is a function of a and b, you can use the methods discussed in the preceding section to find the minimum value of S. Specifically, the first partial derivatives of S are

$$S_a(a, b) = \sum_{i=1}^{n} 2x_i(ax_i + b - y_i)$$

$$= 2a\sum_{i=1}^{n} x_i^2 + 2b\sum_{i=1}^{n} x_i - 2\sum_{i=1}^{n} x_i y_i$$

$$S_b(a, b) = \sum_{i=1}^{n} 2(ax_i + b - y_i)$$

$$= 2a\sum_{i=1}^{n} x_i + 2nb - 2\sum_{i=1}^{n} y_i.$$

By setting these two partial derivatives equal to 0, you obtain the values for a and b that are listed in the theorem. We leave it to you to apply the Second Partials Test (see Exercise 38) to verify that these values of a and b yield a minimum.

If the x-values are symmetrically spaced about the y-axis, then $\Sigma x_i = 0$ and the formulas for a and b simplify to

$$a = \frac{\sum_{i=1}^{n} x_i y_i}{\sum_{i=1}^{n} x_i^2} \qquad \text{and} \qquad b = \frac{1}{n}\sum_{i=1}^{n} y_i.$$

This simplification is often possible with a translation of the x-values. For instance, if the x-values in a data collection consist of the years 1980, 1981, 1982, 1983, and 1984, you could let 1982 be represented by 0.

EXAMPLE 3   Finding the Least Squares Regression Line

Find the least squares regression line for the points $(-3, 0)$, $(-1, 1)$, $(0, 2)$, and $(2, 3)$.

Solution   The table shows the calculations involved in finding the least squares regression line using $n = 4$.

TECHNOLOGY   Many calculators have "built-in" least squares regression programs. If your calculator has such a program, try using it to duplicate the results of Example 3.

x	y	xy	x^2
-3	0	0	9
-1	1	-1	1
0	2	0	0
2	3	6	4
$\displaystyle\sum_{i=1}^{n} x_i = -2$	$\displaystyle\sum_{i=1}^{n} y_i = 6$	$\displaystyle\sum_{i=1}^{n} x_i y_i = 5$	$\displaystyle\sum_{i=1}^{n} x_i^2 = 14$

Applying Theorem 12.18 produces

$$a = \frac{n\displaystyle\sum_{i=1}^{n} x_i y_i - \displaystyle\sum_{i=1}^{n} x_i \displaystyle\sum_{i=1}^{n} y_i}{n\displaystyle\sum_{i=1}^{n} x_i^2 - \left(\displaystyle\sum_{i=1}^{n} x_i\right)^2} = \frac{4(5) - (-2)(6)}{4(14) - (-2)^2} = \frac{8}{13}$$

and

$$b = \frac{1}{n}\left(\sum_{i=1}^{n} y_i - a\sum_{i=1}^{n} x_i\right) = \frac{1}{4}\left[6 - \frac{8}{13}(-2)\right] = \frac{47}{26}.$$

The least squares regression line is $y = \frac{8}{13}x + \frac{47}{26}$, as shown in Figure 12.76.

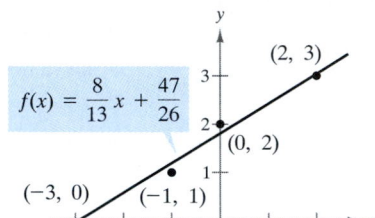

$f(x) = \dfrac{8}{13}x + \dfrac{47}{26}$

Least squares regression line
Figure 12.76

EXERCISES FOR SECTION 12.9

In Exercises 1 and 2, find the minimum distance from the point to the plane $2x + 3y + z = 12$. (*Hint:* To simplify the computations, minimize the square of the distance.)

1. $(0, 0, 0)$ **2.** $(1, 2, 3)$

In Exercises 3 and 4, find the minimum distance from the point to the paraboloid $z = x^2 + y^2$.

3. $(5, 5, 0)$ **4.** $(5, 0, 0)$

In Exercises 5–8, find three positive numbers x, y and z that satisfy the indicated conditions.

5. The sum is 30 and the product is a maximum.

6. The sum is 32 and $P = xy^2z$ is a maximum.

7. The sum is 30 and the sum of the squares is a minimum.

8. The sum is 1 and the sum of the squares is a minimum.

9. *Volume*   The sum of the length and the girth (perimeter of a cross section) of packages carried by a delivery service cannot exceed 108 inches. Find the dimensions of the rectangular package of largest volume that may be sent.

10. *Volume*   The material for constructing the base of an open box costs 1.5 times as much per unit area as the material for constructing the sides. For a fixed amount of money C, find the dimensions of the box of largest volume that can be made.

11. *Volume*   The volume of an ellipsoid

$$\frac{x^2}{a^2} + \frac{y^2}{b^2} + \frac{z^2}{c^2} = 1$$

is $4\pi abc/3$. For a fixed sum $a + b + c$, show that the ellipsoid of maximum volume is a sphere.

12. *Volume*   Show that a rectangular box of maximum volume inscribed in a sphere of radius r is a cube.

13. *Volume and Surface Area*   Show that a rectangular box of given volume and minimum surface area is a cube.

14. Volume Repeat Exercise 9 under the condition that the sum of the perimeters of the two cross sections shown in the figure cannot exceed 108 inches.

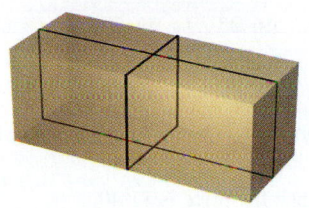

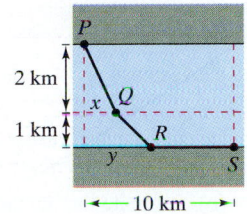

Figure for 14   **Figure for 15**

15. Minimum Cost A water line is to be built from point P to point S and must pass through regions where construction costs differ (see figure). Find x and y such that the total cost C will be minimized if the cost per kilometer in dollars is $3k$ from P to Q, $2k$ from Q to R, and k from R to S.

16. Area A trough with trapezoidal cross sections is formed by turning up the edges of a 10-inch wide sheet of aluminum (see figure). Find the cross section of maximum area.

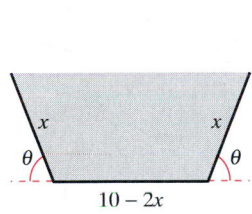

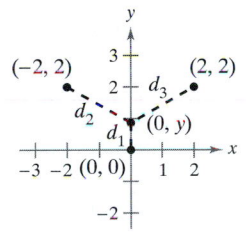

Figure for 16   **Figure for 18**

17. Area Repeat Exercise 16 for a sheet that is w inches wide.

18. Distance A company has retail outlets located at the points $(0, 0)$, $(2, 2)$, and $(-2, 2)$ (see figure). Management plans to build a distribution center located such that the sum of the distances S from the center to the outlets is minimum. From the symmetry of the problem it is clear that the distribution center will be located on the y-axis, and therefore S is a function of the single variable y. Using techniques presented in Chapter 3, find the required value of y.

19. Investigation The retail outlets described in Exercise 18 are located at $(0, 0)$, $(4, 2)$, and $(-2, 2)$ (see figure). The location of the distribution center is (x, y), and therefore the sum of the distances S is a function of x and y.

(a) Write the expression giving the sum of the distances S. Use a computer algebra system to graph S. Does the surface have a minimum?

(b) Use a computer algebra system to obtain S_x and S_y. Observe that solving the system $S_x = 0$ and $S_y = 0$ is very difficult. Therefore, you will approximate the location of the distribution center.

(c) An initial estimate of the critical point is $(x_1, y_1) = (1, 1)$. Calculate $-\nabla S(1, 1)$ with components $-S_x(1, 1)$ and $-S_y(1, 1)$. What direction is given by the vector $-\nabla S(1, 1)$?

(d) The second estimate of the critical point is

$$(x_2, y_2) = (x_1 - S_x(x_1, y_1)t, y_1 - S_y(x_1, y_1)t).$$

If these coordinates are substituted into $S(x, y)$, then S becomes a function of the single variable t. Find the value of t that minimizes S. Use this value of t to estimate (x_2, y_2).

(e) Complete two more iterations of the process in part (d) to obtain (x_4, y_4). For this location of the distribution center, what is the sum of the distances to the retail outlets?

(f) Explain why $-\nabla S(x, y)$ was used to approximate the minimum value of S. In what types of problems would you use $\nabla S(x, y)$?

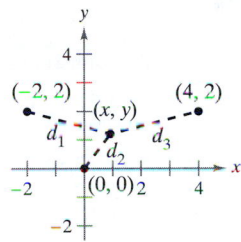

20. Investigation Repeat Exercise 19 for retail outlets located at the points $(-4, 0)$, $(1, 6)$, and $(12, 2)$.

21. Profit A corporation manufactures a product at two locations. The cost of producing x units at location 1 is

$$C_1 = 0.02x_1^2 + 4x_1 + 500$$

and the cost of producing x_2 units at location 2 is

$$C_2 = 0.05x_2^2 + 4x_2 + 275.$$

If the product sells for \$15 per unit, find the quantity that should be produced at each location to maximize the profit $P = 15(x_1 + x_2) - C_1 - C_2$.

22. Hardy-Weinberg Law Common blood types are determined genetically by three alleles A, B, and O. (An allele is any of a group of possible mutational forms of a gene.) A person whose blood type is AA, BB, or OO is homozygous. A person whose blood type is AB, AO, or BO is heterozygous. The Hardy-Weinberg Law states that the proportion P of heterozygous individuals in any given population is

$$P(p, q, r) = 2pq + 2pr + 2qr,$$

where p represents the percent of allele A in the population, q represents the percent of allele B in the population, and r represents the percent of allele O in the population. Use the fact that $p + q + r = 1$ to show that the maximum proportion of heterozygous individuals in any population is $\frac{2}{3}$.

23. Revenue A company manufactures two products. The total revenue from x_1 units of product 1 and x_2 units of product 2 is $R = -5x_1^2 - 8x_2^2 - 2x_1x_2 + 42x_1 + 102x_2$. Find x_1 and x_2 so as to maximize the revenue.

24. Revenue A retail outlet sells two competitive products, the prices of which are p_1 and p_2. Find p_1 and p_2 so as to maximize total revenue, where $R = 500p_1 + 800p_2 + 1.5p_1p_2 - 1.5p_1^2 - p_2^2$.

In Exercises 25–28, (a) find the least squares regression line and (b) calculate *S*, the sum of the squared errors. Use the regression capabilities of a graphing utility to verify your results.

25.

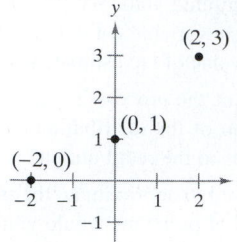

26.

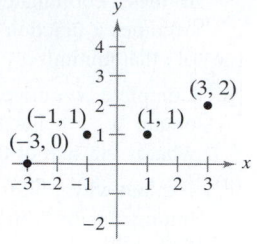

27.

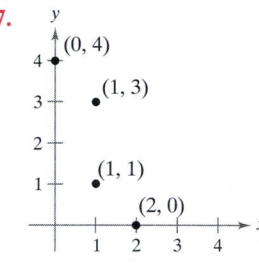

28.

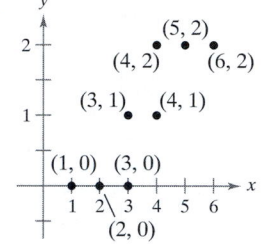

In Exercises 29–32, find the least squares regression line for the points. Use the regression capabilities of a graphing utility to verify your results. Use the graphing utility to plot the points and graph the regression line.

29. $(0, 0), (1, 1), (3, 4), (4, 2), (5, 5)$

30. $(1, 0), (3, 3), (5, 6)$

31. $(0, 6), (4, 3), (5, 0), (8, -4), (10, -5)$

32. $(5, 2), (0, 0), (2, 1), (7, 4), (10, 6), (12, 6)$

33. *Modeling Data* The costs per fine ounce of gold and silver for the years 1990 through 1994 are given in the table. *(Source: U.S. Bureau of Mines)*

Year	1990	1991	1992	1993	1994
Gold (*x*)	$385	$363	$345	$361	$389
Silver (*y*)	$4.82	$4.04	$3.94	$4.30	$5.30

Let *x* and *y* represent the costs per fine ounce of gold and silver.

(a) Use the regression capabilities of a graphing utility to find the least squares regression line for the data.

(b) Use a graphing utility to plot the data and graph the model.

(c) Use the model to approximate the change in the cost of silver for a $1-per-ounce increase in the cost of gold.

34. *Modeling Data* A store manager wants to know the demand for a certain product as a function of price. The daily sales for three different prices of the product are given in the table.

Price (*x*)	$1.00	$1.25	$1.50
Demand (*y*)	450	375	330

(a) Use the regression capabilities of a graphing utility to find the least squares regression line for the data.

(b) Estimate the demand when the price is $1.40.

35. *Modeling Data* An agronomist used four test plots to determine the relationship between the wheat yield (in bushels per acre) and the amount of fertilizer (in hundreds of pounds per acre). The results are given in the table.

Fertilizer (*x*)	1.0	1.5	2.0	2.5
Yield (*y*)	32	41	48	53

Use the regression capabilities of a graphing utility to find the least squares regression line for the data, and estimate the yield for a fertilizer application of 160 pounds per acre.

36. *Modeling Data* The table gives the percent and number (in millions) of women in the work force for selected years. *(Source: Department of Labor)*

Year	1960	1970	1980	1990
Percent (*x*)	37.7	43.3	51.5	57.5
Number (*y*)	23.2	31.5	45.5	56.6

Year	1991	1992	1993	1994
Percent (*x*)	57.3	57.8	57.9	58.8
Number (*y*)	56.9	57.8	58.4	60.2

(a) Use the regression capabilities of a graphing utility to find the least squares regression line for the data.

(b) According to this model, approximately how many women enter the labor force for each one-point increase in the percent of women in the labor force?

37. Find a system of equations whose solution yields the coefficients *a*, *b*, and *c* for the least squares regression quadratic $y = ax^2 + bx + c$ for the points

$(x_1, y_1), (x_2, y_2), \ldots, (x_n, y_n)$

by minimizing the sum

$$S(a, b, c) = \sum_{i=1}^{n} (y_i - ax_i^2 - bx_i - c)^2.$$

38. Use the Second Partials Test to verify that the formulas for *a* and *b* given in Theorem 12.18 yield a minimum.

$$\left[\text{Hint: Use the fact that } n \sum_{i=1}^{n} x_i^2 \geq \left(\sum_{i=1}^{n} x_i \right)^2. \right]$$

In Exercises 39–42, use the result of Exercise 37 to find the least squares regression quadratic for the given points. Use the regression capabilities of a graphing utility to confirm your results. Use the graphing utility to plot the points and graph the least squares regression quadratic.

39. $(-2, 0), (-1, 0), (0, 1), (1, 2), (2, 5)$

40. $(-4, 5), (-2, 6), (2, 6), (4, 2)$

41. $(0, 0), (2, 2), (3, 6), (4, 12)$

42. $(0, 10), (1, 9), (2, 6), (3, 0)$

 43. *Modeling Data*   After a new turbocharger for an automobile engine was developed, the following experimental data were obtained for speed in miles per hour at 2-second intervals.

Time (x)	0	2	4	6	8	10
Speed (y)	0	15	30	50	65	70

(a) Find a least squares regression quadratic for the data. Use a graphing utility to confirm your results.

(b) Use a graphing utility to plot the points and graph the model.

 44. *Modeling Data*   The table gives the world population (in billions) for five different years.   (*Source: U.S. Bureau of the Census*)

Year (x)	1960	1970	1980	1990	1996
Population (y)	3.0	3.7	4.5	5.3	5.8

Let $x = 0$ represent the year 1960.

(a) Use the regression capabilities of a graphing utility to find the least squares regression line for the data.

(b) Use the regression capabilities of a graphing utility to find the least squares regression quadratic for the data.

(c) Use a graphing utility to plot the data and graph the models.

(d) Use both models to forecast the world population for the year 2010. How do the two models differ as you extrapolate into the future?

45. *Modeling Data*   A meteorologist measures the atmospheric pressure P (in kilograms per square meter) at altitude h (in kilometers). The data are shown below.

h	0	5	10	15	20
P	10,332	5583	2376	1240	517

(a) Use the regression capabilities of a graphing utility to find a least squares regression line for the points $(h, \ln P)$.

(b) The result in part (a) is an equation of the form $\ln P = ah + b$. Write this logarithmic form in exponential form.

(c) Use a graphing utility to plot the original data and graph the exponential model in part (b).

46. *Modeling Data*   The endpoints of the interval over which distinct vision is possible are called the *near point* and *far point* of the eye. With increasing age, these points normally change. The table gives the approximate near point y in centimeters for various ages x.

x	10	20	30	40	50
y	7	10	14	22	40

(a) Find a rational model for the data by taking the reciprocal of the near points to generate the points $(x, 1/y)$. Use the regression capabilities of a graphing utility to find a least squares regression line for the revised data. The resulting line has the form

$$\frac{1}{y} = ax + b.$$

Solve for y.

(b) Use a graphing utility to plot the data and graph the model.

(c) Do you think the model can be used to predict the near point for a person who is 60 years old? Explain.

SECTION PROJECT

Building a Pipeline   An oil company wishes to construct a pipeline from its offshore facility A to its refinery B. The offshore facility is 2 miles from shore, and the refinery is 1 mile inland. Furthermore, A and B are 5 miles apart, as indicated in the figure.

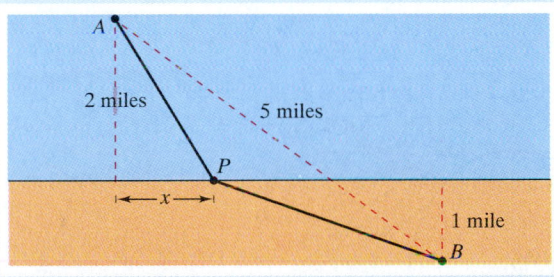

The cost of building the pipeline is \$3 million per mile in the water, and \$4 million per mile on land. Hence, the cost of the pipeline depends on the location of point P, where it meets the shore. What would be the most economical route of the pipeline?

Imagine that you are to write a report to the oil company about this problem. Let x be the distance indicated in the figure. Determine the cost of building the pipeline from A to P, and the cost from P to B. Analyze some sample pipeline routes and their corresponding costs. For instance, what is the cost of the most direct route? Then use calculus to determine the route of the pipeline that minimizes the cost. Explain all steps of your development and include any relevant graphs.

Lagrange Multipliers • Constrained Optimization Problems •
The Method of Lagrange Multipliers with Two Constraints

Lagrange Multipliers

Many optimization problems have restrictions or **constraints** on the values that can
be used to produce the optimal solution. Such constraints tend to complicate optimization problems because the optimal solution can occur at a boundary point of the
domain. In this section, you will study an ingenious technique for solving such problems. It is called the **Method of Lagrange Multipliers.**

To see how this technique works, suppose you want to find the rectangle of
maximum area that can be inscribed in the ellipse given by

$$\frac{x^2}{3^2} + \frac{y^2}{4^2} = 1.$$

Let (x, y) be the vertex of the rectangle in the first quadrant, as shown in Figure 12.77.
Because the rectangle has sides of length $2x$ and $2y$, its area is given by

$$f(x, y) = 4xy. \qquad \text{\color{red}Objective function}$$

You want to find x and y such that $f(x, y)$ is a maximum. Your choice of (x, y) is
restricted to first-quadrant points that lie on the ellipse.

$$\frac{x^2}{3^2} + \frac{y^2}{4^2} = 1 \qquad \text{\color{red}Constraint}$$

Now, consider the constraint equation to be a fixed level curve of

$$g(x, y) = \frac{x^2}{3^2} + \frac{y^2}{4^2}.$$

The level curves of f represent a family of hyperbolas

$$f(x, y) = 4xy = k.$$

In this family, the level curves that meet the given constraint correspond to the hyperbolas that intersect the ellipse. Moreover, to maximize $f(x, y)$, you want to find the
hyperbola that just barely satisfies the constraint. The level curve that does this is the
one that is *tangent* to the ellipse, as shown in Figure 12.78.

To find the appropriate hyperbola, use the fact that two curves are tangent at
a point if and only if their gradient vectors are parallel. This means that $\nabla f(x, y)$
must be a scalar multiple of $\nabla g(x, y)$ at the point of tangency. In the context of constrained optimization problems, this scalar is denoted by λ (the lowercase Greek
letter lambda).

$$\nabla f(x, y) = \lambda \nabla g(x, y)$$

The scalar λ is called a **Lagrange multiplier.** Theorem 12.19 gives the necessary conditions for the existence of such multipliers.

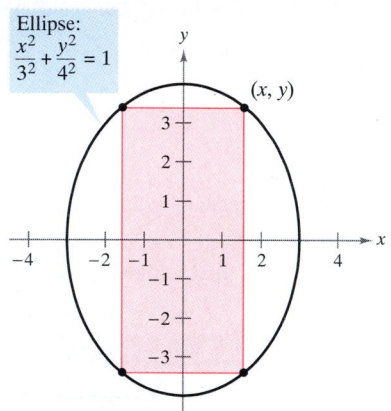

Ellipse:
$\frac{x^2}{3^2} + \frac{y^2}{4^2} = 1$

Objective function: $f(x, y) = 4xy$
Figure 12.77

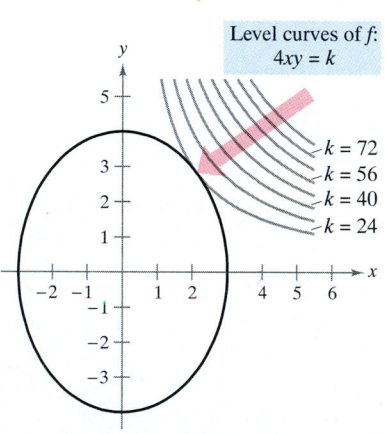

Level curves of f:
$4xy = k$

$k = 72$
$k = 56$
$k = 40$
$k = 24$

Constraint: $g(x, y) = \frac{x^2}{3^2} + \frac{y^2}{4^2} = 1$

Figure 12.78

THEOREM 12.19   Lagrange's Theorem

Let f and g have continuous first partial derivatives such that f has an extremum at a point (x_0, y_0) on the smooth constraint curve $g(x, y) = c$. If $\nabla g(x_0, y_0) \neq \mathbf{0}$, then there is a real number λ such that

$$\nabla f(x_0, y_0) = \lambda \nabla g(x_0, y_0).$$

JOSEPH-LOUIS LAGRANGE (1736-1813)

The Method of Lagrange Multipliers is named after the French mathematician Joseph-Louis Lagrange. Lagrange first introduced the method in his famous paper on mechanics, written when he was just 19 years old.

Proof   To begin, represent the smooth curve given by $g(x, y) = c$ by the vector-valued function

$$\mathbf{r}(t) = x(t)\mathbf{i} + y(t)\mathbf{j}, \qquad \mathbf{r}'(t) \neq \mathbf{0}$$

where x' and y' are continuous on an open interval I. Define the function h as $h(t) = f(x(t), y(t))$. Then, because $f(x_0, y_0)$ is an extreme value of f, you know that

$$h(t_0) = f(x(t_0), y(t_0)) = f(x_0, y_0)$$

is an extreme value of h. This implies that $h'(t_0) = 0$, and, by the Chain Rule,

$$h'(t_0) = f_x(x_0, y_0)x'(t_0) + f_y(x_0, y_0)y'(t_0) = \nabla f(x_0, y_0) \cdot \mathbf{r}'(t_0) = 0.$$

Therefore, $\nabla f(x_0, y_0)$ is orthogonal to $\mathbf{r}'(t_0)$. Moreover, by Theorem 12.12, $\nabla g(x_0, y_0)$ is also orthogonal to $\mathbf{r}'(t_0)$. Consequently, the gradients $\nabla f(x_0, y_0)$ and $\nabla g(x_0, y_0)$ are parallel, and there must exist a scalar λ such that

$$\nabla f(x_0, y_0) = \lambda \nabla g(x_0, y_0). \qquad \blacksquare$$

NOTE   Lagrange's Theorem can be shown to be true for functions of three variables, using a similar argument with level surfaces and Theorem 12.14.

The Method of Lagrange Multipliers uses Theorem 12.19 to find the extreme values of a function f subject to a constraint.

Method of Lagrange Multipliers

Let f and g satisfy the hypothesis of Lagrange's Theorem, and let f have a minimum or maximum subject to the constraint $g(x, y) = c$. To find the minimum or maximum of f, use the following steps.

1. Simultaneously solve the equations $\nabla f(x, y) = \lambda \nabla g(x, y)$ and $g(x, y) = c$ by solving the following system of equations.

$$f_x(x, y) = \lambda g_x(x, y)$$
$$f_y(x, y) = \lambda g_y(x, y)$$
$$g(x, y) = c$$

NOTE   As you will see in Examples 1 and 2, the Method of Lagrange Multipliers requires solving systems of nonlinear equations. This often can require some tricky algebraic manipulation.

2. Evaluate f at each solution point obtained in the first step. The largest value yields the maximum of f subject to the constraint $g(x, y) = c$, and the smallest value yields the minimum of f subject to the constraint $g(x, y) = c$.

Constrained Optimization Problems

At the beginning of this section, we described a problem in which we wanted to maximize the area of a rectangle that is inscribed in an ellipse. Example 1 shows how to use Lagrange multipliers to solve this problem.

EXAMPLE 1   Using a Lagrange Multiplier with One Constraint

Find the maximum value of

$$f(x, y) = 4xy, \qquad x > 0, y > 0$$

subject to the constraint $(x^2/3^2) + (y^2/4^2) = 1$.

NOTE   Example 1 can also be solved using the techniques you learned in Chapter 3. To see how, try to find the maximum value of $A = 4xy$ given that

$$\frac{x^2}{3^2} + \frac{y^2}{4^2} = 1.$$

To begin, solve the second equation for y to obtain

$$y = \tfrac{4}{3}\sqrt{9 - x^2}.$$

Then substitute into the first equation to obtain

$$A = 4x\left(\tfrac{4}{3}\sqrt{9 - x^2}\right).$$

Finally, use the techniques of Chapter 3 to maximize A.

Solution   To begin, let

$$g(x, y) = \frac{x^2}{3^2} + \frac{y^2}{4^2} = 1.$$

By equating $\nabla f(x, y) = 4y\,\mathbf{i} + 4x\,\mathbf{j}$ and $\lambda \nabla g(x, y) = (2\lambda x/9)\mathbf{i} + (\lambda y/8)\mathbf{j}$, you can obtain the following system of equations.

$$4y = \frac{2}{9}\lambda x \qquad\qquad f_x(x, y) = \lambda g_x(x, y)$$

$$4x = \frac{1}{8}\lambda y \qquad\qquad f_y(x, y) = \lambda g_y(x, y)$$

$$\frac{x^2}{3^2} + \frac{y^2}{4^2} = 1 \qquad\qquad \text{Constraint}$$

From the first equation, you obtain $\lambda = 18y/x$, and substitution into the second equation produces

$$4x = \frac{1}{8}\left(\frac{18y}{x}\right)y \quad \Longrightarrow \quad x^2 = \frac{9}{16}y^2.$$

Substituting this value for x^2 into the third equation produces

$$\frac{1}{9}\left(\frac{9}{16}y^2\right) + \frac{1}{16}y^2 = 1 \quad \Longrightarrow \quad y^2 = 8.$$

Thus, $y = \pm 2\sqrt{2}$. Because it is required that $y > 0$, choose the positive value and find that

$$x^2 = \frac{9}{16}y^2 = \frac{9}{16}(8) = \frac{9}{2} \quad \Longrightarrow \quad x^2 = \frac{9}{2} \quad \Longrightarrow \quad x = \frac{3}{\sqrt{2}}.$$

Thus, the maximum of f is

$$f\left(\frac{3}{\sqrt{2}}, 2\sqrt{2}\right) = 4xy = 4\left(\frac{3}{\sqrt{2}}\right)\left(2\sqrt{2}\right) = 24.$$

Note that writing the constraint as

$$g(x, y) = \frac{x^2}{3^2} + \frac{y^2}{4^2} = 1 \qquad \text{or} \qquad g(x, y) = \frac{x^2}{3^2} + \frac{y^2}{4^2} - 1 = 0$$

does not affect the solution—the constant is eliminated when you form ∇g.

EXAMPLE 2   A Business Application

The Cobb-Douglas production function (see Example 5, Section 12.1) for a particular manufacturer is given by

$$f(x, y) = 100x^{3/4}y^{1/4} \qquad \text{Objective function}$$

where x represents the units of labor (at \$150 per unit) and y represents the units of capital (at \$250 per unit). The total cost of labor and capital is limited to \$50,000. Find the maximum production level for this manufacturer.

Solution   From the given function, you have

$$\nabla f(x, y) = 75x^{-1/4}y^{1/4}\mathbf{i} + 25x^{3/4}y^{-3/4}\mathbf{j}.$$

The limit on the cost of labor and capital produces the constraint

$$g(x, y) = 150x + 250y = 50,000. \qquad \text{Constraint}$$

Thus, $\lambda \nabla g(x, y) = 150\lambda \mathbf{i} + 250\lambda \mathbf{j}$. This gives rise to the following system of equations.

$$75x^{-1/4}y^{1/4} = 150\lambda \qquad f_x(x, y) = \lambda g_x(x, y)$$
$$25x^{3/4}y^{-3/4} = 250\lambda \qquad f_y(x, y) = \lambda g_y(x, y)$$
$$150x + 250y = 50,000 \qquad \text{Constraint}$$

By solving for λ in the first equation

$$\lambda = \frac{75x^{-1/4}y^{1/4}}{150} = \frac{x^{-1/4}y^{1/4}}{2}$$

and substituting into the second equation, you obtain

$$25x^{3/4}y^{-3/4} = 250\left(\frac{x^{-1/4}y^{1/4}}{2}\right) \qquad \text{Multiply by } x^{1/4}y^{3/4}.$$
$$25x = 125y.$$

Thus, $x = 5y$. By substituting into the third equation, you have

$$150(5y) + 250y = 50,000$$
$$1000y = 50,000$$
$$y = 50 \text{ units of capital}$$
$$x = 250 \text{ units of labor.}$$

Thus, the maximum production is

$$f(250, 50) = 100(250)^{3/4}(50)^{1/4} \approx 16,719 \text{ product units.}$$

FOR FURTHER INFORMATION For more information on the use of Lagrange multipliers in economics, see the article "Lagrange Multiplier Problems in Economics" by John V. Baxley and John C. Moorhouse in the August–September 1984 issue of *The American Mathematical Monthly*.

Economists call the Lagrange multiplier obtained in a production function the **marginal productivity of money.** For instance, in Example 2 the marginal productivity of money at $x = 250$ and $y = 50$ is

$$\lambda = \frac{x^{-1/4}y^{1/4}}{2} = \frac{(250)^{-1/4}(50)^{1/4}}{2} \approx 0.334$$

which means that for each additional dollar spent on production, 0.334 additional units of the product can be produced.

> **EXAMPLE 3**   **Lagrange Multipliers and Three Variables**

Find the minimum value of

$$f(x, y, z) = 2x^2 + y^2 + 3z^2 \qquad \text{Objective function}$$

subject to the constraint $2x - 3y - 4z = 49$.

Solution   Let $g(x, y, z) = 2x - 3y - 4z = 49$. Then, because

$$\nabla f(x, y, z) = 4x\mathbf{i} + 2y\mathbf{j} + 6z\mathbf{k} \qquad \text{and} \qquad \lambda \nabla g(x, y, z) = 2\lambda \mathbf{i} - 3\lambda \mathbf{j} - 4\lambda \mathbf{k}$$

you obtain the following system of equations.

$$4x = 2\lambda \qquad\qquad f_x(x, y, z) = \lambda g_x(x, y, z)$$
$$2y = -3\lambda \qquad\qquad f_y(x, y, z) = \lambda g_y(x, y, z)$$
$$6z = -4\lambda \qquad\qquad f_z(x, y, z) = \lambda g_z(x, y, z)$$
$$2x - 3y - 4z = 49 \qquad\qquad \text{Constraint}$$

The solution of this system is $x = 3$, $y = -9$, and $z = -4$. Therefore, the optimum value of f is

$$f(3, -9, -4) = 2(3)^2 + (-9)^2 + 3(-4)^2$$
$$= 147.$$

From the original function and constraint, it is clear that $f(x, y, z)$ has no maximum. Thus, the optimum value of f determined above is a minimum.

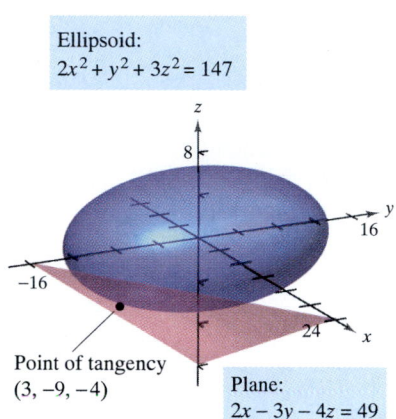

Ellipsoid:
$2x^2 + y^2 + 3z^2 = 147$

Point of tangency
$(3, -9, -4)$

Plane:
$2x - 3y - 4z = 49$

Figure 12.79

At the beginning of this section, we gave a graphical interpretation of constrained optimization problems in two variables. In three variables, the interpretation is similar, except that we use level surfaces instead of level curves. For instance, in Example 3, the level surfaces of f are ellipsoids centered at the origin, and the constraint $2x - 3y - 4z = 49$ is a plane. The minimum value of f is represented by the ellipsoid that is tangent to the constraint plane, as shown in Figure 12.79.

EXAMPLE 4   **Optimization Inside a Region**

Find the extreme values of

$$f(x, y) = x^2 + 2y^2 - 2x + 3 \qquad \text{Objective function}$$

subject to the constraint $x^2 + y^2 \leq 10$.

Solution   To solve this problem, you can break the constraint into two cases.

a. For points *on the circle* $x^2 + y^2 = 10$, you can use Lagrange multipliers to find that the maximum value of $f(x, y)$ is 24—this value occurs at $(-1, 3)$ and at $(-1, -3)$. In a similar way, you can determine that the minimum value of $f(x, y)$ is approximately 6.675—this value occurs at $\left(\sqrt{10}, 0\right)$.

b. For points *inside the circle*, you can use the techniques discussed in Section 12.8 to conclude that the function has a relative minimum of 2 at the point $(1, 0)$.

By combining these two results, you can conclude that f has a maximum of 24 at $(-1, \pm3)$ and a minimum of 2 at $(1, 0)$, as shown in Figure 12.80.

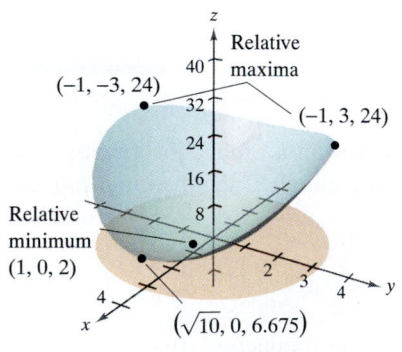

$(-1, -3, 24)$

Relative maxima

$(-1, 3, 24)$

Relative minimum
$(1, 0, 2)$

$\left(\sqrt{10}, 0, 6.675\right)$

Figure 12.80

The Method of Lagrange Multipliers with Two Constraints

For optimization problems involving *two* constraint functions g and h, we introduce a second Lagrange multiplier, μ (the lowercase Greek letter mu), and solve the equation

$$\nabla f = \lambda \nabla g + \mu \nabla h$$

where the gradient vectors are not parallel, as illustrated in Example 5.

EXAMPLE 5   Optimization with Two Constraints

Let $T(x, y, z) = 20 + 2x + 2y + z^2$ represent the temperature at each point on the sphere $x^2 + y^2 + z^2 = 11$. Find the extreme temperatures on the curve formed by the intersection of the plane $x + y + z = 3$ and the sphere.

Solution   The two constraints are

$$g(x, y, z) = x^2 + y^2 + z^2 = 11 \quad \text{and} \quad h(x, y, z) = x + y + z = 3.$$

Using

$$\nabla T(x, y, z) = 2\mathbf{i} + 2\mathbf{j} + 2z\mathbf{k},$$
$$\lambda \nabla g(x, y, z) = 2\lambda x\mathbf{i} + 2\lambda y\mathbf{j} + 2\lambda z\mathbf{k},$$

and

$$\mu \nabla h(x, y, z) = \mu\mathbf{i} + \mu\mathbf{j} + \mu\mathbf{k},$$

you can write the following system of equations.

$2 = 2\lambda x + \mu$	$T_x(x, y, z) = \lambda g_x(x, y, z) + \mu h_x(x, y, z)$
$2 = 2\lambda y + \mu$	$T_y(x, y, z) = \lambda g_y(x, y, z) + \mu h_y(x, y, z)$
$2z = 2\lambda z + \mu$	$T_z(x, y, z) = \lambda g_z(x, y, z) + \mu h_z(x, y, z)$
$x^2 + y^2 + z^2 = 11$	Constraint 1
$x + y + z = 3$	Constraint 2

By subtracting the second equation from the first, you can obtain the following system.

$$\lambda(x - y) = 0$$
$$2z(1 - \lambda) - \mu = 0$$
$$x^2 + y^2 + z^2 = 11$$
$$x + y + z = 3$$

STUDY TIP   The system of equations that arises in the Method of Lagrange Multipliers is not, in general, a linear system, and the solution often requires ingenuity.

From the first equation, you can conclude that $\lambda = 0$ or $x = y$. If $\lambda = 0$, you can show that the critical points are $(3, -1, 1)$ and $(-1, 3, 1)$. (Try doing this—it takes a little work.) If $\lambda \neq 0$, then $x = y$ and you can show that the critical points occur when $x = y = (3 \pm 2\sqrt{3})/3$ and $z = (3 \mp 4\sqrt{3})/3$. Finally, to find the optimal solutions, compare the temperatures at the four critical points.

$$T(3, -1, 1) = T(-1, 3, 1) = 25$$
$$T\left(\frac{3 - 2\sqrt{3}}{3}, \frac{3 - 2\sqrt{3}}{3}, \frac{3 + 4\sqrt{3}}{3}\right) = \frac{91}{3} \approx 30.33$$
$$T\left(\frac{3 + 2\sqrt{3}}{3}, \frac{3 + 2\sqrt{3}}{3}, \frac{3 - 4\sqrt{3}}{3}\right) = \frac{91}{3} \approx 30.33$$

Thus, $T = 25$ is the minimum temperature and $T = \frac{91}{3}$ is the maximum temperature on the curve.

EXERCISES FOR SECTION 12.10

In Exercises 1–4, identify the constraint and level curves of the objective function in the given figure. Use the figure to approximate the indicated extrema, assuming that x and y are positive. Use Lagrange multipliers to verify your result.

1. Maximize $z = xy$
   Constraint: $x + y = 10$

2. Maximize $z = xy$
   Constraint: $2x + y = 4$

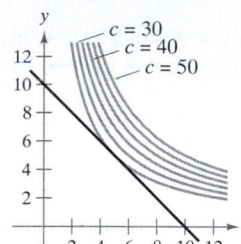
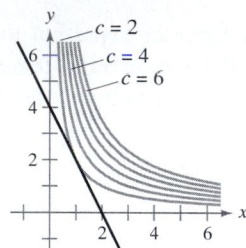

3. Minimize $z = x^2 + y^2$
   Constraint: $x + y - 4 = 0$

4. Minimize $z = x^2 + y^2$
   Constraint: $2x + 4y = 5$

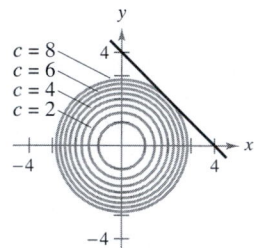
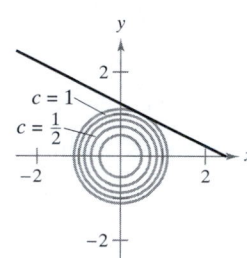

In Exercises 5–12, use Lagrange multipliers to find the indicated extrema, assuming that x and y are positive.

5. Minimize $f(x, y) = x^2 - y^2$
   Constraint: $x - 2y + 6 = 0$

6. Maximize $f(x, y) = x^2 - y^2$
   Constraint: $y - x^2 = 0$

7. Maximize $f(x, y) = 2x + 2xy + y$
   Constraint: $2x + y = 100$

8. Minimize $f(x, y) = 3x + y + 10$
   Constraint: $x^2 y = 6$

9. Maximize $f(x, y) = \sqrt{6 - x^2 - y^2}$
   Constraint: $x + y - 2 = 0$

10. Minimize $f(x, y) = \sqrt{x^2 + y^2}$
    Constraint: $2x + 4y - 15 = 0$

11. Maximize $f(x, y) = e^{xy}$
    Constraint: $x^2 + y^2 = 8$

12. Minimize $f(x, y) = 2x + y$
    Constraint: $xy = 32$

In Exercises 13–16, use Lagrange multipliers to find the indicated extrema, assuming that x, y, and z are positive.

13. Minimize $f(x, y, z) = x^2 + y^2 + z^2$
    Constraint: $x + y + z - 6 = 0$

14. Maximize $f(x, y, z) = xyz$
    Constraint: $x + y + z - 6 = 0$

15. Minimize $f(x, y, z) = x^2 + y^2 + z^2$
    Constraint: $x + y + z = 1$

16. Minimize $f(x, y) = x^2 - 8x + y^2 - 12y + 48$
    Constraint: $x + y = 8$

In Exercises 17–20, use Lagrange multipliers to find the indicated extrema of f subject to two constraints. In each case, assume that x, y, and z are nonnegative.

17. Maximize $f(x, y, z) = xyz$
    Constraints: $x + y + z = 32,$ $x - y + z = 0$

18. Minimize $f(x, y, z) = x^2 + y^2 + z^2$
    Constraints: $x + 2z = 4,$ $x + y = 8$

19. Maximize $f(x, y, z) = xy + yz$
    Constraints: $x + 2y = 6,$ $x - 3z = 0$

20. Maximize $f(x, y, z) = xyz$
    Constraints: $x^2 + z^2 = 5,$ $x - 2y = 0$

In Exercises 21 and 22, use Lagrange multipliers to find any extrema of the function subject to the constraint $x^2 + y^2 \le 1$.

21. $f(x, y) = x^2 + 3xy + y^2$ 22. $f(x, y) = e^{-xy}$

In Exercises 23–26, use Lagrange multipliers to find the minimum distance from the curve or surface to the indicated point. [*Hint:* In Exercise 23, minimize $f(x, y) = x^2 + y^2$ subject to the constraint $2x + 3y = -1$.]

Curve	*Point*
23. Line: $2x + 3y = -1$	$(0, 0)$
24. Circle: $(x - 4)^2 + y^2 = 4$	$(0, 10)$

Surface	*Point*
25. Plane: $x + y + z = 1$	$(2, 1, 1)$
26. Cone: $z = \sqrt{x^2 + y^2}$	$(4, 0, 0)$

In Exercises 27 and 28, find the highest point on the curve of intersection of the surfaces.

27. Sphere: $x^2 + y^2 + z^2 = 36,$ Plane: $2x + y - z = 2$
28. Cone: $x^2 + y^2 - z^2 = 0,$ Plane: $x + 2z = 4$

29. ***Volume***   Find the dimensions of the rectangular package of largest volume subject to the constraint that the sum of the length and the girth cannot exceed 108 inches.

30. *Volume* The material for the base of an open box costs 1.5 times as much as the material for the sides. Find the dimensions of the box of largest volume that can be made for a fixed cost C. (Maximize $V = xyz$ subject to $1.5xy + 2xz + 2yz = C$.)

31. *Cost* A cargo container (in the shape of a rectangular solid) must have a volume of 480 cubic feet. Use Lagrange multipliers to find the dimensions of the container of this size that has minimum cost if the bottom will cost $5 per square foot to construct and the sides and top will cost $3 per square foot to construct.

32. *Surface Area* Use Lagrange multipliers to find the dimensions of a right circular cylinder with volume V_0 cubic units and minimum surface area.

33. *Volume* Use Lagrange multipliers to find the dimensions of a rectangular box of maximum volume that can be inscribed (with edges parallel to the coordinate axes) in the ellipsoid

$$\frac{x^2}{a^2} + \frac{y^2}{b^2} + \frac{z^2}{c^2} = 1.$$

34. *Geometric and Arithmetic Means*

(a) Use Lagrange multipliers to prove that the product of three positive numbers x, y, and z, whose sum has the constant value S, is a maximum when the three numbers are equal. Use this result to prove that

$$\sqrt[3]{xyz} \le \frac{x + y + z}{3}.$$

(b) Generalize the result in part (a) to prove that the product $x_1 x_2 x_3 \cdots x_n$ is a maximum when $x_1 = x_2 = x_3 = \cdots = x_n$, $\sum_{i=1}^{n} x_i = S$, and all $x_i \ge 0$. Use this result to prove that

$$\sqrt[n]{x_1 x_2 x_3 \cdots x_n} \le \frac{x_1 + x_2 + x_3 + \cdots + x_n}{n}.$$

This shows that the geometric mean is never greater than the arithmetic mean.

35. *Refraction of Light* When light waves traveling in a transparent medium strike the surface of a second transparent medium, they tend to "bend" in order to follow the path of minimum time. This tendency is called refraction and is described by **Snell's Law of Refraction,**

$$\frac{\sin \theta_1}{v_1} = \frac{\sin \theta_2}{v_2}$$

where θ_1 and θ_2 are the magnitudes of the angles shown in the figure, and v_1 and v_2 are the velocities of light in the two media. Use Lagrange multipliers to derive this law using the constant $x + y = a$.

36. *Area and Perimeter* A semicircle is on top of a rectangle (see figure). If the area is fixed and the perimeter is a minimum, or if the perimeter is fixed and the area is a maximum, use Lagrange multipliers to verify that the length of the rectangle is twice its height.

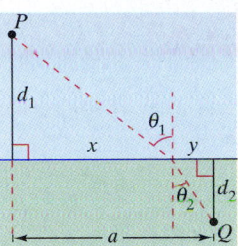

Figure for 35

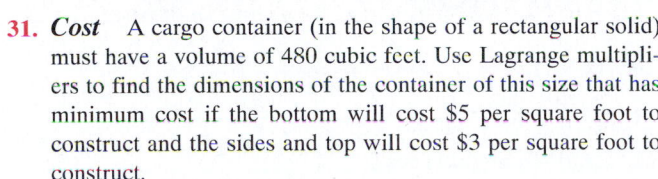

Figure for 36

37. *Hardy-Weinberg Law* Use Lagrange multipliers to maximize $P(p, q, r) = 2pq + 2pr + 2qr$ subject to $p + q + r = 1$. (See Exercise 22 in Section 12.9.)

38. *Temperature Distribution* Let

$$T(x, y, z) = 100 + x^2 + y^2$$

represent the temperature at each point on the sphere $x^2 + y^2 + z^2 = 50$. Find the maximum temperature on the curve formed by the intersection of the sphere and the plane $x - z = 0$.

Production Level **In Exercises 39 and 40, find the maximum production level P if the total cost of labor (at $48 per unit) and capital (at $36 per unit) is limited to $100,000 where x is the number of units of labor and y is the number of units of capital.**

39. $P(x, y) = 100x^{0.25}y^{0.75}$ **40.** $P(x, y) = 100x^{0.6}y^{0.4}$

Cost **In Exercises 41 and 42, find the minimum cost of producing 20,000 units of a product, where x is the number of units of labor (at $48 per unit) and y is the number of units of capital (at $36 per unit).**

41. $P(x, y) = 100x^{0.25}y^{0.75}$ **42.** $P(x, y) = 100x^{0.6}y^{0.4}$

43. *Investigation* Consider the objective function $g(\alpha, \beta, \gamma) = \cos \alpha \cos \beta \cos \gamma$ subject to the constraint that α, β, and γ are the angles of a triangle.

(a) Use Lagrange multipliers to maximize g.

(b) Use the constraint to reduce the function g to a function of two independent variables. Use a computer algebra system to graph the surface represented by g. Identify the maximum values on the graph.

44. *Investigation* Consider the objective function $f(x, y) = ax + by$ subject to the constraint $x^2/64 + y^2/36 = 1$. Assume that x and y are positive.

(a) Use a computer algebra system to graph the constraint. If $a = 4$ and $b = 3$, use the computer algebra system to graph the level curves of the objective function. By trial and error, find the level curve that appears to be tangent to the ellipse. Use the result to approximate the maximum of f subject to the constraint.

(b) Repeat part (a) for $a = 4$ and $b = 9$.

REVIEW EXERCISES FOR CHAPTER 12

In Exercises 1 and 2, use the graph to determine whether z is a function of x and y. Explain.

1.

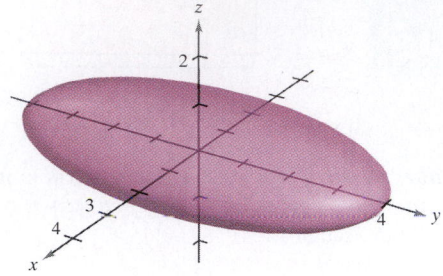

2.

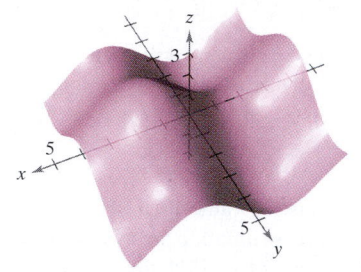

In Exercises 3–6, use a computer algebra system to graph several level curves for the function.

3. $f(x, y) = e^{x^2 + y^2}$

4. $f(x, y) = \ln xy$

5. $f(x, y) = x^2 - y^2$

6. $f(x, y) = \dfrac{x}{x + y}$

In Exercises 7 and 8, use a computer algebra system to graph the function.

7. $f(x, y) = e^{-(x^2 + y^2)}$

8. $g(x, y) = |y|^{1 + |x|}$

In Exercises 9–12, discuss the continuity of the function and evaluate the limit, if it exists.

9. $\displaystyle\lim_{(x, y)\to(1, 1)} \dfrac{xy}{x^2 + y^2}$

10. $\displaystyle\lim_{(x, y)\to(1, 1)} \dfrac{xy}{x^2 - y^2}$

11. $\displaystyle\lim_{(x, y)\to(0, 0)} \dfrac{-4x^2 y}{x^4 + y^2}$

12. $\displaystyle\lim_{(x, y)\to(0, 0)} \dfrac{y + xe^{-y^2}}{1 + x^2}$

In Exercises 13–22, find all first partial derivatives.

13. $f(x, y) = e^x \cos y$

14. $f(x, y) = \dfrac{xy}{x + y}$

15. $z = xe^y + ye^x$

16. $z = \ln(x^2 + y^2 + 1)$

17. $g(x, y) = \dfrac{xy}{x^2 + y^2}$

18. $w = \sqrt{x^2 + y^2 + z^2}$

19. $f(x, y, z) = z \arctan \dfrac{y}{x}$

20. $f(x, y, z) = \dfrac{1}{\sqrt{1 - x^2 - y^2 - z^2}}$

21. $u(x, t) = ce^{-n^2 t} \sin nx$

22. $u(x, t) = c \sin(akx) \cos kt$

In Exercises 23 and 24, find $\partial z / \partial x$ and $\partial z / \partial y$.

23. $x^2 y - 2yz - xz - z^2 = 0$

24. $xz^2 - y \sin z = 0$

In Exercises 25–28, find all second partial derivatives and verify that the second mixed partials are equal.

25. $f(x, y) = 3x^2 - xy + 2y^3$

26. $h(x, y) = \dfrac{x}{x + y}$

27. $h(x, y) = x \sin y + y \cos x$

28. $g(x, y) = \cos(x - 2y)$

Laplace Equation   In Exercises 29–32, show that the function satisfies the Laplace equation

$$\dfrac{\partial^2 z}{\partial x^2} + \dfrac{\partial^2 z}{\partial y^2} = 0.$$

29. $z = x^2 - y^2$

30. $z = x^3 - 3xy^2$

31. $z = \dfrac{y}{x^2 + y^2}$

32. $z = e^x \sin y$

In Exercises 33 and 34, find the indicated derivatives (a) by the Chain Rule and (b) by substitution before differentiating.

33. $u = x^2 + y^2 + z^2$, $\dfrac{\partial u}{\partial r}, \dfrac{\partial u}{\partial t}$

$x = r \cos t, \ y = r \sin t, \ z = t$

34. $u = y^2 - x$, $\dfrac{du}{dt}$

$x = \cos t, \ y = \sin t$

In Exercises 35–38, find the directional derivative in the direction of **v** at the indicated point.

Function	Direction	Point
35. $f(x, y) = x^2 y$	$\mathbf{v} = \mathbf{i} - \mathbf{j}$	$(2, 1)$
36. $f(x, y) = \frac{1}{4}y^2 - x^2$	$\mathbf{v} = 2\mathbf{i} + \mathbf{j}$	$(1, 4)$
37. $w = y^2 + xz$	$\mathbf{v} = 2\mathbf{i} - \mathbf{j} + 2\mathbf{k}$	$(1, 2, 2)$
38. $w = 6x^2 + 3xy - 4y^2 z$	$\mathbf{v} = \mathbf{i} + \mathbf{j} - \mathbf{k}$	$(1, 0, 1)$

In Exercises 39–42, find the gradient and the maximum value of the directional derivative of the function at the indicated point.

39. $z = \dfrac{y}{x^2 + y^2}$, $(1, 1)$ **40.** $z = \dfrac{x^2}{x - y}$, $(2, 1)$

41. $z = e^{-x} \cos y$, $\left(0, \dfrac{\pi}{4}\right)$ **42.** $z = x^2 y$, $(2, 1)$

In Exercises 43–46, find an equation of the tangent plane and parametric equations of the normal line to the surface at the indicated point.

Surface	Point
43. $f(x, y) = x^2 y$	$(2, 1, 4)$
44. $f(x, y) = \sqrt{25 - y^2}$	$(2, 3, 4)$
45. $z = -9 + 4x - 6y - x^2 - y^2$	$(2, -3, 4)$
46. $z = \sqrt{9 - x^2 - y^2}$	$(1, 2, 2)$

In Exercises 47 and 48, find symmetric equations of the tangent line to the curve of intersection of the surfaces at the indicated point.

Surfaces	Point
47. $z = x^2 - y^2$, $z = 3$	$(2, 1, 3)$
48. $z = 25 - y^2$, $y = x$	$(4, 4, 9)$

 In Exercises 49–52, locate and classify any extrema of the function. Use a computer algebra system to graph the function and confirm your analytical results.

49. $f(x, y) = x^3 - 3xy + y^2$

50. $f(x, y) = 2x^2 + 6xy + 9y^2 + 8x + 14$

51. $f(x, y) = xy + \dfrac{1}{x} + \dfrac{1}{y}$

52. $z = 50(x + y) - (0.1x^3 + 20x + 150) - (0.05y^3 + 20.6y + 125)$

Writing   **In Exercises 53 and 54, write a short paragraph about the surface whose level curves (*c*-values evenly spaced) are given. Comment on possible extrema, saddle points, the magnitude of the gradient, etc.**

53.   **54.**

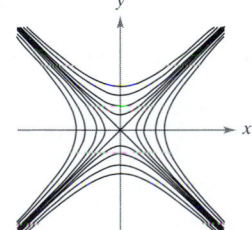   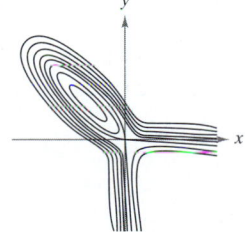

55. *Think About It*   Sketch a graph of a function $z = f(x, y)$ whose derivative f_x is always negative and whose derivative f_y is always negative.

56. *Approximation*   Consider the following approximations centered at $(0, 0)$ for a function $f(x, y)$.

Linear approximation:

$$P_1(x, y) = f(0, 0) + f_x(0, 0)x + f_y(0, 0)y$$

Quadratic approximation:

$$P_2(x, y) = f(0, 0) + f_x(0, 0)x + f_y(0, 0)y + \tfrac{1}{2}f_{xx}(0, 0)x^2 + f_{xy}(0, 0)xy + \tfrac{1}{2}f_{yy}(0, 0)y^2$$

[Note that the linear approximation is the tangent plane to the surface at $(0, 0, f(0, 0))$.]

(a) Find the linear approximation of $f(x, y) = \cos x + \sin y$ centered at $(0, 0)$.

(b) Find the quadratic approximation of $f(x, y) = \cos x + \sin y$ centered at $(0, 0)$.

(c) If $y = 0$ in the quadratic approximation, you obtain the second-degree Taylor polynomial for what function?

(d) Complete the table.

x	y	$f(x, y)$	$P_1(x, y)$	$P_2(x, y)$
0	0			
0	0.1			
0.2	0.1			
0.5	0.3			
1	0.5			

(e) Use a computer algebra system to graph the surfaces $z = f(x, y)$, $z = P_1(x, y)$, and $z = P_2(x, y)$. How does the accuracy of the approximations change as the distance from $(0, 0)$ increases?

In Exercises 57 and 58, find dz.

57. $z = x \sin \dfrac{y}{x}$ **58.** $z = \dfrac{xy}{\sqrt{x^2 + y^2}}$

59. *Error Analysis*   The legs of a right triangle are measured to be 5 centimeters and 12 centimeters, with a possible error of $\tfrac{1}{2}$ centimeter. Approximate the maximum possible error in computing the length of the hypotenuse. Approximate the maximum percent error.

60. *Error Analysis*   To determine the height of a tower, the angle of elevation to the top of the tower was measured from a point 100 feet $\pm \tfrac{1}{2}$ foot from the base. The angle is measured at $33°$, with a possible error of $1°$. Assuming that the ground is horizontal, approximate the maximum error in determining the height of the tower.

61. *Error Analysis* The volume of a right circular cone is $V = \frac{1}{3}\pi r^2 h$. The measured values of r and h for a cone are found to be 2 and 5 inches. Find the approximate error in the volume because of possible error of $\frac{1}{8}$ inch in the measurements of the radius and the height.

62. *Error Analysis* Approximate the error in the lateral surface area of the cone in Exercise 61. (The lateral surface area is given by $A = \pi r \sqrt{r^2 + h^2}$.)

63. *Profit* A corporation manufactures a product at two locations. The cost functions for producing x_1 units at location 1 and x_2 units at location 2 are

$$C_1 = 0.05x_1^2 + 15x_1 + 5400$$
$$C_2 = 0.03x_2^2 + 15x_2 + 6100$$

and the total revenue function is

$$R = [225 - 0.4(x_1 + x_2)](x_1 + x_2).$$

Find the production levels at the two locations that will maximize the profit $P(x_1, x_2) = R - C_1 - C_2$.

64. *Cost* A manufacturer has an order for 1000 units that can be produced at two locations. Let x_1 and x_2 be the numbers of units produced at the two locations. Find the number that should be produced at each to meet the order and minimize cost, if the cost function is

$$C = 0.25x_1^2 + 10x_1 + 0.15x_2^2 + 12x_2.$$

65. *Production Level* The production function for a manufacturer is

$$f(x, y) = 4x + xy + 2y$$

where x is the number of units of labor and y is the number of units of capital. Assume that the total amount available for labor and capital is $2000, and that units of labor and capital cost $20 and $4, respectively. Find the maximum production level for this manufacturer.

66. *Modeling Data* The table gives the drag force y in kilograms for a certain motor vehicle at indicated speeds x in kilometers per hour.

Speed (x)	25	50	75	100	125
Drag (y)	28	38	54	75	102

(a) Use the regression capabilities of a graphing utility to find the least squares regression quadratic for the data.

(b) Use the quadratic to estimate the total drag when the vehicle is moving at 80 kilometers per hour.

67. *Modeling Data* The data in the table give the yield y (in milligrams) of a chemical reaction after t minutes.

t	1	2	3	4
y	1.5	7.4	10.2	13.4

t	5	6	7	8
y	15.8	16.3	18.2	18.3

(a) Use a graphing utility to plot the data. Use the graphing utility to find a linear model for the data and graph the model.

(b) Use a graphing utility to plot the points $(\ln t, y)$. Do these points appear to follow a linear pattern more closely than the plot of the given data in part (a)?

(c) Use a graphing utility to find a linear model for the data $(\ln t, y)$ and obtain the logarithmic model

$$y = a + b \ln t.$$

(d) Use a graphing utility to plot the data and graph the linear and logarithmic models. Which is a better model? Explain.

In Exercises 68 and 69, locate and classify any extrema of the function by using Lagrange multipliers.

68. $z = x^2 y$

Constraint: $x + 2y = 2$

69. $w = xy + yz + xz$

Constraint: $x + y + z = 1$

True or False? **In Exercises 70–72, determine whether the statement is true or false. If it is false, explain why or give an example that shows it is false.**

70. Of all parallelepipeds having a fixed surface area, the cube has the largest volume.

71. The gradient $\nabla f(x_0, y_0)$ is normal to the surface given by $z = f(x, y)$ at the point (x_0, y_0, z_0).

72. The plane $x_0 x + y_0 y + z_0 z = c^2$ is tangent to the sphere $x^2 + y^2 + z^2 = c^2$ at the point (x_0, y_0, z_0).

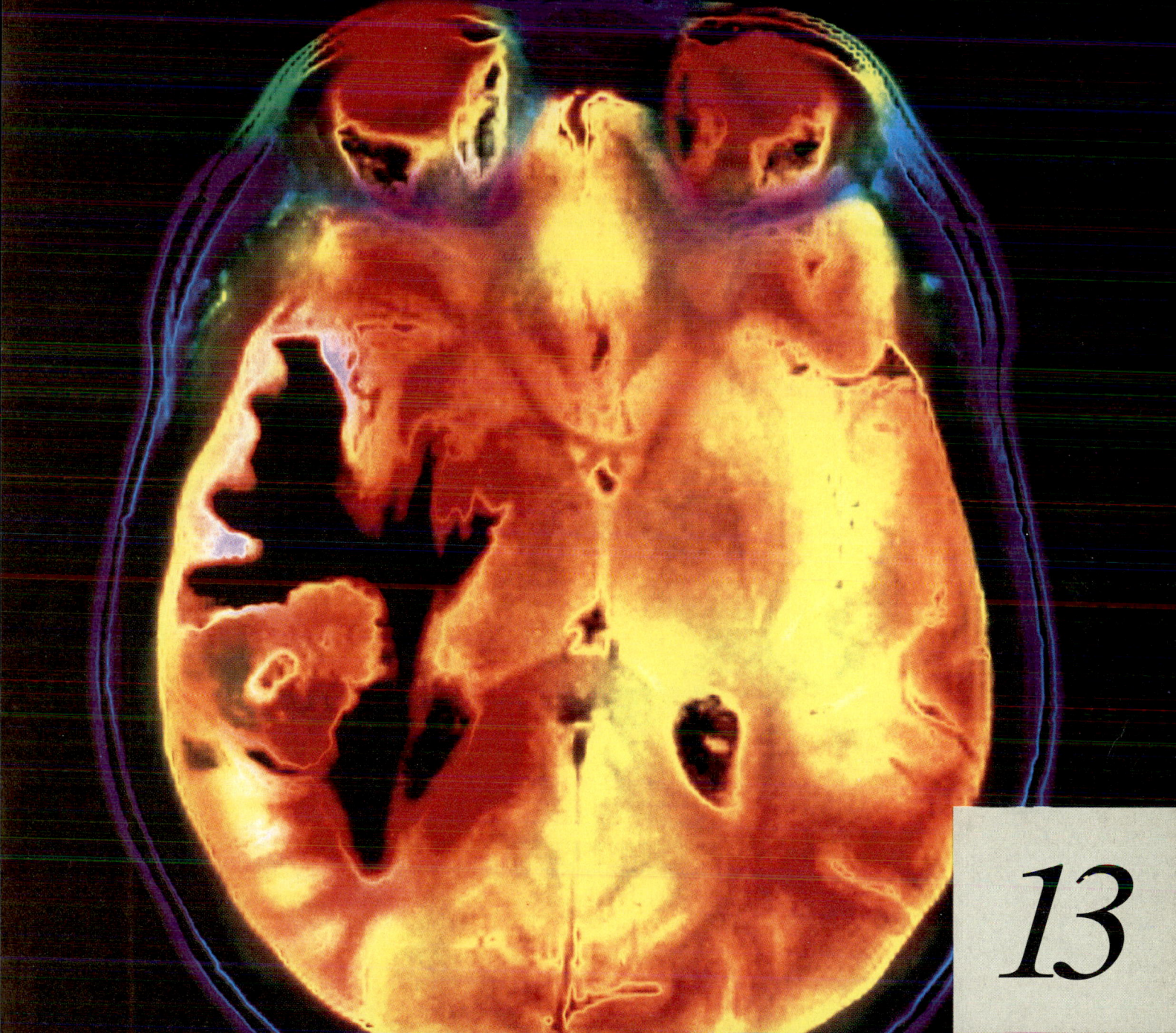

13

Multiple Integration

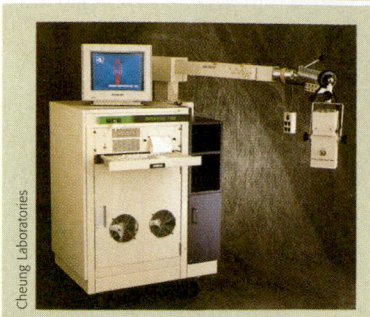

CLI's Hyperthermia Microfocus 1000 introduces a practical way to apply microwaves to tumors, and may bring heat therapy into general use.

Hyperthermia treatment uses elevated temperatures to destroy malignant tissues. Microwaves are employed to raise tumor temperatures to about 107°F. The concept of heat treatment is not a new one, and doctors estimate that application of sufficient heat can increase the effectiveness of both radiation and chemotherapy by a factor of 2. Still the technique is not in wide use today because of difficulties in focusing the energy on the tumor site.

The latest technology, Adaptive Phased Array (APA), is expected to solve the focusing problem. Originally developed to track targets in defense radar systems, APA technology was adapted by the Massachusetts Institute of Technology. In 1996, MIT licensed the technology to Cheung Laboratories, who incorporated it into a device that is able to direct a beam of energy directly into the tumor, minimizing temperature elevation in surrounding areas.

The tumor temperature during treatment is highest at the center and gradually decreases toward the edges. Regions of tissue having the same temperature, or equitherms, can be visualized as closed surfaces that are nested one inside the other. The problem of determining the portion of the tumor that has been heated to an effective temperature reduces to finding the ratio V_T/V, where V is the volume of the entire tumor and V_T is the volume of the portion of the tumor that is heated above temperature T. Using APA technology, the shapes of the equithermal surfaces are determined by the type of applicator that is used, which in turn is determined by the shape of the tumor.

NOTE   Equations for other possible tumor shapes are given in the Section Project in Section 13.7.

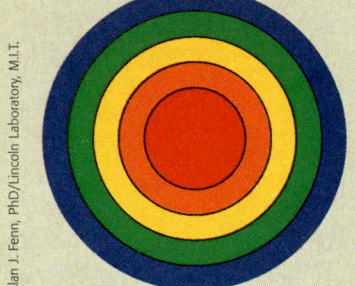 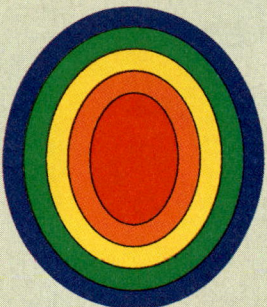

Thermal Patterns Determined by APA Technology

QUESTIONS

1. When treating a spherical tumor, a technician uses a probe to determine that the temperature has reached an appropriate level to about half the radius of the tumor. What is the ratio V_T/V? Is it $\frac{1}{2}$? Explain your reasoning.

2. Consider an ellipsoidal tumor that can be modeled by the equation

$$\frac{x^2}{2.5} + \frac{y^2}{6.5} + \frac{z^2}{2.5} = 1.$$

Consider a sequence of five ellipsoidal equitherms whose major and minor axes increase linearly until the fifth equitherm is the entire tumor. Write an equation for each of the five equitherms. Then find the ratio V_T/V for each of the five equitherms.

The concepts presented here will be explored further in this chapter. For an extension of this application, see the lab series that accompanies this text.

Iterated Integrals • Area of a Plane Region

Iterated Integrals

NOTE In Chapters 13 and 14, you will study several applications of integration involving functions of several variables. Chapter 13 is much like Chapter 6 in that it *surveys* the use of integration to find plane areas, volumes, surface areas, moments, and centers of mass.

In Chapter 12, you saw that it is meaningful to differentiate functions of several variables with respect to one variable while holding the other variables constant. You can *integrate* functions of several variables by a similar procedure. For example, if you are given the partial derivative

$$f_x(x, y) = 2xy$$

then, by considering y constant, you can integrate with respect to x to obtain

$$f(x, y) = \int f_x(x, y)\, dx \qquad \text{Integrate with respect to } x.$$

$$= \int 2xy\, dx \qquad \text{Hold } y \text{ constant.}$$

$$= y \int 2x\, dx \qquad \text{Factor out constant } y.$$

$$= y(x^2) + C(y) \qquad \text{Antiderivative of } 2x \text{ is } x^2.$$

$$= x^2 y + C(y). \qquad C(y) \text{ is function of } y.$$

Note that the "constant" of integration, $C(y)$, is a function of y. In other words, by integrating with respect to x, you are able to recover $f(x, y)$ only partially. The total recovery of a function of x and y from its partial derivatives is a topic you will study in Chapter 14. For now, we are more concerned with extending definite integrals to functions of several variables. For instance, by considering y constant, you can apply the Fundamental Theorem of Calculus to evaluate

$$\int_1^{2y} 2xy\, dx = x^2 y \Big]_1^{2y} = (2y)^2 y - (1)^2 y = 4y^3 - y.$$

x is the variable of integration and y is fixed.

Replace x by the limits of integration.

The result is a function of y.

Similarly, you can integrate with respect to y by holding x fixed. Both procedures are summarized as follows.

$$\int_{h_1(y)}^{h_2(y)} f_x(x, y)\, dx = f(x, y) \Big]_{h_1(y)}^{h_2(y)} = f(h_2(y), y) - f(h_1(y), y) \qquad \text{With respect to } x$$

$$\int_{g_1(x)}^{g_2(x)} f_y(x, y)\, dy = f(x, y) \Big]_{g_1(x)}^{g_2(x)} = f(x, g_2(x)) - f(x, g_1(x)) \qquad \text{With respect to } y$$

Note that the variable of integration cannot appear in either limit of integration. For instance, it makes no sense to write $\int_0^x y\, dx$.

EXAMPLE 1 **Integrating with Respect to y**

Evaluate $\displaystyle\int_{1}^{x} (2x^2 y^{-2} + 2y)\, dy$.

Solution Considering x to be constant and integrating with respect to y produces

$$\int_{1}^{x} (2x^2 y^{-2} + 2y)\, dy = \left[\frac{-2x^2}{y} + y^2\right]_{1}^{x}$$

$$= \left(\frac{-2x^2}{x} + x^2\right) - \left(\frac{-2x^2}{1} + 1\right)$$

$$= 3x^2 - 2x - 1.$$

Notice in Example 1 that the integral defines a function of x and can *itself* be integrated, as shown in the next example.

EXAMPLE 2 **The Integral of an Integral**

Evaluate $\displaystyle\int_{1}^{2} \left[\int_{1}^{x} (2x^2 y^{-2} + 2y)\, dy\right] dx$.

Solution Using the result of Example 1, you have

$$\int_{1}^{2} \left[\int_{1}^{x} (2x^2 y^{-2} + 2y)\, dy\right] dx = \int_{1}^{2} (3x^2 - 2x - 1)\, dx$$

$$= \left[x^3 - x^2 - x\right]_{1}^{2}$$

$$= 2 - (-1)$$

$$= 3.$$

The integral in Example 2 is an **iterated integral.** The brackets used in Example 2 are normally not written. Instead, iterated integrals are usually written simply as

$$\int_{a}^{b}\int_{g_1(x)}^{g_2(x)} f(x, y)\, dy\, dx \quad \text{and} \quad \int_{c}^{d}\int_{h_1(y)}^{h_2(y)} f(x, y)\, dx\, dy.$$

The **inside limits of integration** can be variable with respect to the outer variable of integration. However, the **outside limits of integration** must be constant with respect to both variables of integration. After performing the inside integration, you obtain a "standard" definite integral, and the second integration produces a real number. The limits of integration for an iterated integral identify two sets of boundary intervals for the variables. For instance, in Example 2, the outside limits indicate that x lies in the interval $1 \le x \le 2$ and the inside limits indicate that y lies in the interval $1 \le y \le x$. Together, these two intervals determine the **region of integration R** of the iterated integral, as shown in Figure 13.1.

Because an iterated integral is just a special type of definite integral—one in which the integrand is also an integral—you can use the properties of definite integrals to evaluate iterated integrals.

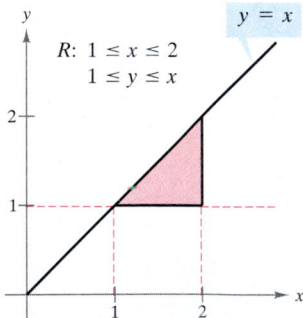

The region of integration for

$$\int_{1}^{2}\int_{1}^{x} f(x, y)\, dy\, dx$$

Figure 13.1

Area of a Plane Region

In the remainder of this section, you will take a new look at an old problem—that of finding the area of a plane region. Consider the plane region R bounded by $a \le x \le b$ and $g_1(x) \le y \le g_2(x)$, as shown in Figure 13.2. The area of R is given by the definite integral

$$\int_a^b [g_2(x) - g_1(x)]\, dx. \qquad \text{Area of } R$$

Using the Fundamental Theorem of Calculus, you can rewrite the integrand $g_2(x) - g_1(x)$ as a definite integral. Specifically, if you consider x to be fixed and let y vary from $g_1(x)$ to $g_2(x)$, you can write

$$\int_{g_1(x)}^{g_2(x)} dy = y \Big]_{g_1(x)}^{g_2(x)} = g_2(x) - g_1(x).$$

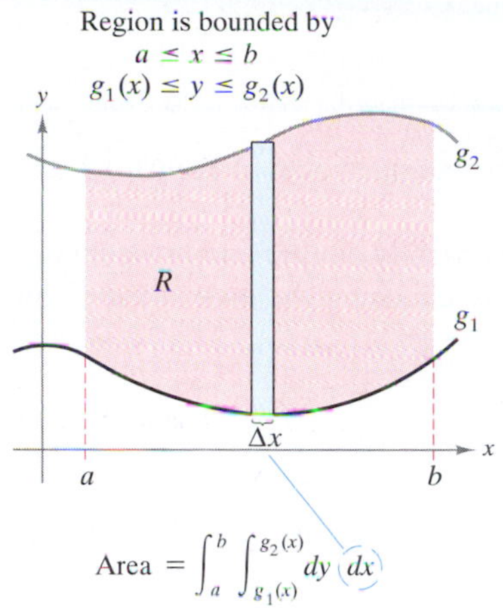

Combining these two integrals, you can write the area of the region R as an iterated integral

$$\int_a^b \int_{g_1(x)}^{g_2(x)} dy\, dx = \int_a^b y \Big]_{g_1(x)}^{g_2(x)} dx \qquad \text{Area of } R$$

$$= \int_a^b [g_2(x) - g_1(x)]\, dx.$$

Placing a representative rectangle in the region R helps determine both the order and the limits of integration. A vertical rectangle implies the order $dy\, dx$, with the inside limits corresponding to the upper and lower bounds of the rectangle, as shown in Figure 13.2. This type of region is called **vertically simple,** because the outside limits of integration represent the vertical lines $x = a$ and $x = b$.

Similarly, a horizontal rectangle implies the order $dx\, dy$, with the inside limits determined by the left and right bounds of the rectangle, as shown in Figure 13.3. This type of region is called **horizontally simple,** because the outside limits represent the horizontal lines $y = c$ and $y = d$. The iterated integrals used for these two types of simple regions are summarized as follows.

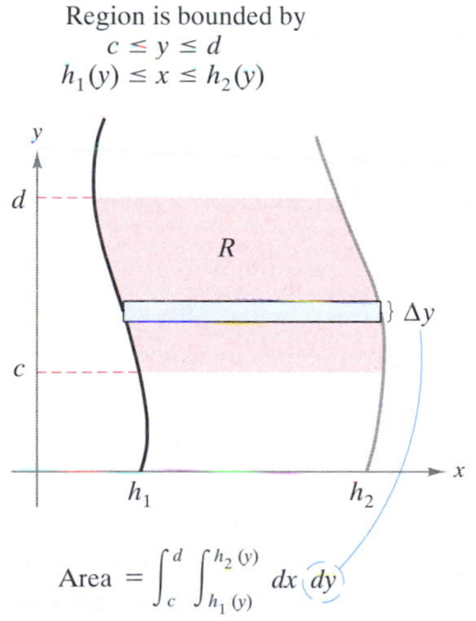

Region is bounded by
$$a \le x \le b$$
$$g_1(x) \le y \le g_2(x)$$

$$\text{Area} = \int_a^b \int_{g_1(x)}^{g_2(x)} dy\, dx$$

Vertically simple region
Figure 13.2

Region is bounded by
$$c \le y \le d$$
$$h_1(y) \le x \le h_2(y)$$

$$\text{Area} = \int_c^d \int_{h_1(y)}^{h_2(y)} dx\, dy$$

Horizontally simple region
Figure 13.3

Area of a Region in the Plane

1. If R is defined by $a \le x \le b$ and $g_1(x) \le y \le g_2(x)$, where g_1 and g_2 are continuous on $[a, b]$, then the area of R is given by

$$A = \int_a^b \int_{g_1(x)}^{g_2(x)} dy\, dx. \qquad \text{Figure 13.2 (vertically simple)}$$

2. If R is defined by $c \le y \le d$ and $h_1(y) \le x \le h_2(y)$, where h_1 and h_2 are continuous on $[c, d]$, then the area of R is given by

$$A = \int_c^d \int_{h_1(y)}^{h_2(y)} dx\, dy. \qquad \text{Figure 13.3 (horizontally simple)}$$

NOTE   Be sure you see that the order of integration of these two integrals is different—the order $dy\, dx$ corresponds to a vertically simple region, and the order $dx\, dy$ corresponds to a horizontally simple region.

If all four limits of integration happen to be constants, the region of integration is rectangular, as shown in Example 3.

EXAMPLE 3 The Area of a Rectangular Region

Use an iterated integral to represent the area of the rectangle shown in Figure 13.4.

Solution The region shown in Figure 13.4 is both vertically simple and horizontally simple, so you can use either order of integration. By choosing the order $dy\,dx$, you obtain the following.

$$\int_a^b \int_c^d dy\,dx = \int_a^b y \Big]_c^d dx$$

$$= \int_a^b (d - c)\,dx$$

$$= \Big[(d - c)x \Big]_a^b$$

$$= (d - c)(b - a)$$

Notice that this answer is consistent with what you know from geometry.

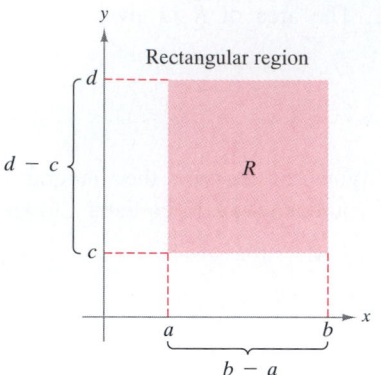

Area $= \int_a^b \int_c^d dy\,dx$

$= (d - c)(b - a)$

Figure 13.4

EXAMPLE 4 Finding Area by an Iterated Integral

Use an iterated integral to find the area of the region bounded by the graphs of

$f(x) = \sin x$ Sine curve forms upper boundary.

$g(x) = \cos x$ Cosine curve forms lower boundary.

between $x = \pi/4$ and $x = 5\pi/4$.

Solution Because f and g are given as functions of x, a vertical representative rectangle is convenient, and you can choose $dy\,dx$ as the order of integration, as shown in Figure 13.5. The outside limits of integration are $\pi/4 \le x \le 5\pi/4$. Moreover, because the rectangle is bounded above by $f(x) = \sin x$ and below by $g(x) = \cos x$, you have

$$\text{Area of } R = \int_{\pi/4}^{5\pi/4} \int_{\cos x}^{\sin x} dy\,dx$$

$$= \int_{\pi/4}^{5\pi/4} y \Big]_{\cos x}^{\sin x} dx$$

$$= \int_{\pi/4}^{5\pi/4} (\sin x - \cos x)\,dx$$

$$= \Big[-\cos x - \sin x \Big]_{\pi/4}^{5\pi/4}$$

$$= 2\sqrt{2}.$$

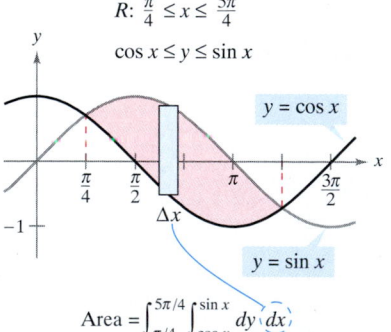

$R: \frac{\pi}{4} \le x \le \frac{5\pi}{4}$

$\cos x \le y \le \sin x$

Area $= \int_{\pi/4}^{5\pi/4} \int_{\cos x}^{\sin x} dy\,dx$

Figure 13.5

NOTE The region of integration of an iterated integral need not have any straight lines as boundaries. For instance, the region of integration shown in Figure 13.5 is *vertically simple* even though it has no vertical lines as left and right boundaries. The quality that makes the region vertically simple is that it is bounded above and below by the graphs of *functions of x*.

One order of integration will often produce a simpler integration problem than the other order. For instance, try reworking Example 4 with the order $dx\,dy$—you may be surprised to see that the task is formidable. However, if you succeed, you will see that the answer is the same. In other words, the order of integration affects the ease of integration, but not the value of the integral.

EXAMPLE 5   Comparing Different Orders of Integration

Sketch the region whose area is represented by the integral

$$\int_0^2 \int_{y^2}^4 dx\,dy.$$

Then find another iterated integral using the order $dy\,dx$ to represent the same area and show that both integrals yield the same value.

Solution   From the given limits of integration, you know that

$$y^2 \le x \le 4 \qquad \text{\color{red}Inner limits of integration}$$

which means that the region R is bounded on the left by the parabola $x = y^2$ and on the right by the line $x = 4$. Furthermore, because

$$0 \le y \le 2 \qquad \text{\color{red}Outer limits of integration}$$

you know that R is bounded below by the x-axis as shown in Figure 13.6(a). The value of this integral is

$$\int_0^2 \int_{y^2}^4 dx\,dy = \int_0^2 x \Big]_{y^2}^4 dy$$

$$= \int_0^2 (4 - y^2)dy$$

$$= \left[4y - \frac{y^3}{3} \right]_0^2 = \frac{16}{3}.$$

To change the order of integration to $dy\,dx$, place a vertical rectangle in the region, as shown in Figure 13.6(b). From this you can see that the constant bounds $0 \le x \le 4$ serve as the outer limits of integration. By solving for y in the equation $x = y^2$, you can conclude that the inner bounds are $0 \le y \le \sqrt{x}$. Therefore, the area of the region can also be represented by

$$\int_0^4 \int_0^{\sqrt{x}} dy\,dx.$$

By evaluating this integral, you can see that it has the same value as the original integral.

$$\int_0^4 \int_0^{\sqrt{x}} dy\,dx = \int_0^4 y \Big]_0^{\sqrt{x}} dx$$

$$= \int_0^4 \sqrt{x}\,dx$$

$$= \frac{2}{3} x^{3/2} \Big]_0^4 = \frac{16}{3}$$

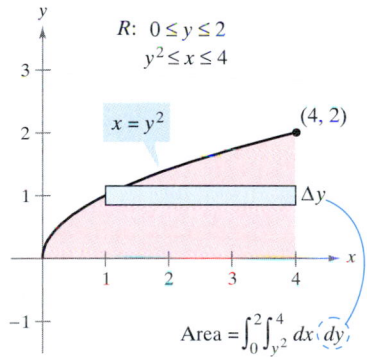

(a)

(b)

Figure 13.6

Sometimes it is not possible to calculate the area of a region with a single iterated integral. In these cases you can divide the region into subregions such that the area of each subregion can be calculated by an iterated integral. The total area is then the sum of the iterated integrals.

EXAMPLE 6 An Area Represented by Two Iterated Integrals

Find the area of the region R that lies below the parabola

$$y = 4x - x^2 \qquad \text{Parabola forms upper boundary.}$$

above the x-axis, and above the line

$$y = -3x + 6 \qquad \text{Line and } x\text{-axis form lower boundary.}$$

as shown in Figure 13.7.

Solution Begin by dividing R into the two subregions R_1 and R_2 shown in Figure 13.7. In both regions, it is convenient to use vertical rectangles, and you have

$$\text{Area} = \int_1^2 \int_{-3x+6}^{4x-x^2} dy \, dx + \int_2^4 \int_0^{4x-x^2} dy \, dx$$

$$= \int_1^2 (4x - x^2 + 3x - 6) \, dx + \int_2^4 (4x - x^2) \, dx$$

$$= \left[\frac{7x^2}{2} - \frac{x^3}{3} - 6x \right]_1^2 + \left[2x^2 - \frac{x^3}{3} \right]_2^4$$

$$= \left(14 - \frac{8}{3} - 12 - \frac{7}{2} + \frac{1}{3} + 6 \right) + \left(32 - \frac{64}{3} - 8 + \frac{8}{3} \right)$$

$$= \frac{15}{2}.$$

The area of the region is $15/2$ square units. Try checking this using the procedure for finding the area between two curves, as presented in Section 6.1.

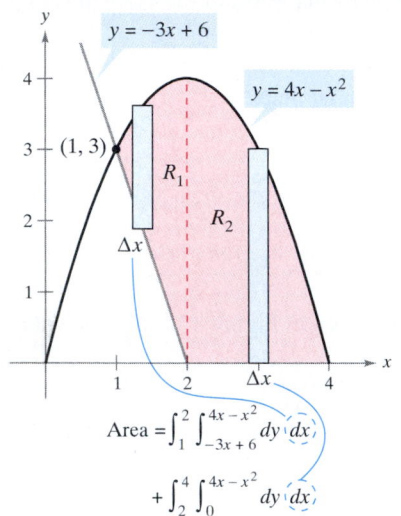

$y = -3x + 6$

$y = 4x - x^2$

$(1, 3)$

R_1

R_2

Δx

Δx

$$\text{Area} = \int_1^2 \int_{-3x+6}^{4x-x^2} dy \, dx$$

$$+ \int_2^4 \int_0^{4x-x^2} dy \, dx$$

Figure 13.7

NOTE In Examples 3 to 6, be sure you see the benefit of sketching the region of integration. We strongly recommend that you develop the habit of making sketches to help determine the limits of integration for all iterated integrals in this chapter.

At this point you may be wondering why you would need iterated integrals. After all, you already know how to use conventional integration to find the area of a region in the plane. (For instance, compare the solution of Example 4 in this section with that given in Example 3 in Section 6.1.) The need for iterated integrals will become clear in the next section. In this section, we have chosen to give primary attention to procedures for finding the limits of integration of the region of an iterated integral and the following exercise set is designed to develop skill in this important procedure.

In Exercises 1–10, evaluate the integral.

1. $\displaystyle\int_0^x (2x - y)\, dy$ **2.** $\displaystyle\int_x^{x^2} \frac{y}{x}\, dy$

3. $\displaystyle\int_1^{2y} \frac{y}{x}\, dx$ **4.** $\displaystyle\int_0^{\cos y} y\, dx$

5. $\displaystyle\int_0^{\sqrt{4-x^2}} x^2 y\, dy$ **6.** $\displaystyle\int_{x^2}^{\sqrt{x}} (x^2 + y^2)\, dy$

7. $\displaystyle\int_{e^y}^{y} \frac{y \ln x}{x}\, dx$ **8.** $\displaystyle\int_{-\sqrt{1-y^2}}^{\sqrt{1-y^2}} (x^2 + y^2)\, dx$

9. $\displaystyle\int_0^{x^3} ye^{-y/x}\, dy$ **10.** $\displaystyle\int_y^{\pi/2} \sin^3 x \cos y\, dx$

In Exercises 11–20, evaluate the iterated integral.

11. $\displaystyle\int_0^1 \int_0^2 (x + y)\, dy\, dx$

12. $\displaystyle\int_0^1 \int_0^x \sqrt{1 - x^2}\, dy\, dx$

13. $\displaystyle\int_1^2 \int_0^4 (x^2 - 2y^2 + 1)\, dx\, dy$

14. $\displaystyle\int_0^1 \int_y^{2y} (1 + 2x^2 + 2y^2)\, dx\, dy$

15. $\displaystyle\int_0^1 \int_0^{\sqrt{1-y^2}} (x + y)\, dx\, dy$

16. $\displaystyle\int_0^2 \int_{3y^2 - 6y}^{2y - y^2} 3y\, dx\, dy$

17. $\displaystyle\int_0^2 \int_0^{\sqrt{4-y^2}} \frac{2}{\sqrt{4 - y^2}}\, dx\, dy$

18. $\displaystyle\int_0^{\pi/2} \int_0^{2\cos\theta} r\, dr\, d\theta$

19. $\displaystyle\int_0^{\pi/2} \int_0^{\sin\theta} \theta r\, dr\, d\theta$

20. $\displaystyle\int_0^{\pi/4} \int_0^{\cos\theta} 3r^2 \sin\theta\, dr\, d\theta$

In Exercises 21–24, evaluate the improper iterated integral.

21. $\displaystyle\int_1^{\infty} \int_0^{1/x} y\, dy\, dx$ **22.** $\displaystyle\int_0^3 \int_0^{\infty} \frac{x^2}{1 + y^2}\, dy\, dx$

23. $\displaystyle\int_1^{\infty} \int_1^{\infty} \frac{1}{xy}\, dx\, dy$ **24.** $\displaystyle\int_0^{\infty} \int_0^{\infty} xye^{-(x^2+y^2)}\, dx\, dy$

In Exercises 25–28, sketch the region R of integration and switch the order of integration.

25. $\displaystyle\int_0^4 \int_0^y f(x, y)\, dx\, dy$ **26.** $\displaystyle\int_0^4 \int_{\sqrt{y}}^2 f(x, y)\, dx\, dy$

27. $\displaystyle\int_{-1}^1 \int_{x^2}^1 f(x, y)\, dy\, dx$ **28.** $\displaystyle\int_{-\pi/2}^{\pi/2} \int_0^{\cos x} f(x, y)\, dy\, dx$

In Exercises 29–36, sketch the region R whose area is given by the iterated integral. Then switch the order of integration and show that both orders yield the same area.

29. $\displaystyle\int_0^1 \int_0^2 dy\, dx$

30. $\displaystyle\int_1^2 \int_2^4 dx\, dy$

31. $\displaystyle\int_0^1 \int_{-\sqrt{1-y^2}}^{\sqrt{1-y^2}} dx\, dy$

32. $\displaystyle\int_0^2 \int_0^x dy\, dx + \int_2^4 \int_0^{4-x} dy\, dx$

33. $\displaystyle\int_0^2 \int_{x/2}^1 dy\, dx$

34. $\displaystyle\int_0^4 \int_{\sqrt{x}}^2 dy\, dx$

35. $\displaystyle\int_0^1 \int_{y^2}^{\sqrt[3]{y}} dx\, dy$

36. $\displaystyle\int_{-2}^2 \int_0^{4-y^2} dx\, dy$

In Exercises 37–42, use an iterated integral to find the area of the region.

37.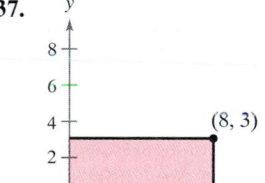

38.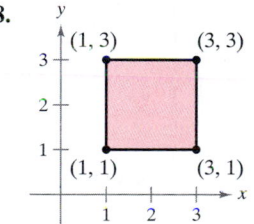

39.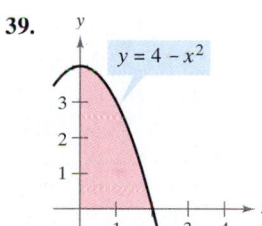

40.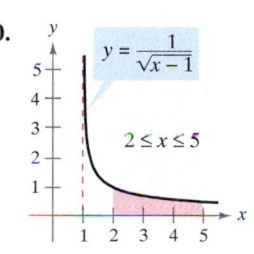

41.

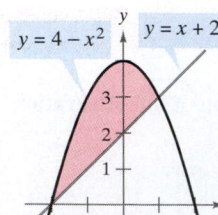

$y = 4 - x^2$ $y = x + 2$

42.

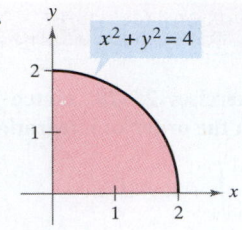

$x^2 + y^2 = 4$

In Exercises 43–48, use an iterated integral to find the area of the region bounded by the graphs of the equations.

43. $\sqrt{x} + \sqrt{y} = 2$, $x = 0$, $y = 0$

44. $y = x^{3/2}$, $y = x$

45. $2x - 3y = 0$, $x + y = 5$, $y = 0$

46. $xy = 9$, $y = x$, $y = 0$, $x = 9$

47. $\dfrac{x^2}{a^2} + \dfrac{y^2}{b^2} = 1$

48. $y = x$, $y = 2x$, $x = 2$

Think About It **In Exercises 49 and 50, give a geometric argument for the given equality. Verify the equality analytically.**

49. $\displaystyle\int_0^5 \int_x^{\sqrt{50-x^2}} x^2 y^2 \, dy \, dx =$

$\displaystyle\int_0^5 \int_0^y x^2 y^2 \, dx \, dy + \int_5^{5\sqrt{2}} \int_0^{\sqrt{50-y^2}} x^2 y^2 \, dx \, dy$

50. $\displaystyle\int_0^2 \int_{x^2}^{2x} x \sin y \, dy \, dx = \int_0^4 \int_{y/2}^{\sqrt{y}} x \sin y \, dx \, dy$

In Exercises 51–54, evaluate the iterated integral. (Note that it is necessary to switch the order of integration.)

51. $\displaystyle\int_0^2 \int_x^2 x\sqrt{1 + y^3} \, dy \, dx$

52. $\displaystyle\int_0^2 \int_x^2 e^{-y^2} \, dy \, dx$

53. $\displaystyle\int_0^1 \int_y^1 \sin x^2 \, dx \, dy$

54. $\displaystyle\int_0^2 \int_{y^2}^4 \sqrt{x} \sin x \, dx \, dy$

In Exercises 55–58, use a symbolic integration utility to evaluate the iterated integral.

55. $\displaystyle\int_0^2 \int_{x^2}^{2x} (x^3 + 3y^2) \, dy \, dx$

56. $\displaystyle\int_0^1 \int_y^{2y} \sin(x + y) \, dx \, dy$

57. $\displaystyle\int_0^4 \int_0^y \dfrac{2}{(x + 1)(y + 1)} \, dx \, dy$

58. $\displaystyle\int_0^a \int_0^{a-x} (x^2 + y^2) \, dy \, dx$

In Exercises 59 and 60, (a) sketch the region of integration, (b) switch the order of integration, and (c) use a symbolic integration utility to show that both orders yield the same value.

59. $\displaystyle\int_0^2 \int_{y^3}^{4\sqrt{2y}} (x^2 y - x y^2) \, dx \, dy$

60. $\displaystyle\int_0^2 \int_{\sqrt{4-x^2}}^{4-x^2/4} \dfrac{xy}{x^2 + y^2 + 1} \, dy \, dx$

In Exercises 61 and 62, use a symbolic integration utility to approximate the iterated integral.

61. $\displaystyle\int_0^2 \int_0^{4-x^2} e^{xy} \, dy \, dx$

62. $\displaystyle\int_0^2 \int_x^2 \sqrt{16 - x^3 - y^3} \, dy \, dx$

63. ***Writing*** Explain what is meant by an iterated integral. How is it evaluated?

64. ***Writing*** Describe regions that are vertically simple and regions that are horizontally simple.

True or False? **In Exercises 65 and 66, determine whether the statement is true or false. If it is false, explain why or give an example that shows it is false.**

65. $\displaystyle\int_a^b \int_c^d f(x, y) \, dy \, dx = \int_c^d \int_a^b f(x, y) \, dx \, dy$

66. $\displaystyle\int_0^1 \int_0^x f(x, y) \, dy \, dx = \int_0^1 \int_0^y f(x, y) \, dx \, dy$

Double Integrals and Volume of a Solid Region • Properties of Double Integrals • Evaluation of Double Integrals

Double Integrals and Volume of a Solid Region

You already know that a definite integral over an *interval* uses a limit process to assign measure to quantities such as area, volume, arc length, and mass. In this section, we use a similar process to define the **double integral** of a function of two variables over a *region in the plane*.

Consider a continuous function f such that $f(x, y) \geq 0$ for all (x, y) in a region R in the xy-plane. The goal is to find the volume of the solid region lying between the surface given by

$$z = f(x, y) \qquad \text{Surface lying above the } xy\text{-plane}$$

and the xy-plane, as shown in Figure 13.8. You can begin by superimposing a rectangular grid over the region, as shown in Figure 13.9. The rectangles lying entirely within R form an **inner partition** Δ, whose **norm** $\|\Delta\|$ is defined as the length of the longest diagonal of the n rectangles. Next, choose a point (x_i, y_i) in each rectangle and form the rectangular prism whose height is $f(x_i, y_i)$, as shown in Figure 13.10. Because the area of the ith rectangle is $\Delta A_i = \Delta x_i \Delta y_i$, it follows that the volume of the ith prism is

$$f(x_i, y_i)\Delta A_i = f(x_i, y_i)\Delta x_i \Delta y_i \qquad \text{Volume of } i\text{th prism}$$

and you can approximate the volume of the solid region by the Riemann sum of the volumes of all n prisms,

$$\sum_{i=1}^{n} f(x_i, y_i)\Delta x_i \Delta y_i \qquad \text{Riemann sum}$$

as shown in Figure 13.11. This approximation can be improved by tightening the mesh of the grid to form smaller and smaller rectangles, as shown in Example 1.

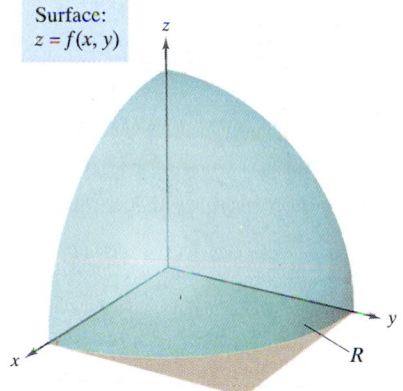

Surface:
$z = f(x, y)$

Figure 13.8

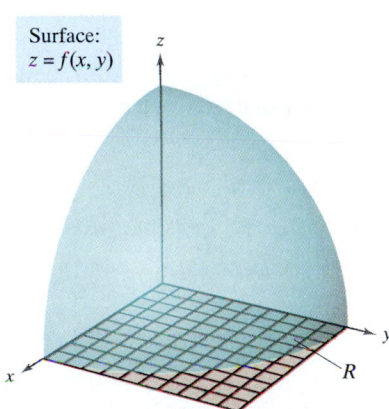

Surface:
$z = f(x, y)$

The rectangles lying within R form an inner partition of R.
Figure 13.9

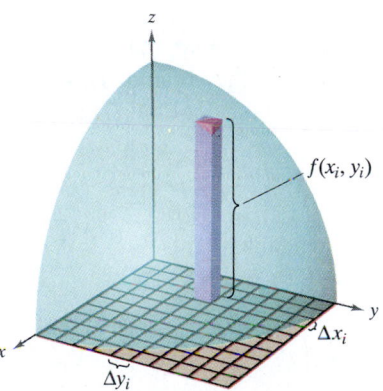

Rectangular prism whose base is Δx_i by Δy_i and whose height is $f(x_i, y_i)$
Figure 13.10

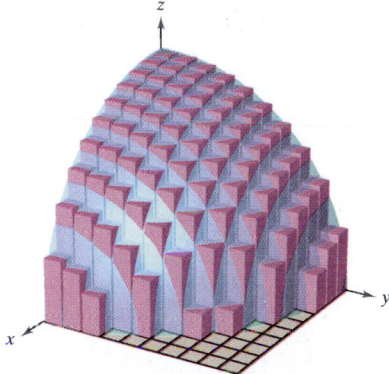

Volume approximated by rectangular prisms
Figure 13.11

EXAMPLE 1   Approximating the Volume of a Solid

Approximate the volume of the solid lying between the paraboloid

$$f(x, y) = 1 - \frac{1}{2}x^2 - \frac{1}{2}y^2$$

and the square region R given by $0 \le x \le 1, 0 \le y \le 1$. Use a partition made up of squares whose edges have a length of $\frac{1}{4}$.

Solution   Begin by forming the specified partition of R. For this partition, it is convenient to choose the centers of the subregions as the points at which to evaluate $f(x, y)$.

$$\left(\tfrac{1}{8}, \tfrac{1}{8}\right) \quad \left(\tfrac{1}{8}, \tfrac{3}{8}\right) \quad \left(\tfrac{1}{8}, \tfrac{5}{8}\right) \quad \left(\tfrac{1}{8}, \tfrac{7}{8}\right)$$

$$\left(\tfrac{3}{8}, \tfrac{1}{8}\right) \quad \left(\tfrac{3}{8}, \tfrac{3}{8}\right) \quad \left(\tfrac{3}{8}, \tfrac{5}{8}\right) \quad \left(\tfrac{3}{8}, \tfrac{7}{8}\right)$$

$$\left(\tfrac{5}{8}, \tfrac{1}{8}\right) \quad \left(\tfrac{5}{8}, \tfrac{3}{8}\right) \quad \left(\tfrac{5}{8}, \tfrac{5}{8}\right) \quad \left(\tfrac{5}{8}, \tfrac{7}{8}\right)$$

$$\left(\tfrac{7}{8}, \tfrac{1}{8}\right) \quad \left(\tfrac{7}{8}, \tfrac{3}{8}\right) \quad \left(\tfrac{7}{8}, \tfrac{5}{8}\right) \quad \left(\tfrac{7}{8}, \tfrac{7}{8}\right)$$

Because the area of each square is $\Delta x_i \Delta y_i = \frac{1}{16}$, you can approximate the volume by the sum

$$\sum_{i=1}^{16} f(x_i\, y_i) \Delta x_i \Delta y_i = \sum_{i=1}^{16} \left(1 - \frac{1}{2}x_i^2 - \frac{1}{2}y_i^2\right)\left(\frac{1}{16}\right) \approx 0.672.$$

This approximation is shown graphically in Figure 13.12. The exact volume of the solid is $\frac{2}{3}$ (see Example 2). You can obtain a better approximation by using a finer partition. For example, with a partition of squares with sides of length $\frac{1}{10}$, the approximation is 0.668.

:   **TECHNOLOGY**   Some three-dimensional graphing utilities are capable of
:   sketching figures such as that shown in Figure 13.12. For instance, the sketch
:   shown in Figure 13.13 was drawn with a computer program. In this sketch, note
:   that each of the rectangular prisms lies within the solid region.

In Example 1, note that by using finer partitions, you can obtain better approximations of the volume. This observation suggests that you could obtain the exact volume by taking a limit. That is

$$\text{Volume} = \lim_{\|\Delta\| \to 0} \sum_{i=1}^{n} f(x_i, y_i) \Delta x_i \Delta y_i.$$

The precise meaning of this limit is that the limit is equal to L if for every $\varepsilon > 0$ there exists a $\delta > 0$ such that

$$\left| L - \sum_{i=1}^{n} f(x_i, y_i) \Delta x_i \Delta y_i \right| < \varepsilon$$

for all partitions Δ of the plane region R (that satisfy $\|\Delta\| < \delta$) and for all possible choices of x_i and y_i in the ith region.

Using the limit of a Riemann sum to define volume is a special case of using the limit to define a **double integral.** The general case, however, does not require that the function be positive or continuous.

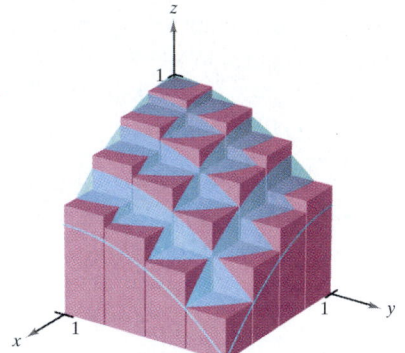

Surface:
$f(x, y) = 1 - \frac{1}{2}x^2 - \frac{1}{2}y^2$

Figure 13.12

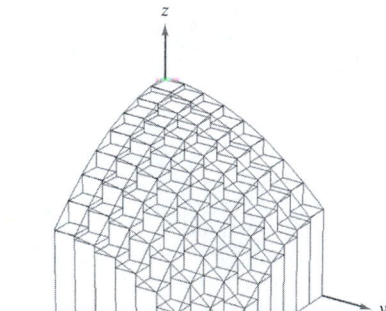

Figure 13.13

The entries in the table represent the depth (in 10-yard units) of earth at the center of each square in the figure below.

x \ y	1	2	3
1	10	9	7
2	7	7	4
3	5	5	4
4	4	5	3

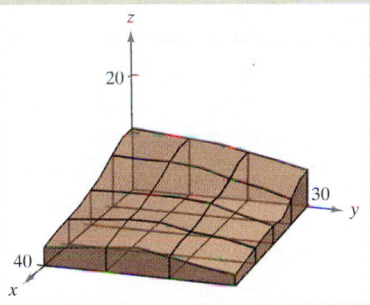

Approximate the number of cubic yards of earth in the first octant. This exploration was submitted by Robert Vojack, Ridgewood High School, Ridgewood, NJ.

Definition of Double Integral

If f is defined on a closed, bounded region R in the xy-plane, then the **double integral of f over R** is given by

$$\iint_R f(x, y)\,dA = \lim_{\|\Delta\| \to 0} \sum_{i=1}^{n} f(x_i, y_i)\Delta x_i \Delta y_i$$

provided the limit exists. If the limit exists, then f is **integrable** over R.

Sufficient conditions for the double integral of f on the region R to exist are that R can be written as a union of a finite number of nonoverlapping (see Figure 13.14) subregions that are vertically or horizontally simple *and* that f is continuous on the region R.

A double integral can be used to find the volume of a solid region that lies between the xy-plane and the surface given by $z = f(x, y)$.

Volume of a Solid Region

If f is integrable over a plane region R and $f(x, y) \geq 0$ for all (x, y) in R, then the volume of the solid region that lies above R and below the graph of f is defined as

$$V = \iint_R f(x, y)\,dA.$$

Properties of Double Integrals

Double integrals share many properties of single integrals.

NOTE   Having defined a double integral, we will occasionally refer to a definite integral as a **single integral.**

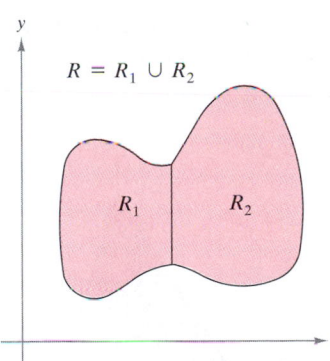

Two regions are nonoverlapping if their intersection is a set that has an area of 0. In this figure, the area of the line segment that is common to R_1 and R_2 is 0.
Figure 13.14

THEOREM 13.1   Properties of Double Integrals

Let f and g be continuous over a closed, bounded plane region R, and let c be a constant.

1. $\displaystyle \iint_R cf(x, y)\,dA = c\iint_R f(x, y)\,dA$

2. $\displaystyle \iint_R [f(x, y) \pm g(x, y)]\,dA = \iint_R f(x, y)\,dA \pm \iint_R g(x, y)\,dA$

3. $\displaystyle \iint_R f(x, y)\,dA \geq 0, \quad$ if $f(x, y) \geq 0$

4. $\displaystyle \iint_R f(x, y)\,dA \geq \iint_R g(x, y)\,dA, \quad$ if $f(x, y) \geq g(x, y)$

5. $\displaystyle \iint_R f(x, y)\,dA = \iint_{R_1} f(x, y)\,dA + \iint_{R_2} f(x, y)\,dA$

where R is the union of two nonoverlapping subregions R_1 and R_2.

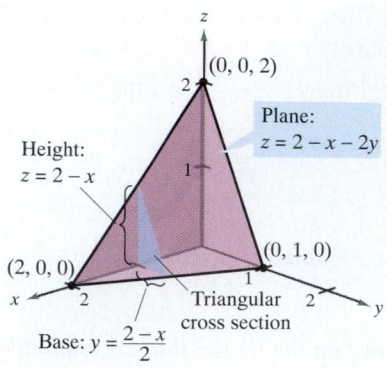

Volume: $\int_0^2 A(x)\, dx$

Figure 13.15

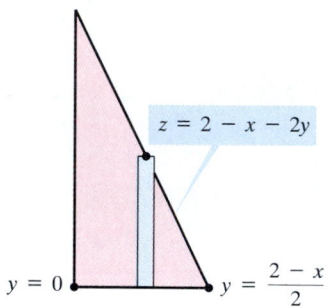

Triangular cross section
Figure 13.16

Evaluation of Double Integrals

Normally, the first step in evaluating a double integral is to rewrite it as an iterated integral. To show how this is done, we use a geometric model of a double integral as the volume of a solid.

Consider the solid region bounded by the plane $z = f(x, y) = 2 - x - 2y$ and the three coordinate planes, as shown in Figure 13.15. Each vertical cross section taken parallel to the yz-plane is a triangular region whose base has a length of $y = (2 - x)/2$ and whose height is $z = 2 - x$. This implies that for a fixed value of x, the area of the triangular cross section is

$$A(x) = \frac{1}{2}(\text{base})(\text{height}) = \frac{1}{2}\left(\frac{2 - x}{2}\right)(2 - x) = \frac{(2 - x)^2}{4}.$$

By the formula for the volume of a solid with known cross sections (Section 6.2), the volume of the solid is

$$\text{Volume} = \int_a^b A(x)\, dx = \int_0^2 \frac{(2 - x)^2}{4}\, dx = -\frac{(2 - x)^3}{12}\bigg]_0^2 = \frac{8}{12} = \frac{2}{3}.$$

This procedure works no matter how $A(x)$ is obtained. In particular, you can find $A(x)$ by integration, as indicated in Figure 13.16. That is, you consider x to be constant, and integrate $z = 2 - x - 2y$ from 0 to $(2 - x)/2$ to obtain

$$A(x) = \int_0^{(2-x)/2} (2 - x - 2y)\, dy$$

$$= \left[(2 - x)y - y^2\right]_0^{(2-x)/2}$$

$$= \frac{(2 - x)^2}{4}.$$

Combining these results, you have the *iterated integral*

$$\text{Volume} = \int_R \int f(x, y)\, dA = \int_0^2 \int_0^{(2-x)/2} (2 - x - 2y)\, dy\, dx.$$

To better understand this procedure, it helps to imagine the integration as two sweeping motions. For the inner integration, a vertical line sweeps out the area of a cross section. For the outer integration, the triangular cross section sweeps out the volume, as shown in Figure 13.17.

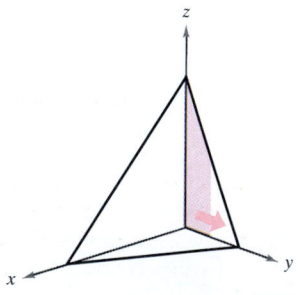
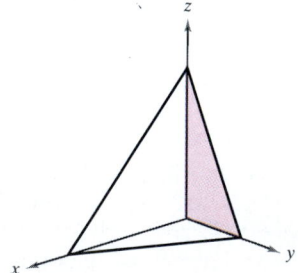
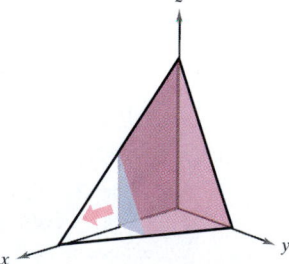
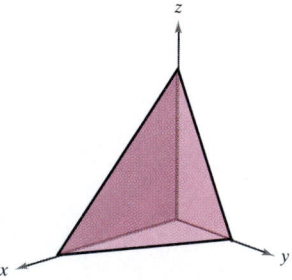

Integrate with respect to y to obtain the area of the cross section.

Integrate with respect to x to obtain the volume of the solid.

Figure 13.17

The following theorem was proved by the Italian mathematician Guido Fubini (1879–1943). The theorem states that if R is a vertically or horizontally simple region and f is continuous on R, the double integral of f on R is equal to an iterated integral.

THEOREM 13.2   Fubini's Theorem

Let f be continuous on a plane region R.

1. If R is defined by $a \le x \le b$ and $g_1(x) \le y \le g_2(x)$, where g_1 and g_2 are continuous on $[a, b]$, then

$$\iint_R f(x, y)\, dA = \int_a^b \int_{g_1(x)}^{g_2(x)} f(x, y)\, dy\, dx.$$

2. If R is defined by $c \le y \le d$ and $h_1(y) \le x \le h_2(y)$, where h_1 and h_2 are continuous on $[c, d]$, then

$$\iint_R f(x, y)\, dA = \int_c^d \int_{h_1(y)}^{h_2(y)} f(x, y)\, dx\, dy.$$

EXAMPLE 2   Evaluating a Double Integral as an Iterated Integral

Evaluate

$$\iint_R \left(1 - \tfrac{1}{2}x^2 - \tfrac{1}{2}y^2\right) dA$$

where R is the region given by $0 \le x \le 1$, $0 \le y \le 1$.

Solution   Because the region R is a simple square, it is both vertically and horizontally simple, and you can use either order of integration. Suppose you choose $dy\, dx$ by placing a vertical representative rectangle in the region, as shown in Figure 13.18. This produces the following.

$$\iint_R \left(1 - \tfrac{1}{2}x^2 - \tfrac{1}{2}y^2\right) dA = \int_0^1 \int_0^1 \left(1 - \tfrac{1}{2}x^2 - \tfrac{1}{2}y^2\right) dy\, dx$$

$$= \int_0^1 \left[\left(1 - \tfrac{1}{2}x^2\right)y - \frac{y^3}{6}\right]_0^1 dx$$

$$= \int_0^1 \left(\frac{5}{6} - \frac{1}{2}x^2\right) dx$$

$$= \left[\frac{5}{6}x - \frac{x^3}{6}\right]_0^1 = \frac{2}{3}$$

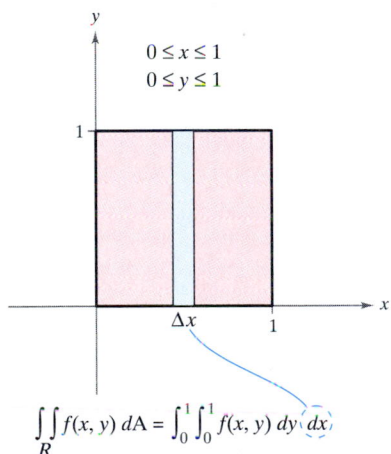

$0 \le x \le 1$
$0 \le y \le 1$

$\iint_R f(x, y)\, dA = \int_0^1 \int_0^1 f(x, y)\, dy\, dx$

The volume of the solid region is $\tfrac{2}{3}$.
Figure 13.18

NOTE   The double integral evaluated in Example 2 represents the volume of the solid region approximated in Example 1. Note that the approximation obtained in Example 1 is quite good $\left(0.672 \text{ vs. } \tfrac{2}{3}\right)$, even though we used a partition consisting of only 16 squares. The error resulted because we used the centers of the square subregions as the points in the approximation. This is comparable to the Midpoint Rule approximation of a single integral.

EXPLORATION

Volume of a Paraboloid Sector
The solid in Example 3 has an elliptical (not a circular) base. Consider the
region bounded by the circular
paraboloid

$$z = a^2 - x^2 - y^2, \quad a > 0$$

and the *xy*-plane. How many ways
do you now know for finding the volume of this solid? For instance, you
could use the disc method to find the
volume as a solid of revolution. Does
each method involve integration?

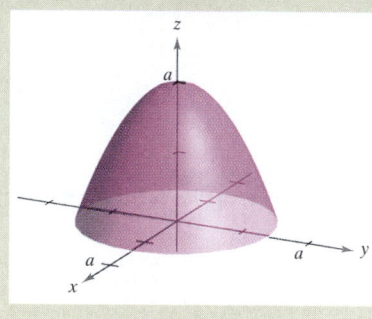

The difficulty of evaluating a single integral $\int_a^b f(x)\, dx$ usually depends on the
function f, and not on the interval $[a, b]$. This is a major difference between single and
double integrals. In the next example, we integrate a function similar to that in
Examples 1 and 2. Notice that a change in the region R produces a much more
difficult integration problem.

EXAMPLE 3 Finding Volume by a Double Integral

Find the volume of the solid region bounded by the paraboloid $z = 4 - x^2 - 2y^2$ and
the *xy*-plane.

Solution By letting $z = 0$, you can see that the base of the region in the *xy*-plane is
the ellipse $x^2 + 2y^2 = 4$, as shown in Figure 13.19. This plane region is both
vertically and horizontally simple, so the order $dy\, dx$ is appropriate.

Variable bounds for y: $\quad -\sqrt{\dfrac{(4 - x^2)}{2}} \le y \le \sqrt{\dfrac{(4 - x^2)}{2}}$

Constant bounds for x: $\quad -2 \le x \le 2$

The volume is given by

$$V = \int_{-2}^{2} \int_{-\sqrt{(4-x^2)/2}}^{\sqrt{(4-x^2)/2}} (4 - x^2 - 2y^2)\, dy\, dx$$

$$= \int_{-2}^{2} \left[(4 - x^2)y - \frac{2y^3}{3} \right]_{-\sqrt{(4-x^2)/2}}^{\sqrt{(4-x^2)/2}} dx$$

$$= \frac{4}{3\sqrt{2}} \int_{-2}^{2} (4 - x^2)^{3/2}\, dx$$

$$= \frac{4}{3\sqrt{2}} \int_{-\pi/2}^{\pi/2} 16 \cos^4 \theta\, d\theta \qquad \textcolor{red}{x = 2\sin\theta}$$

$$= \frac{64}{3\sqrt{2}} (2) \int_{0}^{\pi/2} \cos^4 \theta\, d\theta$$

$$= \frac{128}{3\sqrt{2}} \left(\frac{3\pi}{16} \right) = 4\sqrt{2}\,\pi. \qquad \textcolor{red}{\text{Wallis's Formula}}$$

NOTE In Example 3, note the usefulness of Wallis's Formula to evaluate
$\int_0^{\pi/2} \cos^n \theta\, d\theta$. You may want to review
this formula in Section 7.3.

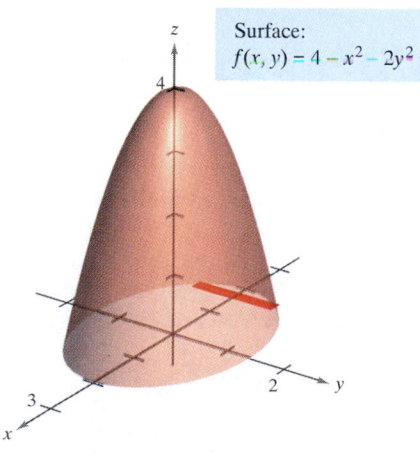

Surface:
$f(x, y) = 4 - x^2 - 2y^2$

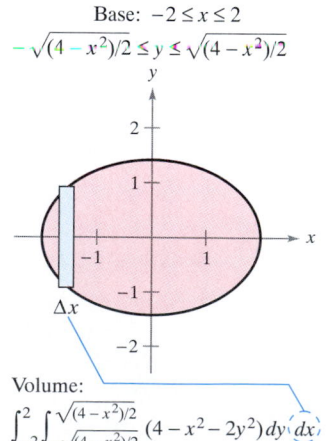

Base: $-2 \le x \le 2$
$-\sqrt{(4-x^2)/2} \le y \le \sqrt{(4-x^2)/2}$

Volume:
$\int_{-2}^{2} \int_{-\sqrt{(4-x^2)/2}}^{\sqrt{(4-x^2)/2}} (4 - x^2 - 2y^2)\, dy\, dx$

Figure 13.19

In Examples 2 and 3, the order of integration was optional, because the regions were both vertically and horizontally simple. Moreover, had you used the order $dx\,dy$, you would have obtained integrals of comparable difficulty. There are, however, some occasions in which one order of integration is much more convenient than the other. Example 4 shows such a case.

EXAMPLE 4   Comparing Different Orders of Integration

Find the volume of the solid region R bounded by the surface

$$f(x, y) = e^{-x^2} \qquad \text{Surface}$$

and the planes $y = 0$, $y = x$, and $x = 1$, as shown in Figure 13.20.

Solution   The base of R in the xy-plane is bounded by the lines $y = 0$, $x = 1$, and $y = x$. The two possible orders of integration are given in Figure 13.21.

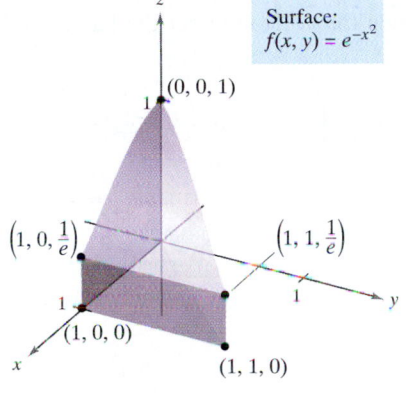

Surface:
$f(x, y) = e^{-x^2}$

Base is bounded by $y = 0$, $y = x$, and $x = 1$.

Figure 13.20

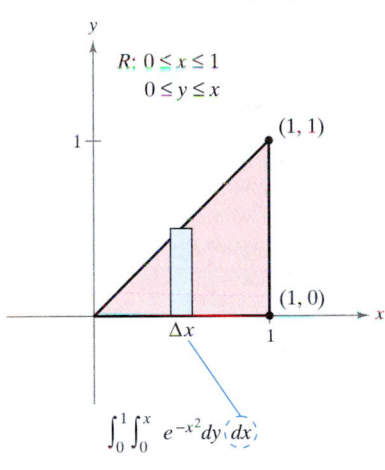
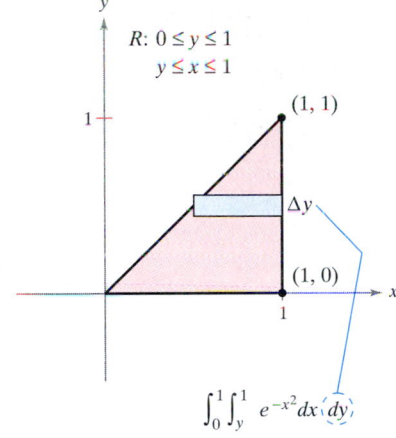

Figure 13.21

By setting up the corresponding iterated integrals, you can see that the order $dx\,dy$ requires the antiderivative $\int e^{-x^2}\,dx$, which is not an elementary function. On the other hand, the order $dy\,dx$ produces the integral

$$\int_0^1 \int_0^x e^{-x^2}\,dy\,dx = \int_0^1 e^{-x^2}y \bigg]_0^x dx$$

$$= \int_0^1 x e^{-x^2}\,dx$$

$$= -\frac{1}{2}e^{-x^2} \bigg]_0^1$$

$$= -\frac{1}{2}\left(\frac{1}{e} - 1\right)$$

$$= \frac{e - 1}{2e}$$

$$\approx 0.316.$$

NOTE   Try using a symbolic integration utility to evaluate the integral in Example 4.

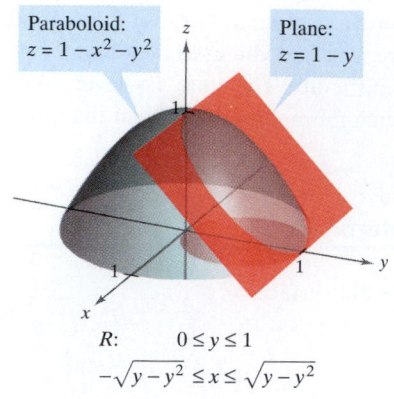

Paraboloid:
$z = 1 - x^2 - y^2$

Plane:
$z = 1 - y$

$R:$ $0 \le y \le 1$
$-\sqrt{y - y^2} \le x \le \sqrt{y - y^2}$

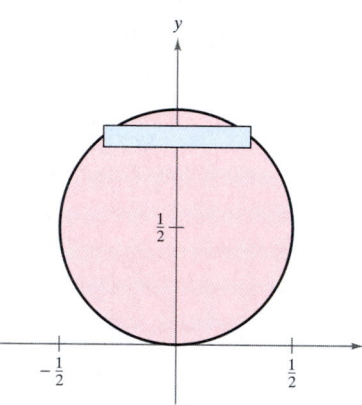

The volume of the solid region is $\pi/32$.
Figure 13.22

EXAMPLE 5   Volume of a Region Bounded by Two Surfaces

Find the volume of the solid region R bounded above by the paraboloid $z = 1 - x^2 - y^2$ and below by the plane $z = 1 - y$, as shown in Figure 13.22.

Solution Equating z-values, you can determine that the intersection of the two surfaces occurs on the right circular cylinder given by

$$1 - y = 1 - x^2 - y^2 \quad \Longrightarrow \quad x^2 = y - y^2.$$

Because the volume of R is the difference between the volume under the paraboloid and the volume under the plane, you have

$$\text{Volume} = \int_0^1 \int_{-\sqrt{y-y^2}}^{\sqrt{y-y^2}} (1 - x^2 - y^2)\, dx\, dy - \int_0^1 \int_{-\sqrt{y-y^2}}^{\sqrt{y-y^2}} (1 - y)\, dx\, dy.$$

$$= \int_0^1 \int_{-\sqrt{y-y^2}}^{\sqrt{y-y^2}} (y - y^2 - x^2)\, dx\, dy$$

$$= \int_0^1 \left[(y - y^2)x - \frac{x^3}{3} \right]_{-\sqrt{y-y^2}}^{\sqrt{y-y^2}} dy$$

$$= \frac{4}{3} \int_0^1 (y - y^2)^{3/2}\, dy$$

$$= \left(\frac{4}{3}\right)\left(\frac{1}{8}\right) \int_0^1 [1 - (2y - 1)^2]^{3/2}\, dy$$

$$= \frac{1}{6} \int_{-\pi/2}^{\pi/2} \frac{\cos^4 \theta}{2}\, d\theta \qquad\qquad\qquad 2y - 1 = \sin\theta$$

$$= \frac{1}{6} \int_0^{\pi/2} \cos^4 \theta\, d\theta = \left(\frac{1}{6}\right)\left(\frac{3\pi}{16}\right) \qquad\qquad \text{Wallis's Formula}$$

$$= \frac{\pi}{32}$$

EXERCISES FOR SECTION 13.2

Approximation In Exercises 1–4, approximate the integral $\int_R \int f(x, y)\, dA$ by dividing the rectangle R with vertices $(0, 0)$, $(4, 0)$, $(4, 2)$, and $(0, 2)$ into eight equal squares and finding the sum

$$\sum_{i=1}^{8} f(x_i, y_i)\Delta x_i \Delta y_i$$

where (x_i, y_i) is the center of the ith square. Evaluate the double integral and compare it with the approximation.

1. $\displaystyle\int_0^4 \int_0^2 (x + y)\, dy\, dx$

2. $\displaystyle\int_0^4 \int_0^2 xy\, dy\, dx$

3. $\displaystyle\int_0^4 \int_0^2 (x^2 + y^2)\, dy\, dx$

4. $\displaystyle\int_0^4 \int_0^2 \frac{1}{(x + 1)(y + 1)}\, dy\, dx$

In Exercises 5–10, sketch the region R and evaluate the double integral $\int_R \int f(x, y)\, dA$.

5. $\displaystyle\int_0^2 \int_0^1 (1 + 2x + 2y)\, dy\, dx$

6. $\displaystyle\int_0^\pi \int_0^{\pi/2} \sin^2 x \cos^2 y\, dy\, dx$

7. $\displaystyle\int_0^6 \int_{y/2}^3 (x + y)\, dx\, dy$

8. $\displaystyle\int_0^1 \int_y^{\sqrt{y}} x^2 y^2\, dx\, dy$

9. $\displaystyle\int_{-a}^a \int_{-\sqrt{a^2 - x^2}}^{\sqrt{a^2 - x^2}} (x + y)\, dy\, dx$

10. $\displaystyle\int_0^1 \int_{y-1}^0 e^{x+y}\, dx\, dy + \int_0^1 \int_0^{1-y} e^{x+y}\, dx\, dy$

In Exercises 11–16, set up an integral for both orders of integration, and use the more convenient order to evaluate the integral over the region *R*.

11. $\displaystyle\int_R\int xy\,dA$

R: rectangle with vertices $(0, 0), (0, 5), (3, 5), (3, 0)$

12. $\displaystyle\int_R\int \sin x \sin y\,dA$

R: rectangle with vertices $(-\pi, 0), (\pi, 0), (\pi, \pi/2), (-\pi, \pi/2)$

13. $\displaystyle\int_R\int \frac{y}{x^2 + y^2}\,dA$

R: triangle bounded by $y = x, y = 2x, x = 2$

14. $\displaystyle\int_R\int \frac{y}{1 + x^2}\,dA$

R: region bounded by $y = 0, y = \sqrt{x}, x = 4$

15. $\displaystyle\int_R\int x\,dA$

R: sector of a circle in the first quadrant bounded by
$y = \sqrt{25 - x^2}, 3x - 4y = 0, y = 0$

16. $\displaystyle\int_R\int (x^2 + y^2)\,dA$

R: semicircle bounded by $y = \sqrt{4 - x^2}, y = 0$

In Exercises 17–26, use a double integral to find the volume of the indicated solid.

17.

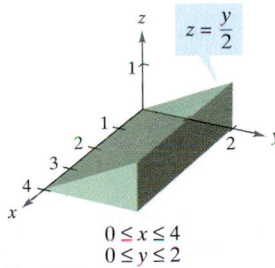

$z = \dfrac{y}{2}$

$0 \le x \le 4$
$0 \le y \le 2$

18.

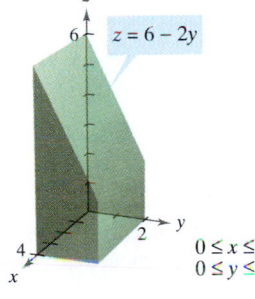

$z = 6 - 2y$

$0 \le x \le 4$
$0 \le y \le 2$

19.

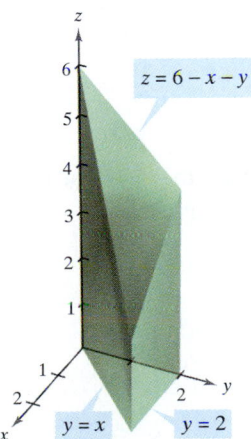

$z = 6 - x - y$

$y = x$ $y = 2$

20.

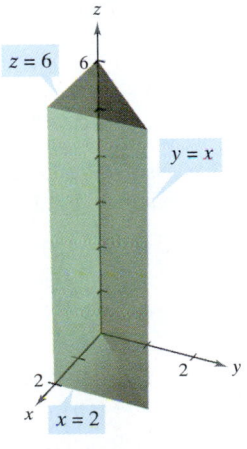

$z = 6$

$y = x$

$x = 2$

21.

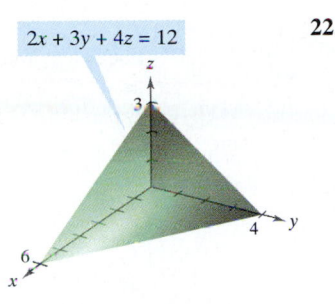

$2x + 3y + 4z = 12$

22.

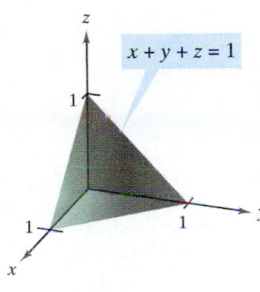

$x + y + z = 1$

23.

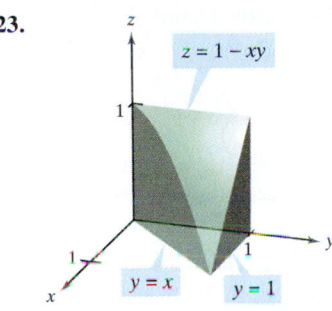

$z = 1 - xy$

$y = x$ $y = 1$

24.

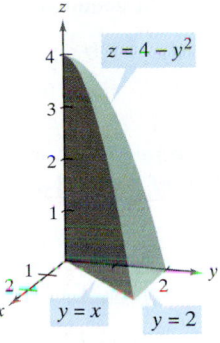

$z = 4 - y^2$

$y = x$ $y = 2$

25. Improper integral

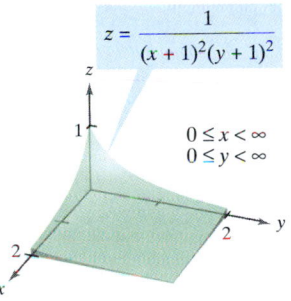

$z = \dfrac{1}{(x + 1)^2(y + 1)^2}$

$0 \le x < \infty$
$0 \le y < \infty$

26. Improper integral

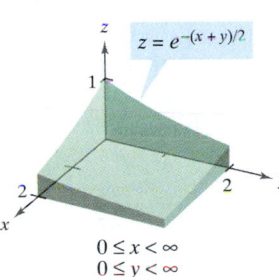

$z = e^{-(x + y)/2}$

$0 \le x < \infty$
$0 \le y < \infty$

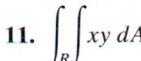 In Exercises 27 and 28, use a symbolic integration utility to find the volume of the solid.

27.

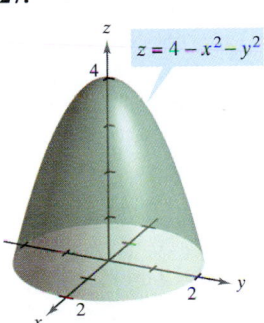

$z = 4 - x^2 - y^2$

28.

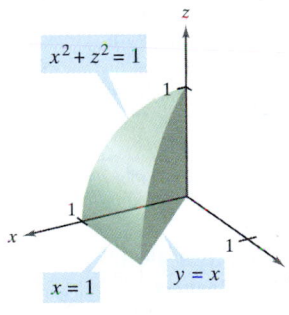

$x^2 + z^2 = 1$

$x = 1$ $y = x$

In Exercises 29–36, set up a double integral to find the volume of the solid bounded by the graphs of the equations.

29. $z = xy, z = 0, y = x, x = 1$, first octant

30. $y = 0, z = 0, y = x, z = x, x = 0, x = 5$

31. $z = 0, z = x^2, x = 0, x = 2, y = 0, y = 4$

32. $x^2 + y^2 + z^2 = r^2$

33. $x^2 + z^2 = 1, y^2 + z^2 = 1$, first octant

34. $y = 1 - x^2, z = 1 - x^2$, first octant

35. $z = x + y, x^2 + y^2 = 4$, first octant

36. $z = \dfrac{1}{1 + y^2}, x = 0, x = 2, y \geq 0$

In Exercises 37 and 38, use Wallis's Formula to find the volume of the solid bounded by the graphs of the equations.

37. $z = x^2 + y^2, x^2 + y^2 = 4, z = 0$

38. $z = \sin^2 x, z = 0, 0 \leq x \leq \pi, 0 \leq y \leq 5$

In Exercises 39–42, use a symbolic integration utility to find the volume of the solid bounded by the graphs of the equations.

39. $z = 4 - x^2 - y^2, z = 0$

40. $x^2 = 9 - y, z^2 = 9 - y$, first octant

41. $z = \dfrac{2}{1 + x^2 + y^2}, z = 0, y = 0, x = 0, y = -0.5x + 1$

42. $z = \ln(1 + x + y), z = 0, y = 0, x = 0, x = 4 - \sqrt{y}$

43. If f is a continuous function such that $0 \leq f(x, y) \leq 1$ over a region R of area 1, prove that

$$0 \leq \int\!\!\int_R f(x, y) \, dA \leq 1.$$

44. Find the volume of the solid in the first octant bounded by the coordinate planes and the plane $(x/a) + (y/b) + (z/c) = 1$, where $a > 0, b > 0$, and $c > 0$.

In Exercises 45–48, evaluate the iterated integral. (Note that it is necessary to switch the order of integration.)

45. $\displaystyle\int_0^1 \int_{y/2}^{1/2} e^{-x^2} \, dx \, dy$

46. $\displaystyle\int_0^1 \int_0^{\arccos y} \sin x \sqrt{1 + \sin^2 x} \, dx \, dy$

47. $\displaystyle\int_0^{\ln 10} \int_{e^x}^{10} \dfrac{1}{\ln y} \, dy \, dx$

48. $\displaystyle\int_0^2 \int_{x^2}^4 \sqrt{y} \cos y \, dy \, dx$

In Exercises 49–52, find the average value of $f(x, y)$ over the region R where

$$\text{Average} = \dfrac{1}{A} \int\!\!\int_R f(x, y) \, dA$$

and where A is the area of R.

49. $f(x, y) = x$

   R: rectangle with vertices $(0, 0), (4, 0), (4, 2), (0, 2)$

50. $f(x, y) = xy$

   R: rectangle with vertices $(0, 0), (4, 0), (4, 2), (0, 2)$

51. $f(x, y) = x^2 + y^2$

   R: square with vertices $(0, 0), (2, 0), (2, 2), (0, 2)$

52. $f(x, y) = e^{x+y}$

   R: triangle with vertices $(0, 0), (0, 1), (1, 1)$

53. *Average Production*   The Cobb-Douglas production function for a company is

$$f(x, y) = 100x^{0.6}y^{0.4}$$

where x is the number of units of labor and y is the number of units of capital. Estimate the average production level if the number of units of labor x varies between 200 and 250 and the number of units of capital y varies between 300 and 325.

54. *Average Profit*   A firm's profit in marketing two products is

$$P = 192x + 576y - x^2 - 5y^2 - 2xy - 5000$$

where x and y represent the numbers of units of the two products. Use a symbolic integration utility to evaluate the double integral yielding the average weekly profit if x varies between 40 and 50 units and y varies between 45 and 60 units.

55. *Think About It*   Let R be a region in the xy-plane whose area is B. If $f(x, y) = k$ for every point (x, y) in R, what is the value of $\int_R\!\!\int f(x, y) \, dA$?

56. *Think About It*   Let R represent a county in the northern part of the United States, and let $f(x, y)$ represent the total annual snowfall at the point (x, y) in R. Give an interpretation of each of the following.

(a) $\displaystyle\int_R\!\!\int f(x, y) \, dA$

(b) $\dfrac{\displaystyle\int_R\!\!\int f(x, y) \, dA}{\displaystyle\int_R\!\!\int dA}$

Probability   A joint density function of the continuous random variables x and y is a function $f(x, y)$ satisfying the following properties.

(a) $f(x, y) \geq 0$ for all (x, y)

(b) $\displaystyle\int_{-\infty}^{\infty} \int_{-\infty}^{\infty} f(x, y) \, dA = 1$

(c) $P[(x, y) \in R] = \displaystyle\int_R\!\!\int f(x, y) \, dA$

In Exercises 57–60, show that that the function is a joint density function and find the required probability.

57. $f(x, y) = \begin{cases} \frac{1}{10}, & 0 \leq x \leq 5, 0 \leq y \leq 2 \\ 0, & \text{elsewhere} \end{cases}$

   $P(0 \leq x \leq 2, 1 \leq y \leq 2)$

58. $f(x, y) = \begin{cases} \frac{1}{4}xy, & 0 \leq x \leq 2, 0 \leq y \leq 2 \\ 0, & \text{elsewhere} \end{cases}$

   $P(0 \leq x \leq 1, 1 \leq y \leq 2)$

59. $f(x, y) = \begin{cases} \frac{1}{27}(9 - x - y), & 0 \le x \le 3, 3 \le y \le 6 \\ 0, & \text{elsewhere} \end{cases}$

$P(0 \le x \le 1, 4 \le y \le 6)$

60. $f(x, y) = \begin{cases} e^{-(x-y)}, & x \ge 0, y \ge 0 \\ 0, & \text{elsewhere} \end{cases}$

$P(0 \le x \le 1, x \le y \le 1)$

61. *Approximation* The base of a pile of sand at a cement plant is rectangular with approximate dimensions of 20 meters by 30 meters. If the base is placed on the xy-plane with one vertex at the origin, the coordinates on the surface of the pile are $(5, 5, 3)$, $(15, 5, 6)$, $(25, 5, 4)$, $(5, 15, 2)$, $(15, 15, 7)$, and $(25, 15, 3)$. Approximate the volume of sand in the pile.

62. *Programming* Consider a continuous function $f(x, y)$ over the rectangular region R with vertices (a, c), (b, c), (a, d), and (b, d) where $a < b$ and $c < d$. Partition the intervals $[a, b]$ and $[c, d]$ into m and n subintervals, so that the subintervals in a given direction are of equal length. Write a program for a graphing utility to compute the sum

$$\sum_{i=1}^{n} \sum_{j=1}^{m} f(x_i, y_j) \, \Delta x_i \Delta y_j \approx \int_a^b \int_c^d f(x, y) \, dA$$

where (x_i, y_j) is the center of a representative rectangle in R.

Approximation In Exercises 63–66, (a) use a symbolic integration utility to approximate the double integral, and (b) use the program in Exercise 62 to approximate the double integral for the indicated values of m and n.

63. $\int_0^1 \int_0^2 \sin \sqrt{x + y} \, dy \, dx$

$m = 4, n = 8$

64. $\int_0^2 \int_0^4 20e^{-x^3/8} dy \, dx$

$m = 10, n = 20$

65. $\int_4^6 \int_0^2 y \cos \sqrt{x} \, dx \, dy$

$m = 4, n = 8$

66. $\int_1^4 \int_1^2 \sqrt{x^3 + y^3} \, dx \, dy$

$m = 6, n = 4$

67. *Writing* From 1963 to 1986, the volume of the Great Salt Lake approximately tripled while its top surface area approximately doubled. Read the article "Relations between Surface Area and Volume in Lakes" by Daniel Cass and Gerald Wildenberg in the November 1990 issue of *The College Mathematics Journal*. Then give examples of solids that have "water levels" a and b such that $V(b) = 3V(a)$ and $A(b) = 2A(a)$, as indicated in the figure at the top of the next column, where V is volume and A is area.

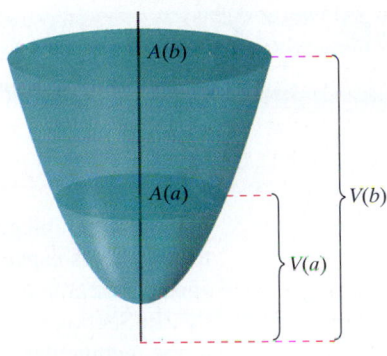

Figure for 67

Think About It In Exercises 68 and 69, determine which value best approximates the volume of the solid between the xy-plane and the function over the region. (Make your selection on the basis of a sketch of the solid and *not* by performing any calculations.)

68. $f(x, y) = 4x$

R: square with vertices $(0, 0)$, $(4, 0)$, $(4, 4)$, $(0, 4)$

(a) -200 (b) 600 (c) 50 (d) 125 (e) 1000

69. $f(x, y) = \sqrt{x^2 + y^2}$

R: circle bounded by $x^2 + y^2 = 9$

(a) 50 (b) 500 (c) -500 (d) 5 (e) 5000

True or False? In Exercises 70 and 71, determine whether the statement is true or false. If it is false, explain why or give an example that shows it is false.

70. The volume of the sphere $x^2 + y^2 + z^2 = 1$ is given by the integral

$$V = 8 \int_0^1 \int_0^1 \sqrt{1 - x^2 - y^2} \, dx \, dy.$$

71. If $f(x, y) \le g(x, y)$ for all (x, y) in R, and both f and g are continuous over R, then

$$\int_R \int f(x, y) \, dA \le \int_R \int g(x, y) \, dA.$$

72. Let $f(x) = \int_1^x e^{t^2} dt$. Find the average value of f on the interval $[0, 1]$.

73. Find $\int_0^\infty \dfrac{e^{-x} - e^{-2x}}{x} dx$. $\left(\text{Hint: Evaluate } \int_1^2 e^{-xy} \, dy.\right)$

74. *Probability* Consider the function

$$f(x, y) = \begin{cases} ke^{-(x+y)/a}, & x \ge 0, y \ge 0 \\ 0, & \text{elsewhere.} \end{cases}$$

Find the relationship between the positive constants a and k such that f is a joint density function of the continuous random variables x and y.

Double Integrals in Polar Coordinates

Double Integrals in Polar Coordinates

Some double integrals are *much* easier to evaluate in polar form than in rectangular form. This is especially true for regions such as circles, cardioids, and petal curves, and for integrands that involve $x^2 + y^2$.

In Section 9.4, you learned that the polar coordinates (r, θ) of a point are related to the rectangular coordinates (x, y) of the point as follows.

$$x = r \cos \theta \quad \text{and} \quad y = r \sin \theta$$

$$r^2 = x^2 + y^2 \quad \text{and} \quad \tan \theta = \frac{y}{x}$$

EXAMPLE 1   Using Polar Coordinates to Describe a Region

Use polar coordinates to describe the regions shown in Figure 13.23.

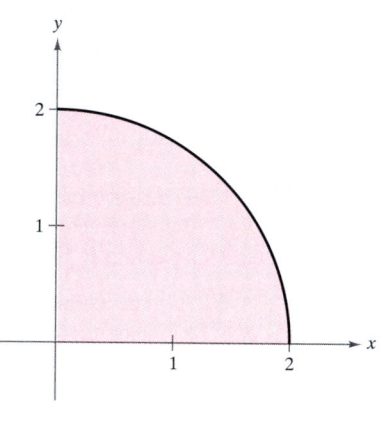

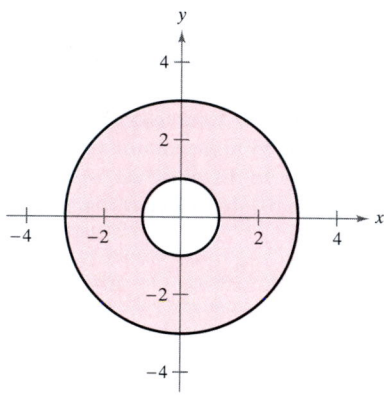

(a)   (b)

Figure 13.23

Solution

a. The region R is a quarter circle of radius 2. It can be described in polar coordinates as

$$R = \{(r, \theta): 0 \le r \le 2, \quad 0 \le \theta \le \pi/2\}.$$

b. The region R consists of all points between the concentric circles of radii 1 and 3. It can be described in polar coordinates as

$$R = \{(r, \theta): 1 \le r \le 3, \quad 0 \le \theta \le 2\pi\}.$$

The regions in Example 1 are special cases of **polar sectors**

$$R = \{(r, \theta): r_1 \le r \le r_2, \quad \theta_1 \le \theta \le \theta_2\} \qquad \text{Polar sector}$$

as shown in Figure 13.24.

Polar sector
Figure 13.24

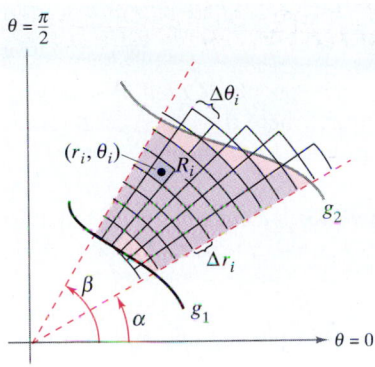

Polar grid is superimposed over region R.
Figure 13.25

To define a double integral of a continuous function $z = f(x, y)$ in polar coordinates, consider a region R bounded by the graphs of $r = g_1(\theta)$ and $r = g_2(\theta)$ and the lines $\theta = \alpha$ and $\theta = \beta$. Instead of partitioning R into small rectangles, use a partition of small polar sectors. On R, superimpose a polar grid made of rays and circular arcs, as shown in Figure 13.25. The polar sectors R_i lying entirely within R form an **inner polar partition** Δ, whose **norm** $\|\Delta\|$ is the length of the longest diagonal of the n polar sectors.

Consider a specific polar sector R_i, as shown in Figure 13.26. It can be shown (see Exercise 49) that the area of R_i is

$$\Delta A_i = r_i \Delta r_i \Delta \theta_i \qquad \text{Area of } R_i$$

where $\Delta r_i = r_2 - r_1$ and $\Delta \theta_i = \theta_2 - \theta_1$. This implies that the volume of the solid of height $f(r_i \cos \theta_i, r_i \sin \theta_i)$ above R_i is approximately

$$f(r_i \cos \theta_i, r_i \sin \theta_i) r_i \Delta r_i \Delta \theta_i$$

and you have

$$\iint_R f(x, y)\, dA \approx \sum_{i=1}^{n} f(r_i \cos \theta_i, r_i \sin \theta_i) r_i \Delta r_i \Delta \theta_i.$$

The sum on the right can be interpreted as a Riemann sum for $f(r \cos \theta, r \sin \theta)r$. The region R corresponds to a *horizontally simple* region S in the $r\theta$-plane, as indicated in Figure 13.27. The polar sectors R_i correspond to rectangles S_i, and the area ΔA_i of S_i is $\Delta r_i \Delta \theta_i$. Thus, the right-hand side of the equation corresponds to the double integral

$$\iint_S f(r \cos \theta, r \sin \theta)r\, dA.$$

From this, you can apply Theorem 13.2 to write

$$\iint_R f(x, y)dA = \iint_S f(r \cos \theta, r \sin \theta)r\, dA$$

$$= \int_\alpha^\beta \int_{g_1(\theta)}^{g_2(\theta)} f(r \cos \theta, r \sin \theta)r\, dr\, d\theta.$$

This suggests the following theorem, whose proof is discussed in Section 13.8.

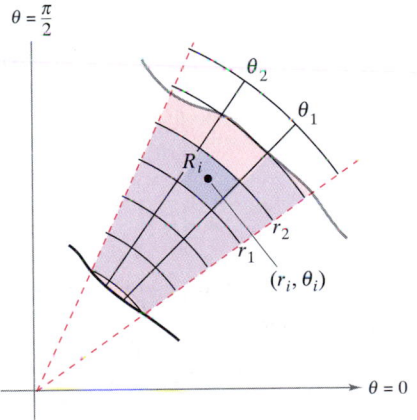

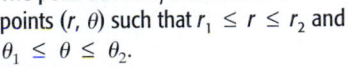

The polar sector R_i is the set of all points (r, θ) such that $r_1 \leq r \leq r_2$ and $\theta_1 \leq \theta \leq \theta_2$.
Figure 13.26

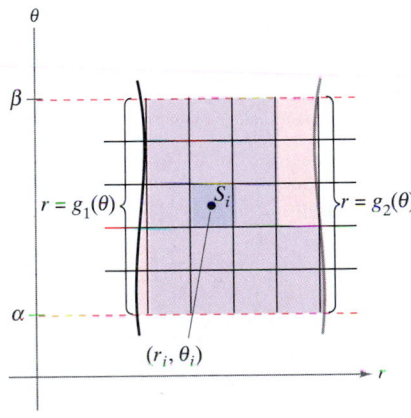

Horizontally simple region S
Figure 13.27

NOTE If $z = f(x, y)$ is nonnegative on R, then the integral in Theorem 13.3 can be interpreted as the volume of the solid region between the graph of f and the region R.

THEOREM 13.3 Change of Variables to Polar Form

Let R be a plane region consisting of all points $(x, y) = (r \cos \theta, r \sin \theta)$ satisfying the conditions $0 \le g_1(\theta) \le r \le g_2(\theta)$, $\alpha \le \theta \le \beta$, where $0 \le (\beta - \alpha) \le 2\pi$. If g_1 and g_2 are continuous on $[\alpha, \beta]$ and f is continuous on R, then

$$\int\int_R f(x, y)\,dA = \int_\alpha^\beta \int_{g_1(\theta)}^{g_2(\theta)} f(r \cos \theta, r \sin \theta)r\,dr\,d\theta.$$

EXPLORATION

Volume of a Paraboloid In the Exploration feature on page 928, you were asked to summarize the different ways you know for finding the volume of the solid bounded by the paraboloid

$$z = a^2 - x^2 - y^2, \quad a > 0$$

and the xy-plane. You now know another way. Use it to find the volume of the solid.

The region R is restricted to two basic types, **r-simple** regions and **θ-simple** regions, as shown in Figure 13.28.

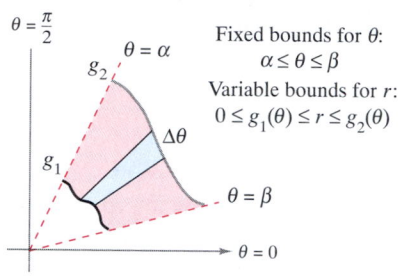

Fixed bounds for θ:
$\alpha \le \theta \le \beta$
Variable bounds for r:
$0 \le g_1(\theta) \le r \le g_2(\theta)$

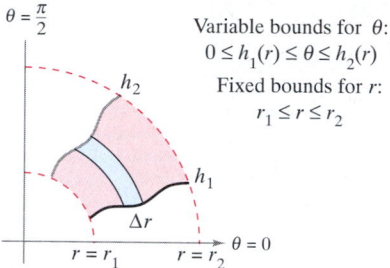

Variable bounds for θ:
$0 \le h_1(r) \le \theta \le h_2(r)$
Fixed bounds for r:
$r_1 \le r \le r_2$

r-Simple region
Figure 13.28

θ-Simple region

EXAMPLE 2 Evaluating a Double Polar Integral

Let R be the annular region lying between the two circles $x^2 + y^2 = 1$ and $x^2 + y^2 = 5$, as shown in Figure 13.29. Evaluate the integral $\int_R\int (x^2 + y)\,dA$.

Solution The polar boundaries are $1 \le r \le \sqrt{5}$ and $0 \le \theta \le 2\pi$. Furthermore, $x^2 = (r \cos \theta)^2$ and $y = r \sin \theta$. Thus, you have

$$\int\int_R (x^2 + y)\,dA = \int_0^{2\pi} \int_1^{\sqrt{5}} (r^2 \cos^2 \theta + r \sin \theta)r\,dr\,d\theta$$

$$= \int_0^{2\pi} \int_1^{\sqrt{5}} (r^3 \cos^2 \theta + r^2 \sin \theta)\,dr\,d\theta$$

$$= \int_0^{2\pi} \left(\frac{r^4}{4} \cos^2 \theta + \frac{r^3}{3} \sin \theta \right)\Big]_1^{\sqrt{5}}\,d\theta$$

$$= \int_0^{2\pi} \left(6 \cos^2 \theta + \frac{5\sqrt{5} - 1}{3} \sin \theta \right)\,d\theta$$

$$= \int_0^{2\pi} \left(3 + 3 \cos 2\theta + \frac{5\sqrt{5} - 1}{3} \sin \theta \right)\,d\theta$$

$$= \left(3\theta + 3 \frac{\sin 2\theta}{2} - \frac{5\sqrt{5} - 1}{3} \cos \theta \right)\Big]_0^{2\pi}$$

$$= 6\pi.$$

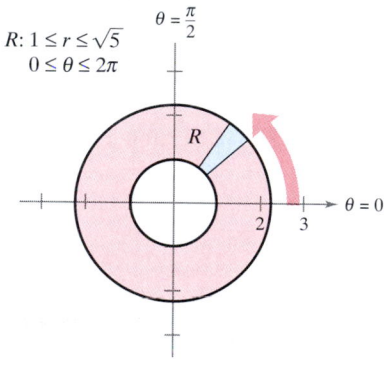

$R: 1 \le r \le \sqrt{5}$
$0 \le \theta \le 2\pi$

r-Simple region
Figure 13.29

In Example 2, be sure to notice the extra factor of r in the integrand. This comes from the formula for the area of a polar sector. In differential notation, you can write

$$dA = r \, dr \, d\theta$$

which indicates that the area of a polar sector increases as you move away from the origin.

EXAMPLE 3   Change of Variables to Polar Coordinates

Use polar coordinates to find the volume of the solid region bounded above by the hemisphere

$$z = \sqrt{16 - x^2 - y^2} \qquad \text{Hemisphere forms upper surface.}$$

and below by the circular region R given by

$$x^2 + y^2 \leq 4 \qquad \text{Circular region forms lower surface.}$$

as shown in Figure 13.30.

Solution   In Figure 13.30, you can see that R has the bounds

$$-\sqrt{4 - y^2} \leq x \leq \sqrt{4 - y^2}, \quad -2 \leq y \leq 2$$

and that $0 \leq z \leq \sqrt{16 - x^2 - y^2}$. In polar coordinates, the bounds are

$$0 \leq r \leq 2 \quad \text{and} \quad 0 \leq \theta \leq 2\pi$$

with height $z = \sqrt{16 - x^2 - y^2} = \sqrt{16 - r^2}$. Consequently, the volume V is given by

$$V = \int\int_R f(x, y) \, dA = \int_0^{2\pi} \int_0^2 \sqrt{16 - r^2} \, r \, dr \, d\theta$$

$$= -\frac{1}{3} \int_0^{2\pi} (16 - r^2)^{3/2} \Big]_0^2 \, d\theta$$

$$= -\frac{1}{3} \int_0^{2\pi} \left(24\sqrt{3} - 64\right) d\theta$$

$$= -\frac{8}{3} \left(3\sqrt{3} - 8\right)\theta \Big]_0^{2\pi}$$

$$= \frac{16\pi}{3} \left(8 - 3\sqrt{3}\right)$$

$$\approx 46.98.$$

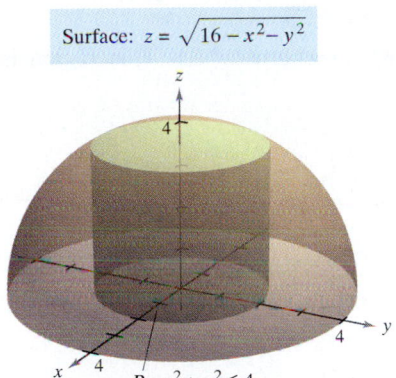

Surface: $z = \sqrt{16 - x^2 - y^2}$

R: $x^2 + y^2 \leq 4$

Figure 13.30

NOTE   To see the benefit of polar coordinates in Example 3, you should try to evaluate the corresponding rectangular double integral.

$$\int_{-2}^{2} \int_{-\sqrt{4-y^2}}^{\sqrt{4-y^2}} \sqrt{16 - x^2 - y^2} \, dx \, dy.$$

TECHNOLOGY   Any symbolic integration utility that can handle double integrals in rectangular coordinates can also handle double integrals in polar coordinates. The reason this is true is that once you have formed the iterated integral, its value is not changed by using different variables. In other words, if you use a symbolic integration utility to evaluate

$$\int_0^{2\pi} \int_0^2 \sqrt{16 - x^2} \, x \, dx \, dy$$

you should obtain the same value as that obtained in Example 3.

Just as with rectangular coordinates, the double integral

$$\int_R \int dA$$

can be used to find the area of a region in the plane.

 EXAMPLE 4   Finding Areas of Polar Regions

Use a double integral to find the area enclosed by the graph of $r = 3\cos 3\theta$.

Solution   Let R be one petal of the curve shown in Figure 13.31. This region is r-simple, and the boundaries are as follows.

$$-\frac{\pi}{6} \le \theta \le \frac{\pi}{6} \qquad \text{Fixed bounds on } \theta$$

$$0 \le r \le 3\cos 3\theta \qquad \text{Variable bounds on } r$$

Thus, the area of one petal is

$$\frac{1}{3}A = \int_R \int dA = \int_{-\pi/6}^{\pi/6} \int_0^{3\cos 3\theta} r\, dr\, d\theta$$

$$= \int_{-\pi/6}^{\pi/6} \frac{r^2}{2}\bigg]_0^{3\cos 3\theta} d\theta$$

$$= \frac{9}{2}\int_{-\pi/6}^{\pi/6} \cos^2 3\theta\, d\theta$$

$$= \frac{9}{4}\int_{-\pi/6}^{\pi/6} (1 + \cos 6\theta)\, d\theta$$

$$= \frac{9}{4}\left[\theta + \frac{1}{6}\sin 6\theta\right]_{-\pi/6}^{\pi/6}$$

$$= \frac{3\pi}{4}.$$

Therefore, the total area is $A = 9\pi/4$.

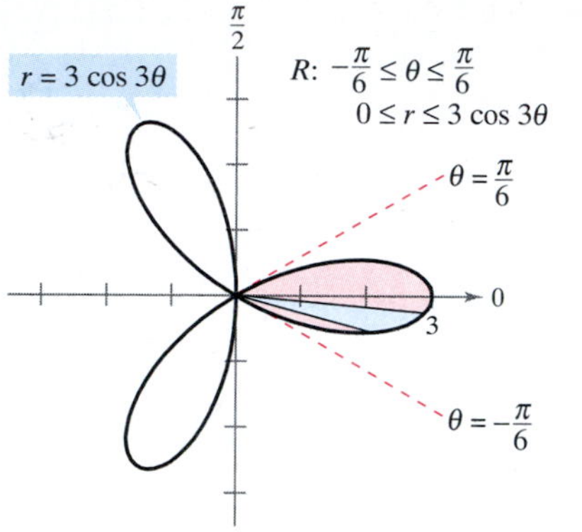

As illustrated in Example 4, the area of a region in the plane can be represented by

$$A = \int_\alpha^\beta \int_{g_1(\theta)}^{g_2(\theta)} r\, dr\, d\theta.$$

If $g_1(\theta) = 0$, you obtain

$$A = \int_\alpha^\beta \int_0^{g_2(\theta)} r\, dr\, d\theta$$

$$= \int_\alpha^\beta \frac{r^2}{2}\bigg]_0^{g_2(\theta)} d\theta$$

$$= \int_\alpha^\beta \frac{1}{2}\,(g_2(\theta))^2\, d\theta$$

which agrees with Theorem 9.13.

$r = 3\cos 3\theta$

$$R: -\frac{\pi}{6} \le \theta \le \frac{\pi}{6}$$
$$0 \le r \le 3\cos 3\theta$$

$$\theta = \frac{\pi}{6}$$

$$\theta = -\frac{\pi}{6}$$

The area of R is $3\pi/4$, and the total area is $9\pi/4$.

Figure 13.31

So far in this section, all of the examples of double integrals in polar form have been of the form

$$\int_{\alpha}^{\beta} \int_{g_1(\theta)}^{g_2(\theta)} f(r \cos \theta, r \sin \theta)\, r\, dr\, d\theta$$

in which the order of integration is with respect to r first. Sometimes you can obtain a simpler integration problem by switching the order of integration, as illustrated in the next example.

EXAMPLE 5   Changing the Order of Integration

Find the area of the region bounded above by the spiral

$$r = \frac{\pi}{3\theta}$$

and below by the x-axis, between $r = 1$ and $r = 2$.

Solution   The region is shown in Figure 13.32. The polar boundaries for the region are

$$1 \le r \le 2 \quad \text{and} \quad 0 \le \theta \le \frac{\pi}{3r}.$$

Hence, the area of the region can be evaluated as follows.

$$A = \int_{1}^{2} \int_{0}^{\pi/3r} r\, d\theta\, dr$$

$$= \int_{1}^{2} r\theta \Big]_{0}^{\pi/3r} dr$$

$$= \int_{1}^{2} \frac{\pi}{3}\, dr$$

$$= \frac{\pi r}{3} \Big]_{1}^{2}$$

$$= \frac{\pi}{3}$$

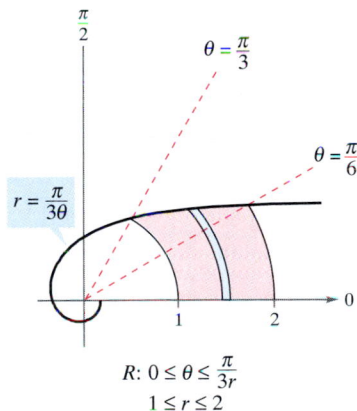

θ-Simple region
Figure 13.32

EXERCISES FOR SECTION 13.3

In Exercises 1–6, evaluate the double integral $\int_R \int f(r, \theta)\,dA$, and sketch the region R.

1. $\displaystyle\int_{0}^{2\pi} \int_{0}^{6} 3r^2 \sin \theta\, dr\, d\theta$

2. $\displaystyle\int_{0}^{\pi/4} \int_{0}^{4} r^2 \sin \theta \cos \theta\, dr\, d\theta$

3. $\displaystyle\int_{0}^{\pi/2} \int_{2}^{3} \sqrt{9 - r^2}\, r\, dr\, d\theta$

4. $\displaystyle\int_{0}^{\pi/2} \int_{0}^{3} re^{-r^2}\, dr\, d\theta$

5. $\displaystyle\int_{0}^{\pi/2} \int_{0}^{1 + \sin \theta} \theta r\, dr\, d\theta$

6. $\displaystyle\int_{0}^{\pi/2} \int_{0}^{1 - \cos \theta} (\sin \theta) r\, dr\, d\theta$

In Exercises 7–12, use a double integral to find the area of the shaded region.

7.

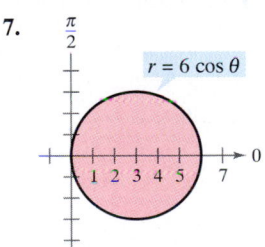

$r = 6 \cos \theta$

8.

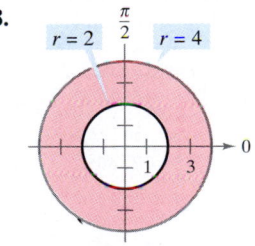

$r = 2 \quad r = 4$

9.

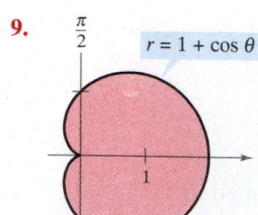

$r = 1 + \cos\theta$

10.

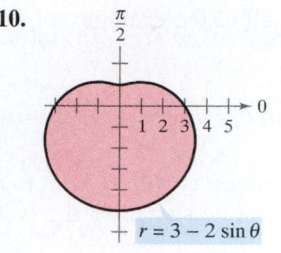

$r = 3 - 2\sin\theta$

11.

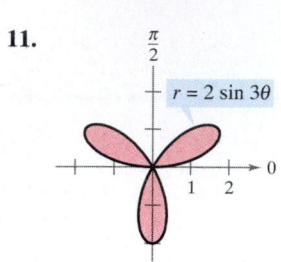

$r = 2\sin 3\theta$

12.

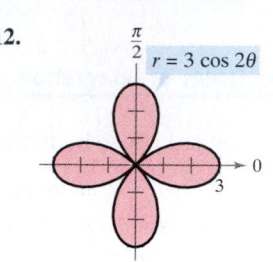

$r = 3\cos 2\theta$

In Exercises 13–18, evaluate the double integral by converting to polar coordinates.

13. $\displaystyle\int_0^a \int_0^{\sqrt{a^2 - y^2}} y \, dx \, dy$

14. $\displaystyle\int_0^a \int_0^{\sqrt{a^2 - x^2}} x \, dy \, dx$

15. $\displaystyle\int_0^3 \int_0^{\sqrt{9 - x^2}} (x^2 + y^2)^{3/2} \, dy \, dx$

16. $\displaystyle\int_0^2 \int_y^{\sqrt{8 - y^2}} \sqrt{x^2 + y^2} \, dx \, dy$

17. $\displaystyle\int_0^2 \int_0^{\sqrt{2x - x^2}} xy \, dy \, dx$

18. $\displaystyle\int_0^4 \int_0^{\sqrt{4y - y^2}} x^2 \, dx \, dy$

In Exercises 19 and 20, combine the sum of the two double integrals into a single double integral by converting to polar coordinates. Evaluate the resulting double integral.

19. $\displaystyle\int_0^2 \int_0^x \sqrt{x^2 + y^2} \, dy \, dx + \int_2^{2\sqrt{2}} \int_0^{\sqrt{8 - x^2}} \sqrt{x^2 + y^2} \, dy \, dx$

20. $\displaystyle\int_0^{5\sqrt{2}/2} \int_0^x xy \, dy \, dx + \int_{5\sqrt{2}/2}^5 \int_0^{\sqrt{25 - x^2}} xy \, dy \, dx$

In Exercises 21–24, use polar coordinates to set up the double integral $\int_R \int f(x, y) \, dA$.

21. $f(x, y) = x + y$, $R: x^2 + y^2 \le 4, x \ge 0, y \ge 0$

22. $f(x, y) = e^{-(x^2 + y^2)}$, $R: x^2 + y^2 \le 4, x \ge 0, y \ge 0$

23. $f(x, y) = \arctan \dfrac{y}{x}$

   $R: x^2 + y^2 \ge 1, x^2 + y^2 \le 4, 0 \le y \le x$

24. $f(x, y) = 9 - x^2 - y^2$, $R: x^2 + y^2 \le 9, x \ge 0, y \ge 0$

Volume **In Exercises 25–30, use a double integral in polar coordinates to find the volume of the solid bounded by the graphs of the equations.**

25. $z = xy, x^2 + y^2 = 1$, first octant

26. $z = x^2 + y^2 + 1, z = 0, x^2 + y^2 = 4$

27. $z = \sqrt{x^2 + y^2}, z = 0, x^2 + y^2 = 25$

28. $z = \ln(x^2 + y^2), z = 0, x^2 + y^2 \ge 1, x^2 + y^2 \le 4$

29. Inside the hemisphere $z = \sqrt{16 - x^2 - y^2}$ and inside the cylinder $x^2 + y^2 - 4x = 0$

30. Inside the hemisphere $z = \sqrt{16 - x^2 - y^2}$ and outside the cylinder $x^2 + y^2 = 1$

31. *Volume*   Find a such that the volume inside the hemisphere $z = \sqrt{16 - x^2 - y^2}$ and outside the cylinder $x^2 + y^2 = a^2$ is one-half the volume of the hemisphere.

32. *Volume*   Use a double integral in polar coordinates to find the volume of a sphere of radius a.

33. *Volume*   Determine the diameter of a hole that is drilled vertically through the center of the solid bounded by the graphs of the equations

$$z = 25e^{-(x^2 + y^2)/4}, z = 0, \quad \text{and} \quad x^2 + y^2 = 16$$

if one-tenth of the volume of the solid is removed.

34. *Machine Design*   The surfaces of a double-lobbed cam are modeled by the inequalities $\frac{1}{4} \le r \le \frac{1}{2}(1 + \cos^2\theta)$ and

$$\frac{-9}{4(x^2 + y^2 + 9)} \le z \le \frac{9}{4(x^2 + y^2 + 9)}$$

where all measurements are in inches.

(a) Use a computer algebra system to graph the cam.

(b) Use a computer algebra system to approximate the perimeter of the polar curve

$$r = \tfrac{1}{2}(1 + \cos^2\theta).$$

This is the distance a roller must travel as it runs against the cam through one revolution of the cam.

(c) Use a symbolic integration utility to find the volume of steel in the cam.

35. *Think About It*   Consider the program you wrote to approximate double integrals in rectangular coordinates (Exercise 62, Section 13.2). If the program is used to approximate the double integral

$$\int_R \int f(r, \theta) \, dA$$

in polar coordinates, how will you modify f when it is entered into the program? Because the limits of integration are constants, describe the plane region of integration.

36. *Approximation* Horizontal cross sections of a piece of ice that broke from a glacier are in the shape of a quarter of a circle with a radius of approximately 50 feet. The base is divided into 20 subregions, as shown in the figure. At the center of each subregion, the height of the ice is measured, yielding the following points in cylindrical coordinates.

$\left(5, \frac{\pi}{16}, 7\right), \left(15, \frac{\pi}{16}, 8\right), \left(25, \frac{\pi}{16}, 10\right), \left(35, \frac{\pi}{16}, 12\right), \left(45, \frac{\pi}{16}, 9\right),$

$\left(5, \frac{3\pi}{16}, 9\right), \left(15, \frac{3\pi}{16}, 10\right), \left(25, \frac{3\pi}{16}, 14\right), \left(35, \frac{3\pi}{16}, 15\right), \left(45, \frac{3\pi}{16}, 10\right),$

$\left(5, \frac{5\pi}{16}, 9\right), \left(15, \frac{5\pi}{16}, 11\right), \left(25, \frac{5\pi}{16}, 15\right), \left(35, \frac{5\pi}{16}, 18\right), \left(45, \frac{5\pi}{16}, 14\right),$

$\left(5, \frac{7\pi}{16}, 5\right), \left(15, \frac{7\pi}{16}, 8\right), \left(25, \frac{7\pi}{16}, 11\right), \left(35, \frac{7\pi}{16}, 16\right), \left(45, \frac{7\pi}{16}, 12\right)$

(a) Approximate the volume of the solid.

(b) Approximate the weight of the solid if ice weighs approximately 56 pounds per cubic foot.

(c) Approximate the number of gallons of water in the solid if there are 7.48 gallons of water per cubic foot.

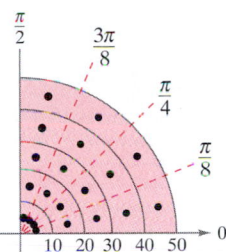

Approximation In Exercises 37 and 38, use a symbolic integration utility to approximate the double integral.

37. $\displaystyle\int_{\pi/4}^{\pi/2} \int_{0}^{5} r\sqrt{1+r^3}\,\sin\sqrt{\theta}\,dr\,d\theta$

38. $\displaystyle\int_{0}^{\pi/4} \int_{0}^{4} 5re^{\sqrt{r\theta}}\,dr\,d\theta$

Think About It In Exercises 39 and 40, determine which value best approximates the volume of the solid between the *xy*-plane and the function over the region. (Make your selection on the basis of a sketch of the solid and *not* by performing any calculations.)

39. $f(x, y) = 15 - 2y$; *R*: semicircle: $x^2 + y^2 = 16, y \geq 0$

(a) 100   (b) 200   (c) 300   (d) −200   (e) 800

40. $f(x, y) = xy + 2$; *R*: quarter circle: $x^2 + y^2 = 9, x \geq 0, y \geq 0$

(a) 25   (b) 8   (c) 100   (d) 50   (e) −30

True or False? In Exercises 41 and 42, determine whether the statement is true or false. If it is false, explain why or give an example that shows it is false.

41. If $\int_R\int f(r, \theta)\,dA > 0$, then $f(r, \theta) > 0$ for all (r, θ) in *R*.

42. If $f(r, \theta)$ is a constant function and the area of the region *S* is twice that of the region *R*, then $2\int_R\int f(r, \theta)\,dA = \int_S\int f(r, \theta)\,dA$.

43. *Probability* The value of the integral

$$I = \int_{-\infty}^{\infty} e^{-x^2/2}\,dx$$

is required in the development of the normal probability density function.

(a) Use polar coordinates to evaluate the double integral.

$$I^2 = \left(\int_{-\infty}^{\infty} e^{-x^2/2}\,dx\right)\left(\int_{-\infty}^{\infty} e^{-y^2/2}\,dy\right)$$
$$= \int_{-\infty}^{\infty}\int_{-\infty}^{\infty} e^{-(x^2+y^2)/2}\,dA$$

(b) Use the result in part (a) to determine *I*.

FOR FURTHER INFORMATION For more information on this problem, see the article "Integrating e^{-x^2} Without Polar Coordinates" by William Dunham in the January 1988 issue of *Mathematics Teacher*.

44. Use the result in Exercise 43 and change of variables to evaluate each of the following integrals. No integration is required.

(a) $\displaystyle\int_{-\infty}^{\infty} e^{-x^2}\,dx$ (b) $\displaystyle\int_{-\infty}^{\infty} e^{-4x^2}\,dx$

45. *Population* The population density of a city is approximated by the model

$$f(x, y) = 4000e^{-0.01(x^2+y^2)}, \quad x^2 + y^2 \leq 49$$

where *x* and *y* are measured in miles. Integrate the density function over the indicated circular region to approximate the population of the city.

46. *Probability* Find *k* such that the function

$$f(x, y) = \begin{cases} ke^{-(x^2+y^2)}, & x \geq 0, y \geq 0 \\ 0, & \text{elsewhere} \end{cases}$$

is a probability density function.

47. *Think About It* Consider the region bounded by the graphs of $y = 2, y = 4, y = x$, and $y = \sqrt{3}x$ and the double integral $\int_R\int f\,dA$. Determine the limits of integration if the region *R* is divided into (a) horizontal representative elements, (b) vertical representative elements, and (c) polar sectors.

48. Repeat Exercise 47 for a region *R* bounded by the graph of the equation $(x - 2)^2 + y^2 = 4$.

49. Show that the area *A* of the polar sector *R* (see figure) is $A = \bar{r}\Delta r\Delta\theta$, where $\bar{r} = (r_1 + r_2)/2$ is the average radius of *R*.

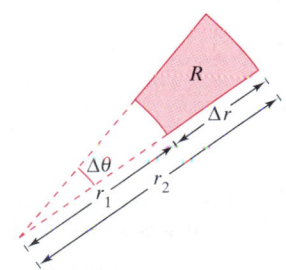

Mass • Moments and Center of Mass • Moments of Inertia

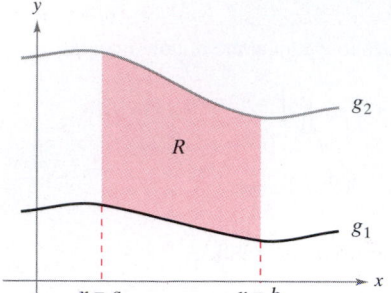

Lamina of constant density ρ
Figure 13.33

Mass

In Section 6.6, we discussed several applications of integration involving a lamina of *constant* density ρ. For example, if the lamina corresponding to the region R, as shown in Figure 13.33, has a constant density ρ, then the mass of the lamina is given by

$$\text{Mass} = \rho A = \rho \iint_R dA = \iint_R \rho \, dA. \qquad \text{Constant density}$$

If not otherwise stated, a lamina is assumed to have a constant density. In this section, however, we extend the definition of the term *lamina* to include thin plates of *variable* density. Double integrals can be used to find the mass of a lamina of *variable* density, where the density at (x, y) is given by the **density function ρ**.

> **Definition of Mass of a Planar Lamina of Variable Density**
>
> If ρ is a continuous density function on the lamina corresponding to a plane region R, then the mass m of the lamina is given by
>
> $$m = \iint_R \rho(x, y) \, dA. \qquad \text{Variable density}$$

NOTE   Density is normally expressed as mass per unit volume. For a planar lamina, however, density is mass per unit surface area

EXAMPLE 1   Finding the Mass of a Planar Lamina

Find the mass of the triangular lamina with vertices $(0, 0)$, $(0, 3)$, and $(2, 3)$, given that the density at (x, y) is $\rho(x, y) = 2x + y$.

Solution   As shown in Figure 13.34, region R has the boundaries $x = 0$, $y = 3$, and $y = 3x/2$ (or $x = 2y/3$). Therefore, the mass of the lamina is

$$m = \iint_R (2x + y) \, dA = \int_0^3 \int_0^{2y/3} (2x + y) \, dx \, dy$$

$$= \int_0^3 \left[x^2 + xy \right]_0^{2y/3} dy$$

$$= \frac{10}{9} \int_0^3 y^2 \, dy$$

$$= \frac{10}{9} \left[\frac{y^3}{3} \right]_0^3$$

$$= 10.$$

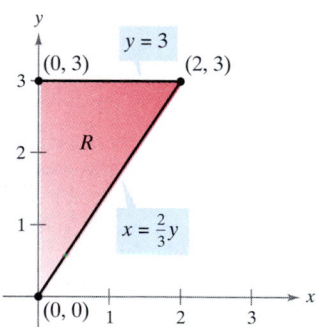

Lamina of variable density $\rho(x, y) = 2x + y$
Figure 13.34

NOTE   In Figure 13.34, note that the planar lamina is shaded so that the darkest shading corresponds to the densest part.

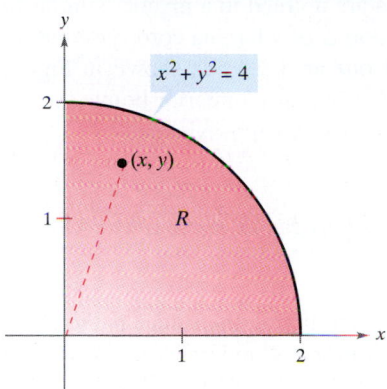

Density at (x, y): $\rho(x, y) = k\sqrt{x^2 + y^2}$

Figure 13.35

EXAMPLE 2   Finding Mass by Polar Coordinates

Find the mass of the lamina corresponding to the first-quadrant portion of the circle

$$x^2 + y^2 = 4$$

where the density at the point (x, y) is proportional to the distance between the point and the origin, as shown in Figure 13.35.

Solution   At any point (x, y), the density of the lamina is

$$\rho(x, y) = k\sqrt{(x - 0)^2 + (y - 0)^2}$$
$$= k\sqrt{x^2 + y^2}.$$

Because $0 \le x \le 2$ and $0 \le y \le \sqrt{4 - x^2}$, the mass is given by

$$m = \int\int_R k\sqrt{x^2 + y^2}\, dA$$
$$= \int_0^2 \int_0^{\sqrt{4-x^2}} k\sqrt{x^2 + y^2}\, dy\, dx.$$

To simplify the integration, you can change to polar coordinates, using the bounds $0 \le \theta \le \pi/2$ and $0 \le r \le 2$. Thus, the mass is

$$m = \int\int_R k\sqrt{x^2 + y^2}\, dA = \int_0^{\pi/2} \int_0^2 k\sqrt{r^2}\, r\, dr\, d\theta$$
$$= \int_0^{\pi/2} \int_0^2 kr^2\, dr\, d\theta$$
$$= \int_0^{\pi/2} \frac{kr^3}{3}\bigg]_0^2 d\theta$$
$$= \frac{8k}{3} \int_0^{\pi/2} d\theta$$
$$= \frac{8k}{3} \Big[\theta\Big]_0^{\pi/2}$$
$$= \frac{4\pi k}{3}.$$

TECHNOLOGY   On many occasions in this text, we have mentioned the benefits of computer programs that perform symbolic integration. Even if you use such a program regularly, you should remember that its greatest benefit comes only in the hands of a knowledgeable user. For instance, notice how much simpler the integral in Example 2 becomes when it is converted to polar form.

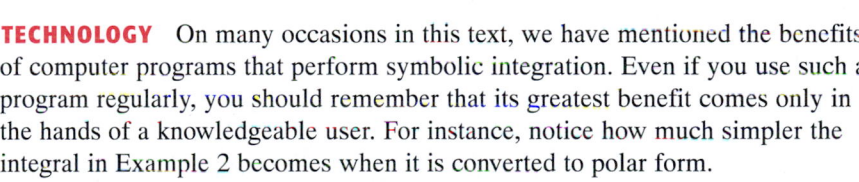

Rectangular Form	*Polar Form*

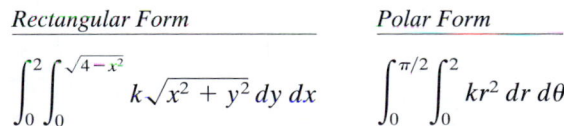

If you have access to software that performs symbolic integration, try using it to evaluate both integrals. Some software programs cannot handle the first integral, but any program that can handle double integrals can evaluate the second integral.

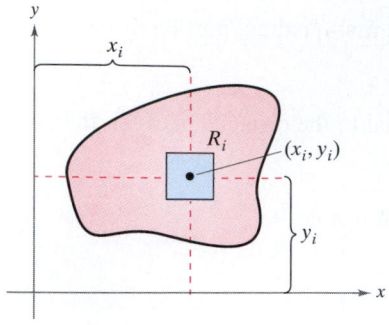

$M_x = \text{(mass)}(y_i)$

$M_y = \text{(mass)}(x_i)$

Figure 13.36

Moments and Center of Mass

For a lamina of variable density, moments of mass are defined in a manner similar to that used for the uniform density case. For a partition Δ of a lamina corresponding to a plane region R, consider the ith rectangle R_i of one area ΔA_i, as shown in Figure 13.36. Assume that the mass of R_i is concentrated at one of its interior points (x_i, y_i). The moment of mass of R_i with respect to the x-axis can be approximated by

$$(\text{Mass})(y_i) \approx [\rho(x_i, y_i)\Delta A_i](y_i).$$

Similarly, the moment of mass with respect to the y-axis can be approximated by

$$(\text{Mass})(x_i) \approx [\rho(x_i, y_i)\Delta A_i](x_i).$$

By forming the Riemann sum of all such products and taking the limits as the norm of Δ approaches 0, you obtain the following definitions of moments of mass with respect to the x- and y-axes.

Moments and Center of Mass of a Variable Density Planar Lamina

Let ρ be a continuous density function on the planar lamina R. The **moments of mass** with respect to the x- and y-axes are

$$M_x = \int_R \int y\rho(x, y)\, dA \quad \text{and} \quad M_y = \int_R \int x\rho(x, y)\, dA.$$

If m is the mass of the lamina, then the **center of mass** is

$$(\bar{x}, \bar{y}) = \left(\frac{M_y}{m}, \frac{M_x}{m}\right).$$

If R represents a simple plane region rather than a lamina, the point $(\bar{x}, \bar{y})$ is called the **centroid** of the region.

For some planar laminas, you can determine the center of mass (or one of its coordinates) using symmetry rather than using integration. For instance, consider the laminas shown in Figure 13.37. Using symmetry, you can see that $\bar{y} = 0$ for the first lamina and $\bar{x} = 0$ for the second lamina.

$R: 0 \le x \le 1$
$-\sqrt{1 - x^2} \le y \le \sqrt{1 - x^2}$

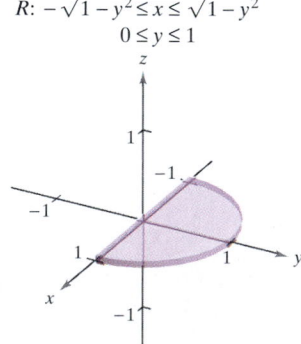

$R: -\sqrt{1 - y^2} \le x \le \sqrt{1 - y^2}$
$0 \le y \le 1$

Symmetric with respect to the x-axis

Symmetric with respect to the y-axis

Figure 13.37

Variable density:
$\rho(x, y) = ky$

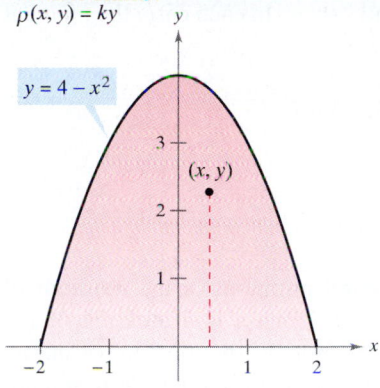

$y = 4 - x^2$

(x, y)

The parabolic region of variable density
Figure 13.38

> **EXAMPLE 3**   **Finding the Center of Mass**

Find the center of mass of the lamina corresponding to the parabolic region

$$0 \le y \le 4 - x^2 \qquad \text{Parabolic region}$$

where the density at the point (x, y) is proportional to the distance between (x, y) and the x-axis, as shown in Figure 13.38.

Solution   Because the lamina is symmetric with respect to the y-axis and

$$\rho(x, y) = ky$$

the center of mass lies on the y-axis. Thus, $\bar{x} = 0$. To find $\bar{y}$, first find the mass of the lamina.

$$\text{Mass} = \int_{-2}^{2} \int_{0}^{4 - x^2} ky \, dy \, dx = \frac{k}{2} \int_{-2}^{2} y^2 \Big]_{0}^{4 - x^2} dx$$

$$= \frac{k}{2} \int_{-2}^{2} (16 - 8x^2 + x^4) \, dx$$

$$= \frac{k}{2} \left[16x - \frac{8x^3}{3} + \frac{x^5}{5} \right]_{-2}^{2}$$

$$= k \left(32 - \frac{64}{3} + \frac{32}{5} \right)$$

$$= \frac{256k}{15}$$

Next, find the moment about the x-axis.

$$M_x = \int_{-2}^{2} \int_{0}^{4 - x^2} (y)(ky) \, dy \, dx = \frac{k}{3} \int_{-2}^{2} y^3 \Big]_{0}^{4 - x^2} dx$$

$$= \frac{k}{3} \int_{-2}^{2} (64 - 48x^2 + 12x^4 - x^6) \, dx$$

$$= \frac{k}{3} \left[64x - 16x^3 + \frac{12x^5}{5} - \frac{x^7}{7} \right]_{-2}^{2}$$

$$= \frac{4096k}{105}$$

Variable
density:
$\rho(x, y) = ky$

$R: -2 \le x \le 2$
$0 \le y \le 4 - x^2$

Center of mass:
$\left(0, \frac{16}{7} \right)$

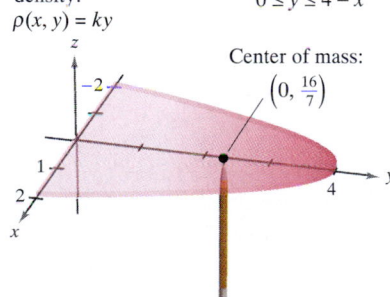

The center of mass of a planar lamina is its balancing point.
Figure 13.39

Thus,

$$\bar{y} = \frac{M_x}{m} = \frac{4096k/105}{256k/15} = \frac{16}{7}$$

and the center of mass is $\left(0, \frac{16}{7} \right)$.

Although you can think of the moments M_x and M_y as measuring the tendency to rotate about the x- or y-axis, the calculation of moments is usually an intermediate step toward a more tangible goal. The use of the moments M_x and M_y in Example 3 is typical—to find the center of mass. Determination of the center of mass is useful in a variety of applications that allow you to treat a lamina as if its mass were concentrated at just one point. Intuitively, you can think of the center of mass as the balancing point of the lamina. For instance, the lamina in Example 3 should balance on the point of a pencil placed at $\left(0, \frac{16}{7} \right)$, as shown in Figure 13.39.

Moments of Inertia

The moments of M_x and M_y used in determining the center of mass of a lamina are sometimes called the **first moments** about the x- and y-axes. In each case, the moment is the product of a mass times a distance.

$$M_x = \int_R \int (y) \underbrace{\rho(x, y)} \, dA \qquad M_y = \int_R \int (x) \underbrace{\rho(x, y)} \, dA$$

Distance   Mass   Distance   Mass
to x-axis   to y-axis

We now look at another type of moment—the **second moment,** or the **moment of inertia** of a lamina about a line. In the same way that mass is a measure of the tendency of matter to resist a change in straight-line motion, the moment of inertia about a line is a *measure of the tendency of matter to resist a change in rotational motion.* For example, if a particle of mass m is a distance d from a fixed line, its moment of inertia about the line is defined as

$$I = md^2 = (\text{mass})(\text{distance})^2.$$

As with moments of mass, you can generalize this concept to obtain the moments of inertia about the x- and y-axes of a lamina of variable density. These second moments are denoted by I_x and I_y, and in each case the moment is the product of a mass times the square of a distance.

$$I_x = \int_R \int (y^2) \underbrace{\rho(x, y)} \, dA, \qquad I_y = \int_R \int (x^2) \underbrace{\rho(x, y)} \, dA,$$

Square of distance   Mass   Square of distance   Mass
to x-axis   to y-axis

The sum of the moments I_x and I_y is called the **polar moment of inertia** and is denoted by I_0.

NOTE For a lamina in the xy-plane, I_0 represents the moment of inertia of the lamina about the z-axis. The term "polar moment of inertia" stems from the fact that the square of the polar distance r is used in the calculation.

$$I_0 = \int_R \int (x^2 + y^2)\rho(x, y) \, dA$$

$$= \int_R \int r^2 \rho(x, y) \, dA$$

EXAMPLE 4   Finding the Moment of Inertia

Find the moment of inertia about the x-axis of the lamina in Example 3.

Solution From the definition of moment of inertia, you have

$$I_x = \int_{-2}^{2} \int_{0}^{4-x^2} y^2 (ky) \, dy \, dx$$

$$= \frac{k}{4} \int_{-2}^{2} y^4 \Big]_{0}^{4-x^2} \, dx$$

$$= \frac{k}{4} \int_{-2}^{2} (256 - 256x^2 + 96x^4 - 16x^6 + x^8) \, dx$$

$$= \frac{k}{4} \left[256x - \frac{256x^3}{3} + \frac{96x^5}{5} - \frac{16x^7}{7} + \frac{x^9}{9} \right]_{-2}^{2}$$

$$= \frac{32{,}768k}{315}.$$

The moment of inertia I of a revolving lamina can be used to measure its kinetic energy. For example, suppose a planar lamina is revolving about a line with an **angular speed** of ω radians per second, as shown in Figure 13.40. The kinetic energy of the revolving lamina is

$$E = \frac{1}{2}I\omega^2.$$ Kinetic energy for rotational motion

On the other hand, the kinetic energy of a mass m moving in a straight line at a velocity v is

$$E = \frac{1}{2}mv^2.$$ Kinetic energy for linear motion

Thus, the kinetic energy of a mass moving in a straight line is proportional to its mass, but the kinetic energy of a mass revolving about an axis is proportional to its moment of inertia.

The **radius of gyration** $\bar{\bar{r}}$ of a revolving mass m with moment of inertia I is defined to be

$$\bar{\bar{r}} = \sqrt{\frac{I}{m}}.$$ Radius of gyration

If the entire mass were located at a distance $\bar{\bar{r}}$ from its axis of revolution, it would have the same moment of inertia and, consequently, the same kinetic energy. For instance, the radius of gyration of the lamina in Example 4 about the x-axis is given by

$$\bar{\bar{y}} = \sqrt{\frac{I_x}{m}} = \sqrt{\frac{32{,}768k/315}{256k/15}} = \sqrt{\frac{128}{21}} \approx 2.47.$$

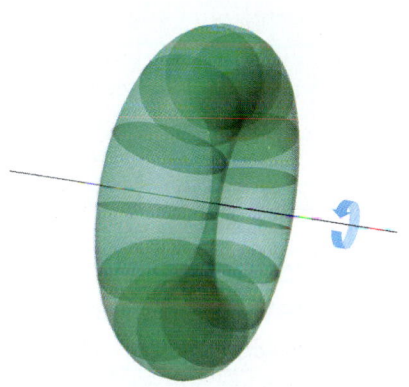

Planar lamina revolving at ω radians per second
Figure 13.40

EXAMPLE 5   Finding the Radius of Gyration

Find the radius of gyration about the y-axis for the lamina corresponding to the region $R: 0 \le y \le \sin x,\ 0 \le x \le \pi$, where the density at (x, y) is given by $\rho(x, y) = x$.

Solution   The region R is shown in Figure 13.41. By integrating $\rho(x, y) = x$ over the region R, you can determine that the mass of the region is π. The moment of inertia about the y-axis is

$$I_y = \int_0^\pi \int_0^{\sin x} x^3 \, dy \, dx$$

$$= \int_0^\pi x^3 y \Big]_0^{\sin x} dx$$

$$= \int_0^\pi x^3 \sin x \, dx$$

$$= \left[(3x^2 - 6)(\sin x) - (x^3 - 6x)(\cos x) \right]_0^\pi$$

$$= \pi^3 - 6\pi.$$

Thus, the radius of gyration about the y-axis is

$$\bar{\bar{x}} = \sqrt{\frac{I_y}{m}} = \sqrt{\frac{\pi^3 - 6\pi}{\pi}} = \sqrt{\pi^2 - 6} \approx 1.97.$$

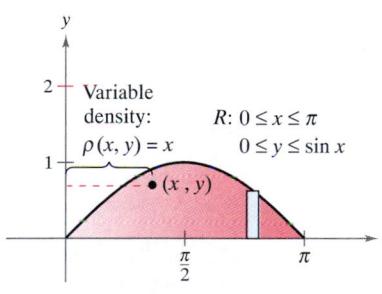

The radius of gyration about the y-axis is approximately 1.97.
Figure 13.41

EXERCISES FOR SECTION 13.4

In Exercises 1–4, find the mass and center of mass of the lamina for the indicated density.

1. R: rectangle with vertices $(0, 0)$, $(a, 0)$, $(0, b)$, (a, b)

　(a) $\rho = k$　(b) $\rho = ky$　(c) $\rho = kx$

2. R: rectangle with vertices $(0, 0)$, $(a, 0)$, $(0, b)$, (a, b)

　(a) $\rho = kxy$　(b) $\rho = k(x^2 + y^2)$

3. R: triangle with vertices $(0, 0)$, $(b/2, h)$, $(b, 0)$

　(a) $\rho = k$　(b) $\rho = ky$　(c) $\rho = kx$

4. R: triangle with vertices $(0, 0)$, $(0, a)$, $(a, 0)$

　(a) $\rho = k$　(b) $\rho = x^2 + y^2$

5. *Translations in the Plane*　Translate the lamina in Exercise 1 to the right five units and determine the resulting center of mass.

6. *Conjecture*　Use the result in Exercise 5 to make a conjecture about the change in the center of mass when a lamina of constant density is translated h units horizontally or k units vertically. Is the conjecture true if the density is not constant? Explain.

In Exercises 7–18, find the mass and center of mass of the lamina bounded by the graphs of the equations for the indicated density or densities. (*Hint*: Some of the integrals are simpler in polar coordinates.)

7. $y = \sqrt{a^2 - x^2}$, $y = 0$,　(a) $\rho = k$　(b) $\rho = k(a - y)y$

8. $x^2 + y^2 = a^2$, $0 \le x$, $0 \le y$,　(a) $\rho = k$　(b) $\rho = k(x^2 + y^2)$

9. $y = \sqrt{x}$, $y = 0$, $x = 4$, $\rho = kxy$

10. $y = x^2$, $y = 0$, $x = 4$, $\rho = kx$

11. $y = \dfrac{1}{1 + x^2}$, $y = 0$, $x = -1$, $x = 1$, $\rho = k$

12. $xy = 4$, $x = 1$, $x = 4$, $\rho = kx^2$

13. $x = 16 - y^2$, $x = 0$, $\rho = kx$

14. $y = 9 - x^2$, $y = 0$, $\rho = ky^2$

15. $y = \sin \dfrac{\pi x}{L}$, $y = 0$, $x = 0$, $x = L$, $\rho = ky$

16. $y = \cos \dfrac{\pi x}{L}$, $y = 0$, $x = 0$, $x = \dfrac{L}{2}$, $\rho = k$

17. $y = \sqrt{a^2 - x^2}$, $0 \le y \le x$, $\rho = k$

18. $y = \sqrt{a^2 - x^2}$, $y = 0$, $y = x$, $\rho = k\sqrt{x^2 + y^2}$

In Exercises 19–22, use a symbolic integration utility to find the mass and center of mass of the lamina bounded by the graphs of the equations for the indicated density.

19. $y = e^{-x}$, $y = 0$, $x = 0$, $x = 2$, $\rho = ky$

20. $y = \ln x$, $y = 0$, $x = 1$, $x = e$, $\rho = k/x$

21. $r = 2 \cos 3\theta$, $-\dfrac{\pi}{6} \le \theta \le \dfrac{\pi}{6}$, $\rho = k$

22. $r = 1 + \cos\theta$, $\rho = k$

Think About It　The center of mass of the lamina of constant density shown in the figure is $\left(2, \frac{8}{5}\right)$. In Exercises 23–26, make a conjecture about how the center of mass $(\bar{x}, \bar{y})$ will change for the nonconstant density $\rho(x, y)$. Explain. (**Make your conjecture** *without* **performing any calculations.**)

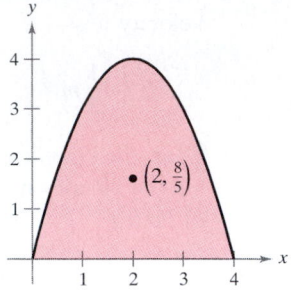

23. $\rho(x, y) = ky$

24. $\rho(x, y) = k|2 - x|$

25. $\rho(x, y) = kxy$

26. $\rho(x, y) = k(4 - x)(4 - y)$

In Exercises 27–32, verify the given moment(s) of inertia and find $\bar{\bar{x}}$ and $\bar{\bar{y}}$. Assume each lamina has a density of $\rho = 1$. (These regions are common shapes used in engineering.)

27. Rectangle

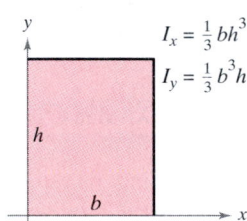

$I_x = \frac{1}{3} bh^3$

$I_y = \frac{1}{3} b^3 h$

28. Right triangle

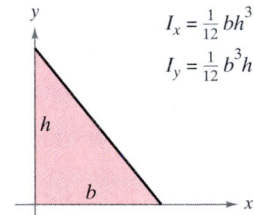

$I_x = \frac{1}{12} bh^3$

$I_y = \frac{1}{12} b^3 h$

29. Circle

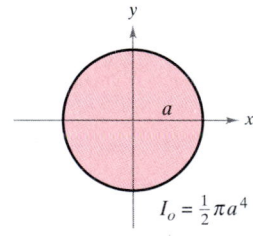

$I_o = \frac{1}{2} \pi a^4$

30. Semicircle

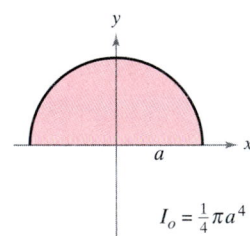

$I_o = \frac{1}{4} \pi a^4$

31. Quarter circle

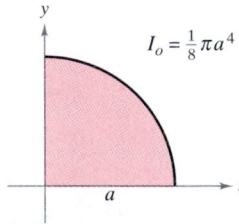

$I_o = \frac{1}{8} \pi a^4$

32. Ellipse

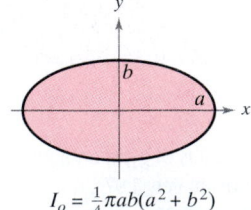

$I_o = \frac{1}{4} \pi ab(a^2 + b^2)$

In Exercises 33–40, find $I_x, I_y, I_0, \bar{\bar{x}}$, and $\bar{\bar{y}}$ for the lamina bounded by the graphs of the equations. Use a symbolic integration utility to evaluate the double integrals.

33. $y = 0$, $y = b$, $x = 0$, $x = a$, $\rho = ky$
34. $y = \sqrt{a^2 - x^2}$, $y = 0$, $\rho = ky$
35. $y = 4 - x^2$, $y = 0$, $x > 0$, $\rho = kx$
36. $y = x$, $y = x^2$, $\rho = kxy$
37. $y = \sqrt{x}$, $y = 0$, $x = 4$, $\rho = kxy$
38. $y = x^2$, $y^2 = x$, $\rho = x^2 + y^2$
39. $y = x^2$, $y^2 = x$, $\rho = kx$
40. $y = x^3$, $y = 4x$, $\rho = ky$

 In Exercises 41–46, set up the double integrals required to find the moment of inertia I, about the indicated line, of the lamina bounded by the graphs of the equations. Use a symbolic integration utility to evaluate the double integrals.

41. $x^2 + y^2 = b^2$, $\rho = k$, line: $x = a$ $(a > b)$
42. $y = 0$, $y = 2$, $x = 0$, $x = 4$, $\rho = k$, line: $x = 6$
43. $y = \sqrt{x}$, $y = 0$, $x = 4$, $\rho = kx$, line: $x = 6$
44. $y = \sqrt{a^2 - x^2}$, $y = 0$, $\rho = ky$, line: $y = a$
45. $y = \sqrt{a^2 - x^2}$, $y = 0$, $x \geq 0$, $\rho = k(a - y)$, line: $y = a$
46. $y = 4 - x^2$, $y = 0$, $\rho = k$, line: $y = 2$

47. Prove the following Theorem of Pappus: Let R be a region in a plane and let L be a line in the same plane such that L does not intersect the interior of R. If r is the distance between the centroid of R and the line, then the volume V of the solid of revolution formed by revolving R about the line is given by $V = 2\pi rA$, where A is the area of R.

Hydraulics In Exercises 48–51, determine the location of the horizontal axis y_a at which a vertical gate in a dam is to be hinged so that there is no moment causing rotation under the indicated loading (see figure). The model for y_a is

$$y_a = \bar{y} - \frac{I_{\bar{y}}}{hA}$$

where $\bar{y}$ is the y-coordinate of the centroid of the gate, $I_{\bar{y}}$ is the moment of inertia of the gate about the line $y = \bar{y}$, h is the depth of the centroid below the surface, and A is the area of the gate.

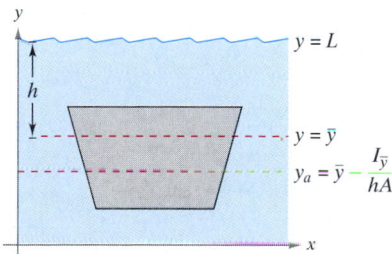

48.

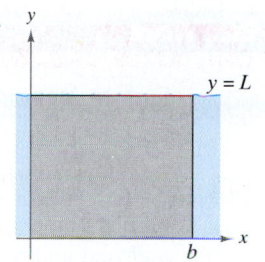

49.

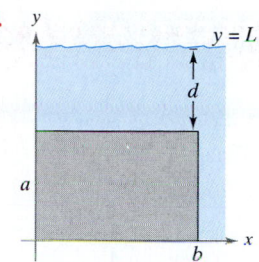

50.

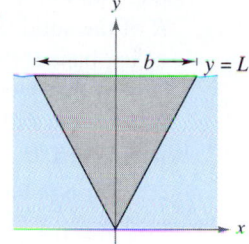

51.

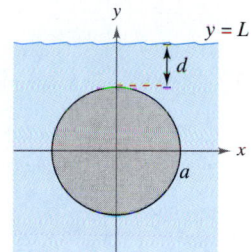

SECTION PROJECT

Center of Pressure on a Sail The center of pressure on a sail is that point (x_p, y_p) at which the total aerodynamic force may be assumed to act. If the sail is represented by a plane region R, the center of pressure is

$$x_p = \frac{\int_R \int xy \, dA}{\int_R \int y \, dA}$$

and

$$y_p = \frac{\int_R \int y^2 \, dA}{\int_R \int y \, dA}.$$

Consider a triangular sail with vertices at $(0, 0)$, $(2, 1)$, and $(0, 5)$. Verify the values of the following three integrals.

$$\int_R \int y \, dA = 10$$

$$\int_R \int xy \, dA = \frac{35}{6}$$

$$\int_R \int y^2 \, dA = \frac{155}{6}$$

Calculate the coordinates (x_p, y_p) of the center of pressure. Sketch a graph of the sail and indicate the location of the center of pressure.

SECTION *13.5* Surface Area

Surface Area

Surface Area

At this point you know a great deal about the solid region lying between a surface and a closed and bounded region R in the xy-plane, as shown in Figure 13.42. For example, you know how to find: the extrema of f on R (Section 12.8), the area of the base R of the solid (Section 13.1), the volume of the solid (Section 13.2), and the centroid of the base R (Section 13.4).

In this section, you will learn how to find the upper **surface area** of the solid. Later, you will learn how to find the centroid of the solid (Section 13.6) and the lateral surface area (Section 14.2).

To begin, consider a surface S given by

$$z = f(x, y) \qquad \text{Surface defined over a region } R$$

defined over a region R. Assume that R is closed and bounded and that f has continuous first partial derivatives. To find the surface area, construct an inner partition of R consisting of n rectangles, where the area of the ith rectangle R_i is $\Delta A_i = \Delta x_i \Delta y_i$, as shown in Figure 13.43. In each R_i let (x_i, y_i) be the point that is closest to the origin. At the point $(x_i, y_i, z_i) = (x_i, y_i, f(x_i, y_i))$ on the surface S, construct a tangent plane T_i. The area of the portion of the tangent plane that lies directly above R_i is approximately equal to the area of the surface lying directly above R_i. That is, $\Delta T_i \approx \Delta S_i$. Hence, the surface area of S is given by

$$\sum_{i=1}^{n} \Delta S_i \approx \sum_{i=1}^{n} \Delta T_i.$$

To find the area of the parallelogram ΔT_i, note that its sides are given by the vectors

$$\mathbf{u} = \Delta x_i \mathbf{i} + f_x(x_i, y_i) \Delta x_i \mathbf{k}$$

and

$$\mathbf{v} = \Delta y_i \mathbf{j} + f_y(x_i, y_i) \Delta y_i \mathbf{k}.$$

From Theorem 10.8, the area of ΔT_i is given by $\|\mathbf{u} \times \mathbf{v}\|$, where

$$
\mathbf{u} \times \mathbf{v} = \begin{vmatrix} \mathbf{i} & \mathbf{j} & \mathbf{k} \\ \Delta x_i & 0 & f_x(x_i, y_i) \Delta x_i \\ 0 & \Delta y_i & f_y(x_i, y_i) \Delta y_i \end{vmatrix}
$$

$$= -f_x(x_i, y_i) \Delta x_i \Delta y_i \mathbf{i} - f_y(x_i, y_i) \Delta x_i \Delta y_i \mathbf{j} + \Delta x_i \Delta y_i \mathbf{k}$$

$$= (-f_x(x_i, y_i)\mathbf{i} - f_y(x_i, y_i)\mathbf{j} + \mathbf{k}) \Delta A_i.$$

Thus, the area of ΔT_i is $\|\mathbf{u} \times \mathbf{v}\| = \sqrt{[f_x(x_i, y_i)]^2 + [f_y(x_i, y_i)]^2 + 1}\ \Delta A_i$, and

$$\text{Surface area of } S \approx \sum_{i=1}^{n} \Delta S_i$$

$$\approx \sum_{i=1}^{n} \sqrt{1 + [f_x(x_i, y_i)]^2 + [f_y(x_i, y_i)]^2}\ \Delta A_i.$$

This suggests the following definition of surface area.

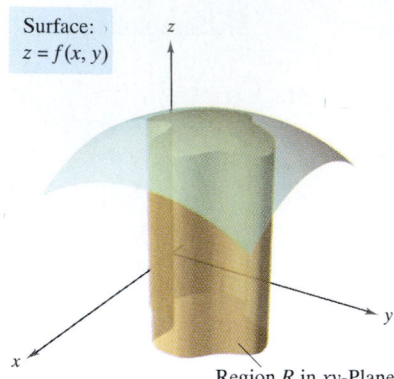

Surface:
$z = f(x, y)$

Region R in xy-Plane

Figure 13.42

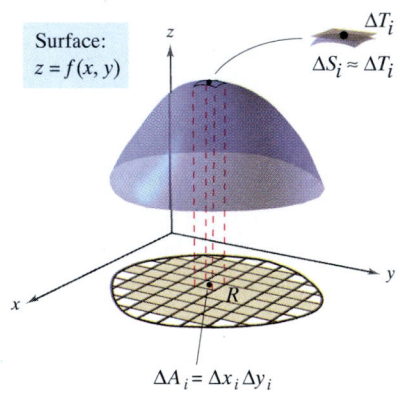

Surface:
$z = f(x, y)$

ΔT_i

$\Delta S_i \approx \Delta T_i$

R

$\Delta A_i = \Delta x_i \Delta y_i$

Figure 13.43

Definition of Surface Area

If f and its first partial derivatives are continuous on the closed region R in the xy-plane, then the **area of the surface** S given by $z = f(x, y)$ over R is given by

$$\text{Surface area} = \iint_R dS$$

$$= \iint_R \sqrt{1 + [f_x(x, y)]^2 + [f_y(x, y)]^2} \, dA.$$

As an aid to remembering the double integral for surface area, it is helpful to note its similarity to the integral for arc length.

Length on x-axis: $\qquad \displaystyle\int_a^b dx$

Arc length in xy-plane: $\qquad \displaystyle\int_a^b ds = \int_a^b \sqrt{1 + [f'(x)]^2} \, dx$

Area in xy-plane: $\qquad \displaystyle\iint_R dA$

Surface area in space: $\qquad \displaystyle\iint_R dS = \iint_R \sqrt{1 + [f_x(x, y)]^2 + [f_y(x, y)]^2} \, dA$

Like integrals for arc length, integrals for surface area are often very difficult to evaluate. However, one type that is easily evaluated is demonstrated in the next example.

EXAMPLE 1   The Surface Area of a Plane Region

Find the surface area of the portion of the plane $z = 2 - x - y$ that lies above the circle $x^2 + y^2 \leq 1$ in the first quadrant, as shown in Figure 13.44.

Solution   Because $f_x(x, y) = -1$ and $f_y(x, y) = -1$, the surface area is given by

$$S = \iint_R \sqrt{1 + [f_x(x, y)]^2 + [f_y(x, y)]^2} \, dA$$

$$= \iint_R \sqrt{3} \, dA$$

$$= \sqrt{3} \iint_R dA.$$

Note that the integral on the right is simply $\sqrt{3}$ times the area of the region R. Thus, the area of S is

$$S = \sqrt{3} \, (\text{area of } R)$$

$$= \sqrt{3} \left(\frac{\pi}{4} \right)$$

$$= \frac{\sqrt{3} \, \pi}{4}.$$

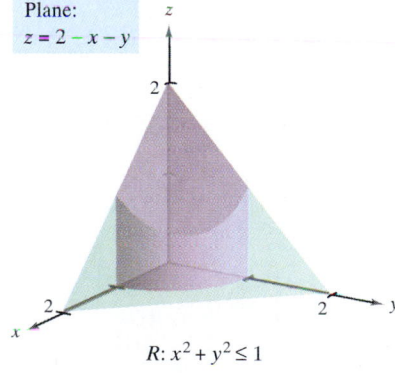

Plane:
$z = 2 - x - y$

$R: x^2 + y^2 \leq 1$

The surface area of the portion of the plane that lies above the quarter circle is $\sqrt{3} \, \pi / 4$.
Figure 13.44

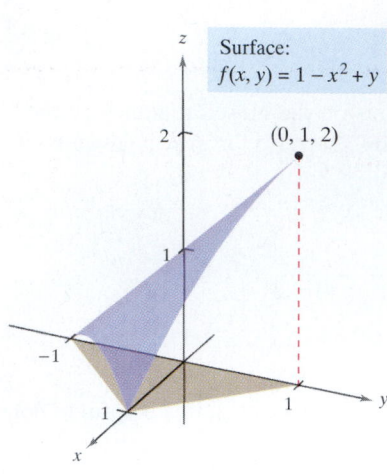

Surface:
$f(x, y) = 1 - x^2 + y$

(0, 1, 2)

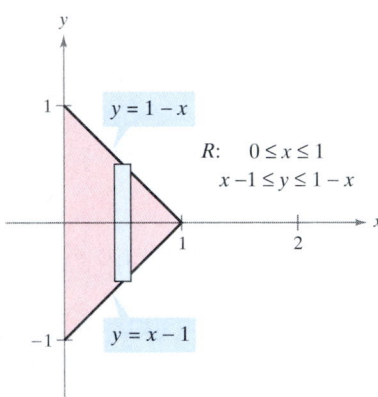

$y = 1 - x$

$R:$ $0 \leq x \leq 1$
  $x - 1 \leq y \leq 1 - x$

$y = x - 1$

Figure 13.45

EXAMPLE 2 **Finding Surface Area**

Find the area of the portion of the surface

$$f(x, y) = 1 - x^2 + y \qquad \text{Surface}$$

that lies above the triangle region with vertices $(1, 0, 0)$, $(0, -1, 0)$, and $(0, 1, 0)$, as shown in Figure 13.45.

Solution Because $f_x(x, y) = -2x$ and $f_y(x, y) = 1$, you have

$$S = \iint_R \sqrt{1 + [f_x(x, y)]^2 + [f_y(x, y)]^2} \, dA = \iint_R \sqrt{1 + 4x^2 + 1} \, dA.$$

In Figure 13.45, you can see that the bounds for R are $0 \leq x \leq 1$ and $x - 1 \leq y \leq 1 - x$. Thus, the integral becomes

$$S = \int_0^1 \int_{x-1}^{1-x} \sqrt{2 + 4x^2} \, dy \, dx$$

$$= \int_0^1 y\sqrt{2 + 4x^2} \,\Big]_{x-1}^{1-x} dx$$

$$= \int_0^1 \left(2\sqrt{2 + 4x^2} - 2x\sqrt{2 + 4x^2}\right) dx \qquad \begin{array}{l}\text{Integration tables (the appendix),}\\ \text{Formula 26 and Power Rule}\end{array}$$

$$= \left[x\sqrt{2 + 4x^2} + \ln\left(2x + \sqrt{2 + 4x^2}\right) - \frac{(2 + 4x^2)^{3/2}}{6}\right]_0^1$$

$$= \sqrt{6} + \ln\left(2 + \sqrt{6}\right) - \sqrt{6} - \ln\sqrt{2} + \frac{1}{3}\sqrt{2}$$

$$\approx 1.618.$$

EXAMPLE 3 **Change of Variables to Polar Coordinates**

Find the surface area of the paraboloid

$$z = 1 + x^2 + y^2 \qquad \text{Paraboloid}$$

that lies above the unit circle, as shown in Figure 13.46.

Solution Because $f_x(x, y) = 2x$ and $f_y(x, y) = 2y$, you have

$$S = \iint_R \sqrt{1 + [f_x(x, y)]^2 + [f_y(x, y)]^2} \, dA = \iint_R \sqrt{1 + 4x^2 + 4y^2} \, dA.$$

You can convert to polar coordinates by letting $x = r\cos\theta$ and $y = r\sin\theta$. Then, because the region R is bounded by $0 \leq r \leq 1$ and $0 \leq \theta \leq 2\pi$, you have

$$S = \int_0^{2\pi} \int_0^1 \sqrt{1 + 4r^2} \, r \, dr \, d\theta = \int_0^{2\pi} \frac{1}{12}(1 + 4r^2)^{3/2} \,\Big]_0^1 d\theta$$

$$= \int_0^{2\pi} \frac{5\sqrt{5} - 1}{12} \, d\theta$$

$$= \frac{\pi\left(5\sqrt{5} - 1\right)}{6} \approx 5.33.$$

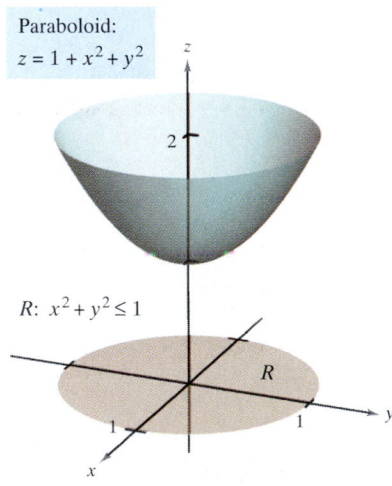

Paraboloid:
$z = 1 + x^2 + y^2$

$R: x^2 + y^2 \leq 1$

R

The surface area of the portion of the paraboloid that lies above the unit circle is approximately 5.33.
Figure 13.46

Hemisphere:
$f(x, y) = \sqrt{25 - x^2 - y^2}$

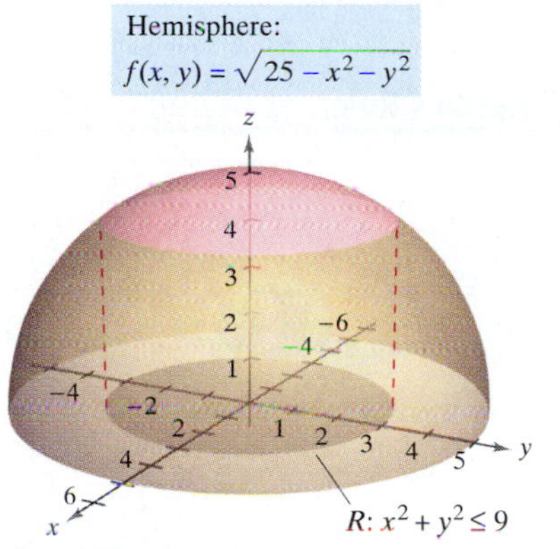

The surface area of the portion of the hemisphere that lies above the circle is 10π.
Figure 13.47

EXAMPLE 4   Finding Surface Area

Find the surface area S of the portion of the hemisphere

$$f(x, y) = \sqrt{25 - x^2 - y^2} \qquad \text{Hemisphere}$$

that lies above the region R bounded by the circle $x^2 + y^2 \leq 9$, as shown in Figure 13.47.

Solution   The first partial derivatives of f are

$$f_x(x, y) = \frac{-x}{\sqrt{25 - x^2 - y^2}} \quad \text{and} \quad f_y(x, y) = \frac{-y}{\sqrt{25 - x^2 - y^2}}$$

and, from the formula for surface area, you have

$$dS = \sqrt{1 + [f_x(x, y)]^2 + [f_y(x, y)]^2}\, dA$$

$$= \frac{5}{\sqrt{25 - x^2 - y^2}}\, dA.$$

Therefore, the surface area is

$$S = \iint_R \frac{5}{\sqrt{25 - x^2 - y^2}}\, dA.$$

You can convert to polar coordinates by letting $x = r\cos\theta$ and $y = r\sin\theta$. Then, because the region R is bounded by

$$0 \leq r \leq 3 \quad \text{and} \quad 0 \leq \theta \leq 2\pi$$

you obtain

$$S = \int_0^{2\pi} \int_0^3 \frac{5}{\sqrt{25 - r^2}}\, r\, dr\, d\theta$$

$$= 5 \int_0^{2\pi} \left. -\sqrt{25 - r^2} \right]_0^3 \, d\theta$$

$$= 5 \int_0^{2\pi} d\theta$$

$$= 10\pi.$$

Hemisphere:
$f(x, y) = \sqrt{25 - x^2 - y^2}$

Figure 13.48

The procedure used in Example 4 can be extended to find the surface area of a sphere by using the region R bounded by the circle $x^2 + y^2 \leq a^2$, where $0 < a < 5$, as shown in Figure 13.48. The surface area of portion of the hemisphere $f(x, y) = \sqrt{25 - x^2 - y^2}$ lying above the circular region can be shown to be

$$S = \iint_R \frac{5}{\sqrt{25 - x^2 - y^2}}\, dA$$

$$= \int_0^{2\pi} \int_0^a \frac{5}{\sqrt{25 - r^2}}\, r\, dr\, d\theta$$

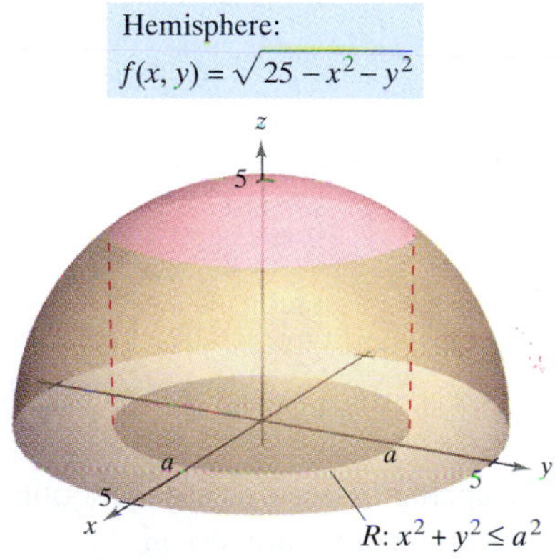

$$= 10\pi \left(5 - \sqrt{25 - a^2} \right).$$

By taking the limit as a approaches 5 and doubling the result, you obtain a total area of 100π. (The surface area of a sphere of radius r is $S = 4\pi r^2$.)

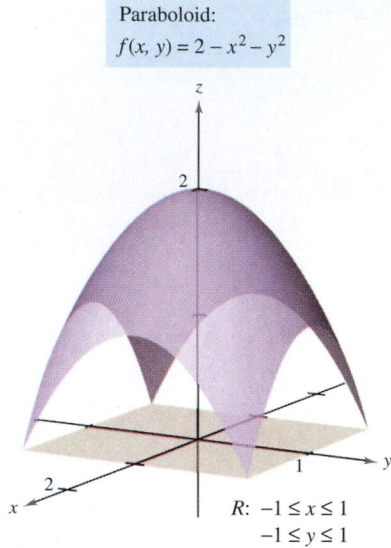

Paraboloid:

$f(x, y) = 2 - x^2 - y^2$

R: $-1 \le x \le 1$
$-1 \le y \le 1$

Figure 13.49

You can use Simpson's Rule or the Trapezoidal Rule to approximate the value of a double integral, *provided* you can get through the first integration. This is demonstrated in the next example.

EXAMPLE 5   Approximating Surface Area by Simpson's Rule

Find the area of the surface of the paraboloid

$$f(x, y) = 2 - x^2 - y^2 \qquad \text{Paraboloid}$$

that lies above the square region bounded by $-1 \le x \le 1$ and $-1 \le y \le 1$, as shown in Figure 13.49.

Solution   Using the partial derivatives

$$f_x(x, y) = -2x \quad \text{and} \quad f_y(x, y) = -2y$$

you have a surface area of

$$S = \iint_R \sqrt{1 + [f_x(x, y)]^2 + [f_y(x, y)]^2} \, dA$$

$$= \iint_R \sqrt{1 + 4x^2 + 4y^2} \, dA.$$

In polar coordinates, the line $x = 1$ is given by $r \cos \theta = 1$ or $r = \sec \theta$, and you can determine from Figure 13.50 that one fourth of the region R is bounded by

$$0 \le r \le \sec \theta \quad \text{and} \quad -\frac{\pi}{4} \le \theta \le \frac{\pi}{4}.$$

Letting $x = r \cos \theta$ and $y = r \sin \theta$ produces

$$\frac{1}{4} S = \frac{1}{4} \iint_R \sqrt{1 + 4x^2 + 4y^2} \, dA$$

$$= \int_{-\pi/4}^{\pi/4} \int_0^{\sec \theta} \sqrt{1 + 4r^2} \, r \, dr \, d\theta$$

$$= \int_{-\pi/4}^{\pi/4} \frac{1}{12} (1 + 4r^2)^{3/2} \Big]_0^{\sec \theta} d\theta$$

$$= \frac{1}{12} \int_{-\pi/4}^{\pi/4} [(1 + 4 \sec^2 \theta)]^{3/2} - 1] \, d\theta.$$

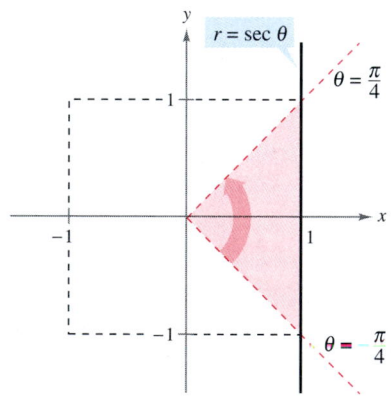

One fourth of the region R is bounded by
$0 \le r \le \sec \theta$ and $-\dfrac{\pi}{4} \le \theta \le \dfrac{\pi}{4}.$
Figure 13.50

Finally, using Simpson's Rule with $n = 10$, you can approximate this *single* integral to be

$$S = \frac{1}{3} \int_{-\pi/4}^{\pi/4} [(1 + 4 \sec^2 \theta)^{3/2} - 1] \, d\theta$$

$$\approx 7.45.$$

TECHNOLOGY   Most computer programs that are capable of performing symbolic integration for multiple integrals are also capable of performing numerical approximation techniques. If you have access to such software, try using it to approximate the value of the integral in Example 5.

EXERCISES FOR SECTION 13.5

In Exercises 1–14, find the area of the surface given by $z = f(x, y)$ over the region R. (*Hint:* Some of the integrals are simpler in polar coordinates.)

1. $f(x, y) = 2x + 2y$

   R: triangle with vertices $(0, 0), (2, 0), (0, 2)$

2. $f(x, y) = 10 + 2x - 3y$

   R: square with vertices $(0, 0), (2, 0), (0, 2), (2, 2)$

3. $f(x, y) = 8 + 2x + 2y$

   $R = \{(x, y): x^2 + y^2 \leq 4\}$

4. $f(x, y) = 10 + 2x - 3y$

   $R = \{(x, y): x^2 + y^2 \leq 9\}$

5. $f(x, y) = 9 - x^2$

   R: square with vertices $(0, 0), (3, 0), (0, 3), (3, 3)$

6. $f(x, y) = y^2$

   R: square with vertices $(0, 0), (3, 0), (0, 3), (3, 3)$

7. $f(x, y) = 2 + x^{3/2}$

   R: rectangle with vertices $(0, 0), (0, 4), (3, 4), (3, 0)$

8. $f(x, y) = 2 + \frac{2}{3}x^{3/2}$

   $R = \{(x, y): 0 \leq x \leq 1, 0 \leq y \leq 1 - x\}$

9. $f(x, y) = \ln |\sec x|$

   $R = \left\{(x, y): 0 \leq x \leq \dfrac{\pi}{4}, 0 \leq y \leq \tan x\right\}$

10. $f(x, y) = 4 + x^2 - y^2$

   $R = \{(x, y): x^2 + y^2 \leq 1\}$

11. $f(x, y) = \sqrt{x^2 + y^2}$

   $R = \{(x, y): 0 \leq f(x, y) \leq 1\}$

12. $f(x, y) = xy$

   $R = \{(x, y): x^2 + y^2 \leq 16\}$

13. $f(x, y) = \sqrt{a^2 - x^2 - y^2}$

   $R = \{(x, y): x^2 + y^2 \leq b^2, b < a\}$

14. $f(x, y) = \sqrt{a^2 - x^2 - y^2}$

   $R = \{(x, y): x^2 + y^2 \leq a^2\}$

In Exercises 15–18, find the area of the surface.

15. The portion of the plane $z = 24 - 3x - 2y$ in the first octant

16. The portion of the paraboloid $z = 16 - x^2 - y^2$ in the first octant

17. The portion of the sphere $x^2 + y^2 + z^2 = 25$ inside the cylinder $x^2 + y^2 = 9$

18. The portion of the cone $z = \sqrt{x^2 + y^2}$ inside the cylinder $x^2 + y^2 = 1$

In Exercises 19–24, write a double integral that represents the surface area of $z = f(x, y)$ over the region R. Use a symbolic integration utility to evaluate the double integral.

19. $f(x, y) = 2y + x^2$

   R: triangle with vertices $(0, 0), (1, 0), (1, 1)$

20. $f(x, y) = 2x + y^2$

   R: triangle with vertices $(0, 0), (2, 0), (2, 2)$

21. $f(x, y) = 4 - x^2 - y^2$

   $R = \{(x, y): 0 \leq f(x, y)\}$

22. $f(x, y) = x^2 + y^2$

   $R = \{(x, y): 0 \leq f(x, y) \leq 16\}$

23. $f(x, y) = 4 - x^2 - y^2$

   $R = \{(x, y): 0 \leq x \leq 1, 0 \leq y \leq 1\}$

24. $f(x, y) = \frac{2}{3}x^{3/2} + \cos x$

   $R = \{(x, y): 0 \leq x \leq 1, 0 \leq y \leq 1\}$

Think About It In Exercises 25 and 26, determine which value best approximates the surface area of $z = f(x, y)$ over the region R. (Make your selection on the basis of a sketch of the surface and *not* by performing any calculations.)

25. $f(x, y) = 10 - \frac{1}{2}y^2$

   R: square with vertices $(0, 0), (4, 0), (4, 4), (0, 4)$

   (a) 16   (b) 200   (c) -100 (d) 72   (e) 36

26. $f(x, y) = \frac{1}{4}\sqrt{x^2 + y^2}$

   R: circle bounded by $x^2 + y^2 = 9$

   (a) -100 (b) 150   (c) 9π (d) 55   (e) 500

In Exercises 27 and 28, use a symbolic integration utility to approximate the double integral that gives the surface area of the graph of f over the region $R = \{(x, y): 0 \leq x \leq 1, 0 \leq y \leq 1\}$.

27. $f(x, y) = e^x$ **28.** $f(x, y) = \frac{2}{5}y^{5/2}$

In Exercises 29–34, set up a double integral that gives the area of the surface on the graph of f over the region R.

29. $f(x, y) = x^3 - 3xy + y^3$

   R: square with vertices $(1, 1), (-1, 1), (-1, -1), (1, -1)$

30. $f(x, y) = e^{-x} \sin y$

   $R = \{(x, y): 0 \leq x \leq 4, 0 \leq y \leq x\}$

31. $f(x, y) = e^{-x} \sin y$

   $R = \{(x, y): x^2 + y^2 \leq 4\}$

32. $f(x, y) = x^2 - 3xy - y^2$

   $R = \{(x, y): 0 \leq x \leq 4, 0 \leq y \leq x\}$

33. $f(x, y) = e^{xy}$

$R = \{(x, y): 0 \le x \le 4, 0 \le y \le 10\}$

34. $f(x, y) = \cos(x^2 + y^2)$

$R = \left\{ (x, y): x^2 + y^2 \le \dfrac{\pi}{2} \right\}$

35. Find the surface area of the solid of intersection of the cylinders $x^2 + z^2 = 1$ and $y^2 + z^2 = 1$ (see figure).

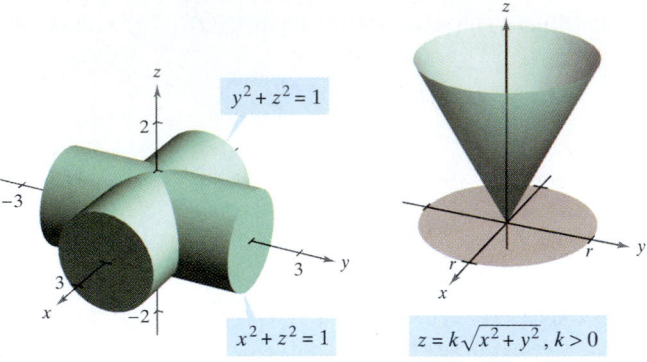

Figure for 35   **Figure for 36**

36. Show that the surface area of the cone

$z = k\sqrt{x^2 + y^2}, \quad k > 0$

over the circular region $x^2 + y^2 \le r^2$ in the xy-plane is $\pi r^2 \sqrt{k^2 + 1}$ (see figure).

37. *Building Design*   A new auditorium is built with a foundation in the shape of $\frac{1}{4}$ of a circle of radius 50 feet. Therefore, it forms a region R bounded by the graph of $x^2 + y^2 = 50^2$ with $x \ge 0$ and $y \ge 0$. The following equations are models for the floor and ceiling.

Floor: $z = \dfrac{x + y}{5}$ Ceiling: $z = 20 + \dfrac{xy}{100}$

(a) Calculate the volume of the room, which is needed to determine the heating and cooling requirements.

(b) Find the surface area of the ceiling.

38. The angle between a plane P and the xy-plane is θ, where $0 \le \theta < \pi/2$. The projection of a rectangular region in P onto the xy-plane is a rectangle whose sides have lengths Δx and Δy, as shown in the figure. Prove that the area of the rectangular region in P is $\sec \theta \, \Delta x \, \Delta y$.

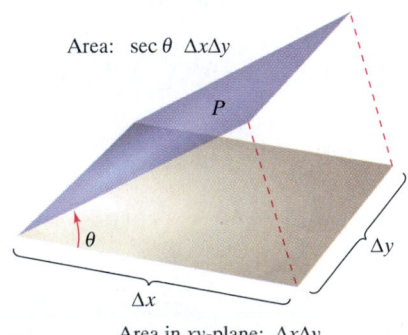

Area: $\sec \theta \ \Delta x \Delta y$

P

θ

Δx

Δy

Area in xy-plane: $\Delta x \Delta y$

39. *Think About It*   Use the result in Exercise 38 to order the planes in ascending order of their surface areas for a fixed region R in the xy-plane. Explain your ordering without doing any calculations.

(a) $z_1 = 2 + x$ (b) $z_2 = 5$

(c) $z_3 = 10 - 5x + 9y$ (d) $z_4 = 3 + x - 2y$

40. *Product Design*   A company produces a spherical object of radius 25 centimeters. A hole of radius 4 centimeters is drilled through the center of the object. Find (a) the volume of the object and (b) the outer surface area of the object.

41. *True or False?*   The surface area of the graph of a function $z = f(x, y)$ over a region R will increase if the graph is shifted k units vertically. Explain.

SECTION PROJECT

Capillary Action   A well-known property of liquids is that they will rise in narrow vertical channels—this property is called "capillary action." The figure shows two plates, which form a narrow wedge, in a container of liquid. The upper surface of the liquid follows a hyperbolic shape given by

$$z = \dfrac{k}{\sqrt{x^2 + y^2}}$$

where x, y, and z are measured in inches. The constant k depends on the angle of the wedge, the type of liquid, and the material that comprises the flat plates.

(a) Find the volume of the liquid that has risen in the wedge. (Assume $k = 1$.)

(b) Find the horizontal surface area of the liquid that has risen in the wedge.

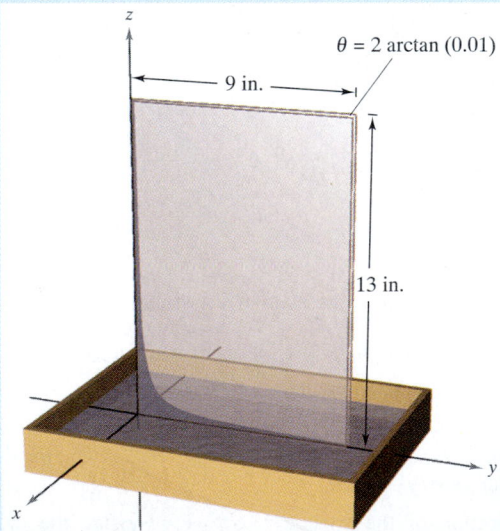

Adaptation of Capillary Action problem from "Capillary Phenomena" by Thomas Greenslade, Jr., PHYSICS TEACHER, May 1992. By permission of the author.

Triple Integrals • Center of Mass and Moments of Inertia

Triple Integrals

The procedure used to define a **triple integral** follows that used for double integrals. Consider a function f of three variables that is continuous over a bounded solid region Q. Then, encompass Q with a network of boxes and form the **inner partition** consisting of all boxes lying entirely within Q, as shown in Figure 13.51. The volume of the ith box is

$$\Delta V_i = \Delta x_i \Delta y_i \Delta z_i. \qquad \text{Volume of } i\text{th box}$$

The **norm** $\|\Delta\|$ of the partition is the length of the longest diagonal of the n boxes in the partition. Choose a point (x_i, y_i, z_i) in each box and form the Riemann sum

$$\sum_{i=1}^{n} f(x_i, y_i, z_i)\Delta V_i.$$

Taking the limit as $\|\Delta\| \to 0$ leads to the following definition.

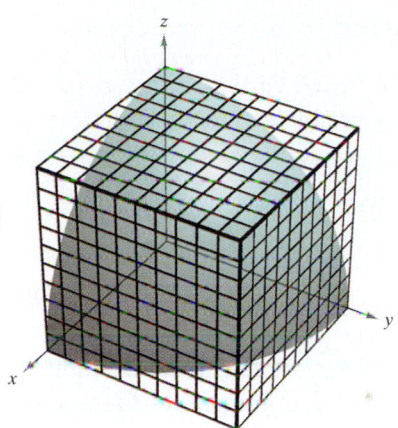

Solid region Q

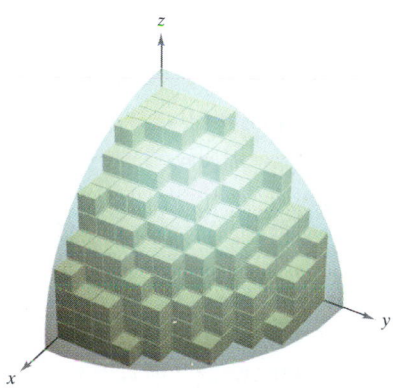

Volume of $Q \approx \sum_{i=1}^{n} f(x_i, y_i, z_i)\, \Delta V_i$

Figure 13.51

Definition of Triple Integral

If f is continuous over a bounded solid region Q, then the **triple integral of f over Q** is defined as

$$\iiint\limits_{Q} f(x, y, z)\, dV = \lim_{\|\Delta\| \to 0} \sum_{i=1}^{n} f(x_i, y_i, z_i)\, \Delta V_i$$

provided the limit exists. The **volume** of the solid region Q is given by

$$\text{Volume of } Q = \iiint\limits_{Q} dV.$$

Some of the properties of double integrals in Theorem 13.1 can be restated in terms of triple integrals

1. $\displaystyle\iiint\limits_{Q} cf(x, y, z)\, dV = c\iiint\limits_{Q} f(x, y, z)\, dV$

2. $\displaystyle\iiint\limits_{Q} [f(x, y, z) \pm g(x, y, z)]\, dV = \iiint\limits_{Q} f(x, y, z)\, dV \pm \iiint\limits_{Q} g(x, y, z)\, dV$

3. $\displaystyle\iiint\limits_{Q} f(x, y, z)\, dV = \iiint\limits_{Q_1} f(x, y, z)\, dV + \iiint\limits_{Q_2} f(x, y, z)\, dV$

where Q is the union of two nonoverlapping solid subregions Q_1 and Q_2. If the solid region Q is simple, the triple integral $\iiint f(x, y, z)\, dV$ can be evaluated with an iterated integral using one of the six possible orders of integration:

$$dx\, dy\, dz \quad dy\, dx\, dz \quad dz\, dx\, dy \quad dx\, dz\, dy \quad dy\, dz\, dx \quad dz\, dy\, dx.$$

EXPLORATION

Volume of a Paraboloid Sector
On pages 928 and 936, you were asked to summarize the ways you know for finding the volume of the solid bounded by the paraboloid

$$z = a^2 - x^2 - y^2, \quad a > 0$$

and the *xy*-plane. You now know one more way. Use it to find the volume of the solid.

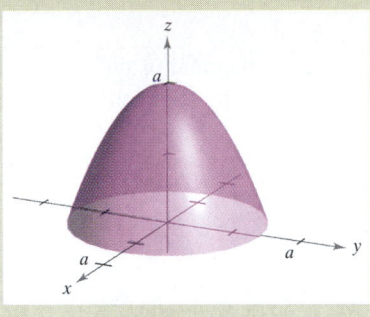

The following version of Fubini's Theorem describes a region that is considered simple with respect to the order *dz dy dx*. Similar descriptions can be given for the other five orders.

THEOREM 13.4   Evaluation by Iterated Integrals

Let *f* be continuous on a solid region *Q* defined by

$$a \le x \le b, \quad h_1(x) \le y \le h_2(x), \quad g_1(x, y) \le z \le g_2(x, y)$$

where $h_1, h_2, g_1,$ and g_2 are continuous functions. Then,

$$\iiint\limits_{Q} f(x, y, z)\, dV = \int_{a}^{b} \int_{h_1(x)}^{h_2(x)} \int_{g_1(x, y)}^{g_2(x, y)} f(x, y, z)\, dz\, dy\, dx.$$

To evaluate a triple iterated integral in the order *dz dy dx*, hold *both* x and y constant for the innermost integration. Then, hold x constant for the second integration.

EXAMPLE 1   Evaluating a Triple Iterated Integral

Evaluate the triple iterated integral

$$\int_{0}^{2} \int_{0}^{x} \int_{0}^{x+y} e^x(y + 2z)\, dz\, dy\, dx.$$

Solution   For the first integration, hold x and y constant and integrate with respect to z.

$$\int_{0}^{2} \int_{0}^{x} \int_{0}^{x+y} e^x(y + 2z)\, dz\, dy\, dx = \int_{0}^{2} \int_{0}^{x} e^x(yz + z^2) \Big]_{0}^{x+y} dy\, dx$$

$$= \int_{0}^{2} \int_{0}^{x} e^x(x^2 + 3xy + 2y^2)\, dy\, dx$$

For the second integration, hold x constant and integrate with respect to y.

$$\int_{0}^{2} \int_{0}^{x} e^x(x^2 + 3xy + 2y^2)\, dy\, dx = \int_{0}^{2} \left[e^x \left(x^2y + \frac{3xy^2}{2} + \frac{2y^3}{3} \right) \right]_{0}^{x} dx$$

$$= \frac{19}{6} \int_{0}^{2} x^3 e^x\, dx$$

$$= \frac{19}{6} \left[e^x(x^3 - 3x^2 + 6x - 6) \right]_{0}^{2}$$

$$= 19\left(\frac{e^2}{3} + 1 \right)$$

NOTE   Example 1 demonstrates the integration order *dz dy dx*. For other orders, you can follow a similar procedure. For instance, to evaluate a triple iterated integral in the order *dx dy dz*, hold both y and z constant for the innermost integration and integrate with respect to x. Then, for the second integration, hold z constant and integrate with respect to y. Finally, for the third integration, integrate with respect to z.

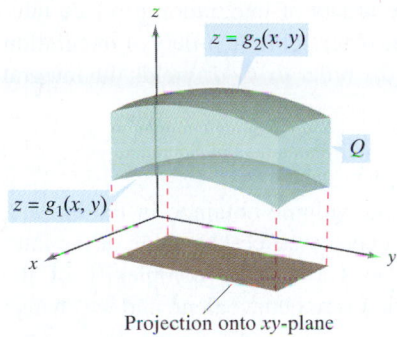

$z = g_2(x, y)$

Q

$z = g_1(x, y)$

Projection onto xy-plane

Solid region Q lies between two surfaces.
Figure 13.52

To find the limits for a particular order of integration, it is generally advisable to first determine the innermost limits, which may be functions of the outer two variables. Then, by projecting the solid Q onto the coordinate plane of the outer two variables, you can determine their limits of integration by the methods used for double integrals. For instance, to evaluate

$$\iiint_Q f(x, y, z)\, dz\, dy\, dx$$

first determine the limits for z, and then the integral has the form

$$\iint \left[\int_{g_1(x, y)}^{g_2(x, y)} f(x, y, z)\, dz \right] dy\, dx.$$

By projecting the solid Q onto the xy-plane, you can determine the limits for x and y as you did for double integrals, as shown in Figure 13.52.

EXAMPLE 2   Using a Triple Integral to Find Volume

Find the volume of the ellipsoidal solid given by $4x^2 + 4y^2 + z^2 = 16$.

Solution   Because x, y, and z play similar roles in the equation, the order of integration is probably immaterial, and you arbitrarily choose $dz\, dy\, dx$. Moreover, you can simplify the calculation by considering only the portion of the ellipsoid lying in the first octant, as shown in Figure 13.53. From the order $dz\, dy\, dx$, you first determine the bounds for z.

$$0 \le z \le 2\sqrt{4 - x^2 - y^2}$$

In Figure 13.54, you can see that the boundaries for x and y are $0 \le x \le 2$ and $0 \le y \le \sqrt{4 - x^2}$, so the volume of the ellipsoid is

$0 \le z \le 2\sqrt{4 - x^2 - y^2}$

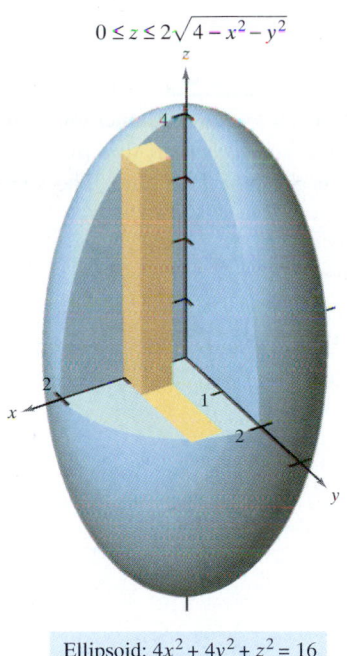

Ellipsoid: $4x^2 + 4y^2 + z^2 = 16$

The volume of the ellipsoid is $64\pi/3$.
Figure 13.53

$$V = \iiint_Q dV$$

$$= 8 \int_0^2 \int_0^{\sqrt{4-x^2}} \int_0^{2\sqrt{4-x^2-y^2}} dz\, dy\, dx$$

$$= 8 \int_0^2 \int_0^{\sqrt{4-x^2}} z \Big]_0^{2\sqrt{4-x^2-y^2}} dy\, dx$$

$$= 16 \int_0^2 \int_0^{\sqrt{4-x^2}} \sqrt{(4 - x^2) - y^2}\, dy\, dx \qquad \text{Integration tables (the appendix), Formula 37}$$

$$= 8 \int_0^2 \left[y\sqrt{4 - x^2 - y^2} + (4 - x^2) \arcsin\left(\frac{y}{\sqrt{4 - x^2}}\right) \right]_0^{\sqrt{4-x^2}} dx$$

$$= 8 \int_0^2 (4 - x^2)\left(\frac{\pi}{2}\right) dx$$

$$= 4\pi \left[4x - \frac{x^3}{3} \right]_0^2$$

$$= \frac{64\pi}{3}.$$

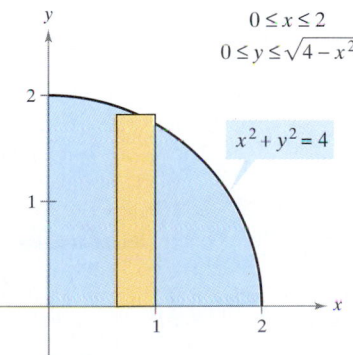

$0 \le x \le 2$
$0 \le y \le \sqrt{4 - x^2}$

$x^2 + y^2 = 4$

Figure 13.54

Example 2 is unusual in that all six possible orders of integration produce integrals of comparable difficulty. Try setting up some other possible orders of integration to find the volume of the ellipsoid. For instance, the order $dx\ dy\ dz$ yields the integral

$$V = 8 \int_0^4 \int_0^{\sqrt{16-z^2}/2} \int_0^{\sqrt{16-4y^2-z^2}/2} dx\ dy\ dz.$$

If you solve this integral, you will obtain the same volume obtained in Example 2. This is always the case—the order of integration does not affect the value of the integral. However, the order of integration often does affect the complexity of the integral. In Example 3, the given order of integration is not convenient, and we change the order to simplify the problem.

EXAMPLE 3 Changing the Order of Integration

Evaluate $\displaystyle\int_0^{\sqrt{\pi/2}} \int_x^{\sqrt{\pi/2}} \int_1^3 \sin y^2\ dz\ dy\ dx$.

Solution Note that after one integration in the given order, you would encounter the integral $2\int \sin(y^2)\ dy$, which is not an elementary function. To avoid this problem, change the order of integration to $dz\ dx\ dy$, so that y is the outer variable. The solid region Q is given by

$$0 \le x \le \sqrt{\frac{\pi}{2}}, \quad x \le y \le \sqrt{\frac{\pi}{2}}, \quad 1 \le z \le 3$$

as shown in Figure 13.55, and the projection of Q in the xy-plane yields the bounds

$$0 \le y \le \sqrt{\frac{\pi}{2}} \quad \text{and} \quad 0 \le x \le y.$$

Therefore, you have

$$V = \int\int\int_Q dV = \int_0^{\sqrt{\pi/2}} \int_0^y \int_1^3 \sin(y^2)\ dz\ dx\ dy$$

$$= \int_0^{\sqrt{\pi/2}} \int_0^y z \sin(y^2) \Big]_1^3 dx\ dy$$

$$= 2\int_0^{\sqrt{\pi/2}} \int_0^y \sin(y^2)\ dx\ dy$$

$$= 2\int_0^{\sqrt{\pi/2}} x \sin(y^2) \Big]_0^y dy$$

$$= 2\int_0^{\sqrt{\pi/2}} y \sin(y^2)\ dy$$

$$= -\cos(y^2) \Big]_0^{\sqrt{\pi/2}}$$

$$= 1.$$

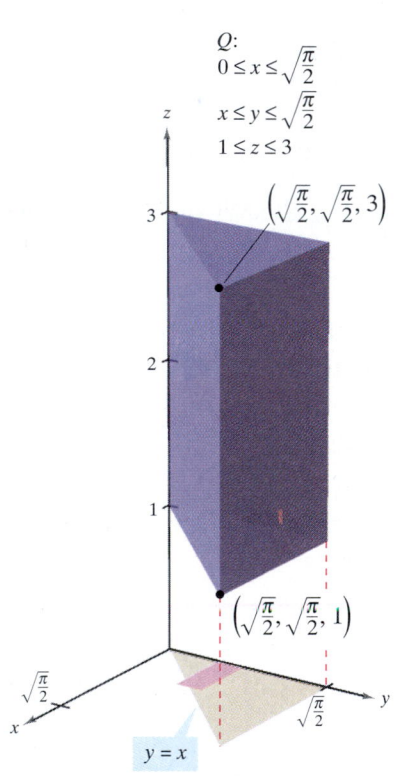

Q:
$0 \le x \le \sqrt{\dfrac{\pi}{2}}$

$x \le y \le \sqrt{\dfrac{\pi}{2}}$

$1 \le z \le 3$

$\left(\sqrt{\dfrac{\pi}{2}}, \sqrt{\dfrac{\pi}{2}}, 3\right)$

$\left(\sqrt{\dfrac{\pi}{2}}, \sqrt{\dfrac{\pi}{2}}, 1\right)$

$y = x$

The volume of the solid region Q is 1.
Figure 13.55

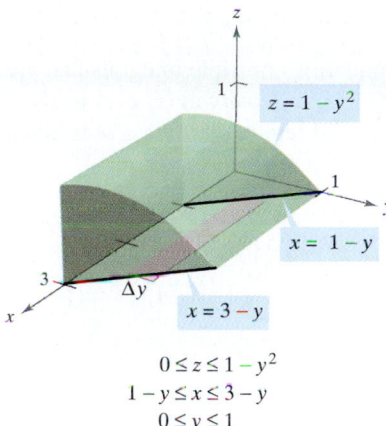

$z = 1 - y^2$

$x = 1 - y$

$x = 3 - y$

$$0 \le z \le 1 - y^2$$
$$1 - y \le x \le 3 - y$$
$$0 \le y \le 1$$

Figure 13.56

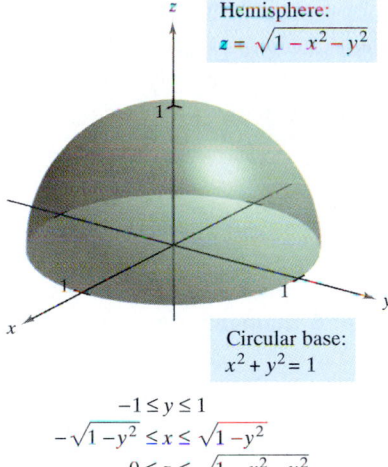

Hemisphere:
$z = \sqrt{1 - x^2 - y^2}$

Circular base:
$x^2 + y^2 = 1$

$$-1 \le y \le 1$$
$$-\sqrt{1 - y^2} \le x \le \sqrt{1 - y^2}$$
$$0 \le z \le \sqrt{1 - x^2 - y^2}$$

Figure 13.57

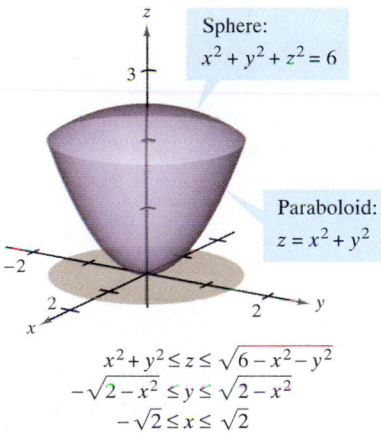

Sphere:
$x^2 + y^2 + z^2 = 6$

Paraboloid:
$z = x^2 + y^2$

$$x^2 + y^2 \le z \le \sqrt{6 - x^2 - y^2}$$
$$-\sqrt{2 - x^2} \le y \le \sqrt{2 - x^2}$$
$$-\sqrt{2} \le x \le \sqrt{2}$$

Figure 13.58

EXAMPLE 4   Determining the Limits of Integration

Set up a triple integral for the volume of each solid region.

a. The region in the first octant bounded above by the cylinder $z = 1 - y^2$ and lying between the vertical planes $x + y = 1$ and $x + y = 3$.

b. The upper hemisphere given by $z = \sqrt{1 - x^2 - y^2}$.

c. The region bounded below by the paraboloid $z = x^2 + y^2$ and above by the sphere $x^2 + y^2 + z^2 = 6$.

Solution

a. In Figure 13.56, note that the solid is bounded below by the xy-plane ($z = 0$) and above by the cylinder $z = 1 - y^2$. Therefore,

$$0 \le z \le 1 - y^2. \qquad \text{\textcolor{red}{Bounds for } } z$$

Projecting the region onto the xy-plane produces a parallelogram. Because two sides of the parallelogram are parallel to the x-axis, you have the following bounds:

$$1 - y \le x \le 3 - y \quad \text{and} \quad 0 \le y \le 1.$$

Therefore, the volume of the region is given by

$$V = \iiint_Q dV = \int_0^1 \int_{1-y}^{3-y} \int_0^{1-y^2} dz \, dx \, dy.$$

b. For the upper hemisphere given by $z = \sqrt{1 - x^2 - y^2}$, you have

$$0 \le z \le \sqrt{1 - x^2 - y^2}. \qquad \text{\textcolor{red}{Bounds for } } z$$

In Figure 13.57, note that the projection of the hemisphere onto the xy-plane is the circle given by $x^2 + y^2 = 1$, and you can use either order $dx \, dy$ or $dy \, dx$. Choosing the first produces

$$-\sqrt{1 - y^2} \le x \le \sqrt{1 - y^2} \quad \text{and} \quad -1 \le y \le 1$$

which implies that the volume of the region is given by

$$V = \iiint_Q dV = \int_{-1}^1 \int_{-\sqrt{1-y^2}}^{\sqrt{1-y^2}} \int_0^{\sqrt{1-x^2-y^2}} dz \, dx \, dy.$$

c. For the region bounded below by the paraboloid $z = x^2 + y^2$ and above by the sphere $x^2 + y^2 + z^2 = 6$, you have

$$x^2 + y^2 \le z \le \sqrt{6 - x^2 - y^2}. \qquad \text{\textcolor{red}{Bounds for } } z$$

The sphere and the paraboloid intersect when $z = 2$. Moreover, you can see in Figure 13.58 that the projection of the solid region onto the xy-plane is the circle given by $x^2 + y^2 = 2$. Using the order $dy \, dx$ produces

$$-\sqrt{2 - x^2} \le y \le \sqrt{2 - x^2} \quad \text{and} \quad -\sqrt{2} \le x \le \sqrt{2}$$

which implies that the volume of the region is given by

$$V = \iiint_Q dV = \int_{-\sqrt{2}}^{\sqrt{2}} \int_{-\sqrt{2-x^2}}^{\sqrt{2-x^2}} \int_{x^2+y^2}^{\sqrt{6-x^2-y^2}} dz \, dy \, dx.$$

: **TECHNOLOGY** Sketch the solid (of
: uniform density) bounded by $z = 0$
: and $z = 1/(1 + x^2 + y^2)$, where
: $x^2 + y^2 \le 1$. From your sketch,
: estimate the coordinates of the center
: of mass of the solid. Now use a
: computer algebra system to verify
: your estimate. What do you observe?

Center of Mass and Moments of Inertia

In the remainder of this section, we discuss two applications of triple integrals that are important in engineering. Consider a solid region Q whose density at (x, y, z) is given by the **density function ρ.** The **center of mass** of a solid region Q of mass m is given by $(\bar{x}, \bar{y}, \bar{z})$, where

$$m = \iiint\limits_{Q} \rho(x, y, z)\, dV \qquad \text{Mass of the solid}$$

$$M_{yz} = \iiint\limits_{Q} x\rho(x, y, z)\, dV \qquad \text{First moment about } yz\text{-plane}$$

$$M_{xz} = \iiint\limits_{Q} y\rho(x, y, z)\, dV \qquad \text{First moment about } xz\text{-plane}$$

$$M_{xy} = \iiint\limits_{Q} z\rho(x, y, z)\, dV \qquad \text{First moment about } xy\text{-plane}$$

and

$$\bar{x} = \frac{M_{yz}}{m}, \quad \bar{y} = \frac{M_{xz}}{m}, \quad \bar{z} = \frac{M_{xy}}{m}.$$

The quantities M_{yz}, M_{xz}, and M_{xy} are called the **first moments** of the region Q about the yz-, xz-, and xy-planes.

The first moments for solid regions are taken about a plane, whereas the second moments for solids are taken about a line. The **second moments** (or **moments of inertia**) about the x-, y-, and z-axes are as follows.

NOTE In engineering and physics, the moment of inertia of a mass is used to find the time required for a mass to reach a given speed of rotation about an axis, as indicated in Figure 13.59. The greater the moment of inertia, the longer a force must be applied for the mass to reach the given speed.

$$I_x = \iiint\limits_{Q} (y^2 + z^2)\rho(x, y, z)\, dV \qquad \text{Moment of inertia about } x\text{-axis}$$

$$I_y = \iiint\limits_{Q} (x^2 + z^2)\rho(x, y, z)\, dV \qquad \text{Moment of inertia about } y\text{-axis}$$

$$I_z = \iiint\limits_{Q} (x^2 + y^2)\rho(x, y, z)\, dV \qquad \text{Moment of inertia about } z\text{-axis}$$

For problems requiring the calculation of all three moments, considerable effort can be saved by applying the additive property of triple integrals and writing

$$I_x = I_{xz} + I_{xy}, \quad I_y = I_{yz} + I_{xy}, \quad \text{and} \quad I_z = I_{yz} + I_{xz}$$

where I_{xy}, I_{xz}, and I_{yz} are as follows.

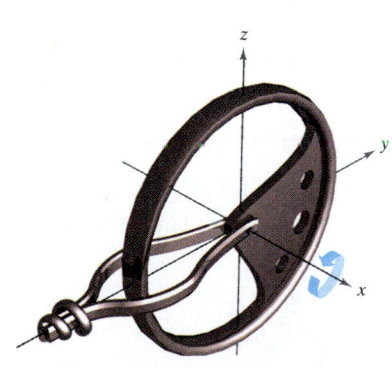

Figure 13.59

$$I_{xy} = \iiint\limits_{Q} z^2\rho(x, y, z)\, dV$$

$$I_{xz} = \iiint\limits_{Q} y^2\rho(x, y, z)\, dV$$

$$I_{yz} = \iiint\limits_{Q} x^2\rho(x, y, z)\, dV$$

EXAMPLE 5   Finding the Center of Mass of a Solid Region

Find the center of mass of the unit cube shown in Figure 13.60, given that the density at the point (x, y, z) is proportional to the square of its distance from the origin.

Solution   Because the density at (x, y, z) is proportional to the square of the distance between $(0, 0, 0)$ and (x, y, z), you have

$$\rho(x, y, z) = k(x^2 + y^2 + z^2).$$

You can use this density function to find the mass of the cube. Because of the symmetry of the region, any order of integration will produce an integral of comparable difficulty.

$$m = \int_0^1 \int_0^1 \int_0^1 k(x^2 + y^2 + z^2)\, dz\, dy\, dx$$

$$= k \int_0^1 \int_0^1 \left[(x^2 + y^2)z + \frac{z^3}{3} \right]_0^1 dy\, dx$$

$$= k \int_0^1 \int_0^1 \left(x^2 + y^2 + \frac{1}{3} \right) dy\, dx$$

$$= k \int_0^1 \left[\left(x^2 + \frac{1}{3} \right)y + \frac{y^3}{3} \right]_0^1 dx$$

$$= k \int_0^1 \left(x^2 + \frac{2}{3} \right) dx$$

$$= k \left[\frac{x^3}{3} + \frac{2x}{3} \right]_0^1 = k$$

The first moment about the yz-plane is

$$M_{yz} = k \int_0^1 \int_0^1 \int_0^1 x(x^2 + y^2 + z^2)\, dz\, dy\, dx$$

$$= k \int_0^1 x \left[\int_0^1 \int_0^1 (x^2 + y^2 + z^2)\, dz\, dy \right] dx.$$

Note that x can be factored out of the two inner integrals, because it is constant with respect to y and z. After factoring, the two inner integrals are the same as for the mass m. Hence, you have

$$M_{yz} = k \int_0^1 x \left(x^2 + \frac{2}{3} \right) dx$$

$$= k \left[\frac{x^4}{4} + \frac{x^2}{3} \right]_0^1$$

$$= \frac{7k}{12}.$$

Therefore,

$$\bar{x} = \frac{M_{yz}}{m} = \frac{7k/12}{k} = \frac{7}{12}.$$

Finally, from the nature of ρ and the symmetry of x, y, and z in this solid region, you have $\bar{x} = \bar{y} = \bar{z}$, and the center of mass is $\left(\frac{7}{12}, \frac{7}{12}, \frac{7}{12} \right)$.

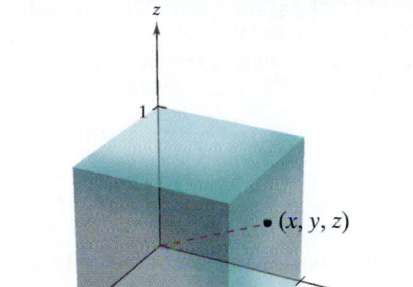

Variable density:
$\rho(x, y, z) = k(x^2 + y^2 + z^2)$
Figure 13.60

EXAMPLE 6   Moments of Inertia for a Solid Region

Find the moments of inertia about the x- and y-axes for the solid region lying between the hemisphere

$$z = \sqrt{4 - x^2 - y^2}$$

and the xy-plane, given that the density at (x, y, z) is proportional to the distance between (x, y, z) and the xy-plane.

Solution   The density of the region is given by $\rho(x, y, z) = kz$. Considering the symmetry of this problem, you know that $I_x = I_y$, and you need to compute only one moment, say I_x. From Figure 13.61, choose the order $dz\, dy\, dx$ and write

$$I_x = \iiint\limits_{Q} (y^2 + z^2)\rho(x, y, z)\, dV$$

$$= \int_{-2}^{2} \int_{-\sqrt{4-x^2}}^{\sqrt{4-x^2}} \int_{0}^{\sqrt{4-x^2-y^2}} (y^2 + z^2)(kz)\, dz\, dy\, dx$$

$$= k \int_{-2}^{2} \int_{-\sqrt{4-x^2}}^{\sqrt{4-x^2}} \left[\frac{y^2 z^2}{2} + \frac{z^4}{4} \right]_{0}^{\sqrt{4-x^2-y^2}} dy\, dx$$

$$= k \int_{-2}^{2} \int_{-\sqrt{4-x^2}}^{\sqrt{4-x^2}} \left[\frac{y^2(4 - x^2 - y^2)}{2} + \frac{(4 - x^2 - y^2)^2}{4} \right] dy\, dx$$

$$= \frac{k}{4} \int_{-2}^{2} \int_{-\sqrt{4-x^2}}^{\sqrt{4-x^2}} \left[(4 - x^2)^2 - y^4 \right] dy\, dx$$

$$= \frac{k}{4} \int_{-2}^{2} \left[(4 - x^2)^2 y - \frac{y^5}{5} \right]_{-\sqrt{4-x^2}}^{\sqrt{4-x^2}} dx$$

$$= \frac{k}{4} \int_{-2}^{2} \frac{8}{5}(4 - x^2)^{5/2}\, dx$$

$$= \frac{4k}{5} \int_{0}^{2} (4 - x^2)^{5/2}\, dx \qquad\qquad x = 2\sin\theta$$

$$= \frac{4k}{5} \int_{0}^{\pi/2} 64\cos^6\theta\, d\theta$$

$$= \left(\frac{256k}{5} \right)\left(\frac{5\pi}{32} \right) \qquad\qquad \text{Wallis's Formula}$$

$$= 8k\pi.$$

Thus, $I_x = 8k\pi = I_y$.

NOTE   In Example 6, notice that the moments of inertia about the x- and y-axes are equal to each other. The moment about the z-axis, however, is different. Does it seem that the moment of inertia about the z-axis should be less than or greater than the moments calculated in Example 6? By performing the calculations, you can determine that $I_z = \frac{16}{3}k\pi$. This tells you that the solid shown in Figure 13.61 has a greater resistance to rotation about the x- or y-axis than about the z-axis.

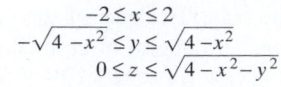

$$-2 \le x \le 2$$
$$-\sqrt{4 - x^2} \le y \le \sqrt{4 - x^2}$$
$$0 \le z \le \sqrt{4 - x^2 - y^2}$$

Hemisphere:
$$z = \sqrt{4 - x^2 - y^2}$$

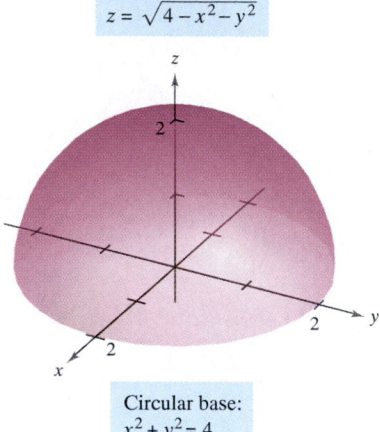

Circular base:
$$x^2 + y^2 = 4$$

Variable density: $\rho(x, y, z) = kz$
Figure 13.61

EXERCISES FOR SECTION 13.6

In Exercises 1–8, evaluate the triple integral.

1. $\displaystyle\int_0^3 \int_0^2 \int_0^1 (x + y + z)\, dx\, dy\, dz$

2. $\displaystyle\int_{-1}^1 \int_{-1}^1 \int_{-1}^1 x^2 y^2 z^2\, dx\, dy\, dz$

3. $\displaystyle\int_0^1 \int_0^x \int_0^{xy} x\, dz\, dy\, dx$

4. $\displaystyle\int_0^4 \int_0^\pi \int_0^{1-x} x \sin y\, dz\, dy\, dx$

5. $\displaystyle\int_1^4 \int_0^1 \int_0^x 2z e^{-x^2}\, dy\, dx\, dz$

6. $\displaystyle\int_1^4 \int_1^{e^2} \int_0^{1/xz} \ln z\, dy\, dz\, dx$

7. $\displaystyle\int_0^9 \int_0^{y/3} \int_0^{\sqrt{y^2 - 9x^2}} z\, dz\, dx\, dy$

8. $\displaystyle\int_0^{\pi/2} \int_0^{y/2} \int_0^{1/y} \sin y\, dz\, dx\, dy$

In Exercises 9 and 10, use a symbolic integration utility to evaluate the triple integral.

9. $\displaystyle\int_0^2 \int_{-\sqrt{4-x^2}}^{\sqrt{4-x^2}} \int_0^{x^2} x\, dz\, dy\, dx$

10. $\displaystyle\int_0^{\sqrt{2}} \int_0^{\sqrt{2-x^2}} \int_{2x^2+y^2}^{4-y^2} y\, dz\, dy\, dx$

In Exercises 11 and 12, use a symbolic integration utility to approximate the triple integral.

11. $\displaystyle\int_0^2 \int_0^{\sqrt{4-x^2}} \int_1^4 \frac{x^2 \sin y}{z}\, dz\, dy\, dx$

12. $\displaystyle\int_0^3 \int_0^{2-(2y/3)} \int_0^{6-2y-3z} ze^{-x^2y^2}\, dx\, dz\, dy$

In Exercises 13–16, sketch the solid whose volume is given by the triple integral and rewrite the integral with the indicated order of integration

13. $\displaystyle\int_0^4 \int_0^{(4-x)/2} \int_0^{(12-3x-6y)/4} dz\, dy\, dx$

Rewrite using the order $dy\, dx\, dz$.

14. $\displaystyle\int_0^4 \int_0^{\sqrt{16-x^2}} \int_0^{10-x-y} dz\, dy\, dx$

Rewrite using the order $dz\, dx\, dy$.

15. $\displaystyle\int_0^1 \int_y^1 \int_0^{\sqrt{1-y^2}} dz\, dx\, dy$

Rewrite using the order $dz\, dy\, dx$.

16. $\displaystyle\int_0^2 \int_{2x}^4 \int_0^{\sqrt{y^2-4x^2}} dz\, dy\, dx$

Rewrite using the order $dx\, dy\, dz$.

In Exercises 17 and 18, list the six possible orders of integration for the triple integral over the solid Q

$$\iiint_Q xyz\, dV.$$

17. $Q = \{(x, y, z): 0 \le x \le 1, 0 \le y \le x, 0 \le z \le 3\}$

18. $Q = \{(x, y, z): 0 \le x \le 2, x^2 \le y \le 4, 0 \le z \le 2 - x\}$

Volume **In Exercises 19–24, use a triple integral to find the volume of the solid bounded by the graphs of the equations.**

19. $x = 4 - y^2, z = 0, z = x$

20. $z = xy, z = 0, x = 0, x = 1, y = 0, y = 1$

21. $x^2 + y^2 + z^2 = a^2$

22. $z = 9 - x^2 - y^2, z = 0$

23. $z = 4 - x^2, y = 4 - x^2$, first octant

24. $z = 9 - x^2, y = -x + 2, y = 0, z = 0, x \ge 0$

Mass and Center of Mass **In Exercises 25–28, find the mass and the indicated coordinates of the center of mass of the solid of indicated density bounded by the graphs of the equations.**

25. Find $\bar{x}$ using $\rho(x, y, z) = k$.

$Q: 2x + 3y + 6z = 12, x = 0, y = 0, z = 0$

26. Find $\bar{y}$ using $\rho(x, y, z) = ky$.

$Q: 2x + 3y + 6z = 12, x = 0, y = 0, z = 0$

27. Find $\bar{z}$ using $\rho(x, y, z) = kx$.

$Q: z = 4 - x, z = 0, y = 0, y = 4, x = 0$

28. Find $\bar{y}$ using $\rho(x, y, z) = k$.

$Q: \dfrac{x}{a} + \dfrac{y}{b} + \dfrac{z}{c} = 1 \ (a, b, c > 0), x = 0, y = 0, z = 0$

Mass and Center of Mass **In Exercises 29 and 30, set up the triple integrals for finding the mass and the center of mass of the solid bounded by the graphs of the equations.**

29. $x = 0, x = b, y = 0, y = b, z = 0, z = b,$

$\rho(x, y, z) = kxy$

30. $x = 0, x = a, y = 0, y = b, z = 0, z = c,$

$\rho(x, y, z) = kz$

Think About It The center of mass of the solid of constant density shown in the figure is $\left(2, 0, \frac{8}{5}\right)$. In Exercises 31–34, make a conjecture about how the center of mass $(\bar{x}, \bar{y}, \bar{z})$ will change for the nonconstant density $\rho(x, y)$. Explain. (Make your conjecture *without* performing any calculations.)

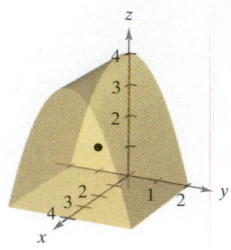

31. $\rho(x, y, z) = kx$ **32.** $\rho(x, y, z) = kz$

33. $\rho(x, y, z) = k(y + 2)$ **34.** $\rho(x, y, z) = kxz^2(y + 2)^2$

Centroid In Exercises 35–40, find the centroid of the solid region bounded by the graphs of the equations or described by the figure. Use a symbolic integration utility to evaluate the triple integrals. (Assume uniform density and find the center of mass.)

35. $z = \dfrac{h}{r}\sqrt{x^2 + y^2}, z = h$

36. $y = \sqrt{4 - x^2}, z = y, z = 0$

37. $z = \sqrt{4^2 - x^2 - y^2}, z = 0$

38. $z = \dfrac{1}{y^2 + 1}, z = 0, x = -2, x = 2, y = 0, y = 1$

39.

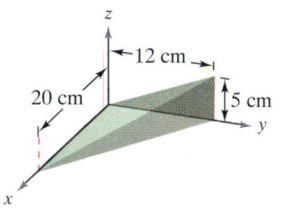

40.

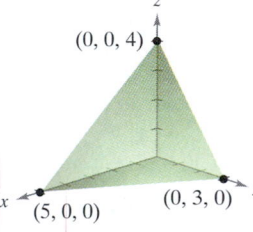

Moments of Inertia In Exercises 41–44, find I_x, I_y, and I_z for the solid of indicated density. Use a symbolic integration utility to evaluate the triple integrals.

41. (a) $\rho = k$ **42.** (a) $\rho(x, y, z) = k$

    (b) $\rho = kxyz$ (b) $\rho(x, y, z) = k(x^2 + y^2)$

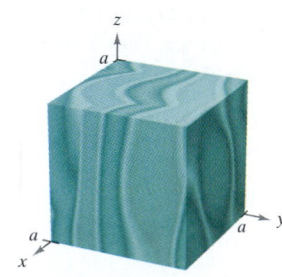

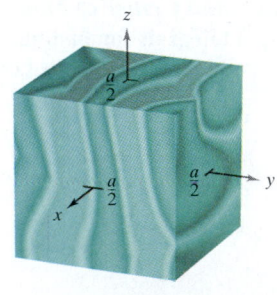

43. (a) $\rho(x, y, z) = k$ **44.** (a) $\rho = kz$

    (b) $\rho = ky$ (b) $\rho = k(4 - z)$

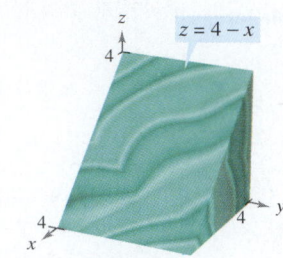

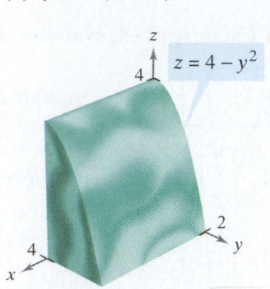

Moments of Inertia In Exercises 45 and 46, verify the moments of inertia for the solid of uniform density. Use a symbolic integration utility to evaluate the triple integrals.

45. $I_x = \frac{1}{12}m(3a^2 + L^2)$ **46.** $I_x = \frac{1}{12}m(a^2 + b^2)$

    $I_y = \frac{1}{2}ma^2$ $I_y = \frac{1}{12}m(b^2 + c^2)$

    $I_z = \frac{1}{12}m(3a^2 + L^2)$ $I_z = \frac{1}{12}m(a^2 + c^2)$

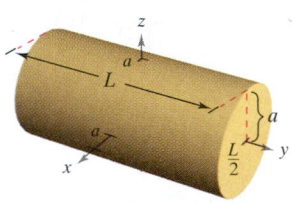

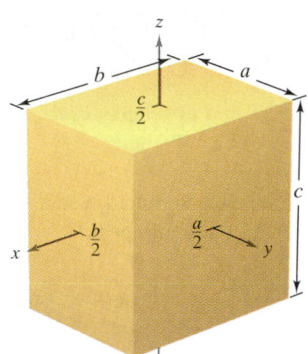

Moments of Inertia In Exercises 47 and 48, set up a triple integral that gives the moment of inertia about the z-axis of the solid Q of density ρ.

47. $Q = \{(x, y, z): -1 \le x \le 1, -1 \le y \le 1, 0 \le z \le 1 - x\}$

    $\rho = \sqrt{x^2 + y^2 + z^2}$

48. $Q = \{(x, y, z): x^2 + y^2 \le 1, 0 \le z \le 4 - x^2 - y^2\}$

    $\rho = kx^2$

49. ***Think About It*** Determine whether the moment of inertia about the y-axis of the cylinder in Exercise 45 will increase or decrease for the nonconstant density $\rho(x, y, z) = \sqrt{x^2 + z^2}$ and $a = 4$.

Triple Integrals in Cylindrical Coordinates • Triple Integrals in Spherical Coordinates

Triple Integrals in Cylindrical Coordinates

Many common solid regions such as spheres, ellipsoids, cones, and paraboloids can yield difficult triple integrals in rectangular coordinates. In fact, it is precisely this difficulty that led to the introduction of nonrectangular coordinate systems. In this section, you will learn how to use *cylindrical* and *spherical* coordinates to evaluate triple integrals.

Recall from Section 10.7 that the rectangular conversion equations for cylindrical coordinates are

$$x = r \cos \theta$$
$$y = r \sin \theta$$
$$z = z.$$

NOTE   An easy way to remember these conversions is to note that the equations for x and y are the same as in polar coordinates and z is unchanged.

In this coordinate system, the simplest solid region is a cylindrical block determined by

$$r_1 \leq r \leq r_2, \quad \theta_1 \leq \theta \leq \theta_2, \quad z_1 \leq z \leq z_2$$

as shown in Figure 13.62. To obtain the cylindrical coordinate form of a triple integral, suppose that Q is a solid region whose projection R onto the xy-plane can be described in polar coordinates. That is,

$$Q = \{(x, y, z) : (x, y) \text{ is in } R, \quad h_1(x, y) \leq z \leq h_2(x, y)\}$$

and

$$R = \{(r, \theta) : \theta_1 \leq \theta \leq \theta_2, \quad g_1(\theta) \leq r \leq g_2(\theta)\}.$$

If f is a continuous function on the solid Q, you can write the triple integral of f over Q as

$$\iiint_Q f(x, y, z) \, dV = \iint_R \left[\int_{h_1(x, y)}^{h_2(x, y)} f(x, y, z) \, dz \right] dA$$

where the double integral over R is evaluated in polar coordinates. That is, R is a plane region that is either r-simple or θ-simple. If R is r-simple, the iterated form of the triple integral in cylindrical form is

$$\iiint_Q f(x, y, z) \, dV = \int_{\theta_1}^{\theta_2} \int_{g_1(\theta)}^{g_2(\theta)} \int_{h_1(r \cos\theta, \, r \sin\theta)}^{h_2(r \cos\theta, \, r \sin\theta)} f(r \cos \theta, r \sin \theta, z) r \, dz \, dr \, d\theta.$$

NOTE   This is only one of six possible orders of integration. The other five are $dz \, d\theta \, dr$, $dr \, dz \, d\theta$, $dr \, d\theta \, dz$, $d\theta \, dz \, dr$, and $d\theta \, dr \, dz$.

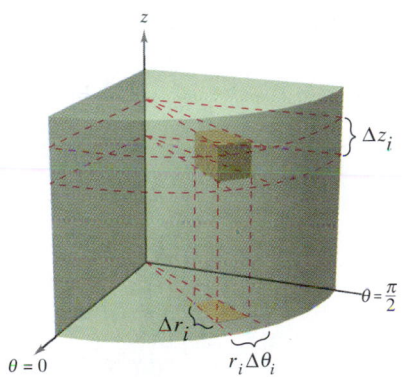

PIERRE SIMON DE LAPLACE (1749–1827)

One of the first to use a cylindrical coordinate system was the French mathematician Pierre Simon de Laplace. Laplace has been called the "Newton of France," and he published many important works in mechanics, differential equations, and probability.

Volume of cylindrical block:
$$\Delta V_i = r_i \, \Delta r_i \, \Delta \theta_i \, \Delta z_i$$
Figure 13.62

EXPLORATION

Volume of a Paraboloid Sector
On pages 928, 936, and 958, you
were asked to summarize the ways
you know for finding the volume of
the solid bounded by the paraboloid

$z = a^2 - x^2 - y^2, \quad a > 0$

and the *xy*-plane. You now know one
more way. Use it to find the volume
of the solid.

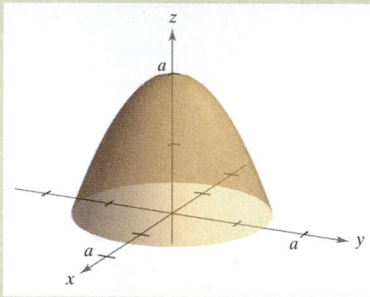

Compare the different methods. What
are the advantages and disadvantages
of each?

To visualize a particular order of integration, it helps to view the iterated integral
in terms of three sweeping motions—each adding another dimension to the solid. For
instance, in order $dr \, d\theta \, dz$, the first integration occurs in the *r*-direction as a point
sweeps out a ray. Then, as θ increases, the line sweeps out a sector. Finally, as *z*
increases, the sector sweeps out a solid wedge, as shown in Figure 13.63.

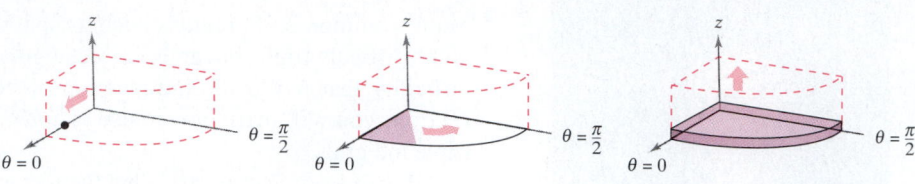

Integrate with respect to *r*.   Integrate with respect to θ.   Integrate with respect to *z*.
Figure 13.63

EXAMPLE 1   Finding Volume by Cylindrical Coordinates

Find the volume of the solid region Q cut from the sphere

$x^2 + y^2 + z^2 = 4$ Sphere

by the cylinder $r = 2 \sin \theta$, as shown in Figure 13.64.

Solution   Because $x^2 + y^2 + z^2 = r^2 + z^2 = 4$, the bounds on *z* are

$$-\sqrt{4 - r^2} \le z \le \sqrt{4 - r^2}.$$

Let R be the circular projection of the solid onto the $r\theta$-plane. Then the bounds on R
are $0 \le r \le 2 \sin \theta$ and $0 \le \theta \le \pi$. Thus, the volume of Q is

$$V = \int_0^\pi \int_0^{2 \sin \theta} \int_{-\sqrt{4-r^2}}^{\sqrt{4-r^2}} r \, dz \, dr \, d\theta$$

$$= 2 \int_0^{\pi/2} \int_0^{2 \sin \theta} 2r\sqrt{4 - r^2} \, dr \, d\theta$$

$$= 2 \int_0^{\pi/2} -\frac{2}{3}(4 - r^2)^{3/2} \Big]_0^{2 \sin \theta} d\theta$$

$$= \frac{4}{3} \int_0^{\pi/2} (8 - 8 \cos^3 \theta) \, d\theta$$

$$= \frac{32}{3} \int_0^{\pi/2} [1 - (\cos \theta)(1 - \sin^2 \theta)] \, d\theta$$

$$= \frac{32}{3} \left[\theta - \sin \theta + \frac{\sin^3 \theta}{3} \right]_0^{\pi/2}$$

$$= \frac{16}{9}(3\pi - 4).$$

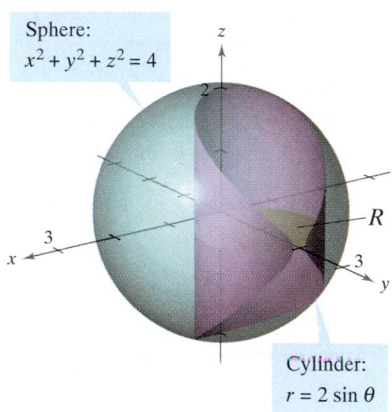

Sphere:
$x^2 + y^2 + z^2 = 4$

Cylinder:
$r = 2 \sin \theta$

The volume of the solid region Q is
$\frac{16}{9}(3\pi - 4)$.
Figure 13.64

$0 \le z \le 2\sqrt{4 - r^2}$

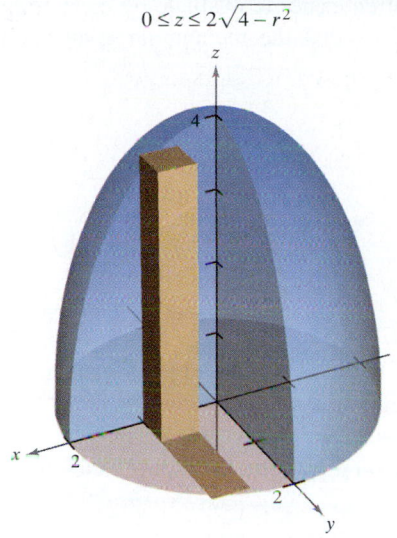

Ellipsoid: $4x^2 + 4y^2 + z^2 = 16$

The mass of the ellipsoidal solid is $16\pi k$.
Figure 13.65

EXAMPLE 2   Finding Mass by Cylindrical Coordinates

Find the mass of the ellipsoidal solid Q given by $4x^2 + 4y^2 + z^2 = 16$, lying above the xy-plane. The density at a point in the solid is proportional to the distance between the point and the xy-plane.

Solution   The density function $\rho(r, \theta, z) = kz$. The bounds on z are

$$0 \le z \le \sqrt{16 - 4x^2 - 4y^2} = \sqrt{16 - 4r^2} = 2\sqrt{4 - r^2}$$

where $0 \le r \le 2$ and $0 \le \theta \le 2\pi$, as shown in Figure 13.65. The mass of the solid is

$$m = \int_0^{2\pi} \int_0^2 \int_0^{\sqrt{16 - 4r^2}} kzr \, dz \, dr \, d\theta$$

$$= \frac{k}{2} \int_0^{2\pi} \int_0^2 z^2 r \Big]_0^{\sqrt{16 - 4r^2}} dr \, d\theta$$

$$= \frac{k}{2} \int_0^{2\pi} \int_0^2 (16r - 4r^3) \, dr \, d\theta$$

$$= \frac{k}{2} \int_0^{2\pi} \left[8r^2 - r^4 \right]_0^2 d\theta$$

$$= 8k \int_0^{2\pi} d\theta = 16\pi k.$$

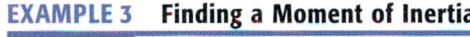

Integration in cylindrical coordinates is useful when factors involving $x^2 + y^2$ appear in the integrand, as illustrated in Example 3.

EXAMPLE 3   Finding a Moment of Inertia

Find the moment of inertia about the axis of symmetry of the solid bounded by the paraboloid $z = x^2 + y^2$ and the plane $z = 4$, as shown in Figure 13.66. The density at each point is proportional to the distance between the point and the z-axis.

Solution   Because the z-axis is the axis of symmetry, and $\rho(x, y, z) = k\sqrt{x^2 + y^2}$, it follows that

$$I_z = \int \int \int_Q k(x^2 + y^2) \sqrt{x^2 + y^2} \, dV.$$

In cylindrical coordinates, $0 \le r \le \sqrt{x^2 + y^2} = \sqrt{z}$. Therefore, you have

$$I_z = k \int_0^4 \int_0^{2\pi} \int_0^{\sqrt{z}} r^2(r)r \, dr \, d\theta \, dz$$

$$= k \int_0^4 \int_0^{2\pi} \frac{r^5}{5} \Big]_0^{\sqrt{z}} d\theta \, dz$$

$$= k \int_0^4 \int_0^{2\pi} \frac{z^{5/2}}{5} d\theta \, dz$$

$$= k \int_0^4 \frac{z^{5/2}}{5} (2\pi) dz = k \left[\left(\frac{2\pi}{5} \right) \left(\frac{2}{7} \right) z^{7/2} \right]_0^4 = \frac{512k\pi}{35}.$$

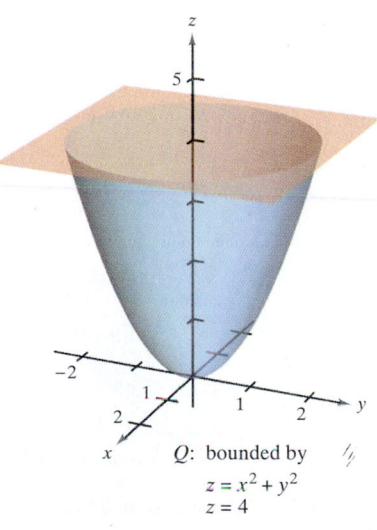

Q: bounded by
$z = x^2 + y^2$
$z = 4$

The moment of inertia about the z-axis is $512k\pi/35$.
Figure 13.66

Triple Integrals in Spherical Coordinates

Triple integrals involving spheres or cones are often easier to evaluate by converting to spherical coordinates. Recall from Section 10.7 that the rectangular conversion equations for spherical coordinates are

$$x = \rho \sin \phi \cos \theta$$
$$y = \rho \sin \phi \sin \theta$$
$$z = \rho \cos \phi.$$

In this coordinate system, the simplest region is a spherical block determined by

$$\{(\rho, \theta, \phi): \rho_1 \le \rho \le \rho_2, \quad \theta_1 \le \theta \le \theta_2, \quad \phi_1 \le \phi \le \phi_2\}$$

where $\rho_1 \ge 0$, $\theta_2 - \theta_1 \le 2\pi$, and $0 \le \phi_1 \le \phi_2 \le \pi$, as shown in Figure 13.67. If (ρ, θ, ϕ) is a point in the interior of such a block, then the volume of the block can be approximated as $\Delta V \approx \rho^2 \sin \phi \, \Delta \rho \, \Delta \phi \, \Delta \theta$ (see Exercise 40).

Using the usual process involving an inner partition, summation, and a limit, you can develop the following version of a triple integral in spherical coordinates for a continuous function f defined on the solid Q.

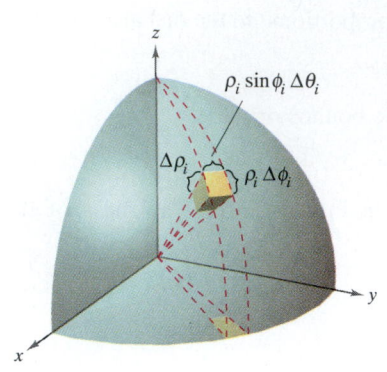

Spherical block:
$\Delta V_i \approx \rho_i^2 \sin \phi_i \, \Delta \rho_i \, \Delta \theta_i \, \Delta \phi_i$
Figure 13.67

$$\iiint\limits_{Q} f(x, y, z) \, dV = \int_{\theta_1}^{\theta_2} \int_{\phi_1}^{\phi_2} \int_{\rho_1}^{\rho_2} f(\rho \sin \phi \cos \theta, \, \rho \sin \phi \sin \theta, \, \rho \cos \phi) \rho^2 \sin \phi \, d\rho \, d\phi \, d\theta.$$

This formula can be modified for different orders of integration and generalized to include regions with variable boundaries.

Like triple integrals in cylindrical coordinates, triple integrals in spherical coordinates are evaluated with iterated integrals. As with cylindrical coordinates, you can visualize a particular order of integration by viewing the iterated integral in terms of three sweeping motions—each adding another dimension to the solid. For instance, the iterated integral

$$\int_0^{2\pi} \int_0^{\pi/4} \int_0^3 \rho^2 \sin \phi \, d\rho \, d\phi \, d\theta$$

(which is used in Example 4) is illustrated in Figure 13.68.

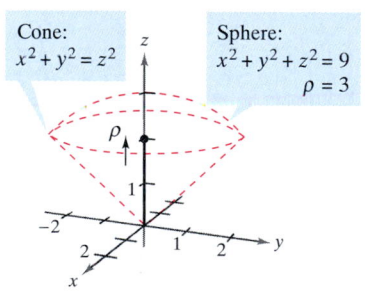

ρ varies from 0 to 3 with
ϕ and θ held constant.
Figure 13.68

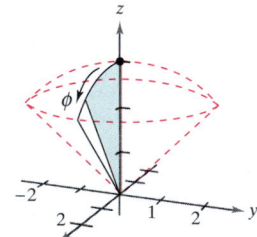

ϕ varies from 0 to $\pi/4$
with θ held constant.

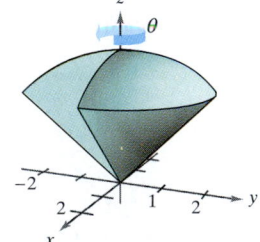

θ varies from 0 to 2π.

NOTE   The Greek letter ρ used in spherical coordinates is not related to density. Rather, it is the three-dimensional analog of the r used in polar coordinates. For problems involving spherical coordinates and a density function, we will use a different symbol to denote density.

EXAMPLE 4   Finding Volume in Spherical Coordinates

Find the volume of the solid region Q bounded below by the upper nappe of the cone $z^2 = x^2 + y^2$ and above by the sphere $x^2 + y^2 + z^2 = 9$, as shown in Figure 13.69.

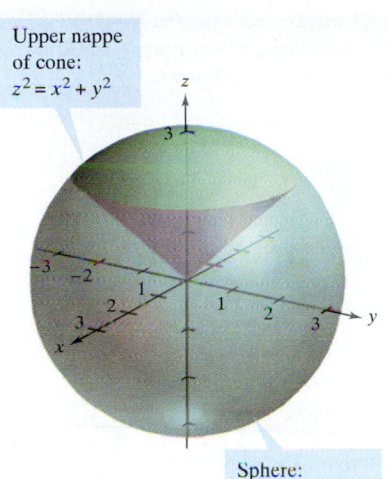

Upper nappe
of cone:
$z^2 = x^2 + y^2$

Sphere:
$x^2 + y^2 + z^2 = 9$

Figure 13.69

Solution   In spherical coordinates, the equation of the sphere is

$$\rho^2 = x^2 + y^2 + z^2 = 9 \quad \Longrightarrow \quad \rho = 3.$$

Furthermore, the sphere and cone intersect when

$$(x^2 + y^2) + z^2 = (z^2) + z^2 = 9 \quad \Longrightarrow \quad z = \frac{3}{\sqrt{2}}$$

and, because $z = \rho \cos \phi$, it follows that

$$\left(\frac{3}{\sqrt{2}}\right)\left(\frac{1}{3}\right) = \cos \phi \quad \Longrightarrow \quad \phi = \frac{\pi}{4}.$$

Consequently, you can use the integration order $d\rho\, d\phi\, d\theta$, where $0 \leq \rho \leq 3$, $0 \leq \phi \leq \pi/4$, and $0 \leq \theta \leq 2\pi$. The volume is

$$V = \iiint_Q dV = \int_0^{2\pi} \int_0^{\pi/4} \int_0^3 \rho^2 \sin \phi\, d\rho\, d\phi\, d\theta$$

$$= \int_0^{2\pi} \int_0^{\pi/4} 9 \sin \phi\, d\phi\, d\theta = 9 \int_0^{2\pi} -\cos \phi \Big]_0^{\pi/4} d\theta$$

$$= 9 \int_0^{2\pi} \left(1 - \frac{\sqrt{2}}{2}\right) d\theta = 9\pi(2 - \sqrt{2}) \approx 16.56.$$

EXAMPLE 5   Finding the Center of Mass of a Solid Region

Find the center of mass of the solid region Q of uniform density, bounded below by the upper nappe of the cone $z^2 = x^2 + y^2$ and above by the sphere $x^2 + y^2 + z^2 = 9$.

Solution   Because the density is uniform, you can consider the density at the point (x, y, z) to be k. By symmetry, the center of mass lies on the z-axis, and you need only calculate $\bar{z} = M_{xy}/m$, where $m = kV = 9k\pi(2 - \sqrt{2})$ from Example 4. Because $z = \rho \cos \phi$, it follows that

$$M_{xy} = \iiint_Q kz\, dV = k \int_0^3 \int_0^{2\pi} \int_0^{\pi/4} (\rho \cos \phi)\rho^2 \sin \phi\, d\phi\, d\theta\, d\rho$$

$$= k \int_0^3 \int_0^{2\pi} \rho^3 \frac{\sin^2 \phi}{2} \Big]_0^{\pi/4} d\theta\, d\rho$$

$$= \frac{k}{4} \int_0^3 \int_0^{2\pi} \rho^3\, d\theta\, d\rho = \frac{k\pi}{2} \int_0^3 \rho^3\, d\rho = \frac{81k\pi}{8}.$$

Therefore,

$$\bar{z} = \frac{M_{xy}}{m} = \frac{81k\pi/8}{9k\pi(2 - \sqrt{2})} = \frac{9(2 + \sqrt{2})}{16} \approx 1.92$$

and the center of mass is approximately $(0, 0, 1.92)$.

EXERCISES FOR SECTION 13.7

In Exercises 1–6, evaluate the triple integral.

1. $\int_0^4 \int_0^{\pi/2} \int_0^2 r \cos\theta \, dr \, d\theta \, dz$

2. $\int_0^{\pi/4} \int_0^2 \int_0^{2-r} rz \, dz \, dr \, d\theta$

3. $\int_0^{\pi/2} \int_0^{2\cos^2\theta} \int_0^{4-r^2} r \sin\theta \, dz \, dr \, d\theta$

4. $\int_0^{\pi/2} \int_0^{\pi} \int_0^2 e^{-\rho^3} \rho^2 \, d\rho \, d\theta \, d\phi$

5. $\int_0^{2\pi} \int_0^{\pi/4} \int_0^{\cos\phi} \rho^2 \sin\phi \, d\rho \, d\phi \, d\theta$

6. $\int_0^{\pi/4} \int_0^{\pi/4} \int_0^{\cos\theta} \rho^2 \sin\phi \cos\phi \, d\rho \, d\theta \, d\phi$

In Exercises 7 and 8, use a symbolic integration utility to evaluate the triple integral.

7. $\int_0^4 \int_0^z \int_0^{\pi/2} re^r \, d\theta \, dr \, dz$

8. $\int_0^{\pi/2} \int_0^{\pi} \int_0^{\sin\theta} (2\cos\phi)\rho^2 \, d\rho \, d\theta \, d\phi$

In Exercises 9–12, sketch the solid region whose volume is given by the integral, and evaluate the integral.

9. $\int_0^{\pi/2} \int_0^3 \int_0^{e^{-r^2}} r \, dz \, dr \, d\theta$

10. $\int_0^{2\pi} \int_0^{\sqrt{3}} \int_0^{3-r^2} r \, dz \, dr \, d\theta$

11. $\int_0^{2\pi} \int_{\pi/6}^{\pi/2} \int_0^4 \rho^2 \sin\phi \, d\rho \, d\phi \, d\theta$

12. $\int_0^{2\pi} \int_0^{\pi} \int_2^5 \rho^2 \sin\phi \, d\rho \, d\phi \, d\theta$

In Exercises 13–16, convert the integral from rectangular coordinates to both cylindrical and spherical coordinates, and evaluate the simplest integral.

13. $\int_{-2}^2 \int_{-\sqrt{4-x^2}}^{\sqrt{4-x^2}} \int_{x^2+y^2}^4 x \, dz \, dy \, dx$

14. $\int_0^2 \int_0^{\sqrt{4-x^2}} \int_0^{\sqrt{16-x^2-y^2}} \sqrt{x^2+y^2} \, dz \, dy \, dx$

15. $\int_{-a}^a \int_{-\sqrt{a^2-x^2}}^{\sqrt{a^2-x^2}} \int_a^{a+\sqrt{a^2-x^2-y^2}} x \, dz \, dy \, dx$

16. $\int_0^1 \int_0^{\sqrt{1-x^2}} \int_0^{\sqrt{1-x^2-y^2}} \sqrt{x^2+y^2+z^2} \, dz \, dy \, dx$

In Exercises 17–22, use cylindrical coordinates to find the indicated characteristic of the cone (see figure).

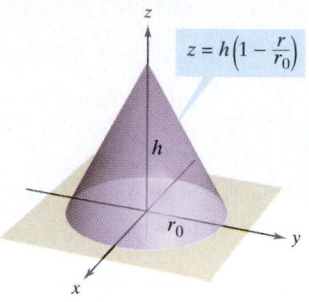

$z = h\left(1 - \dfrac{r}{r_0}\right)$

17. *Volume*   Find the volume of the cone.

18. *Centroid*   Find the centroid of the cone.

19. *Center of Mass*   Find the center of mass of the cone assuming that its density at any point is proportional to the distance between the point and the axis of the cone. Use a symbolic integration utility to evaluate the triple integral.

20. *Center of Mass*   Find the center of mass of the cone assuming that its density at any point is proportional to the distance between the point and the base. Use a symbolic integration utility to evaluate the triple integral.

21. *Moment of Inertia*   Assume that the cone has uniform density and show that the moment of inertia about the z-axis is

$I_z = \frac{3}{10}mr_0^2.$

22. *Moment of Inertia*   Assume that the density of the cone is

$\rho(x, y, z) = k(x^2 + y^2)$

and find the moment of inertia about the z-axis.

Moment of Inertia   **In Exercises 23 and 24, use cylindrical coordinates to verify the given formula for the moment of inertia of the solid of uniform density.**

23. Cylindrical shell: $I_z = \frac{1}{2}m(a^2 + b^2)$

$0 < a \le r \le b, 0 \le z \le h$

24. Right circular cylinder: $I_z = \frac{3}{2}ma^2$

$r = 2a \sin\theta, 0 \le z \le h$

Use a symbolic integration utility to evaluate the triple integral.

Volume   **In Exercises 25–28, use cylindrical coordinates to find the volume of the solid.**

25. Solid inside both $x^2 + y^2 + z^2 = a^2$

and $\left(x - \dfrac{a}{2}\right)^2 + y^2 = \left(\dfrac{a}{2}\right)^2$

26. Solid inside $x^2 + y^2 + z^2 = 16$ and outside $z = \sqrt{x^2 + y^2}$

27. Solid bounded by the graphs of the sphere $r^2 + z^2 = a^2$ and the cylinder $r = a \cos \theta$

28. Solid inside the sphere $x^2 + y^2 + z^2 = 4$ and above the cone $z^2 = x^2 + y^2$

Volume   In Exercises 29 and 30, use spherical coordinates to find the volume of the solid.

29. The torus given by $\rho = 4 \sin \phi$. (Use a symbolic integration utility to evaluate the triple integral.)

30. The solid between the spheres $x^2 + y^2 + z^2 = a^2$ and $x^2 + y^2 + z^2 = b^2$, $b > a$, and inside the cone $z^2 = x^2 + y^2$

Mass   In Exercises 31 and 32, use spherical coordinates to find the mass of the sphere $x^2 + y^2 + z^2 = a^2$ with the indicated density.

31. The density at any point is proportional to the distance between the point and the origin.

32. The density at any point is proportional to the distance of the point from the z-axis.

Center of Mass   In Exercises 33 and 34, use spherical coordinates to find the center of mass of the solid of uniform density.

33. Hemispherical solid of radius r

34. Solid lying between two concentric hemispheres of radii r and R, where $r < R$

Moment of Inertia   In Exercises 35 and 36, use spherical coordinates to find the moment of inertia about the z-axis of the solid of uniform density.

35. Solid bounded by the hemisphere $\rho = \cos \phi$, $\pi/4 \le \phi \le \pi/2$, and the cone $\phi = \pi/4$

36. Solid lying between two concentric hemispheres of radii r and R, where $r < R$

37. *Think About It*   Describe the surface whose equation is a coordinate equal to a constant for each of the coordinates in (a) the cylindrical coordinate system and (b) the spherical coordinate system.

38. *Think About It*   When evaluating a triple integral with constant limits of integration in the cylindrical coordinate system, you are integrating over a part of what solid? What is the solid when you are in spherical coordinates?

39. Find the "volume" of the "four-dimensional sphere"

$$x^2 + y^2 + z^2 + w^2 = a^2,$$

by evaluating

$$16 \int_0^a \int_0^{\sqrt{a^2-x^2}} \int_0^{\sqrt{a^2-x^2-y^2}} \int_0^{\sqrt{a^2-x^2-y^2-z^2}} dw\, dz\, dy\, dx.$$

40. Show that the volume of a spherical block can be approximated by

$$\Delta V \approx \rho^2 \sin \phi\, \Delta \rho\, \Delta \phi\, \Delta \theta.$$

SECTION PROJECT

Wrinkled and Bumpy Spheres   In parts (a) and (b), find the volume of the wrinkled sphere or bumpy sphere. These solids are used as models for tumors.

(a) Wrinkled sphere

$\rho = 1 + 0.2 \sin 8\theta \sin\phi$

$0 \le \theta \le 2\pi, 0 \le \phi \le \pi$

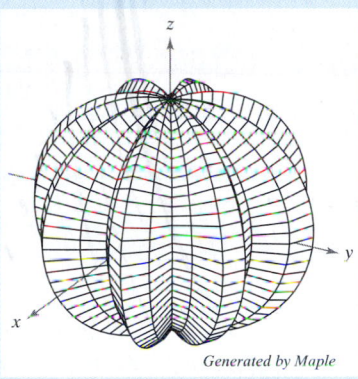

Generated by Maple

(b) Bumpy sphere

$\rho = 1 + 0.2 \sin 8\theta \sin 4\phi$

$0 \le \theta \le 2\pi, 0 \le \phi \le \pi$

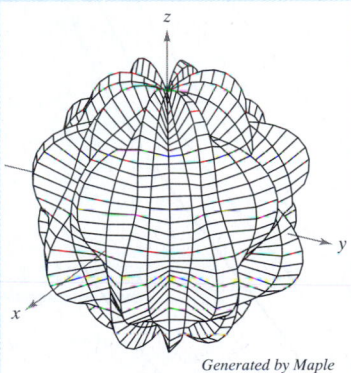

Generated by Maple

FOR FURTHER INFORMATION   For more information, see page 914, or see the article "Heat Therapy for Tumors" by Leah Edelstine-Keshet in the Summer 1991 issue of *The UMAP Journal.*

Jacobians • Change of Variables for Double Integrals

CARL GUSTAV JACOBI (1804–1851)

The Jacobian is named after the German mathematician Carl Gustav Jacobi. Jacobi is known for his work in many areas of mathematics, but his interest in integration stemmed from the problem of finding the circumference of an ellipse.

Jacobians

For the single integral

$$\int_a^b f(x)\, dx,$$

you can change variables by letting $x = g(u)$, so that $dx = g'(u)\, du$, and obtain

$$\int_a^b f(x)\, dx = \int_c^d f(g(u))g'(u)\, du$$

where $a = g(c)$ and $b = g(d)$. Note that the change-of-variables process introduces an additional factor $g'(u)$ into the integrand. This also occurs in the case of double integrals

$$\iint_R f(x, y)\, dA = \iint_S f(g(u, v), h(u, v)) \underbrace{\left| \frac{\partial x}{\partial u} \frac{\partial y}{\partial v} - \frac{\partial y}{\partial u} \frac{\partial x}{\partial v} \right|}_{\text{Jacobian}} du\, dv$$

where the change of variables $x = g(u, v)$ and $y = h(u, v)$ introduces a factor called the **Jacobian** of x and y with respect to u and v. In defining the Jacobian, it is convenient to use the following determinant notation.

Definition of the Jacobian

If $x = g(u, v)$ and $y = h(u, v)$, then the **Jacobian** of x and y with respect to u and v, denoted by $\partial(x, y)/\partial(u, v)$, is

$$\frac{\partial(x, y)}{\partial(u, v)} = \begin{vmatrix} \dfrac{\partial x}{\partial u} & \dfrac{\partial x}{\partial v} \\[2mm] \dfrac{\partial y}{\partial u} & \dfrac{\partial y}{\partial v} \end{vmatrix} = \frac{\partial x}{\partial u} \frac{\partial y}{\partial v} - \frac{\partial y}{\partial u} \frac{\partial x}{\partial v}.$$

EXAMPLE 1 The Jacobian for Rectangular-to-Polar Conversion

Find the Jacobian for the change of variables defined by

$$x = r \cos \theta \qquad \text{and} \qquad y = r \sin \theta.$$

Solution From the definition of a Jacobian, you obtain

$$\frac{\partial(x, y)}{\partial(r, \theta)} = \begin{vmatrix} \dfrac{\partial x}{\partial r} & \dfrac{\partial x}{\partial \theta} \\[2mm] \dfrac{\partial y}{\partial r} & \dfrac{\partial y}{\partial \theta} \end{vmatrix}$$

$$= \begin{vmatrix} \cos \theta & -r \sin \theta \\ \sin \theta & r \cos \theta \end{vmatrix}$$

$$= r \cos^2 \theta + r \sin^2 \theta$$

$$= r.$$

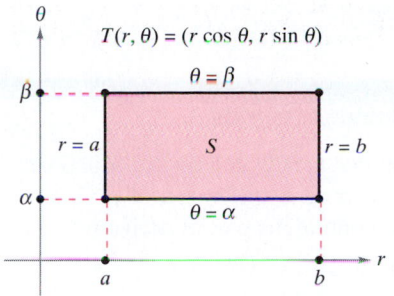

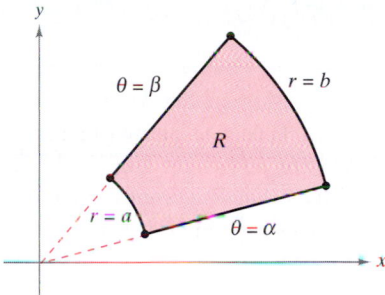

S is the region in the $r\theta$-plane that corresponds to R in the xy-plane.
Figure 13.70

Example 1 points out that the change of variables from rectangular to polar coordinates for a double integral can be written as

$$\iint_R f(x, y)\, dA = \iint_S f(r\cos\theta, r\sin\theta)r\, dr\, d\theta, \qquad r > 0$$

$$= \iint_S f(r\cos\theta, r\sin\theta)\left|\frac{\partial(x, y)}{\partial(r, \theta)}\right| dr\, d\theta$$

where S is the region in the $r\theta$-plane that corresponds to the region R in the xy-plane, as shown in Figure 13.70. This formula is similar to that found on page 935.

In general, a change of variables is given by a one-to-one **transformation** T from a region S in the uv-plane to a region R in the xy-plane, to be given by

$$T(u, v) = (x, y) = (g(u, v), h(u, v))$$

where g and h have continuous first partial derivatives in the region S. Note that the point (u, v) lies in S and the point (x, y) lies in R. In most cases, you are hunting for a transformation for which the region S is simpler than the region R.

EXAMPLE 2   Finding a Change of Variables to Simplify a Region

Let R be the region bounded by the lines

$$x - 2y = 0, \qquad x - 2y = -4, \qquad x + y = 4, \qquad \text{and} \qquad x + y = 1$$

as shown in Figure 13.71. Find a transformation T from a region S to R such that S is a rectangular region (with sides parallel to the u- or v-axis).

Solution   To begin, let $u = x + y$ and $v = x - 2y$. Solving this system of equations for x and y produces $T(u, v) = (x, y)$, where

$$x = \frac{1}{3}(2u + v) \qquad \text{and} \qquad y = \frac{1}{3}(u - v).$$

The four boundaries for R in the xy-plane give rise to the following bounds for S in the uv-plane.

Bounds in the xy-Plane		Bounds in the uv-Plane
$x + y = 1$	⇨	$u = 1$
$x + y = 4$	⇨	$u = 4$
$x - 2y = 0$	⇨	$v = 0$
$x - 2y = -4$	⇨	$v = -4$

The region S is shown in Figure 13.72. Note that the transformation T maps the vertices of the region S onto the vertices of the region R. For instance,

$$T(1, 0) = \left(\tfrac{1}{3}[2(1) + 0], \tfrac{1}{3}[1 - 0]\right)$$
$$= \left(\tfrac{2}{3}, \tfrac{1}{3}\right)$$
$$T(4, 0) = \left(\tfrac{1}{3}[2(4) + 0], \tfrac{1}{3}[4 - 0]\right)$$
$$= \left(\tfrac{8}{3}, \tfrac{4}{3}\right)$$
$$T(4, -4) = \left(\tfrac{1}{3}[2(4) - 4], \tfrac{1}{3}[4 - (-4)]\right)$$
$$= \left(\tfrac{4}{3}, \tfrac{8}{3}\right)$$
$$T(1, -4) = \left(\tfrac{1}{3}[2(1) - 4], \tfrac{1}{3}[1 - (-4)]\right)$$
$$= \left(-\tfrac{2}{3}, \tfrac{5}{3}\right).$$

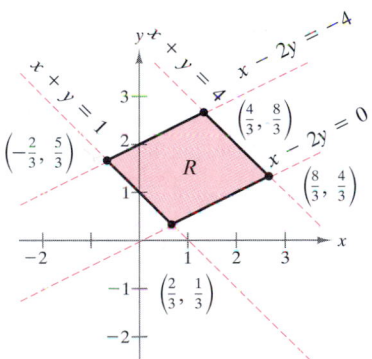

Region R in the xy-plane
Figure 13.71

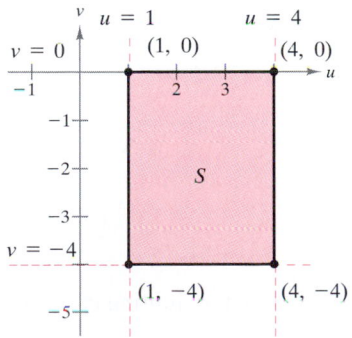

Region S in the uv-plane
Figure 13.72

Change of Variables for Double Integrals

> **THEOREM 13.5** **Change of Variables for Double Integrals**
>
> Let R and S be regions in the xy- and uv-planes that are related by the equations $x = g(u, v)$ and $y = h(u, v)$ such that each point in R is the image of a unique point in S. If f is continuous on R, g and h have continuous partial derivatives on S, and $\partial(x, y)/\partial(u, v)$ is nonzero on S, then
>
> $$\int_R \int f(x, y)\, dx\, dy = \int_S \int f(g(u, v), h(u, v)) \left| \frac{\partial(x, y)}{\partial(u, v)} \right| du\, dv.$$

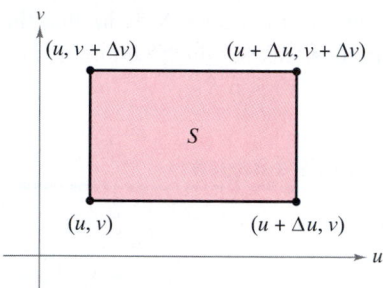

Area of $S = \Delta u \Delta v$
$\Delta u > 0,\ \Delta v > 0$
Figure 13.73

Proof Consider the case in which S is a rectangular region in the uv-plane with vertices (u, v), $(u + \Delta u, v)$, $(u + \Delta u, v + \Delta v)$, and $(u, v + \Delta v)$, as shown in Figure 13.73. The images of these vertices in the xy-plane are shown in Figure 13.74. If Δu and Δv are small, the continuity of g and h implies that R is approximately a parallelogram determined by the vectors $\overrightarrow{MN}$ and $\overrightarrow{MQ}$. Thus, the area of R is

$$\Delta A \approx \|\overrightarrow{MN} \times \overrightarrow{MQ}\|.$$

Moreover, for small Δu and Δv, the partial derivatives of g and h with respect to u can be approximated by

$$g_u(u, v) \approx \frac{g(u + \Delta u, v) - g(u, v)}{\Delta u}$$

and

$$h_u(u, v) \approx \frac{h(u + \Delta u, v) - h(u, v)}{\Delta u}.$$

Consequently,

$$\begin{aligned}
\overrightarrow{MN} &= [g(u + \Delta u, v) - g(u, v)]\mathbf{i} + [h(u + \Delta u, v) - h(u, v)]\mathbf{j} \\
&\approx [g_u(u, v)\Delta u]\mathbf{i} + [h_u(u, v)\Delta u]\mathbf{j} \\
&= \frac{\partial x}{\partial u} \Delta u\, \mathbf{i} + \frac{\partial y}{\partial u} \Delta u\, \mathbf{j}.
\end{aligned}$$

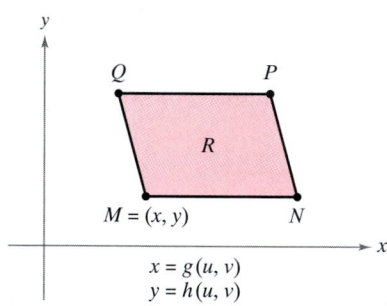

The vertices in the xy-plane are
$M(g(u, v), h(u, v))$,
$N(g(u + \Delta u, v), h(u + \Delta u, v))$,
$P(g(u + \Delta u, v + \Delta v), h(u + \Delta u, v + \Delta v))$,
and $Q(g(u, v + \Delta v), h(u, v + \Delta v))$.
Figure 13.74

Similarly, you can approximate $\overrightarrow{MQ}$ by $\dfrac{\partial x}{\partial v} \Delta v\, \mathbf{i} + \dfrac{\partial y}{\partial v} \Delta v\, \mathbf{j}$, which implies that

$$\overrightarrow{MN} \times \overrightarrow{MQ} \approx \begin{vmatrix} \mathbf{i} & \mathbf{j} & \mathbf{k} \\ \dfrac{\partial x}{\partial u} \Delta u & \dfrac{\partial y}{\partial u} \Delta u & 0 \\ \dfrac{\partial x}{\partial v} \Delta v & \dfrac{\partial y}{\partial v} \Delta v & 0 \end{vmatrix} = \begin{vmatrix} \dfrac{\partial x}{\partial u} & \dfrac{\partial y}{\partial u} \\ \dfrac{\partial x}{\partial v} & \dfrac{\partial y}{\partial v} \end{vmatrix} \Delta u\, \Delta v\, \mathbf{k}.$$

It follows that, in Jacobian notation,

$$\Delta A \approx \|\overrightarrow{MN} \times \overrightarrow{MQ}\| \approx \left| \frac{\partial(x, y)}{\partial(u, v)} \right| \Delta u \Delta v.$$

Because this approximation improves as Δu and Δv approach 0, the limiting case can be written as

$$dA \approx \|\overrightarrow{MN} \times \overrightarrow{MQ}\| \approx \left| \frac{\partial(x, y)}{\partial(u, v)} \right| du\, dv.$$

The next two examples show how a change of variables can simplify the integration process. The simplification can occur in various ways. You can make a change of variables to simplify either the *region R* or the *integrand f(x, y)*, or both.

EXAMPLE 3   Using a Change of Variables to Simplify a Region

Let R be the region bounded by the lines

$$x - 2y = 0, \quad x - 2y = -4, \quad x + y = 4, \quad \text{and} \quad x + y = 1$$

as shown in Figure 13.75. Evaluate the double integral

$$\iint_R 3xy \, dA.$$

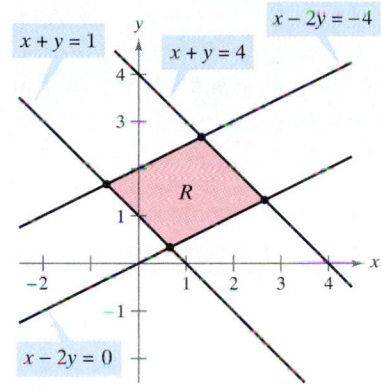

Figure 13.75

Solution   From Example 2, you can use the following change of variables.

$$x = \frac{1}{3}(2u + v) \quad \text{and} \quad y = \frac{1}{3}(u - v)$$

The partial derivatives of x and y are

$$\frac{\partial x}{\partial u} = \frac{2}{3}, \quad \frac{\partial x}{\partial v} = \frac{1}{3}, \quad \frac{\partial y}{\partial u} = \frac{1}{3}, \quad \text{and} \quad \frac{\partial y}{\partial v} = -\frac{1}{3}$$

which implies that the Jacobian is

$$
\begin{aligned}
\frac{\partial(x, y)}{\partial(u, v)} &=
\begin{vmatrix}
\dfrac{\partial x}{\partial u} & \dfrac{\partial x}{\partial v} \\[2mm]
\dfrac{\partial y}{\partial u} & \dfrac{\partial y}{\partial v}
\end{vmatrix} \\[3mm]
&=
\begin{vmatrix}
\dfrac{2}{3} & \dfrac{1}{3} \\[2mm]
\dfrac{1}{3} & -\dfrac{1}{3}
\end{vmatrix} \\[3mm]
&= -\frac{2}{9} - \frac{1}{9} \\[2mm]
&= -\frac{1}{3}.
\end{aligned}
$$

Therefore, by Theorem 13.5, you obtain

$$
\begin{aligned}
\iint_R 3xy \, dA &= \iint_S 3\left[\frac{1}{3}(2u + v)\frac{1}{3}(u - v)\right]\left|\frac{\partial(x, y)}{\partial(u, v)}\right| dv \, du \\[2mm]
&= \int_1^4 \int_{-4}^0 \frac{1}{9}(2u^2 - uv - v^2)\, dv \, du \\[2mm]
&= \frac{1}{9}\int_1^4 \left[2u^2v - \frac{uv^2}{2} - \frac{v^3}{3}\right]_{-4}^0 du \\[2mm]
&= \frac{1}{9}\int_1^4 \left(8u^2 + 8u - \frac{64}{3}\right) du \\[2mm]
&= \frac{1}{9}\left[\frac{8u^3}{3} + 4u^2 - \frac{64}{3}u\right]_1^4 \\[2mm]
&= \frac{164}{9}.
\end{aligned}
$$

EXAMPLE 4 Using a Change of Variables to Simplify an Integrand

Let R be the region bounded by the square with vertices $(0, 1)$, $(1, 2)$ $(2, 1)$, and $(1, 0)$. Evaluate the integral

$$\iint_R (x + y)^2 \sin^2(x - y) \, dA.$$

Solution Note that the sides of R lie on the lines $x + y = 1$, $x - y = 1$, $x + y = 3$, and $x - y = -1$, as shown in Figure 13.76. Letting $u = x + y$ and $v = x - y$, you can determine the bounds for region S in the uv-plane to be

$$1 \le u \le 3 \qquad \text{and} \qquad -1 \le v \le 1$$

as shown in Figure 13.77. Solving for x and y in terms of u and v produces

$$x = \frac{1}{2}(u + v) \qquad \text{and} \qquad y = \frac{1}{2}(u - v).$$

The partial derivatives of x and y are

$$\frac{\partial x}{\partial u} = \frac{1}{2}, \quad \frac{\partial x}{\partial v} = \frac{1}{2}, \quad \frac{\partial y}{\partial u} = \frac{1}{2}, \quad \text{and} \quad \frac{\partial y}{\partial v} = -\frac{1}{2}$$

which implies that the Jacobian is

$$\frac{\partial(x, y)}{\partial(u, v)} = \begin{vmatrix} \dfrac{\partial x}{\partial u} & \dfrac{\partial x}{\partial v} \\[2mm] \dfrac{\partial y}{\partial u} & \dfrac{\partial y}{\partial v} \end{vmatrix} = \begin{vmatrix} \dfrac{1}{2} & \dfrac{1}{2} \\[2mm] \dfrac{1}{2} & -\dfrac{1}{2} \end{vmatrix} = -\frac{1}{4} - \frac{1}{4} = -\frac{1}{2}.$$

By Theorem 13.5, it follows that

$$\iint_R (x + y)^2 \sin^2(x - y) \, dA = \int_{-1}^{1} \int_{1}^{3} u^2 \sin^2 v \left(\frac{1}{2}\right) du \, dv$$

$$= \frac{1}{2} \int_{-1}^{1} (\sin^2 v) \frac{u^3}{3} \Bigg]_{1}^{3} dv$$

$$= \frac{13}{3} \int_{-1}^{1} \sin^2 v \, dv$$

$$= \frac{13}{6} \int_{-1}^{1} (1 - \cos 2v) \, dv$$

$$= \frac{13}{6} \left[v - \frac{1}{2} \sin 2v \right]_{-1}^{1}$$

$$= \frac{13}{6} \left[2 - \frac{1}{2} \sin 2 + \frac{1}{2} \sin(-2) \right]$$

$$= \frac{13}{6} (2 - \sin 2)$$

$$\approx 2.363.$$

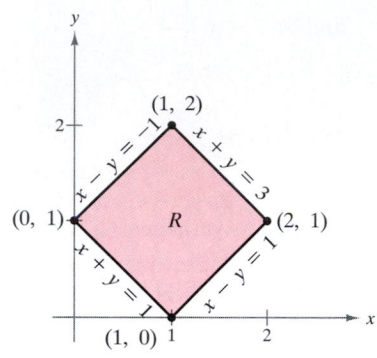

Region R in the xy-plane
Figure 13.76

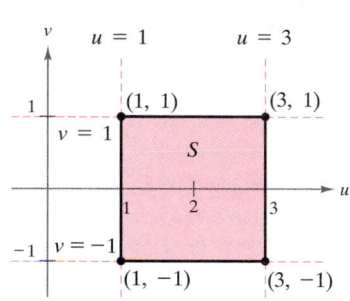

Region S in the uv-plane
Figure 13.77

STUDY TIP In each of the change-of-variables examples in this section, the region S has been a rectangle with sides parallel to the u- or v-axis. Occasionally, a change of variables can be used for other types of regions. For instance, letting $T(u, v) = \left(x, \frac{1}{2}y\right)$ changes the circular region $u^2 + v^2 = 1$ to the elliptical region $x^2 + (y^2/4) = 1$.

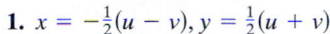
EXERCISES FOR SECTION 13.8

In Exercises 1–8, find the Jacobian $\partial(x, y)/\partial(u, v)$ for the indicated change of variables.

1. $x = -\frac{1}{2}(u - v), y = \frac{1}{2}(u + v)$

2. $x = au + bv, y = cu + dv$

3. $x = u - v^2, y = u + v$

4. $x = u - uv, y = uv$

5. $x = u \cos \theta - v \sin \theta, y = u \sin \theta + v \cos \theta$

6. $x = u + a, y = v + a$

7. $x = e^u \sin v, y = e^u \cos v$

8. $x = \dfrac{u}{v}, y = u + v$

In Exercises 9 and 10, sketch the image S in the uv-plane of the region R in the xy-plane using the given transformations.

9. $x = 3u + 2v$
$y = 3v$

10. $x = 4u + v$
$y = u + 2v$

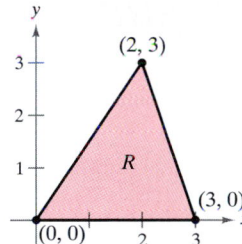

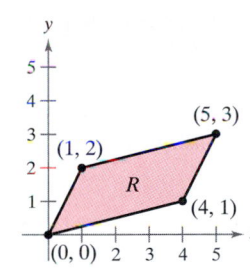

In Exercises 11–16, use the indicated change of variables to evaluate the double integral.

11. $\displaystyle\int_R\!\!\int 4(x^2 + y^2)\, dx\, dy$
$x = \frac{1}{2}(u + v), y = \frac{1}{2}(u - v)$

12. $\displaystyle\int_R\!\!\int 48xy\, dx\, dy$
$x = \frac{1}{2}(u + v), y = \frac{1}{2}(u - v)$

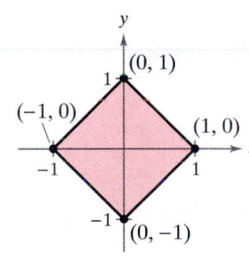

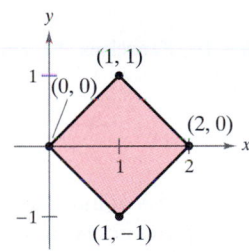

13. $\displaystyle\int_R\!\!\int y(x - y)\, dx\, dy$
$x = u + v, y = u$

14. $\displaystyle\int_R\!\!\int 4(x + y)e^{x-y}\, dy\, dx$
$x = \frac{1}{2}(u + v), y = \frac{1}{2}(u - v)$

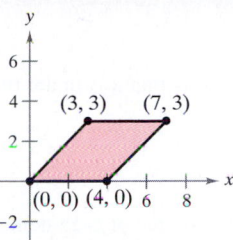

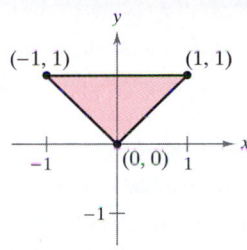

Figure for 13   **Figure for 14**

15. $\displaystyle\int_R\!\!\int e^{-xy/2}\, dy\, dx$

$x = \sqrt{\dfrac{v}{u}}, \; y = \sqrt{uv}$

R: first quadrant region lying between the graphs of $y = \dfrac{1}{4}x$, $y = 2x, y = \dfrac{1}{x}, y = \dfrac{4}{x}$

16. $\displaystyle\int_R\!\!\int y \sin xy\, dy\, dx$

$x = \dfrac{u}{v}, y = v$

R: region lying between the graphs of $xy = 1, xy = 4, y = 1$, $y = 4$

In Exercises 17–22, use a change of variables to find the volume of the solid region lying below the surface $z = f(x, y)$ and above the plane region R.

17. $f(x, y) = (x + y)e^{x-y}$

R: region bounded by the square with vertices $(4, 0)$, $(6, 2)$, $(4, 4)$, $(2, 2)$

18. $f(x, y) = (x + y)^2 \sin^2(x - y)$

R: region bounded by the square with vertices $(\pi, 0)$, $(3\pi/2, \pi/2)$, (π, π), $(\pi/2, \pi/2)$

19. $f(x, y) = \sqrt{(x - y)(x + 4y)}$

R: region bounded by the parallelogram with vertices $(0, 0)$, $(1, 1)$, $(5, 0)$, $(4, -1)$

20. $f(x, y) = (3x + 2y)^2 \sqrt{2y - x}$

R: region bounded by the parallelogram with vertices $(0, 0)$, $(-2, 3)$, $(2, 5)$, $(4, 2)$

21. $f(x, y) = \sqrt{x + y}$

R: region bounded by the triangle with vertices $(0, 0)$, $(a, 0)$, $(0, a)$, where $a > 0$

22. $f(x, y) = \dfrac{xy}{1 + x^2 y^2}$

R: region bounded by the graphs of $xy = 1, xy = 4, x = 1$, $x = 4$ (*Hint:* Let $x = u, y = v/u$.)

23. Consider the region R in the xy-plane bounded by the ellipse

$$\frac{x^2}{a^2} + \frac{y^2}{b^2} = 1$$

and the transformations $x = au$ and $y = bv$.

(a) Sketch the graph of the region R and its image S under the indicated transformation.

(b) Find $\partial(x, y)/\partial(u, v)$.

(c) Find the area of the ellipse.

24. Use the result in Exercise 23 to find the volume of each dome-shaped solid lying below the surface $z = f(x, y)$ and above the elliptical region R. (*Hint:* After making the change of variables indicated by the results in Exercise 23, make a second change of variables to polar coordinates.)

(a) $f(x, y) = 16 - x^2 - y^2, \quad R: \frac{x^2}{16} + \frac{y^2}{9} \leq 1$

(b) $f(x, y) = A \cos\left(\frac{\pi}{2}\sqrt{\frac{x^2}{a^2} + \frac{y^2}{b^2}}\right), \quad R: \frac{x^2}{a^2} + \frac{y^2}{b^2} \leq 1$

In Exercises 25–28, find the Jacobian $\partial(x, y, z)/\partial(u, v, w)$ for the indicated change of variables. If $x = f(u, v, w)$, $y = g(u, v, w)$, and $z = h(u, v, w)$, then the Jacobian of x, y, and z with respect to u, v, and w is

$$\frac{\partial(x, y, z)}{\partial(u, v, w)} = \begin{vmatrix} \dfrac{\partial x}{\partial u} & \dfrac{\partial x}{\partial v} & \dfrac{\partial x}{\partial w} \\[8pt] \dfrac{\partial y}{\partial u} & \dfrac{\partial y}{\partial v} & \dfrac{\partial y}{\partial w} \\[8pt] \dfrac{\partial z}{\partial u} & \dfrac{\partial z}{\partial v} & \dfrac{\partial z}{\partial w} \end{vmatrix}.$$

25. $x = u(1 - v), y = uv(1 - w), z = uvw$

26. $x = 4u - v, y = 4v - w, z = u + w$

27. Spherical Coordinates

$x = \rho \sin \phi \cos \theta, y = \rho \sin \phi \sin \theta, z = \rho \cos \phi$

28. Cylindrical Coordinates

$x = r \cos \theta, y = r \sin \theta, z = z$

REVIEW EXERCISES FOR CHAPTER 13

In Exercises 1 and 2, evaluate the integral.

1. $\displaystyle\int_1^{x^2} x \ln y \, dy$

2. $\displaystyle\int_y^{2y} (x^2 + y^2) \, dx$

In Exercises 3–12, evaluate the multiple integral. Change the coordinate system when convenient.

3. $\displaystyle\int_0^1 \int_0^{1+x} (3x + 2y) \, dy \, dx$

4. $\displaystyle\int_0^2 \int_{x^2}^{2x} (x^2 + 2y) \, dy \, dx$

5. $\displaystyle\int_0^3 \int_0^{\sqrt{9-x^2}} 4x \, dy \, dx$

6. $\displaystyle\int_0^{\sqrt{3}} \int_{2-\sqrt{4-y^2}}^{2+\sqrt{4-y^2}} dx \, dy$

7. $\displaystyle\int_0^h \int_0^x \sqrt{x^2 + y^2} \, dy \, dx$

8. $\displaystyle\int_0^4 \int_0^{\sqrt{16-y^2}} (x^2 + y^2) \, dx \, dy$

9. $\displaystyle\int_{-3}^3 \int_{-\sqrt{9-x^2}}^{\sqrt{9-x^2}} \int_{x^2+y^2}^9 \sqrt{x^2 + y^2} \, dz \, dy \, dx$

10. $\displaystyle\int_{-2}^2 \int_{-\sqrt{4-x^2}}^{\sqrt{4-x^2}} \int_0^{(x^2+y^2)/2} (x^2 + y^2) \, dz \, dy \, dx$

11. $\displaystyle\int_0^a \int_0^b \int_0^c (x^2 + y^2 + z^2) \, dx \, dy \, dz$

12. $\displaystyle\int_0^5 \int_0^{\sqrt{25-x^2}} \int_0^{\sqrt{25-x^2-y^2}} \frac{1}{1 + x^2 + y^2 + z^2} \, dz \, dy \, dx$

In Exercises 13–16, use a symbolic integration utility to evaluate the multiple integral.

13. $\displaystyle\int_{-2}^4 \int_{y^2/4}^{(4+y)/2} (x - y) \, dx \, dy$

14. $\displaystyle\int_{-2}^2 \int_0^{4-y^2} (8x - 2y^2) \, dx \, dy$

15. $\displaystyle\int_{-1}^1 \int_{-\sqrt{1-x^2}}^{\sqrt{1-x^2}} \int_{-\sqrt{1-x^2-y^2}}^{\sqrt{1-x^2-y^2}} (x^2 + y^2) \, dz \, dy \, dx$

16. $\displaystyle\int_0^2 \int_0^{\sqrt{4-x^2}} \int_0^{\sqrt{4-x^2-y^2}} xyz \, dz \, dy \, dx$

Area **In Exercises 17–24, write the limits for the double integral**

$$\int_R \int f(x, y) \, dA$$

for both orders of integration. Compute the area of R by letting $f(x, y) = 1$ and integrating.

17. Triangle: vertices $(0, 0), (3, 0), (0, 1)$

18. Triangle: vertices $(0, 0), (3, 0), (2, 2)$

19. The larger area between the graphs of $x^2 + y^2 = 25$ and $x = 3$

20. Region bounded by the graphs of $y = 6x - x^2$ and $y = x^2 - 2x$

21. Region enclosed by the graph of $y^2 = x^2 - x^4$

22. Region bounded by the graphs of $x = y^2 + 1$, $x = 0$, $y = 0$, and $y = 2$

23. Region bounded by the graphs of $x = y + 3$ and $x = y^2 + 1$

24. Region bounded by the graphs of $x = -y$ and $x = 2y - y^2$

Think About It In Exercises 25 and 26, give a geometric argument for the given equality. Verify the equality analytically.

25. $\displaystyle\int_0^1 \int_{2y}^{2\sqrt{2-y^2}} (x + y)\,dx\,dy = \int_0^2 \int_0^{x/2} (x + y)\,dy\,dx +$
$$\int_2^{2\sqrt{2}} \int_0^{\sqrt{8-x^2}/2} (x + y)\,dy\,dx$$

26. $\displaystyle\int_0^2 \int_{3y/2}^{5-y} e^{x+y}\,dx\,dy = \int_0^3 \int_0^{2x/3} e^{x+y}\,dy\,dx +$
$$\int_3^5 \int_0^{5-x} e^{x+y}\,dy\,dx$$

Volume In Exercises 27–32, use a multiple integral and a convenient coordinate system to find the volume of the solid.

27. Solid bounded by the graphs of $z = x^2 - y + 4$, $z = 0$, $y = 0$, $x = 0$, and $x = 4$

28. Solid bounded by the graphs of $z = x + y$, $z = 0$, $x = 0$, $x = 3$, and $y = x$

29. Solid bounded by the graphs of $z = 0$ and $z = h$, outside the cylinder
$$x^2 + y^2 = 1$$
and inside the hyperboloid
$$x^2 + y^2 - z^2 = 1$$

30. Solid that remains after drilling a hole of radius b through the center of a sphere of radius R ($b < R$)

31. Solid inside the graphs of $r = 2\cos\theta$ and $r^2 + z^2 = 4$

32. Solid inside the graphs of $r^2 + z = 16$ and $r = 2\sin\theta$

Think About It In Exercises 33 and 34, determine which value best approximates the volume of the solid between the xy-plane and the function over the region. (Make your selection on the basis of a sketch of the solid and *not* by performing any calculations.)

33. $f(x, y) = x + y$
R: triangle with vertices $(0, 0)$, $(3, 0)$, $(3, 3)$
(a) $\frac{9}{2}$ (b) 5 (c) 13 (d) 100 (e) -100

34. $f(x, y) = 10x^2y^2$
R: circle bounded by $x^2 + y^2 = 1$
(a) π (b) -15 (c) $\frac{2}{3}$ (d) 3 (e) 15

Probability In Exercises 35 and 36, find k such that the function is a joint density function and find the required probability, where
$$P(a \le x \le b, c \le y \le d) = \int_c^d \int_a^b f(x, y)\,dx\,dy.$$

35. $f(x, y) = \begin{cases} kxye^{-(x+y)}, & x \ge 0, y \ge 0 \\ 0, & \text{elsewhere} \end{cases}$
$P(0 \le x \le 1, 0 \le y \le 1)$

36. $f(x, y) = \begin{cases} kxy, & 0 \le x \le 1, 0 \le y \le x \\ 0, & \text{elsewhere} \end{cases}$
$P(0 \le x \le 0.5, 0 \le y \le 0.25)$

37. Consider the region R in the xy-plane bounded by the graph of the equation
$$(x^2 + y^2)^2 = 9(x^2 - y^2).$$
(a) Convert the equation to polar coordinates. Use a graphing utility to graph the equation.
(b) Use a double integral to find the area of the region R.
(c) Use a symbolic integration utility to determine the volume of the solid over the region R and beneath the hemisphere $z = \sqrt{9 - x^2 - y^2}$.

38. Combine the sum of the two double integrals into a single double integral by converting to polar coordinates. Evaluate the resulting double integral.
$$\int_0^{8/\sqrt{13}} \int_0^{3x/2} xy\,dy\,dx + \int_{8/\sqrt{13}}^4 \int_0^{\sqrt{16-x^2}} xy\,dy\,dx$$

Mass and Center of Mass In Exercises 39 and 40, find the mass and center of mass of the lamina of indicated density bounded by the graphs of the equations. Use a symbolic integration utility to evaluate the multiple integrals.

39. $y = 2x$, $y = 2x^3$, first quadrant
(a) $\rho = kxy$ (b) $\rho = k(x^2 + y^2)$

40. $y = \frac{h}{2}\left(2 - \frac{x}{L} - \frac{x^2}{L^2}\right)$, $\rho = k$, first quadrant

Surface Area In Exercises 41 and 42, find the area of the surface on the graph of the function $f(x, y)$ over the region R.

41. $f(x, y) = 16 - x^2 - y^2$
$R = \{(x, y): x^2 + y^2 \le 16\}$

42. $f(x, y) = 16 - x - y^2$
$R = \{(x, y): 0 \le x \le 2, 0 \le y \le x\}$
Use a symbolic integration utility to evaluate the integral.

43. *Surface Area*   Find the area of the surface of the cylinder $f(x, y) = 9 - y^2$ that lies above the triangle bounded by the graphs of the equations $y = x$, $y = -x$, and $y = 3$.

 44. *Surface Area*   The roof over the stage of an open air theater at a theme park is modeled by

$$f(x, y) = 25\left[1 + e^{-(x^2+y^2)/1000} \cos^2\left(\frac{x^2 + y^2}{1000}\right)\right]$$

where the stage is a semicircle bounded by the graphs of $y = \sqrt{50^2 - x^2}$ and $y = 0$.

(a) Use a computer algebra system to graph the surface.

(b) Use a symbolic integration utility to approximate the number of square feet of roofing required to cover the surface.

Center of Mass   **In Exercises 45–48, find the center of mass of the solid of uniform density bounded by the graphs of the equations.**

45. Solid inside the hemisphere $\rho = \cos \phi$, $\pi/4 \le \phi \le \pi/2$, and outside the cone $\phi = \pi/4$

46. Wedge: $x^2 + y^2 = a^2$, $z = cy$ $(c > 0)$, $y \ge 0$, $z \ge 0$

47. $x^2 + y^2 + z^2 = a^2$, first octant

48. $x^2 + y^2 + z^2 = 25$, $z = 4$ (the larger solid)

Moment of Inertia   **In Exercises 49 and 50, find the moment of inertia I_z of the solid of indicated density.**

49. The solid of uniform density inside the paraboloid

$$z = 16 - x^2 - y^2$$

and outside the cylinder $x^2 + y^2 = 9$, $\quad z \ge 0$

50. $x^2 + y^2 + z^2 = a^2$, density is proportional to the distance from the center

51. *Investigation*   Consider a spherical segment of height h from a sphere of radius a, where $h \le a$, and constant density $\rho(x, y, z) = k$ (see figure).

(a) Find the volume of the solid.

(b) Find the centroid of the solid.

(c) Use the result in part (b) to find the centroid of a hemisphere of radius a.

(d) Find $\lim_{h \to 0} \bar{z}$.

(e) Find I_z.

(f) Use the result in part (e) to find I_z for a hemisphere.

52. *Moment of Inertia*   Find the moment of inertia about the z-axis of the ellipsoid

$$x^2 + y^2 + \frac{z^2}{a^2} = 1$$

where $a > 0$.

In Exercises 53 and 54, give a geometrical interpretation of the triple integral.

53. $\displaystyle\int_0^{2\pi} \int_0^{\pi} \int_0^{6 \sin \phi} \rho^2 \sin \phi \, d\rho \, d\phi \, d\theta$

54. $\displaystyle\int_0^{\pi} \int_0^{2} \int_0^{1+r^2} r \, dz \, dr \, d\theta$

True or False?   **In Exercises 55–58, determine whether the statement is true or false. If it is false, explain why or give an example that shows it is false.**

55. $\displaystyle\int_a^b \int_c^d f(x)g(y) \, dy \, dx = \left[\int_a^b f(x) \, dx\right]\left[\int_c^d g(y) \, dy\right]$

56. If f is continuous over R_1 and R_2, and

$$\int_{R_1} \int dA = \int_{R_2} \int dA$$

then

$$\int_{R_1} \int f(x, y) \, dA = \int_{R_2} \int f(x, y) \, dA.$$

57. $\displaystyle\int_{-1}^{1} \int_{-1}^{1} \cos(x^2 + y^2) \, dx \, dy = 4 \int_0^1 \int_0^1 \cos(x^2 + y^2) \, dx \, dy$

58. $\displaystyle\int_0^1 \int_0^1 \frac{1}{1 + x^2 + y^2} \, dx \, dy < \frac{\pi}{4}$

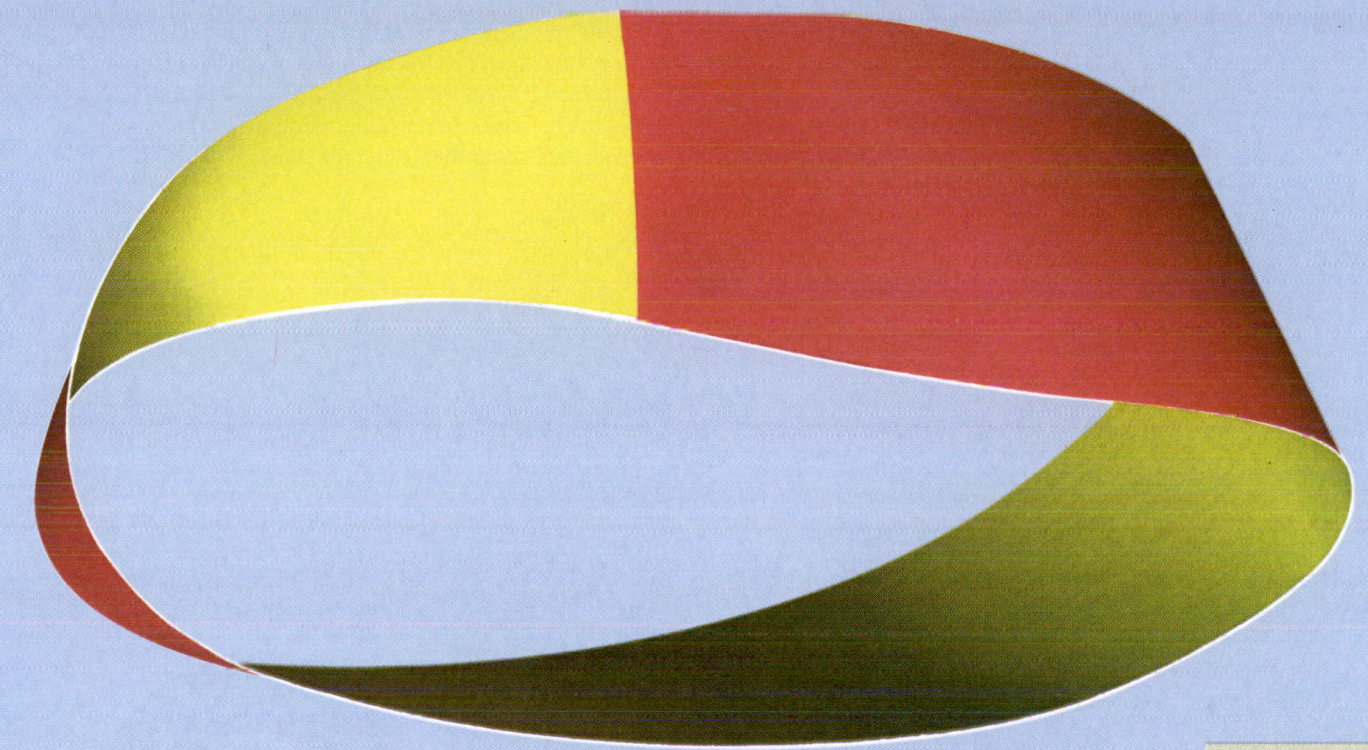

Vector Analysis

Whether mathematics is seen as a science or as an art depends on one's perspective. One mathematician-sculptor, Helaman Ferguson, combines both viewpoints in a unique way. Ferguson's sculptures, which bear such names as *Cosine Wild Sphere* and *Esker Trefoil Torus*, are the concrete embodiments of mathematical concepts that incorporate ideas such as series expansions and vector fields into their creation. Some of the basic images of his work are tori and double tori, Möbius strips, and trefoil knots. One of his techniques is to use three-dimensional computer graphics to model his intended sculptures. The coordinates on the computer screen can then be used to direct the sculpting.

One example of Helaman Ferguson's work, *Umbilic Torus NC*, is shown below. This form can be written as a parametric surface using the following set of parametric equations.

Helaman Ferguson's *The Eight Fold Way*, overlooking the Mathematical Science Research Institute at Berkeley, was created using a tool called the SP-2. This tool has six cables monitored by sensors that permit a computer to continuously calculate tool-tip position (x, y, z) and orientation.

$$x = \sin u \left[7 + \cos\left(\frac{u}{3} - 2v\right) + 2\cos\left(\frac{u}{3} + v\right) \right]$$

$$y = \cos u \left[7 + \cos\left(\frac{u}{3} - 2v\right) + 2\cos\left(\frac{u}{3} + v\right) \right]$$

$$z = \sin\left(\frac{u}{3} - 2v\right) + 2\sin\left(\frac{u}{3} + v\right)$$

$$-\pi \le u \le \pi, \qquad -\pi \le v \le \pi$$

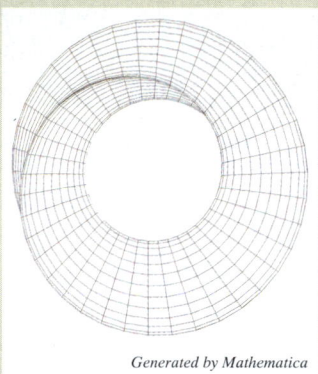

Generated by Mathematica

Mathematica Rendition of *Umbilic Torus NC*

Photograph of *Umbilic Torus NC*

QUESTIONS

1. Explain how the umbilic torus shown above is similar to a Möbius strip. (A Möbius strip is shown in the photograph on page 983.)

2. Use a three-dimensional computer algebra system to graph the torus—you will have to use the *parametric surface* graphing mode.

3. When viewed head-on (along the z-axis), the torus appears nearly circular. Is it possible to alter this appearance so that the shape of the torus is more like that of an elongated ellipse? If so, use a computer algebra system to sketch an "elliptical" torus.

The concepts presented here will be explored further in this chapter. For an extension of this application, see the lab series that accompanies this text.

Jon Ferguson © 1993

Helaman Ferguson holds a Ph.D. in mathematics and works out of his studio in Laurel, Maryland. In addition to exhibiting his works worldwide, he designs algorithms for operating machinery and for scientific visualization.

Vector Fields • Conservative Vector Fields • Curl of a Vector Field • Divergence of a Vector Field

Vector Fields

In Chapter 11, you studied vector-valued functions—functions that assign a vector to a *real number*. There you saw that vector-valued functions of real numbers are useful in representing curves and motion along a curve. In this chapter, you will study two other types of vector-valued functions—functions that assign a vector to a *point in the plane* or a *point in space*. Such functions are called **vector fields,** and they are useful in representing various types of **force fields** and **velocity fields.**

> **Definition of a Vector Field**
>
> Let M and N be functions of two variables x and y, defined on a plane region R. The function $\mathbf{F}$ defined by
>
> $$\mathbf{F}(x, y) = M\mathbf{i} + N\mathbf{j} \qquad \text{Plane}$$
>
> is called a **vector field over R.**
> Let M, N, and P be functions of three variables x, y, and z, defined on a solid region Q in space. The function $\mathbf{F}$ defined by
>
> $$\mathbf{F}(x, y, z) = M\mathbf{i} + N\mathbf{j} + P\mathbf{k} \qquad \text{Space}$$
>
> is called a **vector field over Q.**

NOTE Although a vector field consists of infinitely many vectors, you can get a good idea of what the vector field looks like by sketching several representative vectors $\mathbf{F}(x, y)$ whose initial points are (x, y).

From this definition you can see that the *gradient* is one example of a vector field. For example, if

$$f(x, y) = x^2 + y^2$$

then the gradient of f

$$\nabla f(x, y) = f_x(x, y)\mathbf{i} + f_y(x, y)\mathbf{j}$$
$$= 2x\mathbf{i} + 2y\mathbf{j} \qquad \text{Vector field in the plane}$$

is a vector field in the plane. From Chapter 11, the graphical interpretation of this field is a family of vectors, each of which points in the direction of maximum increase along the surface given by $z = f(x, y)$. For this particular function, the surface is a paraboloid and the gradient tells you that the direction of maximum increase along the surface is the direction given by the ray from the origin through the point (x, y).

Similarly, if

$$f(x, y, z) = x^2 + y^2 + z^2$$

then the gradient of f

$$\nabla f(x, y, z) = f_x(x, y, z)\mathbf{i} + f_y(x, y, z)\mathbf{j} + f_z(x, y, z)\mathbf{k}$$
$$= 2x\mathbf{i} + 2y\mathbf{j} + 2z\mathbf{k} \qquad \text{Vector field in space}$$

is a vector field in space.

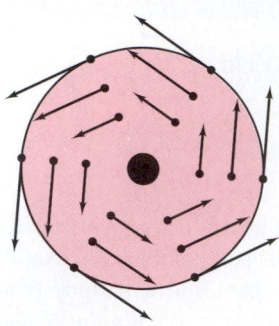

Velocity field

Rotating wheel
Figure 14.1

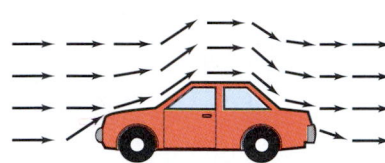

Air-flow vector field
Figure 14.2

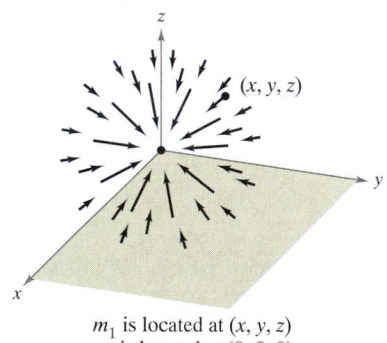

m_1 is located at (x, y, z)
m_2 is located at $(0, 0, 0)$

Gravitational force field
Figure 14.3

Some common *physical* examples of vector fields are **velocity fields, gravitational fields,** and **electric force fields.**

1. *Velocity fields* describe the motions of systems of particles in the plane or in space. For instance, Figure 14.1 shows the vector field determined by a wheel rotating on an axle. Notice that the velocity vectors are determined by the locations of their initial points—the farther a point is from the axle, the greater its velocity. Velocity fields are also determined by the flow of liquids through a container or by the flow or air currents around a moving object, as shown in Figure 14.2.

2. *Gravitational fields* are defined by **Newton's Law of Gravitation,** which states that the force of attraction exerted on a particle of mass m_1 located at (x, y, z) by a particle of mass m_2 located at $(0, 0, 0)$ is given by

$$\mathbf{F}(x, y, z) = \frac{-Gm_1m_2}{x^2 + y^2 + z^2}\mathbf{u}$$

where G is the gravitational constant and $\mathbf{u}$ is the unit vector in the direction from the origin to x, y, z. In Figure 14.3, you can see that the gravitational field $\mathbf{F}$ has the properties that $\mathbf{F}(x, y, z)$ always points toward the origin, and that the magnitude of $\mathbf{F}x, y, z$ is the same at all points equidistant from the origin. A vector field with these two properties is called a **central force field.** Using the position vector.

$$\mathbf{r} = x\mathbf{i} + y\mathbf{j} + z\mathbf{k}$$

for the point (x, y, z), you can express the gravitational field $\mathbf{F}$ as

$$\mathbf{F}(x, y, z) = \frac{-Gm_1m_2}{\|\mathbf{r}\|^2}\left(\frac{\mathbf{r}}{\|\mathbf{r}\|}\right)$$

$$= \frac{-Gm_1m_2}{\|\mathbf{r}\|^2}\mathbf{u}.$$

3. *Electric force fields* are defined by **Coulomb's Law,** which states that the force exerted on a particle with electric charge q_1 located at (x, y, z) by a particle with electric charge q_2 located at $(0, 0, 0)$ is given by

$$\mathbf{F}(x, y, z) = \frac{cq_1q_2}{\|\mathbf{r}\|^2}\mathbf{u}$$

where $\mathbf{r} = x\mathbf{i} + y\mathbf{j} + z\mathbf{k}$, $\mathbf{u} = \mathbf{r}/\|\mathbf{r}\|$, and c is a constant that depends on the choice of units for $\|\mathbf{r}\|$, q_1, and q_2.

Note that an electric force field has the same form as a gravitational field. That is,

$$\mathbf{F}(x, y, z) = \frac{k}{\|\mathbf{r}\|^2}\mathbf{u}.$$

Such a force field is called an **inverse square field.**

Definition of Inverse Square Field

Let $\mathbf{r}(t) = x(t)\mathbf{i} + y(t)\mathbf{j} + z(t)\mathbf{k}$ be the position vector. The vector field $\mathbf{F}$ is an **inverse square field** if

$$\mathbf{F}(x, y, z) = \frac{k}{\|\mathbf{r}\|^2}\mathbf{u}$$

where k is a real number and $\mathbf{u} = \mathbf{r}/\|\mathbf{r}\|$ is a unit vector in the direction of $\mathbf{r}$.

Because vector fields consist of infinitely many vectors, it is not possible to actually create a sketch of the field. Instead, when you sketch a vector field, your goal is to sketch representative vectors that help you visualize the field.

EXAMPLE 1 Sketching a Vector Field

Sketch some vectors in the vector field given by

$$\mathbf{F}(x, y) = -y\mathbf{i} + x\mathbf{j}.$$

Solution You could plot vectors at several random points in the plane. However, it is more enlightening to plot vectors of equal magnitude. This corresponds to finding level curves in scalar fields. In this case, vectors of equal magnitude lie on circles.

$$\|\mathbf{F}\| = c \qquad \text{Vectors of length } c$$
$$\sqrt{x^2 + y^2} = c$$
$$x^2 + y^2 = c^2 \qquad \text{Equation of circle}$$

To begin making the sketch, choose a value for c and plot several vectors on the resulting circle. For instance, the following vectors occur on the unit circle.

Point	Vector
$(1, 0)$	$\mathbf{F}(1, 0) = \mathbf{j}$
$(0, 1)$	$\mathbf{F}(0, 1) = -\mathbf{i}$
$(-1, 0)$	$\mathbf{F}(-1, 0) = -\mathbf{j}$
$(0, -1)$	$\mathbf{F}(0, -1) = \mathbf{i}$

These and several other vectors in the vector field are shown in Figure 14.4. Note in the figure that this vector field is similar to that given by the rotating wheel shown in Figure 14.1.

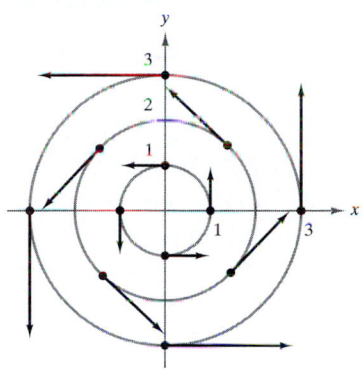

Vector field:
$\mathbf{F}(x, y) = -y\mathbf{i} + x\mathbf{j}$

Figure 14.4

EXAMPLE 2 Sketching a Vector Field

Sketch some vectors in the vector field given by

$$\mathbf{F}(x, y) = 2x\mathbf{i} + y\mathbf{j}.$$

Solution For this vector field, vectors of equal length lie on ellipses given by

$$\|\mathbf{F}\| = \sqrt{(2x)^2 + (y)^2} = c$$

which implies that

$$4x^2 + y^2 = c^2.$$

For $c = 1$, sketch several vectors $2x\mathbf{i} + y\mathbf{j}$ of magnitude 1 at points on the ellipse given by

$$4x^2 + y^2 = 1.$$

For $c = 2$, sketch several vectors $2x\mathbf{i} + y\mathbf{j}$ of magnitude 2 at points on the ellipse given by

$$4x^2 + y^2 = 4.$$

These vectors are shown in Figure 14.5.

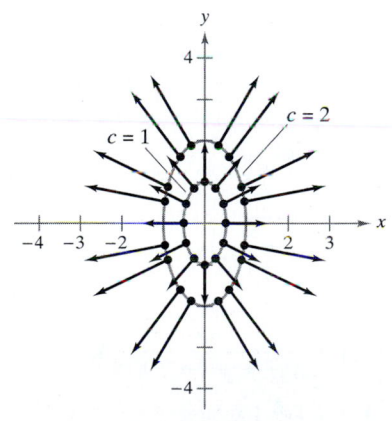

Vector field:
$\mathbf{F}(x, y) = 2x\mathbf{i} + y\mathbf{j}$

Figure 14.5

EXAMPLE 3   Sketching a Velocity Field

Sketch some vectors in the velocity field given by

$$\mathbf{v}(x, y, z) = (16 - x^2 - y^2)\mathbf{k}$$

where $x^2 + y^2 \leq 16$.

Solution   You can imagine that $\mathbf{v}$ describes the velocity of a liquid flowing through a tube of radius 4. Vectors near the z-axis are longer than those near the edge of the tube. For instance, at the point $(0, 0, 0)$, the velocity vector is $\mathbf{v}(0, 0, 0) = 16\mathbf{k}$, whereas at the point $(0, 3, 0)$, the velocity vector is $\mathbf{v}(0, 3, 0) = 7\mathbf{k}$. Figure 14.6 shows these and several other vectors for the velocity field. From the figure, you can see that the speed of the liquid is greater near the center of the tube than near the edges of the tube.

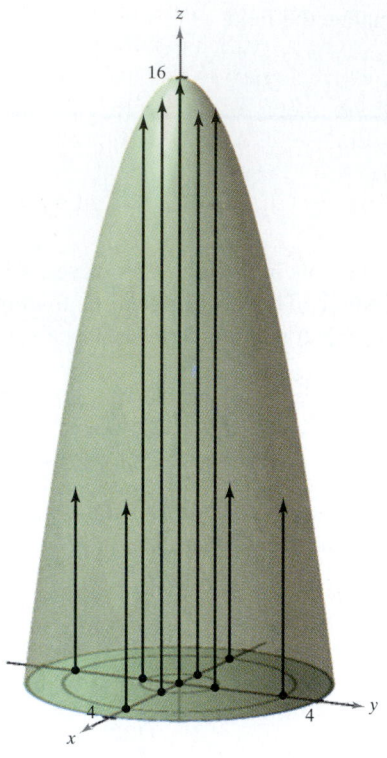

Velocity field:
$\mathbf{v}(x, y, z) = (16 - x^2 - y^2)\mathbf{k}$

Figure 14.6

Conservative Vector Fields

Notice in Figure 14.5 that all the vectors appear to be normal to the level curve from which they emanate. Because this is a property of gradients, it is natural to ask whether the vector field given by $\mathbf{F}(x, y) = 2x\mathbf{i} + y\mathbf{j}$ is the *gradient* for some differentiable function f. The answer is that some vector fields can be represented as the gradients of differentiable functions and some cannot—those that can are called **conservative** vector fields.

> **Definition of Conservative Vector Field**
>
> A vector field $\mathbf{F}$ is called **conservative** if there exists a differentiable function f such that $\mathbf{F} = \nabla f$. The function f is called the **potential function** for $\mathbf{F}$.

EXAMPLE 4   Conservative Vector Fields

a. The vector field given by $\mathbf{F}(x, y) = 2x\mathbf{i} + y\mathbf{j}$ is conservative. To see this, consider the potential function $f(x, y) = x^2 + \frac{1}{2}y^2$. Because

$$\nabla f = 2x\mathbf{i} + y\mathbf{j} = \mathbf{F}$$

it follows that $\mathbf{F}$ is conservative.

b. Every inverse square field is conservative. To see this, let

$$\mathbf{F}(x, y, z) = \frac{k}{\|\mathbf{r}\|^2}\mathbf{u} \quad \text{and} \quad f(x, y, z) = \frac{-k}{\sqrt{x^2 + y^2 + z^2}}$$

where $\mathbf{u} = \mathbf{r}/\|\mathbf{r}\|$. Because

$$\nabla f = \frac{kx}{(x^2 + y^2 + z^2)^{3/2}}\mathbf{i} + \frac{ky}{(x^2 + y^2 + z^2)^{3/2}}\mathbf{j} + \frac{kz}{(x^2 + y^2 + z^2)^{3/2}}\mathbf{k}$$

$$= \frac{k}{x^2 + y^2 + z^2}\left(\frac{x\mathbf{i} + y\mathbf{j} + z\mathbf{k}}{\sqrt{x^2 + y^2 + z^2}}\right) = \frac{k}{\|\mathbf{r}\|^2}\frac{\mathbf{r}}{\|\mathbf{r}\|} = \frac{k}{\|\mathbf{r}\|^2}\mathbf{u}$$

it follows that $\mathbf{F}$ is conservative.

As can be seen in Example 4b, many important vector fields, including gravitational fields, magnetic fields, and electric force fields, are conservative. Most of the terminology in this chapter comes from physics. For example, the term "conservative" is derived from the classic physical law regarding the conservation of energy. This law states that the sum of the kinetic energy and the potential energy of a particle moving in a conservative force field is constant. (The kinetic energy of a particle is the energy due to its motion, and the potential energy is the energy due to its position in the force field.)

The following important theorem gives a necessary and sufficient condition for a vector field *in the plane* to be conservative.

THEOREM 14.1   Test for Conservative Vector Field in the Plane

Let M and N have continuous first partial derivatives on an open disc R. The vector field given by $\mathbf{F}(x, y) = M\mathbf{i} + N\mathbf{j}$ is conservative if and only if

$$\frac{\partial N}{\partial x} = \frac{\partial M}{\partial y}.$$

Proof   To prove that the given condition is necessary for $\mathbf{F}$ to be conservative, you suppose there exists a potential function f such that

$$\mathbf{F}(x, y) = \nabla f(x, y) = M\mathbf{i} + N\mathbf{j}.$$

Then you have

$$f_x(x, y) = M \quad \Longrightarrow \quad f_{xy}(x, y) = \frac{\partial M}{\partial y}$$

$$f_y(x, y) = N \quad \Longrightarrow \quad f_{yx}(x, y) = \frac{\partial N}{\partial x}$$

and, by the equivalence of the mixed partials f_{xy} and f_{yx}, you can conclude that $\partial N/\partial x = \partial M/\partial y$ for all (x, y) in R. The sufficiency of the condition is proved in Section 14.4.   ∎

NOTE   Theorem 14.1 requires that the domain of $\mathbf{F}$ be an open disc. If R is simply an open region, the given condition is necessary but not sufficient to produce a conservative vector field.

EXAMPLE 5   Testing for Conservative Vector Fields in the Plane

a. The vector field given by $\mathbf{F}(x, y) = x^2 y\mathbf{i} + xy\mathbf{j}$ *is not* conservative because

$$\frac{\partial M}{\partial y} = \frac{\partial}{\partial y}[x^2 y] = x^2 \quad \text{and} \quad \frac{\partial N}{\partial x} = \frac{\partial}{\partial x}[xy] = y.$$

b. The vector field given by $\mathbf{F}(x, y) = 2x\mathbf{i} + y\mathbf{j}$ *is* conservative because

$$\frac{\partial M}{\partial y} = \frac{\partial}{\partial y}[2x] = 0 \quad \text{and} \quad \frac{\partial N}{\partial x} = \frac{\partial}{\partial x}[y] = 0.$$

Theorem 14.1 tells you whether a vector field is conservative. It does not tell you how to find a potential function of **F**. The problem is comparable to antidifferentiation. Sometimes you will be able to find a potential function by simple inspection. For instance, in Example 4 you observed that

$$f(x, y) = x^2 + \frac{1}{2}y^2$$

has the property that $\nabla f(x, y) = 2x\mathbf{i} + y\mathbf{j}$.

EXAMPLE 6   Find a Potential Function for F(x, y)

Find a potential function for

$$\mathbf{F}(x, y) = 2xy\mathbf{i} + (x^2 - y)\mathbf{j}.$$

Solution   From Theorem 14.1 it follows that **F** is conservative because

$$\frac{\partial}{\partial y}[2xy] = 2x \quad \text{and} \quad \frac{\partial}{\partial x}[x^2 - y] = 2x.$$

If f is a function whose gradient is equal to $\mathbf{F}(x, y)$, then

$$\nabla f(x, y) = 2xy\mathbf{i} + (x^2 - y)\mathbf{j}$$

which implies that

$$f_x(x, y) = 2xy$$

and

$$f_y(x, y) = x^2 - y.$$

To reconstruct the function f from these two partial derivatives, integrate $f_x(x, y)$ with respect to x and $f_y(x, y)$ with respect to y, as follows.

$$f(x, y) = \int f_x(x, y) \, dx = \int 2xy \, dx = x^2y + g(y)$$

$$f(x, y) = \int f_y(x, y) \, dy = \int (x^2 - y) \, dy = x^2y - \frac{y^2}{2} + h(x)$$

Notice that $g(y)$ is constant with respect to x and $h(x)$ is constant with respect to y. To find a single expression that represents $f(x, y)$, let $g(y) = -\frac{1}{2}y^2 + K$ and $h(x) = K$. Then, you can write

$$f(x, y) = x^2y + g(y) + K$$

$$= x^2y - \frac{y^2}{2} + K.$$

You can check this result by forming the gradient of f—it should be equal to the original function **F**.

NOTE   Notice that the solution in Example 6 is comparable to that given by an indefinite integral. That is, the solution represents a family of potential functions, any two of which differ by a constant. To find a unique solution, you would have to be given an initial condition satisfied by the potential function.

Curl of a Vector Field

Theorem 14.1 has a counterpart for vector fields in space. Before stating that result, we define the **curl of a vector field** in space.

Definition of Curl of a Vector Field

The **curl** of $\mathbf{F}(x, y, z) = M\mathbf{i} + N\mathbf{j} + P\mathbf{k}$ is

$$\text{curl } \mathbf{F}(x, y, z) = \nabla \times \mathbf{F}(x, y, z)$$

$$= \left(\frac{\partial P}{\partial y} - \frac{\partial N}{\partial z} \right)\mathbf{i} - \left(\frac{\partial P}{\partial x} - \frac{\partial M}{\partial z} \right)\mathbf{j} + \left(\frac{\partial N}{\partial x} - \frac{\partial M}{\partial y} \right)\mathbf{k}.$$

NOTE   If **curl F** = **0**, we say that **F** is **irrotational**.

The cross product notation used for curl comes from viewing the gradient ∇f as the result of the **differential operator** ∇ acting on the function f. In this context, you can use the following determinant form as an aid in remembering the formula for curl.

$$\text{curl } \mathbf{F}(x, y, z) = \nabla \times \mathbf{F}(x, y, z)$$

$$= \begin{vmatrix} \mathbf{i} & \mathbf{j} & \mathbf{k} \\ \dfrac{\partial}{\partial x} & \dfrac{\partial}{\partial y} & \dfrac{\partial}{\partial z} \\ M & N & P \end{vmatrix}$$

$$= \left(\frac{\partial P}{\partial y} - \frac{\partial N}{\partial z} \right)\mathbf{i} - \left(\frac{\partial P}{\partial x} - \frac{\partial M}{\partial z} \right)\mathbf{j} + \left(\frac{\partial N}{\partial x} - \frac{\partial M}{\partial y} \right)\mathbf{k}$$

 EXAMPLE 7   Finding the Curl of a Vector Field

Find **curl F** for the vector field given by

$$\mathbf{F}(x, y, z) = 2xy\mathbf{i} + (x^2 + z^2)\mathbf{j} + 2zy\mathbf{k}.$$

Solution   The curl of **F** is given by

$$\text{curl } \mathbf{F}(x, y, z) = \nabla \times \mathbf{F}(x, y, z)$$

$$= \begin{vmatrix} \mathbf{i} & \mathbf{j} & \mathbf{k} \\ \dfrac{\partial}{\partial x} & \dfrac{\partial}{\partial y} & \dfrac{\partial}{\partial z} \\ 2xy & x^2 + z^2 & 2zy \end{vmatrix}$$

$$= \begin{vmatrix} \dfrac{\partial}{\partial y} & \dfrac{\partial}{\partial z} \\ x^2 + z^2 & 2zy \end{vmatrix}\mathbf{i} - \begin{vmatrix} \dfrac{\partial}{\partial x} & \dfrac{\partial}{\partial z} \\ 2xy & 2zy \end{vmatrix}\mathbf{j} + \begin{vmatrix} \dfrac{\partial}{\partial x} & \dfrac{\partial}{\partial y} \\ 2xy & x^2 + z^2 \end{vmatrix}\mathbf{k}$$

$$= (2z - 2z)\mathbf{i} - (0 - 0)\mathbf{j} + (2x - 2x)\mathbf{k}$$

$$= \mathbf{0}.$$

Later in this chapter, we will assign a physical interpretation to the curl of a vector field. But for now, the primary use we make of curl is in the following test for conservative vector fields in space. The test states that for a vector field whose domain is all of three-dimensional space (or an open sphere), the curl is **0** vector at every point in the domain if and only if **F** is conservative. The proof is similar to that given for Theorem 14.1.

THEOREM 14.2   Test for Conservative Vector Field in Space

Suppose that M, N, and P have continuous first partial derivatives in an open sphere Q in space. The vector field given by $\mathbf{F}(x, y, z) = M\mathbf{i} + N\mathbf{j} + P\mathbf{k}$ is conservative if and only if

$$\text{curl } \mathbf{F}(x, y, z) = \mathbf{0}.$$

That is, **F** is conservative if and only if

$$\frac{\partial P}{\partial y} = \frac{\partial N}{\partial z}, \quad \frac{\partial P}{\partial x} = \frac{\partial M}{\partial z}, \quad \text{and} \quad \frac{\partial N}{\partial x} = \frac{\partial M}{\partial y}.$$

From Theorem 14.2, you can see that the vector field given in Example 7 is conservative because **curl** $\mathbf{F}(x, y, z) = \mathbf{0}$. Try showing that the vector field $\mathbf{F}(x, y, z) = x^3y^2z\mathbf{i} + x^2z\mathbf{j} + x^2y\mathbf{k}$ is not conservative—you can do this by showing that its curl is **curl** $\mathbf{F}(x, y, z) = (x^3y^2 - 2xy)\mathbf{j} + (2xz - 2x^3yz)\mathbf{k} \neq \mathbf{0}$.

For vector fields in space that pass the test for being conservative, you can find a potential function by following the same pattern used in the plane (as demonstrated in Example 6).

EXAMPLE 8   Finding a Potential Function for F(x, y, z)

NOTE   Examples 6 and 8 are illustrations of a type of problem called *recovering a function from its gradient*. If you go on to take a course in differential equations, you will study other methods for solving this type of problem. One popular method gives an interplay between successive "partial integrations" and partial differentiations.

Find a potential function for $\mathbf{F}(x, y, z) = 2xy\mathbf{i} + (x^2 + z^2)\mathbf{j} + 2zy\mathbf{k}$.

Solution   From Example 7, you know that the vector field given by **F** is conservative. If f is a function such that $\mathbf{F}(x, y, z) = \nabla f(x, y, z)$, then

$$f_x(x, y, z) = 2xy, \quad f_y(x, y, z) = x^2 + z^2, \quad \text{and} \quad f_z(x, y, z) = 2zy$$

and integrating with respect to x, y, and z separately produces

$$f(x, y, z) = \int M\, dx = \int 2xy\, dx = x^2y + g(y, z)$$

$$f(x, y, z) = \int N\, dy = \int (x^2 + z^2)\, dy = x^2y + z^2y + h(x, z)$$

$$f(x, y, z) = \int P\, dz = \int 2zy\, dz = z^2y + k(x, y).$$

Comparing these three versions of $f(x, y, z)$, you can conclude that

$$g(y, z) = z^2y + K, \quad h(x, z) = K, \quad \text{and} \quad k(x, y) = x^2y + K.$$

Therefore, $f(x, y, z)$ is given by

$$f(x, y, z) = x^2y + z^2y + K.$$

Divergence of a Vector Field

NOTE   Divergence can be viewed as a type of derivative of **F** in that, for vector fields representing velocities of moving particles, the divergence measures the rate of particle flow per unit volume at a point. In hydrodynamics (the study of fluid motion), a velocity field that is divergence free is called **incompressible.** In the study of electricity and magnetism, a vector field that is divergence free is called **solenoidal.**

You have seen that the curl of a vector field **F** is itself a vector field. Another important function defined on a vector field is **divergence,** which is a scalar function.

> ### Definition of Divergence of a Vector Field
>
> The **divergence** of $F(x, y) = M\mathbf{i} + N\mathbf{j}$ is
>
> $$\text{div } \mathbf{F}(x, y) = \nabla \cdot \mathbf{F}(x, y) = \frac{\partial M}{\partial x} + \frac{\partial N}{\partial y}. \qquad \text{Plane}$$
>
> The **divergence** of $F(x, y, z) = M\mathbf{i} + N\mathbf{j} + P\mathbf{k}$ is
>
> $$\text{div } \mathbf{F}(x, y, z) = \nabla \cdot \mathbf{F}(x, y, z) = \frac{\partial M}{\partial x} + \frac{\partial N}{\partial y} + \frac{\partial P}{\partial z}. \qquad \text{Space}$$
>
> If div **F** = 0, then **F** is said to be **divergence free.**

The dot product notation used for divergence comes from considering ∇ as a **differential operator,** as follows.

$$\nabla \cdot \mathbf{F}(x, y, z) = \left[\left(\frac{\partial}{\partial x} \right)\mathbf{i} + \left(\frac{\partial}{\partial y} \right)\mathbf{j} + \left(\frac{\partial}{\partial z} \right)\mathbf{k} \right] \cdot (M\mathbf{i} + N\mathbf{j} + P\mathbf{k})$$

$$= \frac{\partial M}{\partial x} + \frac{\partial N}{\partial y} + \frac{\partial P}{\partial z}$$

EXAMPLE 9   Finding the Divergence of a Vector Field

Find the divergence at $(2, 1, -1)$ for the vector field

$$\mathbf{F}(x, y, z) = x^3 y^2 z\mathbf{i} + x^2 z\mathbf{j} + x^2 y\mathbf{k}.$$

Solution   The divergence of **F** is

$$\text{div } \mathbf{F}(x, y, z) = \frac{\partial}{\partial x}\left[x^3 y^2 z\right] + \frac{\partial}{\partial y}\left[x^2 z\right] + \frac{\partial}{\partial z}\left[x^2 y\right] = 3x^2 y^2 z.$$

At the point $(2, 1, -1)$, the divergence is

$$\text{div } \mathbf{F}(2, 1, -1) = 3(2^2)(1^2)(-1) = -12.$$

There are many important properties of the divergence and curl of a vector field **F** (see Exercises 65–71). One that is used often is described in Theorem 14.3. You are asked to prove this theorem in Exercise 72.

> ### THEOREM 14.3   Relationship Between Divergence and Curl
>
> If $F(x, y, z) = M\mathbf{i} + N\mathbf{j} + P\mathbf{k}$ is a vector field and M, N, and P have continuous second partial derivatives, then
>
> $$\text{div }(\mathbf{curl } \mathbf{F}) = 0.$$

In Exercises 1–6, match the vector field with its graph. [The graphs are labeled (a), (b), (c), (d), (e), and (f).]

(a)
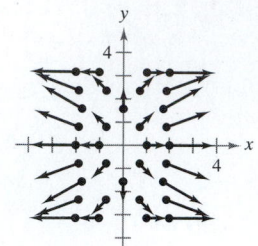

(b)
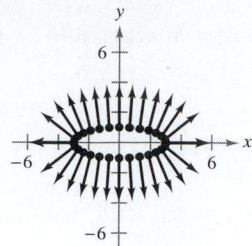

(c)
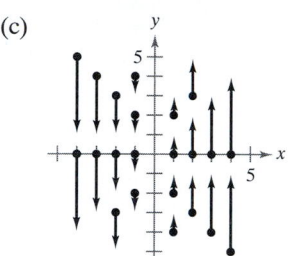

(d)
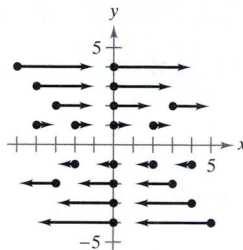

(e)
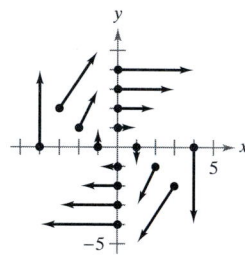

(f)
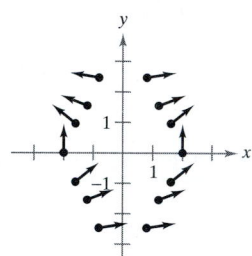

1. $\mathbf{F}(x, y) = x\mathbf{j}$
2. $\mathbf{F}(x, y) = y\mathbf{i}$
3. $\mathbf{F}(x, y) = x\mathbf{i} + 3y\mathbf{j}$
4. $\mathbf{F}(x, y) = y\mathbf{i} - x\mathbf{j}$
5. $\mathbf{F}(x, y) = \langle x, \sin y \rangle$
6. $\mathbf{F}(x, y) = \left\langle \frac{1}{2}xy, \frac{1}{4}x^2 \right\rangle$

In Exercises 7–16, sketch several representative vectors in the vector field.

7. $\mathbf{F}(x, y) = \mathbf{i} + \mathbf{j}$
8. $\mathbf{F}(x, y) = 2\mathbf{i}$
9. $\mathbf{F}(x, y) = x\mathbf{i} + y\mathbf{j}$
10. $\mathbf{F}(x, y) = -x\mathbf{i} + y\mathbf{j}$
11. $\mathbf{F}(x, y, z) = 3y\mathbf{j}$
12. $\mathbf{F}(x, y) = x\mathbf{i}$
13. $\mathbf{F}(x, y) = 4x\mathbf{i} + y\mathbf{j}$
14. $\mathbf{F}(x, y) = \mathbf{i} + (x^2 + y^2)\mathbf{j}$
15. $\mathbf{F}(x, y, z) = \mathbf{i} + \mathbf{j} + \mathbf{k}$
16. $\mathbf{F}(x, y, z) = x\mathbf{i} + y\mathbf{j} + z\mathbf{k}$

In Exercises 17–20, use a computer algebra system to graph several representative vectors in the vector field.

17. $\mathbf{F}(x, y) = \frac{1}{8}(2xy\mathbf{i} + y^2\mathbf{j})$
18. $\mathbf{F}(x, y) = (2y - 3x)\mathbf{i} + (2y + 3x)\mathbf{j}$

19. $\mathbf{F}(x, y, z) = \dfrac{x\mathbf{i} + y\mathbf{j} + z\mathbf{k}}{\sqrt{x^2 + y^2 + z^2}}$
20. $\mathbf{F}(x, y, z) = x\mathbf{i} - y\mathbf{j} + z\mathbf{k}$

In Exercises 21–26, find the gradient vector field for the scalar function. (That is, find the conservative vector field for the potential function.)

21. $f(x, y) = 5x^2 + 3xy + 10y^2$
22. $f(x, y) = \sin 3x \cos 4y$
23. $f(x, y, z) = z - ye^{x^2}$
24. $f(x, y, z) = \dfrac{y}{z} + \dfrac{z}{x} - \dfrac{xz}{y}$
25. $g(x, y, z) = xy \ln(x + y)$
26. $g(x, y, z) = x \arcsin yz$

In Exercises 27–34, determine whether the vector field is conservative. If it is, find a potential function for the vector field.

27. $\mathbf{F}(x, y) = 2xy\mathbf{i} + x^2\mathbf{j}$
28. $\mathbf{F}(x, y) = \dfrac{1}{y^2}(y\mathbf{i} - 2x\mathbf{j})$
29. $\mathbf{F}(x, y) = xe^{x^2 y}(2y\mathbf{i} + x\mathbf{j})$
30. $\mathbf{F}(x, y) = 2xy^3\mathbf{i} + 3y^2x^2\mathbf{j}$
31. $\mathbf{F}(x, y) = \dfrac{x\mathbf{i} + y\mathbf{j}}{x^2 + y^2}$
32. $\mathbf{F}(x, y) = \dfrac{2y}{x}\mathbf{i} - \dfrac{x^2}{y^2}\mathbf{j}$
33. $\mathbf{F}(x, y) = e^x(\cos y\mathbf{i} + \sin y\mathbf{j})$
34. $\mathbf{F}(x, y) = \dfrac{2x\mathbf{i} + 2y\mathbf{j}}{(x^2 + y^2)^2}$

In Exercises 35–38, find the curl of the vector field F at the indicated point.

Vector Field	Point
35. $\mathbf{F}(x, y, z) = xyz\mathbf{i} + y\mathbf{j} + z\mathbf{k}$	$(1, 2, 1)$
36. $\mathbf{F}(x, y, z) = x^2z\mathbf{i} - 2xz\mathbf{j} + yz\mathbf{k}$	$(2, -1, 3)$
37. $\mathbf{F}(x, y, z) = e^x \sin y\mathbf{i} - e^x \cos y\mathbf{j}$	$(0, 0, 3)$
38. $\mathbf{F}(x, y, z) = e^{-xyz}(\mathbf{i} + \mathbf{j} + \mathbf{k})$	$(3, 2, 0)$

In Exercises 39–42, use a computer algebra system to find the curl of the vector field F.

39. $\mathbf{F}(x, y, z) = \arctan\dfrac{x}{y}\mathbf{i} + \ln\sqrt{x^2 + y^2}\,\mathbf{j} + \mathbf{k}$

40. $\mathbf{F}(x, y, z) = \dfrac{yz}{y - z}\mathbf{i} + \dfrac{xz}{x - z}\mathbf{j} + \dfrac{xy}{x - y}\mathbf{k}$

41. $F(x, y, z) = \sin(x - y)\mathbf{i} + \sin(y - z)\mathbf{j} + \sin(z - x)\mathbf{k}$

42. $F(x, y, z) = \sqrt{x^2 + y^2 + z^2}\,(\mathbf{i} + \mathbf{j} + \mathbf{k})$

In Exercises 43–48, determine whether the vector field F is conservative. If it is, find a potential function for the vector field.

43. $F(x, y, z) = \sin y\mathbf{i} - x \cos y\mathbf{j} + \mathbf{k}$

44. $F(x, y, z) = e^z\,(y\mathbf{i} + x\mathbf{j} + \mathbf{k})$

45. $F(x, y, z) = e^z\,(y\mathbf{i} + x\mathbf{j} + xy\mathbf{k})$

46. $F(x, y, z) = 3x^2y^2z\mathbf{i} + 2x^3yz\mathbf{j} + x^3y^2\mathbf{k}$

47. $F(x, y, z) = \dfrac{1}{y}\mathbf{i} - \dfrac{x}{y^2}\mathbf{j} + (2z - 1)\,\mathbf{k}$

48. $F(x, y, z) = \dfrac{x}{x^2 + y^2}\mathbf{i} + \dfrac{y}{x^2 + y^2}\mathbf{j} + \mathbf{k}$

In Exercises 49 and 50, find curl (F × G).

49. $F(x, y, z) = \mathbf{i} + 2x\mathbf{j} + 3y\mathbf{k}$

$\quad G(x, y, z) = x\mathbf{i} - y\mathbf{j} + z\mathbf{k}$

50. $F(x, y, z) = x\mathbf{i} - z\mathbf{k}$

$\quad G(x, y, z) = x^2\mathbf{i} + y\mathbf{j} + z^2\mathbf{k}$

In Exercises 51 and 52, find curl (curl F) = $\nabla \times (\nabla \times F)$.

51. $F(x, y, z) = xyz\mathbf{i} + y\mathbf{j} + z\mathbf{k}$

52. $F(x, y, z) = x^2z\mathbf{i} - 2xz\mathbf{j} + yz\mathbf{k}$

In Exercises 53–56, find the divergence of the vector field F.

53. $F(x, y, z) = 6x^2\mathbf{i} - xy^2\mathbf{j}$

54. $F(x, y, z) = xe^x\mathbf{i} + ye^y\mathbf{j}$

55. $F(x, y, z) = \sin x\mathbf{i} + \cos y\mathbf{j} + z^2\mathbf{k}$

56. $F(x, y, z) = \ln(x^2 + y^2)\mathbf{i} + xy\mathbf{j} + \ln(y^2 + z^2)\mathbf{k}$

In Exercises 57–60, find the divergence of the vector field F at the indicated point.

Vector Field	Point
57. $F(x, y, z) = xyz\mathbf{i} + y\mathbf{j} + z\mathbf{k}$	$(1, 2, 1)$
58. $F(x, y, z) = x^2z\mathbf{i} - 2xz\mathbf{j} + yz\mathbf{k}$	$(2, -1, 3)$
59. $F(x, y, z) = e^x \sin y\mathbf{i} - e^x \cos y\mathbf{j}$	$(0, 0, 3)$
60. $F(x, y, z) = e^{-xyz}\,(\mathbf{i} + \mathbf{j} + \mathbf{k})$	$(3, 2, 0)$

In Exercises 61 and 62, find div (F × G).

61. $F(x, y, z) = \mathbf{i} + 2x\mathbf{j} + 3y\mathbf{k}$

$\quad G(x, y, z) = x\mathbf{i} - y\mathbf{j} + z\mathbf{k}$

62. $F(x, y, z) = x\mathbf{i} - z\mathbf{k}$

$\quad G(x, y, z) = x^2\mathbf{i} + y\mathbf{j} + z^2\mathbf{k}$

In Exercises 63 and 64, find div (curl F) = $\nabla \cdot (\nabla \times F)$.

63. $F(x, y, z) = xyz\mathbf{i} + y\mathbf{j} + z\mathbf{k}$

64. $F(x, y, z) = x^2z\mathbf{i} - 2xz\mathbf{j} + yz\mathbf{k}$

In Exercises 65–71, prove the property for vector fields F and G and scalar function f. (Assume that the required partial derivatives are continuous.)

65. $\text{curl }(F + G) = \text{curl } F + \text{curl } G$

66. $\text{curl}(\nabla f) = \nabla \times (\nabla f) = 0$

67. $\text{div}(F + G) = \text{div } F + \text{div } G$

68. $\text{div}(F \times G) = (\text{curl } F) \cdot G - F \cdot (\text{curl } G)$

69. $\nabla \times [\nabla f + (\nabla \times F)] = \nabla \times (\nabla \times F)$

70. $\nabla \times (fF) = f(\nabla \times F) + (\nabla f) \times F$

71. $\text{div}(fF) = f \,\text{div } F + \nabla f \cdot F$

72. Prove Theorem 14.3.

In Exercises 73–76, let $F(x, y, z) = x\mathbf{i} + y\mathbf{j} + z\mathbf{k}$, and let $f(x, y, z) = \|F(x, y, z)\|$.

73. Show that $\nabla(\ln f) = \dfrac{F}{f^2}$.

74. Show that $\nabla\left(\dfrac{1}{f}\right) = -\dfrac{F}{f^3}$.

75. Show that $\nabla f^n = nf^{n-2}F$.

76. The **Laplacian** is the differential operator

$$\nabla^2 = \nabla \cdot \nabla = \frac{\partial^2}{\partial x^2} + \frac{\partial^2}{\partial y^2} + \frac{\partial^2}{\partial z^2}$$

and Laplace's equation is

$$\nabla^2 w = \frac{\partial^2 w}{\partial x^2} + \frac{\partial^2 w}{\partial y^2} + \frac{\partial^2 w}{\partial z^2} = 0.$$

Any function that satisfies this equation is called **harmonic**. Show that the function $1/f$ is harmonic.

77. *Magnetic Field*   A cross section of the earth's magnetic field can be represented as a vector field in which the center of the earth is located at the origin and the positive y-axis points in the direction of the magnetic north pole. The equation for this field is

$$F(x, y) = M(x, y)\mathbf{i} + N(x, y)\mathbf{j}$$

$$= \frac{m}{(x^2 + y^2)^{5/2}}\,[3xy\mathbf{i} + (2y^2 - x^2)\mathbf{j}]$$

where m is the magnetic moment of the earth. Show that this vector field is conservative.

Piecewise Smooth Curves • Line Integrals • Line Integrals of Vector Fields • Line Integrals in Differential Form

Piecewise Smooth Curves

A classical property of gravitational fields is that, subject to certain physical constraints, the work done by gravity on an object moving between two points in the field is independent of the path taken by the object. One of the constraints is that the **path** must be a piecewise smooth curve. Recall that a plane curve C given by

$$\mathbf{r}(t) = x(t)\mathbf{i} + y(t)\mathbf{j}, \quad a \le t \le b$$

is **smooth** if dx/dt and dy/dt are continuous on $[a, b]$ and not simultaneously 0 on (a, b). Similarly, a space curve C given by

$$\mathbf{r}(t) = x(t)\mathbf{i} + y(t)\mathbf{j} + z(t)\mathbf{k}, \quad a \le t \le b$$

is **smooth** if dx/dt, dy/dt, and dz/dt are continuous on $[a, b]$ and not simultaneously 0 on (a, b). A curve C is **piecewise smooth** if the interval $[a, b]$ can be partitioned into a finite number of subintervals, on each of which C is smooth.

JOSIAH WILLARD GIBBS (1839–1903)

Many physicists and mathematicians have contributed to the theory and applications described in this chapter—Newton, Gauss, Laplace, Hamilton, Maxwell, and many others. However, the use of vector analysis to describe these results is attributed primarily to the American mathematical physicist Josiah Willard Gibbs.

EXAMPLE 1   Finding a Piecewise Smooth Parametrization

Find a piecewise smooth parametrization of the graph C shown in Figure 14.7.

Solution   Because C consists of three line segments C_1, C_2, and C_3, you can construct a smooth parametrization for each segment and piece them together by making the last t-value in C_i correspond to the first t-value in C_{i+1}, as follows.

$$
\begin{aligned}
&C_1: x(t) = 0, && y(t) = 2t, && z(t) = 0, && 0 \le t \le 1 \\
&C_2: x(t) = t - 1, && y(t) = 2, && z(t) = 0, && 1 \le t \le 2 \\
&C_3: x(t) = 1, && y(t) = 2, && z(t) = t - 2, && 2 \le t \le 3
\end{aligned}
$$

Therefore, C is given by

$$
\mathbf{r}(t) = \begin{cases}
2t\mathbf{j}, & 0 \le t \le 1 \\
(t - 1)\mathbf{i} + 2\mathbf{j}, & 1 \le t \le 2 \\
\mathbf{i} + 2\mathbf{j} + (t - 2)\mathbf{k}, & 2 \le t \le 3.
\end{cases}
$$

Because C_1, C_2, and C_3 are smooth, it follows that C is piecewise smooth.

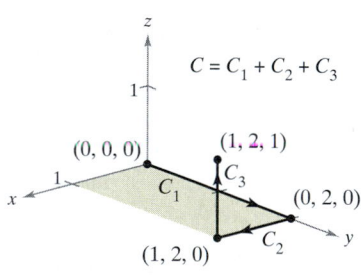

Figure 14.7

Recall that parametrization of a curve induces an **orientation** to the curve. For instance, in Example 1 the curve is oriented such that the positive direction is from $(0, 0, 0)$, following the curve to $(1, 2, 1)$. Try finding parametrization that induces the opposite orientation.

Line Integrals

Up to this point in the text, you have studied various types of integrals. For a single integral

$$\int_a^b f(x)\, dx \qquad \text{Integrate over interval } [a, b].$$

you integrated over the interval $[a, b]$. Similarly, for a double integral

$$\int_R \int f(x, y)\, dA \qquad \text{Integrate over region } R.$$

you integrated over the region R in the plane. In this section, you will study a new type of integral called a **line integral**

$$\int_C f(x, y)\, ds \qquad \text{Integrate over curve } C.$$

for which you integrate over a piecewise smooth curve C. (The terminology is somewhat unfortunate—this type of integral might be better described as a "curve integral.")

To introduce the concept of a line integral, consider the mass of a wire of finite length, given by a curve C in space. The density (mass per unit length) of the wire at the point (x, y, z) is given by $f(x, y, z)$. Partition the curve C by the points $P_0, P_1, \ldots, P_n$, producing n subarcs, as shown in Figure 14.8. The length of the ith subarc is given by Δs_i. Next, choose a point (x_i, y_i, z_i) in each subarc. If the length of each subarc is small, the total mass of the wire can be approximated by the sum

$$\text{Mass of wire } \approx \sum_{i=1}^n f(x_i, y_i, z_i)\, \Delta s_i.$$

If you let $\|\Delta\|$ denote the length of the longest subarc and let $\|\Delta\|$ approach 0, it seems reasonable that the limit of this sum approaches the mass of the wire. This leads to the following definition.

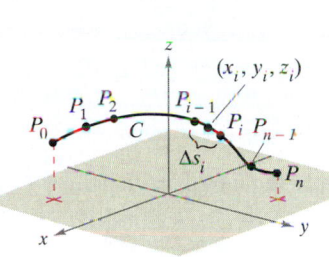

Partitioning of curve C
Figure 14.8

Definition of Line Integral

If f is defined in a region containing a smooth curve C of finite length, then the **line integral of f along C** is given by

$$\int_C f(x, y)\, ds = \lim_{\|\Delta\| \to 0} \sum_{i=1}^n f(x_i, y_i) \Delta s_i \qquad \text{Plane}$$

or

$$\int_C f(x, y, z)\, ds = \lim_{\|\Delta\| \to 0} \sum_{i=1}^n f(x_i, y_i, z_i) \Delta s_i \qquad \text{Space}$$

provided this limit exists.

As with the integrals discussed in Chapter 13, evaluation of a line integral is best accomplished by converting to a definite integral. It can be shown that if f is *continuous*, the limit given above exists and is the same for all smooth parametrizations of C.

To evaluate a line integral over a plane curve C given by $\mathbf{r}(t) = x(t)\mathbf{i} + y(t)\mathbf{j}$, use the fact that

$$ds = \|\mathbf{r}'(t)\| \, dt = \sqrt{[x'(t)]^2 + [y'(t)]^2} \, dt.$$

A similar formula holds for a space curve, as indicated in the following theorem.

THEOREM 14.4   Evaluation of a Line Integral as a Definite Integral

Let f be continuous in a region containing a smooth curve C. If C is given by $\mathbf{r}(t) = x(t)\mathbf{i} + y(t)\mathbf{j}$, where $a \leq t \leq b$, then

$$\int_C f(x, y) \, ds = \int_a^b f(x(t), y(t)) \sqrt{[x'(t)]^2 + [y'(t)]^2} \, dt.$$

If C is given by $\mathbf{r}(t) = x(t)\mathbf{i} + y(t)\mathbf{j} + z(t)\mathbf{k}$, where $a \leq t \leq b$, then

$$\int_C f(x, y, z) \, ds = \int_a^b f(x(t), y(t), z(t)) \sqrt{[x'(t)]^2 + [y'(t)]^2 + [z'(t)]^2} \, dt.$$

Note that if $f(x, y, z) = 1$, the line integral gives the arc length of the curve C, as defined in Section 11.5. That is,

$$\int_C 1 \, ds = \int_a^b \|\mathbf{r}'(t)\| \, dt = \text{length of curve } C.$$

EXAMPLE 2   Evaluating a Line Integral

NOTE   The value of the line integral in Example 2 does not depend on the parametrization of the line segment C (any smooth parametrization will produce the same value). To convince yourself of this, try some other parametrizations, such as $x = 1 + 2t$, $y = 2 + 4t$, $z = 1 + 2t$, $-\frac{1}{2} \leq t \leq 0$, or $x = -t$, $y = -2t$, $z = -t$, $-1 \leq t \leq 0$.

Evaluate

$$\int_C (x^2 - y + 3z) \, ds$$

where C is the line segment shown in Figure 14.9.

Solution   Begin by writing a parametric form of the equation of a line:

$$x = t, \quad y = 2t, \quad \text{and} \quad z = t, \ 0 \leq t \leq 1.$$

Hence, $x'(t) = 1$, $y'(t) = 2$, and $z'(t) = 1$, which implies that

$$\sqrt{[x'(t)]^2 + [y'(t)]^2 + [z'(t)]^2} = \sqrt{1^2 + 2^2 + 1^2} = \sqrt{6}.$$

Thus, the line integral takes the following form.

$$\int_C (x^2 - y + 3z) \, ds = \int_0^1 (t^2 - 2t + 3t) \sqrt{6} \, dt$$

$$= \sqrt{6} \int_0^1 (t^2 + t) \, dt$$

$$= \sqrt{6} \left[\frac{t^3}{3} + \frac{t^2}{2} \right]_0^1$$

$$= \frac{5\sqrt{6}}{6}$$

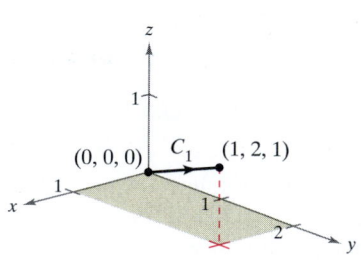

Figure 14.9

Suppose C is a path composed of smooth curves $C_1, C_2, \ldots, C_n$. If f is continuous on C, it can be shown that

$$\int_C f(x, y)\, ds = \int_{C_1} f(x, y)\, ds + \int_{C_2} f(x, y)\, ds + \cdots + \int_{C_n} f(x, y)\, ds.$$

This property is used in Example 3.

EXAMPLE 3   Evaluating a Line Integral Over a Path

Evaluate $\int_C x\, ds$, where C is the piecewise smooth curve shown in Figure 14.10.

Solution   Begin by integrating up the line $y = x$, using the following parametrization.

$$C_1 \colon x = t,\, y = t,\quad 0 \le t \le 1$$

For this curve, $\mathbf{r}(t) = t\mathbf{i} + t\mathbf{j}$, which implies that $x'(t) = 1$ and $y'(t) = 1$. Thus,

$$\sqrt{[x'(t)]^2 + [y'(t)]^2} = \sqrt{2}$$

and you have

$$\int_{C_1} x\, ds = \int_0^1 t\sqrt{2}\, dt = \frac{\sqrt{2}}{2} t^2 \Big]_0^1 = \frac{\sqrt{2}}{2}.$$

Next, integrate down the parabola $y = x^2$, using parametrization

$$C_2 \colon x = 1 - t,\, y = (1 - t)^2,\quad 0 \le t \le 1.$$

For this curve, $\mathbf{r}(t) = (1 - t)\mathbf{i} + (1 - t)^2\mathbf{j}$, which implies that $x'(t) = -1$ and $y'(t) = -2(1 - t)$. Thus,

$$\sqrt{[x'(t)]^2 + [y'(t)]^2} = \sqrt{1 + 4(1 - t)^2},$$

and you have

$$\int_{C_2} x\, ds = \int_0^1 (1 - t)\sqrt{1 + 4(1 - t)^2}\, dt$$

$$= -\frac{1}{8}\left[\frac{2}{3}[1 + 4(1 - t)^2]^{3/2}\right]_0^1$$

$$= \frac{1}{12}(5^{3/2} - 1).$$

Consequently,

$$\int_C x\, ds = \int_{C_1} x\, ds + \int_{C_2} x\, ds = \frac{\sqrt{2}}{2} + \frac{1}{12}(5^{3/2} - 1) \approx 1.56.$$

For parametrizations given by $\mathbf{r}(t) = x(t)\mathbf{i} + y(t)\mathbf{j} + z(t)\mathbf{k}$, it is helpful to remember the form of ds as

$$ds = \|\mathbf{r}'(t)\|\, dt = \sqrt{[x'(t)]^2 + [y'(t)]^2 + [z'(t)]^2}\, dt.$$

This is demonstrated in Example 4.

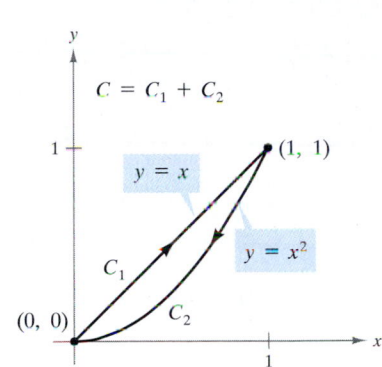

Figure 14.10

EXAMPLE 4 **Evaluating a Line Integral**

Evaluate $\int_C (x + 2) \, ds$, where C is the curve represented by

$$\mathbf{r}(t) = t\mathbf{i} + \frac{4}{3}t^{3/2}\mathbf{j} + \frac{1}{2}t^2\mathbf{k}, \quad 0 \le t \le 2.$$

Solution Because $\mathbf{r}'(t) = \mathbf{i} + 2t^{1/2}\mathbf{j} + t\mathbf{k}$, and

$$\|\mathbf{r}'(t)\| = \sqrt{[x'(t)]^2 + [y'(t)]^2 + [z'(t)]^2} = \sqrt{1 + 4t + t^2}$$

it follows that

$$\int_C (x + 2) \, ds = \int_0^2 (t + 2)\sqrt{1 + 4t + t^2} \, dt$$

$$= \frac{1}{2}\int_0^2 2(t + 2)(1 + 4t + t^2)^{1/2} \, dt$$

$$= \frac{1}{3}\left[(1 + 4t + t^2)^{3/2}\right]_0^2$$

$$= \frac{1}{3}\left(13\sqrt{13} - 1\right)$$

$$\approx 15.29.$$

The next example shows how a line integral can be used to find the mass of a spring whose density varies. In Figure 14.11, note that the density of this spring increases as the spring spirals up the z-axis.

EXAMPLE 5 **Finding the Mass of a Spring**

Find the mass of a spring in the shape of the circular helix

$$\mathbf{r}(t) = \frac{1}{\sqrt{2}}(\cos t\mathbf{i} + \sin t\mathbf{j} + t\mathbf{k}), \quad 0 \le t \le 6\pi$$

where the density of the wire is $\rho(x, y, z) = 1 + z$, as shown in Figure 14.11.

Solution Because

$$\|\mathbf{r}'(t)\| = \frac{1}{\sqrt{2}}\sqrt{(-\sin t)^2 + (\cos t)^2 + (1)^2} = 1$$

it follows that the mass of the spring is

$$\text{Mass} = \int_C (1 + z) \, ds = \int_0^{6\pi}\left(1 + \frac{t}{\sqrt{2}}\right) dt$$

$$= \left[t + \frac{t^2}{2\sqrt{2}}\right]_0^{6\pi}$$

$$= 6\pi\left(1 + \frac{3\pi}{\sqrt{2}}\right)$$

$$\approx 144.47.$$

Density:
$\rho(x, y, z) = 1 + z$

$\mathbf{r}(t) = \frac{1}{\sqrt{2}}(\cos t\,\mathbf{i} + \sin t\,\mathbf{j} + t\,\mathbf{k})$

The mass of the spring is approximately 144.47.

Figure 14.11

Line Integrals of Vector Fields

One of the most important physical applications of line integrals is that of finding the **work** done on an object moving in a force field. For example, Figure 14.12 shows an inverse square force field similar to the gravitational field of the sun. Note that the magnitude of the force along a circular path about the center is constant, whereas the magnitude of the force along a parabolic path varies from point to point.

To see how a line integral can be used to find work done in a force field **F**, consider an object moving along a path C in the field, as shown in Figure 14.13. To determine the work done by the force, you need consider only that part of the force that is acting in the same direction as that in which the object is moving (or the opposite direction). This means that at each point on C, you can consider the projection $\mathbf{F} \cdot \mathbf{T}$ of the force vector **F** onto the unit tangent vector **T**. On a small subarc of length Δs_i, the increment of work is

$$\Delta W_i = (\text{force})(\text{distance})$$
$$\approx [\mathbf{F}(x_i, y_i, z_i) \cdot \mathbf{T}(x_i, y_i, z_i)]\Delta s_i$$

where (x_i, y_i, z_i) is a point in the ith subarc. Consequently the total work done is given by the following integral.

$$W = \int_C \mathbf{F}(x, y, z) \cdot \mathbf{T}(x, y, z) \, ds$$

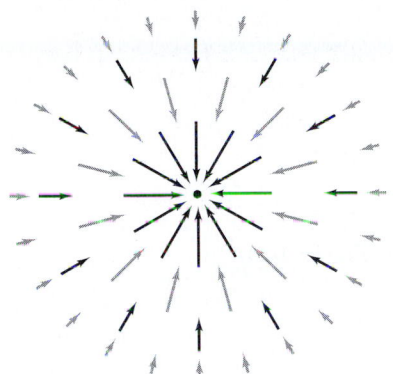

Inverse square force field **F**

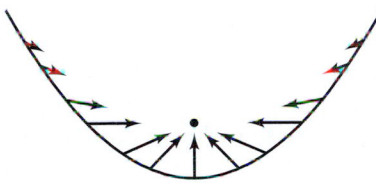

Vectors along a parabolic path in the force field **F**
Figure 14.12

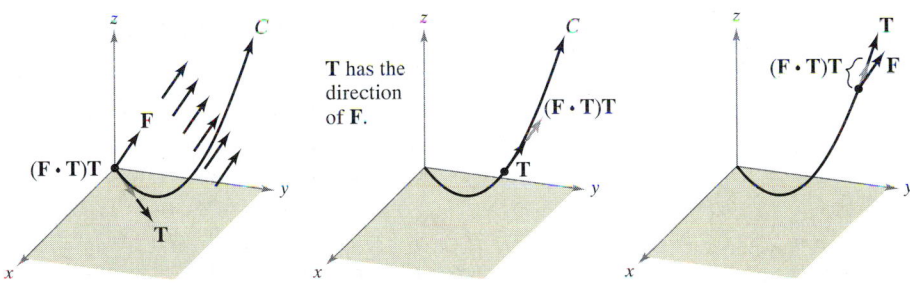

At each point on C, the force in the direction of motion is $(\mathbf{F} \cdot \mathbf{T})\mathbf{T}$.
Figure 14.13

This line integral appears in other contexts and is the basis of the following definition of the **line integral of a vector field.** Note in the definition that

$$\mathbf{F} \cdot \mathbf{T} \, ds = \mathbf{F} \cdot \frac{\mathbf{r}'(t)}{\|\mathbf{r}'(t)\|} \|\mathbf{r}'(t)\| \, dt$$
$$= \mathbf{F} \cdot \mathbf{r}'(t) \, dt$$
$$= \mathbf{F} \cdot d\mathbf{r}.$$

Definition of Line Integral of a Vector Field

Let **F** be a continuous vector field defined on a smooth curve C given by $\mathbf{r}(t), a \leq t \leq b$. The **line integral** of **F** on C is given by

$$\int_C \mathbf{F} \cdot d\mathbf{r} = \int_C \mathbf{F} \cdot \mathbf{T} \, ds = \int_a^b \mathbf{F}(x(t), y(t), z(t)) \cdot \mathbf{r}'(t) \, dt.$$

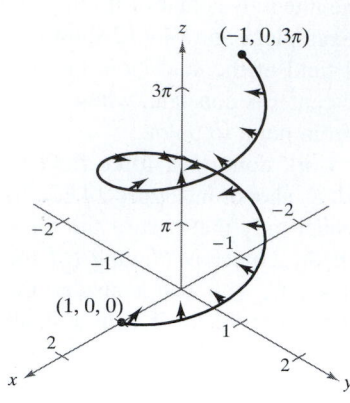

Figure 14.14

EXAMPLE 6   Work Done by a Force

Find the work done by the force field

$$\mathbf{F}(x, y, z) = -\frac{1}{2}x\mathbf{i} - \frac{1}{2}y\mathbf{j} + \frac{1}{4}\mathbf{k} \qquad \text{Force field } \mathbf{F}$$

on a particle as it moves along the helix given by

$$\mathbf{r}(t) = \cos t\mathbf{i} + \sin t\mathbf{j} + t\mathbf{k} \qquad \text{Space curve } C$$

from the point $(1, 0, 0)$ to $(-1, 0, 3\pi)$, as shown in Figure 14.14.

Solution   Because

$$\mathbf{r}(t) = x(t)\mathbf{i} + y(t)\mathbf{j} + z(t)\mathbf{k}$$
$$= \cos t\mathbf{i} + \sin t\mathbf{j} + t\mathbf{k}$$

it follows that $x(t) = \cos t$, $y(t) = \sin t$, and $z(t) = t$. Thus, the force field can be written as

$$\mathbf{F}(x(t), y(t), z(t)) = -\frac{1}{2}\cos t\mathbf{i} - \frac{1}{2}\sin t\mathbf{j} + \frac{1}{4}\mathbf{k}.$$

To find the work done by the force field in moving a particle along the curve C, use the fact that

$$\mathbf{r}'(t) = -\sin t\mathbf{i} + \cos t\mathbf{j} + \mathbf{k}$$

and write the following.

$$W = \int_C \mathbf{F} \cdot d\mathbf{r}$$

$$= \int_a^b \mathbf{F}(x(t), y(t), z(t)) \cdot \mathbf{r}'(t)\, dt$$

$$= \int_0^{3\pi} \left(-\frac{1}{2}\cos t\mathbf{i} - \frac{1}{2}\sin t\mathbf{j} + \frac{1}{4}\mathbf{k}\right) \cdot (-\sin t\mathbf{i} + \cos t\mathbf{j} + \mathbf{k})\, dt$$

$$= \int_0^{3\pi} \left(\frac{1}{2}\sin t \cos t - \frac{1}{2}\sin t \cos t + \frac{1}{4}\right) dt$$

$$= \int_0^{3\pi} \frac{1}{4}\, dt$$

$$= \frac{1}{4}t\Big]_0^{3\pi}$$

$$= \frac{3\pi}{4}$$

NOTE   In Example 6, note that the x- and y-components of the force field end up contributing nothing to the total work. This occurs because *in this particular example* the z-component of the force field is the only portion of the force that is acting in the same (or opposite) direction in which the particle is moving (see Figure 14.15).

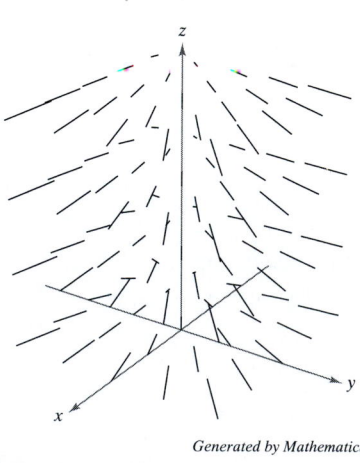

Generated by Mathematica

Figure 14.15

TECHNOLOGY   The computer-generated view of the force field in Example 6 shown in Figure 14.15 indicates that each vector in the force field points toward the z-axis.

For line integrals of vector functions, the orientation of the curve C is important. If the orientation of the curve is reversed, the unit tangent vector $\mathbf{T}(t)$ is changed to $-\mathbf{T}(t)$, and you obtain

$$\int_{-C} \mathbf{F} \cdot d\mathbf{r} = -\int_{C} \mathbf{F} \cdot d\mathbf{r}.$$

EXAMPLE 7   Orientation and Parametrization of a Curve

Let $\mathbf{F}(x, y) = y\mathbf{i} + x^2\mathbf{j}$ and evaluate the line integral $\int_C \mathbf{F} \cdot d\mathbf{r}$ for each of the following parabolic curves (see Figure 14.16).

a. $C_1: \mathbf{r}_1(t) = (4 - t)\mathbf{i} + (4t - t^2)\mathbf{j}, \quad 0 \le t \le 3$
b. $C_2: \mathbf{r}_2(t) = t\mathbf{i} + (4t - t^2)\mathbf{j}, \quad 1 \le t \le 4$

Solution

a. Because $\mathbf{r}_1'(t) = -\mathbf{i} + (4 - 2t)\mathbf{j}$ and

$$\mathbf{F}(x(t), y(t)) = (4t - t^2)\mathbf{i} + (4 - t)^2\mathbf{j}$$

the line integral is

$$\int_{C_1} \mathbf{F} \cdot d\mathbf{r} = \int_0^3 [(4t - t^2)\mathbf{i} + (4 - t)^2\mathbf{j}] \cdot [-\mathbf{i} + (4 - 2t)\mathbf{j}]\, dt$$

$$= \int_0^3 (-4t + t^2 + 64 - 64t + 20t^2 - 2t^3)\, dt$$

$$= \int_0^3 (-2t^3 + 21t^2 - 68t + 64)\, dt$$

$$= \left[-\frac{t^4}{2} + 7t^3 - 34t^2 + 64t \right]_0^3$$

$$= \frac{69}{2}.$$

b. Because $\mathbf{r}_2'(t) = \mathbf{i} + (4 - 2t)\mathbf{j}$ and

$$\mathbf{F}(x(t), y(t)) = (4t - t^2)\mathbf{i} + t^2\mathbf{j}$$

the line integral is

$$\int_{C_2} \mathbf{F} \cdot d\mathbf{r} = \int_1^4 [(4t - t^2)\mathbf{i} + t^2\mathbf{j}] \cdot [\mathbf{i} + (4 - 2t)\mathbf{j}]\, dt$$

$$= \int_1^4 (4t - t^2 + 4t^2 - 2t^3)\, dt$$

$$= \int_1^4 (-2t^3 + 3t^2 + 4t)\, dt$$

$$= \left[-\frac{t^4}{2} + t^3 + 2t^2 \right]_1^4$$

$$= -\frac{69}{2}.$$

The answer in part (b) is the negative of that in part (a) because C_1 and C_2 represent opposite orientations of the same parabolic segment.

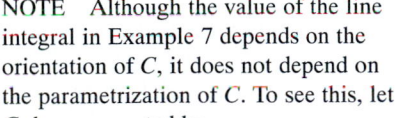

$C_1: \mathbf{r}_1(t) = (4 - t)\mathbf{i} + (4t - t^2)\mathbf{j}$
$C_2: \mathbf{r}_2(t) = t\mathbf{i} + (4t - t^2)\mathbf{j}$

Figure 14.16

NOTE   Although the value of the line integral in Example 7 depends on the orientation of C, it does not depend on the parametrization of C. To see this, let C_3 be represented by

$$\mathbf{r}_3 = (t + 2)\mathbf{i} + (4 - t^2)\mathbf{j}$$

where $-1 \le t \le 2$. The graph of this curve is the same parabolic segment shown in Figure 14.16. Does the value of the line integral over C_3 agree with the value over C_1 or C_2? Why or why not?

Line Integrals in Differential Form

A second commonly used form of line integral is derived from the vector field notation used in the previous section. If $\mathbf{F}$ is a vector field of the form $\mathbf{F}(x, y) = M\mathbf{i} + N\mathbf{j}$, and C is given by $\mathbf{r}(t) = x(t)\mathbf{i} + y(t)\mathbf{j}$, then $\mathbf{F} \cdot d\mathbf{r}$ is often written as $M\,dx + N\,dy$.

$$\int_C \mathbf{F} \cdot d\mathbf{r} = \int_C \mathbf{F} \cdot \frac{d\mathbf{r}}{dt}\,dt$$

$$= \int_a^b (M\mathbf{i} + N\mathbf{j}) \cdot (x'(t)\mathbf{i} + y'(t)\mathbf{j})\,dt$$

$$= \int_a^b \left(M\frac{dx}{dt} + N\frac{dy}{dt} \right) dt$$

$$= \int_C (M\,dx + N\,dy)$$

This **differential form** can be extended to three variables. The parentheses are often omitted, as follows.

$$\int_C M\,dx + N\,dy \qquad \text{and} \qquad \int_C M\,dx + N\,dy + P\,dz$$

Notice how this differential notation is used in Example 8.

NOTE The orientation of C affects the value of the differential form of a line integral. Specifically, if $-C$ has the orientation opposite to that of C, then

$$\int_{-C} M\,dx + N\,dy =$$

$$-\int_C M\,dx + N\,dy.$$

Thus, of the three line integral forms presented in this section, the orientation of C does not affect the form $\int_C f(x, y)\,ds$, but it does affect the vector form and the differential form.

EXAMPLE 8   Evaluating a Line Integral in Differential Form

Let C be the circle of radius 3 given by

$$\mathbf{r}(t) = 3 \cos t\mathbf{i} + 3 \sin t\mathbf{j}, \quad 0 \le t \le 2\pi$$

and evaluate the line integral

$$\int_C y^3\,dx + (x^3 + 3xy^2)\,dy.$$

(See Figure 14.17.)

Solution Because $x = 3 \cos t$ and $y = 3 \sin t$, you have $dx = -3 \sin t\,dt$ and $dy = 3 \cos t\,dt$. Thus, the line integral is

$$\int_C M\,dx + N\,dy$$

$$= \int_C y^3\,dx + (x^3 + 3xy^2)\,dy$$

$$= \int_0^{2\pi} [(27 \sin^3 t)(-3 \sin t) + (27 \cos^3 t + 81 \cos t \sin^2 t)(3 \cos t)]\,dt$$

$$= 81 \int_0^{2\pi} (\cos^4 t - \sin^4 t + 3 \cos^2 t \sin^2 t)\,dt$$

$$= 81 \int_0^{2\pi} \left(\cos^2 t - \sin^2 t + \frac{3}{4} \sin^2 2t \right) dt$$

$$= 81 \int_0^{2\pi} \left[\cos 2t + \frac{3}{4}\left(\frac{1 - \cos 4t}{2} \right) \right] dt$$

$$= 81 \left[\frac{\sin 2t}{2} + \frac{3}{8}t - \frac{3 \sin 4t}{32} \right]_0^{2\pi} = \frac{243\pi}{4}.$$

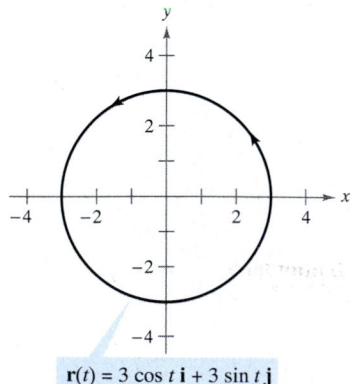

$\mathbf{r}(t) = 3 \cos t\,\mathbf{i} + 3 \sin t\,\mathbf{j}$

Figure 14.17

For curves represented by $y = g(x)$, $a \le x \le b$, you can let $x = t$ and obtain the parametric form

$$x = t \quad \text{and} \quad y = g(t), \quad a \le t \le b.$$

Because $dx = dt$ for this form, you have the option of evaluating the line integral in the variable x or t. This is demonstrated in Example 9.

EXAMPLE 9   Evaluating a Line Integral in Differential Form

Evaluate

$$\int_C y \, dx + x^2 \, dy$$

where C is the parabolic arc given by $y = 4x - x^2$ from $(4, 0)$ to $(1, 3)$, as shown in Figure 14.18.

Solution   Rather than converting to the parameter t, you can simply retain the variable x and write

$$y = 4x - x^2 \quad \Longrightarrow \quad dy = (4 - 2x) \, dx.$$

Then, in the direction from $(4, 0)$ to $(1, 3)$, the line integral is

$$\int_C y \, dx + x^2 \, dy = \int_4^1 \left[(4x - x^2) \, dx + x^2(4 - 2x) \, dx \right]$$

$$= \int_4^1 (4x + 3x^2 - 2x^3) \, dx$$

$$= \left[2x^2 + x^3 - \frac{x^4}{2} \right]_4^1$$

$$= \frac{69}{2}.$$

(See Example 7.)

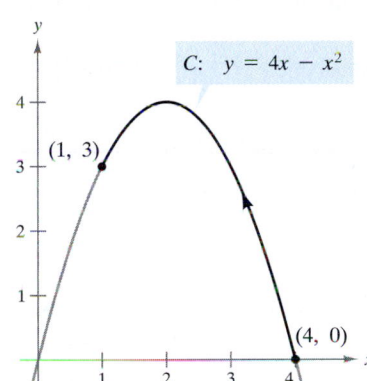

C: $y = 4x - x^2$

$(1, 3)$

$(4, 0)$

The line integral over C from $(4, 0)$ to $(1, 3)$ is $\frac{69}{2}$.

Figure 14.18

EXPLORATION

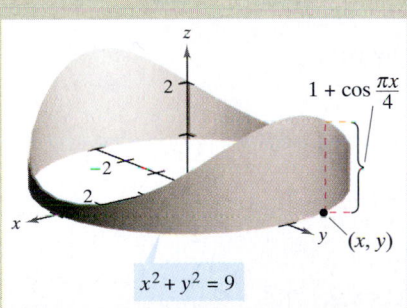

$x^2 + y^2 = 9$

Finding Lateral Surface Area   The figure at the left shows a piece of tin that has been cut from a circular cylinder. The base of the circular cylinder can be modeled by

$$x^2 + y^2 = 9. \qquad \text{Circular base}$$

At any point (x, y) on the base, the height of the object is given by

$$f(x, y) = 1 + \cos \frac{\pi x}{4}.$$

Explain how to use a line integral to find the surface area of the piece of tin.

In Exercises 1–6, find a piecewise smooth parametrization of the path C.

1.

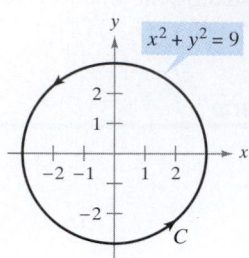

$x^2 + y^2 = 9$

2.

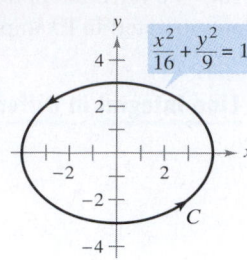

$\dfrac{x^2}{16} + \dfrac{y^2}{9} = 1$

3.

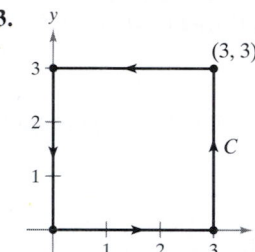

$(3, 3)$

4.

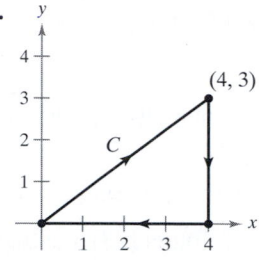

$(4, 3)$

5.

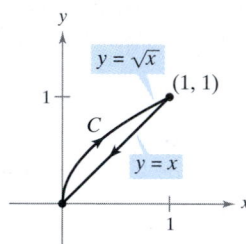

$y = \sqrt{x}$ $(1, 1)$ $y = x$

6.

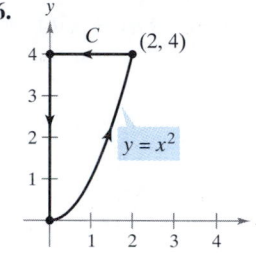

C $(2, 4)$ $y = x^2$

In Exercises 7–10, evaluate the line integral over the indicated path.

7. $\displaystyle\int_C (x - y)\, ds$

   $C\!:\ \mathbf{r}(t) = 4t\mathbf{i} + 3t\mathbf{j}$

   $0 \le t \le 2$

8. $\displaystyle\int_C 4xy\, ds$

   $C\!:\ \mathbf{r}(t) = t\mathbf{i} + (1 - t)\mathbf{j}$

   $0 \le t \le 1$

9. $\displaystyle\int_C (x^2 + y^2 + z^2)\, ds$

   $C\!:\ \mathbf{r}(t) = \sin t\,\mathbf{i} + \cos t\,\mathbf{j} + 8t\mathbf{k}$

   $0 \le t \le \pi/2$

10. $\displaystyle\int_C 8xyz\, ds$

   $C\!:\ \mathbf{r}(t) = 3\mathbf{i} + 12t\mathbf{j} + 5t\mathbf{k}$

   $0 \le t \le 2$

In Exercises 11–14, evaluate

$$\int_C (x^2 + y^2)\, ds$$

along the given path.

11. $C\!:$ x-axis from $x = 0$ to $x = 3$

12. $C\!:$ y-axis from $y = 1$ to $y = 10$

13. $C\!:$ counterclockwise around the circle $x^2 + y^2 = 1$ from $(1, 0)$ to $(0, 1)$

14. $C\!:$ counterclockwise around the circle $x^2 + y^2 = 4$ from $(2, 0)$ to $(0, 2)$

In Exercises 15–18, evaluate

$$\int_C \left(x + 4\sqrt{y}\right) ds$$

along the given path.

15. $C\!:$ line from $(0, 0)$ to $(1, 1)$

16. $C\!:$ line from $(0, 0)$ to $(3, 9)$

17. $C\!:$ counterclockwise around the triangle with vertices $(0, 0)$, $(1, 0)$ and $(0, 1)$

18. $C\!:$ counterclockwise around the square with vertices $(0, 0)$, $(1, 0)$, $(1, 1)$, and $(0, 1)$

Mass   **In Exercises 19 and 20, find the total mass of two turns of a spring with density ρ in the shape of the circular helix**

$$\mathbf{r}(t) = 3\cos t\,\mathbf{i} + 3\sin t\,\mathbf{j} + 2t\mathbf{k}.$$

19. $\rho(x, y, z) = \frac{1}{2}(x^2 + y^2 + z^2)$

20. $\rho(x, y, z) = 2$

In Exercises 21–26, evaluate

$$\int_C \mathbf{F} \cdot d\mathbf{r}$$

where C is represented by $\mathbf{r}(t)$.

21. $\mathbf{F}(x, y) = xy\mathbf{i} + y\mathbf{j}$

   $C\!:\ \mathbf{r}(t) = 4t\mathbf{i} + t\mathbf{j},\quad 0 \le t \le 1$

22. $\mathbf{F}(x, y) = xy\mathbf{i} + y\mathbf{j}$

   $C\!:\ \mathbf{r}(t) = 4\cos t\,\mathbf{i} + 4\sin t\,\mathbf{j},\quad 0 \le t \le \dfrac{\pi}{2}$

23. $\mathbf{F}(x, y) = 3x\mathbf{i} + 4y\mathbf{j}$

   $C\!:\ \mathbf{r}(t) = 2\cos t\,\mathbf{i} + 2\sin t\,\mathbf{j},\quad 0 \le t \le \dfrac{\pi}{2}$

24. $\mathbf{F}(x, y) = 3x\mathbf{i} + 4y\mathbf{j}$

   $C\!:\ \mathbf{r}(t) = t\mathbf{i} + \sqrt{4 - t^2}\,\mathbf{j},\quad -2 \le t \le 2$

25. $F(x, y, z) = x^2 y\mathbf{i} + (x - z)\mathbf{j} + xyz\mathbf{k}$

$C: \mathbf{r}(t) = t\mathbf{i} + t^2\mathbf{j} + 2\mathbf{k}, \quad 0 \le t \le 1$

26. $F(x, y, z) = x^2\mathbf{i} + y^2\mathbf{j} + z^2\mathbf{k}$

$C: \mathbf{r}(t) = \sin t\mathbf{i} + \cos t\mathbf{j} + t^2\mathbf{k}, \quad 0 \le t \le \dfrac{\pi}{2}$

In Exercises 27 and 28, use a symbolic integration utility to evaluate the integral

$$\int_C \mathbf{F} \cdot d\mathbf{r}$$

where C is represented by $\mathbf{r}(t)$.

27. $F(x, y, z) = x^2 z\mathbf{i} + 6y\mathbf{j} + yz^2\mathbf{k}$

$C: \mathbf{r}(t) = t\mathbf{i} + t^2\mathbf{j} + \ln t\mathbf{k}, \quad 1 \le t \le 3$

28. $F(x, y, z) = \dfrac{x\mathbf{i} + y\mathbf{j} + z\mathbf{k}}{\sqrt{x^2 + y^2 + z^2}}$

$C: \mathbf{r}(t) = t\mathbf{i} + t\mathbf{j} + e^t\mathbf{k}, \quad 0 \le t \le 2$

Work In Exercises 29–34, find the work done by the force field **F** on an object moving along the indicated path.

29. $F(x, y) = -x\mathbf{i} - 2y\mathbf{j}$

$C: y = x^3$ from $(0, 0)$ to $(2, 8)$

30. $F(x, y) = x^2\mathbf{i} - xy\mathbf{j}$

$C: x = \cos^3 t, \ y = \sin^3 t$ from $(1, 0)$ to $(0, 1)$

31. $F(x, y) = 2x\mathbf{i} + y\mathbf{j}$

$C:$ counterclockwise around the triangle with vertices $(0, 0)$, $(1, 0)$, and $(1, 1)$

32. $F(x, y) = -y\mathbf{i} - x\mathbf{j}$

$C:$ counterclockwise along the semicircle $y = \sqrt{4 - x^2}$ from $(2, 0)$ to $(-2, 0)$

33. $F(x, y, z) = x\mathbf{i} + y\mathbf{j} - 5z\mathbf{k}$

$C: \mathbf{r}(t) = 2 \cos t\mathbf{i} + 2 \sin t\mathbf{j} + t\mathbf{k}, \quad 0 \le t \le 2\pi$

34. $F(x, y, z) = yz\mathbf{i} + xz\mathbf{j} + xy\mathbf{k}$

$C:$ line from $(0, 0, 0)$ to $(5, 3, 2)$

35. _Work_ Find the work done by a person weighing 150 pounds walking exactly one revolution up a circular helical staircase of radius 3 feet if the person rises 10 feet.

36. _Work_ A particle moves along the path $y = x^2$ from the point $(0, 0)$ to the point $(1, 1)$. The force field **F** is measured at five points along the path and the results are shown in the table. Use Simpson's Rule or a graphing utility to approximate the work done by the force field.

(x, y)	$(0, 0)$	$\left(\frac{1}{4}, \frac{1}{16}\right)$	$\left(\frac{1}{2}, \frac{1}{4}\right)$	$\left(\frac{3}{4}, \frac{9}{16}\right)$	$(1, 1)$
$F(x, y)$	$\langle 5, 0 \rangle$	$\langle 3.5, 1 \rangle$	$\langle 2, 2 \rangle$	$\langle 1.5, 3 \rangle$	$\langle 1, 5 \rangle$

In Exercises 37–40, demonstrate the property that

$$\int_C \mathbf{F} \cdot d\mathbf{r} = 0$$

regardless of the initial and terminal points of C, if the tangent vector $\mathbf{r}'(t)$ is orthogonal to the force field **F**.

37. $F(x, y) = y\mathbf{i} - x\mathbf{j}$

$C: \mathbf{r}(t) = t\mathbf{i} - 2t\mathbf{j}$

38. $F(x, y) = -3y\mathbf{i} + x\mathbf{j}$

$C: \mathbf{r}(t) = t\mathbf{i} - t^3\mathbf{j}$

39. $F(x, y) = (x^3 - 2x^2)\mathbf{i} + \left(x - \dfrac{y}{2}\right)\mathbf{j}$

$C: \mathbf{r}(t) = t\mathbf{i} + t^2\mathbf{j}$

40. $F(x, y) = x\mathbf{i} + y\mathbf{j}$

$C: \mathbf{r}(t) = 3 \sin t\mathbf{i} + 3 \cos t\mathbf{j}$

41. _Investigation_ Determine the value of c such that the work done by the force field

$$F(x, y) = 15[(4 - x^2 y)\mathbf{i} - xy\mathbf{j}]$$

on an object moving along the parabolic path $y = c(1 - x^2)$ between the points $(-1, 0)$ and $(1, 0)$ is a minimum. Compare the result with the work required to move the object along the straight-line path connecting the points.

42. _Think About It_ For each of the following, determine whether the work done in moving an object from the first to the second point through the force field shown in the figure is positive, negative, or zero. Explain your answer.

(a) From $(-3, -3)$ to $(3, 3)$

(b) From $(-3, 0)$ to $(0, 3)$

(c) From $(5, 0)$ to $(0, 3)$

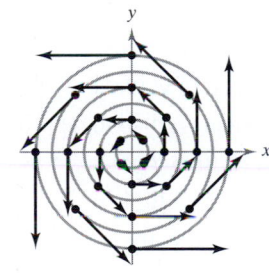

In Exercises 43–46, evaluate the line integral over the path C given by $x = 2t, y = 10t$, where $0 \le t \le 1$.

43. $\displaystyle\int_C (x + 3y^2)\, dy$

44. $\displaystyle\int_C (x + 3y^2)\, dx$

45. $\displaystyle\int_C xy\, dx + y\, dy$

46. $\displaystyle\int_C (y - 3x)\, dx + x^2\, dy$

In Exercises 47–52, evaluate the integral

$$\int_C (2x - y)\, dx + (x + 3y)\, dy$$

along the path.

47. C: x-axis from $x = 0$ to $x = 5$

48. C: y-axis from $y = 0$ to $y = 2$

49. C: line segments from $(0, 0)$ to $(3, 0)$ and $(3, 0)$ to $(3, 3)$

50. C: line segments from $(0, 0)$ to $(0, -3)$ and $(0, -3)$ to $(2, -3)$

51. C: parabolic path $x = t$, $y = 2t^2$, from $(0, 0)$ to $(2, 8)$

52. C: elliptic path $x = 4 \sin t$, $y = 3 \cos t$, from $(0, 3)$ to $(4, 0)$

Lateral Surface Area **In Exercises 53–60, find the area of the lateral surface (see figure) over the curve C in the xy-plane and under the surface $z = f(x, y)$, where**

Lateral surface area $= \displaystyle\int_C f(x, y)\, ds.$

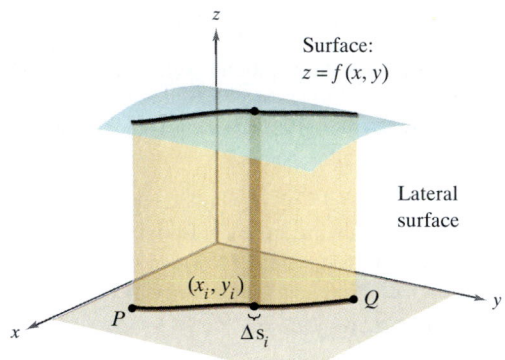

Surface:
$z = f(x, y)$

Lateral surface

(x_i, y_i)

P Δs_i Q

C: Curve in xy-plane

53. $f(x, y) = h$, C: line from $(0, 0)$ to $(3, 4)$

54. $f(x, y) = y$, C: line from $(0, 0)$ to $(4, 4)$

55. $f(x, y) = xy$, C: $x^2 + y^2 = 1$ from $(1, 0)$ to $(0, 1)$

56. $f(x, y) = x + y$, C: $x^2 + y^2 = 1$ from $(1, 0)$ to $(0, 1)$

57. $f(x, y) = h$, C: $y = 1 - x^2$ from $(1, 0)$ to $(0, 1)$

58. $f(x, y) = y + 1$, C: $1 - x^2$ from $(1, 0)$ to $(0, 1)$

59. $f(x, y) = xy$, C: $y = 1 - x^2$ from $(1, 0)$ to $(0, 1)$

60. $f(x, y) = x^2 - y^2 + 4$, C: $x^2 + y^2 = 4$

61. *Engine Design* A tractor engine has a steel component with a circular base modeled by the vector-valued function $\mathbf{r}(t) = 2 \cos t\mathbf{i} + 2 \sin t\mathbf{j}$. Its height is given by $z = 1 + y^2$. All measurements of the component are given in centimeters.

(a) Find the lateral surface area of the component.

(b) If the component is in the form of a shell of thickness 0.2 centimeter, use the result in part (a) to approximate the amount of steel used in its manufacture.

(c) Make a sketch of the component.

62. *Building Design* The ceiling of a building has a height above the floor given by $z = 20 + \frac{1}{4}x$, and one of the walls follows a path modeled by $y = x^{3/2}$. Find the surface area of the wall if $0 \le x \le 40$. (All measurements are given in feet.)

63. *Think About It* Order the surfaces in ascending order of the lateral surface area under the surface and over the curve $y = \sqrt{x}$ from $(0, 0)$ to $(4, 2)$ in the xy-plane. Explain your ordering without doing any calculations.

(a) $z_1 = 2 + x$ (b) $z_2 = 5 + x$

(c) $z_3 = 2$ (d) $z_4 = 10 + x + 2y$

Approximation **In Exercises 64 and 65, determine which value best approximates the lateral surface area over the curve C in the xy-plane and under the surface $z = f(x, y)$. (Make your selection on the basis of a sketch of the surface and *not* by performing any calculations.)**

64. $f(x, y) = y$

C: $y = x^2$ from $(0, 0)$ to $(2, 4)$

(a) 2   (b) 4   (c) 8   (d) 16

65. $f(x, y) = e^{xy}$

C: line from $(0, 0)$ to $(2, 2)$

(a) 54   (b) 25   (c) -250 (d) 75   (e) 100

66. *Investigation* The top outer edge of a solid with vertical sides and resting on the xy-plane is modeled by

$$\mathbf{r}(t) = 3 \cos t\mathbf{i} + 3 \sin t\mathbf{j} + (1 + \sin^2 2t)\mathbf{k}$$

where all measurements are in centimeters. The intersection of the plane $y = b\, (-3 < b < 3)$ with the top of the solid is a horizontal line.

(a) Use a computer algebra system to graph the solid.

(b) Use a computer algebra system to approximate the lateral surface area of the solid.

(c) Find (if possible) the volume of the solid.

True or False? **In Exercises 67–70, determine whether the statement is true or false. If it is false, explain why or give an example that shows it is false.**

67. If C is given by $x(t) = t$, $y(t) = t$, $0 \le t \le 1$, then

$$\int_C xy\, ds = \int_0^1 t^2\, dt.$$

68. If $C_2 = -C_1$, then $\displaystyle\int_{C_1} f(x, y)\, ds + \int_{C_2} f(x, y)\, ds = 0.$

69. The vector functions $\mathbf{r}_1 = t\mathbf{i} + t^2\mathbf{j}$, $0 \le t \le 1$, and $\mathbf{r}_2 = (1 - t)\mathbf{i} + (1 - t)^2\mathbf{j}$, $0 \le t \le 1$, define the same curve.

70. If $\displaystyle\int_C \mathbf{F} \cdot \mathbf{T}\, ds = 0$, then $\mathbf{F}$ and $\mathbf{T}$ are orthogonal.

Fundamental Theorem of Line Integrals • Independence of Path •
Conservation of Energy

Fundamental Theorem of Line Integrals

In the preceding section we pointed out that in a gravitational field the work done by
gravity on an object moving between two points in the field is independent of the path
taken by the object. In this section, you will study an important generalization of this
result—it is called the **Fundamental Theorem of Line Integrals.**
    We begin with an example in which the line integral of a *conservative vector field*
is evaluated over three different paths.

EXAMPLE 1   **Line Integral of a Conservative Vector Field**

Find the work done by the force field

$$\mathbf{F}(x, y) = \frac{1}{2}xy\mathbf{i} + \frac{1}{4}x^2\mathbf{j}$$

on a particle that moves from $(0, 0)$ to $(1, 1)$ along each of the following paths.

a. $C_1: y = x$ **b.** $C_2: x = y^2$ **c.** $C_3: y = x^3$

Solution   (See Figure 14.19.)

a. Let $\mathbf{r}(t) = t\mathbf{i} + t\mathbf{j}$ for $0 \le t \le 1$, so that

$$d\mathbf{r} = (\mathbf{i} + \mathbf{j})\, dt \qquad \text{and} \qquad \mathbf{F}(x, y) = \frac{1}{2}t^2\mathbf{i} + \frac{1}{4}t^2\mathbf{j}.$$

Then, the work done is

$$W = \int_{C_1} \mathbf{F} \cdot d\mathbf{r} = \int_0^1 \frac{3}{4}t^2\, dt = \frac{1}{4}t^3 \bigg]_0^1 = \frac{1}{4}.$$

b. Let $\mathbf{r}(t) = t\mathbf{i} + \sqrt{t}\mathbf{j}$ for $0 \le t \le 1$, so that

$$d\mathbf{r} = \left(\mathbf{i} + \frac{1}{2\sqrt{t}}\mathbf{j}\right) dt \qquad \text{and} \qquad \mathbf{F}(x, y) = \frac{1}{2}t^{3/2}\mathbf{i} + \frac{1}{4}t^2\mathbf{j}.$$

Then, the work done is

$$W = \int_{C_2} \mathbf{F} \cdot d\mathbf{r} = \int_0^1 \frac{5}{8}t^{3/2}\, dt = \frac{1}{4}t^{5/2} \bigg]_0^1 = \frac{1}{4}.$$

c. Let $\mathbf{r}(t) = \frac{1}{2}t\mathbf{i} + \frac{1}{8}t^3\mathbf{j}$ for $0 \le t \le 2$, so that

$$d\mathbf{r} = \left(\frac{1}{2}\mathbf{i} + \frac{3}{8}t^2\mathbf{j}\right) dt \qquad \text{and} \qquad \mathbf{F}(x, y) = \frac{1}{32}t^4\mathbf{i} + \frac{1}{16}t^2\mathbf{j}.$$

Then, the work done is

$$W = \int_{C_3} \mathbf{F} \cdot d\mathbf{r} = \int_0^2 \frac{5}{128}t^4\, dt = \frac{1}{128}t^5 \bigg]_0^2 = \frac{1}{4}.$$

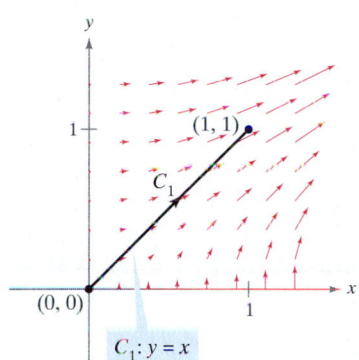

$C_1: y = x$

(a)

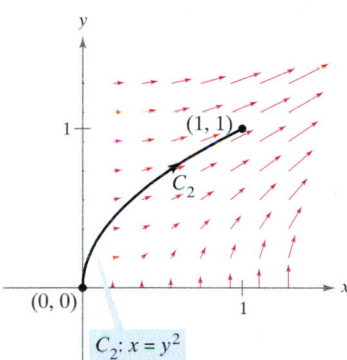

$C_2: x = y^2$

(b)

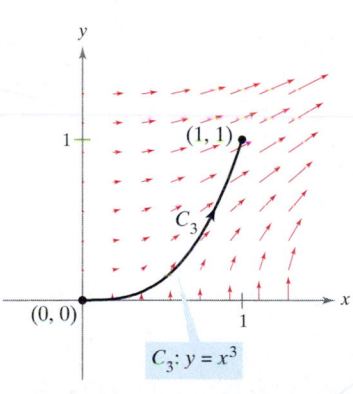

$C_3: y = x^3$

(c)

The work done by a conservative vector field
is the same for all paths.
Figure 14.19

In Example 1, note that the vector field $\mathbf{F}(x, y) = \frac{1}{2}xy\mathbf{i} + \frac{1}{4}x^2\mathbf{j}$ is conservative because $\mathbf{F}(x, y) = \nabla f(x, y)$, where $f(x, y) = \frac{1}{4}x^2y$. In such cases, the following theorem states that the value of $\int_C \mathbf{F} \cdot d\mathbf{r}$ is given by

$$\int_C \mathbf{F} \cdot d\mathbf{r} = f(x(1), y(1)) - f(x(0), y(0))$$

$$= \frac{1}{4} - 0$$

$$= \frac{1}{4}.$$

NOTE   Notice how the Fundamental Theorem of Line Integrals is similar to the Fundamental Theorem of Calculus (page 274), which states that

$$\int_a^b f(x)\, dx = F(b) - F(a)$$

where $F'(x) = f(x)$.

THEOREM 14.5   Fundamental Theorem of Line Integrals

Let C be a piecewise smooth curve lying in an open region R and given by

$$\mathbf{r}(t) = x(t)\mathbf{i} + y(t)\mathbf{j}, \quad a \le t \le b.$$

If $\mathbf{F}(x, y) = M\mathbf{i} + N\mathbf{j}$ is conservative in R, and M and N are continuous in R, then

$$\int_C \mathbf{F} \cdot d\mathbf{r} = \int_C \nabla f \cdot d\mathbf{r} = f(x(b), y(b)) - f(x(a), y(a))$$

where f is a potential function of $\mathbf{F}$. That is, $\mathbf{F}(x, y) = \nabla f(x, y)$.

Proof   We provide a proof only for a smooth curve. For piecewise smooth curves, the procedure is carried out separately on each smooth portion. Because $\mathbf{F}(x, y) = \nabla f(x, y) = f_x(x, y)\mathbf{i} + f_y(x, y)\mathbf{j}$, it follows that

$$\int_C \mathbf{F} \cdot d\mathbf{r} = \int_a^b \mathbf{F} \cdot \frac{d\mathbf{r}}{dt}\, dt = \int_a^b \left[f_x(x, y)\frac{dx}{dt} + f_y(x, y)\frac{dy}{dt} \right] dt$$

and, by the Chain Rule (Theorem 12.6), you have

$$\int_C \mathbf{F} \cdot d\mathbf{r} = \int_a^b \frac{d}{dt}[f(x(t), y(t))]\, dt = f(x(b), y(b)) - f(x(a), y(a)).$$

The last step is an application of the Fundamental Theorem of Calculus.

In space, the Fundamental Theorem of Line Integrals takes the following form. Let C be a piecewise smooth curve lying in an open region Q and given by

$$\mathbf{r}(t) = x(t)\mathbf{i} + y(t)\mathbf{j} + z(t)\mathbf{k}, \quad a \le t \le b.$$

If $\mathbf{F}(x, y, z) = M\mathbf{i} + N\mathbf{j} + P\mathbf{k}$ is conservative and M, N, and P are continuous, then

$$\int_C \mathbf{F} \cdot d\mathbf{r} = \int_C \nabla f \cdot d\mathbf{r}$$

$$= f(x(b), y(b), z(b)) - f(x(a), y(a), z(a))$$

where $\mathbf{F}(x, y, z) = \nabla f(x, y, z)$.

The Fundamental Theorem of Line Integrals states that if the vector field $\mathbf{F}$ is conservative, the line integral between any two points is simply the difference in the values of the *potential* function f at these points.

EXAMPLE 2   Using the Fundamental Theorem of Line Integrals

Evaluate $\int_C \mathbf{F} \cdot d\mathbf{r}$, where C is a piecewise smooth curve from $(-1, 4)$ to $(1, 2)$ and

$$\mathbf{F}(x, y) = 2xy\mathbf{i} + (x^2 - y)\mathbf{j}.$$

(See Figure 14.20.)

Solution   From Example 6 in Section 14.1, you know that $\mathbf{F}$ is the gradient of f where

$$f(x, y) = x^2 y - \frac{1}{2}y^2 + K.$$

Consequently, $\mathbf{F}$ is conservative, and by the Fundamental Theorem of Line Integrals, it follows that

$$\int_C \mathbf{F} \cdot d\mathbf{r} = f(1, 2) - f(-1, 4)$$

$$= \left[1^2(2) - \frac{1}{2}(2^2) \right] - \left[(-1)^2(4) - \frac{1}{2}(4^2) \right]$$

$$= 4.$$

Note that it is unnecessary to include a constant K as part of f, because it is canceled by subtraction.

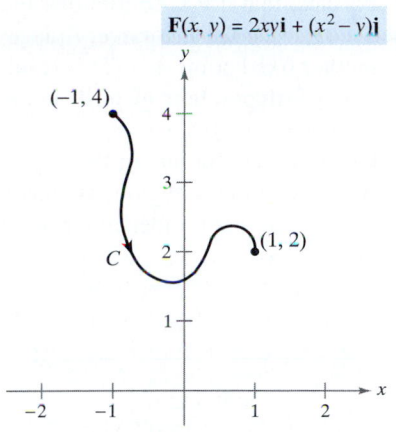

$\mathbf{F}(x, y) = 2xy\mathbf{i} + (x^2 - y)\mathbf{j}$

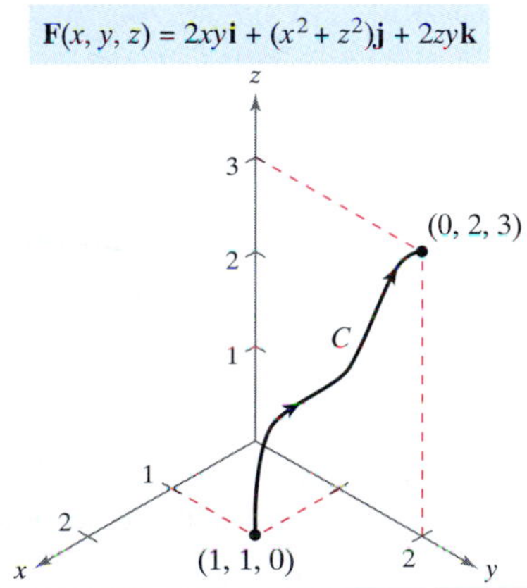

Using the Fundamental Theorem of Line Integrals, $\int_C \mathbf{F} \cdot d\mathbf{r} = 4$.
Figure 14.20

EXAMPLE 3   Using the Fundamental Theorem of Line Integrals

Evaluate $\int_C \mathbf{F} \cdot d\mathbf{r}$, where C is a piecewise smooth curve from $(1, 1, 0)$ to $(0, 2, 3)$ and

$$\mathbf{F}(x, y, z) = 2xy\mathbf{i} + (x^2 + z^2)\mathbf{j} + 2zy\mathbf{k}.$$

(See Figure 14.21.)

Solution   From Example 8 in Section 14.1, you know that $\mathbf{F}$ is the gradient of f where $f(x, y, z) = x^2 y + z^2 y + K$. Consequently, $\mathbf{F}$ is conservative, and by the Fundamental Theorem of Line Integrals, it follows that

$$\int_C \mathbf{F} \cdot d\mathbf{r} = f(0, 2, 3) - f(1, 1, 0)$$

$$= [(0)^2(2) + (3^2)(2)] - [1^2(1) + (0^2)(1)]$$

$$= 17.$$

$\mathbf{F}(x, y, z) = 2xy\mathbf{i} + (x^2 + z^2)\mathbf{j} + 2zy\mathbf{k}$

Using the Fundamental Theorem of Line Integrals, $\int_C \mathbf{F} \cdot d\mathbf{r} = 17$.
Figure 14.21

NOTE   In Examples 2 and 3, be sure you see that the value of the line integral is the same for any smooth curve C that has the given initial and terminal points. For instance, in Example 3, try evaluating the line integral for the curve given by

$$\mathbf{r}(t) = (1 - t)\mathbf{i} + (1 + t)\mathbf{j} + 3t\mathbf{k}.$$

You should obtain

$$\int_C \mathbf{F} \cdot d\mathbf{r} = \int_0^1 (30t^2 + 16t - 1)\, dt$$

$$= 17.$$

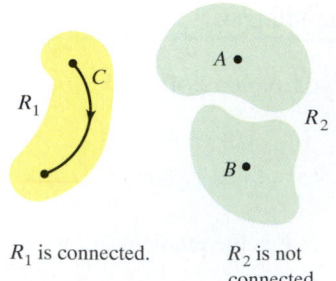

R_1 is connected. R_2 is not connected.

Figure 14.22

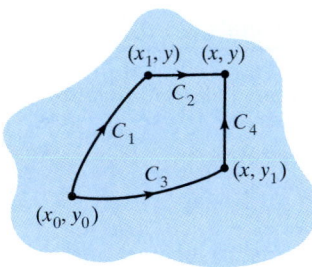

Figure 14.23

Independence of Path

From the Fundamental Theorem of Line Integrals it is clear that if **F** is continuous and conservative in an open region R, the value of $\int_C \mathbf{F} \cdot d\mathbf{r}$ is the same for every piecewise smooth curve C from one fixed point in R to another fixed point in R. This result is described by saying that the line integral $\int_C \mathbf{F} \cdot d\mathbf{r}$ is **independent of path** in the region R.

A region in the plane (or in space) is **connected** if any two points in the region can be joined by a piecewise smooth curve lying entirely within the region, as shown in Figure 14.22. In open regions that are *connected*, the path independence of $\int_C \mathbf{F} \cdot d\mathbf{r}$ is equivalent to the condition that **F** is conservative.

THEOREM 14.6   Independence of Path and Conservative Vector Fields

If **F** is continuous on an open connected region, then the line integral

$$\int_C \mathbf{F} \cdot d\mathbf{r}$$

is independent of path if and only if **F** is conservative.

Proof   If **F** is conservative, then, by the Fundamental Theorem of Line Integrals, the line integral is independent of path. We establish the converse for a plane region R. Let $\mathbf{F}(x, y) = M\mathbf{i} + N\mathbf{j}$, and let (x_0, y_0) be a fixed point in R. If (x, y) is any point in R, choose a piecewise smooth curve C running from (x_0, y_0) to (x, y), and define f by

$$f(x, y) = \int_C \mathbf{F} \cdot d\mathbf{r} = \int_C M\,dx + N\,dy.$$

The existence of C in R is guaranteed by the fact that R is connected. You can show that f is a potential function of **F** by considering two different paths between (x_0, y_0) and (x, y). For the *first* path, choose (x_1, y) in R such that $x \neq x_1$. This is possible because R is open. Then choose C_1 and C_2, as shown in Figure 14.23. Using the independence of path, it follows that

$$f(x, y) = \int_C M\,dx + N\,dy = \int_{C_1} M\,dx + N\,dy + \int_{C_2} M\,dx + N\,dy.$$

Because the first integral does not depend on x, and because $dy = 0$ in the second integral, you have

$$f(x, y) = g(y) + \int_{C_2} M\,dx$$

and it follows that the partial derivative of f with respect to x is $f_x(x, y) = M$. For the *second* path, choose a point (x, y_1). Using reasoning similar to that used for the first path, you can conclude that $f_y(x, y) = N$. Therefore,

$$\nabla f(x, y) = f_x(x, y)\mathbf{i} + f_y(x, y)\mathbf{j}$$
$$= M\mathbf{i} + N\mathbf{j}$$
$$= \mathbf{F}(x, y)$$

and it follows that **F** is conservative.

EXAMPLE 4   Finding Work in a Conservative Force Field

For the force field given by

$$\mathbf{F}(x, y, z) = e^x \cos y\mathbf{i} - e^x \sin y\mathbf{j} + 2\mathbf{k}$$

show that $\int_C \mathbf{F} \cdot d\mathbf{r}$ is independent of path, and calculate the work done by $\mathbf{F}$ on an object moving along a curve C from $(0, \pi/2, 1)$ to $(1, \pi, 3)$.

Solution   Writing the force field in the form $\mathbf{F}(x, y, z) = M\mathbf{i} + N\mathbf{j} + P\mathbf{k}$, you have $M = e^x \cos y$, $N = -e^x \sin y$, and $P = 2$, and it follows that

$$\frac{\partial P}{\partial y} = 0 = \frac{\partial N}{\partial z}$$

$$\frac{\partial P}{\partial x} = 0 = \frac{\partial M}{\partial z}$$

$$\frac{\partial N}{\partial x} = -e^x \sin y = \frac{\partial M}{\partial y}.$$

Hence, $\mathbf{F}$ is conservative. If f is a potential function of $\mathbf{F}$, then

$$f_x(x, y, z) = e^x \cos y$$
$$f_y(x, y, z) = -e^x \sin y$$
$$f_z(x, y, z) = 2.$$

By integrating with respect to x, y, and z separately, you obtain

$$f(x, y, z) = \int f_x(x, y, z)\, dx = \int e^x \cos y\, dx = e^x \cos y + g(y, z)$$

$$f(x, y, z) = \int f_y(x, y, z)\, dy = \int -e^x \sin y\, dy = e^x \cos y + h(x, z)$$

$$f(x, y, z) = \int f_z(x, y, z)\, dx = \int 2\, dz = 2z + k(x, y)$$

By comparing these three versions of $f(x, y, z)$, you can conclude that

$$f(x, y, z) = e^x \cos y + 2z + K.$$

Therefore, the work done by $\mathbf{F}$ along *any* curve C from $(0, \pi/2, 1)$ to $(1, \pi, 3)$ is

$$W = \int_C \mathbf{F} \cdot d\mathbf{r}$$

$$= \left[e^x \cos y + 2z \right]_{(0, \pi/2, 1)}^{(1, \pi, 3)}$$

$$= (-e + 6) - (0 + 2)$$

$$= 4 - e.$$

How much work would be done if the object in Example 4 moved from the point $(0, \pi/2, 1)$ to $(1, \pi, 3)$ and then back to the starting point $(0, \pi/2, 1)$? The Fundamental Theorem of Line Integrals states that there is zero work done. Remember that, by definition, work can be negative. Hence, by the time the object gets back to its starting point, the amount of work that registers positively is canceled out by the amount of work that registers negatively.

A curve, C given by $\mathbf{r}(t)$ for $a \leq t \leq b$, is **closed** if $\mathbf{r}(a) = \mathbf{r}(b)$. By the Fundamental Theorem of Line Integrals, you can conclude that if $\mathbf{F}$ is continuous and conservative on an open region R, the line integral over every closed curve C is 0.

THEOREM 14.7   Equivalent Conditions

Let $\mathbf{F}(x, y, z) = M\mathbf{i} + N\mathbf{j} + P\mathbf{k}$ have continuous first partial derivatives in an open connected region R, and let C be a piecewise smooth curve in R. The following conditions are equivalent.

1. $\mathbf{F}$ is conservative. That is, $\mathbf{F} = \nabla f$ for some function f.

2. $\displaystyle\int_C \mathbf{F} \cdot d\mathbf{r}$ is independent of path.

3. $\displaystyle\int_C \mathbf{F} \cdot d\mathbf{r} = 0$ for every *closed* curve C in R.

NOTE   Theorem 14.7 gives you options for evaluating a line integral involving a conservative vector field. You can use a potential function, or it might be more convenient to choose a particularly simple path, such as a straight line.

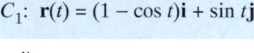

C_1: $\mathbf{r}(t) = (1 - \cos t)\mathbf{i} + \sin t\mathbf{j}$

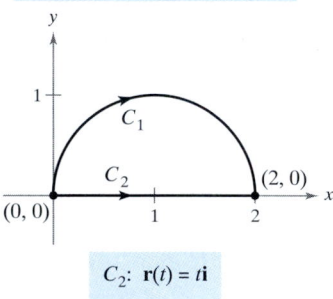

C_2: $\mathbf{r}(t) = t\mathbf{i}$

Figure 14.24

EXAMPLE 5   Evaluating a Line Integral

Evaluate $\int_{C_1} \mathbf{F} \cdot d\mathbf{r}$, where

$$\mathbf{F}(x, y) = (y^3 + 1)\mathbf{i} + (3xy^2 + 1)\mathbf{j}$$

and C_1 is the semicircular path from $(0, 0)$ to $(2, 0)$, as shown in Figure 14.24.

Solution   You have the following three options.

a. You can use the method presented in the preceding section to evaluate the line integral along the *given curve*. To do this, you can use the parametrization $\mathbf{r}(t) = (1 - \cos t)\mathbf{i} + \sin t\mathbf{j}$, where $0 \leq t \leq \pi$. For this parametrization, it follows that $d\mathbf{r} = \mathbf{r}'(t)\, dt = (\sin t\, \mathbf{i} + \cos t\, \mathbf{j})\, dt$, and

$$\int_{C_1} \mathbf{F} \cdot d\mathbf{r} = \int_0^\pi (\sin t + \sin^4 t + \cos t + 3\sin^2 t \cos t - 3\cos^2 t \sin^2 t)\, dt.$$

This integral should dampen your enthusiasm for this option.

b. You can try to find a *potential function* and evaluate the line integral by the Fundamental Theorem of Line Integrals. Using the technique demonstrated in Example 4, you can find the potential function to be $f(x, y) = xy^3 + x + y + K$, and, by the Fundamental Theorem,

$$W = \int_{C_1} \mathbf{F} \cdot d\mathbf{r} = f(2, 0) - f(0, 0) = 2.$$

c. Knowing that $\mathbf{F}$ is conservative, you have a third option. Because the value of the line integral is independent of path, you can replace the semicircular path with a *simpler path*. Suppose you choose the straight-line path C_2 from $(0, 0)$ to $(2, 0)$. Then, $\mathbf{r}(t) = t\mathbf{i}$, where $0 \leq t \leq 2$. Thus, $d\mathbf{r} = \mathbf{i}\, dt$ and $\mathbf{F}(x, y) = (y^3 + 1)\mathbf{i} + (3xy^2 + 1)\mathbf{j} = \mathbf{i} + \mathbf{j}$, so that

$$\int_{C_1} \mathbf{F} \cdot d\mathbf{r} = \int_{C_2} \mathbf{F} \cdot d\mathbf{r} = \int_0^2 1\, dt = t \Big]_0^2 = 2.$$

Of the three options, obviously the third one is the easiest.

MICHAEL FARADAY (1791–1867)

Several philosophers of science have considered Faraday's Law of Conservation of Energy to be the greatest generalization ever conceived by humankind. Many physicists have contributed to our knowledge of this law. Two early and influential ones were James Prescott Joule (1818–1889) and Hermann Ludwig Helmholtz (1821–1894).

Conservation of Energy

In 1840, the English physicist Michael Faraday wrote, "Nowhere is there a pure creation or production of power without a corresponding exhaustion of something to supply it." This statement represents the first formulation of one of the most important laws of physics—the **Law of Conservation of Energy.** In modern terminology, the law is stated as follows: *In a conservative force field, the sum of the potential and kinetic energies of an object remains constant from point to point.*

You can use the Fundamental Theorem of Line Integrals to derive this law. From physics, the **kinetic energy** of a particle of mass m and speed v is $k = \frac{1}{2}mv^2$. The **potential energy** p of a particle at point (x, y, z) in a conservative vector field $\mathbf{F}$ is defined as $p(x, y, z) = -f(x, y, z)$, where f is the potential function for $\mathbf{F}$. Consequently the work done by $\mathbf{F}$ along a smooth curve C from A to B is

$$W = \int_C \mathbf{F} \cdot d\mathbf{r} = f(x, y, z)\Big]_A^B = -p(x, y, z)\Big]_A^B = p(A) - p(B)$$

as indicated in Figure 14.25. In other words, work W is equal to the difference in the potential energies of A and B. Now, suppose that $\mathbf{r}(t)$ is the position vector for a particle moving along C from $A = \mathbf{r}(a)$ to $B = \mathbf{r}(b)$. At any time t, the particle's velocity, acceleration, and speed are $\mathbf{v}(t) = \mathbf{r}'(t)$, $\mathbf{a}(t) = \mathbf{r}''(t)$, and $v(t) = \|\mathbf{v}(t)\|$. Thus, by Newton's Second Law of Motion, $\mathbf{F} = m\mathbf{a}(t) = m(\mathbf{v}'(t))$, and the work done by $\mathbf{F}$ is

$$W = \int_C \mathbf{F} \cdot d\mathbf{r} = \int_a^b \mathbf{F} \cdot \mathbf{r}'(t)\, dt$$

$$= \int_a^b \mathbf{F} \cdot \mathbf{v}(t)\, dt = \int_a^b [m\mathbf{v}'(t)] \cdot \mathbf{v}(t)\, dt$$

$$= \int_a^b m[\mathbf{v}'(t) \cdot \mathbf{v}(t)]\, dt$$

$$= \frac{m}{2} \int_a^b \frac{d}{dt}[\mathbf{v}(t) \cdot \mathbf{v}(t)]\, dt$$

$$= \frac{m}{2} \int_a^b \frac{d}{dt}[\|\mathbf{v}(t)\|^2]\, dt$$

$$= \frac{m}{2} \left[\|\mathbf{v}(t)\|^2\right]_a^b$$

$$= \frac{m}{2} \left[[v(t)]^2\right]_a^b$$

$$= \frac{1}{2}m[v(b)]^2 - \frac{1}{2}m[v(a)]^2$$

$$= k(B) - k(A).$$

Equating these two results for W produces

$$p(A) - p(B) = k(B) - k(A)$$
$$p(A) + k(A) = p(B) + k(B)$$

which implies that the sum of the potential and kinetic energies remains constant from point to point.

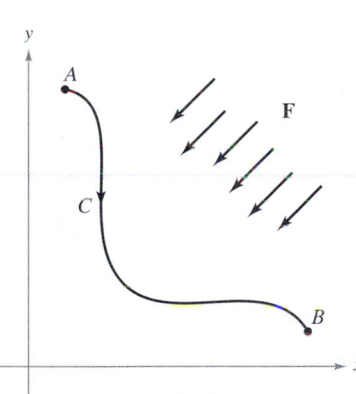

The work done by $\mathbf{F}$ along C is
$$W = \int_C \mathbf{F} \cdot d\mathbf{r} = p(A) - p(B).$$
Figure 14.25

EXERCISES FOR SECTION 14.3

In Exercises 1–4, show that the value of $\int_C \mathbf{F} \cdot d\mathbf{r}$ is the same for both parametric representations of C.

1. $\mathbf{F}(x, y) = x^2\mathbf{i} + xy\mathbf{j}$

(a) $\mathbf{r}_1(t) = t\mathbf{i} + t^2\mathbf{j}, \quad 0 \le t \le 1$

(b) $\mathbf{r}_2(\theta) = \sin\theta\,\mathbf{i} + \sin^2\theta\,\mathbf{j}, \quad 0 \le \theta \le \dfrac{\pi}{2}$

2. $\mathbf{F}(x, y) = (x^2 + y^2)\mathbf{i} - x\mathbf{j}$

(a) $\mathbf{r}_1(t) = t\mathbf{i} + \sqrt{t}\,\mathbf{j}, \quad 0 \le t \le 4$

(b) $\mathbf{r}_2(w) = w^2\mathbf{i} + w\mathbf{j}, \quad 0 \le w \le 2$

3. $\mathbf{F}(x, y) = y\mathbf{i} - x\mathbf{j}$

(a) $\mathbf{r}_1(\theta) = \sec\theta\,\mathbf{i} + \tan\theta\,\mathbf{j}, \quad 0 \le \theta \le \dfrac{\pi}{3}$

(b) $\mathbf{r}_2(t) = \sqrt{t+1}\,\mathbf{i} + \sqrt{t}\,\mathbf{j}, \quad 0 \le t \le 3$

4. $\mathbf{F}(x, y) = y\mathbf{i} + x^2\mathbf{j}$

(a) $\mathbf{r}_1(t) = (2 + t)\mathbf{i} + (3 - t)\mathbf{j}, \quad 0 \le t \le 3$

(b) $\mathbf{r}_2(w) = (2 + \ln w)\mathbf{i} + (3 - \ln w)\mathbf{j}, \quad 1 \le w \le e^3$

In Exercises 5–10, determine whether or not the vector field is conservative.

5. $\mathbf{F}(x, y) = e^x(\sin y\,\mathbf{i} + \cos y\,\mathbf{j})$

6. $\mathbf{F}(x, y) = 15x^2y^2\mathbf{i} + 10x^3y\mathbf{j}$

7. $\mathbf{F}(x, y) = \dfrac{1}{y^2}(y\mathbf{i} + x\mathbf{j})$

8. $\mathbf{F}(x, y, z) = y\ln z\,\mathbf{i} - x\ln z\,\mathbf{j} + \dfrac{xy}{z}\mathbf{k}$

9. $\mathbf{F}(x, y, z) = y^2z\mathbf{i} + 2xyz\mathbf{j} + xy^2\mathbf{k}$

10. $\mathbf{F}(x, y, z) = \sin yz\,\mathbf{i} + xz\cos yz\,\mathbf{j} + xy\sin yz\,\mathbf{k}$

In Exercises 11–24, find the value of the line integral

$$\int_C \mathbf{F} \cdot d\mathbf{r}.$$

(*Hint:* If F is conservative, the integration may be easier on an alternative path.)

11. $\mathbf{F}(x, y) = 2xy\mathbf{i} + x^2\mathbf{j}$

(a) $\mathbf{r}_1(t) = t\mathbf{i} + t^2\mathbf{j}, \quad 0 \le t \le 1$

(b) $\mathbf{r}_2(t) = t\mathbf{i} + t^3\mathbf{j}, \quad 0 \le t \le 1$

12. $\mathbf{F}(x, y) = ye^{xy}\mathbf{i} + xe^{xy}\mathbf{j}$

(a) $\mathbf{r}_1(t) = t\mathbf{i} - \tfrac{3}{2}(t - 2)\mathbf{j}, \quad 0 \le t \le 2$

(b) line segments from $(0, 3)$ to $(0, 0)$, and then from $(0, 0)$ to $(2, 0)$

13. $\mathbf{F}(x, y) = y\mathbf{i} - x\mathbf{j}$

(a) $\mathbf{r}_1(t) = t\mathbf{i} + t\mathbf{j}, \quad 0 \le t \le 1$

(b) $\mathbf{r}_2(t) = t\mathbf{i} + t^2\mathbf{j}, \quad 0 \le t \le 1$

(c) $\mathbf{r}_3(t) = t\mathbf{i} + t^3\mathbf{j}, \quad 0 \le t \le 1$

14. $\mathbf{F}(x, y) = xy^2\mathbf{i} + 2x^2y\mathbf{j}$

(a) $\mathbf{r}_1(t) = t\mathbf{i} + \dfrac{1}{t}\mathbf{j}, \quad 1 \le t \le 3$

(b) $\mathbf{r}_2(t) = (t + 1)\mathbf{i} - \tfrac{1}{3}(t - 3)\mathbf{j}, \quad 0 \le t \le 2$

15. $\displaystyle\int_C y^2\,dx + 2xy\,dy$

(a)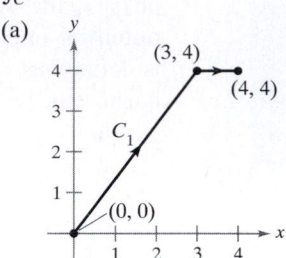

(b)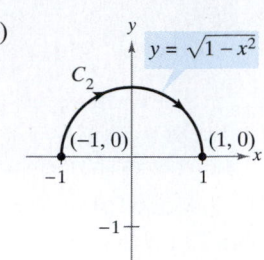

(c)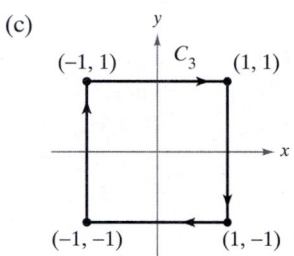

(d)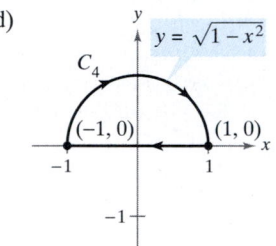

16. $\displaystyle\int_C (2x - 3y + 1)\,dx - (3x + y - 5)\,dy$

(a)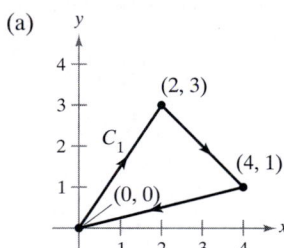

(b)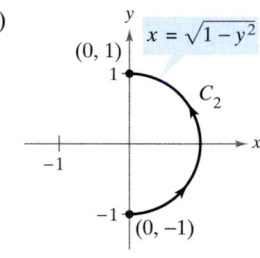

(c)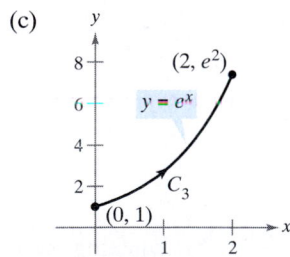

(d)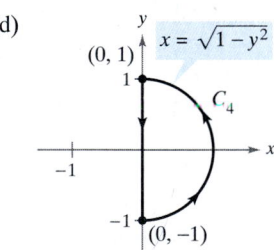

17. $\displaystyle\int_C 2xy\,dx + (x^2 + y^2)\,dy$

(a) C: ellipse $\dfrac{x^2}{25} + \dfrac{y^2}{16} = 1$ from $(5, 0)$ to $(0, 4)$

(b) C: parabola $y = 4 - x^2$ from $(2, 0)$ to $(0, 4)$

18. $\displaystyle\int_C (x^2 + y^2)\, dx + 2xy\, dy$

(a) $\mathbf{r}_1(t) = t^3\mathbf{i} + t^2\mathbf{j}, \quad 0 \le t \le 2$

(b) $\mathbf{r}_2(t) = 2\cos t\mathbf{i} + 2\sin t\mathbf{j}, \quad 0 \le t \le \dfrac{\pi}{2}$

19. $\mathbf{F}(x, y, z) = yz\mathbf{i} + xz\mathbf{j} + xy\mathbf{k}$

(a) $\mathbf{r}_1(t) = t\mathbf{i} + 2\mathbf{j} + t\mathbf{k}, \quad 0 \le t \le 4$

(b) $\mathbf{r}_2(t) = t^2\mathbf{i} + t\mathbf{j} + t^2\mathbf{k}, \quad 0 \le t \le 2$

20. $\mathbf{F}(x, y, z) = \mathbf{i} + z\mathbf{j} + y\mathbf{k}$

(a) $\mathbf{r}_1(t) = \cos t\mathbf{i} + \sin t\mathbf{j} + t^2\mathbf{k}, \quad 0 \le t \le \pi$

(b) $\mathbf{r}_2(t) = (1 - 2t)\mathbf{i} + \pi^2 t\mathbf{k}, \quad 0 \le t \le 1$

21. $\mathbf{F}(x, y, z) = (2y + x)\mathbf{i} + (x^2 - z)\mathbf{j} + (2y - 4z)\mathbf{k}$

(a) $\mathbf{r}_1(t) = t\mathbf{i} + t^2\mathbf{j} + \mathbf{k}, \quad 0 \le t \le 1$

(b) $\mathbf{r}_2(t) = t\mathbf{i} + t\mathbf{j} + (2t - 1)^2\mathbf{k}, \quad 0 \le t \le 1$

22. $\mathbf{F}(x, y, z) = -y\mathbf{i} + x\mathbf{j} + 3xz^2\mathbf{k}$

(a) $\mathbf{r}_1(t) = \cos t\mathbf{i} + \sin t\mathbf{j} + t\mathbf{k}, \quad 0 \le t \le \pi$

(b) $\mathbf{r}_2(t) = (1 - 2t)\mathbf{i} + \pi t\mathbf{k}, \quad 0 \le t \le 1$

23. $\mathbf{F}(x, y, z) = e^z(y\mathbf{i} + x\mathbf{j} + xy\mathbf{k})$

(a) $\mathbf{r}_1(t) = 4\cos t\mathbf{i} + 4\sin t\mathbf{j} + 3\mathbf{k}, \quad 0 \le t \le \pi$

(b) $\mathbf{r}_2(t) = (4 - 8t)\mathbf{i} + 3\mathbf{k}, \quad 0 \le t \le 1$

24. $\mathbf{F}(x, y, z) = y\sin z\mathbf{i} + x\sin z\mathbf{j} + xy\cos x\mathbf{k}$

(a) $\mathbf{r}_1(t) = t^2\mathbf{i} + t^2\mathbf{j}, \quad 0 \le t \le 2$

(b) $\mathbf{r}_2(t) = 4t\mathbf{i} + 4t\mathbf{j}, \quad 0 \le t \le 1$

In Exercises 25–34, evaluate the line integral using the Fundamental Theorem of Line Integrals. Use a symbolic integration utility to verify your results.

25. $\displaystyle\int_C (y\mathbf{i} + x\mathbf{j}) \cdot d\mathbf{r}$

C: smooth curve from $(0, 0)$ to $(3, 8)$

26. $\displaystyle\int_C [2(x + y)\mathbf{i} + 2(x + y)\mathbf{j}] \cdot d\mathbf{r}$

C: smooth curve from $(-1, 1)$ to $(3, 2)$

27. $\displaystyle\int_C \cos x \sin y\, dx + \sin x \cos y\, dy$

C: smooth curve from $(0, -\pi)$ to $\left(\dfrac{3\pi}{2}, \dfrac{\pi}{2}\right)$

28. $\displaystyle\int_C \dfrac{y\, dx - x\, dy}{x^2 + y^2}$

C: smooth curve from $(1, 1)$ to $(2\sqrt{3}, 2)$

29. $\displaystyle\int_C e^x \sin y\, dx + e^x \cos y\, dy$

C: cycloid $x = \theta - \sin\theta, y = 1 - \cos\theta$ from $(0, 0)$ to $(2\pi, 0)$

30. $\displaystyle\int_C \dfrac{2x}{(x^2 + y^2)^2}\, dx + \dfrac{2y}{(x^2 + y^2)^2}\, dy$

C: circle $(x - 4)^2 + (y - 5)^2 = 9$ clockwise from $(7, 5)$ to $(1, 5)$

31. $\displaystyle\int_C (z + 2y)\, dx + (2x - z)\, dy + (x - y)\, dz$

(a) C: line segment from $(0, 0, 0)$ to $(1, 1, 1)$

(b) C: line segments from $(0, 0, 0)$ to $(0, 0, 1)$ to $(1, 1, 1)$

(c) C: line segments from $(0, 0, 0)$ to $(1, 0, 0)$ to $(1, 1, 0)$ to $(1, 1, 1)$

32. Repeat Exercise 31 using the integral

$$\int_C zy\, dx + xz\, dy + xy\, dz.$$

33. $\displaystyle\int_C -\sin x\, dx + z\, dy + y\, dz$

C: smooth curve from $(0, 0, 0)$ to $\left(\dfrac{\pi}{2}, 3, 4\right)$

34. $\displaystyle\int_C 6x\, dx - 4z\, dy - (4y - 20z)\, dz$

C: smooth curve from $(0, 0, 0)$ to $(3, 4, 0)$

Work   In Exercises 35 and 36, find the work done by the force field $\mathbf{F}$ in moving an object from P to Q.

35. $\mathbf{F}(x, y) = 9x^2y^2\mathbf{i} + (6x^3y - 1)\mathbf{j}$

$P(0, 0), Q(5, 9)$

36. $\mathbf{F}(x, y) = \dfrac{2x}{y}\mathbf{i} - \dfrac{x^2}{y^2}\mathbf{j}$

$P(-1, 1), Q(3, 2)$

37. Work   A stone weighing 1 pound is attached to the end of a 2-foot string and is whirled horizontally with one end held fixed. It makes 1 revolution per second. Find the work done by the force $\mathbf{F}$ that keeps the stone moving in a circular path. [*Hint:* Use Force = (mass)(centripetal acceleration).]

38. Work   If $\mathbf{F}(x, y, z) = a_1\mathbf{i} + a_2\mathbf{j} + a_3\mathbf{k}$ is a constant force vector field, show that the work done in moving a particle along any path from P to Q is

$$W = \mathbf{F} \cdot \overrightarrow{PQ}.$$

39. Kinetic and Potential Energy   The kinetic energy of an object moving through a conservative force field is decreasing at a rate of 10 units per minute. At what rate is the potential energy changing?

40. Work   To allow a way of escape for workers in a hazardous job 50 meters above ground level, a slide wire has been installed. It runs from their position to a point on the ground 50 meters from the base of the installation where they are located. Show that the work done by the gravitational force field for a 150-pound man moving the length of the slide wire is the same for the following two paths.

(a) $\mathbf{r}(t) = t\mathbf{i} + (50 - t)\mathbf{j}$

(b) $\mathbf{r}(t) = t\mathbf{i} + \frac{1}{50}(50 - t)^2\mathbf{j}$

41. Work Can you find a path for the slide wire in Exercise 40 such that the work done by the gravitational force field would differ from the amounts of work done for the two paths given? Explain why or why not.

42. Let $\mathbf{F}(x, y) = \dfrac{y}{x^2 + y^2}\mathbf{i} - \dfrac{x}{x^2 + y^2}\mathbf{j}$.

(a) Show that

$$\frac{\partial N}{\partial x} = \frac{\partial M}{\partial y}$$

where

$$M = \frac{y}{x^2 + y^2} \text{ and } N = \frac{-x}{x^2 + y^2}.$$

(b) If $\mathbf{r}(t) = \cos t\,\mathbf{i} + \sin t\,\mathbf{j}$, for $0 \leq t \leq \pi$, find $\int_C \mathbf{F} \cdot d\mathbf{r}$.

(c) If $\mathbf{r}(t) = \cos t\,\mathbf{i} - \sin t\,\mathbf{j}$, for $0 \leq t \leq \pi$, find $\int_C \mathbf{F} \cdot d\mathbf{r}$.

(d) If $\mathbf{r}(t) = \cos t\,\mathbf{i} + \sin t\,\mathbf{j}$, for $0 \leq t \leq 2\pi$, find $\int_C \mathbf{F} \cdot d\mathbf{r}$. Why doesn't this contradict Theorem 14.7?

(e) Show that

$$\nabla\left(\arctan \frac{x}{y}\right) = \mathbf{F}.$$

True or False? In Exercises 43–46, determine whether the statement is true or false. If it is false, explain why or give an example that shows it is false.

43. If C_1, C_2, and C_3 have the same initial and terminal points and $\int_{C_1} \mathbf{F} \cdot d\mathbf{r}_1 = \int_{C_2} \mathbf{F} \cdot d\mathbf{r}_2$, then $\int_{C_1} \mathbf{F} \cdot d\mathbf{r}_1 = \int_{C_3} \mathbf{F} \cdot d\mathbf{r}_3$.

44. If $\mathbf{F} = y\mathbf{i} + x\mathbf{j}$ and C is given by $\mathbf{r}(t) = (4 \sin t)\mathbf{i} + (3 \cos t)\mathbf{j}$, $0 \leq t \leq \pi$, then $\int_C \mathbf{F} \cdot d\mathbf{r} = 0$.

45. If $\mathbf{F}$ is conservative in a region R bounded by a simple closed path and C lies within R, then $\int_C \mathbf{F} \cdot d\mathbf{r}$ is independent of path.

46. If $\mathbf{F} = M\mathbf{i} + N\mathbf{j}$ and $\partial M/\partial x = \partial N/\partial y$, then $\mathbf{F}$ is conservative.

47. f is called harmonic if $\dfrac{\partial^2 f}{\partial x^2} + \dfrac{\partial^2 f}{\partial y^2} = 0$.

Prove that if f is harmonic, then

$$\int_C \left(\frac{\partial f}{\partial y}\,dx - \frac{\partial f}{\partial x}\,dy\right) = 0$$

where C is a smooth closed curve in the plane.

48. Writing Consider the force field shown in the figure.

(a) Give a verbal argument that the force field is not conservative because you can identify two paths that require different amounts of work to move an object from $(-4, 0)$ to $(3, 4)$. Identify two paths and state which requires the greater amount of work.

(b) Give a verbal argument that the force field is not conservative because you can find a close curve C such that

$$\int_C \mathbf{F} \cdot d\mathbf{r} \neq 0.$$

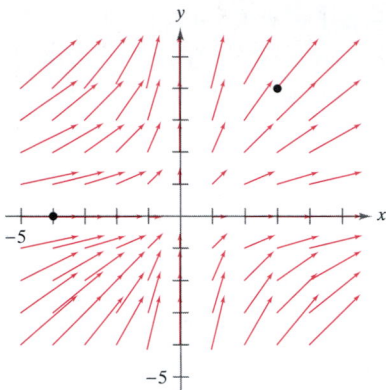

49. Writing The map shows the wind speed vectors at an altitude of 5000 feet over the United States for August 1, 1996. In planning a flight from Dallas to Chicago in a small plane at an altitude of 5000 feet, is the amount of fuel required independent of the flight path? Is the vector field conservative? Explain. *(Source: Ohio State University)*

Green's Theorem • Alternative Forms of Green's Theorem

Green's Theorem

In this section, you will study **Green's Theorem**, named after the English mathematician George Green (1793–1841). This theorem states that the value of a double integral over a *simple connected* plane region R is determined by the value of a line integral around the boundary of R.

A curve C given by $\mathbf{r}(t) = x(t)\mathbf{i} + y(t)\mathbf{j}$, where $a \leq t \leq b$, is **simple** if it does not cross itself—that is, $\mathbf{r}(c) \neq \mathbf{r}(d)$ for all c and d in the open interval (a, b). A plane region R is **simply connected** if its boundary consists of *one* simple closed curve, as shown in Figure 14.26.

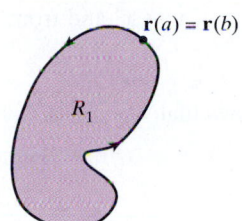

$\mathbf{r}(a) = \mathbf{r}(b)$

R_1

Simply connected

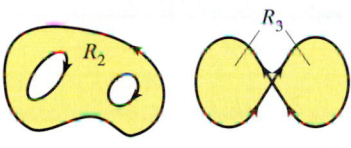

R_2 R_3

Not simply connected

Figure 14.26

THEOREM 14.8   Green's Theorem

Let R be a simply connected region with a piecewise smooth boundary C, oriented counterclockwise (that is, C is traversed *once* so that the region R always lies to the *left*). If M and N have continuous partial derivatives in an open region containing R, then

$$\int_C M\,dx + N\,dy = \iint_R \left(\frac{\partial N}{\partial x} - \frac{\partial M}{\partial y}\right) dA.$$

Proof We give a proof only for a region that is both vertically simple and horizontally simple, as shown in Figure 14.27.

$$\int_C M\,dx = \int_{C_1} M\,dx + \int_{C_2} M\,dx$$

$$= \int_a^b M(x, f_1(x))\,dx + \int_b^a M(x, f_2(x))\,dx$$

$$= \int_a^b [M(x, f_1(x)) - M(x, f_2(x))]\,dx$$

On the other hand,

$$\iint_R \frac{\partial M}{\partial y}\,dA = \int_a^b \int_{f_1(x)}^{f_2(x)} \frac{\partial M}{\partial y}\,dy\,dx$$

$$= \int_a^b M(x, y)\Big]_{f_1(x)}^{f_2(x)}\,dx$$

$$= \int_a^b [M(x, f_2(x)) - M(x, f_1(x))]\,dx.$$

Consequently,

$$\int_C M\,dx = -\iint_R \frac{\partial M}{\partial y}\,dA.$$

Similarly, you can use $g_1(y)$ and $g_2(y)$ to show that $\int_C N\,dy = \iint_R \partial N/\partial x\,dA$. By adding the integrals $\int_C M\,dx$ and $\int_C N\,dy$, you obtain the conclusion stated in the theorem.

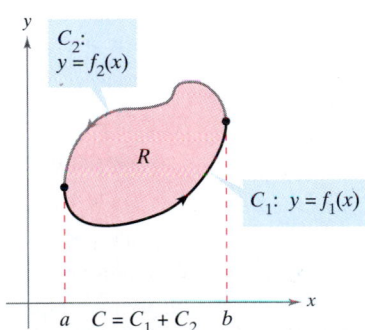

C_2: $y = f_2(x)$

R

C_1: $y = f_1(x)$

a $C = C_1 + C_2$ b

R is vertically simple.

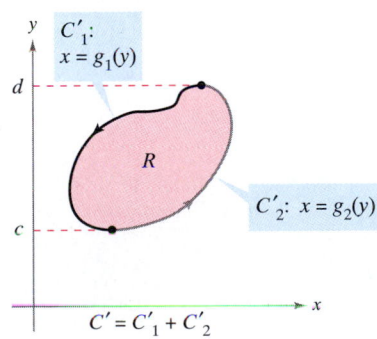

C'_1: $x = g_1(y)$

R

C'_2: $x = g_2(y)$

$C' = C'_1 + C'_2$

R is horizontally simple.

Figure 14.27

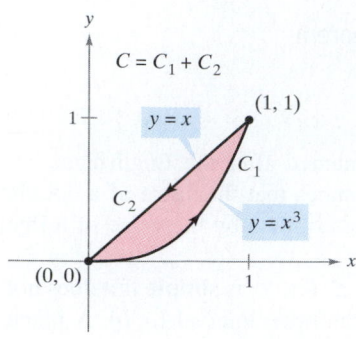

C is simple and closed, and the region R always lies to the left of C.

Figure 14.28

EXAMPLE 1   Using Green's Theorem

Use Green's Theorem to evaluate the line integral

$$\int_C y^3 \, dx + (x^3 + 3xy^2) \, dy$$

where C is the path from $(0, 0)$ to $(1, 1)$ along the graph of $y = x^3$ and from $(1, 1)$ to $(0, 0)$ along the graph of $y = x$, as shown in Figure 14.28.

Solution   Because $M = y^3$ and $N = x^3 + 3xy^2$, it follows that

$$\frac{\partial N}{\partial x} = 3x^2 + 3y^2 \quad \text{and} \quad \frac{\partial M}{\partial y} = 3y^2.$$

Applying Green's Theorem, you then have

$$\int_C y^3 \, dx + (x^3 + 3xy^2) \, dy = \int\int_R \left(\frac{\partial N}{\partial x} - \frac{\partial M}{\partial y} \right) dA$$

$$= \int_0^1 \int_{x^3}^x [(3x^2 + 3y^2) - 3y^2] \, dy \, dx$$

$$= \int_0^1 \int_{x^3}^x 3x^2 \, dy \, dx$$

$$= \int_0^1 3x^2 y \Big]_{x^3}^x \, dx$$

$$= \int_0^1 (3x^3 - 3x^5) \, dx$$

$$= \left[\frac{3x^4}{4} - \frac{x^6}{2} \right]_0^1$$

$$= \frac{1}{4}.$$

Green's Theorem cannot be applied to every line integral. Among other restrictions stated in Theorem 14.8, the curve C must be simple and closed. When Green's Theorem does apply, however, it can save time. To see this, try using the techniques described in Section 14.2 to evaluate the line integral in Example 1. To do this, you would need to write the line integral as

$$\int_C y^3 \, dx + (x^3 + 3xy^2) \, dy =$$

$$\int_{C_1} y^3 \, dx + (x^3 + 3xy^2) \, dy + \int_{C_2} y^3 \, dx + (x^3 + 3xy^2) \, dy$$

where C_1 is the cubic path given by

$$\mathbf{r}(t) = t\mathbf{i} + t^3\mathbf{j}$$

from $t = 0$ to $t = 1$, and C_2 is the line segment given by

$$\mathbf{r}(t) = (1 - t)\mathbf{i} + (1 - t)\mathbf{j}$$

from $t = 0$ to $t = 1$.

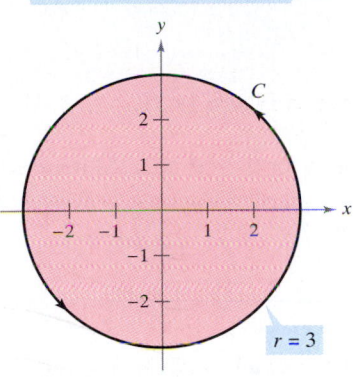

$\mathbf{F}(x, y) = y^3\mathbf{i} + (x^3 + 3xy^2)\mathbf{j}$

The work done by $\mathbf{F}$ as a particle travels once around the circle is $\dfrac{243\pi}{4}$.

Figure 14.29

EXAMPLE 2 Using Green's Theorem to Calculate Work

While subject to the force

$$\mathbf{F}(x, y) = y^3\mathbf{i} + (x^3 + 3xy^2)\mathbf{j}$$

a particle travels once around the circle of radius 3 shown in Figure 14.29. Use Green's Theorem to find the work done by $\mathbf{F}$.

Solution   From Example 1, you know by Green's Theorem that

$$\int_C y^3 \, dx + (x^3 + 3xy^2) \, dy = \int\int_R 3x^2 \, dA.$$

In polar coordinates, using $x = r \cos \theta$ and $dA = r \, dr \, d\theta$, the work done is

$$
\begin{aligned}
W &= \int\int_R 3x^2 \, dA = \int_0^{2\pi}\int_0^3 3(r \cos \theta)^2 \, r \, dr \, d\theta \\
&= 3\int_0^{2\pi}\int_0^3 r^3 \cos^2 \theta \, dr \, d\theta \\
&= 3\int_0^{2\pi} \frac{r^4}{4} \cos^2 \theta \Big]_0^3 \, d\theta \\
&= 3\int_0^{2\pi} \frac{81}{4} \cos^2 \theta \, d\theta \\
&= \frac{243}{8}\int_0^{2\pi} (1 + \cos 2\theta) \, d\theta \\
&= \frac{243}{8}\left[\theta + \frac{\sin 2\theta}{2}\right]_0^{2\pi} \\
&= \frac{243\pi}{4}.
\end{aligned}
$$

When evaluating line integrals over closed curves, remember that for conservative vector fields (those for which $\partial N/\partial x = \partial M/\partial y$), the value of the line integral is 0. This is easily seen from the statement of Green's Theorem:

$$\int_C M \, dx + N \, dy = \int\int_R \left(\frac{\partial N}{\partial x} - \frac{\partial M}{\partial y}\right) dA = 0.$$

EXAMPLE 3 Green's Theorem and Conservative Vector Fields

Evaluate the line integral

$$\int_C y^3 \, dx + 3xy^2 \, dy$$

where C is the path shown in Figure 14.30.

Solution   From this line integral, $M = y^3$ and $N = 3xy^2$. Thus, $\partial N/\partial x = 3y^2$ and $\partial M/\partial y = 3y^2$. This implies that the vector field $\mathbf{F} = M\mathbf{i} + N\mathbf{j}$ is conservative, and because C is closed, you can conclude that

$$\int_C y^3 \, dx + 3xy^2 \, dy = 0.$$

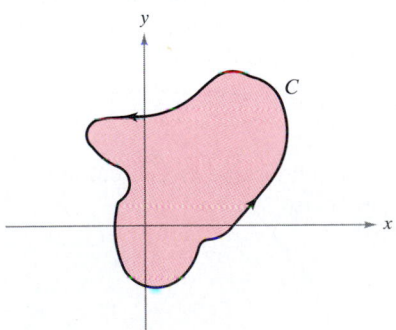

C is closed.

Figure 14.30

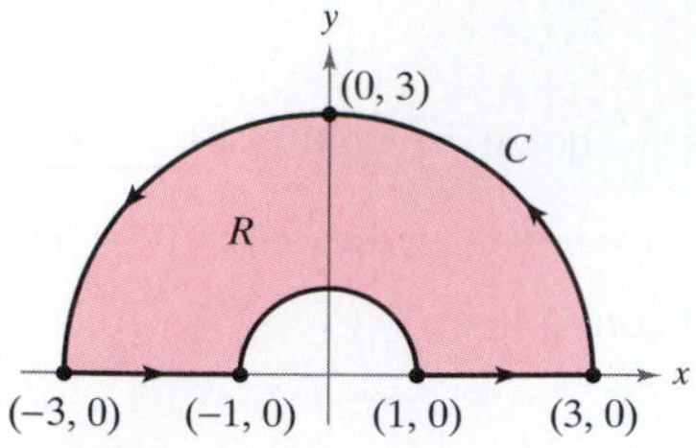

EXAMPLE 4   **Using Green's Theorem for a Piecewise Smooth Curve**

Evaluate

$$\int_C (\arctan x + y^2) \, dx + (e^y - x^2) \, dy$$

where C is the path enclosing the annular region shown in Figure 14.31.

Solution   In polar coordinates, R is given by $1 \le r \le 3$ for $0 \le \theta \le \pi$. Moreover,

$$\frac{\partial N}{\partial x} - \frac{\partial M}{\partial y} = -2x - 2y = -2(r \cos \theta + r \sin \theta).$$

Thus, by Green's Theorem,

$$\int_C (\arctan x + y^2) \, dx + (e^y - x^2) \, dy = \iint_R -2(x + y) \, dA$$

$$= \int_0^{\pi} \int_1^3 -2r(\cos \theta + \sin \theta) r \, dr \, d\theta$$

$$= \int_0^{\pi} -2(\cos \theta + \sin \theta) \frac{r^3}{3} \Big]_1^3 \, d\theta$$

$$= \int_0^{\pi} \left(-\frac{52}{3} \right)(\cos \theta + \sin \theta) \, d\theta$$

$$= -\frac{52}{3} \Big[\sin \theta - \cos \theta \Big]_0^{\pi}$$

$$= -\frac{104}{3}.$$

The figure shows a semicircular annular region R (shaded pink) in the upper half plane with vertices/points labeled $(0, 3)$ at top, $(-3, 0)$, $(-1, 0)$, $(1, 0)$, $(3, 0)$ on the x-axis, with curve C.

C is piecewise smooth.
Figure 14.31

In Examples 1, 2, and 4, Green's Theorem was used to evaluate line integrals as double integrals. You can also use the theorem to evaluate double integrals as line integrals. One useful application occurs when $\partial N / \partial x - \partial M / \partial y = 1$.

$$\int_C M \, dx + N \, dy = \iint_R \left(\frac{\partial N}{\partial x} - \frac{\partial M}{\partial y} \right) dA$$

$$= \iint_R 1 \, dA \qquad \qquad \frac{\partial N}{\partial x} - \frac{\partial M}{\partial y} = 1$$

$$= \text{area of region } R$$

Among the many choices for M and N satisfying the stated condition, the choice of $M = -y/2$ and $N = x/2$ produces the following line integral for the area of region R.

THEOREM 14.9   Line Integral for Area

If R is a plane region bounded by a piecewise smooth simple closed curve C, oriented counterclockwise, then the area of R is given by

$$A = \frac{1}{2} \int_C x \, dy - y \, dx.$$

EXAMPLE 5   Finding Area by a Line Integral

Use a line integral to find the area of the ellipse

$$\frac{x^2}{a^2} + \frac{y^2}{b^2} = 1.$$

Solution   Using Figure 14.32, you can induce a counterclockwise orientation to the elliptical path by letting

$$x = a \cos t \quad \text{and} \quad y = b \sin t, \quad 0 \le t \le 2\pi.$$

Therefore, the area is

$$A = \frac{1}{2} \int_C x \, dy - y \, dx = \frac{1}{2} \int_0^{2\pi} \left[(a \cos t)(b \cos t) \, dt - (b \sin t)(-a \sin t) \, dt \right]$$

$$= \frac{ab}{2} \int_0^{2\pi} (\cos^2 t + \sin^2 t) \, dt$$

$$= \frac{ab}{2} \left[t \right]_0^{2\pi}$$

$$= \pi ab.$$

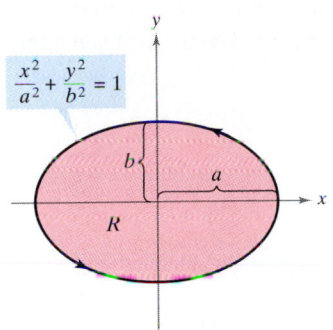

The area of the ellipse is πab.
Figure 14.32

Green's Theorem can be extended to cover some regions that are not simply connected. This is demonstrated in the next example.

EXAMPLE 6   Green's Theorem Extended to a Region with a Hole

Let R be the region inside the ellipse $(x^2/9) + (y^2/4) = 1$ and outside the circle $x^2 + y^2 = 1$. Evaluate the line integral

$$\int_C 2xy \, dx + (x^2 + 2x) \, dy$$

where $C = C_1 + C_2$ is the boundary of R, as shown in Figure 14.33.

Solution   To begin, we introduce the line segments C_3 and C_4, as shown in Figure 14.33. Note that because the curves C_3 and C_4 have opposite orientations, the line integrals over them cancel. Furthermore, you can apply Green's Theorem to the region R using the boundary $C_1 + C_4 + C_2 + C_3$ to obtain

$$\int_C 2xy \, dx + (x^2 + 2x) \, dy = \int\int_R \left(\frac{\partial N}{\partial x} - \frac{\partial M}{\partial y} \right) dA$$

$$= \int\int_R (2x + 2 - 2x) \, dA$$

$$= 2 \int\int_R dA = 2(\text{area of } R)$$

$$= 2[\pi(3)(2) - \pi(1^2)]$$

$$= 10\pi.$$

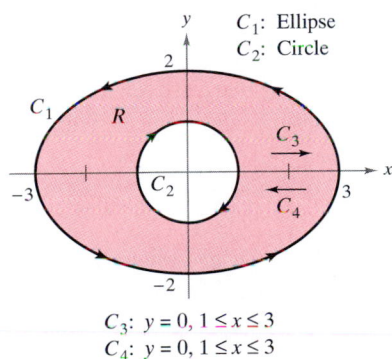

C_1: Ellipse
C_2: Circle

C_3: $y = 0, 1 \le x \le 3$
C_4: $y = 0, 1 \le x \le 3$

Figure 14.33

In Section 14.1, we listed a necessary and sufficient condition for conservative vector fields. There, we proved only one direction of the proof. We now outline the other direction, using Green's Theorem. Let $\mathbf{F}(x, y) = M\mathbf{i} + N\mathbf{j}$ be defined on an open disc R. We want to show that if M and N have continuous first partial derivatives and

$$\frac{\partial M}{\partial y} = \frac{\partial N}{\partial x}$$

then $\mathbf{F}$ is conservative. Suppose that C is a closed path forming the boundary of a connected region lying in R. Then, using the fact that $\partial M/\partial y = \partial N/\partial x$, you can apply Green's Theorem to conclude that

$$\int_C \mathbf{F} \cdot d\mathbf{r} = \int_C M\,dx + N\,dy$$

$$= \int_R\!\!\int \left(\frac{\partial N}{\partial x} - \frac{\partial M}{\partial y} \right) dA$$

$$= 0.$$

This, in turn, is equivalent to showing that $\mathbf{F}$ is conservative (see Theorem 14.7).

Alternative Forms of Green's Theorem

We conclude this section with the derivation of two vector forms of Green's Theorem for regions in the plane. The extension of these vector forms to three dimensions is the basis for the discussion in the remaining sections of this chapter. If $\mathbf{F}$ is a vector field in the plane, you can write

$$\mathbf{F}(x, y, z) = M\mathbf{i} + N\mathbf{j} + 0\mathbf{k}$$

so that the curl of $\mathbf{F}$, as described in Section 14.1, is given by

$$\mathbf{curl\ F} = \nabla \times \mathbf{F} = \begin{vmatrix} \mathbf{i} & \mathbf{j} & \mathbf{k} \\ \dfrac{\partial}{\partial x} & \dfrac{\partial}{\partial y} & \dfrac{\partial}{\partial z} \\ M & N & 0 \end{vmatrix}$$

$$= -\frac{\partial N}{\partial z}\mathbf{i} + \frac{\partial M}{\partial z}\mathbf{j} + \left(\frac{\partial N}{\partial x} - \frac{\partial M}{\partial y} \right)\mathbf{k}.$$

Consequently,

$$(\mathbf{curl\ F}) \cdot \mathbf{k} = \left[-\frac{\partial N}{\partial z}\mathbf{i} + \frac{\partial M}{\partial z}\mathbf{j} + \left(\frac{\partial N}{\partial x} - \frac{\partial M}{\partial y} \right)\mathbf{k} \right] \cdot \mathbf{k}$$

$$= \frac{\partial N}{\partial x} - \frac{\partial M}{\partial y}.$$

With appropriate conditions on $\mathbf{F}$, C, and R, you can write Green's Theorem in the vector form

$$\int_C \mathbf{F} \cdot d\mathbf{r} = \int_R\!\!\int \left(\frac{\partial N}{\partial x} - \frac{\partial M}{\partial y} \right) dA$$

$$= \int_R\!\!\int (\mathbf{curl\ F}) \cdot \mathbf{k}\, dA. \qquad \text{\color{red}First alternative form}$$

The extension of this vector form of Green's Theorem to surfaces in space produces **Stokes's Theorem,** discussed in Section 14.8.

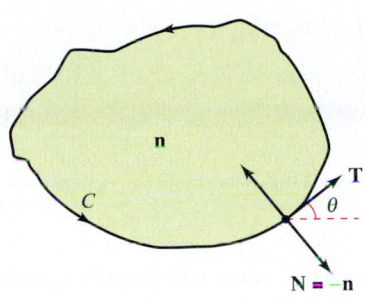

$\mathbf{T} = \cos\theta\mathbf{i} + \sin\theta\mathbf{j}$

$\mathbf{n} = \cos\left(\theta + \dfrac{\pi}{2}\right)\mathbf{i} + \sin\left(\theta + \dfrac{\pi}{2}\right)\mathbf{j}$

$\quad = -\sin\theta\mathbf{i} + \cos\theta\mathbf{j}$

$\mathbf{N} = \sin\theta\mathbf{i} - \cos\theta\mathbf{j}$

Figure 14.34

For the second vector form of Green's Theorem, assume the same conditions for $\mathbf{F}$, C, and R. Using the arc length parameter s for C, you have $\mathbf{r}(s) = x(s)\mathbf{i} + y(s)\mathbf{j}$. Thus, a unit tangent vector $\mathbf{T}$ to curve C is given by

$$\mathbf{r}'(s) = \mathbf{T} = x'(s)\mathbf{i} + y'(s)\mathbf{j}.$$

From Figure 14.34 you can see that the *outward* unit normal vector $\mathbf{N}$ can then be written as

$$\mathbf{N} = y'(s)\mathbf{i} - x'(s)\mathbf{j}.$$

Consequently, for $\mathbf{F}(x, y) = M\mathbf{i} + N\mathbf{j}$, you can apply Green's Theorem to obtain

$$\int_C \mathbf{F} \cdot \mathbf{N}\, ds = \int_a^b (M\mathbf{i} + N\mathbf{j}) \cdot (y'(s)\mathbf{i} - x'(s)\mathbf{j})\, ds$$

$$= \int_a^b \left(M\frac{dy}{ds} - N\frac{dx}{ds} \right) ds$$

$$= \int_C M\, dy - N\, dx$$

$$= \int_C -N\, dx + M\, dy$$

$$= \int\!\!\int_R \left(\frac{\partial M}{\partial x} + \frac{\partial N}{\partial y} \right) dA \qquad \text{Green's Theorem}$$

$$= \int\!\!\int_R \operatorname{div} \mathbf{F}\, dA.$$

Therefore,

$$\int_C \mathbf{F} \cdot \mathbf{N}\, ds = \int\!\!\int_R \operatorname{div} \mathbf{F}\, dA. \qquad \text{Second alternative form}$$

The extension of this form to three dimensions is called the **Divergence Theorem**, discussed in Section 14.7. The physical interpretation of divergence and curl will be discussed in Sections 14.7 and 14.8.

EXERCISES FOR SECTION 14.4

In Exercises 1–4, verify Green's Theorem by evaluating both integrals

$$\int_C y^2\, dx + x^2\, dy = \int\!\!\int_R \left(\frac{\partial N}{\partial x} - \frac{\partial M}{\partial y} \right) dA$$

for the indicated path.

1. C: square with vertices $(0, 0)$, $(4, 0)$, $(4, 4)$, $(0, 4)$

2. C: triangle with vertices $(0, 0)$, $(4, 0)$, $(4, 4)$

3. C: boundary of the region lying between the graphs of $y = x$ and $y = x^2/4$

4. C: circle given by $x^2 + y^2 = 1$

In Exercises 5 and 6, verify Green's Theorem by using a symbolic integration utility to evaluate both integrals

$$\int_C xe^y\, dx + e^x\, dy = \int\!\!\int_R \left(\frac{\partial N}{\partial x} - \frac{\partial M}{\partial y} \right) dA$$

for the indicated path.

5. C: circle given by $x^2 + y^2 = 4$

6. C: boundary of the region lying between the graphs of $y = x$ and $y = x^3$

In Exercises 7–10, use Green's Theorem to evaluate the integral

$$\int_C (y - x)\, dx + (2x - y)\, dy$$

for the indicated path.

7. C: boundary of region lying between the graphs of $y = x$ and $y = x^2 - x$

8. C: $x = 2\cos\theta$, $y = \sin\theta$

9. C: boundary of the region lying inside the rectangle bounded by $x = -5$, $x = 5$, $y = -3$, and $y = 3$, and outside the square bounded by $x = -1$, $x = 1$, $y = -1$, and $y = 1$

10. C: boundary of the region lying inside the circle $x^2 + y^2 = 16$ and outside the circle $x^2 + y^2 = 1$

In Exercises 11–20, use Green's Theorem to evaluate the line integral.

11. $\int_C 2xy \, dx + (x + y) \, dy$

    C: boundary of the region lying between the graphs of $y = 0$ and $y = 4 - x^2$

12. $\int_C y^2 \, dx + xy \, dy$

    C: boundary of the region lying between the graphs of $y = 0$, $y = \sqrt{x}$, and $x = 4$

13. $\int_C (x^2 - y^2) \, dx + 2xy \, dy$

    C: $x^2 + y^2 = a^2$

14. $\int_C (x^2 - y^2) \, dx + 2xy \, dy$

    C: $r = 1 + \cos\theta$

15. $\int_C 2 \arctan \frac{y}{x} \, dx + \ln(x^2 + y^2) \, dy$

    C: $x = 4 + 2\cos\theta, \, y = 4 + \sin\theta$

16. $\int_C e^x \sin 2y \, dx + 2e^x \cos 2y \, dy$

    C: $x^2 + y^2 = a^2$

17. $\int_C \sin x \cos y \, dx + (xy + \cos x \sin y) \, dy$

    C: boundary of the region lying between the graphs of $y = x$, and $y = \sqrt{x}$

18. $\int_C (e^{-x^2/2} - y) \, dx + (e^{-y^2/2} + x) \, dy$

    C: boundary of the region lying between the graphs of the circle $x = 5 \cos\theta, y = 5 \sin\theta$ and the ellipse $x = 2 \cos\theta$, $y = \sin\theta$

19. $\int_C xy \, dx + (x + y) \, dy$

    C: boundary of the region lying between the graphs of $x^2 + y^2 = 1$ and $x^2 + y^2 = 9$

20. $\int_C 3x^2 e^y \, dx + e^y \, dy$

    C: boundary of the region lying between the squares with vertices $(1, 1), (-1, 1), (-1, -1)$, and $(1, -1)$, and $(2, 2)$, $(-2, 2), (-2, -2)$, and $(2, -2)$

Work In Exercises 21–24, use Green's Theorem to calculate the work done by the force **F** on a particle that is moving counterclockwise around the closed path C.

21. $\mathbf{F}(x, y) = xy\mathbf{i} + (x + y)\mathbf{j}$

    C: $x^2 + y^2 = 4$

22. $\mathbf{F}(x, y) = (e^x - 3y)\mathbf{i} + (e^y + 6x)\mathbf{j}$

    C: $r = 2 \cos\theta$

23. $\mathbf{F}(x, y) = (x^{3/2} - 3y)\mathbf{i} + (6x + 5\sqrt{y})\mathbf{j}$

    C: boundary of the triangle with vertices $(0, 0), (5, 0)$, and $(0, 5)$

24. $\mathbf{F}(x, y) = (3x^2 + y)\mathbf{i} + 4xy^2 \mathbf{j}$

    C: boundary of the region lying between the graphs of $y = \sqrt{x}, y = 0$, and $x = 4$

Area In Exercises 25–28, use a line integral to find the area of the region R.

25. R: region bounded by the graph of $x^2 + y^2 = a^2$

26. R: triangle bounded by the graphs of $x = 0, 2x - 3y = 0$, and $x + 3y = 9$

27. R: region bounded by the graphs of $y = 2x + 1$ and $y = 4 - x^2$

28. R: region inside the loop of the folium of Descartes bounded by the graph of

$$x = \frac{3t}{t^3 + 1}, \quad y = \frac{3t^2}{t^3 + 1}$$

In Exercises 29 and 30, use Green's Theorem to verify the line integral formulas.

29. The centroid of the region having area A bounded by the simple closed path C is

$$\bar{x} = \frac{1}{2A} \int_C x^2 \, dy, \quad \bar{y} = -\frac{1}{2A} \int_C y^2 \, dx.$$

30. The area of a plane region bounded by the simple closed path C given in polar coordinates is

$$A = \frac{1}{2} \int_C r^2 \, d\theta.$$

Centroid In Exercises 31–34, use a symbolic integration utility and the result in Exercise 29 to find the centroid of the region.

31. R: region bounded by the graphs of $y = 0$ and $y = 4 - x^2$

32. R: region bounded by the graph of $y = \sqrt{a^2 - x^2}$ and $y = 0$

33. R: region bounded by the graphs of $y = x^3$ and $y = x$, $0 \le x \le 1$

34. R: triangle with vertices $(-a, 0), (a, 0)$, and (b, c), where $-a \le b \le a$

Area In Exercises 35–38, use a symbolic integration utility and the result in Exercise 30 to find the area of the region bounded by the graph of the polar equation.

35. $r = a(1 - \cos\theta)$

36. $r = a \cos 3\theta$

37. $r = 1 + 2 \cos\theta$

    (inner loop)

38. $r = \dfrac{3}{2 - \cos\theta}$

39. *Think About It*   Let

$$I = \int_C \frac{y\,dx - x\,dy}{x^2 + y^2}$$

where C is a circle orientated counterclockwise. Show that $I = 0$ if C does not contain the origin. What is I if C contains the origin?

40. (a) Let C be the line segment joining (x_1, y_1) and (x_2, y_2). Show that

$$\int_C -y\,dx + x\,dy = x_1 y_2 - x_2 y_1.$$

(b) Let $(x_1, y_1), (x_2, y_2), \ldots, (x_n, y_n)$ be the vertices of a polygon. Prove that the area enclosed is

$$\tfrac{1}{2}[(x_1 y_2 - x_2 y_1) + (x_2 y_3 - x_3 y_2) + \cdots + (x_{n-1} y_n - x_n y_{n-1}) + (x_n y_1 - x_1 y_n)].$$

Area   In Exercises 41 and 42, find the area enclosed by the polygon with the given vertices.

41. Pentagon: $(0, 0), (2, 0), (3, 2), (1, 4), (-1, 1)$

42. Hexagon: $(0, 0), (2, 0), (3, 2), (2, 4), (0, 3), (-1, 1)$

43. *Investigation*   Consider the line integral

$$\int_C y^n\,dx + x^n\,dy$$

where C is the boundary of the region lying between the graphs of $y = \sqrt{a^2 - x^2}$ $(a > 0)$ and $y = 0$.

(a) Use a symbolic integration utility to verify Green's Theorem for n an odd integer from 1 through 7.

(b) Use a symbolic integration utility to verify Green's Theorem for n an odd integer from 2 through 8.

(c) For n an odd integer, make a conjecture about the value of the integral.

In Exercises 44 and 45, prove the identity where R is a simply connected region with boundary C. Assume that the required partial derivatives of the scalar functions f and g are continuous. The expressions $D_N f$ and $D_N g$ are the derivatives in the direction of the outward normal vector N of C, and are defined by

$$D_N f = \nabla f \cdot N, \quad D_N g = \nabla g \cdot N.$$

44. Green's first identity:

$$\iint_R (f \nabla^2 g + \nabla f \cdot \nabla g)\,dA = \int_C f D_N g\,ds$$

[*Hint:* Use the alternative form of Green's Theorem and the property div $(f\mathbf{G}) = f$ div $\mathbf{G} + \nabla f \cdot \mathbf{G}$.]

45. Green's second identity:

$$\iint_R (f \nabla^2 g - g \nabla^2 f)\,dA = \int_C (f D_N g - g\,D_N f)\,ds$$

(*Hint:* Use Exercise 44 twice.)

46. Use Green's Theorem to prove that

$$\int_C f(x)\,dx + g(y)\,dy = 0$$

if f and g are differentiable functions and C is a piecewise smooth simple closed path.

47. Let $\mathbf{F} = M\mathbf{i} + N\mathbf{j}$, where M and N have continuous first partial derivatives in a simply connected region R. Prove that if C is simple, smooth, and closed, and $N_x = M_y$, then

$$\int_C \mathbf{F} \cdot d\mathbf{r} = 0.$$

SECTION PROJECT

Hyperbolic and Trigonometric Functions

(a) Sketch the plane curve represented by the vector-valued function $\mathbf{r}(t) = \cosh t\,\mathbf{i} + \sinh t\,\mathbf{j}$ on the interval $0 \le t \le 5$. Show that the rectangular equation corresponding to $\mathbf{r}(t)$ is the hyperbola $x^2 - y^2 = 1$. Verify your sketch by using a graphing utility to graph the hyperbola.

(b) Let $P = (\cosh \phi, \sinh \phi)$ be the point on the hyperbola corresponding to $\mathbf{r}(\phi)$ for $\phi > 0$. Use the formula for area

$$A = \frac{1}{2} \int_C x\,dy - y\,dx$$

to show that the area of the region indicated in the figure is $\tfrac{1}{2}\phi$.

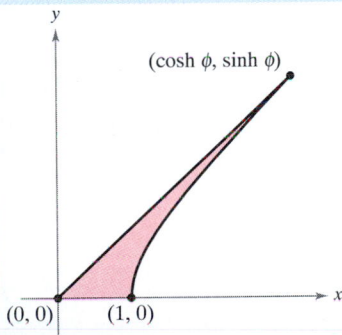

(c) Show that the area of the indicated region is also given by the integral

$$A = \int_0^{\sinh \phi} \left(\sqrt{1 + y^2} - (\coth \phi)y \right) dy.$$

Confirm your answer in part (b) by numerically approximating this integral for $\phi = 1, 2, 4$, and 10.

(d) If θ is the angle between the x-axis and a point (x, y) on the unit circle $x^2 + y^2 = 1$, the area of the corresponding sector is $\tfrac{1}{2}\theta$. That is, the trigonometric functions $f(\theta) = \cos \theta$ and $g(\theta) = \sin \theta$ could have been defined to be the coordinates of that point $(\cos \theta, \sin \theta)$ on the unit circle that determine a sector area of $\tfrac{1}{2}\theta$. Write a short paragraph explaining how you could define the hyperbolic functions in a similar manner, using the "unit hyperbola" $x^2 - y^2 = 1$.

Parametric Surfaces • Finding Parametric Equations for Surfaces •
Normal Vectors and Tangent Planes • Area of a Parametric Surface

Parametric Surfaces

You already know how to represent a curve in the plane or in space by a set of
parametric equations—or, equivalently, by a vector-valued function.

$$\mathbf{r}(t) = x(t)\mathbf{i} + y(t)\mathbf{j}$$ Plane curve

$$\mathbf{r}(t) = x(t)\mathbf{i} + y(t)\mathbf{j} + z(t)\mathbf{k}$$ Space curve

In this section, you will learn how to represent a surface in space by a set of parametric equations—or by a vector-valued function. For curves, note that the vector-valued
function $\mathbf{r}$ is a function of a *single* parameter t. For surfaces, the vector-valued
function is a function of *two* parameters u and v.

> **Definition of Parametric Surface**
>
> Let x, y, and z be functions of u and v that are continuous on a domain D in the
> uv-plane. The set of points (x, y, z) given by
>
> $$\mathbf{r}(u, v) = x(u, v)\mathbf{i} + y(u, v)\mathbf{j} + z(u, v)\mathbf{k}$$ Parametric surface
>
> is called a **parametric surface**. The equations
>
> $$x = x(u, v), \quad y = y(u, v) \quad \text{and} \quad z = z(u, v)$$
>
> are the **parametric equations** for the surface.

If S is a parametric surface given by the vector-valued function $\mathbf{r}$, then S is traced
out by the position vector $\mathbf{r}(u, v)$ as the point (u, v) moves throughout the domain D,
as shown in Figure 14.35.

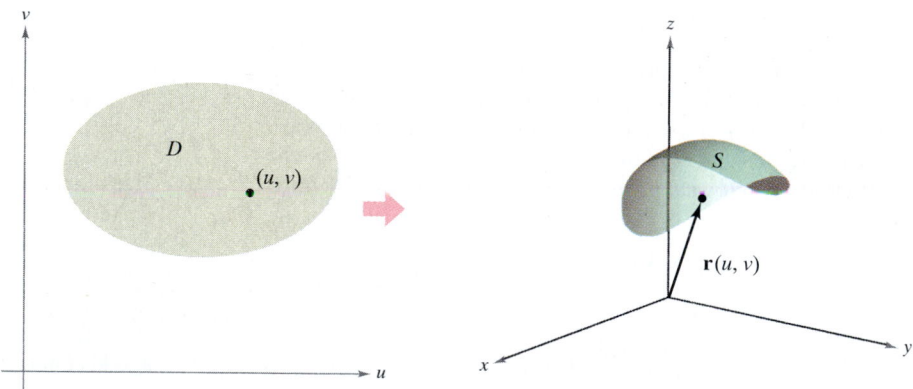

Figure 14.35

TECHNOLOGY Some computer
algebra systems are capable of sketching surfaces that are represented
parametrically. If you have access to
such software, try using it to sketch
some of the surfaces in the examples
and exercises in this section.

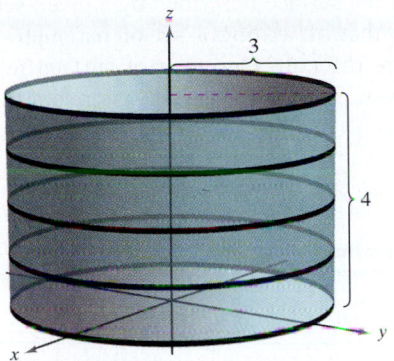

Figure 14.36

EXAMPLE 1   Sketching a Parametric Surface

Identify and sketch the parametric surface S given by

$$\mathbf{r}(u, v) = 3 \cos u\mathbf{i} + 3 \sin u\mathbf{j} + v\mathbf{k}$$

where $0 \le u \le 2\pi$ and $0 \le v \le 4$.

Solution   Because $x = 3 \cos u$ and $y = 3 \sin u$, you know that for each point (x, y, z) on the surface, x and y are related by the equation $x^2 + y^2 = 3^2$. In other words, each cross section of S taken parallel to the xy-plane is a circle of radius 3, centered on the z-axis. Because $z = v$, where $0 \le v \le 4$, you can see that the surface is a right circular cylinder of height 4. The radius of the cylinder is 3, and the z-axis forms the axis of the cylinder, as shown in Figure 14.36.

As with parametric representations of curves, parametric representations of surfaces are not unique. That is, there are many other sets of parametric equations that could be used to represent the surface shown in Figure 14.36.

EXAMPLE 2   Sketching a Parametric Surface

Identify and sketch the parametric surface S given by

$$\mathbf{r}(u, v) = \sin u \cos v\mathbf{i} + \sin u \sin v\mathbf{j} + \cos u\mathbf{k}$$

where $0 \le u \le \pi$ and $0 \le v \le 2\pi$.

Solution   To identify the surface, you can try to use trigonometric identities to eliminate the parameters. After some experimentation, you can discover that

$$\begin{aligned}
x^2 + y^2 + z^2 &= (\sin u \cos v)^2 + (\sin u \sin v)^2 + (\cos u)^2 \\
&= \sin^2 u \cos^2 v + \sin^2 u \sin^2 v + \cos^2 u \\
&= \sin^2 u(\cos^2 v + \sin^2 v) + \cos^2 u \\
&= \sin^2 u + \cos^2 u \\
&= 1.
\end{aligned}$$

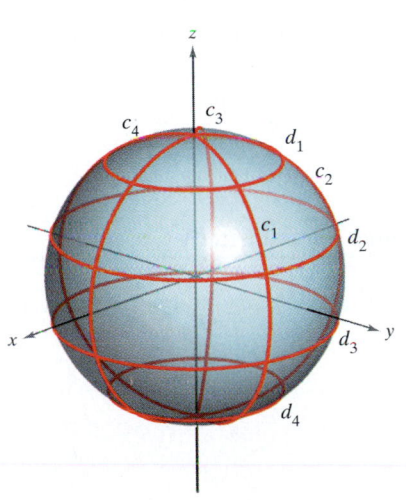

Figure 14.37

Thus, each point on S lies on the unit sphere, centered at the origin, as shown in Figure 14.37. For fixed $u = d_i$, $\mathbf{r}(u, v)$ traces out latitude (or meridian) circles.

$$x^2 + y^2 = \sin^2 d_i, \quad 0 \le d_i \le \pi$$

that are parallel to the xy-plane, and for fixed $v = c_i$, $\mathbf{r}(u, v)$ traces out longitude (or great) circles.

NOTE   To further convince yourself that the vector-valued function in Example 2 traces out the entire unit sphere, recall that the parametric equations

$$x = \rho \sin \phi \cos \theta, \quad y = \rho \sin \phi \sin \theta, \quad \text{and} \quad z = \rho \cos \phi,$$

where $0 \le \theta \le 2\pi$ and $0 \le \phi \le \pi$, describe the conversion from rectangular to spherical coordinates, as discussed in Section 10.7.

Finding Parametric Equations for Surfaces

In Examples 1 and 2, you were asked to identify the surface described by a given set of parametric equations. The reverse problem—that of writing a set of parametric equations for a given surface—is generally more difficult. One type of surface for which this problem is straightforward, however, is a surface that is given by $z = f(x, y)$. You can parametrize such a surface as

$$\mathbf{r}(x, y) = x\mathbf{i} + y\mathbf{j} + f(x, y)\mathbf{k}.$$

EXAMPLE 3   Representing a Surface Parametrically

Write a set of parametric equations for the cone given by

$$z = \sqrt{x^2 + y^2}$$

as shown in Figure 14.38.

Solution   Because this surface is given in the form $z = f(x, y)$, you can let x and y be the parameters. Then the cone is represented by the vector-valued function

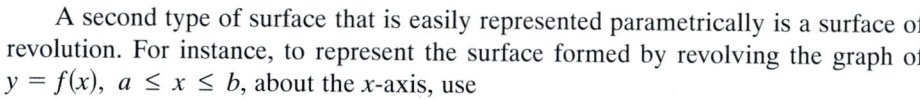

$$\mathbf{r}(x, y) = x\mathbf{i} + y\mathbf{j} + \sqrt{x^2 + y^2}\,\mathbf{k}$$

where (x, y) varies over the entire xy-plane.

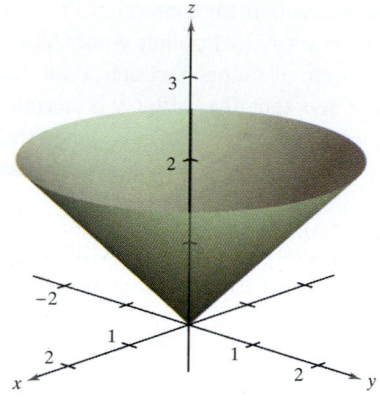

Figure 14.38

A second type of surface that is easily represented parametrically is a surface of revolution. For instance, to represent the surface formed by revolving the graph of $y = f(x)$, $a \leq x \leq b$, about the x-axis, use

$$x = u, \quad y = f(u) \cos v, \quad \text{and} \quad z = f(u) \sin v$$

where $a \leq u \leq b$ and $0 \leq v \leq 2\pi$.

EXAMPLE 4   Representing a Surface of Revolution Parametrically

Write a set of parametric equations for the surface of revolution obtained by revolving

$$f(x) = \frac{1}{x}, \quad 1 \leq x \leq 10$$

about the x-axis.

Solution   Use the parameters u and v as described above to write

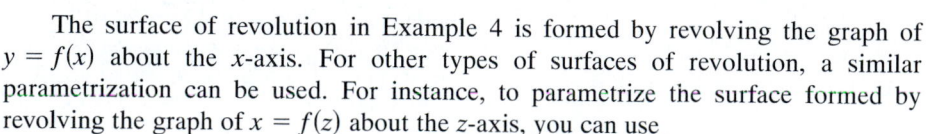

$$x = u, \quad y = f(u) \cos v = \frac{1}{u} \cos v, \quad \text{and} \quad z = f(u) \sin v = \frac{1}{u} \sin v$$

where $1 \leq u \leq 10$ and $0 \leq v \leq 2\pi$. The resulting surface is a portion of *Gabriel's Horn*, as shown in Figure 14.39.

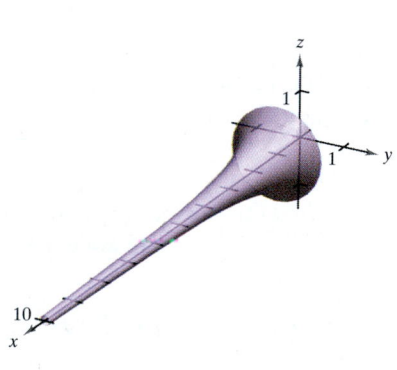

Figure 14.39

The surface of revolution in Example 4 is formed by revolving the graph of $y = f(x)$ about the x-axis. For other types of surfaces of revolution, a similar parametrization can be used. For instance, to parametrize the surface formed by revolving the graph of $x = f(z)$ about the z-axis, you can use

$$z = u, \quad x = f(u) \cos v, \quad \text{and} \quad y = f(u) \sin v.$$

Normal Vectors and Tangent Planes

Let S be a parametric surface given by

$$\mathbf{r}(u, v) = x(u, v)\mathbf{i} + y(u, v)\mathbf{j} + z(u, v)\mathbf{k}$$

over an open region D such that x, y, and z have continuous partial derivatives on D. The **partial derivatives of r** with respect to u and v are defined as

$$\mathbf{r}_u = \frac{\partial x}{\partial u}(u, v)\mathbf{i} + \frac{\partial y}{\partial u}(u, v)\mathbf{j} + \frac{\partial z}{\partial u}(u, v)\mathbf{k}$$

and

$$\mathbf{r}_v = \frac{\partial x}{\partial v}(u, v)\mathbf{i} + \frac{\partial y}{\partial v}(u, v)\mathbf{j} + \frac{\partial z}{\partial v}(u, v)\mathbf{k}.$$

Each of these partial derivatives is a vector-valued function that can be interpreted geometrically in terms of tangent vectors. For instance, if $v = v_0$ is held constant, then $\mathbf{r}(u, v_0)$ is a vector-valued function of a single parameter and defines a curve C_1 that lies on the surface S. The tangent vector to C_1 at the point $(x(u_0, v_0), y(u_0, v_0), z(u_0, v_0))$ is given by

$$\mathbf{r}_u(u_0, v_0) = \frac{\partial x}{\partial u}(u_0, v_0)\mathbf{i} + \frac{\partial y}{\partial u}(u_0, v_0)\mathbf{j} + \frac{\partial z}{\partial u}(u_0, v_0)\mathbf{k}$$

as shown in Figure 14.40. In a similar way, if $u = u_0$ is held constant, then $\mathbf{r}(u_0, v_0)$ is a vector-valued function of a single parameter and defines a curve C_2 that lies on the surface S. The tangent vector to C_2 at the point $(x(u_0, v_0), y(u_0, v_0), z(u_0, v_0))$ is given by

$$\mathbf{r}_v(u_0, v_0) = \frac{\partial x}{\partial v}(u_0, v_0)\mathbf{i} + \frac{\partial y}{\partial v}(u_0, v_0)\mathbf{j} + \frac{\partial z}{\partial v}(u_0, v_0)\mathbf{k}.$$

If the normal vector $\mathbf{r}_u \times \mathbf{r}_v$ is not $\mathbf{0}$ for any (u, v) in D, the surface S is called **smooth** and will have a tangent plane. Informally, a smooth surface is one that has no sharp points or cusps. For instance, spheres, ellipsoids, and paraboloids are smooth, whereas the cone given in Example 3 is not smooth.

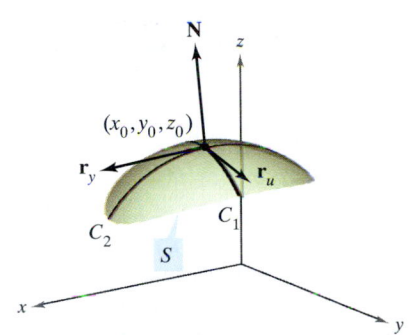

Figure 14.40

Normal Vector to a Smooth Parametric Surface

Let S be a smooth parametric surface

$$\mathbf{r}(u, v) = x(u, v)\mathbf{i} + y(u, v)\mathbf{j} + z(u, v)\mathbf{k}$$

defined over an open region D in the uv-plane. Let (u_0, v_0) be a point in D. A normal vector at the point

$$(x_0, y_0, z_0) = (x(u_0, v_0), y(u_0, v_0), z(u_0, v_0))$$

is given by

$$\mathbf{N} = \mathbf{r}_u(u_0, v_0) \times \mathbf{r}_v(u_0, v_0) = \begin{vmatrix} \mathbf{i} & \mathbf{j} & \mathbf{k} \\ \dfrac{\partial x}{\partial u} & \dfrac{\partial y}{\partial u} & \dfrac{\partial z}{\partial u} \\ \dfrac{\partial x}{\partial v} & \dfrac{\partial y}{\partial v} & \dfrac{\partial z}{\partial v} \end{vmatrix}.$$

NOTE   Figure 14.40 shows the normal vector $\mathbf{r}_u \times \mathbf{r}_v$. The vector $\mathbf{r}_v \times \mathbf{r}_u$ is also normal to S and points in the opposite direction.

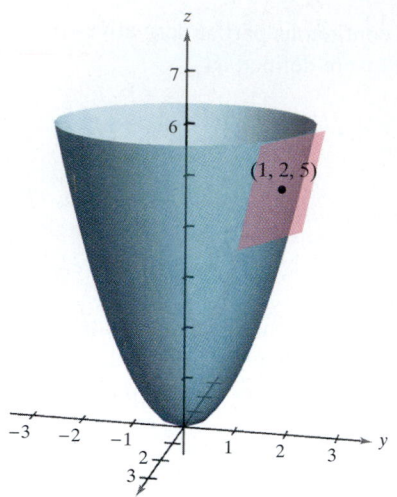

Figure 14.41

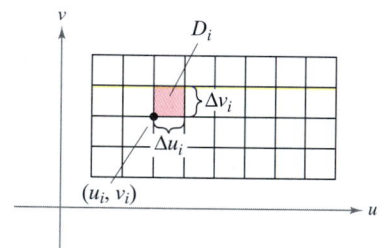

EXAMPLE 5   Finding a Tangent Plane to a Parametric Surface

Find an equation of the tangent plane to the paraboloid given by

$$\mathbf{r}(u, v) = u\mathbf{i} + v\mathbf{j} + (u^2 + v^2)\mathbf{k}$$

at the point $(1, 2, 5)$.

Solution   The point in the uv-plane that is mapped to the point $(x, y, z) = (1, 2, 5)$ is $(u, v) = (1, 2)$. The partial derivatives of $\mathbf{r}$ are

$$\mathbf{r}_u = \mathbf{i} + 2u\mathbf{k} \quad \text{and} \quad \mathbf{r}_v = \mathbf{j} + 2v\mathbf{k}.$$

The normal vector is given by

$$\mathbf{r}_u \times \mathbf{r}_v = \begin{vmatrix} \mathbf{i} & \mathbf{j} & \mathbf{k} \\ 1 & 0 & 2u \\ 0 & 1 & 2v \end{vmatrix} = -2u\mathbf{i} - 2v\mathbf{j} + \mathbf{k}$$

which implies that the normal vector at $(1, 2, 5)$ is $\mathbf{r}_u \times \mathbf{r}_v = -2\mathbf{i} - 4\mathbf{j} + \mathbf{k}$. Thus, an equation of the tangent plane at $(1, 2, 5)$ is

$$-2(x - 1) - 4(y - 2) + (z - 5) = 0$$
$$-2x - 4y + z = -5.$$

The tangent plane is shown in Figure 14.41.

Area of a Parametric Surface

To define the area of a parametric surface, you can use a development that is similar to that given in Section 13.5. Begin by constructing an inner partition of D consisting of n rectangles, where the area of the ith rectangle D_i is $\Delta A_i = \Delta u_i \Delta v_i$, as shown in Figure 14.42. In each D_i let (u_i, v_i) be the point that is closest to the origin. At the point $(x_i, y_i, z_i) = (x(u_i, v_i), y(u_i, v_i), z(u_i, v_i))$ on the surface S, construct a tangent plane T_i. The area of the portion of S that corresponds to D_i, ΔT_i, can be approximated by a parallelogram in the tangent plane. That is, $\Delta T_i \approx \Delta S_i$. Hence the surface of S is given by $\Sigma \, \Delta S_i \approx \Sigma \, \Delta T_i$. The area of the parallelogram in the tangent plane is

$$\|\Delta u_i \mathbf{r}_u \times \Delta v_i \mathbf{r}_v\| = \|\mathbf{r}_u \times \mathbf{r}_v\| \, \Delta u_i \, \Delta v_i$$

which leads to the following definition.

Figure 14.42

Area of a Parametric Surface

Let S be a smooth parametric surface

$$\mathbf{r}(u, v) = x(u, v)\mathbf{i} + y(u, v)\mathbf{j} + z(u, v)\mathbf{k}$$

defined over an open region D in the uv-plane. If each point on the surface S corresponds to exactly one point in the domain D, then the **surface area** of S is given by

$$\text{Surface area} = \iint_S dS = \iint_D \|\mathbf{r}_u \times \mathbf{r}_v\| \, dA$$

where $\mathbf{r}_u = \dfrac{\partial x}{\partial u}\mathbf{i} + \dfrac{\partial y}{\partial u}\mathbf{j} + \dfrac{\partial z}{\partial u}\mathbf{k}$ and $\mathbf{r}_v = \dfrac{\partial x}{\partial v}\mathbf{i} + \dfrac{\partial y}{\partial v}\mathbf{j} + \dfrac{\partial z}{\partial v}\mathbf{k}$.

For a surface S given by $z = f(x, y)$, this formula for surface area corresponds to that given in Section 13.5. To see this, you can parametrize the surface using the vector-valued function

$$\mathbf{r}(x, y) = x\mathbf{i} + y\mathbf{j} + f(x, y)\mathbf{k}$$

defined over the region R in the xy-plane. Using

$$\mathbf{r}_x = \mathbf{i} + f_x(x, y)\mathbf{k} \quad \text{and} \quad \mathbf{r}_y = \mathbf{j} + f_y(x, y)\mathbf{k}$$

you have

$$\mathbf{r}_x \times \mathbf{r}_y = \begin{vmatrix} \mathbf{i} & \mathbf{j} & \mathbf{k} \\ 1 & 0 & f_x(x, y) \\ 0 & 1 & f_y(x, y) \end{vmatrix} = -f_x(x, y)\mathbf{i} - f_y(x, y)\mathbf{j} + \mathbf{k}$$

and $\|\mathbf{r}_x \times \mathbf{r}_y\| = \sqrt{[f_x(x, y)]^2 + [f_y(x, y)]^2 + 1}$. This implies that the surface area of S is

$$\text{Surface area} = \int\int_R \|\mathbf{r}_x \times \mathbf{r}_y\| \, dA$$

$$= \int\int_R \sqrt{1 + [f_x(x, y)]^2 + [f_y(x, y)]^2} \, dA.$$

EXAMPLE 6 Finding Surface Area

NOTE The surface in Example 6 does not quite fulfill the hypothesis that each point on the surface corresponds to exactly one point in D. For this surface, $\mathbf{r}(u, 0) = \mathbf{r}(u, 2\pi)$ for any fixed value of u. However, because the overlap consists of only a semicircle (which has no area), you can still apply the formula for the area of a parametric surface.

Find the surface area of the unit sphere given by

$$\mathbf{r}(u, v) = \sin u \cos v\mathbf{i} + \sin u \sin v\mathbf{j} + \cos u\mathbf{k}$$

where the domain D is given by $0 \le u \le \pi$ and $0 \le v \le 2\pi$.

Solution Begin by calculating $\mathbf{r}_u$ and $\mathbf{r}_v$.

$$\mathbf{r}_u = \cos u \cos v\mathbf{i} + \cos u \sin v\mathbf{j} - \sin u\mathbf{k}$$
$$\mathbf{r}_v = -\sin u \sin v\mathbf{i} + \sin u \cos v\mathbf{j}$$

The cross product of these two vectors is

$$\mathbf{r}_u \times \mathbf{r}_v = \begin{vmatrix} \mathbf{i} & \mathbf{j} & \mathbf{k} \\ \cos u \cos v & \cos u \sin v & -\sin u \\ -\sin u \sin v & \sin u \cos v & 0 \end{vmatrix}$$

$$= \sin^2 u \cos v\mathbf{i} + \sin^2 u \sin v\mathbf{j} + \sin u \cos u\mathbf{k}$$

which implies that

$$\|\mathbf{r}_u \times \mathbf{r}_v\| = \sqrt{(\sin^2 u \cos v)^2 + (\sin^2 u \sin v)^2 + (\sin^2 u \cos u)^2}$$
$$= \sqrt{\sin^4 u + \sin^2 u \cos^2 u}$$
$$= \sqrt{\sin^2 u}$$
$$= \sin u. \qquad \text{sin u > 0 for } 0 \le u \le \pi$$

Finally, the surface area of the sphere is

$$A = \int\int_D \|\mathbf{r}_u \times \mathbf{r}_v\| \, dA = \int_0^{2\pi} \int_0^{\pi} \sin u \, du \, dv$$

$$= \int_0^{2\pi} 2 \, dv = 4\pi.$$

EXAMPLE 7   Finding Surface Area

Find the surface area of the torus given by

$$\mathbf{r}(u, v) = (2 + \cos u) \cos v\mathbf{i} + (2 + \cos u) \sin v\mathbf{j} + \sin u\mathbf{k}$$

where the domain D is given by $0 \le u \le 2\pi$ and $0 \le v \le 2\pi$. (See Figure 14.43.)

Solution   Begin by calculating $\mathbf{r}_u$ and $\mathbf{r}_v$.

$$\mathbf{r}_u = -\sin u \cos v\mathbf{i} - \sin u \sin v\mathbf{j} + \cos u\mathbf{k}$$
$$\mathbf{r}_v = -(2 + \cos u) \sin v\mathbf{i} + (2 + \cos u) \cos v\mathbf{j}$$

The cross product of these two vectors is

$$\mathbf{r}_u \times \mathbf{r}_v = \begin{vmatrix} \mathbf{i} & \mathbf{j} & \mathbf{k} \\ -\sin u \cos v & -\sin u \sin v & \cos u \\ -(2 + \cos u) \sin v & (2 + \cos u) \cos v & 0 \end{vmatrix}$$

$$= -(2 + \cos u)(\cos v \cos u\mathbf{i} + \sin v \cos u\mathbf{j} + \sin u\mathbf{k})$$

which implies that

$$\|\mathbf{r}_u \times \mathbf{r}_v\| = (2 + \cos u)\sqrt{(\cos v \cos u)^2 + (\sin v \cos u)^2 + \sin^2 u}$$

$$= (2 + \cos u)\sqrt{\cos^2 u(\cos^2 v + \sin^2 v) + \sin^2 u}$$

$$= (2 + \cos u)\sqrt{\cos^2 u + \sin^2 u}$$

$$= 2 + \cos u.$$

Finally, the surface area of the torus is

$$A = \int_D\!\!\int \|\mathbf{r}_u \times \mathbf{r}_v\| \, dA = \int_0^{2\pi}\!\!\int_0^{2\pi} (2 + \cos u) \, du \, dv$$

$$= \int_0^{2\pi} 4\pi \, dv = 8\pi^2.$$

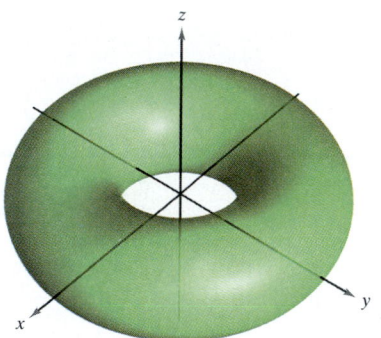

Figure 14.43

If the surface S is a surface of revolution, you can show that the formula for surface area given in Section 6.4 is equivalent to the formula given in this section. For instance, suppose f is a nonnegative function such that f' is continuous over the interval $[a, b]$. Let S be the surface of revolution formed by revolving the graph of f, where $a \le x \le b$, about the x-axis. From Section 6.4, you know that the surface area is given by

$$\text{Surface area} = 2\pi \int_a^b f(x) \sqrt{1 + [f'(x)]^2} \, dx.$$

To represent S parametrically, let $x = u$, $y = f(u) \cos v$, and $z = f(u) \sin v$, where $a \le u \le b$ and $0 \le v \le 2\pi$. Then,

$$\mathbf{r}(u, v) = u\mathbf{i} + f(u) \cos v\mathbf{j} + f(u) \sin v\mathbf{k}.$$

Try showing that the formula

$$\text{Surface area} = \int_D\!\!\int \|\mathbf{r}_u \times \mathbf{r}_v\| \, dA$$

is equivalent to the formula given above.

EXERCISES FOR SECTION 14.5

In Exercises 1–4, match the vector-valued function with its graph. [The graphs are labeled (a), (b), (c), and (d).]

(a)

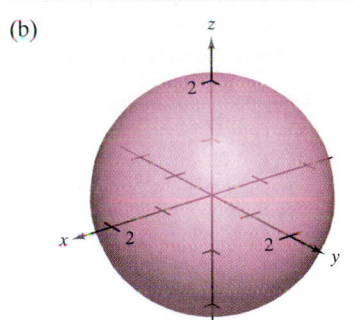

(b)

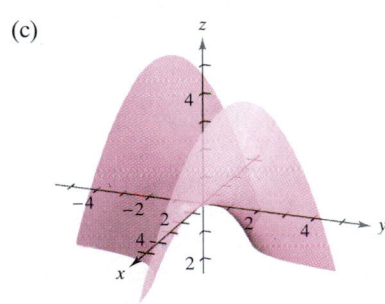

(c)

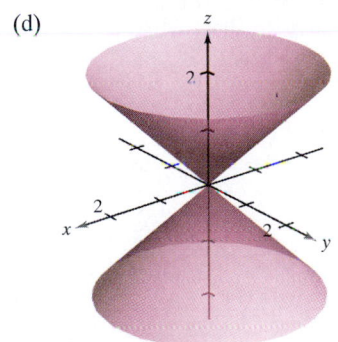

(d)

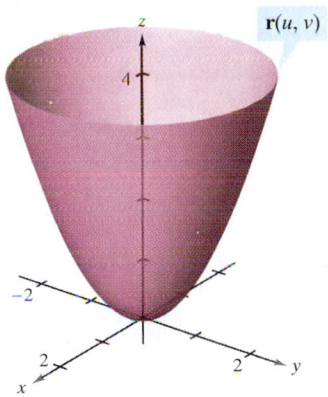

1. $\mathbf{r}(u, v) = u\mathbf{i} + v\mathbf{j} + uv\mathbf{k}$

2. $\mathbf{r}(u, v) = u\cos v\mathbf{i} + u\sin v\mathbf{j} + u\mathbf{k}$

3. $\mathbf{r}(u, v) = 2\cos v\cos u\mathbf{i} + 2\cos v\sin u\mathbf{j} + 2\sin v\mathbf{k}$

4. $\mathbf{r}(u, v) = 4\cos u\mathbf{i} + 4\sin u\mathbf{j} + v\mathbf{k}$

In Exercises 5–8, find the rectangular equation for the surface by eliminating the parameters from the vector-valued function, identify the surface, and sketch its graph.

5. $\mathbf{r}(u, v) = u\mathbf{i} + v\mathbf{j} + \dfrac{v}{2}\mathbf{k}$

6. $\mathbf{r}(u, v) = u\cos v\mathbf{i} + u\sin v\mathbf{j} + u^2\mathbf{k}$

7. $\mathbf{r}(u, v) = 2\cos u\mathbf{i} + v\mathbf{j} + 2\sin u\mathbf{k}$

8. $\mathbf{r}(u, v) = 5\cos v\cos u\mathbf{i} + 5\cos v\sin u\mathbf{j} + 5\sin v\mathbf{k}$

Think About It In Exercises 9–12, determine how the graph of the surface $s(u, v)$ differs from the graph of $\mathbf{r}(u, v) = u\cos v\mathbf{i} + u\sin v\mathbf{j} + u^2\mathbf{k}$ (see figure) where $0 \le u \le 2$ and $0 \le v \le 2\pi$. (It is not necessary to graph s.)

$\mathbf{r}(u, v)$

9. $\mathbf{s}(u, v) = u\cos v\mathbf{i} + u\sin v\mathbf{j} - u^2\mathbf{k}$
   $0 \le u \le 2, \quad 0 \le v \le 2\pi$

10. $\mathbf{s}(u, v) = u\cos v\mathbf{i} + u^2\mathbf{j} + u\sin v\mathbf{k}$
    $0 \le u \le 2, \quad 0 \le v \le 2\pi$

11. $\mathbf{s}(u, v) = u\cos v\mathbf{i} + u\sin v\mathbf{j} + u^2\mathbf{k}$
    $0 \le u \le 3, \quad 0 \le v \le 2\pi$

12. $\mathbf{s}(u, v) = 4u\cos v\mathbf{i} + 4u\sin v\mathbf{j} + u^2\mathbf{k}$
    $0 \le u \le 2, \quad 0 \le v \le 2\pi$

In Exercises 13–18, use a computer algebra system to graph the surface represented by the vector-valued function.

13. $\mathbf{r}(u, v) = 2u\cos v\mathbf{i} + 2u\sin v\mathbf{j} + u^4\mathbf{k}$
    $0 \le u \le 1, \quad 0 \le v \le 2\pi$

14. $\mathbf{r}(u, v) = 2\cos v\cos u\mathbf{i} + 4\cos v\sin u\mathbf{j} + \sin v\mathbf{k}$
    $0 \le u \le 2\pi, \quad 0 \le v \le 2\pi$

15. $\mathbf{r}(u, v) = 2\sinh u\cos v\mathbf{i} + \sinh u\sin v\mathbf{j} + \cosh u\mathbf{k}$
    $0 \le u \le 2, \quad 0 \le v \le 2\pi$

16. $\mathbf{r}(u, v) = 2u\cos v\mathbf{i} + 2u\sin v\mathbf{j} + v\mathbf{k}$
    $0 \le u \le 1, \quad 0 \le v \le 3\pi$

17. $\mathbf{r}(u, v) = (u - \sin u)\cos v\mathbf{i} + (1 - \cos u)\sin v\mathbf{j} + u\mathbf{k}$

$0 \le u \le \pi, \quad 0 \le v \le 2\pi$

18. $\mathbf{r}(u, v) = \cos^3 u \cos v\mathbf{i} + \sin^3 u \sin v\mathbf{j} + u\mathbf{k}$

$0 \le u \le \dfrac{\pi}{2}, \quad 0 \le v \le 2\pi$

19. *Think About It* The four figures are graphs of the surface

$\mathbf{r}(u, v) = u\mathbf{i} + \sin u \cos v\mathbf{j} + \sin u \sin v\mathbf{k},$

$0 \le u \le \dfrac{\pi}{2}, \quad 0 \le v \le 2\pi.$

Match each of the four graphs with the point in space from which the surface is viewed. The four points are $(10, 0, 0)$, $(-10, 10, 0)$, $(0, 10, 0)$, and $(10, 10, 10)$.

(a)
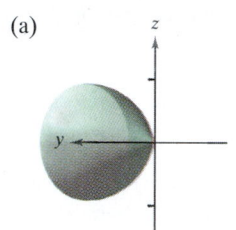

(b)
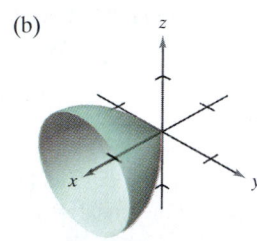

(c)
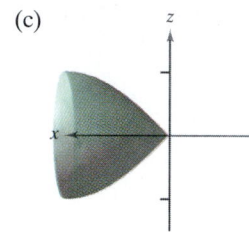

(d)
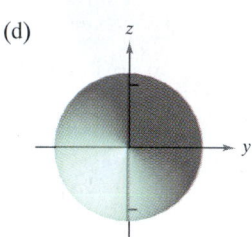

20. Use a computer algebra system to sketch three views of the graph of the vector-valued function

$\mathbf{r}(u, v) = u \cos v\mathbf{i} + u \sin v\mathbf{j} + v\mathbf{k}, \ 0 \le u \le \pi, \ 0 \le v \le \pi$

from the points $(10, 0, 0)$, $(0, 0, 10)$, and $(10, 10, 10)$.

21. *Investigation* Use a computer algebra system to graph the torus

$\mathbf{r}(u, v) = (a + b \cos v)\cos u\mathbf{i} + (a + b \cos v)\sin u\mathbf{j} + b \sin v\mathbf{k}$

for each set of values for a and b, where $0 \le u \le 2\pi$ and $0 \le v \le 2\pi$. Use the results to describe the effect of a and b on the shape of the torus.

(a) $a = 4, \ b = 1$ (b) $a = 4, \ b = 2$
(c) $a = 8, \ b = 1$ (d) $a = 8, \ b = 3$

22. *Investigation* Consider the function in Exercise 16.

(a) Sketch a graph of the function where u is held constant at $u = 1$. Identify the graph.

(b) Sketch a graph of the function where v is held constant at $v = 2\pi/3$. Identify the graph.

(c) Assume that a surface is represented by the vector-valued function $\mathbf{r} = \mathbf{r}(u, v)$. What generalization can you make about the graph of the function if one of the parameters is held constant?

In Exercises 23–30, find a vector-valued function whose graph is the indicated surface.

23. The plane $z = y$

24. The plane $x + y + z = 6$

25. The cylinder $x^2 + y^2 = 16$

26. The cylinder $x^2 + 4y^2 = 16$

27. The cylinder $z = x^2$

28. The ellipsoid $\dfrac{x^2}{9} + \dfrac{y^2}{4} + \dfrac{z^2}{1} = 1$

29. The part of the plane $z = 4$ that lies inside the cylinder $x^2 + y^2 = 9$

30. The part of the paraboloid $z = x^2 + y^2$ that lies inside the cylinder $x^2 + y^2 = 9$

Surface of Revolution In Exercises 31–34, write a set of parametric equations for the surface of revolution obtained by revolving the graph of the function about the indicated axis.

Function	Axis of Revolution
31. $y = \dfrac{x}{2}, \ 0 \le x \le 6$	x-axis
32. $y = \sqrt{x}, \ 0 \le x \le 4$	x-axis
33. $x = \sin z, \ 0 \le z \le \pi$	z-axis
34. $z = 4 - y^2, \ 0 \le y \le 2$	y-axis

Tangent Plane In Exercises 35–38, find an equation of the tangent plane to the surface given by the vector-valued function at the indicated point.

35. $\mathbf{r}(u, v) = (u + v)\mathbf{i} + (u - v)\mathbf{j} + v\mathbf{k}, \ (1, -1, 1)$

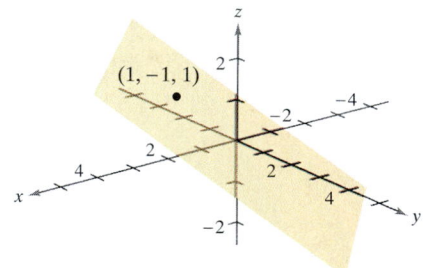

36. $\mathbf{r}(u, v) = u\mathbf{i} + v\mathbf{j} + \sqrt{uv}\ \mathbf{k}, \ (1, 1, 1)$

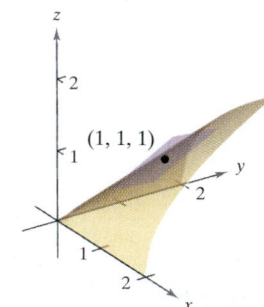

37. $\mathbf{r}(u, v) = 2u \cos v\mathbf{i} + 3u \sin v\mathbf{j} + u^2\mathbf{k}$, $(0, 6, 4)$

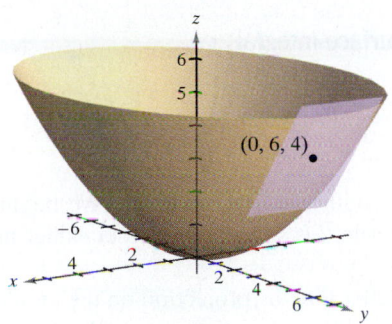

$(0, 6, 4)$

38. $\mathbf{r}(u, v) = u \cosh v\mathbf{i} + u \sinh v\mathbf{j} + u^2\mathbf{k}$, $(-2, 0, 4)$

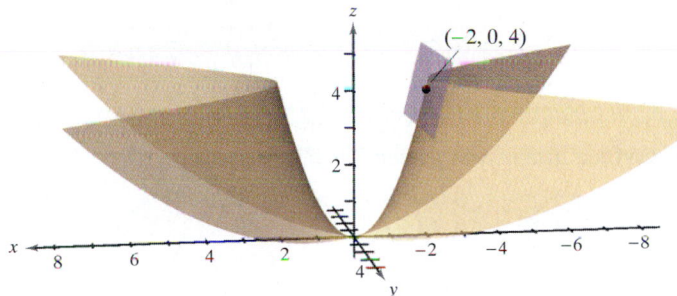

$(-2, 0, 4)$

Area   In Exercises 39–46, find the area of the surface over the indicated region. Use a symbolic integration utility to verify your results.

39. The part of the plane

$$\mathbf{r}(u, v) = 2u\mathbf{i} - \frac{v}{2}\mathbf{j} + \frac{v}{2}\mathbf{k}$$

where $0 \le u \le 2$ and $0 \le v \le 1$

40. The part of the paraboloid

$$\mathbf{r}(u, v) = 2u \cos v\mathbf{i} + 2u \sin v\mathbf{j} + u^2\mathbf{k}$$

where $0 \le u \le 2$ and $0 \le v \le 2\pi$

41. The part of the cylinder

$$\mathbf{r}(u, v) = a \cos u\mathbf{i} + a \sin u\mathbf{j} + v\mathbf{k}$$

where $0 \le u \le 2\pi$ and $0 \le v \le b$

42. The sphere

$$\mathbf{r}(u, v) = a \sin u \cos v\mathbf{i} + a \sin u \sin v\mathbf{j} + a \cos u\mathbf{k}$$

where $0 \le u \le \pi$ and $0 \le v \le 2\pi$

43. The part of the cone

$$\mathbf{r}(u, v) = au \cos v\mathbf{i} + au \sin v\mathbf{j} + u\mathbf{k}$$

where $0 \le u \le b$ and $0 \le v \le 2\pi$

44. The torus $\mathbf{r}(u, v) = (a + b \cos v)\cos u\mathbf{i} + (a + b \cos v)\sin u\mathbf{j} + b \sin v\mathbf{k}$, where $a > b$, $0 \le u \le 2\pi$, and $0 \le v \le 2\pi$

45. The surface of revolution $\mathbf{r}(u, v) = \sqrt{u} \cos v\mathbf{i} + \sqrt{u} \sin v\mathbf{j} + u\mathbf{k}$, where $0 \le u \le 4$ and $0 \le v \le 2\pi$

46. The surface of revolution

$$\mathbf{r}(u, v) = u\mathbf{i} + \sin u \cos v\mathbf{j} + \sin u \sin v\mathbf{k}$$

where $0 \le u \le \pi$ and $0 \le v \le 2\pi$

47. The surface of the dome on a new museum is given by

$$\mathbf{r}(u, v) = 20 \sin u \cos v\mathbf{i} + 20 \sin u \sin v\mathbf{j} + 20 \cos u\mathbf{k}$$

where $0 \le u \le \pi/3$ and $0 \le v \le 2\pi$ and $\mathbf{r}$ is in meters. Find the surface area of the dome.

48. Find a vector-valued function for the hyperboloid

$$x^2 + y^2 - z^2 = 1$$

and determine the tangent plane at $(1, 0, 0)$.

49. Graph and find the area of one turn of the spiral ramp

$$\mathbf{r}(u, v) = u \cos v\mathbf{i} + u \sin v\mathbf{j} + 2v\mathbf{k}$$

where $0 \le u \le 3$, $0 \le v \le 2\pi$.

50. Let f be a nonnegative function such that f' is continuous over the interval $[a, b]$. Let S be the surface of revolution formed by revolving the graph of f, where $a \le x \le b$, about the x-axis. Let $x = u$, $y = f(u)\cos v$, and $z = f(u)\sin v$, where $a \le u \le b$ and $0 \le v \le 2\pi$. Then, S is represented parametrically by

$$\mathbf{r}(u, v) = u\mathbf{i} + f(u)\cos v\mathbf{j} + f(u)\sin v\mathbf{k}.$$

Show that the following formulas are equivalent.

$$\text{Surface area} = 2\pi \int_a^b f(x)\sqrt{1 + [f'(x)]^2}\, dx$$

$$\text{Surface area} = \int\int_D \|\mathbf{r}_u \times \mathbf{r}_v\|\, dA$$

51. *Open-Ended Project*   The parametric equations

$$x = 3 + \sin u[7 - \cos(3u - 2v) - 2\cos(3u + v)]$$
$$y = 3 + \cos u[7 - \cos(3u - 2v) - 2\cos(3u + v)]$$
$$z = \sin(3u - 2v) + 2\sin(3u + v)$$

where $-\pi \le u \le \pi$ and $-\pi \le v \le \pi$, represent the surface shown below. Try to create your own parametric surface using a computer algebra system.

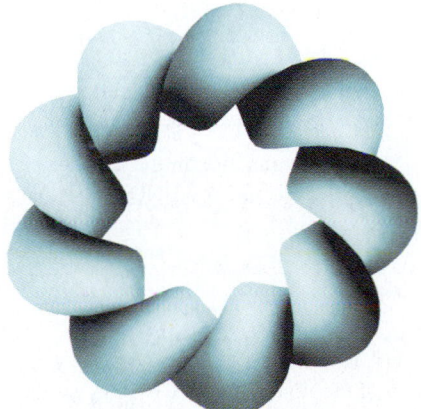

Surface Integrals • Parametric Surfaces and Surface Integrals •
Orientation of a Surface • Flux Integrals

Surface Integrals

The remainder of this chapter deals primarily with **surface integrals.** We begin by considering surfaces given by $z = g(x, y)$. Later in this section we will consider more general surfaces given in parametric form.

Let S be a surface given by $z = g(x, y)$ and let R be its projection on the xy-plane, as shown in Figure 14.44. Suppose that g, g_x, and g_y are continuous at all points in R and that f is defined on S. Employing the procedure used to find surface area in Section 13.5, evaluate f at (x_i, y_i, z_i) and form the sum

$$\sum_{i=1}^{n} f(x_i, y_i, z_i) \, \Delta S_i$$

where $\Delta S_i \approx \sqrt{1 + [g_x(x_i, y_i)]^2 + [g_y(x_i, y_i)]^2} \, \Delta A_i$. Provided the limit as $\|\Delta\|$ approaches 0 exists, the **surface integral of f over S** is defined as

$$\iint_{S} f(x, y, z) \, dS = \lim_{\|\Delta\| \to 0} \sum_{i=1}^{n} f(x_i, y_i, z_i) \, \Delta S_i.$$

This integral can be evaluated by a double integral.

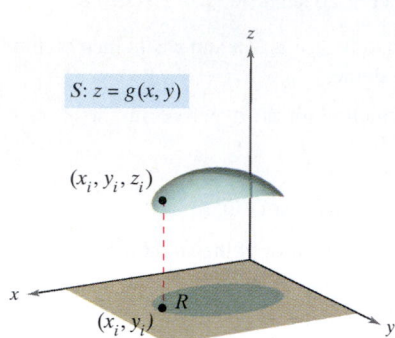

Scalar function f assigns a number to each point of S.
Figure 14.44

THEOREM 14.10   Evaluating a Surface Integral

Let S be a surface with equation $z = g(x, y)$ and let R be its projection on the xy-plane. If g, g_x, and g_y are continuous on R and f is continuous on S, then the surface integral of f over S is

$$\iint_{S} f(x, y, z) \, dS = \iint_{R} f(x, y, g(x, y)) \sqrt{1 + [g_x(x, y)]^2 + [g_y(x, y)]^2} \, dA.$$

For surfaces described by functions of x and z (or y and z), you can make the following adjustments to Theorem 14.10. If S is the graph of $y = g(x, z)$ and R is its projection on the xz-plane, then

$$\iint_{S} f(x, y, z) \, dS = \iint_{R} f(x, g(x, z), z) \sqrt{1 + [g_x(x, z)]^2 + [g_z(x, z)]^2} \, dA.$$

If S is the graph of $x = g(y, z)$ and R is its projection on the yz-plane, then

$$\iint_{S} f(x, y, z) \, dS = \iint_{R} f(g(y, z), y, z) \sqrt{1 + [g_y(y, z)]^2 + [g_z(y, z)]^2} \, dA.$$

NOTE   If $f(x, y, z) = 1$, the surface integral over S yields the surface area of S. For instance, suppose the surface S is the plane given by $z = x$, where $0 \le x \le 1$ and $0 \le y \le 1$. The surface area of S is $\sqrt{2}$ square units. Try verifying that $\int_S \int f(x, y, z) \, dS = \sqrt{2}$.

EXAMPLE 1   Evaluating a Surface Integral

Evaluate the surface integral

$$\iint_S (y^2 + 2yz)\, dS$$

where S is the first-octant portion of the plane $2x + y + 2z = 6$.

Solution   Begin by writing S as

$$z = \frac{1}{2}(6 - 2x - y)$$

$$g(x, y) = \frac{1}{2}(6 - 2x - y).$$

Using the partial derivatives $g_x(x, y) = -1$ and $g_y(x, y) = -\frac{1}{2}$, you can write

$$\sqrt{1 + [g_x(x, y)]^2 + [g_y(x, y)]^2} = \sqrt{1 + 1 + \frac{1}{4}} = \frac{3}{2}.$$

Using Figure 14.45 and Theorem 14.10, you obtain

$$\iint_S (y^2 + 2yz)\, dS = \iint_R f(x, y, g(x, y))\sqrt{1 + [g_x(x, y)]^2 + [g_y(x, y)]^2}\, dA$$

$$= \iint_R \left[y^2 + 2y\left(\frac{1}{2}\right)(6 - 2x - y) \right]\left(\frac{3}{2}\right) dA$$

$$= 3\int_0^3 \int_0^{2(3-x)} y(3 - x)\, dy\, dx$$

$$= 6\int_0^3 (3 - x)^3\, dx$$

$$= -\frac{3}{2}(3 - x)^4 \Big]_0^3$$

$$= \frac{243}{2}.$$

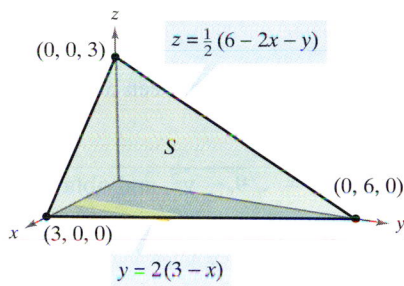

$z = \frac{1}{2}(6 - 2x - y)$

$(0, 0, 3)$

S

$(0, 6, 0)$

$(3, 0, 0)$

$y = 2(3 - x)$

Figure 14.45

An alternative solution to Example 1 would be to project S onto the yz-plane, as shown in Figure 14.46. Then, $x = \frac{1}{2}(6 - y - 2z)$, and

$$\sqrt{1 + [g_y(y, z)]^2 + [g_z(y, z)]^2} = \sqrt{1 + \frac{1}{4} + 1} = \frac{3}{2}.$$

Thus, the surface integral is

$$\iint_S (y^2 + 2yz)\, dS = \iint_R f(g(y, z), y, z)\sqrt{1 + [g_y(y, z)]^2 + [g_x(y, z)]^2}\, dA$$

$$= \int_0^6 \int_0^{(6-y)/2} (y^2 + 2yz)\left(\frac{3}{2}\right) dz\, dy$$

$$= \frac{3}{8}\int_0^6 (36y - y^3)\, dy$$

$$= \frac{243}{2}.$$

Try reworking Example 1 by projecting S onto the xz-plane.

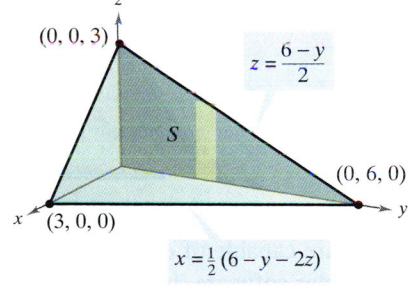

$(0, 0, 3)$

$z = \frac{6 - y}{2}$

S

$(0, 6, 0)$

$(3, 0, 0)$

$x = \frac{1}{2}(6 - y - 2z)$

Figure 14.46

N (up

(horiz

N (dow

Figure

De
Fig

REVIEW EXERCISES FOR CHAPTER 14

In Exercises 1 and 2, sketch several representative vectors in the vector field. Use a computer algebra system to verify your results.

1. $\mathbf{F}(x, y, z) = x\mathbf{i} + \mathbf{j} + 2\mathbf{k}$ 2. $\mathbf{F}(x, y) = \mathbf{i} - 2y\mathbf{j}$

In Exercises 3 and 4, find a three-dimensional vector field that has the potential function f.

3. $f(x, y, z) = 8x^2 + xy + z^2$
4. $f(x, y, z) = x^2 e^{yz}$

In Exercises 5–12, determine if F is conservative. If it is, find the potential function f.

5. $\mathbf{F}(x, y) = \dfrac{1}{y}\mathbf{i} - \dfrac{y}{x^2}\mathbf{j}$

6. $\mathbf{F}(x, y) = -\dfrac{y}{x^2}\mathbf{i} + \dfrac{1}{x}\mathbf{j}$

7. $\mathbf{F}(x, y) = (6xy^2 - 3x^2)\mathbf{i} + (6x^2y + 3y^2 - 7)\mathbf{j}$
8. $\mathbf{F}(x, y) = (-2y^3 \sin 2x)\mathbf{i} + 3y^2(1 + \cos 2x)\mathbf{j}$
9. $\mathbf{F}(x, y, z) = (4xy + z)\mathbf{i} + (2x^2 + 6y)\mathbf{j} + 2z\mathbf{k}$
10. $\mathbf{F}(x, y, z) = (4xy + z^2)\mathbf{i} + (2x^2 + 6yz)\mathbf{j} + 2xz\mathbf{k}$

11. $\mathbf{F}(x, y, z) = \dfrac{yz\mathbf{i} - xz\mathbf{j} - xy\mathbf{k}}{y^2 z^2}$

12. $\mathbf{F}(x, y, z) = \sin z(y\mathbf{i} + x\mathbf{j} + \mathbf{k})$

In Exercises 13–20, find (a) the divergence of the vector field F and (b) the curl of the vector field F.

13. $\mathbf{F}(x, y, z) = x^2\mathbf{i} + y^2\mathbf{j} + z^2\mathbf{k}$
14. $\mathbf{F}(x, y, z) = xy^2\mathbf{j} - zx^2\mathbf{k}$
15. $\mathbf{F}(x, y, z) = (\cos y + y \cos x)\mathbf{i} + (\sin x - x \sin y)\mathbf{j} + xyz\mathbf{k}$
16. $\mathbf{F}(x, y, z) = (3x - y)\mathbf{i} + (y - 2z)\mathbf{j} + (z - 3x)\mathbf{k}$
17. $\mathbf{F}(x, y, z) = \arcsin x\mathbf{i} + xy^2\mathbf{j} + yz^2\mathbf{k}$
18. $\mathbf{F}(x, y, z) = (x^2 - y)\mathbf{i} - (x + \sin^2 y)\mathbf{j}$
19. $\mathbf{F}(x, y, z) = \ln(x^2 + y^2)\mathbf{i} + \ln(x^2 + y^2)\mathbf{j} + z\mathbf{k}$

20. $\mathbf{F}(x, y, z) = \dfrac{z}{x}\mathbf{i} + \dfrac{z}{y}\mathbf{j} + z^2\mathbf{k}$

In Exercises 21–26, evaluate the line integral over the indicated path(s).

21. $\displaystyle\int_C (x^2 + y^2)\, ds$

(a) C: line segment from $(-1, -1)$ to $(2, 2)$
(b) $C: x^2 + y^2 = 16$, one revolution counterclockwise, starting at $(4, 0)$

22. $\displaystyle\int_C xy\, ds$

(a) C: line segment from $(0, 0)$ to $(5, 4)$
(b) C: counterclockwise around the triangle with vertices $(0, 0)$, $(4, 0)$, $(0, 2)$

23. $\displaystyle\int_C (x^2 + y^2)\, ds$

$C: \mathbf{r}(t) = (\cos t + t \sin t)\mathbf{i} + (\sin t - t \cos t)\mathbf{j}, \quad 0 \le t \le 2\pi$

24. $\displaystyle\int_C x\, ds$

$C: \mathbf{r}(t) = (t - \sin t)\mathbf{i} + (1 - \cos t)\mathbf{j}, \quad 0 \le t \le 2\pi$

25. $\displaystyle\int_C (2x - y)\, dx + (x + 3y)\, dy$

(a) C: line segment from $(0, 0)$ to $(2, -3)$
(b) C: counterclockwise around the circle $x = 3 \cos t$, $y = 3 \sin t$

26. $\displaystyle\int_C (2x - y)\, dx + (x + 3y)\, dy$

$C: \mathbf{r}(t) = (\cos t + t \sin t)\mathbf{i} + (\sin t - t \sin t)\mathbf{j}, \; 0 \le t \le \pi/2$

In Exercises 27 and 28, use a symbolic integration utility to evaluate the line integral over the indicated path.

27. $\displaystyle\int_C (2x + y)\, ds$

$\mathbf{r}(t) = a \cos^3 t\,\mathbf{i} + a \sin^3 t\,\mathbf{j}, \quad 0 \le t \le \pi/2$

28. $\displaystyle\int_C (x^2 + y^2 + z^2)\, ds$

$\mathbf{r}(t) = t\mathbf{i} + t^2\mathbf{j} + t^{3/2}\mathbf{k}, \quad 0 \le t \le 4$

In Exercises 29 and 30, find the lateral surface area over the curve C in the xy-plane and under the surface $z = f(x, y)$.

29. $f(x, y) = 5 + \sin(x + y)$
   $C: y = 3x$ from $(0, 0)$ to $(2, 6)$
30. $f(x, y) = 12 - x - y$
   $C: y = x^2$ from $(0, 0)$ to $(2, 4)$

In Exercises 31–36, evaluate

$$\int_C \mathbf{F} \cdot d\mathbf{r}.$$

31. $\mathbf{F}(x, y) = xy\mathbf{i} + x^2\mathbf{j}$
   $C: \mathbf{r}(t) = t^2\mathbf{i} + t^3\mathbf{j}, \quad 0 \le t \le 1$
32. $\mathbf{F}(x, y) = (x - y)\mathbf{i} + (x + y)\mathbf{j}$
   $C: \mathbf{r}(t) = 4 \cos t\,\mathbf{i} + 3 \sin t\,\mathbf{j}, \quad 0 \le t \le 2\pi$

33. $\mathbf{F}(x, y, z) = x\mathbf{i} + y\mathbf{j} + z\mathbf{k}$

$C: \mathbf{r}(t) = 2\cos t\,\mathbf{i} + 2\sin t\,\mathbf{j} + t\mathbf{k}, \quad 0 \le t \le 2\pi$

34. $\mathbf{F}(x, y, z) = (2y - z)\mathbf{i} + (z - x)\mathbf{j} + (x - y)\mathbf{k}$

$C:$ curve of intersection of $x^2 + z^2 = 4$ and $y^2 + z^2 = 4$ from $(2, 2, 0)$ to $(0, 0, 2)$

35. $\mathbf{F}(x, y, z) = (y - z)\mathbf{i} + (z - x)\mathbf{j} + (x - y)\mathbf{k}$

$C:$ curve of intersection of $z = x^2 + y^2$ and $x + y = 0$ from $(-2, 2, 8)$ to $(2, -2, 8)$

36. $\mathbf{F}(x, y, z) = (x^2 - z)\mathbf{i} + (y^2 + z)\mathbf{j} + x\mathbf{k}$

$C:$ curve of intersection of $z = x^2$ and $x^2 + y^2 = 4$ from $(0, -2, 0)$ to $(0, 2, 0)$

In Exercises 37 and 38, use a symbolic integration utility to evaluate the line integral.

37. $\displaystyle\int_C xy\,dx + (x^2 + y^2)\,dy$

$C: y = x^2$ from $(0, 0)$ to $(2, 4)$ and $y = 2x$ from $(2, 4)$ to $(0, 0)$

38. $\displaystyle\int_C \mathbf{F} \cdot d\mathbf{r}$

$\mathbf{F}(x, y) = (2x - y)\mathbf{i} + (2y - x)\mathbf{j}$

$C: \mathbf{r}(t) = (2\cos t + 2t\sin t)\mathbf{i} + (2\sin t - 2t\cos t)\mathbf{j},$ $0 \le t \le \pi$

39. Work Find the work done by the force field $\mathbf{F} = x\mathbf{i} - \sqrt{y}\,\mathbf{j}$ along the path $y = x^{3/2}$ from $(0, 0)$ to $(4, 8)$.

40. Work Find the work done by the engines of a 20-ton aircraft if it climbs 2000 feet while making a 90° turn in a circular arc of radius 10 miles.

In Exercises 41 and 42, use the Fundamental Theorem of Line Integrals to evaluate the integral.

41. $\displaystyle\int_C 2xyz\,dx + x^2z\,dy + x^2y\,dz$

$C:$ smooth curve from $(0, 0, 0)$ to $(1, 4, 3)$

42. $\displaystyle\int_C y\,dx + x\,dy + \frac{1}{z}\,dz$

$C:$ smooth curve from $(0, 0, 1)$ to $(4, 4, 4)$

43. Evaluate the line integral $\displaystyle\int_C y^2\,dx + 2xy\,dy$.

(a) $C: \mathbf{r}(t) = (1 + 3t)\mathbf{i} + (1 + t)\mathbf{j}, \quad 0 \le t \le 1$

(b) $C: \mathbf{r}(t) = t\mathbf{i} + \sqrt{t}\,\mathbf{j}, \quad 1 \le t \le 4$

(c) Use the Fundamental Theorem of Line Integrals where C is a smooth curve from $(1, 1)$ to $(4, 2)$.

44. Area and Centroid Consider the region bounded by the x-axis and one arch of the cycloid with parametric equations $x = a(\theta - \sin\theta)$ and $y = a(1 - \cos\theta)$. Use line integrals to find (a) the area of the region and (b) the centroid of the region.

In Exercises 45–50, use Green's Theorem to evaluate the line integral.

45. $\displaystyle\int_C y\,dx + 2x\,dy$

$C:$ boundary of the square with vertices $(0, 0)$, $(0, 2)$, $(2, 0)$, $(2, 2)$

46. $\displaystyle\int_C xy\,dx + (x^2 + y^2)\,dy$

$C:$ boundary of the square with vertices $(0, 0)$, $(0, 2)$, $(2, 0)$, $(2, 2)$

47. $\displaystyle\int_C xy^2\,dx + x^2y\,dy$

$C: x = 4\cos t, \quad y = 2\sin t$

48. $\displaystyle\int_C (x^2 - y^2)\,dx + 2xy\,dy$

$C: x^2 + y^2 = a^2$

49. $\displaystyle\int_C xy\,dx + x^2\,dy$

$C:$ boundary of the region between the graphs of $y = x^2$ and $y = x$

50. $\displaystyle\int_C y^2\,dx + x^{2/3}\,dy$

$C: x^{2/3} + y^{2/3} = 1$

In Exercises 51 and 52, use a graphing utility to graph the surface represented by the vector-valued function.

51. $\mathbf{r}(u, v) = \sec u \cos v\,\mathbf{i} + (1 + 2\tan u)\sin v\,\mathbf{j} + 2u\,\mathbf{k}$

$0 \le u \le \dfrac{\pi}{3}, \quad 0 \le v \le 2\pi$

52. $\mathbf{r}(u, v) = e^{-u/4}\cos v\,\mathbf{i} + e^{-u/4}\sin v\,\mathbf{j} + \dfrac{u}{6}\mathbf{k}$

$0 \le u \le 4, \quad 0 \le v \le 2\pi$

53. Investigation Consider the surface represented by the vector-valued function

$\mathbf{r}(u, v) = 3\cos v \cos u\,\mathbf{i} + 3\cos v \sin u\,\mathbf{j} + \sin v\,\mathbf{k}.$

Use a computer algebra system when a graph is required.

(a) Graph the surface for $0 \le u \le 2\pi$ and $-\dfrac{\pi}{2} \le v \le \dfrac{\pi}{2}$.

(b) Graph the surface for $0 \le u \le 2\pi$ and $\dfrac{\pi}{4} \le v \le \dfrac{\pi}{2}$.

(c) Graph the surface for $0 \le u \le \dfrac{\pi}{4}$ and $0 \le v \le \dfrac{\pi}{2}$.

(d) Graph and identify the space curve for $0 \le u \le 2\pi$ and $v = \dfrac{\pi}{4}$.

(e) Use a computer algebra system to approximate the area of the surface graphed in part (b).

(f) Use a computer algebra system to approximate the area of the surface graphed in part (c).

54. Evaluate the surface integral $\displaystyle\iint_S z\,dS$ over the surface S:

$$\mathbf{r}(u, v) = (u + v)\mathbf{i} + (u - v)\mathbf{j} + \sin v\,\mathbf{k}$$

where $0 \leq u \leq 2$ and $0 \leq v \leq \pi$.

55. Use a symbolic integration utility to graph the surface S and approximate the surface integral $\displaystyle\iint_S (x + y)\,dS$, where S is the surface S: $\mathbf{r}(u, v) = u \cos v\,\mathbf{i} + u \sin v\,\mathbf{j} + (u - 1)(2 - u)\mathbf{k}$ over $0 \leq u \leq 2$ and $0 \leq v \leq 2\pi$.

In Exercises 56 and 57, verify the Divergence Theorem by evaluating $\displaystyle\int_S \mathbf{F} \cdot \mathbf{N}\,dS$ **as a surface integral and as a triple integral.**

56. $\mathbf{F}(x, y, z) = x\mathbf{i} + y\mathbf{j} + z\mathbf{k}$

Q: solid region bounded by the coordinate planes and the plane $2x + 3y + 4z = 12$

57. $\mathbf{F}(x, y, z) = x^2\mathbf{i} + xy\mathbf{j} + z\mathbf{k}$

Q: solid region bounded by the coordinate planes and the plane $2x + 3y + 4z = 12$

In Exercises 58 and 59, verify Stokes's Theorem by evaluating $\displaystyle\int_C \mathbf{F} \cdot d\mathbf{r}$ **as a line integral and as a double integral.**

58. $\mathbf{F}(x, y, z) = (x - z)\mathbf{i} + (y - z)\mathbf{j} + x^2\mathbf{k}$

S: first-octant portion of the plane $3x + y + 2z = 12$

59. $\mathbf{F}(x, y, z) = (\cos y + y \cos x)\mathbf{i} + (\sin x - x \sin y)\mathbf{j} + xyz\mathbf{k}$

S: portion of $z = y^2$ over the square in the xy-plane with vertices $(0, 0)$, $(a, 0)$, (a, a), $(0, a)$

$\mathbf{N}$ is the upward unit normal to the surface.

SECTION PROJECT

The Planimeter You have learned many calculus techniques for finding the area of a planar region. Engineers use a mechanical device called a *planimeter* for measuring planar areas, which is based on the area formula given in Theorem 14.9 (page 1022). As you can see in the figure, the planimeter is fixed at point O (but free to pivot) and has a hinge at A. The end of the tracer arm AB moves counterclockwise around the region R. A small wheel at B is perpendicular to $\overline{AB}$ and is marked with a scale to measure how much it rolls as B traces out the boundary of region R. You will show in this project that the area of R is given by the length L of the tracer arm $\overline{AB}$ multiplied by the distance D that the wheel rolls.

Assume that point B traces out the boundary of R for $a \leq t \leq b$. Point A will move back and forth along a circular arc around the origin O. Let $\theta(t)$ denote the angle in the figure and let $(x(t), y(t))$ denote the coordinates of A.

(a) Show that the vector $\overrightarrow{OB}$ is given by the vector-valued function

$$\mathbf{r}(t) = [x(t) + L \cos \theta(t)]\mathbf{i} + [y(t) + L \sin \theta(t)]\mathbf{j}.$$

(b) Show that the following two integrals are equal to zero.

$$I_1 = \int_a^b \frac{1}{2}L^2 \frac{d\theta}{dt}\,dt$$

$$I_2 = \int_a^b \frac{1}{2}\left(x\frac{dy}{dt} - y\frac{dx}{dt}\right)dt$$

(c) Use the integral $\displaystyle\int_a^b [x(t) \sin \theta(t) - y(t) \cos \theta(t)]'\,dt$ to show that the following two integrals are equal.

$$I_3 = \int_a^b \frac{1}{2}L\left(y \sin \theta \frac{d\theta}{dt} + x \cos \theta \frac{d\theta}{dt}\right)dt$$

$$I_4 = \int_a^b \frac{1}{2}L\left(-\sin \theta \frac{dx}{dt} + \cos \theta \frac{dy}{dt}\right)dt$$

(d) Let $\mathbf{N} = -\sin \theta\,\mathbf{i} + \cos \theta\,\mathbf{j}$. Explain why the distance D that the wheel rolls is given by

$$D = \int_C \mathbf{N} \cdot \mathbf{T}\,ds.$$

(e) Show that the area of region R is given by $I_1 + I_2 + I_3 + I_4 = DL$.

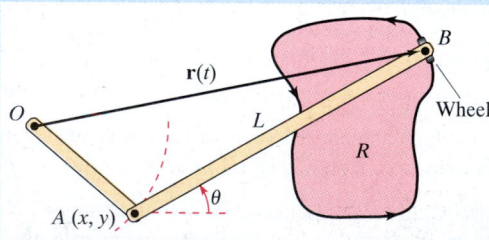

FOR FURTHER INFORMATION For more information about planimeters, see the article by C. L. Strong in the August 1958 issue of *Scientific American*.

15

Differential Equations

A well-publicized example of two species that have a predator/prey relationship is that of the black-footed ferret and the prairie dog. In the western United States, increasing depletion of prairie dog populations led to a dramatic decrease in the population of the black-footed ferrets. In 1986, biologists captured the last 18 ferrets that they could find and put them into a breeding station. Attempts are being made to reintroduce the black-footed ferret populations into natural settings.

Species of animals are introduced or reintroduced into specific habitats for a variety of reasons, both deliberate and accidental. Sometimes such an introduction results in an environmental tragedy in which the alien species threatens or even displaces one or more native species in the region.

In the 1920s, mathematicians Lotka and Volterra independently developed mathematical models that can represent many of the different ways that two species interact with each other. The common ways in which two species interact are (1) as predator and prey, (2) as complimentary species, and (3) as competing species. For instance, consider a large pond that contains a trout population and a bass population. Although these fishes do eat each others' young, they are usually classified as competing populations because they compete for the same food supply. Let x represent the number of bass, let y represent the number of trout, and let t represent the time (in months). Then, the rates of change of each population can be represented by the following system of differential equations. (In this system, a, b, m, and n are positive constants.)

$$\frac{dx}{dt} = ax - bxy \qquad \text{and} \qquad \frac{dy}{dt} = my - nxy$$

These equations have many possible solutions that can be obtained by solving the following differential equation.

$$\frac{dx}{dy} = \frac{dx/dt}{dy/dt} = \frac{ax - bxy}{my - nxy}$$

The particular solutions depend on the initial values of x and y and on the values of a, b, m, and n. Graphs of several solutions are shown at the right. The arrowheads on the graphs denote the directions the populations take over time.

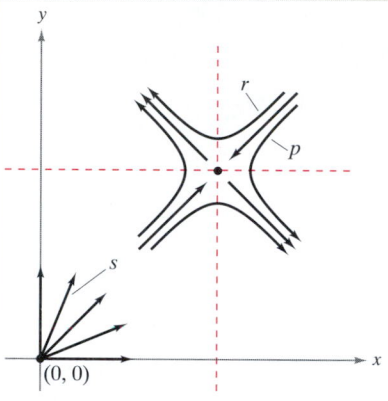

QUESTIONS

1. How would you describe a point of equilibrium for the two populations? Knowing that an equilibrium point must be located where

$$\frac{dx}{dt} = \frac{dy}{dt} = 0$$

   find the coordinates of each equilibrium point shown above.

2. For the curves labeled p, r, and s on the graph, describe what is happening to x and y as t increases. Explain this in terms of bass and trout.

3. From which starting points do the bass "win"? From which do the trout "win"?

4. How might you expect the equations and graphs to change if you were studying two species that had a predator/prey relationship rather than a competing relationship?

5. Solve the differential equation $dx/dy = (ax - bxy)/(my - nxy)$ using a technique you learned in Chapter 5. Then assign different values of a, b, m, and n and graph the resulting solutions. Do you obtain graphs similar to those shown above?

The concepts presented here will be explored further in this chapter. For an extension of this application, see the lab series that accompanies this text.

Exact Differential Equations • Integrating Factors

Exact Differential Equations

In Section 5.6, you studied applications of differential equations to growth and decay problems. In Section 5.7, you learned more about the basic ideas of differential equations and studied the solution technique known as separation of variables. In this chapter, you will learn more about solving differential equations and using them in real-life applications. This section introduces you to a method for solving the first-order differential equation

$$M(x, y)\, dx + N(x, y)\, dy = 0$$

for the special case in which this equation represents the exact differential of a function $z = f(x, y)$.

Definition of an Exact Differential Equation

The equation $M(x, y)\, dx + N(x, y)\, dy = 0$ is an **exact differential equation** if there exists a function f of two variables x and y having continuous partial derivatives such that

$$f_x(x, y) = M(x, y) \qquad \text{and} \qquad f_y(x, y) = N(x, y).$$

The general solution of the equation is $f(x, y) = C$.

From Section 12.3, you know that if f has continuous second partials, then

$$\frac{\partial M}{\partial y} = \frac{\partial^2 f}{\partial y\, \partial x} = \frac{\partial^2 f}{\partial x\, \partial y} = \frac{\partial N}{\partial x}.$$

This suggests the following test for exactness.

THEOREM 15.1   Test for Exactness

Let M and N have continuous partial derivatives on an open disc R. The differential equation $M(x, y)\, dx + N(x, y)\, dy = 0$ is exact if and only if

$$\frac{\partial M}{\partial y} = \frac{\partial N}{\partial x}.$$

Exactness is a fragile condition in the sense that seemingly minor alterations in an exact equation can destroy its exactness. This is demonstrated in the following example.

EXAMPLE 1   Testing for Exactness

NOTE   Every differential equation of the form

$$M(x)\,dx + N(y)\,dy = 0$$

is exact. In other words, a separable variables equation is actually a special type of an exact equation.

a. The differential equation $(xy^2 + x)\,dx + yx^2\,dy = 0$ is exact because

$$\frac{\partial M}{\partial y} = \frac{\partial}{\partial y}\big[xy^2 + x\big] = 2xy \qquad \text{and} \qquad \frac{\partial N}{\partial x} = \frac{\partial}{\partial x}\big[yx^2\big] = 2xy.$$

But the equation $(y^2 + 1)\,dx + xy\,dy = 0$ is not exact, even though it is obtained by dividing both sides of the first equation by x.

b. The differential equation $\cos y\,dx + (y^2 - x\sin y)\,dy = 0$ is exact because

$$\frac{\partial M}{\partial y} = \frac{\partial}{\partial y}\big[\cos y\big] = -\sin y \qquad \text{and} \qquad \frac{\partial N}{\partial x} = \frac{\partial}{\partial x}\big[y^2 - x\sin y\big] = -\sin y.$$

But the equation $\cos y\,dx + (y^2 + x\sin y)\,dy = 0$ is not exact, even though it differs from the first equation only by a single sign.

Note that the test for exactness of $M(x, y)\,dx + N(x, y)\,dy = 0$ is the same as the test for determining whether $\mathbf{F}(x, y) = M(x, y)\mathbf{i} + N(x, y)\mathbf{j}$ is the gradient of a potential function (Theorem 14.1). This means that a general solution $f(x, y) = C$ to an exact differential equation can be found by the method used to find a potential function for a conservative vector field.

EXAMPLE 2   Solving an Exact Differential Equation

Solve the differential equation $(2xy - 3x^2)\,dx + (x^2 - 2y)\,dy = 0$.

Solution   The given differential equation is exact because

$$\frac{\partial M}{\partial y} = \frac{\partial}{\partial y}\big[2xy - 3x^2\big] = 2x = \frac{\partial N}{\partial x} = \frac{\partial}{\partial x}\big[x^2 - 2y\big].$$

The general solution, $f(x, y) = C$, is given by

$$f(x, y) = \int M(x, y)\,dx$$

$$= \int (2xy - 3x^2)\,dx = x^2 y - x^3 + g(y).$$

In Section 14.1, you determined $g(y)$ by integrating $N(x, y)$ with respect to y and reconciling the two expressions for $f(x, y)$. An alternative method is to partially differentiate this version of $f(x, y)$ with respect to y and compare the result with $N(x, y)$. In other words,

$$f_y(x, y) = \frac{\partial}{\partial y}\big[x^2 y - x^3 + g(y)\big] = x^2 + g'(y) = \overbrace{x^2 - 2y}^{N(x, y)}.$$

$$\boxed{g'(y) = -2y}$$

Thus, $g'(y) = -2y$, and it follows that $g(y) = -y^2 + C_1$. Therefore,

$$f(x, y) = x^2 y - x^3 - y^2 + C_1$$

and the general solution is $x^2 y - x^3 - y^2 = C$. Figure 15.1 shows the solution curves that correspond to $C = 1, 10, 100,$ and 1000.

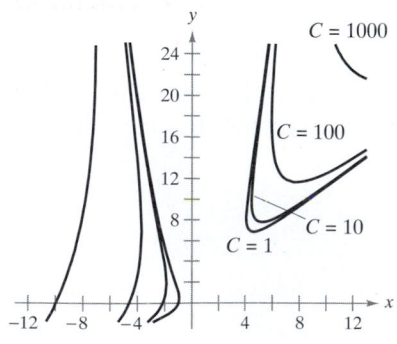

Figure 15.1

EXAMPLE 3   Solving an Exact Differential Equation

Find the particular solution of

$$(\cos x - x \sin x + y^2)\,dx + 2xy\,dy = 0$$

that satisfies the initial condition $y = 1$ when $x = \pi$.

Solution   The differential equation is exact because

$$\underbrace{\frac{\partial}{\partial y}\left[\cos x - x\sin x + y^2\right]}_{\tfrac{\partial M}{\partial y}} = 2y = \underbrace{\frac{\partial}{\partial x}\left[2xy\right]}_{\tfrac{\partial N}{\partial x}}.$$

Because $N(x, y)$ is simpler than $M(x, y)$, it is better to begin by integrating $N(x, y)$.

$$f(x, y) = \int N(x, y)\,dy = \int 2xy\,dy = xy^2 + g(x)$$

$$f_x(x, y) = \frac{\partial}{\partial x}\left[xy^2 + g(x)\right] = y^2 + g'(x) = \overbrace{\cos x - x\sin x + y^2}^{M(x,\,y)}$$

$$\boxed{g'(x) = \cos x - x \sin x}$$

Thus, $g'(x) = \cos x - x \sin x$ and

$$g(x) = \int (\cos x - x\sin x)\,dx$$

$$= x\cos x + C_1$$

which implies that $f(x, y) = xy^2 + x\cos x + C_1$, and the general solution is

$$xy^2 + x\cos x = C.\qquad \text{General solution}$$

Applying the given initial condition produces

$$\pi(1)^2 + \pi\cos\pi = C$$

which implies that $C = 0$. Hence, the particular solution is

$$xy^2 + x\cos x = 0.\qquad \text{Particular solution}$$

The graph of the particular solution is shown in Figure 15.3. Notice that the graph consists of two parts: the ovals are given by $y^2 + \cos x = 0$, and the y-axis is given by $x = 0$.

In Example 3, note that if $z = f(x, y) = xy^2 + x\cos x$, the total differential of z is given by

$$dz = f_x(x, y)\,dx + f_y(x, y)\,dy$$

$$= (\cos x - x\sin x + y^2)\,dx + 2xy\,dy$$

$$= M(x, y)\,dx + N(x, y)\,dy.$$

In other words, $M\,dx + N\,dy = 0$ is called an *exact* differential equation because $M\,dx + N\,dy$ is exactly the differential of $f(x, y)$.

TECHNOLOGY   You can use a graphing utility to graph a particular solution that satisfies the initial condition of a differential equation. In Example 3, the differential equation and initial conditions are satisfied when $xy^2 + x\cos x = 0$, which implies that the particular solution can be written as $x = 0$ or $y = \pm\sqrt{-\cos x}$. On a graphing calculator screen, the solution would be represented by Figure 15.2 together with the y-axis.

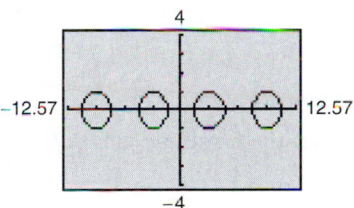

Figure 15.2

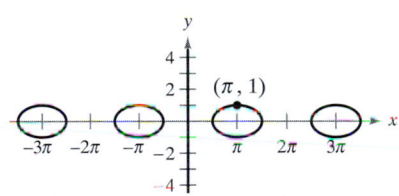

Figure 15.3

Integrating Factors

If the differential equation $M(x, y)\, dx + N(x, y)\, dy = 0$ is not exact, it may be possible to make it exact by multiplying by an appropriate factor $u(x, y)$, which is called an **integrating factor** for the differential equation.

EXAMPLE 4   Multiplying by an Integrating Factor

a. If the differential equation

$$2y\, dx + x\, dy = 0 \qquad \text{\color{red}{Not an exact equation}}$$

is multiplied by the integrating factor $u(x, y) = x$, the resulting equation

$$2xy\, dx + x^2\, dy = 0 \qquad \text{\color{red}{Exact equation}}$$

is exact—the left side is the total differential of $x^2 y$.

b. If the equation

$$y\, dx - x\, dy = 0 \qquad \text{\color{red}{Not an exact equation}}$$

is multiplied by the integrating factor $u(x, y) = 1/y^2$, the resulting equation

$$\frac{1}{y}\, dx - \frac{x}{y^2}\, dy = 0 \qquad \text{\color{red}{Exact equation}}$$

is exact—the left side is the total differential of x/y.

Finding an integrating factor can be difficult. However, there are two classes of differential equations whose integrating factors can be found routinely—namely, those that possess integrating factors that are functions of either x alone or y alone. The following theorem, which we present without proof, outlines a procedure for finding these two special categories of integrating factors.

THEOREM 15.2   Integrating Factors

Consider the differential equation $M(x, y)\, dx + N(x, y)\, dy = 0$.

1. If

$$\frac{1}{N(x, y)}\left[M_y(x, y) - N_x(x, y)\right] = h(x)$$

is a function of x alone, then $e^{\int h(x)\, dx}$ is an integrating factor.

2. If

$$\frac{1}{M(x, y)}\left[N_x(x, y) - M_y(x, y)\right] = k(y)$$

is a function of y alone, then $e^{\int k(y)\, dy}$ is an integrating factor.

STUDY TIP   If either $h(x)$ or $k(y)$ is constant, Theorem 15.2 still applies. As an aid to remembering these formulas, note that the subtracted partial derivative identifies both the denominator and the variable for the integrating factor.

EXAMPLE 5   Finding an Integrating Factor

Solve the differential equation $(y^2 - x)\,dx + 2y\,dy = 0$.

Solution   The given equation is not exact because $M_y(x, y) = 2y$ and $N_x(x, y) = 0$. However, because

$$\frac{M_y(x, y) - N_x(x, y)}{N(x, y)} = \frac{2y - 0}{2y} = 1 = h(x)$$

it follows that $e^{\int h(x)\,dx} = e^{\int dx} = e^x$ is an integrating factor. Multiplying the given differential equation by e^x produces the exact differential equation

$$(y^2 e^x - x e^x)\,dx + 2y e^x\,dy = 0$$

whose solution is obtained as follows.

$$f(x, y) = \int N(x, y)\,dy = \int 2y e^x\,dy = y^2 e^x + g(x)$$

$$f_x(x, y) = y^2 e^x + g'(x) = \overbrace{y^2 e^x - x e^x}^{M(x, y)}$$

$$\boxed{g'(x) = -x e^x}$$

Therefore, $g'(x) = -x e^x$ and $g(x) = -x e^x + e^x + C_1$, which implies that

$$f(x, y) = y^2 e^x - x e^x + e^x + C_1.$$

The general solution is $y^2 e^x - x e^x + e^x = C$, or $y^2 - x + 1 = C e^{-x}$.

In the next example, we show how a differential equation can help in sketching a force field given by $\mathbf{F}(x, y) = M(x, y)\mathbf{i} + N(x, y)\mathbf{j}$.

EXAMPLE 6   An Application to Force Fields

Sketch the force field given by

$$\mathbf{F}(x, y) = \frac{2y}{\sqrt{x^2 + y^2}}\mathbf{i} - \frac{y^2 - x}{\sqrt{x^2 + y^2}}\mathbf{j}$$

by finding and sketching the family of curves tangent to $\mathbf{F}$.

Solution   At the point (x, y) in the plane, the vector $\mathbf{F}(x, y)$ has a slope of

$$\frac{dy}{dx} = \frac{-(y^2 - x)/\sqrt{x^2 + y^2}}{2y/\sqrt{x^2 + y^2}} = \frac{-(y^2 - x)}{2y}$$

which, in differential form, is

$$2y\,dy = -(y^2 - x)\,dx$$
$$(y^2 - x)\,dx + 2y\,dy = 0.$$

From Example 5, we know that the general solution of this differential equation is $y^2 - x + 1 = C e^{-x}$, or $y^2 = x - 1 + C e^{-x}$. Figure 15.4 shows several representative curves from this family. Note that the force vector at (x, y) is tangent to the curve passing through (x, y).

Force field:
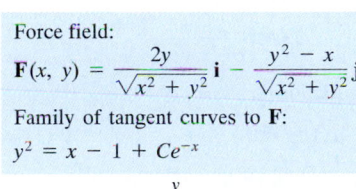
$$\mathbf{F}(x, y) = \frac{2y}{\sqrt{x^2 + y^2}}\mathbf{i} - \frac{y^2 - x}{\sqrt{x^2 + y^2}}\mathbf{j}$$

Family of tangent curves to $\mathbf{F}$:
$$y^2 = x - 1 + C e^{-x}$$

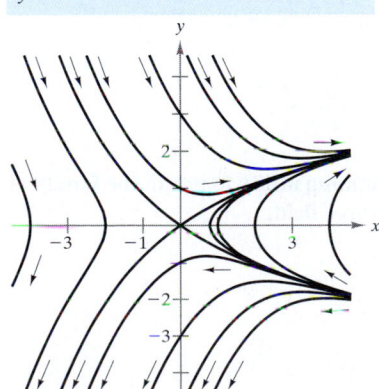

Figure 15.4

EXERCISES FOR SECTION 15.1

In Exercises 1–10, determine whether the differential equation is exact. If it is, find the general solution.

1. $(2x - 3y)\,dx + (2y - 3x)\,dy = 0$

2. $ye^x\,dx + e^x\,dy = 0$

3. $(3y^2 + 10xy^2)\,dx + (6xy - 2 + 10x^2y)\,dy = 0$

4. $2\cos(2x - y)\,dx - \cos(2x - y)\,dy = 0$

5. $(4x^3 - 6xy^2)\,dx + (4y^3 - 6xy)\,dy = 0$

6. $2y^2e^{xy^2}\,dx + 2xye^{xy^2}\,dy = 0$

7. $\dfrac{1}{x^2 + y^2}(x\,dy - y\,dx) = 0$

8. $e^{-(x^2+y^2)}(x\,dx + y\,dy) = 0$

9. $\dfrac{1}{(x - y)^2}(y^2\,dx + x^2\,dy) = 0$

10. $e^y\cos xy\,[y\,dx + (x + \tan xy)\,dy] = 0$

In Exercises 11 and 12, (a) sketch an approximate solution of the differential equation satisfying the initial condition by hand on the direction field, (b) find the particular solution that satisfies the initial condition, and (c) use a graphing utility to graph the particular solution. Compare the graph with the hand-drawn graph of part (a).

Differential Equation	Initial Condition
11. $(2x\tan y + 5)\,dx + (x^2\sec^2 y)\,dy = 0$	$y\left(\tfrac{1}{2}\right) = \pi/4$
12. $\dfrac{1}{\sqrt{x^2 + y^2}}(x\,dx + y\,dy) = 0$	$y(4) = 3$

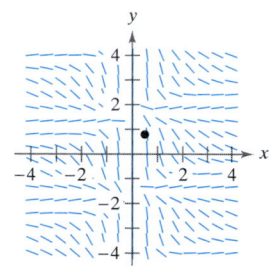

Figure for 11

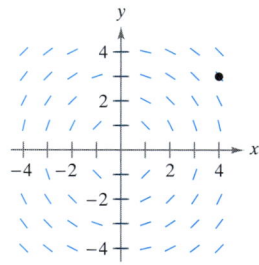

Figure for 12

In Exercises 13–16, find the particular solution that satisfies the initial condition.

Differential Equation	Initial Condition
13. $\dfrac{y}{x - 1}\,dx + [\ln(x - 1) + 2y]\,dy = 0$	$y(2) = 4$
14. $\dfrac{1}{x^2 + y^2}(x\,dx + y\,dy) = 0$	$y(0) = 4$
15. $e^{3x}(\sin 3y\,dx + \cos 3y\,dy) = 0$	$y(0) = \pi$
16. $(x^2 + y^2)\,dx + 2xy\,dy = 0$	$y(3) = 1$

In Exercises 17–26, find the integrating factor that is a function of x or y alone and use it to find the general solution of the differential equation.

17. $y\,dx - (x + 6y^2)\,dy = 0$

18. $(2x^3 + y)\,dx - x\,dy = 0$

19. $(5x^2 - y)\,dx + x\,dy = 0$

20. $(5x^2 - y^2)\,dx + 2y\,dy = 0$

21. $(x + y)\,dx + \tan x\,dy = 0$

22. $(2x^2y - 1)\,dx + x^3\,dy = 0$

23. $y^2\,dx + (xy - 1)\,dy = 0$

24. $(x^2 + 2x + y)\,dx + 2\,dy = 0$

25. $2y\,dx + \left(x - \sin\sqrt{y}\right)dy = 0$

26. $(-2y^3 + 1)\,dx + (3xy^2 + x^3)\,dy = 0$

In Exercises 27–30, use the integrating factor to find the general solution of the differential equation.

27. $(4x^2y + 2y^2)\,dx + (3x^3 + 4xy)\,dy = 0$
    $u(x, y) = xy^2$

28. $(3y^2 + 5x^2y)\,dx + (3xy + 2x^3)\,dy = 0$
    $u(x, y) = x^2y$

29. $(-y^5 + x^2y)\,dx + (2xy^4 - 2x^3)\,dy = 0$
    $u(x, y) = x^{-2}y^{-3}$

30. $-y^3\,dx + (xy^2 - x^2)\,dy = 0$
    $u(x, y) = x^{-2}y^{-2}$

31. Show that each of the following is an integrating factor for the differential equation

    $y\,dx - x\,dy = 0.$

    (a) $\dfrac{1}{x^2}$ (b) $\dfrac{1}{y^2}$ (c) $\dfrac{1}{xy}$ (d) $\dfrac{1}{x^2 + y^2}$

32. Show that the differential equation

    $(axy^2 + by)\,dx + (bx^2y + ax)\,dy = 0$

    is exact only if $a = b$. If $a \neq b$, show that x^my^n is an integrating factor, where

    $$m = -\dfrac{2b + a}{a + b}, \qquad n = -\dfrac{2a + b}{a + b}.$$

In Exercises 33–36, use a graphing utility to graph the family of tangent curves to the given force field.

33. $\mathbf{F}(x, y) = \dfrac{y}{\sqrt{x^2 + y^2}}\mathbf{i} - \dfrac{x}{\sqrt{x^2 + y^2}}\mathbf{j}$

34. $\mathbf{F}(x, y) = \dfrac{x}{\sqrt{x^2 + y^2}}\mathbf{i} - \dfrac{y}{\sqrt{x^2 + y^2}}\mathbf{j}$

35. $\mathbf{F}(x, y) = 4x^2y\mathbf{i} - \left(2xy^2 + \dfrac{x}{y^2}\right)\mathbf{j}$

36. $\mathbf{F}(x, y) = (1 + x^2)\mathbf{i} - 2xy\mathbf{j}$

In Exercises 37 and 38, find an equation for the curve with the specified slope passing through the given point.

Slope	Point
37. $\dfrac{dy}{dx} = \dfrac{y - x}{3y - x}$	$(2, 1)$
38. $\dfrac{dy}{dx} = \dfrac{-2xy}{x^2 + y^2}$	$(0, 2)$

39. Cost   If $y = C(x)$ represents the cost of producing x units in a manufacturing process, the **elasticity of cost** is defined as

$$E(x) = \frac{\text{marginal cost}}{\text{average cost}} = \frac{C'(x)}{C(x)/x} = \frac{x}{y}\frac{dy}{dx}.$$

Find the cost function if the elasticity function is

$$E(x) = \frac{20x - y}{2y - 10x}$$

where $C(100) = 500$ and $x \geq 100$.

40. Euler's Method   Consider the differential equation $y' = F(x, y)$ with the initial condition $y(x_0) = y_0$. At any point (x_k, y_k) in the domain of F, $F(x_k, y_k)$ yields the slope of the solution at that point. Euler's Method gives a discrete set of estimates of the y values of a solution of the differential equation using the iterative formula

$$y_{k+1} = y_k + F(x_k, y_k)\,\Delta x$$

where $\Delta x = x_{k+1} - x_k$.

(a) Write a short paragraph describing the general idea of how Euler's Method works.

(b) How will decreasing the magnitude of Δx affect the accuracy of Euler's Method?

41. Euler's Method   Use Euler's Method (see Exercise 40) to approximate $y(1)$ for the values of Δx given in the table if $y' = x + \sqrt{y}$ and $y(0) = 2$. (Note that the number of iterations increases as Δx decreases.) Sketch a graph of the approximate solution on the direction field in the figure.

Δx	0.50	0.25	0.10
Estimate of $y(1)$			

The value of $y(1)$, accurate to three decimal places, is 4.213.

42. Programming   Write a program for a graphing utility or computer that will perform the calculations of Euler's Method for a specified differential equation, interval, Δx, and initial condition. The output should be a graph of the discrete points approximating the solution.

Euler's Method   In Exercises 43–46, (a) use the program of Exercise 42 to approximate the solution of the differential equation over the indicated interval with the specified value of Δx and the initial condition, (b) solve the differential equation analytically, and (c) use a graphing utility to graph the particular solution and compare the result with the graph of part (a).

Differential Equation	Interval	Δx	Initial Condition
43. $y' = x\sqrt[3]{y}$	$[1, 2]$	0.01	$y(1) = 1$
44. $y' = \dfrac{\pi}{4}(y^2 + 1)$	$[-1, 1]$	0.1	$y(-1) = -1$
45. $y' = \dfrac{-xy}{x^2 + y^2}$	$[2, 4]$	0.05	$y(2) = 1$
46. $y' = \dfrac{6x + y^2}{y(3y - 2x)}$	$[0, 5]$	0.2	$y(0) = 1$

47. Euler's Method   Repeat Exercise 45 for $\Delta x = 1$ and discuss how the accuracy of the result changes.

48. Euler's Method   Repeat Exercise 46 for $\Delta x = 0.5$ and discuss how the accuracy of the result changes.

True or False?   In Exercises 49–52, determine whether the statement is true or false. If it is false, explain why or give an example that shows it is false.

49. The differential equation $2xy\,dx + (y^2 - x^2)\,dy = 0$ is exact.

50. If $M\,dx + N\,dy = 0$ is exact, then $xM\,dx + xN\,dy = 0$ is also exact.

51. If $M\,dx + N\,dy = 0$ is exact, then $[f(x) + M]\,dx + [g(y) + N]\,dy = 0$ is also exact.

52. The differential equation $f(x)\,dx + g(y)\,dy = 0$ is exact.

SECTION | *15.2* | **First-Order Linear Differential Equations**

First-Order Linear Differential Equations • Bernoulli Equations • Applications

First-Order Linear Differential Equations

In this section, you will see how integrating factors help to solve a very important class of first-order differential equations—first-order *linear* differential equations.

> ### Definition of First-Order Linear Differential Equation
>
> A first-order linear differential equation is an equation of the form
>
> $$\frac{dy}{dx} + P(x)y = Q(x)$$
>
> where P and Q are continuous functions of x. This first-order linear differential equation is said to be in **standard form.**

To solve a first-order linear differential equation, you can use an integrating factor $u(x)$, which converts the left side into the derivative of the product $u(x)y$. That is, you need a factor $u(x)$ such that

$$u(x)\frac{dy}{dx} + u(x)P(x)y = \frac{d[u(x)y]}{dx}$$

$$u(x)y' + u(x)P(x)y = u(x)y' + yu'(x)$$

$$u(x)P(x)y = yu'(x)$$

$$P(x) = \frac{u'(x)}{u(x)}$$

$$\ln|u(x)| = \int P(x)\,dx + C_1$$

$$u(x) = Ce^{\int P(x)\,dx}.$$

Because you don't need the most general integrating factor, let $C = 1$. Multiplying the original equation $y' + P(x)y = Q(x)$ by $u(x) = e^{\int P(x)dx}$ produces

$$y'e^{\int P(x)\,dx} + yP(x)e^{\int P(x)\,dx} = Q(x)e^{\int P(x)\,dx}$$

$$\frac{d}{dx}\left[ye^{\int P(x)\,dx}\right] = Q(x)e^{\int P(x)\,dx}.$$

The general solution is given by

$$ye^{\int P(x)\,dx} = \int Q(x)e^{\int P(x)\,dx}\,dx + C.$$

THEOREM 15.3   Solution of a First-Order Linear Differential Equation

An integrating factor for the first-order linear differential equation

$$y' + P(x)y = Q(x)$$

is $u(x) = e^{\int P(x)\,dx}$. The solution of the differential equation is

$$ye^{\int P(x)\,dx} = \int Q(x)e^{\int P(x)\,dx}\,dx + C.$$

STUDY TIP   Rather than memorizing this formula, just remember that multiplication by the integrating factor $e^{\int P(x)\,dx}$ converts the left side of the differential equation into the derivative of the product $ye^{\int P(x)\,dx}$.

EXAMPLE 1   Solving a First-Order Linear Differential Equation

Find the general solution of

$$xy' - 2y = x^2.$$

Solution   The *standard form* of the given equation is

$$y' + P(x)y = Q(x)$$

$$y' - \left(\frac{2}{x}\right)y = x. \qquad \text{Standard form}$$

Thus, $P(x) = -2/x$, and you have

$$\int P(x)\,dx = -\int \frac{2}{x}\,dx = -\ln x^2$$

$$e^{\int P(x)\,dx} = e^{-\ln x^2} = \frac{1}{x^2}. \qquad \text{Integrating factor}$$

Therefore, multiplying both sides of the standard form by $1/x^2$ yields

$$\frac{y'}{x^2} - \frac{2y}{x^3} = \frac{1}{x}$$

$$\frac{d}{dx}\left[\frac{y}{x^2}\right] = \frac{1}{x}$$

$$\frac{y}{x^2} = \int \frac{1}{x}\,dx$$

$$\frac{y}{x^2} = \ln|x| + C$$

$$y = x^2(\ln|x| + C). \qquad \text{General solution}$$

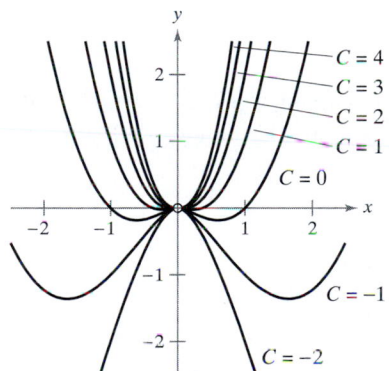

Figure 15.5

Several solution curves (for $C = -2, -1, 0, 1, 2, 3,$ and 4) are shown in Figure 15.5.

EXAMPLE 2   Solving a First-Order Linear Differential Equation

Find the general solution of

$$y' - y \tan t = 1, \qquad -\frac{\pi}{2} < t < \frac{\pi}{2}.$$

Solution   The equation is already in the standard form $y' + P(t)y = Q(t)$. Thus, $P(t) = -\tan t$, and

$$\int P(t)\, dt = -\int \tan t\, dt = \ln |\cos t|$$

which implies that the integrating factor is

$$e^{\int P(t)\, dt} = e^{\ln |\cos t|}$$
$$= |\cos t|. \qquad \text{Integrating factor}$$

A quick check shows that $\cos t$ is also an integrating factor. Thus, multiplying $y' - y \tan t = 1$ by $\cos t$ produces

$$\frac{d}{dt}[y \cos t] = \cos t$$

$$y \cos t = \int \cos t\, dt$$

$$y \cos t = \sin t + C$$

$$y = \tan t + C \sec t. \qquad \text{General solution}$$

Several solution curves are shown in Figure 15.6.

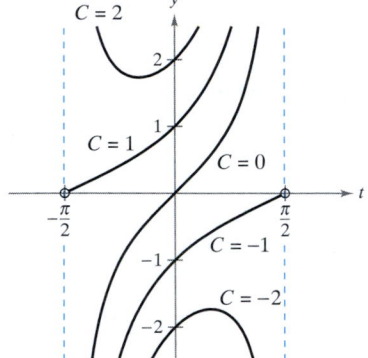

Figure 15.6

Bernoulli Equations

A well-known *nonlinear* equation that reduces to a linear one with an appropriate substitution is the **Bernoulli equation,** named after James Bernoulli (1654–1705).

$$y' + P(x)y = Q(x)y^n \qquad \text{Bernoulli equation}$$

This equation is linear if $n = 0$, and has separable variables if $n = 1$. Thus, in the following development, assume that $n \neq 0$ and $n \neq 1$. Begin by multiplying by y^{-n} and $(1 - n)$ to obtain

$$y^{-n}y' + P(x)y^{1-n} = Q(x)$$
$$(1 - n)y^{-n}y' + (1 - n)P(x)y^{1-n} = (1 - n)Q(x)$$
$$\frac{d}{dx}[y^{1-n}] + (1 - n)P(x)y^{1-n} = (1 - n)Q(x)$$

which is a linear equation in the variable y^{1-n}. Letting $z = y^{1-n}$ produces the linear equation

$$\frac{dz}{dx} + (1 - n)P(x)z = (1 - n)Q(x).$$

Finally, by Theorem 15.3, the *general solution of the Bernoulli equation* is

$$y^{1-n}e^{\int(1-n)P(x)\,dx} = \int (1 - n)Q(x)e^{\int(1-n)P(x)\,dx}\, dx + C.$$

EXAMPLE 3   Solving a Bernoulli Equation

Find the general solution of $y' + xy = xe^{-x^2}y^{-3}$.

Solution   For this Bernoulli equation, let $n = -3$, and use the substitution

$z = y^4$ Let $z = y^{1-n} = y^{1-(-3)}$.

$z' = 4y^3y'$.   Differentiate.

Multiplying the original equation by $4y^3$ produces

$y' + xy = xe^{-x^2}y^{-3}$ Original equation

$4y^3y' + 4xy^4 = 4xe^{-x^2}$ Multiply both sides by $4y^3$.

$z' + 4xz = 4xe^{-x^2}$.   Linear equation: $z' + P(x)z = Q(x)$

This equation is linear in z. Using $P(x) = 4x$ produces

$$\int P(x)\, dx = \int 4x\, dx = 2x^2$$

which implies that e^{2x^2} is an integrating factor. Multiplying the linear equation by this factor produces

$z' + 4xz = 4xe^{-x^2}$ Linear equation

$z'e^{2x^2} + 4xze^{2x^2} = 4xe^{x^2}$ Exact equation

$\dfrac{d}{dx}[ze^{2x^2}] = 4xe^{x^2}$ Write left side as total differential.

$ze^{2x^2} = \int 4xe^{x^2}\, dx$ Integrate both sides.

$ze^{2x^2} = 2e^{x^2} + C$

$z = 2e^{-x^2} + Ce^{-2x^2}$.   Divide both sides by e^{2x^2}.

Finally, substituting $z = y^4$, the general solution is

$y^4 = 2e^{-x^2} + Ce^{-2x^2}$.   General solution

So far you have studied several types of first-order differential equations. Of these, the separable variables case is usually the simplest, and solution by an integrating factor is usually a last resort.

Summary of First-Order Differential Equations

Method	Form of Equation
1. Separable variables:	$M(x)\, dx + N(y)\, dy = 0$
2. Homogeneous:	$M(x, y)\, dx + N(x, y)\, dy = 0$, where M and N are nth-degree homogeneous
3. Exact:	$M(x, y)\, dx + N(x, y)\, dy = 0$, where $\partial M/\partial y = \partial N/\partial x$
4. Integrating factor:	$u(x, y)M(x, y)\, dx + u(x, y)N(x, y)\, dy = 0$ is exact
5. Linear:	$y' + P(x)y = Q(x)$
6. Bernoulli equation:	$y' + P(x)y = Q(x)y^n$

Applications

One type of problem that can be described in terms of a differential equation involves chemical mixtures, as illustrated in the next example.

EXAMPLE 4   A Mixture Problem

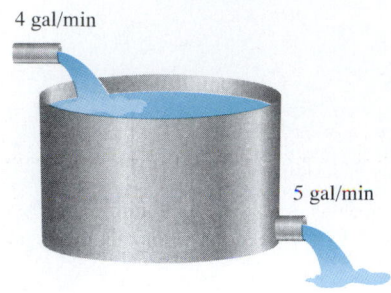

4 gal/min

5 gal/min

Figure 15.7

A tank contains 50 gallons of a solution composed of 90% water and 10% alcohol. A second solution containing 50% water and 50% alcohol is added to the tank at the rate of 4 gallons per minute. As the second solution is being added, the tank is being drained at the rate of 5 gallons per minute, as shown in Figure 15.7. Assuming the solution in the tank is stirred constantly, how much alcohol is in the tank after 10 minutes?

Solution   Let y be the number of gallons of alcohol in the tank at any time t. You know that $y = 5$ when $t = 0$. Because the number of gallons of solution in the tank at any time is $50 - t$, and the tank loses 5 gallons of solution per minute, it must lose

$$\left(\frac{5}{50 - t}\right)y$$

gallons of alcohol per minute. Furthermore, because the tank is gaining 2 gallons of alcohol per minute, the rate of change of alcohol in the tank is given by

$$\frac{dy}{dt} = 2 - \left(\frac{5}{50 - t}\right)y \quad \Longrightarrow \quad \frac{dy}{dt} + \left(\frac{5}{50 - t}\right)y = 2.$$

To solve this linear equation, let $P(t) = 5/(50 - t)$ and obtain

$$\int P(t)\, dt + \int \frac{5}{50 - t}\, dt = -5 \ln|50 - t|.$$

Because $t < 50$, you can drop the absolute value signs and conclude that

$$e^{\int P(t)\,dt} = e^{-5\ln(50-t)} = \frac{1}{(50 - t)^5}.$$

Thus, the general solution is

$$\frac{y}{(50 - t)^5} = \int \frac{2}{(50 - t)^5}\, dt = \frac{1}{2(50 - t)^4} + C$$

$$y = \frac{50 - t}{2} + C(50 - t)^5.$$

Because $y = 5$ when $t = 0$, you have

$$5 = \frac{50}{2} + C(50)^5 \quad \Longrightarrow \quad -\frac{20}{50^5} = C$$

which means that the particular solution is

$$y = \frac{50 - t}{2} - 20\left(\frac{50 - t}{50}\right)^5.$$

Finally, when $t = 10$, the amount of alcohol in the tank is

$$y = \frac{50 - 10}{2} - 20\left(\frac{50 - 10}{50}\right)^5 = 13.45 \text{ gal}$$

which represents a solution containing 33.6% alcohol.

In most falling-body problems discussed so far in the text, we have neglected air resistance. The next example includes this factor. In the example, the air resistance on the falling object is assumed to be proportional to its velocity v. If g is the gravitational constant, the downward force F on a falling object of mass m is given by the difference $mg - kv$. But by Newton's Second Law of Motion, you know that $F = ma = m(dv/dt)$, which yields the following differential equation.

$$m\frac{dv}{dt} = mg - kv \quad \Longrightarrow \quad \frac{dv}{dt} + \frac{k}{m}v = g$$

EXAMPLE 5   A Falling Object with Air Resistance

An object of mass m is dropped from a hovering helicopter. Find its velocity as a function of time t, assuming that the air resistance is proportional to the velocity of the object.

Solution   The velocity v satisfies the equation

$$\frac{dv}{dt} + \frac{kv}{m} = g$$

where g is the gravitational constant and k is the constant of proportionality. Letting $b = k/m$, you can *separate variables* to obtain

$$dv = (g - bv)\, dt$$

$$\int \frac{dv}{g - bv} = \int dt$$

$$-\frac{1}{b}\ln|g - bv| = t + C_1$$

$$\ln|g - bv| = -bt - bC_1$$

$$g - bv = Ce^{-bt}.$$

Because the object was dropped, $v = 0$ when $t = 0$; thus $g = C$, and it follows that

$$-bv = -g + ge^{-bt} \quad \Longrightarrow \quad v = \frac{g - ge^{-bt}}{b} = \frac{mg}{k}\left(1 - e^{-kt/m}\right).$$

NOTE   Notice in Example 5 that the velocity approaches a limit of mg/k as a result of the air resistance. For falling-body problems in which air resistance is neglected, the velocity increases without bound.

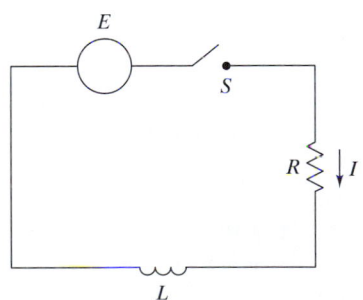

Figure 15.8

A simple electrical circuit consists of electric current I (in amperes), a resistance R (in ohms), an inductance L (in henrys), and a constant electromotive force E (in volts), as shown in Figure 15.8. According to Kirchhoff's Second Law, if the switch S is closed when $t = 0$, the applied electromotive force (voltage) is equal to the sum of the voltage drops in the rest of the circuit. This in turn means that the current I satisfies the differential equation

$$L\frac{dI}{dt} + RI = E.$$

EXAMPLE 6   An Electric Circuit Problem

Find the current I as a function of time t (in seconds), given that I satisfies the differential equation $L(dI/dt) + RI = \sin 2t$, where R and L are nonzero constants.

Solution   In standard form, the given linear equation is

$$\frac{dI}{dt} + \frac{R}{L}I = \frac{1}{L}\sin 2t.$$

Let $P(t) = R/L$, so that $e^{\int P(t)\,dt} = e^{(R/L)t}$, and, by Theorem 15.3,

$$Ie^{(R/L)t} = \frac{1}{L}\int e^{(R/L)t}\sin 2t\, dt$$

$$= \frac{1}{4L^2 + R^2}e^{(R/L)t}(R\sin 2t - 2L\cos 2t) + C.$$

Thus, the general solution is

$$I = e^{-(R/L)t}\left[\frac{1}{4L^2 + R^2}e^{(R/L)t}(R\sin 2t - 2L\cos 2t) + C\right]$$

$$I = \frac{1}{4L^2 + R^2}(R\sin 2t - 2L\cos 2t) + Ce^{-(R/L)t}.$$

EXERCISES FOR SECTION 15.2

True or False?   **In Exercises 1 and 2, determine whether the statement is true or false. If it is false, explain why or give an example that shows it is false.**

1. $y' + x\sqrt{y} = x^2$ is a first-order linear differential equation.

2. $y' + xy = e^x y$ is a first-order linear differential equation.

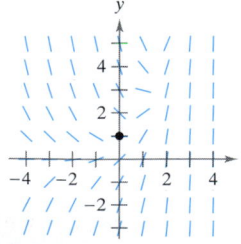

In Exercises 3 and 4, (a) sketch an approximate solution of the differential equation satisfying the initial condition by hand on the direction field, (b) find the particular solution that satisfies the initial condition, and (c) use a graphing utility to graph the particular solution. Compare the graph with the hand-drawn graph of part (a).

Differential Equation	*Initial Condition*
3. $\dfrac{dy}{dx} = e^x - y$	$(0, 1)$
4. $y' + 2y = \sin x$	$(0, 4)$

Figure for 3   **Figure for 4**

In Exercises 5–12, solve the first-order linear differential equation.

5. $\dfrac{dy}{dx} + \left(\dfrac{1}{x}\right)y = 3x + 4$

6. $\dfrac{dy}{dx} + \left(\dfrac{2}{x}\right)y = 3x + 1$

7. $y' - y = \cos x$

8. $y' + 2xy = 2x$

9. $(3y + \sin 2x)\, dx - dy = 0$

10. $(y - 1)\sin x\, dx - dy = 0$

11. $(x - 1)y' + y = x^2 - 1$

12. $y' + 5y = e^{5x}$

In Exercises 13–18, find the particular solution of the differential equation that satisfies the boundary condition.

Differential Equation	*Boundary Condition*
13. $y'\cos^2 x + y - 1 = 0$	$y(0) = 5$
14. $x^3 y' + 2y = e^{1/x^2}$	$y(1) = e$
15. $y' + y\tan x = \sec x + \cos x$	$y(0) = 1$
16. $y' + y\sec x = \sec x$	$y(0) = 4$
17. $y' + \left(\dfrac{1}{x}\right)y = 0$	$y(2) = 2$
18. $y' + (2x - 1)y = 0$	$y(1) = 2$

In Exercises 19–24, solve the Bernoulli differential equation.

19. $y' + 3x^2y = x^2y^3$

20. $y' + 2xy = xy^2$

21. $y' + \left(\dfrac{1}{x}\right)y = xy^2$

22. $y' + \left(\dfrac{1}{x}\right)y = x\sqrt{y}$

23. $y' - y = x^3\sqrt[3]{y}$

24. $yy' - 2y^2 = e^x$

 In Exercises 25–28, (a) use a graphing utility to graph the direction field for the differential equation, (b) find the particular solutions of the differential equation passing through the specified points, and (c) use a graphing utility to graph the particular solutions on the direction field.

Differential Equation	Points
25. $\dfrac{dy}{dx} - \dfrac{1}{x}y = x^2$	$(-2, 4),\ (2, 8)$
26. $\dfrac{dy}{dx} + 2xy = x^3$	$\left(0, \tfrac{7}{2}\right),\ \left(0, -\tfrac{1}{2}\right)$
27. $\dfrac{dy}{dx} + (\cot x)y = x$	$(1, 1),\ (3, -1)$
28. $\dfrac{dy}{dx} + 2xy = xy^2$	$(0, 3),\ (0, 1)$

Electrical Circuits   **In Exercises 29–32, use the differential equation for electrical circuits given by**

$$L\frac{dI}{dt} + RI = E.$$

In this equation, I is the current, R is the resistance, L is the inductance, and E is the electromotive force (voltage).

29. Solve the differential equation given a constant voltage E_0.

30. Use the result of Exercise 29 to find the equation for the current if $I(0) = 0$, $E_0 = 110$ volts, $R = 550$ ohms, and $L = 4$ henrys. When does the current reach 90% of its limiting value?

31. Solve the differential equation given a periodic electromotive force $E_0 \sin \omega t$.

32. Verify that the solution of Exercise 31 can be written in the form

$$I = ce^{-(R/L)t} + \frac{E_0}{\sqrt{R^2 + \omega^2L^2}}\sin(\omega t + \phi)$$

where ϕ, the phase angle, is given by $\arctan(-\omega L/R)$. (Note that the exponential term approaches 0 as $t \to \infty$. This implies that the current approaches a periodic function.)

33. *Population Growth*   When predicting population growth, demographers must consider birth and death rates as well as the net change caused by the difference between the rates of immigration and emigration. Let P be the population at time t and let N be the net increase per unit time resulting from the difference between immigration and emigration. Thus, the rate of growth of the population is given by

$$\frac{dP}{dt} = kP + N, \qquad N \text{ is constant.}$$

Solve this differential equation to find P as a function of time if at time $t = 0$ the size of the population is P_0.

34. *Investment Growth*   A large corporation starts at time $t = 0$ to continuously invest part of its receipts at a rate of P dollars per year in a fund for future corporate expansion. Assume that the fund earns r percent interest per year compounded continuously. Thus, the rate of growth of the amount A in the fund is given by

$$\frac{dA}{dt} = rA + P$$

where $A = 0$ when $t = 0$. Solve this differential equation for A as a function of t.

Investment Growth   **In Exercises 35 and 36, use the result of Exercise 34.**

35. Find A for the following.

 (a) $P = \$100{,}000$, $r = 6\%$, and $t = 5$ years

 (b) $P = \$250{,}000$, $r = 5\%$, and $t = 10$ years

36. Find t if the corporation needs $\$800{,}000$ and it can invest $\$75{,}000$ per year in a fund earning 8% interest compounded continuously.

37. *Intravenous Feeding*   Glucose is added intravenously to the bloodstream at the rate of q units per minute, and the body removes glucose from the bloodstream at a rate proportional to the amount present. Assume $Q(t)$ is the amount of glucose in the bloodstream at time t.

 (a) Determine the differential equation describing the rate of change with respect to time of glucose in the bloodstream.

 (b) Solve the differential equation from part (a), letting $Q = Q_0$ when $t = 0$.

 (c) Find the limit of $Q(t)$ as $t \to \infty$.

38. *Learning Curve*   The management at a certain factory has found that the maximum number of units a worker can produce in a day is 30. The rate of increase in the number of units N produced with respect to time t in days by a new employee is proportional to $30 - N$.

 (a) Determine the differential equation describing the rate of change of performance with respect to time.

 (b) Solve the differential equation from part (a).

 (c) Find the particular solution for a new employee who produced ten units on the first day at the factory and 19 units on the twentieth day.

Mixture In Exercises 39–44, consider a tank that at time $t = 0$ contains v_0 gallons of a solution of which, by weight, q_0 pounds is soluble concentrate. Another solution containing q_1 pounds of the concentrate per gallon is running into the tank at the rate of r_1 gallons per minute. The solution in the tank is kept well stirred and is withdrawn at the rate of r_2 gallons per minute.

39. If Q is the amount of concentrate in the solution at any time t, show that

$$\frac{dQ}{dt} + \frac{r_2 Q}{v_0 + (r_1 - r_2)t} = q_1 r_1.$$

40. If Q is the amount of concentrate in the solution at any time t, write the differential equation for the rate of change of Q with respect to t if $r_1 = r_2 = r$.

41. A 200-gallon tank is full of a solution containing 25 pounds of concentrate. Starting at time $t = 0$, distilled water is admitted to the tank at a rate of 10 gallons per minute, and the well-stirred solution is withdrawn at the same rate.

   (a) Find the amount of concentrate Q in the solution as a function of t.

   (b) Find the time at which the amount of concentrate in the tank reaches 15 pounds.

   (c) Find the quantity of the concentrate in the solution as $t \to \infty$.

42. Repeat Exercise 41, assuming that the solution entering the tank contains 0.05 pound of concentrate per gallon.

43. A 200-gallon tank is half full of distilled water. At time $t = 0$, a solution containing 0.5 pound of concentrate per gallon enters the tank at the rate of 5 gallons per minute, and the well-stirred mixture is withdrawn at the rate of 3 gallons per minute.

   (a) At what time will the tank be full?

   (b) At the time the tank is full, how many pounds of concentrate will it contain?

44. Repeat Exercise 43, assuming that the solution entering the tank contains 1 pound of concentrate per gallon.

In Exercises 45–48, match the differential equation with its solution.

Differential Equation	Solution
45. $y' - 2x = 0$	(a) $y = Ce^{x^2}$
46. $y' - 2y = 0$	(b) $y = -\frac{1}{2} + Ce^{x^2}$
47. $y' - 2xy = 0$	(c) $y = x^2 + C$
48. $y' - 2xy = x$	(d) $y = Ce^{2x}$

In Exercises 49–64, solve the first-order differential equation by any appropriate method.

49. $\dfrac{dy}{dx} = \dfrac{e^{2x+y}}{e^{x-y}}$

50. $\dfrac{dy}{dx} = \dfrac{x+1}{y(y+2)}$

51. $(1 + y^2)\,dx + (2xy + y + 2)\,dy = 0$

52. $(1 + 2e^{2x+y})\,dx + e^{2x+y}\,dy = 0$

53. $y \cos x - \cos x + \dfrac{dy}{dx} = 0$

54. $(x + 1)\dfrac{dy}{dx} = e^x - y$

55. $(x^2 + \cos y)\dfrac{dy}{dx} = -2xy$

56. $y' = 2x\sqrt{1 - y^2}$

57. $(3y^2 + 4xy)\,dx + (2xy + x^2)\,dy = 0$

58. $(x + y)\,dx - x\,dy = 0$

59. $(2y - e^x)\,dx + x\,dy = 0$

60. $(y^2 + xy)\,dx - x^2\,dy = 0$

61. $(x^2 y^4 - 1)\,dx + x^3 y^3\,dy = 0$

62. $y\,dx + (3x + 4y)\,dy = 0$

63. $3y\,dx - (x^2 + 3x + y^2)\,dy = 0$

64. $x\,dx + (y + e^y)(x^2 + 1)\,dy = 0$

SECTION PROJECT

Weight Loss A person's weight depends on both the amount of calories consumed and the energy used. Moreover, the amount of energy used depends on a person's weight—the average amount of energy used by a person is 17.5 calories per pound per day. Thus, the more weight a person loses, the less energy the person uses (assuming that the person maintains a constant level of activity). An equation that can be used to model weight loss is

$$\left(\frac{dw}{dt}\right) = \frac{C}{3500} - \frac{17.5}{3500}w$$

where w is the person's weight (in pounds), t is the time in days, and C is the constant daily calorie consumption.

   (a) Find the general solution of the differential equation.

   (b) Consider a person who weighs 180 pounds and begins a diet of 2500 calories per day. How long will it take the person to lose 10 pounds? How long will it take the person to lose 35 pounds?

   (c) Use a graphing utility to graph the solution. What is the "limiting" weight of the person?

   (d) Repeat parts (b) and (c) for a person who weighs 200 pounds when the diet is started.

FOR FURTHER INFORMATION For more information on modeling weight loss, see the article "A Linear Diet Model" by Arthur C. Segal in the January 1987 issue of *The College Mathematics Journal*.

SECTION | **15.3** | **Second-Order Homogeneous Linear Equations**

Second-Order Linear Differential Equations •
Higher-Order Linear Differential Equations • Applications

Second-Order Linear Differential Equations

In this section and the following section, we discuss methods for solving higher-order linear differential equations.

> ### Definition of Linear Differential Equation of Order n
>
> Let $g_1, g_2, \ldots, g_n$ and f be functions of x with a common (interval) domain. An equation of the form
>
> $$y^{(n)} + g_1(x)y^{(n-1)} + g_2(x)y^{(n-2)} + \cdots + g_{n-1}(x)y' + g_n(x)y = f(x)$$
>
> is called a **linear differential equation of order n**. If $f(x) = 0$, the equation is **homogeneous**; otherwise, it is **nonhomogeneous**.

NOTE   Notice that this use of the term *homogeneous* differs from that in Section 5.7.

We discuss homogeneous equations in this section, and leave the nonhomogeneous case for the next section.

The functions $y_1, y_2, \ldots, y_n$ are **linearly independent** if the *only* solution of the equation

$$C_1 y_1 + C_2 y_2 + \cdots + C_n y_n = 0$$

is the trivial one, $C_1 = C_2 = \cdots = C_n = 0$. Otherwise, this set of functions is **linearly dependent**.

EXAMPLE 1   Linearly Independent and Dependent Functions

a. The functions $y_1(x) = \sin x$ and $y_2 = x$ are linearly independent because the only values of C_1 and C_2 for which

$$C_1 \sin x + C_2 x = 0$$

for all x are $C_1 = 0$ and $C_2 = 0$.

b. It can be shown that two functions form a linearly dependent set if and only if one is a constant multiple of the other. For example, $y_1(x) = x$ and $y_2(x) = 3x$ are linearly dependent because

$$C_1 x + C_2(3x) = 0$$

has the nonzero solutions $C_1 = -3$ and $C_2 = 1$.

The following theorem points out the importance of linear independence in constructing the general solution of a second-order linear homogeneous differential equation with constant coefficients.

> ### THEOREM 15.4   Linear Combinations of Solutions
>
> If y_1 and y_2 are linearly independent solutions of the differential equation $y'' + ay' + by = 0$, then the general solution is
>
> $$y = C_1 y_1 + C_2 y_2$$
>
> where C_1 and C_2 are constants.

Proof   We prove this theorem in only one direction. If y_1 and y_2 are solutions, you can obtain the following system of equations.

$$y_1''(x) + ay_1'(x) + by_1(x) = 0$$
$$y_2''(x) + ay_2'(x) + by_2(x) = 0$$

Multiplying the first equation by C_1, multiplying the second by C_2, and adding the resulting equations together produces

$$[C_1 y_1''(x) + C_2 y_2''(x)] + a[C_1 y_1'(x) + C_2 y_2'(x)] + b[C_1 y_1(x) + C_2 y_2(x)] = 0$$

which means that

$$y = C_1 y_1 + C_2 y_2$$

is a solution, as desired. The proof that all solutions are of this form is best left to a full course on differential equations.

Theorem 15.4 states that if you can find two linearly independent solutions, you can obtain the general solution by forming a **linear combination** of the two solutions.

To find two linearly independent solutions, note that the nature of the equation $y'' + ay' + by = 0$ suggests that it may have solutions of the form $y = e^{mx}$. If so, then $y' = me^{mx}$ and $y'' = m^2 e^{mx}$. Thus, by substitution, $y = e^{mx}$ is a solution if and only if

$$y'' + ay' + by = 0$$
$$m^2 e^{mx} + ame^{mx} + be^{mx} = 0$$
$$e^{mx}(m^2 + am + b) = 0.$$

Because e^{mx} is never 0, $y = e^{mx}$ is a solution if and only if

$$m^2 + am + b = 0. \qquad \text{\textcolor{red}{Characteristic equation}}$$

This is the **characteristic equation** of the differential equation

$$y'' + ay' + by = 0.$$

Note that the characteristic equation can be determined from its differential equation simply by replacing y'' with m^2, y' with m, and y with 1.

EXAMPLE 2   Characteristic Equation with Distinct Real Roots

Solve the differential equation

$$y'' - 4y = 0.$$

Solution   In this case, the characteristic equation is

$$m^2 - 4 = 0 \qquad \text{Characteristic equation}$$

so $m = \pm 2$. Thus, $y_1 = e^{m_1 x} = e^{2x}$ and $y_2 = e^{m_2 x} = e^{-2x}$ are particular solutions of the given differential equation. Furthermore, because these two solutions are linearly independent, you can apply Theorem 15.4 to conclude that the general solution is

$$y = C_1 e^{2x} + C_2 e^{-2x}. \qquad \text{General solution}$$

The characteristic equation in Example 2 has two distinct real roots. From algebra, you know that this is only one of *three* possibilities for quadratic equations. In general, the quadratic equation $m^2 + am + b = 0$ has roots

$$m_1 = \frac{-a + \sqrt{a^2 - 4b}}{2} \qquad \text{and} \qquad m_2 = \frac{-a - \sqrt{a^2 - 4b}}{2}$$

which fall into one of three cases.

1. Two distinct real roots, $m_1 \neq m_2$
2. Two equal real roots, $m_1 = m_2$
3. Two complex conjugate roots, $m_1 = \alpha + \beta i$ and $m_2 = \alpha - \beta i$

In terms of the differential equation $y'' + ay' + by = 0$, these three cases correspond to three different types of general solutions.

THEOREM 15.5   Solutions of $y'' + ay' + by = 0$

The solutions of

$$y'' + ay' + by = 0$$

fall into one of the following three cases, depending on the solutions of the characteristic equation, $m^2 + am + b = 0$.

1. *Distinct Real Roots*   If $m_1 \neq m_2$ are distinct real roots of the characteristic equation, then the general solution is

$$y = C_1 e^{m_1 x} + C_2 e^{m_2 x}.$$

2. *Equal Real Roots*   If $m_1 = m_2$ are equal real roots of the characteristic equation, then the general solution is

$$y = C_1 e^{m_1 x} + C_2 x e^{m_1 x} = (C_1 + C_2 x) e^{m_1 x}.$$

3. *Complex Roots*   If $m_1 = \alpha + \beta i$ and $m_2 = \alpha - \beta i$ are complex roots of the characteristic equation, then the general solution is

$$y = C_1 e^{\alpha x} \cos \beta x + C_2 e^{\alpha x} \sin \beta x.$$

FOR FURTHER INFORMATION For more information on Theorem 15.5, see "A Note on a Differential Equation" by Russell Euler in the 1989 winter issue of the *Missouri Journal of Mathematical Sciences*.

EXAMPLE 3 **Characteristic Equation with Complex Roots**

Find the general solution of the differential equation

$$y'' + 6y' + 12y = 0.$$

Solution The characteristic equation

$$m^2 + 6m + 12 = 0$$

has two complex roots, as follows.

$$\begin{aligned} m &= \frac{-6 \pm \sqrt{36 - 48}}{2} \\ &= \frac{-6 \pm \sqrt{-12}}{2} \\ &= -3 \pm \sqrt{-3} \\ &= -3 \pm \sqrt{3}i \end{aligned}$$

Thus, $\alpha = -3$ and $\beta = \sqrt{3}$, and the general solution is

$$y = C_1 e^{-3x} \cos \sqrt{3}x + C_2 e^{-3x} \sin \sqrt{3}x.$$

NOTE In Example 3, note that although the characteristic equation has two *complex* roots, the solution of the differential equation is *real*.

EXAMPLE 4 **Characteristic Equation with Repeated Roots**

Solve the differential equation

$$y'' + 4y' + 4y = 0$$

subject to the initial conditions $y(0) = 2$ and $y'(0) = 1$.

Solution The characteristic equation

$$m^2 + 4m + 4 = (m + 2)^2 = 0$$

has two equal roots given by $m = -2$. Thus, the general solution is

$$y = C_1 e^{-2x} + C_2 x e^{-2x}. \qquad \text{General solution}$$

Now, because $y = 2$ when $x = 0$, we have

$$2 = C_1(1) + C_2(0)(1) = C_1.$$

Furthermore, because $y' = 1$ when $x = 0$, we have

$$\begin{aligned} y' &= -2C_1 e^{-2x} + C_2(-2xe^{-2x} + e^{-2x}) \\ 1 &= -2(2)(1) + C_2[-2(0)(1) + 1] \\ 5 &= C_2. \end{aligned}$$

Therefore, the solution is

$$y = 2e^{-2x} + 5xe^{-2x}. \qquad \text{Particular solution}$$

Try checking this solution in the original differential equation.

Higher-Order Linear Differential Equations

For higher-order homogeneous linear differential equations, you can find the general solution in much the same way as you do for second-order equations. That is, you begin by determining the n roots of the characteristic equation. Then, based on these n roots, you form a linearly independent collection of n solutions. The major difference is that with equations of third or higher order, roots of the characteristic equation may occur more than twice. When this happens, the linearly independent solutions are formed by multiplying by increasing powers of x, as demonstrated in Examples 6 and 7.

EXAMPLE 5   Solving a Third-Order Equation

Find the general solution of $y''' - y' = 0$.

Solution   The characteristic equation is

$$m^3 - m = 0$$
$$m(m - 1)(m + 1) = 0$$
$$m = 0, 1, -1.$$

Because the characteristic equation has three distinct roots, the general solution is

$$y = C_1 + C_2 e^{-x} + C_3 e^x. \qquad \text{General solution}$$

EXAMPLE 6   Solving a Third-Order Equation

Find the general solution of $y''' + 3y'' + 3y' + y = 0$.

Solution   The characteristic equation is

$$m^3 + 3m^2 + 3m + 1 = 0$$
$$(m + 1)^3 = 0$$
$$m = -1.$$

Because the root $m = -1$ occurs three times, the general solution is

$$y = C_1 e^{-x} + C_2 x e^{-x} + C_3 x^2 e^{-x}. \qquad \text{General solution}$$

 ### EXAMPLE 7   Solving a Fourth-Order Equation

Find the general solution of $y^{(4)} + 2y'' + y = 0$.

Solution   The characteristic equation is as follows.

$$m^4 + 2m^2 + 1 = 0$$
$$(m^2 + 1)^2 = 0$$
$$m = \pm i$$

Because each of the roots $m_1 = \alpha + \beta i = 0 + i$ and $m_2 = \alpha - \beta i = 0 - i$ occurs twice, the general solution is

$$y = C_1 \cos x + C_2 \sin x + C_3 x \cos x + C_4 x \sin x. \qquad \text{General solution}$$

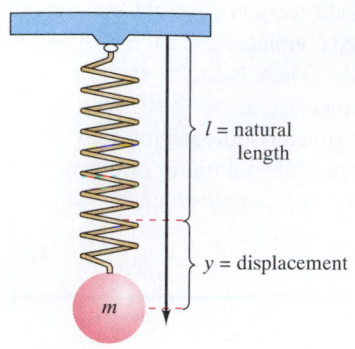

l = natural length

y = displacement

m

A rigid object of mass m attached to the end of the spring causes a displacement of y.
Figure 15.9

Applications

One of the many applications of linear differential equations is describing the motion of an oscillating spring. According to Hooke's Law, a spring that is stretched (or compressed) y units from its natural length l tends to *restore* itself to its natural length by a force F that is proportional to y. That is, $F(y) = -ky$, where k is the **spring constant** and indicates the stiffness of the given spring.

Suppose a rigid object of mass m is attached to the end of a spring and causes a displacement, as shown in Figure 15.9. Assume that the mass of the spring is negligible compared with m. If the object is pulled down and released, the resulting oscillations are a product of two opposing forces—the spring force $F(y) = -ky$ and the weight mg of the object. Under such conditions, you can use a differential equation to find the position y of the object as a function of time t. According to Newton's Second Law of Motion, the force acting on the weight is $F = ma$, where $a = d^2y/dt^2$ is the acceleration. Assuming that the motion is **undamped**—that is, there are no other external forces acting on the object—it follows that $m(d^2y/dt^2) = -ky$, and you have

$$\frac{d^2y}{dt^2} + \left(\frac{k}{m}\right)y = 0. \qquad \text{\color{red}Undamped motion of a spring}$$

EXAMPLE 8   Undamped Motion of a Spring

Suppose a 4-pound weight stretches a spring 8 inches from its natural length. The weight is pulled down an additional 6 inches and released with an initial upward velocity of 8 feet per second. Find a formula for the position of the weight as a function of time t.

Solution   By Hooke's Law, $4 = k\left(\frac{2}{3}\right)$, so $k = 6$. Moreover, because the weight w is given by mg, it follows that $m = w/g = \frac{4}{32} = \frac{1}{8}$. Hence, the resulting differential equation for this undamped motion is

$$\frac{d^2y}{dt^2} + 48y = 0.$$

Because the characteristic equation $m^2 + 48 = 0$ has complex roots $m = 0 \pm 4\sqrt{3}i$, the general solution is

$$y = C_1 e^0 \cos 4\sqrt{3}\,t + C_2 e^0 \sin 4\sqrt{3}\,t = C_1 \cos 4\sqrt{3}\,t + C_2 \sin 4\sqrt{3}\,t.$$

Using the initial conditions, you have

$$\frac{1}{2} = C_1(1) + C_2(0) \quad \Longrightarrow \quad C_1 = \frac{1}{2} \qquad \text{\color{red}}y(0) = \tfrac{1}{2}$$

$$y'(t) = -4\sqrt{3}\,C_1 \sin 4\sqrt{3}\,t + 4\sqrt{3}\,C_2 \cos 4\sqrt{3}\,t$$

$$8 = -4\sqrt{3}\left(\frac{1}{2}\right)(0) + 4\sqrt{3}\,C_2(1) \quad \Longrightarrow \quad C_2 = \frac{2\sqrt{3}}{3}. \qquad \text{\color{red}}y'(0) = 8$$

Consequently, the position at time t is given by

$$y = \frac{1}{2}\cos 4\sqrt{3}\,t + \frac{2\sqrt{3}}{3}\sin 4\sqrt{3}\,t.$$

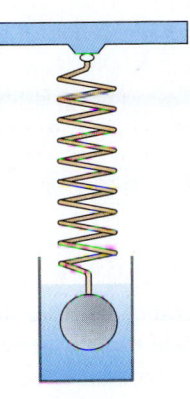

A damped vibration could be caused by friction and movement through a liquid.
Figure 15.10

Suppose the object in Figure 15.10 undergoes an additional damping or frictional force that is proportional to its velocity. A case in point would be the damping force resulting from friction and movement through a fluid. Considering this damping force, $-p(dy/dt)$, the differential equation for the oscillation is

$$m\frac{d^2y}{dt^2} = -ky - p\frac{dy}{dt}$$

or, in standard linear form,

$$\frac{d^2y}{dt^2} + \frac{p}{m}\left(\frac{dy}{dt}\right) + \frac{k}{m}y = 0. \qquad \text{Damped motion of a spring}$$

EXERCISES FOR SECTION 15.3

In Exercises 1–4, verify the solution of the differential equation.

Solution	Differential Equation
1. $y = (C_1 + C_2x)e^{-3x}$	$y'' + 6y' + 9y = 0$
2. $y = C_1e^{2x} + C_2e^{-2x}$	$y'' - 4y = 0$
3. $y = C_1\cos 2x + C_2\sin 2x$	$y'' + 4y = 0$
4. $y = e^{-x}\sin 3x$	$y'' + 2y' + 10y = 0$

In Exercises 5–30, find the general solution of the linear differential equation.

5. $y'' - y' = 0$

6. $y'' + 2y' = 0$

7. $y'' - y' - 6y = 0$

8. $y'' + 6y' + 5y = 0$

9. $2y'' + 3y' - 2y = 0$

10. $16y'' - 16y' + 3y = 0$

11. $y'' + 6y' + 9y = 0$

12. $y'' - 10y' + 25y = 0$

13. $16y'' - 8y' + y = 0$

14. $9y'' - 12y' + 4y = 0$

15. $y'' + y = 0$

16. $y'' + 4y = 0$

17. $y'' - 9y = 0$

18. $y'' - 2y = 0$

19. $y'' - 2y' + 4y = 0$

20. $y'' - 4y' + 21y = 0$

21. $y'' - 3y' + y = 0$

22. $3y'' + 4y' - y = 0$

23. $9y'' - 12y' + 11y = 0$

24. $2y'' - 6y' + 7y = 0$

25. $y^{(4)} - y = 0$

26. $y^{(4)} - y'' = 0$

27. $y''' - 6y'' + 11y' - 6y = 0$

28. $y''' - y'' - y' + y = 0$

29. $y''' - 3y'' + 7y' - 5y = 0$

30. $y''' - 3y'' + 3y' - y = 0$

31. Consider the differential equation $y'' + 100y = 0$ and the solution $y = C_1\cos 10x + C_2\sin 10x$. Find the particular solution satisfying each of the following initial conditions.

(a) $y(0) = 2$, $y'(0) = 0$

(b) $y(0) = 0$, $y'(0) = 2$

(c) $y(0) = -1$, $y'(0) = 3$

32. Determine C and ω such that $y = C\sin\sqrt{3}\,t$ is a particular solution of the differential equation $y'' + \omega y = 0$, where $y'(0) = -5$.

In Exercises 33–36, find the particular solution of the linear differential equation.

33. $y'' - y' - 30y = 0$
$y(0) = 1$, $y'(0) = -4$

34. $y'' + 2y' + 3y = 0$
$y(0) = 2$, $y'(0) = 1$

35. $y'' + 16y = 0$
$y(0) = 0$, $y'(0) = 2$

36. $y'' + 2y' + 3y = 0$
$y(0) = 2$, $y'(0) = 1$

Think About It In Exercises 37 and 38, give a geometric argument to explain why the graph cannot be a solution of the differential equation. It is not necessary to solve the differential equation.

37. $y'' = y'$

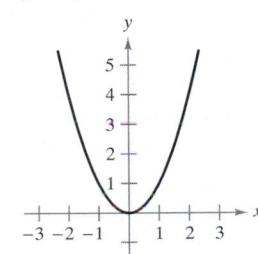

38. $y'' = -\frac{1}{2}y'$

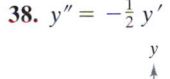
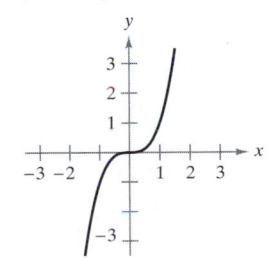

Vibrating Spring In Exercises 39–44, describe the motion of a 32-pound weight suspended on a spring. Assume that the weight stretches the spring $\frac{2}{3}$ foot from its natural position.

39. The weight is pulled $\frac{1}{2}$ foot below the equilibrium position and released.

40. The weight is raised $\frac{2}{3}$ foot above the equilibrium position and released.

41. The weight is raised $\frac{2}{3}$ foot above the equilibrium position and started off with a downward velocity of $\frac{1}{2}$ foot per second.

42. The weight is pulled $\frac{1}{2}$ foot below the equilibrium position and started off with an upward velocity of $\frac{1}{2}$ foot per second.

43. The weight is pulled $\frac{1}{2}$ foot below the equilibrium position and released. The motion takes place in a medium that furnishes a damping force of magnitude $\frac{1}{8}$ speed at all times.

44. The weight is pulled $\frac{1}{2}$ foot below the equilibrium position and released. The motion takes place in a medium that furnishes a damping force of magnitude $\frac{1}{4}|v|$ at all times.

Vibrating Spring In Exercises 45–48, match the differential equation with the graph of a particular solution. [The graphs are labeled (a), (b), (c), and (d).] The correct match can be made by comparing the frequency of the oscillations or the rate at which the oscillations are being damped with the appropriate coefficient in the differential equation.

(a)
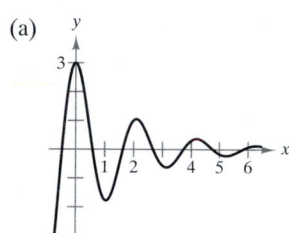

(b)
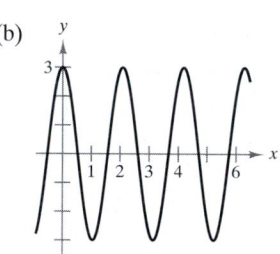

(c)
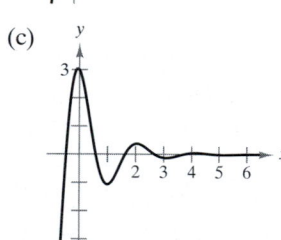

(d)
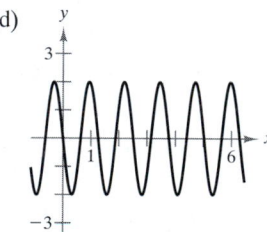

45. $y'' + 9y = 0$

46. $y'' + 25y = 0$

47. $y'' + 2y' + 10y = 0$

48. $y'' + y' + \frac{37}{4}y = 0$

49. If the characteristic equation of the differential equation

$$y'' + ay' + by = 0$$

has two equal real roots given by $m = r$, show that

$$y = C_1 e^{rx} + C_2 x e^{rx}$$

is a solution.

50. If the characteristic equation of the differential equation

$$y'' + ay' + by = 0$$

has complex roots given by $m_1 = \alpha + \beta i$ and $m_2 = \alpha - \beta i$, show that

$$y = C_1 e^{\alpha x} \cos \beta x + C_2 e^{\alpha x} \sin \beta x$$

is a solution.

True or False? In Exercises 51–54, determine whether the statement is true or false. If it is false, explain why or give an example that shows it is false.

51. $y = C_1 e^{3x} + C_2 e^{-3x}$ is the general solution of $y'' - 6y' + 9 = 0$.

52. $y = (C_1 + C_2 x)\sin x + (C_3 + C_4 x)\cos x$ is the general solution of $y^{(4)} + 2y'' + y = 0$.

53. $y = x$ is a solution of $a_n y^{(n)} + a_{n-1} y^{(n-1)} + \cdots + a_1 y' + a_0 y = 0$ if and only if $a_1 = a_0 = 0$.

54. It is possible to choose a and b such that $y = x^2 e^x$ is a solution of $y'' + ay' + by = 0$.

The *Wronskian* of two differentiable functions f and g, denoted by $W(f, g)$, is defined as the function given by the determinant

$$W(f, g) = \begin{vmatrix} f & g \\ f' & g' \end{vmatrix}.$$

The functions f and g are linearly independent if there exists at least one value of x for which $W(f, g) \neq 0$. In Exercises 55–58, use the Wronskian to verify the linear independence of the two functions.

55. $y_1 = e^{ax}$
    $y_2 = e^{bx}, \ a \neq b$

56. $y_1 = e^{ax}$
    $y_2 = xe^{ax}$

57. $y_1 = e^{ax} \sin bx$
    $y_2 = e^{ax} \cos bx, \ b \neq 0$

58. $y_1 = x$
    $y_2 = x^2$

59. Euler's differential equation is of the form

$$x^2 y'' + axy' + by = 0, \quad x > 0$$

where a and b are constants.

(a) Show that this equation can be transformed into a second-order linear equation with constant coefficients by using the substitution $x = e^t$.

(b) Solve $x^2 y'' + 6xy' + 6y = 0$.

60. Solve

$$y'' + Ay = 0$$

where A is constant, subject to the conditions $y(0) = 0$ and $y(\pi) = 0$.

Nonhomogeneous Equations • Method of Undetermined Coefficients • Variation of Parameters

Nonhomogeneous Equations

In the preceding section, we represented damped oscillations of a spring by the *homogeneous* second-order linear equation

$$\frac{d^2y}{dt^2} + \frac{p}{m}\left(\frac{dy}{dt}\right) + \frac{k}{m}y = 0. \qquad \text{Free motion}$$

This type of oscillation is called **free** because it is determined solely by the spring and gravity and is free of the action of other external forces. If such a system is also subject to an external periodic force such as $a \sin bt$, caused by vibrations at the opposite end of the spring, the motion is called **forced,** and it is characterized by the *nonhomogeneous* equation

$$\frac{d^2y}{dt^2} + \frac{p}{m}\left(\frac{dy}{dt}\right) + \frac{k}{m}y = a \sin bt. \qquad \text{Forced motion}$$

In this section, you will study two methods for finding the general solution of a nonhomogeneous linear differential equation. In both methods, the first step is to find the general solution of the corresponding homogeneous equation.

$$y = y_h \qquad \text{General solution of homogeneous equation}$$

Having done this, you try to find a particular solution of the nonhomogeneous equation.

$$y = y_p \qquad \text{Particular solution of nonhomogeneous equation}$$

By combining these two results, you can conclude that the general solution of the nonhomogeneous equation is $y = y_h + y_p$, as stated in the following theorem.

THEOREM 15.6   Solution of Nonhomogeneous Linear Equation

Let

$$y'' + ay' + by = F(x)$$

be a second-order nonhomogeneous linear differential equation. If y_p is a particular solution of this equation and y_h is the general solution of the corresponding homogeneous equation, then

$$y = y_h + y_p$$

is the general solution of the nonhomogeneous equation.

SOPHIE GERMAIN (1776–1831)

Many of the early contributors to calculus were interested in forming mathematical models for vibrating strings and membranes, oscillating springs, and elasticity. One of these was the French mathematician Sophie Germain, who in 1816 was awarded a prize by the French Academy for a paper entitled "Memoir on the Vibrations of Elastic Plates."

Method of Undetermined Coefficients

You already know how to find the solution y_h of a linear *homogeneous* differential equation. The remainder of this section looks at ways to find the particular solution y_p. If $F(x)$ in

$$y'' + ay' + by = F(x)$$

consists of sums or products of x^n, e^{mx}, $\cos \beta x$, or $\sin \beta x$, you can find a particular solution y_p by the method of **undetermined coefficients.** The gist of this method is to guess that the solution y_p is a generalized form of $F(x)$. Here are some examples.

1. If $F(x) = 3x^2$, choose $y_p = Ax^2 + Bx + C$.
2. If $F(x) = 4xe^x$, choose $y_p = Axe^x + Be^x$.
3. If $F(x) = x + \sin 2x$, choose $y_p = (Ax + B) + C \sin 2x + D \cos 2x$.

Then, by substitution, determine the coefficients for the generalized solution.

EXAMPLE 1   Method of Undetermined Coefficients

Find the general solution of the equation

$$y'' - 2y' - 3y = 2 \sin x.$$

Solution   To find y_h, solve the characteristic equation.

$$m^2 - 2m - 3 = 0$$
$$(m + 1)(m - 3) = 0$$
$$m = -1 \quad \text{or} \quad m = 3$$

Thus, $y_h = C_1 e^{-x} + C_2 e^{3x}$. Next, let y_p be a generalized form of $2 \sin x$.

$$y_p = A \cos x + B \sin x$$
$$y_p' = -A \sin x + B \cos x$$
$$y_p'' = -A \cos x - B \sin x$$

Substitution into the original differential equation yields

$$y'' - 2y' - 3y = 2 \sin x$$
$$-A \cos x - B \sin x + 2A \sin x - 2B \cos x - 3A \cos x - 3B \sin x = 2 \sin x$$
$$(-4A - 2B)\cos x + (2A - 4B)\sin x = 2 \sin x.$$

By equating coefficients of like terms, you obtain

$$-4A - 2B = 0 \quad \text{and} \quad 2A - 4B = 2$$

with solutions $A = \frac{1}{5}$ and $B = -\frac{2}{5}$. Therefore,

$$y_p = \frac{1}{5} \cos x - \frac{2}{5} \sin x$$

and the general solution is

$$y = y_h + y_p$$
$$= C_1 e^{-x} + C_2 e^{3x} + \frac{1}{5} \cos x - \frac{2}{5} \sin x.$$

In Example 1, the form of the homogeneous solution

$$y_h = C_1 e^{-x} + C_2 e^{3x}$$

has no overlap with the function $F(x)$ in the equation

$$y'' + ay' + by = F(x)$$

However, suppose the given differential equation in Example 1 were of the form

$$y'' - 2y' - 3y = e^{-x}.$$

Now, it would make no sense to guess that the particular solution were $y = Ae^{-x}$, because you know that this solution would yield 0. In such cases, you should alter your guess by multiplying by the lowest power of x that removes the duplication. For this particular problem, you would guess

$$y_p = Axe^{-x}.$$

EXAMPLE 2   Method of Undetermined Coefficients

Find the general solution of

$$y'' - 2y' = x + 2e^x.$$

Solution   The characteristic equation $m^2 - 2m = 0$ has solutions $m = 0$ and $m = 2$. Thus,

$$y_h = C_1 + C_2 e^{2x}.$$

Because $F(x) = x + 2e^x$, your first choice for y_p would be $(A + Bx) + Ce^x$. However, because y_h *already* contains a constant term C_1, you should multiply the *polynomial part* by x and use

$$y_p = Ax + Bx^2 + Ce^x$$
$$y_p' = A + 2Bx + Ce^x$$
$$y_p'' = 2B + Ce^x.$$

Substitution into the differential equation produces

$$y'' - 2y' = x + 2e^x$$
$$(2B + Ce^x) - 2(A + 2Bx + Ce^x) = x + 2e^x$$
$$(2B - 2A) - 4Bx - Ce^x = x + 2e^x.$$

Equating coefficients of like terms yields the system

$$2B - 2A = 0, \qquad -4B = 1, \qquad -C = 2$$

with solutions $A = B = -\frac{1}{4}$ and $C = -2$. Therefore,

$$y_p = -\frac{1}{4}x - \frac{1}{4}x^2 - 2e^x$$

and the general solution is

$$y = y_h + y_p$$
$$= C_1 + C_2 e^{2x} - \frac{1}{4}x - \frac{1}{4}x^2 - 2e^x.$$

In Example 2, the polynomial part of the initial guess

$$(A + Bx) + Ce^x$$

for y_p overlapped by a constant term with $y_h = C_1 + C_2 e^{2x}$, and it was necessary to multiply the polynomial part by a power of x that removed the overlap. The next example further illustrates some choices for y_p that eliminate overlap with y_h. Remember that in all cases the first guess for y_p should match the types of functions occurring in $F(x)$.

EXAMPLE 3   Choosing the Form of the Particular Solution

Determine a suitable choice for y_p for each of the following.

$y'' + ay' + by = F(x)$	y_h
a. $y'' = x^2$	$C_1 + C_2 x$
b. $y'' + 2y' + 10y = 4 \sin 3x$	$C_1 e^{-x} \cos 3x + C_2 e^{-x} \sin 3x$
c. $y'' - 4y' + 4 = e^{2x}$	$C_1 e^{2x} + C_2 x e^{2x}$

Solution

a. Because $F(x) = x^2$, the normal choice for y_p would be $A + Bx + Cx^2$. However, because $y_h = C_1 + C_2 x$ already contains a linear term, you should multiply by x^2 to obtain

$$y_p = Ax^2 + Bx^3 + Cx^4.$$

b. Because $F(x) = 4 \sin 3x$ and each term in y_h contains a factor of e^{-x}, you can simply let

$$y_p = A \cos 3x + B \sin 3x.$$

c. Because $F(x) = e^{2x}$, the normal choice for y_p would be Ae^{2x}. However, because $y_h = C_1 e^{2x} + C_2 x e^{2x}$ already contains an $x e^{2x}$ term, you should multiply by x^2 to get

$$y_p = Ax^2 e^{2x}.$$

EXAMPLE 4   Solving a Third-Order Equation

Find the general solution of

$$y''' + 3y'' + 3y' + y = x.$$

Solution   From Example 6 in the preceding section, you know that the homogeneous solution is

$$y_h = C_1 e^{-x} + C_2 x e^{-x} + C_3 x^2 e^{-x}.$$

Because $F(x) = x$, let $y_p = A + Bx$ and obtain $y_p' = B$ and $y_p'' = 0$. Thus, by substitution, you have

$$(0) + 3(0) + 3(B) + (A + Bx) = (3B + A) + Bx = x.$$

Thus, $B = 1$ and $A = -3$, which implies that $y_p = -3 + x$. Therefore, the general solution is

$$y = y_h + y_p$$
$$= C_1 e^{-x} + C_2 x e^{-x} + C_3 x^2 e^{-x} - 3 + x.$$

Variation of Parameters

The method of undetermined coefficients works well if $F(x)$ is made up of polynomials or functions whose successive derivatives have a cyclic pattern. For functions such as $1/x$ and $\tan x$, which do not have such characteristics, it is better to use a more general method called **variation of parameters.** In this method, you assume that y_p has the same *form* as y_h, except that the constants in y_h are replaced by variables.

Variation of Parameters

To find the general solution to the equation $y'' + ay' + by = F(x)$, use the following steps.

1. Find $y_h = C_1 y_1 + C_2 y_2$.
2. Replace the constants by variables to form $y_p = u_1 y_1 + u_2 y_2$.
3. Solve the following system for u_1' and u_2'.

$$u_1' y_1 + u_2' y_2 = 0$$
$$u_1' y_1' + u_2' y_2' = F(x)$$

4. Integrate to find u_1 and u_2. The general solution is $y = y_h + y_p$.

EXAMPLE 5   Variation of Parameters

Solve the differential equation

$$y'' - 2y' + y = \frac{e^x}{2x}, \qquad x > 0.$$

Solution   The characteristic equation $m^2 - 2m + 1 = (m - 1)^2 = 0$ has one solution, $m = 1$. Thus, the homogeneous solution is

$$y_h = C_1 y_1 + C_2 y_2 = C_1 e^x + C_2 x e^x.$$

Replacing C_1 and C_2 by u_1 and u_2 produces

$$y_p = u_1 y_1 + u_2 y_2 = u_1 e^x + u_2 x e^x.$$

The resulting system of equations is

$$u_1' e^x + u_2' x e^x = 0$$
$$u_1' e^x + u_2' (x e^x + e^x) = \frac{e^x}{2x}.$$

Subtracting the second equation from the first produces $u_2' = 1/(2x)$. Then, by substitution in the first equation, you have $u_1' = -\frac{1}{2}$. Finally, integration yields

$$u_1 = -\int \frac{1}{2} \, dx = -\frac{x}{2} \quad \text{and} \quad u_2 = \frac{1}{2} \int \frac{1}{x} \, dx = \frac{1}{2} \ln x = \ln \sqrt{x}.$$

From this result it follows that a particular solution is

$$y_p = -\frac{1}{2} x e^x + \left(\ln \sqrt{x} \right) x e^x$$

and the general solution is

$$y = C_1 e^x + C_2 x e^x - \frac{1}{2} x e^x + x e^x \ln \sqrt{x}.$$

EXAMPLE 6   Variation of Parameters

Solve the differential equation

$$y'' + y = \tan x.$$

Solution   Because the characteristic equation $m^2 + 1 = 0$ has solutions $m = \pm i$, the homogeneous solution is

$$y_h = C_1 \cos x + C_2 \sin x.$$

Replacing C_1 and C_2 by u_1 and u_2 produces

$$y_p = u_1 \cos x + u_2 \sin x.$$

The resulting system of equations is

$$u_1' \cos x + u_2' \sin x = 0$$
$$-u_1' \sin x + u_2' \cos x = \tan x.$$

Multiplying the first equation by $\sin x$ and the second by $\cos x$ produces

$$u_1' \sin x \cos x + u_2' \sin^2 x = 0$$
$$-u_1' \sin x \cos x + u_2' \cos^2 x = \sin x.$$

Adding these two equations produces $u_2' = \sin x$, which implies that

$$u_1' = -\frac{\sin^2 x}{\cos x}$$
$$= \frac{\cos^2 x - 1}{\cos x}$$
$$= \cos x - \sec x.$$

Integration yields

$$u_1 = \int (\cos x - \sec x)\, dx$$
$$= \sin x - \ln|\sec x + \tan x|$$

and

$$u_2 = \int \sin x\, dx$$
$$= -\cos x$$

so that

$$y_p = \sin x \cos x - \cos x \ln|\sec x + \tan x| - \sin x \cos x$$
$$= -\cos x \ln|\sec x + \tan x|$$

and the general solution is

$$y = y_h + y_p$$
$$= C_1 \cos x + C_2 \sin x - \cos x \ln|\sec x + \tan x|.$$

EXERCISES FOR SECTION 15.4

In Exercises 1–4, verify the solution of the differential equation.

Solution	Differential Equation
1. $y = 2(e^{2x} - \cos x)$	$y'' + y = 10e^{2x}$
2. $y = \left(2 + \frac{1}{2}x\right)\sin x$	$y'' + y = \cos x$
3. $y = 3\sin x - \cos x \ln|\sec x + \tan x|$	$y'' + y = \tan x$
4. $y = \left(5 - \ln|\sin x|\right)\cos x - x\sin x$	$y'' + y = \csc x \cot x$

In Exercises 5–20, solve the differential equation by the method of undetermined coefficients.

5. $y'' - 3y' + 2y = 2x$

6. $y'' - 2y' - 3y = x^2 - 1$

7. $y'' + y = x^3$
$y(0) = 1, y'(0) = 0$

8. $y'' + 4y = 4$
$y(0) = 1, y'(0) = 6$

9. $y'' + 2y' = 2e^x$

10. $y'' - 9y = 5e^{3x}$

11. $y'' - 10y' + 25y = 5 + 6e^x$

12. $16y'' - 8y' + y = 4(x + e^x)$

13. $y'' + y' = 2\sin x$
$y(0) = 0, y'(0) = -3$

14. $y'' + y' - 2y = 3\cos 2x$
$y(0) = -1, y'(0) = 2$

15. $y'' + 9y = \sin 3x$

16. $y'' + 4y' + 5y = \sin x + \cos x$

17. $y''' - 3y' + 2y = 2e^{-2x}$

18. $y''' - y'' = 4x^2$
$y(0) = 1, y'(0) = 1, y''(0) = 1$

19. $y' - 4y = xe^x - xe^{4x}$
$y(0) = \dfrac{1}{3}$

20. $y' + 2y = \sin x$
$y\left(\dfrac{\pi}{2}\right) = \dfrac{2}{5}$

21. *Think About It*

(a) Explain how, by observation, you know that a particular solution of the differential equation $y'' + 3y = 12$ is $y_p = 4$.

(b) Use the explanation of part (a) to give a particular solution of the differential equation $y'' + 5y = 10$.

(c) Use the explanation of part (a) to give a particular solution of the differential equation $y'' + 2y' + 2y = 8$.

22. *Think About It*

(a) Explain how, by observation, you know that a form of a particular solution of the differential equation $y'' + 3y = 12\sin x$ is $y_p = A\sin x$.

(b) Use the explanation of part (a) to find a particular solution of the differential equation $y'' + 5y = 10\cos x$.

(c) Compare the algebra required to find particular solutions in parts (a) and (b) with that required if the form of the particular solution were $y_p = A\cos x + B\sin x$.

In Exercises 23–28, solve the differential equation by the method of variation of parameters.

23. $y'' + y = \sec x$

24. $y'' + y = \sec x \tan x$

25. $y'' + 4y = \csc 2x$

26. $y'' - 4y' + 4y = x^2e^{2x}$

27. $y'' - 2y' + y = e^x \ln x$

28. $y'' - 4y' + 4y = \dfrac{e^{2x}}{x}$

Electrical Circuits **In Exercises 29 and 30, use the electrical circuit differential equation**

$$\frac{d^2q}{dt^2} + \left(\frac{R}{L}\right)\frac{dq}{dt} + \left(\frac{1}{LC}\right)q = \left(\frac{1}{L}\right)E(t)$$

where R is the resistance (in ohms), C is the capacitance (in farads), L is the inductance (in henrys), $E(t)$ is the electromotive force (in volts), and q is the charge on the capacitor (in coulombs). Find the charge q as a function of time for the electrical circuit described. Assume that $q(0) = 0$ and $q'(0) = 0$.

29. $R = 20, C = 0.02, L = 2$
$E(t) = 12\sin 5t$

30. $R = 20, C = 0.02, L = 1$
$E(t) = 10\sin 5t$

Vibrating Spring **In Exercises 31–34, find the particular solution of the differential equation**

$$\frac{w}{g}y''(t) + by'(t) + ky(t) = \frac{w}{g}F(t)$$

for the oscillating motion of an object on the end of a spring. Use a graphing utility to graph the solution. In the equation, y is the displacement from equilibrium (positive direction is downward) measured in feet, and t is time in seconds (see figure). The constant w is the weight of the object, g is the acceleration due to gravity, b is the magnitude of the resistance to the motion, k is the spring constant from Hooke's Law, and $F(t)$ is the acceleration imposed on the system.

31. $\frac{24}{32}y'' + 48y = \frac{24}{32}(48\sin 4t)$
$y(0) = \frac{1}{4}, y'(0) = 0$

32. $\frac{2}{32}y'' + 4y = \frac{2}{32}(4\sin 8t)$
$y(0) = \frac{1}{4}, y'(0) = 0$

33. $\frac{2}{32}y'' + y' + 4y = \frac{2}{32}(4\sin 8t)$
$y(0) = \frac{1}{4}, y'(0) = -3$

34. $\frac{4}{32}y'' + \frac{1}{2}y' + \frac{25}{2}y = 0$
$y(0) = \frac{1}{2}, y'(0) = -4$

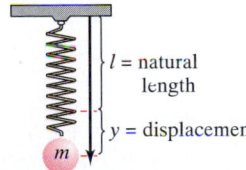

l = natural length

y = displacement

m

Spring displacement

35. Vibrating Spring   Rewrite y_h in the solution for Exercise 31 by using the identity

$$a \cos \omega t + b \sin \omega t = \sqrt{a^2 + b^2} \sin(\omega t + \phi)$$

where $\phi = \arctan a/b$.

36. Vibrating Spring   The figure shows the particular solution of the differential equation

$$\frac{4}{32} y'' + by' + \frac{25}{2} y = 0$$

$$y(0) = \frac{1}{2}, \, y'(0) = -4$$

for values of the resistance component b in the interval $[0, 1]$. (Note that when $b = \frac{1}{2}$, the problem is identical to that of Exercise 34.)

(a) If there is no resistance to the motion ($b = 0$), describe the motion.

(b) If $b > 0$, what is the ultimate effect of the retarding force?

(c) Is there a real number M such that there will be no oscillations of the spring if $b > M$? Explain your answer.

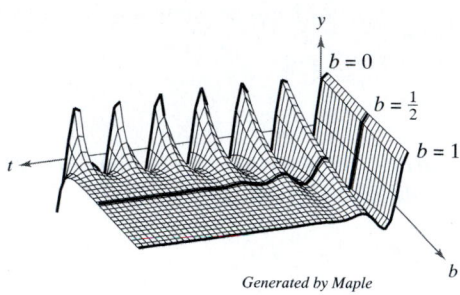

Generated by Maple

37. Parachute Jump   The fall of a parachutist is described by the second-order linear differential equation

$$\frac{w}{g} \frac{d^2 y}{dt^2} - k \frac{dy}{dt} = w$$

where w is the weight of the parachutist, y is the height at time t, g is the acceleration due to gravity, and k is the drag factor of the parachute. If the parachute is opened at 2000 feet, $y(0) = 2000$, and at that time the velocity is $y'(0) = -100$ feet per second, then for a 160-pound parachutist, using $k = 8$, the differential equation is

$$-5y'' - 8y' = 160.$$

Using the given initial conditions, verify that the solution of the differential equation is

$$y = 1950 + 50e^{-1.6t} - 20t.$$

38. Parachute Jump   Repeat Exercise 37 for a parachutist who weighs 192 pounds and has a parachute with a drag factor of $k = 9$.

39. Solve the differential equation

$$x^2 y'' - xy' + y = 4x \ln x$$

given that $y_1 = x$ and $y_2 = x \ln x$ are solutions of the corresponding homogeneous equation.

40. True or False? $y_p = -e^{2x} \cos e^{-x}$ is a particular solution of the differential equation

$$y'' - 3y' + 2y = \cos e^{-x}.$$

SECTION | 15.5 | Series Solutions of Differential Equations

Power Series Solution of a Differential Equation • Approximation by Taylor Series

Power Series Solution of a Differential Equation

We conclude this chapter by showing how power series can be used to solve certain types of differential equations. We begin with the general **power series solution** method.

Recall from Chapter 8 that a power series represents a function f on an interval of convergence, and that you can successively differentiate the power series to obtain a series for f', f'', and so on. These properties are used in the power series solution method demonstrated in the first two examples.

EXAMPLE 1   Power Series Solution

Use a power series to solve the differential equation $y' - 2y = 0$.

Solution   Assume that $y = \Sigma a_n x^n$ is a solution. Then, $y' = \Sigma n a_n x^{n-1}$. Substituting for y' and $-2y$, you obtain the following series form of the differential equation. (Note that, from the third step to the fourth, the index of summation is changed to ensure that x^n occurs in both sums.)

$$y' - 2y = 0$$

$$\sum_{n=1}^{\infty} n a_n x^{n-1} - 2 \sum_{n=0}^{\infty} a_n x^n = 0$$

$$\sum_{n=1}^{\infty} n a_n x^{n-1} = \sum_{n=0}^{\infty} 2 a_n x^n$$

$$\sum_{n=0}^{\infty} (n+1) a_{n+1} x^n = \sum_{n=0}^{\infty} 2 a_n x^n$$

Now, by equating coefficients of like terms, you obtain the **recursion formula** $(n+1)a_{n+1} = 2a_n$, which implies that

$$a_{n+1} = \frac{2a_n}{n+1}, \quad n \geq 0.$$

This formula generates the following results.

a_0	a_1	a_2	a_3	a_4	a_5	$\cdots$
a_0	$2a_0$	$\dfrac{2^2 a_0}{2}$	$\dfrac{2^3 a_0}{3!}$	$\dfrac{2^4 a_0}{4!}$	$\dfrac{2^5 a_0}{5!}$	$\cdots$

Using these values as the coefficients for the *solution* series, you have

$$y = \sum_{n=0}^{\infty} \frac{2^n a_0}{n!} x^n = a_0 \sum_{n=0}^{\infty} \frac{(2x)^n}{n!} = a_0 e^{2x}.$$

In Example 1, the differential equation could be solved easily without using a series. The differential equation in Example 2 cannot be solved by any of the methods discussed in previous sections.

EXAMPLE 2 Power Series Solution

Use a power series to solve the differential equation $y'' + xy' + y = 0$.

Solution Assume that $\displaystyle\sum_{n=0}^{\infty} a_n x^n$ is a solution. Then you have

$$y' = \sum_{n=1}^{\infty} na_n x^{n-1}, \qquad xy' = \sum_{n=1}^{\infty} na_n x^n, \qquad y'' = \sum_{n=2}^{\infty} n(n-1)a_n x^{n-2}.$$

Substituting for y'', xy', and y in the given differential equation, you obtain the following series.

$$\sum_{n=2}^{\infty} n(n-1)a_n x^{n-2} + \sum_{n=0}^{\infty} na_n x^n + \sum_{n=0}^{\infty} a_n x^n = 0$$

$$\sum_{n=2}^{\infty} n(n-1)a_n x^{n-2} = -\sum_{n=0}^{\infty} (n+1)a_n x^n$$

To obtain equal powers of x, adjust the summation indices by replacing n by $n+2$ in the left-hand sum, to obtain

$$\sum_{n=0}^{\infty} (n+2)(n+1)a_{n+2} x^n = -\sum_{n=0}^{\infty} (n+1)a_n x^n.$$

By equating coefficients, you have $(n+2)(n+1)a_{n+2} = -(n+1)a_n$, from which you obtain the recursion formula

$$a_{n+2} = -\frac{(n+1)}{(n+2)(n+1)} a_n = -\frac{a_n}{n+2}, \qquad n \geq 0,$$

and the coefficients of the solution series are as follows.

$$a_2 = -\frac{a_0}{2} \qquad\qquad\qquad a_3 = -\frac{a_1}{3}$$

$$a_4 = -\frac{a_2}{4} = \frac{a_0}{2 \cdot 4} \qquad\qquad a_5 = -\frac{a_3}{5} = \frac{a_1}{3 \cdot 5}$$

$$a_6 = -\frac{a_4}{6} = -\frac{a_0}{2 \cdot 4 \cdot 6} \qquad\qquad a_7 = -\frac{a_5}{7} = -\frac{a_1}{3 \cdot 5 \cdot 7}$$

$$\vdots \qquad\qquad\qquad\qquad\qquad \vdots$$

$$a_{2k} = \frac{(-1)^k a_0}{2 \cdot 4 \cdot 6 \cdots (2k)} = \frac{(-1)^k a_0}{2^k(k!)} \qquad a_{2k+1} = \frac{(-1)^k a_1}{3 \cdot 5 \cdot 7 \cdots (2k+1)}$$

Thus, you can represent the general solution as the sum of two series—one for the even-powered terms with coefficients in terms of a_0 and one for the odd-powered terms with coefficients in terms of a_1.

$$y = a_0\left(1 - \frac{x^2}{2} + \frac{x^4}{2 \cdot 4} - \cdots\right) + a_1\left(x - \frac{x^3}{3} + \frac{x^5}{3 \cdot 5} - \cdots\right)$$

$$= a_0 \sum_{k=0}^{\infty} \frac{(-1)^k x^{2k}}{2^k(k!)} + a_1 \sum_{k=0}^{\infty} \frac{(-1)^k x^{2k+1}}{3 \cdot 5 \cdot 7 \cdots (2k+1)}$$

The solution has two arbitrary constants, a_0 and a_1, as you would expect in the general solution of a second-order differential equation.

Approximation by Taylor Series

A second type of series solution method involves a differential equation *with initial conditions* and makes use of Taylor series, as given in Section 8.10.

EXAMPLE 3   Approximation by Taylor Series

Use a Taylor series to find the series solution of

$$y' = y^2 - x$$

given the initial condition $y = 1$ when $x = 0$. Then, use the first six terms of this series solution to approximate values of y for $0 \le x \le 1$.

Solution   Recall from Section 8.10 that, for $c = 0$,

$$y = y(0) + y'(0)x + \frac{y''(0)}{2!}x^2 + \frac{y'''(0)}{3!}x^3 + \cdots.$$

Because $y(0) = 1$ and $y' = y^2 - x$, you obtain the following.

	$y(0) = 1$
$y' = y^2 - x$	$y'(0) = 1$
$y'' = 2yy' - 1$	$y''(0) = 2 - 1 = 1$
$y''' = 2yy'' + 2(y')^2$	$y'''(0) = 2 + 2 = 4$
$y^{(4)} = 2yy''' + 6y'y''$	$y^{(4)}(0) = 8 + 6 = 14$
$y^{(5)} = 2yy^{(4)} + 8y'y''' + 6(y'')^2$	$y^{(5)}(0) = 28 + 32 + 6 = 66$

Therefore, you can approximate the values of the solution from the series

$$y = y(0) + y'(0)x + \frac{y''(0)}{2!}x^2 + \frac{y'''(0)}{3!}x^3 + \frac{y^{(4)}(0)}{4!}x^4 + \frac{y^{(5)}(0)}{5!}x^5 + \cdots$$

$$= 1 + x + \frac{1}{2}x^2 + \frac{4}{3!}x^3 + \frac{14}{4!}x^4 + \frac{66}{5!}x^5 + \cdots.$$

Using the first six terms of this series, you can compute values for y in the interval $0 \le x \le 1$, as shown in the table at the left.

x	y
0.0	1.0000
0.1	1.1057
0.2	1.2264
0.3	1.3691
0.4	1.5432
0.5	1.7620
0.6	2.0424
0.7	2.4062
0.8	2.8805
0.9	3.4985
1.0	4.3000

EXERCISES FOR SECTION 15.5

In Exercises 1–6, verify that the power series solution of the differential equation is equivalent to the solution found using the techniques in Sections 5.7 and 15.1–15.4.

1. $y' - y = 0$ **2.** $y' - ky = 0$

3. $y'' - 9y = 0$ **4.** $y'' - k^2y = 0$

5. $y'' + 4y = 0$ **6.** $y'' + k^2y = 0$

In Exercises 7–10, use power series to solve the differential equation and find the interval of convergence of the series.

7. $y' + 3xy = 0$ **8.** $y' - 2xy = 0$

9. $y'' - xy' = 0$ **10.** $y'' - xy' - y = 0$

In Exercises 11 and 12, find the first three terms of each of the power series representing independent solutions of the differential equation.

11. $(x^2 + 4)y'' + y = 0$ **12.** $y'' + x^2y = 0$

In Exercises 13 and 14, use Taylor's Theorem to find the series solution of the differential equation under the specified initial conditions. Use n terms of the series to approximate y for the given value of x and compare the result with the approximation given by Euler's Method for $\Delta x = 0.1$.

13. $y' + (2x - 1)y = 0,\ y(0) = 2,\ n = 5,\ x = \frac{1}{2}$,

14. $y' - 2xy = 0,\ y(0) = 1,\ n = 4,\ x = 1$

15. *Investigation* Consider the differential equation $y'' + 9y = 0$ with initial conditions $y(0) = 2$ and $y'(0) = 6$.

(a) Find the solution of the differential equation using the techniques of Section 15.3.

(b) Find the series solution of the differential equation.

(c) The figure shows the graph of the solution of the differential equation and the third-degree and fifth-degree polynomial approximations of the solution. Identify each.

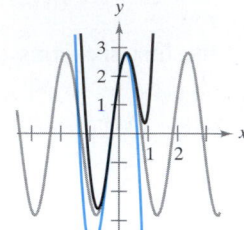

16. Consider the differential equation $y'' - xy' = 0$ with the initial conditions $y(0) = 0$ and $y'(0) = 2$. (See Exercise 9.)

(a) Find the series solution satisfying the initial conditions.

(b) Use a graphing utility to graph the third-degree and fifth-degree series approximations of the solution. Identify the approximations.

(c) Identify the symmetry of the solution.

In Exercises 17 and 18, use Taylor's Theorem to find the series solution of the differential equation under the specified initial conditions. Use *n* terms of the series to approximate *y* for the given value of *x*.

17. $y'' - 2xy = 0$, $y(0) = 1$, $y'(0) = -3$, $n = 6$, $x = \frac{1}{4}$

18. $y'' - 2xy' + y = 0$, $y(0) = 1$, $y'(0) = 2$, $n = 8$, $x = \frac{1}{2}$

In Exercises 19–22, verify that the series converges to the given function on the indicated interval. (*Hint:* Use the given differential equation.)

19. $\sum_{n=0}^{\infty} \frac{x^n}{n!} = e^x$, $(-\infty, \infty)$

Differential equation: $y' - y = 0$

20. $\sum_{n=0}^{\infty} \frac{(-1)^n x^{2n}}{(2n)!} = \cos x$, $(-\infty, \infty)$

Differential equation: $y'' + y = 0$

21. $\sum_{n=0}^{\infty} \frac{(-1)^n x^{2n+1}}{2n+1} = \arctan x$, $(-1, 1)$

Differential equation: $(x^2 + 1)y'' + 2xy' = 0$

22. $\sum_{n=0}^{\infty} \frac{(2n)! x^{2n+1}}{(2^n n!)^2 (2n+1)} = \arcsin x$, $(-1, 1)$

Differential equation: $(1 - x^2)y'' - xy' = 0$

23. Find the first six terms in the series solution of Airy's equation $y'' - xy = 0$.

REVIEW EXERCISES FOR CHAPTER 15

In Exercises 1–4, classify the differential equation according to type and order.

1. $\dfrac{\partial^2 u}{\partial t^2} = c^2 \dfrac{\partial^2 u}{\partial x^2}$

2. $yy'' = x + 1$

3. $y'' + 3y' - 10 = 0$

4. $(y'')^2 + 4y' = 0$

In Exercises 5 and 6, use the given differential equation and its direction field.

5. $\dfrac{dy}{dx} = \dfrac{y}{x}$

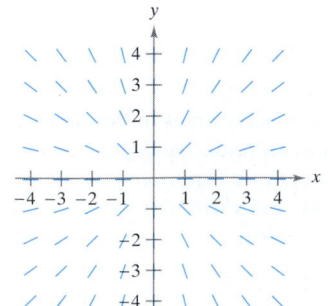

(a) Sketch several solution curves for the differential equation on the direction field.

(b) Find the general solution of the differential equation. Compare the result with the sketches from part (a).

6. $\dfrac{dy}{dx} = \sqrt{1 - y^2}$

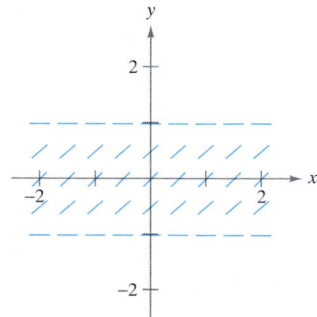

(a) Sketch several solution curves for the differential equation on the direction field.

(b) When is the rate of change of the solution greatest? When is it least?

(c) Find the general solution of the differential equation. Compare the result with the sketches from part (a).

In Exercises 7–10, match the differential equation with its solution.

Differential Equation	Solution
7. $y' - 4 = 0$	(a) $y = C_1 e^{2x} + C_2 e^{-2x}$
8. $y' - 4y = 0$	(b) $y = 4x + C$
9. $y'' - 4y = 0$	(c) $y = C_1 \cos 2x + C_2 \sin 2x$
10. $y'' + 4y = 0$	(d) $y = Ce^{4x}$

In Exercises 11–32, find the general solution of the first-order differential equation.

11. $\dfrac{dy}{dx} - \dfrac{y}{x} = 2 + \sqrt{x}$

12. $\dfrac{dy}{dx} + xy = 2y$

13. $y' - \dfrac{2y}{x} = \dfrac{y'}{x}$

14. $\dfrac{dy}{dx} - 3x^2 y = e^{x^3}$

15. $\dfrac{dy}{dx} - \dfrac{y}{x} = \dfrac{x}{y}$

16. $\dfrac{dy}{dx} - \dfrac{3y}{x^2} = \dfrac{1}{x^2}$

17. $(10x + 8y + 2) \, dx + (8x + 5y + 2) \, dy = 0$

18. $(y + x^3 + xy^2) \, dx - x \, dy = 0$

19. $(2x - 2y^3 + y) \, dx + (x - 6xy^2) \, dy = 0$

20. $3x^2 y^2 \, dx + (2x^3 y + x^3 y^4) \, dy = 0$

21. $dy = (y \tan x + 2e^x) \, dx$

22. $y \, dx - \left(x + \sqrt{xy}\right) dy = 0$

23. $(x - y - 5) \, dx - (x + 3y - 2) \, dy = 0$

24. $y' = 2x\sqrt{1 - y^2}$

25. $x + yy' = \sqrt{x^2 + y^2}$

26. $xy' + y = \sin x$

27. $yy' + y^2 = 1 + x^2$

28. $2x \, dx + 2y \, dy = (x^2 + y^2) \, dx$

29. $(1 + x^2) \, dy = (1 + y^2) \, dx$

30. $x^3 yy' = x^4 + 3x^2 y^2 + y^4$

31. $xy' - ay = bx^4$

32. $y' = y + 2x(y - e^x)$

In Exercises 33–40, find the particular solution of the differential equation that satisfies the boundary condition.

33. $y' - 2y = e^x$

$y(0) = 4$

34. $y' + \dfrac{2y}{x} = -x^9 y^5$

$y(1) = 2$

35. $x \, dy = (x + y + 2) \, dx$

$y(1) = 10$

36. $ye^{xy} \, dx + xe^{xy} \, dy = 0$

$y(-2) = -5$

37. $(1 + y) \ln (1 + y) \, dx + dy = 0$

$y(0) = 2$

38. $(2x + y - 3) \, dx + (x - 3y + 1) \, dy = 0$

$y(2) = 0$

39. $y' = x^2 y^2 - 9x^2$

$y(0) = \dfrac{3(1 + e)}{1 - e}$

40. $2xy' - y = x^3 - x$

$y(4) = 2$

In Exercises 41 and 42, find the orthogonal trajectories of the given family and sketch several members of each family.

41. $(x - C)^2 + y^2 = C^2$

42. $y - 2x = C$

43. *Snow Removal* Assume that the rate of change in the number of miles s of road cleared per hour by a snowplow is inversely proportional to the height h of snow.

(a) Write and solve the differential equation to find s as a function of h.

(b) Find the particular solution if $s = 25$ miles when $h = 2$ inches and $s = 12$ miles when $h = 10$ inches $(2 \le h \le 15)$.

44. *Growth Rate* Let x and y be the sizes of two internal organs of a particular mammal at time t. Empirical data indicate that the relative growth rates of these two organs are equal, and hence we have

$$\frac{1}{x} \frac{dx}{dt} = \frac{1}{y} \frac{dy}{dt}.$$

Solve this differential equation, writing y as a function of x.

45. *Population Growth* The rate of growth in the number N of deer in a state park varies jointly over time t as N and $L - N$, where $L = 500$ is the estimated limiting size of the herd. Write N as a function of t if $N = 100$ when $t = 0$ and $N = 200$ when $t = 4$.

46. *Population Growth* The rate of growth in the number N of elk in a game preserve varies jointly over time t (in years) as N and $300 - N$ where 300 is the estimated limiting size of the herd.

(a) Write and solve the differential equation for the population model if $N = 50$ when $t = 0$ and $N = 75$ when $t = 1$.

(b) Use a graphing utility to graph the direction field of the differential equation and the particular solution of part (a).

(c) At what time is the population increasing most rapidly?

(d) If 400 elk had been placed in the preserve initially, use the direction field to describe the change in the population over time.

47. Slope The slope of a graph is given by $y' = \sin x - 0.5y$. Find the equation of the graph if the graph passes through the point $(0, 1)$. Use a graphing utility to graph the solution.

48. Investment Let $A(t)$ be the amount in a fund earning interest at an annual rate r compounded continuously. If a continuous cash flow of P dollars per year is withdrawn from the fund, the rate of change of A is given by the differential equation

$$\frac{dA}{dt} = rA - P$$

where $A = A_0$ when $t = 0$. Solve this differential equation for A as a function of t.

49. Investment A retired couple plans to withdraw P dollars per year from a retirement account of \$500,000 earning 10% compounded continuously. Use the result of Exercise 48 and a graphing utility to graph the function A for each of the following continuous annual cash flows. Use the graphs to describe what happens to the balance in the fund for each of the cases.

(a) $P = \$40,000$

(b) $P = \$50,000$

(c) $P = \$60,000$

50. Investment Use the result of Exercise 48 to find the time necessary to deplete a fund earning 14% interest compounded continuously if $A_0 = \$1,000,000$ and $P = \$200,000$.

In Exercises 51–54, find the particular solution of the differential equation that satisfies the initial conditions. Use a graphing utility to graph the solution.

Differential Equation	Initial Conditions
51. $y'' - y' - 2y = 0$	$y(0) = 0, y'(0) = 3$
52. $y'' + 4y' + 5y = 0$	$y(0) = 2, y'(0) = -7$
53. $y'' + 2y' - 3y = 0$	$y(0) = 2, y'(0) = 0$
54. $y'' + 2y' + 5y = 0$	$y(1) = 4, y(2) = 0$

In Exercises 55–60, find the general solution of the second-order differential equation.

55. $y'' + y = x^3 + x$

56. $y'' + 2y = e^{2x} + x$

57. $y'' + y = 2\cos x$

58. $y'' + 5y' + 4y = x^2 + \sin 2x$

59. $y'' - 2y' + y = 2xe^x$

60. $y'' + 2y' + y = \dfrac{1}{x^2 e^x}$

In Exercises 61–64, find the particular solution of the differential equation that satisfies the initial conditions.

Differential Equation	Initial Conditions
61. $y'' - y' - 6y = 54$	$y(0) = 2, y'(0) = 0$
62. $y'' + 25y = e^x$	$y(0) = 0, y'(0) = 0$
63. $y'' + 4y = \cos x$	$y(0) = 6, y'(0) = -6$
64. $y'' + 3y' = 6x$	$y(0) = 2, y'(0) = \frac{10}{3}$

Vibrating Spring In Exercises 65 and 66, describe the motion of a 64-pound weight suspended on a spring. Assume that the weight stretches the spring $\frac{4}{3}$ feet from its natural position.

65. The weight is pulled $\frac{1}{2}$ foot below the equilibrium position and released.

66. The weight is pulled $\frac{1}{2}$ foot below the equilibrium position and released. The motion takes place in a medium that furnishes a damping force of magnitude $\frac{1}{8}$ speed at all times.

67. Investigation The differential equation

$$\frac{8}{32} y'' + by' + ky = \frac{8}{32} F(t), \quad y(0) = \frac{1}{2}, \quad y'(0) = 0$$

models the motion of a weight suspended on a spring.

(a) Solve the differential equation and use a graphing utility to graph the solution for each of the assigned quantities for b, k, and $F(t)$.

  (i) $b = 0, k = 1, F(t) = 24 \sin \pi t$

  (ii) $b = 0, k = 2, F(t) = 24 \sin(2\sqrt{2}\, t)$

  (iii) $b = 0.1, k = 2, F(t) = 0$

  (iv) $b = 1, k = 2, F(t) = 0$

(b) Describe the effect of increasing the resistance to motion b.

(c) Explain how the motion of the object would change if a stiffer spring (increased k) were used.

(d) Matching the input and natural frequencies of a system is known as resonance. In which case of part (a) does this occur, and what is the result?

68. Think About It Explain how you can find a particular solution of the differential equation

$$y'' + 4y' + 6y = 30$$

by observation.

In Exercises 69 and 70, find the series solution of the differential equation.

69. $(x - 4)y' + y = 0$

70. $y'' + 3xy' - 3y = 0$

APPENDIX A
Precalculus Review

Real Numbers and the Real Line • Order and Inequalities •
Absolute Value and Distance

Real Numbers and the Real Line

Real numbers can be represented by a coordinate system called the **real line** or x-axis (see Figure A.1). The real number corresponding to a point on the real line is the **coordinate** of the point. As Figure A.1 shows, it is customary to identify those points whose coordinates are integers.

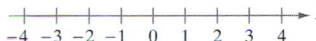

The real line
Figure A.1

The point on the real line corresponding to zero is the **origin** and is denoted by 0. The **positive direction** (to the right) is denoted by an arrowhead and is the direction of increasing values of x. Numbers to the right of the origin are **positive.** Numbers to the left of the origin are **negative.** The term **nonnegative** describes a number that is positive or zero. The term **nonpositive** describes a number that is negative or zero.

Each point on the real line corresponds to one and only one real number, and each real number corresponds to one and only one point on the real line. This type of relationship is called a **one-to-one-correspondence.**

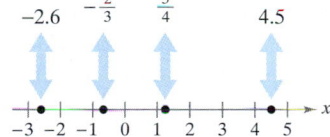

Rational numbers
Figure A.2

Each of the four points in Figure A.2 corresponds to a **rational number**—one that can be expressed as the ratio of two integers. (Note that $4.5 = \frac{9}{2}$ and $-2.6 = \frac{-13}{5}$.) Rational numbers can be represented either by *terminating decimals* such as $\frac{2}{5} = 0.4$, or by *repeating decimals* such as $\frac{1}{3} = 0.333\ldots = 0.\overline{3}$.

Real numbers that are not rational are **irrational.** Irrational numbers cannot be represented as terminating or repeating decimals. In computations, irrational numbers are represented by decimal approximations. Here are three familiar examples.

$$\sqrt{2} \approx 1.414213562$$
$$\pi \approx 3.141592654$$
$$e \approx 2.718281828$$

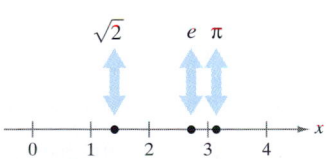

Irrational numbers
Figure A.3

(See Figure A.3.)

Order and Inequalities

One important property of real numbers is that they can be **ordered.** If a and b are real numbers, a is **less than** b if $b - a$ is positive. This order is denoted by the **inequality**

$$a < b.$$

The statement "b is **greater than** a" is equivalent to saying that a is less than b. When three real numbers a, b, and c are ordered such that $a < b$ and $b < c$, we say that b is **between** a and c and $a < b < c$.

Geometrically, $a < b$ if and only if a lies to the *left* of b on the real line (see Figure A.4). For example, $1 < 2$ because 1 lies to the left of 2 on the real line.

The following properties are used in working with inequalities. Similar properties are obtained if < is replaced by ≤ and > is replaced by ≥. (The symbols ≤ and ≥ mean **less than or equal to** and **greater than or equal to,** respectively.)

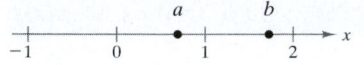

$a < b$ if and only if a lies to the left of b.
Figure A.4

Properties of Inequalities

Let a, b, c, d, and k be real numbers.

1. If $a < b$ and $b < c$, then $a < c$.   Transitive Property

2. If $a < b$ and $c < d$, then $a + c < b + d$.   Add inequalities.

3. If $a < b$, then $a + k < b + k$.   Add a constant.

4. If $a < b$ and $k > 0$, then $ak < bk$.   Multiply by a positive constant.

5. If $a < b$ and $k < 0$, then $ak > bk$.   Multiply by a negative constant.

NOTE   Note that you *reverse the inequality* when you multiply by a negative number. For example, if $x < 3$, then $-4x > -12$. This also applies to division by a negative number. Thus, if $-2x > 4$, then $x < -2$.

A **set** is a collection of elements. Two common sets are the set of real numbers and the set of points on the real line. Many problems in calculus involve **subsets** of one of these two sets. In such cases it is convenient to use **set notation** of the form $\{x:\text{ condition on }x\}$, which is read as follows.

The set of all x such that a certain condition is true.

$$\{ \quad x \quad : \quad \text{condition on } x\}$$

For example, you can describe the set of positive real numbers as

$$\{x:\ x > 0\}. \qquad \text{Set of positive real numbers}$$

Similarly, you can describe the set of nonnegative real numbers as

$$\{x:\ x \geq 0\}. \qquad \text{Set of nonnegative real numbers}$$

The **union** of two sets A and B, denoted by $A \cup B$, is the set of elements that are members of A *or* B *or both*. The **intersection** of two sets A and B, denoted by $A \cap B$, is the set of elements that are members of A *and* B. Two sets are **disjoint** if they have no elements in common.

The most commonly used subsets are **intervals** on the real line. For example, the **open** interval

$$(a, b) = \{x\colon a < x < b\} \qquad \text{Open interval}$$

is the set of all real numbers greater than a and less than b, where a and b are the **endpoints** of the interval. Note that the endpoints are not included in an open interval. Intervals that include their endpoints are **closed** and are denoted by

$$[a, b] = \{x\colon a \le x \le b\}. \qquad \text{Closed interval}$$

The nine basic types of intervals on the real line are shown in the table below. The first four are **bounded intervals** and the remaining five are **unbounded intervals.** Unbounded intervals are also classified as open or closed. The intervals $(-\infty, b)$ and (a, ∞) are open, the intervals $(-\infty, b]$ and $[a, \infty)$ are closed, and the interval $(-\infty, \infty)$ is considered to be both open *and* closed.

Intervals on the Real Line

	Interval Notation	Set Notation	Graph
Bounded open interval	(a, b)	$\{x\colon a < x < b\}$	
Bounded closed interval	$[a, b]$	$\{x\colon a \le x \le b\}$	
Bounded intervals (neither open nor closed)	$[a, b)$	$\{x\colon a \le x < b\}$	
	$(a, b]$	$\{x\colon a < x \le b\}$	
Unbounded open intervals	$(-\infty, b)$	$\{x\colon x < b\}$	
	(a, ∞)	$\{x\colon x > a\}$	
Unbounded closed intervals	$(-\infty, b]$	$\{x\colon x \le b\}$	
	$[a, \infty)$	$\{x\colon x \ge a\}$	
Entire real line	$(-\infty, \infty)$	$\{x\colon x \text{ is a real number}\}$	

NOTE   The symbols ∞ and $-\infty$ refer to positive and negative infinity. These symbols do not denote real numbers. They simply enable you to describe unbounded conditions more concisely. For instance, the interval $[a, \infty)$ is unbounded to the right because it includes *all* real numbers that are greater than or equal to a.

EXAMPLE 1   Liquid and Gaseous States of Water

Describe the intervals on the real line that correspond to the temperature x (in degrees Celsius) for water in

a. a liquid state   **b.** a gaseous state.

Solution

a. Water is in a liquid state at temperatures greater than $0°$ and less than $100°$, as shown in Figure A.5(a).

$$(0, 100) = \{x: 0 < x < 100\}$$

b. Water is in a gaseous state (steam) at temperatures greater than or equal to $100°$, as shown in Figure A.5(b).

$$[100, \infty) = \{x: x \geq 100\}$$

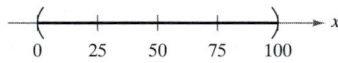

(a) Temperature range of water
     (in degrees Celsius)

(b) Temperature range of steam
     (in degrees Celsius)

Figure A.5

A real number a is a **solution** of an inequality if the inequality is **satisfied** (is true) when a is substituted for x. The set of all solutions is the **solution set** of the inequality.

EXAMPLE 2   Solving an Inequality

Solve $2x - 5 < 7$.

Solution

$$
\begin{aligned}
2x - 5 &< 7 &&\text{Original inequality}\\
2x - 5 + 5 &< 7 + 5 &&\text{Add 5 to both sides.}\\
2x &< 12 &&\text{Simplify.}\\
\frac{1}{2}(2x) &< \frac{1}{2}(12) &&\text{Multiply both sides by } \tfrac{1}{2}.\\
x &< 6 &&\text{Simplify.}
\end{aligned}
$$

The solution set is $(-\infty, 6)$.

NOTE In Example 2, all five inequalities listed as steps in the solution are called **equivalent** because they have the same solution set.

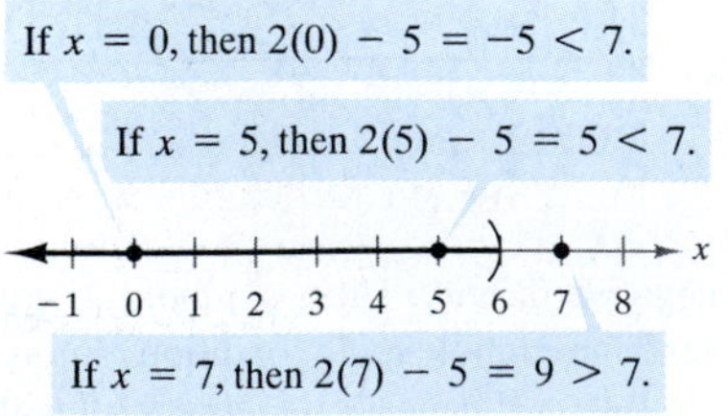

Once you have solved an inequality, check some x-values in your solution set to verify that they satisfy the original inequality. You should also check some values outside your solution set to verify that they *do not* satisfy the inequality. For example, Figure A.6 shows that when $x = 0$ or $x = 5$ the inequality $2x - 5 < 7$ is satisfied, but when $x = 7$ the inequality $2x - 5 < 7$ is not satisfied.

If $x = 0$, then $2(0) - 5 = -5 < 7$.

If $x = 5$, then $2(5) - 5 = 5 < 7$.

If $x = 7$, then $2(7) - 5 = 9 > 7$.

Checking solutions of $2x - 5 < 7$

Figure A.6

EXAMPLE 3   Solving a Double Inequality

Solve $-3 \leq 2 - 5x \leq 12$.

Solution

$$
\begin{array}{llll}
-3 \leq & 2 - 5x & \leq 12 & \text{Original inequality} \\
-3 - 2 \leq 2 - 5x - 2 & \leq 12 - 2 & & \text{Subtract 2.} \\
-5 \leq & -5x & \leq 10 & \text{Simplify.} \\
\dfrac{-5}{-5} \geq & \dfrac{-5x}{-5} & \geq \dfrac{10}{-5} & \text{Divide by } -5 \text{ and reverse both inequalities.} \\
1 \geq & x & \geq -2 & \text{Simplify.}
\end{array}
$$

The solution set is $[-2, 1]$, as shown in Figure A.7.

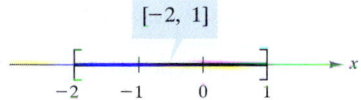

[−2, 1]

Solution set of $-3 \leq 2 - 5x \leq 12$

Figure A.7

The inequalities in Examples 2 and 3 are **linear inequalities**—that is, they involve first-degree polynomials. To solve inequalities involving polynomials of higher degree, use the fact that a polynomial can change signs *only* at its real **zeros** (the numbers that make the polynomial zero). Between two consecutive real zeros, a polynomial must be either entirely positive or entirely negative. This means that when the real zeros of a polynomial are put in order, they divide the real line into **test intervals** in which the polynomial has no sign changes. Thus, if a polynomial has the factored form

$$(x - r_1)(x - r_2) \cdots (x - r_n), \qquad r_1 < r_2 < r_3 < \cdots < r_n$$

the test intervals are

$$(-\infty, r_1), \quad (r_1, r_2), \ldots, \quad (r_{n-1}, r_n), \quad \text{and} \quad (r_n, \infty).$$

To determine the sign of the polynomial in each test interval, you need to test only *one value* from the interval.

EXAMPLE 4   Solving a Quadratic Inequality

Solve $x^2 < x + 6$.

Solution

$$
\begin{array}{ll}
x^2 < x + 6 & \text{Original inequality} \\
x^2 - x - 6 < 0 & \text{Write in standard form.} \\
(x - 3)(x + 2) < 0 & \text{Factor.}
\end{array}
$$

The polynomial $x^2 - x - 6$ has $x = -2$ and $x = 3$ as its zeros. Thus, you can solve the inequality by testing the sign of $x^2 - x - 6$ in each of the test intervals $(-\infty, -2)$, $(-2, 3)$, and $(3, \infty)$. To test an interval, choose any number in the interval and compute the sign of $x^2 - x - 6$. After doing this, you will find that the polynomial is positive for all real numbers in the first and third intervals and negative for all real numbers in the second interval. The solution of the original inequality is therefore $(-2, 3)$, as shown in Figure A.8.

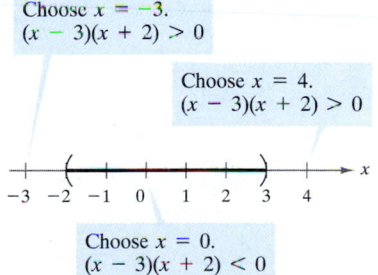

Choose $x = -3$.
$(x - 3)(x + 2) > 0$

Choose $x = 4$.
$(x - 3)(x + 2) > 0$

Choose $x = 0$.
$(x - 3)(x + 2) < 0$

Testing an interval

Figure A.8

Absolute Value and Distance

If a is a real number, the **absolute value** of a is

$$|a| = \begin{cases} a, & \text{if } a \geq 0 \\ -a, & \text{if } a < 0. \end{cases}$$

The absolute value of a number cannot be negative. For example, let $a = -4$. Then, because $-4 < 0$, you have

$$|a| = |-4| = -(-4) = 4.$$

Remember that the symbol $-a$ does not necessarily mean that $-a$ is negative.

Operations with Absolute Value

Let a and b be real numbers and let n be a positive integer.

1. $|ab| = |a|\,|b|$ **2.** $\left|\dfrac{a}{b}\right| = \dfrac{|a|}{|b|}$, $b \neq 0$

3. $|a| = \sqrt{a^2}$ **4.** $|a^n| = |a|^n$

NOTE   You are asked to prove these properties in Exercises 73, 75, 76, and 77.

Properties of Inequalities and Absolute Value

Let a and b be real numbers and let k be a positive real number.

1. $-|a| \leq a \leq |a|$

2. $|a| \leq k$ if and only if $-k \leq a \leq k$.

3. $k \leq |a|$ if and only if $k \leq a$ or $a \leq -k$.

4. *Triangle Inequality:* $|a + b| \leq |a| + |b|$

Properties 2 and 3 are also true if $\leq$ is replaced by $<$.

EXAMPLE 5   Solving an Absolute Value Inequality

Solve $|x - 3| \leq 2$.

Solution   Using the second property of inequalities and absolute value, you can rewrite the original inequality as a double inequality.

$$\begin{aligned} -2 &\leq & x - 3 & \leq 2 & \text{Write as double inequality.} \\ -2 + 3 &\leq & x - 3 + 3 & \leq 2 + 3 & \text{Add 3.} \\ 1 &\leq & x & \leq 5 & \text{Simplify.} \end{aligned}$$

The solution set is $[1, 5]$, as shown in Figure A.9.

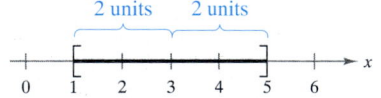

2 units   2 units

Solution set of $|x - 3| \leq 2$

Figure A.9

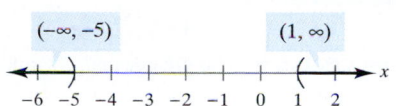

Solution set of $3 < |x + 2|$
Figure A.10

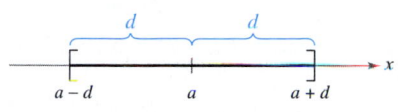

Solution set of $|x - a| \leq d$

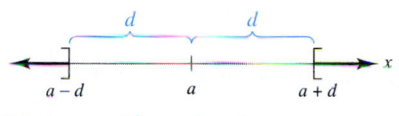

Solution set of $|x - a| \geq d$

Figure A.11

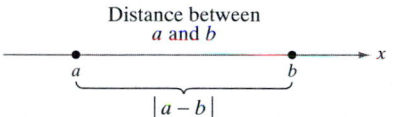

Figure A.12

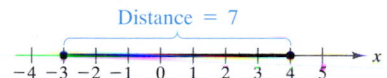

Figure A.13

EXAMPLE 6   A Two-Interval Solution Set

Solve $3 < |x + 2|$.

Solution   Using the third property of inequalities and absolute value, you can rewrite the original inequality as two linear inequalities.

$$3 < x + 2 \quad \text{or} \quad x + 2 < -3$$
$$1 < x \qquad \text{or} \qquad x < -5$$

The solution set is the union of the disjoint intervals $(-\infty, -5)$ and $(1, \infty)$, as shown in Figure A.10.

Examples 5 and 6 illustrate the general results shown in Figure A.11. Note that if $d > 0$, the solution set for the inequality $|x - a| \leq d$ is a *single* interval, whereas the solution set for the inequality $|x - a| \geq d$ is the union of *two* disjoint intervals.

The **distance between two points** a and b on the real line is given by

$$d = |a - b| = |b - a|.$$

The **directed distance from** a **to** b is $b - a$ and the **directed distance from** b **to** a is $a - b$, as shown in Figure A.12.

EXAMPLE 7   Distance on the Real Line

a. The distance between -3 and 4 is

$$|4 - (-3)| = |7| = 7 \quad \text{or} \quad |-3 - 4| = |-7| = 7.$$

(See Figure A.13.)

b. The directed distance from -3 to 4 is $4 - (-3) = 7$.

c. The directed distance from 4 to -3 is $-3 - 4 = -7$.

The **midpoint** of an interval with endpoints a and b is the average value of a and b. That is,

$$\text{Midpoint of interval } (a, b) = \frac{a + b}{2}.$$

To show that this is the midpoint, you need only show that $(a + b)/2$ is equidistant from a and b.

EXERCISES FOR APPENDIX A.1

In Exercises 1–10, determine whether the real number is rational or irrational.

1. 0.7

2. -3678

3. $\dfrac{3\pi}{2}$

4. $3\sqrt{2} - 1$

5. $4.3\overline{451}$

6. $\frac{22}{7}$

7. $\sqrt[3]{64}$

8. $0.\overline{8177}$

9. $4\frac{5}{8}$

10. $\left(\sqrt{2}\right)^3$

In Exercises 11–14, express the repeating decimal as a ratio of integers using the following procedure. If $x = 0.6363\ldots$, then $100x = 63.6363\ldots$. Subtracting the first equation from the second produces $99x = 63$ or $x = \frac{63}{99} = \frac{7}{11}$.

11. $0.\overline{36}$

12. $0.3\overline{18}$

13. $0.\overline{297}$

14. $0.\overline{9900}$

15. Given $a < b$, determine which of the following are true.

(a) $a + 2 < b + 2$

(b) $5b < 5a$

(c) $5 - a > 5 - b$

(d) $\dfrac{1}{a} < \dfrac{1}{b}$

(e) $(a - b)(b - a) > 0$

(f) $a^2 < b^2$

16. Complete the table with the appropriate interval notation, set notation, and graph on the real line.

Interval Notation	Set Notation	Graph
		(graph from -2 to 0) x
$(-\infty, -4]$		
	$\left\{x: 3 \le x \le \frac{11}{2}\right\}$	
$(-1, 7)$		

In Exercises 17–20, verbally describe the subset of real numbers represented by the inequality. Sketch the subset on the real number line, and state whether the interval is bounded or unbounded.

17. $-3 < x < 3$

18. $x \ge 4$

19. $x \le 5$

20. $0 \le x < 8$

In Exercises 21–24, use inequality and interval notation to describe the set.

21. y is at least 4.

22. q is nonnegative.

23. The interest rate on loans is expected to be greater than 3% and no more than 7%.

24. The temperature T is forecast to be above 90° today.

In Exercises 25–44, solve the inequality and graph the solution on the real line.

25. $2x - 1 \ge 0$

26. $3x + 1 \ge 2x + 2$

27. $-4 < 2x - 3 < 4$

28. $0 \le x + 3 < 5$

29. $\dfrac{x}{2} + \dfrac{x}{3} > 5$

30. $x > \dfrac{1}{x}$

31. $|x| < 1$

32. $\dfrac{x}{2} - \dfrac{x}{3} > 5$

33. $\left|\dfrac{x - 3}{2}\right| \ge 5$

34. $\left|\dfrac{x}{2}\right| > 3$

35. $|x - a| < b, \ b > 0$

36. $|x + 2| < 5$

37. $|2x + 1| < 5$

38. $|3x + 1| \ge 4$

39. $\left|1 - \dfrac{2}{3}x\right| < 1$

40. $|9 - 2x| < 1$

41. $x^2 \le 3 - 2x$

42. $x^4 - x \le 0$

43. $x^2 + x - 1 \le 5$

44. $2x^2 + 1 < 9x - 3$

In Exercises 45–48, find the directed distance from a to b, the directed distance from b to a, and the distance between a and b.

45. (number line: $a = -1$, $b = 3$, marks -2 to 4) x

46. (number line: $a = -\frac{5}{2}$, $b = \frac{13}{4}$, marks -3 to 4) x

47. (a) $a = 126, b = 75$

(b) $a = -126, b = -75$

48. (a) $a = 9.34, b = -5.65$

(b) $a = \frac{16}{5}, b = \frac{112}{75}$

In Exercises 49–52, find the midpoint of the interval.

49. (number line: $a = -1$, $b = 3$, marks -2 to 4) x

50. (number line: $a = -5$, $b = -\frac{3}{2}$, marks -6 to 0) x

51. (a) $[7, 21]$

(b) $[8.6, 11.4]$

52. (a) $[-6.85, 9.35]$

(b) $[-4.6, -1.3]$

A blue number indicates that a detailed solution can be found in the Study and Solutions Guide.

In Exercises 53–58, use absolute values to define the interval or pair of intervals on the real line.

53.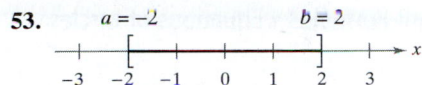

    $a = -2$ $b = 2$

54.

    $a = -3$ $b = 3$

55.

    $a = 0$ $b = 4$

56.

    $a = 20$ $b = 24$

57. (a) All numbers that are at most ten units from 12.

    (b) All numbers that are at least ten units from 12.

58. (a) y is at most two units from a.

    (b) y is less than δ units from c.

59. **Profit**   The revenue from selling x units of a product is

    $R = 115.95x$

    and the cost of producing x units is

    $C = 95x + 750.$

    To make a (positive) profit, R must be greater than C. For what values of x will the product return a profit?

60. **Fleet Costs**   A utility company has a fleet of vans. The annual operating cost of each van is estimated to be

    $C = 0.32m + 2300$

    where C is measured in dollars and m is measured in miles. The company wants the annual operating cost of each van to be less than \$10,000. To do this, m must be less than what value?

61. **Fair Coin**   To determine whether a coin is fair (has an equal probability of landing tails up or heads up), an experimenter tosses it 100 times and records the number of heads x. The coin is declared unfair if

    $$\left|\frac{x - 50}{5}\right| \geq 1.645.$$

    For what values of x will the coin be declared unfair?

62. **Daily Production**   The estimated daily production p at a refinery is

    $|p - 2,250,000| < 125,000$

    where p is measured in barrels of oil. Determine the high and low production levels.

In Exercises 63 and 64, determine which of the two real numbers is greater.

63. (a) π or $\frac{355}{113}$

    (b) π or $\frac{22}{7}$

64. (a) $\frac{224}{151}$ or $\frac{144}{97}$

    (b) $\frac{73}{81}$ or $\frac{6427}{7132}$

65. **Approximation—Powers of 10**   The speed of light is 2.998×10^8 meters per second. Which best estimates the distance in meters that light travels in a year?

    (a) 9.5×10^5 (b) 9.5×10^{15}

    (c) 9.5×10^{12} (d) 9.6×10^{16}

66. **Writing**   The accuracy of an approximation to a number is related to how many significant digits there are in the approximation. Write a definition for significant digits and illustrate the concept with examples.

True or False?   **In Exercises 67–72, determine whether the statement is true or false. If it is false, explain why or give an example that shows it is false.**

67. The reciprocal of a nonzero integer is an integer.

68. The reciprocal of a nonzero rational number is a rational number.

69. Each real number is either rational or irrational.

70. The absolute value of each real number is positive.

71. If $x < 0$, then $\sqrt{x^2} = -x$.

72. If a and b are any two distinct real numbers, then $a < b$ or $a > b$.

In Exercises 73–80, prove the property.

73. $|ab| = |a||b|$

74. $|a - b| = |b - a|$

    $\left[\textit{Hint: } (a - b) = (-1)(b - a)\right]$

75. $\left|\dfrac{a}{b}\right| = \dfrac{|a|}{|b|}, \quad b \neq 0$

76. $|a| = \sqrt{a^2}$

77. $|a^n| = |a|^n, \quad n = 1, 2, 3, \ldots$

78. $-|a| \leq a \leq |a|$

79. $|a| \leq k$, if and only if $-k \leq a \leq k, \quad k > 0.$

80. $k \leq |a|$ if and only if $k \leq a$ or $a \leq -k, \quad k > 0.$

81. Find an example for which $|a - b| > |a| - |b|$, and an example for which $|a - b| = |a| - |b|$. Then prove that $|a - b| \geq |a| - |b|$ for all a, b.

82. Show that the maximum of two numbers a and b is given by the formula

    $\max(a, b) = \frac{1}{2}\left(a + b + |a - b|\right).$

    Derive a similar formula for $\min(a, b)$.

SECTION *A.2* The Cartesian Plane

The Cartesian Plane • The Distance and Midpoint Formulas • Equations of Circles

The Cartesian Plane

An **ordered pair** (x, y) of real numbers has x as its *first* member and y as its *second* member. The model for representing ordered pairs is called the **rectangular coordinate system,** or the **Cartesian plane,** after the French mathematician René Descartes. It is developed by considering two real lines intersecting at right angles (see Figure A.14).

The horizontal real line is usually called the **x-axis,** and the vertical real line is usually called the **y-axis.** Their point of intersection is the **origin.** The two axes divide the plane into four **quadrants.**

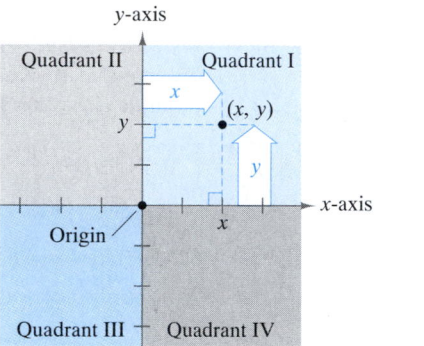

The Cartesian plane
Figure A.14

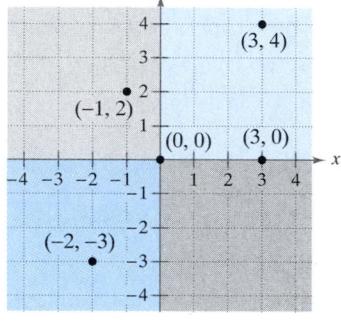

Points represented by ordered pairs
Figure A.15

Each point in the plane is identified by an ordered pair (x, y) of real numbers x and y, called **coordinates** of the point. The number x represents the directed distance from the y-axis to the point, and the number y represents the directed distance from the x-axis to the point (see Figure A.14). For the point (x, y), the first coordinate is the x-coordinate or **abscissa,** and the second coordinate is the y-coordinate or **ordinate.** For example, Figure A.15 shows the locations of the points $(-1, 2)$, $(3, 4)$, $(0, 0)$, $(3, 0)$, and $(-2, -3)$ in the Cartesian plane.

NOTE   The signs of the coordinates of a point determine the quadrant in which the point lies. For instance, if $x > 0$ and $y < 0$, then (x, y) lies in Quadrant IV.

Note that an ordered pair (a, b) is used to denote either a point in the plane *or* an open interval on the real line. This, however, should not be confusing—the nature of the problem should clarify whether a point in the plane or an open interval is being discussed.

The Distance and Midpoint Formulas

Recall from the Pythagorean Theorem that, in a right triangle, the hypotenuse c and sides a and b are related by $a^2 + b^2 = c^2$. Conversely, if $a^2 + b^2 = c^2$, the triangle is a right triangle (see Figure A.16).

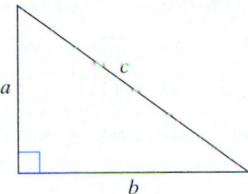

The Pythagorean Theorem:
$a^2 + b^2 = c^2$
Figure A.16

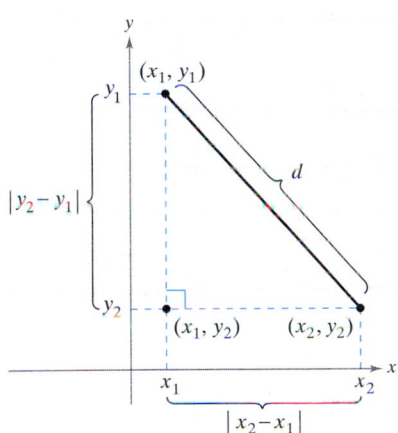

The distance between two points
Figure A.17

Suppose you want to determine the distance d between the two points (x_1, y_1) and (x_2, y_2) in the plane. If the points lie on a horizontal line, then $y_1 = y_2$ and the distance between the points is $|x_2 - x_1|$. If the points lie on a vertical line, then $x_1 = x_2$ and the distance between the points is $|y_2 - y_1|$. If the two points do not lie on a horizontal or vertical line, they can be used to form a right triangle, as shown in Figure A.17. The length of the vertical side of the triangle is $|y_2 - y_1|$, and the length of the horizontal side is $|x_2 - x_1|$. By the Pythagorean Theorem, it follows that

$$d^2 = |x_2 - x_1|^2 + |y_2 - y_1|^2$$
$$d = \sqrt{|x_2 - x_1|^2 + |y_2 - y_1|^2}.$$

Replacing $|x_2 - x_1|^2$ and $|y_2 - y_1|^2$ by the equivalent expressions $(x_2 - x_1)^2$ and $(y_2 - y_1)^2$ produces the following result.

Distance Formula

The distance d between the points (x_1, y_1) and (x_2, y_2) in the plane is given by

$$d = \sqrt{(x_2 - x_1)^2 + (y_2 - y_1)^2}.$$

EXAMPLE 1   Finding the Distance Between Two Points

Find the distance between the points $(-2, 1)$ and $(3, 4)$.

Solution

$$d = \sqrt{[3 - (-2)]^2 + (4 - 1)^2} \qquad \text{Distance Formula}$$
$$= \sqrt{(5)^2 + (3)^2}$$
$$= \sqrt{25 + 9}$$
$$= \sqrt{34}$$
$$\approx 5.83$$

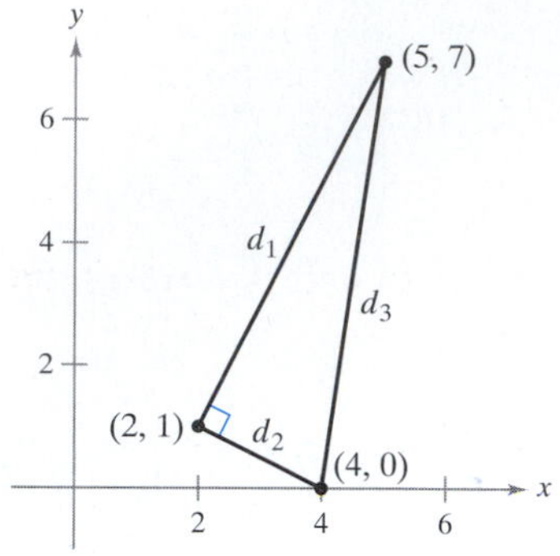

EXAMPLE 2   Verifying a Right Triangle

Verify that the points $(2, 1)$, $(4, 0)$, and $(5, 7)$ form the vertices of a right triangle.

Solution   Figure A.18 shows the triangle formed by the three points. The lengths of the three sides are as follows.

$$d_1 = \sqrt{(5 - 2)^2 + (7 - 1)^2} = \sqrt{9 + 36} = \sqrt{45}$$
$$d_2 = \sqrt{(4 - 2)^2 + (0 - 1)^2} = \sqrt{4 + 1} = \sqrt{5}$$
$$d_3 = \sqrt{(5 - 4)^2 + (7 - 0)^2} = \sqrt{1 + 49} = \sqrt{50}$$

Because

$$d_1{}^2 + d_2{}^2 = 45 + 5 = 50 \qquad \text{Sum of squares of sides}$$

and

$$d_3{}^2 = 50 \qquad \text{Square of hypotenuse}$$

you can apply the Pythagorean Theorem to conclude that the triangle is a right triangle.

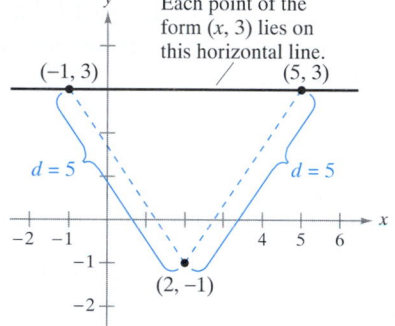

Verifying a right triangle
Figure A.18

EXAMPLE 3   Using the Distance Formula

Find x such that the distance between $(x, 3)$ and $(2, -1)$ is 5.

Solution   Using the Distance Formula, you can write the following.

$$5 = \sqrt{(x - 2)^2 + [3 - (-1)]^2} \qquad \text{Distance Formula}$$
$$25 = (x^2 - 4x + 4) + 16 \qquad \text{Square both sides.}$$
$$0 = x^2 - 4x - 5 \qquad \text{Write in standard form.}$$
$$0 = (x - 5)(x + 1) \qquad \text{Factor.}$$

Therefore, $x = 5$ or $x = -1$, and you can conclude that there are two solutions. That is, each of the points $(5, 3)$ and $(-1, 3)$ lies five units from the point $(2, -1)$, as shown in Figure A.19.

Given a distance, find a point.
Figure A.19

The coordinates of the **midpoint** of the line segment joining two points can be found by "averaging" the x-coordinates of the two points and "averaging" the y-coordinates of the two points. That is, the midpoint of the line segment joining the points (x_1, y_1) and (x_2, y_2) in the plane is

$$\left(\frac{x_1 + x_2}{2}, \frac{y_1 + y_2}{2} \right). \qquad \text{Midpoint Formula}$$

For instance, the midpoint of the line segment joining the points $(-5, -3)$ and $(9, 3)$ is

$$\left(\frac{-5 + 9}{2}, \frac{-3 + 3}{2} \right) = (2, 0)$$

as shown in Figure A.20.

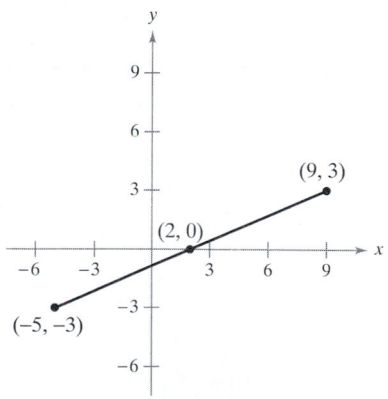

Midpoint of a line segment
Figure A.20

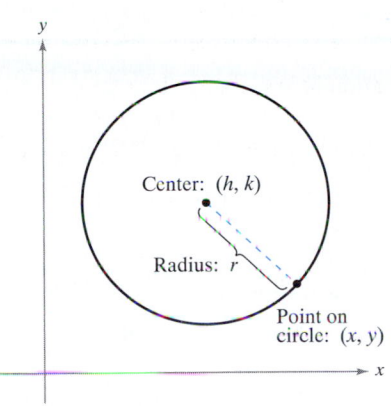

Definition of a circle
Figure A.21

FOR FURTHER INFORMATION
Can you recognize the graph of the equation

$$x^4 + 2x^2y^2 + y^4 -$$

$$5x^2 - 5y^2 = -4?$$

For the solution, see the article "Single Equations Can Draw Pictures" by Keith M. Kendig in the March 1991 issue of *The College Mathematics Journal.*

Equations of Circles

A **circle** can be defined as the set of all points in a plane that are equidistant from a fixed point. The fixed point is the **center** of the circle, and the distance between the center and a point on the circle is the **radius** (see Figure A.21).

You can use the Distance Formula to write an equation for the circle with center (h, k) and radius r. Let (x, y) be any point on the circle. Then the distance between (x, y) and the center (h, k) is given by

$$\sqrt{(x - h)^2 + (y - k)^2} = r.$$

By squaring both sides of this equation, you obtain the **standard form of the equation of a circle.**

Standard Form of the Equation of a Circle

The point (x, y) lies on the circle of radius r and center (h, k) if and only if

$$(x - h)^2 + (y - k)^2 = r^2.$$

The standard form of the equation of a circle with center at the origin, $(h, k) = (0, 0)$, is

$$x^2 + y^2 = r^2.$$

If $r = 1$, the circle is called the **unit circle.**

EXAMPLE 4   Finding the Equation of a Circle

The point $(3, 4)$ lies on a circle whose center is at $(-1, 2)$, as shown in Figure A.22. Find an equation for the circle.

Solution   The radius of the circle is the distance between $(-1, 2)$ and $(3, 4)$.

$$r = \sqrt{[3 - (-1)]^2 + (4 - 2)^2} = \sqrt{16 + 4} = \sqrt{20}$$

You can write the standard form of the equation of this circle as

$$[x - (-1)]^2 + (y - 2)^2 = \left(\sqrt{20}\right)^2$$

$$(x + 1)^2 + (y - 2)^2 = 20. \qquad \text{Standard form}$$

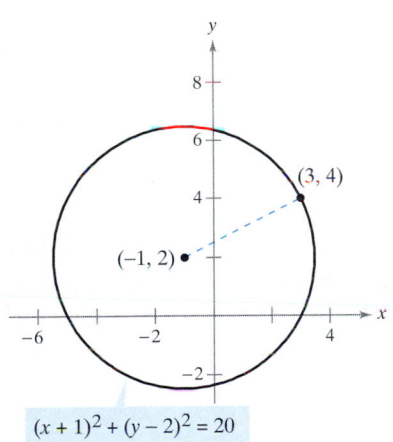

$(x + 1)^2 + (y - 2)^2 = 20$

Standard form of the equation of a circle
Figure A.22

By squaring and simplifying, the equation $(x - h)^2 + (y - k)^2 = r^2$ can be written in the following **general form of the equation of a circle.**

$$Ax^2 + Ay^2 + Dx + Ey + F = 0, \quad A \neq 0$$

To convert such an equation to the standard form

$$(x - h)^2 + (y - k)^2 = p$$

you can use a process called **completing the square.** If $p > 0$, the graph of the equation is a circle. If $p = 0$, the graph is the single point (h, k). If $p < 0$, the equation has no graph.

EXAMPLE 5   Completing the Square

Sketch the graph of the circle whose general equation is

$$4x^2 + 4y^2 + 20x - 16y + 37 = 0.$$

Solution   To complete the square, first divide by 4 so that the coefficients of x^2 and y^2 are both 1.

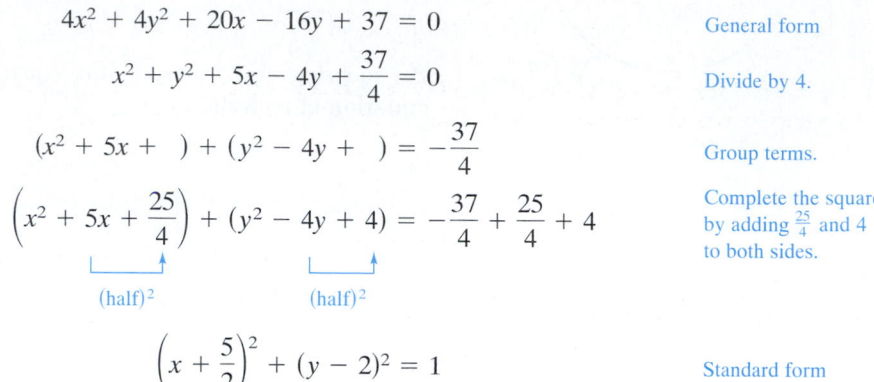

$$4x^2 + 4y^2 + 20x - 16y + 37 = 0 \qquad \text{General form}$$

$$x^2 + y^2 + 5x - 4y + \frac{37}{4} = 0 \qquad \text{Divide by 4.}$$

$$(x^2 + 5x +\ ) + (y^2 - 4y +\ ) = -\frac{37}{4} \qquad \text{Group terms.}$$

$$\left(x^2 + 5x + \frac{25}{4}\right) + (y^2 - 4y + 4) = -\frac{37}{4} + \frac{25}{4} + 4 \qquad \begin{matrix}\text{Complete the square} \\ \text{by adding } \frac{25}{4} \text{ and 4} \\ \text{to both sides.}\end{matrix}$$

$$\underbrace{\qquad}_{(\text{half})^2} \qquad \underbrace{\qquad}_{(\text{half})^2}$$

$$\left(x + \frac{5}{2}\right)^2 + (y - 2)^2 = 1 \qquad \text{Standard form}$$

Note that you complete the square by adding the square of half the coefficient of x *and* the square of half the coefficient of y to both sides of the equation. The circle is centered at $\left(-\frac{5}{2}, 2\right)$ and its radius is 1, as shown in Figure A.23.

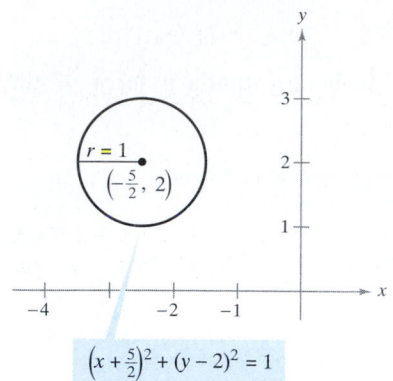

A circle with a radius of 1 and center at $\left(-\frac{5}{2}, 2\right)$

Figure A.23

We have now reviewed some fundamental concepts of *analytic geometry*. Because these concepts are in common use today, it is easy to overlook their revolutionary nature. At the time analytic geometry was being developed by Pierre de Fermat and René Descartes, the two major branches of mathematics—geometry and algebra—were largely independent of each other. Circles belonged to geometry and equations belonged to algebra. The coordination of the points on a circle and the solutions of an equation belongs to what is now called analytic geometry.

It is important to become skilled in analytic geometry so that you can move easily between geometry and algebra. For instance, in Example 4, you were given a geometric description of a circle and were asked to find an algebraic equation for the circle. Thus, you were moving from geometry to algebra. Similarly, in Example 5 you were given an algebraic equation and asked to sketch a geometric picture. In this case, you were moving from algebra to geometry. These two examples illustrate the two most common problems in analytic geometry.

1. Given a graph, find its equation.

2. Given an equation, find its graph.

In the next section, you will review other examples of these two types of problems.

EXERCISES FOR APPENDIX A.2

In Exercises 1–6, (a) plot the points, (b) find the distance between the points, and (c) find the midpoint of the line segment joining the points.

1. $(2, 1), (4, 5)$
2. $(-3, 2), (3, -2)$
3. $\left(\frac{1}{2}, 1\right), \left(-\frac{3}{2}, -5\right)$
4. $\left(\frac{2}{3}, -\frac{1}{3}\right), \left(\frac{5}{6}, 1\right)$
5. $\left(1, \sqrt{3}\right), (-1, 1)$
6. $(-2, 0), \left(0, \sqrt{2}\right)$

In Exercises 7–10, show that the points are the vertices of the polygon. (A rhombus is a quadrilateral whose sides are all of the same length.)

Vertices	Polygon
7. $(4, 0), (2, 1), (-1, -5)$	Right triangle
8. $(1, -3), (3, 2), (-2, 4)$	Isosceles triangle
9. $(0, 0), (1, 2), (2, 1), (3, 3)$	Rhombus
10. $(0, 1), (3, 7), (4, 4), (1, -2)$	Parallelogram

In Exercises 11–14, determine the quadrant(s) in which (x, y) is located so that the condition(s) is (are) satisfied.

11. $x = -2$ and $y > 0$
12. $y < -2$
13. $xy > 0$
14. $(x, -y)$ is in the second quadrant.

15. **Wal-Mart** The number y of Wal-Mart stores for each year x from 1987 through 1996 is given in the table. *(Source: Wal-Mart Annual Report for 1996)*

x	1987	1988	1989	1990	1991
y	980	1114	1259	1399	1568

x	1992	1993	1994	1995	1996
y	1714	1848	1950	1985	1995

Select reasonable scales on the coordinate axes and plot the points (x, y).

16. **Conjecture** Plot the points $(2, 1), (-3, 5)$, and $(7, -3)$ on a rectangular coordinate system. Then change the sign of the x-coordinate of each point and plot the three new points on the same rectangular coordinate system. What conjecture can you make about the location of a point when the sign of the x-coordinate is changed? Repeat the exercise for the case in which the sign of the y-coordinate is changed.

In Exercises 17–20, use the Distance Formula to determine whether the points lie on the same line.

17. $(0, -4), (2, 0), (3, 2)$
18. $(0, 4), (7, -6), (-5, 11)$
19. $(-2, 1), (-1, 0), (2, -2)$
20. $(-1, 1), (3, 3), (5, 5)$

In Exercises 21 and 22, find x such that the distance between the points is 5.

21. $(0, 0), (x, -4)$
22. $(2, -1), (x, 2)$

In Exercises 23 and 24, find y such that the distance between the points is 8.

23. $(0, 0), (3, y)$
24. $(5, 1), (5, y)$

25. Use the Midpoint Formula to find the three points that divide the line segment joining (x_1, y_1) and (x_2, y_2) into four equal parts.

26. Use the result of Exercise 25 to find the points that divide the line segment joining the given points into four equal parts.

   (a) $(1, -2), (4, -1)$
   (b) $(-2, -3), (0, 0)$

In Exercises 27–30, match the equation with its graph. [The graphs are labeled (a), (b), (c), and (d).]

(a)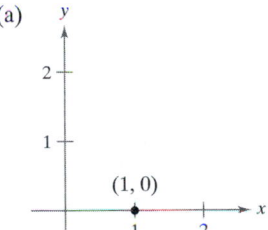

(b)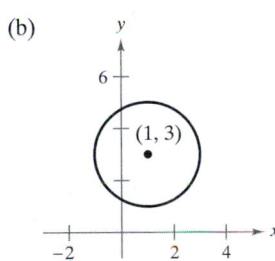

(c)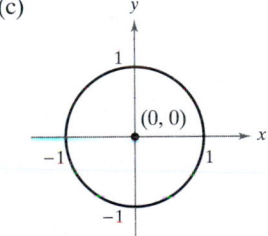

(d)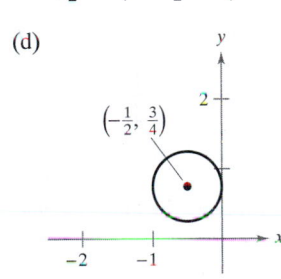

27. $x^2 + y^2 = 1$
28. $(x - 1)^2 + (y - 3)^2 = 4$
29. $(x - 1)^2 + y^2 = 0$
30. $\left(x + \frac{1}{2}\right)^2 + \left(y - \frac{3}{4}\right)^2 = \frac{1}{4}$

In Exercises 31–38, write the equation of the circle in general form.

31. Center: $(0, 0)$
    Radius: 3
32. Center: $(0, 0)$
    Radius: 5
33. Center: $(2, -1)$
    Radius: 4
34. Center: $(-4, 3)$
    Radius: $\frac{5}{8}$

35. Center: $(-1, 2)$

Point on circle: $(0, 0)$

36. Center: $(3, -2)$

Point on circle: $(-1, 1)$

37. Endpoints of diameter: $(2, 5), (4, -1)$

38. Endpoints of diameter: $(1, 1), (-1, -1)$

39. *Satellite Communication* Write an equation for the path of a communications satellite in a circular orbit 22,000 miles above the earth. (Assume that the radius of the earth is 4000 miles.)

40. *Building Design* A circular air duct of diameter D is fit firmly into the right-angle corner where a basement wall meets the floor (see figure). Find the diameter of the largest water pipe that can be run in the right-angle corner behind the air duct.

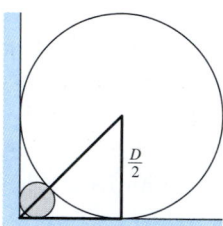

In Exercises 41–48, write the equation of the circle in standard form and sketch its graph.

41. $x^2 + y^2 - 2x + 6y + 6 = 0$

42. $x^2 + y^2 - 2x + 6y - 15 = 0$

43. $x^2 + y^2 - 2x + 6y + 10 = 0$

44. $3x^2 + 3y^2 - 6y - 1 = 0$

45. $2x^2 + 2y^2 - 2x - 2y - 3 = 0$

46. $4x^2 + 4y^2 - 4x + 2y - 1 = 0$

47. $16x^2 + 16y^2 + 16x + 40y - 7 = 0$

48. $x^2 + y^2 - 4x + 2y + 3 = 0$

In Exercises 49 and 50, use a graphing utility to graph the equation. (*Hint:* It may be necessary to solve the equation for y and graph the resulting two equations.)

49. $4x^2 + 4y^2 - 4x + 24y - 63 = 0$

50. $x^2 + y^2 - 8x - 6y - 11 = 0$

In Exercises 51 and 52, sketch the set of all points satisfying the inequality. Use a graphing utility to verify your result.

51. $x^2 + y^2 - 4x + 2y + 1 \leq 0$

52. $(x - 1)^2 + \left(y - \frac{1}{2}\right)^2 > 1$

---

53. Prove that

$$\left(\frac{2x_1 + x_2}{3}, \frac{2y_1 + y_2}{3}\right)$$

is one of the points of trisection of the line segment joining (x_1, y_1) and (x_2, y_2). Find the midpoint of the line segment joining

$$\left(\frac{2x_1 + x_2}{3}, \frac{2y_1 + y_2}{3}\right)$$

and (x_2, y_2) to find the second point of trisection.

54. Use the results of Exercise 53 to find the points of trisection of the line segment joining the following points.

(a) $(1, -2), (4, 1)$ (b) $(-2, -3), (0, 0)$

True or False? **In Exercises 55–58, determine whether the statement is true or false. If it is false, explain why or give an example that shows it is false.**

55. If $ab < 0$, the point (a, b) lies in either the second quadrant or the fourth quadrant.

56. The distance between the points $(a + b, a)$ and $(a - b, a)$ is $2b$.

57. If the distance between two points is zero, the two points must coincide.

58. If $ab = 0$, the point (a, b) lies on the x-axis or on the y-axis.

In Exercises 59–62, prove the statement.

59. The line segments joining the midpoints of the opposite sides of a quadrilateral bisect each other.

60. The perpendicular bisector of a chord of a circle passes through the center of the circle.

61. An angle inscribed in a semicircle is a right angle.

62. The midpoint of the line segment joining the points (x_1, y_1) and (x_2, y_2) is

$$\left(\frac{x_1 + x_2}{2}, \frac{y_1 + y_2}{2}\right).$$

The symbol ⚹ indicates an exercise in which you are instructed to use graphing technology or a symbolic computer algebra system. The solutions of other exercises may also be facilitated by use of appropriate technology.

Angles and Degree Measure • Radian Measure • The Trigonometric Functions •
Evaluating Trigonometric Functions • Solving Trigonometric Equations •
Graphs of Trigonometric Functions

Angles and Degree Measure

An **angle** has three parts: an **initial ray,** a **terminal ray,** and a **vertex** (the point of intersection of the two rays), as shown in Figure A.24. An angle is in **standard position** if its initial ray coincides with the positive x-axis and its vertex is at the origin. We assume that you are familiar with the degree measure of an angle.* It is common practice to use θ (the Greek lowercase *theta*) to represent both an angle and its measure. Angles between $0°$ and $90°$ are **acute,** and angles between $90°$ and $180°$ are **obtuse.**

Positive angles are measured *counterclockwise,* and negative angles are measured *clockwise.* For instance, Figure A.25 shows an angle whose measure is $-45°$. You cannot assign a measure to an angle by simply knowing where its initial and terminal rays are located. To measure an angle, you must also know how the terminal ray was revolved. For example, Figure A.25 shows that the angle measuring $-45°$ has the same terminal ray as the angle measuring $315°$. Such angles are **coterminal.** In general, if θ is any angle, then

$$\theta + n(360), \qquad n \text{ is a nonzero integer}$$

is coterminal with θ.

An angle that is larger than $360°$ is one whose terminal ray has been revolved more than one full revolution counterclockwise, as shown in Figure A.26. You can form an angle whose measure is less than $-360°$ by revolving a terminal ray more than one full revolution clockwise.

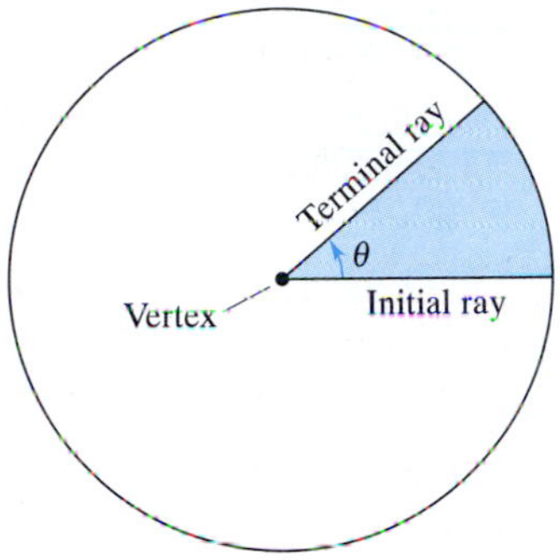

Standard position of an angle
Figure A.24

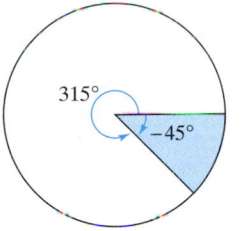

Coterminal angles
Figure A.25

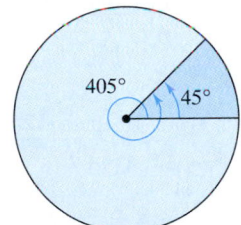

Coterminal angles
Figure A.26

NOTE   It is common to use the symbol θ to refer to both an *angle* and its *measure.* For instance, in Figure A.26, you can write the measure of the smaller angle as $\theta = 45°$.

For a more complete review of trigonometry, see Precalculus, 4th edition, by Larson and Hostetler (Boston, Massachusetts: Houghton Mifflin, 1997).

Radian Measure

To assign a radian measure to an angle θ, consider θ to be a central angle of a circle of radius 1, as shown in Figure A.27. The **radian measure** of θ is then defined to be the length of the arc of the sector. Because the circumference of a circle is $2\pi r$, the circumference of a **unit circle** (of radius 1) is 2π. This implies that the radian measure of an angle measuring $360°$ is 2π. In other words, $360° = 2\pi$ radians.

Using radian measure for θ, the length s of a circular arc of radius r is $s = r\theta$, as shown in Figure A.28.

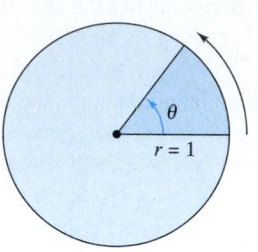

The arc length of the sector is the radian measure of θ.

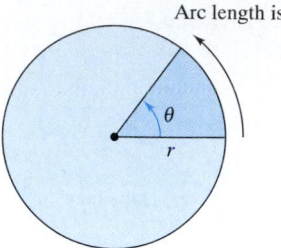

Arc length is $s = r\theta$.

Unit circle
Figure A.27

Circle of radius r
Figure A.28

You should know the conversions of the common angles shown in Figure A.29. For other angles, use the fact that $180°$ is equal to π radians.

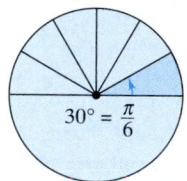
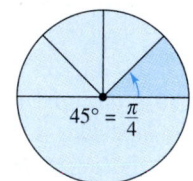
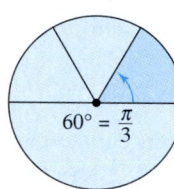
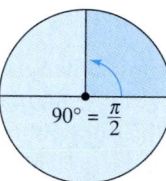
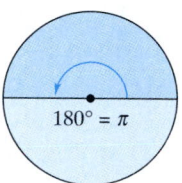
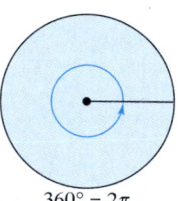

Radian and degree measure for several common angles
Figure A.29

EXAMPLE 1 Conversions Between Degrees and Radians

a. $40° = (40 \text{ deg})\left(\dfrac{\pi \text{ rad}}{180 \text{ deg}}\right) = \dfrac{2\pi}{9}$ radians

b. $-270° = (-270 \text{ deg})\left(\dfrac{\pi \text{ rad}}{180 \text{ deg}}\right) = -\dfrac{3\pi}{2}$ radians

c. $-\dfrac{\pi}{2}$ radians $= \left(-\dfrac{\pi}{2} \text{ rad}\right)\left(\dfrac{180 \text{ deg}}{\pi \text{ rad}}\right) = -90°$

d. $\dfrac{9\pi}{2}$ radians $= \left(\dfrac{9\pi}{2} \text{ rad}\right)\left(\dfrac{180 \text{ deg}}{\pi \text{ rad}}\right) = 810°$

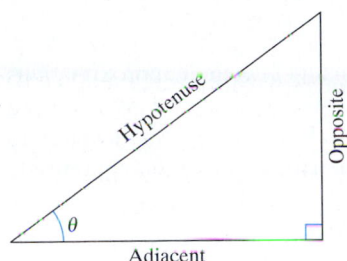

Sides of a right triangle
Figure A.30

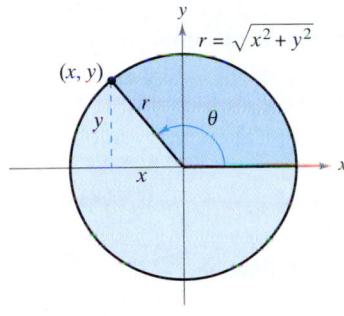

An angle in standard position
Figure A.31

The Trigonometric Functions

There are two common approaches to the study of trigonometry. In one, the trigonometric functions are defined as ratios of two sides of a right triangle. In the other, these functions are defined in terms of a point on the terminal side of an angle in standard position. **We define the six trigonometric functions, sine, cosine, tangent, cotangent, secant,** and **cosecant** (abbreviated as sin, cos, etc.), from both viewpoints.

Definition of the Six Trigonometric Functions

Right triangle definitions, where $0 < \theta < \dfrac{\pi}{2}$ *(see Figure A.30).*

$$\sin\theta = \frac{\text{opposite}}{\text{hypotenuse}} \qquad \cos\theta = \frac{\text{adjacent}}{\text{hypotenuse}} \qquad \tan\theta = \frac{\text{opposite}}{\text{adjacent}}$$

$$\csc\theta = \frac{\text{hypotenuse}}{\text{opposite}} \qquad \sec\theta = \frac{\text{hypotenuse}}{\text{adjacent}} \qquad \cot\theta = \frac{\text{adjacent}}{\text{opposite}}$$

Circular function definitions, where θ *is any angle (see Figure A.31).*

$$\sin\theta = \frac{y}{r} \qquad \cos\theta = \frac{x}{r} \qquad \tan\theta = \frac{y}{x}$$

$$\csc\theta = \frac{r}{y} \qquad \sec\theta = \frac{r}{x} \qquad \cot\theta = \frac{x}{y}$$

The following trigonometric identities are direct consequences of the definitions. (ϕ is the Greek letter *phi*.)

Trigonometric Identities   [*Note that* $\sin^2\theta$ *is used to represent* $(\sin\theta)^2$.]

Pythagorean Identities:

$$\sin^2\theta + \cos^2\theta = 1$$
$$\tan^2\theta + 1 = \sec^2\theta$$
$$\cot^2\theta + 1 = \csc^2\theta$$

Sum or Difference of Two Angles:

$$\sin(\theta \pm \phi) = \sin\theta\cos\phi \pm \cos\theta\sin\phi$$

$$\cos(\theta \pm \phi) = \cos\theta\cos\phi \mp \sin\theta\sin\phi$$

$$\tan(\theta \pm \phi) = \frac{\tan\theta \pm \tan\phi}{1 \mp \tan\theta\tan\phi}$$

Law of Cosines:

$$a^2 = b^2 + c^2 - 2bc\cos A$$

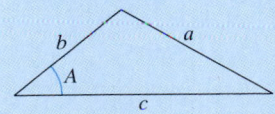

Reduction Formulas:

$$\sin(-\theta) = -\sin\theta \qquad\qquad \sin\theta = -\sin(\theta - \pi)$$
$$\cos(-\theta) = \cos\theta \qquad\qquad \cos\theta = -\cos(\theta - \pi)$$
$$\tan(-\theta) = -\tan\theta \qquad\qquad \tan\theta = \tan(\theta - \pi)$$

Half–Angle Formulas:

$$\sin^2\theta = \frac{1}{2}(1 - \cos 2\theta)$$

$$\cos^2\theta = \frac{1}{2}(1 + \cos 2\theta)$$

Double–Angle Formulas:

$$\sin 2\theta = 2\sin\theta\cos\theta$$

$$\cos 2\theta = 2\cos^2\theta - 1$$
$$= 1 - 2\sin^2\theta$$
$$= \cos^2\theta - \sin^2\theta$$

Reciprocal Identities:

$$\csc\theta = \frac{1}{\sin\theta}$$

$$\sec\theta = \frac{1}{\cos\theta}$$

$$\cot\theta = \frac{1}{\tan\theta}$$

Quotient Identities:

$$\tan\theta = \frac{\sin\theta}{\cos\theta}$$

$$\cot\theta = \frac{\cos\theta}{\sin\theta}$$

Evaluating Trigonometric Functions

There are two ways to evaluate trigonometric functions: (1) decimal approximations with a calculator (or a table of trigonometric values) and (2) exact evaluations using trigonometric identities and formulas from geometry. When using a calculator to evaluate a trigonometric function, remember to set the calculator to the appropriate mode—degree mode or radian mode.

EXAMPLE 2   Exact Evaluation of Trigonometric Functions

Evaluate the sine, cosine, and tangent of $\dfrac{\pi}{3}$.

Solution   Begin by drawing the angle $\theta = \pi/3$ in standard position, as shown in Figure A.32. Then, because $60° = \pi/3$ radians, you can draw an equilateral triangle with sides of length 1 and θ as one of its angles. Because the altitude of this triangle bisects its base, you know that $x = \frac{1}{2}$. Using the Pythagorean Theorem, you obtain

$$ y = \sqrt{r^2 - x^2} = \sqrt{1 - \left(\frac{1}{2}\right)^2} = \sqrt{\frac{3}{4}} = \frac{\sqrt{3}}{2}. $$

Now, knowing the values of x, y, and r, you can write the following.

$$ \sin\frac{\pi}{3} = \frac{y}{r} = \frac{\sqrt{3}/2}{1} = \frac{\sqrt{3}}{2} $$

$$ \cos\frac{\pi}{3} = \frac{x}{r} = \frac{1/2}{1} = \frac{1}{2} $$

$$ \tan\frac{\pi}{3} = \frac{y}{x} = \frac{\sqrt{3}/2}{1/2} = \sqrt{3} $$

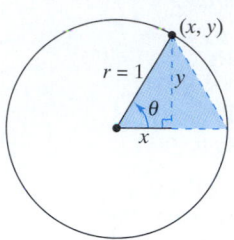

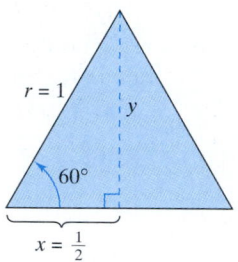

The angle $\pi/3$ in standard position
Figure A.32

NOTE   All angles in this text are measured in radians unless stated otherwise. For example, when we write $\sin 3$, we mean the sine of 3 radians, and when we write $\sin 3°$, we mean the sine of 3 degrees.

The degree and radian measures of several common angles are given in the table below, along with the corresponding values of the sine, cosine, and tangent (see Figure A.33).

Common First Quadrant Angles

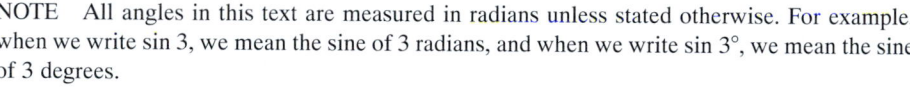

Degrees	0	30°	45°	60°	90°
Radians	0	$\dfrac{\pi}{6}$	$\dfrac{\pi}{4}$	$\dfrac{\pi}{3}$	$\dfrac{\pi}{2}$
sin θ	0	$\dfrac{1}{2}$	$\dfrac{\sqrt{2}}{2}$	$\dfrac{\sqrt{3}}{2}$	1
cos θ	1	$\dfrac{\sqrt{3}}{2}$	$\dfrac{\sqrt{2}}{2}$	$\dfrac{1}{2}$	0
tan θ	0	$\dfrac{\sqrt{3}}{3}$	1	$\sqrt{3}$	Undefined

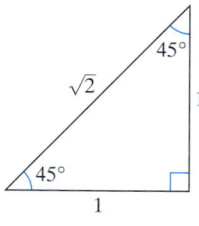

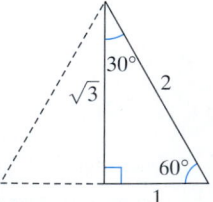

Common angles
Figure A.33

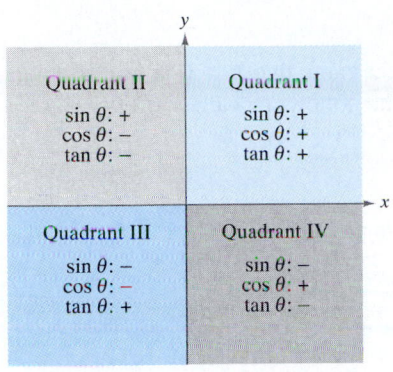

Quadrant signs for trigonometric functions
Figure A.34

The quadrant signs for the sine, cosine, and tangent functions are shown in Figure A.34. To extend the use of the table on the preceding page to angles in quadrants other than the first quadrant, you can use the concept of a **reference angle** (see Figure A.35), with the appropriate quadrant sign. For instance, the reference angle for $3\pi/4$ is $\pi/4$, and because the sine is positive in the second quadrant, you can write

$$\sin\frac{3\pi}{4} = +\sin\frac{\pi}{4} = \frac{\sqrt{2}}{2}.$$

Similarly, because the reference angle for $330°$ is $30°$, and the tangent is negative in the fourth quadrant, you can write

$$\tan 330° = -\tan 30° = -\frac{\sqrt{3}}{3}.$$

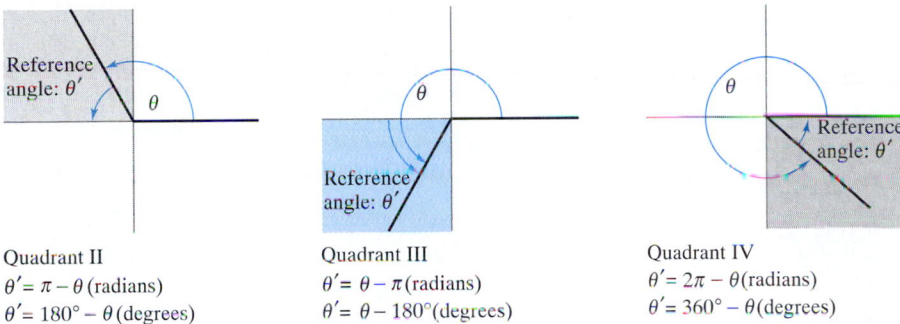

Quadrant II
$\theta' = \pi - \theta$ (radians)
$\theta' = 180° - \theta$ (degrees)

Quadrant III
$\theta' = \theta - \pi$ (radians)
$\theta' = \theta - 180°$ (degrees)

Quadrant IV
$\theta' = 2\pi - \theta$ (radians)
$\theta' = 360° - \theta$ (degrees)

Figure A.35

EXAMPLE 3   Trigonometric Identities and Calculators

Evaluate the trigonometric expression.

a. $\sin\left(-\dfrac{\pi}{3}\right)$ **b.** $\sec 60°$ **c.** $\cos(1.2)$

Solution

a. Using the reduction formula $\sin(-\theta) = -\sin\theta$, you can write

$$\sin\left(-\frac{\pi}{3}\right) = -\sin\frac{\pi}{3} = -\frac{\sqrt{3}}{2}.$$

b. Using the reciprocal identity $\sec\theta = 1/\cos\theta$, you can write

$$\sec 60° = \frac{1}{\cos 60°} = \frac{1}{1/2} = 2.$$

c. Using a calculator, you can obtain

$$\cos(1.2) \approx 0.3624.$$

Remember that 1.2 is given in *radian* measure. Consequently, your calculator must be set in radian mode.

Solving Trigonometric Equations

How would you solve the equation $\sin \theta = 0$? You know that $\theta = 0$ is one solution, but this is not the only solution. Any one of the following values of θ is also a solution.

$$\ldots, -3\pi, -2\pi, -\pi, 0, \pi, 2\pi, 3\pi, \ldots$$

You can write this infinite solution set as $\{n\pi : n \text{ is an integer}\}$.

EXAMPLE 4   Solving a Trigonometric Equation

Solve the equation

$$\sin \theta = -\frac{\sqrt{3}}{2}.$$

Solution   To solve the equation, you should consider that the sine is negative in Quadrants III and IV and that

$$\sin \frac{\pi}{3} = \frac{\sqrt{3}}{2}.$$

Thus, you are seeking values of θ in the third and fourth quadrants that have a reference angle of $\pi/3$. In the interval $[0, 2\pi]$, the two angles fitting these criteria are

$$\theta = \pi + \frac{\pi}{3} = \frac{4\pi}{3} \quad \text{and} \quad \theta = 2\pi - \frac{\pi}{3} = \frac{5\pi}{3}.$$

By adding integer multiples of 2π to each of these solutions, you obtain the following general solution.

$$\theta = \frac{4\pi}{3} + 2n\pi \quad \text{or} \quad \theta = \frac{5\pi}{3} + 2n\pi, \quad \text{where } n \text{ is an integer.}$$

(See Figure A.36.)

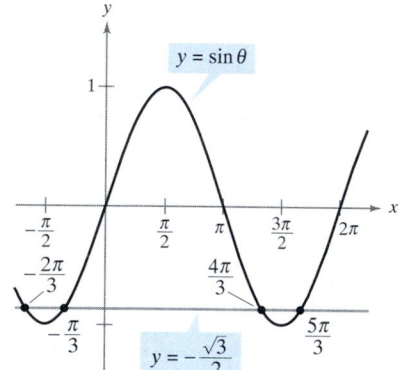

Solution points of $\sin \theta = -\dfrac{\sqrt{3}}{2}$

Figure A.36

EXAMPLE 5   Solving a Trigonometric Equation

Solve $\cos 2\theta = 2 - 3 \sin \theta$, where $0 \le \theta \le 2\pi$.

Solution   Using the double-angle identity $\cos 2\theta = 1 - 2\sin^2 \theta$, you can rewrite the equation as follows.

$$\cos 2\theta = 2 - 3 \sin \theta \qquad \text{Given equation}$$
$$1 - 2\sin^2 \theta = 2 - 3 \sin \theta \qquad \text{Trigonometric identity}$$
$$0 = 2\sin^2 \theta - 3 \sin \theta + 1 \qquad \text{Quadratic form}$$
$$0 = (2\sin \theta - 1)(\sin \theta - 1) \qquad \text{Factor.}$$

If $2 \sin \theta - 1 = 0$, then $\sin \theta = 1/2$ and $\theta = \pi/6$ or $\theta = 5\pi/6$. If $\sin \theta - 1 = 0$, then $\sin \theta = 1$ and $\theta = \pi/2$. Thus, for $0 \le \theta \le 2\pi$, there are three solutions.

$$\theta = \frac{\pi}{6}, \ \frac{5\pi}{6}, \ \text{or} \ \frac{\pi}{2}$$

Graphs of Trigonometric Functions

A function f is **periodic** if there exists a nonzero number p such that $f(x + p) = f(x)$ for all x in the domain of f. The smallest such positive value of p (if it exists) is the **period** of f. The sine, cosine, secant, and cosecant functions each have a period of 2π, and the other two trigonometric functions have a period of π, as shown in Figure A.37.

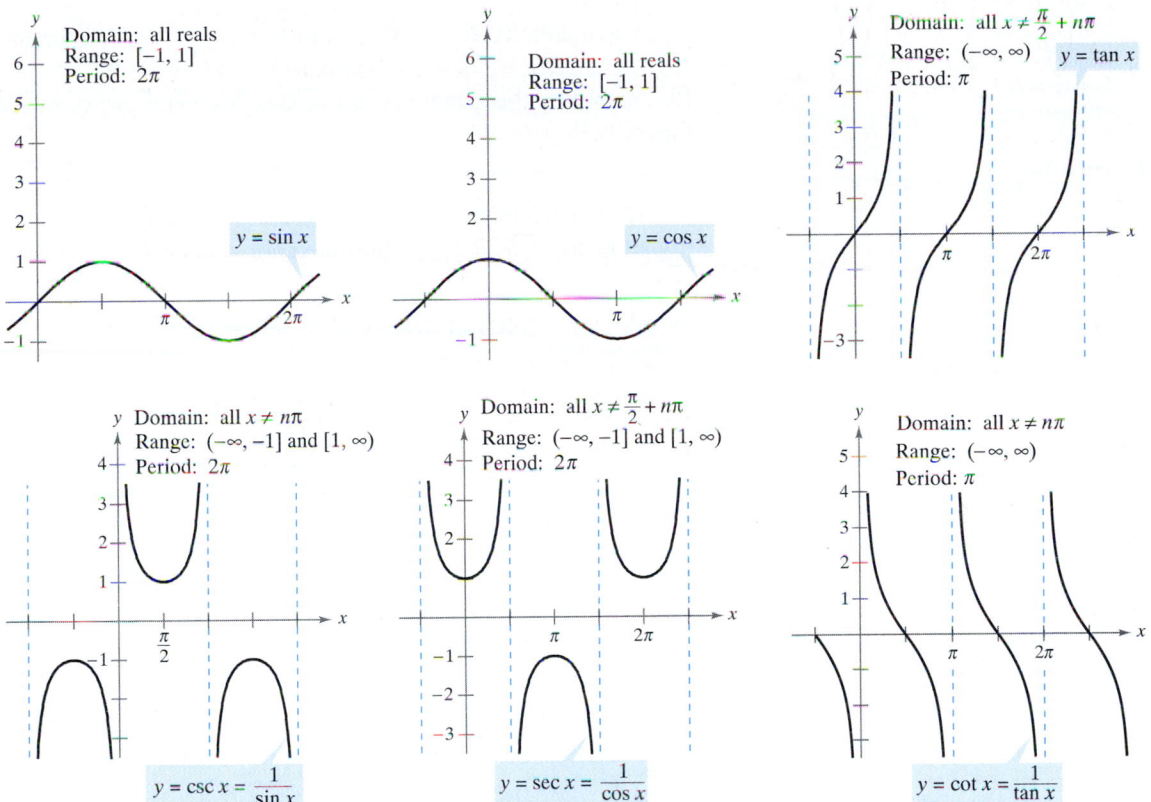

The graphs of the six trigonometric functions
Figure A.37

Note in Figure A.37 that the maximum value of $\sin x$ and $\cos x$ is 1 and the minimum value is -1. The graphs of the functions $y = a \sin bx$ and $y = a \cos bx$ oscillate between $-a$ and a, and hence have an **amplitude** of $|a|$. Furthermore, because $bx = 0$ when $x = 0$ and $bx = 2\pi$ when $x = 2\pi/b$, it follows that the functions $y = a \sin bx$ and $y = a \cos bx$ each have a period of $2\pi/|b|$. The table below summarizes the amplitudes and periods for some types of trigonometric functions.

Function	Period	Amplitude
$y = a \sin bx$ or $y = a \cos bx$	$\dfrac{2\pi}{|b|}$	$|a|$
$y = a \tan bx$ or $y = a \cot bx$	$\dfrac{\pi}{|b|}$	Not applicable
$y = a \sec bx$ or $y = a \csc bx$	$\dfrac{2\pi}{|b|}$	Not applicable

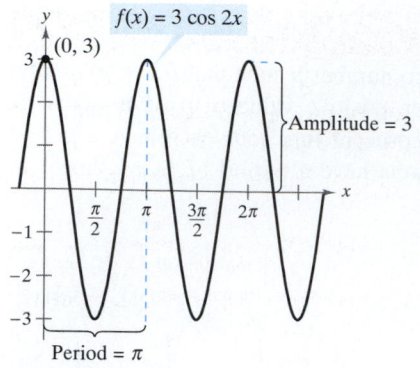

Figure A.38

EXAMPLE 6 Sketching the Graph of a Trigonometric Function

Sketch the graph of $f(x) = 3 \cos 2x$.

Solution The graph of $f(x) = 3 \cos 2x$ has an amplitude of 3 and a period of $2\pi/2 = \pi$. Using the basic shape of the graph of the cosine function, sketch one period of the function on the interval $[0, \pi]$, using the following pattern.

$$\text{Maximum: } (0, 3) \qquad \text{Minimum: } \left(\frac{\pi}{2}, -3\right) \qquad \text{Maximum: } [\pi, 3]$$

By continuing this pattern, you can sketch several cycles of the graph, as shown in Figure A.38.

Horizontal shifts, vertical shifts, and reflections can be applied to the graphs of trigonometric functions, as illustrated in Example 7.

EXAMPLE 7 Shifts of Graphs of Trigonometric Functions

Sketch the graphs of the following functions.

a. $f(x) = \sin\left(x + \frac{\pi}{2}\right)$ **b.** $f(x) = 2 + \sin x$ **c.** $f(x) = 2 + \sin\left(x - \frac{\pi}{4}\right)$

Solution

a. To sketch the graph of $f(x) = \sin(x + \pi/2)$, shift the graph of $y = \sin x$ to the left $\pi/2$ units, as shown in Figure A.39(a).

b. To sketch the graph of $f(x) = 2 + \sin x$, shift the graph of $y = \sin x$ up two units, as shown in Figure A.39(b).

c. To sketch the graph of $f(x) = 2 + \sin(x - \pi/4)$, shift the graph of $y = \sin x$ up two units and to the right $\pi/4$ units, as shown in Figure A.39(c).

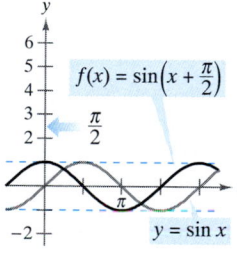
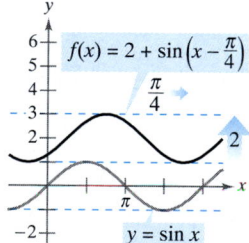

(a) Horizontal shift to the left   (b) Vertical shift upward   (c) Horizontal and vertical shifts

Transformations of the graph of $y = \sin x$

Figure A.39

EXERCISES FOR APPENDIX A.3

In Exercises 1 and 2, determine two coterminal angles (one positive and one negative) for each given angle. Express your answers in degrees.

1. (a) 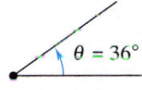   (b)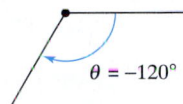

2. (a) 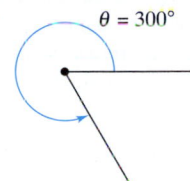   (b)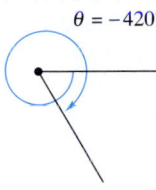

In Exercises 3 and 4, determine two coterminal angles (one positive and one negative) for each given angle. Express your answers in radians.

3. (a) 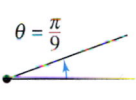   (b)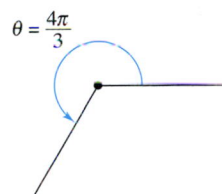

4. (a) 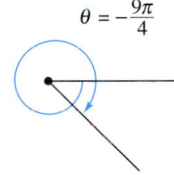   (b)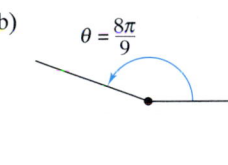

In Exercises 5 and 6, express the angles in radian measure as multiples of π and as decimals accurate to three decimal places.

5. (a) $30°$ (b) $150°$ (c) $315°$ (d) $120°$

6. (a) $-20°$ (b) $-240°$ (c) $-270°$ (d) $144°$

In Exercises 7 and 8, express the angles in degree measure.

7. (a) $\dfrac{3\pi}{2}$ (b) $\dfrac{7\pi}{6}$ (c) $-\dfrac{7\pi}{12}$ (d) -2.367

8. (a) $\dfrac{7\pi}{3}$ (b) $-\dfrac{11\pi}{30}$ (c) $\dfrac{11\pi}{6}$ (d) 0.438

9. Let r represent the radius of a circle, θ the central angle (measured in radians), and s the length of the arc subtended by the angle. Use the relationship $s = r\theta$ to complete the table.

r	8 ft	15 in.	85 cm		
s	12 ft			96 in.	8642 mi
θ		1.6	$\dfrac{3\pi}{4}$	4	$\dfrac{2\pi}{3}$

10. *Angular Speed* A car is moving at the rate of 50 miles per hour, and the diameter of its wheels is 2.5 feet.

  (a) Find the number of revolutions per minute that the wheels are rotating.

  (b) Find the angular speed of the wheels in radians per minute.

In Exercises 11 and 12, determine all six trigonometric functions for the angle θ.

11. (a) 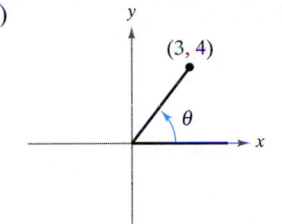   (b)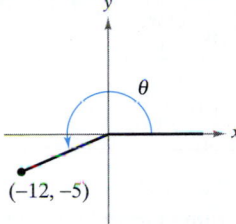

12. (a) 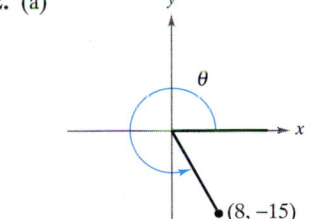   (b)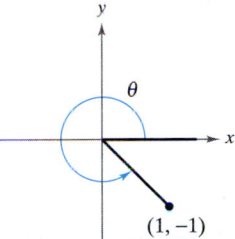

In Exercises 13 and 14, determine the quadrant in which θ lies.

13. (a) $\sin \theta < 0$ and $\cos \theta < 0$

  (b) $\sec \theta > 0$ and $\cot \theta < 0$

14. (a) $\sin \theta > 0$ and $\cos \theta < 0$

  (b) $\csc \theta < 0$ and $\tan \theta > 0$

In Exercises 15–18, evaluate the trigonometric function.

15. $\sin \theta = \frac{1}{2}$

  $\cos \theta = $ ▨

16. $\sin \theta = \frac{1}{3}$

  $\tan \theta = $ ▨

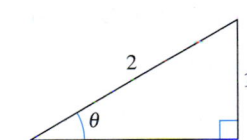   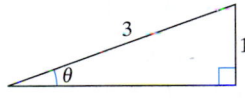

17. $\cos \theta = \frac{4}{5}$

  $\cot \theta = $ ▨

18. $\sec \theta = \frac{13}{5}$

  $\csc \theta = $ ▨

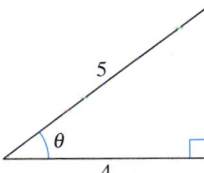   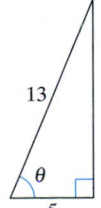

In Exercises 19–22, evaluate the sine, cosine, and tangent of each angle *without* using a calculator.

19. (a) 60°
    (b) 120°
    (c) $\dfrac{\pi}{4}$
    (d) $\dfrac{5\pi}{4}$

20. (a) −30°
    (b) 150°
    (c) $-\dfrac{\pi}{6}$
    (d) $\dfrac{\pi}{2}$

21. (a) 225°
    (b) −225°
    (c) $\dfrac{5\pi}{3}$
    (d) $\dfrac{11\pi}{6}$

22. (a) 750°
    (b) 510°
    (c) $\dfrac{10\pi}{3}$
    (d) $\dfrac{17\pi}{3}$

In Exercises 23–26, use a calculator to evaluate the trigonometric functions to four significant digits.

23. (a) sin 10°
    (b) csc 10°

24. (a) sec 225°
    (b) sec 135°

25. (a) $\tan \dfrac{\pi}{9}$
    (b) $\tan \dfrac{10\pi}{9}$

26. (a) cot(1.35)
    (b) tan(1.35)

In Exercises 27–30, find two solutions of each equation. Express the results in radians $(0 \le \theta < 2\pi)$. Do not use a calculator.

27. (a) $\cos \theta = \dfrac{\sqrt{2}}{2}$
    (b) $\cos \theta = -\dfrac{\sqrt{2}}{2}$

28. (a) $\sec \theta = 2$
    (b) $\sec \theta = -2$

29. (a) $\tan \theta = 1$
    (b) $\cot \theta = -\sqrt{3}$

30. (a) $\sin \theta = \dfrac{\sqrt{3}}{2}$
    (b) $\sin \theta = -\dfrac{\sqrt{3}}{2}$

In Exercises 31–38, solve the equation for θ $(0 \le \theta < 2\pi)$.

31. $2 \sin^2 \theta = 1$

32. $\tan^2 \theta = 3$

33. $\tan^2 \theta - \tan \theta = 0$

34. $2 \cos^2 \theta - \cos \theta = 1$

35. $\sec \theta \csc \theta = 2 \csc \theta$

36. $\sin \theta = \cos \theta$

37. $\cos^2 \theta + \sin \theta = 1$

38. $\cos \dfrac{\theta}{2} - \cos \theta = 1$

39. ***Airplane Ascent*** An airplane leaves the runway climbing at 18° with a speed of 275 feet per second (see figure). Find the altitude a of the plane after 1 minute.

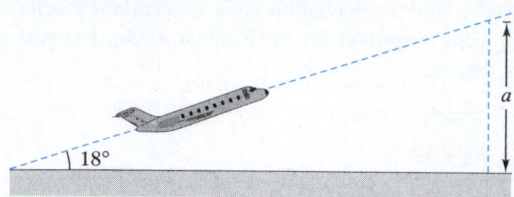

40. ***Height of a Mountain*** In traveling across flat land, you notice a mountain directly in front of you. Its angle of elevation (to the peak) is 3.5°. After you drive 13 miles closer to the mountain, the angle of elevation is 9°. Approximate the height of the mountain.

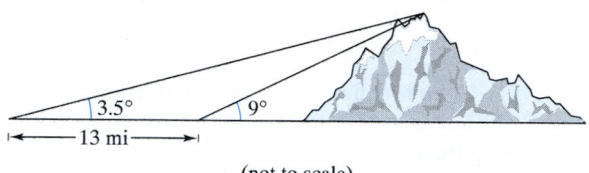

(not to scale)

In Exercises 41–44, determine the period and amplitude of each function.

41. (a) $y = 2 \sin 2x$
    (b) $y = \tfrac{1}{2} \sin \pi x$

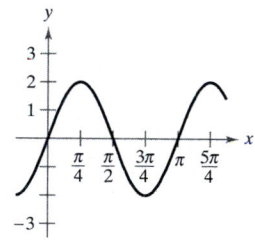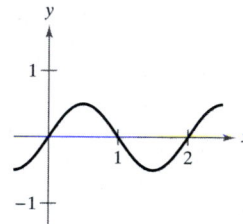

42. (a) $y = \dfrac{3}{2} \cos \dfrac{x}{2}$
    (b) $y = -2 \sin \dfrac{x}{3}$

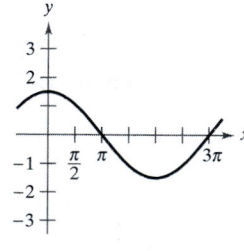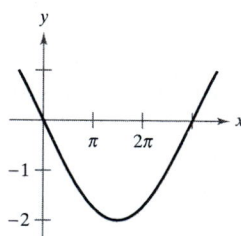

43. $y = 3 \sin 4\pi x$

44. $y = \dfrac{2}{3} \cos \dfrac{\pi x}{10}$

In Exercises 45–48, find the period of the function.

45. $y = 5 \tan 2x$

46. $y = 7 \tan 2\pi x$

47. $y = \sec 5x$

48. $y = \csc 4x$

 Writing   **In Exercises 49 and 50, use a graphing utility to graph each function f on the same set of coordinate axes for $c = -2$, $c = -1$, $c = 1$, and $c = 2$. Give a written description of the change in the graph caused by changing c.**

49. (a) $f(x) = c \sin x$

    (b) $f(x) = \cos(cx)$

    (c) $f(x) = \cos(\pi x - c)$

50. (a) $f(x) = \sin x + c$

    (b) $f(x) = -\sin(2\pi x - c)$

    (c) $f(x) = c \cos x$

In Exercises 51–62, sketch the graph of the function.

51. $y = \sin \dfrac{x}{2}$

52. $y = 2 \cos 2x$

53. $y = -\sin \dfrac{2\pi x}{3}$

54. $y = 2 \tan x$

55. $y = \csc \dfrac{x}{2}$

56. $y = \tan 2x$

57. $y = 2 \sec 2x$

58. $y = \csc 2\pi x$

59. $y = \sin(x + \pi)$

60. $y = \cos\left(x - \dfrac{\pi}{3}\right)$

61. $y = 1 + \cos\left(x - \dfrac{\pi}{2}\right)$

62. $y = 1 + \sin\left(x + \dfrac{\pi}{2}\right)$

Graphical Reasoning   **In Exercises 63 and 64, find a, b, and c such that the graph of the function matches the graph in the figure.**

63. $y = a \cos(bx - c)$

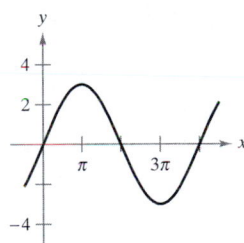

64. $y = a \sin(bx - c)$

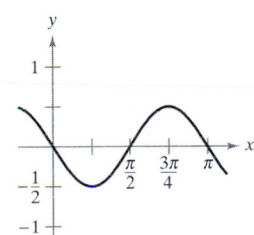

65. *Think About It*   Sketch the graphs of $f(x) = \sin x$, $g(x) = |\sin x|$, and $h(x) = \sin(|x|)$. In general, how are the graphs of $|f(x)|$ and $f(|x|)$ related to the graph of f?

66. *Think About It*   The model for the height h of a Ferris wheel car is

$$h = 51 + 50 \sin 8\pi t$$

where t is measured in minutes. (The Ferris wheel has a radius of 50 feet.) This model yields a height of 51 feet when $t = 0$. Alter the model so that the height of the car is 1 foot when $t = 0$.

 67. *Sales*   Sales S, in thousands of units, of a seasonal product are modeled by

$$S = 58.3 + 32.5 \cos \frac{\pi t}{6}$$

where t is the time in months (with $t = 1$ corresponding to January and $t = 12$ corresponding to December). Use a graphing utility to graph the model for S and determine the months when sales exceed 75,000 units.

68. *Investigation*   Two trigonometric functions f and g have a period of 2, and their graphs intersect at $x = 5.35$.

    (a) Give one smaller and one larger positive value of x where the functions have the same value.

    (b) Determine one negative value of x where the graphs intersect.

    (c) Is it true that $f(13.35) = g(-4.65)$? Give a reason for your answer.

 Pattern Recognition   **In Exercises 69 and 70, use a graphing utility to compare the graph of f with the given graph. Try to improve the approximation by adding a term to $f(x)$. Use a graphing utility to verify that your new approximation is better than the original. Can you find other terms to add to make the approximation even better? What is the pattern? (In Exercise 69, sine terms can be used to improve the approximation and in Exercise 70, cosine terms can be used.)**

69. $f(x) = \dfrac{4}{\pi}\left(\sin \pi x + \dfrac{1}{3}\sin 3\pi x\right)$

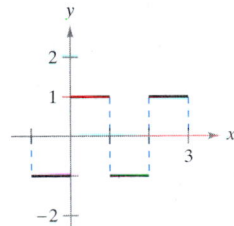

70. $f(x) = \dfrac{1}{2} - \dfrac{4}{\pi^2}\left(\cos \pi x + \dfrac{1}{9}\cos 3\pi x\right)$

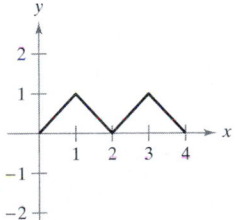

APPENDIX B
Proofs of Selected Theorems

THEOREM 1.2   Properties of Limits (Properties 2, 3, 4, and 5) (page 56)

Let b and c be real numbers, let n be a positive integer, and let f and g be functions with the following limits.

$$\lim_{x \to c} f(x) = L \qquad \text{and} \qquad \lim_{x \to c} g(x) = K$$

2. Sum or difference: $\qquad \lim_{x \to c} [f(x) \pm g(x)] = L \pm K$

3. Product: $\qquad\qquad\;\; \lim_{x \to c} [f(x)g(x)] = LK$

4. Quotient: $\qquad\qquad\; \lim_{x \to c} \dfrac{f(x)}{g(x)} = \dfrac{L}{K}, \qquad$ provided $K \neq 0$

5. Power: $\qquad\qquad\quad\, \lim_{x \to c} [f(x)]^n = L^n$

Proof   To prove Property 2, choose $\varepsilon > 0$. Because $\varepsilon/2 > 0$, you know that there exists $\delta_1 > 0$ such that $0 < |x - c| < \delta_1$ implies $|f(x) - L| < \varepsilon/2$. You also know that there exists $\delta_2 > 0$ such that $0 < |x - c| < \delta_2$ implies $|g(x) - K| < \varepsilon/2$. Let δ be the smaller of δ_1 and δ_2: then $0 < |x - c| < \delta$ implies that

$$|f(x) - L| < \frac{\varepsilon}{2} \quad \text{and} \quad |g(x) - K| < \frac{\varepsilon}{2}.$$

Thus, you can apply the Triangle Inequality to conclude that

$$|[f(x) + g(x)] - (L + K)| \leq |f(x) - L| + |g(x) - K| < \frac{\varepsilon}{2} + \frac{\varepsilon}{2} = \varepsilon$$

which implies that

$$\lim_{x \to c} [f(x) + g(x)] = L + K = \lim_{x \to c} f(x) + \lim_{x \to c} g(x).$$

The proof that

$$\lim_{x \to c} [f(x) - g(x)] = L - K$$

is similar.

To prove Property 3, given that

$$\lim_{x \to c} f(x) = L \quad \text{and} \quad \lim_{x \to c} g(x) = K,$$

you can write

$$f(x)g(x) = [f(x) - L][g(x) - K] + [Lg(x) + Kf(x)] - LK.$$

Because the limit of $f(x)$ is L, and the limit of $g(x)$ is K, you have

$$\lim_{x \to c} [f(x) - L] = 0 \quad \text{and} \quad \lim_{x \to c} [g(x) - K] = 0.$$

Let $0 < \varepsilon < 1$. Then there exists $\delta > 0$ such that if $0 < |x - c| < \delta$, then

$$|f(x) - L - 0| < \varepsilon \quad \text{and} \quad |g(x) - K - 0| < \varepsilon,$$

which implies that

$$|[f(x) - L][g(x) - K] - 0| = |f(x) - L| \, |g(x) - K| < \varepsilon\varepsilon < \varepsilon.$$

Hence,

$$\lim_{x \to c} [f(x) - L][g(x) - K] = 0.$$

Furthermore, by Property 1, you have

$$\lim_{x \to c} Lg(x) = LK \quad \text{and} \quad \lim_{x \to c} Kf(x) = KL.$$

Finally, by Property 2, you obtain

$$\lim_{x \to c} f(x)g(x) = \lim_{x \to c} [f(x) - L][g(x) - K] + \lim_{x \to c} Lg(x) + \lim_{x \to c} Kf(x) - \lim_{x \to c} LK$$

$$= 0 + LK + KL - LK$$

$$= LK.$$

To prove Property 4, note that it is sufficient to prove that

$$\lim_{x \to c} \frac{1}{g(x)} = \frac{1}{K}.$$

Then you can use Property 3 to write

$$\lim_{x \to c} \frac{f(x)}{g(x)} = \lim_{x \to c} f(x) \frac{1}{g(x)} = \lim_{x \to c} f(x) \cdot \lim_{x \to c} \frac{1}{g(x)} = \frac{L}{K}.$$

Let $\varepsilon > 0$. Because $\lim_{x \to c} g(x) = K$, there exists $\delta_1 > 0$ such that if

$$0 < |x - c| < \delta_1, \text{ then } |g(x) - K| < \frac{|K|}{2},$$

which implies that

$$|K| = |g(x) + [|K| - g(x)]| \leq |g(x)| + ||K| - g(x)| < |g(x)| + \frac{|K|}{2}.$$

That is, for $0 < |x - c| < \delta_1$,

$$\frac{|K|}{2} < |g(x)| \quad \text{or} \quad \frac{1}{|g(x)|} < \frac{2}{|K|}.$$

Similarly, there exists a $\delta_2 > 0$ such that if $0 < |x - c| < \delta_2$, then

$$|g(x) - K| < \frac{|K|^2}{2} \varepsilon.$$

Let δ be the smaller of δ_1 and δ_2. For $0 < |x - c| < \delta$, you have

$$\left| \frac{1}{g(x)} - \frac{1}{K} \right| = \left| \frac{K - g(x)}{g(x)K} \right| = \frac{1}{|K|} \cdot \frac{1}{|g(x)|} |K - g(x)| <$$

$$\frac{1}{|K|} \cdot \frac{2}{|K|} \frac{|K|^2}{2} \varepsilon = \varepsilon.$$

Thus, $\lim_{x \to c} \dfrac{1}{g(x)} = \dfrac{1}{K}.$

Finally, the proof of Property 5 can be obtained by a straightforward application of mathematical induction coupled with Property 3.

THEOREM 1.4   The Limit of a Function Involving a Radical (page 57)

Let n be a positive integer. The following limit is valid for all c if n is odd, and is valid for $c > 0$ if n is even.

$$\lim_{x \to c} \sqrt[n]{x} = \sqrt[n]{c}.$$

Proof Consider the case for which $c > 0$ and n is any positive integer. For a given $\varepsilon > 0$, you need to find $\delta > 0$ such that

$$\left| \sqrt[n]{x} - \sqrt[n]{c} \right| < \varepsilon \quad \text{whenever} \quad 0 < |x - c| < \delta,$$

which is the same as saying

$$-\varepsilon < \sqrt[n]{x} - \sqrt[n]{c} < \varepsilon \quad \text{whenever} \quad -\delta < x - c < \delta.$$

Assume $\varepsilon < \sqrt[n]{c}$, which implies that $0 < \sqrt[n]{c} - \varepsilon < \sqrt[n]{c}$. Now, let δ be the smaller of the two numbers.

$$c - \left(\sqrt[n]{c} - \varepsilon \right)^n \quad \text{and} \quad \left(\sqrt[n]{c} + \varepsilon \right)^n - c$$

Then you have

$$-\delta < x - c < \delta$$
$$-\left[c - \left(\sqrt[n]{c} - \varepsilon \right)^n \right] < x - c < \left(\sqrt[n]{c} + \varepsilon \right)^n - c$$
$$\left(\sqrt[n]{c} - \varepsilon \right)^n - c < x - c < \left(\sqrt[n]{c} + \varepsilon \right)^n - c$$
$$\left(\sqrt[n]{c} - \varepsilon \right)^n < x < \left(\sqrt[n]{c} + \varepsilon \right)^n$$
$$\sqrt[n]{c} - \varepsilon < \sqrt[n]{x} < \sqrt[n]{c} + \varepsilon$$
$$-\varepsilon < \sqrt[n]{x} - \sqrt[n]{c} < \varepsilon.$$

THEOREM 1.5   The Limit of a Composite Function (page 58)

If f and g are functions such that $\lim_{x \to c} g(x) = L$ and $\lim_{x \to L} f(x) = f(L)$, then

$$\lim_{x \to c} f(g(x)) = f(L).$$

Proof For a given $\varepsilon > 0$, you must find $\delta > 0$ such that

$$|f(g(x)) - f(L)| < \varepsilon \quad \text{whenever} \quad 0 < |x - c| < \delta.$$

Because the limit of $f(x)$ as $x \to L$ is $f(L)$, you know there exists $\delta_1 > 0$ such that

$$|f(u) - f(L)| < \varepsilon \quad \text{whenever} \quad |u - L| < \delta_1.$$

Moreover, because the limit of $g(x)$ as $x \to c$ is L, you know there exists $\delta > 0$ such that

$$|g(x) - L| < \delta_1 \quad \text{whenever} \quad 0 < |x - c| < \delta.$$

Finally, letting $u = g(x)$, you have

$$|f(g(x)) - f(L)| < \varepsilon \quad \text{whenever} \quad 0 < |x - c| < \delta.$$

> **THEOREM 1.7   Functions That Agree at All But One Point (page 59)**
>
> Let c be a real number and let $f(x) = g(x)$ for all $x \neq c$ in an open interval containing c. If the limit of $g(x)$ as x approaches c exists, then the limit of $f(x)$ also exists and
>
> $$\lim_{x \to c} f(x) = \lim_{x \to c} g(x).$$

Proof Let L be the limit of $g(x)$ as $x \to c$. Then, for each $\varepsilon > 0$ there exists a $\delta > 0$ such that $f(x) = g(x)$ in the open intervals $(c - \delta, c)$ and $(c, c + \delta)$, and

$$|g(x) - L| < \varepsilon \quad \text{whenever} \quad 0 < |x - c| < \delta.$$

Because $f(x) = g(x)$ for all x in the open interval other than $x = c$, it follows that

$$|f(x) - L| < \varepsilon \quad \text{whenever} \quad 0 < |x - c| < \delta.$$

Thus, the limit of $f(x)$ as $x \to c$ is also L.

> **THEOREM 1.8   The Squeeze Theorem (page 62)**
>
> If $h(x) \leq f(x) \leq g(x)$ for all x in an open interval containing c, except possibly at c itself, and if
>
> $$\lim_{x \to c} h(x) = L = \lim_{x \to c} g(x)$$
>
> then $\lim_{x \to c} f(x)$ exists and is equal to L.

Proof For $\varepsilon > 0$ there exist δ_1 and δ_2 such that

$$|h(x) - L| < \varepsilon \quad \text{whenever} \quad 0 < |x - c| < \delta_1$$

and

$$|g(x) - L| < \varepsilon \quad \text{whenever} \quad 0 < |x - c| < \delta_2.$$

Because $h(x) \leq f(x) \leq g(x)$ for all x in an open interval containing c, except possibly at c itself, there exists $\delta_3 > 0$ such that $h(x) \leq f(x) \leq g(x)$ for $0 < |x - c| < \delta_3$. Let δ be the smallest of δ_1, δ_2, and δ_3. Then, if $0 < |x - c| < \delta$, it follows that $|h(x) - L| < \varepsilon$ and $|g(x) - L| < \varepsilon$, which implies that

$$-\varepsilon < h(x) - L < \varepsilon \quad \text{and} \quad -\varepsilon < g(x) - L < \varepsilon$$
$$L - \varepsilon < h(x) \quad \text{and} \quad g(x) < L + \varepsilon.$$

Now, because $h(x) \leq f(x) \leq g(x)$, it follows that $L - \varepsilon < f(x) < L + \varepsilon$, which implies that $|f(x) - L| < \varepsilon$. Therefore,

$$\lim_{x \to c} f(x) = L.$$

THEOREM 1.14   Vertical Asymptotes (page 81)

Let f and g be continuous on an open interval containing c. If $f(c) \neq 0$, $g(c) = 0$, and there exists an open interval containing c such that $g(x) \neq 0$ for all $x \neq c$ in the interval, then the graph of the function given by

$$h(x) = \frac{f(x)}{g(x)}$$

has a vertical asymptote at $x = c$.

Proof Consider the case for which $f(c) > 0$, and there exists $b > c$ such that $c < x < b$ implies $g(x) > 0$. Then for $M > 0$, choose δ_1 such that

$$0 < x - c < \delta_1 \quad \text{implies that} \quad \frac{f(c)}{2} < f(x) < \frac{3f(c)}{2}$$

and δ_2 such that

$$0 < x - c < \delta_2 \quad \text{implies that} \quad 0 < g(x) < \frac{f(c)}{2M}.$$

Now let δ be the smaller of δ_1 and δ_2. Then it follows that

$$0 < x - c < \delta \quad \text{implies that} \quad \frac{f(x)}{g(x)} > \frac{f(c)}{2}\left[\frac{2M}{f(c)}\right] = M.$$

Therefore, it follows that

$$\lim_{x \to c^+} \frac{f(x)}{g(x)} = \infty$$

and the line $x = c$ is a vertical asymptote of the graph of h.

Alternate Form of the Derivative (page 96)

The derivative of f at c is given by

$$f'(c) = \lim_{x \to c} \frac{f(x) - f(c)}{x - c}$$

provided this limit exists.

Proof The derivative of f at c is given by

$$f'(c) = \lim_{\Delta x \to 0} \frac{f(c + \Delta x) - f(c)}{\Delta x}.$$

Let $x = c + \Delta x$. Then $x \to c$ as $\Delta x \to 0$. Thus, replacing $c + \Delta x$ by x, you have

$$f'(c) = \lim_{\Delta x \to 0} \frac{f(c + \Delta x) - f(c)}{\Delta x} = \lim_{x \to c} \frac{f(x) - f(c)}{x - c}.$$

> **THEOREM 2.10   The Chain Rule (page 125)**
>
> If $y = f(u)$ is a differentiable function of u, and $u = g(x)$ is a differentiable function of x, then $y = f(g(x))$ is a differentiable function of x and
>
> $$\frac{dy}{dx} = \frac{dy}{du} \cdot \frac{du}{dx} \quad \text{or, equivalently,} \quad \frac{d}{dx}[f(g(x))] = f'(g(x))g'(x).$$

Proof In Section 2.4, we let $h(x) = f(g(x))$ and used the alternative form of the derivative to show that $h'(c) = f'(g(c))g'(c)$, provided $g(x) \neq g(c)$ for values of x other than c. Now consider a more general proof. Begin by considering the derivative of f.

$$f'(x) = \lim_{\Delta x \to 0} \frac{f(x + \Delta x) - f(x)}{\Delta x} = \lim_{\Delta x \to 0} \frac{\Delta y}{\Delta x}$$

For a fixed value of x, define a function η such that

$$\eta(\Delta x) = \begin{cases} 0, & \Delta x = 0 \\ \dfrac{\Delta y}{\Delta x} - f'(x), & \Delta x \neq 0. \end{cases}$$

Because the limit of $\eta(\Delta x)$ as $\Delta x \to 0$ doesn't depend on the value of $\eta(0)$, you have

$$\lim_{\Delta x \to 0} \eta(\Delta x) = \lim_{\Delta x \to 0} \left[\frac{\Delta y}{\Delta x} - f'(x) \right] = 0$$

and you can conclude that η is continuous at 0. Moreover, because $\Delta y = 0$ when $\Delta x = 0$, the equation

$$\Delta y = \Delta x \eta(\Delta x) + \Delta x f'(x)$$

is valid whether Δx is zero or not. Now, by letting $\Delta u = g(x + \Delta x) - g(x)$, you can use the continuity of g to conclude that

$$\lim_{\Delta x \to 0} \Delta u = \lim_{\Delta x \to 0} [g(x + \Delta x) - g(x)] = 0,$$

which implies that

$$\lim_{\Delta x \to 0} \eta(\Delta u) = 0.$$

Finally,

$$\Delta y = \Delta u \eta(\Delta u) + \Delta u f'(u) \to \frac{\Delta y}{\Delta x} = \frac{\Delta u}{\Delta x} \eta(\Delta u) + \frac{\Delta u}{\Delta x} f'(u), \quad \Delta x \neq 0$$

and taking the limit as $\Delta x \to 0$, you have

$$\frac{dy}{dx} = \frac{du}{dx} \left[\lim_{\Delta x \to 0} \eta(\Delta u) \right] + \frac{du}{dx} f'(u) = \frac{dy}{dx}(0) + \frac{du}{dx} f'(u)$$

$$= \frac{du}{dx} f'(u) = \frac{du}{dx} \cdot \frac{dy}{du}.$$

Concavity Interpretation (page 179)

1. Let f be differentiable at c. If the graph of f is concave upward at $(c, f(c))$, then the graph of f lies *above* the tangent line at $(c, f(c))$ on some open interval containing c.
2. Let f be differentiable at c. If the graph of f is concave downward at $(c, f(c))$, then the graph of f lies *below* the tangent line at $(c, f(c))$ on some open interval containing c.

Proof   Assume that f is concave upward at c. Then, by definition, there exists an interval (a, b) containing c such that f' is increasing on (a, b). The equation of the tangent line to the graph of f at c is given by

$$g(x) = f(c) + f'(c)(x - c).$$

If x is in the open interval (c, b), then the directed distance from point $(x, f(x))$ (on the graph of f) to the point $(x, g(x))$ (on the tangent line) is given by

$$\begin{aligned} d &= f(x) - [\,f(c) + f'(c)(x - c)\,] \\ &= f(x) - f(c) - f'(c)(x - c). \end{aligned}$$

Moreover, by the Mean Value Theorem there exists a number z in (c, x) such that

$$f'(z) = \frac{f(x) - f(c)}{x - c}.$$

Thus, you have

$$\begin{aligned} d &= f(x) - f(c) - f'(c)(x - c) \\ &= f'(z)(x - c) - f'(c)(x - c) \\ &= [\,f'(z) - f'(c)\,](x - c). \end{aligned}$$

The second factor $(x - c)$ is positive because $c < x$. Moreover, because f' is increasing, it follows that the first factor $[\,f'(z) - f'(c)\,]$ is also positive. Therefore, $d > 0$ and you can conclude that the graph of f lies above the tangent line. If x is in the open interval (a, c), a similar argument can be given.   ■

THEOREM 3.10   Limits at Infinity (page 188)

If r is a positive rational number, and c is any real number, then

$$\lim_{x \to \infty} \frac{c}{x^r} = 0.$$

Furthermore, if x^r is defined when $x < 0$, then $\displaystyle\lim_{x \to -\infty} \frac{c}{x^r} = 0$.

Proof   Begin by proving that

$$\lim_{x \to \infty} \frac{1}{x} = 0.$$

For $\varepsilon > 0$, let $M = 1/\varepsilon$. Then, for $x > M$, you have

$$x > M = \frac{1}{\varepsilon} \quad \Rightarrow \quad \frac{1}{x} < \varepsilon \quad \Rightarrow \quad \left| \frac{1}{x} - 0 \right| < \varepsilon.$$

Therefore, by the definition of a limit at infinity, you can conclude that the limit of $1/x$ as $x \to \infty$ is 0. Now, using this result, and letting $r = m/n$, you can write the following.

$$\lim_{x \to \infty} \frac{c}{x^r} = \lim_{x \to \infty} \frac{c}{x^{m/n}}$$

$$= c \left[\lim_{x \to \infty} \left(\frac{1}{\sqrt[n]{x}} \right)^m \right]$$

$$= c \left(\lim_{x \to \infty} \sqrt[n]{\frac{1}{x}} \right)^m$$

$$= c \left(\sqrt[n]{\lim_{x \to \infty} \frac{1}{x}} \right)^m$$

$$= c \left(\sqrt[n]{0} \right)^m$$

$$= 0$$

The proof of the second part of the theorem is similar.

THEOREM 4.2   Summation Formulas (page 253)

1. $\displaystyle\sum_{i=1}^{n} c = cn$ **2.** $\displaystyle\sum_{i=1}^{n} i = \frac{n(n + 1)}{2}$

3. $\displaystyle\sum_{i=1}^{n} i^2 = \frac{n(n + 1)(2n + 1)}{6}$ **4.** $\displaystyle\sum_{i=1}^{n} i^3 = \frac{n^2(n + 1)^2}{4}$

Proof   The proof of Property 1 is straightforward. By adding c to itself n times, you obtain a sum of nc.

To prove Property 2, write the sum in increasing and decreasing order and add corresponding terms as follows.

$$\sum_{i=1}^{n} i = \quad 1 \quad + \quad 2 \quad + \quad 3 \quad + \cdots + (n - 1) + \quad n$$

$$\downarrow \qquad\qquad \downarrow \qquad\qquad \downarrow \qquad\qquad \downarrow$$

$$\sum_{i=1}^{n} i = \quad n \quad + (n - 1) + (n - 2) + \cdots + \quad 2 \quad + \quad 1$$

$$\downarrow \qquad \downarrow \qquad \downarrow \qquad\qquad \downarrow \qquad \downarrow$$

$$2\sum_{i=1}^{n} i = \underbrace{(n + 1) + (n + 1) + (n + 1) + \cdots + (n + 1) + (n + 1)}_{n \text{ terms}}$$

Therefore,

$$\sum_{i=1}^{n} i = \frac{n(n + 1)}{2}.$$

To prove Property 3, use mathematical induction. First, if $n = 1$, the result is true because

$$\sum_{i=1}^{1} i^2 = 1^2 = 1 = \frac{1(1 + 1)(2 + 1)}{6}.$$

Now, assuming the result is true for $n = k$, you can show that it is true for $n = k + 1$, as follows.

$$\sum_{i=1}^{k+1} i^2 = \sum_{i=1}^{k} i^2 + (k + 1)^2$$

$$= \frac{k(k + 1)(2k + 1)}{6} + (k + 1)^2$$

$$= \frac{k + 1}{6}(2k^2 + k + 6k + 6)$$

$$= \frac{k + 1}{6}[(2k + 3)(k + 2)]$$

$$= \frac{(k + 1)(k + 2)[2(k + 1) + 1]}{6}$$

Property 4 can be proved using a similar argument with mathematical induction.

THEOREM 4.8 Preservation of Inequality (page 271)

1. If f is integrable and nonnegative on the closed interval $[a, b]$, then

$$0 \leq \int_a^b f(x)\, dx.$$

2. If f and g are integrable on the closed interval $[a, b]$, and $f(x) \leq g(x)$ for every x in $[a, b]$, then

$$\int_a^b f(x)\, dx \leq \int_a^b g(x)\, dx.$$

Proof To prove Property 1, suppose, on the contrary, that

$$\int_a^b f(x)\, dx = I < 0.$$

Then, let $a = x_0 < x_1 < x_2 < \cdots < x_n = b$ be a partition of $[a, b]$, and let

$$R = \sum_{i=1}^{n} f(c_i)\Delta x_i$$

be a Riemann sum. Because $f(x) \geq 0$, it follows that $R \geq 0$. Now, for $\|\Delta\|$ sufficiently small, you have $|R - I| < -I/2$, which implies that

$$\sum_{i=1}^{n} f(c_i)\Delta x_i = R < I - \frac{I}{2} < 0,$$

which is not possible. From this contradiction, you can conclude that

$$0 \leq \int_a^b f(x)\, dx.$$

To prove Property 2 of the theorem, note that $f(x) \leq g(x)$ implies that $g(x) - f(x) \geq 0$. Hence, you can apply the result of Property 1 to conclude that

$$0 \leq \int_a^b [g(x) - f(x)]\, dx$$

$$0 \leq \int_a^b g(x)\, dx - \int_a^b f(x)\, dx$$

$$\int_a^b f(x)\, dx \leq \int_a^b g(x)\, dx.$$

Properties of the Natural Logarithmic Function (page 312)

$$\lim_{x \to 0^+} \ln x = -\infty \quad \text{and} \quad \lim_{x \to \infty} \ln x = \infty$$

Proof To begin, show that $\ln 2 \geq \frac{1}{2}$. From the Mean Value Theorem for Integrals, you can write

$$\ln 2 = \int_1^2 \frac{1}{x}\, dx = (2 - 1)\frac{1}{c} = \frac{1}{c}$$

where c is in $[1, 2]$. This implies that

$$1 \leq c \quad \leq 2$$

$$1 \geq \frac{1}{c} \quad \geq \frac{1}{2}$$

$$1 \geq \ln 2 \geq \frac{1}{2}.$$

Now, let N be any positive (large) number. Because $\ln x$ is increasing, it follows that if $x > 2^{2N}$, then

$$\ln x > \ln 2^{2N} = 2N \ln 2.$$

However, because $\ln 2 \geq \frac{1}{2}$, it follows that

$$\ln x > 2N \ln 2 \geq 2N\left(\frac{1}{2}\right) = N.$$

This verifies the second limit. To verify the first limit, let $z = 1/x$. Then, $z \to \infty$ as $x \to 0^+$, and you can write

$$\lim_{x \to 0^+} \ln x = \lim_{x \to 0^+} \left(-\ln \frac{1}{x}\right) = \lim_{z \to \infty} (-\ln z) = -\lim_{z \to \infty} \ln z = -\infty.$$

> ### THEOREM 5.8   Continuity and Differentiability of Inverse Functions (page 333)
>
> Let f be a function whose domain is an interval I. If f has an inverse, then the following statements are true.
>
> 1. If f is continuous on its domain, then f^{-1} is continuous on its domain.
> 2. If f is increasing on its domain, then f^{-1} is increasing on its domain.
> 3. If f is decreasing on its domain, then f^{-1} is decreasing on its domain.
> 4. If f is differentiable at c and $f'(c) \neq 0$, then f^{-1} is differentiable at $f(c)$.

Proof   To prove Property 1, first show that if f is continuous on I, and has an inverse, then f is strictly monotonic on I. Suppose that f were not strictly monotonic. Then there would exist numbers x_1, x_2, x_3 in I such that $x_1 < x_2 < x_3$, but $f(x_2)$ is not between $f(x_1)$ and $f(x_3)$. Without loss of generality, assume $f(x_1) < f(x_3) < f(x_2)$. By the Intermediate Value Theorem, there exists a number x_0 between x_1 and x_2 such that $f(x_0) = f(x_3)$. Thus, f is not one-to-one and cannot have an inverse. Hence, f must be strictly monotonic.

Because f is continuous, the Intermediate Value Theorem implies that the set of values of f,

$$\{f(x) : x \in I\},$$

forms an interval J. Assume that a is an interior point of J. From the previous argument, $f^{-1}(a)$ is an interior point of I. Let $\varepsilon > 0$. There exists $0 < \varepsilon_1 < \varepsilon$ such that

$$I_1 = (f^{-1}(a) - \varepsilon_1, f^{-1}(a) + \varepsilon_1) \subseteq I.$$

Because f is strictly monotonic on I_1, the set of values $\{f(x) : x \in I_1\}$ forms an interval $J_1 \subseteq J$. Let $\delta > 0$ such that $(a - \delta, a + \delta) \subseteq J_1$. Finally, if

$$|y - a| < \delta, \text{ then } |f^{-1}(y) - f^{-1}(a)| < \varepsilon_1 < \varepsilon.$$

Hence, f^{-1} is continuous at a. A similar proof can be given if a is an endpoint.

To prove Property 2, let y_1 and y_2 be in the domain of f^{-1}, with $y_1 < y_2$. Then, there exist x_1 and x_2 in the domain of f such that

$$f(x_1) = y_1 < y_2 = f(x_2).$$

Because f is increasing, $f(x_1) < f(x_2)$ holds precisely when $x_1 < x_2$. Therefore,

$$f^{-1}(y_1) = x_1 < x_2 = f^{-1}(y_2),$$

which implies that f^{-1} is increasing. (Property 3 can be proved in a similar way.)

Finally, to prove Property 4, consider the limit

$$(f^{-1})'(a) = \lim_{y \to a} \frac{f^{-1}(y) - f^{-1}(a)}{y - a}$$

where a is in the domain of f^{-1} and $f^{-1}(a) = c$. Because f is differentiable at c, f is continuous at c, and so is f^{-1} at a. Thus, $y \to a$ implies that $x \to c$, and you have

$$(f^{-1})'(a) = \lim_{x \to c} \frac{x - c}{f(x) - f(c)}$$

$$= \lim_{x \to c} \frac{1}{\left(\dfrac{f(x) - f(c)}{x - c}\right)}$$

$$= \frac{1}{\displaystyle\lim_{x \to c} \frac{f(x) - f(c)}{x - c}}$$

$$= \frac{1}{f'(c)}.$$

Hence, $(f^{-1})'(a)$ exists, and f^{-1} is differentiable at $f(c)$.

THEOREM 5.9   The Derivative of an Inverse Function (page 333)

Let f be a function that is differentiable on an interval I. If f has an inverse function g, then g is differentiable at any x for which $f'(g(x)) \neq 0$. Moreover,

$$g'(x) = \frac{1}{f'(g(x))}, \qquad f'(g(x)) \neq 0.$$

Proof From the proof of Theorem 5.8, letting $a = x$, you know that g is differentiable. Using the Chain Rule, differentiate both sides of the equation $x = f(g(x))$ to obtain

$$1 = f'(g(x)) \frac{d}{dx}[g(x)].$$

Because $f'(g(x)) \neq 0$, you can divide by this quantity to obtain

$$\frac{d}{dx}[g(x)] = \frac{1}{f'(g(x))}.$$

THEOREM 5.15   A Limit Involving e (page 352)

$$\lim_{x \to \infty} \left(1 + \frac{1}{x}\right)^x = \lim_{x \to \infty} \left(\frac{x + 1}{x}\right)^x = e$$

Proof Let $y = \lim_{x \to \infty} \left(1 + \frac{1}{x}\right)^x$. Taking the natural logs of both sides, you have

$$\ln y = \ln\left[\lim_{x \to \infty} \left(1 + \frac{1}{x}\right)^x\right].$$

Because the natural logarithmic function is continuous, you can write

$$\ln y = \lim_{x \to \infty} \left[x \ln \left(1 + \frac{1}{x} \right) \right] = \lim_{x \to \infty} \left\{ \frac{\ln [1 + (1/x)]}{1/x} \right\}.$$

Letting $x = \dfrac{1}{t}$, you have

$$\ln y = \lim_{t \to 0^+} \frac{\ln(1 + t)}{t}$$

$$= \lim_{t \to 0^+} \frac{\ln(1 + t) - \ln 1}{t}$$

$$= \frac{d}{dx} \ln x \ \text{ at } \ x = 1$$

$$= \frac{1}{x} \ \text{ at } \ x = 1$$

$$= 1.$$

Finally, because $\ln y = 1$, you know that $y = e$, and you can conclude that

$$\lim_{x \to \infty} \left(1 + \frac{1}{x} \right)^x = e.$$

THEOREM 7.3   The Extended Mean Value Theorem (page 524)

If f and g are differentiable on an open interval (a, b) and continuous on $[a, b]$ such that $g'(x) \neq 0$ for any x in (a, b), then there exists a point c in (a, b) such that

$$\frac{f'(c)}{g'(c)} = \frac{f(b) - f(a)}{g(b) - g(a)}.$$

Proof   You can assume that $g(a) \neq g(b)$, because otherwise, by Rolle's Theorem, it would follow that $g'(x) = 0$ for some x in (a, b). Now, define $h(x)$ to be

$$h(x) = f(x) - \left[\frac{f(b) - f(a)}{g(b) - g(a)} \right] g(x).$$

Then

$$h(a) = f(a) - \left[\frac{f(b) - f(a)}{g(b) - g(a)} \right] g(a) = \frac{f(a)g(b) - f(b)g(a)}{g(b) - g(a)}$$

and

$$h(b) = f(b) - \left[\frac{f(b) - f(a)}{g(b) - g(a)} \right] g(b) = \frac{f(a)g(b) - f(b)g(a)}{g(b) - g(a)}$$

and by Rolle's Theorem there exists a point c in (a, b) such that

$$h'(c) = f'(c) - \frac{f(b) - f(a)}{g(b) - g(a)} g'(c) = 0,$$

which implies that

$$\frac{f'(c)}{g'(c)} = \frac{f(b) - f(a)}{g(b) - g(a)}.$$

THEOREM 7.4   L'Hôpital's Rule (page 524)

Let f and g be functions that are differentiable on an open interval (a, b) containing c, except possibly at c itself. Assume that $g'(x) \neq 0$ for all x in (a, b), except possibly at c itself. If the limit of $f(x)/g(x)$ as x approaches c produces the indeterminate form $0/0$, then

$$\lim_{x \to c} \frac{f(x)}{g(x)} = \lim_{x \to c} \frac{f'(x)}{g'(x)}$$

provided the limit on the right exists (or is infinite). This result also applies if the limit of $f(x)/g(x)$ as x approaches c produces any one of the indeterminate forms ∞/∞, $(-\infty)/\infty$, $\infty/(-\infty)$, or $(-\infty)/(-\infty)$.

You can use the Extended Mean Value Theorem to prove L'Hôpital's Rule. Of the several different cases of this rule, the proof of only one case is illustrated. The remaining cases are left for you to prove.

Proof   Consider the case for which

$$\lim_{x \to c^+} f(x) = 0 \quad \text{and} \quad \lim_{x \to c^+} g(x) = 0.$$

Define the following new functions:

$$F(x) = \begin{cases} f(x), & x \neq c \\ 0, & x = c \end{cases} \quad \text{and} \quad G(x) = \begin{cases} g(x), & x \neq c \\ 0, & x = c \end{cases}.$$

For any x, $c < x < b$, F and G are differentiable on $(c, x]$ and continuous on $[c, x]$. You can apply the Extended Mean Value Theorem to conclude that there exists a number z in (c, x) such that

$$\frac{F'(z)}{G'(z)} = \frac{F(x) - F(c)}{G(x) - G(c)} = \frac{F(x)}{G(x)} = \frac{f'(z)}{g'(z)} = \frac{f(x)}{g(x)}.$$

Finally, by letting x approach c from the right, $x \to c^+$, we have $z \to c^+$ because $c < z < x$, and

$$\lim_{x \to c^+} \frac{f(x)}{g(x)} = \lim_{x \to c^+} \frac{f'(z)}{g'(z)} = \lim_{z \to c^+} \frac{f'(z)}{g'(z)} = \lim_{x \to c^+} \frac{f'(x)}{g'(x)}.$$

The proof for the case where $x \to c^-$ and $x \to c$ are left to the reader.

THEOREM 8.19   Taylor's Theorem (page 602)

If a function f is differentiable through order $n + 1$ in an interval I containing c, then, for each x in I, there exists z between x and c such that

$$f(x) = f(c) + f'(c)(x - c) + \frac{f''(c)}{2!}(x - c)^2 + \cdots + \frac{f^{(n)}(c)}{n!}(x - c)^n + R_n(x)$$

where

$$R_n(x) = \frac{f^{(n+1)}(z)}{(n + 1)!}(x - c)^{n+1}.$$

Proof To find $R_n(x)$, fix x in I ($x \neq c$) and write

$$R_n(x) = f(x) - P_n(x)$$

where $P_n(x)$ is the nth Taylor polynomial for $f(x)$. Then let g be a function of t defined by

$$g(t) = f(x) - f(t) - f'(t)(x - t) - \cdots - \frac{f^{(n)}(t)}{n!}(x - t)^n - R_n(x)\frac{(x - t)^{n+1}}{(x - c)^{n+1}}.$$

The reason for defining g in this way is that differentiation with respect to t has a telescoping effect. For example, you have

$$\frac{d}{dt}[-f(t) - f'(t)(x - t)] = -f'(t) + f'(t) - f''(t)(x - t)$$

$$= -f''(t)(x - t).$$

The result is that the derivative $g'(t)$ simplifies to

$$g'(t) = -\frac{f^{(n+1)}(t)}{n!}(x - t)^n + (n + 1)R_n(x)\frac{(x - t)^n}{(x - c)^{n+1}}$$

for all t between c and x. Moreover, for a fixed x,

$$g(c) = f(x) - [P_n(x) + R_n(x)] = f(x) - f(x) = 0$$

and

$$g(x) = f(x) - f(x) - 0 - \cdots - 0 = f(x) - f(x) = 0.$$

Therefore, g satisfies the conditions of Rolle's Theorem, and it follows that there is a number z between c and x such that $g'(z) = 0$. Substituting z for t in the equation for $g'(t)$ and then solving for $R_n(x)$, you obtain

$$g'(z) = -\frac{f^{(n+1)}(z)}{n!}(x - z)^n + (n + 1)R_n(x)\frac{(x - z)^n}{(x - c)^{n+1}} = 0$$

$$R_n(x) = \frac{f^{(n+1)}(z)}{(n + 1)!}(x - c)^{n+1}.$$

Finally, because $g(c) = 0$, you have

$$0 = f(x) - f(c) - f'(c)(x - c) - \cdots - \frac{f^{(n)}(c)}{n!}(x - c)^n - R_n(x)$$

$$f(x) = f(c) + f'(c)(x - c) + \cdots + \frac{f^{(n)}(c)}{n!}(x - c)^n + R_n(x).$$

THEOREM 9.16 Classification of Conics by Eccentricity (page 689)

The locus of a point in the plane whose distance from a fixed point (*focus*) has a constant ratio to its distance from a fixed line (*directrix*) is a conic. The constant ratio e is the *eccentricity* of the conic.

1. The conic is an ellipse if $0 < e < 1$.

2. The conic is a parabola if $e = 1$.

3. The conic is a hyperbola if $e > 1$.

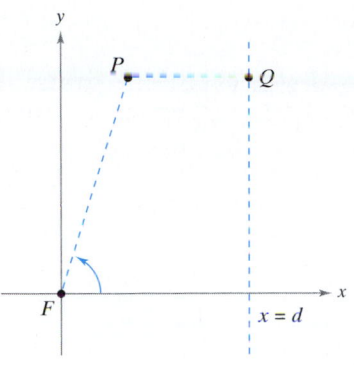

Figure A.40

Proof If $e = 1$, then, by definition, the conic must be a parabola. If $e \neq 1$, then you can consider the focus F to lie at the origin and the directrix $x = d$ to lie to the right of the origin, as shown in Figure A.40. For the point $P = (r, \theta) = (x, y)$, you have $|PF| = r$ and $|PQ| = d - r \cos \theta$. Given that $e = |PF|/|PQ|$, it follows that

$$|PF| = |PQ|e \quad \Longrightarrow \quad r = e(d - r \cos \theta).$$

By converting to rectangular coordinates and squaring both sides, you obtain

$$x^2 + y^2 = e^2(d - x)^2 = e^2(d^2 - 2\,dx + x^2).$$

Completing the square produces

$$\left(x + \frac{e^2 d}{1 - e^2}\right)^2 + \frac{y^2}{1 - e^2} = \frac{e^2 d^2}{(1 - e^2)^2}.$$

If $e < 1$, this equation represents an ellipse. If $e > 1$, then $1 - e^2 < 0$, and the equation represents a hyperbola.

THEOREM 12.6   Chain Rule: One Independent Variable (page 857)

Let $w = f(x, y)$, where f is a differentiable function of x and y. If $x = g(t)$ and $y = h(t)$, where g and h are differentiable functions of t, then w is a differentiable function of t, and

$$\frac{dw}{dt} = \frac{\partial w}{\partial x}\frac{dx}{dt} + \frac{\partial w}{\partial y}\frac{dy}{dt}.$$

Proof Because g and h are differentiable functions of t, you know that both Δx and Δy approach zero as Δt approaches zero. Moreover, because f is a differentiable function of x and y, you know that

$$\Delta w = \frac{\partial w}{\partial x}\Delta x + \frac{\partial w}{\partial y}\Delta y + \varepsilon_1 \Delta x + \varepsilon_2 \Delta y$$

where both ε_1 and $\varepsilon_2 \to 0$ as $(\Delta x, \Delta y) \to (0, 0)$. Thus, for $\Delta t \neq 0$, we have

$$\frac{\Delta w}{\Delta t} = \frac{\partial w}{\partial x}\frac{\Delta x}{\Delta t} + \frac{\partial w}{\partial y}\frac{\Delta y}{\Delta t} + \varepsilon_1 \frac{\Delta x}{\Delta t} + \varepsilon_2 \frac{\Delta y}{\Delta t}$$

from which it follows that

$$\frac{dw}{dt} = \lim_{\Delta t \to 0}\frac{\Delta w}{\Delta t} = \frac{\partial w}{\partial x}\frac{dx}{dt} + \frac{\partial w}{\partial y}\frac{dy}{dt} + 0\left(\frac{dx}{dt}\right) + 0\left(\frac{dy}{dt}\right)$$

$$= \frac{\partial w}{\partial x}\frac{dx}{dt} + \frac{\partial w}{\partial y}\frac{dy}{dt}.$$

APPENDIX C

Basic Differentiation Rules for Elementary Functions

1. $\dfrac{d}{dx}[cu] = cu'$

2. $\dfrac{d}{dx}[u \pm v] = u' \pm v'$

3. $\dfrac{d}{dx}[uv] = uv' + vu'$

4. $\dfrac{d}{dx}\left[\dfrac{u}{v}\right] = \dfrac{vu' - uv'}{v^2}$

5. $\dfrac{d}{dx}[c] = 0$

6. $\dfrac{d}{dx}[u^n] = nu^{n-1}u'$

7. $\dfrac{d}{dx}[x] = 1$

8. $\dfrac{d}{dx}[|u|] = \dfrac{u}{|u|}(u'), \ u \neq 0$

9. $\dfrac{d}{dx}[\ln u] = \dfrac{u'}{u}$

10. $\dfrac{d}{dx}[e^u] = e^u u'$

11. $\dfrac{d}{dx}[\sin u] = (\cos u)u'$

12. $\dfrac{d}{dx}[\cos u] = -(\sin u)u'$

13. $\dfrac{d}{dx}[\tan u] = (\sec^2 u)u'$

14. $\dfrac{d}{dx}[\cot u] = -(\csc^2 u)u'$

15. $\dfrac{d}{dx}[\sec u] = (\sec u \tan u)u'$

16. $\dfrac{d}{dx}[\csc u] = -(\csc u \cot u)u'$

17. $\dfrac{d}{dx}[\arcsin u] = \dfrac{u'}{\sqrt{1 - u^2}}$

18. $\dfrac{d}{dx}[\arccos u] = \dfrac{-u'}{\sqrt{1 - u^2}}$

19. $\dfrac{d}{dx}[\arctan u] = \dfrac{u'}{1 + u^2}$

20. $\dfrac{d}{dx}[\text{arccot } u] = \dfrac{-u'}{1 + u^2}$

21. $\dfrac{d}{dx}[\text{arcsec } u] = \dfrac{u'}{|u|\sqrt{u^2 - 1}}$

22. $\dfrac{d}{dx}[\text{arccsc } u] = \dfrac{-u'}{|u|\sqrt{u^2 - 1}}$

APPENDIX D
Integration Tables

Forms Involving u^n

1. $\displaystyle\int u^n \, du = \frac{u^{n+1}}{n+1} + C, \; n \neq -1$

2. $\displaystyle\int \frac{1}{u} \, du = \ln|u| + C$

Forms Involving $a + bu$

3. $\displaystyle\int \frac{u}{a+bu} \, du = \frac{1}{b^2}(bu - a \ln|a+bu|) + C$

4. $\displaystyle\int \frac{u}{(a+bu)^2} \, du = \frac{1}{b^2}\left(\frac{a}{a+bu} + \ln|a+bu|\right) + C$

5. $\displaystyle\int \frac{u}{(a+bu)^n} \, du = \frac{1}{b^2}\left[\frac{-1}{(n-2)(a+bu)^{n-2}} + \frac{a}{(n-1)(a+bu)^{n-1}}\right] + C, \; n \neq 1, 2$

6. $\displaystyle\int \frac{u^2}{a+bu} \, du = \frac{1}{b^3}\left[-\frac{bu}{2}(2a-bu) + a^2 \ln|a+bu|\right] + C$

7. $\displaystyle\int \frac{u^2}{(a+bu)^2} \, du = \frac{1}{b^3}\left(bu - \frac{a^2}{a+bu} - 2a \ln|a+bu|\right) + C$

8. $\displaystyle\int \frac{u^2}{(a+bu)^3} \, du = \frac{1}{b^3}\left[\frac{2a}{a+bu} - \frac{a^2}{2(a+bu)^2} + \ln|a+bu|\right] + C$

9. $\displaystyle\int \frac{u^2}{(a+bu)^n} \, du = \frac{1}{b^3}\left[\frac{-1}{(n-3)(a+bu)^{n-3}} + \frac{2a}{(n-2)(a+bu)^{n-2}} - \frac{a^2}{(n-1)(a+bu)^{n-1}}\right] + C, \; n \neq 1, 2, 3$

10. $\displaystyle\int \frac{1}{u(a+bu)} \, du = \frac{1}{a} \ln\left|\frac{u}{a+bu}\right| + C$

11. $\displaystyle\int \frac{1}{u(a+bu)^2} \, du = \frac{1}{a}\left(\frac{1}{a+bu} + \frac{1}{a} \ln\left|\frac{u}{a+bu}\right|\right) + C$

12. $\displaystyle\int \frac{1}{u^2(a+bu)} \, du = -\frac{1}{a}\left(\frac{1}{u} + \frac{b}{a} \ln\left|\frac{u}{a+bu}\right|\right) + C$

13. $\displaystyle\int \frac{1}{u^2(a+bu)^2} \, du = -\frac{1}{a^2}\left[\frac{a+2bu}{u(a+bu)} + \frac{2b}{a} \ln\left|\frac{u}{a+bu}\right|\right] + C$

Forms Involving $a + bu + cu^2$, $b^2 \neq 4ac$

14. $\displaystyle \int \frac{1}{a + bu + cu^2}\, du = \begin{cases} \dfrac{2}{\sqrt{4ac - b^2}} \arctan \dfrac{2cu + b}{\sqrt{4ac - b^2}} + C, & b^2 < 4ac \\[3mm] \dfrac{1}{\sqrt{b^2 - 4ac}} \ln \left| \dfrac{2cu + b - \sqrt{b^2 - 4ac}}{2cu + b + \sqrt{b^2 - 4ac}} \right| + C, & b^2 > 4ac \end{cases}$

15. $\displaystyle \int \frac{u}{a + bu + cu^2}\, du = \frac{1}{2c}\left(\ln|a + bu + cu^2| - b \int \frac{1}{a + bu + cu^2}\, du \right)$

Forms Involving $\sqrt{a + bu}$

16. $\displaystyle \int u^n \sqrt{a + bu}\, du = \frac{2}{b(2n + 3)}\left[u^n(a + bu)^{3/2} - na \int u^{n-1}\sqrt{a + bu}\, du \right]$

17. $\displaystyle \int \frac{1}{u\sqrt{a + bu}}\, du = \begin{cases} \dfrac{1}{\sqrt{a}} \ln \left| \dfrac{\sqrt{a + bu} - \sqrt{a}}{\sqrt{a + bu} + \sqrt{a}} \right| + C, & a > 0 \\[3mm] \dfrac{2}{\sqrt{-a}} \arctan \sqrt{\dfrac{a + bu}{-a}} + C, & a < 0 \end{cases}$

18. $\displaystyle \int \frac{1}{u^n \sqrt{a + bu}}\, du = \frac{-1}{a(n - 1)}\left[\frac{\sqrt{a + bu}}{u^{n-1}} + \frac{(2n - 3)b}{2} \int \frac{1}{u^{n-1}\sqrt{a + bu}}\, du \right], \quad n \neq 1$

19. $\displaystyle \int \frac{\sqrt{a + bu}}{u}\, du = 2\sqrt{a + bu} + a \int \frac{1}{u\sqrt{a + bu}}\, du$

20. $\displaystyle \int \frac{\sqrt{a + bu}}{u^n}\, du = \frac{-1}{a(n - 1)}\left[\frac{(a + bu)^{3/2}}{u^{n-1}} + \frac{(2n - 5)b}{2} \int \frac{\sqrt{a + bu}}{u^{n-1}}\, du \right], \quad n \neq 1$

21. $\displaystyle \int \frac{u}{\sqrt{a + bu}}\, du = \frac{-2(2a - bu)}{3b^2}\sqrt{a + bu} + C$

22. $\displaystyle \int \frac{u^n}{\sqrt{a + bu}}\, du = \frac{2}{(2n + 1)b}\left(u^n \sqrt{a + bu} - na \int \frac{u^{n-1}}{\sqrt{a + bu}}\, du \right)$

Forms Involving $a^2 \pm u^2$, $a > 0$

23. $\displaystyle \int \frac{1}{a^2 + u^2}\, du = \frac{1}{a} \arctan \frac{u}{a} + C$

24. $\displaystyle \int \frac{1}{u^2 - a^2}\, du = -\int \frac{1}{a^2 - u^2}\, du = \frac{1}{2a} \ln \left| \frac{u - a}{u + a} \right| + C$

25. $\displaystyle \int \frac{1}{(a^2 \pm u^2)^n}\, du = \frac{1}{2a^2(n - 1)}\left[\frac{u}{(a^2 \pm u^2)^{n-1}} + (2n - 3) \int \frac{1}{(a^2 \pm u^2)^{n-1}}\, du \right], \quad n \neq 1$

Forms Involving $\sqrt{u^2 \pm a^2}$, $a > 0$

26. $\displaystyle \int \sqrt{u^2 \pm a^2}\, du = \frac{1}{2}\left(u\sqrt{u^2 \pm a^2} \pm a^2 \ln \left| u + \sqrt{u^2 \pm a^2} \right| \right) + C$

27. $\displaystyle \int u^2 \sqrt{u^2 \pm a^2}\, du = \frac{1}{8}\left[u(2u^2 \pm a^2)\sqrt{u^2 \pm a^2} - a^4 \ln \left| u + \sqrt{u^2 \pm a^2} \right| \right] + C$

28. $\displaystyle\int \frac{\sqrt{u^2 + a^2}}{u}\, du = \sqrt{u^2 + a^2} - a \ln\left|\frac{a + \sqrt{u^2 + a^2}}{u}\right| + C$

29. $\displaystyle\int \frac{\sqrt{u^2 - a^2}}{u}\, du = \sqrt{u^2 - a^2} - a \operatorname{arcsec} \frac{|u|}{a} + C$

30. $\displaystyle\int \frac{\sqrt{u^2 \pm a^2}}{u^2}\, du = \frac{-\sqrt{u^2 \pm a^2}}{u} + \ln\left|u + \sqrt{u^2 \pm a^2}\right| + C$

31. $\displaystyle\int \frac{1}{\sqrt{u^2 \pm a^2}}\, du = \ln\left|u + \sqrt{u^2 \pm a^2}\right| + C$

32. $\displaystyle\int \frac{1}{u\sqrt{u^2 + a^2}}\, du = \frac{-1}{a} \ln\left|\frac{a + \sqrt{u^2 + a^2}}{u}\right| + C$

33. $\displaystyle\int \frac{1}{u\sqrt{u^2 - a^2}}\, du = \frac{1}{a} \operatorname{arcsec} \frac{|u|}{a} + C$

34. $\displaystyle\int \frac{u^2}{\sqrt{u^2 \pm a^2}}\, du = \frac{1}{2}\left(u\sqrt{u^2 \pm a^2} \mp a^2 \ln\left|u + \sqrt{u^2 \pm a^2}\right|\right) + C$

35. $\displaystyle\int \frac{1}{u^2\sqrt{u^2 \pm a^2}}\, du = \mp \frac{\sqrt{u^2 \pm a^2}}{a^2 u} + C$

36. $\displaystyle\int \frac{1}{(u^2 \pm a^2)^{3/2}}\, du = \frac{\pm u}{a^2\sqrt{u^2 \pm a^2}} + C$

Forms Involving $\sqrt{a^2 - u^2},\ a > 0$

37. $\displaystyle\int \sqrt{a^2 - u^2}\, du = \frac{1}{2}\left(u\sqrt{a^2 - u^2} + a^2 \arcsin \frac{u}{a}\right) + C$

38. $\displaystyle\int u^2\sqrt{a^2 - u^2}\, du = \frac{1}{8}\left[u(2u^2 - a^2)\sqrt{a^2 - u^2} + a^4 \arcsin \frac{u}{a}\right] + C$

39. $\displaystyle\int \frac{\sqrt{a^2 - u^2}}{u}\, du = \sqrt{a^2 - u^2} - a \ln\left|\frac{a + \sqrt{a^2 - u^2}}{u}\right| + C$

40. $\displaystyle\int \frac{\sqrt{a^2 - u^2}}{u^2}\, du = \frac{-\sqrt{a^2 - u^2}}{u} - \arcsin \frac{u}{a} + C$

41. $\displaystyle\int \frac{1}{\sqrt{a^2 - u^2}}\, du = \arcsin \frac{u}{a} + C$

42. $\displaystyle\int \frac{1}{u\sqrt{a^2 - u^2}}\, du = \frac{-1}{a} \ln\left|\frac{a + \sqrt{a^2 - u^2}}{u}\right| + C$

43. $\displaystyle\int \frac{u^2}{\sqrt{a^2 - u^2}}\, du = \frac{1}{2}\left(-u\sqrt{a^2 - u^2} + a^2 \arcsin \frac{u}{a}\right) + C$

44. $\displaystyle \int \frac{1}{u^2\sqrt{a^2-u^2}}\,du = \frac{-\sqrt{a^2-u^2}}{a^2 u} + C$

45. $\displaystyle \int \frac{1}{(a^2-u^2)^{3/2}}\,du = \frac{u}{a^2\sqrt{a^2-u^2}} + C$

Forms Involving $\sin u$ or $\cos u$

46. $\displaystyle \int \sin u\,du = -\cos u + C$

47. $\displaystyle \int \cos u\,du = \sin u + C$

48. $\displaystyle \int \sin^2 u\,du = \frac{1}{2}(u - \sin u \cos u) + C$

49. $\displaystyle \int \cos^2 u\,du = \frac{1}{2}(u + \sin u \cos u) + C$

50. $\displaystyle \int \sin^n u\,du = -\frac{\sin^{n-1} u \cos u}{n} + \frac{n-1}{n}\int \sin^{n-2} u\,du$

51. $\displaystyle \int \cos^n u\,du = \frac{\cos^{n-1} u \sin u}{n} + \frac{n-1}{n}\int \cos^{n-2} u\,du$

52. $\displaystyle \int u \sin u\,du = \sin u - u \cos u + C$

53. $\displaystyle \int u \cos u\,du = \cos u + u \sin u + C$

54. $\displaystyle \int u^n \sin u\,du = -u^n \cos u + n\int u^{n-1} \cos u\,du$

55. $\displaystyle \int u^n \cos u\,du = u^n \sin u - n\int u^{n-1} \sin u\,du$

56. $\displaystyle \int \frac{1}{1 \pm \sin u}\,du = \tan u \mp \sec u + C$

57. $\displaystyle \int \frac{1}{1 \pm \cos u}\,du = -\cot u \pm \csc u + C$

58. $\displaystyle \int \frac{1}{\sin u \cos u}\,du = \ln|\tan u| + C$

Forms Involving $\tan u$, $\cot u$, $\sec u$, $\csc u$

59. $\displaystyle \int \tan u\,du = -\ln|\cos u| + C$

60. $\displaystyle\int \cot u \, du = \ln |\sin u| + C$

61. $\displaystyle\int \sec u \, du = \ln |\sec u + \tan u| + C$

62. $\displaystyle\int \csc u \, du = \ln |\csc u - \cot u| + C$

63. $\displaystyle\int \tan^2 u \, du = -u + \tan u + C$

64. $\displaystyle\int \cot^2 u \, du = -u - \cot u + C$

65. $\displaystyle\int \sec^2 u \, du = \tan u + C$

66. $\displaystyle\int \csc^2 u \, du = -\cot u + C$

67. $\displaystyle\int \tan^n u \, du = \frac{\tan^{n-1} u}{n-1} - \int \tan^{n-2} u \, du, \ n \neq 1$

68. $\displaystyle\int \cot^n u \, du = -\frac{\cot^{n-1} u}{n-1} - \int (\cot^{n-2} u) \, du, \ n \neq 1$

69. $\displaystyle\int \sec^n u \, du = \frac{\sec^{n-2} u \tan u}{n-1} + \frac{n-2}{n-1} \int \sec^{n-2} u \, du, \ n \neq 1$

70. $\displaystyle\int \csc^n u \, du = -\frac{\csc^{n-2} u \cot u}{n-1} + \frac{n-2}{n-1} \int \csc^{n-2} u \, du, \ n \neq 1$

71. $\displaystyle\int \frac{1}{1 \pm \tan u} \, du = \frac{1}{2}\big(u \pm \ln |\cos u \pm \sin u|\big) + C$

72. $\displaystyle\int \frac{1}{1 \pm \cot u} \, du = \frac{1}{2}\big(u \mp \ln |\sin u \pm \cos u|\big) + C$

73. $\displaystyle\int \frac{1}{1 \pm \sec u} \, du = u + \cot u \mp \csc u + C$

74. $\displaystyle\int \frac{1}{1 \pm \csc u} \, du = u - \tan u \pm \sec u + C$

Forms Involving Inverse Trigonometric Functions

75. $\displaystyle\int \arcsin u \, du = u \arcsin u + \sqrt{1 - u^2} + C$

76. $\displaystyle\int \arccos u \; du = u \arccos u - \sqrt{1 - u^2} + C$

77. $\displaystyle\int \arctan u \; du = u \arctan u - \ln \sqrt{1 + u^2} + C$

78. $\displaystyle\int \text{arccot } u \; du = u \text{ arccot } u + \ln \sqrt{1 + u^2} + C$

79. $\displaystyle\int \text{arcsec } u \; du = u \text{ arcsec } u - \ln \left| u + \sqrt{u^2 - 1} \right| + C$

80. $\displaystyle\int \text{arccsc } u \; du = u \text{ arccsc } u + \ln \left| u + \sqrt{u^2 - 1} \right| + C$

Forms Involving e^u

81. $\displaystyle\int e^u \; du = e^u + C$

82. $\displaystyle\int u e^u \; du = (u - 1)e^u + C$

83. $\displaystyle\int u^n e^u \; du = u^n e^u - n \int u^{n-1} e^u \; du$

84. $\displaystyle\int \frac{1}{1 + e^u} \; du = u - \ln (1 + e^u) + C$

85. $\displaystyle\int e^{au} \sin bu \; du = \frac{e^{au}}{a^2 + b^2}(a \sin bu - b \cos bu) + C$

86. $\displaystyle\int e^{au} \cos bu \; du = \frac{e^{au}}{a^2 + b^2}(a \cos bu + b \sin bu) + C$

Forms Involving $\ln u$

87. $\displaystyle\int \ln u \; du = u(-1 + \ln u) + C$

88. $\displaystyle\int u \ln u \; du = \frac{u^2}{4}(-1 + 2 \ln u) + C$

89. $\displaystyle\int u^n \ln u \; du = \frac{u^{n+1}}{(n + 1)^2}[-1 + (n + 1) \ln u] + C, \; n \neq -1$

90. $\displaystyle\int (\ln u)^2 \; du = u\left[2 - 2 \ln u + (\ln u)^2\right] + C$

91. $\displaystyle\int (\ln u)^n \; du = u(\ln u)^n - n \int (\ln u)^{n-1} \; du$

APPENDIX E

Rotation and the General Second-Degree Equation

Rotation of Axes • Invariants Under Rotation

Rotation of Axes

In Section 9.1, you learned that equations of conics with axes parallel to one of the coordinate axes can be written in the general form

$$Ax^2 + Cy^2 + Dx + Ey + F = 0.$$

Horizontal or vertical axes

Here you will study the equations of conics whose axes are rotated so that they are *not* parallel to the x-axis or the y-axis. The general equation for such conics contains an *xy-term.*

$$Ax^2 + Bxy + Cy^2 + Dx + Ey + F = 0$$

Equation in xy-plane

To eliminate this xy-term, you can use a procedure called **rotation of axes.** You want to rotate the x- and y-axes until they are parallel to the axes of the conic. (The rotated axes are denoted as the x'-axis and the y'-axis, as shown in Figure A.41.) After the rotation has been accomplished, the equation of the conic in the new $x'y'$-plane will have the form

$$A'(x')^2 + C'(y')^2 + D'x' + E'y' + F' = 0.$$

Equation in $x'y'$-plane

Because this equation has no $x'y'$-term, you can obtain a standard form by completing the square.

The following theorem identifies how much to rotate the axes to eliminate an xy-term and also the equations for determining the new coefficients $A', C', D', E',$ and F'.

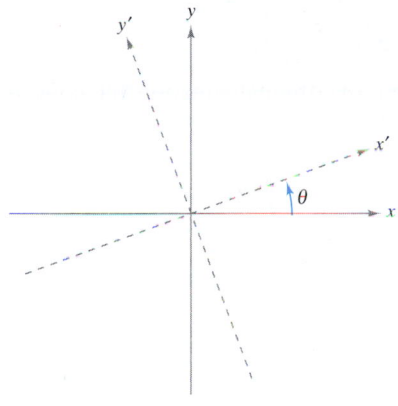

After rotation of the x- and y-axes counterclockwise through an angle θ, the rotated axes are denoted as the x'-axis and y'-axis.
Figure A.41

THEOREM A.1   Rotation of Axes

The general equation of the conic

$$Ax^2 + Bxy + Cy^2 + Dx + Ey + F = 0,$$

where $B \neq 0$, can be rewritten as

$$A'(x')^2 + C'(y')^2 + D'x' + E'y' + F' = 0$$

by rotating the coordinate axes through an angle θ, where

$$\cot 2\theta = \frac{A - C}{B}.$$

The coefficients of the new equation are obtained by making the substitutions

$$x = x'\cos\theta - y'\sin\theta$$

$$y = x'\sin\theta + y'\cos\theta.$$

A51

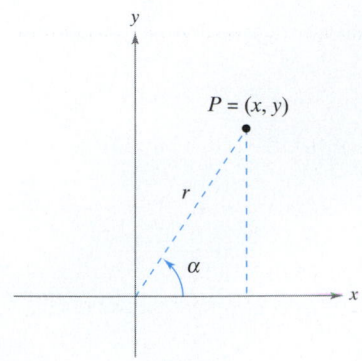

Original: $x = r \cos \alpha$
$\quad\quad\quad\quad y = r \sin \alpha$

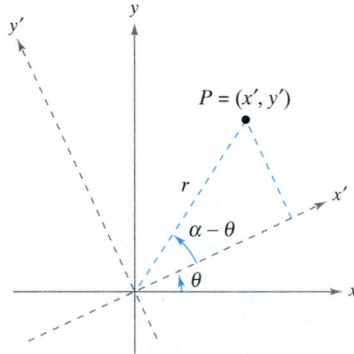

Rotated: $x' = r \cos(\alpha - \theta)$
$\quad\quad\quad\quad y' = r \sin(\alpha - \theta)$

Figure A.42

Proof To discover how the coordinates in the xy-system are related to the coordinates in the $x'y'$-system, choose a point $P = (x, y)$ in the original system and attempt to find its coordinates (x', y') in the rotated system. In either system, the distance r between the point P and the origin is the same, and thus the equations for x, y, x', and y' are those given in Figure A.42. Using the formulas for the sine and cosine of the difference of two angles, you obtain

$$x' = r\cos(\alpha - \theta) = r(\cos \alpha \cos \theta + \sin \alpha \sin \theta)$$

$$= r \cos \alpha \cos \theta + r \sin \alpha \sin \theta = x \cos \theta + y \sin \theta$$

$$y' = r \sin(\alpha - \theta) = r(\sin \alpha \cos \theta - \cos \alpha \sin \theta)$$

$$= r \sin \alpha \cos \theta - r \cos \alpha \sin \theta = y \cos \theta - x \sin \theta.$$

Solving this system for x and y yields

$$x = x' \cos \theta - y' \sin \theta \quad \text{and} \quad y = x' \sin \theta + y' \cos \theta.$$

Finally, by substituting these values for x and y into the original equation and collecting terms, you obtain the following.

$$A' = A \cos^2 \theta + B \cos \theta \sin \theta + C \sin^2 \theta$$

$$C' = A \sin^2 \theta - B \cos \theta \sin \theta + C \cos^2 \theta$$

$$D' = D \cos \theta + E \sin \theta$$

$$E' = -D \sin \theta + E \cos \theta$$

$$F' = F$$

Now, in order to eliminate the $x'y'$-term, you must select θ such that $B' = 0$, as follows.

$$B' = 2(C - A) \sin \theta \cos \theta + B(\cos^2 \theta - \sin^2 \theta)$$

$$= (C - A) \sin 2\theta + B \cos 2\theta$$

$$= B(\sin 2\theta)\left(\frac{C - A}{B} + \cot 2\theta\right) = 0, \quad \sin 2\theta \neq 0$$

If $B = 0$, no rotation is necessary, because the xy-term is not present in the original equation. If $B \neq 0$, the only way to make $B' = 0$ is to let

$$\cot 2\theta = \frac{A - C}{B}, \quad B \neq 0.$$

Thus, you have established the desired results.

EXAMPLE 1   Rotation of a Hyperbola

Write the equation $xy - 1 = 0$ in standard form.

Solution   Because $A = 0$, $B = 1$, and $C = 0$, you have (for $0 < \theta < \pi/2$)

$$\cot 2\theta = \frac{A - C}{B} = 0 \quad \Longrightarrow \quad 2\theta = \frac{\pi}{2} \quad \Longrightarrow \quad \theta = \frac{\pi}{4}.$$

The equation in the $x'y'$-system is obtained by making the following substitutions.

$$x = x' \cos \frac{\pi}{4} - y' \sin \frac{\pi}{4} = x'\left(\frac{\sqrt{2}}{2}\right) - y'\left(\frac{\sqrt{2}}{2}\right) = \frac{x' - y'}{\sqrt{2}}$$

$$y = x' \sin \frac{\pi}{4} + y' \cos \frac{\pi}{4} = x'\left(\frac{\sqrt{2}}{2}\right) + y'\left(\frac{\sqrt{2}}{2}\right) = \frac{x' + y'}{\sqrt{2}}$$

Substituting these expressions into the equation $xy - 1 = 0$ produces

$$\left(\frac{x' - y'}{\sqrt{2}}\right)\left(\frac{x' + y'}{\sqrt{2}}\right) - 1 = 0$$

$$\frac{(x')^2 - (y')^2}{2} - 1 = 0$$

$$\frac{(x')^2}{(\sqrt{2})^2} - \frac{(y')^2}{(\sqrt{2})^2} = 1. \qquad \text{Standard form}$$

This is the equation of a hyperbola centered at the origin with vertices at $\left(\pm\sqrt{2}, 0\right)$ in the $x'y'$-system, as shown in Figure A.43.

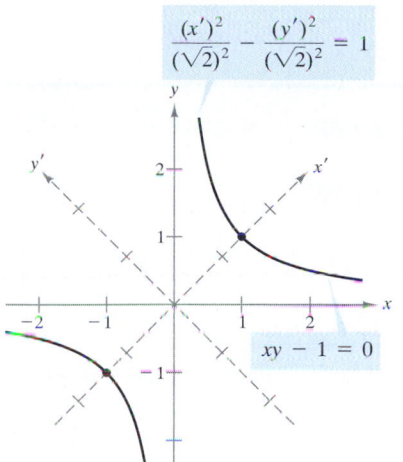

$$\frac{(x')^2}{(\sqrt{2})^2} - \frac{(y')^2}{(\sqrt{2})^2} = 1$$

$$xy - 1 = 0$$

Vertices:
$\left(\sqrt{2}, 0\right), \left(-\sqrt{2}, 0\right)$ in $x'y'$-system
$(1, 1), (-1, -1)$ in xy-system
Figure A.43

EXAMPLE 2   Rotation of an Ellipse

Sketch the graph of $7x^2 - 6\sqrt{3}xy + 13y^2 - 16 = 0$.

Solution   Because $A = 7$, $B = -6\sqrt{3}$, and $C = 13$, you have (for $0 < \theta < \pi/2$)

$$\cot 2\theta = \frac{A - C}{B} = \frac{7 - 13}{-6\sqrt{3}} = \frac{1}{\sqrt{3}} \quad \Longrightarrow \quad \theta = \frac{\pi}{6}.$$

Therefore, the equation in the $x'y'$-system is derived by making the following substitutions.

$$x = x' \cos \frac{\pi}{6} - y' \sin \frac{\pi}{6} = x'\left(\frac{\sqrt{3}}{2}\right) - y'\left(\frac{1}{2}\right) = \frac{\sqrt{3}x' - y'}{2}$$

$$y = x' \sin \frac{\pi}{6} + y' \cos \frac{\pi}{6} = x'\left(\frac{1}{2}\right) + y'\left(\frac{\sqrt{3}}{2}\right) = \frac{x' + \sqrt{3}y'}{2}$$

Substituting these expressions into the original equation eventually simplifies (after considerable algebra) to

$$4(x')^2 + 16(y')^2 = 16$$

$$\frac{(x')^2}{(2)^2} + \frac{(y')^2}{(1)^2} = 1. \qquad \text{Standard form}$$

This is the equation of an ellipse centered at the origin with vertices at $(\pm 2, 0)$ and $(0, \pm 1)$ in the $x'y'$-system, as shown in Figure A.44.

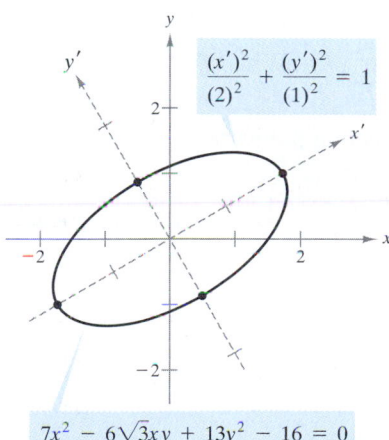

$$\frac{(x')^2}{(2)^2} + \frac{(y')^2}{(1)^2} = 1$$

$$7x^2 - 6\sqrt{3}xy + 13y^2 - 16 = 0$$

Vertices:
$(\pm 2, 0), (0, \pm 1)$ in $x'y'$-system
$\left(\pm\sqrt{3}, \pm 1\right), \left(\pm\frac{1}{2}, \pm\frac{\sqrt{3}}{2}\right)$ in xy-system

Figure A.44

In writing Examples 1 and 2, we chose the equations such that θ would be one of the common angles $30°$, $45°$, and so forth. Of course, many second-degree equations do not yield such common solutions to the equation

$$\cot 2\theta = \frac{A - C}{B}.$$

Example 3 illustrates such a case.

EXAMPLE 3   Rotation of a Parabola

Sketch the graph of $x^2 - 4xy + 4y^2 + 5\sqrt{5}y + 1 = 0$.

Solution   Because $A = 1$, $B = -4$, and $C = 4$, you have

$$\cot 2\theta = \frac{A - C}{B} = \frac{1 - 4}{-4} = \frac{3}{4}.$$

The trigonometric identity $\cot 2\theta = (\cot^2 \theta - 1)/(2 \cot \theta)$ produces

$$\cot 2\theta = \frac{3}{4} = \frac{\cot^2 \theta - 1}{2 \cot \theta}$$

from which you can obtain the equation

$$6 \cot \theta = 4 \cot^2 \theta - 4 \quad \Longrightarrow \quad 4 \cot^2 \theta - 6 \cot \theta - 4 = 0$$

$$(2 \cot \theta - 4)(2 \cot \theta + 1) = 0.$$

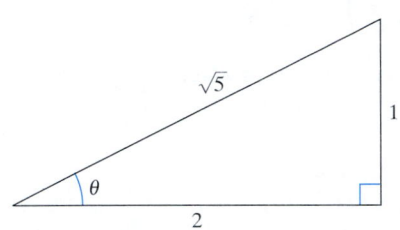

Figure A.45

Considering $0 < \theta < \pi/2$, it follows that $2 \cot \theta = 4$. Thus,

$$\cot \theta = 2 \quad \Longrightarrow \quad \theta \approx 26.6°.$$

From the triangle in Figure A.45, you can obtain $\sin \theta = 1/\sqrt{5}$ and $\cos \theta = 2/\sqrt{5}$. Consequently, you can write the following.

$$x = x' \cos \theta - y' \sin \theta = x'\left(\frac{2}{\sqrt{5}}\right) - y'\left(\frac{1}{\sqrt{5}}\right) = \frac{2x' - y'}{\sqrt{5}}$$

$$y = x' \sin \theta + y' \cos \theta = x'\left(\frac{1}{\sqrt{5}}\right) + y'\left(\frac{2}{\sqrt{5}}\right) = \frac{x' + 2y'}{\sqrt{5}}$$

Substituting these expressions into the original equation produces

$$\left(\frac{2x' - y'}{\sqrt{5}}\right)^2 - 4\left(\frac{2x' - y'}{\sqrt{5}}\right)\left(\frac{x' + 2y'}{\sqrt{5}}\right) + 4\left(\frac{x' + 2y'}{\sqrt{5}}\right)^2 +$$

$$5\sqrt{5}\left(\frac{x' + 2y'}{\sqrt{5}}\right) + 1 = 0$$

which simplifies to

$$5(y')^2 + 5x' + 10y' + 1 = 0.$$

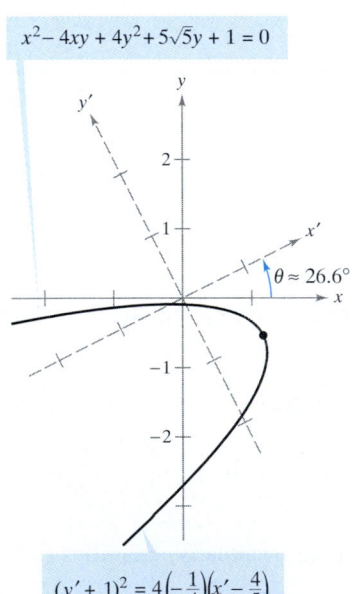

$x^2 - 4xy + 4y^2 + 5\sqrt{5}y + 1 = 0$

$\theta \approx 26.6°$

$(y' + 1)^2 = 4\left(-\frac{1}{4}\right)\left(x' - \frac{4}{5}\right)$

Vertex: $\left(\frac{4}{5}, -1\right)$ in $x'y'$-system

$\left(\frac{13}{5\sqrt{5}}, -\frac{6}{5\sqrt{5}}\right)$ in xy-system

Figure A.46

By completing the square, you can obtain the standard form

$$5(y' + 1)^2 = -5x' + 4$$

$$(y' + 1)^2 = 4\left(-\frac{1}{4}\right)\left(x' - \frac{4}{5}\right). \qquad \text{Standard form}$$

The graph of the equation is a parabola with its vertex at $\left(\frac{4}{5}, -1\right)$ and its axis parallel to the x'-axis in the $x'y'$-system, as shown in Figure A.46.

Invariants Under Rotation

In Theorem A.1, note that the constant term $F' = F$ is the same in both equations. Because of this, F is said to be **invariant under rotation.** Theorem A.2 lists some other rotation invariants. The proof of this theorem is left as an exercise (see Exercise 34).

THEOREM A.2   Rotation Invariants

The rotation of coordinate axes through an angle θ that transforms the equation $Ax^2 + Bxy + Cy^2 + Dx + Ey + F = 0$ into the form

$$A'(x')^2 + C'(y')^2 + D'x' + E'y' + F' = 0$$

has the following rotation invariants.

1. $F = F'$
2. $A + C = A' + C'$
3. $B^2 - 4AC = (B')^2 - 4A'C'$

You can use this theorem to classify the graph of a second-degree equation *with* an xy-term in much the same way you do for a second-degree equation *without* an xy-term. Note that because $B' = 0$, the invariant $B^2 - 4AC$ reduces to

$$B^2 - 4AC = -4A'C' \qquad \text{Discriminant}$$

which is called the **discriminant** of the equation

$$Ax^2 + Bxy + Cy^2 + Dx + Ey + F = 0.$$

Because the sign of $A'C'$ determines the type of graph for the equation

$$A'(x')^2 + C'(y')^2 + D'x' + E'y' + F' = 0$$

the sign of $B^2 - 4AC$ must determine the type of graph for the original equation. This result is stated in Theorem A.3.

THEOREM A.3   Classification of Conics by the Discriminant

The graph of the equation

$$Ax^2 + Bxy + Cy^2 + Dx + Ey + F = 0$$

is, except in degenerate cases, determined by its discriminant as follows.

1. *Ellipse or circle* $B^2 - 4AC < 0$
2. *Parabola* $B^2 - 4AC = 0$
3. *Hyperbola* $B^2 - 4AC > 0$

EXAMPLE 4 Using the Discriminant

Classify the graph of each of the following equations.

a. $4xy - 9 = 0$

b. $2x^2 - 3xy + 2y^2 - 2x = 0$

c. $x^2 - 6xy + 9y^2 - 2y + 1 = 0$

d. $3x^2 + 8xy + 4y^2 - 7 = 0$

Solution

a. The graph is a hyperbola because

$$B^2 - 4AC = 16 - 0 > 0.$$

b. The graph is a circle or an ellipse because

$$B^2 - 4AC = 9 - 16 < 0.$$

c. The graph is a parabola because

$$B^2 - 4AC = 36 - 36 = 0.$$

d. The graph is a hyperbola because

$$B^2 - 4AC = 64 - 48 > 0.$$

EXERCISES FOR APPENDIX E

In Exercises 1–12, rotate the axes to eliminate the xy-term. Give the resulting equation and sketch its graph showing both sets of axes.

1. $xy + 1 = 0$

2. $xy - 4 = 0$

3. $x^2 - 10xy + y^2 + 1 = 0$

4. $xy + x - 2y + 3 = 0$

5. $xy - 2y - 4x = 0$

6. $13x^2 + 6\sqrt{3}xy + 7y^2 - 16 = 0$

7. $5x^2 - 2xy + 5y^2 - 12 = 0$

8. $2x^2 - 3xy - 2y^2 + 10 = 0$

9. $3x^2 - 2\sqrt{3}xy + y^2 + 2x + 2\sqrt{3}y = 0$

10. $16x^2 - 24xy + 9y^2 - 60x - 80y + 100 = 0$

11. $9x^2 + 24xy + 16y^2 + 90x - 130y = 0$

12. $9x^2 + 24xy + 16y^2 + 80x - 60y = 0$

In Exercises 13–18, use a graphing utility to graph the conic. Determine the angle θ through which the axes are rotated. Explain how you used the utility to obtain the graph.

13. $x^2 + xy + y^2 = 10$

14. $x^2 - 4xy + 2y^2 = 6$

15. $17x^2 + 32xy - 7y^2 = 75$

16. $40x^2 + 36xy + 25y^2 = 52$

17. $32x^2 + 50xy + 7y^2 = 52$

18. $4x^2 - 12xy + 9y^2 + (4\sqrt{13} + 12)x - (6\sqrt{13} + 8)y = 91$

In Exercises 19–26, use the discriminant to determine whether the graph of the equation is a parabola, an ellipse, or a hyperbola.

19. $16x^2 - 24xy + 9y^2 - 30x - 40y = 0$

20. $x^2 - 4xy - 2y^2 - 6 = 0$

21. $13x^2 - 8xy + 7y^2 - 45 = 0$

22. $2x^2 + 4xy + 5y^2 + 3x - 4y - 20 = 0$

23. $x^2 - 6xy - 5y^2 + 4x - 22 = 0$

24. $36x^2 - 60xy + 25y^2 + 9y = 0$

25. $x^2 + 4xy + 4y^2 - 5x - y - 3 = 0$

26. $x^2 + xy + 4y^2 + x + y - 4 = 0$

In Exercises 27–32, sketch the graph (if possible) of the degenerate conic.

27. $y^2 - 4x^2 = 0$

28. $x^2 + y^2 - 2x + 6y + 10 = 0$

29. $x^2 + 2xy + y^2 - 1 = 0$

30. $x^2 - 10xy + y^2 = 0$

31. $(x - 2y + 1)(x + 2y - 3) = 0$

32. $(2x + y - 3)^2 = 0$

33. Show that the equation $x^2 + y^2 = r^2$ is invariant under rotation of axes.

34. Prove Theorem A.2.

The symbol ⚛ *indicates an exercise in which you are instructed to use graphing technology or a symbolic computer algebra system. The solutions of other exercises may also be facilitated by use of appropriate technology.*

A blue number indicates that a detailed solution can be found in the Study and Solutions Guide.

APPENDIX F
Complex Numbers

Operations with Complex Numbers • Complex Solutions of Quadratic Equations •
Polar Form of a Complex Number • Powers and Roots of Complex Numbers

Operations with Complex Numbers

Some equations have no real solutions. For instance, the quadratic equation

$$x^2 + 1 = 0 \qquad \text{Equation with no real solution}$$

has no real solution because there is no real number x that can be squared to produce -1. To overcome this deficiency, mathematicians created an expanded system of numbers using the **imaginary unit** i, defined as

$$i = \sqrt{-1} \qquad \text{Imaginary unit}$$

where $i^2 = -1$. By adding real numbers to real multiples of this imaginary unit, we obtain the set of **complex numbers.** Each complex number can be written in the **standard form,** $a + bi$.

Definition of a Complex Number

For real numbers a and b, the number

$$a + bi$$

is a **complex number.** If $b \neq 0$, $a + bi$ is called an **imaginary number,** and bi is called a **pure imaginary number.**

To add (or subtract) two complex numbers, you add (or subtract) the real and imaginary parts of the numbers separately.

Addition and Subtraction of Complex Numbers

If $a + bi$ and $c + di$ are two complex numbers written in standard form, their sum and difference are defined as follows.

Sum: $(a + bi) + (c + di) = (a + c) + (b + d)i$

Difference: $(a + bi) - (c + di) = (a - c) + (b - d)i$

The **additive identity** in the complex number system is zero (the same as in the real number system). Furthermore, the **additive inverse** of the complex number $a + bi$ is

$$-(a + bi) = -a - bi \qquad \text{Additive inverse}$$

Thus, you have

$$(a + bi) + (-a - bi) = 0 + 0i = 0.$$

EXAMPLE 1 Adding and Subtracting Complex Numbers

a. $(3 - i) + (2 + 3i) = 3 - i + 2 + 3i$ Remove parentheses.

$\qquad\qquad\qquad\quad = 3 + 2 - i + 3i$ Group like terms.

$\qquad\qquad\qquad\quad = (3 + 2) + (-1 + 3)i$

$\qquad\qquad\qquad\quad = 5 + 2i$ Standard form

b. $2i + (-4 - 2i) = 2i - 4 - 2i$ Remove parentheses.

$\qquad\qquad\qquad = -4 + 2i - 2i$ Group like terms.

$\qquad\qquad\qquad = -4$ Standard form

c. $3 - (-2 + 3i) + (-5 + i) = 3 + 2 - 3i - 5 + i$

$\qquad\qquad\qquad\qquad\qquad = 3 + 2 - 5 - 3i + i$

$\qquad\qquad\qquad\qquad\qquad = 0 - 2i$

$\qquad\qquad\qquad\qquad\qquad = -2i$

> **NOTE** Notice in Example 1b that the sum of two complex numbers can be a real number.

Many of the properties of real numbers are valid for complex numbers as well. Here are some examples.

Associative Properties of Addition and Multiplication

Commutative Properties of Addition and Multiplication

Distributive Property of Multiplication over Addition

Notice below how these properties are used when two complex numbers are multiplied.

$(a + bi)(c + di) = a(c + di) + bi(c + di)$ Distributive

$\qquad\qquad\quad = ac + (ad)i + (bc)i + (bd)i^2$ Distributive

$\qquad\qquad\quad = ac + (ad)i + (bc)i + (bd)(-1)$ Definition of i

$\qquad\qquad\quad = ac - bd + (ad)i + (bc)i$ Commutative

$\qquad\qquad\quad = (ac - bd) + (ad + bc)i$ Associative

> **STUDY TIP** Rather than trying to memorize the multiplication rule at the right, you can simply remember how the Distributive Property is used to multiply two complex numbers. The procedure is similar to multiplying two polynomials and combining like terms.

EXAMPLE 2   **Multiplying Complex Numbers**

a. $(3 + 2i)(3 - 2i) = 9 - 6i + 6i - 4i^2$ Product of binomials

$$= 9 - 4(-1)$$ $i^2 = -1$

$$= 9 + 4$$ Simplify.

$$= 13$$ Standard form

b. $(3 + 2i)^2 = 9 + 6i + 6i + 4i^2$ Product of binomials

$$= 9 + 4(-1) + 12i$$ $i^2 = -1$

$$= 9 - 4 + 12i$$ Simplify.

$$= 5 + 12i$$ Standard form

Notice in Example 2a that the product of two complex numbers can be a real number. This occurs with pairs of complex numbers of the form $a + bi$ and $a - bi$, called **complex conjugates.**

$$(a + bi)(a - bi) = a^2 - abi + abi - b^2i^2$$

$$= a^2 - b^2(-1)$$

$$= a^2 + b^2$$

To find the quotient of $a + bi$ and $c + di$ where c and d are not both zero, multiply the numerator and denominator by the conjugate of the denominator to obtain

$$\frac{a + bi}{c + di} = \frac{a + bi}{c + di}\left(\frac{c - di}{c - di}\right) = \frac{(ac + bd) + (bc - ad)i}{c^2 + d^2}.$$

EXAMPLE 3   **Dividing Complex Numbers**

$$\frac{2 + 3i}{4 - 2i} = \frac{2 + 3i}{4 - 2i}\left(\frac{4 + 2i}{4 + 2i}\right)$$ Multiply by conjugate.

$$= \frac{8 + 4i + 12i + 6i^2}{16 - 4i^2}$$ Expand.

$$= \frac{8 - 6 + 16i}{16 + 4}$$ $i^2 = -1$

$$= \frac{1}{20}(2 + 16i)$$ Simplify.

$$= \frac{1}{10} + \frac{4}{5}i$$ Standard form

Complex Solutions of Quadratic Equations

When using the Quadratic Formula to solve a quadratic equation, you often obtain a result such as $\sqrt{-3}$, which you know is not a real number. By factoring out $i = \sqrt{-1}$, you can write this number in standard form.

$$\sqrt{-3} = \sqrt{3(-1)} = \sqrt{3}\sqrt{-1} = \sqrt{3}i$$

The number $\sqrt{3}i$ is called the principal square root of -3.

STUDY TIP The definition of principal square root uses the rule

$$\sqrt{ab} = \sqrt{a}\sqrt{b}$$

for $a > 0$ and $b < 0$. The rule is not valid if *both* a and b are negative. For example,

$$\sqrt{-5}\sqrt{-5} = \sqrt{5}i\sqrt{5}i$$
$$= \sqrt{25}i^2$$
$$= 5i^2 = -5$$

whereas

$$\sqrt{(-5)(-5)} = \sqrt{25} = 5.$$

To avoid problems with multiplying square roots of negative numbers, be sure to convert to standard form *before* multiplying.

Principal Square Root of a Negative Number

If a is a positive number, the **principal square root** of the negative number $-a$ is

$$\sqrt{-a} = \sqrt{a}i.$$

EXAMPLE 4 **Writing Complex Numbers in Standard Form**

a. $\sqrt{-3}\sqrt{-12} = \sqrt{3}i\sqrt{12}i = \sqrt{36}i^2 = 6(-1) = -6$

b. $\sqrt{-48} - \sqrt{-27} = \sqrt{48}i - \sqrt{27}i = 4\sqrt{3}i - 3\sqrt{3}i = \sqrt{3}i$

c. $\left(-1 + \sqrt{-3}\right)^2 = \left(-1 + \sqrt{3}i\right)^2$

$$= (-1)^2 - 2\sqrt{3}i + \left(\sqrt{3}\right)^2(i^2)$$

$$= 1 - 2\sqrt{3}i + 3(-1)$$

$$= -2 - 2\sqrt{3}i$$

EXAMPLE 5 **Complex Solutions of a Quadratic Equation**

Solve $3x^2 - 2x + 5 = 0$.

Solution

$$x = \frac{-(-2) \pm \sqrt{(-2)^2 - 4(3)(5)}}{2(3)} \qquad \text{Quadratic Formula}$$

$$= \frac{2 \pm \sqrt{-56}}{6} \qquad \text{Simplify.}$$

$$= \frac{2 \pm 2\sqrt{14}i}{6} \qquad \text{Write in } i\text{-form.}$$

$$= \frac{1}{3} \pm \frac{\sqrt{14}}{3}i \qquad \text{Standard form}$$

Polar Form of a Complex Number

Just as real numbers can be represented by points on the real number line, you can represent a complex number

$$z = a + bi$$

at the point (a, b) in a coordinate plane (the **complex plane**). The horizontal axis is called the **real axis** and the vertical axis is called the **imaginary axis,** as shown in Figure A.47.

The **absolute value** of the complex number $a + bi$ is defined as the distance between the origin $(0, 0)$ and the point (a, b).

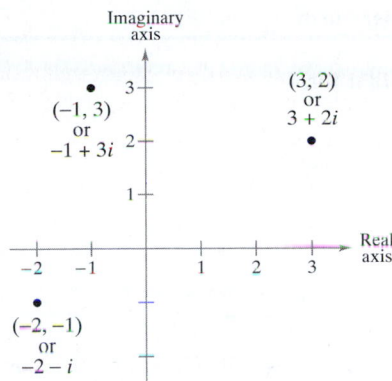

Figure A.47

The Absolute Value of a Complex Number

The **absolute value** of the complex number $z = a + bi$ is given by

$$|a + bi| = \sqrt{a^2 + b^2}.$$

NOTE   If the complex number $a + bi$ is a real number (that is, if $b = 0$), then this definition agrees with that given for the absolute value of a real number.

$$|a + 0i| = \sqrt{a^2 + 0^2} = |a|.$$

To work effectively with *powers* and *roots* of complex numbers, it is helpful to write complex numbers in **polar form.** In Figure A.48, consider the nonzero complex number $a + bi$. By letting θ be the angle from the positive x-axis (measured counterclockwise) to the line segment connecting the origin and the point (a, b), you can write

$$a = r \cos \theta \qquad \text{and} \qquad b = r \sin \theta$$

where $r = \sqrt{a^2 + b^2}$. Consequently, you have

$$a + bi = (r \cos \theta) + (r \sin \theta)i$$

from which you can obtain the **polar form of a complex number.**

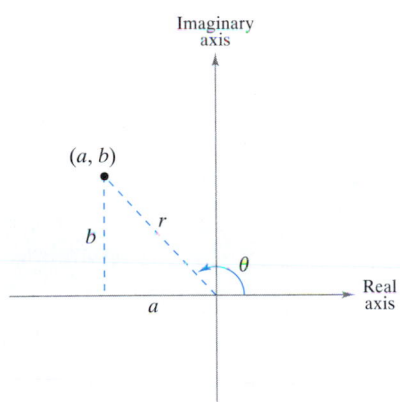

Figure A.48

Polar Form of a Complex Number

The **polar form** of the complex number $z = a + bi$ is

$$z = r(\cos \theta + i \sin \theta)$$

where $a = r \cos \theta$, $b = r \sin \theta$, $r = \sqrt{a^2 + b^2}$, and $\tan \theta = b/a$. The number r is the **modulus** of z, and θ is called an **argument** of z.

NOTE   The polar form of a complex number is also called the **trigonometric form.** Because there are infinitely many choices for θ, the polar form of a complex number is not unique. Normally, θ is restricted to the interval $0 \leq \theta < 2\pi$, although on occasion it is convenient to use $\theta < 0$.

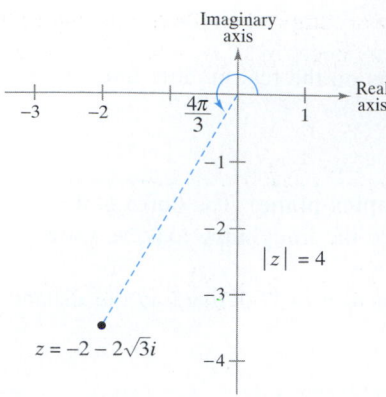

Figure A.49

EXAMPLE 6   Writing a Complex Number in Polar Form

Write the complex number $z = -2 - 2\sqrt{3}i$ in polar form.

Solution   The absolute value of z is

$$r = |-2 - 2\sqrt{3}i| = \sqrt{(-2)^2 + (-2\sqrt{3})^2} = \sqrt{16} = 4$$

and the angle θ is given by

$$\tan \theta = \frac{b}{a} = \frac{-2\sqrt{3}}{-2} = \sqrt{3}.$$

Because $\tan(\pi/3) = \sqrt{3}$ and because $z = -2 - 2\sqrt{3}i$ lies in Quadrant III, you choose θ to be $\theta = \pi + \pi/3 = 4\pi/3$. Thus, the polar form is

$$z = r(\cos \theta + i \sin \theta) = 4\left(\cos \frac{4\pi}{3} + i \sin \frac{4\pi}{3}\right).$$

(See Figure A.49.)

The polar form adapts nicely to multiplication and division of complex numbers. Suppose you are given two complex numbers

$$z_1 = r_1(\cos \theta_1 + i \sin \theta_1) \qquad \text{and} \qquad z_2 = r_2(\cos \theta_2 + i \sin \theta_2).$$

The product of z_1 and z_2 is

$$z_1 z_2 = r_1 r_2 (\cos \theta_1 + i \sin \theta_1)(\cos \theta_2 + i \sin \theta_2)$$
$$= r_1 r_2 [(\cos \theta_1 \cos \theta_2 - \sin \theta_1 \sin \theta_2) + i(\sin \theta_1 \cos \theta_2 + \cos \theta_1 \sin \theta_2)].$$

Using the sum and difference formulas for cosine and sine, you can rewrite this equation as

$$z_1 z_2 = r_1 r_2 [\cos(\theta_1 + \theta_2) + i \sin(\theta_1 + \theta_2)].$$

This establishes the first part of the following rule. Try to establish the second part on your own.

Product and Quotient of Two Complex Numbers

Let $z_1 = r_1(\cos \theta_1 + i \sin \theta_1)$ and $z_2 = r_2(\cos \theta_2 + i \sin \theta_2)$ be complex numbers.

$$z_1 z_2 = r_1 r_2 [\cos(\theta_1 + \theta_2) + i \sin(\theta_1 + \theta_2)] \qquad \text{Product}$$

$$\frac{z_1}{z_2} = \frac{r_1}{r_2} [\cos(\theta_1 - \theta_2) + i \sin(\theta_1 - \theta_2)], \quad z_2 \neq 0 \qquad \text{Quotient}$$

Note that this rule says that to multiply two complex numbers you multiply moduli and add arguments, whereas to divide two complex numbers you divide moduli and subtract arguments.

EXAMPLE 7   Multiplying Complex Numbers in Polar Form

Find the product of the complex numbers.

$$z_1 = 2\left(\cos\frac{2\pi}{3} + i\sin\frac{2\pi}{3}\right) \qquad z_2 = 8\left(\cos\frac{11\pi}{6} + i\sin\frac{11\pi}{6}\right)$$

Solution

$$z_1 z_2 = 2\left(\cos\frac{2\pi}{3} + i\sin\frac{2\pi}{3}\right) \cdot 8\left(\cos\frac{11\pi}{6} + i\sin\frac{11\pi}{6}\right)$$

$$= 16\left[\cos\left(\frac{2\pi}{3} + \frac{11\pi}{6}\right) + i\sin\left(\frac{2\pi}{3} + \frac{11\pi}{6}\right)\right]$$

$$= 16\left[\cos\frac{5\pi}{2} + i\sin\frac{5\pi}{2}\right]$$

$$= 16\left[\cos\frac{\pi}{2} + i\sin\frac{\pi}{2}\right]$$

$$= 16[0 + i(1)] = 16i$$

Check this result by first converting to the standard forms $z_1 = -1 + \sqrt{3}i$ and $z_2 = 4\sqrt{3} - 4i$ and then multiplying algebraically.

EXAMPLE 8   Dividing Complex Numbers in Polar Form

Find the quotient z_1/z_2 of the complex numbers.

$$z_1 = 24(\cos 300° + i\sin 300°) \qquad z_2 = 8(\cos 75° + i\sin 75°)$$

Solution

$$\frac{z_1}{z_2} = \frac{24(\cos 300° + i\sin 300°)}{8(\cos 75° + i\sin 75°)}$$

$$= \frac{24}{8}\left[\cos(300° - 75°) + i\sin(300° - 75°)\right]$$

$$= 3\left[\cos 225° + i\sin 225°\right]$$

$$= 3\left[\left(-\frac{\sqrt{2}}{2}\right) + i\left(-\frac{\sqrt{2}}{2}\right)\right]$$

$$= -\frac{3\sqrt{2}}{2} - \frac{3\sqrt{2}}{2}i$$

Powers and Roots of Complex Numbers

To raise a complex number to a power, consider repeated use of the multiplication rule.

$$z = r(\cos \theta + i \sin \theta)$$

$$z^2 = r^2(\cos 2\theta + i \sin 2\theta)$$

$$z^3 = r^3(\cos 3\theta + i \sin 3\theta)$$

$$\vdots$$

This pattern leads to the following important theorem, which is named after the French mathematician Abraham DeMoivre (1667–1754).

THEOREM A.4 DeMoivre's Theorem

If $z = r(\cos \theta + i \sin \theta)$ is a complex number and n is a positive integer, then

$$z^n = [r(\cos \theta + i \sin \theta)]^n = r^n(\cos n\theta + i \sin n\theta).$$

EXAMPLE 9 Finding Powers of a Complex Number

Use DeMoivre's Theorem to find $\left(-1 + \sqrt{3}i\right)^{12}$.

Solution First convert to polar form.

$$-1 + \sqrt{3}i = 2\left(\cos \frac{2\pi}{3} + i \sin \frac{2\pi}{3}\right)$$

Then, by DeMoivre's Theorem, you have

$$\left(-1 + \sqrt{3}i\right)^{12} = \left[2\left(\cos \frac{2\pi}{3} + i \sin \frac{2\pi}{3}\right)\right]^{12}$$

$$= 2^{12}\left[\cos(12)\frac{2\pi}{3} + i \sin(12)\frac{2\pi}{3}\right]$$

$$= 4096(\cos 8\pi + i \sin 8\pi)$$

$$= 4096.$$

NOTE Notice in Example 9 that the answer is a real number.

Recall that a consequence of the Fundamental Theorem of Algebra is that a polynomial equation of degree n has n solutions in the complex number system. Each solution is an nth root of the equation. The **nth root** of a complex number is defined as follows.

Definition of nth Root of a Complex Number

The complex number $u = a + bi$ is an nth root of the complex number z if

$$z = u^n = (a + bi)^n.$$

To find a formula for an nth root of a complex number, let u be an nth root of z, where

$$u = s(\cos \beta + i \sin \beta) \quad \text{and} \quad z = r(\cos \theta + i \sin \theta).$$

By DeMoivre's Theorem and the fact that $u^n = z$, you have

$$s_n(\cos n\beta + i \sin n\beta) = r(\cos \theta + i \sin \theta).$$

Taking the absolute values of both sides of this equation, it follows that $s^n = r$. Substituting back into the previous equation and dividing by r, you get

$$\cos n\beta + i \sin n\beta = \cos \theta + i \sin \theta.$$

Thus, it follows that

$$\cos n\beta = \cos \theta \quad \text{and} \quad \sin n\beta = \sin \theta.$$

Because both sine and cosine have a period of 2π, these last two equations have solutions if and only if the angles differ by a multiple of 2π. Consequently, there must exist an integer k such that

$$n\beta = \theta + 2\pi k$$

$$\beta = \frac{\theta + 2\pi k}{n}.$$

By substituting this value for β into the polar form of u, you get the following result.

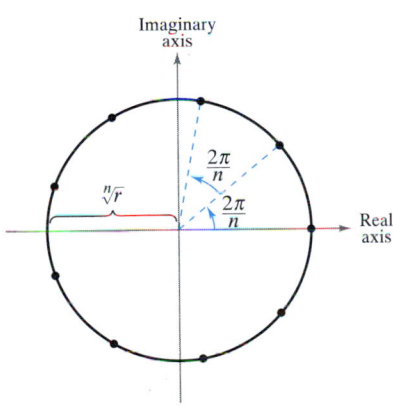

Figure A.50

> **THEOREM A.5 nth Roots of a Complex Number**
>
> For a positive integer n, the complex number $z = r(\cos \theta + i \sin \theta)$ has exactly n distinct nth roots given by
>
> $$\sqrt[n]{r}\left(\cos \frac{\theta + 2\pi k}{n} + i \sin \frac{\theta + 2\pi k}{n} \right)$$
>
> where $k = 0, 1, 2, \ldots, n - 1$.

This formula for the nth roots of a complex number z has a nice geometrical interpretation, as shown in Figure A.50. Note that because the nth roots of z all have the same magnitude $\sqrt[n]{r}$, they all lie on a circle of radius $\sqrt[n]{r}$ with center at the origin. Furthermore, because successive nth roots have arguments that differ by $2\pi/n$, the n roots are equally spaced along the circle.

EXAMPLE 10   Finding the *n*th Roots of a Complex Number

Find the three cube roots of $z = -2 + 2i$.

Solution

Because z lies in Quadrant II, the polar form for z is

$$z = -2 + 2i = \sqrt{8}\,(\cos 135° + i \sin 135°).$$

By the formula for *n*th roots, the cube roots have the form

$$\sqrt[6]{8}\left(\cos \frac{135° + 360°k}{3} + i \sin \frac{135° + 360°k}{3}\right).$$

Finally, for $k = 0$, 1, and 2, you obtain the roots

$$\sqrt{2}\,(\cos 45° + i \sin 45°) = 1 + i$$

$$\sqrt{2}\,(\cos 165° + i \sin 165°) \approx -1.3660 + 0.3660i$$

$$\sqrt{2}\,(\cos 285° + i \sin 285°) \approx 0.3660 - 1.3660i.$$

EXERCISES FOR APPENDIX F

In Exercises 1–24, perform the operation and write the result in standard form.

1. $(5 + i) + (6 - 2i)$

2. $(13 - 2i) + (-5 + 6i)$

3. $(8 - i) - (4 - i)$

4. $(3 + 2i) - (6 + 13i)$

5. $\left(-2 + \sqrt{-8}\right) + \left(5 - \sqrt{-50}\right)$

6. $\left(8 + \sqrt{-18}\right) - \left(4 + 3\sqrt{2i}\right)$

7. $13i - (14 - 7i)$

8. $22 + (-5 + 8i) + 10i$

9. $-\left(\frac{3}{2} + \frac{5}{2}i\right) + \left(\frac{5}{3} + \frac{11}{3}i\right)$

10. $(1.6 + 3.2i) + (-5.8 + 4.3i)$

11. $\sqrt{-6} \cdot \sqrt{-2}$

12. $\sqrt{-5} \cdot \sqrt{-10}$

13. $\left(\sqrt{-10}\right)^2$

14. $\left(\sqrt{-75}\right)^2$

15. $(1 + i)(3 - 2i)$

16. $(6 - 2i)(2 - 3i)$

17. $6i(5 - 2i)$

18. $-8i(9 + 4i)$

19. $\left(\sqrt{14} + \sqrt{10}i\right)\left(\sqrt{14} - \sqrt{10}i\right)$

20. $\left(3 + \sqrt{-5}\right)\left(7 - \sqrt{-10}\right)$

21. $(4 + 5i)^2$

22. $(2 - 3i)^2$

23. $(2 + 3i)^2 + (2 - 3i)^2$

24. $(1 - 2i)^2 - (1 + 2i)^2$

In Exercises 25–32, write the conjugate of the complex number. Multiply the number and its conjugate.

25. $5 + 3i$

26. $9 - 12i$

27. $-2 - \sqrt{5}i$

28. $-4 + \sqrt{2}i$

29. $20i$

30. $\sqrt{-15}$

31. $\sqrt{8}$

32. $1 + \sqrt{8}$

In Exercises 33–46, perform the operation and write the result in standard form.

33. $\dfrac{6}{i}$

34. $-\dfrac{10}{2i}$

35. $\dfrac{4}{4 - 5i}$

36. $\dfrac{3}{1 - i}$

37. $\dfrac{2 + i}{2 - i}$

38. $\dfrac{8 - 7i}{1 - 2i}$

39. $\dfrac{6 - 7i}{i}$

40. $\dfrac{8 + 20i}{2i}$

41. $\dfrac{1}{(4 - 5i)^2}$

42. $\dfrac{(2 - 3i)(5i)}{2 + 3i}$

43. $\dfrac{2}{1 + i} - \dfrac{3}{1 - i}$

44. $\dfrac{2i}{2 + i} + \dfrac{5}{2 - i}$

45. $\dfrac{i}{3 - 2i} + \dfrac{2i}{3 + 8i}$

46. $\dfrac{1 + i}{i} - \dfrac{3}{4 - i}$

A blue number indicates that a detailed solution can be found in the Study and Solutions Guide.

In Exercises 47–54, use the Quadratic Formula to solve the quadratic equation.

47. $x^2 - 2x + 2 = 0$

48. $x^2 + 6x + 10 = 0$

49. $4x^2 + 16x + 17 = 0$

50. $9x^2 - 6x + 37 = 0$

51. $4x^2 + 16x + 15 = 0$

52. $9x^2 - 6x - 35 = 0$

53. $16t^2 - 4t + 3 = 0$

54. $5s^2 + 6s + 3 = 0$

In Exercises 55–62, simplify the complex number and write it in standard form.

55. $-6i^3 + i^2$

56. $4i^2 - 2i^3$

57. $-5i^5$

58. $(-i)^3$

59. $\left(\sqrt{-75}\right)^3$

60. $\left(\sqrt{-2}\right)^6$

61. $\dfrac{1}{i^3}$

62. $\dfrac{1}{(2i)^3}$

In Exercises 63–68, plot the complex number and find its absolute value.

63. $-5i$

64. -5

65. $-4 + 4i$

66. $5 - 12i$

67. $6 - 7i$

68. $-8 + 3i$

In Exercises 69–76, represent the complex number graphically, and find the polar form of the number.

69. $3 - 3i$

70. $2 + 2i$

71. $\sqrt{3} + i$

72. $-1 + \sqrt{3}i$

73. $-2\left(1 + \sqrt{3}i\right)$

74. $\frac{5}{2}\left(\sqrt{3} - i\right)$

75. $6i$

76. 4

In Exercises 77–82, represent the complex number graphically, and find the standard form of the number.

77. $2(\cos 150° + i \sin 150°)$

78. $5(\cos 135° + i \sin 135°)$

79. $\frac{3}{2}(\cos 300° + i \sin 300°)$

80. $\frac{3}{4}(\cos 315° + i \sin 315°)$

81. $3.75\left(\cos \dfrac{3\pi}{4} + i \sin \dfrac{3\pi}{4}\right)$

82. $8\left(\cos \dfrac{\pi}{12} + i \sin \dfrac{\pi}{12}\right)$

In Exercises 83–86, perform the operation and leave the result in polar form.

83. $\left[3\left(\cos \dfrac{\pi}{3} + i \sin \dfrac{\pi}{3}\right)\right]\left[4\left(\cos \dfrac{\pi}{6} + i \sin \dfrac{\pi}{6}\right)\right]$

84. $\left[\dfrac{3}{2}\left(\cos \dfrac{\pi}{2} + i \sin \dfrac{\pi}{2}\right)\right]\left[6\left(\cos \dfrac{\pi}{4} + i \sin \dfrac{\pi}{4}\right)\right]$

85. $\left[\frac{5}{3}(\cos 140° + i \sin 140°)\right]\left[\frac{2}{3}(\cos 60° + i \sin 60°)\right]$

86. $\dfrac{\cos(5\pi/3) + i \sin(5\pi/3)}{\cos \pi + i \sin \pi}$

In Exercises 87–94, use DeMoivre's Theorem to find the indicated power of the complex number. Express the result in standard form.

87. $(1 + i)^5$

88. $(2 + 2i)^6$

89. $(-1 + i)^{10}$

90. $(1 - i)^{12}$

91. $2\left(\sqrt{3} + i\right)^7$

92. $4\left(1 - \sqrt{3}i\right)^3$

93. $\left(\cos \dfrac{5\pi}{4} + i \sin \dfrac{5\pi}{4}\right)^{10}$

94. $\left[2\left(\cos \dfrac{\pi}{2} + i \sin \dfrac{\pi}{2}\right)\right]^8$

In Exercises 95–100, (a) use Theorem A.5 on page A64 to find the indicated roots of the complex number, (b) represent each of the roots graphically, and (c) express each of the roots in standard form.

95. Square roots of $5(\cos 120° + i \sin 120°)$

96. Square roots of $16(\cos 60° + i \sin 60°)$

97. Fourth roots of $16\left(\cos \dfrac{4\pi}{3} + i \sin \dfrac{4\pi}{3}\right)$

98. Fifth roots of $32\left(\cos \dfrac{5\pi}{6} + i \sin \dfrac{5\pi}{6}\right)$

99. Cube roots of $-\dfrac{125}{2}\left(1 + \sqrt{3}i\right)$

100. Cube roots of $-4\sqrt{2}(1 - i)$

In Exercises 101–108, use Theorem A.5 on page A64 to find all the solutions of the equation and represent the solutions graphically.

101. $x^4 - i = 0$

102. $x^3 + 1 = 0$

103. $x^5 + 243 = 0$

104. $x^4 - 81 = 0$

105. $x^3 + 64i = 0$

106. $x^6 - 64i = 0$

107. $x^3 - (1 - i) = 0$

108. $x^4 + (1 + i) = 0$

Answers to Odd-Numbered Exercises

CHAPTER P

Section P.1 *(page 9)*

1. b   **2.** d   **3.** a   **4.** c   **5.** $(0, -2), (-2, 0), (1, 0)$

7. $(0, 0), (-3, 0), (3, 0)$ **9.** $(0, 0)$

11. Symmetric with respect to the y-axis

13. Symmetric with respect to the x-axis

15. Symmetric with respect to the origin

17. Symmetric with respect to the origin

19. $y = -3x + 2$

Symmetry: none

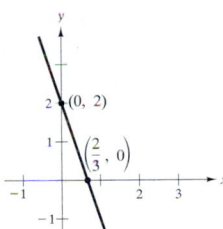

21. $y = \frac{1}{2}x - 4$

Symmetry: none

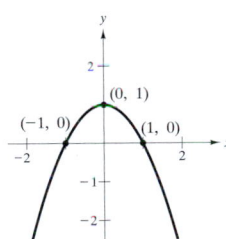

23. $y = 1 - x^2$

Symmetry: y-axis

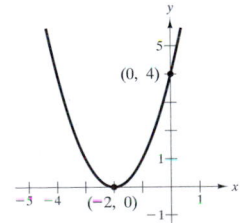

25. $y = x^3 + 2$

Symmetry: none

27. $y = (x + 2)^2$

Symmetry: none

29. $y = \frac{1}{x}$

Symmetry: origin

31.

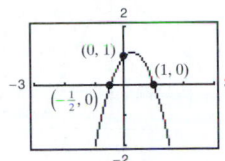

Xmin = -3
Xmax = 5
Xscl = 1
Ymin = -3
Ymax = 5
Yscl = 1

33. $y = -2x^2 + x + 1$

Symmetry: none

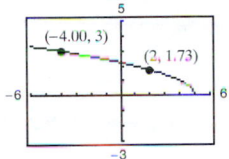

35. $y = \dfrac{5}{x^2 + 1} - 1$

Symmetry: y-axis

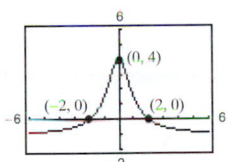

37. $y = \sqrt{5 - x}$

39. $y = (x + 2)(x - 4)(x - 6)$ **41.** $y = x$ **43.** $(1, 1)$

45. $(5, 2)$ **47.** $(-1, -2), (2, 1)$

49. $(-1, -1), (0, 0), (1, 1)$

51. $(-1, -5), (0, -1), (2, 1)$ **53.** $x \approx 3133$ units

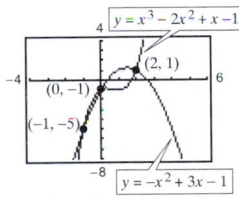

55. (a) $y = 0.0127t^2 + 4.4181t + 36.2896$

(b)

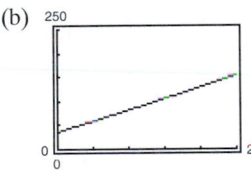

(c) 180.3

57.

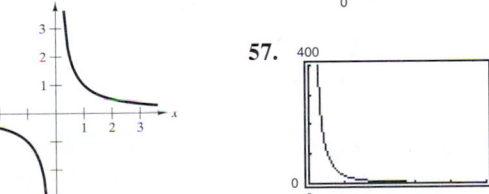

Approximately $\frac{1}{4}$

59. False: $(-1, -2)$ is not a point on the graph of $x = \frac{1}{4}y^2$.

60. True   **61.** True   **62.** True

63. $(1 - K^2)x^2 + (1 - K^2)y^2 + 4K^2x - 4K^2 = 0$

Section P.2 *(page 17)*

1. $m = 1$ **3.** $m = 0$ **5.** $m = -12$

7.

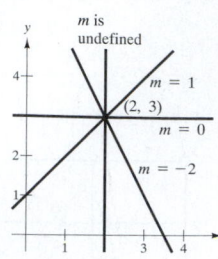

9. $m = 3$

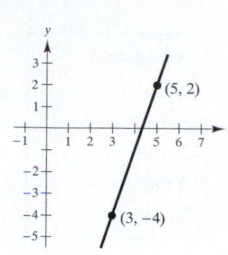

11. $m = -\frac{2}{3}$ **13.** $(0, 1), (1, 1), (3, 1)$

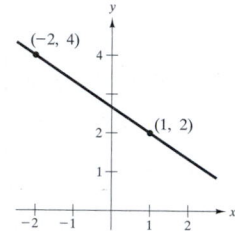

15. $(0, 10), (2, 4), (3, 1)$

17. (a)

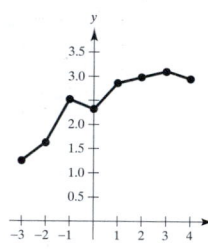

(b) Decreases most rapidly: 1989–1990

Increases most rapidly: 1988–1989

19. Any two points can be used, because the rate of change remains constant.

21. $m = -\frac{1}{5}, (0, 4)$ **23.** m is undefined, no y-intercept

25. $2x - y - 3 = 0$ **27.** $3x + y = 0$

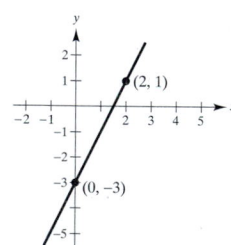

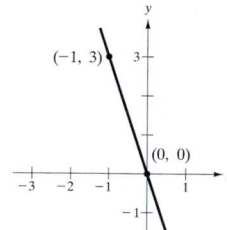

29. $y + 2 = 0$ **31.** $3x - 4y + 12 = 0$

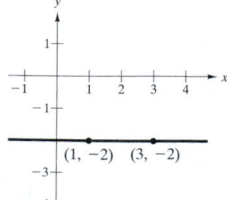

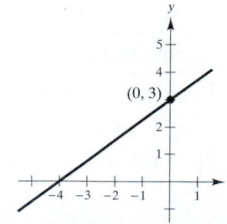

33. $2x - 3y = 0$ **35.** $4x - y + 2 = 0$

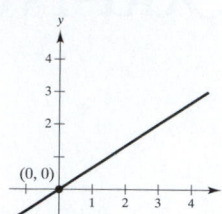

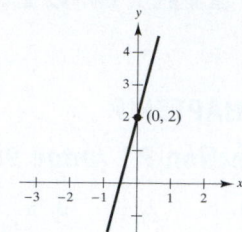

37. $x - 3 = 0$ **39.** $3x + 2y - 6 = 0$ **41.** $x + y - 3 = 0$

43. (a) $2x - y - 3 = 0$ (b) $x + 2y - 4 = 0$

45. (a) $40x + 24y - 53 = 0$ (b) $24x - 40y + 9 = 0$

47. (a) $x - 2 = 0$ (b) $y - 5 = 0$

49.

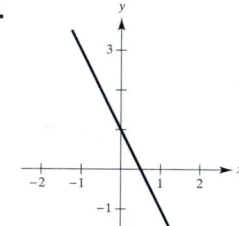

51.

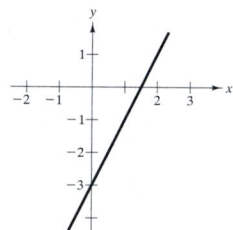

53.

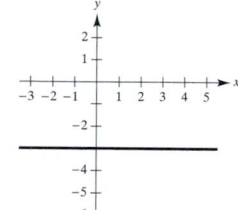

55.

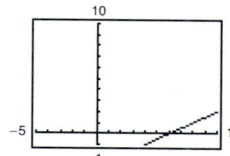

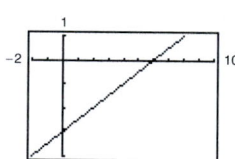

The second gives a more complete graph.

57. $V = 125t + 1540$ **59.** $V = 36{,}400 - 2000t$

61. $y = 2x$ **63.** Not collinear, because $m_1 \neq m_2$

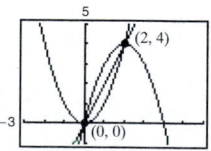

65. $\left(0, \dfrac{-a^2 + b^2 + c^2}{2c}\right)$ **67.** $\left(b, \dfrac{a^2 - b^2}{c}\right)$

69. $5F - 9C - 160 = 0$
$72°F \approx 22.2°C$

71. (a) $W_1 = 12.50 + 0.75x$
$W_2 = 9.20 + 1.30x$

(b)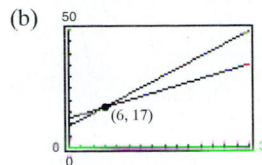

(c) Choose position 1 when less than six units are produced and position 2 otherwise.

73. (a) $x(p) = \dfrac{1330 - p}{15}$

(b)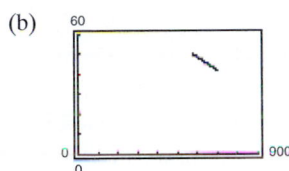

$x(655) = 45$

(c) $x(595) = 49$

75. 2   **77.** $\dfrac{5\sqrt{2}}{2}$ **79.** $2\sqrt{2}$ **81.** Proof   **83.** Proof

85. Proof   **87.** True   **88.** False: If m_1 is positive, then

$$m_2 = -\frac{1}{m_1} \text{ is negative.}$$

Section P.3 *(page 28)*

1. (a) -3 (b) -9 (c) $2b - 3$ (d) $2x - 5$

3. (a) -1 (b) 2   (c) 6   (d) $2t^2 + 4$

5. (a) 1   (b) 0   (c) $-\dfrac{1}{2}$

7. $3x^2 + 3x\,\Delta x + (\Delta x)^2, \ \Delta x \neq 0$

9. $\dfrac{-1}{\sqrt{x-1}\left(1 + \sqrt{x-1}\right)}, \ x \neq 2$

11. $f(x) = 4 - x$
Domain: $(-\infty, \infty)$
Range: $(-\infty, \infty)$

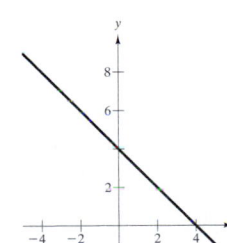

13. $f(x) = \sqrt{x - 1}$
Domain: $[1, \infty)$
Range: $[0, \infty)$

15. $f(x) = \sqrt{9 - x^2}$
Domain: $[-3, 3]$
Range: $[0, 3]$

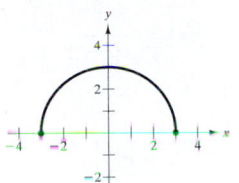

17. $g(t) = 2\sin\pi t$
Domain: $(-\infty, \infty)$
Range: $(-2, 2)$

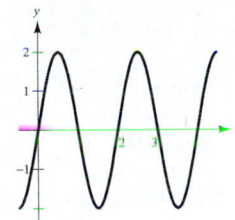

19. y is not a function of x.

21. $f(x) = \begin{cases} -2x + 2, & x < 0 \\ 2, & 0 \le x < 2 \\ 2x - 2, & x \ge 2 \end{cases}$

23. y is not a function of x.   **25.** y is not a function of x.

27. No   **29.** ii, $c = -2$ **30.** i, $c = \frac{1}{4}$ **31.** iv, $c = 32$

32. iii, $c = 3$

33. (a) For each time t there corresponds one depth d.

(b) Domain: $[0, 5]$; Range: $[0, 30]$

(c)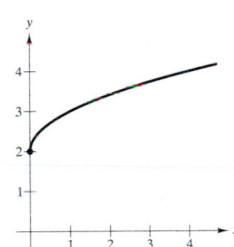

35.
Xmin = -25
Xmax = 25
Xscl = 5
Ymin = -2000
Ymax = 2000
Yscl = 200

37. (a) Vertical translation   (b) Reflection (about the x-axis)

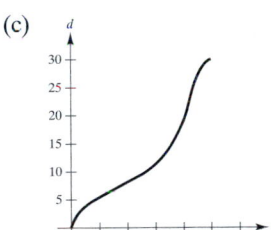

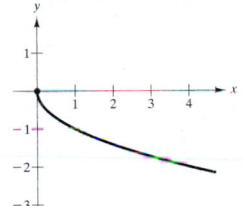

(c) Horizontal translation

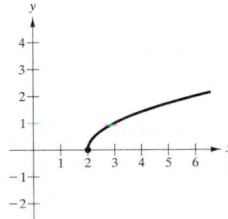

39. (a) $T(4) = 16$, $T(15) = 23$

(b) The change in temperatures would occur 1 hour later.

(c) The temperatures are 1° lower.

41. $(f \circ g)(x) = x$

Domain: $[0, \infty)$

$(g \circ f)(x) = |x|$

Domain: $(-\infty, \infty)$

43. $(f \circ g)(x) = \dfrac{1}{x^2 + 1}$

Domain: $(-\infty, \infty)$

$(g \circ f)(x) = \dfrac{1}{x^2} + 1$

Domain: $(-\infty, 0), (0, \infty)$

45. $(A \circ r)(t) = 0.36\pi t^2$

$A \circ r$ represents the area of the circle at time t.

47. Even   **49.** Odd   **51.** (a) $\left(\frac{3}{2}, 4\right)$ (b) $\left(\frac{3}{2}, -4\right)$

53. Proof   **55.** Proof

57. (a) $f(x) = x^2(4 - x^2)$ (b) $f(x) = x(4 - x^2)$

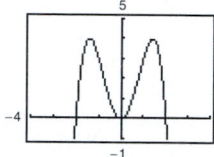 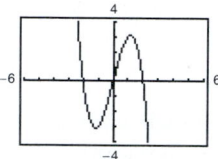

59. (a)   (b) $A = x(50 - x)$

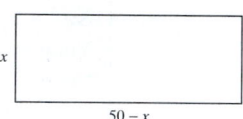

(c)   (d) 25×25 meters

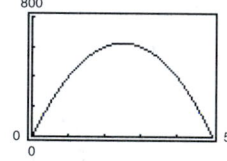

Domain: $(0, 50)$

61. (a)

Height x	Length and width	Volume V
1	$24 - 2(1)$	$1[24 - 2(1)]^2 = 484$
2	$24 - 2(2)$	$2[24 - 2(2)]^2 = 800$
3	$24 - 2(3)$	$3[24 - 2(3)]^2 = 972$
4	$24 - 2(4)$	$4[24 - 2(4)]^2 = 1024$
5	$24 - 2(5)$	$5[24 - 2(5)]^2 = 980$
6	$24 - 2(6)$	$6[24 - 2(6)]^2 = 864$

Guess of maximum volume: 1024 cubic centimeters

(b)   (c) $V = 4x(12 - x)^2$

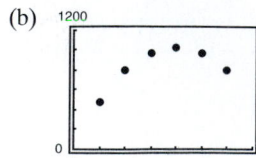

Domain: $(0, 12)$

(d) $4 \times 16 \times 16$ centimeters

V is a function of x.

63. False: if $f(x) = x^2$, then $f(-1) = f(1)$.   **64.** True

65. True   **66.** False: if $f(x) = x^2$, then $f(2x) = 4x^2 \neq 2f(x)$

Section P.4 *(page 34)*

1. Quadratic   **2.** Trigonometric   **3.** Linear

4. No relationship

5. (a) and (b)

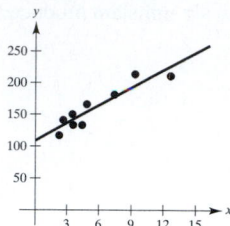

Approximately linear

(c) 136

7. (a) $F(d) = 15.1d + 0.1$

(b)

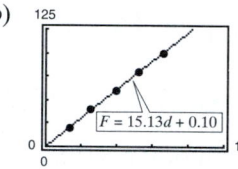

(c) 3.6

9. (a) $y = 2.62x + 2.66$

(b)

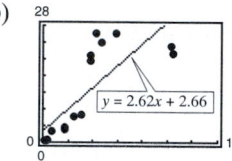

(c) Denmark, France, Italy

11. (a) $y_1 = 0.16t^2 - 2.43t + 13.96$

$y_2 = 0.17t + 0.38$

$y_3 = 0.04t + 0.44$

(b)

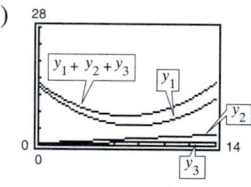

26.66 cents per mile

13. (a) $y_1 = 3.09t + 6.67$

$y_2 = 0.06t^3 - 1.65t^2 + 17.80t - 34.09$

(b)

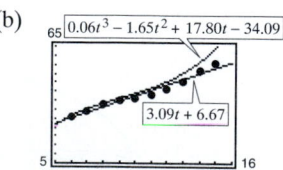

(c) Cubic   (d) $y_3 = 0.17t^2 - 0.44t + 23.82$

(e) Average increase per year in the number of people in HMOs

(f) Linear: 68.5; Cubic: 141.5

15. (a) $y = -1.81x^3 + 14.58x^2 + 16.39x + 10$

(b)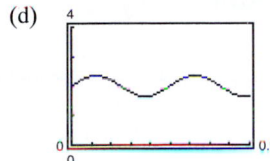

(c) 214

17. (a) Yes. At time t there is one and only one displacement y.

(b) Amplitude: 0.35; Period: 0.5

(c) $y = 0.35 \sin(4\pi t) + 2$

(d)

19. Answers will vary.

Review Exercises for Chapter P *(page 37)*

1. $\left(\frac{3}{2}, 0\right), (0, -3)$ **3.** $(1, 0), \left(0, \frac{1}{2}\right)$ **5.** y-axis symmetry

7. 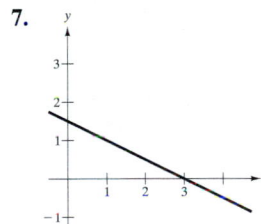   **9.**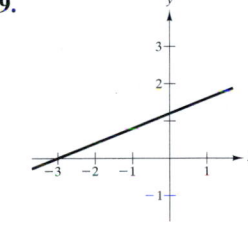

11. 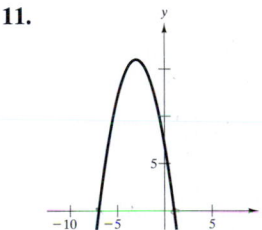   **13.**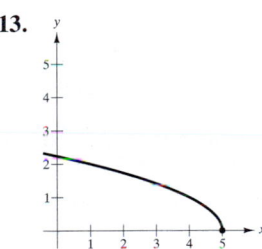

15.
```
Xmin = -5
Xmax = 5
Xscl = 1
Ymin = -30
Ymax = 10
Yscl = 5
```

17. $(4, 1)$ **19.** $y = x^3 - 4x$

21. $m = \frac{3}{7}$

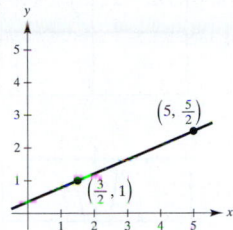

23. $t = \frac{7}{3}$

25. (a) $7x - 16y + 78 = 0$

(b) $5x - 3y + 22 = 0$

(c) $y + 2x = 0$

(d) $x + 2 = 0$

27. $V = 12,500 - 850t$
$9950

29. Not a function

31. Function

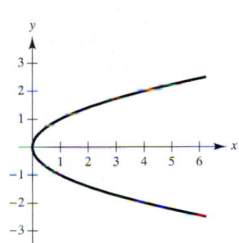  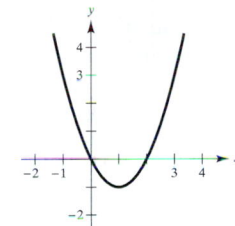

33. (a) Undefined   (b) $\frac{-1}{1 + \Delta x}, \Delta x \neq 0$

35. (a) 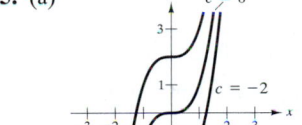 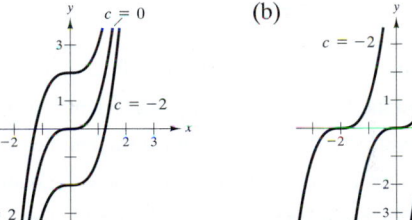   (b)

(c) 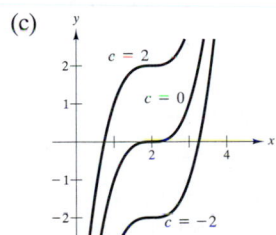   (d)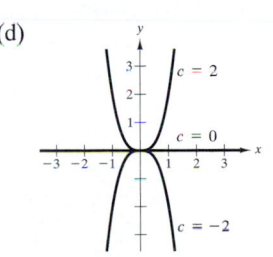

37. (a)

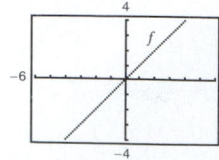 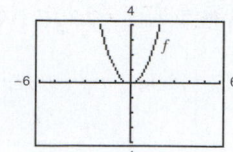

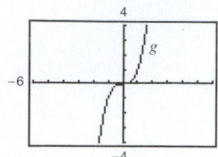 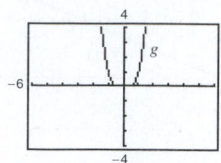

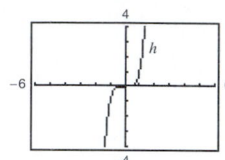 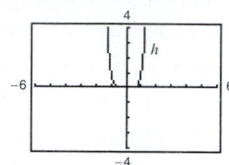

(b) All the graphs pass through the origin. The graphs of the odd powers of x are symmetric with respect to the origin and the graphs of the even powers are symmetric with respect to the y-axis. As the powers increase, the graphs become flatter in the interval $-1 < x < 1$.

39. (a) $A = x(12 - x)$ (c) Maximum area: 36
(b) Domain: $(0, 12)$ 6×6 inches

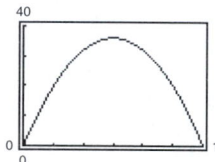

41. (a) Minimum degree: 3; Leading coefficient: negative
(b) Minimum degree: 4; Leading coefficient: positive
(c) Minimum degree: 2; Leading coefficient: negative
(d) Minimum degree: 5; Leading coefficient: positive

43. (a) Yes. For each time t there corresponds one and only one displacement y.
(b) Amplitude: 0.25; Period: 1.1
(c) $y = \frac{1}{4} \cos(5.7t)$

(d)

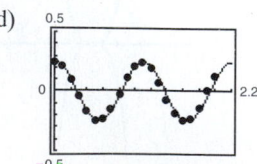

CHAPTER 1

Section 1.1 *(page 46)*

1. Precalculus: 300 feet
3. Calculus: Slope of the tangent line at $x = 2$ is 0.16.
5. Precalculus: $\frac{15}{2}$

7. Precalculus: 24
9. (a)

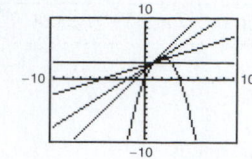

(b) The graphs of y_2 approach the tangent line y_1 at $x = 1$.
(c) 2; $\{0.2, 0.1, 0.01, 0.001\}$

11. (a) 5.66
(b) 6.11
(c) Increase the number of line segments.

Section 1.2 *(page 53)*

1. (a)

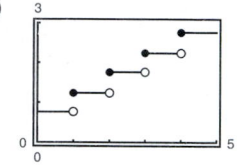

(b)

t	3	3.3	3.4	3.5	3.6	3.7	4
C	1.75	2.25	2.25	?	2.25	2.25	2.25

$$\lim_{t \to 3.5} C(t) = 2.25$$

(c)

t	2	2.5	2.9	3	3.1	3.5	4
C	1.25	1.75	1.75	?	2.25	2.25	2.25

The limit does not exist, because the limits from the right and left are not equal.

3.

x	1.9	1.99	1.999	2.001	2.01	2.1
$f(x)$	0.3448	0.3344	0.3334	0.3332	0.3322	0.3226

$$\lim_{x \to 2} \frac{x - 2}{x^2 - x - 2} \approx 0.3333 \quad \left(\text{Actual limit is } \frac{1}{3}. \right)$$

5.

x	-0.1	-0.01	-0.001	0.001	0.01	0.1
$f(x)$	0.2911	0.2889	0.2887	0.2887	0.2884	0.2863

$$\lim_{x \to 0} \frac{\sqrt{x + 3} - \sqrt{3}}{x} \approx 0.2887 \quad \left(\text{Actual limit is } \frac{1}{2\sqrt{3}}. \right)$$

7.

x	2.9	2.99	2.999	3.001	3.01	3.1
$f(x)$	-0.0641	-0.0627	-0.0625	-0.0625	-0.0623	-0.0610

$$\lim_{x \to 3} \frac{[1/(x + 1)] - (1/4)}{x - 3} \approx -0.0625 \quad \left(\text{Actual limit is } -\frac{1}{16}. \right)$$

9.

x	-0.1	-0.01	-0.001	0.001	0.01	0.1
$f(x)$	0.9983	0.99998	1.0000	1.0000	0.99998	0.9983

$$\lim_{x\to 0}\frac{\sin x}{x}\approx 1.0000 \quad (\text{Actual limit is 1.})$$

11. 1   **13.** 2   **15.** Limit does not exist.

17. Limit does not exist.   **19.** Limit does not exist.

21. $L = 8$. Let $\delta = \dfrac{0.01}{3}\approx 0.0033$

23. $L = 1$. Assume $1 < x < 3$ and let $\delta = \dfrac{0.01}{5} = 0.002$.

25. 5   **27.** -3 **29.** 3   **31.** 0   **33.** 4   **35.** 2

37. It is not obvious from the graph that the limit does not exist at $x = 4$.

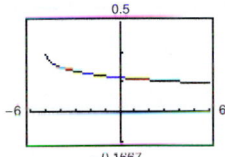

Domain: $[-5, 4), (4, \infty)$
$\lim\limits_{x\to 4} f(x) = \frac{1}{6}$

39. It is not obvious from the graph that the limit does not exist at $x = 9$.

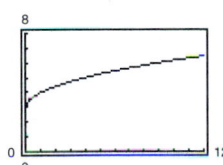

Domain: $[0, 9), (9, \infty)$
$\lim\limits_{x\to 9} f(x) = 6$

41.

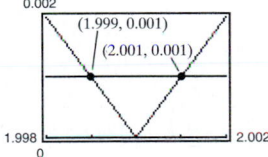

43. False: the existence or nonexistence of $f(x)$ at $x = c$ has no bearing on the existence of the limit of $f(x)$ as $x\to c$.

44. True   **45.** False: see Exercise 13.

46. False: the existence or nonexistence of the limit of $f(x)$ as $x\to c$ has no bearing on the existence of $f(c)$.

47. Answers will vary.   **49.** Proof   **51.** Proof

Section 1.3 *(page 64)*

1.

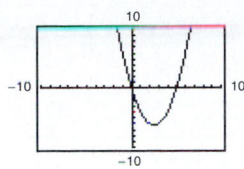

(a) 0   (b) 6

3. 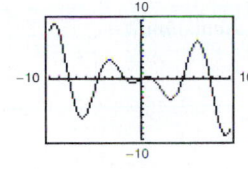   **5.** 16

(a) 0   (b) ≈ 0.52

7. -1 **9.** -4 **11.** 2   **13.** 1   **15.** $\frac{1}{2}$ **17.** -2

19. 1   **21.** -1 **23.** 1   **25.** $\frac{1}{2}$ **27.** -1

29. (a) 15   (b) 5   (c) 6   (d) $\frac{2}{3}$

31. (a) 64   (b) 2   (c) 12   (d) 8

33. (a) 1   (b) 3

$$g(x) = \frac{-2x^2 + x}{x} = -2x + 1, \quad x \neq 0$$

35. (a) 2   (b) 0

$$g(x) = \frac{x^3 - x}{x - 1} = x^2 + x, \quad x \neq 1$$

37. -2

$$f(x) = \frac{x^2 - 1}{x + 1} = x - 1 = g(x), \quad x \neq -1$$

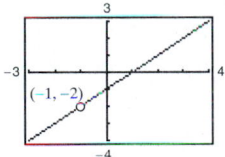

39. 12

$$f(x) = \frac{x^3 + 8}{x + 2} = x^2 - 2x + 4, \quad x \neq -2$$

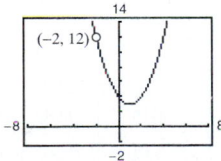

41. $\dfrac{1}{10}$ **43.** $\dfrac{3}{2}$ **45.** $\dfrac{\sqrt{3}}{6}$ **47.** $-\dfrac{1}{4}$

49. 2   **51.** $2x - 2$

53.

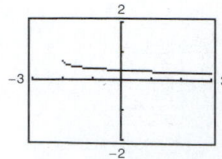

x	-0.1	-0.01	-0.001	0.001	0.01	0.1
$f(x)$	0.358	0.354	0.354	0.354	0.353	0.349

$$\lim_{x \to 0} \frac{\sqrt{x+2} - \sqrt{2}}{x} \approx 0.354 \quad \left(\text{Actual limit is } \frac{1}{2\sqrt{2}}.\right)$$

55.

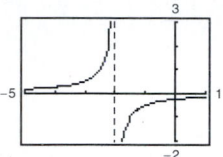

x	-0.1	-0.01	-0.001	0	0.001	0.01	0.1
$f(x)$	-0.263	-0.251	-0.250	?	-0.250	-0.249	-0.238

$$\lim_{x \to 0} \frac{[1/(2+x)] - (1/2)}{x} \approx -0.250 \quad \left(\text{Actual limit is } -\tfrac{1}{4}.\right)$$

57. $\frac{1}{5}$ **59.** 0   **61.** 0   **63.** 0   **65.** 1   **67.** 1

69.

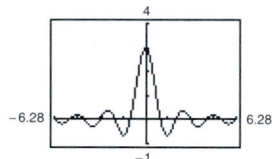

t	-0.1	-0.01	0	0.01	0.1
$f(t)$	2.96	2.9996	?	2.9996	2.96

$$\lim_{t \to 0} \frac{\sin 3t}{t} = 3$$

71.

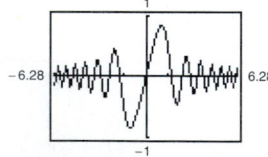

x	-0.1	-0.01	-0.001	0	0.001	0.01	0.1
$f(x)$	-0.1	-0.01	-0.001	?	0.001	0.01	0.1

$$\lim_{x \to 0} \frac{\sin x^2}{x} = 0$$

73. 2   **75.** $-\dfrac{4}{x^2}$ **77.** 4

79. 0   **81.** 0

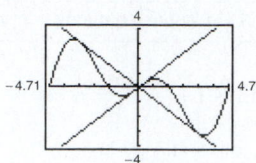   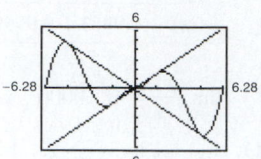

83. 0

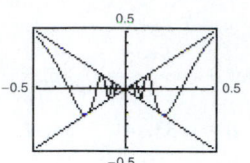

85.

The magnitudes of $f(x)$ and $g(x)$ are approximately equal when x is "close to" 0. Therefore, their ratio is approximately 1.

87. -160 feet per second   **89.** -29.4

91. (a)

(b) Yes. As time increases, S increases at a slower rate.

93. Proof   **95.** Proof

97. False. Let $f(x) = \frac{1}{2}x^2$ and $g(x) = x^2$.

Then $f(x) < g(x)$ for all $x \neq 0$.

However, $\lim\limits_{x \to 0} f(x) = \lim\limits_{x \to 0} g(x) = 0$.

99. Proof   **101.** Proof   **103.** Proof

105. (a) Domain: all real numbers x except $x = 0$ or odd multiples of $\dfrac{\pi}{2}$.

(b)

It is not obvious that $f(0)$ is undefined.

(c) 0.5   (d) $\frac{1}{2}$

107. f and g agree at all but one point if c is a real number such that $f(x) = g(x)$ for all $x \neq c$.

Section 1.4 *(page 75)*

1. (a) The limit does not exist at $x = c$.
   (b) The function is not defined at $x = c$.
   (c) The limit exists, but it is not equal to the value of the function at $x = c$.
   (d) The limit does not exist at $x = c$.

3.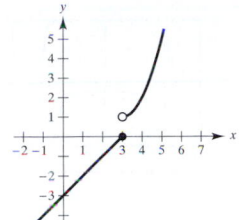

   Not continuous because the $\lim_{x \to 3} f(x)$ does not exist.

5. (a) 1   (b) 1   (c) 1
7. (a) 0   (b) 0   (c) 0
9. (a) 3   (b) -3 (c) Limit does not exist.
11. $\frac{1}{10}$ 13. Limit does not exist   15. Limit does not exist.
17. $-\frac{1}{x^2}$ 19. Limit does not exist.   21. 2
23. Limit does not exist.   25. 3
27. Discontinuous at $x = -2$ and $x = 2$
29. Discontinuous at every integer
31. Continuous for all real x 33. Continuous for all real x
35. Nonremovable discontinuity at $x = 1$
37. Continuous for all real x
39. Removable discontinuity at $x = -2$
    Nonremovable discontinuity at $x = 5$
41. Nonremovable discontinuity at $x = -2$
43. Continuous for all real x
45. Nonremovable discontinuity at $x = 2$
47. Continuous for all real x
49. Nonremovable discontinuities at integer multiples of $\frac{\pi}{2}$
51. Nonremovable discontinuity at each integer
53.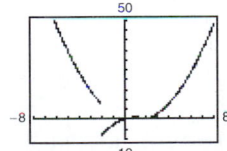

   $\lim_{x \to 0^+} f(x) = 0$
   $\lim_{x \to 0^-} f(x) = 0$
   Discontinuity at $x = -2$
55. $a = 2$ 57. $a = -1$, $b = 1$

59. Continuous for all real x
61. Nonremovable discontinuities at $x = 1$ and $x = -1$
63. Nonremovable discontinuity at each integer.

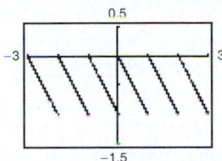

65. Discontinuous at $x = 3$

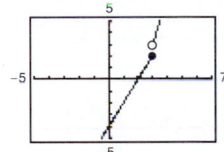

67. Continuous on $(-\infty, \infty)$
69. Continuous on $\dots, (-2\pi, 0), (0, 2\pi), (2\pi, 4\pi), \dots$

71.

   It is not obvious from the graph that the function is discontinuous at $x = 0$.

73. $f(x)$ is continuous on $[2, 4]$.
    $f(2) = -1$ and $f(4) = 3$
    By the Intermediate Value Theorem, $f(c) = 0$ for at least one value of c between 2 and 4.

75. 0.68, 0.6823   77. 0.56, 0.5636   79. $f(3) = 11$
81. $f(2) = 4$
83. Because $V(0) = 0$, $V(5) = 523.6$, and V is continuous, there is at least one real number r, $0 \leq r \leq 5$, such that $V(r) = 275$.
85. The function is discontinuous at every even positive integer. The company must replenish every two months.

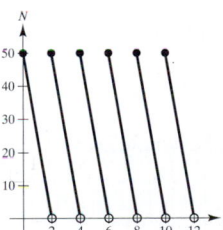

87. Proof   89. Proof   91. True   92. True
93. False: the rational function $f(x) = p(x)/q(x)$ has at most n discontinuities where n is the degree of $q(x)$.
94. False: it is discontinuous at $x = 1$.

95. (a)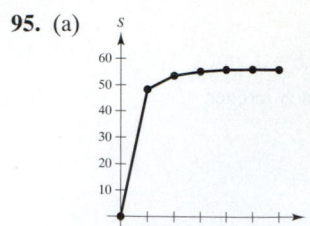

(b) There appears to be a limiting speed, and a possible cause is air resistance.

97. Proof

99. Domain: $[-c^2, 0), \quad (0, \infty)$

Let $f(0) = \dfrac{1}{2c}$.

Section 1.5 *(page 84)*

1. $\displaystyle\lim_{x \to -2^+} \frac{1}{(x+2)^2} = \infty$ **3.** $\displaystyle\lim_{x \to -2^+} \tan \frac{\pi x}{4} = -\infty$

$\displaystyle\lim_{x \to -2^-} \frac{1}{(x+2)^2} = \infty$ $\displaystyle\lim_{x \to -2^-} \tan \frac{\pi x}{4} = \infty$

5.

x	-3.5	-3.1	-3.01	-3.001	-2.999	-2.99
$f(x)$	0.31	1.64	16.6	167	-167	-16.6

x	-2.9	-2.5
$f(x)$	-1.7	-0.36

$\displaystyle\lim_{x \to -3^+} f(x) = -\infty \qquad \lim_{x \to -3^-} f(x) = \infty$

7.

x	-3.5	-3.1	-3.01	-3.001	-2.999	-2.99
$f(x)$	3.8	16	151	1501	-1499	-149

x	-2.9	-2.5
$f(x)$	-14	-2.3

$\displaystyle\lim_{x \to -3^+} f(x) = -\infty \qquad \lim_{x \to -3^-} f(x) = \infty$

9. $x = 0$ **11.** $x = 2, \quad x = -1$ **13.** $x = \pm 1$

15. $x = \dfrac{\pi}{4} + \dfrac{n\pi}{2}$, n an integer   **17.** $t = 0$

19. $x = -2, \quad x = 1$ **21.** No vertical asymptote

23. $t = n\pi$, n a nonzero integer

25.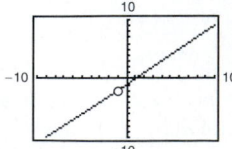

Removable discontinuity at $x = -1$

27.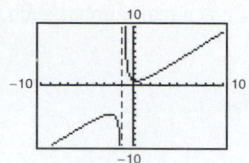

Vertical asymptote at $x = -1$

29. $-\infty$ **31.** ∞ **33.** $\frac{4}{5}$ **35.** $-\infty$ **37.** ∞ **39.** $\frac{1}{2}$

41. ∞ **43.** $-\infty$

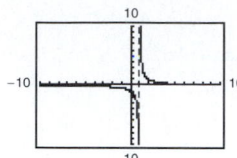   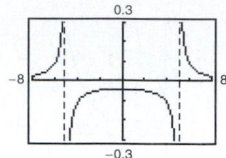

$\displaystyle\lim_{x \to 1^+} f(x) = \infty$

45. ∞

47. (a) $\frac{7}{12}$ foot per second   (b) $\frac{3}{2}$ feet per second   (c) ∞

49. (a) \$176 million   (b) \$528 million

(c) \$1584 million   (d) ∞

51. ∞

53. (a) 850 revolutions per minute

(b) Reverse direction

(c) $L = 60 \cot \phi + 30(\pi + 2\phi)$

Domain: $\left(0, \dfrac{\pi}{2}\right)$

(d)

ϕ	0.3	0.6	0.9	1.2	1.5
L	306.2	217.9	195.9	189.6	188.5

(e) 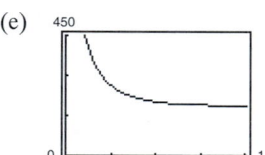   (f) 60π (g) ∞

55. False: let $p(x) = x^2 - 1$.

56. False: let $f(x) = \dfrac{1}{x^2 + 1}$.

57. True

58. False: let $f(x) = \begin{cases} \dfrac{1}{x}, & x \neq 0 \\ 0, & x = 0 \end{cases}$.

59. Let $f(x) = 1/x^2$, $g(x) = 1/x^4$, and $c = 0$.   **61.** Proof

63. (a)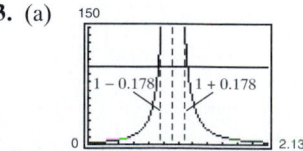

(b)

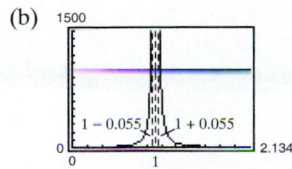

(c) δ decreases for increasing M.

Review Exercises for Chapter 1 *(page 87)*

1. Calculus

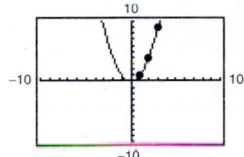

Estimate: 8.261

3.

x	-0.1	-0.01	-0.001	0.001	0.01	0.1
$f(x)$	-0.26	-0.25	-0.250	-0.2499	-0.249	-0.24

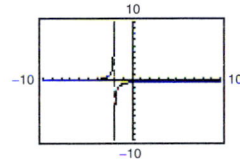

$$\lim_{x \to 0} f(x) \approx 0.25$$

5. (a) -2 (b) -3

7. 7   **9.** 77   **11.** $\frac{10}{3}$ **13.** $-\frac{1}{4}$ **15.** -1 **17.** 75

19. $-\infty$ **21.** $\frac{\sqrt{3}}{2}$ **23.** $-\infty$ **25.** $\frac{1}{3}$ **27.** $-\infty$

29. $\frac{4}{5}$ **31.** ∞

33.

x	1.1	1.01	1.001	1.0001
$f(x)$	0.5680	0.5764	0.5773	0.5773

$$\lim_{x \to 1^+} \frac{\sqrt{2x + 1} - \sqrt{3}}{x - 1} \approx 0.577 \quad \left(\text{Actual limit is } \frac{\sqrt{3}}{3}.\right)$$

35. Nonremovable discontinuity at each integer
Continuous on $(k, k + 1)$ for all integers k

37. Removable discontinuity at $x = 1$
Continuous on $(-\infty, 1) \cup (1, \infty)$

39. Nonremovable discontinuity at $x = 2$
Continuous on $(-\infty, 2) \cup (2, \infty)$

41. Nonremovable discontinuity at $x = -1$
Continuous on $(-\infty, -1) \cup (-1, \infty)$

43. Nonremovable discontinuity at each even integer
Continuous on $(2k, 2k + 2)$ for all integers k

45. $c = -\frac{1}{2}$

47. Nonremovable discontinuity every 6 months

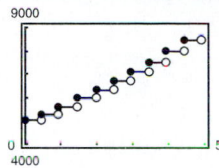

49. $x = 0$ **51.** $x = 10$

53. (a) $14,117.65$ (b) $80,000.00$

(c) $720,000.00$ (d) ∞

55. -39.2 meters per second

57. False: $\displaystyle\lim_{x \to 0^-} \frac{|x|}{x} = -1$ **58.** True   **59.** True

60. False: the existence of the limit of $f(x)$ as $x \to c$ has no bearing on the existence of $f(c)$.

61. True   **62.** False: the limit does not exist.   **63.** True

65. (a) Domain: $(-\infty, 0] \cup [1, \infty)$

(b) $\displaystyle\lim_{x \to 0^-} f(x) = 0$

(c) $\displaystyle\lim_{x \to 1^+} f(x) = 0$

CHAPTER 2

Section 2.1 *(page 98)*

1. (a) $m = 0$ (b) $m = -3$

3.

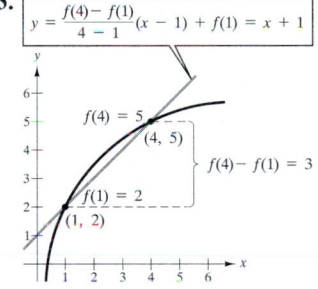

5. $f'(x) = 0$

7. $f'(x) = -5$ **9.** $f'(x) = 4x + 1$ **11.** $f'(x) = 3x^2 - 12$

13. $f'(x) = \dfrac{-1}{(x - 1)^2}$ **15.** $f'(x) = \dfrac{1}{2\sqrt{x - 4}}$

17. (a) Tangent line: $y = 4x - 3$

(b)

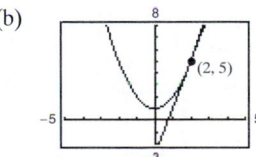

19. (a) Tangent line: $y = 12x - 16$

(b)
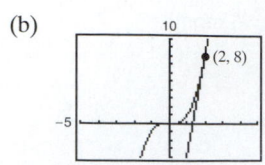

21. (a) Tangent line: $y = 2$

(b)
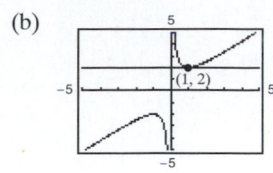

23. $y = 3x - 2$ **25.** $y = 2x + 1$
$y = 3x + 2$ $y = -2x + 9$

27. b   **28.** e   **29.** c   **30.** a   **31.** f   **32.** d

33. (a) -3

(b) 0

(c) The graph is moving downward to the right when $x = 1$.

(d) The graph is moving upward to the right when $x = -4$.

(e) Positive. Because $g'(x) > 0$ on $[3, 6]$, the graph of g is moving upward to the right.

(f) No. Knowing only $g'(2)$ is not sufficient information. $g'(2)$ remains the same for any vertical translation of g.

35.
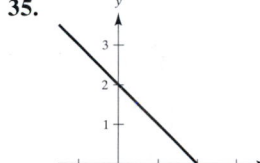

37.
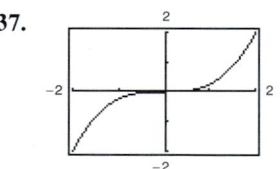

x	-2	-1.5	-1	-0.5	0	0.5	1	1.5	2
$f(x)$	-2	$-\frac{27}{32}$	$-\frac{1}{4}$	$-\frac{1}{32}$	0	$\frac{1}{32}$	$\frac{1}{4}$	$\frac{27}{32}$	2
$f'(x)$	3	$\frac{27}{16}$	$\frac{3}{4}$	$\frac{3}{16}$	0	$\frac{3}{16}$	$\frac{3}{4}$	$\frac{27}{16}$	3

39.
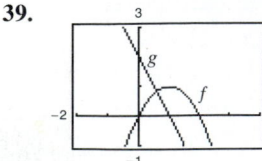

41.
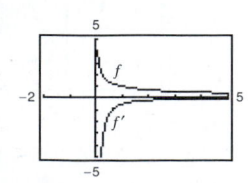

$g(x) \approx f'(x)$

43. (a)
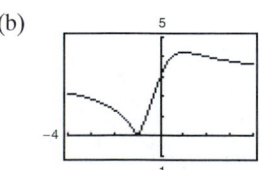

(b) The graphs of S for decreasing values of Δx are secant lines approaching the tangent line to the graph of f at the point $(2, f(2))$.

45. $f(x) = x^2 - 1$

$$f'(2) = \lim_{x \to 2} \frac{f(x) - f(2)}{x - 2}$$

$$= \lim_{x \to 2} \frac{(x^2 - 1) - 3}{x - 2}$$

$$= \lim_{x \to 2} (x + 2) = 4$$

47. $f(x) = x^3 + 2x^2 + 1$

$$f'(-2) = \lim_{x \to -2} \frac{f(x) - f(-2)}{x + 2}$$

$$= \lim_{x \to -2} \frac{(x^3 + 2x^2 + 1) - 1}{x + 2}$$

$$= \lim_{x \to -2} x^2 = 4$$

49. $f(x) = (x - 1)^{2/3}$ **51.** $(-\infty, -3), (-3, \infty)$

$$f'(1) = \lim_{x \to 1} \frac{f(x) - f(1)}{x - 1}$$

$$= \lim_{x \to 1} \frac{(x - 1)^{2/3} - 0}{x - 1}$$

$$= \lim_{x \to 1} \frac{1}{(x - 1)^{1/3}}$$

Limit does not exist.
f is not differentiable at $x = 1$.

53. $(-\infty, -1), (-1, \infty)$ **55.** $(-\infty, 3), (3, \infty)$

57. $(1, \infty)$ **59.** $(-\infty, 0), (0, \infty)$

61. f is not differentiable at $x = 1$.   **63.** $f'(1) = 0$

65. f is differentiable at $x = 2$.

67. (a) $d = \dfrac{3|m + 1|}{\sqrt{m^2 + 1}}$ **69.** False: let $f(x) = |x|$ at $x = 0$.

(b)
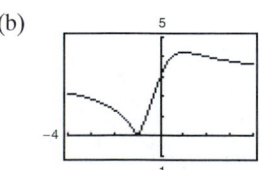

Not differentiable at $m = -1$

70. False: let $f(x) = \begin{cases} x, & x \leq 0 \\ 2x, & x > 0 \end{cases}$ at $x = 0$ **71.** True

73. Proof

Section 2.2 *(page 110)*

1. (a) $\frac{1}{2}$ (b) $\frac{3}{2}$ (c) 2 (d) 3

3. 0 **5.** 1 **7.** $2x$ **9.** $-4t + 3$ **11.** $3t^2 - 2$

13. $2x + \frac{1}{2}\sin x$ **15.** $-\frac{1}{x^2} - 3\cos x$

	Function	Rewrite	Derivative	Simplify
17.	$y = \dfrac{1}{3x^3}$	$y = \dfrac{1}{3}x^{-3}$	$y' = -x^{-4}$	$y' = -\dfrac{1}{x^4}$
19.	$y = \dfrac{1}{(3x)^3}$	$y = \dfrac{1}{27}x^{-3}$	$y' = -\dfrac{1}{9}x^{-4}$	$y' = -\dfrac{1}{9x^4}$
21.	$y = \dfrac{\sqrt{x}}{x}$	$y = x^{-1/2}$	$y' = -\dfrac{1}{2}x^{-3/2}$	$y' = -\dfrac{1}{2x^{3/2}}$

23. -1 **25.** 0 **27.** 4 **29.** 3 **31.** $3x^2 - 3 + \dfrac{8}{x^5}$

33. $2t + \dfrac{4}{t^2}$ **35.** $\dfrac{x^3 - 8}{x^3}$ **37.** $3x^2 + 1$ **39.** $\dfrac{4}{5s^{1/5}}$

41. $\dfrac{2}{\sqrt{x}} - 3\sin x$

43. (a) $2x + y - 2 = 0$

(b)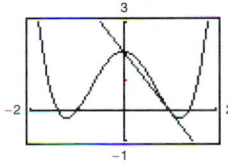

45. (a) $x + 48y - 20 = 0$

(b)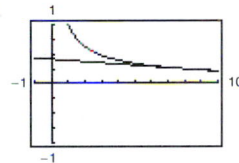

47. $(0, 2), (-2, -14), (2, -14)$

49. No horizontal tangents   **51.** (π, π)

53.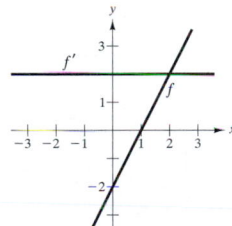

The rate of change of f is constant and therefore f' is a constant function.

55. $y = 2x - 1$ $y = 4x - 4$

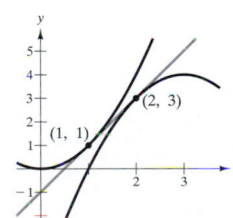   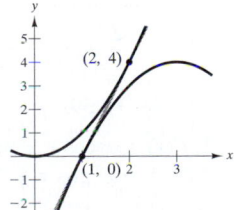

57. $x - 4y + 4 = 0$

59. (a)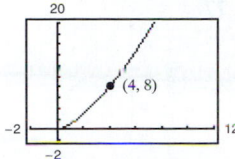

$(3.9, 7.7019)$, $S(x) = 2.981x - 3.924$

(b) $T(x) = 3(x - 4) + 8 = 3x - 4$

(c) It becomes less accurate.

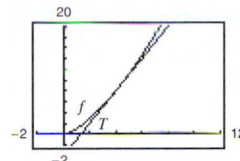

(d)

Δx	-3	-2	-1	-0.5	-0.1	0
$f(4 + \Delta x)$	1	2.828	5.196	6.458	7.702	8
$T(4 + \Delta x)$	-1	2	5	6.5	7.7	8

Δx	0.1	0.5	1	2	3
$f(4 + \Delta x)$	8.302	9.546	11.180	14.697	18.520
$T(4 + \Delta x)$	8.3	9.5	11	14	17

61. False: let $f(x) = x$ and $g(x) = x + 1$.   **62.** True

63. False: $dy/dx = 0$ **64.** True

65. Average rate: 2 **67.** Average rate: $\frac{1}{2}$
Instantaneous rates: Instantaneous rates:
$f'(1) = 2$ $f'(1) = 1$
$f'(2) = 2$ $f'(2) = \frac{1}{4}$

69. (a) A and B (b) Greater

(c)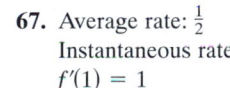

(d) B and C, D and E

71. (a) $s(t) = -16t^2 + 1362$
$v(t) = -32t$

(b) -48 feet per second

(c) $s'(1) = -32$ feet per second
$s'(2) = -64$ feet per second

(d) $t = \dfrac{\sqrt{1362}}{4} \approx 9.226$ seconds

(e) -295.242 feet per second

73. $v(5) = 71$ meters per second
$v(10) = 22$ meters per second

75.

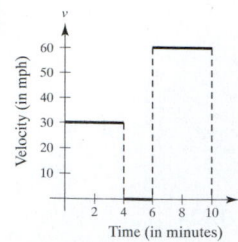

77.

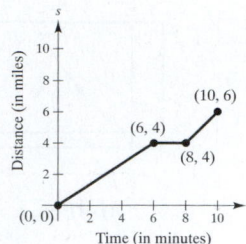

79. (a) $R(v) = 0.167v - 0.02$

(b) $B(v) = 0.006v^2 - 0.024v + 0.460$

(c) $T(v) = 0.006v^2 + 0.143v + 0.440$

(d)

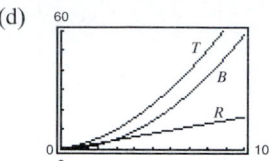

(e) $T'(v) = 0.012v + 0.143$

$T'(40) = 0.623$

$T'(80) = 1.103$

$T'(100) = 1.343$

(f) Stopping distances increase at an increasing rate.

81. 8 meters   **83.** $-\$1.91, -\$1.93$

85. The runner was safe. Explanations will vary.

87. $y = 2x^2 - 3x + 1$ **89.** $y = -9x, y = -\frac{9}{4}x - \frac{27}{4}$

91. $a = \frac{1}{3}, b = -\frac{4}{3}$ **93.** Proof

Section 2.3 *(page 121)*

1. $f'(x) = 2x^2$

$f'(0) = 0$

3. $f'(x) = (x^3 - 3x)(4x + 3) + (2x^2 + 3x + 5)(3x^2 - 3)$

$= 10x^4 + 12x^3 - 3x^2 - 18x - 15$

$f'(0) = -15$

5. $f'(x) = \cos x - x \sin x$

$f'\left(\frac{\pi}{4}\right) = \frac{\sqrt{2}}{8}(4 - \pi)$

Function	Rewrite	Derivative	Simplify
7. $y = \dfrac{x^2 + 2x}{x}$	$y = x + 2$	$y' = 1$	$y' = 1$
9. $y = \dfrac{7}{3x^3}$	$y = \dfrac{7}{3}x^{-3}$	$y' = -7x^{-4}$	$y' = -\dfrac{7}{x^4}$
11. $y = \dfrac{3x^2 - 5}{7}$	$y = \dfrac{1}{7}(3x^2 - 5)$	$y' = \dfrac{1}{7}(6x)$	$y' = \dfrac{6x}{7}$

13. $\dfrac{(2x - 3)(3) - (3x - 2)(2)}{(2x - 3)^2} = -\dfrac{5}{(2x - 3)^2}$

15. $\dfrac{(x^2 - 1)(-2 - 2x) - (3 - 2x - x^2)(2x)}{(x^2 - 1)^2} = \dfrac{2}{(x + 1)^2}, \quad x \neq 1$

17. $\dfrac{\sqrt{x}(1) - (x + 1)[1/(2\sqrt{x})]}{x} = \dfrac{x - 1}{2x^{3/2}}$

19. $6s^2(s^3 - 2)$

21. $\dfrac{(t^2 + 2t + 2)(1) - (t + 1)(2t + 2)}{(t^2 + 2t + 2)^2} = \dfrac{-t^2 - 2t}{(t^2 + 2t + 2)^2}$

23. $(3x^3 + 4x)[(x - 5) \cdot 1 + (x + 1) \cdot 1]$

$\quad + [(x - 5)(x + 1)](9x^2 + 4)$

$\quad = 15x^4 - 48x^3 - 33x^2 - 32x - 20$

25. $\dfrac{(x^2 - c^2)(2x) - (x^2 + c^2)(2x)}{(x^2 - c^2)^2} = -\dfrac{4xc^2}{(x^2 - c^2)^2}$

27. $t(t \cos t + 2 \sin t)$ **29.** $-\dfrac{t \sin t + \cos t}{t^2}$

31. $-1 + \sec^2 x = \tan^2 x$ **33.** $\dfrac{1}{2\sqrt{t}} + 4 \sec t \tan t$

35. $-5x \csc x \cot x + 5 \csc x = 5 \csc x(1 - x \cot x)$

37. $\csc x \cot x - \cos x = \cos x \cot^2 x$

39. $x^2 \cos x + 2x \sin x - 2x \sin x + 2 \cos x$

$\quad = x^2 \cos x + 2 \cos x$

41. $x(x \sec^2 x + 2 \tan x)$

43. $\left(\dfrac{x + 1}{x + 2}\right)(2) + (2x - 5)\left[\dfrac{(x + 2)(1) - (x + 1)(1)}{(x + 2)^2}\right]$

$\quad = \dfrac{2x^2 + 8x - 1}{(x + 2)^2}$

45. $\dfrac{1 - \sin\theta + \theta \cos\theta}{(1 - \sin\theta)^2}$ **47.** $y' = \dfrac{-2 \csc x \cot x}{(1 - \csc x)^2}, \quad -4\sqrt{3}$

49. $h'(t) = \dfrac{\sec t(t \tan t - 1)}{t^2}, \quad \dfrac{1}{\pi^2}$

51. (a) $y = -x + 4$ **53.** (a) $y = -x - 2$

(b) 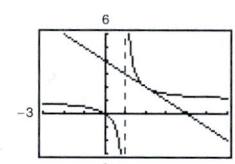   (b)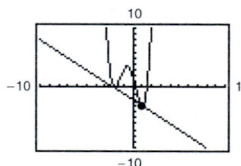

55. (a) $4x - 2y - \pi + 2 = 0$

(b)

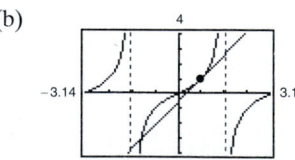

57. $(0, 0), (2, 4)$ **59.** 0 **61.** -10

63. $n = 1, f'(x) = x \cos x + \sin x$

$n = 2, f'(x) = x^2 \cos x + 2x \sin x$

$n = 3, f'(x) = x^3 \cos x + 3x^2 \sin x$

$n = 4, f'(x) = x^4 \cos x + 4x^3 \sin x$

$f'(x) = x^n \cos x + nx^{n-1} \sin x$

65. (a) $-\$38.13$ (b) $-\$10.37$ (c) $-\$3.80$

The costs decrease with increasing order size.

67. 31.55 bacteria per hour   **69.** Proof

71.

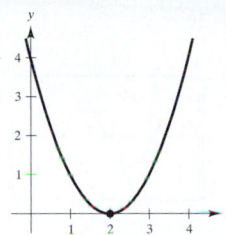

73. $\dfrac{3}{\sqrt{x}}$ **75.** $\dfrac{2}{(x-1)^3}$ **77.** $-3\sin x$

79. $2x$ **81.** $\dfrac{1}{\sqrt{x}}$

83.

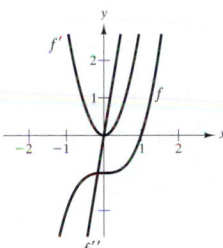

85. (a) $f''(x) = g(x)h''(x) + 2g'(x)h'(x) + g''(x)h(x)$

$f'''(x) = g(x)h'''(x) + 3g'(x)h''(x) +$
$\qquad\qquad 3g''(x)h'(x) + g'''(x)h(x)$

$f^{(4)}(x) = g(x)h^{(4)}(x) + 4g'(x)h'''(x) + 6g''(x)h''(x) +$
$\qquad\qquad 4g'''(x)h'(x) + g^{(4)}(x)h(x)$

(b) $f^{(n)}(x) = g(x)h^{(n)}(x) + \dfrac{n!}{1!(n-1)!}\,g'(x)h^{(n-1)}(x) +$

$\qquad \dfrac{n!}{2!(n-2)!}\,g''(x)h^{(n-2)}(x) + \cdots +$

$\qquad \dfrac{n!}{(n-1)!1!}\,g^{(n-1)}(x)h'(x) + g^{(n)}(x)h(x)$

87. $v(3) = 27$ meters per second

$a(3) = -6$ meters per second per second

The speed of the object is decreasing, but the rate of that decrease is increasing.

89.

x	0	1	2	3	4
$s(t)$	0	57.75	99	123.75	132
$v(t)$	66	49.5	33	16.5	0
$a(t)$	-16.5	-16.5	-16.5	-16.5	-16.5

Average velocity on:

$[0, 1]$ is 57.75

$[1, 2]$ is 41.25

$[2, 3]$ is 24.75

$[3, 4]$ is 8.25

91. (a) $P_1(x) = -\dfrac{\sqrt{3}}{2}\left(x - \dfrac{\pi}{3}\right) + \dfrac{1}{2}$

$P_2(x) = -\dfrac{1}{4}\left(x - \dfrac{\pi}{3}\right)^2 - \dfrac{\sqrt{3}}{2}\left(x - \dfrac{\pi}{3}\right) + \dfrac{1}{2}$

(b)

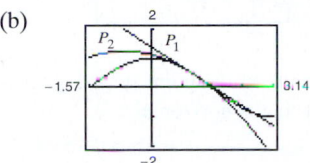

(c) P_2

(d) P_1 and P_2 become less accurate as you move farther from $x = a$.

93. False: $dy/dx = f(x)g'(x) + g(x)f'(x)$ **94.** True

95. True   **96.** True   **97.** True   **98.** True

99. $f'(x) = 2|x|$

$f''(0)$ does not exist.

Section 2.4 *(page 130)*

$y = f(g(x))$	$u = g(x)$	$y = f(u)$
1. $y = (6x - 5)^4$	$u = 6x - 5$	$y = u^4$
3. $y = \sqrt{x^2 - 1}$	$u = x^2 - 1$	$y = \sqrt{u}$
5. $y = \csc^3 x$	$u = \csc x$	$y = u^3$

7. $6(2x - 7)^2$ **9.** $-108(4 - 9x)^3$

11. $\dfrac{2}{3}(9 - x^2)^{-1/3}(-2x) = -\dfrac{4x}{3(9 - x^2)^{1/3}}$

13. $\dfrac{1}{2}(1 - t)^{-1/2}(-1) = -\dfrac{1}{2\sqrt{1 - t}}$

15. $\dfrac{1}{3}(9x^2 + 4)^{-2/3}(18x) = \dfrac{6x}{(9x^2 + 4)^{2/3}}$

17. $(4 - x^2)^{-1/2}(-2x) = -\dfrac{2x}{\sqrt{4 - x^2}}$

19. $-\dfrac{1}{(x - 2)^2}$ **21.** $-2(t - 3)^{-3}(1) = -\dfrac{2}{(t - 3)^3}$

23. $-\dfrac{1}{2(x + 2)^{3/2}}$

25. $x^2[4(x - 2)^3(1)] + (x - 2)^4(2x) = 2x(x - 2)^3(3x - 2)$

27. $x\left(\dfrac{1}{2}\right)(1 - x^2)^{-1/2}(-2x) + (1 - x^2)^{1/2}(1) = \dfrac{1 - 2x^2}{\sqrt{1 - x^2}}$

29. $\dfrac{(x^2 + 1)^{1/2}(1) - x(1/2)(x^2 + 1)^{-1/2}(2x)}{x^2 + 1} = \dfrac{1}{(x^2 + 1)^{3/2}}$

31. $\dfrac{1 - 3x^2 - 4x^{3/2}}{2\sqrt{x}(x^2 + 1)^2}$

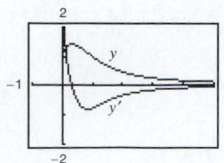

The zero of $g'(x)$ corresponds to the point on the graph of the function where the tangent line is horizontal.

33. $\dfrac{3t(t^2 + 3t - 2)}{(t^2 + 2t - 1)^{3/2}}$

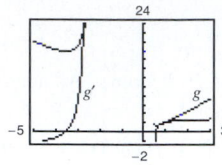

The zero of $g'(x)$ corresponds to the point on the graph of the function where the tangent line is horizontal.

35. $-\dfrac{\sqrt{\dfrac{x + 1}{x}}}{2x(x + 1)}$

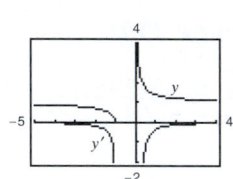

y' has no zeros.

37. $\dfrac{t}{\sqrt{1 + t}}$

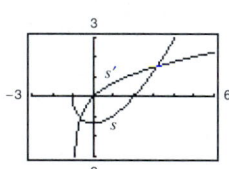

The zero of $s'(t)$ corresponds to the point on the graph of the function where the tangent line is horizontal.

39. $y' = -\dfrac{\pi x \sin \pi x + \cos \pi x + 1}{x^2}$

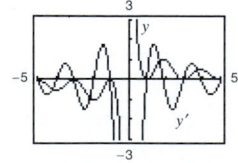

The zeros of y' correspond to the points on the graph of the function where the tangent lines are horizontal.

41. (a) 1   (b) 2   **43.** $-3 \sin 3x$ **45.** $12 \sec^2 4x$

47. $\sin 2\theta \cos 2\theta = \frac{1}{2}\sin 4\theta$ **49.** $\dfrac{1}{2\sqrt{x}} + 2x \cos(2x)^2$

51. $-\sin x \cos(\cos x)$ **53.** $s'(t) = \dfrac{t + 1}{\sqrt{t^2 + 2t + 8}}, \dfrac{3}{4}$

55. $f'(x) = \dfrac{-9x^2}{(x^3 - 4)^2}, -\dfrac{9}{25}$ **57.** $f'(t) = \dfrac{-5}{(t - 1)^2}, -5$

59. $y' = -6 \sec^3(2x) \tan(2x), 0$

61. (a) $9x - 5y - 2 = 0$ **63.** (a) $2x - y - 2\pi = 0$

(b) 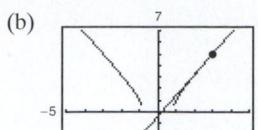   (b)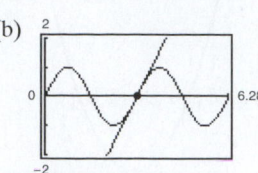

65. 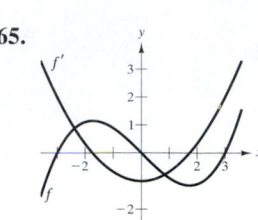   The zeros of f' correspond to the points where the graph of f has horizontal tangents.

67. 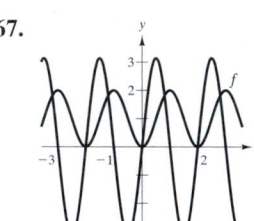   The zeros of f' correspond to the points where the graph of f has horizontal tangents.

69. $12(5x^2 - 1)(x^2 - 1)$ **71.** $2(\cos x^2 - 2x^2 \sin x^2)$

73. (a) 24   (b) Not possible because $g'(h(5))$ is not known.
(c) $\frac{4}{3}$ (d) 162

75. (a) 1.461   (b) -1.016

77. 0.2 radian, 1.45 radians per second   **79.** 0.04224

81.

x	-2	-1	0	1	2	3
$f'(x)$	4	$\frac{2}{3}$	$-\frac{1}{3}$	-1	-2	-4
$g'(x)$	4	$\frac{2}{3}$	$-\frac{1}{3}$	-1	-2	-4
$h'(x)$	8	$\frac{4}{3}$	$-\frac{2}{3}$	-2	-4	-8
$r'(x)$	—	12	1	—	—	—
$s'(x)$	$-\frac{1}{3}$	-1	-2	-4	—	—

83. (a) yes   (b) yes   **85.** Proof   **87.** $\dfrac{2(2x - 3)}{|2x - 3|}, x \neq \dfrac{3}{2}$

89. $-|x| \sin x + \dfrac{x}{|x|} \cos x, x \neq 0$

91. (a) $P_1 = 1$ $P_2 = -\frac{1}{8}(x - \pi)^2 + 1$

(b)

(c) P_2

(d) P_1 and P_2 become less accurate as you move farther from $x = \pi$.

93. False: $y' = \frac{1}{2}(1-x)^{-1/2}(-1)$

94. False: $f'(x) = 2(\sin 2x)(\cos 2x)(2)$ **95.** True
$= 4\sin 2x\cos 2x$

Section 2.5 (page 139)

1. $-\dfrac{x}{y}$ **3.** $-\sqrt{\dfrac{y}{x}}$ **5.** $\dfrac{y-3x^2}{2y-x}$ **7.** $\dfrac{1-3x^2y^3}{3x^3y^2-1}$

9. $\dfrac{4xy-3x^2-3y^2}{2x(3y-x)}$ **11.** $\dfrac{\cos x}{4\sin 2y}$ **13.** $\dfrac{\cos x-\tan y-1}{x\sec^2 y}$

15. $\dfrac{y\cos(xy)}{1-x\cos(xy)}$ **17.** $-\dfrac{y}{x},\ -\dfrac{1}{4}$ **19.** $\dfrac{18x}{(x^2+9)^2 y}$, undefined

21. $-\sqrt[3]{\dfrac{y}{x}},\ -\dfrac{1}{2}$ **23.** $-\sin^2(x+y)$ or $-\dfrac{x^2}{x^2+1},\ 0$

25. $x+2y-6=0$ **27.** $-\dfrac{1}{2}$ **29.** 0

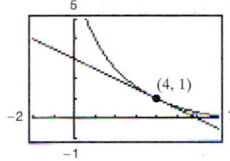

31. (a) $y_1 = \sqrt{16-x^2}$ (b)
$y_2 = -\sqrt{16-x^2}$

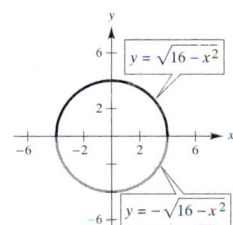

(c) $y' = \mp\dfrac{x}{\sqrt{16-x^2}} = -\dfrac{x}{y}$ (d) $y' = -\dfrac{x}{y}$

33. (a) $y_1 = \dfrac{3}{4}\sqrt{16-x^2}$ (b)
$y_2 = -\dfrac{3}{4}\sqrt{16-x^2}$

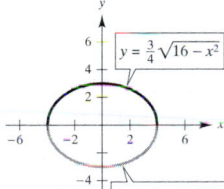

(c) $y' = \mp\dfrac{3x}{4\sqrt{16-x^2}} = -\dfrac{9x}{16y}$ (d) $y' = -\dfrac{9x}{16y}$

35. $\dfrac{10}{x^3}$ **37.** $-\dfrac{16}{y^3}$ **39.** $\dfrac{3y}{4x^2} = \dfrac{3x}{4y}$

41. At $(4,3)$:
Tangent line: $4x+3y-25=0$
Normal line: $3x-4y=0$

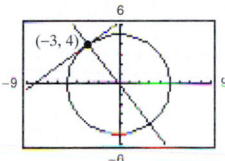

At $(-3,4)$:
Tangent line: $3x-4y+25=0$
Normal line: $4x+3y=0$

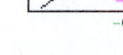

43. Proof   **45.** Horizontal tangents: $(-4,0),(-4,10)$
Vertical tangents: $(0,5),(-8,5)$

47. 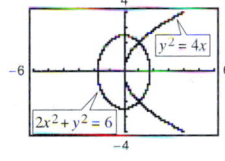   **49.**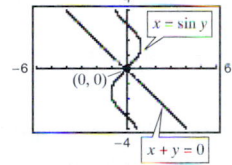

At $(1,2)$:   At $(0,0)$:
Slope of ellipse: -1 Slope of line: -1
Slope of parabola: 1 Slope of sine curve: 1
At $(1,-2)$:
Slope of ellipse: 1
Slope of parabola: -1

51.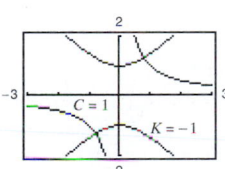

Derivatives: $\dfrac{dy}{dx} = -\dfrac{y}{x},\ \dfrac{dy}{dx} = \dfrac{x}{y}$

53. (a) $4y\dfrac{dy}{dx}-12x^3=0$

(b) $4y\dfrac{dy}{dt}-12x^3\dfrac{dx}{dt}=0$

55. (a) $-\pi\sin\pi y\left(\dfrac{dy}{dx}\right)-3\pi\cos\pi x=0$

(b) $-\pi\sin\pi y\left(\dfrac{dy}{dt}\right)-3\pi\cos\pi x\left(\dfrac{dx}{dt}\right)=0$

57. (a)

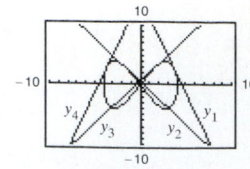

(b) $y_1 = \dfrac{1}{3}\left[\left(\sqrt{7}+7\right)x + \left(8\sqrt{7}+23\right)\right]$

$y_2 = -\dfrac{1}{3}\left[\left(-\sqrt{7}+7\right)x - \left(23 - 8\sqrt{7}\right)\right]$

$y_3 = -\dfrac{1}{3}\left[\left(\sqrt{7}-7\right)x - \left(23 - 8\sqrt{7}\right)\right]$

$y_4 = -\dfrac{1}{3}\left[\left(\sqrt{7}+7\right)x - \left(8\sqrt{7}+23\right)\right]$

(c) $\left(\dfrac{8\sqrt{7}}{7}, 5\right)$

59. Proof

Section 2.6 *(page 146)*

1. (a) $\frac{3}{4}$ (b) 20   **3.** (a) $-\frac{5}{8}$ (b) $\frac{3}{2}$

5. (a) -4 centimeters per second

(b) 0 centimeters per second

(c) 4 centimeters per second

(d) 12 centimeters per second

7. (a) 8 centimeters per second

(b) 4 centimeters per second

(c) 2 centimeters per second

(d) 6.851 centimeters per second

9. (a) Decreases   (b) Increases   **11.** $\dfrac{2(2x^3 + 3x)}{\sqrt{x^4 + 3x^2 + 1}}$

13. (a) 24π square centimeters per minute

(b) 96π square centimeters per minute

15. (a) Proof

(b) When $\theta = \dfrac{\pi}{6}$, $\dfrac{dA}{dt} = \dfrac{\sqrt{3}}{8}s^2$

When $\theta = \dfrac{\pi}{3}$, $\dfrac{dA}{dt} = \dfrac{1}{8}s^2$

(c) If s and $d\theta/dt$ are constant, dA/dt is proportional to $\cos\theta$.

17. (a) $\dfrac{5}{36\pi}$ centimeter per minute

(b) $\dfrac{5}{144\pi}$ centimeter per minute

19. (a) 36 square centimeters per second

(b) 360 square centimeters per second

21. $\dfrac{8}{405\pi}$ foot per minute

23. (a) 12.5%   (b) $\frac{1}{144}$ meter per minute

25. (a) $-\frac{7}{12}$ foot per second

$-\frac{3}{2}$ feet per second

$-\frac{48}{7}$ feet per second

(b) $\frac{527}{24}$ square feet per second

(c) $\frac{1}{12}$ radian per second

27. Rate of vertical change: $\frac{1}{5}$ meter per second

Rate of horizontal change: $-\dfrac{\sqrt{3}}{15}$ meter per second

29. (a) -750 miles per hour   (b) 20 minutes

31. $-\dfrac{28}{\sqrt{10}} \approx -8.85$ feet per second

33. (a) $\frac{25}{3}$ feet per second   (b) $\frac{10}{3}$ feet per second

35. (a) 12 seconds   **37.** Proof

(b) $\dfrac{1}{2}\sqrt{3}$ meter

(c) $\dfrac{\sqrt{5}\,\pi}{120}$ meter per second

39. $v^{0.3}\left(1.3p\,\dfrac{dv}{dt} + v\,\dfrac{dp}{dt}\right) = 0$ **41.** $\dfrac{1}{20}$ radian per second

43. (a) $\frac{1}{2}$ radian per minute   (b) $\frac{3}{2}$ radians per minute

(c) 1.87 radians per minute

45. (a) $\dfrac{dx}{dt} = -600\pi\sin\theta$ (b)

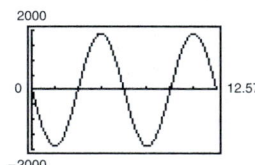

(c) $\theta = 90° + n\cdot 180°;\ \theta = 0° + n\cdot 180°$

(d) -300π centimeters per second;

$-300\sqrt{3}\pi$ centimeters per second

47. $\dfrac{1}{25}\cos^2\theta,\quad -\dfrac{\pi}{4} \le \theta \le \dfrac{\pi}{4}$

49. -0.1808 foot per second per second

51. (a) $m(s) = -1.014\,s^2 + 31.685s - 214.436$

(b) $(-2.028s + 31.685)\dfrac{ds}{dt}$

(c) -2.1 million

Review Exercises for Chapter 2 *(page 150)*

1. $f'(x) = 2x - 2$

3.

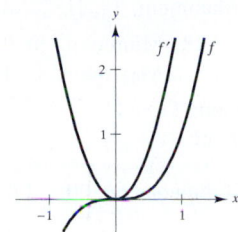 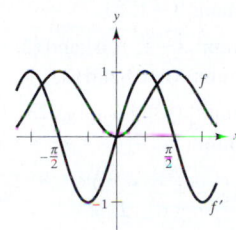

$f' > 0$ where the slopes of tangent lines to the graph of f are positive.

5. $x \neq -1$ **7.** $3x(x - 2)$ **9.** $\dfrac{2(x^3 + 1)}{x^3}$ **11.** $-\dfrac{4}{3t^3}$

13. $\dfrac{-3x^2}{2\sqrt{1 - x^3}}$ **15.** $2(6x^3 - 9x^2 + 16x - 7)$

17. $5\left(x^2 + \dfrac{1}{x}\right)\left(2x - \dfrac{1}{x^2}\right)$ **19.** $-\dfrac{x^2 + 1}{(x^2 - 1)^2}$

21. $\dfrac{6x}{(4 - 3x^2)^2}$ **23.** $-9 \sin(3x + 1)$

25. $-\csc 2x \cot 2x$ **27.** $\frac{1}{2}(1 - \cos 2x) = \sin^2 x$

29. $\sin^{1/2}x \cos x - \sin^{5/2}x \cos x = \cos^3 x\sqrt{\sin x}$

31. $-x \sec^2 x - \tan x$ **33.** $\dfrac{x \cos x - 2 \sin x}{x^3}$

35. $t(t - 1)^4(7t - 2)$

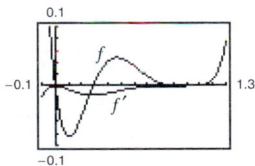

The zeros of f' correspond to the points on the graph of the function where the tangent line is horizontal.

37. $\dfrac{x + 2}{(x + 1)^{3/2}}$ **39.** $\dfrac{5}{6(t + 1)^{1/6}}$

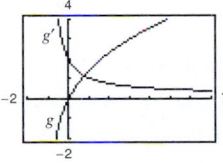 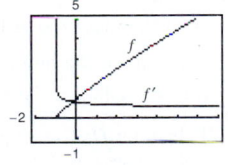

g' is not equal to zero for any x. f' has no zeros.

41. $-\dfrac{\sec^2 \sqrt{1 - x}}{2\sqrt{1 - x}}$ **43.** $-\dfrac{\pi\sqrt{3}}{4}$ **45.** $4 - 4 \sin 2x$

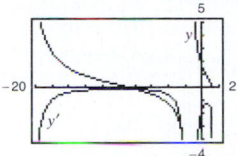

y' has no zeros.

47. $2 \csc^2 x \cot x$ **49.** $\dfrac{2(t + 2)}{(1 - t)^4}$ **51.** $2 \sec^2 x(x \tan x + 1)$

53. $-\dfrac{2x + 3y}{3(x + y^2)}$ **55.** $\dfrac{2y\sqrt{x} - y\sqrt{y}}{2x\sqrt{y} - x\sqrt{x}}$ **57.** $-\dfrac{y \sin x + \sin y}{\cos x - x \cos y}$

59. Tangent line: $3x - y + 7 = 0$
Normal line: $x + 3y - 1 = 0$

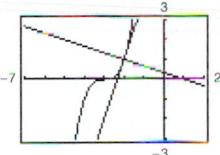

61. Tangent line: $x + 2y - 10 = 0$
Normal line: $2x - y = 0$

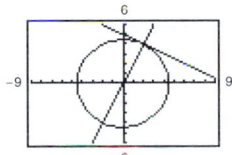

63. Tangent line: $2x - 3y - 3 = 0$
Normal line: $3x + 2y - 11 = 0$

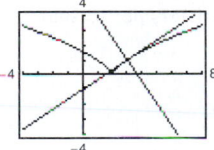

65. (a) $(0, -1), \left(-2, \dfrac{7}{3}\right)$ (b) $(-3, 2), \left(1, -\dfrac{2}{3}\right)$

(c) $\left(-1 +, \sqrt{2}, \dfrac{2\left[1 - 2\sqrt{2}\right]}{3}\right)$

$\left(-1 - \sqrt{2}, \dfrac{2\left[1 + 2\sqrt{2}\right]}{3}\right)$

67. (a) Yes

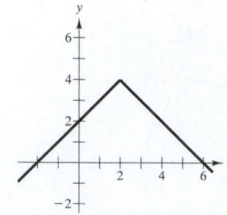

(b) No. The derivatives from the right and left are not equal.

69. $y'' + y = -(2 \sin x + 3 \cos x) + (2 \sin x + 3 \cos x) = 0$

71. (a) -18.667 degrees per hour

(b) -7.284 degrees per hour

(c) -3.240 degrees per hour

(d) -0.747 degree per hour

73. (a) 50 vibrations per second per pound

(b) 33.33 vibrations per second per pound

75. 56 feet per second

77. (a)

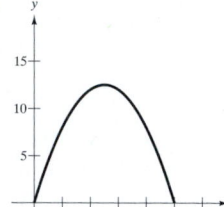

(b) 50   (c) $x = 25$

(d) $y' = 1 - 0.04x$

(e) $y'(25) = 0$

x	0	10	25	30	50
y'	1	0.6	0	-0.2	-1

79. (a) $\dfrac{4\sqrt{6}}{3}$ units per second   (b) $3\sqrt{2}$ units per second

(c) $\dfrac{18\sqrt{5}}{5}$ units per second

81. $\frac{2}{25}$ meter per minute   **83.** 38.34 meters per second

85. (a) $y = 0.14x^2 - 4.43x + 58.4$

(b) 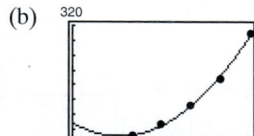   (c)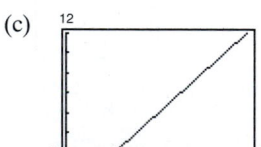

(d) 365 feet

(e) Stopping distance increases at an increasing rate as x increases.

CHAPTER 3

Section 3.1 *(page 160)*

1. $f'(0) = 0$ **3.** $f'(4) = 0$

5. $f'(-2)$ is undefined.   **7.** $x = 0, \ x = 2$

9. $t = \dfrac{8}{3}, \ t = 4$ **11.** $x = 0, \ x = \dfrac{\pi}{3}, \ \pi, \ \dfrac{5\pi}{3}$

13. Minimum: $(2, 2)$ **15.** Minimum: $(0, 0)$ and $(3, 0)$
Maximum: $(-1, 8)$ Maximum: $\left(\frac{3}{2}, \frac{9}{4}\right)$

17. Minimum: $(-1, -4)$ and $(2, -4)$ **19.** Minimum: $(0, 0)$
Maximum: $(0, 0)$ and $(3, 0)$ Maximum: $(-1, 5)$

21. Minimum: $(1, 1)$ **23.** Minimum: $(1, -1)$
Maximum: $(4, 4)$ Maximum: $\left(0, -\frac{1}{2}\right)$

25. Minimum: $\left(\dfrac{1}{6}, \dfrac{\sqrt{3}}{2}\right)$ **27.** Continuous on $\left[0, \dfrac{\pi}{4}\right]$
Maximum: $(0, 1)$ Not continuous on $[0, \pi]$

29. (a) Yes   (b) No   **31.** (a) No   (b) Yes

33. (a) Minimum: $(0, -3)$; maximum: $(2, 1)$

(b) Minimum: $(0, -3)$ (c) Maximum: $(2, 1)$

(d) No extrema

35. (a) Minimum: $(1, -1)$; maximum: $(-1, 3)$

(b) Maximum: $(3, 3)$ (c) Minimum: $(1, -1)$

(d) Minimum: $(1, -1)$

37. 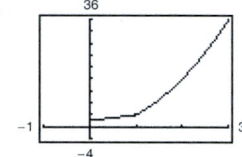   **39.**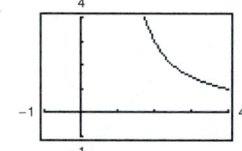

Minimum: $(0, 2)$ Minimum: $(4, 1)$
Maximum: $(3, 36)$

41. (a)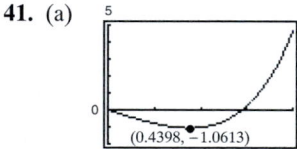

(b) Minimum: $(0.4398, -1.0613)$

43. Maximum: $\left| f''\left(\sqrt[3]{-10 + \sqrt{108}}\right) \right| = f''\left(\sqrt{3} - 1\right) \approx 1.47$

45. Maximum: $\left| f^{(4)}(0) \right| = \frac{56}{81}$

47. Maximum: $P(12) = 72$
No: P is decreasing for $I \geq 12$.

49. The part of the lawn farthest from the sprinkler

51. 0.9553 radian   **53.** True   **54.** True   **55.** True

56. False: let $f(x) = x^2$ and $g(x) = (x - 1)^2$.

57. All real numbers (Slope is zero at nonintegers and undefined at integers.)

Section 3.2 *(page 167)*

1. $f(0) = f(2) = 0$

f is not differentiable on $(0, 2)$.

3. $f'(1) = 0$

5. $f'\left(\dfrac{6 - \sqrt{3}}{3}\right) = 0$

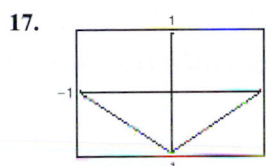

$f'\left(\dfrac{6 + \sqrt{3}}{3}\right) = 0$

7. Not differentiable at $x = 0$

9. $f'\left(-2 + \sqrt{5}\right) = 0$

11. $f'\left(\dfrac{\pi}{2}\right) = 0$ **13.** $f'\left(\dfrac{\pi}{4}\right) = 0$

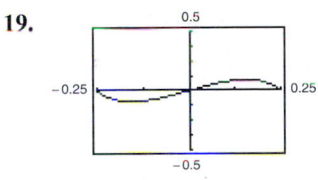

$f'\left(\dfrac{3\pi}{2}\right) = 0$

15. Not continuous on $[0, \pi]$

17.

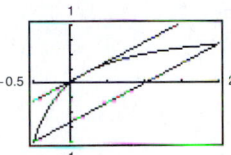

Rolle's Theorem does not apply.

19.

$f'(\pm 0.1533) = 0$

21. (a) $f(1) = f(2) = 64$

(b) Velocity $= 0$ for some t in $[-1, 2]$.

23. No: Let $f(x) = x^2$ on $[-1, 2]$.

25. (a) f is continuous and changes signs in $[-10, 4]$ (Intermediate Value Theorem).

(b) There exist real numbers a and b such that $-10 < a < b < 4$ and $f(a) = f(b) = 2$. Therefore, f' has a zero in the interval by Rolle's Theorem.

(c) (d)

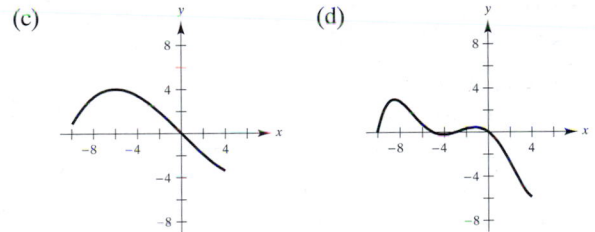

(e) No

27. $f'\left(-\dfrac{1}{2}\right) = -1$ **29.** $f'\left(\dfrac{8}{27}\right) = 1$

31. $f'(3) = \dfrac{1}{2}$ **33.** $f'\left(\dfrac{\pi}{2}\right) = 0$

35. Secant line: $2x - 3y - 2 = 0$

Tangent line: $c = \dfrac{-2 + \sqrt{6}}{2}$, $2x - 3y + 5 - 2\sqrt{6} = 0$

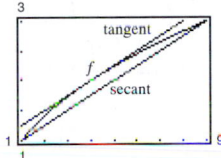

37. Secant line: $x - 4y + 3 = 0$

Tangent line: $c = 4$, $x - 4y + 4 = 0$

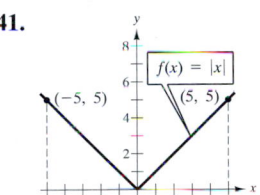

39. The function is discontinuous on $[0, 6]$.

41.

$f(x) = |x|$

$(-5, 5)$ $(5, 5)$

43. (a) -14.7 meters per second   (b) 1.5 seconds

45. By the Mean Value Theorem, there is a time when the speed of the plane must equal the average speed of 454.5 miles per hour. The speed was 400 miles per hour when the plane was accelerating to 454.5 miles per hour and decelerating from 454.5 miles per hour.

47. False: f is not continuous on $[-1, 1]$.

48. False: let $f(x) = \dfrac{x^3 - 4x}{x^2 - 1}$.   **49.** True

50. True   **51.** Proof   **53.** Proof   **55.** Proof

Section 3.3 *(page 176)*

1. Increasing on $(3, \infty)$
Decreasing on $(-\infty, 3)$

3. Increasing on $(-\infty, -2)$ and $(2, \infty)$
Decreasing on $(-2, 2)$

5. Increasing on $(-\infty, 0)$
Decreasing on $(0, \infty)$

7. Critical number: $x = 1$
Increasing on $(-\infty, 1)$
Decreasing on $(1, \infty)$
Relative maximum: $(1, 5)$

9. Critical number: $x = 3$
Increasing on $(3, \infty)$
Decreasing on $(-\infty, 3)$
Relative minimum: $(3, -9)$

11. Critical numbers: $x = -2, 1$
Increasing on $(-\infty, -2)$ and $(1, \infty)$
Decreasing on $(-2, 1)$
Relative maximum: $(-2, 20)$
Relative minimum: $(1, -7)$

13. Critical numbers: $x = -1, 1$
Increasing on $(-\infty, -1)$ and $(1, \infty)$
Decreasing on $(-1, 1)$
Relative maximum: $\left(-1, \frac{4}{5}\right)$
Relative minimum: $\left(1, -\frac{4}{5}\right)$

15. Critical number: $x = 0$
Increasing on $(-\infty, \infty)$
No relative extrema

17. Critical number: $x = 5$
Increasing on $(-\infty, 5)$
Decreasing on $(5, \infty)$
Relative maximum: $(5, 5)$

19. Critical number $x = 0$
Discontinuities: $x = -3, 3$
Increasing on $(-\infty, -3)$ and $(-3, 0)$
Decreasing on $(0, 3)$ and $(3, \infty)$
Relative maximum: $(0, 0)$

21. Critical numbers: $x = 0, 4$
Increasing on $(-\infty, 0)$ and $(4, \infty)$
Decreasing on $(0, 4)$
Relative maximum: $(0, 15)$
Relative minimum: $(4, -17)$

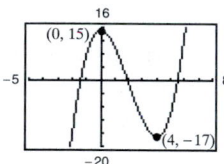

23. Critical number: $x = 1$
Increasing on $(1, \infty)$
Decreasing on $(-\infty, 1)$
Relative minimum: $(1, 0)$

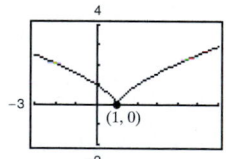

25. Critical numbers: $x = -1, 1$
Discontinuity: $x = 0$
Increasing on $(-\infty, -1)$ and $(1, \infty)$
Decreasing on $(-1, 0)$ and $(0, 1)$
Relative maximum: $(-1, -2)$
Relative minimum: $(1, 2)$

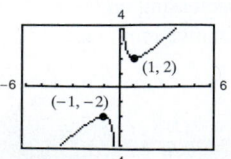

27. Critical numbers: $x = -3, 1$
Discontinuity: $x = -1$
Increasing on $(-\infty, -3)$ and $(1, \infty)$
Decreasing on $(-3, -1)$ and $(-1, 1)$
Relative maximum: $(-3, -8)$
Relative minimum: $(1, 0)$

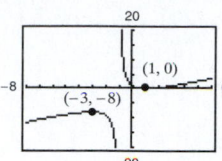

29. Critical numbers: $x = \dfrac{\pi}{6}, \dfrac{5\pi}{6}$

Increasing on $\left(0, \dfrac{\pi}{6}\right), \left(\dfrac{5\pi}{6}, 2\pi\right)$

Decreasing on $\left(\dfrac{\pi}{6}, \dfrac{5\pi}{6}\right)$

Relative maximum: $\left(\dfrac{\pi}{6}, \dfrac{\left[\pi + 6\sqrt{3}\right]}{12}\right)$

Relative minimum: $\left(\dfrac{5\pi}{6}, \dfrac{\left[5\pi - 6\sqrt{3}\right]}{12}\right)$

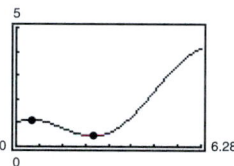

31. Critical numbers: $x = \dfrac{\pi}{2}, \dfrac{7\pi}{6}, \dfrac{3\pi}{2}, \dfrac{11\pi}{6}$

Increasing on $\left(0, \dfrac{\pi}{2}\right), \left(\dfrac{7\pi}{6}, \dfrac{3\pi}{2}\right), \left(\dfrac{11\pi}{6}, 2\pi\right)$

Decreasing on $\left(\dfrac{\pi}{2}, \dfrac{7\pi}{6}\right), \left(\dfrac{3\pi}{2}, \dfrac{11\pi}{6}\right)$

Relative maxima: $\left(\dfrac{\pi}{2}, 2\right), \left(\dfrac{3\pi}{2}, 0\right)$

Relative minima: $\left(\dfrac{7\pi}{6}, -\dfrac{1}{4}\right), \left(\dfrac{11\pi}{6}, -\dfrac{1}{4}\right)$

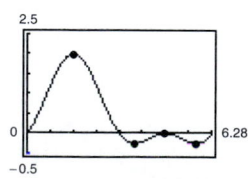

33. (a) $f'(x) = \dfrac{2(9 - 2x^2)}{\sqrt{9 - x^2}}$

(b)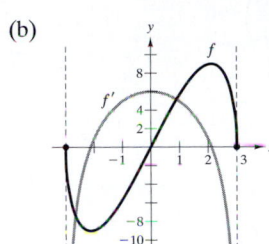

(c) Critical numbers: $x = \pm \dfrac{3\sqrt{2}}{2}$

(d) $f' > 0$ on $\left(-\dfrac{3\sqrt{2}}{2}, \dfrac{3\sqrt{2}}{2}\right)$

$f' < 0$ on $\left(-3, -\dfrac{3\sqrt{2}}{2}\right), \left(\dfrac{3\sqrt{2}}{2}, 3\right)$

35. (a) $f'(t) = t(t \cos t + 2 \sin t)$

(b)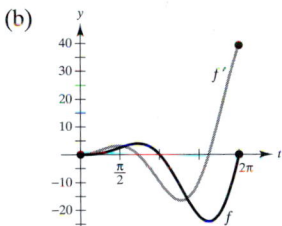

(c) Critical numbers: $x = 2.2889, 5.0870$

(d) $f' > 0$ on $(0, 2.2889), (5.0870, 2\pi)$

$f' < 0$ on $(2.2889, 5.0870)$

37. $g'(0) < 0$ **39.** $g'(-6) < 0$ **41.** $g'(0) > 0$

43. 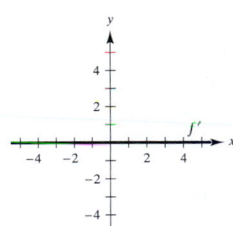   **45.**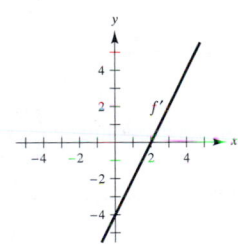

47. 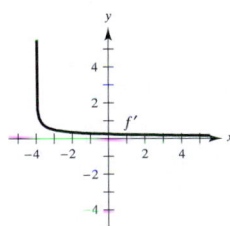   **49.**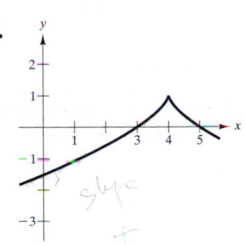

51.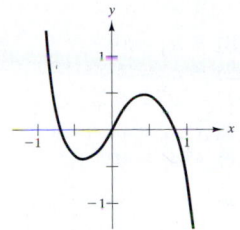

Minimum at the approximate critical number $x = -0.40$.
Maximum at the approximate critical number $x = 0.48$.

53. (a)

x	0.5	1	1.5	2	2.5	3
$f(x)$	0.5	1	1.5	2	2.5	3
$g(x)$	0.48	0.84	1.00	0.91	0.60	0.14

$f(x) > g(x)$

(b)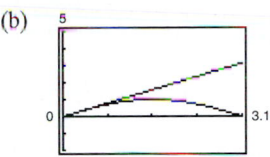

$f(x) > g(x)$

55. $r = \dfrac{2R}{3}$ **57.** Maximum when $R_2 = R_1$

59. (a)

$B = -0.154t^4 + 3.538t^3 - 19.754t^2 + 42.391t + 332.823$

(b)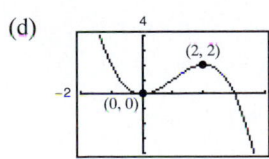

(c) $(12.6, 926.6)$

61. (a) 3

(b) $a_3(0)^3 + a_2(0)^2 + a_1(0) + a_0 = 0$

$a_3(2)^3 + a_2(2)^2 + a_1(2) + a_0 = 2$

$3a_3(0)^2 + 2a_2(0) + a_1 = 0$

$3a_3(2)^2 + 2a_2(2) + a_1 = 0$

(c) $f(x) = -\dfrac{1}{2}x^3 + \dfrac{3}{2}x^2$

(d)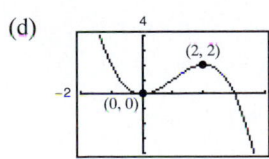

63. (a) 4

(b) $a_4(0)^4 + a_3(0)^3 + a_2(0)^2 + a_1(0) + a_0 = 0$
$a_4(2)^4 + a_3(2)^3 + a_2(2)^2 + a_1(2) + a_0 = 4$
$a_4(4)^4 + a_3(4)^3 + a_2(4)^2 + a_1(4) + a_0 = 0$
$4a_4(0)^3 + 3a_3(0)^2 + 2a_2(0) + a_1 = 0$
$4a_4(2)^3 + 3a_3(2)^2 + 2a_2(2) + a_1 = 0$

(c) $f(x) = \dfrac{1}{4}x^4 - 2x^3 + 4x^2$

(d)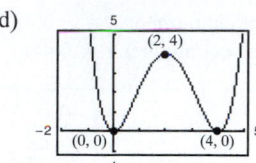

65. True

66. False, let $f(x) = x$, and $g(x) = x$ on the interval $(-\infty, 0)$.

67. False, let $f(x) = x^3$.

68. True   **69.** Proof   **71.** Proof

Section 3.4 *(page 184)*

1. Concave upward: $(-\infty, \infty)$

3. Concave upward: $(-\infty, -2), (2, \infty)$
Concave downward: $(-2, 2)$

5. Concave upward: $(-\infty, -1), (1, \infty)$
Concave downward: $(-1, 1)$

7. Relative maximum: $(3, 9)$

9. Relative minimum: $(5, 0)$

11. Relative maximum: $(0, 3)$
Relative minimum: $(2, -1)$

13. Relative minimum: $(3, -25)$

15. Relative minimum: $(0, -3)$

17. Relative maximum: $(-2, -4)$
Relative minimum: $(2, 4)$

19. No relative extrema, because f is nonincreasing.

21. Relative maximum: $(-2, 16)$
Relative minimum: $(2, -16)$
Point of inflection: $(0, 0)$

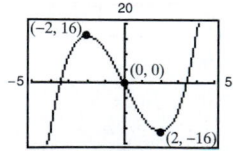

23. Point of inflection: $(2, 8)$

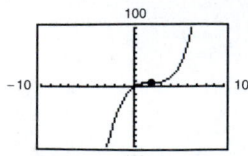

25. Relative minima: $(\pm 2, -4)$
Relative maximum: $(0, 0)$

Points of inflection: $\left(\pm\dfrac{2}{\sqrt{3}}, -\dfrac{20}{9}\right)$

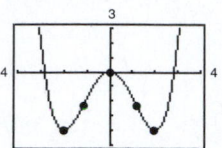

27. Relative minimum: $(1, -27)$
Points of inflection: $(2, -16), (4, 0)$

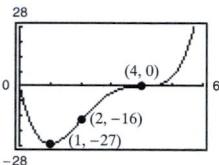

29. Relative minimum: $(-2, -2)$

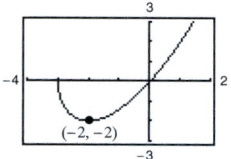

31. Relative minimum: $(3\pi, -1)$
Relative maximum: $(\pi, 1)$
Point of inflection: $(2\pi, 0)$

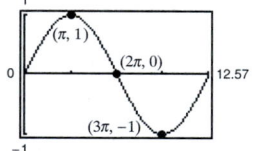

33. Relative minima: $\left(\dfrac{\pi}{2}, 1\right), \left(\dfrac{5\pi}{2}, 1\right)$

Relative maxima: $\left(\dfrac{3\pi}{2}, -1\right), \left(\dfrac{7\pi}{2}, -1\right)$

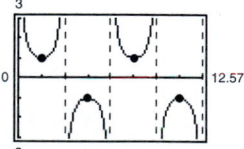

35. Relative minimum: $\left(\dfrac{5\pi}{3}, -2.598\right)$

Relative maximum: $\left(\dfrac{\pi}{3}, 2.598\right)$

Points of inflection: $(\pi, 0)$, $(1.823, 1.452)$, $(4.46, -1.452)$

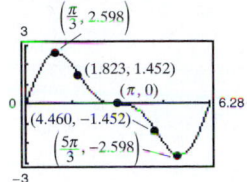

37. (a) $f'(x) = 0.2x(x-3)^2(5x-6)$

$f''(x) = 0.4(x-3)(10x^2 - 24x + 9)$

(b) Relative maximum: $(0, 0)$

Relative minimum: $(1.2, -1.6796)$

Points of inflection: $(0.4652, -0.7048)$,

$(1.9348, -0.9048)$, $(3, 0)$

(c)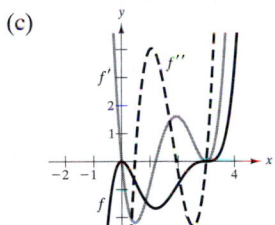

39. (a) $f'(x) = \cos x - \cos 3x + \cos 5x$

$f''(x) = -\sin x + 3\sin 3x - 5\sin 5x$

(b) Relative maximum: $(\pi/2, 1.53333)$

Points of inflection:

$(0.5236, 0.2667)$, $(1.1731, 0.9637)$,

$(1.9685, 0.9637)$, $(2.6180, 0.2667)$

(c)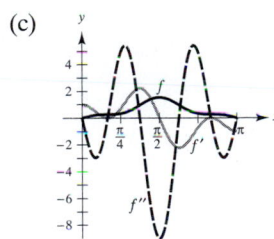

41. 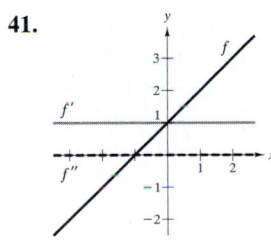   **43.**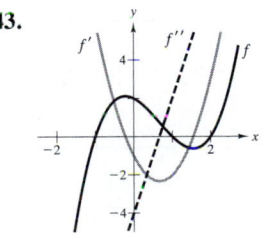

45. (a) 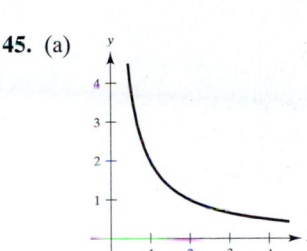   (b)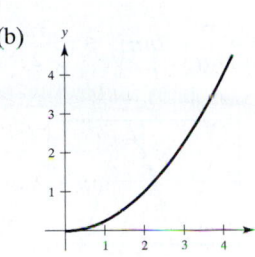

47. 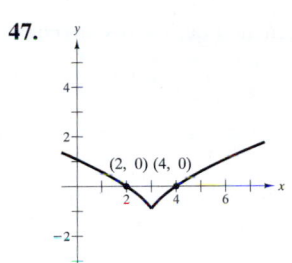   **49.**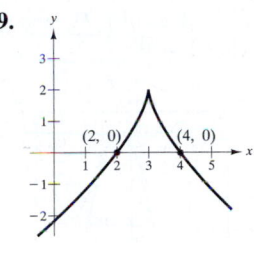

51. $f(x) = \dfrac{1}{2}x^3 - 6x^2 + \dfrac{45}{2}x - 24$

53.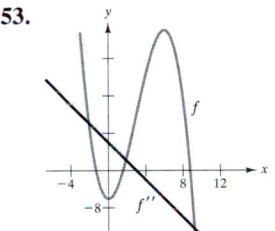

55. (a) $f(x) = (x-2)^n$ has a point of inflection at $(2, 0)$ if n is odd and $n \geq 3$.

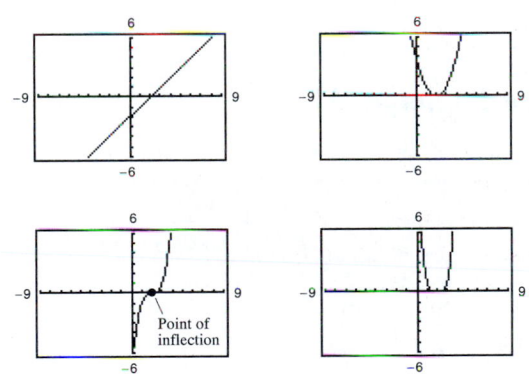

(b) Proof

57. (a) $S'' > 0$ (b) $S' > 0$ (c) $S' = C, S'' = 0$

(d) $S' = 0, S'' = 0$ (e) $S' < 0, S'' > 0$ (f) $S' > 0$

59. (a) $f(x) = \dfrac{1}{32}x^3 + \dfrac{3}{16}x^2$

(b) Two miles from touchdown

61. $x = \left(\dfrac{15 - \sqrt{33}}{16}\right)L \approx 0.578\,L$ **63.** $x = 100$ units

65. $\theta = \dfrac{\pi}{2} + 2n\pi, \quad \theta = \dfrac{3\pi}{2} + 2n\pi$

67.

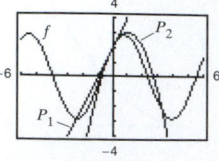

The values of f, P_1, and P_2 and their first derivatives are equal when $x = 0$.

69.

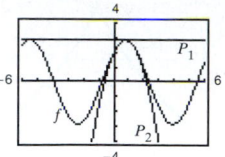

The values of f, P_1, and P_2 and their first derivatives are equal when $x = \pi/4$.

71.

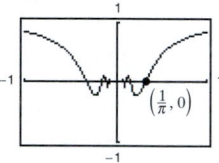

$\left(\frac{1}{\pi}, 0\right)$

73. Proof   **75.** Proof   **77.** True

78. False: 0 is not in the domain of f.

79. False: the maximum value is $\sqrt{13} \approx 3.60555$.

80. True

Section 3.5 *(page 193)*

1. f   **2.** c   **3.** d   **4.** a   **5.** b   **6.** e

7.

x	10^0	10^1	10^2	10^3	10^4
$f(x)$	7	2.2632	2.0251	2.0025	2.0003

x	10^5	10^6
$f(x)$	2.0000	2.0000

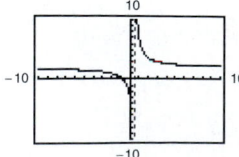

$\lim\limits_{x \to \infty} \dfrac{4x + 3}{2x - 1} = 2$

9.

x	10^0	10^1	10^2	10^3	10^4
$f(x)$	-2	-2.9814	-2.9998	-3.0000	-3.0000

x	10^5	10^6
$f(x)$	-3.0000	-3.0000

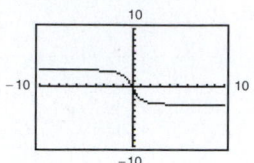

$\lim\limits_{x \to \infty} \dfrac{-6x}{\sqrt{4x^2 + 5}} = -3$

11. $\frac{2}{3}$ **13.** 0   **15.** Limit does not exist.   **17.** -1

19. 2   **21.** 0   **23.** 0   **25.** 1   **27.** 0   **29.** $-\frac{1}{2}$

31.

x	10^0	10^1	10^2	10^3	10^4	10^5	10^6
$f(x)$	1.000	0.513	0.501	0.500	0.500	0.500	0.500

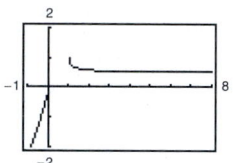

$\lim\limits_{x \to \infty} x - \sqrt{x(x - 1)} = \frac{1}{2}$

33.

x	10^0	10^1	10^2	10^3	10^4	10^5	10^6
$f(x)$	0.479	0.500	0.500	0.500	0.500	0.500	0.500

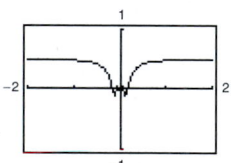

$\lim\limits_{x \to \infty} x \sin \dfrac{1}{2x} = \dfrac{1}{2}$

35.

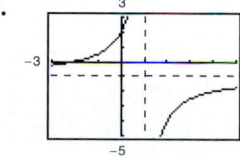

37.

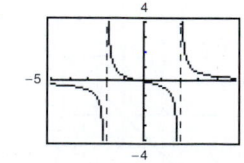

39.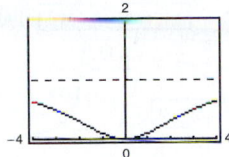

41.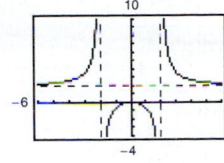

43.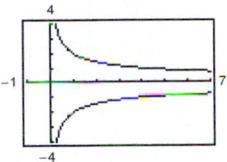

45.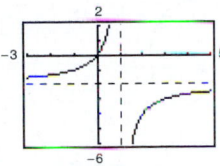

47.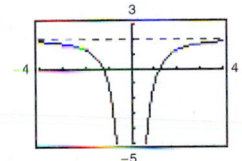

49.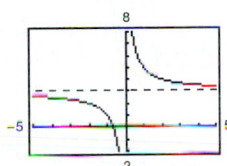

51.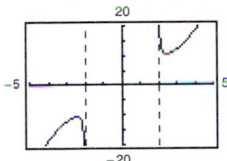

53.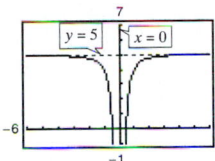

55.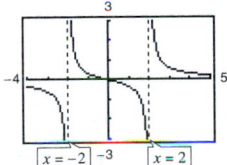

57.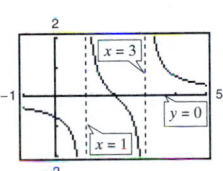

59.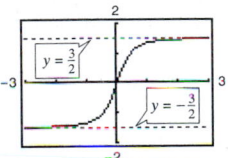

61. (a) 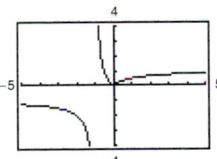   (b)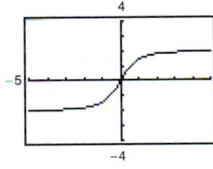

63. (a) 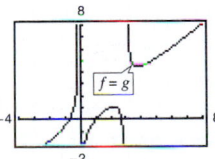   (c)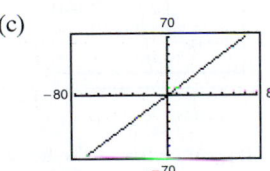

(b) Proof

The slant asymptote $y = x$

65. (a)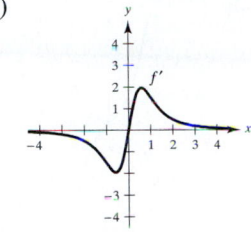

(b) $\lim\limits_{x \to \infty} f(x) = 3, \quad \lim\limits_{x \to \infty} f'(x) = 0$

(c) $y = 3$ is a horizontal asymptote. The rate of increase of the function approaches 0 as the graph approaches $y = 3$.

67. 0.5

69. (a) $T_1 = -0.003t^2 + 0.677t + 26.564$

(b) 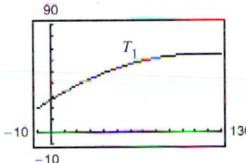   (c)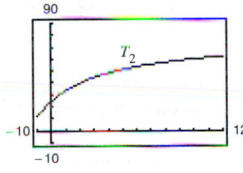

(d) $T_1(0) \approx 26.6°, \quad T_2(0) \approx 24.7°$ (e) 77.5

(f) The limiting temperature is $77.5°$.
T_1 has no horizontal asymptote.

71. False: let $f(x) = \dfrac{2x}{\sqrt{x^2 + 2}}$

$f'(x) > 0$ for all real numbers.

72. False: let $f(x) = \begin{cases} -\frac{1}{8}x^2 + \frac{1}{2}x + 1, & x < 0 \\ \sqrt{x + 1}, & x \geq 0 \end{cases}$

$f''(x) < 0$, $f(x)$ increases without bound.

73. 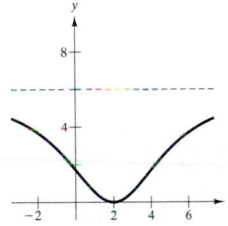   **75.** Proof

Section 3.6 *(page 202)*

1. (a) D   (b) C   (c) A   (d) B

3. 5

5. (a) $f'(x) = 0$ for $x = \pm 2$
$f'(x) > 0$ for $(-\infty, -2), (2, \infty)$
$f'(x) < 0$ for $(-2, 2)$

(b) $f''(x) = 0$ for $x = 0$
$f''(x) > 0$ for $(0, \infty)$
$f''(x) < 0$ for $(-\infty, 0)$

(c) $(0, \infty)$

(d) f' is minimum for $x = 0$.
f is decreasing at the fastest rate.

7.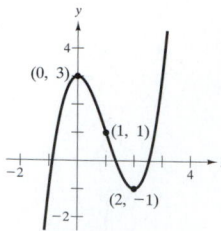

$(0, 3)$
$(1, 1)$
$(2, -1)$

9.

$(0, 2)$

11.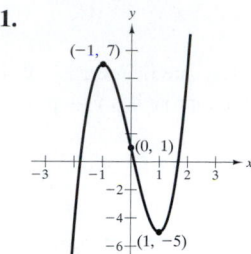

$(-1, 7)$
$(0, 1)$
$(1, -5)$

13.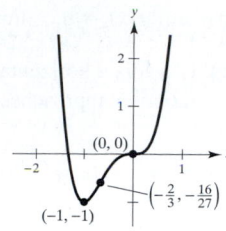

$(0, 0)$
$\left(-\frac{2}{3}, -\frac{16}{27}\right)$
$(-1, -1)$

15.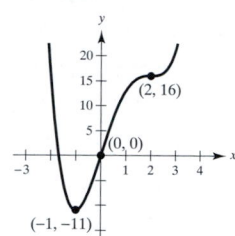

$(2, 16)$
$(0, 0)$
$(-1, -11)$

17.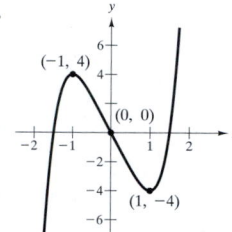

$(-1, 4)$
$(0, 0)$
$(1, -4)$

19.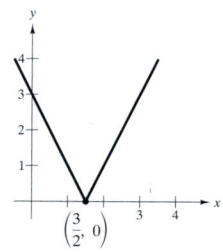

$\left(\frac{3}{2}, 0\right)$

21.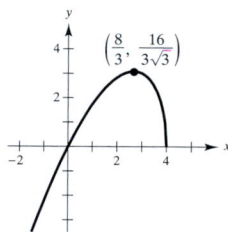

$\left(\frac{8}{3}, \frac{16}{3\sqrt{3}}\right)$

23.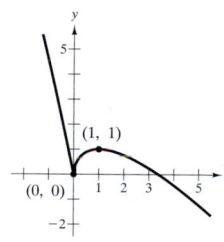

$(1, 1)$
$(0, 0)$

25.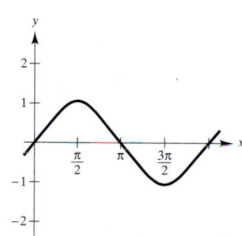

$\frac{\pi}{2}$ π $\frac{3\pi}{2}$

27.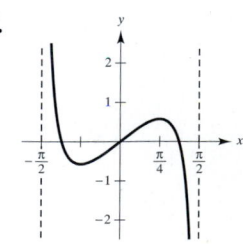

$-\frac{\pi}{2}$ $\frac{\pi}{4}$ $\frac{\pi}{2}$

29.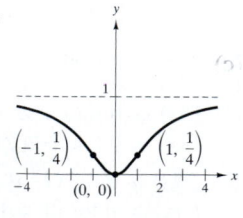

$\left(-1, \frac{1}{4}\right)$ $\left(1, \frac{1}{4}\right)$
$(0, 0)$

31.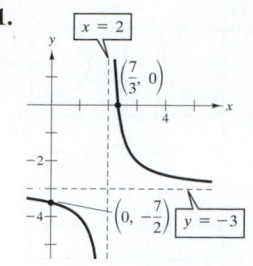

$x = 2$
$\left(\frac{7}{3}, 0\right)$
$\left(0, -\frac{7}{2}\right)$ $y = -3$

33.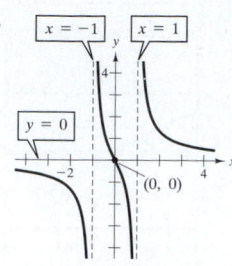

$x = -1$ $x = 1$
$y = 0$
$(0, 0)$

35.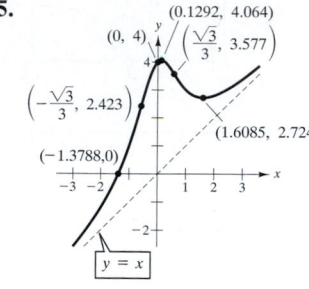

$(0.1292, 4.064)$
$(0, 4)$ $\left(\frac{\sqrt{3}}{3}, 3.577\right)$
$\left(-\frac{\sqrt{3}}{3}, 2.423\right)$
$(1.6085, 2.724)$
$(-1.3788, 0)$
$y = x$

37.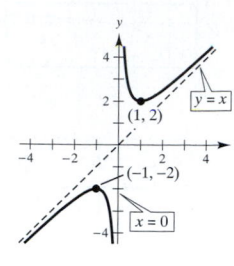

$y = x$
$(1, 2)$
$(-1, -2)$
$x = 0$

39.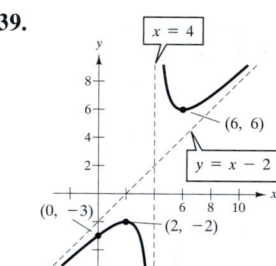

$x = 4$
$(6, 6)$
$y = x - 2$
$(0, -3)$ $(2, -2)$

41.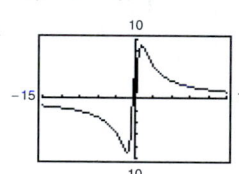

Minimum: $(-1.10, -9.05)$
Maximum: $(1.10, 9.05)$
Points of inflection:
$(-1.84, -7.86), (1.84, 7.86)$
Vertical asymptote: $x = 0$
Horizontal asymptote: $y = 0$

43.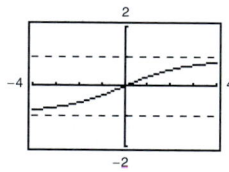

Point of inflection: $(0, 0)$
Horizontal asymptotes: $y = \pm 1$

45.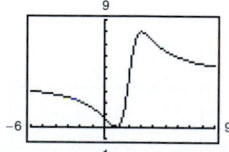

The graph crosses the horizontal asymptote $y = 4$. The graph of f does not cross its vertical asymptote $x = c$ because $f(c)$ does not exist.

47.

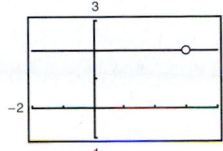

The rational function is not reduced to lowest terms.

49.

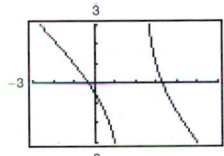

The graph appears to approach the line $y = -x + 1$, which is the slant asymptote.

51. $y = \dfrac{1}{x - 5}$ **53.** $y = \dfrac{3x^2 - 13x - 9}{x - 5}$

55. (a) Rate of change of f changes as a varies. If the sign of a is changed, the graph is reflected through the x-axis.

(b) The locations of the vertical asymptote and the minimum (if $a > 0$) or maximum (if $a < 0$) are changed.

57.

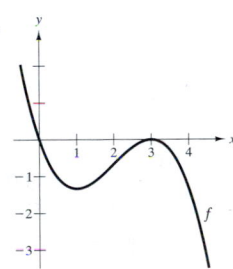

59.

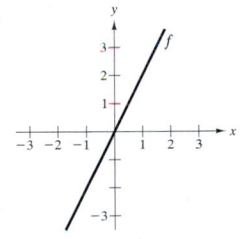

61.

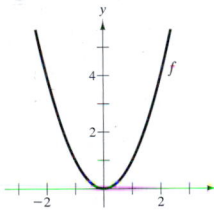

63.

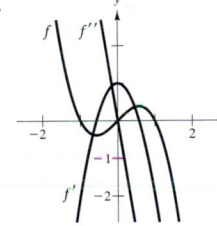

65. $a < 0$ and $b^2 < 3ac$ **67.** $a < 0$ and $b^2 = 3ac$

69. $a < 0$ and $b^2 > 3ac$

71.

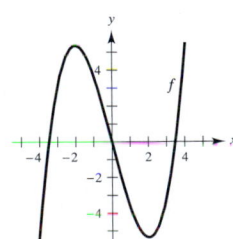

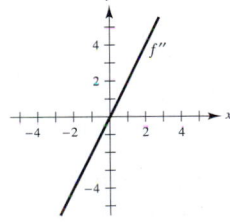

73.

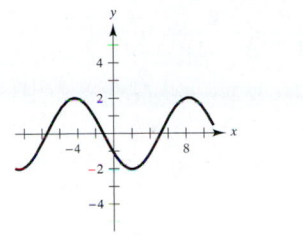

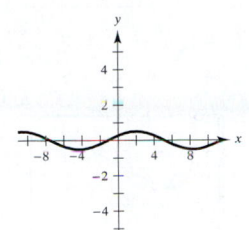

Section 3.7 *(page 210)*

1. (a) and (b)

First Number, x	Second Number	Product, P
10	$110 - 10$	$10(110 - 10) = 1000$
20	$110 - 20$	$20(110 - 20) = 1800$
30	$110 - 30$	$30(110 - 30) = 2400$
40	$110 - 40$	$40(110 - 40) = 2800$
50	$110 - 50$	$50(110 - 50) = 3000$
60	$110 - 60$	$60(110 - 60) = 3000$
70	$100 - 70$	$70(110 - 70) = 2800$
80	$110 - 80$	$80(110 - 80) = 2400$
90	$110 - 90$	$90(110 - 90) = 1800$
100	$110 - 100$	$100(110 - 100) = 1000$

(c) $P = x(110 - x)$

(d)

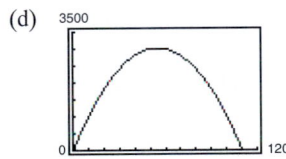

(e) 55 and 55

3. $\sqrt{192}$ and $\sqrt{192}$ **5.** 1 and 1   **7.** $l = w = 25$ feet

9. $l = w = 8$ feet

11. $\left(\dfrac{7}{2}, \sqrt{\dfrac{7}{2}}\right)$ **13.** $x = \dfrac{Q_0}{2}$ **15.** 600×300 meters

17. (a) Proof

(b) $V_1 = 99$ square inches
$V_2 = 125$ square inches
$V_3 = 117$ square inches

(c) $5 \times 5 \times 5$ inches

19. (a) $V = x(s - 2x)^2, \ 0 < x < \dfrac{s}{2}$

Maximum: $V\left(\dfrac{s}{6}\right) = \dfrac{2s^3}{27}$

(b) Increased by a factor of 8

21. Rectangular portion: $\dfrac{16}{\pi + 4} \times \dfrac{32}{\pi + 4}$ feet

23. (a) $L = \sqrt{x^2 + 4 + \dfrac{8}{x - 1} + \dfrac{4}{(x - 1)^2}}, \quad x > 1$

(b)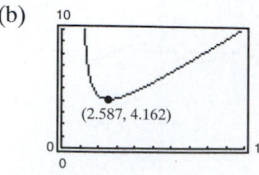

(2.587, 4.162)

Minimum when $x \approx 2.587$

(c) $(0, 0), (2, 0), (0, 4)$

25. Width: $\dfrac{5\sqrt{2}}{2}$; Length: $5\sqrt{2}$

27. Bases: r and $2r$; Altitude: $\dfrac{\sqrt{3}r}{2}$

29. (a) and (b)

Radius, r	Height	Surface Area, S
0.2	$\dfrac{22}{\pi(0.2)^2}$	$2\pi(0.2)\left[0.2 + \dfrac{22}{\pi(0.2)^2}\right] \approx 220.3$
0.4	$\dfrac{22}{\pi(0.4)^2}$	$2\pi(0.4)\left[0.4 + \dfrac{22}{\pi(0.4)^2}\right] \approx 111.0$
0.6	$\dfrac{22}{\pi(0.6)^2}$	$2\pi(0.6)\left[0.6 + \dfrac{22}{\pi(0.6)^2}\right] \approx 75.6$
0.8	$\dfrac{22}{\pi(0.8)^2}$	$2\pi(0.8)\left[0.8 + \dfrac{22}{\pi(0.8)^2}\right] \approx 59.0$
1.0	$\dfrac{22}{\pi(1.0)^2}$	$2\pi(1.0)\left[1.0 + \dfrac{22}{\pi(1.0)^2}\right] \approx 50.3$
1.2	$\dfrac{22}{\pi(1.2)^2}$	$2\pi(1.2)\left[1.2 + \dfrac{22}{\pi(1.2)^2}\right] \approx 45.7$
1.4	$\dfrac{22}{\pi(1.4)^2}$	$2\pi(1.4)\left[1.4 + \dfrac{22}{\pi(1.4)^2}\right] \approx 43.7$
1.6	$\dfrac{22}{\pi(1.6)^2}$	$2\pi(1.6)\left[1.6 + \dfrac{22}{\pi(1.6)^2}\right] \approx 43.6$
1.8	$\dfrac{22}{\pi(1.8)^2}$	$2\pi(1.8)\left[1.8 + \dfrac{22}{\pi(1.8)^2}\right] \approx 44.8$
2.0	$\dfrac{22}{\pi(2.0)^2}$	$2\pi(2.0)\left[2.0 + \dfrac{22}{\pi(2.0)^2}\right] \approx 47.1$

(c) $S = 2\pi r\left(r + \dfrac{22}{\pi r^2}\right)$

(d)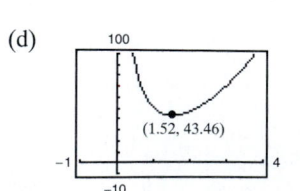

(1.52, 43.46)

(e) $r = \sqrt[3]{\dfrac{11}{\pi}}, h = 2r$

31. $18 \times 18 \times 36$ inches

33. $\dfrac{32\pi r^3}{81}$ 35. $r = \sqrt[3]{\dfrac{9}{\pi}} \approx 1.42$ inches

37. Side of square: $\dfrac{10\sqrt{3}}{9 + 4\sqrt{3}}$

Side of triangle: $\dfrac{30}{9 + 4\sqrt{3}}$

39. $w = 8\sqrt{3}, h = 8\sqrt{6}$ 41. $\theta = \dfrac{\pi}{4}$

43. (a)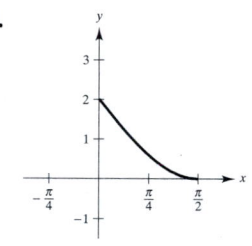

(b) $d = -\dfrac{1}{16}x^4 + x^2$

Maximum when $x = 2\sqrt{2}$

(c) $y = 2\sqrt{2}x - 4$
$y = 2\sqrt{2}x - 8$

Parallel

(d) The tangent lines are parallel.

45. $\dfrac{d\sqrt[3]{I_1}}{\sqrt[3]{I_1} + \sqrt[3]{I_2}}$ units from source 1   47. Proof

49.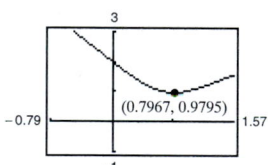

(a) Origin to y-intercept: 2

Origin to x-intercept: $\dfrac{\pi}{2}$

(b) $d = \sqrt{x^2 + (2 - 2\sin x)^2}$

(0.7967, 0.9795)

(c) Minimum distance is 0.9795 when $x \approx 0.7967$.

51. $\theta = \dfrac{2\pi}{3}\left(3 - \sqrt{6}\right) \approx 66°$

53. (a) and (b)

θ	L_1	L_2	$L_1 + L_2$
0.1	$\dfrac{2}{\sin(0.1)}$	$\dfrac{7}{\cos(0.1)}$	≈ 27.1
0.2	$\dfrac{2}{\sin(0.2)}$	$\dfrac{7}{\cos(0.2)}$	≈ 17.2
0.3	$\dfrac{2}{\sin(0.3)}$	$\dfrac{7}{\cos(0.3)}$	≈ 14.1
0.4	$\dfrac{2}{\sin(0.4)}$	$\dfrac{7}{\cos(0.4)}$	≈ 12.7
0.5	$\dfrac{2}{\sin(0.5)}$	$\dfrac{7}{\cos(0.5)}$	≈ 12.1
0.6	$\dfrac{2}{\sin(0.6)}$	$\dfrac{7}{\cos(0.6)}$	≈ 12.0
0.7	$\dfrac{2}{\sin(0.7)}$	$\dfrac{7}{\cos(0.7)}$	≈ 12.3

Minimum length is approximately 12.0 meters.

(c) $L = \dfrac{2}{\sin \theta} + \dfrac{7}{\cos \theta}$

(d)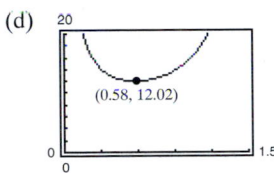

(0.58, 12.02)

Minimum length is approximately 12.02.

(e) $\theta = \arctan \dfrac{\sqrt[3]{98}}{7} \approx 0.5824$ radian

(f) 6.61 meters

55. Rectangle: $\dfrac{3}{2} \times 2$

Circle: $r = 1$

Semicircle: $r = \dfrac{12}{7}$

Calculus was helpful for the rectangle.

Section 3.8 (page 219)

1.

n	x_n	$f(x_n)$	$f'(x_n)$	$\dfrac{f(x_n)}{f'(x_n)}$	$x_n - \dfrac{f(x_n)}{f'(x_n)}$
1	1.7000	-0.1100	3.4000	-0.0324	1.7324
2	1.7324	0.0012	3.4648	0.0003	1.7321

3.

n	x_n	$f(x_n)$	$f'(x_n)$	$\dfrac{f(x_n)}{f'(x_n)}$	$x_n - \dfrac{f(x_n)}{f'(x_n)}$
1	3	0.1411	-0.9900	-0.1425	3.1425
2	3.1425	-0.0009	-1.0000	0.0009	3.1416

5. 0.682   **7.** 1.146, 7.854   **9.** 0.900, 1.100, 1.900

11. -0.489 **13.** 0.569   **15.** 4.493   **17.** 0.74

19. (a)

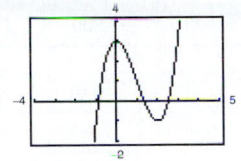

(b) 1.347   (c) 2.532

(d)

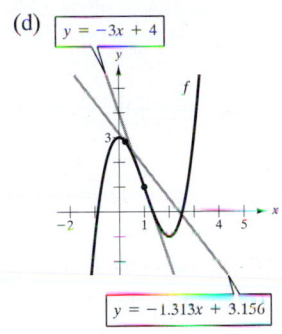

$y = -3x + 4$

$y = -1.313x + 3.156$

(e) If the initial estimate $x = x_1$ is not sufficiently close to the desired zero of a function, the x-intercept of the corresponding tangent line to the function may approximate a second zero of the function.

21. $f'(x_1) = 0$ **23.** $1 = x_1 = x_3 = \ldots$
$0 = x_2 = x_4 = \ldots$

25. $x_{i+1} = \dfrac{x_i^2 + a}{2x_i}$ **27.** 2.646   **29.** 1.565   **31.** Proof

33. 3.141

35. 0.860

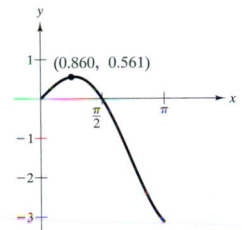

(0.860, 0.561)

37. $(1.939, 0.240)$ **39.** $x \approx 1.563$ miles   **41.** \$384,356

43. False: let $f(x) = \dfrac{x^2 - 1}{x - 1}$ **44.** True   **45.** True

46. True   **47.** $x \approx 11.803$

Section 3.9 (page 226)

1.

x	1.9	1.99	2	2.01	2.1
$f(x)$	3.610	3.960	4	4.040	4.410
$T(x)$	3.600	3.960	4	4.040	4.400

3.

x	1.9	1.99	2	2.01	2.1
$f(x)$	24.761	31.208	32	32.808	40.841
$T(x)$	24.000	31.200	32	32.800	40.000

5.

x	1.9	1.99	2	2.01	2.1
$f(x)$	0.946	0.913	0.909	0.905	0.863
$T(x)$	0.951	0.913	0.909	0.905	0.868

7. $\Delta y = 0.331, \; dy = 0.300$

9. $\Delta y = -0.039, \; dy = -0.040$

11. $6x \, dx$ **13.** $-\dfrac{3}{(2x-1)^2} \, dx$ **15.** $\dfrac{1 - 2x^2}{\sqrt{1 - x^2}} \, dx$

17. $\dfrac{2 \sec^2 x (x^2 \tan x + \tan x - x)}{(x^2 + 1)^2} \, dx$

19. $-\pi \sin\left(\dfrac{6\pi x - 1}{2}\right) dx$ **21.** $\pm \frac{3}{8}$ square inch

23. $\pm 7\pi$ square inches **25.** (a) $\frac{2}{3}\%$ (b) 1.25%

27. (a) $\pm 2.88\pi$ cubic inches (b) $\pm 0.96\pi$ square inches

(c) 1%, $\frac{2}{3}\%$

29. 80π cubic centimeters

31. (a) $\frac{1}{4}\%$ (b) 216 seconds = 3.6 minutes

33. Proof **35.** (a) 0.87% (b) 2.16% **37.** 4961 feet

39. $f(x) = \sqrt{x}, \; dy = \dfrac{1}{2\sqrt{x}} \, dx$

$f(99.4) \approx \sqrt{100} + \dfrac{1}{2\sqrt{100}}(-0.6) = 9.97$

Calculator: 9.97

41. $f(x) = \sqrt[4]{x}, \; dy = \dfrac{1}{4x^{3/4}} \, dx$

$f(624) \approx \sqrt[4]{625} + \dfrac{1}{4(625)^{3/4}}(-1) = 4.998$

Calculator: 4.998

43. $f(x) = x^4, \; dy = 4x^3 \, dx$
$f(0.99) \approx 1^4 + 4(1)(-0.01) = 1 - 4(0.01)$

45. $f(x) = \sec x, \; dy = \sec x \tan x \, dx$
$f(0.03) \approx \sec 0 + \sec 0 \tan 0(0.03) = 1 + 0(0.03)$

47. True **48.** True **49.** True

50. False: if $f(x) = \sqrt{x}$, $x = 1$, and $\Delta x = 3$, then $dy = \frac{3}{2}$ and $\Delta y = 1$.

51. (a) $dA = 2x\Delta x, \; \Delta A = 2x\Delta x + (\Delta x)^2$

(b) and (c)

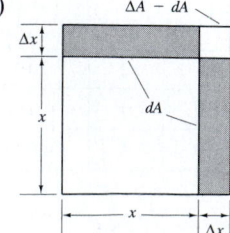

Section 3.10 *(page 232)*

1. (a) Fixed cost

(b)

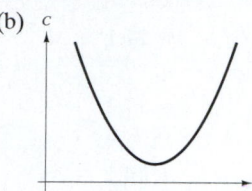

(c) Yes, it occurs when production costs are increasing at their slowest rate.

3. 4500 **5.** 300 **7.** 200 **9.** 200

11. $x = 30$ **13.** $x = 1500$ **15.** $x = 3$ **17.** Proof
$p = 60$ $p = 35$

19. (a)

Order size, x	Price	Profit, P
102	$90 - 2(0.15)$	$102[90 - 2(0.15)] - 102(60) = 3029.40$
104	$90 - 4(0.15)$	$104[90 - 4(0.15)] - 104(60) = 3057.60$
106	$90 - 6(0.15)$	$106[90 - 6(0.15)] - 106(60) = 3084.60$
108	$90 - 8(0.15)$	$108[90 - 8(0.15)] - 108(60) = 3110.40$
110	$90 - 10(0.15)$	$110[90 - 10(0.15)] - 110(60) = 3135.00$
112	$90 - 12(0.15)$	$112[90 - 12(0.15)] - 112(60) = 3158.40$

(b)

Order size, x	Price	Profit, P
⋮	⋮	⋮
146	$90 - 46(0.15)$	$146[90 - 46(0.15)] - 146(60) = 3372.60$
148	$90 - 48(0.15)$	$148[90 - 48(0.15)] - 148(60) = 3374.40$
150	$90 - 50(0.15)$	$150[90 - 50(0.15)] - 150(60) = 3375.00$
152	$90 - 52(0.15)$	$152[90 - 52(0.15)] - 152(60) = 3374.40$
154	$90 - 54(0.15)$	$154[90 - 54(0.15)] - 154(60) = 3372.60$
⋮	⋮	⋮

(c) $P = x[90 - (x - 100)(0.15)] - x(60) = x(45 - 0.15)x,$
  $x \geq 100$

(d) 150 units

(e)

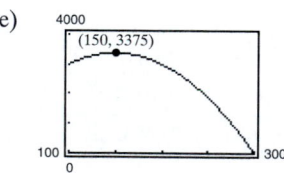

21. $10\sqrt{30} \approx 54.8$ miles per hour

23. Line should run from the power station to a point across the river $3/(2\sqrt{7})$ mile downstream.

25. $y = \dfrac{64}{141} x$, 6.1 miles   **27.** $y \approx 0.3x$, 4.5 miles

29. 8%   **31.** $x \approx 40$ units   **33.** \$30,000

35. (a) $t = 151.25$, May 31

(b) $t = 333.75$, Nov. 30

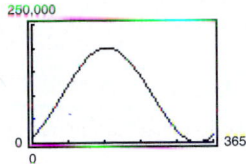

37. (a) 1987

(b) 1994

(c) Least: \$5995 million

Greatest: \$8051 million

(d)

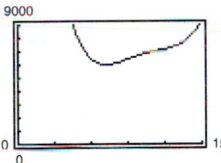

39. (a) Demand function

(b) Cost function

(c) Revenue function

(d) Profit function

41. $\eta = -\dfrac{17}{3}$, elastic   **43.** $\eta = -\dfrac{1}{2}$, inelastic

Review Exercises for Chapter 3 *(page 235)*

1. Let f be defined at c. If $f'(c) = 0$ or if f' is undefined at c, then c is a critical number of f.

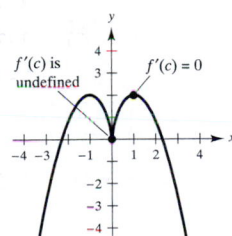

3. Maximum: $(2\pi, 17.57)$

Minimum: $(2.73, 0.88)$

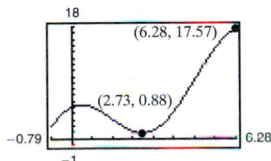

5. (a)

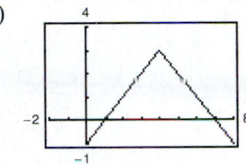

(b) f is not differentiable at $x = 4$.

7. $f'\left(\dfrac{2744}{729}\right) = \dfrac{3}{7}$ **9.** $f'(0) = 1$ **11.** $c = \dfrac{x_1 + x_2}{2}$

13. Critical numbers: $x = 1, \dfrac{7}{3}$

Increasing on $(-\infty, 1), \left(\dfrac{7}{3}, \infty\right)$

Decreasing on $\left(1, \dfrac{7}{3}\right)$

15. Critical numbers: $x = 1, 0$

Increasing on $(1, \infty)$

Decreasing on $(0, 1)$

17. Minimum: $(2, -12)$ **19.** $\left(\dfrac{\pi}{2}, \dfrac{\pi}{2}\right), \left(\dfrac{3\pi}{2}, \dfrac{3\pi}{2}\right)$

21. $\dfrac{2}{3}$ **23.** 0

25. Vertical asymptote: $x = 4$

Horizontal asymptote: $y = 2$

27. Vertical asymptote: $x = 0$

Horizontal asymptote: $y = -2$

29.

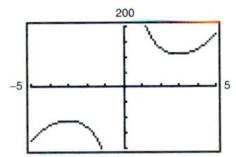

Vertical asymptote: $x = 0$

Relative minimum: $(3, 108)$

Relative maximum: $(-3, -108)$

31.

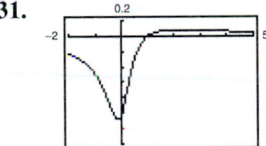

Horizontal asymptote: $y = 0$

Relative minimum: $(-0.155, -1.077)$

Relative maximum: $(2.155, 0.077)$

33.

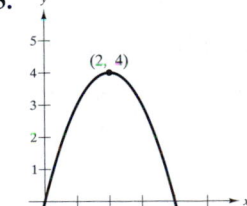

35.

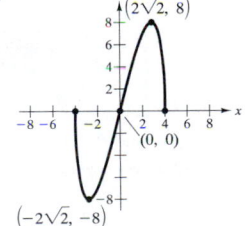

37.

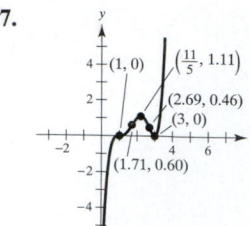

39.

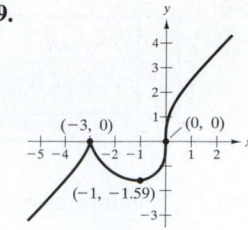

41.

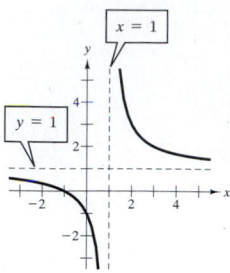

43.

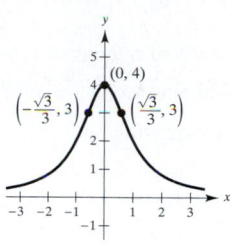

45.

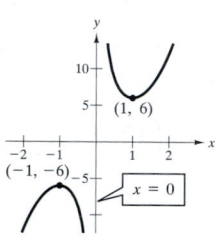

47.

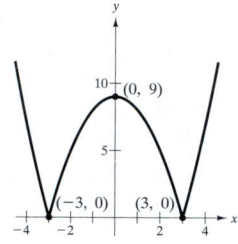

49.

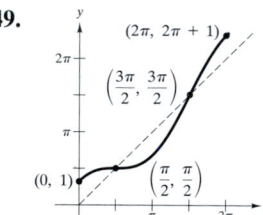

51. $t \approx 4.92 \approx 4{:}55$ p.m.; $d \approx 64$ kilometers

53. $(0, 0), (5, 0), (0, 10)$ **55.** Proof   **57.** 14.05 feet

59. $3(3^{2/3} + 2^{2/3})^{3/2} \approx 21.07$ feet

61. (a) $y = \dfrac{1}{4}$ inch, $v = 4$ inches per second

     (b) Proof   (c) Period: $\dfrac{\pi}{6}$ Frequency: $\dfrac{6}{\pi}$

63. $-0.347, -1.532, 1.879$ **65.** $-1.164, 1.453$

67. $dy = (1 - \cos x + x \sin x)\, dx$

69. (a) $S = -0.081t^3 + 2.920t^2 - 15.546t + 92.325$

     (b)

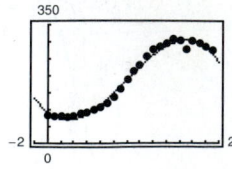

     (c) 1990   (d) 1982

71. $dS = \pm 1.8\pi$ square centimeters, $\dfrac{dS}{S} \times 100 \approx \pm 0.56\%$

     $dV = \pm 8.1\pi$ cubic centimeters, $\dfrac{dV}{V} \times 100 \approx \pm 0.83\%$

73. \$48   **75.** 120   **77.** $x = \sqrt{\dfrac{2Qs}{r}}$

79. Maximum: $(1, 3)$
     Minimum: $(1, 1)$

81. False: let $f(x) = x^3$ and $c = 0$.

82. False: the horizontal asymptotes of $f(x) = 2x/\sqrt{x^2 + 2}$ are $y = 2$ and $y = -2$.

83. Increasing and concave down

CHAPTER 4

Section 4.1 *(page 248)*

1. Proof   **3.** Proof

	Original Integral	*Rewrite*	*Integrate*	*Simplify*
5.	$\displaystyle\int \sqrt[3]{x}\, dx$	$\displaystyle\int x^{1/3}\, dx$	$\dfrac{x^{4/3}}{4/3} + C$	$\dfrac{3}{4} x^{4/3} + C$
7.	$\displaystyle\int \dfrac{1}{x\sqrt{x}}\, dx$	$\displaystyle\int x^{-3/2}\, dx$	$\dfrac{x^{-1/2}}{-1/2} + C$	$-\dfrac{2}{\sqrt{x}} + C$
9.	$\displaystyle\int \dfrac{1}{2x^3}\, dx$	$\dfrac{1}{2}\displaystyle\int x^{-3}\, dx$	$\dfrac{1}{2}\left(\dfrac{x^{-2}}{-2}\right) + C$	$-\dfrac{1}{4x^2} + C$

11. $y = t^3 + C$ **13.** $y = \frac{2}{5}x^{5/2} + C$ **15.** $\frac{1}{4}x^4 + 2x + C$

17. $\dfrac{2}{5} x^{5/2} + x^2 + x + C$ **19.** $\dfrac{3}{5} x^{5/3} + C$ **21.** $-\dfrac{1}{2x^2} + C$

23. $\frac{2}{15}x^{1/2}(3x^2 + 5x + 15) + C$ **25.** $x^3 + \frac{1}{2}x^2 - 2x + C$

27. $\frac{2}{7}y^{7/2} + C$ **29.** $x + C$ **31.** $-2\cos x + 3\sin x + C$

33. $t + \csc t + C$ **35.** $\tan\theta + \cos\theta + C$ **37.** $\tan y + C$

39.   **41.**

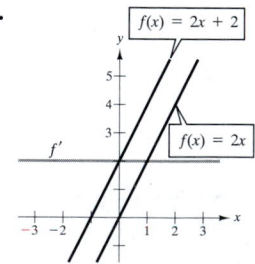

43.

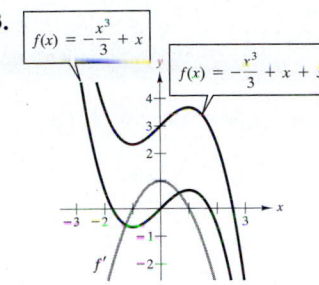

$f(x) = -\dfrac{x^3}{3} + x$

$f(x) = -\dfrac{x^3}{3} + x + 3$

45. $y = x^2 - x + 1$

(g)

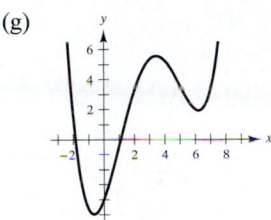

87. Proof

Section 4.2 *(page 260)*

1. 35   **3.** $\dfrac{158}{85}$ **5.** $4c$ **7.** $\displaystyle\sum_{i=1}^{9} \dfrac{1}{3i}$ **9.** $\displaystyle\sum_{j=1}^{8}\left[2\left(\dfrac{j}{8}\right) + 3\right]$

11. $\dfrac{2}{n}\displaystyle\sum_{i=1}^{n}\left[\left(\dfrac{2i}{n}\right)^3 - \left(\dfrac{2i}{n}\right)\right]$ **13.** $\dfrac{3}{n}\displaystyle\sum_{i=1}^{n}\left[2\left(1 + \dfrac{3i}{n}\right)^2\right]$

15. 420   **17.** 2470   **19.** 12,040   **21.** 2930   **23.** $\dfrac{8}{3}$

25. $\dfrac{81}{4}$ **27.** 9   **29.** $\displaystyle\lim_{n\to\infty}\left[8\left(\dfrac{n^2 + n}{n^2}\right)\right] = 8$

31. $\displaystyle\lim_{n\to\infty}\dfrac{1}{6}\left(\dfrac{2n^3 - 3n^2 + n}{n^3}\right) = \dfrac{1}{3}$ **33.** $\displaystyle\lim_{n\to\infty}\left(\dfrac{3n + 1}{n}\right) = 3$

35. $S \approx 0.768$ **37.** $S \approx 0.746$
    $s \approx 0.518$ $s \approx 0.646$

39. (a)   **(b)** Proof   **(c)** Proof

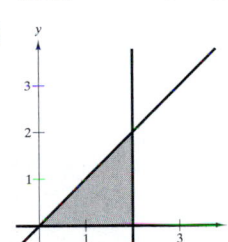

(d) Proof

(e)

n	5	10	50	100
$s(n)$	1.6	1.8	1.96	1.98
$S(n)$	2.4	2.2	2.04	2.02

(f) Proof

47. $y = \sin x + 4$

49. (a)

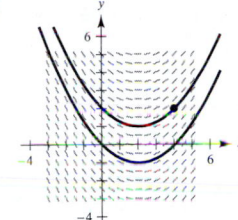

(b) $y = \dfrac{1}{4}x^2 - x + 2$

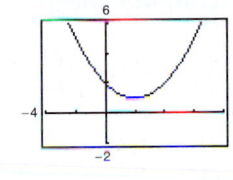

51. $f(x) = x^2 + x + 4$

53. $f(x) = -4x^{1/2} + 3x = -4\sqrt{x} + 3x$

55. (a) $h(t) = \dfrac{3}{4}t^2 + 5t + 12$ **(b)** 69 centimeters

57. 56.25 feet   **59.** $v_0 \approx 187.617$ feet per second   **61.** Proof

63. 5.1 meters   **65.** 320 meters, -32 meters per second

67. (a) $v(t) = 3t^2 - 12t + 9,\ a(t) = 6t - 12$

(b) $(0, 1), (3, 5)$ **(c)** -3

69. $a(t) = \dfrac{-1}{2t^{3/2}},\ s(t) = 2\sqrt{t} + 2$

71. (a) 1.18 meters per second per second   **(b)** 190 meters

73. (a) 300 feet   **(b)** 60 feet per second ≈ 41 miles per hour

75. 7.45 feet per second per second

77. $C(x) = x^2 - 12x + 50$ **79.** $R = x\left(100 - \dfrac{5}{2}x\right)$

$\overline{C}(x) = x - 12 + \dfrac{50}{x}$ $p = 100 - \dfrac{5}{2}x$

81. True   **82.** True   **83.** True   **84.** True

85. (a) -1

(b) No. The slopes of the tangent lines are greater than 2 on $[0, 2]$. Therefore, f must increase more than four units on $[0, 2]$.

(c) No. The function is decreasing on $[4, 5]$.

(d) 3.5. $f'(3.5) \approx 0$

(e) Concave up: $(-\infty, 1), (5, \infty)$
    Concave down: $(1, 5)$

(f) 3

41. $A = 2$

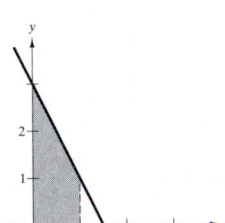

43. $A = \dfrac{7}{3}$

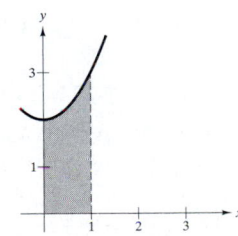

45. $A = 34$

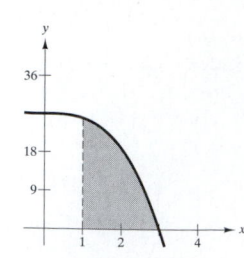

47. $A = \frac{2}{3}$

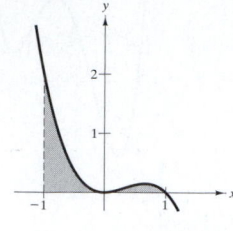

49. $A = 6$

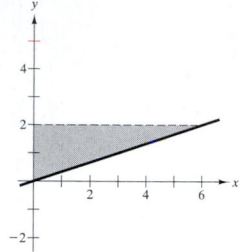

51. $\frac{69}{8}$ **53.** 0.345

55.

n	4	8	12	16	20
Approximate area	5.3838	5.3523	5.3439	5.3403	5.3384

57.

n	4	8	12	16	20
Approximate area	2.2223	2.2387	2.2418	2.2430	2.2435

59. (a)

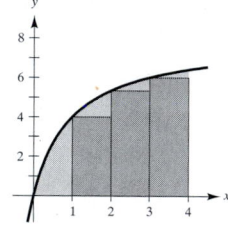

$s(4) = \frac{46}{3}$

(b)

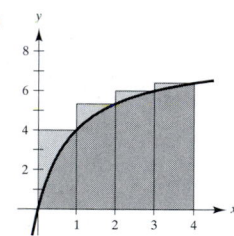

$S(4) = \frac{326}{15}$

(c)

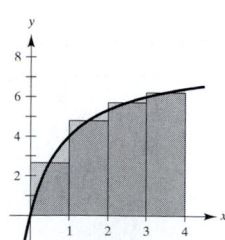

$m(4) = \frac{6112}{315}$

(d) Proof

(e)

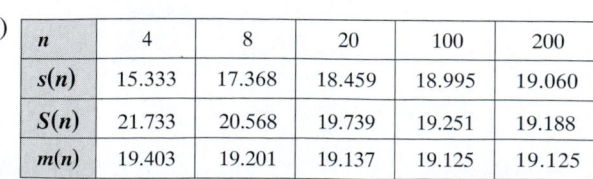

n	4	8	20	100	200
$s(n)$	15.333	17.368	18.459	18.995	19.060
$S(n)$	21.733	20.568	19.739	19.251	19.188
$m(n)$	19.403	19.201	19.137	19.125	19.125

(f) f is an increasing function.

61. b **63.** True **64.** True **65.** Answers will vary.

67. Answer will vary.

69. (a) $y = (-4.09 \times 10^{-5})x^3 + 0.016x^2 - 2.67x + 452.9$

(b) 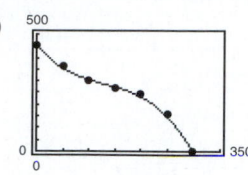 (c) 76,897 square feet

Section 4.3 *(page 271)*

1. $\int_0^5 3\, dx$ **3.** $\int_{-4}^4 \left(4 - |x|\right) dx$ **5.** $\int_{-2}^2 \left(4 - x^2\right) dx$

7. $\int_0^\pi \sin x\, dx$ **9.** $\int_0^2 y^3\, dy$

11. $A = 12$

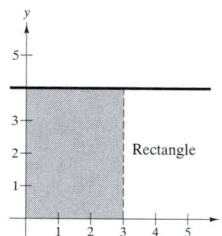

Rectangle

13. $A = 8$

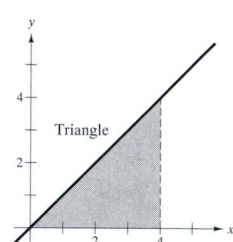

Triangle

15. $A = 14$

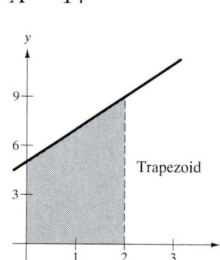

Trapezoid

17. $A = 1$

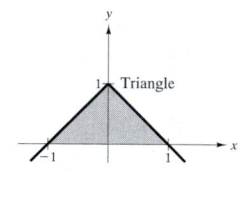

Triangle

19. $A = \dfrac{9\pi}{2}$

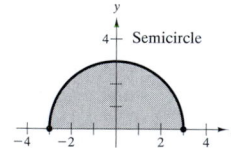

Semicircle

21. (a) 13 (b) -10 (c) 0 (d) 30

23. (a) 8 (b) -12 (c) -4 (d) 30

25. 36 **27.** 0 **29.** $\frac{10}{3}$

31. $\int_{-1}^5 \left(3x + 10\right) dx$ **33.** $\int_0^3 \sqrt{x^2 + 4}\, dx$

35.

n	4	8	12	16	20
$L(n)$	3.6830	3.9956	4.0707	4.1016	4.1177
$M(n)$	4.3082	4.2076	4.1838	4.1740	4.1690
$R(n)$	3.6830	3.9956	4.0707	4.1016	4.1177

37.

n	4	8	12	16	20
$L(n)$	0.5890	0.6872	0.7199	0.7363	0.7461
$M(n)$	0.7854	0.7854	0.7854	0.7854	0.7854
$R(n)$	0.9817	0.8836	0.8508	0.8345	0.8247

39. $\displaystyle\sum_{i=1}^{n} f(x_i)\Delta x > \int_1^5 f(x)\,dx$ **41.** $\displaystyle\sum_{i=1}^{n} f(x_i)\Delta x < \int_1^5 f(x)\,dx$

43. (a) $-\pi$ (b) 4   (c) $-(1+2\pi)$ (d) $3-2\pi$
(e) $5+2\pi$ (f) $23-2\pi$

45. a   **47.** True

48. False: $\displaystyle\int_0^1 x\sqrt{x}\,dx \neq \left(\int_0^1 x\,dx\right)\left(\int_0^1 \sqrt{x}\,dx\right)$ **49.** True

50. True   **51.** False: $\displaystyle\int_0^2 (-x)\,dx = -2.$

52. False: $\displaystyle\int_{-2}^4 x\,dx = 6.$ **53.** 272   **55.** $\dfrac{4\sqrt{2}}{3}$

57. No, $f(x)$ is discontinuous at $x = 4$.   **59.** $f(x) = \dfrac{|x|}{x}$

61. $\frac{1}{3}$ **63.** Proof

Section 4.4 *(page 283)*

1. 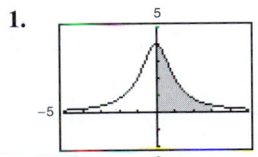   **3.**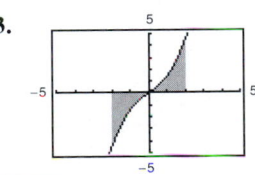

Positive   Zero

5. 1   **7.** $-\frac{5}{2}$ **9.** $-\frac{10}{3}$ **11.** $\frac{1}{3}$ **13.** $\frac{1}{2}$ **15.** $\frac{2}{3}$

17. -4 **19.** $-\frac{1}{18}$ **21.** $-\frac{27}{20}$ **23.** $\frac{9}{2}$ **25.** $\pi + 2$

27. $\dfrac{2\sqrt{3}}{3}$ **29.** 0

31. $\displaystyle\int_0^3 10{,}000(t-6)\,dt = -\$135{,}000$

33. $\dfrac{1}{6}$ **35.** $\dfrac{12\sqrt{3}}{5}$ **37.** 1   **39.** 10   **41.** 6

43. 0.4380, 1.7908   **45.** $\pm\arccos\dfrac{\sqrt{\pi}}{2} \approx \pm 0.4817$

47. Average value $= \dfrac{8}{3}$
$x = \pm\dfrac{2\sqrt{3}}{3} \approx \pm 1.155$
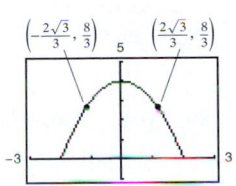

49. Average value $= \dfrac{2}{\pi}$
$x \approx 0.690, x \sim 2.451$
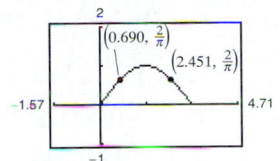

51. -1.5 **53.** 6.5   **55.** 15.5
57. (a) 8   (b) $\frac{4}{3}$ (c) $20, \frac{10}{3}$

59. (a) 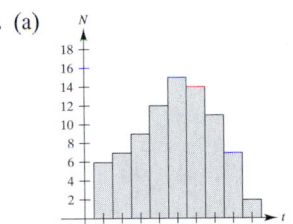   (b) $R'(40) \approx 0.12$
$R'(50) \approx 0.43$
(c) 1.80, 7.35

61. (a) $F(x) = 500\sec^2 x$ (b) 827 newtons

63. (a) 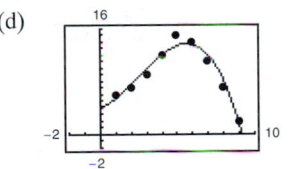   (b) 4980
(c) $N = -0.084t^3 + 0.635t^2 + 0.791t + 4.103$
(d) 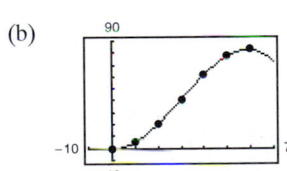   (e) 5129   (f) 12.6

65. (a) $v = (-8.612 \times 10^{-4})t^3 + 0.078t^2 - 0.208t + 0.096$
(b) 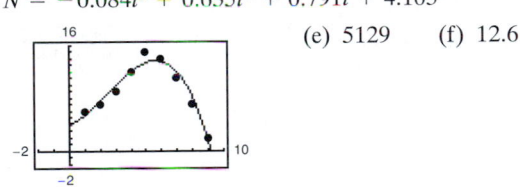   (c) 2457 meters

67. $\frac{1}{2}x^2 + 2x$ **69.** $\frac{3}{4}x^{4/3} - 12$ **71.** $\tan x - 1$
73. $x^2 - 2x$ **75.** $\sqrt{x^4 + 1}$ **77.** $x\cos x$ **79.** 8
81. $\cos x\sqrt{\sin x}$ **83.** $3x^2\sin x^6$

85.

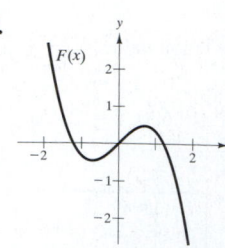

The extrema of F correspond to the zeros of f, and the inflection points of F correspond to the extrema of f.

87. (a) $C(x) = 1000(12x^{5/4} + 125)$ **89.** True   **90.** True

(b) $C(1) = \$137,000$
$C(5) = \$214,721$
$C(10) = \$338,394$

91. False: $f(x) = x^{-2}$ has a nonremovable discontinuity at $x = 0$.

93. Proof

95. (a)

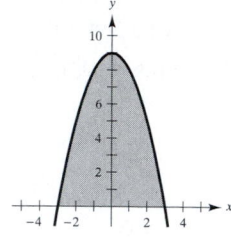

$$A = \int_{-3}^{3} (9 - x^2)\, dx = 36$$

(b) $b = 6, h = 9, A = 36$
(c) $b = 5, h = \frac{25}{4}, A = \frac{125}{6}$

97. 27.37 units

Section 4.5 (page 296)

$\int f(g(x))g'(x)\, dx$	$u = g(x)$	$du = g'(x)$
1. $\int (5x^2 + 1)^2(10x)\, dx$	$5x^2 + 1$	$10x\, dx$
3. $\int \dfrac{x}{\sqrt{x^2 + 1}}\, dx$	$x^2 + 1$	$2x\, dx$
5. $\int \tan^2 x \sec^2 x\, dx$	$\tan x$	$\sec^2 x\, dx$

7. $\dfrac{(1 + 2x)^5}{5} + C$ **9.** $\dfrac{2}{3}(9 - x^2)^{3/2} + C$

11. $\dfrac{(x^3 - 1)^5}{15} + C$ **13.** $-\dfrac{15}{8}(1 - x^2)^{4/3} + C$

15. $-\dfrac{1}{3(1 + x^3)} + C$ **17.** $-\dfrac{1}{4}\left(1 + \dfrac{1}{t}\right)^4 + C$

19. $\sqrt{2x} + C$

21. $\frac{2}{5}x^{5/2} + 2x^{3/2} + 14x^{1/2} + C = \frac{2}{5}\sqrt{x}(x^2 + 5x + 35) + C$

23. $\frac{1}{4}t^4 - t^2 + C$

25. $6y^{3/2} - \frac{2}{5}y^{5/2} + C = \frac{2}{5}y^{3/2}(15 - y) + C$

27. $2x^2 - 4\sqrt{16 - x^2} + C$

29. $-\dfrac{1}{2(x^2 + 2x - 3)} + C$

31. (a)

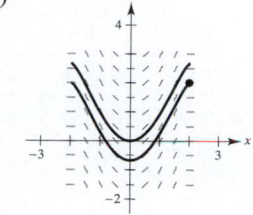

(b) $y = -\frac{1}{3}(4 - x^2)^{3/2} + 2$

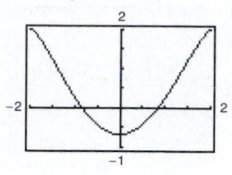

33. $-\dfrac{1}{2}\cos 2x + C$ **35.** $-\sin\dfrac{1}{\theta} + C$

37. $\dfrac{1}{4}\sin^2 2x + C_1$ or $-\dfrac{1}{4}\cos^2 2x + C_2$ or $-\dfrac{1}{8}\cos 4x + C_3$

39. $\frac{1}{5}\tan^5 x + C$ **41.** $\frac{1}{2}\tan^2 x + C$ or $\frac{1}{2}\sec^2 x + C_1$

43. $-\cot x - x + C$ **45.** $f(x) = 2\sin\dfrac{x}{2} + 3$

47. $\frac{2}{15}(x + 2)^{3/2}(3x - 4) + C$

49. $-\frac{2}{105}(1 - x)^{3/2}(15x^2 + 12x + 8) + C$

51. $\dfrac{\sqrt{2x - 1}}{15}(3x^2 + 2x - 13) + C$

53. $-x - 1 - 2\sqrt{x + 1} + C$ or $-\left(x + 2\sqrt{x + 1}\right) + C_1$

55. 0   **57.** 2   **59.** $\dfrac{1}{2}$ **61.** $\dfrac{4}{15}$ **63.** $\dfrac{3\sqrt{3}}{4}$

65. $\dfrac{1209}{28}$ **67.** 4   **69.** $2\left(\sqrt{3} - 1\right)$

71. $\dfrac{10}{3}$ **73.** $\dfrac{144}{5}$

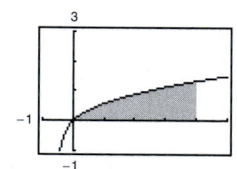

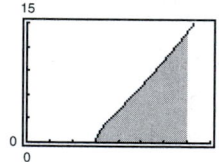

75. 7.38

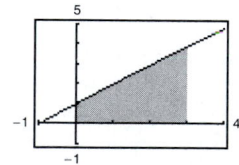

77. $\frac{1}{6}(2x - 1)^3 + C_1 = \frac{4}{3}x^3 - 2x^2 + x - \frac{1}{6} + C_1$
or $\frac{4}{3}x^3 - 2x^2 + x + C_2$
Answers differ by a constant: $C_2 = C_1 - \frac{1}{6}$

79. (a) $\frac{8}{3}$ (b) $\frac{16}{3}$ (c) $-\frac{8}{3}$ (d) 8

81. $2\int_{0}^{4}(6x^2 - 3)\, dx = 232$ **83.** $V(t) = \dfrac{200,000}{t + 1} + 300,000$

$\$340,000$

85. (a) $C(x) = \frac{3}{2}(12x + 1)^{2/3} + 56.35$

(b)

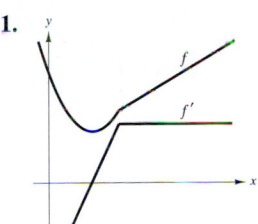

87. (a) 102.352 thousand units

(b) 102.352 thousand units

(c) 74.5 thousand units

89. (a) 1.273 amperes

(b) 1.382 amperes

(c) 0 amperes

91. False: $\int (2x + 1)^2 \, dx = \frac{1}{6}(2x + 1)^3 + C.$

92. False: $\int x(x^2 + 1) \, dx = \frac{1}{4}(x^2 + 1)^2 + C.$

93. True   **94.** True   **95.** True

96. False: $\int \sin^2 2x \cos 2x \, dx = \frac{1}{6}\sin^3 2x + C.$ **97.** Proof

Section 4.6 *(page 304)*

	Exact	Trapezoidal	Simpson's
1.	2.6667	2.7500	2.6667
3.	4.0000	4.2500	4.0000
5.	4.0000	4.0625	4.0000
7.	12.6667	12.6640	12.6667
9.	0.1667	0.1676	1.1667

	Trapezoidal	Simpson's	Graphing utility
11.	3.283	3.240	3.241
13.	0.342	0.372	0.393
15.	0.957	0.978	0.977
17.	0.089	0.089	0.089
19.	0.194	0.186	0.186

21. (a) 0.500   (b) 0.000   **23.** (a) $n = 366$ (b) $n = 26$

25. (a) $n = 130$ (b) $n = 12$

27. (a) $n = 643$ (b) $n = 48$ **29.** Proof

31.

n	$L(n)$	$M(n)$	$R(n)$	$T(n)$	$S(n)$
4	12.7771	15.3965	18.4340	15.6055	15.4845
8	14.0868	15.4480	16.9152	15.5010	15.4662
10	14.3569	15.4544	16.6197	15.4883	15.4658
12	14.5386	15.4578	16.4242	15.4814	15.4657
16	14.7674	15.4613	16.1816	15.4745	15.4657
20	14.9056	15.4628	16.0370	15.4713	15.4657

33.

n	$L(n)$	$M(n)$	$R(n)$	$T(n)$	$S(n)$
4	2.8163	3.5456	3.7256	3.2709	3.3996
8	3.1809	3.5053	3.6356	3.4083	3.4541
10	3.2478	3.4990	3.6115	3.4296	3.4624
12	3.2909	3.4952	3.5940	3.4425	3.4674
16	3.3431	3.4910	3.5704	3.4568	3.4730
20	3.3734	3.4888	3.5552	3.4643	3.4759

35. 0.701   **37.** 10,233.58 foot-pounds

39. (a) 9920 square feet   (b) $10,413\frac{1}{3}$ square feet

41. 89,250 square meters   **43.** 2.477

45. (a) 0   (b) $L'(x) = \frac{1}{x}, L'(1) = 1$ (c) 2.718   (d) Proof

Review Exercises for Chapter 4 *(page 306)*

1.

3. $\frac{2}{3}x^3 + \frac{1}{2}x^2 - x + C$

5. $\frac{1}{2}x^2 - \frac{1}{x} + C$ **7.** $2x^2 + 3\cos x + C$ **9.** $y = 2 - x^2$

11. 240 feet per second

13. (a) 3 seconds   (b) 144 feet   (c) $\frac{3}{2}$ seconds   (d) 108 feet

15. (a) $\sum_{i=1}^{10} (2i - 1)$ (b) $\sum_{i=1}^{n} i^3$ (c) $\sum_{i=1}^{10} (4i + 2)$

17. (a) $S = \frac{5mb^2}{8}, \quad s = \frac{3mb^2}{8}$

(b) $S(n) = \frac{mb^2(n + 1)}{2n}, \quad s(n) = \frac{mb^2(n - 1)}{2n}$

(c) $\frac{1}{2}mb^2$ (d) $\frac{1}{2}mb^2$

19. (a) 13   (b) 7   (c) 11   (d) 50

21. $\frac{1}{7}x^7 + \frac{3}{5}x^5 + x^3 + x + C$ **23.** $\frac{2}{3}\sqrt{x^3 + 3} + C$

25. $\frac{(3x^2 - 1)^5}{30} + C$ **27.** $\frac{1}{4}\sin^4 x + C$

29. $2\sqrt{1 - \cos\theta} + C$ **31.** $\frac{\tan^{n+1} x}{n + 1} + C, \quad n \neq -1$

33. $\frac{1}{3\pi}(1 + \sec\pi x)^3 + C$ **35.** 16   **37.** 0   **39.** 2

41. $\frac{422}{5}$ **43.** $\frac{28\pi}{15}$ **45.** 2

47. 6

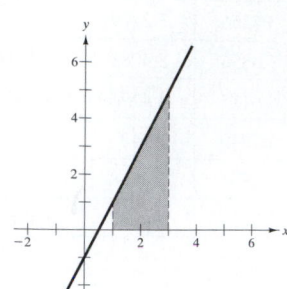

49. $\frac{10}{3}$

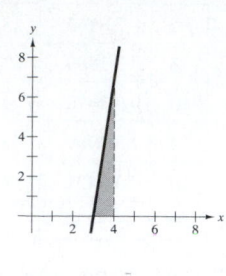

51. $\frac{1}{4}$

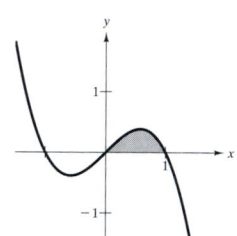

53. 16

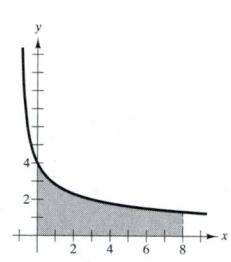

55. $\sqrt{3}$

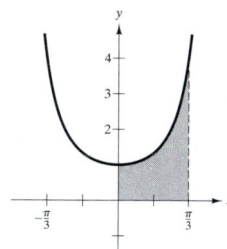

57. Average value $= \frac{2}{5}$, $x = \frac{29}{4}$

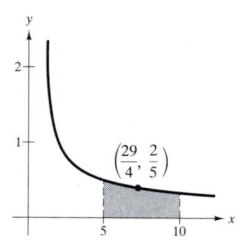

59. Average value $= 2$, $x = 2$

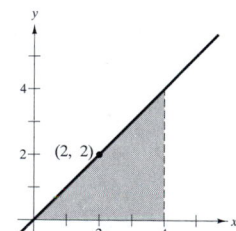

61. Trapezoidal Rule: 0.257
Simpson's Rule: 0.254
Graphing utility: 0.254

63. (a) $C \approx \$9.17$ (b) $C \approx \$3.14$; Savings $\approx \$6.03$

65. 1.6234 liters   **67.** (a) $\dfrac{24{,}300}{M}$ (b) $\dfrac{27{,}300}{M}$

69. (a) $0.025 = 2.5\%$ (b) $0.736 = 73.6\%$

70. False: only constants can be taken through the integral sign.

71. False: $\dfrac{d}{dx}\left[-\dfrac{1}{x^2}\right] \neq \dfrac{1}{x}$.   **72.** True   **73.** True

74. False: $\dfrac{d}{dx}\sec^2 x \neq \tan x$

CHAPTER 5

Section 5.1 *(page 318)*

1.

x	0.5	1.5	2	2.5	3
$\int_1^x (1/t)\, dt$	-0.6932	0.4055	0.6932	0.9163	1.0987

x	3.5	4
$\int_1^x (1/t)\, dt$	1.2529	1.3865

3. b   **4.** d   **5.** a   **6.** c

7. Domain: $x > 0$

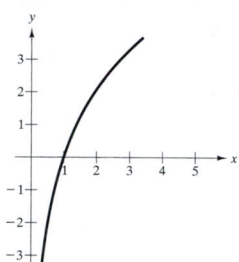

9. Domain: $x > 0$

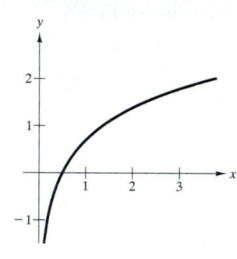

11. Domain: $x > 1$

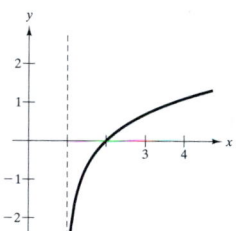

13. (a) 1.7917   (b) -0.4055 (c) 4.3944   (d) 0.5493

15. $\ln 2 - \ln 3$ **17.** $\ln x + \ln y - \ln z$ **19.** $\frac{3}{2}\ln 2$

21. $3[\ln(x+1) + \ln(x-1) - 3\ln x]$

23. $\ln z + 2\ln(z-1)$ **25.** $\ln\dfrac{x-2}{x+2}$ **27.** $\ln\sqrt[3]{\dfrac{x(x+3)^2}{x^2-1}}$

29. $\ln\dfrac{9}{\sqrt{x^2+1}}$ **31.**

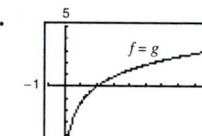

33. $-\infty$

35. $\ln 4$ **37.** 3   **39.** 2   **41.** $\dfrac{2}{x}$ **43.** $\dfrac{4(\ln x)^3}{x}$

45. $\dfrac{2x^2-1}{x(x^2-1)}$ **47.** $\dfrac{1-x^2}{x(x^2+1)}$ **49.** $\dfrac{1-2\ln t}{t^3}$

51. $\dfrac{2}{x\ln x^2} = \dfrac{1}{x\ln x}$ **53.** $\dfrac{1}{1-x^2}$ **55.** $\dfrac{-4}{x(x^2+4)}$

57. $\dfrac{\sqrt{x^2+1}}{x^2}$ **59.** $\cot x$ **61.** $-\tan x + \dfrac{\sin x}{\cos x - 1}$

63. $\dfrac{3 \cos x}{(\sin x - 1)(\sin x + 2)}$ **65.** $\dfrac{2}{x}(\sin 2x + x \cos 2x \ln x^2)$

67. (a) $5x - y - 2 = 0$

(b)

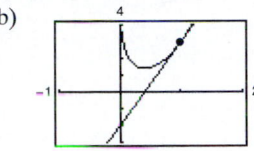

69. $\dfrac{2xy}{3 - 2y^2}$ **71.** $xy'' + y' = x\left(\dfrac{-2}{x^2}\right) + \dfrac{2}{x} = 0$

73. Relative minimum: $\left(1, \frac{1}{2}\right)$

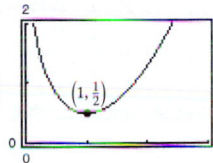

75. Relative minimum: $(e^{-1}, -e^{-1})$

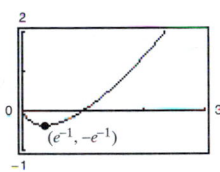

77. Relative minimum: (e, e)

Point of inflection: $\left(e^2, \dfrac{e^2}{2}\right)$

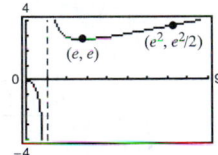

79.

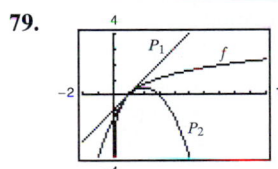

The values of f, P_1, and P_2, and their first derivatives, agree at $x = 1$.

81. $x \approx 0.567$ **83.** $\dfrac{2x^2 - 1}{\sqrt{x^2 - 1}}$

85. $\dfrac{3x^3 - 15x^2 + 8x}{2(x - 1)^3 \sqrt{3x - 2}}$ **87.** $\dfrac{(2x^2 + 2x - 1)\sqrt{x - 1}}{(x + 1)^{3/2}}$

89. $\beta = \dfrac{10}{\ln 10}(\ln I + 16 \ln 10)$

60 decibels

91. (a)

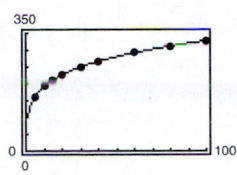

(c)

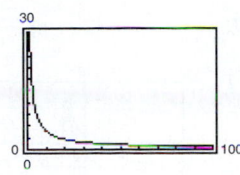

$$\lim_{p \to \infty} T'(p) = 0$$

(b) $p = 10$: $4.75°F$ per pound per square inch

$p = 70$: $0.97°F$ per pound per square inch

93. (a) $h = 0$ is not in the domain of the function.

(b) $h = 0.86 - 6.45 \ln p$

(c)

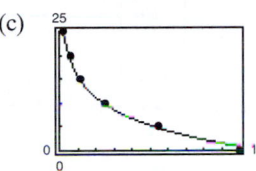

(d) 2.7 kilometers

(e) 0.15 atmosphere

(f) $h = 5$: $\dfrac{dp}{dh} = -0.085$

$h = 20$: $\dfrac{dp}{dh} = -0.009$

As the altitude increases, the pressure decreases at a slower rate.

95. For large values of x, g increases at a higher rate than f in both cases. The natural logarithmic function increases very slowly for large values of x.

(a)

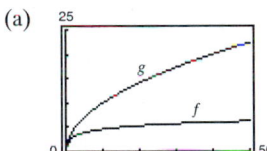

(b)

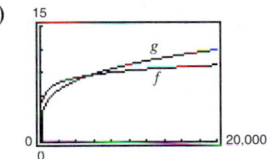

97. False: $\ln x + \ln 25 = \ln 25x$.   **98.** False: $y' = 0$.

Section 5.2 *(page 327)*

1. $\ln|x + 1| + C$ **3.** $-\frac{1}{2} \ln|3 - 2x| + C$

5. $\ln \sqrt{x^2 + 1} + C$ **7.** $\dfrac{x^2}{2} - 4 \ln|x| + C$

9. $\frac{1}{3} \ln|x^3 + 3x^2 + 9x| + C$ **11.** $\frac{1}{3}(\ln x)^3 + C$

13. $2\sqrt{x + 1} + C$ **15.** $x + 6\sqrt{x} + 18 \ln|\sqrt{x} - 3| + C$

17. $2 \ln|x - 1| - \dfrac{2}{x - 1} + C$ **19.** $\ln|\sin \theta| + C$

21. $-\frac{1}{2} \ln|\csc 2x + \cot 2x| + C$ **23.** $\ln|1 + \sin t| + C$

25. $\ln|\sec x - 1| + C$

27. $y = -3 \ln |2 - x| + C$

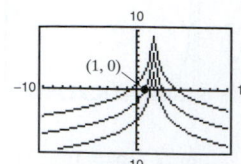

29. $y = -\frac{1}{2} \ln |\cos 2\theta| + C$

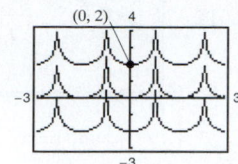

59. $\frac{12}{\pi} \left[2 \ln(\sqrt{3} + 1) - \ln 2 \right] \approx 5.03$

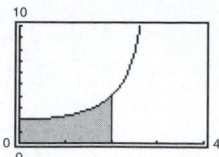

31. (a)

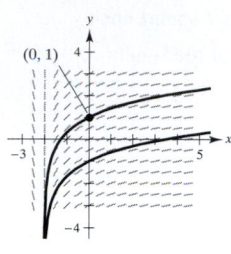

(b) $y = \ln \left| \dfrac{x + 2}{2} \right| + 1$

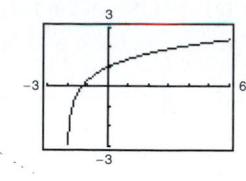

61. $P(t) = 1000(12 \ln |1 + 0.25t| + 1)$ **63.** \$168.27
$P(3) \approx 7715$

65. (a)

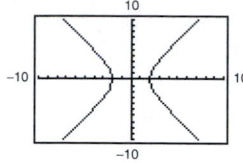

(b) $y^2 = e^{-\ln x + \ln 4} = \dfrac{4}{x}$

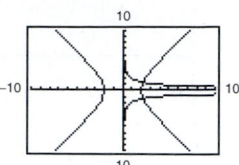

33. $\frac{5}{3} \ln 13 \approx 4.275$ **35.** $\frac{7}{3}$ **37.** $-\ln 3$

39. $\ln \left| \dfrac{2 - \sin 2}{1 - \sin 1} \right| \approx 1.929$

41. $2\left[\sqrt{x} - \ln\left(1 + \sqrt{x}\right) \right] + C$ **43.** $-\sin(1 - x) + C$

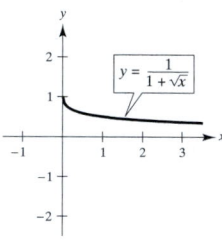

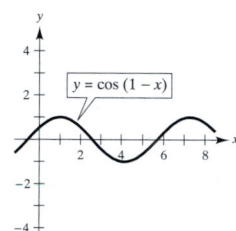

67. False: $\dfrac{1}{2} (\ln x) = \ln x^{1/2}$.   **68.** False: $\dfrac{d}{dx} [\ln x] = \dfrac{1}{x}$.

69. True

70. False: the integrand, $1/x$, has a nonremovable discontinuity in the interval $[-1, 2]$.

Section 5.3 *(page 335)*

1. (a) Proof   **3.** (a) Proof

(b)

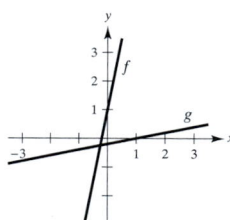

(b)

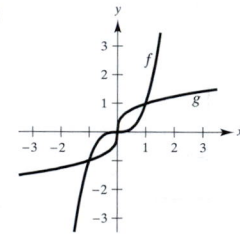

45. $\ln\left(\sqrt{2} + 1\right) - \dfrac{\sqrt{2}}{2} \approx 0.174$ **47.** Proof   **49.** Proof

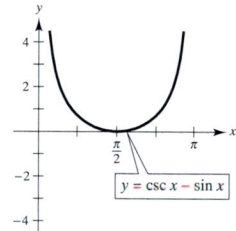

5. (a) Proof   **7.** (a) Proof

(b)

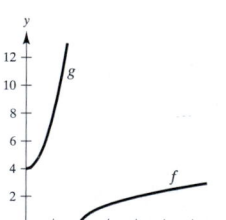

(b)

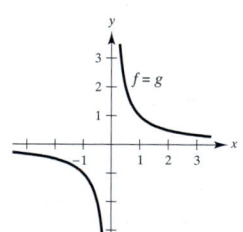

51. $\dfrac{1}{x}$ **53.** 0   **55.** d

57. $\frac{15}{2} + 8 \ln 2 \approx 13.045$

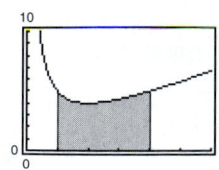

9. c   **10.** b   **11.** a   **12.** d

13. $f^{-1}(x) = \dfrac{x+3}{2}$

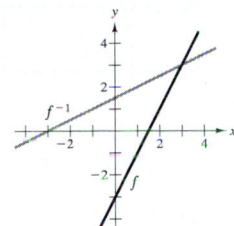

15. $f^{-1}(x) = x^{1/5}$

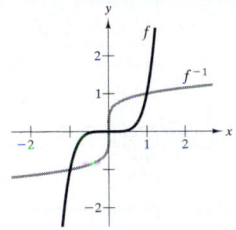

17. $f^{-1}(x) = x^2, \quad x \geq 0$

19. $f^{-1}(x) = \sqrt{4 - x^2}, \quad 0 \leq x \leq 2$

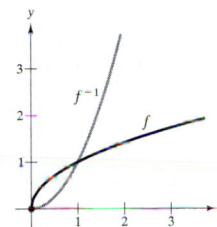

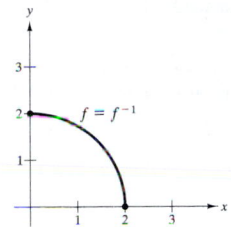

21. $f^{-1}(x) = x^3 + 1$

23. $f^{-1}(x) = x^{3/2}, \quad x \geq 0$

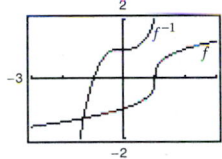

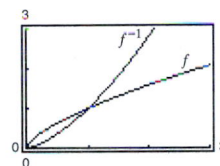

25. $f^{-1}(x) = \dfrac{\sqrt{7}x}{\sqrt{1 - x^2}}, \quad -1 < x < 1$

27. $f^{-1}(x) = \begin{cases} \dfrac{1 - \sqrt{1 + 16x^2}}{2x} & \text{if } x \neq 0 \\ 0 & \text{if } x = 0 \end{cases}$

The graph of f^{-1} is a reflection of the graph of f in the line $y = x$.

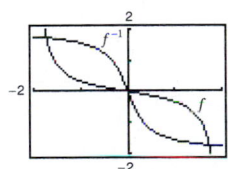

29.

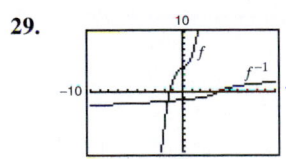

f is one-to-one and has an inverse function.

31.

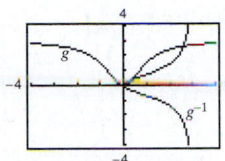

g is not one-to-one and does not have an inverse function.

33.

x	1	2	3	4
$f^{-1}(x)$	0	1	2	4

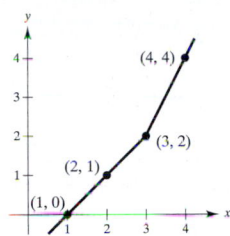

35. (a) Proof

(b) $y = \dfrac{20}{7}(80 - x)$

x: total cost

y: number of pounds of the less expensive commodity

(c) $[62.5, 80]$

(d) 20 pounds

37. Inverse exists.   **39.** Inverse does not exist.

41. One-to-one   **43.** One-to-one

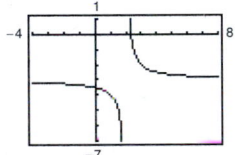

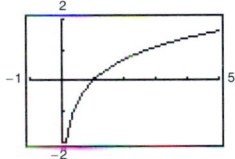

45. One-to-one

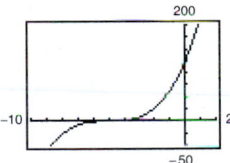

47. Inverse exists.   **49.** Inverse does not exist.

51. Inverse exists.   **53.** $f'(x) = 2(x - 4) > 0$ on $(4, \infty)$

55. $f'(x) = -\dfrac{8}{x^3} < 0$ on $(0, \infty)$

57. $f'(x) = -\sin x < 0$ on $(0, \pi)$

59. Not continuous at $\dfrac{(2n - 1)\pi}{2}$

61. One-to-one

$f^{-1}(x) = x^2 + 2, \quad x \geq 0$

63. One-to-one

$f^{-1}(x) = 2 - x, \quad x \geq 0$

65. $f^{-1}(x) = \sqrt{x} + 3, \quad x \geq 0$

67. $f^{-1}(x) = x - 3, \quad x \geq 0$ **69.** Inverse exists.

71. Inverse does not exist.   **73.** $\dfrac{1}{5}$ **75.** $\dfrac{2\sqrt{3}}{3}$ **77.** $\dfrac{1}{13}$

79. (a) Domain of f: $(-\infty, \infty)$
     Domain of f^{-1}: $(-\infty, \infty)$
   (b) Range of f: $(-\infty, \infty)$
     Range of f^{-1}: $(-\infty, \infty)$

(c)   (d) $f'\!\left(\tfrac{1}{2}\right) = \tfrac{3}{4}, \ (f^{-1})'\!\left(\tfrac{1}{8}\right) = \tfrac{4}{3}$

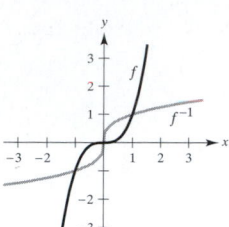

81. (a) Domain of f: $[4, \infty)$ (b) Range of f: $[0, \infty)$
     Domain of f^{-1}: $[0, \infty)$ Range of f^{-1}: $[4, \infty)$

(c)   (d) $f'(5) = \tfrac{1}{2}, \ (f^{-1})'(1) = 2$

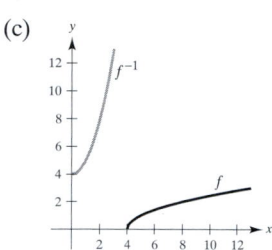

83. $-\dfrac{1}{11}$

85. 32   **87.** 600   **89.** $(g^{-1} \circ f^{-1})(x) = \dfrac{x + 1}{2}$

91. $(f \circ g)^{-1}(x) = \dfrac{x + 1}{2}$ **93.** Proof   **95.** Proof

97. False: let $f(x) = x^2$.   **98.** True   **99.** True

100. False: let $f(x) = \dfrac{1}{x}$.

101. No, let $f(x) = \begin{cases} x, & 0 \leq x \leq 1 \\ 1 - x, & 1 < x \leq 2 \end{cases}$.   **103.** $\sqrt{17}$

Section 5.4 *(page 344)*

1. $\ln 1 = 0$ **3.** $e^{0.6931\ldots} = 2$ **5.** $x = 4$ **7.** $x = e^2$

9.   **11.**

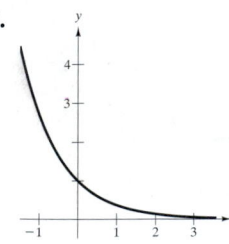 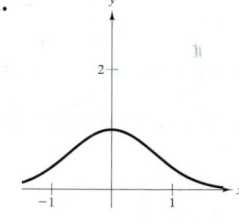

13. (a)   (b)

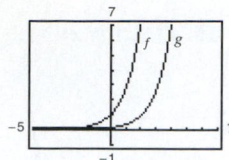 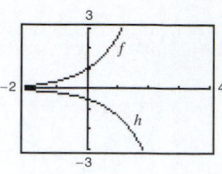

Translation two units   Reflection in the x-axis
to the right   and a vertical shrink

(c)

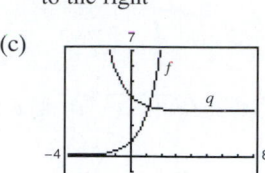

Reflection in the y-axis and
a translation three units upward

15.   **17.**

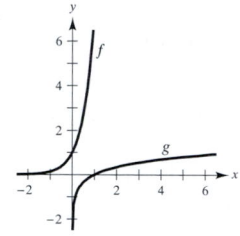 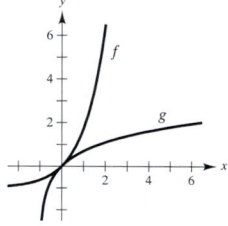

19. c   **20.** d   **21.** a   **22.** b

23. $\displaystyle\lim_{x \to \infty} f(x) = \lim_{x \to \infty} g(x) = e^{0.5}$

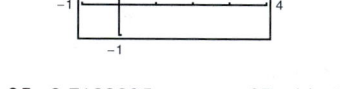

25. $2.7182805 < e$ **27.** (a) 3   (b) -3

29. $2e^{2x}$ **31.** $2(x - 1)e^{-2x+x^2}$

33. $\dfrac{e^{\sqrt{x}}}{2\sqrt{x}}$ **35.** $3(e^{-t} + e^{t})^2(e^{t} - e^{-t})$ **37.** $2x$

39. $\dfrac{2e^{2x}}{1 + e^{2x}}$ **41.** $\dfrac{-2(e^{x} - e^{-x})}{(e^{x} + e^{-x})^2}$ **43.** $x^2 e^{x}$

45. $e^{-x}\!\left(\dfrac{1}{x} - \ln x\right)$ **47.** $2e^{x}\cos x$ **49.** $\dfrac{10 - e^{y}}{xe^{y} + 3}$

51. $3(6x + 5)e^{-3x}$ **53.** Proof

55. Relative maximum: $\left(0, 1/\sqrt{2\pi}\right)$
      Points of inflection: $\left(\pm 1, 1/\sqrt{2\pi e}\right)$

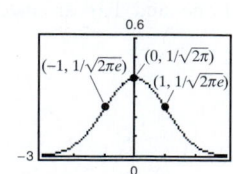

57. Relative minimum: $(0, 1)$

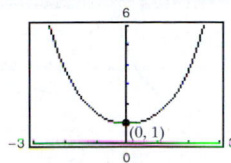

59. Relative minimum: $(0, 0)$
Relative maximum: $(2, 4e^{-2})$
Points of inflection: $\left(2 \pm \sqrt{2}, \left(6 \pm 4\sqrt{2}\right)e^{-(2 \pm \sqrt{2})}\right)$

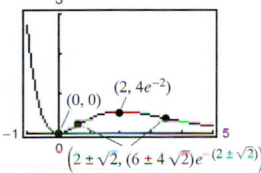

61. $A = \sqrt{2}e^{-1/2}$ **63.** Proof   **65.** 0.567

67. (a)

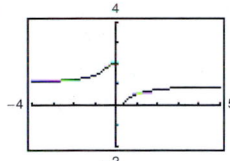

(b) When x increases without bound, $1/x$ approaches zero and $e^{1/x}$ approaches 1. Therefore, $f(x)$ approaches $\frac{2}{1+1} = 1$. Thus, $f(x)$ has a horizontal asymptote at $y = 1$. As x approaches zero from the right, $1/x$ approaches ∞, $e^{1/x}$ approaches ∞, and $f(x)$ approaches 0. As x approaches zero from the left, $1/x$ approaches $-\infty$, $e^{1/x}$ approaches 0, and $f(x)$ approaches 2. The limit does not exist, because the left limit does not equal the right limit. Therefore, $x = 0$ is a nonremovable discontinuity.

69. (a) $\ln P = -0.1499h + 9.3018$

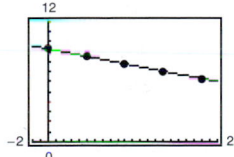

(b) $P = 10{,}957.7e^{-0.1499h}$

(c)

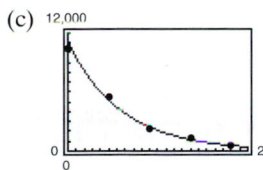

(d) $h = 5$: -776
$\quad\; h = 18$: -111

71.

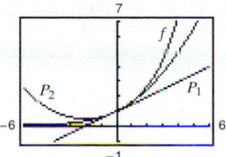

The values of f, P_1, and P_2, and their first derivatives, agree at $x = 0$. The values of the second derivatives of f and P_2 agree at $x = 0$.

73. (a)

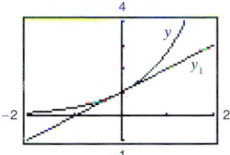

(b)

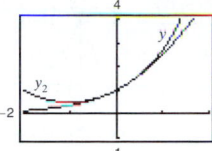

(c)

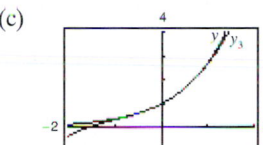

75. $e^{5x} + C$ **77.** $\dfrac{e^2 - 1}{2e^2}$

79. $x - \ln(e^x + 1) + C_1$ or $-\ln(1 + e^{-x}) + C_2$

81. $\dfrac{e}{3}(e^2 - 1)$ **83.** $-\dfrac{2}{3}(1 - e^x)^{3/2} + C$

85. $\ln|e^x - e^{-x}| + C$ **87.** $-\dfrac{5}{2}e^{-2x} + e^{-x} + C$

89. $\dfrac{1}{\pi}e^{\sin \pi x} + C$ **91.** $\ln|\cos e^{-x}| + C$ **93.** $\dfrac{1}{2a}e^{ax^2} + C$

95. $f(x) = \dfrac{1}{2}(e^x + e^{-x})$

97. (a)

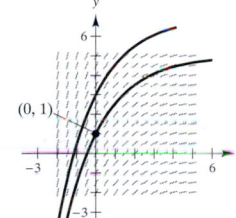

(b) $y = -4e^{-x/2} + 5$

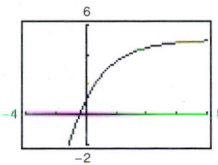

99. $e^5 - 1 \approx 147.413$

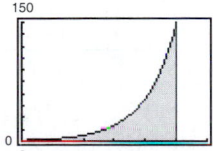

101. $1 - e^{-1} \approx 0.632$ **103.** Proof   **105.** 0.4772

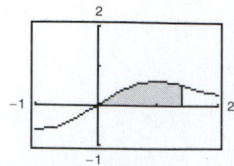

107. (a) $R = 428.76e^{-0.6155t}$

(b) 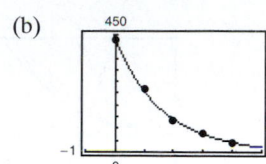   (c) 637 liters

109. (a) 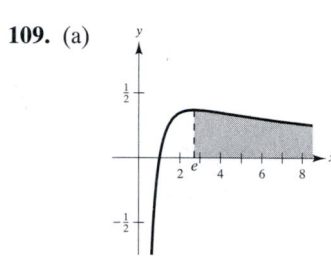   (b) Proof   (c) Proof

Section 5.5 *(page 354)*

1. -3 **3.** 0   **5.** (a) $\log_2 8 = 3$ (b) $\log_3(1/3) = -1$

7. (a) $10^{-2} = 0.01$ (b) $(1/2)^{-3} = 8$

9. (a) $x = 3$ (b) $x = -1$ **11.** (a) $x = \frac{1}{3}$ (b) $x = \frac{1}{16}$

13. (a) $x = -1, 2$ (b) $x = \frac{1}{3}$

15. 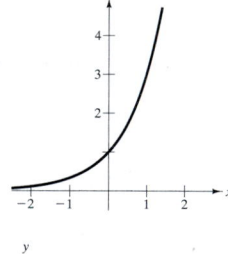   **17.**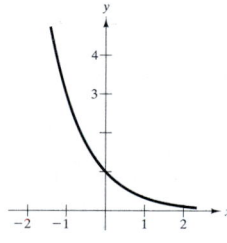

19. 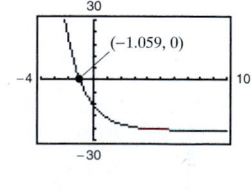   **21.**

23. 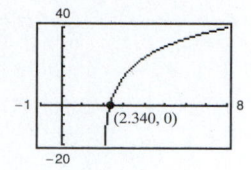   **25.**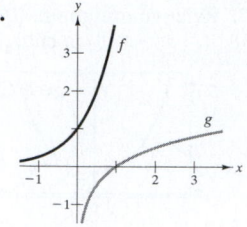

27. (a) False
(b) True: $y = \log_2 x$
(c) True: $2^y = x$
(d) False

29. $(\ln 4)4^x$ **31.** $(\ln 5)5^{x-2}$

33. $t2^t (t \ln 2 + 2)$ **35.** $-2^{-\theta}[(\ln 2)\cos \pi\theta + \pi \sin \pi\theta]$

37. $\dfrac{1}{x(\ln 3)}$ **39.** $\dfrac{x-2}{(\ln 2)x(x-1)}$ **41.** $\dfrac{x}{(\ln 5)(x^2-1)}$

43. $\dfrac{5}{(\ln 2)t^2}(1 - \ln t)$ **45.** $2(1 - \ln x)x^{(2/x)-2}$

47. $(x-2)^{x+1}\left[\dfrac{x+1}{x-2} + \ln(x-2)\right]$

49. $g(x) = x^x, k(x) = 2^x, h(x) = x^2, f(x) = \log_2 x$

51. (a) \$40.64   (b) $C'(1) \approx 0.051P, C'(8) \approx 0.072P$
(c) $\ln 1.05$

53.

n	1	2	4	12
A	\$1410.60	\$1414.78	\$1416.91	\$1418.34

n	365	Continuous
A	\$1419.04	\$1419.07

55.

n	1	2	4	12
A	\$4321.94	\$4399.79	\$4440.21	\$4467.74

n	365	Continuous
A	\$4481.23	\$4481.69

57.

t	1	10	20	30
P	\$95,122.94	\$60,653.07	\$36,787.94	\$22,313.02

t	40	50
P	\$13,533.53	\$8208.50

59.

t	1	10	20	30
P	\$95,132.82	\$60,716.10	\$36,864.45	\$22,382.66

t	40	50
P	\$13,589.88	\$8251.24

61. c

63. (a) 6.7 million cubic feet per acre

(b) $t = 20$: $\dfrac{dv}{dt} = 0.073$

$t = 60$: $\dfrac{dv}{dt} = 0.040$

65. (a)

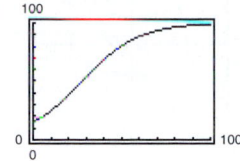

(b) 16.7%   (c) $x \approx 38.8$ or 38,800 egg masses

(d) $x \approx 2.78$ or 27,800 egg masses

67. (a) $y = 271.92e^{0.0919x}$ (b) $y = -254.08 + 418.41 \ln x$

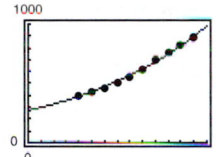   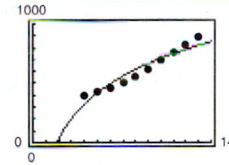

(c) Exponential

(d) Exponential: $\dfrac{dy}{dx} = 157.0$

Logarithmic: $\dfrac{dy}{dx} = 20.9$

Logarithmic

69. $\dfrac{3^x}{\ln 3} + C$ **71.** $\dfrac{7}{\ln 4}$ **73.** $-\dfrac{1}{2 \ln 5}\left(5^{-x^2}\right) + C$

75. $\dfrac{\ln\left(3^{2x} + 1\right)}{2 \ln 3} + C$

77. (a)

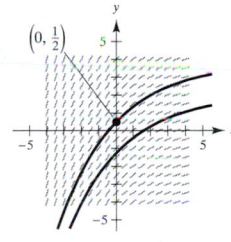

(b) $y = \dfrac{3\left(1 - 0.4^{x/3}\right)}{\ln 2.5} + \dfrac{1}{2}$

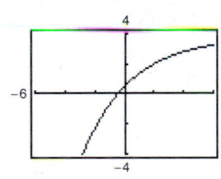

79. (a) 5.67

(b)

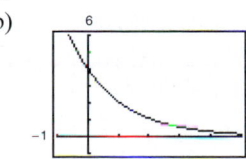

(c) $f(t) = g(t) = h(t)$. No. The definite integrals of two functions over a given interval may be equal when the functions are not equal.

81. $15,039.61   **83.** $y = 1200(0.6^t)$

85. False; e is an irrational number.   **86.** True   **87.** True

88. True   **89.** True   **90.** True   **91.** Proof

Section 5.6 *(page 363)*

1. $y^2 - 5x^2 = C$ **3.** $y = Ce^{(2x^{3/2})/3}$ **5.** $y = C(1 + x^2)$

7. (a)   (b) $y = 6 - 6e^{-x^2/2}$

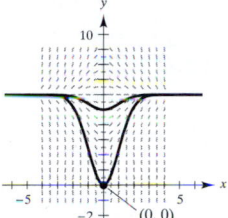   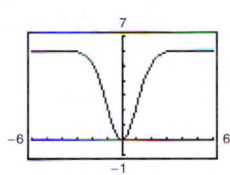

9. $\dfrac{dQ}{dt} = \dfrac{k}{t^2}$ **11.** $\dfrac{dN}{ds} = k(250 - s)$

$Q = -\dfrac{k}{t} + C$ $N = -\dfrac{k}{2}(250 - s)^2 + C$

13. $y = \frac{1}{4}t^2 + 10$ **15.** $y = 10e^{-t/2}$

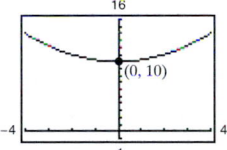   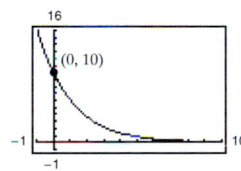

17. $y = \frac{1}{2}e^{0.4605t}$ **19.** $y = 0.6687e^{0.4024t}$

21. Amount after 1000 years: 6.52 grams
Amount after 10,000 years: 0.14 gram

23. Initial quantity: 6.70 grams
Amount after 1000 years: 5.94 grams

25. Initial quantity: 2.16 grams   **27.** 95.81%
Amount after 10,000 years: 1.63 grams

29. Time to double: 11.55 years
Amount after 10 years: $1822.12

31. Annual rate: 8.94%
Amount after 10 years: $1833.67

33. Annual rate: 9.50%   **35.** $112,087.09
Time to double: 7.30 years

37. (a) 10.24 years   **39.** $y \approx 4.22e^{0.0430t}$ **41.** $y \approx 3e^{-0.0091t}$

(b) 9.93 years   9.97 million   2.50 million

(c) 9.90 years

(d) 9.90 years

43. (a) k is the continuous annual percentage rate of change of y.

(b) When $k > 0$ the population is growing, and when $k < 0$ the population is decreasing.

45. 527.06 millimeters of mercury

47. (a) $N \approx 30(1 - e^{-0.0502t})$ (b) 36 days

49. (a) $S \approx 30e^{-1.7918/t}$ **51.** 2014 ($t = 16$)

(b) 20,965 units

(c)
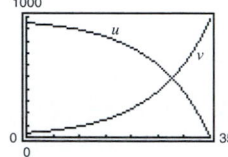

53. (a) 20 decibels (b) 70 decibels
(c) 95 decibels (d) 120 decibels

55. (a) $10^{8.3} \approx 199,526,231.5$ (b) 10^R (c) $\dfrac{1}{I \ln 10}$

57. 22.35°F

59. (a)
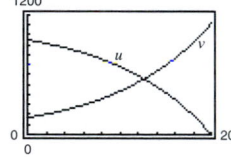

(b) Interest; $t \approx 28$ years

(c) The slopes of the tangent lines to the two curves are equal in magnitude and opposite in sign. $u'(15) = -14.06$, $v'(15) = 14.06$

(d)
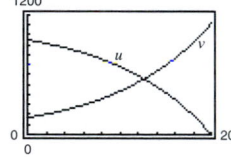

Section 5.7 *(page 374)*

1. Proof   **3.** Proof   **5.** Proof   **7.** Not a solution

9. Solution   **11.** Solution   **13.** Not a solution

15. Solution   **17.** Not a solution   **19.** $k = 0.07$

21. $4y^2 = x^3$

23.
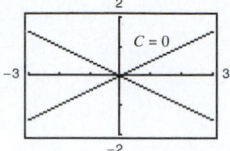 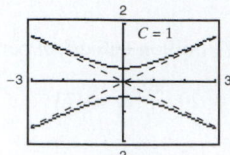
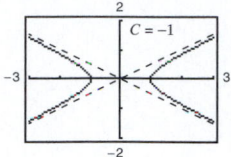 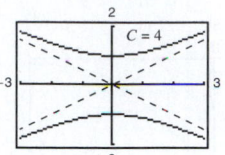
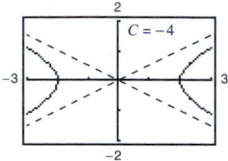

25. $y = 3e^{-2x}$ **27.** $y = 2 \sin 3x - \frac{1}{3} \cos 3x$

29. $y = -2x + \frac{1}{2}x^3$ **31.** $y = x^3 + C$

33. $y = x - \ln x^2 + C$ **35.** $y = -\frac{1}{2} \cos 2x + C$

37. $y = \frac{2}{5}(x - 3)^{5/2} + 2(x - 3)^{3/2} + C$ **39.** $y^2 - x^2 = C$

41. $r = Ce^{0.05s}$ **43.** $y = C(x + 2)^3$ **45.** $y^2 = C - 2\cos x$

47. $y = Ce^{(\ln x)^2/2}$ **49.** $y^2 = 2e^x + 14$ **51.** $y = e^{-(x^2+2x)/2}$

53. $y^2 = 4x^2 + 3$ **55.** $u = e^{(1 - \cos v^2)/2}$ **57.** $P = P_0 e^{kt}$

59. $9x^2 + 16y^2 = 25$ **61.** $f(x) = Ce^{-x/2}$

63. Homogeneous of degree 3   **65.** Not homogeneous

67. Homogeneous of degree 0   **69.** $|x| = C(x - y)^2$

71. $|y^2 + 2xy - x^2| = C$

73. $y = Ce^{-x^2/2y^2}$ **75.** $e^{y/x} = 1 + \ln x^2$ **77.** $x = e^{\sin(y/x)}$

79.   **81.**
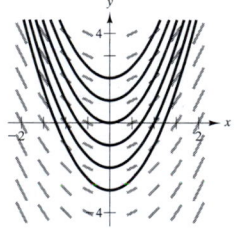   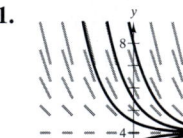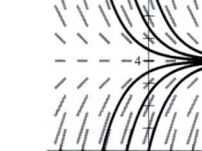

$y = \frac{1}{2}x^2 + C$ $y = 4 + Ce^{-x}$

83. (a) $\dfrac{dy}{dx} = k(y - 4)$ (b) a   (c) Proof

85. (a) $\dfrac{dy}{dx} = ky(y - 4)$ (b) c   (c) Proof

87. (a) $\dfrac{dS}{dt} = kS(L - S)$ (b) $t = 2.7$ months

$$S = \dfrac{100}{1 + 9e^{-0.8109t}}$$

(c)
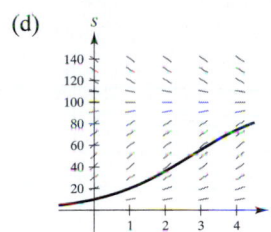

(d)
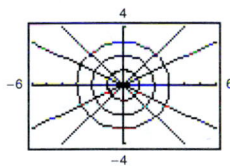

(e) Sales will decrease toward the line $S = L$.

89. 98.9% of the original amount

91. (a) $\dfrac{dv}{dt} = k(W - v)$ (b) $s = 20t + 69.5(e^{-0.2877t} - 1)$

$v = 20(1 - e^{-0.2877t})$

93. Circles: $x^2 + y^2 = C$
Lines: $y = Kx$

95. Parabolas: $x^2 = Cy$
Ellipses: $x^2 + 2y^2 = K$

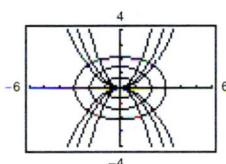

97. Curves: $y^2 = Cx^3$
Ellipses: $2x^2 + 3y^2 = K$

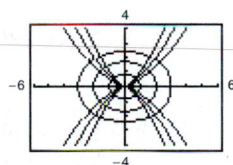

99. False: $y = x^3$ is a solution to $xy' - 3y = 0$, but $y = x^3 + 1$ is not a solution.

100. True   **101.** False: $f(tx, ty) \neq t^n f(x, y)$.   **102.** True

Section 5.8 (page 383)

1.

x	-1	-0.8	-0.6	-0.4	-0.2	0
y	-1.57	-0.93	-0.64	-0.41	-0.20	0

x	0.2	0.4	0.6	0.8	1
y	0.20	0.41	0.64	0.93	1.57

(b)

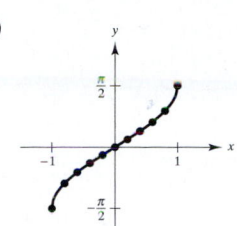

(c)
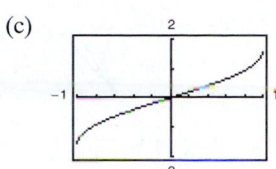

(d) Intercept: $(0, 0)$
Symmetry: origin

3. False: the range of $y = \arccos x$ is $[0, \pi]$.   **5.** $\dfrac{\pi}{6}$ **7.** $\dfrac{\pi}{3}$

9. $\dfrac{\pi}{6}$ **11.** $-\dfrac{\pi}{4}$ **13.** 2.50   **15.** $\arccos\left(\dfrac{1}{1.269}\right) \approx 0.66$

17. The range of $y = \arctan x$ is $-\dfrac{\pi}{2} < y < \dfrac{\pi}{2}$.

19. (a) $\dfrac{3}{5}$ (b) $\dfrac{5}{3}$ **21.** (a) $-\sqrt{3}$ (b) $-\dfrac{13}{5}$

23. $\sqrt{1 - 4x^2}$ **25.** $\dfrac{\sqrt{x^2 - 1}}{|x|}$ **27.** $\dfrac{\sqrt{x^2 - 9}}{3}$

29. $\dfrac{\sqrt{x^2 + 2}}{x}$

31.
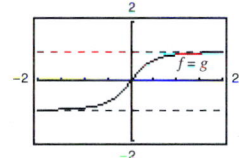

g is the algebraic form of f.
Horizontal asymptotes: $y = -1, y = 1$

33. Proof

35.
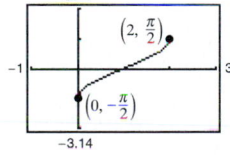

37.
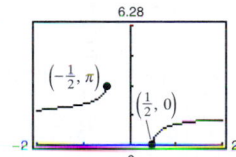

39. $x = \dfrac{1}{3}\left[\sin\left(\dfrac{1}{2}\right) + \pi\right] \approx 1.207$ **41.** $x = \dfrac{1}{3}$

43. $\dfrac{2}{\sqrt{2x - x^2}}$ **45.** $-\dfrac{3}{\sqrt{4 - x^2}}$ **47.** $\dfrac{a}{a^2 + x^2}$

49. $\dfrac{3x - \sqrt{1 - 9x^2}\arcsin 3x}{x^2\sqrt{1 - 9x^2}}$ **51.** $-\dfrac{t}{\sqrt{1 - t^2}}$

53. $\dfrac{1}{1 - x^4}$ **55.** $\arcsin x$

57.

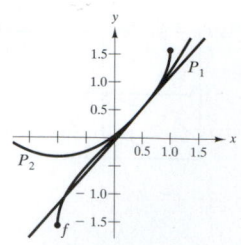

59. Relative maximum: $(1.272, -0.606)$
Relative minimum: $(-1.272, 3.747)$

61. (a) $h(t) = -16t^2 + 256$ **63.** Proof
$t = 4$ seconds

(b) $t = 1$: -0.0520 radian per second
$t = 2$: -0.1116 radian per second

65. $k \le -1$ or $k \ge 1$ **67.** True

68. False: the range is $\left[\dfrac{-\pi}{2}, \dfrac{\pi}{2}\right]$.   **69.** True

70. False: $\arcsin^2 0 + \arccos^2 0 = \left(\dfrac{\pi}{2}\right)^2 \ne 1$.

Section 5.9 *(page 390)*

1. $\dfrac{\pi}{18}$ **3.** $\dfrac{\pi}{6}$ **5.** $\operatorname{arcsec}|2x| + C$

7. $\frac{1}{2}x^2 - \frac{1}{2}\ln(x^2 + 1) + C$ **9.** $\arcsin(x + 1) + C$

11. $\frac{1}{2}\arcsin t^2 + C$ **13.** $\dfrac{\pi^2}{32} \approx 0.308$

15. $\dfrac{\sqrt{3} - 2}{2} \approx -0.134$ **17.** $\dfrac{1}{4}\arctan\dfrac{e^{2x}}{2} + C$ **19.** $\dfrac{\pi}{4}$

21. $\dfrac{\pi}{2}$ **23.** $\ln|x^2 + 6x + 13| - 3\arctan\left(\dfrac{x + 3}{2}\right) + C$

25. $\arcsin\left(\dfrac{x + 2}{2}\right) + C$ **27.** $-\sqrt{-x^2 - 4x} + C$

29. $4 - 2\sqrt{3} + \dfrac{\pi}{6} \approx 1.059$ **31.** $\frac{1}{2}\arctan(x^2 + 1) + C$

33. a and b   **35.** a, b, and c

37. $2\sqrt{e^t - 3} - 2\sqrt{3}\arctan\left(\dfrac{\sqrt{e^t - 3}}{\sqrt{3}}\right) + C$

39. (a)

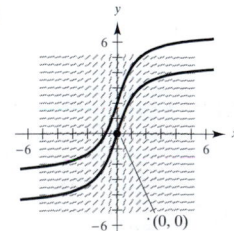

(b) $y = 3\arctan x$

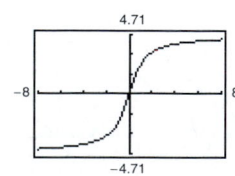

41. $\dfrac{\pi}{8}$ **43.** c

45. (a) Proof   (b) 3.1415918 (c) 3.1415927

47. (a) $y_1 = \arcsin\left(\dfrac{x - 3}{3}\right) + C$

(b) $y_2 = 2\arcsin\sqrt{\dfrac{x}{6}} + C$

(c)

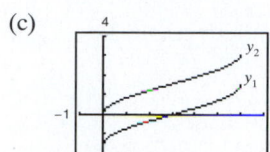

$2\arcsin\sqrt{\dfrac{x}{6}} = \arcsin\left(\dfrac{x - 3}{3}\right) + \dfrac{\pi}{2}$

Domain: $0 \le x \le 6$

49. (a) $v(t) = -32t + 500$

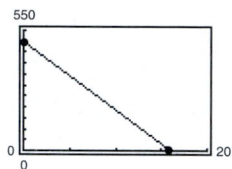

(b) $s(t) = -16t^2 + 500t$, 3906.25 feet

(c) $v(t) = \sqrt{\dfrac{32}{k}}\tan\left[\arctan\left(500\sqrt{\dfrac{k}{32}}\right) - \sqrt{32k}\,t\right]$

(d)

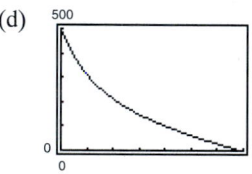

6.86 seconds

(e) 1088 feet

(f) When air resistance is taken into account, the maximum height of the object is not as great.

Section 5.10 *(page 400)*

1. (a) 10.018 (b) -0.964 **3.** (a) $\frac{4}{3}$ (b) $\frac{13}{12}$

5. (a) 1.317 (b) 0.962

7. $\left(\dfrac{e^x - e^{-x}}{e^x + e^{-x}}\right)^2 + \left(\dfrac{2}{e^x + e^{-x}}\right)^2 = 1$

9. $\left(\dfrac{e^x - e^{-x}}{2}\right)\left(\dfrac{e^y + e^{-y}}{2}\right) + \left(\dfrac{e^x + e^{-x}}{2}\right)\left(\dfrac{e^y - e^{-y}}{2}\right) =$
$\dfrac{e^{(x+y)} - e^{-(x+y)}}{2} = \sinh(x + y)$

11. $\left(\dfrac{e^x - e^{-x}}{2}\right)\left[3 + 4\left(\dfrac{e^x - e^{-x}}{2}\right)^2\right] =$

$\dfrac{e^{3x} - e^{-3x}}{2} =$

$\sinh(3x)$

13. $\cosh x = \dfrac{\sqrt{13}}{2}$ **15.** $-2x\cosh(1 - x^2)$ **17.** $\coth x$

$\tanh x = \dfrac{3\sqrt{13}}{13}$

$\operatorname{csch} x = \dfrac{2}{3}$

$\operatorname{sech} x = \dfrac{2\sqrt{13}}{13}$

$\coth x = \dfrac{\sqrt{13}}{3}$

19. $\operatorname{csch} x$ **21.** $\sinh^2 x$ **23.** $\operatorname{sech} t$

25. $\dfrac{y}{x}\left[\cosh x + x(\sinh x)\ln x\right] = \dfrac{x^{\cosh x}}{x}\left[\cosh x + x(\sinh x)\ln x\right]$

27. $-2(\cosh x - \sinh x)^2 = -2e^{-2x}$

29. Relative maxima: $(\pm\pi, \cosh\pi)$
Relative minimum: $(0, -1)$

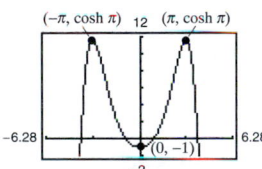

31. Relative maximum: $(1.20, 0.66)$ **33.** Proof
Relative minimum: $(-1.20, -0.66)$

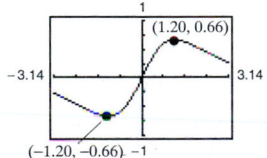

35. $P_1(x) = 0.76 + 0.42(x - 1)$
$P_2(x) = 0.76 + 0.42(x - 1) - 0.32(x - 1)^2$

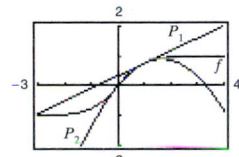

37. $-\dfrac{1}{2}\cosh(1 - 2x) + C$ **39.** $\dfrac{1}{3}\cosh^3(x - 1) + C$

41. $\ln|\sinh x| + C$ **43.** $-\coth\dfrac{x^2}{2} + C$ **45.** $\operatorname{csch}\dfrac{1}{x} + C$

47. $\dfrac{1}{5}\ln 3$ **49.** $\dfrac{\pi}{4}$ **51.** $\dfrac{1}{2}\arctan x^2 + C$ **53.** $\dfrac{3}{\sqrt{9x^2 - 1}}$

55. $|\sec x|$ **57.** $2\sec 2x$ **59.** $2\sinh^{-1}(2x)$

61. $-\dfrac{\sqrt{a^2 - x^2}}{x}$

63. $-\operatorname{csch}^{-1}(e^x) + C = -\ln\left(\dfrac{1 + \sqrt{1 + e^{2x}}}{e^x}\right) + C$

65. $2\sinh^{-1}\sqrt{x} + C = 2\ln\left(\sqrt{x} + \sqrt{1 + x}\right) + C$

67. $\dfrac{1}{4}\ln\left|\dfrac{x - 4}{x}\right| + C$ **69.** $\dfrac{1}{2\sqrt{6}}\ln\left|\dfrac{\sqrt{2}(x + 1) + \sqrt{3}}{\sqrt{2}(x + 1) - \sqrt{3}}\right| + C$

71. $\dfrac{1}{4}\arcsin\left(\dfrac{4x - 1}{9}\right) + C$

73. $-\dfrac{x^2}{2} - 4x - \dfrac{10}{3}\ln\left|\dfrac{x - 5}{x + 1}\right| + C$

75. $8\arctan(e^2) - 2\pi \approx 5.207$ **77.** $\dfrac{5}{2}\ln\left(\sqrt{17} + 4\right) \approx 5.237$

79. $\dfrac{52}{31}$ kilograms

81. If k were increased, the time of descent would increase.

83. Proof   **85.** Proof   **87.** Proof

Review Exercises for Chapter 5 *(page 402)*

1. Vertical asymptote: $x = 0$

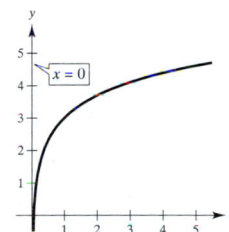

3. $\dfrac{1}{5}\left[\ln(2x + 1) + \ln(2x - 1) - \ln(4x^2 + 1)\right]$

5. $\ln\left(\dfrac{3\sqrt[3]{4 - x^2}}{x}\right)$

7. False: the domain of $f(x) = \ln x$ is the set of all *positive* real numbers.

8. False: $\ln x + \ln y = \ln xy$ **9.** $e^4 - 1 \approx 53.598$

11. $\dfrac{1}{2x}$ **13.** $\dfrac{1 + 2\ln x}{2\sqrt{\ln x}}$ **15.** $\dfrac{x}{(a + bx)^2}$ **17.** $\dfrac{1}{x(a + bx)}$

19. $\dfrac{1}{7}\ln|7x - 2| + C$ **21.** $-\ln|1 + \cos x| + C$

23. $3 + \ln 4$ **25.** $\ln\left(2 + \sqrt{3}\right)$

27. (a) $f^{-1}(x) = 2x + 6$ **29.** (a) $f^{-1}(x) = x^2 - 1, \quad x \geq 0$

(b)

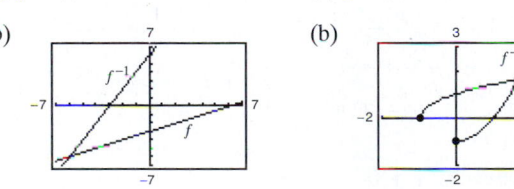

(c) Proof   (c) Proof

31. (a) $f^{-1}(x) = x^3 - 1$ **33.** (a) $f^{-1}(x) = e^{2x}$

(b) 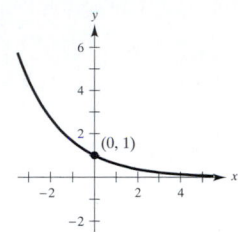   (b)

(c) Proof   (c) Proof

35.   **37.**

39. (a) $\dfrac{1}{2}$ (b) $\dfrac{\sqrt{3}}{2}$ **41.** $-2x$ **43.** $te^t(t + 2)$

45. $\dfrac{e^{2x} - e^{-2x}}{\sqrt{e^{2x} + e^{-2x}}}$ **47.** $3^{x-1} \ln 3$ **49.** $\dfrac{x(2 - x)}{e^x}$

51. $(1 - x^2)^{-3/2}$ **53.** $\dfrac{x}{|x|\sqrt{x^2 - 1}} + \text{arcsec } x$

55. $(\arcsin x)^2$ **57.** $2 - \dfrac{\sinh \sqrt{x}}{2\sqrt{x}}$ **59.** $\dfrac{-y}{x(2y + \ln x)}$

61. (a) ax^{a-1} (b) $(\ln a)a^x$ (c) $x^x(1 + \ln x)$ (d) 0

63. 6.9% **65.** $-\dfrac{1}{6}e^{-3x^2} + C$ **67.** $\dfrac{e^{4x} - 3e^{2x} - 3}{3e^x} + C$

69. $\ln|e^x - 1| + C$ **71.** $\dfrac{1}{2}\arctan(e^{2x}) + C$

73. $\dfrac{1}{2}\arcsin x^2 + C$ **75.** $\dfrac{1}{2}\ln(16 + x^2) + C$

77. $\dfrac{1}{4}\left(\arctan\dfrac{x}{2}\right)^2 + C$ **79.** $\dfrac{1}{2}\ln\left(\sqrt{x^4 - 1} + x^2\right) + C$

81. $-\dfrac{1}{2}(e^{-16} - 1) \approx 0.500$ **83.** $y = \dfrac{x^2}{2} + 3\ln|x| + C$

85. $y = Ce^{x^2}$ **87.** $\dfrac{x}{x^2 - y^2} = C$

89. (a)

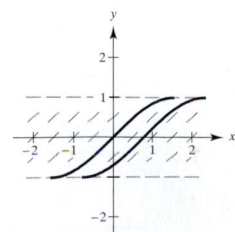

(b) Greatest: $y = 0$
     Least: $y = \pm 1$

(c) $y = \sin(x + C), \quad -\dfrac{\pi}{2} \le x + C \le \dfrac{\pi}{2}$

91. Family of circles: $x^2 + (y - K)^2 = K^2$

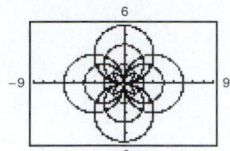

93. (a) 0.60 (b) 0.85 **95.** $y = A\sin\left(\sqrt{\dfrac{k}{m}}\,t\right)$

97. (a) $y = 28e^{0.6 - 0.012s}, s > 50$

(b)

Speed, s	50	55	60	65	70
Miles per gallon, y	28	26.4	24.8	23.4	22.0

CHAPTER 6

Section 6.1 *(page 413)*

1. $-\displaystyle\int_0^6 (x^2 - 6x)\, dx$ **3.** $\displaystyle\int_0^3 (-2x^2 + 6x)\, dx$

5. $-6\displaystyle\int_0^1 (x^3 - x)\, dx$

7.   **9.**

11. d

13. $\dfrac{32}{2}$ **15.** $\dfrac{9}{2}$

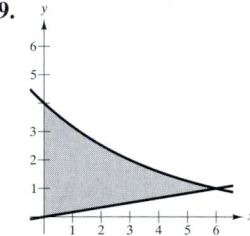   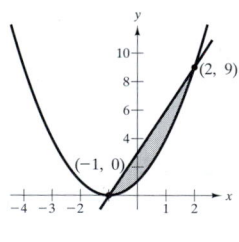

17. 1   **19.** $\dfrac{3}{2}$

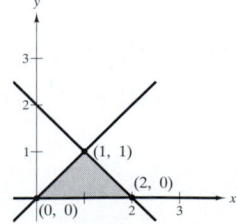   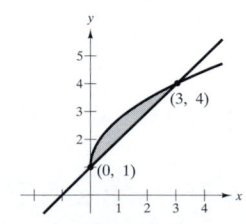

21. $\frac{9}{2}$

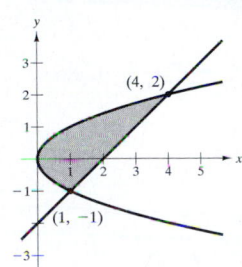

23. 6

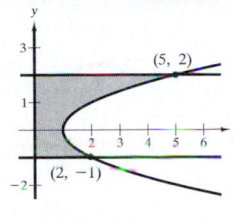

25. $8 \ln 2 \approx 5.545$

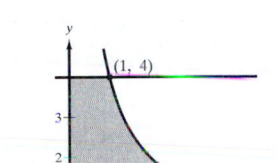

27. $\frac{37}{12}$

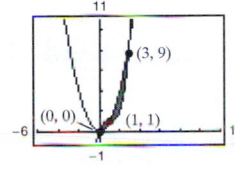

29. $\frac{64}{3}$

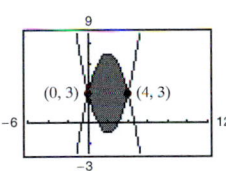

31. 8

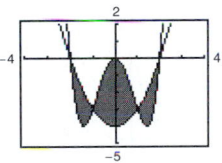

33. $\frac{\pi}{2} - \frac{1}{3} \approx 1.237$

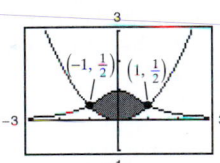

35. ≈ 1.759

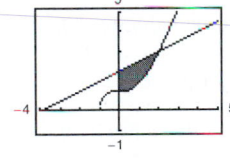

37. $2(1 - \ln 2) \approx 0.614$

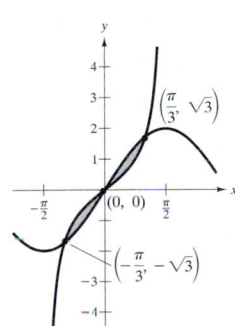

39. $\frac{1}{2}\left(1 - \frac{1}{e}\right) \approx 0.316$

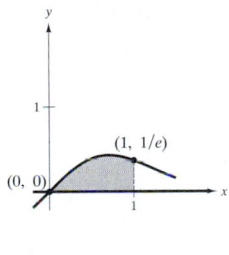

41. 4

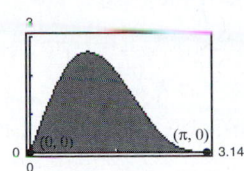

43. ≈ 1.323

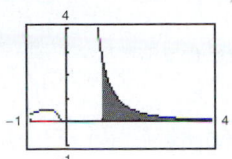

45. (a)

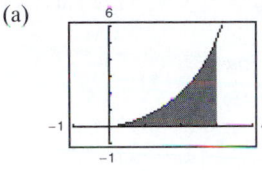

(b) $\displaystyle\int_0^3 \sqrt{\frac{x^3}{4-x}}\, dx$; No   (c) ≈ 4.773

47. $\frac{1}{2} ac$ **49.** $\displaystyle\int_{-2}^1 [x^3 - (3x - 2)]\, dx = \frac{27}{4}$

51. $x^4 - 2x^2 + 1 \le 1 - x^2$ on $[-1, 1]$

$$\int_{-1}^1 [(1 - x^2) - (x^4 - 2x^2 + 1)]\, dx = \frac{4}{15}$$

53. $b = 9\left(1 - \dfrac{1}{\sqrt[3]{4}}\right) \approx 3.330$

55. $\frac{1}{6}$ **57.** \$1.625 billion

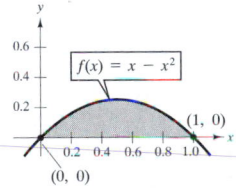

59. (a)

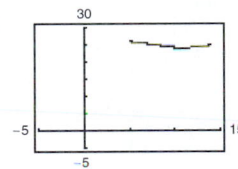

61. \$193,183

(b) 3.16 billion pounds

63. $\frac{16}{3}\left(4\sqrt{2} - 5\right) \approx 3.5$

65. (a) 6.031 square meters

(b) 12.062 cubic meters

(c) 60,310 pounds

67. Consumer surplus = 1600

Producer surplus = 400

69. True   **70.** True

Section 6.2 *(page 423)*

1. $\pi \displaystyle\int_0^1 (-x + 1)^2\, dx = \frac{\pi}{3}$ **3.** $\pi \displaystyle\int_1^4 \left(\sqrt{x}\right)^2 dx = \frac{15\pi}{2}$

5. $\pi \int_0^1 \left[(x^2)^2 - (x^3)^2 \right] dx = \dfrac{2\pi}{35}$ **7.** $\pi \int_0^4 (\sqrt{y})^2 \, dy = 8\pi$

9. $\pi \int_0^1 (y^{3/2})^2 \, dy = \dfrac{\pi}{4}$

11. (a) 8π (b) $\dfrac{128\pi}{5}$ (c) $\dfrac{256\pi}{15}$ (d) $\dfrac{192\pi}{5}$

13. (a) $\dfrac{32\pi}{3}$ (b) $\dfrac{64\pi}{3}$ **15.** 18π

17. $\pi\left(8 \ln 4 - \tfrac{3}{4}\right) \approx 32.49$ **19.** $\dfrac{208\pi}{3}$ **21.** $\dfrac{384\pi}{5}$

23. $\pi \ln 4$ **25.** $\dfrac{3\pi}{4}$ **27.** $\dfrac{\pi}{2}\left(1 - \dfrac{1}{e^2}\right) \approx 1.358$

29. 8π **31.** $\dfrac{\pi^2}{2} \approx 4.935$ **33.** 1.969

35. 49.022 **37.** a

39. (a) $\dfrac{512\pi}{15}$

(b) No, the solid has only been translated horizontally.

41. 18π **43.** Proof

45. $\pi r^2 h\left(1 - \dfrac{h}{H} + \dfrac{h^2}{3H^2}\right)$ **47.** $\dfrac{\pi}{30}$

49. (a) 60π **51.** One-fourth: 32.64 feet

(b) 50π Three-fourths: 67.36 feet

53. (a) ii; right-circular cylinder of radius r and height h

(b) iv; ellipsoid whose underlying ellipse has the equation

$\left(\dfrac{x}{b}\right)^2 + \left(\dfrac{y}{a}\right)^2 = 1$

(c) iii; sphere of radius r

(d) i; right-circular cone of radius r and height h

(e) v; torus of cross-sectional radius r and other radius R

55. (a) $\dfrac{128}{3}$ (b) $\dfrac{32\sqrt{3}}{3}$ (c) $\dfrac{16\pi}{3}$ (d) $\dfrac{32}{3}$

57. (a) $\dfrac{1}{10}$ (b) $\dfrac{\pi}{80}$ (c) $\dfrac{\sqrt{3}}{40}$ (d) $\dfrac{\pi}{20}$

59. $\dfrac{\pi d^3 \sin 70°}{8 \sin 20°}$

61. (a) $\dfrac{2r^3}{3}$ (b) $\dfrac{2r^3 \tan \theta}{3}$, $\lim\limits_{\theta \to 90°} V = \infty$

63. $5\sqrt{1 - 2^{-2/3}} \approx 3.0415$

Section 6.3 *(page 432)*

1. $2\pi \int_0^2 x^2 \, dx = \dfrac{16\pi}{3}$ **3.** $2\pi \int_0^4 x\sqrt{x} \, dx = \dfrac{128\pi}{5}$

5. $2\pi \int_0^2 x^3 \, dx = 8\pi$ **7.** $2\pi \int_0^2 x(4x - 2x^2) \, dx = \dfrac{16\pi}{3}$

9. $2\pi \int_0^2 x(x^2 - 4x + 4) \, dx = \dfrac{8\pi}{3}$

11. $2\pi \int_0^1 x\left(\dfrac{1}{\sqrt{2\pi}} e^{-x^2/2}\right) dx = \sqrt{2\pi}\left(1 - \dfrac{1}{\sqrt{e}}\right) \approx 0.986$

13. $2\pi \int_0^2 y(2 - y) \, dy = \dfrac{8\pi}{3}$

15. $2\pi\left[\int_0^{1/2} y \, dy + \int_{1/2}^1 y\left(\dfrac{1}{y} - 1\right) dy\right] = \dfrac{\pi}{2}$

17. 16π **19.** 64π

21. (a) $\dfrac{128\pi}{7}$ (b) $\dfrac{64\pi}{5}$ (c) $\dfrac{96\pi}{5}$

23. (a) $\dfrac{\pi a^3}{15}$ (b) $\dfrac{\pi a^3}{15}$ (c) $\dfrac{4\pi a^3}{15}$

25. (a) 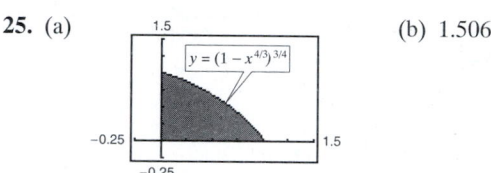   (b) 1.506

27. (a) 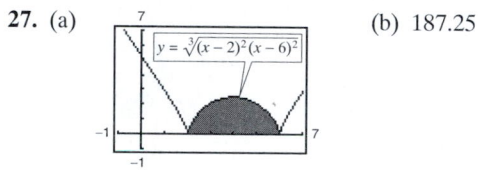   (b) 187.25

29. a, c, b   **31.** d   **33.** Diameter $= 2\sqrt{4 - 2\sqrt{3}} \approx 1.464$

35. Proof   **37.** $2\pi^2 r^2 R$

39. Both integrals yield the volume of the solid generated by revolving the region bounded by the graphs of $y = \sqrt{x - 1}$, $y = 0$, and $x = 5$ about the x-axis.

41. (a) ii; right-circular cone of radius r and height h

(b) v; torus of cross-sectional radius r and other radius R

(c) iii; sphere of radius r

(d) i; right-circular cylinder of radius r and height h

(e) iv; ellipsoid whose underlying ellipse has the equation

$\left(\dfrac{x}{b}\right)^2 + \left(\dfrac{y}{a}\right)^2 = 1$

43. (a) 1,366,593 cubic feet

(b) $d = -0.000561x^2 + 0.0189x + 19.39$

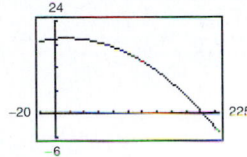

(c) 1,343,345 cubic feet

(d) 10,048,221 gallons

Section 6.4 *(page 442)*

1. 13   **3.** $\frac{2}{3}\left(\sqrt{8} - 1\right) \approx 1.219$ **5.** $5\sqrt{5} - 2\sqrt{2} \approx 8.352$

7. $\frac{33}{16}$

9. (a)

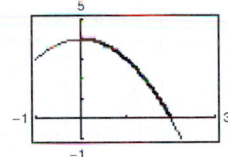

(b) $\int_0^2 \sqrt{1 + 4x^2}\, dx$ (c) ≈ 4.647

11. (a)

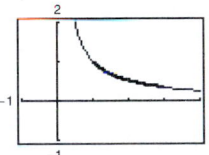

(b) $\int_1^3 \sqrt{1 + \frac{1}{x^4}}\, dx$ (c) ≈ 2.147

13. (a)

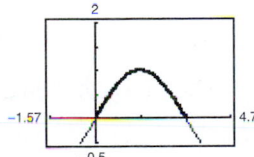

(b) $\int_0^\pi \sqrt{1 + \cos^2 x}\, dx$ (c) ≈ 3.820

15. (a)

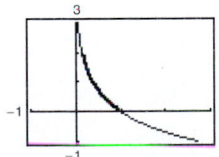

(b) $\int_0^2 \sqrt{1 + e^{-2y}}\, dy$ (c) ≈ 2.221

17. (a)

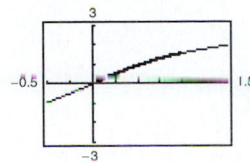

(b) $\int_0^1 \sqrt{1 + \left(\frac{2}{1 + x^2}\right)^2}\, dx$ (c) ≈ 1.871

19. b

21. (a) 64.125   (b) 64.525   (c) 64.666   (d) 64.672

23. (a)   (b) y_1, y_2, y_3, y_4

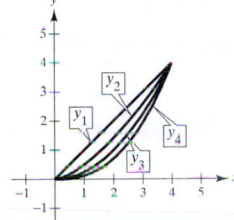

(c) $s_1 \approx 5.657$
$s_2 \approx 5.759$
$s_3 \approx 5.916$
$s_4 \approx 6.063$

25. $\frac{2}{3}$ **27.** $20[\sinh 1 - \sinh(-1)] \approx 47.0$ meters

29. $3\arcsin\frac{2}{3} \approx 2.1892$

31. $2\pi\int_0^3 \frac{1}{3}x^3\sqrt{1 + x^4}\, dx = \frac{\pi}{9}\left(82\sqrt{82} - 1\right) \approx 258.85$

33. $2\pi\int_1^2 \left(\frac{x^3}{6} + \frac{1}{2x}\right)\left(\frac{x^2}{2} + \frac{1}{2x^2}\right) dx = \frac{47\pi}{16}$

35. $2\pi\int_1^8 x\sqrt{1 + \frac{1}{9x^{4/3}}}\, dx =$

$\frac{\pi}{27}\left(145\sqrt{145} - 10\sqrt{10}\right) \approx 199.48$

37. 14.424   **39.** Proof

41. $6\pi\left(3 - \sqrt{5}\right) \approx 14.40$

43. Surface area $= \dfrac{\pi}{27}$ square feet ≈ 16.8 square inches

Amount of glass $= \dfrac{\pi}{27}\left(\dfrac{0.015}{12}\right)$

≈ 0.00015 cubic foot

≈ 0.25 cubic inch

45. Answers will vary.

47. (a) $\dfrac{ds}{dx} = \sqrt{1 + [f'(x)]^2}$

(b) $ds = \sqrt{1 + [f'(x)]^2}\, dx$

$(ds)^2 = \left[1 + \left(\dfrac{dy}{dx}\right)^2\right](dx)^2 = (dx)^2 + (dy)^2$

(c) $s(x) = \int_1^x \sqrt{1 + \frac{9}{4}t}\, dt$

$s(2) = \frac{22}{27}\sqrt{22} - \frac{13}{27}\sqrt{13}$

49. (a) Proof (b) Proof

Section 6.5 *(page 451)*

1. 1000 foot-pounds **3.** 448 newton-meters **5.** c, d, a, b

7. 30.625 inch-pounds $\approx$ 2.55 foot-pounds

9. 87.5 newton-meters

11. 180 inch-pounds = 15 foot-pounds

13. 37.125 foot-pounds

15. (a) 761.905 mile-tons $\approx 8.05 \times 10^9$ foot-pounds

(b) 1454.545 mile-tons $\approx 1.54 \times 10^{10}$ foot-pounds

17. (a) 2.93×10^4 mile-tons $\approx 3.10 \times 10^{11}$ foot-pounds

(b) 3.38×10^4 mile-tons $\approx 3.57 \times 10^{11}$ foot-pounds

19. (a) 2496 foot-pounds (b) 9984 foot-pounds

21. $48{,}000\pi$ kilogram-meters

23. 2995.2π foot-pounds **25.** $20{,}217.6\pi$ foot-pounds

27. 2457π foot-pounds **29.** 337.5 foot-pounds

31. 300 foot-pounds **33.** 168.75 foot-pounds

35. 7987.5 foot-pounds

37. $2000 \ln \frac{3}{2} \approx 810.93$ foot-pounds **39.** $\frac{3k}{4}$

41. 3249.4 foot-pounds **43.** 10,330.3 foot-pounds

Section 6.6 *(page 462)*

1. $\bar{x} = -\frac{6}{7}$ **3.** $\bar{x} = 12$ **5.** (a) $\bar{x} = 17$ (b) $\bar{x} = -3$

7. $x = 6$ feet **9.** $(\bar{x}, \bar{y}) = \left(\frac{10}{9}, -\frac{1}{9}\right)$

11. $(\bar{x}, \bar{y}) = \left(-\frac{7}{8}, -\frac{7}{16}\right)$

13. $M_x = 4\rho$, $M_y = \frac{64\rho}{5}$, $(\bar{x}, \bar{y}) = \left(\frac{12}{5}, \frac{3}{4}\right)$

15. $M_x = \frac{\rho}{35}$, $M_y = \frac{\rho}{20}$, $(\bar{x}, \bar{y}) = \left(\frac{3}{5}, \frac{12}{35}\right)$

17. $M_x = \frac{99\rho}{5}$, $M_y = \frac{27\rho}{4}$, $(\bar{x}, \bar{y}) = \left(\frac{3}{2}, \frac{22}{5}\right)$

19. $M_x = \frac{192\rho}{7}$, $M_y = 96\rho$, $(\bar{x}, \bar{y}) = \left(5, \frac{10}{7}\right)$

21. $M_x = 0$, $M_y = \frac{256\rho}{15}$, $(\bar{x}, \bar{y}) = \left(\frac{8}{5}, 0\right)$

23. $M_x = \frac{27\rho}{4}$, $M_y = -\frac{27\rho}{10}$, $(\bar{x}, \bar{y}) = \left(-\frac{3}{5}, \frac{3}{2}\right)$

25. $A = \int_0^1 (x - x^2)\, dx = \frac{1}{6}$

$M_x = \int_0^1 \left(\frac{x + x^2}{2}\right)(x - x^2)\, dx = \frac{1}{15}$

$M_y = \int_0^1 x(x - x^2)\, dx = \frac{1}{12}$

27. $A = \int_0^3 (2x + 4)\, dx = 21$

$M_x = \int_0^3 \left(\frac{2x + 4}{2}\right)(2x + 4)\, dx = 78$

$M_y = \int_0^3 x(2x + 4)\, dx = 36$

29.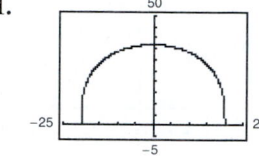

$(\bar{x}, \bar{y}) = (3.0, 126.0)$

31.

$(\bar{x}, \bar{y}) = (0, 16.2)$

33. $(\bar{x}, \bar{y}) = \left(\frac{b}{3}, \frac{c}{3}\right)$ **35.** $(\bar{x}, \bar{y}) = \left(\frac{(a + 2b)c}{3(a + b)}, \frac{a^2 + ab + b^2}{3(a + b)}\right)$

37. $(\bar{x}, \bar{y}) = \left(0, \frac{4b}{3\pi}\right)$

39. (a)

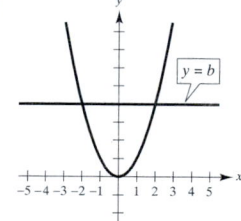

$y = b$

(b) $\bar{x} = 0$ by symmetry

(c) $M_y = \int_{-\sqrt{b}}^{\sqrt{b}} x(b - x^2)\, dx = 0$ because $y = x$ and $y = x^3$ is

an odd function.

(d) $\bar{y} > \frac{b}{2}$, because the area is greater for $y > \frac{b}{2}$.

(e) $\bar{y} = \frac{3}{5}b$

41. (a) $(\bar{x}, \bar{y}) = (0, 12.98)$

(b) $y = (-1.02 \times 10^{-5})x^4 - 0.0019x^2 + 29.28$

(c) $(\bar{x}, \bar{y}) = (0, 12.85)$

43.

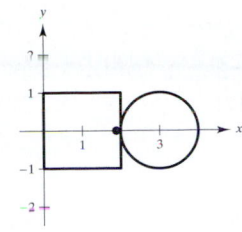

$$(\bar{x}, \bar{y}) = \left(\frac{4 + 3\pi}{4 + \pi}, 0\right)$$

45.

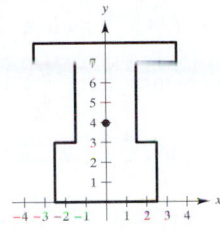

$$(\bar{x}, \bar{y}) = \left(0, \frac{135}{34}\right)$$

47. $(\bar{x}, \bar{y}) = \left(\dfrac{2 + 3\pi}{2 + \pi}, 0\right)$ **49.** $160\pi^2 \approx 1579.14$

51. $\dfrac{128\pi}{3} \approx 134.04$ **53.** $(\bar{x}, \bar{y}) = \left(0, \dfrac{2r}{\pi}\right)$

55. $(\bar{x}, \bar{y}) = \left(\dfrac{n + 1}{n + 2}, \dfrac{n + 1}{4n + 2}\right)$

As $n \to \infty$, $(\bar{x}, \bar{y}) \to \left(1, \dfrac{1}{4}\right)$

Section 6.7 (page 469)

1. 936 pounds   **3.** 748.8 pounds   **5.** 1123.2 pounds

7. 748.8 pounds   **9.** 1064.96 pounds

11. 12,000 kilograms   **13.** 243,000 kilograms

15. 2814 pounds   **17.** 6753.6 pounds   **19.** 94.5 pounds

21. Proof   **23.** 960 pounds

25. 3010.8 pounds   **27.** 6448.7 pounds

29. (a) $\dfrac{3\sqrt{2}}{2} \approx 2.12$ feet

    (b) The pressure increases with increasing depth.

Review Exercises for Chapter 6 (page 471)

1. $\dfrac{4}{5}$ **3.** $\dfrac{\pi}{2}$

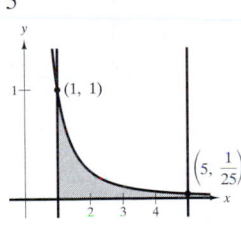
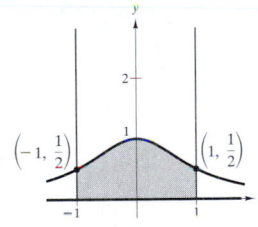

5. $\dfrac{1}{2}$ **7.** $e^2 + 1$

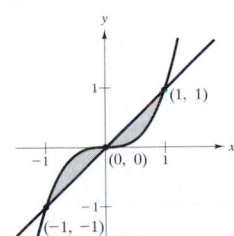
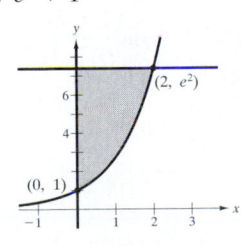

9. $2\sqrt{2}$

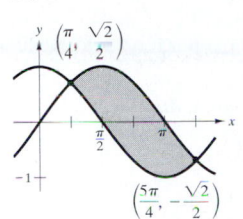

11. $\dfrac{512}{3}$

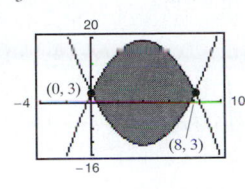

13. $\dfrac{1}{6}$

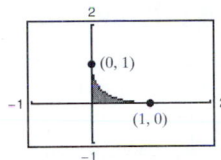

15. $\displaystyle\int_0^2 [0 - (y^2 - 2y)]\, dy = \int_{-1}^0 2\sqrt{x + 1}\, dx = \dfrac{4}{3}$

17. $\displaystyle\int_0^2 \left[1 - \left(1 - \dfrac{x}{2}\right)\right] dx + \int_2^3 [1 - (x - 2)]\, dx =$

$\displaystyle\int_0^1 [(y + 2) - (2 - 2y)]\, dy = \dfrac{3}{2}$

19. Job 1. The salary for job 1 is greater than the salary for job 2 for all the years except the first and tenth years.

21. (a) $\dfrac{64\pi}{3}$ (b) $\dfrac{128\pi}{3}$ (c) $\dfrac{64\pi}{3}$ (d) $\dfrac{160\pi}{3}$

23. (a) 64π (b) 48π **25.** $\dfrac{\pi^2}{4}$

27. $\dfrac{4\pi}{3}(20 - 9\ln 3) \approx 42.359$ **29.** $\dfrac{4}{15}$ **31.** 1.958 feet

33. $\dfrac{8}{15}\left(1 + 6\sqrt{3}\right) \approx 6.076$ **35.** 4018.2 feet   **37.** 15π

39. 50 inch-pounds ≈ 4.167 foot-pounds

41. $104{,}000\pi$ foot-pounds ≈ 163.4 foot-tons

43. 250 foot-pounds   **45.** $a = \dfrac{15}{4}$ **47.** $(\bar{x}, \bar{y}) = \left(\dfrac{a}{5}, \dfrac{a}{5}\right)$

49. $(\bar{x}, \bar{y}) = \left(0, \dfrac{2a^2}{5}\right)$ **51.** $(\bar{x}, \bar{y}) = \left(\dfrac{2(9\pi + 49)}{3(\pi + 9)}, 0\right)$

53. 72,800 pounds (on side walls)
    62,400 pounds (on wall at deep end)
    15,600 pounds (on wall at shallow end)

55. 4992π pounds

CHAPTER 7

Section 7.1 (page 479)

1. b   **3.** c

5. $\displaystyle\int u^n\, du$

$u = 3x - 2,\, n = 4$

7. $\displaystyle\int \frac{du}{u}$

$u = 1 - 2\sqrt{x}$

9. $\displaystyle\int \frac{du}{\sqrt{a^2 - u^2}}$

$u = t,\, a = 1$

11. $\displaystyle\int \sin u\, du$

$u = t^2$

13. $\displaystyle\int e^u\, du$

$u = \sin x$

15. $-\dfrac{1}{5}(-2x + 5)^{5/2} + C$

17. $\dfrac{1}{2}v^2 - \dfrac{1}{6(3v - 1)^2} + C$ **19.** $-\dfrac{1}{3}\ln|-t^3 + 9t + 1| + C$

21. $\dfrac{1}{2}x^2 + x + \ln|x - 1| + C$ **23.** $\ln(1 + e^x) + C$

25. $\dfrac{x}{15}(12x^4 + 20x^2 + 15) + C$ **27.** $\dfrac{1}{4\pi}\sin 2\pi x^2 + C$

29. $-\dfrac{1}{\pi}\csc \pi x + C$ **31.** $\dfrac{1}{5}e^{5x} + C$ **33.** $2\ln(1 + e^x) + C$

35. $\ln|\sec x(\sec x + \tan x)| + C$

37. $\ln(t^2 + 4) - \dfrac{1}{2}\arctan \dfrac{t}{2} + C$

39. $-\dfrac{1}{2}\arcsin(2t - 1) + C$ **41.** $\dfrac{1}{2}\ln\left|\cos \dfrac{2}{t}\right| + C$

43. $3\arcsin \dfrac{x - 3}{3} + C$ **45.** $\dfrac{1}{4}\arctan \dfrac{2x + 1}{8} + C$

47. (a)

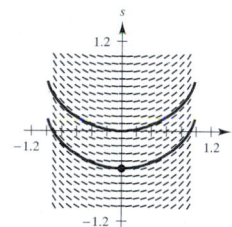

(b) $\dfrac{1}{2}\arcsin t^2 - \dfrac{1}{2}$

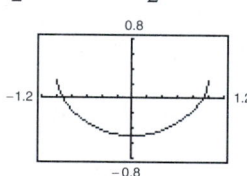

49. $y = \dfrac{1}{2}e^{2x} + 2e^x + x + C$ **51.** $y = \dfrac{1}{2}\arctan \dfrac{\tan x}{2} + C$

53. $\dfrac{1}{2}$ **55.** $\dfrac{1}{2}(1 - e^{-1}) \approx 0.316$ **57.** 4   **59.** $\dfrac{\pi}{18}$

61. $\dfrac{1}{3}\arctan\left(\dfrac{x + 2}{3}\right) + C$ **63.** $\tan\theta - \sec\theta + C$

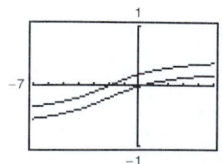

The one graph is a vertical translation of the other.

The one graph is a vertical translation of the other.

65. $a = \sqrt{2},\, b = \dfrac{\pi}{4}$

$-\dfrac{1}{\sqrt{2}}\ln\left|\csc\left(x + \dfrac{\pi}{4}\right) + \cot\left(x + \dfrac{\pi}{4}\right)\right| + C$

67. a   **69.** $\dfrac{4}{3}$ **71.** $a = \dfrac{1}{2}$

73. (a) $\pi(1 - e^{-1}) \approx 1.986$

(b) $b = \sqrt{\ln\left(\dfrac{3\pi}{3\pi - 4}\right)} \approx 0.743$

75. $\dfrac{2}{\arcsin(4/5)} \approx 2.157$ **77.** 1.0320   **79.** True

80. False: if $u = \sin x$, then $du = \cos x\, dx$.

Section 7.2 *(page 487)*

1. (a) ii   (b) iv   (c) iii   (d) i

3. $u = x,\, dv = e^{2x}\, dx$ **5.** $u = (\ln x)^2,\, dv = dx$

7. $u = x,\, dv = \sec^2 x\, dx$ **9.** $-\dfrac{1}{4e^{2x}}(2x + 1) + C$

11. $e^x(x^3 - 3x^2 + 6x - 6) + C$ **13.** $\dfrac{1}{3}e^{x^3} + C$

15. $\dfrac{1}{4}[2(t^2 - 1)\ln|t + 1| - t^2 + 2t] + C$

17. $\dfrac{(\ln x)^3}{3} + C$ **19.** $\dfrac{e^{2x}}{4(2x + 1)} + C$ **21.** $(x - 1)^2 e^x + C$

23. $\dfrac{2(x - 1)^{3/2}}{15}(3x + 2) + C$ **25.** $x\sin x + \cos x + C$

27. $x\arctan x - \dfrac{1}{2}\ln(1 + x^2) + C$

29. $\dfrac{1}{5}e^{2x}(2\sin x - \cos x) + C$ **31.** $y = \dfrac{1}{2}e^{x^2} + C$

33. $y = \dfrac{2}{405}(27t^2 - 24t + 32)\sqrt{2 + 3t} + C$

35. $\sin y = x^2 + C$

37. (a)

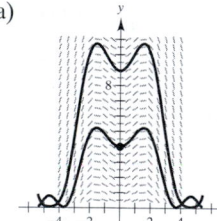

(b) $2\sqrt{y} - \cos x - x\sin x = 3$

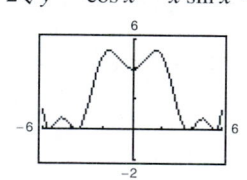

39. $-\dfrac{\pi}{2}$ **41.** $\dfrac{e[\sin(1) - \cos(1)] + 1}{2} \approx 0.909$

43. $\dfrac{\pi}{2} - 1$ **45.** $\dfrac{e^{2x}}{4}(2x^2 - 2x + 1) + C$

47. $(3x^2 - 6)\sin x - (x^3 - 6x)\cos x + C$

49. $x\tan x + \ln|\cos x| + C$

51. $-\dfrac{e^{-4t}}{128}(32t^3 + 24t^2 + 12t + 3) + C$

53. $\frac{1}{13}(2e^{-\pi} + 3) \approx 0.2374$ **55.** $\frac{2}{5}(2x - 3)^{3/2}(x + 1) + C$

57. $\frac{1}{3}\sqrt{4 + x^2}(x^2 - 8) + C$

59. $n = 0$: $x(\ln x - 1) + C$

$n = 1$: $\dfrac{x^2}{4}(2 \ln x - 1) + C$

$n = 2$: $\dfrac{x^3}{9}(3 \ln x - 1) + C$

$n = 3$: $\dfrac{x^4}{16}(4 \ln x - 1) + C$

$n = 4$: $\dfrac{x^5}{25}(5 \ln x - 1) + C$

$\displaystyle\int x^n \ln x \, dx = \dfrac{x^{n+1}}{(n + 1)^2}[(n + 1) \ln x - 1] + C$

61. Proof   **63.** Proof   **65.** Proof

67. $\dfrac{x^4}{16}(4 \ln x - 1) + C$ **69.** $\dfrac{e^{2x}}{13}(2 \cos 3x + 3 \sin 3x) + C$

71. 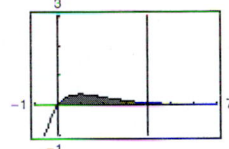   **73.**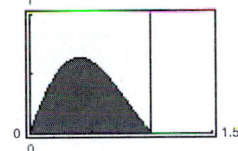

$1 - \dfrac{5}{e^4} \approx 0.908$ $\dfrac{\pi}{1 + \pi^2}\left(\dfrac{1}{e} + 1\right) \approx 0.395$

75. (a) 1   (b) $\pi(e - 2) \approx 2.257$ (c) $\dfrac{(e^2 + 1)\pi}{2} \approx 13.177$

(d) $\left(\dfrac{e^2 + 1}{4}, \dfrac{e - 2}{2}\right) \approx (2.097, 0.359)$

77. $\dfrac{7}{10\pi}(1 - e^{-4\pi}) \approx 0.223$ **79.** \$931,265   **81.** Proof

83. $b_n = \dfrac{8h}{(n\pi)^2} \sin\left(\dfrac{n\pi}{2}\right)$ **85.** Proof

87. (a) No   (b)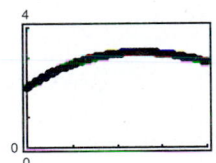

Section 7.3 *(page 496)*

1. (a) $\frac{1}{4}(3 + \cos 4x)$

(b) $2 \cos^4 x - 2 \cos^2 x + 1$

(c) $1 - 2 \sin^2 x \cos^2 x$ (d) $1 - \frac{1}{2}\sin^2 2x$

(e) Four. No: there is often more than one way to rewrite a trigonometric expression.

3. $-\frac{1}{4}\cos^4 x + C$ **5.** $\frac{1}{12}\sin^6 2x + C$

7. $-\frac{1}{3}\cos^3 x + \frac{2}{5}\cos^5 x - \frac{1}{7}\cos^7 x + C$

9. $\frac{1}{12}(6x + \sin 6x) + C$

11. $\frac{1}{8}(2x^2 - 2x \sin 2x - \cos 2x) + C$

13. Proof   **15.** Proof   **17.** $\frac{1}{3}\ln|\sec 3x + \tan 3x| + C$

19. $\frac{1}{15}\tan 5x(3 + \tan^2 5x) + C$

21. $\dfrac{1}{2\pi}\left(\sec \pi x \tan \pi x + \ln|\sec \pi x + \tan \pi x|\right) + C$

23. $\tan^4\left(\dfrac{x}{4}\right) - 2\tan^2\left(\dfrac{x}{4}\right) - 4\ln\left|\cos\dfrac{x}{4}\right| + C$

25. $\frac{1}{2}\tan^2 x + C$ **27.** $\dfrac{\tan^3 x}{3} + C$ **29.** $\dfrac{\sec^6 4x}{24} + C$

31. $\frac{1}{3}\sec^3 x + C$

33. $r = \dfrac{1}{32\pi}(12\pi\theta - 8 \sin 2\pi\theta + \sin 4\pi\theta) + C$

35. $y = \frac{1}{9}\sec^3 3x - \frac{1}{3}\sec 3x + C$

37. (a) 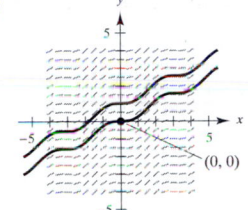   (b) $y = \frac{1}{2}x - \frac{1}{4}\sin 2x$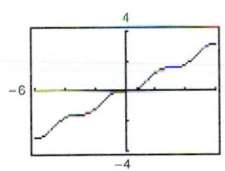

39. $-\frac{1}{10}(\cos 5x + 5 \cos x) + C$ **41.** $\frac{1}{8}(2 \sin 2\theta - \sin 4\theta) + C$

43. $\frac{1}{4}(\ln|\csc^2 2x| - \cot^2 2x) + C$ **45.** $-\cot\theta - \frac{1}{3}\cot^3\theta + C$

47. $\ln|\csc t - \cot t| + \cos t + C$

49. $\ln|\csc x - \cot x| + \cos x + C$ **51.** $t - 2\tan t + C$

53. π **55.** $\frac{1}{2}(1 - \ln 2)$ **57.** $\ln 2$ **59.** $\frac{4}{3}$

61. $\frac{1}{16}(6x + 8 \sin x + \sin 2x) + C$

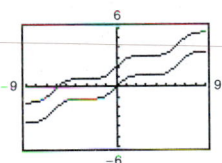

63. $\dfrac{1}{4\pi}\left[\sec^3 \pi x \tan \pi x + \right.$

$\left. \dfrac{3}{2}\left(\sec \pi x \tan \pi x + \ln|\sec \pi x + \tan \pi x|\right)\right] + C$

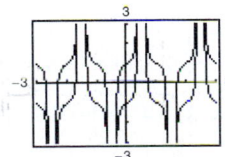

65. $\dfrac{1}{5\pi} \sec^5 \pi x + C$ **67.** $\dfrac{3\sqrt{2}}{10}$ **69.** $\dfrac{3\pi}{16}$

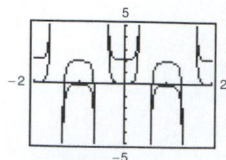

71. (a) $\dfrac{\tan^6 3x}{18} + \dfrac{\tan^4 3x}{12} + C_1, \dfrac{\sec^6 3x}{18} - \dfrac{\sec^4 3x}{12} + C_2$

(b) 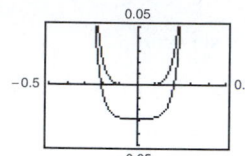   (c) Proof

73. $\dfrac{1}{2}$ **75.** (a) $\dfrac{\pi^2}{2}$ (b) $(\bar{x}, \bar{y}) = \left(\dfrac{\pi}{2}, \dfrac{\pi}{8}\right)$

77. Proof   **79.** Proof

81. $-\dfrac{1}{15} \cos x(3 \sin^4 x + 4 \sin^2 x + 8) + C$

83. $\dfrac{5}{6\pi} \tan \dfrac{2\pi x}{5} \left(\sec^2 \dfrac{2\pi x}{5} + 2\right) + C$

85. (a) Proof   (b) Proof

87. (a) $H(t) = 55.46 - 23.88 \cos \dfrac{\pi t}{6} - 3.34 \sin \dfrac{\pi t}{6}$

(b) $L(t) = 39.34 - 20.78 \cos \dfrac{\pi t}{6} - 4.33 \sin \dfrac{\pi t}{6}$

(c) Summer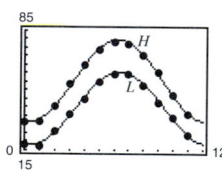

Section 7.4 (page 505)

1. b   **2.** d   **3.** a   **4.** c   **5.** $\dfrac{x}{25\sqrt{25 - x^2}} + C$

7. $5 \ln \left|\dfrac{5 - \sqrt{25 - x^2}}{x}\right| + \sqrt{25 - x^2} + C$

9. $\ln|x + \sqrt{x^2 - 4}| + C$ **11.** $\dfrac{1}{15}(x^2 - 4)^{3/2}(3x^2 + 8) + C$

13. $\dfrac{1}{3}(1 + x^2)^{3/2} + C$ **15.** $\dfrac{1}{2}\left(\arctan x + \dfrac{x}{1 + x^2}\right) + C$

17. $\dfrac{1}{2}x\sqrt{4 + 9x^2} + \dfrac{2}{3} \ln|3x + \sqrt{4 + 9x^2}| + C$

19. $\sqrt{x^2 + 9} + C$ **21.** 2π **23.** $\ln|x + \sqrt{x^2 - 9}| + C$

25. $-\dfrac{(1 - x^2)^{3/2}}{3x^3} + C$ **27.** $-\dfrac{1}{3} \ln \left|\dfrac{\sqrt{4x^2 + 9} + 3}{2x}\right| + C$

29. $-\dfrac{1}{\sqrt{x^2 + 3}} + C$ **31.** $\dfrac{1}{3}(1 + e^{2x})^{3/2} + C$

33. $\dfrac{1}{2}\left(\arcsin e^x + e^x\sqrt{1 - e^{2x}}\right) + C$

35. $\dfrac{1}{4}\left(\dfrac{x}{x^2 + 2} + \dfrac{1}{\sqrt{2}} \arctan \dfrac{x}{\sqrt{2}}\right) + C$

37. $x \operatorname{arcsec} 2x - \dfrac{1}{2} \ln|2x + \sqrt{4x^2 - 1}| + C$

39. $\arcsin\left(\dfrac{x - 2}{2}\right) + C$

41. $\sqrt{x^2 + 4x + 8} - 2 \ln|\sqrt{x^2 + 4x + 8} + (x + 2)| + C$

43. (a) and (b) $\sqrt{3} - \dfrac{\pi}{3} \approx 0.685$

45. (a) and (b) $9(2 - \sqrt{2}) \approx 5.272$

47. $\dfrac{1}{2}(x - 15)\sqrt{x^2 + 10x + 9}$
$+ 33 \ln|\sqrt{x^2 + 10x + 9} + (x + 5)| + C$

49. $\dfrac{1}{2}\left(x\sqrt{x^2 - 1} + \ln|x + \sqrt{x^2 - 1}|\right) + C$

51. πab **53.** $6\pi^2$

55. $\ln\left[\dfrac{5(\sqrt{2} + 1)}{\sqrt{26} + 1}\right] + \sqrt{26} - \sqrt{2} \approx 4.367$ **57.** Proof

59. (a)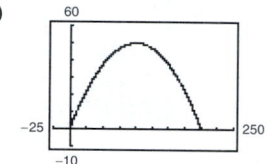

(b) 200   (c) $100\sqrt{2} + 50\ln\left(\dfrac{\sqrt{2} + 1}{\sqrt{2} - 1}\right) \approx 229.559$

61. $\dfrac{\pi}{32}[102\sqrt{2} - \ln(3 + 2\sqrt{2})] \approx 13.989$

63. $\left(\dfrac{4(3\pi + 7)}{3\pi + 4}, \dfrac{104}{5(3\pi + 4)}\right) \approx (4.89, 1.55)$

65. (a) 187.2π pounds   (b) $62.4\pi d$ pounds

67. (a) Proof

(b) $y = -12 \ln\left(\dfrac{12 - \sqrt{144 - x^2}}{x}\right) - \sqrt{144 - x^2}$

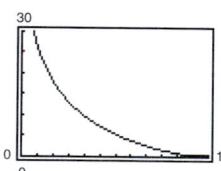

(c) $x = 0$ (d) 5.2 meters

69. True   **70.** False: $\displaystyle\int \dfrac{\sqrt{x^2 - 1}}{x} \, dx = \int \tan^2 \theta \, d\theta$

71. False: $\displaystyle\int_0^{\sqrt{3}} \dfrac{dx}{(1 + x^2)^{3/2}} = \int_0^{\pi/3} \cos \theta \, d\theta.$

72. True   **73.** Proof

Section 7.5 *(page 515)*

1. $\dfrac{A}{x} + \dfrac{B}{x-10}$　**3.** $\dfrac{A}{x} + \dfrac{Bx+C}{x^2+10}$　**5.** $\dfrac{A}{x} + \dfrac{B}{x^2} + \dfrac{C}{x-10}$

7. $\dfrac{1}{2}\ln\left|\dfrac{x-1}{x+1}\right| + C$　**9.** $\ln\left|\dfrac{x-1}{x+2}\right| + C$

11. $\dfrac{3}{2}\ln|2x-1| - 2\ln|x+1| + C$

13. $5\ln|x-2| - \ln|x+2| - 3\ln|x| + C$

15. $x^2 + \dfrac{3}{2}\ln|x-4| - \dfrac{1}{2}\ln|x+2| + C$

17. $\dfrac{1}{x} + \ln|x^4 + x^3| + C$　**19.** $\ln\left|\dfrac{x^2+1}{x}\right| + C$

21. $\dfrac{1}{6}\left[\ln\left|\dfrac{x-2}{x+2}\right| + \sqrt{2}\arctan\left(\dfrac{x}{\sqrt{2}}\right)\right] + C$

23. $\dfrac{1}{16}\ln\left|\dfrac{4x^2-1}{4x^2+1}\right| + C$

25. $\ln|x+1| + \sqrt{2}\arctan\left(\dfrac{x-1}{\sqrt{2}}\right) + C$

27. $\ln 2$　**29.** $\dfrac{1}{2}\ln\left(\dfrac{8}{5}\right) - \dfrac{\pi}{4} + \arctan 2 \approx 0.557$

31. $y = 3\ln|x-3| - \dfrac{9}{x-3} + 9$

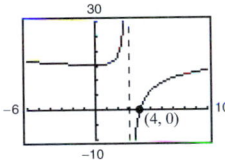

33. $y = \dfrac{\sqrt{2}}{2}\arctan\dfrac{x}{\sqrt{2}} - \dfrac{1}{2(x^2+2)} + \dfrac{5}{4}$

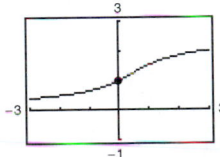

35. $y = \ln|x-2| + \dfrac{1}{2}\ln|x^2 + x + 1| -$

$\sqrt{3}\arctan\left(\dfrac{2x+1}{\sqrt{3}}\right) - \dfrac{1}{2}\ln 13 +$

$\sqrt{3}\arctan\dfrac{7}{\sqrt{3}} + 10$

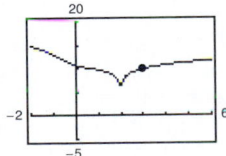

37. $y = \ln|x-1| - \arctan x +$

$\arctan 2 + 6$

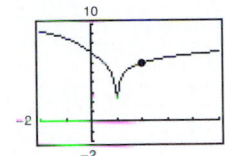

39. $\ln\left|\dfrac{\cos x}{\cos x - 1}\right| + C$

41. $\ln\left|\dfrac{-1+\sin x}{2+\sin x}\right| + C$　**43.** $\dfrac{1}{5}\ln\left|\dfrac{e^x-1}{e^x+4}\right| + C$

45. Proof　**47.** Proof　**49.** c

51. (a) $2\pi\left(\arctan 3 - \dfrac{3}{10}\right) \approx 5.963$　**53.** $x = \dfrac{n[e^{(n+1)kt}-1]}{n+e^{(n+1)kt}}$

(b) $(\bar{x}, \bar{y}) \approx (1.521, 0.412)$

55. $\dfrac{\pi}{8}$　**57.** $\dfrac{1/12}{x} + \dfrac{1/10}{x-1} + \dfrac{111/140}{x+4} + \dfrac{1/42}{x-3}$

Section 7.6 *(page 521)*

1. $-\dfrac{1}{2}x(2-x) + \ln|1+x| + C$

3. $\dfrac{1}{2}\left[e^x\sqrt{e^{2x}+1} + \ln\left(e^x + \sqrt{e^{2x}+1}\right)\right] + C$

5. $-\dfrac{\sqrt{1-x^2}}{x} + C$

7. $\dfrac{1}{16}(6x - 3\sin 2x\cos 2x - 2\sin^3 2x\cos 2x) + C$

9. $-2\left(\cot\sqrt{x} + \csc\sqrt{x}\right) + C$

11. $x - \dfrac{1}{2}\ln(1 + e^{2x}) + C$　**13.** $\dfrac{1}{16}x^4(4\ln x - 1) + C$

15. (a) and (b) $e^x(x^2 - 2x + 2) + C$

17. (a) and (b) $\ln\left|\dfrac{x+1}{x}\right| - \dfrac{1}{x} + C$　**19.** $\dfrac{1}{2}e^{x^2} + C$

21. $\dfrac{1}{2}\left\{(x^2+1)\text{arcsec}(x^2+1) - \ln\left[(x^2+1) + \sqrt{x^4+2x^2}\right]\right\} + C$

23. $\dfrac{1}{9}x^3(-1 + 3\ln x) + C$　**25.** $\dfrac{\sqrt{x^2-4}}{4x} + C$

27. $\dfrac{2}{9}\left(\ln|1-3x| + \dfrac{1}{1-3x}\right) + C$

29. $e^x\arccos(e^x) - \sqrt{1-e^{2x}} + C$

31. $\dfrac{1}{2}(x^2 + \cot x^2 + \csc x^2) + C$　**33.** $\arctan(\sin x) + C$

35. $\dfrac{\sqrt{2}}{2}\arctan\left(\dfrac{1+\sin\theta}{\sqrt{2}}\right) + C$　**37.** $-\dfrac{\sqrt{2+9x^2}}{2x} + C$

39. $(t^4 - 12t^2 + 24)\sin t + (4t^3 - 24t)\cos t + C$

41. $\dfrac{1}{4}\left(2\ln|x| - 3\ln|3 + 2\ln|x||\right) + C$

43. $\dfrac{3x-10}{2(x^2-6x+10)} + \dfrac{3}{2}\arctan(x-3) + C$

45. $\dfrac{1}{2}\ln|x^2 - 3 + \sqrt{x^4 - 6x^2 + 5}| + C$

47. $-\dfrac{1}{3}\sqrt{4-x^2}\,(x^2+8) + C$

49. $\dfrac{2}{1+e^x} - \dfrac{1}{2(1+e^x)^2} + \ln(1+e^x) + C$　**51.** Proof

53. Proof   **55.** Proof

57. $y = -\dfrac{2\sqrt{1-x}}{\sqrt{x}} + 7$

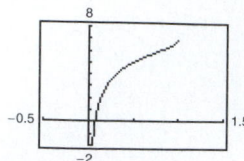

59. $y = \sqrt{2 - 2x - x^2} - \sqrt{3}\ln\left|\dfrac{\sqrt{3} + \sqrt{2 - 2x - x^2}}{x+1}\right| + \sqrt{3}\ln\left(\sqrt{3} + \sqrt{2}\right)$

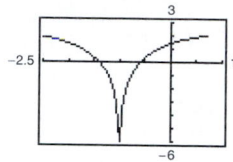

61. $y = -\csc\theta + \sqrt{2} + 2$

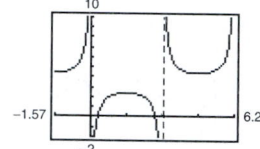

63. $\dfrac{1}{\sqrt{5}}\ln\left|\dfrac{2\tan(\theta/2) - 3 - \sqrt{5}}{2\tan(\theta/2) - 3 + \sqrt{5}}\right| + C$

65. $\ln 2$ **67.** $\frac{1}{2}\ln(3 - 2\cos\theta) + C$ **69.** $2\sin\sqrt{\theta} + C$

71. $\frac{40}{3}$ **73.** 1919.145 foot-pounds

75. (a) $V = 80\ln\left(\sqrt{10} + 3\right) \approx 145.5$ cubic feet
$W = 11{,}840\ln\left(\sqrt{10} + 3\right) \approx 21{,}530.4$ pounds

(b) $(0, 1.19)$

77. (a) $k = \dfrac{30}{\ln 7} \approx 15.42$

(b)

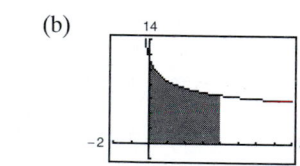

Section 7.7 *(page 530)*

1.

x	-0.1	-0.01	-0.001	0.001	0.01	0.1
$f(x)$	2.4132	2.4991	2.500	2.500	2.4991	2.4132

2.5

3.

x	1	10	10^2	10^3	10^4	10^5
$f(x)$	0.9900	90,483.7	3.7×10^9	4.5×10^{10}	0	0

0

5. $\frac{1}{3}$ **7.** $\frac{1}{4}$ **9.** $\frac{5}{3}$ **11.** 3   **13.** 0   **15.** 2

17. $n = 1: 0$
$n = 2: \frac{1}{2}$
$n \geq 3: \infty$

19. $\frac{2}{3}$ **21.** 1   **23.** $\frac{3}{2}$ **25.** ∞ **27.** 1   **29.** 0

31. (a) $0 \cdot \infty$ (b) 0   **33.** (a) $0 \cdot \infty$ (b) 1

(c) 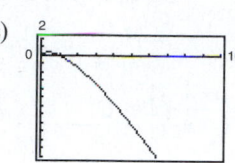   (c)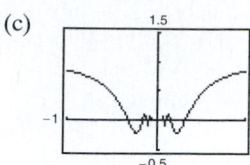

35. (a) Not indeterminate   (b) 0

(c)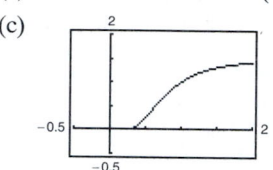

37. (a) ∞^0 (b) 1   **39.** (a) 1^∞ (b) e

(c) 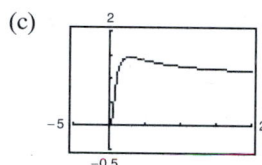   (c)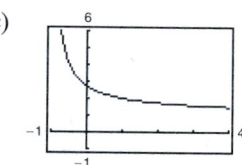

41. (a) $\infty - \infty$ (b) $-\frac{3}{2}$ **43.** (a) $\infty - \infty$ (b) ∞

(c) 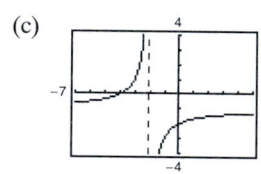   (c)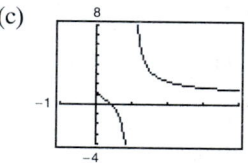

45. (a)   **47.** (a)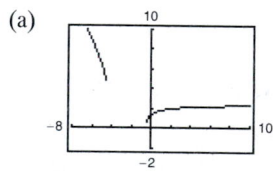

(b) $\frac{1}{2}$ (b) $\frac{5}{2}$

49. (a) $f(x) = x^2 - 25, g(x) = x - 5$
(b) $f(x) = (x - 5)^2, g(x) = x^2 - 25$
(c) $f(x) = x^2 - 25, g(x) = (x - 5)^3$

51. 0   **53.** 0   **55.** 0

57.

x	10	10^2	10^4	10^6	10^8	10^{10}
$\dfrac{(\ln x)^4}{x}$	2.811	4.498	0.720	0.036	0.001	0.000

59. Horizontal asymptote: $y = 1$

Relative maximum: $(e, e^{1/e})$

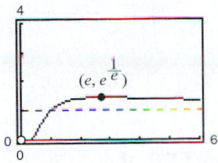

61. Horizontal asymptote: $y = 0$

Relative minimum: $\left(1, \dfrac{2}{e}\right)$

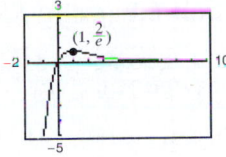

63. Limit is not of the form $0/0$ or ∞/∞.

65. Limit is not of the form $0/0$ or ∞/∞.

67. (a) 1   (b) Proof

69. $v = 32t + v_0$

(c)

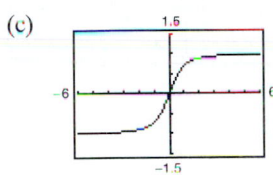

71. $\dfrac{3}{4}$ **73.** $c = \dfrac{2}{3}$ **75.** $c = \dfrac{\pi}{4}$

77. False: L'Hôpital's Rule does not apply, because $\lim\limits_{x \to 0}(x^2 + x + 1) \neq 0$.

78. False: $y' = \dfrac{e^x(x-2)}{x^3}$.   **79.** True

80. False: let $f(x) = x$ and $g(x) = x + 1$.

81. (a) $\dfrac{1}{2}\sin\theta - \dfrac{1}{2}\sin\theta\cos\theta$ (c) $\dfrac{\sin\theta - \sin\theta\cos\theta}{\theta - \sin\theta\cos\theta}$

(b) $\dfrac{1}{2}\theta - \dfrac{1}{2}\sin\theta\cos\theta$ (d) $\dfrac{3}{4}$

83. Proof   **85.** Proof

Section 7.8 (page 540)

1. Infinite discontinuity at $x = 0$; 4

3. Infinite discontinuity at $x = 1$; Diverges

5. Infinite limit of integration; 1

7. Infinite discontinuity at $x = 0$; Diverges

9. Diverges   **11.** 2   **13.** 1   **15.** $\dfrac{1}{2}$ **17.** π

19. $\dfrac{\pi}{4}$ **21.** Diverges   **23.** Diverges   **25.** 6   **27.** $-\dfrac{1}{4}$

29. Diverges   **31.** $\ln(2 + \sqrt{3})$ **33.** 0   **35.** $p > 1$

37. Proof   **39.** Diverges   **41.** Converges

43. Converges   **45.** Diverges   **47.** Converges

49. $\dfrac{1}{s}$ **51.** $\dfrac{2}{s^3}$ **53.** $\dfrac{s}{s^2 + a^2}$ **55.** $\dfrac{s}{s^2 - a^2}$

57. (a) 1   (b) $\dfrac{\pi}{2}$ (c) 2π

59.   Perimeter $= 6$

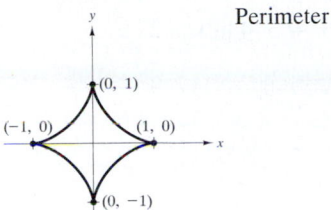

61. (a) $\Gamma(1) = 1$, $\Gamma(2) = 1$, $\Gamma(3) = 2$ (b) Proof

(c) $\Gamma(n) = (n-1)!$

63. (a) Proof   (b) $P = 43.53\%$ (c) $E(x) = 7$

65. (a) \$757,992.41   (b) \$837,995.15   (c) \$1,066,666.67

67. $\dfrac{k\left(\sqrt{a^2 + 1} - 1\right)}{a^2\sqrt{a^2 + 1}}$

69. (a) $\dfrac{1}{6}$ (b) $\dfrac{1}{24}$ (c) $\dfrac{1}{60}$ **71.** False: let $f(x) = \dfrac{1}{x + 1}$.

72. False: $\displaystyle\int_0^\infty \dfrac{1}{\sqrt{x+1}}\,dx$ diverges and $\lim\limits_{x \to \infty} \dfrac{1}{\sqrt{x+1}} = 0$.

73. True   **74.** True

Review Exercises for Chapter 7 (page 542)

1. $\dfrac{e^{2x}}{13}(2\sin 3x - 3\cos 3x) + C$

3. $\dfrac{1}{3\pi}\sin(\pi x - 1)[\cos^2(\pi x - 1) + 2] + C$

5. $\dfrac{2}{3}\left[\tan^3\left(\dfrac{x}{2}\right) + 3\tan\left(\dfrac{x}{2}\right)\right] + C$ **7.** $\dfrac{3\sqrt{4 - x^2}}{x} + C$

9. $\dfrac{1}{4}[6\ln|x - 1| - \ln(x^2 + 1) + 6\arctan x] + C$

11. $x + \dfrac{9}{8}\ln|x - 3| - \dfrac{25}{8}\ln|x + 5| + C$

13. $\tan\theta + \sec\theta + C$ **15.** $-\dfrac{1}{x}(1 + \ln 2x) + C$

17. $\dfrac{1}{2}\left(4\arcsin\dfrac{x}{2} + x\sqrt{4 - x^2}\right) + C$

19. $\dfrac{3}{2}\ln(x^2 + 1) - \dfrac{1}{2(x^2 + 1)} + C$ **21.** $16\arcsin\dfrac{x}{4} + C$

23. $\dfrac{1}{2}\ln|x^2 + 4x + 8| - \arctan\left(\dfrac{x + 2}{2}\right) + C$

25. $\dfrac{1}{8}(\sin 2\theta - 2\theta\cos 2\theta) + C$ **27.** $\dfrac{1}{2}(2\theta - \cos 2\theta) + C$

29. $2\sqrt{1 - \cos x} + C$ **31.** $\sin x \ln(\sin x) - \sin x + C$

33. $\dfrac{1}{16}\left[(8x^2 - 1)\arcsin 2x + 2x\sqrt{1 - 4x^2}\right] + C$

35. $\dfrac{4}{3}[x^{3/4} - 3x^{1/4} + 3\arctan(x^{1/4})] + C$

37. $y = 6\ln|x - 1| - \dfrac{8x - 7}{2(x - 1)^2} + \dfrac{1}{2}x^2 + 3x + \dfrac{1}{2}$

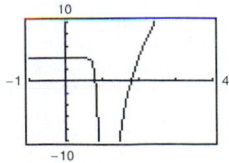

39. $y = \frac{1}{2} \arctan \frac{x}{2} + \ln(x^2 + 4) + 4 \ln|x - 2| -$

$\frac{1}{2} \arctan \frac{3}{2} - \ln 13 + 2$

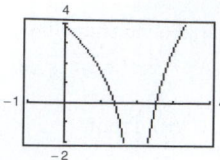

41. $y = -\frac{\sqrt{5}}{5} \ln \left| \frac{\cos \theta + \sqrt{5}(\sin \theta - 1)}{\cos \theta - \sqrt{5}(\sin \theta - 1)} \right| + 0.3$

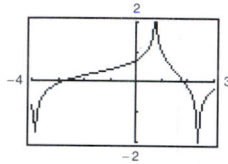

43. $y = \ln \left| \frac{1 + \sin \theta + \cos \theta}{1 + \cos \theta} \right|$ **45.** $y = \frac{1}{2} \ln(3 - 2\cos \theta)$

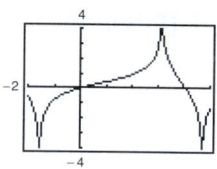   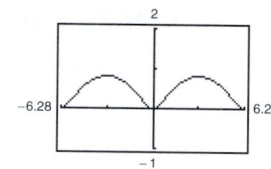

47. $y = \frac{3}{2} \ln \left| \frac{x - 3}{x + 3} \right| + C$

49. $y = x \ln |x^2 + x| - 2x + \ln |x + 1| + C$

51. (a) and (b) $-\frac{\sqrt{4 + x^2}}{4x} + C$

53. (a), (b), and (c) $\frac{1}{3} \sqrt{4 + x^2}(x^2 - 8) + C$

55. $\frac{1}{5}$ **57.** $\frac{1}{2}(\ln 4)^2 \approx 0.961$ **59.** π **61.** $\frac{128}{15}$

63. $(\bar{x}, \bar{y}) = \left(0, \frac{4}{3\pi} \right)$

65. (a) $e - 1 \approx 1.72$ (b) 1

(c) $\frac{1}{2}(e - 1) \approx 0.86$ (d) 1.46 (Simpson's Rule)

67. 3.82   **69.** 0   **71.** ∞ **73.** 1

75. $1000e^{0.09} \approx 1094.17$ **77.** Converges; $\frac{32}{3}$ **79.** Diverges

81. (a) \$6,321,205.59   (b) \$10,000,000

83. (a) 0.4581   (b) 0.0135

85. Proof

87. False: $\int \frac{\ln x^2}{x} dx = \frac{1}{2} \int (\ln x^2) \frac{2}{x} dx = \frac{1}{2} \int u \, du$.

88. True

89. False: $\int_{-1}^{1} \sqrt{x^2 - x^3} \, dx = \int_{-1}^{1} |x| \sqrt{1 - x} \, dx$.

90. True   **91.** Proof

CHAPTER 8
Section 8.1 *(page 555)*

1. 2, 4, 8, 16, 32   **3.** $-\frac{1}{2}, \frac{1}{4}, -\frac{1}{8}, \frac{1}{16}, -\frac{1}{32}$

5. $-1, -\frac{1}{4}, \frac{1}{9}, \frac{1}{16}, -\frac{1}{25}$ **7.** $3, \frac{9}{2}, \frac{27}{6}, \frac{81}{24}, \frac{243}{120}$

9. 3, 4, 6, 10, 18   **11.** 32, 16, 8, 4, 2   **13.** d

14. a   **15.** c   **16.** b

17. 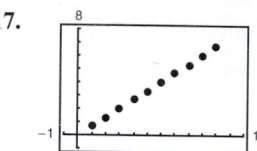   **19.**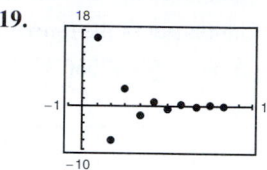

21.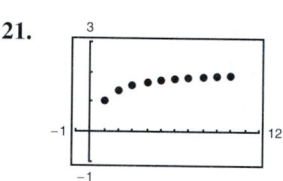

23. 14, 17; add 3 to previous term

25. $\frac{3}{16}, -\frac{3}{32}$; multiply the previous term by $-\frac{1}{2}$

27. $10 \cdot 9$ **29.** $n + 1$ **31.** $\frac{1}{(2n + 1)(2n)}$

33. $3n - 2$ **35.** $n^2 - 2$ **37.** $\frac{n + 1}{n + 2}$ **39.** $\frac{(-1)^{n-1}}{2^{n-2}}$

41. $\frac{n + 1}{n}$ **43.** $\frac{n}{(n + 1)(n + 2)}$

45. $\frac{(-1)^{n-1}}{1 \cdot 3 \cdot 5 \cdots (2n - 1)} = \frac{(-1)^{n-1}2^n n!}{(2n)!}$

47. 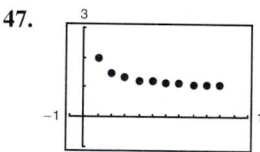   **49.**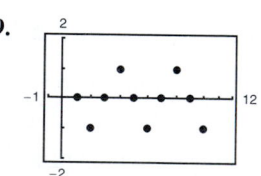

Converges to 1   Diverges

51. Diverges   **53.** Converges to $\frac{3}{2}$ **55.** Converges to 0

57. Converges to 0   **59.** Diverges   **61.** Converges to 0

63. Converges to 0   **65.** Converges to e^k

67. Monotonic, bounded   **69.** Not monotonic, bounded

71. Not monotonic, bounded   **73.** Monotonic, bounded

75. Not monotonic, bounded

77. (a) Proof

(b)

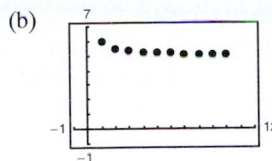

Limit = 5

79. (a) Proof

(b)

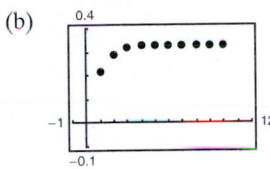

Limit = $\dfrac{1}{3}$

81. (a) $a_n = 10 - \dfrac{1}{n}$

(b) A monotonic, bounded sequence must converge (Theorem 8.5).

(c) $a_n = \dfrac{3n}{4n + 1}$

(d) An unbounded sequence does not converge.

83. (a) No

(b)

n	1	2	3	4	5
A_n	\$9086.25	\$9173.33	\$9261.24	\$9349.99	\$9439.60

n	6	7	8	9	10
A_n	\$9530.06	\$9621.39	\$9713.59	\$9806.68	\$9900.66

85. (a) $\$2,500,000,000(0.8)^n$

(b)

Year	1	2
Budget	\$2,000,000,000	\$1,600,000,000

Year	3	4
Budget	\$1,280,000,000	\$1,024,000,000

(c) Converges to 0

87. (a) $a_n = 697.32 + 59.69n$

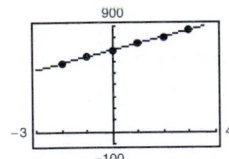

(b) \$1294

89. $S_6 = 240,\ S_7 = 440,\ S_8 = 810,\ S_9 = 1490,\ S_{10} = 2740$

91. (a) $a_9 = a_{10} = \dfrac{1,562,500}{567}$

(b) Decreasing

(c) Factorials increase more rapidly than exponentials.

93. 1, 1.4142, 1.4422, 1.4142, 1.3797, 1.3480; Converges to 1

95. (a) 1, 1, 2, 3, 5, 8, 13, 21, 34, 55, 89, 144

(b) 1, 2, 1.5, 1.6667, 1.6, 1.6250, 1.6154, 1.6190, 1.6176, 1.6182

(c) Proof

(d) $\rho = \dfrac{1 + \sqrt{5}}{2} \approx 1.6180$

97. True   **98.** True   **99.** True   **100.** True

101. 1.4142, 1.8478, 1.9616, 1.9904, 1.9976

$\displaystyle\lim_{n \to \infty} a_n = 2$

Section 8.2 *(page 564)*

1. 1, 1.25, 1.361, 1.424, 1.464

3. 3, -1.5, 5.25, -4.875, 10.3125

5. 3, 4.5, 5.25, 5.625, 5.8125

7. $\displaystyle\lim_{n \to \infty} a_n = 1 \neq 0$ **9.** $\displaystyle\lim_{n \to \infty} a_n = 1 \neq 0$

11. Geometric series: $r = \dfrac{3}{2} > 1$

13. Geometric series: $r = 1.055 > 1$

15. $\displaystyle\lim_{n \to \infty} a_n = \dfrac{1}{2} \neq 0$

17. Geometric series: $r = \dfrac{3}{4} < 1$

19. Geometric series: $r = 0.9 < 1$

21. Telescoping series: $a_n = \dfrac{1}{n} - \dfrac{1}{n + 1}$

23. c; 3   **24.** b; 3   **25.** a; 3   **26.** d; 3

27. (a) 20

(b)

n	5	10	20	50	100
S_n	8.1902	13.0264	17.5685	19.8969	19.9995

(c)

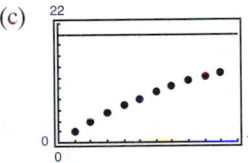

(d) The terms of the series decrease in magnitude relatively slowly, and the sequence of partial sums approaches the sum of the series relatively slowly.

29. (a) $\frac{40}{3}$

(b)

n	5	10	20	50	100
S_n	13.3203	13.3333	13.3333	13.3333	13.3333

(c)

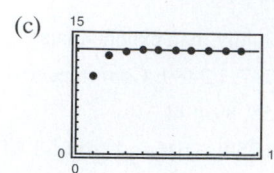

(d) The terms of the series decrease in magnitude relatively rapidly, and the sequence of partial sums approaches the sum of the series relatively rapidly.

31. 2   **33.** $\frac{2}{3}$ **35.** $\frac{10}{9}$ **37.** $\frac{9}{4}$ **39.** $\frac{3}{4}$ **41.** 3

43. $\frac{1}{2}$ **45.** $\sum_{n=0}^{\infty} \frac{4}{10}(0.1)^n = \frac{4}{9}$ **47.** $\sum_{n=0}^{\infty} \frac{3}{40}(0.01)^n = \frac{5}{66}$

49. Diverges   **51.** Converges   **53.** Diverges

55. Converges   **57.** Diverges   **59.** Diverges

61. (a) x

(b) $f(x) = \dfrac{1}{1-x}, \quad |x| < 1$

(c)

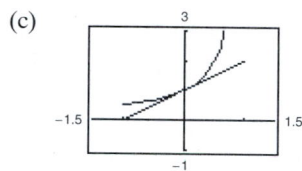

63.

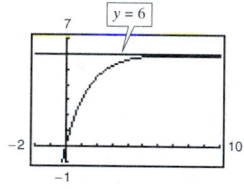

Horizontal asymptote: $y = 6$

The horizontal asymptote is the sum of the series.

65. The required terms for the two series are $n = 100$ and $n = 5$, respectively. The second series converges at a faster rate.

67. $80,000(1 - 0.9^n)$ **69.** 152.42 feet   **71.** $\frac{1}{8}$

73. (a) 126 square inches   (b) 128 square inches

75. (a) $5,368,709.11   (b) $10,737,418.23   (c) $21,474,836.47

77. (a) $16,415.10   (b) $16,421.83

79. (a) $118,196.13   (b) $118,393.43

81. Proof   **83.** Proof   **85.** $3,623,993.23

87. $\sum_{n=0}^{\infty} 1, \sum_{n=0}^{\infty}(-1)$ (Answer is not unique.)

89. False: $\sum_{n=1}^{\infty} \frac{1}{n}$ diverges.   **90.** True

91. False: the series must begin with $n = 0$ in order for the limit to be $a/(1 - r)$.

92. True   **93.** Answers will vary.

Section 8.3 (*page 571*)

1. Diverges   **3.** Converges   **5.** Converges

7. Diverges   **9.** Diverges   **11.** Converges

13. $p > 1$ **15.** Diverges   **17.** Diverges

19. Converges   **21.** Converges   **23.** a, diverges

24. d, diverges   **25.** b, converges   **26.** c, converges

27. No. For some the terms decrease toward 0 too slowly for the series to converge.

29. (a)

M	2	4	6	8
N	4	31	227	1674

(b) No. Because the magnitude of the terms of the series are approaching zero, it requires more and more terms to increase the partial sum by 2.

31. Proof

33. $R_6 \approx 0.0015$ **35.** $R_{10} \approx 0.0997$
$S_6 \approx 1.0811$ $S_{10} \approx 0.9818$

37. $R_4 \approx 5.6 \times 10^{-8}$ **39.** $N \geq 7$ **41.** $N \geq 2$
$S_4 \approx 0.4049$

43. No. Because $\sum_{n=1}^{\infty} \frac{1}{n}$ diverges, $\sum_{n=10,000}^{\infty} \frac{1}{n}$ also diverges. The convergence or divergence of a series is not determined by the first finite number of terms of the series.

45. (a) Proof

(b) $\sum_{n=2}^{\infty} \frac{1}{n^{1.1}} = 0.4665 + 0.2987 + 0.2176 +$

$0.1703 + 0.1393 + \cdots$

$\sum_{n=2}^{\infty} \frac{1}{n \ln n} = 0.7213 + 0.3034 + 0.1803 +$

$0.1243 + 0.0930 + \cdots$

(c) $n \geq 3.431 \times 10^{15}$

47. $\sum_{n=1}^{\infty} \frac{1}{n}$ **49.** Diverges   **51.** Converges

53. Converges   **55.** Diverges

57. Diverges   **59.** Converges

Section 8.4 *(page 578)*

1. (a)

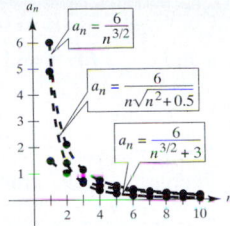

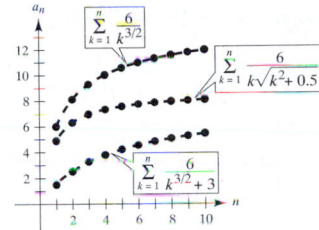

(b) $\sum_{n=1}^{\infty} \dfrac{6}{n^{3/2}}$; Converges

(c) Magnitudes of terms are less than magnitudes of terms of *p*-series. Therefore, series converges.

(d) The smaller the magnitudes of the terms, the smaller the magnitudes of the terms of the sequence of partial sums.

3. Converges   **5.** Diverges   **7.** Converges

9. Diverges   **11.** Converges   **13.** Converges

15. Diverges   **17.** Diverges   **19** Converges

21. Diverges   **23.** Converges   **25.** Diverges

27. Diverges   **29.** Diverges; *p*-Series Test

31. Converges; Direct Comparison Test with $\sum_{n=1}^{\infty} \left(\dfrac{1}{3}\right)^n$

33. Diverges; *n*th-Term Test   **35.** Converges; Integral Test

37. Proof   **39.** Diverges   **41.** Converges   **43.** Proof

45. (a) Proof

(b)

n	5	10	20	50	100
S_n	1.1839	1.2087	1.2212	1.2287	1.2312

(c) $0.1226,\ \dfrac{\pi^2}{8} - \dfrac{1}{(2 \cdot 1 - 1)^2} - \dfrac{1}{(2 \cdot 2 - 1)^2}$

(d) 0.0277

47. False: let $a_n = \dfrac{1}{n^3}$ and $b_n = \dfrac{1}{n^2}$.   **48.** True   **49.** True

50. False: let $a_n = \dfrac{1}{n}$, $b_n = \dfrac{1}{n^2}$, and $c_n = \dfrac{10}{n}$.   **51.** Proof

53. $\sum_{n=1}^{\infty} \dfrac{1}{n^2}$, $\sum_{n=1}^{\infty} \dfrac{1}{n^3}$ **55.** (a) Proof   (b) Proof

57.

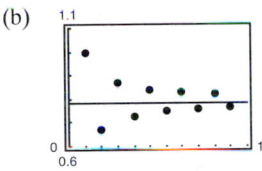

Because $0 < a_n < 1$, $0 < a_n^2 < a_n < 1$.

Section 8.5 *(page 586)*

1. b   **2.** d   **3.** c   **4.** a

5. (a)

n	1	2	3	4	5
S_n	1.0000	0.6667	0.8667	0.7238	0.8349

n	6	7	8	9	10
S_n	0.7440	0.8209	0.7543	0.8131	0.7605

(b)

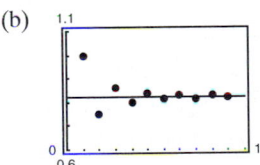

(c) The points alternate sides of the horizontal line $y = \pi/4$ that represents the sum of the series. The distance between the successive points and the line decreases.

(d) The distance in part (c) is always less than the magnitude of the next term of the series.

7. (a)

n	1	2	3	4	5
S_n	1.0000	0.7500	0.8611	0.7986	0.8386

n	6	7	8	9	10
S_n	0.8108	0.8312	0.8156	0.8280	0.8180

(b)

(c) The points alternate sides of the horizontal line $y = \pi^2/12$ that represents the sum of the series. The distance between the successive points and the line decreases.

(d) The distance in part (c) is always less than the magnitude of the next term of the series.

9. Converges   **11.** Converges   **13.** Diverges

15. Converges   **17.** Diverges   **19.** Diverges

21. Converges   **23.** Converges   **25.** Converges

27. Converges

29. (a) 7 terms (Note that the sum begins with $N = 0$.)

     (b) 0.368

31. (a) 3 terms (Note that the sum begins with $N = 0$.)

     (b) 0.842

33. (a) 1000 terms   (b) 0.693

35. 7   **37.** Converges absolutely

39. Converges conditionally   **41.** Diverges

43. Converges conditionally   **45.** Converges absolutely

47. Converges absolutely   **49.** Converges conditionally

51. Converges absolutely   **53.** Proof

55. $\displaystyle\sum_{n=1}^{\infty} (-1)^n \frac{1}{n^p}$ for $0 < p \le 1$

57. (a) $-1 < x < 1$ (b) $x = -1$

59. False: let $a_n = \dfrac{(-1)^n}{n}$.   **60.** True

Section 8.6 *(page 594)*

1. Proof   **3.** Proof   **5.** d   **6.** c   **7.** a   **8.** b

9. (a) Proof

     (b)

n	5	10	15	20	25
S_n	9.2104	16.7598	18.8016	19.1878	19.2491

     (c)

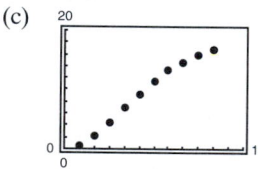

     (d) 19.26

     (e) The more rapidly the terms of the series approach 0, the more rapidly the sequence of partial sums approaches the sum of the series.

11. Diverges   **13.** Converges   **15.** Converges

17. Diverges   **19.** Converges   **21.** Diverges

23. Converges   **25.** Converges   **27.** Diverges

29. Converges   **31.** Proof   **33.** Converges

35. Converges   **37.** Diverges   **39.** Converges

41. Converges; Alternating Series Test

43. Converges; p-Series Test   **45.** Diverges; nth-Term Test

47. Diverges; Ratio Test

49. Converges; Limit Comparison Test with $b_n = \dfrac{1}{2^n}$

51. Converges; Direct Comparison Test with $b_n = \dfrac{1}{2^n}$

53. Converges; Ratio Test   **55.** Converges; Ratio Test

57. Converges; Ratio Test   **59.** a and c   **61.** a and b

63. $\displaystyle\sum_{n=0}^{\infty} \frac{n+1}{4^{n+1}}$ **65.** (a) 9   (b) -0.7769

67. No: the series $\displaystyle\sum_{n=1}^{\infty} \frac{1}{n + 10,000}$ diverges.   **69.** Proof

Section 8.7 *(page 604)*

1. d   **2.** c   **3.** a   **4.** b

5. (a)

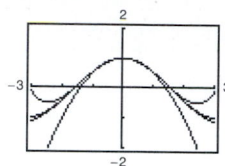

     (b) $f^{(2)}(0) = -1$ $P_2^{(2)}(0) = -1$

        $f^{(4)}(0) = 1$ $P_4^{(4)}(0) = 1$

        $f^{(6)}(0) = -1$ $P_6^{(6)}(0) = -1$

     (c) $f^{(n)}(0) = P_n^{(n)}(0)$

7. $1 - x + \frac{1}{2}x^2 - \frac{1}{6}x^3$ **9.** $1 + 2x + 2x^2 + \frac{4}{3}x^3 + \frac{2}{3}x^4$

11. $x - \frac{1}{6}x^3 + \frac{1}{120}x^5$ **13.** $x + x^2 + \frac{1}{2}x^3 + \frac{1}{6}x^4$

15. $1 - x + x^2 - x^3 + x^4$

17. $1 - (x-1) + (x-1)^2 - (x-1)^3 + (x-1)^4$

19. $(x-1) - \frac{1}{2}(x-1)^2 + \frac{1}{3}(x-1)^3 - \frac{1}{4}(x-1)^4$

21. (a) $P_3(x) = x + \dfrac{1}{3}x^3$ (b) $P_5(x) = x + \dfrac{1}{3}x^3 + \dfrac{2}{15}x^5$

     (c) $Q_3(x) = 1 + 2\left(x - \dfrac{\pi}{4}\right) + 2\left(x - \dfrac{\pi}{4}\right)^2 + \dfrac{8}{3}\left(x - \dfrac{\pi}{4}\right)^3$

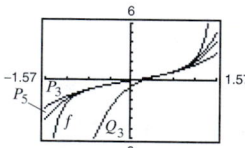

23. (a)

x	0	0.25	0.50	0.75	1.00
$\sin x$	0	0.2474	0.4794	0.6816	0.8415
$P_1(x)$	0	0.25	0.50	0.75	1.00
$P_3(x)$	0	0.2474	0.4792	0.6797	0.8333
$P_5(x)$	0	0.2474	0.4794	0.6817	0.8417
$P_7(x)$	0	0.2474	0.4794	0.6816	0.8415

(b)

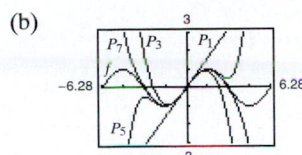

(c) As the distance increases, the polynomial approximation becomes less accurate.

25. (a) $P_3(x) = x + \frac{1}{6}x^3$

(b)

x	-0.75	-0.50	-0.25	0	0.25
$f(x)$	-0.848	-0.524	-0.253	0	0.253
$P_3(x)$	-0.820	-0.521	-0.253	0	0.253

x	0.50	0.75
$f(x)$	0.524	0.848
$P_3(x)$	0.521	0.820

(c)

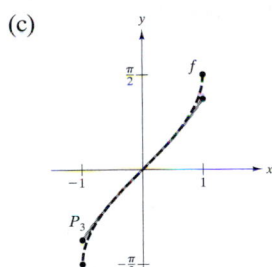

27.

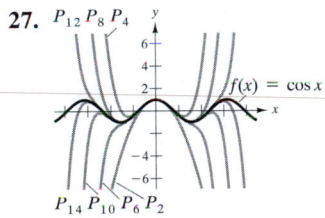

29. 0.6042   **31.** 0.1823   **33.** $R_4 \le 2.03 \times 10^{-5}$

35. $R_3 \le 7.82 \times 10^{-3}$ **37.** 3   **39.** 9; 0.4055

41. $-0.3936 < x < 0$

43. (a) $f(x) \approx P_4(x) = 1 + x + \frac{1}{2}x^2 + \frac{1}{6}x^3 + \frac{1}{24}x^4$

$g(x) \approx Q_5(x) = x + x^2 + \frac{1}{2}x^3 + \frac{1}{6}x^4 + \frac{1}{24}x^5$

$Q_5(x) = x\,P_4(x)$

(b) $g(x) \approx P_6(x) = x^2 - \frac{x^4}{3!} + \frac{x^6}{5!}$

(c) $g(x) \approx P_4(x) = 1 - \frac{x^2}{3!} + \frac{x^4}{5!}$

45. Proof   **47.** Proof

Section 8.8 *(page 613)*

1. $R = 1$ **3.** $R = \frac{1}{2}$ **5.** $R = \infty$ **7.** $(-2, 2)$

9. $(-1, 1]$ **11.** $(-\infty, \infty)$ **13.** $x = 0$ **15.** $(-4, 4)$

17. $(0, 10]$ **19.** $(0, 2]$ **21.** $(0, 2c)$ **23.** $\left(-\frac{1}{2}, \frac{1}{2}\right)$

25. $(-\infty, \infty)$ **27.** $(-1, 1)$ **29.** $x = 3$

31. (a) $(-2, 2)$ (b) $(-2, 2)$ (c) $(-2, 2)$ (d) $[-2, 2)$

33. (a) $(0, 2]$ (b) $(0, 2)$ (c) $(0, 2)$ (d) $[0, 2]$

35. c   **36.** a   **37.** b   **38.** d

39. (a) For $f(x)$: $(-\infty, \infty)$; For $g(x)$: $(-\infty, \infty)$

(b) Proof   (c) Proof

(d) $f(x) = \sin x$
$g(x) = \cos x$

41. Proof

43. (a) Proof   (b) Proof

(c)

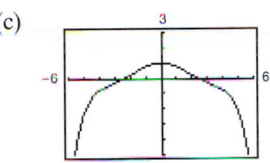

(d) 0.92

45. $f(x) = \cos x$

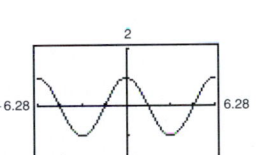

47. $f(x) = \dfrac{1}{1 + x}$

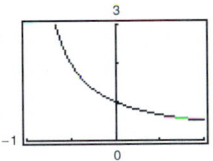

49. (a) $\frac{8}{5}$ (b) $\frac{8}{11}$

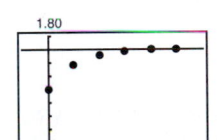

(c) The alternating series converges more rapidly. The partial sums of the series of positive terms approach the sum from below. The partial sums of the alternating series alternate sides of the horizontal line representing the sum.

(d)

M	10	100	1000	10,000
N	4	9	15	21

51. False: let $a_n = \dfrac{(-1)^n}{n2^n}$.   **52.** True

53. True   **54.** True

Section 8.9 *(page 620)*

1. $\displaystyle\sum_{n=0}^{\infty} \frac{x^n}{2^{n+1}}$ **3.** $\displaystyle\sum_{n=0}^{\infty} \frac{(-1)^n x^n}{2^{n+1}}$

5. $\displaystyle\sum_{n=0}^{\infty} \frac{(x-5)^n}{(-3)^{n+1}}$ **7.** $-3 \displaystyle\sum_{n=0}^{\infty} (2x)^n$

$(2, 8)$ $\left(-\frac{1}{2}, \frac{1}{2}\right)$

9. $-\dfrac{1}{11} \displaystyle\sum_{n=0}^{\infty} \left[\frac{2}{11}(x+3)\right]^n$ **11.** $\dfrac{3}{2} \displaystyle\sum_{n=0}^{\infty} \left(\frac{x}{-2}\right)^n$

$\left(-\dfrac{17}{2}, \dfrac{5}{2}\right)$ $(-2, 2)$

13. $\displaystyle\sum_{n=0}^{\infty} \left[\frac{1}{(-2)^n} - 1\right] x^n$ **15.** $2 \displaystyle\sum_{n=0}^{\infty} x^{2n}$

$(-1, 1)$ $(-1, 1)$

17. $2 \displaystyle\sum_{n=0}^{\infty} x^{2n}$ **19.** $\displaystyle\sum_{n=1}^{\infty} n(-1)^n x^{n-1}$

$(-1, 1)$ $(-1, 1)$

21. $\displaystyle\sum_{n=0}^{\infty} \frac{(-1)^n x^{n+1}}{n+1}$ **23.** $\displaystyle\sum_{n=0}^{\infty} (-1)^n x^{2n}$

$(-1, 1]$ $(-1, 1)$

25. $\displaystyle\sum_{n=0}^{\infty} (-1)^n (2x)^{2n}$

$\left(-\dfrac{1}{2}, \dfrac{1}{2}\right)$

27.

x	0.0	0.2	0.4	0.6	0.8	1.0
S_2	0.000	0.180	0.320	0.420	0.480	0.500
$\ln(x+1)$	0.000	0.182	0.336	0.470	0.588	0.693
S_3	0.000	0.183	0.341	0.492	0.651	0.833

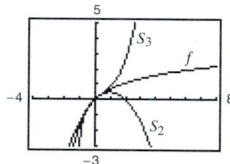

29. c **30.** d **31.** a **32.** b

33. $f(x) = \arctan x$ is an odd function (symmetric to the origin).

35. 0.245 **37.** 0.125

39. $\displaystyle\sum_{n=0}^{\infty} nx^{n-1}$

$(-1, 1)$

41. $E(n) = 2$. Because the probability of obtaining a head on a single toss is $\frac{1}{2}$, it is expected that, on average, a head will be obtained in two tosses.

43. Proof **45.** (a) Proof (b) 3.14 **47.** $\ln\frac{3}{2} \approx 0.4055$

49. $\ln\frac{7}{5} \approx 0.3365$ **51.** $\arctan\frac{1}{2} \approx 0.4636$

53. The series in Exercise 50 converges to its sum at a slower rate because its terms approach 0 at a much slower rate.

55. -0.6931

Section 8.10 *(page 630)*

1. $\displaystyle\sum_{n=0}^{\infty} \frac{(2x)^n}{n!}$ **3.** $\dfrac{\sqrt{2}}{2} \displaystyle\sum_{n=0}^{\infty} \frac{(-1)^{n(n+1)/2}}{n!} \left(x - \frac{\pi}{4}\right)^n$

5. $\displaystyle\sum_{n=0}^{\infty} \frac{(-1)^n(x-1)^{n+1}}{n+1}$ **7.** $\displaystyle\sum_{n=0}^{\infty} \frac{(-1)^n(2x)^{2n+1}}{(2n+1)!}$

9. $1 + \dfrac{x^2}{2!} + \dfrac{5x^4}{4!} + \cdots$ **11.** $\displaystyle\sum_{n=0}^{\infty} (-1)^n(n+1)x^n$

13. $\dfrac{1}{2}\left[1 + \displaystyle\sum_{n=1}^{\infty} \frac{(-1)^n \, 1 \cdot 3 \cdot 5 \cdots (2n-1)x^{2n}}{2^{3n}n!}\right]$

15. $1 + \dfrac{x^2}{2} + \displaystyle\sum_{n=2}^{\infty} \frac{(-1)^{n+1}1 \cdot 3 \cdot 5 \cdots (2n-3)x^{2n}}{2^n n!}$

17. $1 + \dfrac{x^2}{2} + \dfrac{x^4}{2^2 2!} + \dfrac{x^6}{2^3 3!} + \cdots$ **19.** $\displaystyle\sum_{n=0}^{\infty} \frac{(-1)^n(2x)^{2n+1}}{(2n+1)!}$

21. $\displaystyle\sum_{n=0}^{\infty} \frac{(-1)^n x^{3n}}{(2n)!}$ **23.** $\displaystyle\sum_{n=0}^{\infty} \frac{(-1)^n x^{2n}}{(2n+1)!}$ **25.** $\displaystyle\sum_{n=0}^{\infty} \frac{x^{2n+1}}{(2n+1)!}$

27. $\dfrac{1}{2}\left[1 + \displaystyle\sum_{n=0}^{\infty} \frac{(-1)^n(2x)^{2n}}{(2n)!}\right]$ **29.** Proof

31. $P_5(x) = x + x^2 + \dfrac{1}{3}x^3 - \dfrac{1}{30}x^5 + \cdots$

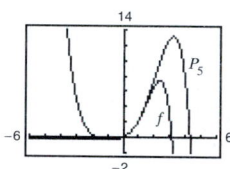

33. $P_5(x) = x - \dfrac{1}{2}x^2 - \dfrac{1}{6}x^3 + \dfrac{3}{40}x^5 + \cdots$

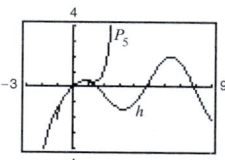

35. $P_4(x) = x - x^2 + \dfrac{5}{6}x^3 - \dfrac{5}{6}x^4 + \cdots$

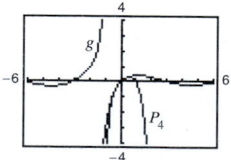

37. a; $y \approx x \sin x$ **38.** b; $y \approx x \cos x$

39. c; $y \approx xe^x$ **40.** d; $y \approx x^2\left(\dfrac{1}{x-1}\right)$

41. $\displaystyle\sum_{n=0}^{\infty} \frac{(-1)^{(n+1)}x^{2n+3}}{(2n+3)(n+1)!}$ **43.** 0.6931 **45.** 7.3891

47. 0   **49.** 0.9461   **51.** 0.7040

53. 0.2010   **55.** 0.3413

57. $P_5(x) = x - 2x^3 + \frac{2}{3}x^5$

$\left[-\frac{3}{4}, \frac{3}{4}\right]$

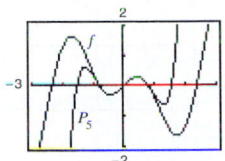

59. $P_5(x) = (x - 1) - \frac{1}{24}(x - 1)^3 + \frac{1}{24}(x - 1)^4 - \frac{71}{1920}(x - 1)^5$

$\left[\frac{1}{4}, 2\right]$

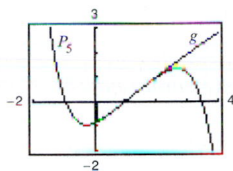

61. Proof

63. (a)

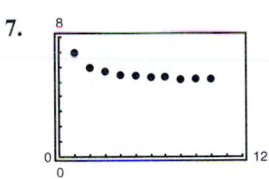

(b) Proof   (c) $\displaystyle\sum_{n=0}^{\infty} 0x^n = 0 \neq f(x)$

65. Proof

Review Exercises for Chapter 8 *(page 632)*

1. $a_n = \dfrac{1}{n!}$ **3.** a   **4.** c   **5.** d   **6.** b

7.

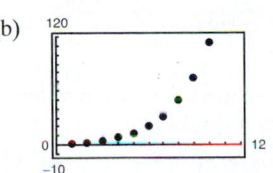

Converges to 5

9. Converges to 0   **11.** Diverges

13. Converges to 0   **15.** Converges to 0

17. (a)

n	1	2	3	4	5
A_n	\$5062.50	\$5125.78	\$5189.85	\$5254.73	\$5320.41

n	6	7	8
A_n	\$5386.92	\$5454.25	\$5522.43

(b) \$8218.10

19. (a)

k	5	10	15	20	25
S_k	13.2	113.3	873.8	6648.5	50,500.3

(b)

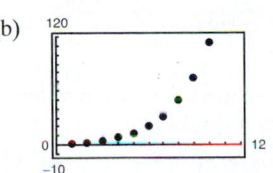

(c) Diverges

21. (a)

n	5	10	15	20	25
S_n	0.4597	0.4597	0.4597	0.4597	0.4597

(b)

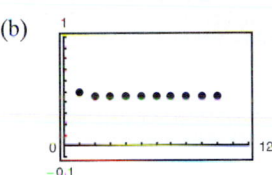

(c) Converges

23. 3   **25.** $\frac{1}{2}$ **27.** $\frac{1}{11}$ **29.** $45\frac{1}{3}$ meters   **31.** \$5087.14

33. (a) Proof

(b)

n	5	10	15	20	25
S_n	2.8752	3.6366	3.7377	3.7488	3.7499

(c)

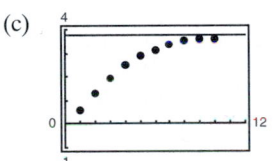

(d) 3.75

35. Diverges   **37.** Converges   **39.** Converges

41. Diverges   **43.** Diverges   **45.** Diverges

47. (a)

N	5	10	20	30	40
$\displaystyle\sum_{n=1}^{N} \frac{1}{n^p}$	1.4636	1.5498	1.5962	1.6122	1.6202
$\displaystyle\int_{N}^{\infty} \frac{1}{x^p}\,dx$	0.2000	0.1000	0.0500	0.0333	0.0250

(b)

N	5	10	20	30	40
$\displaystyle\sum_{n=1}^{N} \frac{1}{n^p}$	1.0367	1.0369	1.0369	1.0369	1.0369
$\displaystyle\int_{N}^{\infty} \frac{1}{x^p}\,dx$	0.0004	0.0000	0.0000	0.0000	0.0000

The series in part (b) converges more rapidly. This is evident from the integrals that give the remainders of the partial sums.

49. $(-10, 10)$ **51.** $[1, 3]$ **53.** Converges only at $x = 2$

55. $\dfrac{\sqrt{2}}{2} \displaystyle\sum_{n=0}^{\infty} \dfrac{(-1)^{n(n+1)/2}}{n!}\left(x - \dfrac{3\pi}{4}\right)^n$ **57.** $\displaystyle\sum_{n=0}^{\infty} \dfrac{(x \ln 3)^n}{n!}$

59. $-\displaystyle\sum_{n=0}^{\infty}(x+1)^n$ **61.** $\displaystyle\sum_{n=0}^{\infty} \dfrac{2}{3}\left(\dfrac{x}{3}\right)^n$

63. $\ln \dfrac{5}{4} \approx 0.2231$ **65.** $e^{1/2} \approx 1.6487$

67. $\cos \dfrac{2}{3} \approx 0.7859$

69. The series for Exercise 37 converges to its sum at a slower rate because its terms approach 0 at a slower rate.

71. $1 + 2x + 2x^2 + \dfrac{4}{3}x^3$ **73.** $f(x) = \dfrac{3}{3-2x};\ \left(-\dfrac{3}{2}, \dfrac{3}{2}\right)$

75. $\displaystyle\sum_{n=0}^{\infty} \dfrac{(-1)^n x^{2n+1}}{(2n+1)(2n+1)!}$ **77.** $\displaystyle\sum_{n=0}^{\infty} \dfrac{(-1)^n x^{n+1}}{(n+1)^2}$

79. 0   **81.** Proof   **83.** 0.996   **85.** 0.560

87. (a) 4   (b) 6   (c) 5   (d) 10

CHAPTER 9

Section 9.1 *(page 647)*

1. h   **2.** a   **3.** e   **4.** b

5. f   **6.** g   **7.** c   **8.** d

9. Vertex: $(0, 0)$

Focus: $\left(-\dfrac{3}{2}, 0\right)$

Directrix: $x = \dfrac{3}{2}$

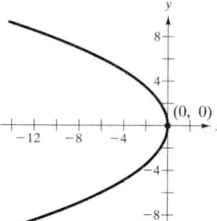

11. Vertex: $(-3, 2)$

Focus: $\left(-\dfrac{13}{4}, 2\right)$

Directrix: $x = -\dfrac{11}{4}$

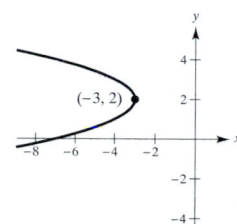

13. Vertex: $(-1, 2)$

Focus: $(0, 2)$

Directrix: $x = -2$

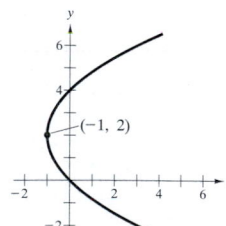

15. Vertex: $(-2, 2)$

Focus: $(-2, 1)$

Directrix: $y = 3$

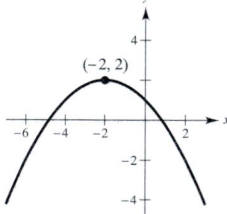

17. Vertex; $\left(\dfrac{1}{4}, -\dfrac{1}{2}\right)$

Focus: $\left(0, -\dfrac{1}{2}\right)$

Directrix: $x = \dfrac{1}{2}$

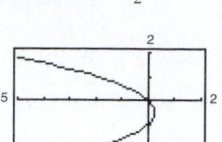

19. Vertex: $(-1, 0)$

Focus: $(0, 0)$

Directrix: $x = -2$

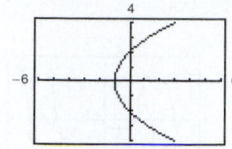

21. $y^2 - 4y + 8x - 20 = 0$ **23.** $x^2 - 24y + 96 = 0$

25. $x^2 + y - 4 = 0$ **27.** $5x^2 - 14x - 3y + 9 = 0$

29. $\dfrac{9}{4}$ meters   **31.** $y = 2ax_0 x - ax_0^2$

33. (a) Proof

(b) Tangent lines: $2x + y - 1 = 0$

$$2x - 4y - 1 = 0$$

Point of intersection: $\left(\dfrac{1}{2}, 0\right)$ (on the directrix)

35. $x_0 = \dfrac{2\sqrt{3}}{3}$; Distance from hill: $\dfrac{2\sqrt{3}}{3} - 1$

37. $\dfrac{16\left(4 + 3\sqrt{3} - 2\pi\right)}{3} \approx 15.536$ square feet

39. (a) $y = \dfrac{1}{180}x^2$

(b) $10\left[2\sqrt{13} + 9\ln\left(\dfrac{2 + \sqrt{13}}{3}\right)\right] \approx 128.4$ meters

41. Center: $(0, 0)$

Foci: $\left(\pm\sqrt{3}, 0\right)$

Vertices: $(\pm 2, 0)$

$e = \dfrac{\sqrt{3}}{2}$

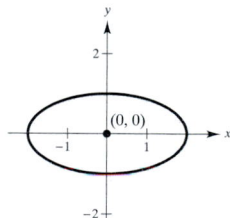

43. Center: $(1, 5)$

Foci: $(1, 9)\ (1, 1)$

Vertices: $(1, 10)\ (1, 0)$

$e = \dfrac{4}{5}$

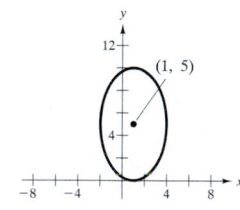

45. Center: $(-2, 3)$

Foci: $(-2, 3 \pm \sqrt{5})$

Vertices: $(-2, 6), (-2, 0)$

$e = \dfrac{\sqrt{5}}{3}$

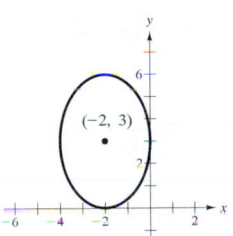

47. Center: $\left(\dfrac{1}{2}, -1\right)$

Foci: $\left(\dfrac{1}{2} \pm \sqrt{2}, -1\right)$

Vertices: $\left(\dfrac{1}{2} \pm \sqrt{5}, -1\right)$

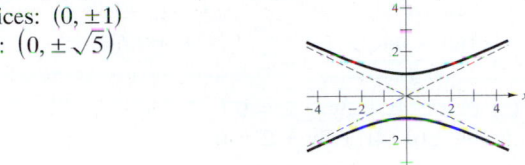

49. Center: $\left(\dfrac{3}{2}, -1\right)$

Foci: $\left(\dfrac{3}{2} - \sqrt{2}, -1\right), \left(\dfrac{3}{2} + \sqrt{2}, -1\right)$

Vertices: $\left(-\dfrac{1}{2}, -1\right), \left(\dfrac{7}{2}, -1\right)$

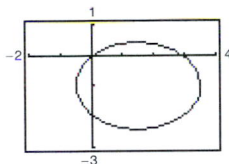

51. $\dfrac{x^2}{9} + \dfrac{y^2}{5} = 1$ **53.** $\dfrac{(x-3)^2}{9} + \dfrac{(y-5)^2}{16} = 1$

55. $\dfrac{x^2}{16} + \dfrac{7y^2}{16} = 1$

57.

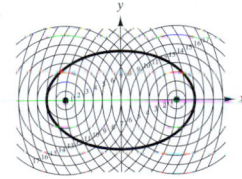

59. The tacks should be placed 1.5 feet from the center. The string should be $2a = 5$ feet long.

61. Proof   **63.** $e \approx 0.9672$

65. Minor-axis endpoints: $(-6, -2), (0, -2)$
Major-axis endpoints: $(-3, -6), (-3, 2)$

67. $\left(0, \dfrac{25}{3}\right)$

69. (a) Area $= 2\pi$

(b) Volume $= \dfrac{8\pi}{3}$

Surface area $= \dfrac{2\pi\left(9 + 4\sqrt{3}\pi\right)}{9} \approx 21.48$

(c) Volume $= \dfrac{16\pi}{3}$

Surface area $= \dfrac{4\pi\left[6 + \sqrt{3}\ln\left(2 + \sqrt{3}\right)\right]}{3} \approx 34.69$

71. 22.10   **73.** 40

75. Center: $(0, 0)$
Vertices: $(0, \pm 1)$
Foci: $(0, \pm\sqrt{5})$

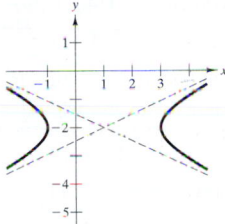

77. Center: $(1, -2)$
Vertices: $(-1, -2), (3, -2)$

Foci: $(1 \pm \sqrt{5}, -2)$

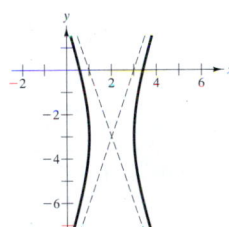

79. Center: $(2, -3)$
Vertices: $(1, -3), (3, -3)$
Foci: $(2 \pm \sqrt{10}, -3)$

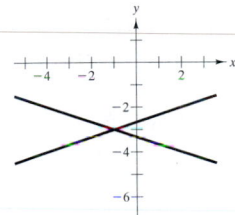

81. Degenerate hyperbola
Graph is two lines intersecting at $(-1, -3)$.

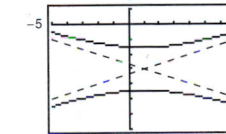

83. Center: $(1, -3)$
Vertices: $(1, -3 \pm \sqrt{2})$
Foci: $(1, -3 \pm 2\sqrt{5})$

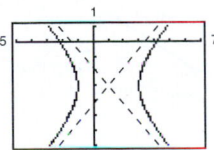

85. Center: $(1, -3)$

Vertices: $(-1, -3), (3, -3)$

Foci: $(1 \pm \sqrt{10}, -3)$

87. $\dfrac{x^2}{1} - \dfrac{y^2}{9} = 1$ **89.** $\dfrac{y^2}{9} - \dfrac{(x-2)^2}{9/4} = 1$

91. $\dfrac{y^2}{4} - \dfrac{x^2}{12} = 1$ **93.** $\dfrac{(x-3)^2}{9} - \dfrac{(y-2)^2}{4} = 1$

95. $\dfrac{(x-6)^2}{9} - \dfrac{(y-2)^2}{7} = 1$ **97.** $\left(y - \dfrac{3}{2}\right)^2 - \dfrac{(x+3)^2}{5/4} = 1$

99. Proof

101. $x = \dfrac{-90 + 96\sqrt{2}}{7} \approx 6.538$

$y = \dfrac{160 - 96\sqrt{2}}{7} \approx 3.462$

103. (a) $(4, 6)$: $4x - 3y + 2 = 0$
$(4, -6)$: $4x + 3y + 2 = 0$

(b) $(4, 6)$: $3x + 4y - 36 = 0$
$(4, -6)$: $3x - 4y - 36 = 0$

105. Proof   **106.** False: see the definition of a parabola.

107. True   **108.** True

109. False: the equation of an ellipse is second degree in x and y.

110. False: $y^2 - x^2 + 2x + 2y = 0$ yields two intersecting lines.

111. True   **112.** True   **113.** Proof

115. Hyperbola   **117.** Parabola   **119.** Parabola

121. Ellipse   **123.** Ellipse

Section 9.2 *(page 659)*

1. (a)

t	0	1	2	3	4
x	0	1	$\sqrt{2}$	$\sqrt{3}$	2
y	1	0	-1	-2	-3

(b) and (c)

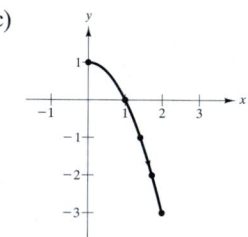

(d) $y = 1 - x^2, \quad x \geq 0$

3. $2x - 3y + 5 = 0$ **5.** $y = (x - 1)^2$

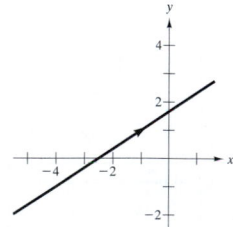   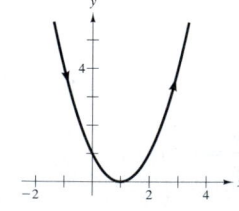

7. $y = \dfrac{1}{2}x^{2/3}$

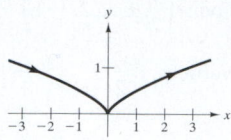

9. $y = \dfrac{x+1}{x}$

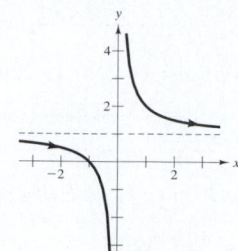

11. $y = \dfrac{|x-4|}{2}$

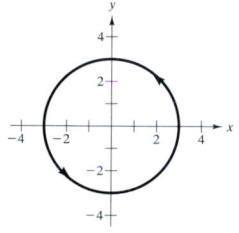

13. $y = \dfrac{1}{x}, \quad |x| \geq 1$

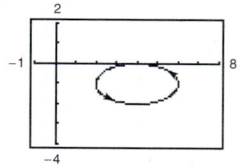

15. $x^2 + y^2 = 9$

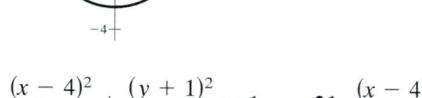

17. $\dfrac{x^2}{16} + \dfrac{y^2}{4} = 1$

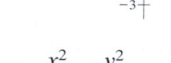

19. $\dfrac{(x-4)^2}{4} + \dfrac{(y+1)^2}{1} = 1$

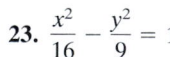

21. $\dfrac{(x-4)^2}{4} + \dfrac{(y+1)^2}{16} = 1$

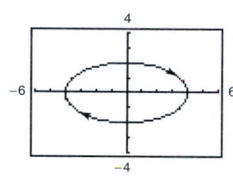

23. $\dfrac{x^2}{16} - \dfrac{y^2}{9} = 1$

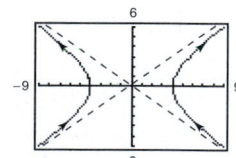

25. $y = \ln x$

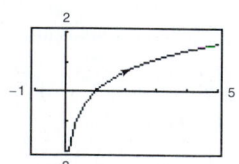

27. $y = \dfrac{1}{x^3}, \quad x > 0$

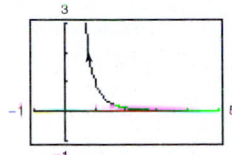

29. Each curve represents a portion of the line $y = 2x + 1$.

Domain	Orientation	Smooth
(a) $-\infty < x < \infty$	Up	Yes
(b) $-1 \le x \le 1$	Oscillates	No, $\dfrac{dx}{d\theta} = \dfrac{dy}{d\theta} = 0$ when $\theta = 0, \pi$
(c) $0 < x < \infty$	Down	Yes
(d) $0 < x < \infty$	Up	Yes

31. (a) and (b) represent the parabola $y = 2(1 - x^2)$ for $-1 \le x \le 1$. The curve is smooth. The orientation is from right to left in part (a) and from left to right in part (b).

33. (a)

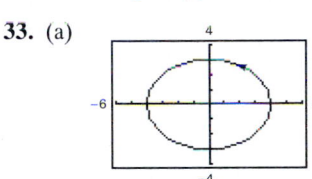 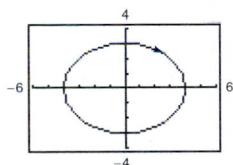

(b) The orientation is reversed.

(c) The orientation is reversed.   (d) Answers will vary.

35. $y - y_1 = \dfrac{y_2 - y_1}{x_2 - x_1}(x - x_1)$ **37.** $\dfrac{(x - h)^2}{a^2} + \dfrac{(y - k)^2}{b^2} = 1$

39. $x = 5t$
$y = -2t$
(Solution is not unique.)

41. $x = 2 + 4\cos\theta$
$y = 1 + 4\sin\theta$
(Solution is not unique.)

43. $x = 5\cos\theta$
$y = 3\sin\theta$
(Solution is not unique.)

45. $x = 4\sec\theta$
$y = 3\tan\theta$
(Solution is not unique.)

47. $x = t$
$y = 3t - 2$
$x = t - 3$
$y = 3t - 11$
(Solution is not unique.)

49. $x = t$
$y = t^3$
$x = \tan t$
$y = \tan^3 t$
(Solution is not unique.)

51.

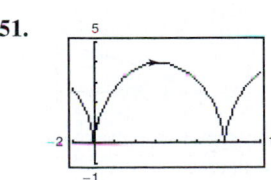

Not smooth when $\theta = 2n\pi$

53.

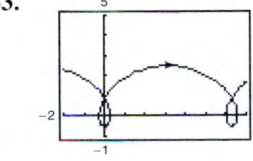

55.

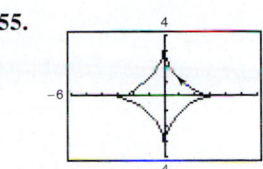

57.

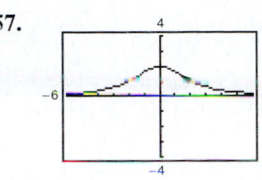

Not smooth when $\theta = \dfrac{1}{2}n\pi$

59. d   **60.** a   **61.** b   **62.** c

63. $x = a\theta - b\sin\theta$ **65.** True
$y = a - b\cos\theta$

66. False: because neither x nor y can be negative, the graph consists of only the portion of the line $y = x$ in the first quadrant.

67. False: let $x = t^2$ and $y = t$.

68. False: let $x = \sin t$ and $y = \cos t$.

69. (a) $x = \left(\dfrac{440}{3}\cos\theta\right)t \qquad y = 3 + \left(\dfrac{440}{3}\sin\theta\right)t - 16t^2$

(b)

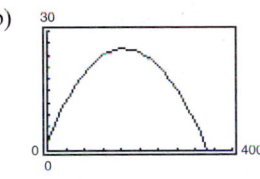

(c)

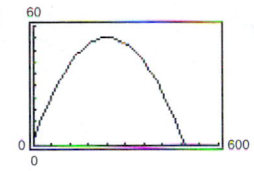

Not a home run   Home run

(d) $19.4°$

71. $t = 2: 3^2 + 4^2 = 5^2$
$t = 3: 8^2 + 6^2 = 10^2$

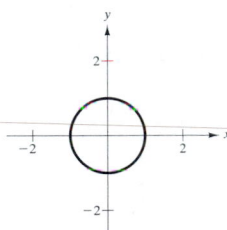

Section 9.3 *(page 668)*

1. $\dfrac{dy}{dx} = \dfrac{3}{2}, \qquad \dfrac{d^2y}{dx^2} = 0$

3. $\dfrac{dy}{dx} = 2t + 3, \qquad \dfrac{d^2y}{dx^2} = 2$

At $t = -1, \dfrac{dy}{dx} = 1, \qquad \dfrac{d^2y}{dx^2} = 2$

5. $\dfrac{dy}{dx} = -\cot\theta, \qquad \dfrac{d^2y}{dx^2} = -\dfrac{\csc^3\theta}{2}$

At $\theta = \dfrac{\pi}{4}, \dfrac{dy}{dx} = -1, \qquad \dfrac{d^2y}{dx^2} = -\sqrt{2}$

7. $\dfrac{dy}{dx} = 2\csc\theta, \qquad \dfrac{d^2y}{dx^2} = -2\cot^3\theta$

At $\theta = \dfrac{\pi}{6}, \dfrac{dy}{dx} = 4, \qquad \dfrac{d^2y}{dx^2} = -6\sqrt{3}$

9. $\dfrac{dy}{dx} = -\tan\theta,$ $\dfrac{d^2y}{dx^2} = \dfrac{\sec^4\theta\,\csc\theta}{3}$

At $\theta = \dfrac{\pi}{4}, \dfrac{dy}{dx} = -1,$ $\dfrac{d^2y}{dx^2} = \dfrac{4\sqrt{2}}{3}$

11. $\left(-\dfrac{2}{\sqrt{3}}, \dfrac{3}{2}\right), 3\sqrt{3}x - 8y + 18 = 0$

$(0, 2), y - 2 = 0$

$\left(2\sqrt{3}, \dfrac{1}{2}\right), \sqrt{3}x + 8y - 10 = 0$

13. (a) and (d)

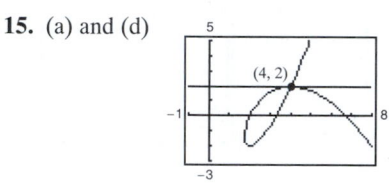

(b) At $t = 2, \dfrac{dx}{dt} = 2, \dfrac{dy}{dt} = 4,$ and $\dfrac{dy}{dx} = 2$

(c) $y = 2x - 5$

15. (a) and (d)

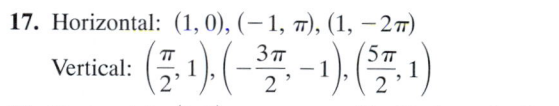

(b) At $t = -1, \dfrac{dx}{dt} = -3, \dfrac{dy}{dt} = 0,$ and $\dfrac{dy}{dx} = 0$

(c) $y = 2$

17. Horizontal: $(1, 0), (-1, \pi), (1, -2\pi)$

Vertical: $\left(\dfrac{\pi}{2}, 1\right), \left(-\dfrac{3\pi}{2}, -1\right), \left(\dfrac{5\pi}{2}, 1\right)$

19. Horizontal: $(1, 0)$ **21.** Horizontal: $(0, -2), (2, 2)$
    Vertical: none   Vertical: none

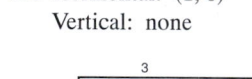 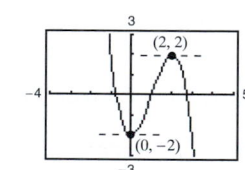

23. Horizontal: $(0, 3), (0, -3)$ **25.** Horizontal: $(4, 0), (4, -2)$
    Vertical: $(3, 0), (-3, 0)$ Vertical: $(2, -1), (6, -1)$

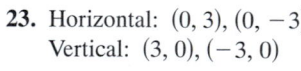

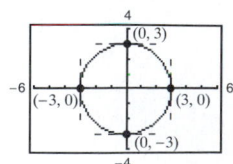 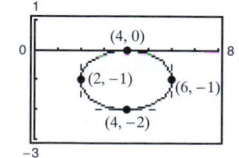

27. Horizontal: none
    Vertical: $(1, 0), (-1, 0)$

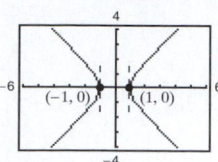

29.

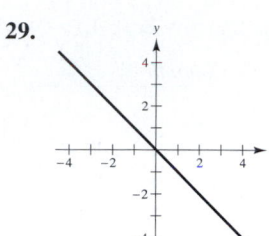

31. $\sqrt{2}(1 - e^{-\pi/2}) \approx 1.12$ **33.** $2\sqrt{5} + \ln(2 + \sqrt{5}) \approx 5.916$

35. $\dfrac{1}{12}\left[\ln(\sqrt{37} + 6) + 6\sqrt{37}\right] \approx 3.249$ **37.** $6a$ **39.** $8a$

41. (a) 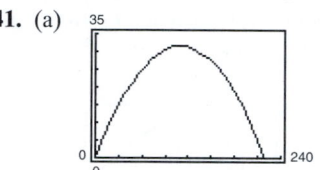   (b) 219.2 feet

                        (c) 230.8 feet

43. (a)

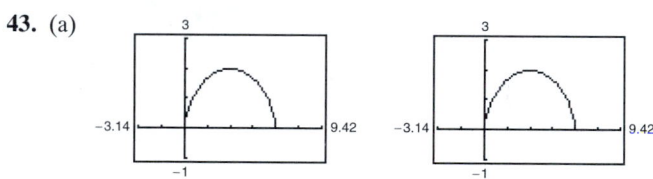

(b) The average speed of the particle on the second path is twice the average speed of the particle on the first path.

(c) 4π

45. (a) $32\pi\sqrt{5}$ (b) $16\pi\sqrt{5}$ **47.** 32π **49.** $\dfrac{12\pi a^2}{5}$

51. $2\pi r^2(1 - \cos\theta)$ **53.** $\left(\dfrac{3}{4}, \dfrac{8}{5}\right)$ **55.** 36π **57.** $\dfrac{3\pi}{2}$

59. d   **60.** b   **61.** f   **62.** c   **63.** a   **64.** e

65. (a)

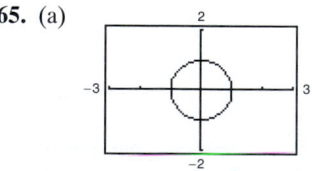

Circle of radius 1

(b) As t increases from -20 to 0, the speed increases, and as t increases from 0 to 20, the speed decreases.

67. (a) $x = \cos 70° (150 - 375t)$
 $y = \sin 70° (150 - 375t)$
 $x = -\cos 45° (190 - 450t)$
 $y = \sin 45° (190 - 450t)$

(b) $d = \sqrt{[-\cos 45°(190 - 450t) - \cos 70°(150 - 375t)]^2 + [\sin 45°(190 - 450t) - \sin 70°(150 - 375t)]^2}$

(c)

The minimum distance is 7.59 miles when $t = 0.4145$.

69. False: $\dfrac{d^2y}{dx^2} = \dfrac{\dfrac{d}{dt}\left[\dfrac{g'(t)}{f'(t)}\right]}{f'(t)} = \dfrac{f'(t)g''(t) - g'(t)f''(t)}{[f'(t)]^3}$.

70. False: the graph of $y = x^{2/3}$ does not have a horizontal asymptote at the origin.

Section 9.4 *(page 678)*

1.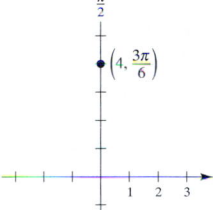

$(0, 4)$

3.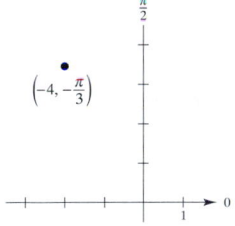

$(-2, 3.464)$

5.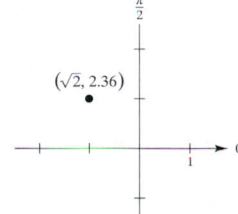

$(-1.004, 0.996)$

7.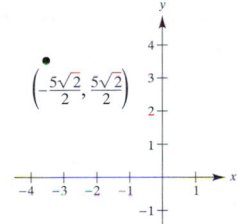

9.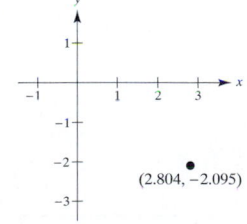

11. $\left(\sqrt{2}, \dfrac{\pi}{4}\right), \left(-\sqrt{2}, \dfrac{5\pi}{4}\right)$ **13.** $(5, 2.214), (-5, 5.356)$

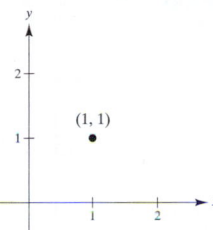   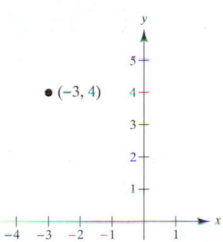

15. $(3.606, -0.588)$ **17.** $(2.833, 0.490)$

19. (a)   (b)

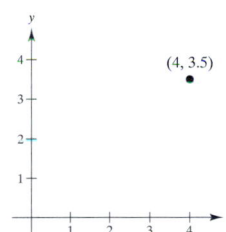   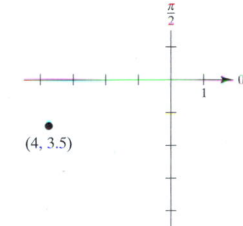

21. $r = a$ **23.** $r = 4 \csc\theta$

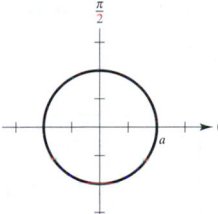   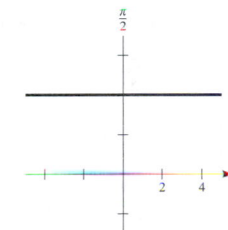

25. $r = \dfrac{-2}{3\cos\theta - \sin\theta}$ **27.** $r = 9 \csc^2\theta \cos\theta$

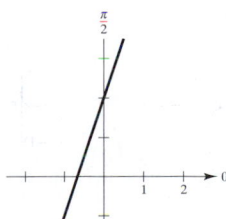   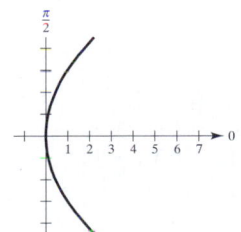

29. $x^2 + y^2 = 9$

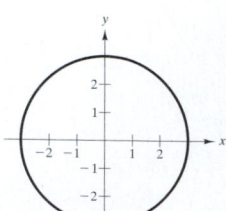

31. $x^2 + y^2 - y = 0$

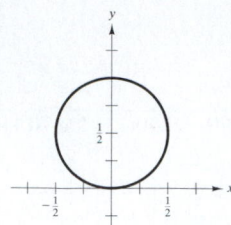

53. $(0, 0), (1.4142, 0.7854), (1.4142, 2.3562)$

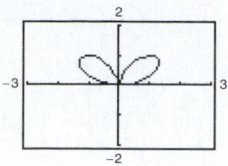

55. $(7, 1.5708), (3, 4.7124)$

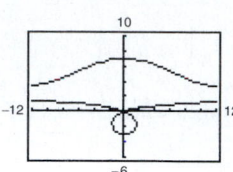

57. $\theta = 0$

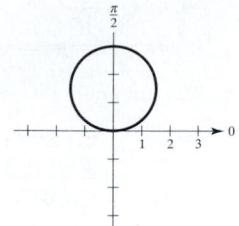

33. $\sqrt{x^2 + y^2} = \arctan \dfrac{y}{x}$

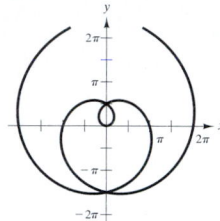

35. $x - 3 = 0$

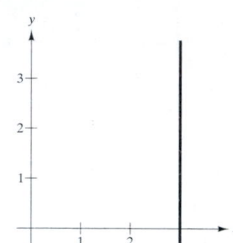

59. $\theta = \dfrac{\pi}{6}, \dfrac{\pi}{2}, \dfrac{5\pi}{6}$

61. $\theta = 0, \dfrac{\pi}{2}$

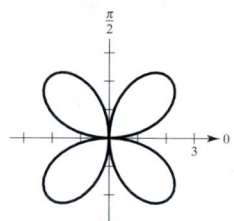

37. $(x - h)^2 + (y - k)^2 = h^2 + k^2$

Center: (h, k)

Radius: $\sqrt{h^2 + k^2}$

39. $2\sqrt{5}$ **41.** 5.6

43. $\dfrac{dy}{dx} = \dfrac{2\cos\theta(3\sin\theta + 1)}{6\cos^2\theta - 2\sin\theta - 3}$

$\left(5, \dfrac{\pi}{2}\right), \dfrac{dy}{dx} = 0$

$(2, \pi), \dfrac{dy}{dx} = -\dfrac{2}{3}$

$\left(-1, \dfrac{3\pi}{2}\right), \dfrac{dy}{dx} = 0$

63.

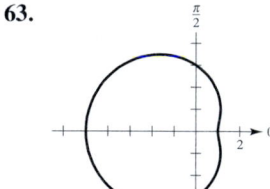

65.

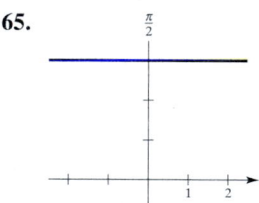

45. (a) and (b)

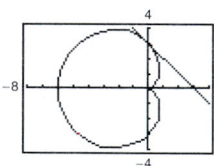

(c) -1

47. (a) and (b)

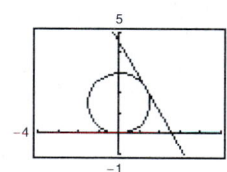

(c) $-\sqrt{3}$

67.

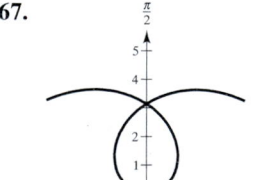

69.

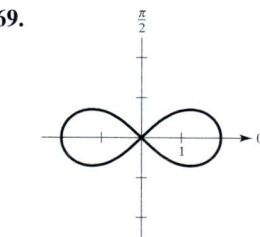

49. Horizontal: $\left(2, \dfrac{\pi}{2}\right), \left(\dfrac{1}{2}, \dfrac{7\pi}{6}\right), \left(\dfrac{1}{2}, \dfrac{11\pi}{6}\right)$

Vertical: $\left(\dfrac{3}{2}, \dfrac{\pi}{6}\right), \left(\dfrac{3}{2}, \dfrac{5\pi}{6}\right)$

51. $\left(5, \dfrac{\pi}{2}\right), \left(1, \dfrac{3\pi}{2}\right)$

71. $0 \le \theta < 2\pi$

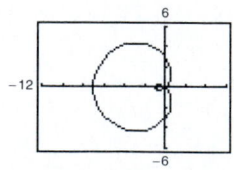

73. $0 \le \theta < 2\pi$

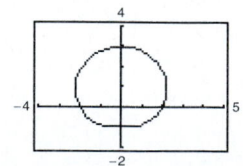

75. $-\pi < \theta < \pi$

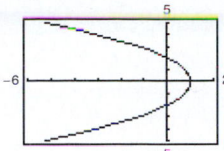

77. $0 \le \theta < 4\pi$

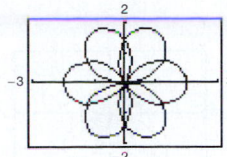

79. $0 \le \theta < \pi/2$

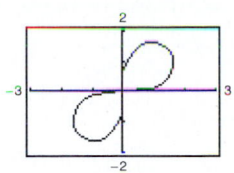

81.

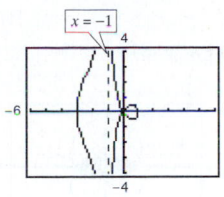

83.

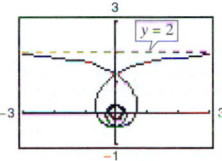

85. (a)

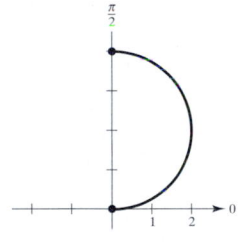

(b)

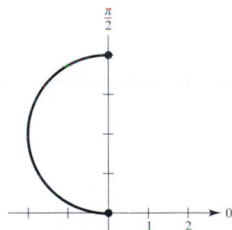

(c)

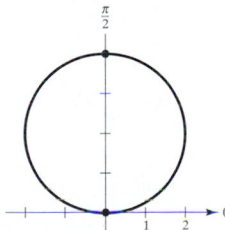

87. Proof

89. (a) $r = 2 - \sin\left(\theta - \dfrac{\pi}{4}\right) = 2 - \dfrac{\sqrt{2}(\sin\theta - \cos\theta)}{2}$

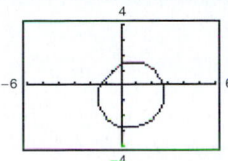

(b) $r = 2 + \cos\theta$

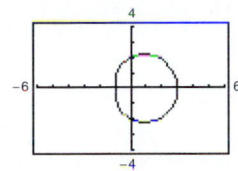

(c) $r = 2 + \sin\theta$ (d) $r = 2 - \cos\theta$

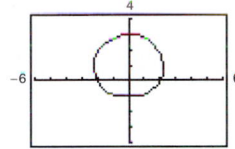
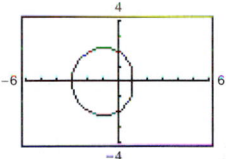

91. (a)

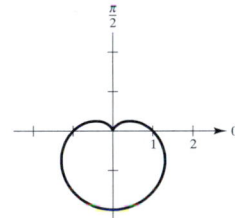

(b)

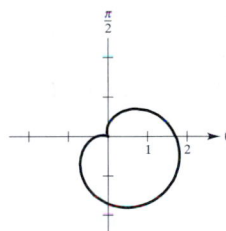

93. $\psi = \dfrac{\pi}{2}$ **95.** $\psi = 0$

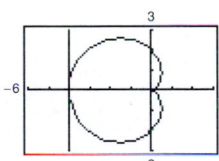
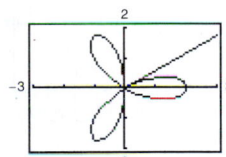

97. $\psi = \dfrac{\pi}{3}$

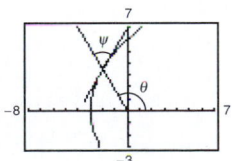

99. $0 \le \theta \le 9\pi$

100. True   **101.** True   **102.** True   **103.** True

Section 9.5 *(page 687)*

1. 16π 3. $\dfrac{\pi}{3}$ 5. $\dfrac{\pi}{8}$ 7. $\dfrac{3\pi}{2}$

9. $\dfrac{2\pi - 3\sqrt{3}}{2}$

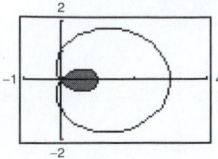

11. $\pi + 3\sqrt{3}$

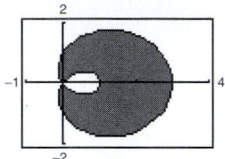

13. $\left(1, \dfrac{\pi}{2}\right), \left(1, \dfrac{3\pi}{2}\right), (0, 0)$

15. $\left(\dfrac{2 - \sqrt{2}}{2}, \dfrac{3\pi}{4}\right), \left(\dfrac{2 + \sqrt{2}}{2}, \dfrac{7\pi}{4}\right), (0, 0)$

17. $\left(\dfrac{3}{2}, \dfrac{\pi}{6}\right), \left(\dfrac{3}{2}, \dfrac{5\pi}{6}\right), (0, 0)$ 19. $(2, 4), (-2, -4)$

21. $\left(2, \dfrac{\pi}{12}\right), \left(2, \dfrac{5\pi}{12}\right), \left(2, \dfrac{7\pi}{12}\right), \left(2, \dfrac{11\pi}{12}\right)$

$\left(2, \dfrac{13\pi}{12}\right), \left(2, \dfrac{17\pi}{12}\right), \left(2, \dfrac{19\pi}{12}\right), \left(2, \dfrac{23\pi}{12}\right)$

23. $(0.581, \pm 2.607), (2.581, \pm 1.376)$

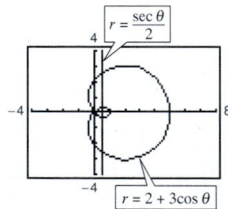

25. $(0, 0), (0.935, 0.363), (0.535, -1.006)$
The graphs reach the pole at different times (θ-values).

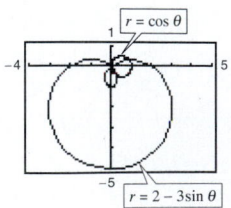

27. $\dfrac{4}{3}\left(4\pi - 3\sqrt{3}\right)$

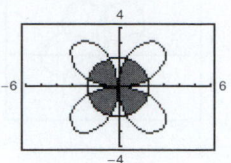

29. $11\pi - 24$

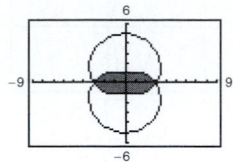

31. $\dfrac{2}{3}\left(4\pi - 3\sqrt{3}\right)$

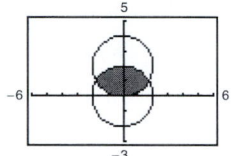

33. $\dfrac{5\pi a^2}{4}$ 35. $\dfrac{a^2}{2}(\pi - 2)$

37. (a) $(x^2 + y^2)^{3/2} = ax^2$

(b)

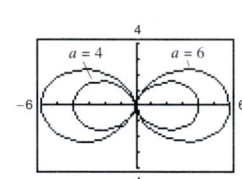

(c) $\dfrac{15\pi}{2}$

39. The area enclosed by the function is $\dfrac{\pi a^2}{4}$ if n is odd and is $\dfrac{\pi a^2}{2}$ if n is even.

41. $2\pi a$ 43. 8

45.

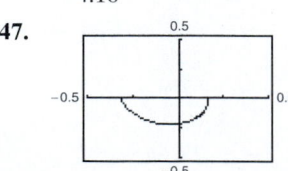

≈ 4.16

47.

≈ 0.71

49.

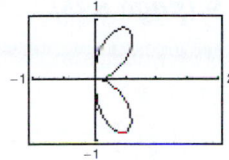

≈ 4.39

51. 4π **53.** $\dfrac{2\pi\sqrt{1+a^2}}{1+4a^2}(e^{\pi a}-2a)$

55. 21.87 **57.** $4\pi^2 ab$

59. (a) 16π

(b)

θ	0.2	0.4	0.6	0.8	1.0	1.2	1.4
A	6.32	12.14	17.06	20.80	23.27	24.60	25.08

(c) $\dfrac{1}{4}$: $\theta \approx 0.4$ (d) $\dfrac{1}{4}$: $\theta \approx 0.42$

$\dfrac{1}{2}$: $\theta \approx 1.5$ $\dfrac{1}{2}$: $\theta \approx 1.57$

$\dfrac{3}{4}$: $\theta \approx 2.7$ $\dfrac{3}{4}$: $\theta \approx 2.73$

(e) No: when solving the equation, the radius squared divides from both sides.

60. False: the area is given by $\displaystyle\int_0^{\pi/2} \sin^2\theta\, d\theta$.

61. False: the graphs of $f(\theta)=1$ and $g(\theta)=-1$ coincide.

62. False: if $f(\theta)=0$ and $g(\theta)=\sin 2\theta$, there is only one point of intersection.

63. True: the area enclosed by the first is $\pi/2$ and the area enclosed by the second is $\pi/4$.

Section 9.6 *(page 694)*

1.

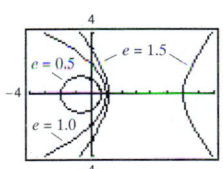

3.

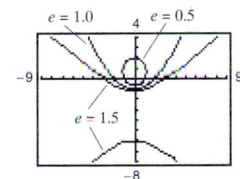

5. (a) Ellipse

As $e \to 1^-$, the ellipse becomes more elliptical, and as $e \to 0^+$, it becomes more circular.

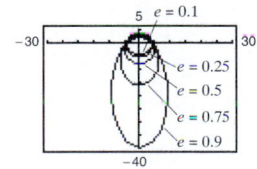

(b) Parabola

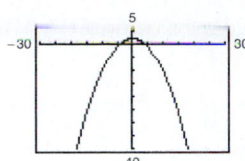

(c) Hyperbola

As $e \to 1^+$, the hyperbola opens more slowly, and as $e \to \infty$, they open more rapidly.

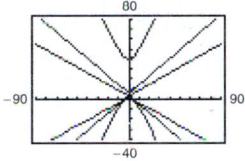

7. c   **8.** f   **9.** a   **10.** e   **11.** b   **12.** d

13. Parabola   **15.** Ellipse

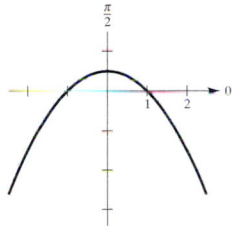

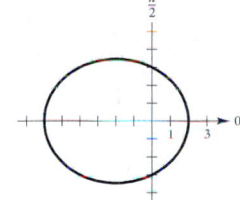

17. Ellipse   **19.** Hyperbola

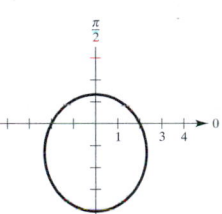

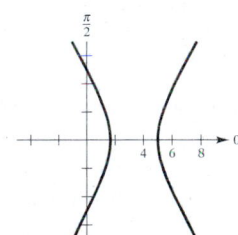

21. Hyperbola

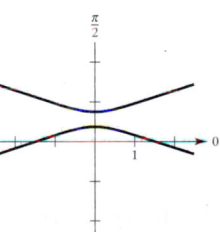

23.

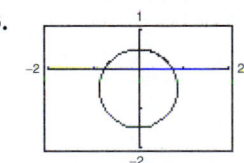

25.

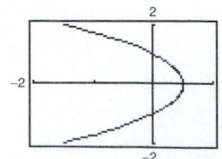

27. Rotated $\pi/4$ radians counterclockwise

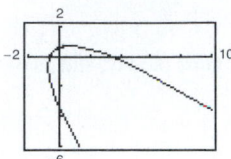

29. Rotated $\pi/6$ radians clockwise

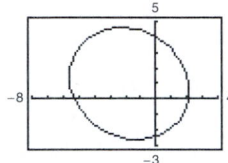

31. $r = \dfrac{5}{5 + 3\cos\left(\theta + \dfrac{\pi}{4}\right)}$

33. $r = \dfrac{1}{1 - \cos\theta}$ **35.** $r = \dfrac{1}{2 + \sin\theta}$ **37.** $r = \dfrac{2}{1 + 2\cos\theta}$

39. $r = \dfrac{2}{1 - \sin\theta}$ **41.** $r = \dfrac{16}{5 + 3\cos\theta}$

43. $r = \dfrac{9}{4 - 5\sin\theta}$ **45.** Proof

47. $r^2 = \dfrac{9}{1 - (16/25)\cos^2\theta}$ **49.** $r^2 = \dfrac{-16}{1 - (25/9)\cos^2\theta}$

51. 10.88   **53.** $r = \dfrac{345{,}996{,}000}{43{,}373 - 40{,}627\cos\theta}$; 11,004 miles

55. $r = \dfrac{92{,}931{,}075.2223}{1 - 0.0167\cos\theta}$

Perihelion: 91,404,618 miles
Aphelion: 94,509,382 miles

57. $r = \dfrac{5.537 \times 10^9}{1 - 0.2481\cos\theta}$

Perihelion: 4.436×10^9 kilometers
Aphelion: 7.364×10^9 kilometers

59. (a) 9.341×10^{18} square kilometers; 21.867 years

(b) 0.8995 radians, larger

(c) (a) 2.559×10^9 kilometers, 1.17×10^8 kilometers per year

(b) 4.119×10^9 kilometers, 1.88×10^8 kilometers per year

61. Proof

Review Exercises for Chapter 9 *(page 696)*

1. d   **2.** b   **3.** a   **4.** c

5. Circle
Center: $\left(\dfrac{1}{2}, -\dfrac{3}{4}\right)$
Radius: 1

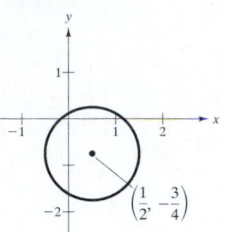

7. Hyperbola
Center: $(-4, 3)$
Vertices: $\left(-4 \pm \sqrt{2}, 3\right)$

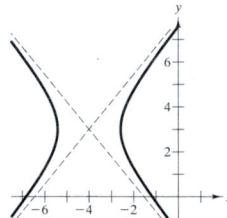

9. Ellipse
Center: $(2, -3)$
Vertices: $\left(2, -3 \pm \dfrac{\sqrt{2}}{2}\right)$

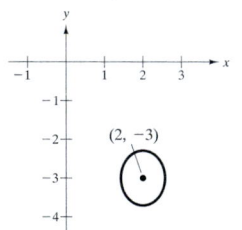

11. $y^2 - 4y - 12x + 4 = 0$

13. $\dfrac{(x - 2)^2}{25} + \dfrac{y^2}{21} = 1$ **15.** 15.87   **17.** $4x + 4y - 7 = 0$

19. (a) 192π cubic feet   (b) 7057.3 pounds

(c) 4.212 feet   (d) 429.105 square feet

21. (a) $\dfrac{dy}{dx} = -\dfrac{3}{4}$; Horizontal tangents: none

(b) $y = \dfrac{-3x + 11}{4}$

(c)

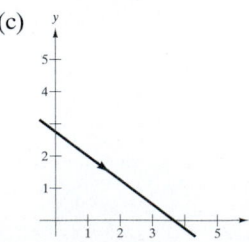

23. (a) $\dfrac{dy}{dx} = -2t^2$; Horizontal tangents: none

(b) $y = 3 + \dfrac{2}{x}$

(c)

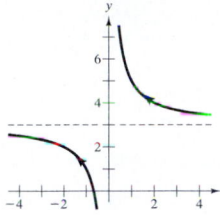

25. (a) $\dfrac{dy}{dx} = \dfrac{(t-1)(2t+1)^2}{t^2(t-2)^2}$; Horizontal tangents: $\left(\dfrac{1}{3}, -1\right)$

(b) $y = \dfrac{4x^2}{(5x-1)(x-1)}$

(c)

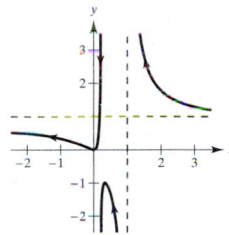

27. (a) $\dfrac{dy}{dx} = -\dfrac{5}{2}\cot\theta$; Horizontal tangents: $(3, 7), (3, -3)$

(b) $\dfrac{(x-3)^2}{4} + \dfrac{(y-2)^2}{25} = 1$

(c)

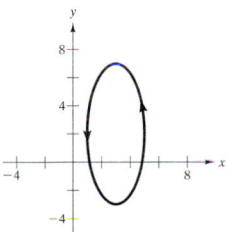

29. (a) $\dfrac{dy}{dx} = -4\tan\theta$; Horizontal tangents: none

(b) $x^{2/3} + (y/4)^{2/3} = 1$

(c)

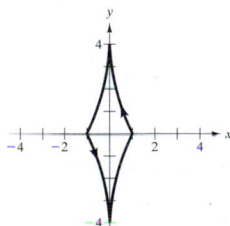

31. (a) and (c)

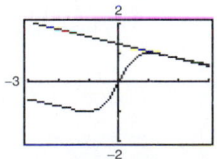

(b) $\dfrac{dx}{d\theta} = -4, \dfrac{dy}{d\theta} = 1,$

$\dfrac{dy}{dx} = -\dfrac{1}{4}$

33. $x = 3 + 5t$
   $y = 2 - 4t$

35. $x = -3 + 4\cos\theta$
   $y = 4 + 3\sin\theta$

37.

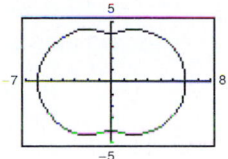

39. Proof   **41.** $\dfrac{\pi^2 r}{2}$

43.

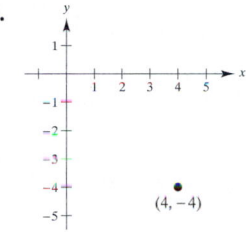

$\left(4\sqrt{2}, \dfrac{7\pi}{4}\right), \left(-4\sqrt{2}, \dfrac{3\pi}{4}\right)$

45. $x^2 + y^2 - 3x = 0$ **47.** $(x^2 + y^2 + 2x)^2 = 4(x^2 + y^2)$

49. $(x^2 + y^2)^2 = x^2 - y^2$ **51.** $y^2 = x^2\left(\dfrac{4-x}{4+x}\right)$

53. $r = a\cos^2\theta\sin\theta$ **55.** $r^2 = a^2\theta^2$

57. Circle   **59.** Line

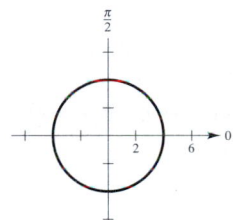

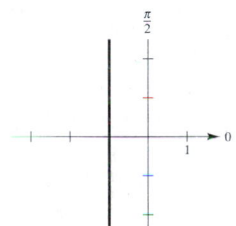

61. Cardioid

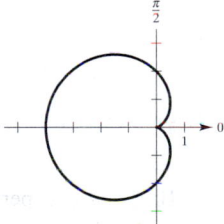

63. Limaçon

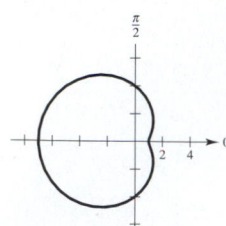

65. Rose curve

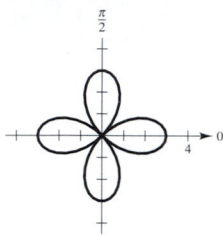

67. Rose curve

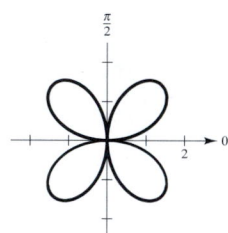

69. Parabola

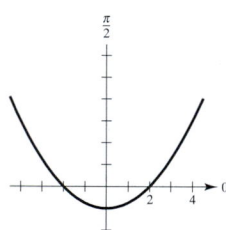

71.

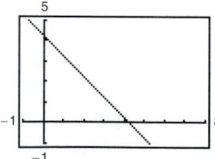

73.

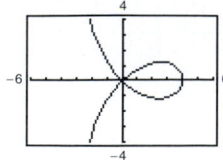

75. (a) $\pm \dfrac{\pi}{3}$

    (b) Vertical: $(-1, 0), (3, \pi), \left(\frac{1}{2}, \pm 1.318\right)$

        Horizontal: $(-0.686, \pm 0.568), (2.186, \pm 2.206)$

    (c)

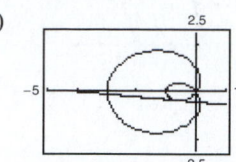

77. Proof   **79.** $\arctan\left(\dfrac{2\sqrt{3}}{3}\right) \approx 49.1°$

81.

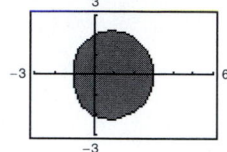

$$A = 2\left(\frac{1}{2}\right)\int_0^{\pi} (2 + \cos\theta)^2\, d\theta \approx 14.14$$

83.

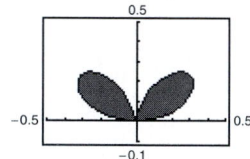

$$A = 2\left(\frac{1}{2}\right)\int_0^{\pi/2} \sin^2\theta \cos^4\theta\, d\theta \approx 0.10$$

85.

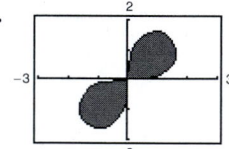

$$A = 2\left(\frac{1}{2}\right)\int_0^{\pi/2} 4 \sin 2\theta\, d\theta \approx 4$$

87.

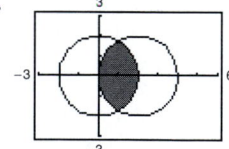

$$A = 2\left(\frac{1}{2}\int_0^{\pi/3} 4\, d\theta + \frac{1}{2}\int_{\pi/3}^{\pi/2} 16\cos^2\theta\, d\theta\right) \approx 4.91$$

89. $8a$ **91.** $r = 10 \sin\theta$

93. $r = \dfrac{4}{1 - \cos\theta}$ **95.** $r = \dfrac{5}{3 - 2\cos\theta}$

CHAPTER 10
Section 10.1 *(page 708)*

1. $\langle 4, 2 \rangle$

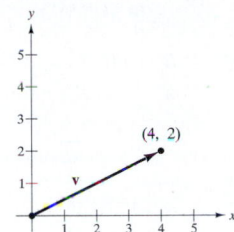

3. $\langle -7, 0 \rangle$

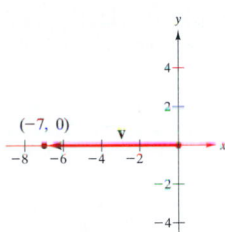

5. $\langle 4, 3 \rangle$

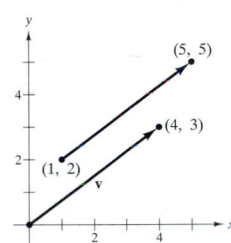

7. $\langle -4, -3 \rangle$

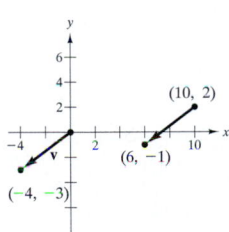

9. $\langle 0, 4 \rangle$

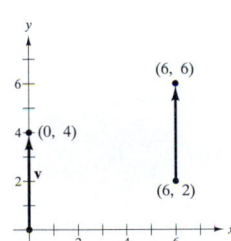

11. $\langle -1, \frac{5}{3} \rangle$

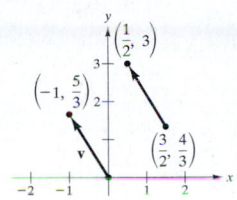

13. (a) $\langle 4, 6 \rangle$ (b) $\langle -6, -9 \rangle$

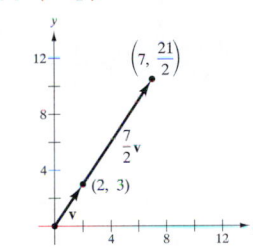   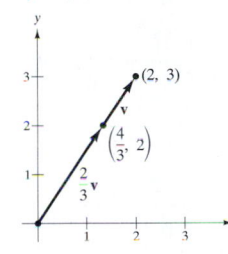

   (c) $\langle 7, \frac{21}{2} \rangle$ (d) $\langle \frac{4}{3}, 2 \rangle$

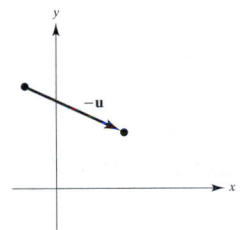   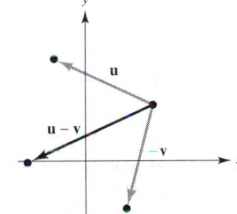

15.   **17.**

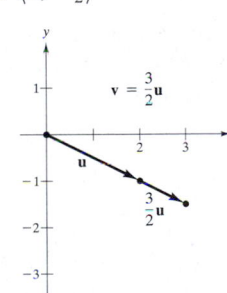   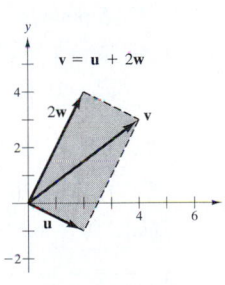

19. $\langle 3, -\frac{3}{2} \rangle$ **21.** $\langle 4, 3 \rangle$

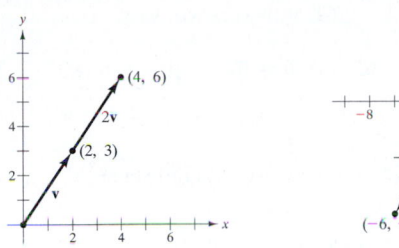

23. $a = 1, b = 1$ **25.** $a = 1, b = 2$ **27.** $a = \frac{2}{3}, b = \frac{1}{3}$

29. $(3, 5)$ **31.** 5   **33.** $\sqrt{61}$ **35.** 4

37. (a) $\sqrt{2}$ (b) $\sqrt{5}$ (c) 1   (d) 1   (e) 1   (f) 1

39. (a) $\sqrt{5}/2$ (b) $\sqrt{13}$ (c) $\sqrt{85}/2$
    (d) 1   (e) 1   (f) 1

41. Proof   **43.** $\langle 2\sqrt{2}, 2\sqrt{2}\rangle$ **45.** $\langle 1, \sqrt{3}\rangle$

47. (a) $\pm\dfrac{1}{\sqrt{10}}\langle 1, 3\rangle$ (b) $\pm\dfrac{1}{\sqrt{10}}\langle 3, -1\rangle$

49. (a) $\pm\dfrac{1}{5}\langle -4, 3\rangle$ (b) $\pm\dfrac{1}{5}\langle 3, 4\rangle$ **51.** $\langle 3, 0\rangle$

53. $\langle -\sqrt{3}, 1\rangle$ **55.** $\left(\dfrac{3+\sqrt{2}}{\sqrt{2}}\right)\mathbf{i} + \left(\dfrac{3}{\sqrt{2}}\right)\mathbf{j}$

57. $(2\cos 4 + \cos 2)\mathbf{i} + (2\sin 4 + \sin 2)\mathbf{j}$

59. $-\dfrac{\sqrt{2}}{2}\mathbf{i} + \dfrac{\sqrt{2}}{2}\mathbf{j}$ **61.** Answers will vary.

63. $1.33, 132.5°$ **65.** (a) $\theta = 0°$ (b) $\theta = 180°$ (c) No

67. (a) Direction: $\alpha = 11.8°$
     Magnitude: 440.2 N

     (b) $M = \sqrt{(275 + 180\cos\theta)^2 + (180\sin\theta)^2}$

$$\alpha = \arccos\left(\frac{36\cos\theta + 55}{\sqrt{3960\cos\theta + 4321}}\right)$$

     (c)

θ	0°	30°	60°	90°	120°
M	455.0	440.2	396.9	328.7	241.9
α	0°	11.8°	23.1°	33.2°	40.1°

θ	150°	180°
M	149.3	95.0
α	37.1°	0°

     (d)

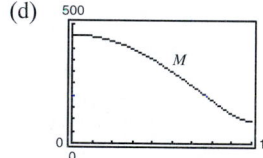 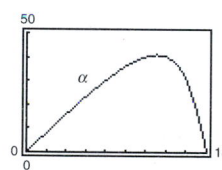

     (e) M decreases because the forces change from acting in the same direction to acting in opposite directions as θ increases from 0° to 180°.

69. $71.3°, 228.5$ pounds   **71.** $(-4, -1), (6, 5), (10, 3)$

73. (a) Tension ≈ 1.1547 pounds   (b) $T = \sec\theta$
       $\|\mathbf{u}\| \approx 0.5774$ pound $\|\mathbf{u}\| = \tan\theta$

     (c)

θ	0°	10°	20°	30°	40°
T	1	1.0154	1.0642	1.1547	1.3054
$\|\mathbf{u}\|$	0	0.1763	0.3640	0.5774	0.8391

θ	50°	60°
T	1.5557	2
$\|\mathbf{u}\|$	1.1918	1.7321

(d)

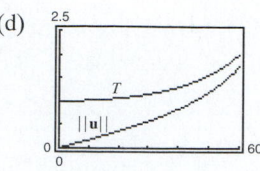

     (e) Both are increasing functions for $0° \le \theta \le 60°$.

     (f) $\displaystyle\lim_{\theta\to\pi/2^-} T = \infty$, $\displaystyle\lim_{\theta\to\pi/2^-} \|\mathbf{u}\| = \infty$

75. Horizontal: 1193.43 feet per second
     Vertical: 125.43 feet per second

77. 38.3° north of west   **79.** $T_2 = 157.316$
     882.9 kilometers per hour $T_3 = 3692.482$

81. Proof   **83.** Proof   **84.** True   **85.** True   **86.** True

87. False: $a = b = 0$ **88.** False: $\|a\mathbf{i} + b\mathbf{j}\| = \sqrt{2}|a|$

89. True

Section 10.2 *(page 717)*

1. 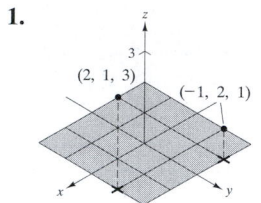   **3.**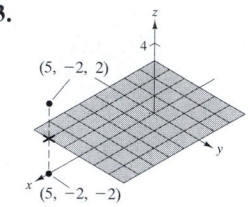

5. $A(2, 3, 4)$ **7.** $(-3, 4, 5)$ **9.** $(10, 0, 0)$ **11.** 0
    $B(-1, -2, 2)$

13. 3 units below the xy-plane, to the right of the xz-plane and in front of the yz-plane, *or* 3 units below the xy-plane, to the left of the xz-plane and behind the yz-plane

15. To the right of the xz-plane and behind the yz-plane *or* to the left of the xz-plane and in front of the yz-plane

17. $3, 3\sqrt{5}, 6$ **19.** $6, 6, 2\sqrt{10}$
    Right triangle   Isosceles triangle

21. $(0, 0, 5), (2, 2, 6), (2, -4, 9)$

23. $\left(\dfrac{3}{2}, -3, 5\right)$ **25.** $(x - 0)^2 + (y - 2)^2 + (z - 5)^2 = 4$

27. $(x - 1)^2 + (y - 3)^2 + (z - 0)^2 = 10$

29. Center: $(1, -3, -4)$ **31.** Center: $\left(\dfrac{1}{3}, -1, 0\right)$
    Radius: 5   Radius: 1

33. (a) $\langle -2, 2, 2\rangle$ **35.** (a) $\langle -3, 0, 3\rangle$

    (b) 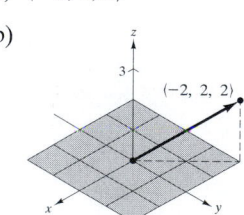   (b)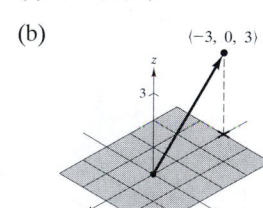

37. (a) and (c)   (b) $\langle 4, 1, 1 \rangle$

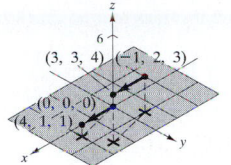

39. $(3, 1, 8)$

41. (a)   (b)

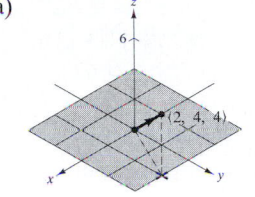 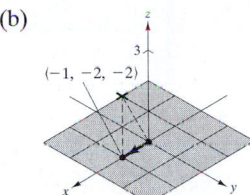

(c)   (d)

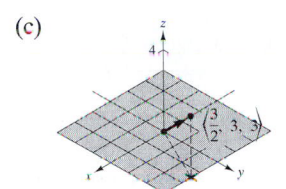 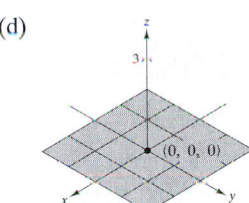

43. $\langle -1, 0, 4 \rangle$ **45.** $\langle 6, 12, 6 \rangle$ **47.** $\langle \frac{7}{2}, 3, \frac{5}{2} \rangle$

49. a and b   **51.** a   **53.** Collinear   **55.** Not collinear

57. Proof   **59.** 0   **61.** $\sqrt{14}$ **63.** $\sqrt{34}$

65. (a) $\frac{1}{3} \langle 2, -1, 2 \rangle$ (b) $-\frac{1}{3} \langle 2, -1, 2 \rangle$

67. (a) $\frac{1}{\sqrt{38}} \langle 3, 2, -5 \rangle$ (b) $-\frac{1}{\sqrt{38}} \langle 3, 2, -5 \rangle$

69. Answers will vary.   **71.** $\pm \frac{5}{3}$ **73.** $\langle 0, \frac{10}{\sqrt{2}}, \frac{10}{\sqrt{2}} \rangle$

75. $\langle 1, -1, \frac{1}{2} \rangle$

77. $\langle 0, \sqrt{3}, \pm 1 \rangle$ **79.** $(2, -1, 2)$

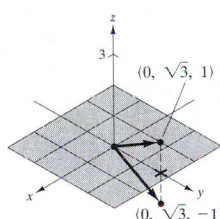

81. (a)   (b) Proof   (c) $a = b = 1$

(d) Proof

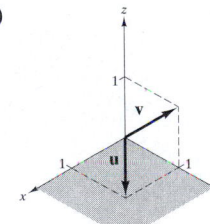

83. (a) $T = \dfrac{8L}{\sqrt{L^2 - 18^2}}$

(b)

L	20	25	30	35	40	45	50
T	18.4	11.5	10	9.3	9.0	8.7	8.6

(c)

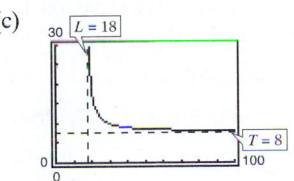

(d) Proof   (e) 30 inches

85. $\dfrac{\sqrt{3}}{3} \langle 1, 1, 1 \rangle$

87. Tension in $\overline{AB}$: 202.919 N
Tension in $\overline{AC}$: 157.909 N
Tension in $\overline{AD}$: 226.521 N

89. $\left(x - \dfrac{4}{3}\right)^2 + (y - 3)^2 + \left(z + \dfrac{1}{3}\right)^2 = \dfrac{44}{9}$

Section 10.3 *(page 727)*

1. (a) -6 (b) 25   (c) 25   (d) $\langle -12, 18 \rangle$ (e) -12

3. (a) 2   (b) 29   (c) 29   (d) $\langle 0, 12, 10 \rangle$ (e) 4

5. (a) 1   (b) 6   (c) 6   (d) $\mathbf{i} - \mathbf{k}$ (e) 2

7. \$17,139.05, total revenue

9. 20   **11.** $\dfrac{\pi}{2}$ **13.** $\arccos\left(-\dfrac{1}{5\sqrt{2}}\right) \approx 98.1°$

15. $\arccos\left(\dfrac{\sqrt{2}}{3}\right) \approx 61.9°$ **17.** $\arccos\left(-\dfrac{8\sqrt{13}}{65}\right) \approx 116.3°$

19. Answers will vary.   **21.** Neither   **23.** Orthogonal

25. Neither   **27.** Orthogonal   **29.** Proof

31. (a) $\theta = \dfrac{\pi}{2}$ (b) $0 < \theta < \dfrac{\pi}{2}$ (c) $\dfrac{\pi}{2} < \theta < \pi$

33. Proof

35. $\cos\alpha = \dfrac{1}{3}$ **37.** $\cos\alpha = 0$

$\cos\beta = \dfrac{2}{3}$ $\cos\beta = \dfrac{3}{\sqrt{13}}$

$\cos\gamma = \dfrac{2}{3}$ $\cos\gamma = -\dfrac{2}{\sqrt{13}}$

39. Magnitude: 124.310 pounds
$\alpha = 29.48°$
$\beta = 61.39°$
$\gamma = 96.53°$

41. $\arccos\left(\dfrac{1}{\sqrt{3}}\right) \approx 54.7°$

43. $\alpha = 90°$
$\beta = 45°$
$\gamma = 45°$

45. (a) $\left\langle \frac{5}{2}, \frac{1}{2} \right\rangle$ (b) $\left\langle -\frac{1}{2}, \frac{5}{2} \right\rangle$

47. (a) $\left\langle 0, \frac{33}{25}, \frac{44}{25} \right\rangle$ (b) $\left\langle 2, -\frac{8}{25}, \frac{6}{25} \right\rangle$

49. Answers will vary. **51.** $\langle 0, 0 \rangle$

53. (a) $\mathbf{u}$ and $\mathbf{v}$ are parallel. (b) $\mathbf{u}$ and $\mathbf{v}$ are orthogonal.

55. $\langle 4, 3 \rangle, \langle -4, -3 \rangle$ **57.** $\langle 2, 0, 3 \rangle, \langle -2, 0, -3 \rangle$

59. (a) 8282.2 pounds (b) 30,909.6 pounds

61. 425 foot-pounds **63.** 72 **65.** Proof **67.** Proof

Section 10.4 (page 735)

1. $-\mathbf{k}$

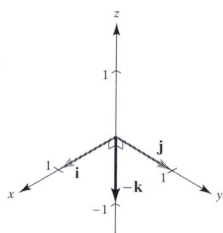

3. $\mathbf{i}$

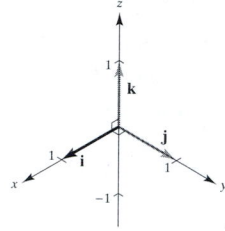

5. $-\mathbf{j}$

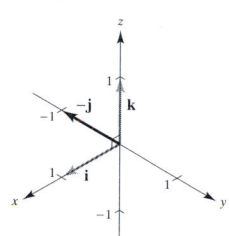

7. $\langle -1, -1, -1 \rangle$

9. $\langle 0, 0, 54 \rangle$ **11.** $\langle -2, 3, -1 \rangle$

13.

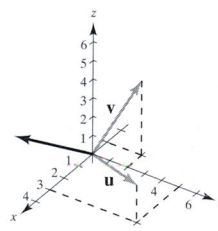

15.

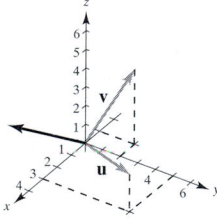

17. $\left\langle -70, -23, \frac{57}{2} \right\rangle$

$\left\langle \frac{-140}{\sqrt{24,965}}, \frac{-46}{\sqrt{24,965}}, \frac{57}{\sqrt{24,965}} \right\rangle$

19. $\left\langle \frac{71}{20}, -\frac{11}{5}, \frac{5}{4} \right\rangle$ **21.** Answers will vary.

$\left\langle \frac{-71}{\sqrt{7602}}, \frac{-44}{\sqrt{7602}}, \frac{25}{\sqrt{7602}} \right\rangle$

23. 1 **25.** $6\sqrt{5}$ **27.** $2\sqrt{83}$ **29.** $\frac{3\sqrt{13}}{2}$ **31.** $\frac{9\sqrt{6}}{2}$

33. 1 **35.** 6 **37.** 2 **39.** 75

41. $10 \cos 40° \approx 7.66$ foot-pounds

43. (a) $\overrightarrow{AB} = -\frac{5}{4}\mathbf{j} + \mathbf{k}$

$\mathbf{F} = -200(\cos \theta \mathbf{j} + \sin \theta \mathbf{k})$

(b) $25(8 \cos \theta + 10 \sin \theta)$

(c) $25\left(4\sqrt{3} + 5\right)$

(d) $\theta = 51.34°$; The vectors are orthogonal.

(e)

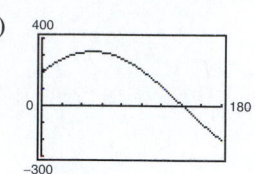

$\theta \approx 141.34°, \overrightarrow{AB}$ is parallel to $\mathbf{F}$.

45. Proof **47.** Proof **49.** Proof **51.** Proof

53. Proof

55. Find the cross product of the vectors $\langle x_2 - x_1, y_2 - y_1, z_2 - z_1 \rangle$ and $\langle x_3 - x_1, y_3 - y_1, z_3 - z_1 \rangle$.

57. Proof

Section 10.5 (page 744)

1. (a)

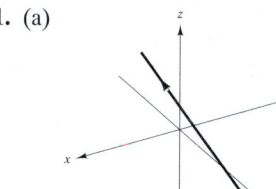

(b) $P = (1, 2, 2), Q = (10, -1, 17), \overrightarrow{PQ} = \langle 9, -3, 15 \rangle$ (There are many correct answers.) The components of the vector and the coefficients of t are proportional because the line is parallel to $\overrightarrow{PQ}$.

(c) $\left(-\frac{1}{5}, \frac{12}{5}, 0 \right), (7, 0, 12), \left(0, \frac{7}{3}, \frac{1}{3} \right)$

Parametric Equations	Symmetric Equations	Direction Numbers
3. $x = t$	$x = \frac{y}{2} = \frac{z}{3}$	1, 2, 3
$y = 2t$		
$z = 3t$		
5. $x = -2 + 2t$	$\frac{x+2}{2} = \frac{y}{4} = \frac{z-3}{-2}$	2, 4, −2
$y = 4t$		
$z = 3 - 2t$		
7. $x = 1 + 3t$	$\frac{x-1}{3} = \frac{y}{-2} = \frac{z-1}{1}$	3, −2, 1
$y = -2t$		
$z = 1 + t$		

Parametric Equations	Symmetric Equations	Direction Numbers

9. $x = 5 + 17t$ $\dfrac{x-5}{17} = \dfrac{y+3}{-11} = \dfrac{z+2}{-9}$ $17, -11, -9$

$y = -3 - 11t$

$z = -2 - 9t$

11. $x = 2$

$y = 3$

$z = 4 + t$

13. a, b   **15.** $(2, 3, 1)$, $\cos\theta = \dfrac{7\sqrt{17}}{51}$ **17.** Nonintersecting

19. $(7, 8, -1)$

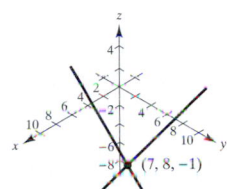

21. (a) $P = (0, 0, -1)$, $Q = (0, -2, 0)$, $R = (3, 4, -1)$

$\overrightarrow{PQ} = \langle 0, -2, 1 \rangle$, $\overrightarrow{PR} = \langle 3, 4, 0 \rangle$

(There are many correct answers.)

(b) $\overrightarrow{PQ} \times \overrightarrow{PR} = \langle -4, 3, 6 \rangle$

The components of the cross product are proportional to the coefficients of the variables in the equation. The cross product is parallel to the normal vector.

23. $x - 2 = 0$ **25.** $2x + 3y - z = 10$

27. $x - y + 2z = 12$ **29.** $3x + 9y - 7z = 0$

31. $4x - 3y + 4z = 10$ **33.** $z = 3$ **35.** $x + y + z = 5$

37. $7x + y - 11z = 5$ **39.** $y - z = -1$ **41.** Orthogonal

43. $83.5°$ **45.** Parallel

47. 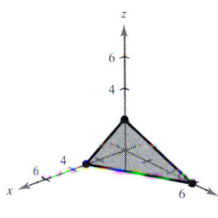   **49.**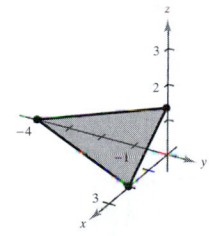

51. 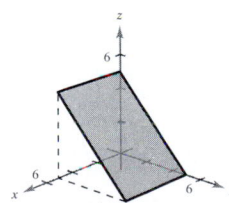   **53.**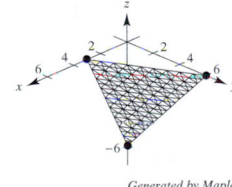

Generated by Maple

55. 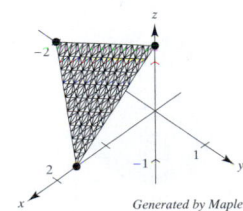   **57.** $x = 2$

$y = 1 + t$

$z = 1 + 2t$

Generated by Maple

59. $(2, -3, 2)$ **61.** Nonintersecting   **63.** $\dfrac{6\sqrt{14}}{7}$

65. $\dfrac{2\sqrt{26}}{13}$ **67.** $149\dfrac{\sqrt{17}}{17} \approx 36.14$

69. (a)

Year	1970	1975	1980	1985	1990
z (approx.)	213.6	173.7	144.7	121.1	85.3

Year	1991	1992	1993
z (approx.)	81.9	81.4	83.6

(b) Decrease

(c)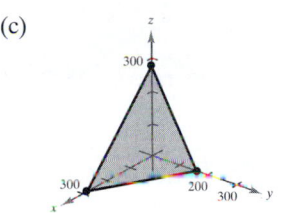

71. (a) Sphere

$x^2 + y^2 + z^2 - 6x + 4y - 10z + 22 = 0$

(b) Planes

$4x - 3y + z = 10 \pm 4\sqrt{26}$

73. $\arccos\frac{1}{65} \approx 89.1°$ **75.** True

76. False: they may be nonintersecting. (See Exercise 61.)

77. Proof

Section 10.6 *(page 756)*

1. c   **2.** e   **3.** f   **4.** b   **5.** d   **6.** a

7. Plane   **9.** Right circular cylinder

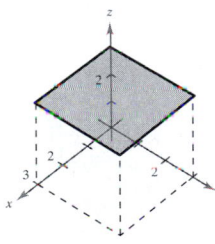   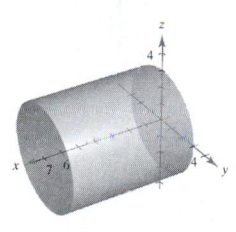

11. Parabolic cylinder   **13.** Elliptic cylinder

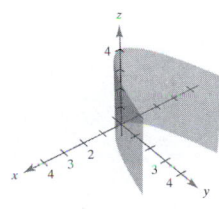   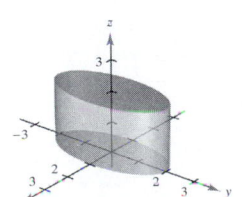

15. Cylinder

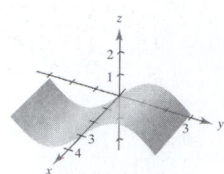

17. (a) $(20, 0, 0)$
(b) $(10, 10, 20)$
(c) $(0, 0, 20)$
(d) $(0, 20, 0)$

39.

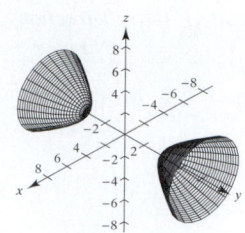

41.

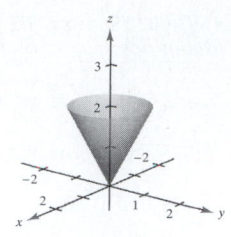

19. Ellipsoid

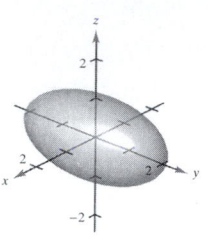

21. Hyperboloid of one sheet

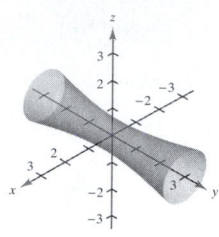

43.

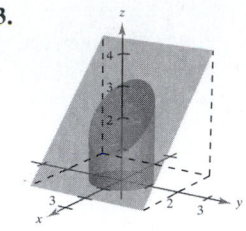

45. $x^2 + z^2 = 4y$

23. Elliptic paraboloid

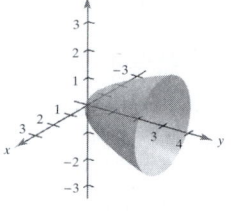

25. Hyperbolic paraboloid

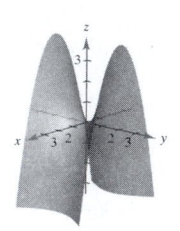

47. $4x^2 + 4y^2 = z^2$ **49.** $y^2 + z^2 = \dfrac{4}{x^2}$ **51.** $y = \sqrt{2z}$

53. $\dfrac{128\pi}{3}$

55. (a) Major axis: $4\sqrt{2}$ (b) Major axis: $8\sqrt{2}$
  Minor axis: 4 Minor axis: 8
  Foci: $(0, \pm 2, 2)$ Foci: $(0, \pm 4, 8)$

57. $\dfrac{x^2}{3963^2} + \dfrac{y^2}{3963^2} + \dfrac{z^2}{3942^2} = 1$

59. $x = at, y = -bt, z = 0;$
$x = at, y = bt + ab^2, z = 2abt + a^2b^2$

27. Elliptic cone

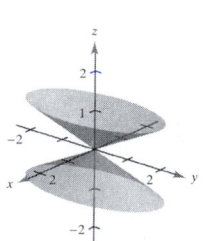

29. Ellipsoid

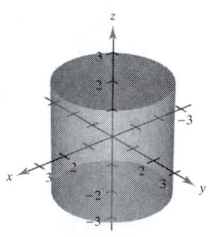

Section 10.7 (page 763)

1. $\left(5, \dfrac{\pi}{2}, 1\right)$ **3.** $\left(2, \dfrac{\pi}{3}, 4\right)$ **5.** $\left(2\sqrt{2}, -\dfrac{\pi}{4}, -4\right)$

7. $(5, 0, 2)$ **9.** $\left(1, \sqrt{3}, 2\right)$ **11.** $\left(-2\sqrt{3}, -2, 3\right)$

13. $x^2 + y^2 = 4$ **15.** $x - \sqrt{3}y = 0$

31.

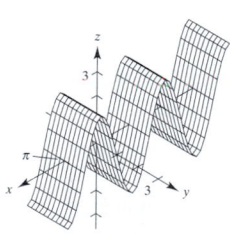

33.

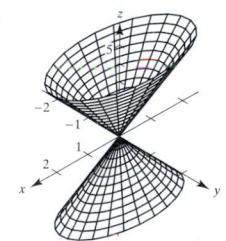

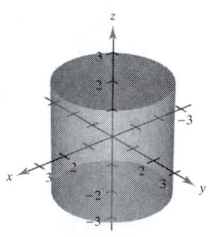

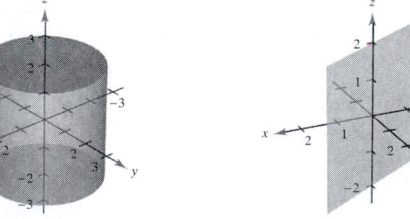

17. $x^2 + y^2 - 2y = 0$ **19.** $x^2 + y^2 + z^2 = 4$

35.

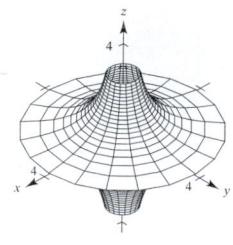

37.

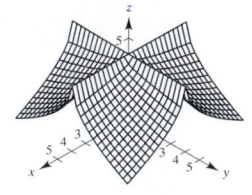

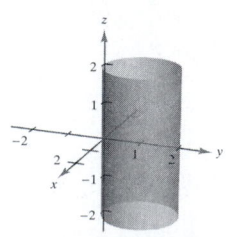

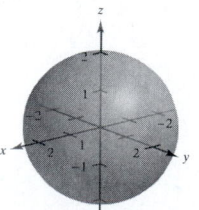

21. $\left(4, 0, \dfrac{\pi}{2}\right)$ **23.** $\left(4\sqrt{2}, \dfrac{2\pi}{3}, \dfrac{\pi}{4}\right)$ **25.** $\left(4, \dfrac{\pi}{6}, \dfrac{\pi}{6}\right)$

27. $\left(\sqrt{6},\ \sqrt{2},\ 2\sqrt{2}\right)$ **29.** $(0, 0, 12)$ **31.** $\left(\dfrac{5}{2}, \dfrac{5}{2}, -\dfrac{5\sqrt{2}}{2}\right)$

33. (a) Answers will vary.   (b) $(5.385, -0.927, 1.190)$

35. $x^2 + y^2 + z^2 = 4$ **37.** $3x^2 + 3y^2 - z^2 = 0$

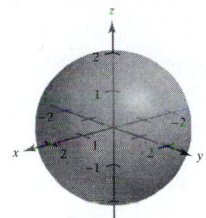   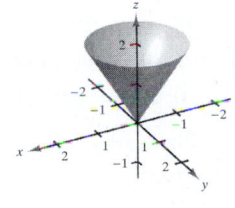

39. $x^2 + y^2 + (z - 2)^2 = 4$ **41.** $x^2 + y^2 = 1$

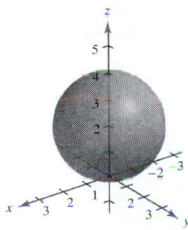   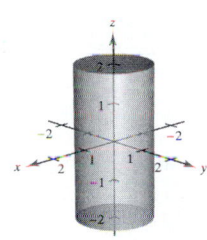

43. $\left(4, \dfrac{\pi}{4}, \dfrac{\pi}{2}\right)$ **45.** $\left(2\sqrt{13}, -\dfrac{\pi}{6}, \arccos\left[\dfrac{3}{\sqrt{13}}\right]\right)$

47. $\left(13, \pi, \arccos\left[\dfrac{5}{13}\right]\right)$ **49.** $\left(10, \dfrac{\pi}{6}, 0\right)$

51. $\left(3\sqrt{3}, -\dfrac{\pi}{6}, 3\right)$ **53.** $\left(4, \dfrac{7\pi}{6}, 4\sqrt{3}\right)$

Rectangular	Cylindrical	Spherical
55. $(4, 6, 3)$	$(7.211, 0.983, 3)$	$(7.810, 0.983, 1.177)$
57. $(4.698, 1.710, 8)$	$\left(5, \dfrac{\pi}{9}, 8\right)$	$(9.434, 0.349, 0.559)$
59. $(-7.071, 12.247, 14.142)$	$(14.142, 2.094, 14.142)$	$\left(20, \dfrac{2\pi}{3}, \dfrac{\pi}{4}\right)$
61. $(3, -2, 2)$	$(3.606, -0.588, 2)$	$(4.123, -0.588, 1.064)$
63. $\left(\dfrac{5}{2}, \dfrac{4}{3}, -\dfrac{3}{2}\right)$	$(2.833, 0.490, -1.5)$	$(3.206, 0.490, 2.058)$
65. $(-3.536, 3.536, -5)$	$\left(5, \dfrac{3\pi}{4}, -5\right)$	$(7.071, 2.356, 2.356)$
67. $(2.804, -2.095, 6)$	$(-3.5, 2.5, 6)$	$(6.946, 5.641, 0.528)$

69. d   **70.** e   **71.** c   **72.** a   **73.** f   **74.** b

75. (a) $r^2 + z^2 = 16$ (b) $\rho = 4$

77. (a) $r^2 + (z - 1)^2 = 1$ (b) $\rho = 2\cos\phi$

79. (a) $r = 4\sin\theta$ (b) $\rho = \dfrac{4\sin\theta}{\sin\phi} = 4\sin\theta\csc\phi$

81. (a) $r^2 = \dfrac{9}{\cos^2\theta - \sin^2\theta}$ (b) $\rho^2 = \dfrac{9\csc^2\phi}{\cos^2\theta - \sin^2\theta}$

83.

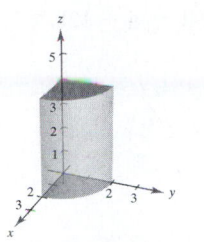

85.

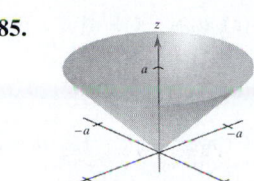

87.

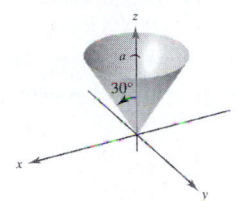

89. Rectangular: $0 \le x \le 10$
$0 \le y \le 10$
$0 \le z \le 10$

91. Spherical: $4 \le \rho \le 6$ **93.** Ellipse

Review Exercises for Chapter 10 *(page 765)*

1. (a) $\mathbf{u} = 3\mathbf{i} - \mathbf{j}$
$\mathbf{v} = 4\mathbf{i} + 2\mathbf{j}$
(b) $2\sqrt{5}$ (c) 10 (d) $10\mathbf{i}$ (e) $2\mathbf{i} + \mathbf{j}$ (f) $\mathbf{i} - 2\mathbf{j}$

3. $(-5, 4, 0)$

5. Above the xy-plane and to the right of the xz-plane *or* below the xy-plane and to the left of the xz-plane

7. $(x - 3)^2 + (y + 2)^2 + (z - 6)^2 = \dfrac{225}{4}$

9. Center: $(2, 3, 0)$
Radius: 3

11. (a) $\mathbf{u} = -\mathbf{i} + 4\mathbf{j}$ (b) 3 (c) $24\mathbf{i} + 6\mathbf{j} + 12\mathbf{k}$
$\mathbf{v} = -3\mathbf{i} + 6\mathbf{k}$

(d) $4x + y + 2z = 20$ (e) $x = 4 - t, y = 4 + 4t, z = 0$

13. Orthogonal **15.** $\theta = \arccos\left(\dfrac{\sqrt{2} + \sqrt{6}}{4}\right) = 15°$

17. π **19.** $-2\sqrt{2}\mathbf{i} + 2\sqrt{2}\mathbf{j}$

21. $\dfrac{3}{\sqrt{26}}\mathbf{i} - \dfrac{9}{\sqrt{26}}\mathbf{j} + \dfrac{12}{\sqrt{26}}\mathbf{k}$ **23.** $\sqrt{14}$ **25.** Proof

27. $\left\langle -\frac{15}{14}, \frac{5}{7}, -\frac{5}{14} \right\rangle$ **29.** Proof **31.** 4

33. Tension in $\overline{AB}$: 21.7 N **35.** $100 \sec 20° \approx 106.4$ pounds
Tension in $\overline{BC}$: 17.7 N

37. (a) $x = 1, y = 2 + t, z = 3$ (b) None

39. (a) $x = t, y = -1 + t, z = 1$ (b) $x = y + 1, z = 1$

41. $x + 2y = 1$ **43.** $\dfrac{8}{7}$ **45.** $\dfrac{\sqrt{3}}{3}$

47. 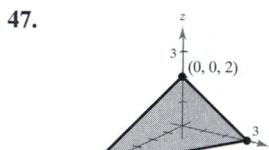 **49.**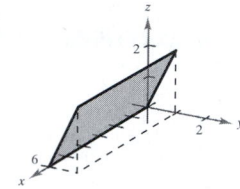

51. 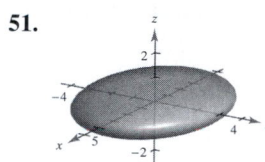 **53.**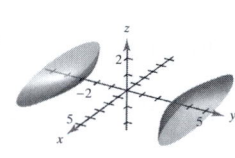

$\dfrac{x^2}{16} + \dfrac{y^2}{9} + z^2 = 1$

55. (a) $x^2 + y^2 - 2z + 2 = 0$

(b) $4\pi \approx 12.6$ cubic centimeters

(c) $\dfrac{225\pi}{64} \approx 11.0$ cubic centimeters

57. (a) $\left(4, \dfrac{3\pi}{4}, 2\right)$ (b) $\left(2\sqrt{5}, \dfrac{3\pi}{4}, \arccos\left[\dfrac{\sqrt{5}}{5}\right]\right)$

59. (a) $r^2 \cos 2\theta = 2z$ (b) $\rho = 2 \sec 2\theta \cos \phi \csc^2 \phi$

CHAPTER 11

Section 11.1 *(page 774)*

1. $(-\infty, 0) \cup (0, \infty)$ **3.** $(0, \infty)$

5. $[0, \infty)$ **7.** $(-\infty, \infty)$

9. (a) $\dfrac{1}{2}\mathbf{i}$ (b) $\mathbf{j}$ (c) $\dfrac{1}{2}(s + 1)^2\mathbf{i} - s\mathbf{j}$

(d) $\dfrac{1}{2}\Delta t(\Delta t + 4)\mathbf{i} - \Delta t\mathbf{j}$

11. (a) $\ln 2\mathbf{i} + \dfrac{1}{2}\mathbf{j} + 6\mathbf{k}$ (b) Not possible

(c) $\ln(t - 4)\mathbf{i} + \dfrac{1}{t - 4}\mathbf{j} + 3(t - 4)\mathbf{k}$

(d) $\ln(1 + \Delta t)\mathbf{i} - \dfrac{\Delta t}{1 + \Delta t}\mathbf{j} + 3\Delta t\mathbf{k}$

13. $\sqrt{1 + t^2}$ **15.** $t^2(5t - 1)$ The dot product is a scalar.

17. b **18.** c **19.** d **20.** a

21. (a) $(-20, 0, 0)$ (b) $(10, 20, 10)$

(c) $(0, 0, 20)$ (d) $(20, 0, 0)$

23. 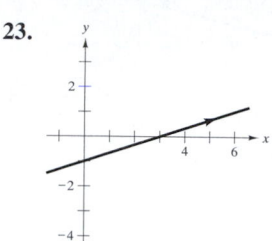 **25.**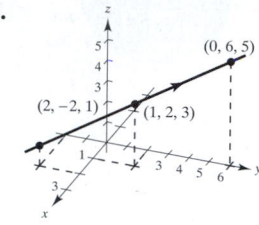

27. 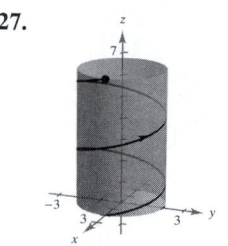 **29.**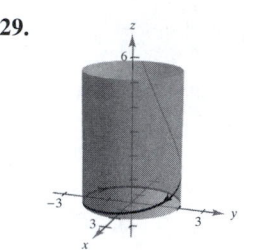

31. 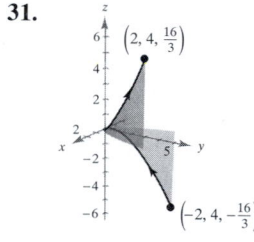 **33.** Parabola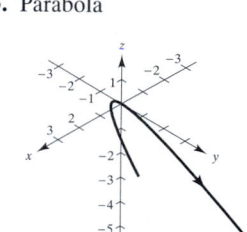

35. Helix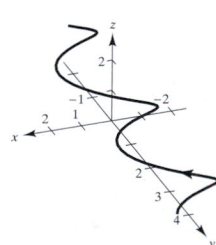

37.

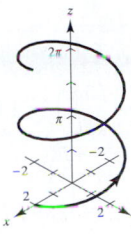

(a) The helix is translated two units back on the x-axis.

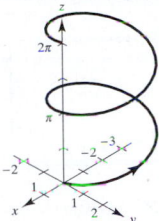

(b) The height of the helix increases at a faster rate.

(c) The orientation of the graph is reversed.

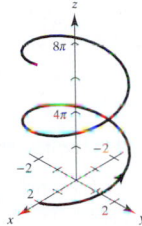

(d) The axis of the helix is the x-axis.

(e) The radius of the helix is increased from 2 to 6.

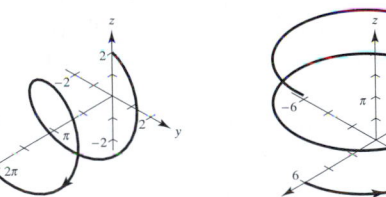

39. $\mathbf{r}(t) = t\mathbf{i} + (4 - t)\mathbf{j}$ **41.** $\mathbf{r}(t) = 5\cos t\mathbf{i} + 5\sin t\mathbf{j}$

43. $\mathbf{r}(t) = \langle 2 - 2t, 3 + 5t, 8t \rangle$

45. $\mathbf{r}_1(t) = t\mathbf{i}, \quad 0 \le t \le 4$
$\mathbf{r}_2(t) = (4 - 4t)\mathbf{i} + 6t\mathbf{j}, \quad 0 \le t \le 1$
$\mathbf{r}_3(t) = (6 - t)\mathbf{j}, \quad 0 \le t \le 6$

47. $\mathbf{r}_1(t) = t\mathbf{i} + t^2\mathbf{j}, \quad 0 \le t \le 2$
$\mathbf{r}_2(t) = (2 - t)\mathbf{i}, \quad 0 \le t \le 2$
$\mathbf{r}_3(t) = (4 - t)\mathbf{j}, \quad 0 \le t \le 4$

49. $\mathbf{r}(t) = t\mathbf{i} - t\mathbf{j} + 2t^2\mathbf{k}$ **51.** $\mathbf{r}(t) = 2\sin t\mathbf{i} + 2\cos t\mathbf{j} + 4\sin^2 t\mathbf{k}$

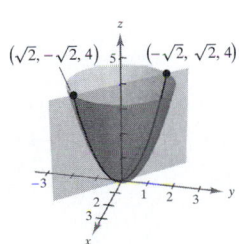

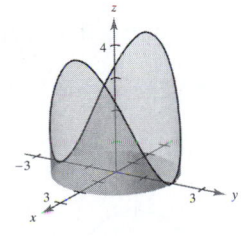

53. $\mathbf{r}(t) = (1 + \sin t)\mathbf{i} + \sqrt{2}\cos t\mathbf{j} + (1 - \sin t)\mathbf{k}$
and
$\mathbf{r}(t) = (1 + \sin t)\mathbf{i} - \sqrt{2}\cos t\mathbf{j} + (1 - \sin t)\mathbf{k}$

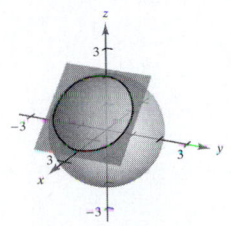

55. $\mathbf{r}(t) = t\mathbf{i} + t\mathbf{j} + \sqrt{4 - t^2}\mathbf{k}$

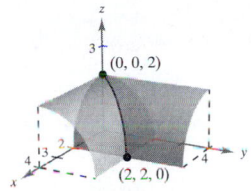

57. $2\mathbf{i} + 2\mathbf{j} + \frac{1}{2}\mathbf{k}$ **59.** $\mathbf{0}$ **61.** Limit does not exist.

63. $(-\infty, 0), (0, \infty)$ **65.** $[-1, 1]$

67. $\left(-\dfrac{\pi}{2} + n\pi, \dfrac{\pi}{2} + n\pi\right)$ **69.** Proof

71. Proof   **73.** True

74. False: $x = y = z = t^3$ represents a line.

Section 11.2 *(page 783)*

1. $\mathbf{r}(2) = 4\mathbf{i} + 2\mathbf{j}$

$\mathbf{r}'(2) = 4\mathbf{i} + \mathbf{j}$

3. $\mathbf{r}\left(\dfrac{\pi}{2}\right) = \mathbf{j}$

$\mathbf{r}'\left(\dfrac{\pi}{2}\right) = -\mathbf{i}$

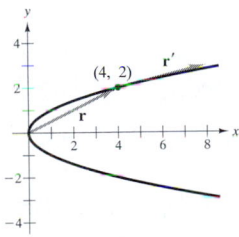

$\mathbf{r}'(t_0)$ is tangent to the curve at t_0.

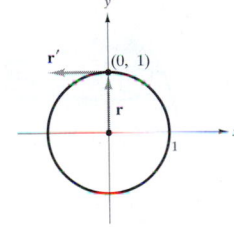

$\mathbf{r}'(t_0)$ is tangent to the curve at t_0.

5. (a) and (b)

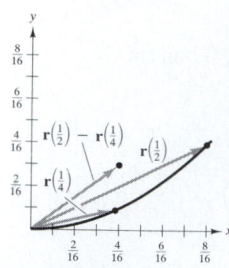

(c) The vector

$$\frac{\mathbf{r}\left(\frac{1}{2}\right) - \mathbf{r}\left(\frac{1}{4}\right)}{\frac{1}{2} - \frac{1}{4}}$$

approximates the tangent vector $\mathbf{r}'\left(\frac{1}{4}\right)$.

7. $\mathbf{r}\left(\dfrac{3\pi}{2}\right) = -2\mathbf{j} + \left(\dfrac{3\pi}{2}\right)\mathbf{k}$

$\mathbf{r}'\left(\dfrac{3\pi}{2}\right) = 2\mathbf{i} + \mathbf{k}$

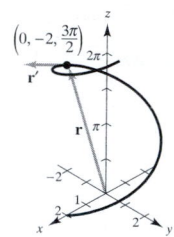

9. $\dfrac{\mathbf{r}'\left(-\frac{1}{4}\right)}{\left\|\mathbf{r}'\left(-\frac{1}{4}\right)\right\|} = \dfrac{1}{\sqrt{4\pi^2 + 1}}\left(\sqrt{2}\,\pi\mathbf{i} + \sqrt{2}\,\pi\mathbf{j} - \mathbf{k}\right)$

$\dfrac{\mathbf{r}''\left(-\frac{1}{4}\right)}{\left\|\mathbf{r}''\left(-\frac{1}{4}\right)\right\|} = \dfrac{1}{2\sqrt{\pi^4 + 4}}\left(-\sqrt{2}\,\pi^2\mathbf{i} + \sqrt{2}\,\pi^2\mathbf{j} + 4\mathbf{k}\right)$

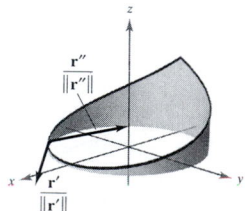

11. $6\mathbf{i} - 14t\mathbf{j} + 3t^2\mathbf{k}$ **13.** $-3a\sin t\cos^2 t\mathbf{i} + 3a\sin^2 t\cos t\mathbf{j}$

15. $-e^{-t}\mathbf{i}$ **17.** $\langle \sin t + t\cos t, \cos t - t\sin t, 1\rangle$

19. (a) $\mathbf{i} + 3\mathbf{j} + 2t\mathbf{k}$ (b) $2\mathbf{k}$ (c) $8t + 9t^2 + 5t^4$

(d) $-\mathbf{i} + (9 - 2t)\mathbf{j} + (6t - 3t^2)\mathbf{k}$

(e) $8t^3\mathbf{i} + (12t^2 - 4t^3)\mathbf{j} + (3t^2 - 24t)\mathbf{k}$

(f) $\dfrac{10 + 2t^2}{\sqrt{10 + t^2}}$

21. $\theta(t) = \arccos\left(\dfrac{-7\sin t\cos t}{\sqrt{9\sin^2 t + 16\cos^2 t}\,\sqrt{9\cos^2 t + 16\sin^2 t}}\right)$

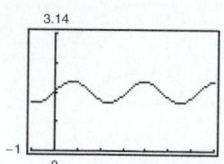

Maximum: $\theta\left(\dfrac{\pi}{4}\right) = \theta\left(\dfrac{5\pi}{4}\right) \approx 1.855$

Minimum: $\theta\left(\dfrac{3\pi}{4}\right) = \theta\left(\dfrac{7\pi}{4}\right) \approx 1.287$

Orthogonal: $\theta\left(\dfrac{n\pi}{2}\right) = \dfrac{\pi}{2}$

23. $(-\infty, 0),\ (0, \infty)$ **25.** $\left(\dfrac{n\pi}{2}, \dfrac{(n+1)\pi}{2}\right)$ **27.** $(-\infty, \infty)$

29. $(-\infty, 0),\ (0, \infty)$ **31.** $\left(-\dfrac{\pi}{2} + n\pi, \dfrac{\pi}{2} + n\pi\right)$, n is an integer.

33. $\mathbf{r}'(t) = 3\mathbf{i} - 2t\mathbf{j}$

35. The three components of $\mathbf{u}$ are increasing functions of t at $t = t_0$.

37. $t^2\mathbf{i} + t\mathbf{j} + t\mathbf{k} + \mathbf{C}$ **39.** $\ln t\mathbf{i} + t\mathbf{j} - \frac{2}{5}t^{5/2}\mathbf{k} + \mathbf{C}$

41. $(t^2 - t)\mathbf{i} + t^4\mathbf{j} + 2t^{3/2}\mathbf{k} + \mathbf{C}$ **43.** $\tan t\mathbf{i} + \arctan t\mathbf{j} + \mathbf{C}$

45. $2e^{2t}\mathbf{i} + 3(e^t - 1)\mathbf{j}$ **47.** $600\sqrt{3}t\mathbf{i} + (-16t^2 + 600t)\mathbf{j}$

49. $\left(\dfrac{2 - e^{-t^2}}{2}\right)\mathbf{i} + (e^{-t} - 2)\mathbf{j} + (t + 1)\mathbf{k}$

51. $4\mathbf{i} + \dfrac{1}{2}\mathbf{j} - \mathbf{k}$ **53.** $a\mathbf{i} + a\mathbf{j} + \dfrac{\pi}{2}\mathbf{k}$

55. Proof   **57.** Proof   **59.** Proof   **61.** Proof

63. False: let $\mathbf{r}(t) = \cos t\mathbf{i} + \sin t\mathbf{j} + \mathbf{k}$, then $\dfrac{d}{dt}\left[\|\mathbf{r}(t)\|\right] = 0$, but $\|\mathbf{r}'(t)\| = 1$.

64. False: $D_t[\mathbf{r}(t) \cdot \mathbf{u}(t)] = \mathbf{r}(t) \cdot \mathbf{u}'(t) + \mathbf{r}'(t) \cdot \mathbf{u}(t)$

Section 11.3 (page 791)

1. $\mathbf{v}(1) = 3\mathbf{i} + \mathbf{j}$ **3.** $\mathbf{v}(2) = 4\mathbf{i} + \mathbf{j}$
$\mathbf{a}(1) = \mathbf{0}$ $\mathbf{a}(2) = 2\mathbf{i}$

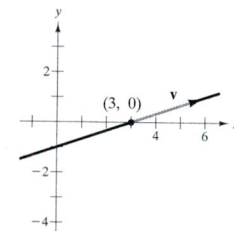

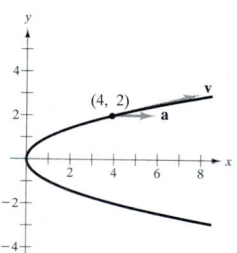

5. $\mathbf{v}\left(\dfrac{\pi}{4}\right) = -\sqrt{2}\mathbf{i} + \sqrt{2}\mathbf{j}$

$\mathbf{a}\left(\dfrac{\pi}{4}\right) = -\sqrt{2}\mathbf{i} - \sqrt{2}\mathbf{j}$

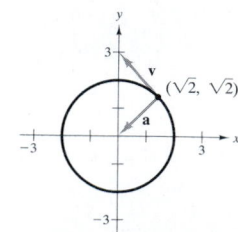

7. $\mathbf{v}(\pi) = 2\mathbf{i}$
$\mathbf{a}(\pi) = -\mathbf{j}$

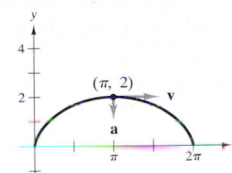

9. $\mathbf{v}(t) = \mathbf{i} + 2\mathbf{j} + 3\mathbf{k}$ **11.** $\mathbf{v}(t) = \mathbf{i} + 2t\mathbf{j} + t\mathbf{k}$
$s(t) = \sqrt{14}$ $s(t) = \sqrt{1 + 5t^2}$
$\mathbf{a}(t) = \mathbf{0}$ $\mathbf{a}(t) = 2\mathbf{j} + \mathbf{k}$

13. $\mathbf{v}(t) = \mathbf{i} + \mathbf{j} - \dfrac{t}{\sqrt{9 - t^2}}\mathbf{k}$

$s(t) = \sqrt{\dfrac{18 - t^2}{9 - t^2}}$

$\mathbf{a}(t) = \dfrac{-9}{(9 - t^2)^{3/2}}\mathbf{k}$

15. $\mathbf{v}(t) = 4\mathbf{i} - 3\sin t\mathbf{j} + 3\cos t\mathbf{k}$
$s(t) = 5$
$\mathbf{a}(t) = -3\cos t\mathbf{j} - 3\sin t\mathbf{k}$

17. (a) $x = 1 + t$ (b) $(1.100, -1.200, 0.325)$
$\quad\quad y = -1 - 2t$
$\quad\quad z = \dfrac{1}{4} + \dfrac{3}{4}t$

19. $\mathbf{v}(t) = t(\mathbf{i} + \mathbf{j} + \mathbf{k})$
$\mathbf{r}(t) = \dfrac{t^2}{2}(\mathbf{i} + \mathbf{j} + \mathbf{k})$
$\mathbf{r}(2) = 2(\mathbf{i} + \mathbf{j} + \mathbf{k})$

21. $\mathbf{v}(t) = \left(\dfrac{t^2}{2} + \dfrac{9}{2}\right)\mathbf{j} + \left(\dfrac{t^2}{2} - \dfrac{1}{2}\right)\mathbf{k}$

$\mathbf{r}(t) = \left(\dfrac{t^3}{6} + \dfrac{9}{2}t - \dfrac{14}{3}\right)\mathbf{j} + \left(\dfrac{t^3}{6} - \dfrac{1}{2}t + \dfrac{1}{3}\right)\mathbf{k}$

$\mathbf{r}(2) = \dfrac{17}{3}\mathbf{j} + \dfrac{2}{3}\mathbf{k}$

23. $\mathbf{r}(t) = 44\sqrt{3}t\mathbf{i} + (10 + 44t - 16t^2)\mathbf{j}$

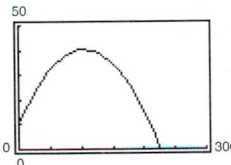

25. $v_0 = 40\sqrt{6}$ feet per second, 78 feet

27. (a) 54.1 feet per second   (b) 22.0 feet   (c) 2.0 seconds

29. $v_0 = 28.78$ feet per second, $\theta = 58.28°$ **31.** $1.91°$

33. (a) 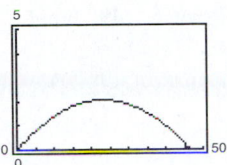   Maximum height: 2.1 feet
Range: 46.6 feet

(b) 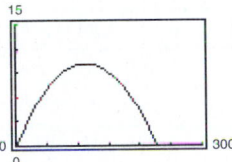   Maximum height: 10.0 feet
Range: 227.8 feet

(c) 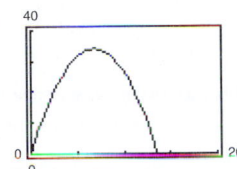   Maximum height: 34.0 feet
Range: 136.1 feet

(d) 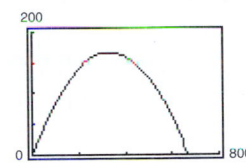   Maximum height: 166.5 feet
Range: 666.1 feet

(e) 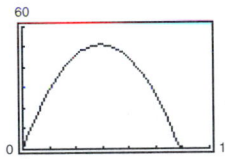   Maximum height: 51.0 feet
Range: 117.9 feet

(f) 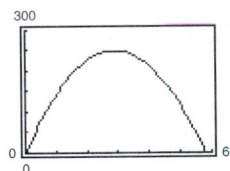   Maximum height: 249.8 feet
Range: 576.9 feet

35. $\mathbf{r}(t) = \left(40\sqrt{2}t\right)\mathbf{i} + \left(40\sqrt{2}t - 16t^2\right)\mathbf{j}$
Direction: $8\sqrt{2}(5\mathbf{i} + 2\mathbf{j})$
Speed: $8\sqrt{58} \approx 60.9$ feet per second

37. Maximum height: 129.1 meters
Range: 886.3 meters

39. $\mathbf{v}(t) = b\omega[1 - \cos \omega t)\mathbf{i} + \sin \omega t\mathbf{j}]$
$\mathbf{a}(t) = b\omega^2(\sin \omega t\mathbf{i} + \cos \omega t\mathbf{j})$
(a) $\|\mathbf{v}(t)\| = 0$ when $\omega t = 0, 2\pi, 4\pi, \ldots$
(b) $\|\mathbf{v}(t)\|$ is maximum when $\omega t = \pi, 3\pi, \ldots$

41. $\mathbf{v}(t) = -b\omega \sin \omega t\mathbf{i} + \omega t \cos \omega t\mathbf{j}$
$\mathbf{v}(t) \cdot \mathbf{r}(t) = 0$

43. $\mathbf{a}(t) = -b\omega^2(\cos \omega t\mathbf{i} + \sin \omega t\mathbf{j}) = -\omega^2\mathbf{r}(t)$

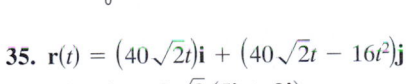

45. $8\sqrt{10}$ feet per second   **47.** Proof   **49.** Proof

51. (a) $\mathbf{v}(t) = -6\sin t\mathbf{i} + 3\cos t\mathbf{j}$
$\|\mathbf{v}(t)\| = 3\sqrt{3\sin^2 t + 1}$
$\mathbf{a}(t) = -6\cos t\mathbf{i} - 3\sin t\mathbf{j}$

(b)

t	0	$\pi/4$	$\pi/2$	$2\pi/3$	π
Speed	3	$3\sqrt{10}/2$	6	$3\sqrt{13}/2$	3

(c)

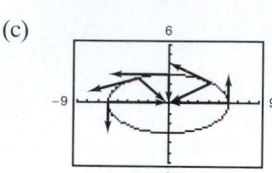

(d) The speed is increasing when the angle between $\mathbf{v}$ and $\mathbf{a}$ is in the interval $[0, \pi/2)$ and decreasing when the angle is in the interval $(\pi/2, \pi]$.

Section 11.4 (page 800)

1. $\mathbf{T}(0) = \dfrac{\sqrt{2}}{2}(\mathbf{i} + \mathbf{k})$ **3.** $\mathbf{T}(0) = \dfrac{\sqrt{5}}{5}(2\mathbf{j} + \mathbf{k})$

$x = t$ $x = 2$
$y = 0$ $y = 2t$
$z = t$ $z = t$

5. $\mathbf{T}\left(\dfrac{\pi}{4}\right) = \dfrac{1}{2}\langle -\sqrt{2}, \sqrt{2}, 0\rangle$

$x = \sqrt{2} - \sqrt{2}t$
$y = \sqrt{2} + \sqrt{2}t$
$z = 4$

7. $\mathbf{T}(3) = \frac{1}{19}\langle 1, 6, 18\rangle$

$x = 3 + t$
$y = 9 + 6t$
$z = 18 + 18t$

9. Tangent line: $x = 1 + t$
$y = t$
$z = 1 + \dfrac{1}{2}t$

$\mathbf{r}(1.1) \approx \langle 1.1, 0.1, 1.05\rangle$

11. $1.2°$

13. $\mathbf{v}(t) = 4\mathbf{i}$
$\mathbf{a}(t) = \mathbf{0}$
$\mathbf{T}(t) = \mathbf{i}$
$\mathbf{N}(t)$ is undefined. The path is a line and the speed is constant.

15. $\mathbf{v}(t) = 8t\mathbf{i}$
$\mathbf{a}(t) = 8\mathbf{i}$
$\mathbf{T}(t) = \mathbf{i}$
$\mathbf{N}(t)$ is undefined. The path is a line and the speed is variable.

17. $\mathbf{T} = \dfrac{\sqrt{2}}{2}(\mathbf{i} - \mathbf{j})$ **19.** $\mathbf{T} = \dfrac{\sqrt{2}}{2}(-\mathbf{i} + \mathbf{j})$

$\mathbf{N} = \dfrac{\sqrt{2}}{2}(\mathbf{i} + \mathbf{j})$ $\mathbf{N} = -\dfrac{\sqrt{2}}{2}(\mathbf{i} + \mathbf{j})$

$a_T = -\sqrt{2}$ $a_T = \sqrt{2}e^{\pi/2}$

$a_N = \sqrt{2}$ $a_N = \sqrt{2}e^{\pi/2}$

21. $\mathbf{T} = (\cos \omega t_0)\mathbf{i} + (\sin \omega t_0)\mathbf{j}$
$\mathbf{N} = (-\sin \omega t_0)\mathbf{i} + (\cos \omega t_0)\mathbf{j}$
$a_T = \omega^2$
$a_N = \omega^3 t_0$

23. $\mathbf{T}(t) = -\sin(\omega t)\mathbf{i} + \cos(\omega t)\mathbf{j}$
$\mathbf{N}(t) = -\cos(\omega t)\mathbf{i} - \sin(\omega t)\mathbf{j}$
$a_T = 0$
$a_N = a\omega^2$

25. $\|\mathbf{v}(t)\| = a\omega$ The speed is constant because $a_T = 0$.

27. $\mathbf{r}(2) = 2\mathbf{i} + \dfrac{1}{2}\mathbf{j}$

$\mathbf{T}(2) = \dfrac{\sqrt{17}}{17}(4\mathbf{i} - \mathbf{j})$

$\mathbf{N}(2) = \dfrac{\sqrt{17}}{17}(\mathbf{i} + 4\mathbf{j})$

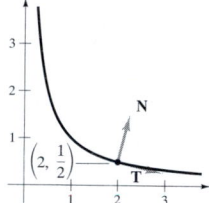

29. (a) $t = \dfrac{1}{2}$: $a_T = \dfrac{\sqrt{2}\pi^2}{2}, a_N = \dfrac{\sqrt{2}\pi^2}{2}$

$t = 1$: $a_T = 0, a_N = \pi^2$

$t = \dfrac{3}{2}$: $a_T = -\dfrac{\sqrt{2}\pi^2}{2}, a_N = \dfrac{\sqrt{2}\pi^2}{2}$

(b) $t = \dfrac{1}{2}$: Increasing
$t = 1$: Maximum
$t = \dfrac{3}{2}$: Decreasing

31. $\mathbf{T}(t) = \dfrac{\sqrt{14}}{14}(\mathbf{i} + 2\mathbf{j} - 3\mathbf{k})$

$\mathbf{N}(t)$ is undefined.
a_T is undefined.
a_N is undefined.

33. $\mathbf{T} = \dfrac{\sqrt{6}}{6}(\mathbf{i} + 2\mathbf{j} + \mathbf{k})$

$\mathbf{N} = \dfrac{\sqrt{30}}{30}(-5\mathbf{i} + 2\mathbf{j} + \mathbf{k})$

$a_T = \dfrac{5\sqrt{6}}{6}$

$a_N = \dfrac{\sqrt{30}}{6}$

35. $T = \frac{1}{5}(4i - 3j)$

$N = -k$

$a_T = 0$

$a_N = 3$

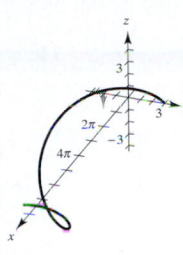

37. $T\left(\frac{\pi}{2}\right) = \frac{\sqrt{17}}{17}(-4i + k)$

$N\left(\frac{\pi}{2}\right) = -j$

$B\left(\frac{\pi}{2}\right) = \frac{\sqrt{17}}{17}(i + 4k)$

39. (a) $4\sqrt{625\pi^2 + 1} \approx 314$ miles per hour

(b) $a_T = 0$, $a_N = 1000\pi^2$

41. $a_T = \frac{-32(v_0 \sin\theta - 32t)}{\sqrt{v_0{}^2 \cos^2\theta + (v_0 \sin\theta - 32t)^2}}$

$a_N = \frac{32v_0 \cos\theta}{\sqrt{v_0{}^2 \cos^2\theta + (v_0 \sin\theta - 32t)^2}}$

43. (a) Centripetal component is quadrupled.

(b) Centripetal component is halved.

45. 4.83 miles per second   **47.** 4.67 miles per second

49. Proof   **51.** Proof

Section 11.5 *(page 811)*

1. $4\sqrt{10}$

3. $6a$

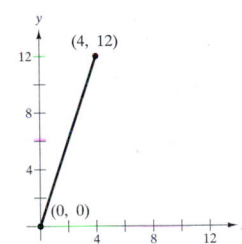

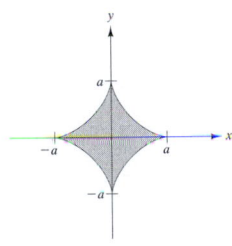

5. $2\sqrt{14}$

7. $2\pi\sqrt{a^2 + b^2}$

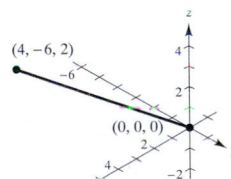

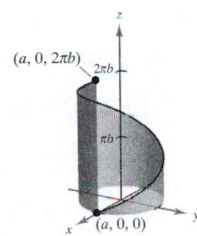

9. 8.37

11. (a) $2\sqrt{21} \approx 9.165$

(b) 9.529

(c) Increase the number of line segments.

(d) 9.571

13. (a) $s = \sqrt{5}\,t$

(b) $r(t) = 2\cos\frac{s}{\sqrt{5}}i + 2\sin\frac{s}{\sqrt{5}}j + \frac{s}{\sqrt{5}}k$

(c) $s = \sqrt{5}$: $(1.081, 1.683, 1.000)$

$s = 4$: $(-0.433, 1.953, 1.789)$

15. $K = 0$ **17.** $K = \frac{2}{5}$ **19.** $K = 0$ **21.** $\frac{\sqrt{2}}{2}$

23. $\frac{1}{4}$ **25.** $\frac{1}{a}$ **27.** $\frac{\sqrt{2}}{2}e^{-t}$ **29.** $\frac{1}{\omega t}$

31. $\frac{\sqrt{5}}{(1 + 5t^2)^{3/2}}$ **33.** $\frac{3}{25}$ **35.** $K = 0$, $\frac{1}{K}$ is undefined.

37. $K = \frac{4}{17^{3/2}}$, $\frac{1}{K} = \frac{17^{3/2}}{4}$ **39.** $K = \frac{1}{a}$, $\frac{1}{K} = a$

41. (a) $\left(x - \frac{\pi}{2}\right)^2 + y^2 = 1$

(b) Because the curvature is not as great, the radius of the curvature is greater.

43. $(x - 1)^2 + \left(y - \frac{5}{2}\right)^2 = \left(\frac{1}{2}\right)^2$

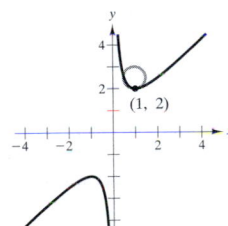

45. $(x + 2)^2 + (y - 3)^2 = 8$

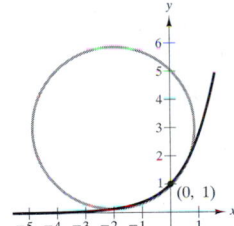

47. (a) $(1, 3)$ (b) 0 **49.** (a) $K \to \infty$ as $x \to 0$ (b) 0

51. $(1, 3)$ **53.** Proof

55. (a) $K = \dfrac{2|6x^2 - 1|}{(16x^6 - 16x^4 + 4x^2 + 1)^{3/2}}$

(b) $x = 0$: $\quad x^2 + \left(y + \dfrac{1}{2}\right)^2 = \dfrac{1}{4}$

$x = 1$: $\quad x^2 + \left(y - \dfrac{1}{2}\right)^2 = \dfrac{5}{4}$

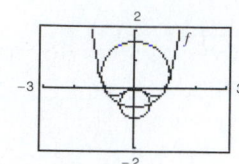

(c)

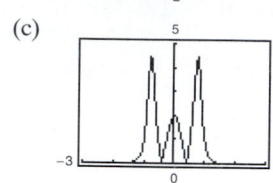

The curvature tends to be greatest near the extrema of the function and decreases as $x \to \pm\infty$. However, f and K do not have the same critical numbers.

Critical numbers of f: $x = 0, \pm\dfrac{\sqrt{2}}{2} \approx \pm 0.7071$

Critical numbers of K: $x = 0, \pm 0.7647, \pm 0.4082$

57. (a) 12.25 units (b) $\dfrac{1}{2}$

59. At a smooth relative extremum $K = |y''|$. Yes. $y = x^4$ has a curvature of 0 at its minimum $(0, 0)$. The curvature is positive for any other point on the curve.

61. $(-2, 3)$ **63.** $\dfrac{3}{2\sqrt{2(1 + \sin\theta)}}$ **65.** $\dfrac{2}{|a|}$

67. (a) 0 (b) 0 **69.** $\dfrac{1}{4}$ **71.** Proof

73. $K = \dfrac{1}{4a}\left|\csc\dfrac{\theta}{2}\right|$

Minimum: $K = \dfrac{1}{4a}$

There is no maximum.

75. 2420 pounds **77.** Proof **79.** Proof

81. Proof **83.** Proof

85. (a) a (b) πa (c) $K = \pi a$

Review Exercises for Chapter 11 *(page 815)*

1. (a) All reals except $n\pi$, n is an integer

(b) Continuous except at $t = n\pi$, n is an integer

3. (a) $(0, \infty)$ (b) Continuous for all $t > 0$

5. (a) $\mathbf{i}$

(b) $-3\mathbf{i} + 4\mathbf{j} + \dfrac{8}{3}\mathbf{k}$

(c) $(2c - 1)\mathbf{i} + (c - 1)^2\mathbf{j} + \dfrac{1}{3}(1 - c)^3\mathbf{k}$

(d) $2\,\Delta t\mathbf{i} + \Delta t(\Delta t + 2)\mathbf{j} - \dfrac{1}{3}\Delta t[(\Delta t)^2 + 3\Delta t + 3]\mathbf{k}$

7. $1 - t - \sin t$ No: the dot product is a scalar.

9.

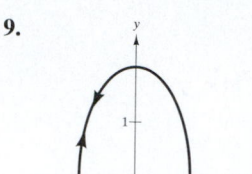

11.

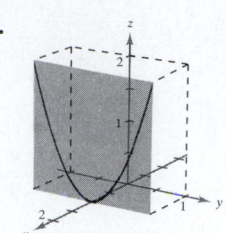

13.

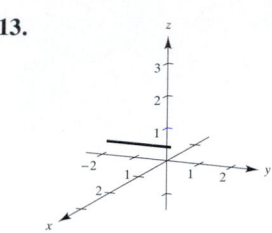

15.

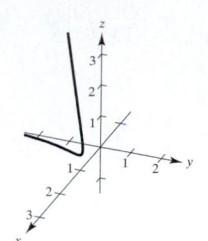

17. $\mathbf{r}_1(t) = 4t\mathbf{i} + 3t\mathbf{j}, \quad 0 \le t \le 1$

$\mathbf{r}_2(t) = 4\mathbf{i} + (3 - t)\mathbf{j}, \quad 0 \le t \le 3$

$\mathbf{r}_3(t) = (4 - t)\mathbf{i}, \quad 0 \le t \le 4$

19. $\mathbf{r}(t) = \langle -2 + 7t, -3 + 4t, 8 - 10t \rangle$

21. $x = t, y = -t, z = 2t^2$ **23.** $4\mathbf{i} + \mathbf{k}$

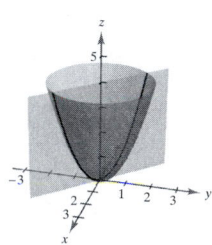

25. (a) $3\mathbf{i} + \mathbf{j}$ (b) $\mathbf{0}$ (c) $4t + 3t^2$

(d) $-5\mathbf{i} + (2t - 2)\mathbf{j} + 2t^2\mathbf{k}$ (e) $\dfrac{10t - 1}{\sqrt{10t^2 - 2t + 1}}$

(f) $\left(\dfrac{8}{3}t^3 - 2t^2\right)\mathbf{i} - 8t^3\mathbf{j} + (9t^2 - 2t + 1)\mathbf{k}$

27. $x = -\sqrt{2} - \sqrt{2}t$

$y = \sqrt{2} - \sqrt{2}t$

$z = \dfrac{3\pi}{4} + t$

29. $x(t)$ and $y(t)$ are increasing functions at $t = t_0$, and $z(t)$ is a decreasing function at $t = t_0$.

31. $x(t) = t, y(t) = 16 + 8t, z(t) = 2 + \dfrac{1}{2}t$

$\mathbf{r}(4.1) \approx \langle 0.1, 16.8, 2.05 \rangle$

33. $\sin t\mathbf{i} + (t \sin t + \cos t)\mathbf{j} + \mathbf{C}$

35. $\dfrac{1}{2}\left(t\sqrt{1 + t^2} + \ln|t + \sqrt{1 + t^2}|\right) + \mathbf{C}$

37. $\mathbf{r}(t) = (t^2 + 1)\mathbf{i} + (e^t + 2)\mathbf{j} - (e^{-t} + 4)\mathbf{k}$

39. $\frac{32}{3}\mathbf{j}$

41. $\mathbf{v}(t) = \langle -3\cos^2 t \sin t, 3\sin^2 t \cos t, 3 \rangle$
$\|\mathbf{v}(t)\| = 3\sqrt{\sin^2 t \cos^2 t + 1}$
$\mathbf{a}(t) = \langle 3\cos t(3\sin^2 t - 1), 3\sin t(2\cos^2 t - \sin^2 t), 0 \rangle$

43. 152 feet   **45.** 34.9 meters per second

47. $\mathbf{v} = 5\mathbf{i}$
$\|\mathbf{v}\| = 5$
$\mathbf{a} = \mathbf{0}$
$\mathbf{a} \cdot \mathbf{T} = 0$
$\mathbf{a} \cdot \mathbf{N}$ does not exist.
$K = 0$

49. $\mathbf{v} = \mathbf{i} + \dfrac{1}{2\sqrt{t}}\mathbf{j}$
$\|\mathbf{v}\| = \dfrac{\sqrt{4t + 1}}{2\sqrt{t}}$
$\mathbf{a} = -\dfrac{1}{4t\sqrt{t}}\mathbf{j}$
$\mathbf{a} \cdot \mathbf{T} = \dfrac{-1}{4t\sqrt{t}\sqrt{4t+1}}$
$\mathbf{a} \cdot \mathbf{N} = \dfrac{1}{2t\sqrt{4t+1}}$
$K = \dfrac{2}{(4t+1)^{3/2}}$

51. $\mathbf{v} = e^t\mathbf{i} - e^{-t}\mathbf{j}$
$\|\mathbf{v}\| = \sqrt{e^{2t} + e^{-2t}}$
$\mathbf{a} = e^t\mathbf{i} + e^{-t}\mathbf{j}$
$\mathbf{a} \cdot \mathbf{T} = \dfrac{e^{2t} - e^{-2t}}{\sqrt{e^{2t} + e^{-2t}}}$
$\mathbf{a} \cdot \mathbf{N} = \dfrac{2}{\sqrt{e^{2t} + e^{-2t}}}$
$K = \dfrac{2}{(e^{2t} + e^{-2t})^{3/2}}$

53. $\mathbf{v} = \mathbf{i} + 2t\mathbf{j} + t\mathbf{k}$
$\|\mathbf{v}\| = \sqrt{1 + 5t^2}$
$\mathbf{a} = 2\mathbf{j} + \mathbf{k}$
$\mathbf{a} \cdot \mathbf{T} = \dfrac{5t}{\sqrt{1 + 5t^2}}$
$\mathbf{a} \cdot \mathbf{N} = \dfrac{\sqrt{5}}{\sqrt{1 + 5t^2}}$
$K = \dfrac{\sqrt{5}}{(1 + 5t^2)^{3/2}}$

55. $\dfrac{\sqrt{5}\pi}{2}$ **57.** 4.56 miles per second

59. The curvature changes abruptly from zero to a nonzero constant.

61. (a)

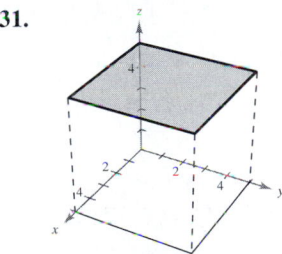

(b) $\sqrt{4\pi^2 + 1} + \dfrac{1}{2\pi}\ln\left(\sqrt{4\pi^2 + 1} + 2\pi\right) \approx 6.77$

(c) $K = \dfrac{\pi(\pi^2 t^2 + 2)}{(\pi^2 t^2 + 1)^{3/2}}$

$K(0) = 2\pi, K(1) = \dfrac{\pi(\pi^2 + 2)}{(\pi^2 + 1)^{3/2}} \approx 1.04$

$K(2) = \dfrac{2\pi(2\pi^2 + 1)}{(4\pi^2 + 1)^{3/2}} \approx 0.51$

(d)

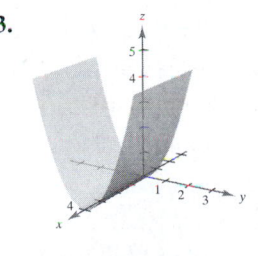

(c) 0

(f) As the graph spirals outward, the curvature decreases.

CHAPTER 12

Section 12.1 *(page 827)*

1. z is a function of x and y.   **3.** z is not a function of x and y.

5. (a) $\dfrac{3}{2}$ (b) $-\dfrac{1}{4}$ (c) 6   (d) $\dfrac{5}{y}$ (e) $\dfrac{x}{2}$ (f) $\dfrac{5}{t}$

7. (a) 5   (b) $3e^2$ (c) $\dfrac{2}{e}$ (d) $5e^y$ (e) xe^2 (f) te^t

9. (a) $\frac{2}{3}$ (b) 0   **11.** (a) $\sqrt{2}$ (b) $3\sin 1$

13. (a) 4   (b) 6   **15.** (a) $2x + \Delta x$ (b) -2

17. Domain: $\{(x, y): x^2 + y^2 \leq 4\}$
Range: $0 \leq z \leq 2$

19. Domain: $\{(x, y): -1 \leq x + y \leq 1\}$
Range: $-\dfrac{\pi}{2} \leq z \leq \dfrac{\pi}{2}$

21. Domain: $\{(x, y): y < -x + 4\}$
Range: all real numbers

23. Domain: $\{(x, y): x \neq 0, y \neq 0\}$
Range: all real numbers

25. Domain: $\{(x, y): y \neq 0\}$
Range: $z > 0$

27. Domain: $\{(x, y): x \neq 0, y \neq 0\}$
Range: $|z| > 0$

29. (a) $(20, 0, 0)$ (b) $(-15, 10, 20)$
(c) $(20, 15, 25)$ (d) $(20, 20, 0)$

31.

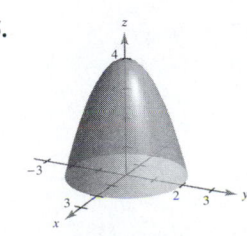

33.

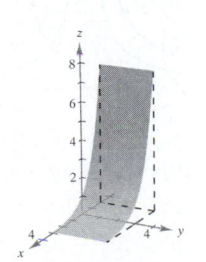

35.

37.

39. 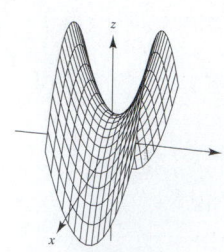   **41.**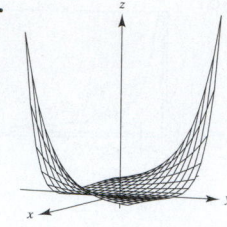

43. (a)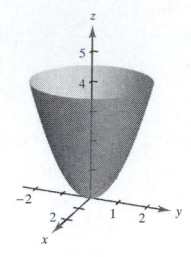

(b) g is a vertical translation of f two units upward.

(c) g is a horizontal translation of f two units to the right.

(d) g is a reflection of f in the xy-plane followed by a vertical translation 4 units upward.

(e) 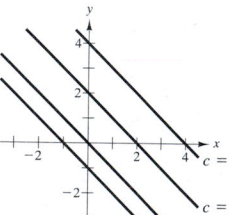   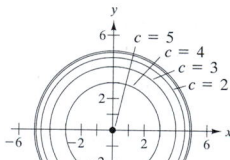

$z = f(1, y)$ $z = f(x, 1)$

45. c   **46.** d   **47.** b   **48.** a

49. Lines: $x + y = c$

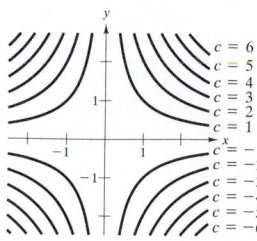

51. Circles centered at $(0, 0)$
Radius ≤ 5

53. Hyperbolas: $xy = c$

55. Circles passing through $(0, 0)$

Centered at $\left(\dfrac{1}{2c}, 0\right)$

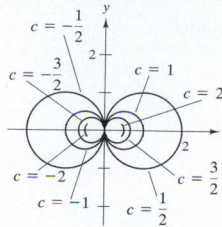

57. No: $z = e^{-(x^2 + y^2)}$

59. The surface may be shaped like a saddle. For example, let $f(x, y) = xy$. The graph is not unique: any vertical translation will produce the same level curves.

61.

Tax Rate	Inflation Rate		
	0	**0.03**	**0.05**
0	\$2593.74	\$1929.99	\$1592.33
0.28	\$2004.23	\$1491.34	\$1230.42
0.35	\$1877.14	\$1396.77	\$1152.40

63.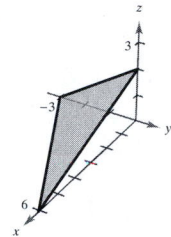

65.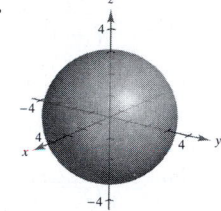

67.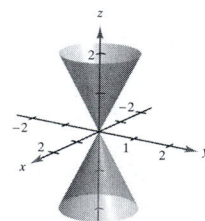

69. (a) 243 board-feet
(b) 507 board-feet

71.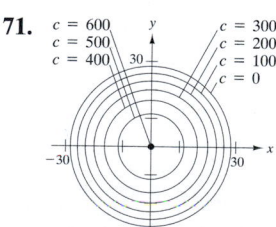

$c = 600$
$c = 500$
$c = 400$
$c = 300$
$c = 200$
$c = 100$
$c = 0$

73. $C = 0.75xy + 0.80(xz + yz)$

75. (a) $k = \dfrac{520}{3}$

(b) $P = \dfrac{520T}{3V}$

The level curves are lines.

77. (a) C (b) A (c) B

79. (a) The boundaries between colors represent level curves.

   (b) No: the colors represent intervals of different lengths.

   (c) Use more colors.

81. False: let $f(x, y) = 4$.   **82.** True

83. False: let $f(x, y) = xy^2$. $f(ax, ay) = a^3 f(x, y)$.   **84.** True

Section 12.2 *(page 837)*

1. 2   **3.** 15   **5.** 5, continuous

7. -3, continuous for $x \neq y$

9. 0, continuous for $xy \neq -1$, $y \neq 0$, $\left| \dfrac{x}{y} \right| \leq 1$

11. 1, continuous   **13.** $2\sqrt{2}$, continuous for $x + y + z \geq 0$

15. 1, continuous

17. Continuous except at $(0, 0)$; the limit does not exist.

19.

(x, y)	$(1, 0)$	$(0.5, 0)$	$(0.1, 0)$	$(0.01, 0)$	$(0.001, 0)$
$f(x, y)$	0	0	0	0	0

$y = 0$: 0

(x, y)	$(1, 1)$	$(0.5, 0.5)$	$(0.1, 0.1)$	$(0.01, 0.01)$	$(0.001, 0.001)$
$f(x, y)$	1/2	1/2	1/2	1/2	1/2

$y = x$: $\frac{1}{2}$

Limit does not exist.

Continuous except at $(0, 0)$

21.

(x, y)	$(1, 1)$	$(0.25, 0.5)$	$(0.01, 0.1)$
$f(x, y)$	$-1/2$	$-1/2$	$-1/2$

(x, y)	$(0.0001, 0.01)$	$(0.000001, 0.001)$
$f(x, y)$	$-1/2$	$-1/2$

$x = y^2$: $-\frac{1}{2}$

(x, y)	$(-1, 1)$	$(-0.25, 0.5)$	$(-0.01, 0.1)$
$f(x, y)$	1/2	1/2	1/2

(x, y)	$(-0.0001, 0.01)$	$(-0.000001, 0.001)$
$f(x, y)$	1/2	1/2

$x = -y^2$: $\frac{1}{2}$

Limit does not exist.

Continuous except at $(0, 0)$

23. 0   **25.** Limit does not exist.

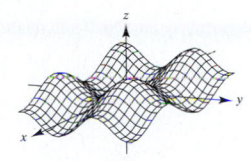
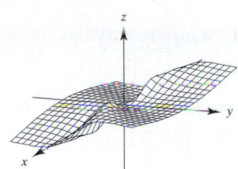

27. Limit does not exist.   **29.** 1   **31.** 0

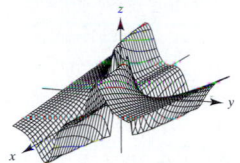

33. Continuous except at $(0, 0, 0)$

35. Continuous   **37.** Continuous

39. Continuous for $y \neq \dfrac{3x}{2}$ **41.** (a) $2x$ (b) -4

43. (a) $2 + y$ (b) $x - 3$ **45.** Proof

47. No: the existence of $f(2, 3)$ has no bearing on the existence of the limit as $(x, y) \to (2, 3)$.

49. True   **50.** False: let $f(x, y) = \dfrac{xy}{x^2 + y^2}$.

51. False: let $f(x, y) = \begin{cases} \ln(x^2 + y^2), & x \neq 0, y \neq 0 \\ 0, & x = 0, y = 0. \end{cases}$

52. True

Section 12.3 *(page 846)*

1. $f_x = (4, 1) < 0$ **3.** $f_y = (4, 1) > 0$

5. $f_x(x, y) = 2$ **7.** $\dfrac{\partial z}{\partial x} = \sqrt{y}$

   $f_y(x, y) = -3$ $\dfrac{\partial z}{\partial y} = \dfrac{x}{2\sqrt{y}}$

9. $\dfrac{\partial z}{\partial x} = 2xe^{2y}$ **11.** $\dfrac{\partial z}{\partial x} = \dfrac{2x}{x^2 + y^2}$

   $\dfrac{\partial z}{\partial y} = 2x^2 e^{2y}$ $\dfrac{\partial z}{\partial y} = \dfrac{2y}{x^2 + y^2}$

13. $\dfrac{\partial z}{\partial x} = \dfrac{-2y}{x^2 - y^2}$ **15.** $h_x(x, y) = -2xe^{-(x^2 + y^2)}$

   $\dfrac{\partial z}{\partial y} = \dfrac{2x}{x^2 - y^2}$ $h_y(x, y) = -2ye^{-(x^2 + y^2)}$

17. $f_x(x, y) = \dfrac{x}{\sqrt{x^2 + y^2}}$ **19.** $\dfrac{\partial z}{\partial x} = 2\sec^2(2x - y)$

   $f_y(x, y) = \dfrac{y}{\sqrt{x^2 + y^2}}$ $\dfrac{\partial z}{\partial y} = -\sec^2(2x - y)$

21. $\dfrac{\partial z}{\partial x} = ye^y \cos xy$

$\dfrac{\partial z}{\partial y} = e^y(x \cos xy + \sin xy)$

23. $f_x(x, y) = 1 - x^2$
$f_y(x, y) = y^2 - 1$

25. $f_x(x, y) = 2$
$f_y(x, y) = 3$

27. $f_x(x, y) = \dfrac{1}{2\sqrt{x + y}}$

$f_y(x, y) = \dfrac{1}{2\sqrt{x + y}}$

29. $g_x(1, 1) = -2$
$g_y(1, 1) = -2$

31. $\dfrac{\partial z}{\partial x} = -1$

$\dfrac{\partial z}{\partial y} = 0$

33. $\dfrac{\partial z}{\partial x} = \dfrac{1}{4}$

$\dfrac{\partial z}{\partial y} = \dfrac{1}{4}$

35. $\dfrac{\partial z}{\partial x} = -\dfrac{1}{4}$

$\dfrac{\partial z}{\partial y} = \dfrac{1}{4}$

37.

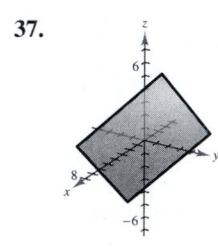

39. $-\dfrac{1}{2}$

41. 18

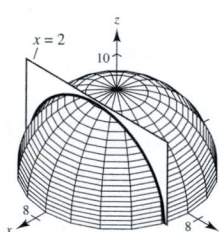

43. $x = -6, y = 4$ **45.** $x = 1, y = 1$

47. $\dfrac{\partial^2 z}{\partial x^2} = 2$

$\dfrac{\partial^2 z}{\partial y^2} = 6$

$\dfrac{\partial^2 z}{\partial y \partial x} = \dfrac{\partial^2 z}{\partial x \partial y} = -2$

49. $\dfrac{\partial^2 z}{\partial x^2} = \dfrac{y^2}{(x^2 + y^2)^{3/2}}$

$\dfrac{\partial^2 z}{\partial y^2} = \dfrac{x^2}{(x^2 + y^2)^{3/2}}$

$\dfrac{\partial^2 z}{\partial y \partial x} = \dfrac{\partial^2 z}{\partial x \partial y} = \dfrac{-xy}{(x^2 + y^2)^{3/2}}$

51. $\dfrac{\partial^2 z}{\partial x^2} = e^x \tan y$

$\dfrac{\partial^2 z}{\partial y^2} = 2e^x \sec^2 y \tan y$

$\dfrac{\partial^2 z}{\partial y \partial x} = \dfrac{\partial^2 z}{\partial x \partial y} = e^x \sec^2 y$

53. $\dfrac{\partial^2 z}{\partial x^2} = \dfrac{2xy}{(x^2 + y^2)^2}$

$\dfrac{\partial^2 z}{\partial y^2} = \dfrac{-2xy}{(x^2 + y^2)^2}$

$\dfrac{\partial^2 z}{\partial y \partial x} = \dfrac{\partial^2 z}{\partial x \partial y} = \dfrac{y^2 - x^2}{(x^2 + y^2)^2}$

55. $\dfrac{\partial z}{\partial x} = \sec y$

$\dfrac{\partial z}{\partial y} = x \sec y \tan y$

$\dfrac{\partial^2 z}{\partial x^2} = 0$

$\dfrac{\partial^2 z}{\partial y^2} = x \sec y(\sec^2 y + \tan^2 y)$

$\dfrac{\partial^2 z}{\partial y \partial x} = \dfrac{\partial^2 z}{\partial x \partial y} = \sec y \tan y$

57. $\dfrac{\partial z}{\partial x} = \dfrac{y^2 - x^2}{x(x^2 + y^2)}$

$\dfrac{\partial z}{\partial y} = \dfrac{-2y}{x^2 + y^2}$

$\dfrac{\partial^2 z}{\partial x^2} = \dfrac{x^4 - 4x^2y^2 - y^4}{x^2(x^2 + y^2)^2}$

$\dfrac{\partial^2 z}{\partial y^2} = \dfrac{2(y^2 - x^2)}{(x^2 + y^2)^2}$

$\dfrac{\partial^2 z}{\partial y \partial x} = \dfrac{\partial^2 z}{\partial x \partial y} = \dfrac{4xy}{(x^2 + y^2)^2}$

59. $\dfrac{\partial w}{\partial x} = \dfrac{x}{\sqrt{x^2 + y^2 + z^2}}$

$\dfrac{\partial w}{\partial y} = \dfrac{y}{\sqrt{x^2 + y^2 + z^2}}$

$\dfrac{\partial w}{\partial z} = \dfrac{z}{\sqrt{x^2 + y^2 + z^2}}$

61. $F_x(x, y, z) = \dfrac{x}{x^2 + y^2 + z^2}$

$F_y(x, y, z) = \dfrac{y}{x^2 + y^2 + z^2}$

$F_z(x, y, z) = \dfrac{z}{x^2 + y^2 + z^2}$

63. $H_x(x, y, z) = \cos(x + 2y + 3z)$
$H_y(x, y, z) = 2\cos(x + 2y + 3z)$
$H_z(x, y, z) = 3\cos(x + 2y + 3z)$

65. $f_{xyy}(x, y, z) = f_{yxy}(x, y, z) = f_{yyx}(x, y, z) = 0$

67. $f_{xyy}(x, y, z) = f_{yxy}(x, y, z) = f_{yyz}(x, y, z) = z^2e^{-x}\sin yz$

69. $\dfrac{\partial^2 z}{\partial x^2} + \dfrac{\partial^2 z}{\partial y^2} = 0 + 0 = 0$

71. $\dfrac{\partial^2 z}{\partial x^2} + \dfrac{\partial^2 z}{\partial y^2} = e^x \sin y - e^x \sin y = 0$

73. $\dfrac{\partial^2 z}{\partial t^2} = -c^2 \sin(x - ct) = c^2 \dfrac{\partial^2 z}{\partial x^2}$

75. $\dfrac{\partial z}{\partial t} = -e^{-t}\cos\dfrac{x}{c} = c^2 \dfrac{\partial^2 z}{\partial x^2}$

77. (a) $\dfrac{\partial C}{\partial x} = 183, \dfrac{\partial C}{\partial y} = 237$

(b) The fireplace-insert stove results in the cost increasing at a faster rate, because the coefficient of y is greater in magnitude than the coefficient of x.

79. An increase in either charge for food and housing or tuition will cause a decrease in the number of applicants.

81. $\dfrac{\partial T}{\partial x} = -2.4°$ per meter, $\dfrac{\partial T}{\partial y} = -9°$ per meter

83. (a) $U_x = -10x + y$

(b) $U_y = x - 6y$

(c) $U_x(2, 3) = -17, U_y(2, 3) = -16$

The person should consume one more unit of product y, because the rate of decrease of satisfaction is less for y.

(d)

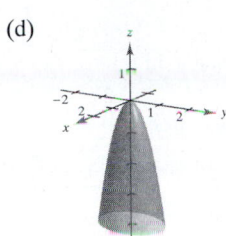

85. (a) $f_x(x, y) = \dfrac{y(x^4 + 4x^2y^2 - y^4)}{(x^2 + y^2)^2}$

$f_y(x, y) = \dfrac{x(x^4 - 4x^2y^2 - y^4)}{(x^2 + y^2)^2}$

(b) $f_x(0, 0) = 0$, $f_y(0, 0) = 0$

(c) $f_{xy}(0, 0) = -1$, $f_{yx}(0, 0) = 1$

(d) f_{xy} or f_{yx} or both are not continuous at $(0, 0)$.

86. False: let $z = x + y + 1$.   **87.** True

88. True   **89.** True

Section 12.4 *(page 855)*

1. $dz = 6xy^3 dx + 9x^2y^2 dy$ **3.** $dz = \dfrac{2}{(x^2 + y^2)^2}(x\, dx + y\, dy)$

5. $dz = (\cos y + y \sin x)\, dx - (x \sin y + \cos x)dy$

7. $dw = 2z^3 y \cos x\, dx + 2z^3 \sin x\, dy + 6z^2 y \sin x\, dz$

9. $dw = \dfrac{1}{z - 2y}dx + \dfrac{z + 2x}{(z - 2y)^2}dy - \dfrac{x + y}{(z - 2y)^2}dz$

11. (a) $f(1, 2) = 4, f(1.05, 2.1) = 3.4875, \Delta z = -0.5125$

(b) $dz = -0.5$

13. (a) $f(1, 2) = 0.90930, f(1.05, 2.1) = 0.90637, \Delta z = -0.00293$

(b) $dz = 0.00385$

15. (a) $f(1, 2) = -5, f(1.05, 2.1) = -5.25, \Delta z = -0.25$

(b) $dz = -0.25$

17. 0.094 **19.** -0.012

21. $dA = h\, dl + l\, dh$

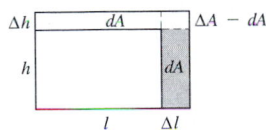

23.

Δr	Δh	dV	ΔV	$\Delta V - dV$
0.1	0.1	4.7124	4.8391	0.1267
0.1	-0.1	2.8274	2.8264	-0.0010
0.001	0.002	0.0565	0.0565	0.0001
-0.0001	0.0002	-0.0019	-0.0019	0.0000

25. 10%   **27.** ± 0.24 square inch   **29.** 7%

31. (a) 107.3 feet   (b) ± 8.27 feet

33. $L \approx 8.096 \times 10^{-4} \pm 6.6 \times 10^{-6}$ microhenrys

35. (a)

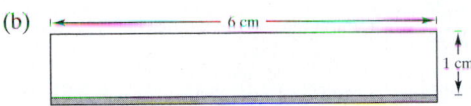

(b)

(c) Height

(d) $dl = 0.01$, $dh = 0$: $dA = 0.01$

$dl = 0$, $dh = 0.01$: $dA = 0.06$

37. Answers will vary.

39. $\varepsilon_1 = \Delta x$ **41.** $\varepsilon_1 = y\Delta x$

$\varepsilon_2 = 0$ $\varepsilon_2 = 2x\Delta x + (\Delta x)^2$

(Answer is not unique.)   (Answer is not unique.)

43. Proof

Section 12.5 *(page 863)*

1. $2(e^{2t} - e^{-2t})$ **3.** $e^t \sec(\pi - t)[1 - \tan(\pi - t)]$

5. $2 \cos 2t$ **7.** $4e^{2t}$ **9.** $3(2t^2 - 1)$ **11.** $\dfrac{24t}{(1 - 4t^2)^{5/2}}, 0$

13. $\dfrac{\partial w}{\partial s} = 4s, 8$ **15.** $\dfrac{\partial w}{\partial s} = 2s \cos 2t, 0$

$\dfrac{\partial w}{\partial t} = 4t, -4$ $\dfrac{\partial w}{\partial t} = -2s^2 \sin 2t, -18$

17. $\dfrac{\partial w}{\partial r} = 0$ **19.** $\dfrac{\partial w}{\partial r} = 0$

$\dfrac{\partial w}{\partial \theta} = 8\theta$ $\dfrac{\partial w}{\partial \theta} = 1$

21. $\dfrac{3y - 2x + 2}{2y - 3x + 1}$ **23.** $-\dfrac{x + y(x^2 + y^2)}{y + x(x^2 + y^2)}$

25. $\dfrac{\partial z}{\partial x} = \dfrac{-x}{z}$ **27.** $\dfrac{\partial z}{\partial x} = \dfrac{-\sec^2(x + y)}{\sec^2(y + z)}$

$\dfrac{\partial z}{\partial y} = \dfrac{-y}{z}$ $\dfrac{\partial z}{\partial y} = -1 - \dfrac{\sec^2(x + y)}{\sec^2(y + z)}$

29. $\dfrac{\partial z}{\partial x} = \dfrac{-x}{y + z}$ **31.** $\dfrac{\partial z}{\partial x} = -\dfrac{ze^{xz} + y}{xe^{xz}}$

$\dfrac{\partial z}{\partial y} = \dfrac{-z}{y + z}$ $\dfrac{\partial z}{\partial y} = -e^{-xz}$

33. $\dfrac{\partial w}{\partial z} = \dfrac{yw - xy - xw}{xz - yz + 2w}$ **35.** $\dfrac{\partial w}{\partial x} = \dfrac{y \sin xy}{z}$

$\dfrac{\partial w}{\partial y} = \dfrac{-xz + zw}{xz - yz + 2w}$ $\dfrac{\partial w}{\partial y} = \dfrac{x \sin xy - z \cos yz}{z}$

$\dfrac{\partial w}{\partial x} = \dfrac{-yz - zw}{xz - yz + 2w}$ $\dfrac{\partial w}{\partial z} = -\dfrac{y \cos yz + w}{z}$

37. 1   **39.** 0   **41.** $\dfrac{\sqrt{2}}{10}(15 + \pi)$ square meters per hour

43. $\dfrac{dV}{dt} = 1536\pi$ cubic inches per minute

$\dfrac{dS}{dt} = \dfrac{36\pi}{5}\left(20 + 9\sqrt{10}\right)$ square inches per minute

(Surface area includes base.)

45. $28m$ square centimeters per second

47. (a) $x = 32\sqrt{2}t$
$y = 32\sqrt{2}t - 16t^2$

(b) $\alpha = \arctan\left(\dfrac{y}{x + 50}\right)$

$= \arctan\left(\dfrac{32\sqrt{2}t - 16t^2}{32\sqrt{2}t + 50}\right)$

(c) $\dfrac{d\alpha}{dt} = \dfrac{-16\left(8\sqrt{2}t^2 + 25t - 25\sqrt{2}\right)}{64t^4 - 256\sqrt{2}t^3 + 1024t^2 + 800\sqrt{2}t + 625}$

(d)

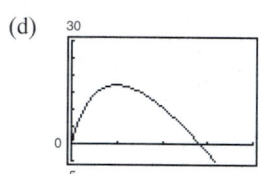

No. The rate of change of α is greatest when the projectile is closest to the camera.

(e) α is maximum when $t = 0.98$.
No: the projectile is at its maximum height when $t = \sqrt{2} \approx 1.41$ seconds

49. Proof   **51.** Proof   **53.** Proof   **55.** Proof

Section 12.6 *(page 874)*

1. $\dfrac{\sqrt{3} - 5}{2}$ **3.** $\dfrac{5\sqrt{2}}{2}$ **5.** $-\dfrac{7}{25}$ **7.** $-e$ **9.** $\dfrac{2\sqrt{6}}{3}$

11. $\dfrac{(8 + \pi)\sqrt{6}}{24}$ **13.** $\sqrt{2}(x + y)$

15. $\left(\dfrac{2 + \sqrt{3}}{2}\right)\cos(2x - y)$ **17.** $-7\sqrt{2}$ **19.** $\dfrac{7\sqrt{19}}{19}$

21. (a) Answers will vary.

(b) $-\dfrac{2}{5}\mathbf{i} + \dfrac{1}{10}\mathbf{j}$

(c) $\dfrac{2}{5}\mathbf{i} - \dfrac{1}{10}\mathbf{j}$
The direction opposite that of the gradient

23. (a)

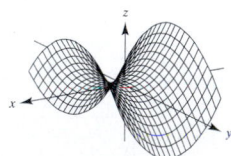

(b) $D_{\mathbf{u}}f(4, -3) = 8\cos\theta + 6\sin\theta$

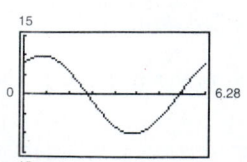

(c) $\theta \approx 2.21$, $\theta \approx 5.36$
Directions in which there is no change in f.

(d) $\theta \approx 0.64$, $\theta \approx 3.79$
Directions of greatest rate of change in f.

(e) 10; magnitude of the greatest rate of change

(f)

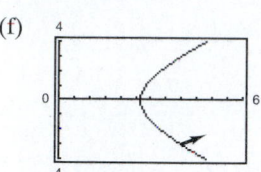

Orthogonal to the level curve

25. $\tan y\mathbf{i} + x\sec^2 y\mathbf{j}$, $\sqrt{17}$ **27.** $\dfrac{2}{3(x^2 + y^2)}(x\mathbf{i} + y\mathbf{j})$, $\dfrac{2\sqrt{5}}{15}$

29. $\dfrac{x\mathbf{i} + y\mathbf{j} + z\mathbf{k}}{\sqrt{x^2 + y^2 + z^2}}$, 1   **31.** $\dfrac{x\mathbf{i} + y\mathbf{j} + z\mathbf{k}}{(1 - x^2 - y^2 - z^2)^{3/2}}$, 0

33.

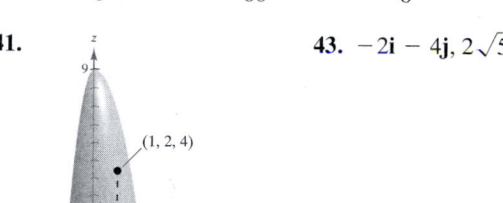

35. (a) $\dfrac{2 + 3\sqrt{3}}{12}$ (b) $\dfrac{3 - 2\sqrt{3}}{12}$

37. (a) $-\dfrac{1}{5}$ (b) $-\dfrac{11\sqrt{10}}{60}$ **39.** $\dfrac{\sqrt{13}}{6}$

41.

43. $-2\mathbf{i} - 4\mathbf{j}$, $2\sqrt{5}$

45. $6\mathbf{i} + 8\mathbf{j}$ **47.** $-\dfrac{1}{2}\mathbf{j}$

49. $\dfrac{\sqrt{257}}{257}(16\mathbf{i} - \mathbf{j})$ **51.** $\dfrac{\sqrt{85}}{85}(9\mathbf{i} - 2\mathbf{j})$

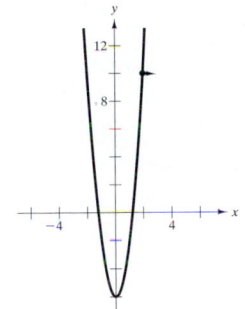

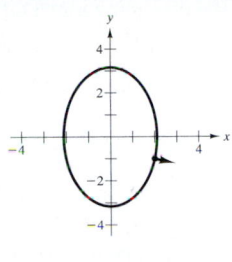

53. $\dfrac{1}{625}(7\mathbf{i} - 24\mathbf{j})$ **55.** $y^2 = 10x$

57. (a)

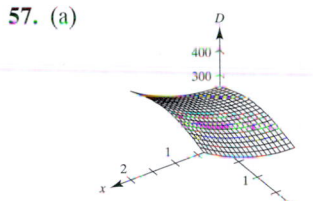

(b) Graph $-D$.   (c) 315 feet   (d) 60.0

(e) 55.5   (f) $60.0\mathbf{i} + 55.5\mathbf{j}$

59. A   **61.** True

62. False: $D_{\mathbf{u}} f(x, y) = \sqrt{2} > 1$ where $\mathbf{u} = \cos\dfrac{\pi}{4}\mathbf{i} + \sin\dfrac{\pi}{4}\mathbf{j}$.

63. True   **64.** True   **65.** $f(x, y, z) = e^x \cos y + \dfrac{1}{2}z^2 + C$

Section 12.7 *(page 883)*

1. $\dfrac{\sqrt{3}}{3}(\mathbf{i} + \mathbf{j} + \mathbf{k})$ **3.** $\dfrac{\sqrt{2}}{10}(3\mathbf{i} + 4\mathbf{j} - 5\mathbf{k})$

5. $\dfrac{\sqrt{2049}}{2049}(32\mathbf{i} + 32\mathbf{j} - \mathbf{k})$ **7.** $\dfrac{\sqrt{113}}{113}\left(-\mathbf{i} - 6\sqrt{3}\mathbf{j} + 2\mathbf{k}\right)$

9. $\dfrac{\sqrt{3}}{3}(\mathbf{i} - \mathbf{j} + \mathbf{k})$ **11.** $6x + 2y + z = 35$

13. $2x - y + z = 2$ **15.** $10x - 8y - z = 9$

17. $2x - z = -2$ **19.** $3x + 4y - 25z = 25(1 - \ln 5)$

21. $x - 4y + 2z = 18$ **23.** $x + y + z = 1$

25. $2x + 4y + z = 14$ **27.** $3x + 2y + z = -6$

$\dfrac{x - 1}{2} = \dfrac{y - 2}{4} = \dfrac{z - 4}{1}$ $\dfrac{x + 2}{3} = \dfrac{y + 3}{2} = \dfrac{z - 6}{1}$

29. $x - y + 2z = \dfrac{\pi}{2}$

$\dfrac{x - 1}{1} = \dfrac{y - 1}{-1} = \dfrac{z - (\pi/4)}{2}$

31. (a) Line: $x = 1, y = 1, z = 1 - t$
Plane: $z = 1$

(b) Line: $x = -1, y = 2 + \dfrac{6}{25}t, z = -\dfrac{4}{5} - t$
Plane: $6y - 25z - 32 = 0$

(c)

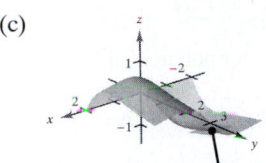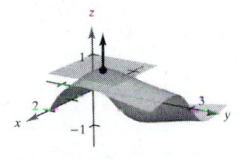

(d) At $(1, 1, 1)$ the tangent plane is parallel to the xy-plane, implying that the surface is level. At $\left(-1, 2, -\dfrac{4}{5}\right)$, the function does not change in the x-direction.

33. (a) $\dfrac{x - 2}{1} = \dfrac{y - 1}{-2} = \dfrac{z - 2}{1}$ (b) $\dfrac{\sqrt{10}}{5}$, not orthogonal

35. (a) $\dfrac{x - 3}{4} = \dfrac{y - 3}{4} = \dfrac{z - 4}{-3}$ (b) $\dfrac{16}{25}$, not orthogonal

37. (a) $\dfrac{y - 1}{1} = \dfrac{z - 1}{-1}, x = 2$ (b) 0, orthogonal

39. (a) $x = 1 + t$
$y = 2 - 2t$
$z = 4$
$\theta \approx 48.2°$

(b)

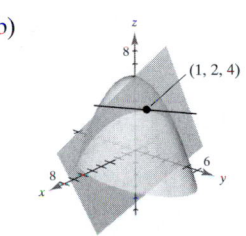

41. $86.0°$ **43.** $77.4°$

45. $(0, 3, 12)$

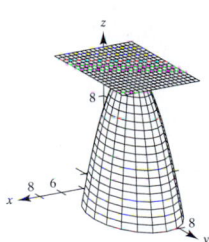

The function is maximum.

47. $x = 4e^{-4kt}, y = 3e^{-2kt}, z = 10e^{-8kt}$

49. Proof   **51.** Proof

53. (a) $P_1(x, y) = 1 + x - y$

(b) $P_2(x, y) = 1 + x - y + \frac{1}{2}x^2 - xy + \frac{1}{2}y^2$

(c) If $x = 0$, $P_2(0, y) = 1 - y + \frac{1}{2}y^2$

This is the second-degree Taylor polynomial for e^{-y}. If $y = 0$, $P_2(0, y) = 1 + x + \frac{1}{2}x^2$

This is the second-degree Taylor polynomial for e^x.

(d)

x	y	$f(x, y)$	$P_1(x, y)$	$P_2(x, y)$
0	0	1	1	1
0	0.1	0.9048	0.9000	0.9050
0.2	0.1	1.1052	1.1000	1.1050
0.2	0.5	0.7408	0.7000	0.7450
1	0.5	1.6487	1.5000	1.6250

(e)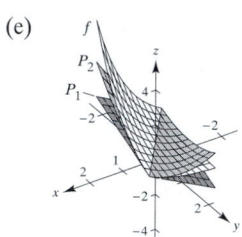

55. Proof

Section 12.8 *(page 892)*

1. Relative minimum: $(1, 3, 0)$ **3.** Relative minimum: $(0, 0, 1)$

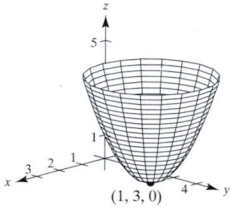 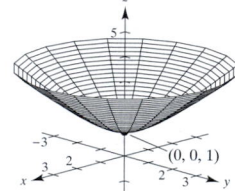

5. Relative minimum: $(-1, 3, -4)$

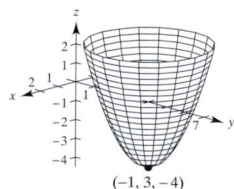

7. Relative minimum: $(-1, 1, -4)$

9. Relative maximum: $(8, 16, 74)$

11. Relative minimum: $(1, 2, -1)$

13. Saddle point: $(1, -2, -1)$ **15.** Saddle point: $(0, 0, 0)$

17. Saddle point: $(0, 0, 0)$
Relative minimum: $(1, 1, -1)$

19. There are no critical numbers.

21. Relative maximum: $(-1, 0, 2)$
Relative minimum: $(1, 0, -2)$

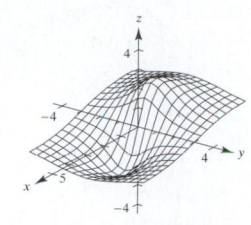

23. Relative maximum: $(0, 0, 1)$
Saddle points: $(0, 2, -3)$, $(\pm\sqrt{3}, -1, -3)$

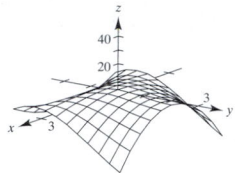

25. Insufficient information   **27.** Saddle point

29.   **31.**

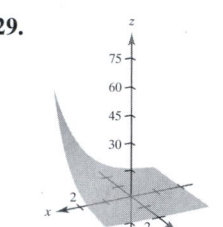

No extrema   Saddle point

33. $-4 < f_{xy}(3, 7) < 4$

35. Absolute maximum: $(0, 1, 10)$
Absolute minimum: $(1, 2, 5)$

37. Absolute maxima: $(\pm 2, 4, 28)$
Absolute minimum: $(0, 1, -2)$

39. Absolute maxima: $(2, 1, 6)$, $(-2, -1, 6)$
Absolute minima: $\left(-\frac{1}{2}, 1, -\frac{1}{4}\right)$, $\left(\frac{1}{2}, -1, -\frac{1}{4}\right)$

41. Absolute maxima: $(2, 2, 16)$, $(-2, -2, 16)$
Absolute minima: $(x, -x, 0)$, $|x| \leq 2$

43. Absolute maximum: $(1, 1, 1)$
Absolute minimum: $(0, 0, 0)$

45. Saddle point: $(0, 0, 0)$
Test fails

47. Absolute minima: $(1, a, 0)$, $(b, -4, 0)$
Test fails

49. Absolute minimum: $(0, 0, 0)$
Test fails

51. Relative minimum: $(0, 3, -1)$

53. Point A is a saddle point.

55. (a)

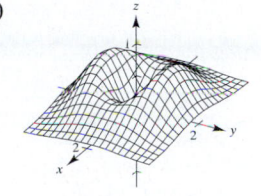

(b)

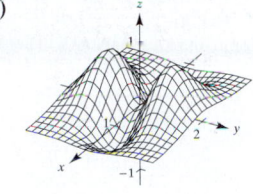

Minimum: $(0, 0, 0)$ Minima: $(\pm 1, 0, -e^{-1})$
Maxima: $(0, \pm 1, 2e^{-1})$ Maxima: $(0, \pm 1, 2e^{-1})$
Saddle points: $(\pm 1, 0, e^{-1})$ Saddle point: $(0, 0, 0)$

(c) $\alpha > 0$ $\alpha < 0$
Minimum: $(0, 0, 0)$ Minima: $(\pm 1, 0, \alpha e^{-1})$
Maxima: $(0, \pm 1, \beta e^{-1})$ Maxima: $(0, \pm 1, \beta e^{-1})$
Saddle points: $(\pm 1, 0, \alpha e^{-1})$ Saddle point: $(0, 0, 0)$

57. False: let $f(x, y) = |1 - x - y|$ at the point $(0, 0, 1)$.

58. True   **59.** False: let $f(x, y) = x^4 - 2x^2 + y^2$.

60. True

Section 12.9 *(page 898)*

1. $\dfrac{6\sqrt{14}}{7}$ **3.** 6   **5.** 10, 10, 10   **7.** 10, 10, 10

9. $36 \times 18 \times 18$ inches   **11.** Proof   **13.** Proof

15. $x = \dfrac{\sqrt{2}}{2} \approx 0.707$ kilometer

$y = \dfrac{3\sqrt{2} + 2\sqrt{3}}{6} \approx 1.284$ kilometers

17. Each edge of $w/3$ inches is turned up $60°$ from the horizontal.

19. (a) $S = \sqrt{x^2 + y^2} + \sqrt{(x + 2)^2 + (y - 2)^2} +$
$\sqrt{(x - 4)^2 + (y - 2)^2}$

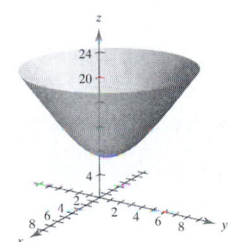

The surface has a minimum.

(b) $S_x = \dfrac{x}{\sqrt{x^2 + y^2}} + \dfrac{x + 2}{\sqrt{(x + 2)^2 + (y - 2)^2}} +$

$\dfrac{x - 4}{\sqrt{(x - 4)^2 + (y - 2)^2}}$

$S_y = \dfrac{y}{\sqrt{x^2 + y^2}} + \dfrac{y - 2}{\sqrt{(x + 2)^2 + (y - 2)^2}} +$

$\dfrac{y - 2}{\sqrt{(x - 4)^2 + (y - 2)^2}}$

(c) $-\dfrac{1}{\sqrt{2}}\mathbf{i} - \left(\dfrac{1}{\sqrt{2}} - \dfrac{2}{\sqrt{10}}\right)\mathbf{j}$

$\theta \approx 186.0°$

(d) $(x_2, y_2) \approx (0.05, 0.90)$

(e) $(x_4, y_4) \approx (0.06, 0.45)$

(f) $-\nabla S(x, y)$ gives the direction of greatest rate of decrease of S. Use $\nabla S(x, y)$ when finding a maximum.

21. $x_1 = 275, x_2 = 110$ **23.** $x_1 = 3, x_2 = 6$

25. (a) $y = \dfrac{3}{4}x + \dfrac{4}{3}$ (b) $\dfrac{1}{6}$

27. (a) $y = -2x + 4$ (b) 2

29. $y = \dfrac{37}{43}x + \dfrac{7}{43}$ **31.** $y = -\dfrac{175}{148}x + \dfrac{945}{148}$

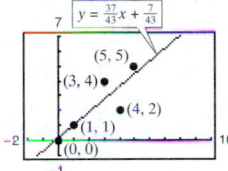

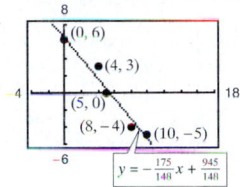

33. (a) $y = 0.03x - 6.29$

(b)

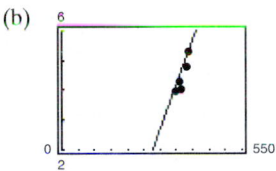

(c) \$0.03

35. $y = 14x + 19$
41.4 bushels per acre

37. $a\displaystyle\sum_{i=1}^{n} x_i^4 + b\sum_{i=1}^{n} x_i^3 + c\sum_{i=1}^{n} x_i^2 = \sum_{i=1}^{n} x_i^2 y_i$

$a\displaystyle\sum_{i=1}^{n} x_i^3 + b\sum_{i=1}^{n} x_i^2 + c\sum_{i=1}^{n} x_i = \sum_{i=1}^{n} x_i y_i$

$a\displaystyle\sum_{i=1}^{n} x_i^2 + b\sum_{i=1}^{n} x_i + cn = \sum_{i=1}^{n} y_i$

39. $y = \dfrac{3}{7}x^2 + \dfrac{6}{5}x + \dfrac{26}{35}$ **41.** $y = x^2 - x$

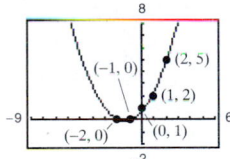

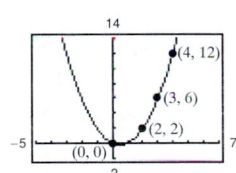

43. (a) $y = -0.22x^2 + 9.66x - 1.79$

(b)

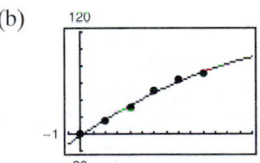

45. (a) $\ln P = -0.1499h + 9.3018$

(b) $P = 10{,}957.7e^{-0.1499h}$

(c)

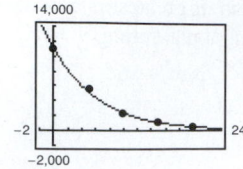

43. (a) $g\left(\dfrac{\pi}{3}, \dfrac{\pi}{3}, \dfrac{\pi}{3}\right) = \dfrac{1}{8}$ (b)

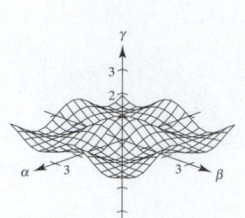

Section 12.10 *(page 908)*

1.

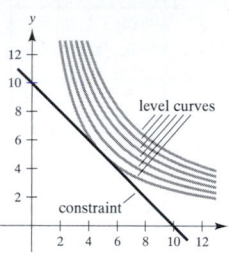

3.

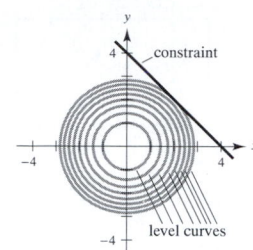

$f(5, 5) = 25$ $\qquad$ $f(2, 2) = 8$

5. $f(2, 4) = -12$ $\quad$ **7.** $f(25, 50) = 2600$ $\quad$ **9.** $f(1, 1) = 2$

11. $f(2, 2) = e^4$ $\quad$ **13.** $f(2, 2, 2) = 12$ $\quad$ **15.** $f\left(\frac{1}{3}, \frac{1}{3}, \frac{1}{3}\right) = \frac{1}{3}$

17. $f(8, 16, 8) = 1024$ $\quad$ **19.** $f\left(3, \frac{3}{2}, 1\right) = 6$

21. Maxima: $f\left(\dfrac{\sqrt{2}}{2}, \dfrac{\sqrt{2}}{2}\right) = \dfrac{5}{2}$

$\qquad\qquad f\left(-\dfrac{\sqrt{2}}{2}, -\dfrac{\sqrt{2}}{2}\right) = \dfrac{5}{2}$

$\qquad$ Minima: $f\left(-\dfrac{\sqrt{2}}{2}, \dfrac{\sqrt{2}}{2}\right) = -\dfrac{1}{2}$

$\qquad\qquad f\left(\dfrac{\sqrt{2}}{2}, -\dfrac{\sqrt{2}}{2}\right) = -\dfrac{1}{2}$

23. $\dfrac{\sqrt{13}}{13}$ $\qquad$ **25.** $\sqrt{3}$

27. $x = \dfrac{10 + 2\sqrt{265}}{15}$

$\quad y = \dfrac{5 + \sqrt{265}}{15}$

$\quad z = \dfrac{-1 + \sqrt{265}}{3}$

29. $36 \times 18 \times 18$ inches $\qquad$ **31.** $\sqrt[3]{360} \times \sqrt[3]{360} \times \frac{4}{3}\sqrt[3]{360}$ feet

33. $\dfrac{2\sqrt{3}a}{3} \times \dfrac{2\sqrt{3}b}{3} \times \dfrac{2\sqrt{3}c}{3}$ $\qquad$ **35.** Proof $\qquad$ **37.** $\frac{2}{3}$

39. $P\left(\dfrac{3125}{6}, \dfrac{6250}{3}\right) = 147{,}314$

41. $\quad x = 50\sqrt{2}$

$\qquad y = 200\sqrt{2}$

$\qquad$ cost = \$13,576.45

Review Exercises for Chapter 12 *(page 910)*

1. Not a function

3.

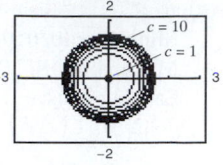

5.

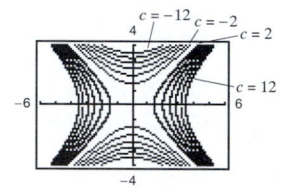

7.

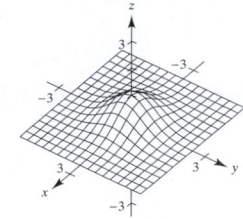

9. Continuous except at $(0, 0)$
$\quad$ Limit: $\frac{1}{2}$

11. Continuous except at $(0, 0)$
$\quad$ Limit does not exist.

13. $f_x(x, y) = e^x \cos y$
$\quad f_y(x, y) = -e^x \sin y$

15. $f_x(x, y) = e^y + ye^x$
$\quad f_y(x, y) = xe^y + e^x$

17. $f_x(x, y) = \dfrac{y(y^2 - x^2)}{(x^2 + y^2)^2}$

$\quad f_y(x, y) = \dfrac{x(x^2 - y^2)}{(x^2 + y^2)^2}$

19. $f_x(x, y, z) = \dfrac{-yz}{x^2 + y^2}$

$\quad f_y(x, y, z) = \dfrac{xz}{x^2 + y^2}$

$\quad f_z(x, y, z) = \arctan \dfrac{y}{x}$

21. $u_x(x, t) = cne^{-n^2 t}\cos nx$
$\quad u_t(x, t) = -cn^2 e^{-n^2 t}\sin nx$

23. $\dfrac{\partial z}{\partial x} = \dfrac{2xy - z}{x + 2y + 2z}$

$\quad \dfrac{\partial z}{\partial y} = \dfrac{x^2 - 2z}{x + 2y + 2z}$

25. $f_{xx}(x, y) = 6$
$\quad f_{yy}(x, y) = 12y$
$\quad f_{xy}(x, y) = f_{yx}(x, y) = -1$

27. $f_{xx}(x, y) = -y \cos x$
$\quad f_{yy}(x, y) = -x \sin y$
$\quad f_{xy}(x, y) = f_{yx}(x, y) = \cos y - \sin x$

29. Proof $\qquad$ **31.** Proof

33. $\dfrac{\partial u}{\partial r} = 2r$

$\quad \dfrac{\partial u}{\partial t} = 2t$

35. 0   **37.** $\frac{2}{3}$ **39.** $\langle -\frac{1}{2}, 0 \rangle, \frac{1}{2}$

41. $\left\langle -\frac{\sqrt{2}}{2}, -\frac{\sqrt{2}}{2} \right\rangle, 1$

43. $4x + 4y - z = 8$

$\frac{x-2}{4} = \frac{y-1}{4} = \frac{z-4}{-1}$

45. Tangent plane: $z = 4$
Normal line: $x = 2, y = -3, z = 4 + t$

47. $\frac{x-2}{1} = \frac{y-1}{2}, z = 3$

49. Relative minimum: $\left(\frac{3}{2}, \frac{9}{4}, -\frac{27}{16} \right)$

Saddle point: $(0, 0, 0)$

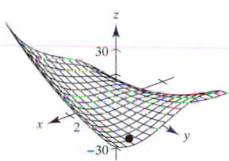

51. Relative minimum: $(1, 1, 3)$

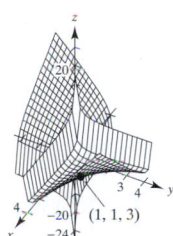

53. The level curves are hyperbolas. The critical point $(0, 0)$ may be a saddle point or an extrema.

55.

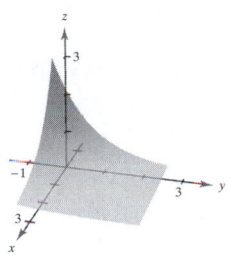

57. $\left(\sin\frac{y}{x} - \frac{y}{x}\cos\frac{y}{x} \right) dx + \left(\cos\frac{y}{x} \right) dy$

59. 0.6538 centimeter, 5.03%   **61.** π cubic inches

63. $x_1 = 94, x_2 = 157$ **65.** $f(49.4, 253) = 13,189.8$

67. (a) $y = 2.29t + 2.34$ (b)

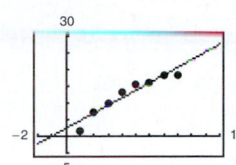
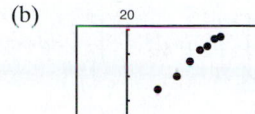

More closely linear

(c) $y = 1.54 + 8.37 \ln t$ (d)

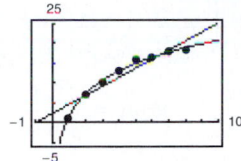

Logarithmic model is a better fit.

69. Maximum: $f\left(\frac{1}{3}, \frac{1}{3}, \frac{1}{3} \right) = \frac{1}{3}$ **70.** True

71. False: $\nabla \mathbf{f}(x_0, y_0, z_0)$ is normal.   **72.** True

CHAPTER 13

Section 13.1 *(page 921)*

1. $\frac{3x^2}{2}$ **3.** $y\ln(2y)$ **5.** $\frac{4x^2 - x^4}{2}$ **7.** $\frac{y}{2}\left[(\ln y)^2 - y^2 \right]$

9. $x^2(1 - e^{-x^2} - x^2 e^{-x^2})$ **11.** 3   **13.** $\frac{20}{3}$ **15.** $\frac{2}{3}$

17. 4   **19.** $\frac{\pi^2}{32} + \frac{1}{8}$ **21.** $\frac{1}{2}$ **23.** Diverges

25. $\displaystyle\int_0^4 \int_0^y f(x, y)\, dx\, dy = \int_0^4 \int_x^4 f(x, y)\, dy\, dx$

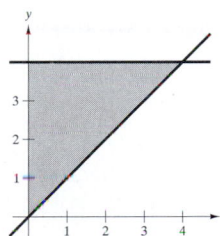

27. $\displaystyle\int_{-1}^1 \int_{x^2}^1 f(x, y)\, dy\, dx = \int_0^1 \int_{-\sqrt{y}}^{\sqrt{y}} f(x, y)\, dx\, dy$

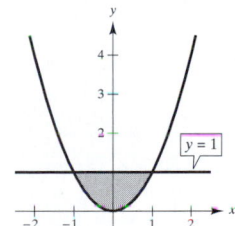

29. $\int_0^1 \int_0^2 dy\,dx = \int_0^2 \int_0^1 dx\,dy = 2$

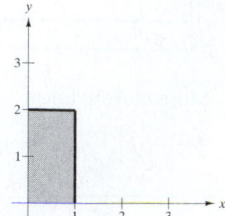

31. $\int_0^1 \int_{-\sqrt{1-y^2}}^{\sqrt{1-y^2}} dx\,dy = \int_{-1}^1 \int_0^{\sqrt{1-x^2}} dy\,dx = \dfrac{\pi}{2}$

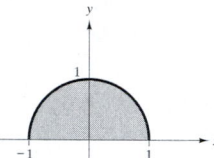

33. $\int_0^2 \int_{x/2}^1 dy\,dx = \int_0^1 \int_0^{2y} dx\,dy = 1$

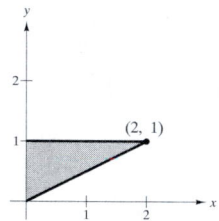

35. $\int_0^1 \int_{y^2}^{\sqrt[3]{y}} dx\,dy = \int_0^1 \int_{x^3}^{\sqrt{x}} dy\,dx = \dfrac{5}{12}$

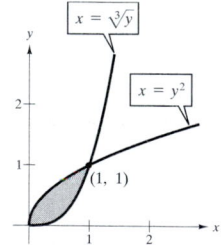

37. 24   **39.** $\dfrac{16}{3}$ **41.** $\dfrac{9}{2}$ **43.** $\dfrac{8}{3}$ **45.** 5   **47.** πab

49. Common region of integration is the sector of the circle shown in the figure.

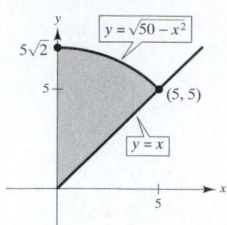

Value of the integrals: $\dfrac{15{,}625\pi}{24}$

51. $\dfrac{26}{9}$ **53.** $\dfrac{1}{2}(1 - \cos 1) \approx 0.230$

55. $\dfrac{1664}{105}$ **57.** $(\ln 5)^2$

59. (a)

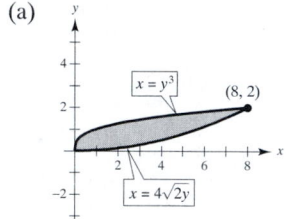

(b) $\int_0^2 \int_{y^3}^{4\sqrt{2y}} (x^2 y - xy^2)\,dx\,dy = \int_0^8 \int_{x^2/32}^{\sqrt[3]{x}} (x^2 y - xy^2)\,dy\,dx$

(c) $\dfrac{67{,}520}{693}$

61. 20.5648

63. Integration of a function of several variables. Integrate with respect to one variable while holding the other variables constant.

65. True   **66.** False: let $f(x, y) = x$.

Section 13.2 *(page 930)*

1. 24 (approximation is exact)

3. Approximation: 52; Exact: $\dfrac{160}{3}$

5. 8   **7.** 36

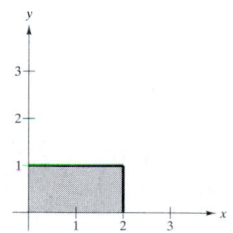   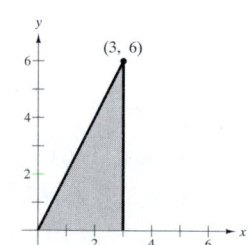

9. 0

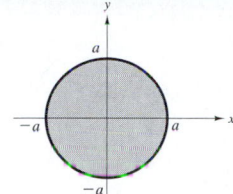

11. $\displaystyle\int_0^3\int_0^5 xy\,dy\,dx = \frac{225}{4}$

$\displaystyle\int_0^5\int_0^3 xy\,dx\,dy = \frac{225}{4}$

13. $\displaystyle\int_0^2\int_x^{2x} \frac{y}{x^2+y^2}\,dy\,dx = \ln\frac{5}{2}$

$\displaystyle\int_0^2\int_{y/2}^y \frac{y}{x^2+y^2}\,dx\,dy + \int_2^4\int_{y/2}^2 \frac{y}{x^2+y^2}\,dx\,dy = \ln\frac{5}{2}$

15. $\displaystyle\int_0^3\int_{4y/3}^{\sqrt{25-y^2}} x\,dx\,dy = 25$

$\displaystyle\int_0^4\int_0^{3x/4} x\,dy\,dx + \int_4^5\int_0^{\sqrt{25-x^2}} x\,dy\,dx = 25$

17. 4   **19.** 8   **21.** 12   **23.** $\frac{3}{8}$ **25.** 1   **27.** 8π

29. $\displaystyle\int_0^1\int_0^x xy\,dy\,dx = \frac{1}{8}$ **31.** $\displaystyle\int_0^2\int_0^4 x^2\,dy\,dx = \frac{32}{3}$

33. $\displaystyle 2\int_0^1\int_0^x \sqrt{1-x^2}\,dy\,dx = \frac{2}{3}$

35. $\displaystyle\int_0^2\int_0^{\sqrt{4-x^2}} (x+y)\,dy\,dx = \frac{16}{3}$

37. 8π **39.** 8π **41.** 1.2315   **43.** Proof

45. $1 - e^{-1/4} \approx 0.221$ **47.** 9   **49.** 2   **51.** $\frac{8}{3}$

53. 25,645.24   **55.** kB **57.** $\frac{1}{5}$ **59.** $\frac{7}{27}$

61. 2500 cubic meters   **63.** (a) 1.784   (b) 1.788

65. (a) 11.057   (b) 11.041   **67.** Answers will vary.

69. a   **70.** False: $V = 8\displaystyle\int_0^1\int_0^{\sqrt{1-y^2}} \sqrt{1-x^2-y^2}\,dx\,dy$.

71. True   **73.** $\ln 2$

Section 13.3 *(page 939)*

1. 0   **3.** $\dfrac{5\sqrt{5}\,\pi}{6}$

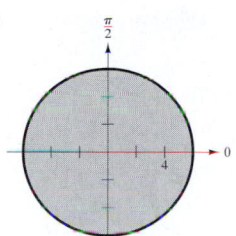   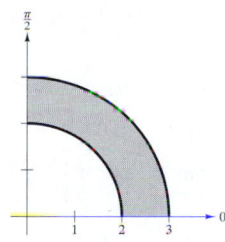

5. $\dfrac{9}{8} + \dfrac{3\pi^2}{32}$

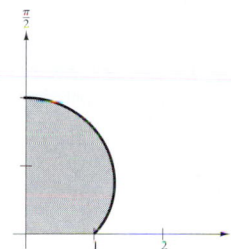

7. 9π **9.** $\dfrac{3\pi}{2}$ **11.** π **13.** $\dfrac{a^3}{3}$ **15.** $\dfrac{243\pi}{10}$ **17.** $\dfrac{2}{3}$

19. $\displaystyle\int_0^{\pi/4}\int_0^{2\sqrt{2}} r^2\,dr\,d\theta = \frac{4\sqrt{2}\,\pi}{3}$

21. $\displaystyle\int_0^{\pi/2}\int_0^2 r^2(\cos\theta + \sin\theta)\,dr\,d\theta$ **23.** $\displaystyle\int_0^{\pi/4}\int_1^2 r\theta\,dr\,d\theta$

25. $\dfrac{1}{8}$ **27.** $\dfrac{250\pi}{3}$ **29.** $\dfrac{64}{9}(3\pi - 4)$

31. $2\sqrt{4 - 2\sqrt[3]{2}}$ **33.** 1.2858

35. Insert a factor of r.   Sector of a circle

37. 56.051   **39.** c

41. False: let $f(r, \theta) = r - 1$ and let R be a sector where $0 \le r \le 6$ and $0 \le \theta \le \pi$.

42. True   **43.** (a) 2π (b) $\sqrt{2}\pi$ **45.** 486,788

47. (a) $\displaystyle\int_2^4\int_{y/\sqrt{3}}^y f\,dx\,dy$

(b) $\displaystyle\int_{2/\sqrt{3}}^2\int_2^{\sqrt{3}x} f\,dy\,dx + \int_2^{4\sqrt{3}}\int_x^{\sqrt{3}x} f\,dy\,dx + \int_{4\sqrt{3}}^4\int_x^4 f\,dy\,dx$

(c) $\displaystyle\int_{\pi/4}^{\pi/3}\int_{2\csc\theta}^{4\csc\theta} f r\,dr\,d\theta$

49. Proof

Section 13.4 (page 948)

1. (a) $m = kab, \left(\dfrac{a}{2}, \dfrac{b}{2}\right)$ (b) $m = \dfrac{kab^2}{2}, \left(\dfrac{a}{2}, \dfrac{2b}{3}\right)$

   (c) $m = \dfrac{ka^2b}{2}, \left(\dfrac{2a}{3}, \dfrac{b}{2}\right)$

3. (a) $m = \dfrac{kbh}{2}, \left(\dfrac{b}{2}, \dfrac{h}{3}\right)$ (b) $m = \dfrac{kh^2b}{6}, \left(\dfrac{b}{2}, \dfrac{h}{2}\right)$

   (c) $m = \dfrac{khb^2}{4}, \left(\dfrac{7b}{12}, \dfrac{h}{3}\right)$

5. (a) $\left(\dfrac{a}{2} + 5, \dfrac{b}{2}\right)$ (b) $\left(\dfrac{a}{2} + 5, \dfrac{2b}{3}\right)$

   (c) $\left(\dfrac{2(a^2 + 15a + 75)}{3(a + 10)}, \dfrac{b}{2}\right)$

7. (a) $m = \dfrac{k\pi a^2}{2}, \left(0, \dfrac{4a}{3\pi}\right)$

   (b) $m = \dfrac{ka^4}{24}(16 - 3\pi), \left(0, \dfrac{a}{5}\left[\dfrac{15\pi - 32}{16 - 3\pi}\right]\right)$

9. $m = \dfrac{32k}{3}, \left(3, \dfrac{8}{7}\right)$ 11. $m = \dfrac{k\pi}{2}, \left(0, \dfrac{\pi + 2}{4\pi}\right)$

13. $m = \dfrac{8192k}{15}, \left(\dfrac{64}{7}, 0\right)$ 15. $m = \dfrac{kL}{4}, \left(\dfrac{L}{2}, \dfrac{16}{9\pi}\right)$

17. $m = \dfrac{k\pi a^2}{8}, \left(\dfrac{4\sqrt{2}a}{3\pi}, \dfrac{4a(2 - \sqrt{2})}{3\pi}\right)$

19. $m = \dfrac{k}{4}(1 - e^{-4}), \left(\dfrac{e^4 - 5}{2(e^4 - 1)}, \dfrac{4}{9}\left[\dfrac{e^6 - 1}{e^6 - e^2}\right]\right)$

21. $m = \dfrac{k\pi}{3}, (1.12, 0)$ 23. $\bar{y}$ will increase.

25. $\bar{x}$ and $\bar{y}$ will both increase.

27. $\bar{\bar{x}} = \dfrac{\sqrt{3}b}{3}, \bar{\bar{y}} = \dfrac{\sqrt{3}h}{3}$ 29. $\bar{\bar{x}} = \dfrac{a}{2}, \bar{\bar{y}} = \dfrac{a}{2}$

31. $\bar{\bar{x}} = \dfrac{a}{2}, \bar{\bar{y}} = \dfrac{a}{2}$

33. $I_x = \dfrac{kab^4}{4}$ 35. $I_x = \dfrac{32k}{3}$

   $I_y = \dfrac{kb^2a^3}{6}$ $I_y = \dfrac{16k}{3}$

   $I_0 = \dfrac{3kab^4 + 2ka^3b^2}{12}$ $I_0 = 16k$

   $\bar{\bar{x}} = \dfrac{\sqrt{3}a}{3}$ $\bar{\bar{x}} = \dfrac{2\sqrt{3}}{3}$

   $\bar{\bar{y}} = \dfrac{\sqrt{2}b}{2}$ $\bar{\bar{y}} = \dfrac{2\sqrt{6}}{3}$

37. $I_x = 16k$

   $I_y = \dfrac{512k}{5}$

   $I_0 = \dfrac{592k}{5}$

   $\bar{\bar{x}} = \dfrac{4\sqrt{15}}{5}$

   $\bar{\bar{y}} = \dfrac{\sqrt{6}}{2}$

39. $I_x = \dfrac{3k}{56}$

   $I_y = \dfrac{k}{18}$

   $I_0 = \dfrac{55k}{504}$

   $\bar{\bar{x}} = \dfrac{\sqrt{30}}{9}$

   $\bar{\bar{y}} = \dfrac{\sqrt{70}}{14}$

41. $2k\displaystyle\int_{-b}^{b}\int_{0}^{\sqrt{b^2 - x^2}}(x - a)^2\,dy\,dx = \dfrac{k\pi b^2}{4}(b^2 + 4a^2)$

43. $\displaystyle\int_{0}^{4}\int_{0}^{\sqrt{x}}kx(x - 6)^2\,dy\,dx = \dfrac{42{,}752k}{315}$

45. $\displaystyle\int_{0}^{a}\int_{0}^{\sqrt{a^2 - x^2}}k(a - y)(y - a)^2\,dy\,dx = ka^5\left(\dfrac{7\pi}{16} - \dfrac{17}{15}\right)$

47. Proof   49. $\dfrac{a(3L - 2a)}{3(2L - a)}$ 51. $-\dfrac{a^2}{4L}$

Section 13.5 (page 955)

1. 6   3. 12π 5. $\dfrac{3}{4}\left[6\sqrt{37} + \ln(\sqrt{37} + 6)\right]$

7. $\dfrac{4}{27}(31\sqrt{31} - 8)$ 9. $\sqrt{2} - 1$ 11. $\sqrt{2}\pi$

13. $2\pi a(a - \sqrt{a^2 - b^2})$ 15. $48\sqrt{14}$ 17. 20π

19. $\displaystyle\int_{0}^{1}\int_{0}^{x}\sqrt{5 + 4x^2}\,dy\,dx = \dfrac{27 - 5\sqrt{5}}{12}$

21. $\displaystyle\int_{-2}^{2}\int_{-\sqrt{4 - x^2}}^{\sqrt{4 - x^2}}\sqrt{1 + 4x^2 + 4y^2}\,dy\,dx =$

   $\displaystyle\int_{0}^{2\pi}\int_{0}^{2}r\sqrt{1 + 4r^2}\,dr\,d\theta = \dfrac{\pi}{6}(17\sqrt{17} - 1)$

23. $\displaystyle\int_{0}^{1}\int_{0}^{1}\sqrt{1 + 4x^2 + 4y^2}\,dy\,dx = 1.8616$

25. e 27. 2.0035

29. $\displaystyle\int_{-1}^{1}\int_{-1}^{1}\sqrt{1 + 9(x^2 - y)^2 + 9(y^2 - x)^2}\,dy\,dx$

31. $\displaystyle\int_{-2}^{2}\int_{-\sqrt{4 - x^2}}^{\sqrt{4 - x^2}}\sqrt{1 + e^{-2x}}\,dy\,dx$

33. $\displaystyle\int_{0}^{4}\int_{0}^{10}\sqrt{1 + e^{2xy}(x^2 + y^2)}\,dy\,dx$ 35. 16

37. (a) 30,415.74 cubic feet   (b) 2081.53 square feet

39. z_2, z_1, z_4, z_3; The greater the angle between the plane and the xy-plane, the greater the surface area.

41. False: the surface area will remain the same for any vertical translation of a surface.

Section 13.6 *(page 965)*

1. 18   **3.** $\dfrac{1}{10}$ **5.** $\dfrac{15}{2}\left(1 - \dfrac{1}{e}\right)$ **7.** $\dfrac{729}{4}$

9. $\dfrac{128}{15}$ **11.** 2.44167

13. $\displaystyle\int_0^3 \int_0^{(12-4z)/3} \int_0^{(12-4z-3x)/6} dy\, dx\, dz$

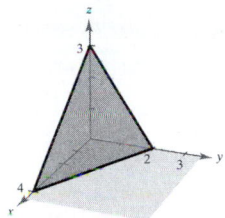

15. $\displaystyle\int_0^1 \int_0^x \int_0^{\sqrt{1-y^2}} dz\, dy\, dx$

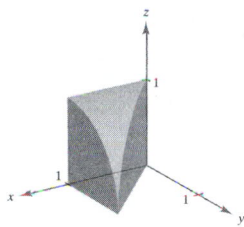

17. $\displaystyle\int_0^1 \int_0^x \int_0^3 xyz\, dz\, dy\, dx$, $\displaystyle\int_0^1 \int_y^1 \int_0^3 xyz\, dz\, dx\, dy$,

$\displaystyle\int_0^1 \int_0^3 \int_0^x xyz\, dy\, dz\, dx$, $\displaystyle\int_0^3 \int_0^1 \int_0^x xyz\, dy\, dx\, dz$,

$\displaystyle\int_0^3 \int_0^1 \int_y^1 xyz\, dx\, dy\, dz$, $\displaystyle\int_0^1 \int_0^3 \int_y^1 xyz\, dx\, dz\, dy$

19. $\dfrac{256}{15}$ **21.** $\dfrac{4\pi a^3}{3}$ **23.** $\dfrac{256}{15}$

25. $m = 8k$ **27.** $m = \dfrac{128k}{3}$
$\bar{x} = \dfrac{3}{2}$ $\bar{z} = 1$

29. $m = k\displaystyle\int_0^b \int_0^b \int_0^b xy\, dz\, dy\, dx$

$M_{yz} = k\displaystyle\int_0^b \int_0^b \int_0^b x^2 y\, dz\, dy\, dx$

$M_{xz} = k\displaystyle\int_0^b \int_0^b \int_0^b xy^2\, dz\, dy\, dx$

$M_{xy} = k\displaystyle\int_0^b \int_0^b \int_0^b xyz\, dz\, dy\, dx$

31. $\bar{x}$ will be greater than 2, and $\bar{y}$ and $\bar{z}$ will be unchanged.

33. $\bar{x}$ and $\bar{z}$ will be unchanged, and $\bar{y}$ will be greater than 0.

35. $\left(0, 0, \dfrac{3h}{4}\right)$ **37.** $\left(0, 0, \dfrac{3}{2}\right)$ **39.** $\left(5, 6, \dfrac{5}{4}\right)$

41. (a) $I_x = \dfrac{2ka^5}{3}$ (b) $I_x = \dfrac{ka^8}{8}$

$\qquad I_y = \dfrac{2ka^5}{3}$ $\qquad I_y = \dfrac{ka^8}{8}$

$\qquad I_z = \dfrac{2ka^5}{3}$ $\qquad I_z = \dfrac{ka^8}{8}$

43. (a) $I_x = 256k$ (b) $I_x = \dfrac{2048k}{3}$

$\qquad I_y = \dfrac{512k}{3}$ $\qquad I_y = \dfrac{1024k}{3}$

$\qquad I_z = 256k$ $\qquad I_z = \dfrac{1048k}{3}$

45. Proof   **47.** $\displaystyle\int_{-1}^1 \int_{-1}^1 \int_0^{1-x} (x^2 + y^2)\sqrt{x^2 + y^2 + z^2}\, dz\, dy\, dx$

49. Increase

Section 13.7 *(page 972)*

1. 8   **3.** $\dfrac{52}{45}$ **5.** $\dfrac{\pi}{8}$ **7.** $\pi(e^4 + 3)$

9. $\dfrac{\pi}{4}(1 - e^{-9})$ **11.** $\dfrac{64\sqrt{3}\pi}{3}$

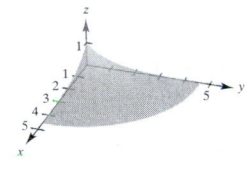 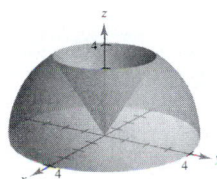

13. Cylindrical: $\displaystyle\int_0^{2\pi} \int_0^2 \int_{r^2}^4 r^2 \cos\theta\, dz\, dr\, d\theta = 0$

Spherical: $\displaystyle\int_0^{2\pi} \int_0^{\arctan(1/2)} \int_0^{4\sec\phi} \rho^3 \sin^2\phi \cos\theta\, d\rho\, d\phi\, d\theta +$

$\displaystyle\int_0^{2\pi} \int_{\arctan(1/2)}^{\pi/2} \int_0^{\cot\phi\csc\phi} \rho^3 \sin^2\phi \cos\phi\, d\rho\, d\phi\, d\theta = 0$

15. Cylindrical: $\displaystyle\int_0^{2\pi} \int_0^a \int_a^{a + \sqrt{a^2 - r^2}} r^2 \cos\theta\, dz\, dr\, d\theta = 0$

Spherical: $\displaystyle\int_0^{\pi/4} \int_0^{2\pi} \int_{a\sec\phi}^{2a\cos\phi} \rho^3 \sin^2\phi \cos\theta\, d\rho\, d\theta\, d\phi = 0$

17. $\dfrac{\pi r_0^2 h}{3}$ **19.** $\left(0, 0, \dfrac{h}{5}\right)$ **21.** Proof   **23.** Proof

25. $\dfrac{2a^3}{9}(3\pi - 4)$ **27.** $\dfrac{2a^3}{9}(3\pi - 4)$ **29.** $16\pi^2$

31. $k\pi a^4$ **33.** $\left(0, 0, \dfrac{3r}{8}\right)$ **35.** $\dfrac{k\pi}{192}$

37. (a) r constant: right circular cylinder
θ constant: plane
z constant: plane

(b) ρ constant: sphere
θ constant: plane
ϕ constant: cone

39. $\dfrac{1}{2}\,\pi^2 a^4$

Section 13.8 (page 979)

1. $-\dfrac{1}{2}$ **3.** $1 + 2v$ **5.** 1 **7.** $-e^{2u}$

9.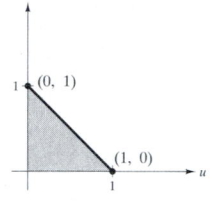

11. $\dfrac{8}{3}$ **13.** 36

15. $(e^{-1/2} - e^{-2})\ln 8 \approx 0.9798$

17. $12(e^4 - 1)$ **19.** $\dfrac{100}{9}$ **21.** $\dfrac{2}{5}a^{5/2}$

23. (a)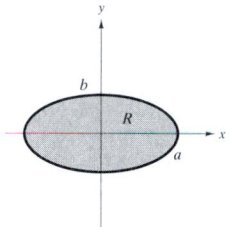

(b) ab (c) πab

25. $u^2 v$ **27.** $-\rho^2 \sin\phi$

Review Exercises for Chapter 13 (page 980)

1. $x - x^3 + x^3 \ln x^2$ **3.** $\dfrac{29}{6}$ **5.** 36

7. $\dfrac{h^3}{6}\left[\ln\left(\sqrt{2} + 1\right) + \sqrt{2}\right]$ **9.** $\dfrac{324\pi}{5}$

11. $\dfrac{abc}{3}(a^2 + b^2 + c^2)$ **13.** $\dfrac{27}{5}$ **15.** $\dfrac{8\pi}{15}$

17. $\displaystyle\int_0^3 \int_0^{(3-x)/3} dy\, dx = \int_0^1 \int_0^{3-3y} dx\, dy = \dfrac{3}{2}$

19. $\displaystyle\int_{-5}^3 \int_{-\sqrt{25-x^2}}^{\sqrt{25-x^2}} dy\, dx = \int_{-5}^{-4}\int_{-\sqrt{25-y^2}}^{\sqrt{25-y^2}} dx\, dy +$

$\displaystyle\int_{-4}^4 \int_{-\sqrt{25-y^2}}^{3} dx\, dy + \int_4^5 \int_{-\sqrt{25-y^2}}^{\sqrt{25-y^2}} dx\, dy =$

$\dfrac{25\pi}{2} + 12 + 25 \arcsin\dfrac{3}{5} \approx 67.36$

21. $4\displaystyle\int_0^1 \int_0^{x\sqrt{1-x^2}} dy\, dx = 4\int_0^{1/2} \int_{\sqrt{(1-\sqrt{1-4y^2})/2}}^{\sqrt{(1+\sqrt{1-4y^2})/2}} dx\, dy = \dfrac{4}{3}$

23. $\displaystyle\int_2^5 \int_{x-3}^{\sqrt{x-1}} dy\, dx + 2\int_1^2 \int_0^{\sqrt{x-1}} dy\, dx = \int_{-1}^2 \int_{y^2+1}^{y+3} dx\, dy = \dfrac{9}{2}$

25. Integration over the common region R shown in the figure

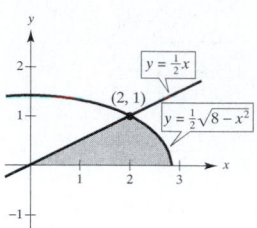

27. $\dfrac{3296}{15}$ **29.** $\dfrac{\pi h^3}{3}$ **31.** $\dfrac{32}{3}\left(\dfrac{\pi}{2} - \dfrac{2}{3}\right)$

33. c **35.** $k = 1,\quad 0.070$

37. (a) $r = 3\sqrt{\cos 2\theta}$

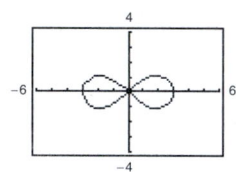

(b) 9 (c) $3(3\pi - 16\sqrt{2} + 20) \approx 20.392$

39. (a) $m = \dfrac{k}{4},\ \left(\dfrac{32}{45}, \dfrac{64}{55}\right)$

(b) $m = \dfrac{17k}{30},\ \left(\dfrac{936}{1309}, \dfrac{784}{663}\right)$

41. $\dfrac{\pi}{6}\left(65\sqrt{65} - 1\right)$ **43.** $\dfrac{1}{6}\left(37\sqrt{37} - 1\right)$

45. $\left(0, 0, \dfrac{1}{4}\right)$ **47.** $\left(\dfrac{3a}{8}, \dfrac{3a}{8}, \dfrac{3a}{8}\right)$ **49.** $\dfrac{833k\pi}{3}$

51. (a) $\dfrac{1}{3}\pi h^2 (3a - h)$ (b) $\left(0, 0, \dfrac{3(2a-h)^2}{4(3a-h)}\right)$

(c) $\left(0, 0, \dfrac{3}{8}a\right)$ (d) a (e) $\dfrac{\pi}{30}h^3(20a^2 - 15ah + 3h^2)$

(f) $\dfrac{4}{15}\pi a^5$

53. Volume of a torus formed by a circle of radius 3, centered at $(0, 3, 0)$, and revolved about the z-axis

55. True **56.** False: $\displaystyle\int_0^1 \int_0^1 x\, dx\, dy \neq \int_1^2 \int_1^2 x\, dx\, dy$

57. True

58. True: $\displaystyle\int_0^1 \int_0^1 \dfrac{1}{1 + x^2 + y^2} dx\, dy < \int_0^1 \int_0^1 \dfrac{1}{1 + x^2} dx\, dy = \dfrac{\pi}{4}$.

CHAPTER 14

Section 14.1 *(page 994)*

1. c　　**2.** d　　**3.** b　　**4.** e　　**5.** a　　**6.** f

7.

9.

11.

13.

15.

17.

19.

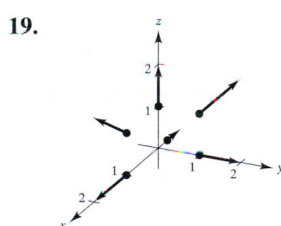

21. $(10x + 3y)\mathbf{i} + (3x + 20y)\mathbf{j}$　　**23.** $-2xye^{x^2}\mathbf{i} - e^{x^2}\mathbf{j} + \mathbf{k}$

25. $\left[\dfrac{xy}{x + y} + y\ln(x + y)\right]\mathbf{i} + \left[\dfrac{xy}{x + y} + x\ln(x + y)\right]\mathbf{j}$

27. Conservative: $f(x, y) = x^2y + K$

29. Conservative: $f(x, y) = e^{x^2y} + K$

31. Conservative: $f(x, y) = \frac{1}{2}\ln(x^2 + y^2) + K$

33. Not conservative　　**35.** $2\mathbf{j} - \mathbf{k}$　　**37.** $-2\mathbf{k}$

39. $\dfrac{2x\mathbf{k}}{x^2 + y^2}$　　**41.** $\cos(y - z)\mathbf{i} + \cos(z - x)\mathbf{j} + \cos(x - y)\mathbf{k}$

43. Not conservative　　**45.** Conservative: $f(x, y, z) = xye^z + K$

47. Conservative: $f(x, y, z) = \dfrac{x}{y} + z^2 - z + K$

49. $6x\mathbf{j} - 3y\mathbf{k}$　　**51.** $z\mathbf{j} + y\mathbf{k}$　　**53.** $12x - 2xy$

55. $\cos x - \sin y + 2z$　　**57.** 4　　**59.** 0

61. $2z + 3x$　　**63.** 0　　**65.–77.** Proof

Section 14.2 *(page 1002)*

1. $3\cos t\mathbf{i} + 3\sin t\mathbf{j}, \quad 0 \le t \le 2\pi$

3. $\mathbf{r}(t) = \begin{cases} t\mathbf{i}, & 0 \le t \le 3 \\ 3\mathbf{i} + (t - 3)\mathbf{j}, & 3 \le t \le 6 \\ (9 - t)\mathbf{i} + 3\mathbf{j}, & 6 \le t \le 9 \\ (12 - t)\mathbf{j}, & 9 \le t \le 12 \end{cases}$

5. $\mathbf{r}(t) = \begin{cases} t\mathbf{i} + \sqrt{t}\mathbf{j}, & 0 \le t \le 1 \\ (2 - t)\mathbf{i} + (2 - t)\mathbf{j}, & 1 \le t \le 2 \end{cases}$

7. 10　　**9.** $\dfrac{\sqrt{65}\,\pi}{6}(3 + 16\pi^2)$　　**11.** 9　　**13.** $\dfrac{\pi}{2}$

15. $\dfrac{19\sqrt{2}}{6}$　　**17.** $\dfrac{19}{6}\left(1 + \sqrt{2}\right)$

19. $\dfrac{2\sqrt{13}\,\pi}{3}(27 + 64\pi^2) \approx 4973.8$　　**21.** $\frac{35}{6}$　　**23.** 2

25. $-\frac{17}{15}$　　**27.** 249.49　　**29.** -66　　**31.** 0　　**33.** $-10\pi^2$

35. 1500 foot-pounds　　**37.** Proof　　**39.** Proof

41. Parabolic path: minimum when $c = \dfrac{1}{4}$; $W = 119.5$
Straight-line path: $W = 120$

43. 1010　　**45.** $\frac{190}{3}$　　**47.** 25　　**49.** $\frac{63}{2}$　　**51.** $\frac{316}{3}$

53. $5h$　　**55.** $\dfrac{1}{2}$　　**57.** $\dfrac{h}{4}\left[2\sqrt{5} + \ln\left(2 + \sqrt{5}\right)\right]$

59. $\dfrac{1}{120}\left(25\sqrt{5} - 11\right)$

61. (a) $12\pi \approx 37.70$ square centimeters

　　(b) $\dfrac{12\pi}{5} \approx 7.54$ cubic centimeters

　　(c)

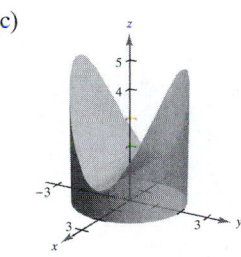

63. z_3, z_1, z_2, z_4;　The greater the height of the surface over the curve $y = \sqrt{x}$, the greater the lateral surface area.

65. b　　**67.** False: $\displaystyle\int_C xy\,ds = \sqrt{2}\int_0^1 t^2\,dt.$

68. False: the orientation of C does not affect the form $\int_C f(x, y)\,ds$.

69. False: the orientations are different.

70. False: see Exercise 32.

Section 14.3 *(page 1016)*

1. $\frac{11}{15}$ **3.** $-\ln(2 + \sqrt{3}) = -\frac{1}{2}\ln(7 + 4\sqrt{3})$

5. Conservative **7.** Not conservative **9.** Conservative

11. (a) 1 (b) 1 **13.** (a) 0 (b) $-\frac{1}{3}$ (c) $-\frac{1}{2}$

15. (a) 64 (b) 0 (c) 0 (d) 0

17. (a) $\frac{64}{3}$ (b) $\frac{64}{3}$ **19.** (a) 32 (b) 32

21. (a) $\frac{2}{3}$ (b) $\frac{17}{6}$ **23.** (a) 0 (b) 0

25. 24 **27.** -1 **29.** 0 **31.** (a) 2 (b) 2 (c) 2

33. 11 **35.** 30,366 **37.** 0

39. Increasing: 10 units per minute

41. No, the force field is conservative.

43. False: it would be true if **F** were conservative.

44. True **45.** True

46. False: the requirement is $\dfrac{\partial M}{\partial y} = \dfrac{\partial N}{\partial x}$. **47.** Proof

49. No: fuel consumption is dependent on wind speed and direction.

Section 14.4 *(page 1025)*

1. 0 **3.** $\frac{32}{15}$ **5.** 19.99 **7.** $\frac{4}{3}$ **9.** 56 **11.** $\frac{32}{3}$

13. 0 **15.** 0 **17.** $\frac{1}{12}$ **19.** 8π **21.** 4π **23.** $\frac{225}{2}$

25. πa^2 **27.** $\frac{32}{3}$ **29.** Proof **31.** $\left(0, \frac{8}{5}\right)$

33. $\left(\dfrac{8}{15}, \dfrac{8}{21}\right)$ **35.** $\dfrac{3\pi a^2}{2}$ **37.** $\pi - \dfrac{3\sqrt{3}}{2}$

39. $I = -2\pi$ when C is a circle that contains the origin.

41. $\frac{19}{2}$

43. (a) $n = 1$: 0 (b) $n = 2$: $-\dfrac{4}{3}a^3$ (c) 0

$n = 3$: 0 $n = 4$: $-\dfrac{16}{15}a^5$

$n = 5$: 0 $n = 6$: $-\dfrac{32}{35}a^7$

$n = 7$: 0 $n = 8$: $-\dfrac{256}{315}a^9$

45. Proof **47.** Proof

Section 14.5 *(page 1035)*

1. c **2.** d **3.** b **4.** a

5. $y - 2z = 0$ **7.** $x^2 + z^2 = 4$

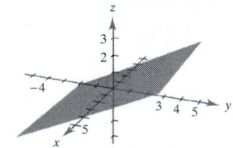 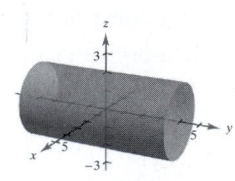

9. The paraboloid is reflected (inverted) through the xy-plane.

11. The height of the paraboloid is increased from 4 to 9.

13. 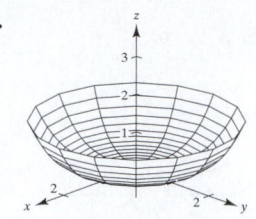 **15.**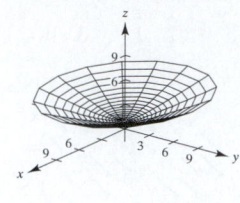

17.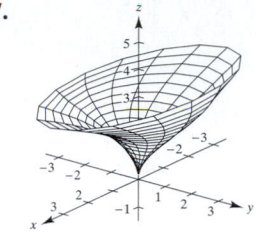

19. (a) $(-10, 10, 0)$ (b) $(10, 10, 10)$
(c) $(0, 10, 0)$ (d) $(10, 0, 0)$

21. (a) 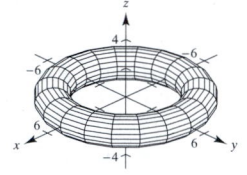 (b)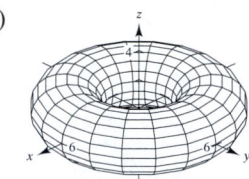

(c) 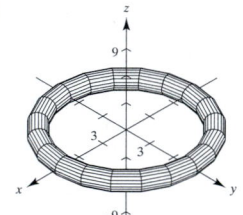 (d)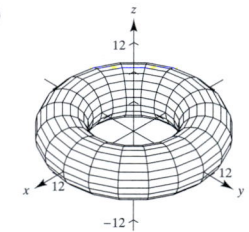

The radius of the generating circle that is revolved about the z-axis is b, and its center is a units from the axis of revolution.

23. $\mathbf{r}(u, v) = u\mathbf{i} + v\mathbf{j} + v\mathbf{k}$

25. $\mathbf{r}(u, v) = 4\cos u\mathbf{i} + 4\sin u\mathbf{j} + v\mathbf{k}$

27. $\mathbf{r}(u, v) = u\mathbf{i} + v\mathbf{j} + u^2\mathbf{k}$

29. $\mathbf{r}(u, v) = v\cos u\mathbf{i} + v\sin u\mathbf{j} + 4\mathbf{k}, \quad 0 \le v \le 3$

31. $x = u, y = \dfrac{u}{2}\cos v, z = \dfrac{u}{2}\sin v, \quad 0 \le u \le 6, \quad 0 \le v \le 2\pi$

33. $x = \sin u \cos v, y = \sin u \sin v, z = u$
$0 \le u \le \pi, 0 \le v \le 2\pi$

35. $x - y - 2z = 0$ **37.** $4y - 3z = 12$ **39.** $2\sqrt{2}$

41. $2\pi ab$ **43.** $\pi ab^2\sqrt{a^2 + 1}$

45. $\dfrac{\pi}{6}\left(17\sqrt{17} - 1\right) \approx 36.177$ **47.** 400π

49. $2\pi\left[\frac{3}{2}\sqrt{13} + 2\ln\left(3 + \sqrt{13}\right) - 2\ln 2\right]$

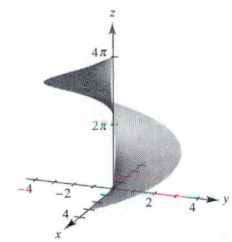

51. Answers will vary.

Section 14.6 *(page 1048)*

1. 0   **3.** 10π **5.** $\dfrac{27\sqrt{6}}{2}$ **7.** $\dfrac{391\sqrt{17} + 1}{240}$

9. -11.47 **11.** $6\sqrt{5}$ **13.** 8   **15.** $\dfrac{19\sqrt{2}\pi}{4}$

17. $\dfrac{32\pi}{3}$ **19.** 486π **21.** $-\dfrac{4}{3}$ **23.** $\dfrac{243\pi}{2}$ **25.** 20π

27. $\frac{5}{2}$ **29.** $\frac{364}{3}$ **31.** Proof   **33.** $2\pi a^3 h$ **35.** $64\pi\rho$

37. (a)

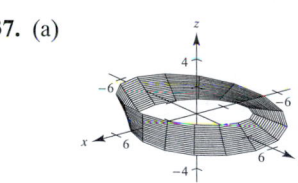

(b) If a normal vector at a point P on the surface is moved around the Möbius strip once, it will point in the opposite direction.

(c)

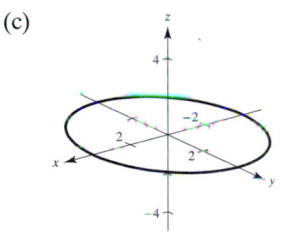

    Circle

(d) Construction

(e) A strip with a double twist and twice as long as the Möbius strip.

39. 0

Section 14.7 *(page 1056)*

1. a^4 **3.** 18   **5.** $3a^4$ **7.** 0   **9.** 32π **11.** 0

13. 2304   **15.** $\dfrac{128\pi}{3}$ **17.** Proof   **19.** 0   **21.** Proof

23. Proof   **25.** Proof

Section 14.8 *(page 1063)*

1. $-xy\mathbf{i} - \mathbf{j} + (yz - 2)\mathbf{k}$ **3.** $\left(2 - \dfrac{1}{1 + x^2}\right)\mathbf{j} - 8x\mathbf{k}$

5. $z(x - 2e^{y^2 + z^2})\mathbf{i} - yz\mathbf{j} - 2ye^{x^2 + y^2}\mathbf{k}$ **7.** 2π **9.** 0

11. 1   **13.** 0   **15.** 0   **17.** $\dfrac{8}{3}$ **19.** $\dfrac{a^5}{4}$ **21.** 0

23. Proof   **25.** Proof

Review Exercises for Chapter 14 *(page 1064)*

1. 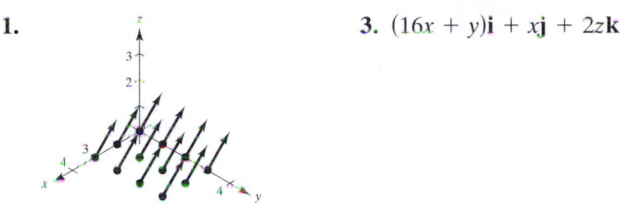   **3.** $(16x + y)\mathbf{i} + x\mathbf{j} + 2z\mathbf{k}$

5. Not conservative

7. Conservative: $f(x, y) = 3x^2y^2 - x^3 + y^3 - 7y + K$

9. Not conservative

11. Conservative: $f(x, y, z) = \dfrac{x}{yz} + K$

13. (a) div $\mathbf{F} = 2x + 2y + 2z$ (b) curl $\mathbf{F} = \mathbf{0}$

15. (a) div $\mathbf{F} = -y\sin x - x\cos y + xy$

     (b) curl $\mathbf{F} = xz\mathbf{i} - yz\mathbf{j}$

17. (a) div $\mathbf{F} = \dfrac{1}{\sqrt{1 - x^2}} + 2xy + 2yz$

     (b) curl $\mathbf{F} = z^2\mathbf{i} + y^2\mathbf{k}$

19. (a) div $\mathbf{F} = \dfrac{2x + 2y}{x^2 + y^2} + 1$ (b) curl $\mathbf{F} = \dfrac{2x - 2y}{x^2 + y^2}\mathbf{k}$

21. (a) $6\sqrt{2}$ (b) 128π **23.** $2\pi^2(1 + 2\pi^2)$

25. (a) $\dfrac{35}{2}$ (b) 18π **27.** $\dfrac{9a^2}{5}$

29. $\dfrac{\sqrt{10}}{4}(41 - \cos 8) \approx 32.528$ **31.** $\dfrac{5}{7}$ **33.** $2\pi^2$

35. $\dfrac{64}{3}$ **37.** $\dfrac{4}{3}$ **39.** $\dfrac{8}{3}\left(3 - 4\sqrt{2}\right) \approx -7.085$

41. 12   **43.** (a) 15   (b) 15   (c) 15

45. 4   **47.** 0   **49.** $\dfrac{1}{12}$

51.

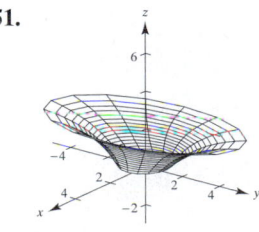

53. (a) 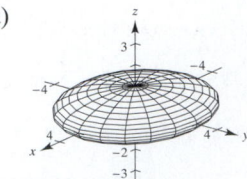   (b)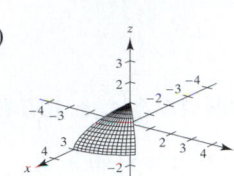

(c) 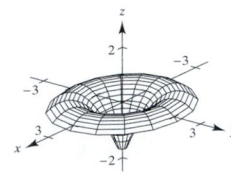   (d)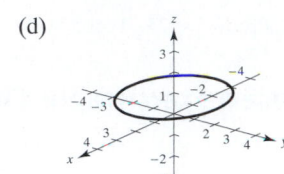

(e) 14.436   (f) 4.269

55.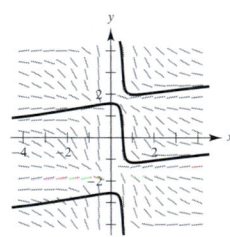

0

57. 66   **59.** $\dfrac{2a^6}{5}$

CHAPTER 15

Section 15.1 *(page 1074)*

1. $x^2 - 3xy + y^2 = C$ **3.** $3xy^2 + 5x^2y^2 - 2y = C$

5. Not exact   **7.** $\arctan \dfrac{x}{y} = C$ **9.** Not exact

11. (a) Answers will vary.

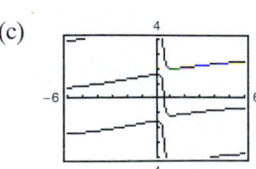

(b) $x^2 \tan y + 5x = \dfrac{11}{4}$

(c)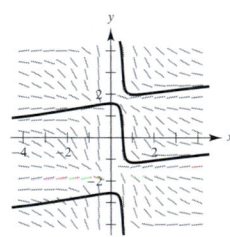

13. $y \ln(x - 1) + y^2 = 16$ **15.** $e^{3x} \sin 3y = 0$

17. Integrating factor: $\dfrac{1}{y^2}$ **19.** Integrating factor: $\dfrac{1}{x^2}$

$\dfrac{x}{y} - 6y = C$ $\dfrac{y}{x} + 5x = C$

21. Integrating factor: $\cos x$

$y \sin x + x \sin x + \cos x = C$

23. Integrating factor: $\dfrac{1}{y}$

$xy - \ln y = C$

25. Integrating factor: $\dfrac{1}{\sqrt{y}}$ **27.** $x^4y^3 + x^2y^4 = C$

$x\sqrt{y} + \cos\sqrt{y} = C$

29. $\dfrac{y^2}{x} + \dfrac{x}{y^2} + C$ **31.** Proof

33. $x^2 + y^2 = C$ **35.** $2x^2y^4 + x^2 = C$

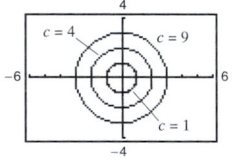   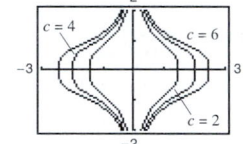

37. $x^2 - 2xy + 3y^2 = 3$ **39.** $C = \dfrac{5\left(x^2 + \sqrt{x^4 - 1{,}000{,}000x}\right)}{x}$

41.

Δx	0.50	0.25	0.10
Estimate	3.7798	3.9875	4.1207

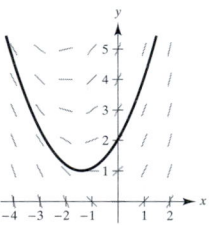

43. (a)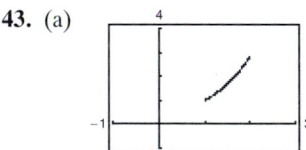

(b) $3y^{2/3} - x^2 = 2$

(c)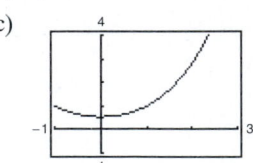

45. (a)

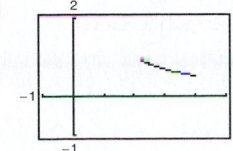

(b) $y^2(2x^2 + y^2) = 9$

(c)

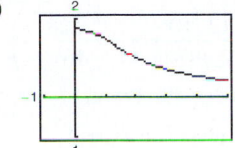

47. (a)

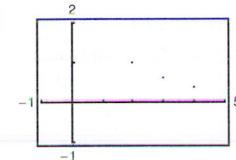

(b) $y^2(2x^2 + y^2) = 9$

(c)

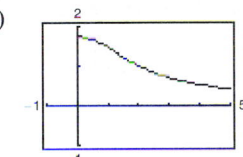

Less accurate

49. False: $\dfrac{\partial M}{\partial y} = 2x, \quad \dfrac{\partial N}{\partial x} = -2x.$

50. False: $ydx + xdy = 0$ is exact, but $xydx + x^2dy = 0$ is not exact.

51. True **52.** True

Section 15.2 *(page 1082)*

1. False: $y' + xy = x^2$ is linear. **2.** True

3. (a) Answers will vary. (b) $y = \dfrac{1}{2}(e^x + e^{-x})$

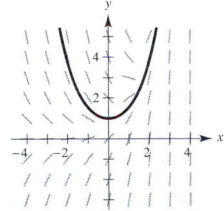

(c)

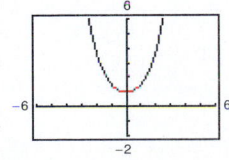

5. $y = x^2 + 2x + \dfrac{C}{x}$ **7.** $y = \tfrac{1}{2}(\sin x - \cos x) + Ce^x$

9. $y = -\dfrac{1}{13}(3\sin 2x + 2\cos 2x) + Ce^{3x}$

11. $y = \dfrac{x^3 - 3x + C}{3(x - 1)}$ **13.** $y = 1 + 4e^{-\tan x}$

15. $y = \sin x + (x + 1)\cos x$ **17.** $xy = 4$

19. $\dfrac{1}{y^2} = Ce^{2x^3} + \dfrac{1}{3}$ **21.** $y = \dfrac{1}{Cx - x^2}$

23. $y^{2/3} = Ce^{2x/3} - \dfrac{1}{4}(4x^3 + 18x^2 + 54x + 81)$

25. (a)

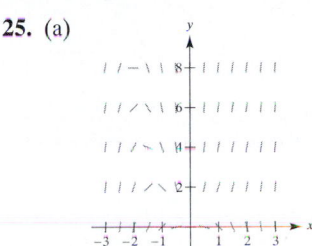

(b) $(-2, 4):\quad y = \dfrac{1}{2}x(x^2 - 8)$

$(2, 8):\quad y = \dfrac{1}{2}x(x^2 + 4)$

(c)

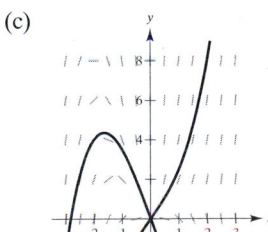

27. (a)

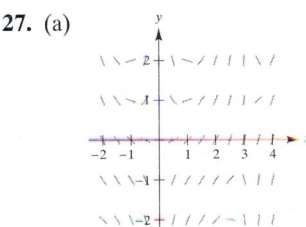

(b) $(1, 1):\quad y = \cos 1\,\csc x - x\cot x + 1$

$(3, -1):\quad y = (3\cos 3 - 2\sin 3)\csc x - x\cot x + 1$

(c)

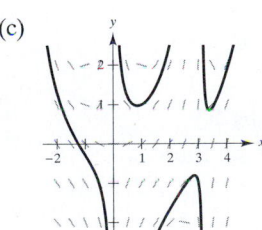

29. $I = \dfrac{E_0}{R} + Ce^{-Rt/L}$

31. $I = Ce^{-(R/L)t} + \dfrac{E_0}{R^2 + \omega^2 L^2}(R\sin\omega t - \omega L\cos\omega t)$

33. $P = -\dfrac{N}{k} + \left(\dfrac{N}{k} + P_0\right)e^{kt}$

35. (a) \$583,098.01   (b) \$3,243,606.35

37. (a) $\dfrac{dQ}{dt} = q - kQ$

    (b) $Q = \dfrac{q}{k} + \left(Q_0 - \dfrac{q}{k}\right)e^{-kt}$

    (c) $\dfrac{q}{k}$

39. Proof

41. (a) $Q = 25e^{-t/20}$

    (b) $-20\ln\left(\dfrac{3}{5}\right) \approx 10.2$ minutes

    (c) 0

43. (a) $t = 50$ minutes

    (b) $100 - \dfrac{25}{\sqrt{2}} \approx 82.32$ pounds

45. c   **46.** d   **47.** a   **48.** b   **49.** $2e^x + e^{-2y} = C$

51. $x + xy^2 + \frac{1}{2}y^2 + 2y = C$

53. $y = Ce^{-\sin x} + 1$ **55.** $x^2 y + \sin y = C$

57. $x^3 y^2 + x^4 y = C$ **59.** $y = \dfrac{e^x(x-1) + C}{x^2}$

61. $x^4 y^4 - 2x^2 = C$ **63.** $3\arctan\dfrac{x}{y} - y = C$

Section 15.3 *(page 1091)*

1. Proof   **3.** Proof   **5.** $y = C_1 + C_2 e^x$

7. $y = C_1 e^{3x} + C_2 e^{-2x}$ **9.** $y = C_1 e^{x/2} + C_2 e^{-2x}$

11. $y = C_1 e^{-3x} + C_2 x e^{-3x}$ **13.** $y = C_1 e^{x/4} + C_2 x e^{x/4}$

15. $y = C_1 \sin x + C_2 \cos x$ **17.** $y = C_1 e^{3x} + C_2 e^{-3x}$

19. $y = e^x\left(C_1 \sin\sqrt{3}x + C_2 \cos\sqrt{3}x\right)$

21. $y = C_1 e^{(3+\sqrt{5})x/2} + C_2 e^{(3-\sqrt{5})x/2}$

23. $y = e^{2x/3}\left(C_1 \sin\dfrac{\sqrt{7}x}{3} + C_2 \cos\dfrac{\sqrt{7}x}{3}\right)$

25. $y = C_1 e^x + C_2 e^{-x} + C_3 \sin x + C_4 \cos x$

27. $y = C_1 e^x + C_2 e^{2x} + C_3 e^{3x}$

29. $y = C_1 e^x + e^x(C_2 \sin 2x + C_3 \cos 2x)$

31. (a) $y = 2\cos 10x$ (b) $y = \frac{1}{5}\sin 10x$

    (c) $y = -\cos 10x + \frac{3}{10}\sin 10x$

33. $y = \frac{1}{11}(e^{6x} + 10e^{-5x})$

35. $y = \dfrac{1}{2}\sin 4x$

37. y'' and y' are not equal for $x < 0$. $y'' > 0$ for all x, but $y' < 0$ for $x < 0$.

39. $y = \frac{1}{2}\cos 4\sqrt{3}t$

41. $y = \dfrac{2}{3}\cos 4\sqrt{3}t - \dfrac{\sqrt{3}}{24}\sin 4\sqrt{3}t$

43. $y = \dfrac{e^{-t/16}}{2}\left(\cos\dfrac{\sqrt{12{,}287}\,t}{16} + \dfrac{\sqrt{12{,}287}}{12{,}287}\sin\dfrac{\sqrt{12{,}287}\,t}{16}\right)$

45. b   **46.** d   **47.** c   **48.** a   **49.** Proof

51. False: the general solution is $y = C_1 e^{3x} + C_2 x e^{3x}$.

52. True   **53.** True

54. False: the solution $y = x^2 e^x$ requires that $m = 1$ is a triple root of the characteristic equation. Because the characteristic equation is quadratic, $m = 1$ can be at most a double root.

55. Proof   **57.** Proof

59. (a) Proof   (b) $y = \dfrac{C_1}{x^3} + \dfrac{C_2}{x^2}$

Section 15.4 *(page 1099)*

1. Proof   **3.** Proof   **5.** $y = C_1 e^x + C_2 e^{2x} + x + \frac{3}{2}$

7. $y = \cos x + 6\sin x + x^3 - 6x$

9. $y = C_1 + C_2 e^{-2x} + \frac{2}{3}e^x$

11. $y = (C_1 + C_2 x)e^{5x} + \frac{3}{8}e^x + \frac{1}{5}$

13. $y = -1 + 2e^{-x} - \cos x - \sin x$

15. $y = \left(C_1 - \dfrac{x}{6}\right)\cos 3x + C_2 \sin 3x$

17. $y = C_1 e^x + C_2 x e^x + \left(C_3 + \dfrac{2x}{9}\right)e^{-2x}$

19. $y = \left(\dfrac{4}{9} - \dfrac{1}{2}x^2\right)e^{4x} - \dfrac{1}{9}(1 + 3x)e^x$

21. (a) $y''_p = 0$ and $3y_p = 12$ (b) $y_p = 2$ (c) $y_p = 4$

23. $y = (C_1 + \ln|\cos x|)\cos x + (C_2 + x)\sin x$

25. $y = \left(C_1 - \dfrac{x}{2}\right)\cos 2x + \left(C_2 + \dfrac{1}{4}\ln|\sin 2x|\right)\sin 2x$

27. $y = (C_1 + C_2 x)e^x + \dfrac{x^2 e^x}{4}(\ln x^2 - 3)$

29. $q = \frac{3}{25}(e^{-5t} + 5te^{-5t} - \cos 5t)$

31. $y = \frac{1}{4}\cos 8t - \frac{1}{2}\sin 8t + \sin 4t$

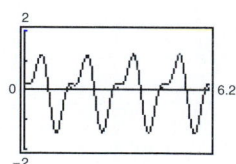

33. $y = \left(\frac{9}{32} - \frac{3}{4}t\right)e^{-8t} - \frac{1}{32}\cos 8t$

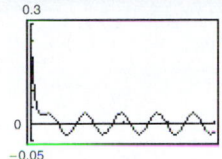

35. $y = \frac{\sqrt{5}}{4}\sin\left(8t - \arctan\frac{1}{2}\right)$ **37.** Proof

$\qquad = \frac{\sqrt{5}}{4}\sin(8t - 0.4636)$

39. $y = C_1 x + C_2 x \ln x + \frac{2}{3}x(\ln x)^3$

Section 15.5 (page 1103)

1. Proof   **3.** Proof   **5.** Proof

7. $y = a_0 \sum_{k=0}^{\infty} \frac{(-3)^k}{2^k k!} x^{2k}$

Interval of convergence: $(-\infty, \infty)$

9. $y = a_0 + a_1 \sum_{k=0}^{\infty} \frac{x^{2x+1}}{2^k(k!)(2k+1)}$

Interval of convergence: $(-\infty, \infty)$

11. $y = a_0\left(1 - \frac{x^2}{8} + \frac{x^4}{128} - \cdots\right) +$

$\qquad a_1\left(x - \frac{x^3}{24} + \frac{7x^5}{1920} - \cdots\right)$

13. Taylor's Theorem: $y = 2 + \frac{2x}{1!} - \frac{2x^2}{2!} - \frac{10x^3}{3!} +$

$\qquad \frac{2x^4}{4!} + \cdots$

$\qquad y\left(\frac{1}{2}\right) \approx 2.547$

Euler's Method: $y\left(\frac{1}{2}\right) \approx 2.672$

15. (a) $y = 2(\cos 3x + \sin 3x)$

(b) $y = 2\left[\sum_{n=0}^{\infty} \frac{(-1)^n(3x)^{2n}}{(2n)!} + \sum_{n=0}^{\infty} \frac{(-1)^n(3x)^{2n+1}}{(2n+1)!}\right]$

(c)

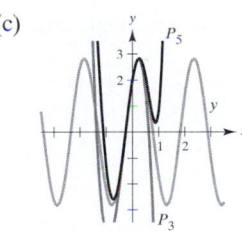

17. $y = 1 - \frac{3x}{1!} + \frac{2x^3}{3!} - \frac{12x^4}{4!} + \frac{16x^6}{6!} - \frac{120x^7}{7!} + \cdots$

$\qquad y\left(\frac{1}{4}\right) \approx 0.253$

19. Proof   **21.** Proof

23. $y = a_0 + a_1 x + \frac{a_0}{6}x^3 + \frac{a_1}{12}x^4 + \frac{a_0}{180}x^6 + \frac{a_1}{504}x^7$

Review Exercises for Chapter 15 (page 1104)

1. Type: partial   **3.** Type: ordinary
Order: 2   Order: 2

5. (a)   (b) $y = Cx$

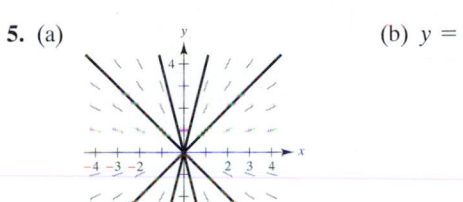

7. b   **8.** d   **9.** a   **10.** c

11. $y = x \ln x^2 + 2x^{3/2} + Cx$ **13.** $y = C(1-x)^2$

15. $y^2 = x^2 \ln x^2 + Cx^2$

17. $5x^2 + 8xy + 2x + \frac{5}{2}y^2 + 2y = C$

19. $xy - 2xy^3 + x^2 = C$

21. $y = e^x(1 + \tan x) + C \sec x$

23. $x^2 - 2xy - 10x - 3y^2 + 4y = C$ **25.** $y^2 = 2Cx + C^2$

27. $y^2 = x^2 - x + \frac{3}{2} + Ce^{-2x}$ **29.** $\frac{y-x}{1+xy} = C$

31. $y = \frac{bx^4}{4-a} + Cx^a$ **33.** $y = 5e^{2x} - e^x$

35. $y = x \ln|x| - 2 + 12x$ **37.** $\ln|1+y| = (\ln 3)e^{-x}$

39. $y = \frac{3(1 + e^{2x^3+1})}{1 - e^{2x^3+1}}$

41. Family of circles: $x^2 + (y-K)^2 = K^2$

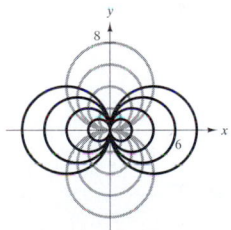

43. (a) $\frac{ds}{dh} = \frac{k}{h}$ (b) $s = 25 - \frac{13\ln(h/2)}{\ln 5}, \quad 2 \le h \le 15$

$\qquad s = k \ln h + C$

45. $N = \frac{500}{1 + 4e^{-0.2452t}}$

47. $y = \dfrac{2}{5}(\sin x - 2 \cos x) + \dfrac{9}{5}e^{-x/2}$

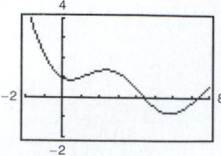

49. (a) Balance increases.

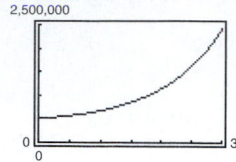

(b) Balance remains $500,000.

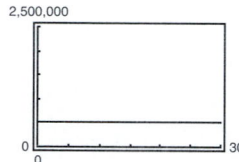

(c) Balance is depleted in 17.9 years.

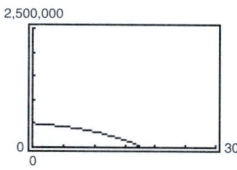

51. $y = e^{2x} - e^{-x}$

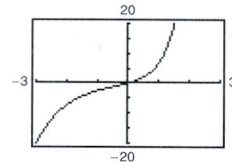

53. $y = \dfrac{3}{2}e^x + \dfrac{1}{2}e^{-3x}$

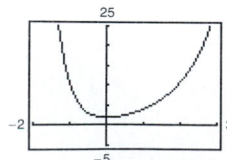

55. $y = C_1 \sin x + C_2 \cos x - 5x + x^3$

57. $y = (C_1 + x)\sin x + C_2 \cos x$

59. $y = \left(C_1 + C_2 x + \dfrac{1}{3}x^3\right)e^x$

61. $y = \dfrac{11}{5}\,(2e^{3x} + 3e^{-2x}) - 9$

63. $y = \dfrac{17}{3}\cos 2x - 3 \sin 2x + \dfrac{1}{3}\cos x$

65. $y = \dfrac{1}{2}\cos\left(2\sqrt{6}\,t\right)$

67. (a) (i) $y = \dfrac{1}{2}\cos 2t + \dfrac{12\pi}{\pi^2 - 4}\sin 2t + \dfrac{24}{4 - \pi^2}\sin \pi t$

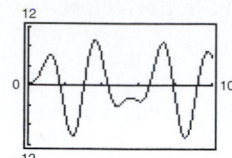

(ii) $y = \dfrac{1}{2}\left[\left(1 - 6\sqrt{2}t\right)\cos\left(2\sqrt{2}t\right) + 3 \sin\left(2\sqrt{2}t\right)\right]$

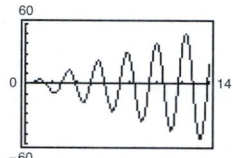

(iii) $y = \dfrac{e^{-t/5}}{398}\left[199 \cos \dfrac{\sqrt{199}t}{5} + \sqrt{199}\sin \dfrac{\sqrt{199}t}{5}\right]$

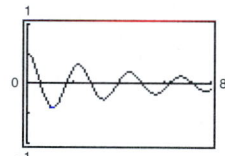

(iv) $y = \tfrac{1}{2}e^{-2t}(\cos 2t + \sin 2t)$

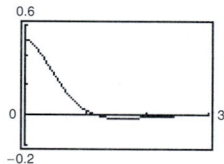

(b) The object would come to rest more quickly. It might not oscillate at all, as in part (iv).

(c) The object would oscillate more rapidly.

(d) Part (ii). The amplitude becomes increasingly large.

69. $y = a_0 \displaystyle\sum_{n=0}^{\infty} \dfrac{x^n}{4^n}$

APPENDIX A

Section A.1 *(page A8)*

1. Rational   **3.** Irrational   **5.** Rational   **7.** Rational

9. Rational   **11.** $\frac{4}{11}$ **13.** $\frac{11}{37}$

15. (a) True   (b) False   (c) True   (d) False
(e) False   (f) False

17. x is greater than -3 and less than 3.

 The interval is bounded.

19. x is no more than 5.

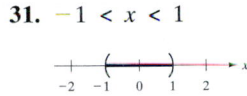 The interval is unbounded.

21. $y \geq 4, [4, \infty)$ **23.** $0.03 < r \leq 0.07, (0.03, 0.07]$

25. $x \geq \frac{1}{2}$ **27.** $-\frac{1}{2} < x < \frac{7}{2}$

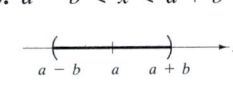

29. $x > 6$ **31.** $-1 < x < 1$

33. $x \geq 13, x \leq -7$ **35.** $a - b < x < a + b$

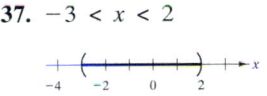   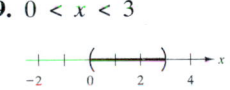

37. $-3 < x < 2$ **39.** $0 < x < 3$

41. $-3 \leq x \leq 1$ **43.** $-3 \leq x \leq 2$

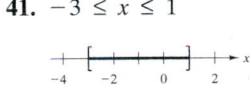

45. $4, -4, 4$ **47.** (a) $-51, 51, 51$ (b) $51, -51, 51$

49. 1   **51.** (a) 14   (b) 10   **53.** $|x| \leq 2$

55. $|x - 2| > 2$ **57.** (a) $|x - 12| \leq 10$ (b) $|x - 12| \geq 10$

59. $x \geq 36$ units   **61.** $x \leq 41$ or $x \geq 59$

63. (a) $\frac{355}{112} > \pi$ (b) $\frac{22}{7} > \pi$ **65.** b

67. False: the reciprocal of 2 is $\frac{1}{2}$, which is not an integer.

68. True   **69.** True   **70.** False: $|0| = 0$.   **71.** True

72. True   **73.** Proof   **75.** Proof   **77.** Proof

79. Proof

81. $|-3 - 1| > |-3| - |1|$
$|3 - 1| = |3| - |1|$

Section A.2 *(page A15)*

1. $d = 2\sqrt{5}$ **3.** $d = 2\sqrt{10}$

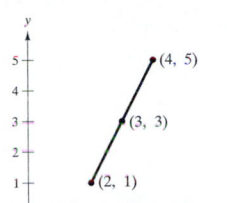   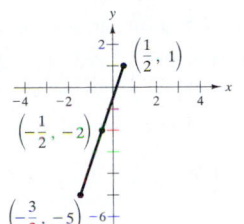

5. $8\sqrt{8 - 2\sqrt{3}}$

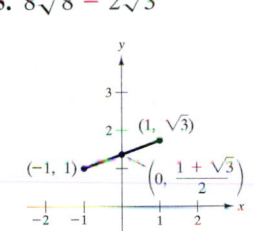

7. Right triangle:
$d_1 = \sqrt{45}, d_2 = \sqrt{5}$
$d_3 = \sqrt{50}$
$(d_1)^2 + (d_2)^2 = (d_3)^2$

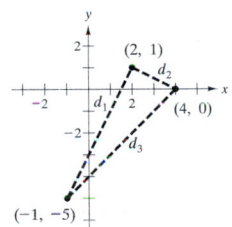

9. Rhombus: the length of each side is $\sqrt{5}$.

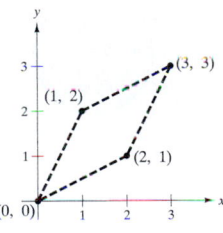

11. Quadrant II   **13.** Quadrants I and III

15. $x = 0 \leftrightarrow 1990$

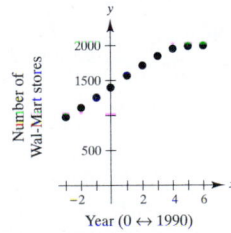

17. $d_1 = 2\sqrt{5}, d_2 = \sqrt{5}, d_3 = 3\sqrt{5}$
Collinear, because $d_1 + d_2 = d_3$

19. $d_1 = \sqrt{2}, d_2 = \sqrt{13}, d_3 = 5$
Not collinear, because $d_1 + d_2 > d_3$

21. $x = \pm 3$ **23.** $y = \pm\sqrt{55}$

25. $\left(\dfrac{3x_1 + x_2}{4}, \dfrac{3y_1 + y_2}{4}\right)$ $\left(\dfrac{x_1 + x_2}{2}, \dfrac{y_1 + y_2}{2}\right)$

$\left(\dfrac{x_1 + 3x_2}{4}, \dfrac{y_1 + 3y_2}{4}\right)$

27. c   **28.** b   **29.** a   **30.** d   **31.** $x^2 + y^2 - 9 = 0$

33. $x^2 + y^2 - 4x + 2y - 11 = 0$

35. $x^2 + y^2 + 2x - 4y = 0$

37. $x^2 + y^2 - 6x - 4y + 3 = 0$ **39.** $x^2 + y^2 = 26{,}000^2$

41. $(x - 1)^2 + (y + 3)^2 = 4$ **43.** $(x - 1)^2 + (y + 3)^2 = 0$

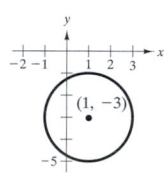

45. $\left(x - \tfrac{1}{2}\right)^2 + \left(y - \tfrac{1}{2}\right)^2 = 2$ **47.** $\left(x + \tfrac{1}{2}\right)^2 + \left(y + \tfrac{5}{4}\right)^2 = \tfrac{9}{4}$

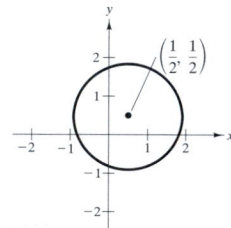

49. 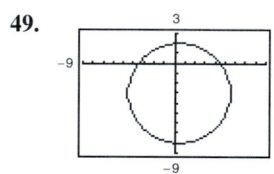   **51.**

53. Proof   **55.** True   **56.** False: the distance is $|2b|$.

57. True   **58.** True   **59.** Proof   **61.** Proof

Section A.3 *(page A25)*

1. (a) $396°, -324°$ (b) $240°, -480°$

3. (a) $\dfrac{19\pi}{9}, -\dfrac{17\pi}{9}$ (b) $\dfrac{10\pi}{3}, -\dfrac{2\pi}{3}$

5. (a) $\dfrac{\pi}{6}, 0.524$ (b) $\dfrac{5\pi}{6}, 2.618$

(c) $\dfrac{7\pi}{4}, 5.498$ (d) $\dfrac{2\pi}{3}, 2.094$

7. (a) $270°$ (b) $210°$ (c) $-105°$ (d) $-135.6°$

9.

r	8 ft	15 in.	85 cm	24 in.	$\dfrac{12{,}963}{\pi}$ mi
s	12 ft	24 in.	63.75π cm	96 in.	8642 mi
θ	1.5	1.6	$\dfrac{3\pi}{4}$	4	$\dfrac{2\pi}{3}$

11. (a) $\sin\theta = \tfrac{4}{5}$ $\csc\theta = \tfrac{5}{4}$ (b) $\sin\theta = -\tfrac{5}{13}$ $\csc\theta = -\tfrac{13}{5}$
$\cos\theta = \tfrac{3}{5}$ $\sec\theta = \tfrac{5}{3}$ $\cos\theta = -\tfrac{12}{13}$ $\sec\theta = -\tfrac{13}{12}$
$\tan\theta = \tfrac{4}{3}$ $\cot\theta = \tfrac{3}{4}$ $\tan\theta = \tfrac{5}{12}$ $\cot\theta = \tfrac{12}{5}$

13. (a) Quadrant III   (b) Quadrant IV

15. $\dfrac{\sqrt{3}}{2}$ **17.** $\dfrac{4}{3}$

19. (a) $\sin 60° = \dfrac{\sqrt{3}}{2}$ (b) $\sin 120° = \dfrac{\sqrt{3}}{2}$
$\cos 60° = \dfrac{1}{2}$ $\cos 120° = -\dfrac{1}{2}$
$\tan 60° = \sqrt{3}$ $\tan 120° = -\sqrt{3}$

(c) $\sin\dfrac{\pi}{4} = \dfrac{\sqrt{2}}{2}$ (d) $\sin\dfrac{5\pi}{4} = -\dfrac{\sqrt{2}}{2}$
$\cos\dfrac{\pi}{4} = \dfrac{\sqrt{2}}{2}$ $\cos\dfrac{5\pi}{4} = -\dfrac{\sqrt{2}}{2}$
$\tan\dfrac{\pi}{4} = 1$ $\tan\dfrac{5\pi}{4} = 1$

21. (a) $\sin 225° = -\dfrac{\sqrt{2}}{2}$ (b) $\sin(-225°) = \dfrac{\sqrt{2}}{2}$
$\cos 225° = -\dfrac{\sqrt{2}}{2}$ $\cos(-225°) = -\dfrac{\sqrt{2}}{2}$
$\tan 225° = 1$ $\tan(-225°) = -1$

(c) $\sin\dfrac{5\pi}{3} = -\dfrac{\sqrt{3}}{2}$ (d) $\sin\dfrac{11\pi}{6} = -\dfrac{1}{2}$
$\cos\dfrac{5\pi}{3} = \dfrac{1}{2}$ $\cos\dfrac{11\pi}{6} = \dfrac{\sqrt{3}}{2}$
$\tan\dfrac{5\pi}{3} = -\sqrt{3}$ $\tan\dfrac{11\pi}{6} = -\dfrac{\sqrt{3}}{3}$

23. (a) 0.1736   (b) 5.759   **25.** (a) 0.3640   (b) 0.3640

27. (a) $\theta = \dfrac{\pi}{4}, \dfrac{7\pi}{4}$ (b) $\theta = \dfrac{3\pi}{4}, \dfrac{5\pi}{4}$

29. (a) $\theta = \dfrac{\pi}{4}, \dfrac{5\pi}{4}$ (b) $\theta = \dfrac{5\pi}{6}, \dfrac{11\pi}{6}$

31. $\theta = \dfrac{\pi}{4}, \dfrac{3\pi}{4}, \dfrac{5\pi}{4}, \dfrac{7\pi}{4}$ **33.** $\theta = 0, \dfrac{\pi}{4}, \pi, \dfrac{5\pi}{4}$

35. $\theta = \dfrac{\pi}{3}, \dfrac{5\pi}{3}$ **37.** $\theta = 0, \dfrac{\pi}{2}, \pi$ **39.** 5099 feet

41. (a) Period: π (b) Period: 2   **43.** Period: $\frac{1}{2}$
Amplitude: 2   Amplitude: $\frac{1}{2}$ Amplitude: 3

45. Period: $\dfrac{\pi}{2}$ **47.** Period: $\dfrac{2\pi}{5}$

49. (a) Change in amplitude   (b) Change in period

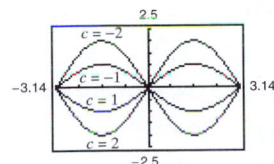 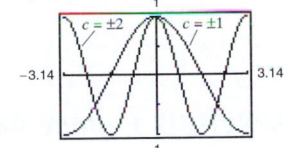

(c) Horizontal translation

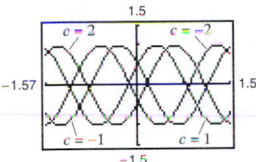

51. 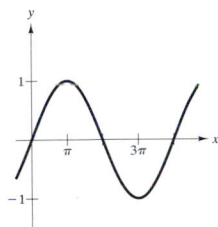   **53.**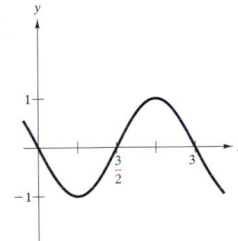

55. 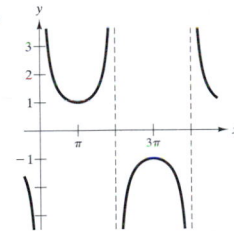   **57.**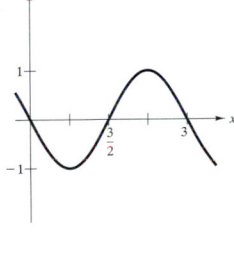

59. 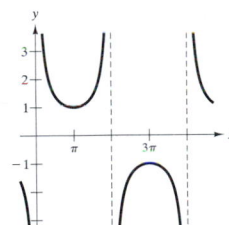   **61.**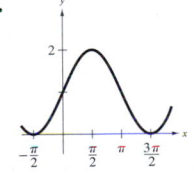

63. $a = 3, b = \dfrac{1}{2}, c = \dfrac{\pi}{2}$

65.

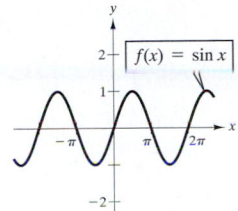   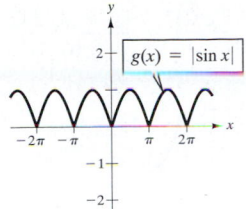

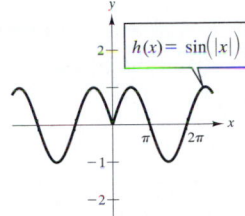

The graph of $|f(x)|$ will reflect any parts of the graph of $f(x)$ below the x-axis about the x-axis. The graph of $f(|x|)$ will reflect the part of the graph of $f(x)$ left of the y-axis about the x-axis.

67.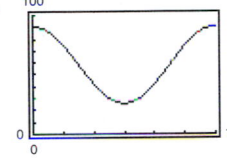

January, November, December

69. $f(x) = \dfrac{4}{\pi}\left(\sin \pi x + \dfrac{1}{3}\sin 3\pi x + \dfrac{1}{5}\sin 5\pi x + \cdots\right)$

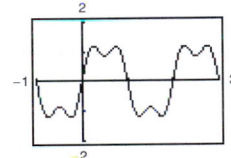

APPENDIX E *(page A56)*

1. $\dfrac{(y')^2}{2} - \dfrac{(x')^2}{2} = 1$ **3.** $\dfrac{(x')^2}{1/4} - \dfrac{(y')^2}{1/6} = 1$

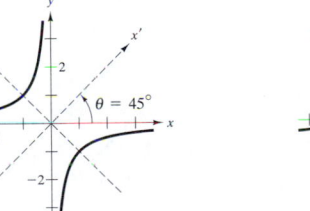   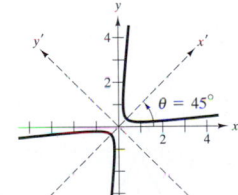

5. $\dfrac{(x' - 3\sqrt{2})^2}{16} - \dfrac{(y' - \sqrt{2})^2}{16} = 1$

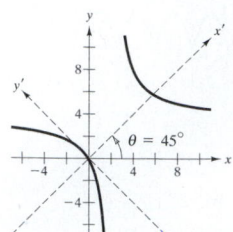

7. $\dfrac{(x')^2}{3} + \dfrac{(y')^2}{2} = 1$

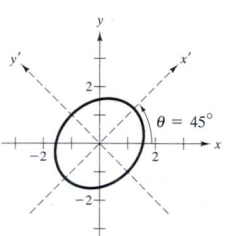

9. $x' = -(y')^2$

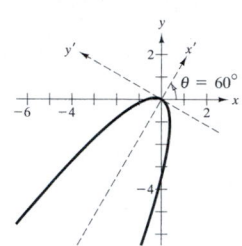

11. $y' = \dfrac{(x')^2}{6} - \dfrac{x'}{3}$

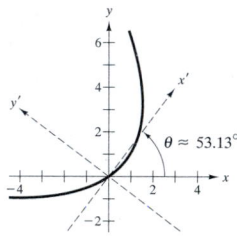

13. $\theta = 45°$

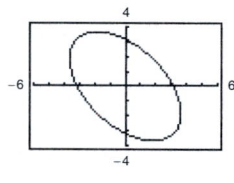

15. $\theta \approx 26.57°$

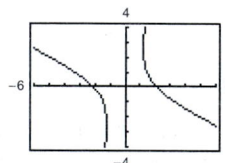

17. $\theta \approx 31.72°$

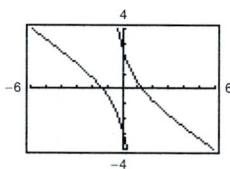

19. Parabola **21.** Ellipse **23.** Hyperbola **25.** Parabola
27. Two lines **29.** Two parallel lines

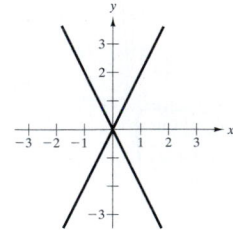

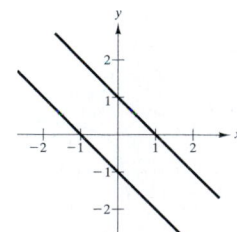

31. Two lines **33.** Proof

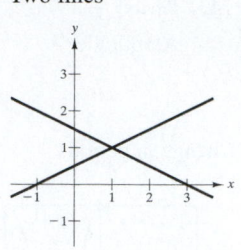

APPENDIX F *(page A66)*

1. $11 - i$ **3.** 4 **5.** $3 - 3\sqrt{2}i$ **7.** $-14 + 20i$
9. $\frac{1}{6} + \frac{7}{6}i$ **11.** $-2\sqrt{3}$ **13.** -10 **15.** $5 + i$
17. $12 + 30i$ **19.** 24 **21.** $-9 + 40i$ **23.** -10
25. 34 **27.** 9 **29.** 400 **31.** 8 **33.** $-6i$
35. $\frac{16}{41} + \frac{20}{41}i$ **37.** $\frac{3}{5} + \frac{4}{5}i$ **39.** $-7 - 6i$
41. $-\frac{9}{1681} + \frac{40}{1681}i$ **43.** $-\frac{1}{2} - \frac{5}{2}i$ **45.** $\frac{62}{949} + \frac{297}{949}i$
47. $1 \pm i$ **49.** $-2 \pm \frac{1}{2}i$ **51.** $-\frac{5}{2}, -\frac{3}{2}$ **53.** $\frac{1}{8} \pm \dfrac{\sqrt{11}}{8}i$
55. $-1 + 6i$ **57.** $-5i$ **59.** $-375\sqrt{3}i$ **61.** i
63. 5 **65.** $4\sqrt{2}$

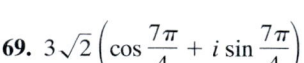

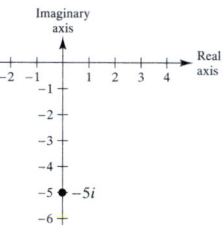

67. $\sqrt{85}$ **69.** $3\sqrt{2}\left(\cos\dfrac{7\pi}{4} + i\sin\dfrac{7\pi}{4}\right)$

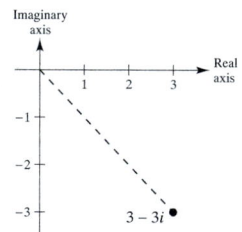

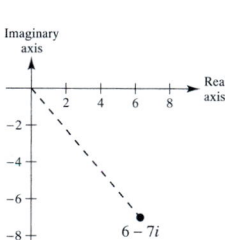

71. $2\left(\cos\dfrac{\pi}{6} + i\sin\dfrac{\pi}{6}\right)$ **73.** $4\left(\cos\dfrac{4\pi}{3} + i\sin\dfrac{4\pi}{3}\right)$

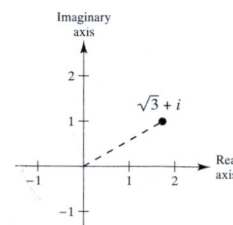

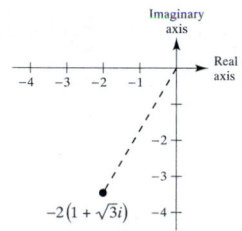

75. $6\left(\cos\dfrac{\pi}{2} + i\sin\dfrac{\pi}{2}\right)$

77. $-\sqrt{3} + i$

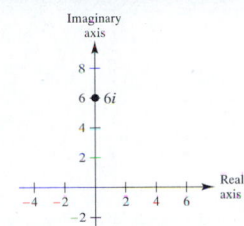

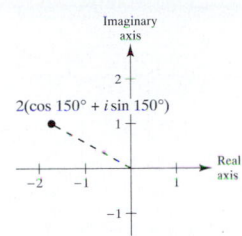

79. $\dfrac{3}{4} - \dfrac{3\sqrt{3}}{4}i$

81. $\dfrac{-15\sqrt{2}}{8} + \dfrac{15\sqrt{2}}{8}i$

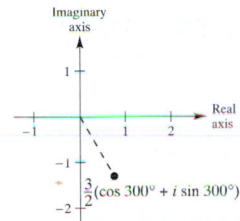

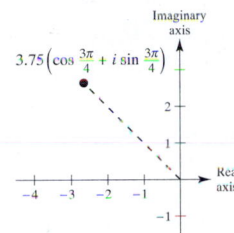

83. $12\left(\cos\dfrac{\pi}{2} + i\sin\dfrac{\pi}{2}\right)$ **85.** $\dfrac{10}{9}(\cos 200° + i\sin 200°)$

87. $-4 - 4i$ **89.** $-32i$ **91.** $-128\sqrt{3} - 128i$ **93.** i

95. (a) $\sqrt{5}(\cos 60° + i\sin 60°)$
   $\sqrt{5}(\cos 240° + i\sin 240°)$

(b)

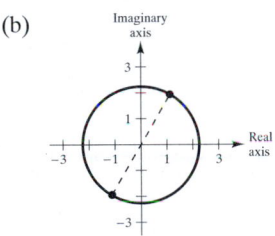

(c) $\dfrac{\sqrt{5}}{2} + \dfrac{\sqrt{15}}{2}i, \ -\dfrac{\sqrt{5}}{2} - \dfrac{\sqrt{15}}{2}i$

97. (a) $2\left(\cos\dfrac{\pi}{3} + i\sin\dfrac{\pi}{3}\right)$

   $2\left(\cos\dfrac{5\pi}{6} + i\sin\dfrac{5\pi}{6}\right)$

   $2\left(\cos\dfrac{4\pi}{3} + i\sin\dfrac{4\pi}{3}\right)$

   $2\left(\cos\dfrac{11\pi}{6} + i\sin\dfrac{11\pi}{6}\right)$

(b)

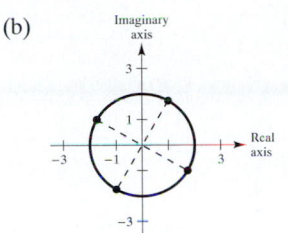

(c) $1 + \sqrt{3}i, \ -\sqrt{3} + i, \ -1 - \sqrt{3}i, \ \sqrt{3} - i$

99. (a) $5\left(\cos\dfrac{4\pi}{9} + i\sin\dfrac{4\pi}{9}\right)$

   $5\left(\cos\dfrac{10\pi}{9} + i\sin\dfrac{10\pi}{9}\right)$

   $5\left(\cos\dfrac{16\pi}{9} + i\sin\dfrac{16\pi}{9}\right)$

(b)

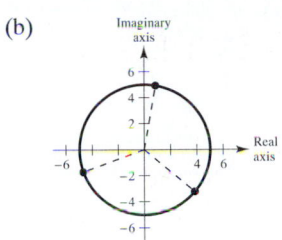

101. $\cos\dfrac{\pi}{8} + i\sin\dfrac{\pi}{8}$

   $\cos\dfrac{5\pi}{8} + i\sin\dfrac{5\pi}{8}$

   $\cos\dfrac{9\pi}{8} + i\sin\dfrac{9\pi}{8}$

   $\cos\dfrac{13\pi}{8} + i\sin\dfrac{13\pi}{8}$

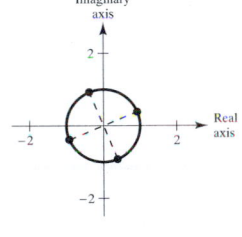

103. $3\left(\cos\dfrac{\pi}{5} + i\sin\dfrac{\pi}{5}\right)$

   $3\left(\cos\dfrac{3\pi}{5} + i\sin\dfrac{3\pi}{5}\right)$

   $3(\cos \pi + i\sin \pi)$

   $3\left(\cos\dfrac{7\pi}{5} + i\sin\dfrac{7\pi}{5}\right)$

   $3\left(\cos\dfrac{9\pi}{5} + i\sin\dfrac{9\pi}{5}\right)$

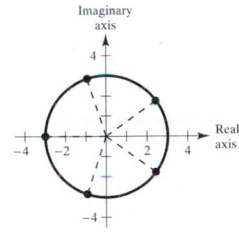

105. $4\left(\cos\dfrac{\pi}{2} + i\sin\dfrac{\pi}{2}\right)$

   $4\left(\cos\dfrac{7\pi}{6} + i\sin\dfrac{7\pi}{6}\right)$

   $4\left(\cos\dfrac{11\pi}{6} + i\sin\dfrac{11\pi}{6}\right)$

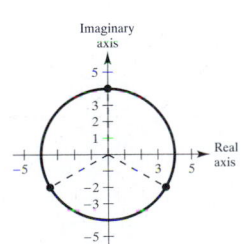

107. $\sqrt[6]{2}(\cos 105° + i\sin 105°)$

   $\sqrt[6]{2}(\cos 225° + i\sin 225°)$

   $\sqrt[6]{2}(\cos 345° + i\sin 345°)$

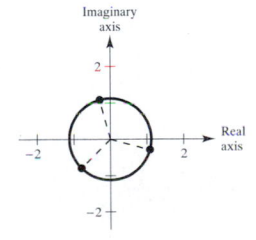

INDEX

Increments, 849
Indefinite integral, 242
Indefinite integral (or antiderivative) of a
  vector-valued function, 781
Independent of path, 1012
Independent variable, 20, 819
Indeterminate form, 60, 189, 523
Index of summation, 252
Inelastic demand, 234
Inequality (inequalities), A2
  and absolute value, A6
    equivalent, A4
    linear, A5
    properties of, A2
    solution of, A4
Inertia, moment of, 946, 962
  polar, 946
Infinite discontinuity, 533
Infinite interval, A3
Infinite limit, 79
  from the left, 79
  properties of, 83
  from the right, 79
Infinite series (or series), 558
  alternating, 581
  convergence of, 558
  divergence of, 558
  geometric, 560
  harmonic, 576
  *n*th partial sum of, 558
  p-series, 570
  properties of, 562
  sequence of partial sums of, 558
  sum of, 558
  telescoping, 559
  term of, 558
Infinity, A3
  limit at, 187
Inflection point, 181
Initial condition, 246, 367
Initial point, 701
Initial ray of an angle, A17
Initial value of exponential growth and
  decay models, 359
Inner partition, 923, 957
  polar, 935
Inner product, 498, 720
Inner radius, 419
Inscribed rectangle, 256
Inside limits of integration, 916
Instantaneous velocity, 109
Integrability, and continuity, 266
Integrable, 925
Integrable function, 266
Integral
  definition of, 266
  double, 923, 925
  flux, 1044

improper, 533
indefinite, 242
iterated, 916
line, 997
surface, 1038
triple, 957
Integral Test, 568
Integration, 242, 533
  completing the square, 386
  of even and odd functions, 295
  factor, 1072
  guidelines for, 324
  of a hyperbolic function, 394
  involving inverse hyperbolic function,
    398
  involving inverse trigonometric
    functions, 385
  involving logarithmic functions, 321
  involving secant and tangent, 493
  involving sine and cosine, 490, 495
  limits of, 916
  Log Rule, 321
  lower limit of, 266
  by partial fraction, 508
  by parts, 481
    summary, 486
    tabular method, 486
  of power series, 611
  region of, 916
  rules, 243, 244
  rules for exponential functions, 342
  by substitution, 287
  summary of formulas, 1062
  by tables, 517
  of trigonometric functions, 325, 326,
    332
  by trigonometric substitution, 499
  upper limit of, 266
  of a vector-valued function, 781
Integration formulas, special, 503
Integration rules, 243
Intercept of a graph, 5
Interest, compound, 352
Interior point, 831, 837
Intermediate Value Theorem, 74
Intersection of two sets, A2
Interval
  of convergence, 607
  midpoint of, A7
  partition of, 265
  on the real line, A3
Inverse cosecant function, 377
  derivative of, 380
  graph of, 378
Inverse cosine function, 377
  derivative of, 380
  graph of, 378

Inverse cotangent function, 377
  derivative of, 380
  graph of, 378
Inverse function, 329
  continuity of, 333
  derivative of, 333
  existence of, 331
  graph of, 330
  guidelines for finding, 332
  horizontal line test for, 331
  reflective property of, 330
Inverse hyperbolic cosecant function, 396
  derivative of, 398
  graph of, 397
  integrals involving, 398
Inverse hyperbolic cosine function, 396
  derivative of, 398
  graph of, 397
  integrals involving, 398
Inverse hyperbolic cotangent function, 396
  derivative of, 398
  graph of, 397
  integrals involving, 398
Inverse hyperbolic functions, 396
Inverse hyperbolic secant function, 396
  derivative of, 398
  graph of, 397
  integrals involving, 398
Inverse hyperbolic sine function, 396
  derivative of, 398
  graph of, 397
  integrals involving, 398
Inverse hyperbolic tangent function, 396
  derivative of, 398
  graph of, 397
  integrals involving, 398
Inverse properties, 379
Inverse secant function, 377
  derivative of, 380
  graph of, 378
Inverse sine function, 377
  derivative of, 380
  graph of, 378
Inverse square field, 986
Inverse tangent function, 377
  derivative of, 380
  graph of, 378
Inverse trigonometric function(s), 377
  derivative of, 380
  integration involving, 385
Involute of a circle, 697
Irrational number, A1
Irrotational, 991
Isobars, 822
Isotherm, 822
Isothermal surfaces, 825
Iterated integral, 915, 916
Iteration, 215
*i*th term of a sum, 252

ALGEBRA

Factors and Zeros of Polynomials

Let $p(x) = a_n x^n + a_{n-1} x^{n-1} + \cdots + a_1 x + a_0$ be a polynomial. If $p(a) = 0$, then a is a *zero* of the polynomial and a solution of the equation $p(x) = 0$. Furthermore, $(x - a)$ is a *factor* of the polynomial.

Fundamental Theorem of Algebra

An nth degree polynomial has n (not necessarily distinct) zeros. Although all of these zeros may be imaginary, a real polynomial of odd degree must have at least one real zero.

Quadratic Formula

If $p(x) = ax^2 + bx + c$, and $0 \le b^2 - 4ac$, then the real zeros of p are $x = (-b \pm \sqrt{b^2 - 4ac})/2a$.

Special Factors

$$x^2 - a^2 = (x - a)(x + a) \qquad\qquad x^3 - a^3 = (x - a)(x^2 + ax + a^2)$$
$$x^3 + a^3 = (x + a)(x^2 - ax + a^2) \qquad\qquad x^4 - a^4 = (x^2 - a^2)(x^2 + a^2)$$

Binomial Theorem

$$(x + y)^2 = x^2 + 2xy + y^2 \qquad\qquad (x - y)^2 = x^2 - 2xy + y^2$$
$$(x + y)^3 = x^3 + 3x^2 y + 3xy^2 + y^3 \qquad\qquad (x - y)^3 = x^3 - 3x^2 y + 3xy^2 - y^3$$
$$(x + y)^4 = x^4 + 4x^3 y + 6x^2 y^2 + 4xy^3 + y^4 \qquad\qquad (x - y)^4 = x^4 - 4x^3 y + 6x^2 y^2 - 4xy^3 + y^4$$
$$(x + y)^n = x^n + nx^{n-1}y + \frac{n(n-1)}{2!}x^{n-2}y^2 + \cdots + nxy^{n-1} + y^n$$
$$(x - y)^n = x^n - nx^{n-1}y + \frac{n(n-1)}{2!}x^{n-2}y^2 - \cdots \pm nxy^{n-1} \mp y^n$$

Rational Zero Theorem

If $p(x) = a_n x^n + a_{n-1} x^{n-1} + \cdots + a_1 x + a_0$ has integer coefficients, then every *rational zero* of p is of the form $x = r/s$, where r is a factor of a_0 and s is a factor of a_n.

Factoring by Grouping

$$acx^3 + adx^2 + bcx + bd = ax^2(cx + d) + b(cx + d) = (ax^2 + b)(cx + d)$$

Arithmetic Operations

$$ab + ac = a(b + c) \qquad \frac{a}{b} + \frac{c}{d} = \frac{ad + bc}{bd} \qquad \frac{a + b}{c} = \frac{a}{c} + \frac{b}{c}$$

$$\frac{\left(\dfrac{a}{b}\right)}{\left(\dfrac{c}{d}\right)} = \left(\frac{a}{b}\right)\left(\frac{d}{c}\right) = \frac{ad}{bc} \qquad \frac{\left(\dfrac{a}{b}\right)}{c} = \frac{a}{bc} \qquad \frac{a}{\left(\dfrac{b}{c}\right)} = \frac{ac}{b}$$

$$a\left(\frac{b}{c}\right) = \frac{ab}{c} \qquad \frac{a - b}{c - d} = \frac{b - a}{d - c} \qquad \frac{ab + ac}{a} = b + c$$

Exponents and Radicals

$$a^0 = 1, \quad a \ne 0 \qquad (ab)^x = a^x b^x \qquad a^x a^y = a^{x+y} \qquad \sqrt{a} = a^{1/2} \qquad \frac{a^x}{a^y} = a^{x-y} \qquad \sqrt[n]{a} = a^{1/n}$$

$$\left(\frac{a}{b}\right)^x = \frac{a^x}{b^x} \qquad \sqrt[n]{a^m} = a^{m/n} \qquad a^{-x} = \frac{1}{a^x} \qquad \sqrt[n]{ab} = \sqrt[n]{a}\,\sqrt[n]{b} \qquad (a^x)^y = a^{xy} \qquad \sqrt[n]{\frac{a}{b}} = \frac{\sqrt[n]{a}}{\sqrt[n]{b}}$$